THE ASTRONOMY AND ASTROPHYSICS ENCYCLOPEDIA

THE ASTRONOMY AND ASTROPHYSICS ENCYCLOPEDIA

Edited by

Stephen P. Maran

Foreword by Carl Sagan

VNR VAN NOSTRAND REINHOLD
NEW YORK

CAMBRIDGE UNIVERSITY PRESS
Cambridge Melbourne Sydney

Library of Congress Catalog Card Number 91-23241
ISBN 0-442-26364-3

Manufactured in the United States of America

Published in the USA and its possessions by Van Nostrand Reinhold
115 Fifth Avenue
New York, New York 10003

Distributed in Canada by Nelson Canada
1120 Birchmount Road
Scarborough, Ontario M1K 5G4, Canada

16 14 14 13 12 11 10 9 8 7 6 5 4 3 2 1

Published outside North America by the Press Syndicate of the University of Cambridge
The Pitt Building, Trumpington Street, Cambridge, CB2 1RP
10 Stamford Road, Oakleigh, Victoria 3166, Australia

British Library Cataloging in publication data available

ISBN 0 521 41744 9 hardback
Cambridge University Press

Library of Congress Cataloging-in-Publication Data

The Astronomy and astrophysics encyclopedia / edited by Stephen P. Maran.
p. cm.
Includes bibliographical references and index.
ISBN 0-442-26364-3
1. Astronomy—Encyclopedias. 2. Astrophysics—Encyclopedias.
I. Maran, Stephen P.
QB14.A873 1991
520′.3–dc20 91-23241
CIP

The Astronomy and Astrophysics Encyclopedia

Editorial Advisory Board

Contributors

Luis A. Aguilar, Instituto de Astronomía
Universidad Nacional Autonoma de México
Ensenada, B. C., Mexico
Star Clusters, Globular, Galactic Tidal Interactions

Syun-Ichi Akasofu, Geophysical Institute
University of Alaska, Fairbanks
Fairbanks, AK, U.S.A.
Earth, Magnetosphere and Magnetotail

Michael Allison
Goddard Institute for Space Studies
New York, NY, U.S.A.
Planetary Atmospheres, Dynamics

Johannes Andersen
Copenhagen University Observatory
Tølløse, Denmark
Binary Stars, Eclipsing, Determination of Stellar Parameters

William J. H. Andrewes, Collection of Historical Scientific Instruments
Harvard University
Cambridge, MA, U.S.A.
Time and Clocks

Robert Antonucci, Department of Physics
University of California, Santa Barbara
Santa Barbara, CA, U.S.A.
Active Galaxies and Quasistellar Objects, Blazars

Eugene H. Avrett
Harvard-Smithsonian Center for Astrophysics
Cambridge, MA, U.S.A.
Sun, Atmosphere

W. Ian Axford
Max-Planck-Institut für Aeronomie
Katlenburg-Lindau, Germany
Heliosphere

Marc A. Azzopardi
Observatoire de Marseille
Marseille, France
Stars, Wolf-Rayet Type

Donald C. Backer, Department of Astronomy
University of California, Berkeley
Berkeley, CA, U.S.A.
Pulsars, Millisecond

Daniel N. Baker, Laboratory for Extraterrestrial Physics
NASA/Goddard Space Flight Center
Greenbelt, MD, U.S.A.
Earth, Geomagnetic Storms

Norman H. Baker, Department of Astronomy
Columbia University
New York, NY, U.S.A.
Stellar Evolution, Pulsations

Sallie Baliunas
Harvard-Smithsonian Center for Astrophysics
Cambridge, MA, U.S.A.
Stars, Activity and Sunspots

Joshua E. Barnes, Institute for Astronomy
University of Hawaii
Honolulu, HI, U.S.A.
Galaxies, Disk, Evolution

John D. Barrow, Astronomy Centre
University of Sussex
Brighton, East Sussex, England, U.K.
Cosmology, Theories

Peter Barthel, Kapteyn Astronomical Institute
University of Groningen
Groningen, The Netherlands
Active Galaxies and Quasistellar Objects, Interrelations of Various Types

Gibor Basri, Department of Astronomy
University of California, Berkeley
Berkeley, CA, U.S.A.
Stars, T Tauri

Alan H. Batten, Dominion Astrophysical Observatory
National Research Council of Canada
Victoria, BC, Canada
Binary and Multiple Stars, Types and Statistics

Siegfried J. Bauer, Institute for Meteorology and Geophysics
University of Graz
Graz, Austria
Planetary Atmospheres, Escape Processes

Wendy Hagen Bauer, Department of Astronomy
Wellesley College
Wellesley, MA, U.S.A.
Stars, Red Supergiant

William A. Baum, Department of Astronomy
University of Washington
Seattle, WA, U.S.A.
Galaxies, Elliptical

J. Kelly Beatty
Sky & Telescope
Cambridge, MA, U.S.A.
Outer Planets, Space Missions

Steven V. W. Beckwith, Department of Astronomy
Cornell University
Ithaca, NY, U.S.A.
Stars, Circumstellar Disks

Jacob D. Bekenstein, Racah Institute of Physics
Hebrew University of Jerusalem
Jerusalem, Israel
Gravitational Theories

Roger A. Bell, Astronomy Program
University of Maryland
College Park, MD, U.S.A.
Stars, Temperatures and Energy Distributions

Claude Bertout
Institut d'Astrophysique
Paris, France
Protostars

Edmund Bertschinger, Department of Physics
Massachusetts Institute of Technology
Cambridge, MA, U.S.A.
Dark Matter, Cosmological

Michael S. Bessell, Mount Stromlo and Siding Spring Observatories
Australian National University
Woden, ACT, Australia
Magnitude Scales and Photometric Systems

John W. Bieber, Bartol Research Institute
University of Delaware
Newark, DE, U.S.A.
Interplanetary Magnetic Field

John H. Bieging, Department of Astronomy
University of Arizona
Tucson, AZ, U.S.A.
Stars, Young, Continuum Radio Observations

Giovanni F. Bignami, Istituto di Fisica Cosmica
Consiglio Nazionale delle Ricerche
Milan, Italy
Gamma-Ray Sources, Galactic Distribution

Bruno Binggeli, Astronomical Institute
University of Basel
Binningen, Switzerland
Virgo Cluster

David C. Black
Lunar and Planetary Institute
Houston, TX, U.S.A.
Planetary Systems, Formation, Observational Evidence

Jacques Blamont
Centre Nationale d'Etudes Spatiale
Paris, France
Observatories, Balloon

Victor M. Blanco
Cerro Tololo Inter-American Observatory
La Serena, Chile
Galactic Bulge

Roger D. Blandford
California Institute of Technology
Pasadena, CA, U.S.A.
Active Galaxies and Quasistellar Objects, Accretion

Robert C. Bless, Department of Astronomy
University of Wisconsin, Madison
Madison, WI, U.S.A.
Telescopes, Detectors and Instruments, Ultraviolet

Leo Blitz, Astronomy Program
University of Maryland
College Park, MD, U.S.A.
Galactic Structure, Interstellar Clouds

Albert Boggess
NASA/Goddard Space Flight Center
Greenbelt, MD, U.S.A.
Hubble Space Telescope

Michael J. Bolte, Dominion Astrophysical Observatory
National Research Council of Canada
Victoria, BC, Canada
Star Clusters, Globular

Roger M. Bonnet
European Space Agency
Paris, France
Scientific Spacecraft and Missions

Arnold I. Boothroyd, Kellogg Radiation Laboratory
California Institute of Technology
Pasadena, CA, U.S.A.
Stars, Carbon

Alan P. Boss, Department of Terrestrial Magnetism
Carnegie Institution of Washington
Washington, DC, U.S.A.
Solar System, Origin

Gregory D. Bothun, Department of Physics
University of Oregon
Eugene, OR, U.S.A.
Protogalaxies

Stephen P. Boughn, Observatory
Haverford College
Haverford, PA, U.S.A.
Background Radiation, Microwave;
Gravitational Radiation

Jérôme Bouvier
Canada-France-Hawaii Telescope Corporation
Kamuela, HI, U.S.A.
Stars, Young, Rotation

Stuart Bowyer, Department of Astronomy
University of California, Berkeley
Berkeley, CA, U.S.A.
Background Radiation, Ultraviolet

Joseph M. Boyce
National Aeronautics and Space Administration
Washington, DC, U.S.A.
Mars, Space Missions

Hale Bradt, Department of Physics
Massachusetts Institute of Technology
Cambridge, MA, U.S.A.
X-Ray Astronomy, Space Missions

Robert Brandenberger, Department of Physics
Brown University
Providence, RI, U.S.A.
Cosmology, Inflationary Universe

John C. Brandt, Laboratory for Atmospheric and Space Physics
University of Colorado
Boulder, CO, U.S.A.
Comets, Plasma Tails

Michel Breger, Institute of Astronomy
University of Vienna
Vienna, Austria
Stars, δ Scuti and Related Types

Asgeir Brekke, Auroral Observatory
University of Tromsø
Tromsø, Norway
Earth, Aurora

Donald E. Brownlee, Department of Astronomy
University of Washington
Seattle, WA, U.S.A.
Interplanetary Dust, Collection and Analysis

Guenter E. Brueckner, Space Sciences Division
Naval Research Laboratory
Washington, DC, U.S.A.
Solar Physics, Space Missions

Frederick C. Bruhweiler, Department of Physics
Catholic University of America
Washington, DC, U.S.A.
Interstellar Medium, Local

Marc W. Buie
Space Telescope Science Institute
Baltimore, MD, U.S.A.
Pluto and Its Moon

Leonard F. Burlaga, Laboratory for Extraterrestrial Physics
NASA/Goddard Space Flight Center
Greenbelt, MD, U.S.A.
Interplanetary Medium, Shock Waves and Traveling Magnetic Phenomena

W. Butler Burton, Sterrewacht
University of Leiden
Leiden, The Netherlands
Galactic Structure, Large Scale

P. Brendan Byrne
Armagh Observatory
Armagh, Northern Ireland, U.K.
Stars, Red Dwarfs and Flare Stars

Carla Cacciari
Osservatorio Astronomico di Bologna
Bologna, Italy
Star Clusters, Globular, Variable Stars

John J. Caldwell, Department of Physics
York University
North York, ON, Canada
Planetary and Satellite Atmospheres

Claude R. Canizares, Center for Space Research
Massachusetts Institute of Technology
Cambridge, MA, U.S.A.
Supernova Remnants, Evidence for Nucleosynthesis

Ray G. Carlberg, Department of Astronomy
University of Toronto
Toronto, ON, Canada
Galaxies, Spiral, Nature of Spiral Arms

Bernard J. Carr, School of Mathematical Sciences
Queen Mary and Westfield College, University of London
London, England, U.K.
Cosmology, Population III

Michael H. Carr
U.S. Geological Survey
Menlo Park, CA, U.S.A.
Mars, Surface Features and Geology

George R. Carruthers
Naval Research Laboratory
Washington, DC, U.S.A.
Sounding Rocket Experiments, Astronomical

T. Richard Carson, Department of Physics and Astronomy
University of St. Andrews
St. Andrews, Scotland, U.K.
Stars, Interior, Radiative Transfer

Andrea Carusi, Istituto di Astrofisica Spaziale, Reparto di Planetologia
Consiglio Nazionale delle Ricerche
Rome, Italy
Asteroid and Comet Families

Joseph P. Cassinelli, Department of Astronomy
University of Wisconsin, Madison
Madison, WI, U.S.A.
Stars, Winds

Clark R. Chapman
Planetary Science Institute
Tucson, AZ, U.S.A.
Asteroids

Gary A. Chapman, San Fernando Observatory
California State University, Northridge
Sylmar, CA, U.S.A.
Sun, Atmosphere, Photosphere

Robert D. Chapman
Bendix Field Engineering Corporation
Columbia, MD, U.S.A.
Comets, Types

Roger A. Chevalier, Department of Astronomy
University of Virginia
Charlottesville, VA, U.S.A.
Supernova Remnants, Evolution and Interaction with the Interstellar Medium

Guido Chincarini, Osservatorio di Brera
Universitá di Milano
Milan, Italy
Superclusters, Dynamics and Models

Stephen P. Christon, Applied Physics Laboratory
Johns Hopkins University
Laurel, MD, U.S.A.
Mercury, Magnetosphere

You-Hua Chu, Department of Astronomy
University of Illinois
Urbana, IL, U.S.A.
Nebulae, Wolf-Rayet

Edward L. Chupp, Department of Physics
University of New Hampshire
Durham, NH, U.S.A.
Telescopes, Detectors and Instruments, Gamma Ray

Barry G. Clark
National Radio Astronomy Observatory
Socorro, NM, U.S.A.
Radio Telescopes, Interferometers and Aperture Synthesis

David H. Clark
Science and Engineering Research Council
Surnden, England, U.K.
Supernovae, Historical

Donald D. Clayton, Department of Physics and Astronomy
Clemson University
Clemson, SC, U.S.A.
Cosmology, Cosmochronology

Thomas L. Cline, Laboratory for High Energy Astrophysics
NASA/Goddard Space Flight Center
Greenbelt, MD, U.S.A.
Gamma Ray Bursts, Observed Properties and Sources

Marshall H. Cohen, Department of Astronomy
California Institute of Technology
Pasadena, CA, U.S.A.
Active Galaxies and Quasistellar Objects, Superluminal Motion

Stephen Cole
American Geophysical Union
Washington, DC, U.S.A.
Mercury and Venus, Space Missions

Barney J. Conrath, Laboratory for Extraterrestrial Physics
NASA/Goddard Space Flight Center
Greenbelt, MD, U.S.A.
Jupiter, Atmosphere

Arthur N. Cox
Los Alamos National Laboratory
Los Alamos, NM, U.S.A.
Stars, Pulsating, Theory

Donald P. Cox, Department of Physics and Astronomy
University of Wisconsin, Madison
Madison, WI, U.S.A.
Interstellar Medium, Hot Phase

David L. Crawford, Kitt Peak National Observatory
National Optical Astronomy Observatories
Tucson, AZ, U.S.A.
Telescopes, Large Optical

Kyle M. Cudworth, Yerkes Observatory
University of Chicago
Williams Bay, WI, U.S.A.
Galactic Structure, Globular Clusters

Louis A. D'Amario, Jet Propulsion Laboratory
California Institute of Technology
Pasadena, CA, U.S.A.
Interplanetary Spacecraft Dynamics

Francesca D'Antona
Osservatorio Astronomico di Roma
Monteporzio, Italy
Stellar Evolution, Low Mass Stars

Gary Da Costa
Anglo-Australian Observatory
Epping, NSW, Australia
Star Clusters, Globular, Stellar Populations

Conard C. Dahn, Flagstaff Station
U.S. Naval Observatory
Flagstaff, AZ, U.S.A.
Stars, Distances and Parallaxes

Anton Dainty, Earth Resources Laboratory
Massachusetts Institute of Technology
Cambridge, MA, U.S.A.
Moon, Seismic Properties

Alexander Dalgarno
Harvard-Smithsonian Center for Astrophysics
Cambridge, MA, U.S.A.
Interstellar Clouds, Chemistry

Klaas de Boer
Sternwarte der Universität Bonn
Bonn, Germany
Stars, Horizontal Branch

Jean-Pierre De Greve, Astrofysich Instituut
Vrije Universiteit Brussel
Brussels, Belgium
Binary Stars, Semidetached

Valérie de Lapparent-Gurriet, Institut d'Astrophysique de Paris
Centre Nationale de la Recherche Scientifique
Paris, France
Voids, Extragalactic

Shuji Deguchi, Nobeyama Radio Observatory
National Astronomical Observatory of Japan
Nagano, Japan
Masers, Interstellar and Circumstellar

Armand H. Delsemme, Department of Physics and Astronomy
University of Toledo
Toledo, OH, U.S.A.
Comets

Pierre Demarque, Department of Astronomy
Yale University
New Haven, CT, U.S.A.
Star Clusters, Stellar Evolution

John R. Dickel, Department of Astronomy
University of Illinois
Urbana, IL, U.S.A.
Supernova Remnants, Observed Properties

Robert L. Dickman, Department of Physics and Astronomy
University of Massachusetts
Amherst, MA, U.S.A.
Bok Globules

Roger E. Diehl, Jet Propulsion Laboratory
California Institute of Technology
Pasadena, CA, U.S.A.
Interplanetary Spacecraft Dynamics

Stanislav G. Djorgovski
California Institute of Technology
Pasadena, CA, U.S.A.
Star Clusters, Globular, Mass Segregation

LeRoy E. Doggett, Nautical Almanac Office
U.S. Naval Observatory
Washington, DC, U.S.A.
Mercury and Venus, Transits

George A. Doschek
Naval Research Laboratory
Washington, DC, U.S.A.
Solar Activity, Solar Flares

Laurance R. Doyle, SETI Institute
NASA/Ames Research Center
Moffett Field, CA, U.S.A.
Planetary Rings

Alan Dressler
Observatories of the Carnegie Institution of Washington
Pasadena, CA, U.S.A.
Galaxies, Properties in Relation to Environment

Sandra L. Dueck, Jet Propulsion Laboratory
California Institute of Technology
Pasadena, CA, U.S.A.
Mars, Space Missions

George A. Dulk, Department of Astrophysical, Planetary and Atmospheric Sciences
University of Colorado
Boulder, CO, U.S.A.
Sun, Radio Emissions

René Dumont
Observatoire de Bordeaux
Floirac, France
Zodiacal Light and Gegenschein

Richard H. Durisen, Department of Astronomy
Indiana University
Bloomington, IN, U.S.A.
Binary and Multiple Stars, Origin

Palmer Dyal
NASA/Ames Research Center
Moffett Field, CA, U.S.A.
Planetary Magnetism, Origin

Michael G. Edmunds, Department of Physics
College of Cardiff, University of Wales
Cardiff, Wales, U.K.
Galaxy, Dynamical Models

Suzan Edwards, Five College Astronomy Department
Smith College
Northampton, MA, U.S.A.
Stars, Pre-Main Sequence, Winds and Outflow Phenomena

Heinrich E. Eichhorn, Department of Astronomy
University of Florida
Gainesville, FL, U.S.A.
Star Catalogs, Historic

David Eichler, Department of Physics
Ben Gurion University
Beer-Sheva, Israel
Neutrinos, Cosmic

Jean A. Eilek, Department of Physics
New Mexico Institute of Mining and Technology
Socorro, NM, U.S.A.
Active Galaxies and Quasistellar Objects, Jets

Jonathan H. Elias
Cerro Tololo Inter-American Observatory
La Serena, Chile
Telescopes, Detectors and Instruments, Infrared

Moshe Elitzur, Department of Physics and Astronomy
University of Kentucky
Lexington, KY, U.S.A.
Masers, Interstellar and Circumstellar, Theory

Bruce G. Elmegreen
IBM Watson Research Center
Yorktown Heights, NY, U.S.A.
Galaxies, Spiral, Structure

Debra Meloy Elmegreen
Vassar College Observatory
Poughkeepsie, NY, U.S.A.
Galaxies, Spiral, Structure

James P. Emerson, Department of Physics
Queen Mary and Westfield College, University of London
London, England, U.K.
Stars, Young, Millimeter and Submillimeter Observations

Oddbjørn Engvold, Institute of Theoretical Astrophysics
University of Oslo
Oslo, Norway
Telescopes and Observatories, Solar

Neal J. Evans II, Department of Astronomy
University of Texas at Austin
Austin, TX, U.S.A.
Interstellar Clouds, Molecular

Giuseppina Fabbiano
Harvard-Smithsonian Center for Astrophysics
Cambridge, MA, U.S.A.
Galaxies, X-Ray Emission

Anthony P. Fairall, Department of Astronomy
University of Cape Town
Rondebosch, South Africa
Superclusters, Observed Properties

Donald J. Faulkner, Mount Stromlo and Siding Spring Observatories
Australian National University
Weston Post Office, ACT, Australia
Star Clusters, Globular, Mass Loss

Michael W. Feast
South African Astronomical Observatory
Cape, South Africa
Stars, Pulsating, Overview

Eric D. Feigelson, Department of Astronomy
Pennsylvania State University
University Park, PA, U.S.A.
Stars, Pre-Main Sequence, X-Ray Emission

Robert A. Fesen, Department of Physics and Astronomy
Dartmouth College
Hanover, NH, U.S.A.
Supernovae, General Properties

Carl E. Fichtel, Laboratory for High Energy Astrophysics
NASA/Goddard Space Flight Center
Greenbelt, MD, U.S.A.
Background Radiation, Gamma Ray

Gerald J. Fishman
NASA/Marshall Space Flight Center
Huntsville, AL, U.S.A.
Gamma-Ray Astronomy, Space Missions

M. Pim Vatter FitzGerald, Department of Physics
University of Waterloo
Waterloo, ON, Canada
Galactic Structure, Optical Tracers

F. Michael Flasar, Laboratory for Extraterrestrial Physics
NASA/Goddard Space Flight Center
Greenbelt, MD, U.S.A.
Saturn, Atmosphere; Uranus and Neptune, Atmospheres

Gilles Fontaine, Département de Physique
Université de Montréal
Montreal, PQ, Canada
Stars, White Dwarf, Structure and Evolution

Wendy Laurel Freedman
Observatories of the Carnegie Institution of Washington
Pasadena, CA, U.S.A.
Hubble Constant; Stars, Cepheid Variable, Period-Luminosity Relation and Distance Scale

Michael J. Gaffey, Department of Geology
Rensselaer Polytechnic Institute
Troy, NY, U.S.A.
Asteroids, Earth Crossing

Catharine D. Garmany, Joint Institute for Laboratory Astrophysics
University of Colorado
Boulder, CO, U.S.A.
Stellar Associations, OB Type

Robert F. Garrison, David Dunlap Observatory
University of Toronto
Richmond Hill, ON, Canada
Stars, Hertzsprung-Russell Diagram

Mark S. Giampapa, National Solar Observatory
National Optical Astronomy Observatories
Tucson, AZ, U.S.A.
Stars, Magnetism, Observed Properties

Gerard F. Gilmore
Institute for Astronomy
Cambridge, England, U.K.
Solar Neighborhood

Riccardo Giovanelli, Arecibo Observatory
National Astronomy and Ionosphere Center
Arecibo, Puerto Rico
Clusters of Galaxies, Radio Observations

Giuliano Giuricin, Dipartimento di Astronomia
Universitá degli Studi di Trieste
Trieste, Italy
Galaxies, Local Supercluster

Christoph K. Goertz, Department of Physics and Astronomy
University of Iowa
Iowa City, IA, U.S.A.
Planetary Magnetospheres, Jovian Planets

Matthew P. Golombek, Jet Propulsion Laboratory
California Institute of Technology
Pasadena, CA, U.S.A.
Planetary Tectonic Processes, Terrestrial Planets

James L. Gooding
NASA/Lyndon B. Johnson Space Center
Houston, TX, U.S.A.
Meteorites, Composition, Mineralogy, and Petrology

Stephen T. Gottesman, Department of Astronomy
University of Florida
Gainesville, FL, U.S.A.
Galaxies, Barred Spiral

Christos D. Goudas, Department of Physics
University of Patras
Patras, Greece
Interstellar Medium, Stellar Wind Effects

John A. Graham, Department of Terrestrial Magnetism
Carnegie Institution of Washington
Washington, DC, U.S.A.
Magellanic Clouds

David F. Gray, Department of Astronomy
University of Western Ontario
London, ON, Canada
Stars, Atmospheres, Turbulence and Convection

J. Mayo Greenberg, Huygens Laboratory
University of Leiden
Leiden, The Netherlands
Interstellar Medium, Chemistry, Laboratory Studies

Donald A. Gurnett, Department of Physics and Astronomy
University of Iowa
Iowa City, IA, U.S.A.
Planetary Radio Emissions

Harm J. Habing, Sterrewacht
University of Leiden
Leiden, The Netherlands
Stars, Evolved, Circumstellar Masers

Margherita Hack, Dipartimento di Astronomia
Universitá degli Studi di Trieste
Trieste, Italy
Binary Stars, Atmospheric Eclipses

Douglas S. Hall, Dyer Observatory
Vanderbilt University
Nashville, TN, U.S.A.
Binary Stars, RS Canum Venaticorum Type

Alan W. Harris, Jet Propulsion Laboratory
California Institute of Technology
Pasadena, CA, U.S.A.
Planetary Rotational Properties

Hugh C. Harris, Flagstaff Station
U.S. Naval Observatory
Flagstaff, AZ, U.S.A.
Stars, BL Herculis, W Virginis, and RV Tauri Type

William E. Harris, Department of Physics
McMaster University
Hamilton, ON, Canada
Star Clusters, Globular, Extragalactic

Martin O. Harwit, National Air and Space Museum
Smithsonian Institution
Washington, DC, U.S.A.
Background Radiation, Infrared

Robert L. Hawkes, Department of Physics
Mount Allison University
Sackville, NB, Canada
Meteors, Shower and Sporadic

Timothy M. Heckman, Department of Physics and Astronomy
Johns Hopkins University
Baltimore, MD, U.S.A.
Galaxies, Nuclei

Alan Hedin, Laboratory for Atmospheres
NASA/Goddard Space Flight Center
Greenbelt, MD, U.S.A.
Planetary Atmospheres, Solar Activity Effects

Wulff D. Heintz, Department of Physics and Astronomy
Swarthmore College
Swarthmore, PA, U.S.A.
Stars, Masses, Luminosities, and Radii

William Herbst, Van Vleck Observatory
Wesleyan University
Middletown, CT, U.S.A.
Stellar Associations, R- and T-Type

Paul L. Hertz
Naval Research Laboratory
Washington, DC, U.S.A.
Star Clusters, Globular, X-Ray Sources

James E. Hesser, Dominion Astrophysical Observatory
National Research Council of Canada
Victoria, BC, Canada
Star Clusters, Globular

Dieter Heymann, Department of Geology and Geophysics
Rice University
Houston, TX, U.S.A.
Meteorites, Origin and Evolution

Ronald W. Hilditch, University Observatory
University of St. Andrews
St Andrews, Scotland, U.K.
Binary Stars, Detached

Jack G. Hills, Theoretical Astrophysics Group
Los Alamos National Laboratory
Los Alamos, NM, U.S.A.
Missing Mass, Galactic

Robert M. Hjellming
National Radio Astronomy Observatory
Socorro, NM, U.S.A.
Stars, Radio Emission

Paul T. P. Ho
Harvard-Smithsonian Center for Astrophysics
Cambridge, MA, U.S.A.
Galactic Center

Paul Hodge, Department of Astronomy
University of Washington
Seattle, WA, U.S.A.
Galaxies, Local Group

David Hollenbach, Space Science Division
NASA/Ames Research Center
Moffett Field, CA, U.S.A.
Shock Waves, Astrophysical

Joseph V. Hollweg, Space Science Center
University of New Hampshire
Durham, NH, U.S.A.
Interplanetary Medium, Wave-Particle Interactions

Thomas E. Holzer, High Altitude Observatory
National Center for Atmospheric Research
Boulder, CO, U.S.A.
Interplanetary Medium, Solar Wind

Lon L. Hood, Lunar and Planetary Laboratory
University of Arizona
Tucson, AZ, U.S.A.
Planetary Interiors, Terrestrial Planets

James R. Houck, Department of Astronomy
Cornell University
Ithaca, NY, U.S.A.
Infrared Astronomy, Space Missions

William B. Hubbard, Lunar and Planetary Laboratory
University of Arizona
Tucson, AZ, U.S.A.
Planetary Interiors, Jovian Planets

John P. Huchra
Harvard-Smithsonian Center for Astrophysics
Cambridge, MA, U.S.A.
Galaxies, Starburst

Hugh S. Hudson, Center for Astrophysics and Space Science
University of California, San Diego
La Jolla, CA, U.S.A.
Sun, Solar Constant

David W. Hughes, Department of Physics
University of Sheffield
Sheffield, England, U.K.
Meteoroid Streams, Dynamical Evolution

David G. Hummer, Joint Institute for Laboratory Astrophysics
University of Colorado
Boulder, CO, U.S.A.
Stars, Atmospheres, Radiative Transfer

Roberta M. Humphreys, Department of Astronomy
University of Minnesota
Minneapolis, MN, U.S.A.
Stars, High Luminosity

Colin M. Humphries
Royal Observatory, Edinburgh
Edinburgh, Scotland, U.K.
Telescopes, Wide Field

Deidre A. Hunter
Lowell Observatory
Flagstaff, AZ, U.S.A.
Galaxies, Irregular

James H. Hunter, Jr., Department of Astronomy
University of Florida
Gainesville, FL, U.S.A.
Galaxies, Barred, Spiral

Piet Hut
Institute for Advanced Study
Princeton, NJ, U.S.A.
Star Clusters, Globular, Gravothermal Instability

Icko Iben, Jr., Department of Astronomy
Pennsylvania State University
University Park, PA, U.S.A.
Nebulae, Planetary, Origin and Evolution

Wing-Huen Ip
Max-Planck-Institut für Aeronomie
Katlenburg-Lindau, Germany
Mercury and the Moon, Atmospheres

Frank P. Israel, Sterrewacht
University of Leiden
Leiden, The Netherlands
H II Regions

Kenneth A. Janes, Department of Astronomy
Boston University
Boston, MA, U.S.A.
Star Clusters, Open

Istvan Jankovics, Konkoly Observatory
Hungarian Academy of Sciences
Budapest, Hungary
Pre-Main Sequence Objects: Ae Stars and Related Objects

Edward B. Jenkins, Department of Astrophysical Sciences
Princeton University
Princeton, NJ, U.S.A.
Interstellar Medium, Chemical Composition

Jack R. Jokipii, Lunar and Planetary Laboratory
University of Arizona
Tucson, AZ, U.S.A.
Cosmic Rays, Propagation

Burton F. Jones, Lick Observatory
University of California, Santa Cruz
Santa Cruz, CA, U.S.A.
Astrometry

Christine Jones
Harvard-Smithsonian Center for Astrophysics
Cambridge, MA, U.S.A.
Clusters of Galaxies, X-Ray Observations

Stephen W. Kahler
Air Force Geophysics Laboratory
Hanscom Air Force Base, MA, U.S.A.
Solar Activity, Coronal Mass Ejections

Franz D. Kahn, Department of Astronomy
University of Manchester
Manchester, England, U.K..
H II Regions, Dynamics and Evolution

James B. Kaler, Department of Astronomy
University of Illinois
Urbana, IL, U.S.A.
Nebulae, Planetary, Energetics and Radiation;
Nebulae, Planetary, Origin and Evolution

Jonathan I. Katz, Department of Physics
Washington University
St. Louis, MO, U.S.A.
Stars, Neutron, Physical Properties and Models

Steven D. Kawaler, Department of Physics and Astronomy
Iowa State University
Ames, IA, U.S.A.
Stars, Rotation, Observed Properties

Demosthenes Kazanas, Laboratory for High Energy Astrophysics
NASA/Goddard Space Flight Center
Greenbelt, MD, U.S.A.
Radiation, High-Energy Interaction with Matter

William C. Keel, Department of Physics and Astronomy
University of Alabama
Tuscaloosa, AL, U.S.A.
Galaxies, Binary and Multiple, Interactions

Scott J. Kenyon
Harvard-Smithsonian Center for Astrophysics
Cambridge, MA, U.S.A.
Stars, Symbiotic

Yoji Kondo, Laboratory for Astronomy and Solar Physics
NASA/Goddard Space Flight Center
Greenbelt, MD, U.S.A.
Ultraviolet Astronomy, Space Missions

Arieh Königl, Astronomy and Physics Center
University of Chicago
Chicago, IL, U.S.A.
Jets, Theory of

William L. Kraushaar, Department of Physics
University of Wisconsin, Madison
Madison, WI, U.S.A.
Background Radiation, Soft X-Ray

Lawrence M. Krauss, Center for Theoretical Physics & Department of Astronomy
Yale University
New Haven, CT, U.S.A.
Neutrinos, Supernova

Stamatios M. Krimigis, Applied Physics Laboratory
Johns Hopkins University
Laurel, MD, U.S.A.
Interplanetary Medium, Solar Cosmic Rays

Kevin Krisciunas
Joint Astronomy Centre
Hilo, HI, U.S.A.
Telescopes, Historical

Julian H. Krolik, Department of Physics and Astronomy
Johns Hopkins University
Baltimore, MD, U.S.A.
Active Galaxies and Quasistellar Objects, Central Engine

Max Kuperus, Astronomical Institute
University of Utrecht
Utrecht, The Netherlands
Magnetohydrodynamics, Astrophysical

George Lake, Department of Astronomy
University of Washington
Seattle, WA, U.S.A.
Galaxies, Formation

Kurt Lambeck, Research School of Earth Sciences
Australian National University
Canberra, ACT, Australia
Earth, Figure and Rotation

Philippe L. Lamy
Laboratoire d'Astronomie Spatiale
Marseille, France
Comets, Dust Tails

Louis J. Lanzerotti
AT&T Bell Laboratories
Murray Hill, NJ, U.S.A.
Earth, Magnetosphere, Space Missions

Richard B. Larson, Department of Astronomy
Yale University
New Haven, CT, U.S.A.
Star Clusters, Globular, Formation and Evolution

Myron Lecar
Harvard-Smithsonian Center for Astrophysics
Cambridge, MA, U.S.A.
N-Body Problem

Martin Alan Lee, Space Science Center
University of New Hampshire
Durham, NH, U.S.A.
Shock Waves, Collisionless, and Particle Acceleration

David Leisawitz, Department of Astronomy
Pennsylvania State University
University Park, PA, U.S.A.
Interstellar Medium, Dust, Large Scale Galactic Properties

Kam-Ching Leung, Department of Physics and Astronomy
University of Nebraska
Lincoln, NE, U.S.A.
Binary Stars, Contact

Anny-Chantal Levasseur-Regourd, Service d'Aéronomie
Centre National de la Recherche Scientifique
Verrières-le-Buisson, France
Interplanetary Dust, Remote Sensing

Joel S. Levine, Atmospheric Sciences Division
NASA/Langley Research Center
Hampton, VA, U.S.A.
Earth, Atmosphere

Walter H. G. Lewin, Space Research Center
Massachusetts Institute of Technology
Cambridge, MA, U.S.A.
X-Ray Bursters

James W. Liebert, Steward Observatory
University of Arizona
Tucson, AZ, U.S.A.
Stars, Low Mass and Planetary Companions

Richard E. Lingenfelter, Center for Astrophysics and Space Sciences
University of California, San Diego
La Jolla, CA, U.S.A.
Cosmic Rays, Origin

Sarah Lee Lippincott, Sproul Observatory
Swarthmore College
Swarthmore, PA, U.S.A.
Binary Stars, Astrometric and Visual

William C. Livingston, National Solar Observatory
National Optical Astronomy Observatories
Tucson, AZ, U.S.A.
Solar Magnetographs

Thomas Lloyd Evans
South African Astronomical Observatory
Cape, South Africa
Stars, Cepheid Variable

George Lovi, Hayden Planetarium
American Museum of Natural History
New York, NY, U.S.A.
Constellations and Star Maps

Frank J. Low, Steward Observatory
University of Arizona
Tucson, AZ, U.S.A.
Observatories, Airborne

Paul D. Lowman, Jr., Geology and Geomagnetism Branch
NASA/Goddard Space Flight Center
Greenbelt, MD, U.S.A.
Moon, Lunar Bases; Moon, Space Missions

R. Earle Luck, Department of Astronomy
Case Western Reserve University
Cleveland, OH, U.S.A.
Stars, Chemical Composition

Janet G. Luhmann, Institute of Geophysics and Planetary Physics
University of California, Los Angeles
Los Angeles, CA, U.S.A.
Venus, Magnetic Fields

Jonathan Lunine, Lunar and Planetary Laboratory
University of Arizona
Tucson, AZ, U.S.A.
Satellites, Ices and Atmospheres

Marcos E. Machado
Observatorio Nacional de Fisica Cosmica
San Miguel, Argentina
Solar Activity, Sunspots and Active Regions, Theories

Barry F. Madore, Infrared Processing and Analysis Center
California Institute of Technology
Pasadena, CA, U.S.A.
Stars, Cepheid Variable, Period-Luminosity Relation and Distance Scale

Loris A. Magnani, Arecibo Observatory
National Astronomy and Ionosphere Center
Arecibo, Puerto Rico
Interstellar Medium, Dust, High Galactic Latitude

Paul D. Maley
NASA/Lyndon B. Johnson Space Center
Houston, TX, U.S.A.
Artificial Satellites, Observational Phenomena

Richard N. Manchester, Australia Telescope National Facility
Commonwealth Scientific and Industrial Research Organisation
Marsfield, NSW, Australia
Supernova Remnants and Pulsars, Galactic Distribution

Stephen P. Maran
Chevy Chase, MD, U.S.A.
Cosmology, Observational Tests;
Telescopes, Next Generation

John T. Mariska, E. O. Hulburt Center for Space Research
Naval Research Laboratory
Washington, DC, U.S.A.
Solar Activity, Solar Flares, Theories

Mikhail Marov
Keldysh Institute of Applied Mathematics
Moscow, U.S.S.R.
Mars, Atmosphere

Laurence A. Marschall, Department of Physics
Gettysburg College
Gettysburg, PA, U.S.A.
Coordinates and Reference Systems

John S. Mathis, Department of Astronomy
University of Wisconsin, Madison
Madison, WI, U.S.A.
Interstellar Medium, Dust Grains

Italo Mazzitelli, Istituto di Astrofisica Spaziale
Consiglio Nazionale delle Ricerche
Frascati, Italy
Stellar Evolution, Low Mass Stars

Patrick J. McCarthy
Observatories of the Carnegie Institution of Washington
Pasadena, CA, U.S.A.
Galaxies, High Redshift

George E. McCluskey, Jr., Division of Astronomy, Department of Mathematics
Lehigh University
Bethlehem, PA, U.S.A.
Binary Stars, Observations of Mass Loss and Transfer in Close Systems

Alfred S. McEwen
U.S. Geological Survey
Flagstaff, AZ, U.S.A.
Io, Volcanism and Geophysics

Christopher McKee, Department of Physics
University of California, Berkeley
Berkeley, CA, U.S.A.
Shock Waves, Astrophysical

D. Harold McNamara, Department of Physics and Astronomy
Brigham Young University
Provo, UT, U.S.A.
Stars, Cepheid Variable, Dwarf

Harry Y. McSween, Jr., Department of Geological Sciences
University of Tennessee
Knoxville, TN, U.S.A.
Meteorites, Classification

Stephen J. Meatheringham, Mount Stromlo and Siding Spring Observatories
Australian National University
Weston Post Office, ACT, Australia
Nebulae, Planetary, Extragalactic

H. Jay Melosh, Lunar and Planetary Laboratory
University of Arizona
Tucson, AZ, U.S.A.
Moon, Origin and Evolution

David R. Merritt, Department of Physics and Astronomy
Rutgers University
Piscataway, NJ, U.S.A.
Galaxies, Elliptical, Dynamics

Peter Mészáros, Department of Astronomy
Pennsylvania State University
University Park, PA, U.S.A.
Radiation, Scattering and Polarization

Peter Meyer, Enrico Fermi Institute and Department of Physics
University of Chicago
Chicago, IL, U.S.A.
Cosmic Rays, Observations and Experiments

Peter M. Millman, Herzberg Institute of Astrophysics
National Research Council of Canada
Ottawa, ON, Canada
Earth, Impact Craters

David G. Monet, Flagstaff Station
U.S. Naval Observatory
Flagstaff, AZ, U.S.A.
Astrometry, Techniques and Telescopes

Ronald L. Moore, Space Science Laboratories
NASA/Marshall Space Flight Center
Huntsville, AL, U.S.A.
Solar Activity, Sunspots and Active Regions, Observed Properties

Mark R. Morris, Department of Astronomy
University of California, Los Angeles
Los Angeles, CA, U.S.A.
Interstellar Medium, Galactic Center

David Moss, Department of Mathematics
University of Manchester
Manchester, England, U.K.
Stars, Magnetism, Theory

Jeremy R. Mould
California Institute of Technology
Pasadena, CA, U.S.A.
Galaxies, Dwarf Spheroidal

Michael J. Mumma, Laboratory for Extraterrestrial Physics
NASA/Goddard Space Flight Center
Greenbelt, MD, U.S.A.
Comets, Atmospheres

Richard Mushotzky, Laboratory for High Energy Astrophysics
NASA/Goddard Space Flight Center
Greenbelt, MD, U.S.A.
Active Galaxies and Quasistellar Objects, X-Ray Emission

Philip C. Myers
Harvard-Smithsonian Center for Astrophysics
Cambridge, MA, U.S.A.
Molecular Clouds and Globules, Relation to Star Formation

Andrew F. Nagy, Department of Atmospheric and Oceanic Science
University of Michigan
Ann Arbor, MI, U.S.A.
Planetary Atmospheres, Ionospheres

James Nemec, Department of Astronomy
University of British Columbia
Vancouver, BC, Canada
Stars, RR Lyrae Type

Ken'ichi Nomoto, Department of Astronomy
University of Tokyo
Tokyo, Japan
Supernovae, Type I, Theory and Interpretation

John E. Norris, Mount Stromlo and Siding Spring Observatories
Australian National University
Weston Post Office, ACT, Australia
Star Clusters, Globular, Chemical Composition

Laurence J. November, National Solar Observatory
National Optical Astronomy Observatories
Sunspot, NM, U.S.A.
Filters, Tunable Optical

Robert W. O'Connell, Department of Astronomy
University of Virginia
Charlottesville, VA, U.S.A.
Galaxies, Stellar Content

Christopher P. O'Dea
Space Telescope Science Institute
Baltimore, MD, U.S.A.
Intracluster Medium

Stephen L. O'Dell, Space Science Laboratory
NASA/Marshall Space Flight Center
Huntsville, AL, U.S.A.
Radio Sources, Emission Mechanisms

Augustus Oemler, Jr., Department of Astronomy
Yale University
New Haven, CT, U.S.A.
Clusters of Galaxies, Component Galaxy Characteristics

Takaya Ohashi, Department of Physics
University of Tokyo
Tokyo, Japan
X-Ray Astronomy, Space Missions

Edward Olszewski, Steward Observatory
University of Arizona
Tucson, AZ, U.S.A.
Stellar Associations and Open Clusters, Extragalactic

Jonathan F. Ormes, Laboratory for High Energy Astrophysics
NASA/Goddard Space Flight Center
Greenbelt, MD, U.S.A.
Cosmic Rays, Space Investigations

Frank Q. Orrall, Institute for Astronomy
University of Hawaii
Honolulu, HI, U.S.A.
Solar Activity

Patrick S. Osmer
National Optical Astronomy Observatories
Tucson, AZ, U.S.A.
Quasistellar Objects, in Galaxy Clusters and Superclusters

Bernard E. J. Pagel
Nordic Institute for Theoretical Atomic Physics
Copenhagen, Denmark
Cosmology, Big Bang Theory

R. Bruce Partridge, Department of Astronomy
Haverford College
Haverford, PA, U.S.A.
Background Radiation, Microwave

Jay M. Pasachoff, Hopkins Observatory
Williams College
Williamstown, MA, U.S.A.
Sun, Eclipses

John Peacock
Royal Observatory, Edinburgh
Edinburgh, Scotland, U.K.
Gravitational Lenses

Stanton J. Peale, Department of Physics
University of California, Santa Barbara
Santa Barbara, CA, U.S.A.
Satellites, Rotational Properties

Manuel Peimbert, Instituto de Astronomía
Universidad Nacional Autonoma de México
México D.F., Mexico
Interstellar Medium, Chemical Composition, Galactic Distribution

Carolyn Collins Petersen
Denver, CO, U.S.A.
Uranus and Neptune, Satellites

Bradley M. Peterson, Department of Astronomy
Ohio State University
Columbus, OH, U.S.A.
Quasistellar Objects, Absorption Lines

Jeffrey B. Plescia, Jet Propulsion Laboratory
California Institute of Technology
Pasadena, CA, U.S.A.
Jupiter and Saturn, Satellites

Kenneth A. Pounds, Department of Physics
University of Leicester
Leicester, England, U.K.
X-Ray Astronomy, Space Missions

Richard H. Price, Department of Physics
University of Utah
Salt Lake City, UT, U.S.A.
Black Holes, Theory

Carlton Pryor, Department of Physics and Astronomy
Rutgers University
Piscataway, NJ, U.S.A.
Star Clusters, Globular, Binary Stars

Peter J. Quinn, Mount Stromlo and Siding Spring Observatories
Australian National University
Woden Post Office, ACT, Australia
Galaxies, Elliptical, Origin and Evolution

Reuven Ramaty, Laboratory for High Energy Astrophysics
NASA/Goddard Space Flight Center
Greenbelt, MD, U.S.A.
Sun, High-Energy Particle Emissions

Joanna M. Rankin, Department of Physics
University of Vermont
Burlington, VT, U.S.A.
Pulsars, Observed Properties

Kavan U. Ratnatunga, Space Data and Computing Division
NASA/Goddard Space Flight Center
Greenbelt, MD, U.S.A.
Galactic Structure, Stellar Kinematics

Gail A. Reichert
Universities Space Research Association
Greenbelt, MD, U.S.A.
Active Galaxies, Seyfert Type

Ronald J. Reynolds, Department of Physics
University of Wisconsin, Madison
Madison, WI, U.S.A.
Interstellar Medium

Edward J. Rhodes, Jr., Department of Astronomy
University of Southern California
Los Angeles, CA, U.S.A.
Sun, Oscillations

George H. Rieke, Steward Observatory
University of Arizona
Tucson, AZ, U.S.A.
Active Galaxies and Quasistellar Objects, Infrared Emission and Dust

Hans Ritter
Max-Planck-Institut für Physik und Astrophysik
Garching bei München, Germany
Binary Stars, Cataclysmic

William W. Roberts, Jr., Department of Applied Mathematics
University of Virginia
Charlottesville, VA, U.S.A.
Galactic Structure, Spiral, Interstellar Gas, Theory

Brian J. Robinson, Division of Radiophysics
Commonwealth Scientific and Industrial Research Organisation
Epping, NSW, Australia
Galactic Structure, Spiral, Observations

Alex W. Rodgers, Mount Stromlo and Siding Spring Observatories
Australian National University
Woden Post Office, ACT, Australia
Spectrographs, Astronomical

Luis F. Rodríguez, Instituto de Astronomía
Universidad Nacional Autonoma de México
México D.F., Mexico
Nebulae, Cometary and Bipolar

Colin A. Ronan
Hastings, East Sussex, England, U.K.
Calendars

Robert Rosner, Department of Astronomy and Astrophysics
University of Chicago
Chicago, IL, U.S.A.
Stars, High-Energy Photon and Cosmic Ray Sources

Janet Rountree, Technical Services Staff
Bolling Air Force Base
Washington, DC, U.S.A.
Stars, Spectral Classification

Christopher T. Russell, Institute of Geophysics and Planetary Physics
University of California, Los Angeles
Los Angeles, CA, U.S.A.
Venus, Magnetic Fields

Barbara S. Ryden
Canadian Institute for Theoretical Astrophysics
Toronto, ON, Canada
Cosmology, Galaxy Formation

I. Juliana Sackmann, Kellogg Radiation Laboratory
California Institute of Technology
Pasadena, CA, U.S.A.
Stars, Carbon

Robert E. Samuelson, Laboratory for Extraterrestrial Physics
NASA/Goddard Space Flight Center
Greenbelt, MD, U.S.A.
Planetary Atmospheres, Clouds and Condensates

William C. Saslaw, Department of Astronomy
University of Virginia
Charlottesville, VA, U.S.A.
Cosmology, Clustering and Superclustering

Blair D. Savage, Department of Astronomy
University of Wisconsin, Madison
Madison, WI, U.S.A.
Interstellar Medium, Galactic Corona

Malcolm P. Savedoff, Department of Physics and Astronomy
University of Rochester
Rochester, NY, U.S.A.
Stellar Evolution

GertJan Savonije, Astronomical Institute
University of Amsterdam
Amsterdam, The Netherlands
Binary Stars, X-Ray, Formation and Evolution

Roberto Scaramella
Scuola Internazionale Superiore de Studi Avanzati
Miramare-Trieste, Italy
Superclusters, Dynamics and Models

Bradley E. Schaefer, Laboratory for High Energy Astrophysics
Goddard Space Flight Center
Greenbelt, MD, U.S.A.
Gamma-Ray Bursts, Optical Flashes

Detlef Schönberner, Institut für Theoretische Physik & Sternwarte
Universität Kiel
Kiel, Germany
Stars, R Coronae Borealis

Richard D. Schwartz, Department of Physics
University of Missouri, St. Louis
St. Louis, MO, U.S.A.
Herbig-Haro Objects and Their Exciting Stars

Nicholas Z. Scoville, Department of Astronomy
California Institute of Technology
Pasadena, CA, U.S.A.
Interstellar Medium, Galactic Molecular Hydrogen

Alvin Seiff
San Jose State University Foundation
San Jose, CA, U.S.A.
Venus, Atmosphere

Patrick Seitzer
Space Telescope Science Institute
Baltimore, MD, U.S.A.
Cameras and Imaging Detectors, Optical Astronomy

Kristen Sellgren, Department of Astronomy
Ohio State University
Columbus, OH, U.S.A.
Nebulae, Reflection

Maurice M. Shapiro, Department of Physics and Astronomy
University of Maryland
College Park, MD, U.S.A.
Neutrino Observatories

Peter A. Shaver
European Southern Observatory
Garching bei München, Germany
Interstellar Medium, Radio Recombination Lines

Neil R. Sheeley, Jr.
Naval Research Laboratory
Washington, DC, U.S.A.
Sun, Atmosphere, Corona

Gregory A. Shields, Department of Astronomy
University of Texas at Austin
Austin, TX, U.S.A.
Active Galaxies and Quasistellar Objects, Emission Line Regions

Steven N. Shore
Computer Sciences Corporation
Greenbelt, MD, U.S.A.
Stars, Atmospheres

Frank H. Shu, Department of Astronomy
University of California, Berkeley
Berkeley, CA, U.S.A.
Protostars

William L. H. Shuter, Department of Physics
University of British Columbia
Vancouver, BC, Canada
Interstellar Medium, Galactic Atomic Hydrogen

Steven B. Simon, Department of Geophysical Sciences
University of Chicago
Chicago, IL, U.S.A.
Moon, Rock and Soil

Edward M. Sion, Department of Astronomy and Astrophysics
Villanova University
Villanova, PA, U.S.A
Stars, White Dwarf, Observed Properties

Arne Slettebak, Department of Astronomy
Ohio State University
Columbus, OH, U.S.A.
Stars, Be-Type

Raymond N. Smartt, National Solar Observatory
National Optical Astronomy Observatories
Sunspot, NM, U.S.A.
Coronagraphs, Solar

Graeme H. Smith, Lick Observatory
University of California, Santa Cruz
Santa Cruz, CA, U.S.A.
Star Clusters, Mass and Luminosity Functions

Malcolm G. Smith
Joint Astronomy Centre
Hilo, HI, U.S.A.
Quasistellar Objects, Statistics and Distribution

Myron A. Smith, Division of Astronomical Sciences
National Science Foundation
Washington, DC, U.S.A.
Stars, Beta Cephei Pulsations;
Stars, Nonradial Pulsation in B-Type

Ronald L. Snell, Five College Radio Observatory
University of Massachusetts
Amherst, MA, U.S.A.
Infrared Sources in Molecular Clouds

Herschel B. Snodgrass, Department of Physics
Lewis and Clark College
Portland, OR, U.S.A.
Sun, Magnetic Field

Lewis E. Snyder, Department of Astronomy
University of Illinois
Urbana, IL, U.S.A.
Radio Astronomy, Receivers and Spectrometers

Sabatino Sofia, Center for Solar and Space Research
Yale University
New Haven, CT, U.S.A.
Sun, Interior and Evolution

Baruch T. Soifer, Division of Physics, Mathematics, and Astronomy
California Institute of Technology
Pasadena, CA, U.S.A.
Galaxies, Infrared Emission

Daniel S. Spicer, Space Data and Computing Division
Goddard Space Flight Center
Greenbelt, MD, U.S.A.
Plasma Transport, Astrophysical

Henk C. Spruit
Max-Planck-Institut für Physik und Astrophysik
Garching bei München, Germany
Accretion

Paul D. Spudis
Lunar and Planetary Institute
Houston, TX, U.S.A.
Moon, Geology

Steven W. Stahler, Department of Physics
Massachusetts Institute of Technology
Cambridge, MA, U.S.A.
Stars, Population III;
Stellar Evolution, Pre-Main Sequence

Gary Steigman, Department of Physics
Ohio State University
Columbus, OH, U.S.A.
Antimatter in Astrophysics

Robert F. Stein, Department of Physics and Astronomy
Michigan State University
East Lansing, MI, U.S.A.
Sun, Atmosphere, Chromosphere

Robert E. Stencel, Center for Astrophysics and Space Astronomy
University of Colorado
Boulder, CO, U.S.A.
Stars, Circumstellar Matter

Robert A. Stern
Lockheed Palo Alto Research Laboratory
Palo Alto, CA, U.S.A.
Stars, Atmospheres, X-Ray Emission

S. Alan Stern, Center for Astrophysics and Space Astronomy
University of Colorado
Boulder, CO, U.S.A.
Comets, Nucleus Structure and Composition

Daniel R. Stinebring, Department of Physics
Oberlin College
Oberlin, OH, U.S.A.
Pulsars, Binary

Alan Stockton, Institute for Astronomy
University of Hawaii
Honolulu, HI, U.S.A.
Quasistellar Objects, Host Galaxies

Robert G. Stone, Laboratory for Extraterrestrial Physics
Goddard Space Flight Center
Greenbelt, MD, U.S.A.
Radio Astronomy, Space Missions

Robert G. Strom, Department of Planetary Sciences
University of Arizona
Tucson, AZ, U.S.A.
Mercury, Geology and Geophysics

Curtis J. Struck-Marcell, Department of Physics
Iowa State University
Ames, IA, U.S.A.
Galaxy, Chemical Evolution

Linda L. Stryker, Department of Arts and Sciences
Arizona State University, West
Phoenix, AZ, U.S.A.
Stars, Blue Stragglers

George W. Swenson, Jr., Everitt Laboratory
University of Illinois
Urbana, IL, U.S.A.
Radio Telescopes and Radio Observatories

Mark V. Sykes, Steward Observatory
University of Arizona
Tucson, AZ, U.S.A.
Interplanetary Dust, Dynamics

Victor Szebehely, Department of Aerospace Engineering and Engineering Mechanics
University of Texas at Austin
Austin, TX, U.S.A.
Three-Body Problem

Paula Szkody, Department of Astronomy
University of Washington
Seattle, WA, U.S.A.
Binary Stars, Polars (AM Herculis Type)

Ronald E. Taam, Department of Physics and Astronomy
Northwestern University
Evanston, IL, U.S.A.
Stellar Evolution, Binary Systems

Kenneth L. Tanaka
U.S. Geological Survey
Flagstaff, AZ, U.S.A.
Planetary Volcanism and Surface Features

Jean-Louis Tassoul, Département de Physique
Université de Montréal
Montreal, PQ, Canada
Stellar Evolution, Rotation

Guillermo Tenorio-Tagle
Max-Planck-Institut für Astrophysik
Garching bei München, Germany
Star Formation, Propagating

Susan Terebey, Infrared Processing and Analysis Center
California Institute of Technology
Pasadena, CA, U.S.A.
Interstellar Clouds, Collapse and Fragmentation

Yervant Terzian, Department of Astronomy
Cornell University
Ithaca, NY, U.S.A.
Nebulae, Planetary

Paul J. Thomas, Department of Physics and Astronomy
University of Wisconsin, Eau Claire
Eau Claire, WI, U.S.A.
Satellites

Peter C. Thomas
Cornell University
Ithaca, NY, U.S.A.
Satellites, Minor

M. Nafi Toksöz, Earth Resources Laboratory
Massachusetts Institute of Technology
Cambridge, MA, U.S.A.
Moon, Seismic Properties

Monica Tosi
Osservatorio Astronomico di Bologna
Bologna, Italy
Galaxies, Chemical Evolution

Laurence M. Trafton, Department of Astronomy
University of Texas at Austin
Austin, TX, U.S.A.
Planetary Atmospheres, Structure and Energy Transfer

Virginia Trimble, Department of Physics
University of California, Irvine
Irvine, CA, U.S.A.
Cosmology, Observational Tests

James W. Truran, Department of Astronomy
University of Illinois
Urbana, IL, U.S.A.
Stellar Evolution, Massive Stars

Bruce T. Tsurutani, Jet Propulsion Laboratory
California Institute of Technology
Pasadena, CA, U.S.A.
Comets, Solar Wind Interactions

R. Brent Tully, Institute for Astronomy
University of Hawaii
Honolulu, HI, U.S.A.
Distance Indicators, Extragalactic

Barry E. Turner
National Radio Astronomy Observatory
Charlottesville, VA, U.S.A.
Interstellar Medium, Molecules

Neil Turok, Department of Physics
Princeton University
Princeton, NJ, U.S.A.
Cosmology, Cosmic Strings

Anne B. Underhill, Department of Geophysics and Astronomy
University of British Columbia
Vancouver, BC, Canada
Binary Stars, Spectroscopic

Arthur R. Upgren, Van Vleck Observatory
Wesleyan University
Middletown, CT, U.S.A.
Stars, Proper Motions, Radial Velocities, and Space Motions

Giovanni B. Valsecchi, Istituto di Astrofisica Spaziale, Reparto di Planetologia
Consiglio Nazionale delle Ricerche
Rome, Italy
Comets, Dynamical Evolution

James A. Van Allen, Department of Physics and Astronomy
University of Iowa
Iowa City, IA, U.S.A.
Interplanetary and Heliospheric Space Missions

Michiel van der Klis, Astronomical Institute
University of Amsterdam
Amsterdam, The Netherlands
X-Ray Sources, Quasiperiodic Oscillators

Peter O. Vandervoort, Astronomy and Astrophysics Center
University of Chicago
Chicago, IL, U.S.A.
Stellar Orbits, Galactic

Mahendra S. Vardya
Tata Institute of Fundamental Research
Bombay, India
Stars, Long Period Variable

Gerrit L. Verschuur
Bowie, MD, U.S.A.
Galactic Structure, Magnetic Fields

Paolo Vettolani, Istituto di Radioastronomia
Consiglio Nazionale delle Ricerche
Bologna, Italy
Superclusters, Dynamics and Models

Robert V. Wagoner, Department of Physics
Stanford University
Stanford, CA, U.S.A.
Cosmology, Nucleogenesis

Robert M. Walker, McDonnell Center for the Space Sciences
Washington University
St. Louis, MO, U.S.A.
Meteorites, Isotopic Analyses

René A. M. Walterbos, Department of Astronomy
New Mexico State University
Las Cruces, NM, U.S.A.
Andromeda Galaxy

Wayne H. Warren Jr., National Space Science Data Center
Goddard Space Flight Center
Greenbelt, MD, U.S.A.
Star Catalogs and Surveys

Michael G. Watson, Department of Physics and Astronomy
University of Leicester
Leicester, England, U.K.
Black Holes, Stellar, Observational Evidence

Ronald F. Webbink, Department of Astronomy
University of Illinois
Urbana, IL, U.S.A.
Binary Stars, Theory of Mass Loss and Transfer in Close Systems

John P. Wefel, Department of Physics and Astronomy
Louisiana State University
Baton Rouge, LA, U.S.A.
Cosmic Rays, Acceleration

Volker Weidemann, Institut für Theoretische Physik & Sternwarte
Universität Kiel
Kiel, Germany
Stellar Evolution, Intermediate Mass Stars

Achim Weiss
Max-Planck-Institut für Physik und Astrophysik
Garching bei München, Germany
Stellar Evolution, Massive Stars

Martin Weisskopf
NASA/Marshall Space Flight Center
Huntsville, AL, U.S.A.
Telescopes, Detectors and Instruments, X-Ray

Paul R. Weissman, Jet Propulsion Laboratory
California Institute of Technology
Pasadena, CA, U.S.A.
Comets, Oort Cloud

William J. Welch, Radio Astronomy Laboratory
University of California, Berkeley
Berkeley, CA, U.S.A.
Stars, Young, Masers

Francois Wesemael, Département de Physique
Université de Montréal
Montreal, PQ, Canada
Stars, White Dwarf, Structure and Evolution

Michael J. West
Canadian Institute for Theoretical Astrophysics
Toronto, ON, Canada
Clusters of Galaxies

J. Craig Wheeler, Department of Astronomy
University of Texas at Austin
Austin, TX, U.S.A.
Supernovae, New Types

Ewen A. Whitaker, Lunar and Planetary Laboratory
University of Arizona
Tucson, AZ, U.S.A.
Moon, Eclipses, Librations, and Phases

David A. Williams, Department of Mathematics, Astrophysics Group
UMIST
Manchester, England, U.K.
Interstellar Extinction, Galactic

Beverley J. Wills, Department of Astronomy & McDonald Observatory
University of Texas at Austin
Austin, TX, U.S.A.
Quasistellar Objects, Spectroscopic and Photometric Properties

Rogier A. Windhorst, Department of Physics and Astronomy
Arizona State University
Tempe, AZ, U.S.A.
Radio Sources, Cosmology

George L. Withbroe, Space Physics Division
National Aeronautics and Space Administration
Washington, DC, U.S.A.
Sun, Coronal Holes and Solar Wind

Adolf N. Witt, Ritter Observatory
University of Toledo
Toledo, OH, U.S.A.
Diffuse Galactic Light

Lincoln Wolfenstein, Department of Physics
Carnegie Mellon University
Pittsburgh, PA, U.S.A.
Neutrinos, Solar

Sidney C. Wolff
National Optical Astronomy Observatories
Tucson, AZ, U.S.A.
Stars, Magnetic and Chemically Peculiar

Stanford E. Woosley, Department of Astronomy and Astrophysics
University of California, Santa Cruz
Santa Cruz, CA, U.S.A.
Supernovae, Type II, Theory and Interpretation

Rosemary F. G. Wyse, Department of Physics and Astronomy
Johns Hopkins University
Baltimore, MD, U.S.A.
Galactic Structure, Stellar Populations

Donald K. Yeomans, Jet Propulsion Laboratory
California Institute of Technology
Pasadena, CA, U.S.A.
Comets, Historical Apparitions

Judith S. Young, Department of Physics and Astronomy
University of Massachusetts
Amherst, MA, U.S.A.
Galaxies, Molecular Gas in

William J. Zealey, Department of Physics
University of Wollongong
Woolongong, NSW, Australia
Stars, Young, Jets

Jack Zirker, National Solar Observatory
National Optical Astronomy Observatories
Sunspot, NM, U.S.A.
Sun

Herbert A. Zook
NASA/Lyndon B. Johnson Space Center
Houston, TX, U.S.A.
Meteoroids, Space Investigations

Maria T. Zuber, Geodynamics Branch
NASA/Goddard Space Flight Center
Greenbelt, MD, U.S.A.
Venus, Geology and Geophysics

Foreword

Astronomy is the study of everything—with the sole exception of the planet Earth and what lies in it and on it. Because the Universe is so vast and the Earth so tiny, the exception is a trivial one, although understandably important for us. In its infancy and early childhood, astronomy was necessarily restricted to that tiny volume of the Cosmos that could be made out with the naked eye—the Sun, the Moon, five planets, some stars and an occasional meteor or comet. Only a few thousand stars are bright enough to be seen unaided, even by individuals with excellent eyesight. You could estimate the brightness and relative positions of the stars and planets, the Sun and Moon, look for regularities in their motion, and try to predict when a given configuration of bodies might repeat—perhaps in the far future. We were painfully ignorant even of what these celestial bodies were made of, and why they moved. We invested them with mystical and superstitious fancies of which contemporary astrology is a living fossil. Progress was slow in part because certain views were unacceptable to the religious authorities. Nevertheless, the regularities in the apparent motions in the heavens were so striking that, over the millennia, they called out for an explanation, an understanding of the celestial clockwork that made it all go. This endeavor—of astronomers in many nations and historical epochs—by fits and starts, down many blind alleys, with much heated debate and correction of errors has led not only to modern astronomy and astrophysics, but also to nothing less than the scientific world view, and, for better and for worse, our contemporary civilization.

The findings and perspectives of modern astronomy would not only dazzle those ancients who began our subject; they would, I think, have elicited disbelief: We observe objects so far away that their light set out on its intergalactic voyage before the Earth was formed; and this light reveals that the laws of nature are the same there as here. Our airplanes fly through the upper air collecting microscopic remnants of long dead comets. Meteorites are identified: this one comes from an asteroid, this one comes from Mars; this little meteoritic grain was formed billions of years ago in the dark between the stars. Every atom but primordial hydrogen and helium is traced to generative processes in the hearts of stars. We bounce radiowaves off planets to map volcanic mountains and cratered plains, and look for hidden oceans. Radio telescopes in Massachusetts and Argentina, supported by small contributions from a hundred thousand people all over the Earth, are searching for possible signals from alien intelligences on the planets of other stars. We postulate a connection between mass extinctions of life on Earth and the motions of comets in an invisible cloud far beyond the outermost planet, stirred up by passing stars. From the concentration and motion of stars in distinct galaxies, the presence of invisible and massive black holes is deduced. Radiowaves filling all of space are interpreted as the remnant of an immense explosion which initiated the present incarnation of the Universe; that cosmic background radiation is so uniform in all directions as to challenge astronomers who are finding that the matter in the Universe on the largest scales we can see is arranged into great non-uniform bubble-like structures. The Universe seems destined to expand forever—unless, some astronomers say, the Universe is mainly composed of a kind of matter that does not exist on Earth. We are beginning to understand the lives and deaths of worlds, stars, galaxies, and perhaps the Universe itself.

The entire electromagnetic spectrum from short gamma rays to long radiowaves is being examined to see what information the Universe is sending us. We have peppered the tops of the Earth's mountains with optical and infrared telescopes, have covered over valleys with wire to make immense radio telescopes, and have observed the neutrinos from an exploding star in a satellite galaxy of the Milky Way from observatories in mines far beneath the surface of the Earth. We have carried telescopes in airplanes, balloons, and Earth satellites above the obscuring atmosphere of the Earth. We have sent robotic ships to dozens of previously unknown worlds. We have landed on the Moon and two planets and brought back lunar samples. As I write, there is a spacecraft in orbit around Venus, mapping the surface hidden beneath its opaque clouds; another, having encountered Halley's Comet, is journeying to another comet; two others are on their way to peer down at the poles of the Sun and to drop an entry probe into the atmosphere of the giant planet Jupiter. And four artifacts of the human species are beyond the orbit of Pluto, seeking the boundary between the solar system and the interstellar medium, inexorably on their way to the realm of the stars.

This is an age of ferment in astronomy. Many old paradigms are being shaken. New insights into the intricacy and elegance of the Cosmos are being uncovered. These are signs of tempestuous adolescence in a subject rapidly maturing. Since astronomy does not, in the short run, put bread on the average person's table, it might seem surprising that it has been supported at so high a level. But astronomy excites our sense of wonder, answers many of the deepest questions that humans have ever asked, inspires young people to pursue science and provides a useful antidote for the anthropocentric conceit. It would be cost-effective at twice the price.

Carl Sagan

David Duncan Professor of Astronomy and Space Sciences,
and Director, Laboratory for Planetary Studies,
Cornell University.

Preface

This volume represents an attempt to provide an authoritative summary of current knowledge of astronomy and astrophysics, including the exploration of the solar system, at the outset of the final decade of the Twentieth Century. In selecting the 403 article topics, the editors have emphasized the subjects of recent and current investigation. The intended readers are scientifically literate lay persons, teachers, science writers and editors, and also professional scientists who need a ready source of introductions to astronomical subjects in which they themselves are not specialists.

The *Encyclopedia* represents the product of an extensive collaboration. The Associate Editors, Professors Hunten, Margon, and Salpeter, reviewed the overall plan for the book including the initial list of about 480 article titles, nominated the two dozen equally distinguished members of the Editorial Advisory Board, and provided much other help including reviewing some manuscripts. The Board members and some additional consultants reviewed and amended the topic lists within their individual areas of expertise, and nominated candidate and alternate authors on the basis of recognized accomplishments in research. Of the final 403 articles, 399 were written by these expert-nominated authors, and each of them has been carefully read by an advisor, associate editor, or other consultant or referee. Of the remaining four articles, three on planetary exploration were written by established science writers who regularly cover that subject (J. Kelly Beatty, technical editor of *Sky & Telescope*, Stephen Cole, managing editor of *EOS*, and free lance writer Carolyn Collins Petersen), and one article on telescopes was written by the Editor.

Of the original manuscripts for this book, one (originally drafted as a sample article by Professor Virginia Trimble) was written in late 1988, two in early 1991, and the remaining 400 in 1989 and 1990. A few articles have dual or multiple authors. At least one author of every article received and returned galley proofs in 1991, and a great many authors took that opportunity to bring their articles up to date on the latest developments. Galleys of every article were also carefully read and marked by the Editor: I have not hesitated to clarify the English, make minor corrections and additions, and add useful bibliographical citations, sometimes with further expert advice. For any error that may have inadvertently been introduced, I apologize to authors and readers alike.

All articles but one have detailed bibliographies, in most cases giving the titles of the listed articles and papers. The authors were encouraged to emphasize review articles and monographs in these sections, headed "Additional Reading." Some also cited papers reporting individual research results. All of the *Encyclopedia* articles are cross-referenced to other appropriate entries. Accordingly, the reader can think of any given entry in the *Encyclopedia* as a resumé atop a pyramid of information; the mid-levels of the pyramid consist of the other cross-referenced entries, which appear under the heading "See also;" the base of the pyramid consists of the works cited in the bibliographies attached to the given and cross-referenced entries.

It is a pleasure to acknowledge the collegial attitude that our authors, editors, and advisors took to the preparation of the individual manuscripts. Many of the authors contacted the writers of related entries to coordinate the respective contents of their articles, and many also called members of the Editorial Advisory Board and myself for direction beyond the author's guidelines and sample article that were provided to them. Many an author made valuable suggestions in areas beyond the scope of his or her own article, and they were much appreciated.

I thank Robert N. Ubell, the Managing Editor, and his capable associates, Elaine Cacciarelli and Barbara D. Sullivan, for their immense effort in soliciting and acquiring so many manuscripts, and the associated reviews, over such a short interval. Special thanks are due to a number of past and present editors at the publisher, Van Nostrand Reinhold, for encouraging this *Encyclopedia*: Eric Rosen convinced me to begin the project, Charles Hutchinson shepherded it during a critical period when it might well have folded, Marjan Bace took the vital step of engaging Robert Ubell Associates to assist in the elaborate task of corresponding near-simultaneously with more than 600 prospective and actual authors, advisors, and consultants, Judith R. Joseph, now President and Chief Executive Officer of Van Nostrand Reinhold, took charge at a critical juncture, and finally Steve Chapman, Robert Esposito, and Alberta Gordon, nursed the *Encyclopedia* to completion. Thanks also to Dr. Simon Mitton of Cambridge University Press, who took a lively interest in the book, and eventually arranged for the Press to become the co-publisher.

The expert copy editors and typesetters at Science Typographers, Inc., under the direction of Sarah Roesser, successfully accomplished the daunting task of combining the many emendations to the manuscripts and galleys that were made by authors, advisors, and the Editor. On the principle that every important step in the editorial process should be tested by a second party, I was very fortunate to persuade Sally Scott Maran, of the Board of Editors at *Smithsonian* magazine, to check the more than 1000 page proofs, to verify that the thousands of corrections by authors and editors had been faithfully entered into the final *Encyclopedia* wherever possible. Our children, Elissa, Enid, and Michael Maran, made, collated and stapled a comparable number of photocopies of proofs. Finally, a special note of thanks to the couriers, clerks, and foreign associates of the Federal Express Corporation, who expedited an enormous number of our shipments, even making a valiant attempt to track down an author who had last been heard from in South America and who was eventually traced to College Park, Maryland. To the best of my knowledge, no overnight letter nor package was lost; when one arrived a day late, the Corporation cheerfully canceled the charge.

This book is dedicated to the thousands of astronomers in every part of the world who have worked alone and together since ancient times to gather the knowledge that is described in the pages that follow.

Stephen P. Maran
Chevy Chase, Maryland

Accretion

Henk C. Spruit

Accretion is understood as the accumulation of mass onto a static object (such as a star) from its surroundings, due to the gravitational pull of the object. Stars, for example, are believed to form by the accretion of mass onto a protostellar object, a process that takes an estimated 10^5–10^6 years. The time scales on which objects grow by accretion can be very different. Neutron stars in x-ray binaries receive mass from their companions on time scales of 10^8–10^9 years; protoplanets and stars grow on a much shorter time scale. The shortest time scale for mass accretion is the free fall time scale $(GM/r^3)^{-1/2}$, where M is the total mass and r is the distance between the gas and the object accreting it. If the process is this fast, one does not speak of accretion but of gravitational collapse. Such fast processes typically happen in systems in which a previously existing balance between gravitational force and an internal pressure fails catastrophically. Examples are the collapse of a stellar core that initiates a supernova explosion and certain stages in the contraction of a protostellar cloud. In the remainder of this entry we confine ourselves to accretion, that is, to processes by which a central object grows on a long time scale compared to the free fall time.

Accreting objects are numerous in the universe and are of very different size and appearance. Some of them are among the most spectacular objects presently known. Thus the accreting massive black holes believed to exist in the centers of some galaxies are the most powerful energy sources in the universe. The strongest x-ray sources in our own galaxy are binaries in which a neutron star accretes mass from a normal companion star. Other examples are protostars and protoplanets. The physics of disk galaxies is also closely connected with that of accretion disks.

The oldest example of an accreting system that has occupied the attention of astrophysicists is the cloud from which both the sun and the planets formed. Though many early ideas on the formation of the solar system were very different, the current picture of a gaseous disk from which the sun and planets formed roughly simultaneously by accretion processes is also an old one; it goes back to Laplace. Though only remnants of the process, such as the planets and the Oort cloud of comets, remain today, this is an important example because it shows that star formation involves *disk-like* structures. One can estimate the thickness and radial extent of the star forming disk from the present dimensions in the solar system, and one can learn about physical processes in the protosolar disk by studying meteorites. Disk-like structures have also been observed more directly around some young stars.

With some exceptions (to be discussed) accretion is thought to occur generally in the form of disks. Though observations also point in this direction, the most compelling reasons are more theoretical. In a typical accretion situation, the size of the mass accreting object is very small compared with the distance from which mass accretes. Therefore, a small cloud of gas falling toward the object will usually miss it unless its initial velocity happened to be pointing very precisely in the direction of the object. In the more likely case where the initial velocity does not point exactly to the object, the cloud will instead go into an orbit around it, like a comet around the sun, and the gas will never end up on the object unless something else happens. This stability of orbits in the gravitational potential of a point mass is expressed by the conservation of angular momentum. In the absence of external forces the angular momentum of the cloud is constant in time and because orbits that touch the central object have an angular momentum near zero, the cloud cannot accrete as a whole onto the central object in the absence of external forces. It was realized early on, however, that this argument does not exclude that *parts* of the cloud accrete. If the cloud is assumed to be viscous, there are internal forces in it that reduce the angular momentum of one part while at the same time increasing that of another. As a result, an orbiting viscous cloud spreads over the entire plane of the orbit. That is, the cloud spreads into a disk. To a good approximation, the gas in the disk moves around the central object in circular Keplerian orbits. It can be shown that, if one waits long enough, the viscous friction between neighboring orbits causes almost all the gas in such a disk to accrete onto the central object while almost all the angular momentum is carried off to large distances by a vanishing fraction of the mass. The orientation of the disk plane is determined by the initial direction of the gas falling toward the central object. In close binaries, for example, the disks are formed by gas falling directly from the surface of the companion star. This gas moves in the same direction as the orbital motion of the companion, so that the disk forms in the orbital plane.

The assumption of a viscosity leads naturally to a picture of accretion in the form of disks. Conversely, observations indicating accretion in the form of disks have stimulated the assumption that something like a viscosity operates in accreting gas. Such models were first developed in the context of the formation of the solar system. More recent developments were stimulated by the discovery in the 1970s of x-ray binaries, which were soon recognized to be objects in which mass from an ordinary star is transferred in a disk-like fashion to a neutron star. To enable quantitative calculations for such objects, a viscous model was introduced by Nicholaj J. Shakura and Rashid A. Sunyaev and it has been the main basis for the interpretation of observations ever since. A severe problem with this hypothesis, however, has remained largely unsolved, namely: What causes the gas to behave as if it were highly viscous? The kinematic viscosity deduced from observations of cataclysmic variables, for example, is 13–15 orders of magnitude larger than the microscopic value due to collisions between the particles of the gas. Popular ideas involve turbulent viscosities due to some sort of hydrodynamic or hydromagnetic turbulence driven by the shear flow in the disk. Because turbulence is a problem of notorious difficulty, no convincing calculations of the efficiency or even the existence of such processes in disks exist at present. Other processes that do not involve any viscosity or small scale turbulent motions are also being investigated; in particular, angular momentum loss through a magnetically driven wind from the disk and angular momentum transport by spiral shock waves.

As the accreting gas gets closer to the central object, gravitational binding energy is released. Nearly half of this goes into the kinetic energy of orbital motion (for nearly Keplerian orbital motion); the other half is dissipated as heat (in the case of a viscous disk). This heat makes the disk luminous. Because the energy dissipated increases in inverse proportion to the distance to the central object, the inner parts of a disk are the hottest and most luminous. If the central object is a relativistic object (neutron star or black hole), that is, if the escape speed from its surface is comparable to the speed of light, a very high luminosity is possible with only a small rate of mass accretion. In relativistic terms, a part of the rest mass of the accreting gas is converted into radiation energy during the accretion process. The efficiency of this conversion can be substantial, about 30% for accretion onto a neutron star or black hole. This is a hundred times higher than in the nuclear burning of hydrogen into helium. For this reason, the nuclear explosions taking place in the gas accreted onto the surface of neutron stars are not very spectacular compared with the energy released during the accretion.

In the theory of accretion onto compact objects (white dwarfs, neutron stars, and black holes) the so-called Eddington luminosity L_E and an associated critical accretion rate $\dot{M}_E$ play important roles. If $\dot{M}$ is the mass accretion rate (g s^{-1}), κ the opacity (cm^2 g^{-1}) of the accreting gas, c the speed of light, R the radius of the accreting object, and M its mass, then

$$L_E = 4\pi cGM/\kappa, \qquad \dot{M}_E = 4\pi cR/\kappa.$$

When $\dot{M}$ approaches $\dot{M}_E$, the pressure of radiation near the central object becomes an important factor influencing the accretion process. For $\dot{M}$ larger than about $\dot{M}_E$ this pressure may expell gas as a radiation driven wind along the axis while accretion is still taking place in the disk plane. It can be shown that although gas can, in principle, be accreted at a rate much larger than $\dot{M}_E$ (especially by black holes), the resulting luminosity cannot exceed L_E by a large factor. In fact, the most luminous x-ray binaries have a luminosity near L_E (about 10^{38} erg s^{-1} or one million times the solar luminosity for a one solar mass accreter), and this may also be the case for quasars (but this is less certain because the mass of the central object is not well known in this case).

Several accreting objects also show evidence of strong *outflows*. Perhaps the most detailed observations to date are those of the CO outflows, jets, and Herbig–Haro objects associated with many protostellar objects. These observations show two oppositely directed outflows and sometimes inside this a well collimated jet in the same direction, a central luminous object at the origin of the flows, and indications of a disk or torus in a plane perpendicular to the flows. Very similar associations, but on a vastly larger scale, occur in the nuclei of active galaxies, and jets have also been found associated with some x-ray binaries. Because of this association, jets are believed to be due to a physical process intrinsic to accretion. Radiation pressure driven outflows may be caused by accretion near the Eddington rate. However, jets are also seen from objects that probably accrete well below the Eddington value. Also, highly collimated flows and relativistic flow speeds are hard to explain this way. A promising mechanism is the acceleration and collimation of a flow from the disk surface by magnetic fields anchored in the disk (a "magnetic slingshot" mechanism). Detailed calculations already exist showing how high collimation and relativistic speeds can be obtained.

Accretion often involves more than a disk. If the central object is magnetic the magnetic forces can be strong enough to disrupt the disk. Inside a critical distance (the magnetosphere radius) the gas is then dominated by magnetic forces and consequently flows down along the field lines. This happens in many accreting white dwarfs (the so-called polars and intermediate polars) and neutron stars (the x-ray pulsars). Outside the magnetosphere radius, accretion may still take place via a disk. In such magnetic field-guided accretion the gravitational energy is not released until the gas hits the surface of the central object at high speed. This takes place in the form of a stationary shock between the infalling gas and the surface of the object.

Another situation is that of a star moving through a cloud or through the stellar wind blown off by a nearby star (massive x-ray binaries). The flow of gas relative to the star is usually supersonic, but by its gravitational field it is still able to capture some of the gas flowing around it. The cross section of this process is πR_a^2, that is, the star catches all gas that passes within a distance R_a, the so-called (Bondi) accretion radius. If v is the star's speed with respect to the gas, it is of the order $R_a = GM/v^2$.

Accretion seems to be accompanied by the production of cosmic rays (highly energetic particles) in several objects. At least some active galactic nuclei radiate a significant fraction of their luminosity in the form of gamma rays near 1 MeV. To explain this, some theories assume that (part of) the accretion energy is first spent in the acceleration of fast particles, which then generate the observed gamma rays. Two galactic x-ray binaries, Her X-1 and Cyg X-3, may prehaps be sources of extremely energetic particles, which are observed at the Earth's surface as a component of cosmic rays. Standard accretion disks consist of a thermal plasma and do not account for the generation of such fast particles. Apparently, a part of the accretion energy can go into the production of a highly nonthermal plasma, even when a thermal disk is also present. Current theories involve the same processes as those invoked for cosmic rays in general (first order Fermi acceleration at shock fronts or direct acceleration in electric fields).

Additional Reading

Frank, J., King, A. R., and Raine, D. J. (1985). *Accretion Power in Astrophysics*. Cambridge University Press, Cambridge.

Lada, C. J. (1982). Energetic outflows from young stars. *Scientific American* **247** (No. 1) 74.

MacKeown, P. K. and Weekes, T. C. (1985). Cosmic rays from Cyg X-3. *Scientific American* **253** (No. 5) 40.

Meyer, F., Duschl, W., Frank, J., and Meyer-Hofmeister, E., eds. (1989). *Theory of Accretion Disks*. Kluwer Academic Publishers, Dordrecht.

Pringle, J. E. (1981). Accretion disks in astrophysics. *Ann. Rev. Astr. Ap.* **19** 137.

Shaham, J. (1987). The oldest pulsars in the universe. *Scientific American* **266** (No. 2) 34.

See also **Active Galaxies and Quasistellar Objects, Accretion; Binary Stars, Theory of Mass Loss and Transfer in Close Systems; Galaxies, Spiral; Protostars, Theory; Stars, Circumstellar Disks; Stars, Pre-Main Sequence, Winds and Outflow Phenomena.**

Active Galaxies, Seyfert Type

Gail A. Reichert

Seyfert galaxies are galaxies having bright compact nuclei and exhibiting in their spectra broad high-excitation emission lines arising from a wide range of ionization states. First systematically studied by Carl K. Seyfert in 1943, they are one of several classes of galaxies that show signs of intense and violent activity within their nuclei.

DISCOVERY

Examples of Seyfert galaxies have been known since the early 1900s. However, it was not until the middle 1960s, when it was realized that these galaxies might be the lower luminosity cousins of the mysterious quasars, that Seyfert galaxies began to attract widespread interest. At about the same time, the pioneering work by B. E. Markarian and collaborators greatly expanded the list of known Seyfert galaxies, and showed that nonthermal activity in galaxies is far from uncommon.

Searches for Seyfert galaxies usually rely on the fact that the continuum emission from the active nucleus differs markedly from the thermal emission seen from stars. Instead, the nuclear continuum extends to far higher (and lower) photon energies, and is often characterized as a power law. Markarian and his collaborators searched for galaxies with compact nuclei that were comparatively bright in the ultraviolet. Subsequent spectroscopy showed that about 10% of these "UV excess" galaxies were Seyfert galaxies. The search criteria now also include unusual radio, infrared, and x-ray continuum properties as compared to normal galaxies, as well as strong emission lines. Each type of survey yields its own characteristic sample of Seyfert galaxies. Today, more than 600 Seyfert galaxies have been identified, and more continue to be found.

Seyfert galaxies form a continuous sequence with quasistellar objects (QSOs) in their properties, and share many of their characteristics with other types of active galaxies. Indeed, all active galaxies may actually be the same kind of object, involving essentially the same physical phenomena but perhaps under different external and/or observational conditions. The term *active galactic nuclei* (or AGN) is commonly used to mean all active galaxies, including QSOs. Although the discussion here will concern only Seyfert galaxies per se, much of it will also apply for AGN in general.

SEYFERT GALAXY TYPES

Seyfert galaxies are separated into two basic types, according to the relative widths of emission lines arising from permitted transitions in hydrogen, helium, and so forth, and those arising from forbidden transitions. In Seyfert type 1 (hereafter abbreviated Sy 1) galaxies, the permitted lines are broad, typically with full widths at half maximum (FWHM) of several thousand to ten thousand kilometers per second, whereas the forbidden lines are much narrower, with typical FWHM of several hundred kilometers per second. In Seyfert type 2 (hereafter Sy 2) galaxies, the permitted and forbidden emission lines have comparable widths, similar to the forbidden line widths in Sy 1 galaxies.

The original scheme has since been expanded to distinguish Seyferts whose permitted lines show both narrow and broad components as type 1.*N*, with *N* ranging from 1 to 9. In these objects, the narrow components of the profiles resemble the lines in Sy 2 galaxies whereas the broad components resemble the lines in Sy 1s. The number *N* increases with the relative strengths of the narrow components. For example, Sy 1.5 galaxies clearly show both narrow and broad components, whereas in Sy 1.9s the broad components are quite weak. It is not clear whether the Seyfert types 1.*N* are subclasses of the first Seyfert type or separate classes intermediate between Seyfert types 1 and 2. A few objects have also been observed to change in their Seyfert types. Finally, a class of narrow line Sy 1 galaxies has been identified. These objects have permitted lines that are not much broader than the forbidden lines (FWHM ~ 1000–2000 km s^{-1}) and appear to be Sy 1s at the low end of the distribution in velocity width.

COMPARISON OF PROPERTIES BY TYPE

Is the separation into Seyfert types physically meaningful? The first step in answering this question is to compare other observable properties, such as radio brightness and size, nuclear luminosity, continuum spectrum, and so forth, and determine how these vary with type.

Emission Line Spectra

Overall, the emission line spectra of the various Seyfert types are very similar. They show the same atomic transitions and similar ratios of the emission line fluxes. However, there are subtle differences. For example, the Sy 1.8s and 1.9s resemble Sy 2s in the ratios of their narrow line fluxes, whereas the Sy 1.5s are more like Sy 1s. Sy 1.8s and 1.9s also tend to show large broad Hα to broad Hβ flux ratios, much greater than the ratios predicted by normal recombination theory. The large ratios indicate that dust may play an important role in reddening the broad line spectra of these objects. In contrast, the broad line ratios for Sy 1s and 1.5s tend to show little if any reddening.

Sy 1 galaxies also tend to show higher ionization transitions than most Sy 2s. Ionized species as high as [Fe x]λ6375 are seen in Sy 1s. Broad Fe II emission from many blended multiplets appears to be universally present in Seyferts with broad line components (including the narrow line Sy 1s), but so far has been observed in only a handful of Sy 2s. Broad Fe II emission is also observed in other classes of broad line AGN. There is some evidence that broad line radio galaxies tend to show weaker Fe II emission, but some of this may be due to increased line blending.

The forbidden lines have different velocity widths, such that lines of higher critical density (i.e., density at which the level collisionally deexcites) and/or ionization energy tend to be broader. Line width appears to correlate with ionization energy in Sy 1s and with critical density in Sy 2s, although there are many exceptions.

Luminosity of the Active Nucleus

The nuclei of Sy 1s and 1.5s tend to be more luminous than Sy 2 nuclei. At the highest luminosities there are no Sy 2s. Practically all QSOs are broad line objects.

Optical/Ultraviolet Continuum

The presence of a featureless, nonthermal optical/UV continuum is a defining characteristic of Seyfert galaxies. However, the nonthermal continuum is much weaker in Sy 2 galaxies than in Sy 1s. Sy 2s therefore tend to have redder colors than Sy 1s, and so are often missed from surveys selecting for blue colors or ultraviolet excess.

Host Galaxy

When the host galaxies of Seyfert nuclei can be classified, they are nearly always spiral or "barred" spiral galaxies. Many Seyfert nuclei are found in interacting galaxies and/or in galaxies with peculiar morphology. In contrast, strong radio galaxies are nearly all ellipticals.

Radio Brightness and Size

Seyfert galaxies tend to be faint radio sources, both fainter and less extended than the sources in radio galaxies. Sy 2 galaxies tend to be brighter radio sources, relative to their optical and ultraviolet emission, than Sy 1s. Sy 2s also tend to be more extended radio sources than Sy 1s. Sy 1.5s appear to be intermediate between Sy 1s and Sy 2s in both brightness and extent.

The Seyfert radio sources that are spatially resolved are nearly always linear in structure. A small number of Seyferts have been observed with high enough spatial resolution that the emission from the narrow lines can be mapped. In these cases, there is a very strong tendency for the radio and narrow line emission to be coaligned.

Infrared Emission

Seyfert galaxies are typically strong infrared emitters, but their infrared properties remain poorly understood. Most presently available data have insufficient spectral and spatial resolution to separate the nonthermal emission from the active nucleus from other sources, for example, thermal photospheric emission from stars in the host galaxy and thermal emission from dust heated either by the active nucleus or by the intense bursts of star formation often found in active galaxies. Present data indicate that the infrared-visible continuum spectra of Sy 1s are generally dominated by nonthermal emission from the active nucleus, whereas the infrared spectra of Sy 2s may be dominated by thermal emission from extended regions of hot dust.

Seyfert galaxies found in infrared surveys tend to be more heavily reddened than those found by UV excess. Dust opacities are many times lower in the infrared than in the visible or ultraviolet, so that infrared emission is far less affected by reddening. Many more previously unknown Sy 2s (which tend to be more heavily reddened than Sy 1s) have been discovered via infrared surveys than Sy 1s. Hence the "true" proportion of Sy 2s to broad line Seyferts is roughly 4:1, far higher than the 1:3 ratio found for UV excess surveys. Infrared surveys selected on the basis of "warm" (i.e., color temperatures of approximately hundreds of degrees kelvin) infrared colors also tend to be biased in favor of narrow line Sy 1s, which generally have warmer infrared colors than most Sy 1s.

X-Ray Emission

Active galaxies are almost always strong x-ray emitters relative to their optical brightness, and x-ray surveys have led to the identification of many previously unknown AGN. However, almost all of the Seyferts detected in x-rays show broad components to their emission lines at some level (although in some cases the broad components are extremely weak, weaker than those found in Sy 1.9s). Bona fide Sy 2s tend to be much weaker x-ray emitters than broad line Seyferts, typically by factors of $\sim 30-\gtrsim 100$. However, the x-ray spectra of the few that have been measured do not differ appreciably from the spectra of broad line Seyferts.

Although x-ray spectra do not appear to differ systematically by Seyfert type, the spectra of low luminosity objects tend to show the effects of photoelectric absorption by large amounts of intervening material. In some cases the absorbing material appears to cover only a fraction of the x-ray emitter. The spectra for higher luminosity objects rarely show such absorption. There is some evidence that suggests that the probability of intervening absorption increases as luminosity decreases, rather than the covered fraction or the intervening column density.

Optical Polarization

Optical polarization observations provide some of the clearest evidence linking Sy 1s and 2s. Only a small fraction of Seyfert galaxies emit polarized light, typically at the level of a few percent of the total emission. Polarized Sy 2s tend to have relatively higher polarization that is usually independent of wavelength, indicating that it is caused by electron scattering. Sy 1s generally have lower polarizations due to a variety of causes. In general, polarization from an unresolved source indicates intrinsic asymmetry. The distribution in polarization angle also appears to be bimodal. Sy 2s show polarization that is perpendicular to the radio structure, whereas Sy 1s show polarization that is parallel. Seyferts classified 1.*N* are polarimetrically similar to Sy 1s, as are low polarization QSOs.

Optical spectra in polarized light have been obtained for a few of the brightest Sy 2 galaxies, for example, NGC 1068. The polarized spectra of several show broad components in the permitted lines, Fe II emission, and featureless nonthermal continua. These Sy 2 galaxies must therefore contain "hidden" Sy 1 nuclei. Except for light scattered and polarized by external regions of hot electrons, the nuclei must be totally blocked from our view.

Variability

Active galaxies can vary substantially, both in continuum output and in the fluxes of the broad emission lines. Continuum variability on time scales as short as 10 min has been reported, whereas broad line fluxes typically vary on longer time scales of $\gtrsim 1$ week to a few months. Narrow line fluxes have been observed to vary in only one object, over a period of a few years. The relative levels of broad to narrow line components can change dramatically. Some objects have been observed to change in Seyfert classification from Sy 1s to 1.9s or even 2s, and vice versa. There are suggestions that the continua of lower luminosity objects *may* vary on faster time scales than those of higher luminosity objects, but the issue is far from settled.

Space Densities

The true relative space densities of the various Seyfert types are difficult to determine. UV-excess and x-ray surveys tend to be biased against Sy 2s in favor of broad line Sys, whereas infrared surveys tend to be biased in favor of Sy 2s and narrow line Sy 1s. Sy 1.9s can also be misclassified as Sy 2s if the spectra are of low signal/noise so that the broad components are not detected. Unbiased emission line surveys suggest that there are roughly as many Sy 1.8s and 1.9s as there are Sy 1s and 1.5s, and eight times as many Sy 2s.

PHYSICAL INTERPRETATIONS OF THE DIFFERENCES BETWEEN SEYFERT TYPES

What causes the observed differences between the various Seyfert types? The general picture is far from clear, because our understanding of the physical nature(s) of AGN remains limited. Several possibilities have been proposed:

1. *The various types of Seyfert galaxies may actually be fundamentally different kinds of objects.* This picture is unlikely because there are many basic similarities, not only between the different Seyfert types, but also between Seyfert galaxies and other classes of AGN.

2. *The different Seyfert types may actually be intrinsically the same kind of object, viewed under different external and/or observational conditions.* For example, Sy 2s may simply lack the denser, higher velocity regions of gas responsible for the broad emission line components seen in Sy 1s to 1.9s. Alternatively, the broad emission line (BLR) and continuum regions may also be present in Sy 2s, but obscured from the line of sight. Some Sy 2s are known to show otherwise "hidden" Sy 1 nuclei in polarized light and would appear as Sy 1s if viewed from a different direction. Obscuration (and reddening by associated dust) may also explain the differences in x-ray emission between Sy 1s and 2s, the wide range in infrared continuum shapes observed for Sy 1s, and the elongated shapes of the narrow emission line regions in Sy 2s. The problems with this hypothesis are the differences in Sy 1 and 2 radio properties, the presence of weak featureless continua (which should be blocked if the BLR is blocked) in Sy 2s, and the existence of Seyferts that appear to change in Seyfert type without accompanying changes in reddening or obscuration.

3. *The various Seyfert types may represent different evolutionary stages in the lifetime of a single kind of object.* For example, Sy 1s might evolve into Sy 2s via triggering of bursts of star formation that are left behind when the type 1 nucleus switches off. Alternatively, Sy 2 nuclei might be triggered by interactions between galaxies and might evolve to Sy 1s by expelling the dust from the BLR. The problems with these scenarios are again the differences in radio properties and the fact that some objects can change Seyfert type on short time scales of months. The relative numbers of Sy 2s versus Sy 1s also suggest that the active nuclei must spend most of their time switched off. If this were the case, however, we might expect the narrow line regions in Sy 1s and 2s to show very different excitation properties, which are not observed.

Additional Reading

Osterbrock, D. E. (1989). *Astrophysics of Gaseous Nebulae and Active Galactic Nuclei*. University Science Books, Mill Valley, CA.

Shipman, H. L. (1980). *Black Holes, Quasars, and the Universe*, 2nd ed. Houghton Mifflin, Boston, MA.

Weedman, D. W. (1977). Seyfert galaxies. *Annual Reviews of Astronomy and Astrophysics* **15** 69.

Weedman, D. W. (1986). *Quasar Astronomy*. Cambridge University Press, Cambridge.

Weymann, R. J. (1969). Seyfert galaxies *Scientific American*, **220** (No. 1) 28.

See also **Active Galaxies and Quasistellar Objects, Emission Line Regions; Active Galaxies and Quasistellar Objects, Infrared Emission and Dust; Active Galaxies and Quasistellar Objects, Interrelations of Various Types; Active Galaxies and Quasistellar Objects, X-Ray Emission.**

Active Galaxies and Quasistellar Objects, Accretion

Roger D. Blandford

Quasars, which were first discovered in 1963, are active regions in the centers of galaxies that are so luminous that they outshine the surrounding stars. Many astronomers have concluded that the source of their prodigious power is a massive black hole that attracts surrounding gas by its gravitational force and liberates gravitational energy as radiation. In the context of the black hole model for quasars, this inward drift of matter is known as accretion and the rate and the manner of the accretion in individual active galactic nuclei is a major factor in determining the type of object we observe.

ACCRETION ONTO BLACK HOLES

There is a natural limit, known as the Eddington limit and named after the famous astronomer, Sir Arthur Eddington, to the luminosity L that can be radiated by a compact object of mass M. This limit arises because both the attractive gravitational force acting on an electron–ion pair and the repulsive force due to radiation pressure decrease inversely with the square of the distance from the black hole. When the luminosity exceeds the Eddington limit, which is given by

$$L_{\text{Edd}} = \frac{4\pi G M m_p c}{\sigma_T},$$

the gas will be blown away by the radiation. (In this equation, G is the gravitational constant, m_p is the mass of a proton, c is the speed of light, and σ_T is the Thomson cross section or the effective area of an electron when it is illuminated by radiation.) Note that the Eddington limit is independent of distance from the compact object. Numerically, if we express the mass in units of the mass of the sun $M_\odot$ and the luminosity in units of the luminosity of the sun $L_\odot$, then

$$L_{\text{Edd}} = 30{,}000\left(\frac{M}{M_\odot}\right)L_\odot.$$

A bright quasar has a luminosity of about 10^{13} $L_\odot$, and so if it is to continue to attract gas to power itself, the central mass must exceed about 3×10^{8} $M_\odot$.

Most galaxies display evidence for central activity and so they may also contain black holes although they need not be as massive as the black holes in bright quasars. Indeed, we know that the maximum mass of a hypothetical black hole in the center of our Milky Way galaxy is only 3×10^{6} $M_\odot$. Correspondingly, most galaxies that contain massive black holes do not necessarily accrete gaseous fuel at a sufficient rate to maintain their luminosities at the Eddington value.

ACCRETION DISKS

Stars and gas are observed to rotate about the centers of galaxies. This implies that if and when gas accretes toward a black hole, the centrifugal force acting upon the gas will increase in importance relative to the gravitational force. The gas is then expected to settle into a rotating disk, known as an accretion disk, just like the gas in some mass transfer binary stars. If the gravitational field were spherically symmetrical, then there would be no reason for the orbiting gas to move in any particular plane. However, the stars in the centers of galaxies are not spherically distributed and probably define a preferred plane into which the disk can settle. In addition, the black hole itself would probably be spinning rapidly and general relativistic effects may twist the accretion disk into its equatorial plane. This need not coincide with that defined by the stars.

The orbital period of the gas in an accretion disk will change with radius, just like the orbital periods of the planets in the solar system. This implies that adjacent rings of gas will rub against each other and be subject to friction which will allow the gas to move toward the central black hole. Several sources of this decelerating frictional force have been suggested. It may be caused by turbulent motions of the gas; alternatively, it has been attributed to magnetic field lines that are stretched between one ring and the next. Other possibilities, which are more likely to operate in the outer parts of accretion disks, include the development of gas clouds, bars, and spiral arms. Many active galactic nuclei produce a pair of jets—two collimated outflows that carry gas away from the nucleus to the outer parts of the galaxy and beyond. It is widely believed that these jets are launched perpendicular to the central accretion disks. Other objects exhibit outflowing winds. It is possible that the creation of either jets or a wind might also produce a reaction force on the gas in the disk allowing the gas to sink inward toward the central black hole.

Whatever its origin, this frictional force is responsible for heating the gas in the disk which can then radiate. The source of the radiant energy is ultimately gravitational and up to about 10^{20} erg of energy may be released for every gram of gas that is accreted onto a black hole. (This is several hundred times more efficient than the nuclear processes occurring in stars.) Most of this energy will be released fairly close to the black hole, within a radius of typically 10^{15} cm for a massive black hole in a quasar. In order to fuel a bright quasar, gas must accrete at a rate of up to 10 $M_\odot$ yr^{-1}.

If there is enough gas around the black hole, then the escaping photons will be absorbed and reemitted several times before they escape. The characteristic frequencies of the radiation can be calculated from Stefan's law just as is done for a stellar atmosphere. For an active galactic nucleus, this turns out to be in the ultraviolet part of the spectrum, which is where most observed objects appear to be most luminous. However, not all this ultraviolet radiation need escape. Some of it will be intercepted by dense clouds and converted into the emission lines by which active galaxies are frequently recognized. More of it may be intercepted by dust grains in the outer parts of the disk and transformed into infrared radiation.

Accretion disks are probably endowed with a magnetic field and their orbital velocity is necessarily supersonic. It is therefore expected that they are embedded in very hot, though transparent, coronae, analogous to the solar corona. The magnetic field lines will be twisted and torn by the motion of the disk and this may lead to the acceleration of relativistic electrons, which some astronomers believe emit the x-rays and gamma rays that are seen coming from some active galaxies.

ORIGIN OF GAS

The large fueling rates required by the most energetic quasars demand a more copious source than normal stars, evolving in the body of the surrounding galaxy. Some astronomers believe that direct collisions between individual stars in the nucleus of the galaxy is responsible for releasing the gas. Another possibility, particularly relevant to lower power objects, is that stars that pass too close to the black hole itself may be torn apart by the tidal gravitational force exerted by the black hole.

However, recent observations suggest that quasar activity is actually triggered by interactions between galaxies. In some instances, a small galaxy makes a direct hit on a larger galaxy and is ingested and falls to the center of the larger galaxy. In other collisions, the incident galaxy may only strike a glancing blow, and there will only be a small transfer of mass. In fact, no mass transfer is necessary, and in most cases, just the gravitational perturbation due to the incident galaxy can be sufficient to trigger the formation of spiral arms and bars in the galaxy surrounding the active nucleus which may, in turn, drive the gas inward.

FUTURE PROSPECTS

The previous description of the workings of a quasar is still largely conjectural. This is mainly because it is not possible to resolve the smallest regions where most of the energy is released. Indeed, it is proving to be very hard to produce clear and unambiguous proof that massive black holes are present. In addition, it is still quite uncertain what is the source of the accreting gas and by what mechanism does it settle toward the black hole.

Fortunately, observations scheduled over the next five years may test the accreting black hole model and provide answers to these difficult questions. Very long baseline interferometry, performed with the VLBA and from space, should reveal finer detail in radio maps of nearby and distant active galactic nuclei and may even be able to trace the outer parts of accretion disks. The Hubble space telescope, with its unprecedented resolution at optical wavelengths, should be able to trace the central velocity dispersions of stars in nearby galaxies and thereby measure the central mass which ought to be a fuel-starved black hole in most instances. However, the greatest progress in our understanding may be less direct and come from observing gas in the outer parts of galaxies either accreting onto or flowing away from their nuclei. Only when we understand the accretion process in physical terms will we be able to account for the evolution of quasars, Seyfert galaxies, and radio galaxies.

Additional Reading

Balick, B. and Heckman, T. M. (1982). Extranuclear clues to the origin and evolution of activity in galaxies. *Ann. Rev. Astron. Ap.* **20** 431.

Meyer, F., Duschl, W. J., Frank, J., and Meyer-Hofmeister, E., eds. (1989). *Theory of Accretion Disks*. Kluwer Academic Publishers, Dordrecht.

Frank, J. H., King, A. R., and Raine, D. J. (1986). *Accretion Power in Astrophysics*. Cambridge University Press, Cambridge.

Shapiro, S. L., and Teukolsky, S. A. (1983). *Black Holes, White Dwarfs, and Neutron Stars: The Physics of Compact Objects*. Wiley, New York.

See also **Accretion; Active Galaxies and Quasistellar Objects, Jets; Black Holes, Theory; Galaxies, Nuclei**

Active Galaxies and Quasistellar Objects, Blazars

Robert Antonucci

Blazars are members of the family of active galactic nuclei and quasars, defined specifically by their strong optical polarization and variability. These unique defining properties seemed mysterious and even paradoxical in the 1960s and 1970s, but now there is a growing consensus that their behavior and their role among quasars is qualitatively understood. Many of the modern ideas started to emerge during an important meeting in 1978 (the Pittsburgh Conference on BL Lac Objects). This is a good place to pick up the historical thread.

BL Lac objects have historically been defined as point-like sources of optical radiation that show little or no line emission, and strong and variable brightness and polarization. Pittsburgh meeting participants made it clear that some nearby objects exhibit all of these properties, along with narrow emission lines of considerable strength. Because they did not seem fundamentally different from the original BL Lacs, they were generally accepted as members of the class. This was especially reasonable in light of the fact that the narrow emission line equivalent widths (strengths of the lines compared with that of the continuum) vary inversely with continuum flux. Without this unification, an object's class would sometimes be a function of time!

Optically violently variable (OVV) quasars presented a similar situation. They were defined as broad emission line objects which otherwise showed the characteristics of BL Lacs. In fact, in their contributions to the Proceedings, Joseph Miller and collaborators showed that very high signal-to-noise ratio spectroscopy of known BL Lacs sometimes reveals broad emission lines. Furthermore, some OVVs clearly look like BL Lacs when their continua are in high brightness states. These facts are closely related. Since 1978, several of the BL Lacs discussed by the meeting participants have shown broad emission lines when observed carefully in low states. Therefore, it is no longer possible to distinguish BL Lacs and OVVs in a rigorous way and the two classes were merged under the name *blazars*. (Of course, this does not imply that all blazars are intrinsically exactly the same.)

The old-fashioned view of blazars was that their high polarization and tiny sizes (from variability arguments) meant that they were bare quasars, with the fundamental energy generation process being observed directly, perhaps within a few gravitational radii of supermassive black holes. In this picture, ordinary quasars are surrounded by gas that reprocesses and depolarizes the radiation and damps out the variability.

Roger Blandford and Martin Rees presented a very different idea at Pittsburgh, an idea which has since had many successes and which prevails among most researchers today. The high polarization and "power law" spectra could naturally be produced by synchrotron radiation, as in the Crab nebula. However, Blandford and Rees pointed out the very severe constraints on any such model that result from the rapid optical variability and high observed luminosities. The variability seems to require that even the luminous sources are intrinsically tiny (light-days or less). However, the polarization requires that both the optical depth to electron scattering and the optical depth to the synchrotron self-absorption process must be low; the reason is that both of these processes destroy polarization. A source satisfying all of these constraints basically cannot be as luminous as those observed!

Now all of the constraints would be greatly alleviated if we made one assumption: Suppose the synchrotron sources are not stationary, but are moving in bulk at relativistic speeds toward Earth. (This idea is called the beam model.) Then two things happen. Because the radiation is "beamed forward" by special relativistic aberration, the observed fluxes are greatly boosted. Therefore, the luminosities in the rest frames are much less than was otherwise thought. Also, with the emitting volume moving toward Earth and nearly keeping up with its own past images, the rapid observed variability is partially an illusion. The variations have been compressed in time. In the rest frames they are substantially slower, so the sources can be rather larger than in a stationary model.

After Blandford and Rees' paper was written, the variability constraints became even stronger. Papers by Chris Impey and collaborators and by P. A. Holmes and collaborators reported studies of variability in the infrared. This is where blazars put out most of their energy. Now, independent of the emission mechanism, radiation from black hole accretion is generally not expected to vary on time scales shorter than the travel time of light across the event horizon of a maximally accreting black hole. Yet infrared monitoring showed such enormous apparent luminosities and such rapid variability that even this conservative expectation was violated in at least five cases!

The assumption that all blazars are moving relativistically toward Earth may seem ad hoc or even crazy. In fact, it is very reasonable. Blazars invariably have very bright compact radio cores, and these cores often show very strong evidence for such a scenario. It was well known since the work of Fred Hoyle, Geoffrey Burbidge, and Wallace Sargent in the 1960s that a stationary synchrotron model for compact radio sources was not tenable. Radio variability seems to require extremely compact sources, and from these sizes and the observed radio fluxes, the surface brightnesses can be calculated. These turned out to be far above the "Compton limit" at 10^{12} K in brightness temperature. A stationary synchrotron source must emit fantastically large and observationally excluded inverse-Compton x-ray emission in order to have such a high brightness temperature. Therefore, relativistic motion in the line of sight had already been invoked. The idea was confirmed when superluminal (apparent faster-than-light) motion of the milliarcsecond-scale radio jets was discovered.

The "time compression" of the observed variability was also verified by James Condon and B. Dennison. They showed that if the sources were really as small as naively expected from the radio variability data, they should have such small angular sizes that they should show interstellar scintillation (twinkling), and they do not!

Blandford and Rees supplied an astrophysical context for synchrotron sources undergoing relativistic bulk motion in the line of sight. They suggested that the sources were simply the bases of the jets of normal double radio galaxies and quasars that happened to point in our direction. After all, some of these objects must be oriented in that way. Beaming of radiation by the aberration effect referred to earlier boosts the radio core fluxes in such objects, so they should be greatly over-represented in flux-limited surveys.

M. Orr and I. Browne adopted a simplified version of Blandford and Rees' idea. They postulated that all blazars, other core-dominant radio sources, and normal double sources all have similar relativistic bulk speeds, that the motions are along straight lines, and that the jets are linear in shape (rather than, say, conical). They concluded that such a simple model was consistent with a variety of source count data. Finally, Orr and Browne gave the name *unified scheme* to the hypothesis that flat-spectrum core-dominant sources are just normal doubles seen end-on. (The flat-spectrum core-dominant sources are just a slightly larger superset of blazars.)

The hypothesis that blazars are double radio sources seen along their jet (symmetry) axes obviously predicts that the double lobes should be seen projected as halos on the strong radio cores. It was just becoming possible in the early 1980s to achieve the required dynamic range in interferometer maps that was needed to detect such halos. (Remember that the blazar radio cores are tremendously strong.) Several groups discovered significant diffuse radio emission around many sources; this includes work by R. T. Schilizzi and A. G. de Bruyn with the Westerbork telescope, and Wardle and collaborators and James Ulvestad and collaborators with the NRAO Very Large Array.

The author and Ulvestad carried out an exhaustive blazar mapping program with the VLA and discovered substantial diffuse radio emission in almost all cases. The emission had qualitatively the right power, morphology, and projected linear size for the unified scheme. They critically examined various counter arguments in the literature, and then showed that if the beam model is qualitatively correct, the unified scheme must be, too. Suppose the beam model is correct and the core radio flux is beamed into a small solid angle that includes the direction to Earth. Suppose also that the large diffuse sources discovered in association with blazars emit isotropically. (This is very likely for the large, diffuse, and often two-sided halos.) Some blazars have sufficient flux *in the diffuse radio halos alone* to qualify for inclusion in the flux-limited radio catalogs. Therefore, under our two hypotheses, blazars *not* directed at Earth would still be in the catalogs, but classified as something else. The only candidates are the normal double sources. In fact, statistically, many or most normal doubles would have to be misdirected blazars!

Two exciting recent developments need to be mentioned. First, according to the unified scheme, normal double quasars should show much lower speeds in their cores than blazars do (although they should still be superluminal). Sensitive, very long baseline interferometry experiments are now being carried out, and the speeds are, in fact, coming in at 1–5c rather than the 5–10c typical of blazars.

The second recent development also seems to be a great success for the beam model and the unified scheme. Luminous double radio sources have two lobes that are generally fairly similar in flux, but jets that are very dissimilar in flux. This is at first sight unexpected because the jets appear to be the source of energy feeding the lobes. In the beam model, the jet radiation asymmetry is nicely explained as the result of beaming of the radiation from the jet closer to the line of sight toward us and beaming of the far jet radiation away from us. This does not require the axis to be *very* close to the line of sight as the blazar phenomenology does. Now the exciting new development is that Robert Laing and collaborators have discovered that in almost every case, *one* of the radio lobes is depolarized by passage through a magnetoionic medium (or "Faraday screen"), so that the depolarized lobe would be past the screen and the polarized lobe would be on its near side. (The Faraday screen would then probably be associated with the host galaxy.) The near side determined in this way is essentially always the side with the strong radio jet! This seems to mean that the jet radiation is beamed forward. Other interpretations are still possible but most researchers feel that the discovery of Laing and collaborators is a tremendous boost for the beam model.

Finally, there is evidence that the normal double quasars and broad-line radio galaxies that lie *very close to the sky plane* are observed and classified as narrow-line radio galaxies, at least in some cases. The optical continuum sources and broad emission line regions are apparently obscured by opaque tori composed of dust clouds. The evidence comes from optical spectropolarimetry and from some statistical tests which seem to show that too few objects classified as quasars lie very close to the sky plane. These arguments are summarized and the appropriate references given in a recent review paper by the present author, which discusses orientation effects in radio-quiet objects as well.

Additional Reading

Antonucci, R. (1989). *Evidence for and Against Relativistic Beaming in Active Galactic Nuclei. Fourteenth Texas Symposium on Relativistic Astrophysics*. Academy of Sciences Press, New York.

Antonucci, R. and Ulvestad, J. (1985). Extended radio emission and the nature of blazars. *Astrophys. J.* **294** 158.

Hoyle, F., Burbidge, G., and Sargent, W. (1966). On the nature of the quasi-stellar sources. *Nature* **209** 751.

Orr, M. and Browne, I. (1982). Relativistic beaming and quasar statistics. *MNRAS* **200** 1067.

Wolfe, A. M., ed. (1978). *Pittsburgh Conference on BL Lac Objects* (Physics and Astronomy Department, University of Pittsburgh). Includes papers by J. Miller, H. French, and S. Hawley; J. Miller and H. French; and R. Blandford and M. Rees.

See also **Active Galaxies and Quasistellar Objects, Jets; Active Galaxies and Quasistellar Objects, Superluminal Motion; Galaxies, Nuclei.**

Active Galaxies and Quasistellar Objects, Central Engine

Julian H. Krolik

Objects with a great variety of names—QSOs (or quasars), blazars, Seyfert galaxies, radio galaxies, and sometimes LINERs (low ionization nuclear emission line galaxies)—are all grouped into the category *active galactic nuclei* (AGN) because they share a basic set of common properties: very small spatial extent (on the galactic scale), luminosity comparable to or greater than that of an entire galaxy, and substantial power radiated in frequency bands where stars emit very little if at all. In addition to this set subscribed to by all AGN, many show evidence for bulk motion at relativistic speeds. Somewhere inside each object there must be a system responsible for the tremendous amounts of energy released; because they share so many basic characteristics, it is generally thought that in each of the different varieties of active galaxy this "central engine" is built according to a basic design that is common to all. The "specifications" for this central engine are exactly this list of common properties, and we begin by briefly elaborating on them.

At present, observations only give upper limits on the sizes of these objects. Atmospheric "seeing" limits angular resolution of ground-based telescopes to ~ 1 arcsecond, corresponding to ~ 100 parsec (pc) in even the nearest AGN. Some AGN are strongly variable; in these, causality limits the size to the distance light can travel in a characteristic variability time. This limit is often considerably less than 1 parsec, but systematic studies of AGN variability are still in their infancy.

Active galactic nuclei can be found over a very wide range of luminosity. The all-time record is $\sim 10^{48}$ erg s^{-1}, or more than 10^4 times brighter than an average galaxy, but luminosities this large are quite rare. At redshifts around 2, AGN with luminosities $\sim 10^{46}$ erg s^{-1} existed in $\sim 1\%$ of galaxies, whereas at the present epoch a few percent of all galaxies contain AGN with luminosities $\sim 10^{44}$ erg s^{-1}. It is possible that somewhat weaker AGN are still more common.

Perhaps most remarkable of all, whereas stars emit nearly all of their power in a frequency band a mere factor of 3 wide, and the range of stellar temperatures broadens that range for a galaxy by no more than another factor of 3, most AGN produce roughly equal amounts of power per logarithmic frequency band all the way from the mid-infrared to hard x-rays—a span of 10^7 in frequency. The exceptions (radio galaxies) produce such a large ratio of very low frequency (radio) power to optical that they, too, could hardly be stars.

Whatever constitutes the central engine, it almost certainly must have a very large mass. Two arguments lead to this conclusion. First, because the force due to radiation pressure falls off with distance from the source in exactly the same inverse square fashion as gravity, there is a critical luminosity to mass ratio beyond which a self-gravitating and radiating structure cannot exist. This is called the Eddington luminosity, and is $\simeq 4\times10^4$ in units of solar luminosities per solar mass. From this argument we infer that the central engines of active galaxies must have a mass at least $\sim 10^6(L/10^{44}$ erg s$^{-1})$ $M_\odot$.

Second, the total active lifetime of an AGN must be at least $\sim 10^8$ yr. This is the minimum mean active lifetime derived from the observed frequency of AGN if all galaxies are occasionally active. It is possible that only a few percent of all galaxies have ever been active, but in that case the observed frequency of AGN means that they must have been active throughout the lifetime of the universe, $\sim 1\text{–}2\times10^{10}$ yr. Thus the minimum total energy released by an average AGN is $\sim 10^{60}$ erg. It is possible to estimate the minimum accumulated mass in the central engine by supposing that this energy was derived from processing some sort of "fuel." Chemical fuels release $\sim 10^{-9}$ of their rest-mass energy when burnt; nuclear reactions release $\sim 10^{-3}$. Only the conversion of gravitational potential energy into heat when matter falls into a relativistically deep potential well produces energy with efficiency approaching unity in rest-mass units: Accretion onto a neutron star releases a fraction $\simeq 0.1\text{–}0.2$, accretion onto a maximally rotating black hole can release up to $\simeq 0.29$. Even with these high efficiencies, the minimum accumulated mass for a typical AGN is still $\sim 10^7$ $M_\odot$, and if only a small fraction of galaxies ever become active, the minimum accumulated mass could be much greater than this.

It is the difficulty in understanding how such large masses are brought so close to the centers of galaxies that has led most astronomers to believe that the basic power source for an AGN is accretion into a relativistically deep potential, probably a massive black hole. Although a dense cluster of neutron stars cannot absolutely be ruled out, it seems less likely: If typical AGN are less than a few light-days across (as the variability would in some cases suggest), the cluster would have to be so dense that stellar collisions would cause collapse to a black hole in less than the minimum active lifetime of $\sim 10^8$ yr.

The principal hurdle in bringing so much mass so close to the center of a galaxy is dumping the matter's angular momentum. Average stars in galaxies have 10^5 times more angular momentum than the maximum permitted for accretion onto a central black hole, and it is hard to identify mechanisms that efficiently remove angular momentum from material orbiting in galaxies. By contrast, energy can be lost comparatively easily by radiation. For this reason, it is generally thought that material approaches the central black hole along trajectories in which the energy is the minimum consistent with an orbit of that angular momentum. These trajectories taken together form a flat disk. At large distances from the center, it is possible that global disturbances in the gravitational field of the host galaxy remove angular momentum from the accreting matter; at small distances, friction between material on neighboring orbits may cause a slow outward transport of angular momentum and an associated slow sifting inward of the matter in the disk.

Energy can be released and transformed into radiation in a variety of ways. Within the disk, the same friction causing the angular momentum transport also causes local heating. The energy source, of course, is the slow fall of material in the gravitational field of the central object. This heat can then be lost by thermal radiation. Because the greatest amount of gravitational potential energy is lost in the innermost rings of the disk, this inner region dominates the total power radiated, and its typical temperature ($\sim 10^5$ K) forces most of the photons radiated by the disk to emerge in the ultraviolet.

Nonthermal mechanisms are possible also. Indeed, to explain the very broad range of photon energies seen, they are probably required. These generally involve populations of relativistic electrons (and sometimes positrons) in which the numbers of particles with a particular energy are proportional to a power of the energy. Relativistic particles are often distributed in energy in this way because the only characteristic energy scale relevant to them is the particle rest mass. Relativistic electrons can create new photons by the synchrotron mechanism and they can also multiply the energy of already-existing photons by large factors as a result of inverse Compton scattering. Many suggestions have been made about how to produce such large quantities of very energetic electrons, but no consensus currently exists. Some, but not all, of these suggestions depend on accretion into a relativistic gravitational potential.

Electromagnetic effects are also likely to play a role in the energy release: The characteristic scale of magnetic field strength near the edge of the black hole is $\sim 10^4(L/L_E)(L/10^{46}$ erg s$^{-1})^{-1/2}$ G, though electric fields should (in most places) be efficiently shorted out by the high densities of ionized plasma. Magnetic fields threading the surface of the black hole and coupled to the external plasma can allow the rotational energy in the black hole to be tapped, and energy stored in the field itself can be released nonthermally if regions with oppositely directed field can be brought together to reconnect.

Our present understanding of the generation of bulk relativistic motions is even cruder. Possibilities include acceleration of plasma on magnetic field lines attached to a rotating black hole, hydrodynamic acceleration inside funnels formed by general relativistic dynamical effects along the rotation axis of a black hole, or acceleration by radiation pressure, but it is quite possible that the correct answer is something else altogether.

In sum, measured against the specifications for a central engine, massive black holes do better than any other model yet proposed. They are certainly sufficiently compact, having radii $\sim 10^{-5}(M/10^8$ $M_\odot)$ pc; very large luminosities can be produced with a minimum mass in fuel consumed; the high energies afforded by the depth of their potential wells help in the creation of relativistic particles that can radiate over a broad range of frequencies; and they potentially provide sites for the bulk acceleration of matter to relativistic speeds. However, there are few statements that can be made on this subject with great confidence.

Additional Reading

Begelman, M. C. (1989). Physics of the central engine. In *IAU Symposium 134: Active Galactic Nuclei*, p. 141. Kluwer Academic Publishers, Dordrecht.

Rees, M. J. (1984). Black hole models of active galactic nuclei. *Ann. Rev. Astron. Ap.* **22** 471.

Rees, M. J. (1990). Black holes in galactic centers. *Scientific American* **263** (No. 5) 56.

See also **Active Galaxies and Quasistellar Objects, Jets; Active Galaxies and Quasistellar Objects, X-Ray Emission; Cosmic Rays, Acceleration; Quasistellar Objects, Statistics and Distribution; Radiation, High-Energy Interaction with Matter; Radio Sources, Emission Mechanisms.**

Active Galaxies and Quasistellar Objects, Emission Line Regions

Gregory A. Shields

Twenty-five thousand light-years away, behind the stars of Sagittarius, the nucleus of our galaxy lies hidden by intervening gas and dust. Peering through the dust with instruments sensitive to radio and infrared radiation, astronomers are finding there a bizarre scene. Clouds of ionized gas, heated by ultraviolet radiation of unknown origin, orbit a black hole with the mass of three million suns. Striking as it is, this activity is dwarfed by the powerful active galactic nuclei (AGN) observed in some other galaxies. Drawing their energy from a source no bigger than the solar system, the brightest AGN far outshine their host galaxy, with its hundreds of billions of stars. Their radiation, spanning all bands of the electromagnetic spectrum, consists of a "continuum" and "emission lines." The continuum, with its energy smoothly distributed over all wavelengths, may be radiated by cosmic rays or by a hot disk of gas orbiting the black hole. The emission lines, occurring at discrete wavelengths in the infrared, optical, and ultraviolet, represent photons of light emitted by atoms in hot, ionized clouds of gas. The clouds' location, chemical composition, physical conditions, and motion can be derived by careful observation and theoretical analysis. Such information may help to answer the following basic questions: Do AGN really contain giant black holes? Does the energy derive from gas falling into these holes? What is the source of the gas?

Galaxies containing an active nucleus are called *active galaxies*. These include Seyfert galaxies, BL Lac objects, radio galaxies, and QSOs. In 1943, Carl Seyfert described a class of spiral galaxies, otherwise normal, that have brilliant, starlike nuclei. The spectrum of a Seyfert nucleus shows a continuum whose energy distribution, polarization, and lack of absorption lines imply that it is not ordinary starlight. The continuum now is known to extend from radio to gamma ray frequencies. Superimposed on this "nonthermal" continuum are broad emission lines at the wavelengths familiar from nebulae in our galaxy but somewhat smeared out in wavelength. QSOs are believed to be active galaxies in which the image of the host galaxy is lost in the glare of the brilliant nucleus. AGN occur over a huge range in luminosity, from 10^{14} $L_{\odot}$ for the brightest QSOs to only 10^{6} $L_{\odot}$ for the dimmest known Seyfert nucleus (fainter than the brightest normal stars), where $L_{\odot}$ denotes the luminosity of the Sun. The character of the continuum and emission lines of AGN is preserved over this entire range of luminosity.

The emission lines of AGN fall into two categories according to the line profile (see Quasistellar Objects, Spectroscopic and Photometric Properties). The profile describes how the intensity of the line radiation varies with wavelength offset from the central wavelength. The light is spread out in wavelength because the motion of the gas in various parts of the emitting region causes a Doppler shift. The blue (short wavelength) wing of the profile corresponds to photons emitted by gas moving toward the observer, and vice versa. The profile's breadth in wavelength gives the speed of the gas. Typically, AGN show a combination of "broad lines," with widths up to $\sim 10{,}000$ km s^{-1} and "narrow lines" with widths ~ 1000 km s^{-1}. These are attributed to two distinct regions in the nucleus. The broad-line region (BLR) involves fast-moving clouds of dense gas near the central engine, whereas the narrow-line region (NLR) involves more remote clouds with lower velocities and densities. Both regions are believed to be powered by the nonthermal continuum.

THE NARROW-LINE REGION

The relative intensities of the narrow lines resemble those of planetary nebulae. The intensity ratios of certain pairs of lines give the gas temperature and density (atoms per cm^3). Typical values for the NLR are $T \approx 1\text{–}2 \times 10^4$ K and $N \approx 10^3\text{–}10^6$ cm^{-3}. The elements are spread over a wide range of ionization states from O^0 to Fe^{+9}. The most prominent lines are from H, He, C, N, O, Mg, Si, S, and Fe, once or several times ionized. In planetary nebulae, the gas is heated and ionized by "photoionization," in which ultraviolet photons from the central star eject electrons from atoms in the surrounding gas. A similar model explains the narrow lines of AGN, but here the ionizing radiation is the high-frequency extrapolation of the continuum observed in the optical and ultraviolet. Computer models of the physical processes in the photoionized clouds, called *photoionization models*, can explain the line intensities if the gas has a normal chemical composition. The NLR gas typically extends to distances of a few hundred light-years from the continuum source. The emitting clouds fill only a small fraction of this volume; and there is uncertainty about what if anything, fills the rest of the volume. The mass of gas in the NLR is $\sim 10^5$ $M_{\odot}$, only a tiny fraction of the combined mass of the stars in the same region. Arguments involving the asymmetry of the line profiles (unequal brightness of the red and blue wings) suggest that the gas is moving outward, but its origin is uncertain. Some observations indicate that within the NLR, denser gas moves at higher velocities.

THE BROAD-LINE REGION

The BLR is interesting because of its proximity to the center of the nucleus. Do its high velocities represent orbital motion around the black hole, infall toward it, or outflow? Does this gas actually fuel the energy source, or is it merely a by-product? Can its motion be used to weigh the black hole?

The BLR gas is very dense by nebular standards. The absence of forbidden emission lines such as [O III] $\lambda 5007$ implies $N \geq 10^8$ cm^{-3}. On the other hand, the presence in the spectrum of intercombination lines such as [C III] $\lambda 1909$ implies $N \leq 10^{10}$ cm^{-3}. Remarkably, the implied value of $\sim 10^9$ cm^{-3} applies over the whole range of AGN luminosities. Gas at these densities radiates efficiently, so that only a few solar masses of gas are needed to produce the broad-line emission.

The line intensities can be fairly well explained by photoionization models. The ionization structure of the emitting clouds consists of a surface layer highly ionized by the ultraviolet continuum and a deeper layer ionized by x-rays. This layer emits the Balmer lines and continuum of hydrogen and lines of O I, Mg II, Ca II, and Fe II. The complex physics of x-ray ionization and radiative transfer in optically thick lines requires elaborate computer programs. The continuum of AGN may be beamed in certain directions; and this contributes to the uncertainty as to whether or not the observed continuum, extrapolated to high frequencies, contains enough energy to power the BLR. Also uncertain is the fraction of the continuum intercepted by the BLR; some arguments suggest values $\sim 10^{-1}$, with larger values in lower luminosity AGN.

Photoionization models involve the "ionization parameter," $U \equiv \phi_i / Nc$, where ϕ_i (photons cm^{-2} s^{-1}) is the flux of ionizing radiation striking a typical cloud and c is the speed of light. Larger values of U correspond to higher degrees of ionization in the gas, and values $U \sim 10^{-2}$ best reproduce the observed spectrum. This value varies little over the entire range of AGN luminosities, which in turn implies a uniform value of ϕ_i. The radius of the BLR, R_B, and the central ionizing luminosity, L_i, are related to ϕ_i through the inverse square law. This gives $R_B \approx 1$ ly for a typical QSO, and R_B must vary roughly as $L_i^{1/2}$ over the range of AGN luminosities. The gas emitting the broad lines occupies a tiny fraction of the volume within R_B. The traditional picture is a swarm of clouds, only $\sim 10^{-4}$ ly thick, immersed in some hot medium whose pressure holds the clouds intact.

An alternative measure of R_B has been used in recent years. The continuum of some AGN varies in brightness on time scales of days to years. The emission lines should change the brightness accordingly, but with a delay corresponding to the time required for light to cross the BLR. Observations confirm that more luminous AGN have larger R_B; but in a given AGN, this method gives values of R_B as much as 10 times smaller than the "U method" described previously. This discrepancy is a puzzle.

The nature of the BLR's motion is unclear. The clouds are opaque at the wavelengths of some emission lines, and simple arguments then imply that the line profile should be asymmetrical in one way for outflow and in the reverse way for infall. Generally, the observed profiles are symmetrical. Another argument involves time variations. For outflow, the blue wing of the line should change before the red wing, and conversely for infall. Observed variations usually are symmetrical. These observations argue against pure infall or outflow but allow gas moving in elliptical or circular orbits. One picture is that the line emission comes, not from clouds, but from the surface of an "accretion disk" of gas extending from near the hole's horizon to radii ≥ 1 ly. (The horizon, or boundary, of a black hole of mass 10^8 $M_\odot$ has a radius comparable with the Earth's orbit around the Sun.) The AGN derive their energy from the inner disk, but radiation intercepted by the outer disk could produce the lines. This model gives a unified explanation of the lines and continuum, but it has trouble fitting the line profiles in detail. In another model, the broad lines come from interstellar clouds that fall near the hole just as comets orbit near the Sun. A more exotic theory suggests that the broad lines are emitted by the gaseous debris of stars ripped apart by gravity as they pass too close to the black hole ("tidal disruption").

If the broad-line clouds are orbiting a black hole, or falling into it, the mass of the hole is given by

$$M = R_B v^2 / G,$$

where v is the clouds' velocity. The masses implied are reasonable for the values of R_B derived from variability, 10^8–10^9 $M_\odot$ for QSOs and 10^6–10^7 $M_\odot$ for Seyfert galaxies. Such values are consistent with the Eddington limit, which specifies, for a given luminosity, the minimum mass that can accrete gas by overpowering the outward force of radiation pressure. Nearby galactic nuclei, such as M31, show evidence for a black hole of mass $\sim 10^7$ $M_\odot$; but in these cases, there is no AGN energy source.

CONCLUSIONS

The emission lines of AGN currently provide one of our few clues to the conditions prevailing at distances $\sim 10^{-1}$–10^3 ly from the central object. The lines hold the promise of measuring the mass of the black hole and determining the nature of its fuel supply. Gradually, we are unraveling the secrets of the emission lines, but today there are more questions than answers.

Additional Reading

Burbidge, G. and Burbidge, E. M. (1967). *Quasi-Stellar Objects*. W. H. Freeman, San Francisco.

Davidson, K. and Netzer, H. (1979). The emission lines of quasars and similar objects. *Rev. Mod. Phys.* **51** 715.

Miller, J. S. and Osterbrock, D. E. eds. (1984). *Active Galactic Nuclei and Quasi-Stellar Objects*. University Science Books, San Francisco.

Osterbrock, D. E. (1989). *Astrophysics of Gaseous Nebulae and Active Galactic Nuclei*. University Science Books, Mill Valley, Calif.

Shaffer, D. B. and Shields, G. A. (1980). Why all the fuss about quasars? *Astronomy Magazine* **8** (No. 10) 6.

Strittmatter, P. A. and Williams, R. E. (1976). The line spectra of quasi-stellar objects. *Ann. Rev. Astron. Ap.* **14** 307.

Weedman, D. W. (1986). *Quasar Astronomy*. Cambridge University Press, Cambridge.

See also **Active Galaxies, Seyfert Type; Galactic Center; Quasistellar Objects, Spectroscopic and Photometric Properties.**

Active Galaxies and Quasistellar Objects, Infrared Emission and Dust

George H. Rieke

"Normal" galaxies have fairly similar populations of stars in their nuclei, which produce similar nuclear spectra and colors. Some galaxies—roughly 1–2% of large spirals and ellipticals—harbor a powerful compact nuclear source that emits copiously outside the spectral region occupied by normal stellar radiation (e.g., in the radio, infrared, ultraviolet, and x-ray).

Historically, these active galactic nuclei and QSOs have been discovered by detecting excess radiation in the blue and ultraviolet above that to be expected from "normal" galaxies. However, an abnormal excess in the infrared is at least as ubiquitous among these sources as a blue or ultraviolet one; they are described as blue-excess objects only because of the greater ease of making sensitive observations with optical rather than infrared detectors.

These infrared excesses can arise from two underlying causes: The source that powers the nuclear activity may emit a nonthermal spectrum that is observed directly in the infrared, or the output of this source may be absorbed by the surrounding interstellar dust and reemitted thermally in the infrared. In the first case, the infrared region usually accounts for a significant fraction of the total source luminosity and therefore can reveal fundamental aspects of the nuclear activity. In the second case, an energetically important infrared component requires that the interstellar material has altered the output spectrum at the wavelengths where the energy has been absorbed, typically in the blue and ultraviolet; the infrared properties help reveal the underlying characteristics of the nuclear source in these other spectral regions as well as the placement and characteristics of the interstellar medium around the nuclear source.

OBSERVATIONAL BEHAVIOR

Figure 1 shows the behavior in the infrared of typical representatives of various types of active nucleus and QSO. Although the

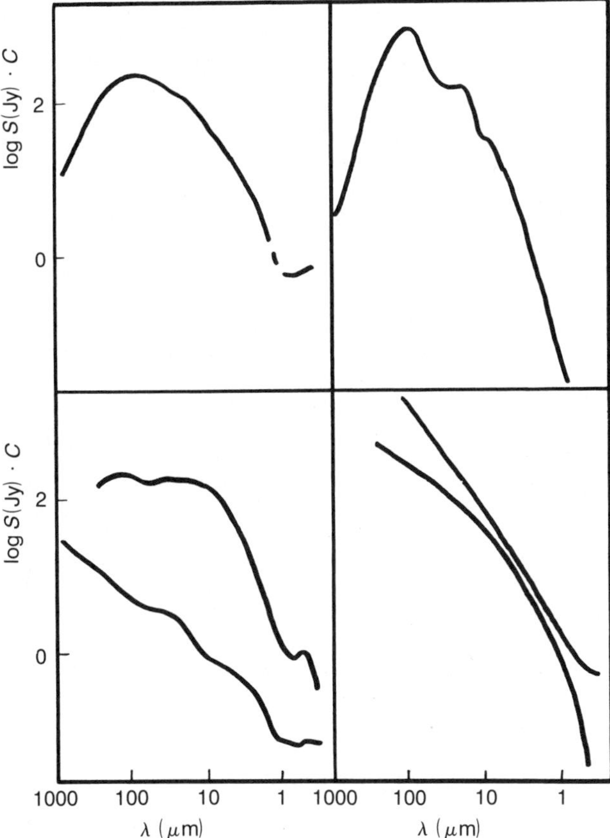

Figure 1. Infrared radiation of active galactic nuclei and QSOs. Spectral energy distributions extend from a wavelength of 1 mm (1000 μm) to the visible (0.5 μm). Fluxes are given in units of Jy ($= 10^{-26}$ W m^{-2} Hz) and have been offset arbitrarily for clarity. Stellar contributions to the infrared fluxes have been removed. The upper left panel shows the type 1 Seyfert galaxy NGC 4151 and the upper right panel shows the type 2 Seyfert NGC 1068. The lower left panel shows the QSOs 13349 + 2438 (upper curve) and 3C 273 (lower curve). The lower right panel shows the blazars 3C 279 (upper curve) and 0235 + 164 (lower curve). In these figures a power law with index -1, ν^{-1}, would be a diagonal line at 45° extending from the upper left of the frame downward.

characteristics of these broad classes are discussed extensively elsewhere, we give here brief definitions with emphasis on specific aspects of interest from the infrared perspective.

1. Type 1 Seyfert galaxies have very broad (~ 5000 km s^{-1}), high-excitation emission lines that are generated in highly disturbed gas very close (within $\sim$ a light-year) to their nuclei. Their blue and ultraviolet excesses are typically variable on time scales of a few weeks, and they have strong x-ray fluxes. Type 1.5 Seyfert galaxies have additional line components of moderate (~ 1000 km s^{-1}) width that originate from gas farther from the nucleus.

2. Type 2 Seyfert galaxies have moderate width, high-excitation emission lines that are similar to the additional line components of type 1.5 Seyfert galaxies. Type 2 galaxies have weaker blue and UV excesses than type 1 galaxies, they are less variable, and their x-ray fluxes are weaker. Both type 1 and 2 Seyfert galaxies have total nuclear luminosities of 10^9–10^{12} $L_\odot$ (where $L_\odot$ is the luminosity of the Sun).

3. QSOs appear to be related to type 1 Seyfert galaxies; the most fundamental difference is their much greater luminosities, which are typically 10^{12}–10^{14} $L_\odot$. A minority of QSOs have very strong radio emission ("radio loud").

4. Blazars are defined by extremely short variability time scales, of the order of a day. They can be strongly polarized and have optical spectra that approximate power laws with indices of about -1; for example, their spectra go as S_ν (W m^{-2} Hz) $= C\nu^\alpha$, where C is a constant and $\alpha \sim -1$. They are frequently strong x-ray sources and, so far as is currently known, always strong high-frequency radio sources. They have apparent luminosities of 10^{11}–10^{14} $L_\odot$, although a variety of arguments suggest that the emission from these sources is beamed strongly and we see only those where we fall in the beam. As a result, the luminosity estimates are strongly dependent on the detailed model of the source.

As indicated by Fig. 1, these four classes of active source also tend to have distinctive infrared behavior. For example, virtually all type 1 Seyfert galaxies have a strong near-infrared excess that seems to fall very steeply toward the optical, becoming weak or undetectable at 1 μm and shorter wavelengths. This component is variable, but with smaller amplitude than the ultraviolet variability. These galaxies have modest far-infrared emission, unless they are embedded in a galaxy that has strong far-infrared emission from the disk probably associated with star formation in the outer parts of the galaxy.

Virtually all Seyfert 2 galaxies have a strong infrared excess rising steeply toward longer wavelengths from about 3 μm, with few if any confirmed observations of variability. Far infrared emission from the galaxy disk is relatively common, and there is evidence that the far-infrared luminosities of the host galaxies tend to be larger than those from a similar set of "normal" galaxies. This suggested association between galaxy-wide properties and the presence of an active nucleus is not well understood.

QSOs frequently have a spectrum that falls steeply toward 1 μm, where there is an inflection with a much flatter spectrum extending to the blue and ultraviolet. In most cases, the variability damps out dramatically from the blue and visible to the near infrared. QSO far-infrared spectra often show only weak excesses above a $\nu^{-0.7}$–ν^{-1} power law extending from the near infrared (such power laws connect smoothly between the near IR and radio for radio-loud QSOs). Out to the longest wavelength where there are abundant observations, 120 μm, differences between radio-loud and radio-quiet QSOs are not apparent.

Blazars generally are characterized by very nearly power law spectra extending from the optical all the way to the submillimeter- or millimeter-wave regions, where the infrared joins onto the strong high-frequency radio emission. The strong polarization and rapid variability also seem to extend from the optical into the infrared with little attenuation.

There are two important infrared-discovered variants to the patterns of behavior described previously. First, a census of the most luminous objects in the local part of the Universe shows them to be roughly equally divided into traditional QSOs and sources such as 13349+2438 that are relatively unremarkable optically, but have very large and powerful far-infrared emission. Current research is examining whether these latter objects are QSOs so heavily dust embedded that virtually all their power is absorbed and reradiated thermally. Second, although about 10% of the high-frequency extragalactic radio sources have invisible or extremely faint visible counterparts, nearly all these objects are detectable in the near infrared and appear (from their rapid variability and strong polarization) to be a form of blazar with a very steep spectral cutoff from the near infrared to the visible. Because only a fraction of the total sample of high-frequency radio sources are blazars, the infrared variant is relatively common among this source type, accounting for perhaps 20–30% of them.

INTERPRETATIONS

The infrared emission of blazars is the most straightforward to interpret. Because it follows the rapid variations and strong polarization of the optical light very closely, it is clearly an extension of the optical synchrotron emission and, at least in the near infrared, is probably generated by the same population of relativistic electrons and in the same source region and magnetic field as the optical emission. The continuity of the power law spectra from the optical through the infrared provides further support for this view. The sharp spectral cutoffs in a significant portion of this source type suggest that the electron acceleration mechanism is operative only up to a very sharply defined upper energy limit.

For the type 2 Seyfert galaxies, it is generally agreed that the infrared emission is dominated by thermal reradiation of the nuclear emission by circumnuclear dust. In the case of the archetype Seyfert 2, NGC 1068, the diameter of the nuclear source has been measured at 2–10 μm and is in direct support of this conclusion. For other Seyfert 2 galaxies, the evidence is circumstantial, but convincing. For example, the spectral energy distributions are accurately reproduced by thermal radiation models and there is evidence in the optical emission line ratios for reddening by dust.

The type of circumstantial evidence found for Seyfert 2 galaxies is weaker for Seyfert 1s and QSOs, and the nature of their infrared emission has been controversial for 20 years. Opinion is now leaning toward reradiation by dust as the dominant emission in most cases. The strongest argument is derived from the behavior during variability. The distance, d, of heated dust grains from the source of ultraviolet heating energy can be shown to be

$$d^2 = \frac{\varepsilon_{UV}}{\varepsilon_{IR}} \left[\frac{L_{UV}}{16\pi\sigma T^4} \right], \tag{1}$$

where ε_{IR} is the infrared emission efficiency of the grains, ε_{UV} is the ultraviolet absorption efficiency, T is the grain temperature, σ is the Stefan–Boltzmann constant, and L_{UV} is the ultraviolet luminosity of the source. For any except highly contrived source geometries, this distance must be small enough that the heated dust can respond in the observed variability time scale to changes in L_{UV} propagating outward from the nucleus at the speed of light, or

$$d < c\tau_{IR}, \tag{2}$$

where c is the speed of light and τ_{IR} is the time scale for infrared variations. For plausible grain emission and absorption efficiencies, these relations and the blackbody law that relates grain temperature to emission wavelength together predict the variability behavior over the infrared spectrum as a function of the ultraviolet luminosity. Qualitatively, variations are expected to occur more and more slowly as the central source luminosity is increased and to have decreasing amplitude with increasing wavelength in the infrared. The behavior of most of the sources is in good agreement with these predictions, suggesting that much of their infrared is thermal reradiation. However, there are exceptions, such as the archetype QSO 3C 273, whose infrared variations are too fast for a thermal (reradiation by dust) source and which therefore must have a strong nonthermal component in its infrared spectrum. The conclusion (consistent with other lines of evidence) is that the outputs of most Seyfert 1s and QSOs are dominated in the near infrared by heated dust, and that cooler dust is likely to be an important component of their mid-infrared emission. From the arguments given previously, the hottest dust lies within a few light-months to light-years of the nucleus and may be involved in the matter that is accreting into these regions to replenish the fuel for their nuclear sources.

If dust absorbs a large portion of the visible and ultraviolet output of Seyfert galaxies and QSOs to emit it in the infrared, one could expect to find significant modifications due to this same dust in the characteristics observed in the visible and ultraviolet. An intriguing hypothesis is that the differences between type 1 and type 2 Seyfert galaxies, and many of those observed among members of the same class, arise solely because of the effects of circumnuclear matter. Reddening of the nucleus by dust then hides the broad-line regions in type 2 galaxies, and absorption by gas eliminates the low-energy x-rays. In dramatic support for this possibility, it has been shown that the archetypical type 2 Seyfert, NGC 1068, has the very broad emission lines characteristic of a type 1 Seyfert when viewed in scattered light that avoids the absorption path directly along the line of sight. A few other type 2 galaxies show similar behavior. Additional ones show broad components in their infrared lines, also indicating visible obscuration of the regions where the broad lines are produced. However, it has been impossible to show that this "unified model of Seyfert galaxies" holds for more than a small subset of type 2 galaxies.

It is also conceivable that some active nuclei might be so heavily dust embedded that they have escaped our attention in traditional searches. An example would be the very powerful infrared objects such as 13349+2438 that overlap with traditional QSOs in luminosity. It has been suggested that these sources could be QSOs that are at a different evolutionary state than the traditional ones. The association of many of these objects with interacting galaxies suggests that this stage might be triggered by galaxy collisions. Most of these objects have some signature of an active nuclear source in their emission line spectra. However, it is difficult with existing tools to separate quantitatively the contributions of the nuclear source from those of the dramatic enhancements in the rate of star formation that are also known to be triggered by galaxy interactions. If such a distinction can be achieved, we may find that circumnuclear dust has been hiding the formative stages in the lives of active nuclei and QSOs.

Additional Reading

Lawrence, A. (1987). Classification of active galaxies and the prospect of a unified phenomenology. *Publ. Astronom. Soc. Pacific* **99** 309.

Rieke, G. H. and Rieke, M. S. (1979). Infrared emission of extragalactic sources. *Ann. Rev. Astron. Ap.* **17** 477.

Soifer, B. T., Houck, J. R., and Neugebauer, G. (1987). The *IRAS* view of the extragalactic sky. *Ann. Rev. Astron. Ap.* **25** 187.

See also **Active Galaxies, Seyfert Type; Active Galaxies and Quasistellar Objects, Blazars; Galaxies, Starburst; Quasistellar Objects, Spectroscopic and Photometric Properties.**

Active Galaxies and Quasistellar Objects, Interrelations of Various Types

Peter Barthel

Discovering and investigating the various properties of QSOs and active galaxies such as radio galaxies, Seyfert galaxies, and BL Lacertae objects over the past three decades, astronomers noted that different classes of objects share properties, even in a quantitative sense. This led them to propose interrelating schemes for seemingly different objects. Some of these schemes had to be abandoned; other ones survived, sometimes after minor or major revision. This entry deals with such interrelating and/or unifying schemes. Because knowledge of the similarities and dissimilarities of the various types of QSOs and active galaxies is essential to understanding possible interrelations, I will first briefly introduce the subject and review the basic, relevant properties of the different classes of objects.

Normal galaxies such as our Milky Way galaxy or the Andromeda galaxy (M31) emit the combined radiation of some hundred billion stars as the bulk of their radiation. The evolution of such normal galaxies is a gradual one, governed by the evolution of these very many stars, which are born, go through various phases of stellar life, and finally die. Unless major disturbances from outside take place, the structure and radiation of a normal galaxy will not change during a time span of a hundred million years.

Active galaxies produce more than just the radiation of stars. Radio and x-ray radiation from nonthermal processes, but also optical line radiation originating in hot gas clouds are among the signs of this activity. The activity usually originates in the centers of these galaxies: hence the acronym AGN, for active galactic nuclei. Species in the active galaxy "zoo" include Seyfert galaxies, radio galaxies, BL Lacertae (BL Lac) objects, and QSOs. The QSO or quasar class can be subdivided into weak radio sources and strong radio sources (the latter group is often referred to as quasistellar radio sources, or QSRs). Taken with more or less effort, optical images of BL Lac objects, Seyfert galaxies, and (nearby) radio galaxies generally show the underlying galaxy associated with the active nucleus. Although the actual galaxy is not observed in most quasars, all but a few astronomers nowadays classify these objects as distant and ultraluminous AGN.

Following the selection criteria of their discoverer Carl Seyfert, Seyfert galaxies are characterized by having small, bright nuclei (optical) and strong emission lines in their optical spectrum. Such emission lines are emitted by hot, ionized gas clouds, which in turn require the presence of an intense flux of ionizing ultraviolet photons originating in the galactic nucleus. Broad emission lines originate in chaotic moving, dense, hot gas (the broad-line region, BLR), located within a few light-years from the nucleus of the galaxy. Narrow emission lines are produced in a more tenuous hot gas further out; this narrow-line region (NLR) stretches out to distances of several thousand light-years, and, in Seyfert galaxies that are not too far away, the NLR can sometimes be resolved with optical telescopes. Early in the study of Seyfert galaxies, it appeared that two subclasses in the population could be separated, based on the relative width or strength of the narrow emission lines with respect to the broad emission lines. These lines are of comparable strength in Seyfert type 1 galaxies, whereas the broad lines are considerably less luminous than the narrow lines in type 2 Seyferts. Most Seyfert galaxies produce stronger radio emission than normal galaxies; however, the radio luminosity of a typical Seyfert galaxy is in the range of 0.1–1% of an average radio galaxy or QSR. The radio emission generally emanates from within the dimensions of the optical galaxy.

Radio galaxies display strong radio emission, most of which originates in two giant radio lobes that straddle the associated optical galaxy. The radio sources can be of gargantuan dimension; radio galaxy 3C 236 is the largest known object in the universe, having a (projected!) linear size of 13 million light-years. The associated galaxies are of (giant) elliptical type in radio galaxies of low and moderate luminosity, but they often have peculiar optical appearances in the most luminous cases. Most notable in the latter cases are spatial elongations of hot emission line gas in the direction of the extended radio emission and the presence of morphological peculiarities such as faint tails and wisps. Luminous radio galaxies often display strong narrow emission lines. Based on the radio luminosity, the radio galaxy population can be divided into two subclasses: Above a certain luminosity the radio sources have a linear, edge-brightened, simple double-lobed morphology, whereas most nearby radio galaxies that are below this "break luminosity" tend to have more or less complex, edge-darkened morphologies. Whereas (two-sided, at both sides of the nucleus) radio jets are generally observed in the latter group, the luminous radio galaxies seldom display jets.

BL Lacertae or BL Lac objects are elliptical galaxies with very bright, variable nuclei, displaying strong, variable radio emission, the morphology of which is dominated by a compact radio core. Emission lines are not seen; the redshift, and thereby distance of a BL Lac object, is inferred from stellar absorption features in the faint extended optical emission of the elliptical galaxy itself.

An important subpopulation of quasistellar objects (QSOs or quasars) is formed by the quasistellar radio sources (QSRs), the radio-loud QSOs. Historically, quasars have been characterized by a stellar appearance and the presence of strong, broad, redshifted emission lines. The optical spectra also display narrow, redshifted emission lines. These redshifts, resulting from the general expansion of the universe, indicate that quasars are the most distant objects in the observable universe. QSRs are powerful radio sources, all exceeding the previously mentioned break point in radio luminosity. Broadly speaking, they have either a compact core-halo radio structure (with dominant, variable core emission) or an extended double-lobed morphology (with dominant lobe emission). QSRs of the latter morphology resemble the luminous radio galaxies, although the QSRs are smaller. A further difference with powerful radio galaxies is that many QSRs display radio jets, but always at one side of the nucleus only. Many of the compact QSRs vary rapidly in their optical light; such optically violent variable (OVV) QSRs are usually combined with the BL Lac objects into the so-called blazar class. Not all radio-quiet QSOs are radio silent; sensitive radio telescopes detect weak, compact radio emission in a considerable fraction of the QSO population. The radio luminosity of these QSOs is two to four orders of magnitude weaker than in typical, otherwise similar QSRs. The fraction of QSRs among QSOs is about ten percent.

It will be clear that characteristic AGN properties, such as a bright nucleus, powerful radio lobes, or strong emission lines, are shared by members of different types. With growing databases, the notion developed that the various AGN types could be interrelated. The qualitative similarities between the Seyfert 1 class and the QSOs and the quantitative continuity in their properties were already recognized in the early 1970s. Observations of radio jets were reported from 1977 onward, and the identification of BL Lacs with AGN in which jets were pointing at us was subsequently suggested in 1978. This suggestion attempted for the first time to attribute widely differing properties of active galaxies to the effects of their orientation. Several more were to follow, which will be discussed, in roughly chronological order.

Using the radio astronomical very long baseline interferometry (VLBI) technique, by the end of the 1970s three QSRs and one Seyfert type 1 galaxy had been found to display superluminal velocities in their nuclear radio jets. The accepted explanation for this phenomenon is based on relativistic motion in a radio jet pointing nearly at the observer. Matter moving at nearly the speed of light in a direction close to the line of sight will almost overtake its own radiation; time intervals will be compressed, creating the illusion of transverse speeds in excess of the speed of light. This relativistic beaming model is attractive, because it also explains the apparent one-sidedness of the radio jets, as well as the observed core dominance in superluminal and core-halo QSRs in general, as a result of Doppler boosting of relativistically approaching material. Expanding on this picture, a proposal that the radio-loud QSRs and the radio-quiet QSOs could be interrelated through orientation, in the sense that QSRs are QSOs with jets pointing in our direction, was put forward. This unified scheme had to be dismissed a few years later, after close examination of the weak radio emission of the QSOs. A subsequent proposal, however, unifying the compact, core-halo QSRs and the extended, double-lobed QSRs through orientation, has found rather widespread approval. Defining the source axis as the line connecting the two radio lobes (and passing through the nucleus), the important parameter in this scheme is the angle θ of this source axis with respect to the line of sight, toward the observer. One prediction of this scheme, namely that the lobes of a QSR should be seen in projection on the strong, boosted core, when seen end-on (at small angle θ), has been successful. Furthermore, the relative numbers of core-halo and double-lobed QSRs in radio source catalogs appear reasonably consistent with this scheme. The model implicitly assumes that the axes of quasistellar radio sources are randomly oriented in space. This assumption may not be valid, as we will see later on. The alternative to this QSR unified scheme would be the picture where intrinsically small (young?) QSRs have stronger and more variable radio core emission. To many astronomers these and related lines of thought appear somewhat contrived, however. A combination of the effects of orientation and source intrinsic effects, such as

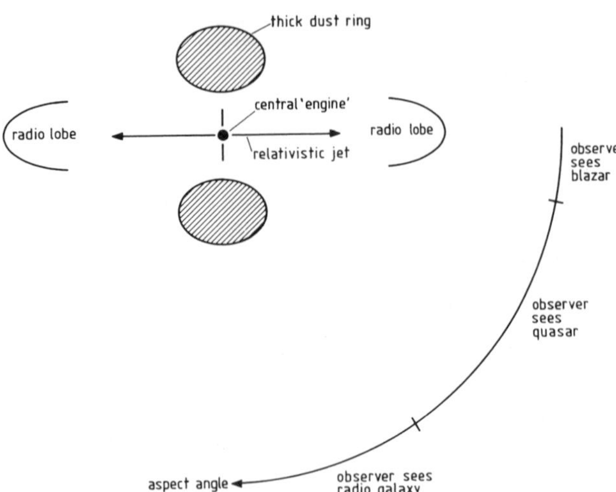

Figure 1. Radio galaxies, quasistellar radio sources, and blazars may be the same objects viewed from different directions.

individual source evolution, is likely of course. The relative importance of these two effects will no doubt be determined in the next few years.

Considerable progress in understanding the interrelation between the classes of Seyfert galaxies was made using the technique of spectropolarimetry, that is, spectroscopy of the polarized light component. This technique has revealed several cases of type 2 Seyfert galaxies with obscured type 1 regions; the polarized light spectrum was found to display a strong optical continuum and broad emission lines, which are typical type 1 characteristics. The angle of polarization in general, and the case of the prototypical Seyfert 2 galaxy NGC 1068 in particular, where nuclear light reflected off a dust cloud was actually measured, strongly argue for the presence of an obscuring dust torus (doughnut) around a bright continuum and broad emission line nucleus. Seen from above or below, this nucleus is directly visible; seen from the side, the nucleus is only indirectly visible through reflection (causing the polarization) by gas and dust above and below the torus. Evidence for dust obscuration through correlations with x-ray emission has also been reported, but it is not yet clear how general these phenomena in Seyfert galaxies are. However, there is no doubt that anisotropic radiation exists, because an optical image of the proposed bi-conical radiation field escaping from the presumed torus has already been obtained for one Seyfert 2 galaxy.

As described previously, the unification of compact core-halo QSRs and extended double-lobed QSRs via the effects of jet flow speeds near the speed of light and orientation was quite successful in explaining a number of observed properties. Continued VLBI monitoring of moving radio components in QSR nuclei, however, revealed many more cases of superluminal motion, not only in compact but also in large double-lobed QSRs. Using sophisticated radio imaging techniques, it was also found that radio jets in QSRs generally occur at one side of the nucleus only. In the framework of relativistic beaming, this would imply that many if not all QSRs are oriented more or less toward us. Stated otherwise: QSRs whose relativistic beams are perpendicular to our line of sight in effect hide or masquerade themselves as other sorts of objects. Based on the facts that powerful radio galaxies exist everywhere in the (history of the) universe where QSRs exist, and that the former are on average a factor of 2 larger in projected size and have comparable narrow emission line luminosities (hence comparable level of activity in the NLR), unification of these objects was proposed. As in the Seyfert case, a dusty torus perpendicular to the radio axis could hide the QSR broad emission line region as well as the bright continuum radiation, when observed from the side (edge-on). See Fig. 1. Such a configuration would naturally explain the extension of the emission line gas as discovered in several powerful distant radio galaxies, as due to cones of ionizing radiation emitted by a hidden energy source and escaping along the radio axis. Also, the strong excesses as far-infrared wavelengths reported in these radio galaxies could well be due to reradiation by obscuring dust. Many observational facts are consistent with this unified model, but the case has to be fully established. The alternative picture would be to identify a luminous radio galaxy with a burned-out QSR. A QSR in which the violent nuclear activity has (temporarily) stopped would look like a powerful radio galaxy, but the questions as to the existence of truly randomly oriented QSRs or the origin of the apparent radio morphological asymmetries remain. As mentioned already, the relative importance of orientation and object evolution will be a subject of detailed study in the coming years.

Although additional effects of orientation are not ruled out, the class of radio-quiet QSOs is most likely evolutionary linked to another, recently recognized, class of active galaxies, namely powerful infrared galaxies. These objects were discovered by the infrared astronomy satellite IRAS and produce more than 10^{12} $L_\odot$ in the far-infrared part of the electromagnetic spectrum. These luminosities are comparable to QSO luminosities. Because these infrared galaxies also display QSO spectral characteristics and their optical morphologies resemble those of some nearby QSOs, identification with dust-enshrouded QSOs was proposed. Once all the dust has been blasted away, a brilliant QSO should appear. Equally important is the fact that optical images of these QSOs-in-the-making suggest galaxy interaction as the ultimate origin of the nuclear activity.

In a way that is complementary to the unification of radio-loud QSRs with powerful radio galaxies, the association of BL Lac objects with favorably oriented radio galaxies of low and moderate luminosity has been suggested. BL Lac objects occur rather close to our galaxy. Unifying them with the powerful radio galaxies which are rare in the local universe is therefore not possible. Lower luminosity radio galaxies do exist in larger numbers locally, and because their radio lobe luminosities are comparable to the luminosity of the halo emission in BL Lac objects, unification of these classes of objects is likely. The absence of emission lines in BL Lacs is no surprise in this picture, because the lower luminosity radio galaxies also lack those lines. Moreover, the optical emission of BL Lac objects is strongly dominated by the (polarized) nonthermal jet component. Note that the OVV QSRs, which together with the BL Lacs make up the blazar class, do have emission lines. Compared to other QSRs, the effects of beaming are most pronounced in OVV QSRs, which are therefore regarded as QSRs with jets closest to the sight line.

For the previous interrelating schemes to hold, it is imperative that the properties of the host galaxies as well as their local environments are undistinguishable. Investigations to test the various predictions concerning host galaxies and environments are currently in progress.

Important steps have been made in the past 10 years to explain the observed AGN diversity as a result of the combined effects of beaming, projection, and anisotropic radiation/obscuration in a small number of intrinsically different, evolving classes of objects. The effects of orientation and dust obscuration appear far more important than previously thought. The puzzle is not yet solved, but many pieces of the picture seem to be falling into place.

Additional Reading

Antonucci, R. (1989). Evidence for and against relativistic beaming in active galactic nuclei. *Proceedings 14th Texas Symposium on Relativistic Astrophysics*. New York Academy of Sciences, New York.

Barthel, P. D. (1989). Is every quasar beamed? *Ap. J.* **336** 606.

Blandford, R. D., Begelman, M. C., and Rees, M. J. (1982). Cosmic jets. *Scientific American* **246** (No. 5) 84.

Sanders, D. B., Scoville, N. Z., and Soifer, B. T. (1988). IRAS 14348–1447, an ultraluminous pair of colliding, gas-rich galaxies: The birth of a quasar? *Science* **239** 625.

Zensus, J. A. and Pearson, T. J., eds. (1987). *Superluminal Radio Sources*. Cambridge University Press, Cambridge.

See also **Active Galaxies and Quasistellar Objects, Jets; Active Galaxies and Quasistellar Objects, Superliminal Motion; Galaxies, Radio Emission.**

Active Galaxies and Quasistellar Objects, Jets

Jean A. Eilek

Fundamental considerations of the energy supply in large-scale extragalactic radio sources ("radio galaxies") led astronomers to predict that these sources should contain jets or beams that connect the nucleus of the parent galaxy to the large-scale radio lobes. These predictions were made shortly after the discovery of the sources; their proof had to wait for the advent of radio interferometers. Interferometers produced much higher quality radio images than those available from early radio telescopes; the jets were indeed found to be a major feature of all radio galaxies.

RADIO GALAXIES

Strong, double radio sources were first identified with external galaxies in the 1950s, and in fact the search for optical identifications of radio sources led to the discovery of quasars in the 1960s. Over the following couple of decades, data were accumulated on large numbers of radio galaxies, and a basic picture of these sources emerged. Radio galaxies generally consist of two fairly symmetric radio-luminous "lobes"; the parent galaxy lies midway between the lobes. The parent galaxy is almost always a giant elliptical, at least at nearby redshifts; it can also be the dominant central galaxy in a cluster of galaxies. At high redshift, double radio sources are found to be associated with quasars (for which the classification of the parent galaxy is still a matter of discussion) and with distorted distant galaxies. Not all active galaxies produce large and powerful radio jets, however. Active nuclei in spirals are much weaker radio sources, which may involve small jets or may only lead to uncollimated core sources. Most QSOs are not strong radio sources; only about 1 in 10 produces the dramatic radio jets which are the topic of this entry.

The radio structure is typically a few to 10 times the size of the parent galaxy [i.e., 50–200 kiloparsecs (kpc)]; however, smaller (down to 10 kpc) and much larger (up to 1 Mpc) examples are known. The spectrum and polarization of the radio emission suggest that the radiation mechanism is incoherent synchrotron emission, produced by relativistic electrons spiraling in magnetic fields. Knowledge of the emission mechanisms allows us to estimate the minimum energy contained in the radio lobes (that part of it in relativistic electrons and the magnetic field). This minimum energy, divided by the luminosity, predicts a very short lifetime. Thus, unless we are seeing these sources at a very special epoch, there must be either a great deal of energy hidden in the lobes, or a continuous resupply of energy to the lobes. One way of doing this is through a "pipeline" from the galactic nucleus; jets were therefore predicted fairly soon after radio galaxies were discovered.

Although optical jets had been known in M87 (a nearby elliptical galaxy) and in 3C 273 (a quasar) for some time, our understanding of jets really grew only after interferometer observations found that jets are common in radio galaxies and in radio-strong quasars. There are now enough data (several hundred radio jets are currently known) to justify studying the basic physics of the jets and to test models against observations.

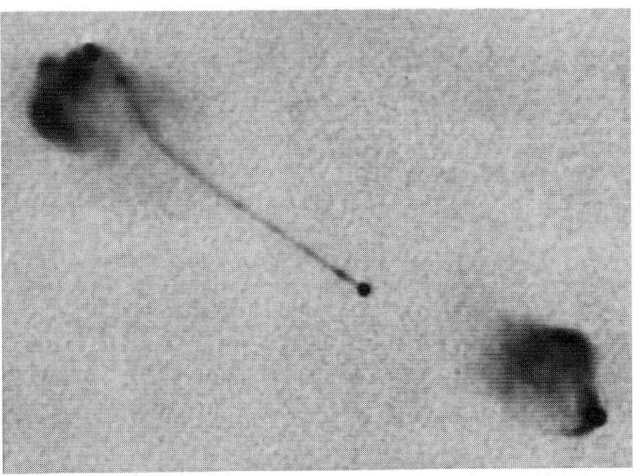

Figure 1. Very Large Array (VLA) image of the radio emission from the quasar 3C 175. The radio source is approximately 200 Mpc end-to-end. The bright central core coincides with the optical quasar, which is at a redshift of 0.77. Several knots of enhanced emission can be seen in the jet. The "snowy" background in the image is the noise level of the observation. (*Courtesy of A. H. Bridle*.)

OBSERVATIONS OF RADIO JETS

This section summarizes important trends in the data. Two examples are shown, Fig. 1 (3C 175, a 200-kpc-long source, associated with a quasar) and Fig. 2 (M87, a 3-kpc-long jet in the core of a nearby galaxy). Nearly all of our information on jets comes from radio observations; only a few jets have been detected at other frequencies (optical and x-ray).

Almost all of our direct information on jets relates to the morphology and luminosity of the jets. The jets are very well collimated structures, with opening angles of no more than a few degrees; jets in more powerful sources tend to have smaller opening angles. The jets often remain straight over their entire length. On the other hand, they can show small wiggles (which do not change the jet direction significantly), or they may show gradual or sharp bends (which do change the jet direction). These bends, small or large, generally do not disrupt the jet; if the jet is a flow, this flow can turn corners.

In higher-power radio sources, the jets end in "hot spots" (small, bright regions) at the outer edges of the radio lobes. If the jet is a flow—most likely supersonic in high-power sources—the hot spot might be where it meets the ambient medium and decelerates through a shock transition. The radio lobes must then be backflow of the shocked jet material (perhaps mixed with the surrounding extragalactic gas).

In lower-power radio sources, the jets broaden and brighten as they enter the lobes; the surface brightness of the lobes then decays going away from the galaxy (this is opposite to the surface brightness in high-power sources). The broadening and brightening can be either sudden or gradual. If the jet is a flow—perhaps transonic or subsonic in these sources—the broadening and brightening may represent deceleration and the onset of turbulence.

Most jets are barely resolved or unresolved, so little can be said in general about substructure within the jets. The few jets that have been resolved, however, show dramatic substructure: bright knots, what appear to be twisted helices of emission, or surface emission (e.g., the jet in M87, Fig. 2). It seems likely that complex substructure will prove to be the rule rather than the exception.

The basic working model of jets is that they represent an outflow of matter and energy from the nucleus. However, there has been no direct measurement of an outflow of jet material (because no

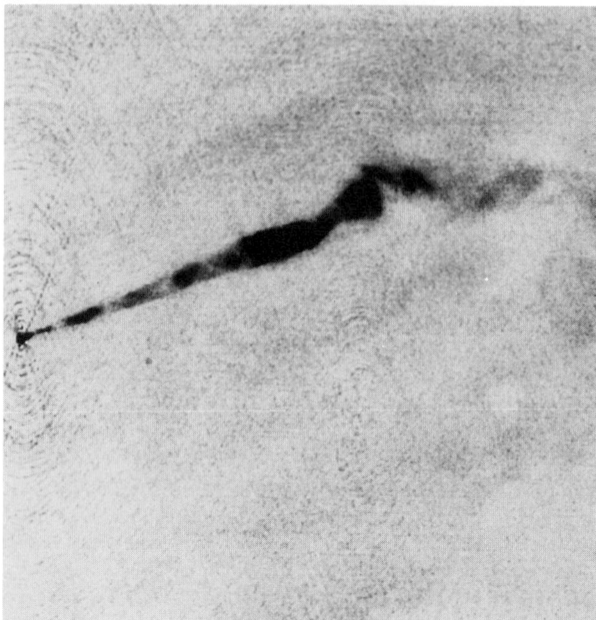

Figure 2. VLA image of the radio emission from the inner part of the galaxy M87. The jet shown here is about 3 kpc long; it emanates from a source in the core of the galaxy, seen at the left of this image. This jet sits in the center of a nearby elliptical galaxy, which is in the Virgo cluster at a distance of approximately 20 megaparsecs. Complex substructure can be seen in the jet, as well as a sudden transition from well-collimated to more disordered flow. Other radio data (not presented here) show that the jet sits inside a larger envelope of diffuse radio emission. The "rings" around the core and the "shadows" above and below the jet are artifacts of the telescope response and the data reduction process. (*Courtesy of F. N. Owen.*)

emission lines are seen, Doppler shifts cannot be measured). Proper motion of bright knots has been detected. Many parsec-scale nuclear jets have been observed by very long baseline interferometry and motion of bright knots is commonly detected. The knots move away from the nucleus, at apparent velocities (projected on the sky) of 2 to 10 times the speed of light, c. This is believed to result from a speed close to c in a jet lying close to the observer's line of sight; light travel effects can account for the apparent superlight speeds. Proper motion has also been searched for from the kiloparsec-scale jet in M87. The data up to 1990 rule out velocities greater than c (again projected on the sky); detection of sublight speed proper motion has not yet been confirmed in this source. It is important to note that these are not necessarily detections of a material flow speed; they could equally well be measuring a pattern speed (such as a traveling wave).

Another indirect measure of jet speed is possible: the measurement of side-to-side brightness ratios for the two jets in a particular source. High-power jets are often asymmetric, with sidedness ratios ranging from a few to several tens. This is especially true in quasar jets, but is also seen in high-power radio galaxies.

One possible explanation of this (although not the only one) is that the sidedness is due to relativistic beaming of the emitted radiation. A pair of jets, each moving at a velocity $v = \beta c$, of equal intrinsic luminosity, will have an apparent brightness ratio given by $(1 - \beta \cos \theta)^x$, where x lies between 2 and 3 (depending on the spectrum of the radio emission) and θ is the angle between the jet and the observer's line of sight. Thus, for $\beta \cos \theta \simeq 1$ (i.e.: for jets moving with $v \simeq c$ and lying close to the line of sight), strong side-to-side asymmetries are seen.

This can, in principle, be used to estimate the average jet velocity in a set of radio galaxies or quasars. If one has a set of jets whose orientation (relative to the line of sight) is known to be random (the "parent sample"), their average sidedness will provide a measure of their average velocity. This type of analysis has been attempted, using different sets of observations, and the mean Lorentz factor of the jets $[\gamma = (1 - \beta^2)^{-1/2}]$ has been estimated at between 2 and 10. This analysis is not yet conclusive, however; several problems remain. First, it is not yet clear that a proper parent sample exists, so that the statistical arguments may not be conclusive. Second, it has not been established that the detailed jet structures seen in high-quality images are consistent with highly relativistic bulk flow. Third, the deprojected length of some of the one-sided quasar jets must be several megaparsecs, which is larger even than clusters of galaxies; this may be an awkward consequence of the model. Thus the question of sidedness is tantalizing, but has not yet produced conclusive evidence on the jet speed.

THE PHYSICS OF RADIO JETS

Work on physical models of the jets is hampered by the fact that none of the parameters that are likely to be important can be measured directly. We have little information on the density, internal energy, composition, velocity, or magnetic field in the jet material. (Note that the synchrotron emissivity, which we can measure, is a nonlinear function of the density and energy of the relativistic electrons, and of the magnetic field. Early work hoped that detection of Faraday rotation would provide independent information on the gas density and magnetic field; however, almost all Faraday rotation detected so far turns out to come from thermal gas surrounding the radio source.) We can, however, list some general considerations that any model of the jets must satisfy.

The total energy supplied to the lobes, over the lifetime of the source, must be at least as large as the current minimum energy contained in the lobes plus the net energy radiated over the lifetime of the source (perhaps 10^{60} erg; this estimate depends on the details of the model as well as on the estimated lifetime). The momentum flux transported by the jet must be large enough to push the extragalactic gas aside and establish the current size of the radio source. The direction of the flow must remain nearly constant over the age of the source (at least 10^7 yr, perhaps much longer).

The internal state of the plasma in the radio jet and lobes must be consistent with the observed synchrotron luminosity. The relativistic electrons are probably transported out from the galactic nucleus, and may also be reaccelerated locally in the jet or lobes. These sources are, in fact, distant laboratories for studying charged-particle acceleration. The same mechanisms that accelerate cosmic rays, and the energetic particles in solar flares and supernova remnants, are likely to be operating in radio galaxies. Shocks and turbulence in the radio source plasma are currently thought to be the most likely mechanisms for accelerating the particles.

Models of the sources must also maintain the magnetic field strength necessary to produce the observed radio synchrotron emission. Because the plasma is highly conducting, induction will reduce magnetic fields in the nuclear plasma to very low levels as this plasma expands into the jet and lobes. It is very likely, therefore, that turbulence in the plasma maintains a dynamo, which amplifies the magnetic field in the jet and lobes. If so, this will involve the same processes that operate to maintain the magnetic field in the Sun and in most of the planets in the solar system.

Most current models of radio jets assume the jets can be described in terms of directed fluid flow, similar to buoyant smoke plumes or to supersonic jet-engine exhausts. Numerical simulations, combined with analytic modeling, have had quite a bit of success in reproducing the morphology of the sources. This work suggests the high-power sources are driven by jets that are very supersonic and much less dense than the medium into which they propagate. Lower-power sources are thought to be transonic or subsonic flows; turbulence develops in the flow, which entrains the surrounding gas and decelerates the flow.

Alternative models have been proposed for these jets, however. It has been suggested that the jets are magnetohydrodynamic rather than hydrodynamic phenomena. In these models, the jet morphology and propagation are controlled as much or more by magnetic and electric forces as by the inertia and turbulent stresses that govern fluid flow. In addition, it has been proposed that the jets are collisionless particle beams—where only electric and magnetic forces on charged particles are important in the jet dynamics.

At this point, there are more competing models of jets than there are ways to discriminate between them. The next major advance in this field might involve a more specific comparison of the data with the models, than has been done up to now, in order to verify or disprove some of the theories described previously. When this is accomplished, the jets should be important tools for studying other astrophysical problems. They will provide probes of their local environment, through their interaction with the extragalactic gas, and with the local gravitational potential. But, perhaps their most interesting use will be to study the central engine in active galactic nuclei; why do some (but not all) active nuclei produce these jets, and how do they do so?

Additional Reading

Begelman, M. C., Blandford, R. D., and Rees, M. J. (1984). Theory of extragalactic radio sources. *Rev. Mod. Phys.* **56** 255.

Bridle, A. H. and Perley, R. P. (1984). Extragalactic radio jets. *Ann. Rev. Astron. Ap.* **22** 319.

De Young, D. S. (1976). Extended extragalactic radio sources. *Ann. Rev. Astron. Ap.* **14** 447.

Hughes, P., ed. (1991). *Beams and Jets in Astrophysics*. Cambridge University Press, Cambridge.

Kellermann, K. I. and Owen, F. N. (1988). Radio galaxies and quasars. In *Galactic and Extragalactic Radio Astronomy*, G. L. Verschuur and K. I. Kellermann, eds. Springer-Verlag, New York, p. 563.

Miley, G. (1980). The structure of extended extragalactic radio sources. *Ann. Rev. Astron. Ap.* **18** 165.

See also **Active Galaxies and Quasistellar Objects, Superluminal Motion; Galaxies, Radio Emission; Jets, Theory of.**

Active Galaxies and Quasistellar Objects, Superluminal Motion

Marshall H. Cohen

When compact radio sources are examined on a scale of a milliarcsecond (mas), they frequently are seen to contain moving components. This is described as an internal proper motion μ, which is expressed in millarcseconds per year. It may be converted to an "apparent transverse velocity" v with the formula $v = \mu D(1+z)$, where D is an appropriate distance, z is the redshift, and the factor $(1+z)$ scales the velocity to the value a comoving observer (i.e., one at rest relative to the source) would measure. When $v > c$ (the velocity of light), the motion is called superluminal; when $v < c$, it is subluminal.

This phenomenon has a tiny angular scale, and very long baseline interferometry (VLBI) at radio wavelengths is the only technique capable of seeing it today. Hints of superluminal motion were seen as early as 1969, two years after the invention of VLBI; but the first definitive measurements were not made until 1971, when two teams, one headed by Irwin Shapiro and the other by Marshall Cohen and Kenneth Kellermann, saw changes in the quasars 3C 273 and 3C 279. These observations followed by four months the accidental discovery, by Shapiro's group, that those objects had simple double structure. For both sources the separation increased in four months, with $v > c$.

The velocity v is not limited to c by the special theory of relativity, because it may be due to phase or retardation effects. The usual description has two bright spots, one (A) stationary, and the other (B) moving with velocity βc at an angle θ to the line of sight. In unit time B moves a distance βc, whereas its transverse distance from A increases by $\beta c \sin\theta$. But the elapsed time for the observer to see the beginning and end of this motion is only $(1-\beta\cos\theta)$ because the velocity of light is finite. Hence the apparent transverse velocity is

$$v = \frac{\beta c \sin\theta}{1-\beta\cos\theta} \tag{1}$$

and v has a maximum value $\beta\gamma c$ when $\csc\theta = \gamma = (1-\beta^2)^{-1/2}$. Equation (1) is a general formula and can apply when B is a "light echo" with $\beta = 1$ as well as to luminous "bullets" with $\beta < 1$, or to regions successively excited by a wave. In 1967, Martin Rees used Eq. (1) to explain rapid flux variations in active nuclei, and at that time he predicted superluminal motion, four years before it was seen!

A different kinematic model could involve motion that is simulated by independent fluctuations in intensity of a few stationary spots. This explanation, however, is incapable of explaining two key features of the observations: (a) all well-determined motions are expansions, never contractions; and (b) in some cases spots have been observed to more than double their separation.

GENERAL BEHAVIOR

The best-studied source is the quasar 3C 345, for which a series of maps are shown in Fig. 1. The increasing separations are apparent, even though these images have lower dynamic range than is becoming available today. Figure 2 shows an "expansion diagram" for component C2 of 3C 345. The data are well fit by a straight line, indicating a uniform proper motion of 0.47 mas yr^{-1}. With $z = 0.595$ and $q_0 = 0.5$, standard cosmology gives $v = 9.3h^{-1}c$, where $h = H_0/100$ is the reduced Hubble constant in kilometers per second per 100 megaparsecs.

Figures 1 and 2 represent the classic form of superluminal motion. The morphology is characterized by a linear structure with a rather stable, particularly compact, component at one end, and one or more moving components. The end component is called the "core" and is thought to be close to the central engine. In one case (3C 345), absolute astrometry was able to show that the core is stationary on the sky. A core generally has a flat radio spectrum, whereas the other (moving) components have a steep spectrum. The moving components usually evolve and grow weak at short centimeter wavelengths, with a time scale of 2–10 yr.

A number of other sources have similarly shown this classic behavior, but there are numerous complications.

1. Many sources display more than one traveling component and their speeds may be different. In two well-studied cases (3C 273 and 3C 345), the more distant components have the higher speeds. In 3C 345 the new component C4, born near the core around 1980, was observed to accelerate in 1982.

2. Several sources have stationary components in addition to moving components. The most famous case is 4C 39.25, which had been a stable double for a decade. Around 1980, however, a new component appeared near one of the original ones and began traveling superluminally toward the other, which has remained stationary but is fading. None of the three components appears to be a core in the sense described previously, and it is not at all certain that any of the three is truly stationary on the sky.

3. A source with more than two components is usually not straight but shows curvature or even a kink. A major question then has been whether the components move ballistically, on straight

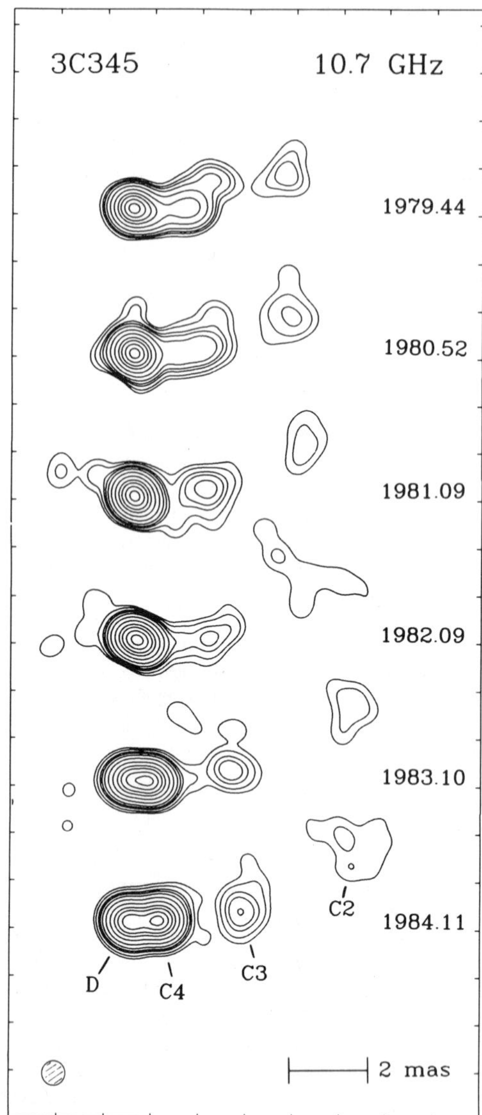

Figure 1. A sequence of maps of 3C 345 made at 10.7 GHz. The smoothing beam is a circular Gaussian of half-width 0.6 milliarcseconds (mas), shown in the lower left corner. [*Reproduced from Biretta, Moore, and Cohen (1986)* Ap. J.]

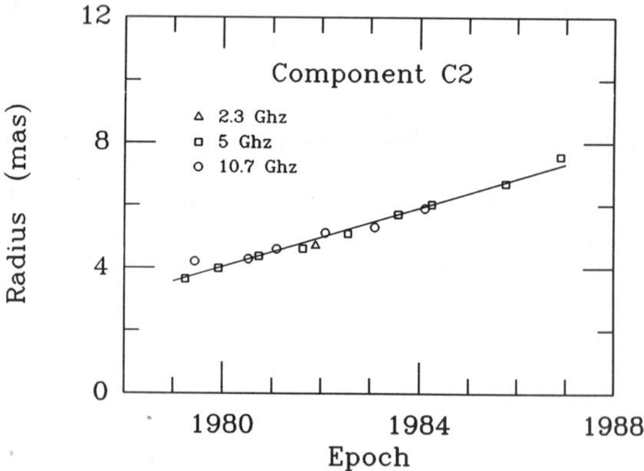

Figure 2. Expansion diagram for component C2 of 3C 345. The line represents a uniform proper motion of 0.47 mas yr^{-1}. The points, each corresponding to the centroid of C2 at the labeled frequency, cluster tightly about the line, showing that the location of the centroid is nearly independent of frequency.

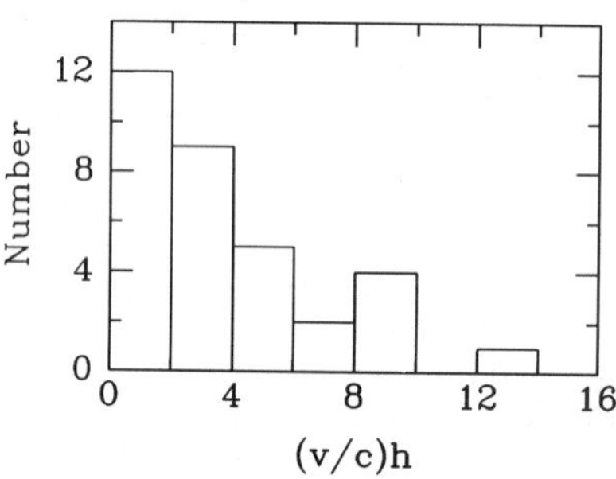

Figure 3. Histogram of superluminal velocities for 33 objects. The apparent transverse velocity v is normalized by c, the velocity of light, and is inversely proportional to the reduced Hubble constant h. [*Reproduced from Cohen, in BL Lac Objects*, Lecture Notes in Physics **334**, *Springer-Verlag, Berlin (1988), p. 13.*]

lines, or move on curved tracks and perhaps even follow each other like beads on a string. The present indication in both 3C 273 and 3C 345 is that the motion is along a curved track. In 3C 345 the new components C5 and C4 have had curved tracks, but they were at different position angles; that is, they were not like beads on a fixed string. Component C3, however, may be following the track taken earlier by C2. There evidently is a variety of possible motions, and it will take many years of observation to sort them out.

DISTRIBUTION OF SUPERLUMINAL VELOCITIES

Proper motions (or upper limits) have now been established in about 33 objects, of which 23 are superluminal, 2 are subluminal, and 8 have upper limits that generally require $v < 1.5h^{-1}c$. Figure 3 shows the histogram of their apparent transverse velocities; for those sources with more than one moving component only the fastest is plotted. The sample is inhomogeneous and incomplete, but most of the objects were chosen because they are "active"; that is, they are variable and strong, and Fig. 3 is probably representative of core-dominated objects. Thus it is significant that the median velocity corresponds to angles much less than 60°. The median velocity for all objects in Fig. 3 is $3.3h^{-1}c$. From Eq. (1) the median "maximum allowed angle" (for $\beta = 1$) is approximately $34h$ °, which for $0.5 < h < 1$ means that most of the objects are pointed close to our line of sight.

Why is there such a strong selection effect? Three possibilities have been suggested.

Obscuration

Many Seyfert galaxies are known to have a central torus that blocks optical light and allows a clear view of the central region only from a restricted solid angle around the axis. This idea is geometrically simple, but it is difficult to make it work for superluminal sources, because they are observed at radio wavelengths. Dust is transparent at these wavelengths, and blocking the radiation with plasma requires far more electron density than is customarily ascribed to the narrow-line region.

Relativistic Beaming

The moving bright spot must have velocity close to 1; for example, for $v = 10c$, $\beta \geq 0.995$. If the radiating material itself is moving with the same or a similar velocity, then it has a Lorentz factor $\gamma \geq 10$; that is, it is relativistic. Its radiation is necessarily beamed into a cone of half-angle γ^{-1} rad, and this automatically gives a strong preference to sources beamed at us. This explanation is usually favored, because, in addition to explaining the angular anisotropy, it explains in a simple way the one-sided morphology. Furthermore, it uses the bulk relativistic motion that appears to be needed in any event to explain the rapid time variations, the lack of inverse Compton x-rays, and the polarization asymmetry seen in the large-scale jets.

Wide Cones

Statistical difficulties associated with the thin-beam model have prompted speculation of multiple beams in wide cones. If a source has many narrow beams, all within a wide cone, then there could be a substantial probability that a source oriented at random still has one of its beams aimed close to our line of sight. We would see the Doppler-boosted radiation from that beam but not from the others, and the superluminal jet would look thin when projected on the sky. The beams do not have to coexist like multiple searchlights, but could follow each other; for example, they could be successive shock waves with different trajectories. A general objection to this picture is that in some cases our line of sight should be inside the cone, and then the superluminal motions should be projected into many directions. The source should have a two-dimensional appearance, whereas these sources always have a linear appearance. In support of the wide-cone hypothesis, however, is the observation that the two new components in 3C 345 emerged from the core at rather different position angles, and have been following different tracks.

STATISTICS

The synchrotron radiation from a beam with bulk relativistic motion has an angular pattern given by δ^n, where $n \approx (2-\alpha)$, α is the spectral index of the radiation, and δ is the Doppler factor $\gamma^{-1}(1-\cos\theta)^{-1}$. A source with $\gamma = 10$ thus has an enormous front-to-back ratio, so that an intrinsically weak source aimed at us may look strong, whereas an intrinsically strong one aimed away will be weak. When this strong boosting is combined with an intrinsic luminosity function (many more weak sources than strong ones), the result is that a flux-limited sample of beamed sources will consist nearly entirely of objects at small angles. When this calculation is carried through in detail, it turns out that the expected distribution of superluminal velocities has a strong peak near the maximum possible velocity, $v_{max} = \beta\gamma c$. Thus the histogram in Fig. 3 may reflect the distribution of Lorentz factors, rather than revealing the angular distribution of superluminal sources.

A more direct way to study these objects is to select a sample on the basis of some "aspect-independent" quantity. The superluminal motions will then be oriented at random, and so the measured distribution of velocities should be a direct test of the beaming hypothesis. Several experiments along this line are currently under way. The aspect-independent quantity is taken to be the flux density in the outer lobes; nearly everyone thinks that the outer lobes radiate isotropically. This work is difficult and slow, because the central components are generally weak. The first results from these tests are in crude agreement with the beaming hypothesis: The velocities observed so far are smaller, on average, than those of the strong core-dominated sources.

COSMOLOGY

The superluminal objects include quasars at large redshifts, and they may contain a "standard velocity," that is, the velocity of light. Thus, almost from their discovery, there have been attempts to use them for cosmology. The early work used a "light echo" model, but it now is clear that this picture is too simple. It does not easily allow for the observed morphology, especially the asymmetry around the core. The same objections have been raised against models that use excitations along magnetic field lines. More recent work has used models with relativistic beams. If two-jet objects could be found they should be near the plane of the sky, and then from Eq. (1) $v \approx c$; that is, we have recovered the standard velocity. Unfortunately, there are as yet no reliable two-sided motions. A different method uses the highest velocities that are measured; these are limited to $v_{max} \approx \gamma c$ and so the top bin in Fig. 3 implies $\gamma H_0 \approx 900$ km sec^{-1} Mpc^{-1}. If there were a reliable physical model from which the Lorentz factor could be found, then the Hubble constant could be calculated. So far, though, it has been the other way around; H_0 is used to estimate the Lorentz factor. Values of γ found this way generally agree with those needed to explain the variability and the spectrum.

Additional Reading

Kellermann, K. I. and Pauliny-Toth, I. I. K. (1981). Compact radio sources. *Ann. Rev. Astron. Ap.* **19** 373.

Pearson, T. J. and Zensus, J. A. (1987). Introduction. In *Superluminal Radio Sources*, J. A. Zensus and T. J. Pearson, eds. Cambridge University Press, Cambridge, p. 1.

Porcas, R. W. (1987). Summary of superluminal sources. In *Superluminal Radio Sources*, J. A. Zensus and T. J. Pearson, eds. Cambridge University Press, Cambridge, p. 12.

Readhead, A. C. S. (1982). Radio astronomy by very-long-baseline interferometry. *Scientific American* **246** (No. 6) 38.

See also **Active Galaxies, Seyfert Type; Active Galaxies and Quasistellar Objects, Blazars; Active Galaxies and Quasistellar Objects, Emission Line Regions.**

Active Galaxies and Quasistellar Objects, X-Ray Emission

Richard Mushotzky

X-ray emission from active galactic nuclei (AGN) is one of the major signatures of these objects and thus it is necessary to understand how it is created in order to study the physical conditions in these objects. It is thought that the x-ray emission comes from those regions that are very close to the source of energy, frequently presumed to be a massive black hole. This conclusion is based on general theoretical grounds and observations that indicate that the fastest time scale for variability most often occurs in the x-ray band. Most known forms of active galaxies (e.g., quasars, Seyfert galaxies, broad-line radio galaxies, and BL Lac objects) have been well studied in the x-ray band. Located in the centers of "normal" galaxies, AGN often outshine them, radiating from 10^6–10^{15} times the bolometric luminosity of the Sun from a region smaller than a light-month in radius. Their spectra can be roughly characterized by a power law distribution with equal energy per decade of frequency from the mid-infrared ($\nu \sim 10^{12}$ Hz or $\lambda \sim 3$ cm) to the hard ($\nu \sim 5\times10^{19}$ Hz or $E \sim 200$ keV) x-ray (where λ is the wavelength, ν is the frequency of the radiation, and E is its energy).

This entry concentrates on the x-ray emission from quasars and Seyfert 1 galaxies and excludes BL Lac objects, Seyfert 2 nuclei, and other types (see Additional Reading).

X-RAY SPECTRA

The observed spectrum is a signature of the mechanism via which the energy is radiated. Although x-ray spectroscopy of AGN is in its infancy (past instruments having had spectral resolution of only $E/dE \leq 10$), some general results are clear. The x-ray spectra of AGN that have broad optical emission lines, in particular Seyfert 1 galaxies, are dominated in the 2–20 keV band by a simple power law continuum, $F(\nu) \sim \nu^{-\alpha}$ erg cm^{-2} s^{-1} Hz^{-1}. The spectral index α is most commonly ~ 0.7, a value that does not seem to be a function of luminosity and, in a given object, does not seem to change much when the source intensity varies by a factor of ~ 10.

More than 10% of the bolometric luminosity of an AGN is emitted in x-rays with energies greater than 2 keV. This seems to be the only spectral domain where the radiation is dominated by "nonthermal" processes. That is, the physical mechanism for generation of the x-ray photons is not due to "thermal" processes such as recombination or blackbody radiation; instead the effects of Compton scattering and/or electron–positron pairs must be very important. However, recent observations indicate that there may be a substantial contribution to the 20–50 keV band radiation from "thermal" reprocessing of the "nonthermal" radiation.

At energies less than 2 keV, the x-ray spectra of many AGN deviate from the simple power law description. About one-third of all Seyfert galaxies show a deficit of photons at low energies. This deficit occurs most often in low-luminosity [$L(x) < 5 \times 10^{43}$ erg cm^{-2} s^{-1}] objects and has the characteristic signature of absorption of x-rays by cold material in the line of sight. The inferred column densities are on the order of 10^{22}–10^{23} atoms cm^{-2}, consistent with the absorbing material being the same clouds responsible for the emission of the broad optical lines characteristic of AGN. However, recent observations have indicated that in some objects the absorber is not a simple "cold" ($T < 10^5$ K) slab but must be either partially ionized or have "holes" in it, which some of the x-rays leak through. There are several possible locations for the absorbing material with the prime candidates being the clouds that are responsible for the broad optical emission lines and the outer parts of the accretion disk itself.

In approximately one-half of the AGN that do not show strong absorption, there is a "soft x-ray excess." That is, the flux at energies less than 2 keV exceeds what would be predicted by an extension of the higher-energy continuum. The poor spectral resolution of past experiments was inadequate to determine the nature of this excess and there are several viable explanations for this excess emission. It may be: (1) radiation from an accretion disk, (2) the extension into the x-ray band of a "synchrotron" emission observed in the infrared band, (3) a thermal photoionized plasma associated with the gas confining the broad-line clouds, or (4) the signature of reprocessing of the radiation from the central engine of the AGN by a thermal accretion disk or "cold" accreting matter.

The next generation of x-ray experiments, to be launched in the early 1990s, should have sufficient spectral resolution and sensitivity to determine the nature of these "soft x-ray excesses" and resolve these outstanding problems.

TIME VARIABILITY

The general condition of causality states that the fastest time scale over which an object can vary (in the absence of special and general relativistic effects) is the light travel time, $\delta t \sim R/c$, where R is the "size" of the object and c is the speed of light. If the energy for the x-ray radiation comes from the innermost regions around a massive black hole, $R \sim 10GM_{\rm BH}/c^2$, where $M_{\rm BH}$ is the mass of the black hole in solar masses. Defining $M_7 = 10^7\ M_\odot$, one finds that $\delta t \sim 500(M_{\rm BH}/M_7)$ s. Another fundamental consideration is the efficiency with which matter can be converted into energy. The physics of such conversion requires that $\delta t \geq \eta\, \delta L(x)/(5 \times 10^{44})$ where η is the efficiency with which matter is converted into energy via the release of gravitational energy (nominally 10–40%) and $\delta L(x)$ is the change in the x-ray luminosity of the source in a time δt.

So far AGN have *only* been observed to vary strongly ($> 50\%$) on time scales of less than 1 day in the x-ray domain. Such "rapid" variability has not been seen in the radio, infrared, optical, or ultraviolet bands. This seems to verify that x-rays do indeed come from the smallest regions in AGN. Analysis of a large number of Seyfert 1 galaxies with data from the Exosat satellite indicates that whereas "rapid" variability ($\Delta t < 1/2$ day) seems to be a common property of these objects, there is a wide variety in the nature of the variability. For example, compare the properties of three low-luminosity Seyfert 1 galaxies: NGC 4051, which is continuously and always variable and is well described by a process with no characteristic time scale; NGC 4151, which shows quiescent periods but also flares with a one-day time scale; and NGC 6814, which shows "quasiperiodic" behavior with a ~ 3–4 h time scale. The origin of these phenomena is not understood at present and we do not know if these are fundamentally different behaviors or just different aspects of the same phenomenon.

Frequently, the x-ray variability is rather dramatic, such as a reduction in flux in NGC 6814 by a factor of > 5 in less than 300 s, or an increase in the flux in MCG 6−30−15 (a Seyfert 1 galaxy) by more than 50% in less than 2000 s and its fall by more than a factor of 2 in less than 1000 s. With a few exceptions there has been no detection of spectral variability during these events. This lack of significant spectral changes rules out a "thermal" origin for the x-ray flux and confirms the spectral result that nonthermal processes dominate.

There seems to be a global pattern to the variability time scales with more luminous objects varying on rather longer time scales than less luminous objects; however, this conclusion is currently controversial. The data are consistent with most objects having variability time scales that scale linearly with the luminosity, which might be expected if they were accreting at a similar ratio of the Eddington limit. This limit is the maximum luminosity an accreting object with a steady-state spherical flow can have for a given mass $L \sim 1.4 \times 10^{38}$ erg s^{-1} $M_\odot$.

CORRELATIONS OF THE X-RAY WITH OTHER WAVELENGTHS

Because most AGN detected in the x-ray band have a very low flux (~ 0.005 photons cm^{-2} s^{-1} in the 2–10 keV range), their spectral and temporal properties are poorly determined. However, one can use the large numbers of x-ray detections and the large numbers of upper limits for other objects to correlate with their optical and radio properties, to determine the evolution of AGN with luminosity and redshift to derive scaling laws between the different spectral bands, and to obtain the properties of AGN which are at very high redshift or which are very faint.

The distribution of the number, N, of x-ray-detected AGN versus x-ray flux, S, called the $\log N$–$\log S$ diagram, shows that AGN have evolved in either number or luminosity with cosmic epoch. Luminosity evolution seems to be more likely. However, comparison of the evolution of AGN in the x-ray and optical bands indicates that the x-ray luminosity has changed less rapidly with cosmic time than the optical.

Detailed comparison of the soft x-ray with the optical luminosity shows that the ratio of the two is a function of the optical luminosity with $L_x/L_{\rm opt} \sim L_{\rm opt}^{-1/4}$. Thus the more optically luminous an object, the smaller is the soft x-ray fraction of the total luminosity. A similar study involving x-ray and radio luminosity shows that the two are strongly correlated with the relationship being linear. These two correlations seem to be additive with the best fit of the form $\log L_x = \log(AL_{\rm opt}^{3/4} + BL_{\rm radio})$, where A and B are constants. This may indicate that the x-ray emission from radio bright AGN is due to two physically separate components even though the spectral data do not require it.

Additional Reading

Maraschi, L., Maccacaro, T., and Ulrich, M.-H., eds. (1989). *Workshop on BL Lac Objects*. Springer-Verlag, Berlin.

Tanaka, Y., ed. (1988). *International Symposium on the Physics of Neutron Stars and Black Holes*, see Section 5, Extragalactic sources: Active galactic nuclei and cosmic x-ray background. Universal Academy Press, Tokyo.

Turner, T. J. and Pounds, K. A. (1989). The EXOSAT spectral survey of AGN. *Monthly Not. Roy. Acad. Sci.* **240** 833.

See also **Active Galaxies, Seyfert Type; Active Galaxies and Quasistellar Objects, Accretion.**

Andromeda Galaxy

René A. M. Walterbos

The Andromeda galaxy is the largest member of the local group of galaxies. At a distance of 700 kpc or 2.3 million light-years, it is also the closest spiral galaxy to our own (a kiloparsec is 3260 light-years). It is often referred to as M31 or Messier 31, because it is number 31 in the famous catalog of Messier (1730–1817). The Andromeda galaxy has played a major role in developing our understanding of galactic structure and evolution. In the 1920s, Edmund Hubble discovered Cepheid variable stars in M31, which established its extragalactic nature, because the distance derived for the Cepheids from the period–luminosity relation placed M31 well beyond the bounds of our Milky Way. This discovery proved that galaxies other than our own exist. In 1943, Walter Baade was able to resolve the central part of M31 into thousands of red stars. These were clearly different from the bluer, more luminous stars that can be seen in the spiral arms. His observations established the concept of two different stellar populations, types I and II. Type I stars are young and luminous and are currently still forming in most galaxies. Type II stars, on the other hand, are old and much fainter. The discovery of two stellar populations provided important input for models of stellar evolution.

Progress in observational techniques has been rapid and today catalogs exist of the distributions of H II regions, OB associations, globular clusters, open clusters, planetary nebulae, novae, and other objects in M31. In addition, detailed observations exist of the various phases of the interstellar medium: atomic and molecular hydrogen gas, dust, regions of ionized gas (H II regions and supernova remnants), and relativistic electrons and the magnetic field. Both space- and ground-based telescopes operating over the entire wavelength range from radio to x-ray have been used in these observations.

STRUCTURE

As in most spiral galaxies, the stellar distribution in M31 has two major components, a thin disk with superposed on it the spiral arms, and a spheroidal component which is strongly concentrated to the center and is flattened with an axial ratio of about 0.6. The light distribution of the disk falls off more slowly like an exponential function with a scale length of about 5 kpc. Table 1 gives a summary of the properties of M31 and of our galaxy. Figure 1 shows an optical image of M31. The light from the stellar disk is bluer at larger distances from the center. This can be caused by a decrease in the average metallicity of the disk stars with radius, or by an increase in the fraction of relatively younger stars with radius. At about 18 kpc from the center, the stellar disk seems to bend out of its principal plane, a phenomenon known as a "warping." A similar warp occurs in the gas distribution in M31, as is apparent from radio observations of atomic hydrogen. The warp in the gas disk occurs at slightly larger distances from the center.

An as-yet unsolved problem concerns the shape of the spiral arm pattern of M31. Spiral arm segments show up clearly in various tracers of young objects, such as interstellar gas and dust, regions of ionized gas, O and B stars, and open clusters. However, analyses of the distribution of these tracers do not result in a unique picture of the spiral arm structure. Both a one-armed leading spiral and a two-armed trailing spiral structure have been proposed. The problem is that the small angle between our line of sight and the principal plane of M31 makes it difficult to disentangle the spiral arms. Furthermore, the arms may be distorted by the gravitational interaction with the close elliptical companion, M32.

In addition to these visible components, M31, like other spiral galaxies, has an invisible halo of unknown matter. The dark halo is inferred from the rotational speed of gas that orbits the galaxy. The velocity of the gas stays constant at 230 km^{-1} s out to at least 35 kpc from the center. The amount of mass that is required to explain this is much larger than can be accounted for by the visible matter; the discrepancy is at least a factor 2, possibly as much as a factor 10.

Table 1. Properties of Andromeda and Milky Way Galaxies

Parameter	*Andromeda*	*Milky Way*
Distance to center	690 kpc	8 kpc
Hubble type	Sb I–II	Sbc II
Total luminosity (in solar units)	$3.1 \times 10^{10}\ L_\odot$	$1.8 \times 10^{10}\ L_\odot$
Luminosity of spheroidal component	$7.7 \times 10^{9}\ L_\odot$	$2.8 \times 10^{9}\ L_\odot$
Velocity dispersion of stars in bulge	155 km s^{-1}	130 km s^{-1}
Luminosity of disk component	$2.4 \times 10^{10}\ L_\odot$	$1.6 \times 10^{10}\ L_\odot$
Optical diameter at brightness of 25 mag arcsec^{-2}	30 kpc	25 kpc
Scale length of disk	5–6 kpc	4–5 kpc
Number of globular clusters	300–400	130–160
Mass of atomic hydrogen gas (in solar units)	$3.9 \times 10^{9}\ M_\odot$	$4 \times 10^{9}\ M_\odot$
Atomic gas disk extends to at least:	25–30 kpc radius	20–25 kpc radius
Infrared luminosity emitted by dust	$2.6 \times 10^{9}\ L_\odot$	$12 \times 10^{9}\ L_\odot$

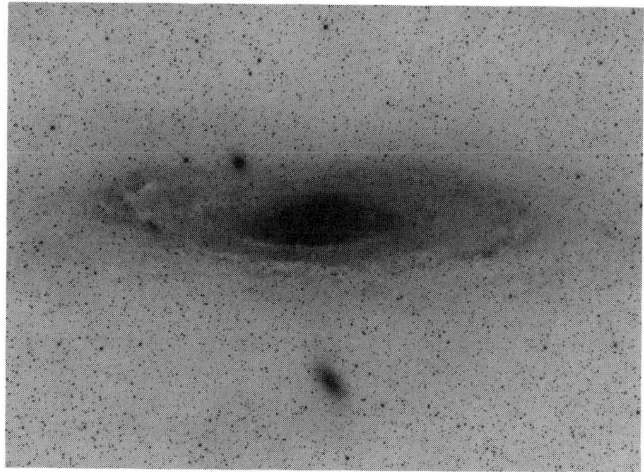

Figure 1. The Andromeda galaxy photographed in blue light with the Burrell–Schmidt telescope at Kitt Peak. Two elliptical companion galaxies, M32 (top) and NGC 205 (bottom) are visible. North is to the bottom right in this figure, as in Fig. 2. (*Courtesy National Optical Astronomy Observatories.*)

STELLAR POPULATIONS

In Baade's picture, the spheroidal component contains old stars with low abundances of metals (elements heavier than helium), whereas the disk contains younger stars with relatively high metal abundances. The metal abundance increases with time, because the material is continuously enriched by the processes of stellar evolution. Therefore, young stars will be more metal rich than old stars. Recent studies show that the distribution of stellar populations in M31 is more complicated. Stars in the outer parts of the spheroid indeed have metal abundances about 10 times lower than our sun, but there is considerable spread. The central part of the spheroid contains metal-rich stars, and the spheroidal component may consist of a metal-poor halo and a central stellar bulge that has stars of varying metal abundances, some of them more metal rich than our sun. The disk contains stars in a range of ages and metal abundances. Observations with the Hubble space telescope will aid considerably in unraveling the stellar populations in M31 and other galaxies, since with this instrument much fainter stars can be detected and much more crowded regions close to the center can be resolved than with ground-based observations.

The Andromeda galaxy has more globular clusters than the Milky Way, which is possibly related to its larger stellar spheroid. Some 300 have been identified; the true number may be even larger. The cluster system in M31 resembles the galactic cluster system in many ways, but some differences exist. In particular, detailed spectroscopic studies indicate differences in the stellar populations of some clusters that were unexpected and are not yet fully understood. In addition, M31 has some blue compact clusters that must be much younger than the canonical globular cluster that is as old as the universe.

INTERSTELLAR MEDIUM AND STAR FORMATION

All tracers of star-forming regions and the interstellar medium are concentrated in an annulus between 8–12 kpc from the center. The annular structure results from projection of various segments of the complicated spiral arm pattern, and is probably not a true

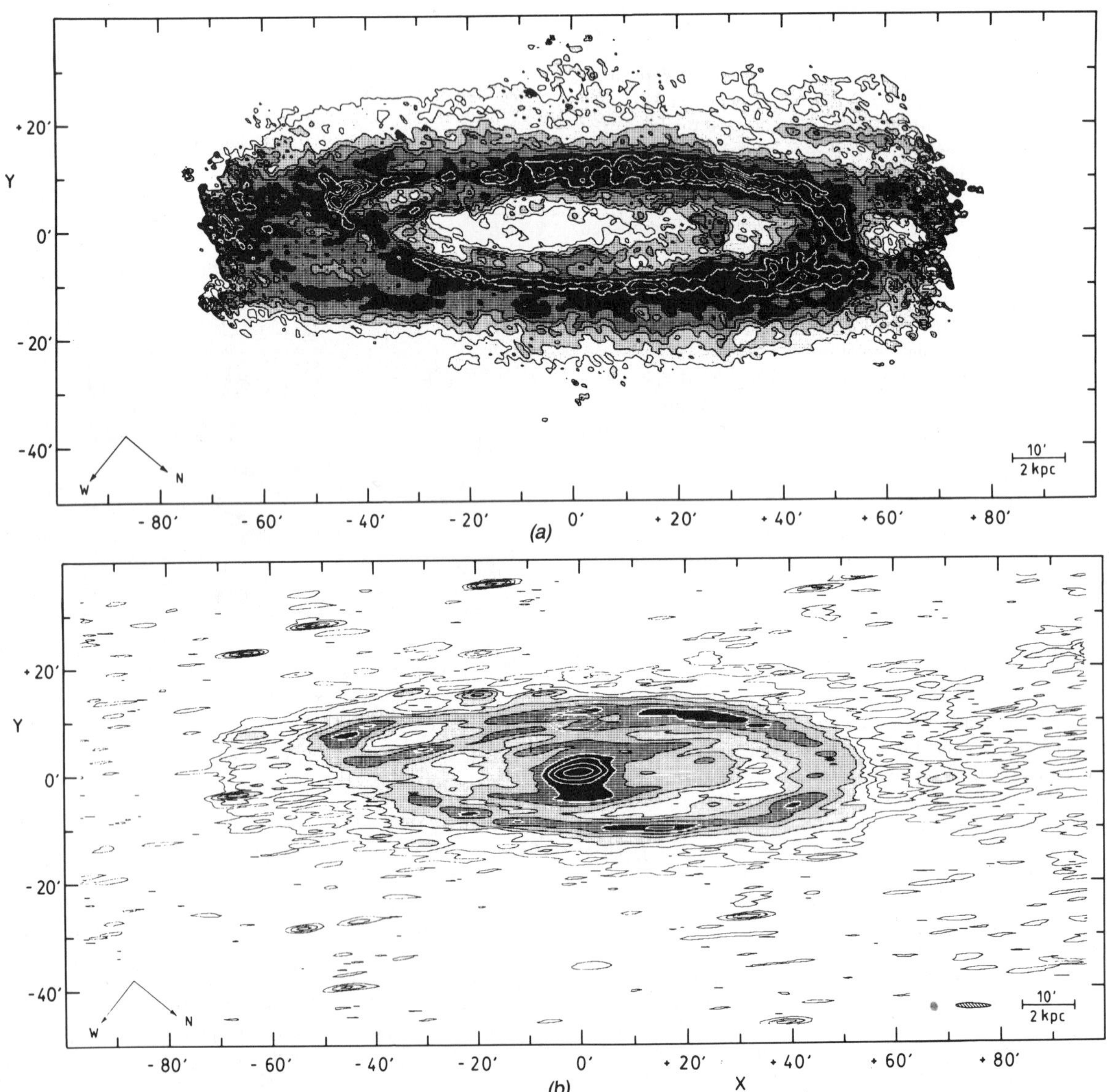

Figure 2. (*a*) The atomic hydrogen distribution in M31 as mapped with the Westerbork synthesis radio telescope in the Netherlands. The hydrogen distribution peaks around 10 kpc from the center. It extends further along the *X* direction than shown in this picture. (*b*) The infrared emission from dust as mapped by the *IRAS* satellite at a wavelength of 12 μm. The infrared emission also peaks around 10 kpc, but is bright in the nuclear region as well.

torus. Figure 2 shows the distribution of atomic hydrogen gas and infrared emission from dust. The amount of atomic hydrogen gas in M31 is comparable to that in our galaxy. However, molecular gas, as inferred from observations of the CO molecule, has lower surface densities than in our galaxy. Observations of regions of ionized gas and far-infrared emission from dust indicate that the overall level of star formation is also down from our galaxy by a factor of 5–10. In spite of this relatively low level of activity, detailed observations of the massive O and B stars in the spiral arms can and have been done. The massive stars appear to be similar to those in our galaxy, although there are some indications that stars above 60 $M_\odot$ may be absent in M31. The atomic hydrogen distribution is characterized by numerous holes and shell-like structures. These result from stellar winds from young stars and supernova explosions that deposit energy in the interstellar medium.

Information on the magnetic field structure in M31 has been obtained from radio observations of the synchrotron radiation that is emitted by fast-moving electrons spiraling around the field lines. The magnetic field is oriented along the spiral arms with a large degree of ordering. Some 30 supernova remnants have been discovered in M31 from optical images taken in various emission lines. Most of them are found in the spiral arms which indicates that they result from massive stars that exploded. About half of these have recently also been detected at radio wavelengths. The most recent supernova explosion occurred in 1885 close to the center of M31. This was not a massive star that exploded, but a supernova originating in a close double-star system.

THE CENTRAL REGION

The central region of the Andromeda galaxy differs in several respects from that of our galaxy. The most conspicuous difference concerns the lack of interstellar matter in the inner kiloparsecs in M31. No atomic or molecular hydrogen has been detected here, nor is there any evidence for ongoing star formation. This is very different from the situation in our galaxy, where active star formation is taking place at the center. Optical images do show small dust patches in the inner region; this dust is also detected in far-infrared images obtained with the *IRAS* satellite. Furthermore, ionized gas has been found but the mass involved is small, perhaps a few thousand solar masses. The structure of the ionized gas is intriguing, however. It shows spiral arms in a plane that is tilted from the principal plane of M31's disk. The size of this spiral pattern is about 1 kpc. It is also visible in radio continuum images. The gas presumably originates from mass loss by old stars. The hot central stars of planetary nebulae and collisions between gas clouds are responsible for ionizing the gas. The overall level of radio continuum emission from the central region of M31 is 10 times weaker than that from the central region of our galaxy. The emission is nonthermal in nature. Furthermore, there is no evidence for a compact, unresolved radio source in the center of M31, contrary to our galaxy.

High-resolution optical imagery of M31 done from a balloon-borne telescope in the mid-1970s shows a distinct nuclear component in the inner 1 or 2 pc of the light profile. Recent findings provide evidence for the presence of a black hole at the center with a mass of 10 or 100 million solar masses. The evidence comes from high-resolution spectroscopy of the center, which shows a large rotational velocity and velocity dispersion very close, at about 1 to 2 pc, from the nucleus. The presence of a black hole is usually postulated in galaxies with active nuclei; the M31 observations provide strong evidence that normal galaxies may also have black holes at their centers.

With the upcoming new generation of space- and ground-based instruments, M31 will once again be one of the prime targets for observations. Much remains to be learned, and much will be learned in the next decade.

Additional Reading

Beck, R. and Wielebinski, R. (1981). Radio waves from M31 and M33. *Sky and Telescope* **61** 495.
Hodge, P. W. (1981). *Atlas of the Andromeda Galaxy*. University of Washington Press, Seattle.
Hodge, P. W. (1981). The Andromeda galaxy. *Scientific American* **244** (No. 1) 92.

See also **Galaxies, Infrared Emission; Galaxies, Local Group.**

Antimatter in Astrophysics

Gary Steigman

All matter comes in particle–antiparticle pairs. As Paul Dirac demonstrated in 1928, this is required by any theory that is consistent with quantum mechanics, the special theory of relativity and causality. Particles and their antiparticles have the same mass and lifetime. Electrically charged particles, such as, for example, the electron and the proton, have antiparticles—the positron and the antiproton—with equal but opposite electrical charges. Other, electrically neutral, particles such as the photon are their own antiparticles; such particles are called self-conjugate. Five years after Dirac's prediction, the first antiparticle—the positron—was discovered by Carl D. Anderson in 1933. There then followed a hiatus of some 22 years before the antiproton was produced and detected at an accelerator at Berkeley by Owen Chamberlain and collaborators in 1955. Subsequent accelerator experiments at higher and higher energies provided convincing confirmation that all particles do indeed come in pairs. These experiments suggest further that particles have associated with them certain "quantum numbers" such as electrical charge, baryon number, lepton number, and so forth, which seem to be conserved in all reactions. If these conservation "laws," inferred from the experimental data, are exact —inviolable—then all matter is restricted to appear (creation) or disappear (annihilation) only as particle–antiparticle pairs. That is, if an electron is created in a high-energy collision, then so must a positron be created in the same collision. In addition, an electron can only disappear if it annihilates with a positron. This apparent symmetry in the laws of physics led Maurice Goldhaber in 1956 to speculate on the antimatter content of the universe. At the same time, Geoffrey Burbidge and Fred Hoyle considered some of the possible astrophysical consequences of antimatter.

In considering the problem of the amount and possible astrophysical role(s) of antimatter in the universe, it is useful to distinguish between two general categories of questions. First, we may inquire: Must the universe be symmetric? That is, do the known laws of physics require that there be exactly equal numbers of particles and antiparticles in the universe (Must the universe be symmetric?)? In contrast to the somewhat philosophical nature of this question, we may pose a more practical question: Is the universe symmetric? That is, empirically, based on terrestrial experiments and astronomical observations, is there evidence that the universe contains exactly equal amounts of matter and antimatter (Is the universe symmetric?)? In addressing this latter question, it will be necessary to consider the possible astrophysical role of antimatter.

SEARCHING FOR ANTIMATTER

Antimatter is trivially easy to detect. For a detector, the most rudimentary device will suffice. The sample and the detector are placed in contact and, if the detector disappears (matter and antimatter annihilate on contact), the sample was antimatter. Because antimatter cannot survive in the presence of ordinary matter (atoms and molecules made of neutrons and protons and electrons), it is clear that there are no macroscopic amounts of antimatter on the Earth (antiparticles are, of course, produced in high-energy collisions at accelerators and when cosmic rays impact atmospheric nuclei). Aside from the Earth, only the solar system and the galactic cosmic rays provide a sample of the universe that may be subjected to such a direct test. The Apollo series of lunar landings and the probes to Venus provided direct evidence that the Moon and Venus are made of ordinary matter. In a somewhat less direct manner, the solar wind that sweeps through the solar system establishes that there are no antiplanets in the solar system. The reason is that, were any of the planets made of antimatter, annihilation of the solar wind particles that strike them would have turned them into the strongest gamma ray sources in the sky (gamma rays—high-energy photons—are among the products of matter–antimatter annihilations).

If the solar system resulted from the collapse of a gaseous, presolar nebula, then any antimatter initially present would have annihilated before any solid bodies condensed because the annihilation rate is faster than the collapse rate. In this case, the solar system should consist entirely of ordinary matter. However, because condensed bodies of antimatter could survive indefinitely in the environment of the solar system, the capture of antiplanets could have led to the presence of macroscopic amounts of antimatter in the solar system. For this reason, the space probes and the solar wind have provided valuable—direct—information: There are no macroscopic amounts of antimatter in the solar system.

The only direct sample of extrasolar system material is provided by the galactic cosmic rays (the high-energy nuclei of atoms, mainly hydrogen and helium). Perhaps the debris of exploding stars (supernovae) or the accelerated nuclei of interstellar gas atoms, cosmic rays bring samples of the material content of the Galaxy. However, as they traverse the Galaxy, cosmic rays occasionally collide with interstellar gas nuclei and may be transformed. For example, in such encounters, cosmic ray nuclei of carbon, nitrogen, and oxygen can be broken up into lighter nuclei such as lithium, beryllium, and boron. From time to time, such cosmic ray–gas collisions are sufficiently energetic to produce a proton–antiproton pair (just as in collisions at a high-energy accelerator) or an electron–positron pair (the source of the positrons discovered by Anderson in 1933). Because any antiprotons in cosmic rays may be "secondary" (produced in collisions), they do not provide an unambiguous signal for the existence of "primary" sources of antimatter (e.g., antistars) in the Galaxy. Indeed, the very small flux of antiprotons found in cosmic rays is consistent with a secondary origin.

In contrast to antiprotons, no secondary antialpha particles (the nuclei of antihelium atoms) should be present in the cosmic rays. The discovery of even one antialpha particle in the cosmic rays would provide compelling evidence for the presence of macroscopic amounts of antimatter in the Galaxy. None has ever been found. From the limits to the cosmic ray flux of antialphas, it may be inferred that the fraction of antimatter in those parts of the Galaxy probed by the observed cosmic rays is less than 1 part in 10,000 ($\lesssim 10^{-4}$).

INDIRECT SEARCHES

The solar system and the cosmic rays provide the only sample of material in the universe that can be examined directly. To search further afield for antimatter, it is necessary to rely on indirect evidence.

The conclusion, from the absence of antimatter in cosmic rays, that the Galaxy is not matter–antimatter symmetric, receives further support from observations of "Faraday rotation." Polarized light (or radio waves) traversing the magnetized interstellar plasma will have its plane of polarization rotated. The amount of rotation (the "rotation measure" RM) depends on the wavelength of the light and on the column density of electrons (the number of electrons per square centimeter) along the line of sight through the Galaxy. For positrons along the line of sight, the sense of rotation would be opposite. There would be no significant net Faraday rotation observed if typical lines of sight through the Galaxy intersected roughly equally many regions and "antiregions." Indeed, by comparing the RM with the "dispersion measure" DM, which depends on the sum of the electron and positron (if any) column densities, it is confirmed that $N(e^-) - N(e^+) \approx N(e^-) + N(e^+ -)$. Faraday rotation provides "another nail in the coffin" of a symmetric Galaxy.

Because annihilation provides unmistakable evidence for antimatter, ordinary matter is an excellent probe for the presence of antimatter. Observations of the products of annihilation could provide indirect evidence for antimatter; the absence of annihilation products can lead to bounds on antimatter. The primary products of nucleon–antinucleon annihilation are pions ($\pi^{\pm}, \pi^0$); for annihilation in nonrelativistic collisions, approximately 5 or 6 pions (equal numbers of π^+ and π^- due to electrical charge conservation and approximately the same number of π^0) are produced. The charged pions decay very quickly to muons ($\mu^{\pm}$) and μ-neutrinos ($\nu_\mu, \bar{\nu}_\mu$) and the muons decay to electrons, positrons, μ-neutrinos, and e-neutrinos ($\nu_e, \bar{\nu}_e$). The neutral pions decay primarily to a pair of gamma rays. The end products of a typical nucleon–antinucleon annihilation are $\sim 3e^{\pm}$ which carry off $\sim 1/6$ of the total available energy ($2M_N c^2$), equal numbers of $\nu_e, \bar{\nu}_e$ which carry off comparable energy, twice as many $\nu_\mu, \bar{\nu}_\mu$ which carry off $\sim 1/3$ of the total annihilation energy, and ~ 3 high-energy (~ 200 MeV) gamma rays which carry off $\sim 1/3$ of the total annihilation energy.

Neutrinos are so weakly interacting that it is unlikely that any evidence for (or against) antimatter in the universe can be inferred from annihilation neutrinos. Neither are the annihilation $e^{\pm}$ pairs of much value as a probe for antimatter. The reason is that the cosmic rays contain primary electrons as well as secondary (produced in cosmic-ray–gas collisions) electrons and positrons. It is hopeless to separate a possible annihilation component from this background. Furthermore, $e^{\pm}$ are "tied" to local magnetic fields and lose energy rapidly via Compton scattering and synchrotron radiation. Therefore, annihilation secondary $e^{\pm}$ pairs are fated to die where they are born. Indeed, it is this fact that led to proposals that the strongest extragalactic sources observed (quasistellar objects, active galactic nuclei, etc.) might be powered by annihilation. However, rather than provide a panacea for the energy woes of these sources, annihilation merely exacerbates their already strained energy budgets. The problem is that to produce the observed radio/optical radiation from ~ 100 MeV $e^{\pm}$ requires such enormous magnetic fields ($B \gtrsim 10^4$–10^8 G) that most of the energy in such sources would reside in the magnetic fields.

Annihilation gamma rays provide the most useful probe for antimatter in the universe. Any site in the universe where matter and antimatter mix will be a gamma ray source. Because gamma rays are unaffected by magnetic fields and are scattered or absorbed only in environments of extremely high density (column density $\gtrsim$ several grams per centimeter squared), the distribution of observed gamma rays yields information on their sources. In particular, individual galactic and extragalactic sources can be identified and a diffuse galactic component can be separated from an isotropic (extragalactic) background. Because there are alternate (to annihilation) sources for the production of gamma rays (e.g., decay of cosmic-ray-produced π^0), the observations can be used to constrain the fraction, f, of the material in various environments that could be antimatter.

In our galaxy, the gamma ray emissivity per hydrogen atom restricts f to less than a part in 10^{15} ($f \lesssim 10^{-15}$). This tiny upper limit is not surprising when it is realized that, in the gaseous interstellar medium, an antiatom will survive for only ~ 300 yr before annihilating. Indeed, for a collapsing protogalactic gas cloud, the annihilation rate always exceeds the collapse rate.

Although antimatter cannot survive in a gaseous form in the interstellar medium, any condensed object made of antimatter (e.g., antistars) could survive almost indefinitely because most of the mass, being in the interior, is shielded from annihilation. Furthermore, because in such cases annihilation would be limited to the surface layers, antistars would not necessarily be strong sources of gamma rays. Still, the lack of galactic gamma ray sources constrains the distance to the nearest antistar to be more than 10 ly. More constraining, however, is the galactic gamma ray background which limits the possible number of antistars to $\lesssim 10^7$ ($f \lesssim 10^{-4}$). The lack of gamma rays from M31 suggests for that galaxy, too, $f \lesssim 10^{-3}$.

Although the evidence is overwhelming that the Galaxy contains no significant antimatter, could every second galaxy in the universe be an antigalaxy? To address this issue, consider clusters of galaxies and, in particular, the hot intracluster gas observed through its x-ray emission (via thermal bremsstrahlung radiation). Because the x-ray emission is due to collisions between electrons and nuclei (or positrons and antinuclei!), the observed x-ray emission can be related to the expected gamma ray emission if some of the intracluster gas were made of antimatter. Such a comparison of x-ray and gamma ray observations restricts the antimatter fraction to $f \lesssim 10^{-5}$. At least on the scale of rich clusters, every second galaxy is not an antigalaxy.

The problems with annihilation-driven strong sources (QSOs, AGN, etc.) have already been mentioned. The gamma rays provide yet another constraint. In annihilation, $\sim 10^4$ gamma rays are produced for every erg of energy carried off by the electron–positron pairs. Strong radio, infrared, and/or optical sources should, if annihilation powered, be strong gamma ray sources. The absence of observed gamma rays excludes antimatter as the fuel for these objects.

The presence of a smoothly distributed, hot, intergalactic gas is controversial at present. It is the absence of any absorption that requires such a gas—if present—to be very hot ($T \gtrsim 10^8$ K). Using the observations of the isotropic x-ray background to constrain the density of such a gas and the gamma ray background to bound its possible antimatter content, $f \lesssim 10^{-7}$. Either there is no antimatter in the intergalactic medium or there is no intergalactic medium.

Direct or indirect evidence could have revealed the presence of antimatter in the universe. On scales from the solar system, to the Galaxy, to clusters of galaxies, and beyond, the absence of evidence suggests the universe is not symmetric. Keeping in mind that the absence of evidence is not evidence of absence, it would nevertheless be churlish to ignore the data which strongly suggest a clear answer to the question, Is the universe symmetric? The observations suggest an asymmetric universe with little—if any—antimatter on macroscopic scales.

A SYMMETRIC COSMOLOGY?

Could the universe be symmetric between matter and antimatter and the absence of evidence be due to a very large scale separation between matter and antimatter? To address this issue, we turn to a consideration of the cosmological implications of a symmetric universe.

At very early times when the universe was very hot and very dense, neither molecules nor atoms, nor even nuclei, existed. The "matter" at such early epochs consisted of baryons (e.g., neutrons and protons or their constituent quarks) and leptons (e.g., electrons, muons, tauons, and their corresponding neutrinos), their antiparticle counterparts and photons, gluons, $W^{\pm}, Z^0$ bosons. The early universe was a hot soup of all fundamental particles. As the universe expands and cools, annihilation occurs reducing the number of massive particles in every "comoving volume" (a volume in the universe that expands with the universal expansion). If the universe were matter-antimatter symmetric and fully mixed, annihilation would have been so efficient that the density of matter in the universe today (when the cosmic background radiation is at a temperature of ~ 3 K) would be some 10 billion times smaller than what is observed. An initially symmetric (and well-mixed) universe would, today, be an empty universe. The observed mass density provides a strong clue that the early universe was not totally symmetric. At very early times, when particle–antiparticle pairs were abundant, a very small asymmetry ($\sim 10^{-10}$–10^{-9}) could account for the presently observed universe.

Suppose, instead, that total annihilation was avoided because—somehow—matter and antimatter were separated early on. To account for the observed matter density, it is necessary that such a separation have occurred when the universe was less than ~ 1 ms old. However, at that time, the amount of matter that could have been in causal contact (i.e., within the "horizon") is less than $10^{-6}\ M_{\odot}$. Because antimatter today (if present at all) must be separated from matter on scales at least as large as that of clusters of galaxies ($M_{CL} \sim 10^{15}\ M_{\odot}$), macroscopic separation could not have occurred in the early universe. Similarly, it is easy to see that purely statistical fluctuations are entirely inadequate to avoid the cosmological annihilation catastrophe.

In recent years the idea of "inflation" has come to play an increasingly important role in scenarios for the evolution of the universe. The effect of inflation (if it does, indeed, occur) is to cause a small (causally connected) region of the universe to expand exponentially ("inflate"). After inflation ends, it is conceivable that some regions contain an excess of matter and others contain an excess of antimatter. If these regions were at least the size of clusters, it could be that the observable universe is—on average—symmetric. This, however, would require extreme "fine-tuning" of inflation, the point being that it is "natural" in inflation to have one, small (causally connected) region expand to encompass the entire observable universe. An extremely delicate balance would be required to inflate up to scales at least as large as clusters but not as large as the observable universe. Antimatter may be out there but, beyond the horizon!

BARYON ASYMMETRY AND GRAND UNIFIED THEORIES

The evidence is overwhelmingly in favor of the conclusion that the universe is not matter–antimatter symmetric. Must the universe be symmetric? The experimental results that particles are created and annihilated in pairs suggested that there were some quantum numbers (generalized charges) that are conserved (e.g., baryon number, lepton number, etc.). However, grand unified theories, which are attempts to unify the strong and electroweak interactions, suggest that at very high energies (or, in the early universe, at very high temperatures), these conservation "laws" are violated. If these theories are correct, then, as Andrei Sakharov outlined in 1967, there is a recipe for generating an asymmetric universe from an initially symmetric universe. First, of course, there need to be interactions that violate baryon and lepton number conservation. The grand unified theories predict this will occur during the very early evolution of the universe (how early is model dependent). In addition, some other symmetries (charge conjugation and parity) must be violated as well. This, too, occurs in many grand unified theories. Sakharov's major contribution was to realize that another ingredient was crucial to the success of this recipe for an asymmetric universe. These symmetry violating processes must occur out of equilibrium. Here the expansion of the universe is indispensable. The symmetry violating interactions occurring very early in the evolution of the universe will, due to the dilution and cooling of the universe as it expands, have dropped out of equilibrium, leaving behind a tiny remnant: the matter–antimatter asymmetry of the universe. This relic, the legacy of the earliest epochs in the evolution of the universe, is (most likely) responsible for the observed matter–antimatter asymmetry of the present universe.

Additional Reading

Adair, R. (1988). A flaw in a universal mirror. *Scientific American* **258** (No. 2) 50.

Goldhaber, M. (1956). *Science* **124** 218.

Sakharov, A. (1967). *JETP Lett.* **5** 24.

Steigman, G. (1969). Antimatter and cosmology. *Nature* **224** 477.

Steigman, G. (1973). The case against antimatter in the universe. In *Cargèse Lectures in Physics* **6** 505.

Steigman, G. (1976). Observational tests of antimatter cosmologies. *Ann. Rev. Astron. Ap.* **14** 339.

See also **Background Radiation, Gamma Ray; Cosmic Rays, Observations and Experiments; Cosmology, Big Bang Theory.**

Artificial Satellites, Observational Phenomena

Paul D. Maley

Figure 1. A trail of star-like points marks the path of the Japanese EGP satellite on October 24, 1986, as photographed from Japan. Flashes of 3.5 magnitude are seen as sunlight glints or mirrored surfaces on the exterior of the spacecraft. Flashes are one way a satellite can manifest itself to astronomers. (*Courtesy NASDA.*)

When the first artificial Earth satellite was launched in 1957, it attracted worldwide media attention. Astronomers had very little reason to worry about it because it was an isolated curiosity. But before long satellites were being launched with lifetimes of hundreds and thousands of years. At the end of 1989, the growing population had reached nearly 6700 trackable satellites, rockets, and pieces of debris in orbit. In addition, there is an untold number of smaller objects 10 cm or less in size that cannot be tracked by normal radar methods.

Satellites appear as star-like bodies that receive their illumination from sunlight and then reflect it to the observer on Earth. It is to this property that we owe most of the observational phenomenology. There are several types of manifestations: (1) normal nightly flyovers visible to the unaided eye, telescope, or other recording device; (2) reentry into the Earth's atmosphere; (3) expulsion of material and subsatellites while the satellite is in orbit; (4) orbital debris; and (5) phenomena associated with ascent into space. Astronomical terms can also be applied to artificial satellites. For example, spacecraft have been seen to transit the Moon, undergo eclipses as they pass in and out of the Earth's shadow, and exhibit light curves like variable stars.

There have been a number of developments in the field over its first 33 years. Most recently, satellites have been found to cause interference in areas such as research into gamma ray sources, flare stars, photographic sky surveys, the Moon, and instrument testing. Invention of low-light television has made it possible to tape and rebroadcast satellite phenomena seldom (or never) seen by the general public such as rendezvous and reentry. Some previously reported "unidentified flying objects" (UFOs) have been found to be caused by sunlight reflecting from ascending rocket plumes. Theoretically predicted phenomena involving artificial vehicles entering the atmosphere at high speed have been verified since the launch of the Space Shuttle in 1982.

Shortly after the beginning of the space age, there were concentrated efforts to monitor and track satellite behavior using amateur astronomers around the world in programs such as "Moonwatch" and the "Western Satellite Research Network." But by the time of the landing of the last astronauts on the Moon, interest and funding for these projects had been terminated. It is probably accurate to say that optical and visual artificial Earth satellite observing went "underground" as an adjunct to amateur astronomy. On the other hand, radio frequency monitoring has progressed uninterrupted. Although not observational, this medium has generated substantial benefits from the deciphering of telemetry from covert Soviet probes which continues today based on the tireless efforts of the famed Kettering Group. Because much information on the development of the Soviet space program has been kept out of the literature, it was only natural that space detectives would enter the market to fill the void when previous organized satellite tracking programs were terminated.

As phenomena, active satellites in geostationary orbit, for example, are being considered as test targets for evaluating nightly seeing conditions with 1-m size telescopes, but for the most part their presence is transparent to astronomers. These bodies are confined to areas within 10° north or south of the celestial equator (depending on the observer's latitude) and easily reveal themselves by lack of significant proper motion with respect to the rotation of the Earth. The introduction of large-aperture Dobsonian-type reflecting telescopes means that it is now possible for geostationary satellites to be detected by greater numbers of astronomers. Seen through instruments of 32 cm aperture or larger, for example, they can be viewed by reflected sunlight almost the full duration of each night as 14th or 15th magnitude objects. During the equinoxes they can brighten to magnitude +9 or +10.

Some satellites orbiting between 800 and 2000 km have incorporated lamps that intentionally cast short moderate intensity flashes to the ground. Their purpose is geodetic calibration. The lamps would be turned on in darkness while passing over a location where they could be photographed. Other designs incorporate arrays of highly polished mirrors such as in the Japanese EGP spacecraft (see Fig. 1), which can be used in laser ranging technology.

THE SPACE SHUTTLE

Still closer to the Earth in the 200–600 km zone, manned spacecraft such as the reusable Space Shuttle have provided some ground-based observers spectacular glimpses of these unique satellites. Most missions have been launched into orbits that track no farther north or south than 28.5° latitude. But one or two flights per year have been sent into inclinations high enough for passage over most of the population centers of the world. As a phenomenon, the Space Shuttle has been the brightest artificial Earth satellite ever launched. Under its normal orientation the white painted wings and upper body combine with the shiny payload bay radiators to produce a traveling point of light that often reaches a visual magnitude of −4, competing with the planet Venus.

Persons situated near its final descent trajectory have witnessed a remarkable nighttime reentry phenomena as the spacecraft interacts with increasingly denser layers of the atmosphere. As the shuttle passes across the sky, one can easily detect a chemiluminescent glow in its wake. The region around the shuttle becomes illuminated forming a contrail-like channel 10° or more long; and the glowing area can remain visible for as long as 80–90 s after the shuttle passes by. The phenomenon is much like a fireball (very bright meteor), except that the shuttle remains intact as it descends toward its landing site. A similar description has been given for the reentry of Soviet Soyuz manned capsules.

Another less well known phenomenon accompanying the reentry is the auditory perception of the double sonic boom—a feature that has been commonly heard by television viewers as the shuttle is about to land. Sonic booms have been perceived on three missions prior to this writing while the shuttle was flying over central Texas at altitudes as high as 55–65 km. There has also been a hissing or whistling phenomenon that is very much analogous to some fireball reports.

The giant hyperbolic fuel tank, which initially remains attached to the shuttle as it begins its climb toward orbit, has also created fireworks as it has been visually observed from the Hawaiian Islands on its destructive reentry over the Pacific Ocean. The huge aluminum container was not intended to survive. Instead, it dissolves into a myriad of small blue and green pieces some as bright as −2 to −4 visual magnitude. Two interesting phenomena are then observed. First, a short chemiluminescent trail accompanies the descent. This is a by-product of friction with the atmosphere and excitation created by the high-velocity reentry that is initiated by the shuttle as it plunges toward the ground at 24 times the speed of sound. Second, some of the residue from the aluminum tank which has disintegrated into very small grains remains aloft as aluminum oxide particles. The material lingers at an altitude of around 80 km as a long stringy cloud-like feature that is broken in places. When reentry occurs in predawn hours, it remains invisible until sunrise. Then the aluminum cloud may be seen for up to several hours under direct solar illumination. Whereas the preceding discussions have related to night sky conditions, we now have an event that can be spotted in a clear daytime sky (Fig. 2).

Another unique nighttime phenomenon associated with most every shuttle mission is the water dump cloud. Waste and supply water tanks are filled with nonrecyclable liquid by-products generated by the shuttle's fuel cells and astronaut-related activity which must be periodically expelled. When this occurs near the day/night terminator, a cloud as much as 5° in length may be seen without optical aid streaming tenuously from the star-like shuttle as it streaks across the sky. The cloud is composed of ice crystals of varying size and dimension which quickly decay out of orbit; however, in late 1987, the shuttle crew from mission 61-A did report seeing the stream of particles from an earlier water dump, appearing like a "meteor shower," collide harmlessly with the shuttle hours after it was released.

Figure 2. Spacecraft phenomena are normally only seen at night. However, when a large satellite reenters the atmosphere, it can leave a residual silvery cloud of small particles that is suspended at high altitudes. This photograph is of the remnants of the Space Shuttle External Tank 45 minutes after its reentry on April 6, 1984. The Mauna Loa volcano can be seen erupting in the foreground. (*Photograph by the author from Mauna Kea, Hawaii.*)

SATELLITE PREDICTION AND OBSERVATION

Satellites must have orbits that are well enough known to be able to permit their future prediction and observation. There are a number of software programs designed for personal computers that can perform this task. Forecasting is not a problem for most orbit configurations. Yet there are some regimes where the generation of accurate look angles are difficult, if not impossible. The process of falling out of orbit as described previously is the most obvious. In the last few days in orbit a satellite may be affected by unstable behavior or by shifting of its frontal surface area. The drag coefficient then markedly changes as it plows through the atmosphere, thus causing confusion in propagating orbital parameters that are not very fresh. Another situation applies to satellites that have extremely eccentric Earth orbits such as the scientific spacecraft *Granat* launched in late 1989 whose perigee is around 2000 km but whose apogee extends out to 200,000 km. This results in the orbit being severely influenced by the Moon as well as the Earth and complicates not only the tracking process but the ability to forecast in advance just where the satellite will appear. Satellites, like the PAGEOS orbited in the 1960s, are not always metal shells but have been designed in balloon form. When the balloon loses its shape and collapses, fragments become limp, unformed masses that are "tossed about" by effects of changes in the solar flux levels that vary during the solar cycle. As the atmosphere becomes more dense toward the high end of the cycle, the balloon pieces encounter a moving ocean of heavier drag so that forecasts of the future orbital positions are badly degraded.

The majority of satellites that are in altitudes low enough to contact significant atmospheric friction will burn up as they reenter, most of the time going unnoticed and having virtually no impact on the environment. Other reentry processes are sporadically described as behaving very similar to fireballs. On very rare occasions surviving fragments are recovered.

Much can be learned about satellites through a program of periodic optical examination. A single passage can yield data on a satellite's reflectance and its rotational characteristics. Over months and years we have used a simple technique known as visual photometry to acquire information on groupings of spacecraft using optical instruments as small as 7×35 binoculars. Families of satellites exist similar in nature to asteroidal families, classified by their orbital parameters. Also many satellites can be "typed" like variable stars and information gathered on their behavior. More advanced studies can be made with much larger telescopes on such bodies as tiny calibration subsatellites ejected from certain parent spacecraft. Another area to probe is the size of a satellite's solar arrays which can be a measure of the electrical power needed for its operation. Satellite structural designs are evolutionary because a common purpose such as navigation or geodesy is served by new generations of vehicles. Changes in reflectivity from one member of a family to another can signal an improvement in technology that had thus far remained unannounced.

Even when satellites are on their way to orbit, there have been instances of unusual phenomena accompanying them. Mass hysteria has been the result in the Soviet Union and in several other countries after Soviet rocket launches which were reported as UFO phenomena. Colorful displays of rocket fuel plumes illuminated by very favorable conditions could be seen for hundreds of kilometers by unwary citizens.

Chemical canisters are sometimes launched aboard rockets or ejected from orbiting spacecraft. Their contents are intended to interact with the magnetosphere so scientists can monitor their progress over time. A 1980 project called AMPTE released elements such as lithium and barium at very high altitudes (100,000 km) producing reddish clouds of visual magnitude +4. Other lower-altitude chemical releases after sunset have formed clouds several degrees across and much brighter. These too have been detected without optical aid; but often the public is not informed until after reports to the media have been received. This same situation exists for the reentry of satellites and rocket bodies which sometimes provide highly visible light shows across metropolitan areas. The reason that the times and places of reentries are not advertised in advance is due to the very unreliable nature of the prediction process.

Low-light level video recording beginning in the early 1980s has made it possible to transcribe satellite phenomena. We have used the video medium to broadcast such interesting satellite events as the rendezvous of the Space Shuttle *Columbia* and the Long Duration Exposure Facility payload on January 25, 1990, on a national television network. The approach of the two craft was only visible to the unaided eye from a very narrow geographical region of about 300 km^2, but through the television medium was transmitted afterward so that millions of people could observe the event from the comfort of their homes.

SPACE DEBRIS INTERFERENCE

Historically, the presence of the Earth satellite population has had no appreciable impact on ground/space-based astronomy. Earliest manifestations were the streaking of long-exposure photographs of large areas of the sky in the hours just after sunset and immediately before sunrise. The moving lines traced out by satellites have been documented on all-sky survey plates but are mainly just bothersome, making otherwise pristine photographs appear fouled. Other satellites leave their marks through pinpoint flashes or flares that undulate across the frame.

The first indication that satellite passages could directly impact the celestial discovery process was identified in 1987. Sporadically observed flashes in the night sky believed by the discoverers to be evidence of bursts of energy in the visual spectrum from a gamma ray source were shown to be caused by sunlight glinting from the metallic surfaces of artificial Earth satellites. The primary piece of evidence was a very bright flash that had been photographed in the location where this "source" was presumed to be. This photograph was found to coincide with the passage of a defunct Soviet spacecraft across the flash point. Flashes lasting less than 1 s in duration could be correlated to the passage of still other spacecraft during this episode. Glints reaching naked-eye brightness have therefore been seen at least as far as 36,000 km from the Earth. This highlights the fact that there is so far no effective way to cleanse the near-Earth environment of orbital debris except by waiting for the process of natural decay. During periods of high solar activity, the process of sweeping out this debris is accelerated as the atmosphere becomes increasingly dense, and satellites encounter greater resistance when traveling through it.

A second discovery that relates to a flash photographed on the dark side of the waxing crescent Moon now seems to be related to the same phenomenon—sun reflecting from a bright solar panel of a satellite that had been inoperative for two years. At first glance the occurrence of a natural lunar event such as outgassing and ionization or even a meteoritic impact might be supposed. But after evaluating the orbits of nearly 5700 space objects, there is sufficient evidence that points to a disabled U.S. weather satellite being the culprit.

Further interference is being found in flare star projects where satellites have transited fields of view and induced simulated flares into photographs. There have also been optical surveys of flare stars that have encountered similar sightings of a "new star" flaring up and then disappearing after seconds or minutes. Some of these cases may be due to satellites.

CONSIDERING SPACE AS AN ENVIRONMENT

Future designs of large Earth-orbiting space structures are subjects for consideration. Diagrams of the U.S. Space Station show very large solar arrays, portions of which are as expansive as an eight-story building. Reflected light from a giant array could be extraordinarily brilliant. Others have wanted to build artistic configurations which have no material purpose or benefit that would potentially reflect so much sunlight to the ground as to be considered "environmental light pollution." A continuing challenge will be to find ways to purge (as well as designs to minimize future debris) existing debris rather than let meandering pieces of space junk pose permanent collision and light pollution hazards.

At least one space project being designed is taking into account these considerations. A plan to launch two large mylar balloons tethered together during 1992 has been formulated so that its orbital lifetime will not exceed six months while still enabling its mission to be accomplished. If the project is approved the twin balloons would be observable from the ground with the unaided eye. The primary goal would be to test the material strength of tether lines in the space environment and to study free-flying tether dynamics. Concern regarding the sanctity of near-Earth space has apparently affected military applications. A tendency for a certain type of bright Soviet satellite to be commanded to destruct has been halted, this after a continuous pattern over a number of years. It has been shown that low-orbit explosions generate thousands of fragments which, though very small, could disable an active satellite upon contact.

Artificial satellites are much more than casual phenomena. There now exists a definite connection between them and sporadic interference with ongoing scientific research. The satellite population has become an active part of the nightly landscape. It is time that the public, plus casual and serious astronomers, become cognizant of their full effect on the celestial environment.

Additional Reading

Johnson, N. L. (1989). *The Soviet Year in Space 1988.* Teledyne Brown Engineering, Colorado Springs.

Maley, P. D. (1985). Earth based observation of the STS-11 Space Shuttle mission on orbit and during reentry. *Acta Astronautica* **12** (No. 10) 755.

Maley, P. D. (1987). Specular satellite reflection and the 1985 March 19 optical outburst in Perseus. *Astrophys. J.* (*Lett.*) **317** L39.

Maley, P. D. (1991). Space debris and a flash on the moon. *Icarus.*

Oberg, J. E. (1988). UFO's in the mind's eye. *Sky and Telescope* **75** 572.

Schaefer, B. E., Barber, M., Brooks, J. J., DeForrest, A., Maley, P. D., McCleod, N. W., McNeil, R., Noymer, A. J., Presnell, A. K., Schwartz, R., and Whitney, S. (1987). The Perseus flasher and satellite glints. *Astrophys. J.* **320** 398.

See also **Meteors, Shower and Sporadic.**

Asteroid and Comet Families

Andrea Carusi

Objects in the solar system are usually divided into two broad classes, depending upon their size. "Major" bodies include the Sun and the planets. "Minor" bodies, in their turn, are grouped into several subclasses, mainly discriminated by their dynamical and physical properties: satellites of planets, asteroids, comets, meteoroids, and dust.

The two major subclasses, asteroids and comets, include an enormous number of objects: observed asteroids, with well-determined orbits, permitting their recovery practically at each orbital period, amount to almost 4000, but their actual number for sizes greater than 1 km is thought to be greater by at least one order of magnitude; the number of observed comets, both of long and short period, is not that high (around 1200), but cometary theories suggest that the total number of comets bound to the Sun is in the range 10^{12}–10^{14}. In addition, both populations are not evenly distributed in size, location, physical, and dynamical properties. It is, therefore, natural to make further subdivisions within each population, with the goal of a better understanding of their internal structure, origin, and evolution from the birth of the system to the present epoch.

With the exception of comet P/Halley (P/ means periodic, i.e., of more than one apparition), no asteroid or comet has been observed closely by a space probe: All the information available on these objects, up to now, comes from ground observations and Earth-orbiting satellites. Important characteristics, such as sizes, shapes, surface features, and compositions, are known for only a small fraction of objects. This is the principal reason why the subdivisions of the asteroid and comet populations have been made mainly on dynamical grounds.

ASTEROID FAMILIES

The distribution of asteroids in the solar system varies dramatically with the distance from the Sun (see Fig. 1). Most objects have orbits with a semimajor axis between 2.2 and 3.3 AU. These constitute the "main belt" of asteroids. A small number of bodies revolve around the Sun on closer orbits, sometimes crossing those of the inner planets (Mars, Earth, Venus, and Mercury). Other small groups of asteroids reside on the other side of the main belt, at a distance larger than 3.3 AU from the Sun.

Inside the main belt, the distribution of asteroids is marked by distinct gaps, discovered by Daniel Kirkwood in 1867. The values of semimajor axes at which these gaps are located correspond to orbits resonant with that of Jupiter.

The most pronounced Kirkwood gaps occur at 2.06, 2.50, 2.82, and 3.28 AU and correspond to the resonances 1/4, 1/3, 2/5, and 1/2, respectively (e.g., an asteroid at 2.06 AU would have a period 1/4 of Jupiter's). Less pronounced gaps occur at 2.96 and 3.03 AU (resonances 3/7 and 4/9). In contrast, there are three groups of asteroids, beyond the outer border of the main belt, that are located around three more resonances with Jupiter's motion. They are

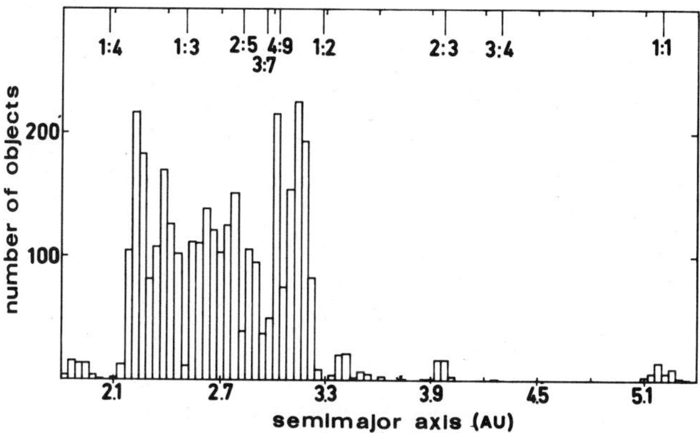

Figure 1. Distribution of numbered asteroids according to their orbital semimajor axis (last numbered asteroid in this plot: 3495). The "main belt" extends from roughly 2.2–3.3 AU. On top of the figure are marked the positions of the principal resonances with Jupiter's motion. Note the correspondence of these with the Kirkwood gaps, in the main belt, and with the groups of Hildas (2:3), Thule (3:4), and Trojan asteroids (1:1).

named the Hilda, Thule, and Trojan groups, at 3.97, 4.29, and 5.2 AU (2/3, 3/4, and 1/1 resonances). It appears, therefore, that the spatial distribution of asteroids is of two types: In the main belt the objects are mainly located between resonances that produce deep, narrow gaps; outside the main belt they are grouped near resonances, and the space between them is almost empty.

In a series of articles, between 1918 and 1933, K. Hirayama noted that many asteroids that cluster at similar values of semimajor axis also have similar values of orbital eccentricity and inclination. He called these clusters "families," with the implication that these asteroids are genetically related. The search for families was carried out by Hirayama looking at asteroid distributions on bidimensional (e.g., semimajor axis vs. eccentricity) maps (with the exception of a work published in 1919). His particular family identifications were not all supported by subsequent quantitative estimates, but the general idea that families exist has been confirmed.

The importance of this discovery is evident: If an asteroid family formed by the fragmentation of a single, disrupted parent body, we may have the unique opportunity to look "inside" a planetary body by simply analyzing the composition of various members of the family. On the other hand, statistics of the occurrence of asteroid families would give some indication of the rate of disruption of asteroids in the belt; this, in turn, would provide a quantitative check of validity for theories of the collisional evolution.

For these reasons many investigations have been performed since the time of Hirayama, searching for other families. In 1951, Dirk Brouwer was the first to use a planetary theory to compute the so-called "proper elements," which are intended to eliminate as much as possible the dynamical "noise" introduced by the accumulation of planetary perturbations in time (mainly of Jupiter). As a matter of fact, the clusters identified by Hirayama as families appear even more concentrated if proper elements are used; Brouwer found a number of additional families, not recognized by Hirayama.

In 1969, J. R. Arnold repeated the search on a larger sample of asteroids, using for the first time an electronic computer and a statistical method to identify possible families. He also found families in addition to those of Hirayama, but the agreement between his and Brouwer's families was poor. Further searches have been performed more recently by B. A. Lindblad and R. B. Southworth, Andrea Carusi and E. Massaro, and James G. Williams, and Yoshihide Kozai.

In the meantime, new physical data on asteroids have become available thanks to large-scale observational campaigns and many researchers have tried to interface these data with the dynamically identified families.

Unfortunately, the six classifications of asteroids in families do not agree completely. As shown in 1982 in a comparative review by Carusi and Giovanni B. Valsecchi, there may be several reasons for the disagreement: different methods for computing proper elements, different criteria for accepting or rejecting family members, different sizes of the available data sample, and/or different computational techniques. In fact, only three families have been identified by all authors: Themis, Eos, and Koronis. These are also the first families recognized by Hirayama in 1918, and they have systematically grown in time, as new asteroids were being discovered. The three concentrations appear very clearly in semiaxis–eccentricity and semiaxis–inclination plots (see Fig. 2). Two more Hirayama families have been given a wide consensus: Maria and Flora. The first of these, however, has not grown with improved statistical sampling at the same rate of the other three, whereas Flora has been frequently divided in a number of subfamilies. A sixth Hirayama family, Phocaea, has always been identified, but it seems to represent a group of objects isolated for dynamical reasons (secular resonances), not genetic ones.

The search for families is very much complicated by the background structure of the asteroid belt. One of the major problems is the definition of the "background." The presence of the Kirkwood

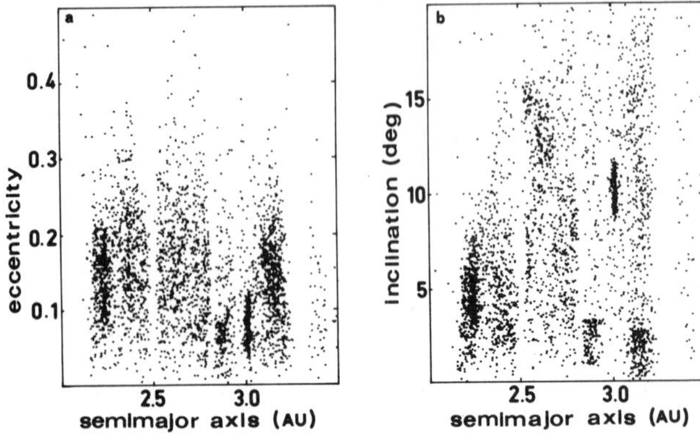

Figure 2. (*a*) Distribution of numbered asteroids (up to number 3495) in the plane semimajor axis–orbital eccentricity; (*b*) distribution of the same asteroids in the plane semimajor axis–orbital inclination. The Hirayama families Themis, Eos, and Koronis are the three clusters at 3.2, 3.0, and 2.9 AU. The diffuse cloud between 2.2 and 2.3 is the Hirayama family Flora. The plots have been drawn using osculating orbital elements.

gaps tends to mask the "original" distribution of asteroids. Also, theories for the computation of proper elements often fail close to the gaps, because oscillations of orbital parameters are not completely predictable. Moreover, in recent years numerous selection effects have received attention: among others, the distribution of observatories on the Earth, the prevailing observing condition in the northern hemisphere (where most observatories were located until recently), and the distribution of asteroids in inclination.

It has also been shown that many proposed "families" have no meaning from a genetic point of view, because the supposed composition of their members is inconsistent with the appurtenance to a single primeval body.

Notwithstanding these difficulties, it is commonly thought that a precise dynamical classification of asteroids in families is a prerequisite for any physical interpretation of the evolution of the asteroid population. The classification still needs to be refined, using more than one planetary theory and several different techniques.

COMET FAMILIES

Completely different problems arise from an analysis of the distribution of comets. Unlike asteroids, these objects are less confined within the solar system: They are subject to strong perturbations, especially by the giant planets, which can drastically change their orbital paths. The majority of comets is thought to reside in a wide reservoir with no precise limits, well outside the planetary region called the Oort Cloud. From time to time passing stars (or other perturbing mechanisms) remove comets from the Oort Cloud, forcing some paths through the planetary region where they can interact with the gravitational fields of the major planets (long-period comets). By a sequence of close encounters with the planets, comets may then be confined in the inner regions of the solar system, where they rapidly decay due to the disruptive action of the Sun's heat (short-period comets).

Families like those proposed in the asteroidal case do not exist among comets. We do know of a few cases of splitting of cometary nuclei, but the minor fragments decay very rapidly, usually disappearing within one revolution. The only known cases of comets with a possible common origin are given by the Kreutz group (long-period, Sun-grazing comets with orbits that are exceptionally similar), and, among short-period comets, the case shown quite recently of P/Neujmin 3 and P/Van Biesbroeck, which are probably the fragments of a comet that split around 1850.

In some respect meteor streams could be thought of as "cometary families," because meteoritic material comes from the gradual disintegration of comets. This terminology, however, is never used.

The existence of other associations among short-period comets has been proposed looking at the aphelion distances: They tend to group at distances from the Sun corresponding to the semiaxes of the major planets (the so-called "planetary families of comets"). However, it has been recently shown that the associations with Saturn, Uranus, and Neptune are fictitious and are due to the existence of librations about high-order resonances with Jupiter.

Among short-period comets there are groups of objects that present similar dynamical characters. It is not yet clear, however, to what extent members of one group may evolve so as to become members of another group.

One of these groups is formed by the so-called "Halley-type" comets: These generally have periods between 20 and 200 years and orbits of high eccentricity. Their main characteristics are a low probability of gravitational interactions with planets and the presence of retrograde orbits. High-order resonances with Jupiter's motion have been found among these comets.

Other groups may be tentatively identified with comets librating around low-order resonances with Jupiter (mainly the 1/2 resonance), or with comets having low-velocity encounters with Jupiter, leading often to temporary satellite captures and major orbital variations. All these groupings, however, denote similar orbital evolutions and are not indicative of a common physical origin.

If asteroid families are formed by the disruption of a common parent body, it is understandable why similar clusterings do not exist among comets. Collisions have very low probabilities in the Oort Cloud, and even less among comets observed in the planetary region at present. Collisions may have been common in the early stages of accumulation of planets, when comets were still orbiting in the planetary region, but subsequent orbital evolution must have completely obliterated any family structure formed in that way.

Additional Reading

Carusi, A. and Valsecchi, G. B. (1982). On asteroid classifications in families. *Astron. Ap.* **115** 327.

Carusi, A. and Valsecchi, G. B. (1985). Dynamical evolution of short-period comets. In *Interplanetary Matter*, Z. Ceplecha and P. Pecina, eds. Astronomical Institute of the Czechoslovak Academy of Sciences, Prague, p. 21.

Gradie, J. C., Chapman, C. R., and Williams, J. G. (1979). Families of minor planets. In *Asteroids*, T. Gehrels, ed. University of Arizona Press, Tucson, p. 359.

Valsecchi, G. B., Carusi, A., Knezevic, Z., Kresak, L., and Williams, J. G. (1989). Identification of asteroid dynamical families. In *Asteroids II*, R. P. Binzel, T. Gehrels, and M. S. Matthews, eds. University of Arizona Press, Tucson, p. 368.

See also **Asteroids; Comets, Dynamical Evolution; Comets, Oort Cloud.**

Asteroids

Clark R. Chapman

Asteroids are a population of small planetary bodies, ranging in size from 950 km diameter (Ceres) down to countless boulders and meteoroids. Most asteroids orbit in roughly circular orbits in the main asteroid belt, located beyond the orbit of Mars, from about 2.2–3.2 AU. However, any heliocentrically orbiting body that fails to show cometary activity is termed an asteroid. The Aten asteroids have semimajor axes less than the Earth's, while 2060 Chiron is in the outer solar system. Actually, Chiron has shown some comet-like activity, and it may well be that many other asteroids beyond the main belt as well as nearly dead comets on Earth-crossing orbits (which belong to the so-called Apollo group) may contain volatiles that occasionally give rise to cometary behavior.

Approximately 4000 asteroids have been cataloged; when their orbits are sufficiently well known, they are given numbers and the discoverer may select a name. Most asteroids have relatively modest orbital eccentricities (about 0.15) and inclinations (averaging about 10°). Our knowledge of the main-belt population is incomplete below about 40 km diameter. Of the estimated 1000 Earth-approaching asteroids over 1 km diameter, less than 10% have been cataloged so far.

Very few asteroids ever show resolvable disks in ground-based telescopes, so most of what we know about their properties is based on photometric, spectral, and positional data from their unresolved, star-like images. Naturally, discovery and measurement of smaller, darker, and more distant asteroids are more difficult, so data must be corrected for such observational biases before inferences are made about statistical properties of the population.

A rather different set of biases affects the infrared catalog produced in the mid-1980s by the *Infrared Astronomical Satellite* (*IRAS*). Its data obtained in several bandpasses record the thermal emission from asteroids. Because darker asteroids are warmer than high-albedo asteroids at the same distance from the Sun, the *IRAS* catalog gives weight to dark objects, which predominate from the middle of the main belt outward. From a combination of infrared radiometric brightness and optical magnitude, it is possible to solve for the albedo (with only minor modeling uncertainties) and hence for the diameter of an asteroid. This method of determining albedos and sizes has been calibrated by occasional well-observed occultations of stars by asteroids. So far, there is at least crude information on the sizes and albedos of nearly half of the cataloged asteroids.

Another nonoptical observational technique is radar. Several dozen asteroids have been observed, chiefly by Steven J. Ostro using the Arecibo radar installation. Radar provides independent data that help constrain size, spin, orbital motion, and surface properties. Very high radar reflectivities for a few asteroids have demonstrated that they are probably of metallic composition. Several promising observational techniques have yet to bear much fruit, including ultraviolet measurements from Earth-orbiting satellites, radio, and speckle interferometry. The Hubble Space Telescope has the capacity to revolutionize observations in the ultraviolet and visible, but in the initial time allocation, the asteroid community received no observing time. No spacecraft has yet visited an asteroid, but a powerful array of instruments aboard the Galileo Jupiter mission may focus in on the small asteroid Gaspra en route to Jupiter.

The distributions of orbital properties of asteroids between the orbits and Mars and Jupiter is far from homogeneous. The Kirkwood gaps (associated with ratios of small integers between asteroid orbital period and Jupiter's period) were recognized long ago. Early this century, clusters of orbits in a-e-i space (a = semimajor axis, e = eccentricity, i = inclination) were recognized by K. Hirayama; about a dozen families are generally recognized and thought to represent the collisional disruption of a larger precursor body. More than 100 smaller families have also been proposed, but the methodology of family identification has been questioned and the properties of members of many of the proposed families are difficult to understand. There are also gaps in other projections of orbital phase space, generally explicable by Jupiter's gravitational influences.

The masses and bulk densities of asteroids are generally not known, but the few that have been measured appear to have densities similar to ordinary rocks and meteorites. The size distribution of asteroids crudely follows a power law with an index that results in most of the mass being concentrated in the largest bodies. The total mass of all the asteroids (about 5% of that of the Moon) is only about three times the mass of Ceres alone. In detail, the size distribution shows departure from a simple power law, probably reflecting the role of gravitational cohesion and different bulk properties in affecting the outcomes of interasteroidal collisions. Asteroids are sufficiently numerous within the confines of the toroidal region of the solar system in which they orbit, and they have sufficiently high relative velocities (typically 5 km s^{-1}), so that most asteroids must have suffered major, catastrophic collisions during the last few billion years. The larger ones may have reaccreted from the fragments, but most of the original population has been fragmented into the power-law-like size distribution we observe today.

Asteroid shapes and spins also reflect their history of collisional evolution. These properties are inferred from photometric observations made around the world. A few small asteroids spin once every couple hours, but the most common period is 8 or 9 hours, and several others spin anomalously slowly (once every few weeks). There are some differences in average spin periods for asteroids of different sizes and compositions. There has been considerable interest in some large, rapidly rotating objects that have unusually large light curve amplitudes. Pole direction, sense of rotation, and three-dimensional body shape are properties that can be inferred, with some difficulty, for those asteroids that have been precisely observed over many years. Some asteroids are roughly spherical, but more commonly they have shapes similar to natural rock fragments.

The visible and infrared spectra of sunlight reflected from asteroid surfaces exhibit a variety of different colors and absorption bands, which have led to considerable understanding of the gross mineralogy of asteroids. Laboratory measurements of meteorite spectra help provide a context for understanding asteroid compositions. When excellent spectra are obtained throughout the visible and IR (with high precision, good spectral resolution, and good time resolution as the asteroid rotates), considerable insight can be gained about the detailed surface mineralogy. For most asteroids, however, there is much less precise data, for example, eight-color spectrophotometry in the visible and near-IR alone. Thus a compositional "taxonomy" has been established, which enables many hundreds of asteroids to be placed into one of about a dozen different classes.

When corrected for observational bias, the spectral data show that most of the asteroids are of the very low-albedo C type, exhibiting relatively neutral reflectance spectra, except for an ultraviolet absorption and, sometimes, a 3-μm water-of-hydration band. These asteroids look very much like laboratory spectra of the CM carbonaceous chondritic meteorites and are almost certainly similar in composition to those meteorites. There are several other low-albedo spectral groups, including the reddish P types, and the even redder D types, which tend to predominate beyond the main asteroid belt. Within the belt, a small number of asteroids show C-like spectra (termed B, G, and F), which may reflect degrees of metamorphism of C-like material. The presence or absence of the 3-μm band indicates whether the surface minerals have been subjected to hydrous alteration, perhaps by thermal heating of the icy interiors of these primitive C-type bodies.

A large fraction of inner-belt asteroids are of the higher-albedo S type, characterized by reddish spectra, a deep UV charge transfer band, and one or more near-infrared bands indicating the presence of the iron-bearing minerals pyroxene and olivine. A straightforward interpretation of these spectra likens these asteroids to the stony-iron meteorites. However, the same bulk mineralogy is also characteristic of the ordinary chondritic meteorites, which are the most common kind of meteorite in our museums. Only a couple of tiny asteroids have been found in Earth-approaching orbits that look like ordinary chondritic meteorites, and none have been found in the main belt. Because of the commonness of ordinary chondrites and the fact that there are known dynamical mechanisms to deliver meteorite fragments from near the 3:1 resonance in the main belt, where S types are common, many cosmochemists expect that the S class contains ordinary chondrite parent bodies, perhaps with their spectra slightly altered by regolithic processes.

The resolution of this problem could solve a number of problems in understanding the origin and evolution of the asteroids and concerning their relationship to meteorites.

A small number of asteroids exhibit spectra indicative of highly differentiated assemblages. A few asteroids (termed A type) are pure olivine, like the mantle of the Earth. Indeed, it is surprising that so few A-type asteroids occur, because olivine mantles should make up a large fraction of the bulk of differentiated asteroids. Some moderate-albedo, slightly reddish spectra were interpreted more than a decade ago as possibly indicating pure nickel–iron, although the featureless spectra were not very diagnostic. More recently, several of these M-type objects have been observed by radar and found to have very high radar reflectivities, almost certainly confirming that they are made out of pure metal. Presumably they are the metallic cores of thermally differentiated bodies, collisionally stripped of overlying crustal and mantle rocks. There is one example of an intact differentiated body: Vesta is covered with basaltic lava, but somehow has managed to avoid major disruptive impacts.

One of the most significant aspects of asteroid compositions is that they vary with distance from the Sun. At the inner edge of the main belt, there are a number of E-type asteroids (high-albedo objects interpreted as being made mostly of enstatite). S types predominate in the inner one-third of the main belt. C types predominate in the outer two-thirds. P types predominate in the Cybele and Hilda regions beyond the main belt, and D types predominate among the Trojans in Jupiter's orbit. It is worth noting here that recent assessments of the population of Trojans suggest that they are nearly as numerous as the main-belt population. Perhaps there are other asteroid populations still farther out in the solar system.

This compositional gradient reflected by asteroids from 1–5 AU has generally been thought to reflect the variation in composition of solar nebular condensates. In this sense, the asteroids would represent a tableau of primordial conditions, relatively unmodified by subsequent thermal, collisional, or dynamical processes. (Of course, a minority of asteroids must have been melted to yield the relatively rare differentiated types, and others have certainly been collisionally destroyed or dynamically ejected from the belt.) More recently, several researchers have emphasized the degree of apparent thermal alteration indicated by many asteroid taxonomic classes (especially so if the S types are taken to be stony-irons rather than primitive ordinary chondrites). It is possible that the compositional gradient is dominated by thermal processing, or even by differential collisional processing. It has even been suggested that part of the inferred gradient may actually be due to asteroids being dynamically transported into parts of the asteroid belt from elsewhere in the inner solar system, so the asteroid belt is kind of a "dumping ground" for stray planetesimals. In conclusion, it remains uncertain to what degree the asteroids are a relatively pristine remnant of primordial planetesimals.

Additional Reading

Binzel, R. P., Gehrels, T., and Matthews, M. S., eds. (1989). *Asteroids II.* University of Arizona Press, Tucson. This is the definitive technical book on asteroids.

Chapman, C. R. (1990). Asteroids. In *The New Solar System*, 3rd ed. J.K. Beatty and A. Chaikin, eds. Sky Publishing Corp., Cambridge, MA and Cambridge University Press, New York, p. 231. This is an up-to-date illustrated review article.

Cunningham, C. J. (1988). *Introduction to Asteroids.* Willmann-Bell, Inc. This paperback volume provides a history of asteroid studies.

Gehrels, T. and Matthews, M. S., eds. (1979). *Asteroids.* University of Arizona Press, Tucson.

See also **Asteroid and Comet Families; Asteroids, Earth-Crossing; Meteorites, Origin and Evolution.**

Asteroids, Earth-Crossing

Michael J. Gaffey

The Earth-crossing asteroids are a population of small rocky or metallic bodies whose orbits approach that of the Earth. These "Earth-approaching" or "near-Earth" asteroids are of special interest for several reasons. They are major contributors to the flux of meteorites falling to the surface of the Earth. Because of their short dynamical lifetimes, the Earth-crossing objects must be continuously replenished from long-lived reservoirs, thus providing samples from beyond the Earth–Moon region. Collisions with the near-Earth asteroids have caused many of the large impact structures on the Earth, Moon, and other terrestrial planets. Such impacts are a major geologic process especially on the airless bodies. Large impacts may have had profound consequences for biological evolution such as the proposed impact-related extinctions at the end of the Cretaceous period. And finally, because the Earth-approaching asteroids include objects that can be reached with relatively small propulsion energies, such bodies could provide relatively cheap raw materials to support large-scale space manufacturing and transportation systems.

DYNAMICAL TYPES

Three general groups (Apollo, Amor, and Aten) of Earth-crossing or Earth-approaching asteroids are recognized based upon the semimajor axis (a), perihelion (q), and aphelion (Q) of the asteroid's orbit. Figure 1 shows an example of each of these types of orbits projected onto the ecliptic plane.

The *Apollo* objects ($a > 1$ AU, $q < 1.017$ AU) have orbits that lie mostly outside that of the Earth but that have perihelion distances smaller than the Earth's aphelion distance (1.017 AU). The Apollo asteroids are either presently on Earth-crossing orbits or undergo periodic variations in their orbital elements which make them Earth-crossing on a short time scale. For example, an asteroid with a perihelion distance of 0.99 AU will not be Earth-crossing if its perihelion is near the ecliptic longitude of the Earths' perihelion (0.983 AU). However, the longitude of perihelion of such an orbit will precess rapidly, so that shallow Apollo orbits ($q > 0.983$ AU)

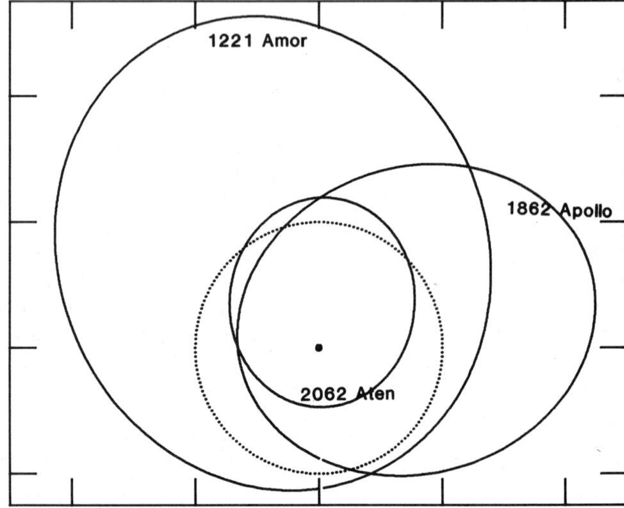

Figure 1. The orbits of three Earth-approaching asteroids (1862 Apollo, 1221 Amor, and 2062 Aten) projected onto the plane of the Earth's orbit (ecliptic plane). The asteroid orbits are shown as solid lines, and the Earth's orbit is shown as a dotted circle (radius = 1 AU) with the Sun (small solid symbol) at the center. Tick marks along the margins are at 1-AU intervals.

can vary between Earth-crossing and noncrossing many times during their dynamical lifetimes. About one-third of the known Apollo asteroids are such intermittent Earth crossers.

The *Amor* objects ($a > 1$ AU, $q = 1.017–1.3$ AU) follow orbits that lie entirely outside that of the Earth but that approach the orbit of the Earth when they are near perihelion. The upper bound of 1.3 AU on the perihelion distance for this class was arbitrarily chosen near a minimum of the radial frequency distribution of perihelion distances. The boundary between Apollo and Amor objects is also somewhat arbitrary, because orbital perturbations can convert Apollo objects into Amors, and vice versa. An apparent Amor object may be an Apollo during a noncrossing period. These two classes probably represent a continuous population with a portion of the relatively long-lived Amors evolving into shorter-lived Apollos which are subsequently lost due to collision or ejection.

The *Aten* objects ($a < 1$ AU, $Q > 0.983$ AU) follow orbits that lie mostly inside that of the Earth but that cross the orbit of the Earth when they are near aphelion. As is the case for the Apollo objects, it is probable that the short-lived Aten population is maintained by replenishment from the Amor and Apollo populations. A fourth orbital type may exist which follows orbits that lie entirely within that of the Earth, but none have been discovered to date.

THE NUMBER OF EARTH-APPROACHING ASTEROIDS

The number of known Earth-approaching asteroids has increased rapidly in recent years. The first Earth crosser (1862 Apollo) was discovered in 1932, and the first Aten (2062 Aten) was discovered in 1976. In 1979, the known population of 46 objects included 3 Atens, 23 Apollos, and 20 Amors. By February 1989, the known population had increased to 118 objects with 7 Atens, 53 Apollos, and 58 Amors. Four of the Earth-approaching asteroids (2 Apollos and 2 Amors) discovered in the past remain lost because their original orbits were not sufficiently accurate to permit recovery during subsequent oppositions and they have not yet been rediscovered.

Active near-Earth asteroid search programs carried out at several observatories around the world on a few nights each month have yielded about half of the Earth-approaching asteroids discovered in the last decade. The additional new objects have been chance discoveries in other observational programs, such as deep sky photography. Because of a greater awareness among the general astronomical community, such chance detections are now often communicated rapidly so that the fleeting object can be relocated and tracked in time to determine a good orbit.

The total population of Earth-approaching asteroids has been estimated based upon discovery statistics from search programs, upon terrestrial impact rates, and upon models of the source mechanisms that replenish this transient population. These estimates suggest that there are approximately 500 Apollo, 1500 Amor, and 100 Aten bodies brighter than absolute visual magnitude 18. That magnitude limit corresponds to diameters of about 1.7 and 0.9 km for the low- and high-albedo asteroid types, respectively.

The number of Earth-approaching objects increases approximately as the inverse square of the diameter change. Thus down to a magnitude limit of 20.5 (170 and 90 m, respectively), the population of Apollo, Amor, and Aten objects should be about 100 times higher (e.g., 50,000 Apollos). The number of bodies in each class increases with decreasing size, until nongravitational forces (e.g., Poynting–Robertson drag, Yarkovsky effect) begin to significantly deplete the population. There is a continuous size distribution between the large Earth-approaching asteroids and the sand- and dust-sized particles that briefly flare as faint meteors as they enter the Earth's atmosphere. However, the size at which one distinguishes between asteroids and meteoroids is arbitrary.

SIZES AND PHYSICAL PROPERTIES

The individual sizes and albedos (surface reflectances) for about one-quarter of the known near-Earth asteroids have been determined using the infrared radiometric technique. Three, all Amor objects [1036 Ganymed, 38.5 km; 433 Eros, 22 km; 3552 1983SA (presently nameless), 18.7 km], have effective diameters larger than 10 km. The smallest for which a radiometric diameter is available is 1915 Quetzalcoatl with a size near 0.5 km. The derived geometric visual albedos of the known Earth-approaching asteroids range from about 1–53%, most being near 20%. The radiometric data suggest that the surfaces of some of these small bodies are covered with thick (> 10 cm) layers of particulate material (regolith), whereas others are bare rock.

Color and/or spectral data are available for about one-third of the known Apollo, Amor, and Aten objects. Reflectance spectra indicate various mixtures of silicates (olivine, pyroxene), nickel–iron metal, clay minerals, and opaque phases (probably carbonaceous compounds and iron oxides) generally similar to the surface materials on the main-belt asteroids. Types identified in the near-Earth population include: S (metal + olivine + pyroxene), M (metal), V (basaltic), E (enstatite, an iron-free silicate), Q (olivine + pyroxene + metal; not to be confused with the abbreviation for aphelion), and the dark C/D/F (clay minerals or silicates + carbonaceous material).

The known population of near-Earth asteroids is dominated by the high-albedo types in contrast to the main belt where the dark objects outnumber the brighter by a factor of at least 4. It is not yet clear what the actual distribution of compositional types is within the Earth-approaching population, because there is a strong observational bias to discover and to observe the brighter objects, which selects for the high-albedo bodies in any size interval.

METEORITES FROM THE EARTH-CROSSING ASTEROIDS

By definition, any meteorite that falls to Earth is an Apollo object, even though the size of the original meteoroid is generally below any reasonable size limit for asteroids. It seems likely that many meteorites are fragments knocked off Earth-approaching asteroids by collisions.

Most meteor showers occur when the Earth encounters dense concentrations of debris in the orbit of a comet. Recent analysis of radar meteor data has shown that several Earth-approaching asteroids are also the parent bodies of meteor showers. Some near-Earth asteroids (e.g., 3200 Phaethon, a parent object of the large Geminid shower) are probably "dead" cometary nuclei, crusted over with silicate dust and carbonaceous material left behind by the evaporation of the near-surface ice. Several Earth-approaching asteroids have anomalous properties suggestive of a cometary origin.

However, some of the shower-producing near-Earth asteroids show surface mineralogies at odds with our models of the nature of the rocky component of cometary nuclei. It is probable that these showers represent debris ejected at relatively low velocities by a comparably recent impact onto the asteroid. Fragments large enough to survive atmospheric entry should occasionally fall during such showers.

One of the major unsolved problems in the asteroid–meteorite relationship is why the most abundant ($> 85\%$) meteorite type (the ordinary chondrites) are apparently absent from the main belt. The presence of the Q-type objects, which are very much like ordinary chondrites, in the near-Earth population provides a partial solution. However, the source of the Q asteroids themselves remains in considerable dispute.

LIFETIMES AND SOURCES OF NEAR-EARTH ASTEROIDS

The Earth-approaching asteroids have lifetimes much shorter than the age of the solar system. Their orbits are subjected to gravitational perturbations by the planets, so that on a time scale of 10^7–10^8 yr most will either impact one of the major planets or will be expelled from the solar system. In general, the members of this population that are subject to the strongest perturbations (i.e., those that more closely approach a larger planet, those with low inclinations, those in more eccentric orbits) have the shortest lifetimes.

An extant short-lived population requires continuous replenishment from some external reservoir. Two major sources have been proposed to maintain the near-Earth population: the main asteroid belt and the comets. The long-term orbital evolution of inner-belt asteroids on Mars-approaching orbits could contribute significantly to the Amor population. The more rapid transfer of asteroids from the chaotic zone near the 3:1 Kirkwood gap at 2.50 AU in the main asteroid belt could supply at least half of the Earth-approaching population. Dynamical models also suggest that at least half of the needed replenishment could be derived from the capture of short-period comets whose surface ices have evaporated leaving a layer of silicate dust and carbonaceous material, similar to most of the surface of the nucleus of Halley's comet as seen by the *Giotto* spacecraft. The actual relative contributions from these potential reservoirs is still uncertain.

IMPACTS FROM THE EARTH-APPROACHING ASTEROIDS

In the ecliptic projection shown in Fig. 1, the orbits of 1862 Apollo and 2062 Aten (an Apollo and Aten, respectively) intersect the Earth's orbit implying that collisions would be frequent. However, the Earth-approaching asteroids have orbital inclinations ranging up to 68° and therefore are generally located a significant distance above or below the ecliptic when crossing the Earth's orbit. Moreover, a collision with the Earth can only occur when the asteroid intersects the Earth's orbit at a time when the Earth is also at that same point in its orbit.

The probability that during its lifetime, any individual Earth-approaching object will actually collide with the Earth is a function of how strongly the asteroid's orbit is subject to gravitational perturbations by the major planets. This probability is increased if the asteroid can closely approach the Earth, Venus, Jupiter, and to a lesser extent, Mars. Such close encounters can greatly modify the orbit of the asteroid and can change it toward or away from a collision orbit. Although the probability of a close encounter can be accurately assessed, the resultant orbit cannot generally be calculated with any accuracy, so that long-term (i.e., beyond the next close encounter) risk estimation for individual bodies is not possible. Earth-crossing asteroids which are in resonant orbits that prevent close encounters can have relatively long lifetimes and low collision probabilities.

Considerations from calculations of close encounters and from the terrestrial cratering record suggest that about one-quarter of the Earth-crossing asteroids will eventually terminate their existence by impacting one of the planets, most likely Earth. The other three-quarters will be expelled from the solar system by gravitational interactions, most probably with Jupiter.

Additional Reading

Davis, J. K. (1985). Is 3200 Phaethon a dead comet? *Sky and Telescope* **70** 317.

McFadden, L. A., Gaffey, M. J., and McCord, T. B. (1985). Near-Earth asteroids: Possible sources from reflectance spectroscopy. *Science* **229** 160.

Shoemaker, E. M., Williams, J. G., Helin, E. F., and Wolfe, R. F. (1979). Earth-crossing asteroids: orbital classes, collision rates with Earth, and origin. In *Asteroids*, T. Gehrels and M. S. Matthews, eds. University of Arizona Press, Tucson, p. 253.

Wetherill, G. W. (1979). Apollo objects. *Scientific American* **240** (No. 3) 54.

Wetherill, G. W. (1988). Where do the Apollo objects come from? *Icarus* **76** 1.

See also **Asteroid and Comet Families; Asteroids; Earth, Impact Craters; Meteorites, Origin and Evolution.**

Astrometry

Burton F. Jones

Astrometry (measuring the stars) is the part of astronomy that deals with the positions and changes in positions of celestial objects. It is the oldest branch of astronomy, with roots in pre-Greek civilization. Astrometry provided the information for the calculation of eclipses for many ancient cultures. It was the accurate astrometry of Tycho Brahe that provided the data that Johannes Kepler needed for the calculation of planetary orbits, thus ushering in the scientific revolution. Today, the ancient science of astrometry is undergoing a revolution as modern technology is brought to bear on the problems of position determination.

PRINCIPLES

Because the directions to most celestial objects can be determined much more accurately than their distances, the term position in astrometry (and in astronomy in general) is used to refer to the direction to celestial objects. To be meaningful, the direction must be specified in some well-defined coordinate system, easily accessible to all observers.

Traditionally and currently, the Earth and its rotation have been used to provide the celestial reference frame. The celestial poles and equator are the projection of the Earth's rotation axis and equator onto the sky. The *vernal equinox* is the intersection of the Sun's apparent path through the sky (the ecliptic) with the celestial equator as the Sun moves from south to north. The position of a celestial object is given by two angles. The *declination* is the angle between the object and the celestial equator along a great circle going through the celestial poles. Declination is usually measured in degrees north (positive) or south (negative) of the celestial equator. The *right ascension* is the angle measured on the equator between the vernal equinox and the intersection of the celestial equator with the great circle going through the object and the celestial poles. Right ascension is commonly given in units of time, because, historically, it was measured as the time between the meridian passages of the vernal equinox and the celestial object. Twenty-four hours of right ascension are equal to 360°. These two angles, the right ascension and the declination, serve to uniquely define the direction to the object.

To be useful, astronomers need a coordinate system that approaches, as closely as possible, an inertial system. Unfortunately, the Earth's rotation axis is not stationary. It changes its direction due to tidal torques provided by both the Moon and Sun (luni–solar precession and nutation). Moreover, the plane of the Earth's orbit also changes due to the gravitational effects of the planets (planetary precession). Thus the right ascension and declination as defined previously are constantly changing: The right ascension and declination of a star determined tonight will be different from the right ascension and declination determined last night. However, we know how the Earth's axis and orbital plane are changing and will change. Thus, knowing the position of a star in the system defined by the Earth's axis and orbital plane on any date, we can predict the position a star would have in the coordinate system defined by the Earth's axis and orbital plane on any other date. The process of transforming from the coordinate system of one date to that of another is frequently called *precessing* the coordinates. What limits how closely we can approach an inertial system is how well we can determine the changing orientations of the Earth's axis and orbital plane.

When giving positions, the date defining the coordinate system must be given. This date is referred to as the *equator and equinox* (or simply the equinox) of the coordinate system. Although any date could be used, the most common dates are January 1, 1950 or, more recently, January 1, 2000. Thus observations reduced to the coordinate system defined by the Earth's rotational axis and orbital plane in 1950 are referred to as equator and equinox, 1950.0.

One must be careful to distinguish between the date of the coordinate system and the date of the observation. The date of the observation is called the *epoch* of the observation. For example, if I directly determine the position of an object on June 5, 1989, the equator and equinox are 1989.42 and the epoch is also 1989.42. If I correct for precession and nutation to January 1, 2000, the equator and equinox will be 2000.0, but the epoch will still be 1989.42.

If we make a series of observations of the position of a star and reduce the positions to a common equator and equinox, we will find in general that the positions at the different epochs will be different. The differences will be due to several causes. Even though we have corrected for precession and nutation, our imperfect knowledge of these effects will lead to some error in our corrections. Of more interest is the positional shift due to our changing perspective as the Earth orbits the Sun. The amount of this shift depends on the distance of the star, and the shift is periodic, with a period of one year. Each star will appear to move in a small ellipse, which is the projection of the Earth's orbit on the plane of the sky at the star's position and distance. The semimajor axis of this ellipse is simply the angular size of the radius of the Earth's orbit as seen from the star and is called the *parallax* of the star. It is customarily measured in seconds of arc. The common unit of astronomical distance, the parsec (*par*allax *sec*ond), has its historical roots in the parallactic shifts of stars. One parsec (206,265 AU or approximately 3.086×10^{18} cm) is the distance of a star with a parallax of 1 arcsecond. Another source of positional change is due to the motions of the object and the Sun. Both will have some motion in an inertial reference frame, and these motions will cause the object to change its position. The resulting time rate of change of position is called the *proper motion* of the object. It depends on the relative tangential speeds of the object and the Sun and the distance of the object. Proper motion units are usually given as arcseconds per year on a great circle, although the unit of proper motion in right ascension is frequently given as seconds of time on a small circle parallel to the celestial equator.

METHODS

The position of a star can be determined from first principles, directly in the system defined by the Earth's rotation, or in a secondary way by referring its position to other stars with known positions. Typically, direct positional determinations are measured using a transit instrument called a *meridian circle*, constrained to move on a north–south line, the observer's meridian. A meridian circle can measure declinations by measuring the angular distances of objects from the zenith as they cross the meridian. This information combined with the latitude of the observer gives the declination (the latitude can be determined by observing a circumpolar star as it crosses the meridian above and below the pole). Such an instrument utilizes the Earth's rotation to measure right ascension differences. It is equipped with instrumentation to determine the times of passage of stars through the meridian. This can be so simple as a cross-hair in the eyepiece and a clock, or some electronic device. Measurement of the times of passage of stars through the meridian gives the difference in right ascension between the stars. Observations of the Sun directly, or other solar system objects, can be used to determine the vernal equinox. Although simple in principle, great care must be made in the design and calibration of transit instruments, and corrections must be made for errors in the construction of the instruments, as well as for refraction introduced by the Earth's atmosphere.

One can use stars with known right ascension and declinations as primary standards to measure the positions of large numbers of stars in a restricted region of the sky. This is most often done by utilizing photography. The rectangular positions of stars with known coordinates are measured along with the rectangular positions of stars of interest on a photographic plate. The primary standards are used to determine the transformation from the rectangular positions to right ascension and declination.

For many investigations of proper motion and parallax, the coordinates of the stars are not calculated. For such investigations, what is important is the change in position as a function of time, not the positions themselves. In these cases, the rectangular coordinates are measured, either from photographic plates or, more recently, by utilizing a photoelectronic device (charge coupled device, Ronchi ruling, and phototubes) directly in the focal plane of the telescope. After allowance is made for the differences in orientation of the detector and the scale of the telescope at different epochs (determined using the measures themselves), the reduced measures give directly the positional changes of the object. These methods, however, give only the proper motions and parallax relative to the mean proper motion and parallax of the stars used in the reduction. Thus one must either have some object with a known proper motion and parallax in the field or some model must be used to correct the relative proper motion and parallax to absolute. Although in some investigations only relative proper motions are needed, it is nearly always the case that one wants the absolute parallax.

To bypass the problems of uncertain precession or models, there are several programs now underway that are attempting to measure proper motions not in the reference frame provided by the Earth and its movements, but by using faint galaxies as a reference frame. Faint galaxies are so far away that their proper motions will be undetectable, thus providing objects with known zero proper motion in an inertial frame.

Recently, radio telescopes have been used for accurate determination of positions. The radio telescope observations of declinations are absolute, whereas those of right ascension are relative. Absolute radio determinations of right ascension are hampered by the lack of point radio sources in the solar system. The tie-in of radio right ascensions to the optical system is either through radio observation of the few objects in a fundamental catalog that are radio emitters, or through the determination of optical positions for objects (mainly quasars) that are radio sources. Unfortunately, most suitable radio sources are faint optically, so that the determination of high-precision optical positions for them is difficult.

USES OF ASTROMETRIC DATA

It is fair to say that astrometric data provide the foundation for much of modern astrophysics. Our knowledge of the cosmic distance scale, and hence the luminosities of cosmic sources, rests either directly or indirectly on astrometric data. Much of our knowledge on the kinematics, rotation, and structure of our galaxy comes from astrometric measurement.

One of the main uses of astrometry is in the determination of stellar distances, primarily through the determination of parallaxes. Stellar distances are necessary for the determination of stellar luminosities. Although the direct determination of stellar distances through parallax measurement is limited to the solar neighborhood, it is these distances that provide the foundation for the series of steps that lead to our knowledge of the distances and brightnesses of cepheids, RR Lyrae stars, and the other standard candles and high-luminosity stars that are necessary for the determination of the distances of external galaxies.

Our knowledge of stellar masses and the mass–luminosity relation rests on astrometric measurement. Astrometric observations of binary stars can, in the best cases, give the individual apparent orbits of both stars around their center of mass and the distance to the system. After correction for projection effects, the true orbits can be obtained, and the resulting knowledge of the period and semimajor axes of the orbits give directly the masses of the two stars through Kepler's third law.

One of the main observational predictions of the theory of stellar evolution and structure is the luminosity of a star with a given mass and chemical composition as a function of age. It is astrometry that ultimately provides this body of observational data against which to compare theory.

Proper motions of stars in the vicinity of a cluster are useful for the determination of membership of stars in the cluster. Because clusters are gravitationally bound, it is a good assumption that they move through space together. For faint stars in particular, proper motion data is the most efficient and bias-free method of selecting cluster members. Although in a cluster such as the Pleiades, the bright stars are easy to pick out because their surface density is so much higher than that of field stars, the percentage of stars that are cluster members drops rapidly with magnitude, so that only a small fraction of faint stars in the region of a cluster are members, whereas the vast majority are field stars either background or foreground to the cluster. In the case of nearby clusters with rich observational histories, the proper motions can reveal the internal motion of stars in the cluster.

FUTURE DIRECTION: THE QUEST FOR ACCURACY

The positional changes of stars, due either to their proper motions or their parallactic shift, are small. The volume of space available for observation of parallax depends upon the cube of the error of the observation of the parallax. Thus, as the errors of observation decrease, the volume of space available for observation increases dramatically, and direct parallax determination of uncommon stars becomes possible.

Improved methods of photographic astrometry, in plate handling and measuring, have decreased the error of relative positional determination. Today, with care, photographic techniques can yield errors as small as 2–3 milliarcseconds (mas) in the determination of a parallax. Even so, photographic determination of stellar parallaxes is time consuming, and the rate of new parallax determinations is low. Because of this, new techniques are being explored. The most promising uses electronic detectors directly in the focal plane. These new ground-based techniques give promise of lowering the errors in the parallax below the milliarcsecond level.

Astrometry, like much of astronomy, benefits greatly from space observations. The accuracy of a positional measurement is compromised by the atmosphere in many ways. Refraction caused by the Earth's atmosphere changes the position of the stars and causes the light to be smeared out in a little spectrum. Turbulence causes the image to be smeared out and to wander, degrading the telescope performance and making it more difficult to determine the centroid of the image position.

In the near future, the Hubble Space Telescope will be used to measure parallaxes with unprecedented accuracy. In an attempt to provide high-accuracy parallaxes and proper motions for a large number of mostly bright stars, in 1989 the European Space Agency launched an astrometric satellite, called HIPPARCOS, to measure the positions and proper motions of 100,000, mostly bright, stars to an accuracy of 2 mas in position and parallax, and 2 mas yr^{-1} in proper motion. The satellite was placed in an unsuitable orbit and its performance remains to be seen. The more distant future will undoubtedly see astrometric telescopes in space obtaining accuracies of 10^{-5} arcseconds. These accuracies will allow astronomers to determine the motions and distances of stars of all types in our galaxy and to determine the motion of nearby galaxies.

Additional Reading

Counselman, C. C., III (1976). Radio astrometry. *Ann. Rev. Astron. Ap.* **14** 197.

Eichhorn, H. (1974). *Astronomy of Star Positions*. Frederick Ungar Publishing Company, New York.

Monet, D. G. (1988). Recent advances in optical astrometry. *Ann. Rev. Astron. Ap.* **26** 413.

van Altena, W. F. (1983). Astrometry. *Ann. Rev. Astron. Ap.* **21** 131.

See also **Astrometry, Techniques and Telescopes; Coordinates and Reference Systems; Stars, Proper Motions, Radial Velocities and Space Motions; Time and Clocks.**

Astrometry, Techniques and Telescopes

David G. Monet

Astrometry is the measurement and interpretation of the apparent positions of celestial objects. Topics include the definition and measurement of celestial coordinate systems, the measurement of position, proper motion, and parallax, and the detection of otherwise invisible objects through gravitational perturbations of the position of visible stars. Because the scientific importance of astrometry depends critically on the accuracy of the observations, astrometry is greatly affected by advances in observational technology. Astrometric research includes the development and testing of new detector systems as well as the usage of tried and proven systems to provide the fundamental data required by other branches of astronomy. Given the limitations of available technology, astrometry is primarily carried out at optical and radio wavelengths, but recent advances in array detector fabrication offer the prospect of infrared astrometry.

Astrometric programs such as the determination of a celestial coordinate system require measurement of stellar positions over the entire sky, whereas programs such as the determination of a particular star's parallax require only relative measurements in the small area of sky adjacent to the program star. Currently available instruments for measuring small angles are about a factor of 10 more accurate than those for large-angle measurement, but the next generation of astrometric instruments, particularly those using interferometers, may offer similar accuracies for both types of observations.

LARGE-ANGLE SYSTEMS

The workhorse for the measurement of the positions of bright stars at optical wavelengths is the meridian circle (also known as a transit telescope or transit circle), Fig. 1. This telescope moves only along the celestial meridian (the great circle including both celestial poles and the observer's zenith) and the stars drift through its field of view. Some measurement schemes rely on a human observer using a motorized micrometer to record the position of the star and its time of transit, but an increasing number are equipped with impersonal detector systems (photomultiplier tubes, array scanners, etc.) that make observations automatically. Measurements of a ruled and calibrated circle mounted on the horizontal telescope axis are used in the declination computation. Celestial observations are augmented by observations of the local nadir, artificial stars, and other calibration sources to monitor telescope performance. Meteorological measurements are used to compensate for

Figure 1. The U.S. Naval Observatory 6-in. transit circle is typical of modern instruments used for fundamental measurement of stellar positions. (*Official U.S. Navy Photograph.*)

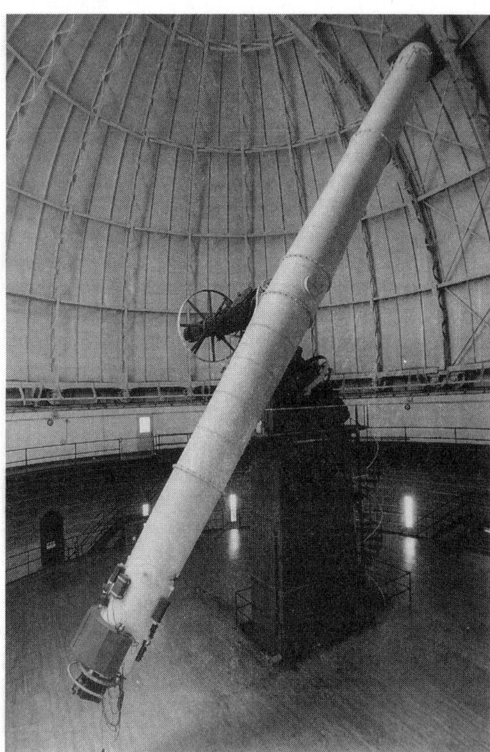

Figure 2. The 40-in. refracting telescope of the Yerkes Observatory is the world's largest and, like most long focal length refractors, has a long and distinguished history of measuring stellar proper motions and parallaxes. (*Yerkes Observatory photograph.*)

refraction and other atmospheric effects. Reduction and analysis of the observations to produce a catalog of apparent positions is a complex and exacting process. Revisions to and augmentation of the algorithms is an ongoing process, as is the recalibration of the various encoders and sensors. The *Fifth Fundamental Catalog* contains coordinates derived from fundamental observations for approximately 1500 stars accurate to about 20 milliarcseconds.

Other optical instruments have been developed and have produced important results. Astrolabes provide accurate measurement of the time of transit, and vertical circles provide accurate measurement of declination. Zenith tubes provide very accurate measurements in a narrow band of declination near the observer's zenith.

Radio interferometric astrometry has produced positions accurate to 1 milliarcsecond and beyond. Such data are collected by the Very Large Array (VLA) and by the network of Very Long Baseline Interferometry (VLBI) antennas. Although still in its infancy, optical interferometric astrometry has already produced preliminary results with accuracies near 10 milliarcseconds. Observing at optical wavelengths is more difficult for two reasons: The mechanical tolerances are substantially tighter and laser metrology systems are required in an optical interferometer and atmospheric phenomena such as refraction and seeing are far larger problems at optical wavelengths. In addition, the typical radio object is a distant galaxy or quasar, while the typical optical object is a nearby star. Hence the observed optical coordinates require systematic correction terms to transform them into an inertial coordinate system.

SMALL-ANGLE SYSTEMS

Since the early 1900s, astronomical photographs have been taken with the intention of measuring stellar positions and this type of observing continues today. A modern photographic astrometry program takes high-sensitivity fine-grain photographic plates with a long focal length telescope (Fig. 2) and measures them with a computer-controlled microdensitometer. A rotating sector, neutral density spot, or other device is used to selectively dim bright stars and thereby extend the dynamic range of the photographic plate. The current *Yale Parallax Catalog* lists some 7500 photographically determined stellar parallaxes and the best accuracies approach 2 milliarcseconds. Disadvantages of this technology include the need for wet-process photographic chemistry, the relatively poor quantum efficiency of the photographic emulsion, and the lack of real-time data analysis and quality control.

Usage of a charge coupled device (CCD) instead of the photographic plate offers many advantages. Its quantum efficiency approaches 80%, its geometry is defined during fabrication, and the data are digitized immediately. Preliminary results have demonstrated accuracies of 1 milliarcsecond, and better, on stars as faint as 20th magnitude. Disadvantages include the relatively small physical size of currently available devices (about 5 cm^2) and the need to archive large quantities of digital data.

The Multichannel Astrometric Photometer (MAP) developed at the Allegheny Observatory uses a Ronchi ruling in the image plane to modulate the starlight and an array of photomultiplier tubes to detect these signals. The metrology is defined by the ruling which can be quite accurately fabricated. First results show an accuracy in the range of 1–2 milliarcseconds. This instrument is well suited to the selective dimming required for bright stars but has a relatively low overall quantum efficiency.

Variations on the preceding approaches have been developed for use by telescopes that scan the sky. For CCDs, the charge image is shifted at the sidereal rate and the exposure time is the time it takes the star to drift the length of the chip. Detectors based on detection of the modulation of starlight as stars drift past a stationary Ronchi ruling have been constructed. The principal advantage of scan-mode observing is the ability to map large portions of the sky. Although exciting preliminary results have been demonstrated, neither technique has produced milliarcsecond-class accuracies yet.

Because meridian circles and other large-angle astrometric detectors measure only one object at a time, only a relatively few objects define the coordinate system. Astrographs are used to extend the coordinate system to many more and fainter stars and galaxies. These telescopes have large fields of view (typically 4–25 square degrees), are designed to have smooth and predictable focal plane geometry, and produce coordinates with accuracies in the range of 10–50 milliarcseconds. Very wide field instruments such as Schmidt telescopes can be used for astrometry but accuracies are rarely better than 100 milliarcseconds.

An important part of astrometry is the study of double and multiple star systems. When both components of a double star system are observable, astrometric and spectroscopic data can be combined to determine the mass of each component. If perturbations in the position of an apparently single star are detected, the existence and some physical properties of the unseen companion can be inferred. Such an unseen companion might be a brown dwarf star, a black hole, or a planet orbiting the visible star. Three techniques are available for this type of astrometry: visual measurement with a filar micrometer, measurement of photographic plates or CCDs, and speckle interferometry. Speckle interferometry (the collection and analysis of many short exposures during which the atmospheric turbulence is constant) offers exceptional accuracy and allows resolving systems at the telescope diffraction limit rather than at the atmospheric seeing limit. Speckle astrometry is being done at optical and infrared wavelengths.

Radio astronomy yields important astrometric information. The VLA, VLBI, and other radio telescope systems can measure proper motions as small as 0.1 milliarcseconds per year. Several radio sources show motion at this level and, assuming standard cosmological models, many show apparently superluminal motions that require relativistic models for explanation. Through the study of maser sources, the distance of galactic objects such as the galactic center can be measured directly.

SPACE ASTROMETRY AND THE FUTURE

The European *HIPPARCOS* satellite will produce milliarcsecond-class positions for some 100,000 stars during its 2.5-yr mission lifetime. It measures large angles very precisely, and its observations will produce a self-consistent, inertial coordinate system that can be transformed to the equatorial system, the radio system, or any other through the observation of appropriate bridge objects. Because the *HIPPARCOS* mission is short, its measurements of proper motions will be relatively imprecise and the accuracy of its catalog will slowly degrade in the coming decades. Data from the *HIPPARCOS* star mapper will be used by the *TYCHO* experiment to produce coarse positions (typically 30 milliarcseconds) for some 500,000 stars.

Astrometry is an important part of the Hubble Space Telescope (HST) project: Two instruments have enhanced astrometric capabilities. The Fine Guidance Sensors were designed to make milliarcsecond-class measures on stars from 4th to 17th magnitude, and the Wide field and Planetary Camera CCD detectors will make differential astrometric measurements on objects as faint as 28th magnitude, when the telescope is repaired.

Future space-based astrometric instruments promise significant improvements over currently available ground-based performance. Proposals have been prepared for instruments with 10-microarcsecond class accuracy using radio interferometric, optical interferometric, and Ronchi ruling technologies. Such instruments could resolve perturbations from planetary systems around nearby stars as well as directly measure the distances to a wide variety of astrophysically interesting objects. Ground-based astrometric instruments are in transition from technology demonstration to catalog production modes of operation. It is still not clear whether the current 1-milliarcsecond accuracy is set by the atmosphere or whether advanced technologies used at sites with superlative seeing can make significant improvements.

Astrometry continues in a state of transition. Traditional methods provide the fundamental data essential to the science. New ideas and technologies are being incorporated into advanced astrometric detector systems. Although many may not survive, some will demonstrate significant increases in both accuracy and productivity and will become the standard systems in the future. Great astrometric problems such as the establishment of an inertial coordinate system, the measurement of the distances of Cepheid variables and other stars needed for the cosmological distance scale determination, and the detection of black holes and planets around other stars have yet to be solved, but important intermediate steps are being taken.

Additional Reading

Eichhorn, H. (1974). *Astronomy of Star Positions*. Frederick Ungar Publishing Company, New York.

Monet, D. G. (1988). Recent advances in optical astrometry. *Ann. Rev. Astron. Ap.* **26** 413.

Murray, C. A. (1983). *Vectorial Astrometry*. Adam Hilger, Bristol.

Podobed, V. V. (1962). *Fundamental Astrometry*. University of Chicago Press, Chicago.

van Altena, W. F. (1983). Astrometry. *Ann. Rev. Astron. Ap.* **21** 131.

van de Kamp, P. (1967). *Principles of Astrometry*. W. H. Freeman, San Francisco.

See also **Astrometry; Coordinates and Reference Systems; Time and Clocks.**

B

Background Radiation, Gamma Ray

Carl E. Fichtel

It is now known that the diffuse celestial radiation extends well into the gamma ray region, at least to approximately 200 MeV, and that it is isotropic at least on a coarse scale. The degree of isotropy that has already been shown to exist, taken together with the spectrum being different from that of the galactic diffuse radiation, strongly supports this diffuse radiation being extragalactic in origin. The intensity of the radiation is rather weak and its level has already caused the rejection of the steady-state theory of the universe wherein matter and antimatter are continuously created in equal amounts in an expanding universe. For this theory to have been correct, the radiation would have had to be several factors of 10 more intense.

The intensity, energy spectrum, and the degree of isotropy that has been established have, in fact, eliminated many theories that have either involved a diffuse celestial gamma radiation or been designed to explain it. Among the still possible explanations are that it is the sum of radiation from active galaxies or it is from the interaction of matter and antimatter at the boundaries of superclusters of galaxies. Future measurements may very well be able to separate these two alternatives. It is also possible that part of the diffuse emission, but probably not all because of the spectral shape, may be the result of the quantum-mechanical decay of small black holes created by inhomogeneities in the early universe. These theories and the details of the experimental tests that can be made will be considered after a brief review of the current status of the knowledge of this radiation.

THE EXISTING OBSERVATIONAL PICTURE

The first indication that the diffuse radiation extended above the x-ray region into at least the low-energy gamma ray region came from instruments flown in the *Ranger 3* and *5* Moon probes. Subsequently, there have been a large number of measurements in the low-energy (approximately 0.1–20 MeV) region, establishing its general nature. In the high-energy gamma ray realm, the first measurements came from *Explorer 11*. Although they gave only an upper limit, they provided the refutation of the steady-state theory mentioned previously. The first suggestion of a high-energy component came from a telescope on OSO-3, and the results from the SAS-2 gamma ray telescope provided measurements on the isotropy and the energy spectrum from 35 MeV to about 200 Mev, above which energy the intensity falls below the galactic diffuse radiation even well away from the galactic plane.

Although the intensity of the measured gamma radiation in the low-energy region is stronger compared to the galactic radiation than the high-energy gamma radiation, there is the considerable problem of background radiation created in the surrounding material and even in the instrument itself. The difficulty is sufficiently severe that early results consisted principally of upper limits and uncertain reported positive results. The background radiation is now better understood, atlhough still high. The *Apollo 15* and *16* gamma ray telescopes were on booms of variable lengths allowing the background radiation to be carefully measured as a function of distance from the spacecraft. The final results from these missions are now generally accepted as being a good general representation of the radiation level. There remains, however, some question about the exact shape in the 1/2–10-MeV region.

In the high-energy region, the nature of the instrumentation and the physical interactions of both gamma rays and cosmic particles causes the background radiation to be very low relative to the measured radiation. However, the separation of the extragalactic diffuse radiation from the galactic diffuse emission is a challenge. Several different types of analyses have been made, starting with the most simple one of studying the intensity and energy spectrum as a function of galactic latitude and continuing with careful comparisons to the matter column density and even galaxy counts. All of the studies give approximately the same result for the magnitude and energy spectrum in this portion of the energy range.

Figure 1 shows the energy spectrum of the diffuse radiation. Because of the large number of experimental points, only a few of the results are included to avoid confusion. References to the considerable number of other results, including the early ones, may be found in the articles listed at the end of this entry. The level of the galactic radiation is also shown for comparison. Notice that there is a general trend toward a steepening as the energy increases from the low-energy x-ray region into the gamma ray region. However, in addition, there appears to be a "bump" in the region around 1 MeV. This feature, should it be real, is important in trying to interpret the nature of this radiation. The concern, in spite of the now much-improved instrument design and background radiation subtraction techniques, is that this is just the region where a significant background is known to exist. Early experiments indeed saw a very large background in this region.

The degree of spatial isotropy is not yet precisely known, but the information that does exist is important in constraining the possible interpretations. The x-ray emission through about 100 keV is known to be isotropic to within about 5%, and the low-energy gamma radiation is uniform to within 20%. At high gamma ray energies, the center-to-anticenter ratio for intermediate galactic latitude radiation is one to within 20%, and the extragalactic diffuse radiation perpendicular to the plane relative to that at

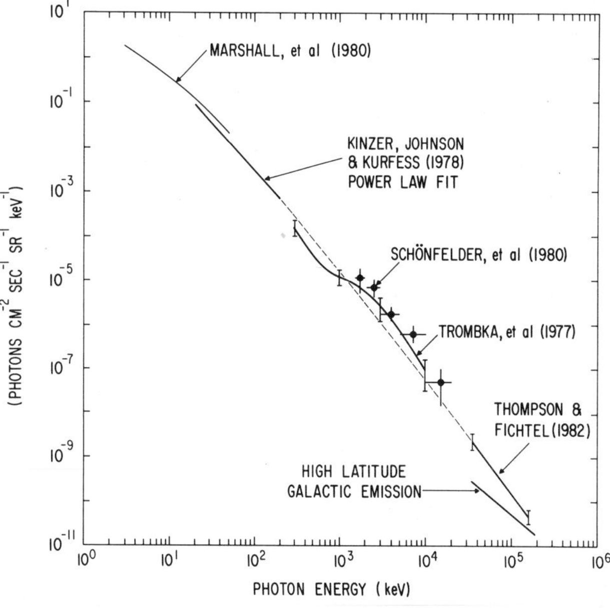

Figure 1. The diffuse extragalactic gamma radiation. The high-energy high-latitude galactic emission is shown for comparison. The dashed line is shown only to guide the eye. [*From Trombka and Fichtel (1983).*]

intermediate latitudes is also consistent with isotropy to within 20%. Thus, although more precise measurements are desired, no evidence for a major anisotropy exists. As an example of the significance of the results that already exist, a spherical galactic halo origin is clearly eliminated.

POSSIBLE ORIGINS OF THE DIFFUSE RADIATION

As noted, the current knowledge of the extragalactic diffuse gamma rays has already significantly constrained the possible explanations of this radiation. The three possibilities for the origin of the diffuse gamma radiation, or part of it, that will be pursued here are the annihilation of nucleons and antinucleons at the boundaries of superclusters of galaxies of matter and antimatter, active galaxies, and primordial black holes.

In the first concept, the proposed baryon nonconserving forces in the grand unified field theory revived interest in considering a baryon symmetric universe, containing superclusters of matter and others of antimatter. The low level of the extragalactic diffuse radiation, itself, has effectively eliminated regions of baryons and regions of antibaryons on a smaller scale, because in that case the gamma ray intensity would be much larger than observed. Although the nucleon–antinucleon annihilation spectrum has a maximum at about 70 MeV, when it is integrated over cosmological time and the redshifts are properly considered, a spectrum similar to the one in Fig. 1 is found with a bump at about 1 MeV corresponding to the largest appropriate redshift. In this model the x-ray emission, including the hard x-rays, is presumed to be due to a different source. The intensity level is also consistent with reasonable choices for the parameters.

On the largest scale, this theory predicts a smooth distribution over the sky; however, a crucial test of this theory (in addition to a precise measure of the energy spectrum) would be the detection of enhancements of the radiation in the direction of boundaries between close superclusters of galaxies. The excess diffuse radiation associated with these boundaries would be at the higher energies. Calculations suggest that with future high-sensitivity gamma ray telescopes, there is hope of seeing these ridges if they exist. Because there are very few ways to test whether our universe is baryon symmetric on this scale, the search for these ridges is of considerable interest for cosmological theory.

Although the information on gamma ray emission from active galaxies is very limited at present, a Seyfert galaxy, a quasar, and a radio galaxy have all been seen. It is, of course, very speculative to assume that they are typical. However, if this assumption is made or even that they are a bit exceptional, and if x-ray data and gamma ray upper limits on other galaxies are used as support of the spectral shape that is seen, then a cosmological integration using parameters developed from other areas of astronomy shows that active galaxies could explain the diffuse radiation. An important test of this theory will be whether enough additional active galaxies are seen and their spectra are appropriate as the increased sensitivity of future gamma ray telescopes permit the detection of weaker sources. The appeal of this model is that active galaxies are certainly expected to produce gamma rays, and the intensity level is not unreasonable; so the only question that remains is whether the spectra and level are compatible with the observed diffuse intensity.

The third possible origin to be mentioned here is that associated with the death of primordial black holes. It has been postulated that these primordial black holes can evaporate by the tunneling process at an accelerating rate that ends in an explosive gamma ray burst. These events are predicted to be significantly harder in energy spectrum than the low-energy gamma ray bursts that have been observed. The primordial black hole bursts when integrated over cosmological time might produce the diffuse gamma ray background, or at least part of it. The observed spectral shape may create a difficulty, but the most significant test of this model would be the observation of high-energy gamma ray bursts of the appropriate type. Attempts to detect them so far have been unsuccessful; however, in view of the uncertain mass spectrum and other features, further searches are certainly appropriate.

SUMMARY

An extragalactic diffuse gamma radiation appears to exist with reasonably well-defined characteristics. There are at least three possible explanations for this radiation, each of which has a specific experimental test. In view of the desire to know whether the universe as a whole is baryon symmetric, to understand the nature of active galaxies, and to know if primordial black holes exist and what their properties are if they do, there is certainly considerable fundamental astrophysics involved with the resolution of the matters associated with the understanding of the diffuse extragalactic radiation. Hopefully, this future scientific work will add to the important contributions to science that the study of the extragalactic diffuse radiation has already made.

Additional Reading

Bignami, G. F., Fichtel, C. E., Hartman, R. C., and Thompson, D. J. (1979). *Ap. J.* **232** 649.

Fichtel, C. E. and Trombka, J. I. (1981). *Gamma Ray Astrophysics, New Insight into the Universe.* NASA SP-453.

Kinzer, R. L., Johnson, W. N., and Kurfess, J. D. (1978). *Ap. J.* **222** 370.

Marshall, F., Boldt, E., Holt, S., Miller, R., Mushotzky, R., Rose, L., Rothschild, R., and Serlemitsos (1980). *Ap. J.* **235** 4.

Page, D. N. and Hawking, S. W. (1976). *Ap. J.* **206** 1.

Ramaty, R. and Lingenfelter, R. E. (1982). *Ann. Rev. Nucl. Part. Sci.* **32** 235.

Schonfelder, V., Graml, F., and Penningsfeld, F. P. (1980). *Ap. J.* **240** 350.

Stecker, F. W., Morgan, D. L., and Bredekamp, J. (1971). *Phys. Rev. Lett.* **27** 1469.

Thompson, D. J. and Fichtel, C. E. (1982). *Astron. Ap.* **109** 352.

Trombka, J. I., Dyer, C. S., Evans, L. G., Bielefeld, M. J., Seltzer, S. M., and Metzger, A. E. (1977). *Ap. J.* **212** 925.

Trombka, J. I. and Fichtel, C. E. (1983). *Phys. Rep.* **97** 173.

White, R. S. (1987). *Enc. Phys. Sci. Tech.* **763**.

Wolfendale, A. W. (1983). *Quart. J. Roy. Soc. Astron. Soc.* **24** 226.

See also **Antimatter in Astrophysics; Gamma Ray Astronomy, Space Missions.**

Background Radiation, Infrared

Martin Harwit

Even on the darkest nights, a diffuse glow permeates the sky. Part of this glow condenses into stellar images in the focal plane of large telescopes, but a component remains which is genuinely diffuse. Some diffuse sources are nebular patches on the sky, gas clouds in the vicinity of stars, or entire galaxies at extreme distances. Still other sources of radiation produce a totally isotropic background having one and the same surface brightness in whatever direction we look. An isotropic background flux is the most difficult to measure accurately.

INFRARED BACKGROUND MEASUREMENTS

Any warm body emits radiation—glows. At room temperature and below, this radiated energy is emitted at infrared wavelengths that span the range of 1–1000 μm. Ambient temperature telescopes,

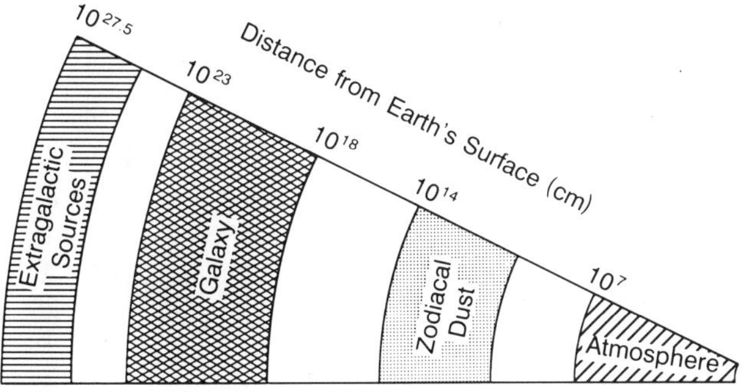

Figure 1. Sources of infrared background radiation and their distances from the center of the Earth.

therefore, emit infrared radiation, and that emission would normally not be distinguishable from a diffuse cosmic glow were it not for the possibility of cooling telescopes and detectors to extremely low temperatures, around 2 K. That way, practically all infrared radiation incident on the astronomer's detectors must be arriving at the telescope from space, rather than from the telescope itself.

The most ambitious instrument constructed to detect the diffuse infrared background is a space telescope cooled to just such low temperatures. It is the *Cosmic Background Explorer*, (*COBE*), built under the auspices of the U.S. National Aeronautics and Space Administration, NASA. In Earth orbit, above any glow from the atmosphere, it has been able to peer out into the universe to detect faint glows across most of the infrared wavelength domain. Before the advent of *COBE* in late 1989, such measurements could only be undertaken in short-lived rocket observations using similarly cooled telescopes.

The extreme cooling of telescopes is not strictly necessary in observations conducted at short infrared wavelengths, around 1 μm, and some such observations can be undertaken from mountaintop observatories, although the Earth's atmosphere then produces a limiting foreground glow.

SOURCES OF DIFFUSE INFRARED RADIATION

An astronomer's telescope receives light not only from the star or galaxy being observed, but from a wide variety of other sources also present in the telescope's field of view. Those sources are at vastly differing distances from the telescope (see Fig. 1) and contribute radiation through a wide selection of different mechanisms.

Airglow

Airglow emanating from the terrestrial atmosphere enters the telescope from the lowest few hundred kilometers above Earth's surface. For the astronomer this radiation is a nuisance that must be weeded out to reveal the genuine astronomical sources beyond the atmosphere.

At the shortest wavelengths, ranging out to about 4 μm, collisionally excited OH radicals at high altitudes emit radiation at scores of well-defined wavelengths, each wavelength corresponding to a particular transition of a collisionally excited OH radical from one vibrationally and rotationally excited state to a lower one. At wavelengths between 1 and 1000 μm, the atmosphere also emits over broad wavelength ranges, through emission by water vapor, carbon dioxide, methane, N_2O, ozone, and other atmospheric constituents. In much of this wavelength regime, the atmosphere is also opaque, so that astronomical observations must be carried out from high altitudes or from space.

Zodiacal Light

The Sun is enclosed in a cloud of finely dispersed dust that orbits the Sun in a disk-shaped volume symmetrically distributed about the ecliptic plane. This zodiacal cloud contains roughly micrometer-sized grains that scatter light at visible and near-infrared wavelengths. At wavelengths longward of 3 μm, the thermal emission from these grains begins to dominate scattered sunlight. For observations on diffuse astronomical sources, this foreground emission can represent a serious contaminant that needs to be subtracted before the intrinsic surface brightness of more distant astronomical nebulosity can be assessed.

The uncertainty of whether infrared radiation from the telescope, from the atmosphere, or from zodiacal dust has been adequately subtracted from flux levels recorded by infrared detectors makes the detection of an isotropic, cosmic background flux particularly difficult. For patchy, diffuse glow emanating, say, from the galactic plane, a reference signal can always be obtained at higher galactic latitudes; but isotropic background radiation can only be determined through an absolute flux level measurement, independent of any other reference region in the sky.

Galactic Starlight

As we leave the solar system and penetrate further out into the Milky Way, we find four other sources of diffuse radiation, stars, dust, gaseous—atomic, ionic, and molecular—constituents, and electrons both at thermal energies and at the high energies of cosmic rays. The glow emanating from these broadly distributed sources is called the *diffuse galactic emission*.

1. Unresolved stars produce a diffuse glow. Most of this radiation lies in the visible part of the spectrum and in the near infrared where stars emit much of their energy. The component of this light that reaches us from remote regions preferentially lies in the infrared, because much of the ultraviolet and visible radiation is absorbed in interstellar dust clouds along the way.

2. The interstellar dust heated by starlight appears to consist of two quite distinct components. The first consists of small grains in the size range around 10^{-5} cm (0.1 μm). This dust has an equilibrium temperature that is much cooler than the illuminating star and, by Planck's radiation law, correspondingly reemits the absorbed energy at much longer wavelengths—around 10 μm when the absorbing dust cloud lies relatively close to the illuminating stars, and 100 μm when the dust is at larger distances. The emitted radiation has a continuum spectrum, somewhat akin to a blackbody spectrum except peaked toward shorter wavelengths. This spectral shift comes about because small grains radiate with systematically decreasing efficiency at wavelengths substantially longer than the grain diameter.

A second component formerly thought to be dust recently has been identified with macromolecules, possibly polyaromatic hydrocarbon (PAH) molecules. Upon absorbing a single quantum of starlight, parts of these molecules are excited to energies equivalent to temperatures of several hundred degrees kelvin. The molecular heat capacities are so small that the tiny amounts of energy carried by a single absorbed quantum can raise temperatures to these heights, causing the PAHs to emit at substantially shorter wavelengths than larger grains. Being small, PAHs also act more like molecules than like macroscopic objects, and they emit radiation at rather well-defined energies, in contrast to the continuum radiation emitted by larger grains. Characteristic PAH wavelengths lie at 3.3, 6.2, 7.7, 8.6, and 11.3 μm. Because this radiation is precipitated by the absorption of single photons of radiation, its spectrum tends to be independent of a PAH grain's distance from the illuminating star or stars. As a result, PAH emission can be observed in emission from clouds of dust at high galactic latitudes, from regions that might be as far as 100 pc from the galactic plane, producing a diffuse though patchy glow straddling the entire plane.

3. Gas clouds between the stars heated and ionized by starlight emit radiation when their atomic, ionic, or molecular constituents become collisionally excited and cascade down to the ground state emitting quanta of radiation at well-defined wavelengths characterizing the active atom or molecule.

In hot, ionized gases, electrons undergoing close encounters with ions are electrostatically deflected in their trajectories and emit radiation through *free–free transitions*; that is, the electron is bound to the ion neither in its initial approach nor after deflection. Free–free emission is observed in diffuse ionized regions throughout the galactic plane and is most readily identified at radio wavelengths, though the emission takes place at shorter infrared wavelengths as well.

4. Electrons accelerated to relativistic energies constitute part of the *cosmic ray particle* component traversing interstellar space. They scatter—collide with—radiation in what is termed *Compton scattering*, a process in which the scattered energy can be substantially changed from the incident energy. When the electron loses significant amounts of energy to the quantum of incident radiation, in Compton scattering, we speak of *inverse Compton scattering* and the associated energy increase of the quanta of radiation expresses itself in a shortening of the wavelength. In particular, it can lead to the conversion of radio waves into infrared radiation. That kind of a diffuse infrared glow appears primarily to be emitted in active galaxies and quasars exhibiting powerful radio as well as x-ray emission in a compact nuclear region, but it may be present in galactic emission as well.

Highly energetic electrons gyrating in the interstellar magnetic field can also produce *synchrotron radiation* at infrared wavelengths, that is, radiation associated with the accelerations the electrons experience in their motions across the magnetic field.

Extragalactic Emission

Beyond the Milky Way, stretching to the cosmic horizon, lie perhaps a hundred billion galaxies and millions of quasars. These sources appear to be the primary contributors to the *extragalactic background light*.

Most of these sources are believed to emit infrared radiation by virtue of the various processes enumerated previously, but there may be exceptions among objects whose sources of energy are not well understood. The most extreme among these are the quasars, which appear to be galaxies in whose nuclei violent processes hundreds of times more powerful than the emission from normal galaxies take place.

A number of other extragalactic contributors to the diffuse infrared radiation have been postulated but never observed.

1. Individual stars and individual globular clusters may be tidally removed from galaxies during intergalactic collisions and left to drift in extragalactic space. If so, their stellar emission could be a significant contributor to the near-infrared diffuse background.

2. The earliest galaxies to have been formed may all have had their origin in one brief outburst when the universe was still young. Many stars formed at that time could have been massive and highly luminous. Such massive stars are thought to be the primary contributors to the formation of heavy elements, such as carbon, oxygen, nitrogen, and possibly iron, and their injection into the interstellar medium through supernova explosions. Such elements are found in appreciable abundances in all stars observed today, suggesting that all known stars formed in the wake of an earlier generation of massive—and therefore luminous—precursors. Granted these assumptions, we would expect a diffuse background flux due to early galaxies whose radiation, though originally at visible and ultraviolet wavelengths, would now be observed redshifted into the near infrared.

3. Dust ejected into intergalactic space, shortly after the first galaxies formed, would have radiated at wavelengths around 100 μm, and the redshifted flux received today would most likely be found at submillimeter wavelengths.

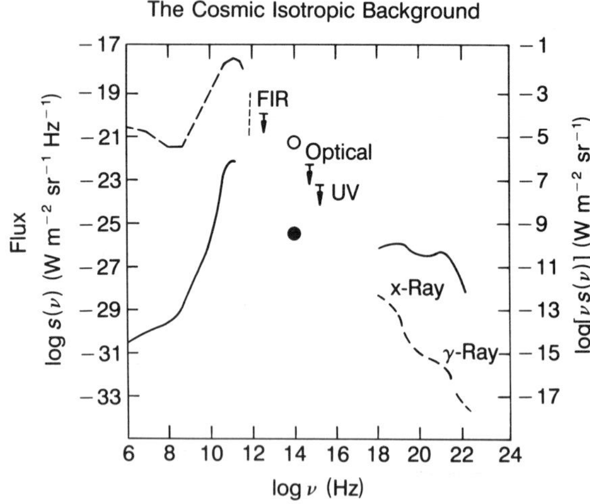

Figure 2. Spectrum of the isotropic background radiation. The solid curve is to be used with the scale on the right and indicates flux levels integrated over a logarithmic frequency interval. It permits ready comparison of the microwave and x-ray background energies. The dashed curve, to be used with the scale on the left, shows the flux received per unit frequency interval. The circles indicate recent measurements in the infrared. The open circle corresponds to the dashed line, and closed circle corresponds to the solid curve. Upper limits to any observed flux are indicated by arrows. A wavelength of 100 μm corresponds to a frequency of 3×10^{12} Hz, and a 10-fold decrease in wavelength corresponds to a 10-fold increase in frequency.

MICROWAVE BACKGROUND RADIATION

In addition to all of the previously mentioned sources of background radiation, there is the ubiquitous *microwave background radiation*, a remnant of the earliest hot phases of the universe (see Background Radiation, Microwave). A portion of this radiation stretches from the radio region into the infrared domain. As seen in Fig. 2, this cosmic background radiation contains far more radiant energy than sources at all other wavelengths combined.

Additional Reading

Bowyer, S. and Leinert, C., eds. (1989). *Galactic and Extragalactic Background Radiation*. IAU Symposium 139. Kluwer, Dordrecht.

Clegg, P. (1978). Cosmic background: Measurements of the spectrum. In *Infrared Astronomy*, G. Setti and G. G. Fazio, eds., NATO Advanced Studies Institutes Series (Series C, Mathematical and Physical Sciences). D. Reidel, Dordrecht, p. 181.

Harrison, E. (1987). *Darkness at Night*. Harvard University Press, Cambridge, MA.

Harwit, M. (1978). Infrared astronomical background radiation. In *Infrared Astronomy*, G. Setti and G. G. Fazio, eds., NATO Advanced Studies Institutes Series (Series C, Mathematical and Physical Sciences). D. Reidel, Dordrecht, p. 173.

See also **Background Radiation, Microwave; Diffuse Galactic Light; Galaxies, Infrared Emission; Infrared Astronomy, Space Missions; Zodiacal Light and Gegenschein.**

Background Radiation, Microwave

S. P. Boughn and R. B. Partridge

The microwave background radiation (MBR) is a relic of a hot, early phase in the history of the universe. It fills the universe and is thus detected as a uniform background. Its discovery by radio astronomers in 1964 provided the second major piece of evidence in support of the big bang theory of the universe. The first, Edmund Hubble's discovery of the expansion of the universe, may have been more dramatic but studies of the MBR have yielded much more detailed information about the universe, ranging from the physical conditions a few minutes after the big bang to galaxy formation. The current temperature of the MBR together with observed helium abundances of very old stars have even been used to constrain the number of possible families of elementary particles, a result of major importance to particle physics.

Between the wavelengths of 1 mm and 10 cm (the microwave region), the overall brightness of the sky is dominated by the MBR. Its spectrum closely approximates that emitted by a blackbody at a temperature of 2.74 K and it is very nearly isotropic; that is, it has the same intensity in all directions. These two properties of the MBR, which will be discussed in more detail, provide the primary evidence of its cosmological origin. If the radiation is left over from a hot, dense phase of the early history of the universe, the spectrum should be that of a blackbody, whereas virtually every other astrophysical source of microwave radiation has a spectrum that is quite distinct from a blackbody. Similarly, if the radiation is left over from an earlier epoch of the universe, it should fill the universe uniformly and therefore be isotropic. Relatively local foreground sources of microwaves, such as the Sun or the Galaxy, on the other hand, are not isotropic. By eliminating other possible sources for the MBR, astrophysicists have come to accept the cosmological interpretation.

Although the MBR is the cooled relic of thermal radiation that was present when the universe was very young (<100 s), the microwave photons observed today have not propagated directly to us from that epoch. The matter in the early universe was a dense plasma consisting primarily of electrons, protons (hydrogen nuclei), and alpha particles (helium nuclei) in thermodynamic equilibrium with the radiation. In such an environment radiation is scattered many times in a small distance, especially by free electrons, so the universe was opaque. As the universe expanded and cooled to a temperature 10,000 K, the electrons began to combine with the protons and alpha particles to form neutral hydrogen and helium atoms. However, at that epoch the density of ionized matter was still high enough to keep the universe opaque. Finally, when the temperature dropped to about 3000 K (about 300,000 years after the big bang), so few free electrons were left that the universe became transparent and the thermal radiation was able to travel more or less unimpeded from that point on. This is known as the epoch of "decoupling" or of last scattering. The peak wavelength of the radiation was then about 0.6 μm in the red portion of the visible spectrum. As the universe continued to expand, the radiation continued to cool, so that today it is observed as microwaves with a temperature of 2.74 K.

THE DISCOVERY

The serendipitous discovery of the microwave background radiation by Arno A. Penzias and Robert W. Wilson in 1964 followed several "forgotten" predictions and measurements that could have led to the discovery but did not. In 1941, Walter S. Adams observed absorption lines from both the ground and excited states of interstellar cyanogen (CN) molecules. From the ratio of the intensities of these two lines, A. McKellar deduced an excitation temperature of 2.3 K. In retrospect, these molecules were almost surely being excited by the MBR and thus the preceding value constitutes a measurement of the temperature of the background radiation. However, the hot big bang model had not yet been introduced and, of course, the proper interpretation was not made. It is interesting to note that refined observations of interstellar CN absorption lines have recently yielded some of the most accurate determinations of the temperature of the MBR (see the following section).

In 1946, George Gamow proposed a hot, radiation-dominated, early universe in which heavy elements were produced by neutron capture. Two of his colleagues, Ralph A. Alpher and Robert

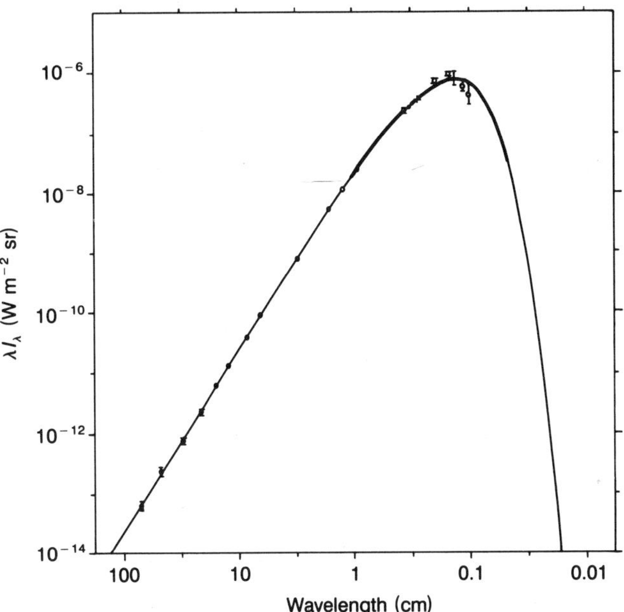

Figure 1. The intensity of the microwave background radiation as a function of wavelength. The solid curve is a theoretical blackbody spectrum with a temperature of 2.74 K. Ground-based measurements are indicated by ●, balloon measurements by ○, and CN measurements by +. Error bars are included unless they are smaller than the symbol. The heavy line between wavelengths of 1 and 0.05 cm represent the measurements of the *COBE* satellite. The errors in these measurements are much smaller than the width of the line.

Herman, deduced that the relict radiation would today be at a temperature of 5 K. Although this work was well publicized at the time, neither Penzias and Wilson nor Robert H. Dicke's group at Princeton were aware of it. In 1964, Dicke independently proposed a hot, early universe, not to produce heavy elements but rather to destroy them at the end of every cycle of a hypothesized oscillating universe. Members of his research group were already building a microwave radiometer to search for the relict radiation when they heard of the excess signal observed by Penzias and Wilson with their 7-cm wavelength radiometer at the Bell Telephone Laboratories in Holmdel, NJ. In 1965, the discovery was reported by Penzias and Wilson and interpreted by the Princeton group.

THE SPECTRUM

Measurements of the intensity of the MBR are indicated in Fig. 1 along with the spectrum of a 2.74 K blackbody. It is clear that from ~1 mm to ~50 cm, a factor of 500 in wavelength, the data are remarkably consistent with a blackbody spectrum, which is essential to the cosmological interpretation as mentioned previously. The longer-wavelength measurements were made with ground-based microwave radiometers. At shorter wavelengths ($\lambda \lesssim$ 3 mm), the atmosphere becomes more emissive and as a consequence direct observations in this region of the spectrum must be made with high-altitude balloon-borne, rocket, or satellite detectors. It should be noted that one of the most accurate determinations was deduced from interstellar CN absorption at 2.6 mm. They appear to be ruled out by measurements made by the *COBE* satellite and reported in 1990.

The first quantity of cosmological significance coming from these measurements is simply the temperature of the radiation, which is $T = 2.74 \pm 0.02$ K (an accuracy rare in cosmology). Initially, in 1965, it was an estimate of the baryon density in the universe and the primordial helium content that allowed Peebles to estimate the current temperature of the MBR. Now that the temperature is so well determined, this argument can be reversed to yield a value for the baryon density of the universe, a crucial cosmological datum. A similar calculation involving T and the abundances of light elements limits the number of possible neutrino families to ≤ 4, an important result for particle physics (this limit could be made tighter still if the lifetime of the free neutron were known to higher precision). Another cosmological parameter that comes directly from the value of T is the ratio of the number of microwave photons to baryons, or entropy per baryon in the universe. This ratio is on the order of 10^9, and until recently there was no compelling explanation for it. It is regarded as a considerable success of the new grand unified theories of elementary particles that they make possible a natural explanation of the value of this parameter.

ISOTROPY

Within a few years after the discovery of the MBR, it was shown to be isotropic to within less than 1%. Although this isotropy was an essential test for a cosmological interpretation as mentioned previously, astrophysicists knew that the MBR had to be anisotropic at some level and began immediately to search for such anisotropies. It is generally accepted that, on the scale of galaxies and larger, the lumpiness of matter in the universe is due to aggregation by gravity. If this is so, the mass in the universe at the epoch of decoupling cannot have been perfectly smooth but rather must have exhibited small fluctuations that would later grow in amplitude to become galaxies and clusters of galaxies. A microwave map of the sky is a snapshot of the universe at this epoch more than 10 billion years ago and thus provides us with a unique opportunity to see these "gravitational seeds" from which galaxies grew as spatial fluctuations in the MBR intensity.

Dozens of isotropy measurements have been made with many different instruments, including ground-based radiometers only a few centimeters across, the 1 mile Very Large Array in New Mexico, a variety of high-altitude balloon-borne radiometers, and even radiometers operating aboard satellites. Because of diffraction effects in radio telescopes, the smaller instruments can only be used to search for large-scale anisotropy, whereas the very large telescopes are capable of investigating very small scales. There is still no convincing evidence that any deviation from isotropy has been detected with the exception of the "dipole anisotropy," which will be discussed in the following paragraph.

Figure 2 is a map of nearly the entire sky made with a balloon-borne radiometer with a ruby maser receiver operating at a wavelength of 15.6 mm. Figure 2 is plotted in galactic coordinates and the strip across the center of the map is the microwave emission from our own Milky Way. The 2.74 K isotropic part of the MBR has been subtracted, so only the deviations are represented in the map. Signal levels correspond to temperatures that range from -0.004 to $+0.004$ or $\pm 0.1\%$ of the MBR. Aside from the plane of the Milky Way, the only other obvious feature in the map is the "dipole anisotropy" which is generally characterized by the hot area to the right and cold area to the left (the remaining bumpiness in the map is due to instrumental noise). It is generally accepted that this feature is due to the motion of the Earth with respect to the "comoving frame" of the universe in which the MBR is at rest. The MBR is Doppler-shifted to shorter wavelengths in the direction of the Earth's motion and thus appears hotter, whereas in the direction opposite to the Earth's motion the radiation is shifted toward longer wavelengths and appears cooler. This anisotropy certainly has nothing to do with the "gravitational seeds" out of which galaxies form, but it is of fundamental cosmological importance nonetheless. From the magnitude of the "dipole anisotropy" (about 0.1% of T) and the motion of the Earth with respect to the Local Group of galaxies, it is found that the Local Group as a whole has the rather large velocity of 600 km s^{-1}. This number is being actively exploited to tell us about the large-scale distribution of matter in the universe, because the velocity of the Local Group seems to be at least in part produced by the gravitational acceleration caused by nearby clumps of matter (such as the Virgo cluster of galaxies). Such arguments can also be used to set constraints on the density parameter.

There is no convincing evidence of any other anisotropy of the MBR. The upper limits on the deviations from isotropy on angular scales from an arcminute to 90° are on the order of or less than 1 part in 10^4. However, even these null results have provided important evidence in evaluating cosmological models and theories of galaxy formation. For example, the limit on the "quadrupole anisotropy" (90° scale) places important constraints on anisotropic expansion of the universe and, in turn, provides strong support for the isotropic Friedmann model of the big bang. Limits on anisotropy on the scale of a few arcminutes are already in conflict with many standard models of galaxy formation (especially those not including some form of "dark matter") and such models are no longer considered viable.

The lack of a measurable anisotropy on the scale of several degrees and larger has caused astrophysicists discomfort for some time. According to the standard big bang theory, the microwave radiation incident from parts of the sky that are further than a few degrees apart was emitted (at the time of decoupling) from regions of the universe that were not causally connected; that is, there had not been time from the beginning of the universe for light signals to have been propagated between them. How then could these regions possibly have the same temperature to within less than 1 part in 10^4? It is considered one of the strengths of the inflationary theory of the universe that it offers a natural resolution to this puzzle: The present universe as a whole was inflated from a tiny causally connected region very early in the history of the universe.

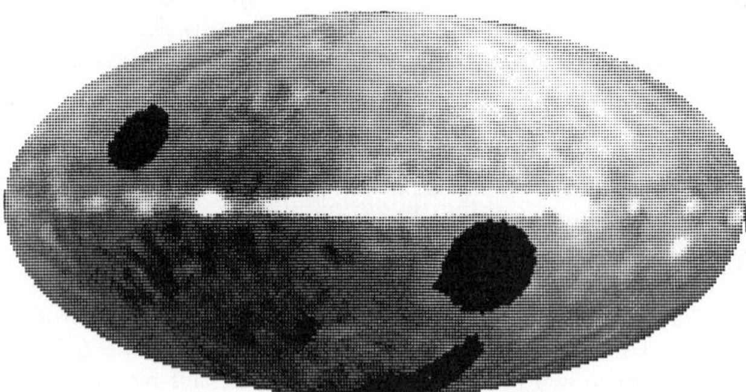

Figure 2. Microwave map of the sky at a wavelength of 1.6 cm. The darkest regions are 0.004 K cooler than the average and the brightest regions are 0.004 K hotter than the average. The strip down the middle is emission from the Milky Way. The blank patches were not observed.

FUTURE DIRECTIONS

Although no investigations have yet found any asymmetry of the MBR (other than the dipole), astrophysicists are very optimistic. Satellites including NASA's *COBE* mission that was launched in November 1989 and the Soviet Union's Relict II mission to be launched in the 1990s promise to improve the sensitivity to large-scale anisotropy, whereas several ground-based missions seek to improve the situation at smaller scales. The goal is a sensitivity of 10^{-6} of T. Astrophysicists agree that the lumpiness of the MBR *must* be larger than this level if we are to preserve our most basic ideas about cosmology and galaxy formation.

Additional Reading

Dicke, R. H., Peebles, P. J. E., Roll, P. G., and Wilkinson, D. T. (1965). Cosmic blackbody radiation. *Ap. J.* **142** 414.

Ferris, T. (1983). *The Red Limit*. Quill, New York.

Harrison, E. R. (1981). *Cosmology: The Science of the Universe*. Cambridge University Press, Cambridge.

Mather, J. C. et al. (1990). *Ap. J. Lett*. **354** L37.

Penzias, A. A. and Wilson, R. W. (1965). A measurement of excess antenna temperature at 4080 Mc/s. *Ap. J.* **142** 419.

Silk, J. (1989). *The Big Bang*. W. H. Freeman, New York.

Weinberg, S. (1984). *The First Three Minutes: A Modern View of the Origin of the Universe*. Bantam Books, New York.

Wilkinson, D. T. (1986). Anisotropy of the cosmic blackbody radiation. *Science* **232** 1517.

Wilkinson, D. T. and Peebles, P. J. E. (1989). Discovery of the 3°K radiation. *The Cosmic Microwave Background: 25 Years Later*, N. Mandolesi and N. Vittorio, eds. Kluwer Publishing Company, Dordrecht.

See also **Cosmology, Big Bang Theory; Cosmology, Inflationary Universe; Cosmology, Observational Tests; Galaxies, Local Group.**

Background Radiation, Soft X-Ray

William L. Kraushaar

"Soft x-rays" means x-radiation that is easily absorbed. Here we shall be discussing the x-ray energy region between 80 and 2000 eV. Adjacent at lower energies is the "extreme ultraviolet" and adjacent at higher energies are "hard x-rays." A beam of 2000 eV x-rays is attenuated to half-intensity by an air path of about 1.5 cm, whereas an air path of only 0.01 cm is sufficient to attenuate to half-intensity a beam of 80 eV x-rays. The detection technology of soft x-rays is awkward because of this extreme susceptibility to absorption and, of course, all astronomical observations must be carried out well above the Earth's atmosphere. Given the x-ray energy, E (in electron volts), the wavelength, λ (in angstroms), can be calculated from $\lambda = 12{,}400/E$.

Very early in the history of x-ray astronomy, it was noted that in addition to the radiation coming from discrete astronomical objects that are largely confined to the plane of our galaxy (the Milky Way), there was also a diffuse component that appeared to be isotropic. These early measurements were generally made at energies greater than 2000 eV where interstellar absorption (absorption by the gas between the stars of the Milky Way) is not significant. The isotropy of the detected radiation implied, therefore, that the diffuse background radiation above 2000 eV had an origin unrelated to the discrete objects in our galaxy and most likely originated in the vast regions beyond. This interpretation of the diffuse background radiation above 2000 eV has persisted over the years and is discussed further in other sections of this encyclopedia. When the first measurements of diffuse soft x-rays near 250 eV were made (here interstellar absorption *is* an important effect), there were pronounced deviations from isotropy. An appreciation of the possible significance of this anisotropy requires familiarity with a few of the details of the interstellar absorption process.

SOFT X-RAY MEASUREMENTS

All of the soft x-ray background measurements made during the decade since discovery have been made with detectors of modest-to-poor energy resolution. Survey data such as those used to prepare Fig. 1 have been obtained with proportional counters. Thin filters in front of a counter window, sometimes fabricated as part of the window itself, have been used to limit the response to a fairly narrow band of x-ray energies. The soft x-ray band names and definitions are shown in the two left-hand columns of Table 1. Maps of the entire sky are available for all except the Be band. The most complete data set, of which the C band map shown in the figure is a part, has a spatial resolution of about 6°. Be, B, and C refer to beryllium, boron, and carbon as filter materials.

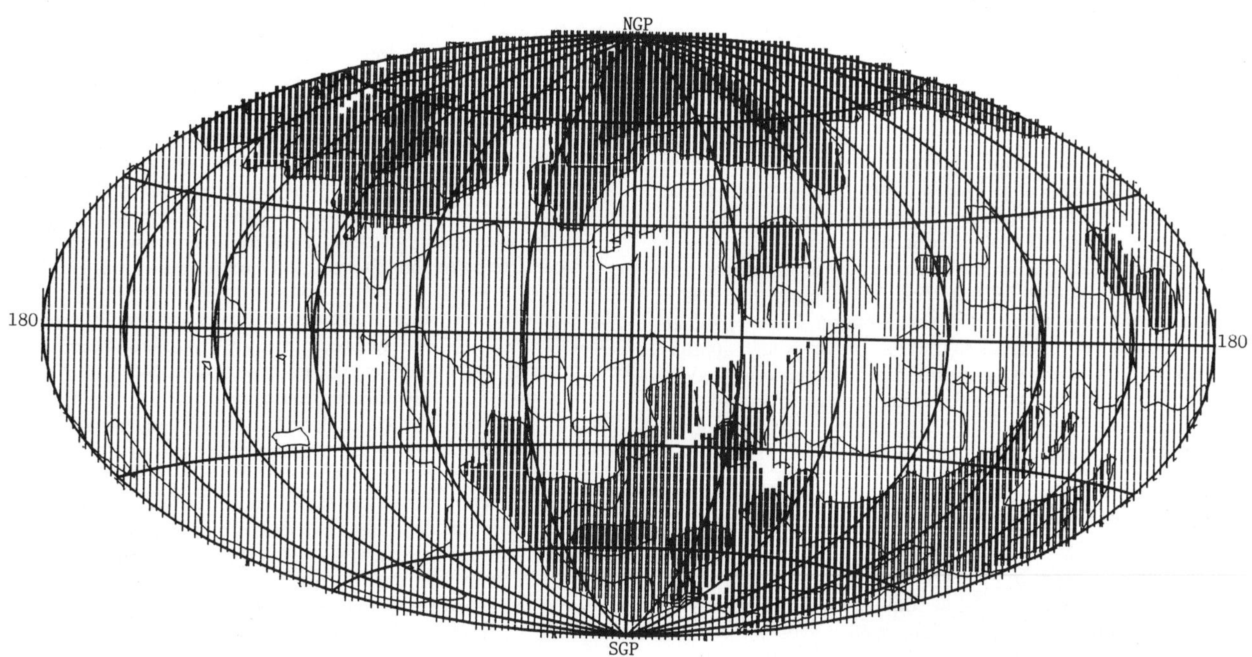

Figure 1. All-sky map in galactic coordinates for the C band (130–284 eV) soft x-ray background. The galactic center (l = 0, b = 0) is in the center of the map and galactic longitude increases from right to left, that is, from l = 180°, to l = 360° (or 0°), to l = 180°. The north and south galactic poles are at b = 90° and b = −90°. The latitude–longitude grid is in 30° intervals. The spatial resolution is 7° FWHM. The contour interval is 50 (arbitrary) units of intensity and the contour nearest l = 180°, b = 30° is for an intensity level of 100 units. Uncontoured blank areas indicate missing data or areas where the background emission was obscured by strong discrete sources.

Table 1

Band Name	Energy Bounds (eV)	Mean Free Path	
		H atoms cm^{-2}	*pc*
Be	78–111	1.0×10^{19}	3
B	130–188	5.9×10^{19}	20
C	160–284	1.2×10^{20}	40
M1	440–930	1.7×10^{21}	550
M2	600–1100	2.7×10^{21}	900
I	770–1500	4.6×10^{21}	1500
J	1100–2200	1.4×10^{22}	5000

INTERSTELLAR X-RAY ABSORPTION

The physical process by which soft x-rays are attenuated by the interstellar medium is photoelectric absorption, usually with the ejection of a *K*-shell electron. Hydrogen and helium are responsible for almost all of the absorption at energies less than 250 eV but at higher x-ray energies, carbon, nitrogen, oxygen, iron (*L*-shell electron ejection), neon, and magnesium together make the dominant contribution. Interstellar absorption has usually been estimated quantitatively from a measured knowledge of only the atomic hydrogen column density. This is because column densities of atomic hydrogen can be and have been measured in all directions, using the hydrogen hyperfine 21-cm radio transition in emission. The column density of most other elements can only be measured in optical or ultraviolet absorption, using a bright star for the background source. X-ray absorption as inferred from 21-cm column density measurements relies, therefore, on a knowledge of the abundances of the other elements, relative to hydrogen. Furthermore, 21-cm-measured column densities extend through the whole Galaxy and do not terminate on a star. Interstellar absorption mean free paths, expressed as the column density of atomic hydrogen, are shown in Table 1. The mean free paths (in parsecs) assume an atomic hydrogen density of 1 cm^{-3}. Recent evidence indicates that absorption by interstellar gas associated with ionized hydrogen may be comparable in some directions with that associated with atomic hydrogen. The detailed spatial distribution of ionized hydrogen is not known. It is not simply related to that of atomic hydrogen.

The 21-cm studies have shown that the galactic atomic hydrogen is distributed in a flat disk some 200 pc or so thick and 15,000 pc in radius. The solar system is located about midway through the disk and 8000 pc from the galactic center. Lines of sight near the perpendicular to the galactic plane (galactic latitude, $|b|$, of about 90° or toward a galactic pole) traverse generally a minimum column density of $N_m = 10^{20}$ H atoms cm^{-2}, although there are a few special directions with only half that amount. The interstellar gas distribution is quite irregular so that only in very rough approximations does the column density conform to the expected value, $N_m/\sin|b|$. In the galactic plane the column density, Sun to outside the Galaxy, is of order 10^{22} H atoms cm^{-2}, sufficient to absorb completely in all but the I and J bands.

THE SOFT X-RAY SKY

Among the objects evident in all-sky soft x-ray surveys are stars of many types, accreting binary systems and supernova remnants of various ages. In preparing maps for background studies, these more or less discrete objects are usually removed, if possible, to leave maps of the background alone. The map shown in Fig. 1 is for the C band and is in galactic coordinates. A prominent feature is the tongue of enhanced intensity that dips down from near the north galactic pole along galactic longitudes near 15°. It appears also on the B, M1, M2, and I band maps with a tendency for the intensity peak to move toward the galactic plane at higher x-ray energies. An apparently related but not quite coincident feature in the nonthermal radio sky is known as the North Polar Spur. The synchrotron radio emission is thought to originate in a shell at about 100 pc. It is likely that the x-rays are emitted from a region of hot gas interior to the radio-emitting shell. The connected tongue that runs along longitude 300° also has a radio noise counterpart and the total feature is known as Loop I. In those parts of the sky not dominated by Loop I, there is, in the C and B bands, a marked tendency for the intensity to be larger at high galactic latitudes. The intensity near the galactic poles is two or three times as great as it is near the galactic plane. The intensity has a negative correlation with the column density of hydrogen, though with considerable scatter. The finite intensity in the galactic plane is clear evidence for nearby emission.

Whereas the Loop I x-ray feature remains prominent on the M1, M2, and I band all-sky soft x-ray maps, the intensity enhancement at high galactic latitudes is lacking. Instead, there is a large irregular high-intensity feature about 40° in extent near the galactic center. Although there are a few other enhanced regions, possibly relic supernova remnants, the overall picture is one of increasing isotropy as the x-ray energy increases from the M1, to M2, to I, to J band.

DISTANCE TO THE DIFFUSE EMISSION SOURCES

The B and C band observations might easily be taken to suggest a strong extragalactic or galactic halo source subject to interstellar absorption (to explain the negative correlation of intensity with hydrogen column density) superimposed on a local diffuse source (to explain the intensity in the galactic plane). This seemingly attractive suggestion has serious problems, however, outlined as follows.

1. Shadowing of soft x-rays by the absorbing interstellar gas of external galaxies should be observable, has been searched for, but has not been seen.
2. Intensity fall-off with increasing amounts of galactic gas is observed but only in an overall average sense. Many individual features in the gas distribution, where deep absorption is predicted, show no decrease whatever in x-ray intensity.
3. The steepness of the intensity fall-off with gas column density should vary inversely with the absorption mean free path. Yet the C and B band intensities (and the Be band intensity also, where data are available) appear to fall off as though their mean free paths were all the same and about twice the predicted value for the C band. Extreme clumping of the gas of the interstellar medium could result in these anomalous mean free paths, but the required clumping is not seen.

Extragalactic intensities or even upper limits are often of cosmological interest. A power law spectrum $11E^{-1.4}$ photons (cm^2 s sr keV)$^{-1}$ (E is the photon energy in kiloelectron volts) is a good representation of the extragalactic intensity from 2–6 keV. For the C band the intensity upper limit for x-rays arriving from beyond the absorbing galactic gas is about three times that predicted from the extrapolated extragalactic power law. This upper limit does not include a correction factor estimated to be between 2 and 4 to allow for absorption due to gas associated with ionized hydrogen. (The upper limit measurement was carried out in directions near to but not coincident with that of the Small Magellanic Cloud.) The corresponding intensity upper limit for the M2 band is about twice that predicted from the extrapolated extragalactic power law. No correction factor is appropriate here.

As far as the galactic emission is concerned, shadowing by interstellar clouds a known distance away would serve to give direct indicators of distance to the soft x-ray emitting gas. This shadowing has not been observed and the best distance indicator comes from an indirect argument, as follows. If there were no absorbing material between us and the emission source, the C and

B band maps would reflect just variations in the source strength and so, except for minor effects of source energy spectra, the intensities would track and the maps would look much the same. For small but variable (in direction) amounts of intervening absorber, say 10^{20} H atoms cm^{-2} or so, the amount of absorption and its variation would be very different in the B and C bands. Features, if due to absorption, should be twice as deep in the B band as in the C band. This is not observed and so we conclude that the bulk of the galactic emission must be separated from us by less than 10^{20} H atoms cm^{-2}. With an atomic hydrogen density of 1 cm^{-3}, this would correspond to a distance of less than 30 pc. Note, however, that the solar system is in a region of density appreciably less than the galactic plane average of 1 cm^{-3}.

EMISSION MECHANISMS

The energy spectrum of the isotropic extragalactic radiation between 2 and 6 keV is well represented by a power law. The nature of the source and its emission mechanism is uncertain. If extrapolated into the soft x-ray region, this power law spectrum would provide a significant fraction of the measured intensity in the J, I, M2, and M1 bands. Its extrapolated contribution to the C and B bands would be negligible except in those few special directions where the 21-cm H data would predict extraordinarily small absorption of an extragalactic source.

M-dwarf stars, which have known x-ray emission properties and a very large space density, may contribute an apparent diffuse intensity of as much as 25% of that measured in the M bands. The contribution predicted in the lower energy bands is only a few percent.

For the remaining emission source, the available evidence favors radiation from very hot regions of interstellar gas. Ultraviolet measurements have shown that O VI, the oxygen ionization state favored at 0.3 million degrees, exists along the lines of sight to many stars. Furthermore, in the M1 band evidence for emission lines of O VII, uniquely characteristic of gas at 2 million degrees kelvin, have been observed from the North Polar Spur and from the general background radiation. Emission lines have not yet been observed in the other soft x-ray bands.

Additional Reading

Bochkarev, N. G. (1987). The structure of the local interstellar medium, and the origin of the soft x-ray background. *Soviet Astron.* **31** (No. 1) 20.

Cox, D. P. and Reynolds, R. J. (1987). The local interstellar medium. *Ann. Rev. Astron. Ap.* **25** 303.

McCammon, D., Burrows, D. N., Sanders, W. T., and Kraushaar, W. L. (1983). The soft x-ray background. *Ap. J.* **269** 107.

McCammon, D., and Sanders, W. T. (1990). The soft x-ray background and its origins. *Ann. Rev. Astron. Ap.* **28** 657.

Tanaka, Y. and Bleeker, J. A. M. (1977). The diffuse soft x-ray sky. *Space Science Rev.* **20** 815.

See also **Background Radiation, Gamma Ray; Background Radiation, X-Ray; Interstellar Medium, Hot Phase.**

Background Radiation, Ultraviolet

Stuart Bowyer

The ultraviolet band of the spectrum is generally defined as extending from the end of the soft x-ray band at 100 Å to the cutoff imposed by the transmission of the atmosphere at ~ 3000 Å. It is best divided into two separate bands: the extreme ultraviolet (EUV) from 100– ~ 1000 Å and the far ultraviolet (FUV) from 1000–3000 Å. If one considers the cosmic ultraviolet background, each of these bands is sampling vastly different regions; the EUV background is expected to be produced by emission from the hot component of the local interstellar medium, whereas the FUV background could be produced by a variety of mechanisms in our galaxy and may have contributions from a variety of exotic and cosmological processes. Studies of the EUV background are in their infancy but show great promise. The FUV background has been the subject of a substantial amount of work from the beginning of space research. The subset of the FUV band from ~ 1300–2000 Å has been studied most extensively. Scattering from the very intense, diffuse hydrogen Lyman α line at 1216 Å, which would severely compromise any observational data, can be eliminated through the use of crystal filters that block all radiation below ~ 1300 Å. Longward of 2000 Å, zodiacal light contributes strongly to the background in a complex and as yet poorly determined manner, and it is extremely difficult to separate the stellar contribution from the truly diffuse flux. It is a statement of the difficulty of this work that despite considerable effort substantial progress in our understanding of this background has only recently been achieved. This effort has clearly been worthwhile; recent results provide information on such diverse topics as the character of the long-prophesied but never-established galactic halo, star formation rates in the universe, and the character of dust in our galaxy.

THE COSMIC FAR ULTRAVIOLET BACKGROUND

From the beginnings of space research, attempts were made to measure the cosmic FUV background. This work was strongly motivated by the hope that in this waveband a true extragalactic flux could be detected and characterized. Theoretical speculation as to possible sources for this radiation was unconstrained by the available data and included such diverse processes as emission from a lukewarm intergalactic medium, emission from hot gas produced in a protogalaxy collapse phase in the early universe, the summed emission from a star formation burst phase in young galaxies, and photons from the electromagnetic decay of real or hypothetical exotic particles that were produced, or may have been produced, in the early universe.

It was the expectation that all of the problems that had bedeviled attempts to measure the cosmic extragalactic flux in the optical band would be overcome: The zodiacal light would be absent because the Sun's radiation falls rapidly in the ultraviolet, the measurements would be made above the atmospheric airglow, and the difficulty of separating the diffuse background radiation from that produced by stars would be reduced or even eliminated because the UV-producing early stars would be limited to the galactic plane.

The first 20 years of measurements, carried out by a number of groups in at least five major countries, indicated that the flux was uniform across the celestial sphere and hence was cosmological in origin. Estimates of the intensity of this flux, however, varied by *three orders of magnitude* with no clustering around a mean. In retrospect, there are a number of reasons for these discrepant results; perhaps it is most useful to state that these results provide empirical evidence that measurements of this type are intrinsically difficult.

A turning point in the study of this background occurred in 1980 when data obtained with a FUV channel of a telescope flown as part of the Apollo–Soyuz mission were analyzed and published. These data exhibited a correlation between intensity of the background and galactic neutral hydrogen column as derived from 21-cm radio measurements. Although these initial results were criticized on a variety of grounds, they were quickly confirmed by independent rocket and satellite measurements. The intensities obtained in these experiments were consistent within a factor of 2, and they varied from roughly 200–1500 photons cm^{-2} s^{-1} sr^{-1} $Å^{-1}$ depending upon the galactic hydrogen column. These results showed that the vast majority of the FUV flux was connected with processes in our Milky Way galaxy and was not, in fact, extragalactic in origin. Nonetheless, all of the data sets were consistent with a small part of this flux being isotropic. However, there was no way to determine whether this component was due to residual airglow processes at satellite altitudes, due to processes occurring within the galaxy, or was extragalactic in origin.

In the first half of the 1980s, a variety of processes was suggested as possible sources for the galactic component of this radiation, including starlight scattered by dust, molecular hydrogen emission, line emission from a hot (~ 10^5 K) interstellar gas, two-photon recombination radiation, and others. A variety of extragalactic processes was suggested as the source of the small isotropic component. Given the character of the existing data, there was no way to establish which, if any, of the processes mentioned previously were contributing to the FUV background. Two experiments carried out in the second half of the 1980s proved to be especially productive in establishing the sources of this radiation.

The first of these was a FUV imaging detector flown as a piggyback experiment at the focal plane of a 1-m EUV/FUV telescope developed as a collaborative effort between the United States and Germany. Data from this imager were subjected to a power spectrum analysis in a search for any component of the FUV background flux that was correlated with angular separation. This search was motivated by the fact that galaxies are known to cluster, and the scale of this clustering is well characterized by a two-point correlation function derived from optical data. A component of the FUV background was, in fact, found with the same angular correlation as that found in the optical, leading to the conclusion that this component is indeed extragalactic in origin and is the summed flux of galaxies at great distances. Because of the cosmological redshift, the FUV flux emitted by galaxies is shifted out of the observed FUV band at distances corresponding to about one-third of a Hubble time.

The detection of a cosmological component has an important astrophysical consequence. The intensity of this flux (about 50 photons cm^{-2} s^{-1} sr^{-1} $Å^{-1}$) is consistent only with a relatively low and constant star formation rate in galaxies for the time scales indicated. It also limits so-called star burst galaxies to be less than 10% of the total population of galaxies in this period.

A second major experiment that provided a variety of new data relevant to this field was a FUV diffuse spectrometer which flew in 1986 on the last shuttle flight before the *Challenger* tragedy. This instrument was sensitive from about 1350–1850 Å with about 16 Å resolution. Astronomical data were obtained from eight different directions in the celestial sphere with a wide range of neutral hydrogen column densities.

Several important results came from the study of the data obtained with this instrument. Emission lines from a collisionally excited hot (~ 10^5 K) gas in the interstellar medium were discovered. Combining these results with absorption data obtained with the International Ultraviolet Explorer proved the existence of the long-postulated, but unproven, hot galactic halo. The mass infall rate of ~ 10 $M_{\odot}$ yr requires that this halo must be continually replenished, which provides direct experimental confirmation of the "galactic fountain" model of the galactic halo gas. Data obtained with this instrument on FUV radiation scattered from dust in directions with relatively high column densities of neutral hydrogen showed that this dust has a relatively low albedo ($\omega \lesssim 0.2$) and that it scatters FUV radiation almost isotropically, indicating the dust itself is relatively small and roughly spherical. There are strong indications that at least traces of this dust are present everywhere, including regions in directions of very low hydrogen column density.

The cosmic FUV flux from 1400–1850 Å is shown in Fig. 1. The lower curve shows the spectrum in a direction of low neutral hydrogen column. A variety of processes contributes to this flux; the most intense spectral lines are emitted by the hot interstellar medium. The upper curve shows the spectrum in a direction of higher neutral hydrogen column. The majority of this radiation is starlight scattered by dust in the interstellar medium, but other processes also contribute to this flux.

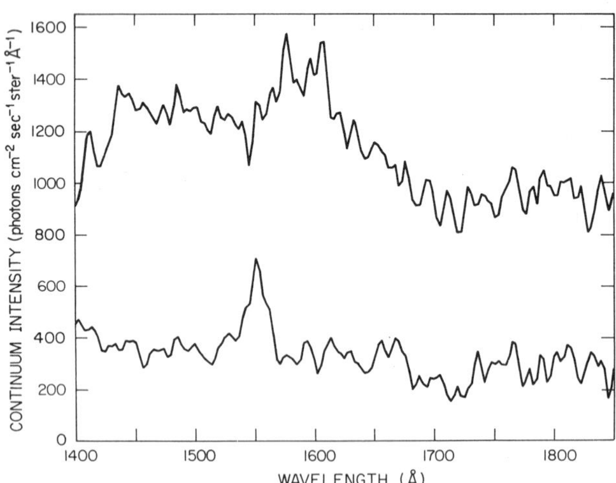

Figure 1. The cosmic FUV flux from 1400–1850 Å. The lower curve shows the spectrum in a direction of low neutral hydrogen column. The broad line at 1550 Å is the C IV from hot interstellar gas. The upper curve shows the spectrum in a direction of higher neutral hydrogen column. The majority of this radiation is starlight scattered by interstellar dust, but other processes also contribute to this flux.

THE COSMIC EXTREME ULTRAVIOLET BACKGROUND

Early attempts at measurements of the cosmic EUV background were made with broadband filters which included the resonantly scattered solar lines of He II at 304 Å and He I at 584 Å. These lines, whose study is important in its own right, dominated the data and made estimates of a truly cosmic flux impossible. The first useful estimates of the cosmic diffuse EUV background below 500 Å were obtained with an EUV telescope flown on the Apollo–Soyuz mission. These data were also obtained with a broadband filter, but they had an important advantage over previous results, in that a small field of view was employed and a very extensive data set was obtained. The geocoronal ionized helium responsible for scattering the He II 304 Å flux is confined to a few Earth radii by the Earth's magnetic field. In the antisolar direction, the Earth's shadow leaves the helium in darkness, greatly reducing the scattered intensity. By analyzing the extensive data obtained in the antisolar direction, significant upper limits to this flux were obtained.

Several broadband measurements of the EUV background have been made at wavelengths longer that 500 Å. In addition to measurements in the 500–800 Å band, several groups have obtained upper limits to the diffuse flux from 912–1080 Å with instruments with relatively narrow bandpasses.

Two relatively coarse spectrographic measurements of the cosmic EUV background have been carried out. High galactic latitude observations were obtained with 40 Å resolution during the interplanetary cruise phase of the Voyager 2 mission near the orbit of Saturn. The EUV part of these observations was dominated by the solar He I 584 Å flux resonantly scattered from neutral helium flowing through the solar system. The effects of this bright He I line, along with several other sources of background, were removed from the raw spectra by analysis, and upper limits to the EUV background were derived.

Spectroscopy of the diffuse EUV background below 500 Å is in its infancy. Grazing incidence optics must be used in order to obtain any reasonable throughput, and grazing incidence spectro-

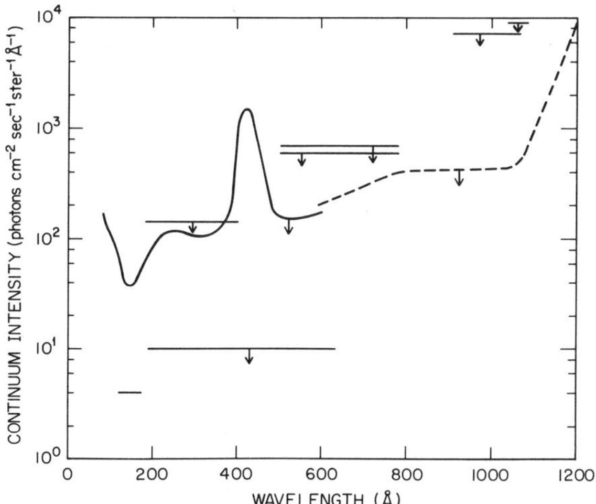

Figure 2. The cosmic EUV flux from 100–1000 Å. Most of the measurements have been made with broadband detectors and are shown as solid horizontal bars. Most are upper limits to the diffuse flux. The curved dotted line is an upper limit to the diffuse flux obtained with a low-resolution (~ 40 Å) spectrometer. The solid curved line was also obtained with a low-resolution (~ 20 Å) spectrometer; some lines or line complexes may have been detected in this measurement.

meters from diffuse radiation are technically difficult. An instrument that was flown on a sounding rocket in 1986 obtained upper limits on the EUV flux from 90–700 Å with about 20 Å resolution.

The existing data on the diffuse cosmic EUV background are shown in Fig. 2.

CONCLUSIONS

Investigations of the FUV background were initiated in the early phases of space astronomy. Substantial efforts were devoted to this work with disparate results. The discovery that most of this flux is produced within our galaxy opened a new chapter in the study of this radiation. Subsequent work in this field has established the existence and character of the hot galactic halo, has provided limits on the star formation rate in the universe for the last third of a Hubble time, and has defined the character of a component of interstellar dust. Future work in this field will explore questions such as: Why is this flux only roughly, and not exactly, correlated with neutral hydrogen column? Is this correlation the same in all directions of the sky? What is the true extragalactic background level? What is the correlation of the FUV background with the infrared background? and higher-order questions. Answers to these questions will certainly lead us to new insights about our galaxy and the universe.

The exploration of the cosmic EUV background is in its infancy, and the existing data are only suggestive of the underlying processes believed to be in operation. Another order of magnitude in sensitivity and spectral resolution may be required for these studies to provide definitive results.

Additional Reading

Davidsen, A., Bowyer, S., and Lampton, M. (1974). Cosmic far ultraviolet background, *Nature* **247** 513.

Paresce, F. and Jakobsen, P. (1980). The diffuse ultraviolet background, *Nature* **288** 119.

See also **Diffuse Galactic Light; Interstellar Medium, Galactic Corona; Zodiacal Light and Gegenschein.**

Background Radiation, X-Ray

Elihu Boldt

Although x-ray astronomy certainly flourished during the initial 15 years of its history, research on the CXB (cosmic x-ray background) remained comparatively dormant. During this early period experiments concentrated on well-isolated bright sources, particularly compact objects of high astrophysical interest such as neutron stars in binary stellar systems. It remained for the first two *High-Energy Astronomy Observatories* (*HEAO-1* and *HEAO-2*) to significantly remedy the relatively poor observational situation that existed before concerning the CXB. The all-sky survey carried out over a broad band of photon energies (from 0.1 keV to over 0.5 MeV) with the *HEAO-1* mission involved newly developed experiments especially designed to unambiguously distinguish the x-ray sky background from that due to other causes. The grazing incidence focusing x-ray telescope flown on the *HEAO-2* mission, usually known as the *Einstein* observatory, brought the power of focusing optics to x-ray astronomy. For soft x-rays (≤ 3 keV), the imaging detectors at the focus of this telescope were used to resolve a substantial portion of the background into discrete faint sources.

Basic measured characteristics of the cosmic x-ray background are reviewed here, including recent results from autocorrelation studies of surface brightness fluctuations examined with the *HEAO-1* and the *Einstein* observatory (*HEAO-2*). Prospects for addressing some key outstanding issues with future experiments are discussed as regards possible weak large-scale anisotropies (e.g., dipole) and spectral tests to help discriminate among candidate scenarios for the sources.

OBSERVED CHARACTERISTICS

The x-ray sky (at photon energies greater than about 3 keV) is dominated by a remarkably isotropic extragalactic CXB exhibiting an optically thin thermal-type spectrum characterized by a photon energy of 40 keV corresponding to a temperature $T = 46 \times 10^7$ K, much as the sky in the microwave band is dominated by an isotropic thermal CMB (cosmic microwave background), albeit characteristic of a blackbody at 3 K. Discrete x-ray sources resolvable with the imaging telescope of the *Einstein* observatory account for about 20% of the CXB (evaluated at 3 keV); they number about 10 per square degree. The upper limit to arcminute scale fluctuations in the apparent surface brightness of the unresolved background observed with this telescope implies that most of this residual CXB is either diffuse (e.g., due to a hot intergalactic plasma) or arises from a discrete source population having a number density on the sky exceeding one per square arcminute, much more than can be accounted for by quasars. An upper limit to the autocorrelation function for surface brightness fluctuations on scales ≥ 5 arcminutes has also been derived from *Einstein* observatory data; it sets an upper bound on the correlation length for the possible dominant sources of the residual CXB that is only 0.3% of the characteristic distance scale (c/H_0) associated with the Hubble constant (H_0) for the expansion of the universe ($c \equiv$ speed of light; $1/H_0 = 10$–20 billion years).

The broadband (3–50 keV) all-sky study of the background carried out with the gas proportional chambers of the *HEAO-1* A2 experiment resolved only the brightest foreground sources; in toto these resolved sources account for about 1% of the CXB (at 3 keV). Fluctuations in the surface brightness of the CXB on scales $\geq 3°$ observed with this experiment are consistent with random variations in the expected number of unresolved foreground sources. The *HEAO-1*-derived upper limit to the autocorrelation function for these CXB surface brightness fluctuations on scales $\geq 3°$ sets an upper bound on the correlation length for x-radiating rich clusters of galaxies which is consistent with the correlation found in the optical. For those unresolved AGN (active galactic nuclei) within the present epoch (i.e., at redshifts $z \leq 0.1$) contributing to the foreground fluctuations in surface brightness of the CXB the *HEAO-1* limit implies an upper bound on their correlation length $\approx 0.6\%(c/H_0)$.

LARGE-SCALE ISOTROPY

If the proper frame of the CXB were at rest relative to the proper frame of the CMB, then our own velocity (v) relative to this universal system would induce a CXB dipole anisotropy (i.e., Compton–Getting effect) in the direction of the CMB apex having an amplitude $(\alpha+3)v/c$ that varies from $\approx 0.4\%$ at ≤ 10 keV (where the effective CXB energy spectral index $\alpha \approx 0.4$) to $\approx 0.6\%$ at ≥ 100 keV (where the effective $\alpha \approx 1.6$). However, fluctuations in CXB surface brightness observed with the *HEAO-1* amount to $\geq 0.3\%$ for bands in the sky which are ≤ 1 sr (i.e., ≤ 3000 square degrees) in solid angle. Within the uncertainty imposed by such fluctuations, the weak large-scale anisotropy of the CXB is consistent with that implied by the CMB. If we could resolve out foreground sources at a level corresponding to ≥ 10 per square degree (i.e., comparable to the point source sensitivity level of the Einstein Observatory or better) over the whole sky, then large-scale surface brightness fluctuations in the residual CXB might well be suppressed by at least an order of magnitude, thereby permitting a sufficiently precise determination of the CXB Compton–Getting dipole anisotropy. Possible structure of the CXB on intermediate angular scales (e.g., such as might be associated with extragalactic objects like the "great attractor" or even with our galaxy) could, however, constitute a fundamental complication that would still have to be addressed. At ≥ 100 keV the best available all-sky data are from the *HEAO-1* A4 scintillation counter experiment. The limitation in measuring weak large-scale CXB anisotropies with this experiment arises from variable radioactivity induced in the scintillators by the radiation environment of the *HEAO-1* orbit, in spite of the precautions employed. Future high-energy studies of the CXB will have to avoid or further minimize this complication.

SPECTRAL ISSUES

It is now well established that the apparently thermal-type spectrum of the CXB is significantly different from the power law x-ray spectrum characteristic of typical bright AGN (active galactic nuclei). This has led to the notion that perhaps there is an evolution of AGN whereby those at the largest redshifts have x-ray spectra that differ from the canonical spectrum observed for those within the present epoch. However, such spectral evolution would not be necessary if most of the CXB were due to a hot IGM (intergalactic medium). Although this hot IGM would have to be of such magnitude as to dominate the baryonic matter content of the universe, cooling by the CMB would restrict it to redshifts within $z = 6$. Given this redshift constraint, an acceptable IGM model for the observed CXB spectrum would have to demand that the perturbation permissible from canonical AGN contributions be less than currently estimated.

A Residual Background?

Including the contribution of foreground extragalactic sources unresolved with the *HEAO-1*, the *total* 3–100 keV CXB spectrum observed may be nicely described by optically thin thermal bremsstrahlung radiation corresponding to a hot plasma at $T = 46 \times 10^7$ K (i.e., $kT = 40$ keV, where k is the Boltzmann constant). A good thermal fit characterized by $kT = 40$ keV was first established for the band 3–50 keV with gas proportional chamber data from the *HEAO-1* A2 experiment. Data from the scintillation counters of the *HEAO-1* A4 experiment were then used to determine that a thermal fit was similarly valid up to 100 keV. The temperature associated with this CXB spectrum is an order of magnitude higher

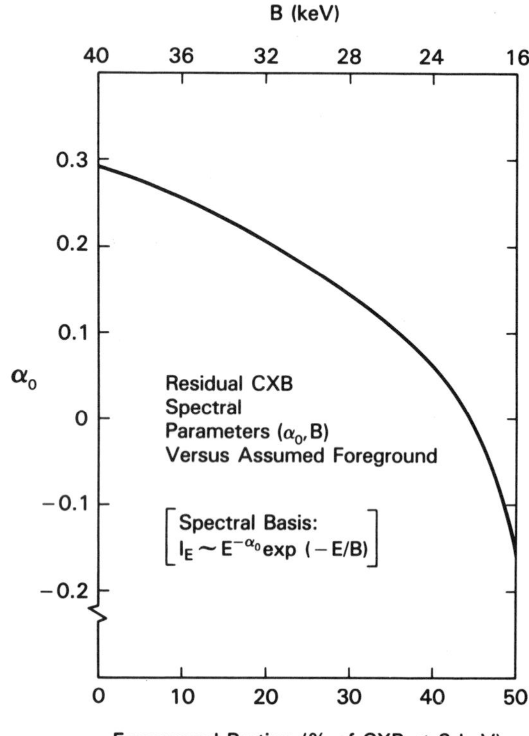

Figure 1. Parameters α_0 and B characterizing the indicated spectral fit to the residual CXB (3–50 keV) as a function of the assumed percentage of AGN foreground (at 3 keV).

than that for the x-radiating thermal plasmas associated with rich clusters of galaxies. Furthermore, the thermal form of the 3–100 keV CXB spectrum is apparently distinct from the nonthermal power law x-ray spectra characteristic of the brightest AGN observed with the *HEAO-1* over the same band. Subtracting the estimated contributions of sources making up such known extragalactic populations (i.e., a foreground amounting to about 40% of the CXB, at 3 keV) yields a residual CXB energy spectrum that is remarkably well described by a simple exponential function characterized by an e-folding energy of 23 keV; this spectrum is significantly flatter below about 10 keV than what is expected from a thin thermal plasma. Hence the spectral paradox posed by the total CXB appears to become appreciably more so by our attempting to isolate a *residual* CXB. With this procedure, however, we have sharpened our picture of the very particular sort of spectrum required for major sources of the CXB.

The principal portion of the subtracted foreground discussed above arises from AGN with canonical power law spectra characterized by an energy spectral index $\alpha \approx 0.7$. To exhibit the spectral consequences of subtracting various different estimated amounts of such AGN foreground from the CXB measured with the *HEAO-1* A2 experiment, the residual energy spectrum (3–50 keV) corresponding to each assumed foreground level has been fitted with the functional form

$$I_E \equiv dI/dE \propto E^{-\alpha_0} \exp(-E/B),$$

where E is the photon energy and α_0 and B are parameters determined from the spectral fit. In doing so the simplifying assumption is made that all foreground AGN have canonical power law spectra (i.e., $\propto E^{-0.7}$) over this band. A graphical representation of α_0 and B thereby obtained for the residual CXB is displayed in Fig. 1 as a function of the AGN foreground level at 3 keV. For zero foreground, we recover $\alpha_0 = 0.29$ and $B = 40$ keV corresponding to the thermal bremsstrahlung spectrum characterized by

$kT = 40$ keV that describes the total CXB. We note that for an AGN contribution exceeding about 30% (at 3 keV), $\alpha_0 < 0.2$ and $B < 30$ keV; this limit on α_0 would imply that the candidate "thin thermal" sources of the residual CXB have $kT > 200$ keV in their proper frame and, coupled with the limit on B, that they are located at redshifts $z = [(kT/B) - 1] > 6$, beyond the highest-redshift quasars as well as beyond a possible hot x-radiating intergalactic plasma.

If due to discrete objects, the residual CXB spectrum could readily arise from the Comptonized thermal emission characteristic of extremely compact sources whose accretion powered luminosity is mainly in x-radiation, at the maximum permissible level. These compact x-ray sources of the residual CXB could very well be high-redshift objects that represent an early stage in the evolution of AGN and *not* a new ad hoc population; this sort of *spectral* evolution would be inherent to the underlying physical processes involved. Unlike canonical AGN (e.g., those at low redshifts), their radiation in the IR, optical, and UV would be relatively small. In summary, AGN spectral evolution is not only an attractive simple solution to the severe spectral paradox associated with a residual CXB but could provide us with strong evidence that redshift is indeed a direct measure of the "arrow of time."

An Alternate View

Is the pronounced difference between the spectrum of the residual CXB and that of foreground sources compelling evidence for AGN spectral evolution? To explore this question, we consider the alternate possibility that there exists an as-yet unknown broadband x-ray spectral form for AGN which is essentially independent of cosmological epoch and can be understood to account for the puzzling spectrum of the entire CXB by suitably integrating the redshifted contributions of all such sources. This universal spectrum would have to be significantly flatter than the canonical ($\alpha = 0.7$) power law over a substantial portion of the x-ray band. In particular, as shown in Fig. 2, the redshifted spectra of such principal sources of the CXB must superpose to a composite spectrum characterized by $\alpha \approx 0.4$ over the band 3–10 keV and $\alpha \approx 0.7$ over the band 10–20 keV. We already know, though, that the spectra for essentially all present-epoch AGN are well described by $\alpha = 0.7$ over the band 3–10 keV. If the principal sources of the CXB are indeed to be representative of a universal AGN spectral form, they should exhibit spectra which, in the reference frame of emission, matches the canonical one over the 3–10-keV band. To ensure that this spectral component manifests itself to the observer mainly *below* 3 keV, however, most of the CXB would have to arise from sources of redshifts $z > 2$ [i.e., E(observed) = E(emitted)/$(1 + z)$]. In fact, our knowledge of the CXB spectrum below 3 keV is still relatively uncertain, and the possibility of a universal AGN spectrum cannot as yet be ruled out.

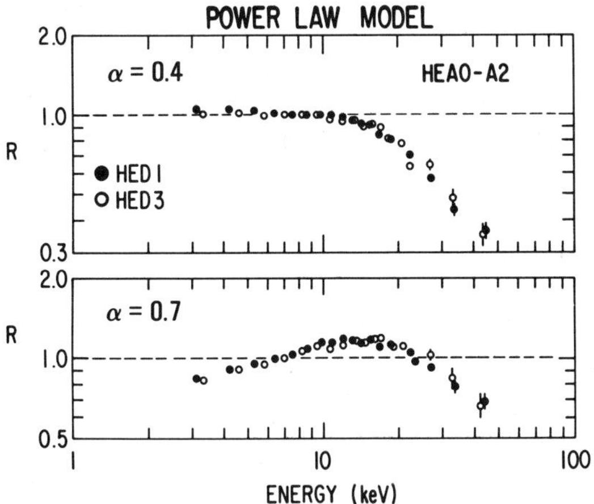

Figure 2. The ratio (R) as a function of energy of the counts observed for the *total* CXB to that predicted by convolving, with the detector response function, power law spectra characterized by spectral indices $\alpha = 0.4$ and $\alpha = 0.7$. Different symbols distinguish data obtained with the HED-1 and HED-3 high-energy detectors of the *HEAO-1* A2 experiment. These units were xenon-filled multianode proportional chambers. Statistical errors are shown when larger than the size of the symbols.

OUTLOOK

If the CXB is mainly due to AGN (i.e., not diffuse and not due to other discrete objects such as star-forming galaxies), then these sources must undergo substantial evolution in their luminosity whether or not spectral evolution is involved. As such, much of the CXB would arise from AGN at $z > 2$. Hence we have the possibility of a redshift test for the spectral evolution of these sources relative to present-epoch AGN. However, we note that future observations of such faint sources to be made possible by the powerful high-resolution imaging x-ray telescope of the *AXAF* (*Advanced X-Ray Astrophysics Facility*) will be restricted to photon energies less than 10 keV. Furthermore, if most of the CXB does indeed arise from faint discrete x-ray sources, then observations with *AXAF* must *necessarily* yield a significant sample of sources that exhibit a spectral index $\alpha \leq 0.4$ (over the 3–10-keV band) just to be compatible with the known CXB spectrum (see Fig. 2), regardless of AGN spectral evolution. However, if these same sources exhibit steeper spectra (corresponding to $\alpha \approx 0.7$) below 3 keV this could constitute real evidence for a universal AGN spectrum (i.e., absence of spectral evolution). On the other hand, if the flat spectra of high-redshift sources of the CXB are found to persist well below 3 keV, this would suggest that spectral evolution is a fundamental aspect of the AGN phenomenon.

Finally, should we fail to find a population of point sources that dominate the CXB, then we can use "fast" moderate-resolution x-ray optics for arcminute mapping of the weak surface brightness variations that could be characteristic of a thermal background that is intrinsically diffuse. The study of such variations might provide us with a direct indication of the large-scale gravitational fluctuations that trace the distribution of all matter (i.e., dark as well as visible).

Additional Reading

Barcons, X. and Fabian, A. (1989). The small-scale autocorrelation function of the x-ray background. *Monthly Notices Roy. Astron. Soc.* **237** 119.

Boldt, E. (1987). The cosmic x-ray background. *Phys. Rep.* **146** 215.

Leiter, D. and Boldt, E. (1982). Spectral evolution of active galactic nuclei. *Ap. J.* **260** 1.

Marshall, F., Boldt, E., Holt, S., Miller, R., Mushotzky, R., Rose, L., Rothschild, R., and Serlemitsos, P. (1980). The diffuse x-ray background spectrum from 3 to 50 keV. *Ap. J.* **235** 4.

Persic, M., DeZotti, G., Boldt, E., Marshall, F., Danese, L., Franceschini, A., and Palumbo, G. (1989). The autocorrelation properties of fluctuations in the cosmic x-ray background. *Ap. J.* (*Lett.*) **336** L47.

Schwartz, D. and Tucker, W. (1988). Production of the diffuse x-ray background spectrum by active galactic nuclei. *Ap. J.* **332** 157.

Tucker, W. (1984). *The Star Splitters* (*The High Energy Astronomy Observatories*). NASA, SP-466, Washington, DC.

See also **Active Galaxies and Quasistellar Objects, X-Ray Emission; Clusters of Galaxies, X-Ray Observations; X-Ray Astronomy; Space Missions.**

Binary and Multiple Stars, Origin

Richard H. Durisen

Binary and multiple stars are systems of two or several gravitationally bound stars orbiting their common center of mass. Since the first observational demonstration of their existence in 1803, it has become increasingly apparent that the preponderance ($>2/3$) of well-observed stars are members of such systems, whereas single stars like the Sun are less common, if not rare. Binary and multiple stars are remarkably diverse and can undergo complex evolutions from one type to another. The question of origin concerns not this evolution of types, but the way in which binary and multiple stars form in the first place as gravitationally bound systems.

OBSERVATIONAL CONSTRAINTS

Statistical compilations of data on unevolved main-sequence binaries provide important constraints on origin theories. Binaries range in scale from contact systems, with periods of a day or less, up to common proper motion pairs, with periods of a million years and separations of 0.1 pc. Although selection effects introduce considerable uncertainty, orbit periods seem to be distributed smoothly between these extremes with a median near 10 years, corresponding to solar system–scale separations. Systems with periods greater that 10 days tend to have eccentric orbits, whereas orbits with shorter periods are usually circularized by tidal interactions. Despite some exceptions, the binary fraction and other statistics do not vary drastically with spectral type along the main sequence. Although observers do not entirely agree, it is usually said that the widest binaries have random mass ratios of components, whereas closer systems have more nearly equal masses. Multiple systems are less common than binaries by a factor of a few and tend to be hierarchical, in that triples consist of a close pair with a wide companion, quadruples of two widely separated close pairs, and so on.

Important data relevant to origin are currently being gathered from observations of young stellar objects (YSOs) in star-forming regions. High-mass stars tend to form together as loosely bound associations in giant molecular cloud (GMC) complexes (e.g., Orion). Low-mass stars form out of the numerous dense clumps in GMCs, and also in smaller dark clouds (e.g., Taurus-Auriga), by a more quiescent process.

The visible T Tauri stars, the best studied YSOs, are 1–3 $M_{\odot}$ stars with ages of about 10^6 yr. As a class, they are slow rotators yet often have circumstellar disks accompanied by winds and bipolar outflows. About a dozen binaries with determined orbits are known among T Tauri stars, including systems with and without disks. Low-mass protostars still embedded in dense clumps almost always show spectroscopic evidence for circumstellar disks of solar system dimensions. Many of these disks are "active" in that their temperature structures imply an enormous rate of unexplained energy redistribution. Various infrared imaging techniques reveal apparent doubles among YSOs which remain to be confirmed as true binaries.

THEORETICAL CONSTRAINTS

Ignoring rotation, a star forms when a clump of a molecular cloud becomes too dense and cold for thermal gas pressure to resist self-gravity and gravitational collapse ensues. Clumps subject to collapse are typically a fraction of a light-year across and have stellar masses. The mechanisms for reaching the onset of collapse probably differ between high- and low-mass stars. The nature of the subsequent collapse, lasting about 10^5 yr, is best understood for low-mass stars.

Due to efficient cooling by dust, the collapsing gas remains cold until a central portion becomes opaque to far-infrared radiation and heats to provide pressure support. This "equilibrium core" only contains a small percentage of the clump's mass and quickly recollapses when it reaches the dissociation and ionization temperatures of hydrogen. A final equilibrium core of a few main-sequence radii forms in force balance and spends about 10^5 yr accreting the rest of the collapsing gas.

This picture is complicated by the fact that many dense clumps have measurable rotation rates. In most cases, the measured rates provide sufficient angular momentum that centrifugal balance should cause the collapsing mass to produce a rapidly rotating system of solar system dimensions, not stellar dimensions. Given that most stars are binaries or multiples with median separations of solar system scale, it seems evident that clumps most often solve their "angular momentum problem" by forming a binary or multiple star; that is, the clump rotational angular momentum is converted to binary orbital angular momentum. But how does this happen?

ORIGIN THEORIES

There are at least six theories of binary origin: cluster disintegration, capture, conucleation, fission, fragmentation, and disk instability. Despite almost two centuries of study, there is still no consensus about which of these mechanisms predominates.

Cluster Disintegration

Loosely bound groups of stars can form close binaries and hierarchical systems when they disintegrate into stable systems by ejecting stars. Only 10% of stars are estimated to form in bound groups, and so the process is inefficient. Perhaps some high-mass binaries originate in this way, but it cannot be a significant mechanism overall.

Capture

A bound binary system can form by three-body encounters of unbound stars or by tidal dissipation during an encounter between two unbound stars. Both mechanisms require high stellar densities to operate efficiently. Capture has been shown to be important for forming compact binaries in globular clusters. In star-forming regions, however, the capture time is too long to produce a significant fraction of binaries, even if solar system–sized disks are used for the collision cross section. Tidal interactions between precollapse clumps could produce bound systems but would at best be limited to explaining the widest binaries.

Conucleation

Conucleation theories suppose that mechanisms at the molecular cloud level lead to bound systems of precollapse clumps that later become binary or multiple stars. In addition to tidal interaction, possible mechanisms include phase transitions, thermal instabilities, and magnetic effects; but few suggestions along these lines have been investigated in enough detail to evaluate. Until recently, observations have been limited in resolution to the size of typical clumps, which are themselves about the size of the widest binaries. Improvements in observational resolution may soon discern directly whether binary or multiple clumps actually exist.

Fission

The fission theory proposes that a two- (or more) body system can form by the breakup of a single equilibrium object through hydrodynamic instabilities or successive bifurcations of form. Equilibrium in this context means that forces due to gravity, gas pressure, and rotation are balanced. The self-gravitating equilibrium states of

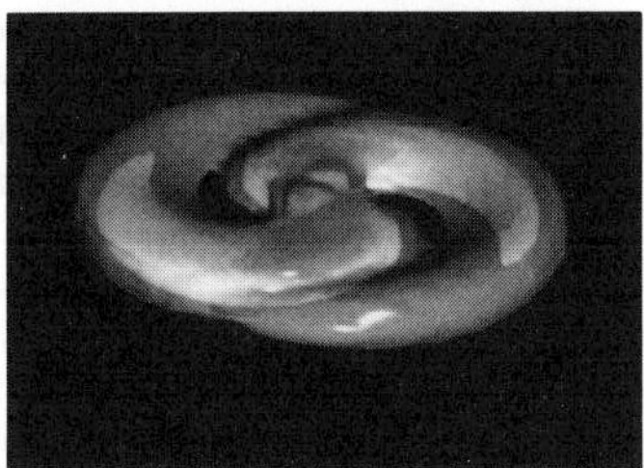

Figure 1. Computer hydrodynamic simulation of fission. An axisymmetric rotating star with $\beta = 0.33$ is given a slight initial disturbance toward producing a binary star. After two and one-half rotations, as shown here, the disturbance grows into two spiral arms that almost separate at the center. Rotation is counterclockwise in this view from above the equatorial symmetry plane, and so the arms are trailing. Gravitational torques in the arms therefore transport angular momentum from the center toward the ends of the arms. Because of this, binary formation proceeds no further than shown here; the arms never completely pinch through at the center.

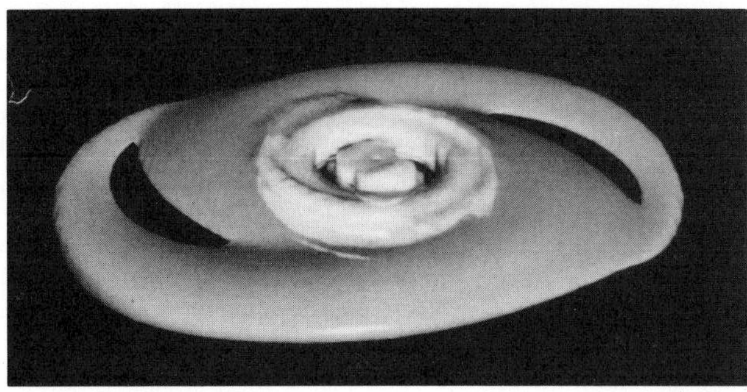

Figure 2. The same simulation as in Fig. 1 at five rotations. The bulk of the gas in the center has formed a single, barlike equilibrium star. A small amount of mass with the excess angular momentum has formed a detached ring. The material remaining outside the ring falls into it during the next few rotations, and so the end state is a ring around a bar. The ring shows no evidence of condensing into a single object during the simulation. (*Both figures are reproduced, with permission, from a 1989 videotape made by R. H. Durisen, S. Yang, R. Grabhorn, and J. B. Yost at the University of Illinois National Center for Supercomputing Applications.*).

incompressible fluids attracted the interest of many great mathematical physicists over the last two centuries. The theory has been extended to realistically compressible fluids, resembling stellar gases, only during the last two decades through computer modeling. Until the 1980s, fission was the most popular theory of binary origin, especially for contact systems.

A useful parameter in fission theory is β, the ratio of total rotational energy to total gravitational energy. For $\beta > 0.14$, equilibrium states with symmetry about the rotation axis (axisymmetric) "bifurcate" into forms resembling American footballs tumbling end over end. The axisymmetric states continue to higher β, but the nonaxisymmetric forms also exist. Similar bifurcations of more complex shapes occur as β increases further. Some bifurcations are stable and occur smoothly; others lead to violent instabilities. In either case, as β increases, a binary or multiple star could result, if a surface distortion of high enough complexity pinches the star into pieces.

The parameter β will increase as the equilibrium core of a protostar accretes collapsing gas and as a pre–main-sequence star contracts to the main sequence. Unfortunately, observations alone seem to preclude the latter possibility for fission. T Tauri stars are such slow rotators that they can not even achieve the first significant bifurcation at $\beta = 0.14$ during subsequent contraction. On the theoretical side, hydrodynamic computer simulations show that the fission instabilities tested so far cause shedding of a ring, not a binary (see Fig. 1 and 2). Even if the ring condenses into a star later, the system would have a very low mass ratio and could not explain the bulk of close binaries. The fatal problem is that compressible fluids produce trailing spiral arms when distorted at high β, and the gravitational torques caused by the arms transport orbital angular momentum outward.

Fragmentation

Fragmentation is also a process of hydrodynamic breakup due to nonaxisymmetric distortions; but it is distinguished from fission by occurrence during collapse, under nonequilibrium conditions. It has been extensively studied by the same computer techniques as fission and has been demonstrated to work, given the right initial conditions.

A dense clump of high β that has uniform density and rigid rotation fragments into a wide few-body system. Secondary collapses of the individual pieces can then produce a hierarchical multiple star. A similar clump with a moderate initial β can produce a binary of intermediate separation. Fragmentation can thus explain binary and multiple star formation for Earth-orbit separations and larger, when precollapse conditions are relatively uniform. To say that closer binaries evolve from these by orbital decay is problematic. For periods of 10 days and greater, observed binary orbits are eccentric on the main sequence. Dissipative orbital decay would probably have circularized them.

Even the formation of wider systems by fragmentation is disputed, at least for low-mass stars, because these are thought by some to reach the brink of collapse by a slow process of magnetic diffusion, producing a precollapse clump that is much denser at the center than the edges. Both analytic calculations and computer simulations show that the collapse of such a clump forms a single star plus disk, without fragmentation during collapse.

Disk Instability

Centrally condensed precollapse clumps form single stars with disks. Embedded protostars and classic T Tauri stars have such disks, with masses that range from 0.01 $M_\odot$ to a few $\times 0.1$ $M_\odot$. The detection of binaries among T Tauri stars suggests that binary formation occurs during this stage or earlier, that is , when a disk is likely to be present. Disks that are detected in YSOs are often "active," indicating global processes of considerable power. For these reasons, there has been an intensive recent effort to study nonaxisymmetric instabilities in disks around stars.

Most instabilities that have been found so far are localized. They may have influence on planet formation or angular momentum transport, but probably not on binary formation. However, one particular instability holds some promise. Global one-armed spiral disturbances in a disk can sling the central star off the center of mass. They grow rapidly whenever the disk becomes sufficiently massive, typically a few tenths of the central star's mass. It remains to be shown, however, that a binary can result and that mass ratios close to unity can be achieved.

ASSESSMENT

Scientists always hope that a single physical phenomenon will have a single cause, but the evidence suggests that more than one mechanism for binary and multiple star formation may be necessary. Precollapse tidal capture or conucleation may be needed to explain the widest systems, where mass ratios seem random. Fragmentation works well for a wide range of separations and particularly for hierarchical multiples, but only when the precollapse conditions are smooth. It does not work well for close systems. Fission seems to produce only low-mass ratios, if that, but global disk instabilities hold some promise for creating binaries of moderate mass ratio with close to moderate separations. In all cases, evolution subsequent to seed binary or multiple formation, due to accretion or orbital decay, could complicate the picture.

Given that binary and multiple star formation is such a clean resolution of the angular momentum problem, Nature may indeed arrive there by more than one route. Conditions just prior to the collapse of dense molecular clumps are potentially critical for determining the route. The distribution of main-sequence binary properties may some day be traced back to the rotation speed and central concentration distributions of precollapse clumps. Apart from theoretical studies of disk instabilities and of mechanisms prior to the onset of collapse, future advances are likely to come from observations of star-forming regions at higher angular resolution.

Additional Reading

Aitken, R. G. (1964). *The Binary Stars*. Dover Publications, New York.

Boss, A. P. (1985). Collapse and formation of stars. *Scientific American* **252** (No. 1) 40.

Boss, A. P. (1988). Binary stars: formation by fragmentation. *Comments on Ap.* **12** 169.

Durisen, R. H. and Tohline, J. E. (1985). Fission of rapidly rotating fluid systems. In *Protostars and Planets II*, D. C. Black and M. S. Matthews, eds., p. 534. University of Arizona Press, Tucson.

Shu, F. H., Adams, F. C., and Lizano, S. (1987). Star formation in molecular clouds: observation and theory. *Ann. Rev. Astron. Ap.* **25** 23.

Tohline, J. E. (1982). Hydrodynamic collapse. *Fundamentals of Cosmic Physics* **8** 1.

See also **Binary and Multiple Stars, Types and Statistics; Infrared Sources in Molecular Clouds; Interstellar Clouds, Collapse and Fragmentation; Molecular Clouds and Globules, Relation to Star Formation; Protostars; Protostars, Theory; Star Clusters, Globular, Binary Stars; Stars, Circumstellar Disks; Stars, T Tauri; Stars, Young, Rotation; Stellar Evolution, Binary Systems; Stellar Evolution, Rotation.**

Binary and Multiple Stars, Types and Statistics

Alan H. Batten

A binary system is a pair of stars revolving in a stable orbit about a common center of mass. The definition could be extended to include pairs of stars traveling together through space without any discernible relative orbital motion—the common-proper-motion pairs. Binary systems are usually classified by the observational means that we use to recognize them:

Visual binaries are instantly seen to be double stars through the telescope—quite a small one often suffices. Although some of these pairs are chance alignments of stars at very different distances from us, many are genuine binary systems.

Spectroscopic binaries cannot be seen as double even through a large telescope, but either two spectra can be seen with the aid of a spectrograph or periodic motions of features in the one spectrum (that result from the orbital motion of the star) betray the presence of a fainter companion.

Eclipsing binaries are recognized by the variation of their light caused by mutual eclipses of the two components. The light variations of these stars are characteristic and usually easily distinguished from those of intrinsically variable stars.

This classification came about for historical reasons. Sir William Herschel (1738–1822) first demonstrated the existence of visual binary stars (and coined the term "binary") from observations made between 1782 and 1803. Spectroscopic binaries were recognized about 100 years later, when E. C. Pickering (1846–1919) and H. C. Vogel (1841–1907) independently published, in 1889, the first demonstrations of the existence of such systems. John Goodricke (1764–1786) first proposed that mutual eclipses of two stars in a binary system could explain the variations (which he discovered in 1782 to be periodic) in the light of Algol. His explanation did not find general acceptance, however, until the discoveries of 1889 were announced.

Classification by methods of observation is not fully satisfactory for several reasons. First, binaries may belong to more than one class: Many visual binaries, for example, can also be observed spectroscopically, and *all* eclipsing binaries are potentially spectroscopic. Second, subdivisions inevitably arise as new observing techniques enable intermediate types to be discovered. Thus many authorities would add *astrometric* and *interferometric* binaries to the three classes given previously. Astrometric binaries are pairs in which the companion cannot be seen directly, but its presence may be inferred from the periodic motion of the brighter star, as seen projected on the plane of the sky. The first astrometric binary (Sirius) was discovered in 1834 by Friedrich Wilhelm Bessel (1784–1846), although the companion was subsequently directly observed by Alvan Clark (1832–1897). Interferometric binaries are pairs that may or may not be directly resolvable with the aid of a telescope, but which are best measured by studying the interference patterns produced by the two stars. J. A. Anderson (1876–1959) is usually credited with the first interferometric observations (of Capella in 1920) but earlier pioneering observations were made in 1895 by Karl Schwarzschild (1873–1916). In the last two decades, the increasing importance of interferometry (especially of speckle interferometry) and the development of photometric methods of measuring radial velocities have blurred the distinction between visual and spectroscopic binaries.

The strongest objection to the preceding classification scheme, however, is that it does not correspond to any obvious physical differences between the classes. We have come to recognize that the most important distinction is between *wide* pairs, in which the separation of the two components is so great that they do not affect each other's evolution, and *close* pairs, in which the two stars interact in many ways and change the course of each other's development. Even these two classes do not have a well-defined common boundary, but most visual binaries are *wide*, and probably most spectroscopic and eclipsing pairs are *close*, some even being pairs of stars in contact. Probably many unusual types of objects can be explained as the results of interactions between the components of close binary systems. For example, all cataclysmic variables (novae and other classes of erupting stars) appear to be binary. Loss of mass from the system, or transfer of mass between the two components, has made them evolve very differently from single stars, or members of wide binary systems. For this reason, the study of the incidence of binaries within special groups of stars has become of interest. All stars having an enhanced abundance of the element barium, to give another example, also appear to be members of binary systems, but the reason for this is not yet fully clear. It is still uncertain whether or not there are as many binaries among the old (Population II) stars found near the galactic center and in globular clusters, as there are among the younger (Population I) stars found in the solar neighborhood.

STATISTICS OF BINARIES

Many authors have estimated that a half or more of all stars are binary. Put another way, more stars are to be found in binary systems than in single isolation. This is certainly true if the common-proper-motion pairs are included as binaries. We cannot even rule out that the Sun belongs to such a pair, although it seems unlikely. There is a great disparity, however, between the numbers of binaries that are known to exist and the numbers for which we have detailed information—such as a knowledge of their orbits. For example, the *Index Catalogue of Double Stars*, in its original form published in 1961, lists around 65,000 pairs, but the most recent catalog of orbital elements of visual binaries (1983) contains orbital elements for fewer than 850 pairs. Similarly, the fourth edition of the *General Catalogue of Variable Stars* (1988) contains nearly 25,000 stars, of which perhaps around 4000 are eclipsing binaries, but only about 200 were listed in a 1970 catalog of the orbital elements of such systems. As mentioned previously, all eclipsing binaries are potentially spectroscopic binaries, and there must be many times more of the latter that do not display eclipses; yet we currently know orbital elements for only about 1500 spectroscopic binaries.

Such small samples may not be representative of the whole population of binaries, yet if we want to know the properties of that population, we have no choice but to begin with those few members that are well studied. All discussions of the distribution of periods, orbital eccentricities, ratios of the masses of the two components, and so on are inevitably tentative. We have (not always consciously) selected the systems that are well studied and we do not know how to make full allowance for that fact.

For example, the observed distribution of the ratios of the masses of the two components of binary systems shows a strong peak at unity. There are, indeed, some physical reasons for supposing that the mechanisms of formation of binary systems will favor the production of two nearly equal components. Stars of equal mass, however, are usually of equal luminosity—and pairs of nearly equal stars are much easier to discover than pairs whose components are very different. There is still no general agreement on the significance of the peak in the observed distribution. Again, there is much evidence for the existence of a period–eccentricity relation for visual binaries: Higher orbital eccentricities are associated with longer orbital periods. Observational selection has contributed to this. Stars in elliptical orbits move more quickly (relative to each other) when they are near their point of closest mutual approach, than do stars in circular orbits with the same average separation. The longer the period, the more noticeable this effect. Thus the attention of observers is drawn preferentially to eccentric long-period orbits, rather than to circular ones. On the other hand, most spectroscopic binaries have circular orbits, and we know that tidal and other forces act within such systems to make orbits circular quickly, even if they were not originally so. No orbits with an eccentricity greater than 0.7 (and only a few with eccentricity greater than 0.6) have yet been found for periods less than 10 days. So there are both real and spurious period–eccentricity relations, and we have not yet completely sorted them out.

MULTIPLE SYSTEMS

Because binary systems are evidently very common, it should come as no surprise that many of them are even more complex. Triple, quadruple, quintuple, and even sextuple systems are known, and possibly systems with even more components are waiting to be recognized. Usually, a triple system consists of a close pair attended by a distant companion, and the ratio of the two orbital periods is large (usually several hundred or thousand, although in λ Tauri the ratio is only 8). This can be understood from general considerations of stability: The close pair acts gravitationally like a mass point, on the distant companion, and is itself too tightly bound to be disrupted by the latter's tidal force. Similarly, quadruple systems usually consist of two close pairs separated by a relatively large distance. An extreme example is A.D.S. 9537, which is a common-proper-motion pair, showing as yet no perceptible orbital motion, each of whose components is an eclipsing binary consisting of a pair of stars in contact, with a period of less than 1 day. However, a class of multiple systems is recognized (the so-called Trapezium systems) in which the mutual separations of individual stars are more nearly comparable. The class takes its name from its prototype: The Trapezium in the constellation of Orion. It is believed that the orbital planes of multiple systems are usually nearly coplanar. This again is probably a requirement for the stability of the system, but the observational data are meager, and some systems are known for which there is reason to believe that the different orbital planes are highly inclined to each other.

The incidence of triple systems has not yet been studied enough for there to be general agreement on what proportion of binaries is triple. Obviously, not all binaries are triple, nor all triples quadruple, and so on. The higher multiples must be rarer than the lower. Observational selection again makes it difficult for us to determine just how much rarer. The more complex a system, the more difficult it is to observe and to interpret. This is particularly true of spectroscopic multiples: Four or five sets of spectral features blended together may be impossible to resolve and to measure properly. Only recently have we become sure that λ Tauri *is* triple, and another triple, V389 Cygni, defied all attempts at interpretation until a decade ago. Sometimes, a third star may be so distant from the close pair that the physical association of the three must long remain doubtful. In still other systems, the existence of a close companion of one member of a relatively wide pair is inferred only from the anomalous masses and luminosities that would otherwise be deduced for the components. It seems certain, however, that, at some level, higher multiplicity becomes so rare that only one example of such multiplicity will be found in a galaxy. The next level of multiplicity should be nonexistent. This conclusion seems to be invalidated by the existence of star clusters, in which hundreds or thousands of stars may be found together in a relatively small volume. They are not bound to each other in closed orbits, however. Even though most binaries were probably formed within clusters, a star cluster is *not* simply a still more complex multiple system. The border region between multiple systems and small clusters (containing only about 10 or 20 members) is still relatively poorly studied.

Additional Reading

Abt, H. A. (1977). The companions of sunlike stars. *Scientific American* **236** (No. 4) 96.

Abt, H. A. (1983). Normal and abnormal binary frequencies. *Ann. Rev. Astron. Ap.* **21** 343.

Evans, D. S. (1968). Stars of higher multiplicity. *Quart. J. Roy. Astron. Soc.* **9** 388.

McAlister, H. A. (1988). Seeing stars with speckle interferometry. *American Scientist* **76** (No. 2) 166.

See also **Binary and Multiple Stars, Origin; Binary Stars, Astrometric and Visual; Binary Stars, Eclipsing; Determination of Stellar Parameters; Binary Stars, Spectroscopic; Stellar Evolution, Binary Systems.**

Binary Stars, Astrometric and Visual

Sarah Lee Lippincott

Visual and astrometric binaries are gravitationally bound pairs of stars identified observationally through positional measurements made on the plane of the sky. It is well known that stars are gregarious and that a large proportion of stars are found in pairs or in multiple systems. Their importance lies in a number of areas:

1. The analysis of a visual binary leads to the only direct method for the evaluation of stellar mass, one of the most important parameters of the physical universe. In some cases the mass

of the system is determined; in other situations the individual masses or only the mass ratio of the components is determined.

2. Stellar duplicity plays an important role in the study of the genesis and evolution of stars; it also provides the potential for the study of stellar mass loss.
3. The spatial arrangement of the orbital planes of wide visual binaries in clusters or in small regions in the Galaxy may play a role in interacting galactic dynamics.

The particular interest in astrometric binaries is in the solutions to a number of problems:

1. Luminosity function of late spectral type dwarfs.
2. Missing mass
3. Frequency of planet-like objects and brown dwarfs

Visual and astrometric binaries may appear in multiple systems in hierarchical form with space distributions that allow for a pair to be treated as a two-body system. There is no obvious preference for particular magnitude differences except for selection effects, and a wide range of spectral types are found in combination. The apparent orbit is that which appears on the plane of the sky; the major axis of this ellipse is foreshortened with respect to the true orbit by the inclination of the true orbit to the plane of the sky. The periods of revolution of visual binaries range from a little under two years to centuries, whereas their separations on the plane of the sky range upward from less than one second of arc. There is no intrinsic physical significance to these limits, which result from observational selection effects.

Astrometric binaries are those physical pairs of stars which because of their small angular separation and/or their large differences in magnitude are not discovered as visual doubles. They appear on photographic plates as single images, called the photocenters. They may be revealed as binary systems through the detection of variable proper motion (interpreted as Keplerian motion) from astrometric studies of parallax and proper motion. Other optical techniques and detectors can make visible the close companions in systems previously identified through analyses of variable proper motion. This technique is also used to survey nearby stars to discover duplicity.

METHODS OF OBSERVATIONS AND ORBITAL ANALYSES

Visual binary stars have been observed for almost two centuries. The two components of a system are referred to as primary and secondary or by the designations A and B. Observations are generally characterized by measurements of B relative to A in polar coordinates: ρ, the separation, and Θ, the position angle referred to a specific equator and equinox on the plane of the sky. Visual observation programs most frequently make use of the filar micrometer to measure these quantities. Other programs are designed to inspect all stars systematically for duplicity. The commission on double stars of the International Astronomical Union authorized a depository for visual binary and speckle-interferometric observations (currently with over 70,000 entries) which is maintained and kept up to date in Washington, DC, and duplicated elsewhere.

The greatest problem interfering with the ability to resolve close pairs from Earth-based telescopes is the turbulence in our atmosphere, or "seeing effects." Bad seeing causes the stellar image to rapidly displace itself in the focal plane of the telescope by many times the diameter of the diffraction image of the telescope optics. The trained eye has the ability to capitalize on moments of good seeing to take advantage of the quality of the optics, and by so doing the observer is able to make measurements of separation or to make discoveries of duplicity of close visual binaries. Current techniques are able to "beat" the seeing. Charged-coupled devices (solid-state electronic detectors, CCDs) provide an attractive alternative to photography but are currently limited to a small field of view.

The components of visual binaries whose angular separations on the plane of the sky are several arcseconds or more are also measured on photographic plates. These measurements of the components may be made relative to one another as in the visual mode or each star may be referred to a background of so-called fixed stars appearing on the photograph. The two-dimensional measurements must be analyzed with algorithms that consider the times of observation, the transference of the apparent orbit to the true orbit through the geometry of the system, and the dynamical properties of Keplerian motion. If each component is measured against a reference frame, proper motion and parallax are included in the solution. The resulting quantities that have further physical interest are the period of revolution and the scale of the orbit, given as the semiaxis major in arcseconds. These values along with the parallax of the system provide the mass of the system,

$$\sum M_{\odot} = M_{\mathrm{A}} + M_{\mathrm{B}} = a^3/\pi^3 P^2.$$

Here, the unit of mass is that of the Sun, a is the semiaxis major in arcseconds, P is the period in years, and π is the trigonometric parallax in arcseconds. The accuracy of the sum of the masses is usually limited by the accuracy of the parallax.

In order to determine the individual masses, each component of the system must be measured against a reference frame of fixed stars close to the line of sight. For this purpose, it has been practical to accumulate a series of photographic plates exposed with a long-focus refractor or an astrometric reflector in order to achieve a large-scale portrayal of the field. The images on the photographic plate may be 3 arcseconds in diameter with less than good seeing. Therefore, generally only binaries with separations greater than this photographic resolving power limit are measured individually against the reference star frame. (There are currently a limited number of programs that employ charged coupled devices as receivers instead of photographic plates for positional determinations.) The photographic plates are measured in rectangular coordinates and the data on the two target stars undergo a combined least squares solution, merging the two coordinates when appropriate. The solution yields values for the proper motion of the system, and one value for the parallax. In addition, orbital parameters from the visual observation analysis are used for the determination of the orbital scale of each component. Here, each component revolves about the inertial point of the system and the orbits mimic the relative orbit with different scales and an orientation reversal of 180° for the orbit of the more massive star. The ratio of the scales of the two orbits yields the sought-after value of the mass ratio. The mass ratio can be well determined and gives good insight into the relative differences in masses of stars of differing spectral types, colors, and magnitudes. The individual masses are also determined from the scale as given by the semiaxis major of the relative orbit. But this value, a, in astronomical units, is dependent on the value for the trigonometric parallax of the binary system. If the binary is unresolved on the photograph, the plates are measured and analyzed as before, only it is then the photocenter that is measured and the orbital motion analyzed is that of the photocentric orbit. The resulting scale of this orbit, α, is a modification of that of the relative orbit by the difference in luminosity of the two components and their mass ratio:

$$\alpha/a = \mathbf{B} - \beta,$$

where $\mathbf{B} = M_{\mathrm{B}}/(M_{\mathrm{A}} + M_{\mathrm{B}})$, M refers to mass, and $\beta = L_{\mathrm{B}}/(L_{\mathrm{A}} + L_{\mathrm{B}})$, where L refers to luminosity. Here, the ratio depends on the accuracy of determination of the blending of the images.

DISCOVERY AND CONFIRMATION OF ASTROMETRIC BINARIES

The nearby stars within about 25 pc (about 80 ly) of the Sun, the region currently limited to the Earth-based trigonometric parallax observations of sufficient accuracy, are predominantly main-sequence stars of solar and later spectral types. Therefore, the fainter unseen or astrometric companions in this region are likely to be of later spectral type, lower-mass stars and perhaps brown dwarfs or planets. Alternately, they may be at the other end of the evolutionary scale, that is, degenerate stars, such as white dwarfs. Discovery of astrometric binaries has long been a by-product of proper motion and parallax studies of relatively nearby stars. Variable proper motion accounted for by Keplerian motion will yield computed geometric and dynamical orbital elements with the true scale of the relative orbit being substituted by the photocentric orbital scale, as in the case of blended photographic images. If the difference in magnitude of the components is greater than 5 magnitudes, the image is formed solely by the brighter component and therefore the scale of the orbit is that of the visible component around the center of mass of the system. Until the astrometric binary is resolved by another method, only limits on the individual masses and the sum of the masses can be given. The mass of the visible component can be estimated from the mass–luminosity relation, which provides plausible working values or limits on the masses.

There are a number of programs with techniques devised to enhance the resolution of binaries of too small angular separation to be seen or recorded directly, primarily because of turbulence in the Earth's atmosphere. These programs vary in optical analysis technique, scanning or "staring" detection, and detecting device. Binary star astrometry programs which yield significant contributions are generated by speckle interferometry. With the largest telescopes we can expect a resolving power of 0.02 arcseconds. This technique is suitable for a difference in magnitude of up to 5 or 6. Current speckle interferometry programs are shifting to CCDs as receivers. The CCD data can be digitized, eliminating the transfer of information via a series of photographic processes. There are a variety of optical devices and techniques being tested and used. Observations made in the infrared have the increased potential for the discovery of very low luminosity stars or brown dwarf companions.

AN EXAMPLE OF THE COMBINATION OF PHOTOGRAPHIC ASTROMETRIC AND OTHER OPTICAL TECHNIQUES FOR BINARY STAR ANALYSIS

μ Cassiopeiae has long been an intriguing astrometric binary to the astrophysicist in the study of stellar evolution. This fifth magnitude star is classified as a Population II high-velocity subdwarf. It is likely that μ Cas came into being before the formation of the galactic disk, giving it a chemical composition that may be characteristic of the early stages of the formation of our Galaxy. A determination of the helium abundance is of interest in evolutionary studies of the Galaxy. Once the mass and the luminosity of the star are known, its helium abundance can be determined when the abundance of "metals" (heavier elements) is known. Hence the determination of mass of the components of this astrometric binary is all important to the question of the primordial helium abundance in the Galaxy. Proper motion studies from two observatories have shown Keplerian motion. Observations over the years have provided the photocentric orbit from the Sproul Observatory as shown in Fig. 1. Missing from this study alone is the important knowledge of the scale of the relative orbit, or simply, the separation of the two components and their difference in magnitude. Assuming a large magnitude difference, the orbit is virtually that of the primary star around the common center of mass. The relative orbit of the B–A components has an orientation difference of 180°.

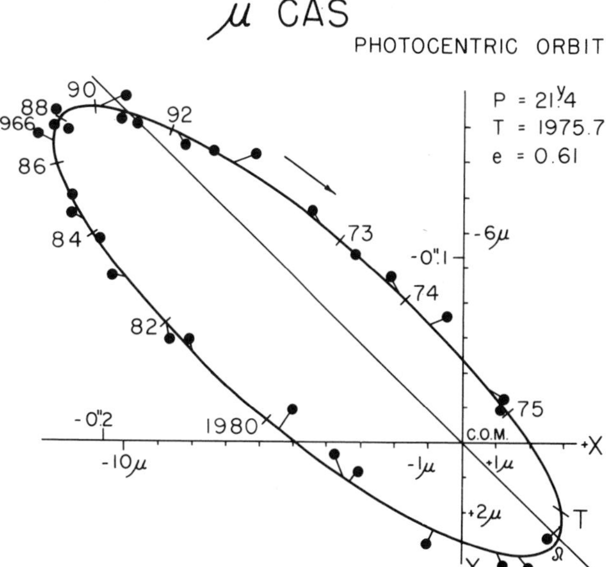

Figure 1. The photocentric orbit of μ Cassiopeiae, with yearly average residuals determined from photographs taken with the Sproul Observatory refractor. The proper motion and parallax, simultaneously computed with the orbital elements, have been removed, leaving the inertial point of the system as the origin. [*Reproduced by permission from* Astrophysical Journal ***248*** *1055 (1981).*]

Since the middle 1960s, many attempts have been made to "see" the companion at various wavelengths: photographically, visually, and with many kinds of optical techniques that are used to resolve close binaries. These observations, with increasing accuracy as time goes by, are used to measure the separation and position angle of the components for comparison with the concurrent photocentric orbital positions. In addition, their difference in magnitude is measured. By now, with continued speckle-interferometric observations and with direct-imaging CCD cameras, there is sufficient accuracy for improved photocentric orbit and parallax determinations. The required highly accurate consistent astrometric positions over a long interval of time are difficult to maintain; however, with forthcoming technical advances on the ground and in space, improvements are in sight. Other binary stars that relate to important astrophysical questions also are given due attention by the observers.

OBSERVATIONS FROM SPACE

With programs designed for the *HIPPARCOS* and *Hubble Space Telescope* satellite observatories, visual and astrometric binaries play a significant role. In preparation for the accumulation of data from space by *HIPPARCOS*, a catalog of the components of double and multiple stars (CCDM) was compiled. The CCDM lists the positions of the individual components aimed at a positional accuracy of 0.1 arcsecond, resulting in a database that will be of great value for the future, and on which we can build. For the purpose of identification for the *HIPPARCOS* program, coordinates are required for individual components where the separations are greater than 3 arcseconds and for the photocenter where the separations are less than 3 arcseconds. The *HIPPARCOS* program will provide binary star observations Θ and ρ for separations from 0.2–5 arcseconds. The positions, parallaxes, and proper motions for 100,000 stars including visual and astrometric binaries brighter than the 13th magnitude were to be determined with a positional accuracy of 0.002 arcsecond. As the launch of *HIPPARCOS* in 1989 placed it in a nonoptimum orbit, the final performance of the satellite remains to be seen.

The *Hubble Space Telescope* (*HST*) is also concerned with the identification of binary stars. Ground-based observations with high-resolution cameras have been systematically surveying potential *HST* guide stars for duplicity. (Double stars will in many cases be unsuitable guide stars.) There is potential for astrometric studies of binaries and the discovery of faint low-mass companions particularly among the nearby stars with the Wide Field/Planetary Camera of the *HST*. The Fine Guidance System and also the Faint Object Camera can be used as well. Positional accuracy for a bright star is expected to be in the range of 0.001 arcsecond. Stars as faint as magnitude 28 may be detected. There is provision for determining the position angles and separations of close binaries where the separation is greater than 0.03 arcsecond. After the April, 1990 launch of the *HST*, spherical aberration was discovered in the telescope; corrective optics to be installed in the future are needed to attain the full performance of *HST*.

Additional Reading

Couteau, P. (1981). *Observing Double Stars.* MIT Press, Cambridge, MA.

Kafatos, M. C., Harrington, R. S., and Maran, S. P., eds. (1986). *Astrophysics of Brown Dwarfs.* Cambridge University Press, Cambridge.

McAlister, H. A. (1985). High resolution measurements of stellar properties, *Ann. Rev. Astron. Ap.* **23** 59.

Monet, D. G. (1989). Recent advances in optical astronomy. *Ann. Rev. Astron. Ap.* **26** 413.

van de Kamp, P. (1981). *Stellar Paths.* D. Reidel Publishing Company, Dordrecht.

See also **Stars, Distances and Parallaxes; Stars, Low Mass and Planetary Companions; Stars, Masses, Luminosities and Radii.**

Binary Stars, Atmospheric Eclipses

Margherita Hack

A small group of eclipsing binaries (binary stars whose orbital plane contains or forms a small angle with the line of sight of the observer, so that an eclipse of one star by the other can occur) exhibits atmospheric eclipses. The typical members of the group, also called ζ Aurigae stars from the name of their prototype, consist of a cool giant or supergiant star and a hot companion, which periodically passes behind the larger star and shines through its very extended, rarefied atmosphere. This gives us the opportunity to observe the absorption lines produced by the cool atmosphere of the larger star in the continuous spectrum of the companion. In fact, the latter, when approaching the conjunction, before disappearing behind the cool giant, and then reappearing after conjunction, moves at various angular distances from the cool star photosphere and acts like a probe of the cool atmosphere, permitting us to study its physical properties at various heights above the photosphere. The well-known binaries belonging to this group are ζ Aurigae, 31 Cygni, 32 Cygni, 22 Vulpeculae, and VV Cephei, with orbital periods ranging from 249 days to 20 yr (Table 1). A different case, unique among all binaries, is the ε Aurigae system, which has the longest known period of all the eclipsing binaries: 27.1 yr. Whereas in the previously mentioned cases we observe the eclipse of the hot companion star by the cool giant or supergiant (the eclipse of the cool giant by the hot companion giving no measurable effects), in the case of ε Aurigae the observable photometric and spectroscopic phenomena occur when the supergiant is eclipsed by a companion whose nature was very obscure before the observations of the most recent eclipse (1982–1984) which, for the first time, was observed over the ultraviolet, optical, and infrared ranges.

Table 1. Physical Parameters of the Well-Known Atmospheric Eclipse Binaries

Name	*Orbit Period (day)*	*Spectral Type*	*Mass* ($M_\odot$)	*Radius* ($R_\odot$)
22 Vul	249.111	(1) G7 II	4.2	40
		(2) B9 V	3.0	2.9
ζ Aur	972.162	(1) K4 Ib	8	154
		(2) B8 V	5.8	4.3
32 Cyg	1147.8	(1) K5 Ia, b	19:	215
		(2) B4 IV–V	10:	3
31 Cyg	3784.3	(1) K1 Ib	9.2	169
		(2) B4 V	4.2	3.2
VV Cep	7450	(1) M2 Ib		1600
		(2) A0		13
ε Aur	9890	(1) A9 Ia		214
		(2) B: (binary?)		
		+ disk or rings		1221

In Table 1, the spectral types of the components of each binary system are listed on successive lines, along with the corresponding estimates of their masses and radii. It should be noted that the table lists the result of many contemporary astronomers who have contributed to this field.

THE ζ AURIGAE AND VV CEPHEI SYSTEMS

The common and most interesting property of the eclipses of the hot companion by the cool supergiant is that the length of the atmospheric phases is generally not constant, and the ingress and egress of one and the same eclipse do not have equal lengths either; that is, the extent of the atmosphere of the cool supergiant is variable and is not the same at ingress and egress. This is not surprising for an extended, very rarefied atmosphere. We know, for example, that the outer atmosphere of our Sun—the chromosphere and the corona, which have much smaller radial extension than their analogs in cool supergiants—undergoes both irregular and cyclical variations, the latter being related to the 11-yr solar cycle.

The common property of the eclipses in the ζ Aurigae and VV Cephei systems is that the resonance doublet of singly ionized calcium at 3933 and 3968 Å shows enhanced sharp absorption cores several weeks before the geometrical eclipse starts and continues to do so several weeks after it has ended. (A resonance line is a spectral line produced by a transition between two atomic energy levels in which the lower level is the ground state of the atom.) These cores are characteristic of the absorption in the outer atmosphere of the cool supergiant. They are often split into two or more components with different radial velocities, generally more negative than the orbital radial velocity derived from the photospheric absorption lines (i.e., the excited lines from neutral metallic atoms). This indicates that the outer atmosphere is composed of two or more discrete layers that are moving toward the observer, often at velocities larger than escape velocity, and therefore establishes the existence of mass loss from the star. Large differences are found between different eclipses and also between the ingress and egress phases of one and the same eclipse, which is further proof of the variable and unstable structure of the outer atmosphere.

The observations of the ultraviolet spectral range (1200–3200Å) made with the *International Ultraviolet Explorer* (*IUE*) satellite have not only permitted a better determination of the spectral type of the companion star and a more complete study of the outer atmosphere of the cool supergiant star, but have also shown that the degree of interaction between the surroundings of the two members of the binary system is much larger than was previously suspected. In fact, in the far ultraviolet the emission from the cool star is completely negligible and the spectrum of the hot star is dominant, whereas in the optical spectral range the opposite occurs. Furthermore, the only observable absorption lines formed in a rarefied medium are the strongest resonance lines of abundant ions. The resonance lines of neutral and singly ionized calcium and the resonance doublet of neutral sodium are the only resonance lines present in the optical range. In the ultraviolet, on the contrary, we have many such lines: low ionization lines such as those of neutral oxygen and nitrogen; singly ionized carbon, magnesium, aluminum, silicon, and iron; and high ionization ions like those of four-times ionized nitrogen and three-times ionized carbon and silicon. Hence it is possible to measure the degree of ionization and stratification effects in the atmosphere, that is, where ions of different degrees of ionization are formed.

The new and unexpected results from the *IUE* studies of ζ Aurigae and VV Cephei systems are the following: the presence in the spectrum of high ionization lines like those of four-times ionized nitrogen and three-times ionized carbon which are not usually found in the chromospheric spectra of cool supergiants and which, when present, are much weaker than those observed in ζ Aur, 31 Cyg, and VV Cep. When the hot companion is totally eclipsed, many emission lines appear, perfectly matching the absorption lines produced by the atmosphere of the cool supergiant in the continuum of the hot companion, proving that they are actually formed in the extended atmosphere, which is too rarefied to emit appreciably in the continuum, but which is optically thick in the corresponding lines.

The other and still more unexpected result was found for the first time by Robert D. Chapman and then by Imad A. Ahmad, Robert E. Stencel, and others, in ζ Aur, 31 Cyg, 32 Cyg, and 22 Vul. It consists of evidence of shock wave interactions between the slower and denser wind from the cool star and the faster, thinner wind from the hot one (where by "wind" astronomers mean the flow of gas continuously escaping from the star). The observational evidence consists of the appearance of an absorption component that is shifted longward by several tens of kilometers per second in some chromospheric lines, an effect which is not manifest before the hot star arrives directly in front of the cool supergiant. This phenomenon is illustrated in the accompanying spectrum of Mg II profiles in ζ Aurigae (Fig. 1). These longward-shifted components are attributed to absorption in an accretion column inside the shock wave behind the hot star. These components should be present as long as the line of sight to the hot star passes through the shock cone, whose width is about 20°, as shown in the accompanying schematic diagram of the shock zone around the hot star (Fig. 2). This width is proportional to the square root of the stellar wind temperature and to the inverse of the relative velocity of the companion star in the wind. This effect has been detected in the Mg^{+}, N^{4+}, O^{3+}, and Si^{3+} resonance doublets. (An accretion column is a spatially restricted flow of accreting matter that is channeled toward the compact star in an interacting binary system.)

Unfortunately, no ultraviolet observations are available for the eclipse of VV Cep that was just ending in April 1978 when *IUE* began its regular observing program. However, *IUE* has permitted us to determine that the spectral type of the companion is an early A, and not a very hot O-type star as the optical observations had suggested. In spite of its rather low surface temperature, about 9000 K, and the low temperature of the M supergiant (about 3000 K), the resonance lines of three-times ionized carbon and silicon are present in VV Cephei. They are never observed either in the chromospheric spectra of single M-type supergiants or in the photospheric and chromospheric spectra of A-type stars.

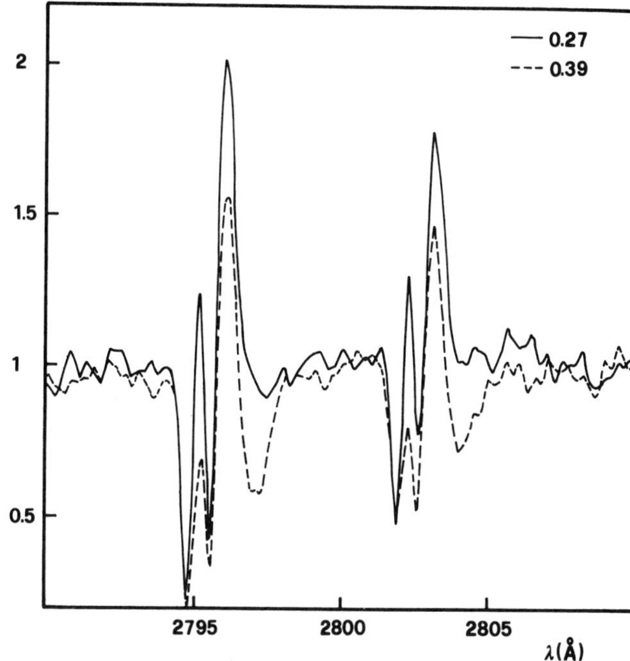

Figure 1. The singly ionized magnesium (Mg II) line profiles in the spectrum of ζ Aurigae at phase 0.27 (counted from the epoch of the eclipse of the hot star) and at phase 0.39 when the redshifted absorptions due to the accretion column inside the shock cone can be seen. [*From Hack and Stickland (1987).*]

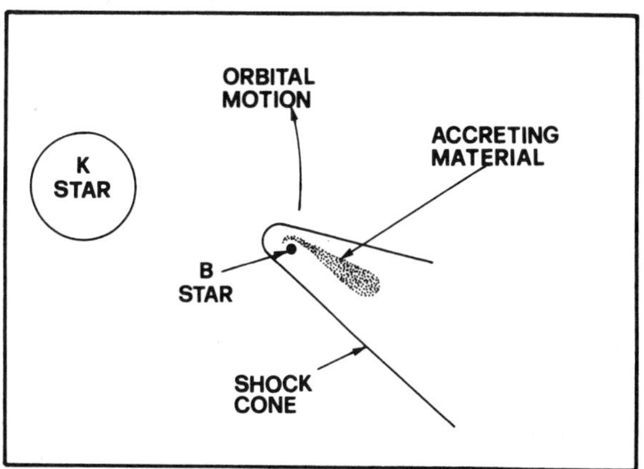

Figure 2. A schematic diagram showing the shock zone around the hot star in the ζ Aurigae system. [*Adapted from Hack and Stickland (1987).*]

ε *AURIGAE*

Optical observations of past eclipses of ε Aurigae show a very peculiar behavior. The primary A9 supergiant star undergoes a total eclipse (with a flat minimum lasting 330 days) during which its spectrum remains visible. The depth of the eclipse (0.8 magnitude) indicates that the flux is reduced to about one-half. The two partial phases (ingress and egress) last about 140 days each; hence, in all, the eclipse lasts about 610 days. No trace of the spectrum of the secondary is visible at any time. These facts were explained by assuming that the companion is surrounded by an opaque disk that covers half of the disk of the primary star. The thick disk can satisfactorily explain the photometric eclipse, but not the spectroscopic observations. In fact, the three eclipses observed with sufficiently high spectroscopic resolution (in 1929–1930, 1956–1957, and 1982–1984) all show the appearance of a "shell spectrum," that is, an array of spectral lines that are formed in a medium slightly cooler and more rarefied than the photosphere of the primary star. The shell spectrum is redshifted by about +15 km s^{-1} before the start of the photometric eclipse and is violet-shifted by about −35 km s^{-1} after the end of the photometric eclipse. This proves that a rarefied gaseous envelope is associated with the eclipsing body and is more extended than it. The nature of the companion was partly clarified by the observations made during the last eclipse in the ultraviolet with the *IUE* satellite and in the infrared from mountaintop observatories. The ultraviolet observations indicate that the depth of the eclipse becomes negligible in the far ultraviolet ($\lambda < 1500$ Å); hence a hot body is in front at the epoch of the eclipse. The depth of the eclipse is small in the far infrared ($\lambda = 20$ μm); hence a cool dusty body, whose temperature is estimated at about 500 K, is in front at the epoch of the eclipse.

Therefore, the mysterious eclipsing body in the ε Aurigae system is a hot star surrounded by a thick disk or ring and a gaseous rarefied envelope. However, if the mass of the primary star is about 15 $M_{\odot}$, as suggested by its high luminosity, then the orbital parameters indicate that the companion should also be a very massive star, and if so it would not be easy to understand why we do not see any trace of its spectrum in the optical range. A solution was suggested by Jack J. Lissauer and Dana E. Backman in 1984: The companion might be itself a binary consisting of two hot close stars. In this case, the luminosity of the binary companion could be sufficiently low to explain why it is not detectable in the optical range, but the total mass of the binary companion would be compatible with the orbital data. Only radial-velocity observations of the ultraviolet spectrum of the companion will permit us to derive its mass and to solve the problem. High-resolution spectral observations are necessary in order to reach this goal. The ultraviolet flux of ε Aurigae is too faint to be measured with *IUE* in the high-resolution mode. Only the powerful *Hubble Space Telescope* may be able to give us the correct answer.

The structure of the thick eclipsing body in ε Aurigae was recently investigated by S. Ferluga using the synthetic light curve method. This method consists of making several assumptions about the structure of the body, computing the resultant light curve, and modifying the assumed structure until the best agreement is obtained with the light curves that were observed during 1983 and 1984 and in the two previous eclipses. The main results are the following: The eclipsing body is a system of rings encircling the secondary and forming a thin disk with a central aperture. During the last eclipse it had the following characteristics: a large central aperture with radius = 1.6 AU (1 AU or astronomical unit equals about 1.5×10^{8} km), a semitransparent outer edge of the disk (at radius = 5.9 AU), and a transparent annular zone (radius 3.1 AU, width 0.8 AU) splitting the main opaque ring (radius = 5.0 AU) into two concentric bands. Secular variability (in this structure) has been investigated by applying the same method to the eclipses of 1927 and 1956. However, only minor features appear to be variable, like the relative extension of concentric bands and the sharpness of the external edge of the disk. The result is unequivocal in the sense that variations of the parameters larger than about 10% give synthetic light curves that are in much larger disagreement with the observed light curves than the observational errors (see Fig. 3).

CONCLUSION

Atmospheric eclipses have permitted us to probe the atmospheres of cool supergiants at different heights above the stellar photosphere in a way that had previously been possible only for the Sun. Observations from space of their ultraviolet spectra, where a wealth of atomic and ionic resonance lines are present, have revealed the complex physical structure of these atmospheres and the interaction between the winds of the two components, and given a better understanding of the character of the companion star.

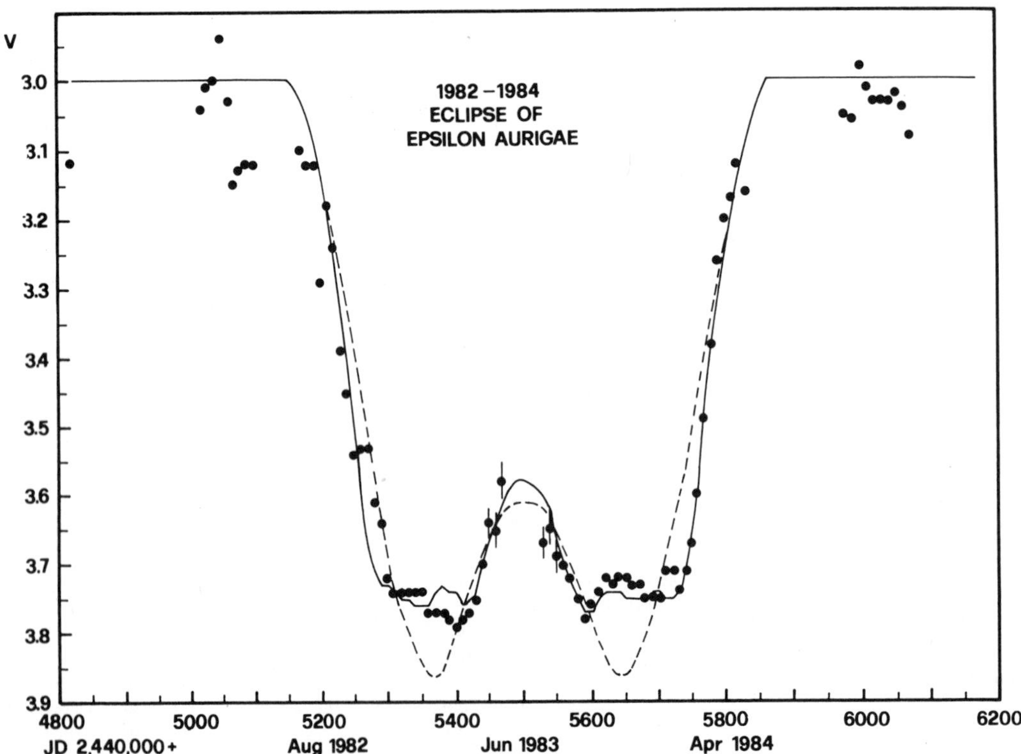

Figure 3. Light curve of the 1982–1984 eclipse of ε Aurigae. Dots: visual observations by Schmidtke. Dashed line: single-ring model; the unsatisfactory fit with the observations is very evident. Solid line: split-ring model described in the text; the agreement with the observations is excellent except at ingress; this is due to the intrinsic variability of the hot companion. [*Courtesy of S. Ferluga; based on work published in* Astron. Ap. **238** *270 (1990)*.]

Additional Reading

Ferluga, S. (1989). The rings of Epsilon Aurigae. *Scienza Dossier* (No. 34) 12.

Hack, M. (1984). Epsilon Aurigae. *Scientific American* **251** (No. 4) 98.

Hack, M. and Stickland, D. (1987). Atmospheric eclipses by supergiants in binary systems. In *Exploring the Universe with the IUE Satellite*, Y. Kondo, ed. D. Reidel Publishing Company, Dordrecht, p. 445.

Wright, K. O. (1970). The Zeta Aurigae stars. *Vistas in Astron.* **12** 147.

See also **Stars, Atmospheres; Stars, Circumstellar Matter.**

Binary Stars, Cataclysmic

Hans Ritter

Stars that undergo prominent and erratic temporal brightness variations have always been of special interest to astronomers. This is particularly true for a class of eruptive variable stars for which Robert P. Kraft, in 1962, coined the collective name cataclysmic variables (CVs). The objects that we consider now as belonging to the CVs are: classical novae, recurrent novae, dwarf novae, and nova-like objects. In every case, they are binary systems with one member being a white dwarf star.

The first major breakthrough in the understanding of CVs came through the photometric study of the old nova DQ Her by Merle F. Walker in 1954 and the pioneering spectroscopic studies of old novae and dwarf novae by Kraft in 1960–1965, which showed convincingly that all CVs probably are close binary systems with orbital periods that are among the shortest known. Another major step forward in revealing the structure of cataclysmic binaries (CBs) came through extensive application of high-speed photometry by R. Edward Nather, Edward L. Robinson, Brian Warner, and co-workers in the early 1970s. As more structural details were unveiled by subsequent observations, it became clearer that CBs are objects of high astrophysical interest. Detailed investigations of CBs now play an important role in constraining theories in fields as diverse as close binary evolution, structure of accretion flows around magnetized and nonmagnetic compact stars, and nuclear astrophysics, to mention just a few.

PHENOMENOLOGICAL CLASSIFICATION OF CVs

The brightness variations observed among CVs are characterized by a wide range of amplitudes and time scales. Yet the variability of most of the CVs follows one of a few characteristic patterns. This, in turn, led to a coarse phenomenological classification of CVs into the following main classes and subclasses.

Novae

This class consists of two subclasses: the classical novae and the recurrent novae.

Classical novae (N) are distinguished from the other CVs by the fact that they have undergone one nova outburst in historical times. The main characteristics of the visual light curve during a nova outburst are shown schematically in the top panel of Fig. 1. Years to decades after their outburst, when they have returned to near-minimum light, these stars are sometimes simply referred to as old novae.

Recurrent novae (Nr) are a very small, yet heterogeneous group of stars. Observationally, they are distinguished from the classical novae by having undergone more than one outburst in historic

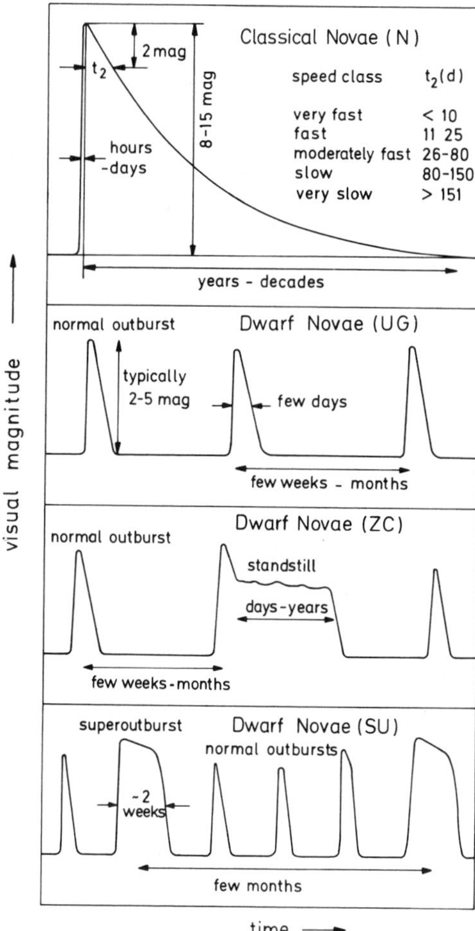

Figure 1. Main characteristics of the visual light curves (schematic) of classical novae (top panel) and of dwarf novae of the U Gem, Z Cam, and SU UMa types (lower three panels).

times. The shortest recurrence time scales observed are of the order of a few decades.

Theoretical arguments suggest that the classical nova outburst is also a recurrent phenomenon, however, with a much longer recurrence time scale of approximately 10^4–10^6 yr.

Dwarf Novae

As the name implies, dwarf novae show less spectacular outbursts than novae but they erupt much more frequently. According to the characteristic patterns seen in the visual light curves, shown schematically in the lower three panels of Fig. 1, one distinguishes three subclasses of dwarf novae, that is, the U Gem stars (UG), the Z Cam stars (ZC), and the SU UMa stars (SU), where U Gem, Z Cam, and SU UMa are the names of the prototypes that define the corresponding subclasses.

U Gem stars show outbursts of 2–5 magnitudes in amplitude at quasiregular intervals of a few weeks to a few months. The outburst maximum is reached within a few hours to as much as a day, the maximum lasting typically a few days, and the decay from outburst to minimum is typically one to a few days.

Z Cam stars differ from U Gem stars in that the decline from outburst is occasionally interrupted about 1 magnitude below maximum brightness. The star may then remain at this intermediate brightness, that is, in the so-called standstill, for extended periods (days to years) before returning to minimum again.

SU UMa stars are distinguished from U Gem stars by showing a second type of outburst, the so-called superoutbursts, which last longer, typically two weeks, occur less frequently, typically every few months, and are brighter by about 1 magnitude than the normal outbursts of U Gem stars.

Nova-Like Objects

This class comprises all the remaining CVs, that is, systems that are structurally cataclysmic binaries, but that are not known to undergo prominent outbursts. Currently, the following subtypes are distinguished: UX UMa stars, VY Scl stars, AM Her stars, and DQ Her stars; the latter two classes are collectively known as magnetic CVs. Again the names of the four classes are derived from their corresponding prototypes.

UX UMa stars (UX) resemble dwarf novae during outburst or Z Cam stars during standstill. In fact, they may be regarded as Z Cam stars trapped in permanent standstill.

VY Scl stars (VY) are similar to UX UMa stars but show, in addition, occasional minima during which they resemble dwarf novae at minimum. Thus they may be regarded as Z Cam stars in standstill for most of the time. They are sometimes, therefore, referred to as anti-dwarf novae.

Magnetic systems. It is common practice to include them among the nova-like systems, although the defining characteristic is not their photometric behavior, which resembles that of VY Scl stars, but rather the direct or indirect evidence for the presence of a strongly magnetized white dwarf. Consequently, it is not surprising that this class is not disjoint from the others defined previously.

Finally, it is important to note that most old novae resemble UX UMa stars. However, a few exhibit dwarf nova-like outbursts (e.g., GK Per, or Nova Per 1901, and WY Sge, or Nova Sge 1783).

BINARY MODEL OF CVs

Distinct as CVs of the various types just described may appear phenomenologically, detailed spectroscopic and photometric observations have shown that CVs of all types invariably conform to the same basic binary model. The key ingredients of this model are a white dwarf primary star and a low-mass nondegenerate secondary that fills its critical Roche volume and spills matter through the inner Lagrangian point toward the primary. Now, CBs can be further subdivided into magnetic and nonmagnetic systems depending on whether the accretion flow is dominated anywhere by the white dwarf's magnetic field (magnetic CBs) or not (nonmagnetic CBs).

Nonmagnetic CBs In about 80% of the known CBs, the white dwarf's magnetic field is so weak that the transferred matter forms a gaseous disk, usually referred to as an accretion disk, around the primary which extends all the way down to its surface. At the point where the matter flowing from the secondary collides with the disk's outer rim, a shock front is formed which gives rise to what is known observationally as the hot or bright spot. In these systems most of the light is accretion luminosity liberated in the disk and the hot spot.

Magnetic CBs In the remaining 20% of the known CBs, the white dwarf's magnetic moment is so high, that is, of order 10^{33}–10^{34} G cm^3, that the accretion flow is strongly influenced by the magnetic field. In about half of the magnetic systems, the field is strong enough not only to prevent the formation of a disk but also to synchronize the spin of the white dwarf with the orbital revolution. These systems are usually referred to either as AM Her stars or polars. In the rest of the magnetic systems (about 10% of the CBs), the white dwarf's rotation is not synchronized with the orbital motion. Magnetic accretion on the poles of the nonsynchronously rotating white dwarf gives rise to a lighthouse effect, that is, to brightness variations at the spin period, analogous to what is observed in x-ray pulsars. These systems are usually referred to either as DQ Her stars or intermediate polars. Depending on the size of the magnetosphere, that is, on the magnetic moment and on the accretion rate, it may or may not be surrounded by a residual accretion disk.

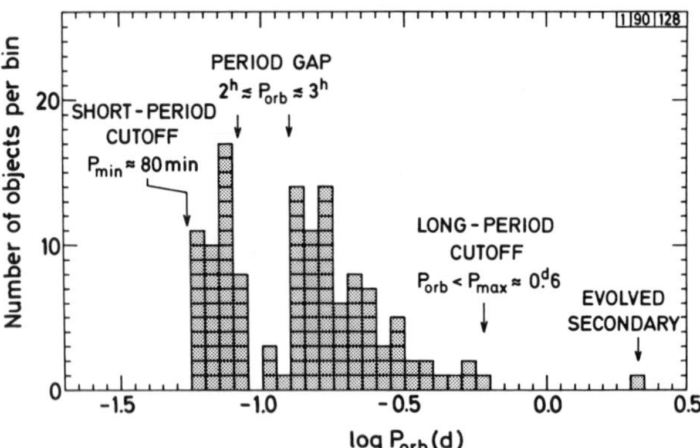

Figure 2. Distribution of observed orbital periods of cataclysmic binaries.

PHYSICAL PARAMETERS

Orbital Periods

The orbital period is the only physical parameter that is accurately known for a relatively large number of CVs. Although the most recent edition of the *General Catalogue of Variable Stars* lists some 600 stars as CVs, at present the orbital periods of only about 130 CBs are reasonably well known. The remarkable period distribution is shown in Fig. 2. Its main characteristics are:

1. With one exception, all periods are shorter than $P_{max} \approx 0^{d}.6$.
2. There is a cutoff at a minimum period of $P_{min} \approx 80$ min.
3. There is a remarkable deficiency of systems in the period range $2^{h} \lesssim P \lesssim 3^{h}$. This is usually referred to as the period gap.

Secondary Components

Observations show rather convincingly that in CBs with $P \lesssim 0^{d}.6$ the secondary is a nondegenerate low-mass star near the main sequence. Because of Roche geometry the secondary's mass correlates strongly with the orbital period. Typically,

$$M_2 \approx 0.060\ M_{\odot}\ P_{orb}^{1.3}\ (\text{h}) \tag{1}$$

for systems in the period range 80 min $\lesssim P_{orb} \lesssim 10$ h. From this it follows that $0.1\ M_{\odot} \lesssim M_2 \lesssim 1\ M_{\odot}$. Here $M_{\odot}$ is the solar mass.

White Dwarf Primary

Determination of the masses of CBs, and in particular of the white dwarf component, has been and still is a notoriously difficult task. Therefore, the resulting masses are always subject to some debate. Nevertheless, it seems clear that the average white dwarf observed in a CB is significantly more massive, that is, $\langle M_{WD} \rangle \approx (0.9 \pm 0.1)$ $M_{\odot}$, than an average single white dwarf with $\langle M_{SWD} \rangle \approx (0.6 \pm 0.1)$ $M_{\odot}$.

Mass Ratio

In all CB systems, as far as known, the mass ratio $M_1/M_2 \gtrsim 1$. This and the obvious longevity suggests that CBs are stable against runaway mass transfer on the thermal and the dynamical time scale. This, in turn, together with the fact that the primary is a white dwarf, that is, that $M_1 \lesssim 1.4\ M_{\odot}$, explains the long-period cutoff of the period distribution at $P_{max} \approx 0^{d}.6$.

Mass Transfer Rates

The mass transfer rate $\dot{M}$ is one of the most important and at the same time one of the least well known quantities characterizing a CB. Despite the enormous difficulties encountered in determining $\dot{M}$, there is some consensus that $\dot{M}$ is generally very low. Typical values are $10^{-11}\ M_{\odot}\ \mathrm{yr}^{-1} \lesssim \dot{M} \lesssim 10^{-10}\ M_{\odot}\ \mathrm{yr}^{-1}$ in systems below the period gap ($P \lesssim 2^{\mathrm{h}}$) and $10^{-9}\ M_{\odot}\ \mathrm{yr}^{-1} \lesssim \dot{M} \lesssim 10^{-8}\ M_{\odot}\ \mathrm{yr}^{-1}$ in systems above the gap ($P \gtrsim 3^{\mathrm{h}}$).

THE OUTBURSTS

Dwarf Nova Outbursts

Detailed observations, in particular of eclipsing systems, have unambiguously shown that dwarf nova outbursts are due to a brightening of the accretion disk which, in turn, is a consequence of increased mass flow through the disk. Whether that increased mass flow is caused by mass transfer bursts from the secondary star or by a thermal instability of the disk has been the subject of a long debate, though in recent years disk instability models have become more powerful in explaining the observed phenomena.

Novae and Recurrent Novae

There is a general consensus that the outbursts of classical novae and of at least some of the recurrent novae are best explained in terms of a thermonuclear explosion of hydrogen at the base of the layer of hydrogen-rich gas accumulated on the white dwarf before the outburst. Because the white dwarf seems to be scarcely affected by such an explosion, the outbursts of classical novae can be and very probably are repetitive.

LONG-TERM EVOLUTION

The stability of CBs against mass transfer implies that the observed mass transfer must be driven by an external force. In the vast majority of CBs, this can only occur via systemic loss of orbital angular momentum. Viable mechanisms are gravitational radiation and/or a magnetically coupled stellar wind from the secondary ("magnetic braking"). Many of the observed properties of CBs (such as the minimum period of approximately 80 min, the period gap and the period distribution of the AM Her and DQ Her systems) can be explained with models where the evolution is driven in this way. The corresponding evolutionary time scales are of the order of few 10^{8} yr to few 10^{9} yr. Nuclear evolution of the secondary can drive a significant mass transfer only if its mass is $M_2 \gtrsim 1 M_{\odot}$, that is, if the system's orbital period is $P \gtrsim 10^{\mathrm{h}}$.

GALACTIC DISTRIBUTION, SPACE DENSITY, AND BIRTH RATE

Because even the nearest CVs are too far away to have their distances reliably determined from parallaxes, and because absolute magnitudes of CVs are very poorly known in general, the distances of only few CVs are reliably known. Therefore, the estimates of the space density of CVs are also very uncertain. The currently best data yield a space density of dwarf novae in the solar vicinity of approximately 10^{-6} pc^{-3}. Proper motions, the distribution of the systemic radial velocities and an apparent scale height of dwarf novae perpendicular to the galactic plane of approximately 100–200 pc are all consistent with CVs belonging to the old disk population. The same conclusion is also reached from the observed spatial distribution of novae in our galaxy and in the Large Magellanic Cloud. On the other hand, the novae in M31 (Andromeda galaxy) are predominantly found in its bulge. Based on the observed nova rate ($\sim$ 3.5 yr^{-1}), one estimates a total rate of 50–100 novae yr^{-1} in our galaxy. This value taken together with theoretical estimates of the nova recurrence time scale suggests that the intrinsic space density of novae (and thus of CVs) is much higher than observed, namely of order 10^{-4} pc^{-3}. Several other lines of argument end up with a similar value for the intrinsic space density. If that many CVs have been formed in the past 10^{10} yr or so, the corresponding birth rate of CVs in the Galaxy must be of order 10^{-14} pc^{-3} yr^{-1}. To put this number in perspective, we note that the birth rates of single white dwarfs and of Type I supernovae are estimated, respectively, as approximately 10^{-12} pc^{-3} yr^{-1} and 10^{-13} pc^{-3} yr^{-1}.

Additional Reading

Bode, M. F. and Evans, A., eds. (1989). *Classical Novae*. Wiley, Chichester.

King, A. R. (1988). The evolution of compact binaries. *Quart. J. Roy. Astron. Soc.* **29** 1.

Pringle, J. E. and Wade, R. A., eds. (1985). *Interacting Binary Stars*. Cambridge University Press, Cambridge.

Ritter, H. (1986). Secular evolution of cataclysmic binaries. In *The Evolution of Galactic X-Ray Binaries*, J. Trümper, W. H. G. Lewin, and W. Brinkmann, eds. NATO ASI Series C, Vol. 167. D. Reidel Publishing Company, Dordrecht, p. 207.

Ritter, H. (1990). *Catalogue of Cataclysmic Binaries, Low-Mass X-Ray Binaries and Related Objects*, 5th ed. *Astron. Ap. Suppl. Ser.* **85** 1179.

Warner, B. (1976). Observations of dwarf novae. In *Structure and Evolution of Close Binary Systems*, P. Eggleton, J. Whelan, and S. Mitton, eds. IAU Symposium, No. 73. D. Reidel Publishing Company, Dordrecht, p. 85.

See also **Accretion; Binary Stars, Observations of Mass Loss and Transfer in Close Systems; Binary Stars, Polars (AM Herculis Type); Stars, White Dwarf, Observed Properties.**

Binary Stars, Contact

Kam-Ching Leung

Generally, eclipsing binary stars are divided superficially into three groups. They are classified traditionally according to the shape of their light variations or light curves; namely the Algol type, the β Lyrae type, and the W Ursae Majoris (W UMa) type. The W UMa systems have the following characteristics: short periods (about a quarter of a day), spectral type mostly of G to K, continuous light variation in their light curves with equal or almost equal minima, and most of them showing double-line (usually rather broad) spectra. It was believed that their components are joined together in the shape of a dumbbell or a peanut. The term *contact binary* has been synonymous with W UMa stars since it was first introduced by Gerard P. Kuiper in 1941.

The configuration of a binary system is derived from analyses of the light curve (i.e., photometric solution of the binary light curve). The orbital parameters for a close binary (i.e., excluding visual binaries) are deduced from the radial velocity curve of the system. The absolute dimensions of a binary system are determined from the combined results of a photometric and spectroscopic analysis. There are two kinds of stellar atmospheres (excluding degenerate and exotic objects such as white dwarfs, neutron stars, black holes, etc.) defined by the mechanism of energy transportation in the outer layer of the star: radiative atmospheres and convective atmospheres. Coincidentally, our Sun is at about the dividing point with respect to its temperature or mass. Stars with earlier spectral type, or hotter than the Sun, have radiative atmospheres, and stars with spectra similar to or later than the Sun, have convective atmospheres. The W UMa system consists of two late-type components (i.e., both have convective stellar atmospheres). It was generally believed that a contact system of this type would have a convective common envelope. In the late 1960s, Leon B. Lucy had some success in constructing models of contact systems with convective common envelopes.

The problem then was to determine if all W UMa systems are contact systems. It was difficult to solve this problem in the 1960s. The major difficulty was that the classical methods of photometric analyses (e.g., Russell–Merrill, Kopal, etc.) for deriving the configurations of a close binary star were based on the assumption that the components (stars) were ellipsoidal in shape. Obviously, this cannot be the true shape of the stars in very close systems. However, an alternative approach would have been formidable or prohibitively difficult without the current general availability of fast computers. Usually, a system was "defined" as in contact if the sum of the fractional radii of the components was 0.75 or larger. There was no direct way to judge whether a system was in actual contact. However, if one thinks about this seriously, for a typical short-period system with a mass ratio of 2 to produce nearly equal eclipsing depths, can only mean that there must be an energy exchange between the components. Therefore, the stars have to be in contact.

In the beginning of the 1970s, several groups around the world were developing more realistic models utilizing zero-velocity surfaces or Roche-type geometry for stellar surfaces for deriving more accurate configurations for close binary systems. Such methods of analysis were developed simultaneously by Lucy; Graham Hill and John B. Hutchings; Robert E. Wilson and Edward J. Devinney; and Stephan W. Mochnachi and Noel A. Doughty; and by some others in later years. At present, the most flexible and most utilized computing codes are those developed by Wilson and Devinney in 1971 and being kept current with additional features by Wilson. With the advance and general availability of fast computers, close binary stars research entered a new phase of rapid development. Now, it is very well known that not all the systems with W UMa type light curves are contact, and that there are contact systems among almost all spectral types (a contact system with a star of spectral type M has not been discovered). The current stage of research of contact binary stars is no longer restricted to the late-type W UMa systems.

The term *contact system* in the modern sense means that the binary stellar surfaces of both components are in contact with the same equal potential surface (taking into account gravity and orbital rotation) which is loosely called the *common Roche surface*. That is, the two components are "physically" touching (critical/point contact at the inner Lagrangian point) or joining (overcontact) with each other. The data show that some systems are at or near critical contact, whereas most of the others are at a variety of degrees of overcontact.

TYPES OF CONTACT BINARY SYSTEMS

Contact binary systems can be classified into five distinct categories according to the structural characteristics of their atmospheres or their common envelopes. The contact common envelopes are classified as follows.

Late-Type Contact Systems or W UMa Systems

These have rather short periods of the order of about a quarter of one day. They consist of late-type stars, mostly of G and K spectra, and the common envelopes are convective. They are low-mass binary systems (see the previous discussion). In 1965, Leendert Binnendijk suggested that these systems can be further divided into two subclasses of A and W types. A-type systems have their primary eclipses at transit, whereas the W-type systems have their primary eclipses at occultation. Generally, the A-type systems have deeper contact, slightly longer periods, and earlier spectra. There is observational evidence that the A-type systems may be evolved, but the evolutionary stage of the W-type systems are less certain. The interpretation of these two distinct types of late-type contact systems remains a real challenge.

Early-Type Contact Systems

These systems consist of early-type stars with spectra mostly of types O and B. Because they consist of very hot stars, their contact common envelopes are believed to be radiative. They have longer periods (in terms of days generally) and are very massive systems compared with the W UMa systems. The majority of the members of this group was discovered by the author and his collaborators. One may say that this group is just the opposite of the late-type contact systems mentioned previously, that is, their common envelopes are radiative instead of convective.

Mixed- or Intermediate-Type Contact Systems

Sometimes systems of this group are called contact systems with large temperature differences between the components. They generally consist of stars of intermediate spectral types with a hotter component of spectral type A (or F) and a cool companion of spectral type G or later. This type of system can also be described as binaries consisting of early-type stars in contact with late-type stars. Their light curves resemble β Lyrae light curves with considerably shallower secondary minima. With modern photometric analyses, they were found to have contact configurations. They show large temperature differences (typically several thousand degrees) between the components as a result of the large differences between the depths of eclipses. The systems consist of stars with radiative atmospheres in contact with stars with convective atmospheres. The common envelopes of such systems may most likely be quite complicated, due to the large temperature differences at the interface/neck. How can these systems maintain stability under such large temperature differentials at the interfaces? So far, there is no reliable theoretical model for contact systems of this type.

Some researchers seriously question the reliability of the contact solutions derived from the modern analysis for these intermediate-type systems. At present, no one has been able to provide an answer for this tough problem. Certainly more investigations of this group of "contact" systems are needed before any intelligent response can be put forward.

Supergiant Contact Systems

Recently, late-type (G and K) binary stars with periods of a hundred days or longer were found to be contact systems by the author and his collaborators. These systems consist of cool supergiant or giant stars with very deep convective atmospheres. Thus the common convective envelopes must be very large or deep. Their extremely long period suggests that these systems may be the result of very advanced case B mass loss (i.e., very far from the terminal-age sequence in an Hertzprung–Russell diagram). Because of the problem of deep convection atmospheres, constructing model atmospheres for late-type giants and supergiants has been very difficult. It is very easy to see that modeling a common convective atmosphere for late-type supergiant contact systems would be an impossible task with our present knowledge of convection. No doubt in the coming decade these systems will be a great challenge for theorists.

Double Contact Systems

The term *double contact systems* was introduced by Wilson in 1979. In general, most binary systems are in synchronized rotation; that is, the orbital period is the same as the rotational period of each component. In the case of nonsynchronized rotation, the rotation period of one or both components is different from the binary orbital period. Usually, it is the rotational period of one component that is faster than the orbital rotation as a result of gaining mass and angular momentum during the mass exchange phase of close binary evolution. Therefore, we have to include the distortion due to stellar rotation in calculating the equipotential surfaces of a binary system. As a result, we can define another critical equipotential surface (beyond which a star would break apart because of rotational instability). Thus in a binary system a star will have two different critical potential surfaces: the regular one mentioned at the beginning of this entry, and the one just described. A double contact system could mean that one of its components is contacting the regular critical potential surface, whereas the other is contacting its critical rotational potential surface; that is, each component is in contact with a different critical surface. Thus we have a double contact system! Wilson suggested several members belonging to this group. It is important to note that in a double contact system the components *are not in physical contact* with each other, whereas the components of the four types of contact systems described previously *are in actual physical contact* with each other.

CONTACT BINARY SYSTEMS COME IN TWO KINDS: ZERO-AGE OR EVOLVED

Where do contact binary stars come from? Are they born as Siamese twins (in contact) or do they evolve into contact? There are at least two easy ways to determine whether a system is at zero-age or is evolved. For the systems in star clusters, the locations of the components in the cluster H–R diagram will indicate the age of the systems. Fortunately, there are about four or five systems that are members of open (galactic) clusters. The second method is to compare the absolute dimensions (radii and masses) of the components with the radii and masses of zero-age main-sequence stars. If the radius of the component(s) is larger than the radius of the zero-age star for the corresponding mass, the system is evolved. The observational data indicate that both kinds of contact systems exist. However, less than a handful of zero-age contacts have been discovered. The evolved contacts outnumber the zero-age contacts by at least a factor of 10. Most of the evolved contact systems are located near the main sequence and the terminal-age main sequence. Therefore, they are mostly the result of case A mass loss of close binary evolution. Very few systems are believed to be the result of case B mass loss. It is believed that this statistic is subject to a very strong selection effect, because observers have a strong bias toward shorter-period systems. The interior structures could be very different between the zero-age contacts and the evolved contacts, even for the same spectral type or mass.

Lucy and Frank H. Shu and collaborators produced models of contact binary systems from fairly different basic physical assumptions. It is generally believed that these are preliminary models only, and that more realistic models will be developed as we understand better the dynamical processes involved in contact atmospheres.

Additional Reading

Leung, K. C., ed. (1988). *Critical Observations Versus Physical Models for Close Binary Stars*. Gordon and Breach, New York, p. 93.

Pringle, J. E. and Wade, R. A., eds. (1985). *Interacting Binary Stars*. Cambridge University Press, Cambridge.

Sahade, J. and Wood, F. B. (1978). *Interacting Binary Stars*. Pergamon Press, Oxford.

Wilson, R. E. (1979). Eccentric orbit generalization and simultaneous solution of binary star light and velocity curves. *Ap. J.* **234** 1054.

Wilson, R. E., Van Hamm, W., and Pettera, L. (1985). RZ Scuti as a double contact binary. *Ap. J.* **289** 748.

See also **Binary Stars, Theory of Mass Loss and Transfer in Close Systems; Stellar Evolution, Binary Systems.**

Binary Stars, Detached

Ronald W. Hilditch

Binary systems of stars are referred to as detached binaries if the components are well separated, that is, if their radii are typically less than about one-quarter of the mean distance between them. The stars are still close enough that duplicity, even at very close astronomical distances, may be recognized only by their spectra demonstrating Doppler-shifted lines arising from orbital motion, and perhaps also mutual eclipses of the two stars. Orbital periods may range from a few hours, for systems containing very small, low-mass M dwarfs, up to perhaps several hundred days, where there is still sufficient room for stars to expand up to red giant dimensions and still not interact with their companions. At such long orbital periods, orbital velocities are at the limit of detectability (~ 2 km s^{-1}) and the probability of observable eclipses is minimal. This is the transition region from close binaries to speckle-interferometric/astrometric/visual binaries where the two stars may be seen separately.

It is usual to regard the term *detached binaries* to mean those systems in which the components have not yet evolved sufficiently far beyond the initial main-sequence stage to have started any mass exchange or mass loss process. The mutual gravitational field around two bodies is not spherically symmetric as for a single body, except at very large distances, and the standard Roche model for describing that gravitational field identifies two limiting volumes, one around each mass center, interior to which matter may be said to belong to one or the other body. These so-called Roche lobes define upper limits to the sizes of stars in binaries, and any evolutionary expansion by a star will ultimately result in matter being transferred to the companion or being lost from the system. If mechanisms exist to lose angular momentum from the orbit, then the Roche lobes may shrink downwards onto the stellar surfaces and also cause mass exchange processes to occur. For detached systems, such phenomena have not (yet) occurred, and hence we can study the properties of these stars without any need to consider the complex problems of binary star evolution. Thus the masses, radii, and luminosities in solar (or SI) units determined for stars in detached binaries are seen as strict observational tests of theories of stellar structure for stars on the zero-age main sequence and of the initial stages of stellar evolution perhaps as far as the red giant branch.

It is a combination of chance statistics on the orientations of binary orbits to our lines of sight and of observational practicalities (e.g., available telescope time and weather statistics) that dictate that most well-studied detached binaries have orbital periods of a few days. Although some systems are known to be eclipsing detached binaries with orbital periods measured in many tens of days, it takes very particular combinations of the preceding factors to allow the fundamental parameters of such systems to be determined accurately.

Thus detached binaries are close, spectroscopic, eclipsing systems for which accurate fundamental data are obtainable by means of analyses of the radial velocity variations due to orbital motion and of the brightness changes caused by mutual eclipses and other geometrical and physical effects. In practice, modern-day accuracy of the order of 1–2% in all quantities is achievable for many binary stars observed with the best instruments and with the data analyzed in an appropriately rigorous manner. In favorable systems, detailed investigations may also be made of the stellar atmospheric properties of each component because the companion can act during an eclipse as a means of scanning across the unresolved stellar surface. Thus empirical data on the degree of limb darkening, gravity darkening (the dependence of emergent flux on local surface gravity), and the mutual irradiation of the facing hemispheres of the two stars may be obtained. Furthermore, for binaries in eccentric orbits with components not too well separated, the phenomenon of apsidal motion—the precession of the orbit in its own plane due to the gravitational quadrupole moment induced by rotation and tides and to the general-relativistic term—may be observed. The rate of apsidal rotation depends upon the distribution of matter within each star and hence may act as a further test of stellar structure theory. Only recently have much of the disagreements (factors of 3) between observed and theoretical apsidal rotation rates been largely resolved.

FUNDAMENTAL PARAMETERS

Accurate determinations of masses, radii, and luminosities of the components of detached binaries require data of high precision, free from systematic errors, and where there are no complications resulting from intrinsic variability of either star or of circumstellar matter. Radial-velocity data must be secured from spectrograms of high signal-to-noise ratio and high spectral resolution ($R > 25{,}000$, where R is the ratio of the wavelength to the smallest resolved spectral element) which are obtained in integration times of less than 1–2% of the orbital period. These requirements are made to ensure that orbital motion during the integration time does not degrade the spectral resolution, and that the spectral resolution is sufficient to ensure that spectral lines from both components are well separated at most orbital phases, else blending of the absorption lines will result in systematic underestimates of the semiamplitudes of the velocity curves. [Note that the semiamplitudes enter the equation for determining minimum masses ($m_{1,2} \sin^3 i$) at the third power.] The best lines for radial-velocity measurement are those with residual intensities approximately 10–40% below the continuum level of the spectrum which originate in the stellar photosphere and are not contaminated by possible extended envelopes or circumstellar matter. Thus for most systems, the broad Balmer-series hydrogen lines are unusable for accurate velocity determinations because velocity separations in excess of 600 km s^{-1} are required to resolve such profiles adequately—a figure that is reached only rarely among very massive systems. Hydrogen lines are also readily contaminated by circumstellar material. Equally, other strong features in spectra of cooler stars, such as the ionized calcium H and K lines and the G band of the CH molecule among others, should be avoided. Whether the means of velocity measurement is by line-profile fitting of individual absorption lines of accurately known rest wavelength or by cross-correlation of the program star spectra against spectra of radial-velocity standard stars, the same requirements apply. It was the advent of profile-fitting comparators around 1970 that finally demonstrated unequivocally that pair blending of spectral lines could be a source of significant systematic error. The greatly improved efficiency of spectrographs around that time, which allowed higher-resolution studies per unit integration time, placed much of our empirical knowledge of stellar masses and dimensions on a much more accurate foundation.

Acquisition of photometry to a precision of better than ± 0.01 magnitude has also aided in the general improvement. High-quality light curves defined by large numbers of observations per unit phase interval (~ 10 observations per 0.01 in orbital phase) are required at least at one well-defined wavelength. The data set is even better when multichannel photometers provide such light curves in, say, four well-defined passbands simultaneously. Photometry obtained in the Strömgren–Crawford *uvbyβ* system with four- or six-channel photometers has enabled astronomers to determine interstellar reddening corrections and stellar effective temperatures for O to F stars much more precisely than hitherto. More work with simultaneous optical/infrared photometers is required particularly for binaries of spectral classes G to M.

The consequences of high-precision data are that the techniques used for their analysis must be at least as sophisticated. Although analytical procedures for the study of eclipse curves due to spherical stars in circular orbits have been well established for more than 50 years, they have been largely superseded by numerical codes. Such codes are capable not only of allowing accurate physical representations of the nonspherical shapes of stars in the Roche model, and the associated gravitational and radiative interactions between the two stars, but also, and very importantly, of the use by interpolation of tables of fluxes and limb-darkening coefficients as functions of temperature and surface gravity calculated by sophisticated stellar model–atmosphere codes. Thus the light-curve synthesis codes, first introduced in 1970, are used routinely in studies of light curves of detached binaries so that the most accurate stellar radii, orbital parameters, and luminosity ratios are determined. Knowledge of the effective temperature of one component via well-calibrated multicolor photometry and MK spectral classifications permits the determination of separate luminosities for both components. Such photometric results, in combination with the analyses of radial-velocity curves, then provide masses, radii, and luminosities in solar units for each component.

In the analyses of both radial-velocity curves and light curves, it is possible to solve by least squares procedures for many parameters, some of which are of only second-order importance. A good knowledge of the relative importance of the various parameters is essential toward making effective use of such codes, and it cannot be emphasized too strongly that the independent data sets of spectroscopy and photometry should be used in combination to obtain the model that represents best all of the data.

COMPARISONS OF OBSERVATIONS WITH THEORY

Chemical Compositions and Evolution Tracks

The extensive review by Daniel M. Popper of stellar parameters derived from binary stars in 1980 has been extended substantially in the last 9 years due to the work of several groups. The results of 79 detached binaries (cf. with 48 in 1980) are shown in the mass–radius and mass–luminosity diagrams (Figs. 1 and 2) together with the zero-age main-sequence (ZAMS) loci according to two model sequences covering different mass ranges. In the mass–radius plane, where evolution results in radius increases only, the observational data define a lower envelope that is generally well matched by the zero-age main-sequence models with helium content $Y = 0.25$–0.28 and heavy element content $Z = 0.02$–0.03. For early-type stars (O to early-A) comparisons of theoretical evolution tracks and the individual stars show very good agreement with initial chemical compositions of hydrogen:metals of $X{:}Z = 0.70$–$0.72{:}0.02$–0.03. ("Metals" means elements heavier than hydrogen.) For later-A and early-F stars, discrepancies arise with some models in the sense that helium deficiencies or heavy element excesses are required to provide agreement. For later-F and G stars, the good agreement returns. One ZAMS system, VZ CVn, does stand out significantly from other systems as having both components too cool, and hence luminosities too small, for their ZAMS masses. As yet, there is no resolution of this particular problem, nor of the "A-star" discrepancy. There is also good overall agreement in the mass–luminosity plane between these ZAMS loci and the observational data.

Detailed investigations of the longer-period detached systems allows strict comparisons with theoretical evolutionary tracks to test the differential evolution of the two stars in a binary resulting from their differing masses. Because both stars were formed at the same time, such comparisons are crucial tests provided that observed systems may be found at unambiguous stages of evolution. Very close to the zero-age main sequence, the evolution tracks are too close together for many masses and ages for realistic discrimination between models. Toward the end of the main-sequence phase for more massive stars, the S-shaped kink in evolution tracks also causes difficulties of interpretation. But, in between such stages, and particularly for binary systems wide enough to allow evolution up the red giant branch before any mass transfer

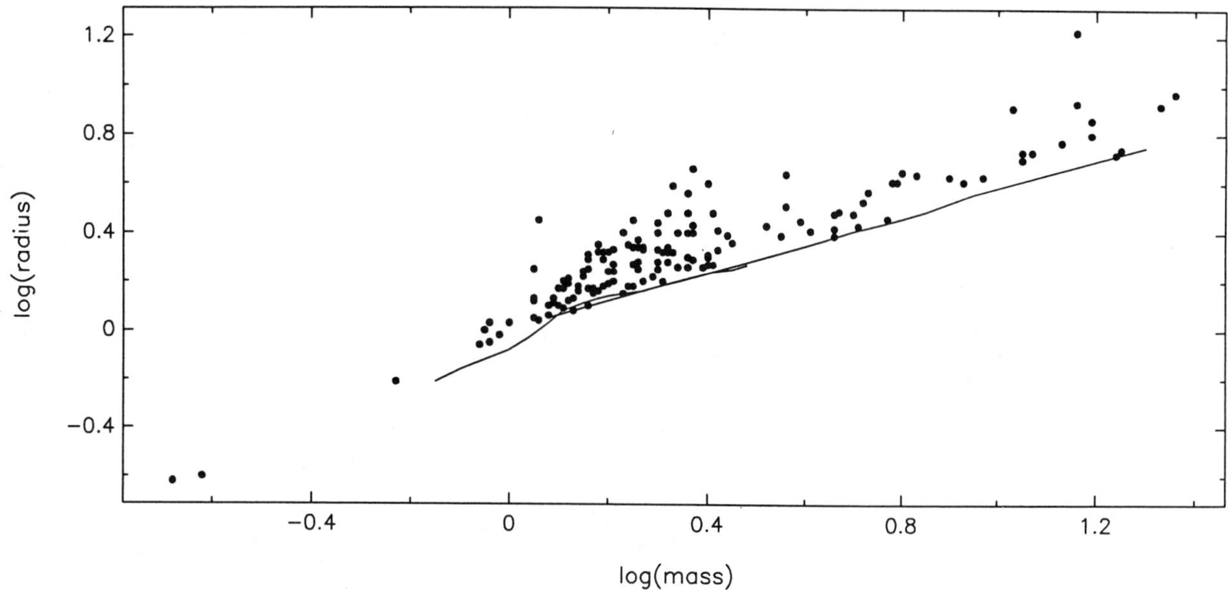

Figure 1. The masses and radii in solar units of 158 stars in 79 detached binary systems compiled from the 1980 review by Daniel M. Popper and many subsequent publications in astronomical journals. Also shown are the loci of theoretical zero-age main-sequence models covering two ranges of mass.

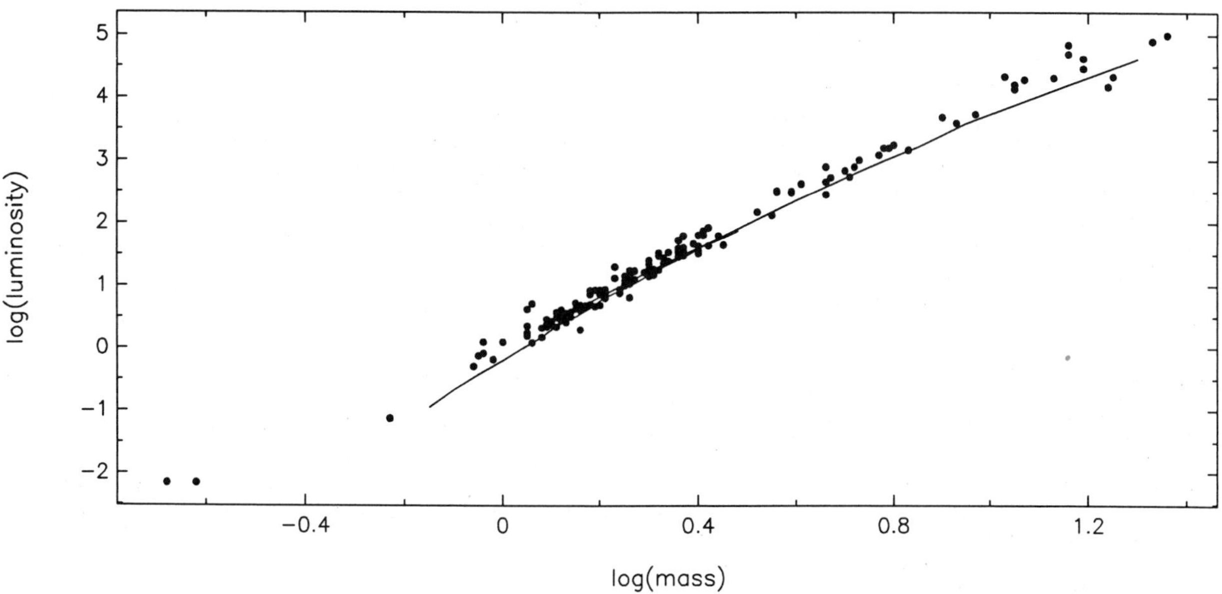

Figure 2. The masses and luminosities in solar units of the same 158 stars and zero-age main-sequence models as in Fig. 1.

events, the differential evolution test is strong. It is pleasing to note that in most well-studied systems the agreement between ages derived for each star in a binary is very good, even when the mass differences are as large as 6 $M_{\odot}$ (e.g., V380 Cyg).

The important G-type eclipsing binary HD 27130, located in the Hyades cluster, and therefore of known distance and age, also demonstrates good agreement with modern solar models at the enhanced metallicity appropriate for the Hyades stars and an age of 4×10^8 yr. The primary component with 1.08 ± 0.02 $M_{\odot}$, and only one-tenth of the age of the Sun, does match the theoretical model for the Sun at that age, with due allowance for the difference in mass and chemical composition.

Despite these pleasing overall agreements of models and data at intermediate masses, it is clear from Figs. 1 and 2, that at both ends of the empirical mass range, the data are extremely sparse. It is in both regions that the theoretical models are also far less certain.

Above 15 $M_{\odot}$, there are only four stars with well-determined masses (better than about 5%) and none at all above 40 $M_{\odot}$. It is at such masses that the phenomenon of overshooting in the convective cores of high-mass stars, and substantial mass loss by means of stellar winds (at rates of $\sim10^{-6}$ $M_{\odot}$ yr^{-1}), can alter substantially the durations of main-sequence stages. There are substantial theoretical problems here whose solution would be aided by high-quality observational data. The inherent paucity of high-mass stars and random orientations of binary star orbits conspire to limit the number of accessible binaries for which masses and other properties may be determined. Because tidal forces usually enforce synchronism between axial rotation periods and orbital periods, the rotational velocities of these high-mass stars in detached systems are often of the order of 200 km s^{-1}, resulting in broad diffuse spectral lines that are difficult to measure accurately. Furthermore, the extended nature of the atmospheres of hot stars also renders it difficult to define meaningful effective temperatures; and the maximum of the energy output for stars at temperatures around 30,000 K is in the far ultraviolet (< 1000 Å) —a region of the electromagnetic spectrum that has not so far been open to exploration. So our empirical calibration of temperatures via optical and near-ultraviolet characteristics is severely limited.

The same paucity of data is apparent also below about 0.7 $M_{\odot}$. It is not at all clear why astronomers have so far not found substantial numbers of detached eclipsing binaries composed of two low-mass stars. But excellent data from visual binaries in recent years have provided masses and luminosities for very low-mass stars, though not of course their radii determined geometrically. Again there are substantial theoretical problems in this area of stellar structure, due most notably to the calculation of opacity at lower temperatures—many transitions of complex atoms, molecules—and also the difficulties of determining the extent of convection zones in low-mass stars. The comparison of empirical results and theoretical models is much needed, not just for binary star evolution including the properties of the low-mass companions in cataclysmic variables and low-mass x-ray sources, but also for understanding the properties of degenerate, hydrogen-rich objects (brown dwarfs) whose masses are too low to sustain nuclear fusion in their interiors.

Synchronism between rotation and orbital periods again ensures that in these low-mass systems, these late-type stars with convective envelopes are rotating much more rapidly than their field-star counterparts—50–100 km s^{-1} rather than 2 km s^{-1} in the case of the Sun. The consequence of such rapid rotation is enhanced magnetic activity in the form of star spots, active chromospheres and coronae, which can reduce the precision with which light curves are defined and stellar radii and orbital parameters are determined; they do of course tell us more about magnetic activity in such stars.

Apsidal Motion

Apsidal periods for binary stars, the time taken for one complete precession period of the orbit, can range over many decades, with observed or inferred values from about 40 years up to several thousand years. Detailed knowledge of the parameters of a binary system, including rotational velocities, are required to provide accurate determinations of the apsidal constant k_2, a measure of the density distribution within a star. For example, the radii of both stars enter the expression for k_2 at the fifth power. Because only the mean apsidal constant for both components is determinate, it is necessary to study systems with closely similar components. For many years, disagreements between observationally derived constants and those from theoretical models were as large as a factor of 3. It was finally recognized only a few years ago that evolution across the main-sequence band results in very large changes in the internal density distributions of stars, with the stars becoming much more centrally condensed and therefore acting more like classical point masses. Apsidal constants calculated as functions of age across the main sequence now compare much more favorably with most systems with well-determined apsidal motion, the discrepancies having been reduced to 20% or less. There is still a tendency for observed rates to suggest higher central concentrations than the models.

One system, DI Her, has an apsidal rotation rate that is far lower than either the expected classical or general-relativistic contributions. Considerable discussion has led to the most recent suggestion that the system is young enough (5×10^7yr) for both components not to have their rotation axes perpendicular to the orbital plane. An additional (negative) term is then required in the classical model which resolves the problem, with rotation speeds typical of ordinary B stars (150 km s^{-1}) but rather extreme angles of inclination of about 70° to the orbital angular momentum vector. Though the major problems about understanding apsidal rotation rates have been removed, there are still some systematic deviations and occasional large discrepancies for individual systems. It is to be hoped that new main-sequence models with the latest opacities will resolve the remaining problems of apsidal motion.

SUMMARY

The last 20 years have seen major advances in the empirical and theoretical foundations of stellar structure and evolution. It is a delight to see the detailed interagreement between the masses, radii, temperatures, luminosities, apsidal constants, and stages of evolution of the components of a particular binary system and the theoretical evolutionary models for detached systems incorporating all the latest understandings and numerical implementations of nuclear physics, quantum mechanics, thermodynamics, and radiative transfer. One hopes that such agreements may be extended to both higher and lower masses than currently available, not only to improve the subject of stellar structure and evolution, but also that of the chemical evolution of galaxies. It is the high-mass stars that interact significantly with the interstellar medium and change the initial chemical composition for future generations of stars, whereas low-mass stars dominate numerically the stellar population in the solar neighborhood.

Additional Reading

Anderson, J. and Clausen, J. V. (1989). Absolute dimensions of eclipsing binaries. XV. EM Carinae. *Astron. Ap.* **213** 183.

Chiosi, C. and Maeder, A. (1986). The evolution of massive stars with mass loss. *Ann. Rev. Astron. Ap.* **24** 329.

Hilditch, R. W. and Bell, S. A. (1987). On OB-type close binary stars. *Monthly Notices Roy. Astron. Soc.* **229** 529.

Maeder, A. and Renzini, A., eds. (1984). *Observational Tests of the Stellar Evolution Theory.* D. Reidel Publishing Company, Dordrecht.

Popper, D. M. (1980). Stellar masses. *Ann. Rev. Astron. Ap.* **18** 115.

VandenBerg, D. A. (1985). Evolution of 0.7–3.0 $M_{\odot}$ stars having $-1.0 \leq [Fe/H] \leq 0.0$. *Ap. J. (Suppl.)* **58** 711.

See also **Binary Stars, Spectroscopic; Binary Stars, Theory of Mass Loss and Transfer in Close Systems; Stars, Masses, Luminosities and Radii.**

Binary Stars, Eclipsing, Determination of Stellar Parameters

Johannes Andersen

The properties of a star are specified by a set of physical parameters. Present overall characteristics are described by such fundamental parameters as mass, radius, surface temperature (or, equivalently if the radius is known, luminosity), chemical composition, and rotation. These quantities are in principle observable. Age, internal structure, evolutionary stage and history, and so on, are other properties of the star of great interest. However, except for the Sun, these parameters are not directly observable. Instead, they are inferred from theoretical models computed to fit the observable fundamental parameters. Not all of these can be well determined in all stars. Hence it is important that fundamental parameters be

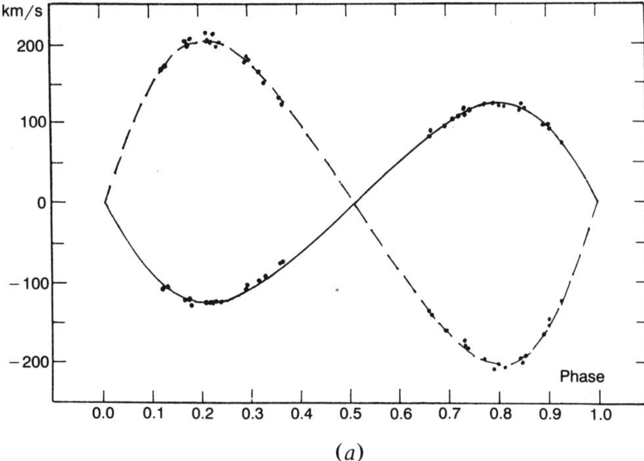

(*a*)

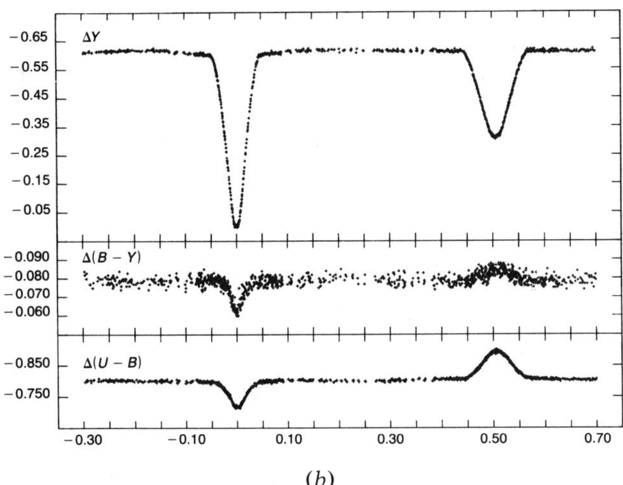

(*b*)

Figure 1. Radial-velocity curves (*a*) and light and color curves in the Strömgren *uvby* system (*b*) for the B-type eclipsing binary GG Lupi. Such data can yield component masses and radii to ≈ 1% accuracy.

determined accurately and completely for at least *some* stars, so their relationships can be established and the theoretical models tested as far as possible.

Binary stars with orbits seen so nearly edge-on that the component stars eclipse each other during the orbital cycle offer unique opportunities for a complete determination of the fundamental parameters of the component stars. Hence spectroscopic and photometric observations of eclipsing binaries (see Fig. 1) form much of the empirical foundation for our current knowledge of the masses, radii, and evolution of both single and binary stars. In recent years, advances in the quality and quantity of such data have been combined with improvements in their analysis, with particular emphasis on the *consistency* between complementary data sets. The results have led to significant insights in stellar evolution theory.

STELLAR MASSES

The most fundamental parameter of a star, and that upon which its other properties depend most strongly, is its *mass*. Except for the Sun, where we inhabit a convenient test particle, the Earth, the mass of a star can only be directly measured by its gravitational attraction on a companion star. Therefore, the discussion of mass determinations in eclipsing binaries merits special attention. For stars close enough to exhibit eclipses, only the line-of-sight component (radial velocity) of the orbital motion is seen; it is measured from the Doppler shift of spectral lines in the stars. From the orbital period and the two radial velocity amplitudes (Fig. 1*a*), the masses and separation of the two stars can be determined from Kepler's laws, except for the projection factor $\sin i$, where i is the inclination of the orbital plane relative to the sky and can be determined from the light curve of an eclipsing system. If only one star is seen in the spectrum (single-line binary), the individual masses cannot be found unless the mass ratio can be fixed from other data.

As the masses depend on the cube of the observed velocity variations, great care is necessary in order to obtain accurate results. Observations of adequate spectral resolution and signal/noise ratio are an obvious first requirement. Given such data, the principal difficulty arises from line blending in the double-lined spectra. First, rotation may broaden spectral features so as to include nearby lines (mostly in late-type stars), leading to different effective wavelengths than observed in single stars. Second, lines with broad wings may not be fully resolved when double, causing systematic errors in the measured velocities. This is most important in early-type (OBA) stars, where velocities from the extended, Stark-broadened wings of Balmer and diffuse He I lines typically lead to masses underestimated by ≈ 35% and 10%, respectively. Third, in line-rich spectra, lines of one star may blend with different lines of the other star at double-lined phases. These effects differ from line to line and from star to star. In some stars, they are best dealt with by using a small number of very carefully selected lines. In others, it is preferable to average over a large number of lines by cross-correlating entire regions of the binary spectrum with that of a reference star.

In both cases it is important to verify that the radial velocities measured represent the motion of the centers of mass of the two stars. If gas clouds or streams exist in the system, or if the shapes and light distribution of the stars are significantly distorted by their nearby companions, this may not be true. One therefore prefers spectral lines of purely photospheric origin and corrects for the possible effects of distortion of the stars as computed from a physical model of the binary system which also fits the observed light curve.

With proper precautions, the masses of stars in selected detached, double-lined and eclipsing binary systems can now be determined to an accuracy of ≈ 0.5%. Whereas *visual binaries* also supply useful data, mostly for later-type stars than eclipsing binaries, no other current technique can match this precision. Summaries of mass determinations for various types of stars are included in the Additional Reading and in the entry on *detached binaries*, which also shows the observed mass–radius and mass–luminosity relations. Applications of such accurate data will be discussed in the following section.

STELLAR RADII

Whereas the radii of single stars can also be estimated by various indirect methods, the light variations observed in eclipsing binaries (Fig. 1*b*) provide direct, potentially very accurate, geometric evidence of the sizes of the component stars. From an analysis of the light curve, the radii of the two stars in units of their separation, (the ratio of) their surface temperatures, the orbital inclination, and the eccentricity and orientation of the orbit and the possible variation of the latter (apsidal motion) can in principle be determined. Combining these parameters with the spectroscopic determination of the orbital radius yields the absolute radii of the two stars. From the temperatures (or, more basically, surface fluxes) and the radii, the luminosity ratio follows. For accurate results, the light curves must contain a sufficient number (≈ 1000) of precise (≈ 0.5%) photometric observations in two or more intermediate-width passbands (≈ 25 nm); wide-band photometric data are generally inadequate.

Modern analysis methods generate synthetic light curves, computed with a realistic physical model of the gravitational and rotational deformation of the stars, the limb darkening and gravity brightening of their disks, and the amounts of light reflected from them. Spectroscopic data on the component masses and rotations are essential input to the model. By systematically varying the radii, temperatures, and inclination (and some secondary parameters mentioned previously), one searches for the synthetic light curve that best matches the observations. However, the problem is highly nonlinear, and in systems of relatively low inclination (partial eclipses), one often finds that a wide range of parameter sets produce essentially identical light curves. In particular, the ratio of the two radii is often poorly constrained, and one must invoke other (e.g., spectroscopic) data in order to define the "best" solution with useful accuracy. In favorable systems with deep eclipses (inclination $\approx 90°$; Fig. 1*b*), the radii can be determined to better than 1%.

In some interacting binaries with recent or ongoing mass transfer, circumstellar gas streams and/or an accretion disk may distort the light curve and prevent accurate determination of the stellar radii. Strongly spotted binary stars may suffer from similar problems. However, if the orbital inclination is known and the *rotational* broadening of the spectral lines of the stars can be measured, the radii follow directly from the rotation rate and orbital period if the rotations are synchronized with the orbital motion. Finally, the radius of a Roche lobe-filling star is in principle a function of the mass ratio, but the latter is often poorly determined in just such systems.

STELLAR SURFACE TEMPERATURES

As in single stars, surface temperatures of eclipsing binary components are generally determined from color indices in a well-calibrated photometric system. Complications arise because the observed colors refer to the combined light of both stars. The best approach is to obtain also the light curves in the same photometric system; from the combined indices and the luminosity ratio in each color, the individual color indices and temperatures can then be found. Important consistency checks are possible in eclipsing binaries: The eclipse depths indicate the central surface flux ratio, and the color of the light lost in eclipse can also be derived directly from the data. Moreover, the ratio of the central surface fluxes (related to temperature and color via an empirically calibrated scale) should be consistent with the observed color difference.

CHEMICAL COMPOSITION

Together with its initial mass, the chemical composition of a star determines the course of its evolution; it also places it among the stellar populations in the Galaxy. Generally, the chemical composition of a binary star is determined just as for a single star, by suitable photometry or high-resolution spectroscopy. However, the presence of two sets of blended and rotationally broadened spectral lines is an added difficulty. Most chemical analyses of stars concern their *metal abundance*, the content of elements heavier than helium. However, for main-sequence binaries with accurately known masses, radii, luminosities, and metal content, the *helium abundance* may also be derived by comparison with stellar evolution models, even for stars without observable spectral lines of He. This possibility of determining the initial He abundance even for old, unevolved binary stars is a feature of particular interest.

EVOLUTIONARY STAGE, AGE

Apart from pulsational phenomena or such surface features as star spots or magnetic fields, the fundamental stellar parameters discussed previously comprise about as much basic information as we can obtain for any star. Our model of its evolutionary history must be built on this basis. Detached binary stars for which these data are complete and accurate at the 1% level can be fitted into the framework of stellar evolution models to a precision unmatched for any single star except the Sun. Such binary stars can be located within the main-sequence band to within $\approx 5\%$ of the width of the main sequence. Highly significant differences between the masses and radii of "main-sequence" stars of a given spectral type are found, depending on their (composition and) precise degree of evolution. Table 1 summarizes the observational results on masses and radii at the beginning and end of the main-sequence phase of evolution. The data are not yet adequate for a similar table for evolved stars.

Table 1. Mass and radius (in solar units) for Population I stars at the beginning (ZAMS) and end (TAMS) of their main-sequence evolution, based on recent data

	ZAMS		*TAMS*	
Spectral Type	*Mass*	*Radius*	*Mass*	*Radius*
O8	20 :	6 :	35 :	20 :
B2	8.0	3.5	11	8
B6	4.0	2.3	7	6
A0	2.3	1.7	3.4	4.5
A4	1.8	1.6	2.4	4.3
A8	1.5	1.4	2.2	4.0
F2	1.4	1.3	2.0	3.5
F6	1.3	1.2	1.5	2.5

Such accurate data also provide significant tests of and constraints on theoretical models of stellar evolution: Models for the observed mass and composition should have the observed radius and luminosity for the age of the binary. The age is generally not independently known, but if the masses are considerably different, a simultaneous fit to *both* stars in a binary provides a nontrivial test. For stars with masses up to 1.2 $M_{\odot}$, standard evolutionary models using the latest opacities have recently been shown to pass these tests with flying colors.

For only slightly more massive stars, 1.5–2.5 $M_{\odot}$, very recent work shows that nonstandard models incorporating convective core overshooting are needed. Even though only a modest amount of overshooting is required to match both the binary data and a wide range of other observations, the implications are far from trivial: The interior structure and future course of evolution of these stars are significantly different from the predictions of standard models, and the ages estimated for moderately evolved stars are considerably larger.

SPECIAL OBJECTS

Some types of star are the unique products of *binary* as distinct from single-star evolution and are found only in certain types of binary system such as Algol or RS CVn systems, cataclysmic variables, x-ray binaries, and so on. As these types of binary are treated elsewhere, we shall not discuss their properties in detail. We do note, however, that our understanding of the nature and origin of also these objects, however incomplete it may be, ultimately rests on data on their fundamental parameters determined primarily from the eclipsing members of each class, using the methods discussed previously.

Additional Reading

Andersen, J., Clausen, J. V., Gustafsson, B., Nordström, B., and VandenBerg, D. A. (1988). Absolute dimensions of eclipsing binaries. XIII. AI Phoenicis: a case study in stellar evolution. *Astron. Ap.* **196** 128.

Andersen, J., Clausen, J. V., and Nordström, B. (1980). Determination of absolute dimensions of main-sequence binaries. In *Close*

Binary Stars: Observations and Interpretation, M. J. Plavec, D. M. Popper, and R. K. Ulrich, eds. D. Reidel Publishing Company, Dordrecht, p. 81.

Batten, A. H., ed. (1989). Algols. *Space Sci. Rev.* **50** 1.

Hilditch, R. W., King, D. J., and McFarlane, T. M. (1988). The evolutionary state of contact and near-contact binary stars. *Monthly Notices Roy. Astron. Soc.* **231** 341.

Popper, D. M. (1980). Stellar masses. *Ann. Rev. Astron. Ap.* **18** 115.

Popper, D. M. (1984). Error analysis of light curves of detached eclipsing binary systems. *Astronom. J.* **89** 132.

See also **Binary Stars, Detached; Stars, Masses, Luminosities and Radii.**

Binary Stars, Observations of Mass Loss and Transfer in Close Systems

George E. McCluskey, Jr.

As the second half of the 20th century has progressed, astronomers have determined that mass loss is an important factor in the birth, evolution, and death of stars. This is particularly true for close binary stars. The term *close binary star* is derived from the fact that if the two components of a binary system are sufficiently close to one another, then the evolution of one or both components is significantly affected by the existence of the other. The primary result of the interaction between the two stars is the transfer of mass from one to the other and the loss of mass from the binary system into interstellar space. We present here the observational evidence that has proven that mass transfer and mass loss in close binary stars is a critical factor in their evolution and leads to such exotic systems as x-ray binaries and the relativistic jet binary SS 433.

It will be useful to introduce a simple and commonly used scheme by which we can classify close binaries. The classification uses the concept of the critical Roche lobe (hereafter RL). All binary systems are surrounded by an equipotential energy surface in the shape of a three-dimensional, asymmetrical figure 8. The two lobes of the 8 touch at a point between the two stars on the line joining their centers. Each lobe contains one of the stars. These lobes are the critical Roche lobes of the system. The size of the RL increases in proportion to the separation of the stars and the more massive star has a larger lobe than the less massive star. If the stars are very far apart relative to their radii, the RL is also much larger than its star. If the stars are separated by only a few radii of the larger star, then the RL of the larger star will not be much larger than its star. It is well known from observation and from the theory of stellar evolution, that stars undergo dramatic increases of their radii at certain stages of their evolution. If a star expands sufficiently and if the distance between the two stars is relatively small, then at some point the evolving star will fill and overflow its RL. It is at this point that mass transfer and/or mass loss can occur very rapidly. If both stars are smaller than their RLs, the system is called detached. If one star fills its RL and the other does not, the system is called semidetached. If both stars fill their RL, the system is called contact. If one or both stars are overflowing their RL, surrounding the system with gas, it is called a common envelope binary.

OBSERVATIONAL EVIDENCE

Ground-Based Observations

Until 1941, essentially all astronomers believed that close binaries were uncomplicated. From 1941 onward, Otto Struve and collaborators began to realize that many close binaries have gas streams between the components and circumbinary material that escapes from the system. By the mid-1950s general acceptance of the existence of mass transfer and mass loss had taken place, and from 1966 onward, major observational and theoretical programs have been improving our understanding of gas flow in close binaries.

Two important diagnostic tools that are used to detect and analyze mass flow in close binaries are photometry and spectroscopy. In essence, photometry can detect the existence of light added or subtracted to or from the normal light of the two stars by streams and/or envelopes of gas in the system. Spectroscopy can detect the presence of absorption lines and/or emission lines due to this gas and also yields important information concerning its physical properties.

Mass transfer and mass loss give rise to numerous complications of the light and velocity curves of a close binary. Increases, decreases, or irregular variations in the orbital periods of close binaries are often detected. Most of these changes can only be explained by mass transfer and/or mass loss. The existence of time-dependent variations in the shapes of light and velocity curves is indicative of unsteady gas flow. In strongly interacting close binaries, the depth and shape of the eclipses may vary with time and the shapes of light and velocity curves depend on the wavelength of observation. The stronger these effects are, the stronger is the mass flow.

Any gas flowing between or around the stars or out of the system will give rise to absorption and/or emission lines. Because the gas will often have velocities significantly different from the orbital velocities of the stellar components, the Doppler effect will cause these lines to appear at wavelengths different from those of the stellar lines. In addition, because the temperature and density of this gas will generally differ from those of the stellar atmospheres, lines not present in the spectrum of either star will appear. The detection of emission lines normally indicates the presence of relatively extensive low-density gas, exactly the property of most gas flows in close binaries. All of these effects are seen in common envelope systems and at least some of them are seen in some contact and semidetached systems. Detached systems show these effects weakly if at all.

By the late 1960s, the existence of disk or disk-like structures arising from mass transfer was well documented observationally. In addition, the cataclysmic variables, which include novae and dwarf novae, appeared to be close binaries with the outburst related in some way to mass transfer.

By 1970, ground-based astronomy had provided a firm observational base strongly indicating that mass transfer and mass loss play a major role in the evolution of close binaries. Theoretical work led to the conclusion that Roche lobe overflow is the dominant mechanism for rapid mass transfer and mass loss in close binaries and also plays a prominent role in many systems with a low rate of mass flow. With the launching of numerous Earth-orbiting astronomical observatories, new observational data related to mass transfer and mass flow in close binaries became available to astronomers for the first time.

Observatories in Space

The advantage of space-based observatories is that astronomers are able to observe regions of the electromagnetic spectrum that are inaccessible from the Earth's surface due to absorption by the atmosphere. Ground-based observatories are limited to the optical and radio regions of the spectrum, whereas gamma rays, x-rays, ultraviolet radiation, and much infrared and microwave radiation are unobservable. With the appropriate detectors, the entire electromagnetic spectrum can be seen from observatories in space.

The greatest efforts of space astronomy have been devoted to the ultraviolet and x-ray regions of the spectrum. The gamma ray region has also received considerable attention. These short wavelength regions of the electromagnetic spectrum correspond to the highest energies and have attracted the most attention.

The ultraviolet and x-ray wavelengths are extremely important for studying mass transfer and mass loss in close binaries. If both components are normal stars (as opposed to degenerate objects, i.e., white dwarfs, neutron stars, and black holes), the energies acquired by the atoms and ions making up the gas flow can attain values equivalent to temperatures of 30,000–200,000 K. The sources of this energy include the radiation, gravitational, and magnetic fields of the stars. At these temperatures, the gas radiates and absorbs primarily in the ultraviolet region of the spectrum and consequently this is where the most useful information concerning mass transfer and mass loss can be obtained. If the components of a close binary include a white dwarf or neutron star or black hole, far higher energies can be attained due to the incredibly strong gravitational fields of these objects. These objects can generate powerful x-rays and even gamma rays in addition to ultraviolet radiation.

Many nondegenerate binaries have been observed in the ultraviolet and x-ray regions of the spectrum by a number of satellites. Two of the most productive satellites have been the *International Ultraviolet Explorer*, launched in 1978 and still operating well in 1990, and the *Einstein x-ray observatory (HEAD 2)* satellite which operated from 1978 to 1981. Many other astronomy satellites have also contributed to the study of close binaries as well as to essentially all areas of astronomy.

The strongly interacting close binary β Lyrae has been observed in the ultraviolet by many satellites. This system shows strong emission lines due to an extensive envelope of gas surrounding the stars. Ions such as Si IV, C IV, and N V are present, indicating ionization temperatures considerably above that of the stars. Our understanding of the physical processes occurring in this circumbinary plasma is still incomplete. Absorption lines of the preceding ions indicate that this plasma is leaving the system at 200–400 km s^{-1}. The existence of gas streams is clearly indicated by the ultraviolet spectrum of β Lyrae.

A number of other close binaries have been found to have strong ultraviolet emission lines. These systems are common envelope binaries. Emission lines of lesser intensity have been discovered in a number of semidetached binaries such as U Cephei. Also found in these semidetached close binaries are absorption lines of the ions Si IV, C IV, and N V which are associated with the detached component. The energy source for these lines is postulated to be a high-temperature accretion region near the detached component where mass transferred from the contact companion is falling onto this star. These lines show small, but significant nonorbital velocities indicating that gas flow is present. Even semidetached systems such as U Sagittae which appear quiescent in ground-based study show the clear-cut effects of mass transfer in ultraviolet spectra. Essentially all semidetached contact and common envelope binaries are undergoing mass transfer and mass loss. Even some detached systems have been found to show mild mass flow phenomena. It is clear that ultraviolet photometry and spectroscopy have shed light on a heretofore unknown treasure of plasma phenomena.

The various x-ray observatories launched beginning in 1970 (earlier hints were provided by suborbital rocket experiments carrying x-ray detectors) have also discovered a wealth of mass transfer and mass loss phenomenology. In the early 1970s, x-ray binaries were detected. These consist of a nondegenerate star paired with a neutron star or, in a few cases, a black hole. The powerful (10^{35}–10^{38} erg s^{-1}) x-rays emitted by these systems can only arise from the energy released when gas from the companion falls into the intense gravitational field of the degenerate object. Some gas also escapes into interstellar space. The massive x-ray binary Cygnus X-1 is a very strong case for a black hole. It is paired with an O-type supergiant which is the source of the mass being transferred and lost. The very existence of neutron stars or black holes in these x-ray binary systems is also proof of earlier mass flow because as these stars evolved, they must have lost mass, eventually becoming supernovae and losing still more mass. The neutron star or black hole is the stellar remnant of the supernova.

The x-rays detected from cataclysmic binaries are also indicative of the accretion of mass onto a degenerate object, which is a white dwarf.

In 1978, a combination of ground-based optical and radio observations with satellite x-ray data led to the discovery of a unique close binary, SS 433. Spectroscopic analysis showed that SS 433 is a close binary with a neutron star or black hole and an O star that is losing mass rapidly. Gas flows toward the neutron star or black hole forming a thick disk. The compact star is too small (radius less than 20 or 30 km) to accrete the gas at the rate it arrives. The disk behaves like a tube of toothpaste open at both ends and squeezed in the middle. Gas flows out of the system in two oppositely directed jets at slightly more than one-fourth the speed of light. No other stellar-sized object is known in which bulk masses of gas attain relativistic velocities.

It is clear that mass transfer and mass loss play a critical role in the evolution of close binaries and can lead to some very interesting systems. Astronomers look forward with great anticipation to further research with more sophisticated astronomical space observatories in the coming years, especially the *Hubble Space Telescope* which is the largest space observatory ever launched. We can be certain that many fascinating discoveries await us.

Additional Reading

Batten, A. H. (1973). *Binary and Multiple Systems of Stars*. Pergamon Press, London.

Gursky, H. and van den Heuvel, E. P. J. (1975). X-ray emitting double stars. *Scientific American* **232** (No. 3) 24.

Margon, B. (1980). The bizarre spectrum of SS 433. *Scientific American* **243** (No. 4) 54.

Shaham, J. (1987). The oldest pulsars in the universe. *Scientific American* **246** (No. 2) 50.

Sahade, J. and Wood, F. B. (1978). *Interacting Binary Stars*. Pergamon Press, London.

See also **Binary and Multiple Stars, Types and Statistics; Binary Stars, Theory of Mass Loss and Transfer in Close Systems; Binary Stars, X-Ray, Formation and Evolution; Ultraviolet Astronomy, Space Missions.**

Binary Stars, Polars (AM Herculis Type)

Paula Szkody

The *polars* are a class of cataclysmic variable stars whose major distinguishing characteristic is the presence of strong (10%–30%) circular polarization. They are also commonly referred to as AM Herculis stars, after the prototype of the class (AM Her) which was the first and brightest system found. The polarization is the result of mass transfer onto one or more magnetic poles of a white dwarf from a close binary companion.

The magnetic field strength of the white dwarf, combined with the separation of the two stars (which is related to the orbital period), determines the characteristics of the accretion of the system. In the polars, the field strengths are all near 3×10^7 G. Because the orbital periods are all less than 4 hr, this field strength is sufficient to control the flow of material as soon as it leaves the Roche lobe of the secondary star. The transferred material follows the field lines until it encounters a shock close to the surface of the white dwarf. This shock slows the material and cools it by emitting thermal bremsstrahlung emission in x-rays and cyclotron radiation in the optical (which results in circular polarized light). Some of this radiation escapes to space and some heats the white dwarf surface, creating a hot accretion zone on the white dwarf. Because the magnetic field dominates the flow out to the secondary, the spin of the white dwarf is locked to the orbital revolution period. Thus all observed variations from the system are at the orbital time scale.

Related systems containing white dwarfs with magnetic poles but with different observational characteristics are the *intermediate polars* (also called DQ Herculis stars after their prototype). These systems generally have longer orbital periods (larger separation between the two stars) and are thought to contain white dwarfs with lower magnetic field strengths (smaller than 10^6 G). In this situation, the magnetic field only controls the flow near the white dwarf. An outer disk forms, but the inner portion (close to the white dwarf) is channeled to the magnetic pole. Because the spin of the white dwarf is not locked to the orbit, there is a rotating beacon effect as the accretion pole of the white dwarf sweeps across the line of sight. This results in a large pulse in the x-ray and optical light curves at the spin period (or, alternatively, at the beat period between the spin and orbit for the optical pulse, if the accretion column light interacts with something fixed in the orbital reference frame). The presence of this pulse (on time scales of minutes), with a longer orbital period (hours), provides the primary identification of a DQ Her system.

The polars and intermediate polars represent relatively new classes of cataclysmic variables, because AM Her was not discovered until 1976 and the DQ Her systems as a class in the early 1980s (although DQ Her and its spin and binary periods were already recognized by Merle F. Walker around 1955). The addition of further members to these classes has been mainly the result of optical identifications of strong x-ray sources found by the *HEAO* and *EXOSAT* satellites.

AM HERCULIS SYSTEMS

Table 1 provides a summary of the AM Herculis systems that are known to date, together with orbital periods, magnetic field strengths, and stellar characteristics. Some common trends that must be explained in terms of evolution emerge from this table. First, it is apparent that all of the magnetic field strengths cluster about 10^7 G. Second, most of the orbital periods are below 2 h, with a large number of objects (6) having an orbital period of 114–115 min.

Besides the evolutionary aspect, major current problems under study for the polars involve the energy balance and geometry of the accretion column and its associated flow. Space observations of hard and soft x-ray radiation have shown that a simplified theoretical picture of the accretion column heating the white dwarf does not provide the correct observed balance. In several systems, there is too much soft x-ray emission for the observed hard x-ray emission. Contributing to this problem are the large corrections that must be made because all parts of the x-ray spectrum are not observed, because local and interstellar absorption can absorb much of the x-ray spectrum, and because the reflectance (albedo) of the white dwarf surface can make a large difference in how much of the energy is absorbed by the white dwarf.

The geometry of the accretion column can be studied through analysis of the orbital light and polarization curves, using light from the x-ray through the infrared. This is especially useful in the case of eclipsing systems. Because the magnetic pole of the system is typically inclined to the rotation axis, both the inclination of the system to the line of sight and the inclination of the pole must be determined. In several cases, the geometry reveals a large hump in the light curves, indicating the self-eclipse of the pole behind the white dwarf during part of the orbit. In three systems, the secondary star eclipses the pole. Studies of this type reveal a much more complicated picture than a simple circular accretion zone on the white dwarf. The observations often suggest extended, asymmetric regions that extend above the white dwarf. Also, large temporal changes have been found in the x-ray and optical light curves, indicating the emission may change from two poles to one and the geometry of the accretion column itself may change.

The optical emission line spectra of the polars typically show abnormally strong high excitation lines of He II at 4686 Å and a blend of carbon and nitrogen lines at 4640 Å. The lines generally have a very complex structure, with a broad component having motion that identifies it with the accretion stream, whereas a sharp narrow component, 180° out of phase with the broad one, is traced to the irradiation of the secondary by the accretion pole. Furthermore, narrow components that may originate in other parts of the stream can be identified in several systems. Understanding the location of these components helps to map out the geometry of the material as it flows along the field lines.

At unpredictable times, the radiation from the accretion column can disappear (causing a reduction in brightness by several magnitudes) and the white dwarf and its companion may be viewed directly. These "low states" are extremely useful for the determination of the field strength of the white dwarf (from Zeeman features) and of the spectral type of the secondary star (from TiO bands), which then yields a distance to the system. Studies during the low state have shown that the mass transfer greatly diminishes but some remnant radiation from the hot white dwarf usually remains.

Table 1. Known AM Herculis Systems

Object	*X-Ray Name*	*Period (min)*	*V (mag)*	*Polar Field (MG)*	*White Dwarf Mass ($M_\odot$)*
EF Eri	2A0311 − 227	81	14–18		0.4
DP Leo	E1114 + 182	90	17–22	44	0.4
VV Pup		100	14–18	32	1.0
V834 Cen	E1405 − 451	102	14–17	22	
Grus V1		109	18–21		
MR Ser	PG1550 + 191	114	15–17	36	0.6
BL Hyi	H0139 − 68	114	14–17	30	
ST LMi		114	15–17	30	0.6
	E1048 + 543	115	18–20		
	E0234 − 523	115	19–23		
AN UMa		115	15–19	35	
	E0333 − 255	127	17–21	55	> 0.9
AM Her	4U1809 + 50	186	12–15	20	0.6
	H0538 + 608	195	15–18	41	
	V1500 Cyg	201	2–22		> 0.9
QQ Vul	E2003 + 225	223	15–18		0.6
	E0329 − 261	228	17–19		

DQ HERCULIS SYSTEMS

Table 2 summarizes the known DQ Herculis systems. It is generally apparent that this group has longer orbital periods than the polars. Because the general trend of close binary evolution is toward decreasing orbital period, it has been speculated that DQ Hers may evolve into AM Hers. However, difficulties with the observed period distributions suggest alternative individual evolutionary scenarios. The DQ Hers show a wide range of spin periods. AE Aqr and DQ Her have very fast periods (on the order of 1 min), the majority have spin periods of minutes and a few have long periods near 1 h. The true spin period is identified from the x-ray pulse period, whereas the optical can show either the spin or the beat period between the spin and the orbit. In most cases, the same period is evident in the x-ray and the optical, which implies the pulsations originate from the white dwarf. In the few cases where the beat period is present in the optical, the implication is that the pulsations originate from x-rays that are reprocessed by the hot spot (where the mass transfer stream intersects the outer disk) or possibly by the secondary. However, the strength of the optical pulsations compared to the known x-ray fluxes implies that the reprocessing area must receive an abnormally large amount of the x-rays. It is possible that the x-rays are beamed (for an unknown reason) or there is a large amount of flux in the unobserved region of the extreme ultraviolet.

Of the DQ Her stars, only BG CMi shows a weak circular polarization, primarily evident in the infrared. This would be consistent with a lower magnetic field strength for this group as

Table 2. Known DQ Herculis Systems

Object	*X-Ray Name*	*Spin P (min)*	*Orbital P (h)*	*Beat P (min)*	*V (mag)*	*White Dwarf Mass ($M_\odot$)*
AE Aqr		0.6	9.8		11	0.9
DQ Her		1.2	4.7		15	0.5
GK Per	A0327 + 43	5.9	48		13	0.5–1
V1223 Sgr	4U1849 − 31	12.4	3.4	14.1	13	0.5
AO Psc	H2235 − 035	13.4	3.6		13	1.0
BG CMi	3A0729 − 10	15.2	3.2		15	0.6–1
SW UMa		15.9	1.4		17	0.5
FO Aqr	H2215 − 086	20.9	4.0	22.9	13	0.5–1.4
TV Col	2A0526 − 328	31.9	5.5		14	0.5–1.4
	H0542 − 407	32.0	6.2		16	
EX Hya	4U1228 − 29	67.0	1.6		13	0.8

compared to the polars. The observed rate of spin period changes for some DQ Hers also gives a low value for the magnetic field, but there are large uncertainties in the associated parameters of accretion rate, mass of the white dwarf and the geometry of the accretion. Thus the true magnetic field strengths of the white dwarfs are not known.

The x-ray light curves generally show quasisinusoidal variations at the rotation period of the white dwarf, implying that the accretion zone is large in extent. The x-ray spectral studies indicate that a large amount of local absorption may be present, whereas the shape of the modulations indicate that the emission regions are extended and may have a ring or arc structure. The actual extent of the accretion disk (and even its existence in the short period systems) is difficult to determine.

FUTURE WORK

The past decade has seen the emergence of two new classes of cataclysmic variables containing magnetic white dwarfs (the AM Her and DQ Her systems). Although a dozen members are now known in each category, there are many questions that remain to be explored concerning the evolution and the physics of the accretion flow under high field strength configurations. The resolution of these problems will come from space-based observations of the soft x-ray and extreme ultraviolet (with the *ROSAT*, *EUVE*, and *AXAF* satellites) which will explore the accretion zone close to the white dwarf and from the interpretation of these data with adequate theoretical models.

Additional Reading

Liebert, J. L. and Stockman, H. S. (1985). The AM Herculis magnetic variables. In *Cataclysmic Variables and Low-Mass X-Ray Binaries*, J. Patterson and D. Q. Lamb, eds. D. Reidel, Dordrecht, p. 151.

Liller, W. (1977). The story of AM Herculis. *Sky and Telescope* **53** 351.

Szkody, P. and Cropper, M. (1988). Cataclysmic variables. In *Multiwavelength Astrophysics*, F. A. Cordova, ed. Cambridge University Press, Cambridge, p. 109.

Wade, R. A. and Ward, M. (1985). Cataclysmic variables: Observational overview. In *Interacting Binary Stars*, J. E. Pringle and R. A. Wade, eds. Cambridge University Press, Cambridge, p. 129.

Warner, B. (1983). The intermediate polars. In *Cataclysmic Variables and Related Objects*, M. Livio and G. Shaviv, eds. D. Reidel, Dordrecht, p. 155.

See also **Binary Stars, Cataclysmic; Binary Stars, X-Ray, Formation and Evolution.**

Binary Stars, RS Canum Venaticorum Type

Douglas S. Hall

This group of peculiar binary stars was defined by the author in 1975. The group is named after the prototype RS Canum Venaticorum (abbreviated RS CVn), a star of magnitude 7.9 and the 11th variable discovered in the constellation Canes Venatici (the hunting dogs). RS CVn is an eclipsing binary with an orbital period of 4.798 days and a spectral type of F5 V + K0 IV*e* (see Fig. 1).

The definition is based on three observable characteristics: The orbital period should be between one day and two weeks; the hotter of the two stars should be of spectral type F or G; and the H and K lines of singly ionized calcium (Ca II) should show strong emission in their cores at all orbital phases, that is, not only during an eclipse. At the same time, two subgroups were defined, the short-period and the long-period RS CVn binaries, although it was not clear whether the divisions in orbital period at 1 day and at 14 days would prove to be significant. Many peculiar characteristics, discussed later, are common to the RS CVn binaries, but the three mentioned previously constituted the formal definition.

One can find references in the literature dating back to the late 1940s that showed a few astronomers were aware of a small group of binaries whose spectroscopic and physical characteristics seemed to set them apart. Subsequently, however, those binaries (which we now know as RS CVn binaries) generally lost their distinction and throughout the 1950s, 1960s, and most of the 1970s their peculiar characteristics were universally misinterpreted. The principal confusion was with the semidetached Algol-type binaries, in which one star fills its Roche lobe and is transferring luminous gaseous material down onto its companion star. The resulting circumstellar material (in the form of gas streams, circumstellar disks, hot spots, and circumbinary shells) was invoked (incorrectly) to explain the peculiar observational properties of the RS CVn binaries as well.

The discovery of the RS CVn binaries was intimately related to the discovery of starspots, an interesting episode of astronomical history. The very first physical mechanism proposed to explain the photometric variability of a variable star (other than a nova or supernova) was the suggestion by Ismael Boulliau in 1667 that dark spots distributed nonuniformly in stellar longitude could explain the 11-month periodicity of *o* Ceti (Mira Ceti). Starspots became the favored physical mechanism invoked to explain *all* subsequently discovered variable stars that were not obviously eclipsing like Algol. This continued for the next 250 years, until the correct mechanisms were identified one by one and found *not* to be related to dark spots. Then, rather abruptly around the turn of this century, the very idea of starspots became anathema. In scores of articles over the next 50 years, starspots were not even mentioned as astronomers groped to explain peculiar light variations in spotted stars, one of which was RS CVn itself. The first mention of

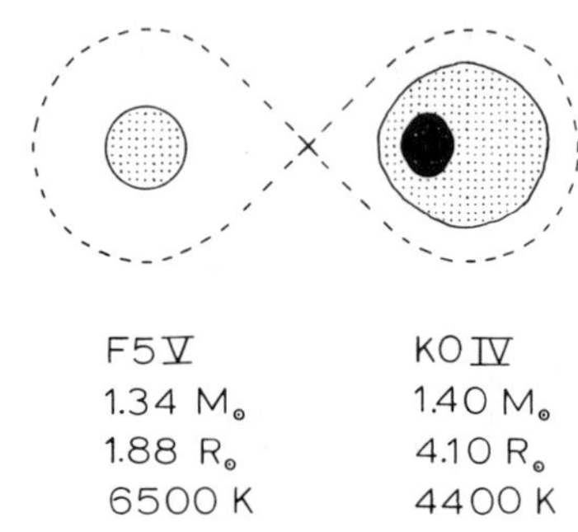

Figure 1. The two stars in RS CVn as they appear projected on the sky. Note that both, even the evolved star on the right, lie comfortably within their Roche lobes. The dark spot on the synchronously rotating K0 subgiant produces the wave (seen in Fig. 2) as it turns into and out of view.

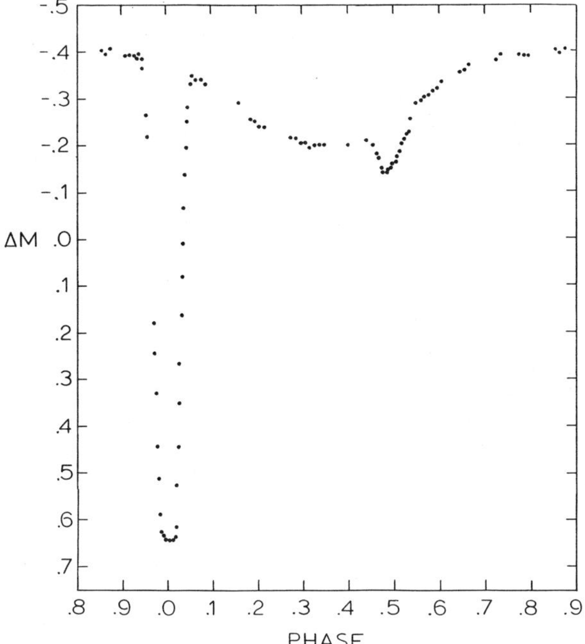

Figure 2. The 1964 light curve of RS CVn in magnitudes versus orbital phase. The conspicuous wave had its minimum at phase 0.34, when the dark spot was facing toward Earth. Note that the shallow secondary eclipse, when the smaller F5 star transits across the spotted star, is asymmetric.

starspots in connection with a truly spotted star was by Gerald E. Kron, in an article dated 1947; there was a nearly 20-year lull before starspots were mentioned again, by Cuno Hoffmeister in 1965 and by P. F. Chugainov in 1966; and the majority of astronomers continued not to "believe in spots" until the RS CVn group emerged in 1975 (see Fig. 2).

CHROMOSPHERIC ACTIVITY

The RS CVn binaries have many peculiar properties, most of them analogous to the panoply of features usually referred to as solar activity but most far more extreme than their counterparts on the Sun's surface.

1. Starspots can blanket up to 50% of a hemisphere of the active star in an RS CVn binary, whereas sunspots cover only about 0.2% of a solar hemisphere even during a sunspot maximum.
2. In the core of the Ca II H and K absorption lines, there is a sharp emission line that can rise well *above* the continuum, the continuous part of the spectrum. On the Sun the corresponding emission lines, which are chromospheric in origin, reach only 1% of the way to the continuum.
3. Similar emission lines in the far ultraviolet spectrum of an RS CVn binary, produced in the transition region between the chromosphere and the corona, are more intense than the corresponding solar lines by a comparable factor.
4. Hα, a strong absorption line in the solar spectrum, sporadically appears in emission in many RS CVn binaries and often appears as incipient emission, partially filling in the absorption.
5. Rapid (time scale of minutes) solar-like flares occur and are especially conspicuous (making the star brighten as much as 10-fold in the optical continuum) if the active star is much fainter than the Sun.
6. Relative to the spectral energy distribution of normal stars, in which the photosphere is essentially at one temperature, RS CVn binaries show excess light in the near infrared, because the spotted and unspotted portions of the photosphere comprise a two-temperature source.
7. They also show excess ultraviolet light which can be attributed to the more active chromosphere.
8. Emission at soft x-ray wavelengths is about 10^4 times stronger than in the Sun, because of a corona that is about 10^7 K compared to the Sun's 10^6 K.
9. Flare-like radio emission at centimeter wavelengths can be stronger by a factor of 10^7 compared to similar radio flaring observed in the Sun.
10. Magnetic fields of several *thousand* G have been measured by specialized techniques or deduced indirectly (e.g., from observed radio emission), far greater than the Sun's global magnetic field of only 1 or 2 G.
11. Differential rotation (which means parts of a stellar atmosphere at different latitudes actually rotate at different rates) inferred from systematic photometry is reminiscent of that in the Sun, though the difference in rotation rate per degree of latitude can be 300 times less (i.e., approaching rigid-body rotation) and in some stars might have the opposite sign.
12. Long-term cycles analogous to the 11-yr solar cycle, suggested by long-term photometric studies of starspot coverage and mean brightness levels, appear to show the same approximate time scale.
13. Mass loss in a stellar wind, inferred from orbital period changes and also from mass determinations and evolutionary considerations, is about 10^4 times greater than the $10^{-13}\ M_\odot\ \text{yr}^{-1}$ solar wind.

Nowadays these characteristics are usually referred to collectively as chromospheric activity. Stars displaying them more strongly than the Sun does are called chromospherically active. It is important to realize that the phenomenon of chromospheric activity occurs in stars at several quite different stages of evolution and is found in both single and binary stars. We find it in the following 10 groups of stars:

1. RS CVn binaries of the short-period, regular, and long-period subgroups
2. Stars like AY Ceti, 29 Dra, and HD 185510 which might be considered RS CVn binaries but which, because of their white dwarf secondaries, fail the formal definition by a technicality
3. BY Dra variables, some of which are in binaries and some of which are single
4. Flare star (or UV Ceti) variables, which are found generally to overlap with the BY Dra variables
5. Solar-type (mostly G) single dwarfs
6. The pre–main-sequence T Tau variables
7. The contact W UMa binaries
8. FK Com stars
9. Single, rapidly rotating giants with chromospheric activity not strong enough to qualify them as FK Com stars
10. The cool contact component in semidetached Algol-type binaries

The physical mechanism responsible for extreme chromospheric activity is understood only in broad strokes. Inference from observational material points to two essential ingredients: a deep convective outer envelope and rapid rotation. A 1977 study by Bernard W. Bopp and Francis C. Fekel was decisive, because these authors examined the BY Dra variables which, unlike the RS CVn binaries, include both single and binary stars. So we came to know that binarity per se is not directly responsible for chromospheric activity. A star rotating synchronously in a binary with a short orbital period will rotate faster than the same star would if single. Although quantitative measures of some signatures of chromospheric activity appear to correlate well with rotational period alone, others differ little over the full range of observed rotational periods, from less than a day to over a year. Equatorial rotational velocity (units of kilometers per second) is a useful parameter that includes stellar radius (hence depth of convective zone) as well as rotational period. A rotational velocity of about 5 km s^{-1} seems to be a threshold for the onset of extreme chromospheric activity.

Most theoretical models of chromospheric activity involve some form of stellar dynamo. A typical dynamo model will include as parameters some sort of rotation rate and some scale to measure convection depth, both of which were mentioned previously, but also differential rotation, which was not. It is therefore curious, and probably important, that the RS CVn binaries have much weaker differential rotation (as a function of latitude) but *not* correspondingly weaker chromospheric activity. It may be that differential rotation as a function of radial depth down into the active star is the effective parameter, and that the active star in an RS CVn binary is a nearly rigid rotator only on its surface. HR 1362 and HD 181943 are two chromospherically active single giant stars with quite long rotation periods (slightly over 1 yr) which provide support for this possibility. They probably evolved from rapidly rotating early-type main-sequence stars, slowed down as they expanded to leave the main sequence, but preserved their rapidly rotating cores and are experiencing strong differential rotation at the core–envelope interface.

EVOLUTION

The RS CVn binaries were defined in such a way that they do not necessarily comprise a single evolutionary state. Nevertheless, most of them do. The average mass of the active star is 1.3 $M_\odot$ and the mass ratio of the two stars is typically not far from unity. Presumably, one star was slightly more massive than the other, left the main sequence first, is now a subgiant, but has not yet filled up its Roche lobe and hence is not expected to be experiencing bona fide mass transfer as the semidetached Algol-type binaries are. If it is now slightly less massive than the other star, its enhanced stellar wind was probably the cause. Whereas a 1.3 $M_\odot$ main-sequence star would rotate slowly, like the Sun with its 25-day period, this star rotating synchronously in a binary with an orbital period of only a few days would be a rapid rotator. Moreover, as a subgiant, it would now have a deep convective envelope. Eventually, it should fill up its Roche lobe and begin mass transfer, which should occur on the catastrophic dynamical time scale (days) if it is still the more massive of the two stars but should occur on the more leisurely thermal time scale (about 10^7 yr) or on the glacially slow nuclear time scale (about 10^{10} yr) if it is the less massive. A few of the known RS CVn binaries (e.g., RZ Cnc, RT Lac, and AR Mon) apparently are in a slow mass transfer stage now.

The short-period and long-period RS CVn binaries are not profoundly different. The former are somewhat less massive (active star around 0.9 $M_\odot$) and less evolved, perhaps not evolved at all. The latter are somewhat more massive (active star around 2 $M_\odot$) and, because of their larger separations, the more massive star can expand to a giant (as opposed to a subgiant) without filling up its Roche lobe.

STARSPOTS

A typical starspot in an RS CVn binary is a large dark region covering an appreciable fraction of one hemisphere of the surface of the chromospherically active star; the record-holder is II Peg, with its spot covering 50% in 1986. A spot is darker because it is cooler than the surrounding photosphere by about 1000 or 2000 K, as measured by spectrophotometry and by multi-bandpass photometry. As the spotted star rotates, it produces a roughly sinusoidal (wave-like) light variation, the period of which indicates the star's rotation rate. This has been called the photometric distortion wave, or simply the "wave." A very high percentage (around 90%) of known RS CVn binaries show photometrically detectable waves.

When the two stars in a binary are relatively close together, we expect (and, indeed, we find) synchronous rotation. That means the star's rotation period equals the binary's orbital period. If the spotted star undergoes differential rotation, then spots at different latitudes will rotate at slightly different rates and cannot exactly equal the binary's orbital period even if the average rotation is synchronous. As a result, when the light curve is plotted versus orbital phase, the wave will move or "migrate" in phase as time goes on. The time required for the wave to migrate one complete orbital cycle is called the "migration period." We have the relation

$$1/P_{\mathrm{migr}} = 1/P_{\mathrm{rot}} - 1/P_{\mathrm{orb}}.$$

Different values of P_{rot} are seen at different times in the same star; from this we deduce that spots are occurring at different latitudes and that differential rotation is present. Systematic photometry can show how long a given spot maintains its identity. From this we infer spot lifetimes, which seem to range from somewhat less than 1 yr to several years but rarely longer than a decade. This is a curious time scale, longer than the lifetime of a large sunspot group (1 month) but shorter than the solar cycle (11 yr).

There is growing suspicion that chromospherically active stars have a four-sector structure containing two preferential longitude zones 180° apart in which a spotted region can develop, rotate at the rate dictated by its latitude, and then dissolve when its lifetime is over.

ROTATION PERIODS AND ORBITAL PERIODS

Photometry of spotted stars is very useful in determining accurate rotation periods. Results show that most stars in binaries rotate synchronously. In circular orbits, synchronism means $P_{\mathrm{rot}} = P_{\mathrm{orb}}$. In eccentric orbits we must talk about pseudosynchronism, a situation where the star matches its rotation rate more nearly with the effective orbital rate at periastron. Thus, when $e > 0$, $P_{\mathrm{rot}} < P_{\mathrm{orb}}$. According to most theories, synchronization or pseudosynchronization occurs on a time scale inversely proportional to the star's relative radius, r, raised to some large power, where

$$r = R/a.$$

Here R is the star's radius and a is the semimajor axis of the binary orbit. Moreover, convective stars synchronize more quickly than radiative stars. Thus asynchronism should be found only in stars that have a very small r and/or are very young. Only about 10% of known RS CVn binaries are not rotating synchronously or pseudosynchronously. In virtually all of the exceptions, the spotted star is a giant that apparently has recently evolved off the main sequence, slowed its rotation rate to conserve angular momentum, and not had time to resynchronize. A short orbital period by itself is *not* a good predictor of synchronization, unless stars of equal radius are compared. For example, the dwarf in HR 45088 (with $P_{\mathrm{orb}} = 7.0$ days) rotates *asynchronously*, whereas the giant in HR 7428 (with $P_{\mathrm{orb}} = 108.6$ days) rotates *synchronously*.

Accurate times of minimum brightness in those RS CVn binaries that eclipse can show how the orbital period varies. In those that show variable orbital periods, the majority (like RT And, AR Lac, and SZ Psc) are speeding up, presumably because of mass loss in the enhanced stellar wind. Several (RS CVn itself is a good example) show period changes that alternate between decreases and increases on a time scale of decades. The most likely explanation is magnetic cycles that cause the active star to expand and contract. If spin-orbit coupling by tidal forces occurs on a sufficiently short time scale, the resulting changes in rotational angular momentum produce corresponding changes in the orbital angular momentum and hence the orbital period.

Additional Reading

Baliunas, S. L. and Vaughan, A. H. (1985). Stellar activity cycles. *Ann. Rev. Astron. Ap.* **23** 379.

Hall, D. S. (1983). Starspots. *Astronomy* **11** (No. 2) 66.

Hall, D. S. (1987). Variability in chromospherically active stars—at all time scales. *Publications of the Astronomical Institute of the Czechoslovak Academy of Science* **70** 77.

Linsky, J. L. (1980). Stellar chromospheres. *Ann. Rev. Astron. Ap.* **18** 439.

Linsky, J. L. and Stencel, R. E., eds. (1987). *Cool Stars, Stellar Systems, and the Sun*. Springer, Berlin.

Strassmeier, K. G., Hall, D. S., Zeilik, M., Nelson, E., Eker, Z., and Fekel, F. C. (1988). A catalogue of chromospherically active binary stars. *Astron. Ap. Suppl.* **72** 291.

See also **Stars, Activity and Starspots; Stars, Atmospheres, Turbulence and Convection; Stars, Atmospheres, X-Ray Emission.**

Binary Stars, Semidetached

Jean-Pierre De Greve

Semidetached binary stars are systems in which one of the components fills its critical Roche lobe. In the literature these systems are identified with the most abundant group of this type, the Algol systems. Photometrically, these systems show deep primary and shallow secondary eclipses. They usually consist of a massive and unevolved primary (a main-sequence star) with a low-mass evolved secondary (a subgiant or giant star). Some classical examples of semidetached stars are u Her, β Per (Algol), V356 Sgr, and μ^1 Sco. Up to now some 260 systems are known of this type. This implies that one in a thousand stars brighter than a given magnitude is an Algol system. The other semidetached systems are classified into other groups, according to additional characteristics. For example, low-mass x-ray binaries are also semidetached systems, with the emission of x-rays powered by mass transfer from the Roche lobe filling component toward a compact star (neutron star or black hole). Another prominent example is the group of the cataclysmic variables, short-period systems, where the late-type component fills its Roche lobe and is transferring mass to a white dwarf star.

HISTORY

The prototype of the semidetached systems is Algol (β Persei). Its brightness variations were first discovered in 1667 by the Italian astronomer Geminiano Montanari (1633–1687), but their explanation, as a result of eclipses of one component of a double system by the other with a periodicity of 2.87 days, was suggested only in 1783 by John Goodricke, and finally confirmed by spectroscopic observations at the end of the nineteenth century.

The contradiction of the existence of Algol systems with our knowledge of stellar evolution was formulated in 1950 by P. P. Parenago as the so-called Algol paradox: "The more massive a star, the faster it evolves. Why then is the less massive star in these binaries the more evolved component?" The answer was an ingenious hypothesis; based on the existence of a critical common equipotential surface around the two components, the Roche volume, intersecting in a saddle point, called the first Lagrangian point. Equating each volume to a sphere leads to the definition of corresponding critical radii, the Roche radii. With that knowledge, J. A. Crawford proposed in 1955 that the now less massive star was originally the more massive one. While the stars are evolving, they grow larger and larger. At a given moment, the fastest evolving component (the more massive one) will exceed its Roche radius and lose mass toward its companion. The large and rapid mass transfer causes a reversal of the mass ratio and leads to a system such as Algol. In the following discussion, the process of mass transfer in a semidetached system will be abbreviated as RLOF (Roche lobe overflow). As usual, the now more massive (and most luminous) component—the mass-accreting star—is called the primary.

OBSERVATIONS

The conclusion for a semidetached nature of an observed close binary system depends on the evidence that one component (up to now the less massive one) is close to the limit of dynamical instability. Supporting data come from the combined analysis of light curves and velocity curves. Additionally, features in the spectra, such as emission effects in the Balmer lines, point to circumstellar matter in the system. Ultraviolet spectra of hot components of semidetached systems, mostly taken by the *IUE* satellite (*International Ultraviolet Explorer*), show signs of activities such as gas streams and circumstellar matter. They are revealed by the abnormal and variable strength of resonance lines such as Si IV. Such variations as well as those of the C IV resonance lines are, for example, found in the system U Sge. In the infrared region, distortions of the continuum in comparison with that of free gaseous components are also signs of mass transfer activity.

U Cep, a classical semidetached binary, has been long known for its high level of activity in the optical. It consists of a B7 V and a G8 III star, with respective masses of 4.4 and 2.9 $M_\odot$. The period is 2.49 days. From the interpretation of the radial velocities of the circumstellar absorption lines emanates the picture of a gas stream encircling the B star. According to this model, most of the matter leaves the system after orbiting 3/4 of the B star. The G star also shows a strong carbon deficiency.

β Lyrae is another well-known active system. In this system, the underluminous secondary is probably embedded in a thick disk. The W Serpentis stars, a class of semidetached systems with intermediate periods (13–605 days), consist of a cool star with a hot companion. The latter is surrounded by an accretion disk. In most cases the spectrum in the ultraviolet shows strong emission lines on a hot continuum, whereas in the optical a cool continuum is observed. Examples are W Ser, SX Cas, RX Cas, RW Per, and V367 Cyg. Of these systems, SX Cas has been studied intensively, resulting in a model with a hot spot on the hot component, fed by an accretion ring.

GENERAL CHARACTERISTICS

As a whole, the group of observed semidetached systems shows the following statistical properties:

1. There is a single peaked distribution of the mass ratio ($q^{-1} = M_1/M_2$). The pronounced maximum is found in the interval 0.2–0.3.
2. There is no distinction between low- and high-mass systems with regards to the observational characteristics.
3. After correction for selection effects, the period distribution is shifted to smaller values than observed for detached systems, indicating a loss of angular momentum (and mass) during the mass transfer.
4. There is a general trend to low-mass ratios at longer periods, in accordance with the expectations of the theory of mass transfer, although many serious discrepancies are noticed.
5. The mass ratio q increases with increasing mass of the components. This is a natural result of the decrease of the fraction of transferred matter with increasing initial mass. In turn, this is a consequence of the proportional increase of the initial convective core with increasing mass.
6. Except for massive systems with a short period and a high-mass ratio (e.g., SX Aur, IU Aur, δ Pic, V Pup, μ^1 Sco), all

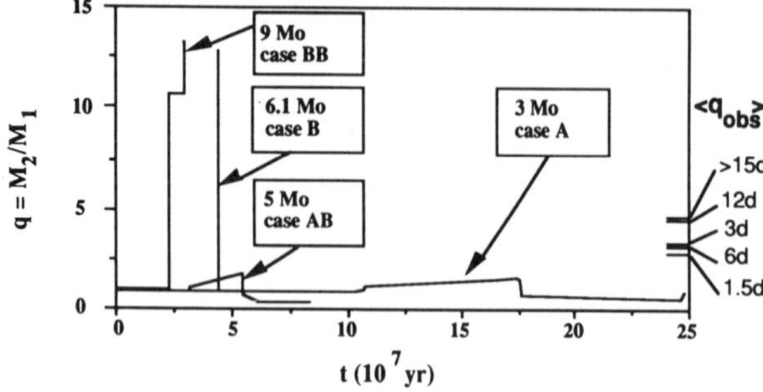

Figure 1. Evolution of the mass ratio with time for different cases of mass transfer and different systems. The mass of the primary star is indicated. Average mass ratios of groups of observed semidetached systems are shown at the right. The groups have periods in a 20% range around respectively 1.5, 3, 6, and 12 days and periods larger than 15 days, respectively. (*By permission from ESO.*)

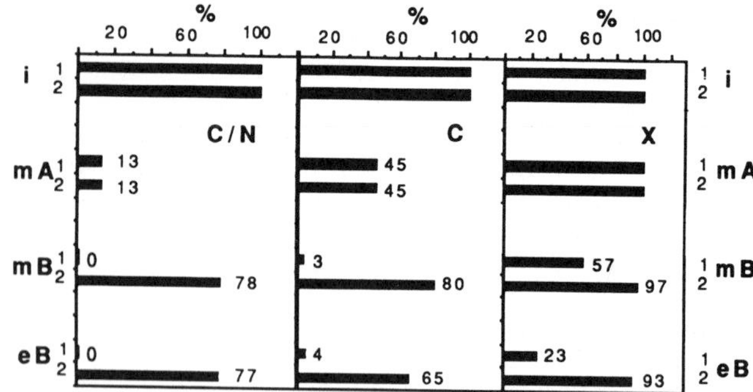

Figure 2. Proportional variation of the surface abundances of carbon-to-nitrogen (C/N), carbon (C), and hydrogen (X), during case AB of mass transfer of a system with masses 9.2 $M_{\odot}$ + 8.5 $M_{\odot}$, P = 4.48 days, assuming thermohaline mixing in the gainer when hydrogen-depleted layers are accreted by this star. The number 1 refers to the initially more massive star; the number 2 refers to the mass-gaining star. The following phases are given: i = initial state, mA = mid case A, mB = mid case B, eB = end case B.

mass-losing stars are oversized with respect to their mass. Generally, the mass-gaining stars are not undersized.

7. From a sample of 244 candidate stars emerges the following "average" system: P = 7.5 days, M(primary) = 3.3 $M_{\odot}$, R(primary) = 2.3 $M_{\odot}$, q = M(secondary)/M(primary) = 0.28.

THEORETICAL MODELS

The semidetached systems that we observe are interactive binaries that have reversed their mass ratio. To compute their evolution, one must adopt a number of hypotheses and simplifications regarding the mass transfer. The most important one is the conservation of mass and orbital angular momentum. Most of the theoretical modeling has been based on it. It leads to a predicted evolution of the mass ratio with time, as shown in Fig. 1. The variation depends on the initial values of the masses and the time that the mass transfer starts (determined by the initial period).

When the initial period is very short (1–2 days), the mass ratio remains bound to values smaller than 2 as a result of a reversal of the mass transfer. For longer initial periods, large values (of the order of tens) are reached near the end of the mass transfer (in the conservative case!). However, the expected final values show a clear discrepancy with the values of observed semidetached binaries. In Fig. 1, the average ratios are given, derived for groups of observed semidetached systems, with periods in a 20% interval around respectively 1.5, 3, 6, and 12 days and for all periods larger than 15 days. They are shown at the right of the figure. This correlation between q and period indicates a nonconservative character of the mass transfer process (although the weak dependence of q on P may indicate that this process does have a conservative part, probably during the slow phase), or that a system at the end of the process remains merely unobserved as a semidetached system. Except for very few individual systems such as AS Eri or TV Cas, for which specific models have been proposed, all evolutionary computations were performed for a general description of the evolution and the dimensions of the Algol binaries. The main reason is the need for a very accurate determination of the absolute dimensions of the individual systems. Moreover, the nonconservative mass transfer introduces additional unknown parameters into the treatment of the mass transfer process and the derivation of the initial dimensions.

The moment that the original less massive component will swell up after accretion and in turn start losing mass depends on the initial masses and period, and the way the mass transfer is treated (conservative or nonconservative). Especially if the period is short (so-called mid or late case A) and the initial mass ratio close to 1, the evolution of the system takes a different course. Successive phases of semidetached mass transfer of comparable duration occur, with opposite directions of the mass stream. The occurrence of such evolution is basically determined by the competition between two mechanisms:

1. The increase of the central hydrogen content of the mass-accreting star, as a result of the growing convective core. The star gets younger [it evolves closer to the zero-age main sequence (ZAMS), apart from a radius excess due to disruption of thermal equilibrium during the fast phase.]
2. Nuclear burning accelerates as the accreting star grows more and more massive. Hence, although fresh hydrogen is mixed in during accretion, the core hydrogen content will decrease faster and faster. If the initial period is shorter than 1–2 days, the second effect dominates and reversed mass transfer occurs while both components are still main-sequence stars. However, this type of semidetached binary may be quite rare, because the number of early-case A progenitors is extremely small for spectral types B and A. Among the observed semidetached systems, SX Aurigae seems to be the best candidate.

Analyses of high-dispersion spectra also reveal carbon (C) and nitrogen (N) anomalies in the chemical composition of the surface. Computations of close binary evolution predict that the mass transfer affects the surface composition of the components, as nuclear processed layers appear at the surface of the mass loser and are (partly) accreted by the other component. After accretion, mixing of such modified matter with underlying layers may result in a different chemical composition at the surface of the two stars.

Figure 2 shows typical percentage fractions of hydrogen, carbon, and carbon-to-nitrogen, expected at various stages of the mass transfer. Verification of these predictions form an important key to understanding the details of the mass transfer process. They will also enhance our understanding of the inner structure of the stars. For a number of systems rough quantitative estimates of the C and C/N abundances are readily obtained from high-dispersion ultraviolet spectra of the mass-gaining star. They reveal abundances lower than the cosmic values (i.e., the initial surface abundances), but larger than expected if the transferred matter is accreted without any mixing. This agrees with the predictions of the theoretical models (cf. Fig. 2).

CONCLUSION

Semidetached stars form the observable evidence of evolution of close binaries by mass transfer. Their importance in astrophysics comes first of all from the fact that many groups of interesting binaries have reached their present state passing through a semidetached phase. But they also offer an opportunity to look into the interior of a star, as the mass-losing star exposes nuclearly processed layers. Furthermore, as it is now readily accepted that the mass transfer is nonconservative, these systems contribute to the replenishment and modification of the interstellar medium. Until now, the importance of the latter has barely been addressed quantitatively, because of a lack of research on their space density or galactic distribution.

Additional Reading

Batten, A. H. (1988). Some aspects of a century of spectroscopic-binary studies. *Publ. Astron. Soc. Pacific* **100** 160.

Batten, A. H. (1989). Algols. *Space Sci. Rev.* **50** 1.

De Greve, J. P. (1986). Semidetached systems: Evolutionary viewpoints and observational constraints. *Space Sci. Rev.* **43** 139.

Plavec, M. (1973). Evolutionary aspects of circumstellar matter in binary systems. In *Extended Atmospheres and Circumstellar Matter in Spectroscopic Binary Systems*, A. H. Batten, ed. D. Reidel Publishing Company, Dordrecht, p. 216.

Sahade, J. and Wood, F. B. (1978). *Interacting Binary Stars*. Pergamon Press, Oxford.

See also **Stellar Evolution, Binary Systems.**

Binary Stars, Spectroscopic

Anne B. Underhill

Spectroscopic binaries are stars that appear to be single in the telescope, but are observed to be moving in a periodic manner backward and forward along the line of sight. This motion is inferred from the fact that their spectral lines periodically shift longward and shortward in wavelength. Such wavelength shifts are a result of the Doppler effect.

When a star has a component of motion away from an observer, the apparent wavelength of each line in the spectrum of the star shifts longward by an amount $\Delta\lambda$,

$$\Delta\lambda(t) = \lambda(t) - \lambda_0. \tag{1}$$

Here $\lambda(t)$ is the wavelength of the spectral line measured at time t and λ_0 is the wavelength of the line in a source that is stationary relative to the observer. A negative shift, that is, a shift to a shorter wavelength than λ_0, indicates a component of velocity along the line of sight toward the observer. The component of velocity along the line of sight is often called the radial velocity of the star. It is

$$v(t) = (c/\lambda_0)\,\Delta\lambda(t) \text{ km s}^{-1}. \tag{2}$$

In Eq. (2), c denotes the speed of light, 299,790 km s^{-1}.

Single-line spectroscopic binaries, SB1, show the spectrum of one star only. Double-lined spectroscopic binaries, SB2, show the spectra of two stars superposed. The two sets of spectral lines are observed to move half a period out of phase with each other. When the spectrum of star 1 has its largest displacement longward, the spectrum of star 2 has its largest displacement shortward. In the case of SB2, the brightnesses of the two stars are comparable in the observed spectral region, that is, usually within a factor of 1–5. A few stars have been observed to show the spectra of three stars at one time.

MOTION IN AN ELLIPSE

Two bodies that are moving relative to each other under the action of a force that varies as r^{-2}, where r is the distance separating the bodies, will move along a conic section. In our case, the active force is the gravitational attraction of one star for the other. Periodic motion is along an ellipse. Each star moves along an ellipse with the center of gravity of the two stars at one focus. When no perturbing forces act, the periodic motion of each star is such that the areal velocity is constant. The areal velocity is $\frac{1}{2}r^2\,d\theta/dt$, where θ is the angle at time t measured in the plane of the orbit from the line of apsides.

The plane of the orbit of a pair of stars, in general, is inclined at an angle i to the plane of the sky. Consideration of the geometry of the situation shows that the instantaneous radial velocity of a star moving in an elliptical orbit is given by

$$v(t) = v_0 + K[e\cos\omega + \cos(\theta+\omega)]. \tag{3}$$

Here v_0 is the line-of-sight component of velocity of the system with respect to the Sun, that is, the systemic velocity, K is the semiamplitude of the radial-velocity changes of the star, e is the eccentricity of the ellipse, and ω is the longitude of periastron. The geometry of binary star orbits is fully described in books by Aitken and Smart (see Additional Reading).

From the geometry of the orbit, we find

$$K = \frac{2\pi a\sin i}{P(1-e^2)^{1/2}}. \tag{4}$$

Here a is the semimajor axis of the ellipse in linear measure. By fitting the observed radial velocities as a periodic function of time to Eq. (3), one can solve for the orbital elements P, e, a, and ω. This solution is carried out by a numerical method that improves a set of preliminary elements found by one of the graphical methods explained in the Aitken and Smart books.

ORBITAL ELEMENTS AND STELLAR MASSES

The projected semimajor axis of the orbit of a binary star can be found from Eq. (4) once the values of P, e, and K have been determined from the observations. When P is expressed in days and K in kilometers per second, we have

$$a\sin i = 1.3751\times10^4 KP(1-e^2)^{1/2}\text{ km}. \tag{5}$$

This length is sometimes expressed in terms of $R_\odot$, the solar radius. From Kepler's third law we know that

$$M_1 + M_2 = (a_1+a_2)^3/P^2. \tag{6}$$

Here a_1 and a_2 are the semimajor axes of the relative orbits of stars 1 and 2 expressed in astronomical units, M_1 and M_2 are the masses expressed in solar masses, and P is the period expressed in years.

When the spectra of both stars have been observed, we can find a_1/a_2, which is equal to K_1/K_2. In the case that one sees only one spectrum, one can find the value of the mass function $f(m)$:

$$f(m) = \frac{(M_2\sin i)^3}{(M_1+M_2)^2} = 1.0385\times10^{-7}K_1^3P(1-e^2)^{3/2}\,M_\odot. \tag{7}$$

When two spectra are observed, minimum masses for both stars can be determined:

$$M_1 \sin^3 i = 1.0385 \times 10^{-7} K_2 (K_1 + K_2)^2 P (1 - e^2)^{3/2} \, M_\odot \quad (8)$$

and

$$M_2 \sin^3 i = 1.0385 \times 10^{-7} K_1 (K_1 + K_2)^2 P (1 - e^2)^{3/2} \, M_\odot. \quad (9)$$

In Eqs. (7), (8), and (9), P is in days and K is in kilometers per second.

For those double-lined spectroscopic binaries for which the inclination, i, is known from the solution of a photometric orbit or from measured polarization changes, absolute values of the mass can be found. In the case of single-lined spectroscopic binaries, masses may be inferred from the mass function in terms of an adopted value for the parameter $q = M_2/M_1$. If q and i can be estimated from external information, one can then find M_1 and M_2.

RESULTS FROM SPECTROSCOPIC BINARIES

Orbital elements for 1469 spectroscopic binaries are contained in the most recent catalog of spectroscopic binaries, by A. H. Batten, J. M. Fletcher, and D. G. MacCarthy. Each system is discussed critically there. Information like this is the source of most of the presently available information about the masses of stars. P. Harmanec (1988) has updated the list of stars by D. M. Popper (1980) which have known masses, radii, and luminosities.

Few spectroscopic binaries behave exactly as the theory for two mass points attracting each other gravitationally indicates they should. Precision stellar masses are obtained only from well-behaved systems; see the article by Popper, listed in Additional Readings.

Problems may arise because of unresolved line blends in the spectra and because the selected stellar spectral lines do not move with the center of mass of the star. Streams of gas in the system are sometimes detected.

Examples of radial-velocity curves of the stars in the double-lined spectroscopic binary V444 Cygni are shown in Figs. 1 and 2. This system has a circular orbit, thus $e = 0.00$. In Figs. 1 and 2 the observed radial velocity is plotted as a function of the radial-velocity phase in the orbit. The radial-velocity phase is $(T - T_0)/P$ less

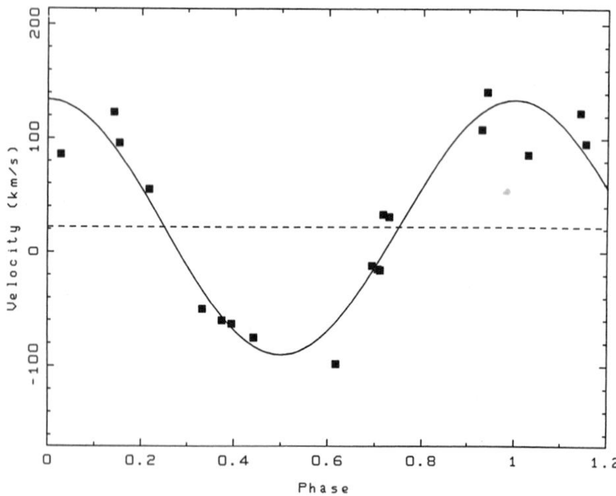

Figure 1. The radial velocity of the O6 star in V444 Cygni shown by the O III absorption line at 5592 Å. The radial-velocity phase is plotted as abscissa; see text. [*Reproduced courtesy of the Publications of the Astronomical Society of the Pacific.*]

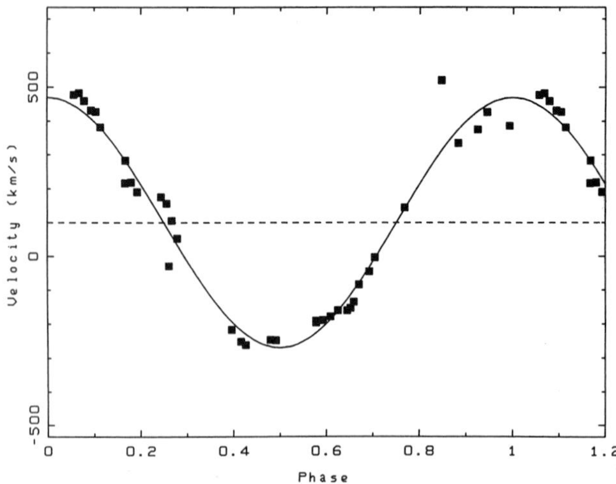

Figure 2. The radial velocity of the WN5 star in V444 Cygni shown by the He II emission line at 5411 Å. The radial-velocity phase is plotted as abscissa; see text. [*Reproduced courtesy of the Publications of the Astronomical Society of the Pacific.*]

an integral number of cycles. Here T_0 is a Julian day at which the star reached maximum positive velocity and T is the Julian day of the observation. In Fig. 1, T_0 = JD2447048.0869, whereas in Fig. 2 it is JD2447045.7662. These values of T_0 were found by the computer from the observations. Because the orbit is circular, they should differ by $\frac{1}{2}P$. In the case of V444 Cygni, $P = 4.212424$ day, but the difference is 2.3207 day. The center of light for the He II emission line leads the position of the Wolf–Rayet star in the presumed circular orbit. The systemic velocity found by the computer is shown by a dashed line in Figs. 1 and 2.

From Figs. 1 and 2 one sees for the radial-velocity amplitude K that K(WR star) ≥ K(O star). Hence we infer that the Wolf–Rayet star is less massive than the O6 star in the system. In the case of V444 Cygni and of other systems that include a Wolf–Rayet star, the K_{WR} depends on the line measured!

The systemic velocity indicated by the O6 star is +22 km s^{-1} whereas that indicated by the He II lines of the Wolf–Rayet star is +100 km s^{-1}. This longward shift of the He II emission lines is significant. Such shifts for the He I and He II emission lines in Wolf–Rayet spectroscopic binaries have been known for many years. Their interpretation is not agreed upon.

An interpretation of these discrepancies with the theory of spectroscopic binaries in the case of V444 Cygni can be found in the paper by A. B. Underhill, S. Yang, and G. M. Hill (1988). The resolution of discrepancies with straightforward theory may lead to new information about the stars in binary systems.

Additional Reading

Aitken, R. G. (1935). *Binary Stars*, 2nd ed., Chap. 6. McGraw Hill, New York.

Batten, A. H., Fletcher, J. M., and MacCarthy, D. G. (1989). *Eighth Catalogue of the Orbital Elements of Spectroscopic Binary Systems. Pub. Dominion Ap. Obs.* **17**.

Harmenec, P. (1988). *Bull. Astron. Inst. Czechoslovakia* **39** 329.

Popper, D. M. (1980). Stellar masses. *Ann. Rev. Astron. Ap.* **18** 115.

Smart, W. M. (1949). *A Textbook on Spherical Astronomy*, 4th ed., Chap. 14. Cambridge University Press, Cambridge.

Underhill, A. B., Yang, S., and Hill, G. M. (1988). *Pub. Astron. Soc. Pacific* **100** 741.

See also **Binary and Multiple Stars; Types and Statistics; Stars, Masses, Luminosities and Radii.**

Binary Stars, Theory of Mass Loss and Transfer in Close Systems

Ronald F. Webbink

In a close binary star system, two stars orbit so close to each other that, through tides, stellar winds, or even irradiation, one star may profoundly influence, and be influenced by, the evolution of its companion star. Objects and phenomena are produced in close binaries that are never found among isolated, single stars—extraordinarily low-mass giant-branch stars in Algol binaries, x-ray binaries, novae, and many other peculiar stars.

In order to see how the evolution of a star may be altered by the presence of a companion, it is useful first to consider the gravitational effects of each star on the other. To do so, one takes advantage of the fact that stars have internal mass distributions that are highly concentrated toward their centers—to a first approximation their gravitational fields are those of point masses.

ROCHE LOBES

Consider two point masses, M_1 and M_2, separated by a distance A, and in circular orbit about their common center of mass. A test particle with velocity $\mathbf{v}$ in a coordinate frame corotating with the binary experiences an acceleration

$$\frac{d\mathbf{v}}{dt} = -\frac{GM_1}{r_1^3}\mathbf{r}_1 - \frac{GM_2}{r_2^3}\mathbf{r}_2 - \boldsymbol{\Omega}\times(\boldsymbol{\Omega}\times\mathbf{r}) - 2\boldsymbol{\Omega}\times\mathbf{v},$$

where $\mathbf{r}_1$ and $\mathbf{r}_2$ are vectors from masses 1 and 2, respectively, to the test particle, $\mathbf{r}$ is a vector from the center of mass of the binary to the test particle, $\boldsymbol{\Omega} = [G(M_1 + M_2)/A^3]^{1/2}\hat{\mathbf{z}}$ is the angular velocity vector of the rotating coordinate frame (in the $+\hat{\mathbf{z}}$ direction), and G is the universal gravitational constant. The scalar (dot) product of this acceleration with the particle velocity $\mathbf{v}$ is the time rate of change of the kinetic energy per unit mass of the test particle in the corotating frame. Because Coriolis forces (represented by the final term in the preceding force equation) do no work on the particle, the time integral of the scalar product is a function of position only; that is, the kinetic energy per unit mass of the test particle (but not its momentum) can be derived from the potential

$$\Psi = -\frac{GM_1}{r_1} - \frac{GM_2}{r_2} - \frac{G(M_1+M_2)}{A^3}\frac{\rho^2}{2},$$

where ρ is the distance of the test particle from the orbital axis of the binary. Surfaces on which Ψ is constant correspond to surfaces of constant potential energy, known as *Roche equipotentials*. Real stars, though not point masses, are so centrally concentrated that Roche equipotentials are excellent approximations of their equilibrium shapes in binary systems with circular orbits.

There exist five points (the *Lagrangian points*), all in the orbital plane, and conventionally denoted as $L_1, \ldots, L_5$, where the gradient of this potential vanishes (see Fig. 1). At L_1, the *inner Lagrangian point*, the gravitational acceleration due to the lesser mass, augmented by centrifugal terms, is balanced by that of the greater mass. The equipotential passing through this point, the *inner critical surface*, is the innermost one common to both stars. The volume about each mass bounded by this equipotential is the *Roche lobe* of that mass. At the two points L_2 and L_3, the *outer Lagrangian points*, the combined gravity of the two orbiting masses is balanced by centrifugal forces. The equipotentials passing through these points, the *outer critical surfaces*, open to infinity. The equilateral points, L_4 and L_5, are of interest in solar system dynamics (marking the positions of the Trojan asteroids in the Sun–Jupiter system, for example), but are of minor interest among interacting binary stars.

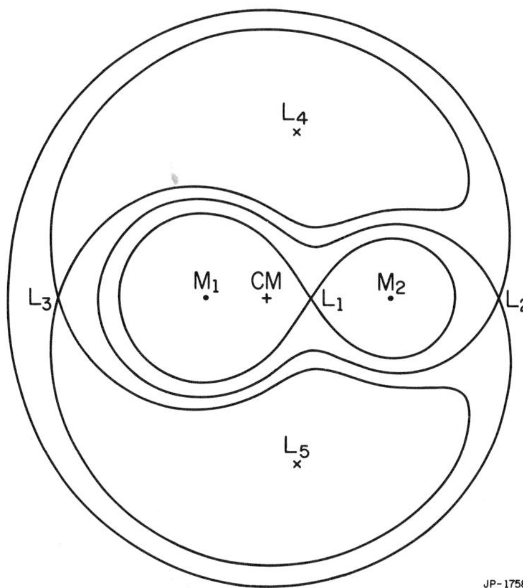

Figure 1. A cross section through the orbital plane of a close binary system, showing the topology of the inner and outer critical equipotential surfaces and the locations of the Lagrangian points, $L_1, \ldots, L_5$. In this example, the mass M_1 is twice as great as M_2, and the center of mass of the system is marked *CM*.

The existence of common closed equipotential surfaces encompassing both stars opens the possibility of mass exchange between those stars. If a star just fills its Roche lobe, matter at the L_1 points is neutrally stable: Matter driven beyond that point by expansion of that star, or by contraction of its Roche lobe, becomes dynamically unstable and will cascade toward the companion star.

The size of the Roche lobe of a star is itself affected by mass exchange between stars in a binary system and by mass and angular momentum losses from the system. Measured relative to the orbital separation, it is a function only of the ratio of masses of the two stars in the system; the larger the relative mass of a star, the larger is its Roche lobe in terms of the orbital separation. However, the separation itself, $A = J^2(M_1 + M_2)/GM_1^2M_2^2$, is affected by mass exchange or systemic losses: It is an increasing function of orbital angular momentum (J), a decreasing function of total mass, and at constant total mass and orbital angular momentum is minimized when the stellar masses are equal. The Roche lobe of the donor star in a binary system thus tends to decrease in size as that star loses mass, until the star has been reduced to a mass slightly smaller than that of its companion; thereafter, further mass transfer leads to Roche lobe expansion, provided that angular momentum losses are negligible.

MASS TRANSFER

The evolution of isolated single stars can be characterized as consisting of long phases of expansion, driven by nuclear burning either in the core or in outwardly advancing shells, punctuated by occasional phases of rapid contraction (e.g., upon helium ignition in low- and intermediate-mass stars). In a close binary system, this expansion is limited by the Roche lobe enclosing that star; further expansion of a star drives a mass flow beyond the L_1 point to its companion star. Thus, in a collection of binary stars with various orbital periods or separations, mass transfer is most likely to be initiated not during the longest-lived evolutionary phases of the donor stars, but during those phases in which they undergo their most dramatic expansion: when they cross the Hertzsprung gap (i.e., expand from the main sequence to the base of the giant branch), ascend the giant branch, or ascend the asymptotic giant branch. Because stellar lifetimes decrease rapidly with increasing mass, it is invariably the case that, in any binary consisting of two normal (i.e., nondegenerate) stars, it is the more massive component that will first expand to fill its Roche lobe. Indeed, even among completely unevolved, zero-age main-sequence stars, the mass–radius relationship is such that the more massive component of a binary is the first to fill its lobe.

The rate at which mass flows from a lobe-filling star to its companion depends primarily upon the internal structure of the lobe-filling star. Just as there are three time scales characterizing the structure of individual stars—dynamical (sound travel), thermal (heat diffusion), and nuclear (fusion)—so these same three time scales may characterize the mass transfer process in a close binary system. *Dynamical* time scale mass transfer occurs when the adiabatic expansion of the interior of a mass-losing star pushes the newly exposed stellar surface beyond the confines of that star's Roche lobe. The radius of the star may thus substantially exceed that of its Roche lobe, and mass in this outer envelope thus flows freely, at sonic velocity (i.e., on a dynamical time scale) to the companion star, restrained only by the gravitational constriction of the flow near the L_1 point. This circumstance arises most often if the donor star (component 1 here) is not significantly less massive than the accreting star, and it possesses a deep convective envelope, or if it is degenerate. Mass transfer rates then arise which are dimensionally of order $\dot{M}_1 \sim -M_1\Omega[\ln(R_1/R_L)]^3$, where Ω is again the orbital angular frequency of the binary and R_L is the Roche lobe equivalent radius of the donor star. *Thermal* time scale mass transfer occurs when thermal relaxation (rather than hydrostatic readjustment alone) drives the expansion of the donor star beyond its Roche lobe. In this case, the mass loss process is self-regulating, as expansion driven by thermal relaxation competes against contraction driven by rapid mass loss. Mass transfer on this time scale, at rates dimensionally of order $\dot{M}_1 \sim -R_1\mathscr{L}_1/GM_1$, where $\mathscr{L}_1$ is the luminosity of the donor star, tends to occur when the donor star has a radiative envelope and is of comparable or somewhat greater mass than the accreting star. *Nuclear* time scale mass transfer occurs when a star continues to fill its Roche lobe only by virtue of its evolutionary expansion. This circumstance generally requires that the donor star be significantly less massive than the accretor, so that mass exchange produces an expansion of both the binary separation and the Roche lobe of the donor. Mass transfer rates are then of order $\dot{M}_1 \sim -M_1\dot{R}_1/R_1$. In other cases, slow mass transfer may also be driven by angular momentum losses from the binary system (e.g., through magnetically coupled stellar winds or by the emission of gravitational waves). Provided that the conditions for stability against dynamical and thermal time scale mass loss are fulfilled, the resultant transfer rates are of order $\dot{M}_1 \sim M_1\dot{J}/J$.

The response of the companion star to the influx of accreted matter depends in a similar way on the intrinsic time scales of that star. In general, the thermal and nuclear time scales of the accretor are longer than those of the donor, whereas its dynamical time scale is somewhat shorter. Thus the accretor is driven further from thermal equilibrium than the donor during rapid mass transfer and tends to expand rapidly where a donor contracts. However, from a physical standpoint, the two conditions (mass loss and mass accretion) are not altogether parallel: Matter being accreted is generally heated by the accretion process to much higher temperatures (and entropies) than those preexisting in that star's photosphere, and it may also carry a substantial angular momentum content, especially if the accretor resides well within its Roche lobe. The accretion of high-entropy gas exaggerates the tendency of the accretor to expand as mass accumulates and may produce a *contact* binary—a binary star in which both components fill their respective Roche lobes. The angular momentum of this gas imparts the accretor with very rapid rotation—enough in principle to bring that star to the point of nonaxisymmetric instability if the accretor increases in mass by a large fraction. Real binaries evidently avoid this instability by restoring much of the angular momentum content of accreted matter to the orbit on the accretion time scale, but how they do so is not yet understood.

BINARIES WITH COMMON ENVELOPES

The situation often arises in the course of mass transfer that both components of a binary simultaneously fill their Roche lobes. It is useful to distinguish two circumstances that may give rise to this configuration, depending upon the time scale on which contact is achieved.

If contact arises from thermal or nuclear time scale evolution, we may expect that frictional energy dissipation in the outer envelopes of the two stars is strong enough to enforce approximate synchronism, so that the Roche model remains an adequate description of the dynamical state of the system. A common envelope may then develop which extends beyond the bounds of the inner critical surface to an equipotential lying within the *outer critical surface* (the equipotential surface passing through the L_2 point). Matter can still flow from each component to the other in different parts of the common envelope, but far from the L_1 point that envelope is very near hydrostatic equilibrium. Near L_1, small differences in the structure of the envelopes of the two stars suffice to drive substantial mass fluxes in one or both directions between those stars. These currents, even if producing no net mass flow in either direction, can result in a net transfer of energy, as manifested by the near equality of surface temperatures of the components of W Ursae Majoris binaries. However, the details of the transfer mechanism remain elusive and controversial at present.

In dynamical time scale mass exchange, mass flow rates between components become so high that little accretion energy can be radiated from the system. A common envelope is then expected to form, but on so rapid a time scale that dissipation cannot enforce synchronism within that envelope. The Roche model thus fails, although the notion of a centrifugally bounded envelope may still be physically meaningful. The cores of the two stars find themselves embedded within a more slowly rotating envelope, which acts as a dissipative medium. Energy is extracted from the orbital motion of the embedded cores, causing them to spiral toward each other. This dissipated orbital energy inflates the envelope. In most systems containing at least one degenerate core, enough energy is in principle available from the orbit in this way to unbind the common envelope. The result is a relatively compact binary containing one, or even two, degenerate components. It is widely accepted that cataclysmic variables, for example, must originate in this way.

MASS TRANSFER REMNANTS

The properties of a remnant binary produced by mass exchange also depend upon the time scale for that exchange. Except where a donor star is able to regain thermal and dynamical equilibrium, and thus enter a phase of mass transfer driven by nuclear evolution itself, the transfer process is so rapid as to abort core evolution in that star when it first fills its Roche lobe. The donor remnant thus depends strongly on the evolutionary status of that star at the onset of mass transfer, and only weakly on the properties of its companion. Mass transfer typically continues at a high rate until a strong composition gradient or active nuclear-burning shell is reached, at which point it stops abruptly. (If the star initially possessed more than one nuclear energy source, a second, generally slower phase of mass transfer may subsequently occur as the inner source exhausts a central core.) In the alternative case, where thermal equilibrium can be restored in the donor before the envelope has been entirely stripped away, the properties of the donor remnant depend strongly on the final orbital angular momentum of the binary, which ultimately limits the size of its Roche lobe. Although the initial conditions required of such a binary may be more restrictive than most (a low-mass initial binary of nearly equal component masses, small enough in orbital separation to encounter mass transfer before the donor had first developed a degenerate core), these systems are the most commonly observed interacting binaries (the Algol-type binaries) on account of their long lifetimes in the interactive state.

WIND ACCRETION

In addition to the wide variety of binary systems that arise from mass transfer by Roche lobe overflow, capture of the stellar wind of one star by its companion, especially if that companion is a degenerate or compact object, can give rise to observable phenomena (e.g., massive x-ray binaries if the accretors are neutron stars or black holes, or symbiotic stars if the accretors are white dwarf stars). This process, also known as *Bondi–Hoyle accretion*, occurs by virtue of the gravitational focusing of the wind by the companion star. That fraction of the stellar wind passing within an *accretion radius* of the companion is accreted by that star. This accretion radius is of order $R_A \sim GM_2 / v_{\mathrm{rel}}^2$, where M_2 is the mass of the accretor and v_{rel} is the velocity of the wind in the rest frame of the accretor. (It is assumed that the relative velocity, v_{rel}, is supersonic; if it is subsonic, the velocity scale appropriate to the accretion radius is the sound velocity in the stellar wind.) To the extent that wind velocities are typically of the order of the escape velocity from the surface of the star giving rise to the wind, and that the escape velocity from a source that must lie within its own Roche lobe is invariably greater than the orbital velocity of the accreting star in the binary system, the fraction of a stellar wind actually accreted by a companion star is never large. The dominant dynamical effect of wind mass loss on a binary system is an expansion of the binary orbit. If the wind carries away the same angular momentum per unit mass as that characterizing its source in its binary orbit (*Jeans-mode* mass loss), the orbital separation of the binary varies inversely as the total mass of the binary, and the orbital period varies inversely as the square of the total mass.

Additional Reading

Paczyński, B. (1971). Evolutionary processes in close binary systems. *Ann. Rev. Astron. Ap.* **9** 183.

Pringle, J. E. and Wade, R. A., eds. (1985). *Interacting Binary Stars.* Cambridge University Press, Cambridge.

Smith, R. C. (1984). The theory of contact binaries. *Quart. J. Roy. Astron. Soc.* **25** 405.

Thomas, H.-C. (1977). Consequences of mass transfer in close binary systems. *Ann. Rev. Astron. Ap.* **15** 127.

Truemper, J., Lewin, W. H. G., and Brinkmann, W., eds. (1986). *The Evolution of Galactic X-Ray Binaries.* D. Reidel Publishing Company, Dordrecht.

See also **Accretion; Binary Stars, Cataclysmic; Binary Stars, Contact; Stellar Evolution, Binary Systems.**

Binary Stars, X-Ray, Formation and Evolution

GertJan Savonije

There are roughly 100 bright ($> 10^{36}$ erg s^{-1}) x-ray sources in the Galaxy, most of which are binary systems. These so-called x-ray binaries are gravitationally bound pairs of stars. In each pair, one member is a compact star (a neutron star or, in a few cases, even a black hole). The inferred intrinsic x-ray brightness of the sources is rather large: 100 or more times the total solar luminosity. This means that there must be a powerful energy source at work in these binaries.

ACCRETION ONTO A COMPACT STAR

One of the most efficient ways to produce energy is the accretion of matter onto a compact object. In x-ray binary systems there is a large potential supply of matter that can be accreted, namely, the less compact companion star. If the neutron star is in a sufficiently close orbit, its gravitational attraction may suck the outer layers from the companion. Such tidal mass transfer is called Roche lobe overflow. The Roche lobe is a critical surface such that matter exterior to it is not bound to the star, but is attracted by the nearby companion. Also, when the companion star is massive, it produces a strong stellar wind and the neutron star may accrete part of the ejected wind material even while its companion is confined inside its Roche lobe. This accretion may be sufficient to power a bright x-ray source. When the accreting matter falls in the deep potential well of a neutron star, an accretion energy of about $0.1c^2$ (where c is the speed of light) is liberated per gram, which is much more than could be generated by nuclear fusion (more than two orders of magnitude greater, in the case of fusion of hydrogen to helium). The (kinetic) energy of the accreting matter is dissipated when it strikes the neutron star's surface, heating the infalling gas to high temperatures so that it radiates x-rays. When the accreting matter originates from Roche lobe overflow, it has so much angular momentum with respect to the neutron star that it cannot fall straight onto its surface, but forms an accretion disk around the neutron star. As a result of dissipational process(es), disk matter then spirals inward to the neutron star. When the neutron star is young and possesses a strong magnetic field, its accretion disk cannot extend inwards to the stellar surface, but is limited to the region exterior to where the magnetic field dominates the flow. The region where the magnetic field is dominant is called the magnetosphere. The extension of the magnetosphere may be very many stellar radii. If the strong magnetic field has essentially a dipole structure, the accretion flow will be forced to follow magnetic field lines and will hit the stellar surface only near the magnetic poles. If the axis of the dipole magnetic field is tilted with respect to the rotation axis of the neutron star, then the rapidly spinning neutron star is an x-ray pulsar, as the two hot magnetic poles function as a kind of x-ray lighthouse.

CLASSIFICATION OF X-RAY BINARIES

One distinguishes low-mass x-ray binaries (LMXBs), in which the mass of the neutron star's companion is smaller than about 2 $M_\odot$, and masive x-ray binaries (MXBs), in which the companion is a massive star (with a mass larger than, say 10 $M_\odot$) . X-ray binaries in which the mass donor has a mass intermediate between these values are rare. Because massive stars evolve on short time scales (less than about 10^7 yr), MXBs must be rather young objects. The massive x-ray binaries often contain x-ray pulsars and in many cases they show periodic eclipses of the x-ray source. This reveals their obital periods which are typically a few days.

The LMXBs on the other hand, are much older systems (up to about 10^{10} yr) and rarely contain x-ray pulsars or show x-ray eclipses. In contrast to the MXBs that are situated close to the galactic plane (as are all massive stars), the LMXBs have a much larger average distance from the galactic plane, and several of them have been found in globular clusters. The LMXBs have in general small orbital periods, which are usually of the order of a few hours, but which are sometimes much smaller (about 10 min). Thus LMXBs are in general rather compact binary systems.

FORMATION AND EVOLUTION OF MASSIVE X-RAY BINARIES

X-ray binaries distinguish themselves from other binary systems in that one of the stars is a neutron star or black hole. Therefore, they must be rather evolved systems, because compact stars are the end points of normal stellar evolution. It is well known that a neutron star (and in some cases perhaps a black hole) can be formed by core collapse in a massive star that had an original mass larger than about 7 $M_{\odot}$. The core collapse occurs when the burnt-out core becomes gravitationally unstable. The gravitational energy liberated by the collapsing core blows the remaining mass (which forms most of the stellar mass) into space in a violent (Type II) supernova explosion. The more massive a star, the faster its nuclear burning proceeds, and hence the faster it completes its evolution. Therefore, the more massive component in a binary will reach the supernova stage first. The supernova explosion disrupts the binary if more than half of the total system mass is expelled. Apparently, something has prevented this catastrophe in x-ray binaries.

The Importance of Mass Transfer

When the hydrogen in the core of the more massive star in the progenitor system of an MXB is depleted by nuclear fusion reactions, the star commences to expand at a fast rate and will eventually reach its Roche lobe. During the ensuing phase of Roche lobe overflow, the massive star loses almost its entire envelope on a thermal time scale. The mass transfer is very violent as a result of a positive feedback mechanism whereby the orbit, and with it the Roche lobe, shrinks as mass flows from the more massive to the less massive component. (The orbital shrinkage is due to conservation of angular momentum.) This shrinkage intensifies the Roche lobe overflow, which in turn accelerates the shrinkage. Only when the star is peeled to virtually its bare helium core does the mass tranfer come to a stop. The companion star will now have become the more massive star in the binary as it has accreted all or most of the lost mass.

Although it is now the less massive component, the helium star remnant remains the faster evolving star. Helium fusion in the much hotter helium star is much quicker than hydrogen fusion in the rejuvenated core of the companion. The further evolution of the helium star proceeds at an ever faster pace due to the copious emission of neutrinos that fly directly from the dense center of the star out into space. Long before the massive companion has finished hydrogen burning in its core, the helium star will have built up the catastrophic structure of a presupernova star and explode. The core may at the same time collapse into a neutron star. Because the helium star lost most of its envelope, the supernova explosion will not disrupt the binary, but will result in a young neutron star in an eccentric orbit about the still-unevolved companion. For a few million years, the system is in a quiet stage and tidal effects will tend to circularize the orbit.

The X-Ray Phase

After a few million years, the unevolved companion star finally expands beyond its Roche lobe and sends back the matter received in the previous phase of mass transfer. Some time before this, its stellar wind may have become sufficiently strong to turn the still young neutron star into a bright x-ray source. Once Roche lobe overflow has started in the companion star, the mass transfer will soon (within a thermal time scale) become so fast that it will smother the x-ray source. In other words, so much matter is accreted that the x-rays can no longer penetrate it and are absorbed and emitted as softer (longer wavelength) radiation. By then, the x-ray source has disappeared from the sky. The total duration of the x-ray emission phase is expected to be only a few times 10^4 yr, in rough agreement with the statistics of massive x-ray binaries.

Binary Radio Pulsars

We have seen that in MXBs the x-ray phase terminates after the second phase of Roche lobe overflow develops whereby the x-ray source is quenched. The continued expansion of the companion star and mass transfer will probably give rise to a spiralling-in of the neutron star toward its companion, whereby the latter star will lose its envelope. Depending on the mass of the remnant helium core, the binary may or may not be disrupted in the second supernova explosion. When disruption occurs, two runaway neutron stars will be ejected, and one or both may act as a radio pulsar. When the binary star remains bound, a binary radio pulsar such as the observed systems PSR 1913+16 and PSR 2303+46 is formed.

THE FORMATION OF LOW-MASS X-RAY BINARIES

It is somewhat surprising that binaries such as LMXBs exist in which a neutron star is accompanied by a low-mass star. In the previous section we have seen that neutron stars are born in supernova explosions and that such explosive events tend to disrupt a binary system if less than half of the system mass survives the explosion. Disruption of the binary must certainly be the case, however, if the companion has a mass smaller or comparable to the neutron star, because the neutron star contains only a small fraction (the collapsed core) of the presupernova star's mass.

The neutron star in an LMXB must therefore have been formed in a different way. One possibility for a less violent birth is accretion-induced collapse of a white dwarf star. Subrahmanyan Chandrasekhar showed in the 1930s that white dwarfs cannot have a mass larger than a certain critical value (the Chandrasekhar mass), which is about 1.4 $M_{\odot}$. For a more massive white dwarf, the relativistically degenerate electron gas that supplies almost all the pressure can no longer balance the self-gravity and the white dwarf collapses to a much denser state, a neutron star. This gives an alternative way to produce a neutron star in a binary: by transferring mass onto a white dwarf until its mass exceeds the Chandrasekhar mass and it collapses to a neutron star. However, numerical calculations indicate that this happens only if the white dwarf is already massive ($> 1.2\ M_{\odot}$) before mass transfer begins. Furthermore, it requires a high rate of mass accretion, as otherwise the white dwarf will be totally disrupted by explosive thermonuclear burning. It needs considerable fine tuning to meet these requirements, and we expect accretion-induced collapse to happen only rarely in binaries. This would be consistent with the relatively small number of LMXBs in the Galaxy (about 50) compared to the large number of cataclysmic variables (CVs). CVs are related binary systems in which mass transfer from a low-mass star to a white dwarf occurs. But in CVs the white dwarf is less massive and the accretion rate much lower than required for accretion-induced collapse.

Evolution by Inward Spiraling

By adopting the posibility of accretion-induced collapse of a white dwarf, we have created a new problem: White dwarfs are the remnants of giant stars with dimensions much larger than the average orbital separation of LMXBs or CVs. How can the binary have accommodated such a giant star? The binary system orbits must have been much larger in the past! The current idea is that the progenitor systems were indeed very wide binaries (with orbital periods of the order of one hundred days) and with extreme mass ratios. The more massive star evolves to the giant stage, whereas its companion remains practically unevolved. When the giant star expands toward its Roche lobe, tidal effects become important and orbital angular momentum is fed into the giant's envelope in an "attempt" to keep its rotation rate synchronous with the orbital motion. Because the giant star has a large moment of inertia, it absorbs a substantial fraction of the orbital angular momentum, and the companion star plunges into the giant's envelope. This effect is accelerated by violent mass transfer from the giant, which reduces the orbital separation even more. The details of the following spiral-in process are not well understood, but the dissipated orbital energy is amply sufficient to "boil away" the giant's whole envelope. However, because a large fraction of the dissipated energy may be radiated away and not be available for mass ejection, the outcome is not clear. Nevertheless, observations of so-called "double core" planetary nebulae (with binary central stars) indicate that the giant star in such a system is indeed stripped of most of its envelope, and that the outcome of the spiral-in process is a compact detached binary (with orbital period of the order of one day) consisting of a hot dwarf (the dense core of the giant) and a low-mass star. The more massive post–spiral-in systems may then evolve into LMXBs and the others may evolve into the more common CVs.

Other Formation Mechanisms of LMXBs

There are a few x-ray binaries with low-mass donor stars for which accretion-induced collapse cannot explain the compact star's existence. One of these is A0620-00, which is believed to contain a black hole with a mass larger than 5 $M_{\odot}$. This system may have evolved through a spiral-in phase, but with a much more massive (super)giant star, the core of which may have imploded to a black hole after most of the envelope had already been ejected, so that the binary was not disrupted. It should be mentioned that possibly the neutron stars in LMXBs are formed by direct collapse at the end of the spiral-in phase as well. This seems a viable alternative to the accretion-induced collapse hypothesis.

A group of LMXBs with a completely different formation mechanism are the globular cluster x-ray binaries. The known number of these x-ray binaries (about 10) is disproportionately large compared to the number of LMXBs in the galactic disk, if we take into account the small amount of mass involved in globular clusters, compared to the mass of the stars in the disk. Apparently, the formation mechanism of LMXBs in globular clusters is different and more efficient than in the galactic disk. It is believed to be related to the high star density in globular clusters. This makes the tidal capture of a free "fossil" (old) neutron star by a main sequence or (sub)giant star in a close encounter a realistic possibility during the lifetime of the cluster.

EVOLUTION OF LMXBs

The further evolution of the progenitor LMXB depends critically on the orbital period of the system when it emerges from the spiral-in phase. If the orbital period is less than a critical value (which depends on the stellar masses), the evolution of the system is dominated by orbital angular momentum losses due to gravitational radiation and stellar wind losses ("magnetic braking"). These losses cause the binary orbit to shrink on a time scale short compared to the nuclear time scale of the unevolved low-mass star. After a long period of time, the orbit has shrunk so much that a still-unevolved star fills its Roche lobe and begins a phase of mass tranfer. If the companion is a white dwarf, it may after a while undergo an accretion-induced collapse. After that the system becomes a LMXB.

The systems that emerge from the spiral-in phase with an orbital period longer than the critical value cannot be captured by angular momentum losses (these losses decrease with increasing orbital period) so that the unevolved star is able to continue its nuclear evolution. Therefore, the star expands slowly to its Roche lobe and begins to transfer mass. Contrary to the systems with smaller orbital periods, mass transfer is now driven by the expansion caused by the nuclear evolution of the unevolved star. The bifurcation in the evolution after spiral-in occurs for orbital periods in the range of the observed post–spiral-in systems. There are observational indications that both types of evolution occur.

Evolution Towards Ultra-Compact Binaries

Let us first discuss the binaries that are captured by angular momentum losses and that shrink to very small dimensions. The mass tranfer rate by Roche lobe overflow is much smaller than for MXBs, so that the x-ray source is long lived and not smothered. The continued angular momentum losses cause the orbit to become very compact. After a long period of mass transfer (on the order of 10^9 yr), the lobe-filling star has lost an appreciable fraction of its mass and its thermonuclear burning dies out. After that it becomes degenerate and begins to expand in reaction to further mass loss. This forces the orbital period to pass through a minimum (of about 40–60 min for hydrogen-rich, lobe-filling stars). The mass transfer has now become very slow, so that the orbit reexpands very slowly on the order of a Hubble time.

Evolution Towards Millisecond Binary Radio Pulsars

The wider post–spiral-in systems for which the angular momentum losses are unimportant evolve to very wide binaries as the mass flows to the more massive neutron star. The Roche-lobe-filling star expands together with the Roche lobe to large dimensions and ascends the giant branch in the Hertzsprung–Russell diagram. The giant star slowly depletes its envelope both by mass loss and by nuclear burning in a shell just exterior to the degenerate helium core. Such a system shows fairly high mass transfer rates and presumably lives for some 10^8 yr as a bright x-ray source. But in the end the hydrogen-rich envelope gets depleted and the expansion stops quite abruptly on a thermal time scale. Meanwhile the neutron star has been spun up by accretion to a rotation period of order 1 ms. Now that mass transfer has stopped, the rapidly spinning neutron star, which still has a residual magnetic field, will appear as a millisecond radio pulsar. The radio pulsar is accompanied by a low-mass white dwarf star (the degenerate helium core remnant of the giant) in a very wide circular orbit. Several of these peculiar systems have been discovered, such as PSR 1957+20 (period 1.6 ms), PSR 1855+09 (5.4 ms), and PSR 1953+29 (6.1 ms).

Additional Reading

Gursky, H. and van den Heuvel, E. P. J. (1975). X-ray emitting double stars. In *New Frontiers in Astronomy*. W. H. Freeman, San Francisco, p. 268.

Lewin, W. H. G. and van den Heuvel, E. P. J. (1983). *Accretion-driven Stellar X-Ray Sources*. Cambridge University Press, Cambridge.

Verbunt, F. (1989). The origin and evolution of x-ray binaries and low magnetic field radio pulsars. In *Neutron Stars: Their Birth, Evolution, Radiation and Winds*, W. Kundt, ed. Kluwer Academic Publishers, Dordrecht.

See also **Binary Stars, Theory of Mass Loss and Transfer in Close Systems; Star Clusters, Globular, X-Ray Sources; Stellar Evolution, Binary Systems.**

Black Holes, Stellar, Observational Evidence

M. G. Watson

X-ray binaries provide the astrophysical setting in which we have the best chance of finding evidence for stellar mass black holes. In these systems the normally elusive black holes can become observable through accretion processes that lead to prodigious fluxes of x-rays being produced. Unfortunately, a major difficulty remains, namely that of distinguishing the type of compact object present, that is, between a black hole or a neutron star in the binary system. Black holes have only three basic properties that can be observed outside the event horizon: mass, charge, and angular momentum. Coupled with the effective lack of a surface, this means that black holes cannot display many of the detailed phenomena associated with neutron stars. Nevertheless, the basic phenomena associated with accretion onto both black holes and neutron stars depend primarily on the depth of the gravitational potential well and the size of the emission region, parameters that do not differ markedly between, say, a 1 $M_\odot$ neutron star and a black hole with mass ~ 5–10 $M_\odot$. Thus, although a number of observational signatures have been proposed to distinguish between black holes and neutron stars, none of these has turned out so far to be completely reliable.

The most reliable method we have of distinguishing between a black hole and neutron star in an x-ray bright binary system is through the determination of the mass of the compact object. This method works because there is a strict upper limit to the mass of a neutron star of ~ 3 $M_\odot$ based on causality arguments: Any compact object with a mass larger than this must be completely gravitationally collapsed; that is, it must be a black hole. (The actual value of the upper limit mass for neutron stars is not precisely known because of uncertainties concerning the equation of state of neutron star matter; current estimates range from 2–2.7 $M_\odot$.) Such mass determinations have been made for several x-ray binaries. To date, such studies have revealed three systems that may contain a black hole: Cyg X-1, LMC X-3, and A0620–00.

OBSERVATIONAL SIGNATURES OF BLACK HOLE SYSTEMS

The possibility of finding black holes in x-ray binaries was foreseen by the Russian astronomer Ya. B. Zeldovich before the existence of x-ray binaries themselves had been clearly demonstrated. Since then comprehensive observational evidence has been accumulated for the existence in our galaxy of some hundred or more x-ray binary systems powered by accretion onto a compact object. A large fraction of these are believed to be neutron star systems, primarily because they are x-ray pulsars with characteristic modulation of their x-ray flux at the spin period of the magnetized neutron star, or because they have x-ray bursts, believed to involve a thermonuclear flash on the neutron star surface. Indeed, in perhaps a few dozen x-ray binaries, the evidence for the existence of an accreting neutron star is overwhelming: Here we have reasonably accurate estimates of several of the key properties of the neutron stars such as masses, radii, and surface magnetic fields.

For the remainder of the x-ray binaries, the nature of the compact object is not definitely known, although the general similarity of many of these systems to known neutron star binaries (e.g., x-ray spectra and variability characteristics) would argue that the majority also involve a neutron star. For a few x-ray binaries there has long been speculation that the compact object was a black hole. Cygnus X-1 was the first such system. Early x-ray observations showed that it lacked the periodic signal characteristic of x-ray pulsars, but instead had very rapid aperiodic flickering that extended down to very short time scales. Rapid aperiodic flickering thus quickly became a possible signature of a black hole, even though on theoretical grounds there was no reason why both neutron stars and black holes should not vary on similar short time scales. Dynamical studies of Cygnus X-1 did, in fact, indicate that the system contained a black hole. Thus other observational properties of Cygnus X-1, such as the shape of the x-ray spectrum and its bimodal spectral behavior, also became accepted as possible black hole signatures. More recently, the catalog of black hole observational signatures has been expanded to include the ultrasoft x-ray spectral shapes that are characteristic of several of the better black hole candidates (e.g., A0620–00).

These observational signatures are, at best, very indirect evidence for the presence of a black hole in an x-ray binary. Their unreliability can be illustrated by the example of the x-ray binary Circinus X-1 which was long regarded as a good black hole candidate on the basis of its rapid aperiodic variability and bimodal spectral behavior. Recent observations with the European x-ray astronomy satellite *EXOSAT*, however, showed that Circinus X-1 also produces x-ray bursts, clearly indicating that it is a neutron star system.

Table 1. Black Hole Candidates

Property	*Cyg X-1*	*LMC X-3*	*A0620 – 00*
Binary period P (days)	5.6	1.7	0.32
Velocity semiamplitude K_c (km s^{-1})	76 ± 1	235 ± 11	457 ± 8
Mass function $f(M/M_\odot)$	0.25 ± 0.01	2.3 ± 0.3	3.18 ± 0.16

THE BEST BLACK HOLE CANDIDATES

The evidence for black holes in three x-ray binaries is nevertheless convincing. In all three cases the evidence is based on detailed spectroscopic observations of the optical counterpart to the x-ray source that measure the amplitude of its radial velocity variations K_c through each binary period. Measuring the radial velocity amplitude enables the optical mass function $f(M)$ to be determined which can then be related, via Kepler's laws, to the masses M_C and M_X of the optical companion star and the compact object, and the inclination of the binary system to the line of sight i. For a circular orbit with period P, we have

$$f(M) = \frac{PK_c^3}{2\pi G} = \frac{(M_X \sin i)^3}{(M_X + M_C)^2}.$$

The binary periods, measured radial-velocity amplitudes, and derived mass functions for the three x-ray binaries that are good black hole candidates are given in Table 1. The mass function $f(M)$ provides an absolute lower limit for the compact object mass, that is, $M_X \geq f(M)$. Thus in the case of A0620–00 the large mass function already suggests a very massive compact object, whereas for LMC X-3 and Cyg X-1 the case also depends on determining reliable values for a the companion star mass M_C and the orbital inclination i. For all three systems the evidence also relies on the radial velocities measured correctly reflecting the orbital motion of the companion star. These issues are now briefly addressed for each of the three binaries.

Cygnus X-1

The first x-ray binary to be seriously considered as a black hole candidate was Cyg X-1. The optical companion (HDE 226868) is an O9.7 supergiant. Such a star would normally have a mass over 30 $M_\odot$, giving a lower limit to the compact object mass of 7 $M_\odot$. This is not a firm conclusion, however, because of the strong possibility that the companion is significantly undermassive for its spectral type. It turns out that a firm lower limit to the compact object mass can be set without recourse to specific arguments about the nature of the companion star. This method, due to Paczynski, uses the observed (dereddened) flux of the companion and the absence of x-ray eclipses, combined with the measured mass function, to provide a lower limit to the compact object mass that depends only on the distance to Cyg X-1. For a distance of 2 kpc or greater, the compact object mass exceeds 3.4 $M_\odot$. Present measurements of Cyg X-1 do not provide indisputable evidence that its distance must be greater than 2 kpc, although all the indications are that it must be. There is thus a chance, albeit a slim one, that the compact object mass in the Cyg X-1 system could be low enough for it to be a neutron star.

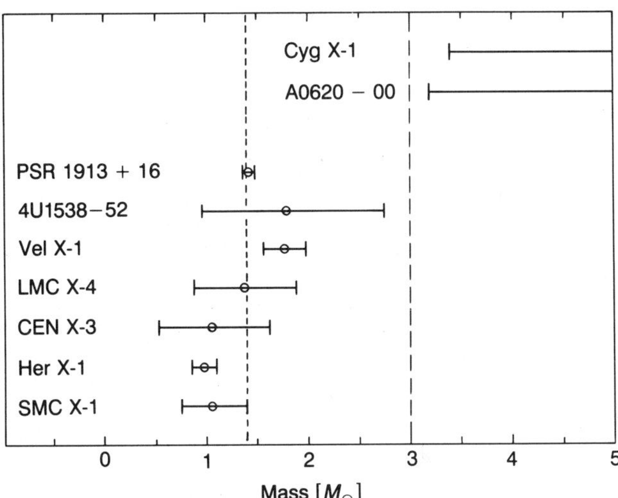

Figure 1. Mass determinations for the compact object star in eight x-ray binary systems, plus the binary radio pulsar PSR 1913+16. The masses are all consistent with the compact object being a neutron star with the exception of A0620−00 and Cyg X-1. The dashed lines indicate the canonical neutron star mass (1.4 $M_\odot$) and the upper limit neutron star mass based on causality arguments (3 $M_\odot$).

LMC X-3

The optical star in the LMC X-3 system is of spectral type B3 V. A straightforward application of Paczynski's method to this system yields a compact object mass of 6 $M_\odot$, without any substantial uncertainty about the distance because LMC X-3 is known to be a member of the Large Magellanic Cloud. Doubts do exist, however, about possible distortion to the radial velocity measurements and the extent to which the optical light from the companion star might be diluted by the accretion disk. Both factors could lower the compact object mass estimate substantially, and heavy neutron star models are therefore viable.

A0620−00

The most recent, and perhaps most convincing, black hole candidate is the x-ray transient system A0620–00. In quiescence the optical counterpart of this binary (V616 Mon) is a faint K7 dwarf. If conservative assumptions are made about the mass of the K dwarf and the orbital inclination, the measured mass function yields a firm lower limit to the compact object mass in this system of $M_X \geq 3.2\ M_\odot$. In contrast with the other systems, systematic effects that might distort this limit substantially are not believed to be important. Thus A0620–00 is the best black hole candidate that we have to date.

SUMMARY

The present evidence for the existence of black holes of stellar mass is summarized in Fig. 1 which shows the current best estimates of the compact object mass for a number of systems believed to contain a neutron star, together with the estimates for A0620–00 and Cyg X-1. The masses for A0620–00 and Cyg X-1 exceed the 3 $M_\odot$ maximum mass considered possible for neutron stars and are thus good candidates for black hole systems. Although the evidence is good, there are still a number of uncertainties in the interpretation of the observations and ad hoc explanations of the data, such as those involving triple systems, that can be invoked. Even if these uncertainties can be resolved, the question of what constitutes definite proof of the existence of stellar mass black holes, given their very special and remarkable nature, remains a very difficult one to answer.

Additional Reading

Bradt, H. V. and McClintock, J. E. (1983). The optical counterparts of compact galactic x-ray sources. *Ann. Rev. Astron. Ap.* **21** 13.

Lewin, W. H. G. and van den Heuvel, E. P. J., eds. (1983). *Accretion-Driven Binary X-Ray Sources*, Cambridge University Press, Cambridge.

McClintock, J. E. (1986). Black holes in x-ray binaries. In *The Physics of Accretion onto Compact Objects*, K. O. Mason, M. G. Watson, and N. E. White, eds. Springer, Heidelberg, p. 211.

McClintock, J. E. (1988). Do black holes exist? *Sky and Telescope* **75** 28.

See also **Accretion; Binary Stars, X-Ray, Formation and Evolution; Black Holes, Theory.**

Black Holes, Theory

Richard H. Price

A black hole is a region of space in which the pull of gravity is so strong that nothing can emerge from it. The name "hole" is used because it is not an object in the usual sense, but more like a void; it is "black" because not even light can overcome the pull of gravity to escape from this void. Some of the features of black holes were first envisioned by 18th-century thinkers, but what we mean by black hole today has exotic and nonclassical features that can be completely understood only with Albert Einstein's relativistic theory of gravity. Evidence is now strong that these exotic voids do exist and have detectable astrophysical signatures.

THE HORIZON

The idea of an overpowering gravitational pull is one that can be quantified with classical mechanics. Outside a spherical body of mass M, at a distance r from the center of the body, a baseball or bullet, or quantum of light, if it is to avoid falling back down, must be moving faster than the escape velocity $\sqrt{2GM/r}$, where G is the universal gravitational constant. Let us suppose that there is an object in the universe so massive and compact that the escape velocity from its surface is twice the speed of light. Light then could not escape from the surface of this body or from the nearby surrounding region. There would be an imaginary boundary surface with a radius four times the radius of the body; outside that surface light could escape, inside it could not. This surface is called the horizon, because a distant observer could not get light from (and hence could not see) inside it. This then is a black hole: a region with a horizon. It can be of any mass, because the horizon defining relation simply requires that the horizon radius be proportional to the mass. For a hole of 1 $M_\odot$, the horizon radius is 3 km; for a 10^9 $M_\odot$ hole, the radius is 3×10^9 km.

This picture of a black hole becomes perplexing when relativity theory is mixed in. For any particle or signal to proceed outward through the horizon would mean that it is moving faster than outward directed light, and relativity forbids faster-than-light motion. The horizon then not only hides its interior, but acts as a one-way membrane. Particles, signals, interstellar matter, curious astronauts, and so forth, can move inward into the hole, but never again outward. It is a surface of no return. (There is a slight exception: Stephen Hawking has shown that quantum effects should cause a hole to act as a thermal source of radiation. For astrophysical objects the effective temperature is less than 10^{-7} K and Hawking radiation is of interest only in principle.) It must be remembered that despite the crucial role it plays, the horizon, the black hole "surface," is not really there. It is a demarcation surface, not a material surface or a surface in which gravitational fields are dramatically stronger than just outside or just inside the horizon. The doomed astronaut falling into the hole would feel no singularly large gravitational fields at the horizon to warn him or her of this fateful error. The astronaut would experience his or her passage through the horizon as a smooth event.

As is typical in relativity, the observation of events is dramatically different for different observers. The astronaut's viewpoint that there is nothing special at the horizon is in marked disagreement with the observations by an astronomer watching the astronaut's progress. This astronomer can receive no signal from the interior of the black hole and hence cannot see the astronaut after he or she has crossed the horizon. But as the astronomer watches, the astronaut does not suddenly disappear from the telescope. Rather, there is a "time dilation," an effect most familiar in connection with the moving clock of the twin paradox of relativity. For black holes the time dilation is due primarily not to motion, but to gravitation, and is infinite right at the horizon. The astronomer then sees the astronaut's inward progress increasingly slowed down, so that the horizon is approached, but never quite reached.

BLACK HOLE PROPERTIES

The one-way property of the horizon has a crucial consequence for the interior of the hole. If nothing can move outward at the horizon, the implication (confirmed by the mathematics of general relativity) is that interior to the horizon nothing can even remain stationary; everything must be falling inward. A consequence is that the astrophysical "body" cannot be a stationary object. It must be collapsing, and it cannot halt its collapse, according to Einstein's theory, until it has become a singularity, such as a point mass. Such a singularity in a physical process is unacceptable to theorists, who assume that quantum mechanical effects, which have not yet been fully combined with general relativity, will come into play when conditions in the interior become sufficiently dramatic and will prevent the collapse from forming a true singularity.

Whether a true singularity is reached or not is in some sense astrophysically irrelevant. Because the collapse is happening inside the (nonsingular) horizon, the singularity can have no influence on the rest of the universe outside the horizon. Theorists think of the black hole singularity as shielded or clothed by the horizon and distinguish it from a "naked" singularity, one that can be in principle investigated arbitrarily closely by external observers. It is strange and interesting that attempts to construct theoretical models of the formation of a naked singularity have failed. There appears to be a conspiracy in nature, or in the theory, always to form a horizon around a singularity.

The action of the horizon in hiding the black hole interior makes characterizing black holes extremely simple. For the type of hole (Schwarzschild) discussed so far, there is only a single parameter: the black hole mass. More generally, a black hole can have spin, or angular momentum (Kerr hole), electrical charge (Reissner–Nørdstrom hole), or all three (Kerr–Newman hole). But these three features are the only qualities of holes. This near featurelessness of black holes is sometimes referred to as their lack of "hair."

As astrophysical objects, holes would arise as the end point of the collapse of less exotic bodies. In Einstein's theory it is also mathematically possible to have holes without collapse. In these objects the interior of the hole also acts as the interior of another hole, forming a bridge (Einstein–Rosen bridge) between two locations in the universe or joining two otherwise disconnected universes. These bridges are useful to science fiction as cosmic shortcuts but are more mathematical than astrophysical entities. The bridges do not form in any process; they must be present in the initial structure of the universe. Furthermore, the bridges are transitory; they either pinch off due to their own structure or are unstable to any perturbation.

BLACK HOLE OBSERVABILITY

Because nothing can get out of a black hole, it might seem that black holes would be impossible to observe. But the gravitational pull, as well as the electrical charge and angular momentum of the hole, *do* influence the hole's astrophysical environment and produce observable effects. (The gravitational, electrical, and angular momentum fields do not "escape" from the black hole; these fields are always there, linking the hole and its exterior.)

A black hole can be observable, for example, through the effect of its pull on another object in a binary system. A black hole can also pull matter inward from its surroundings; the accreting matter will be compressionally heated and, while still outside the horizon, can emit observable radiation. A black hole can bend the path of passing photons and thereby act as a gravitational lens distorting the observed pattern of background stars. A black hole in a much less obvious way can act to anchor magnetic field lines, and the hole's rotation can thereby produce an astrophysical dynamo.

Observational interest at present is focused on two classes of objects as black hole candidates. Several compact x-ray sources are thought to be stellar mass black holes that heat accreting matter to x-ray emitting temperatures. Quasars and active galactic nuclei may be powered by very massive (10^8–10^{10} $M_\odot$) holes, either through accretion or a black hole dynamo. Black holes may also be important sources of the gravitational waves that are being sought with detectors of ever increasing sensitivity. The detection of a black hole with such a device would be a breakthrough combining technology and theory and would be one of the major events in the astronomy of the 1990s.

Additional Reading

Hawking, S. W. (1977). The quantum mechanics of black holes. *Scientific American* **236** (No. 1) 34.

Penrose, R. (1972). Black holes. *Scientific American* **226** (No. 5) 38.

Price, R. H. and Thorne, K. S. (1988). The membrane paradigm for black holes. *Scientific American* **258** (No. 4) 69.

Shapiro, S. L. and Teukolsky, S. A. (1983). *Black Holes, White Dwarfs, and Neutron Stars* Wiley, New York.

Thorne, K. S. (1974). The search for black holes. *Scientific American* **231** (No. 6) 32.

See also **Active Galaxies and Quasistellar Objects, Central Engine; Black Holes, Stellar, Observational Evidence; Gravitational Theories.**

Bok Globules

Robert L. Dickman

Bok globules are small, relatively dense interstellar clouds that are spherical or nearly spherical in appearance. They are named after the Dutch-American astronomer Bart J. Bok (1906–1983), who with Edith Reilly in 1947, first pointed out their potential role as star formation sites.

Bok and Reilly distinguished between two types of globules: "small globules," the minuscule, virtually opaque condensations (Fig. 1) often seen silhouetted against the bright, ionized emission nebulae called H II regions; and "large globules," isolated, more massive clouds that usually possess a visibly differentiated internal structure (Fig. 2). Despite their name, and despite the fact that they are the most morphologically regular of interstellar clouds, globules are rarely perfectly spherical; ellipsoids are quite common (see Fig. 2), and windblown, "cometary" shapes are often exhibited, particularly by the outer layers of the clouds. Partly because they are much more easily studied than their smaller counterparts, the large globules have continued to be the subject of studies by astronomers since the seminal article of Bok and Reilly, and they are, for the most part, the subject of this entry. By contrast, the small globules remain objects about which relatively little is known.

Astronomers now know that the vast majority of stars formed in the Galaxy are born in the objects known as giant molecular clouds (GMCs; discussed in the following section), rather than in globules. Nevertheless, Bok's oft-repeated suspicions that globules are places where the mysterious process of star birth occurs were correct. And whereas Bok globules are certainly not the most prolific star formation sites in the Galaxy, they have nevertheless remained of considerable importance to astrophysicists. The reason for this lies in the enormous complexity of GMCs: Star formation is often an immensely vigorous process, involving the energetic ejection of matter from newborn stars, and, in the most massive GMCs, the formation of luminous, hot stars that ionize and disrupt the clouds out of which they recently formed. As a result, the study of star formation is all too frequently complicated by the need to disentangle the threads of multiple recent, vigorous stellar births in order to obtain a picture of the events that precede the formation of a *single* star. Such a task is immensely difficult, and it is in this connection that Bok globules are particularly attractive objects of study: They rarely form more than one star at a time,and any newly formed stars usually have masses low enough so that the conditions associated with the star formation process are not strongly distorted or masked. Bok globules are thus the simplest star-forming structures in the Galaxy, and in many respects offer the best prospects for understanding the still obscure process of star birth.

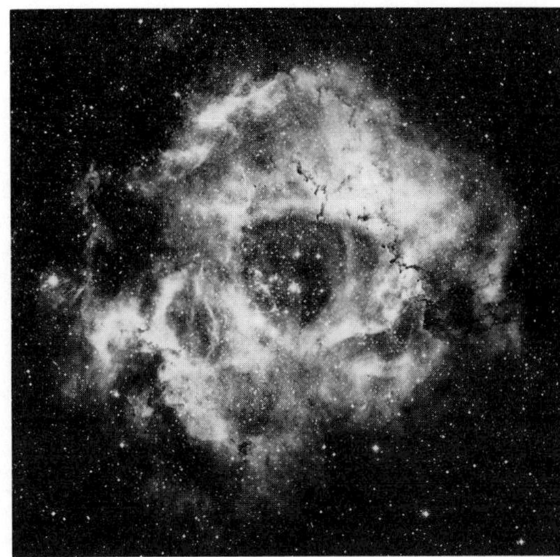

Figure 1. Small Bok globules are visible as dark spots projected against the glowing, ionized hydrogen gas of the Rosette nebula in the constellation Monoceros. The gas is ionized by a cluster of luminous, massive young stars, some of which are visible in the cavity that they have begun to evacuate at the center of the nebula. The globules are typically about 1/3 ly across, and most show strong departures from sphericity. If, as other indirect arguments suggest, the globules are actually embedded within the hot, nebular gas, their nonspherical shapes are readily explained as a result of the energetic expansion of the ionized gas and the erosion of neutral globule material by the ionizing radiation of the central cluster. This Hale Observatory photograph was made with the 48-in. Schmidt Telescope on Mt. Palomar.

BACKGROUND

Bok globules are a type of interstellar molecular cloud. In order to define this term, we begin by noting that roughly 90% of the mass in the disk of a spiral galaxy like our own Milky Way consists of stars. The remainder is spread between the stars in the form of an *interstellar medium* (ISM). Hydrogen gas is by far the most common constituent of the ISM, followed by helium (which by mass is about three times less abundant); all other elements, such as carbon, oxygen, and so forth, are far less common than hydrogen, by factors of a thousand or more. Although other physical states can occur, the ISM in the galactic disk tends to take one of two forms: a "warm" (approximately 100 K), diffuse state (densities of only about one particle per cubic centimeter or less) in which the hydrogen is in atomic form, and a cold (10 K), dense state (exceeding a few hundred particles per cubic centimeter) in which the hydrogen is molecular (H_2). Interstellar matter in the latter state tends to clump into well-defined *molecular clouds*, the largest of which are many hundreds of light-years across and have masses several million times that of the Sun. These giant molecular clouds (or GMCs) are the most prolific star formation sites in the Galaxy, and the largest GMCs dominate the mass distribution of the interstellar medium; by contrast, globules are the smallest and least massive molecular clouds in the Galaxy.

Were globules (and indeed all molecular clouds) made solely of hydrogen and helium, they would be virtually undetectable from the surface of the Earth—the clouds are far too cold to excite the helium gas to emit spectral lines, and whereas molecular hydrogen gas at temperatures of 10 K does emit certain characteristic spectral lines, these lie in the infrared portion of the electromagnetic spectrum blocked by the Earth's atmosphere, and are thus invisible from the ground.

This situation is ameliorated by two factors, both stemming from the fact that the gas within molecular clouds is not quite pure. All interstellar clouds in the galactic disk also possess traces of dust, tiny solid particles akin to soot. Because the dust attenuates and selectively reddens starlight, interstellar clouds not too far from the Sun (typically within a few thousand light-years) can often be detected because they block, wholly or in part, the light of the stars behind them. An opaque Bok globule lying in a rich star field will appear as a striking, nearly circular "hole" in the heavens, as in Fig. 2. Indeed, long before the nature of dense interstellar clouds was understood by astronomers, this clear-cut visual signature of their presence led to the compilation of catalogs of such objects. Although radio surveys using emission lines from carbon monoxide have proven to be more efficient at locating the largest molecular clouds in the Galaxy, globules, easily overlooked in the surveys, are still best found by studying long-exposure images of the sky. Because the widespread distribution of dusty molecular clouds in the plane of the Galaxy prohibits such images from probing more than a few thousand light-years beyond the Earth, astronomers do not have a very clear understanding of how many large globules exist in the Galaxy. Several years before his death, Bok estimated the galactic population of large globules to be some 25,000; no better estimate yet exists.

The second mitigating factor for detecting and studying the properties of Bok globules is that under virtually the same physical conditions that lead to the formation of a molecular ISM, the rare trace gases within these clouds—elements such as oxygen, carbon, nitrogen, and sulfur—begin to form surprisingly complex chemical compounds. As a result, the H_2 and helium admixture of molecular clouds is leavened by minute traces of compounds like carbon monoxide (CO), hydrogen cyanide (HCN), formaldehyde (H_2CO), carbon monosulfide (CS), ammonia (NH_3), sulfur monoxide (SO), and ethyl alcohol (CH_3OH). Although these trace chemical species comprise but a tiny fraction of the mass of a typical molecular cloud (0.01% or less), they produce spectral lines that are easily observed through the Earth's atmosphere and that are present in numbers large enough to be readily detected (usually by radio

Figure 2. A large Bok globule in the constellation Cassiopeia seen in a digital CCD image obtained with the 36-in. telescope of Kitt Peak National Observatory by T. H. Jarrett of the University of Massachusetts. The globule is markedly ellipsoidal rather than perfectly spherical, a not-uncommon occurrence. Its resemblance to a ghostly smoke ring is caused by dust particles in the relatively diffuse outer layers of the cloud; the dust scatters starlight, producing the bright rim. Background stars easily shine through the outer layers of the globule (where there is relatively little gas and dust), but the dense core of the cloud is essentially opaque. This globule is believed to be about 1800 ly from Earth, is about 3 ly across, and has a total mass about 15 times that of the Sun.

telescopes operating at millimeter wavelengths); the emission line of the carbon monoxide molecule at a wavelength of 2.6 mm is especially useful in this connection. All molecular clouds are thus "molecular" from two standpoints, both in consisting mainly of molecular hydrogen and in possessing a rich array of trace molecules.

PHYSICAL CONDITIONS WITHIN GLOBULES

The faint traces of dust and exotic chemicals in molecular clouds do more than render these interstellar structures observable. They also make possible their detailed scientific study. Virtually everything presently known about the distribution, elemental composition, velocity, and temperature of the gas and dust within interstellar clouds—including Bok globules—comes from analyzing observations of dust and trace molecular species. From such studies the following picture emerges.

The typical large globule is a few light-years across, although diameters as large 10–15 ly are sometimes encountered. A globule without an embedded newborn star to heat it is extremely cold—observations of molecules such as CO indicate gas temperatures of only about 10 K (−441°F); far-infrared observations of the dust within globules (studies that usually must be made from high-flying aircraft, balloons, or satellites) likewise imply temperatures in the same range. Even when newly formed stars are embedded in globules, gas and dust temperatures show little departure from these frigid values; except in their immediate vicinity, the low-mass stars that form in globules generally emit too little energy to substantially heat their nursery sites. Observations of various molecular species, as well as measurements of the gradient in starlight extinction produced by the dust in Bok globules, all indicate that both the gas and dust within these objects are centrally condensed—although whether these two basic constituents of the clouds remain mixed in equal proportions is unknown (some investigators have suggested that over time heavy dust particles may, under the action of gravity, preferentially settle to the centers of globules). Although astronomers habitually refer to molecular clouds like Bok globules as "dense" objects, the "high" central densities of more than 10^6 H_2 molecules per cubic centimeter that are frequently inferred to exist in the cores of globules without embedded stars represent a vacuum many orders of magnitude better than any ever attained in a laboratory on Earth.

Observations of molecular spectral lines permit the gas motions within interstellar clouds to be studied in detail via the Doppler effect: The width of a spectral line reflects how fast gas along the line of sight is moving, and by mapping shifts in the central frequency of a spectral line across a cloud, the presence of rotation can be unambiguously discerned. Observations indicate that the gas within a typical large Bok globule is in a largely turbulent, mildly supersonic state, although a few globules are known in which the gas moves at or below the local sound speed (about 300 m s^{-1}). Theoretical difficulties exist in understanding why such highly dissipative motions are not quickly damped out. Remarkably few globules showing coherent rotation are known, despite the fact that several careful surveys for such motions have been carried out. The comparative rarity of rotation may be tied to the state of the interstellar magnetic field within globules, a subject that has only recently begun to receive systematic attention.

Several methods exist for estimating globule masses. Although the techniques are not individually very reliable (each involves oversimplifications and assumptions that are difficult to justify rigorously), in most cases the various methods concur, typically yielding masses for large globules that are in the range 10–100 times that of the Sun. Given the mass of a globule, together with the other physical data noted previously, it is then possible to evaluate the cloud's stability against disruption or collapse. Most large globules are found to be close to a state of overall equilibrium.

STAR FORMATION IN BOK GLOBULES

Star formation begins when a localized region within an otherwise stable molecular cloud becomes unstable to contraction under its own gravity. Exactly how this occurs is unclear, but a mechanism of *local* condensation in a context of *global* stability implies that star formation must be an inherently inefficient process. Indeed, only a small percentage of the total mass of a molecular cloud may be actively engaged in the star formation process at any given time. Given the range of masses noted previously, this implies that within Bok globules the formation of stars much more massive than the Sun is expected to be very unlikely. This is unfortunate, because such objects are relatively easy to detect, and a search in the early 1970s for stars with masses in excess of about 3 $M_\odot$ which had formed in large globules was, in fact, entirely negative. The overall inefficiency of the star birth process also implies that the formation of more than one or two stars at a time is unlikely in even the largest globules. This further complicates the search for newly formed stars in such clouds and raises the question of how astronomers have managed to succeed in confirming that globules are star birth sites. In order to answer this question, we must consider how astronomers detect newly formed stars.

Youthful, low-mass stars usually go through a period of variability before settling down to stable maturity; such *T Tauri stars*, if associated with Bok globules, are fairly conclusive signposts of recent star birth activity. However, even before a newly formed low-mass star actually emerges from its birth cloud and becomes optically visible, its presence can be inferred in several ways. First, it can be detected as a near- or mid-infrared source. All newborn stars are initially enshrouded in a cocoon of gas and dust, placental matter from the parent cloud made dense in the gravitational collapse process that led to the formation of the star itself. Although the photosphere of the newborn star has a temperature of several thousand degrees, energy emitted from the stellar surface is intercepted by the dusty shell, which then reradiates it in the infrared portion of the electromagnetic spectrum. Infrared surveys have, in fact, revealed the presence of recently formed stars deep in the central regions of certain large globules.

One of the most significant discoveries of the 1980s was the finding that virtually all newborn stars go through a (probably episodic) phase of energetic *mass outflow*, in which cloud material is driven from the immediate vicinity of a newly formed star. Although the mechanisms responsible for this process are as yet obscure, outflows are often collimated into a pair of oppositely directed "bipolar" beams; these may help maintain the puzzling levels of turbulence observed in globules and other molecular clouds. Outflows can be rather easily detected by radio observations, and studies of the centers of several large Bok globules have revealed the unmistakable star formation signature of these vigorous stellar winds. Outflows are sometimes also associated with *Herbig–Haro* objects, small, bright blobs or streamers of hot gas which are presumably ejected by the young stars that power the outflows. Even if a newborn star is too feeble to be detectable as an embedded infrared point source, and even if no associated outflow is discernible (because of an unfavorably oriented bipolar ejection geometry, for example), the luminous knots that mark the presence of a Herbig–Haro object therefore also signal that the star formation process has occurred. One of Bok's last scientific discoveries was the detection in 1978 of a Herbig–Haro object at the edge of the southern globule 210-6a.

Additional Reading

Blitz, L. (1982). Giant molecular cloud complexes in the Galaxy. *Scientific American* **246** (No. 4) 84.

Bok, B. J. and Bok, P. F. (1981). *The Milky Way*, 5th ed. Harvard University Press, Cambridge, MA.

Clemens, D. P. and Barvainis, R. (1988). A catalog of small, optically selected molecular clouds: optical, infrared and millimeter properties. *Astrophys. J. (Suppl.)* **68** 257.

Dickman, R. (1977). Bok globules. *Scientific American* **236** (No. 6) 66.

Lada, C. J. (1984). Bart Jan Bok. *Mercury* **8** (No. 2) 35.

Reipurth, B. (1984). Bok globules. *Mercury* **8** (No. 2) 50.

Rodriguez, L. (1984). Cosmic jets: bipolar outflows in the universe. *Astronomy* **12** 66.

See also **Interstellar Clouds, Giant Molecular; Interstellar Medium; Stellar Evolution, Low Mass Stars.**

C

Calendars

Colin Ronan

A calendar is a method of arranging days in a system for regulating everyday life, for religious usage, and for historical and scientific purposes. In practice this means determining the length of the year, its date of commencement, and deciding on suitable subdivisions. Because the year and its seasons are governed by the behavior of the sun and moon, calendar determination requires knowledge of astronomy. Various calendars have been devised by different civilizations throughout the world, though now modern communication and travel have made it necessary for all nations to adopt the same standard system. This is based on the Gregorian calendar, developed in Western Europe in the late sixteenth century A.D. from a system devised in ancient Rome. Indeed, the word "calendar" is itself derived from the Latin *calendarium*, an account book, the word itself being derived from *calendae*, the calends, which was the first day of the Roman month, and the day on which accounts were due and also proclamation day for forthcoming feasts and various civic events.

THE SOLAR CALENDAR

Devising an annual calendar which will keep pace with the seasons —vitally necessary for agriculture—requires the determination of the time taken by the Sun to complete an apparent circuit of the heavens. Such a purely solar calendar for civil purposes was devised first in ancient Egypt, not later, it seems, than the period of the Old Kingdom (circa 3400 to 2475 B.C.). The motivation was agricultural, being linked to the annual flooding of the Nile, an event of vital importance to the Egyptian farmer. This flooding occurred at approximately the same time each year, and was seen to coincide nearly enough with Sirius rising just before dawn (heliacal rising). The Egyptian administration therefore settled on a calendar of 365 days, dividing it into 12 months of 30 days each with 5 extra or *intercalary* days added at the end.

The sidereal year, which is that indicated by the heliacal rising of Sirius, was longer than 365 days; it amounted to 365.25636 days. So also is the tropical year—the time taken for the Sun to travel from equinox to equinox—which is the period in which the annual cycle of the seasons recurs, since it is that period during which the Sun completes its apparent journeys north and south of the celestial equator. This has a value of 365.24219 days. It is clear, therefore, that 365.25 days would be a more precise, though not exact, expression for the length of the year. In consequence the Egyptian civil year would be 1 day short every 4 years, and this would amount to 10 days in 40 years—a noticeable error. The administration therefore concluded that the calendar would only become properly in step again after 365×4 or 1460 years. This became known as the *Sothic cycle*, Sothis being the Egyptian name for Sirius. In China also, during Shang times (the sixteenth to eleventh centuries B.C.), a year of 365.25 days was adopted.

Many civilizations—those in Meso-America, for instance—used a 365-day calendar, but it was the Egyptian civil year of 365.25 days which Julius Caesar introduced into Rome in 45 B.C. To keep close to the seasons, an extra day was intercalated between February 23 and 24, once every four years. Since on this *Julian calendar* February had 28 days, February 23 appeared 6 days before the calendae of March, and was known as the *sexto-calendae*. When the intercalary day arrived, it came a day later, and was known as the *bis-sexto-calendae*; hence the name *bissextile* for what are now generally known as leap years, because in such years the date of any fixed festival leaps forward one whole day.

The Julian calendar of 365.25 days was too long; it exceeded the tropical year by 0.00791 days or a little over 11 minutes. This discrepancy amounted to almost eight days in 1000 years and led to the calendar reform promulgated by Pope Gregory XIII in 1582. The *Gregorian calendar* omitted 10 days to bring the year into step with the seasons and thus with the equinoxes, and so brought the vernal equinox back to 21 March. A length of 365.2422 days was adopted for the length of the tropical year (equinox to equinox); this was 0.0078 days shorter than the Julian year. The difference amounts to 3.14 days every 400 years, and Gregory decreed that 3 out of 4 centennial years should not be leap years. In practice a centennial year is only a leap year when exactly divisible by 400. This ensures that the calendar only moves out of step with the equinoxes by one day every 3000 years.

The year continued to begin on January 1, having been altered from March to January in the Roman calendar in 153 B.C. However, in Britain December 25 was New Year's Day until the fourteenth century, when it became March 25 until the Gregorian calendar was adopted. In the nineteenth century Friedrich Wilhelm Bessel introduced a method of calculating the beginning of the year from the moment the "mean sun" (a fictitious sun orbiting the heavens at a uniform rate) reaches a right ascension of 18 h 40 min. This *Besselian year* is used in astrometry; thus epoch 1980.0 is taken to begin at the start of that Besselian year.

For religious reasons, there were delays in adopting the Gregorian calendar in some countries. In the United States and Great Britain this occurred only in 1752, and in Russia not until 1918.

THE LUNAR CALENDAR

In very early times every society seems to have adopted a lunar calendar rather than the astronomically more sophisticated solar calendar, and such a calendar has become incorporated into the feasts and festivals of many religions. The Islamic and Jewish calendars are of this kind. Such a calendar of months (a word derived from "moon") is independent and incommensurate with the year. The month may be defined essentially in two ways; one relates to the moon's orbital motion around the earth, the other to its cycle of phases. The former demands some astronomical expertise, but the latter, the *synodic* month, is a cycle of phases and is simple to observe and measure. It was used in earliest times, and its duration is 29.531 days.

Since 12 synodic months total 354.372 days, short of the tropical year by 10.87 days, long cycles were once calculated to try to find some way of reconciling the lunar and solar calendars. The most successful of these was the *Metonic cycle*, first devised in the second half of the fifth century B.C. by the astronomer Meton of Athens, who worked with Euctemon, another astronomer. Taking the synodic month as 29.5 days, and the tropical year as 365.25 days, Meton arranged his cycle on the assumption—already known before his time—that 19 years (6939.75 days) correspond closely to 235 synodic months or *lunations* (6932.50 days). He therefore devised a cycle of "full" 30-day and "hollow" 29-day months. Together these totaled 6,940 days, having 125 full months (3750 days) and 110 hollow months (3190 days). The advantage of the cycle was that it gave a definite rule for the intercalation of months into a lunar calendar to keep in step with the seasons; in addition it gave an improved long-term value for the tropical year. It was adopted by the Seleucid rulers in Mesopotamia as the basis of their reckoning. Around the time of the Council of Nicea (circa 325 A.D.)

it was used in the calendars of the Christian church and also by the Jews; it even influenced Indian astronomical reckoning. In the next century the astronomer Callippus improved the precision of the system by adopting a period of four Metonic cycles with a different arrangement for full and hollow months; its use was primarily confined to astronomers.

THE DAYS OF THE WEEK

In both solar and lunar calendars, it is convenient to subdivide the month into smaller intervals. The groups of days may vary in number, as also the time at which a day is supposed to start. Most primitive peoples reckoned from dawn to dawn, a method adopted in Babylonia and in Greece, but Egypt started the day at midnight, while the Jews and later the Italians reckoned from sunset to sunset. For convenience in observing the skies, astronomers from about the time of Ptolemy (second century A.D.), until as comparatively recently as 1925, reckoned the day to commence at noon. In the West each day was divided into 24 hours, 12 during daylight and 12 at night; these were known as "unequal hours" because periods of daylight varied with the annual procession of the seasons. In Babylonia, however, they used 12 hours to cover the whole period.

The day count varied in different places. In West Africa some peoples used a 4-day grouping; in central Asia it was 5 days; in Assyria 6 days; in Babylonia 7; in ancient Rome there was a 9-day "week"; whilst the Egyptians adopted a 10-day grouping and the Chinese a sexagenary cycle. The seven-day week with which we are now familiar may have arisen partly due to a grouping of 28 days for the visible phases of the moon (at New Moon the moon being invisible) and partly because of the Babylonian belief that seven was a sacred number; the latter due possibly to the existence of seven naked-eye "planets" (Mercury, Venus, Mars, Jupiter, Saturn, the Moon, and the Sun.) Certainly the names of the days of the week have connections with the seven planets, though in English-speaking countries these have been modified by Saxon usage. At all events, by the first century B.C. the Romans had adopted the seven-day week, then already in use among the Jews.

THE DATE OF EASTER

The Jewish religious calendar is lunar and the Christian church was forced therefore to find rules for calculating the date of Easter which celebrates the Resurrection—its most important festival—according to the civil solar calendar in use at any particular time. The Crucifixion was the basic date to be determined; this occurred three days before the Jewish Passover. The beginning of Passover falls at full moon, that is, on the 14th day of the first lunar month (Nisan) of the Jewish religious year. In practice this works out as the 14th day of Nisan, which falls either on or just after the vernal equinox. To simplify matters the churches of eastern Mediterranean countries used to celebrate Easter Day on the third day after the 14th day of Nisan, on whatever day of the week that happened to be; they now use a Sunday. The Western churches complicated the issue because they wanted to ensure that Easter Day fell on a Sunday during the week of Passover. Moreover, fixing the date for the Crucifixion was a matter of some difficulty and disagreement. Partly because of this and partly because of interpreting day counts based on the Jewish practice of reckoning day from sunset to sunset, Easter in the Eastern church is celebrated at a different time than in the Western churches.

In the West, once the date for the Crucifixion was fixed, two problems remained to be solved. One was to determine the days of the week for any date in any year; the second to determine the dates of full moon in any year. The first was achieved by assigning seven letters, *A* through *G*, consecutively to the days of the week. These *Dominical Letters* commenced on January 1. For the second, a full-moon cycle similar to the Metonic cycle was used. But such *Golden Numbers*, introduced in 530 A.D., proved to introduce errors, and were replaced officially at the time of the Gregorian calendar by a system of *Epacts* (Greek *epagein*—to intercalate), which did not depend on the Metonic cycle but on the more precise Gregorian calendar. Today, special computer programs have eased all the previous difficulties of calculation.

JULIAN DAYS

For determining the periods between historical events and for some astronomical work, it is convenient to have a sequential day-numbering system independent of weeks, months, and leap years. Such a system was devised by Joseph Justus Scaliger in 1582, and named by him after his father Julius Caesar Scaliger. Choosing a cycle of 7,980 years based on multiplying together various calendrical periods, this useful *Julian Day* numbering method begins on January 1, 4713 B.C.

Additional Reading

The calendar. In *Explanatory Supplement to the Astronomical Ephemeris and the American Ephemeris and Nautical Almanac* (third impression with amendments). U.S. Government Printing Office, Washington, D.C., 1974, p. 407.

Coyne, G. V., Hoskin, M. A., and Pederson, O., ed. (1983). *Gregorian Reform of the Calendar*. The Vatican Observatory, Vatican City.

Harvey, O. L. (1983). *Calendar Conversions by Way of the Julian Day Number*. American Philosophical Society, Philadelphia.

Ronan, C. A. et al. (1974). Calendar. *Encyclopaedia Britannica, Macropaedia* **3**, 15th edition. Encyclopaedia Britannica, Chicago, p. 595.

See also **Time and Clocks.**

Cameras and Imaging Detectors, Optical Astronomy

Patrick Seitzer

The last 10 years have seen a tremendous revolution in detectors for optical astronomy. In the mid-seventies, the majority of astronomers were using photographic plates for their observations. Today electronic detectors are used in almost all applications. They offer tremendous advantages in linearity, dynamic range, and repeatability over the photographic emulsion. The prime disadvantage is their small size (existing devices range up to a few centimeters on each side) as compared with a photographic plate (which can be up to 20 in. on a side).

Photography continues to play a major role in wide-area imaging. For example, sky surveys with Schmidt telescopes are still done with photographic plates. In addition, they offer a convenient long-term storage format (glass plates) that, in terms of information density, equals or exceeds that of electronic media.

CHARGE COUPLED DETECTORS

The most widely used detector is the charge coupled device (CCD). This device was invented at Bell Laboratories in 1970. Initial applications were for shift registers and computer memories.

As used in astronomy, a CCD is a two-dimensional grid of picture elements (pixels) on a very thin silicon wafer. Every pixel is sensitive to light: As light is incident on the detector, electron-hole pairs are produced. Deposited on the wafer is a structure of electrodes that, with the application of the proper voltages and the sequencing of these voltages between electrodes, can be used to control the positions of the packets of electrons and move them around the grid. An exposure is started by clearing the device of all charge and then opening the shutter. When the exposure is complete, the shutter is closed and the device is read out to an external computer by shifting the charge out row by row. All of the changes on the pixels in each row are shifted vertically into what is called the serial register located at one edge of the grid. Then this register is shifted pixel by pixel out one end to an amplifier and associated electronic circuitry. The amount of charge in each pixel is measured and converted to a digital number. Once this row has been completely read out, another row of pixels is shifted into the serial register. The final output in the computer consists of a two-dimensional array of numbers corresponding to the charge levels (and hence the amount of incident light) in the original pixels during the exposure. This digital picture is also known as an image.

Typical pixel sizes for current CCDs range from 7–30 μm. In general, the larger the pixel, the more charge it can hold before overflowing into neighboring pixels (an effect known as saturation). The capacity of a 30-μm-square pixel on some existing CCDs can exceed 500,000 electrons. Thus brighter objects can be imaged or longer exposures taken. Formats available commercially today range in size from 385×576 pixels to 1024×1024 pixels and even 2048×2048 pixels. Plans are underway to produce detectors with as many as 4096×4096 pixels, for a total of over 16 million pixels! Because each pixel typically has 2 bytes of digital information when represented in the computer, this means more than 32 megabytes of data every time the device is read out. (But for comparison, a large photographic plate can have the resolution of tens of thousands of pixels on each side.)

There are two major types of CCDs made by semiconductor manufacturers today.

1. The frame transfer CCD. The entire area of the device is sensitive to light, and a readout of one exposure must be complete before another exposure can be started. Astronomical observations almost exclusively use this type of CCD.

2. Interline transfer CCD. Each column of pixels is split into two pixels: a light-sensitive one and a second, covered one. Now during an exposure charge can be shifted rapidly from the first sensitive pixel into the second covered one. The rows of pixels can now be shifted out without closing the shutter, an important consideration for high-speed imaging. But the result is that only about half of the area is light sensitive, which reduces the usefulness of this sort of CCD for astronomy. It does find major applications in commercial imaging.

The prime advantages of CCDs are as follows.

1. They are very linear over a wide range of incident illumination levels. Typical nonlinearities in a carefully designed CCD system are significantly less than 1%.

2. When cooled to temperatures near −100°C, the dark current is very small and exposures of one hour or greater can be done with no significant problems.

3. Sensitivity over a very wide range of wavelength from less than 4000 Å to greater than one micron. The peak quantum efficiencies reported from various CCDs can range up to 80–90% in the 6000–8000-Å range, where the natural response of the silicon is greatest.

Many of the most astrophysically interesting observations are made in the ultraviolet portion of the spectrum (between 3200 and 4500 Å). Unfortunately, the response of a typical CCD tends to fall off very rapidly below 5000 Å to only a few percent or less due to absorption of photons in the silicon. Various techniques have been used to improve the response of CCDs in this region. The first involves thinning the CCD to cut down the thickness of the silicon and then illuminating the device from the back side (a typical CCD is illuminated from the front through the electrode structure). These thinned chips (as compared with the standard thick CCD) can have peak quantum efficiencies in the range of 10–50% around 4000 Å.

A second technique involves coating a thick CCD with a selection of chemicals that absorb photons in the ultraviolet but reradiate near 5000 Å. Such coating was done to the eight 800×800 CCDs that form the heart of the wide field and planetary camera on the Hubble Space Telescope and allow observations out to wavelengths near 2000 Å. The disadvantages of these coatings is that they are not nearly as efficient as thinning the CCD.

Even thinned CCDs can suffer from decreased ultraviolet (UV) sensitivity due to surface charging effects. To overcome this problem, UV flooding has been successfully implemented for some devices, which involves flooding the CCD with UV light near 2500 Å. Although very effective, the increased response is lost if the CCDs are allowed to warm (typically, to warmer than −50°C), and the flood must be repeated. More permanent techniques such as ion implantation or application of a thin film (flash gate) offer ways to obtain stable long-term UV sensitivity.

4. In a properly designed and operating CCD camera, the response of the system is very stable over long time scales. This is important for very accurate photometry and calibration, since calibration images are typically taken at a different time than the primary science images. For example, the response of the CCD is not the same for all pixels. Even the best CCDs have sensitivity variations of a few percent from pixel to pixel. This can be calibrated by measuring the response of the CCD to a perfectly spatially uniform light source and dividing this response into the original science picture on a pixel-by-pixel basis (a technique known as flat fielding). Careful calibration can lead to relative measurements that are accurate to better than 0.1% of the incident illumination.

In spite of these impressive advantages over other detectors, CCDs suffer from a number of problems:

1. When the charge is clocked out through the readout amplifier, electronic noise is added. For early CCDs this could be as much as several hundred electrons per pixel. The best current detectors have readout noise of around five electrons per pixel. Still, even this small number puts a limit to the signal-to-noise achievable in a given exposure time. Clearly, if one is trying to detect a signal of just a few electrons, a readout noise of an equivalent amount will seriously impact the observation.

2. The transfer of charge from one pixel to the next is not perfect. In the best of CCDs perhaps 99.999% of the charge will be transferred. For small-format devices, this is excellent. But consider the situation in a 4096×4096 detector. The charge packet in the far corner of the array will have to be shifted 4096 times to reach the serial register, and then another 4096 times to reach the output amplifier. With even 99.999% charge transfer efficiency at each step, only 92% of the original charge will be read out. Thus an image will appear progressively smeared.

3. Silicon detectors are very sensitive to high-energy particles (cosmic rays), which, upon interacting with silicon atoms, result in electrons being left in the detector. The number of cosmic rays detected in any CCD depends on the type of CCD (thinned devices being less sensitive than thick devices). Typical rates for a 512×512 CCD are 1–2 cosmic rays detected over the whole array per minute. Because these events occur at random over the area of the detector, one can identify them in the final digital images by comparing in the computer multiple exposures of the same field taken in sequence. The cosmic-ray rates can be much higher (by factors of 5–20 or more) for CCDs flown on orbiting space telescopes.

Despite these disadvantages, the overwhelming advantages of increased sensitivity and stability mean that CCDs are the detectors of choice for many astronomical observations today.

PHOTON COUNTING DETECTORS

Photon counting detectors can use a CCD as the heart of the detector system, but they operate in a fundamentally different way. Their prime advantage is the lack of readout noise, which makes them the choice for applications involving very low signal levels (such as high-dispersion spectroscopy or imaging through very narrow filters). The first implementation of such a detector system was by Alec Boksenberg, who used a television camera instead of a CCD.

Most photon counters couple an image intensifier (either an image-intensifier tube or a microchannel plate), used as the first stage of the system, to a CCD as the final detector. As a result of a single photon striking the front surface of the intensifier, a spot of several thousand electrons is produced at the back end. This can either impact the CCD directly or can cause an intermediate phosphor to generate photons that then illuminate the CCD. Most photon counters use interline transfer CCDs operated in a continuous-readout mode, so a complete picture is produced every tenth of a millisecond. The number of events (not the individual photons or electrons produced by the back of the image intensifier) is counted and recorded in a digital memory.

The major advantages are the very high time resolution obtained and the zero readout noise. Disadvantages are the low count rate (typically less than 1–5 counts per pixel per second before the device starts to suffer from coincidence losses), and the fragility of the front end of the system. Whereas it is nearly impossible to destroy a CCD as a result of overillumination, image intensifiers are very susceptible to permanent damage from bright sources. Nonetheless, photon counters have proved themselves to be complementary to CCDs and to excel in several areas of modern observational astronomy.

Additional Reading

Boksenberg, A. (1975). In *Image Processing Techniques in Astronomy*, C. de Jager and H. Nieuwenhuizen, eds. D. Reidel, Boston.

Boyle, W. and Smith, G. (1970). Charge coupled semiconductor devices. *Bell Syst. Tech. J.* **49** 587.

Janesick, J. R., Elliott, T., Collins, S., Blouke, M. M., and Freeman, J. (1987). Scientific charge-coupled devices. *Opt. Eng.* **26** 692.

MacKay, C. D. (1986). Charge-coupled devices in astronomy. *Ann. Rev. Astron. Ap.* **24** 255.

Robinson, L. B., ed. (1988). *Instrumentation for Ground-Based Optical Astronomy*. Springer, New York.

See also **Telescopes, Detectors and Instruments, Ultraviolet; Telescopes, Large Optical; Telescopes, Wide Field.**

Clusters of Galaxies

Michael J. West

Clusters of galaxies are gravitationally bound systems of tens, hundreds, or even thousands of galaxies. They are among the largest known objects in the universe, having typical dimensions of roughly 10–20 million light years across. Observations suggest that most galaxies belong to groups of various sizes, although it is estimated that only 5–10% of all galaxies in the universe reside in richly populated clusters. Thus, while very rich clusters of galaxies are certainly impressive behemoths in the cosmic menagerie, they are relatively rare. Nevertheless, thousands of clusters have been revealed by telescopes, and there are no doubt countless others as yet undetected.

The study of clusters of galaxies has yielded a wealth of astronomical information and has provided valuable insights into such diverse areas as cosmology, galaxy formation and evolution, high-energy astrophysics, and the origin of the large-scale structure of the universe. Clusters can be readily detected out to quite great distances, and hence can serve as beacons for probing the universe on large scales. For this reason, they play an essential role in astronomers' quest to determine the ultimate fate of the universe. Furthermore, because looking at greater astronomical distances means looking further back into the past, distant clusters may provide important clues as to what conditions were like in the early universe and may allow astronomers to actually observe how galaxies and clusters evolve with time.

CLASSIFICATION

Clusters of galaxies exhibit a wide range of properties. Just as it has proven advantageous for biologists to classify plants and animals based on certain shared characteristics or features, so too astronomers find it useful to recognize different morphological classes of clusters. A number of different schemes have been devised for classifying clusters.

The *richness* of a cluster is a measure of how many galaxies comprise it, which is determined by counting the number of bright galaxies within a given distance of the cluster center. Thus, astronomers classify a cluster as either *poor* or *rich* depending on whether it has relatively few or many member galaxies. In addition to the observable cluster galaxies, there are no doubt also many smaller member galaxies which are too faint to be seen at the great distances of most clusters from earth.

A cluster may also be classified as *regular* or *irregular*, depending on its overall appearance. Regular clusters, which are usually quite rich, show a smooth, centrally concentrated galaxy distribution and an overall symmetric shape. The closest example of a very rich, regular cluster is the Coma cluster which lies at a distance of approximately 70 Megaparsecs (1 Mpc = 3.26 million light years) from earth. Irregular clusters, on the other hand, have a generally amorphous appearance, usually showing little overall symmetry or central concentration. They are often composed of several distinct clumps of galaxies referred to as *subclusters*. The majority of clusters are irregular. The nearby Virgo cluster, at a distance of ~ 15 Mpc, is a good example of this type of cluster.

Clusters can also be classified according to their galaxy content. The galaxy populations of rich regular clusters tend to be composed primarily of elliptical and S0 (lenticular) galaxies, with only a small fraction of spirals. In the Coma cluster, for example, as many as 80% of the brightest galaxies may be ellipticals, with the few spirals that are found relegated to the outer regions of the cluster. Irregular clusters, on the other hand, have a more even mixture of different galaxy types. Nearly half of all galaxies in the Virgo cluster are spirals. Thus, astronomers sometimes classify clusters according to the fraction of spiral galaxies that they possess, making a distinction between spiral-rich and spiral-poor clusters.

More sophisticated cluster classification schemes have also been proposed. One example is the Rood–Sastry classification scheme, which differentiates between six basic cluster morphologies: cD-dominated (cD, see next section), binary (B), linear (L), core-halo (C), flattened (F), and irregular (I). A cluster is assigned to a particular morphological classification depending on its dominant characteristic feature. A very flattened cluster, for instance, would belong to Rood–Sastry class F, while a cluster that is dominated by two large galaxies at its center would be a B cluster in this system.

There is a good deal of overlap between various classification schemes, which indicates that different cluster properties are correlated with one another. For instance, very rich clusters are almost always regular in appearance and spiral poor in terms of their galaxy population. It is important to emphasize, however, that cluster classification schemes are merely a crude but useful means of trying to search for gross similarities between clusters. Clusters, of course, exhibit a continuum of properties and recognize no sharp divisions between the different morphological classes proposed by astronomers.

Several catalogs listing the locations and some of the basic properties of thousands of clusters have been compiled. The most widely used are those of George Abell and Fritz Zwicky, who each identified clusters using somewhat different criteria. These catalogs were compiled several decades ago from laborious searches of photographic surveys of the sky. More recent developments in cluster cataloging include the use of computers for automated cluster searches, which are faster and much less prone to observer bias. Catalogs of clusters are useful because they provide the foundations for systematic studies of clusters and their properties. Numerous detailed studies of individual clusters have been published in the astronomical literature, with more and more such studies appearing every year.

STRUCTURE

The number of galaxies found in a rich cluster is typically two orders of magnitude greater than the number that would be found on average in some randomly selected region of space of comparable volume. The distribution of these galaxies usually shows a clear tendency to be centrally condensed, i.e., to have a high concentration of galaxies near the cluster center which decreases further out. The central densities in rich clusters may reach 10,000 or more times the mean density of galaxies in the universe. Clusters of galaxies do not have sharp, well-defined boundaries; rather, their galaxy distributions gradually fade into the cosmic background of galaxies. Their high densities and degree of central concentration provide compelling evidence that most rich clusters are genuine gravitationally bound, dynamical systems rather than chance groupings of unrelated galaxies.

Cluster shapes range from spherical to quite flattened, with most clusters showing some degree of elongation. Statistical studies suggest that clusters are intrinsically flatter on average than elliptical galaxies. There is also intriguing observational evidence that the long axes of neighboring clusters may tend to point at one another, even when the clusters are separated by quite large distances, which may be telling us something quite interesting about the way in which clusters have formed.

As mentioned earlier, the galaxy population in rich clusters of galaxies appears in general to differ from that found over most of the sky. In the central regions of rich, regular clusters, elliptical and S0 galaxies are by far the most common galaxy types, with only a very small fraction of spiral galaxies found. This is the converse of the situation outside of clusters, where the majority of galaxies in isolation or in small groups are spirals. Why different galaxy populations are found in and out of clusters is a matter of some debate. Some astronomers interpret this difference as implying that spiral galaxies have somehow been transformed into S0 or elliptical galaxies within clusters. Others argue, however, that the different galaxy populations reflect intrinsic differences between the galaxy-formation process in different environments, with conditions in rich clusters favoring the formation of elliptical and S0 galaxies over spirals.

Also lurking in the central regions of many rich clusters are the largest known galaxies in the universe, called cD *galaxies*. These supergiant galaxies are unique to the dense environments of rich clusters. A cD galaxy is characterized by a very extended envelope of diffuse stellar light surrounding what may be an otherwise normal elliptical galaxy nucleus. The visible extent of a cD galaxy may be as much as 10–100 times that of a normal galaxy. Although the densities of the cD envelopes are quite low, they extend to such great distances that the total luminosity associated with these galaxies is extremely large. Many cD galaxies exhibit more than one nucleus. Whether cD galaxies represent simply the extreme end of the normal distribution of galaxy luminosities, or are instead a unique type of galaxy which has formed or evolved under special conditions in clusters, has still not been fully resolved. The fact that many cD galaxies have multiple nuclei suggests that they may have grown through mergers of smaller galaxies.

For decades, most observations of clusters of galaxies were done in the visible wavelengths of light. In recent years, however, great emphasis has also been placed on observing clusters at other wavelengths, such as x-ray and radio, as these can provide different insights into the structure and properties of clusters and their constituent galaxies. Many clusters are now known to be strong emitters of x-rays, which indicates that, in addition to the visible galaxies, they also possess an *intracluster medium* of hot, rarefied gas at temperatures of $\sim 10^7$–10^8 K. The amount of mass in the form of gas in rich clusters is probably comparable to the total mass in the form of visible galaxies. Spectral analysis indicates that this gas contains iron and other heavy elements which must have been produced in stars, suggesting that much of the intracluster gas may have once been located within individual galaxies and was subsequently liberated. Radio observations of the hydrogen gas content of cluster galaxies also show that those galaxies in the highest density regions are systematically lower in gas, suggesting that the gas has been removed from these galaxies.

DYNAMICS

The smooth, centrally condensed distribution of galaxies within rich, regular clusters suggests that these clusters have reached a state of *dynamical equilibrium*, sometimes referred to as *relaxation*. Irregular clusters, on the other hand, are probably unrelaxed systems, having not yet reached a steady state. Thus, it seems likely that different cluster morphologies correspond to different stages of dynamical evolution. Computer simulations of the formation and evolution of clusters of galaxies support this notion.

Galaxies in a cluster move under the influence of gravity in orbits about their common center of mass. Although it is not possible to observe the true spatial velocities of galaxies, redshift measurements make it possible to determine the radial velocities (i.e., the component of velocity along the line of sight) of galaxies in clusters. Such measurements show that the member galaxies are all moving with different speeds about the cluster. A cluster can be characterized by its *velocity dispersion*, which is a measure of the typical speed with which the galaxies are moving in the cluster. The velocity dispersion of a rich cluster is typically ~ 1000 km s^{-1}. The velocity dispersion in many clusters shows a systematic decrease as the distance from the cluster center increases. In the Coma cluster, for instance, the velocity dispersion near the outskirts of the cluster has fallen by roughly a factor of 2 from its peak value at the cluster center.

Given the observed size of a cluster and the velocity dispersion of its constituent galaxies, it is straightforward to compute the *crossing time* of the system, which is simply the time required for a galaxy moving with a speed equal to the cluster velocity dispersion to travel from one side of the cluster to the other. Crossing times in the inner regions of rich clusters are generally on the order of 10^9 years. Assuming the age of the universe to be between 10 and 20 billion years, this means that galaxies in the cores of clusters have had sufficient time to make several traversals of the system. This suggests that clusters of galaxies are well mixed and provides further evidence that rich clusters are dynamically stable systems, since the galaxies would otherwise long ago have dispersed if the cluster was not held together by gravity.

Assuming that clusters of galaxies are in dynamical equilibrium, it is possible to determine their masses using the *virial theorem*. The virial theorem relates the total mass of a self-gravitating system of stars or galaxies to its size and velocity dispersion,

$$M = \frac{V^2 R}{G},$$

where M is the total mass of the cluster, R is its characteristic size, V is the cluster velocity dispersion, and G is the gravitational constant. The essence of the virial theorem is quite simple. Galaxies in a cluster are in motion because they are accelerated by the total gravitational force that they experience from every atom that is present in the cluster. Therefore, the greater the total mass of the cluster, the greater the gravitational tug felt by each galaxy, and, consequently, the greater the speed with which the galaxies move in their orbits. Thus, by observing the velocity dispersion and characteristic size of a cluster, astronomers can infer the total mass that is present in that system. Typical masses of rich clusters of galaxies obtained using the virial theorem are $\sim 10^{15}$ solar masses.

Another way that one can determine the mass of a cluster is by simply counting the number of galaxies that it contains. Astronomers have a fairly good idea of how much mass there is in a typical galaxy, and so by adding up the contribution from each galaxy that is observed within a cluster one gets an estimate of the total cluster mass. This number should probably be doubled to also include the intracluster gas.

When Fritz Zwicky used these two methods to compute the mass of the Coma cluster in a seminal study over 50 years ago, he made one of the most important discoveries in modern astronomy. What Zwicky found was that the virial mass that he had computed for the Coma cluster was some 400 times greater than the mass that could be accounted for in the form of visible galaxies! A subsequent study of the Virgo cluster by Sinclair Smith in 1936 suggested that the total mass of that cluster was roughly 200 times the amount of the observable luminous matter. Thus was born one of the great puzzles of modern astronomy, which has come to be known as the *missing mass problem*. This discrepancy between the virial mass and visible mass has since been found in many other clusters as well, although modern estimates are that the total cluster mass is more likely 10 times the mass contained in galaxies and gas. The missing mass problem is one of the most fascinating discoveries to come out of the study of clusters of galaxies and remains one of the major unsolved problems of modern astronomy. What it may be telling us is that as much as 90% of the matter in rich clusters (and perhaps the universe) is in a form that is invisible and can be detected only by the gravitational influence that it exerts on visible matter such as galaxies and stars. In a sense, then, the term "missing" mass is really a misnomer—it is the light, rather than mass, that is truly missing! Virial mass determinations tell us beyond any doubt just how much mass must be present in a cluster because we can see its gravitational influence on the motions of the cluster galaxies, but the fact is that we are unable to observe this dark matter directly. What the dark matter may be is still pure speculation at present. There have been many suggestions, ranging from faint low-mass stars, to black holes, to exotic forms of matter predicted by particle physics theories.

There are several interesting dynamical processes that can occur in the dense environments of clusters of galaxies. In an individual galaxy, the odds of two stars colliding are extremely small because of the vast spaces between them; the average distance between stars in a galaxy is millions of times greater than the size of a typical star. However, the situation is very different in clusters of galaxies, where the high density of galaxies means that the separation between neighboring galaxies is often only a few times greater than their sizes. Frequent collisions between galaxies are therefore very likely to occur in clusters. Such collisions can have several important consequences. If two galaxies pass near each other at relatively low speeds, their mutual gravitational attraction may cause them to merge together, resulting in a single, larger galaxy. Two galaxies can also pass right through one another, sweeping out each other's interstellar gas in the process (even though their stars will rarely collide), which may be heated to high temperatures. The large gravitational forces exerted as two galaxies pass very near one another can also remove loosely bound stars from the outer regions of each of them in a process known as *tidal stripping*. The halos of cD galaxies may be composed largely of tidally stripped material captured from other galaxies. It is also possible that many of the largest galaxies in clusters may have grown by devouring smaller neighboring galaxies in a process referred to as *galactic cannibalism*.

Since gravity is a long-range force, distant encounters between galaxies in a cluster that do not involve direct contact can still have very important consequences for the overall cluster evolution. Gravitational interactions between galaxies allow energy to be exchanged between them. During such *two-body interactions*, more massive galaxies will tend to give up some of their energy of motion to smaller ones and as a consequence will slowly sink to the center of the cluster, while the energy gained by the less massive galaxies will cause them to move in orbits which take them further away from the cluster center. Given enough time, this mechanism can lead to *mass segregation* between the high-mass and low-mass galaxies. Estimates vary, but in general the two-body relaxation time for an average galaxy in a rich cluster is expected to be quite long, roughly 10^{11} years or more. A related effect is *dynamical friction*, which occurs when a massive galaxy moves through a background of smaller mass galaxies or dark matter. As the heavy galaxy moves through the cluster, its large gravitational tug will deflect background objects from their original orbits such that they form a wake behind the path traveled by the massive galaxy. This then sets up a sort of feedback mechanism whereby the wake that has been created by the passage of the massive galaxy now exerts a gravitational tug back on it, causing this galaxy to slow down in its orbit and eventually spiral in towards the cluster center. The efficiency of dynamical friction is proportional to galaxy mass, and thus the time scale for this process can be significantly shorter than the standard two-body relaxation time for very massive galaxies, perhaps as short as $\sim 10^9$ years. Observationally, the extent to which mass segregation has already occurred in rich clusters is debated at present. Although there is no doubt that the very largest galaxies are invariably found at the centers of clusters, the overall amount of mass segregation which has already occurred appears to be marginal. The fact that more pronounced mass segregation is not observed tells us something very interesting about the distribution of the dark matter in clusters. If all the missing mass was bound to individual galaxies (in the form of dark halos, for example), then the time scale for relaxation would be much less than the age of the universe, and thus one would expect to see strong mass segregation in clusters. The fact that this is not observed indicates that most of the dark matter cannot be attached to individual galaxies, but rather must be distributed smoothly throughout the entire cluster.

Additional Reading

Bahcall, N. A. (1977). Clusters of galaxies. *Ann. Rev. Astron. Ap.* **15** 505.

Dressler, A. (1984). The evolution of galaxies in clusters. *Ann. Rev. Astron. Ap.* **22** 185.

Gorenstein, P. and Tucker, W. (1978). Rich clusters of galaxies. *Scientific American* **239** (No. 5), 110.

Shu, F. H. (1982). *The Physical Universe*. University Science Books, Mill Valley.

Struble, M. F. and Rood, H. J. (1988). Diversity among Galaxy Clusters. *Sky and Telescope* **75** 16.

See also **Clusters of Galaxies, Component Galaxy Characteristics; Clusters of Galaxies, Radio Observations; Clusters of Galaxies, X-Ray Observations; Dark Matter, Cosmological; Galaxies, Properties in Relation to Environment.**

Clusters of Galaxies, Component Galaxy Characteristics

Augustus Oemler, Jr.

A few percent of galaxies are members of rich clusters, containing hundreds of galaxies in volumes a few megaparsecs across. The existence of such clusters was recognized early in this century, and they became important tools in pioneering studies of galaxies by Edwin Hubble, Fritz Zwicky, and others. Because all cluster members are at the same distance from us, one redshift (which determines the distance by Hubble's law) suffices to provide the distance to all. Cluster members are, therefore, the easiest galaxies to study, and much of what we know about galaxies is derived from them. However, what is learned about cluster galaxies can be applied to galaxies in general only if the former are typical. It was soon realized that they are not: Cluster galaxies differ in a number of ways from those which lie outside of rich clusters, in what is often termed the field. Although this fact limits their usefulness as exemplars of the general galaxy population, it provides a most important tool for understanding the processes of galaxy formation and evolution. The differences between cluster and field galaxies are presumably due to the differences in the environments in which they occur. This, by itself, is significant, and shows that galaxies have a sensitivity to their surroundings that stars, for example, lack. If one can understand how the environment has shaped cluster galaxies, one will have gone a long way towards understanding the processes governing the formation and evolution of all galaxies.

THE GALAXY CONTENT OF CLUSTERS

Morphology

The most obvious difference between cluster and field galaxies is in their morphological types. The majority of field galaxies are gas-rich spirals and irregulars; gas-poor ellipticals and S0s are outnumbered by about 2 to 1. In contrast, the centers of rich clusters are dominated by the latter types. The few spirals seen are probably chance superpositions: field galaxies or outlying cluster members lying along the line of sight to the cluster center. The shift in galaxy populations is not abrupt: There is a continuum from spiral-dominated to S0-dominated populations as one moves from the most remote field to the densest cluster centers. The S0-to-spiral ratio also depends on the nature of the cluster: populous, dense, symmetric clusters have fewer spirals than do poorer, more irregular clusters. In addition to this dependence on the global cluster properties, it has been shown that galaxy populations are sensitive to the local environment: The ratio of S0s to spirals increases with the number of neighboring galaxies.

There is reasonably good evidence for a continual variation in environmental sensitivity as one moves along the Hubble sequence of galaxy types, from ellipticals, through S0s, to spirals and irregulars. As the density of the environment increases, irregulars disappear first, followed by late-type spirals. The abundance of S0s rises as spirals disappear, but only at the highest densities does the abundance of ellipticals begin to increase. This suggests that the effect of environment is to skew the distribution of galaxies along the Hubble sequence towards one or the other end, but such an interpretation is by no means unavoidable. As important as the differences between cluster and fields populations are the similarities: Cluster and field galaxies are both drawn from the same Hubble sequence; cluster ellipticals look approximately like field ellipticals, and cluster Sc galaxies like field Scs.

Galaxy Luminosities

If environment influences what kind of galaxies form, it might also influence how bright they are. Furthermore, the luminosity function of ellipticals, at one end of the Hubble sequence, differs markedly from that of irregulars, at the other end. For both these reasons, one might expect the luminosity function of cluster galaxies to differ from that of field galaxies. If such differences exist, however, they are quite subtle. The gross shape and characteristic luminosity of the cluster-galaxy luminosity function is almost indistinguishable from that of field galaxies. The small differences which do exist are attributable to processes involving the brightest few galaxies, as discussed below.

THE STRUCTURE OF CLUSTER GALAXIES

In the cores of populous clusters, the average separation between galaxies is less than 10 times their size. One would, therefore, expect frequent close encounters between galaxies, during which their mutual gravitational forces could affect their internal structure. The tidal forces of encounters can remove the outer envelopes of the galaxies. In the rarer event of interpenetrating collisions, the process of dynamical friction can remove enough of the orbital energy of the galaxies to cause them to merge.

Direct evidence for tidal perturbation of cluster galaxies is difficult to find. Optical evidence of tidal interactions is much rarer in clusters than in small groups. This is not in itself unexpected. It can be shown that most tidal stripping of galaxy envelopes should have occurred early in the history of the cluster. Furthermore, tidal encounters become less effective when the relative velocities of the galaxies are much higher than the velocities with which their stars orbit within them; in most rich clusters this ratio is of order 2–10. The only clusters in which many tidal encounters can be seen are that rare subset with irregular structure and a large population of spirals. This suggests that these clusters are, in fact, quite young, and have yet to reach equilibrium. In their present state, the relative velocities of neighboring galaxies may be sufficiently low to permit effective tidal encounters. There is, indeed, very little evidence that significant tidal stripping ever occurred in most clusters. A comparison of the structure of elliptical galaxies in clusters and the field shows little sign that the outer envelopes of the former have been tidally truncated.

There is somewhat better evidence for the occurrence of mergers, driven by dynamical friction. The brightest member of a cluster, which is always a giant elliptical and usually located in the cluster center, often has a number of satellite galaxies located within its outer envelope. Observations of the velocities and positions of the satellites, and of tidal distortions due to interaction with the brightest cluster member, suggest that some fraction of these satellites may be merging with the central galaxy. Such a process will cause the central galaxy to grow at the expense of the smaller galaxies. There is also some evidence of the results of this process in the luminosity functions of the galaxies in clusters with a dominant central galaxy.

Although the dominant cluster member is always an elliptical, it is not always a normal one. W. W. Morgan and his collaborators first pointed out that some are sufficiently unusual to warrant a special class, which they named cD. cD galaxies are characterized by elliptical-like centers on which are superimposed very extended, low surface brightness envelopes. The total size and luminosity of some envelopes is remarkable, with diameters greater than 2 Mpc (10 times that of a large elliptical) and total luminosities 10 times that of the underlying central galaxy. These galaxies are clearly products of the cluster environment in a way unique for galaxies. They only occur as brightest members of clusters. Their envelopes are so extended that the outer parts must be considered part of the cluster as a whole, rather than of the central galaxy. Most telling, the total luminosity of the envelope is a unique function of the total cluster richness.

It is most likely that cD galaxies began as normal luminous ellipticals which, because of their mass, sank to the center of their clusters during the period of cluster formation. Once there, they were in a favorable position to accrete material from within the cluster. This material might be smaller galaxies, which were captured by dynamical friction, or it might be tidal debris stripped from other galaxies. Neither possibility is entirely consistent with what is known about cluster galaxies. Material from mergers will be distributed over a much smaller area in the cluster center than that covered by the cD envelopes. Tidal debris would be expected to have a broader distribution, but, as mentioned earlier, there is little evidence that cluster galaxies have lost much of their outer envelopes to tidal stripping.

THE CONTENTS OF CLUSTER GALAXIES

The difference between the galaxy populations of the field and of clusters is essentially a difference between galaxies which contain significant amounts of gas and young stars and those which do not. This is probably the only difference between the spirals which dominate the field population and the S0s which dominate clusters. The lack of young stars is confirmed by detailed study of the colors and spectral energy distributions of cluster galaxies, which show that most contain only old stars. A similar difference in the gas content is also well documented. This persists when one examines each Hubble type separately. S0s seldom contain much gas, but field S0s contain substantially more than do those in clusters. Those spirals which survive in clusters are much depleted in gas compared to their field counterparts. Interestingly, this gas depletion seems to be confined to the outer parts of a galaxy, beyond the region in which star formation normally occurs. In the inner parts, the gas content appears to be normal, which is why the galaxies are still able to make the new stars which label them as spirals.

One cannot find, among the now-quiescent elliptical and S0 cluster galaxies, much evidence for recent star formation. However, examination of clusters at earlier epochs shows that clusters contained many more star-forming galaxies in the recent past. Because of the finite velocity of light, distant clusters are observed at earlier times, which allows one to examine the properties of cluster galaxies over about the last third of the age of the universe. The fraction of cluster members whose colors indicate recent star formation has decreased from about 25% five billion years ago to a few percent today.

We do not yet know whether these blue galaxies are identical to the spirals seen in nearby clusters. Their distribution within the cluster is similar, and there are some indications that they are disk galaxies, rather than star-forming ellipticals. However, their spectral energy distributions suggest that a significant fraction may have undergone large bursts of star formation shortly before the epoch at which they are observed. This is in contrast to today's spirals, in which star formation appears to have decreased at a slow and steady rate throughout their lifetimes. Large bursts of star formation are rare among galaxies today, except for those which are interacting. These observation suggest a fundamental evolution in the cluster environment, or in the properties of galaxies, over the last third of the age of the universe.

WHY ARE CLUSTER GALAXIES DIFFERENT?

There are three possible routes by which the population of cluster galaxies might have come to differ from that in the field.

1. Only elliptical and S0 galaxies were formed in the cores of clusters.
2. The spiral galaxies which formed in clusters were of a type which exhausted their gas more rapidly than the typical field galaxy; thus evolving, on their own, into S0s.
3. Cluster populations were originally identical to those in the field, but have been transformed by the cluster environment.

The existing evidence is contradictory. The discovery of a blue population in distant clusters would seem to rule out possibility 1, but it is possible that those galaxies have faded into obscurity as their starbursts died, leaving an unchanging primordial population of E/S0s. The observation that the dependence of galaxy type on local density extends to regions beyond the cores of rich clusters favors 1 and 2, and suggests that the density of the environment out of which the galaxy formed is the proximate cause of its morphological type. However, models of the formation of clusters indicate that the environment near a galaxy at its birth is only loosely related to its environment today, and early differences may not have been great enough to affect the morphological type of a forming galaxy.

The strong correlation of global cluster properties and galaxy types favors possibility 3, but effective mechanisms to alter the properties of galaxies have been difficult to find. It has been suggested that interactions between galaxies, and between galaxies and the hot gaseous intracluster medium, could remove the gas from the disks of spirals and transform them into S0s. Unfortunately, the observations of gas in cluster galaxies mentioned earlier show that gas does not get stripped from those regions of the galaxies in which star formation usually occurs. Furthermore, the existence of population differences outside of the cores of clusters cannot be explained easily by stripping mechanisms. It is possible that the evolution of cluster galaxies and their present properties are the result of a number of independent processes, acting in concert over the entire epoch of galaxy formation and evolution, to produce the populations that we observe today.

Additional Reading

Bahcall, N. A. (1977). Clusters of galaxies. *Ann. Rev. Astron. Ap.* **15** 505.

Dressler, A. (1984). The evolution of galaxies in clusters. *Ann. Rev. Astron. Ap.* **22** 185.

Haynes, M. P. (1988). Morphology and environment. In *The Minnesota Lectures on Clusters of Galaxies and Large Scale Structure, Astronomical Society of the Pacific Conference Series No. 3*, J. M. Dickey, ed. Astronomical Society of the Pacific, San Francisco. p. 71.

Sandage, A. (1990). Properties of galaxies in groups and clusters. In *Clusters of Galaxies*, W. Oegerle, L. Danly, and M. Fitchett, eds. Cambridge University Press, Cambridge.

Strom, S. E. and Strom K. E. (1979). The evolution of disk galaxies. *Scientific American* **240** (No. 4) p. 72.

See also **Clusters of Galaxies; Galaxies, Formation; Galaxies, Properties in Relation to Environment.**

Clusters of Galaxies, Radio Observations

Riccardo Giovanelli

A large fraction of the mass of rich clusters of galaxies that emits electromagnetic radiation is in the form of a diffuse (density 10^{-3} atoms cm^{-3}) intracluster medium (ICM), macroscopically at rest in the cluster gravitational potential well. This gas is hot (10^8 K) and especially conspicuous in x-rays, which are produced by thermal bremsstrahlung.

Individual galaxies in clusters are, of course, radio emitters, as are those outside of clusters; however, the location of cluster galaxies in a high-density environment brings about peculiar characteristics which are usually attributed to their interaction with the ICM or to the exceptional character of the objects dwelling in the cluster cores (e.g., cD galaxies). The radio properties of individual galaxies, including radio galaxies associated with cDs, are described elsewhere in this volume. Here we discuss the characteristics of the radio emission arising from the interaction of radio sources with the ICM, as well as the global radio properties associated with the ICM itself.

CLUSTER HALO AND RELIC SOURCES

A handful of clusters of galaxies are known to harbor diffuse sources of radio emission which cannot be attributed to single galaxies in the cluster. The best studied of these sources is associated with the Coma cluster. Cluster *halo* sources are concentric with the distribution of galaxies, have very low surface brightness, and are thus difficult to distinguish from a population of distributed discrete, weak sources, or from the outskirts of strong, centrally located radio galaxies. The mean properties of halo sources are derived from a few clusters where clear identification and distinction from other cluster radio sources have been made. Halo sources appear to have steep spectral index (i.e., the radio flux drops with increasing frequency ν with a power larger than -1, typically as $\nu^{-1.2}$), a moderately high radio luminosity (about 10^{31}–10^{32} erg s^{-1} Hz^{-1} at a wavelength of 21 cm or 10^{41} erg s^{-1} over the whole radio spectrum) and large size (diameter of about 1 Mpc). Clusters that contain radio halo sources have very high x-ray luminosities; their galaxian population appears compact and dynamically evolved and is characterized by a high velocity dispersion. These clusters have a rich, widely distributed ICM and lack a single dominating central cD galaxy.

Relic sources are also found in clusters. Their properties resemble those of halo sources: They are extended, diffuse, without an optical counterpart, and have steep radio spectra. They are, however, not as extended as halo sources and are not centrally located in the cluster. It is postulated that relic radio sources are ejecta of now-quiescent radio galaxies that have moved away from the scene. It has also been suggested that a cluster halo source is produced by the collective radio emission of a superposition of relic radio sources in the cluster.

The steep nonthermal spectra of these radio sources suggests that the emission process is synchrotron radiation by a cluster population of relativistic electrons in an intracluster magnetic field that needs not be stronger than 1 μG.

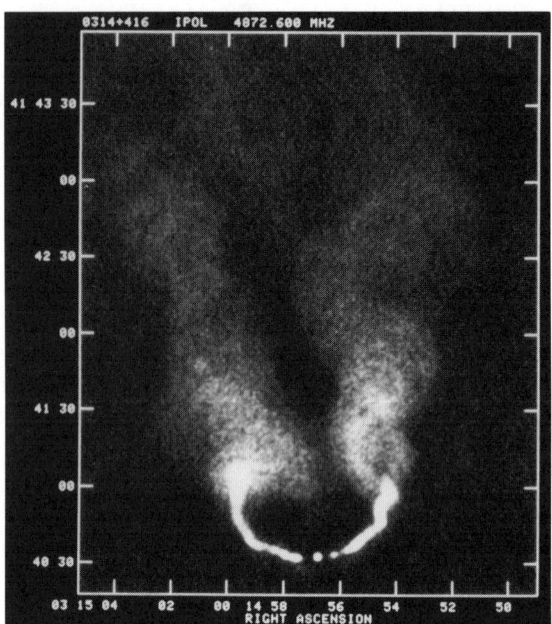

Figure 1. Total intensity map of the radio source in NGC 1265 in the Perseus cluster. The image was obtained by C. O'Dea and F. Owen at a frequency of 1413 MHz with the Very Large Array radio telescope of the National Radio Astronomy Observatory.

ACTIVE RADIO GALAXIES AND THE INTRACLUSTER MEDIUM

A "standard" radio galaxy has a compact radio source associated with the active nucleus of an optical galaxy; part of the radio emission is observed in the form of radio *lobes*, extended regions of emission diametrically opposed with respect to and quite distant from the compact radio source. Narrow *jets* originating in the central compact source extend out to the lobes and are the conduits by which energy is carried from an active region deep in the core of the central source to the lobes. The energy is carried by a mixture of relativistic and thermal gas, which outlines the jets. In clusters of galaxies, the rapid motion of the central source and the interaction of the gas outflow with the ICM are thought to be responsible for the observed spectacular departures from alignment of the radio lobes. As the angle between the direction from the central source to the lobes progressively departs from 180°, the terminology describing the radio source morphology varies from *wide-angle tails* (WATs) to *narrow-angle tails* (NATs) to *head-tail* sources, the latter of striking cometary appearance. Figure 1 illustrates an archetype of the NAT category, the source associated with the elliptical galaxy NGC 1265 in the Perseus cluster. The conventional model to explain head-tail or NAT sources pictures them as conventional radio galaxies moving at high velocity through the ICM. As plasma beams are quasicontinuously ejected from the active nucleus of the galaxy, they are bent either by direct ram pressure or, if there is a significant interstellar medium in the galaxy, the latter forms a cocoon around the plasma beams and the pressure gradients created by the motion of the galaxy through the ICM ultimately cause the bending of the plasma beams. The ram pressure acting on the ejected gas is $P = n_{\text{icm}} v_g^2$, where n_{icm} is the ICM density and v_g the velocity of the galaxy relative to it. While the basic idea of plasma ejecta interacting with the ICM is still the main driver in explanations of the peculiar cluster radio source morphology, many aspects of the interpretation are still quite uncertain, and are related to the poorly known circumstances associated with the process of ejection, with the difficulty of deprojecting two-dimensional images into three-dimensional representations and with the effects of inhomogeneities in the ICM.

INTERSTELLAR GAS DEFICIENCY AND CONTINUUM DISK EMISSION IN CLUSTER SPIRALS

Normal spiral galaxies are characterized by disks containing a fair fraction of their mass in the form of interstellar gas. The neutral component of that gas, primarily hydrogen, extends much farther out from the center of the galaxy than the stellar component. This gas is endowed with a large specific angular momentum, which supports it in equilibrium to the outer reaches of the galaxy's potential well. As a result, this gas is quite vulnerable to external dynamical and thermal perturbations. As a spiral galaxy travels at high speed through the denser regions of a cluster, its interaction with the ICM can sweep a large fraction of its interstellar gas off its outer disk. The mechanics of the interaction are difficult to model; the sweeping efficiency depends on the parameters that determine the ICM ram pressure, the angle between the galaxy's vector of motion and that defining the inclination of its disk, the degree of clumpiness of the interstellar gas, the galaxy's total mass, and the degree of thermal shielding provided by its gaseous corona. Thermal conduction between the cold interstellar gas and the hot ICM can, in fact, play as important a role as ram pressure in the sweeping episode.

Interstellar neutral hydrogen is easily detected in external galaxies by means of its line emission at 21 cm. Observations of the H I line have shown that a spiral galaxy moving through the inner parts of a dense cluster core can have most of its interstellar neutral hydrogen removed from the disk. The total neutral hydrogen mass of a normal spiral galaxy ranges between 1 and 30 billion solar masses. A single high-velocity passage through the cluster core can free much of that gas from the galaxy, transferring it to the ICM. A spiral galaxy is called H I-deficient when its H I mass is at least 2–3 times smaller than that found on the average in noncluster spirals of the same type and size. On the other hand, radio observation of the 2.6-mm line of CO in H I-deficient spirals indicate that the molecular gas content of those galaxies is little

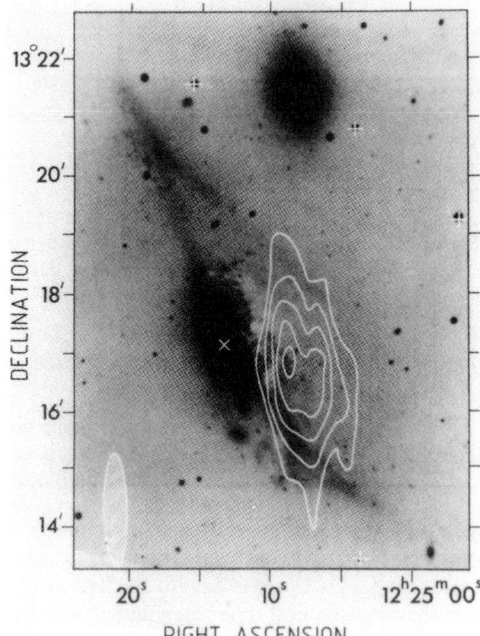

Figure 2. Contours of radio emission of the displaced disk of NGC 4438, superimposed on an optical image of the galaxy. Also in the picture (top) is the companion galaxy NGC 4435. The radio observations were made with the Westerbork Synthesis Radio Telescope at 1.4 GHz by C. Kotanyi and R. Ekers.

affected by the sweeping events that deplete its neutral gas reservoir. The explanation of this difference may be in the different degree of clumpiness and radial distribution in the galactic disk of the two components of the interstellar medium: The diffuse, more peripheral H I presents a larger cross-section to ram pressure forces than the clumpy, more centrally located molecular gas and it is less tightly bound to the galaxy than the molecular gas. It has been suggested that the removal of the neutral component of the interstellar medium has a delayed, quenching effect on the star-formation rate in the swept galaxies. This "interruption of fertility" does eventually result in an aging of the stellar population, an obliteration of the spiral pattern, and the conversion from spiral to lenticular morphology. It has been proposed that the marked morphological segregation of galaxy types observed in cluster cores, whereby ellipticals and lenticulars dominate the core population, may in part be due to this "secular," environment-driven process.

Disks of spiral galaxies also harbor a diffuse population of relativistic electrons, which are produced continuously by localized stellar sources. The interaction of these electrons with a galactic magnetic field, supported by the interstellar gas, produces synchrotron radiation, providing the main contribution to the radio emission of normal spiral galaxies. The ram pressure produced by the transit of a spiral galaxy through the ICM can produce large-scale displacements of its interstellar medium, which will carry along field and relativistic electrons. Thus, images of radio continuum disks can appear displaced from the stellar disks of galaxies moving through the ICM. Figure 2 illustrates the point for the galaxy NGC 4438, which is known to be moving at high speed through the core of the Virgo cluster.

COOLING FLOWS IN CLUSTER ELLIPTICALS

The ICM radiates mainly in the x-ray energy band, via thermal bremsstrahlung. This process provides the primary means of cooling for the intracluster gas. The cooling time of the ICM, that is, the time necessary for the temperature of the gas to drop significantly from its equilibrium value, can be approximated by

$$t_{\text{cool}} = 8.5\times 10^{10}\left[n_{\text{icm}}/10^{-3}\ \text{cm}^{-3}\right]^{-1}\left[T_{\text{icm}}/10^{8}\ \text{K}\right]^{1/2}\ \text{yr},$$

where n_{icm} and T_{icm} are the density and temperature of the ICM. In most of the ICM, t_{cool} is comparable with or longer than the age of the universe, and therefore no substantial cooling is expected to have occurred. However, in the central parts of the cluster, where the densities are higher, cooling may be important. The cooling gas will then flow towards the center of the cluster and coalesce onto massive galaxies sitting at the bottom of the gravitational potential well. Once in the cooling flow, the gas may rapidly cool to temperatures where hydrogen recombination can occur, and the 21-cm line of the neutral hydrogen may then be detected in emission, or in absorption if the cluster core harbors a strong radio continuum source. The likelihood of the presence of a cooling flow in a cluster is estimated by the inspection of the observed x-ray parameters of the ICM at the cluster core, which yield an indication of n_{icm} and of T_{icm}. Early searches for cooling flows in the 21-cm line led to negative results. More recent measurements with the Arecibo telescope, however, have provided very encouraging evidence of this interesting phenomenon.

THE COSMIC MICROWAVE BACKGROUND RADIATION AND THE ICM

The ICM is not completely transparent to radio waves. In fact, the optical depth for scattering of microwave photons of a gas with electron density n_e is

$$\tau = \int \sigma_T n_e\, dl,$$

where the integration is performed along the line of sight through the ICM and σ_T is the Thompson electron scattering cross-section, which is on the order of 10^{-28} m^2. For a typical cluster, τ is between 0.001 and 0.01. Thus, a small fraction of the photons from any radio source behind a cluster will be scattered off the line of sight by the cluster's ICM. The cosmic microwave background radiation (MBR) is a nearly isotropic bath of radio photons well described by a blackbody temperature of 2.7 K, a "fossil" relic of the Big Bang. Because of its nearly isotropic character, MBR photons can be scattered into the line of sight from any direction. Because they are much "cooler" than the ICM electrons (2.7 versus 10^8 K), they can gain energy, that is, be "heated" by the interaction with the ICM electrons (the process is known as inverse Compton scattering). The net result of the transit of the MBR photons through a cluster is that the blackbody curve that describes their spectrum is slightly shifted. The shift of the spectral energy density of the MBR $F(\nu)$ at the frequency ν is described by

$$\Delta F(\nu) = \chi\nu\frac{d}{d\nu}\left\{\nu^4\frac{d}{d\nu}\left[\nu^{-3}F_{\text{bb}}(\nu)\right]\right\},$$

where $\chi = (kT_{\text{icm}}\tau/m_e c^2)$, k is the Boltzmann constant, m_e the electron's mass, c the speed of light, and F_{bb} is the blackbody curve at 2.7 K. Because the number of MBR photons is conserved in transit through the ICM, the heating of the MBR by this process translates into a decrease of the number of photons on the low-frequency side of the blackbody curve, and an increase on the high-frequency side. Because the measurements are generally done on the low-frequency side (the Rayleigh–Jeans domain) of the blackbody curve, the effect of the cluster on that spectral region is an at first sight paradoxical apparent cooling of the MBR: The MBR flux in the line of sight to the cluster is somewhat lower than in the directions around it. However, if the energy budget is estimated over all frequencies, on both sides of the peak of the energy distribution curve, it is found that the total energy carried by MBR photons after transit through the cluster has increased.

This effect was first proposed by Ya. B. Zeldovich and Rashid A. Sunyaev in 1972. The shift for the densest and hottest clusters can be described as a temperature shift in the Rayleigh–Jeans part of the MBR spectrum of about -0.1 mK. This is a very tiny effect, a $\Delta T/T_{\mathrm{mbr}}$ on the order of -10^{-4}. It has however been successfully measured in a few clusters at frequencies of about 20 GHz. Testing the Sunyaev–Zeldovich effect has important cosmological applications. Among them is the possibility of obtaining an estimate of the Hubble constant in a manner completely independent from that of more traditional methods and the uncertainties associated with local calibrators. Although these measurements are difficult and still relatively inaccurate, the potential of this line of radio astronomical work is exceptionally attractive.

Additional Reading

Fabian, A. C., ed. (1987). *Cooling Flows in Clusters and Galaxies.* Reidel, Dordrecht.

Haynes, M. P., Giovanelli, R., and Chincarini, G. L. (1984). Environmental effects on the H I content of galaxies. *Ann. Rev. Astron. Ap.* **22** 445.

O'Dea, C. and Owen, F. N. (1987). Astrophysical implications of the multifrequency VLA observations of NGC 1265. *Ap. J.* **301** 841.

O'Dea, C. and Uson, J. M., eds. (1986). *Radio Continuum Processes in Cluster of Galaxies.* NRAO, Green Bank.

Sarazin, C. L. (1986). X-ray emission from clusters of galaxies. *Rev. Mod. Phys.* **58** 1.

Uson, J. M. and Wilkinson, D. T. (1988). The microwave background radiation, In *Galactic and Extragalactic Radio Astronomy*, G. L. Verschuur and K. I. Kellermann, eds. Springer-Verlag, Berlin.

See also **Background Radiation, Microwave; Clusters of Galaxies; Clusters of Galaxies, X-Ray Observations; Galaxies, Radio Emission; Intracluster Medium.**

Clusters of Galaxies, X-Ray Observations

Christine Jones

The luminous material in clusters of galaxies falls primarily into two forms—the visible galaxies and the hot, x-ray–emitting intracluster medium. A visible light image of a cluster shows an overdense region of galaxies. The richest, densest clusters contain predominantly elliptical and lenticular (S0) galaxies, while in less dense clusters up to half the galaxies are spirals. Observations of the relative velocities of galaxies (the velocity dispersion) in rich clusters result in mass-to-light ratios of ~250 in solar units—the mass-to-light ratio of the Sun is one. Thus with a mass-to-light ratio of ~8 for individual galaxies, only about 3% of the total cluster mass is contained within the visible galaxies.

X-ray observations provide a different view from that obtained at visible wavelengths. Although emission from individual galaxies in the cluster is sometimes seen, the primary source of x-ray emission is thermal bremsstrahlung from a hot, intracluster medium (ICM). X-ray luminosities of clusters range from 10^{42}–10^{45} erg s^{-1} with gas temperatures of 10^7–10^8 degrees kelvin, comparable to the equivalent temperatures as measured by the velocity dispersions for the galaxies in the cluster. In the cores of rich clusters, the mass of gas is ~10% of the total cluster mass. The hot intracluster medium is a major observed luminous component of clusters with a mass equal to or greater than that in the stellar component of the galaxies. Thus, understanding the formation of clusters, and the galaxies within them, requires knowledge of the origin and evolution of the hot intracluster gas.

The visible galaxies and the x-ray emitting hot gas do not comprise most of the cluster mass. Instead, most of the mass in rich clusters is "dark matter." Although this material is not directly observable at any wavelength and its nature remains unknown, x-ray and visible light observations are used to determine the amount and distribution of the dark matter.

CLUSTER MORPHOLOGY AND STRUCTURE

Present epoch clusters display a wide variety of properties both in visible light and at x-ray wavelengths, which can be understood in a framework of cluster evolution (dynamical evolution). Rich clusters with low x-ray luminosities, cool intracluster gas (~3×10^7 degrees kelvin), and low velocity dispersions have longer dynamical time scales and show more substructure. The very x-ray–luminous clusters with hot gas (10^8 degrees kelvin) and high velocity dispersions have shorter dynamical time scales and are relaxed systems.

One property that does not fit neatly into this scenario is the presence of a massive, centrally located galaxy. It had been suggested that the importance of a central galaxy was directly related to the evolutionary stage of a cluster. However, x-ray observations show that there is a class of clusters whose properties are those of dynamically young systems which nevertheless contain a massive, centrally located galaxy. More recent theoretical studies of cluster dynamics argue that the importance of the central galaxy is determined at early stages of cluster evolution. Thus, the suggestion has been made that a second parameter—the importance of a central galaxy—be added to the dynamical time scale to generate a two-dimensional cluster classification system as described in Table 1 and Fig. 1.

Figure 1 illustrates the classification system by comparing cluster optical images with their x-ray emission shown as isointensity contours. The clusters on the top (A1367—left and A262—right) are examples of unevolved systems with long dynamical time scales. The clusters shown at the bottom (A2256—left and A85—right) are dynamically more evolved systems. The clusters on the left do not have a central, dominant galaxy, whereas those at the right have a bright galaxy around which the x-ray emission is clearly centered and concentrated. The clusters with bright galaxies at the centers of their x-ray emission comprise nearly 40% of those clusters surveyed with *Einstein* satellite observations, which, in

Table 1. Cluster Morphology and Dynamical Evolution

	No Dominant Central Galaxy	*Dominant Central Galaxy*
Less evolved	Low L_x (<10^{44} erg s^{-1}) Cool ICM (few $\times 10^7$ K) Low velocity dispersion High spiral fraction (>30%) Low central galaxy density Low central gas density (no cooling flows) Examples: A1367, A1314	Low L_x (<10^{44} erg s^{-1}) Cool ICM (few $\times 10^7$ K) Low velocity dispersion High spiral fraction (>30%) Low central galaxy density High central gas density (cooling flows) Examples: Virgo, A262
More evolved	High L_x (>10^{44} erg s^{-1}) Hot ICM (10^8 K) High velocity dispersion Low spiral fraction (<30%) High central galaxy density Low central gas density (no cooling flows) Examples: Coma, A2256	High L_x (>10^{44} erg s^{-1}) Hot ICM (10^8 K) High velocity dispersion Low spiral fraction (<30%) High central galaxy density High central gas density (cooling flows) Examples: Perseus, A85

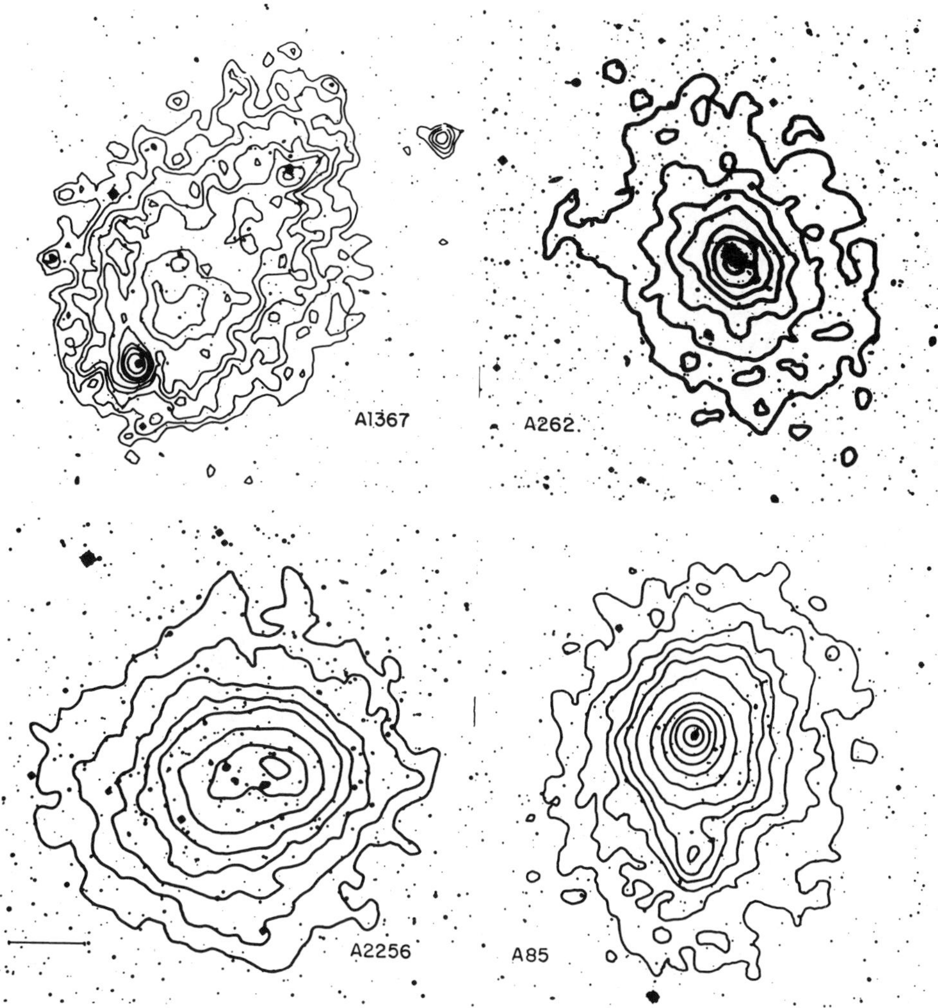

Figure 1. X-ray isointensity contours obtained from *Einstein* satellite observations are plotted on the optical photographs for four Abell clusters of galaxies (A1367, A262, A2256, and A85). These illustrate the cluster properties outlined in Table 1.

this sample, is twice the incidence of clusters selected optically as having a central, dominant galaxy.

Related to the morphology of clusters is the presence of substructure. X-ray images of the emission from the intracluster gas are particularly effective for determining cluster structure because the gas effectively maps the distribution of the total underlying mass. Based on an x-ray imaging survey of about 150 rich clusters with the *Einstein* observatory, substructure is a common phenomenon. About 30% of the clusters have multiple peaks, of which two-thirds are double and one-third are more complex. The fraction of clusters showing substructure in their x-ray emission agrees with that determined from optical studies of galaxy distributions. These multiple-peaked structures are evidence for a still-evolving cluster potential. Figure 2 shows two examples of double and complex systems.

The remaining 70% of clusters show single peaks in their x-ray surface brightness distributions. Of these clusters, roughly half have bright galaxies at their x-ray centers. These clusters tend to have high gas densities in the central regions, so the gas can radiate its energy and cool on relatively short (billion-year) time scales. Of those clusters with single peaks, about 40% have cooling times less than the age of the universe. For half of these, the estimated rates at which the gas cools and flows toward the cluster center exceed 100 $M_{\odot}$ yr^{-1}. Although often referred to as "cooling flows," little matter is flowing and the flow rate is slow (velocities of tens of kilometers per second), so that the gas generally remains in a near-equilibrium state.

Perhaps the best-studied example of a cooling flow cluster is the Perseus cluster. Observed temperatures for gas in the Perseus cluster range from the cluster mean ($\sim 10^8$ degrees Kelvin) to a factor of 10 lower. The observation of gas with temperatures below the cluster mean is the strongest evidence for mass deposition. For the Perseus cluster, estimates of the cooling rates for the gas at each observed temperature yield consistent mass deposition rates of ~ 200 $M_{\odot}$ yr^{-1}. For the different temperature regimes, the cooling times of the gas span a range up to several billion years, suggesting that cooling flows are long lived. Detailed analysis of x-ray observations of the Perseus cluster shows that the gas is not all flowing to the center of the cluster, but is cooling out over a wide range of radii. Although high mass deposition rates can contribute significantly to the mass of the central galaxy, the precise fate of the cooling gas remains uncertain. Optical observations of the central galaxy show that any new star formation can produce only a very few stars with masses exceeding one solar mass.

The large mass deposition rates inferred for many clusters have been controversial. A variety of suggestions have been made to attempt to reduce the calculated mass deposition rates to values which would not significantly affect the mass balance of the central galaxies. Models to reheat the cooling gas through thermal conduc-

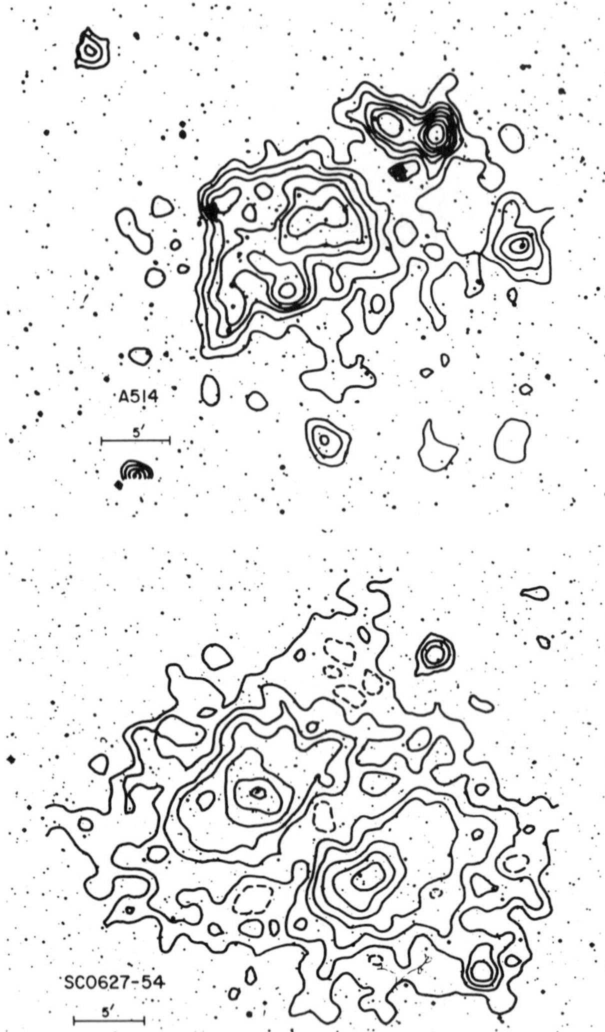

Figure 2. X-ray isointensity contours are shown for the "double" cluster SC0627-54 and the more complex cluster A514 superposed on optical photographs. The x-ray emission that traces the gravitational potential of the clusters shows the subclustering of these systems. Approximately 20% of the clusters which could be classified through the x-ray observations have two components, whereas 10% of clusters appear to be more complex.

tion from the outer, hotter gas, or through supernovae, galaxy motions, and relativistic particles have not been successful in reducing the inferred cooling rates. In conclusion, either cooling flows deposit a significant amount of mass around central cluster galaxies, or curtailing a large cooling flow requires a heat source capable of providing energy up to 10^{44} erg s^{-1} for a total of 10^{62} erg over the lifetime of a cluster.

CLUSTER MASS DETERMINATIONS

The x-ray emitting gas, while itself an important mass component of clusters, also can be used to trace the total underlying mass, which determines the gravitational potential. The intracluster medium relaxes on a relatively short time scale to hydrostatic equilibrium, which means that any flow velocities in the gas are small. Thus the equation for a hydrostatic gas can be used to determine the cluster gravitational mass. This application of the x-ray observations requires the determination of both the gas density, which can be determined from the observed x-ray surface brightness, and the temperature distributions. Gas temperature distributions have been obtained for only a few systems and indicate the presence of dark matter in agreement with optical determinations. Future x-ray satellite observations will provide spatially resolved measurements of the gas temperature for many clusters and will allow the accurate determination of cluster mass distributions.

THE ORIGIN OF THE INTRACLUSTER MEDIUM

In addition to providing information on the structure and morphology of clusters and the total mass distribution, the study of the intracluster medium is important in its own right. Because the ICM mass is equal to or greater than that in stars, it is of particular importance to determine the origin of this large fraction of the known baryonic mass.

One of the most important results related to the origin of the ICM was the discovery of emission lines from iron and other heavy elements in the energy spectrum of the hot gas. Cluster x-ray spectra show both that the x-ray emission is thermal in origin and that the heavy element abundances are between 20 and 50% of the solar values. Because heavy elements can be produced only through thermonuclear reactions in stars or by supernovae, the discovery that the intracluster medium was rich in heavy elements requires that material processed through stars be ejected into the ICM.

While the enriched material must come from the galaxies (in the absence of a very early stellar population outside of galaxies), recent studies have suggested that the bulk of the ICM of a rich cluster could not have originated within the galaxies because the ICM mass is up to several times larger than the mass of the galactic stellar component. Thus, a considerable fraction of the ICM in rich clusters must be left over from the formation of galaxies.

FUTURE OBSERVATIONS

This entry has highlighted some of the properties of clusters of galaxies in which x-ray observations have been particularly useful. However, the study of x-ray emission from clusters has only begun. Other important contributions will arise as more powerful observatories become available. Detailed x-ray studies of distant clusters, combined with radio observations, have the potential for determining a precise value of the scale of the universe, the Hubble constant, whose value has eluded astronomers for decades. Measurements of the abundance of heavy elements in the intracluster medium for clusters at different look-back times will provide information on the formation of galaxies and nucleosynthesis in stars. Mass determinations in clusters may elucidate the nature of the dark matter. Through the unique view of clusters of galaxies provided by x-ray observations, we can determine fundamental properties of galaxies, clusters, and the universe.

Additional Reading

Fabian, A. (1988). *Cooling Flows in Clusters of Galaxies.* Kluwer Academic, Dordrecht.

Forman, W. and Jones, C. (1982). X-ray-imaging observations of clusters of galaxies. *Ann. Rev. Astron. Ap.* **20** 547.

Sarazin, C. L. (1986). X-ray emission from clusters of galaxies. *Rev. Mod. Phys.* **58** 1.

See also **Galaxies, X-Ray Emission; Intracluster Medium.**

Comets

Armand H. Delsemme

WHAT IS A COMET?

Historically, a comet is a heavenly body with a fuzzy head and a long tail that makes a transient appearance in the sky. Near the center of its head a starlike image is often visible; traditionally it was called the comet "nucleus." At large distances from the Sun, the tail and the nebulous head disappear but the nucleus keeps a star-like appearance. Because the nucleus is the only permanent feature of the comet, a consensus emerged in the 1950s that it had to be a small solid body.

The passage of comet Halley in 1986 has finally confirmed that a *comet nucleus* was a small body of irregular shape, in the size range of about 10 km, made of a cold conglomerate of silicate dust, water ice, and frozen volatiles that are mostly made of organic compounds. The larger volatility of cometary material is the major difference between cometary nuclei and small asteroids. Another difference is that cometary orbits are often more eccentric than those of asteroids, which brings them closer to the Sun, where their ice and their frozen gases sublimate and drag dust away.

Repelled by the radiation pressure of the solar light, the dust makes spectacular dust tails, and the gaseous atmosphere (which is really an exosphere) expands quasiisotropically and is steadily lost to space. Some of its molecules or molecular fragments are ionized by the ultraviolet light of the Sun and become able to interact with the solar wind, not by direct collisions of particles but through the solar wind's magnetic field. This interaction is the source of the plasma tail, which is more bluish and fainter than the dust tail, and hence is less spectacular. Finally, the latent heat absorbed by the sublimation process keeps the nucleus much cooler than a blackbody, which explains the nucleus's survival after close passages to the Sun.

Nevertheless, the very appearance of the tail and the head is the signature of cometary decay. As soon as they come into the inner solar system, comets decay quickly, because they disappear completely (or at least they lose their cometary appearance) in scores of centuries or millenia, that is, in durations extremely short as compared to the age of the solar system. Since there are still many comets around, there must be a permanent source that replenishes the cometary supply of the inner solar system. This source is the *Oort cloud*, an unobserved reservoir of pristine comets bound to the Sun, occupying a spherical volume whose radius is roughly 1000 times as large as that of the planetary system. Its existence was deduced from orbital statistics.

ORBITAL STATISTICS

Jan Oort established in 1950 that two or three comets were being discovered every year, whose original orbits were in the same very narrow range of binding energy. Original orbits are orbits as they were before planetary perturbations. These "new" comets had aphelia all in the range of 30,000–50,000 AU (this includes a modern correction). Had these objects been on stable orbits, their previous perihelion passages would of course have brought them already across the planets, which would have scattered their binding energies considerably, hence hiding their common origin. Therefore they were really entering the planetary system for the first time. This implied that their previous perihelion was beyond Neptune; but their orbit had been somewhat perturbed later, when they were far away from the Sun, back in the outer shell of the Oort cloud. Over there the mass distributed in the galactic disk does not attract comets and the Sun equally. This differential attraction, usually referred to as the galactic tide, competes with the action of individual stars passing nearby and explains the observed depletion of the aphelia of new comets, in the galactic polar caps as well as in a galactic equatorial belt.

Influenced by the gravitational pull of the giant planets, the orbits of new comets evolve rapidly. If not slung out of the solar system, their periods diminish from a few million years down to tens of thousands or even a thousand years, becoming what we call long-period comets. They come from all directions, as new comets do, whereas the orbital distribution of short-period comets has a sharply different symmetry. Short-period comets (with periods less than 200 years) move mostly on direct paths around the Sun, close to the plane of the ecliptic (inclinations of 15–20° are typical).

Because the distribution of inclinations remains well preserved during the diffusion of cometary orbits, short-period comets cannot arise from a parent population with an all-sky distribution like the Oort cloud, but from a source that already has a flattened distribution, like the belt of comets proposed by Gerard P. Kuiper in 1951. The existence of the short-period comets is now considered the best proof of the existence of the *Kuiper belt*. As exemplified by the dust disk discovered around Beta Pictoris, the Kuiper belt is likely to extend beyond Neptune outward to 500 or 1000 AU or even much more. It would be a remnant of the solar accretion disk, surviving at those distances where planets never formed. It is also likely to be the primeval source of the Oort cloud. The scattering of its outer fringe by galactic tides has produced and is still replenishing the Oort-cloud outer shell (because this shell still is periodically depleted by encounters with large molecular clouds).

So, the two ends of the remnants of our accretion disk have remained gravitationally unstable: The inner end is perturbed by chaotic motions induced by the resonances with the giant planets and yields the short-period comets, whereas the outer end is perturbed by the tides perpendicular to the galactic disk, feeding the Oort cloud. The cloud's outer shell produces a steady trickle of new comets, whose orbits in turn decay into the long-period comets.

ELEMENTAL ABUNDANCES IN COMETS

Numerous population-I stars present in our galactic vicinity have, by and large, the same elemental proportions as the Sun's. These proportions are known as the cosmic (or solar) abundances. In those primitive meteorites called *chondrites*, the most volatile elements are depleted in various amounts, but there are about 50 heavier elements that are in solar proportions. As a matter of fact, their correlation with solar abundances is so good, and the error bars so much smaller than those for the solar atmosphere, that the abundances of the type C I carbonaceous chondrites (complemented by solar and other data for the most volatile elements) have become the standards for cosmic abundance tables.

To compare elemental abundances in comets with cosmic abundances, the dust fraction and the volatile fraction must be added. The 1986 passage of comet Halley confirmed that the elemental abundances were about the same in its volatile fraction as in those of several bright comets of the 1970s. Before the passage of comet Halley, all that was known for sure about the cometary dust was that it showed the emission bands of silicates in the thermal infrared and that it had a very low albedo, suggesting the presence of carbon to darken the silicates. The mass spectrometry results of the *Vega 1* spacecraft give for the first time the chemical composition of dust in a comet. Figure 1 compares the elemental abundances in the Sun and in comet Halley (sum of gas plus dust). Except hydrogen (depleted by a factor of 500) and presumably the nonmeasured inert gases, all the elements seem close to solar abundances in comets. As a matter of fact, the only elements that seem to differ from solar abundances by more than their error bars are silicon (enriched by a factor of 2) and iron (depleted by a factor of 2). The error bars include in particular the uncertainty in the dust-to-gas ratio. The iron-to-silicon ratio is therefore 0.25 solar, which is particularly puzzling, because iron and silicon are major constituents. If not a fluke in the (unique) experiment, this anomalous ratio reminds us of the well-documented variations of Fe/Si in the different types of chondrites. In particular, this ratio goes from 0.90 in C I carbonaceous chondrites, down to 0.45 in some ordinary and some enstatite chondrites; the mean ratio for a homogenized Earth would also be close to 0.45, whereas recent solar data give 1.30 ± 0.25 in the photosphere. Such a variable iron/silicon ratio, which seems even lower for comets, implies the existence of not yet completely elucidated fractionation processes in the accretion disk that gave birth to planets and comets; otherwise, the close-to-cosmic elemental abundances in comet Halley and presumably in most unevolved comets imply that they are made of a primitive material even less differentiated than chondrites.

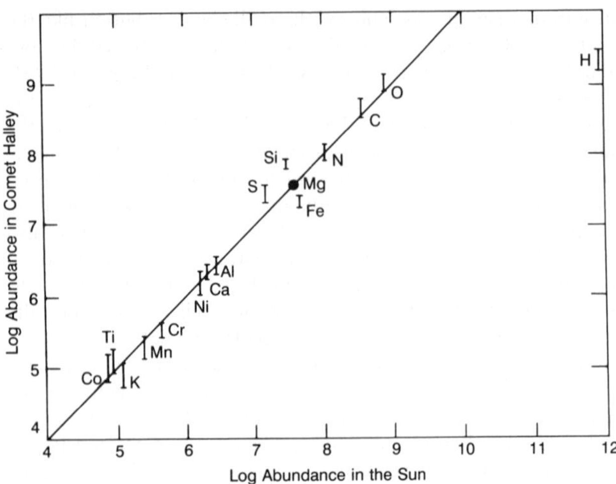

Figure 1. The elemental abundances in comet Halley (sum of gas and dust) are compared with solar abundances of the same elements. Magnesium is used to normalize the two scales, because in comet Halley silicon and iron seem to be at variance with solar abundances (this is also true in many chondrites).

Table 1. Chemical Species Identified So Far in Comets

1. By visual, ultraviolet, infrared or radio frequency spectroscopy

(a) Organics:

Hydrocarbons:	C, C_2, C_3, CH (also, isotopic ratio C_{12}/C_{13})
With nitrogen:	CN, HCN, CH_3CN (also, isotopic ratio N_{14}/N_{15})
With oxygen:	HCO, H_2CO, CO_2

(b) Inorganics:

Ammonia group:	NH, NH_2, NH_3, NH_4
Water group:	H, O, OH, H_2O (also, ortho/para ratio in H_2O)
Sulfur group:	S, S_2

(c) Metals:

Na, K, Ca, V, Mn, Fe, Co, Ni, Cu (in Sun-grazing comets)

(d) Ions:

Organic: C^+, CH^+,	CO^+, CO_2^+
With N: N_2^+,	With S: H_2S^+
With O: OH^+,	Metal: Ca^+

2. From peaks in mass spectra during spacecraft flybys

H_3O^+, $(H_2CO)_n^+$; CH_4^+, CH_3^+, $C_3H_3^+$;
NH_4^+, NH_3^+; S_2^+, CS^+, CS_2^+;
In CHON grains: numerous fragments of organics

CHEMICAL COMPOSITION

The 79 microscopic grains of comet Halley dust analyzed by the impact ionization mass spectrometer of the spacecraft *Vega 1* show a surprisingly large chemical diversity, in spite of the fact that their summation yields the quasisolar abundances of the metals listed in Table 1. Thirty-seven grains are carbon rich (mean 53% C, 12% O, 10% metals); 18 grains are oxygen rich (mean 54% O, 10% C, 15% metals) 10 grains are Mg and Si rich (mean 81% metals, but 5% Fe, 2% C, and 2% O); finally 11 grains are iron rich (mean 33% Fe), but 9% Mg, 5% Si, and 4% O. Cometary dust seems to be an extreme example of the same kind of sedimentation that is well known for the matrix of chondrites. However, the diversity of cometary grains is much larger, suggesting the mixing of interstellar grains coming from different sources (reduced in carbon stars as well as oxidized in oxygen stars, for instance). Altogether, the dust contained an inorganic fraction (67% by mass) with 52% silicates, 6% FeS, 3% reduced carbon, 1% sulfur, and 5% water, as well as an organic fraction containing 16% unsaturated hydrocarbons, 12% organics with O and or N and S, and another 5% water. A long list of specific compounds is proposed to explain all the molecular fragments observed, including heterocycles with nitrogen and oxyheterocycles with nitrogen, and also aldehydes, acids, etc. Only a cometary mission of the CRAF type (comet rendezvous and asteroid flyby) could improve our knowledge enough to bring positive identifications that are final. In the *volatile* fraction of comet Halley, the major molecular compounds were 78–80% water, 4%–5% of $CO+CO_2$, 7–8% of other CO-producing molecules by photodissociation (like formaldehyde H_2CO and formic acid HCO$\cdot$OH); 5–6% with N (N_2, HCN, NH_3, N_2H_4, etc.); 0.2% with S (H_2S, CS_2, S_2, etc.) and 2–3% hydrocarbons (C_2H_2, CH_4, C_3H_2, etc.).

When dust and gas are added together, the cometary composition turns out to be (in mass fraction) 45% water, 28% stony material, mostly siliceous, and 27% organic material (16% refractory organics, 11% volatile organics). This large abundance of organic material shows that the cosmic abundance of carbon is well preserved in comets.

PRIMORDIAL SIMILARITY

Most of the historical comets have displayed a large diversity of transient phenomena. However, our best data on the physical and chemical nature of the nucleus were derived from the combined ground-based and spacecraft observations of a single comet, namely from the 1986 passage of comet Halley. Can these results be extended with any confidence to the rest of the comets? The answer is a cautious probable yes. The reason is that the diversity in cometary phenomena seems to come more from different consequences of their aging process than from intrinsic differences in composition. Let us consider some of the consequences of cometary decay that seem at least partially understood.

Nuclear Size The number of new comets versus their absolute magnitude shows a unimodal distribution, with a maximum near magnitude 5.5. This indicates a surprising size homogeneity, in the range of 10 km in diameter. In contrast, the distribution by number of long-period and short-period comets shows no maximum; the number of objects of fainter magnitudes grows indefinitely. This suggests a general dimming produced by activity fading, complicated by repeated observed splitting into smaller pieces.

Brightness Laws The brightness dependence on heliocentric distance also depends on the cometary decay. The brightness variation is shallower for new comets, and steeper and steeper for older and older comets. It has been interpreted by the shift of the major vaporizations to shorter and shorter heliocentric distances in aging comets, because the more volatile gases are exhausted first.

Dust Tails The diversity of the dust-tail shapes comes from the diversity of the grain sizes, as is easily established: Their kinematics in the tail is influenced by an acceleration that depends on their size, because it is due to the radiation pressure of solar light. What is the origin of the variation in the grains' size distribution? Again, it is a matter of differential aging of the dust that leaves the nucleus. Grains are dragged away in fragile clusters; however, there is a sintering of the clusters when the solar heat is large enough. This happens either on the nucleus or before the grains leave the nucleus vicinity. The processing of the outer layers by heat may also produce the polymerization of an organic glue. Finally, the complete vaporization of the smallest grains may also occur in Sun-grazing comets, completely disturbing their grain size distribution. It must be concluded that the dust size distribution as observed in cometary dust tails is not intrinsic or primordial.

Plasma Tails The ionization of cometary molecules is not a direct function of the ionization rates only, but of the variable fraction confined in a narrow magnetic tube. Outside of the tube, a large and variable amount of the ions is blown away by the solar wind, too fast to be observable. This implies that, in a first approximation, the intensity of the plasma tail depends not only on the total production rate of the comet, but also on a rather high negative power (−4 to −5) of the heliocentric distance. This explains the variable onset of the plasma tail, and its quasidisappearance for old short-period comets.

CONCLUSION

The full diversity of the transient phenomena observed in comets seems to derive from their decay process. Preserved in the Kuiper belt and in the Oort cloud by the deep cold of interstellar space, comets lose their volatile material as soon as they come into the inner solar system. Their mass is only known by order of magnitude (10^{17} g) as is their density (0.1–1.0 g cm^{-3}). The best data by far come from comet Halley. Its large-scale homogeneity is testified to by the quasicosmic abundances of most of its elements (Fig. 1); the microscopic diversity of its grains, as well as the detection of isotopic anomalies, suggests that the individuality of its primeval grains has been surprisingly well preserved, promising many more clues on the origin of the solar system when a spacecraft makes a rendezvous with the proper comet.

Additional Reading

Delsemme, A. H. (1987). Diversity and similarity of comets. European Space Agency, Paris, Publication No. ESA SP-278, p. 19.

Delsemme, A. H. (1989). Whence come comets? *Sky and Telescope* **77** 260.

Grewing, M., Praderie, F., and Rheinhard, R., eds. (1988). *Exploration of Halley's Comet*. Springer-Verlag, Berlin.

Newburn, R. N. et al., eds. (1990). *Comets in the Post-Halley Era.* Kluwer Academic, Dordrecht.

See also **Comets, Nucleus Structure and Composition; Comets, Oort Cloud; Interplanetary Dust, Collection and Analysis; Meteorites, Classification.**

Comets, Atmospheres

Michael J. Mumma

The atmosphere of a comet (the coma) is an ephemeral one. Its origin is the frozen gases that form the matrix of the cometary solid body, the nucleus. At distances far from the Sun, the nucleus is cold and inert, and there is no significant gaseous atmosphere surrounding it. Warming as the comet approaches the Sun, the surface layers of the nucleus begin to release their volatiles, which pass from the condensed phase directly into the gaseous phase. The distance at which this sublimation first becomes significant depends on the chemical composition of the snows and ices in the nucleus. The more volatile of these ices, for example, carbon monoxide (CO) and carbon dioxide (CO_2), begin to vaporize at distances beyond 10 AU from the Sun, but water ice (H_2O) only begins to sublime significantly within about 3 AU. Comet Halley showed its first significant sublimation of the 1986 apparition at about 6 AU, and it was later found to contain all three of these ices in significant amounts.

Once the ices have begun to vaporize, they immediately enter the space surrounding the nucleus. Because cometary nuclei are relatively lightweight, the comet's gravitational field is too weak to bind the sublimed gases to it and so they stream off into space. For example, even a fairly large nucleus, with radius ~10 km and density ~0.4 g cm^{-3} (i.e., mass ~1.6×10^{18} g), has an escape velocity of only ~4.7 m s^{-1} at its surface, far lower than the gas thermal velocity of ~300 m s^{-1} at 1 AU. The kinetic energy of a water molecule exceeds the gravitational binding energy by a factor of ~4000, in this example. In this respect, the coma resembles the upper atmospheres of the major planets, including Earth, where atoms and molecules whose kinetic energy exceeds the local gravitational potential energy may escape into interplanetary space. This atmospheric region is called the exosphere in planets, and so the atmospheres of comets are really cometary exospheres, whose lower boundary (or exobase) is at the surface of the nucleus.

Telescopic and space-based investigations of comet Halley confirmed that the gas and dust are produced almost exclusively from the sunlit side of the nucleus, and this is understood as being a consequence of the extreme difference in day–night surface temperatures on the slowly rotating cometary nucleus. Thus the volatiles and dust flow primarily into the sunward-facing hemisphere of the coma. The volatiles sublimed directly from the nucleus (the parent molecules) are subject to destructive processes in the coma. They may undergo chemical reactions, may break down into simpler fragments (the daughter species), or may become electrically charged (the ions). A partial listing of some species found in the coma of comet Halley is given in Table 1.

Table 1. Major Constituents in the Coma of Comet Halley

	*Production Rate**	*Comment*
	Gases	
Parent species		
Water (H_2O)	100	
Formaldehyde (H_2CO)	4.5	Variable in nucleus
Carbon monoxide (CO)	5	Directly sublimed
	10	Secondary source
Carbon dioxide (CO_2)	3	Variable in nucleus
Methane (CH_4)	1–2	Estimate
Ammonia (NH_3)	0.2–0.3	Variable in nucleus
Hydrogen cyanide (HCN)	0.1	
Carbonyl sulfide (OCS)	0.7 ?	Tentative
Daughter species		
Hydrides (OH, CH, NH)		
Dihydrides (CH_2, NH_2)		
Carbon (C_2, C_3)		
Nitrogen (CN)		
Atomic (H, C, O, N, S, . . .)		
Ions		
Water group (H_3O^+, H_2O^+, OH^+)		H_3O^+ dominant
Organic (CO_2^+, CO^+, C^+, CH^+, . . .)		
	Grains	
Carbon, hydrogen, oxygen, nitrogen (CHON)		
Siliceous (silicates, FeS, C, S, . . .)		

*Relative to water.

DUST IN THE COMETARY COMA

The escaping gases sweep up particles of dust with which the ices are seeded and carry these into the coma. Quite a lot of dust is present in the nucleus; for example, the nucleus of comet Halley released about equal amounts of gas and dust each second. At 1 AU from the Sun, gas was produced at a rate of about 30 metric tons per second, while dust was lost at a rate of about 24 metric tons per second. Because most of the dust grains were microscopic in size (grain diameters of 0.1–0.2 μm were typical for comet Halley), the number released from the nucleus was enormous. Their production rate would exceed 4×10^{21} s^{-1} at 1 AU, if every grain had a mass of 5×10^{-15} g, a typical value. The dust is accelerated (mainly) radially away from the nucleus by the outward flow of gas, and it diffuses more slowly in the transverse direction. If gas and dust production is produced only from certain patches on the surface of the nucleus, then the dust entrained in the outflowing gas will appear only above those active regions, and will be confined to structures called jets. These often persist as recognizable structures for enormous distances, approaching or even exceeding 100,000 km in length. In the inner coma, the jets show a narrow cone-like shape, which is a consequence of the supersonic gas flow. Seventeen such jets were identified near the nucleus in images of comet Halley acquired by the *Giotto* spacecraft, and the apex angle of the cones was typically $\sim 10°$. Although they initially flow outward in straight lines, the jets become progressively more curved because the radiation pressure of sunlight accelerates the individual dust grains away from the Sun, just as gravity accelerates a stream of water downward, forcing it to fall in a curve toward the earth. This physical explanation of the curvature of dust jets is often called the fountain model. The strength of the effect is dependent on the ratio of the grain's surface area to its mass. Thus, small, fluffy grains are more strongly affected than are large, compact grains, and this leads to spatial segregation by size in the cometary dust tail.

A major discovery of the spacecraft missions to comet Halley was that cometary dust consists of several distinct compositional types, including the newly discovered CHON (carbon-hydrogen-oxygen-nitrogen) particles, and the previously known siliceous grains. If their relative abundances varied from vent to vent, this would lead to jets of different composition—and these may have been detected already in the form of jets of CN and C_2 in the coma of comet Halley. The CN jets seem to be distinct from siliceous jets, suggesting that carbonaceous and siliceous grains are distributed differently within the nucleus. If so, this would need to be cosmogonic and could be evidence for gravitational diffusion of individual cometesimals from chemically distinct regions of the presolar nebula, prior to their aggregation into the final cometary nucleus.

THE KINETIC TEMPERATURE OF THE NEUTRAL COMA

The freshly sublimed gases are cold by ordinary standards, being about 200 K at 1 AU from the Sun. As they expand outward from the nucleus they cool adiabatically, causing their kinetic temperature to decrease to about 20 K at a distance of about 100 km from the surface. The lost thermal energy goes into the bulk motion of the gases and the flow becomes supersonic. There is reason to believe that some of the gases may recondense into icy grains in this region, liberating their latent heat of sublimation and possibly reheating the cometary atmosphere locally in doing so, but this has not yet been confirmed by direct measurement. The outward-flowing gas also is subject to the pressure of sunlight, and so it too experiences a fountain effect, but in this case the fountain is not as clearly recognized since the gases form a much more uniform cloud and are not confined in jets. [The jets of cyanogen (CN) discussed later are associated with production from dust grains, and not with directly sublimed material.] At larger distances from the nucleus (500–1000 km), the coma temperature increases again (to $\sim$ 50–100 K) as molecular dissociation heats the coma, and it remains at about this level through the rest of the coma. The dissociation of the parent molecules heats the middle and outer coma, because the dissociation fragments are produced with significant excess kinetic energy and collisions between these fast fragments and other molecules in the coma heat the cooler gas. The increased temperature and the increased bulk outflow velocity predicted by theoretical models have been confirmed by infrared spectroscopy of the water molecule, which was observed directly for the first time in comet Halley.

THE SIZE OF THE COMA

The gaseous molecules are destroyed by the combined effects of photodissociation and photoionization. Their exact survival times are strongly dependent on their chemical identities because the rates at which these processes proceed depend on the quantum mechanical structure of the individual molecule. For easily dissociated species such as carbonyl sulfide (OCS) and formaldehyde (H_2CO) the lifetime against photodestruction at 1 AU is about 1 h, while for water (H_2O) it is about one day. The ionization time for hydrogen atoms is about 10 days. The outflow velocity near the nucleus is about 300 m s^{-1}, increasing to about 1 km s^{-1} at several thousand kilometers from the nucleus. Thus, the cloud of OCS molecules would extend to only about several thousand kilometers, while the water-vapor cloud would extend to several hundred thousand kilometers from the nucleus. The hydrogen atoms receive enormous kinetic energy as a consequence of their liberation by the dissociation of H_2O and hydroxyl (OH). The combined effect of their high velocities (about 20 km s^{-1} for the fastest atoms) and long lifetimes is to produce a hydrogen coma which can exceed 30 million km in radius, about 5,000 times larger in radius than the Earth. This was discovered in comet Tago–Sato–Kosaka (1969 IX) using the OAO-2 space observatory. The visual observer, however, sees the cometary coma primarily in the sunlight scattered by molecular fragments such as C_2, C_3, NH_2, H_2O^+, and CN and in the solar continuum scattered by dust, and these typically extend outward from the nucleus about 100,000 km.

IONS

The dominant ions may be divided into three mass groups: the water group (H_2O^+, H_3O^+, OH^+), the sulphur/carbon monoxide group (CO^+, S^+), and the carbon dioxide group (CO_2^+, CS^+). The dominant ion in Halley's comet was found to be H_3O^+ within $\sim$ 20,000 km of the nucleus, confirming the predicted importance of ion-molecule chemistry in the inner coma. The motion of ionized atoms and molecules is controlled by the local magnetic field. The origin of this field is the interplanetary magnetic field carried by the solar wind, whose interaction with the cometary atmosphere causes the field lines to drape around the cometary coma in a "hairpin" configuration. The ions follow this configuration. Brightness variations, called rays or streamers, are often seen, and the brightest of these can be traced from the plasma-tail region back into the inner coma—even approaching the nucleus itself. It is not yet known whether the brightness variations are caused by changes in density or in excitation. In comet Halley, the draped field exhibited a bow shock at distances of $\sim$ 200,000 km from the nucleus, and penetrated the coma to within $\sim$ 4700 km from the nucleus. The outflowing gas prevented further penetration, and resulted in a region of zero field strength within this region, the diamagnetic cavity. This boundary also marks the ionopause, outside of which the ions are accelerated away from the coma by the magnetic field, causing their densities to decrease. Ultimately, the ions are swept away from the coma and form the plasma tail of the comet.

DISTRIBUTED SOURCES OF GAS

Besides the gas that sublimes directly from the nucleus, it appears that some gas production is associated with grains in the coma. Jet structures of CN were discovered in telescopic images of comet Halley. Because CHON grains (i.e., carbon-hydrogen-oxygen-nitrogen) were found in the coma of this comet by instruments on the *Giotto* spacecraft, it is tempting to identify the CN in the jets as a breakdown product of CHON grains. However, it is possible that the grains are acting as sites where other gas molecules are converted into CN molecules by catalytic chemical reactions, rather than themselves decomposing into CN and other gases.

Certain spectra of the ion and neutral mass spectrometers on the *Giotto* spacecraft showed periodic peaks in the mass spectra which have been identified with a polymeric form of formaldehyde, and the same instruments also indicate a distributed source for CO in the inner coma, within 10,000 km of the nucleus. It seems clear that some gas, at least, is produced from grains. Some grains undergo disintegration as they move outward through the coma, as measured directly by the particle detection instruments on *Giotto*. It is not yet known how this disintegration occurs. Vaporization of a somewhat refractory glue is one possibility.

THE COMA AND THE NUCLEUS

One of the most important aspects of the study of the coma is the insight that it can provide into the nature of the cometary nucleus itself, and thus into the relationship between these most primitive bodies and the origins of our solar system. It is possible to measure remotely many of the directly sublimed parent volatiles, and also the carbonaceous and siliceous refractory grains. In earlier comets, several important species (e.g., CO and S_2) were detected at ultraviolet wavelengths, and HCN was detected at radio wavelengths. In Halley, both CO and HCN were again detected remotely, and the application of infrared spectroscopy from the Earth and from space provided detections of the siliceous grain features, and the first definitive detections of water, carbon dioxide, formaldehyde, and the "organic grain" feature in the atmospheres of comets. Infrared spectroscopic observations of cometary water provided a direct measure of the outflow velocity in the coma, of the kinetic temperature in the middle coma, of the nuclear spin temperature of water, and of the asymmetry of ejection of water from the cometary nucleus. Many other parent volatiles are detectable as a consequence of improved instrumentation, and we can expect that remote spectroscopic investigations with space-borne and ground-based observatories will enable us to establish compositions for a large number of these primitive objects. Future spacecraft will investigate individual comets in great detail, and the remote observations will permit extension of these detailed studies to a statistically significant sample of comets. In this way, we can establish their significance as records of the processes that attended the formation of our solar system.

Additional Reading

Bailey, M. E., Clube, S. V. M., and Napier, W. M. (1990). *The Origin of Comets*. Pergamon Press, Oxford.

Balsiger, H., Fechtig, H., and Geiss, J. (1988). A close look at Halley's comet. *Scientific American* **259** (No. 9) 96.

Grewing, M., Praderie, F., and Reinhard, R., eds. (1988). *Exploration of Halley's Comet*. Springer-Verlag, Berlin.

Newburn, R. L., Neugebauer, M., and Rahe, J., eds. (1990). *Comets in the Post-Halley Era*. Kluwer Academic, Dordrecht.

Wilkening, L. L., ed. (1982). *Comets*. University of Arizona Press, Tucson.

See also **Comets, Nucleus Structure and Composition; Comets, Solar Wind Interactions.**

Comets, Dust Tails

Philippe L. Lamy

Historically, comets have been perceived as long objects with diffuse tails. Their slow motion in the sky, their ever-changing form, and their variety of aspect have always fascinated and frightened people. The Egyptians and the Greeks imagined flaming hair; chronicles of the Middle Ages mention an axe, a scimitar, a sword, and a dagger.

Astronomers now recognize two types of cometary tails: plasma tails and dust tails. The composition and evolution of these two types of tails are drastically different. This entry addresses the tenuous cloud of dust particles expelled by the sublimating gas from the nucleus and repelled by the radiation pressure of sunlight in the antisolar direction to form the so-called dust tail. This tail itself is seen as sunlight reflected off the dust grains—like the zodiacal light—and becomes fainter and fainter further away from the coma as the dust disperses through interplanetary space.

The shape of the tail itself results from the gravitational solar attraction, the different radiation pressure forces experienced by grains of different sizes, the cometary activity, and the viewing geometry as the tail is seen from the Earth. These mechanisms essentially account for the typical fan-shaped, often curved and featureless tail, including some peculiarities such as the antitail.

Our understanding of cometary dust tails is rather recent, as our curiosity did not go beyond a mere classification according to their forms (swords...) until the fifteenth century. It seems that Frascator and Peter Apianus were the first, almost simultaneously and independently in the years 1530–1540, to characterize the fundamental behavior of cometary tails, namely, that their direction is opposite to that of the Sun. A remark in the annals of the T'ang dynasty on the 837 A.D. apparition of comet Halley reveals that the Chinese had noticed earlier a related property of the tails. It states the constant rule that the tail extends toward the west when a comet rises in the morning and toward the east when it rises in the evening. The sixteenth century also witnessed the first explanations of tails, and Jerome Cardan proposed an optical theory, wherein the tails were produced by refraction of solar light off the head of the comet, indeed satisfying the antisolar orientation. Johannes Kepler (1571–1630) should probably be credited for the concept of a repulsive force due to solar light, which explained not only the general direction of the tails but also their curvature and their concavity directed toward the place the comet left. Isaac Newton proposed a mechanical theory, describing the motion of particles ejected from the head and accelerated in the antisolar direction. This idea was pursued by Heinrich Olbers, who formulated such a theory based on a repulsive force inversely proportional to the square of the heliocentric distance (r^{-2}); he thought this force to be electrical. After having observed the 1835 apparition of the comet Halley, Friedrich Wilhelm Bessel worked out an analytical development in an effort to explain the form of its tail. He went one step further than Olbers, introducing two competing forces varying as r^{-2}, a repulsive one (as Olbers did) and an attractive one, the solar gravity. He attributed the origin of the repulsive force to the surrounding ether. Feodor A. Bredichin (1831–1904) improved on the work of Bessel and introduced the key concepts of synchrones and syndynes (or syndynames), which are fundamental, and now commonly used, in the understanding of cometary dust tails. Two parameters are needed to properly define them, namely, the ratio β of the radiation pressure force to the solar gravitational force for a given grain and the time elapsed between its ejection and observation. A synchrone, then, is the geometrical locus of the grains ejected from the nucleus at the same time and having any value of β, whereas a syndyne is the locus of the grains leaving the nucleus continuously and having a particular value of β. The evidence for the presence of dust in comets came from spectroscopic observations performed after 1860 and Svante

Arrhenius in 1900 proposed that indeed tails are made of dust particles repelled by radiation pressure exerted by the solar light. It was not until 1968 that the next breakthrough came with the dynamical-photometric model devised by Michael L. Finson and Ronald F. Probstein, which may be considered the foundation of modern studies of dust tails.

FORMATION OF DUST TAILS

As a cometary nucleus approaches the Sun, it becomes active as its ices start to sublime; the expanding gas drags dust particles away. Once released from the nucleus, those particles are subjected to two opposite forces, both being radial (along the direction to the Sun) and inversely proportional to the square of the heliocentric distance (r^{-2}):

1. The gravitational force, which is attractive and proportional to the mass of the particle.
2. The radiation pressure force, which is repulsive and is created by the corpuscular nature of light: Each time a photon interacts with a dust grain, it transfers to it an impulse or momentum in the direction of its initial velocity.

Contrary to the gravitational attraction, the radiation pressure force is proportional to the cross-section of the grain, at least in a first approximation. Therefore, the smaller the particles, the larger the repulsive force. Exact calculations show, however, that the strongest effect takes place when the size of the particle is comparable to the wavelengths of the visible light, that is, in the range of 0.5 μm.

Cometary astrophysicists use the ratio β of these two forces to characterize the behavior of a dust particle. This ratio strongly varies with the size of a dust particle: As its radius decreases from millimeters ($\beta = 0.0003$), β increases to reach a maximum value β_{max} at a radius of approximately 0.2 μm and decreases rapidly beyond that. β_{max} itself is determined by the composition of the grains, ranging from 0.5 for silicates to values in excess of 1 for absorbing materials such as metals, oxides, and graphite ($\beta_{max} \simeq 4$).

The radiation pressure force induces a differential motion of the dust particles with respect to the nucleus (which is insensitive to this force); this leads to a progressive separation of the particles that further depends upon their size and composition. This results in an efficient spatial segregation or sorting of the grains. The large millimetric ones evolve slowly, remaining rather close to the nucleus, whereas the submicronic ones separate rapidly and spread far away in the tail.

The formation and geometry of a dust tail may be well understood by combining two kinds of grain trajectories (Fig. 1):

1. Grains of the same size and composition (same β) emitted continuously from the nucleus form a syndyne (or syndyname); it is tangent to the radial direction at its origin (the nucleus) and departs more and more from it in the direction opposed to the motion of the comet, as the radiation pressure force is less and less effective. But the particles do not have the same size and composition and are therefore not subjected to the same repulsive force; the tail is the superposition of various syndynes.
2. Grains of various size and composition hence subjected to the same repulsive force, and ejected at a given time (in reality during a short time interval) form a synchrone. As the comet emits particles more or less continuously, the tail may be viewed as a superposition of various synchrones.

While the two points of view of a dust tail, syndyne and synchrone, are equally valid, it turns out that the most common structures seen in tails, sometimes called streamers, correspond well to synchrones; hence the terminology "synchronic bands."

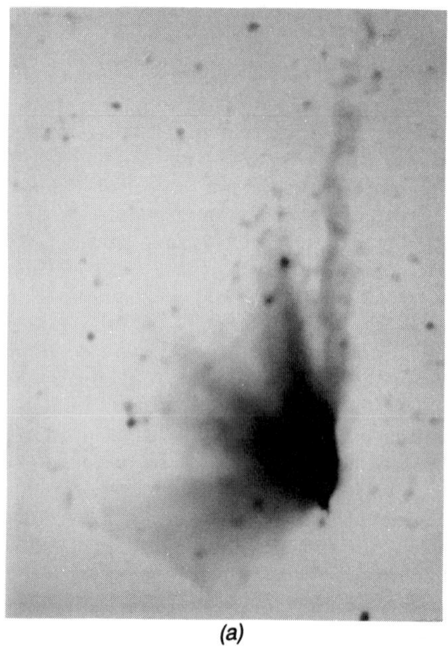

(a)

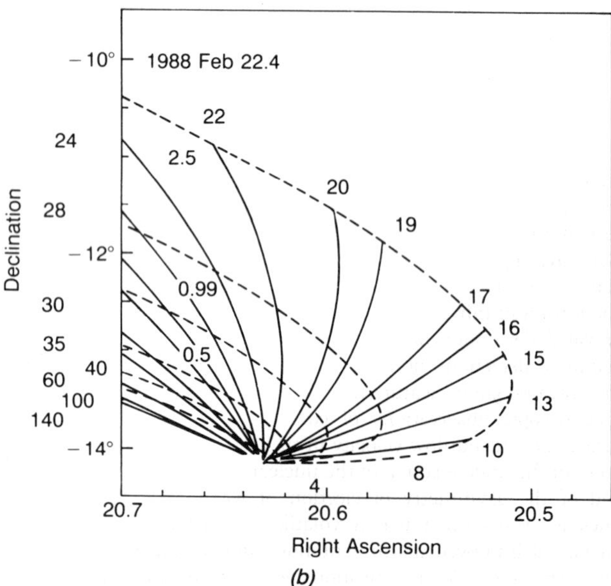

(b)

Figure 1. The dust tail of comet Halley on February 22.4, 1986. (*a*) An image obtained with the wide-field CCD camera of the European Southern Observatory by H. Pedersen and R. Vio. The "smoky" plasma tail points to the antisolar direction, whereas the broad, fan-shaped dust tail extends over a sector angle of approximately 130°. Note that the coma and inner part of the tail are saturated, creating an artificial spike in the solar direction. (*b*) The corresponding graph of the synchrones (solid lines) labeled in days elapsed since ejection and the syndynes (dashed lines) labeled in values of β. A simple comparison shows that the various components of the dust tail, the streamers, are synchronic formations.

Those are associated with short periods of enhanced dust ejection, the outbursts, which appear to be the dominant effect for controlling the distribution of dust in the tail. In essence, dust tails are primarily of synchronic nature or formation.

When a grain leaves the nucleus, it has initially the same total energy. In the absence of radiation pressure, it would therefore follow the same orbit as the nucleus around the Sun. But radiation pressure adds an extra energy to the grain, putting it in a different orbit of higher total energy. In the case of a new comet that has generally a parabolic orbit, this means that all the ejected grains are injected on unbound, hyperbolic orbits, and consequently will leave the solar system. In the case of short-period comets which have elliptic orbits, the fate of the particles depends upon the level of extra energy added by the radiation pressure, that is, upon their size and composition. They may be injected on bound elliptic or on unbound hyperbolic orbits.

While the actual orbital calculations of the trajectory of the grains show good agreement with the observed tails (taking into account the perspective or projection effects to be described later), it corresponds to the idealized situation of a two-dimensional planar motion; that is, the synchrones and syndynes lie in the orbital plane of the comet: The tail has no thickness. Actually, the dust grains leaving the nucleus are accelerated in the coma by the expanding gas to some "terminal velocity," which may be regarded as an ejection velocity. The component of this velocity perpendicular to the cometary orbital plane causes the trajectory to depart from this plane and creates a full three-dimensional tail. The terminal velocity is a function of grain size: It increases as the size decreases and reaches a maximum value for submicronic grains, for instance, 0.7 km s^{-1} when comet P/Halley was at 1 AU postperihelion. This is to be compared and vectorially added to the orbital velocity of the nucleus, typically 50–100 km s^{-1}, for rigorous calculations of the trajectories of the grain. Their out-of-plane expansion remains therefore reduced in comparison with the extent in the orbital plane.

Finally, the dust ejection from the nucleus is not necessarily isotropic and observations of comet P/Halley in 1986 have indeed revealed that the emission is concentrated in a few small active regions on the solar-illuminated hemisphere of the nucleus.

Although a priori small, the effects of the ejection velocity and of the anisotropy of dust emission have perceptible effects, especially for the large submillimetric grains.

THE FORM OF DUST TAILS

If tails are intrinsic cometary phenomena as described above, it is important to realize that our perception of them is biased, even spoiled, as we actually seem them as projected on the plane of sky, presently from the Earth. A key parameter is the position of the Earth with respect to the orbital plane of the comet, as the tail mostly spreads over it. The cometocentric latitude of the Earth, that is, the angle between the comet-Earth vector and the comet orbital plane, gives a very good feeling on how a tail is seen. If this angle is large, the tail is seen in its full extent. The perspective effect becomes more and more important as it decreases to the point where the Earth crosses the orbital plane of the comet, resulting in an edge-on view of the tail. This is only one part of the story. The appearance of a tail is also a question of contrast above the local background. This involves both the intrinsic brightness of the tail (via the level of dust production) and the brightness of the atmospheric (twilight), interplanetary (zodiacal light), galactic (Milky Way) backgrounds and the moonlight.

The viewing geometry is also responsible for a particular shape of the tail known as the antitail or anomalous tail. Although the antisolar direction of dust tails has been constantly emphasized, it happens occasionally that a comet displays a tail oriented in the opposite direction, named an antitail for that reason. These structures are perfectly understood in the synchrones-syndynes framework and result from grains subjected to very low radiation pressure forces (practically, large, submillimeter grains). Consequently, their trajectories are strongly curved, and considerably separated from the antisolar direction. A simple perspective effect allows these grains to be seen, in projection onto the plane of the sky, "trailing" the coma, that is, in the solar direction. There is also a tendency for the old synchrones to pile up toward the earliest ejection times, resulting in a sharp edge or even an apparent

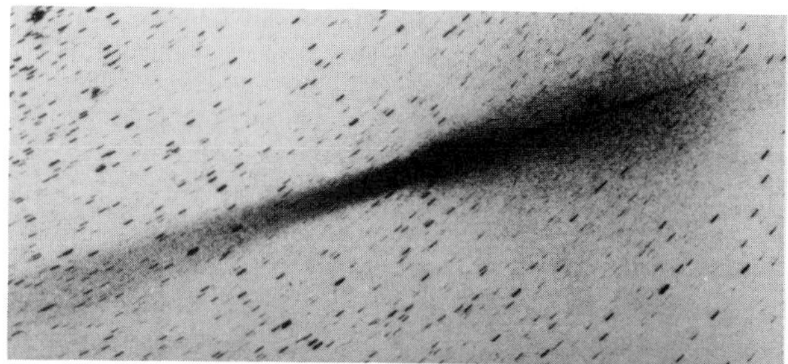

Figure 2. Comet Arend–Roland 1957 III on April 25, 1957 photographed by R. Forgelquist (Uppsala) when the Earth crossed the cometary orbital plane. Three phenomena are well illustrated: the tail, the antitail, and superimposed on it, the sunward spike.

reinforcement (particularly if the Earth is close to the orbital plane of the comet), sometimes called a spike (we shall, however, reserve this name for a different structure). Here again, this is entirely due to the geometry of projection; there is a continuity from this sharp edge, through the fuzzy antitail to the normal tail.

Since an unusual geometry is required, the antitail phenomenon is quite rare. Comets Arend–Roland 1957 III (Fig. 2), Tago–Sato–Kosaka 1969 IX and Kohoutek 1973 XII are among the best recent examples. Periodic comets do also occasionally exhibit an antitail, as seen in comets P/d'Arrest 1976 XI and P/Encke 1977 XI, for example.

STRUCTURES IN DUST TAILS

Plasma and dust tails are often distinguished by their structures, the latter ones being usually qualified as structureless. This may often be the case, but dust tails sometimes display a wealth of structures. Besides their esthetic value, these structures are very informative about physical properties of dust in comets.

Streamers

Streamers appear as discrete reinforcements in or outside the main body of a dust tail; they originate from the nucleus and follow theoretical synchrones very well (Fig. 1). Henceforth, they are correctly associated with short periods of enhanced dust ejection from the nucleus, such as outbursts or explosive phenomena such as splitting of the nucleus (e.g., comet West 1976 VI). Streamers have been displayed by a relatively large number of comets.

Spikes

When the motion of dust grains is considered in three-dimensional space, it is a well-known fact of celestial mechanics that a particular grain ejected at anomaly f will again cross the orbital plane of the nucleus at the anomaly $f + \pi$, that is, at the second node of its orbit. In 1977, H. Kimura and C.-P. Liu calculated that a dust shell formed preperihelion will collapse into a flat ellipse postperihelion, at the second node. The size distribution of particles leads to a set of overlapping ellipses whose distance to the nucleus depends upon the radiation pressure force. Large grains, which are subjected to a weak force, form an ellipse approximately centered on the nucleus, half of it toward the Sun, the other half away from the Sun. Smaller grains form ellipses further away, in the antisolar direction. Kimura and Liu called the locus of the centers of all the ellipses the neckline. When the Earth is close to the comet orbital plane, all these ellipses appear edge-on, sometimes gaining enough contrast to be seen against the surrounding tail or antitail. This results in a sharp spike appearing either in the solar or the antisolar directions, or both (e.g., comet Bennett 1970). The terminology "sunward spike" and "antisolar spike" for the manifestation of this common neckline, is certainly more appropriate. As a result of the above mechanism of formation, the sunward spike is shorter than the antisolar one, and is composed of large, submillimetric particles.

Striae

At first glance, striae may appear like streamers and have often been misinterpreted as such in the past. They are systems of discrete bands or rays whose direction distinctly differs from synchrones and does not converge to the nucleus; they usually do not last very long, a fact which renders their observation quite difficult. They have been positively identified as such in five comets. Comet West 1976 V represents by far the best-documented example and has been studied in detail by Serge Koutchmy and the author and by Zdenek Sekanina and J. A. Farrell. Although the two pairs of authors agree on the basic characteristics of the striae, Koutchmy and I advocate an electromagnetic interaction, whereas Sekanina and Farrell favor a fragmentation of grains in the tail. Probably the question will not be settled until new comets exhibit striae; it is hoped that these will be well observed.

MODERN THEORY OF DUST TAILS

Why do astrophysicists build models of cometary dust tails? Certainly they do so to understand their various shapes, a task essentially involving the calculation of grain trajectories as seen above. But Finson and Probstein realized that the photometric information contained in calibrated images of dust tails (i.e., the distribution of brightness in tails) could be analyzed to retrieve a wealth of information on cometary physics. They treated the motion of dust particles as a hypersonic, collision-free flow and expressed the intensity of light scattered by the dust in terms of three parametric functions: the size distribution, the time-dependent dust production rate, and the time- and size-dependent dust ejection velocity. These functions are determined by fitting the calculated surface-density distribution with the observed isophotes of the tail by trial and error. Practically, the procedure is simplified assuming zero ejection velocity, and, as such, has been successfully applied to several comets. The Finson–Probstein method contains several limitations inherent to its approximations; various attempts have been proposed as alternatives or improvements, such as a local linearization of the equations or a general, closed-form solution. However, the most successful alternative is a complete numerical Monte Carlo simulation as pioneered by Kimura and Liu in 1977: The photometric images of the dust tail are reconstructed by grid calculations involving some 10^6–10^7 grains and properly sampling the required parametric functions. It represents a full three-dimensional treatment wherein the trajectory of each grain is rigorously calculated using the appropriate initial velocity; hence the most general case of anisotropic dust ejection, even with a rotating nucleus, may be investigated. This, of course, requires extensive computing power best handled by supercomputers using vectorial calculation and parallel processors. This approach has successfully explained the sunward and antisolar spikes. But more generally, it allows the retrieval of:

1. The size distribution function of the dust particles, that is, the number of particles per unit size interval.
2. The temporal evolution of the size and mass production rate of the dust over long time periods, because one image of a tail may display over a month of cometary activity.
3. The size- (and possibly time-) dependent ejection or "terminal" velocity of the grains, a parameter which has never been directly measured and is often estimated from theoretical considerations.

Additional Reading

Finson, M. L. and Probstein, R. F. (1968). Theory of dust comets. *Ap. J.* **154** 327.

Kimura, H. and Liu, C. P. (1977). The structure of dust tails of comets. *Chinese Astron.* **1** 235.

Lamy, P. L. (1990). The dust tail of comet Halley. In *Comet Halley—Investigations, Results and Interpretations.* Ellis Horwood Ltd., London.

Sekanina, Z. (1974). The prediction of anomalous tails of comets. *Sky and Telescope* **47** 374.

Whipple, F. L. (1978). Comets. In *Cosmic Dust*, J. A. M. McDonnell, ed. John Wiley and Sons, New York, p. 1.

Whipple, F. L. (1985). *The Mystery of Comets.* Smithsonian Institution Press, Washington, D.C.

See also **Comets, Historical Apparitions; Comets, Types; Zodiacal Light and Gegenschein.**

Comets, Dynamical Evolution

Giovanni B. Valsecchi

In the study of comets, the problem of their dynamical evolution is of special importance because we know that the comets that we observe cannot have been in the orbits in which we see them for very long. There are two reasons for that: their physical aging (due to solar warming) on those orbits proceeds at a very high rate, and most of the observable orbits have a short dynamical lifetime. Therefore we need to identify reservoirs able to supply observable comets at the observed rate (often implicitly assumed to be typical), together with dynamical processes connecting the reservoirs to the observed orbits.

Planetary encounters play a major role in governing the dynamical evolution of comets when they move within the planetary region, whereas outside that region evolution is dominated by the vertical component of the galactic tidal potential and by encounters with passing stars and giant molecular clouds. As a consequence of these processes, cometary orbits are often said to be chaotic. There is not yet a single generally accepted definition of a chaotic orbit, and it is possible to argue that those of comets are, in fact, "open" orbits, in the sense that, sooner or later, they will become unbound from the solar system. Here we consider chaotic those orbits occupying a region in the phase space in which the following phenomenon occurs: Mass points whose evolutions start arbitrarily close to each other after a sufficiently long time become separated by an exponentially growing distance. Generally speaking, a body in a chaotic orbit may experience various regimes of motion, because sooner or later it will go through all the connected chaotic regions of phase space. Thus we can find comets over a range of distances from the Sun much greater than those spanned by other solar-system bodies like planets and asteroids.

Reservoirs for long- and short-period comets are thought to exist in the outer solar system; those reservoirs that have gained wider acceptance in recent years are the classical Oort cloud and its inner core, and the trans-Neptunian belt. In the following we will examine in more detail the main characteristics of these reservoirs and especially the dynamical processes that link them to long- and short-period comets. The discussion also shows the chain of arguments that allows us to infer the existence of the reservoirs, which are not observed directly.

Figure 1 shows various regions of the solar system relevant to this discussion. It is a log-log plot of aphelion distance Q versus the perihelion distance q, both in astronomical units; the triangle in which $q > Q$ is of course by definition a forbidden region, while

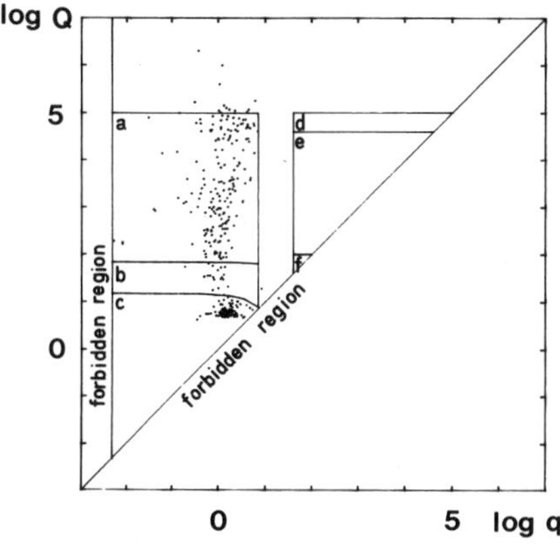

Figure 1. Log-log plot of aphelion distance Q versus perihelion distance q of observed cometary orbits. The forbidden region on the left represents orbits that cross the surface of the Sun; the forbidden triangle on the lower right represents orbits with $q > Q$, excluded by definition. Region (a) is that of observed long-period comets (the fact that there are some comets with aphelia outside the commonly accepted Oort cloud may be due to observational errors); region (b) is that of Halley-type comets; region (c) contains the Jupiter family. The three regions on the right, (d), (e) and (f) represent, respectively, the Oort cloud, its inner core, and the trans-Neptunian belt.

orbits with $q < 0.0465$ AU lead to collision with the Sun (some of the members of the Kreutz group of sun-grazing comets have been actually observed in such orbits).

The points in Fig. 1 represent the orbits of all the short-period comets at their last observed perihelion passage, as given in the 1986 edition of Marsden's Catalogue, and 220 original orbits of long-period comets. The original orbits of long-period comets are obtained by computing their motion backwards in time taking into account planetary perturbations and, when known, nongravitational forces, up to a distance from the Sun large enough to allow ignoring planetary perturbations; at this point the orbit is shifted from heliocentric to barycentric (i.e., it is computed so that the focus is no longer placed in the Sun, but at the barycenter of the solar system), and it can then be considered as the orbit along which the comet has approached the planetary region.

Regions (a)–(f) in the diagram can be described as follows: Regions (d)–(f), corresponding to large perihelion distances (essentially unobservable orbits), are (d) the classical Oort cloud, (e) its inner core, and (f) the trans-Neptunian belt; regions (a)–(c) are (a) the observed long-period (orbital period $P > 200$ yr), (b) Halley-type (20 yr $< P <$ 200 yr), and (c) Jupiter-family comets ($P < 20$ yr). The dot between the trans-Neptunian belt (f) and the Jupiter family (c) corresponds to a peculiar asteroid, 2060 Chiron. Numerical studies of its orbital evolution have shown that its most probable fate is to end in an orbit like those of Jupiter-family comets in a few tens of thousand years. Because of this typically cometary dynamical behavior, it has been put in the figure. Recently Chiron has displayed what may be cometary activity.

THE OORT CLOUD AND ITS INNER CORE

In 1950, in a famous paper, Jan Oort analyzed the distribution of $1/a$ (i.e., the inverse of the orbital semimajor axis a) in a sample of 19 comet orbits known with a precision sufficient for the computation of a reliable original orbit, finding a definite concentration (10 orbits out of 19) in the range $0 < 1/a < 0.00005$ AU^{-1}; as an explanation for this he proposed the existence of a comet cloud extending to distances of the order of 100,000 AU from the Sun as the source of long-period comets.

Since Oort's original proposal, the concept of the comet cloud has gained widespread acceptance, as it appears to explain very well the immediate origin of the long-period comets that we observe; the dimensions of the cloud are currently thought to be a little smaller, both because many more original orbits of comets of good quality are now available, and because there is a better understanding of the processes that shape the cloud itself. Oort thought of a range between 50,000 and 150,000 AU, whereas more recent estimates put it rather closer to the Sun, at about 50,000 AU.

There are three causes of perturbations of cometary orbits in the cloud. The first, stellar passages, was proposed by Oort as the cause of both the decoupling of the perihelia of comets from the planetary region, and of the later reinjection of comets in that region. Encounters with passing stars take place much more often than encounters with giant molecular clouds, but the latter are significant perturbers of the comet cloud; although their frequency and strength is not well known, the effect of giant molecular clouds is to remove a substantial fraction of the cometary population of the cloud at each encounter, possibly leading over the age of the solar system to a complete depletion of the comet cloud. For that reason an inner core of the Oort cloud has been hypothesized, more massive and heavily populated than the classical cloud. Comets residing in the core would normally not be much perturbed by star passages; in the case of an encounter with a giant molecular cloud, with the consequent stripping of comets from the outer cloud, comets of the core would similarly be perturbed into outer orbits, thus replenishing the Oort cloud and guaranteeing its survival.

In recent years a third perturbing process has been identified, which turns out to be possibly the most effective: the vertical component of the galactic tidal field. Oort cloud cometary orbits have such large semimajor axes that they are significantly perturbed by the galactic field; the z component of this field, in particular, modifies the angular momentum. Thus the perihelion distances (especially of orbits with inclinations near 45° or 135° with respect to the galactic plane) change, with an overall effect slightly larger than that of stellar perturbations.

NEW AND OLD LONG-PERIOD COMETS

The injection of the perihelia of outer Oort cloud comets well into the planetary region, in orbits that make the comets observable, is due essentially to the same three factors (discussed above) that govern their evolution in the cloud.

The perturbations that cause the reduction of the perihelion distance q also affect the angular momentum (which for nearly parabolic orbits is proportional to $\sqrt{q}$). Orbital energy (which is proportional to $1/a$) changes much less, so the value of $1/a$ of an Oort cloud comet making its first passage through the planetary region is very close to the one that the comet had while residing in the Oort cloud. However, every cometary passage among the planets yields gravitational perturbations that change $1/a$ by amounts several times larger than the typical value of $1/a$ of comets in the cloud. Thus, after the first passage through the planetary region, comets from the Oort cloud are either ejected from the solar system or are transferred to orbits of much smaller semimajor axes.

Therefore, the simple examination of the original value of $1/a$ can tell us if a comet is presumably making its first passage through the planetary region, after a long residence in the Oort cloud, or if it is likely to be making a subsequent passage; in the first case we speak of dynamically new comets in Oort's sense, in the second of dynamically old ones. It is not totally excluded, however, although rather improbable, that a seemingly new comet is in fact making its second planetary passage, because it can happen that its $1/a$ during the first passage was not changed much.

SHORT-PERIOD COMETS: HALLEY TYPE AND JUPITER FAMILY

The dynamical routes leading comets from outer solar system reservoirs to short-period orbits are a bit more complicated than those just described. We first distinguish between Halley-type and Jupiter-family comets on the basis of their orbital period, as noted earlier, because these two categories probably have very different dynamical histories.

Halley-type comets are most likely former long-period comets that during a passage within the planetary region have undergone a close encounter with Jupiter, the consequence of which has been a drastic reduction of the orbital period. This encounter can take place on the first passage or in a subsequent one, and a rather large change of orbital energy is required to produce a Halley-type orbit (more than an order of magnitude larger than the typical planetary perturbations suffered from long-period comets at each perihelion passage), thus implying an encounter distance of the order of a few hundredths of an astronomical unit. Halley-type comets could therefore be considered as an extreme case of "old" comets, differing from these essentially because of the shorter period; in fact, among them we find also many comets in highly inclined or even retrograde orbits, something quite common for long-period comets, but extremely rare among Jupiter-family ones.

Most of the known short-period comets belong to the Jupiter family; the origin of the name for this group of objects is due to the clustering of the aphelion distances of their orbits around the orbit of Jupiter. Until recently it was thought that short-period comet orbits could be grouped in families, each one related to an outer planet, because of the apparent clustering of aphelion distances at the orbital radii of the four giant planets; a critical analysis of the situation, however, has shown that only the Jupiter family can be recognized on this basis, and that the aphelion clusterings in the case of other planets lack both statistical significance and dynamical plausibility, because Jupiter is by far the strongest perturber of all short-period comets.

The orbits of Jupiter-family comets are quite different from those of the Halley type: the eccentricities and especially the inclinations are much smaller, and the perihelion distances are in general larger. As a result, these comets encounter Jupiter at much lower relative velocities, so that the encounters are more effective in changing their orbits. While Halley-type orbits remain essentially unaltered for millennia, those of Jupiter-family comets can change drastically within centuries or even decades. Some of these comets have passed from orbits totally exterior to that of Jupiter to orbits totally interior to it in this century, just before being discovered. In an extreme case, that of comet *P*/Oterma, in less than 30 years the orbit passed from totally outside to totally inside that of Jupiter, and then, after three revolutions of the comet on the inner orbit, it was ejected again outside Jupiter's orbit. The encounters of such comets happen at a velocity relative to Jupiter so low that they can be captured by the planet as temporary satellites for a few years.

Although there is not yet a consensus on the location of the reservoir from which Jupiter-family comets come, it is probably not the same as for long-period and Halley-type comets, that is, it is not the classical Oort cloud. Both the inner cloud and the trans-Neptunian belt are possible candidates, but their relative efficiency in providing appropriate final possible orbits is not yet known well. In both cases, the process would involve repeated close encounters with the four giant planets in succession, that is, a comet starting in an orbit close to that of Neptune would, after a series of encounters with that planet, have its perihelion distance decreased enough to make encounters with Uranus possible. At this point, encounters with Uranus could start to control the evolution of the comet, until it would be "passed" to Saturn and, by a similar process, finally to Jupiter. At any stage of this evolution, the comet could as well be ejected out of the solar system by an encounter with the planet controlling its motion, and in fact this fate is very probable if the comet has a rather eccentric orbit. Further studies, involving sophisticated computer modeling, will in the near future clarify the details of these processes.

Additional Reading

Fernandez, J. A. (1985). The formation and dynamical survival of the comet cloud. In *Dynamics of Comets: Their Origin and Evolution*, A. Carusi and G. B. Valsecchi, eds. D. Reidel, Dordrecht, p. 45.

Marsden, B. G. (1985). Nongravitational forces on comets: The first fifteen years. In *Dynamics of Comets: Their Origin and Evolution*, A. Carusi and G. B. Valsecchi, eds. D. Reidel, Dordrecht, p. 45.

Marsden, B. G. (1986). *Catalogue of Cometary Orbits*, 5th ed., Smithsonian Astrophysical Observatory, Cambridge.

Marsden, B. G. and Roemer, E. (1982). Basic information and references. In *Comets*, L. L. Wilkening, ed. University of Arizona Press, Tucson, p. 707.

Rickman, H. and Froeschlé, C. (1988). Cometary dynamics. In *Long Term Evolution of Planetary Systems*, R. Dvorak and J. Henrard, eds. Kluwer Academic Publishers, Dordrecht, p. 243.

See also **Asteroid and Comet Families; Comets, Oort Cloud.**

Comets, Historical Apparitions

Donald K. Yeomans

Only a few comets each century are so visually impressive as to be termed "great comets." The right set of circumstances must occur to make a comet easily visible to the unaided, or naked, eye. Far from the Sun, the nuclei of comets, which consist mostly of water ice, are inactive, and their modest sizes are not nearly large enough to allow naked-eye observations from Earth. However, when near the Sun, the icy cometary surfaces can vaporize and throw off quantities of gas and dust forming the enormous atmospheres, or comae, for which comets are known. It is the fluorescing of these gases, and particularly the reflection of sunlight from the minute dust particles in the comet's atmospheres, that can make these objects so visually impressive. However, this activity by itself does not insure that a comet will become an impressive, naked-eye object as seen from the Earth. Often, an active comet can join the ranks of the great apparitions by making a reasonably close approach to the Earth from such a direction as to allow its tail to be viewed broadside in a dark, or nearly dark, sky.

HISTORICAL APPARITIONS

Because great comets travel on highly elliptic orbits about the Sun, they appear only briefly as obvious naked eye objects when in the neighborhood of the Sun and Earth. Their brief and unpredictable appearances, or apparitions, suggested to the ancient Greeks that comets were atmospheric phenomena. In the Greek philosophy, true celestial bodies were permanent objects whose circular motions were predictable from night to night. Aristotle wrote of comets as combustible emanations rising from the Earth's surface to the region of fire surrounding the airy atmosphere. Because of Aristotle's enormous authority, comets were considered terrestrial phenomena well into the sixteenth century.

Perhaps because of their unexpected, and often awesome appearances, comets were feared as portents of coming disasters. They were often considered as presaging, or marking, the deaths of great rulers (Fig. 1). This concern with cometary apparitions was nearly universal, occurring as it did in widely diverse cultures that did not interact with one another. Mainly because of the astrological importance of comets to the ruling emperors, the ancient Chinese

Figure 1. This Roman silver coin (denarius) was struck during the reign of the first emperor Augustus (27 B.C.–14 A.D.). The head of Augustus appears on the obverse with the partly obliterated inscription "Caesar Augustus." The reverse shows a stylized comet and the inscription "DIVVS IVLIVS" (Divine Julius). After the assassination of Caesar, his soul was believed to have risen to take the form of the great comet of 44 B.C. (*Collection of D. K. Yeomans.*)

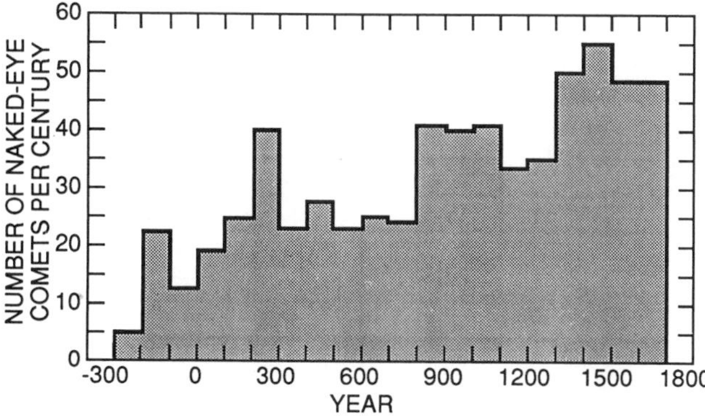

Figure 2. Histogram showing the number of naked-eye comets reported for each century between 301 B.C. and 1700 A.D.

Figure 3. A nineteenth century lithograph of the great comet Donati as seen during October 1858 over the Notre Dame Cathedral in Paris.

looked for comets and recorded their positions with an accuracy unknown in Europe until the fifteenth century. Figure 2 is a histogram showing the number of naked-eye comets per century reliably recorded from 301 B.C. until 1700 A.D. Although most of the sightings were made by the ancient Chinese, the Japanese, Korean, and European records were consulted as well. If one considers the more recent period from 800 through 1700 A.D., an interval for which there are more complete records, one can determine an average rate of approximately 43 naked-eye comets per century or one apparition every two to three years. Thus, naked-eye comets can be noted in a dark sky fairly frequently if one makes diligent, continuous searches. However, the vast majority of these naked-eye comets would not have been immediately obvious to casual observers; they could not be considered great comets.

THE GREAT COMETS

Although applying the appellation "great comet" to a particular cometary apparition is necessarily a subjective process, Table 1 is an attempt to list the most impressive cometary apparitions from 373 B.C. through 1976 A.D. With the single exception of periodic comet Halley, all the tabulated comets are nonperiodic in the sense that they passed through the inner solar system either for the first time ever or that the intervals between their returns are measured in thousands or millions of years. The vast majority of the great comets have made relatively few passages near the Sun, so they still retain much of the ices with which they formed during the solar's system's origin some 4.5 billion years ago.

From Table 1, it is apparent that all but a few of the great comets had perihelion and perigee distances well inside one astronomical unit; perihelion and perigee are the times when a comet comes closest to the Sun and Earth, respectively and one astronomical unit is the mean distance between the Sun and Earth, or approximately 150 million kilometers. A number of these great comets owe their impressive appearances to the rapid outpouring of their gases and dust as they passed close to the Sun's surface. The comets of 1680, 1843, 1882, 1965 (and possibly those of 373–372 B.C. and 1106) were so-called sun-grazer comets. The sun grazers of 373–372 B.C., 1882, and 1965 were observed to split into pieces near their perihelia. Comets passing close to the Sun can appear so bright as to be easily noticed within a few degrees of the daytime Sun itself (i.e., the comets of 1222, 1402, 1680, 1743, 1843, 1910, 1927). Often a comet that is not a sun grazer owes its greatness to a close-Earth approach, a circumstance that can render a moderately active comet with a tail of modest linear dimensions into an extraordinary spectacle as seen from the Earth. Comet Halley's closest known approach to the Earth in April 837 presented an awesome sight, with the Chinese recording a tail stretching more than halfway across the entire sky. Although comets 1858 Donati and 1976 West did not make particularly close Earth approaches, the enormous extent of their tails was seen to best advantage on Earth because they were viewed broadside. That is, when their tails were seen to their best advantage, they were stretched out in a

Table 1. Great Comets in History *

First Date Reported		Observation Interval	Perihelion		Perigee		Brightness Maximum		Notes
			Date	Distance	Date	Distance	Date	Mag.	
Julian calendar									
B.C. dates									
373–372	Winter								1
164	Oct.	30	11/12	0.58	9/29	0.11	9/29	1	Halley
147	Aug. 6	32	6/28	0.43	8/4	0.15			
87	July	35	8/6	0.59	7/27	0.44	7/27	2	Halley
12	Aug. 26	56	10/10	0.59	9/10	0.16	9/10	1	Halley
A.D. Dates									
66	Jan. 31	69	1/26	0.59	3/20	0.25	3/20	1	Halley
141	Mar. 27	30	3/22	0.58	4/22	0.17	4/22	−1	Halley
178	Sep.	80							2
191	Oct.								2
218	May	40	5/17	0.58	5/30	0.42	5/30	0	Halley
240	Nov. 10	40	11/10	0.37	11/30	1.00	11/20	1–2	
295	May	30	4/20	0.58	5/12	0.32	5/12	0	Halley
374	Mar. 4	30	2/16	0.58	4/2	0.09	4/2	−1	Halley
390	Aug. 7	40	9/5	0.92	8/18	0.10	8/18	−1	2
400	Mar. 19	30	2/25	0.21	3/31	0.08	3/19	0	
442	Nov. 10	100	12/15	1.53	12/7	0.58	12/7	1–2	
451	June 10	60	6/28	0.58	6/30	0.49	6/30	0	Halley
530	Aug. 29	30	9/27	0.58	9/3	0.28	9/3	1–2	Halley
565	July 22	100	7/15	0.82	9/13	0.54	9/13	0–1	
568	July 28	100	8/27	0.87	9/25	0.09	9/25	0	
607	Mar.–Apr.	30	3/15	0.58	4/19	0.09	4/19	−2	Halley
684	Sep. 6	33	10/2	0.58	9/7	0.26	9/7	1–2	Halley
760	May 17	50	5/20	0.58	6/3	0.41	6/3	0	Halley
770	May 26	60	6/5	0.58	7/10	0.30	7/10	1–2	
837	Mar. 22	46	2/28	0.58	4/11	0.03	4/11	−3	Halley [3]
891	May 12	50							2
905	May 18	25	4/26	0.20	5/25	0.21	5/23	0	2
912	July		7/18	0.58	7/16	0.49	7/19	0	Halley
962	Jan. 28	64	12/28/61	0.63	2/24	0.35	2/21	1	
989	Aug. 12	30	9/5	0.58	8/20	0.39	8/20	1–2	Halley
1066	Apr. 3	60	3/20	0.58	4/24	0.10	4/24	−1	Halley
1106	Feb. 4	40							4
1132	Oct. 5	20	8/30	0.74	10/7	0.04	10/7	−1	
1145	Apr. 15	81	4/18	0.58	5/12	0.27	5/12	0	Halley
1222	Sep. 3	35	9/28	0.58	9/6	0.31	9/24	1–2	Halley [5]
1240	Jan. 27	64	1/21	0.67	2/2	0.36	2/2	0	
1264	July 17	80	7/20	0.82	7/29	0.18	7/29	0	6
1301	Sep. 1	61	10/25	0.58	9/23	0.18	9/23	1–2	Halley
1378	Sep. 26	15	11/10	0.58	10/3	0.12	10/3	1	Halley [7]
1402	Feb. 8	70	3/21	0.38	2/19	0.71	3/12	−3	8
1456	May 27	42	6/9	0.58	6/19	0.45	6/19	0	Halley
1468	Sep. 18	81	10/7	0.85	10/2	0.67	10/2	1–2	
1471	Dec. 25	58	3/1/72	0.49	1/23	0.07	1/23	−3	
1531	Aug. 5	34	8/26	0.58	8/14	0.44	8/27	1	Halley
1532	Sep. 2	119	10/18	0.52	9/21	0.67	10/13	−1	
1533	June 27	81	6/15	0.25	8/2	0.42	6/27	0	9
1556	Feb. 27	73	4/22	0.49	3/13	0.08	3/14	−2	
1577	Nov. 1	79	10/27	0.18	11/10	0.63	11/8	−3	
Gregorian calendar									
1607	Sep. 21	35	10/27	0.58	9/29	0.24	9/29	1–2	Halley
1618	Nov. 16	67	11/8	0.40	12/6	0.36	11/29	0–1	
1664	Nov. 17	75	12/5	1.03	12/29	0.17	12/29	−1	

Table 1. *(continued)*

First Date Reported		Observation Interval	Perihelion		Perigee		Brightness Maximum		Notes
			Date	Distance	Date	Distance	Date	Mag.	
1665	Mar. 27	24	4/24	0.11	4/4	0.57	4/20	−1	10
1668	Mar. 3	27	2/28	0.07	3/5	0.80	3/8	1–2	
1680	Nov. 23	80	12/18	0.01	11/30	0.42	12/29	1–2	11
1682	Aug. 15	40	9/15	0.58	8/31	0.42	8/31	0–1	Halley
1686	Aug. 12	30	9/16	0.34	8/16	0.32	8/27	1–2	
1743	Nov. 29	45	3/1	0.22	2/27	0.83	2/20	−3	12
1769	Aug. 15	100	10/8	0.12	9/10	0.32	9/22	0	13
1807	Sep. 9	90	9/19	0.65	9/27	1.15	9/20	1–2	
1811	Apr. 11	260	9/12	1.04	10/16	1.22	10/20	0	
1843	Feb. 5	48	2/27	0.006	3/6	0.84	3/7	1	14
1858	Aug. 20	80	9/30	0.58	10/11	0.54	10/7	0–1	
1861	May 13	90	6/12	0.82	6/30	0.13	6/27	0	13
1865	Jan. 17	36	1/14	0.03	1/16	0.94	1/24	1	15
1874	June 10	50	7/9	0.68	7/23	0.29	7/13	0–1	
1882	Sep. 1	135	9/17	0.008	9/16	0.99	9/8	−2	16
1901	Apr. 12	38	4/24	0.24	4/30	0.83	5/5	1	
1910	Jan. 13	17	1/17	0.13	1/18	0.86	1/30	1–2	17
1910	Apr. 10	80	4/20	0.59	5/20	0.15	5/20	0–1	Halley
1927	Nov. 27	32	12/18	0.18	12/12	0.75	12/8	1	18
1965	Oct. 3	30	10/21	0.008	10/17	0.91	10/14	2	19
1970	Feb. 10	80	3/20	0.54	3/26	0.69	3/20	0–1	20
1976	Feb. 5	55	2/25	0.20	2/29	0.79	3/1	0	21

*The first tabular entry gives the approximate date when the comet was first reported as a naked-eye object. The following entry is the approximate observation interval (in days) during which the comet remained a naked-eye object. The next two entries give, respectively, the month and day of perihelion passage, and the distance between the comet and Sun at that time. Next follows the approximate date when the comet passed perigee and the minimum distance from Earth, the date when the comet appeared brightest in a darkened sky and the apparent magnitude at that time. The perihelion and perigee distances are given in astronomical units. An object with an apparent magnitude of 6 is just visible to the naked eye in a clear, dark sky. Compared to a comet whose apparent magnitude is 6, a magnitude 5 comet is 2.5 times brighter and a magnitude 4 comet is 2.5 × 2.5 = 6.3 times brighter still, etc. The bright star Vega has an apparent magnitude of 0 while the brightest star in the sky (Sirius) has an apparent magnitude of −1.5. The planet Jupiter appears at magnitude −2.7 when at its brightest.

[1] Reported by the Greek historian Ephorus to have split into two pieces.
[2] The Chinese reported that the tail reached more than 70°.
[3] The closest approach to the Earth that comet Halley has ever made. On Apr. 13, the comet's tail was more than 90° in length.
[4] This comet passed very close to the Sun and is perhaps the progenitor of the sun-grazing comets of 1882 and 1965 or that of 1843.
[5] Korean observers reported the comet was visible during the daylight hours on Sept. 9.
[6] On July 26, Chinese observers reported the tail as reaching 100°.
[7] The Chinese observers reported cloudy weather from Oct. 11 until Nov. 9, at which time the comet had passed behind the Sun.
[8] In mid-March, the comet entered solar conjunction and there were reports that it was a daylight object for eight days.
[9] Comet was discovered emerging from solar conjunction.
[10] Last observed on Apr. 20 as it approached solar conjunction.
[11] Discovered with aid of a telescope on Nov. 14. On Dec. 18, it was observable 2° from Sun at noon.
[12] Visible in broad daylight only 12° from Sun on Feb. 27.
[13] Tail reported as longer than 90° near Earth close approach.
[14] On the date of perihelion, this sun-grazing comet was observed in broad daylight nearly 1° from the Sun.
[15] Comet observed in southern hemisphere.
[16] The Great September comet was a brilliant object that was observed very close to the Sun, and split into at least four separate nuclei near perihelion. This object and comet Ikeya–Seki in 1965 are believed to be members of the same family of sun-grazing comets.
[17] Comet was easily observed on Jan. 17 only 4.5° from Sun. This comet is often confused with the later apparition of comet Halley in mid-1910.
[18] On Dec. 18, comet was observed in broad daylight only 5° from the Sun. At the end of December the tail was reported to be nearly 40 degrees in length.
[19] Sun-grazing comet Ikeya–Seki split into two or possibly three pieces near perihelion. Toward the end of October, the impressive tail reached lengths in excess of 45°.
[20] The tail of comet Bennett reached 10° in mid-March.
[21] Comet West's impressive broad tail reached a length of 30° on Mar. 8. Near perihelion, the comet split into four pieces.

direction opposite to that of the Sun and the Sun-comet-Earth angle was nearly 90°. Comet 1858 Donati, with its impressive tails, was called the most beautiful comet ever by those fortunate enough to have seen it (see Fig. 3).

The great comet of 1811, whose influence was said to have created a memorable vintage that year, did not make a particularly close approach to the Sun or to the Earth. However, it was an intrinsically active comet whose perihelion distance was just outside the Earth's orbit and whose highly inclined orbit kept it well above the Earth's orbital plane when it was at its brightest. Thus, it was easily sighted in the more populous northern hemisphere, remaining in view to the naked eye over a period spanning 260 days.

The icy nuclei of comets are amongst the most diminutive solar system bodies and most often they pass unnoticed through the Sun's realm. But occasionally the right set of circumstances can come together to allow the atmosphere and tail of a great comet to briefly appear on our celestial stage and demand the rapt attention of an Earth-bound audience.

Additional Reading

Hasegawa, I. (1980). Catalogue of ancient and naked-eye comets. *Vistas Astron.* **24** 59.

Hellman, C. D. (1971). *The Comet of 1577; Its Place in the History of Astronomy*. AMS Press, New York. (Reprint of the 1944 edition with addenda, errata, and a supplement to the appendix.)

Hughes, D. W. (1987). The criteria for cometary remarkability. *Vistas Astron.* **30** 145.

Marsden, B. G. (1989). *Catalogue of Cometary Orbits*, 6th ed. Smithsonian Astrophysical Observatory, Cambridge.

Yeomans, D. K. and Kiang, T. (1981). The long term motion of comet Halley. *Mon. Not. R. Astron. Soc.* **197** 633.

Yoke, Ho Peng (1964). Ancient and medieval observations of comets and novae in Chinese sources. *Vistas Astron.* **5** 127.

See also **Comets, Dust Tails; Comets, Types.**

Comets, Nucleus Structure and Composition

S. Alan Stern

The cometary nucleus is the actual solid body of each comet. Typically only a few kilometers across, cometary nuclei are too small to be resolved using Earth-based telescopes. Our knowledge of cometary nuclei must therefore be pieced together from Earth-based optical observations of the huge coma and tail which originate from the nucleus, Earth-based radar observations of those comets that pass very close to earth, and in situ spacecraft observations.

It is widely believed that most comets are located in the Oort cloud, a vast halo of comets that orbit the Sun far beyond the known planets. According to standard theory, comets are natural products of solar-system formation, which have been ejected to their present, distant orbits by gravitational perturbations during the epoch of giant planet formation. It is believed that this scenario resulted in the formation of the Oort cloud, which is the vast assemblage of comets orbiting the Sun at many hundreds of times the distance of Pluto. Having been stored so far from the Sun in this reservoir, comets almost certainly contain a record of the events which occurred during (and perhaps even before) the period of planetary formation.

Mankind's knowledge of comets has progressed considerably in the twentieth century. At present, almost 800 comets have been observed from Earth. Over 20 comets have been studied by satellite observatories operating in Earth orbit. Two comets,

Figure 1. Comet Halley's nucleus as photographed by the *Giotto* spacecraft. (*Reproduced by permission of European Space Agency, 1986.*)

Giacobini–Zinner and Halley, have been investigated by in situ observations by interplanetary spacecraft.

STRUCTURE

The concept of a discrete nucleus was suggested by the nineteenth century mathematicians Simon de Laplace of France and Friedrich Wilhelm Bessel of Germany. However, the modern nucleus model, often called the dirty snowball model, was first proposed by the contemporary American astronomer Fred L. Whipple in 1950. In Whipple's model, the cometary nucleus is a gravitationally and/or mechanically bound structure consisting of water ice, dust, and much smaller amounts of other ices. It has been confirmed by ground- and space-based studies, including the Halley flybys (Fig. 1). The Whipple model is also referred to as the dirty iceball, snowy mudball, or icy conglomerate model. The degree of nuclear heterogeneity is presently under debate.

To date, Halley is the only comet whose nucleus has been imaged. However, the radar cross sections of several other comets have been measured. Based on the Halley images, photometric measurements of cometary light curves, and cometary radar measurements, it has been shown that most cometary nuclei range are shaped like elongated triaxial ellipsoids. Halley itself has such a shape, much like a potato or a peanut. Some researchers have speculated that more complex shapes, including contact binaries, prolate "pancakes," and even fractal assemblages, may be possible.

The minimum and maximum sizes of comets in our solar system are unknown. However, published radii measurements for those nuclei studied optically in the infrared range and by radar give axial dimensions in the 1–8 km range. Halley's nucleus has dimensions $8.2 \times 8.4 \times 16$ km. The outer solar system objects Chiron and Schwassmann–Wachmann 1 may be very large comets; each displays periodic activity attributed to volatile release. Chiron is estimated to be 50–100 km in radius; the radius of Schwassmann–Wachmann 1 is thought to be near 30–40 km.

No direct detection of the mass of a comet has yet been made. Such a measurement would require a slow flyby or (preferably) a rendezvous of a spacecraft with a comet. However, indirect arguments based on perturbations of cometary orbits caused by the jetting forces of nonsymmetric sublimation (i.e., "the rocket effect") have led to general agreement that typical cometary masses lie in the range 10^{15}–10^{18} g. (Ice sublimation is the change from solid to gas, without passing through the liquid state; as is readily observed after a snowfall on a dry winter day.) Dynamical observations and nuclear models are in agreement that nuclear densities are typically in the range 0.2–1.2 g cm^{-3}; however, it is not impossible that some comets may have very fluffy interiors, with densities as low as 0.03 g cm^{-3}. Evidence for very low cometary densities comes from samples of cometary dust collected near Halley and in the Earth's high upper atmosphere, as well as from studies of the trajectories of fireballs in the Earth's upper atmosphere. Densities less than 0.8 g cm^{-3}, if confirmed by spacecraft determination of cometary masses, would indicate that comets contain substantial voids, either at the macroscopic scale (chambers) or at the microscopic scale (porosity). Low packing densities in cometary nuclei imply low thermal conductivities. Low densities and material strengths, if such prevail, would help explain the numerous observed instances of cometary splitting and breakup. If a density range between 0.2 and 1.2 g cm^{-3} is adopted, a spherical comet of radius 3 km would have a mass in the range 2.3×10^{16} -1.4×10^{17} g.

The rotation periods of almost two dozen cometary nuclei have now been estimated. These rotation periods range from ~4 h to ~7 day, with an average somewhat less than 1 day. The rotation period of comet Halley is still controversial, despite photographs made from space probes; Halley's nucleus appears to be spinning and tumbling in a complicated way. Scientific opinion differs as to whether the rotation period is 2.2 or 7.4 day; it may be that the 2.2-day period is the rotation and the 7.4-day period is the tumbling. Presently, there are not enough data to accurately determine the distribution of cometary rotation periods. However, it is becoming increasingly clear that cometary rotation periods are systematically slower than asteroid rotation periods.

Cometary nuclei are now known to be dark bodies. Based on *Giotto* and *Vega* flyby imaging, comet Halley's surface albedo (reflectance) is calculated to be just 2.7%. This means that Halley's surface absorbs over 97% of all light falling on it. This result is consistent with the general prevalence of dark bodies in the outer solar system, such as the outermost asteroids and the darker regions of Saturn's moon Iapetus, which is as dark as the blackest velvet. Low surface albedo may be the result of intrinsically dark materials, or radiation darkening of volatiles and organic chemicals, or microscopic surface structure effects, or a combination of all three. Studies of comet nucleus colors have shown that different comets display different surface colors, ranging from grey to dark red; these color differences may be due to differing dust/ice ratios in the surface, or to a combination of surface composition and radiation exposure history.

COMPOSITION

The firm detection of the long-suspected primary constituent of comets, water (H_2O) ice, came only in 1986 using sophisticated infrared spectroscopy techniques. Prior to this time, the major evidence for water ice in comets was circumstantial. Two lines of evidence were available. First, comets were known to become active at the point in their orbits where water ice begins vigorously to sublime. Second, the photodissociation products of water, H and OH, were observed to be prevalent in comets. Based on this evidence, water has long been suspected to be the major volatile constituent of comets.

In addition to water ice, refractory dust, and organic solids, CO and CO_2 have been found to be minor constituents in comets (each at the 5–15% level). These *parent molecules* were detected from Earth-based and satellite instruments by the technique of fluorescence spectroscopy. Fluorescence spectroscopy is the study of emissions from cometary gases that are stimulated by sunlight. Evidence for nitrogen (N_2), argon (Ar), ammonia (NH_3), cyanogen (CN), HCN, formaldehyde (H_2CO), and a number of complex hydrocarbons have also been found. Molecules and atoms of these and other volatile species may be trapped in the nuclear water-ice matrix, forming a so-called clathrate-hydrate snow.

Silicate and carbonaceous dust also exists in cometary nuclei. Much of this dust is believed to be a relic of the solar nebula; some of the dust may actually be of interstellar origin. The dust-to-ice ratio in comets appears to vary among the observed population as a whole, and with time for some comets; the dust-to-ice ratio is known to range for different comets at least from 0.1–3.0 by mass. To date, however, no systematic differences in the dust-to-gas ratio of old and new comets have been observed (old comets are those that have made many passages by the Sun). The dust particles sampled by various detectors on the Halley flyby spacecraft consisted of fine (10^{-20}–10^{-8} g) particulates. The particles detected were composed of organic solids, silicates, and graphite; the silicate grains were found to contain the refractory elements aluminum (Al), manganese (Mn), magnesium (Mg), titanium (Ti), and calcium (Ca), as well as sulfur (S). An unexpectedly large number of low-mass particles, with mass 10^{-17}–10^{-10} g, were detected. In situ observations of Halley gave a dust-to-gas ratio near 2.0. In some comets, photometric and polarimetric studies have revealed typical dust particle radii of order 1 μm. The presence of still larger (millimeter to centimeter scale) grains in comets has been proposed in order to account for observations at long infrared wavelengths (e.g., 20–100 μm, the so-called "thermal infrared" where emission depends strongly on the temperature of the material) and radar observations. The European spacecraft *Giotto* was apparently struck by a millimeter-sized grain just before its closest approach to Halley. The presence of organic substances has been detected through observations of infrared spectral features due to both C═O (carbon-oxygen bonds) and C—H (carbon-hydrogen bonds), and by means of the mass spectrometry performed by spacecraft near Halley. Discoveries of organic compounds with a series of regular mass peaks between 45 and 120 amu (atomic mass units) in Halley's inner coma indicate that complex organic polymers known as CHON particles (containing a carbon-hydrogen-oxygen-nitrogen composition) are present in the nucleus.

ORIGIN OF COMETARY NUCLEI

Theories of the origin of comets are, as might be expected, closely tied to theories of the Oort cloud. The presence of H_2O ice, and the extremely volatile ices CO (carbon monoxide), CH_4 (methane), and S_2 (diatomic sulfur) leave little doubt that comets were formed at very low temperatures, perhaps as low as 15–25 K. The most likely site for cometary formation is in the Uranus–Neptune zone of the solar nebula, or in a region of the nebula exterior to that. All comets may not have been formed at the same distance from the Sun. The range of formation distances may be reflected in compositional differences between comets. However, except for CO and CO_2 abundances and dust-to-ice ratio variations, no major compositional differences have been identified between comets.

LIFE CYCLE

The life cycle of a comet is outlined in Fig. 2. Comets spend most of their lives in the Oort cloud, where the ambient temperature is ≈ 4 K. During their long residence in the cloud, cometary surfaces are exposed to several kinds of evolutionary processes, including erosion by energetic collisions with dust grains in the interstellar medium, heating by passing luminous stars and nearby supernovae, and irradiation by cosmic rays. In the inner portions of the Oort cloud, where orbit velocities are higher and the density of objects is greater, collisions between comets and small rocky or icy debris may also be important. These mechanisms may modify the outer 5–50 m of cometary nuclei in the Oort cloud.

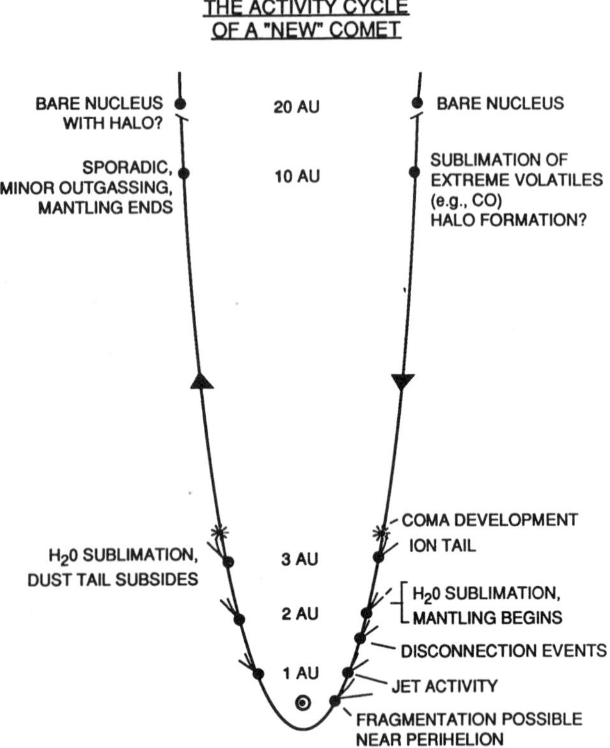

Figure 2. The cycle of cometary activity over an orbit. (*Reproduced by permission from Alan Stern, Ph.D. Thesis, University of Colorado, Boulder, CO, © 1989.*)

After a comet's orbit is perturbed so that its perihelion falls within the planetary system, it may either be returned to the cloud, ejected from the solar system on a hyperbolic orbit, or captured into a short-period orbit (short-period comet orbits are defined as those less than 200 years in length).

The physics of a comet approaching the Sun is controlled by the effects of increasing solar radiation acting on a dusty-ice surface. Primary among these effects is the sublimation of surface volatiles. Sublimation of trace constituents such as CO and CO_2 may begin as far as 30 AU from the Sun. Once a comet approaches within ~ 2.8 AU of the Sun, water sublimation becomes important. This process leads to the nucleus becoming enveloped in a cloud of escaping vapor and dust (dust escapes by being entrained in the sublimation outflow); this cloud is the comet's atmosphere; it is known as the *coma*. In the most well-studied comet, Halley, sublimation activity was observed to arise from three sources: gas leaking through the mantled surface, gas evolving from sublimating dust grains in the coma, and gas originating from localized, highly directed vents on the nuclear surface. Although such jets may be a common feature of active cometary nuclei, it is not known how many jets a typical comet develops, nor over how many orbits a typical jet persists.

In addition to the primary volatile H_2O, other chemical species have been observed in the coma. Some of these species are created by chemical reactions in the coma; others (called parent or mother species) are directly present in the nucleus. Known parent species include CO, CO_2, S_2, S_8, NH_3, HCN, H_2, CO, C_2, and C_3. The most abundant of the volatile species after water are CO and CO_2, which may together comprise as much as one-third as much mass as the water ice. Literally dozens of so-called daughter products, resulting from kinetic and photochemical reactions between parent species, have been observed in cometary comae.

The effect of solar radiation and the solar wind on the coma is (a) to drive complex photochemistry, (b) to dissociate molecules and ionize both molecules and atoms, and (c) to form and control the dynamics of the cometary *ion tail*. This tail results because atoms and molecules in the coma ionized by solar radiation are efficiently swept into a nearly linear, anti-sunward-pointing ionized structure by electromagnetic coupling to the solar wind. Cometary dust typically decouples from the gas and forms a separate, more arcuate, anti-sunward-pointing *dust tail*. Complex plasma interactions in the ion tail, including tail disconnection, shock formation, helical disturbances, and other wave phenomena, have been widely observed. The coma and cometary dust tail persist throughout perihelion passage and disperse only after water sublimation ceases to be important.

Water sublimation rates may reach many tons per second during a comet's approach to perihelion. Active comets may shrink by an average of a few meters in radius due to water sublimation over the course of each perihelion passage.

As a comet draws away from the Sun after a perihelion passage, the rate of water sublimation diminishes, and the ability of the outflowing gases to entrain dust becomes increasingly small. Eventually, there is not enough gas flow to lift dust away from the nucleus, so the surface dust remains behind as the ice sublimes (additionally, some cometary particles are too large to be lifted off the surface even at perihelion). This causes the buildup of a nonvolatile lag deposit, or *mantle*. This residue is thought to be partly responsible for the very low albedo and suspected very low thermal conductivities of cometary nuclei. After many orbits, the increasingly thick mantle impedes or even prevents the sublimation of ice. For this reason, some comets may actually "choke" themselves off and come to resemble asteroids.

FUTURE PROSPECTS FOR STUDY

Plans for future comet studies include continued ground-based optical and radar observations, as well as orbital observations in the 1990s from the *Hubble Space Telescope* (*HST*), various Spacelab flights (including the 1990 mission of the ASTRO ultraviolet observatory), and the collection of cometary debris by facilities mounted on the Soviet Mir and U.S./International space stations. In the late 1990s, the NASA Comet Rendezvous/Asteroid Flyby (CRAF) mission will be launched; according to plan, this mission will make detailed studies of the nucleus and coma of comet Kopff, beginning in the year 2000. NASA and the European Space Agency are considering a successor mission to CRAF that would return a number of samples from a comet, sometime early in the 21st century. [Editor's note: In 1990, both the *HST* and the *ASTRO-1* mission made cometary observations.]

Additional Reading

Bailey, M. E., Clube, S. V. M., and Napier, W. M. (1986). The origin of comets. *Vistas in Astronomy* **29** 53.

Brandt, J. C. and Chapman, R. D. (1981). *Introduction to Comets.* Cambridge University Press, New York.

Wilkening, L. L., ed. (1982). *Comets.* University of Arizona Press, Tucson.

See also **Comets, Dynamical Evolution; Comets, Oort Cloud.**

Comets, Oort Cloud

Paul R. Weissman

It is common to think of the solar system as ending at the orbit of the outermost planet, Pluto, at a distance of ~ 40 AU, or about 6×10^9 km from the Sun. But the Sun's gravitational sphere of influence is much larger, extending out well over 10^5 AU, or about two light years. And that space is not empty; it is "filled" with a vast cloud of some 10^{12}–10^{13} cometary nuclei, preserved rem-

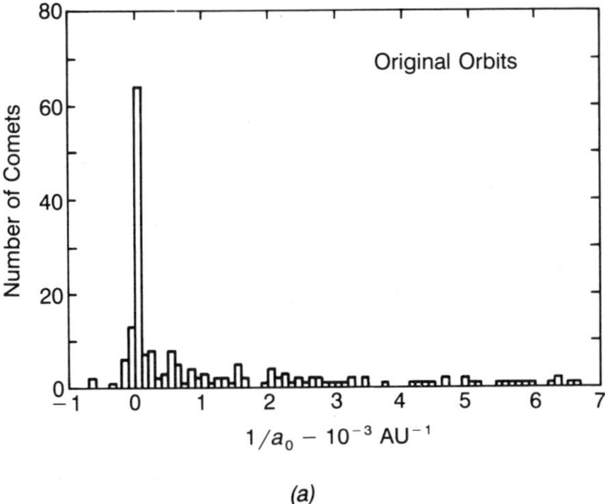

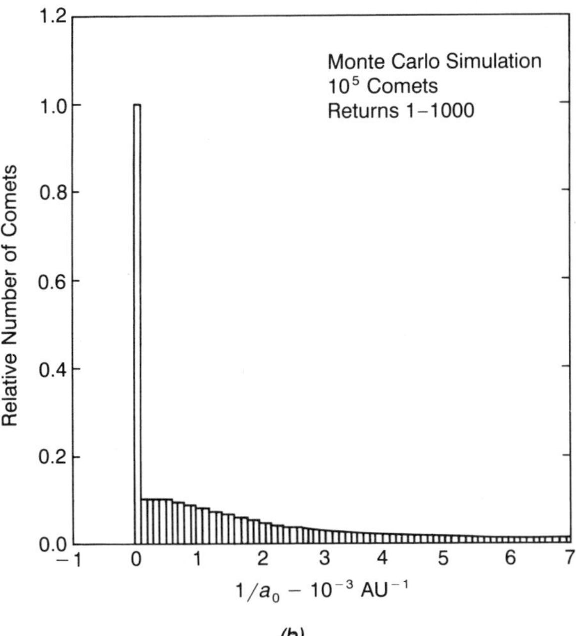

Figure 1. (*a*) Observed distribution of $1/a_0$, the original inverse semimajor axes for long-period comets. The sharp spike of comets at near-zero but bound energies are the dynamically new comets from the Oort cloud. The few hyperbolic comets likely only appear to be so as the result of errors in their orbit determinations. [*Orbital data from Marsden (1989)*.] (*b*) Estimated distribution of inverse semimajor axes based on a Monte Carlo simulation model which includes the effects of planetary, stellar, and nongravitational perturbations, and physical loss due to collisions, random disruption, and loss of all volatiles. [*From Weissman (1990)*.]

nants of the formation of the Sun and planetary system. This is the Oort cloud, as first suggested by Dutch astronomer Jan Oort in 1950.

Oort was attempting to explain the unusual distribution of orbital energy for the observed long-period comets, as shown in Fig. 1*a* (plotted as a function of the original inverse semimajor axis, $1/a_0$, which is equivalent to orbital energy). The distribution is marked by a sharp spike of comets at near-zero but bound energies, representing orbits with semimajor axes between 10^4 AU and infinity; a low continuous distribution of more tightly bound, less eccentric orbits; and a few apparently hyperbolic orbits. Earlier dynamical calculations had shown that Jupiter would perturb the orbits of comets passing through the planetary system so as to randomly spread them in energy, giving rise to the low continuous distribution. But what then was the explanation for the spike of comets?

Oort recognized that the spike had to be the source of the long-period comets, a vast spherical cloud of comets at distances greater than 10^4 AU from the Sun, but still gravitationally bound to it. Oort showed that comets in the Oort cloud are so far from the Sun that distant perturbations from random passing stars can change their orbital elements and occasionally send the comets back into the planetary system. On their first pass through the planetary system, Jupiter's random perturbations eject roughly half the "new" comets to interstellar space, while capturing the other half to more tightly bound, less eccentric orbits. On subsequent returns the comets continue to random walk in orbital energy until they are either ejected, collide with a planet or the Sun, or are physically destroyed by one of several poorly understood physical mechanisms. The presence of a few apparently hyperbolic comets in Fig. 1*a*, those with negative values of $1/a_0$, is most likely the result of small errors in their orbit determination, or unmodeled nongravitational forces resulting from jetting of volatiles on the surfaces of the cometary nuclei (which make the orbits appear more eccentric than they actually are).

It is possible to simulate the dynamical evolution of comets in the Oort cloud using a computer-based Monte Carlo simulation model. The results of one such simulation for 10^5 hypothetical comets are shown in Fig. 1*b*, where the comets are perturbed by the major planets, random passing stars, and nongravitational forces, and removed by collisions, random disruption (splitting), and loss of all volatiles. A fairly good match to the observed distribution in Fig. 1*a* is obtained. By "tuning" such a model to improve the fit, some insight into the possible physical and dynamical loss mechanisms can also be obtained.

As a result of such simulations, it is found that 65% of all long-period comets are dynamically ejected from the solar system on hyperbolic orbits, 27% are randomly disrupted (10% on the first perihelion passage), and the remainder are lost by a variety of processes such as loss of all volatiles or collision with the Sun and planets. The average ejection velocity is 0.6 km s^{-1}. The average long-period comet with perihelion < 4 AU makes five passages through the planetary region before arriving at some end state, with a mean lifetime of 6×10^5 yr between the first and last passages.

To explain the observed flux of dynamically new long-period comets, Oort estimated that the population of the cometary cloud was 1.9×10^{11} objects. More recent dynamical modeling using computer simulations has produced somewhat higher estimates, by about an order of magnitude. These result in part from higher estimates of the flux of long-period comets (brighter than absolute magnitude, $H_{10} = 11$) through the planetary system, and in part from a recognition of the role of the giant planets in blocking the diffusion of cometary orbits back into the planetary region. Comets perturbed to perihelia near the orbits of Jupiter and Saturn will likely be hyperbolically ejected before they can diffuse to smaller perihelia and be observed. Thus, the terrestrial planets region is undersupplied in long-period comets.

Since first proposed in 1950, Oort's vision of a cometary cloud gently stirred by perturbations from distant passing stars has evolved considerably. Additional perturbers have been recognized: giant molecular clouds (GMCs) in the galaxy which were unknown before 1970, and the galactic gravitational field itself, in particular the tidal field of the galactic disk. In addition, random stars will occasionally pass directly through the Oort cloud, ejecting a substantial number of the comets and severely perturbing the orbits of others. As a result, it is now estimated that the mean dynamical lifetime of comets in the cloud is only about 60% of the age of the solar system (though some estimates are even shorter), and that the Oort cloud must somehow be replenished, either by capture of comets from interstellar space, or from a more populous inner Oort cloud reservoir, comets in orbits closer to the Sun which are pumped up to replace the lost comets.

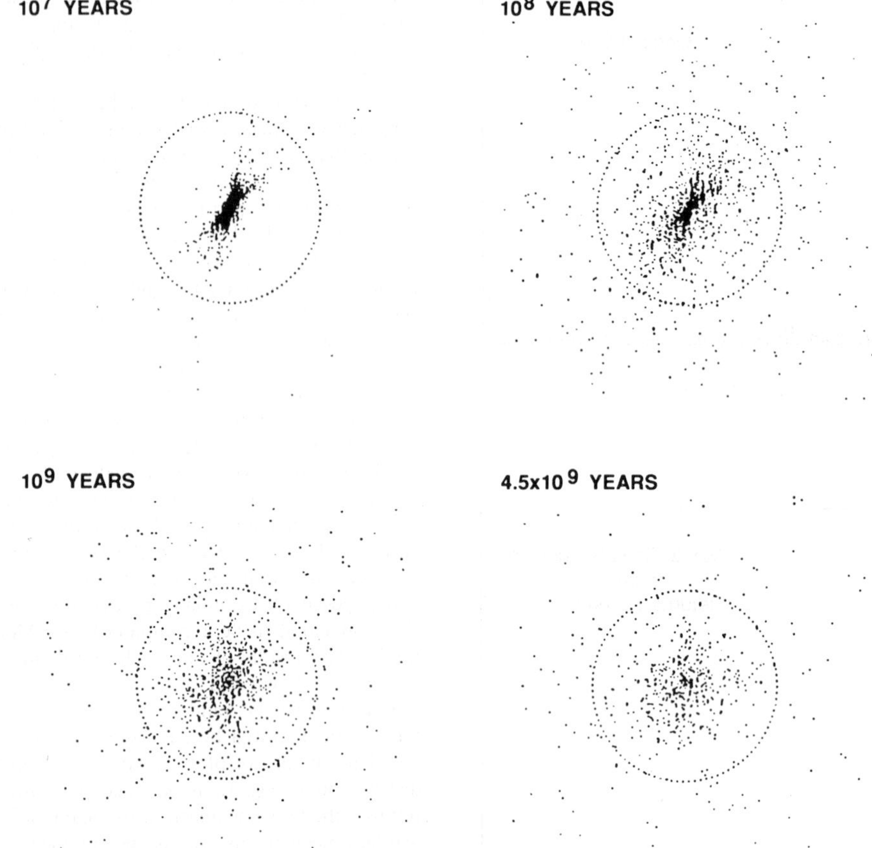

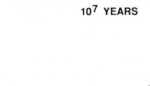

Figure 2. Dynamical evolution of a hypothetical cloud of comets ejected out of the Uranus–Neptune zone, viewed at several times during the history of the solar system, under a combination of galactic, stellar, and planetary perturbations (projected onto a plane perpendicular to the galactic plane). The dotted circle is at a radius of 2×10^4 AU, the boundary between the inner and outer Oort clouds. [*From Duncan et al. (1987).*]

It now appears that the galactic disk is the major perturber on the Oort cloud, though stars and GMCs still play an important role in repeatedly randomizing the cometary orbits. Galactic tidal perturbations peak at galactic latitudes of $\pm 45°$ and go to zero at the galactic equator and poles. The observed distribution of galactic latitudes of cometary aphelion directions mimics that dependence. Although a lack of comet discoveries near the galactic equator could be the result of observational selection effects (confusion with galactic nebulae), the lack of comets near the poles appears to confirm the importance of the galactic disk perturbations on the Oort cloud.

The galactic field also sets the limits on the outer dimensions of the Oort cloud. The cloud is a prolate spheroid with the long axis oriented toward the galactic nucleus. Maximum semimajor axes are about 10^5 AU for direct orbits (relative to galactic rotation) oriented along the galactic radius vector, decreasing to about 8×10^4 AU for orbits perpendicular to the galactic radius vector, and increasing to 1.2×10^5 AU for retrograde orbits (opposite to galactic rotation). Most comets in the Oort cloud are in highly eccentric orbits, with the mean eccentricity being equal to 0.7.

The question of how to replenish the Oort cloud appears to be settled in favor of a massive inner Oort cloud. Capture of hyperbolic comets is a highly unlikely mechanism, particularly at the Sun's random velocity of 20 km s^{-1} relative to the local group of stars. On the other hand, it has been shown that icy planetesimals formed in the Uranus-Neptune zone would be perturbed by the forming protoplanets, and by stars and the galactic field, to form an inner Oort cloud reservoir, with a population between five and ten times that of the outer cloud. As comets are stripped away from the outer cloud by close stellar and GMC encounters, the same perturbations will pump up the comets in the inner cloud to replace them. A computer simulation of the dynamical evolution of the inner and outer Oort cloud is shown in Figure 2.

Current best estimates are that the outer, dynamically active Oort cloud has a population $\sim 10^{12}$ comets, and the inner cloud a population of $\sim 10^{13}$ comets. The population estimates depend on the assumption that the currently observed long-period comet flux through the planetary system is equivalent to the long-term average flux. If the current flux is enhanced due to a recent perturbation on the Oort cloud, then the population estimate of the cloud is too high, and vice versa. Close or penetrating passages by random stars typically cause variations in the flux from the outer Oort cloud by a factor of 2–3, with variations up to a factor of 10 occasionally.

Even more extreme variations in the flux are possible if a star passes directly through the cloud, in particular the inner Oort cloud. A star passage at 3×10^3 AU from the Sun would perturb a shower of $\sim 0.6 \times 10^9$ comets into Earth-crossing orbits, raising the expected impact rate by a factor of 300 or more, and lasting 2 to 3 Myr. However, such events are rare, occurring perhaps once every 3–5×10^8 yr. Nevertheless, cometary showers dominate the cometary contribution to cratering on the terrestrial planets. Such showers have been suggested as a possible cause of biological extinction events on the Earth.

In fact, several hypotheses were put forth in 1984 to suggest that cometary showers could be caused periodically, resulting in alleged periodic extinctions seen in the fossil record on the Earth. These hypotheses involved (1) a dwarf companion star in a distant, 26-Myr-period orbit; (2) a 10th planet circulating in the inner Oort cloud; or (3) the solar system's epicyclic motion above and below the galactic plane (with a half-period of 32–33 Myr), all causing periodic comet showers. However, each of these hypotheses has a variety of dynamical problems associated with it, and no evidence in support of any of them has been found. In addition, the observed cratering records on the Earth and Moon do not show the expected number of craters that would result from intense periodic comet showers every 26–32 Myr.

The total mass of the Oort cloud is not well known because of the high uncertainty in the bulk density of cometary nuclei and in the nucleus size distribution. Estimates for the total cloud mass range from about 15–1000 Earth masses, the upper limit being greater than twice the total mass of the planetary system. A best guess, taking an average nucleus mass of 4×10^{16} g, is ~ 50 Earth masses. Over the history of the solar system, the Oort cloud has lost between 40% and 80% of its population, so the original mass must have been a factor of 2–5 larger.

An interesting recent development has been the addition of a third component to the inner and outer zones of the Oort cloud. Dynamical studies have shown that the short-period comets likely do not result from the dynamical evolution of long-period comets, but rather may come from a flattened belt or ring of comets beyond Neptune. This belt is estimated to be ~ 300 times more dynamically efficient for producing short-period comets than direct evolution of long-period comets from the outer Oort cloud. The number of comets in the Kuiper belt, named for Gerard Kuiper, who first suggested its existence in 1951, is on the order of 10^8–10^9, with an estimated total mass of 0.01–0.1 Earth masses. The belt of comets is presumably a remnant of the solar nebula's primordial accretion disk: icy planetesimals beyond the orbit of Neptune which were never absorbed into planets or ejected to Oort cloud distances.

The most common loss mechanism for comets in the Oort cloud is ejection to interstellar space, either due to close stellar and GMC perturbations, or as a result of Jupiter perturbations during a comet's pass through the planetary system. Many more comets were likely ejected during the formation of the solar system. If all stars produce interstellar comets in similar numbers, then the density of comets in interstellar space may be as much as 10% of that in the outer Oort cloud. No comet on a clearly interstellar trajectory has been observed passing through the planetary region. However, because the planetary system is so small compared with the mean spacing between stars, this does not set a very strict limit on the space density of interstellar comets.

Additional Reading

Duncan, M., Quinn, T., and Tremaine, S. (1987). The formation and extent of the solar system comet cloud. *Astron. J.* **94** 1330.

Duncan, M., Quinn, T., and Tremaine, S. (1988). The origin of short-period comets. *Ap. J.* **328** (*Lett.*) L69.

Heisler, J. and Tremaine, S. (1986). The influence of the galactic tidal field on the Oort comet cloud. *Icarus* **65** 13.

Marsden, B. G. (1989). *Catalogue of Cometary Orbits*, 6th ed. Smithsonian Astrophysical Observatory, Cambridge, MA.

Oort, J. H. (1950). The structure of the cometary cloud surrounding the solar system and a hypothesis concerning its origin. *Bull. Astron. Inst. Neth.* **11** 91.

Weissman, P. R. (1990). The Oort cloud. *Nature* **344** 825.

See also **Asteroid and Comet Families; Comets, Dynamical Evolution; Interstellar Clouds, Giant Molecular.**

Comets, Plasma Tails

John C. Brandt

The long plasma tails of comets are the dramatic product of the comet's interaction with the solar wind, the fully ionized tenuous plasma flowing away from the sun at approximately 400 km s^{-1}. The plasma tails are distinct from the dust tails, which are composed of dust particles released from the nucleus and pushed in the antisolar direction by the sun's radiation pressure. The two types of tails are often present in the same comet. They can be distinguished on color photographs because the plasma tail is blue (see below) and the dust tail is yellow because it shines in sunlight reflected off dust particles.

The serious study of plasma tails began with the development of photographic techniques early in the twentieth century. The basic physical ideas involving the interaction with the solar wind were not developed until the 1950s and 1960s when the solar wind began to be understood. Indeed, the discovery of the solar wind is credited to Ludwig Biermann in the early 1950s and was invoked by him to explain certain properties of plasma tails.

The study of plasma tails was greatly changed in 1985 and 1986 because of the space missions to two comets. These missions not only provided in situ data on the cometary plasma, but the interpretation of this data led to advances in theoretical understanding.

HISTORICAL STUDIES

Plasma tails are easily photographed with wide-angle imaging systems on blue-sensitive emulsions. Because the early emulsions had good blue sensitivity, substantial numbers of good-quality images have been obtained on most comets since the turn of the century (Fig. 1). The comet's blue appearance arises because a major constituent of the plasma tail is ionized carbon monoxide (CO^+), which has strong bands near 4200 Å. Spectroscopic studies have established that the ions present in the tail are mainly H_2O^+, OH^+, CO^+, and CO_2^+.

From the photographs, the basic dimensions and structure of the plasma tails were derived. Lengths are typically a few tenths of astronomical units, with a range from approximately 10^7 to over 10^8 km. Widths are roughly $\frac{1}{10}$ or less of the length. Plasma tails usually are quite straight and are oriented about 5° from the antisolar direction in the direction opposite the comet's motion. This lag occurs because the tail acts like a wind sock in the solar wind and because of the aberration effect due to the comet's motion.

Considerable fine structure was noted in the photographs, much of it in the form of linear structures called rays or streamers. In addition, the plasma structures are constantly in a state of change. Many features, such as condensations and kinks, show motion away from the nucleus. The extent of the changes is such that the entire plasma structure of a comet can change from one day to the next. Thus, observations at a single site are usually not suitable for studying the evolution of plasma tails. This was known to Edward E. Barnard around the turn of the century and later incorporated into the approach to recording the plasma tails taken by the Large-Scale Phenomena Network of the International Halley Watch in the 1980s. Observers widely distributed in longitude provided the opportunity to follow the evolution with closely spaced observations.

One of the major features of plasma tail evolution is the disconnection event, or DE. Here the entire plasma tail is disconnected from the head region, drifts away in the antisolar direction, and is replaced by a new plasma tail. This is a common phenomenon and is believed to be part of an evolutionary cycle for plasma tails.

Understanding of plasma tail phenomena prior to the space missions was based on the theoretical work of Hannes Alfvén and

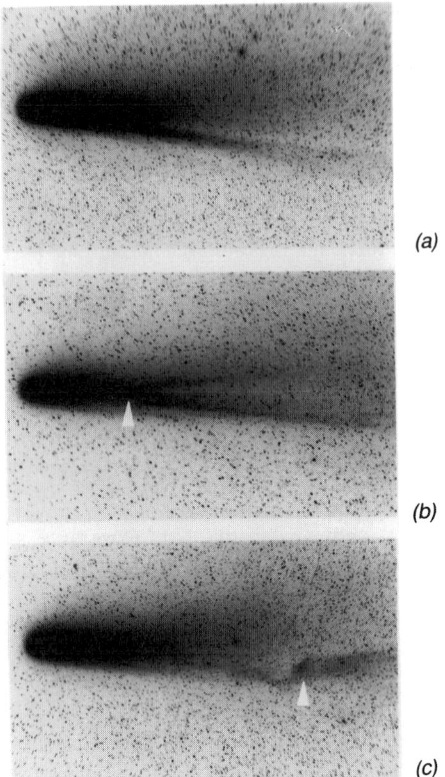

Figure 1. The plasma tail of Halley's comet as photographed on three successive days showing dramatic changes in appearance including a major disconnection event (DE). These are negative prints. The homogeneous dust tail lies above and the plasma tail below. The photographs were taken on UT dates of (*a*) March 19.50, (*b*) 20.42, and (*c*) 21.44, 1986. The tail has a normal appearance on March 19, but a DE (arrow) appears on March 20 and is quite obvious (arrow) on March 21. The DE on March 21 is approximately 6.3×10^6 km from the nucleus. (*Photographs taken by William Liller as part of the International Halley Watch's Large-Scale Phenomena Network's Island Network, Maunga Orito, Easter Island, Chile.*)

Biermann. All of the atmospheric phenomena for comets begin with the sublimation of ices (mostly water ice) from the nucleus. The freshly sublimated gases cannot be held by the nucleus (unlike most planetary atmospheres) because the surface gravity of a comet is far too low. The gases expand away from the nucleus to form a roughly spherical cloud of gases around the nucleus called the coma. The gas is composed of neutral molecules (mostly water) and some of these molecules are ionized by the sun's ultraviolet radiation and by the solar wind. This forms the cometary ionosphere, which becomes an obstacle in the solar wind. The solar wind, which carries the solar magnetic field with it, cannot easily penetrate the ionospheric obstacle, and the solar magnetic field lines become embedded in the comet's ionosphere. The flow is greatly slowed near the comet and relatively unimpeded well away from the comet. This causes the field lines to wrap or drape around the comet to form a magnetic tail structure composed of two oppositely polarized magnetic lobes separated by a current sheet. The plasma tail is visible because the ionized cometary molecules are trapped on the solar magnetic field lines when the field lines become embedded in this ionosphere, and they shine by fluorescence of the solar radiation field. Because the comet is an obstacle in a supersonic and super-Alfvénic solar-wind flow, a bow shock is formed on the sunward side. The shock slows the solar wind and allows it to flow smoothly around the comet.

The deceleration of the solar-wind flow upstream of the shock (and elsewhere) occurs through the "pick-up" process. Neutral

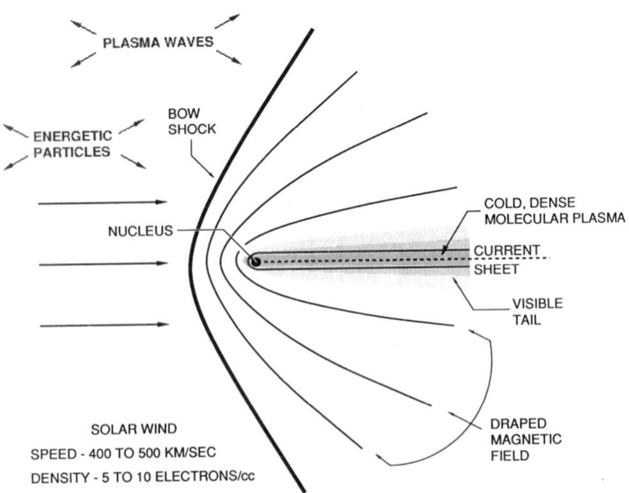

Figure 2. Schematic view of magnetic field and plasma structures in comets.

molecules from the comet freely pass through the solar-wind magnetic field, but when ionized they are picked up by the field lines and spiral around them. The heavy molecular ions, which are initially essentially at rest, are accelerated to the solar-wind flow speed while simultaneously decelerating the flow itself. Only a small percentage of heavy ions added to the flow produces significant decelerations.

The plasma tail is normally a feature attached to the head region of the comet and should not be considered as a cloud of independent particles drifting in the antisolar direction. The exception is the disconnection event (Fig. 1*b* and 1*c*). The DEs are probably triggered by rapid changes in solar-wind conditions at the comet. The leading candidates are rapid changes in solar-wind magnetic field polarity such as when comets cross sector boundaries or in the solar-wind momentum flux such as when comets encounter the high-speed streams.

SPACE MISSIONS

On September 11, 1985, the *International Cometary Explorer* (*ICE*) passed through the plasma tail of comet Giacobini–Zinner. During March of 1986, five spacecraft passed on the sunward side of comet Halley at distances ranging from 600 km to approximately 7×10^5 km. In addition, the *ICE* spacecraft was a distant probe of comet Halley at a sunward distance of 3×10^7 (1 million = 10^6) km. On all of the spacecraft sent to comets in 1985 and 1986, there were approximately 50 experiments. Almost all of them worked and about half of them were devoted to measurements of plasma parameters, including magnetic field; plasma speeds, temperatures, and densities; plasma waves; and high-energy particles. A schematic summary of the results is given in Fig. 2.

The results from the space missions confirmed the broad outlines of our physical picture. The solar-wind magnetic field is captured by the comet and draped around the comet to form the bilobed magnetic tail structure. The existence of a current sheet separating the regions of opposite magnetic polarity was verified and the plasma near it was found to be cold and dense. This plasma, composed of electrons and ionized molecules, is the gas we record in photographic images as it shines by fluorescence excited by sunlight. The deceleration occurs by the pick-up process and the bow shock described above.

There were also surprises from the missions. Plasma waves were found over extended regions and at high levels of intensity. The waves are thought to be excited by the process of "picking up" cometary ions by the solar-wind magnetic field. Energetic ions were also found around the comets at energies higher than could be produced by the pick-up process. An acceleration mechanism is needed. Stochastic mechanisms and magnetic reconnection are candidates.

During the week of the spacecraft encounters with comet Halley, March 6–14, 1986, there were at least one minor and two major DEs. The major DE of March 8.5 shows a close association with a sector boundary crossing and, when taken together with the minor DE on March 6, is entirely consistent with the sector-boundary front-side–reconnection theory. The analysis for this period is relatively straightforward because the spacecraft was close to the comet and the solar-wind conditions are known. The other 20–30 DEs in comet Halley during the 1985–1986 apparition [Fig. 1(*b*) and 1(*c*)] require detailed studies (which are in progress) before the single or dominant DE mechanism can be definitely identified.

The ongoing nature of the investigations for plasma structures in comets must be stressed. Even though the encounters took place in 1985 and 1986, several years may be required before the final data reductions and analysis are complete. A part of this process will be utilization of data from the *Archive* of the International Halley Watch.

FUTURE DIRECTIONS

The field of cometary plasmas in general, and plasma tails in particular, has received a tremendous boost from the in situ and ground-based data obtained during the 1985–1986 apparition of Halley's comet and the 1985 apparition of comet Giacobini–Zinner. These data provide a solid base for developing a physical understanding of cometary plasmas and for testing models.

Yet we must understand the limitations of these data, which refer generally to two comets; direct tail measurements, for example, refer specifically to one comet at one time and at one heliocentric distance. Additional missions with in situ measurements are required to be sure that models are widely applicable.

Plasmas make up 90–95% or more of the matter in the universe and an understanding of plasma physics is clearly important. Comets and comet tails are a unique plasma physics laboratory where the macroscopic and microscopic perspectives converge to provide a powerful investigative technique.

Additional Reading

Balsiger, H., Fechtig, H., and Geiss, J. (1988). A close look at Halley's comet. *Scientific American* **250** (No. 3) 96.

Brandt, J. C. and Chapman, R. D. (1991). *The New Comet Book.* W. H. Freeman and Co., New York.

Chapman, R. D. and Brandt, J. C. (1984). *The Comet Book.* Jones and Bartlett Publishers, Boston.

Grewing, M., Praderie, F., and Reinhard, R., eds. (1988). *The Exploration of Halley's Comet.* Springer-Verlag, Berlin.

Newburn, R. L., Neugebauer, M., and Rahe, J., eds. (1990). *Comets in the Post-Halley Era.* Kluwer Academic Publishers, Dordrecht.

Wilkening, L. L., ed. (1982). *Comets.* The University of Arizona Press, Tucson.

See also **Comets, Solar Wind Interactions; Interplanetary Magnetic Field; Interplanetary Medium, Solar Wind; Interplanetary Medium, Wave–Particle Interactions.**

Comets, Solar-Wind Interactions

Bruce T. Tsurutani

Most cometary nuclei are <20 km in diameter, but when they approach the Sun, sublimation of the water ice leads to a production rate of 10^{28}–10^{31} atoms and molecules per second traveling radially away from the nucleus at typical speeds of ~ 1 km s^{-1}. The time scale for photoionization or charge exchange (with solar-wind protons) is $\sim 10^6$ s; thus the neutrals propagate approximately 10^6 km from the nucleus before being ionized. This cloud of neutrals can be optically detected, giving comets an enormous scale size of 10^6–10^7 km. Far from the nucleus, the neutrals do not interact with the low-density, ~ 5 particles cm^{-3}, solar-wind plasma. However, once ionized, the strong solar-wind electric fields ($\mathbf{V}_{sw} \times \mathbf{B}_0 / c$) can lead to rapid acceleration of the ions. The solar wind velocity V_{sw} is typically 400 km s^{-1} at 1 AU, and the interplanetary magnetic field B_0 is typically 5 nT. c is the speed of light. If the interplanetary magnetic field is orthogonal to $\mathbf{V}_{sw}$, the electric force is $V_{sw} B_0 / c$ and the ions are accelerated to form a gyrating ring in velocity space. The ion guiding center moves with the solar wind and the ion has a gyro velocity V_{sw} relative to the convected magnetic field. If, on the other hand, the interplanetary magnetic field is parallel to $\mathbf{V}_{sw}$, then the newly formed ions experience no solar-wind electric field. However, the ions comprise a beam with velocity ~ 400 km s^{-1} relative to the solar-wind plasma. An ion (beam) instability occurs that leads to the generation of electromagnetic cyclotron waves, whose fields rotate about the ambient magnetic field in the same sense as electrons (right-hand polarized). As the cometary ions overtake these waves, an anomalous Doppler shift will occur and the ions will sense the waves at the ion gyrofrequency with the same sense of rotation (left hand). The particles will be scattered by the waves and accelerated up to the solar-wind velocity. The above interactions, the "pickup" of cometary ions by the solar wind, is the basic starting point of the solar-wind–comet interaction.

Prior to 1985, our knowledge about the solar-wind–comet interaction was very limited. In 1951, Ludwig Biermann observed that the cometary plasma (mainly CO^+) tails were oriented in basically an antisolar direction and hypothesized the existence of the solar wind prior to spacecraft measurements. Hannes Alfvén (1957) predicted that cometary magnetic tails would be oriented along the comet-sun line, arguing that the interplanetary fields would be draped around the cometary plasma region. Other plasma structures, such as a collisionless bow shock, a solar-wind–cometary neutral interaction layer, a solar-wind magnetic-field-free region close to the nucleus and a collisional shock inside the discontinuity to the magnetic-field-free region (ionopause) were only hypothesized. It was not until 1985–1986, when a fleet of spacecraft encountered two comets, that there was any possibility of testing these ideas. In September 1985, the *International Cometary Explorer* (*ICE*) went through the tail of comet Giacobini–Zinner 7800 km downstream of the nucleus. In March 1986, the European *Giotto*, Soviet *Vega 1* and *Vega 2*, and the Japanese *Sakigake* and *Suisei* spacecraft encountered comet Halley. *Giotto* went within 600 km of the nucleus, providing the wonderful images of the nucleus that have been published worldwide. The plasma, field, and particle measurements made aboard these spacecraft not only allowed testing of past theoretical ideas, but produced many surprises.

PLASMA TURBULENCE

The most striking of the new findings was the extremely high level of turbulence surrounding comets, particularly that about Giacobini–Zinner. This turbulence is in the form of ion acoustic waves, electromagnetic whistlers, electron plasma oscillations, broad band electrostatic noise, and low frequency plasma waves generated by the anisotropic distribution of cometary pickup H_2O group ions (O^+, OH^+, H_2O^+, H_3O^+) and their photoelectrons. The VLF, ELF, and LF (10^5 to 10^{-2} Hz) waves had the highest amplitudes ever detected by the *ICE* spacecraft after seven years in the solar wind and one year in the earth's geotail. The low-frequency magnetosonic waves often had amplitudes equal to the ambient magnetic field ($B_\omega / B_0 \approx 1$). The waves had peak power at their generation frequency $\sim 0.2\ \Omega_{ion}$ (ion gyrofrequency), but because of wave steepening and the generation of large amplitude upstream whistler-mode waves at $f \sim 10\ \Omega_{ion}$, the wave spectrum had significant power at both higher and lower frequencies. The total wave power has been estimated to be $\sim 5 \times 10^{15}$ W (for watt = joule per

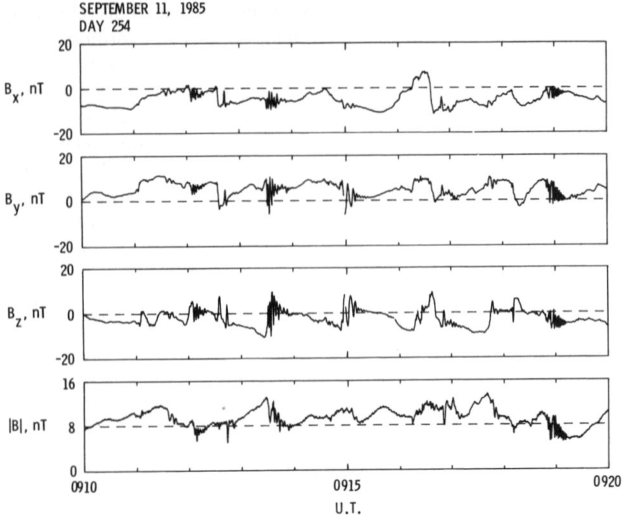

Figure 1. Example of steepened ~ 100-s period ($\tau_{H_2O^+}$) magnetosonic waves and the occurrence of whistler-mode wave packets at the leading (compressive) edges. The trailing portions of the waves appear to be linearly polarized. This is only an artifact of the steepening process. [*This figure is Figure 4 in Tsurutani et al.*, J. Geophys. Res. **92** *11074 (1987).]*

second) at a small comet such as Giacobini–Zinner. It is unclear how this energy is reabsorbed by the plasma, but it is speculated that it can go into heating solar-wind ions and may be involved in the stochastic (Fermi) acceleration of the cometary ions to the observed hundreds of kiloelectronvolts kinetic energy.

NONLINEAR WAVES AND WAVE-WAVE INTERACTIONS

The very high growth rate of the low-frequency right-hand waves associated with the pickup of cometary ions and the continuous feeding of this free energy source into the solar wind produce wave amplitudes comparable to the background magnetic field. Wave steepening occurs, broadening the spectrum by generating high-frequency precursor whistlers and forming an elongated low frequency "linearly polarized" trailing portion of the wave (see Fig. 1). The cometary experience allows one to study the development and evolution of such highly nonlinear phenomena.

Higher-frequency whistler waves driven by the cometary ions are susceptible to the decay instability, supplying energy to backward-traveling R-mode waves with slightly lower k as well as forward-traveling longitudinal waves. This "inverse cascading" continues, forming a wide band spectrum. We note, however, that because of intense wave-particle interactions the low-frequency magnetosonic waves do not similarly decay. Self-consistent wave-wave and wave-particle interaction studies are needed to understand the overall picture. These studies are just beginning to be developed for the case of comets.

MASS LOADING AND THE BOW SHOCK

The pickup of the cometary ions in the solar wind [both direct, when $\mathbf{B}_0 \perp \mathbf{V}_{sw}$, and indirect when $\mathbf{B}_0 \| \mathbf{V}_{sw}$ (via plasma instabilities)] increases the mass density of the streaming plasma, and from momentum conservation, leads to gradual solar-wind deceleration. However, this is only possible if the loaded mass is below some critical value. Before this critical value is reached, a collisionless reverse shock forms and moves upstream to direct the solar-wind flow around the comet. Calculations have predicted that the shock should have a subsolar magnetosonic Mach number of ~ 2 and should occur when the cometary number density is ≈ 1% of the solar-wind density.

The observed location and strength of the bow shocks at comets Halley and Giacobini–Zinner are in good agreement with theory, although this is perhaps more clear at Halley than at Giacobini–Zinner. At Halley on the inbound pass, a sharp transition with a ~ 25% decrease in flow velocity, a magnetic field increase of 50–100% and a solar-wind helium thermal pressure increase by ~ 100% were observed. (It should be noted however, that the solar-wind proton thermal pressure increase is far less than the helium increase and the cometary proton and H_2O-group ion temperatures are essentially unchanged.) The subsolar standoff distance is 3.9×10^5 km assuming an axisymmetric paraboloid for the shape of the shock. The magnetosonic Mach number (the solar wind flow speed divided by the magnetosonic wave speed) has been calculated to be ~ 1.6. Although the very thick (~ 10^4 km) nature of the bow "wave" at Giacobini–Zinner put the existence of a shock at that comet into question, a full Rankine–Hugoniot analysis of the boundary has established that at least the inbound pass contained a shock with magnetosonic Mach number 1.5. Calculations indicate the outbound transition may have been subsonic. However, because of the presence of the cometary turbulence, the calculated normal direction of the shock surface may be in significant error. Taking this potential inaccuracy into account, another scenario indicates that the outbound structure may have been a perpendicular shock with a magnetosonic Mach number of 1.6.

COMETOPAUSE

This boundary, observed by the *Vega* plasma and energetic particle experiments, is a broad region (~ 10^4 km) separating the cometary sheath region from the magnetized cometary plasma region. The former is dominated by solar-wind plasma and the latter by compressed interplanetary magnetic fields and cometary ions.

The explanation for this boundary layer and whether it is a permanent cometary feature or is associated with the solar wind are not agreed upon. Assuming that it is a permanent feature, one proposed scenario has collisions between the incoming solar wind plasma and the (radially) outflowing neutrals as the mechanism. The position where this occurs has been approximately calculated by equating the total momentum transfer collision mean free path to the distance from the nucleus. Since the collision cross section is different for different ion/neutral species, a chemical separation is expected. Another possible explanation for this phenomenon is a charge exchange avalanche between the incoming solar-wind protons and the outflowing cometary neutrals. Both mechanisms can explain the increase of magnetic fields cometward of the boundary region. The rapid deceleration of the solar-wind plasma and the cooling due to charge exchange with cometary neutrals will lead to a rapid increase in the magnetic field, forming a region where the predominant pressure is magnetic in nature. However the *Giotto* plasma and energetic ion scientists do not find similar features in their data. It is therefore thought by them that the features in the *Vega* data may be associated with disturbed solar wind conditions. Additional cometary encounters will be needed to resolve this question.

MAGNETIC TAIL AND PLASMA SHEET

The only observations of the cometary magnetotail and plasma sheet were made by *ICE*. The diameter of the magnetotail of comet Giacobini–Zinner was ~ 10,000 km at a distance 7800 km downstream. The tail is composed of two opposite directed lobes bounding a ~ 1500-km-thick plasma sheet. Peak lobe field strengths were ~ 60 nT. In the center of the tail lobes, the field was nearly aligned along the aberrated comet-sun line. Flaring of 20–40° is present in the outer regions of the tail lobes. The observations are in good agreement with the model of Alfvén.

In the lobes the electron β, the ratio of the electron thermal pressure to the magnetic pressure, is 0.05–0.10, indicating that this region is dominated by the magnetic pressure. The plasma sheet has an electron β of 2–4, indicating dominance of the plasma pressure in this region of space. It is significant to note that this latter region is not magnetically field-free, but has a strength of ~7 nT throughout. One possible explanation for this observation is that magnetic fields draped around the cometary ionosphere slipped into the outer regions of the ionosphere and have been dragged back into the plasma sheet by the solar wind. Calculations indicate that the Giacobini–Zinner ionosphere could have been susceptible to such field slippage (while the Halley ionosphere was not). Another possible explanation is that *ICE* crossed the tail well downstream of the antisolar extent of the ionosphere and that what was observed was the remnants of magnetic reconnection which occurred much closer to the nucleus. It should be noted that the cometary plasma sheet is considerably different from those observed in planetary magnetotails. At comets, the electron temperature is quite low, 1.6×10^4 K, and the density is high, 650 cm^{-3}.

MAGNETIC FIELD-FREE REGION

The only spacecraft that entered the cometary magnetic field-free region was *Giotto*. Deep inside the ionosphere of comet Halley, *Giotto* found that there was a region with essentially zero magnetic field (however, the exact upper limit is not available because of the large spacecraft fields which had to be subtracted from the measurements). The ion temperatures were as low as 340 K with outflowing velocities of ~1 km s^{-1}.

The boundary between this cavity and the magnetic tail region in comet Halley was quite thin, involving a field drop from 40 to ~0 nT in ~30 km. The location of this boundary can be estimated by equating the drag force on the ions imposed by the neutrals with the $\mathbf{J}\times\mathbf{B}$ electromagnetic force, where $\mathbf{J}$ is the current density. A calculated location of 4700 km from the nucleus is in good agreement with the observations. More recent calculations have included plasma pressure gradients and mass loading effects. These results give even better agreement with observations.

INNER SHOCK

The existence of a standing shock inside the ionopause was hypothesized prior to the spacecraft encounters. The effect of this reverse shock is to decelerate and divert the supersonically flowing plasma as it approaches the ionopause. The shock could be collisional, dominated by ion–ion interactions, or perhaps is "electrostatic" in nature. Measurements of comet Halley gave no evidence for the existence of such a shock. The ion densities appeared to be sufficiently high to allow plasma pileup and a rapid charge recombination (neutralization) process to occur, easing the physical necessity for a shock.

A fast forward Mach ~1.2 shock was detected outside the ionopause of comet Halley and propagating away from the nucleus. Because of the nature of the shock and its location, it could not have been created by the mechanism stated above. It is speculated that it was generated by a sudden pulse of plasma density/velocity in the ionosphere. The shock clearly was propagating outward from the nucleus of Halley and was not a standing shock.

CONCLUSIONS

We have just had a brief glimpse of the solar-wind interaction with comets. The joint NASA/Germany Comet Rendezvous and Asteroid Flyby mission will be launched in the mid 1990s and will orbit a short period comet early the next century. The mission will allow scientists to examine this interaction in detail over a solar distance range from 4.5 AU to perihelion at 1.5 AU. During this period, the comet will go from virtually no activity to full activity. The formation of the various regions of cometary space can be carefully monitored by the orbiting spacecraft.

Additional Reading

Alfvén, H. (1957). On the theory of plasma tails. *Tellus* **9** 92.

Biermann, L. (1951). Kometenschweife und solare korpuskularstrahlung. *Z. Ap.* **29** 274.

Galeev, A. A. (1987). Encounter with comets: Discoveries and puzzles in cometary plasma physics. *Astron. Ap.* **187** 12.

Goldstein, B. E., Altwegg, K., Balsiger, H., Fuselier, S. A., Ip, W.-H., Meier, A., Neugebauer, M., Rosenbauer, H., and Schwenn, R. (1989). Observations of a shock and a recombination layer at the contact surface of comet Halley. *J. Geophys. Res.* **94** 17251.

Mendis, D. A., Flammer, K. R., Reme, H., Sauvaud, J. A., d'Uston, C., Cotin, F., Cros, A., Anderson, K. A., Carlson, C. W., Curtis, D. W., Larson, D. E., Lin, R. P., Mitchell, D. L., Korth, A., Richter, A. K. (1989). On the global nature of the solar wind interaction with comet Halley. *Ann. Geophys.* **7** 99.

Neugebauer, M. (1990). Spacecraft observations of the interaction of active comets with the solar wind. *Rev. Geophys.* **28** 231.

Tsurutani, B. T., Thorne, R. M., Smith, E. J., Gosling, J. T., and Matsumoto, H. (1987). Steepened magnetosonic waves at comet Giacobini–Zinner. *J. Geophys. Res.* **92** 11074.

A special issue of the *ICE* Giacobini–Zinner encounter first results is given in *Science* **232** (1986), the comet Halley first results in *Nature*, **34** (1986), and a comprehensive review of all cometary science in *Astron. Ap.* **187** (1987).

See also **Comets, Plasma Tails; Interplanetary Medium, Shock Waves and Traveling Magnetic Phenomena; Interplanetary Medium, Wave-Particle Interactions.**

Comets, Types

Robert D. Chapman

The periods of revolution of known comets range from the short 3.3-yr period of comet Encke, through the 1000-yr period of comet 1887 II, to the indeterminately long periods of the parabolic comets. The numbers of known comets as of 1979, as a function of their orbital periods, are plotted in Fig. 1. Three groupings of comets can

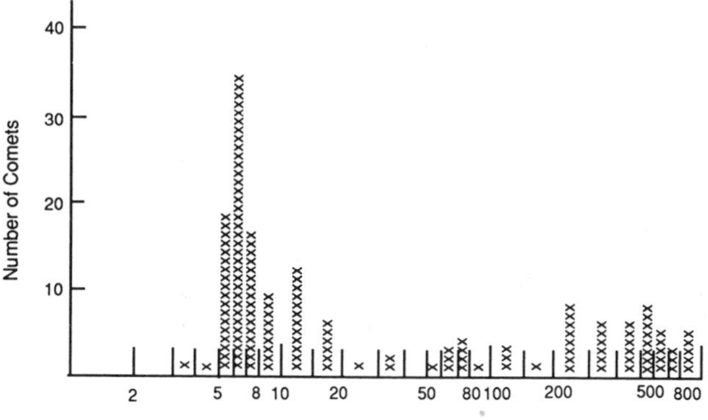

Figure 1. Plot of numbers of comets as a function of period of revolution around the Sun, as of 1979. The large number of near-parabolic and parabolic comets with indeterminately long periods are not shown on the figure. The three groupings of comets discussed in the text are obvious in the figure.

be identified. The first group consists of the short-period comets, with periods in the range 3.3 yr $< P <$ 20 yr. The second group is the intermediate-period comets, with periods in the range 20 yr $< P <$ 200 yr. Finally comes the group of long-period comets with $P >$ 200 yr. This latter group contains the parabolic and nearly parabolic comets with indeterminate periods. Some researchers combine the short-period and intermediate-period groups and call all comets with $P <$ 200 yr short-period comets; however, the orbital characteristics of the intermediate-period comets are different from those of the short-period comets. Two other identifiable types of comets are interesting: the sun grazers and extra-solar comets.

This entry will discuss four aspects of comets: the characteristics of the groups mentioned above, the characteristics of examples of some of the groups, the lifetimes of short-period comets, and the origins of short-period comets.

SHORT-PERIOD COMETS

The orbits of short-period comets share properties in common with the orbits of the planets. The inclinations of the orbits, for instance, are low, with 48% in the range $0° \leq i < 10°$, 37% in the range $10° \leq i < 20°$, and 14% in the range $20° \leq i < 60°$. The overwhelming majority move in direct orbits with inclinations less than 20°. The major difference between short-period cometary orbits and planetary orbits lies in the larger eccentricities of the cometary orbits. The distribution of periods, as can be seen from Fig. 1, has a peak between 6 and 7 yr. Most of these comets have their aphelia near the orbit of Jupiter.

Encke's comet is an interesting example of a short-period comet, though its short 3.3-yr period makes it somewhat atypical. The comet was first discovered over two centuries ago, and has made over 50 appearances since. In 1818 the German mathematician Johann F. Encke attempted to calculate the orbital parameters of a comet discovered by the great observer Jean L. Pons. He used a mathematical technique invented by Isaac Newton, which assumed the orbit of the comet to be a parabola. Encke was not able to fit Pons's observations using Newton's methodology. He then turned to a method first devised by Carl Friedrich Gauss to calculate the orbital parameters of minor planets. He found that the comet moves in an orbit inclined by 12° to the ecliptic, and with an eccentricity of 0.846. Once Encke determined the orbital parameters of the comet, he was able to show that it had been seen at three earlier appearances.

Since comet Encke's orbit carries it across the orbits of Venus, the Earth, Mars, and outward nearly to Jupiter, planetary perturbations have a significant effect on its motion. Encke continued to study its motion during several passes and found that, even after accounting for all planetary perturbations, the period exhibited a secular decrease of about 0.1 day per revolution. He attempted to explain the acceleration as being due to a uniform resisting medium. In honor of these pioneering dynamical studies, the comet was named after Encke; it is one of the relatively few comets not named for its discoverer.

Today, comet researchers invoke a different mechanism to explain the nongravitational force observed on comet Encke and a number of other comets. Solar radiation causes the nuclear ices to sublimate preferentially on the side facing the Sun. Since the nucleus rotates, the point that has been heated by sunlight moves a small distance while the sublimation process takes place. The result is a reaction force with a component along the comet's orbit. A little reflection will show that the nongravitational effect can lead either to a secular acceleration or to a deceleration depending on the direction of rotation of the nucleus. The average radius of the nucleus of a short-period comet is too small to be measured directly. A much-used technique for estimating the sizes of comet nuclei makes use of their measured brightnesses and estimated reflectivity properties. Estimated sizes are in the 1–2-km range. Each time a comet passes perihelion, sublimation and other processes cause a few meters of the nucleus to become coma and tail material and to be lost to the comet. Armand Delsemme has estimated that a typical short-period comet will disappear after about $i_0 = 200–100$ perihelion passages, or about 1200 yr. The lifetimes of these short-period comets are very small compared to the age of the solar system, and the fact that they are observed today means that there must be a mechanism to resupply the population. The resupply mechanism will be addressed in the next section.

We have probably witnessed the final passes of at least one comet. Comet Biela, with a period of 6.6 yr, was observed as comet 1772, 1806 I, 1826 II, 1832 III, 1846 II, and 1852 III. At the 1846 return, the nucleus of the comet split into two parts, both of which became fully formed comets. During the remainder of the passage the two comets exhibited unusual activity. In 1852, both comets reappeared, but were separated by about 2×10^6 km. Once again they were unusually active. The comets were predicted to be favorably situated relative to the Sun and Earth at the 1866 return. However, they did not appear, and have never been seen since.

LONG-PERIOD COMETS

The distribution of the inclinations of the long-period comets, by contrast to that of the short-period comets, is a random distribution. In addition, if one plots the direction of the aphelia of the orbits of all known long-period comets, the points are distributed at random on the celestial sphere.

The frequency distribution of the reciprocal of the semimajor axes $(1/a)$ of the orbits of long-period comets shows a marked peak between 10^{-4} and 10^{-5} inverse astronomical units. This peak represents the inner edge of the Oort cloud, a cloud surrounding the Sun and containing an estimated 2×10^{12} comets. Oort calculated, in the 1950s, that a star will occasionally come close enough to the Sun that it will pass through the cloud and will cause some of the comets to fall in toward the inner solar system, producing the observed population of parabolic comets.

The current working hypothesis on the origin of the short-period comets is that a fraction of these parabolic comets are captured into the inner solar system and become short-period comets due to the influence of Jupiter and the other giant planets. For the number of short-period comets to remain constant, the rate of their destruction must equal the rate of capture of long-period comets. Delsemme has analyzed the problem in detail, and has found that the capture rate is adequate to replenish the supply of short-period comets if 1000–3000 comets pass perihelion between 4 and 6 AU per year. Such comets would be virtually unobservable, so that the hypothesis remains plausible, but untested. The long-period comets have estimated nuclear sizes significantly larger than those of the short-period comets.

INTERMEDIATE PERIOD COMETS

There are so few intermediate-period comets known that it is not possible to draw statistically meaningful conclusions on the characteristics of their orbits. What we can say is that 4 of the 13 known intermediate-period comets move in retrograde orbits, and the absolute values of their orbital inclinations range from 17° to 85°. Despite their small numbers, they are clearly a different population than the short-period comets, and should be identified as such.

Comet Halley is one of the most famous examples of an intermediate-period comet, with a period of 76 years. It moves in a retrograde direction around an orbit inclined by 18° to the ecliptic. Using historical records, researchers have established that comet Halley has been seen at 28 appearances, beginning in 86 B.C. Of course, comet Halley was studied in minute detail by both spacecraft- and ground-based observatories during its 1985–1986 passage by the Sun. Much new information on the nature of the comet was gathered at that passage. For instance, the comet's nucleus was found to be elongated, with a long axis of 15 km, and it is one of the darkest bodies in the solar system with an albedo of about 0.04. If the reflectivity of the nuclei of the short-period comets is as low as is the case for Halley's comet, then our estimates of their sizes are too small. (The reflectivity method led to an initial estimate of the size of Halley's nucleus of 5 km.)

SUN-GRAZING COMETS

The *Solar Maximum Mission* (*SMM*) spacecraft was launched in 1980 with a suite of scientific instruments designed to study the intense activity on the Sun during the sunspot maximum. One instrument, the coronagraph, has provided cometary researchers with the discovery of at least 10 sun-grazing comets, as well.

Sun-grazing comets, as the name implies, are comets that come very close to the Sun at perihelion. (Their perihelion distances are in the range $0.0055 < q < 0.067$ AU.) The first of the *SMM* discoveries was comet Howard–Koomen–Michels (1979 XI), which was first observed as it approached the Sun. Within a few hours of discovery, the head of the comet had disappeared behind the occulting disk of the *SMM* coronagraph, which has a projected radius on the sky of 2.5 solar radii. Soon thereafter, the solar corona brightened significantly. The comet was not seen again. The preperihelion observations covered a sufficient arc during the short time the comet was followed to permit orbital elements to be derived with sufficient accuracy that researchers should have been able to locate it after perihelion. There is little doubt that the comet disintegrated as it passed perihelion.

Sun-grazing comets had been known long before *SMM*. Their proximity to the Sun causes them to be among the historically brightest comets. The Great Comet of 1843 (1843 I), for instance, actually could be observed in broad daylight when it was near perihelion. The orbital elements of eight of the sun grazers known before *SMM* are all nearly identical. They are members of what has come to be known as the Kreutz group of comets. A reasonable hypothesis is that the Kreutz group comets are all fragments of a very large parent comet which passed close to the Sun some time in the past and was disrupted into a number of fragments, that we observe today as the individual sun grazers. Several sun grazers (Comet 1883 II and Comet 1965 VIII, Ikeya–Seki) have been observed to split as they passed perihelion, adding credibility to the hypothesis.

EXTRASOLAR COMETS

The current theory of the origin of the Oort cloud suggests that about 10^{14} comets were ejected from the solar system during the formation process. If most of the stars in the solar neighborhood have Oort clouds, formed by the same process as the Sun's cloud, then there may be as many as 10^{13} comets per cubic parsec in the solar neighborhood. Given a reasonable hypothesis for their velocity distribution, there should be a flux of 0.6 comets per year within 2 AU of the Sun. Such comets should exhibit clearly hyperbolic orbits. A number of researchers, including the present author, have addressed the question of the probability of detection of such comets. We have concluded that the probability of detecting an individual hyperbolic comet within 2 AU of the Sun is very roughly 0.07. Therefore, about four extrasolar comets should have been observed in the last century, certainly not a large number.

The largest orbital eccentricity listed for any comet in 1979 was $e = 1.004$, and there were about 80 comets listed with $1.000 < e < = 1.004$. A careful analysis of the motions of the majority of those comets shows that they were moving in elliptical orbits relative to the center of mass of the solar system when they were beyond Neptune. For the remainder of the comets, nongravitational forces appear to account for the observed excess hyperbolic eccentricities. In short, there is as yet no evidence that a comet moving in a truly hyperbolic has ever been observed.

As observing technology improves, researchers may finally discover a comet of extra-solar origin. Such a discovery would provide a significant data point in our attempts to understand the origin of the solar system.

Additional Reading

Brandt, J. C. and Chapman, R. D. (1981). *Introduction to Comets*. Cambridge University Press, Cambridge.

Marsden, B. G. (1989). *Catalog of Cometary Orbits*, 6th ed. Smithsonian Astrophysical Observatory, Cambridge, MA.

McGlynn, T. A. and Chapman, R. D. (1989). On the non-detection of extrasolar comets. *Ap. J. (Letters)* **346** L105.

Wilkening, L. L., ed. (1982). *Comets*. University of Arizona Press, Tucson.

See also **Asteroid and Comet Families; Comets, Dynamics; Comets, Historical Apparitions; Comets, Oort Cloud.**

Constellations and Star Maps

George Lovi

Since antiquity, watchers of the starry firmament have observed that the star locations appear "fixed" relative to each other. Aided by vivid imaginations, a number of cultures created patterns among the stars; these patterns came to be called *constellations*, which in a loose sense became regarded as star "pictures." These were often influenced by the legends, heroes/heroines, deities, creatures, and artifacts of the respective cultures. Such constellations that have survived to the present retain some astronomical value, not as fanciful sky pictures but as arbitrary divisions of the face of the sky; these divisions can facilitate finding one's way around the sky and referring to particular locations, much as similar political divisions —countries, states, and so on—do on Earth maps. To help chart the starry sky, various forms of star "maps" have been constructed over the centuries and millenia, whether they be flat two-dimensional renditions like terrestrial maps or three-dimensional models such as globes, armillary spheres, or even modern-day planetarium projectors. Unlike terrestrial maps, which portray an actual physical entity—the Earth's surface or portions thereof—star maps record the *directions* in relation to us of stars and manifold other celestial objects (which are intrinsically at enormously varying spatial distances) as projected upon a tremendous, larger-than-the-universe, imaginary, but mathematically real, *celestial sphere*.

Although constellations have been relatively widespread among the world's peoples, those still recognized today are of fundamentally Western origin, though created during distinctly separate historical eras with the earliest tracing back to the Mesopotamian peoples of five or more millenia ago. It is there that today's oldest constellations, such as the Lion, Bull, and Strong Man (modern Leo, Taurus, and Hercules), among others, trace their earliest antecedents; for example, Hercules—Herakles to the Greeks—has been linked to the Mesopotamian strong man Gilgamesh and even to the biblical Samson.

Of particular significance to even those early peoples was a special band around the sky some 16° wide, within which the Sun, Moon, and "wandering stars" (planets) confine their movements. This became known as the *zodiac*, whose central line is the *ecliptic*—the Sun's apparent celestial path or Earth's orbital plane extended out to the sky. By the first millenium B.C., when astronomy flourished during the Greek civilization, the zodiac—which has remained inextricably intertwined with astrology even to this day—was divided into 12 equal 30° sectors or "signs." These were named for the zodiacal constellations which then approximately coincided with each sector; since then the precession of the equinoxes has shifted each zodiacal constellation roughly one sign length eastward.

Largely because of the factors enumerated in the previous paragraph, the ecliptic became the primary celestial reference circle to ancient and medieval Western cultures (but not Oriental cultures, which were early adoptors of today's equatorial system, which parallels the apparent rotation of the sky). The ecliptic became the basis of an all-sky coordinate grid of celestial longitude and latitude, which represent, respectively, angular distance parallel to the ecliptic and perpendicular to it. There was also a strong tendency to list the locations of celestial objects by their zodiacal sign and number of longitude degrees within each one, even if these celestial objects were located at some distance above and below the zodiacal band, as these signs were extended to the ecliptic poles to cover the entire sky. This is how Claudius Ptolemy stipulated stellar longitudes in the star catalogue in his epochal *Almagest* of about 150 A.D, as did other early astronomers.

The ancient Greeks were the most influential in establishing today's roster of traditional constellations, with Aratus of Soli (ca. 300 B.C.) being of particular import. His epic work, the *Phaenomena*, has a running commentary on the then-recognized constellations; it was largely based on a same-titled work by Eudoxus of Cnidus of a century previous. All this, along with the extensive body of Greek legends associated with these star patterns, later influenced Ptolemy when he compiled his *Almagest* catalogue, which established the constellations known as the "classical 48" until the Renaissance some $1\frac{1}{2}$ millenia later.

With the reawakening of Western culture and its corollary Age of Exploration, a new postclassical constellation-forming period commenced. Particularly significant was the filling of the large blank area around the south celestial pole; this area could not be seen from the lands of classical Western culture. The first 12 such patterns were created by the Dutch explorer Pieter Dirckszoon Keyser during a 1590s East Indies voyage; these were mostly exotic birds and creatures.

Keyser's constellations were "officially" introduced by Johann Bayer in his famous *Uranometria* star atlas of 1603, a singularly influential work of celestial cartography. Largely based on the then unprecedentedly accurate pretelescopic star positions obtained by Tycho Brahe during previous decades, the *Uranometria* is best known for its high artistic merit in delineating the mythological constellation figures; it essentially inaugurated the ensuing $2\frac{1}{2}$-century pictorial star-map period. However, Bayer's major astronomical legacy was introducing the star-labeling scheme, still used today, of assigning lower-case Greek letters to the brighter naked-eye stars of a constellation, roughly (but far from strictly) in order of brightness.

Oddly enough, such fanciful Bayer-style artistic augmentations had some astronomical usefulness back then, because astronomers still sometimes referred to a sky area as being "in Orion's belt" or whatever, continuing a tradition that went back to the time of Ptolemy.

Two other astronomers were primarily responsible for creating the remaining postclassical constellations that are presently recognized. Johannes Hevelius's 1687 *Firmamentum Sobiescianum* atlas introduced seven new ones, including Lynx, Leo Minor, Canes Venatici, and Sextans—named for both animate and inanimate objects. The second, that of Nicolas Louis de Lacaille, introduced, in the 1750s, 14 new far-southerly patterns to supplement Keyser's. Totally eschewing traditional people and creatures, they mostly represent scientific/technical artifacts, for example, Telescopium, Microscopium, Horologium, and others, all composed of relatively faint stars, because the brighter ones were by then already incorporated into previously defined constellations.

As the mid-nineteenth century approached, pictorial star maps rapidly lost favor, and were disdainfully dismissed by astronomers like John Herschel, who referred to constellation figures as "men and monsters scribbled over celestial globes and maps." By the latter part of that century, star maps became strictly nonartistic utilitarian tools; yet with the growth of large-aperture telescopic observing, a peculiar problem arose: what are the "limits" of each constellation? For example, the rapid discoveries of new variable stars, which continued to be named according to their constellations, resulted in considerable confusion early in the twentieth century. The solution was a network of official constellation boundaries, introduced in 1930 by the International Astronomical Union, that ran strictly north-south, east-west (for equinox 1875.0, since at that time Benjamin A. Gould had already created such boundaries for far-southerly groups). Additionally, this IAU project established the present definitive roster of 88 constellations.

Figure 1. The prominent constellation Orion as portrayed in the classical *Uranometria* star atlas of Johann Bayer, published in 1603, the most famous compilation of pictorial star maps. (*Courtesy U. S. Naval Observatory*.)

Although these IAU boundaries are usually found on detailed star-atlas–type maps, popular star charts have since the last century usually delineated the constellations with arbitrary straight lines connecting the principal stars; some of these have attempted to be somewhat pictorial.

The most recent major star-mapping breakthrough, in the past quarter century, has been the use of electronic computers and plotters to compile and plot star maps with unprecedented speed and accuracy. The two major computer-plotted atlases in this period are the *Smithsonian Astrophysical Observatory Star Atlas* of the 1960s and *Uranometria 2000.0* of the late 1980s. Despite the computer's substantial advantages, one problem that has yet to be completely solved is the optimum placement of labeling and certain other auxiliary graphics. In the production of *Uranometria 2000.0*, the noted celestial cartographer Wil Tirion, one of its authors, added these features by hand; the earlier Smithsonian work is devoid of any labeling within the body of its maps.

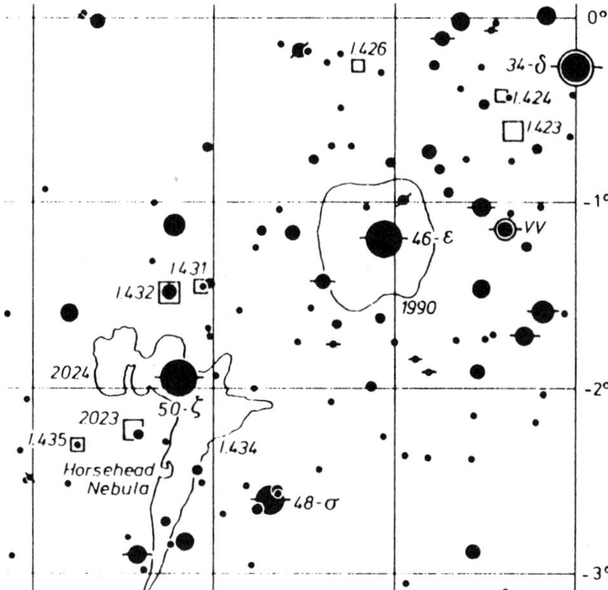

Figure 2. The area of Orion's belt, shown below center on the Bayer plate, as portrayed in the 1987–88 *Uranometria 2000.0* star atlas, a computer-plotted two-volume work, to about magnitude $9\frac{1}{2}$. (*Courtesy Willmann-Bell, Inc.*)

Additional Reading

Davis, G. A. (1959). The origin of the ancient constellations. *Sky and Telescope* **18** 424.

Gingerich, O. (1974). Astronomical maps. In *Encyclopaedia Britannica*. Encyclopaedia Britannica, Inc., Chicago.

Lovi, G. (1987). Uranography yesterday and today. In *Uranometria 2000.0* **1**, W. Tiron, B. Rappaport, and G. Lovi, eds. Willmann-Bell, Inc., Richmond, Va.

Lovi, G. and Tirion, W. (1989). *Men, Monsters, and the Modern Universe*. Willmann-Bell, Inc., Richmond, VA

Ridpath, I. (1988). *Star Tales*. Universe Books, New York.

Staal, J. D. W. (1988). *The New Patterns in the Sky*. McDonald & Woodward, Blacksburg, VA.

Warner, D. J. (1979). *The Sky Explored, Celestial Cartography 1500–1800*. Alan R. Liss, New York.

See also **Coordinates and Reference Systems; Star Catalogs, Historical.**

Coordinates and Reference Systems

Laurence A. Marschall

Astronomical constants and reference systems are the standard "weights and measures" of astronomy. Astronomy requires, in addition to the standards of mass, length, time, etc. used in the other physical sciences, a set of specialized reference quantities that can be used to specify observing techniques and to compare measurements of celestial bodies made by different observers and at different times. Astronomical standards include:

1. Units describing the dynamical behavior of objects in the solar system. These so-called "astronomical constants" are useful, among other things, in calculating accurate ephemerides, or timetables, for the positions of objects in the sky.
2. A stable coordinate reference frame, defined by objects whose relative motions are assumed to be negligible, against which the changing positions of celestial bodies can be measured and followed from one time period to another.

CONSTANTS

The fundamental system of astronomical constants is a set of defined values, observed quantities, and values derived from other constants by the application of fundamental physical and geometrical relations. These constants are agreed upon by an international governing body, the International Astronomical Union (IAU), based upon the needs of the astronomical community and the best available measurements. The current system was adopted by the IAU in 1976, and the values of certain quantities have been modified subsequently as better determinations have become available.

Astronomical constants are, for the most part, consistent with those of other sciences, incorporating the *Système Internationale*, or SI, in which distance, mass, and time are expressed in units of meters, kilograms, and seconds, which are defined by properties of light, standard masses, and atomic clocks. However, because of the specific requirements of astronomy—the large distances and long time periods used in many calculations, as well as the need to specify special quantities not encountered in a laboratory—the IAU System of Astronomical Constants considerably modifies and extends the SI. In the IAU system, for instance, the fundamental unit of mass is taken to be the mass of the Sun, and the units of time are the day (86,400 s), and the Julian century (36,525 days).

The astronomical unit of length, sometimes called the unit distance or simply the astronomical unit, is defined as that length for which the Gaussian gravitational constant k takes on the value of 0.01720209895 when the conventional astronomical units of length, mass, and time are used to calculate it (k^2 is essentially the constant in Newton's universal law of gravitation). This astronomical unit of length is virtually identical with the semimajor axis of the Earth's orbit. Its current value in SI units is $1.4959787066 \times 10^{11}$ m. By adopting a value of k and measuring the period of the Earth's orbit around the Sun, we have effectively specified the value of the astronomical unit. Thus the astronomical unit is often considered a "derived constant," whereas k is a "defining constant." The speed of light is also taken as a defining constant. Its standard value is 299, 792, 458 m s^{-1}, from which we can derive distance units such as light-seconds and light-years.

Among other constants unique to the IAU system are the equatorial radius of the Earth (6,378,137 m), the ratio of the mass of the Moon to the mass of the Earth (0.012300034), and the ratios of the masses of the planets to the mass of the Sun.

Several astronomical constants describe the dynamical properties of the Earth. Among these are the precession constant, the nutation constant, and the aberration constant. Precession is the term given to a conical motion of the Earth's axis of rotation around a direction perpendicular to the *ecliptic* (the plane of the Earth's orbit). Precession causes the *equinoxes* (the intersections between the celestial equator and the ecliptic) to move along the ecliptic. A complete cycle takes about 26,000 yr. Nutation describes an elliptical "nodding" motion of the Earth's axis, with a period of about 18.6 yr, that is superimposed on precession. The aberration constant essentially measures the rate at which the equinoxes move, whereas the nutation constant measures the effect of nutation in changing the inclination of the Earth's axis to the ecliptic (the *obliquity*).

Aberration is a third effect, referring to a cyclic displacement of the apparent positions of stars due to the changing orbital velocity of the Earth. The effect resembles that observed by running through

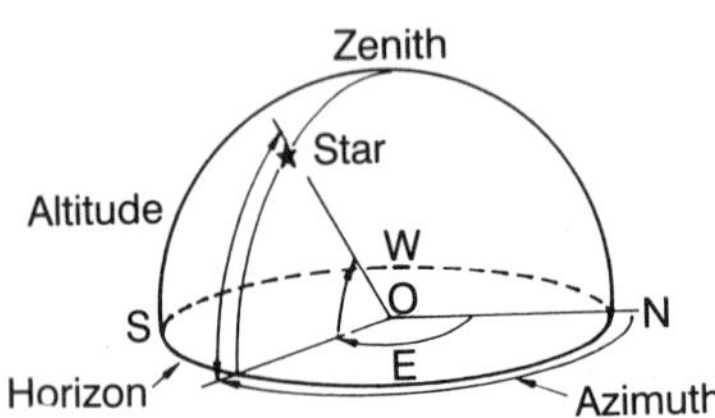

Figure 1. The horizon coordinate system. [*By permission from Paul E. Trejo,* Introductory Astronomy, *Kendall/Hunt Publishing Company, Dubuque, Iowa (1986)*.]

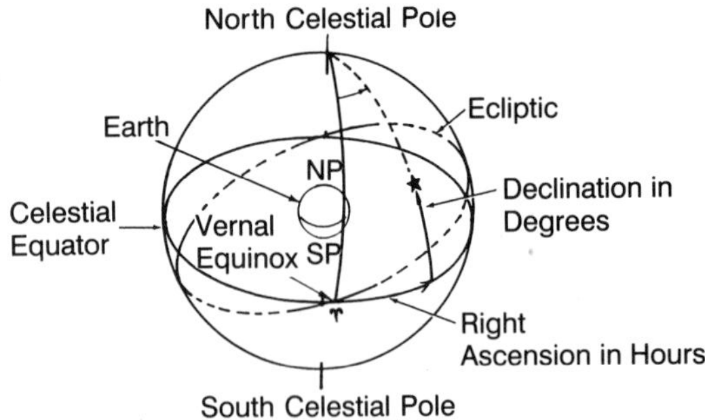

Figure 2. The equatorial coordinate system. [*By permission from Paul E. Trejo,* Introductory Astronomy, *Kendall/Hunt Publishing Company, Dubuque, Iowa (1986)*.]

a rainstorm. Raindrops, which descend vertically when an observer is stationary, seem to slant toward the observer who is running. Furthermore, the amount of shift depends on the ratio of raindrop speed to the speed of the runner. Stellar aberration similarly shifts the direction of starlight towards the direction in which the Earth is traveling and by an amount proportional to the ratio of the earth's orbital velocity to the velocity of light. Thus stars trace out small ellipses in the sky as the Earth orbits. The aberration constant, which measures the semimajor axis of the aberrational ellipse, can be derived from measurements of the Earth's orbital velocity and from the defined speed of light.

A more complete tabulation of astronomical constants, along with explanatory notes, is provided yearly in *The Astronomical Almanac*, published by the U.S. Government Printing Office.

COORDINATE SYSTEMS

Since the fundamental data of positional astronomy are angular measurements made from the Earth, astronomical reference systems effectively provide a two-dimensional grid on the celestial sphere against which the angular coordinates of celestial objects can be measured. Common astronomical coordinate systems all resemble the system of latitude and longitude on the Earth. They are defined by diametrically opposed poles, a primary circle halfway between the poles, and a reference position on the primary circle. One coordinate measures angular distance from the primary circle (similar to latitude), and another measures angular distance around the primary circle from a reference point to a great circle through the poles intersecting the object in question. A number of coordinate systems are in common use.

In the *horizon system* (see Fig. 1), the primary circle is the observer's horizon, and the reference point is the due north point on the horizon. One pole, directly overhead, is the *zenith*, the one diametrically opposite, the *nadir*. The angle above or below the horizon is called the *altitude*, while the other coordinate, conventionally measured eastward from the north point, is called the *azimuth*. The great circle through the zenith, north, and south points is called the *celestial meridian*. Direct measurements of star positions are commonly made in the horizon system, and meridian altitudes and crossing times are the fundamental data determining precise star positions for reference purposes. But because the altitudes and azimuths of objects depend on the observer's terrestrial location and the time of day, horizon coordinates are commonly transformed into one of the other coordinate systems described below.

In the *equatorial system* (see Fig. 2), the poles are defined by the Earth's axis, the primary circle is the celestial equator, and the reference point on the celestial equator is the *vernal equinox* (the intersection of equator and ecliptic on the northward passage of the Sun). *Right ascension*, measured in hours (1 h = 15°) eastward from the vernal equinox, and *declination*, measured in degrees north or south of the celestial equator, are the two coordinates defined by the system. Because of precession the origin of this system continuously changes, so that equatorial coordinates of stars are generally tabulated for a standard reference date or *epoch*, currently the beginning of the year 2000 (see below).

In the *ecliptic system*, useful for the description of planetary motion, the primary circle is defined by the ecliptic, and the reference point is the vernal equinox again. Ecliptic latitude is measured north or south of the ecliptic, and ecliptic longitude is measured eastward from the equinox point.

In the *galactic system*, commonly used for studies of objects in the Milky Way galaxy, the primary circle is the plane of our galaxy and the reference point is in the direction of the galactic center. Since 1958, the precise positions of the galactic equator and galactic center have been defined primarily by observations of radio emission from atomic hydrogen. The current position of the north galactic pole is near right ascension 12^h52^m, declination 27°8′ and the galactic center is located (in the constellation of Sagittarius) at about right ascension 17^h45^m, declination −28°56′.

REFERENCE SYSTEMS

The coordinate systems defined above are geocentric; that is, they are defined with respect to axes fixed to the constantly moving Earth. Ideally, these coordinates should be related to a stationary or "inertial" frame of reference. Long-term observations of a set of standard stars can be used to compile a fundamental catalog of star positions and motions, implicitly defining a reference system against which the positions of other objects can be referred. The FK4 catalog, containing values for 3522 stars, provided the basic stellar reference frame adopted by the IAU in 1976. The FK5 catalog, completed in the late 1980s, extended the sample to fainter stars with improved accuracy. The fundamental reference system for the present is therefore the system of the FK5. Positions are defined for the current standard epoch, the first second of the year 2000, referred to as J2000.0.

Another related reference system is being defined by the *HIPPARCOS* satellite, a dedicated astrometric satellite launched in 1989. Though it failed to reach its intended high orbit, the satellite is operating smoothly and is conducting a survey of 120,000 stars, deriving relative positions and motions to high precision.

The most precise astronomical reference system, however, is that being defined by radio astronomy. Very long baseline interferometry (VLBI) measurements of distant extragalactic radio sources yield measurements of the relative positions of the sources along with the separation of the receiving antennas and the orientation of the axis of the Earth. Because of the intrinsic precision of VLBI methods and the extreme distance of the radio sources, this is a more nearly inertial reference system than that defined by stellar observations. Determining how to convert measurements made in the stellar reference frame to those defined in the radio reference frame is currently one of the more challenging problems of positional astronomy.

Additional Reading

Carter, W. E. and Robertson, D. S. (1986). Studying the Earth by Very-Long-Baseline Interferometry. *Scientific American* **255** (No. 5) 46.

Eichhorn, H. (1974). *Astronomy of Star Positions*. Frederick Ungar, New York.

Fricke, W. (1972). Fundamental systems of positions and proper motions. *Ann. Rev. Astron. Ap.* **10** 101.

Kulikov, K. A. (1964). *Fundamental Constants of Astronomy*. Israel Program for Scientific Translations, Jerusalem.

Taff, L. G. (1981). *Computational Spherical Astronomy*. John Wiley and Sons, New York.

The Astronomical Almanac (published yearly). U.S. Government Printing Office, Washington.

See also **Earth, Figure and Rotation; Star Catalogs and Surveys; Time and Clocks.**

Coronagraphs, Solar

Raymond N. Smartt

DESIGN PRINCIPLES

A coronagraph is a specialized astronomical telescope designed to observe, outside of a total solar eclipse, the faint solar corona surrounding the Sun. The primary concern in the design is to prevent the direct solar light, and as far as possible its scattered and diffracted components, from reaching the final image plane. Otherwise the coronal light will be overwhelmed by the million times brighter background of light from the Sun itself.

In its classical form, (see Fig. 1) as invented by Bernard Lyot, the primary objective of a coronagraph is a singlet objective lens made of high-quality optical glass with surfaces specially polished to minimize scattering of the sunlight passing through it. To reduce aberrations, one or both surfaces are usually aspherics. A normal achromatic doublet objective, even if cemented, would scatter far too much light. An occulting disk, a piece of metal with a very smooth circular edge, is located in the primary image plane to block the solar image completely. For this, the occulting disk is made a percent or so larger than the diameter of the Sun's image. Since the angular size of the Sun varies during the year, several occulting disks are usually employed so that the coronal light can be observed as close to the limb of the Sun as possible, consistent with some small motion of the image due to atmospheric seeing and possible small guiding errors of the coronagraph. Beyond the primary focal plane, a field lens forms an image of the primary objective. Since diffraction of the solar light at the rim of the objective constitutes a large source of stray light, a circular aperture (Lyot stop) is located at the image of the objective with a slightly smaller opening, such that the bright diffracted rim of light is blocked from passing to the final image plane. A further lens, or lens system, forms the final image. High imaging quality can be achieved at specific wavelengths by optimizing the secondary optical system to compensate for the chromatic aberrations produced by a singlet objective, but the occulting disk image is not color-free. Multiple reflections within the primary objective can constitute an additional source of stray light. These will come to a focus in the vicinity of the Lyot stop and can be blocked by locating a tiny occulting disk in this plane. To minimize the contribution of scattered light from the sky, coronagraphs are located at high-altitude sites where the brightness of the sky, even at less than one minute of arc above the limb, can be only a few millionths that of the disk of the Sun on extremely clear days.

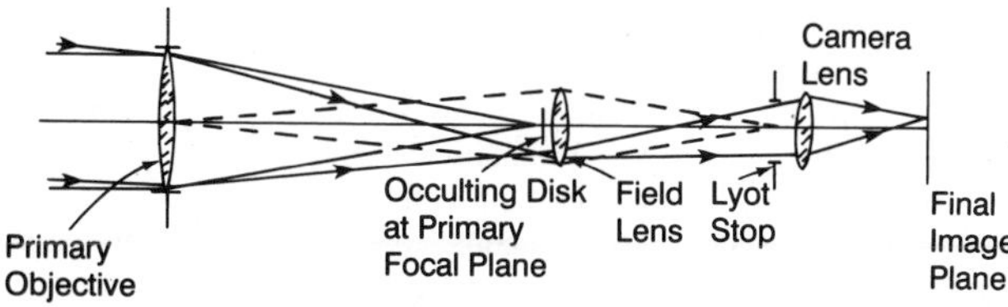

Figure 1. Diagram showing the key features of a coronagraph used for observing the solar corona.

GROUND-BASED CORONAGRAPHS

To observe images of the corona from the ground, filters are used with bandwidths that match closely the emission spectral line widths, thus optimizing the coronal signal relative to the sky background over this spectral interval. For this, dichroic interference filters can be used, but filters stable in wavelength with time and accommodating larger angular fields, such as a Lyot birefringent filter, are generally preferred. Birefringent filters do require interference filters for spectral blocking, but their broader bandwidths reduce their performance constraints. Alternatively, spectrographs are commonly used for spectral line studies. Both types of measurements can be recorded photographically, but diode arrays are more convenient for studies of rapidly varying coronal activity over small fields of view. The performance requirements of coronagraphs with regard to scattered light are reduced if the field of view is limited, commonly to one quadrant of the coronal field, or even smaller, with provision for selecting any part of the coronal field of interest. A full annular field of view can nevertheless be achieved with high-quality performance. Even under relatively poor sky conditions, the brightness of the emission corona can be measured accurately using a coronal photometer based on a polarization spectral-chopping technique.

Coronagraphs designed to observe the white-light electron-scattered K-corona are constructed according to the same principles as given above, except a broad spectral range of at least several hundred angstroms is required. A different detection technique is essential since the spectral brightness of the K-corona, even close to the limb, is usually only a few millionths that of the solar disk. The method used takes advantage of the fact that the K-corona is linearly polarized with the electric vector tangential to the limb. This polarized signal can be differentiated from that due to the sky by a polarizing-analyzer chopping technique. With precise calibration, the contribution of the F-component (photospheric light scattered by interplanetary dust) of the corona, at heights above the limb typically greater than about 8 min of arc, can be determined.

SPACE-BASED CORONAGRAPHS

Coronagraphs have been operated successfully from stratospheric balloon platforms, rockets, and satellites. The designs used are essentially those of ground-based systems, except, without a sky background, external occulting is generally employed. This reduces instrumental scattered light, since the objective is then shadowed; otherwise it would constitute the dominant source of scattering. External occulting then provides higher sensitivity to very faint coronal light that, for ground-based observations with some residual sky background, would mostly be below the detection limit. External occulting has possible disadvantages, however; vignetting reduces the irradiance and also the spatial resolution of the inner corona, and observations close to the limb are precluded. Since extremely high pointing accuracy is not easily achieved in space systems, most designs have substantial over-occulting. Both circular and straight occulting edges have been used; straight edges apply where only a small part of the coronal field is observed. The occulter is more efficient if it consists of a series of plates with their edges located such that the first is followed by a second that lies just within the shadow of the first, and so on. Three edges in a series have been used successfully. The design is such that the second edge blocks most of the diffracted light from the first edge from reaching the objective, while the third edge blocks most of the residual diffracted light from the second edge. Of several types of designs, tests indicate that a slightly tapered, threaded rod is the most efficient. In addition, its performance is relatively insensitive to guiding errors. Internal occulting is used at an image plane of the external occulter to remove residual diffracted and scattered light.

REFLECTING CORONAGRAPHS

The recent development of "superpolished" mirrors has allowed the use of mirror objectives in coronagraphs. This development has several advantages. Principally, achromaticity gives high optical quality independent of wavelength and allows simultaneous multi-wavelength observations. Further, a mirror can have a wide spectral coverage and large apertures are feasible. In a fully reflecting design, an annular-shaped mirror close to the focal plane serves as an inverse occulter. An off-axis configuration is used because a clear-aperture entrance pupil is required for optimum coronagraph performance. A miniature rocket-borne reflecting coronagraph has been successfully operated, as has a small ground-based instrument. Larger instruments are planned.

Additional Reading

Fisher, R. R., Lee, R. H., MacQueen, R. M., and Poland, A. I. (1981). New Mauna Loa coronagraph systems. *Appl. Opt.* **20** 1094.

Koutchmy, S. (1988). Space-borne coronagraphy. *Space Sci. Rev.* **47** 95.

Lyot, B. (1939). A study of the solar corona and prominences without eclipses. *Mon. Not. Roy. Astron. Soc.* **99** 580.

Smartt, R. N. (1982). Solar corona photoelectric photometer using mica etalons. *SPIE J.* **331** 442.

Smartt, R. N., Dunn, R. B., and Fisher, R. R. (1981). Design and performance of a new emission-line coronagraph. *SPIE J.* **288** 395.

See also **Sun, Atmosphere, Corona; Telescopes and Observatories, Solar.**

Cosmic Rays, Acceleration

John P. Wefel

Particle acceleration is commonplace in astrophysical environments ranging from the Earth's magnetosphere, to interplanetary space, to the Sun and other stars, to interstellar space, and to the nuclei of active galaxies. These environments cover many orders of magnitude in size and are characterized by vastly different physical conditions, for example, temperature and density, and yet all produce energetic charged particles. The accelerated particles have been studied directly, within our solar system by spacecraft, and indirectly, via secondary radiations produced by the particles, principally radio synchrotron and gamma ray emission.

The galactic cosmic rays present a particular challenge in understanding the particle acceleration process. Since the discovery by Victor F. Hess during 1911–1912 that the cosmic rays were extraterrestrial, and the subsequent measurements which finally established that these high-energy particles came from the galaxy and not from the Sun, it has been recognized that a powerful "cosmic accelerator" must exist to energize the cosmic rays. Enrico Fermi (in 1949) was one of the first to propose a model for the acceleration of galactic cosmic rays based upon particle scattering in a turbulent, magnetized, conducting medium. The difficulty, however, was to find suitable astrophysical environments in which the Fermi process might operate.

Galactic cosmic rays have long been connected with supernovae, at least since 1934, when Walter Baade and Fritz Zwicky pointed out that supernovae can provide the power source necessary to maintain the cosmic rays in the galaxy. Following the (apparent) failure of the Fermi mechanism, attention focused upon developing acceleration models connected to supernova explosions or to the remnants remaining after the event, that is, a pulsar or the expanding cloud of debris. These models, however, suffered from the fact that the energetic particles would be decelerated in the subsequent expansion of the debris.

In the last decade, the galactic cosmic ray acceleration problem has come much closer to a solution through particle acceleration at shock waves. This mechanism involves the Fermi process acting at the shock front between an expanding supernova blast wave and the interstellar medium, with the energy coming from the supernova. The "new" shock acceleration models appear to be able to explain the acceleration of the bulk of the cosmic rays, at least up to energies of 10^{14}–10^{15} eV.

OBSERVATIONAL CONSTRAINTS

The cosmic rays consist of all of the elements in the periodic table, up to uranium. Hydrogen is the dominant element, followed by helium; but, relative to the metals such as iron, H and He are underabundant in cosmic rays compared to material of solar system composition. Thus, the "cosmic accelerator" must accelerate all of the elements, and the injection/acceleration or the environment must provide the needed composition.

The cosmic ray source abundances, the relative composition derived from the measured abundances by correcting for the secondary contribution due to nuclear fragmentation in the interstellar medium, resemble closely (for atomic number $Z \geq 6$) the relative composition observed among solar flare energetic particles. Both sets of abundances show ordering, relative to solar system composition, by the first ionization potential of the element. Such ordering could indicate conditions of high temperatures in the matter that is accelerated.

Cosmic rays have power-law energy spectra $[(dJ/dE) \propto E^{-\gamma}]$, illustrated in Fig. 1 for protons, with a spectral index $\gamma \sim 2.7$ over 5–6 decades in energy. Below a few gigaelectronvolts per nucleon the spectra flatten, due largely to effects of the outflowing solar wind on the incoming cosmic rays. At still higher energies, the all-particle spectrum shows a change in slope, as illustrated in Fig. 2, at 10^6–10^7 GeV (10^{15}–10^{16} eV) total energy, becoming steeper up to the highest measured energies, 10^{19}–10^{20} eV. Thus, the acceleration process or processes must account for the $\gamma = 2.7$ power law for the bulk of the cosmic rays as well as for the transition above 10^{15}–10^{16} eV.

The cosmic radiation is extremely isotropic. The anisotropy is 10^{-3}–10^{-4} below 10^{13} eV, and it gradually increases with increasing energy. At the highest energies ($\geq 10^{19}$ eV), the anisotropy indicates that the particles are extragalactic, coming from the general direction of the Virgo supercluster, and in this energy region ($> 10^{10}$ GeV in Fig. 2) the spectrum appears to change as well. However, the bulk of the cosmic rays are local, accelerated in and confined to our galaxy.

Assuming that the cosmic rays fill the galaxy, the energy density in energetic particles is of order 1 eV cm^{-3}. The mean confinement time of the cosmic rays, as measured by the radioactive isotope ^{10}Be, is ~ 15 million years, and cosmic rays have been present at about the same intensity for at least the last 2 billion years. This requires replenishment of the cosmic rays at a rate of a few $\times 10^{40}$ erg s^{-1}. This is consistent with the mean energy generation rate for supernovae, assuming that only a few percent of the total supernova energy is channeled into particle acceleration.

The cosmic rays propagate in the galaxy, traversing a mean amount of matter that is energy dependent and undergoing nuclear

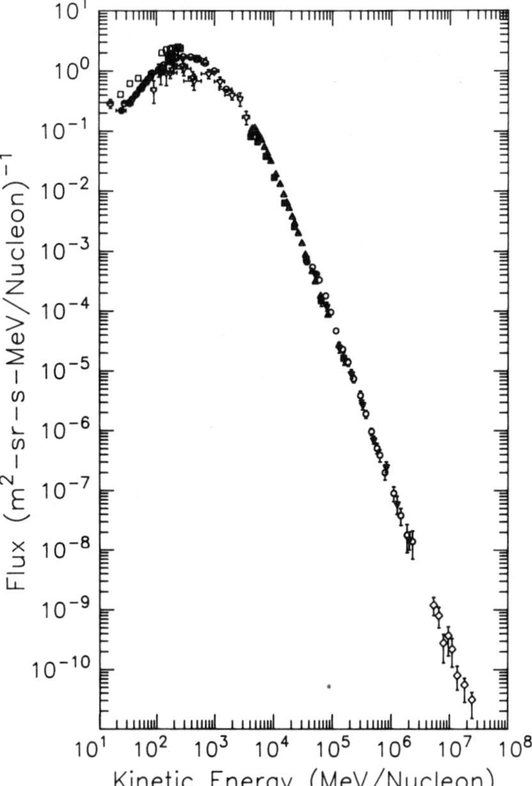

Figure 1. The differential energy spectrum for protons compiled from measurements reported in the literature.

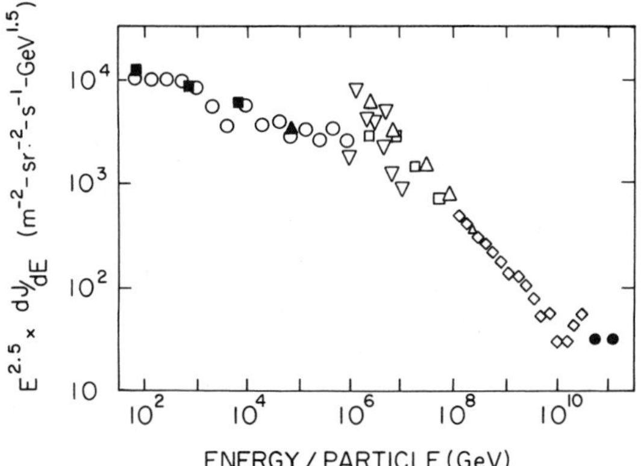

Figure 2. The all-particle cosmic ray differential energy spectrum at high energies. The ordinate is multiplied by $E^{2.5}$ and the abscissa is the total energy per particle. The region from 10^2–10^4 GeV corresponds to 10^5–10^7 MeV per nucleon in Fig. 1.

interactions and ionization energy loss. The ratio of predominantly secondary species (produced by nuclear interactions) to primary elements decreases with increasing energy above a few gigaelectron-volts per nucleon approximately as a power law with spectral index of ~ 0.5. This propagation energy dependence must be subtracted from the measured spectra to obtain the source spectral index, $\gamma \sim 2.2$, which is what must be supplied by the acceleration mechanism.

Continuous acceleration models, in which the cosmic rays are accelerated at a constant rate all during their propagation in the galaxy, can be eliminated. In such models the secondaries created early are accelerated along with the primaries. This leads to a secondary-to-primary ratio that is constant or increasing with energy, contrary to the observations. Thus, cosmic ray acceleration must take place mainly in discrete events from which the high-energy particles emerge to begin their propagation/confinement in the galaxy.

SHOCK ACCELERATION

Although supernova shocks provide the dominant power input to the galaxy, interstellar shock waves are also produced by novae, expanding H II regions, stellar winds and, possibly, a galactic wind, all of which have been suggested as possible cosmic ray accelerators. Shock waves meet the conditions for the Fermi process. In the shock rest frame, an observer sees material converging from both the upstream direction (high speed ejecta) and the downstream side (the material into which the shock is moving). The material on both sides is magnetized and turbulent and scatters charged particles. The particles travel back and forth across the shock boundary, gaining energy upon each scattering, until they finally escape. Simplified model calculations have shown that this process produces a power-law energy spectrum with index $\gamma = (2+R)/(R-1)$, where R is the compression ratio of the shock. For strong shocks, $R = 3$–4 which gives $\gamma = 2$–2.5.

A supernova explosion typically ejects 1–10 solar masses of material at velocities of 3000–10,000 km s^{-1}. After several hundred years, the ejecta will have swept up about its own mass of interstellar material, will be several parsecs in radius, and will start to decelerate. Observations of remnants at this stage show radio synchrotron emission from relativistic electrons whose energy spectrum is inferred to have $\gamma \sim 2.2$. As the remnant continues to expand, the shock weakens and particle acceleration may decrease. Note that the material in and around the shock is a mixture of interstellar matter and supernova envelope, likely to be rich in metals relative to hydrogen and helium.

The maximum particle energy from shock acceleration is determined by the particle gyroradius in the local magnetic field. When the particle obtains sufficient energy so that its gyroradius is larger than the acceleration region, the particle is effectively lost from the acceleration process. For supernovae, this energy is a few times 10^{14} eV, still below the steepening of the spectrum in Fig. 2. For stellar wind termination shocks the maximum energy can be somewhat higher, but the total energy available from stellar wind shocks is insufficient to maintain the cosmic ray energy density. A galactic wind termination shock could accelerate particles to $\sim 10^{17}$ eV, and may be the source of the highest-energy cosmic rays.

Shock acceleration models appear to offer a plausible explanation for the acceleration of cosmic rays but are not without problems. The configuration of the magnetic field, the turbulence spectrum, the nonlinear effects of the accelerated particles on the shock evolution, particle energy losses during acceleration, and injection of particles for acceleration are all areas that require further investigation.

DISCRETE SOURCES

There are discrete objects that are known to accelerate particles. These include pulsars and x-ray binary systems, for example, Cygnus X-3 and Hercules X-1. The presence of energetic charged particles is inferred from the secondary radiation, gamma rays in the case of the highest-energy particles. While the acceleration mechanisms in these objects are still not well defined, escape of some of the accelerated particles into the interstellar medium could contribute to the cosmic rays, particularly at high energies greater than 10^{15} eV. Similarly, extragalactic radio sources, active galactic nuclei, accretion disks around massive black holes, such as may exist at the center of our galaxy or of other galaxies, and accretion onto neutron stars may also provide energetic particles.

Additional Reading

Axford, W. I. (1981). The acceleration of galactic cosmic rays. In *Origin of Cosmic Rays*, G. Setti, G. Spada, and A. W. Wolfendale, eds. Reidel, Dordrecht. p. 339.

Blandford, R. and Eichler, D. (1987). Particle acceleration at astrophysical shocks: A theory of cosmic ray origin. *Phys. Rep.* **154** 1.

Ginzburg, V. L. and Syrovatskii, S. I. (1964). *The Origin of Cosmic Rays*. Pergamon Press, Oxford.

Hillas, A. M. (1984). The origin of ultra-high-energy cosmic rays. *Ann. Rev. Astron. Ap.* **22** 425.

Koch-Miramond, L. and Lee, M. A., eds. (1984). Particle acceleration processes, shockwaves, nucleosynthesis and cosmic rays. *Adv. Space Res.* **4** 1.

See also **Cosmic Rays, Origin; Cosmic Rays, Propagation; Magnetohydrodynamics, Astrophysical; Supernova Remnants, Evolution and Interaction with the Interstellar Medium.**

Cosmic Rays, Observations and Experiments

Peter Meyer

SHORT HISTORY OF COSMIC RAY OBSERVATIONS

The intense research late in the nineteenth century on the phenomenon of radioactivity prepared the way for the discovery of the cosmic radiation. Ionization chambers were developed to detect and measure with ever-increasing sensitivity the ionizing radiations emitted by radioactive substances. It was soon noted that even in the absence of all radioactive materials, a residual amount of radiation passed through the chambers and the question arose whether this radiation may originate from the surface of the Earth, in the atmosphere, or possibly even beyond. To explore this question, a number of manned balloon flights, carrying ionization chambers, were made by the Austrian physicist Victor F. Hess. In 1912, Hess published his result, for which he later received the Nobel Prize: The radiation intensity increases with altitude. The source of the radiation must therefore be sought in the overlying atmosphere or outside the Earth. With no knowledge of its nature, Hess gave it the name Höhenstrahlung ("radiation from above").

It took almost four decades of research to establish that the primary cosmic radiation, before it interacts with the matter of the atmosphere, consists mostly of atomic nuclei of high kinetic energy. These years were extremely rich in discoveries. Elementary particle physics and high-energy physics were born, using the energetic cosmic ray particles as projectiles to study interactions, and to produce new particles. Numerous particles were discovered, including positrons, muons, pions, and many more. New detection instruments were invented and used for cosmic ray measurements: Geiger counters, nuclear emulsions, and cloud chambers were among the first with scintillation counters, Čerenkov counters, spark chambers, solid-state detectors, multiwire proportional chambers, transition radiation detectors, and plastic track detectors coming later. With these refined observational tools it was possible to study the cascades of secondary particles that are generated after a high-energy nucleus impinges on the atmosphere. Similar detectors are used to determine the identity of the primary particles. In the 1950s, when large particle accelerators were developed, most of high-energy physics moved to the accelerator laboratory, and research in cosmic rays concentrated largely on the astrophysical questions of their nature, their origin, and their acceleration.

The high-energy primary nuclei interact in the first few g cm^{-2} of the atmosphere, which altogether is about 1000 g cm^{-2} thick, and produce large numbers of secondary, tertiary, etc. particles and energetic photons. Therefore, the primary radiation could only be studied when it became possible to bring detectors near the top of the atmosphere or above it. A famous balloon flight using Geiger counters first established that the primaries are mostly protons.

Cosmic ray particles reach the solar system and the earth from all directions. Galactic magnetic fields deflect the trajectories of the charged particles with the result that their direction of arrival is unrelated to the direction toward their source. To investigate their origin one must rely on other characteristic features: their composition and their energy spectra. Major technical innovations opened the door to modern cosmic ray research. Reliable high-altitude balloons, earth satellites, and deep space probes made it possible to place instruments near the top of the atmosphere or outside the atmosphere and even outside the magnetosphere of the Earth. Equally important was the simultaneous development of modern solid-state electronics, which is ideally suited for space vehicles because of its low mass, low power consumption, and high stability.

OBSERVATIONS OF NUCLEAR ABUNDANCES

Elemental Composition

Balloon flights in 1949 carrying nuclear emulsions brought the first evidence that, besides the overwhelmingly abundant protons (almost 90% of the cosmic rays), heavier nuclei are present in the primary cosmic radiation. The relative abundance distribution of the elements in the cosmic radiation, as it is measured today with modern instruments, is shown in Fig. 1. Normalization is at the element Si. The solid circles represent measurements at about 200 MeV per nucleon; the open circles are measurements around 2 GeV per nucleon. The open diamonds are the best estimate of the abundance distribution of the elements in the solar system.

This figure represents a major milestone in particle astrophysics. First, it shows the striking similarity between cosmic ray abundances and solar system abundances, leading immediately to the important conclusion that the cosmic ray nuclei, just as the solar system nuclei, were produced by nucleosynthesis in the interiors of stars and only afterwards accelerated to their high energies.

A second conclusion follows from the differences between cosmic ray abundances and solar system abundances. These are particularly noticeable in the group of the light elements, Li, Be, and B, and in the subiron group from Sc to Mn. The filling of the abundance valleys in the solar system distribution by the cosmic rays is due to collisions of the source nuclei with the interstellar medium, leading to nuclear disintegration or spallation, with the secondary fragments produced with almost the same velocity as that of the parent nucleus. The ratio of secondary to primary intensity provides a measure of the average amount of material that a cosmic ray particle traverses before being lost from the galaxy. This is called the escape mean free path and amounts to about 7 g cm^{-2} of material at the energies to which Fig. 1 applies.

In recent years it was discovered that the intensity ratio of secondary nuclei to primary nuclei decreases with increasing energy, indicating that the escape mean free path decreases at high energies. The highest energy at which the abundance of individual nuclear species could be measured is about 1 TeV per nucleon (1000 GeV per nucleon) in an experiment carried by the space shuttle. At those energies the escape mean free path is only 1 g cm^{-2}. Several models attempt to describe this behavior.

Ultraheavy Nuclei

The unique position of Fe (iron) as the stablest of all elements in the periodic system of the elements is manifest by its high relative abundance in the cosmic rays and in the solar system. Beyond Fe the abundances drop rapidly and by many orders of magnitude. In spite of their extreme scarcity, ultraheavy nuclei (UH nuclei) have

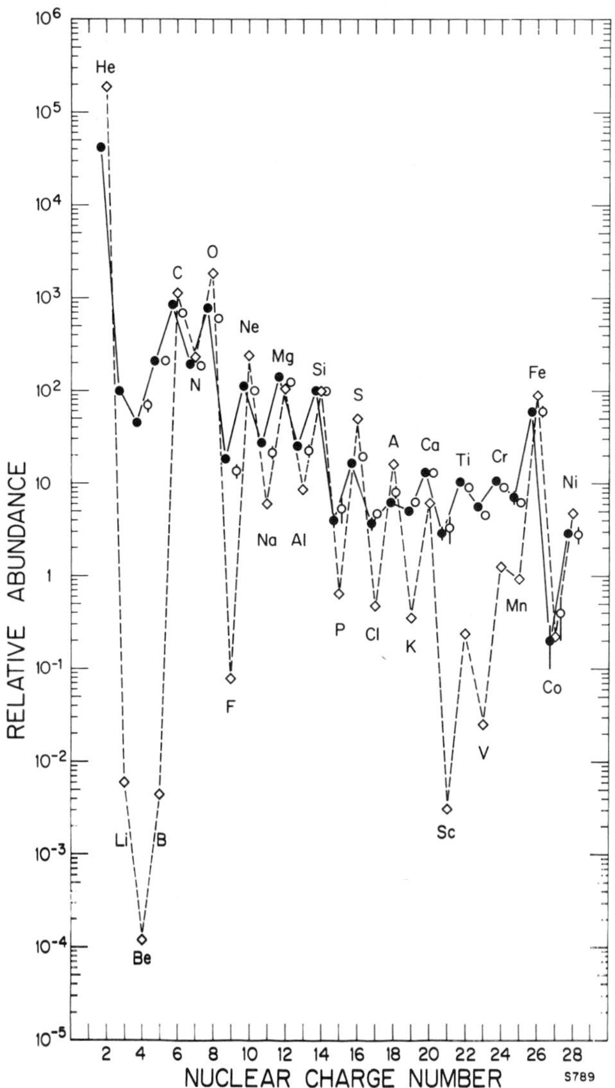

Figure 1. The relative abundance distribution of the elements in the cosmic radiation and in the solar system (normalized to Si = 100) from He to Ni (solid circles, 70–280 MeV per nucleon; open circles, 1000–2000 MeV per nucleon; open diamonds, solar system abundance distribution). [*Reproduced with permission from J. A. Simpson (1983).* Ann. Rev. Nucl. Part. Sci. *33 by Annual Reviews, Inc.*].

been discovered in the cosmic radiation, first through the identification of tracks they had left in natural minerals of meteorites. Shortly afterwards balloon experiments using large-area nuclear emulsions and plastic track detectors provided rough indications of their contemporary flux and composition. More recent satellite experiments with electronic detectors (1978) were able to separate most of the abundant elements with even atomic number. They established the existence of the entire periodic table of the elements in the cosmic rays. Since the UH nuclei are probably produced in nucleosynthesis processes different from those that produced nuclei of lower mass, this knowledge contains significant information on the origin of the elements. Attaining full charge resolution of the UH nuclei is still a task for the future.

Isotopes

In the past 10–15 years it has become possible to resolve the isotopic abundance distribution for several nuclear species. Differences in isotopic abundances for a given element in the cosmic rays and in the solar system must be due to nuclear phenomena rather than atomic effects. They therefore are important in selecting from alternate types of nucleosynthesis processes responsible for the observed sample of isotopes. Although the determination of accurate isotopic abundances is still in its beginning, a number of important results have already been obtained. Cosmic ray Ne and Mg have been observed to have overabundances of their neutron-rich isotopes ^{22}Ne, ^{25}Mg, and ^{26}Mg when compared to the solar system isotopic abundances. This evidence is of considerable importance in understanding the nature of the cosmic ray sources. The relative abundance of the radioactive isotope ^{10}Be has provided a measurement of the average time cosmic rays are contained in the Galaxy. Experiments are planned that will yield much more precise data on isotopic abundances for a wide range of elements.

OBSERVATIONS OF ELECTRONS AND POSITRONS

The presence of energetic electrons spiraling around galactic magnetic fields was first deduced from the observation of a background of radio frequency synchrotron radiation from the galactic disc. Balloon experiments in 1960 established the presence of a flux of cosmic ray electrons, amounting to about 1% of the flux of protons. Not much later, through the use of a magnet spectrometer in a balloon-borne instrument, it became possible to separate the negatively charged electrons from the positively charged positrons. Negative electrons were found to be about 10 times more abundant than positrons at energies between a few hundred megaelectronvolts and a few gigaelectronvolts. Positrons and electrons are expected in about equal amounts if they originate from interstellar collisions of cosmic ray nuclei that produce $\pi^{\pm}$ mesons which subsequently decay via $\mu^{\pm}$ mesons to electrons or positrons. The fact that negative electrons were found to be more abundant than positrons is proof that part of the electrons must have a different source such as the Crab nebula, or other supernova remnants which are known to produce electrons of high energy. The shape of the electron energy spectrum is of particular interest. In contrast to the nuclei, electrons are not lost through collisions with interstellar particles, but lose energy by synchrotron radiation as they spiral about magnetic fields, and by Compton collisions with photons of starlight and the universal blackbody radiation. These energy-loss processes are reflected in the spectral shape. Hence, the electron spectrum is a probe of interstellar electromagnetic fields.

OBSERVATIONS OF ANTIPARTICLES

Antiprotons of high energy are expected in the galaxy as a product of high-energy nuclear collisions. Measurements with a balloon-borne superconducting magnet spectrometer have established the existence of an antiproton flux. The observed antiproton intensity slightly exceeds that predicted by calculation of their production in nuclear collisions. Speculation on the existence of primary antiprotons, however, is premature until further, more accurate experiments are carried out.

Antinuclei with atomic number $Z \geq 2$ have been searched for but not found in several experiments. The discovery of a single antioxygen or antiiron nucleus would indeed be a spectacular result, because it would signify the existence of macroscopic regions of antimatter.

THE ENERGY SPECTRA OF THE NUCLEAR COMPONENTS

Spectral measurements of individual nuclear species extend to about 1 TeV per nucleon. The primary nuclei He, C, O, and Mg are observed to follow similar spectral forms to the highest measured energies. These can be described by power laws of the form $(dJ/dE)_{\text{prim}} \sim E^{-\gamma}$, γ having the value of 2.6 to 2.7. Spectra for a

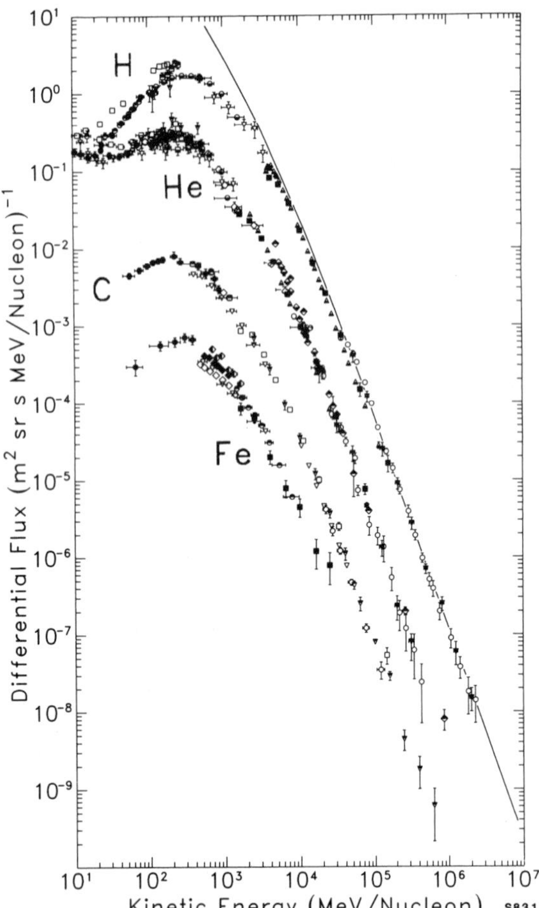

Figure 2. The differential energy spectra of the primary cosmic ray H, He, C, and Fe at Earth. [*Reproduced with permission from J. A. Simpson (1983).* Ann. Rev. Nucl. Part. Sci. **33** *by Annual Reviews, Inc.*].

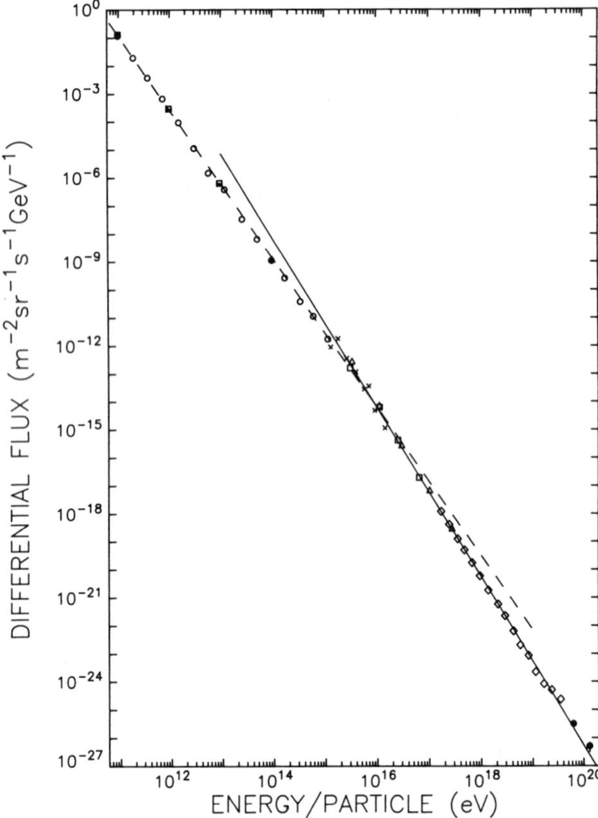

Figure 3. The differential energy spectrum of all cosmic ray particles at Earth. This spectrum can be approximated by two power laws with a break around 10^{15}–10^{16} eV.

few primary elements are shown in Fig. 2. The flattening at low energies is due to the local effect of modulation in the solar wind.

Due to the decreasing escape mean free path with energy the spectra of the pure secondary nuclei are all steeper than the primary spectra. It is observed that at high energies their spectra follow power laws $(dJ/dE)_{\text{sec}} \sim E^{-(\gamma+\delta)}$, where $\delta = 0.6 \pm 0.1$. For the same reason the spectra at the cosmic ray sources must be flatter than the observed primary spectra, $(dJ/dE)_{\text{source}} \sim E^{-(\gamma-\delta)}$, having an exponent of about 2.1. This number is significant. The most promising and extensively studied mechanism for cosmic ray acceleration is collisionless shocks, produced, for example, in the expanding shell of a supernova remnant. The shock acceleration model predicts a power law for the spectrum of the accelerated particles with an exponent around 2.

The energy spectrum of all cosmic rays together has been measured over a much wider range than that of individual species. This is achieved through the analysis of giant air showers where secondaries, tertiaries, and so on propagate all the way down to mountain altitude, or even to sea level. The density and extent of the shower provides a measure of the energy of the incident particle. Figure 3 is a compilation of the all-particle spectrum that reaches to the incredible energy of 10^{20} eV, by far the largest energy of any known radiation. Around 10^{16} eV the power-law spectrum exhibits a steepening whose origin is not understood. This spectral change may be related to a transition to different cosmic ray sources and composition. At low energies the flux is dominated by protons. Nothing is known about the composition or the acceleration mechanisms at the highest energies. Shock acceleration is not likely to be effective at energies in excess of 10^{14}–10^{15} eV. While the bulk of the cosmic rays is undoubtedly of galactic origin, an extragalactic origin of the highest-energy particles is frequently postulated in the absence of any other explanation. The transition from galactic to extragalactic sources may be the cause for the change in slope of the all-particle spectrum.

FUTURE INVESTIGATIONS

A substantial observational program in cosmic ray research lies ahead. The elemental composition must be measured at higher energies than has so far been possible, extending into the air-shower region. High resolution composition measurements in the UH regime of nuclei are expected to illuminate several questions of cosmic ray origin. Measurements of isotopic abundances of all elements up to and beyond the iron group are needed to better understand the origin of the elements and the material of which cosmic rays are made. An extension of the isotopic analysis into the regime of the UH nuclei is an important experimental challenge. Finally, detailed measurements of the antiproton and positron spectra, and more sensitive searches for heavier antinuclei, will address fundamental questions of astrophysics and cosmology.

Additional Reading

Mewaldt, R. A., Stone, E. C., and Wiedenbeck, M. E. (1982). Samples of the Milky Way. *Scientific American* **247** (No. 6) 100.

Rossi, B. (1964). *Cosmic Rays*. McGraw-Hill, New York.

Shapiro, M. M., ed. (1983). *Composition and Origin of Cosmic Rays*. Reidel, Dordrecht.

Simpson, J. A. (1983). Elemental and isotopic composition of the galactic cosmic rays. *Ann. Rev. Nucl. Part. Sci.* **33** 323.

Sokolsky, P. (1989). *Introduction to Ultrahigh Energy Cosmic Ray Physics*. Addison-Wesley, Reading, MA.

See also **Antimatter in Astrophysics; Cosmic Rays, Space Investigations; Radiation, High-Energy Interaction with Matter.**

Cosmic Rays, Origin

Richard E. Lingenfelter

The question of the origin of cosmic rays has been the focal point of cosmic ray studies since their discovery nearly a century ago, and a wide variety of sources, ranging from solar to cosmological, have been suggested over the years. Although the question is still not fully answered, the observational constraints now at least allow us to strongly restrict the possibilities. In this entry we will concentrate on the question of the sites and energetics of cosmic ray sources, leaving the details of their acceleration and propagation to other entries.

The existence of cosmic radiation was first suggested in 1900 by Charles T. R. Wilson, following his discovery of atmospheric ionization, when he proposed that extraterrestrial gamma rays might be responsible for the ionization. Twelve years later, with balloon-borne ionization detectors showing that the ionization rate increased with altitude and had no diurnal variation, Victor F. Hess demonstrated the extraterrestrial and extrasolar origin of the ionizing radiation, which soon came to be known as "cosmic rays." These rays were generally thought to be gamma rays until the variation of their flux with geomagnetic latitude was discovered in 1927. This clearly established that they were not gamma rays, but charged particles deflected by the magnetic field. Subsequent observations from balloons and satellites have shown that the cosmic rays are mostly relativistic protons with a significant fraction of heavier nuclei and electrons. The energy spectrum and anisotropy of the cosmic rays have now been measured over more than 10 decades in energy all the way up to $\sim 10^{20}$ eV.

Studies of the observed cosmic ray energy spectrum, anisotropy, and composition, together with related gamma ray and radio observations, now suggest that the bulk of the cosmic rays observed below $\sim 10^{19}$ eV are of galactic origin, while those at higher energies are primarily of nearby extragalactic origin. These suggestions, nonetheless, are still debatable.

GALACTIC COSMIC RAYS

The observational constraints on the cosmic rays with energies $< 10^{19}$ eV strongly suggest a galactic origin for the bulk of the cosmic rays and a self-consistent argument can be made.

The simplest argument for a galactic origin of at least some of the cosmic rays comes from the cosmic ray electrons, which have been observed to have a power-law energy spectrum of $E^{-2.6 \pm 0.05}$ extending up to energies of $\sim 10^{12}$ eV. Because such relativistic electrons suffer energy losses by Compton scattering on the 2.7-K microwave background radiation, the highest-energy electrons have a lifetime against energy loss of only $\sim 10^6$ yr. Thus they cannot have traveled more than ~ 0.3 Mpc, even if they traveled in straight lines. Since the overall anisotropy of the cosmic rays at that energy is 0.5×10^{-3}, their source is probably no more than 0.1 kpc away, because the anisotropy is essentially the ratio of the rectilinear distance divided by the total distance traveled. Thus the cosmic ray electrons clearly must be of galactic origin, and that would suggest a similar origin for protons of similar energies.

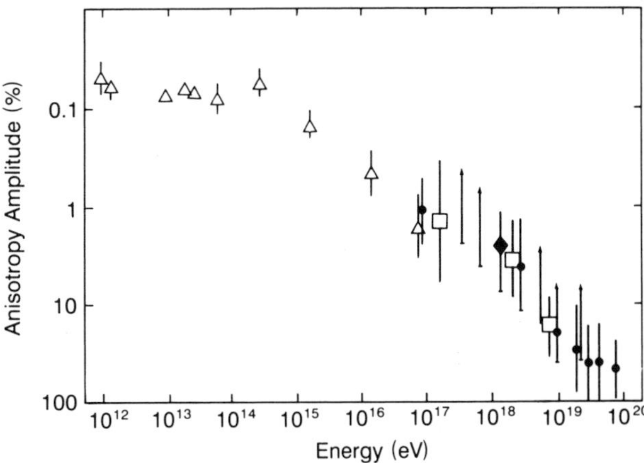

Figure 1. Cosmic ray anisotropy, expressed in terms of the amplitude of the first harmonic, versus cosmic ray energy. (*Adapted from Hillas, 1984.*)

Similar estimates of the volume of space in which the bulk of the cosmic rays are contained and of their lifetime within that containment volume can be made from a number of different observations. A containment volume of at least galactic disk dimensions with roughly the local energy density is required to account for the observed diffuse galactic flux of high energy (> 50 MeV) gamma rays, if they result primarily from the decay of pions produced by nuclear interactions of the cosmic rays with the interstellar gas.

A minimum containment volume several times that of the galactic disk is required by measurements of cosmic ray nuclei of secondary origin, produced by spallation of the primary cosmic rays in nuclear interactions with ambient matter. The cosmic ray abundances of secondaries ^{2}H, ^{3}He, Li, Be, and B all require that the bulk of the comic rays must have gone through a mean column depth $x \sim \rho c \tau$ of ~ 5 g cm^{-2}. Moreover, the cosmic ray abundances of radioactive secondaries, ^{10}Be, ^{26}Al, and ^{36}Cl, with half lives of 2.3×10^6, 1.0×10^6, and 0.43×10^6 yr, further require that these interactions must have taken place over a mean time $\tau \geq 2 \times 10^7$ yr. Thus the mean density of the ambient matter in which the cosmic rays are contained and interact is $\rho \leq 0.5 \times 10^{-24}$ g cm^{-3}. This density is significantly less than the $\sim 2 \times 10^{-24}$-g cm^{-3} local mean density of gas in the disk, implying a minimum cosmic ray containment volume ~ 4 or more times that of the galactic disk.

Such a size is also consistent with the measured anisotropy $\delta \sim 0.5 \times 10^{-3}$ for the bulk of the cosmic rays (Fig. 1). Because the anisotropy is essentially $\delta \sim r/c\tau$, the ratio of the rectilinear distance r divided by the total distance traveled $c\tau$, then the mean distance $r \sim \delta x/\rho$, which for the above values gives a distance ~ 2 kpc, or several times the scale height of the disk gas.

This very small anisotropy of the bulk of the cosmic rays can be understood in terms of diffusive propagation resulting from scattering of the charged cosmic rays by irregularities in the galactic magnetic field. In the simplest diffusion treatment, the mean size of the irregularities is $\lambda \sim \delta^2 c\tau \sim \delta^2 x/\rho$, which, for the above values of δ, x, and ρ, gives $\lambda \sim 1$ pc. This size is typical of interstellar distances in the disk.

Diffusive scattering should be effective as long the cosmic ray gyroradius is less than the mean λ of 1 pc. Such a gyroradius, $r_g(\text{pc}) \sim E_{15}/B_{\mu\text{G}}$, where E_{15} is the energy in units of 10^{15} eV, corresponds to a cosmic ray proton energy of $\sim 3 \times 10^{15}$ eV in a ~ 3-μG galactic magnetic field, and could thus account for the observed breaks in both the cosmic ray energy spectrum (Fig. 2) and anisotropy (Fig. 1) at $10^{15} \sim 10^{16}$ eV. As the energy increases, the diffusive scattering becomes less efficient, the anisotropy of the cosmic rays increases, and their containment time decreases inversely with anisotropy $\tau(E) \sim \tau/c\delta(E)$. Thus cosmic rays pro-

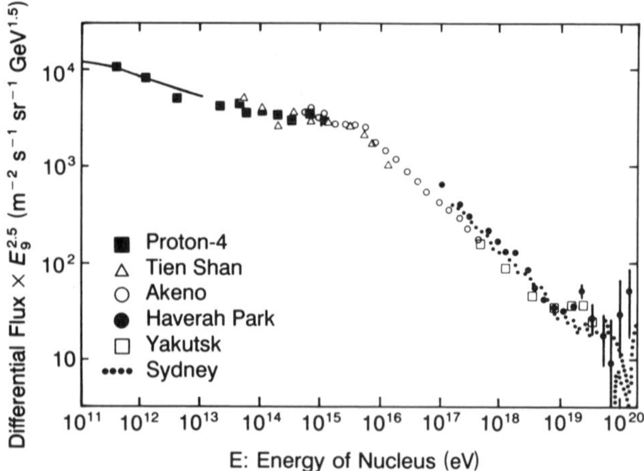

Figure 2. Cosmic ray energy spectrum multiplied by $E^{2.5}$ to better show the spectral variations. (*Adapted from Hillas, 1984.*)

duced with a source energy spectrum $Q(E)$ would be observed within the containment volume with an equilibrium energy spectrum $N(E) \sim Q(E)/\delta(E)$. The observed energy spectrum and anisotropy are, in fact, quite consistent with such a relationship for a single power-law source spectrum of $E^{-2.47}$ all the way from 10^{11} to 10^{19} eV.

This observed correlation between the energy spectrum and anisotropy also provides the only direct evidence for a galactic origin of the bulk of the cosmic rays. For with the exception of the indirect evidence of the cosmic ray electrons, none of the other observations places any upper bound on the size of the containment volume that would require a galactic origin. The limits on the isotropic background of high energy (>50 MeV) gamma rays are not inconsistent with cosmic rays of the local energy density filling the entire Universe and interacting with the intergalactic gas to produce pion decay gamma rays, as long as the density of the gas was less than 3×10^{-2} of that required to close the Universe. Although only a negligible amount of matter could be traversed in such densities by the extragalactic cosmic rays even in the Hubble time, the bulk of the 5 g cm^{-2} of matter that locally observed cosmic rays must have passed through could still have been traversed in the galactic disk, just as if they originated in the disk. The inverse correlation between the observed energy spectrum and anisotropy, however, is expected only within a containment volume from which cosmic rays are escaping, and not for cosmic rays filling all of space. The maximum size of the containment volume for the bulk of the cosmic rays is limited by the minimum anisotropy of $\sim 3\times10^{-4}$ to $r \le \delta c/H_0 \le 2$ Mpc, since the cosmic ray containment time can not exceed the Hubble time c/H_0. Thus the bulk of the observed cosmic rays must be contained in a volume no larger than that of the Local Group of galaxies, in which our Galaxy makes up $\sim \frac{1}{4}$ of the observable mass.

One further argument for an even more restricted galactic disk origin of the bulk of the cosmic rays is the remarkable similarity in magnitude of the cosmic ray energy density and the magnetic field energy density in the galactic disk, both of which are of the order of 10^{-12} erg cm^{-3}. This would be expected if the cosmic rays are produced within the galactic disk and are contained by the galactic magnetic field. For if the energy density, or pressure, of the cosmic rays exceeded that of the magnetic field they could not be contained by it. Such a similarity would not be expected if the cosmic rays originated outside of the galactic disk. Furthermore, the turbulent energy density of the interstellar gas in the galactic disk is also of the same order of magnitude as that of the cosmic rays and the magnetic field. This suggests a coupling and possible equipartition of energy between the three, supporting suggestions that the cosmic rays could be accelerated by stochastic processes in the galactic disk, as will be discussed further. There is, however, one further energy density that is also of the same order of magnitude, that of the 2.7-K microwave background radiation, and this puzzling similarity has not yet found any ready explanation with either a galactic or an extragalactic origin of the cosmic rays.

Independent of the size of the galactic cosmic ray containment volume or the cosmic ray lifetime within it, however, a straightforward determination can be made of the combined cosmic ray luminosity of all galactic sources required to maintain the measured, local cosmic ray energy density of $\sim 10^{-12}$ erg cm^{-3}. The determination of this power depends only on the measured energy density and average amount of matter traversed by the cosmic rays, and the total mass of interstellar matter in which the cosmic rays can interact.

In particular, the galactic cosmic ray luminosity is $L_G \sim wV/\tau$, the total cosmic ray energy in whatever galactic volume V the local energy density w fills, divided by the mean cosmic ray containment time τ within that volume. However, the mean time τ is simply $\sim x/\rho c$, where x is the average amount of matter per unit area that the cosmic rays have traversed, and the average density of that matter seen by the cosmic rays, ρ, is equal to M_g/V, the total mass of gas M_g in the containment volume V. Thus, the galactic cosmic ray luminosity is simply $L_G \sim wcM_g/x$. Taking a mean amount of matter $x \sim 5$ g cm^{-2} traversed by cosmic rays, determined from the relative abundances of secondaries produced by nuclear spallation, and a total mass of interstellar gas $M_g \sim 10^{10}$ $M_\odot \sim 2\times10^{43}$ g, equal to $\sim 10\%$ of the mass of the Galaxy, then $L_G \sim 10^{41}$ erg s^{-1}. This is the total rate of cosmic ray production in the Galaxy, required to maintain the local cosmic ray energy density $\sim 10^{-12}$ erg cm^{-3}, independent of both the containment volume and time.

There are a variety of galactic sites that have been suggested as the source of the cosmic rays, but they most likely all derive their energy in one way or another from supernovae. Galactic supernovae are estimated to occur on the average about once every ~ 30 yr and release some $\sim 10^{51}$ erg just in the kinetic energy of their expanding ejecta, corresponding to a time-averaged energy release of $\sim 10^{42}$ erg s^{-1}. This is an order of magnitude greater than that required in cosmic rays, and thus various models have been suggested for cosmic ray acceleration in the initial explosion of the supernova, in supernova shock waves as they subsequently expand into the interstellar medium, and later yet in stochastic scattering by turbulent irregularities resulting from the dissipation of the supernova energy in the interstellar medium. Supernovae may also leave behind a comparable amount of rotational energy in rapidly spinning (<3-ms period) magnetic neutron stars which could generate an enormous electric field and directly accelerate cosmic rays. Subsequent accretion of gas onto such compact objects can also lead to the acceleration of particles to energies of at least 10^{18} eV, as is evident from observations of Cygnus X-3, Hercules X-1, and other ultrahigh energy sources.

EXTRAGALACTIC COSMIC RAYS

The arguments for an extragalactic origin of the highest-energy cosmic rays are simpler and more straightforward. Unlike the cosmic rays at much lower energies, which are nearly isotropic, those at the highest energies are extremely anisotropic. As can be seen in Fig. 1, the measured anisotropy approaches unity at energies $>10^{19}$ eV. Although there is no clear data on the composition of these cosmic rays, at such energies the gyroradius of protons in the galactic magnetic field of a few microgauss becomes comparable to the dimensions of the Galaxy. Thus, such cosmic ray protons should suffer little deflection in the magnetic field and travel in nearly straight lines, so that their arrival directions should point close to the direction of their origin.

The energy-weighted mean direction of the highest-energy cosmic rays is not in the direction of the inner part of the Galaxy, or even close to the galactic disk, as might be expected if these cosmic rays came from a galactic source. Instead, these highest energy cosmic rays come from a mean direction within 10° of the North Galactic Pole. This is close to the direction of the Virgo supercluster of galaxies, the center of which lies at a distance of ~ 20 Mpc, and the direction is even closer to that of the mean of the galaxies weighted by their mass divided by their distances squared. Estimates of the magnitudes of intergalactic magnetic fields suggest that these cosmic rays would also not be significantly deflected over such distances. Thus their arrival directions seem to point to a nearby extragalactic origin.

Photopion production by interactions of these cosmic rays with the 2.7-K microwave background radiation further constrains their possible sources. In the rest frame of these ultrarelativistic particles the microwave photons are blueshifted by the particle Lorentz factor to energies of the order of 200 MeV, sufficient to produce pions which can carry away a significant fraction of the particle energy. Thus the maximum distance that the highest-energy cosmic rays could have traveled through the microwave radiation without losing most of their energy is only about 30 Mpc.

An estimate of the cosmic ray luminosity required for such an extragalactic source can be made from their energy spectrum. For anisotropy is not the only difference between the highest-energy cosmic rays and those at energies $<10^{19}$ eV; their energy spectrum is also quite different, supporting the possibility that they might have a separate origin. As can be seen in Fig. 2, for several decades below this energy the spectrum roughly follows a power law in energy, $E^{-\gamma}$ with an index $\gamma \sim 3.1 \pm 0.1$, but above 10^{19} eV the spectrum abruptly flattens to a shape that can be crudely approximated by a power-law index $\gamma \sim 2.4 \pm 0.2$.

If such a power-law spectrum extended on down to much lower energies ($\sim 10^9$ eV) typical of the bulk of the cosmic rays, they would make up only $\sim 10^{-4 \pm 1}$ of the flux at those energies, and have a local energy density $w \sim 10^{-16 \pm 1}$ erg cm^{-3}. If these cosmic rays fill a volume of radius $R \sim 20$ Mpc, comparable to the distance of the Virgo supercluster, with a containment time $10^8 < \tau < 10^{10}$ yr between the light-travel time and the Hubble time, then the cosmic ray luminosity of the supercluster would have to be $L_{SC} \sim 4\pi R^3 w/3\tau \sim 3\times 10^{45 \pm 2}$ erg s^{-1}. The lower luminosity, divided among the $\sim 3\times 10^3$ galaxies in the cluster, would require an average galactic luminosity of $\sim 10^{40}$ erg s^{-1}, only $\sim 10\%$ of that required for our Galaxy, if it is the source of bulk of the cosmic rays, as was discussed above. The mean value is comparable to the bolometric luminosity of the Seyfert galaxies, NGC 4151 and NGC 1068, which lie in the cluster. The upper bound would require local quasistellar objects.

Additional Reading

Cesarsky, C. J. (1980). Cosmic ray confinement in the Galaxy, *Ann. Rev. Astron. Ap.* **18** 289.

Hillas, A. M. (1984). The origin of ultra-high-energy cosmic rays. *Ann. Rev. Astron. Ap.* **22** 425.

Setti, G., Spada, G.; and Wolfendale, A. W., ed. (1981). *Origin of Cosmic Rays, IAU Symposium No. 94*. Reidel, Dordrecht.

Shapiro, M. M., ed. (1986). *Cosmic Radiation in Contemporary Astrophysics*, Reidel, Dordrecht.

Shapiro, M. M. and Wefel, J. P., ed. (1988). *Genesis and Propagation of Cosmic Rays*. Reidel, Dordrecht.

Simpson, J. A. (1983). Elemental and isotopic composition of the galactic cosmic rays. *Ann. Rev. Nucl. Part. Sci.* **33** 323.

See also **Background Radiation, Microwave; Cosmic Rays, Acceleration; Cosmic Rays, Propagation; Magnetohydrodynamics, Astrophysical; Supernova Remnants, Evolution and Interaction with the Interstellar Medium.**

Cosmic Rays, Propagation

J. R. Jokipii

Essential to an understanding of most areas of high-energy astrophysics is an understanding of the physics of the acceleration and propagation of energetic charged particles or cosmic rays. Since in all cases of interest the energy change and spatial transport are closely coupled, they will both be referred to as parts of a general transport process. In all cases of interest the ambient thermal plasma is sufficiently rarified that collisions with particles are extremely rare. Moreover, the effects of gravity on the energetic particles are negligible. The number of cosmic ray particles (and hence their mass and momentum) is much less than that of the background plasma. Hence, the transport is governed by the effects of ambient electric and magnetic fields, determined by the thermal plasma. Collisions need only be considered in cases where the production of secondary particles by the infrequent nuclear interactions are of interest. This latter aspect of the problem is considered briefly later in this entry.

The acceleration and transport of charged particles in astrophysics may then be described quite succinctly by the equation of motion for a particle of charge q, momentum $\mathbf{p}$, and velocity $\mathbf{w}$

$$\dot{\mathbf{p}} = q\left[\mathbf{E} + \frac{\mathbf{w}\times\mathbf{B}}{c}\right], \tag{1}$$

where the electric and magnetic fields $\mathbf{E}$ and $\mathbf{B}$ are determined by the state of the background plasma, and where c is the speed of light. In any given situation, the basic problem is to call upon our knowledge of the structure and dynamics of the plasma to determine the electric and magnetic fields to be used in Eq. (1) and to determine the initial and boundary conditions for the accelerated particles. In nearly all cases of interest, the background plasma satisfies the hydromagnetic approximation, in which the electric field is determined by the magnetic field and the fluid flow velocity relative to the observer by the relation $\mathbf{E} = -\mathbf{U}\times\mathbf{B}/c$. Hence the statement that the particle transport is governed by the magnetic field.

Because the equation is impossible to solve in general, a number of approximations have been utilized. Beginning with the seminal work of Enrico Fermi, these all recognize the fact that the plasma and magnetic field are in general irregular and turbulent, so that a statistical treatment is required. The most important and generally used approximation is the diffusion approximation. This is based chiefly on the observation that energetic particles in many important cases are observed to have a very nearly isotropic pitch angle distribution relative to the local plasma. This isotropy is a consequence of the "scattering" of the particles by small-scale, irregular fluctuations in the ambient magnetic field, and may be derived from a statistical analysis of the equation of motion. Space does not permit a discussion of this scattering theory here. In many situations the scattering is sufficiently rapid that the distribution of cosmic rays may be taken to be nearly isotropic, the anisotropies (in the local fluid frame) being generated by spatial gradients of the density. The basic, physical effects to be included are (a) *diffusion*, which describes the random walk of the cosmic rays through the random magnetic field irregularities; (b) *convection* of the cosmic rays with the background fluid flow; (c) *drift* of the cosmic ray guiding centers caused by the gradient and curvature of the large-scale magnetic field; (d) *energy change* of the cosmic rays due to the expansion or compression of the background fluid and its magnetic field; and (e) *acceleration* due to possible random motions of the magnetic irregularities (denoted second-order Fermi acceleration). The basic transport equation, which combines all these effects, may be written (with names of the various physical effects noted next to the corresponding terms)

$$\frac{\partial f}{\partial t} = \frac{\partial}{\partial x_i}\left[\kappa_{ij}\frac{\partial f}{\partial x_j}\right] \qquad \text{(diffusion)}$$

$$- U_i \frac{\partial f}{\partial x_i} \qquad \text{(convection)}$$

$$- V_{Di} \frac{\partial f}{\partial x_i} \qquad \text{(drift)}$$

$$+ \frac{1}{3}\frac{\partial U_i}{\partial x_i}\frac{\partial f}{\partial \ln p} \qquad \text{(adiabatic energy change)}$$

$$+ \frac{1}{2}\frac{\partial}{\partial p}\left[D_{pp}\frac{\partial f}{\partial p}\right] \qquad \text{(second-order Fermi acceleration)}$$

$$+ Q(x_i, p, t) \qquad \text{(local source density)}, \qquad (2)$$

where $f(\mathbf{r}, p, t)$ is the omnidirectional particle distribution as a function of position r, momentum magnitude p, and time t. $\mathbf{U}$ is the background convection velocity, κ_{ij} is the diffusion tensor, which may be obtained from the spectrum of magnetic irregularities in the turbulence, the drift velocity $\mathbf{V_D}$ is determined by the large-scale magnetic field $\mathbf{V_D} = (pcw/3q)\nabla \times (\mathbf{B}/B^2)$, D_{pp} is the rate of acceleration due to the random motion of the magnetic irregularities (relative to $\mathbf{U}$), and Q is the local source function. Another way of viewing Eq. (2), which shows that it should be in fact a reasonable approximation in widely varied circumstances, is to note from Liouville's theorem that a homogeneous, isotropic distribution in an arbitrary static magnetic field is in a steady state. Equation (2) essentially describes the first-order consequences of gradients in the distribution function and plasma flow velocity. Hence, even if we do not know some of the transport coefficients (such as κ_{ij}) accurately, the general form of Eq. (2) may be adequate for many purposes.

Equation (2) contains nearly all of the cosmic ray acceleration and transport mechanisms discussed in recent years and will be the basis of the present discussion. Examples include the diffusion of cosmic rays in the interstellar gas or in supernova remnants, the basic theory of acceleration at hydromagnetic shocks or the venerable second-order Fermi acceleration. Of course, in many cases, some of the terms in Eq. (2) may be omitted to simplify the analysis.

PARAMETERS

The most important applications of Eq. (2) have been to the solar wind and to the interstellar medium. Before considering detailed modeling, it is useful to consider the general magnitude of the basic parameters, which have been found to give reasonable agreement with observations. The typical cosmic ray is a proton with an energy of 1 GeV. For such a particle we expect

	Parameter	*Solar Wind*	*Interstellar Gas*
Magnetic field	$\mathbf{B}$	5×10^{-5} G8	3×10^{-6} G
Diffusion coefficient	κ	10^{22} cm^2 s^{-1}	10^{28} cm^2 s^{-1}
Fluid velocity	$\mathbf{U}$	5×10^{7} cm s^{-1}	10^{7} cm s^{-1}
Drift velocity	$\mathbf{V_d}$	10^{8} cm s^{-1}	10^{7} cm s^{-1}
Diffusion time	$\frac{L^2}{\kappa}$	10^4 s	10^7 yr
Convection time	$\frac{L}{U}$	10^5 s	10^7 yr
Drift time	$\frac{L}{V_d}$	10^5 s	10^{10} yr

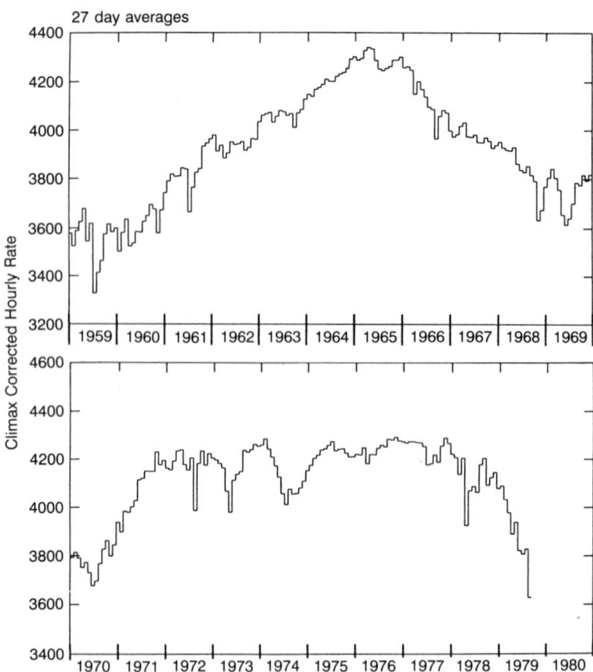

Figure 1. The Climax neutron monitor counting rate, which measures the intensity of $\approx$ 1-GeV protons at Earth, for two sunspot cycles. (*Data from J. A. Simpson, private communication.*)

Clearly, we expect diffusion, convection, and drift to play significant roles in the solar wind, whereas drift may clearly be neglected for the bulk of the cosmic rays in the Galaxy.

APPLICATIONS AND OBSERVATIONSS

Solar Wind

The most sophisticated application of Eq. (2) has been to the transport of galactic cosmic rays in the solar wind, which is subject to in situ observational testing. Here the solar system is regarded as residing in a constant, isotropic bath of galactic cosmic rays. The radially outflowing solar wind acts to partially exclude these particles from the inner solar system, "modulating" the cosmic ray intensity with a basic 11-yr period in antiphase with the sunspot cycle (see Fig. 1). In this case, all the terms in the equation except for the D_{pp} term play important roles. Indeed, the full form of Eq. (2) (without the D_{pp} term) was first written down by Eugene N. Parker in response to the challenges presented by solar modulation. For particle energies greater than approximately 1 MeV, the D_{pp} term is small and may be safely neglected. Hence the solar modulation may be regarded as a balance between the inward random walk or diffusion, the outward convection by the solar wind, gradient and curvature drifts caused by the large-scale structure of the interplanetary magnetic field, and the adiabatic cooling due to the radial expansion of the solar wind.

Sophisticated three-dimensional numerical solutions have been obtained, utilizing the presently accepted picture of the solar wind and its entrained magnetic field, (the heliosphere). In this picture, the solar wind flows out to some termination radius (not yet known, but certainly greater than 50 AU). At this boundary the cosmic ray intensity takes on its interstellar value. The heliospheric model is quite accurate in the solar equatorial regions, where there are many direct observations, but the values of the parameters such as ambient magnetic field and flow velocity in the polar regions are quite uncertain. This uncertainty notwithstand-

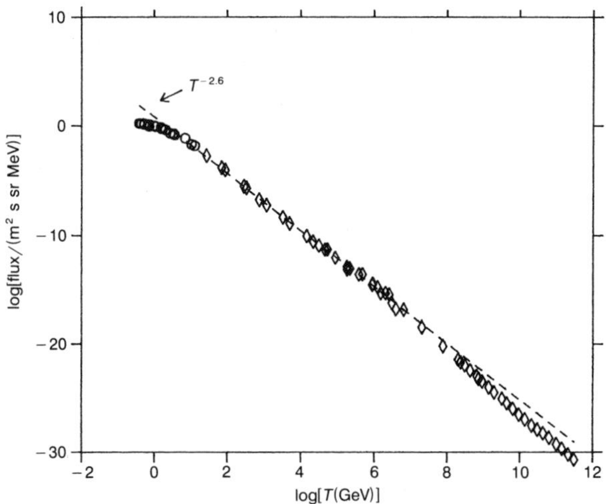

Figure 2. The galactic cosmic ray spectrum for energies greater than 500 MeV. The dashed line is a power law with index -2.6. Note the change in slope at about 10^{15} eV. The turn over of the spectrum at low energies is due to solar effects on the intensity observed at Earth. [*Data compiled from Meyer (1969) and Linsley (1980).*]

ing, the calculated properties of the cosmic rays (radial and latitudinal gradients, energy spectrum, and time variations) are in basic agreement with Earth-based and spacecraft observations of cosmic rays.

In summary, the modulation of cosmic rays by the solar wind provides a reasonably complete verification of the basic transport equation, with the exception of the Fermi acceleration term, which is unimportant in this case.

Interstellar Medium

The observed energy spectrum of galactic cosmic rays is shown in Fig. 2. The part of the spectrum corresponding to energies between 10^8 and 10^{15} eV is thought to originate in supernova explosions in the Galaxy, and to arrive at the Earth after propagating for some 10^6 years in the interstellar medium. The cosmic ray transport in the irregular magnetic field, governed by the basic transport equation, establishes a quasisteady equilibrium where the sources are very nearly balanced by losses. The ultra-high-energy particles (energies $\gtrsim 10^{16}$ eV) have a different origin and may come from outside the Galaxy. Only those galactic particles with energies between about 10^9 and 10^{15} eV are considered, to avoid the part of the spectrum that is seriously distorted by the solar wind. These "unmodulated" galactic cosmic rays are observed to be highly isotropic in arrival directions (with a relative anisotropy $\delta < 10^{-3}$ at $\approx 10^{12}$ eV). Furthermore, evidence from observed γ rays and synchrotron emission indicates that they are uniformly distributed throughout the galactic disk. Finally, observations of unstable isotopes in meteorites (caused by nuclear interactions of impacting cosmic rays) show that the intensity of these cosmic rays has been constant to within about 50%, at the solar system, for approximately the last 10^9 yr. The energy spectrum in interstellar space is apparently very smooth.

Although we could use Eq. (2) with appropriate parameters and boundary conditions, it is adequate to illustrate the propagation of these cosmic rays and their confinement to the Galaxy in terms of a simpler model. The galaxy is taken to be a leaky box containing cosmic rays of species i with a density $n_i(T)$. The loss of these particles is described by a mean leakage time τ (which is related to the diffusion coefficient κ by $\tau \approx L^2/\kappa$, where L is a characteristic scale of the galactic confinement region). If there is a source of particles $Q_i(T)$, conservation of particles is then described by the equation

$$\frac{\partial n_i}{\partial t} = -\frac{n_i(T)}{\tau} + Q_i(T). \tag{3}$$

Observations show that the mean trapping time τ for cosmic rays in the Galaxy is about a few times 10^7 yr, much less than galactic time scales, so the present distribution of cosmic rays reflects a steady state between sources and losses, and we have the simple results from Eq. (3):

$$n_i(T) \approx \frac{L^2}{\kappa(T)} Q_i(T) \tag{4}$$

The equilibrium energy spectrum depends on both the source spectrum and the energy dependence of the transport. In particular, because the leakage time decreases with increasing energy, the spectrum is steepened by the loss process.

The observed secondary cosmic ray nuclei as a function of energy per nucleon provide valuable constraints on the transport in the Galaxy. Secondaries are produced by spallation of heavier cosmic ray nuclei by the rare collisions with the ambient interstellar gas particles. Observations show that for energies above roughly 5 GeV per nucleon, the ratio of secondaries to primaries decreases monotonically with increasing energy. One observes, approximately, for the ratio of the secondaries to the primaries from which they have been created, as a function of energy T,

$$\frac{n_s}{n_p} \approx T^{-0.4} \tag{5}$$

above 2 GeV per nucleon. This is generally interpreted to be a result of an energy-dependent transport and loss from the Galaxy.

Because the cross sections for nuclear interactions at energies $\gtrsim 2$ GeV are roughly independent of energy, the dependence given in Eq. (5) would be produced if the leakage time were approximately proportional to $T^{-0.4}$ above ~ 1–2 GeV.

Galactic cosmic rays are currently believed to be accelerated by collisionless shock waves. Subject only to a few quite reasonable restrictions, which essentially amount to assuming a planar shock on scales of the particle gyroradius, and the validity of the basic transport equation (2), one finds the spectrum of accelerated particles above $T \gg 1$ GeV per nucleon to be

$$Q(T) \approx AT^{-(2+\varepsilon)}, \tag{6}$$

where $\varepsilon = 0$ if the shock is strong, and increases as the shock strength decreases. This shape is independent of the particle propagation parameters, insofar as the basic assumptions are satisfied. Quite naturally, then, from Eq. (4) we would expect the observed $\sim T^{-2.6}$ primary nucleon energy spectrum from this source if $\varepsilon \approx 0.2$ (moderately strong shock), and a leakage time corresponding to that obtained above from the secondary to primary ratio.

We note that although this source spectrum is located at the shock fronts, over the $\gtrsim 10^7$-yr lifetime of a typical cosmic ray particle, one expects that much of the interstellar medium will be traversed by shocks. Hence, it is reasonable to assume a smooth source distribution in the disk.

STATUS

The current diffusive theory of cosmic ray propagation, summarized in Eq. (2), has proved to be successful in explaining many observed features of cosmic rays in the Galaxy and the solar system. It may be used with confidence in other situations as well.

Additional Reading

Axford, W. I. (1981). Acceleration of cosmic rays by shock waves. In *Proceedings of the 17th International Cosmic Ray Conference* **12**, p. 155.

Cesarsky, C. J. (1980). Cosmic-ray confinement in the Galaxy. *Ann. Rev. Astron. Ap.* **18** 289.

Drury, L. (1983). An introduction to the theory of diffusive shock acceleration of energetic particles in tenuous plasmas. *Rep. Prog. Phys.* **46** 973.

Jokipii, J. R. (1971). Propagation of cosmic rays in the solar wind. *Rev. Geophys. Space Phys.* **9** 27.

Linsley, J. (1980). Very high energy cosmic rays. In *IAU Symposium 94, Origin of Cosmic Rays*, G. Setti, G. Spada, and R. A. Wolfendale, eds. Reidel, Dordrecht, p. 53.

Meyer, P. (1969). Cosmic rays in the Galaxy. *Ann. Rev. Astron. Ap.* **7** 1.

Toptygin, I. N. (1985). *Cosmic Rays in Interplanetary Magnetic Fields*. Reidel, Dordrecht.

Völk, H. J. (1976). Cosmic-ray propagation in interplanetary space. *Rev. Geophys. Space Phys.* **13** 547.

See also **Cosmic Rays, Acceleration; Heliosphere; Interplanetary Medium, Shock Waves and Traveling Magnetic Phenomena; Supernova Remnants, Evolution and Interaction with the Interstellar Medium.**

Cosmic Rays, Space Investigations

Jonathan F. Ormes

In 1948 it was discovered that the cosmic rays contained highly energetic (relativistic) ions that include all the elements up through iron in the periodic table, and the study of the astrophysics of galactic cosmic rays (GCR) was born. These balloon-borne studies showed the general similarity between the material in the galactic cosmic rays and solar system material. Inferences about the general nature of the spectra and composition came from ground-based neutron monitors, balloon-borne ion chambers, and air-shower arrays. These studies showed that the particles were coming from beyond the solar system. The first detailed studies of these particles were made on small satellites and balloons. The direct observations from space of GCR began in the 1960s with *Interplanetary Monitoring Platform* (*IMP*) and *Orbiting Geophysical Observatory* (*OGO*) experiments. These experiments were capable of studying particles with kinetic energies up to approximately 1 GeV n^{-1} (here GeV is gigaelectronvolt and n^{-1} means per nucleon). In this entry the instrumentation and methods used to observe galactic cosmic rays above 1 GeV n^{-1} will be described.

Balloon-borne experiments have played an important role in this discipline and are responsible for many of the pioneering observations. In recent years satellite observations have become increasingly important. Satellites offer two very significant advantages. First of all, they carry instrumentation outside the atmosphere, where properties of the particles can be measured without the need to compensate for the effects of the overlying atmosphere. Second, and most important, the missions can be of extended duration (a year or more), whereas balloon payloads can be maintained at high altitudes for typically one day. The added observing time makes possible studies of the rarer species and more subtle signatures of astrophysical effects.

Observations of GCR from space now include high-precision measurements of the energy spectra of the more abundant species, element identification of nuclei throughout most of the periodic table, and exploratory observations of the isotopic composition of the more abundant nuclei. This entry will describe the following experiments:

1. Measurements of the proton and helium spectra and the all-particle spectrum (calorimetry): *Proton* satellites (USSR).
2. Measurements of ultraheavy ($Z > 26$) abundances in cosmic rays: *Skylab*, *Ariel VI* (UK), and *HEAO-3*.
3. Measurements of cosmic ray isotopes: ISEE-3 isotope experiments and the *HEAO-3* Danish-French experiment.
4. Measurements of high-energy spectra using transition radiation detectors: *Spacelab 2* cosmic ray nuclei experiment.

Papers on these kinds of experiments appear regularly in the journal *Nuclear Instruments and Methods* and in books constituting the *Proceedings* of the International Cosmic Ray Conferences that are held biannually.

CALORIMETRY

The proton and helium spectra and the all-particle spectrum were the first to be measured by direct observations from space. These experiments were done on the *Proton* satellites flown by the Soviet Union in the early 1960s.

The first major space experiment to explore the cosmic ray composition and spectra directly at high energy was an instrument known as SEZ-14 which was flown on the *Proton* satellites. A cross section of the instrumentation is shown in Fig. 1. The observations were based on a technique in which the total energy of a highly relativistic particle is measured by counting the secondary particles produced in a cascade of nuclear and electromagnetic interactions.

The results from this series of satellites were quite surprising. They indicated that while the spectrum of all primary cosmic rays could be fit with a single power law between 60 and 10^5 GeV, the proton spectrum showed pronounced steepening at $1-2\times10^3$ GeV. Later, it was demonstrated that the number of backscattered particles from the calorimeter increased with the log(E). The presence of these backscattered particles was sufficient to confuse the identity of single charged particles, artificially steepening the slope of the proton spectrum, but not the all-particle spectrum.

Notwithstanding these difficulties, these pioneering measurements were the first to observe the high-energy component of cosmic rays with direct observations from space.

ULTRAHEAVY NUCLEI EXPERIMENTS

The next major thrust in measurements from space focused on the measurement of elements in the transiron part of the periodic table. Early results from an exposure on *Skylab* gave indications that actinides were present in a high proportion.

The balloon and *Skylab* experiments used large area arrays of thin plastic sheets as a detecting medium. The very highly ionizing particles pass through the plastic sheet, leaving behind a trail of damaged polymer bonds. When the sheets are placed in a solution saturated with hydroxyl radicals, conical pits are etched out of these damaged sites. The etch rate and hence the size and shape of the conical pits had been discovered to be a strong function of the ionization loss of the incident particles. The size and geometry of the etch pits were used to determine the identity of the incident nucleus. The ionization loss rate is given by the Bethe–Block

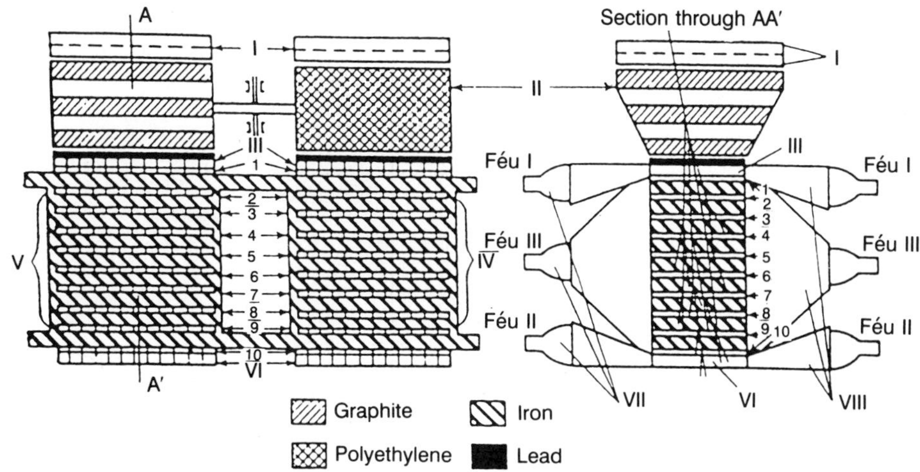

Figure 1. Schematic of the SEZ-14 instrument flown on the *Proton* satellites. (*I*) Charge detectors (proportional counters), (*II*) filters, (*III*) interaction detectors, (*IV*) and (*V*) energy detectors, (*VI*) scintillation counters, (*VII*) photomultipliers, and (*VIII*) diffusers. (*Reprinted by permission of N. L. Grigorov.*)

formula:

$$dE/dx \propto (Z^2/\beta^2) f(\beta\gamma),$$

where Z is the charge of the incident nucleus, β is its velocity in units of the speed of light, and $f(\beta\gamma)$ represents the relativistic increase in ionization loss at energies above several gigaelectronvolts.

The next generation of detectors to study the transiron elements were those flown by the British on a satellite known as *Ariel VI* and by a Washington University (St. Louis)–California Institute of Technology (Caltech)–University of Minnesota collaboration on the third in a series of *High Energy Astronomy Observatory* (*HEAO*) satellites. These detectors were electronic counter experiments in which a combination of ionization chambers and Čerenkov detectors were used to determine velocity and charge. The functional dependence of Čerenkov light on velocity and charge is given by

$$C = C_0 Z^2 [1 - (1/\beta^2 n^2)].$$

The index of refraction n can be chosen, within some restrictive limits, to cover a specific velocity domain of interest, $\beta > 1/n$. Just above threshold, the response is very sensitive to β. Once β is known, Z can be determined and the species identified.

The *Ariel VI* detector had an unusual spherical geometry. It is shown in Fig. 2. The detector consisted of a thin spherical shell of doped acrylic Čerenkov detector suspended concentrically within a thin aluminum shell. The complete volume was filled with a mixture of scintillating noble gases, which responded proportionally to the ionization loss.

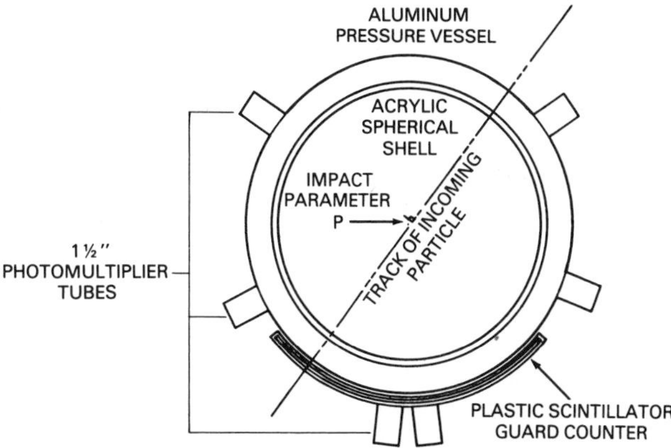

Figure 2. Schematic cross section through the *Ariel VI* ultraheavy cosmic ray detector. (*Reprinted by permission of P. H. Fowler.*)

The next major satellite investigations of GCR were planned by NASA as part of the *HEAO* series. A very nice history of this program can be found in Tucker's *The Star Splitters*. Two cosmic ray investigations were part of the payload carried on the third satellite, *HEAO-C*, renamed *HEAO-3* following the launch.

The *HEAO-3* heavy nuclei experiment was of a more conventional design than that of the *Ariel VI* experiment. It consisted of two large ionization chambers mounted back to back, with a large Čerenkov counter between. The instrument measured the elemental composition of nuclei from chlorine ($Z = 17$) through at least uranium ($Z = 92$).

The relative abundances of the elements in the cosmic rays observed by the *Ariel VI* and *HEAO-3* experiments are similar to the relative abundances of the elements in solar system material (SSM), but they are not identical. These experiments did not find the excess of actinides initially reported by the *Skylab* investigators.

The differences in abundances between GCR and SSM may contain important clues as to how elements originate. It was observed on *HEAO-3* that the elements that are more difficult to ionize are less abundant in the GCR than in SSM. This behavior is difficult to understand if cosmic rays are produced by violent events such as supernova explosions. Under these extremely energetic conditions, the atoms should be completely stripped of all their electrons. On the other hand, if cosmic rays originate in a comparatively low-energy environment having a temperature of about 10,000 K, the ionization potential may parametrize an important selection effect. These conditions are found on the edges of bubbles carved in interstellar space by hot young stars, around red dwarf stars known for their frequent, intense flares (flare stars), or in the wakes of shock waves plowing through the interstellar medium. If this is true, then the material sample represented by the GCR is probably part of the interstellar medium.

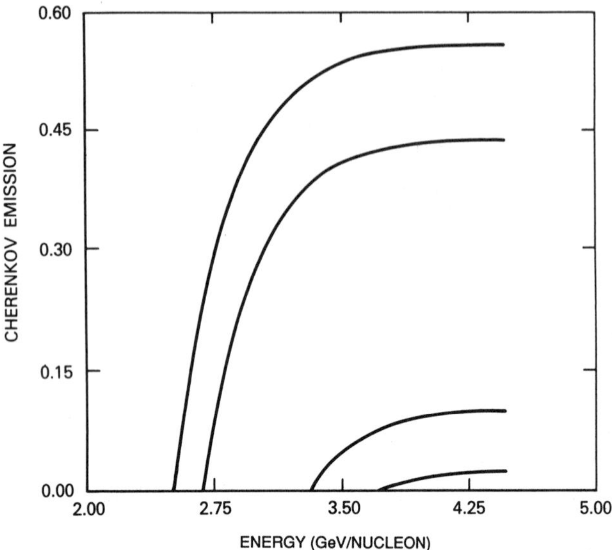

Figure 3. Response curves for the Čerenkov detectors used on the Danish–French cosmic ray instrument.

ISOTOPE MEASUREMENTS

Studying the isotopic composition of GCR is another powerful method for investigation of these ideas. Here again, low-energy cosmic ray experiments on balloons, deep space missions (*Pioneer*, *Voyager*) and near-Earth monitors of solar and galactic cosmic rays made the pioneering observations. Early balloon results indicated that the isotopic composition of GCR is different from solar system material. Subsequent studies using solid-state detectors on the *ISEE-3* spacecraft, in orbit outside the Earth's magnetosphere, have confirmed the composition differences and have established their magnitude for the most abundant species (e.g., the isotopes of Ne).

The first attempt to make measurements at high energies came on a *HEAO-3* experiment built by the Danish Space Research Institute and the Center for Nuclear Studies, Saclay, France. This experiment utilized a set of five Čerenkov detectors, including two with very low index of refraction silica aerogel counters, in conjunction with a flash tube hodoscope used to measure the trajectory of the particles through the instrument. The two outermost Čerenkov detectors were made of glass ($n = 1.5$) and were used for charge determination. The requirement of a consistent response in these two detectors was used to eliminate all those particles which had interacted in the material in the detector. The three inner Čerenkov detectors were chosen to match the spectrum observable in the satellite's orbit and were for velocity determination. They were made of teflon ($n = 1.33$), aerogel blocks ($n = 1.053$), and an aerogel sand ($n = 1.012$). These three indices of refraction were chosen to cover a reasonable energy range, as shown in Fig. 3.

This experiment was able to obtain excellent charge and velocity (and hence momentum) resolution. High precision spectra were obtained covering the nuclei $4 \leq Z \leq 28$ and over the energy range from 0.5 to 20 GeV n^{-1}.

Using the Earth's magnetic field as a filter, isotopic composition information was also obtained. At each point on the satellite's orbit there is a range of energies for which particles of a given A/Z ratio are allowed to reach the spacecraft, whereas particles of a different A/Z ratio are not allowed (e.g., their trajectories, when traced back, hit the Earth). This experiment indicated that the isotopic composition anomalies found at lower energies persist into the trans-gigaelectronvolt energy range. The excellent charge resolution and energy spectra from this experiment have firmly established the spectral difference between the secondary and primary cosmic rays hinted at by prior balloon experiments.

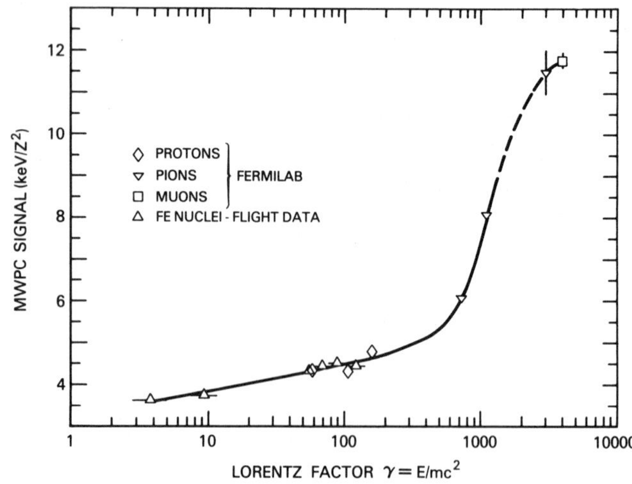

Figure 4. Response curve for the CRNE transition radiation detectors. (*Reprinted by permission of* The Astrophysical Journal.)

TRANSITION RADIATION DETECTORS

The study of the spectra of individual elements at still higher energies required the introduction of a large-area, lightweight detector to obtain the necessary collection power. A new detector, a transition radiation detector, was incorporated into an experiment designed and built at the University of Chicago and flown on the *Spacelab-2* mission. Known as the cosmic ray nuclei (CRN) experiment, its purpose was to extend knowledge of the charge composition of cosmic rays to energies in the range 50–5000 GeV n^{-1}, well beyond the range of the Danish–French experiment. The experiment orbited for the eight days of a space shuttle mission in 1985. CRN employed scintillation detectors for charge measurement and proportional counters for measuring particle trajectories. It used a gas Čerenkov detector for energy measurements in the range from 40–150 GeV n^{-1}.

The transition radiation detector was used for energy measurements above 500 GeV n^{-1}. It utilized the radiation of x-rays produced at the boundary of two media with different indices of refraction by the passage of a charged particle. The radiation is proportional to the Lorentz factor γ of the particle and the square of the charge of the particle. The production of x-rays is improbable at any single interface, so the detector is made up of several layers, each consisting of thousands of interfaces. The detector is made of thin polyolefin fiber threads of 4.5-μm and 21-μm diameter, which are also used for the insulation of ski jackets. The x-rays generated in the fiber bundles are detected in the multiwire proportional counters filled with high Z gas. The response for singly charged particles as a function of Lorentz factor is shown in Fig. 4.

This new detector system has extended knowledge of the spectra of individual nuclear components of the cosmic rays to higher energies. CRN has shown that the energy dependence of the secondary nuclei to primary nuclei ratio extends to the highest energies yet measured.

THE FUTURE

The experiments described have traced 25 years of space-based investigations of the galactic cosmic rays. A few basic detectors have been used in a variety of configurations for different objectives. Great progress has been made, but much more remains to be done. Determination of the isotopic composition of galactic cosmic rays for rarer isotopes will be the focus of the next generation of instruments. Studies of the spectra of antiprotons and searches for other heavier antinuclei will be important in defining the limitations on matter-antimatter symmetry in the Universe. The positrons and antiprotons, secondaries from the collisions of cosmic ray protons and interstellar matter, will be compared to the spallation products of heavier nuclei to establish the relationship between the origin of cosmic ray protons and heavier nuclei. These investigations will require magnetic spectrometers for measurements above 1 GeV n^{-1} and new large arrays of high precision solid-state detectors for measurements at lower energies. The 1990s should see some of these new investigations, currently on the drawing boards, come to fruition.

Additional Reading

Engelmann, J. J., Goret, P., Juliusson, E., Koch-Miramond L., Lund, N., Masse, P., Rasmussen, I. L., and Soutoul, A. (1985). Source energy spectra of heavy cosmic ray nuclei as derived from the French–Danish experiment on *HEAO-3*. *Astron. Ap.* **148** 12.

Fleisher, R. L., Price, P. B., and Walker, R. M. (1975). *Nuclear Tracks in Solids: Principles and Applications*. University of California Press, Berkeley.

Grigorov, N. L., Murzin, V. S., and Rapoport, I. D. (1958). Method of measuring particle energies above 10^{11} eV. *Zh. Eksper. Teor. Fiz.* (*Sov. Phys.-JETP*) **34** 506.

Jelley, J. V. (1958). *Čerenkov Radiation and its Applications*. Pergamon Press, London.

Rossi, B. (1952). *High Energy Particles*. Prentice-Hall, Englewood Cliffs, NJ.

Tucker, Wallace H. (1984). The star splitters. *NASA Report No. SP*-466.

See also **Antimatter in Astrophysics; Cosmic Rays, Observations and Experiments; Supernova Remnants, Evolution and Interaction with the Interstellar Medium.**

Cosmology, Big Bang Theory

Bernard E. J. Pagel

"Big Bang" is the name given (originally by Sir Fred Hoyle with semifacetious intent) to theories which assume that the whole observable universe has expanded in a finite time (according to our usual time reckoning) from an earlier state of much higher density. The scope of such theories is limited at present by quantum effects that are expected to become significant (and cause existing concepts of a smooth space-time to break down) at some density not exceeding the *Planck density* ($c^5/\hbar G^2 = 5\times10^{93}$ g cm^{-3}, where c is the speed of light, $\hbar$ is Planck's action constant divided by 2π, and G is the gravitational constant), corresponding to the *Planck time* $\sqrt{G\hbar/c^5} = 5\times10^{-44}$ s, so that even the boldest extrapolations of current physical ideas into the past are confined to epochs later than this.

Observations (the microwave background radiation and the abundances of light elements) strongly suggest the existence of very high temperatures as well as densities in the past. Attention is therefore confined here to "hot Big Bang" theories, which envisage a "universal fireball" at early times, dominated by radiation and relativistic particles. These were pioneered by George Gamow, Ralph Alpher, and Robert Herman in the late 1940s. Application of such theories can test high-energy physics and basic properties of matter at energies greatly exceeding the 10,000 GeV or so of the largest man-made accelerator currently planned (the Superconducting Super Collider) if we can interpret the data and, conversely, some of their predictions can be tested by laboratory experiments with accelerators.

BIG-BANG COSMOLOGICAL MODELS

Development of consistent cosmological models became possible only after that of the general relativity theory, which accounts for the mutual effects of mass energy and space-time and can thus deal with both finite and infinite amounts of matter, avoiding difficulties with boundary conditions that arise in cosmology with classical dynamics. However, many properties of cosmological models can be qualitatively and even quantitatively described in Newtonian terms, which makes them easier to visualize.

Most models are based on the *cosmological principle*, which asserts that the universe (considered on a sufficiently large scale) is homogeneous and isotropic. Mass energy in the universe is approximated as a smooth fluid called the *substratum*, populated by imaginary *fundamental observers*, each one being at rest relative to the substratum at his position. Homogeneity implies that the universe presents the same large-scale aspect, at any given time, to all fundamental observers, and this in turn permits them to agree on a universal cosmic time which can be standardized by observing a property of the universe (e.g., average density or the radiation temperature). Isotropy implies that, to a fundamental observer, the universe presents the same large-scale aspect in all directions, and this in itself implies homogeneity if all fundamental observers are equivalent, as well as providing a local standard of rest. Anisotropic (e.g., rotating) models have been studied theoretically, but are so severely constrained by the observed isotropy of the microwave background (apart from an effect that can be attributed to motion of the Earth at a few hundred kilometers per second relative to the substratum) that they need not be considered further. The real universe is, of course, lumpy on scales up to at least that of superclusters of galaxies; but the microwave background temperature is isotropic (apart from the effect of the Earth's motion) to a few parts in 100,000 or less, implying that homogeneity is a good approximation on large-enough scales unless the Earth is in a unique central position—a proposition that is unthinkable after Copernicus.

The cosmological principle restricts the evolution of the substratum to one of three possibilities: It may be static, or it may be expanding or contracting with the relative velocity of any two points, at some instant of cosmic time, being directed along the line (or geodesic curve) joining them and proportional to the distance between them. The expanding case is the one corresponding to observation as expressed in Edwin P. Hubble's law of redshifts (1929). Thus the relative position of any two points of the substratum can be expressed by "comoving" coordinates (a distance, and two angles specifying a direction) that remain fixed during expansion (or contraction), but with a universal scale factor $R(t)$ that depends on time. In addition, space may be curved in one of three different ways: spherical (a three-dimensional analog of the surface of a balloon) , flat, i.e., Euclidean; and hyperbolic (analogous to a saddle-shaped surface). Mathematically, these possibilities are embodied in an equation first clearly formulated by Howard P. Robertson (in the USA) and by Arthur G. Walker (in the UK), which relates four-dimensional intervals in space-time between two events to the corresponding differences in time and in spatial coordinates. A consequence of this equation is that when a light signal is emitted by one fundamental observer at time t and received by another at a later time t_0 the wavelength received is scaled up relative to the wavelength emitted in the ratio $(1+z) = R(t_0)/R(t)$. Thus expansion leads to a redshift z caused by a dilation of space; when small, this is given by the usual Doppler formula as it appears in Hubble's law

$$cz = c(\lambda - \lambda_1)/\lambda_1 = V = H_0 D,$$

where V is the recession velocity, D the distance and λ, λ_1 are, respectively, the observed and the emitted (or laboratory) wavelength of light or any other electromagnetic waves. H_0 is the *Hubble constant* (actually a function of time; the subscript zero refers to some epoch of observation such as the present), which is estimated from observation to be somewhere between about 50 and 100 km s^{-1} Mpc^{-1} (1 Mpc = 3.26 million light-years) corresponding to a Hubble time $1/H_0$ between about 2×10^{10} and 10^{10} years, respectively.

The Robertson–Walker equation permits a threefold infinity of cosmological (or "world") models differing in the sign (or absence) of curvature and in the way in which the scale factor depends on time. These properties depend in turn on the form of the field equation of general relativity (involving, or not, an arbitrary "cosmological" constant Λ, that acts—when positive—as though producing a repulsive force that increases with distance from the observer), on the density of mass-energy, and on its equation of state. Two important limiting cases of the latter are (i) radiation and relativistic particles (particles traveling at, or almost at, the speed of light), dominant in the early stages, for which pressure makes a significant contribution to the mass-energy density; and (ii) effectively pressure-free "cold" matter which dominates today.

The first relativistic cosmological model was that of Albert Einstein (1917), who envisaged a static, finite spherical universe in which the gravitational attraction of matter is balanced by repulsion arising from a positive Λ. Willem de Sitter produced in the same year an alternative, flat model, also with positive Λ, that is empty of matter, and was later interpreted as an (exponentially) expanding universe with a Hubble constant that is independent of time. Explicitly nonstatic models were studied first by Alexander Friedmann in the USSR (1922) and independently by Georges Lemaître in Belgium (1927). Sir Arthur Eddington showed theoretically in 1929 that Einstein's static model is unstable and Hubble's announcement in the same year ruled nonexpanding models out of court. Lemaître first formulated explicitly the idea of an initial state of enormous density (called by him "the primeval atom"), which is nowadays referred to as the Big Bang, and Friedmann calculated the properties of models with $\Lambda = 0$, an assumption which is quite commonly made today, at least for the present stage of expansion, although there is a hypothesis with certain attractions known as the inflationary universe scenario, which holds that this phase was preceded in the first 10^{-35} s or so by a very rapid expansion according to de Sitter's formula. The fate of a Friedmann universe depends on whether the kinetic energy of expansion is, or is not, sufficient for it to continue indefinitely despite the deceleration produced by the self-gravitation of the matter within it (see Fig. 1), analogously to the fate of a projectile that is, or is not, thrown with enough velocity to escape from the Earth. This condition is expressed by the value of a dimensionless parameter Ω, which is the ratio of the actual average density to the so-called *critical density* $3H^2/8\pi G = 1.9\times10^{-29}h^2$ g cm^{-3}. Here h is another dimensionless parameter expressing the uncertainty in the Hubble constant (or its variation with time); it is 0.5 if $H = 50$ km s^{-1} Mpc^{-1} and 1 if $H = 100$ km s^{-1} Mpc^{-1}. Ω less than 1 (low density) leads to an infinite hyperbolic universe that will expand forever with Ω decreasing. $\Omega = 1$ (critical density) gives an infinite flat universe that will just barely expand forever with $\Omega = 1$ always (Einstein–de Sitter model). Ω greater than 1 (high density) leads to a closed spherical universe that will eventually stop expanding and recontract, possibly to reach another Big Bang (or Big Crunch) and perhaps start the process over again (oscillating universe). The appropriate Friedmann model to describe the actual universe can be specified (if such an assumption is valid) by measuring two parameters, H_0 and Ω_0, at the present time.

There are several ways of estimating Ω_0. One is to find a dimensionless deceleration parameter q_0 from the *Hubble diagram* relating redshifts to apparent magnitudes of, for example, the brightest galaxies in clusters treated as standard candles. ($q_0 = 0.5\Omega_0$ if $\Lambda = 0$.) There are difficulties due to evolutionary effects on the luminosities of the galaxies, but the observations imply that q_0 is not more than a few and it could even be slightly negative, that is, $\Lambda > 0$. Another method relies on departures from a uniform Hubble flow caused by the lumpy distribution of matter; depending on what kinds of galaxies are assumed to represent the distribution of gravitating matter, one obtains values between about 0.15 and 1.0. A third method comes from certain arguments relating to Big Bang nucleosynthesis: According to the standard model, the abundances of deuterium and helium limit the contribution to Ω_0 from normal baryonic matter (protons and neutrons) to below about 0.1, but an arbitrary additional contribution from nonbaryonic matter in the form of exotic elementary particles not yet discovered is not excluded. Quasiaesthetic considerations (and the inflationary scenario) lead many scientists to believe that Ω is very close to 1, in which case the age of the universe according to the Friedmann model is $\frac{2}{3}$ of the Hubble time. Other estimates of the age come from colors and luminosities of the oldest stars, which give an age of about 1.4×10^{10} years, and from abundances of radioactive elements which give a less precise number of the order of 10^{10} years. Stellar age dating seems to require H_0 to be nearer to 50 rather than 100 km s^{-1} Mpc^{-1} if this cosmological model applies.

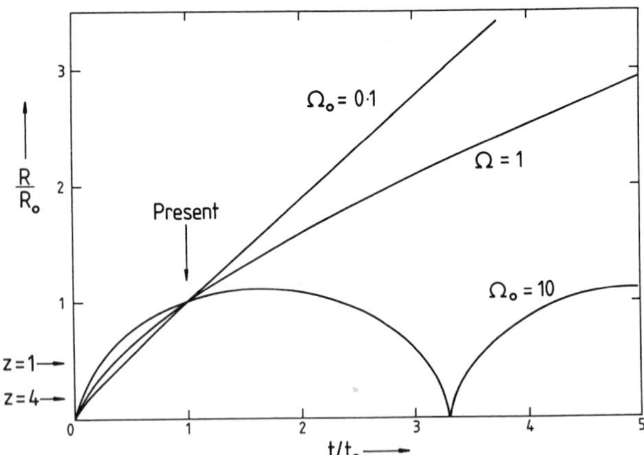

Figure 1. Variation of the scale factor $R(t)$ with cosmic time $[R_0 = R(t_0)]$ according to three Friedmann models with different values of Ω_0 at a chosen epoch such as the present. $\Omega_0 = 0.1$ refers to an open hyperbolic model in which Ω is initially just below unity and steadily decreases and expansion proceeds "coasting" forever. After a brief initial phase, any two fundamental observers recede from each other at constant velocity. $\Omega = 1$ refers to the "critical" Friedmann (or Einstein–de Sitter) model in which $\Omega = 1$ at all times and the scale factor varies as the $\frac{2}{3}$ power of the time throughout the pressure-free phase. This is a flat, open model which just barely manages to expand forever. $\Omega_0 = 10$ refers to a closed spherical model in which Ω starts out slightly greater than 1 and increases during expansion; in such cases the expansion is eventually halted and followed by a recollapse, which may lead to oscillations.

HORIZONS

Because the universe has a finite age, there is a limit to the distance over which two fundamental observers can communicate, which is of the order of the product of the speed of light with the Hubble time and is known as the *particle horizon*. This distance is thus roughly as many light-years as the age of the universe in years and in Friedmann models more and more of the universe becomes visible as time goes on. The converse of this effect leads to the paradox that at early times such as that of the last scattering of the microwave background radiation, which took place at an epoch corresponding to a redshift of the order of 1000, regions that we now see at large angular separations cannot have ever been in causal contact with one another, and yet they manage to have closely equal temperatures. The inflationary scenario provides a possible solution to this "horizon problem." In models with positive Λ one can have accelerated, rather than decelerated, expansion, and in that case (and also in oscillating models) galaxies can also

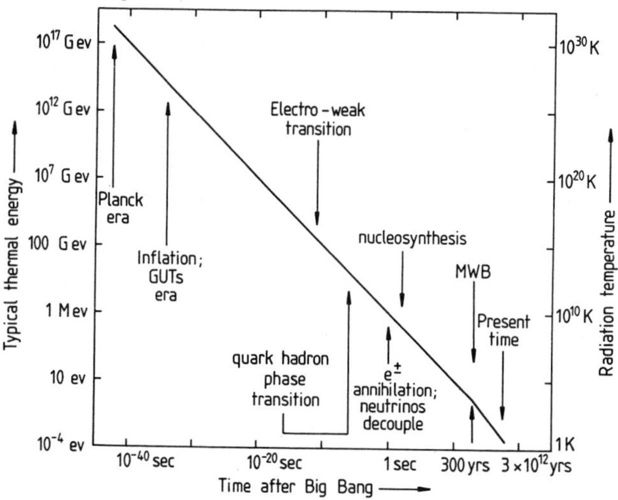

Figure 2. Schematic thermal history of the universe showing some of the major episodes envisaged in the standard model. GUTs is short for grand unification theories and MWB is short for (the last scattering of) the microwave background radiation. The universe is dominated by radiation and relativistic particles up to a time near that corresponding to MWB and by matter (possibly including nonbaryonic matter) thereafter.

disappear from view, giving rise to another sort of horizon known as an event horizon. When both sorts of horizon exist, there are parts of the universe that remain forever unobservable (absolute horizon).

THERMAL HISTORY OF THE UNIVERSE

Our neighborhood and presumably the whole universe is now filled with blackbody radiation at a temperature of 2.7 °C. above absolute zero—interpreted as a consequence of adiabatic expansion from higher temperatures at earlier times which follow the law

$$T = T_0(1+z),$$

where T is the radiation temperature at some past epoch and T_0 is its value now. Thus the energy density of radiation varies as $(1+z)^4$, whereas the density of matter varies only as $(1+z)^3$, leading to the result that radiation dominated over matter at redshifts above 1000 or so when the universe was less than about 100,000 years old and the temperature more than a few thousand degrees (see Fig. 2). Before more or less the same epoch, matter (mostly hydrogen) would have been ionized and closely coupled to radiation by electron scattering, leading to thermal equilibrium, whereas afterwards protons combined with electrons to make neutral hydrogen, which is transparent, and the radiation underwent its last scattering from the cosmic "photosphere"; the latter forms an opaque screen through which it is impossible to see. Only in the subsequent matter-dominated era would it have been possible for galaxies to form, presumably as a consequence of growing density enhancements arising from small fluctuations that had occurred at much earlier epochs. The comoving density of quasars—which may be associated with newly formed galaxies—is found to increase up to a redshift of 2 or 3, with a possible decline somewhere beyond that, and the largest measured redshifts are between 4 and 5.

THE BIG BANG, PARTICLE PHYSICS, AND PRIMORDIAL NUCLEOSYNTHESIS

At the earliest times, according to theory, the temperature was still higher, reaching typical thermal energies equal to the rest-mass energy mc^2 of elementary particles—electrons and positrons (0.5 MeV), protons and antiprotons (1 GeV), quarks and antiquarks, etc.—leading to copious production of these particles together with neutrinos, mesons, W and Z bosons (100 GeV), and, at a very early stage, presumably particles associated with grand unification theories (10^{15} GeV). As time went on and the universe cooled down; particles whose rest-mass energy now exceeded the thermal energy would have annihilated with their antiparticles to make gamma-ray photons leaving a small excess or ordinary matter over antimatter. After about a second, at a thermal energy just below 1 MeV, electrons would have annihilated with positrons leaving a sea of photons and neutrinos (and antineutrinos) with a small admixture of protons, electrons, and neutrons. After 100 s, at a temperature corresponding to 0.1 MeV, nuclear reactions would first become possible as a result of the capture of neutrons by protons to make deuterium nuclei; these would have been destroyed at higher temperatures by photodisintegration. Deuterium formation then led rapidly through further nuclear reactions to primordial nucleosynthesis, in which it appears that 23% of nuclear matter ended up as ordinary helium (^{4}He) with 77% hydrogen and small traces of deuterium (^{2}H), light helium (^{3}He), and the commoner isotope of lithium (^{7}Li). Observed abundances of these elements are well explained by the standard theory assuming that there are just three species of light neutrinos and antineutrinos, corresponding to three (and no more) families of quarks and leptons, and that the contribution of baryons to Ω_0 is between $0.01/h_0^2$ and $0.02/h_0^2$. Standard Big Bang nucleosynthesis theory assumes a homogeneous universe (in this phase, Ω is close to 1 in all Friedmann models and Λ is unimportant even if nonzero); but there are also nonstandard models assuming the existence either of density fluctuations or of unstable particles whose decay products modified the outcome of primordial synthesis by further nuclear reactions. Such theories may be able to account for the abundances even if there is a larger baryonic contribution to Ω_0; but the standard model seems to give the best fit to the data. The ratio of the number of baryons to the number of photons, on which the predicted primordial abundances primarily depend, is found from application of the standard model to be about 3×10^{-10}. This ratio, unchanged since the epoch of electron-positron annihilation, represents the excess of ordinary matter over antimatter that may have been originally set up at (or close to) the grand unification stage, and its reciprocal is closely related to a thermodynamic property of the universe—the entropy per baryon (not counting entropy of any black holes). Its value appears so far as an ad hoc parameter, but it may be expected to emerge (when understanding improves) as a consequence of fundamental properties of elementary particles and their interactions.

Additional Reading

Harwit, M. (1988). *Astrophysical Concepts*, 2nd ed. Springer, New York.

Hawking, S. W. (1988). *A Brief History of Time*. Bantam Press, London.

Peebles, P. J. E. (1980). *The Large-Scale Structure of the Universe*. Princeton University Press, Princeton, NJ.

Rowan-Robinson, M. (1981). *Cosmology*, 2nd ed. Clarendon Press, Oxford.

Rowan-Robinson, M. (1985). *The Cosmological Distance Ladder*. W. H. Freeman, New York.

Silk, J. (1989). *The Big Bang*, rev. ed. W. H. Freeman, New York.

Weinberg, S. (1977). *The First Three Minutes*. Basic Books, New York.

See also **Antimatter in Astrophysics; Background Radiation, Microwave; Cosmology, Clustering and Superclustering; Cosmology, Cosmochronology; Cosmology, Galaxy Formation; Cosmology, Inflationary Universe; Cosmology, Nucleogensis; Cosmology, Observational Tests; Cosmology, Theories; Dark Matter, Cosmological; Gravitational Theories; Neutrinos, Cosmic; Quasistellar Objects, Statistics and Distribution; Star Clusters, Stellar Evolution.**

Cosmology, Clustering and Superclustering

William C. Saslaw

On large scales, the probability that galaxies occupy a given size volume of space is not random, like the toss of a coin, but is related to the presence of nearby galaxies. Positions of galaxies depend on one another. Details of this dependence may provide important clues to the origin and evolution of the universe. Modern systematic searches for these clues started in the 1930s with Edwin P. Hubble's galaxy counts, accelerated in the 1950s with the development of new statistical techniques, and surged in the 1970s and 1980s as computers became powerful enough to handle large amounts of data and calculate complex simulations of clustering physics.

There are two main descriptions of galaxy clustering. The first is essentially pictorial, derived by searching the sky for filaments, voids, and overdense regions. Galaxies in projected high-density regions having similar redshift distances are likely to form a physically related cluster. If such a cluster has been bound gravitationally for a large fraction of the present Hubble time, it evolves fairly independently of its surrounding galaxies. Occasionally several contiguous large clusters form a supercluster. This may result partly from the initial positioning of clusters and partly from their subsequent motions.

Most galaxies are not in large physically bound clusters. They are, however, clustered in a more general statistical sense. Rather than examining individual clusters, this second description looks for statistical departures from a Poisson distribution (which is the uniform random spatial distribution of objects whose positions are entirely independent of one another, i.e., totally uncorrelated). In a Poisson distribution there will be some regions where galaxies are strongly clustered just by chance, and other regions which are unusually empty, also by chance. The observed numbers and statistics of such regions may then be compared with those expected from various theories.

STATISTICAL MEASURES OF CLUSTERING

The most useful and informative statistics can be measured objectively from observations and related analytically to physical processes of clustering and computer experiments. Low-order correlation functions are one example. The two-point correlation function $\xi(r)$ in its simplest form for a homogeneous isotropic system is defined by

$$n(r)\,dr = 4\pi\bar{n}r^2[1+\xi(r)]\,dr.$$

Here $n(r)\,dr$ is the average number of galaxies between radial distance r and $r+dr$ from any given galaxy. The overall average number density of galaxies in the entire system, or over a very large volume of the universe, is $\bar{n}$. For a random Poisson distribution, $\xi(r)=0$; so $n(r)$ is determined just by $\bar{n}$ and geometry. Therefore $\xi(r)$ helps measure departures from the Poisson state. Higher-order correlation functions use the relative positions of three or more galaxies for a more refined description that, unfortunately, is harder to measure observationally. The first accurate observations of $\xi(r)$ for galaxies, made by H. Totsuji and T. Kihara in 1969, gave a power law of the form $\xi(r) \approx 50\ r_{\mathrm{Mpc}}^{-1.8}$ on scales $0.2 \lesssim r \lesssim 20$ Mpc (for a Hubble constant of 50 km s^{-1} Mpc^{-1}). Large clusters of galaxies are observed to have a similar two-point correlation function if each cluster is represented by a single point, but this result is much more uncertain. On small scales $\xi(r) \gg 1$, so the observed clustering is highly nonlinear; that is, correlations dominate for $r \lesssim 10$ Mpc.

Another observed simple objective clustering statistic is the distribution function $f(N)$. This is the probability for finding N galaxies in a volume of size V, or projected onto the sky in an area of size A. If the distribution is statistically homogeneous then it will not depend significantly on the shape of the volumes or areas, provided they are sufficiently large and numerous to give a fair average sample. Recent analyses of the area counts of galaxies show that they have a distribution of the form

$$f(N) = \frac{\bar{N}(1-b)}{N!}[\bar{N}(1-b)+Nb]^{N-1}\exp\{-\bar{N}(1-b)-Nb\},$$

where $\bar{N}=\bar{n}V$ is the average number in a volume V for an average number density $\bar{n}$. The quantity b is a measure of clustering and is related to gravitational correlations. The observed value of b is 0.70 ± 0.05 for galaxies whose separations are typically 1–10 Mpc. For large clusters with separations of $\sim$10–50 Mpc, $b=0.3\pm0.1$. For a sample of faint radio sources with separations $\gtrsim 50$ Mpc, $b=0.0$, which is a random Poisson distribution. The $f(N)$ distribution for $N=0$ gives the probability that a region is a void with no galaxies at all.

Other statistics applied to galaxy clustering include minimal spanning trees (the shortest line connecting all the galaxies in a sample), topological patterns formed by contour maps of regions with the same density, and multifractal analyses (a single fractal dimension does not adequately describe galaxy clustering), which are related to how the average number density of galaxies around a given galaxy changes with distance from the galaxy. These other statistics also yield valuable information. Unlike $\xi(r)$ and $f(N)$, however, they have not yet been related generally to an underlying dynamical theory. Some specialized computer experiments have examined their behavior.

THEORIES OF CLUSTERING

To understand the observed statistics we need to know the initial conditions for clustering as well as a physical theory for its subsequent evolution. Initial conditions may indicate properties of the early universe before galaxies formed and perhaps even close to the Big Bang. Some possibilities are that galaxy clustering started from a random Poisson distribution, or from a state with local clustering or from large-scale coherent structures. No clear observational evidence for any particular initial state has been found. On small scales the clustering processes themselves tend to destroy this evidence, while on large scales it is difficult to detect.

Different types and distributions of dark matter may also be important for forming galaxies and clusters. For example, massive neutrinos, other weakly interacting massive particles, cosmic strings, quark nuggets, or other currently speculative objects of various high-energy theories may influence galaxy clustering if they exist in sufficient quantity.

All known forms of matter gravitate and gravitation promotes clustering. Therefore, astronomers have developed analytical theories and examined many computer simulations to describe the gravitational clustering of galaxies. Results for different models are then compared with $\xi(r)$ and $f(N)$. The models generally differ in their initial conditions, the role of dark matter, and the time available for clustering.

Computer simulations calculate the gravitational orbits of many thousands of particles, each one representing a galaxy, in the background of the expanding universe and any dark matter present. The orbits are found either by integrating the thousands of equations of motion—each with thousands of terms—directly, or by averaging the gravitational forces in different ways to simplify the problem. Averaging sacrifices detailed information in order to include a larger number of galaxies.

Computer models which start with strong structure on scales of tens of megaparsecs frequently do not agree with the observed correlation and distribution functions. Those that do, often agree only for a short span of their evolution. On the other hand, models with fairly homogeneous initial distributions, such as an uncorre-

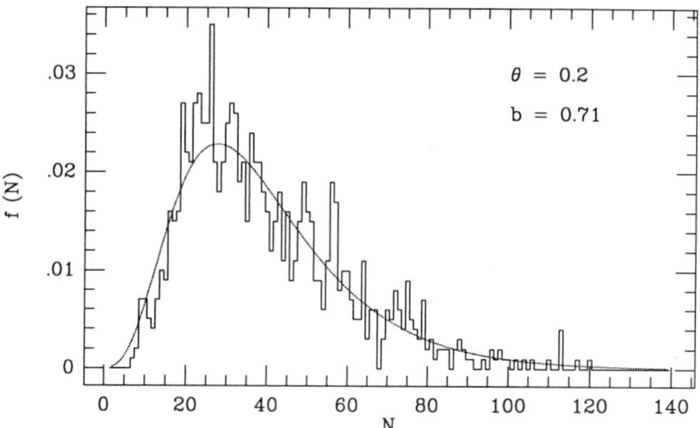

Figure 1. An example of a computer N-body simulation of galaxy clustering (*courtesy M. Itoh and S. Inagaki*) and its related distribution function (analysis by E. Lufkin). The simulation contains 4000 mass points, each representing a galaxy, starting from an initial uniform Poisson distribution and evolving gravitationally in an expanding universe whose density $\Omega_0 = 1$ just produces closure. The upper figure shows positions of about 2000 galaxies projected onto the "sky" after the radius of the universe has increased by a factor of about 8 from its initial size. The lower histogram shows the distribution function $f(N)$ for this same sample. It gives the probability that N galaxies occur in a conical volume whose projected radius on the sky is 0.2 rad. The dotted curve is the theoretical gravitational $f(N)$ distribution described in the text with $b = 0.71$. Observations of actual galaxies in the sky show they also have a distribution very similar to this theoretical curve.

lated Poisson state, evolve gravitationally to agree reasonably well with the observations and remain in agreement as they continue to evolve. In other words, they relax into the observed state and remain there rather than just pass through it. This may make them more aesthetically pleasing, although it does not guarantee they are correct. For example, some models in which galaxies have formed and clustered very recently may conflict with the uniformity of the cosmic microwave background.

Gravitational theory predicted the observed form of $f(N)$ given earlier for relaxed statistically homogeneous clustering of point masses in a slowly expanding universe. The value of b is essentially the ratio of gravitational correlation energy (representing departures from a uniform Poisson distribution) to the kinetic energy of galaxies' peculiar velocities (representing departures from the Hubble flow). Computer experiments such as the example in Fig. 1 show that this relaxed state is a very good description for universes which start with no or little large-scale correlation, have $\Omega_0 \approx 1$, and have expanded by more than several times their initial radius. As differences with these conditions become greater, agreement with the observed $f(N)$ decreases. These conditions also lead to two-point correlation functions in reasonable agreement with the observations provided, for example, that clustering started at redshift $z \approx 8$ in $\Omega_0 = 1$ models and $z \approx 30$ in $\Omega_0 = 0.1$ models. Therefore, gravitational clustering starting from fairly simple initial conditions seems likely to account for the objective statistical evidence now available. These include large underdense regions and filamentary structures, some of which could have formed just by chance concentration of independently clustering regions or their boundaries. When more subtle statistics are developed further and related to dynamical evolution, perhaps they will reveal clear evidence for other processes such as primordial explosions, or large-scale initial structures.

Additional Reading

Hut, P. and MacMillan, S., eds. (1986). *The Use of Supercomputers in Stellar Dynamics*. Springer-Verlag, Berlin.

Itoh, M., Inagaki, S., and Saslaw, W. C. (1988). Gravitational clustering of galaxies: Comparison between thermodynamic theory and N-body simulations. *Ap. J.* **331** 45.

Oort, J. H. (1983). Superclusters. *Ann. Rev. Astron. Ap.* **21** 373.

Peebles, P. J. E. (1980). *The Large-Scale Structure of the Universe*. Princeton University Press, Princeton, NJ.

Rubin, V. C. and Coyne, G. V., eds. (1988). *Large-Scale Motions in the Universe*. Princeton University Press, Princeton, NJ.

Saslaw, W. C. (1987). *Gravitational Physics of Stellar and Galactic Systems*. Cambridge University Press, Cambridge.

See also **Clusters of Galaxies; Galaxies, Formation; Superclusters, Dynamics and Models.**

Cosmology, Cosmic Strings

Neil Turok

One of the most active areas of current research in physics and astronomy is the search for a theory of the formation of structure in the universe.

Historically, this is a result of the success of two different theories and the attempt to combine them. In astronomy, the Hot Big Bang model of the universe has three big successes. It successfully explains the expansion of the universe, the relic microwave background radiation, and the abundances of the light elements today. The weakness of the standard Hot Big Bang model is that it says nothing about how structure in the universe (galaxies and cluster of galaxies) could have originated.

In high-energy physics, the idea that the underlying theory of particles and their interactions has a high degree of symmetry which is broken at low energies forms the basis of the Weinberg–Salam model of the electroweak interactions. Over the last decade many predictions of this model have been confirmed, the discovery of the W and Z particles being the most dramatic. Based on the idea of symmetry breaking, theories which unify all the forces except gravity (grand unified theories), and theories including gravity (superstring theories) have been developed. Unfortunately, at present there are many different theories, and few ways of testing them.

One idea, which emerged from particle physics in the early 1980s, was that the same process which broke the symmetry between the particles and forces might break the spatial symmetry of the universe, producing the structure we see today. This is physically a very reasonable idea. In fact, similar processes happen in everyday substances. Most liquids are quite homogeneous and isotropic, which is not surprising, because there is nothing in the description of atoms and their interactions that singles out a particular direction or place in space as different from any other. Cool the liquid, however, and it freezes. The crystal structure of the

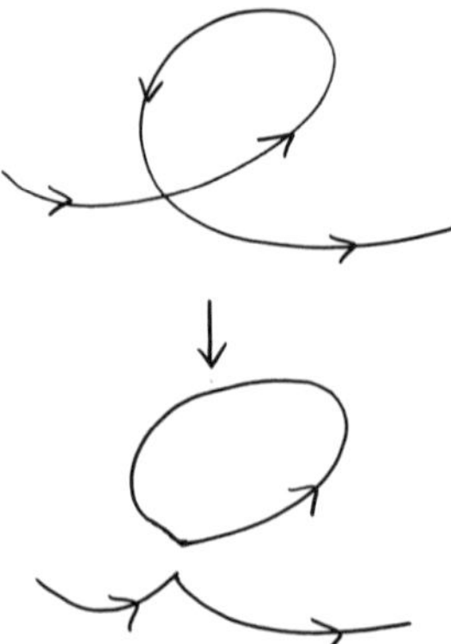

Figure 1. A long string intersects itself and reconnects the other way, breaking off a loop. In unified theories both directional strings (shown here) and nondirectional strings are predicted.

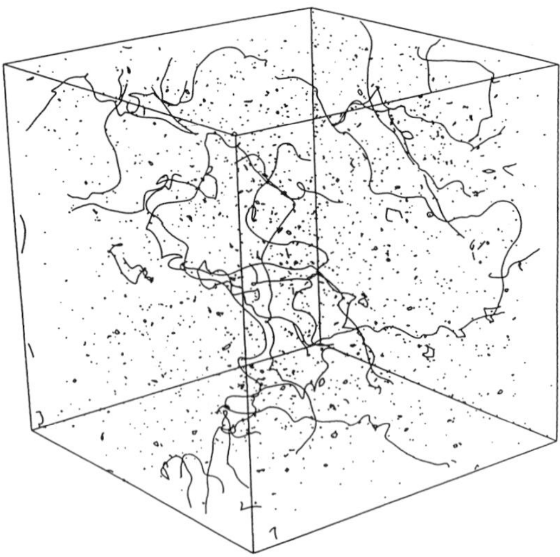

Figure 2. A cubical box of cosmic strings evolved on a computer by A. Albrecht and the author. To see the perspective in the picture, note that the vertical edge of the box to the center left is nearest you; the edge to the center right is furthest from you. The strings oscillate at speeds close to the speed of light, and reconnect the other way if they cross.

solid picks a particular direction; but in different regions different directions are chosen. The result is that (unless the process happens very slowly) the solid is formed full of defects where there is a mismatch between neighboring crystalline regions.

Cosmic strings are very similar to these defects. They are predicted to occur by some grand unified theories and superstring theories. In the very early universe, at high temperature the fields in these theories are random, just as the atoms of a liquid. The density is quite uniform. As the universe cools below a certain temperature, some of the fields "freeze." As they do so, defects are formed, just as in solids.

In different theories, the defects may be at points, along lines, or in sheets. Cosmic strings are the linelike defects: They have special properties which make them well suited to forming structure later in the universe. For topological reasons they cannot have ends; they must form closed loops or continue on forever. The only way to change the length of a string is if it crosses itself and reconnects the other way (Fig. 1), chopping off a loop. This means that if one starts with some strings which wander right across the universe, there is no way to get rid of them completely. At best one can progressively chop more and more of the long string off into loops, and the loops can then radiate away (see Fig. 1). Thus some of the strings formed at very early times (around 10^{-34} s after the Big Bang in most models predicting strings) survive right up to today.

Cosmic strings are very thin: approximately 10^{-15} the radius of a proton! But they are also very massive. 1 meter of cosmic string weighs approximately 10^{20} kg. This is not quite enough for them to form black holes: The dimensionless number measuring their coupling to gravity is $G\mu/c^2$. Here G is Newton's constant, μ is the mass per unit length of the string, and c the speed of light. This would have to be of order unity for the strings to form black holes. Typically, in grand unified theories predicting cosmic strings, one finds $G\mu/c^2 \approx 10^{-6}$.

When cosmic strings form, some of the string is in the form of closed loops. Most, however, is in the form of very long strings, which wander right across the universe. They are infinite if the universe is infinite, or wrap right around the universe and close back on themselves if it is finite. The universe is filled with a random, tangled network of strings. The motion of the string is quite complicated. However, a remarkable feature of it is that it does not depend at all on the mass per unit length of the string. The tension of the string equals its mass per unit length times c^2, so waves propagate on the string at the speed of light. This makes the cosmic string theory very predictive: The distribution of the strings is fully specified at any time.

As the universe expands, the long strings chop themselves up into loops. This is a complicated process, but is well described by a simple scaling theory. According to this theory, at any time after the network is produced it has a characteristic scale, and the long strings look like random walks with this scale. The long strings continually chop off a distribution of loops, characterized by this scale. Furthermore, this scale grows linearly with time. The result is that the total density in a long string remains a fixed fraction (approximately 10^{-4}) of the total density in the universe from the time when strings form right up to the time when density fluctuations start to grow around them, at around 10^{12} s.

Computer simulations such as those shown in Fig. 2 have confirmed the scaling theory, and have led to predictions of the distribution of the strings as the universe evolves. Sharpening these predictions is a difficult task: The biggest computers can only evolve a network for a factor 100 or so in time, and one must evolve them from 10^{-34} s to the time when matter starts clustering around them, about 10^{12} s, in order to compare to observations. So developing an analytic understanding of the evolution of the network is an important area of current research.

In the simplest picture of how structure forms around the strings, one simply associates one loop of string with one object. A loop 10 pc long has a mass of 10^8 solar masses, and accretes a galaxy mass by today (remember the growth factor of 10^4 mentioned earlier). Likewise a loop of 10 kpc accretes 10^{15} solar masses, the mass of a cluster of galaxies. Of course, as it clusters matter around it, it also clusters the loops accreting galaxies.

Whilst the mass of the accreted objects depends on μ, the *pattern* in which the galaxies and clusters are laid down does not (this is only true on large scales, where nonlinear effects are not important). Thus the accretion pattern produced by the strings is a clear prediction with no adjustable parameters. Remarkably, the pattern formed by cosmic string loops closely matches the observed pattern of giant galaxy clusters on the sky. Unfortunately, there are still large uncertainties as to whether the loop distribution measured in string simulations is the correct scaling distribution. More detailed calculations of large-scale structure produced by cosmic strings are only just beginning: There are hints that the "wakes" produced by long strings do produce galaxy "sheets," "bubbles," and "filaments," just as the observations indicate.

The most exciting aspect of the theory is that there are other completely independent ways of observing the effects of cosmic strings. The most direct way is in the pattern of anisotropy produced in the observed microwave background radiation. If a light

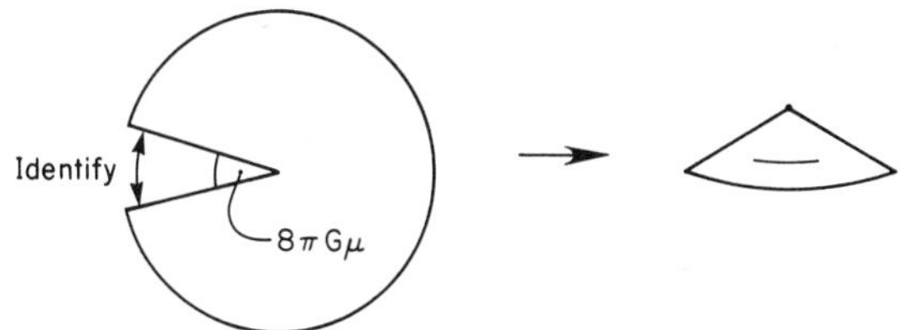

Figure 3. The space-time around a stationary straight string is conical: If one removes a segment with opening angle $8\pi G\mu/c^2$ from a flat piece of paper and sticks the two cut edges together, one makes a cone. Light rays passing on either side of the string (the apex of the cone) follow straight lines on the original flat piece of paper and so are bent towards each other. This leads to the gravitational lensing effect discussed in the text.

ray passes behind a moving string on its path to us, it is shifted towards the blue relative to a ray passing in front of the string. This is a sort of "gravitational slingshot" effect. The result is that if cosmic strings exist, and observations of the microwave background radiation become sensitive to 1 part in 10^5 (which they are close to doing), one should see stripes in the temperature pattern on the sky! This signal is quite unambiguous and would leave little doubt that cosmic strings actually exist.

The second direct effect is gravitational lensing. Light is bent as it passes on either side of a string (Fig. 3). Consequently, a galaxy observed behind a string would be seen as a double image. There are several "lensed" galaxies that have been seen so far, but there is no firm evidence that a cosmic string is responsible: A massive galaxy or dust cloud can produce the same effect.

The third test is the gravity wave background. The loops chopped off of the network radiate away into gravity waves: This process is actually crucial to the consistency of the theory because otherwise the small loops would build up and eventually come to dominate the density of the universe. The gravity wave background produced by the radiation from loops has been calculated, and amounts today to 1 part in 1000 of the density of the microwave background density. Amazingly, there are stringent limits on even this low a density in gravity waves. They come from accurate timing measurements on a millisecond pulsar. Basically, a gravity wave passing through us would move us relative to the pulsar, and cause the frequency of the signals from the pulsar to be Doppler shifted. A random background of gravity waves would therefore lead to "noise" in the pulsar timing data. After various corrections for the evolution of the pulsar, which are believed to be well understood, no such effect is observed. The present status is that the simplest cosmic string theory, explained earlier, is very close to being ruled out by these measurements.

Cosmic strings do not have to be so simple. It has also been found that in some simple grand unified theories the cosmic strings are actually superconducting—they can carry enormous electric currents (10^{21} A!) with zero resistance. This leads to a number of fascinating possibilities. Electric currents can build up on such strings via a "dynamo" effect, and the strings then build up a large magnetic field around them. This leads to a number of new ways for observing strings directly, and to a variety of novel astrophysical phenomena.

There are other equally speculative theories of the origin of large-scale structure in the universe. The virtue of the simplest cosmic string theory is that it makes quite clear testable predictions. If it is wrong, we shall know soon.

Additional Reading

Albrecht, A., Brandenberger, R., and Turok, N. (1987). Cosmic strings and cosmic structure. *New Scientist* **114** (No. 1556) 40.

Albrecht, A. and Turok, N. (1989). Evolution of cosmic string networks. *Phys. Rev. D* **40** 973.

Kibble, T. W. B. (1976). Topology of cosmic domains. *J. Phys. A* **9** 1387.

Vilenkin, A. (1985). Cosmic strings and domain walls. *Phys. Rep.* **121** 263.

Turok, N. (1989). Phase transitions as the origin of large scale structures in the universe. In *Particle Physics and Astrophysics: Current Viewpoints*, H. Mitter and F. Widder, eds. Springer-Verlag, Berlin.

See also **Cosmology, Clustering and Superclustering; Cosmology, Inflationary Universe; Gravitational Lenses; Gravitational Radiation.**

Cosmology, Cosmochronology

Donald D. Clayton

Cosmochronology may be defined as the scientific attempt to determine the age of the universe by determining the ages of some of its key components. In an expanding universe one may expect that galaxies formed some 10^9 yr after the Big Bang. So the age of the universe can be inferred from the ages of galaxies. The age of our own Milky Way galaxy is in fact the objective of almost all methods, because the Galaxy is the only cosmologically significant object on which astronomers can obtain a large amount of detailed data. A correct model of the universal expansion requires that the age of the Galaxy fit sensibly to the inverse of the Hubble constant. So, for example, if $q = \frac{1}{2}$, a sensible galactic age would be

$$\text{galaxy age} = \frac{2}{3}\frac{1}{H_0} - 10^9 \text{ yr},$$

where the last term, which is about 5–15% the first term, depending on the correct value of the Hubble constant H_0, represents the estimate of the time required to form galaxies.

Clearly we hope to have several methods to independently determine the age of the Galaxy. There are in fact four reliable methods:

1. The ages of the oldest clusters of stars (globular clusters).
2. The ages of the oldest white-dwarf stars (cooling times).
3. The ages of galactic solar-type stars in conjunction with the astronomically observed concentration of radioactive thorium in their surfaces.
4. Radioactivity age of the elements in the solar system.

This article will detail only the "nuclear cosmochronologies," numbers 3 and 4. However, for the sake of completeness, the results at present of the first two methods would seem to be:

1. The oldest globular clusters appear to be 15–17 × 10^9 years old, although ages outside this range cannot strictly be excluded at the present time.
2. The oldest white dwarfs, which are also the faintest white dwarfs, appear to be 9–11 × 10^9 years old, although ages outside this range may not be strictly excluded. This is a relatively new method having good future prospects.

The reader should be warned that the two ages above need not agree. In the first place, the globular clusters are distant objects that are probably as old as the Galaxy itself, whereas the very faint white dwarfs can be studied only in the galactic disk, near the position of the Sun, which could be a considerably later aspect of galactic structure than the early phase that formed globular clusters. In the second place, both age estimates involve formidable observing problems for astronomers, plus formidable theoretical problems concerning the physical structure of the stars studied.

For comparison's sake the nuclear cosmochronologies currently seem to yield similar ages:

3. The inferred extent of thorium decay in the oldest solar-type stars in the neighborhood of the Sun appears to indicate that they are $9–14 \times 10^9$ years old.
4. The radioactive dating techniques suggest that the chemical elements in the solar system are $10–15 \times 10^9$ years old.

These last two ages also apply to the formation of the galactic disk structure near the Sun rather than to the Galaxy itself. From even this simple survey it can be seen that no unique age can be ascribed to the Galaxy; but neither is there any glaring or irresolvable conflict comparable to the controversy over the Hubble constant itself.

RADIOACTIVITY AGES

Use of radioactivity to date events has a venerable history. It has measured, for example, ages of terrestrial fossils, ages of terrestrial and lunar rocks, the age of the Earth itself and of meteorites that fall onto the Earth, and, of special interest here, the age of the chemical elements. This last application is conceived for a universe that began with only hydrogen and helium and that has fused the heavier elements from them as a by-product of the thermonuclear evolution of the interiors of stars. If the heavy elements were synthesized within stars, it becomes sensible to ask when that occurred.

The property of radioactivity that enables these applications is that all radioactive decays are quantum mechanical transitions, all of which occur probabilistically. For each radioactive nucleus there exists a transition probability λ, which defines the chance that the nucleus will radioactively decay to its daughter nucleus during the next small time step. Unlike things of direct human experience, that probability is independent of how long the radioactive nucleus has already existed. That is, radioactivity as a process has no memory. Formally, one writes that the probability for decay in the next infinitesimal time interval dt is equal to the product $\lambda\, dt$, independent of the age of the radioactive nucleus. But each radioactive nucleus has a different transition probability λ.

A direct consequence is that if a large number $N(t)$ of radioactive nuclei exist at some time, the number of them that will decay in the next interval dt is given by the product $dN = -N(t)\lambda\, dt$, where the minus sign means a "reduction in the number." From techniques of integral calculus follows the famous law of exponential decay,

$$N(t) = N(t_0)e^{-\lambda(t-t_0)},$$

where t_0 is the initial time when the number was $N(t_0)$. One says that if there exist no sources of production of additional radioactive nuclei, the number of each declines exponentially. The famous concept, half-life, can be defined as the time required for an exponentially declining population to be halved. Its value $t_{1/2} = \ln 2/\lambda$. The value of each half-life is determined by measurement of a population of identical radioactive nuclei. It is intellectually stimulating that a fundamental quantum process without memory can be used as a clock by counting the population. There is no way of predicting when a single radioactive nucleus will decay.

The first application to nuclear cosmochronology was advanced in 1929 by Lord Rutherford (Ernest Rutherford), famous pioneer of nuclear radioactivity. Shortly after the determination that uranium has two naturally occurring radioactive isotopes, ^{235}U and ^{238}U, having different half-lives $t_{1/2}(235) = 0.704$ Gyr and $t_{1/2}(238) = 4.47$ Gyr (where Gyr $= 10^9$ yr), and having the lighter shorter-lived isotope less abundant in the ratio $^{235}U/^{238}U = 0.00725$, Rutherford suggested that they had initially been produced in equal abundance but that time had caused a much greater decline of the ^{235}U population. Application of this argument gives an age of 6 Gyr for uranium, about 1.5 Gyr older than the Earth itself; however, the argument is based on incorrect assumptions about the nature of the astrophysical system. Nonetheless, it conveys the history and spirit of radioactivity ages. In a nutshell Rutherford's observation was that radioactive nuclei could not always have existed. He also pioneered the idea of daughter chronologies—that the age can be calculated from the ratio of abundances of the radioactive parent to that of the stable daughter to which it decays. In geochronological applications it is necessary to know independently something of the initial abundances within the object under study (a rock, an antler, the Earth, etc.), because it is the formation time of that object that is sought. In nuclear cosmochronology one seeks the age of the elements themselves. The independent information needed for that problem is knowledge of *what the natural abundances would have been were the parents stable instead of radioactive*. The ratio of the abundance of ^{238}U, for example, to the abundance it would have had were it stable is called the *remainder*, $r(^{238}U)$. Values of the remainders must be determined from nucleosynthesis theory, independent of the chronology. This determination consists of calculating the production rate of a radioactive nucleus relative to those of stable nuclei. Those remainders, along with a similar determination from nucleosynthesis theory of what the daughter abundances would have been were their parents stable, contain the age information.

The logistical process can be summarized schematically:

$$(\text{natural abundances}) + (\text{nucleosynthesis theory}) \rightarrow (\text{remainders}),$$

$$(\text{remainders}) + (\text{astrophysical model}) \rightarrow (\text{age of elements}).$$

The second of these two equations notes explicitly that the age calculated is model dependent. General astrophysical theory must provide a model setting for the production of new nuclei and for their incorporation into the matter destined eventually for our solar system, for it is the abundances in the solar system that are the basic data.

The useful radioactive nuclei are given in Table 1 in decreasing order of half-life. Also given there are the stable daughters to which they decay and recent evaluations from nucleosynthesis theory of relevant production ratios. Cosmoradiogenic chronologies, those based on daughter abundances, list the remainder at solar birth of the parent.

The first five table entries occur naturally in the Earth and are the useful ones for determining galactic age. The last five entries are "extinct radioactivities," detectable only by their radiogenic daughters in meteorites, and which yield interesting constraints on the nucleosynthesis rate shortly prior to solar formation. For the extinct radioactivities the entry gives the abundance ratio inferred to have existed in the early solar system. This information is more useful for the general astrophysics of solar system formation than for the age of the elements, for which their half-lives are too short.

Table 1. Useful Radioactive Nuclei

Nucleus	*Half-life*	*Daughter*	*Production Ratio*
^{87}Rb	48.0 Gyr	^{87}Sr	$r(87) = 0.92^{+0.05}_{-0.07}$
^{187}Re	42.8 Gyr	^{187}Os	$r(187) = 0.90^{+0.01}_{-0.03}$
^{232}Th	14.1 Gyr	^{208}Pb	$p(232)/p(238) = 1.65 \pm 0.15$ $r(232)$ unmeasurable
^{238}U	4.47 Gyr	^{206}Pb	$r(238)$ very uncertain
^{235}U	0.704 Gyr	^{207}Pb	$p(235)/p(238) = 1.34 \pm 0.2$ $r(235) = 0.12 \pm 0.05$
^{244}Pu	82 Myr	(^{232}Th)	$^{244}Pu/^{238}U = 0.007$
^{129}I	16 Myr	^{129}Xe	$^{129}I/^{129}Xe = 1.5 \times 10^{-4}$
^{107}Pd	6.5 Myr	^{107}Ag	$^{107}Pd/^{107}Ag = 3 \times 10^{-5}$
^{53}Mn	3.7 Myr	^{53}Cr	$^{53}Mn/^{53}Cr = 3 \times 10^{-4}$
^{26}Al	0.75 Myr	^{26}Mg	$^{26}Al/^{26}Mg = 4 \times 10^{-5}$

CHEMICAL EVOLUTION OF SOLAR NEIGHBORHOOD

Certain features of chemical evolution in the solar neighborhood require special emphasis within the framework of cosmochronology. First, the presolar nucleosynthesis of the elements is now believed to be distributed in time, rather than occurring at one point in time as Rutherford initially assumed. This requires that the presolar gas concentration of a radioactive parent be governed by a production term as well as the decay rate term

$$\frac{dN}{dt} = p(t) - \lambda N.$$

Second, these $N(t)$ are not total interstellar abundances, which depend on the variable and unknown total mass of interstellar gas $M_G(t)$, but are rather concentrations within solar matter (normally defined per 10^6 Si atoms). Thus $N_A(t) = X_A(t)$ in usual astronomical nomenclature. Third, the foregoing requires that $p(t)$ be not actually the galactic production rate, but technically the *birthdate spectrum* of only those nuclei that will ultimately appear in the solar system where the measurements will be made. Furthermore, it is the birthdate spectrum of the coproduced stable nuclei, rather than of the surviving radioactive ones. Fourth, the conceptual requirement of a birthdate spectrum of the interstellar nuclei when (and where) the Sun formed requires that the total galactic production rate in the solar neighborhood be adjusted for nuclei locked up within stars and for the interplay between the rate of growth of total mass in the solar neighborhood to the total rate of star formation there. If the accretion of new disk mass at the Sun's galactocentric distance continues well after the birth of globular clusters, the mean age of solar system nuclei may be substantially less than half the age of the globular clusters. In other words, a rather complete description of the growth and chemical evolution of the galactic disk is a prerequisite for relating the mean age of solar nuclei to the actual age of the Galaxy. It is also necessary to decide whether the solar composition is truly typical of a well-mixed interstellar medium. Some indications (e.g., oxygen richness) exist that suggest that the Sun may in fact be atypical. If one assumes the interstellar gas in a galactic annulus containing the Sun to be well mixed, that composition is described by a set of coupled differential equations relating the star formation rate to the mass and composition of the gaseous medium. This is an elaborate computer program. However, if the star formation rate is taken to be proportional to the mass of gas in the same given galactic annulus, instructive analytic solutions of the entire problem exist (see Additional Reading). These analytic solutions reveal the interplay among the conventional astronomical observables of chemical evolution, the rate of mass growth of the solar annulus, and the nuclear cosmochronologies for solar material.

A unique method for nonsolar matter finds its natural context also within the chemical evolution of the solar neighborhood. It stems from the observed concentration of Th in solar-type stars (G dwarfs), which can themselves have any age up to 15 Gyr. Astronomical challenges are the measurement and interpretation of the Th line strength and that of a stable nucleus (e.g., Nd) in these faint dwarfs and a calculation of the ages of those stars from their luminosities and colors. The latter is quite uncertain. Given those quantities, however, the method works as follows. The Th/Nd ratio declines with time in the evolving interstellar medium, so that later-forming G dwarfs form with smaller initial ratios. After the stars form, however, the Th/Nd ratios in their atmospheres decline even faster than that in the interstellar gas, because the latter ratio is held up by fresh nucleosynthesis whereas the Th decays exponentially in the stellar atmospheres. As a consequence, the trend of Th/Nd in their atmospheres *today* versus the ages of the stars gives the age of the galactic disk in the solar neighborhood. This method currently suggests that age to be 9–14 Gyr; however, major uncertainties clouding this method are the actual ages of the dwarf stars, the line strength interpretation, and the relative rates of nucleosynthesis of Th and the comparison element. The latter is an important unsolved problem in the chemical evolution of the Galaxy, which is itself a general topic that has often been underrated in cosmochronological calculations.

RECENT RESULTS

Each method of cosmochronology continuously evolves. New observations and new developments in the theory of nucleosynthesis continuously improve knowledge at the same time that they illuminate the uncertainties. At the present time this writer judges the best three nuclear cosmochronological techniques to be the Th/Nd ratio in G dwarf stars, the ^{238}U/^{232}Th ratio in the solar system, and the radiogenic ^{187}Os. The status of the last two is summarized below.

^{238}U/^{232}Th The production ratio in *r*-process nucleosynthesis is dominated by the relative numbers of nonfissioning radioactive parents of these nuclei. ^{238}U has 3.1 parents ($A = 238$, 242, 246, and 10% of 250), whereas ^{232}Th has 5.9 parents ($A = 232$, 236, 240, 244, 91% of 248, and 97% of 252). If the fission branches in the *r*-process are identical to those in the laboratory, and if each parent had equal intrinsic abundance, then the production ratio is $p(232)/p(238) = 1.89$. Nucleosynthesis calculations, though uncertain, suggest a slightly smaller production ratio. The relative abundances of U and Th must be obtained from meteorites, where differing ratios are found in differing meteoritic types. If the abundance ratio from Type-I carbonaceous chondrites is adopted, the ratio of remainders for these two long-lived nuclei at the time the solar system formed was $r(238)/r(232) = 0.73 \pm 0.07$. That is, the fraction of ^{238}U nuclei surviving until the formation of the solar system was 73% of the fraction of longer-lived ^{232}Th that survived the same history.

To see what this remainder ratio requires for an age for the solar neighborhood, one must take an astrophysical model. A popular simple model is to assume gradual conversion of initial gas to stars at a rate proportional to the remaining gas. In that case the disk age is 10 ± 1.5 Gyr. However, if infall increased the disk mass by a factor of 2 or more over an infall epoch greater than 3 Gyr, the same remainder ratio implies a disk age between 10 and 15 Gyr. To each model of galactic evolution corresponds a "best range of ages" for the remainder ratio. To obtain a more specific answer requires a stricter specification of the nature of the history of the galactic disk. That in turn is a problem for conventional astronomy, which thereby strongly impacts what might at first glance seem to be a nuclear technique rather than an astronomical one.

^{187}Re/^{187}Os With careful neutron cross-section measurements and the application of *s*-process theory it has been possible to show that the cosmoradiogenic portion of ^{187}Os, defined as that part owing its existence to ^{187}Re decay during presolar history, is 47 ± 5% of total ^{187}Os. That argument, first advanced by this writer in 1964, places quantitative bedrock under one of the powerful methods of determining the time of the beginning of disk nucleosynthesis. The major impediment to its application lies in uncertainty over the appropriate ^{187}Re half-life. That uncertainty arises from the expectation that the isotope decays faster within stars than in the laboratory. The latest study of this suggests that the half-life is reduced from 62 to 44 Gyr by this effect (which is in itself novel). With these assumptions, however, the simple constant-mass model of the solar neighborhood gives a galactic age 12^{+4}_{-2} Gyr, whereas models allowing for several-gigayear growth of the disk mass yield 17 ± 3 Gyr. Again, the detailed age depends upon the detailed model.

Summary of Methods

Table 2 lists the cosmochronological techniques described in this entry. Both the most likely result and a range estimate are given. A

Table 2. Cosmochronological Estimates of Galactic Age.

Method	*Best Value (Gyr)*	*Major Uncertainty*
Globular clusters	15–17	Evolution without mass loss?
Th in G dwarfs	9–14	Ages of the G stars. Evolution of interstellar Th/Nd ratio.
Faintest white dwarfs	9–12	Luminosity function of white dwarfs. Cooling rate at low luminosity.
$^{235}U/^{238}U$	8–18	Production ratio. History of disk growth.
$^{238}U/^{232}Th$	9–16	Production ratio. History of disk growth. Th/U abundance ratio.
$^{187}Os/^{187}Re$	11–18	^{187}Re decay rate in stars. History of disk growth. ^{187}Os excited state.
$^{87}Sr/^{87}Rb$	8–25	Neutron cross section ratio. Branching in the *s*-process.
$^{207}Pb/^{235}U$	7–20	*r*-process production $207 < A < 235$. Pb neutron cross sections. Pb abundance.
$^{207}Pb/^{206}Pb$	9–20	*r*-process production $207 < A < 235$. Pb neutron cross sections, abundance.

third column gives the major uncertainties that plague a precise result from that method. Those with the smallest age spread are clearly more reliable. That estimate of uncertainty is somewhat subjective, representing this writer's evaluation of the scientific literature. It will be clear that no definite answer is yet available; but the most comfortable compromise may be a Galaxy that began about 15 Gyr ago and a solar neighborhood disk that had matured to the point of having most of its mass in place about 12 Gyr ago. The future holds plenty of room for improvements and/or conflicts.

Additional Reading

Butcher, H. R. (1987). Thorium in G-dwarf stars as a chronometer for the galaxy. *Nature* **328** 127.

Clayton, D. D. (1988). Nuclear cosmochronology within analytic models of the chemical evolution of the solar neighborhood. *Mon. Not. R. Astron. Soc.* **234** 1.

Fowler, W. A. and Hoyle, F. (1960). Nuclear cosmochronology. *Ann. Phys.* **10** 280.

Sandage, A. (1986). The population concept, globular clusters, subdwarfs, ages, and the collapse of the Galaxy. *Ann. Rev. Astron. Ap.* **24** 421.

Winget, D. E., Hansen, C. J., Liebert, J., Van Horn, H. M., Fontaine, G., Nather, R. E., Kepler, S. O., and Lamb, D. Q. (1987). An independent method for determining the age of the universe. *Ap. J. Lett.* **315** L77.

See also **Cosmology, Observational Tests; Galactic Structure, Globular Clusters; Galaxy, Chemical Evolution; Star Clusters, Stellar Evolution; Stars, White Dwarf, Structure and Evolution.**

Cosmology, Galaxy Formation

Barbara Ryden

Cosmology, the study of the evolution of the universe as a whole, is closely linked with the study of the evolution and formation of galaxies. The properties of galaxies today depend, in part, on the physical conditions in the very early universe. Various cosmological models for the evolution of the universe make definite predictions of the masses, binding energies, angular momenta, spatial distribution, and other properties of galaxies. Comparison of the predicted with the observed values for galaxies can act as a discriminant among competing cosmological models.

The general cosmological framework in which the formation of galaxies must be explained is the standard Hot Big Bang model. Edwin P. Hubble's discovery of a distance-redshift relation for galaxies is evidence for a uniformly expanding universe. Four decades after Hubble's discovery of the expansion of the universe, Arno Penzias and Robert Wilson discovered the cosmic microwave background (CMB), evidence for the Hot Big Bang; the universe, initially very hot, is cooling as it expands. To create galaxies in an expanding universe, there must be some force, working against the universal expansion, which pulls the mass of a galaxy together into a bound lump. In most theories of galaxy formation, this force is taken to be gravity, working on an initial density perturbation. Limits on the early height of the perturbations are imposed by the observed smoothness of the CMB. Observations of the CMB show that its temperature fluctuations $\Delta T/T$ are smaller than a few times 10^{-5} over all angular scales smaller than the dipole. To achieve such a smooth microwave background, the universe must have been quite homogeneous at the time of recombination (at a redshift $z_{\rm rec} \sim 1000$), when the universe became transparent to radiation. A successful theory of galaxy formation must thus explain the transition between the smooth universe at the time of recombination and the lumpy universe (where the lumps are galaxies) seen today.

GRAVITATIONAL THEORY

Suppose that the universe has a density ρ which varies throughout space, but which has a mean value ρ_b. The gravitational instability scenario for galaxy formation states that galaxies form in regions where ρ is larger than ρ_b; that is, where the overdensity $\delta = (\rho - \rho_b)/\rho_b$ is sufficiently greater than zero. These regions will collapse under their own self-gravity; if they are then dense and hot enough for the gas within them to dissipate energy by bremsstrahlung and radiative recombination, the gas will fall to the center of the collapsed region, fragmenting to form stars, and thus forming the visible portions of galaxies that we see today.

The rate at which the density perturbations δ grow with time under the influence of gravity was first computed by James Jeans. He found that in a static, nearly uniform universe, perturbations longer than a critical length, known as the Jeans length, grow exponentially with time. Perturbations smaller than this size are stabilized by the pressure gradients which build up during the attempted collapse.

Jeans assumed a static universe; in an expanding universe, the growth of the perturbations is slowed by the universal expansion. If the universe expands, perturbations grow at most as a power of the time t, instead of exponentially. In a flat ($\Omega = 1$) matter-dominated universe, for instance, $\delta \propto t^{2/3}$. The growth of the perturbations can thus be followed from their initially small values to the time when $\delta \approx 1$, when they collapse to form galaxies. The question remains, however: What was the cause of the initial small density perturbations?

Gaussian Perturbations

In one promising theory, the density perturbations are formed during an inflationary era in the early universe, and take the form of an isotropic homogeneous gaussian density field. Inflation was originally proposed as a means of making the universe flat (i.e., setting the density parameter Ω extremely close to one) and of making the universe homogeneous on large scales. Although inflation flattens out previously existing perturbations, during the process of expansion, it stamps a new set of perturbations on the universe. Inflation occurs when a field ϕ undergoes a phase transition from a false vacuum to a true vacuum of lower energy; during the transition, the universe expands exponentially. As a result, microscopic quantum fluctuations in ϕ are expanded, along with the general expansion of the universe, to cosmological length scales. The perturbations in the field ϕ lead, via distortions of the space-time metric, to perturbations δ in the mass-energy density of the universe. As a necessary consequence of the quantum origins of the perturbations, the density is a gaussian random field; that is, the probability distribution function for δ takes the form $f(\delta) \propto \exp(-\delta^2/2\sigma^2)$. The simplest inflationary scenarios predict a scale-invariant distribution of perturbation sizes. An alternate way of expressing this result is to say that the power spectrum $P(k)$ of the gaussian field δ has the Harrison–Zel'dovich form $P(k) \propto k$.

When the universe stops inflating (when it is $\sim 10^{-30}$ s old, in most inflationary scenarios), it enters a period when the majority of the energy of the universe is in the form of radiation. During this radiation-dominated period, density fluctuations with wavelengths smaller than the cosmological horizon size remain nearly constant in amplitude. Fluctuations which are larger than the horizon, however, are able to grow at the rate $\delta \propto t$.

As the radiation-dominated epoch continues, perturbations of longer and longer wavelength fall within the horizon and have their growth frozen. Thus, during this epoch, the shape of the power spectrum $P(k)$ is continuously modified. As the radiation redshifts away its energy, the massive particles in the universe contribute a larger and larger fraction of the mass-energy density of the universe. Finally, at a redshift $z_{eq} \approx 2.5 \times 10^4 \Omega h^2$ (h is Hubble's constant in units of 100 km s^{-1} Mpc^{-1}), the contributions of radiation and matter are equal, and the universe enters a matter-dominated phase, when all density perturbations inside and outside the horizon are free to grow. At z_{eq}, the power spectrum of the inflation-born perturbations, as modified during the radiation-dominated era, remains $P(k) \propto k$ at the smallest wave numbers (corresponding to long wavelengths), but falls off at the rate $P(k) \propto k^{-3} \log^2 k$ as k goes to infinity. The turnover in the power spectrum occurs gradually around wave number $k_{eq} = 0.5 \Omega h^2$ Mpc, corresponding to the horizon size at the redshift z_{eq}.

In standard cosmological models, the universe has been matter dominated from z_{eq} until the present. If Ω is presently nearly equal to one, a natural result of inflation, then most of the mass must be in the form of dark matter. The fate of density perturbations after z_{eq} depends on whether the dark matter is cold, having negligibly small streaming velocities, or is hot, having relativistic streaming velocities at the time when galactic-sized masses first enter the horizon.

Cold dark matter (or CDM) may consist of particles (e.g., photinos) which are massive enough to render their thermal velocities negligible, or of particles (e.g., axions) which are created with negligible velocities. In either case, the velocities are too small to smooth out density fluctuations, and the power spectrum $P(k)$ is unmodified by smoothing at large k; see Fig. 1 for the shape of the CDM power spectrum. A CDM-dominated universe has density fluctuations on all length scales from subgalactic sizes upward. In the CDM scenario, then, galaxies form by a hierarchical process of collisions and mergers; the smallest and densest objects collapse first, then subsequent collisions gradually build up objects the size of galaxies. In recognition of this process, the CDM scenario is referred to as a "hierarchical clustering," or "bottom-up" model. The CDM model has been extensively studied with analytic calculations and n-body simulations. The main successes of CDM are that it produces objects with the proper masses, density profiles, and angular momenta to match observed galaxies. The principal failures of CDM are that it fails to produce sufficient large-scale structure; it cannot match the observed correlation function for clusters of galaxies or the streaming velocities reported on scales as large as $\sim 60h^{-1}$ Mpc.

Hot dark matter (HDM), by contrast with CDM, smooths out the smallest density fluctuations by its free streaming velocity. The leading candidate for the HDM particle is a massive neutrino. If the neutrino has a mass of 30 eV, thus providing enough mass to

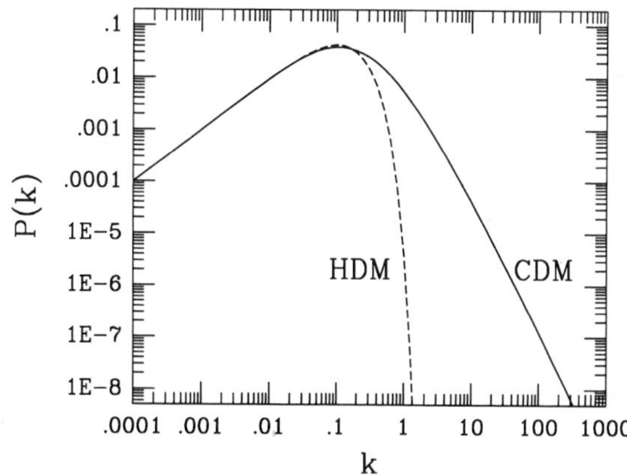

Figure 1. The power spectrum of primordial density perturbations in a universe dominated by CDM (solid line) or by HDM (dashed line). The comoving wavenumber k is measured in units of Ωh^2 Mpc^{-1}, as measured at the present time; the amplitude of the spectrum is in arbitrary units.

make $\Omega = 1$, then cosmological neutrinos will not cool sufficiently to have nonrelativistic thermal velocities until the universe has a comoving horizon size of 4 Mpc, containing a mass $\sim 10^{15}$ $M_\odot$, roughly that of a supercluster of galaxies. Perturbations smaller than this size will be smoothed away; the resulting HDM power spectrum is shown in Fig. 1. In the HDM scenario, the first objects which form are superclusters; these collapse first along their short axes to form "pancakes," which then fragment to form galaxies. For this reason, the HDM scenario is referred to as a "pancaking" or "top-down" scenario. The main success of HDM is that a top-down scenario more easily explains the observed streaming velocities and correlations on large scale. The principal, perhaps fatal, failure of HDM is that it requires galaxies to have formed very recently, at $z \lesssim 1$; as astronomers look deeper into space, they find galaxies at considerably earlier redshifts than this.

In order to evade the flaws of both the CDM and HDM scenarios, astrophysicists have speculated about universes filled with warm dark matter (intermediate between hot and cold in its properties), or with a mixture of hot dark matter and cold dark matter, or with unstable dark matter, in which heavy cold particles decay into light hot particles. All of these speculations are variations on the theme of gravitational collapse of gaussian density fields.

Cosmic Strings

Galaxies might form by the collapse of nongaussian density perturbations. One possible source of such perturbations are the topological defects known as cosmic strings. As the universe cools from its initial high temperature, symmetry breaking can result in the formation of stable topological defects, which have high energy density and may act as seeds for galaxy formation. Depending on the form of the symmetry, the topological defects can be monopoles (pointlike defects), cosmic strings (linear defects), or domain walls (planar defects). Monopoles and domain walls have undesirable side effects, cosmologically speaking; thus, attention has been focused on cosmic strings as a possible source of seeds for galaxy formation.

Analytic and numerical computations of the evolution of cosmic string networks indicate that at any given time after the strings form, there will be approximately one infinite string, reaching from horizon to horizon. In addition, the number of strings with radii in the interval $R \to R + dR$ will be $n(R)\,dR \sim R^{-5/2}\,dR$. These string loops of all sizes act as seeds for the gravitational accretion of matter. Computer simulations have been run of the accretion of mass onto cosmic strings embedded in a background of CDM,

HDM, or simply ordinary gas. Of these models, the one which most closely resembles the observed universe is the model which combines strings with hot dark matter. The strings provide the seeds for galaxies; the HDM provides the longer wavelength fluctuations which are required to explain the large-scale structure (on scales greater than 10 Mpc).

EXPLOSION THEORY

The difficulty of simultaneously explaining the existence of galaxies and of large-scale structure in gravitational collapse models has led to the consideration of models in which the mass destined to become galaxies is shoved into place by forces other than gravity. One such model is the explosion theory. If a large amount of energy is injected into a small region of the early universe—by a powerful explosion, for instance—then a blast wave will expand outward from that region. The blast sweeps up a shell of matter; this shell is gravitationally unstable, and fragments into galaxies. Thus, if these explosions went off at random positions in the early universe, today galaxies would exist on thin shells surrounding the voids which were evacuated by the explosion. In fact, this is what is seen; redshift surveys show a "bubbly" or "spongy" distribution of galaxies, containing voids which are typically $\sim 30h^{-1}$ Mpc across.

Thus, the explosion theory provides a plausible explanation for the observed large-scale distribution of galaxies. A drawback of the theory, however, is the large energies which it requires. Creating a void 30 Mpc across requires an explosion with energy $\sim 10^{64}$ erg. Procuring such a large energy requires exotic mechanisms, such as, for instance, a radiating superconducting cosmic string.

PRESENT STATUS: FUTURE TRENDS

None of the cosmological scenarios sketched above is completely satisfactory at describing the known properties of galaxies. The borderland between cosmology and the study of galaxy formation is, and will probably long remain, a fruitful realm for research. On one hand, further observations of galaxies will give us more clues about how they must have formed; on the other hand, theoretical advances in cosmology will hint at new mechanisms for making galaxies. In the end, a better knowledge of galaxy formation will improve our understanding of cosmology; a better knowledge of cosmology will improve our understanding of galaxy formation.

Additional Reading

Blumenthal, G. R., Faber, S. M., Primack, J. R., and Rees, M. J. (1984). Formation of galaxies and large-scale structure in the universe. *Nature* **311** 517.

Krauss, L. M. (1986). Dark matter in the Universe. *Scientific American* **255** (No. 6) 58.

Ostriker, J. P. (1988). Explosive origins of large-scale structures. In *Large Scale Structures of the Universe*, J. Audouze et al., eds. Reidel, Dordrecht, p. 321.

Rees, M. J. (1987). The emergence of structure in the universe: Galaxy formation and dark matter. In *Three Hundred Years of Gravitation*, S. W. Hawking and W. Israel, eds. Cambridge University Press, Cambridge, p. 459.

Silk, J. (1987). Galaxy formation: Confrontation with observations. In *Observational Cosmology*, A. Hewitt et al., eds. Reidel, Dordrecht, p. 391.

See also **Background Radiation, Microwave; Cosmology, Big Bang Theory; Cosmology, Cosmic Strings; Cosmology, Inflationary Universe; Dark Matter, Cosmological; Voids, Extragalactic.**

Cosmology, Inflationary Universe

Robert H. Brandenberger

The inflationary universe is a cosmological model in which there is a period during the very early universe when the volume of space expands exponentially. A period of inflation may arise when matter is described by particle physics (instead of by an ideal gas) in the early universe. Inflationary models provide potential solutions to some of the problems of standard cosmological models. They also provide a mechanism which produces energy density fluctuations which can seed galaxies and other structures in the universe.

MODELS

The inflationary universe may arise when one couples particle physics and classical general relativity, in particular when one describes matter in terms of fields. It is known that field theory describes the interaction of particles very well at high energies and temperatures when the ideal gas description of matter breaks down. Hence, the cosmological models obtained in this way will give a better description of the early universe than the classical models.

When describing matter in terms of fields, it is possible to obtain equations of state with negative pressure. For a scalar field (a field which does not change under Lorentz transformations) $\phi(\mathbf{x}, t)$, three terms contribute to the energy density ρ and pressure p—the kinetic energy density, the spatial gradient energy, and the contribution of a potential energy term $V(\phi)$:

$$\rho = \frac{1}{2}\dot{\phi}^2 + \frac{1}{2}R^{-2}(t)(\nabla\phi)^2 + V(\phi),$$

$$p = \frac{1}{2}\dot{\phi}^2 - \frac{1}{6}R^{-2}(t)(\nabla\phi)^2 - V(\phi).$$

Here, $R(t)$ is the scale factor of the universe. If $\phi(\mathbf{x}, t)$ is homogeneous and static and $V(\phi)$ is positive, the equation of state is $p = -\rho$. Hence, using the Einstein equations for a Friedmann universe, one obtains exponential increase of the scale factor (the de Sitter phase).

In some models, ϕ is homogeneous but time dependent. In this case, the equation of state is no longer $p = -\rho$. However, it is still possible for such a cosmological model to solve the homogeneity and flatness problem (see the following), provided $p < -\frac{1}{3}\rho$. This is the condition for $R(t)$ to increase more rapidly than the Hubble radius $H^{-1}(t)$, where

$$H(t) = \dot{R}(t)/R(t).$$

Models with $-\rho < p < -\frac{1}{3}\rho$ are said to give rise to generalized inflation.

Scalar fields arise in many particle physics models. They are often used to break gauge symmetries. In this case, a potential of the form of curve (a) in Fig. 1 is required. New inflationary universe models take such a potential and assume that at very

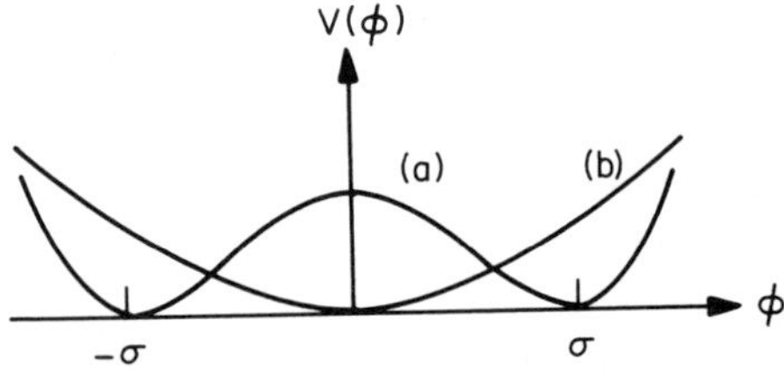

Figure 1. A sketch of the potential $V(\phi)$ postulated in new inflationary universe models (a) and for chaotic inflation (b).

early times $\phi(\mathbf{x}) \simeq 0$. $\phi(\mathbf{x})$ will slowly roll towards one of the minimum energy values $\pm\sigma$. If the potential is very flat, then it will take the field a long time to reach $\pm\sigma$. During most of the rolling phase, $p \simeq -\rho$ and the universe expands exponentially.

Constraints stemming from the magnitude of density perturbations produced during inflation requires the coupling constant which determines the flatness of $V(\phi)$ to be so small that the initial condition $\phi(\mathbf{x}) \simeq 0$ can no longer be justified. This is taken into account in the chaotic inflationary universe. In this case, either potential (curve a or b of Fig. 1) can be chosen. The part of the universe inside our horizon is assumed to stem from a region in space where initially $\phi(\mathbf{x})$ was homogeneous over a few Hubble volumes and in addition $\phi \gg m_{\text{pl}}$ (where m_{pl} stands for the Planck mass). In this case, $\phi(\mathbf{x})$ will slowly roll towards $\phi = \sigma$ (or $\phi = 0$ in the case of curve b in Fig. 1), its motion being damped by the expansion of the universe. Again, $p \simeq -\rho$ and inflation occurs.

HOMOGENEITY AND FLATNESS PROBLEMS

Inflationary models provide a potential explanation of the flatness of the universe and of its homogeneity on large scales. Homogeneity follows from the observed isotropy of the microwave background radiation to less than 1 part in 10^4. The flatness can be quantified by comparing the present energy density ρ with the energy density ρ_c for a flat ($K = 0$) universe. The ratio $\Omega = \rho/\rho_c$ is most likely between 0.1 and 1.

In classical cosmology, the isotropy of the microwave background has no causal explanation, because the radiation detected in different directions on the sky comes from patches of the last scattering surface, which at last scattering t_{rec} could not have been in causal contact. Similarly, in classical cosmology $\Omega = 1$ is an unstable fixed point of an expanding universe. Thus, having Ω close to 1 today implies that Ω had to be tuned to almost $\Omega = 1$ with an extreme accuracy at early times.

As illustrated in Fig. 2, in inflationary models the horizon expands exponentially during the period of inflation. Provided inflation lasts for more than a few Hubble expansion times $H^{-1}(t)$, the horizon at t_{eq} is larger than the surface at t_{rec} from which the microwave radiation reaches us.

Similarly, inflation solves the flatness problem. During the period of inflation, $|\Omega - 1|$ decreases exponentially. Thus, Ω in the range between 0.1 and 1 today no longer requires unnatural fine tuning in the very early universe.

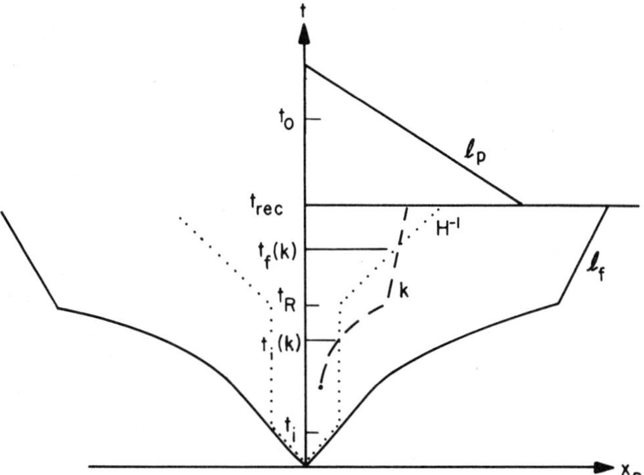

Figure 2. A sketch of the cosmology of inflationary universe models. The phase of exponential expansion lasts from t_i to t_R. Before and after, the universe is dominated by radiation. The horizon $l_f(t)$ expands exponentially in the de Sitter phase. This provides a solution of the homogeneity problem: at recombination t_{rec}, the horizon is much larger than the domain l_p for which we observe isotropy of the microwave background. Also shown in this sketch of physical distance x_p versus time t is how comoving scales k exit the Hubble radius at time $t_i(k)$ during the de Sitter phase and reenter at $t_f(k)$ during the period of radiation dominance. t_0 is the present time.

INFLATION AND DENSITY PERTURBATIONS

Inflationary models provide a possible mechanism for generating the primordial density perturbations which can lead to structure formation. The simplest inflationary models give a scale invariant spectrum of linear, adiabatic random phase perturbations. Scale invariant means that the root-mean-square mass perturbation on a fixed comoving scale has a value which is independent of the scale when measured at the time when the scale equals the Hubble radius.

The basic mechanism is illustrated in Fig. 2. All comoving scales originate inside the Hubble radius (causal horizon) during the de Sitter phase. Hence, it is possible to obtain a causal explanation of the origin of perturbations.

Quantum fluctuations in the de Sitter phase which arise on all comoving scales are frozen in as classical perturbations on scale k when the wavelength k^{-1} equals the Hubble radius at time $t_i(k)$. Between $t_i(k)$ and $t_f(k)$, the time when the wavelength "reenters" the Hubble radius, the mass perturbations grow according to linear gravitational perturbation theory and increase by a constant factor. Since the amplitude of the perturbations is k independent at $t_i(k)$, a scale-invariant spectrum is generated.

The amplitude of the spectrum of density perturbations depends on the coupling constants in the particle physics model. In order not to give a too-large amplitude, self-coupling constants in $V(\phi)$ typically must be tuned to be smaller than 10^{-12}. This is a serious problem which affects most inflationary models.

RECENT RESULTS

The inflationary universe has become a paradigm in search of a concrete realization. There have been attempts to obtain inflation in supergravity and superstring models, in theories with modified gravity or with extra dimensions. Extended inflation is a new attempt to use a dynamical Brans–Dicke scalar field to obtain generalized inflation.

Chaotic inflation is at present the most developed model. It predicts that on scales much larger than our present horizon, the universe will consist of many very different, almost disconnected miniuniverses.

In models in which more than one scalar field is important during the period of exponential expansion, it is possible to obtain deviations from scale invariance, isothermal perturbations, and even non-Gaussian fluctuations.

Additional Reading

Guth, A. and Steinhardt, P. (1984). The inflationary universe. *Scientific American* **250** (No. 5) 116.

Linde, A. (1984). The inflationary universe. *Rep. Prog. Phys.* **47** 925.

Linde, A. (1987). Particle physics and inflationary cosmology. *Physics Today* **40** 61.

Linde, A. (1989). *Inflation and Quantum Cosmology*. Academic Press, Boston.

Turner, M. (1985). The inflationary paradigm. In *Proceedings of the Cargese School on Fundamental Physics and Cosmology*, J. Audouze and J. Tran Thanh Van, eds. Editions Frontieres, Gif-sur-Yvette.

See also **Background Radiation, Microwave; Cosmology, Observational Tests; Cosmology, Theories.**

Cosmology, Nucleogenesis

Robert V. Wagoner

The matter that fills the universe today retains a significant imprint of its history: the relative abundances of the wide variety of atomic nuclei. These abundances have been determined in many different sites; either directly (by analyzing meteorites, cosmic rays, etc.) or, most commonly, indirectly (via spectra of stars, interstellar gas, other galaxies, quasars, etc.). Using these data, astrophysicists can become "cosmic archaeologists," sifting through these ashes for clues to the nature of the universe's past. It is now clear that most of the heavier elements were produced by nuclear reactions within stars and in other locations during the recent history of the universe. However, it is also clear that the bulk of the lightest elements (hydrogen, helium, and lithium, with the exception of the isotope ^{6}Li) were probably formed throughout the universe, during an epoch in the remote past. The consequences of this process of nucleosynthesis, which produces our deepest direct probe of the early universe, are the subject of this entry. As we shall see, the agreement between the observed and calculated abundances of these lightest nuclei provides strong evidence for the validity of the standard Hot Big Bang model of the universe, which leads to an estimate of the amount of ordinary matter in the present universe as well as constraints on new types of elementary particles.

Although the concept that our universe may have expanded from a state of much higher density was developed in the period around 1930 (mostly via Lemaître's "primeval atom"), the knowledge of nuclear reactions necessary to calculate the abundances produced during such an era was not yet available. The first such calculations were carried out in the late 1940s by Ralph Alpher, George Gamow, and Robert Herman, who argued that the temperature was also very high at that time. Although the correct physical description of this era was developed by 1953, the nuclear abundances were not recalculated until the 1960s (apparently because of the great success of stellar nucleosynthesis and the realization that only the lightest nuclei could be produced in the expanding and cooling "primeval fireball"). The discovery of the cosmic microwave radiation in 1965 reinvigorated the Hot Big Bang model, and soon thereafter nuclear-reaction network calculations had produced unambiguous predictions. In the simplest ("standard") model, the resulting abundances [of ^{1}H (protons), ^{2}H (deuterium), ^{3}He, ^{4}He, and ^{7}Li] depended only on the total density of baryons (i.e., nuclei) at a given temperature.

Various developments during the subsequent two decades strengthened the argument that these nuclei were indeed relics of such a remote epoch. Around 1970, calculations began to show that those light nuclei whose observed abundances were too high to fit a fireball-production model (^{6}Li, ^{9}Be, ^{10}B, and ^{11}B) were precisely those produced by collisions of cosmic ray nuclei with the interstellar gas. Meanwhile, the belief that stellar interiors or other galactic sites could produce all the heavier nuclei continued to grow.

In 1973, the *Copernicus* satellite detected deuterium in the nearby interstellar gas. The resulting abundance determination was the first not subject to the large uncertainties of chemical fractionation, and was of critical importance. Deuterium is difficult to produce after the Big Bang era is over, and stars easily consume it. Therefore, it was argued, the Big Bang deuterium abundance must have been at least as great as the abundance presently observed. The predicted Big Bang abundance is a decreasing function of the assumed density of baryons. The deuterium measurement therefore implied an upper limit on the amount of "ordinary" (baryonic) matter in the universe today.

A third important discovery was that there are galaxies with a significant deficiency of heavy elements but an almost normal abundance of helium. It was natural to assume that these were systems in which stellar processing had not yet occurred to a great extent, so that the addition to the primordial helium (mostly ^{4}He) was small. Thus a more relevant value was also obtained for this abundance.

The next major advance began with the discovery (in 1982) of old stars which appeared to have formed from gas containing about 10 times less lithium than the previous "cosmic" abundance. It could be argued that this new determination was more likely to represent the primordial abundance. At about the same time, improved nuclear-reaction rate data led to an increase in the predicted abundance of ^{7}Li. It was remarkable that the observed and predicted primordial abundances then came into agreement for that value of the cosmic density which also produced agreement for the other nuclear abundances.

PRODUCTION OF NUCLEI IN THE EARLY UNIVERSE

The framework within which one calculates the primordial synthesis of nuclei is provided by a model of the structure and dynamics of the universe as a whole. We shall first discuss the results obtained within the standard Big Bang model, which represents the simplest extrapolation of our present knowledge into the past. This model is essentially constructed from the assumptions that (a) general relativity governs the expansion of the universe, (b) the properties of the universe were independent of position (homogeneity) and direction (isotropy) at any particular time in the deep past (before structures like galaxies formed), and (c) only the known particles were present. A key property of the early universe which emerges from this model is its expansion rate,

$$(dV/dt)/V = [24\pi G\rho]^{1/2}, \tag{1}$$

where V is any small volume containing a fixed number of nucleons (neutrons plus protons), and ρ is the *total* density of matter (including radiation). A more familiar property is the cooling of the gas due to this expansion, with the cube of its temperature T^3 proportional to $1/V$, in turn proportional to the density of these nucleons (which provide most of the mass of ordinary matter). But because the density of relativistic particles (photons, neutrinos, and electron-positron pairs) was proportional to T^4, they dominated during the epoch of interest, which began when the temperature was about 10^{11} K. (The present temperature of the relic photons is 3 K.)

It is extremely fortunate that the rates of the reactions among these particles were faster than the expansion rate (1) at this time (10^{-2} s after the Big Bang). This meant that they were in equilibrium with each other, so that their properties (such as density) were fixed, independent of the previous history of the universe! The sole exceptions were the conserved quantities: the number of baryons (nucleons), electron-leptons (electrons plus electron neutrinos minus their antiparticles), plus other flavors of leptons (muon and tau neutrinos); they must be specified. In the standard model all three lepton numbers are taken to be so small that the corresponding neutrinos are nondegenerate, with roughly equal numbers of leptons and antileptons.

The first event of importance during this era occurred when the temperature had dropped almost to 10^{10} K, at which time the reactions $\nu_e + n \rightleftharpoons p + e^-$ and $e^+ + n \rightleftharpoons p + \bar{\nu}_e$, which had kept the neutrons and protons at a relative abundance

$$n/p = \exp[-15.01/T_9], \qquad (T_9 = T/10^9\ \text{K}), \tag{2}$$

dropped out of equilibrium. Subsequently this ratio remained almost fixed, until the next major event—the emergence of other nuclei. This is shown in Fig. 1, which depicts the evolution of the nuclear abundances during the expansion of a particular model. It should be noted that then (and now) there were 10^9–10^{10} photons per nucleon.

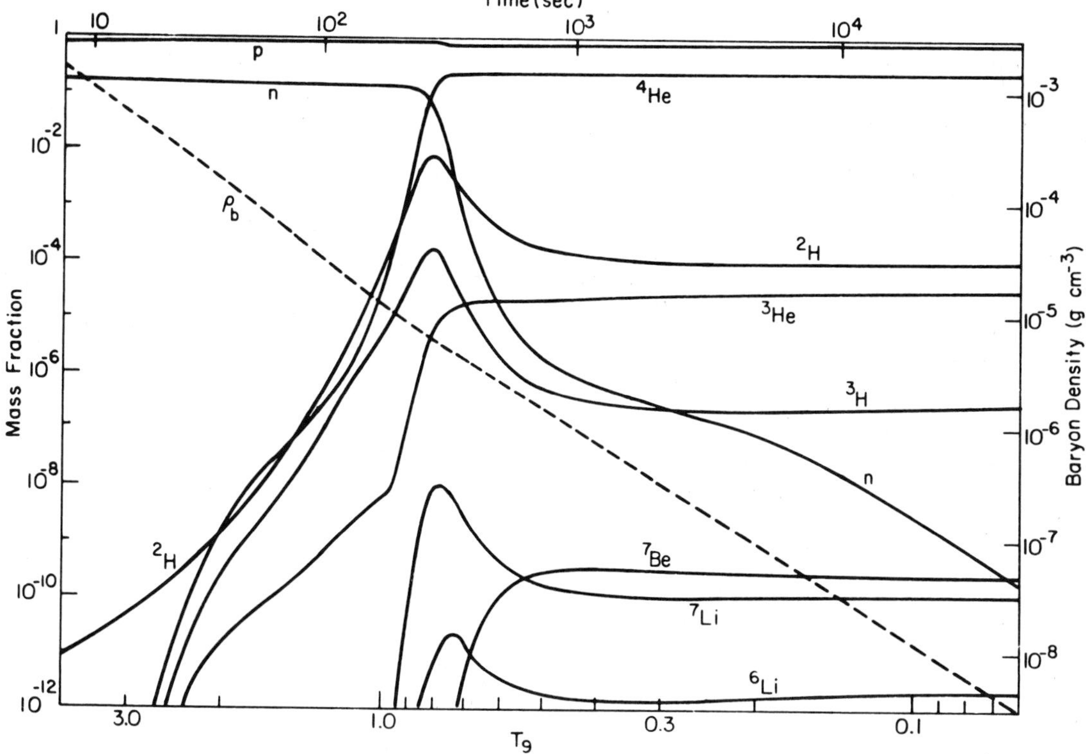

Figure 1. The evolution of the nuclear abundances with temperature (in units of 10^9 K) and time is shown within a particular standard model. The evolution of the total density of nuclei (baryons), the value of which, at any particular temperature, characterizes the model, is also shown (dashed line).

Since the average energy of the photons was decreasing as the universe cooled, eventually their ability to dissociate deuterium into nucleons decreased to the point (near $T = 10^9$ K) where a sufficient abundance of ^{2}H existed to allow further nuclear reactions to proceed quickly. Roughly 30 such reactions among the nuclei shown played a role in determining the final abundances. Thanks to the heroic efforts of many nuclear physicists, essentially all of the relevant nuclear cross sections have been measured, so that we can have confidence in these predicted abundances. This cannot yet be said about any process occurring earlier in the history of the universe.

COMPARISON WITH OBSERVED ABUNDANCES

The abundances which emerge from this process are compared with the range of observed abundances which might represent their primordial values in Fig. 2. Agreement can be achieved for those standard models in the range indicated. Let us first explore the consequences of the abundance of ^{4}He, which has the most precisely determined mass fraction: 0.24 ± 0.02.

If the expansion rate had differed from Eq. (1), the neutrons and protons would have ceased to be in equilibrium at a different temperature, producing a different "frozen-out" relative abundance [from Eq. (2)]. Because almost all neutrons are eventually incorporated into ^{4}He, its abundance is directly affected. Thus powerful constraints can be put upon other theories of gravitation, the amount of anisotropy, the degree of neutrino degeneracy [which can also change Eq. (2)], and any other factor which affects the expansion rate. In particular, if more types of neutrino [or any other particle which contributed to the density in Eq. (1)] existed, more helium would have been produced. This constraint has very recently been spectacularly confirmed at the Stanford Linear Collider and at CERN, where the decay rate of the Z^0 particle pro-

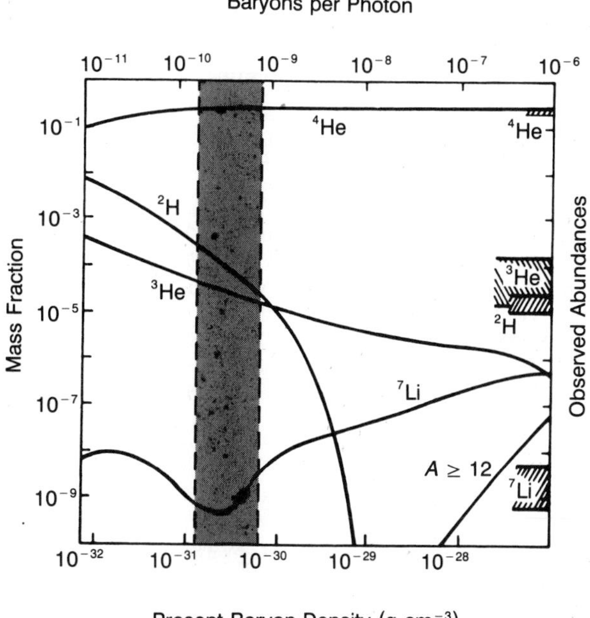

Figure 2. The final abundances of nuclei produced by the range of standard models is compared with their observed abundances. The band indicates those models that are consistent with the allowed values of "primordial" abundances, which can be deduced from the observational data.

duced by electron–positron collisions has shown that there are only three types of light neutrinos, consistent with the limit obtained from the helium abundance.

Another major consequence, which follows mainly from the other abundances, is the pair of upper and lower limits on the present density of ordinary (baryonic) matter. The range shown in Fig. 2 corresponds to at most about 20% of that required to produce a universe with no spatial curvature (favored by many on aesthetic grounds). But more importantly, this upper limit may turn out to be less than the measured total density of matter, which can be determined by its gravitational influences on observed matter (galaxies, etc.). If so, it would indicate that a new form of matter pervades the universe.

The desire to avoid this conclusion has recently motivated two other modifications of the standard model, which attempt to produce an agreement with observed abundances for larger values of the baryon density. The first invokes a first-order phase separation during the quark–hadron transition that formed the protons and neutrons, which could have produced large inhomogeneities in their abundances. However, this model has great difficulty avoiding overproduction of ^{7}Li. The second involves late-decaying massive particles, whose electromagnetic and hadronic showers strongly affect most products of the previous standard nucleosynthesis. The resulting abundances can be made consistent with those observed, but the predicted excess of ^{6}Li relative to ^{7}Li may prove to be a fatal problem.

FUTURE

The quest for truly primordial abundances is the key challenge for the future. Mapping the space and time (redshift) dependence of the amounts of these light nuclei with the largest ground-based telescopes as well as from space (with the *Hubble Space Telescope*, initially) will be required.

On the theoretical side, we must be wary of being channeled into a parochial view of the scope of possible universes. (Could a very early generation of the stars ("Population III") have produced the helium and lithium, as well as the cosmic microwave radiation?) But the standard model has thus far withstood the powerful test provided by these nuclear relics.

Additional Reading

Boesgaard, A. M. and Steigman, G. (1985). Big Bang nucleosynthesis: Theories and observations. *Ann. Rev. Astron. Ap.* **23** 319.

Schramm, D. N. and Steigman, G. (1988). Particle accelerators test cosmological theory. *Scientific American* **258** (No. 6) 66.

Schramm, D. N. and Wagoner, R. V. (1977). Element production in the early universe. *Ann. Rev. Nucl. Sci.* **27** 37.

Silk, J. (1989). *The Big Bang*. W. H. Freeman and Co., New York.

Wagoner, R. V. and Goldsmith, D. W. (1983). *Cosmic Horizons*. W. H. Freeman and Co., San Francisco.

See also **Cosmology, Big Bang Theory; Cosmology, Population III; Dark Matter, Cosmological; Stars, Chemical Composition; Supernova Remnants, Evidence for Nucleosynthesis.**

Cosmology, Observational Tests

Virginia Trimble and Stephen P. Maran

Astronomers make observational tests of cosmology to determine the large-scale properties of the universe so as to distinguish between alternative concepts, or models, that attempt to account for its nature, history, and future. The comparison process is often described as observational testing of the models, which are generally based on Albert Einstein's theory of general relativity. Because it appears that the universe is expanding, these observational tests include measurements to determine the rate at which the universe is now expanding and the rate at which this expansion changes with time. Other observational tests include attempts to determine the density of matter in the universe, the abundances of light elements and isotopes thought to be created in the Big Bang, measurements of the microwave background radiation, and studies of the large-scale distribution of galaxies in space. These tests are intended to yield estimates of the age and size of the universe and the curvature of space, to provide information on the distribution and properties of unseen "dark matter," and to elucidate how the universe may have come to take its present form, and how that form will change with time.

The first, and for years the only, discovery in modern observational cosmology was Edwin Hubble's recognition in the 1920s of a linear relationship between redshifts seen in the spectra of galaxies outside our own cluster (the Local Group) and the distances of those galaxies from us. The relationship

$$cz = H_0 d$$

is called Hubble's law. Here z is the redshift ($=\Delta\lambda/\lambda$, a measure of the amount that light of wavelength λ is shifted toward longer wavelengths), c is the speed of light, and d is distance. H_0, the proportionality factor, is called Hubble's constant, although it can vary with time. H_0 has units of velocity divided by distance, or reciprocal time, and is typically given in kilometers per second per megaparsec. One megaparsec (Mpc) equals approximately 3.086×10^{24} cm, or about 3.26 million light-years.

Most scientists quickly interpreted Hubble's law to mean overall, uniform expansion of the observable universe. The redshifts result from the expansion of space-time, rather than from the motion of objects through space, and so are not quite the same as ordinary Doppler shifts of spectral lines. In particular, there is no unique relationship between redshift and relative velocity, as there is in Doppler shifts. Hubble's law does not, as it might first seem, imply the existence of a center for the expansion, let alone our presence at that center. Rather, an observer anywhere in the universe will see redshifts proportional to distances from himself, and the same value of H_0 at a given time.

Since Hubble's work, the majority of cosmologists has thought in terms of a universe expanding out of a hot, dense initial state over a finite (but long) period of time. Such an expansion is describable by Einstein's equations in the form written down, also in the 1920s, by Alexander Friedmann in the USSR, Georges Lemaître in Belgium, and Howard P. Robertson and Arthur Walker in the US. A model for such a universe is called a Big Bang model or evolutionary cosmology. Wholly different concepts include the steady-state cosmology and sometimes involve claims that the redshifts of galaxies have a noncosmological origin.

For about the next 35 years, observational cosmology consisted largely of a search for accurate values of two numbers, the present expansion rate H_0, and its acceleration or deceleration, denoted q_0. During that period, the best estimates of H_0 gradually dropped from about 500 to about 50 km s^{-1} Mpc^{-1} (corresponding to a characteristic time scale or "age of the universe" of 2–20 billion years, respectively), while q_0 has been uncertain even in sign.

The situation changed rapidly after, and largely because of, the 1965 discovery by Arno Penzias and Robert Wilson of a universal background of microwave radiation at a temperature of about 3 K. This radiation, predicted in 1948–49 by Ralph Alpher and Robert Herman, working with George Gamow, is interpreted as a first-hand relic of the hot, dense early universe. Its properties allow us to ask, and partially to answer, a wide range of questions about processes that occurred after the Big Bang 10–20 billion years ago. In particular, the agreement between the observed amounts of helium, deuterium, and lithium in the universe and calculations based on a Hot Big Bang model is an impressive indicator of the correctness of our standard picture.

PARAMETERS

It is convenient to label the positions of galaxies by coordinates that remain attached to them (comoving coordinates), multiplied by a factor $R(t)$ that describes the expansion of the universe. Hubble's constant H is then simply

$$H = \dot{R}(t)/R(t) = [dR/dt]/R(t),$$

and H_0 is its present value. The "dot" notation, for example, $\dot{R}(t)$, for time derivatives is standard. The acceleration or deceleration of the expansion can be stated in dimensionless form as

$$q = -\ddot{R}(t)R(t)/[\dot{R}(t)]^2$$

Again, q_0 is the present value, positive if the expansion is slowing down, negative if the expansion is speeding up. During the expansion, the density of rest-mass energy drops as R^{-3}, and that of radiation energy as R^{-4}. The age of the universe, that is, the time elapsed since densities were arbitrarily (extremely) high, is $1/H_0$ for $q = 0$ and $\frac{2}{3}H_0$ for $q = \frac{1}{2}$. In addition, if ordinary gravitation is the only important force on a cosmological scale, a universe with $q \leq \frac{1}{2}$ will expand and cool forever, while one with $q > \frac{1}{2}$ will eventually recontract, heating us all out of existence as the temperature of the background radiation increases, long before we are crushed by the increasing density.

The above considerations come just from application of calculus and conservation of mass energy. To go further, we apply general relativity to express how $R(t)$ responds to the presence throughout the universe of an average mass-energy density ρ and a uniform pressure P:

$$\ddot{R} = \frac{-4\pi G}{3} R\left(\rho + \frac{3P}{c^2} + \frac{c^2\Lambda}{8\pi G}\right)$$

The new parameter, Λ, is Einstein's famous cosmological constant, which he is said, late in life, to have deeply regretted including in his equations. Originally introduced to permit a static universe, it may seem unnecessary in an expanding one, but is nevertheless part of the most general equation. Λ can be either positive or negative. Thus it can increase or decrease the effect of gravitating matter. But, since ρ decreases as R grows and Λ does not, Λ, if nonzero, will eventually dominate universal dynamics over sufficiently long times and distances.

The pressure term can be dropped at this point as it is negligible after the first 10^5 yr or so of expansion. Then the equation of motion has a first integral of the form

$$\dot{R}^2 = \frac{8\pi G}{3} R\left(\rho + \frac{c^2\Lambda}{8\pi G}\right) - kc^2,$$

where the integration constant k can have values $+1$, -1, or 0. For $k = \pm 1$, the four-dimensional space-time is curved like a sphere ($k = +1$) or saddle (hypersphere, $k = -1$), and $R(t)$ represents the radius of curvature of the space. For $k = 0$, the universe is the flat Euclidean space of high school geometry.

Thus we have the five parameters H, q, ρ, Λ, and k to describe the behavior of $R(t)$. Mercifully, they are not all independent. If we express the density in the convenient dimensionless form

$$\Omega = 8\pi G\rho/3H^2,$$

then

$$\Lambda = 3H^2(\Omega/2 - q)$$

and

$$kc^2 = H^2R^2(3\Omega/2 - q - 1).$$

If $\Lambda = 0$, then $q = \Omega/2$, and the dynamical future of the universe can be found from its present density, with the dividing point between eternal expansion and eventual recontraction occurring at $\rho = 3H^2/8\pi G$, called the closure or critical density ρ_c. For this critical case, $k = 0$. Thus a universe that just barely expands forever has Euclidean (flat) geometry. It is important to realize that ρ includes all forms of mass energy present in the universe, and thus must include invisible matter, not just the stars, galaxies, and radiation we easily detect.

All these equations and considerations apply only to a homogeneous (same in all places) and isotropic (same in all directions) universe. This is clearly a bad description on small scales, since the distributions of planets, stars, galaxies, and clusters of galaxies are all inhomogeneous. But the uniformity around the sky both of the 3-K radiation and of Hubble's law (for sufficiently large distances) suggests that homogeneity and isotropy are reasonable assumptions on very large scales of at least 100–6000 Mpc. (There are theorems within general relativity that guarantee that local measurements of, for instance, density will be cosmologically meaningful under these circumstances.) Additional parameters are needed to describe a universe in which the expansion is accompanied by overall rotation or shear. Such models have not been explored in as much detail as the purely expansionary ones, and can be severely constrained by the isotropy of the 3-K radiation.

If we expand the idea of observational cosmology to cover processes occurring in the early universe, then a number of additional measurable quantities become of interest. These include the precise temperature of the microwave radiation; its deviations either from a blackbody spectrum or from exact isotropy; the abundances of helium, deuterium, and lithium (before nuclear reactions in stars began modifying them); the properties of the largest structures (superclusters of galaxies and voids between them) in the universe and of the largest deviations from smooth Hubble expansion; and the nature of the primordial density fluctuations that eventually formed galaxies.

OBSERVATIONS

Distance Scale and H

From Hubble's time to the present, cosmology has been bedeviled by the difficulty of establishing distances to astronomical objects beyond the range of parallax measurements. The general approach has been for nearby stars and star clusters, where geometrical methods are applicable, to be used to calibrate more distant, brighter ones (for which apparent brightness or apparent angular size plus known brightness or size imply distance), and for these, in turn, to be used to calibrate whole galaxies and clusters of galaxies. The large drop in the estimated value of H_0 from 500 to 50–100 km s^{-1} Mpc^{-1} between 1930 and 1965 occurred in several steps as astronomers recognized, first, that a bright sort of Cepheid variable star had been mistaken for a fainter sort, and second, that small star clusters and the gas they illuminate had been mistaken for single stars. The redshifts of galaxies, also required for Hubble's constant H_0, are readily measured to much higher accuracy than are the distances.

Since about 1965, two lines of investigation, associated closely with the names of Allan Sandage and Gerard de Vaucouleurs, have persisted in yielding two different values of H_0, near 50 and 100 km s^{-1} Mpc^{-1}, respectively. The two distance scales differ even within our own galaxy and are separated by a full factor of 2 when they reach out to where the universal expansion dominates over local, peculiar motions of galaxies (there being also some disagreement about whether this occurs at about 10 or 100 Mpc). Thus, there is immediately a factor of at least 2 uncertainty in the estimated age of the universe and of 4 in the brightness and other properties of distant galaxies and quasars. Hopes have been expressed that study of Cepheids in the Virgo cluster of galaxies with the Hubble Space Telescope (after its optics are corrected), or sufficiently careful modeling of the behavior of supernovae in still more distant galaxies, may resolve the prolonged discrepancy.

Other Parameters

The distance scale enters into all the other parameters in one way or another, to the detriment of precise measurement. The least-well-determined parameter is Λ, which does not dominate any directly measurable effect. Thus it is constrained only by its relationship to the other parameters.

The geometrical parameters, k and R, are not much better off. To determine them directly, we need to measure lengths, surface areas, or volumes at known, very large distances and to compare them with Euclidean values. However, this requires that accurate distances be known and that the existence of objects of known linear size at large distance be established, and neither requirement is satisfied. Nevertheless, we know that the current radius of curvature of the universe, R_0, must be quite large compared with c/H_0 (3000–6000 Mpc), and it could be infinite. If $\Lambda = 0$, then universes with $k = 0$ or $k = -1$ are infinite in volume and the part we can observe is infinitesimal. A $k = +1$ universe has finite volume proportional to $R^3(t)$, and we might be observing at least a few percent of it.

The deceleration parameter q_0 can be measured by looking at the value of H in the past. This means extending the Hubble relation out far enough so that the linear form breaks down. But the only distance indicator we have, apart from the redshift itself, is the apparent brightness of galaxies. At the redshifts required to see deviations from linearity, galaxies are half or less of their present age. Thus apparent brightness can only be turned into a distance if we understand clearly how the real brightnesses vary with time. As Fig. 1 indicates, the differences in cz (the product of the speed of light and the redshift) versus apparent brightness due to plausible models of brightness evolution are larger than those due to the likely range of q_0. Rapid improvement in this situation is not expected.

The density parameter Ω is rather better known, partly because the value of H cancels in the ratio of ρ_c to ρ contributed by any particular kind of object. Measured values of Ω increase with the size of the system whose dynamics are investigated, from less than 0.01 for single galaxies to 0.1–0.3 for large clusters and superclusters. Significant material may well exist between these structures, however, and $\Omega = 1$, favored by some theories, cannot be excluded observationally.

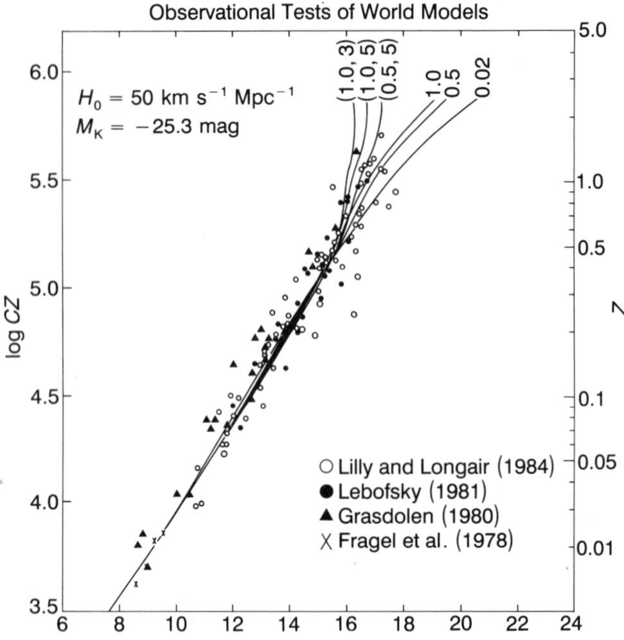

Figure 1. Hubble diagram (plot of log cz versus apparent infrared magnitude as a distance indicator). Points are observations of galaxies. Three curves to the right are predictions of cosmological models with $q_0 = 0.02$, 0.5, and 1.0, and no allowance for evolution of galaxy brightness with time. Three curves to the left are predictions of cosmological models with $q_0 = 0.5$ or 1, and two likely models of galactic luminosity evolution. Clearly the evolutionary effects are stronger than the cosmological ones, and existing data are equally well fit by a range of models, both open and closed. (*Reproduced from Sandage, 1988. Reproduced with permission, from the* Annual Reviews of Astronomy and Astrophysics, ***26****. © 1988 by Annual Reviews Inc.*)

The density contributed by ordinary (baryonic) matter, made of protons, neutrons, and electrons, is separately limited to roughly $\Omega_b = 0.015$–0.15 if the helium, deuterium, and lithium we now see are relics of a homogeneous Hot Big Bang. Thus a density as large

Table 1. Current Status of Cosmological Parameters

Quantity	*Likely Value*	*Significance*	*Barriers to Improve Measurement*
H_0	50–100 km s^{-1} Mpc^{-1}	Time and distance scales in the universe	Uncertainty in distance scale
q_0	-1–$+1$	Dynamical future of expanding universe	Evolution of galaxy brightnesses
Ω	0.1–1.0	Dynamical future; existence of dark matter	Size and velocity properties of largest structures
k	$+1, 0, -1$	Geometry of the universe	Absence of standard meter sticks
R	$>$ 10,000 Mpc	Radius of curvature of the universe	Absence of standard meter sticks
Λ	$-H^2/c^2$ to $+H^2/c^2$	Distant past and future of universal expansion	No observational test known
Age	$\geq 10^{10}$ yr	Constrains combinations of H, q, R, Λ, Ω; upper limit to star and galaxy ages	Uncertainties in models of formation and evolution of stars and galaxies
T_0	2.735 ± 0.06 K	Present and past temperatures; nucleosynthesis	Reducing systematic errors of measurement
$\Delta T/T$	$\leq 3 \times 10^{-5}$ on most angular scales	Fluctuations in early universe; constraints on galaxy formation	Contributions from galactic plane emission and radio sources
He/H	0.24 ± 0.015	Density of baryonic material	Effects of nuclear reactions in stars
D/H	10^{-5}	Density of baryonic material	Effects of nuclear reactions in stars
Li/H	10^{-9}	Density of baryonic material	Effects of nuclear reactions in stars
Largest structures	$\geq$ 170 Mpc	Constrains models of galaxy formation	Selection effects in surveys
Largest deviations from Hubble's law	$\sim$ 1000 km s^{-1}	Constrains models of galaxy formation	Selection effects in surveys; uncertainties in distance scale

as ρ_c requires either the existence of a nonbaryonic component or a highly inhomogeneous early universe. If, on the other hand, Ω (or q_0) were very much larger than 1, its effects would show in a Hubble diagram like Fig. 1, unless a carefully tuned value of Λ compensated.

Finally, limits can be set to the present age of the universe both from the oldest stars (in globular clusters) in our galaxy and from the abundances of radioactive nuclides, including U^{235}, U^{238}, Th^{232}, Rb^{87}, and Sr^{87}. Values found are invariably in excess of 10×10^9 years, ruling out, for instance, the combination $H=100$ km s^{-1} Mpc^{-1}, $q_0=\frac{1}{2}$, $\Lambda=0$, whose age is only 6.7×10^9 years, and marginally ruling out any combination of large H and $\Lambda=0$. Just how much larger than 10×10^9 years the age must be depends on how much time is required for the formation of galaxies and the first stars, on the precise chemical compositions of globular cluster stars, and (yet again!) on precise distances. Values from at least 12 to 17×10^9 years are currently defensible and defended.

STATUS OF OBSERVATIONAL COSMOLOGY

Table 1 summarizes our present understanding of the measurable quantities. From the point of view of a professional astronomer, the lingering factor of two uncertainties are a persistent frustration, particularly because they bracket the dividing line between open (infinite volume, expands forever) and closed (finite volume, eventually recontracts) cosmological models. From a philosophical point of view, however, it should probably be regarded as a triumph that we have a self-consistent model of what has happened over the past $10–20\times10^9$ years in a volume larger by a factor of 10^{40} or more than our own solar system.

Additional Reading

Gott, J. R., III, Gunn, J. E., Schramm, D. N., and Tinsley, B. M. (1976). Will the Universe expand forever? *Scientific American* **234** (No. 3) 62.

Harrison, E. R. (1981). *Cosmology: The Science of the Universe.* Cambridge University Press, Cambridge.

Sandage, A. R. (1988). Observational tests of world models. *Ann. Rev. Astron. Ap.* **26** 561.

Silk, J. (1989). *The Big Bang*, rev. ed. W. H. Freeman and Company, New York.

Trimble, V. (1987). Existence and nature of dark matter in the Universe. *Ann. Rev. Astron. Ap.* **25** 425.

Zeldovich, Ya. B. and Novikov, I. D. (1983). *Relativistic Astrophysics* **2**. *The Structure and Evolution of the Universe.* University of Chicago Press, Chicago.

See also **Background Radiation, Microwave; Cosmology, Big Bang Theory; Cosmology, Galaxy Formation; Cosmology, Inflationary Universe; Dark Matter, Cosmological; Hubble Constant; Superclusters, Dynamics and Models.**

Cosmology, Population III

Bernard J. Carr

The term "Population III" has been used to describe two types of stars: (1) the ones which form out of the pristine gas left over after cosmological nucleosynthesis and generate the first metals; and (2) the ones which have been hypothesized to provide the dark matter in galactic halos. Stars of the first kind definitely exist, but may not warrant a special name. Those of the second kind may not exist, because galactic halos could also be composed of some sort of elementary particle, but they certainly warrant a special name if they do, and they could have many interesting cosmological consequences. Population III stars of either kind could be pregalactic, but they might also have formed during the first phase of galaxy formation.

FORMATION OF POPULATION III STARS

In the most conservative cosmological scenario, the first stars form in the process of galaxy formation: As each protogalaxy cools and collapses, it fragments first into a spheroidal distribution of Population II stars, and then—if there is any gas left over—into a rotationally supported disk of Population I stars. The problem with this picture is that, in both of the standard scenarios for the origin of cosmological structure, the first bound objects would be much smaller than galaxies. For example, in the *hierarchical clustering* scenario the first bound clouds have a mass of about $10^6\ M_\odot$ and bind at a redshift of order 100. Larger bound objects—like galaxies and clusters of galaxies—would then build up through a process of gravitational clustering. A currently popular version of this model is the "cold dark matter" scenario, in which the density of the universe is dominated by some cold elementary particle like the photino or axion. In the *pancake* scenario, the first objects to appear are of cluster or supercluster scale and they form at a rather low redshift. This applies, for example, in the "hot dark matter" picture, in which the universe's mass is dominated by hot particles like neutrinos with nonzero rest mass. However, one still expects these pancakes to fragment into clouds of mass $10^8\ M_\odot$ and these clouds must then cluster in order to form galaxies. In both scenarios, therefore, an appreciable fraction of the universe must go into subgalactic clouds before galaxies themselves form.

The question then arises of what happens to these clouds. In some circumstances, one expects them to be disrupted by collisions with other clouds because their cooling time is too long for them to collapse before coalescing. However, there is usually some subgalactic mass range in which the clouds survive. In this case, they could face various possible fates. They might just fragment into ordinary stars and form objects like globular clusters. On the other hand, the conditions of star formation could have been very different at early times and several alternatives have been suggested.

1. The first stars could have been smaller than at present because of the enhanced formation of molecular hydrogen at early epochs.
2. They could have been larger than at present because the lack of metals or the effects of the microwave background would increase the fragment mass.
3. There may have been a mixture of small and large stars; for example, angular momentum effects could lead to a disk of small stars around a central very massive star, or massive stars could form in the core of the cloud and low-mass stars in the outer regions.
4. The first clouds may not fragment at all, but might collapse directly to supermassive black holes or remain in purely gaseous form and become Lyman-α clouds.

This indicates that, although there is clearly considerable uncertainty as to the fate of the first clouds, they could well fragment into stars that are very different from the ones forming today. They certainly need to be very different if they are to produce much dark matter. Note that the appellation Population III is sometimes assigned to the first clouds rather than the first stars. However, in this case, all the stars which they spawn must also be called Population III, and this can lead to semantic confusion if the clouds fragment bimodally. It is therefore more sensible to reserve the term Population III for the stars.

It is not necessarily required that Population III stars be pregalactic. Some of the arguments for their having a different initial mass function (IMF) would also apply if they formed protogalactically, and this gives rise to a less radical hypothesis, in which the Population III objects form during the first phases of protogalactic collapse. In this case, the Population III stars or their remnants would be confined to galaxies, whereas they would be spread throughout space in the pregalactic case.

POPULATION III AS THE FIRST METAL PRODUCERS

Since heavy elements can only be generated through stellar nucleosynthesis, the existence of stars of type (1) is inevitable. However, the stars warrant a special name only if they are qualitatively different from ordinary Population II stars. For example, it would not be justified if the first metal-producing stars were merely the ones at the high-mass end of the Population II mass spectrum. For in this case they would generate the first metals simply because they evolve fastest. The introduction of a new term would only be warranted if the first metal-producing stars formed at a distinct epoch or if the IMF of the first stars was bimodal (i.e., with a distinct population of high- and low-mass stars forming in different locations).

If one studies the abundances of metal-poor stars in our own galaxy, there is no compelling reason for supposing the first stars were distinct from Population II. For example, field halo stars with $Z < 0.1 Z_{\odot}$ have enhancements in the ratios of O, Mg, Si, and Ca to Fe by a factor of 3 relative to the Sun, but this is naturally explained by the fact that these elements are preferentially produced by the sort of massive stars which would complete their evolution on the time scale (10^8 yr) associated with the formation of the galactic halo. Thus, abundance data itself does not require the existence of Population III stars.

The best evidence for a distinct population of stars would be a lower cutoff in the metallicity distribution of Population II stars. If the first stars had the same IMF as today, with a lower cutoff at about 0.1 $M_{\odot}$, one might expect stars smaller than 0.8 $M_{\odot}$ (whose lifetime exceeds the age of the Universe) to display arbitrarily low metallicity. At one time, it seemed there was a metallicity $Z_{\min}$ of order 10^{-5} below which no stars were found. If this were true, it would suggest that the first stars had an IMF with a lower cutoff above 0.8 $M_{\odot}$. For only then could they produce the minimum enrichment $Z_{\min}$ without surviving until the present epoch. This would imply that the first stars had a different IMF from ordinary Population II stars. Unfortunately, the evidence for such a cutoff is now in dispute: The Z distribution for Population II stars extends well below 10^{-5} and there exists one object with $Z = 6 \times 10^{-7}$. In any case, the number of low-Z objects is not necessarily incompatible with the assumption that the IMF has always been the same; so the first stars may not have been qualitatively different from Population II stars. Thus the introduction of the term Population III may be unnecessary in this context.

POPULATION III AS DARK MATTER PRODUCERS

The success of the standard Big Bang picture in explaining the light element abundances only applies if the baryon density is about 10% of the critical density. Since the theory of inflation requires the total density to have the critical value, this suggests that there must be much nonbaryonic dark matter (most of it unclustered). On the other hand, visible material only has about 1% of the critical density, so it seems that there must also be some baryonic dark matter. It is possible that this is in the form of a hot intergalactic medium, but there could also be enough of it to explain the dark matter in galactic halos. The dark baryons in halos cannot be in the form of ordinary gas, or else they would generate too may x-rays. They must therefore have been processed into some dark form through a first generation of pregalactic or protogalactic stars.

In principle, there are many mass ranges in which stars could produce dark remnants. For example, stars smaller than 0.1 $M_{\odot}$ would always be dim enough to explain galactic halos and those smaller than 0.08 $M_{\odot}$ (jupiters; also called brown dwarfs) would never even ignite their nuclear fuel. Stars in the range 0.8–4 $M_{\odot}$ would leave white dwarf remnants, whereas those between 8 $M_{\odot}$ and some mass M_{BH} (probably about 50 $M_{\odot}$) would leave neutron star remnants. In either case, the remnants would eventually cool

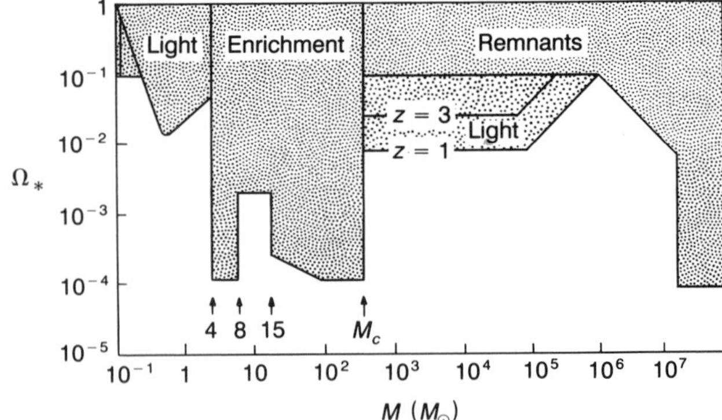

Figure 1. Constraints on the density of Population III stars of mass M. Ω_* is the density in units of the critical cosmological value; $\Omega_* = 0.1$ corresponds to the density associated with galactic halos. The shaded region is excluded, the light constraint depending on the redshift (z) at which the stars burn.

and become dark. Stars more massive than M_{BH} would leave black holes. The ones larger than 100 $M_{\odot}$ are termed very massive objects (VMOs), and are particularly interesting because they could collapse entirely (without any metal ejection) due to an instability encountered in their oxygen-burning phase. This would apply for VMOs larger than $M_c \approx 200$ $M_{\odot}$. Stars larger than 10^5 $M_{\odot}$ are termed supermassive objects (SMOs), and would collapse directly to black holes due to relativistic instabilities even before nuclear burning, at least if they were metal-free. It must be stressed that the existence of VMOs and SMOs is entirely speculative and they are invoked primarily for the purpose of making dark matter.

Although stars can in principle produce dark remnants, various constraints require that the dark matter in galactic halos can only be baryonic if it comprises jupiters or the black hole remnants of VMOs. These constraints are summarized in Fig. 1. Low-mass stars are excluded by source count limits, other stellar remnants by nucleosynthesis and background light constraints, and supermassive black holes by dynamical considerations. At first sight, it might seem rather unlikely that Population III clouds would fragment into such objects with high efficiency, but we have seen that there are theoretical reasons for expecting the first stars to be larger or smaller than at present.

In fact, there are circumstances where dark stars form profusely even at the present epoch. Direct observational evidence that gas can be turned into low-mass stars with high efficiency may come from x-ray observations of cooling flows in the cores of rich clusters. These suggest that 90% of the gas is being turned dark, possibly as a result of the high pressure. Since such cooling flows are confined to the central galaxies in clusters, they could not explain the dark matter in galactic halos. However, one could expect analogous high-pressure flows to occur at earlier cosmological epochs, and these would have been on much smaller scales than clusters. This conclusion pertains in either the hierarchical clustering or pancake scenarios. One could envisage forming dark clusters of jupiters which then agglomerate to form galactic halos. Although VMOs are certainly rare at the present epoch, massive stars do seem to form efficiently in starburst galaxies, and they may have been more abundant in the past. VMOs would certainly have had more exciting cosmological consequences than jupiters.

Note that the formation epoch is very important for the relative distribution of baryonic and nonbaryonic dark matter. If Population III stars form before galaxies, one might expect their remnants to be distributed throughout the universe, with the ratio of the baryonic and nonbaryonic densities being the same everywhere. If they form at the same time as galaxies, one would expect the remnants to be confined to halos with the baryonic component probably dominating.

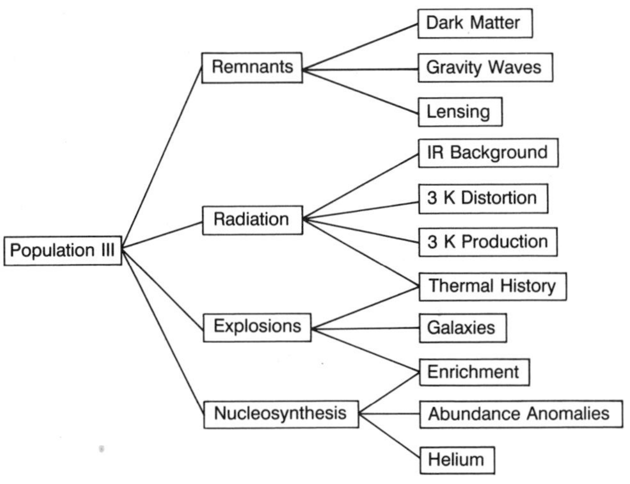

Figure 2. Cosmological consequences of Population III stars.

WHAT POPULATION III STARS CAN EXPLAIN

In this section we will discuss some of the cosmological consequences of Population III stars. We will mostly focus on the VMO scenario, but the last three effects could be important in a more general context. Although Population III stars can explain certain cosmological problems, it would be stressed that this does not provide unequivocal evidence for their existence, because they are not the only explanation. Figure 2 summarizes the effects.

Infrared Background The most direct evidence for a population of VMOs would come from the detection of the background light they generate. Each VMO has the Eddington luminosity and a mass-independent surface temperature of about 10^5 K. In the absence of dust absorption, one would expect a background peaking in the infrared with a density of order 10% that of the microwave background. The detection of such a background has in fact been reported at around 2 μm, although this has not yet been verified. In any case, comparison with the upper limits on the background radiation density in the IR band shows that VMOs with the density required to explain galactic halos would have to burn at a redshift exceeding 30—at least in the absence of dust—and this would imply that they were necessarily pregalactic.

Microwave Distortions If dust were present, the radiation from VMOs would be reprocessed into the far-infrared or submillimeter range, where the limits on the background density are weaker. In the latter case, the radiation would merely appear as a distortion of the microwave background. The dust could either be confined to galaxies (if they cover the sky), or it could be pregalactic in origin. There have, in fact, been two claimed detections of such a background. Data from the *Infrared Astronomical Satellite* (*IRAS*) may indicate a background at 100 μm with 10% of the microwave density, although this may be due to zodiacal emission. More recently, a significant distortion in the spectrum of the microwave background radiation has been reported in the 400–700-μm waveband (cf. the microwave background peak at 1400 μm). This follows a rocket experiment by a team from Nagoya and Berkeley. It corresponds to a submillimeter excess similar to that expected in the VMO-plus-dust scenario. However, preliminary results from the *Cosmic Background Explorer* (*COBE*) satellite have not confirmed either the 100-μm or the 400–700-μm excess.

Generation of 3-K Background Some people have proposed that the entire microwave background is dust-grain-thermalized starlight. This is possible in principle, but the grains would have to form at very high redshift ($z > 100$) or be very elongated in order to thermalize at long wavelengths. An alternative scheme is to propose that the black hole remnants of the stars generate the 3-K background through accretion. Of course, any scheme which envisages the background deriving from Population III stars or their remnants must also require that the early universe was cold or tepid (with the primordial photon-to-baryon ratio being much less than its present value of 10^9). Stars would, in fact, naturally generate a ratio of order 10^9. However, it is now more popular to assume that the ratio arises as a result of baryosynthesis in the very early universe.

Helium Production Because pulsations lead to mass-shedding of material convected from its core, a VMO is expected to return helium to the background medium during core-hydrogen burning. The net yield depends sensitively on the mass loss fraction. The yield will be low if the mass loss is very large or very small, but it is optimized if the envelope is ejected during hydrogen-shell burning. In this case, the fraction of mass returned as new helium is $\Delta Y = 0.25(1 - Y_i)^2(1 - Y_i/2)^{-1}$, where Y_i is the initial (primordial) helium abundance. This would have profound cosmological implications. If $Y_i = 0.24$, corresponding to the conventional primordial value, $\Delta Y = 0.16$, so helium would be substantially overproduced if much of the universe went into VMOs. In this case, the only remaining black hole candidate for the dark matter would be SMOs in the mass range 10^5–10^6 $M_\odot$. On the other hand, if $Y_i = 0$, corresponding to no primordial production, $\Delta Y = 0.25$, which is tantalizingly close to the standard primordial value. This raises the question of whether the Population III VMOs invoked to produce the dark matter might not also generate the helium which is usually attributed to cosmological nucleosynthesis. It must be stressed, however, that the hot Big Bang model also predicts the observed abundances for deuterium, helium-3, and lithium. Although one can conceive of astrophysical ways of generating these elements in a cold universe, they are somewhat contrived.

Dynamical Effects of Halo Holes The most interesting constraint on the mass of any holes in our own galactic halo is associated with their tendency to puff up the galactic disc. A detailed calculation of this effect suggests that halo holes could actually be responsible for all of the observed disk puffing, providing the halo objects have a mass of 2×10^6 $M_\odot$. While one cannot definitely identify halo holes as the explanation for disk heating (e.g., spiral density waves might also work), it is interesting that the hole mass required can be specified so precisely. Note that this argument does not require that the halo object be a single hole; even a cluster of smaller holes—or indeed a cluster of jupiters—would suffice. From a theoretical point of view, the halo objects are more likely to be clusters because too many supermassive black holes would be dragged into the galactic nucleus by dynamical friction. Clusters would be destroyed by collisions within the galactocentric radius where dynamical friction is operative, providing they had a radius larger than 1 pc. Further evidence that galactic halos comprise objects with mass of order 10^6 $M_\odot$ may come from gravitational lensing effects. Evidence for the jupiters themselves could come from microlensing.

Gravity Waves from Black Holes The formation of a population of pregalactic black holes would be expected to generate bursts of gravitational radiation. The characteristic period of the waves would be in the range 10^{-2}–10^4 s, depending on the holes' mass and formation redshift. If the holes are numerous enough to make up galactic halos, one would expect the bursts to overlap in time to form a background of waves. This background could be detectable by laser interferometry if the mass is below 10^2 $M_\odot$ or by the Doppler tracking of interplanetary spacecraft if it is in the range 10^5—10^{10} $M_\odot$. The prospects of detecting the gravitational radiation would be even better if the holes formed in binaries, because one would then get both continuous waves as the binaries spiral inward due to gravitational wave emission and a final burst of waves when the components finally merge. This would increase the amplitude of the waves and extend the spectrum to longer periods. One could also hope to observe the mergers which are occurring in nearby galaxies at the present epoch.

Heavy Element Production For some purposes it would be advantageous to have a slight pregalactic enrichment (e.g., to explain the small grain abundance required to produce alleged distortions in the 3-K spectrum), so the question arises of whether Population III stars could generate just a small amount of metals. A rather plausible way of doing this would be to suppose that the formation of the first metal-producing stars is suspended once enough of them have formed to reionize the universe. This would naturally generate a metallicity of order 10^{-6}. If one also wants dark-matter-producing stars, one must either suppose that they form later or fine-tune the IMF.

Thermal History The light generated by Population III stars could have an important effect on the thermal history of the universe even in the conventional Big Bang scenario. During its main-sequence phase, each star (or cluster of stars) would be surrounded by an H II region. The fraction of the universe in such regions would progressively grow, because of both the increasing number of stars and the increasing size of the individual H II regions; even a small density of Population III stars would soon reionize the whole universe. Such reionization could have important cosmological implications (e.g., in reducing anisotropies in the 3-K background). After the H II regions have merged, the universe would maintain a high degree of ionization until the stars cease burning, and even thereafter if they leave black hole remnants which heat the universe through accretion. It is not clear whether more conventional sources (like quasars) can ionize the intergalactic medium early enough.

Pregalactic Explosions Stars in the mass range $4\text{–}10^2\ M_\odot$ should produce explosive energy with an efficiency $\varepsilon = 10^{-5}\text{–}10^{-4}$; larger ones may explode with comparable efficiency if they eject their envelopes during hydrogen-shell burning. This explosive release could have an important effect on the large-scale structure of the universe. One would expect the shock wave generated by each exploding star (or cluster of stars) to sweep up a shell of gas. Under suitable circumstances, this shell could eventually fragment into more stars. If the new stars themselves explode, one could then initiate a bootstrap process in which the shells grow successively larger until they overlap. This mechanism has been proposed in two contexts: (1) as a means to boost the fraction of the universe being processed through pregalactic stars and (2) as a way of producing the giant voids and filaments, whose existence is indicated by observational data. An upper limit to the final shell size in all circumstances is $\sqrt{\varepsilon}$ times the current horizon size. This is 60 Mpc for $\varepsilon = 10^{-4}$ and a Hubble parameter of 100, which is just about large enough to explain the largest voids.

CONCLUSION

We have seen that one must distinguish between metal-producing and dark-matter-producing Population III stars. The first must exist, but only warrant a special name if there is a lower cutoff in the metallicity distribution of Population II stars, and it is not clear that this is the case. The second may not exist, but, if they do, they certainly warrant a separate name. They would have to be either jupiters or black holes. The detection of microwave distortions would favor the black hole option, but the claim that cooling flows make low-mass stars may favor the jupiter option. In principle, both kinds of Population III stars could derive from a single mass spectrum, but that would require the IMF to be finely tuned.

Additional Reading

Ashman, K. M. and Carr, B. J. (1988). Pregalactic cooling flows and baryonic dark matter. *Mon. Not. R. Astron. Soc.* **234** 219.

Beers, T. C., Preston, G. W., and Schectman, S. A. (1985). A search for stars of very low metal abundance. *Astron J.* **90** 2089.

Bond, H. E. (1981). Where is Population III? *Ap. J.* **248** 606.

Bond, J. R., Carr, B. J., and Hogan, C. J. (1986). Spectrum and anisotropy of the cosmic infrared background. *Ap. J.* **306** 428.

Carr, B. J. and Lacey, C. G. (1987). Dark clusters in galactic halos? *Ap. J.* **316** 23.

Carr, B. J., Bond, J. R., and Arnett, W. A. (1984). Cosmological consequences of Population III stars. *Ap. J.* **277** 445.

Cayrel, R. (1987). And if Population III were Population II? *Astron. Ap.* **168** 81.

Fabian, A. C., Nulsen, P. E. J., and Canizares, C. R. (1984). Cooling flows in clusters of galaxies. *Nature* **310** 733.

Hartquist, T. W. and Cameron, A. G. W. (1977). Pregalactic nucleosynthesis. *Ap. Space Sci.* **48** 145.

Kashlinsky, A. and Rees, M. J. (1983). Formation of Population III stars and pregalactic evolution. *Mon. Not. R. Astron. Soc.* **205** 955.

McDowell, J. C. (1986). The light from Population III stars. *Mon. Not. R. Astron. Soc.* **223** 763.

Pagel, B. E. J. (1987). Galactic chemical evolution. In *The Galaxy*, G. Gilmore and B. Carswell, eds. Reidel, Dordrecht, p. 341.

Rees, M. J. (1978). Origin of pregalactic microwave background. *Nature* **275**, 35.

Truran, J. W. (1984). Nucleosynthesis. *Ann. Rev. Nucl. Part. Sci.* **34** 53.

White, S. D. M. and Rees, M. J. (1978). Core condensation in heavy halos: A two-stage theory for galaxy formation and clustering. *Mon. Not. R. Astron. Soc.* **183** 341.

See also **Background Radiation, Microwave; Cosmology, Big Bang Theory; Cosmology, Galaxy Formation; Gravitational Radiation.**

Cosmology, Theories

John D. Barrow

The twentieth century has seen cosmology transformed from metaphysics into a branch of physics, and the laws governing fundamental forces and elementary particles have been wedded to astronomical observations to produce a description of the past and present states of the visible universe.

Prior to the creation of the general theory of relativity in 1915 by Albert Einstein, no attempts had been made to produce a mathematical description of the entire universe. Einstein's new relativistic theory of gravitation allowed consistent mathematical models of entire universes—even those with infinite size—to be formulated. Einstein's equations describe how these universes will change in time and from place to place. The simplest possible universe that can arise is one that is unchanging in time and uniform from place to place. This *static* universe was first proposed by Einstein in 1917 as a manifestation of the centuries-old prejudice that the universe as a whole be unchanging. In order to achieve this static state Einstein had to modify his original equations by the addition of a small constant term (dubbed the "cosmological constant") that was allowed, but not required, by the internal consistency of the theory. Subsequently, in the 1920s, it was shown by Willem de Sitter, Alexander Friedmann, and Georges Lemaître that such static solutions are of a very special sort that would not arise in practice; the slightest deviation from perfect uniformity would cause the universe either to expand or contract as a whole. Following this discovery attention focused upon universes that expand in time.

In the late 1920s Edwin Hubble discovered that the light from distant galaxies is shifted in the direction of the red end of the spectrum of visible light by an amount that is directly proportional to their distance away from us. This redshifting of the spectrum is characteristic of the Doppler shift produced by a receding source of radiation. These observations established the expanding-universe theory as the basic paradigm of twentieth-century cosmology.

The standard theory of the expanding universe is a reconstruction of its past history and is usually called the *Hot Big Bang theory* (a term invented by Fred Hoyle), because the expansion implies that the universe was hotter and denser in the past.

The expansion and the attractive nature of gravity imply that the expansion must have begun at some finite past time (about 15 billion years ago) if the laws of physics and the theory of general relativity apply unchanged at all times in the past. However, it is known that general relativity must cease to be a good description of the universe when it is less than 10^{-43} s from its apparent beginning and its density exceeds the Planck value of 10^{96} g cm^{-3}. To extend the Hot Big Bang theory into these first instants of time we require a *quantum theory of gravitation*. The search for such a theory, and hence a new quantum cosmology, is at present the greatest unsolved problem in physics.

The only viable alternative to the Hot Big Bang model that has been suggested is the *steady-state theory*, proposed in 1948 by Thomas Gold, Hermann Bondi, and Hoyle. This is an expanding universe that remains the same at all times on the average. Whereas the density of matter falls as the Hot Big Bang models expand, and all the matter was apparently created at some finite past time, the steady-state model proposed that there is continuous creation of matter at a rate that exactly counterbalances the natural dilution of the density by the expansion.

In the steady-state theory the universe was predicted to be the same on the average at all times as well as in all places. During the 1950s observational evidence against this theory mounted. Astronomers discovered more astronomical radio sources at large distances than nearby. Since radio waves travel at the speed of light, the distant sources must have emitted their radiation earlier than those nearby. This shows that the average properties of the universe change with the passage of time. In 1965, the discovery of the 3-K *microwave background radiation*, which is expected as a relic of an earlier hotter and denser phase of cosmic evolution, confirmed the Hot Big Bang theory, and the steady-state theory ceased to be a plausible cosmological model.

The expanding universe exhibits a number of remarkable properties and the explanation of these properties requires detailed extensions of the Hot Big Bang theory.

The expansion of the universe is extremely uniform and isotropic over the largest observed dimensions. The temperature of the microwave radiation is the same in every direction to within a few parts in 100,000. Hence the simplest cosmological models, which assume isotropic and uniform expansion, are an excellent approximation to the structure of the real universe. Two particular problems remain for the theory to explain? Why is the universe expanding in such a uniform and isotropic fashion when there would seem to be so many ways for it to be irregular and chaotic? What is the origin of the nonuniformities in the density that now exist in the form of galaxies and clusters of galaxies?

In the early 1970s the favorite theory for the origin of the large-scale uniformity of the universe was the "chaotic cosmology" of Charles Misner. It proposed that no matter how the universe began there always arise natural processes that smooth out the irregularities and anisotropies as the expansion proceeds. So long as the expansion lasts long enough, the universe will appear uniform and isotropic to observers like ourselves living about 15 billion years after the expansion began in a chaotic state. Unfortunately, this idea could not explain the observed regularity of the universe regardless of the starting state, and in 1981 the author and Richard A. Matzner showed that any universal smoothing process would produce far more heat radiation than is observed in the universe.

All expanding-universe theories regard galaxies as islands of above-average density that began as very small enhancements. During the expansion history of the universe the gravitational force amplifies these overdense regions at the expense of the underdense ones. The question that cosmological theories must answer is this: What is the source and pattern of the original small overdensities that start the process and how intense are they?

Expanding universes need not continue to expand forever. They do so if their average density is below a critical value. But if their average density exceeds this value then at some time in the future the expansion will be reversed into contraction. Redshifted radiation from distant sources will become blueshifted and the universe will collapse toward a "Big Crunch" of ever-increasing density and temperature.

One of the most puzzling properties of our universe is that it is now expanding at a rate very close to the critical divide separating the ever-expanding universes from the recollapsing ones. It is a puzzle because as the universe expands it tends to deviate steadily from the critical divide because of the attractive force of gravity. In order that we be as close to the divide as is observed today, the universe must have begun expanding extraordinarily close to the divide originally.

Since 1981 cosmologists have focused attention upon a refinement of the expanding Hot Big Bang theory—called the *inflationary-universe theory* by its originator, Alan Guth. It proposes that for some finite period of time during the first instants of its expansion the dominant form of matter in the universe exhibited gravitational repulsion rather than attraction. This requires the sum of the matter density plus the pressure stresses exerted in the three directions of space to be negative. (Current theories of elementary particles predict that such forms of matter can arise in the high-density environment expected in the first moments of the universe.) For this period the universe will then accelerate rather than decelerate (as it does at all times in the standard Hot Big Bang theory). The consequences of this acceleration are dramatic. The universe is driven very close to the critical divide, rather than away from it, as in the standard decelerating models. The universe tends to become uniform and isotropic regardless of how it started out and a particular type of seed irregularity is created everywhere in the universe. It remains to be seen whether this form of irregularity can successfully explain the existence of galaxies.

The inflationary-universe theory is the most promising current edition of the Hot Big Bang theory. But it leaves many important questions unanswered. It does not predict on which side of the critical divide we lie—only that we should be close to it. It does not tell us whether the universe is finite or infinite, or whether it had a beginning in time. The answer to these questions must at least await the creation of a quantum theory of the universe as a whole.

Additional Reading

Barrow, J. D. (1988). *The World within the World*. Oxford University Press, Oxford.

Barrow, J. D. (1988). The inflationary Universe: Modern developments. *Q. J. R. Astron. Soc.* **29** 101.

Barrow, J. D. and Tipler, F. J. (1986). *The Anthropic Cosmological Principle*. Oxford University Press, Oxford.

Guth, A. and Steinhardt, P. (1984). The inflationary Universe. *Scientific American* **250** (No. 5) 116.

Harrison, E. R. (1981). *Cosmology: The Science of the Universe*. Cambridge University Press, Cambridge.

Hawking, S. W. and Israel, W. (1987). 300 *Years of Gravity*. Cambridge University Press, Cambridge.

Munitz, M., ed. (1957). *Theories of the Universe—From Babylonian Myth to Modern Science*. Free Press, New York.

North, J. D. (1965). *The Measure of the Universe: A History of Modern Cosmology*. Oxford University Press, Oxford.

Silk, J. (1989). *The Big Bang*, rev. ed. W. H. Freeman, New York.

Weinberg, S. (1977). *The First Three Minutes*. Basic Books, New York

See also **Cosmology, Big Bang Theory; Cosmology, Inflationary Universe; Gravitational Theories.**

D

Dark Matter, Cosmological

Edmund Bertschinger

Dark matter is mass that does not emit or reflect detectable electromagnetic radiation, yet is detectable by its gravitational effect on other, luminous, matter. Perhaps 90% or more of all the matter in the universe is dark. Dark matter has been inferred to exist in galaxies and on larger scales in the universe, but not in the solar system. The nature and total amount of dark matter are unknown, although there are constraints from astronomical observations and particle physics experiments. The abundance, distribution, and nature of dark matter are outstanding questions in modern cosmology.

The total abundance of dark matter has important implications for the evolution of the universe. If the mean density is large enough, dark matter can close the universe, causing the universal Hubble expansion eventually to halt and reverse. For cosmological purposes, the most convenient way to express the abundance of some type of mass labeled i is by the ratio of the mean mass density ρ_i of that substance to the mean mass density ρ_{crit} required to close the universe:

$$\Omega_i \equiv \frac{\rho_i}{\rho_{\text{crit}}} = \frac{8\pi G \rho_i}{3H_0^2}; \qquad \rho_{\text{crit}} = 1.9 \times 10^{-29} h^2 \text{ g cm}^{-3}$$

The critical density depends on the gravitational constant G and the Hubble constant $H_0 = 100h$ km s^{-1} Mpc^{-1}. The Hubble constant is poorly known, but nearly all modern estimates give $0.5 < h < 1.0$. If the total density parameter Ω from all types of matter exceeds 1, the universe is closed and will eventually collapse. If $\Omega < 1$, the universe will continue expanding forever. The inflationary universe model of Big Bang cosmology predicts that $\Omega = 1$ to high precision, but this idea has not been confirmed by observations. Observational estimates yield a total $\Omega \approx 0.1 - 1$, with some preference for smaller values ($\Omega \approx 0.2$).

To place the abundance question in perspective, it is useful to compare the mean density of dark matter with that of luminous matter—stars and gas in galaxies and galaxy clusters—$\Omega_{\text{lum}} \approx 0.01h^{-1}$. It is plausible that most of the ordinary, baryonic matter (with atomic nuclei made of baryons, i.e., protons and neutrons) in the universe does not emit radiation detectable using present technology. For example, planets, brown dwarfs, cold white dwarfs, neutron stars, and intergalactic gas are difficult to detect at large distances, although they are not dark in principle. Baryonic matter in these forms could increase the total baryonic density parameter to $\Omega_b \approx 0.05$–0.10. Support for this possibility comes from the theory of primordial nucleosynthesis, which predicts the abundance of the light isotopes of hydrogen, helium, and lithium produced during the first three minutes after the Big Bang. Excellent agreement with measurements is obtained for $\Omega_b = (0.02 \pm 0.01)h^{-2}$.

OBSERVATIONAL EVIDENCE FOR DARK MATTER

The first measurement of dark matter in the Galaxy was made by Jan Oort in 1932, who concluded that visible stars near the sun could account for only about half the mass implied by the velocities of stars perpendicular to the disk of our Galaxy. In 1933, Fritz Zwicky applied a similar dynamical argument to clusters of galaxies, noting that observed galaxies accounted for 10% or less of the mass needed to gravitationally bind clusters, given the large velocities of galaxies in a cluster. For a self-gravitating system in equilibrium, the mass is $M \approx RV^2/G$, where R is the characteristic size of the system and V is the characteristic velocity of stars or other test bodies in the system.

The most straightforward and extensive mass measurements have been made for spiral galaxies, for which V is the circular rotation speed at radius R and M is the mass interior to R. The rotation curve $V(R)$ has been measured for hundreds of spirals, using the Doppler shift of the optical Hα line or the radio 21-cm line of hydrogen. In almost all cases, $V(R)$ is nearly constant outside of the galactic nucleus, indicating a mass increasing linearly with radius or a density decreasing with the inverse square of the radius. Since the luminosity density typically decreases exponentially with radius, the mass-to-light ratio becomes large in the outer parts, implying that spiral galaxies are embedded in halos of dark matter. For our own Galaxy, with $V \approx 220$ km s^{-1}, M could be as large as 10^{12} solar masses if the halo extends to 100 kpc. Adopting a mean separation of $5h^{-1}$ Mpc, Ω in bright spirals alone is ~ 0.03.

Gravitational mass measurements have also been performed for elliptical galaxies, small groups, and rich clusters of galaxies. These measurements are less certain than those for spirals, largely because of the uncertainty of the distribution of stellar or galactic orbits in the systems analyzed. X-ray emission from hot gas in hydrostatic equilibrium in clusters should allow more precise determinations once accurate gas temperature measurements become available. The mass measurements of ellipticals, groups, and clusters confirm the existence of dark matter and increase the estimated total Ω in galaxies and clusters to ~ 0.1–0.2. Similar results follow from the cosmic virial theorem, a statistical method based on the relative velocities of all close pairs of galaxies.

Because galaxies and clusters occupy a small fraction of the volume of the universe, measurements on larger length scales are needed to obtain the total mean density in dark matter. Unfortunately, equilibrium structures larger than galaxy clusters do not exist, so that large-scale gravitational mass density measurements cannot be based on the simple formula $M \approx RV^2/G$. Instead, cosmologists apply the linear theory of gravitational instability in an expanding universe, supposing that the mass density fluctuations have small amplitude on large scales. The mass density is written $\rho = \bar{\rho} + \delta\rho$, where $\bar{\rho}$ is the mean density and $\delta\rho/\bar{\rho}$ is the spatially varying relative density fluctuation. When smoothed on the scale of superclusters of galaxies, $\delta\rho/\bar{\rho}$ should be, according to theory, related to the "peculiar" velocity field—the velocity remaining after the Hubble velocity of uniform cosmological expansion is subtracted—with a constant of proportionality depending on Ω. Measurements of Ω based on this relation have yielded values in the range 0.2 to 1, with a preference for small values. However, this technique suffers from a major problem. The net density contrast $\delta\rho/\bar{\rho}$ must be known, but the density on large scales is dominated by the unseen dark matter. In practice the assumption is usually made that on large scales dark and luminous matter are distributed similarly, so that $\delta\rho/\bar{\rho} = \delta n_g/\bar{n}_g$, where $\bar{n}_g + \delta n_g$ is the smoothed galaxy density. However, there is no empirical evidence supporting this assumption, and there are sound theoretical arguments suggesting that the galaxy distribution should be *biased* with respect to the dark matter distribution. In the simplest theoretical model, the galaxy distribution has a density contrast larger by a factor b, called the bias parameter, than the matter distribution: $\delta n_g/\bar{n}_g = b(\delta\rho/\bar{\rho})$. If $b \approx 2.5$, then the *apparent* Ω could be ~ 0.2, whereas the true $\Omega = 1$. This possibility is favored by theorists who advocate the inflationary-universe model, but presently is a conjecture neither confirmed nor refuted by observations.

It is possible to measure Ω on still larger scales, by computing the rate of deceleration of the Hubble expansion using observations of cosmologically distant objects. There are a variety of methods for accomplishing this, but all those employed to date suffer from large uncertainties of the structure and cosmological evolution of the objects studied.

THEORETICAL ISSUES

Many important theoretical questions are raised by the existence of dark matter. Perhaps the most obvious are: What is it? Is it baryonic? This possibility is marginally allowed if $\Omega \approx 0.1$, being consistent with primordial nucleosynthesis for $h = 0.5$ and with most dynamical determinations. However, the isotropy of the cosmic microwave background radiation imposes theoretical constraints on baryonic dark matter models that are difficult to satisfy.

If the dark matter is nonbaryonic, it probably consists of elementary particles without electromagnetic or strong interactions; for otherwise it should have been detected by now. There is no shortage of possible candidates proposed by particle physicists, although none except the neutrino are known to exist. Most of these dark matter candidates undergo weak nuclear interactions, so it should be possible to detect them in laboratory experiments of sufficient sensitivity. A key implication of the dark matter hypothesis is that these particles should be abundant in every laboratory on the Earth, with a flux $\sim 10^2$ cm^{-2} s^{-1} if the particle has mass comparable to a proton. Many experiments are underway to try to detect these particles.

From an astrophysical point of view, most of the properties of the dark matter are irrelevant. The one significant detail is the temperature of the dark matter distribution. Cold dark matter (CDM) particles have negligible random velocities before the epoch of galaxy formation, while hot dark matter (HDM) is hot enough to evaporate (erase by free streaming) galaxy-scale primordial density perturbations. Cosmological scenarios with hot and cold dark matter differ in that, in the former, galaxy formation occurs only after the fragmentation of cluster- or supercluster-sized sheets of collapsed matter ("pancakes"), whereas in the CDM model, galaxy formation and clustering proceeds hierarchically, with small objects merging to form larger ones. The latter scenario appears to be more consistent with the relative ages of galaxies and superclusters, but no model is entirely successful. Another problem with HDM is that it cannot cluster enough to provide the dark matter in dwarf galaxies. The best-known HDM candidate is a neutrino with mass $\sim 20h^{-2}$ eV c^{-2}, for example, the tau neutrino, whose experimental mass limit allows this possibility. The most widely discussed CDM candidates are axions, invoked to solve problems in the theory of quantum chromodynamics, and the lightest supersymmetric particle, which is predicted to be stable. Theories of unstable dark matter or of two or more types of dark matter have been advanced occasionally but they are not as appealing as the simple models with one stable dark matter particle.

Additional Reading

Faber, S. M. and Gallagher, J. S. (1979). Mass and mass-to-light ratios of galaxies *Ann. Rev. Astron. Ap.* **17** 135.

Krauss, L. M. (1986). Dark matter in the universe. *Scientific American* **255** (No. 6) 58.

Peebles, P. J. E. (1986). The mean mass density of the Universe. *Nature* **321** 27.

Rubin, V. C. (1983). Dark matter in spiral galaxies. *Scientific American* **248** (No. 6) 96.

Trimble, V. (1987). Existence and nature of dark matter in the Universe. *Ann. Rev. Astron. Ap.* **25** 425.

Tucker, W. and Tucker, K. (1988). *The Dark Matter*. William Morrow and Co., New York.

See also **Cosmology, Galaxy Formation; Cosmology, Theories.**

Diffuse Galactic Light

Adolf N. Witt

To the unaided human eye, the Milky Way appears as an irregular band of hazy light stretching across the sky in a great circle. A small telescope will resolve most of this hazy glow into light from numerous faint stars, as was first discovered by Galileo Galilei in 1610. However, a fraction of about 25% of the total integrated light of the Milky Way at visible wavelengths remains truly diffuse, even in the largest telescope. This component is referred to as the diffuse galactic light (DGL). The principal cause of the DGL is scattering of starlight by dust grains occurring throughout the interstellar space in our Milky Way galaxy.

The existence of the DGL and its possible significance was not recognized until shortly after Robert J. Trumpler's 1930 discovery of general interstellar extinction produced by the same dust grains. The first steps leading to the correct prediction, detection, and surprisingly accurate measurement of the DGL were undertaken by Otto Struve and his collaborators, C. T. Elvey, Franklin E. Roach, Louis G. Henyey, and Jesse L. Greenstein at the Yerkes Observatory between 1933 and 1941. It was from these measurements that the first meaningful constraints for the scattering properties of interstellar grains were derived.

OBSERVATION OF THE DGL

The difficulty of observing the DGL lies in the fact that it is only one minor component among several sources, contributing to the integrated brightness of the night sky for the ground-based observer. Some of the other components are the integrated light of stars in our galaxy; the zodiacal light, produced by interplanetary dust particles scattering sunlight; the permanent airglow, resulting from the emissions of light by ions, atoms, and molecules in the upper atmosphere of the Earth; and finally, light scattered by molecules and dust in the lower Earth atmosphere. With the exception of the direct starlight, all other components are diffuse in nature, and it is difficult to separate the DGL from this background.

The attempts to measure the DGL intensity in the past have followed mainly one of two approaches: the wide-field sky photometry method and the narrow-field approach designed to eliminate contributions from direct starlight at the outset. Wide-field sky photometry has the advantage that it can be carried out with very small telescopes; it has the disadvantage that it includes direct light from stars in the measured signal. The subtraction of the direct starlight from the total signal often is the source of major uncertainties in the final result. Narrow-field photometry avoids this problem, but it requires the use of a fairly large telescope to gather enough photons from preselected star-free fields, typically not more than one are minute in size.

For the separation of the DGL from the other diffuse components, one takes advantage of the strong concentration of the DGL toward the Milky Way plane. Observations can be timed such that all measurements are carried out at a constant zenith distance, essentially allowing the apparent motion of the sky to sweep the Milky Way plane across the viewing direction. Diffuse light components, such as the air glow and atmospheric scattering, which depend mainly on zenith distance, are thus kept at a constant level. Furthermore, if one schedules observations in seasons when the Milky Way section of interest lies close to the antisolar direction, then also the zodiacal light represents only a background which varies slowly with position. A photometric sweep across the Milky Way with these constraints will then produce a diffuse light signal, consisting of a high but essentially constant background underlying a spatially strongly varying DGL signal, which can then be separated.

In Table 1 typical values for the intensity of the different components of sky brightness at visible wavelengths are given for direc-

Table 1. Approximate Values for the Components of Sky Brightness*

	Spectral Region	
Source	*360 nm*	*430 nm*
Airglow and zodiacal light	140	80
Total star background	160	170
Diffuse galactic light	55	50

*Intensities are in S_{10} units (see text) for regions near the galactic equator.

tions near the galactic equator. The units used here are those most frequently employed in the photometry of the night sky, numbers of the tenth-magnitude stars per square degree which would produce the same integrated intensity.

In recent years, photometry of the DGL has increasingly been done from space. This meant that the wavelength coverage could be extended into the ultraviolet and much of the atmospheric components of the night sky light could be eliminated. The zodiacal light, having a spectrum similar to that of the Sun, also becomes insignificant at wavelengths below 200 nm. But, on the negative side, the fields of view of these ultraviolet studies have been larger than is desirable, leading to uncertain corrections for the direct starlight included in the measurements. Still, reasonably accurate data exist now for the intensity of the DGL to wavelengths as short as 150 nm for directions close to the plane of the Milky Way.

Important progress in observing the DGL at visible wavelengths has recently been made with an all-sky photometric survey by the *Pioneer 10* deep-space probe, which carried out these observations at distances beyond 3 AU from the Sun, where zodiacal light is negligible at all wavelengths in the antisolar direction. The results from this survey form an important data set for the combined intensity of integrated starlight and DGL, sampled with a 2.3-square-degree field of view. Accurate photometry of the stars and star counts can be done from Earth at the *Pioneer 10* wavelengths, thus offering an excellent opportunity for absolute DGL photometry.

INTERPRETATION OF THE DGL

Scattering Parameters of Interstellar Dust

The intensity of the DGL along a given line of sight in the Galaxy depends on the number, brightness, and spatial distribution of the sources of illumination, that is, stars, as well as on the number, distribution, and scattering properties of dust grains in that direction. Such information is generally not available except in a statistical sense. One proceeds, therefore, by constructing a highly idealized model for the optically observable section of Galaxy, incorporating this statistical information. With such a model, the generation and transfer of DGL can then be predicted, using the scattering properties of the dust grains as free, adjustable parameters. In an actual application, the process is inverted, meaning that model fits are produced for the observed brightness and distribution of the DGL and acceptable sets of scattering parameters are identified.

Given the uncertainties inherent in the problem, one is normally limited to the simplest description of the dust-scattering properties which are physically still reasonable. The ratio of the light scattered in all directions by a dust particle to the total light intercepted by the particle is specified by the albedo. As some of the light is normally absorbed in this process, the albedo ranges between zero and unity and is usually wavelength dependent as well. The scattered light is generally not thrown equally in all directions but is scattered asymmetrically, mostly in the forward direction for wavelength-sized grains. In addition to the albedo, a second parameter is therefore required to measure the asymmetry of the scattering pattern. It is convenient to choose the average cosine of the scattering angle, weighted over the pattern of scattered light, to measure this asymmetry. For forward-scattering grains, this parameter, g, also has a value between zero and unity. The wavelength dependence of g provides information about the size of grains most effective in contributing to the scattered light at a given wavelength.

Given that the interstellar grains along a line of sight include particles with greatly different chemical compositions, carbon grains and silicate grains among them, and with a wide spectrum of sizes, ranging from 1–1000 nm in diameter, scattering parameters derived from the DGL are the effective averages for such a composite mixture of particles, not the parameters of any grain type in particular. The average nature of the parameters is further enhanced by the highly idealized galaxy models, which at best describe some average galactic conditions. To make the application of such models meaningful, DGL data from as many regions of the Milky Way as possible should be included in the analysis.

Model Dependencies

It is difficult to assess how well the idealized Galaxy models reproduce the actual radiative transfer occurring in the Milky Way. The fact that a significant fraction of the illuminating radiation seen by a dust grain in the galactic plane is previously scattered light implies that multiple scattering must be included in realistic models. This is done easily, provided the assumed distribution of stars and dust is sufficiently simple, symmetric, and uniform. In the actual Milky Way, dust is concentrated in clouds of various sizes and geometries which occupy only a relatively small fraction of interstellar space. Comparisons of models with uniform dust layers and models where the dust is concentrated into clouds, but which have the same statistical properties otherwise, show the cloud models to produce less observable DGL for a given albedo and asymmetry parameter. One may conclude, therefore, that values for the dust albedo and the asymmetry parameter derived from simple models in which the sun is located in the central plane of a plane-parallel slab with similar uniform distributions of stars and dust with distance from the plane are generally too low. Additional model uncertainties arise when the distributions of illuminating sources and scattering dust with distance from the plane are no longer similar. This problem is real, because at longer wavelengths in the red the illumination is increasingly due to cooler giant stars, which have a galactic distribution extending well outside the dust layer, while in the ultraviolet the illumination of the same dust layer comes almost exclusively from hot O and B stars, which, aside from having a highly nonuniform distribution, are usually much more closely confined to the galactic plane than the dust. The application of uniform models to either one of these cases will lead to an albedo too low in both instances, and an asymmetry parameter too low in the first case and too high in the second case. The uncertainties in the derived scattering properties arising from observational difficulties and from model inadequacies are probably comparable.

OTHER CONTRIBUTIONS

While light scattering by interstellar dust is clearly the dominant continuum source of DGL in the visible and through much of the ultraviolet spectral regions, other diffuse emission processes contribute to the diffuse galactic background. In the red, at a wavelength of 656.3 nm, diffuse Balmer line radiation is detectable from many extended Milky Way regions, produced by the recombination of ionized hydrogen in interstellar space. Another source apparently contributing to the red diffuse background at wavelengths beyond 600 nm is dust fluorescence. In this process, some of the light energy absorbed by dust grains in the blue and ultraviolet parts of the spectrum is reemitted in a broad emission band, similar to one seen in the laboratory for photoluminescence by hydrogenated amorphous carbon.

In the ultraviolet, fluorescence by molecular hydrogen has recently been observed from interstellar space. Here, molecular hydrogen absorbs radiation shortward of 110-nm wavelength and isotropically reemits ultraviolet photons with wavelengths as long as 165 nm, thus enhancing the diffuse background between 100 and 200 nm. Finally, in the far-ultraviolet region between 95 and 127 nm, a diffuse emission line background has been detected from the Milky Way. This radiation has been interpreted as resulting from fluorescence of heavy elements interacting with the far-ultraviolet radiation field in interstellar space.

All these diffuse emission processes would be difficult to distinguish from dust scattered light in purely photometric measurements. Accordingly, DGL observations in the affected spectral regions may not be useful for the determination of dust-scattering properties.

SOME RESULTS

The analysis of existing measurements of the DGL suggests that the dust albedo in interstellar space ranges from about 0.6–0.7 in the visible-wavelength region to a minimum of about 0.35 near 220 nm. At wavelengths shortward of 220 nm there is evidence for a renewed increase of the albedo to a value of about 0.5 near 150 nm. The minimum near 220 nm coincides with a broad feature in the interstellar extinction spectrum, demonstrating that this feature results from an absorption process.

The asymmetry parameter g has a value near 0.7 in the visible-wavelength region, indicating that the scattering is dominated by particles with diameters about equal to the light wavelength. Similar values for g have been derived for the ultraviolet, but these results are in conflict with conclusions drawn from the observation of other interstellar scattering systems. More work is clearly needed.

Additional Reading

Bailey, M.E. and Williams, D. A., eds. (1988). *Dust in the Universe.* Cambridge University Press, Cambridge.

Martin, P. G. (1978). *Cosmic Dust: Its Impact on Astronomy.* Clarendon Press, Oxford.

Roach, F. E. and Gordon, J. L. (1973). *The Light of the Night Sky.* Reidel, Dordrecht.

Savage, B. D. and Mathis, J. S. (1979). Observed properties of interstellar dust. *Ann. Rev. Astron. Ap.* **17** 73.

See also **Background Radiation, Ultraviolet; Interstellar Medium, Dust Grains.**

Distance Indicators, Extragalactic

R. Brent Tully

There are two principal reasons for the great interest in the extragalactic distance scale. In the first place, the distances of objects must be known if we are to have a correct understanding of their properties, such as their masses or the energy that they produce. In the second place, the scale of the universe is linked with the age of the universe, and there is some difficulty in reconciling several different kinds of observations in the context of the standard world model. This review elaborates upon certain aspects of the entry Cosmology, Observational Tests.

DISCOVERY OF AN EXPANDING UNIVERSE

It was first firmly established that galaxies were independent self-gravitating islands of stars at large distances in 1925, when Edwin Hubble found that the Andromeda nebula (now called the Andromeda galaxy) and Messier 33 (M33) contained a certain kind of pulsating star called a Cepheid variable. It had been known from observations of the same kind of objects in the Small Magellanic Cloud made by Henrietta Leavitt in 1912 that the pulsation periods are closely correlated with the luminosities of the stars. Those that brighten and dim over a hundred days are intrinsically brighter by a factor of 20 than those that perform similar oscillations over a few days. The critical perception by Hubble was to appreciate that the pulsating objects in the separate systems were basically the same, but dimmed in Andromeda and M33 because of the very much greater distances.

Just a few years later, in 1929, Hubble made a second remarkable discovery that the spectra of almost all galaxies are displaced from our expectations based on laboratory experiments in the sense that the wavelengths of familiar spectral features are displaced toward longer values. Because observations at the time were only in the optical passband of the electromagnetic spectrum and the color red is at the long-wavelength limit of this band, the effect came to be called the redshift. The corollary of Hubble's discovery was that this redshift is larger for galaxies that are fainter and, hence, farther away. A description of the linear relationship between redshift and distance is given by the equation

$$H_0 = cz/d,$$

where $z = \Delta\lambda/\lambda$ is the spectral displacement, d is the distance of the object, c is the velocity of light, and H_0 is the Hubble constant. The subscript 0 indicates that we are referring to the present epoch. The standard interpretation of this redshift posits that the universe is in expansion, so systems that are farther away from an observer are rushing away with higher velocity, whence cz is a relative velocity.

The value of the Hubble constant is a measure of the scale of the universe, as it is the link between the observed relative velocities, or redshifts, of galaxies and their separations. Indeed, once the Hubble constant has been determined reliably, then the distance of a galaxy can be found by calculating the quotient of the galaxy's redshift divided by the Hubble constant. The smaller the value of the Hubble constant, the greater the distance between galaxies and, hence, the larger the observable universe. The inverse of the Hubble constant, $1/H_0$, is a measure of the age of the universe, because taking the simplest case, if objects with known separation are moving apart at constant velocity, the time when they were at the same point is defined.

HISTORICAL DEVELOPMENT OF THE MEASUREMENT OF H_0

The generally accepted value of the Hubble constant was revised radically during the first two decades after World War II, from a value of around 500 kilometers per second per megaparsec (km s^{-1} Mpc^{-1} where 1 Mpc = 3.3 million light-years = 3.1×10^{24} cm) to a value that, over the last two decades, has been argued to lie between 50 and 100 km s^{-1} Mpc^{-1}. One of the reasons for the drastic downward revision was the determination that the Cepheids observed in Andromeda, M33, and the Small Magellanic Cloud are, in fact, not exactly the same as the stars with similar behavior that are close by and provided the distance calibration. Hubble's idea was right, but observers were fooled by a quirk that two kinds of stars have similar properties. Moreover, in more distant galaxies where the pulsating Cepheids could not be seen, astronomers had been confusing star clusters and emission nebulae with individual stars, resulting in underestimated distances.

ALTERNATIVE DISTANCE ESTIMATORS

The example of the Cepheid variable was provided as a means of estimating distances. The correlation between the pulsation period and intrinsic luminosity emitted by such objects is so tight that the root-mean-square (rms) scatter corresponds to only a 10% uncertainty in distance and, if detailed information is available on a few dozen stars in a single galaxy, a distance with an internal error of only a few percent is possible. Systematic errors because of problems with intervening obscuration and scale calibration should be restricted to less than 10%. Unfortunately, Cepheid stars are not sufficiently bright to be easily detected in any but the nearest galaxies. There are other stars called RR Lyrae variables that also have well-delineated luminosities, but they are even fainter and more difficult to find at large distances. (Objects with such well-defined properties can be called "standard candles.")

About 10 galaxies within 4 Mpc have accurately known distances through the pulsating star methods. However, this limited domain is not relevant for the determination of the Hubble constant, because the nearest galaxies have relative motions that are dominated by local gravitational influences rather than the overall expansion of the universe. It is usually considered necessary to study objects beyond 10 Mpc to have a hope of measuring H_0. Until recently, there were no well-calibrated standard candles that were useful at these distances. All that were available were less-precise distance estimators, such as upper limits on the luminosities of the brightest stars in a single galaxy, or morphological characteristics of galaxies related to their intrinsic brightness, or a tendency for clusters of old stars to have integrated luminosities that might be the same from galaxy to galaxy.

Supernovae provide interesting possibilities as distance estimators, because they are so bright and can be seen at great distances. There is a particular kind, called Type Ib, that are especially promising as distance estimators because they appear to be restricted to a narrow range in luminosity at maximum light. The progenitors of the explosive event are taken to be white dwarf stars that are accreting mass from a close companion in a binary star system. If the white dwarf gathers enough mass from the companion to exceed the Chandrasekhar limit of 1.4 solar masses, then the inward pull of gravity overcomes the outward pressure provided by the lattice of degenerate matter that has supported the white dwarf. The subsequent free-fall collapse of the core to the state of a neutron star releases energy that is transferred to shocks that blow off the outer envelope. The precise mass of the progenitor at the time of the explosion might explain the limited range of observed luminosity maxima.

Though there is growing evidence that Type Ib supernovae are good standard candles, the problem remains that they must be calibrated. If a strictly empirical approach is taken, then good distances to a few nearby galaxies with Type Ib events must be known by alternative techniques. However, the situation in this regard is not satisfactory because supernovae are so rare. Alternatively, the luminosity of the supernova event can be reckoned from theoretical calculations. Such considerations lead to low values for H_0, implying a relatively large and old universe. However, these results are contradicted somewhat by results that will be discussed next. This contradiction means that either there is a problem with the theoretical supernova calculations or with the train of the ensuing discussion.

TULLY–FISHER, FABER–JACKSON, AND PLANETARY NEBULA METHODS

It can be argued that a value for H_0 at the level of 20% uncertainty is at hand. There appear to be several precision distance estimators that are giving consistent results and suggest that H_0 is on the high side of the range of modern values. The technique with the longest history as a precision distance tool is based on the Tully–Fisher relation. This method draws upon an empirically observed correlation between a measure of the internal motion of galaxies and their luminosities (see Fig. 1). The physical basis for the correlation is related to the fact that both motions and luminosities are tied to the mass of a system: More-massive galaxies are more luminous and rotate faster than less-massive galaxies. A simple measure of the internal motion of a spiral galaxy is given by the width of the neutral-hydrogen spectral line at wavelength 21 cm. This line is emitted by cold interstellar gas moving in near-circular orbits. The global width of this line is an observable directly related to the mass of the system and is unaffected by distance. The luminosity measured at optical or near-infrared bands depends on the number of ordinary stars and fades as the square of the distance.

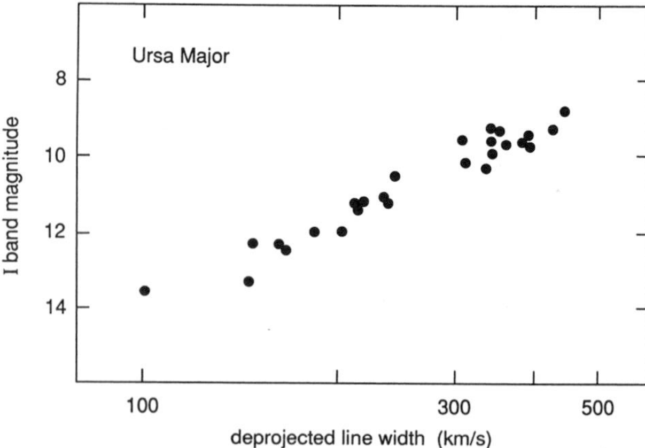

Figure 1. The Tully–Fisher relation between I-band (near-infrared) luminosities and neutral hydrogen profile widths for spiral galaxies in the Ursa Major cluster. Only luminosities are distance dependent, so a relative distance is given by the vertical displacement of an object relative to the mean regression fit to these data.

Depending on the authority, the luminosity-line width correlation has a dispersion that might be as low as to correspond to an rms uncertainty of only 15% or as high as to correspond to an uncertainty of 60%. If errors are anything like the optimistic expectation, then the true dispersion cannot be calibrated by other methods, which are inferior, and can only be ascertained by drawing samples from a few nearby clusters, where there are many galaxies at essentially the same distance. The optimistic expectation is sustained by studies of several well-defined samples. However, the dispersion has been found to be substantial in the direction of the Virgo cluster. This cluster has historically played a key role in the study of the distance scale because it is nearby and contains hundreds of members. At present, there is an ongoing debate. One viewpoint holds that we can measure distances accurately and that many of the objects in the line of sight toward the Virgo cluster are actually in the background. Hence, the observed scatter is large because of our inability to discriminate in favor of true members when the sample is chosen. The other viewpoint holds that the luminosity-line width distance estimator breaks down in the cluster environment, presumably because galaxy properties are altered, so our distance estimates are unreliable.

There appears to be a way to choose between these possibilities. Overwhelmingly, the background contaminants should be gas-rich spiral systems, as these kinds of galaxies are the dominant population outside the environment of rich clusters. Gas-poor ellipticals in the region should lie in the cluster, and distance estimates to these objects should not suffer from the contamination problem. In recent developments, methods have been found to establish good distances to elliptical galaxies.

The Faber–Jackson relation for gas-poor systems has the same physical basis as the Tully–Fisher relation for gas-rich systems. However, in lieu of measuring the motions of gas in near-circular orbits, one observes the spectral dispersion caused by the random motions of stars in a system with little global angular momentum (e.g., an elliptical galaxy). This measure related to the mass of the system is again correlated with the global luminosity, though the dispersion about the mean relation is greater and provides a less accurate distance estimate. However, it was subsequently found that the Faber–Jackson relation could be modified by incorporation of a third observable, a measure of the dimension of the system, and the improved relationship can provide distance estimates with rms uncertainties of 15%. There is a dearth of nearby ellipticals, but it appears possible to calibrate the zero point of the relationship using a reasonably reliable distance to a small group of galaxies in Leo that contains two ellipticals. In this manner, a distance is found for the Virgo cluster of 15 Mpc, which is a value in good agreement with the distance derived from gas-rich systems if the background contamination hypothesis is accepted.

Finally, a promising new procedure is based on properties of a phase of stellar death that produces planetary nebulae. The stars eject matter that is subsequently illuminated by the star. Observations of this envelope are made in a spectral emission line of oxygen. There appears to be an abrupt high-luminosity cutoff in the nebular emission corresponding to a central star with about two-thirds of a solar mass. The limit is presumably due to the combined statistics of the evolution time scale for the planetary nebula phase, which decreases sharply for more massive stars, and the number of stars at a given mass which, again, decreases toward high mass. This cutoff can be established in galaxies at moderate distances and the displacement in luminosity with respect to the calibration gives a distance. To date, distances have been established to about a dozen galaxies with this method, and the agreement with the other precision estimates is excellent. Most recently, this method has been used to derive a distance to the Virgo cluster of 15 Mpc.

SUMMARY

There has been an ongoing dispute over the value of the Hubble constant, which provides a measure of the distances between galaxies. Modern estimates range between 50 and 100 km s^{-1} Mpc^{-1}. Now there are consistent measurements that place the Virgo cluster at a distance of 15 Mpc$\pm$10%. Some of the earlier dispute must be attributed to the background contamination problem. If the presently postulated distance for the cluster is sustained, then a value for the Hubble constant can be determined, since the Virgo cluster has a velocity (corrected for flow motions) of 1300 km $s^{-1}\pm$10%. Hence, $H_0 = V/d = 87$ km s^{-1} $Mpc^{-1}\pm$14%. Perhaps the greatest additional uncertainty is in the corrections necessary to account for flow patterns and random motions. However, if techniques compatible with those that lead to the Virgo cluster distance are used for other samples, consistent results are found.

The corresponding characteristic age for the universe is $1/H_0 =$ 11 ± 3 billion years (estimated 95% probability). In the standard model with $\Lambda = 0$, if the density is equal to the critical value for closure (preferred by some theories), the age of the universe would be 7 ± 2 billion years, and if the density is equal to 10% of the critical value (comparable with observations), the universe would be 10 ± 3 billion years old. This estimate is in only marginal agreement with values of 13–18 billion years for the oldest star clusters in the Galaxy, which cannot be older than the universe. We are not compelled to consider more elaborate models of the universe, given the uncertainties, but the situation is not satisfactory. If the proposed distance scale is correct, then Type Ib supernovae are less bright than anticipated by theory by almost a factor of 2.

Additional Reading

Madore, B. F. and Tully, R. B., eds. (1986). *Galaxy Distances and Deviations from Universal Expansion*. Reidel, Dordrecht.

Rowan-Robinson, M. (1985). *The Cosmological Distance Ladder*. W. H. Freeman, New York.

van den Bergh, S. (1975). The extragalactic distance scale. In *Galaxies and the Universe*, A. Sandage, M. Sandage, and J. Kristian, eds. University of Chicago Press, Chicago, p. 509.

van den Bergh, S. and Pritchet C. J. (1988). *The Extragalactic Distance Scale* (Astronomical Society of the Pacific Conference Series No. 4). Brigham Young University Press, Provo, UT.

See also **Cosmology, Observational Tests; Hubble Constant; Virgo Cluster.**

E

Earth, Atmosphere

Joel S. Levine

The Earth's atmosphere begins at the surface and eventually merges with the interplanetary medium. About 99% of the total mass of the atmosphere resides in the troposphere (surface to 10–15 km) and stratosphere (10–15 to 50 km). The three major permanent gases in the atmosphere account for about 99.96% of the total volume of the atmosphere: nitrogen (78.08% by volume), oxygen (20.95% by volume), and argon (0.93% by volume). Ironically, it is the gases in very minute or trace quantities that control the chemistry of the atmosphere and the temperature of our planet via the greenhouse effect. It is believed that the atmosphere originally formed via the release of gases that had been trapped in the Earth's interior during its formation some 4.6×10^9 years ago. Over geological time and up to the present time, the chemical composition of the atmosphere has changed significantly, primarily as a result of the biogeochemical cycling of the elements, through the presence of life, a significant source of gases to the atmosphere.

STRUCTURE OF THE ATMOSPHERE

The Earth's atmosphere extends to several thousand kilometers above the surface, where it eventually merges with the medium of interplanetary space. The total mass of the atmosphere (5.1×10^{21} g) is very small compared with the total mass of the oceans (1.39×10^{24} g) and the total mass of the Earth (5.98×10^{27} g). Yet this thin gaseous envelope that surrounds our planet has been, and is today, of critical importance to life. Beginning about 4.6×10^9 years ago with the process of chemical evolution (i.e., the synthesis of complex organic molecules, the precursors of living systems, from the simple gases in the prebiological paleoatmosphere), there has been a continuing interaction between life and the atmosphere. Life is protected from biologically lethal solar ultraviolet radiation by the presence of atmospheric ozone, which is produced via photochemical processes from oxygen, a biogenic gas. Over the history of the Earth, living organisms have significantly modified, and to this day, still continue to modify the chemical composition of the atmosphere.

The troposphere extends from the surface to about 10–15 km, depending on latitude. About 80–85% of the total mass of the atmosphere resides in the troposphere. The different regions in the atmosphere are defined by their respective temperature gradients. Temperature decreases with altitude through the troposphere. The troposphere is in direct contact with the biosphere, and hence, regulates or modulates the transfer of gases and particulates in and out of the biosphere. Almost all of the water vapor in the atmosphere is found in the troposphere (see Table 1) where its distribution is controlled by the evaporation-condensation cycle. The troposphere is the region of weather phenomena, that is, cloud formation and precipitation. Most of the remaining mass of the atmosphere is located in the stratosphere (10–15 to 50 km), directly above the troposphere. About 90% of the ozone in the atmosphere is located in the stratosphere (see Table 1). The absorption of biologically lethal solar ultraviolet radiation (200–300 nm) by ozone makes the stratosphere of great importance to the biosphere and to life. The temperature of the atmosphere increases with altitude through the stratosphere.

Above the stratosphere is the mesosphere (50–90 km), which is a region of decreasing atmospheric temperature with altitude. The thermosphere (90–500 km), a region of increasing atmospheric temperature with altitude, is the next higher region. The exosphere, which begins at about 500 km and eventually merges with the interplanetary medium, is a region of constant temperature with altitude. Light atmospheric gases in the exosphere, such as hydrogen and helium, may escape from the Earth.

Superimposed on the structure of the neutral atmosphere are several regions of high concentrations of electrons and charged atoms and molecules resulting from the ionization of neutral gases by high-energy solar (x-rays and ultraviolet) radiation and cosmic rays. The regions of charged particles are found in four layers collectively termed the ionosphere: the D-layer (below about 90 km), the E-layer (90–120 km), the F-1 layer (a daytime feature centered at about 150 km) and the F-2 layer (centered at about 300 km). The magnetosphere or radiation belts, another region of charged particles, mostly of solar origin and contained by the Earth's magnetic field, extends out to several planetary radii. With two notable exceptions—water vapor and ozone—the chemical composition of the atmosphere is fairly constant up to about 100 km, although the number density of the atmosphere (molecules per cubic centimeters) decreases exponentially with altitude.

Table 1. Major and Trace Gases in the Atmosphere

Gases	*Concentration*
Major Gases	
Nitrogen (N_2)	78.08% by volume
Oxygen (O_2)	20.95% by volume
Argon (Ar)	0.93% by volume
Selected Trace Gases	
Water vapor (H_2O)	0 to 1–2% by volume
Carbon dioxide (CO_2)	350 ppmv
Ozone (O_3)	
In troposphere	0.02–0.1 ppmv
In stratosphere	0.1–10 ppmv
Methane (CH_4)	1.7 ppmv
Nitrous oxide (N_2O)	0.31 ppmv
Freon-12 (CF_2Cl_2)	0.3 ppbv
Freon-11 ($CFCl_3$)	0.2 ppbv

The concentrations of atmospheric gases are given in either percent by volume, which is the same as parts per hundred by volume, or parts per million by volume (ppmv), parts per billion by volume (ppmb), or parts per trillion by volume (pptv).

COMPOSITION OF THE ATMOSPHERE

The atmosphere of our planet contains several hundred different gases. These gases have very diverse origins. Some of these gases were originally trapped in the interior during the formation of our planet 4.6×10^9 years ago. These trapped gases were released by the process of "volatile outgassing," which led to the formation of the atmosphere very early in the Earth's history. Some atmospheric gases result from volcanic activity and the biogeochemical cycling of elements between the biosphere and the atmosphere. Other atmospheric gases are formed by the combustion of fossil fuels and the burning of living and dead biomass (trees, vegetation, grass, etc.), and by different industrial, technological, and agricultural activities. Finally, some gases are chemically transformed from one compound to another by atmospheric chemical and photochemical processes.

The major gases and some selected "trace" gases of the atmosphere are listed in Table 1. Inspection of this table indicates that about 99.96% by volume of the atmosphere is due to the presence of the three major permanent constituents of the atmosphere: nitrogen (N_2)—78.08% by volume, oxygen (O_2)—20.95% by volume, and argon (Ar)—0.93% by volume. (Atmospheric water vapor has a variable concentration that ranges from a very small fraction of a percent to 1–2% by volume, depending on local meteorological conditions.) The remaining constituents of the atmosphere are trace gases, the atmospheric abundances of which are too low to be measured in percent. Instead, the concentrations of the trace gases are measured in terms of "mixing ratio," rather than in percentage by volume. The surface mixing ratio is the number of atoms or molecules of a trace gas at the surface of the Earth, divided by the total number of atmospheric molecules (nitrogen and oxygen and argon and all the others) at the surface in a 1-cm^3 volume. The major permanent gases in the atmosphere, nitrogen, oxygen, and argon, are measured in percent by volume, which in terms of mixing ratio is the same as parts per hundred by volume (pphv). Trace gases are measured in mixing ratio units of parts per million by volume (ppmv), parts per billion by volume (ppbv), and parts per trillion by volume (pptv).

From an astronomical perspective, the Earth's nitrogen-oxygen atmosphere is very unique. The atmospheres of Venus and Mars are predominantly carbon dioxide (96% by volume for Venus and 95% for Mars), with smaller amounts of nitrogen (4% for Venus and 3% for Mars; with argon accounting for about 2% of the atmosphere of Mars). The atmospheres of Jupiter and Saturn are predominantly molecular hydrogen (H_2) (89% for Jupiter and 94% for Saturn), with helium accounting for the remainder of these atmospheres. On Jupiter and Saturn, carbon, nitrogen, and oxygen are primarily present in the form of saturated trace hydrides (methane, ammonia, and water vapor) at approximately the solar ratios of carbon, nitrogen, and oxygen. In addition to photosynthetic oxygen, our atmosphere contains several other important trace gases of biological origin, including nitrous oxide, ammonia, methane, hydrogen sulfide, and dimethyl sulfide.

ORIGIN OF THE ATMOSPHERE

It is generally believed that during the period of volatile outgassing in the Earth's early history, the mixture of trapped gases that came through the crust was not unlike that emitted by present-day volcanoes. This mixture most probably was composed of water vapor (80% by volume), carbon dioxide (about 12% by volume), sulfur dioxide (about 6% by volume) and nitrogen (about 1–2% by volume). Clearly the composition of the present atmosphere (Table 1) bears very little resemblance to the composition of the outgassed volatiles. The bulk of the H_2O that outgassed from the interior condensed out of the atmosphere, forming the oceans. Only small amounts of H_2O remained in the atmosphere, with almost all of it confined to the troposphere. At the surface, the H_2O concentration is variable, ranging from a fraction of a percent to 1–2% by volume. At the top of the troposphere, H_2O has a mixing ratio in the parts per million by volume range.

Most of the CO_2 that outgassed over the Earth's history formed sedimentary carbonate rocks [calcite, $CaCO_3$, and dolomite, $CaMg(CO_3)_2$] after dissolution in the ocean. The mixing ratio of CO_2 in the present atmosphere is about 350 ppmv (0.035% by volume). It has been estimated that the preindustrial (around the year 1860) level of atmospheric CO_2 was about 280 ppmv (0.028%). For each CO_2 molecule presently in the atmosphere, there are about 10^5 CO_2 molecules incorporated as carbonates in sedimentary rocks. All of the carbon presently in sedimentary rocks outgassed from the interior of the Earth and was at one time in the atmosphere in the form of CO_2. Hence, the early atmosphere may have contained orders of magnitude more CO_2 than the present atmosphere.

Sulfur dioxide is chemically active and is quickly transformed to other forms of sulfur, particularly sulfate aerosols, which fall out of the atmosphere. Hence, sulfur dioxide and sulfate aerosols have very short residence times in the atmosphere.

Molecular nitrogen is chemically inert (unlike SO_2), non-water-soluble (unlike CO_2), and noncondensable (unlike H_2O). Hence most of the outgassed nitrogen accumulated in the atmosphere over geological time to become the most abundant constituent (78.08% by volume). Argon, the third most abundant permanent constituent of the atmosphere (0.93% by volume), resulted from the radiogenic decay of potassium-40 in the Earth's crust.

It is important to note that oxygen, the second most abundant constituent of the atmosphere (20.95% by volume), is not released via volatile outgassing or volcanic activity. Oxygen evolved as a result of photosynthetic activity. Accompanying and directly controlled by the buildup of atmospheric oxygen was the evolution of ozone (O_3), which is produced from oxygen as a result of atmospheric photochemical and chemical transformations. The development of the stratospheric ozone layer had a very significant impact on the evolution of life on our planet. Throughout most of its 4.6×10^9 year history, there was an insufficient level of ozone in the atmosphere to shield the Earth's surface from the biologically lethal ultraviolet radiation emitted from the Sun. About 600×10^6 years ago, as oxygen built up in the atmosphere (as a result of photosynthetic activity), ozone reached sufficient levels in the atmosphere to effectively absorb incoming solar ultraviolet radiation. At this point, for the first time in history, life could safely leave the ocean and go on land. Before the development of the atmospheric ozone layer, life was restricted to a depth several meters below the ocean surface, since this quantity of seawater can absorb solar ultraviolet radiation and offered the only protection to early life.

GLOBAL ATMOSPHERIC CHANGE

As already noted, the surface of our planet and the myriad of life forms that reside there, including the human species, are shielded by atmospheric ozone from the lethal ultraviolet radiation (200–300 nm) emitted by the Sun. About 90% of the total atmospheric content of ozone is located in the stratosphere (10–15 to 50 km). Ground-based, aircraft, and satellite measurements have indicated an alarming trend in the level of ozone in the stratosphere over the Antarctic over the last two decades. These measurements indicate that for a 4–6-week period, beginning in late September (the beginning of the spring in the southern hemisphere), the total amount of ozone has decreased by more than 50% over the last decade. The measurements indicated that the size of the Antarctic ozone hole, which was about the size of the continental United States in 1987, has increased each year over this time period. From 4–6 weeks after the first appearance of the Antarctic ozone hole in late September, ozone from the southern hemisphere midlatitudes is transported to the South Pole by the general circulation patterns of the atmosphere and replenishes the missing ozone. Then the following spring the ozone hole reappears.

Is there evidence for stratospheric ozone depletion outside of the Antarctic stratosphere? The Ozone Trends Panel formed by NASA in October 1986 in collaboration with the National Oceanic and Atmospheric Administration, the Federal Aviation Administration, the World Meteorological Organization, and the United Nations Environment Program was charged with the reanalysis and reassessment of the data on ozone abundances from the 30-year-old network of ground-based Dobson spectrophotometers concentrated in the northern hemisphere, and, since the late 1970s, from satellite instruments. In March 1988, the Panel reported their findings: Global levels of stratospheric ozone had fallen several percent between 1969 to 1986. Between 30 and 64°N, from 1969 to 1986, stratospheric ozone decreased between 1.7 and 3.0% with larger decreases in the winter and at high latitudes, and smaller decreases in the summer. The observed decreases in stratospheric ozone in the Antarctic and northern hemisphere are attributable to increasing atmospheric levels of atomic chlorine and chlorine oxide in the atmosphere. These chlorine compounds result from chlorofluorocarbons (CFCs), a family of man-made gases used primarily in aerosol spray cans, such as Freon-11 and Freon-12. Freon-11 and Freon-12 are also used in refrigeration and air-conditioning systems and in styrofoam containers. There is evidence to suggest that concentrations of ozone in the troposphere, which contains the remaining 10% of atmospheric ozone, may have been increasing by about 1% per year over the past 20 years. This means that the decreases in stratospheric ozone may exceed those for the total column content, which is the sum of tropospheric and stratospheric ozone. In the troposphere, ozone is chemically formed from the combustion products of fossil fuel burning, including automobile exhaust gases. It is ironic, but while ozone in the stratosphere is beneficial to human health and survival, tropospheric ozone is a pollutant and harmful to human, animal, and plant life in high concentrations.

Although ozone in the stratosphere appears to be decreasing, measurements indicate that some trace gases in the troposphere are increasing with time. Increasing trace gases in the troposphere, in addition to ozone, include carbon dioxide (CO_2), methane (CH_4), nitrous oxide (N_2O), and the chlorofluorocarbons, such as Freon-11 ($CFCl_3$), and Freon-12 (CF_2Cl_2). Atmospheric levels of Freon-11 and Freon-12 are each increasing about 5% per year; methane is increasing about 1–2% per year; carbon dioxide is increasing about 0.4% per year, and nitrous oxide is increasing about 0.3% per year. All six of these gases are greenhouse gases. Greenhouse gases have the ability to absorb or trap infrared or heat energy emitted by the Earth's surface. The greenhouse gases quickly reemit or release the absorbed heat energy with approximately 50% of the reemitted energy redirected back towards the Earth's surface. This results in additional heating of the surface by the energy that would have been lost to space were it not for the trapping by the greenhouse gases. Hence, the action of the greenhouse gases results in an additional warming of the Earth's surface.

Theoretical climate calculations indicate that a doubling of atmospheric carbon dioxide would increase the global atmospheric surface temperature by as much as 4.5°C. A doubling of nitrous oxide would raise the Earth's surface temperature by 0.3–0.6°; a surface temperature increase of about 0.5°C would result from a doubling of Freon-11, and another 0.5°C increase would result from a doubling of Freon-12; a 0.2–0.4°C temperature increase would result from a doubling of methane. Theoretical calculations also indicate that a global heating of several degrees would be amplified by a factor of 2–4 at the poles. Such a temperature increase at the poles would result in the melting of large amounts of surface ice and snow. This melting, coupled with the thermal expansion of seawater, resulting from increased global temperatures, would raise the height of the ocean surface and lead to global flooding. The sources of the greenhouse gases that may lead to global warming and flooding are varied. Major sources of atmospheric carbon dioxide are the burning of fossil fuels (50–75% of the total CO_2) and the burning of biomass, that is, trees, grasslands, agricultural stubble (25–50%). Nitrous oxide is produced by bacterial activity in soils and by biomass burning. Sources of methane include bacterial activity in wetlands, enteric fermentation in ruminants, and biomass burning. As already noted, Freon-11 and Freon-12 sources are aerosol spray cans, refrigeration and air-conditioning systems, and styrofoam containers.

Additional Reading

Brasseur, G. and Solomon, S. (1984). *Aeronomy of the Middle Atmosphere: Chemistry and Physics of the Stratosphere and Mesosphere*. Reidel, Dordrecht.

Goody, R. M. and Walker, J. C. G. (1972). *Atmospheres*. Prentice-Hall, Englewood Cliffs, NJ.

Levine, J. S., ed. (1985). *The Photochemistry of Atmospheres: Earth, the Other Planets, and Comets*. Academic Press, Orlando, FL.

Levine, J. S. (1989). Planetary atmospheres. In *Encyclopedia of Astronomy and Astrophysics*, R. A. Meyers, ed. Academic Press, San Diego, p. 361.

Walker, J. C. G. (1977). *Evolution of the Atmosphere*. Macmillan, New York.

Wayne, R. P. (1985). *Chemistry of Atmospheres*. Clarendon Press, Oxford.

See also **Earth, Magnetosphere and Magnetotail; Planetary and Satellite Atmospheres.**

Earth, Aurora

Asgeir Brekke

The aurora or the polar light is a magnificient, colorful, heavenly display that rewards those who defy the long and dark polar nights. The gracefully waving curtains, brilliantly sweeping arcs, and magically folding bands have for generations inspired the imaginations of people living in the Arctic region. A dancing bevy of veiled ghosts often came to their minds, and for many tribes in the polar regions the aurora was related to the realm of their dead relatives and friends. At the west coast of Norway, for instance, there was an old saying when people talked unflatteringly about an old maid: "She is so old that she will soon be taken by the northern lights." Among the Eskimos in Greenland the northern lights were thought to be caused by a ball game among the spirits of their dead friends.

Maybe it was more common in Scandinavia than anywhere else to relate the aurora to the weather and use it as a prognostic sign for weather forecasting. In fact, there are names in the Scandinavian languages like "weather light," "wind light," etc., which all relate to the northern lights.

At lower latitudes the aurora was more often regarded as an ominous sign threatening people with war, famine, plagues, and death. It was not uncommon in medieval times for both the imperial and the clerical authorities to use a magnificient red aurora to compel their subjects to obedience, subject them to humiliation, and remind them to pay their taxes and tithes.

Some of the first attempts to explain the aurora can be found in the literature by Aristotle (384–322 B.C.), and even the Vikings had their own theories, published in the old Norse chronicle *The King's Mirror*, probably written about 1250 A.D.

The name aurora borealis, which can be directly translated to "the northern dawn," is a deceptive designation of a light phenomenon that does occur mainly in the north, but has nothing whatsoever to do with the dawn. The reason for this confusing nomenclature is the fact that Galileo Galilei (1564–1642), in his writing, mentioned a northern light which he had seen in 1618, and called it "boreale aurora." When Pierre Gassendi (1592–1655) wrote about the same phenomenon in 1621 he coined the phrase "aurora borealis." Since then this has been accepted as a scientifically acceptable form, although the Swedish physicist Anders Celsius (1701–1744) argued strongly for the more proper designation "lumine boreali" as a direct translation of the name "Nordurljos," which was used for the first time in *The King's Mirror*.

POSITION OF THE AURORA

One of the first important discoveries made in auroral physics was by Herman Fritz (1830–1893), a Swiss physicist. In his (1881) book *Das Polarlicht* he showed that the occurrence of the northern lights has a maximum zone around the polar region, close to 67°N. This zone has since been called the auroral zone. Similar results had earlier been reached by the American physicist Elias Loomis

(1811–1889) in 1860; for some reason, however, his work was not given the same attention.

In the Antarctic region the records are very few but a similar maximum zone of the "aurora australis" has also been established. One should note, however, that the auroral zone is a statistical concept only and does not represent the true location of the aurora at any given instant.

AURORAL OVAL

From the large network of auroral observatories established in the International Geophysical Year (IGY) 1956–1957, one of the most outstanding results was the concept of the auroral oval as originally presented by the Russian physicist Yasha Feldstein in 1963. In contrast to the auroral zone the auroral oval is supposed to characterize the instantaneous distribution of the aurora above the globe at any time. The auroral oval can be thought of as an irregular and dynamic, oval-shaped pattern, centered at the geomagnetic pole and oriented with respect to the Sun. The geomagnetic poles are the two theoretical points through which a magnetic axis would point if the Earth's magnetic field could be described as the field of one single gigantic dipole magnet centered close to the center of the Earth. Therefore the oval is centered neither around the geographic nor the magnetic pole but around these fictitious imaginary poles. The discovery of the auroral oval reveals that the Earth's magnetic field is a very important agent in determining the form and position of the auroras. The diameter of the auroral oval is on the average about 4000 km. During very disturbed conditions, however, it widens and becomes larger than 6000 km, and during periods of prolonged quiescence it may shrink to about 3000 km.

One of the main efforts in auroral physics during the last 30 years has been to establish the concept of the auroral oval as a reference frame for future analysis. Until about 1972 this research was founded on ground-based techniques like photography and photometry, all depending strongly on the capricious weather conditions in the polar regions. In 1972, however, the first satellite equipped with a photometric scanner was able to map the auroral oval from above and to confirm the picture already obtained by ground-based techniques.

Since then the technology of space research has advanced tremendously, and today it is even possible to take snapshots of the whole auroral oval a few minutes apart by optical imaging from satellites (see Fig. 1). These images demonstrate that the aurora is there all the time within a circumpolar illuminated band like a saint's glory on the top of the Earth.

SEQUENCE OF AURORAL DISPLAY

The Earth is rotating under the auroral oval and therefore an observer at an auroral zone station, where the aurora is most frequently seen at midnight, will observe a diurnal variation in the auroral activity and intensity according to his position with respect to the oval. By watching a typical auroral display during evening hours, one finds that the display often commences with a quiet, homogeneous arc-shaped aurora on the northern sky. As the time goes on, the arc moves southward and finally reaches the zenith. Sometimes the arc subsides and moves back northward, but at other times it breaks up in the zenith and an explosive-like display can be observed. Large and bright forms or fragments of forms appear to move in irregular patterns across the sky. After half an hour or so, the violent motion calms down and a faint glow is spread across the sky. As time goes on toward dawn, irregular faint patches and forms fluctuate in intensity.

ALTITUDE OF THE AURORA

At the turn of the century little was known about the altitude of the auroral displays. Numerous reports existed on the aurora being observed in front of mountains and even right down to sea level.

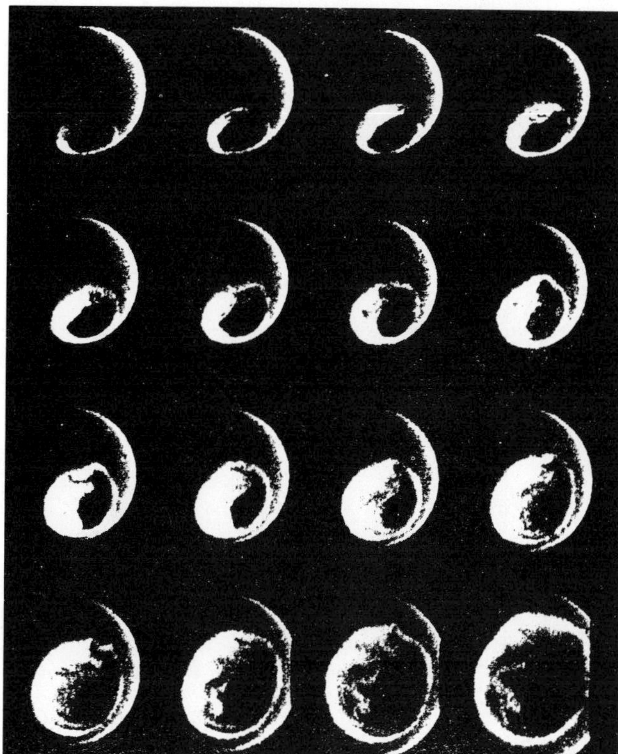

Figure 1. A series of images of the northern auroral oval on November 10, 1989. The observations are made at 130.4 nm and the color-coding is therefore "false." Notice the presence of the complete ring even at the daylit part of the hemisphere. (*By permission from L. A. Frank and J. D. Craven, University of Iowa.*)

Such tales were generally related by people without scientific training as good observers, and were most probably illusionary.

In 1892 the first known successful picture of an aurora was taken by the German physicist Martin Brendel (1862–1939) at Bossekop in Finnmark, Norway. This made it possible for the Norwegian physicist Carl Fredrik Mülerts Størmer (1874–1957) to try out a triangulation technique using two separate cameras in the mountains of Finnmark to measure the height of the aurora. From 1910–1952 he measured by photographic triangulation the height of 18,500 auroral points altogether. The distribution of his measurements by height shows a maximum of 1000 points at 100 km (see Fig. 2). Only 150 and 200 points were observed at 90 and 130 km, respectively. To date Størmer's work represents the most thorough efforts ever made to determine the altitude of the aurora, and it has formed and will continue to form a very important reference frame for many auroral discussions.

OPTICAL SPECTRUM OF THE AURORA

The optical spectrum of the aurora is characterized by numerous emission lines and bands from neutral as well as ionized atomic and molecular states of nitrogen and oxygen. In addition, lines from sodium and helium are occasionally observed, and hydrogen lines are regular features in proton aurora.

The auroral spectrum was probably first observed properly by the Swedish physicist Anders Ångström (1814–1874) in 1867. Since then many attempts have been made to identify the various lines and bands in auroral spectra.

The most prominent feature in the auroral spectrum is the yellow-green line at 5577 Å. It remained a mystery for many decades, but was finally identified in 1926 as a forbidden transition between the ^3P and ^1S states of atomic oxygen. The hydrogen lines

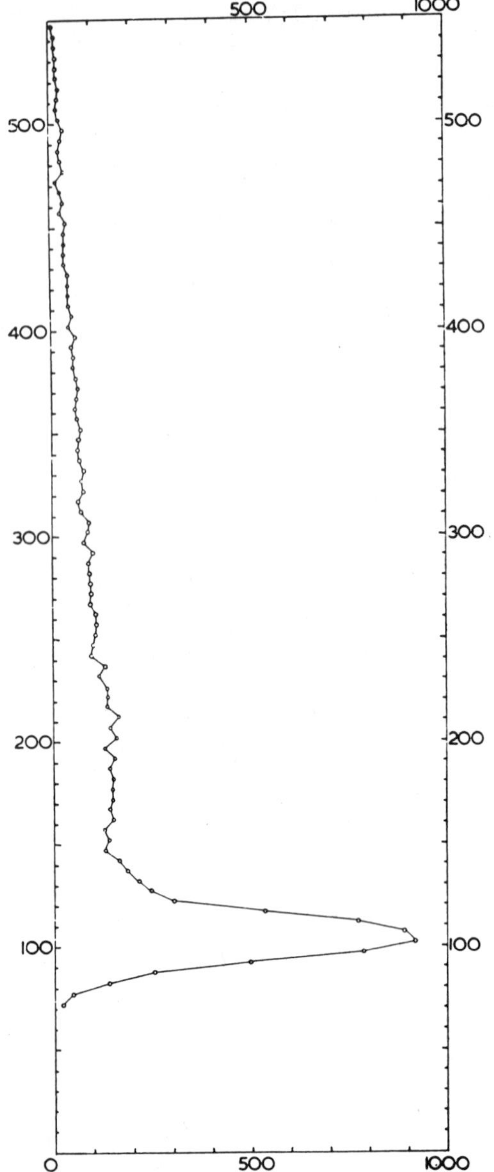

Figure 2. The distribution of the numbers of auroral height measurements (horizontal scale) with altitude in kilometers (vertical scale) obtained by Størmer showing a clear maximum at about 100 km. [*From Størmer, C. (1955).* The Polar Aurora, *Oxford.*]

were identified by the Norwegian physicist Lars Vegard (1880–1963) in 1939, and later in 1948 he found that these lines were Doppler shifted. These observations demonstrated clearly that protons were penetrating the polar atmosphere, causing part of the light emissions in aurora. Most of the auroras, though, are caused by precipitating electrons.

PHYSICAL MECHANISM CREATING THE AURORAL EMISSIONS

As stated earlier, an auroral display most typically takes place about 100 km above the Earth's surface. It is caused by energetic particles penetrating the Earth's atmosphere from interplanetary space. The fundamental clue to the formation of the aurora is therefore the interaction of a charged particle beam and the neutral atmosphere, a mechanism similar to that which causes light emissions in a discharge tube. The important difference, however, is that in the auroral laboratory there are no wall effects. The auroral process takes place in a free space in nature without any pollutant effects. Therefore, a study of the auroral mechanisms is a study of the mechanisms in a gigantic discharge in nature itself. Because the aurora takes place at such high altitudes, where the density and pressure of the neutral gas is extremely low, emissions from such metastable states as the ^1S state in oxygen can occur without being quenched by collisions with the ambient particles.

One of the most important objectives of auroral research has been to understand the interaction between a charged particle beam and the neutral atmosphere. For obvious reasons this research was forced to rely upon ground-based measurements until the late 1950s, when the first rockets were launched into auroral displays. Such rocket measurements have revealed that strong fluxes of precipitating electrons with energies below 100 keV and less strong fluxes of protons are present in aurora.

The energy distribution of the precipitating particles shows a monotonously decreasing background flux of particles between 100 eV and 100 keV. Overlying this background flux is often a strong peak in the flux of electrons between 5 and 10 keV, indicating that on their way toward the upper atmosphere the electrons have passed through an accelerating potential drop of a few kilovolts. These energy spectra can be expressed by the following formula:

$$N(E)\,dE = E^{\gamma}\exp(-E/\alpha).$$

Here $N(E)$ is the number of electrons with energies between E and $E + dE$, γ is a constant between zero and unity, and α is the e-folding energy, usually about 10 keV for the lowest auroral forms and 1 keV for the highest ones.

The excitation of the auroral spectra emitted from atmospheric atoms and molecules can be attributed to four classes of processes:

1. direct excitation by primary particles or secondary electrons,
2. thermal collisions involving ionized or excited atoms and molecules,
3. excitation by heated ambient electrons,
4. discharge mechanisms and heating by electric fields.

The excitation of atmospheric atoms and molecules during auroras is most directly performed by the primary particles and the secondary electrons, which have enough energy to create secondary excitation and ionization. Ionization by tertiary electrons may also occur.

Experimental results indicate that only about 30% of the total number of the ions and electrons are produced directly by fast primary electrons, and 70% are produced by secondary and higher-order processes. The number of ion pairs produced is about three for every 100 eV of energy. As far as can be judged from available data, relative excitation and ionization rates are fairly independent of the energy of the primary particles.

CROSS SECTION FOR EXCITATION AND IONIZATION

The fact that the primary particles are not very effective in exciting the particular molecules and atoms that are causing auroral emissions can be understood from a study of the most relevant energy levels of the excited particles. For atomic oxygen, for instance, the ionizing potential is 13.55 eV and in no transition level for auroral oxygen emissions is the energy above the ground state more than 15 eV. The energy level above the ground state for the ^1S state is 4.17 eV for the auroral green line at 5577 Å, the most dominant emission in the aurora. The cross section for excitation decreases as the inverse of the energy of the primary particle. For the nitrogen molecule the ionization potential is 15.5 eV and in no transition is the upper level more than 20 eV above the ground state.

The excitation and ionization cross section for the 5577-Å transition in oxygen peaks at 20 eV, and the ionization cross section for the nitrogen molecule peaks at about 100 eV. The excitation cross section of the $O(^1S)$ state is of the order of 2.5×10^{-18} cm^2 and much less than the cross section for the $N_2(A^3\Sigma)$ state, which has a peak value of 2.5×10^{-17} cm^2. Even so, the yellow-green line at 5577 Å is much stronger than the Vegard–Kaplan band in the aurora, which originates from the $N_2(A^3\Sigma)$ state. Both emissions, however, are due to so-called forbidden transitions.

AURORA AND BIRKELAND'S CURRENTS

As the aurora is caused mostly by intense electron beams entering the atmosphere from interplanetary space, there are electrical currents related to auroral displays. These currents form loops which create magnetic field variations that can be observed on the ground. This was first observed during the winter of 1741–1742 by Olof Peter Hiorter (1696–1750), an assistant to Celsius. Since then the magnetic field fluctuations at high latitudes were used as indicators of auroral activity, even if the aurora cannot be seen due to overcast or sunlight.

The Norwegian physicist Kristian Olaf Birkeland (1867–1917) postulated at about the turn of the twentieth century that the particle beams creating the auroral displays are entering the atmosphere from the outside and follow the Earth's magnetic field lines toward the poles. In doing so they carry with them the electrical currents which have to be closed in loops. One part of the loop is therefore a current in the Earth's atmosphere parallel to the Earth's surface along or adjacent to the auroral arcs. This current is continued along magnetic field lines toward and away from space, where they are closed in the magnetosphere far away from the Earth. The problem of the exact closure of the inward and outward field-aligned or Birkeland's currents in space is still unsolved, and has formed the basis for several satellite experiments. The exact relationship between these currents and the auroral forms is also an open question, especially because the downward current can either be formed by up-going negative ions or electrons or down-going positive ions, and the up-going current can be carried by the charged particles in the opposite motions. Large positive-ion beams reaching into space from the polar regions during auroral displays have in fact been observed.

Additional Reading

Brekke, A. and Egeland, A. (1983). *The Northern Light*. Springer, Berlin.

Eather, R. H. (1980). *Majestic Lights*. American Geophysical Union, Washington, DC.

Jones, A. V. (1974). *Aurora*. Reidel, Dordrecht.

Omholt, A. (1971). *The Optical Aurora*. Springer, Berlin.

See also **Earth, Geomagnetic Storms; Earth, Magnetosphere and Magnetotail.**

Earth, Figure and Rotation

Kurt Lambeck

The Earth's shape and rotation are time-honored subjects in the study of our planet. Both have their roots in nineteenth-century mathematics and both have received renewed attention with the development of modern technologies in astronomy and geophysics. Both subjects are of intrinsic scientific interest whose understanding requires a knowledge of physics, mathematics, and the Earth sciences. They also form essential parts of other areas of science. The charting of a satellite around the Earth requires a detailed knowledge of the Earth's shape and gravity field and the high-precision maneuvering of planetary space probes requires precise information on the orientation (the rotation) of the Earth in space. In both examples the needs could not be met by the classical terrestrial and astronomical methods, and it is the space methods themselves that are responsible for the improved knowledge of the gravity field and rotation of the planet.

SHAPE AND GRAVITY OF THE EARTH

When geophysicists describe the shape of the Earth they refer not so much to the geometric shape or topography as to the shape of its constant gravitational surfaces. For a fluid body, the geometric shape tends to match the gravitational shape. The shape and gravity of the planet are therefore two closely linked quantities. Gravity measured on the Earth's surface varies from place to place for several reasons. First, the distance of the point of measurement from the center of mass of the planet is not the same everywhere. Second, the apparent force of gravity at any point rotating with the Earth is the sum of the attraction of the remainder of the planet and of the centrifugal force. This latter component varies with the latitude of the measurement point. Third, any irregularities in the mass distribution within the Earth produce spatial variations in gravity. Geophysicists measure the gravity or shape of the Earth as a means of deducing this internal density structure and constraining models of crustal evolution or mantle dynamics, thereby contributing to the understanding of the dynamic processes that have shaped the planet through time.

The centrifugal component represents only about 0.3% of the gravitational force. Yet it has played a major role in shaping the Earth. At the equator the centrifugal force is a maximum, but of opposite sign to the attraction. At the poles the centrifugal force vanishes. If the planet responded as a fluid to the latitude-dependent force, then the Earth's shape would be an ellipsoid of revolution with its short axis directed along the rotation axis. Indeed the Earth's shape in the first approximation is such a shape, with a difference in equatorial and polar radii of about 21 km. This indicates that when subject to forces over long periods of time the planet responds essentially as if it were a fluid.

If the response of the Earth to the centrifugal force were complete, then the planet would be in a state of hydrostatic equilibrium in which all constant density surfaces also approximate ellipsoids of revolution. The figure of the Earth would not be very interesting. Upon closer observation the shape of the planet deviates from the idealized equilibrium figure by a few hundred meters and a more precise definition of the shape is required. This is the geoid, a level or equipotential surface that closely approximates the ocean surface. At sea the geoid can be determined directly from gravity measurements and on land the shape can be inferred from gravity measurements and heights, measured by spirit leveling, above sea level.

Traditionally the gravity measurements were laboriously made using gravity meters and global coverage of the data was limited. But even with such incomplete data, it was realized that these observations pointed to a planet that was out of equilibrium, with lateral variations in its interior.

Over the past few decades a powerful new method has become available for measuring the Earth's gravity field or gravitational potential. The motion of a satellite about the Earth is affected by the gravitational attraction of the planet. For a spherically symmetric planet this motion would be an ellipse following Kepler's laws. For an oblate spheroid this ellipse would precess in space. The careful monitoring of a satellite's motion reveals small perturbations superimposed upon the precessing elliptic motion, and this permits the gravitational field to be deduced if one accounts correctly for other forces, including solar radiation pressure, drag forces from a tenuous atmosphere at satellite elevations, and the gravitational attraction of the Sun and Moon. Early results obtained soon after the launch of the first satellites confirmed the conclusions drawn from the terrestrial measurements. A vigorous

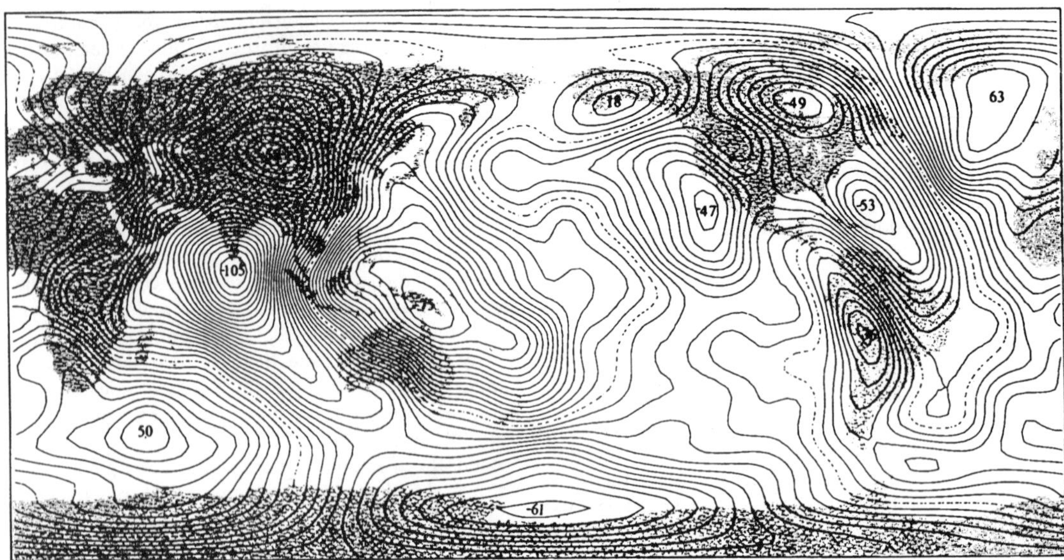

Figure 1. The shape of the Earth. The contours represent the departures of the geoid from a best-fitting ellipsoid. The dashed line indicates zero geoid height, and the contour interval is 5 m. [*From Lambeck, K.,* Geophysical Geodesy, the Slow Deformations of Earth, *Oxford University Press (1988).*]

debate developed as to whether these gravity or geoid anomalies reflected conditions early in the Earth's history, as appears to be the case for the Moon, or whether they reflected dynamic processes occurring within the mantle. These early results were to play an important role in concluding that the Earth's mantle is a convecting region and that the Earth is a mobile planet indeed.

As the accuracy of tracking of satellites has increased and the understanding of the other forces has improved, so has the spatial resolution of the geoid improved (Fig. 1). Further improvements for very high-resolution measurements of the geoid have come from radar altimeter observations of satellite height above the ocean surface from which the shape of the geoid (relative to the satellite orbit) can be deduced directly. As a result, the gravity field is now one of the best-known geophysical quantities of the Earth.

The interpretation of the field is actually more of a problem. The existence of gravity or geoid anomalies indicates either (1) that lateral density anomalies occur, in which case regions of the planet must be sufficiently strong to support such anomalies, or (2) the anomalies are supported dynamically by convection currents. It is now widely accepted that for the longer-wavelength variations, as illustrated in Fig. 1, the latter interpretation is correct. However, the estimation of the density anomalies and the mapping of these convection currents remain an insoluble problem. They key difficulty is the nonuniqueness of the interpretation of a potential field. An infinite number of density distributions can, in principle, give rise to the same gravity anomaly. Only a few of these will be plausible within a framework set by other geophysical observations and models, but what is plausible to one geophysicist may be improbable to another! Other supporting geophysical observations are required and one of the exciting recent developments has been the three-dimensional mapping of the Earth's interior by seismic methods.

Shorter-wavelength variations in gravity and the geoid reflect mainly lateral variations in the density of the crust and uppermost mantle. When used in conjunction with seismic and geological data they permit models of tectonic history to be established, from which conclusions may be drawn about crustal evolution and the forces acting on the crust or, about the likelihood of hydrocarbon or mineral deposits.

ROTATION OF THE EARTH

At first glance, changes in the rotation of the Earth may seem to be relatively simple: response of an oblate planet to the gravitational torques exerted upon it by the Sun and Moon. This variation is the precession in which the rotation axis traces out a circular path against the star background about the pole of the ecliptic. Superimposed upon this are minor oscillations, the nutations, produced by the periodic fluctuations in these gravitational torques. If the Earth were rigid this would be a mathematical problem of little physical interest. The actual problem is both more complex and compelling.

As part of the precession and nutation the Earth's rotation axis is not quite fixed relative to the crust. It is also predicted to trace out a periodic path about some average position fixed in the crust. This is the nutation predicted by Leonhard Euler in 1765. For a rigid Earth the period of this motion should be about 305–310 days, and is determined directly by the oblateness. For over a century, distinguished mathematicians and astronomers sought in vain for observational evidence for this term, only to have an amateur astronomer point out that it actually occurs with a period of about 430 days and an amplitude of about 0.1 arcsec. The correction from 310 days is now well understood in terms of the Earth's departure from rigidity due to its fluid core, anelastic mantle, and oceans.

The Chandler wobble, as it is now known, represents a free oscillation of the Earth. If it persists, as it has for the past 100 years or more, it must be continually excited to overcome the damping effects of internal friction. The nature of this excitation mechanism and where the rotational energy is dissipated remain unresolved questions whose solutions must lead through solid Earth physics, seismology, geomagnetism, oceanography, and meteorology. Is the wobble excited by aperiodic changes in the spatial distribution of air mass, or is it excited by earthquakes? Is the wobble energy dissipated in the oceans or in the solid Earth? Answers to these questions have important consequences for other areas of geophysics.

The traditional observation of polar motion has used the methods of optical astrometry, by measuring the zenith distance of known stars at the time of meridian transit. The accuracy of these results is relatively low compared with the methods of very long baseline ratio interferometry (VLBI). The new record is still too short to resolve questions about the Chandler wobble, but it is sufficiently promising to give the subject a new lease on life.

In addition to the 430-day wobble of the rotation axis relative to the crust, an annual oscillation of comparable amplitude is also observed, and is readily explained in terms of a seasonal redistribution of mass in the atmosphere, hydrosphere, and oceans. A long-period oscillation of decade time scale has also been reported and attributed to the dynamics of the core. But this may be a figment of the noisy astronomical data and its existence remains to be verified when long series of new observations become available. Better established is a small secular drift of the pole in a westerly direction at a rate of about 0.002 yr^{-1}. This is generally attributed to the interaction between the melting of the ice sheets in the Late Pleistocene time and a deformable Earth.

A third aspect of the Earth's rotation concerns the angular velocity about the instantaneous rotation axis. Astronomers have traditionally observed the time of transit of stars across their meridian; the interval between successive transits of the same star determines the length of the day. Successive observations indicate that the quantity varies perceptibly from day to day, and that the time kept by the Earth is not uniform when compared with highly accurate clocks. Typically, the length of the day fluctuates by about 1 part in 10^8, or about 0.001 s. If this change persists, however, the integrated amount by which the Earth is slow or fast becomes very significant.

The observed quantity is the difference τ between the time kept by the Earth, universal time (UT), and a uniform reference time, atomic time (AT) or, by definition

$$\tau = -(\mathrm{UT} - \mathrm{AT}),$$

and the change in length of day Δ(l.o.d.) is

$$\frac{\Delta(\mathrm{l.o.d.})}{\mathrm{l.o.d.}} = -\frac{d\tau}{dt}.$$

Atomic time has only been available since about 1955, and before that the reference time had to be established from the astronomical observations themselves, using a time scale that satisfied the equations of motion of the Moon and planets (ephemeris time, ET). ET was relatively poorly determined and, in consequence, only changes in the length of the day that persisted for long periods of time ($\geq$ 5–10 yr) could be observed. After the introduction of atomic clocks many short-period oscillations were noted. The spectrum of the proportional change in the length of the day is illustrated schematically in Fig. 2. At very low frequencies the dominant change is a secular slowing down of the Earth such that the length of the day is increasing at a rate of about 2 ms per century. Superimposed upon this are irregular oscillations of characteristic time scales of several decades and amplitudes of up to 1 part in 10^7. Seasonal terms are also pronounced, with amplitudes of about 1 part in 10^8, as are shorter-period terms of about 28 and 14 days and a host of irregular fluctuations with time constants as short as a day.

The explanation for these various irregularities again lies in the Earth's departures from being a rigid body or from forces acting on the planet. The full explanation again demands inquiry over the whole range of Earth sciences. One cause is the tidal deformation of the Earth. Periodic changes in the lunar orbit as well as in the Earth's orbit introduce tidal deformations in the planet's oceans as well as in its solid part—the Earth tides. The concomitant changes in the inertia tensor modify the planet's rotation, producing periodic changes in the length of the day at 14 and 28 days (the lunar tides) and at six months (the principal solar tide). A more insidious result is that the lag in the Earth's tidal response to the gravitational attraction of the Moon or Sun produces a small secular torque that slows the planet down. The effect is very small from day to day, but over centuries it becomes more significant.

The seasonal changes are primarily the result of angular-momentum exchange between the solid Earth and the atmosphere. Winds exert torques on the planet and modify its rotation by measurable amounts. Much of the high-frequency signal in the length-of-day spectrum is also of meteorological origin. The large variations on time scales of decades cannot be attributed to the redistribution of mass or angular momentum of the outer fluid parts of the Earth; the angular momentum of the atmosphere relative to the Earth is wholly inadequate to produce changes in the length of the day approaching 1 part in 10^7. The cause must be sought in relative motions between the core and mantle, with the coupling between the two being generally attributed to electromagnetic interactions.

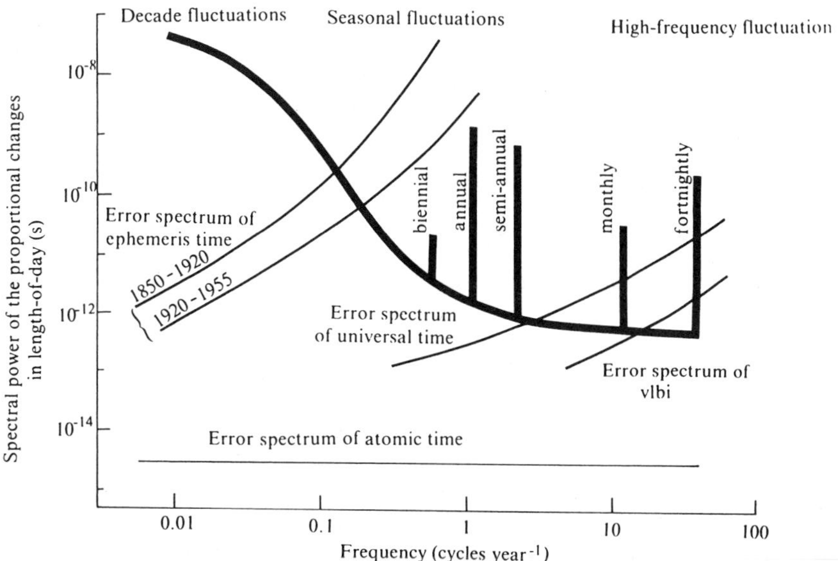

Figure 2. The schematic spectrum of fluctuations in the length of the day and observational error spectra. [*From Lambeck, K.,* Geophysical Geodesy, the Slow Deformations of Earth, *Oxford University Press (1988).*]

The irregular nature of many of the fluctuations in the Earth's rotation makes it difficult to predict future changes and the motion needs to be continually monitored. In turn, the observations provide constraints on a wide range of other geophysical problems.

TIME DEPENDENCE OF THE EARTH'S SHAPE

If the deformations of the planet play an important part in the body's rotation, then what changes in shape can also be expected? The tidal deformations of the solid planet are one example in which the surface is displaced radially by up to 20–30 cm with a range of tidal frequencies. The irregular rotation itself changes the centrifugal force, which further deforms the Earth by amounts that can be measured with modern instrumentation. On longer time scales the Earth is deforming in response to the unloading of the last great ice sheets and the redistribution of the melt water into the oceans. One measurable effect is a change in the Earth's oblateness through a nonuniform precession of satellite orbits about the Earth and through changes in the planetary rotation. More locally the effect is a rebound of the crust beneath the formerly glaciated regions of Fennoscandia, currently at maximum rates of nearly 1 cm yr^{-1}. This indeed offers a good illustration of a mobile planet.

Additional Reading

Jeffreys, H. (1970). *The Earth*, (5th ed.). Cambridge University Press, Cambridge.

Kaula, W. M. (1967). *Satellite Geodesy*. Blaisdell, Waltham.

Lambeck, K. (1980). *The Earth's Variable Rotation*. Cambridge University Press, Cambridge.

Lambeck, K. (1988). *Geophysical Geodesy, the Slow Deformations of the Earth*. Oxford University Press, Oxford.

See also **Coordinates and Reference Systems; Time and Clocks.**

Earth, Geomagnetic Storms

Daniel N. Baker

The outermost part of the Earth's atmospheric envelope is called the magnetosphere. This region extends to about 10 Earth radii ($R_\oplus$; 1 $R_\oplus = 6375$ km) on the dayside of the planet and stretches in an extended magnetotail, millions of kilometers long, on the nightside of the Earth (Fig. 1). As the term magnetosphere implies, this part of the terrestrial atmosphere is dominated by the Earth's magnetic field and is filled with a tenuous gas of ionized particles (a *plasma*) whose properties have been measured by direct in situ probes.

Geomagnetic storms are major disturbances in the magnetosphere. As the name suggests, such storms manifest themselves by global changes (for periods of hours to days) in the magnetic (and electric) fields surrounding the Earth. Storms also produce profound changes in the energy distributions and spatial locations of the plasmas filling the magnetosphere. Moderate magnetic storms may occur relatively frequently (every month or so), but really large storms due to major solar disturbances usually occur at intervals of many years. Geomagnetic storms often produce huge disturbances in the polar ionosphere, and intense auroral luminosity can reach over much of the high-latitude portions of the Earth. Low-latitude magnetometers on the Earth's surface can register changes of up to ~1% of the normal ambient field. Such changes in the magnetic field can have significant effects on navigation, resource exploration, and other human activities.

Geomagnetic storms have been known since ancient times. Their effects were noted by astute observers using magnetized needles suspended by thin threads. Today geomagnetic storms are

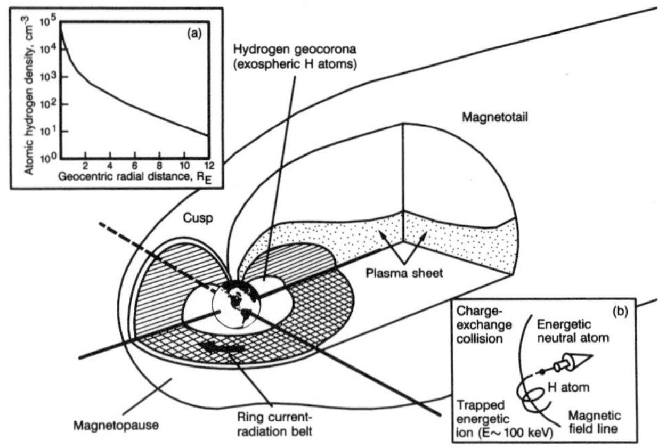

Figure 1. A simplified, cutaway diagram of the Earth's magnetosphere showing the magnetopause, the magnetotail, the plasma sheet in the central part of the magnetotail, and the ring current region. The Sun is far off to the left of this picture and the solar wind plasma flows strongly toward the right in the diagram, confining the magnetosphere to its cometary shape. Surrounding the Earth is an exospheric region (the hydrogen geocorona) of neutral H atoms. (*a*) The radial dependence of exospheric hydrogen atom density which determines charge exchange lifetimes. (*b*) A schematic diagram showing a charge-exchange collision with an exospheric hydrogen atom that converts an energetic ring current ion into an energetic neutral atom. (*Adapted by permission from Roelof and Williams, 1988.*)

documented by systematic measurements from worldwide networks of sensitive ground-based magnetometer stations. The physical causes of geomagnetic storms have been greatly illuminated in recent years by detailed measurements from Earth-orbiting spacecraft.

OBSERVED PHASES OF MAGNETIC STORMS

The geomagnetic storm is conveniently described in terms of the horizontal (H) component of the Earth's surface magnetic field at low latitudes. The dominant contribution to the surface field is due to the dynamo-produced intrinsic field which emerges from the molten iron-nickel core of the Earth. Near the Earth, this intrinsic field is approximately described as a magnetic dipole such as would arise from a large bar magnet near the Earth's center, slightly tilted with respect to the planetary rotation axis. The equatorial field strength from this dipole is $\sim \frac{1}{3}$ G (actually, 3.11×10^{-1} G $= 3.11 \times 10^{-4}$ T). The magnetic field from the Earth's core can change quite dramatically, sometimes even reversing polarity, but these changes only occur over very long time scales ($\sim 10^6$ yr). Thus, for most practical purposes, the intrinsic field of the Earth can be considered constant.

The measured magnetic field, however, is subject to quite rapid and substantial changes due to effects external to the Earth's surface. These external magnetic effects are produced by large-scale electrical currents flowing in and around the magnetosphere. The magnetosphere is formed and maintained by the flow of the hot, continuously expanding solar corona (the solar wind). In confining the magnetosphere into its elongated cometary shape, the ever-changing solar wind contorts and distorts the size and shape of the magnetosphere. The embedded magnetic field in the solar wind flow (the interplanetary magnetic field, IMF) interacts and interconnects in complex ways with the Earth's magnetic field, transferring varying amounts of mass and energy from the solar wind into the magnetosphere.

The outer surface of the magnetosphere is called the magnetopause and is formed by a surface current system first described by Sydney Chapman and V. C. A. Ferraro. This Chapman–Ferraro current arises due to the flowing solar wind plasma and its interaction with the physically distinct, magnetized plasma of the Earth's magnetosphere. The magnetopause separates and defines these different plasma regimes, and the currents flowing in the magnetopause layer have a clear effect on the magnetic fields measured on the Earth's surface. As will be discussed in more detail below, other global scale magnetospheric features, namely, the ring current, magnetotail current, and ionospheric current systems, have an even more profound effect on the surface magnetic fields.

Sudden Commencement

Many magnetic storms begin when the quiet day value of the geomagnetic surface field is abruptly increased by perhaps 10–30 nT (1 nT $=10^{-9}$ T). This initial rise in the field magnitude is called the storm sudden commencement (SSC). The field rises from its quiet value to its new elevated level in a short period (see Fig. 2*a*). The SSC is understood to result from a very sudden compression of the magnetosphere due to the passage of an interplanetary shock wave or other major solar wind pressure discontinuity. The dayside magnetopause position is largely determined by a balance of the solar wind dynamic pressure ($P = nmV^2$) and the Earth's magnetic field pressure ($B^2/8\pi$). Here n is the solar wind density, m is the mass of the solar wind ions (mostly protons), and V is the solar wind speed. Given that the Earth's field falls off like a dipole with equatorial distance (r), $B = B_0/r^3$, where $B_0 = 3.11 \times 10^{-4}$ T. The subsolar distance (R_S) where these pressures balance is given by

$$R_S = \left(\frac{B_0^2}{4\pi nmV^2} \right)^{1/6} \tag{1}$$

As the solar wind pressure increases, the magnetosphere is compressed and this increases the magnetic field at low latitudes on the Earth's surface. The magnitude of the SSC is proportional to the quantity $\Delta W = (P_2^{1/2} - P_1^{1/2})$, where P_1 is the solar wind pressure before the SSC and P_2 is the pressure afterward.

Considered globally, the SSC takes 2–3 min to develop worldwide. This is understood to be due to the time it takes the compression magnetohydrodynamic (MHD) wave to pass through the magnetosphere to the ground and establish a new quasiequilibrium configuration of the magnetosphere.

Initial Phase

The initial phase of the magnetic storm is the period of 1–10 h (varying from storm to storm) in which the H component of the surface field remains elevated by some tens of nanotesla over the quiet value (cf. Fig. 2*a*). As noted above, the increase of the H component is due to the compression of the magnetosphere. De-

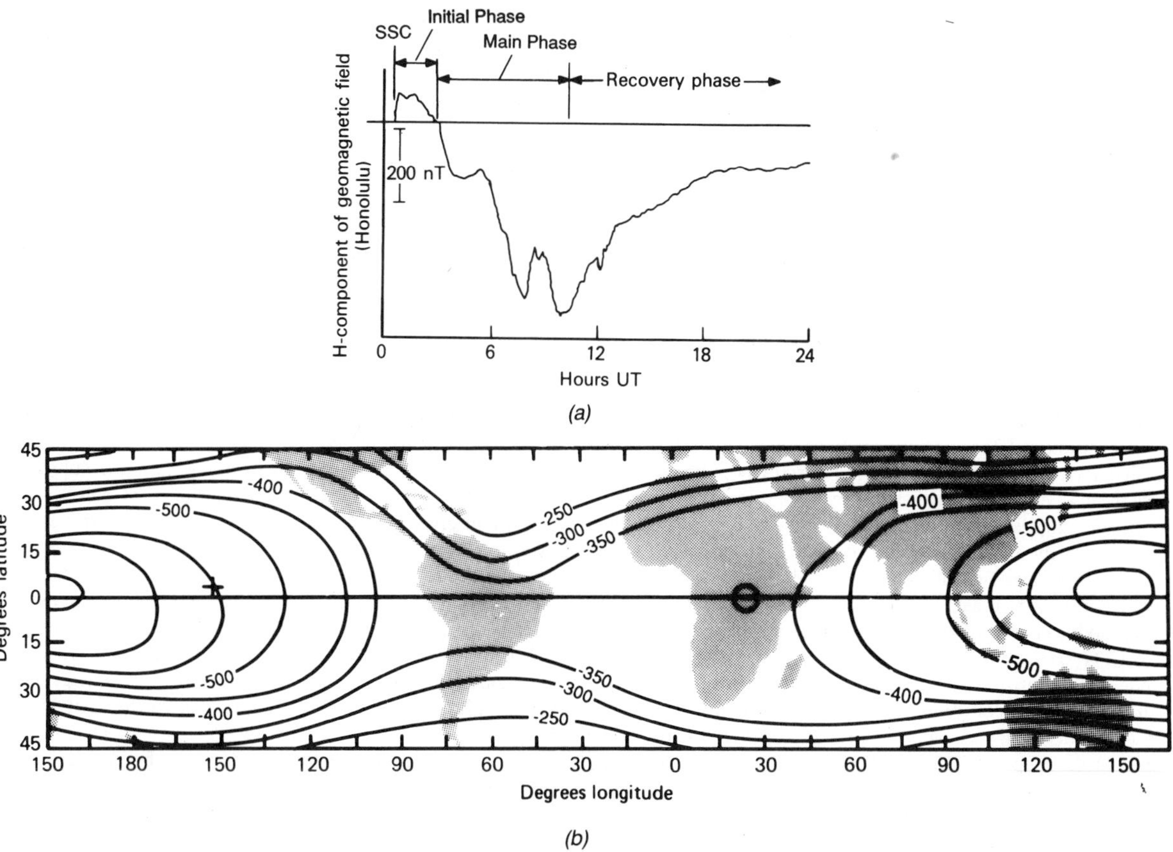

Figure 2. (*a*) The time profile of the low-latitude surface perturbation of the Earth's magnetic field as measured at Honolulu for the great magnetic storm of 13 September 1957. The SSC, initial phase, main phase, and recovery phase of the storm are indicated. (*b*) Isointensity contours (in nanotesla) of the surface magnetic perturbations during the 13 September 1957 geomagnetic storm. Contours are shown in 50-nT increments during the height of the ring current development at 1000 UT. (*Adapted by permission from Akasofu and Chapman, 1972.*)

tailed calculations show that the change in surface field strength, ΔB, is related to the subsolar magnetopause position R_S by

$$\frac{\Delta B}{B_0} = 0.6\left(\frac{R_S}{R_\oplus}\right)^3 \tag{2}$$

During large magnetic storms after very strong solar-flare–induced shock waves in the solar wind have passed the Earth's orbit, the magnetopause can be pushed in to a geocentric distance as small as $R_S \sim 5\ R_\oplus$. Under such circumstances, the usual ΔB of 10–20 nT can be increased to $\Delta B \sim 150$ nT. In these cases, spacecraft normally within the confines of the magnetosphere can be exposed to the unimpeded solar wind flow and massive disruptions of the Earth's polar ionosphere can take place.

Main Phase

The main phase of a geomagnetic storm is the period of strongly *decreasing* surface field strength (Fig. 2*a*). Thus, in contrast to the initial phase, which is caused by a compression of the magnetospheric field, the main phase is caused by an inflation of the magnetosphere due to strong, rapid introduction of hot, magnetically trapped plasma particles. This new plasma population is confined predominantly in a typical geocentric radial range of 2–5 $R_\oplus$ (see Fig. 1). Early observations in the magnetosphere showed that the ring current is comprised mainly of ions in the tens to hundreds of kiloelectronvolt range in kinetic energy.

In carefully considering the MHD equations describing the pressure gradients and magnetic fields resulting from an axially symmetric ring of plasma particles around the Earth, Alex J. Dessler and Eugene N. Parker found the remarkable result that the surface magnetic field change (a *decrease* in this case) due to the ring current is

$$\frac{\Delta B}{B_0} = -\frac{2E}{3E_M} \tag{3}$$

Here E is the total energy contained in the ring current plasma population and E_M is the total energy in the dipole magnetic field exterior to the Earth's surface. The numerical value of E_M is $\sim 8 \times 10^{24}$ erg. This Dessler–Parker relation has been shown to hold for any distribution of trapped particles in the magnetospheric field irrespective of where the particles are located. Thus, all that matters for the surface field decrease in the storm main phase (see Fig. 2*b*) is the total energy content of the particles, not where they are positioned in radial distance.

Detailed studies in the satellite era have shown how the storm main phase actually develops. The ring current population of hot plasma is produced in an intense series of explosive episodes of magnetic reconfiguration in the Earth's magnetotail region (see Fig. 1). In each of these shorter events (called *magnetospheric substorms*), there is rapid addition of solar wind plasma and magnetic energy at the dayside magnetopause (controlled largely by the IMF direction) followed by enhanced global magnetospheric convective transport. Greatly enhanced cross-magnetotail currents on the nightside of the Earth in the plasma-sheet region lead to an effective storage of magnetic energy in the magnetotail. This briefly stored magnetic energy is tapped repeatedly during substorms by the process of magnetic reconnection. This reconnection rapidly converts magnetic energy into hot, jetting plasmas that are injected into the outer parts of the ring current region. The magnetic reconnection, plasma heating, and particle acceleration occurring during substorms in the magnetotail appear to be closely analogous to flare processes on the Sun. Strong, global electric fields during the storm main phase complete the ring current formation process by driving the substorm-energized particles deeper into the inner magnetosphere.

As noted above, the ring current plasma population appears to be concentrated in the geocentric radial range of 2–5 $R_\oplus$. This is quite understandable because the hot, high-pressure gas of the ring current population tends to strongly inflate the magnetic field lines on which it resides. Thus, the ring current plasma must be contained relatively close to the Earth where the magnetic field is strong enough to restrain the plasma pressure. This has been directly verified by spacecraft measurements. Generally speaking, the larger the magnetic storm, the closer to the Earth must be the ring current.

RECOVERY PHASE

The recovery phase of a magnetic storm is the period in which the disturbance is turning off and the magnetosphere is returning toward its normal state (see Fig. 2*a*). The recovery usually exhibits two separate time scales, namely, a rapid initial recovery with characteristic time scale of a few hours and then a much slower recovery lasting days to weeks. The problem in the recovery phase is to understand how to cool or largely eliminate the hot, intense plasma population that constitutes the ring current. As shown by Eq. (3), the ring current particles will continue to contribute to the magnetic field perturbation until they are completely depleted of their energy or else are completely voided from the magnetospheric trapping region.

It was concluded from early studies that radiative cooling of the ring current plasma was much too slow a process to match the observed decay times. Thus the process of charge exchange emerged as the most viable kind of interaction to explain the ring current decay. During charge exchange, a fast (energetic) ion interacts with a cold neutral atom from the exospheric cloud of hydrogen which surrounds the Earth (see Fig. 1 inset). In this process, the energetic ring current ion can pick up an electron from the cold hydrogen atom. The energetic ion then becomes a fast neutral atom which can rapidly move (unimpeded) across the magnetic field lines and escape the magnetosphere, and the low-energy hydrogen ion (proton) is left behind, trapped in the magnetic field. This can efficiently cool, and thus dissipate, the ring current.

RECENT RESULTS

Although it was generally known through careful ground-based work and clever theoretical inferences that the ring current must consist dominantly of ions with energies of tens to hundreds of kiloelectronvolt, until recently little has been known for sure about the actual elemental composition and detailed energy distribution of the ring current population. New plasma and energetic particle detection techniques [as recently utilized on the *Active Magnetospheric Particle Tracer Explorers* (*AMPTE*) spacecraft, for example] have shown clearly which elements and which charge states of ions are predominant. The new measurements show substantial variability from storm to storm, but for the most part the main ring current intensity increase occurs in the 20–200-keV range. Moreover, in all but the largest storms, protons (H^+) with energies $\gtrsim 50$ keV dominate the main phase ring current energy density. Singly charged oxygen ions (O^+) of ionospheric origin may contribute $\sim 25\%$ of the energy density in moderate storms. Generally speaking, high charge states of oxygen (indicative of a solar wind source) contribute a relatively small fraction of the total energy density, but $O^{>5+}$ plays an increasingly larger role further out in geocentric radial distance.

New measurement techniques are also becoming important in understanding the recovery phase of magnetic storms. Using sensitive energetic neutral atom (ENA) sensor systems, it is possible to observe directly the fast atoms produced in the charge exchange process. This ENA technique therefore allows the "imaging" of the ring current energetic particles as they escape from the confines of the Earth's magnetic field. Given suitable spacecraft orbits from which to image the neutral atoms from the charge exchange process, it is possible to build up a global picture of the ring current population, and it is also possible to follow the decay of the ring current in time as the storm recovers. Present results suggest that the fast and slow time scale portions of the storm recovery phase are due, respectively, to the O^+ (rapid charge exchange) and to the H^+ (slower charge exchange) parts of the total ring current population.

FUTURE DIRECTIONS

Geomagnetic storms reach a peak of occurrence frequency and strength during the period around, and immediately following, the maximum in the sunspot number in each solar cycle. During this period (e.g., in 1991) there is the highest probability of strong solar flares that, in turn, drive intense shock waves through the interplanetary medium. These shock waves initiate the magnetic storm sequence as described above. The practical implications of magnetic storms in terms of human navigation, remote communications, near-Earth spacecraft operations, and resource exploration is enormous. The nature of the solar wind magnetosphere mass, momentum, and energy transfer process during geomagnetic storms is rather readily accessible to direct in situ investigation by satellite sensors. During the 1990s there will be an unprecedented international effort to measure all regions of the magnetosphere concurrently, using dozens of spacecraft. This program of observation should clarify most remaining issues concerning geomagnetic storms, giving both scientific and practical benefits.

As is evident, the physical processes which occur in geomagnetic storms are prototypical of plasma processes occurring in other planetary and remote astrophysical systems, for example, as would take place when a cosmic shock wave, or blast wave, encounters an isolated, distinct plasma region (e.g., a planetary or stellar magnetosphere). These remote objects are largely or completely inaccessible to direct in situ measurements, and thus the geomagnetosphere becomes the best and most accessible prototype to study in order to gain insight into general cosmic plasma interactions. A substantial element of the theoretical and observational space plasma physics program, now and in the future, will be devoted to gaining a more general understanding of the kinds of basic processes occurring in magnetic storms.

Additional Reading

Akasofu, S.-I. and Chapman, S. (1972). *Solar Terrestrial Physics.* Oxford University Press, London.

Chapman, S. and Ferraro, V. C. A. (1931). A new theory of magnetic storms. *Terr. Magn. Atmos. Electr.* **36** 77 and 171.

Dessler, A. J. and Parker, E. N. (1959). Hydromagnetic theory of geomagnetic storms. *J. Geophys. Res.* **64** 2239.

Parker, E. N. (1968). Dynamical properties of the magnetosphere. In *Physics of the Magnetosphere*, R. L. Caravillano, J. F. McClay, and H. R. Radoski, eds. Reidel, Dordrecht, p. 3.

Roelof, E. C. and Williams, D. J. (1988). The terrestrial ring current: From in situ measurements to global images using energetic neutral atoms. *Johns Hopkins APL Tech. Dig.* **9** 144.

See also **Earth, Aurora; Earth, Magnetosphere and Magnetotail; Interplanetary Medium, Shock Waves and Traveling Magnetic Phenomena; Interplanetary Medium, Solar Wind.**

Earth, Impact Craters

Peter M. Millman†

The international space program has made it possible for scientists to map the topographic features on the solid surfaces of some 30 extraterrestrial worlds in our solar system. The majority of these worlds have no appreciable atmosphere, and hence no clouds obscure the surface detail. Our knowledge in this field is recorded by remotely controlled cameras on spacecraft such as the *Voyagers*, the *Vikings*, or the *Sputniks*. By far the most common type of feature found is an assortment of circular crater forms of various sizes and ages. A few of these have been identified as volcanic in origin, but the vast majority are impact scars resulting from the collision of small solid objects with either planets or their natural satellites in our solar system. The Earth has been subjected to this bombardment for more that 4.5 billion years, but since the Earth has an active atmosphere and much of its solid surface is covered by water, evidence left by space impacts is often destroyed or obscured and it is only comparatively recently that the historical importance of these events has been recognized. The moon's surface, seen through a telescope, is a good example of an unprotected planetary surface exposed directly to space for several thousand million years.

HISTORY

The first feature on Earth, definitely identified as a space scar, is a roughly circular crater some 1200 m in diameter, 170 m deep, and with a rim averaging some 50 m above the desert in northern Arizona, USA. This feature, just south of Highway 66 near Winslow, was known for years as Coon Butte, and later as Meteor Crater. It is now officially named Barringer Crater after Daniel Moreau Barringer, a mining engineer who made a detailed study of the crater, assisted by his friend Benjamin C. Tilghman. Together, they suggested its extraterrestrial origin in a communication to the Academy of Natural Sciences of Philadelphia on 5 December, 1905. This was published in detail in the *Proceedings* of this Academy.

Barringer faced a great deal of opposition to the impact theory which, in 1905, seemed improbable to a generation of scientists brought up on the universal importance of volcanism. However, he and four of his sons persevered in a continued study of the subject. In 1927 the Odessa crater in Texas was recognized as a possible impact scar, and in 1933 the British geographer, Leonard J. Spencer, published a paper listing seven possible impact sites, all but one of which had meteoritic material associated with them. Over the years additional examples were added and, in the early 1950s, Canada's Dominion Astronomer, Carlyle S. Beals, organized a comprehensive program of searching for impact sites and for studying suspect locations by using various different geophysical techniques. To investigations of surface geology and mineralogy were added mapping of the magnetic and gravitational fields, seismic studies, and deep core drilling. In addition, new methods for detecting ancient severe shocking in rocks became available through the microscopic examination of thin sections of samples from the area being studied. It is now possible to separate with some certainty the features which have a space-impact origin from those formed by purely terrestrial activity. By 1970 over 60 space-impact sites had been identified on Earth. This subject now involves a large group of scientists in a number of countries, and various lists of probable and possible sites have been published.

LOCATIONS

One of the most recent and convenient of these lists is by two Canadian experts in this field, Richard A. F. Grieve and P. B. Robertson. They table 116 sites, together with the name, geographic coordinates, original diameter of feature, and age, where known. They also plot the location of each site on a world map. In the general discussion that follows I have updated this list of 116 sites by an additional 6, verified since 1987, making a total of 122 terrestrial sites. Since at least 8 of these sites give evidence of more than one crater formed at the time of impact, we now have a total of 314 individual impact features available for study.

The well-authenticated impact sites occur worldwide, as can be seen from Table 1. The number of sites recognized in any area of

†Dr. Millman died on December 11, 1990. —Editor.

Table 1. Locations of Terrestrial Impact Sites

North America	
Canada	25
United States	17
Europe	31
Australia	17
Asia	16
Africa	11
South America	5
Total	122

the earth depends primarily on three basic factors—the total size of the area, the nature of the terrain, and the extent of the scientific search. In the table North America, with the largest number of sites, has been listed under two headings because each country involved has important differences in the average type of terrain and has used somewhat different search techniques. In the last two decades the greatest advances in the recognition of new sites have been in Europe and Asia, thanks to a fast-growing interest in the USSR. Both Canada and the USSR have prime search areas because of their size and the quantities of earth's most ancient rocks that are exposed on the surface in these countries.

PHYSICAL STRUCTURE

It is instructive to examine the general physical nature of these features as we progress from the smallest to the largest examples. Under a diameter of 9 m they are essentially pits or holes in the ground, with a shape largely dependent on the geology of the surface and the effects due to weathering since their formation. Here the impacting body, up to a mass of approximately one metric ton (10^6 g) will normally remain within the pit, especially if the impacting body, or meteoroid, is iron. For larger features, up to a diameter of about 90 m, we find fragmentation of both meteoroid and surface rocks, with the crater approaching the shape of a uniform bowl. Above a diameter of roughly 90 m, where the impacting meteoroid will have retained more than half of the original velocity with which it approached the Earth in space, the feature becomes a true explosion crater. The energy of impact is great enough to melt and fuse together some of the mass involved and we may find rock breccia, silica glass, and small metallic spherules. The resulting crater is bowl shaped, with a base below the level of the surrounding terrain and an upturned rim somewhat above this level. These features, up to a diameter of 2 km in sedimentary rock and 4 km in crystalline rock, are termed simple impact features. The Barringer crater, noted earlier, belongs in this class. Above these sizes we have complex impact features, where there is a central uplift on the floor of the crater. In craters greater than a diameter of 24 km in sedimentary rock and 30 km in crystalline rock, the central uplift assumes the appearance of a ring structure, which may have a multiring characteristic if the diameter is greater than about 100 km. These basic types are confirmed by what we observe on extraterrestrial worlds in the solar system, once we have corrected for the variations in gravity on the surfaces of these planetary bodies.

IMPACTING BODIES

The impacting bodies themselves will be, in general, fragments of larger bodies which were originally small minor planets. They may be entirely metallic nickel-iron, or stone with metallic content, or a mixture of both. The energy available to create an impact feature varies with the mass of the impacting body multiplied by the square of its velocity in space relative to the Earth. Typical relative velocities range from 15–25 km s^{-1}, and at these speeds the meteoroid, gram for gram, acts as an explosive substance many times more powerful than TNT. An iron meteoroid with a mass between one and two megatons (10^6 tons) would produce the Barringer crater if moving 15 km s^{-1} relative to the Earth. This mass of nickel-iron corresponds to a body approximately 70 m in diameter. If moving 25 km s^{-1} it would have to be only 50 m in diameter to produce the same crater. During the last century fragments totaling over 30 tons of nickel-iron meteorite have been collected in an area within 8 km of the Barringer crater, in spite of the large quantities that have been oxidized since it fell to earth some 25,000 years ago. The impacting masses that created the largest examples in our list of terrestrial features must have been in the mass range of 10^{11}–10^{13} tons, with diameters of several kilometers.

AGES

The determination of the age of a terrestrial impact feature may be difficult unless we have a detailed knowledge of the local geology and its history throughout past ages. For 110 out of 122 sites information about age or age limits is available, and for some 60 sites the uncertainty in the age is less than 20 percent of its value. The two oldest features identified to date are just under 2000 million years old, and the original craters in both cases are estimated to have been 140 km in diameter, rim to rim. The ages at some 80 sites are 50 million years or older. Listed below is more detailed information about nine typical examples from the list of terrestrial impact features, starting with the youngest and progressing with age to one of the two oldest.

EXAMPLES

Sikhote Alin (46°07′N, 134°40′E), age < 50 yr, diameter (largest crater) 26.5 m. On a bright frosty morning some 500 km northeast of Vladivostok, at about 10:30 a.m. local time, 12 February 1947, citizens of the Maritime Territory in the USSR were startled by an extremely brilliant fireball moving across the sky from north to south and leaving a heavy smoke trail that remained visible for two hours. After the fireball disappeared behind the hills, explosions like the firing of heavy-caliber guns were heard. Numerous craters were found in a heavily wooded area of about 2 km^2, 24 craters with diameters in the range 26.5–9 m, and 98 smaller craters with diameters of 9–0.5 m. Hundreds of fragments of nickel-iron with a total weight of 23 tons were collected in and near these craters.

Barringer (35°02′N, 111°01′W), age 25,000 yr diameter, 1.2 km. This is the best-preserved impact crater of the simple class. It is well worth a visit, and there is an excellent small museum

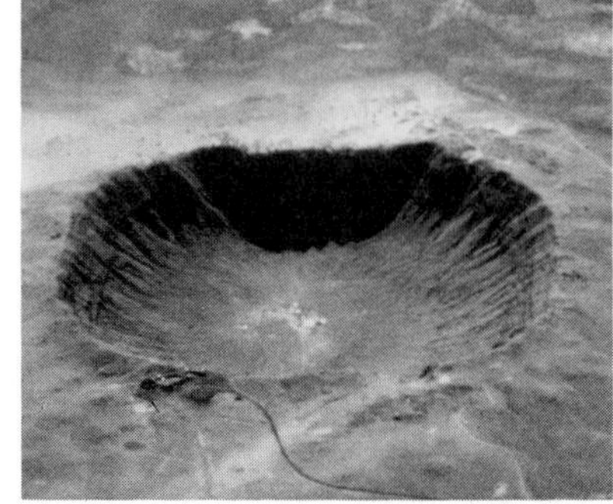

Figure 1. The Barringer Crater, Arizona, as seen from the air looking south near noon on 29 January 1967. The crater is roughly circular with a diameter of 1.2 km, but is somewhat polygonal in form as has been noted in many lunar impact scars. (*Photo by P. M. Millman.*)

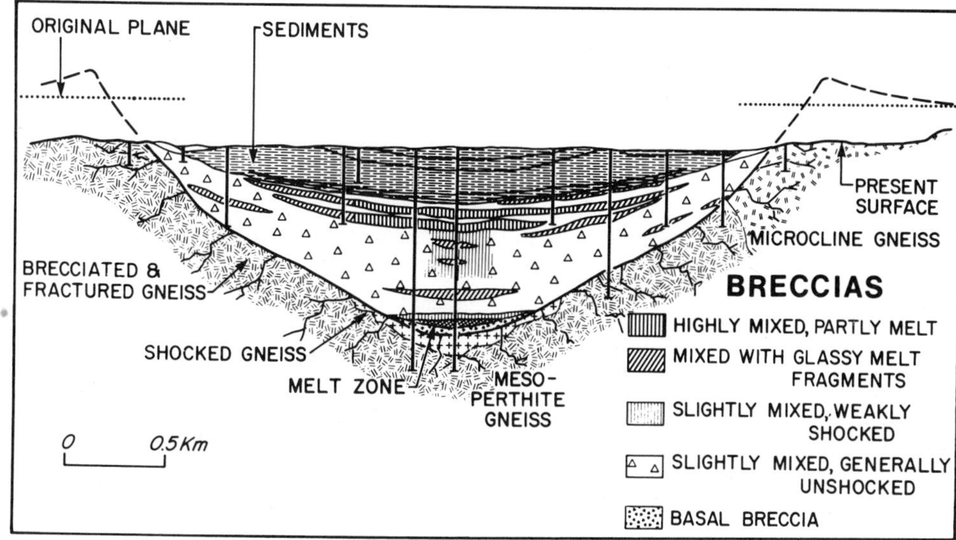

Figure 2. A vertical cross-section through the center of the Brent Crater, drawn to scale. The locations and depths of the twelve core drill holes are shown. (*Courtesy of Geological Survey of Canada.*)

located on the north rim. Through the use of a magnetic pick up large quantities of small nickel-iron spherules have been recovered in the area within a few kilometers of the crater.

Bosumtwi (06°32′N, 001°25′W) age 1.3 million yr (Ma), diameter 10.5 km. This circular crater in Ghana, sometimes referred to as Ashanti, has a floor roughly 400 m below the rim. The rim rises from 90 to 180 m above the surroundings and a lake, 8 km in diameter with a depth of over 70 m, is in the crater. Neither volcanic rocks nor meteoritic material have been found in this area.

Ries (48°53′N, 010°37′E), age 15 Ma, diameter 24 km. The circular form of this large complex feature stands out clearly on a geological map of the area, a region extensively studied by geologists. The original rim-to-rim diameter was of a size near the dividing line between central uplift craters and those showing a ring structure. It is possible that the green-glassy tektites called moldavites, found in Bohemia over 200 km to the east, are a fused splash from the impact that formed the Ries.

Montagnais (42°53′N, 064°13′W), age 50 Ma, diameter 45 km. This is the only impact feature currently recognized on the ocean floor. It lies on the Scotian Shelf, some 200 km south of Halifax at a depth of 115 m under the surface of the Atlantic Ocean. The associated rocks, and rock breccia, show clear evidence of strong shocking at the time of the impact. This feature is unique in exhibiting the long-term effects of the marine environment on an impact site.

Gosses Bluff (23°50′S, 132°19′E), age 142 Ma, diameter 22 km. Today this feature is seen as an area of a few square kilometers near the center of Australia, enclosed by a ragged ridge of hills roughly 200 m above the plain. These hills have a shape between a circle and a polygon. Scientific study has shown that the resemblance to a crater is purely accidental, and the ridge simply marks the location of sandstones resistant to erosion in the central uplift of a very old impact feature of much larger dimensions. The original crater has been eroded away in the millions of years since the impact.

Manicouagan (51°23′N, 068°42′W), age 210 Ma, diameter 100 km. Situated some 200 km north of the Gulf of St. Lawrence in the Province of Quebec, this locality has been flooded in recent years by industrial damming to the south. As a result the circular-ring form of this feature shows very clearly on satellite photographs of the earth's surface.

Brent (46°05′N, 078°29′W), age 450 Ma, diameter 3.8 km. This crater in Ontario is a simple explosion feature and small in comparison to its age. Without the aid of an aerial photographic survey, it is unlikely that it would have been discovered. The exploration at Brent, which included 12 deep core drill holes, has been the most extensive of any carried out on impact sites in Canada. As will be seen from fig. 2 the crater has been filled with very old sedimentary rocks.

Sudbury (46°36′N, 081°11′W), age 1850 Ma, diameter 140 km. This feature, some 60 km north of Georgian Bay on Lake Huron, is now elongated east–west, 60×25 km in size. It has lost all resemblance to its original form. Sudbury is well known for its nickel mines, and the area exhibits many evidences of severe shocking at the time of the impact. It is generally believed that the nickel mined here did not arrive in the impacting body, but that the shock was powerful enough to release magma from the earth's interior.

Additional Reading

Barringer, D. M. and Tilghman, B. C. (1905). Coon mountain and its crater. *Proc. Acad. Nat. Sci. Philadelphia* **57** 861.

Beals, C. S., Innes, M. J. S., and Rottenberg, J. A. (1963). Fossil meteorite craters. In *The Moon, Meteorites and Comets*, B. M. Middlehurst and G. P. Kuiper, eds. University of Chicago Press, Chicago, pp. 235.

Dence, M. R., Grieve, R. A. F., and Robertson P. B. (1977). Terrestrial impact structures: Principal characteristics and energy considerations. In *Impact and Explosion Cratering*. Pergamon, New York, pp. 247.

Grieve, R. A. F. and Robertson, P. B. (1987). Terrestrial impact structures. In *Geological Survey of Canada*, Map 1658A, scale 1:63 000 000.

Krinov, E. L. (1966). *Giant Meteorites*, J. S. Romankiewicz, trans. Pergamon, New York.

Mark, K. (1987). *Meteorite Craters*. The University of Arizona Press, Tucson, AZ.

Millman, P. M. (1971). The space scars of earth. *Nature* **232** 161.

Nininger, H. H. (1956). *Arizona's Meteorite Crater*. World Press, Denver, CO.

See also **Meteorites, Classification; Moon, Geology.**

Earth, Magnetosphere and Magnetotail

Syun-Ichi Akasofu

The Earth is surrounded by a vast region of space, in which magnetic fields and ionized gases (called plasmas) play key roles. This particular region is called the magnetosphere and is shaped mainly by the interaction between the solar wind, a flow of hot plasma from the sun, and the Earth's magnetic field. It is also this interaction which generates the power for auroral displays.

The magnetosphere is a comet-shaped (a long cylinder with a blunt nose) region around the Earth, which is carved in the solar wind. Like a comet, the magnetosphere has an extensive tail, called the magnetotail. The Earth is located near the "head" of this comet-shaped cavity; the distance from the nose of the cavity to the Earth is about 10 Earth radii, and the tail is inferred to extend to a distance of more than 1000 Earth radii. At the nose of the magnetopause, a simple pressure balance relationship holds approximately. The ram pressure of the solar wind, $p_s = 2nmV^2$, is balanced by the magnetic pressure, $B_m^2/8\pi$, where V, n, and m denote speed, the number density, and mass of the solar wind particles, and B_m is the magnitude of the magnetic field just inside the boundary. At this distance, the Earth's atmosphere consists mainly of both neutral and ionized hydrogen atoms at a density of about 10 cm^{-3} or less.

The magnetosphere is not an empty cavity; it is filled with different plasma regimes with different characteristics. Both the solar wind and the ionosphere are sources of those plasmas. The plasma regimes are divided into the ionosphere, the plasmasphere, the Van Allen belts, the ring current belt (the storm-time Van Allen belt), the plasma sheet, and the plasma layers along the boundary of the magnetosphere. Figure 1 shows the noon–midnight cross section of the magnetosphere, in which some of the plasma regimes are shown. It has been found only recently that the ionosphere is a very significant plasma source in all these plasma regimes. The magnetosphere has a bow shock, because the solar wind flow is supersonic; the region between the bow shock and the boundary of the magnetosphere (the magnetopause) is called the magnetosheath.

The solar wind is a magnetized plasma. Thus, some of its magnetic field lines reconnect with some of those which originate from the Earth. In fact, a bundle of the magnetic field lines from

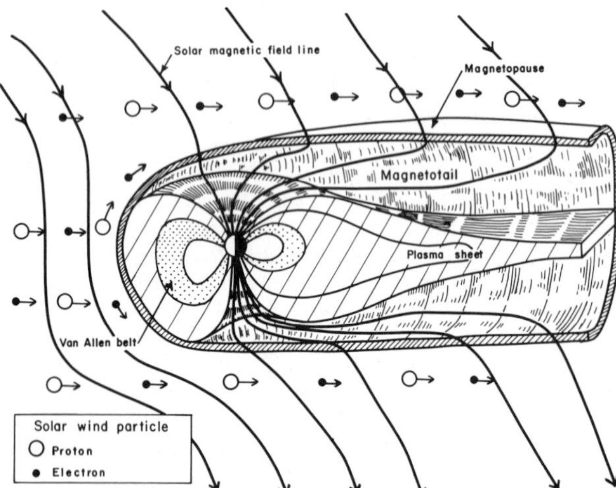

Figure 1. A schematic cross section of the noon–midnight cross section of the magnetosphere, with some of the plasma regimes. The magnetic field lines of the solar wind are reconnected with the Earth's magnetic field lines across the boundary of the magnetosphere (the magnetopause).

both the northern and southern polar regions is reconnected with those in the solar wind across the magnetopause (see Fig. 1.) The solar wind blows across the reconnected field lines along the magnetopause in the antisolar direction. This situation is similar to that in a magnetohydrodynamic generator process, powering a great variety of magnetospheric processes in the magnetosphere, including the aurora. The power P in megawatts is given by

$$\text{power (MW)} = 20V\ (\text{km s}^{-1}) \times B_m^2(nT) \times \sin^4(\theta/2),$$

where θ is the polar angle of the solar wind magnetic field (with respect to the Earth's dipole axis).

Much of the current generated by the generator flows across the magnetotail, from the dawn side to the dusk side. As a result, the magnetotail has two solenoidal loop currents, one in the northern half of the tail and the other in the southern half, which are separated by the plasma sheet. The resulting two bundles of the magnetic flux occupy significant parts of the tail, which are called the northern and southern high-latitude lobes, respectively. The two bundles of the magnetic flux are nearly antiparallel in the magnetotail. For this reason, it had long been thought that the magnetotail was an ideal location for spontaneous and explosive magnetic reconnection. It has, however, become apparent that magnetic reconnection does not occur spontaneously and explosively, annihilating the energy contained in the antiparallel field system. Within a time scale of a few hours, the total amount of energy released from the system is the same as that introduced into it by the generator.

There are complex current systems in the magnetosphere, which are all powered by the generator. In particular, intense field-aligned currents flow between the magnetopause and the ionosphere, downward toward the ionosphere in the dawn sector and upward away from the ionosphere in the dusk sector, together with the connecting ionospheric currents, called the auroral electrojets. Much of the energy generated by the generator is dissipated as Joule heat in the ionosphere.

One of the most interesting findings in magnetospheric physics is that in the upward field-aligned current region (carried by downward-moving electrons), there occurs a potential drop of a few kilovolts, enough to accelerate the current-carrying electrons to cause ionizations and excitations in the lower ionosphere. The aurora results from the emission of radiations by such processes. The aurora emits the radiations over a wide wavelength region, from x-rays to infrared radiations. Auroral processes produce also radio emissions from ultra-low-frequency (ULV) radiations to kilometric radiations.

In the solar wind, V, B, and θ are always varying in time, causing the power of the generator to vary in turn. The magnetosphere responds specifically to an increase of the generator power, which lasts for a few hours. The magnetospheric phenomena in response to such an enhanced genertor power are called collectively the magnetospheric substorm. Among the substorm phenomena the only visible phenomenon is the auroral substorm during which the aurora undergoes a systematic global change. The direct cause of the auroral substorm is enhanced field-aligned currents (which in turn result from an increase of the generator power). During an auroral substorm, the plasma sheet also undergoes dynamical variations. Some plasma escapes from the magnetotail in the antisolar direction in the form of plasmoids. A great variety of electromagnetic and/or electrostatic waves are generated around and within the magnetosphere. One of the most intense radio emissions is the auroral kilometric radiations (AKR), which are generated in the auroral acceleration region. It is believed that the emissions are intense enough to be detected even from a point outside the solar system.

The power of the generator often increases considerably as a shock wave (generated by intense solar activity) collides with the magnetosphere after it traverses the distance of 1 AU from the Sun in about 40 hours. Solar activities that are responsible for such shock waves are solar flares, coronal mass ejections, and/or sudden disappearance of filaments. The shock wave is associated with an increased V and B, so that the power can be greatly enhanced, if θ is $\sim 180°$. A greatly enhanced current in the magnetosphere during such an occasion causes an intense magnetic disturbance which we identify as a geomagnetic storm.

It has been found by deep space probes that magnetized planets, such as Mercury, Jupiter, Saturn, Uranus, and Neptune, have magnetospheres. The size of the jovian magnetosphere is much greater than that of the Sun. Thus, it can be said that the magnetosphere and its tail result wherever the interaction between a magnetized supersonic plasma flow and a magnetized planet (or any magnetized celestial body) can occur. Thus, it is expected that similar interactions occur in a variety of astrophysical conditions. However, the Earth's magnetosphere is the only magnetosphere which can be frequently accessible by spacecraft, enabling us to make in situ observations. Such multisatellite observations have enabled us to clarify the theoretical idea of spontaneous and explosive magnetic reconnection. Spontaneous and explosive magnetic reconnection has been proposed to occur in some astrophysical conditions, including solar and stellar atmospheres, pulsar environments, comet-shaped galaxies, etc. The discovery of the field-aligned potential in a very rarified collisionless plasma (the number density is $0.1 \sim 1.0\ \mathrm{cm}^{-3}$) may also be important in considering acceleration processes of charged particles in similar astrophysical conditions, although it was believed in astrophysics that such a potential drop could not exist in a rarified plasma that is permeated by magnetic field lines.

Additional Reading

Akasofu, S.-I. and Chapman, S. (1972). *Solar-Terrestrial Physics*. Oxford University Press, Oxford.

Akasofu, S.-I. and Kamide, Y., eds. (1987). *The Solar Wind and the Earth*. Terra Sci. Pub. Co., Tokyo.

Akasofu, S.-I. and Kan, J. R. (1981). *Physics and Auroral Arc Formation* (Geophysical Monograph 25). American Geophysical Union, Washington, DC.

Hones, E. W., Jr., ed. (1984). *Magnetic Reconnection in Space and Laboratory Plasmas* (Geophysical Monograph 30). American Geophysical Union, Washington, DC.

Lui, A. T. Y., ed. (1987). *Magnetotail Physics*. Johns Hopkins University Press, Baltimore.

Moore, T. E. and Waite, J. H., Jr. (1988) *Modeling Magnetospheric Plasma* (Geophysical Monograph 44). American Geophysical Union, Washington, DC.

Potemra, T. A. (1984). *Magnetospheric Currents* (Geophysical Monograph 28). American Geophysical Union, Washington, DC.

Tsurutani, B. T. and Stone, R. G. (1985). *Collisionless Shocks in the Heliosphere: Reviews of Current Research* (Geophysical Monograph 35). American Geophysical Union, Washington, DC.

See also **Earth, Aurora; Earth, Geomagnetic Storms; Earth, Magnetosphere, Space Missions; Magnetohydrodynamics, Astrophysical; Plasma Transport, Astrophysical.**

Earth, Magnetosphere, Space Missions

Louis J. Lanzerotti

One of the great exploration urges of humankind has been to escape the force of gravity and to investigate the environment above the Earth's surface. Jules Verne in the nineteenth century stimulated the imagination of many people on several continents with his fanciful depictions of a space flight to the moon. Also in the nineteenth century, the American writer Edward E. Hale proposed in his short story "The Brick Moon" a navigation satellite in order to determine longitude. Konstantin Tsiolkovsky in the Soviet Union and Robert Goddard in the United States provided much of the engineering advocacy for propulsion vehicles that could break the shackles of gravitational attraction. The push in both of these nations to construct and fly Earth-orbiting satellites as a portion of each country's contribution to the International Geophysical Year (1956–1957) resulted in the launching of *Sputnik I* on October 4, 1957, and the onset of the space age. James Van Allen and his colleagues of the University of Iowa, flying Geiger–Mueller (GM) charged particle detectors on the first US spacecraft *Explorers I* and *III*, discovered the trapped (by the Earth's magnetic field) radiation environment which bears his name. This discovery, and the essentially contemporaneous measurements of electrons by Sergei N. Vernov of Moscow State University on the *Sputnik III* Soviet satellite, established the existence of the wondrously complex plasma physics environment which envelopes the planet Earth and stimulated a new and vigorous research field.

EARLY MAGNETOSPHERE STUDIES

Prior to the launching of *Sputnik I* and the radiation-belt discovery of *Explorer I*, there had been extensive discussions in the U.S. in the early 1950s (and undoubtedly in the Soviet Union as well) as to the scientific objectives amenable to investigation by Earth-orbiting satellites and the appropriate engineering developments for the instrumentation to be carried. The foundations for much of these investigations had been laid by balloon-borne and, after World War II, rocket-borne instruments carried aloft to study, at high altitudes, such phenomena as the ionosphere, cosmic rays, solar ultraviolet radiations, and the geomagnetic field. Van Allen's instrument on *Explorer I* was an evolutionary development from instruments that he and his students had earlier flown successfully on rockets and on rockoons—small rockets which were launched after first being borne aloft by high-altitude balloons.

The coupling of a momentous discovery about the Earth's space environment with the capability of following up quickly on the discovery by the launching of instrumented spacecraft, as well as balloons and suborbital rockets, at what appears with hindsight very frequent intervals (with a fair number of failures) revealed rather rapidly the overall basic morphology of the Earth's magnetosphere. In the United States the magnetospheric spacecraft were principally flown by the National Aeronautics and Space Administration (NASA, organized in 1958) and were numbered in the *Pioneer* and the *Explorer* series. These spacecraft were launched from the Earth's surface into a variety of orbital paths. There were low-altitude orbits which had inclinations relative to the equator ranging from near 0 to near 90° (polar orbits). Orbits with apogees at high altitudes provided measures of the geomagnetic tail and discoveries of the Earth's magnetopause, the bow shock and, of great importance, the magnetic fields and plasmas in interplanetary space—the solar wind.

As the research activities (and years) progressed in the 1960s and 1970s, the instrumentation carried by the NASA spacecraft increased in sophistication, in type with respect to geophysical parameters measured, and in number carried on a given probe. The number of investigators and their parent institutions increased rapidly. The instrumentation carried for charged particle studies evolved from the original simple GM tubes to solid-state detectors, Faraday cages, Langmuir probes, and channel electron multipliers. The energy range of particles capable of being detected slowly increased, with the development of new instrumentation technologies, to cover plasma particles with energies from near zero electron volts (eV) to many tens of mega-electron-volts. Capabilities for measuring ion species and, in the 1980s, ionic charge states, have been developed. The Earth's magnetic field and its temporal changes were measured, and electromagnetic and electrostatic plasma waves were discovered, measured, and studied with the development and launch into the magnetosphere of magnetic-field– and electric-field–measuring antennae with the plasma-measuring sensors.

Several spacecraft in the NASA *Orbiting Geophysical Observatory* (*OGO*) series of Earth orbiters (1960s) carried especially comprehensive sets of instrumentation to map, and to begin the physical understanding of, the space plasma environment. Particular scientific interests and accomplishments included detailed studies of particle-wave interactions, mechanisms of auroral particle acceleration and wave generation, and the time, space, and composition dependencies of the magnetosphere cold plasma populations. Several spacecraft in the *Interplanetary Monitoring Platform* (*IMP*) subset of the *Explorer* series were designed with instrumentation, and flown in appropriate orbits, to investigate both the solar wind (which controls magnetosphere dynamics) and the Earth's magnetotail at distances of several 10s of Earth radii, and even to the orbit of the Moon (64 earth-radii distances) in the magnetotail. Instrumentation on these spacecraft provided data and understanding of the dynamics of the magnetotail boundary regions and of the plasma sheet deep within the tail.

In addition to NASA magnetosphere satellites carrying instrumentation provided by investigators from many research institutions [universities, primarily, but also government (both NASA and some non-NASA) and industrial laboratories], single university groups constructed, instrumented, saw launched, and even acquired the telemetry from their own magnetosphere satellites. The physics department of the University of Iowa was notable in this respect. Their spacecraft and instrumentation were designed to study processes ranging from auroral particle acceleration at low altitudes to solar wind particle penetration into the magnetospheric cusp at high altitudes. Furthermore, particularly in the 1960s and somewhat less so into the 1970s, instrumentation on otherwise primarily military- and communications-oriented (industrial as well as NASA) spacecraft provided important data and the physical understanding of aspects of the magnetosphere plasma environment.

In the late 1960s and the 1970s, in addition to the U.S. and the U.S.S.R., several European nations, as well as Canada, began to contribute strongly to studies of the Earth's space plasma environment by providing instrumentation for NASA spacecraft and, eventually, constructing and flying their own satellites. Important spacecraft in the 1980s dedicated to studies primarily of the aurora have been the *Viking* mission (Sweden, with important contributions from six other nations) and *EXOS-D* (Japan).

COORDINATED MAGNETOSPHERE MISSIONS

The mappings of the basic morphological characteristics of the plasmas and of the magnetic and electric fields of the Earth's magnetosphere (the Earth's geospace environment), carried out by individual spacecraft missions, provide a type of "static" picture of the morphology of space. However, the morphology changes from satellite pass to satellite pass through a given region of geospace. In order to understand the dynamics of the magnetosphere, correlated measurements of plasmas and fields are needed at a variety of locations. In the first two decades of magnetosphere research, such intersatellite correlations were largely carried out by fortuitous conjunctions of spacecraft orbital locations and instrument availability. Also of importance, from the early days of space flight, have been correlations of space-obtained parameters with ground-based, balloon, and rocket measurements.

A joint NASA/European Space Agency (ESA) program was the first multisatellite mission specifically targeted for a coordinated study of large-scale magnetosphere plasma processes. This *International Sun-Earth Explorer* (*ISEE*) venture consisted of three spacecraft. Two closely orbiting spacecraft (launched in 1977) were configured to study the spatial and temporal characteristics of, especially, magnetosphere plasma boundaries and structures such as the magnetopause and bow shock. The third spacecraft (1978), positioned at the Sun-Earth libration point (the so-called L-1 point, about 1% of the distance from the Earth to the Sun), measured, continually, the solar wind incident on the magnetosphere.

This magnetosphere program was followed by several other multisatellite missions. The NASA Dynamics Explorer mission (1981) consisted of one low-altitude polar-orbiting spacecraft and one in a highly elliptical orbit with apogee initially at about four to five Earth radii altitude over the northern polar regions. Coordinated measurements were made of particles and waves incident on the upper atmosphere and its response to these energy inputs. The *Active Magnetosphere Particle Tracer Explorer* (*AMPTE*, 1984) was a three-spacecraft joint effort between the U.S., the United Kingdom (U.K.), and the Federal Republic of Germany (F.R.G.). *AMPTE* was specifically designed to conduct coordinated measurements of artificial ion releases in the solar wind and the magnetotail and to study their entry into the magnetosphere. The *AMPTE* mission also was designed for coordinated, multisatellite studies of the magnetosphere and of the outer magnetosphere regions near the equatorial plane.

As these specifically designed, coordinated missions were carried out, research progress on studies of magnetosphere processes continued to rely on the serendipitous locations of other flying magnetosphere missions to provide even more global coverage of the geospace environment. Researchers devoted special efforts to defining intervals of time when particularly good conjunctions of spacecraft (U.S. and foreign) would exist for addressing research problems related to the dynamics of a specific region of geospace. Attempts were always made to ensure appropriate and coordinated ground-based measurements that could contribute key data as well as help untangle the knotty ambiguities introduced by mixing space and time variations in a data set. Nevertheless, the absence, for a variety of reasons, of specific coordinated spacecraft mission sets has resulted in recent years in a complete lack of solar wind data for considerable portions of each month. The venerable IMP 8 (*Explorer 50*) magnetosphere spacecraft, launched in 1973, has carefully been nurtured so that, in the appropriate portions of its orbit, it can continue to measure key parameters of the interplanetary medium: the plasma conditions and the magnetic fields.

FUTURE MAGNETOSPHERE MISSIONS

In general, although not completely so (and improvements continue to be made), the sensor and data-handling techniques exist to design and build instruments which can provide the requisite measurements on essentially all aspects of the geospace plasma environment. A major problem lies in instrumenting sufficient spacecraft and in placing them in appropriate orbits in order to address some of the key problems of magnetosphere dynamics. These problems include the tracing of particle and energy flows from the solar wind to the atmosphere, studying the details of plasma processes, and investigating the origin and loss of plasmas near the Earth. To accomplish these goals, the International Solar Terrestrial Physics (ISTP) program is being developed for the early to mid-1990s by an international consortium of space agencies consisting of NASA, ESA, and the Institute of Space and Astronautical Science (ISAS) of Japan. There are plans for magnetosphere spacecraft from the U.S.S.R. to participate as well. In addition to the space-based measurements at key points in geospace, certain complementary ground-based arrays will be deployed at several locations around the world.

The principal NASA/ESA/ISAS spacecraft in the ISTP program, shown in Fig. 1, will all have maneuvering capabilities for large orbit modifications. The *Ulysses* spacecraft (an ESA/NASA collaboration), while not a part of the program, is shown because, launched in October 1990, it will be making measurements of the interplanetary environment over the polar regions of the sun during the ISTP endeavor.

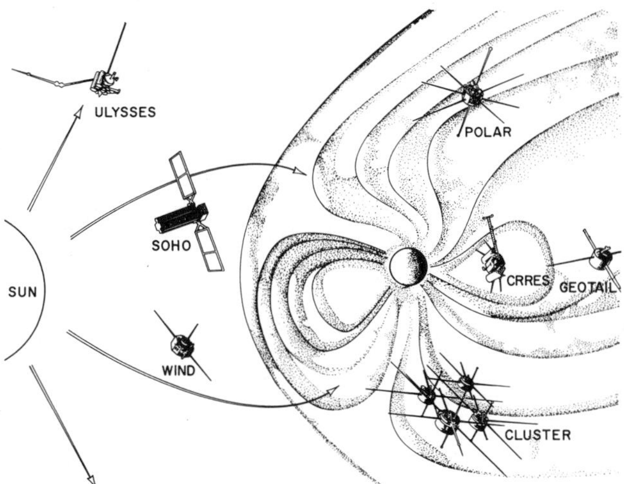

Figure 1. Schematic illustration of the principal spacecraft which will comprise the major program in magnetosphere studies for the 1990s: the International Solar-Terrestrial Physics program, a collaborative effort of NASA, ESA, and ISAS.

The *Wind* and *Polar* spacecraft are primarily NASA's responsibility. *Wind* will monitor the solar wind upstream from the Earth, and *Polar* will measure plasma flows from the ionosphere into geospace and energy and plasma flows from the magnetosphere into the high-latitude atmosphere (forming, for example, the aurora). *CRRES* (launched in July 1990) is a joint NASA/US Air Force spacecraft designed, for the ISTP portion of its mission lifetime, to monitor the plasmas in the equatorial regions of the magnetosphere.

The *Geotail* spacecraft (built by ISAS and containing some NASA-sponsored instrumentation) will orbit from near Earth to flybys of the moon. It will study plasma storage and release mechanisms in the tail of the magnetosphere.

The *Cluster* spacecraft set (ESA responsibility with some NASA-sponsored instrumentation) will consist of four small satellites positioned in an halo orbit to investigate small-scale spatial and temporal structures in the solar wind and (especially) the high-latitude magnetosphere. *Soho* (an ESA project with some US instrumentation) will investigate, from the Earth-Sun L-1 libration point, the physics of the solar interior, solar corona, and some aspects of the solar wind.

Extensive and elaborate data handling, manipulation, dissemination, and analysis efforts are required, and are being planned, in order to achieve the optimum scientific returns from the set of spacecraft and the ground-based elements of the ISTP program.

In the early 1990s, during the ISTP program, there will continue to be more limited, specifically targeted missions directed to discrete magnetosphere problems. In addition, some measurements are planned for opportunities on the space shuttle which will contribute to new understandings of the Earth's space environment.

CONCLUSION

The capability of sending instruments into the geospace environment has revolutionized understanding of the environment and of the connections between the sun, this environment, and processes occurring on the surface of the Earth. Further research advances will come from the extensive use of closely coordinated, intensively instrumented spacecraft sent to specifically designated regions of the magnetosphere. These satellites will return data that will lead to fundamental plasma physics understandings of the processes that control the transfer of energy and mass throughout the magnetosphere.

Additional Reading

Akasofu, S.-I. (1989). The dynamic aurora. *Scientific American* **260** (No. 5) 90.

Bird, J. (1988). The upper atmosphere: Threshold of space (NASA Publication No. NP-105). NASA, Washington, DC.

Goldstein, M. L. and Schmidt, R., eds. (1988). *The* Cluster *Mission*. European Space Agency Special Publication No. ESA SP-1103. European Space Agency, Noordwijk, The Netherlands.

Hess, W. N. (1968). *The Radiation Belt and Magnetosphere*. Blaisdell, Waltham, MA.

Krimigis, S. M., Haerendel, G., McEntire, R. W., Paschmann, G., and Bryant, D. A. (1982). The active magnetospheric particle tracer explorers (AMPTE) program. *Eos* **63** 843.

Newell, H. E., Jr. (1980). *Beyond the Atmosphere*. NASA, Washington, DC.

Thompson, T. D., ed. (1989). *TRW Space Log*. TRW Space and Technology Group, Redondo Beach, CA.

Van Allen, J. A. (1983). *Origins of Magnetospheric Physics*. Smithsonian Institution Press, Washington, DC.

Viking Science Team (1986). The Viking Program. *Eos* **67** 793.

See also **Earth, Magnetosphere and Magnetotail.**

F

Filters, Tunable optical

Laurence J. November

Tunable optical filters are devices which provide a wavelength-adjustable passband over a two-dimensional imaging field. Most of the narrow-passband tunable filter designs are based either on amplitude interference or polarization extinction. Amplitude-interference devices use diffraction or variable refraction (e.g., a grating or prism spectrograph), or selective transmission and reflection, as a function of wavelength (e.g., an interferometer or thin-film filter). These devices are "tuned" in wavelength by changing an appropriate angle or effective path length. Polarization-extinction devices operate by the selective absorption of polarized light in a polarization analyzer because of the variation of retardation with wavelength in birefringent media. Birefringent media show different indices of refraction for light in two polarization states. Polarization-extinction devices are usually tuned by varying the retardation of the medium.

SPECTROGRAPHS

A grating or prism is commonly used to separate the wavelengths of incident light in angle by diffraction or variable refraction. The spectrum, as produced by a spectrograph, is the convolution of an image with the response function of the spectrograph in the dispersion direction. If the image consists of discrete objects or the spectrum of narrow emission features, a slitless spectrograph can be regarded as a spectral imaging device. It produces overlapping images at each spectral feature. In a conventional spectrograph, the input source is a linear slit. In this case spatial variation occurs only in one axis, whereas spectral variation occurs only in the perpendicular axis of the resulting spectrogram. Hale's spectroheliograph is a conventional spectrograph in which the image is stepped perpendicular to the entrance slit while a photographic plate is synchronously stepped at an exit slit. The resulting "spectroheliogram" provides a two-dimensional image at the selected wavelength. This instrument constitutes a tunable optical filter with low throughput, as it samples only one spatial line at a time.

In contrast, the double-grating spectrograph produces a two-dimensional image in a selectable spectral passband. The configuration transmits a two-dimensional image in a single passband, via two gratings with an intermediate slit. The system has high throughput, but the center of the spectral passband varies over the image in the dispersion direction. A modified form uses a multiple-slit intermediate mask to give multiple images at the exit from adjacent spectral positions. However, its spectrum is inherently undersampled.

MICHELSON INTERFEROMETER

The classical form of the Michelson interferometer divides the incident light amplitude into two beams at a beam splitter, and reflects and recombines the beams with differing optical path lengths to give by interference, modulated transmission as a function of wavelength. The Michelson interferometer is an imaging device; however, a path difference of many wavelengths, which is desirable for high spectral resolution, can be obtained only over a limited angular field of view. Several modified designs have been developed to increase the range of the angular acceptance. The Michelson interferometer is the basis of the Fourier transform spectrograph, which gives a complete spectral sample as the path-length difference is stepped through a range of values. The numerical Fourier transform of the resulting modulated intensity recovers the source spectrum. If used in an imaging mode, this device produces the equivalent of a tunable filter that is stepped in wavelength at its spectral resolution. The Fourier transform spectrograph has wide use in laboratory applications because of its high intrinsic throughput and spectral resolution. The sampling time required to make a complete spectral measurement poses a limitation in practical imaging applications.

FABRY–PEROT FILTER

Interference filters are formed by vacuum deposition of a succession of high- and low-index of refraction thin-film coatings onto an optical glass substrate. The succession of partial reflections causes most of the light amplitude to be reflected. The coatings can be adjusted to give a single fixed passband at a selected wavelength. Two high-reflectance coatings with a fixed separation is called an etalon, which is characterized by a comb-function spectrum consisting of widely spaced narrow passbands. The passband of a Fabry–Perot interferometer (adjustable etalon) is tuned by changing the cavity optical size. The width of the passband depends upon the surface reflectance and cavity uniformity. The "finesse," a term useful for describing all channeled spectra, is the ratio of the interorder separation to the passband width. In imaging applications the Fabry–Perot is limited in field angle. For example, for surfaces separated by 2000 wavelengths in air, an off-axis ray of 1.8° gives an effective path-length difference of one wave. However, the acceptance angle increases with the index of the cavity. The limited field can also be alleviated by increasing the aperture of the filter.

Fabry–Perot filters can be cascaded in series or used in a double-pass optical configuration to give a system with very high effective finesse. They are stable within the dimensional stability of the plate spacing and refractive index of the cavity medium. The transmission function is approximately Lorentzian in shape. The Fabry–Perot is tunable by various methods. The dimensions of the spacing can be changed precisely by using piezoelectric crystals. Alternatively, the optical thickness can be changed by varying the gas pressure for a cavity type, or the voltage of an electro-optical material for a solid etalon, like lithium niobate ($LiNbO_3$). Liquid crystals are being tested as a spacing material because they exhibit a sizable index change with a low applied voltage.

LYOT–ÖHMAN FILTER

Polarization-extinction filters use a birefringent and hence phase-retarding medium between polarizers. The birefringent medium is oriented with respect to the entrance polarizer, so that the light is entirely extinguished by the analyzing polarizer at some wavelengths. In the simplest design, due to Bernard Lyot and Yngve Öhman, a single birefringent crystal is oriented with its ordinary axis at 45° with respect to the entrance polarizer and following analyzer; the ordinary axis is in the optic axis normal direction. This arrangement results in the incoming polarization state being equally divided in amplitude between the two principal (orthogonal) "ordinary" and "extraordinary" polarization states that propagate differentially in phase through the crystal. The polarization state at the exit of the crystal is a function of wavelength, varying from a linear ±45° polarization state (with respect to the ordinary axis of the crystal) through circular polarization states. The light is transmitted or extinguished by the analyzer periodically in wavelength with a sinusoidal form; this spectral period is called the element's "free-spectral range." A number of Lyot elements stacked in series, each having twice the free-spectral range of the previous, give a *sync-function* passband of width one half the free-spectral range of the thickest element, which is repeated in each free-spectral range of the thinnest element. A polarizing beam-splitter with differing path lengths in two reflection channels can be used as a substitute for a birefringent crystal.

OTHER BIREFRINGENT FILTERS

Numerous other designs have been found that give a single passband over some free-spectral range. A more desirable form of passband has been demonstrated by using multiple redundant "contrast elements" in series, often of Lyot design, but made with light-leaking partial polarizers. One form of the Evans "split-element" consists of three oriented birefringent crystals having the thickness ratios 1:4:1, between two polarizers. The split-element scheme provides a passband with the same width as that provided by an equivalent-thickness Lyot element, but having all the even passbands extinguished and thus twice the finesse. Solc filters consist of stacks of equal-thickness crystals, each crystal with its ordinary axis rotated relative to adjacent crystals, between two polarizers. The Solc "fanned" and "folded" solutions each give a single narrow passband with an equivalent width of the same-thickness Lyot element, but the free-spectral range between passbands increases as the crystal elements are sliced thinner and thinner. The finesse of a Solc element using always two polarizers scales asymptotically as the number of crystal elements used in the stack. Other general types of solutions have been found that are not as well studied as these.

All of the birefringent filter designs are tuned by changing the optical path length and hence the retardation through the birefringent material. One wavelength of phase difference gives tuning through one free-spectral range. A simple design uses a sliding crystal wedge for each subelement to effect a total change of length in the crystal. Alternatively, electro-optic birefringent materials, such as potassium dihydrogen phosphate (KD*P), lithium niobate, or liquid crystals, can be used. Most commonly, tunable Lyot elements use an achromatic quarter-wave plate, following the crystal, oriented with its principal axes at 45° with respect to the ordinary axis of the crystal. The quarter-wave plate has the effect of mapping the elliptical polarized states following the crystal into linear polarization states, so that any phase of the output light might be extinguished simply by rotating the following analyzer. Rotating the analyzer through 180° changes the phase of the spectral transmission through one free-spectral range. Two designs are commonly used for making achromatic wave plates: (1) Two different materials of opposite spectral variation of birefringence can be adjusted in relative thickness to give less than 5% variation through the visible. (2) A multiple-element stack with elements of varying orientation (following a Solc design) can be made to give any desired degree of spectral compensation.

Birefringent crystals exhibit a natural compensation for off-axis path-length changes. The effective birefringence in the crystal decreases away from the optic axis in at least one plane to compensate naturally for the increased path between the two parallel surfaces. A large anisotropy between the rays in the plane of, and in the plane perpendicular to, the ordinary axis is generally exhibited by uniaxial crystals. It is usual to "field widen" an element if it is longer than about 1000 waves in phase retardation. A Lyot element is field widened by dividing the crystal in half in the optic axis direction. The later half is rotated 90° and an (achromatic) one-half wave plate is inserted between the two crystal halves. The polarization-retardance properties of a field-widened element are unchanged, but the natural field compensation effect of the crystal becomes approximately equal for all the inclined rays. Other designs have field widened variants. Because birefringent filters are single–polarization-state devices they can never transmit more than half the light of an unpolarized signal.

MAGNETO-OPTICAL FILTER

Resonance scattering in an atomic gas in a magnetic field produces strong retardation variations in wavelength near a transition. A heated atomic gas cell with an applied magnetic field between crossed polarizers rejects all wavelengths except near transitions. The geometry of the field, its strength, the density and temperature of the gas, the length of the cell, and the entrance polarization state are free parameters that can be changed to adjust the spectral transmission characteristic. Although the magneto-optical filter has limited applicability in general spectral tuning, it has been used successfully for precise Doppler and magnetic field measurements around particular atomic transitions.

ACOUSTO-OPTICAL FILTER

Bragg diffraction, like diffraction from a grating, occurs with acoustic waves in an optical medium; the acoustic waves give a periodic index variation. Although the simple Bragg cell cannot usually be used for spectral imaging, a modified form given by acoustic waves in a birefringent crystal gives an excellent tunable filter. Light is introduced polarized parallel to the extraordinary axis of a birefringent crystal that is attached to a high-frequency acoustic modulator. The diffracted light has an output angle and polarization state that is a function of wavelength. The condition that the output polarization state be parallel to the ordinary axis of the crystal is very restrictive. The acoustic wave direction and diffraction angle can be adjusted to optimize the angular acceptance field, and the filter is tuned by changing the acoustic wave frequency. The acousto-optical filter has very favorable field properties with a narrow passband and exhibits a wide range of spectral tunability. Its applicability in astronomical problems is only beginning to be realized.

Additional Reading

Atherton, P. D., Reay, N. K., Ring, J., and Hicks, T. R. (1981). Tunable Fabry–Perot filters. *Opt. Eng.* **20** (6) 806.

Chang, I. C. (1981). Acousto-optic tunable filters. *Opt. Eng.* **20** (6) 824.

Evans, J. W. (1949). The birefringent filter. *J. Opt. Soc. Am.* **39** (3) 229.

Gunning, W. J. (1981). Electro-optically tuned spectral filters: A review. *Opt. Eng.* **20** (6) 837.

Solc, I. (1965). Birefringent chain filters. *J. Opt. Soc. Am.* **55** (6) 621.

Title, A. M. and Rosenberg, W. J. (1981). Tunable birefringent filters. *Opt. Eng.* **20** (6) 815.

Vanasse, G. A. and Sakai, H. (1973). Fourier spectroscopy. In *Progress in Optics*, E. Wolf, ed. North-Holland, Amsterdam, Vol. 6, p. 259.

See also **Telescopes and Observatories, Solar.**

G

Galactic Bulge

Victor M. Blanco

Spiral galaxies similar to our own exhibit reddish-hued bulge-like brightness concentrations around their dynamic centers. Near the centers of such galaxies, of which the Andromeda galaxy (M31) is a prototype, the surface brightness increases sharply towards the center, where, in the case of M31, a star-like nucleus is observed. Intervening interstellar dust clouds, which are concentrated in the galactic plane, where the Sun is located at a distance of about 7 or 8 kpc from the galactic center, diminish any visual light radiated from the nucleus of our galaxy by about 10^{12} times or more. As a result, the nuclear region of the Milky Way and its surroundings can only be observed at infrared or longer wavelengths. Much of what is known about the inner part of the Milky Way's bulge has been gained from such observations or by assuming it to be identical to the inner parts of the bulges of prototype galaxies such as M31, where the central regions are observable without the hindrance of interstellar dust. This assumption is supported by a remarkable similarity in the bulges of the Milky Way and of M31 of the radial distribution of surface brightness at a wavelength of 2.2 μm. Spectroscopic observations of M31 show that red giant stars of spectral types K and M are the source of the 2.2-μm radiation. Such stars are also found in large numbers in a few areas of the Milky Way bulge where the line of sight is not too affected by interstellar dust. One such area, called Baade's Window, located about 4° south of the galactic center direction, has played an important role in the exploration of the Milky Way bulge. Following a current trend, we here designate as the bulge of our galaxy the volume surrounding the nucleus and extending outward to about 1 kpc from the center, where it merges with the Galaxy's halo and disk. Whether the bulge is simply the inner part of the halo or of the disk has been a matter of controversy, but recent work shows the bulge to contain a stellar population unlike any found elsewhere in the Galaxy.

The strong concentration of globular clusters towards the center of the Milky Way led Walter W. Baade to conclude in 1944 that the galactic bulge contained a stellar population similar to that found in those clusters. Baade designated those stars as belonging to Population II to differentiate them from the type of stars found in galactic spiral arms, which were assigned to stellar Population I. Baade's conjecture was later supported by the discovery of RR Lyrae variable stars and planetary nebulae in Baade's Window and elsewhere in the bulge. These are archetype Population II objects. The RR Lyrae stars in Baade's Window show a pronounced density maximum as the line of sight passes closest to the galactic center. The apparent magnitudes of the stars near that maximum provide at present the most reliable way of determining the distance of the Sun from the galactic center. Baade's stellar populations concept eventually grew to include differences in stellar ages, chemical compositions, kinematics, and space distributions and to be of major importance in the development of a generally accepted hypothesis for the formation of the Galaxy.

Briefly, the hypothesis calls for a density fluctuation in a lumpy, turbulent, protogalactic hydrogen-helium gas medium with some angular momentum to become unstable against collapse under its own gravitation. In theory, a fluctuation larger than the Jeans length, $L_J = S(\pi/G\rho)^{1/2}$, where G is the gravitational constant, S is the speed of sound in the gas and ρ its density, can suffer gravitational collapse if the gas loses kinetic energy as a result of internal collisions and subsequent radiation. If the kinetic energy generated by the collapse is radially dissipated, the collapse continues until stopped by rotation in a plane normal to the angular-momentum vector while continuing in directions perpendicular to that plane. Presumably, as the collapse proceeds stars are first formed in a large centrally condensed spheroid and eventually in a galactic disk. As these stars evolve, heavier elements—so-called metals—are built by nuclear reactions in their interiors and eventually ejected. Baade's Population II stars are found in the galactic halo and supposedly were formed early in the collapse and reflect the space distribution, kinematics, and a chemical composition low in heavy elements, as expected for the earlier stellar generations. Population I stars and the interstellar dust are younger higher-metallicity objects confined to the galactic disk. According to this hypothesis the galactic bulge is expected to contain old Population II stars as well as younger Population I objects.

Soon after Baade's seminal 1944 work, spectroscopic observations showed that the integrated light from the bulges of the Milky Way and of M31 was principally radiated by stars rich in heavy elements, and thus unlike Population II red giant stars. Individual stars in both of those bulges have now been segregated and observed spectroscopically. The spectra confirm the integrated light results. In Baade's Window, red giant stars show a wide range of metallicity but on the average, unlike red giant stars in globular clusters, they have about twice the heavy-element content of prototype Population I stars in the solar vicinity. They are also less luminous, show a much lesser frequency of variability, and from stellar evolution theory, their average mass is about one solar mass. As far as we know these stars have no counterparts outside the bulge. Presumably, similar red giant stars populate the bulges of other spiral galaxies, a conjecture supported by Albert E. Whitford's finding in 1978 of a close spectrophotometric similarity in the integrated light from Baade's Window and from the bulges of other spiral galaxies.

Other objects found in the bulge of the Milky Way include OH/IR masser sources. Some of these sources possibly are planetary nebulae precursors. The *Infrared Astronomical Satellite* (*IRAS*) has shown numerous point sources of 12-μm radiation in the Milky Way. In Baade's Window these sources have been identified with Mira variables and OH/IR sources. The faintest and reddest of the 12-μm point sources are concentrated in what appears to be a box-like volume included between galactic latitudes $\pm 7°$ and galactic longitudes $\pm 10°$. Such box-like bulges may be found in other spiral galaxies. The 12-μm point sources are also concentrated in the galactic disk.

The Milky Way bulge must contain numerous faint stars. So far unobserved, they are the principal contributors to the total bulge mass. This mass can be estimated from the rotational velocity of a circumnuclear gas disk rich in neutral atomic hydrogen whose 21-cm spectral line is readily observed. From about 50 to about 700 pc from the center, the gas disk rotates in practically Keplerian circular orbits. The rotational velocity at a given galactocentric distance R can be used to estimate the total interior mass. An alternative way of estimating the bulge mass is based on the distribution of 2.2-μm surface brightness throughout the bulge. Although the total mass of the red giant stars is insignificant, they are the source of most of the 2.2-μm radiation. If the ratio by volume between the frequency of red giants and the mass of the faint objects does not depend on R, then the ratio $M/L_{2.2\ \mu m}$ between the total mass contributing to the radiation observable from a given bulge area and the observed 2.2-μm luminosity of the area should be constant. Rotational velocities of the circumnuclear gas disk, where known with some accuracy, and 2.2-μm surface brightness measurements have been used to check for the constancy and to calibrate it. The results indicate that $M/L_{2.2\ \mu m} \approx 1$, in solar units. The 2.2-μm brightness distribution is known in the Milky Way bulge from close to the galactic center out to about 1 kpc. Inward from about 200 pc a sharp increase in the 2.2-μm luminosity is caused by a cluster of red giant stars whose center seems to agree with the center of the Milky Way. Within 2 pc of the center violent gas motions are observed, and the bulge merges with the galactic nucleus.

The 2.2-μm surface-brightness data yield total masses comparable to those derived from purely kinematic measurements and confirm that within 500 pc the Milky Way bulge contains about 10^{10} solar masses. Of this total, some 10^7 solar masses are found within about 1 pc from the center, as indicated by the kinematics of doubly ionized neon near the nucleus. The radial velocity of the neon gas is observable because of a 12.8-μm emission. Besides neutral atomic hydrogen, the interstellar gas in the inner part of the bulge has molecular components that are concentrated in massive clouds whose motions deviate appreciably from circular rotation. These clouds may have been expelled from the nucleus, where poorly understood energetic processes are found. The total mass of the gas within the bulge must be much less than 10^{10} solar masses. A plausible assumption of the distribution of stellar masses combined with the frequency of red giant stars observed in bulge windows indicates that most of the total mass of the bulge is in the form of stars.

Gerard de Vaucouleurs found an empirical dimensionless formula often called the $R^{1/4}$ law, which in general, but not always, describes the variation of surface brightness with galactocentric distance R in the bulges and halos of spiral galaxies. It has been shown that a spherically symmetrical stellar population with a uniform mass-to-luminosity ratio will follow the above relationship if its density ρ obeys the relationship

$$\rho(R/R_e) \propto (R/R_e)^{-7/8} e^{-7.67(R/R_e)^{1/4}}$$

where R_e is the galactocentric radius containing half the total luminosity of the system. This relationship holds for an oblate spheroidal density distribution if R is replaced by $\alpha = [x^2 + z^2(a/c)^2]^{1/2}$, where x and z are, respectively, distances from the center along the galactic plane and perpendicular to it, and a and c are the major and minor axes of the spheroid. The said density distribution may be assumed to apply to our galactic halo. Combined with an analogous expression applicable to the galactic disk, one may estimate the frequency of stars observable in a given sky area. Such estimates are in reasonable agreement with observations in several sample sky areas for the halo. Applied to the bulge, however, the density distributions fail to predict the large total mass and the density distribution derived kinematically or from the 2.2-μm luminosity. Instead one finds that $\rho \propto R^{-x}$, where, except very close to the galactic center, $x = 1.8$ for $R \leq 50$ pc and increases to 3.5 at the bulge periphery. This suggests that the Milky Way bulge is more massive than expected from the $R^{1/4}$ law. Our galaxy is, however, not unique. The $R^{1/4}$ law is not universal and indeed galaxies similar to our own are known. The bulge may not have the simple spheroidal symmetry expected from the above density relationships; departures from circular motion of the bulge gas may be possibly explained by a prolate spheroidal density distribution.

In summary, and in regard to the controversy over whether the galactic bulge is part of the galactic halo or disk, observations show that within the bulge are found objects characteristic of both the halo and disk and also a unique high-metallicity stellar population. We also find the bulge to be much more massive than expected from projections of disk and halo densities.

Additional Reading

Frogel, J. A. (1988). The galactic nuclear bulge and the stellar content of spheroidal systems. *Ann. Rev. Astron. Ap.* **26** 51.

Genzel, R. and Townes, C. H. (1987). Physical conditions, dynamics and mass distribution in the center of the Galaxy. *Ann. Rev. Astron. Ap.* **25** 377.

Mould, J. R. (1986). The bulge of the Galaxy and M31. In *Stellar Populations*, C. A. Norman, A. Renzini, and M. Tosi, eds. Space Telescope Science Institute, p. 9.

Oort, J. H. (1977). The galactic center. *Ann. Rev. Astron. Ap.* **15** 295.

Whitford, A. E. (1985). The stellar population of the galactic bulge. *Publ. Astron. Soc. Pacific.* **97** 205.

See also **Andromeda Galaxy; Galactic Structure, Stellar Populations; Galaxies, Formation.**

Galactic Center

Paul T. P. Ho

In the past decade, great progress has been made in the study of our own galactic center. Because the center is obscured in the optical wavelengths by intervening dust, it is only in the past few years that detailed information on the structure and nature of the center of the Milky Way has been revealed by new methods. Infrared and radio observations, which are relatively unaffected by dust, have shown that there is a clustering of stars toward the center of the Galaxy, that there is a clearing of gas and dust within the central parsec, that there is a point-like nonthermal radio source in the center, and that the kinematics around this point source seem to be consistent with motions around a massive black hole. The possibility that there is a central massive black hole in the Milky Way is particularly exciting, because this would be the nearest example of such an object. In the next decade, there lies the possibility of resolving such an object, and studing its energetics, its possible origin, and how matter may be funneled toward the center.

The center of the Milky Way is now reckoned to lie in the direction of Sagittarius at approximately 8.5 kpc from the Sun. The exact center of the system has been difficult to locate, because we ourselves reside within the plane of the Galaxy with some 30 mag of visual extinction between us and the galactic center. With the advent of radio and infrared techniques, astronomers have succeeded in penetrating this veil of dust. The plethora of phenomena that have been revealed include massive molecular clouds, supernovae, stellar clusters, magnetized filaments and jets, a molecular ring, and a possible massive black hole. In the near infrared, the spatial distribution of the starlight indicates a concentration of stars with a density that varies with the radial distance as $r^{-1.8}$, close to the expectations for a relaxed isothermal cluster. If these stars, mostly red giants, reflect the galactic gravitational potential, then the center of the stellar distribution locates the center of the Galaxy to a precision of about 1 pc. In the radio, a point-like source known as Sgr A*, has been found very near the center of the infrared stellar distribution. Because of its many unusual properties, including size, spectrum, luminosity, and proper motion, as well as the observed gas motions in its vicinity, Sgr A* has been suggested to be a massive black hole. If true, this object would be the nearest example of the supposed engines of active galactic nuclei. The possibility of spatially resolving the accretion disk around a massive black hole is enormously exciting.

THE INNER 100 PARSECS

At the 100-pc scale, the galactic center is characterized by a number of giant molecular clouds. There is a concentration of

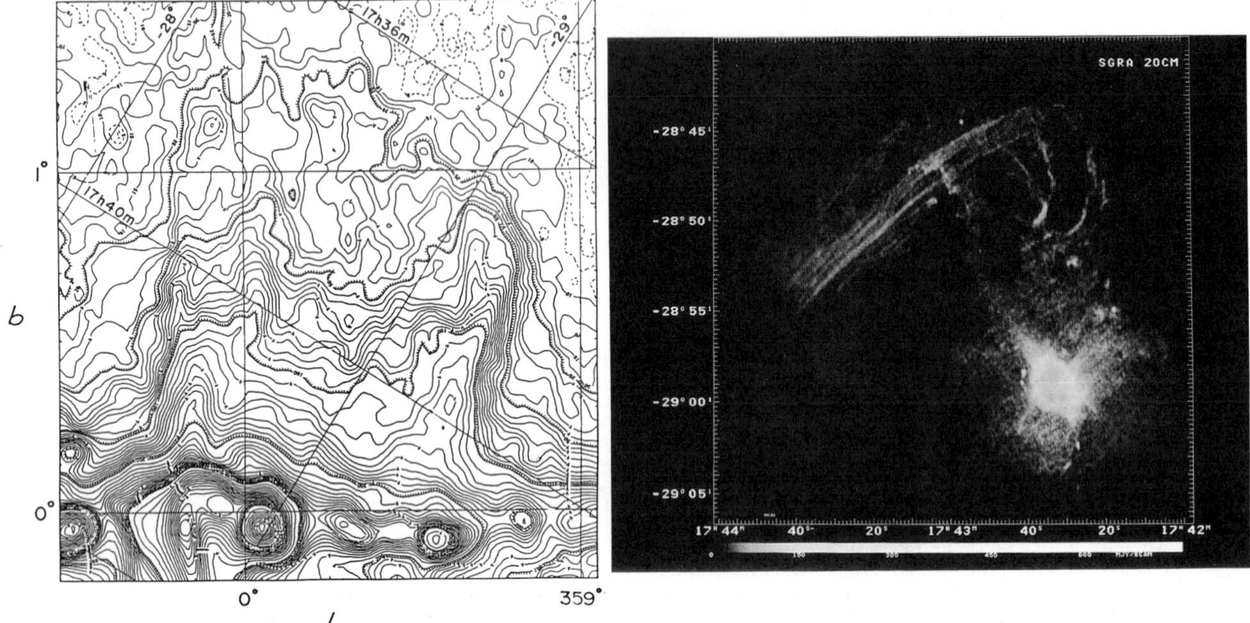

Figure 1. Left—radio continuum map of the large scale structure around the galactic center, showing the large Ω-shaped lobe. Right—higher-resolution picture of the arc-like feature to the northeast of the center. The arc-like feature appears to be composed of many very fine and parallel filaments. The morphology and the fact that the emission is synchrotron radiation strongly suggests the presence of a poloidal magnetic field perpendicular to the disk of the Galaxy. The arc visible in this map actually continues in the direction perpendicular to the disk and returns to the disk in the southwestern part to form an Ω-shaped structure. These nonthermal filaments are associated in places with thermal H II regions and molecular material indicating interactions with the local interstellar medium. It is unclear whether such features are due to ejection phenomena from the nucleus of the Galaxy.

gaseous matter toward the center of the Galaxy, with perhaps 100 times more mass per unit area of the galactic disk than at large radii in the disk. The largest of these clouds are Sagittarius A and Sagittarius B2, with masses of 10^5–10^6 $M_{\odot}$. The Sagittarius B2 cloud located at galactic longitude 0.65° is a large complex with vigorous star formation activities as indicated by numerous H II regions, infrared sources, and interstellar masers. This complex is also where most of the interstellar molecular lines have been discovered, no doubt because of the large column density of gaseous matter in this cloud. The Sagittarius A cloud located at longitude 0.0° is very close to the galactic center, and recent observations suggest that indeed parts of this cloud may be within 10 pc of the center. Star formation activities within the Sagittarius A cloud are, however, relatively restrained, with perhaps only a handful of OB stars. There are many more smaller molecular clouds within the central 100 pc. The most outstanding features of all the galactic center molecular clouds are their large spectral linewidths and temperatures (20–50 km s^{-1}; 50 K) as compared to the typical galactic disk material (5–10 km s^{-1}; 5–10 K). Whether this is due to larger turbulence or star formation activities or to tidal effects is unclear at this time. However, it is safe to surmise that the formation and evolution of the gaseous material must be very different in the galactic center than elsewhere in the Galaxy.

One of the most exciting and intriguing observations in the past decade has been the discovery of very fine filaments or strands in the radio arc-like structure seen at longitude 0.1° (see Fig. 1). Apparently originating from the galactic center, this arc is extended over 20′ (50 pc) and is part of an even more extensive 1° (250-pc) Ω-shaped "lobe" perpendicular to the plane of the Galaxy. In the arc structure, there are a number of very fine (~5″; 0.2 pc) filaments that are parallel for the most part and partially intertwined in a helical fashion. From their radio spectrum, the filaments and the arc structure appear to be producing synchrotron emission. The filamentary structure of these features and their linearity suggest strongly that a galactic poloidal magnetic field must be important. If these arcs, filaments, and lobes are parts of magnetic "loops," they are certainly reminiscent of similar, much smaller loops seen ejected at the surface of the Sun. Whether such magnetic events are associated with or can be produced at the galactic nucleus remain interesting conjectures at the moment.

THE INNER 10 PARSECS

Within the inner 10 pc, a number of interesting objects can be seen in radio wavelengths. In the radio continuum, there are two dominant objects known as Sagittarius A East and Sagittarius A West (see Fig. 2). The Sgr A East object is most likely a supernova remnant extended on a scale of 3′ (8 pc). Its radiation is clearly synchrotron emission from its radio spectrum. There is evidence for a second and much fainter ring-like structure to the south and southeast which is also emitting synchrotron emission. This fainter source may be a second supernova remnant also at the galactic center. Given the higher stellar density toward the center of the Galaxy, it is not surprising that there are a number of radio supernova remnants. Superposed on Sgr A East is the Sgr A West source, which is extended on the order of 1.5′. Low-frequency radio observations have now shown Sgr A West to be in absorption against Sgr A East, thereby placing Sgr A West in the foreground. The morphology consists of a number of thin arms emanating from the center, very much like a minispiral. These arms have been seen both in the radio and in the near-infrared continuum.

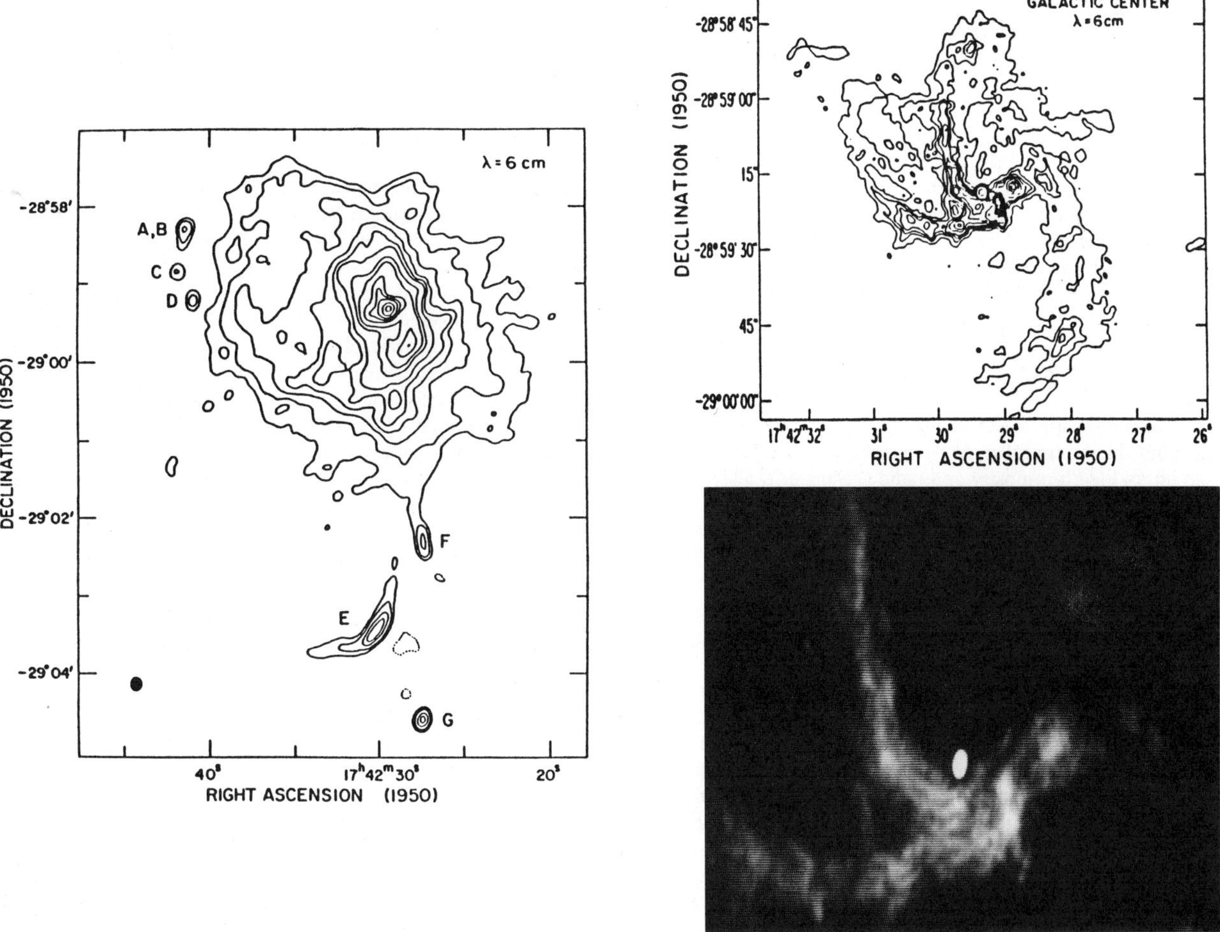

Figure 2. Left—radio-continuum picture of the central region which shows Sgr A East, Sgr A West, a possible second supernova remnant to the south formed in part by features E and F, and H II regions to the east and south, marked A, B, C, D, and G. Right—a close-up view in the radio continuum of the Sgr A West source. Note the intricate and thin spiral-like arms emanating from the center. The strong point-like source in the center is offset to the north from the arm which runs east–west, and has been suggested to be a candidate supermassive black hole. The thermal ionized gas composing the spiral-like arms appears to move in bound orbits around the central source. It is the kinematics of these ionized streamers that provide the strongest argument for a compact massive object. A neutral ring lies immediately outside of the thermal Sgr A West features. A larger molecular complex surrounds this whole region with evidence of interactions with the supernova remnant Sgr A East and of inflow toward the neutral ring.

The radiation from this source is apparently due to thermal bremsstrahlung. The photoionization process is confirmed by recombination lines seen both in the radio and in the infrared, although the source of ionization of the gas in the minispiral is unclear at this time. The consensus seems to be that these are gaseous arms without embedded stars and that they are somehow ionized by a central source. Spectroscopy of the recombination lines has allowed the kinematics of the ionized gas to be measured. The motion detected is consistent with orbital motions of the individual streamers in a gravitational potential. The motions of the various streamers suggest that they are not in the same orbital plane.

Near the center of Sgr A West is a point-like radio continuum source. Careful astrometry has shown that this source is not associated with any near-infrared sources. Very-long-baseline interferometry (VLBI) has resolved this source and it appears to be somewhat elliptical. This point-like source has been suggested to be possibly a massive black hole, as discussed in the next section.

The radio continuum complex Sgr A East and Sgr A West are surrounded by molecular material. The Sgr A molecular cloud is immediately to the south of the continuum complex, and another molecular cloud is immediately to the east of Sgr A East. An important question has been whether these molecular clouds are as close to the galactic center as their projected distances. In other words, can the molecular gas and the continuum complex be actually separated along the line of sight? The argument has been that in order for the molecular clouds to resist tidal effects from the central gravitational potential, they must be very dense, $>10^4$ cm^{-3}. The density of the molecular material material may in fact be at least that high so that self-gravitation and stability against the central tidal forces seem quite reasonable. Recent kinematic and morphological data at 10″ resolution suggest that the molecular gas is in contact with the central supernova remnants. The actual impact of Sgr A East on the ambient matter is indicated both by a dense ridge of molecular gas on the east and southeast edge, and an apparent episode of star formation behind the boundary/shock front. The morphology of the ridge suggests that the molecular material has been compressed by the supernova remnants.

There is a suggestion that part of the neutral gas in the south is flowing toward the galactic center in a thin streamer. Whether such gas flows supply a central "engine" is not obvious. From various molecular line observations, it appears that the inflowing material forms a circumnuclear ring of material immediately outside of Sgr A West. Parts of this ring correspond very well in position and velocity with the ionized arm-like structures of Sgr A West. This suggests that the ionized minispiral may be the inner surfaces of this neutral ring or shell of material or perhaps may consist of parts of the ring that are now falling in toward the center.

THE INNER PARSEC, A MASSIVE BLACK HOLE?

The most interesting object in the inner parsec is the nonthermal radio point-like source Sgr A*. It is neither a star nor a pulsar, based on its high luminosity and radio spectrum. It also cannot be a young supernova or supernova remnant, because its expansion motions are <13 km s^{-1}. There are several pieces of evidence which seem to suggest that it might be a black hole. The most direct evidence is the kinematics measured in a variety of lines for the ionized streamers of Sgr A West. The observed velocities are high, up to 300 km s^{-1}, vary smoothly, and can be fitted by orbital motions. Assuming that the streamers are in bound orbits, the deduced enclosed mass is on the order of $10^6\ M_\odot$. One important aspect of the ionized streamers is that one of them passes within 0.25 pc of Sgr A*. The large enclosed mass within such a small area could be in a compact object or be distributed in a centrally condensed stellar cluster. However, the density distribution of a cluster that satisfies the required kinematics must scale with radius as $r^{-2.7}$, much steeper than is observed in the near-infrared. Although extinction effects and the contribution of dark stars to the mass distribution are difficult to eliminate, there is no evidence at the moment for such a steep stellar mass distribution. Hence a compact massive object of $10^6\ M_\odot$ seems to be required by the kinematics. We note that although the kinematics are consistent with Sgr A* being the massive object, the positional uncertainty of the kinematical center is on the order of 5″.

The indirect evidence for a compact massive object has to do with the properties of Sgr A* itself. Its high luminosity is confined to a small area of 0.002″ (13 AU), which is about 100 times larger than the Schwarzschild radius of a massive black hole of a few million solar masses. The source seems to be elongated with $\sim 2{:}1$ axial ratio, as might be expected for a tilted disk. In addition, Sgr A* appears to be at rest, as might be expected for a supermassive object at the bottom of a potential well. After corrections for the solar peculiar motion with respect to the local standard of rest and the galactic rotation of the solar system around the galactic center, the proper motion of Sgr A* is less than 40 km s^{-1}. If Sgr A* is an ordinary object near the center of the galactic gravitational potential well, it should have a larger proper motion due to rotational motions. Hence the upper limit to its proper motion is consistent with the possibility that Sgr A* is associated with the accretion disk around a supermassive black hole.

All of this is not to suggest that the case for a supermassive object at the center of the Milky Way is conclusive. Indeed, there are many problems remaining with such an interpretation. First, Sgr A* is not associated with any infrared object. If the black hole is actively accreting, its luminosity should be detected in the infrared. Second, if Sgr A* is a massive black hole, accretion of stars should leave behind a cusp in the stellar distribution perhaps as large as 20″. Such a cusp has not been detected. Finally, the infrared stellar cluster IRS16 is projected to be within 0.5″ of Sgr A*, and yet it is not tidally disrupted. This seems difficult to explain unless these objects are significantly separated along the line of sight.

PROSPECTUS

There are many interesting and exciting questions which have been raised concerning the center of the Milky Way. Is there a supermassive black hole? The kinematic data seem to suggest so, and Sgr A* appears to be a possible candidate. To improve our understanding, we need to set more stringent limits on the proper motions of Sgr A*. If it is indeed at rest, this would be a very strong argument that it is the most massive object in the nuclear region. With higher-resolution radio continuum observations, with the new Very Long Baseline Array, and with space VLBI, which will improve the angular resolution, we should be able to resolve and study the structure of Sgr A*. Is this a disk? Can we see ejection phenomena? With an improvement of a factor of 100 in angular resolution, we can resolve the Schwarzschild radius of a massive black hole. In the near-infrared, better measurements of the stellar density distribution using the new-generation array cameras would be especially useful to constrain the possible radial dependences and also perhaps to see the cusp around a black hole. There is the question of possible funnelling of gaseous material into the nuclear region. The galactic center is the nearest example where we can study with such high spatial resolution the flow of gaseous material. The mechanism, the morphology, the time scale, the kinematics, and the energetics of such flows are open questions. Development of spectroscopic and interferometric techniques will provide the increased angular resolution needed to study the gaseous components at a scale comparable to that available for stellar components. Finally, the magnetic loops seen at 100-pc scales imply the great importance of magnetic fields in the galactic center region. Studies of the field strengths and morphologies are difficult, but may be increasingly possible in the submillimeter and far infrared, where dust emission becomes optically thin and polarization measurements can be interpreted with less ambiguity. In any event, it is clear that the study of the galactic center will remain a valuable paradigm for the more exotic and more energetic active nuclei in distant galaxies and quasars. Great progress can be made in the next decade.

Additional Reading

Backer, D. C., ed. (1987). *The Galactic Center* (Proceedings of AIP Conference No. 155 in Honor of Charles H. Townes). American Institute of Physics, New York.

Genzel, R. and Townes, C. H. (1987). Physical conditions, dynamics, and mass distribution in the center of the Galaxy. *Ann. Rev. Astron. Ap.* **25** 377.

Lo, K. Y. (1986). The galactic center: Is it a massive black hole? *Science* **233** 1394.

Morris. M., ed. (1989). *The Center of the Galaxy* (International Astronomical Union Symposium No. 136). Kluwer, Boston.

Oort, J. H. (1977). The galactic center. *Ann. Rev. Astron. Ap.* **15** 295.

Riegler, G. R. and Blanford, R. D., eds. (1982). *The Galactic Center* (Proceedings of AIP Conference No. 83). American Institute of Physics, New York.

See also **Active Galaxies and Quasistellar Objects, Central Engine; Interstellar Medium, Galactic Center.**

Galactic Structure, Globular Clusters

Kyle M. Cudworth

Globular clusters were originally defined as rich, compact, nearly spherical, groups of hundreds of thousands (or even millions) of stars. Work in the past few decades has shown that the stars in globular clusters are among the oldest stars in the Galaxy, with ages greater than 10^{10} years. As a consequence, for most astronomers the working definition of a globular cluster now concentrates more on the age of a cluster than on its richness, and a few are quite sparse. About 140 of these clusters are now known in our Milky Way galaxy. The brightness and distinctive appearance of globulars make them relatively easy to detect at large distances (except in directions where dust very severely absorbs starlight), so it is likely that most that exist in our Galaxy have been discovered. Furthermore, globular clusters are found in the galactic halo, well above and below the thin disk of the Galaxy that contains most stars and the younger open clusters. (The galactic halo should not be thought of as a shell, but rather as a roughly spherical volume of space within which globular clusters and some old stars are found.) While globulars are strongly concentrated toward the center of the Galaxy, some are found at very large distances from the galactic center. These characteristics are major reasons why globulars are key objects for the study of distant parts of the Galaxy. Not surprisingly, globular clusters are also seen in and around other galaxies.

The stars in globular clusters have been found to differ in chemical composition from most stars in the galactic disk in that globular cluster stars are depleted in heavy elements (metal poor) by factors ranging from at least 2 up to 200. In most clusters all stars have very similar chemical compositions, but the composition differs from cluster to cluster.

Because the spatial distribution and chemical composition of the globulars are thus distinctly different from those of most stars, these clusters reveal a different aspect of galactic structure than ordinary stars, and because the clusters are the oldest identifiable objects in the Galaxy, these differences contain information regarding the formation and early evolution of the Galaxy. Globular clusters thus provide much, probably most, of the basic observational data on which any understanding of the early epochs of our Galaxy must be based. We seek to learn more about these early stages in our Galaxy's history in order to understand how the Galaxy came to have its current structure and other characteristics, and we expect that much of what we learn about our Galaxy's history will also be applicable to other galaxies as well.

CLASSIC STUDIES

In an early (1915–1919) use of globular clusters in what we would now call a study of galactic structure, Harlow Shapley derived distances for many globulars, and found that the distribution of globulars was centered at about 15 kpc (kiloparsecs, 1 kpc = 1000 pc = approximately 3260 light-years) away from the Sun in the direction of the constellation Sagittarius. Shapley based his cluster distances upon the brightnesses of individual stars in each cluster when possible, and on the size and brightness of each cluster as a whole when individual stars could not be studied. Since many of the globulars that Shapley studied are out of the dusty plane of the Galaxy, the distances that he found were not too severely affected by the lack of a correction for the absorbing effects of dust. Shapley went on to argue that such massive objects as globular clusters were most likely to be centered around the galactic center, and that thus the center of the Milky Way was 15 kpc away from the Sun toward Sagittarius. Similar studies in the 1970s and 1980s with absorption corrections and with much better data from which to determine cluster distances have yielded distances to the center of the distribution of about 8 instead of 15 kpc.

While Shapley's conclusions remained controversial for a few years, they were eventually accepted by essentially all astronomers, and this technique is still considered one of the primary means of determining the distance to the center of the Galaxy. This classic work by Shapley was crucial in that it was the first study of the structure of our Galaxy to indicate a center well away from the Sun. This shift away from a heliocentric (Sun-centered) Galaxy may even be likened to the Copernican shift away from a geocentric (Earth-centered) solar system (and universe) that had occurred a few centuries earlier, but the shock from Shapley's studies was significantly less severe from either an astronomical or a philosophical/religious point of view.

In the 1940s, Walter Baade developed the concept of stellar populations to describe the differences between the stars commonly found in the thin Galactic disk (Population I) and those distributed spherically about the galactic center (Population II). Globular clusters were and are the primary example of Population II, or the halo population. We now understand the fundamental difference between these populations to be age, and that the differences in the types of stars, their chemical composition, and their distribution in the Galaxy, are all related to their ages. Further work has broken the division down into more populations, including a distinction between extreme Population II and intermediate Population II that was noted in the 1950s and has been reemphasized in the 1980s.

TWO POPULATIONS OF GLOBULAR CLUSTERS: GALACTIC EVOLUTION

Most globular clusters are in the extreme Population II or halo class, having stars metal-poor relative to the Sun by factors of at least 6 or 7 (and in most cases greater than 10), and being distributed essentially spherically about the galactic center. Halo clusters are often found as far as 10 kpc or more above or below the galactic plane. These clusters, like everything else in the Galaxy, are in orbit around the galactic center. Although many millions of years are required for a cluster to complete one orbit, we can learn a good deal about the shapes of the orbits from the current locations and velocities of the clusters. Such investigations have revealed that the orbits of halo globulars are not at all circular, but are generally quite elongated, and are oriented essentially at random. Clusters in such orbits participate only slightly, if at all, in the general rotation of the Galaxy that is the dominant motion of Population I stars in the galactic disk. Some halo globulars even show a motion opposite to galactic rotation.

In contrast to the halo objects, about 20% of the globular clusters in our Galaxy are less metal poor and are found within about 1 or 2 kpc of the galactic plane (compared to most Population I stars lying within 0.4 kpc of the plane), and thus belong to intermediate Population II, often called the thick-disk population. These clusters are also in orbit around the galactic center, of course, but their orbits are more nearly circular and are oriented near the galactic plane. These clusters are moving in the same direction as galactic rotation, though at a rate that is slightly slower than the rotation of the thin disk of Population I stars.

Nearly all of the thick-disk globulars lie closer to the galactic center than does the Sun (i.e., within about 8 kpc of the center), but halo clusters are found out to much larger distances, though also with a strong concentration toward the center. Globular clusters of both types are thus found in large numbers in the general direction of the galactic center, but only halo clusters are found well away from the center. An individual globular can usually be assigned unambiguously to one population or the other on the basis of its chemical composition or velocity.

All of these differences presumably reflect changes that happened in the Galaxy very early in its history, as stars and star clusters formed first from a large, nearly spherical protogalactic cloud of gas. This cloud would have contracted under its own gravitational pull and its small initial rotation would have sped up as the cloud contracted and caused the contraction to end in a disk rather than a small sphere. According to a commonly accepted picture, stars and clusters that formed a little later would have been formed in a thick disk and would have a higher metal abundance, as the chemical composition of the gas from which stars could form was enriched by the products of nuclear reactions in the most massive of the earliest stars.

A major issue in current research on globular clusters is the question of whether there are significant age differences from cluster to cluster, and especially between the two populations of globulars. Such age differences could tell us how long it took the Galaxy to contract from the original spherical cloud to a thick disk, and how long the thick-disk phase lasted. It presently appears that the difference in ages is no more than a few billion years (out of a total age of about 15 billion years), implying fairly rapid changes in the early history of the Milky Way. Our instruments for obtaining appropriate data, and our ability to infer reliable ages from those data, are both improving rapidly, however, so these conclusions regarding ages and age differences could change in the next few years.

THE OUTER GALACTIC HALO

As noted earlier, halo globulars exist much further from the galactic center than the Sun's position. The outer limit for normal halo clusters seems to be about 30–40 kpc from the galactic center, though a few globulars are also found between about 60 and 100 kpc from the center. Most of these outer halo clusters are less dense than typical globulars and most show some pecularities in their constituent stars that may indicate slightly younger ages. A few dwarf spheroidal galaxies are also found at distances similar to those of the outer halo globulars. These systems contain old metal-poor stars similar to those in globular clusters, and have masses comparable to the most massive globulars, but are different in that the star density within each system is very low compared even to the low density outer halo globulars.

It is generally believed that the outer-halo globulars and the dwarf spheroidal galaxies are in orbit around the Milky Way, and are not merely passing by. As they are the most distant objects associated with our Galaxy, their motions should reflect the effects of the gravitational pull of the entire mass of the Milky Way. Observations of their velocities in recent years have thus yielded some of the best available derivations of the mass of our Galaxy: about 5×10^{11} times the mass of the Sun. Since this is considerably more mass than can be accounted for by adding up the masses of all the stars and gas clouds we observe directly, this is part of the evidence for unseen dark matter (sometimes called missing mass) in the Galaxy.

Because of their importance in questions of galactic evolution and the study of the outer parts of the Galaxy, globular clusters continue to be investigated intensively by many astronomers today. Questions such as the distance scale, chemical compositions, and ages of globulars are all interrelated and are receiving considerable attention from both observational and theoretical perspectives. We can thus look forward to increasing knowledge and, one hopes, a better understanding of these issues in the coming years.

Additional Reading

Armandroff, T. E. (1989). The properties of the disk system of globular clusters. *Astron. J.* **97** 375.

Bok, B. J. (1981). The Milky Way galaxy. *Scientific American* **244** (No. 3) 92.

Freeman, K. C. and Norris, J. (1981). The chemical composition, structure, and dynamics of globular clusters. *Ann. Rev. Astron. Ap.* **19** 319.

Harris, W. E. and Racine, R. (1979). Globular clusters in galaxies. *Ann. Rev. Astron. Ap.* **17** 241.

King, I. R. (1985). Globular clusters. *Scientific American* **252** (No. 6) 78.

Zinn, R. (1985). The globular cluster system of the Galaxy. IV. The halo and disk subsystems. *Ap. J.* **293** 424.

See also **Galactic Structure, Large Scale; Galactic Structure, Stellar Populations; Galaxy, Chemical Evolution; Star Clusters, Globular; Star Clusters, Globular, Chemical Composition; Star Cluster, Globular, Extragalactic; Star Clusters, Globular, Variable Stars; Stars, RR Lyrae Type; Stellar Orbits, Galactic.**

Galactic Structure, Interstellar Clouds

Leo Blitz

The disk-like nature of the Milky Way has long been known from its optical appearance on the celestial sphere. However, the extinction of starlight by ubiquitous interstellar dust makes it impossible to untangle the detailed large-scale structure of the Milky Way by analyzing its visible radiation. The discovery of spectral lines from atomic hydrogen (H I) and carbon monoxide (CO) in the radio portion of the spectrum has, however, made it possible to observe these gas tracers everywhere in the Galaxy because the radio radiation from these species is unimpeded by the presence of the dust. The H I and CO are both pervasive, but are distributed differently from one another; their distribution and kinematics serve as the best indicators of what the Milky Way looks like and how the material that comprises it is distributed.

GALACTIC COORDINATES

The first task in determining the structure of the Milky Way is to define as precisely as possible the direction of the center and the orientation of the midplane of stars, which make up about 95% of the mass of the Galaxy. Because both the stars and the gas orbit the center in response to the same gravitational field, and because the Galaxy is transparent to the 21-cm radio line of H I, surveys of atomic hydrogen can be used to determine accurately the orientation of the Galaxy as a whole. The current set of galactic coordinates was devised from atomic hydrogen surveys with the direction of the galactic center defined as the origin and zero latitude defined as the midplane of the hydrogen layer. More recent measurements of the position of the galactic center in the infrared portion of the spectrum have shown that the determination of the center from atomic hydrogen lines is in error by no more than 3 arcminutes. The determination of the orientation of the galactic plane from the atomic hydrogen observations has not yet been improved.

ROTATION CURVE

The Milky Way rotates differentially, rather than like a solid body; material orbiting the nucleus close to the center has a shorter rotation period than material farther out. This had been determined from observations of the kinematics of stars close to the Sun as early as the 1920s, and also was inferred from observations of other spiral galaxies. The functional dependence of the rotational velocity of the Galaxy on the distance from the center is known as the rotation curve, and it is a fundamental measurement from which much of the basic structure of the Milky Way can be inferred. Different techniques are required to measure the rotation curve inside and outside the Sun's orbit around the galactic center, but both methods require observations of the interstellar gas clouds.

To a first level of approximation, the orbits of the gas clouds about the center of the Galaxy are circular. For gas in circular orbits, the *apparent* velocity of a gas cloud with respect to the Sun (or more properly, with respect to the standard of rest defined by the stars within a few hundred parsecs of the Sun) is uniquely determined by its distance from the center of the Milky Way. When one observes the gas along a line of sight that passes inside the solar orbit, gas is found at all velocities up to a terminal velocity. Gas reaches the terminal velocity at the point along the line of sight that lies closest to the center of the Galaxy (see Fig. 1). The distance of this point from the center can be determined from simple geometry. Thus, by observing the atomic hydrogen along many directions that pass inside the orbit of the Sun, it is possible to obtain the rotation curve of the Galaxy out to the solar distance.

Beyond the solar distance, it is possible to obtain the rotation curve only from independent measurements of the distances and velocities of a large number of objects. The best measurements to

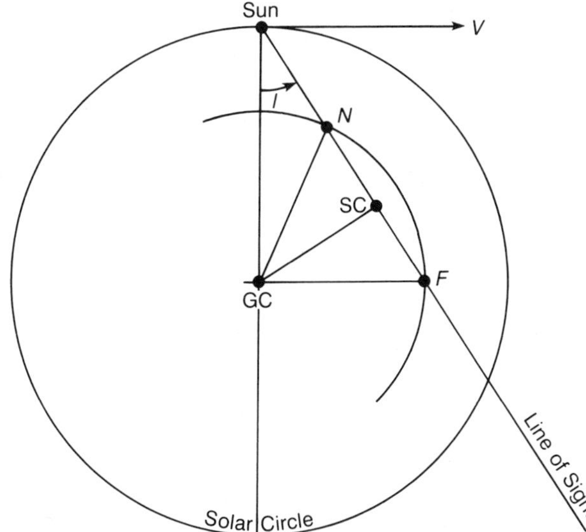

Figure 1. Diagram showing how the geometry of the Galaxy and measured velocities can be used to determine the rotation curve of the inner Milky Way. The Sun moves in a nearly circular orbit in the direction of the arrow. The solar circle is a circle drawn about the Sun's position centered on the galactic center. An observation along the line of sight, at galactic longitude l, of either CO or H I will show a spectral line that has a large range of radial velocities, from zero at the solar circle to a maximum value at the point SC, the place along the line of sight closest to the galactic center. The distance of SC from the galactic center is just the Sun's distance from the center times the sine of angle l. By making observations along different lines of sight, that is, at different angles l, it is possible to derive the rotation curve inside the solar circle.

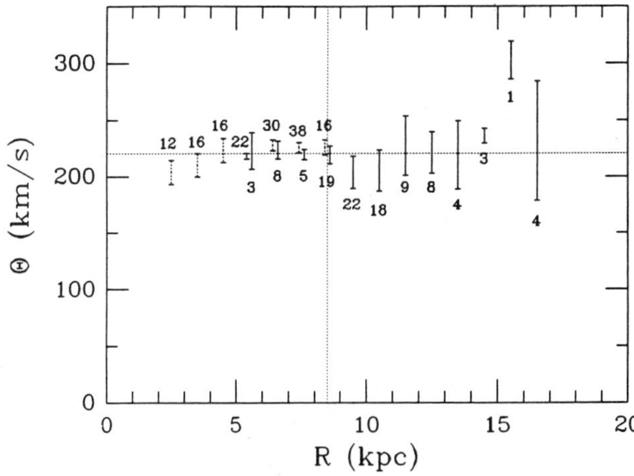

Figure 2. The rotation curve of the Milky Way out to twice the solar radius. The circular velocity Θ is plotted against galactocentric radius R. The dotted lines are the values of Θ and R for the Sun. The extent of the bars shows the uncertainties in the measured values of Θ for each R; dashed bars are from the H I data; solid bars are from the CO data. The rotation curve itself is the smoothest function that traverses the vertical bars. In this case it is adequately represented by a straight line essentially coincident with the horizontal dotted line in the figure. Beyond the Sun's distance from the galactic center, the rotation curve should decrease if all of the mass in the Galaxy were in the form of stars, gas, and dust.

date come from observations of very young bright clusters of stars and the giant molecular clouds out of which they formed. Such stars are always found to be in close proximity to their parent clouds, and thus a distance measurement to the stars is also a measurement of the distance to the clouds; the differences in the respective distances to the stars and the clouds are generally to small to be measured. The velocity of the molecules in the cloud (CO is generally used because it is the most abundant, easily detectable molecule) can be measured to great accuracy because radio techniques are used to observe the spectral lines. The rotation curve thus obtained can be measured out to a distance about twice that of the Sun from the galactic center. Apparently, there are few stars at larger distances.

A composite rotation curve for the Milky Way from both methods is shown in Fig. 2. The rotation velocity of the gas and stars is quite constant out to the last measured point, as is the case for nearly all other spiral galaxies, implying that there is a great deal of "dark matter" in the Galaxy. By analyzing how the atomic hydrogen layer thickens with distance from the galactic center, it is possible to show that the dark matter cannot be confined to the disk and must therefore be in a spheroidal distribution about the center of the Galaxy.

LARGE-SCALE STRUCTURE

Once the galactic coordinate system is established and the rotation curve is measured, it is then possible to use the information to determine the detailed large-scale distribution of matter in the Galaxy. The extent of the galactic disk is, for example, best determined from studies of the atomic and molecular gas clouds. The measured radial velocity of a gas cloud and its galactic longitude (that is, its position in the Galaxy) are uniquely determined by the distance of the cloud from the galactic center and the rotation curve. Observations of the velocity of the H I and CO can therefore be used to determine the distribution and extent of the gas disk of the Galaxy. For atomic hydrogen, the disk extends to approximately three times as far as the Sun's distance from the center. The molecular gas layer defined by the CO is less extensive, and is observed only about twice as far out as the Sun. However, because all stars form from molecular clouds, and because the migration of stars to larger distances from the galactic center is very slow, the extent of the molecular hydrogen disk is approximately the size of the disk of stars. This has recently been confirmed from observations of carbon stars, which can be seen to large distances from the galactic center in certain directions; very few stars are seen at larger distances than the molecular clouds.

In 1957, Frank J. Kerr showed that in the outermost parts of the Milky Way, the atomic hydrogen layer is not confined to the plane of the Galaxy, but exhibits a pronounced warp; one edge is tilted down and the opposite edge is tilted up like the brim of a hat. In 1983, it was shown that the outer edge of the Galaxy is scalloped, that is, there is an undulation of the hydrogen layer around the perimeter of the disk that exhibits about 10 complete cycles. Warps are common features of spiral galaxies, but the sensitivity of present day-instruments make such scalloping difficult to detect in other galaxies.

The distribution of the atomic hydrogen shows that it is not precisely in circular rotation, but that there are systematic deviations of 10–20% from circularity over the entire face of the Galaxy. These have been very recently shown to be the result of the response of the gas to an oval distortion of the stellar spheroid in the inner portions of the Milky Way. That is, the bulge of stars in the inner Galaxy is not round as previously thought, but has a shape similar to that of an American football. This bar-like pattern rotates like a solid body around the center, and similar structures have been both observed directly and inferred indirectly from observations of other galaxies. There is also evidence from the velocities and the distribution of both the H I and the CO that there is a more pronounced stellar bar in the inner core of the Milky Way, but no self-consistent dynamical model has yet been devised which adequately describes the observations.

SPIRAL STRUCTURE

It had long been hoped that the distribution of the atomic and molecular gas clouds in the Milky Way would reveal where the spiral arms are located, and thus what a face-on view of the Milky Way would look like. In some other galaxies, both the atomic hydrogen and the CO are concentrated in the arms of grand design spirals. These are spiral galaxies with a beautiful, symmetric spiral structure. Large-scale surveys of the Milky Way have not yielded an unambiguous picture of the spiral structure of the inner Galaxy. This is in part due to limitations in the analysis techniques, but most probably also due to a semichaotic distribution of the gas in spiral arms. Although there do appear to be identifiable spiral arms in the gas in the inner parts of the Milky Way, there are also small segments and fragments that complicate the picture enormously.

In the part of the Milky Way that is more distant from the galactic center than the Sun, the situation is much clearer. Long, connected spiral arms have been identified, some of which are 20 kpc or more in length: greater than the radius of the Galaxy. The best picture to emerge from these analyses is that the Milky Way has four spiral arms in the outer parts, but only three are clearly visible because the fourth is on the far side of the Galaxy beyond the galactic center. The fourth spiral arm is thus confused in the measurements with foreground gas. Following the arms inward they appear to be somewhat chaotic at the Sun's distance from the galactic center, with the chaos increasing toward the galactic center. The Milky Way thus appears to be a multiarmed spiral with the clearest spiral structure occurring in the outermost regions.

THE GALACTIC CENTER

The center of the Milky Way was the first source of cosmic radio waves detected, as it was discovered by Karl Jansky in 1932. Detailed observations of the structure of the center of the Milky Way have been made primarily at infrared and radio wavelengths because of the enormous extinction of visible light along the line of sight to the center. An analysis of the motions of the neutral gas clouds at the center of the Milky Way suggests that within the central few parsecs there is an unusual mass concentration of about a million solar masses. No similar concentration is observed elsewhere in the Milky Way, and several groups of astronomers have suggested that a massive black hole at the center of the Galaxy is the most reasonable explanation.

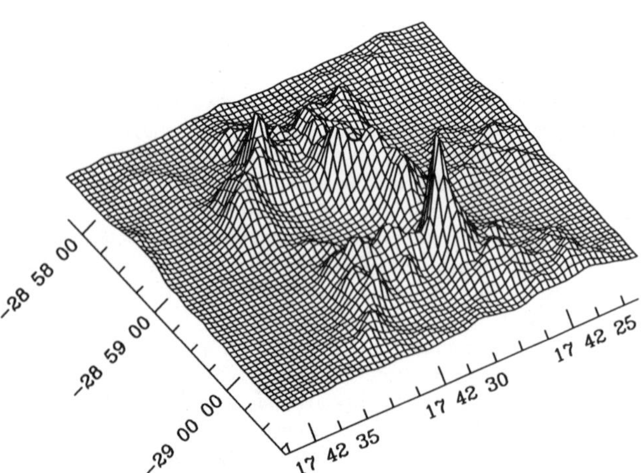

Figure 3. A graphical representation of the emission from the HCN (hydrogen cyanide) molecule within 3 pc of the galactic center. HCN traces the location of the much more abundant hydrogen molecule. Note the ring-like morphology of the gas that is rotating about the galactic center. (*Courtesy of Reinhard Genzel.*)

An outstanding feature of the nucleus of the Milky Way is a ring of molecular clouds in circular rotation about the center. This ring of dense clouds may be an accretion disk of material that feeds interstellar matter into the core of the Milky Way. The ring is shown in Fig. 3, and is itself embedded in a large concentration of molecular clouds associated with the central regions of the Galaxy. Recent radio observations at different wavelengths show astonishing loop-like structures at the galactic center similar in appearance to the prominences on the surface of the Sun, but far larger in scale.

Observations of the interstellar gas clouds in the central region of the Milky Way have completely revised the understanding of the structure and physical processes at the galactic center. Analysis of the observations is far from complete and new observations are likely to turn up additional surprises.

Additional Reading

Bok, B. J. (1981). The Milky Way Galaxy. *Scientific American* **244** (No. 3) 92.

Bok, B. J. and Bok, P. F. (1981). *The Milky Way*, 5th ed.

Blitz, L., Fich, M., and Kulkarni, S. (1983). The new Milky Way. *Science* **220** 1233.

Gilmore, G. and Carswell, B., eds. (1987). *The Galaxy*. Reidel, Dordrecht.

Lo, K. Y. (1986). The galactic center. *Science* **233** 1394.

Shuter, W. L. H., ed. (1983). *Kinematics, Dynamics, and Structure of the Milky Way*. Reidel, Dordrecht.

See also **Galactic Center; Missing Mass, Galactic.**

Galactic Structure, Large Scale

W. B. Burton

The Milky Way is one of the larger and more luminous spiral galaxies in the local supercluster. Our embedded perspective hinders study of some global properties of our own galaxy, notably concerning details of its spiral structure and decomposition of its space motions, but facilitates others, notably study of the global warp and of details of the interstellar medium.

LENGTH SCALES IN THE MILKY WAY

$R_\odot$, The Distance of the Sun from the Galactic Center

Discussions of the morphology of the Galaxy depend in almost all cases on the values of the galactic constants $R_\odot$ and $\Theta_\odot$. The particular values of these constants adopted involve a number of assumptions. A review by Frank J. Kerr and Donald Lynden-Bell summarized the arguments which in 1985 led IAU Commission 33 to recommend usage for most purposes of the standard values $R_\odot = 8.5$ kpc and $\Theta_\odot = 220$ km s^{-1}, replacing the values of 10 kpc and 250 km s^{-1} that were in wide prior use. Many galactic quantities scale simply by the appropriate power of $R_\odot$, but this scaling is complicated for quantities based on kinematic distances by vagaries in the velocity-to-distance transformation.

$R_\odot$ is the distance of the Sun from the galactic center. Several independent methods, incorporating a range of observational techniques, yield a reasonable consensus. One may distinguish between direct measurement of the parallax of the center and less-direct geometrical and kinematic distance determinations. In all cases, systematic errors are probably more important than internal ones.

The measurement of $R_\odot$ with the fewest intervening assumptions is that provided by very long baseline interferometry observations of the statistical parallax of proper motions of H_2O maser features associated with the Sgr B2 molecular complex, which is probably located within a few hundred parsecs of the true galactic center. The internal accuracy of the proper motion method is currently limited by the number of maser sources which can be observed.

Geometrical measurements of $R_\odot$ involve finding the distribution centroid of objects whose distances can be determined at great distance. It is assumed, of course, that the centroid of these visible tracers coincides with the true galactic barycenter, even though the total galactic mass is dominated by invisible matter distributed over a much larger volume.

Harlow Shapley's classical determination of the scale of the Galaxy was based on distances to globular clusters determined from Cepheids. The globular cluster method is limited by the small number of clusters and by their distribution over a large volume. Cepheids remain excellent distance indicators. Observations of Cepheids in the Magellanic Clouds show a very small internal dispersion in distances derived from the period-luminosity (P-L) relationship. Although the intrinsic internal scatter of the P-L relationship is small, calibration of the zero point of the Cepheid period-luminosity relation involves many intervening steps. The Cepheid distance scale is based on distances to galactic clusters containing Cepheids, whose distances in turn depend on main-sequence fitting to the Hyades and other local open clusters and ultimately on the parallaxes of nearby stars. An important problem that remains is of the influence of metallicity on the P-L relationship. Metallicity may vary with galactocentric distance, but in a way not yet fully specified.

An important geometrical determination of $R_\odot$ involves finding the centroid of the distribution of field RR Lyrae variables, which can be identified in the galactic bulge in several windows of relatively low, but certainly not negligible, obscuration. As with Cepheids, the intrinsic scatter in RR Lyrae absolute magnitudes is small, but the true RR Lyrae brightness and its possible dependence on metallicity remain problematic. Charge coupled device (CCD) detectors extend the power of the geometrical methods by allowing photometry in crowded galactic-bulge fields.

Mira variables are strongly concentrated toward the galactic center. Their promise as distance indicators has been enhanced by the discovery of an infrared period-luminosity relation of small intrinsic dispersion; the P-L zero point must yet be calibrated accurately. Long-period Miras generally have dust shells radiating in the far infrared. The large sample of variable sources detected in the galactic bulge by the *Infrared Astronomical Satellite* (*IRAS*) holds promise for accurate determination of the shape and centroid of that region. Accurate photometry in the infrared would alleviate some of the problems of the large and variable interstellar extinction that plague optical work.

A quite direct determination of $R_\odot$ involves finding distances to OH maser sources associated with infrared variable stars. A linear dimension follows from the light-travel time across a circumstellar shell, corresponding to the time lag between the maser responses of the near and far sides of the shell to variability in the central star. The angular dimension can be measured independently using radio interferometry and the distance can then be derived. $R_\odot$ follows from comparing distances to the OH/IR sources with kinematic distances which depend on $R_\odot$.

Kinematic determinations of $R_\odot$ involve restrictive assumptions. The most important of these requires well-behaved kinematic symmetry; this assumption does not allow for streaming motions that might be associated with spiral arms or other structural irregularities. The principal kinematic determination of $R_\odot$ involves identifying objects of known distance whose radial velocity of 0 km s^{-1} indicates that they are located on the circle $R = R_\odot$. Stellar tracers used in this way include Cepheids and OB stars: as is the case with the use of optical tracers in geometric programs, absolute magnitudes and absorption must be well known. H II regions are particularly useful for kinematic measures of the distance scale. Radial velocities of molecular clouds associated with the H II regions can be measured accurately in radio emission lines; distances to the H II regions are determined spectrophotometrically.

$\Theta_\odot$, *The Circular Rotation Velocity at* $R_\odot$; *the Galactic Rotation Curve*

The galactic rotation curve $\Theta(R)$ gives the variation with R of the velocity of an object in a circle of radius R moving such that centrifugal force balances the gravitational attraction of the Galaxy. $\Theta_\odot$ is the circular velocity of $R_\odot$, and defines the velocity of the local standard of rest (LSR). Knowledge of the axially symmetric rotation curve $\Theta(R)$ of the Milky Way is required for most studies of galactic morphology, and is widely used to derive kinematic distances from observed radial velocities. The acceleration due to the Galaxy's gravity is given by $\Theta^2 = R \cdot K$; thus the rotation curve is fundamental to the construction of models of the galactic mass distribution.

The Sun itself has a motion with respect to the LSR, usually specified as the standard solar motion of 20 km s^{-1} toward $\alpha, \delta = 18^h + 30°$ (epoch 1900). A systematic error of between 5 and 10 km s^{-1} cannot be ruled out, and in fact is suggested by several lines of evidence.

The constant $\Theta_\odot$ has been determined from several rather indirect methods. The motion of the LSR has been measured with respect to the ensemble of globular clusters in the galactic halo. The modest flattening of this ensemble implies that it undergoes slow rotation. RR Lyrae stars and halo field stars have been similarly used. In each case additional information on the degree of rotation of the halo itself is required. The LSR motion has also been determined with respect to galaxies in the Local Group. The LSR motion with respect to H I at the distant edges of the galactic gas layer in the first and second quadrants yields a value of 220 km s^{-1}, but a similar study of the southern-hemisphere data yields a substantially higher value; the difference focuses attention on the assumption of total circular symmetry.

Assuming axial symmetry and that the angular velocity $\Theta(R)/R$ decreases smoothly with R, differential rotation results in a Doppler-shifted radial velocity with respect to the LSR along a line of sight in the galactic plane at longitude l of $v = R_\odot[\Theta(R)/R - \Theta_\odot/R_\odot]|\sin(l)|$. If the rotation curve $\Theta(R)$ and the constants $R_\odot$ and $\Theta_\odot$ are known, then distances can be attributed to observed radial velocities.

The rotation curve is determined differently in the inner Galaxy than in the outer, and special methods are necessary in the bulge region. The inner-Galaxy rotation curve has been determined from 21-cm observations of atomic hydrogen, which is a good tracer for this purpose because, being diffusely spread and largely transparent, it can be observed essentially everywhere in the galactic disk, and because it has a low (≈ 5 km s^{-1}) intrinsic velocity dispersion. If the Galaxy rotates as assumed, then the extreme velocity along a line of sight through the inner galaxy is contributed by material at the point closest to the galactic center. Here the line of sight is tangent to a galactocentric circle and $R = R_\odot|\sin(l)|$ is at its minimum for that line of sight and $\Theta(R)$, and thus $|v|$, at their maxima. This terminal velocity may be found from the suitably defined cutoff value on H I spectra observed on lines of sight along the galactic equator in the inner Galaxy. The inner-Galaxy rotation curve is then $\Theta(R_\odot|\sin(l)|) = |v_t| + \Theta_\odot|\sin(l)|$. The rotation curve derived from the terminal velocity method formally represents only motions from the locus of tangent points, that is, the circle passing through the Sun and the galactic center.

The terminal-velocity method cannot be used in the outer Galaxy, because the amplitude of the radial velocity increases smoothly with increasing distance from the Sun along lines of sight at $R > R_{\odot}$. Determination of Θ ($R > R_{\odot}$) requires independently measured velocities and distances. Recent work combines spectrophotometric distances to O- and B-type stars in H II regions, which invariably have associated molecular emission whose radial velocity can be measured in radio emission lines. Unlike the situation in the terminal-velocity method, derivation of the rotation curve in this way is not restricted to a particular locus and can therefore in principal yield the generalized galactic velocity *field*, rather than the azimuthally smoothed rotation curve. Substantial uncertainties in the derived velocity field remain; optical obscuration limits the number of H II regions visible at large R, and in any case H II regions occur in relatively small numbers in the outer Galaxy.

The kinematics for the gas component within the bulge region, at $R < 3$ kpc, must be considered separately because in that region deviations from circular rotation are of approximately the same amplitude as the rotational motions themselves. The noncircular motions are revealed in several ways, but most obviously by the prevalence of H I and CO emission at positive velocities at $l < 0°$ and at negative velocities at $l > 0°$, where such emission is forbidden in terms of simple rotation. Recognition of the forbidden H I contribution at negative velocities is complicated at $b = 0°$ by absorption against continuum sources concentrated near the galactic core. Application of the terminal-velocity method under the associated assumption of circular rotation has resulted in a putative rotation curve that shows a rapid rise at small R and a definite peak between 1 and 2 kpc. Such a curve is almost certainly not correct. Absorption spectra reveal that the noncircular motions in the inner few kiloparsecs of the Galaxy are in the sense of expansion. It is not clear, however, if this signature should be interpreted as a genuine expansion or as motions in closed orbits. The total gravitational potential of the bulge region may be triaxial in form, and under such circumstances families of stable gas orbits have been predicted that show some of the characteristics of those observed; systematic noncircular motions are also expected from flow along a bar-shaped potential.

In addition to the pervasive noncircular motions in the inner few kiloparsecs, there are several examples of isolated, rather large-scale structures showing unambiguous noncircular motion. The most dramatic of these is the 3-kpc arm, observed in emission and absorption in H I as well as in molecular lines. The total gas mass involved in this feature is about $10^8\ M_{\odot}$; it is some few kiloparsecs in extent, and displays a net motion away from the galactic center of some 53 km s^{-1}. The mechanism responsible remains unknown; it is in particular not clear if the 3-kpc arm is transient or if it represents a permanent flow of gas in closed orbits.

The terminal-velocity analysis of H I and CO data between R = 3 kpc and $R_{\odot}$, and of combined optical/radio data at $R_{\odot} < R < 2 R_{\odot}$, show that the rotation curve of the Galaxy is approximately flat, $\Theta(R) = 220$ km s^{-1}, between 3 and 16 kpc. Deviations from this constant value exist. Over much of the disk, the irregularities can be attributed to systematic streaming motions associated with localized concentrations of mass. An increase in tangential velocity outwards across a spiral arm is observed commonly in external galaxies, where the perspective for such studies is better than that afforded by our own Galaxy.

Although over much of the Galaxy the deviations from circular velocity are of the order of a few percent or less of the rotation velocity, these noncircular motions considerably complicate analysis of galactic spectral-line data. A local error of 5 km s^{-1} in Θ can cause a large error in a kinematic distance, especially in regions where the radial velocity changes only slowly with distance from the Sun; such an error can also cause an ambiguity in the distance, if the velocity-to-distance transformation becomes locally multivalued.

Because of the problems of determining distances, as well as because of problems stemming from our embedded perspective, it has proven difficult to determine convincingly a number of important aspects of the structure of the Milky Way, including whether or not it displays a "grand design" of spiral structure, if the spiral arms are trailing or winding with respect to rotation, what pitch angle and spacing characterize the arms, what relative motions and spacings characterize the distributions of stars and gas, etc. Many aspects of such questions are more profitably pursued in external systems.

There are several disturbing indications that the assumption of global circular symmetry may not be valid in the Galaxy at large. For example, irregularities at the few-percent level in the inner-Galaxy rotation curve derived from H I or CO data in the first galactic quadrant are somewhat differently placed than those in the curve derived from fourth-quadrant data. These irregularities no doubt represent the kinematic response to local density fluctuations. If no other perturbing influence were present, then one would expect that the two curves would agree in the mean. In fact, there is a systematic difference of about 7 km s^{-1} between the two curves over the region $3 < R < 8$ kpc. Furthermore, the mean velocity of emission observed in the direction of the galactic center, as well as that observed in the direction of the anticenter, differ from 0 km s^{-1} by about 7 km s^{-1}. An error in the determination of the LSR could be relevant in this regard, and cannot yet be ruled out.

There are other kinematic properties of the Galaxy which cannot be reconciled with the assumption of circular rotation, or with that assumption qualified by the presence of localized streaming and an error in the LSR. Thus the kinematic cutoff at $b = 0°$ corresponding to the far outskirts of the Galaxy occurs at velocities some 25 km s^{-1} more extreme in the fourth quadrant than in the first. It is not yet certain if this lopsidedness is of kinematic or of structural origin; in either case, it indicates violation of axial symmetry. Several nearby spirals, notably M101, are structurally lopsided in the distribution both of starlight and of gas emission. One notes also that the warped outer-Galaxy gas layer extends to higher angular distances in the northern-hemisphere data than at comparable distances from the line of nodes in the southern data. It seems more plausible to view these global outer-Galaxy asymmetries as structural asymmetries rather than as kinematic ones, because the north/south differences in the inner-Galaxy rotation curve and in the outer-Galaxy velocity field are not large enough to account for them kinematically. Nevertheless, we note that just as is the case in the bulge region of the Galaxy, where stable low-latitude orbits can exist in the bulge potential, the total gravitational potential of the Galaxy (which remains of largely unknown composition and of poorly known shape) may harbor stable, noncircular orbits.

The period of revolution of the LSR around the galactic center is $2\pi\ R_{\odot}/\Theta_{\odot} = 2.4\ 10^8$ yr, about 1% or 2% of the age of the galactic disk. Because Θ is approximately constant over much of the Galaxy, the length of the galactic year scales with R; the classic winding dilemma poses the question of how structure in the disk can be maintained against the shearing forces of such strong differential rotation. This dilemma applies to spiral-arm structures throughout the Galaxy, as well as to the global warp dominating the outer Galaxy.

Radial Distributions in the Galactic Disk; Radial Scale Lengths

The morphology of the gaseous disk of the Galaxy can be ascertained from a number of tracers of the interstellar medium; with more difficulty, some aspects of the stellar morphology may be determined. For the cases of H I and CO and other tracers observed as spectral lines, conversion of the longitude-velocity distributions to radial distributions can be done using the velocity-to-distance transformation inherent in the rotation curve. For components traced in continuum data, including starlight, dust emission, nonthermal radio emission, and gamma radiation, kinematic distances are lacking. A geometrical unfolding process is used instead.

The radial distribution of H I can be followed from the galactic nucleus to about $R = 25$ kpc. This total extent is substantially greater than that revealed by any other known constituent of the galactic disk. The midplane volume density of H I gas is approximately constant over the range 3–15 kpc, at about 0.4 cm^{-3}; quantitative evaluation of this density involves many uncertainties. Interior to 3 kpc, the H I density is substantially less than elsewhere in the inner Galaxy. This relative deficiency in the inner few kiloparsecs is a global characteristic that is shared by other tracers of the interstellar medium. In the inner 2 or 3 kpc of the Galaxy (but excluding the rich molecular complexes in the innermost few hundred parsecs), there are few H II regions or molecular clouds (and thus evidently little star formation), and generally only weak emission from molecules and dust, as well as weak gamma and synchrotron radiation. The molecular gas distribution in the inner few kiloparsecs is much less heavily clumped than it is elsewhere in the Galaxy. The vertical thickness of the H I and CO gas layer in the inner few kiloparsecs can be measured with some accuracy. The full thickness at half maximum of the H I is about 90 pc, less than half the thickness elsewhere in the gas layer.

Radial distribution profiles have been examined in some detail in external galaxies, in particular in the H I and the CO emission. It is interesting that those systems which reveal a central depression are morphologically similar to the Milky Way in other ways. Galaxies which show a central depression in CO include M31 and NGC 891, which, like the Galaxy, are classified as Sb systems. Sc systems, including M51, M101, NGC 6946, and IC 342, do not show a central depression in integrated molecular emission. The physical mechanism which is responsible for the central deficiency and for the distinction between those systems which do show it and those which do not has not yet been identified. It is interesting that the radial extent of the central depression in our Galaxy, as well as of the major noncircular motions, coincides with the extent of the stellar bulge component.

Population I tracers other than H I are largely confined to an annulus extending from about 3 to about 7 kpc. CO and dust emission occur sparsely in the second and third quadrants, where $R > R_{\odot}$. Population I tracers including molecular clouds, pulsars, supernova remnants, O- and B-stars, and H II regions (all associated with aspects of star formation) are decreasing rapidly in space density as R increases beyond $R_{\odot}$. At the peak of the molecular annulus near $R = 5$ or 6 kpc, the surface density of H I, derived from observations of CO emission, is about 10 $M_{\odot}$ pc^{-2}. Unlike the situation pertaining to other tracers, the H I surface density remains approximately constant over the inner Galaxy: In the region $3 < R < 15$ kpc the H I surface density is about 8 $M_{\odot}$ pc^{-2}. The ratio of H I to H_2 surface densities increases from about 5 near the Sun to 150 at $R = 20$ kpc. The total mass of H I in the outer Galaxy at $R > 8.5$ kpc is about 5.3×10^9 $M_{\odot}$; the total outer-Galaxy mass of H_2 clouds is some 5.8×10^8 $M_{\odot}$. Derivation of densities, especially those of H_2, involves a number of controversial assumptions.

Recent work has extended the information on molecular material and the star formation properties in the outer Galaxy by observing CO emission in the direction of *IRAS* sources selected on the basis of having infrared radiation characteristics consistent with those of molecular clouds with embedded heat sources. Molecular clouds detected using infrared finding charts can be traced to about 20 kpc from the galactic center. Failure to detect CO emission at $R > 20$ kpc seems to represent a structural limit, not a sensitivity one. Because the clouds selected in this way all have embedded heat sources, evidently star formation is occurring in the far outer Galaxy. The infrared properties of these clouds imply, however, that relatively fewer very massive stars are being formed than in the inner Galaxy.

There are several additional lines of evidence that suggest that the physical or chemical conditions governing the appearance of the interstellar medium in the outer Galaxy differ from those pertaining in the solar vicinity or in the inner Galaxy. One notes in particular the breakdown at large R of the correlation between H I intensities and far-infrared emissivities; these quantities are tightly correlated over the inner Galaxy. It is not yet clear which physical differences between the inner and outer Galaxy are responsible for this. For example, it is not certain that the relative lack of dust emission indicates a lower dust particle density. The emission characteristics of dust particles may well change across the Galaxy, either because of the softening of the ambient interstellar radiation field at larger R or because of changes in the emission properties of individual dust particles following changes across the Galaxy in their size and/or composition.

THE SHAPE OF THE GALACTIC DISK

The Shape of the Inner Galactic Disk

The thickness and degree of flatness of the galactic disk are important aspects of its shape. For transgalactic tracers revealing kinematic information, notably H I and molecular gas, the linear thickness and the deviation of the centroid of the emission from the galactic equator can be measured at the tangent-point locus, where there is no distance ambiguity. Studies of this sort show that most of the inner-Galaxy H I and molecular structures are confined to a remarkably thin and flat layer. The typical thickness of the H I layer to half-density points is about 200 pc at $R < 8$ kpc. The vertical distribution of densities is approximately Gaussian in form, with conspicuous low-intensity deviations from this form extending to higher distances. The vertical thickness of the cold, dense, clumped gas in the molecular annulus is about half that of the H I layer. The molecular clouds comprise material available for star formation; thus the youngest stars are also confined to a similarly thin layer.

The layer thickness of tracers accessible throughout the Galaxy but only in continuum radiation can be derived by unfolding techniques. The distribution of dust observed at 60 and 100 μm has a z-thickness equivalent to that of the H I gas. In the inner Galaxy, far-infrared and H I emissivities are generally tightly correlated.

Over the range of radii $2 < R < 8$ kpc, deviations of the emission centroid of the galactic disk from the equator $b = 0°$ are remarkably small. For the H I gas layer and for the dust emitting at 100 μm, these deviations are less than about 30 pc; for the molecular cloud ensemble, the deviations from flatness are typically even less.

The gas layer at $R < 2$ kpc deviates systematically from the galactic equator. Most of the atomic as well as molecular gas lying between a few hundred parsecs and 2 kpc from the nucleus lies in a disk tilted some 20° with respect to $b = 0°$. This tilted disk is the plane of symmetry of the kinematics as well as of the distribution of gas in that region. The dynamic situation is not yet clear. If the motions are along closed streamlines, no net flux of matter is involved; otherwise, a net flux of some 4 $M_{\odot}$ per year must be accounted for.

The thinness and flatness of the inner Galaxy disk are consistent with the systematic motions in the z-direction measured near the Sun of only a few kilometers per second; it is difficult to measure z-motions elsewhere in the Galaxy. It is expected that these motions would increase toward smaller R in order to maintain the observed constant layer thickness against the total mass density. The line-of-sight random motions in the ensemble of molecular clouds are typically 4 km s^{-1}; although their motions are small, their masses of some 10^3–10^6 $M_{\odot}$ make the clouds effective scattering agents. It is likely that molecular clouds play an important role in the increase of stellar random motions and layer thickness observed with increasing stellar age.

The Shape of the Warped Outer Galactic Disk

It was evident in the early Leiden and Sydney surveys of 21-cm radiation that were made in the 1950s that the gas layer in the outer parts of the Milky Way is systematically warped from the equator $b=0°$. Studies of external galaxies have subsequently revealed that large-scale warps are a common aspect of the H I morphology in the outer parts of spiral galaxies. The two nearest large spirals are both warped: The shape of the hydrogen layer in M31 resembles that of the gas layer in our own Galaxy, and the warp in M33 is more severe. Warps are most easily studied in systems that are viewed more or less edge on and with the line of nodes of the warp oriented more or less along the line of sight. Such an orientation is presented by few external galaxies; the warp in the Milky Way is, however, viewed in this favorable way.

The H I gas layer in the Milky Way remains quite flat until about 11 kpc from the galactic center. The amplitude of the warp grows linearly until about $R=15$ kpc; at this distance, the warp amplitude is approximately equal in the hemispheres of the Galaxy accessible from the northern and southern data. At larger radii, the behavior of the warp is not symmetric. The amplitude continues to increase in the northern data until a mean-layer z-height of about 3 kpc is reached at $R=24$ kpc, at the sensitivity limits of current observations. The gas layer revealed in the southern data, on the other hand, after reaching a maximum excursion of about 1 kpc below the equator at $R=15$ kpc, folds back to almost reach the galactic equator again at the largest distances. NGC 3729 is one of the several external galaxies which show a similar floppy warp, with the inclination angle of the warp flattening at the largest radii. The southern warp differs from the northern one also in extending to more extreme velocities. Whether this difference is of kinematic or spatial origin, the larger amplitude of the northern warp and the floppy aspect of the southern one remain incompatible with axial symmetry.

The line of nodes of the warp may be specified by locating the galactic azimuths at which the gas layer crosses $b=0°$. The regularity of the warp is indicated by the fact that the plane crossings in the northern- and southern-hemisphere data are separated by approximately 180° of azimuth, as well as by the fact that the maximum excursions occur about 90° from the line of nodes. It is remarkable that the line of nodes of the galactic warp is quite straight. Evidently the shape of the warp is stable against the shearing forces of differential rotation and against precession.

Our Galaxy shares another characteristic commonly observed in warped systems, namely, the flare of the gas layer to larger thickness in the outer parts. The thickness of the H I layer increases steadily from the onset of the warp near 11 kpc, where the vertical thickness is about 320 pc (measured as the full width where the emission is half its maximum value), until the outermost regions where the thickness is substantially more than a kiloparsec. The outer-Galaxy molecular-cloud ensemble partakes in the flaring. The molecular thickness is systematically less than the H I thickness over the entire Galaxy, but the relative difference is less in the outer Galaxy than in the inner, or in the solar vicinity, where the H I thickness is typically twice the molecular one. Thus at $R=15$ kpc, the full H I thickness is 600 pc; the cloud-ensemble thickness is 510 pc.

There is no indication that the line-of-sight velocity dispersion of any of the gas components increases with increasing R. Although the vertical dispersion cannot be measured at large R, observations of face-on external galaxies show little variation of the z-dispersion with R. The flaring nature of the disk, the evidently constant z-dispersion, and the flat rotation curve imply together that the dark matter in the Milky Way cannot be confined strongly to a disk.

Although many other galaxies have been identified as warped, no other galaxy reveals its warp in such detail as the Milky Way. The responsible dynamics are not yet clear. The approximate azimuthal symmetry and regularity of the warp suggest that it is a persistent, relaxed, global phenomenon, and not, for example, the consequence of a tidal interaction or merger with a companion system. In any case, warped, totally isolated galaxies have been observed. The large vertical thickness of the total gravitational potential directs attention to the possibility of families of stable orbits which may be both out of plane and noncircular.

Because the H I surface density is rather constant over the range $3<R<15$ kpc, and because both the H I and molecular distributions show a deficiency at $R<3$ kpc, a radial scale length measured with respect to the galactic nucleus (as used to describe the stellar morphology) is not applicable for the gas morphologies. The radial scale length describing the rate at which the H I surface density falls off in the Galaxy beyond 13 kpc is 4.0 kpc. The radial scale length of the surface density of molecular-cloud masses beyond 13 kpc is 1.5 kpc; an abrupt falloff relative to the H I situation also characterizes the radial distribution of H II regions. (The radial scale lengths of the midplane densities are smaller than those of the surface densities because of the flaring of the warped H I and CO disk.) That the distribution of clouds with embedded heat sources, presumably indicating formation of stars of modest mass, ends more abruptly at large R than that of H I suggests that the physical environment is different in the outermost Galaxy than elsewhere.

The radial scale length of the optical surface brightness of the Milky Way has been derived from photometric observations made with the *Pioneer 10* spacecraft when it was far enough out in the solar system to be relatively uncontaminated by zodiacal light. This analysis resulted in a scale length of the stellar disk, referred to the galactic center, of 5.0 ± 0.5 kpc. The stellar and gaseous scale lengths of our Galaxy as well as its stellar color and integrated magnitude, deficient central gas distribution, star formation characteristics, and rotation curve, are consistent with a Hubble classification of Sb.

Additional Reading

Burton, W. B. (1988). The structure of our Galaxy derived from observations of neutral hydrogen. In *Galactic and Extragalactic Radio Astronomy*, G. L. Verschuur and K. I. Kellermann, eds. Springer-Verlag, New York.

Fich, M., ed. (1988). *The Mass of the Galaxy* (Workshop proceedings). Canadian Institute for Theoretical Astrophysics, Toronto.

Fich, M., Blitz, L., and Stark, A. A. (1989). The rotation curve of the Milky Way to 2 $R_{\odot}$. *Ap. J.* **342** 272.

Gilmore, G. and Carswell, B., eds. (1987). *The Galaxy*. Reidel, Dordrecht. (See especially the chapters by M. W. Feast and P. C. van der Kruit.)

Gilmore, G., King, I., and van der Kruit, P. C. (1989). In *The Milky Way as a Galaxy*, R. Buser and I. King, eds. Geneva Observatory.

Kerr, F. J. and Lynden-Bell, D. (1986). Review of galactic constants. *Mon. Not. R. Astron. Soc.* **221** 1023.

van der Kruit, P. C. (1986). The distribution of luminosity in the disk of the Galaxy derived from the *Pioneer 10* background experiment. *Astron. Ap.* **157** 230.

Wouterloot, J. G. A., Brand, J., Burton, W. B., and Kwee, K. K. (1990). *IRAS* sources beyond the solar circle: Distribution in the galactic warp. *Astron. Ap.* **230** 21.

See also **Galactic Bulge; Galactic Structure, Interstellar Clouds; Galactic Structure, Optical Tracers; Stars, RR Lyrae Type.**

Galactic Structure, Magnetic Fields

Gerrit L. Verschuur

The Galaxy is permeated by magnetic fields whose structures vary on scales from parsecs to galactic dimensions. Close to the Sun the field follows the axis of the local spiral arm, but on a larger scale a field of 1.6 microgauss (μG) appears to be circular about the galactic center.

Of all the forces in the astronomical universe the most subtle and pervasive is magnetism. Reaching their tentacles over vast volumes of space, magnetic fields thread their way between the stars, permeate clouds of gas and dust, guide the flow of matter in spiral arms, and reach into dense molecular clouds to preside over star birth. Yet the role of magnetic fields in determining the evolution of the Galaxy, their role in shaping spiral structure, or in setting the scene for star birth, is far from understood. One reason lies in the fact that magnetic fields are notoriously difficult to observe. Evidence for their presence throughout the Galaxy is strong, but experiments to measure their strength and direction are difficult and time consuming. Another reason is that basic theories of galactic structure, interstellar processes, and star formation were initially more easily worked out by ignoring the presence of magnetic fields. However, as the observational data improve, the fields are beginning to be taken into account.

The study of galactic magnetic fields invariably involves the observation of polarization (linear or circular) of some form of electromagnetic radiation (e.g., light or radio waves). The nature of polarization may be illustrated by considering an analogy. Radiation carrying energy through space is like a wave running along a stretched rope. If the rope is flicked up and down at one end, a linearly polarized wave is created and its plane of polarization is said to be vertical. If it is flicked sideways a horizontally polarized wave is created. A third alternative is to twirl one end of the rope in a circular fashion to generate a circularly polarized wave. By measuring the state of polarization of light and radio waves in particular, astronomers have been able to interpret the data to derive information about the presence of magnetic fields in the source of the radiation or, in some cases, along the path between the source and the Sun.

Optical Polarization of Starlight In 1949 W. Albert Hiltner and John S. Hall attempted to observe the polarization of starlight by watching one member of a binary star move in front of the other. They hoped to observe changes in the light intensity consistent with a model that predicted that the light was polarized along the edge of the star being eclipsed. Instead, they discovered that nearby comparison stars were polarized and that the degree of polarization was correlated with the amount of interstellar dust through which the star was shining. Theories were soon developed that required slightly elongated dust particles to become aligned under the influence of the interstellar magnetic field. The particles then interact with starlight to produce a net polarization parallel to the field direction.

Optical polarization data for thousands of stars plotted on a map of the sky reveal the transverse component of the magnetic field in the dust clouds responsible for the polarization. Because the stellar distances are known, we can be certain that the fields revealed in the data are closer than the stars. The data show that the stars are most highly polarized around galactic longitude $l = 140°$, where the polarization vectors are parallel to the galactic plane. Relatively nearby stars (300 pc distant) are as highly polarized as stars several kiloparsecs away, so most of the polarization must be produced relatively locally. The optical polarization vectors observed in other parts of the sky indicate that the local field runs from $l = 50°$ to $l = 230°$ (i.e., normal to the line of sight at $l = 140°$), which is along the axis of the Orion spiral arm in which the Sun is embedded. Models for the interaction between dust grains and fields to account for the polarization require field strengths of 10–20 μG.

Radio Polarization In the early 1960s it was discovered that radio waves from the Milky Way and from extragalactic radio sources were linearly polarized. This not only confirmed the idea that the synchrotron process was the cause of the radio waves, but presented an opportunity to study the galactic magnetic field in a new way. According to the synchrotron theory, radio waves should be as much as 70% polarized in a direction normal to the field direction. However, the fields in space are not perfectly ordered, and as a result the polarization will be less because polarization vectors within the source overlap and cancel. Also, the observed plane of polarization is different at widely spaced radio wavelengths as a consequence of Faraday rotation. If the polarized wave passes through a cloud of electrons on its way to the Sun, it propagates as two oppositely circularly polarized waves, one rotating clockwise, the other counterclockwise. They will bear a certain phase relation to one another determined by the plane of polarization. In the ionized medium, however, one wave travels slightly slower than the other, so that by the time the two emerge from the cloud their relative phases will have shifted. At this point the two components couple and travel as a single wave, but the change in phase difference acts to produce a rotation of the plane of polarization with respect to the original wave. This phenomenon is known as Faraday rotation. For example, the polarization at the source would have been normal to the magnetic field direction, but the wave that reaches Earth will be polarized at a completely different angle.

Fortunately the amount of Faraday rotation can be measured because it is a function of wavelength. A rotation measure, RM, is defined as the amount of rotation in radians per unit of wavelength squared (rad m^{-2}) and is given by $RM = 8 \times 10^5 n_e B_p L$, where n_e is the average electron density along the line of sight, B_p is the magnetic field in Gauss parallel to the line of sight (positive for a field toward and negative for a field away from the observer), and L is the distance in parsecs through the medium doing the rotating. Observations of the galactic background polarization at at least three wavelengths allow the Faraday rotation to be determined. This, in turn, allows the transverse component of the field direction in the source to be derived. Figure 1 shows the galactic magnetic field directions derived from data obtained at five wavelengths in each part of the sky. The distance to the source of the polarized radiation can be estimated by making a model that accounts for the degree of intrinsic polarization and the way polarization drops off with wavelength. A "depolarization" parameter is determined, which depends on the field structure variations within the source and in intervening space.

Estimates of the field strength involved in producing the observed rotation measures of a few to 20 rad m^{-2} require assumptions about path length and mean electron density along the path to the source. Given the inherent uncertainties, fields of 5–10 μG are determined, less than required to produce optical polarization.

Interpretation of radio background polarization places the typical source region 500 pc from the Sun, in particular for the patch of high polarization seen just above the galactic plane at around $l = 140°$. This is the same region where the optical polarization data show a highly ordered field. The patch lies on a great circle on the sky defined by a series of highly polarized patches, an effect consistent with the Sun being located inside a field structure oriented along the axis of the local Orion spiral arm.

Several other patches of the radio sky show higher polarization than would be expected for a uniform field aligned along this arm. For example, a tongue of bright radio emission called the North Polar Spur (believed to be the remnants of an old supernova) is highly polarized where it projects above the galactic plane around $l = 30°$ (Fig. 1). Within the spur the field directions are fairly complex. To add to the difficulty of interpreting the data, the depth to which one can observe polarization increases at shorter wavelengths. At 10 cm it is possible to observe polarization originating at kiloparsec distances, whereas at 75 cm, typically used in these studies, the polarization originates about 500 pc away.

The galactic center region is unique in showing radio polarization that indicates a field normal to the galactic plane in a radio arc near the center. The rotation measure in this arc is very large ($\sim$ 1000 rad m^{-2}) and shows positional variations that may be due to highly twisted fields in regions with high electron density. The fact that the field is normal to the galactic disk (poloidal) is taken by some as evidence for a primeval origin of the magnetic field, which has been wound up as the Galaxy rotates and is now seen in concentrated form near the galactic center. The field is

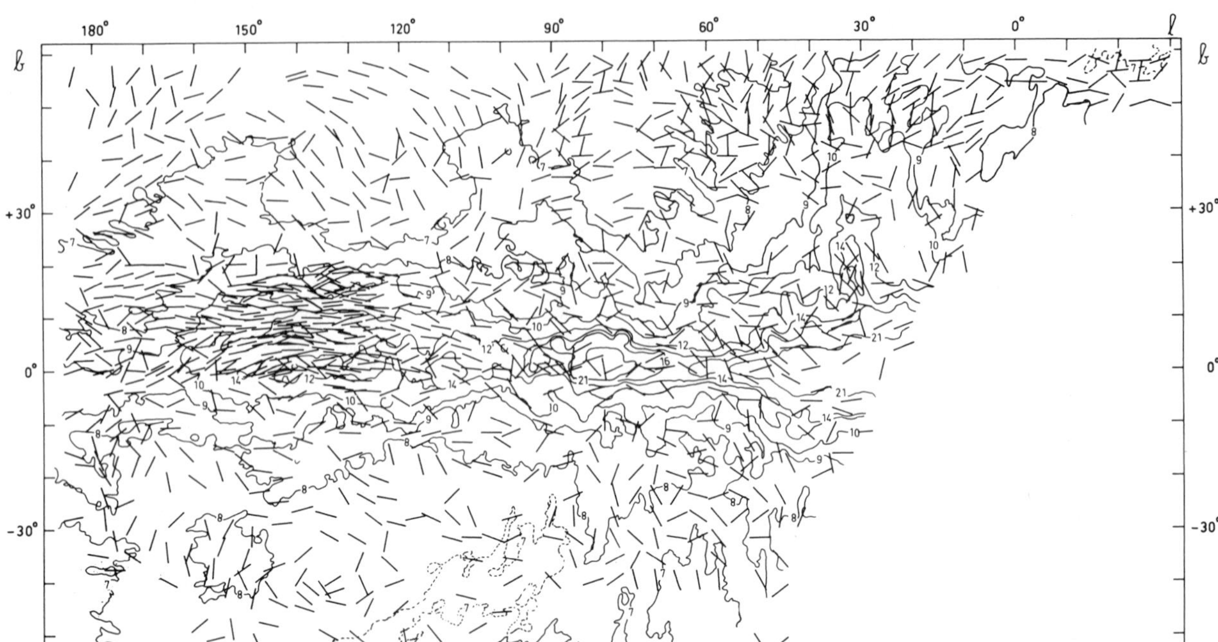

Figure 1. The direction of the perpendicular component of the galactic magnetic field as derived from polarization measurements of the galactic radio background emission. The small lines indicating field directions are superimposed on a contour map indicating the strength of the background emission at 75-cm wavelength. The coordinates are galactic longitude (horizontal) and latitude (vertical). The North Polar Spur emerges from the galactic plane around l = 35°. The highly organized field pattern around l = 140° is very striking. (*Courtesy T. A. Th. Spoelstra.*)

supposed to diverge from the center and then follow the plane of the galactic disk so that in the solar neighborhood it is planar (toroidal), as is observed. In such a model the field direction would be expected to show reversals along the galactic plane.

Radio Source Faraday Rotation Data The study of the distribution of radio source rotation measures gives a clear indication of the field direction along the line of sight on a scale comparable with the depth of the Galaxy in the direction of the sources. Figure 2 shows the data for 543 radio sources. An open circle indicates a negative rotation measure, that is, an intervening mean field directed away from the observer, and a filled circle indicates a field directed toward the observer. Although some of the Faraday rotation may occur within the radio sources themselves, the rotation measures are a strong function of galactic latitude, so they are largely produced in the Galaxy.

The data in Fig. 2 have been interpreted to show that within a few kpc of the Sun the field is circular and directed toward $l = 90°$. Between $l = 315°$ and 45° anomalous positive rotation measures indicate that the field is directed toward the Sun. This region is associated with the North Polar Spur, which may have distorted the field and stamped its own signature onto the rotation measure distribution.

The data also suggest that the general field is oriented along the direction of galactic rotation outside the Sagittarius spiral arm, which is located about 2 kpc inside the Sun's location, and in the opposite direction inside that arm. An additional magnetic field structure of unknown origin and associated with enhanced numbers of electrons in space is required to account for the anomalously large rotation measure in a region below the galactic plane between $l = 60°$ and 140°.

In order to derive field strengths from the rotation measures, assumptions about mean electron densities and path lengths in clouds along the line of sight have to be made, a highly uncertain task. However, observations of pulsars allow this mean interstellar field strength to be determined without making any assumptions.

Pulsar Rotation Measure Data The radio waves from pulsars are linearly polarized. In passing through interstellar clouds of ionized material their plane of polarization suffers Faraday rotation and this has been measured for some 200 pulsars. In addition, the pulses experience an effect known as dispersion, a measure of the delay in arrival time of the pulses as a function of wavelength. The dispersion measure, DM, depends on the total number of electrons between the pulsar and the telescope and is given by $\mathrm{DM} = n_e L$, where n_e is the average electron density, and L is the path length. If both rotation measure and dispersion measure are derived for a pulsar, the two quantities allow the field strength to be directly estimated from $B_p = \mathrm{RM/DM}$. Observations of pulsar rotation and dispersion measures thus allow the mean interstellar magnetic field strength along the path to the pulsars to be unambiguously derived. The mean field is found to be 1.6 μG and applies to a region within a few kiloparsecs of the Sun.

When plotted on a map similar to Fig. 2, the pulsar data indicate that the mean field is directed toward $l = 96°$, consistent with the radio source rotation measure data, but apparently at odds with the notion that the local field is directed along the Orion spiral arm toward $l = 50°$. However, both the radio source and pulsar rotation measures are sensitive to fields stretching over greater depths of space with scale lengths of many kiloparsecs. On the large scale, then, the field is concentric about the galactic center, whereas close to the Sun it is directed along the local Orion arm. The pulsar data also show that the field reverses about 600 pc inside the Sun's orbit, in the region of the Sagittarius spiral arm.

Origin of the Fields Despite the obvious existence of galactic magnetic fields, their origin remains a mystery. One school of thought invokes a dynamo on a galactic scale to create an underlying field which is then pushed hither and thither and amplified by

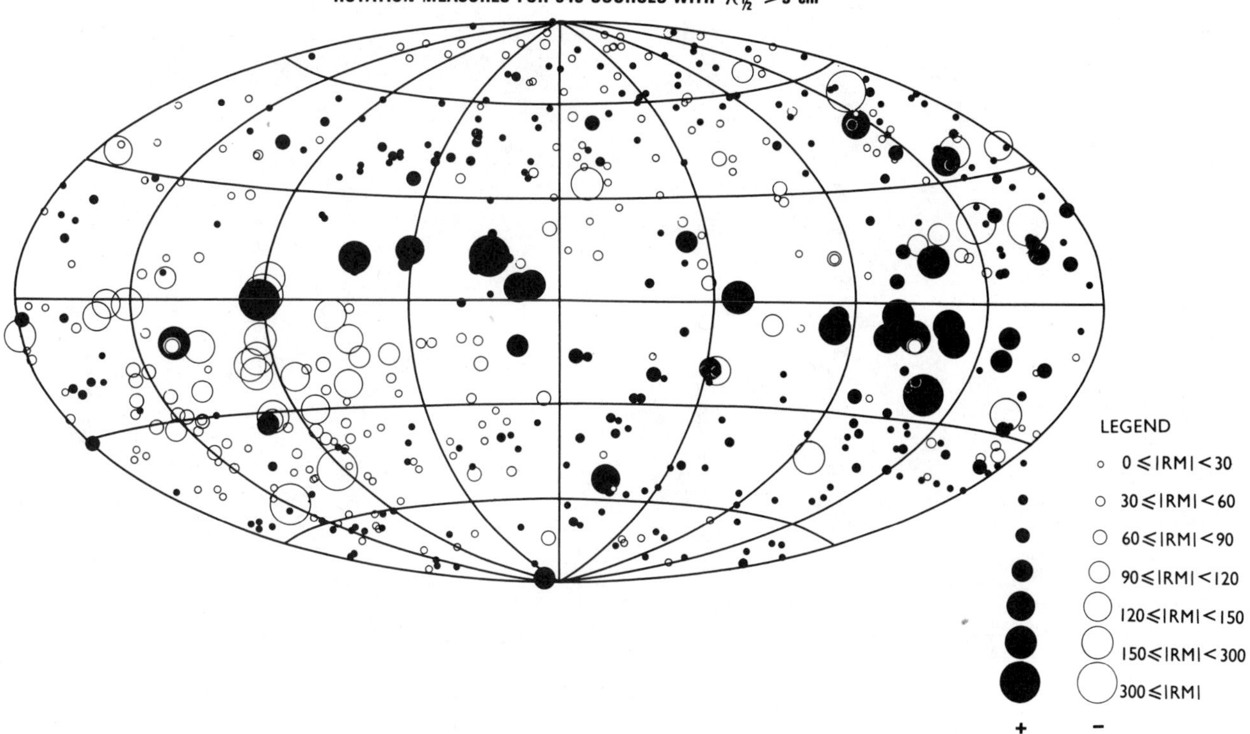

Figure 2. The rotation measure data for 543 radio sources plotted on a map of the entire sky. A positive rotation measure indicates a mean field toward the observer. Galactic longitude runs from right to left, starting at $l = 180°$ with lines drawn every 30° in longitude. The galactic center, $l = 0°$, $b = 0°$, is at the center of the map. Galactic latitude is measured vertically with grid lines shown at 30° intervals. (*Courtesy M. Simard-Normandin and P. P. Kronberg.*)

localized events such as supernova explosions. Another assumes that a primeval field was wound up as the Galaxy formed. This "seed" field may later have been amplified by the dynamo effect.

Additional Reading

Beck, R. and Gräve, R., eds. (1987). *Interstellar Magnetic Fields* (Proceedings of a Workshop). Springer-Verlag, Heidelberg.

Heiles, C. (1976). The interstellar magnetic field. *Ann. Rev. Astron. Ap.* **14** 1.

Verschuur, G. L. (1979). Observations of the galactic magnetic field. *Fundam. Cosmic Phys.* **5** 113.

See also **Radiation, Scattering and Polarization; Stars, Neutron, Physical Properties and Models.**

Galactic Structure, Optical Tracers

M. Pim Vatter FitzGerald

BACKGROUND INFORMATION

Definitions

A composite map of spiral structure detected by optically luminous objects and the radio CO emission is shown in the last two figures of this article. In some places the names of spiral features, as indicated in the last figure will be used. Some terms are herewith defined:

(*Optical*) *spiral tracers* are (optically observable) stars or extended objects that lie nearly exclusively along galactic spiral segments.

Spiral arms are long continuous features dominating the appearance of a galaxy.

Spiral segments are curved features that make a galaxy appear spiral.

Spiral structure is the spiral appearance of a galaxy.

Here we examine optical spiral tracers; that is, those objects that would trace out its spiral features when viewed by eye from the outside. Because we live in the galactic plane, it is very difficult to "see" any great distance due to intervening interstellar dust. This is effectively demonstrated by intercomparing Fig. 1 with any external spiral galaxy viewed "edge-on." This may be contrasted with face-on views of M 31 (the Andromeda galaxy), a tightly wound Sb I–II spiral and Fig. 2. Though smaller, the Milky Way galaxy is quite similar to M 31; consequently it is useful to use M 31 to give us clues as to which types of objects (a) define optical spiral segments visually, (b) follow those optical segments, and (c) follow spiral segments, but not necessarily those defined visually.

That our galaxy is spiral was suspected as soon as the existence of other galaxies was established in 1924. However, detailed proof of its spiral structure has been elusive and controversial, partly because it is tightly wound and it is difficult to determine accurate internal distances. The first optical work credited with showing evidence of spiral segments was a study of OB stars by J. J. Nassau and William W. Morgan in 1951; radio observations of the radial velocity profiles of interstellar H I gave the first convincing evidence. In H I profiles by Frank J. Kerr and Gart Westerhout, map distances depend upon the *adopted galactic rotation curve* and relative distances may be strongly distorted by *streaming motions*. Because of the problems of dust, inaccurate distances, and streaming motions it is wise to examine M 31 to see which objects are the best spiral tracers. Unless specified otherwise, in all the diagrams in this article we assume streaming motions are zero. Further we adopt a distance to the galactic center of $R_0 = 8.5$ kpc and a

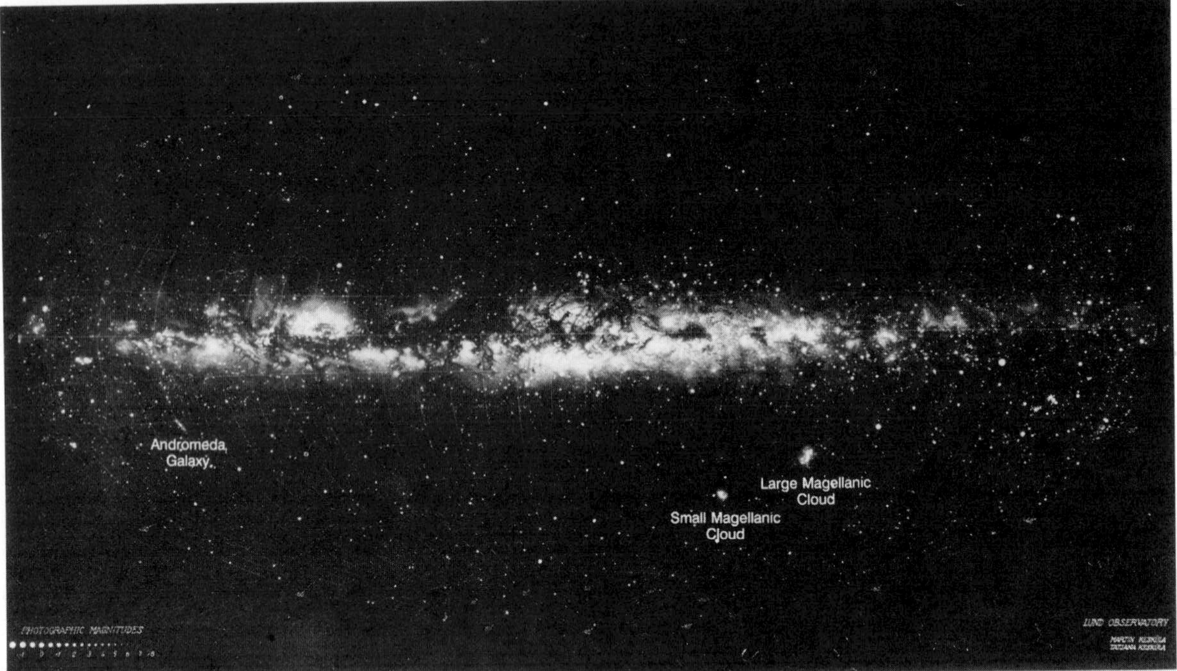

Figure 1. Composite view of the Milky Way (Sbc II). The direction to the center of the Milky Way is at the center of this Aitoff projection. One can clearly see the bulge associated with the galactic center in Sagittarius, but the actual center is totally obscured by intervening dust.

Figure 2. M 31 (NGC 224) the Andromeda nebula (SB I–II). Our sister galaxy, seen inclined at 11°7. Although this galaxy is obviously also spiral, it is difficult to trace individual arms in this blue (B) photograph. (*The Hubble Atlas of Galaxies, Carnegie Institute of Washington, Publication 618, 1961.*)

velocity at the solar circle of $\Theta(R) = \Theta(R_0) = 220$ km s^{-1}. Different values of R_0 and $\Theta(R)$ produce global scaling distortions; streaming motions produce (sometimes severe) local distortions. These simplifications allow straightforward (though quite simplistic) intercomparison of results without detailed discussion of local streaming, distortion, and dynamics.

Evidence from Spiral Structure in M 31, Andromeda

Here we do not discuss the spiral structure of M 31 as such, except to use it as an indicator of what tracers might be useful in the Milky Way. We see any *external* galaxy projected onto the plane of the sky; hence, the relative positions of objects are undistorted by streaming and distance determination problems.

Individual stars are difficult to study in M 31; nevertheless it does appear that OB stars and type I Cepheids prefer spiral segments; those Cepheids with $P > 12.5$ days have the better correlation.

The objects that most obviously trace optical spiral structure in M 31 are complexes containing one or more OB star, such as OB associations (star clouds), open clusters with at least one OB star, and H II regions; single OB stars prefer spiral segments but are not exclusively confined to them. Optically detected dust is associated with spiral structure, but dust clouds are poor spiral tracers. In many galaxies dust clouds tend to lie to the inside of the optical arms, and in some to either side of the obvious optical arm.

In M 31 gas, in particular neutral hydrogen (H I) and carbon monoxide (CO),, does show obvious spiral structure. However the H I is quite strongly concentrated to a ring region, clearly obvious in the infrared. CO shows streaming motions of up to 30 km s^{-1}; in the Milky Way such motions lead to quite incorrect kinematic distances.

DISTANCE DETERMINATION METHODS

Before we can turn to our study of the Milky Way it is essential to review the methods by which we find distance; unlike M 31 (and all external galaxies) we cannot tell the relative positions of objects by comparing their directions in the sky. We must find individual accurate distances. Four distinct methods will be mentioned.

Geometric and Statistical

These methods depend upon the observed parallactic or proper motion and radial velocity of a star or stars. Unfortunately no spiral tracer lies close enough to have an accurate trigonometric or moving cluster parallax (except possibly the Scorpio–Centaurus association), and no spiral tracer is in sufficient local abundance to have a well-determined absolute magnitude by mean or statistical parallax.

Dynamic

In this entry *dynamic* refers to distance-determination methods for individual objects that depend on changes in their diameters. In principle any spherically symmetric object whose radius changes secularly or periodically with time can have its distance (and hence M_V) determined. The required observations are a function of time and consist of the variation in radial velocity and either (a) angular diameter or (b) apparent magnitude and temperature. Planetary nebulae, novae, and supernovae have distances determined by the former, and variable stars by the latter.

Because supernovae of all types are very infrequent and only those of type II lie in spiral arms, this technique is not useful in practice for them. However distances and absolute magnitudes are found for the type I Cepheids by the latter method. These stars are spiral tracers, but because of the time consuming nature of the observations required, the technique is useful only for calibration of the Cepheid period–luminosity law.

Kinematic

In this entry *kinematic* refers to distance-determination methods making use of the rotation of the Milky Way. They are extremely important for mapping spiral structure and are essential for non-stellar objects. The principle of kinematic methods is easily illustrated by considering Fig. 3. It is fairly easy to show that the observed radial velocity of the point at R is

$$V_{\mathrm{LSR}} = R_0\left(\frac{\Theta}{R} - \frac{\Theta_0}{R_0}\right)\sin l. \qquad (1)$$

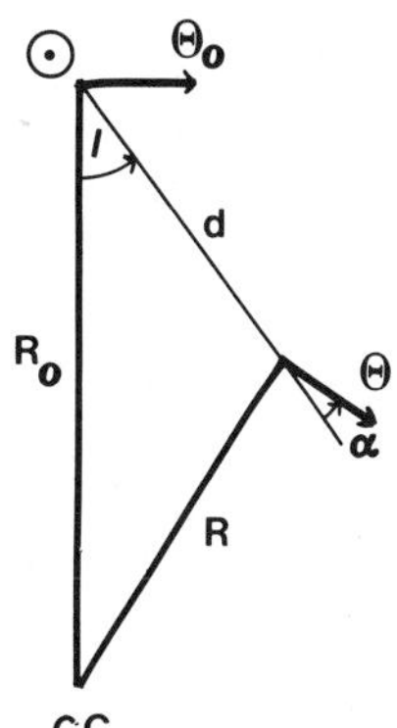

Figure 3. Circular galactic rotation. The local standard of rest (LSR) is imagined to lie in the galactic plane distant R_0 from the galactic center (GC) and to move at a speed Θ_0. A star, or cloud, lies in position $(l, b) = (l, 0°)$, R from the center and moves with velocity Θ. In each case the orbits are assumed circular with no streaming motion or other such effects.

In order to intercompare the spiral pictures obtained by different tracers, we will assume (for the purposes of this entry alone) that

$$\Theta = \Theta_0 \quad \text{for all values of } R,$$

whence Eq. (1) reduces to

$$V_{\mathrm{LSR}} = \Theta_0\left(\frac{R_0}{R} - 1\right)\sin l. \qquad (2)$$

Observations of external galaxies including M 31 indicate that most rotation curves are very close to smooth and often flat, except near the center. It is likely that local deviations are not uniform at any given R, and consequently a smooth rotation curve must usually be assumed. Local deviations from the curve will produce local errors in the distances obtained for all objects at the same position. Streaming motions can produce severe errors, which may be different for different types of objects. Deviations from the flat curve and from our adopted values of

$$(R_0, \Theta_0) = (8.5\ \mathrm{kpc},\ 220\ \mathrm{km\ s^{-1}}) \qquad (3)$$

will produce general scaling errors in our distribution that also do not affect relative positions. Streaming can introduce more severe problems. We adopt this simple method of finding kinematic distances because it allows ease of calculation and simple intercomparison of spiral segments traced by various types of object. A better way to intercompare kinematic spiral tracers with each other is to plot V_{LSR} versus l, with intensity (corresponding to density of the material) shown by shading or contour lines. However, it is as unreliable to transform positional diagrams into kinematic tracers as vice versa. Therefore we choose to map in the plane of the Galaxy. The worst problem with our technique is that it does not allow for the complexity of streaming motions. A lesser but real problem may be that the assumption of a flat uniform curve may give too much distortion to be able to compare the kinematic distances with those found by distance moduli methods.

Kinematic methods can be used for any object or cloud for which a radial velocity is measurable. The method is particularly important for large distances, because optical tracers are generally obscured by interstellar dust. Unfortunately distances are poor or unobtainable kinematically near $l = 0, 180°$ ($\sin l = 0$) and $l = 90, 270°$ ($R_0/R \simeq 1$) and are bivalued for $|l| < 90°$.

Distance Moduli

The corrected distance modulus $V_0 - M_V$ of an object is found by intercomparing its apparent and absolute magnitude. To get the distance d we must correct for the interstellar extinction A_V (in magnitudes):

$$[V_0 - M_V] = [V - M_V - A_V] = 5\log\frac{d}{10}. \qquad (4)$$

Essentially the method boils down to measuring an object's apparent magnitude (V), color ($B - V$), and some other property from which its absolute magnitude M_V and intrinsic color $(B - V)_0$ may be inferred. The visual extinction is given by

$$A_V = \mathscr{R}_V[(B - V) - (B - V)_0], \qquad (5)$$

where $\mathscr{R}_V = 3.2$ needs to be determined in practice. Unfortunately, the errors in finding individual stellar distance moduli are high. The normal best standard deviations (σ_{M_V}) are given in magnitudes here:

Spectroscopic Parallax [$\sigma_{M_V} = 0.7$] This is the method generally used to find the distance for a single star. It depends on an accurate Morgan–Keenan (MK) spectral class and a well-calibrated relationship between that class and the intrinsic M_V and $(B-V)_0$. Unfortunately this calibration, based largely on stars in clusters fitted by zero age main sequence (ZAMS) cluster fitting (see following text) has various determinations and no uniform standard has been adopted. Thus in addition to errors from random sources (especially undetected binaries, $\sigma_{M_V}=0.7$), systematic errors of equivalent size are also present. Thus spectroscopic parallaxes are not reliable for resolving spiral structure, except on a statistical basis. This includes distances found for OB stars $[\sigma_{M_V} \geq 0.7]$, supergiants $[\sigma_{M_V} \geq 1.0]$, and small H II regions $[\sigma_{M_V} \geq 1.2]$. Errors of 0.7–1.25 magnitudes correspond to a factor of 1.4–1.8 in distance.

Type I Cepheids $[\sigma_{M_V}=0.12]$ Type I Cepheids obey a very well determined period-luminosity (PL) or period-luminosity-color (PLC) relation, with the period a reliable predictor of mean absolute magnitude. Consequently we expect accurate Cepheid distances and well mapped spiral features, especially for long period Cepheids (from evidence from M 31 and the evolutionary ages of Cepheids).

Open Cluster (ZAMS) Fitting Parallax $[\sigma_{M_V} \geq 0.15]$ This method has the greatest potential for good distance determination after the Cepheids. However, disagreement about calibration of the ZAMS, intrinsic colors, and methods for dereddening, can give distances different by a factor of 1.4 for some clusters with apparently well-determined ZAMS; metallicity is also an important and ill-determined factor. Such errors are both systematic and random in nature and worst for OB clusters (and ultimately affect the Cepheid PLC law and MK calibrations).

Cluster fitting parallax is used for associations $[\sigma_{M_V} \geq 0.15]$, open clusters $[\sigma_{M_V} \geq 0.15]$, and large H II regions $[\sigma_{M_V} \geq 0.3]$ (which usually contain several OB and other associated stars, often on the ZAMS, but often with spectra and magnitudes contaminated by the H II region itself).

MILKY WAY

From M 31 we can conclude that the best Milky Way spiral tracers are large H II regions, OB associations, OB clusters, and long period Cepheids. These are followed by shorter period Cepheids, small H II regions, and possibly dust or other optical tracers as yet insufficiently observed in M 31. In the radio, H I and possibly CO are quite reliable, as is anomalous OH. However, for optical spiral structure no tracer appears superior to the optical tracers themselves, not even H I. However, interstellar dust prevents us from seeing optical tracers beyond 2–6 kpc, requiring the use of radio and/or kinematic methods to locate spiral tracers in intermediate to distant parts of the Milky Way.

Stellar Tracers

OB Stars Spectroscopic distances to OB stars are not sufficiently accurate to locate an individual star in a spiral segment. Nevertheless, if such stars are spatially concentrated to spiral segments then, even with moderately large errors, more stars in the observed distribution should lie at the position of the segment. G. Klare and T. Neckel have determined the distribution of 6200 OB stars, and they find that spiral segments are detected in positional conformity with the results of other techniques.

Supergiants from O to M (mainly O and B) These supergiants were studied by Roberta M. Humphreys and they conform to the spiral structure revealed by the OB stars, but do not define it as well, possibly because of larger uncertainties in M_V $[\sigma_{M_V} \simeq 1.2]$. The most obvious feature is the Carina arm.

Type I Cepheids The distribution of type I Cepheids has been considered by Gustav Tammann and C. Kim. Tammann suggested that Cepheids with $P>15$ days should be well confined to their regions of birth (presumably spiral arms), whereas those with $15>P>11.25$ should be less well concentrated and those with $11.25>P$, poorly concentrated (long, intermediate, and short period, respectively). Figure 4 is adapted from Kim. Only the Carina feature $l \simeq 290°$ is well delineated and separation of the long, intermediate, and short period Cepheids does not greatly improve

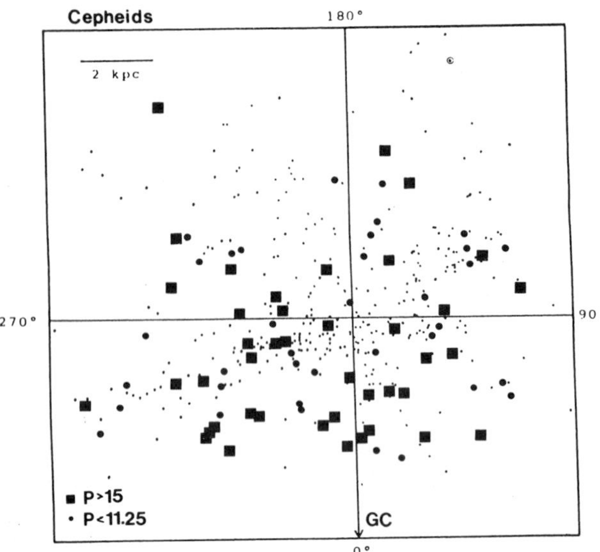

Figure 4. Distribution of 446 type I Cepheids on the projected galactic plane. The data are from Kim [Ap. Space Sci. *136 101 (1987)*] with Cepheids designated by period: long ■; intermediate (●); short (·).

the picture. Both Kim's and Tammann's figures are consistent with spiral structure as mapped for OB stars and OB clusters (Fig. 7), but are not convincing by themselves. Consistent with indications from M 31, short period Cepheids also show concentrations to spiral features, though they are not seen to as great a distance as the longer period objects.

It is tempting to see if kinematic distances to Cepheids would indicate better spiral structure, but a plot for 30 Cepheids with $P>9$ days was not helpful. On the other hand, an intercomparison of kinematic and PLC distances for Cepheids with $P>9$ days (Fig. 5*a*) shows kinematic and PLC distances to be in moderate agreement $[\sigma(d_k-d_m)=0.42$ kpc] in comparison to the short period Cepheids $\sigma(d_k-d_m)=0.86$ kpc, with a similar number of omitted objects.

We can say in summary that the distribution of long (and short) period Cepheids is consistent with spiral structure from other tracers, but disappointing if taken alone.

Wolf–Rayet Stars These stars are tantalizing because it is certain that at least some of them are very young, and hence should lie in spiral segments. They are easy to detect in surveys, yet do not show evident concentration to spiral segments probably because of a large dispersion in absolute magnitude. If the distances could be much better obtained, these stars may prove useful spiral tracers because they can be seen at large distance and are easily detected.

Objects Associated with OB Stars

As expected these are the most useful optical spiral tracers. OB associations do seem to lie along the arms, as do OB clusters. H II regions are not quite as good, probably because of large random errors in distance modulus.

OB Associations OB associations have been studied by Humphreys. They do suggest spiral segments (especially the Local and Perseus), generally consistent with those found by other stellar based methods.

OB Clusters OB clusters have been studied by many researchers and in their diagrams there is little doubt that spiral features are detected. Kenneth A. Janes et al. disagree: The disagreement lies in the implicit definition of what constitutes the detection of spiral structure. This entry assumes it exists, and

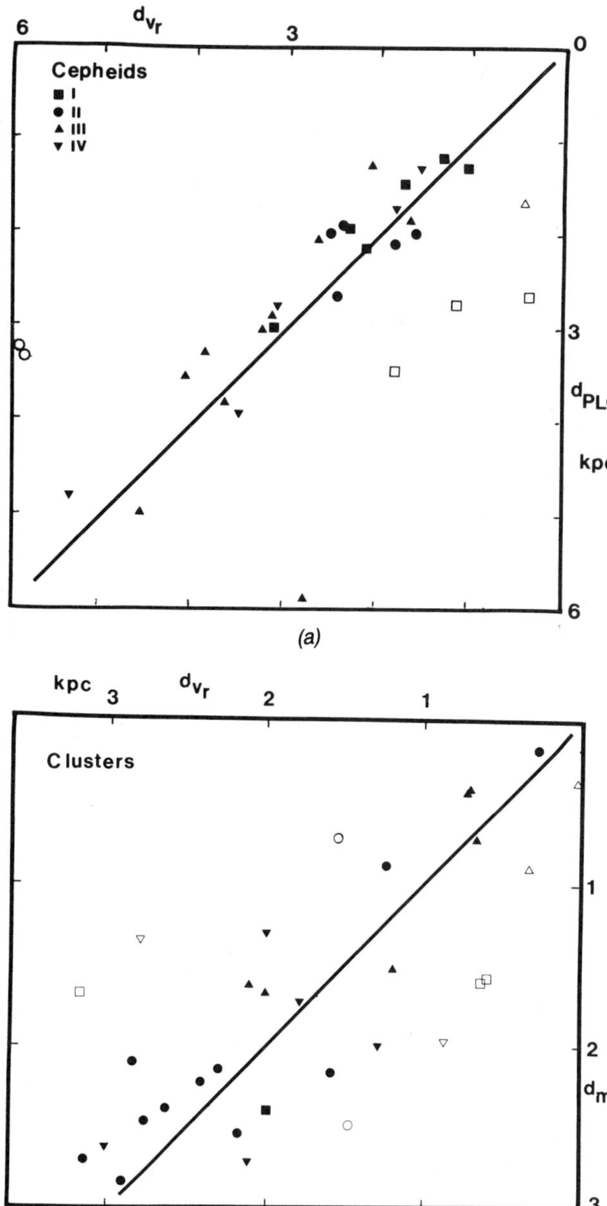

Figure 5. Cepheid dynamical versus PLC distances. (*a*) Objects with $P < 9$ days and within 10° of $l = 0, 90, 180, 270°$ are excluded. The slope of the relation is not significantly different from 1.0, indicating good systematic agreement. If the objects indicated by open symbols are omitted, $\sigma(d_k - d_m) = 0.42$ kpc, which indicates agreement good enough to map spiral structure, provided anomalous objects are excluded. ■: first quadrant; ○: second quadrant; △: third quadrant; ▽: fourth quadrant. (*b*) Same as (*a*), but for OB clusters.

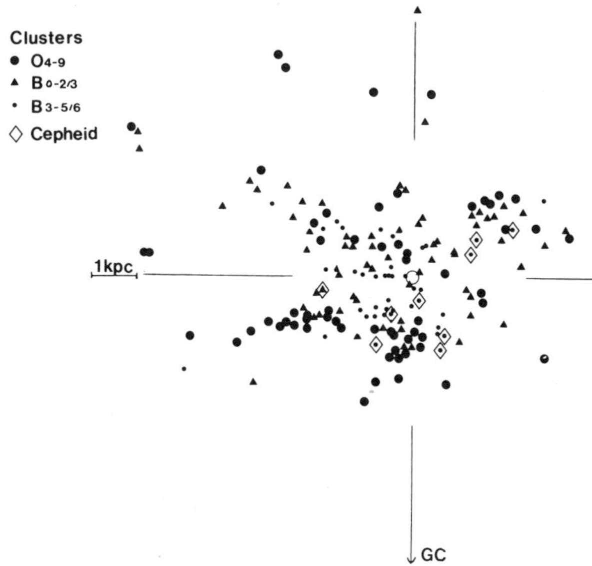

Figure 6. Distribution of OB2/3 and B4-5/6 clusters. The distances to clusters are largely taken from Vogt and Moffat [Astron. Ap. **39** *477 (1975)*] and Moffat et al. [Astron. Ap. Suppl. **38** *197 (1979)*] with some additions. Most objects have their distance determined by essentially the same methods. ⊕: O clusters; △: B0-B2/3 clusters; ●: B3-B5/6 clusters; ◇: B2-5 clusters with Cepheids.

looks for strong consistency; Janes et al. appear to require optical spiral structure be mapped on a large scale, before accepting proof it is there. Both views are defensible. Several of the research figures are consistent with spiral structure, and we note the following important points:

1. Janes et al. give a figure of nearby OB clusters that clearly indicates concentrations of OB clusters. These are well determined, but do not constitute, by themselves, good evidence for spiral structure.
2. Becker and Fenkart give a figure that shows clear spiral segments, probably because of more uniform distance reduction methods.
3. Figure 6 shows the distribution of OB clusters based largely on work done using the techniques employed by N. Vogt and Anthony Moffat. Figure 7 shows the same clusters presented in the manner as presented by Klare and Neckel.

In the latter figure, spiral segments are quite obvious, and clusters as late as B5/6 adhere to the general picture, supporting the finding that Cepheids with $P > 9$ days show spiral structure. Of all the

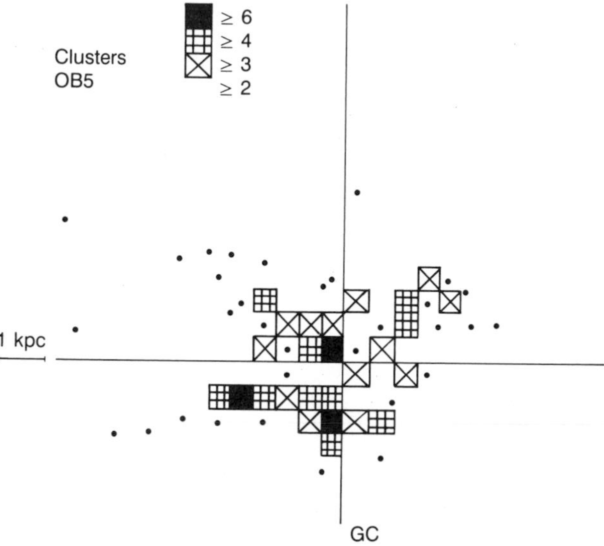

Figure 7. Distribution of O to B5/6 clusters. Each square of 0.25 per side indicates the number of OB-6 clusters projected into the plane at that position. In this diagram there are many regions shown blank containing up to 2 OB-6 clusters each. Number of stars/$(0.25\ \text{kpc})^2$ is shown in the figure. ■: $n > 6$ clusters; ⊗: $6 \geq n > 4$; ⊕: $4 \geq n > 3$; o: $3 \geq n > 2$; omitted: $2 \geq n$.

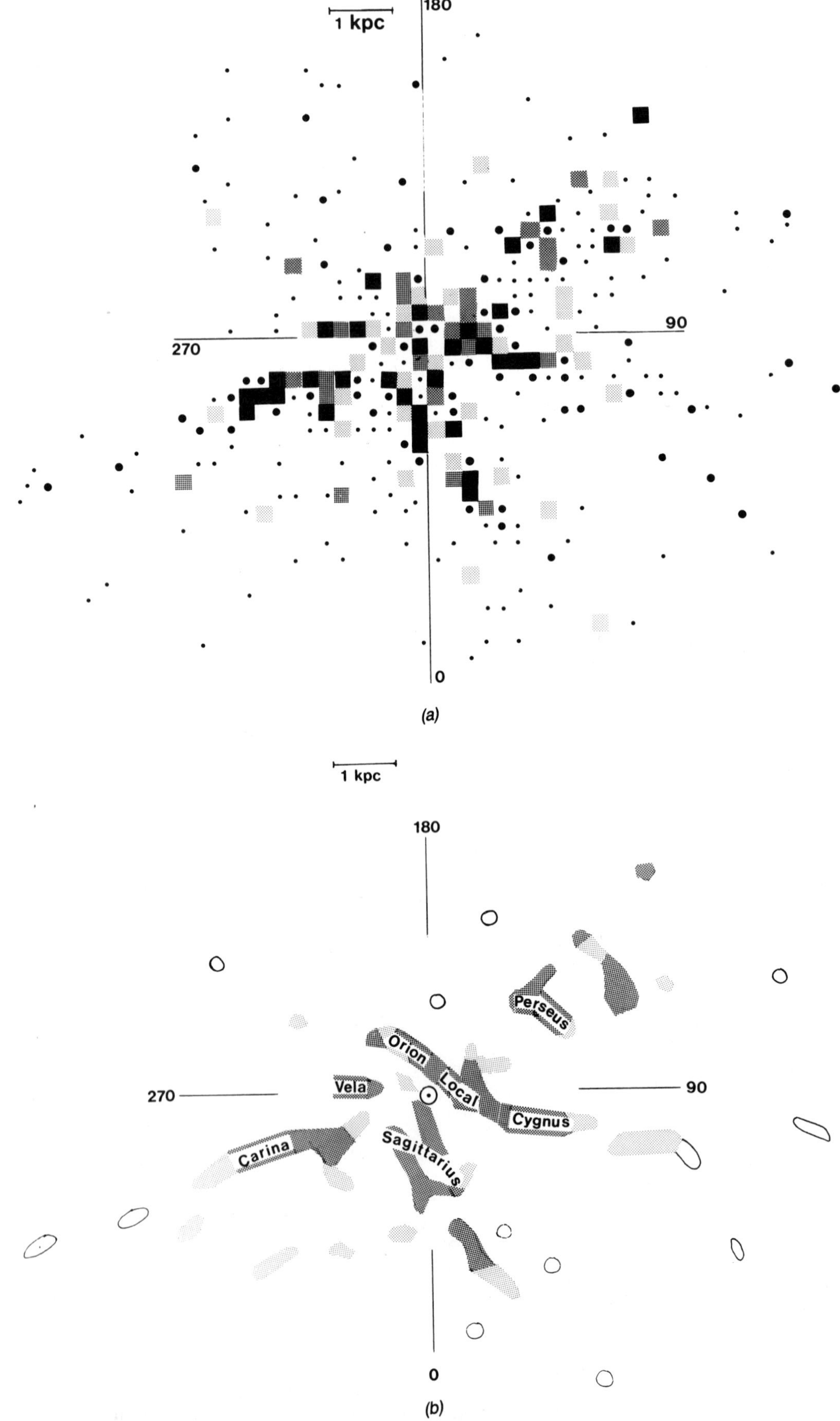

Figure 8. Composite map of optical spiral tracers. (*a*) The amount of shading indicates the cumulative abundance of tracers in any given region. The shaded scale indicates 6, 5, 4, 3, or 2 objects, with OB stars assigned weight 0.1. (*b*) The traditional names of the features.

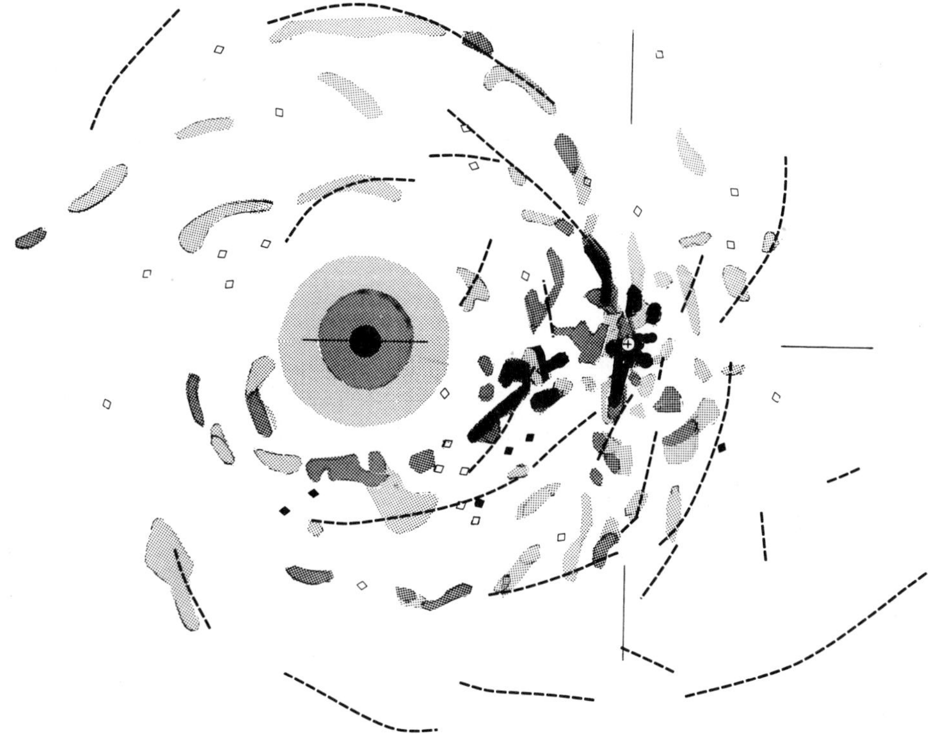

Figure 9. Possible optical appearance of our galaxy. Only regions with several objects are drawn in. Many spiral tracers lie in "empty" regions of the diagram. The local features are based on Fig. 8 and kinematic distances to CO and OH; the more distant features are based exclusively on kinematic distances of CO and OH. The center region is shown schematically. The location of the maximum H I densities are shown as dashed lines.

optical modulus tracers the OB clusters seem best. They still do not show tight correlation to spiral segment patterns, partly because simple spiral structure is not a feature of our system, and also because of still-present uncertainty in the distances. In comparison, the BA clusters show no evidence of spiral structure whatsoever.

The agreement between dynamical and ZAMS fitting for clusters, $\sigma(d_k - d_m) = 0.38$ kpc, is similar to that for Cepheids $P > 9$ days (Fig. 5*b*). Unfortunately, kinematic distances for the OB clusters do *not* show spiral structure. One tentative conclusion that may be drawn with respect to the OB stars and Cepheids found in the field compared to those found in clusters is that field objects do not exclusively form in spiral segments, whereas clusters do. On the other hand, OB clusters formed in spiral segments may move more rapidly from their place of origin than field stars.

Large and Small H II Regions Large H II regions usually contain several exciting OB stars, as well as others later than B2, thus allowing a good spectroscopic or ZAMS parallax. On the other hand, small H II regions frequently contain only one OB star. Consequently, only the less reliable (spectroscopic) distances are available. On the whole the H II region distribution is consistent with, but not as well determined as, the OB cluster distribution.

The distances to H II regions may also be obtained kinematically by Hα interferometry in the optical, but the surveys are far from complete.

Another method of determining kinematic distances to H II regions is to measure the velocity of the associated CO. Leo Blitz et al. have done this for the Sharpless H II regions; one of their figures also shows that the kinematic distances of H II regions based on CO not only show spiral structure in the outer Milky Way, but also conform to, or lie slightly to the inside of, the H I 21-cm spiral structure. The agreement is not perfect, but roughly the same spiral structure is revealed by CO, H II, and distance moduli techniques.

Other Optical Tracers Other optical tracers have been studied; and two are worth particular mention:

The interstellar absorption lines of Ca and Na superimposed on stellar spectra show clear evidence of both galactic rotation and spiral structure in the second quadrant. This early promise was not so convincing in other quadrants as shown by James J. Rickard in the southern hemisphere.

Interstellar dust definitely shows association with spiral arms in some other galaxies, though not very well in M 31. Assorted studies of dust clouds near the Sun are not inconsistent with the OB cluster picture.

NONOPTICAL TRACERS

We have already alluded to CO associated with H II regions and to H I. In addition, the anomalous lines of OH are good radio tracers. The kinematic mapped distances [based on Eq. (2)] of CO, H I, and anomalous OH clouds do not overlap perfectly, almost certainly due to a combination of slightly different locations of the various clouds and to the effects of streaming motions.

SUMMARY AND MAP OF LOCAL STRUCTURE

It should be obvious from the preceding discussion that optical spiral segments are present in the vicinity of the Sun, though intercomparisons of the various objects and methods used to map such structures are not in agreement. Indeed, it is quite probable that different types of object lie in different parts of a segment or sometimes in different segments. In addition, kinematic distances are affected by streaming motions. In an attempt to "map" the local optical segments in as unbiased a manner as possible, Fig. 8*a* provides a composite diagram similar to those of Figure 7. In this diagram the most heavily shaded regions indicate the highest concentration of spiral tracers. The diagram is based upon distance moduli estimates for OB clusters and associations, OB stars, and Cepheids of all periods, and the mean distances for H II regions (moduli, kinematic H II, and CO). Figure 9 shows how the Milky Way might appear optically to the outside observer. The region of the Sun is enhanced in brightness because more tracers have been observed locally; distant regions are based solely on OH and CO, which are believed to trace the optical galactic regions. H I arms do not directly conform to the optical arms.

Additional Reading

Arp, H. (1964). Spiral structure in M 31. *Ap. J.* **139** 1045.

Baade, W. (1951). *Publ. Univ. Michigan Observatory* **10** 7–17.

Becker, W. and Contopoulos, G., eds. (1970). *The Spiral Structure of Our Galaxy*. D. Reidel, Dordrecht, pp. 51, 69, 126, 205, 236.

Binney, J. and Tremaine, S. (1987). *Galactic Dynamics*. Princeton University Press, Princeton, N.J., pp. 339–342.

Blitz, L., Fich, M., and Kulkarni, S. (1983). The new Milky Way. *Science* **220** 1233.

Casili, F., Combes, F., and Stark, A. A. (1987). Mapping of a molecular complex in a northern spiral arm of M 31. *Astron. Ap.* **173** 43.

Clemens, D. P., Sanders, D. B., and Scoville, N. Z. (1988). The large-scale distribution of molecular gas in the first galactic quadrant. *Ap. J.* **327** 139.

Efremov, Y. N. (1980). Cepheids and structure of the spiral arm in the Andromeda nebula. *Sov. Astron. Lett.* **6** 152, 184.

FitzGerald, M. P. (1968). The distribution of interstellar reddening material. *Astron. J.* **73** 983.

Gingerich, O. (1983). The discovery of spiral arms in the Milky Way. In *International Astronomical Union Symposium* **106**, p. 59.

Habing, H. J. (1983). In *International Astronomical Union Symposium* **106**, p. 451.

Herbst, W. (1975). R associations III. Local optical spiral structure. *Astron. J.* **80** 503.

Hodge, P. W. (1979). The open star clusters of M 31 and its spiral structure. *Astron. J.* **84** 744; **85** 376.

Hubble, E. (1958). *Realm of the Nebulae*. Dover, New York. Reprinted from the 1936 edition, Yale University Press, New Haven, Conn., pp. 27–28.

Humphreys, R. M. (1976). A model for the local spiral structure of the Galaxy. *Publ. Astron. Soc. Pacific* **88** 647.

Janes, K. A., et al. (1988). Properties of the open cluster system. *Astron. J.* **95** 771.

Kerr, F. J. and Westerhout, G. (1965). Distribution of interstellar hydrogen. In *Galactic Structure*, A. Blaauw and M. Schmidt, eds., p. 167.

Kim, C. (1987). Distribution of classical Cepheids in the galactic plane. *Ap. Space Sci.* **136** 101.

Klare, G. and Neckel, T. (1967). *Z. Astrophysik* **66** 45.

Nassau, J. J. and Morgan, W. W. (1950). *Sky and Telescope* **9** 243.

Pellet, A., Astier, N., Viale, A., Courtès, G., Maucherat, A., Monnet, G., and Simien, F. (1978). A survey of H II regions in M 31, *Astron. Ap. Suppl.* **31** 439.

Rickard, J. J. (1974). Interstellar lines in the southern hemisphere. *Astron. Ap.* **31** 47.

Ryden, B. S. and Stark, A. A. (1986). Molecules in galaxies I. CO observations in a spiral arm of M 31. *Ap. J.* **305** 823.

Solomon, P. M. and Rivolo, A. R. (1989). A face-on view of the first galactic quadrant in molecular clouds. *Ap. J.* **329** 919.

Stark, A. A. (1983). Distribution and motion of CO in M 31. In *International Astronomical Union Symposium* **106**, p. 445.

Turner, B. E. (1979). Galactic emission of OH at 1720 MHz. *Astron. Ap. Suppl.* **27** 1.

Vogt, N. and Moffat, A. F. J. (1975). Galactic structure based on young southern open star clusters. *Astron. Ap.* **39** 477.

Unwin, S. C. (1980). Neutral hydrogen in the Andromeda nebula —I. H I emission in the southwest. *Monthly Not. Roy. Astron. Soc.* **190** 551.

Unwin, S. C. (1980). Neutral hydrogen in the Andromeda nebula —II. H I emission in the northeast. *Monthly Not. Roy. Astron. Soc.* **192** 243.

Walker, A. R. (1988). Calibration of the Cepheid period–luminosity relation. *Astron. Soc. Pacific Conf. Series* **4** 89.

See also **Andromeda Galaxy; Galaxies, Spiral, Nature of Spiral Arms; Galaxies, Spiral Structure.**

Galactic Structure, Spiral, Interstellar Gas, Theory

William W. Roberts, Jr.

Because of the winding dilemma associated with material arms, a wave interpretation of large-scale spiral structure is necessary for disk-shaped galaxies such as our own Milky Way. Such a wave interpretation was first provided in the linear density wave theory, initiated by C. C. Lin and F. H. Shu over 25 years ago. The present-day linear modal theory for spiral density waves, when merged with the theory for the nonlinear dynamics of the interstellar gas, not only makes strides in solving the dilemma of why spiral arms do not necessarily wind up but also supplies a mechanism for forming those stars which delineate spiral arms. In this theory, the luminosity of a spiral arm is believed to originate primarily from the very young, newly forming stars whose ages are only about 1/1000 of the age of the Galaxy and whose mass makes up only a small percentage of the bulk of the galactic mass, and the spiral arm itself is believed to be a spiral wave—a *nonlinear galactic shock wave* in the interstellar gas—that is capable of triggering from the gas the formation of the young stars selectively along the wave crest.

What was thought of two decades ago as a fairly uniform galactic gaseous medium is now seen to be highly nonuniform, with numerous 1–20-pc-scale clumps or clouds of relatively cool, dense gas at one end of the spectrum, 20–100-pc-scale giant molecular clouds [GMCs] and diffuse atomic hydrogen clouds in the middle, and massive hundred parsec to several kiloparsec scale aggregations of GMCs and diffuse cloud complexes at the other, all embedded in a highly rarefied, partly ionized intercloud medium. Present-day high-resolution observations of the atomic and molecular gas components in global grand design spirals indicate that the largest of these local nonuniformities, the GMC complexes and aggregations, are almost exclusively located in spiral arms. The spiral arms often show a very high resolution into knots that comprise these aggregations and associations of clouds, active star formation regions of newly forming protostars, young luminous stars, and giant H II regions. Such concentrations of GMC complexes and aggregations in spiral arms may also be the case for our own Galaxy. However, their exact distances and their global distribution within our Milky Way system are known with less certainty because of the fact that the bird's eye view we enjoy of the spiral structure in extragalactic systems is largely lacking from the vantage point of our location within the disk of the Milky Way system.

On the global scale in grand design spirals, the clumpy, cloudy gaseous interstellar medium is found to exhibit strong nonlinear contrasts in arm-to-interarm density with peak-to-mean values measured on the order of 2:1–4:1 and strong systematic velocity streaming motions with measured magnitudes on the order of 30–80 km s^{-1} across the major spiral arms. Striking examples are the extragalactic systems: M31, M51, and M81. Prominent narrow dust lanes also delineate the striking nonlinear character of the atomic and molecular interstellar gas and provide excellent optical tracers of the major spiral arms. In view of the observed highly

Table 1. Physical Mechanisms and Dynamical Processes

- Galactic gravity. Gaseous self-gravitational effects.
- Orbital dynamics of interstellar gas clouds. Finite cloud cross sections; representative cloud collisional mean free paths.
- Inelastic, energy-dissipating cloud-cloud collisions.
- Birth of protostars. Time delay before active period of star formation. Stellar associations.
- Participation of gas clouds in subsequent star formation delayed for a refractory period.
- Supernova explosions; gas-star interactions. Replenishment of random kinetic energy.

nonuniform and multiphase gaseous interstellar medium, a basic problem stands out; namely, what physical mechanisms and dynamical processes can organize the locally clumpy and cloudy interstellar gas into its strongly coherent nonlinear distribution on the global scale and can trigger star formation from the cloudy gas along a grand design of spiral structure in an orderly fashion and yet simultaneously provide for the formation of spurs, feathers, arm branchings, and secondary features on intermediate scales?

DOMINANT PHYSICAL MECHANISMS AND DYNAMICAL PROCESSES

The basic problem to be addressed spans the hierarchy of galactic scales: the locally clumpy, cloudy interstellar medium, giant molecular clouds, and star formation on local scales; spurs, feathers, arm branchings, and secondary features on intermediate scales; and the grand design of spiral structure on the global scale. Dominant physical mechanisms and dynamical processes are outlined in Table 1, with particular emphasis on the self-gravitational effects, dissipative effects, and collisional dynamics of cloudy gaseous galactic disks.

The self-gravitational response of the gas driven by an underlying two-armed spiral density wave mode computed in the modal density wave theory is shown in Fig. 1 for one representative model galaxy with a 2% gas mass fraction. The cloudy gaseous galactic disk consists in part of a system of N finite-cross-section clouds (taken to be 10,000), initially distributed randomly over a two-dimensional disk out to a maximum radius $R_{\max}$ (taken as 12 kpc) and given local circular velocities plus small peculiar velocities (amounting to a one-dimensional dispersion, v_d, of 6 km s^{-1}) and in part of a system of young stellar associations that form from the clouds. The clouds and young stellar associations interact gravitationally with each other, and their motions are governed gravitationally by their own self-gravity as well as by the background modal-spiral–perturbed gravitational field which attains a maximum amplitude of 12% that of the central axisymmetric force field.

Displayed in Fig. 1 through a photographic intensity map at the sample time epoch of 480 million years (Myr) during the computations are the computed global distributions of the system of gas clouds (represented by patches) and the system of young to middle-aged stellar associations active with supernova events during the past 60 Myr (represented by white dots). Most striking is the strongly peaked gas density distribution that traces global gaseous spiral arms. A galactic shock formed within the gaseous component imparts distinctly nonlinear characteristics to these global gaseous spiral-wave arms. The regions of most active star formation lie along the ridge of the gas density distribution tracing the galactic shock. Consequently, the systems of gas clouds and young stellar associations triggered from the clouds both exhibit aggregations of giant complexes along the galactic shock and global spiral-wave arm structure. The most prominent aggregations of young stellar associations are strongly correlated with the regions of highest gas density adjacent to the galactic shock, with few associations which are not adjacent to clouds.

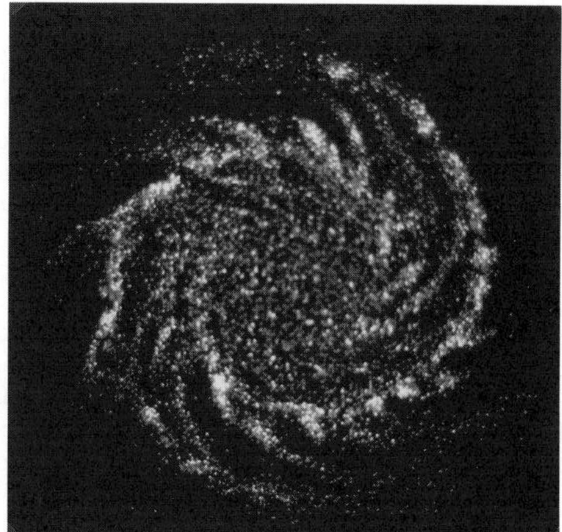

Figure 1. Photographic intensity map of the self-gravitating distributions of gas clouds (patches) and young stellar associations (white dots) born from the gas, superposed at one sample time epoch (480 Myr), for a representative modal-spiral galaxy model with 2% gas mass fraction, derived from the modal density wave theory. The modal-spiral perturbation-force amplitude is 6–12% that of the axisymmetric field. White dots include all stellar associations currently undergoing supernova (SN) events as well as all post-SN associations active during the past 60 Myr.

During the short periods of time (≤ 20 Myr) corresponding to their active star formation stages, the young stellar associations assume the mass of those clouds from which they form; after this active phase, the associations redeposit their mass back into clouds through supernova events. Consequently, the total mass attributed to and interchanged between these two "mobile" systems—gas clouds and young stellar associations—is taken together in the definition of *total gas mass*, and both systems are taken to constitute the total mobile, self-gravitating, *gaseous component*. The ratio of total gas mass (for these mobile systems) to underlying basic state mass (prescribed as a fixed background component) is defined as the *gas mass fraction* and is one of the constants by which this cloudy galactic disk model is characterized.

Self-gravitational effects of the gaseous component (both the gas clouds and young stellar associations) play an important multifold role. On the large scale (e.g., 10–50 kpc) gaseous self-gravity acts to enhance the overall collective gravitational field driving the gaseous response and thus helps enhance and maintain the global spiral structure. On local scales, (e.g., up to hundreds of parsecs) gaseous self-gravity aids the gathering and assembling of clouds into the massive complexes and aggregations. Striking is the local raggedness and patchiness of the computed distributions of gas clouds and young stellar associations formed from the gas. On intermediate scales (e.g., hundreds of parsecs to several kiloparsecs) gaseous self-gravity helps in the formation of spurs, feathers, arm branchings, and secondary features. These intermediate-scale features continually break apart and reform as the loosely associated aggregations, and giant complexes of clouds continually disassemble and reassemble over time. Such transient features on these local and intermediate scales give rise to disorder and chaotic activity within the global spiral structure and tend to blur the global coherence.

NONLINEAR GAS DYNAMICAL EFFECTS

Nonlinear gas dynamical effects dominate the cold gaseous component in galactic disks. Nonlinear characteristics of the gas, including the galactic shock formed in the gas, are evident in Fig. 2. Displayed are the computed variations of selected physical quantities at the sample time epoch 500–520 Myr in the gaseous self-gravitating modal-spiral–galaxy model (of Fig. 1). Plotted with respect to spiral phase around a representative half-annulus (at 10 kpc) in the model galactic disk are cloud number density, components of velocity perpendicular and parallel to spiral equipotential loci, computed velocity dispersion among gas clouds, and the computed distribution of young stellar associations currently active with supernova events (SN rate). First and foremost is the strong gaseous galactic shock that has developed and is being maintained on the global scale, despite local stochastic variations and perturbations. This is evidenced in part by the strong, persistent cloud-density and SN-density enhancements just downstream of 180° spiral phase (i.e., at the potential minimum of the modal spiral) and in part by the corresponding large-scale systematic motions exhibited by the cloud system with magnitudes of 40–60 km s^{-1} (bottom two panels), representing strong systematic perturbations from purely circular rotation. The density distribution of this self-gravitating cloud system (middle panel) is seen to be strongly peaked with peak-to-mean values on the order of 3:1–4:1 and arm-to-interarm contrasts typically 6:1–8:1, with arm thicknesses on the order of 1–2 kpc. The sharp deceleration of gas from supersonic to subsonic speeds, reflected in the $u_\perp$ velocity component just preceding 180° spiral phase with much more gradual characteristic rise downstream (bottom panel) appears as a striking galactic shock manifestation in these self-gravitational computations. The characteristic skewness in the $u_\perp$ velocity component as well as the characteristic asymmetry in the $u_\parallel$ velocity component together delineate the galactic shock structure that is formed. Such skewness is less apparent in the density distribution, with the density rise occurring over the broad shock width of a number of collisional mean free paths.

It is the dissipative character of the cold, cloudy gaseous component that makes possible these strong nonlinear effects, which largely distinguish the interstellar gas from the stellar component. First, dissipative cloud-cloud collisions serve to maintain a low-velocity-dispersion gaseous component. Only such a cold component is able to respond with sharp nonlinear characteristics. The presence of a cold, cloudy, dissipative gaseous component can also promote the physical regimes necessary for unstable, growing global density wave spiral modes. Indeed, in real galaxies, it is likely to be the balance achieved between such moderately to rapidly growing global modes on the one hand and the dissipative nonlinear gaseous response on the other that allows the emergence of finely tuned, coherent grand designs of global spiral structures. Without the presence of a cold and dissipative gaseous component, real galactic disks would be hard pressed to produce and exhibit any such sharp, clear-cut structures on global scales. Likewise, on local and intermediate scales, the presence of a cold, dissipative gaseous component is essential for the formation of giant aggregations of cloud complexes, corresponding active star formation regions, prominent spurs, feathers, arm branchings, and secondary features. Only such a cold, dissipative component can provide the appropriate environment for the effective assembling of the giant massive cloud complexes (GMCs) and thereby produce the fertile beds for star formation activity.

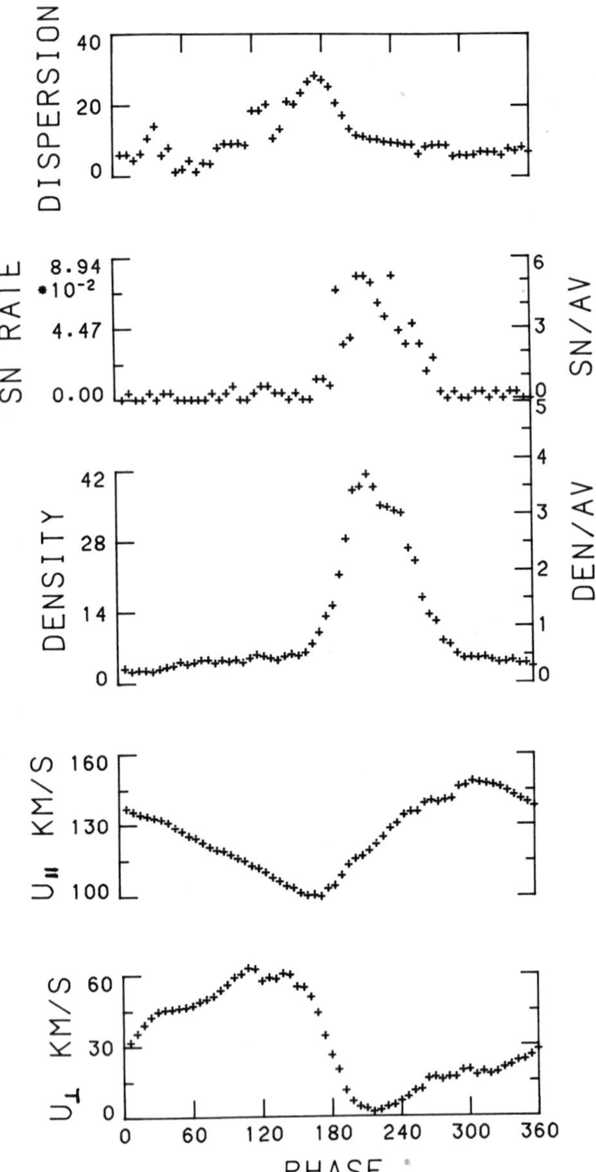

Figure 2. Spatial variation of supernovae (SN rate) and of the number density, velocity dispersion, and velocity components $u_\perp$ and $u_\parallel$ perpendicular and parallel to spiral equipotential contours for the gas cloud system, plotted versus spiral phase about a representative half-annulus (1000 pc wide) at 10 kpc in the modal-spiral galaxy model (Fig. 1). The distributions span a representative epoch (500–520 Myr) during the evolution of the model galaxy, with the (+) values at each phase time averaged over 20 intervals, each of 1-Myr width.

COMPARISONS WITH OBSERVATIONS

These theoretical-computational results appear to be in good agreement with recent high-resolution observational studies of measured observational tracers along and across the spiral arms in several selected grand design spirals. For example, Very Large Array observations of the ratio continuum emission in M81 show nonthermal radio emission spiral arms that are patchy and well resolved, with widths of 1–2 kpc. The H I gas, the nonthermal radio emission from the arms, the dust and narrow dust filaments, the young stars, and the set of giant radio H II regions are each distributed across a broad spiral compression zone that starts near the measured position of the spiral velocity shock front in the H I gas and extends 1–2 kpc downstream from the shock. These features are in good agreement with the theory presented herein, which places important emphasis on the cloudy nature of the interstellar medium.

There also appears to be good agreement with the results of recent observations of CO emission from M51 that reveal density wave–galactic shock streaming motions with magnitudes on the order of 60–80 km s^{-1} in the molecular gas which are coincident with the prominent dust lanes along the major spiral arms. These results appear as strong evidence that the spiral density wave-galactic shock in M51 assembles preexisting molecular clouds into giant associations and triggers the collapse of suitably primed clouds, leading to the formation of stars.

On the basis of the theory, we would expect the GMCs in grand design spiral galaxies to be strongly concentrated in the spiral arms. For M51, the CO emission is found to be strongly peaked in the global spiral arms, with average arm-to-interarm integrated CO brightness ratios of 2.4 for a selected inner arm and 3.0 for a selected outer arm. A narrow dust lane traces closely the inner CO arm, with the dust lane and the CO arm having similar widths of about 300 pc. For M31, the CO distribution is also found to be strongly concentrated to the spiral arms. High-resolution CO observations reveal GMCs in M31 similar in size and molecular hydrogen mass to those in the solar neighborhood of our Milky Way system. These clouds are found to be closely associated with H II regions and are active sites of star formation, in good agreement with the theory.

Additional Reading

Bertin, G., Lin, C. C., Lowe, S. A., and Thurstans, R. P. (1989). Modal approach to the morphology of spiral galaxies: I. Basic structure and astrophysical viability. *Ap. J.* **338** 78, and Modal approach to the morphology of spiral galaxies: II. Dynamical mechanisms. *Ap. J.* **338** 104.

Kaufman, M., Bash, F. N., Hine, B., Rots, A. H., Elmegreen, D. M., and Hodge, P. W. (1989). A comparison of spiral tracers in M81. *Ap. J.* **345** 674.

Lada, C. J., Margulis, M., Sofue, Y., Nakai, N., and Handa, T. (1988). Observations of molecular and atomic clouds in M31. *Ap. J.* **328** 143.

Miller, R. H. (1976). Validity of disk galaxy simulations. *J. Comput. Phys.* **21**(4) 400.

Roberts, W. W., Lowe, S. A., and Adler, D. S. (1990). Simulations of cloudy, gaseous galactic disks. In *Galactic Models*, J. R. Buchler, S. T. Gottesman, and J. H. Hunter, eds. *Ann. N.Y. Acad. Sci.* **596** 130.

Vogel, S. N., Kulkarni, S. R., and Scoville, N. Z. (1988). Star formation in giant molecular associations synchronized by a spiral density wave. *Nature* **334** 402.

See also **Galactic Structure, Interstellar Clouds; Galactic Structure, Large Scale; Interstellar Medium, Galactic Atomic Hydrogen.**

Galactic Structure, Spiral, Observations

Brian Robinson

The bright band of the Milky Way across the sky has long suggested that we live inside a disk of stars, gas, and dust. Flat, rotating disks of stars can be seen in photographs of other galaxies. When edge-on they reveal a thin, flat, dusty disk with rotational motions of some hundreds of kilometers per second. When face-on, the young stars, ionized hydrogen regions, giant molecular clouds, and dust delineate a large-scale structure with several trailing spiral arms. Since the 1950s evidence has been accumulating that young stars and gas clouds in our galaxy have an arm-like distribution. But determining the connection of the fragments of arms into a unique spiral pattern remains a challenge.

From the position of the solar system inside the disk of the Galaxy, and well out from the center, the appearance is essentially that of an edge-on galaxy and it is difficult to infer the overall structure. A further problem is the large amount of dust in the disk (as in any spiral galaxy), which attenuates starlight rapidly. Between the solar system and the galactic center blue light suffers 25 magnitudes of absorption. So, in the disk, optical tracers can only be seen at distances corresponding to about one tenth of the overall size of the Galaxy.

Infrared, millimeter, and radio waves are not attenuated by the fine dust particles, and at these wavelengths measurements can be made right across the Galaxy. In the disk a line of sight will intercept several arm fragments. To determine an overall pattern, we need to be able to recognize arms at different distances and to measure those distances. Making reliable measurements of the distances is the most difficult observational problem.

THE DISK OF THE GALAXY

Within a radius of about 8 kpc from the center of the Galaxy the dust and molecular clouds lie in a thin, flat disk, with the same ratio of thickness to diameter as a black vinyl phonograph record. The thinness of the disk is most clearly seen in the infrared observations by the *Infrared Astronomical Satellite* (*IRAS*) satellite (Fig. 1) and in observations of the 2.6-mm spectral line of carbon monoxide. The molecular disk has a thickness of 130 pc between its half-intensity points. Further than 10 kpc from the center, the disk thickness increases and its centroid deviates by up to 1.6 kpc from the plane of the inner disk.

The thin disk of dust and molecules is prima facie evidence that our galaxy is a member of the class of spiral galaxies classified as Sb, Sc, or Sd. Elliptical, S0, Sa and irregular galaxies are not so flat, while E, S0, and Sa galaxies are not so dusty.

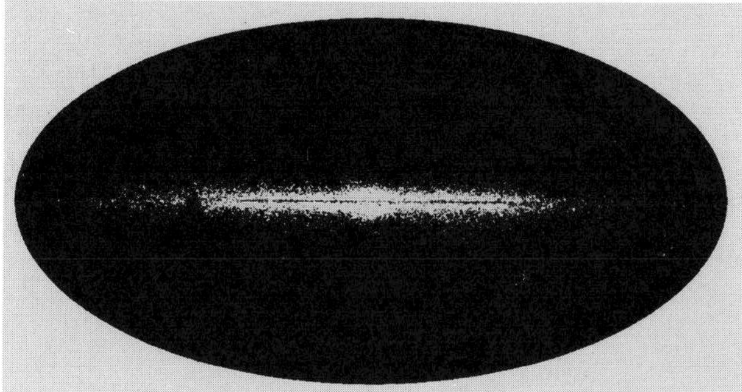

Figure 1. Composite *IRAS* infrared image of our galaxy. White shows 12–25-μm emission from older stars and their envelopes. The thin dark band is a superimposed negative of the 60- and 100-μm emission from cool dusty matter.

In our Galaxy the long-lived stellar populations and atomic hydrogen clouds occupy a thicker disk. In Fig. 1 the bright areas show the distribution of 12–25-μm emission from older stars and their envelopes, as measured by the *IRAS* satellite. The thin dark band is a superimposed negative of the *IRAS* 60- and 100-μm emission from the cool dusty matter in which young, massive stars have formed.

THE CENTRAL BULGE

The *IRAS* infrared observations show that there is a central bulge to our galaxy. The bulge is clearly seen on the composite image in Fig. 1. Long-period variable stars in the bulge have also been studied at optical and radio wavelengths.

The extent of the bulge is very similar to those of Sb galaxies like NGC 891 or NGC 4565. Sc galaxies have smaller nuclear bulges, while most Sa galaxies have large, amorphous central regions without dust.

ASSOCIATIONS OF O AND B STARS

Close to the solar system the young O and B stars in clusters or associations are the best spiral tracers. Such stars have ages of only a few million years. Their ultraviolet radiation ionizes the surrounding gas to form H II regions. The distance of an OB cluster or association is derived from known absolute magnitudes and colors of the embedded stars, and from an estimate of the intervening absorption by dust. The arm fragments near the Sun derived from OB associations are shown in Fig. 4.

DIFFERENTIAL GALACTIC ROTATION

Spiral galaxies rotate rapidly, ensuring the stability of the flat disk. Our galaxy shows clear evidence for rotation, with rotational speeds of about 200 km s^{-1} at and beyond the distance of the solar system from the galactic center. Velocities near 200 km s^{-1} are typical of Sc galaxies; Sb galaxies have rotation speeds more like 250 km s^{-1}, and Sa galaxies have speeds above 300 km s^{-1}.

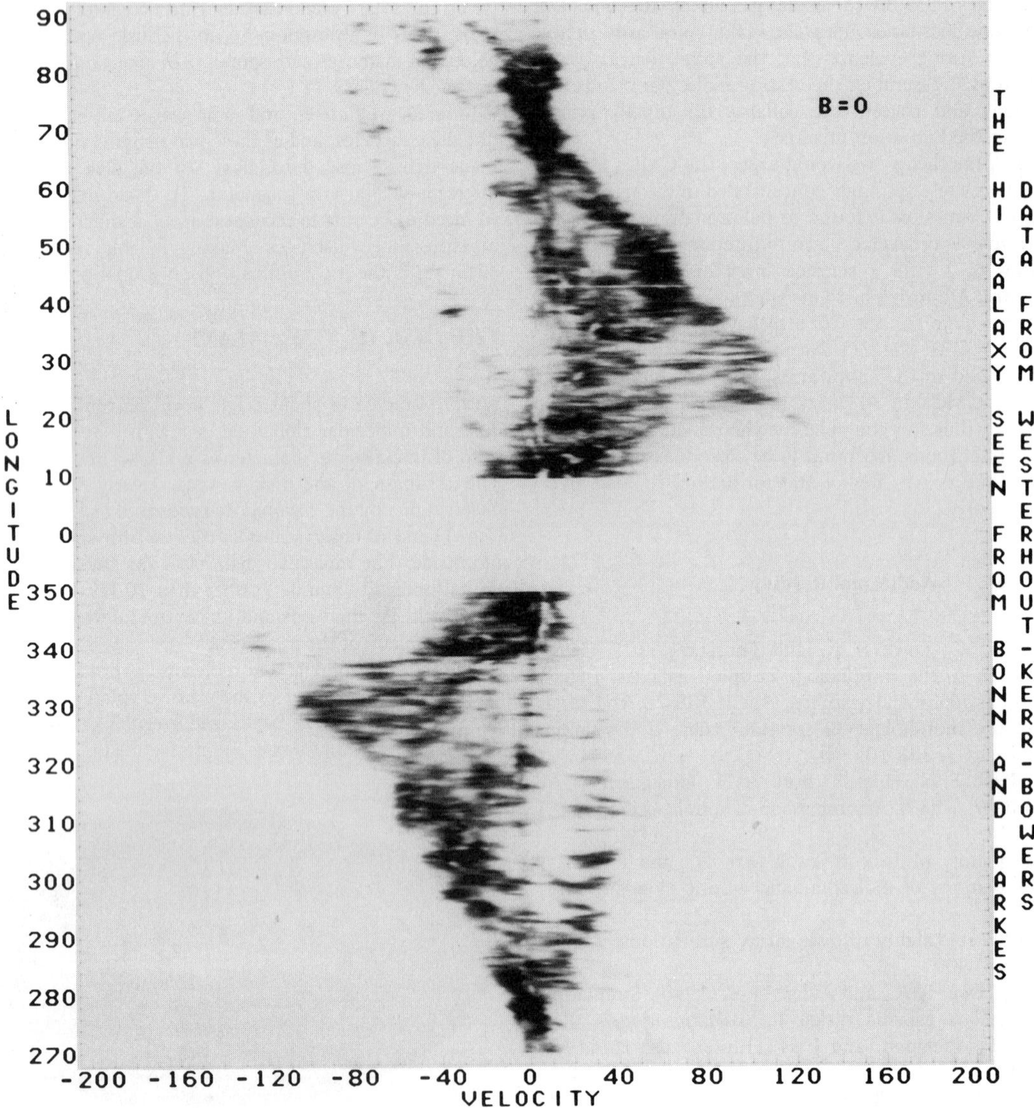

Figure 2. Observed Doppler shift (expressed as a radial velocity) for 21-cm spectral-line emission from atomic hydrogen, plotted as a function of galactic longitude. The brightness of the 21-cm emission is shown on a grey scale.

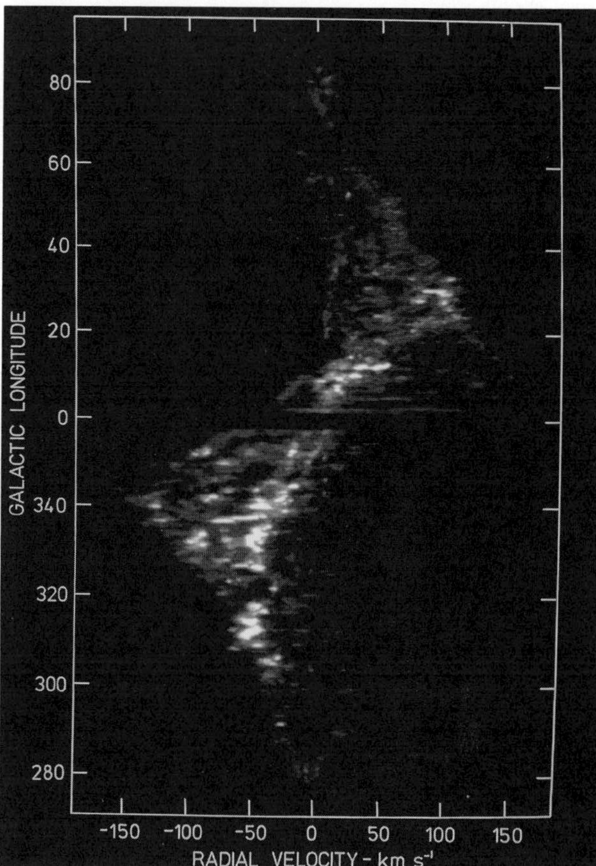

Figure 3. Observed Doppler shift (expressed as a radial velocity) for J = 1–0 carbon monoxide spectral-line emission as a function of galactic longitude. The lighter colors show regions of greater CO brightness.

Atomic hydrogen (H) and carbon monoxide (CO) spectral line emission show a marked change in Doppler shift as the galactic longitude l increases (l is the angle of the line of sight in the plane of the Galaxy measured from the direction of the galactic center). Figures 2 and 3 show the Doppler shifts for H and CO, respectively, as functions of l. For gas moving in a circle of radius R the maximum Doppler shift is observed at a galactic longitude where the line of sight is tangential to the circle. If R_0 is the distance of the solar system from the center, the tangential direction has galactic longitude l given by $\sin l = R/R_0$. The Doppler shift (expressed as a radial velocity) is the vector sum of the motion of the gas at radius R and the orbital motion of the solar system at R_0.

Figures 2 and 3 show a high degree of symmetry between the side of the Galaxy observed from the northern hemisphere ($0° < l < 90°$) and that observed from the southern hemisphere ($270° < l < 360°$). The symmetry shows that to a first approximation the gas is moving in circular orbits with the orbital angular velocity increasing closer to the galactic center, reflecting an increase of stellar density in the central regions.

KINEMATIC DISTANCES

The change of angular velocity with radius provides a means of measuring the normalized radius R/R_0 from the radial velocity of atomic hydrogen, carbon monoxide, or recombination-line emission. This can provide a "kinematic distance" for the gas.

For $R < R_0$, the radial velocity $V(R)$ for motion on a circle of radius R varies as

$$V(R) = V_{\max} \times \frac{R_0}{R} \times \sin l, \qquad (1)$$

where $V_{\max}$ is the maximum radial velocity at the tangent point, observed at galactic longitude $l_T = \sin^{-1}(R/R_0)$.

For $R < R_0$ the line of sight cuts the same radius at two different distances. Often the observed width of the gas layer helps decide whether the gas is at the "near" or "far" point and permits the assignment of an unambiguous kinematic distance. For recombination-line observations an association with OB stars or optical H II regions would indicate the near distance, and absorption of the continuum radiation from the H II region at velocities near $V_{\max}$ would indicate a far distance.

When $R > R_0$, there is no tangent point. To a first approximation the gas is found to be moving at a constant velocity $V_0 \approx 200$ km s^{-1} at all $R > R_0$. Then

$$V(R) = V_0 \left[\frac{R_0}{R} - 1 \right] \sin l \qquad (2)$$

For directions within 10° or so of the galactic center or anticenter the $\sin l$ term in Eqs. (1) or (2) crowds the loci of constant R and kinematic distances cannot be found.

TANGENTIAL DIRECTIONS

From the offset position of the solar system we would expect to see more spiral arm tracers when the line of sight passes along a spiral arm than when it traverses an interarm region.

Observations in the radio continuum at meter wavelengths have located marked jumps in the strength of the integrated synchrotron radiation as the Galaxy is scanned in galactic longitude. The directions of these jumps define edges in the longitudinal distribution of high-energy electrons, and are identified as directions where the telescope looks lengthwise along a spiral arm. These edge directions are found at approximately l = 283°, 310°, 328°, 339°, 31°, and 50°.

The distribution of ionized-hydrogen regions along the galactic plane, based on observations at 6-cm wavelength, shows tangential concentrations near l = 283°, 310°, 328°, 24°, 30°, and 49°.

Carbon monoxide observations show that the warm giant molecular clouds are about five times more abundant within spiral arms than in interarm regions. (The less-massive, colder molecular clouds are distributed homogeneously throughout the zone $0.4 < R/R_0 < 1.0$.) In the northern hemisphere the CO tangent points are at l = 33° and l = 50°. In the southern hemisphere they are at l = 282°, 309°, 326°, and 336°. The so-called "3-kpc expanding arm" is tangential at l = 342°. These tangential points can be seen clearly as peaks in the CO brightness along the tangential velocity locus in Fig. 3. Between these longitudes there are pronounced holes in the CO emissivity, particularly between the l = 282°, l = 309°, and l = 326° tangent points.

A PLAN VIEW OF THE GALAXY

Figure 4 presents an artist's impression of our galaxy as it might be seen by an observer in another galaxy. The crosses near the solar system show the position of OB star clusters and associations, based on photometric distances. We lie on the inside of a spur termed the Orion arm. About 2 kpc further out the OB stars define a portion of an outer arm, the Perseus arm. About 2 kpc closer to the galactic center than the Orion arm there is evidence for a section of an inner arm, the Sagittarius arm.

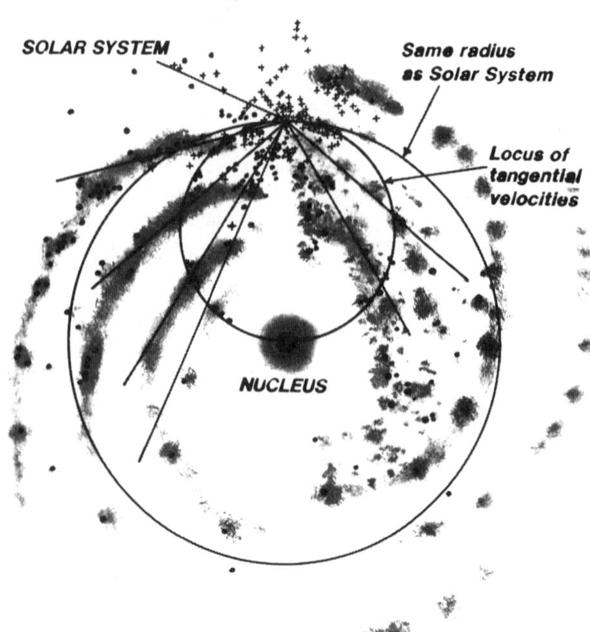

Figure 4. Artist's impression of the plan view of our galaxy as it would be seen from the outside. Crosses mark OB star clusters and associations and dots mark ionized hydrogen regions. The fuzzy bands show the distribution of gas: Inside the circle through the Sun the bands are CO observations of giant molecular clouds; outside this circle the bands show atomic hydrogen clouds. The stellar distances are photometric, whereas the ionized hydrogen, molecular clouds, and atomic hydrogen are located by kinematic distances assuming circular orbits about the galactic center.

The dots show the location of ionized-hydrogen regions with kinematic distances deduced from recombination-line radial velocities. The majority of the points lie on our side of the galactic center, as a result of constraints on sensitivity and angular resolution in surveys at radio wavelengths. However, some very bright regions can be detected on the far side of the Galaxy.

Within the circle $R = R_0$ (the distance of the solar system from the nucleus) the fuzzy bands given an impression of the distribution of giant molecular clouds derived from the 2.6-mm CO line observations in Fig. 3. The distances are again kinematic distances based on the observed tangential velocities [Eq. (1)]. The bands drawn for $R > R_0$ are derived from 21-cm hydrogen-line observations, using a constant-velocity rotation curve [Eq. (2)].

The straight lines radiating from the solar system are the tangential directions discussed in the previous section. From simple geometry these are perpendicular to a radius from the center at points which lie on a circle drawn through the center and the solar system. The tangential directions define a series of benchmarks to which any pattern of spiral structure must be fitted.

The observations sketched in Fig. 4 can be fitted with equiangular spiral arms with pitch angles between 10° and 15°. However, the overall pattern depends on how the various arm fragments are joined together, particularly between the left side (southern-hemisphere observations) and the right side (northern-hemisphere observations). Kinematic distances cannot be defined near $l = 0°$ and $l = 180°$, whereas noncircular motions (see next section) are comparable with or exceed the projected rotational velocities in these directions. Some authors argue for a pitch angle as small as 6° (like an Sa galaxy), whereas others argue for a pitch angle as large as 25° (like an Sc galaxy).

NONCIRCULAR MOTIONS

The artist's impression in Fig. 4 is based on the first approximation of circular orbits derived from the locus of maximum radial velocity in Fig. 2 and 3 (observed for $R < R_0$) and a constant velocity of rotation for $R > R_0$. Observations in certain areas reveal a variety of noncircular motions described as shocks, warps, shears, rolling motions, and streaming motions, as well as a component of random motions of the order of 8 km s^{-1}. The magnitude of the noncircular velocities make kinematic distances uncertain and make it difficult to link features from one side of the Galaxy to the other.

SUMMARY

The basic observational data lead to the following picture. Our galaxy has a flat, thin disk of dust and giant molecular clouds similar to those of Sb or Sc galaxies. The disk is warped about 1° in galactic latitude in its outer parts. Within the disk the gas, young stars, and ionized-hydrogen regions near and beyond the solar system rotate to a first approximation in circular orbits with velocities of about 200 km s^{-1}. This velocity is typical of Sc galaxies. There is a central bulge of evolved stars typical of an Sb galaxy. Spiral arm fragments are shown by various tracers, the separation between arms being about 25% of the distance from the Sun to the center. Kinematic distances based on circular orbits suggest an equiangular spiral structure with pitch angles of 10°–15°, like most Sc galaxies; Sb galaxies are more tightly wound. Shocks, warps, shears, and streaming motions lead to uncertainties in kinematic distances and some authors have suggested pitch angles as low as 6° or as large as 25°.

Additional Reading

Beichman, C. A. (1987). The *IRAS* view of the Galaxy and the solar system. *Ann. Rev. Astron. Ap.* **25** 521.

Blitz, L. (1982). Giant molecular cloud complexes in the Galaxy. *Scientific American* **246** (No. 4) 84.

Bok, B. J. and Bok, P. F. (1981). *The Milky Way*, 5th ed. Harvard University Press, Cambridge, MA.

Burton, W. B. (1976). The morphology of hydrogen and of other tracers in the Galaxy. *Ann. Rev. Astron. Ap.* **14** 275.

Kerr, F. J. (1969). The large-scale distribution of hydrogen in the Galaxy. *Ann. Rev. Astron. Ap.* **7** 39.

Sandage, A. (1961). *The Hubble Atlas of Galaxies.* Carnegie Institution of Washington, Washington, DC.

Scoville, N. and Young, J. S. (1984). Molecular clouds, star formation and galactic structure. *Scientific American* **250** (No. 4) 42.

See also **Galactic Structure, Interstellar Clouds; Galactic Structure, Large Scale; Galaxies, Spiral, Structure.**

Galactic Structure, Stellar Kinematics

Kavan U. Ratnatunga

Stellar kinematics is the study of the orbital motions of stars in our Milky Way Galaxy. The motions of the Sun and other galactic stars are governed by the gravitational potential of the overall distribution of mass in the Galaxy. Therefore, stellar kinematics is also a method for probing the structure and evolution of the Milky Way.

The mean circular motion of stars in the solar neighborhood, belonging to the galactic disk population, defines a coordinate frame called the local standard of rest (LSR). The coordinate frame is realized from the positions and proper motions of 1553 stars in a fundamental catalog called the FK5. The kinematics of any stellar population in the Galaxy are defined by the net streaming motion of that population relative to the LSR and by the dispersion (root-mean-square velocity) with respect to the mean motion.

The circular velocity of the LSR at the Sun, which is 8.5 kpc (kiloparsecs) from the center of the Galaxy, is about 220 km s^{-1}, a value adopted by the International Astronomical Union in 1985. This velocity implies that the LSR takes about 240 million years to complete one orbit about the galactic center. (A velocity of 1 km s^{-1} corresponds to 1.023 pc per million years.) The Sun has a so-called peculiar velocity with respect to the LSR of 17 km s^{-1} towards the star μ Herculis, which has coordinates $l = 53°$, $b = +25°$, where l and b are the galactic longitude and latitude, respectively.

The quantities observed for each star that are studied in stellar kinematics are the line-of-sight velocity and the proper-motion vector. (In the study of galactic kinematics the term line-of-sight velocity is now used instead of its traditional name, radial velocity, to avoid confusion, because "radial" also is used to refer to the direction from the galactic center.) The line-of-sight velocity, taken as positive when directed away from the Sun, is determined by measuring the Doppler shift of spectral lines. The proper motion is the motion of the star in angular coordinates on the sky, determined relative to much more distant stars or galaxies. The transverse velocity of a star, that is, its motion in kilometers per second in the plane perpendicular to the line of sight, is determined from the proper motion when the distance to the star can be estimated. The orientations of stellar proper-motion vectors constitute a useful distance-independent quantity for analysis of stellar kinematics. The observed stellar kinematics also reflect the peculiar motion of the Sun. A precise correction for the orbital and rotational velocities of the Earth at the time of observations is implicit.

EVOLUTION

Stellar velocity dispersion increases slowly with time due to infrequent interactions with stars or large molecular clouds. At birth stars have kinematics typical of the interstellar medium from which they are formed. Some of the youngest stars are observed as moving groups with residual streaming motion. Relaxed older populations have larger velocity dispersion, and are at equilibrium in the radial direction (towards the galactic center) and in the vertical direction (perpendicular to the galactic plane), with a net motion which is only in the azimuthal direction (circular galactic rotation). The mean azimuthal velocity with respect to the LSR is known as asymmetric drift. It is observed to be linearly proportional to the square of the velocity dispersion in the radial direction.

Stars are formed with masses from about 100 to 0.1 $M_{\odot}$. The more massive stars have brighter intrinsic magnitudes and shorter lifetimes. The most luminous stars we observe are therefore much younger and formed in the last few 100 million years. Low mass stars with stellar lifetimes much longer than the current age of the universe are observed from all ages. In view of the wide range of stellar lifetimes, kinematics, intrinsically a function of age, appears like a function of spectral class (stellar temperature and luminosity), which is determined by the mass and state of evolution of a star. As the ages of stars are not directly observable, empirical descriptions of stellar kinematics and galactic structure are often based on spectral class.

The mean metallicity, meaning the collective abundance of elements heavier than hydrogen and helium in the interstellar medium, is continuously enriched by heavier elements created during the evolution and death of stars. Since stars form from the interstellar medium, the metallicity of stars depends on their epochs of formation in the sense that the oldest stars generally have lowest metallicity. Consequently, because older stellar populations have larger velocity dispersions, an observed increase in velocity dispersion is correlated with a decrease in metallicity. The interrelationships between age, kinematics, metallicity, and spatial density distribution of stellar populations give evidence from which we can attempt to derive the evolution and the structure of the Galaxy.

REPRESENTATION

There are a number of distinct stellar populations in the Galaxy. The very youngest stars, with a velocity dispersion typically of order 10 km s^{-1}, are concentrated in the spiral arms of the Galaxy. They have too short a lifetime to leave the neighborhood where they are born. The older stars, with a velocity dispersion typically of order 20 km s^{-1}, form a smooth disk distribution with a radius of order 15 kpc. The oldest and lowest-metallicity stars, with a velocity dispersion typically of order 120 km s^{-1}, form a spheroidal distribution called the halo, with a radius of order 30 kpc.

The velocity distribution is approximately Gaussian in any line of sight. The velocity dispersion σ of stars is found to be anisotropic with a larger dispersion radially (i.e., toward the galactic center) than vertically (perpendicular to the galactic plane), an observation with important dynamical implications. The azimuthal velocity dispersion (toward galactic rotation) is typically intermediate to the radial and vertical dispersions.

The observed anisotropy in velocity dispersion was represented in 1905 by Jacobus Kapteyn as the sum of two isotropic star streams; this is known as the drift hypothesis. Astronomers now adopt Karl Schwarzschild's 1907 representation of the full velocity distribution by a single velocity ellipsoid. The principal axes of the velocity ellipsoid are oriented along the cardinal directions of the galactic coordinate frame, except for the youngest stars. The changes in the velocity dispersions and the orientation of velocity ellipsoid away from the galactic plane, that is, the extent to which it is tilted towards the galactic center, is a current topic of investigation.

Table 1 summarizes the typical kinematic parameter estimates for the different stellar populations. These values depend on which stars are included in each discrete population, and therefore are only illustrative. There has been recent discussion of whether the Galaxy can be separated statistically into discrete kinematic components, or if this division is only a convenient representation for a smooth distribution as could be expected from a continuous process of star formation. The existence of discrete components suggest events over a period short compared to the history of our Galaxy, such as disk formation by a rapid collapse, bursts of star formation, and/or the accretion of a satellite galaxy.

Table 1. Representative Kinematics of Discrete Components

	Age (approximate) (billion years)	*Metallicity (Fe / H) (dex.)*	*Density Scale (Height) (kpc)*	*Velocity Dispersion* σ *(Radial) (km* s^{-1}*)*	σ *(Azimuthal) (km* s^{-1}*)*	σ *(Vertical) (km* s^{-1}*)*	*Mean Drift Velocity (km* s^{-1}*)*
Young disk stars	0.3	+0.1	0.1	10	9	6	1
Thin disk stars	5.0	−0.1	0.2	25	15	10	5
Old disk stars	10.0	−0.5	0.6	60	45	30	35
Halo stars	15.0	−1.5	3.5	140	100	70	190

SAMPLES

The study of kinematics requires the isolation of the separate stellar populations. Most of the stars in the solar neighborhood (i.e., for any spectral class the stars of the brightest apparent magnitude) belong to the galactic thin disk population. Disk kinematics has therefore been extensively investigated.

Study of the halo is more challenging. Only about 1 in 500 stars in the solar neighborhood belong to the halo population; they must be identified spectroscopically by their low metallicity or kinematically by their higher proper motion. However, only some galactic orbits bring halo stars to the solar neighborhood and nearby stars may be a biased sample. Kinematic study of the halo requires fainter (more distant) stars that are situated well outside the disk.

Globular clusters are the best-studied distant members of the galactic halo. Field stars that do not belong to any globular cluster represent another distinct constituent of the halo.

Selection of distant intrinsically bright field halo stars from among the more numerous intrinsically faint nearby disk stars with the same faint apparent magnitude is not simple. With recent advances in automated surveys to fainter magnitudes, more samples of stars can now be located in the outer regions of the galactic halo.

RR Lyrae stars can be identified by their variation in brightness. However, stellar atmospheric motions in these variable stars make them difficult candidates for measurement of the center-of-mass line-of-sight velocity needed for kinematic study.

Field blue horizontal branch (BHB) stars are located by color selection. These stars have the advantage of a relatively narrow range of intrinsic magnitude that leads to better distance estimates. However, since we know that BHB stars are observed mainly in low-metallicity globular clusters, an unknown metallicity-dependent selection bias is probably present.

K-giant stars are representative of the general field population and are found from sky surveys made with an objective prism, which form low-resolution spectra of all stars in the field of view. Fig. 1 shows the line-of-sight velocity and metallicity distribution of K-giants towards a high-latitude field SA127, in a direction opposite to galactic rotation. The broader distribution illustrates the larger velocity dispersion of the more-metal-weak giants, which are up to 25 kpc from the Sun. The large observed mean velocity of 150 km s^{-1} indicates that the halo is at most slowly rotating. The more-metal-strong giants in this sample are on average of 3 kpc from the galactic plane and have a rotation velocity almost as large as the LSR. These giants probably belong to the old disk population, which has also been called a thick disk population.

THEORY

The density distribution of stars moving in the galactic potential depends on their kinematics. Stars with larger vertical velocity dispersion can reach greater distances from the galactic plane and have a more extended density distribution. The density law can be derived from a solution of the Poisson and collisionless Boltzmann equations. (The mathematical theory of stellar dynamics which governs the interrelationships between density and kinematics is beyond the scope of this entry.)

The analysis of stellar kinematics is used to probe the galactic potential. The vertical distribution of a single isothermal population of stars (velocity dispersion independent of distance from the galactic plane) can be used to derive the component of the gravitational force perpendicular to the galactic plane, which in turn gives the potential and the mass density on the galactic plane. Some studies suggest a discrepancy of about a factor of 2 in the disk mass density near the Sun. This appeared to imply that there exists a currently unknown contributor, spatially distributed like the stars of the disk population. More recent determinations have shown the derived mass density to be consistent, within the error of estimate, with the observed total mass of all known constituents. The discrepancy between these studies remains unresolved and controversial. This gives impetus to isolate a pure tracer population with well-defined selection properties, often poorly known in many of the samples that have been studied.

The circular velocity as a function of galactic radius is known as the rotation curve. Inside the solar radius the rotation curve is well determined from radio observations of the interstellar medium.

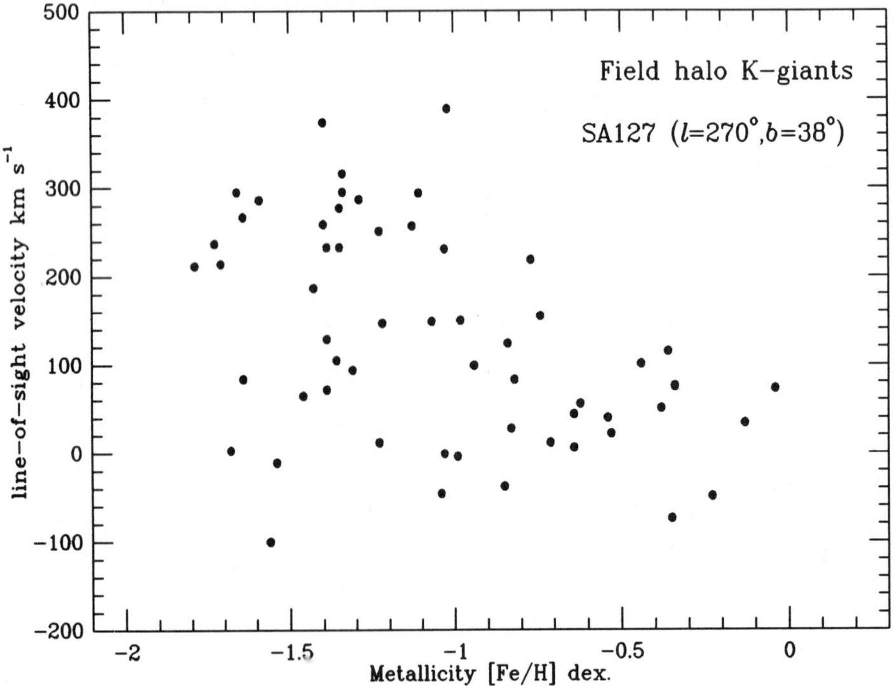

Figure 1. The line-of-sight velocity and metallicity distribution of field K-giants toward a high galactic latitude field SA127.

Stellar kinematics is a useful probe of the rotation curve outside the solar radius, and/or away from the galactic plane. The almost-constant rotation curve that is derived for the Galaxy suggests the presence of an extended dark halo like those that are similarly inferred in external galaxies.

ANALYSIS

The evaluation of the space motion of a star requires data on the line-of-sight velocity, proper motion, galactic coordinates, and distance to the star. For most stars line-of-sight velocity can be measured to an accuracy of better than 1 km s^{-1}, but large catalogs are difficult to compile, since individual spectroscopic observations are required of each star. Proper motions can be derived from observations of changes in the direction coordinates to a star, typically observed over a 25-yr interval either on photographic plates or from transit observations. An accuracy of the order of 1 arc seconds per century is typical. This translates to an error of 47.4 km s^{-1} for a star at 1 kpc from the Sun. Until observations from space provide improved accuracy, proper motion studies are necessarily limited to the solar neighborhood.

The distances to the nearest stars are determined from measurements of their parallaxes. The error in parallax measurement is typically 0.01 arcsecond, which gives an uncertainty larger than 10% for stars over 10 pc from the Sun. Distances to more remote stars are estimated from their spectroscopic and photometric properties calibrated using stars with measured parallax. Distance to a cluster of stars can also be determined from the apparent directional convergence of proper motion vectors of cluster members. However, systematic errors have plagued these distance calibrations, as illustrated by the increase in the adopted "standard distance" to the Hyades cluster from 35 to 45 pc over the 40 years from 1935 to 1975.

Because estimating space velocity requires the combination of many observables with varied error distributions, it is more prudent to study stellar kinematics using the directly observed quantities. Conceptually, rather than taking observations to a theoretical frame of reference by mixing observables, a theoretical Galaxy model is projected to evaluate, within the selection limits, the expected distributions of line-of-sight velocities and proper motions, for comparison with the catalogued data. Complementary maximum-likelihood statistical techniques, which allow for incompleteness and selection biases, are used to derive the kinematic parameters needed to model the Galaxy. These approaches for kinematic analysis have only been possible with the recent growth of computational power.

FUTURE

Recent improvements in computer technology and in electronic detectors for observations at the telescope are giving the study of stellar kinematics a great boost. This follows a roughly 20-yr period of only modest progress after the technological limit of conventional photographic observations was reached in the early 1960s. Astrometric measurements from space represent a new frontier in stellar kinematics. For example, the European Space Agency satellite, *HIPPARCOS*, is now expected to measure the proper motions and the parallaxes of more than 100,000 stars to an accuracy of 0.002 s of arc. The science mission endangered by *HIPPARCOS'* failure to reach its original orbit when launched in August 1989, has had a remarkable recovery. We are truly at the dawn of a new era in the study of galactic structure by stellar kinematics.

Additional Reading

Binney, J. and Tremaine, S. (1987). *Galactic Dynamics*. Princeton University Press, Princeton.

Blaauw, A. and Schmidt, M., eds. (1965). *Galactic Structure*. University of Chicago Press, Chicago.

Mihalas, D. and Binney, J. (1981). *Galactic Astronomy, Structure and Kinematics*. W. H. Freeman and Co., San Francisco.

Smart, W. M. (1968). *Stellar Kinematics*. Longmans Green and Co., London.

Trumpler, R. J. and Weaver, H. F. (1953). *Statistical Astronomy*. University of California Press, Berkeley.

See also **Galactic Structure, Stellar Populations; Galaxy, Dynamical Models; Star Clusters, Globular; Stars, RR Lyrae Type; Stellar Orbits, Galactic.**

Galactic Structure, Stellar Populations

Rosemary F. G. Wyse

A stellar population will here mean a group of stars that is characterized by several properties. These properties are (i) its structure—the spatial distribution; (ii) the motions of its member stars (the kinematics)—both the mean streaming motions, which will essentially be rotation, because we assume the Galaxy is neither expanding nor contracting, and the random motions about these means, (iii) the chemical enrichment of its member stars—both the overall enrichment (metallicity) and the abundances of specific elements, such as oxygen and iron; (iv) the ages of the stars—both the age of the oldest stars and the spread in ages are important; and last (v) the formation mechanism—different physical processes may give rise to similar stellar properties. The Milky Way galaxy offers a unique opportunity to identify stellar populations and determine their characteristic properties, due to our ability to measure three-dimensional quantities in a model-independent way—we can analyze the light from individual stars, unlike studies of external galaxies where the properties of many stars, all those along the line-of-sight, must be analyzed.

Most models of a protogalaxy envisage a mixture of gas and dark matter, with the gas being transformed into stars as the galaxy evolves. The dark matter is assumed present as a means of explaining the flat rotation curves of disk galaxies (constancy of rotational velocity with distance from the center). However, there is no generally accepted theory of galaxy formation. The more of the defining properties of a stellar population listed above that are known, the more one can infer about disk galaxy formation and evolution. The kinematic properties of stars in a galaxy are related to their spatial distribution through the properties of the gravitational potential of the galaxy. Most of the mass of galaxies is believed to be composed of dark matter, that is, objects that are not ordinary stars, and the gravitational potential is in general unknown, to be determined by analysis of the motions of test particles such as stars. The study of galactic structure is strongly motivated by this fact, given the uniqueness of our location within the Milky Way. The defining properties of stellar populations are dependent on the ratios of several timescales—the time for gas to be consumed through star formation; the time for gas to radiate binding energy and dissipate, sinking deeper into the galactic potential well; the collapse time of the initial protogalaxy under its self-gravity. These of course are not the only parameters that determine the nature of a stellar population and they are probably themselves controlled by initial conditions such as the density and angular momentum content of the protogalaxy, but they are the parameters that one can hope to constrain by study of stellar populations and galactic structure.

CONCEPT

The concept of stellar populations was essentially introduced by Walter Baade in 1944, when he succeeded in resolving into individual stars the central regions of our companion spiral galaxy the Andromeda nebula (M31), as well as its elliptical galaxy satellites

M32 and NGC 205. The only information that Baade had available for these stars was their location in the color-magnitude (HR) diagram, since the available technology could not obtain kinematic data for stars in external galaxies and the theory of stellar nucleosynthesis had not yet been established for chemical abundance data to be meaningful. Thus his sole criterion for the assignment of systems of stars to different populations was their HR diagrams. This limited information led to the simplest scheme—Population I stars were those whose HR diagram resembled that of the open clusters in the disk of our galaxy, whereas the red giant branch of Population II stars occupied the same locus in the HR diagram as the galactic globular cluster giants. Baade assigned those stars that lie outside the thin disks of spiral galaxies—in globular clusters, elliptical galaxies, and the stellar halo components of disk galaxies, to Population II.

The kinematics of disk stars in our galaxy had earlier led Bertil Lindblad to model the disk as consisting of a superposition of many subsystems with different mean rotational velocities and random velocities, the subsystems becoming rounder in shape as the mean rotation velocity decreased. These ideas were combined with the newly understood concept of chemical evolution at the 1957 Vatican Symposium, where a scheme of stellar populations was agreed upon that included as discrete classes, between those of Baade, Intermediate Population II and Disk Population.

The term Population II became commonly understood to refer to old, metal-poor stars on low-angular-momentum orbits. This followed from the seminal work of Olin Eggen, Donald Lynden-Bell, and Allan Sandage in 1962, whose detailed observations of the galactic subdwarf stars, observed passing through the solar neighborhood at high velocity, revealed them to be metal poor and members of a system with little or no net rotation, but with large amplitude random motions. This work provided evidence for the existence of a smooth relationship between kinematics and chemistry from thin disk stars through to the extreme subdwarfs. Assuming that the chemical enrichment of a star is a monotonic and universal function of time alone, as thought in the 1960s, implies that the properties of stellar populations are fully determined by the epoch at which those stars formed. The interpretation—and indeed reality—of this correlation has been debated over the years.

THREE POPULATIONS?

Modern data have revealed, perhaps not surprisingly, that the Galaxy is a complicated system and much current research aims to understand it. We now know that the different stellar populations are not a one-parameter sequence, as had been suggested by the smooth correlations that were evident between kinematics and chemistry, and between chemistry and time. Scatter and discontinuities in such relationships may be indicative of such exotic phenomena as mergers between our galaxy and another.

There are most probably three major stellar populations in the Galaxy. This is a controversial statement. In order of decreasing total mass, these stellar populations are the thin disk, the thick disk, and the stellar halo. The characteristic properties of these components are as follows.

Thin Disk

The radial light distribution follows an exponential law, so that the intensity has fallen from its central value by a factor of e by ~ 4 kpc from the galactic center. The scale height of the thin disk is $\lesssim 350$ pc (for the oldest stars of this component, which contain most of the mass, and less for the youngest stars) resulting in a very flattened—hence, thin—spatial distribution. The total blue-band luminosity is $\sim 10^{10}$ solar luminosities ($L_\odot$). This component contains about 10% by mass of gas, with the gas fraction being an increasing function of galactocentric radius, equalling about $\frac{1}{3}$ at the solar distance.

The Sun is a typical youngish thin-disk star and moves in an approximately circular orbit, with rotational velocity around the Galaxy of ~ 220 km s^{-1}. The Sun is $\sim 5\times10^9$ yr old, and there exist stars being born today; the age of the oldest thin disk star is of great interest and (thus!) is very controversial; an age of 12×10^9 yr has recently been obtained for a cluster of stars in the thin disk. The older thin-disk stars have higher random motions, lower rotational velocities, and lower metallicities than the younger stars, on average. The existence of inherent scatter in the age-metallicity relationship for disk stars has only recently been unequivocally established, although its presence was suspected in earlier samples; scatter is expected in many theories of disk evolution and its amplitude is obviously a crucial constraint. The presence of radial metallicity gradients, with the outer regions of the disk being of lower metallicity than the solar neighborhood at the same age, must all be predicted by successful theories.

Thick Disk

This stellar component was first detected in star counts towards the south galactic pole, subsequent to the discovery of thick disks in external spiral galaxies. The existence of the thick disk was controversial in the early 1980s, when only star-count data were available. Spectroscopic data have provided indisputable evidence, but the origins and evolutionary status of the thick disk remain subject to debate.

The thick disk has a scale height of ~ 1 kpc, at least at the solar radius, and of the order of 2% of local stars belong to this component (the other ~ 98% being members of the thin disk, with the stellar halo accounting for merely a fraction of a percent). The total luminosity of the thick disk can at present best be constrained by analogy with external galaxies. The implications are that the thick disk dominates over the stellar halo out to many kiloparsecs away from the plane, and probably has a total luminosity several times that of the metal-poor stellar halo, or a few $\times 10^9$ $L_\odot$. The detailed values of the descriptive parameters remain poorly determined, however, and await the large statistical spectral surveys currently being completed.

The corresponding vertical velocity dispersion for a scale height of ~ 1 kpc, ~ 45 km s^{-1}, has now been observed in many samples. Typical thick disk stars appear to be on high-angular-momentum orbits lagging behind the Sun by only ~ 40 km s^{-1}. The thick disk is apparently kinematically distinct from the subdwarf system to an adequate approximation. The thick disk is probably kinematically distinct from the galactic old thin disk, though the data remain inadequate for robust conclusions. This is obviously an extremely important point, with major ramifications for the formation mechanism of the thick disk.

The mean metallicity of thick disk stars, which dominate 1–2 kpc above the galactic plane, is about one-quarter of the solar metallicity. Provided they are present in sufficient numbers, these stars have a major effect on the chemical evolution of the thin disk. In particular, they can provide a simple, self-consistent solution to the G-dwarf problem, which is the observed paucity of metal-poor G-dwarfs in the solar neighborhood, relative to the predictions of the most naive model of chemical evolution. Since G-dwarfs live for of order of the age of the Galaxy, all G-dwarfs ever born should still be around today, and counts of G-dwarfs in a representative volume will constrain the integrated chemical evolution of that volume. One of the complications of the G-dwarf problem is that it is easy to think of solutions by adopting more sophisticated models. The discovery of the thick disk allows one to retain simplicity by the identification of the thick disk as the reservoir for the missing metal-poor stars. These stars were not found in the earlier small surveys of stars in the solar neighborhood, most probably because of their relatively large scale height and low local space-density normalization.

At least the metal-poor tail of the thick disk is comparable in age to the younger members of globular cluster system, $\sim 14\times10^9$ yr, leaving little room for a hiatus in star formation between the formation of the stellar halo and the formation of a disk, as envisaged in some models of galaxy formation. The age spread is unknown at present, although some results imply that there exist thick disk stars as young as the oldest thin disk stars.

Stellar Halo

We assume that the stellar halo is represented in the solar neighborhood by the high-velocity, metal-poor subdwarfs, the classical (Extreme) Population II of Baade (there are metal-rich stars in the nuclear bulge which may pose a problem). As mentioned above, only $\sim 0.1\%$ of local stars belong to the stellar halo, which has total luminosity $\sim 10^9\ L_\odot$. This component apparently follows a de Vaucouleurs light profile with half of its light being contained within a radius of 3 kpc. The central metal-rich bulge, as detected in the infrared, is more centrally concentrated.

The shape of the stellar halo is important for its implications for the early stages of galaxy collapse and star formation, the interpretation of the kinematics of high-velocity stars, and the shape of the underlying dark matter that generates the gravitational potential in which these stars move. The kinematics of the subdwarfs leads to the expectation of a flattened shape, although the classical Population II was round. A recent determination of the shape of the stellar halo from stars counts has reconciled expectation and reality, however, deriving an axis ratio of approximately 2:1.

The mean metallicity of the subdwarfs is well established at around $\frac{1}{30}$ of the solar value. Several elements are enhanced, when normalized to iron and compared with the solar ratio, with, in particular, oxygen being overabundant by a factor of ~ 3. Analysis of this element ratio, as a function of iron abundance, can lead to constraints on the time scale of stellar halo formation, with the result that a metallicity of $\frac{1}{10}$ solar was reached in the relatively short time of $\sim 10^9$ yr. Indeed, recent analyses of certain unusual products of stellar nucleosynthesis have allowed an estimate that it only took $\sim 10^8$ yr to enrich up to $\frac{1}{100}$ of solar metallicity. These results imply rapid star formation—remember that the freefall collapse time of a typical protogalaxy is $\sim 3\times 10^8$ yr, which is a lower limit to the possible collapse time of the Galaxy, if it formed in a coherent collapse. A rapid collapse was proposed by Eggen, Lynden-Bell, and Sandage, fully 25 years prior to the inferences from studies of the elemental abundances.

Additional Reading

Eggen, O. J., Lynden-Bell, D., and Sandage, A. (1962). Evidence from the motions of old stars that the Galaxy collapsed. *Ap. J.* **136** 748.

Gilmore, G., Wyse, R. F. G., and Kuijken, K. (1989). Kinematics, chemistry and structure of the Galaxy. *Ann. Rev. Astron. Ap.* **27** 555.

Norman, C. A., Renzini, A., and Tosi, M., eds. (1986). *Stellar Populations.* Cambridge University Press, Cambridge.

O'Connell, D. J. K., ed. (1958). *Stellar Populations* (The Vatican Symposium 1957). North-Holland, Amsterdam.

See also **Galactic Structure, Large Scale; Galaxies, Formation; Galaxy, Dynamical Models.**

Galaxies, Barred Spiral

Stephen T. Gottesman and James H. Hunter, Jr.

EARLY OBSERVATIONS

Nonstellar nebulous objects, some of which are visible to the naked eye in the night sky, have been recorded for several centuries. In the 1840s and 1850s the Earl of Rosse, using his great 6-ft telescope, found that several had a spiral form. In 1918 Heber D. Curtis drew attention to a second type of spiral: "its main characteristic is a band of matter extending diametrically across the nucleus and inner parts of the spiral." Owing to their shape, he called these ϕ-type spirals. Edwin P. Hubble, in his 1926 study of the properties of galaxies, called these systems barred spirals, which he described as having the form of a θ. He found that these were less abundant than normal spirals, but (with the exception of their bar) they appeared to have similar properties of the normal systems, a view supported by the photometric studies of Erik B. Holmberg some 30 years later.

Initially, Hubble saw these as distinct but parallel families with a thin population of mixed types lying between the two. Barred spirals are observed to possess two main forms: those in which the arms are tangent to an external ring (r) at which the bar ends, and those for which the arms begin directly from the ends of the bar (s). The same phenomena are present, though less obvious, in normal spirals. Thus, by the late 1950s it was clear that Hubble's ordering was too coarse and that instead one should consider a classification volume incorporating the main families, transition groups, and the r and s varieties. Gerard de Vacouleurs' synthesis accomplished this end. In Fig. 1, a cross section of this volume is shown, illustrating the transition between the normal (A) and the barred (B) families, and encompassing the r and s varieties. Owing to the continuum of forms, de Vaucouleurs suggested there was nothing abnormal about the barred species and proposed that A-type systems be called ordinary spirals. Furthermore, de Vaucouleurs found that ordinary and barred systems were about equally abundant and accounted for almost 50% of the easily classifiable (bright) systems. However, if one adds to the barred family systems with oval distortions, barred galaxies are the dominant shape. Indeed, it has been suggested that the bar-forming tendency is a more significant characteristic of disk systems than the tendency to form spirals.

PHYSICAL DESCRIPTION

Allan Sandage and others characterize several zones when describing barred spirals: (1) the nuclear bulge, (2) a region concentric with the nucleus that has the shape of a convex lens seen in projection (located between the bulge and disk; it is most prominent in the early-type barred systems), (3) a luminous ring of matter that, when present, lies at the outer edge of the lens and, as the r subvariety progresses, develops into the spiral arms (in some cases an outer ring is seen as well, aligned either perpendicular or parallel to the bar axis), and (4) the outer envelope of the galaxy. In addition, there is always present the defining characteristic, a bar that extends across the center and the lens (if present) terminating at its edge or at the innermost regions of the spiral arms. The bar phenomenon is associated with disk systems; barred elliptical galaxies do not exist. Bars are associated with high-angular-momentum structures and rotate much faster than do ellipticals. Furthermore, bars appear to rotate in the plane of the disk, in the same sense as galactic rotation, and as rigid bodies. However, the rate at which they tumble (the pattern speed) is uncertain.

The existence of bars correlates strongly with the presence of grand design spiral structure in disk systems. Bars, as they are not axially symmetric, appear to be an extremely effective mechanism for driving density waves (both gaseous and stellar), particularly if the luminosity profile of the bar is flat rather than exponential. Several high-resolution studies of barred systems, which are rich in atomic hydrogen, have been made in order to investigate the dynamics of these galaxies. Hydrogen is a ubiquitous component of the interstellar medium and its distribution correlates well with spiral arms. Therefore, it is a medium that reveals the structure and kinematics of disk systems against which dynamical models can be tested. In the inner parts of these galaxies, the isopleths of radial velocity are twisted and skew the apparent line of nodes. In these regions, the bar effects the gas orbits most strongly. Streaming motions also are seen in association with the spiral arms. It is assumed that the arms are driven by the bars, and the direction of radial motions is sensitive to the radius at which the pattern corotates with the disk. Inside the corotation radius, radial motions

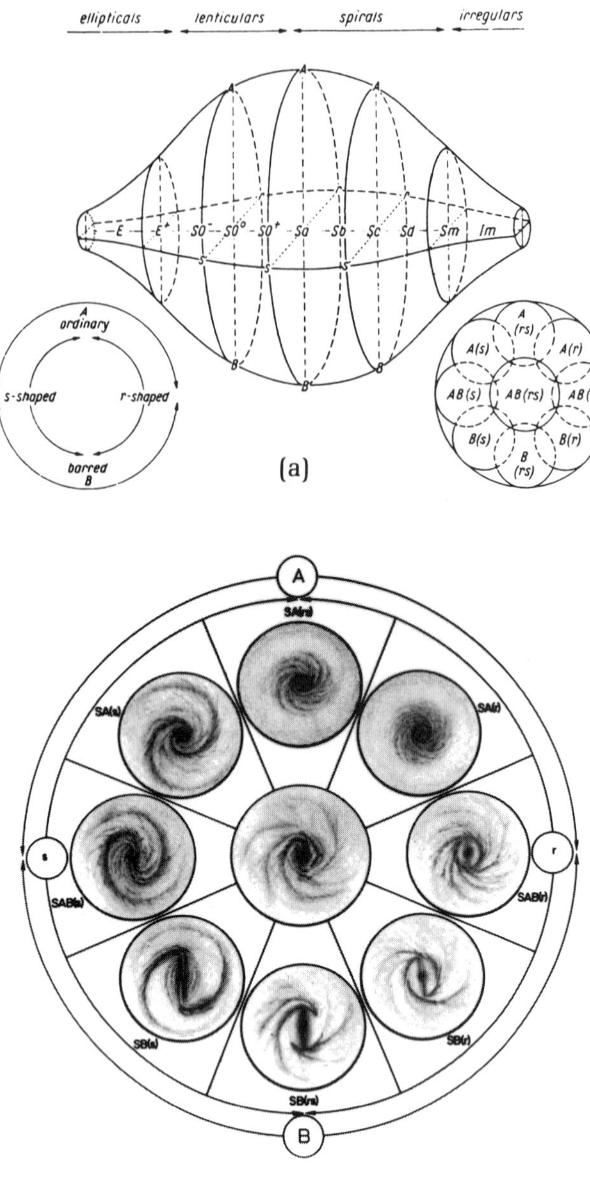

Figure 1. (*a*) The classification volume of de Vaucouleurs showing the relationships between the basic forms. (*b*) A cross section of the volume at about type Sb. Note especially the continuity of the varieties from r through s of the types A and B.

will be inwards, whereas beyond this critical radius the gas gains momentum at the expense of the pattern. Structural features may help locate this domain; many bars show narrow and strategically located dust lanes, frequently along their leading edges. Often at or near the bar–spiral-arm transition region these dust lanes abruptly shift to the trailing edge of the arm. The dust lanes appear to delineate shock regions and the abrupt change from leading to trailing phenomena may indicate passage across a resonance, such as corotation. Also, the location of inner rings may indicate the radius of corotation; for they too are expected to occur at or near resonances. Therefore, there is observational evidence that corotation happens at or just beyond the end of the bar.

Unlike other triaxial phenomena in galaxies, bars have very large axial ratios, which can approach 5:1. However, the luminosity of the bar is typically only a few percent of the luminosity of the disk, though it can be as large as 20%. Nonetheless, the distribution of light is not smooth. As we have noted, dust is seen often in the form of narrow lanes; in other cases it may be seen as patches symmetrically located at the ends of the bars. Also, the cusp region, at the ends of the bar, is often the locus of significant activity, ionized hydrogen or, particularly for early-type systems, lobes in the optical (stellar) emission. Qualitatively, these phenomena can be understood in terms of the prominent families of stellar orbits (and, by association, gas streamlines); where they cross, shocks will be generated in the gas. These appear to be regions of observed dust and/or enhanced star formation. In gas-free systems, the density of stellar orbits generates the luminosity distribution. Below, we will enlarge upon these points.

Many, but not all, of the observed galaxies show either an absence of H I in the bar region, or a very severe reduction in the surface density of the gas. Also, many barred systems show abnormal amounts of stellar activity in their nuclei. A natural interpretation is that the gas in the central regions has lost angular momentum to the bar (the bar is inside corotation) and has been swept into the nucleus, where bursts of star formation occur. Often, CO emission is observed, confirming the association of a dense interstellar component with star formation. However, other galaxies that are centrally evacuated do not show these phenomena; perhaps nuclear star bursts are episodic. In contrast, some of these galaxies have gaseous bars. The critical difference appears to be revealed by dynamical models. If models can be constructed successfully employing a material bar, evacuation occurs. If an oval distortion of the disk will suffice, gaseous bars are formed. Therefore, it is extremely important that multicolor photometry be used to place constraints upon the mass distribution of the bar.

ORBITAL CONSIDERATIONS: PARTICLE DYNAMICS

The morphology of barred galaxies can be understood qualitatively from the (linear) theory of small oscillations, superimposed upon circular orbits (their guiding centers). Because most of the mass of a galactic disk resides in its stars, the disk consists principally of stars, orbiting about its center. In most respects, the ubiquitous interstellar gas is not of great dynamical importance, because it contributes but a small fraction of the global gravitational potential.

In the absence of a bar, small disturbances of a star, moving initially in a circular orbit, oscillate about its circular-orbit guiding center at the epicyclic frequency in the plane of the disk, and with a more rapid frequency normal to the plane. Both frequencies can be calculated at each radial location in the disk plane from the (known) gravitational disk potential. The epicyclic frequency distribution is an important parameter in determining the morphology of barred spiral galaxies. If the angular velocity of stars orbiting at radius r in the plane of the background and axisymmetric disk is defined by $\Omega(r)$, the epicyclic frequency is defined as

$$\kappa(r) = 2\Omega(r)\left[1 + \frac{1}{2}\frac{d\ln\Omega(r)}{d\ln r}\right]^{1/2}$$

For a Keplerian potential, in which $\ln\Omega \propto -1.5\ln r$, $\kappa = \Omega$. In disk galaxies, Ω usually does not diminish this steeply with radius; hence, usually $\kappa(r) > \Omega(r)$. (For the Sun, $\kappa \sim 1.3\Omega$, meaning that a star undergoes ~ 1.3 radial oscillations per orbital period about the galactic center.)

Bar formation occurs naturally in rotating systems undergoing gravitational collapse. This prediction, initially made from linear theory, has been confirmed by a wide range of numerical experiments. An intriguing characteristic of the numerical bars is that, in addition to their being very robust figures (difficult to destroy), the individual stellar orbits behave collectively, so that each bar rotates at a particular angular rate. Therefore, not surprisingly, a second quantity of fundamental importance to the morphology of barred galaxies is the angular velocity, or pattern speed Ω_p of the bar figure as it rotates rigidly in the central portion of the disk, thereby disturbing the nearly circular orbits of the stars.

Of particular interest are those radial distances from the center of a disk where resonances occur between integer multiples n of the angular velocities of stars relative to the bar pattern speed and the epicyclic frequency at that radius, namely,

$$\Omega(r) = \Omega_p \pm \frac{\kappa(r)}{n}$$

The values of r where these resonances occur are called Lindblad radii, in honor of the Swedish astronomer Bertil Lindblad (1895–1965). The (+) sign denotes an inner resonance of order n, while the corresponding outer resonances bears the (−) sign. Stars orbiting near the inner resonances move more rapidly than (and therefore overtake) the bar potential, whereas those near outer resonances orbit more slowly than the bar pattern. The most important of the resonances occur when $n = 2$ (because most galaxies are two-armed systems), whence $\Omega(r) = \Omega_p \pm \kappa(r)/2$. These particular resonances conventionally are called the inner and outer Lindblad resonances, or ILR and OLR. Higher-order resonances usually are designated as inner (+) and outer (−) $n/1$ resonances. In addition to the ILR and OLR, a third resonance of fundamental importance in barred galaxies is the corotation resonance, defined by

$$\Omega(r) = \Omega_p$$

Stars orbiting at the corotation radius have the same angular velocity as the bar.

Among the important predictions of linear theory are that the departures from circular motions become greatest near the principal resonances, and that orbits change their alignments by 90° at the ILR, the OLR, and at the corotation resonance. An interesting class of models is one in which no ILRs exist. (The shape of rotation curves, derived from high-resolution radio observations of neutral hydrogen in selected barred spiral galaxies, show that no ILRs exist in these galaxies if their corotation radii are $\lesssim 1.25$ bar radii.) In the absence of ILRs, linear theory shows that the stellar orbits are aligned parallel to, and therefore reinforce, the bar within the corotation radius, and that the orbits are elongated perpendicular to the bar between corotation and the OLR. Thus, the latter class of orbits are aligned in a sense that would reinforce spiral structure between corotation and the OLR.

The gravitational potentials of generic bars often are represented by expansions of the form $\Sigma_n A_n(r)\cos(n\theta)$, where θ is an angular coordinate in the disk plane, n is an even integer, and the coefficients $A_n(r)$ are the multipole amplitudes, which diminish rapidly with r. Usually only a few of the lowest-order terms are included (e.g., $n = 2$ and $n = 4$), with the 2θ component being the dominant one. In addition to the bar, a spiral potential may be included between corotation and the OLR.

Extensive numerical calculations have been carried out on two-dimensional models in which the motions are restricted to the disk plane. Particular generic characteristics that have been unveiled by these nonlinear studies are as follows.

(1) When a strong bar is present, there are two main families of stable orbits elongated along the bar that have quite different shapes. The inner family of orbits, which are oval shaped and sometimes have loops at their ends, are found between the central region and the inner 4/1 resonance. However, between the inner 4/1 resonance and corotation, the orbits are shaped like parallelograms, with the longer sides parallel to the bar. In some models, the 4/1 family of parallelogram-like orbits is dominant, and gives a boxy appearance to the bar, whereas this family is of little importance in other models.

(2) Orbits between the outer 4/1 resonance and the OLR support a strong, imposed spiral potential. Between corotation and the outer 4/1 resonance, the orbits support the spiral only if it is weak. However, in strong bars and spirals, most orbits are stochastic in this region. Consequently, a strong spiral cannot be reinforced by stellar orbits between corotation and the outer 4/1 resonance. As illustrated in Fig. 2, barred spiral galaxies are observed to have both oval- and parallelogram-shaped bars. Thus, it seems clear that each of the principal families of stable, periodic orbits within corotation can dictate the appearances of bars in real galaxies. Patches of dust, ionized gas and stars, situated at the ends of some bars, plausibly may be associated with loops at the ends of certain oval orbits. Self-consistent model bars, as well as nearly self-consistent, two-dimensional models of barred spiral galaxies, have been constructed from orbit theory. Families of stable, periodic orbits form the backbone structures of these models.

(3) For systems with very strong bars, stochastic orbits may become important. Stars on such orbits fill a zone which closely resembles the lenses that are seen in many early-type galaxies.

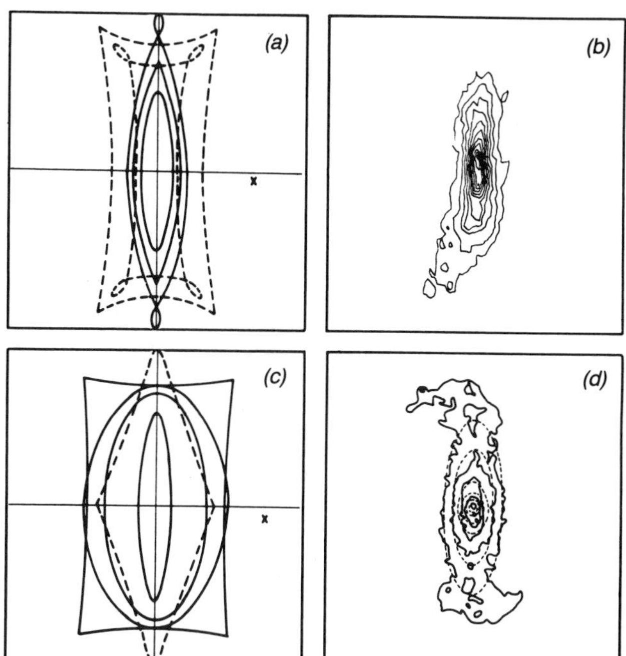

Figure 2. (*a*) and (*c*) illustrate parallelogram and oval orbits, respectively, in a strong, bar potential. (*b*) and (*d*) show I-band photometry of the box- and oval-shaped bars in NGC 4731 and NGC 1300, respectively [*(a) and (c) are from G. Contopoulos, 1988, reproduced with permission from* Annals of the New York Academy of Sciences ***536*** *6. (b) is from observations by E. Costa and G. Fitzgibbons. (d) is from M. England, 1989, reproduced with permission from* The Astrophysical Journal ***344*** *674.*]

GAS DYNAMICS

Gas flows in barred spiral galaxies have been studied extensively, and many gaseous features observed in barred galaxies are replicated by theoretical models. For example, the dust lanes observed on the leading edges of the bars in galaxies such as NGC 1300 may be associated with shock fronts, which appear at those locations in gas dynamical models with strong bars. However, bars alone cannot excite the tightly wound, spiral gaseous arms that are observed in many barred galaxies. Thus, it is necessary to invoke an additional, nonaxisymmetric forcing mechanism, such as an imposed stellar spiral potential, to explain the gas morphology in these systems. Models of the gas flows in selected barred spirals, which have been observed in neutral hydrogen at high resolution, show that their bars must have pattern speeds which place corotation only slightly beyond the ends of the bars. Although it is a minor mass component, interstellar gas may play an important role in the structure of strongly barred galaxies. Since gas streams cannot cross, gas flows may smoothly connect the stellar spiral structure in the stochastic region between corotation and the outer 4/1 resonance. This conjecture has been borne out by preliminary calculations.

Although much progress has been made in modeling barred spiral galaxies, fundamental questions about these systems remain unanswered. In particular, we do not yet understand in detail how collapsing, rotating systems of gas and stars evolve into the configurations that we observe, which include both disks and bars.

Additional Reading

Contopoulos, G. and Grosbol, P. (1989). Orbits in barred galaxies. *Astron. Ap. Rev.* **1** 261.

Hunter, J. H., Jr. (1990). Model gas flows in selected barred spiral galaxies. In *Galactic Models* (Proceedings of the Fourth Florida Workshop in Nonlinear Astronomy), J. R. Buchler, S. T. Gottesman, and J. H. Hunter, Jr., eds. *Ann. N.Y. Acad. Sci.* **596** 174.

Kormandy, J. (1982). Observations of galaxy structure and dynamics. In *Morphology and Dynamics of Galaxies*, L. Martinet and M. Mayor, eds. Geneva Observatory, Sauverny, Switzerland, p. 115.

Sandage, A. (1961). *The Hubble Atlas of Galaxies*. Carnegie Institution of Washington, Washington, DC.

Sandage, A. (1975). Classification and stellar content of galaxies obtained from direct photography. In *Galaxies and the Universe*, A. Sandage, M. Sandage, and J. Kristian, eds. University of Chicago Press, Chicago, p. 1.

See also **Galaxies, Disk, Evolution; Galaxies, Spiral, Structure; Stellar Orbits, Galactic.**

Galaxies, Binary and Multiple, Interactions

William C. Keel

The large size of many galaxies in comparison to their separation means that tidal effects during close passages can significantly affect the galaxies' morphology, star formation, and gas content. An important fraction of luminous galaxies is found in binary systems or small groups, the properties of which can be used to probe the masses and dynamical history of the constituent galaxies. Close encounters are highly inelastic; galaxies can merge with one another in relatively short times. This process has been associated with the formation of elliptical galaxies from collisions of disk systems and with the formation of luminous galaxies from smaller precursors. There is direct observational support for links between interactions and such energetic and short-lived phenomena as infrared-bright galaxies, starbursts, and active nuclei of galaxies.

BINARY GALAXIES

Early studies of the projected distribution of galaxies on the sky showed a clear excess of close pairs above the number to be expected for a population randomly distributed in space. Erik Holmberg at Lund Observatory in Sweden compiled the first extensive catalog of such pairs, and noted correlations between several properties of galaxies in pairs. Such correlations suggest evolutionary connections between the galaxies, resulting either from common environment during galaxy formation or from mutual influences during the history of the galaxies. Specifically, Holmberg found that galaxies in a pair tend to have similar morphological (Hubble) types and, to a degree larger than expected from the type correlation, to have similar optical colors (the Holmberg effect). Later analysis of the binary-galaxy population, incorporating radial-velocity information, has confirmed that these are bound systems, and not simply chance encounters of unrelated galaxies. There are too few pairs with evidence of physical association and the large velocity differences to be expected from chance encounters of unbound systems; these flyby systems are found mainly in clusters, where the high galaxy density and velocity dispersion both favor chance encounters.

The dynamical properties of binary galaxies can be used to trace the extended mass distribution in galaxy halos to very large radii, much as is done with rotation curves. Attempts to do so in detail have been hampered by the difficulty of isolating pure samples of physical pairs of galaxies. The most important source of interlopers in statistically selected catalogs is not foreground or background systems, which can be weeded out using redshift data. Rather, it is the large number of galaxies in the vicinity of a given galaxy (on the scale of a group, for example), which can appear, in projection, artificially close to it. Statistics of the clustering of galaxies (such as the two-point correlation function) do not clearly distinguish pairs from the general clustering trend among galaxies; so there is no physical scale at which pairs can be distinguished from two members of a group that happen to lie (or appear to lie) unusually close together. High-reliability subsets of the binary-galaxy population can be identified by using very stringent selection criteria or requiring evidence of tidal disturbance, but these techniques bias the sample toward various subsets of the whole population of binaries and are not satisfactory for overall statistical use.

INTERACTIONS AND MERGERS

Because the mass in galaxies is widely distributed, some of a pair's orbital energy can be converted into the energy of orbits of stars and gas clouds within the galaxies. This leads to changes in the morphology of galaxies during and after a close passage due to tidal disturbances (see Fig. 1), and to decay of the mutual orbit followed by actual merger of the galaxies.

Morphological disturbances are most pronounced in disk galaxies; on the other hand, the larger random stellar velocities at each point in an elliptical smear tidal features into broad fan-like structures. The dynamically cold situation in disks, with much smaller velocity dispersion, allows production of coherent structures on very large scales, in some cases greater than the initial size of the disk. These are in turn most pronounced after direct (prograde), rather than retrograde, encounters, because the companion galaxy moves slowly in the rotating reference frame of the disk and so builds up a large total tidal impulse in part of the affected disk.

Detailed numerical simulation of individual interacting systems, matching the overall forms and radial-velocity fields of the galaxies, has proven quite successful in understanding the spectacular bridges and tails produced in close tidal encounters. The interacting systems so modeled include ellipticals as well as disk systems; in these cases the recent orbital history of each pair is well understood. In principle, comparisons of observations and simulations could yield estimates of the amount of mass in an extensive unseen halo; this test is not yet sensitive enough to make it a useful probe of halo matter.

There has been more success in exploiting the differing dynamics of stars in disks, and in elliptical galaxies or bulges in tracing relatively faint remnants of accreted companion galaxies that once had disks. These remnants are often seen as sets of ripples superimposed on the smooth background light of the more massive, relatively undisturbed system. Tidal debris around disk galaxies is best seen in polar rings, which represent material accreted from a companion that remains in near-polar orbit over the existing disk. Such orbits decay very slowly compared to those nearer the disk plane, and the material is not spread widely by the differential precession that would affect matter in an annulus close to the disk

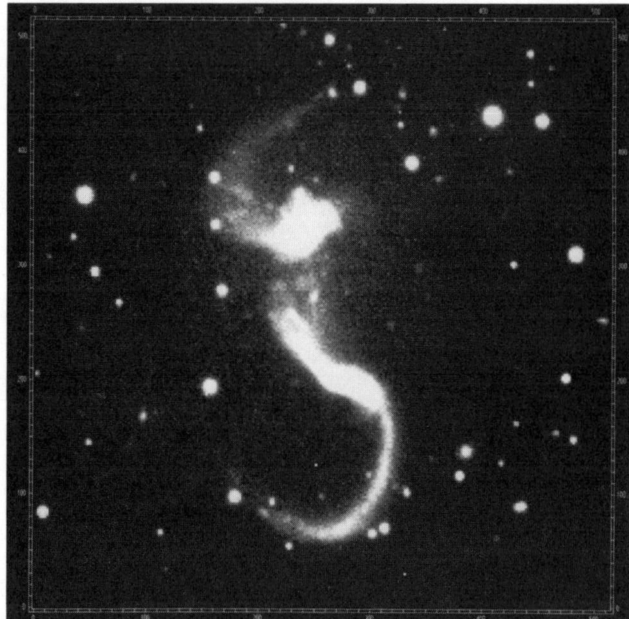

Figure 1. Tidal distortions in interacting spirals are well developed in the southern system MCG −3–25–19 = ESO 0942 − 19, shown in a red-light CCD image from the ESO 3.6-m telescope. The very thin tidal tails are indicative of the response of a stellar disk with small local velocity dispersion, and may remain visible after the main parts of such galaxies have merged.

plane. These structures are best seen in S0 systems, in which there is little interstellar material to interfere with the motion of the accreted matter.

The transfer of energy responsible for tidal distortion in a galaxy's stellar distribution can also have a dramatic impact on the orbital evolution of a galaxy pair. In addition, an effect known as dynamical friction can accelerate the decay of an orbit when one galaxy approaches within the stellar distribution of the other. This is the statistical result of gravitational scattering of stars by the intruding galaxy. The whole aggregate of collisions, even if they are individually elastic in the intruder's reference frame, saps the intruder's orbital energy, because of the asymmetry in stellar distribution produced in the wake of the intruder. This loss of energy is faster for larger intruder masses and at smaller distances from the larger galaxy's core. For an initially grazing encounter of two marginally bound galaxies, a merger can result in only a few initial orbital periods. The past history of the interacting-galaxy population is therefore very difficult to ascertain, because there should have been many more systems that have already merged than we now see as pairs for any plausible primordial distribution of orbital parameters.

The identity of the merged remnants is important in understanding the dynamical history and formation of galaxies in general. Broad statistical considerations suggest that most present luminous galaxies have undergone at least one merger of comparable systems at some time. Mergers with companions of much lower mass may not leave strong traces, but a merger of near equals should produce a dynamically distinct result. A substantial body of work suggests that the product of a merger of disk systems would look much like an elliptical galaxy; the burst of star formation often triggered during such an event can sweep most of the interstellar matter out of the remnant. Some ellipticals in fact show evidence, such as a core in retrograde rotation of having swallowed another system, but it remains unclear whether most ellipticals grew in this way. Protracted, piece-by-piece formation of bright galaxies through accretion and mergers of smaller predecessors may fit well with the high star formation rates inferred in some high-redshift galaxies (such as the most luminous radio galaxies in the 3C [Third Cambridge] catalog) while preserving the range of epochs for galaxy formation required by cosmologies dominated by cold dark matter in the amounts implied by galactic rotation curves.

INTERACTIONS AND STAR FORMATION

Considerable evidence links tidal disturbances of galaxies to increases in their rates of star formation. The fraction of strongly interacting galaxies with starburst nuclei, for example, is much higher than that of similar noninteracting galaxies with such nuclei. Statistical examination of large samples of systems indicates that both nuclear and disk-wide star formation rates are higher in the interacting systems, though the median enhancement is rather small and the most spectacular effects are confined to a small percentage of the systems. It must be stressed that most interacting galaxies do not show strong starburst responses to interactions, but that those that do have such high (temporary) luminosities that they are found very efficiently by flux-limited surveys such as the *Infrared Astronomical Satellite* (*IRAS*) far-infrared survey or the Markarian ultraviolet-excess survey.

The enhancement of star formation in some interacting systems is widely thought to result from the behavior of interstellar clouds during cloud-cloud collisions, the rate of which is greatly increased during interactions even when clouds from one galaxy never contact those from the other. More detailed understanding is limited by lack of direct checks on just how a molecular cloud responds to a collision. The existence of many disk systems with strong tidal disturbances but very low star-forming rates implies that several parameters are involved. The situation is clearly more complicated than a simple triggering by even strong tidal disturbance.

Interaction-induced star formation is not dominant at the present epoch; only a few percent of the population of massive stars in nearby galaxies is due to such induced processes. Extrapolation back in time is important to see whether interactions have ever been important in driving the evolution of galaxies in general (as opposed to just the few with strong bursts), which is not yet possible with certainty. If large numbers of today's bright galaxies are merger remnants, much of the total star formation in them might have been induced. On the other extreme, if the merger rate has been rather low, there might not have been many more close interactions in the past than are seen now.

Considerable recent work has focused on the nature of a population of galaxies largely found as far-infrared sources during the *IRAS* survey which have bolometric luminosities up to 10^{12} times that of the Sun, most of which is radiated in the 25–100-μm range. It is not clear in many instances what the original source of energy is; grains might be heated by starbursts or dust-shrouded active nuclei, and in fact optical and near-infrared spectroscopy finds each of these in some systems. Many of these systems are in strongly interacting or merging systems, with fractions estimated from 50–100% in various surveys. An evolutionary scheme has been introduced involving mergers inducing starbursts which then feed a central massive object via mass loss and disruption of red-supergiant envelopes, thus linking all these phenomena in a causal sense. However, many physical details remain to be clarified; star formation in violent environments is not understood. A competing picture, of energy release directly from plasma processes in cloud collisions during direct impacts and mergers, has had some success in accounting for the occurrence and luminosity of these systems. A mix of physical processes may well be operating in these infrared-bright systems.

TRIGGERING OF ACTIVE GALACTIC NUCLEI

Several lines of evidence point to a connection between galaxy interactions and nuclear activity, defined here as phenomena in addition to those produced by stars or stellar remnants and thus including quasistellar objects (QSOs), Seyfert nuclei, radio galaxies, and their relatives. Indeed, the earliest round of extragalactic radio-source identifications showed a suspicious number of interacting or colliding systems, but the original interaction model for such sources went out of vogue once interferometric maps showed the sources to be symmetric about individual galaxies.

Some studies show that optically selected Seyfert nuclei occur preferentially in galaxies with companions. However, the exact significance of this result is difficult to assess, since other studies with a somewhat different way of selecting a control sample show only a marginal excess of companions around Seyferts. This perhaps emphasizes the crucial role played by comparison samples in observational studies of interactions. A stronger result is shown by the converse experiment, in which large, statistically selected samples of interacting systems are observed spectroscopically and the number of Seyferts found in this way is compared with control samples. There is a clear excess of Seyfert nuclei in pairs with small projected separation but little tidal distortion, and there is an equally clear deficit of Seyfert nuclei among strongly distorted systems. These conflicting trends may imply a temporal sequence, in which case the activity must be short-lived compared to the lifetime of tidal features.

Studies of the structure and environment of radio galaxies also indicate that most high-luminosity radio galaxies show structural evidence of recent interactions or mergers; the dust lane of Centaurus A prefigured this conclusion long ago. Signs of interactions frequently seen in these objects include close companion galaxies, tidal tails, "shells," and kinematic irregularities. These symptoms are largely confined to the systems with the most powerful radio emission, suggesting that there may be a distinction in cause between low- and high-luminosity radio galaxies. This distinction parallels the structural difference reflected in the Fanaroff–Riley classification of the radio sources themselves: More powerful sources are more likely to have undergone a recent interaction, and the associated extended radio lobes generally have prominent hot spots and well-defined edges (expected for supersonic jets).

The situation for QSOs is more difficult to assess, because most are so distant that structural information on surrounding galaxies or companion systems is not yet available. For objects at redshifts below about 0.4, there are strong indications that more companions or instances of tidal distortion in the host galaxies are present than might otherwise be expected. Such a trend appears to continue that found for Seyfert nuclei, but the fact that many of the Seyfert nuclei are of rather low luminosity, and the apparent occurrence of some QSOs in very disturbed host galaxies, may argue for two different mechanisms of production.

The existence of triggering of active nuclei via interactions must probe both the mechanism of fueling for the central engines of these objects and the transport of mass and angular momentum across large radial distances during interactions. Unraveling the connection is both one of the most promising and one of the least tractable problems in studying active nuclei. Several physical mechanisms likely to be operating have been explored in some detail. The nucleus may be fueled either by interstellar gas, or by gas liberated during tidal disruption of stars venturing too close to a massive core object. If the fueling is in the form of diffuse gas, it most likely originates within the host galaxy, but has been moved from farther out in the disk; this form of fueling must somehow overcome a large angular-momentum barrier in order to reach the proximity needed to be accreted by the central source, or captured into an accretion disk. Fueling via disruption of stars appears to require either that massive binary black holes are common, or that substantial bursts of star formation produce many red supergiants with loosely bound atmospheres. This is the extreme case of the general problem of "feeding the monster."

Regardless of just how active galactic nuclei are fueled during interactions, the fact that such fueling can be recognized in large samples of systems has several interesting consequences. Individual encounters (or mergers, for that matter) last a very short time compared to a galaxy's lifetime. Thus, any episodes of nuclear activity triggered by encounters must be comparably short-lived, or we would not observe a significant excess of active nuclei in visibly interacting systems. Simple estimates of this lifetime suggest that individual episodes of activity last on the order of 10^8 yr. This implies that the luminosity function of these objects is to be interpreted as a statistical property of a population of short episodes, not as the distribution of individual objects that might change with cosmic epoch. Furthermore, the existence of triggering suggests that most luminous galaxies have the prerequisites for nuclear activity (such as a central massive object) and need only the proper conditions for its expression. Thus, most bright galaxies would have spent a few percent of their history with active nuclei, so that fossil quasars, for example, should be rather common.

DENSE GROUPS OF GALAXIES

The rapid time scales for merging of close neighbors indicate that strong interactions should mostly be two-galaxy processes; one would seldom expect additional objects to become involved during the relatively brief interaction between two systems. There is, however, a substantial population of groups that are so dense that their very existence may pose a challenge to current understanding of the origin and dynamical evolution of groups of galaxies. In some cases tidal interaction can be observed among four or five galaxies in a single group.

If we observe these groups at a typical time (that is, if they are not looser groups accidentally viewed in angle or time so as to appear unusually dense), the time scale for many of these groups to merge into single objects is much less than a Hubble time. The presence of tidal features in at least some of these suggest that the galaxies are about as close together as they appear. There is a two-sided puzzle posed by the expected mergers of these systems: What are the remnants of the historical population of these groups, and why are we just now seeing the dregs of an originally much larger population of these groups?

Additional Reading

Arp, H. C. (1966). *Atlas of Peculiar Galaxies*. California Institute of Technology, Pasadena. (Also appeared in *Astrophys. J. Suppl. Ser.* **14** 1).

Arp, H. C. and Madore, B. F. (1987). *A Catalogue of Southern Peculiar Galaxies and Associations*. Cambridge University Press, Cambridge.

Binney, J. and Tremaine, S. (1987). *Galactic Dynamics*. Princeton University Press, Princeton, Chap. 7.

Keel, W. C. (1989). Crashing galaxies, cosmic fireworks. *Sky and Telescope* **77** 18.

Schweizer, F. (1986). Colliding and merging galaxies. *Science* **231** 227.

Toomre, A. and Toomre, J. (1973). Violent tides between galaxies. *Scientific American* **229** (No. 6) 39. [Reprinted in P. Hodge, ed. (1984). *The Universe of Galaxies*. Freeman, San Francisco, p. 55.]

See also **Active Galaxies, Seyfert Type; Active Galaxies and Quasistellar Objects, Central Engine; Galaxies, Infrared Emission; Galaxies, Properties in Relation to Environment.**

Galaxies, Chemical Evolution

Monica Tosi

The study of the chemical evolution of galaxies includes the various processes which have led to the present chemical abundances observed in the stars and the interstellar gas of galaxies. This study started essentially in 1963 with a seminal paper by Maarten Schmidt.

The evolution of a galaxy of any type is due to several complicated and interconnected processes. For the sake of simplicity astronomers tend to divide them into three major categories and to treat the photometric, dynamical, and chemical evolution separately. However, to fully understand the actual evolution of galaxies we will need to take all of them into account at the same time.

OVERVIEW

The main goal in modeling the chemical evolution of galaxies is to provide a correct interpretation of the element abundance distributions observed in their stars and interstellar media. In addition, comparison between observed galactic properties and theoretical predictions has frequently put important constraints on the theories of star formation, stellar nucleosynthesis, evolution, etc.

The main processes governing chemical evolution are summarized in Fig. 1, which is adapted from fundamental review by Beatrice Tinsley. As indicated in the sequence of the figure, when the protogalactic cloud reaches a critical density which may vary from one galaxy to another, it starts to fragment and form stars of various masses (ranging essentially from 0.1 to 100 times the mass of the Sun). These stars evolve and burn their nuclear fuel, synthesizing in their interiors chemical elements of progressively higher atomic weight. The original chemical composition of the stars, mostly hydrogen and helium, changes in favor of heavier elements like carbon, oxygen, and iron. During their life and mainly at their death, stars eject most of their initial mass, thus polluting the circumstellar medium with the new elements produced in their interiors. From this medium, modified in mass and metallicity, new stars form, with an initial chemical composition and, therefore, with a subsequent evolution somewhat different from that of the previous stellar generation. In principle this cycle may last forever, leading to a continuous decrease of the abundance of hydrogen and to a corresponding increase of that of heavier elements. However, as it has been found that in every stellar generation the fraction of formed stars is a decreasing function of their mass (i.e., in a single generation there are more low- than high-mass stars) most of the gas remains trapped in low-mass, long-lived stars and cannot contribute to the cycle for a long time. Besides, a dying star never returns all its mass back to the medium and always leaves a dead remnant (white dwarf, neutron star, or black hole, depending on the initial mass) unable to contribute to the productive cycle. After a few stellar generations, the situation can remain frozen for several billion years (or even forever through the exhaustion of the gas reservoir).

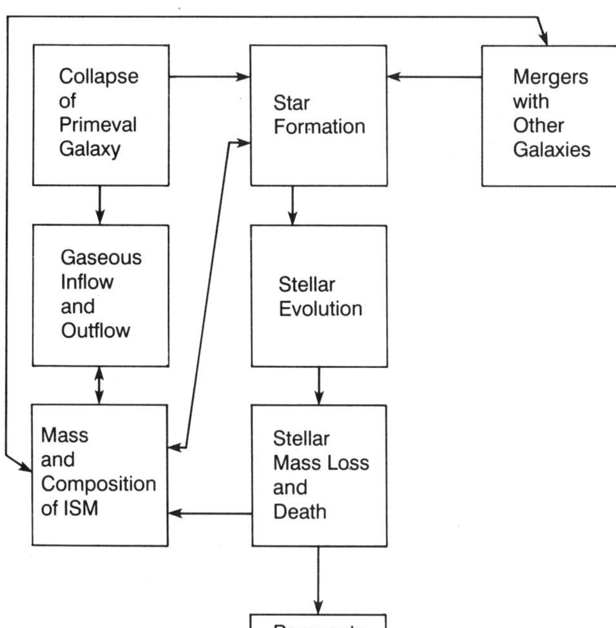

Figure 1. Scheme of the main processes involved in the chemical evolution of galaxies.

Meanwhile, other phenomena can affect the cycle and thus the chemical evolution of a galactic region. Some gas can fall in from outside, either a residual of the collapse of the protogalactic cloud, or intergalactic gas trapped by the gravitational force of the galaxy. Some gas, on the other hand, can also leave the region, for instance, when supernova explosions produce winds effective enough to sweep away the surrounding medium. Finally, the presence of nearby galaxies can provoke strong interactions that may even produce a merger of two or more systems and in any case that will significantly alter the star formation rate of the galaxy.

Modelers of galactic chemical evolution take into account all these possible phenomena by treating them as free parameters of the problem. Modelers borrow from stellar evolution theory the stellar lifetimes and the amount of elements synthesized or burnt by stars of different mass, "cook" all the ingredients together, and search for the recipe that provides objects as similar as possible to actual galaxies. The most important parameters for building a model of a galactic region are the star formation rate (SFR, i.e., the amount of gas mass which goes into stars per unit time), the initial mass function (IMF, i.e., the fraction of stars of each mass formed in a stellar generation), the amount and the rate of gas flows in and out of the region, and the amount of newly formed elements ejected by stars (stellar yields). Some parameters of the stellar evolution theory are still rather uncertain and may consequently be treated as variables in the chemical evolution models. The free parameters are therefore quite numerous, and the only case where they do not outnumber the observational constraints is that of our own galaxy. For external galaxies more observational data are still necessary to properly constrain models.

MODELS FOR THE VARIOUS TYPES OF GALAXIES

As is well known, galaxies are usually classified in three major types: ellipticals, spirals, and irregulars. This classification was originally based only on morphological characteristics, but it actually corresponds to profound differences in chemical evolution as well. The main features characterizing these three types of galaxies that a chemical evolution model must account for are the following.

(a) Elliptical galaxies are more metal rich than the Sun, have very little or no gas left, and show an average stellar population older than that around the Sun. All these facts suggest that they are evolved objects, with no, or very little, recent activity of star formation.

(b) Spiral galaxies show two distinct populations: (1) metal poor, old stars, associated with a gas-poor halo; (2) stars of any age and with metallicity of the same order of magnitude as the Sun, imbedded in a significant amount of gas (ranging roughly between 2 and 20% of the total disk mass) in the disk. (1) is generally interpreted in terms of a first rapid phase of star formation that has generated the halo population. Most of the gas, however, must have escaped this star formation activity, and has concentrated into a disk where the star formation seems to be in a sort of stationary regime.

(c) Irregular galaxies have a much higher gas fraction (mass in interstellar gas relative to the total mass in gas and stars) than the other galaxies, have chemical abundances significantly lower than solar, and show a large fraction of young stars, thus indicating that they are still in an early stage of evolution.

The most up-to-date scenarios for the evolution of the different types of galaxies can be summarized as follows.

Elliptical Galaxies

All the models that are in agreement with the observational features of ellipticals predict, right after galaxy formation, a rapid very strong espisode of star formation, which consumes almost all the gas. After a certain amount of time, which increases with the galactic mass but never exceeds 1 billion years, the explosion of a large number of supernovae in a relatively short time triggers galactic winds, which remove all the residual gas. This prevents any further star formation and explains the average age of the stars presently observed. Despite the rapidity of these events, the star formation activity has been so high until this point that the system has already reached a metallicity even larger than solar (it has taken 10 times longer for our galaxy to produce the Sun, with its metallicity). After this major episode of galactic wind, the only gas available in the galaxy is that which is returned little by little by dying stars. Since there is no diluting medium, this gas is very metal rich. It can reach a present mass around 10^9–10^{10} solar masses (i.e., $\frac{1}{1000}-\frac{1}{100}$ the mass of the galaxy) and is presumably recognizable in the hot gaseous coronae sometimes observed around ellipticals.

Spiral Galaxies

Given the old age of all the halo stars of our galaxy and nearby spirals, we can easily presume that early star formation activity in these galaxies was even more rapid than in ellipticals. The presence of the disk, however, strongly modifies the situation, because the major gas reservoir lies in this case in a potential well deep enough to protect it from the supernova winds in the halo. A steady state of moderate star formation can then start in the disk, with more activity in the central, denser regions and less in the outer ones. In fact, the SFR presently derived in nearby spirals from indicators like supernovae and pulsars is roughly exponentially decreasing with increasing distance from the galactic center. As time goes by, this difference in the SFR of different regions produces relevant effects on their properties: Central regions show lower values of gas-to-total-mass ratios and larger chemical abundances, whereas in the outermost layers of the disk the fraction of gas is still up to 50% and the metallicity is lower than solar. The immediate results of this phenomenon are the radial abundance gradients of heavy elements observed in the disks of spirals. These gradients are always negative (i.e., any element heavier than hydrogen is more abundant in the inner than in the outer regions), but their slope varies from galaxy to galaxy and from element to element. No tight correlation has been found between the variation with time of the SFR and the galaxy morphological subtype, but there is some indication that in early-type spirals like M31 the SFR has slowly decreased, whereas in late-type objects like M33 it has been practically constant since galaxy formation.

Another important parameter in the chemical evolution of spiral galaxies is infall. An infall of gas from outside the system was first suggested by Richard Larson in 1972 to explain why so many galaxies have not yet exhausted their gas through star formation processes. The observational evidence of high velocity clouds, gas condensations falling toward our galaxy, supported his idea of gas replenishment. Currently infall is still the best way to interpret not only the gas survival in galactic disks, but also the steep abundance gradients of some spirals and local features like the so-called G-dwarf problem. In fact, if the infall density is roughly uniform and its metallicity low, as would be reasonable in the case of intergalactic gas attracted in the disk potential well, its ratio to the SFR increases toward the outer regions, thus diluting them more efficiently and leading to a steepening of the metallicity gradient. Moreover, the observational evidence of gas clouds falling into a galaxy, which was formerly restricted to the Milky Way, is now found in other spirals as well (e.g., M101). There are only about 10 nearby spirals studied in enough detail at the appropriate wavelengths to put reliable constraints on the models. For these objects it is found that a conspicuous amount of infall (around 1–2 solar masses per year) is predicted for the more massive galaxies like M101 and NGC 6946, but very little or none is required for the smaller ones like M33 and NGC 2403. This is naturally consistent with the idea that the infall is due to diffuse gas captured by the gravitational force of the disk.

Irregular Galaxies

Gas flows seem to be very important also in the evolution of these small diffuse objects, the irregular galaxies. In this case, however, the main flow is outwards and the galaxy tends to lose most of its gas, because the potential well of an irregular is low enough to allow a strong galactic wind triggered by supernova explosions. In fact, the models predict stronger winds for the lower mass irregulars. The reason these gas outflows seem to be necessary in the evolution of irregulars is that these galaxies show such low metal abundances that we are led to conclude that most of the enriched gas ejected by the dying stars is not retained by the galaxies. The low mass and metallicity of these galaxies have also suggested to some theorists that star formation cannot have been continuous as in spiral galaxies, but has been confined to short-lived episodes or bursts, or that it has increased with time and has only recently reached the present value. There is no definite conclusion on this subject, but certainly small galaxies like the closest dwarf irregulars in the Local Group cannot have sustained a star formation rate of the intensity observed now for their entire lifetimes. There are even some galaxies, called blue compact galaxies because of their appearance, that show incredibly strong star formation activity but have metal abundances so low as to suggest that they are perhaps now experiencing their first burst (e.g., the galaxy n.18 in Fritz Zwicky's catalogue).

Additional Reading

Diaz, A. I. and Tosi, M. (1986). The origin of nitrogen and the chemical evolution of spiral galaxies. *Astron. Ap.* **158** 60.

Gilmore, G., Wyse, F. G., and Kuijken, K. (1989). Kinematics, chemistry, and structure of the Galaxy. *Ann. Rev. Astron. Ap.* **27** 555.

Larson, R. B. (1972). Infall of matter in galaxies. *Nature* **236** 21.

Matteucci, F. (1984). The chemical enrichment of galaxies. *ESO Messenger* (No. 36) 17.

Schmidt, M. (1963). The rate of star formation. II. The rate of formation of stars of different mass. *Ap. J.* **137** 758.

Tinsley, B. M. (1980). Evolution of the stars and gas in galaxies. *Fundam. Cosmic Phys.* **5** 287.

See also **Galactic Structure, Stellar Populations; Galaxies, Disk, Evolution; Galaxies, Formation.**

Galaxies, Disk, Evolution

Joshua E. Barnes

Galaxies come in a bewildering variety of shapes and sizes, but a large majority—perhaps 80%—possess a *disk* of some kind. Galactic disks are thin, basically circular distributions of stars, gas, and dust; this material moves on nearly circular orbits about a common center. Many disks exhibit beautiful spiral patterns as a result of this rotation, and some have pronounced bars crossing their centers. Other disks, however, are nearly featureless, and can only be identified by a characteristic falloff of brightness with radius.

The evolution of disk galaxies is inextricably bound up with the highly controversial problem of galaxy formation. This entry focuses on the history of disks in galaxies such as the Milky Way, as a way of distinguishing the present subject matter from larger issues. Progress in this field has come by combining detailed studies of the motions, compositions, and ages of stars in the solar neighborhood with less-detailed knowledge of the overall structure of the Milky Way and global observations of other galaxies. This approach, while productive, is fraught with uncertainties. Presently one cannot offer a definitive account of the evolution of disk galaxies.

The mere existence of a galactic disk has two basic implications. First, the gas from which the disk formed must have settled into circular orbits before the disk stars were born; once a star is formed its future path is determined entirely by gravitational forces, and gravity cannot circularize a random distribution of stellar orbits. Second, since the disk formed, the gravitational field has not undergone any sudden, dramatic changes, which would disrupt the circular pattern of stellar orbits. We cannot, however, rule out the possibility that the mass distribution, and hence the gravitational field, has evolved *slowly*.

AGES, COLORS, AND STAR FORMATION

The more massive a star, the brighter it shines, the bluer its color, and the shorter its lifetime. Thus it is relatively easy to tell the age of a system in which all stars formed at the same time by observing the colors of the brightest stars still on the main sequence. Stellar associations and open clusters in the disk of the Milky Way yield ages between 3×10^6 and 6×10^9 yr, and the oldest disk stars are at least 10^{10} years old. These widely ranging ages imply that star formation in the Milky Way started when the Universe was less than half its present age and continues up to the present time.

Observations of other galaxies show that the Milky Way is hardly unique in this respect. The broadband colors of a galaxy depend largely on the rate of star formation averaged over the last $\sim10^8$ yr; higher rates of star formation yield bluer colors. Late-type galaxies (type Sc and Irr) indeed have rather blue colors, suggesting a constant rate of star formation. In very early-type disk galaxies (type S0), star formation has largely ceased, although it is hard to tell how long ago this occurred. Finally, the intermediate colors of galaxies like the Milky Way (type Sb) are consistent with a present star formation rate ~3 times lower than the average rate over the lifetime of the galaxy.

How long can galaxies continue to form stars at these rates? In all but the most extreme cases, the gas is $\lesssim15\%$ of the total mass in stars. This suggests that late-type galaxies are literally about to run out of gas, ending their phase of star formation. It seems unlikely, however, that we find ourselves at such a unique moment in cosmic evolution. Alternatively, galaxies may accrete gas from their surroundings, their present gas content representing a rough balance between income and expenditure. This accretion hypothesis solves several problems in galactic evolution, but at present there is little *direct* evidence that galaxies such as the Milky Way are accreting significant amounts of gas.

CHEMICAL EVOLUTION

Some of the gas invested in stars is returned to the galactic reservoir as the stars age, having been enriched in metals (elements heavier than H and He). Massive stars very rapidly become Type II supernovae, spewing a wide range of elements back into the clouds from which they form, whereas close, intermediate-mass binaries may evolve into Type I supernovae, favoring production of iron-group elements. Modeling galactic chemical evolution is in principle just a matter of bookkeeping, but in practice uncertainties in the physics of evolved stars make detailed predictions difficult. Some simple models, however, can at least be ruled out.

Simplest is the closed-box model, in which gas neither enters nor leaves the galaxy; as the metals must be built up over time, the closed-box model predicts large numbers of metal-poor, low-mass disk stars. In fact, only 2% of the low-mass stars in the solar neighborhood have less than a quarter of the solar fraction of metals, compared to the 44% predicted by the closed-box model. This is known as the G-dwarf problem; the scarcity of metal-poor disk stars indicates that a closed-box model is inappropriate for our galaxy.

One obvious solution to the G-dwarf problem is to supply metals from the outside. In addition to the disk, the Milky Way has a spheroidal component, which is older than the disk and massive enough to have contaminated the protodisk with a significant quantity of metals. This disk-spheroid model also explains why the metal fraction of a disk typically increases towards the center, because that is where the bulk of the mass in the spheroid component is found.

Paradoxically, another way to solve the G-dwarf problem is to slowly but steadily build the disk from metal-*poor* gas. In this case the metal fraction soon reaches roughly the present value, and then remains constant. By the present epoch most stars will have formed during the phase in which the metal fraction is constant, and metal-poor stars will be rare.

Which solution is preferred? Data for F stars show the fraction in metals increasing with time up until $\sim3\times10^9$ years ago, and then leveling off. If real, this leveling off indicates that our disk has only just reached the constant-metals phase, suggesting a compromise between the disk-spheroid and accretion solutions. These two hypotheses may be complementary sides of the same story; both challenge the assumption that galactic disks are closed systems.

RANDOM VELOCITIES

The Sun and nearby disk stars share a common orbital motion about the galactic center, but in addition each has a small random velocity, reflecting the fact that their orbits are not perfect circles. On the average, older stars have larger random velocities; for stars less than 10^9 years old, rms velocities toward or away from the galactic center are $\sigma_R\simeq10$ km s^{-1}, while for the oldest disk stars, we find $\sigma_R\simeq40$–60 km s^{-1}.

The most likely explanation for the trend of random velocity with age is that stars are born on nearly circular orbits, and are subsequently deflected onto more random orbits by fluctuations in the galactic gravitational field. This led Lyman Spitzer, Jr. and Martin Schwarzschild to postulate the existence of giant molecular clouds, long before such clouds were detected. Present calculations indicate that clouds of mass $10^6\ M_\odot$ can produce random velocities of up to ~30 km s^{-1}, but not the higher velocities seen in the oldest disk stars. In addition, the Spitzer–Schwarzschild mechanism predicts that random velocities grow rather slowly, roughly as $\sigma \propto t^{0.25}$, and the observations are better fit by $\sigma \propto t^{0.5}$. Another source of fluctuations is needed; transient spiral structure, to be discussed next, may fill the bill.

SPIRAL STRUCTURE

Many different kinds of spiral structure are seen in disk galaxies. Most photogenic are the grand-design two-armed spiral galaxies such as M51, but far more common are ragged or flocculent spirals made up of many short arms. The diversity of spiral galaxies is paralleled by the diversity of theories of spiral structure. Grand-design spirals are often discussed in terms of the Lin–Shu theory (after Chia-Chiao Lin and Frank H. Shu), which views the spirals as slowly turning wave patterns maintaining their form for many rotation periods. However, classic grand-design spirals like M51 often have close companions, and it is possible that such spirals are actually excited by tidal interactions. Flocculent spirals, on the other hand, are generally thought to evolve over time, with individual spiral arms constantly forming and dissolving.

Computer models of rotating disks can produce spiral patterns similar to those seen in real galaxies. In these models, thousands of particles represent the disk; each particle moves in the net gravitational field produced by all the others. If the particles start out in nearly circular orbits with small random velocities, striking multi-armed spiral patterns soon develop. These spirals result from the gravitational amplification of small fluctuations in a disk that rotates differentially (i.e., not like a solid body). As a result of these ever-changing spiral patterns, particles acquire increasingly large random velocities. After a few rotation periods the random velocities become large enough to shut off the gravitational amplifier, and the spiral-making activity dies away.

Transient spiral structure can in principle provide the fluctuating gravitational field needed to generate the random motions of old disk stars, and a theoretical analysis even predicts $\sigma \propto t^{0.5}$. However, the spiral activity seen in the simplest computer models lasts only a few rotation periods, whereas in real galaxies it persists more than 10 times as long. Computer experiments show that spiral structure can be maintained by adding stars to the disk on nearly circular orbits, consistent with the above discussion of random velocities. Moreover, the kind of spiral pattern produced depends on the rate at which stars are injected; high rates produce open, well-defined patterns typical of late-type spirals, and slower injection results in weaker, tightly wound spirals like those seen in early-type disk galaxies. These results support the view that accretion provides a disk galaxy with the shot in the arm needed to promote vigorous development of spiral structure, the type of the resulting spiral depending largely on the rate of accretion.

DEATH AND REBIRTH OF DISK GALAXIES

Complementing the mechanisms which build up galactic disks are those which destroy them. According to the accretion hypothesis, spiral galaxies are susceptible to starvation: If the inflow of raw material for new stars is cut off, the spiral soon fades, leaving a smooth disk resembling an S0 galaxy. Indeed, S0 galaxies are generally found in high-density regions where starvation is likely. A disk galaxy that has the misfortune to fall into a rich cluster may be swept clean of interstellar material by the ram pressure of the hot, low-density gas pervading such clusters. Alternatively, the overpressure of the cluster gas may compress molecular clouds within the galaxy, provoking a burst of rapid star formation. This process may account for some unusually blue galaxies observed in high-redshift clusters.

Finally, instead of accreting gas, a disk galaxy may ingest a companion. Computer simulations show that interactions between galaxies often result in mergers, the outcome depending on the mass ratio of the colliding systems. Large disk galaxies can swallow small companions, of less than ~10% their mass, with only minor damage: Random motions of disk stars increase, and the disks become thicker. A number of galaxies, including the Milky Way, are reported to have thick disks which may have been produced in this way. Mergers between disk galaxies of comparable mass have a very different outcome. So violent is the interaction that neither disk survives; such mergers may in fact produce elliptical galaxies.

Disk galaxies are thus rather fragile and delicate objects. The evidence, while fragmentary and largely circumstantial, suggests that these galaxies grow best in quiet, undisturbed locations where their disks can develop slowly without outside perturbations. When such galaxies become involved with others, they run the risk of violent transformation. But from the wreckage of such cosmic accidents a new disk galaxy may arise, given only time and a sufficient supply of raw materials.

Additional Reading

Binney, J. and Tremaine, S. (1987). *Galactic Dynamics*. Princeton University Press, Princeton, Sects. 6.3, 6.4, 7.5, 9.1, and 9.2.

Gilmore, G., Wyse, R. F. G., and Kuijken, K. (1989). Kinematics, chemistry, and structure of the Galaxy. *Ann. Rev. Astron. Ap.* **27** 555.

Gunn, J. E. (1982). The evolution of galaxies. In *Astrophysical Cosmology*, H. A. Brück, G. V. Coyne, and M. S. Longair, eds. Pontificia Academia Scientiarum, Città del Vaticano, p. 233.

Tinsley, B. M. (1980). Evolution of the stars and gas in galaxies. *Fundam. Cosmic Phys.* **5** 287.

See also **Galactic Structure, Stellar Kinematics; Galaxies, Binary and Multiple, Interactions; Galaxies, Chemical Evolution; Galaxies, Formation; Galaxies, Molecular Gas in; Galaxies, Properties in Relation to Environment; Galaxies, Spiral Structure; Galaxies, Starburst.**

Galaxies, Dwarf Spheroidal

Jeremy Mould

Physically small and intrinsically faint elliptical galaxies are known as dwarfs. The smallest and faintest of these are referred to as dwarf spheroidal galaxies. Low surface brightness is a characteristic of dwarf spheroidals, but in this and other respects the distinction from dwarf ellipticals appears to be simply a matter of degree.

Location Ten galaxies define the class. Seven of these are satellites of the Milky Way. They are known by the names of their home constellations: Draco, Ursa Minor, Carina, Leo I and II, Sculptor, and Fornax. The other three are satellites of M31, and are known as Andromeda I–III. Positions of all of these galaxies are given in Table 1. The radial distances of these objects from their parent galaxies are similar to those of the globular cluster systems of these galaxies. No dwarf spheroidals are known outside the Local Group of galaxies, probably because the surface brightness of dwarf spheroidals is typically ~1% of that of the night sky, and because individual giant stars in these galaxies would fall below the limit of the Palomar Observatory Sky Survey beyond the Local Group. Within the Local Group, however, satellite status seems to be a real characteristic of dwarf spheroidal galaxies.

Luminosity The absolute visual magnitudes of dwarf spheroidals range from $M_V = -8$ to $M_V = -13$ mag. Distances assumed in calculating these values are given in Table 1. The sources for the data in Table 1 are numerous, and cannot be cited here in full. None of their known properties except their satellite status distinguish dwarf spheroidals from the dwarf ellipticals surveyed in the Virgo cluster of galaxies. In particular, dwarf ellipticals have been found in Virgo as faint as the brightest dwarf spheroidal in the Local Group (Fornax). Neither the Virgo dwarf elliptical sample nor the local dwarf spheroidal sample is well adapted to defining the luminosity *distribution* of galaxies at these magnitudes. The Virgo sample is very incomplete (but still seems to be rising in number) at $M_V \geq -12$; the local sample is too small to define any distribution.

Size The effective radius of a galaxy is defined to be the radius which contains half the light. The effective radii of dwarf spheroidals range from 0.2–1.0 kpc. The ellipticity of a galaxy is $1-b/a$, where b is the minor, and a is the major axis of an elliptical isophote. The ellipticities of dwarf spheroidals range from 0 to 0.55. These data, from the review by Paul Hodge, are contained in column 6 of Table 1. The ellipticity of Ursa Minor is high for an elliptical galaxy, but flatter dwarf ellipticals are known in the Virgo cluster. The term "spheroidal" is truly a misnomer.

Mass Measurement of the Doppler velocities of stars in dwarf spheroidal galaxies allows an estimate to be made of the mass required to bind them gravitationally within a measured volume. Although the luminosities of dwarf spheroidals span almost 2 orders of magnitude, the masses measured by Marc Aaronson and his colleagues cover a smaller range from 7×10^5–13×10^6 $M_\odot$. The mass-to-light ratios of the two smallest dwarf spheroidals,

Table 1. The Dwarf Spheroidal Galaxies*

Name (1)	*RA (1950)* (2)	δ (3)	*Distance (kpc)* (4)	M_V *(mag)* (5)	$1-b/a$ (6)	r_c *(arcmin)* (7)	r_t *(arcmin)* (8)	M/L_V $(M_\odot/L_\odot)$ (9)	*AGB (%)* (10)	*AC (No.)* (11)	*MS* (12)	$M_{\text{H I}}$ $(10^3\ M_\odot)$ (13)	$\log Z/Z_\odot$ (14)
Andromeda I	$0^h43.0^m$	$+37°44'$	790	-10.9	0	—	~ 3	—	< 20	—	—	—	-1.4
Andromeda II	$1^h13.5^m$	$+33°09'$	—	$-11:$	0	—	—	—	> 0	—	—	—	—
Andromeda III	$0^h32.6^m$	$+36°14'$	—	$-11:$	0.3	—	—	—	—	—	—	—	—
Sculptor	$0^h57.6^m$	$-33°58'$	80	-11.1	0.35	11.9	53	6	5	3	Y?	130 :	-1.85
Fornax	$2^h37.84^m$	$-34°44.4'$	140	-12.6	0.35	16	50	1.4	25	7	Y	< 8.8	-1.4
Carina	$6^h40.4^m$	$-50°55'$	100	-9.4	0.41	10.7	33.8	1.4	70	9	Y	—	-1.9
Leo I	$10^h05.77^m$	$+12°33.2'$	220	-11.4	0.31	4.5	13.9	≤ 1.5	15	12	—	< 7.2	$-1.7:$
Leo II	$11^h10.83^m$	$+22°26.1'$	230	-10.2	0.01	2.5	9.6	≤ 13	15	4	—	< 11	-1.9
Ursa Minor	$15^h08.2^m$	$+67°23.0'$	65	-8.5	0.55	11.1	59	100	—	7	N	< 0.28	-2.2
Draco	$17^h19.21^m$	$+57°57.5'$	75	-8.5	0.29	6.5	26	50	—	5	N	< 0.068	-2.1

*See text for definitions of tabulated quantities.

Draco and Ursa Minor, appear to be much greater than those of the larger dwarfs. The data are summarized in column 9 of Table 1. Such large mass-to-light ratios seem to require the existence of dark matter in Draco and Ursa Minor, if we suppose that matter is distributed like light. The true distribution of mass in these galaxies, however, is quite unknown.

Density Within the effective radius the mean density is lowest in the Fornax galaxy: $\rho_e \simeq 0.003\ M_\odot\ \text{pc}^{-3}$. For comparison $\rho_e \simeq 0.03\ M_\odot\ \text{pc}^{-3}$ in the Large Magellanic Cloud. For Draco $\rho_e \simeq 0.1\ M_\odot\ \text{pc}^{-3}$, a density comparable to the density due to stars in the solar neighborhood.

Stellar Content The dwarf spheroidal companions of the Milky Way have an interesting dual character. On the one hand, galaxies such as Ursa Minor have color magnitude diagrams almost indistinguishable from those of globular clusters. On the other hand, there are systems like Carina, in which there is a considerable stellar population of intermediate age. The other five dwarf spheroidals are a blend of these characteristics—some are more like Ursa Minor, others are more like Carina. As shown in column 10 of Table 1 the fraction of the light in each galaxy that comes from a population of age between 1 and 10 billion years ranges from 5% for Sculptor up to 70% for Carina. Indicators of an intermediate age population include the presence of asymptotic giant branch (AGB) carbon stars, anomalous Cepheid variable stars, and an overluminous main sequence turnoff. Table 1 contains the latest count of anomalous Cepheids (AC) (column 11), and indicates whether or not a main sequence (MS) turnoff, more luminous and younger than that of globular clusters, is seen (column 12).

The Interstellar Medium Stringent upper limits have been set on the mass of neutral hydrogen in six of the seven Local Group dwarf spheroidal galaxies (see column 13 of Table 1). These upper limits are $\sim 10^4\ M_\odot$ for Fornax and Leo I and II. For Sculptor the data of Gillian Knapp and her colleagues show a small positive perturbation in the baseline at the now accurately known heliocentric velocity (107 km s^{-1}). If this detection is real, Sculptor may contain $\sim 1.3\times10^5\ M_\odot$ of neutral hydrogen. The upper limits on H I in Draco and Ursa Minor are much lower: 68 and 280 $M_\odot$, respectively. Mass loss from dying stars will readily furnish an interstellar medium exceeding these limits in a billion years. However, given the measured wind velocities from the envelopes of red giants, and the shallow gravitational potential of dwarf spheroidals, it is likely that such material is able to escape the galaxy as a wind on a considerably shorter time scale.

Chemical Composition It is generally supposed that the chemical composition of the stellar systems of Population II is $\sim 75\%$ hydrogen by mass, $\sim 25\%$ helium, and less than 1% "heavy elements" (i.e., everything else). If the heavy-element mass fraction is denoted by Z, $Z/Z_\odot$ ranges from $\frac{1}{30}$ to $\frac{1}{300}$ in dwarf spheroidals. Individual values are given in column 14 of Table 1. There is also evidence for a modest variation in Z from star to star in these galaxies. More luminous dwarf spheroidals have higher values of $\langle Z\rangle$, and in fact the review by Aaronson shows a linear relation between absolute magnitude M_V and metallicity (i.e., $\log\langle Z\rangle$). This result is loosely termed the "mass-metallicity relation" for elliptical galaxies and smoothly connects all dwarf ellipticals studied to date. There is a plausible extension to giant elliptical galaxies.

Star Formation History In the chemical enrichment history of a galaxy a generation of massive stars produces heavy elements in millions of years by means of Type II supernovae, but a generation of intermediate mass stars takes characteristically a billion years to return its products to the interstellar medium. Although the spread in Z seen in all dwarf spheroidals (and also in the massive globular cluster ω Centauri) may have its origin in the first few generations of massive stars, it is clear that an isolated galaxy with an escape velocity as low as 10–20 km s^{-1} cannot retain its interstellar medium for many such generations. Ejection of the gas before more than a few percent of it has been turned into stars can account for the low metallicity of dwarf spheroidal galaxies.

Ejection of the interstellar medium after a few million years is inconsistent, however, with the existence of intermediate age populations in a number of dwarf spheroidals. A more acceptable possibility is to suppose that dwarf galaxies are produced by the aggregation of many gas-rich clouds (perhaps similar to the Lyman-α clouds known to exist at earlier epochs). Each merger event stimulates a little more star formation, and the galaxy grows, while at the same time the interstellar medium is enriched in heavy elements. This accumulation provides an alternative basis for a mass-metallicity relation. In this scenario we have to suppose that, for the first few billion years of a galaxy's history, gravity wins over the disruptive effect of massive star formation, but that eventually the balance is reversed, and the interstellar medium is ejected for good.

Structure and Dynamics Classically, the idealized structure of a dwarf spheroidal galaxy is an isothermal sphere whose gravitational potential is truncated by the tidal field of the Milky Way. Core radii and tidal radii of such models, which are known as King models (after Ivan R. King), are given in columns 7 and 8 of Table 1. More recently it has been noted that an exponential surface brightness distribution can also yield an acceptable fit: This is not a strong constraint on the structure of a dwarf galaxy, because the visible galaxy covers rather few scale lengths. Galaxies with exponential disks are generally supported against gravitational collapse by rotation. Rotation has only been detected, however, in the Fornax galaxy. Substructure in the star distribution has also been detected in Fornax.

Dark Matter At the present time the nature of the dark matter which constitutes most of the mass of *spiral* galaxies is quite unknown. Conceivably it is baryonic (e.g., Jupiters); alternatively, it could be self-gravitating elementary particles of either known or unknown form. *If* the particles are massive neutrinos, for example, the Pauli exclusion principle of quantum mechanics prescribes a maximum number density for a given velocity dispersion, σ. A lower limit on the mass of the neutrino follows from the mass that they contribute to bind the galaxy. This lower limit scales like $\sigma^{-3/4}$. Therefore dwarf galaxies place the strongest lower limit. If the measured velocity dispersion of Draco and Ursa Minor is due to the presence of such dark matter, the mass of such hypothetical neutrinos must exceed ~ 500 eV. Such large masses violate a number of other constraints.

Additional Reading

Aaronson, M. (1986). The older stellar population in dwarf galaxies. In *Star Forming Dwarf Galaxies and Related Objects*, D. Kunth, T. X. Thuan, and J. Tran Thanh Van, eds. Editions Frontieres, Gif-sur-Yvette.

Aaronson, M. (1986). Local group dwarf galaxies: The red stellar population. In *Stellar Populations*, C. Norman, A. Renzini, and M. Tosi, eds. Space Telescope Science Institute Symposium Series, Vol. 1. Cambridge University Press, Cambridge, p. 45.

Aaronson, M. and Olszewski, E. (1987). The search for dark matter in Draco and Ursa Minor: A three year progress report. In *Dark Matter in the Universe*, J. Kormendy and G. R. Knapp eds. Reidel, Dordrecht, p. 153.

Hodge, P. (1971). Dwarf galaxies. *Ann. Rev. Astron. Ap.* **9** 35.

Knapp, G., Kerr, F., and Bowers, P. (1978). Upper limits to the H I content of the dwarf spheroidal galaxies. *Astron. J.* **83**, 360.

Sandage, A., Bingelli, B., and Tammann, G. (1985). Morphological and physical characteristics of the Virgo cluster: First results from the Las Campanas photographic survey. In *The Virgo Cluster of Galaxies, ESO Conference and Workshop Proceedings No. 20*, O.-G. Richter and B. Bingelli, eds. European Space Observatory, Garching bei Munchen, p. 239.

See also **Galaxies, Elliptical; Galaxies, Local Group.**

Galaxies, Elliptical

William A. Baum

Ellipticals are galaxies that appear ellipsoidal in form, have no disks, and are devoid of features such as spiral arms, bars, or dust lanes. Because such features would be associated with recent or ongoing star formation, their absence indicates that nearly all of the stars in elliptical galaxies must be somewhat old.

BASIC PROPERTIES

The majority of bright galaxies in large clusters are ellipticals, but fewer than 15% of galaxies in the general field are ellipticals. Classical ellipticals (E galaxies), with their bright compact nuclei and steep brightness gradients, range from absolute visual magnitude $M_V \sim -23$ down to $M_V \sim -16$ mag. The very brightest of them, called cD galaxies, are found at the centers of clusters, and are among the most luminous galaxies in the Universe.

Dwarf ellipticals (dE galaxies) include an assortment of morphological types and range from $M_V \sim -19$ down at least to $M_V \sim -12$ mag, where their numbers are rising steeply and where surveys become seriously incomplete. They probably extend to (and may overlap) the range of globular star clusters, which commences at $M_V \sim -10$ mag. Dwarfs differ greatly from one another in compactness.

Seen on the plane of the sky, some E galaxies are quite round, and others are elongated. Although various classification schemes have been devised, the degree of elongation is commonly designated by Edwin P. Hubble's subclass, $10(1-b/a)$, where a and b refer to major and minor axes. Thus, an E0 galaxy is round, and an E7 (the most elongated subclass) has a projected axis ratio $b/a \sim 0.3$. In the elongated galaxies, the ellipticity is typically a function of the isophotal level. In some, the position angles of the major axes of the isophotes are also a function of the isophotal level; that is, such galaxies possess an isophotal twist. Moreover, isophotes sometimes depart from pure ellipses in the sense of being slightly rectangular ("boxy").

The three-dimensional shape of a galaxy has to do with the distribution of stellar velocities within it. In disk galaxies (spirals and their featureless S0 cousins), the angular momentum due to Keplerian rotation dominates over random motions, and the resulting galaxy is an oblate spheroid. But in E galaxies the angular momentum is *not* dominant, and the three-dimensional shape of the system is maintained mainly by the dispersion of stellar velocities within it. In principle, the velocity dispersion can be anisotropic, so that an E galaxy can be a prolate ellipsoid or even a triaxial one. Unfortunately, the three-dimensional shapes cannot be directly observed.

Since E galaxies are not primarily supported by Keplerian rotation, it is not possible to calculate individual masses from their rotation curves in the manner used for tilted spirals. Internal velocity dispersions give only limited information. If one is willing to assume that clusters of galaxies (which often consist mainly of ellipticals) are in gravitational equilibrium, the mass of a whole cluster of galaxies can be estimated by application of the virial theorem to the observed dispersion of galactic velocities within the cluster, but the inferred mass tends to be puzzlingly large. The mass–luminosity ratio of the stellar population in giant ellipticals therefore remains quite uncertain. It might even be different in a cluster environment than in the general field.

Various formulas have been used to fit the observed radial distribution of brightness of E galaxies in the plane of the sky, but the one most often used today is the de Vaucouleurs law (after Gerard de Vaucouleurs). It says that the logarithm of the surface brightness (usually expressed on a stellar magnitude scale) is an approximately linear function of the $\frac{1}{4}$-power of the radial distance from the nucleus. This is a purely empirical finding that has no direct theoretical basis. Also, the goodness of fit varies somewhat from object to object.

At the very center of a large cluster of galaxies there is typically an unusually luminous elliptical surrounded by a faint but extensive envelope pervading the heart of the cluster. These cD galaxies are suspected of having grown by mergers and by the cannibalization of nearby dwarfs.

STELLAR POPULATIONS

In the 1940s, elliptical galaxies were assigned by Walter Baade to his Population II, by which he meant specifically that the color-magnitude diagram of their stellar population should be essentially the same as that for globular star clusters of the Milky Way. Owing to their high random velocities and their nearly spherical distribution in the Milky Way, globular star clusters were known to have formed during a very short interval before the Milky Way had settled into a disk and before the interstellar medium had become enriched with heavy elements ("metals"). It was therefore assumed that elliptical galaxies had similarly formed during an early burst of star formation.

But integrated photoelectric photometry soon showed (and spectroscopy later confirmed) that giant elliptical galaxies differ greatly in spectral energy distribution from globular star clusters. Therefore, the dominant stellar population of giant ellipticals cannot be similar to that of globular clusters. There was thus already good evidence in the 1950s that star formation in giant ellipticals must have continued long enough for a high degree of metal enrichment to permeate the stellar population. It is only with the accumulation of further evidence in recent years, however, that the concept of temporally distributed star formation in ellipticals has finally gained general acceptance.

Judged from integrated colors and spectra, some dwarf ellipticals must be at least partly similar in stellar content to giant ellipticals, while other dwarfs consist more of stars like those in globular star clusters. There are also a few dwarf ellipticals near enough for telescopic resolution of the brightest stars, but the resolved stars do not necessarily belong to the population that dominates the integrated light or the mass. Such is clearly the case for M32, the compact metal-rich companion of the Andromeda galaxy. On the other hand, several of the very tenuous metal-poor galaxies classed as dwarf spheroidals are near enough that the color-magnitude diagrams of their stellar populations can be definitively identified.

Owing to the fact that no *giant* elliptical happens to lie near us, none has ever been resolved well enough to permit photometry of individual stars. Incipient resolution of the nearest ellipticals can barely be achieved under the very best Earth-based observing conditions. Some improvement can be expected with telescopes in space. Using a form of noise analysis on such images, one can infer the nature of the brightest stars, the number of them per unit surface brightness, and the relative distance of nearby ellipticals. Noise parameters, taken together with known properties of the integrated light, should enhance our knowledge of the population mix.

It has been known for a long time that giant ellipticals tend to be slightly redder in their inner regions than in their outskirts, suggesting a gradient in the metallicity of the stellar population. Gradients have now been measured spectroscopically for several elements, and the gradient for the prominent magnesium feature around 5175 Å has been extended to faint levels in the outskirts by narrow-band CCD photometry. On reasonable assumptions, those data indicate the inner regions to be more metal rich than stars in the solar neighborhood, while the halos are only a little less so. In other words, the halo population of ellipticals is much more metal rich than that of the Milky Way. On the other hand, globular star clusters that have been detected in the halos of nearby ellipticals do not seem (from their colors) to be so metal rich; so the globulars must have formed at an earlier time than many of the individual stars that now populate the same halo regions.

It should be noted that the strength of an absorption feature such as the 5175-Å magnesium band is dependent upon stellar surface gravity as well as metallicity. However, the temperature of the giant branch of the H–R diagram (or color-magnitude diagram) is largely controlled by metallicity, and it is the giant branch that would be expected to dominate the integrated light. For subgiants just above the main-sequence turnoff, surface gravity differences (which are correlated with age differences) play a role, but the turnoff stars probably do not contribute strongly to the integrated light.

MODELS

No existing theoretical model for the formation and evolution of giant elliptical galaxies is able to explain all of their observed properties, but some form of inhomogeneous dissipative collapse appears to be indicated, and star formation (though not active today) was evidently not limited to a single early burst. It presumably required a long time to build up the high observed metallicity. Mergers may also have played a role. Any successful theory of giant ellipticals must take account of the following observed properties: (1) They have high metallicities but low metallicity gradients, resulting in relatively metal-rich halos. (2) They are inferred to have triaxial figures supported by anisotropic velocity dispersion, rather than by rotation. (3) Globular clusters in giant ellipticals have lower metallicities than halo stars in the same regions. (4) The distribution of globular clusters in giant ellipticals is less centrally concentrated than is the main body of stars.

Additional Reading

Burstein, D. (1985). Observational constraints on the ages and abundances of old stellar populations. *Publ. Astron. Soc. Pac.* **97** 89.

de Vaucouleurs, G. (1987). General historical introduction. In *Structure and Dynamics of Elliptical Galaxies* (Proceedings of IAU Symposium 127), T. de Zeeuw, ed. Reidel, Dordrecht, p. 3.

Dressler, A. (1984). The evolution of galaxies in clusters. *Ann. Rev. Astron. Ap.* **22** 185.

O'Connell, R. W. (1986). Analysis of stellar populations at large lookbacks. In *Spectral Evolution of Galaxies*, C. Chiosi and A. Renzini, eds. Reidel, Dordrecht, p. 321.

Pickles, A. (1987). Population synthesis of composite systems. In *Structure and Dynamics of Elliptical Galaxies* (Proceedings of IAU Symposium 127), T. de Zeeuw, ed. Reidel, Dordrecht, p. 203.

See also **Galaxies, Dwarf, Spheroidal; Galaxies, Elliptical, Dynamics; Galaxies, Elliptical, Origin and Evolution.**

Galaxies, Elliptical, Dynamics

David Merritt

Elliptical galaxies are the second largest class of galaxies by number, comprising roughly 30% of all observed galaxies, and a larger fraction of galaxies in dense environments such as galaxy clusters. Compared with spiral galaxies, elliptical galaxies are very regular and smooth in appearance, and contain very little gas, dust, or young stars.

Until 1975, most astronomers viewed elliptical galaxies as oblate spheroids flattened by rotation. In this model, elliptical galaxies were thought to be similar to spiral galaxies in their dynamics and formation history, the principal difference being the efficiency of star formation during collapse of the protogalactic cloud (low efficiency in spirals, high in ellipticals). However, that year saw the publication of the first accurate measurements of the rotation velocities of elliptical galaxies, derived from absorption-line spectra of the stars. These studies, and a large body of observational work since that time, have indicated that the majority of bright elliptical galaxies rotate much too slowly for their shapes to be determined by rotation. It is now believed that the motions of the stars in most elliptical galaxies are essentially random; the shapes of these galaxies are determined by the large-scale anisotropy of the stellar motions, that is, the degree to which random velocities are different in different directions. Since rotation is not very important in these galaxies, there is no compelling reason to assume that they are oblate, and it is now generally believed that elliptical galaxies are triaxial, that is, that their figures are ellipsoids (possibly slowly rotating) with three unequal axes. Since 1975, theoretical and observational work on the dynamics of elliptical galaxies has focused on the following problems: determining their three-dimensional shapes, understanding the character of stellar orbits in triaxial potential wells, constructing self-consistent triaxial models, searching for correlations between the kinematical and morphological parameters of elliptical galaxies, deriving their masses and mass distributions, and resolving the structure of their cores.

MORPHOLOGY

To first order, the appearance of an elliptical galaxy on the sky can be described in terms of its surface brightness profile and its apparent shape. Most elliptical galaxies are well described by an empirical surface brightness law first proposed by Gerard de Vaucouleurs:

$$\Sigma(R) = \Sigma_e \exp\{-7.67[(R/R_e)^{1/4} - 1]\},$$

where R_e is the effective radius and corresponds to the radius that encloses half of the total integrated luminosity of the galaxy, and Σ_e is the surface brightness at R_e. Deviations from this law are often correlated with a galaxy's environment. For instance, dwarf companions to larger galaxies often show surface brightness profiles that fall off more rapidly than de Vaucouleurs's law, an effect that may be attributable to tidal truncation of the envelopes of these galaxies. The brightest elliptical galaxies, called cD galaxies, have more extensive envelopes than predicted by de Vaucouleurs's law. These galaxies are always located at the centers of galaxy clusters, and their envelopes are thought to consist of tidal debris from other cluster galaxies.

The isophotal contours of most elliptical galaxies are usually elliptical to a high degree of accuracy. However, the elongation and orientation of these isophotal ellipses often varies with position. In contrast to surface brightness, ellipticity shows no characteristic dependence on radius: Flattening sometimes increases, sometimes decreases, and sometimes remains roughly constant with radius. The same is true with regard to the orientation of the isophotal major axis as a function of position. The origin and significance of these isophotal twists is not well understood. One possibility is that they result simply from the galaxy's triaxial form, because a set of nested, triaxial ellipsoids with differing ellipticities will show twisted isophotes if viewed from a direction that does not lie along one of the symmetry axes. Alternatively, the twists may be intrinsic, resulting from tidal interactions with neighboring galaxies.

STELLAR DYNAMICS OF ELLIPTICAL GALAXIES

Galaxies are "collisionless" systems: Each star moves along its orbit under the influence of the smooth gravitational potential of the whole galaxy, and hardly ever comes close enough to another star for its motion to be significantly perturbed by the encounter. The basic time scale that governs the dynamical evolution of a collisionless system is just the time for a typical star to cross it in its orbit, called the "dynamical" or "crossing" time; it is given roughly by

$$T_{\rm dyn} \approx \left(GM_{\rm gal}/R_{\rm gal}^3\right)^{-1/2}$$

$$\approx 5\times10^7\ {\rm yr}\left(\frac{M_{\rm gal}}{10^{11}\ M_\odot}\right)^{-1/2}\left(\frac{R_{\rm gal}}{10\ {\rm kpc}}\right)^{3/2}$$

where $M_{\rm gal}$ and $R_{\rm gal}$ are the galaxy's mass and radius, $M_\odot$ is the mass of the Sun, and G is the gravitational constant. A typical large elliptical galaxy has a dynamical time near the center of about 10^8 yr. Because this value is much shorter than the age of the universe ($\sim 10^{10}$ yr), elliptical galaxies are thought to be well "relaxed"; that is, the spatial distribution of stars in these galaxies should long since have reached a smooth unchanging state, and the potential well through which each star moves should be nearly fixed in time.

Computer modeling has shown that most of the stellar orbits in nonrotating, triaxial potentials fill volumes with one of two characteristic shapes. "Box" orbits densely fill regions similar to rectangular parallelepipeds; the basic character of the motion is up and down along the long axis of the galaxy. "Tube" orbits fill roughly doughnut-shaped regions; these orbits circulate around the short or the long axis of the galaxy and avoid the center. Tube orbits are the only orbits present in axisymmetric (oblate or prolate) potentials; it is the box orbits that are uniquely associated with triaxial potentials, and that permit self-consistent triaxial galaxies to exist. Box orbits have the important additional property that a star on such an orbit eventually passes arbitrarily close to the center of the galaxy. This could be important if—as recent observations suggest—some elliptical galaxies contain massive objects (such as black holes) in their cores, since the massive objects will perturb the orbits and induce slow changes in the galaxy's shape.

In addition to the box and tube orbits, which are called "regular," some fraction of the orbits in most triaxial potentials are found to be "irregular," or "stochastic." Stochastic orbits have no well-defined shape; instead they traverse first one, then another, volume, with the transition occurring nearly randomly. It is at present uncertain whether the slow diffusion of these stochastic orbits implies a slow evolution of the structure of elliptical galaxies, on a time scale longer than the dynamical time but shorter than the age of the universe.

Important as orbital studies are for understanding the internal structure of elliptical galaxies, the interpretation of observational data generally involves considerably less-detailed models than the sort described above. This is because the dynamical information available observationally—the rotation velocity and the velocity dispersion of the stars, projected along lines of sight through the galaxy—is not nearly sufficient to constrain a unique orbital model, even if the three-dimensional shape of the galaxy were known. Fortunately, many interesting questions about the dynamics of elliptical galaxies can be answered, at least in part, with the types of observational data currently available. One such question, mentioned above, concerns the degree to which elliptical galaxies are supported by rotation as opposed to velocity anisotropies. Simple calculations based on an *isotropic* model, in which the random component of the stellar motion is the same in all directions at a given point, indicate that the rotation velocity of an elliptical galaxy with an axis ratio of 1:2 should be roughly comparable to its velocity dispersion along the line of sight. In the early 1970s, advances in photon-detection systems and digital data processing permitted the first accurate determinations of these quantities in a number of elliptical galaxies. The results showed clearly that the majority of bright ellipticals were rotating too slowly, by a factor of about 2, for their flattenings to be explained by rotation; thus the velocity distributions in these galaxies had to be strongly anisotropic. In general, the degree of rotational support is observed to increase as galaxy luminosity decreases, with the least-luminous elliptical galaxies having rotation velocities consistent with that expected for isotropic oblate rotators. More recent work has uncovered significant rotation around the apparent long axis of several elliptical galaxies, a result that is easiest to understand if these galaxies are strongly triaxial or prolate.

Much observational work has concentrated on measuring the dependence of stellar velocity dispersion on radius in a large sample of galaxies. These data can be used in one of two ways. If one assumes that the variation of mass density with radius is known for a galaxy—for instance, by equating the mass density to some fixed multiple (the mass-to-light ratio) of the luminosity density, which is easily measured—then one can use the velocity dispersion profile to understand how the basic character of the stellar motion varies with radius. For instance, if the stellar orbits are predominantly radial (i.e., boxes), then the component of their motion along the line of sight falls off more rapidly with radius than if the orbits are mostly circular (i.e., tubes). Alternatively, if one makes an assumption about the orbital character—for instance, that the stellar motions are approximately isotropic—then the variation of velocity dispersion with radius constrains the distribution of mass in the galaxy, since the typical orbital velocity at every radius depends on the amount of matter producing the gravitational acceleration. Unfortunately, it is impossssible to determine both the orbital character and the mass distribution given only the observed velocity dispersions, and this fact has greatly hampered the interpretation of dynamical data. Fortunately, other techniques can sometimes be used to place independent constraints on the mass-to-light ratio; these include measurement of the rotation velocity of the gaseous disks that appear in a small fraction of elliptical galaxies, and observations at x-ray wavelengths of the hot gas that is sometimes present in galactic potential wells. At present, however, there is no elliptical galaxy for which the mass distribution or the character of the stellar orbits has been well determined. It is not yet certain, for instance, whether elliptical galaxies are typically surrounded by the massive dark matter halos that are known to be prevalent around spiral galaxies, although the x-ray data strongly suggest the existence of such halos around a few elliptical galaxies.

CORES AND NUCLEI

Observations of the very centers of elliptical galaxies are hampered by distortions induced by motions in the earth's atmosphere, which limit angular resolution to about one second of arc, regardless of the size of the telescope. It was not until about 1985 that careful observations verified the existence of cores in most elliptical galaxies, that is, regions near the center where the surface brightness levels off to a nearly constant value. (Note that de Vaucouleurs's law predicts an ever-rising central brightness.) Typical core radii, that is, radii at which the surface brightness equals half its central value, range from about 1 kpc for the brightest elliptical galaxies to less than 100 pc for the fainter ones. Many elliptical galaxies are also observed to have unresolved, pointlike nuclei, with much smaller radii and much higher surface brightnesses. The first elliptical galaxy for which useful dynamical information about its core was obtained was M87, the brightest member of the Virgo galaxy cluster. Because M87 is relatively nearby and very large, its core is easily resolved. Early observations of the core of M87 revealed stellar random velocities that rose steeply toward the center, an effect that was initially attributed to the gravitational influence of a massive central object, perhaps a black hole. It was later pointed out that other models were equally consistent with the data, including models in which the stellar velocities are very anisotropic, or the core is elongated along the line of sight. However, recent studies of several other nearby elliptical galaxies make a much stronger case for massive central objects. For instance, stars within a few parsecs of the center of M32, a dwarf elliptical galaxy in the Local Group, are observed to rotate about the center with a velocity of about 100 km s^{-1}. The mass required to produce such large rotational velocities can be derived from the classical equation relating centripetal acceleration to force:

$$\frac{V^2}{R} = \frac{GM}{R^2},$$

where V is the velocity of a star in orbit around a point mass M. The implied mass for the central object in M32 is of order 10^7 solar masses; because of the high rotation, this estimate is not strongly dependent on the unknown anisotropies. A number of other nearby galaxies (including our own) appear to have dark massive objects in their nuclei, with masses ranging from 10^6 to 10^9 solar masses. The nature and origin of the central mass concentrations in elliptical galaxies is still obscure. One possibility is that many elliptical galaxies were once quasars, and that the massive objects are the black holes that once powered their phenomenal radio and optical emission. Alternatively, the nuclei of these galaxies may contain dense clusters of stellar remnants such as neutron stars.

On a somewhat larger scale, the presence of significant rotation in the cores of elliptical galaxies is now believed to be a common phenomenon. Observations of nearby elliptical galaxies show that the cores of $\frac{1}{4}$ to $\frac{1}{2}$ of these galaxies exhibit strong rotation, often around a different axis than that of the rest of the galaxy. These rapidly rotating cores are thought to be the remains of dwarf galaxies that have spiraled to the center of the larger galaxies, remaining substantially intact during the descent; their large rotation velocities are all that remains of the original orbital velocity of the dwarf.

PARAMETER CORRELATIONS

In spite of their complexity, elliptical galaxies appear to obey certain laws relating their kinematical and morphological properties. The existence of these laws is important, because they suggest that the process of galaxy formation—which is still very poorly understood—somehow imposes constraints on the present-day form of galaxies, preferring certain final configurations over others. One such correlation, mentioned above, is between the luminosity and rotation velocity of elliptical galaxies: Luminous elliptical galaxies exhibit less rotation, measured in terms of the amount required for rotational support, than faint elliptical galaxies. The dependence is not very tight, however. A much tighter correlation is seen between the total luminosity L and the central velocity dispersion σ of elliptical galaxies:

$$L \propto \sigma^4$$

called the Faber–Jackson relation (after Sandra M. Faber and Robert E. Jackson). This relation is important because it allows one to estimate the intrinsic luminosity of a galaxy from its observed velocity dispersion, and hence to calculate the galaxy's approximate distance. The Faber–Jackson relation, combined with Fish's law (after Robert Fish), which states that the average surface brightness of elliptical galaxies is roughly constant, implies that all elliptical galaxies have roughly the same mass-to-light ratio. Additional correlations are observed between the parameters that characterize the cores of elliptical galaxies. For instance, the core radius r_c depends on total luminosity through the approximate relation

$$r_c \propto L^{5/4}$$

These relations suggest that elliptical galaxies, in spite of their individual complexity, constitute a basically one-parameter family, the single parameter being the total luminosity or the total mass.

Additional Reading

Binney, J. and Tremaine, S. (1987). *Galactic Dynamics.* Princeton University Press, Princeton.

de Zeeuw, T., ed. (1987). *Structure and Dynamics of Elliptical Galaxies. Proceedings of IAU Symposium 127.* Reidel, Dordrecht.

Martinet, L. and Mayor, M., eds. (1982). *Morphology and Dynamics of Galaxies.* Geneva Observatory, Geneva.

Merritt, D., ed. (1989). *Dynamics of Dense Stellar Systems.* Cambridge University Press, Cambridge.

Mihalas, D. and Binney, J. (1981). *Galactic Astronomy.* W. H. Freeman, San Francisco.

See also **Galaxies, Elliptical; Galaxies, Elliptical, Origin and Evolution.**

Galaxies, Elliptical, Origin and Evolution

Peter J. Quinn

During the 1920s, astronomers began to realize that many of the diffuse nebulae that they sketched and photographed through their telescopes actually were distant galaxies similar to our own Milky Way. These distant galaxies were grouped into two broad categories: the spirals and the ellipticals. Both of these categories can be further divided into subtypes, giving a classification scheme known as the Hubble sequence. Spirals are disk-like systems consisting of stars, gas, and dust. They possess global features such as spiral arms as well as young, active, star-forming regions. Ellipticals, on the other hand, appear to be ellipsoidal distributions of mainly old stars, lacking in cold gas and spectacular features like spiral arms. To the two main galaxy families we can add the S0 galaxies, which are similar to spirals in the sense that they have a flattened disk of stars, but which are also similar to ellipticals in that they lack a young stellar population. There are also irregular galaxies which cannot easily be classified as either spiral or elliptical. For bright, isolated galaxies, the relative abundance of the various galaxy types is elliptical:S0:Spiral:Irregular: 13:22:61:4. The exact mix of galaxy types depends on the environment in which we choose to sample them. In the densest parts of rich clusters of galaxies, for instance, spiral and irregular galaxies are almost completely absent.

In this entry we will examine the origin and evolution of elliptical galaxies. In asking why ellipticals are the way they are, we will contrast them to the observed structural and kinematic makeup of spirals and ask what mechanisms operating during galaxy formation and evolution could have produced these two very different types of galaxies. The following ideas are a synthesis of the work of many researchers, based on clues from the observations of ellipticals and insights gained from computer models of cosmological galaxy formation and mergers of galaxies.

WHEN DID ELLIPTICALS FORM?

In the mid-1960s, researchers discovered that the Earth is being bombarded with microwave radiation from deep space. This radiation is very uniform in intensity over the whole sky and is now generally believed to be the relic radiation from the Big Bang. We can only set an upper limit on the order of 1 part in 10,000 for the fluctuations in the microwave radiation temperature on scales less than an arcminute. This means that the perturbations in the matter density that eventually resulted in galaxies had to grow from dimples of 0.01% to their current density (about 1000 times the mean density of the Universe) over the time in which the Universe had expanded by a factor of 10,000. Just exactly when galaxies appeared as dense systems of gas and stars in this period between the epoch corresponding to a redshift of 10,000 and now is still an open question. One constraint on the epoch of galaxy formation comes from the time it would take a large galaxy to collapse. In other words, a galaxy must be at least one collapse time old or else it would not be a recognizable galaxy today. For a galaxy like the Milky Way, the collapse time is approximately twice the free-fall time from the outer edge of its dark halo (50–100 kpc), which is about 2×10^9 yr. This corresponds to a collapse redshift of less than 4. This seems reasonable, as the most distant quasars we have yet detected have redshifts of about 4. So, sometime between a time when the Universe was about one quarter of its current size and now, most large galaxies began to form. However, there is evidence that some galaxies are undergoing major changes at even smaller redshifts. The types of galaxies that are prevalent in large clusters of galaxies (which are now mostly ellipticals) seem to change at a redshift of about 0.5 (6×10^9 years ago for a Hubble constant of 50 km s^{-1} Mpc^{-1}). Beyond (earlier than) redshift 0.5, there appear to be more blue, star-forming galaxies then at lower redshifts. So the general population of galaxies in clusters seems to have not been in any kind of equilibrium until fairly recently in the Universe. Indeed, we see galaxies very close to us that are undergoing major changes either due to massive bursts of star formation or collisions with other galaxies. Hence we should not necessarily think of galaxy building as happening at a given time and then turning off. Rather, galaxies may be continually modified during their history. If the processes that formed the first systems of stars and gas are the same as those we see operating at smaller redshifts (e.g., mergers and tidally induced star formation), then both formation and evolution can be considered as part of the same ongoing process. Because the stars in ellipticals are at least as old as the stars in the spheroidal bulges and old disks of spiral galaxies, we can assume that the components of ellipticals were formed at a similar time to the oldest components of spirals.

WHERE DID ELLIPTICALS FORM?

One of the most important observations relevant to the formation of galaxies of various morphological types is that the properties of galaxies reflect their environments. The mix of spirals and ellipticals depends on the local density of galaxies, in that denser regions have a higher proportion of ellipticals. Hence the mechanisms that were responsible for forming different types of galaxies could not have been purely internal. There had to be some external influence that chose the preferred type of galaxy in a given environment. There are two important questions raised by the discovery of a "morphology-density" connection.

Where did the morphology-density relationship develop? Because a similar morphology-density relationship exists for compact groups as well as for clusters, and because clusters have a longer collapse time than small groups, we are led to conclude that clusters of galaxies probably inherited their morphological mix from smaller-scale environments that are now incorporated into clusters.

When did the morphology-density relationship develop? We know from the previous section that galaxies take a few times 10^9 yr to collapse. Compact groups of galaxies have a mean density lower than that of a galaxy, so their collapse time is longer. This means that the initial galaxy had already formed before the environment around it had collapsed. If galaxies do not change much after they are formed, then the morphology-density relationship is a consequence of the initial, pre–group-collapse environment. In that case the mean mass density of a region in which a galaxy forms must drive the resultant galaxy morphology. However, if the collapse and evolution of the group environment has an influence on the morphology of the galaxies it contains, then the morphology-density relationship evolves as the group does. We occasionally find ellipticals in the field, outside rich groups and clusters. This could be an example of a group environment that has evolved to complete collapse, that is, where all the constituent members have been accumulated into one large elliptical galaxy. Indeed, close encounters between galaxies in groups very often result in mergers, because the internal motions in the galaxies and their orbital motions are similar, making tidal coupling between the motions of the galaxies and the motions of their constituent stars very efficient.

We are now left in a situation where we are not sure whether ellipticals were formed as a consequence of the same process that formed the region that is now a compact group of galaxies (that is, gravitational collapse of a region that is more dense than the mean of the Universe around it) or as a consequence of the evolution of the group environment after collapse. In order to sort out the relative importance of initial conditions and evolution for the current appearance of ellipticals, we need to examine some dynamical properties of ellipticals that may contain clues.

HOW DID ELLIPTICALS FORM?

A great deal of progress has been made in the area of galaxy formation from the study of dynamical evolution in cosmology using supercomputers and *N*-body models. These models allow us to ask questions about the evolution of the structure of objects that form through purely dissipationless collapse, that is, where gravity is the only force responsible for moving matter around. In reality, galaxies contain neutral gas and plasma that can interact electromagnetically as well as gravitationally. However, we believe that the dark matter that dominates the size and mass of a galaxy evolved in a dissipationless manner, as it is probably not made of normal matter. One important property of a galaxy is the total amount of angular momentum it contains. This is predicted by *N*-body models and probably does not depend too much on what type of matter is in the model (i.e., dissipative or nondissipative), because the angular momentum is produced by tidal interactions between protogalaxies before they collapse. If the dissipative material in a protogalaxy loses energy and shrinks, it will eventually form a disk that is held up by its angular momentum. If we measure the amount of angular momentum in a galaxy like the Milky Way we find that it agrees with the angular momentum content in a protogalaxy as determined from the *N*-body models. So it appears that spiral galaxies have retained most of the angular momentum that they acquired as protogalaxies. Elliptical galaxies are similar in size to spiral galaxies of comparable luminosity. Hence the two types of galaxies presumably collapsed by about the same amount due to the protogalactic gas cooling by radiation. However, whereas spirals are today rapidly rotating because of all their angular momentum, ellipticals hardly rotate at all. So at some point elliptical galaxies lost more than 90% of their angular momentum, whereas spirals did not. This loss of angular momentum has to be related to the process that made ellipticals different from spirals and hence to some property of the local environment.

How can we remove angular momentum from a protoelliptical? Angular momentum is removed by applying a torque; that is, something has to pull on the matter that formed the elliptical and hence slow down its orbital motion. Such a process is very common in dynamical collapse. It is called dynamical friction. Consider a cloud of matter which is not completely homogeneous; that is, it contains some lumps. If we start this cloud of matter off being a little more dense than the surrounding Universe, then it will eventually stop expanding with the Hubble flow and begin collapsing. As the lumps of matter move through the cloud they create a wake of matter behind them similar to the wake caused by a boat passing through water. This wake is formed by particles that are deflected by the gravitational attraction of a lump and eventually converge behind it. The wake slows the lump down and hence removes some of its kinetic energy and angular momentum. Eventually the lumps will settle to the center of the cloud and form a low-angular-momentum, tightly bound system.

Our protogalaxy consists of both dissipationless dark matter and normal matter that can potentially form stars. If the normal dissipative matter in protoellipticals was in the form of lumps, then we would be in a good position to produce a low-angular-momentum galaxy. The dark matter could do the job of absorbing the unwanted angular momentum. By implication this means that the normal matter in the disks of spirals would have to remain quite smoothly distributed during the collapse process. Now all we need is a process that preferentially makes lumpy protogalaxies in dense environments and we would have the basis of a theory for the formation of ellipticals. Again, the *N*-body models have been useful in pointing out an effect that was already known from analytic studies.

Consider two identical regions of the early Universe that are each destined to become a galaxy. These regions are higher in density than the Universe at large and contain lumps of matter of various sizes (consisting of gas, dark matter, and possibly some stars). We will call these lumpy, overdense regions protogalaxies. One protogalaxy is located in a region of the Universe that has a lower-than-average density and the other in a region that has above-average density, that is, one protogalaxy is on top of a "hill" of the density field and the other is in a "valley". The protogalaxy in the low-density region will evolve into a galaxy in a low-density environment, and the other will evolve into a galaxy within a group of galaxies. It can be shown that the final density of a system after gravitational collapse scales as the third power of its initial density with respect to the Universe at large. If the mean density of the regions within which our two protogalaxies are collapsing varies by a factor of only 2 (the height of the density hill over that of the density valley), then the final protogalaxies will have densities that differ by a factor of 8. So protogalactic systems in regions that will become groups of galaxies will generally have higher densities than those in the field. At some point the lumps in the protogalaxies will begin to form stars. We know from studies of external galaxies that the ability of gas to turn into stars is related to the density of the gas. So it would be natural to expect that many of the lumps in protogalaxies in dense regions will consist of stars as well as gas. Lumps of stars are efficiently decelerated by dynamical friction and hence the final stellar system formed by the accumulation of the lumps will have a low specific angular momentum. Therefore, via the action of lumps of stars forming in dense protogalaxies, we can account for both the low-angular-momentum content of ellipticals and their preference for high-density environments.

As noted before, small dense groups of galaxies are also excellent places for galaxies to tidally interact and eventually merge into a single system. The merging process is dynamically similar to the formation of a galaxy from lumpy initial conditions. Two galaxies are brought together because of an exchange of energy and angular momentum between orbital and internal motions. This exchange takes place between either the luminous parts of the galaxies and their dark halos or, in the final stages of the merger, between the luminous components. *N*-body models have again showed that disks are rather fragile and tend not to survive mergers between galaxies of comparable mass. When disk galaxies collide, the final result is an elliptical galaxy, where again the stars have lost angular momentum during the merger. Merging between galaxies in small groups thus adds to the morphology-density relationship, as it removes disk-like systems from high-density environments. In this way we can think of mergers going on today in small groups as being a continuation of the elliptical-making process that began at higher redshifts.

A COMPARISON WITH BASIC PROPERTIES

If ellipticals form through a process of agglomeration of smaller systems, then based on what we know of dissipationless merging, we can compare the theory with some basic properties of ellipticals.

Shape Whether the protoelliptical lumps are themselves small regular galaxies or some type of irregular systems of stars and gas, the final distribution of stars after multiple mergers will be very similar to a triaxial system which will, in projection, look similar to an elliptical. If two roughly spherical lumps of stars collide on a nearly radial orbit, then the final system will be roughly prolate, with the long axis pointing in the direction of the initial orbital motion. This shape will be supported by anisotropic velocities, not rotation. Some of the initial orbital motion will be maintained along the collision axis. Hence triaxial aggregations of stars are a natural consequence of a sequence of dissipationless mergers.

Size and mass Ellipticals span a range of luminosities from 10^8–10^{12} solar luminosities. Some ellipticals could have been formed from lumpy protogalaxies and then have remained untouched, whereas others could have suffered major mergers late in their history. It is possible that large ellipticals may have experienced more late merging than small ellipticals. The progenitor lumps of small ellipticals may no longer exist in their original form today, but we know that systems as small as 10^6 solar luminosities exist (globular clusters and dwarf spheroidals) and are commonly found around large galaxies. Spiral galaxies like the Milky Way have luminosities of several times 10^{11} solar luminosities. Because these disk galaxies cannot have suffered major mergers (as these would have destroyed the disks), we know that individual protogalaxies could be at least as large as the Milky Way. Although we have potential lumps that span the range of luminosity of current ellipticals, the distribution of lump sizes that made up a given elliptical is still unclear. Numerical simulations are currently not capable of following the evolution of normal and dark matter over a range of 10^6 in mass that would be necessary to follow the buildup of a large galaxy from very small lumps. If low-mass ellipticals need to have their angular momentum lowered as much as large ellipticals seem to, then the protoelliptical lumps must have been smaller than 10^8 solar luminosities.

Densities The mean and peak luminosity density of ellipticals is higher than that of spirals of a comparable luminosity. Two possible processes arise from the above theory to contribute to this higher density. First, we believe that ellipticals arise in environments where the density of collapsed objects is higher than in the regions where spirals generally form. These deeper potential wells will lead to denser stellar systems after gaseous dissipation and dynamical friction have done their work. Second, since the protoelliptical lumps have lost kinetic energy to the dark matter through dynamical friction, then the binding energy per unit mass of the final amalgamated system may be higher than that of the system of progenitor lumps.

Stellar motions Because mergers naturally give rise to systems with shapes similar to ellipticals, the permitted equilibrium orbits are also those that we suspect are present in ellipticals. We know that large ellipticals have very small net rotational motion. So the dynamical support provided by the mean motion of the stars is insignificant for the system. This is clearly not the case in spirals, where the whole disk maintains its shape due to the centripetal acceleration provided by organized rotation of disk stars. Yet, despite the inconsequential size of the mean motion in ellipticals, this motion is generally aligned with the structural axes and is organized throughout the elliptical. In other words, the shape of the elliptical is aligned with the small mean motion in the sense that the long axis of the galaxy is usually perpendicular to the mean rotation axis. Also, the magnitude of the mean motion varies with radius in a manner similar to the way that circular orbit velocities vary with radius for the elliptical. This organization of mean motion and correlation with structural axes has been seen in N-body models of mergers. When small dense systems are dropped into initially spherical, nonrotating systems, the final system is flattened and brought into rotation by the matter and kinetic energy deposited by the small system during the merger. In this way, lumps falling into galaxies can organize them in a way that appears consistent with ellipticals. This would strongly suggest that ellipticals have had a long history of lumpy accretions.

ELLIPTICAL EVOLUTION

Many of the kinematic properties (e.g., mean motion alignment and distribution) and peculiarities (shells, counterrotating cores, and peculiar outer isophotes) suggest that many ellipticals have had a long history of bombardment by small systems of stars and gas. This rain of material is probably the tail end of a process that was more intense at earlier times. As shells, counterrotation, and other features of ellipticals have been shown to result from mergers, it is reasonable to suppose that ellipticals evolve primarily by merger processes. The fraction of ellipticals with shells suggests that all large ellipticals went through at least one shell-forming event in their history. The dynamical processes occurring in shell and counterrotating-core formation are similar to those seen in models of lumpy protogalaxy collapse, and so the evolution of ellipticals via mergers is nothing more than a continuation of the galaxy-building process that began at higher redshifts.

CONCLUSIONS

Although mergers of mostly stellar lumps can account for most of the structural and kinematic properties of large ellipticals, we are still unable to explore this process across the entire range of masses appropriate for ellipticals. This is basically due to the limitations of current computer hardware. Our ability to form a complete theory of elliptical formation is limited by our poor knowledge of the star formation process and the history of chemical enrichment in a galaxy that it implies. Although we can incorporate gas as well as dissipationless matter into our models, we lack a basic understanding of how the transition from the dissipative gas to the dissipationless stellar system occurs. The motivation for further N-body studies of dissipationless formation is that we may be able to place useful constraints on star formation physics by completely understanding the dissipationless processes. For example, certain spectra of lump sizes and densities may be required to fit the kinematics and morphology of ellipticals across the range of elliptical masses. These spectra, when compared to the input cosmological spectra of density perturbations, may reveal some density constraints on star formation. Similarly, the point in the mass range of ellipticals where dissipationless merging of purely stellar systems no longer accounts for the gross kinematic properties of ellipticals (such as specific-angular-momentum content) may mark the transition between systems that are formed purely by primordial processes (dissipation and dynamical friction) and those that have been affected by mostly dissipationless mergers late in their evolution.

A guiding light for future theoretical research on the formation and evolution of ellipticals will surely be observations of galaxies at high redshifts, presumably in the process of formation. Recently, new techniques based on low-frequency radio observations have been extremely successful at finding galaxies at high redshifts ($z \sim 3$–4). These galaxies are indeed very complex and lumpy, as our theoretical picture would predict. A detailed study of these lumpy, high-redshift galaxies using instruments like the Hubble Space Telescope will provide the essential physical input for future developments in our theory of elliptical galaxy formation.

Additional Reading

Faber, S. M., ed. (1986). *Nearly Normal Galaxies, from the Planck Time to the Present*. (Proceedings of the Eighth Santa Cruz Summer Workshop in Astronomy and Astrophysics). Springer-Verlag, Berlin.

Fall, S. M. and Lynden-Bell, D., eds. (1981). *The Structure and Evolution of Normal Galaxies*. Cambridge University Press, Cambridge.

Kormendy, J. and Djorgovski, S. (1989). Surface photometry and the structure of elliptical galaxies. *Ann. Rev. Astron. Ap.* **27**.

Zurek, W. H., Quinn, P. J., and Salmon, J. K. (1988). Rotation of halos in open and closed universes: Differentiated merging and natural selection of galaxy types. *Ap. J.* **330** 519.

See also **Galaxies, Elliptical, Dynamics; Galaxies, Formation; Galaxies, Properties in Relation to Environment.**

Galaxies, Formation

George Lake

There are several different ways to define galaxy formation. We leave the problem of generating small fluctuations at a very early epoch or "seeding" at later times to the entries on cosmology. Here we will be concerned with the collapse of a gas cloud and the initial burst of star formation. This definition leads to some ambiguity. Galaxy formation can be the final assembly of the objects that we see today, or it can refer to the formation of the first stars or the formation of a fiducial fraction of the stars. It was once assumed that a large fraction of the stars formed in an initially rapid collapse. In this case, galaxy formation is defined by all of the above happening synchronously. This would produce extremely bright events at high redshift which would be unmistakable. By this criterion, we have not clearly seen any galaxies forming. However, if we adopt either half of the above definitions, we find that galaxy formation is an ongoing process and that it is relatively easy to find nearby examples of galaxies in formation.

THE REDSHIFT OF FORMATION

From direct observations, it is not clear when galaxies formed. The local disk of the Milky Way is an example of continuous galaxy formation. We find that the present rate of star formation is not much smaller than the lifetime average. Generally, the current star formation rate in late-type spiral galaxies multiplied by a Hubble time is approximately the observed disk mass. Similarly, in giant elliptical galaxies with cooling flows, the cooling flow rate multiplied by the Hubble time is roughly the mass of the galaxy. Some nearby objects with high ratios of gas mass to stellar luminosity ($\gtrsim 5$ in solar units) have been labeled protogalaxies. Extreme examples are Malin-1 and the Giovanelli–Haynes object, but there are several others (for example, DDO 170 and DDO 154). These objects are fully assembled but have had little star formation.

On the other hand, we find elliptical galaxies at relatively high redshifts ($z \lesssim 0.8$) where the 4000-Å break in the spectrum shows that the bulk of the stars in these systems were formed much earlier. These galaxies must have formed before $z \sim 2$. There are objects with $z \lesssim 3.5$ seen around radio sources. One of these, at $z = 3.4$, is observed to have a high ratio of 2 μm to optical flux, which is interpreted as indicating an old stellar population. These unusual objects, which are found in surveys of radio galaxies or in searches for very red objects, may not be representative of the formation of an average galaxy. Nonetheless, it is striking that old stellar populations exist at high redshift.

The epoch of galaxy "assembly" can be calculated if we assume that the event was isolated in both space and time. Consider the evolution of a "top hat" fluctuation, a spherical region of space with an excess density embedded in a flat Universe (one with exactly the critical density). Such a system behaves like a miniature closed Universe and turns around when its density is $9\pi^2/16$ times the critical density, $\rho_{crit} = 3H^2/(8\pi G)$, where H is the expansion rate or Hubble constant. If pressure is unimportant, it will collapse by a factor of 2 to virialize. Thus its final density is $4.5\pi^2\rho_{crit@turnaround}$. Hence, the density of an object that formed without dissipation uniquely specifies its redshift of formation. For a typical disk galaxy, the redshift of turnaround for dissipationless formation is $(1+z_{turn})_{no\ dissipation} \sim 40$. Once we know the additional collapse factor owing to dissipation, **C**, we can correct this number to find the true epoch of formation: $(1+z_{turn}) = \mathbf{C}^{-1}(1+z_{turn})_{no\ dissipation}$. There are two ways to find **C**. If galaxies formed as part of a dissipationless clustering hierarchy, then their surface brightnesses would match a smooth extrapolation of the properties of groups and clusters. In reality, their surface brightnesses are roughly 100 times the extrapolated value, which implies that $\mathbf{C} \sim 10$. (A more rigorous determination matches the luminosity density within galaxies to that derived from the two-point correlation function of galaxies.)

The second method of finding **C** uses the angular momentum of galaxies. The dissipational collapse factor needed for a tidally torqued protogalaxy to produce a centrifugally supported disk again tells us that $\mathbf{C} \sim 10$. Therefore, we find $z_{turn} \sim 3$. The maximum half-mass radius of an average disk galaxy was ~ 100 kpc and was achieved at $z \sim 4$.

The preceding arguments assumed that the dimensionless cosmological density parameter $\Omega = 1$ where $\Omega = \rho/\rho_{crit}$. If $\Omega = 0.1$, this does not change the evaluation of the collapse factor owing to dissipation, but it does change the formation redshifts. In such a model, galaxy formation occurred at a redshift of ~ 6. For the sake of simplicity, we will continue to assume that the Universe is flat.

We calculated the above numbers for disk galaxies. We run into difficulties if we instead consider elliptical galaxies. Using the density argument, one concludes that elliptical galaxies have dissipated by a factor of 20. However, the angular momentum of ellipticals is exactly what is expected from tidal torques without any subsequent dissipation. For this reason, understanding the origin of the Hubble sequence of galaxy shapes has been a stumbling block for theories of galaxy formation.

THE FORMATION OF THE HUBBLE SEQUENCE

Ten years ago, one popular picture was that ellipticals were products of dissipationless collapse at high redshift, whereas spirals formed later with considerable dissipation. Here, ellipticals and bulges are formed in an early epoch of Compton cooling, whereas disks result from radiative cooling at a later time. For some time, this was the standard model of dissipational galaxy formation. In this picture, the problem was split into two halves, where the morphological types are cooled by different physical processes from perturbations of vastly different amplitudes and owe their luminosity functions to two independent processes.

It would be preferable to discover a control parameter that causes a bifurcation leading to disks and spheroids. Over 50 years ago, Sir James Jeans noted that the flattest spheroids had the maximum flattening of the Maclaurin sequence. He proposed that angular momentum was the control parameter and the dynamical instability of rapidly rotating systems was the physical mechanism that separated disk galaxies and spheroids. A variant of this scheme links initial angular momentum to "overdensity," and hence the clustering environment. Unfortunately, the range of spins measured in *N*-body experiments is not sufficiently broad to explain the full Hubble sequence. These simulations and other calculations also show that the spin has a relatively weak overdensity dependence. Nevertheless, the Hubble sequence is a sequence of angular momentum and, until recently, most schemes focused on stretching the range of initial angular momentum to produce the final range.

The key to the shift away from this picture has been the realization that, to first order, the Hubble sequence is a velocity or mass sequence. The characteristic velocity dispersion of an elliptical galaxy is ~ 250 km s^{-1}, which implies a circular velocity of ~ 425 km s^{-1}, whereas a typical spiral galaxy has a rotation velocity of only ~ 180 km s^{-1}. Recent *N*-body simulations have shown how the final virial velocity of a system determines its morphology. If the dark matter has a characteristic velocity dispersion at the epoch of galaxy formation, then more-massive objects underwent "cold collapses," and the less-massive collapses were "warm." Gas settles gently into circular orbits in warm collapses, leading to a disk galaxy. In cold collapses, the gas undergoes violent relaxation leading to strong density gradients and the outward transport of angular momentum. The dense inner region quickly makes stars, whereas the gas in the outer parts takes a long time to cool. After 10 dynamical times, roughly half of the gas (with approximately a third of the specific angular momentum) forms a dense slowly rotating bulge.

The final outcome of these simulations is that the more-massive elliptical galaxies have a half-light radius that is half that of a spiral galaxy and a specific angular momentum that is down by a factor of 3 from that of a spiral. This is an excellent match to the observations and shows that it is possible for disks and ellipticals to form from a continuous fluctuation spectrum.

Disks form in these experiments with an angular momentum distribution similar to that in the initial state. This validates our early calculation for the redshift of formation, z_{turn}. The observed differences between spirals and ellipticals also fix the velocity dispersion of the dark matter at turnaround, σ_{turn}. Numerous simulations have shown that z_{turn}, σ_{turn}, and the mass of the protogalaxy uniquely determine the density profile and core radius of the dark matter. This prediction of the density distributions is qualitatively borne out by current observations and may prove to be a stringent test of the theory.

The characteristic velocity dispersion needed to separate disks from spheroids results in a pressure that is more important for the perturbations that become disks. As a result, we expect that disk formation will be delayed by comparison to that of the spheroids. Indeed, the simulations show a rapid early formation of spheroids (during the violent relaxation phase which occurs on a collapse time scale), whereas disks are formed more slowly by continual infall.

Alternative approaches to galaxy formation emphasize environmental factors. These are the ongoing addition and removal of mass: accretion, cannibalism, merging, and stripping. The merger hypothesis proposes to make disks and collide them to make ellipticals. Accretion advocates envision forming spheroids early and slowly depositing disk material around them. There are composite schemes where systems that undergo a lot of merging become ellipticals, and relatively undisturbed coherent collapses turn into disks. These schemes require that star formation is carefully timed to take advantage of the merger dynamics. There is no doubt that all of these effects have varying degrees of importance.

OBSERVING THE EPOCH OF FORMATION

Evidence is rapidly emerging that supports the notion that galaxies formed at a $z \sim 3$. In deep surveys, a population of blue galaxies emerges at B-magnitudes greater than 22. These objects are so numerous that it is difficult to explain them as anything other than a cosmologically distant population of star-forming galaxies. They must have a redshift $z > 0.8$, as an irregular galaxy at lower redshift would be too red. They cannot have $z > 3.5$, as this would place the Lyman continuum break in the B-band and they would again become too red. More recently, a large population of spatially extended objects with "flat red" spectra have been discovered. If we attribute the red population to the Lyman break and note that no "ultrared" objects have been found, we conclude that the process of galaxy formation started at $z \sim 4$.

A recent study of $z > 4$ quasars indicates that the illusive Gunn–Peterson trough has been found. (The trough is caused by the Lyman absorption of intervening hydrogen, which lies at all redshifts up to that of the quasar.) At lower redshifts, absorption-line systems with low neutral fractions are seen. Since it is now believed that significant ionization sources other than quasars are needed to ionize these clouds, the output from young galaxies starting at $z \lesssim 4$ is an appealing option.

The absorption-line systems at lower redshifts may also probe galaxy formation. If the clouds with high optical depths (neutral column densities $\gtrsim 10^{18}$ cm^{-2}) are normally associated with galaxies, it would imply a covering fraction of order unity to a radius of 50 kpc. This is in good agreement with our estimated size of protogalaxies.

Also seen are Lyman-α clouds, which have lower column densities. These clouds may be a part of the galaxy-formation process, either small-scale gravitational collapse in the hierarchical growth of galaxies or condensations from thermal instabilities in the protogalaxy. An alternative proposal is that they are intergalactic, confined by an explosively heated medium. Searches for He^+ ($\lambda 304$) and He I ($\lambda 584$) absorption features are the key to determining where these clouds fit in the formation process.

SUMMARY

The formation of galaxies was a gradual process beginning at a redshift of ~ 4 and continuing today. There are numerous observations of galaxies forming. Current and average rates of star formation are known in disk galaxies. Extended emission around radio galaxies has been seen to a redshift of 3.4. Protogalaxies have been "discovered" in the form of large numbers of blue objects in deep surveys, flat-red extended sources, fuzz around high-redshift radio galaxies and nearby gas-rich, low-surface-brightness objects. Relating all these observations to systematic properties of the distribution of dark and luminous matter as well as the cosmological conditions proposed for galaxy formation is the hard work for the coming decade.

Additional Reading

Berry, M. (1978). *Principles of Cosmology and Gravitation*. Cambridge University Press, Cambridge.

Binney, J. and Tremaine, S. (1988). *Galactic Dynamics*. Princeton University Press, Princeton.

Faber, S. (1982). Galaxy formation via hierarchical clustering and dissipation: the structure of disk systems *and* Galaxy formation via hierarchical clustering and dissipation: the structure of spheroids. In *Astrophysical Cosmology*, H. A. Bruck, G. V. Coyne, and M. S. Longair, eds. Pontificia Academia Scientarum, Vatican City, 191 *and* 219.

Gunn, J. E. (1982). The evolution of galaxies. In *Astrophysical Cosmology*, H. A. Bruck, G. V. Coyne, and M. S. Longair, eds. Pontificia Academia Scientarum, Vatican City, p. 233.

Hodge, P. (1986). *Galaxies*. Harvard University Press, Cambridge.

Kaiser, N. and Lasenby, A. N., eds. (1988). *The Post-Recombination Universe*. Kluwer, Dordrecht.

Sandage, A. (1961). *The Hubble Atlas of Galaxies*. Carnegie Institute of Washington, Washington, DC.

See also **Cosmology, Galaxy Formation; Cosmology, Observational Tests; Galaxies Elliptical, Origin and Evolution; Galaxies, High Redshift; Galaxies, X-Ray Emission; Protogalaxies; Quasi-stellar Objects, Absorption Lines.**

Galaxies, High Redshift

Patrick J. McCarthy

The search for distant galaxies was first motivated by the desire to use them in the classical cosmological tests for determining the value of the deceleration parameter (q_0). It quickly became clear that the evolution of galaxies must be taken into account before cosmological questions could be addressed. Thus the emphasis of high-redshift galaxy research shifted to empirically quantifying the evolution of galaxies and identifying the epoch of galaxy formation. If the latter can be achieved, it will provide an important constraint on competing models of galaxy formation and the growth of structure in the Universe. Nearly all galaxies known to have redshifts greater than unity were recognized on the basis of their unusually powerful radio emission. The optical properties of these galaxies are closely linked to their radio properties, making them less useful for cosmological and evolutionary studies. Some important inferences regarding the formation of massive galaxies can, however, be drawn from this highly unusual population of galaxies.

FINDING GALAXIES AT HIGH REDSHIFT

The night sky, when viewed at the faintest levels in optical light, is crowded with galaxies. The most difficult aspect of high-redshift galaxy research is to distinguish distant, intrinsically luminous galaxies from inherently faint foreground galaxies. Our lack of an a priori knowledge of how distant, and hence young, galaxies should appear compounds the problem.

Redshift surveys of complete samples of faint galaxies have unveiled information concerning the properties of galaxies out to redshifts of $z = 0.75$ or so. The redshifts of these galaxies are determined both from stellar absorption lines (primarily ionized calcium) and from strong emission lines from ionized gas (usually doubly ionized oxygen). The shape of the luminosity function of galaxies ensures that most faint galaxies will be objects of more or less average luminosity at modest distances, making it unlikely that galaxies with $z > 1$ will be found in random surveys.

Rich clusters of galaxies, readily identified on photographic plates because they stand out from the more uniform distribution of field galaxies, have been identified out to redshifts of 0.9. These observations have shown significant evolution in the properties of cluster galaxies in the last $4\text{–}8 \times 10^9$ years, but foreground confusion problems have prevented the detection of clusters beyond $z = 1$, if any exist.

The most productive method of isolating distant galaxies has been identifying the optical counterparts of bright radio sources. High-luminosity radio sources can be readily distinguished from

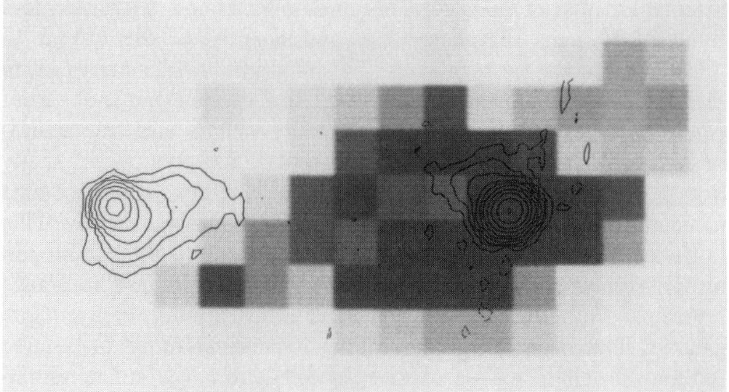

Figure 1. Lyman-α (gray scale) and radio continuum (contours) images of the distant radio galaxy 3C 326.1 ($z = 1.8$). This was one of the first distant galaxies found to be surrounded by a giant cloud of ionized gas. The Lyman-α emitting cloud is roughly 100 kpc (300,000 ly) across and is aligned with the axis of the double-lobed radio source. The Lyman-α image was taken at the Lick Observatory by the author and Prof. Hyron Spinrad. The radio continuum image was made with the Very Large Array by the author and Dr. Wil van Breugel.

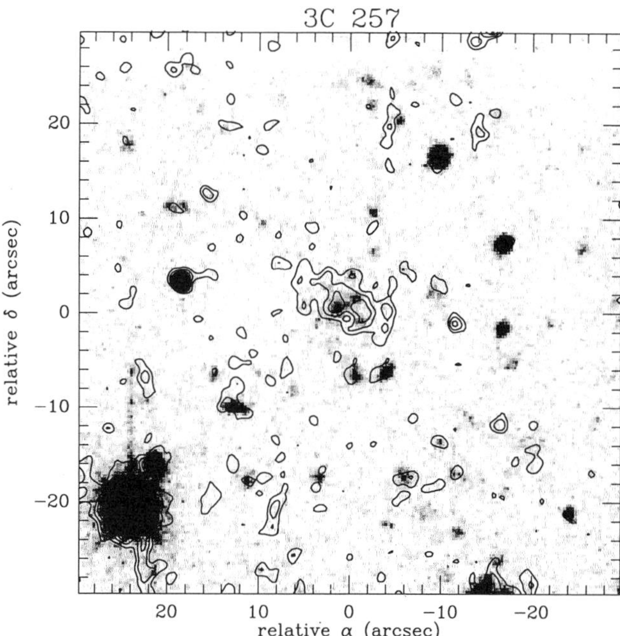

Figure 2. Optical continuum (gray scale) and Lyman-α (contours) images of the radio galaxy 3C 257, one of the most distant galaxies currently known ($z = 2.5$). This galaxy has the clumpy and elongated structure that is typical of radio galaxies with redshifts greater than 1. The images were taken with the 4-m Mayall telescope on Kitt Peak by Hyron Spinrad and Mark Dickinson.

low-luminosity sources on the basis of their double-lobed morphology. Bright radio sources are less crowded than faint optical galaxies, and thus confusion is not as likely. The culmination of high-redshift radio galaxy identification in the era of photographic astronomy was Rudolph L. Minkowski's 1960 measurement of $z = 0.46$ for the galaxy associated with the radio source designated 3C 295. The advent of digital detectors allowed other observers to carry on the approach pioneered by Minkowski to much fainter limits. The first galaxy with a redshift greater than unity, 3C 368, was observed by Hyron Spinrad at Lick Observatory in 1982. In the past five years galaxies identified with relatively strong radio sources have been found out to redshifts of 2, providing the first glimpse of galaxies when the Universe was less than $\frac{1}{2}$ of its current age. The redshifts of these faint galaxies are determined from their strong emission lines of ionized oxygen, carbon, and hydrogen. Recently, astronomers at the University of Hawaii and Johns Hopkins University have extended the search to somewhat fainter radio fluxes and have located galaxies with redshifts approaching 4, a distance at which the Universe was between 15 and 30% of its current age.

PROPERTIES OF HIGH-REDSHIFT RADIO GALAXIES

Only recently has comparison between the optical and radio properties of distant galaxies become possible. Multiwavelength investigations have shown that strong correlations exist between the two. High-luminosity radio galaxies are often associated with large emission-line nebulae, some extending > 100 kpc (300,000 ly). For redshifts greater than ~ 0.5, these giant emission-line regions are found exclusively along the axis defined by the double-lobed structure of the radio source (see Fig. 1). These emission-line regions are extremely luminous and thus require a large input of energy to maintain their highly ionized state. One potential source of energy is the active nucleus itself. A number of researchers have recently suggested that high-redshift radio galaxies are closely linked to radio-loud quasars and may only differ in their angle to the line of sight to the observer, quasars being the pole-on subset of radio galaxies. A more striking correlation between radio and optical properties at high redshifts concerns the rest-frame ultraviolet continuum, which is believed, but not conclusively proven, to be the light of massive young stars. Powerful radio galaxies with $z > 1$ often have ultraviolet continuum emission distributed in highly lumpy linear structures extending 30 kpc or more (see Fig. 2). These elongated continuum structures are in turn closely aligned with their radio lobes. While the origin of this radio/optical alignment is unclear, it is strong evidence that the high-redshift *radio* galaxies are qualitatively different from present-day massive galaxies (often identified as the descendants of powerful radio galaxies) in a fundamental manner. Thus, by selecting galaxies on the basis of their radio emission, we have found a population of objects that are not likely to be representative of most galaxies, either now or in the distant past.

All hope of learning about galaxy evolution and formation from these unusual objects is not lost. The most natural part of the electromagnetic spectrum in which to study high-redshift galaxies is the near infrared (wavelengths, $\lambda \sim 1$–$3\ \mu$m). As we observe galaxies at higher and higher redshifts from our fixed visual window we observe them deeper into the rest-frame ultraviolet, where the light is dominated by the contribution from a minority population of massive stars. We must shift our observing window to the infrared to examine the bulk of the stars, which emit their light in the rest-frame visual and near infrared. Observers on Mauna Kea in Hawaii have found that most distant radio galaxies have well-developed old stellar populations, even at redshifts corresponding to ages of only a few $\times 10^9$ years. This is strong evidence that these galaxies formed at very high redshifts, greater than 5 or 10, depending on the value of the deceleration parameter q_0. Thus these highly unusual objects tell us that there were massive stellar systems very early in the history of the Universe. If this same conclusion could be shown to hold for the majority of massive galaxies, then certain cosmological models, the cold dark matter model in particular, would be in serious difficulty. The matter is not so simple. The latest generation of infrared arrays have allowed us to obtain images in the rest-frame visual and near infrared. The images show that the light at long wavelengths, believed to be the light of old stars, is also strongly aligned with the radio axis. This alignment is not seen in present-day radio galaxies, and its implications are not yet clear.

THE FUTURE OF THE DISTANT PAST

The future of high-redshift galaxy research is likely to lie with objects that are less extreme than the powerful radio galaxies. The latter have provided us with our first glimpse of galaxies in the early universe and have given us important clues concerning the formation and evolution of galaxies, but they are not well suited to addressing the development of ordinary galaxies. The near infrared, both from the ground and, ultimately, from space, will likely be our most productive window on galaxies and their stellar populations in the distant past.

Additional Reading

Bergeron, J. and Kunth, D., eds. (1988). *High Redshift and Primeval Galaxies*. Editions Frontieres, Paris.

Chambers, K. C. (1989). Radio galaxies at $z > 2$. In *The Hubble Symposium on the Evolution of Galaxies*, R. G. Kron, ed. *Publ. Astron. Soc. Pacific Conference Series* **10** 373.

Frenk, C. et al., eds. (1989). *The Epoch of Galaxy Formation*. Kluwer, Dordrecht.

McCarthy, P. and van Breugel, W. (1989). Emission line properties of high redshift radio galaxies. In *Extranuclear Activity in Galaxies*, E. J. A. Meuers and R. A. E. Fosbury, eds. European Southern Observatory, Garching, p. 55.

Spinrad, H. (1986). Faint galaxies and cosmology. *Publ. Astron. Soc. Pac.* **98** 269.

van Breugel, W. and McCarthy, P. (1989). Relations between radio and optical properties in redshift radio galaxies. In *Extranuclear Activity in Galaxies*, E. J. A. Meuers and R. A. E. Fosbury, eds. European Southern Observatory, Garching, p. 227.

See also **Active Galaxies and Quasistellar Objects, Interrelations of Various Types; Galaxies, Formation; Protogalaxies.**

Galaxies, Infrared Emission

Baruch T. Soifer

Infrared emission from galaxies is comprised of three major components. The first is emission from stellar photospheres that peaks at 1–3 μm, and dominates the energy output of most galaxies. Emission lines due to fine structure transitions in atomic and ionic constituents of interstellar gas contribute measurable emission at some infrared wavelengths as well, as do molecular rotational lines, and in some cases molecular vibration-rotation lines are seen. Such lines can contribute as much as a few percent of the total infrared luminosity of galaxies. By far the dominant component of infrared emission from galaxies at wavelengths longer than 3 μm is that due to thermal radiation by dust in the wide variety of environments found in galaxies. Mechanisms for producing infrared emission in active galaxies and quasistellar objects are discussed in another entry.

Dust is a ubiquitous component of the matter surrounding the stars and constituting the interstellar medium. This dust absorbs the photons radiated at shorter wavelengths by stars (or other energy sources), and reradiates this energy in the infrared. The wavelengths at which this energy is radiated are dependent on the environment in which the dust is found. In a steady state the dust is at a temperature where it radiates an amount of energy equal to the amount of energy it absorbs at shorter wavelengths. This temperature ranges from ~ 1000 K for dust in circumstellar environments, to < 20 K for dust in interstellar space, with corresponding wavelengths of peak emission ranging from ~ 3 μm for the hot dust to > 150 μm for the coldest dust. The dust grains that radiate via this steady-state emission are comparatively large, with particle radii ~ 0.1 μm. There is also a significant fraction of interstellar dust composed of much smaller grains, with radii less than 0.001 μm, that contribute significantly to the output of infrared radiation from galaxies. This smaller dust does not radiate in a steady state, but rather each individual photon that is absorbed (these particles can be thought of equally well as large molecules) heats the absorbing grain to a substantial temperature of many hundreds of degrees for a very short time. The dust at such a high temperature radiates energy very effectively, and cools rapidly. This non-steady-state emission process is important to the production of emission at the shorter infrared wavelengths (i.e., $\lambda < 30$ μm).

Almost from the beginning of infrared astronomy (mid 1960s), galaxies have been known to produce copious amounts of infrared radiation, well in excess of that expected from the stellar photospheres in these galaxies. Two of the galaxies that are brightest in the infrared, the nearby irregular galaxy, M82, and the closest Seyfert 2 galaxy, NGC 1068, were discovered to be bright infrared sources, emitting far more luminosity at infrared wavelengths than in the visible, almost as soon as infrared observations began. Such galaxies were observed from ground-based and airborne telescopes over the entire infrared spectrum. Nearly all such galaxies that were found to be bright infrared sources were chosen based on peculiar properties at other wavelengths.

It was not until the launch of the *Infrared Astronomical Satellite* (*IRAS*) in 1983, with its program of performing an unbiased all-sky survey at 12, 25, 60, and 100 μm, that astronomers were able to assess the significance of infrared emission in the energy budget of galaxies. By conducting a survey of objects based on their infrared brightness, astronomers were able, for the first time, to make an unbiased evaluation of the fraction of galaxies that are bright at infrared wavelengths, and thereby to determine the fraction of the total luminosity of galaxies that is emitted in the infrared. Based on this survey it was found that infrared emission compromises $\sim 20\%$ of the radiant energy in the local universe (i.e., the universe within a radius of ~ 100 Mpc or 3×10^8 light years). At the highest luminosities, the infrared bright galaxies represent the most populous constituent of the local universe, having a higher space density than quasars at the same total power.

It was found that the kinds of galaxies that are bright in the infrared are very dusty galaxies, typically spiral galaxies. Three local galaxies, the Milky Way, the great nebula in Andromeda (M31), and the nearest external spiral galaxy, M33, represent three different kinds of galaxies based on their infrared brightness. M31 is a comparatively dust-free galaxy and less than 10% of its total luminosity emerges in the infrared. M33 is a face-on spiral galaxy whose infrared luminosity is $\sim 25\%$ of the total energy output. Finally, the Milky Way is a comparatively bright galaxy in the infrared, radiating roughly half its total luminosity in the infrared.

Figure 1 illustrates the energy distributions of a variety of galaxies from the radio ($\log \nu = 9$) through to the visible ($\log \nu = 15$), including the infrared. The infrared luminosity ranges from a few percent of the total stellar luminosity in the central 4′ of M31 to roughly 98% of the total energy output of the galaxy Arp 220.

The environments that produce the infrared radiation are found to depend on the level of infrared radiation being produced in the galaxy. In quiescent galaxies like M31, the infrared emission is dominated by the interstellar dust absorbing starlight from the ambient radiation field and reemitting it in the infrared. In galaxies like M33 or the Milky Way the vast majority of the infrared emission is emitted by dust in the environments of giant molecular cloud complexes, where many massive, high-luminosity stars are currently forming or have recently been formed. These complexes are still very dusty, and the dust is very efficient at absorbing the ultraviolet radiation emitted by the hot, massive, young stars that dominate the luminosity output of these regions. This absorbed radiation is then reemitted in the infrared. In this way the amount of infrared luminosity from such galaxies is a measure of the total luminosity in recently formed stars, and by assuming a relation between the luminosity and mass of such stars, the rate at which stars have formed over the last $\sim 10^8$ years can be calculated.

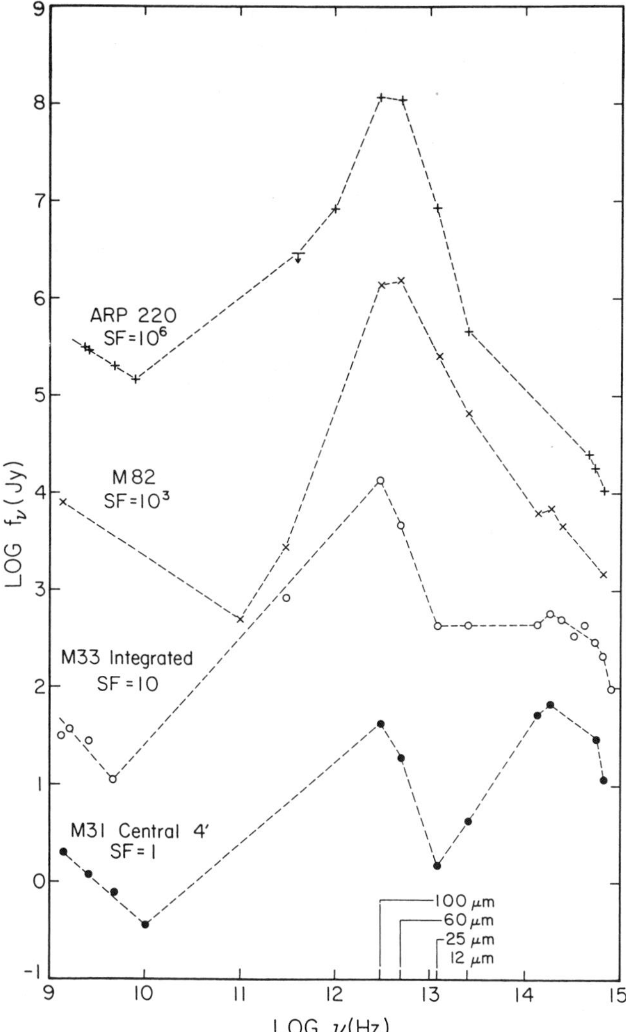

Figure 1. The flux density (W m^{-2} Hz × 10^{-26}) versus frequency (logarithmic scales) for four galaxies representing a range of relative infrared-to-visible brightness. The central 4′ of M31 is much like an elliptical galaxy, M33 represents a normal spiral galaxy, M82 represents an active star-forming galaxy, and Arp 220 is an extremely luminous infrared galaxy. The thermal infrared data, from *IRAS* observations, are plotted from $\log \nu = 12.5$ to $\log \nu = 13.5$; near-infrared and optical data are plotted from $\log \nu = 14$ to $\log \nu = 15$. Radio observations are plotted from $\log \nu = 9$ to $\log \nu = 11$. (*Reproduced from Soifer, Houck, and Neugebauer, 1987. Reproduced, with permission, from the* Annual Reviews of Astronomy and Astrophysics ***25***, *© 1987 by Annual Reviews, Inc.*)

The amount of infrared radiation emitted by galaxies depends to a great extent on the amount of dust that they contain. This in turn corresponds to the amount of interstellar gas, since typically there is 100–200 times more gas than dust in interstellar material. For a galaxy like M31, the amount of dust responsible for the far-infrared radiation is roughly $10^6\ M_\odot$ (1 $M_\odot$ is the mass of the Sun or 2×10^{33} g), corresponding to roughly $10^8\ M_\odot$ of interstellar gas. In a galaxy like the Milky Way, approximately $10^7\ M_\odot$ of dust produce the infrared emission, with a corresponding amount of gas of $10^9\ M_\odot$. In the most luminous infrared galaxies, there is more than an order of magnitude more gas and dust than is found in the Milky Way.

Because infrared emission from galaxies is a process that requires the presence of both dust and higher-energy photons to heat the dust, the infrared emission extends only as far as the optical limits of the galaxies, with a decreasing brightness that is roughly equivalent to that seen in the starlight from the galaxies. In detail it appears that the infrared emission from galaxies decreases somewhat faster than the starlight in the galaxies. This is consistent with the fact that there are less "metals" (elements heavier than helium, which are needed to form dust grains) in the outer parts of galaxies than in the inner regions. In infrared-bright galaxies of high luminosity it appears that much of the emission is generated in the central regions of the galaxy, within 1–2 kpc from the center (as compared to a typical radius for a spiral galaxy of 10 kpc).

A process that has been shown by *IRAS* observations to be important in causing production of infrared bright galaxies is the interaction of two gas-rich galaxies. Such interacting galaxies are identified by such visible phenomena as overlapping disks, multiple nuclei, gravitationally induced tails of stars streaming out from the body of the galaxy, etc. Interacting galaxies comprise an increasing fraction of the infrared bright galaxies as the luminosity of the galaxies increases. Indeed, among galaxies with infrared luminosities $> 10^{11}\ L_\odot$, or 10 times that of the Milky Way (1 $L_\odot$ is the total luminosity of the Sun, or $\sim 4 \times 10^{33}$ erg s^{-1}), interacting systems are the majority. This shows that the process of galaxy interaction/merging is an important one for triggering very luminous galaxies in the infrared. It is expected that in such colliding systems the molecular gas is in some way compressed, either by global compression associated with the interaction of the interstellar media of the two galaxies, or perhaps via more direct cloud-cloud collisions. This compressed molecular gas forms stars quite rapidly and thereby produces the strong infrared emission seen in such galaxies.

The most luminous galaxies found in the infrared emit 100–1000 times more luminosity in the infrared than the Milky Way, or 10^{12}–$10^{13}\ L_\odot$. This is as much power as is emitted by the most luminous objects in the universe, the quasars. The objects that emit this amount of infrared radiation are all quite peculiar, being found in massive, strongly interacting/merging galaxy systems. Because all these galaxies are strongly interacting, it is believed that the interaction has triggered the process that produces the luminosity. Young, massive stars are believed to power a significant fraction of the luminosity in these galaxies. It is also possible that the bulk of the luminosity from these galaxies is not powered by stars, but rather by quasars enshrouded in so much dust that their optical signatures are hidden from view, and the major signature of the quasar is the bolometric luminosity that emerges in the infrared. This picture, if correct, provides an explanation as to how quasars form, in collisions of two gas-rich normal galaxies. The subsequent evolution of the nucleus of the merged galaxy would disrupt the cloud in which it forms, and lead to the appearance of a visible quasar.

Elliptical galaxies, previously thought to be virtually devoid of dust, have been shown to have a significant amount of interstellar dust, via the detection of infrared emission in excess of that expected from the stars in these systems. *IRAS* observations of these galaxies have shown that there is 10^4–$10^5\ M_\odot$ of dust in many of these galaxies. The source of the interstellar dust is believed to be material from old stars in the later stages of evolution. These stars are ejecting their outer envelopes in the form of gas; dust then condenses in the outflowing material. Typically in these galaxies it takes roughly 10^8 yr for the evolving stars to eject as much dust as is observed via their far-infrared emission. This in turn requires a mechanism for removing this interstellar material, since the time to accumulate it is only 1% of the age of the galaxies. One possible removal mechanism is the formation of new stars.

Additional Reading

Lonsdale, C. J., ed. (1987). *Star Formation in Galaxies* (NASA Conference Publication No. 2466). U.S. Government Printing Office, Washington, DC.

Rieke, G. H. and Lebofsky, M. (1979). Infrared emission of extragalactic sources. *Ann. Rev. Astron. Ap.* **17** 477.

Soifer, B. T., Beichman, C. A., and Sanders, D. B. (1989). An infrared view of the universe. *American Scientist* **77** 46.

Soifer, B. T., Houck, J. R., and Neugebauer, G. (1987). The *IRAS* view of the extragalactic sky. *Ann. Rev. Astron. Ap.* **25** 187.

Telesco, C. M. (1989). Enhanced star formation and infrared emission in the centers of galaxies. *Ann. Rev. Astron. Ap.* **26** 343.

See also **Active Galaxies and Quasistellar Objects, Infrared Emission and Dust; Galaxies, Binary and Multiple, Interactions; Galaxies, Molecular Gas in; Galaxies, Stellar Content; Interstellar Medium, Dust Grains.**

Galaxies, Irregular

Deidre A. Hunter

Even within the great family of normal spiral galaxies there is a large variety of structures. As one looks along the Hubble sequence of spirals towards later and later type, one sees the nucleus of the galaxy become less pronounced, and from Sc- to Sm-type galaxies the spiral arms become more ratty. Eventually one arrives at the class of Magellanic-type irregular galaxies, which Hubble described as "lack[ing] both dominating nuclei and rotational symmetry." The Magellanic irregular galaxies, designated as Im galaxies, are the end of the line, at least in terms of the Hubble sequence. However, morphologically, the transition from spiral to Im galaxies is a fairly smooth one. The Magellanic irregular galaxies, as their name implies, are chaotic in appearance, lacking the symmetrical spiral patterns (see Fig. 1). A galaxy can look chaotic because it has been disrupted by a collision with another galaxy. However, there exists a class of normal, noninteracting, intrinsically irregular systems, which comprise $\frac{1}{3}$–$\frac{1}{2}$ of all galaxies.

There is another class of irregular galaxies with an appearance that is very different from that of the Im galaxies: the amorphous irregulars. These galaxies are very smooth in appearance and are not resolved into the many luminous star clusters that give the Im galaxies their distinctive jumbled appearance (see Fig. 2). In fact, they more closely resemble elliptical galaxies, except that, unlike elliptical galaxies, they are blue, contain much neutral hydrogen gas, and are actively forming massive stars. The amorphous irregulars are rare compared to Im galaxies.

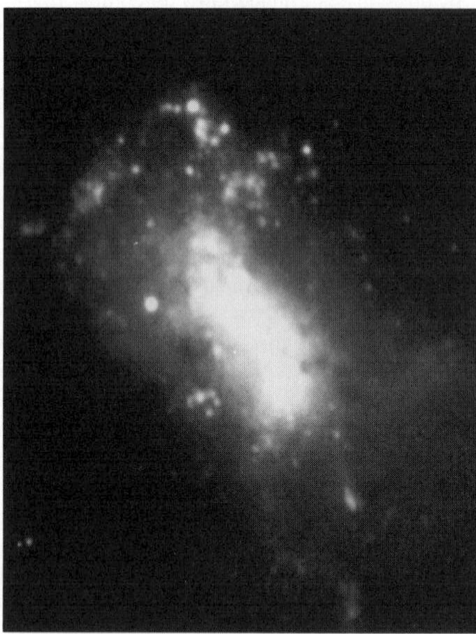

Figure 1. NGC 4449, a Magellanic-type irregular galaxy. This image was obtained with a charge-coupled device on a 0.9-m telescope at Kitt Peak National Observatory through a filter that passes the blue starlight.

Figure 2. The amorphous irregular galaxy NGC 1800. Unlike the Magellanic irregulars, the amorphous systems have a smooth optical appearance. This image was obtained with a video camera on the 2.1-m telescope at Kitt Peak National Observatory through a blue filter.

THE MAGELLANIC IRREGULARS

Global Properties

Compared to spiral galaxies, the Im galaxies are smaller, less-massive systems. Consequently, they are also less luminous and have lower rotational velocities. The optical light that they emit is generally bluer in color, which indicates the relative importance of a hot, young stellar component. A larger fraction of the mass of Im galaxies is in the form of gas, the fuel for star formation, and this gas contains relatively less of the heavy elements (that is, heavier than He), the products of stellar evolution. These latter two characteristics indicate that the irregulars are less evolved than spirals in the sense that less of their mass has been locked up in or processed through stars. Because they lack the spiral density waves that are the cause of the spiral arms, the irregulars are also somewhat simpler systems.

A typical Im galaxy has an absolute blue magnitude M_B of -17 (range -13 to -20) and a mass of $10^9\ M_\odot$ (range $\sim 10^8$–$10^{10}\ M_\odot$, where $M_\odot$ is the solar mass). Average surface brightnesses within an isophotal diameter of 25.0 mag arcsec^{-2} are 22–24 mag arcsec^{-2}. Abundances of oxygen generally range from somewhat less than to $\frac{1}{8}$ of the solar value. A typical late-type spiral galaxy has a maximum rotation velocity of ~ 175 km s^{-1}, and most irregulars have values less than half this. Diameters to an isophote of 26.6 mag arcsec^{-2} are 5–50 kpc. Star formation rates per unit area are 10^{-9}–$10^{-11}\ M_\odot$ pc^{-2} and the timescales to exhaust the current gas supply at the current rate of turning the gas into stars is 10^9–10^{11} yr.

Stellar Content

As a group the irregulars are generally the bluest of the normal galaxies, having average color indices of $U-B = -0.3$ and $B-V =$

0.4, although there is a continuum of normal irregulars from a $U-B=-6$ to $U-B=0$. The blue color indices of irregulars correlate well with other indicators of young stellar populations and are therefore primarily indicative of the presence of massive stars. The global B–V color indices cover a smaller range than U–B, as is expected if the B and V light are primarily contributed by older, intermediate-age stars. Near-infrared colors of irregulars are even more homogeneous, with mean values of J–H = 0.6 and H–K = 0.2. These color indices are nearly the same as the averages of all types of spirals and ellipticals, which probably reflects the basic similarities in old and intermediate-age red stars.

A few irregular galaxies are near enough so that the most luminous stars can be individually resolved. Color-magnitude diagrams for these systems are similar and show pronounced blue supergiant branches and more sparsely populated red supergiant branches. Thus, evolved massive stars are an important component of the stellar population in irregular galaxies. The normal and comparatively uniform properties of young, massive stars in irregulars is demonstrated by other information as well: (1) the optical stellar luminosity functions for irregulars are similar to one another, (2) Wolf–Rayet stars, the descendents of very massive stars, are found in many irregulars, and (3) ultraviolet spectroscopy of luminous star clusters reveal normal mixes of hot massive stars.

The luminous young stars in irregulars are often seen to be superimposed on a spatially more extended, partially resolved sheet of red stars. The spatial smoothness of this sheet is consistent with an older stellar population in which the effects of individual star-forming events have been averaged out. Precise ages are difficult to determine, but the smooth sheet of red stars must originate from stellar populations having ages of ≥ 1 Gyr (1 Gyr = 10^9 yr).

The Interstellar Medium

The Im galaxies are very rich in pristine atomic hydrogen and He gas; typically 20–50% of the mass of an irregular is in this form. Within the galaxies the gas is lumpy, which is expected because gas density enhancements are required to form stars. What is unusual is that in Im galaxies the gas often reaches far beyond the extent of the stars in the galaxy. In one galaxy, NGC 4449, the H I extends to 10 times the optical radius of the galaxy. In spiral galaxies, on the other hand, generally less than 20% of the H I lies beyond the optical galaxy. Because the density of the gas declines radially, the fact that star formation has not occurred beyond some radius in the gas indicates that there is a threshold gas density below which clouds do not condense and form stars.

Another component of the interstellar medium of a galaxy is molecular gas. It is from clouds composed mostly of H_2 that stars appear to form. The symmetric H_2 molecule cannot be directly observed in quiescent clouds. However, it has been found that the easily observed CO molecule can be used as a tracer of H_2. A proportionality constant is applied to the observed CO flux to infer the quantity of H_2 that is present. Observations of CO emission in Im galaxies, however, show that the CO fluxes are much lower relative to the star formation activity compared to spirals. The most likely explanation at present is that the proportionality constant used to deduce the H_2 content from the CO observation is inappropriate to Im galaxies, where the CO flux is low but the H_2 content is not. This conclusion is based on CO observations of individual molecular clouds in the nearest Im galaxy, the Large Magellanic Cloud. The velocity dispersion in a cloud is also a measure of the mass of the cloud, and a plot of the velocity dispersion against the CO flux for clouds in the Large Magellanic Cloud shows a linear relationship that parallels but is offset from that for Milky Way clouds. This is interpreted as evidence for a different proportionality factor necessary to convert CO fluxes to H_2 masses, and is consistent with theoretical expectations based on the overall lower abundance of heavy elements in Im galaxies.

Dust is another component of the interstellar medium. In optical pictures of spiral galaxies one can see dark patches and lanes that are the result of obscuration by concentrations of dust. Such dark patches, however, are less prevalent in optical images of irregular galaxies. Generally the dust reradiates in the far-infrared wavelength region the starlight that heats it. From infrared fluxes one finds that the global dust-to-gas ratios are lower in Im galaxies than in spirals. Furthermore, the characteristic temperature of the dust that radiates at 60–100 μm is warmer and that which radiates at 12–25 μm is cooler in Im galaxies than in spirals. This is at least partially the result of the higher ultraviolet surface brightnesses and lower opacities of the irregulars. The grain characteristics, however, can also affect the temperature of the dust. All else being the same, graphite will be warmer than silicate materials and small grains will be warmer than large grains. If the average composition of the grains depends on the metallicity of the galaxy as a whole, this could be an additional factor in producing the observed properties of the more metal-poor irregulars.

Global galactic kinematics also influence the state of the galactic interstellar medium. Most spiral galaxies rotate differentially, shearing the interstellar medium. Irregular galaxies rotate very slowly, often as a solid body, like a record on a turntable. Thus, there is little, if any, shear. In addition, noncircular and random motions seem to be significant in some Im galaxies. Furthermore, bar structures in the distributions of the older stars, seen in visible light, are frequently present and do not necessarily lie at the center of rotation of the galaxy. This adds additional perturbations to the velocity fields, and it has been suggested that gas flows at the ends of the bars may enhance the formation of large star forming complexes there.

Star Formation

In the preceding section we saw that there are differences between irregulars and spirals in terms of various aspects of the interstellar medium. The heavy-element abundance is lower, the dust-to-gas ratios are lower, the CO content is lower, and the shear due to rotation is lower in irregulars. Because stars are forming from the interstellar medium, we might expect these environmental differences to have observable consequences on the star formation process. For example, theoretical models suggest that the temperatures and sizes of the natal clouds may influence the masses of stars that are formed; lower abundances of heavy elements may lower the efficiency with which a cloud of gas turns into stars, and a lower quantity of dust relative to gas and smaller dust grains may be necessary for the formation of very massive stars. However, an observational comparison of the optical and ultraviolet properties, sizes, luminosities, and morphologies of giant H II regions in Im galaxies with those in spirals shows that they are very similar. Thus, the differences in interstellar-medium characteristics between spirals and Ims have not led to obvious differences in the nature of individual star-forming complexes.

Most irregular galaxies are too distant for us to readily distinguish any but the very brightest individual stars. Therefore, we must find a means other than counting stars to learn about how fast a galaxy is forming them. The most massive stars ($\geq 10\ M_\odot$) are the shortest lived, so they act as tracers of the most recent star formation activity. The prodigious fluxes of ultraviolet photons from these stars ionize the natal gas around the stars. By relating the intensity of the optical atomic line cooling radiation produced by the ionized gas to the number of stellar ultraviolet photons needed to maintain the ionized gas, one can infer the number of massive stars that have formed recently (in the past 10 million years). However, stars of masses down to about 0.1 $M_\odot$ also form, and the lower-mass stars are more numerous, but they are cooler and do not ionize their natal gas. To find the total number and mass of stars recently formed, one must combine the number of massive stars observed with a function that describes the number of lower-mass stars expected for a given number of massive stars. This latter, called the initial mass function, is an empirical relationship that is determined from studies of the Milky Way and several nearby galaxies.

If one uses the above method to measure the rate at which galaxies are converting gas into stars, one finds that Im galaxies have star formation rates that are comparable to those of spiral galaxies relative to their size. This implies that spiral density waves, which spirals have and Im galaxies do not, are not necessary for vigorous production of stars. In the past, the spiral density waves were believed to play the key role in triggering galactic star formation. As the wave passes through the disk, it compresses the interstellar gas into dense clouds from which stars would precipitate. The Im galaxies, however, show that star formation can take place without spiral density waves.

Our empirical picture of star formation in Im galaxies is like that of a pot of oatmeal bubbling on the stove: Star formation moves around the galaxy with time. One region will form stars, deplete the gas locally, and die down; then later another region will become active. However, the bubbling is not entirely random. The star formation rate is not only constant over the lifetime of the galaxy in all but a few peculiar systems; but, on average over long periods of time, it is constant as a function of radius within the galaxy.

When density waves were deposed as the sole mechanism for initiating star formation, there remained the question of what does trigger a region to form stars in Im galaxies. One model that addresses that question, the stochastic self-propagating star formation model (SSPSF), depends on the newly formed stars themselves. Massive stars affect the interstellar medium that surrounds them. They ionize the gas, emit tremendous ionic "winds," and eventually explode as supernovae. The energy dumped back into the interstellar medium must have some effect on the galaxy as either a positive or negative feedback. In the Large Magellanic Cloud and the nearby spiral M31, one can see instances where clusters of massive stars have blown kiloparsec-sized holes in the disk gas. In these cases the gas around the hole has been compressed and appears to have been induced to make stars. The SSPSF model argues that this mechanism allows star formation to propagate around the galaxy as one region afteranother ignites its neighbor. Thus, there is a causal relationship between a young star forming region and its older neighbors. However, although star-induced star formation has been seen to occur in a few cases, whether it occurs on a scale and at a frequency necessary to be the dominant mechanism in a galaxy's star formation activity is not now known.

AMORPHOUS IRREGULARS

Although the amorphous irregulars differ morphologically from the Im galaxies, in terms of most other global properties the two classes of irregulars are quite similar. Furthermore, because the amorphous irregulars are blue and contain ionized gas, we know that they are forming stars at the present time. Many amorphous irregulars, however, differ from the Im irregulars in yet another way: the spatial distribution of the star forming regions. In Im galaxies the star forming regions are generally distributed over the disk of the galaxy. In many amorphous irregulars, on the other hand, the star formation is concentrated to a single supergiant region located at the center of the galaxy. In NGC 1140, for example, the massive stars in the central star forming complex have produced a region of ionized gas that has a luminosity 100 times that of the supergiant H II region 30 Doradus in the Large Magellanic Cloud and contains thousands of O stars.

Why are the amorphous irregulars both so similar and so different compared to the Im galaxies? One model has been proposed that explains the amorphous irregulars as a result of a gentle interaction between two galaxies. Several billion years after the interaction the gas will have piled up at the center, giving rise to a central concentration of the star formation, while the stars are relatively unaffected by the interaction. In this model amorphous galaxies originally must have been Im galaxies, because the global properties are so similar. A crucial test of this model, mapping the H I distribution in these galaxies, has not been completed.

This entry was drawn in part from Hunter, D. A. and Gallagher, J. S. (1989). *Science* **243** 1557 (copyright 1989 by the AAAS).

Additional Reading

Gallagher, J. S. and Hunter, D. A. (1984). Structure and evolution of irregular galaxies. *Ann. Rev. Astron. Ap.* **22** 37.

Hunter, D. A. and Gallagher, J. S. (1986). Stellar populations and star formation in irregular galaxies. *Publ. Astron. Soc. Pac.* **98** 5.

Hunter, D. A. and Gallagher, J. S. (1989). Star formation in irregular galaxies. *Science* **243** 1557.

Sandage, A. (1961). *The Hubble Atlas of Galaxies.* The Carnegie Institution of Washington, Washington, DC.

See also **Star Formation, Propagating.**

Galaxies, Local Group

Paul Hodge

Our Milky Way galaxy is a member of a small group of galaxies that forms a modest density enhancement in the universe of galaxies. This group, which includes about 25 galaxies, is called the Local Group and is similar to many other loose clusters of galaxies in nearby extragalactic space. Its importance comes from the fact that all of its members are near enough to us to resolve well into their individual stellar and interstellar components, and thus we can study them in great detail. This fact allows the Local Group to be the testing ground for many of our ideas, for example, about the distance scale, stellar populations, and galaxy evolution.

PROPERTIES OF THE GROUP

Number of Members

If the local group were to be observed from a distant galaxy, it would seem to include only seven or so members, because there are only about that many that would be conspicuous from such a vantage point. However, there are many small, faint members, and a total census would have to include at least 25 galaxies. Table 1 lists them and gives certain of their vital statistics.

The true number of members remains unknown, and there are three reasons for this. First, there are parts of the sky, especially those areas obscured by the Milky Way dust, that have not yet been searched for members. We know that no large spiral galaxy member lies hidden, because we could detect such a galaxy by its radio emission (especially its neutral hydrogen emission), even if its optical image were completely absorbed by Milky Way dust. However, an elliptical galaxy, perhaps one of low luminosity like the Sculptor dwarf, would not be easy to find because it would emit virtually no radio radiation. Considering the size of the area obscured by the Milky Way, we would expect no more than one or two hidden elliptical galaxies that might have been missed by searches thus far.

The second reason that the list may be incomplete is that there might be objects, like the extremely inconspicuous Ursa Minor

Table 1. Galaxies of the Local Group (listed in order of right ascension)

Name	*RA(1950) Dec*	*Type*	*Apparent Mag. (B)*	*Distance (10^6 ly)*	*Diameter (10^3 ly)*
IC 10	00 17.6 +59 02	Irr	11.7	4.0	6
NGC 147	00 30.4 +48 14	E5	10.4	2.2	10
And III	00 32.6 +36 14	E5	—	2.2	3
NGC 185	00 36.1 +48 04	E3	10.1	2.2	6
NGC 205	00 37.6 +41 25	E5	8.6	2.2	10
M32	00 40.0 +40 36	E2	9.0	2.2	5
M31	00 40.0 +41 00	Sb	4.4	2.2	200
And I	00 43.0 +37 44	E3	14.4	2.2	2
SMC	00 51.0 −73 10	Irr	2.8	0.3	15
Sculptor	00 57.5 −33 58	E3	9.1	0.2	1
Pisces	01 01.0 +21 47	Irr	15.5	3.0	0.5
IC 1613	01 02.3 +01 51	Irr	10.0	2.5	12
And II	01 13.5 +33 09	E2	—	2.2	2
M33	01 31.1 +30 24	Sc	6.3	2.5	45
Fornax	02 37.5 −34 44	E3	8.5	0.5	3
LMC	05 24.0 −69 50	Irr	0.6	0.2	20
Carina	06 40.5 −50 55	E4	—	0.3	0.5
Leo A	09 56.5 +30 59	Irr	12.7	5.0	7
Leo I	10 05.8 +12 33	E3	11.8	0.6	1
Sextans I	10 10.3 −01 26	E	?	0.3	3
Leo II	11 10.8 +22 26	E0	12.3	0.6	0.5
GR8	12 56.7 +14 25	Irr	14.6	4.0	0.2
Ursa Minor	15 08.2 +67 18	E5	—	0.3	1
Draco	17 19.4 +57 58	E3	—	0.3	0.5
Milky Way	17 42.5 −28 59	Sbc	—	0.03	130
SagDIG	19 27.1 −17 47	Irr	15.6	4.0	5
NGC 6822	19 42.1 −14 53	Irr	9.3	1.7	8
DDO 210	20 44.2 −13 02	Irr	15.3	3.0	4
IC 5152	21 59.6 −51 32	Irr	11.7	2.0	5
Tucana	22 38.5 −64 41	?	?	?	?
Pegasus	23 26.1 +14 29	Irr	12.4	5.0	8
WLM	23 59.4 −15 45	Irr	11.3	2.0	7

dwarf, that are simply too faint and sparse to have been found. The most distant dwarf elliptical galaxies are And I, II, and III, which might not have been found easily if their discoverer, Sidney van den Bergh, had not been specifically searching the Andromeda area for them. Other such galaxies, in other parts of the sky and perhaps even a little farther away, might still await discovery. If these types of galaxies are approximately uniformly spaced in the Local Group, there could be as many as 50–100 of these objects within its boundaries. However, it is believed that there is a higher than average density of them near our galaxy, because of its large mass, and that there may be only a few undiscovered examples in the more distant parts of the group.

The third reason that the number is uncertain is the fact that the "boundaries" of the group are not clearly defined. There are several small galaxies, mostly irregular galaxies, that lie at distances of about 1 Mpc (3 million light years), and it is not always clear whether they are members of the group or merely field galaxies.

Types

Within the Local Group there are examples of all three main types of galaxies: spirals, ellipticals, and irregulars. The three spirals (the Milky Way, the Andromeda galaxy, and M33) are the most luminous galaxies of the group. The Magellanic Clouds and other irregular galaxies are also fairly bright. There are no giant, luminous elliptical galaxies, however, even though such galaxies are often conspicuous members of more populous clusters of galaxies. The elliptical members include some of intermediate brightness, like M32, and eight extremely faint dwarf ellipticals.

Size

The Local Group is very small, when compared to the famous galaxy clusters like those in Virgo and Coma, which span hundreds of millions of light years. From our perspective, the group has a diameter of approximately 3 million light years; that is, the most distant certain members are about that distance from the Milky Way galaxy. We are not yet sure, however, just where to draw the boundaries of the group, especially because the distances to some of the more distant dwarf irregular galaxies are not yet reliably known.

Dynamics

The best way to decide on membership in the Local Group has been to measure the velocities of all nearby galaxies and then to see which seem to be moving together in space. If a certain galaxy has a velocity that is very different from the rest, then it is probably an interloper. Our fellow members of the group do not seem to be participating in the Hubble flow (the general expansion of the universe), because they are all gravitationally held together, at least loosely, in the group. Various tests of the stability of the group have been made over the years, with somewhat uncertain results. It appears likely from these studies that the group is a fairly stable dynamical entity, but that it is held together principally by its interstellar dark matter. Most of the visible mass of the group is contained in just two members: the Milky Way and the Andromeda galaxy, and these are falling towards each other at a velocity of 300 km s^{-1}. The group is probably not collapsing, but rather these two galaxies are probably in highly elongated orbits around the group center of mass, which is somewhere between them. The velocities of all members are probably balanced by the distribution of dark matter in the group, about which we know very little.

Member of the Virgo Cloud

The Local Group is not a simple, isolated clump of galaxies, but is one of a large number of groups that belong, at least peripherally, to a giant complex of galaxies called the Virgo cloud. The center of the Virgo cloud is the Virgo cluster, a large, massive, irregular cluster of many hundreds of galaxies. Studies of the velocities of expansion of the universe in our neighborhood show that the Local Group is falling toward the Virgo cloud with a velocity of a few hundred kilometers per second, which is probably caused by the large gravitational field of the Virgo galaxies. We are thus gravitationally attached to Virgo as one of its outlying members.

DESCRIPTION OF THE MEMBERS

Spiral Galaxies

The Milky Way galaxy and the Andromeda galaxy (frequently referred to as M31) are the most luminous and massive members of the group. M31 is a Hubble Sb type, with a luminous, large bulge of older stars, surrounded by a less-luminous disk of gas, dust, and younger stars, arranged in spiral arms. Its diameter is about 200,000 ly and its mass is approximately 700 billion times that of the sun. Among the many objects that have been studied within it are about 300 globular clusters, 400 open clusters, numerous dust clouds, gas clouds, stellar associations, supernovae remnants, and other components.

The Milky Way galaxy is of Hubble type Sb/Sc, with a somewhat less-conspicuous central bulge and a brighter disk and looser arms than M31. We do not know its total luminosity, because we cannot see it from a distant perspective and because so much of its visible light is obscured by the dust in the disk in which we are enveloped, but indirect evidence suggests that it may be roughly twice as bright as M31. The total mass is approximately 500 billion suns, though this is uncertain because of the unknown distribution of dark matter in our galaxy.

M33 is a Hubble type Sc galaxy; it is smaller and fainter than the other two spirals, being only about as bright visually as a few billion suns. It contains many blue luminous stars in its complex, thick spiral arms, and has several spectacular glowing gas clouds (giant H II regions).

Elliptical Galaxies

There are only four moderately bright elliptical galaxies in the Local Group and all four are companions to the Andromeda galaxy. Two very close companions are M32 and NGC 205, both of which are seen superimposed on the outer parts of M31. M32 is a nearly circular galaxy with a population of exclusively very old stars, whereas NGC 205 is more elongated in shape and contains a small but remarkable population of young stars, with accompanying dust and gas. The other two companions, NGC 147 and NGC 185, are somewhat fainter and more distant from M31.

All other known elliptical members of the local group are classed as dwarf ellipticals; they are very low in density, faint in luminosity, and small in size, with typical brightnesses being about a million suns and typical diameters being only about 10 thousand light years. These objects contain primarily old stars, though some show evidence of star formation that took place not too long ago (only 7 or 8 billion years ago, compared to the oldest stars, with ages of 15 billion years).

Irregular Galaxies

The two best studied irregular galaxies in the Local Group are the two Magellanic Clouds, galaxies that are so bright that they were well known to the early explorers and were named after the great navigator Ferdinand Magellan. They are in the southern skies and can only be observed well from below the equator. The biggest of the two, the Large Magellanic Cloud, is only about 150 thousand light years away (some astronomers cite a distance of 160,000 or even 170,000 ly) and the Small Magellanic Cloud is not much farther. They can be studied in great detail because of this proximity. We have information about nearly every kind of star and interstellar object in these two galaxies and they have played an important role in helping our understanding of such fundamental issues as stellar evolution and the extragalactic distance scale. The Large Cloud has a truly giant H II region as one of its most conspicuous features, a huge complex of massive stars, stars being formed, gas, and dust, called 30 Doradus, popularly known as the Tarantula nebula. In 1987 a bright supernova explored near it, causing a sensation among astronomers by being the first nearby supernova in over 300 years. This star, Supernova 1987a, was observed in optical and radio wavelengths, as well as in x-rays, and even its neutrinos were detected, making it a veritable bonanza for our understanding of the supernova phenomenon.

The numerous other irregular members are dwarf galaxies, some only a thousand or so light years across. They show a surprising variety in their structure and in their histories, with some apparently having had recent bursts of star formation and others having been relatively quiet for most of their recent lives. Some are sufficiently far that we do not yet have good distances for them, and we are still uncertain as to whether they are members of the group. But the nearer ones are well studied and they continue to give astronomers clearer and clearer pictures of the nature of galaxy evolution and the behavior of stars in a variety of different environments.

Additional Reading

Baade, W. (1963). *Evolution of Stars and Galaxies*. Harvard University Press, Cambridge, MA.

Hodge, P. W. (1986). *Galaxies*. Harvard University Press, Cambridge, MA.

Hodge, P. W. (1989). Populations in Local Group Galaxies. *Ann. Rev. Astron. Ap.* **27** 139.

Hubble, E. P. (1936). *The Realm of the Nebulae*. Yale University Press, New Haven, CT.

Iwanowska, W. (1989). The Local Group. In *From Stars to Quasars*, S. Grudzinska and B. Krygier, eds. Nicolas Copernicus University, Torun, p. 55.

Mould, J. (1988). Review of the Local Group distance scale. In *The Extragalactic Distance Scale*, S. van den Bergh and C. J. Pritchet, eds. Astronomical Society of the Pacific, San Francisco, p. 32.

van den Bergh, S. (1979). Our galaxy as a member of the Local Group. In *The Large-Scale Characteristics of the Galaxy*, W. B. Burton, ed. Reidel, Dordrecht, p. 577.

See also **Andromeda Galaxy; Galaxies, Dwarf Spheroidal; Magellanic Clouds; Virgo Cluster.**

Galaxies, Local Supercluster

Giuliano Giuricin

A HISTORICAL PERSPECTIVE

The large-scale surveys of the last twenty years have verified the existence of large agglomerations of galaxies. The most conspicuous groupings can contain several clusters of galaxies, which explains why they have been given the name "superclusters."

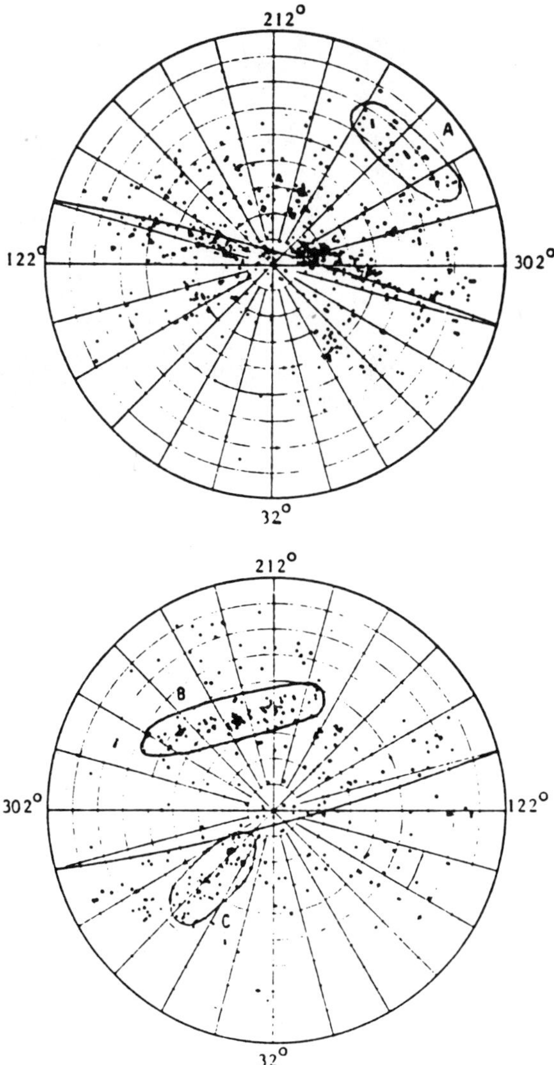

Figure 1. The distribution of galaxies brighter than the thirteenth photographic magnitude. The upper panel shows the northern galactic hemisphere, the lower the southern galactic hemisphere. The galactic poles are at the centers; the circles are at intervals of 10° in latitude; the galactic longitude is shown at the circumference. Galaxies in regions A, B, and C are probably beyond the Local Supercluster. The flat curve that passes close to the poles is the so-called supergalactic equator. (*This figure is reproduced from the* Annual Review of Astronomy and Astrophysics, *vol. 21, 1983, by kind permission of J. H. Oort and Annual Reviews Inc.*)

Some evidence of the existence of our own supercluster (dubbed the Local, or Virgo, Supercluster), where our Galaxy and the Local Group of galaxies reside, dates back to the earliest surveys of extragalactic nebulae made by William and John Herschel in the nineteenth century. However, the reality of the Local Supercluster has become generally accepted only in recent years.

The surveys by the Herschels showed an excess of bright nebulae in the northern galactic hemisphere, a fact that reflects the outlying position of our galaxy with respect to nearby galaxies. In the 1920s, J. H. Reynolds and Knut Lundmark noted a remarkable concentration of the largest nebulae along a great circle of the sphere, but drew no conclusion from this. (At that time several astronomers were not yet convinced that many nebulae were external galaxies.)

The French astronomer Gérard de Vaucouleurs of the University of Texas at Austin was the first to describe and define the Local Supercluster. He proposed that the concentration of the brightest galaxies along a great circle of the sphere was the trace of a flat supersystem (the Local Supercluster) centered in the Virgo cluster, which has a diameter of a few megaparsecs (Mpc) and is about 10 Mpc away (1 Mpc ≈ 3.26 million light-years). [In the present article we adopt a value of 100 km s^{-1} Mpc^{-1} for the Hubble constant in the relationship (Hubble's law) between the redshift observed in a galaxy's spectrum and the galaxy's distance; a change in the value of the Hubble constant alters the spatial scale.] The Local Supercluster was found to contain also several tens of galaxy groups, along with single galaxies scattered among them.

THE COMPONENTS OF THE LOCAL SUPERCLUSTER

A first rough idea of the structure of the Local Supercluster can be obtained by inspecting the two-dimensional distribution of bright galaxies (i.e., the location of galaxies projected onto the globe of the sky, without any indication of the distance of each). The distribution of ~1280 galaxies (brighter than about the thirteenth photographic magnitude), taken from the Harvard surveys of Harlow Shapley and Adelaide Ames (published in the 1930s), clearly reveals the Virgo cluster (at galactic longitude $l \sim 280°$ and galactic latitude $b \sim 75°$) and the Ursa Major cloud (which extends from a region near the galactic pole to $l \sim 140°$, $b \sim 50°$) in the northern hemisphere (see Fig. 1). Galaxies in the region marked by contour A probably form a system that does not belong to the Local Supercluster because they have, in general, higher redshifts. The axis corresponding to the dense chain of galaxies that contains the Virgo cluster, called the supergalactic equator by de Vaucouleurs, is shown by a curve in Fig. 1. The supergalactic longitude is counted along the supergalactic equator from its intersection with the galactic equator at $l \sim 137°$. The southern galactic hemisphere is less populous than the northern; the two most striking features, denoted by contours B (the so-called south galactic chain) and C, deviate from the supergalactic equator and may be considered to be distant, separate superclusters. If we omit these structures, there is some concentration along the supergalactic equator. In the zone below 20° galactic latitude, the relative lack of galaxies is due to the absorption of light by the disk of our own galaxy.

A much better description of the Local Supercluster is made possible if we know the three-dimensional distribution of nearby galaxies, as derived from their positions as well as from their redshifts. In this case, one often simply assumes that the redshifts measure distances fairly well (according to Hubble's law); this is true if there are no large systematic deviations from Hubble's law. Two basic sources of information for the positions and radial velocities of bright (nearby) galaxies in the whole sky are the catalogs of bright galaxies published in the 1980s, the *Revised Shapley–Ames Catalog of Bright Galaxies* by Allan Sandage and Gustav A. Tammann and the *Nearby Galaxy Catalog* (NBG) by R. Brent Tully. From the system of polar supergalactic coordinates [supergalactic longitude (SGL), supergalactic latitude (SGB), distance] it is useful to derive orthogonal Cartesian supergalactic coordinates (SGX, SGY, SGZ): the Sun is at the origin; the SGX axis, definedby the intersection of the galactic and supergalactic planes, is aligned toward SGL = 0°, SGB = 0°; the SGY axis is roughly in the direction of the Virgo cluster (it is aligned toward SGL = 90°, SGB = 0°); and the SGZ axis is perpendicular to the supergalactic plane and points in the direction of SGB = 90°. The positive SGY axis is only 6° removed from the north galactic pole, and the SGX-SGZ plane is almost coincident with the plane of the Galaxy. Our galaxy thus would be seen almost face-on from the center of the Local Supercluster. Figure 2 shows the projection of all (~1260) northern galaxies of the NBG catalog onto the SGX-SGZ plane. The viewer is located in the plane of the supercluster's equator. The disk of the Local Supercluster stretches across the middle of the diagram. The Virgo cluster is the densest concentra-

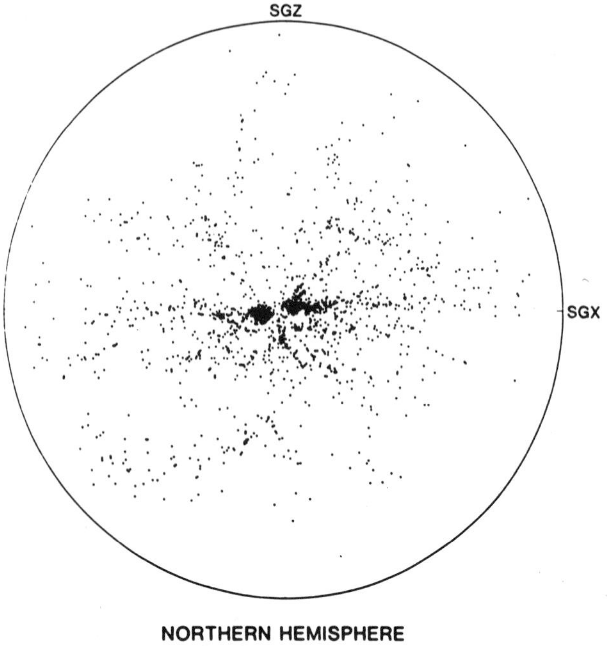

SGZ

SGX

NORTHERN HEMISPHERE

Figure 2. The projection of all galaxies in the NBG catalog in the northern galactic hemisphere onto the SGZ-SGX plane in supergalactic coordinates. The radius of the outer boundary is 30 Mpc. Three-dimensional positions have been computed assuming a uniformly expanding universe and a Hubble constant of 100 km s^{-1} Mpc^{-1}. The Virgo cluster is just to the left of center and the Canes Venatici or Ursa Major cloud (which contains the Milky Way) is just to the right of center. (*This figure is reproduced by kind permission of R. B. Tully and* The Astrophysical Journal.)

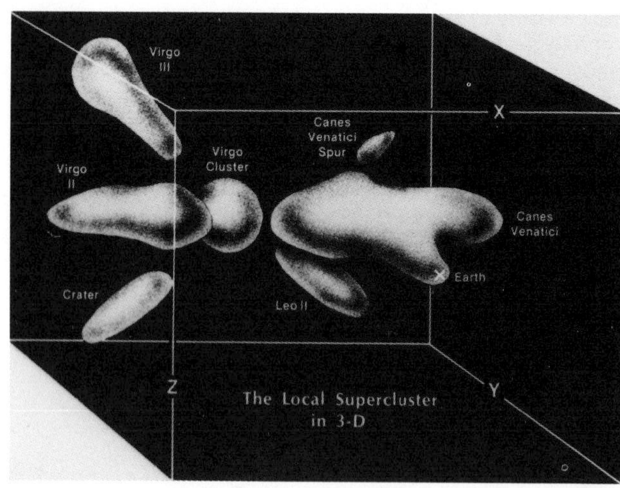

Figure 3. A schematic picture of the Local Supercluster in three dimensions, by Rob Hess. (*Reproduced, by kind permission, from an article of R. B. Tully in* Sky and Telescope). The location of the Earth is shown.

tion of galaxies, somewhat to the left with respect to the center of the figure; to the left of the Virgo cluster is the "southern extension" (also called the Virgo II cloud) and to its right is the Ursa Major cloud (also called the Canes Venatici cloud), which is the largest cloud in the Local Supercluster and the one in which we reside. There are also some clouds located off the supergalactic plane. A schematic picture of the Local Supercluster in three dimensions is shown in Fig. 3: Some major clouds (Virgo III, Crater, Canes Venatici spur, Leo II) lie off the supergalactic plane, which is essentially formed by the Virgo cluster and the Virgo II and Canes Venatici clouds.

In general, it is possible to distinguish three components of the Local Supercluster: the Virgo cluster, the "disk" (i.e., the galaxies distributed along a thin layer coinciding with the supergalactic plane), and the "halo" (galaxies that are distributed roughly spherically, considerably off the plane). Of the NBG galaxies, roughly 40% belong to the halo component; of the remaining 60%, one-third belong to the Virgo cluster and two-thirds to the disk component, which contains the two extended clouds (Canes Venatici and Virgo II). Taken together, these clouds display a structure that is flattened in the ratio SGX:SGY:SGZ = 6:3:1. The thickness of the disk of the Local Supercluster (the short axis) is very thin (~ 1 Mpc). The vast majority of (disk and halo) bright galaxies reside in a small number of clouds; the number density of bright galaxies falls off as $1/d^2$, where d is the distance from the Virgo cluster. Collectively, these clouds occupy a very small fraction of the volume of the Local Supercluster (most of the space is empty). Within the clouds, on the smallest scale we can identify groups (which are mostly bound systems of galaxies) that together contain about 70% of the galaxies of the NBG catalog, if the Virgo members are included. Remarkably, only ~ 1% of the galaxies appear to be isolated (outside clouds).

The internal motions of galaxies in most of the various large clouds of the Local Supercluster are small (less than 100 km s^{-1} along the line of sight). Hence, individual member galaxies can have crossed only a relatively small portion of the cloud in the billions of years since the galaxies came into being, which suggests that little mixing can have occurred in the supercluster as a whole. This fact gives insight into evolutionary history that is simply not obtainable at scales smaller than those of superclusters (i.e., in groups and clusters, where the original distribution of matter tends to be smeared out by evolutionary mixing).

Elliptical and lenticular galaxies have a tendency to be located in the high-density regions of the Local Supercluster, whereas early-type spiral galaxies tend to be found in regions of intermediate density, and late-type spiral and irregular galaxies in regions of low density.

MOTIONS OF GALAXIES IN THE LOCAL SUPERCLUSTER

A better approach to the mapping of the distribution of the galaxies in space is to estimate galaxy distances from a variety of redshift-independent distance indicators. The two most widely used distance indicators are the stellar velocity dispersion in an elliptical galaxy and the width of the neutral-hydrogen emission line (observed at the 21-cm radio wavelength) in a spiral galaxy. The use of redshift-independence distance indicators has the advantage of avoiding the assumption that galaxies strictly obey Hubble's law and allowing us to map the (non-Hubble) peculiar motions of galaxies. A galaxy's peculiar motion is found by subtracting from its observed radial velocity the fraction of its recession velocity attributable to cosmic expansion (according to Hubble's law and a redshift-independent estimate of the galaxy's distance) within a velocity reference system (e.g., that defined by the cosmic microwave background).

In recent years, spectroscopy and photometry of large samples of galaxies showed that, in the cosmic microwave background rest frame, the peculiar motions of the galaxies of the Local Supercluster (including our Local Group) and of a few nearby superclusters can best be fitted by a flow (mostly at a velocity of ~ 600 km s^{-1}) towards a great mass concentration (dubbed "The Great Attractor") centered on the galactic coordinates $l \sim 310°$, $b \sim 10°$ (in the Centaurus region) at a distance of ~ 40 Mpc. There seems to be a large concentration of galaxy clusters in the vicinity of the Great Attractor but, unfortunately, much of this grouping lies behind the obscuring dusty plane of the Milky Way.

Additional Reading

Burns, J. O. (1986). Very large structures in the universe. *Scientific American* **255** (No. 1) 38.

Dressler, A. (1987). The large scale streaming of galaxies. *Scientific American* **257** (No. 3) 46.

Gregory, S. A. and Thompson, L. A. (1982). Superclusters and voids in the distribution of galaxies. *Scientific American* **246** (No. 3) 106.

Oort, J. H. (1983). Superclusters. *Ann. Rev. Astron. Ap.* **21** 373.

Silk, J., Szalay, A. S., and Zeldovich, Y. B. (1983). The large scale structure of the universe. *Scientific American* **249** (No. 4) 72.

See also **Galaxies, Local Group; Hubble Constant; Superclusters, Observed Properties; Virgo Cluster.**

Galaxies, Molecular Gas in

Judith S. Young

Much of the beauty in the Milky Way and other spiral galaxies depends on star formation, an ongoing process that began when galaxies began to form more than 10 billion years ago. Although most of the stars in the night sky are billions of years old, there are a small number of young, high-mass stars that formed recently (i.e., within the past 10 million years) from clouds of interstellar gas and dust. These young stars heat the gas clouds from which they formed, and produce an abundance of glowing nebulae that are commonly found in spiral galaxies. These stars finally explode as supernovae, the best recyclers of all, and return the processed elements from their interiors back into the galaxy to be incorporated into future generations of stars. Recent observations of star-forming regions made with radio telescopes have revealed that the birth sites for future generations of stars are giant clouds of molecular gas, typically more than 100 light years across and containing a mass of gas more than one million times the mass of the sun. These clouds of gas, which indicate the potential for future star formation in a galaxy, can be studied in detail in our own Milky Way galaxy. In other galaxies we can determine the large-scale distribution of molecular gas in relation to past and present star formation. Thus, studies of molecular gas in galaxies will help provide an understanding of the large scale processes that influence the evolution of galaxies.

Because stars form in molecular clouds, most studies of molecular gas in galaxies have been confined to studies of galaxies in which young stars are present, that is, primarily spiral and irregular galaxies. These studies have been made with radio telescopes operating at millimeter wavelengths. The radiation from the molecules in the giant gas clouds arises from changes in the rotation of molecules as a whole. The most abundant molecule in the giant molecular clouds is molecular hydrogen (H_2), with numerous trace molecules, the most important of which is carbon monoxide (CO). Determinations of the molecular gas content and distribution in galaxies are based on observations of the abundance of the CO molecule. It is only within the past 15 years that sensitive radio receivers have been built that can detect the emission from molecules at millimeter wavelengths. Prior to the discovery of molecules in the interstellar medium of our galaxy, our knowledge of the gas content of galaxies was based on observations of atomic hydrogen. Studies of these atomic gas clouds in galaxies were very important, but led to an incomplete view of the gas content and evolution of galaxies.

MOLECULAR GAS DISTRIBUTIONS IN GALAXIES

Ever since pioneering studies of the atomic gas content of galaxies in the 1950s and 1960s, it has been known that most spiral galaxies have similar distributions of atomic gas. Thus, with the discovery of the molecular component of the interstellar medium, it was initially expected that molecular gas distributions among spiral galaxies might also be similar. However, this expectation turned out to be incorrect: There are large differences from galaxy to galaxy in the distribution and total mass of molecular gas, and elucidating these differences has led to a new understanding of star formation in galaxies.

The distribution of molecular gas in the Milky Way galaxy has been studied in detail since 1975. Molecular clouds were found to be extremely abundant in the central 500–1000 ly out from the galactic center, but their number fell off at larger radii. It was indeed surprising when a ring of molecular clouds was discovered in our galaxy, with a peak at about 15,000 ly out from the galactic center. Following the study of the molecular gas distribution in the Milky Way, a number of questions arose that could best be addressed by observations of other galaxies. For example, is the ring of molecular clouds in the Milky Way a common feature of other galaxies? Does the abundance and distribution of molecular gas in a galaxy depend on the form of a galaxy? Does the total luminosity of a galaxy depend on the quantity of molecular gas in a galaxy? Why are spiral arms apparent in spiral galaxies? Specifically, do more stars form in the arms because there is more gas, or do stars form more efficiently in spiral arm locations in galaxies?

Studies of molecular clouds in the *disks* of luminous, nearby face-on spiral galaxies have revealed that the molecular gas distributions are unlike the distributions of neutral atomic hydrogen gas. That is, the molecular gas is concentrated in the centers of most galaxies, with distributions that decrease with increasing radii from the galactic center. Thus, most of the molecular gas in spiral galaxies is confined to the inner half of the optical disk of the galaxy. In contrast, the distributions of atomic hydrogen gas are relatively flat as a function of radius in spiral galaxies, and these gas distributions extend well beyond the optical edges of galaxies. In the most luminous spiral galaxies, there is more molecular than atomic gas in the inner parts of the galaxy, and more atomic than molecular gas in the outer parts. Because atomic gas is probably the reservoir of material out of which molecular clouds form, these results indicate that the formation of molecular clouds is most efficient in the inner disks of spiral galaxies.

Within spiral galaxies where the amount of molecular gas exceeds that of atomic gas, it has been discovered that there is a one-to-one proportionality between the distribution of molecular gas and the distribution of luminosity in young stars. These observations have been interpreted to indicate that the rate of star formation, deduced from the luminosity in young stars, depends simply on the amount of molecular gas present. This implies that more star formation occurs when more gas is present, and that it does not matter whether the cloud is located in the interior of a galaxy or in the outer parts.

Recent observations of face-on spiral galaxies have revealed that the *spiral arms* have different properties than the disk. Specifically, molecular clouds are found everywhere in the inner disks of spiral galaxies, both on the arms *and* in the interarm regions, whereas young stars are found primarily on the spiral arms. This indicates that the rate of star formation per unit molecular gas mass, or the star formation efficiency, is higher on the spiral arms than in the interarm regions of a galaxy. Thus, the formation of high-mass stars is enhanced in the arm regions of a spiral galaxy, leading to the spiral appearance of the galaxy.

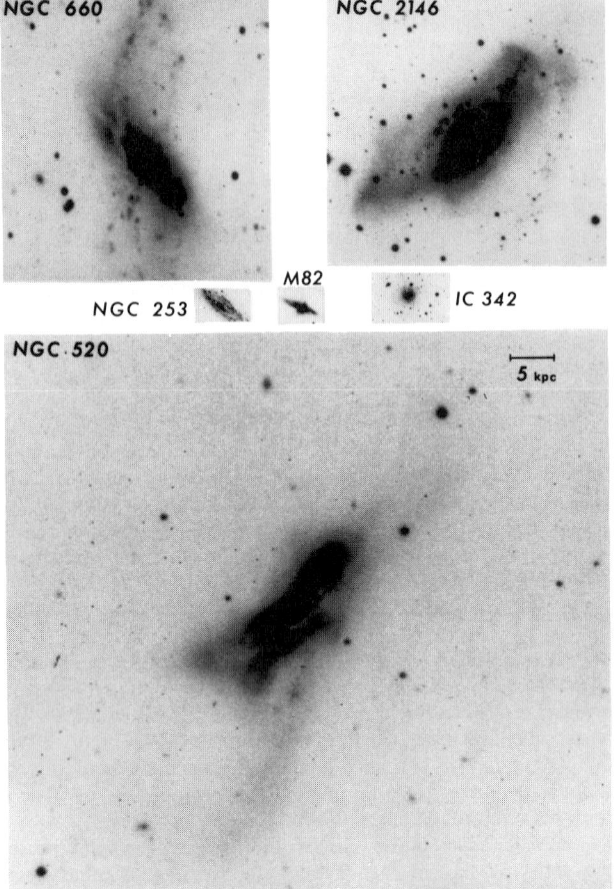

Figure 1. Red images of six galaxies, displayed as if they were all the same distance away from us. The scale is indicated in the lower panel, where 5 kpc = 15,000 ly. For comparison, the Sun is ~ 8 kpc from the center of the Milky Way, showing that some galaxies are much smaller than our own.

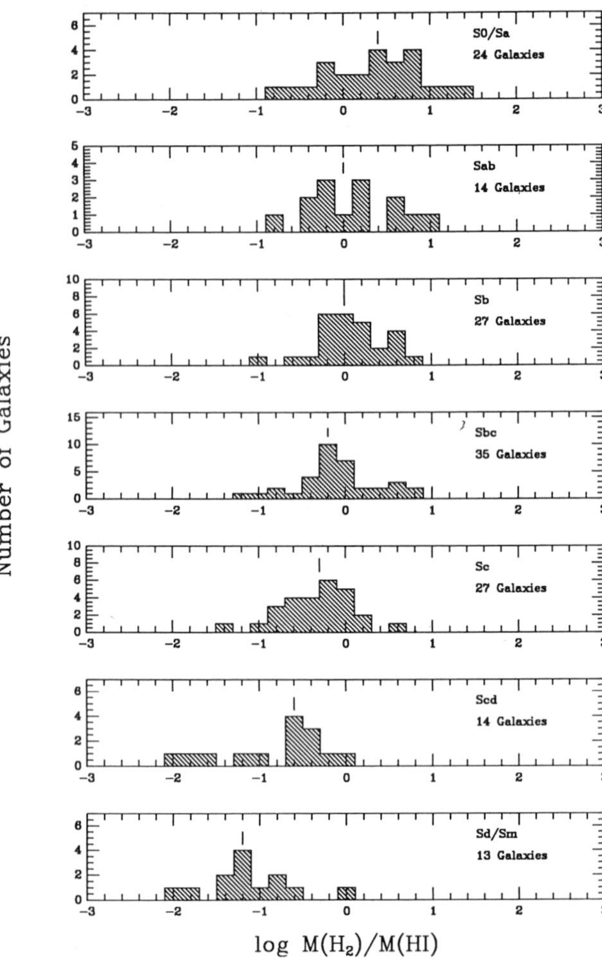

Figure 2. Ratio of molecular to atomic gas mass, $M(H_2)/M(\text{H I})$, as a function of morphological type for spiral galaxies. The S0 and Sa galaxies are bulge dominated, whereas the Sc, Scd, Sd, and Sm galaxies have little or no bulge. The vertical tick mark in each panel represents the median value for that type. Although there is a considerable range in the H_2/H I ratio within each type, the mean value clearly decreases from type S0/Sa to Sd/Sm.

EFFICIENCY OF STAR FORMATION IN GALAXIES

Through a survey by the *Infrared Astronomical Satellite* (*IRAS*), the infrared emission has now been measured from galaxies over the entire sky. This emission is from dust that is heated by young, high-mass stars embedded in molecular clouds. The luminosity of a galaxy in the infrared, therefore, provides a measure of the rate of formation of high-mass stars. Figure 1 illustrates images of six galaxies with a wide range of absolute size and total star formation rate. The nearby, well-known, actively star-forming galaxies M82 and NGC 253 are quite small on an absolute scale.

One of the contributions of the *IRAS* survey is that many distant galaxies with high rates of star formation were identified and subsequently found to have suffered from collisions with other galaxies. This result led to studies of the rate and efficiency of star formation in galaxies in different environments. Specifically, when the most isolated galaxies are compared with strongly interacting galaxies, it is found that both samples exhibit similar ranges of molecular gas mass, whereas the interacting galaxies display more stellar luminosity for a given supply of gas. Thus, the interacting galaxies have ~ 5 times higher ratios of young stellar luminosity to molecular gas mass than the isolated galaxies. This indicates that the environment of a galaxy has a strong influence on the efficiency of formation of high-mass stars.

Studies have also been conducted comparing the rates of high-mass star formation with the molecular gas contents in different types of spiral and irregular galaxies. Spiral galaxies exhibit a variety of forms, or morphologies, including the flat disk of the galaxy (with the spiral arms) and the spheroidal bulge centered on the galaxy. Galaxies of type Sa exhibit large bulges and tightly wound arms; galaxies of type Sc exhibit small bulges and loosely wound arms; galaxies of type Sb are intermediate in terms of bulge size and spiral arms. The bulges of spiral galaxies are generally composed of old stars, which have red colors, whereas the spiral arms have blue colors indicative of the presence of young stars. Because of the large bulges in Sa galaxies, these galaxies have red colors, and because of the small bulges and prominent spiral arms in Sc galaxies, these galaxies have blue colors. This color difference in galaxies as a function of type has led to the perception that there is less star formation in Sa galaxies than in Sc galaxies, when in fact the color difference can be attributed largely to the presence of the bulge and not to the absence of star formation.

If there is star formation occurring in Sa galaxies, then these galaxies should contain sufficient quantities of the prerequisite molecular gas out of which stars form. This has been found to be the case: Sa galaxies have the same range of molecular gas masses as Sc galaxies. In fact, Sa galaxies also have the same range of star formation rates as Sc galaxies. Thus, when investigating the rate of star formation per unit molecular gas mass in the disks of Sa versus Sc galaxies, it appears that the two types of galaxies are indistinguishable. That is, Sa and Sc galaxies have similar mean values of the yield of young stars per unit mass of molecular gas. This indicates that the process of star formation is generally not affected by global properties, such as the galaxy morphology, but is a local process primarily sensitive to the amount of molecular gas present. The result that star formation is a local process and depends primarily on the amount of molecular gas present was previously discussed for the yield of young stars per unit mass of molecular gas within the disks of individual galaxies. The exceptions to this rule are found in spiral arms and in interacting galaxies, where the yield in young stars per unit mass of molecular gas is enhanced.

One explanation for the difference in the efficiency of formation of high-mass stars in interacting versus isolated galaxies and in spiral arms versus the disks of spiral galaxies, is that the formation of high-mass stars is enhanced when more molecular gas is accumulated in a small region, be it on a spiral arm or in the center of an interacting galaxy. The formation of these high-mass stars could result directly from collisions between giant molecular clouds or from the overcoming of the magnetic or turbulent forces that generally support the clouds against collapse.

RATIO OF MOLECULAR TO ATOMIC GAS MASS IN GALAXIES

Ever since the discovery of atomic gas in the universe and studies of the gas content of spiral galaxies, it has been recognized that the mass of atomic gas per unit stellar luminosity in Sa galaxies was about 5 times lower than the atomic gas mass to luminosity ratio for Sc galaxies. This led to the perception that Sa galaxies had less gas than Sc galaxies, when in fact they simply have larger stellar luminosities due to the presence of the central bulge.

In order to determine whether or not Sa galaxies truly have less gas than Sc galaxies, it is necessary to measure the total gas content of spiral galaxies, that is, molecular *plus* atomic gas. For a sample of ~ 200 galaxies, it has recently been found that although the atomic gas content per unit luminosity increases from Sa through Sc galaxies, the molecular gas content of these same galaxies per unit luminosity remains constant. Thus, there is a change in the ratio of molecular to atomic gas in spiral galaxies as a function of morphological type: Sa galaxies have ~ 4 times more molecular than atomic gas, whereas Sc and Sd galaxies have ~ 5 times less molecular than atomic gas as shown in Figure 2. Overall, the total gas content of galaxies is fairly constant, with Sa galaxies having only a factor of 2 less total gas mass than Sc galaxies. The principal result as a function of morphological type is that the phase of the gas changes with galaxy form. Because the morphology of a galaxy is a reflection of the mass distribution, and if atomic gas is the reservoir out of which molecular clouds form, this result indicates that the conversion of atomic to molecular gas is most efficient (i.e., leads to the highest molecular to atomic gas ratios) in galaxies with centrally concentrated mass distributions, where more gas is likely to be collected in the gravitational potential.

SUMMARY

Molecular gas is found to be an important constituent of galaxies because it is the reservoir of gas out of which stars form. In the disks of spiral galaxies and among isolated spiral galaxies, stars form in proportion to the amount of molecular gas present, independent of the morphology of the galaxy. Within these galaxies, the property that does depend on galaxy form is the ratio of molecular to atomic gas.

Because the formation of a molecular cloud from an atomic gas cloud is a large scale process, it is reasonable that gravity on a large scale is important. This is consistent with the observation that the molecular to atomic gas ratio in spiral galaxies changes with galaxy morphology, because morphology is a reflection of the large scale mass distribution. Star formation, on the other hand, is a small scale process relative to the size of a giant molecular cloud, and in the disks of isolated galaxies star formation appears to be a local process that is insensitive to the large scale gravitational field. Only in interacting galaxies or in spiral arms does the enhanced conversion of gas into stars appear to operate.

Additional Reading

Hubble, E. (1936). *The Realm of the Nebulae*. Dover, New York.

Schweizer, F. (1986). Colliding and merging galaxies. *Science* **231** 227.

Scoville, N. (1989). Molecular gas in spiral galaxies. In *Evolutionary Phenomena in Galaxies*, J. E. Beckman and B. E. J. Pagel, eds. Cambridge University Press, Cambridge, p. 63.

Scoville, N. and Young, J. S. (1984). Molecular clouds, star formation and galactic structure. *Scientific American* **250** (No. 4) 42.

See also **Galaxies, Binary and Multiple, Interactions; Galaxies, Infrared Emission; Interstellar Clouds, Giant Molecular.**

Galaxies, Nuclei

Timothy M. Heckman

Unlike the nucleus of a cell, the nucleus of a galaxy is not a precisely defined entity or distinct subcomponent. Rather, it is simply the central-most part of a galaxy. The region referred to as "the nucleus" is roughly the innermost 1% of a galaxy. Most galaxies are rather symmetric in form with the density of stars decreasing smoothly from the center (nucleus) outward. Thus, the nucleus is not only the center of the galaxy, it is also the region of highest density. As such, the nucleus can also be thought of as the "bottom" of the galaxy: gas clouds or stars that move too slowly within the galaxy can be pulled inward by gravity toward the nucleus. This may have some interesting consequences, as described below.

The nuclei of galaxies have been extensively investigated by astronomers for at least two reasons. The first, more prosaic reason is that the nucleus is usually the brightest part of a galaxy (because the density of stars is highest there). This means that the nucleus is the most easily studied part of a galaxy. The second and more exciting reason is that the nuclei of galaxies are often the sites of qualitatively unusual energetic phenomena that are observed nowhere else. These are the so-called active nuclei.

Galactic nuclei (like galaxies themselves) are composed of stars and interstellar matter (mostly gas, plus small dust grains). To explain active galactic nuclei, some additional object must be present. Because the fundamental nature of this object remains mysterious, it is often referred to by deliberately vague and fanciful terms like "the monster" or "central engine."

STARS AND INTERSTELLAR MATTER

Galaxies can be broadly classified into early-type galaxies, which consist predominantly of old stars, and late-type galaxies, which contain old stars, young stars, and cool interstellar gas clouds out of which young stars are formed. This pattern is repeated in the nuclei of early- and late-type galaxies. The nuclei of early-type galaxies apparently formed the great majority of their stars billions of years ago, and in so doing depleted the raw material needed to make new stars (cool, dense gas clouds). The properties of the nuclei of late-type galaxies are consistent with a rate of forming new stars that has been almost constant since the time the galaxies themselves formed. A spectacular exception is the class of starburst nuclei, which are apparently undergoing short-lived episodes of star formation at rates much higher than the past average rate.

The most unusual property of the stars and gas in the nuclei of galaxies is the chemical composition. Astronomers refer to all the chemical elements heavier than hydrogen and helium as metals. The sun has about 2% of its mass in the form of metals, and this solar metal abundance is typical of the chemical composition of the bulk of the stars and gases in bright galaxies like our Milky Way. In contrast, the metal abundance in the nuclei of bright galaxies is apparently two or three times higher than that of the sun. Such high metal abundances are essentially unique to the nuclei of galaxies. Indeed, many galaxies show a steady decrease in metal abundance from the nucleus outward. The reason for the high content of metals in galactic nuclei is not entirely clear. Because metals are formed by nuclear reactions inside stars, the high metal abundance in galactic nuclei means that the material we see today has been extensively processed by previous generations of stars. One possibility is that the strong gravitational field in the nucleus has enabled it to retain a relatively larger fraction of the metals expelled by dying stars in the form of winds or supernova explosions than was possible in the outer parts of the galaxy.

The most conspicuous form of interstellar gas in the nuclei of galaxies is ionized gas (gas consisting of free electrons and the corresponding ions—atoms with one or more of their normal complement of electrons missing). In the nuclei of late-type galaxies, emission from this gas can be very bright, and its properties imply that the gas is kept in its ionized state by energetic photons emitted from hot young stars. Such regions of gas ionized by young stars are commonly found throughout late-type galaxies and not just in the nucleus.

The emission from ionized gas in the nuclei of early-type galaxies is usually weak. Surprisingly, the strength of this emission does not appear to be related to the relative numbers of young stars (which are scarce in such nuclei in any case). Moreover, the nature of the gas is inconsistent with ionization by normal young stars. The detailed properties of these so-called LINERs (low ionization nuclear emission-line regions) can be explained if either they are ionized by a source of photons that is much hotter than ordinary stars or ionized by shock waves resulting from high speed collisions between gas clouds or from explosions. LINERs are often taken as evidence that the nuclei of early-type galaxies are commonly in a state of low-level activity: a dormant, but still living monster may lurk at the heart of most such galactic nuclei.

The recent birth of the field of extragalactic millimeter-wave astronomy has led to the discovery that most of the interstellar matter in the nuclei of many spiral galaxies is in the form of molecular gas. Throughout the spiral disk of our own Milk Way galaxy, molecular gas is intimately related to the process of star formation. Thus, it is not surprising that the nuclei of galaxies that are actively forming stars are rich in molecular gas. Indeed, some starburst nuclei may contain as much molecular gas as an entire normal spiral galaxy.

TYPES OF ACTIVE OR UNUSUAL NUCLEI

As already noted, an active nucleus is one in which processes are observed that cannot be readily explained by the mere presence of normal stars and interstellar gas clouds. By this definition, a starburst nucleus is not a truly active nucleus, but we will discuss such objects in this section because they are rare and can be highly energetic.

LINERs

LINERs are the most common type of active nucleus, and may in fact be present at a very low level in the nucleus of every early-type galaxy. Their spectra are characterized by weak emission lines that have been significantly broadened by the Doppler effect, indicating high speed gas motions (typically a few hundred to a few thousand kilometers per second). LINERs usually contain a compact source of radio synchrotron emission that is qualitatively similar to (but much weaker than) the radio sources seen in radio galaxies and quasars.

Seyfert Galaxies

Seyfert galaxies are usually spiral galaxies whose nuclei are exceptionally bright. A few percent of spiral galaxies contain a Seyfert nucleus. The spectrum of the nucleus shows Doppler-broadened emission lines whose widths are similar to those in LINERs but whose strengths are much greater. The gas is in a highly ionized state, requiring the presence of a source of photons of much greater energies than can be produced by ordinary stars. Direct evidence for this ionization source is provided by the strong ultraviolet and/or x-ray continuum emission observed from Seyfert nuclei.

Radio Galaxies

Radio galaxies are usually elliptical or elliptical-like galaxies that are strong sources of radio sychrotron emission (emission produced by electrons moving nearly at the speed of light while spiraling around magnetic field lines). A few percent of bright elliptical galaxies are classified as radio galaxies. Although most of the radio emission arises from twin radio "lobes" located far outside the radio galaxy, there is convincing evidence that the lobes are powered by matter that has been expelled from the active nucleus of the galaxy: Narrow radio-emitting channels or jets link the distant radio lobes to a compact radio source in the nucleus.

Quasars

Quasars were originally defined to be star-like (quasistellar) objects with large redshifts. Today they are believed by the great majority of astronomers to be the highly powerful nuclei of distant active galaxies. Quasars share many properties in common with Seyfert nuclei (strong, broad emission lines and powerful ultraviolet and x-ray emission). The subclass of "radio-loud" quasars, which are strong radio emitters, is closely related to radio galaxies in their observed properties.

Starburst Nuclei

Starburst nuclei can rival Seyfert nuclei or even some of the less powerful quasars in terms of their total power output. However, unlike Seyferts or quasars, the properties of starburst nuclei can be adequately explained by young stars (albeit a highly unusual *number* of such stars). Starburst nuclei often radiate most strongly in the infrared portion of the electromagnetic spectrum. This infrared emission comes from dust grains that have been heated to temperatures of several tens to several hundreds of degrees kelvin by the ultraviolet light produced by the hot young stars. The presence of the dust is not surprising, because dust is found to be closely associated with cool, dense molecular clouds of the kind that are apparently present with great abundance in starburst nuclei.

MONSTERS AND DEAD QUASARS IN GALACTIC NUCLEI

Currently, the most popular theory holds that the monster that powers active galactic nuclei is a supermassive black hole, a region of high density within which the escape velocity exceeds the speed of light. The mass of the black hole must be at least several million times the mass of the sun for a Seyfert nucleus, and several billion times the mass of the sun for a powerful quasar. Energy would be produced by the supermassive black hole as its powerful gravitational field compresses and heats infalling gas, causing the gas to emit highly energetic photons before it falls into the hole and vanishes. Recall that galactic nuclei are at the bottom of the galaxy, a favorable location for accreting material to "feed" the monster. Although there are a variety of plausible lines of indirect evidence that favor this model, the case is by no means clear.

The difficulty in proving this model rests largely with the fact that powerful active nuclei are located so far away that the direct presence of the supermassive black hole cannot be unambigously detected. However, there are at least two pieces of evidence that suggest that dormant supermassive black holes may reside at the centers of many nuclei that are not presently in a highly active state. The first is the LINER phenomenon, which may be the result of a "starved monster"—a supermassive black hole that is producing very little energy because it is receiving only a slow trickle of food in the form of infalling gas. The second is the fact that the most powerful active nuclei were evidently much more common in the distant past than they are today. Many more galactic nuclei may contain the essential equipment for producing a quasar (i.e., a supermassive black hole) than is evident from the scarcity of highly active nuclei in the present universe.

Motivated by such ideas, astronomers have recently searched the nuclei of nearby galaxies (including the Milky Way) for the presence of a "dead quasar" (a supermassive black hole that is presently producing little or no light). The basic technique is to measure the motions of stars in the nucleus, and then to use Newton's laws of motion and law of gravity to "weigh" the mass contained in the center of the nucleus. The presence of a supermassive black hole would be revealed by stellar velocities that increase rapidly toward the center of the nucleus (because of the strong gravitational field of the black hole) and by a calculated mass that is far in excess of the mass that could be contained in normal stars. The interpretation of such observations is a subtle and difficult task. Nevertheless, there is now good enough evidence to strongly suggest that the nuclei of several of the nearest galaxies may well contain supermassive black holes with masses that are several millions to several tens of millions times that of the sun.

Additional Reading

Filippenko, A. V. (1988). Indirect evidence for massive black holes in nearby galactic nuclei. In *Supermassive Black Holes*, M. Kafatos, ed. Cambridge University Press, New York, p. 104.

Heckman, T. M. (1986). Optical emission-line gas in the nuclei of normal galaxies: Implications for nuclear activity. *Pub. Astr. Soc. Pacific* **98** 159.

Keel, W. C. (1985). Low luminosity active galactic nuclei. In *Astrophysics of Active Galaxies and Quasi-Stellar Objects*, J. S. Miller, ed. University Science Books, Mill Valley, CA, p. 1.

O'Connell, R. W. (1986). Nuclei of normal galaxies. *Pub. Astr. Soc. Pacific* **98** 163.

Rees, M. J. (1990). Black holes in galactic centers. *Scientific American* **263** (No. 5) 56.

Richstone, D. O. (1988). Evidence for massive black holes in galaxies. In *Supermassive Black Holes*, M. Kafatos, ed. Cambridge University Press, New York, p. 87.

See also **Active Galaxies, Seyfert-Type; Active Galaxies and Quasistellar Objects, Central Engine; Active Galaxies and Quasistellar Objects, Interrelations of Various Types.**

Galaxies, Properties in Relation to Environment

Alan Dressler

Astronomers study variations in galaxy properties with environment because they seek to understand the process of galaxy formation, which, for the vast majority of galaxies, occurred long ago. The dependence of galaxy type on environment, first noted by Edwin Hubble and Milton Humason in 1931, provides an opportunity to learn about events that took place in a largely unobservable past.

THE ENVIRONMENT

Galaxies cluster; although a small fraction live in splendid isolation, the average galaxy is found in a low-density, marginally bound group. The typical density of such a group is only 1 galaxy for a volume of the order of 20 cubic megaparsecs (a sphere with diameter ~10 million light-years); therefore, encounters between galaxies are relatively rare. However, the sky is punctuated by rich clusters of several hundred galaxies in regions 100–1000 times as dense as this. These rich clusters, which contain only ~5–10% of all galaxies, are surrounded by extensive plateaus of modestly enhanced galaxy density, called superclusters, that reach like vast, lacy bridges to neighboring clusters.

The existence of groups and clusters is itself an important clue to the conditions of galaxy formation. Gravity, which unlike electromagnetism is a one-pole force, relentlessly gathers matter together. After the universe had cooled sufficiently from the Big Bang for the first segments of galaxies to form, fluctuations in density, perhaps arising from quite specific physical processes, were continually amplified by gravity. Only energy released from star formation could have temporarily resisted this process of "hierarchical clustering" by which smaller lumps aggregate into larger and larger ones. Models of stellar structure and energy generation show that a star more than 100 times as massive as the Sun would blow itself apart, so it is understood why hierarchical clustering ended with billions of stars per galaxy mass rather than a few supermassive stars. What physics prevented the coalescence of cluster galaxies into a single supergalaxy?

The answer appears to lie in the play-off between the time scale for collapse of the protogalaxy and the cooling time of its largely gaseous mass. Galaxies of the size we see today are the largest objects that could have radiated away significant amounts of their binding energies by dissipation in the gaseous component. As also happens in the birth of a star, dissipation enables a galaxy to contract to a higher density than specified by its initial size and temperature. The densities and temperatures associated with protoclusters, on the other hand, result in cooling times much longer than their dynamical times. Consequently, galaxies retain their individual identities and the effect of clustering is to collect galaxies into high-density regions rather than to build massive supergalaxies. Thus, the existence of galaxy groups and clusters is indicative of the role of gas dynamics and hierarchical clustering in the process of galaxy formation.

VARIATIONS OF PROPERTIES WITH ENVIRONMENT: IMPLICATIONS FOR GALAXY FORMATION AND EVOLUTION

The most fundamental property of a galaxy is its mass; however, astronomers usually substitute absolute luminosity because its determination requires only a single photometric measurement and a redshift (as an estimate of the galaxy's distance). (Studies of the internal dynamics of galaxies indicate a range in mass-to-light ratio of about a factor of 5 compared to a range in luminosity of the order of 1000, moderately justifying this substitution.) The number distribution of galaxy luminosities (the *luminosity function*) is thus used in place of the more fundamental mass distribution.

How does the luminosity function vary with environment? In cataloging thousands of rich clusters, George Abell found little or no variation from cluster to cluster; subsequent investigations determined a similar luminosity function for "field galaxies," those in low-density regions. That galaxies in what are today strikingly different environments could have formed with approximately the same distribution in luminosity (roughly equated to mass) suggests that the process of galaxy formation took place at an early epoch, when environmental differences were considerably smaller. This is an argument that galaxies formed before groups and clusters were well developed.

The second major property of a galaxy is its morphology. Averaged over all environments, spiral and irregular galaxies account for ~ 70% of all galaxies, the balance being elliptical and S0 galaxies. In the typical environment (that of loose groups), elliptical and S0 galaxies account for only 10–20% of the population. Hubble and Humason, and later (1936) Fritz Zwicky, noted the startling difference that rich clusters are *mainly* composed of elliptical and S0 galaxies. In addition to their unusual population, rich clusters are also notable for their unusual environment: Cluster galaxies exist in a relatively crowded space 1000 times more dense than the locale of a typical galaxy. Moreover, the core volume of a rich cluster is usually permeated by extremely hot plasma with a pressure much higher than that of a galaxy's interstellar gas. It is reasonable to speculate that some property of this extreme cluster environment favored the formation or evolution of less-common galaxy types—a clue to discovering the processes that produced morphological variation.

At such densities, interactions with the plasma and even direct encounters with other cluster members are likely to be important events in the life of a galaxy. In 1951, Walter Baade and Lyman Spitzer suggested that S0 galaxies might be formed when spiral galaxies cross in the dense environment of a cluster, stripping each other of their interstellar gas and losing their ability to form new stars. To be effective, this process requires a relatively rapid encounter (as would be common in a rich cluster) because a low-speed collision would probably result in the merging or disruption of the original spirals. A related idea, published by James E. Gunn and Richard Gott in 1972, held that a rapidly moving spiral would be stripped of its gas by ram pressure from the hot intracluster plasma.

In the mid-1970s Alar Toomre suggested a model for the formation of ellipticals that, like these S0 production models, imagined a metamorphosis of spiral galaxies in later life, this time by disruptive, largely nondissipative mergers with other galaxies. However, the prevalence of ellipticals in rich, dense clusters (where the encounter velocities are ~ 1000 km s^{-1}) is somewhat problematical in this model because at these speeds, much greater than the orbital speeds of stars in the galaxies, mergers are less likely than more elastic encounters. This suggests that if mergers were responsible for forming cluster ellipticals, they occurred relatively early, when the rich cluster was a collection of poorer, less-dense groups, each with a lower characteristic speed of member galaxies. Furthermore, the fact that the stellar density at the center of an elliptical galaxy is typically much higher than in a spiral suggests that a significant amount of dissipation must have occurred, indicating that the process occurred at an earlier time when galaxies were more gaseous. Both of these arguments suggest that if mergers were important in forming elliptical and perhaps S0 galaxies, they might have occurred so early as to be considered part of the formation process.

Elliptical galaxies are very similar to the bulges of disk galaxies. It is therefore natural to extend the merger picture to suggest that the spheroidal components of disk galaxies also formed by mergers. However, a counterargument holds that the chaotic addition of even a small amount of mass would disrupt the thin disk found in most spiral galaxies. Although a possible aid in explaining the evolution of S0 galaxies, the fact that many spirals with relatively large bulges still retain thin disks might be hard to understand unless these disks themselves formed *after* the bulges. If disk disruption is a serious problem for the model of building bulges by accretion, the importance of mergers in the formation of ellipticals is brought into question as well.

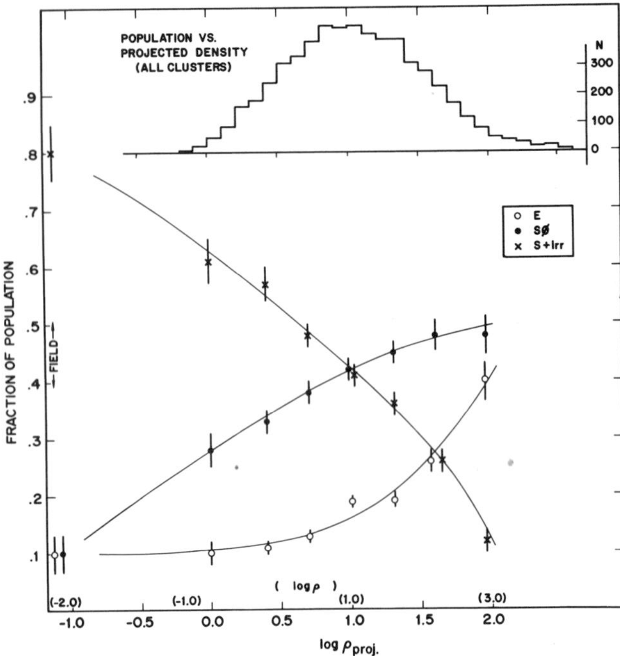

Figure 1. The morphology–density relation [*reproduced from Dressler (1981) Ap. J.* **236** *351*]. The fractions of E, S0, and spiral plus irregular galaxies are shown as functions of the projected local density (in galaxies per square megaparsec) and true density (in galaxies per cubic megaparsec). The upper histogram shows the number distribution of galaxies found in these environments, for a sample of over 6000 galaxies in 55 rich clusters. The fraction of spiral galaxies falls steadily for increasing local density, compensated by a corresponding rise in the fraction of elliptical and S0 galaxies. Identical trends have been found to hold for galaxies in poorer groups, but the dependence weakens or disappears in groups where the crossing time is comparable to the age of the universe.

Perhaps the best evidence for an environmentally induced merger process is the unique presence of giant galaxies in clusters. William W. Morgan identified a class that he called cD galaxies which are surrounded by extensive stellar envelopes. Many of these have extraordinary luminosities and masses, well above those found for galaxies in low-density environments. The view is commonly held (and supported by some admittedly disputed evidence) that cD galaxies have cannibalized neighboring galaxies, accreting by nondissipative mergers unfortunate companions that stray into their domain. Here again, however, both observation and theory suggest that this process was more important in the past when clusters were coalescing from smaller associations of galaxies with lower relative speeds.

The remarkably tight dependence of galaxy type on local density shown in Fig. 1 can be interpreted as evidence that local processes, perhaps early in the lifetime of the galaxy, played an important role in determining galaxy type. Over four orders of magnitude of increasing local density, the fraction of spirals decreases steadily as the fraction of S0s and ellipticals increases. This means that most elliptical and S0 galaxies exist *outside* of rich clusters, even though they are more *prevalent* in rich clusters. A mechanism like ram-pressure stripping by a hot plasma does not then offer a complete model for what makes a galaxy a dormant S0 instead of a star-forming spiral, because it accounts for only a small fraction of the population. Low-velocity encounters between spirals, as would be typical in the lower-density regions where most S0s are found, would be more likely to result in a merger than the mere removal of interstellar gas. However, a better description of the relation between the luminous baryonic component of a galaxy and its massive, nonluminous halo is needed to evaluate fully such models.

As in the case of the luminosity function, the very slow change in the morphological mix over orders of magnitude in density suggests that differentiation took place at an early epoch before the density contrast was very large. The morphology–density relation could even be evidence for a process that formed galaxies of different types *ab initio*, with little or no alteration later in the galaxy's life. In this picture, the formation of high-density spheroids, such as the bulges of disk and elliptical galaxies, is the result of more efficient star formation in denser environments. Crosstalk between the early fluctuations that formed galaxies and the much longer wave perturbations that formed clusters would then account for the fact that denser galaxies are more commonly found in denser environments. However, an elliptical or a disk galaxy with a dominant spheroid could form even in a globally low-density region if there had been a small-scale fluctuation of unusually high amplitude. Here again, the nonluminous halo and variations in *its* properties (e.g., density, angular momentum) with environment could play the key role.

Recent work suggests that the luminosity functions for different galaxy types are different but that each may show little or no variation with environment. If so, it is even more curious that these luminosity functions play off against the morphology–density relation in such a way as to make the summed luminosity function nearly universal. However, it is important to remember that the mass-to-light ratio for spirals is significantly lower than that of elliptical or S0 galaxies. Thus, the universality of the luminosity function may be little more than an accident masking a more fundamental dependence of the mass function on environment.

STUDYING ENVIRONMENTAL INFLUENCES IN THE PAST

In the 1980s astronomers began observational studies of the dependence of galaxian properties on environment *as a function of cosmic time*. Clusters of galaxies can be recognized at great distances corresponding to lookback times of 5–10 billion years, and the identification of these clusters as the ancestors of present-day clusters allows for a statistical comparison of their populations. It has not yet been possible to determine morphological types for these distant galaxies (due to limitations in image detail imposed by the atmosphere on observations with Earth-based telescopes) and observations with the *Hubble Space Telescope* are eagerly awaited to correct this deficiency. In the meantime, however, it has been possible through spectrophotometry to detect in distant galaxies the level of star formation, closely related to galaxy type at the present epoch.

The surprising result is that, in contrast to the relatively dormant elliptical and S0 galaxies found in today's clusters, rich clusters at $z \sim 0.5$ contained a much higher fraction of galaxies with vigorous star formation. This implies that the ancestors of some elliptical and S0 galaxies (or the component parts that preceded them) were much more active in star formation only a few billion years ago and/or that strong bursts of star formation were once common in the spiral galaxies of these clusters. It is not yet understood to what extent this is related to the cluster environment rather than just a reflection of galaxy evolution with cosmic time. It is certain, however, that adding the dimension of time to the observational data base will help distinguish between alternate models for the evolution of different galaxy types as a function of environment.

Additional Reading

Binggeli, B., Sandage, A., and Tammann, G. A. (1988). The luminosity function of galaxies. *Ann. Rev. Astron. Ap.* **26** 509.

Dressler, A. (1984). The evolution of galaxies in clusters. *Ann. Rev. Astron. Ap.* **22** 185.

Giovanelli, R. and Haynes, M. P. (1986). Morphological segregation in the Perseus–Pisces supercluster. *Ap. J.* **300** 77.

Gott, J. R. (1977). Recent theories of galaxy formation. *Ann. Rev. Astron. Ap.* **15** 235.

Schweizer, F. (1986). Colliding and merging galaxies. *Science* **231** 227.

Strom, S. E. and Strom, K. M. (1979). The evolution of disk galaxies. *Scientific American* **240** (No. 4) 72.

See also **Clusters of Galaxies, Component Galaxy Characteristics; Galaxies, Disk, Evolution; Galaxies, Elliptical, Origin and Evolution; Galaxies, Formation; Intracluster Medium.**

Galaxies, Radio Emission

Ulrich Klein

HISTORY

Almost six decades have passed since the first detection of cosmic radio waves by Karl Jansky in the early 1930s, the origin of which he first associated with the Galactic center (1933) and later with the plane of the Milky Way (1935). The first attempt to detect radio waves of the nearest external spiral galaxy similar to our own (M 31 in the constellation of Andromeda) was reported by Grote Reber in 1944, but the first undoubted detection of M 31 was that by Robert Hanbury Brown and Cyril Hazard in 1951. The limitation to observations at low frequencies at that time, however, did not allow detailed investigation of the distribution of radio emission, even in the closest spiral galaxies. This was partly due to the limited sensitivity of telescopes and receiver equipment, but was also due, in particular, to the low angular resolution, which as a rule of thumb is given by

$$\theta = 70\frac{\lambda}{D} \quad \text{(in degrees)},$$

where D is the telescope aperture and λ is the wavelength (measured in the same units as D and related to the frequency ν via the speed of light c, by $\lambda = c/\nu$). In the case of Reber's observations, the beam width was approximately 12° at wavelength 1.9 m, whereas the (optical) angular extent of M 31 is about $2.^\circ5 \times 0.^\circ7$. Hence, although the early radio surveys of the Milky Way galaxy already had provided some detailed information about the (projected) distribution in our home galaxy, such studies of external galaxies had to await modern high-resolution aperture synthesis and single-dish telescopes equipped with high-frequency receivers with good sensitivity and stability.

Detailed measurements of galaxies other than those in the Local Group become feasible starting in the early 1970s with the operation of the Westerbork Synthesis Radio Telescope (WSRT) in the Netherlands, the Effelsberg 100-m telescope in western Germany, and the Very Large Array (VLA) in the United States (New Mexico). Measurements of global properties, such as flux densities and the positions of radio sources associated with galaxies had been conducted earlier with telescopes of various kinds, mainly interferometers.

EMISSION PROCESSES

Thermal Radiation

Galaxies are continuously forming stars from the available gas reservoir. A small fraction of these newly born stars consists of stars with large masses (≥ 8 solar masses) and correspondingly high surface temperatures, so that the bulk of their emission is at wavelengths shorter than 912 Å. This energetic ultraviolet radiation is capable of ionizing the surrounding gas, forming so-called H II regions in which free (i.e., unbound) protons and electrons are present. The electrons move on hyperbolic paths between the protons and the corresponding accelerations give rise to the emission of radiation (called thermal free–free radiation, or bremsstrahlung). According to the typical velocities involved, this emission is observed in the centimeter and decimeter wavelength range and its intensity I_ν decreases only slightly as a function of frequency ν,

$$I_\nu \sim \nu^{-0.1}.$$

This thermal emission (and the H II regions) is seen predominantly in the spiral arms of galaxies in which the current star formation rate is highest.

Nonthermal Radiation

A massive star ends its life as a star (i.e., the hydrogen-burning phase) in the form of a supernova, which releases considerable energy ($\approx 10^{51}$ ergs) into the interstellar medium, along with radiation, fast particles (e.g., protons and electrons), heavier nuclei, and probably also magnetic fields. The fast particles may eventually assume relativistic speeds (i.e., $v \approx c$) by continuous (re)acceleration in the interstellar medium, which is filled to a fairly large degree by supernova "bubbles," produced by the huge energy releases of these explosions. When these fast particles encounter a magnetic field, they are forced to follow helical paths around it so that they are continuously accelerated, giving rise to so-called synchrotron radiation, which also shows up in the radio spectral region. This radiation mechanism is substantially different from that of thermal radiation, in that it is caused (mainly) by relativistic electrons and its intensity decreases rapidly as a function of frequency,

$$I_\nu \sim B_\perp^{1+\alpha_{\mathrm{nth}}},$$

where $B_\perp$ is the component of the magnetic field perpendicular to the line of sight and α_{nth} is the spectral index. (The subscript "nth" denotes nonthermal radiation.) Furthermore, this emission is partly polarized, that is, the emitted waves are oscillating in a preferential direction that is perpendicular to the magnetic field that gives rise to this radiation. However, the polarization direction of the received wave is, in general, different from that of the wave when it left the source, due to the effect of Faraday rotation in which the wave plane is rotated while penetrating a magnetized plasma. The amount of Faraday rotation is wavelength-dependent and is given by

$$\Delta\chi = \mathrm{RM} \cdot \lambda^2,$$

where λ is the wavelength (in meters) and RM is the so-called rotation measure, given by

$$\mathrm{RM} = (8.1 \times 10^5) \int_0^L n_e B_\parallel \, dl,$$

where n_e is the thermal electron density of the medium (in cm^{-3}), $B_\parallel$ is the magnetic field component parallel to the line of sight, and the integral is taken over the whole line of sight L (in parsecs). We therefore have a powerful tool to detect magnetic fields in galaxies! Karl O. Kiepenhéuer (1950) was the first to consider the process of synchrotron radiation as the predominant mechanism producing the observed low-frequency emission of the Milky Way.

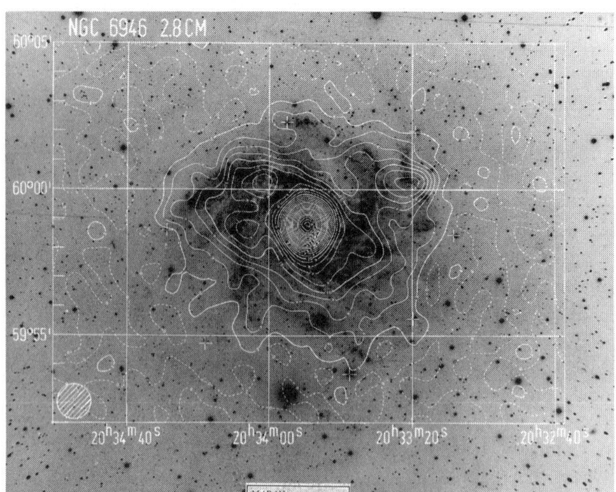

Figure 1. Map of the radio continuum emission of the spiral galaxy NGC 6946 at 2.8-cm wavelength, obtained with the Effelsberg 100-m telescope. The contours represent isointensities of radio emission. They are superimposed onto an optical photograph reproduced from Halton Arp's *Atlas of Peculiar Galaxies*. The hatched circle in the lower left indicates the beam size; coordinates are right ascension and declination, epoch 1950. [*Reproduced from Klein et al. (1982). Reproduced, with permission, from* Astron. Ap. ***108***. *©1982 by* Astronomy and Astrophysics.]

In general, the radio continuum emission in a normal spiral galaxy is a mixture of the two components described here (thermal and nonthermal), and it takes observations made at two or more frequencies to disentangle them, either globally or, if the quality of the radio maps is sufficient, even within different locations in the galaxy.

A TYPICAL GALAXY IN THE RADIO WINDOW

Face-On View

Figure 1 presents a typical example of what a spiral galaxy looks like in the radio window when seen face-on. Shown here are isointensity contours of the radio emission of the spiral galaxy NGC 6946 taken at $\lambda = 2.8$ cm ($\nu = 10.7$ GHz) with the Effelsberg 100-m telescope and superimposed onto an optical photograph of the galaxy. Because the observations were taken at a fairly high radio frequency, thermal emission shows up as enhanced radiation in some locations that are coincident with giant H II complexes, whereas the underlying synchrotron emission decreases more or less smoothly from the center of the galaxy toward the outer (optical) edge. This decrease is best described by an exponential law of the observed intensity I_ν at frequency ν,

$$I_\nu(R) = I_0 e^{-R/R_0},$$

where R is the galactocentric radius and R_0 is the scale length of the distribution. In general, R_0 is always larger than the scale length of the visible light. If mapped with high-resolution aperture synthesis instruments (e.g., WSRT and VLA), which emphasize the small-scale structure, such a galaxy exhibits enhanced emission in the spiral arms, where the young massive stars are located. However, the distribution of the synchrotron emission is always found to be smoother than that of the thermal emission, probably reflecting the diffusive propagation of the relativistic particles away from their sites of origin. Therefore, spiral galaxies in general are more extended in the radio continuum than at optical wavelengths. If we map the distribution of the spectral index

$$\alpha_{\nu_1,\nu_2} = \frac{\ln(S_{\nu_1}/S_{\nu_2})}{\ln(\nu_1/\nu_2)}$$

between any two frequencies ν_1, ν_2 across the galaxy, we can investigate the varying relative proportion of thermal and nonthermal emission as a function of location in the galaxy because we are looking everywhere at the sum of a thermal flux with a "flat" spectrum and a nonthermal flux with a "steep" spectrum, that is,

$$S_\nu(\mathbf{r}) = S_{\text{th}}(\mathbf{r})\nu^{-0.1} + S_{\text{nth}}(\mathbf{r})\nu^{-\alpha_{\text{nth}}},$$

where $S_{\text{th}}(\mathbf{r})$ is the thermal contribution at a certain frequency at location $\mathbf{r}$ in the galaxy and $S_{\text{nth}}(\mathbf{r})$ is the corresponding nonthermal flux. This tells us where the sites of recent formation of massive stars are and how the cosmic rays propagate away from these regions, assuming that they, too, are produced by the aftermaths of stellar death. Flat spectra ($\alpha \approx 0.1, \ldots, 0.5$), which indicate a high contribution of thermal emission, are found coincident with H II regions and central starburst regions of galaxies, whereas steeper spectra ($\alpha \approx 0.5, \ldots, 0.9$) are seen away from these, especially between the spiral arms and at the outer peripheries of galaxies.

Observations of the linearly polarized component of the emission allows to trace the magnetic field in galaxies. If measurements are made at several wavelengths, these can be corrected for Faraday rotation, thus recovering the intrinsic field direction. After initial attempts, (commencing in the mid-1970s) to apply this technique to the spiral galaxies M 51, M 81, and M 31, using the WSRT, this kind of study was resumed with the 100-m telescope (in the early 1980s) and later with the VLA. This field of research is one of the most important in the investigation of galaxies, given the significance and impact of magnetic fields on various physical processes in the interstellar medium (and probably the intergalactic medium, too!). The large-scale structure of magnetic fields in spiral galaxies is known for about a dozen nearby objects, and two substantially different field configurations appear to prevail: an axisymmetric field, in which the field lines run inward everywhere in the galaxy's disk (note that these field lines must be closed above and below the galaxy's plane to avoid magnetic monopoles!), and a bisymmetric field, in which the field enters the disk on one side and leaves it on the opposite side, probably closing in intergalactic space. It is noteworthy that these two scenarios have been postulated by theoreticians modeling the galactic dynamo and that the researchers studying the dynamo mechanism in disk galaxies have become increasingly interested in the prolific observations of polarization in galaxies. Magnetic field strengths can be estimated by assuming energy equipartition between the magnetic field and the cosmic rays in a galaxy (which are coupled). Typical field strengths obtained are in the range of a few microgauss (μG) up to 10 μG, with higher values only encountered in the dense and energetic central regions of starburst galaxies like M 82.

Edge-On View

If we look at a spiral galaxy edge-on in the radio window, we can study the distribution of radio emission away from the galaxy's plane, in the so-called z direction. Figure 2 presents a typical example, the radio continuum map of the spiral galaxy NGC 4631, obtained with the VLA at $\lambda = 20$ cm and superimposed onto an optical picture. One striking difference between radio and optical emissions is immediately obvious: The radio emission extends much further from the plane than does the optical. If the same

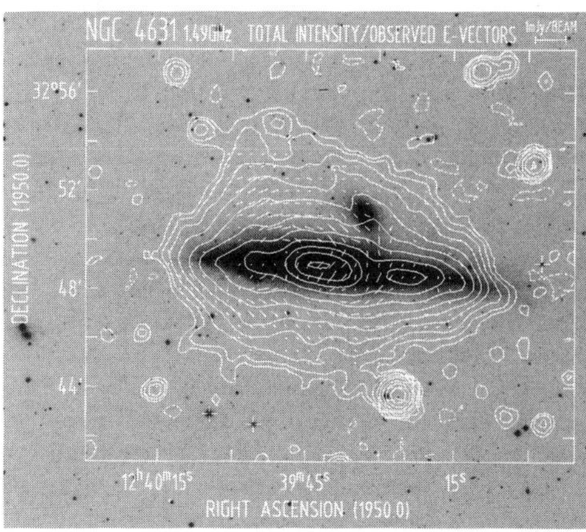

Figure 2. Radio map of the edge-on spiral galaxy NGC 4631 at 20-cm wavelength, observed with the VLA. Radio contours are superimposed onto an optical picture. The vectors indicate the orientation of the plane of the electric field of the linearly polarized radiation, which was simultaneously obtained in the observation. Note the large extent of the radio emission away from the galaxy's plane, a feature that appears to be fairly unique in spiral galaxies. [*Reproduced from Hummel et al. (1988). Reproduced, with permission, from* Astron. Ap. ***197***. *©1988 by* Astronomy and Astrophysics.]

galaxy is investigated at a higher radio frequency, this "halo" of radio emission is found to have a much smaller extent in z. This behavior is explicable in terms of the processes of propagation and energy loss of relativistic particles during their lifetime. What we see here is the superposition of a thin radio disk with predominantly thermal emission and a thicker one with purely nonthermal emission. The relativistic particles lose energy via synchrotron radiation due to their interaction with the magnetic field that is also present outside the (optically visible) disk of the galaxy. They are evidenced by the presence of polarized radio emission at large z distances and are marked in Fig. 1 by the "vectors," which represent the direction of the electric field of the received radiation. This would be perpendicular to the magnetic field direction if there were no Faraday rotation. The latter can be assumed to be negligible outside the disk, so that in NGC 4631 the magnetic field seems to be oriented primarily away from the disk at larger z [a few kiloparsecs (kpc)]. More detailed measurements are needed to show what happens to the field at the interface between the disk and the halo.

The existence of extended radio halos such as the one visible in Fig. 2 was postulated more than 30 years ago (e.g., by Iosif S. Sklovskij in 1952), yet it has been proven only for two or three nearby edge-on galaxies. Most galaxies possess what is usually referred to as a thick radio disk, with z-extents of a few kiloparsecs, whereas in NGC 4631 radio emission can be traced as far out as $z \approx 7.5$ kpc. The spectrum of the radio emission steepens as a function of z, reflecting the energy and possibly the adiabatic losses of the relativistic electrons as they escape from the disk (note that there is hardly any reacceleration of such particles once they have left the disk and its violent interstellar medium).

DWARF IRREGULAR GALAXIES

Dwarf irregular galaxies, such as the Magellanic Clouds, exhibit a rather chaotic appearance at optical wavelengths. A similar structure is found at radio wavelengths, most likely due to the patchy distribution of star-forming complexes. These galaxies obviously lack the exponential nonthermal disk that governs the radio images of massive spiral galaxies. Furthermore, the synchrotron emission from such stellar systems is rather weak, which is one reason for the lack of detailed studies of its distribution in dwarf galaxies. The overall scenario that is presently favored to explain the radio properties of dwarf galaxies is that most of them are forming stars at lower rates than massive galaxies and that the containment of cosmic rays in them is also significantly reduced as compared to normal spiral galaxies. This could be due to a substantially different magnetic field configuration, but that possibility will have to be investigated by future studies (with superior sensitivity) of polarized radio emission in dwarf galaxies. This picture is supported by the properties of a special class of dwarf galaxies that are vigorously forming stars at present, the so-called blue compact dwarf galaxies. These galaxies are undergoing bursts of star formation, yet their nonthermal radio emission is comparatively low. If in these galaxies, too, cosmic rays are produced in supernova events, then the deficiency of synchrotron emission can be explained only in terms of different properties of cosmic ray propagation and magnetic field.

GLOBAL RADIO PROPERTIES OF GALAXIES

Normal spiral galaxies (as opposed to radio galaxies) emit monochromatic radio luminosities at 5 GHz of up to approximately 10^{22} W Hz^{-1}, whereas for dwarf galaxies at the same frequency the lowest luminosities are of the order of approximately 5×10^{17} W Hz^{-1}. This emission accounts for only a small fraction of the total energy output of a galaxy, which is generally dominated by the far-infrared emission (i.e., thermal radiation from dust). The total radio emission shows a good correlation with that radiated in other spectral domains, such as blue light or x-ray, but the tightest correlation by far is that with the far-infrared emission from dust. The radio emission is of predominantly nonthermal origin down to even shorter (centimeter) wavelengths, whereas in the far-infrared we measure the reradiation of the energetic Lyman continuum photons that are produced by massive OB stars and that are efficiently absorbed by dust grains and transformed into longer-wavelength radiation. Because of the obviously entirely different and independent radiation processes (including their origins) this close correlation is still awaiting a thorough and comprehensive theoretical interpretation.

The spectral indices of spiral galaxies, measured between 408 MHz and 10.7 GHz, show a remarkably small scatter around a mean value of $\langle\alpha\rangle = 0.75$. After separating the thermal content of the radiation, the mean value becomes somewhat higher, $\langle\alpha_{\mathrm{nth}}\rangle = -0.88$, representing the pure synchrotron spectrum of a typical galaxy. The dispersion of this quantity is not very large either, clearly showing that the same overall and unique process is controlling cosmic ray acceleration and/or propagation in all massive spiral galaxies. The frequency spectrum of synchrotron radiation in galaxies is connected with the energy distribution $N(E)$ of the relativistic particles,

$$N(E)\,dE = AE^{-\gamma}\,dE,$$

where A is a constant and γ is the so-called energy spectral index, both of which can be measured directly near the Earth. The latter is related to the frequency spectral index α_{nth} via

$$\gamma = 2\alpha_{\mathrm{nth}} + 1$$

The value of γ measured in the Earth's immediate vicinity is $\gamma \approx 2.6$, in remarkably good agreement with $\alpha_{\mathrm{nth}} = 0.88$, the global value obtained for a sample of normal spiral galaxies.

Dwarf galaxies appear to have global radio properties that are markedly different from those of normal spirals. Only a few of them have been thoroughly investigated, and their radio spectra hint at rather steep synchrotron spectra and a much smaller proportion of synchrotron emission relative to thermal emission. This probably reflects the lack of confinement of cosmic rays in dwarf galaxies, as described above.

CURRENT AND FUTURE INVESTIGATIONS

There are a great many open questions (partially addressed here) that require scrutiny, both observationally and by theoretical modeling. The following questions are intriguing issues: Why do some galaxies possess extended radio coronae and others don't? What gives rise to axisymmetric magnetic field configurations in some spiral galaxies and bisymmetric fields in others? How is this related to galaxian dynamics? What determines the confinement of cosmic rays (which appears to be so different in massive and dwarf galaxies)? Which parameters control the conspicuously tight correlation of nonthermal radio emission and thermal far-infrared emission? Related to this, where and how are relativistic particles produced, accelerated, and reaccelerated, and how do they propagate? (This issue, though discussed in many papers, is far from being thoroughly understood.) The key to the resolution of these questions lies in the development of radio astronomical techniques to facilitate improved observations, in the intercomparison of data obtained in various spectral domains, and in progress in the theories of magnetic field production and maintenance and of cosmic ray production and propagation (possibly facilitated by magnetohydrodynamical modeling of this constituent of galaxies).

Additional Reading

Beck, R. (1986). Interstellar magnetic fields. *IEEE Trans. Plasma Science* **PS-14** 740.

Hummel, E., Lesch, H., Wielebinski, R., and Schlickeiser, R. (1988). The radio halo of NGC 4631: Ordered magnetic fields far above the plane. *Astron. Ap.* **198** L29.

Klein, U., Beck, R., Buczilowski, U. R., and Wielebinski, R. (1982). A survey of the distribution of 2.8 cm radio continuum in nearby galaxies. *Astron. Ap.* **108** 176.

Sofue, Y., Fujimoto, M., and Wielebinski, R. (1986). Global structure of magnetic fields in spiral galaxies. *Ann. Rev. Astron. Ap.* **24** 459.

van der Kruit, P. C. and Allen, R. J. (1976). The radio continuum morphology of spiral galaxies. *Ann. Rev. Astron. Ap.* **14** 417.

Verschuur, G. L. and Kellermann, K. I. (1988). *Galactic and Extragalactic Radio Astronomy*. Springer-Verlag, Berlin.

See also **Clusters of Galaxies, Radio Observations; Cosmic Rays, Origin; Galactic Structure, Magnetic Fields; H II Regions; Radio Sources, Emission Mechanisms.**

Galaxies, Spiral, Nature of Spiral Arms

Ray Carlberg

The beautiful spiral patterns that occur in disk galaxies are now understood to be primarily a result of gravitational forces between stars and gas in a shearing, rotating disk. The path toward this simple idea has had many twists, turns, and blind alleys.

Two previous, incorrect theories are that spirals are galactic analogues of a spinning garden sprinkler spraying from the ends of a rotating central object or that magnetic fields trap the gas so that the two are wound up together in spiral patterns.

Attempts to understand spiral structure on the basis of gravitational forces between very large numbers of stars have a long history, but limited success precisely because there are so many stars in galaxies. In fact, there are so many stars in galaxies that any single star in a galaxy could be removed and the motion of every other star would be almost the same, which is very different from the situation in star clusters. This feature has the important consequence that the motions of stars in galaxies are controlled by the collective gravity of all the stars, and do not depend on the details of any star's position. This type of system, called a collisionless fluid, allows one to ignore the details of the individual stars, but requires complex techniques of analysis for adding together the joint motions and gravitational effects of all the stars.

The fact that stars cannot interact as individuals immediately leads to a simple explanation of the fact that all spirals appear to be "trailing," that is, they lag the direction of rotation with increasing radius in the disk. The explanation is that gravitating systems attempt to increase their gravitational energy, or at least become more concentrated, if at all possible. For a disk of stars an increase in gravitational energy of the central region can only occur if the stars move closer to the center. To move stars inward requires that they lose angular momentum, which can be accomplished if there is a torque directed opposite to their direction of rotation. A trailing spiral pattern supplies just such a torque because the stars in the spiral wave near the center are pulled backward by the stars further out in the spiral. Consequently the central gravitational energy is increased at the expense of a decrease in the energy of stars further out in the disk.

The *density wave theory* introduced by Chia-Chiao Lin and Frank H. Shu began the modern era of understanding spirals in terms of gravitational dynamics. Using a simplified set of equations they showed that the solutions were waves that maintained their spiral shape, rotating through the stars and gas in a galactic disk. In the inner part of the disk the stars move faster than the spiral and overtake it; in the outer part of the disk the spiral pattern moves faster than the stars and the relative velocities are reversed. For the gas clouds in the disk this idea predicts that they will be compressed and form dark dust lanes on the inside of the trailing spiral patterns, as is usually the case. More recently detailed investigations of the relative locations of the spiral arms as seen in the stars as compared to the arms seen in the molecular gas provide additional evidence that spiral arms are waves that move through the gas and stars in the disk.

The simple version of the density wave theory requires that the same pattern rotates around and around for the lifetime of the galaxy. However, analysis shows that only in special situations can these density waves persist. The more likely result is that the inner part of the wave travels toward the center, the outer part travels outward, and, in the absence of a source of new waves, the spiral arm dies away. Resolving the question "Where do spiral arms come from?" remains an issue of current investigation.

Real galaxies provide only a limited amount of information about spiral waves. The main problem is that a typical galaxy has a rotation period of 100 million years, so we cannot simply observe the galaxy to see whether the wave pattern remains steady or dies away. Computer simulations have done much to alleviate this problem. Although the equations of motion for stars in a disk were written down by Isaac Newton, it is not simple to solve them for a system of billions of stars. Therefore, computer studies usually have tried to mimic several thousand stars with the motion of one "superstar." If the superstar becomes too massive then the studies start seeing the motions of other stars as individuals, rather than the collective effect of the collisionless fluid that is an essential feature of disk galaxies. For these reasons, it is only recently that computer simulations have been large enough and accurate enough to provide a lot of valuable information regarding the motions of stars and gas within a galaxy disk.

The results of one simulation incorporating 30,000 superstars and 10,000 super gas clouds are shown in, Fig. 1. It is immediately obvious that these model galaxies have strong resemblances to real galaxies: the spirals are relatively open, the top one is nearly bisymmetric but not quite, and it is possible to produce model galaxies with many, very fine, spiral arms. It is also quite easy to make the barred spirals that are the other part of Edwin Hubble's tuning fork diagram of galaxy classification.

Part of the process that is responsible for the spirals in these model galaxies is quite well understood. In the models a gradual inward drift of stars in the central region can be detected exactly as

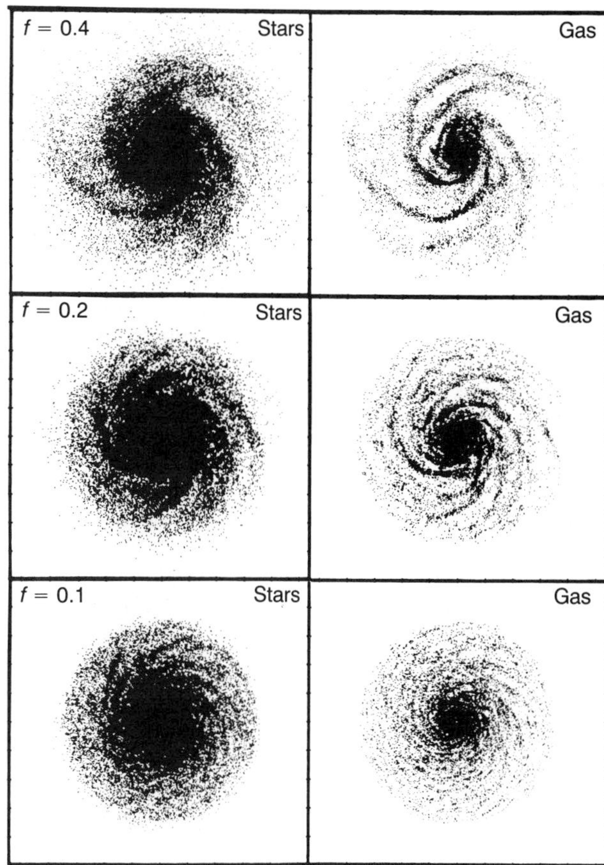

Figure 1. Spiral arms in the stellar and gas components of disks with increasing ratios of halo to disk mass. [*Reprinted from* The Astrophysical Journal ***298*** *486 (1986) by permission of The University of Chicago Press.*]

predicted by the idea that explains trailing spirals. However, measurements of the motions of the spirals in these galaxies find that the patterns are not permanent: they come and go, but at almost any time spirals of the same general appearance are present.

A good partial understanding of these transient spiral patterns was developed prior to the computer models, and the process has been dubbed *swing amplification*. The same effect that makes a trailing spiral wind inward and disappear does exactly the opposite to leading spiral waves. A small leading spiral will travel outward and grow enormously, by factors of up to a hundred in intensity, into a strong training pattern. That is, spirals do not preserve their shape, but are bent backward by the rotation of the disk although not nearly as fast as the motion of the stars alone gives. The spirals continue to move through the stars, thus preserving the ability to predict the characteristics of the dust lanes. However, once a very large trailing spiral has been created it then continues to slowly wind up and fade away. Careful measurements of a large number of simulations confirm that this idea explains many of the details of the changes in the spiral features in these simulations and can nicely explain the trend of increasing number of spiral arms seen in the diagram.

The keys to understanding the amplification is the rotation of the disk and the idea of pumping a swing. To increase the amplitude of motion on a garden swing, it is crucial that the pushes be applied at the correct times, say at both the forward and backward ends of the swing. Such a tuning of an applied force to an oscillator is known as a resonance. In a galaxy the situation is a little more complicated than a swing. Stars do not travel around the disk in perfect circles, but orbit on precessing ellipses that can be conveniently thought of as the combination of motion around a circle combined with a small "epicycle" (as in the Ptolemaic theory of the solar system). Stars on orbits nearer the center go around the circle (not the epicycle) in a shorter time than those further out, an effect that will destroy any material feature within the disk of stars. The epicycles themselves generally have longer periods with increasing radius in the disk. However, it turns out that for the mass distributions of most galaxies the combination of the motion around the circle and half of the epicyclic frequency does not vary much with radius. That is, if a two-armed spiral pattern is created with a pattern speed that matches this combination of frequencies, than the stars on their slightly elliptical orbits will be able to respond in tune with this oscillation; that is, a wide range of radii will be near a resonance of the applied spiral wave so that it should grow in intensity. Ordinarily this spiral wave amplifier does not have a "feedback" circuit available, so the wave is excited, grows into the large visible spiral, and then fades away as the shearing forces overwhelm the amplifying forces.

A question only partially answered is "Where do the leading spirals that trigger the whole process originate?" Within the computer models the answer is fairly readily evident because these models contain 10,000 gas clouds, similar to the number in real galaxies. Quite frequently these clouds are positioned in such a way that chance alignments effectively create leading spiral patterns. Furthermore, the remnants of previous trailing spirals as they wind up leave small gaps and debris from spiral arms that can by chance line up to create a small leading spiral that keeps the process continually recurring. This continually recurring process can be damped down since the spirals disturb the star orbits, causing them to become less circular and therefore less easy to organize into spiral patterns. However, as long as the galaxy has a supply of young stars and gas in which the spirals form, the spirals can keep recurring.

In the end a complete theoretical description of all the details of spiral patterns will involve a tremendous amount of complicated physics of stars, gas, magnetic fields, and gravity. A very good approximate description of spiral structures is a description of many stars moving about as described by Newton's dynamics and theory of gravity. Stars moving together in nearly circular orbits are strong amplifiers of regions of higher star density, and they create the visible trailing spirals that we observe.

Additional Reading

Bertin, G. (1980). On the density wave theory for normal spiral galaxies. *Phys. Rep.* **61** 1.

Carlberg, R. and Freedman, W. (1985). Dissipative models of spiral galaxies. *Astrophys. J.* **298** 486.

Hohl, F. (1971). Numerical experiments with a disk of stars. *Astrophys. J.* **168** 343.

Lin, C. C. and Shu, F. H. (1964). On the spiral structure of disk galaxies. *Astrophys. J.* **140** 646.

Toomre, A. (1977). Theories of spiral structure. *Ann. Rev. Astron. Astrophys.* **15** 437.

Toomre, A. (1981). What amplifies the spirals? In *The Structure and Evolution of Normal Galaxies*, S. M. Fall and D. Lynden-Bell, eds. Cambridge University Press, Cambridge, p. 111.

See also **Galactic Structure, Spiral, Interstellar Gas, Theory; Galaxies, Spiral, Structure.**

Galaxies, Spiral, Structure

Bruce G. Elmegreen and Debra Meloy Elmegreen

Spiral galaxies are flattened stellar systems that rotate in a nearly circular fashion. They generally contain stars with a wide range of ages. The gas and youngest stars are in a flattened disk, older stars occupy a slightly thicker disk, and the oldest stars tend to reside in a more three-dimensional distribution, which may include a central bulge, a system of globular clusters, and a low-luminosity halo. This division of stars into distinct *populations* was first recognized by Walter Baade in 1944.

Spirals form when compressional waves propagate through the disk, growing in length and amplitude because of self-gravitational forces. The waves appear as spirals because the angular rotation rate of the disk varies approximately inversely with distance from the center. In many galaxies, spiral waves become so strong that they drive shock fronts into the gas, trigger new stars to form, and impart large random motions to the older disk stars at critical resonance points. Then the spiral structure can dominate the appearance, internal dynamics, and evolution of a galaxy.

The optical appearance of spiral galaxies is extremely diverse. A classification system was developed in the 1940s by Edwin Hubble and expanded in the 1960s by Allan Sandage, and by Gerard and Antoinette de Vaucouleurs. Along the Hubble sequence from type Sa to Sb, Sc and Sd, the bulge gets smaller and the spiral arms become more open. Two parallel sequences, SBa to SBd and SABa to SABd, describe galaxies with central bars and ovals, respectively. Because Hubble originally envisioned an evolutionary sequence, the Sa and Sb galaxies are termed *early* type and the Sc and Sd galaxies are termed *late* type. Early type galaxies contain proportionately more old stars and less gas and star-forming activity than late type galaxies; presumably the early type galaxies formed stars more rapidly in the past.

The spiral structure itself varies from galaxy to galaxy, even within a Hubble type. A classification system based on the regularity of the spiral arms was introduced in the 1960s by Sidney van den Bergh and revised in the 1980s by Debra and Bruce Elmegreen. Approximately 10% of spiral galaxies contain only a single, bisymmetric spiral that extends in a *grand design* from the edge of the galactic bulge to the outer limit of the perceptible disk. Most galaxies have a *multiple-arm* spiral structure, or even a highly chaotic or *flocculent* spiral-like structure. Multiple arm galaxies represent 60% of the early and intermediate Hubble types with bars and intermediate types without bars. Flocculent spirals represent 60% of early Hubble types without bars.

An example of a grand design galaxy, Messier 81, is shown in Fig. 1, rectified by computer to a face-on orientation and enhanced in detail by subtraction of the average radial light profile. The spiral pattern could be a quasistationary wave mode, as discussed by Chia-Chiao Lin, Giuseppe Bertin, and collaborators. Such a mode comprises both inward and outward propagating spiral wave packets. Interference between these wave packets may produce the gentle oscillation of the arm amplitudes seen in the figure. Wave modes are thought to grow by the transfer of energy and angular momentum from the inner to the outer region at the radius where the spiral pattern corotates with the disk. The trailing spiral components propagate away from this radius, outward and inward, until they either lose their energy by resonant interactions with the peculiar (noncircular) motions of stars or they reflect or refract off regions where a high stellar velocity dispersion makes propagation impossible. The energy-losing resonances are called *Lindblad resonances*, after Bertil Lindblad; they occur where the period of a star's random excursion around its circular orbit equals the time spent between the main spiral arms. The reflection point occurs in the inner region, near the bulge. After reflection, the wave returns to the corotation radius as either a leading wave packet or an open trailing wave packet; there it stimulates more of the tightly wound trailing components to reinforce the original trailing waves. Theory predicts that an initially small spiral perturbation should double in amplitude approximately once in each orbit time. Eventually the growth of the spiral wave mode is limited by energy dissipation in shocked gas and by stars scattering at the Lindblad resonances.

Highly symmetric spirals like those in Messier 81 can also form when a galaxy is strongly perturbed by a close encounter with another galaxy. This mechanism was illustrated in an atlas of peculiar galaxies by Halton Arp (1966) and it was modeled theoret-

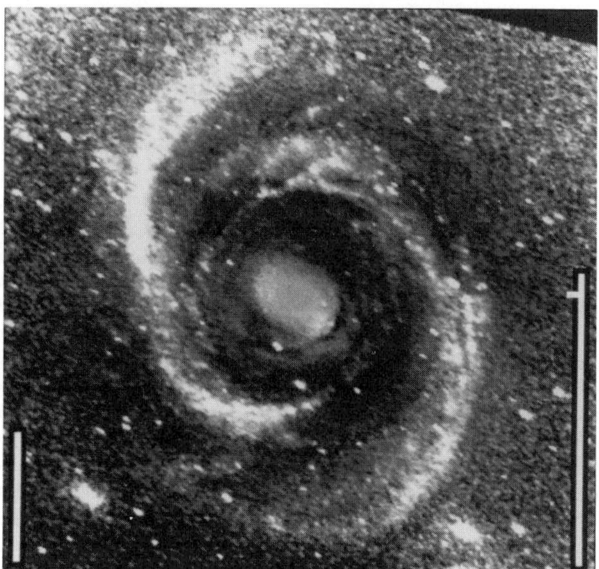

Figure 1. Blue band image of M81, rectified to a face-on orientation and with the average radial light profile subtracted. The calibration bar on the left represents 100 pixels; the bar on the right has a length equal to the isophotal radius at 25 magnitudes per square arc second. (From *Elmegreen, Elmegreen, and Seiden, 1989. Reprinted with permission from* The Astrophysical Journal.)

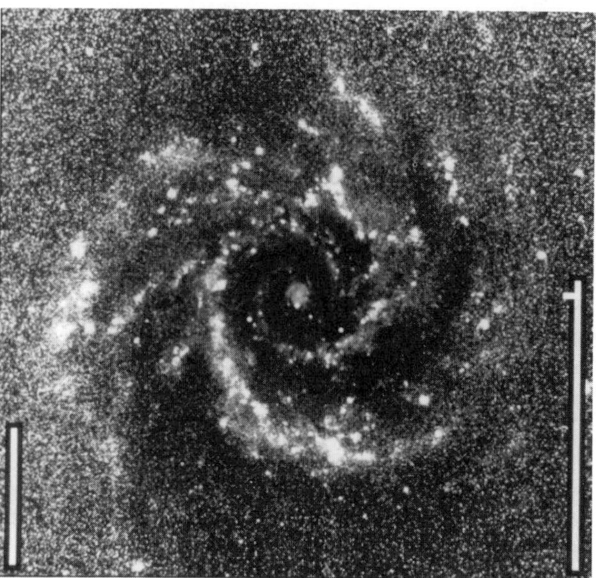

Figure 2. Blue band image of M74, as in Fig. 1. [*By permission from B. G. Elmegreen*, in Galactic Models, *New York Academy of Sciences, New York (1990)*.]

ically by Alar and Juri Toomre (1972). Tidal forces during such an encounter can perturb the stars so much that their previously near-circular orbits become elongated, especially in the outer parts. This elongation may be initially along a common line of apses, but under the influence of galactic shear, the elongations at different radii rotate at different rates, causing the stars to bunch together in a symmetric spiral pattern. Self-gravity then amplifies this pattern, and it propagates into the inner disk, possibly reflecting off the bulge. The outer tidal arms may also stimulate the growth of an inner spiral wave mode, which then persists for many more orbits than the initial tidal wave.

An example of a *multiple-arm* galaxy is Messier 74, shown with computer enhancement in Fig. 2. The symmetric spiral in the inner region is surrounded by numerous, apparently independent, long-arm spirals in the outer region. Theory suggests that the inner symmetric spiral could be a single wave mode or a constructive superposition of several wave modes with different angular speeds, or it could be a rapidly evolving wave packet. The outer spirals could be the constructively interfering parts of several overlapping wave modes, or they could be random and independent wave packets triggered by local gravitational instabilities in the gas and stars.

A different theory, developed by Philip E. Seiden and Humberto Gerola in 1978, suggests that some of the outer structure in multiple-arm galaxies could result entirely from star formation. Gaseous instabilities first trigger star formation, and then supernova explosions and shell-like compressions following star formation trigger more star formation, leading to chain reactions that are swept back into spirals arcs by shear.

Flocculent galaxies contain many more, and much shorter, spiral arms than multiple arm galaxies. Such short arms could form by the same mechanisms as the longer arms in multiple arm galaxies, with the larger number of arms causing the chaotic appearance, or they could differ because of a lack of global wave stimulation, or an inability to amplify global waves in flocculent galaxies. An example is Messier 63 (not shown).

Bars and oval distortions may drive spiral waves in some galaxies. Among galaxies with early Hubble types, those with bars are twice as likely to have symmetric spirals in their outer disks as those without bars. Among intermediate and late-type galaxies, however, those with bars have approximately the same proportions of grand design, multiple arm, and flocculent spirals as those without bars. This dependence of the bar-spiral correlation on Hubble type may result from a variation of bar length and strength with Hubble type, that is, the bars in early-type galaxies tend to be larger and stronger than the bars in late-type galaxies. Perhaps early-type bars end near the corotation resonance, where they can drive outward-moving trailing waves that fill the outer disk. The bars in late-type galaxies may end far inside this resonance, possibly at an inner Lindblad resonance.

An important quantity to determine is the angular rate of the spiral pattern, which equals the angular rate of the stars at the corotation resonance, in the modal theory. Stars move faster than this pattern rate inside corotation and slower outside corotation. Recent work suggests that the angular pattern speed can sometimes be determined from optical tracers of wave-orbit resonances. In the most symmetric spirals, the far outer extent of the spiral should be the outer Lindblad resonance, where outward moving wave packets convert their energy into random stellar motions. The inner extent may be an inner Lindblad resonance if there is a bar or oval there, because such ovals can reflect an inward-moving wave. If the inner galaxy is symmetric, or if there is a strong bulge, then the inner extent of the spiral need not be a resonance but merely a reflection point. The corotation resonance occurs at the radius where gaseous shocks and wave-triggered star formation end, and in some cases, where large patches of interarm star formation occur. The 4:1 resonance, where the time spent between the main spiral arms equals twice the period of the peculiar stellar motions (rather than one times this period, as for a Lindblad resonance), may contain small spiral-like spurs midway between the main arms. Once any one resonance location is identified in a galaxy, all of the others follow from the radial distribution of circular velocities.

Considerable progress has been made in understanding galactic spirals, but neither the theoretical framework nor the observations of galaxies are sufficiently detailed at the present time to determine the origin and evolution of spirals on a case-by-case basis. Part of the problem is that the basic state of a galaxy is uncertain, that is, the spatial and temporal distributions of the gas and stars, and of their velocity dispersions, are poorly known, as are the magnitudes and histories of environmental perturbations. Progress requires detailed observations at optical, infrared, and radio wavelengths of all types of spirals in a variety of environments. Fully nonlinear theories, including gas and stars, with or without external perturbations, are also needed.

Additional Reading

Arp, H. (1966). Atlas of peculiar galaxies. *Ap. J. Suppl.* **14** 1.

Athanassoula, E. (1984). The spiral structure of galaxies. *Phys. Rep.* **114** 319.

Baade, W. (1944). The resolution of Messier 32, NGC 205, and the central region of the Andromeda nebula. *Ap. J.* **100** 137.

Bertin, G. (1980). On the density wave theory of normal spiral galaxies. *Phys. Rep.* **61** 1.

de Vaucouleurs, G., de Vaucouleurs, A., and Corwin, H. G., Jr. (1976). *Second Reference Catalog of Bright Galaxies*. University of Texas Press. Austin.

Elmegreen, B. G. (1989). Grand design, multiple arm and flocculent spiral galaxies. In *Galactic Models*, J. R. Buchler, S. T. Gottesman, and J. H. Hunter, eds. Academy of Science, New York.

Elmegreen, B. G., Elmegreen, D. M., and Seiden, P. E. (1989). Spiral arm amplitude variations and pattern speeds in the grand design galaxies M51, M81, and M100. *Ap. J.* **343** 602.

Elmegreen, D. M. and Elmegreen, B. G. (1987). Arm classes for spiral galaxies. *Ap. J.* **314** 3.

Gerola, H. and Seiden, P. E. (1978). Stochastic star formation and the structure of galaxies. *Ap. J.* **223** 129.

Hubble, E. P. (1926). Extra-galactic nebulae. *Ap. J.* **26** 321.

Lin, C. C. and Shu, F. H. (1964). On the spiral structure of disk galaxies. *Ap. J.* **140** 646.

Lindblad, B. (1958). *Stockholm Observatory Annals* **20** No. 6.

Sandage, A. (1961). The Hubble atlas of galaxies. Publication 618, Carnegie Institute of Washington.

Toomre, A. (1977). Theories of spiral structure. *Ann. Rev. Astron. Ap.* **15** 437.

Toomre, A. and Toomre, J. (1972). Galactic bridges and tails. *Ap. J.* **178** 623.

van den Bergh, S. (1960). A preliminary luminosity classification of late type galaxies. *Ap. J.* **131** 215.

See also **Galactic Structure, Large Scale; Galactic Structure, Spiral, Observations; Galaxies, Spiral, Nature of Spiral Arms.**

Galaxies, Starburst

John P. Huchra

Starburst galaxies are galaxies presently undergoing an intense period of star formation. Because all late type galaxies (spiral and Magellanic irregular galaxies) are forming stars at some rate, to qualify as a true starburst, the current rate of star formation must greatly exceed the rate calculated if the total mass of the galaxy were turned into stars at a constant rate over the age of the galaxy (which is usually taken to be nearly the age of the universe, 10–20 billion years).

In 1968, Beatrice Tinsley showed that the properties of normal galaxies could be explained by models in which their stellar populations are formed by either constant or declining star formation rates over the age of the universe. However, in 1973, Leonard Searle, Wallace Sargent, and William Bagnuolo noted that the optical colors of the bluest galaxies (which had been discovered in surveys for compact and peculiar galaxies, by Fritz Zwicky and in surveys for objects with ultraviolet excesses, by Guillermo Haro and B. E. Markarian) were bluer than could be explained by models with even constant star formation. Subsequent models by the author and by Richard Larson and Tinsley in the late 1970s showed that the observed properties of such galaxies could easily be explained by superposing a burst of star formation on an older population. In typical starburst galaxies, 1–10% of the mass of the galaxy is involved in a star formation episode that lasts ~0.1% of the age of the universe.

Starburst galaxies fall into three basic classes: (1) extragalactic H II regions, first identified as such by Sargent and Searle in 1970; (2) clumpy irregular galaxies, a subset of the Magellanic irregulars; (3) starburst nuclei, often just called starburst galaxies. In all of these objects, the energy output is dominated by recently formed hot young stars (of spectral classes O and B and ages less than or equal to 10 million years). The presence of hot young stars is evidenced at optical wavelengths by the blue color of such galaxies and by the presence of strong emission lines from gas ionized by such stars (see Fig. 1). Recent star formation is also usually evidenced in the infrared by strong thermal emission from warm dust that has been heated by the hot stars. (The exception to this might occur if the metal content of the interstellar medium in the galaxy is so low that dust cannot form effectively.) The *Infrared Astronomical Satellite* (*IRAS*) whole-sky survey has detected nearly 250,000 high galactic latitude sources at a wavelength of 60 μm, most of which are galaxies and a large fraction (10–30%) of which could be classified as starburst galaxies. In some extreme cases, a signature of starburst activity can also be seen at radio wavelengths as extended synchrotron emission from young supernova remnants, produced as massive young stars reach the end points of their evolution.

The three classes of starburst galaxies differ in their morphology, average luminosity, spectroscopic properties, and probably in the mechanisms responsible for inducing their bursts of star formation. Extragalactic H II regions are generally of low luminosity ($\frac{1}{100}$ of our galaxy's), very low metal abundance ($\frac{1}{30}$–$\frac{1}{10}$ of the solar abundance ratios), and "compact" morphology. As the name implies, they can best be described as isolated giant H II regions. In the most extreme cases, Markarian 36 and Markarian 116 = I Zw 18, the majority of the stars formed (by mass) have been formed in the last few tens of millions of years. The clumpy irregular galaxies are Magellanic irregular galaxies that are forming stars in a majority of the volume of the galaxy at the same time. NGC 4214 (Fig. 2*a*) and NGC 4449 are examples of irregular galaxies that qualify as starbursts. These galaxies have intermediate luminosities ($\frac{1}{10}$–$\frac{1}{2}$ of our galaxy's), and somewhat low metal abundances (~$\frac{1}{2}$ solar). The starburst nuclei galaxies, which is somewhat of a misnomer because star formation is often proceeding vigorously well outside the galaxy's nucleus, are usually spiral galaxies and are very often members of interacting or disturbed systems. They are luminous (1–10 times our galaxy) and metal-rich. Examples of this type of starburst galaxy include NGC 7714 = Markarian 538, NGC 3690 = Markarian 171 (Fig. 2*b*), and M 82. These galaxies contain large amounts of warm and hot dust and thus emit strongly in the far infrared. In many (for example, NGC 6240 and NGC 3690) strong OH maser emission has been detected at the 18-cm radio wavelength, another indicator of violent star formation activity.

Several different mechanisms have been proposed to explain bursts of star formation in galaxies. The most likely cause is dynamical interaction. A majority of the starburst galaxies uncovered in early surveys of galaxy colors and in such surveys as that done by the *IRAS* satellite are in interacting systems and appear to be kinematically disturbed (see Fig. 2*a*). Dynamical interactions can cause star formation by shocking the gas and even by gas transfer. Other processes that can be responsible for extreme star formation rates in galaxies are exceptionally strong spiral density waves (the spiral shock patterns that Lin proposed to explain the very regular spiral patterns of many bright spiral galaxies) or the infall of intergalactic gas clouds (as opposed to another galaxy). It is probable that all of these mechanisms operate.

Markarian 36

E_g H II Region

(a)

II Zw 70

E_g H II Region

(b)

NGC 4214

Irregular I–Clumpy Irr

He I

(c)

NGC 3310

Starburst

[O I]

(d)

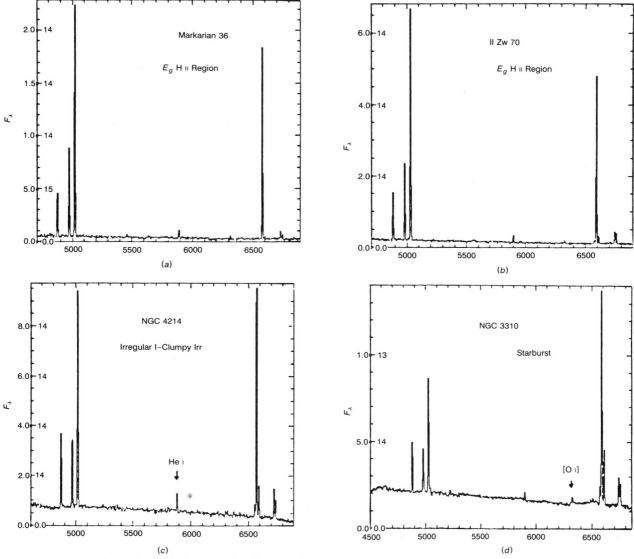

Figure 1. Optical spectra of starburst galaxies of different types: (*a*) Markarian 36 and (*b*) II Zwicky 70, which are both classified as extragalactic H II regions. (*c*) The clumpy irregular galaxy NGC 4214. (*d*) The starburst spiral galaxy NGC 3310. All the spectra are dominated by strong emission lines of hydrogen, oxygen, nitrogen, sulfur, and sometimes helium.

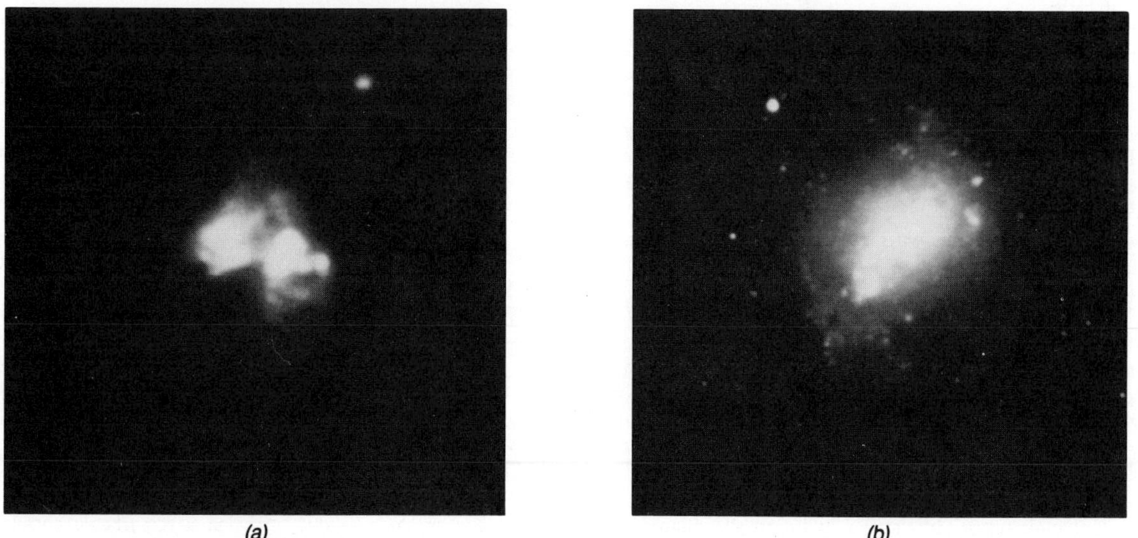

(a) (b)

Figure 2. Optical images of two starburst galaxies: (*a*) the pair NGC 3690 + IC 694 = Markarian 171 = Holmberg 256. (*b*) NGC 4214, a clumpy irregular galaxy. (*Both photos courtesy of R. Schild, Smithsonian Astrophysical Observatory.*)

Additional Reading

Kunth, D., Thuan, T. X., and Tran Than Van, J., eds. (1985). *Star-Forming Dwarf Galaxies*. Editions Frontières, Gif sur Yvette.
Lonsdale-Persson, C. J., ed. (1987). Star formation in galaxies. NASA Conference Publication No. 2466.
Shields, G. A. (1990). Extragalactic H II regions. *Ann. Rev. Astron. Ap.* **28** 525.
Thuan, T. X., Montmerle, T., and Tran Than Van, J., eds. (1987). *Starbursts and Galaxy Evolution*. Editions Frontières, Gif sur Yvette.
Vorontsov-Vel'yaminov, B. A. (1987). *Extragalactic Astronomy*. Harwood Academic Publishers, London.
See also **Galaxies, Binary and Multiple, Interactions; Galaxies, Infrared Emission; Galaxies, X-Ray Emission.**

Galaxies, Stellar Content

Robert W. O'Connell

Stars can have lifetimes that exceed the age of the universe. Therefore, the stellar content of a galaxy represents a fossilized record of its entire evolutionary history, and this is the main motivation for studying the populations of stars in other galaxies. Thanks to the enormous progress made in stellar physics since the 1920s, stars are the best understood constituent of galaxies. They are also responsible for the vast majority of the observational characteristics of galaxies. So, even if it should turn out that the "dark matter" is a more important constituent in terms of mass, the analysis of the stellar component will remain fundamental to understanding galaxies.

The stellar content of galaxies is studied in two distinct ways. In nearby galaxies, the properties of individual luminous stars can be studied with large telescopes. The statistical properties of large samples of individual stars can also be analyzed in the *Hertzsprung–Russell* (H–R) diagram (see Fig. 1). To date, this approach has been limited mainly to the galaxies of the Local Group, but new generations of ground- and space-based telescopes are rapidly expanding its scope. The study of individual stars also plays a fundamental role in measuring the distances to galaxies. In fact, observations of Cepheid variable stars in Local Group galaxies were used by Edwin Hubble to demonstrate the very existence of external galaxies and to provide the first good estimate of the scale of the universe.

In more distant galaxies, individual stars are too difficult to observe, and one must study instead the *integrated*, or composite, light of the stellar component. Although there are limitations to this technique, nonetheless it can be applied in principle to any galaxy in the observable universe—even to very distant protogalaxies still in the process of formation. Data obtained across the entire electromagnetic spectrum, from x-ray to radio wavelengths, are now used in this kind of analysis.

A galaxy may consist of many generations of stars. To clarify their measurable properties, consider the evolution of a single such generation. Stars form over a large range of masses, typically from 0.1–100 $M_\odot$. The distribution over mass is defined by the *initial mass function*, which may vary with physical conditions. The longest-lived stage of stellar evolution is the main sequence phase, in which the star burns hydrogen into helium. In this phase, more massive stars are brighter and hotter (and therefore bluer) than less massive stars. Stars over 10 $M_\odot$ are luminous enough to be detected individually throughout the Local Group and beyond. They also emit copious amounts of far-ultraviolet radiation, which is capable of ionizing the interstellar hydrogen gas around them. The ionized, or H II, regions are readily recognizable by their emission lines. However, massive stars evolve quickly off the main

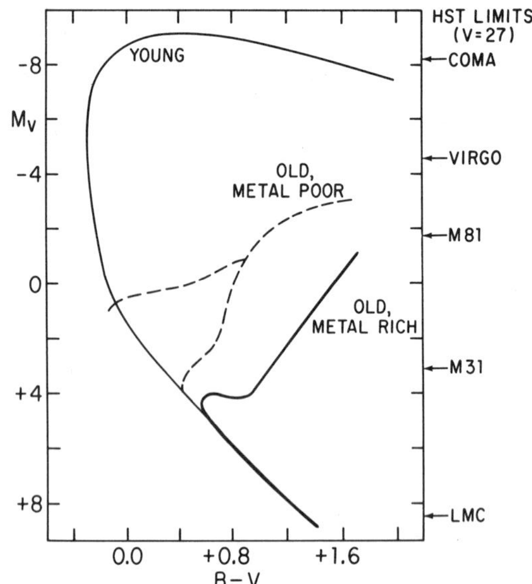

Figure 1. A Hertzsprung–Russell diagram for stellar population types. The absolute magnitude of stars is plotted against their $(B-V)$ color. Brighter stars are higher in the diagram; bluer stars are further to the left. The locus of the young stellar generation contains very bright stars, and its main sequence (the vertical section) is also blue. This corresponds to Baade's Population I and is typical of spiral disks. Old populations contain only low mass stars, which are considerably fainter than Population I. The properties of these depend on their metal abundance, as indicated. Baade's pure Population II would correspond to the old, metal poor locus. Elliptical galaxies and spiral bulges are mixtures of old stars with a range of metal abundances. The design capabilities of the *Hubble Space Telescope* (*HST*) to study individual stars at various distances are indicated on the right-hand side. In the Large Magellanic Cloud (LMC) the *HST* could resolve individual stars fainter than the main sequence turnoff of the old population. In the Andromeda galaxy (M31), the largest galaxy in the Local Group, it will just reach the old, main sequence turnoff. In the Virgo and Coma clusters of galaxies (approximately 20 and 140 million parsecs distant, respectively), individual Population I stars could be studied. [*Reprinted from* Stellar Populations *(Copyright 1986) with permission from Cambridge University Press.*]

sequence, in times of only about 10 million years, to become even shorter-lived giants or supergiants. Most will explode as supernovae, leaving neutron star or black hole remnants.

Low mass stars like the sun are initially much fainter and cooler and have main sequence lifetimes of 10 billion years or more. However, once they leave the main sequence, they brighten to become cool, red giants before they fade to white dwarf remnants. Thus, the progression of a stellar generation is from a bright, blue phase at early times, with many individually bright stars and much ionizing radiation, to a red phase at later times, dominated by the light of cool, low mass giant stars. Evolution in the integrated properties of the generation is much more rapid at early times. Chemical composition also plays a second-order role: for a given age, stars with a smaller abundance of "metals" (elements heavier than helium) will be slightly hotter.

STELLAR POPULATIONS AND GALAXY MORPHOLOGY

Early photographic studies revealed a remarkable correlation between the stellar content of galaxies and their morphology. The disks of spiral galaxies are blue and exhibit considerable structure in the form of individual bright stars, star clusters or associations, and H II regions. On the other hand, elliptical and S0 galaxies and the bulges of spirals are red and amorphous in appearance, with little evidence of resolution into stars and usually no H II regions or dust clouds. Spectroscopic studies by Milton Humason, William W. Morgan, and others also showed a systematic trend in the integrated spectrum of galaxies from one end of the Hubble classification sequence to the other: Elliptical galaxies and large-bulge spirals have cool, red giant spectra whereas Sc and irregular have the blue spectra typical of hot stars and also exhibit emission lines.

The concept of *stellar populations* was crystallized by Walter Baade in 1944. At Mt. Wilson Observatory, taking advantage of the dark skies provided by the wartime blackout, he was able to resolve the brightest stars in several Local Group elliptical galaxies and the bulge of the Andromeda Galaxy. He found them to be cool, low mass, red giants, apparently of the same kind that are found in the globular star clusters in our galaxy. He designated this type of system Population II, to be contrasted with the blue, easily resolved Population I that is characteristic of spiral arms and open star clusters in our galaxy. Subsequent development of the theory of stellar evolution (described above) led to the realization that the distinction between the population types is primarily one of age: Population I represents young, recently formed generations, whereas Population II consists of old generations whose main sequences have burned down to low mass stars.

Parameters other than age are also included in the modern concept of stellar populations. Metal abundance is second in importance. Baade's original system was oversimplified in grouping the elliptical galaxies and globular clusters together. Both may be old, but it has since been demonstrated that the globular clusters in our galaxy are predominantly metal-poor systems (with abundances as low as 1% of those in the Sun), whereas galaxy centers can have metal abundances 2–4 times higher than that of the Sun. Kinematics is also an important population parameter, although observational difficulties limit its application for stars in other galaxies.

HISTORY OF STAR FORMATION IN GALAXIES

Much of the subsequent work on the stellar content of galaxies has concentrated on inferring the detailed time dependence of star formation and metal abundance in galaxies. Ellipticals and spiral bulges evidently completed most of their star formation long ago. Either because of the efficiency of this process or of mechanisms like galactic winds, which sweep out residual gas, little cool interstellar gas remained to form more recent generations of stars. Although it seems clear these systems are old (over 5 billion years), it is a matter of controversy whether they completed star formation as long ago as the globular star clusters (15 billion years).

By contrast, appreciable amounts of cool interstellar matter (typically 10% of the total mass) remain in irregular galaxies and spiral disks, and this fuels ongoing star formation. (It is not completely understood why different galaxy types should differ so much in their current cool gas content.) Studies of these objects indicate that most experienced constant or declining mean star formation rates during the last 10 billion years and probably have an initial mass function like that in the galactic disk. Again, it is unclear exactly how old the disks are; some studies of white dwarf statistics and chemical abundances suggest an age well below 15 billion years for the disk of our galaxy. Most galaxies should consume all their remaining cool gas within another 10 billion years.

Some information on the chemical history of galaxies has also been gleaned. Globular star clusters are found in all types of bright galaxies, and many are very old and have very low metal abundances. Both spiral and elliptical galaxies also exhibit internal gradients in metal abundance, with the central regions having higher metallicities. There is also a dependence of the mean abundance of a galaxy on its mass: more massive galaxies have higher mean metallicities. All these features can be understood in a picture where there is self-enrichment of the gas in protogalaxies by star formation. Globular clusters are representative of the earliest generations of stars formed directly from primordial gas. Massive stars rapidly synthesize heavier elements in their interiors and release them through stellar winds or supernova explosions. The enriched gas collapses inward, so that stars formed later or nearer the center of the protogalaxy have higher abundances. Enrichment ceases when supernova explosions become so frequent that the gas is heated to high temperatures and escapes (a galactic wind). This would happen at an earlier stage in lower mass galaxies, leading to the observed mass-abundance correlation.

RECENT DEVELOPMENTS

New observational techniques are permitting better study of stellar remnant populations (white dwarfs, neutron stars, and black holes) and their relationship to their parent galaxies. For example, compact binary systems including neutron stars or black holes can be copious emitters of x-rays, and such objects can be detected individually in other galaxies. Far-ultraviolet observations of elliptical galaxies have apparently detected the "post-asymptotic–giant-branch" phase of evolution that immediately precedes the formation of white dwarf stars. The relationship between a galaxy's far-UV spectrum and its metal abundance is an important clue to the little-understood physics of advanced stellar evolution. The possibility that the "dark matter" consists of remnants or very low mass stars has not yet been ruled out, so these advances in our ability to study intrinsically faint stars are welcome from that standpoint as well.

It is becoming evident that the history of galaxies can be punctuated by periods of intense star formation, often called *starbursts*. Interactions between galaxies, for instance tidal encounters, often play an important role in triggering starbursts. The best known nearby example is the unusual galaxy M82, in which a burst converting about 10 $M_\odot$ of gas into stars each year was probably initiated by a close encounter with its spiral neighbor, M81. The most spectacular instances are distinct interacting systems identified by the *Infrared Astronomical Satellite*, where the star formation rate may reach 1000 $M_\odot$ yr^{-1}. The supernova rate in such objects (including M82) appears to be high enough to drive hot gas plumes out of the galaxies and into the intergalactic medium.

Unlike most areas of science, where earlier history can only be inferred, astronomers can directly witness evolutionary processes as they occurred billions of years ago simply by observing distant galaxies at large "lookback" times. In the last 10 years important progress has been made in techniques for such studies. Observations of distant clusters of galaxies indicate that the stellar populations of galaxies were undergoing rapid evolution as recently as a few billion years ago. This unexpectedly dramatic change is named the Butcher–Oemler effect, after its discoverers, Harvey Butcher and Augustus Oemler. Once again, interactions between galaxies or with the surrounding gaseous medium are apparently largely responsible. Galaxy mergers, in particular, can have drastic consequences, producing wholesale transformations of the stellar component and morphology of galaxies. A merger could convert a pair of spirals to an elliptical galaxy, for example. It is unclear if mergers play a role in the evolution of most galaxies, but they certainly influence some.

The most distant galaxies detected to date have redshifts higher than 3, meaning that they are seen at over 75% of the lookback time to the Big Bang, and appear to be still in the protogalaxy phase. Research on these remarkable systems holds the promise of providing, at last, an understanding of the processes by which galaxies form.

Additional Reading

Hodge, P. (1981). The Andromeda galaxy. *Scientific American* **244** (No. 1) 89.

Mihalas, D. and Binney, J. (1981). *Galactic Astronomy*. W. H. Freeman, San Francisco.

Morgan, W. and Osterbrock, D. (1969). On the classification of the forms and the stellar content of galaxies. *Astron. J.* **74** 515.

Norman, C., Renzini, A., and Tosi, M. (1986). *Stellar Populations*. Cambridge University Press, Cambridge.

Sandage, A. (1986). The population concept, globular clusters, subdwarfs, ages, and the collapse of the galaxy. *Ann. Rev. Astron. Ap.* **24** 421.

See also **Galactic Structure, Stellar Populations; Galaxy, Chemical Evolution; Star Clusters, Globular, Stellar Populations; Star Clusters, Stellar Evolution; Stellar Evolution.**

Galaxies, X-Ray Emission

Giuseppina Fabbiano

The study of the x-ray emission of normal galaxies is a very recent part of astronomy. This work has been made possible by the sensitive x-ray imaging observations of the *Einstein* (*HEAO 2*) satellite, launched by NASA in November 1978. Before then, with the exclusion of the bright x-ray sources associated with Seyfert nuclei, only four galaxies were known to emit x-rays: the Milky Way, M 31 (Andromeda), and the Magellanic Clouds. The *Einstein* satellite observed over 200 galaxies during its $2\frac{1}{2}$-yr life span. Some were detected with enough detail to allow a study of their x-ray morphology, spectra, and individual sources, and to make comparisons with optical, infrared, and radio data. These observations have shown that normal galaxies of all morphological types are spatially extended sources of x-ray emission with luminosities in the range of 10^{38}–10^{42} erg s^{-1}. Although this is only a small fraction of the total energy output of a galaxy, x-ray observations are uniquely suited to study phenomena that are otherwise elusive. These include the end products of stellar evolution (supernova remnants and compact remnants, such as neutron stars, white dwarfs, and black holes), the hot component of the interstellar medium, and active nuclear regions.

SPIRAL AND IRREGULAR GALAXIES

Normal stars are responsible for the optical emission of galaxies. However, their integrated x-ray emission is only a small fraction of the x-ray emission of a normal galaxy. Observations of the Milky Way and of the Local Group galaxies suggest that a good fraction of the x-ray emission of late-type (spiral and irregular) galaxies is due to a collection of a relatively small number of individual bright sources, such as close accreting binaries with a compact companion, and supernova remnants, with luminosities ranging from $\sim 10^{35}$ erg s^{-1} up to a few times 10^{38} erg s^{-1}.

Only a few very bright individual x-ray sources can be detected in the *Einstein* images of more distant galaxies, which typically appear as extended x-ray emission regions, because at that distance individual sources could not be resolved with the *Einstein* instruments. However, we believe that the x-ray emission of these galaxies is due to sources akin to those detected in the Local Group. The x-ray spectra of these galaxies are consistent with the hard spectra expected from binary x-ray sources, and the x-ray luminosities are linearly correlated with the emission in the optical *B* band, suggesting that the x-ray emission is mostly due to sources constituting a constant fraction of the stellar population.

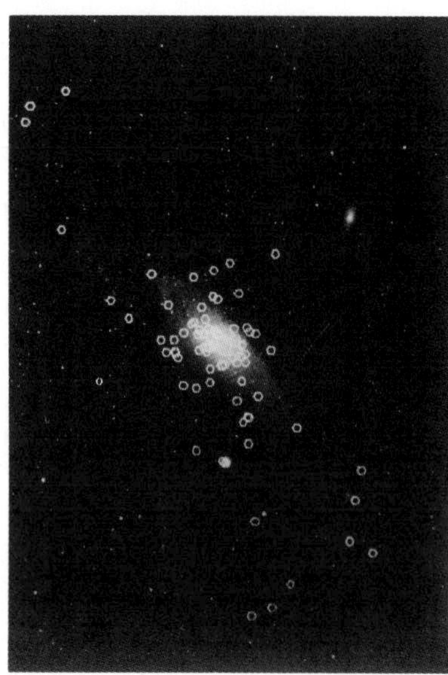

Figure 1. The circles show the positions of the x-ray sources of M 31, superimposed onto an optical photograph. (*Courtesy of L. Van Speybroeck. Reproduced, with permission, from the* Annual Review of Astronomy and Astrophysics, *27 © 1989 by* Annual Reviews, Inc.) Notice the clustering of sources in the bulge.

In the pre-*Einstein* era, the x-ray sources of the Milky Way were classified as young Population I, or spiral arm, sources with massive early-type star counterparts and Population II, or bulge sources, with low-mass stellar counterparts. The *Einstein* imaging observations of spiral galaxies have led us to modify this classification and to gain new insight into the evolution of binary x-ray sources. We can now identify a "spiral arm," a "bulge," and a "disk" component of the x-ray emitting population.

The presence of spiral arm and bulge (and globular cluster) x-ray sources is immediately demonstrated by the x-ray images of nearby galaxies. Bright point sources are detected in the spiral arms of M 31 (see Fig. 1) and M 33. In particular, one of the M 33 sources has a variable light curve that is similar to those of some massive Galactic x-ray binaries. A bulge component of the x-ray emitting population is also evident in M 31; these sources have properties similar to those of low-mass x-ray binaries in the Galaxy. Statistical analyses of the sample of spirals observed with *Einstein* indicate that x-ray bulge emission is present in all bulge-dominated spirals. Disk x-ray sources are suggested by the close resemblance of the radial profile of the x-ray surface brightness of a few face-on spirals (M 83, M 51, and NGC 6946) with that of the optical light of their exponential disk. This implies that a good fraction of low-mass x-ray binaries may originate from the evolution of binary systems belonging to the disk stellar population, rather than from dynamical evolution (capture of a low-mass companion by a compact object in a dense environment and/or disruption of globular clusters), as has been suggested to explain galactic low-mass binaries.

Some of the sources detected in spiral galaxies have x-ray luminosities well above the Eddington limit for accretion onto a one-solar-mass compact object, which is $\sim 1.3 \times 10^{38}$ erg s^{-1}. One of these sources is the supernova SN 1980k detected in NGC 6946 approximately 35 days after maximum light. The variability reported for some bright sources in M 101 suggests point-like objects, possibly bright accretion binaries. If these sources are truly single objects, they could indicate the presence of massive black holes in these galaxies. It is, however, possible that the distances of some galaxies might have been overestimated, making these sources appear more luminous that they are in reality.

Another source of x-ray emission that has been searched for in spiral galaxies is diffuse thermal emission from a hot phase of the interstellar medium. Supernovae release $\sim 10^{42}$ erg s^{-1} in a galaxy, and it has been suggested that hot gaseous coronae, or galactic fountains, could be produced and should be visible in soft x-rays in the *Einstein* range. There is evidence of soft thermal diffuse emission both in the galactic plane and in the large Magellanic Cloud (LMC), and perhaps in M 33. However, this type of emission has not been detected in more distant galaxies. The lack of intense diffuse soft x-ray emission could imply that most of the supernova energy is radiated in the unobservable far ultraviolet. The only reported instance of this type of soft x-ray emission in a spiral galaxy is in the edge-on galaxy NGC 4631, where this component could have an x-ray luminosity of 5×10^{39} erg s^{-1}, which represents $\sim 13\%$ of the total emission in the *Einstein* band.

STARBURST GALAXIES AND NUCLEAR OUTFLOWS

Bluer starburst, often interacting, galaxies tend to have enhanced x-ray emission when compared with galaxies with redder, more normal, colors. The bulk of the x-ray emission of these galaxies can be understood in terms of a number of young supernova remnants and massive x-ray binaries, with X-ray luminosity possibly enhanced by the low metallicity of the accreting gas, similar to those observed in the Magellanic Clouds.

There are galaxies in which the starburst activity is confined to the nuclear regions. Starburst nuclei studied in x-rays include the Milky Way galactic center region, and the nuclei of M 82, NGC 253, M 83, NGC 6946, IC 342, and NGC 3628. A common characteristic of the emission spectrum of these nuclei is their intense far-infrared emission, indicative of dusty nuclear regions heated by newly formed early-type stars. The x-ray emitting regions are extended (whenever they are observed with high enough spatial resolution) and in M 82 there is evidence of a population of bright individual sources. To explain this emission requires, in different cases, different amounts of evolved sources (supernova remnants and x-ray binaries) superimposed on the integrated stellar emission from a young stellar population.

An unexpected result of the *Einstein* observations of these nuclei has been the discovery of extended emission components, suggestive of gaseous bipolar outflows from the nuclear regions, in the edge-on galaxies M 82 and NGC 253. These outflows, if generally associated with violent star formation activity, could be responsible for the formation and enrichment of a large part of the gaseous intracluster medium.

ELLIPTICAL AND S0 GALAXIES

A hot gaseous component dominates the x-ray emission of x-ray-luminous early-type galaxies. These galaxies can be ~ 100 times brighter in x-rays than spiral galaxies of similar optical magnitude, where the x-ray emission instead is due to evolved stellar sources. These x-ray bright galaxies also tend to have x-ray spectra different from those of binary sources, and sometimes show distortions of their x-ray surface brightness, relative to the optical images (Fig. 2). The latter suggest that the x-ray emission is not due to the stellar population and is consistent with the interaction of the hot galactic halo with a surrounding cluster gas.

X-ray observations have revealed the presence of a long-sought interstellar medium in early-type galaxies, whose apparent absence had been explained by invoking galactic winds to remove the gaseous by-products of stellar evolution. The amount of hot gas present in these galaxies can be as high as 10^{10} solar masses, significantly more than the amount of cold interstellar medium seen in neutral hydrogen and in the infrared, which is well below the amount of interstellar medium seen in spiral galaxies. Not all E and S0 galaxies, however, may be able to retain their hot interstellar medium. The *Einstein* survey has shown that there is a wide

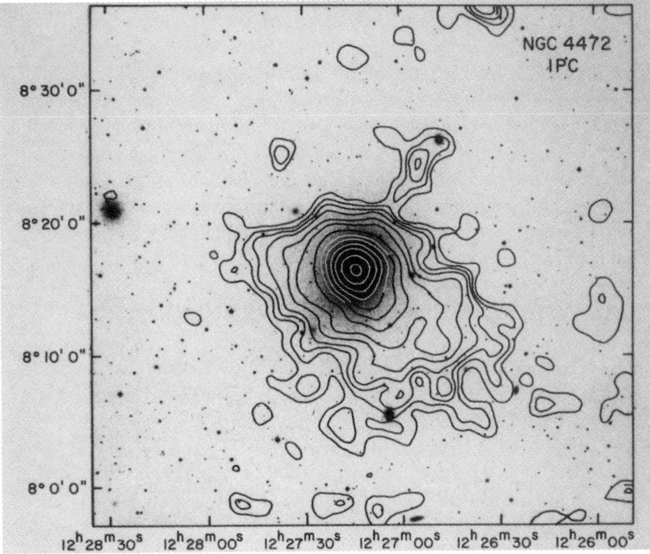

Figure 2. The x-ray map of NGC 4472, an elliptical galaxy in the Virgo cluster, obtained with the imaging proportional counter of the *Einstein* observatory. Notice the asymmetrical halo. (*Reproduced, with permission, from the* Annual Review of Astronomy and Astrophysics, 27 *© 1989 by Annual Reviews, Inc.*)

spread of x-ray luminosities in early-type galaxies of similar optical luminosity. The lowest x-ray luminosities can be explained easily by the integrated emission of bulge-type x-ray sources in these galaxies and do not require any additional gaseous component. The ability (or lesser ability) to retain a hot gaseous halo might be the result of several factors, including large amounts of dark matter in the galaxy, the amount of supernovae present, and the interaction with a surrounding hot cluster medium.

In an x-ray-luminous galaxy with a large hot gaseous halo, the gas is so dense that it will cool in a time shorter than the galaxy's lifetime and then accrete to the galaxy's core, giving rise to "cooling flows." These cooling flows would have interesting consequences: One could be the formation of new stars from matter detaching from the flows; another could be the accretion of gas into the nucleus and the consequent fueling of nuclear sources. There is some evidence of the latter, in that powerful radio sources, connected with nuclear activity, tend to be found in x-ray-bright, gas-rich galaxies. The presence of a hot gaseous halo is also determinant in the formation of extended radio lobes. In the range of radio core power (i.e., of the intensity of the nuclear source) of the radio galaxy Centaurus A, powerful radio lobes more extended than the optical size of the parent galaxy are only found in relatively gas-poor (i.e., x-ray-dim) galaxies.

One of the potentially very important results of x-ray observations of early-type galaxies is the possibility of measuring their masses. The method generally used employs the equation of hydrostatic equilibrium, in which the gas pressure and the gravitational pull balance. Combining this with the ideal gas law, one obtains

$$M(<r_{\text{gas}}) = -\left(\frac{r_{\text{gas}}kT_{\text{gas}}}{G\mu m_{\text{H}}}\right)\left(\frac{d\log\rho_{\text{gas}}}{d\log r} + \frac{d\log T_{\text{gas}}}{d\log r}\right).$$

Four quantities must therefore be measured to determine the mass within a certain radius r_{gas}: the radius itself, the temperature T_{gas} at that radius, and the temperature and gas-density (ρ_{gas}) gradients at that radius. The uncertainty in the mass measurement will reflect the uncertainties in the determination of these quantities. Applied to M 87, the dominant galaxy at the center of the Virgo cluster, this method reveals a large amount of dark matter. However, when this method is applied to more normal early-type galaxies, which are more than 100 times less x-ray-luminous that

M 87, the uncertainties are very large and the presence of large dark halos cannot be demonstrated firmly, although it is suggested in some cases. X-ray measurements with the German satellite *ROSAT* (launched June 1, 1990) and with the future satellites AXAF and XMM will allow accurate mass determinations in the x-ray-bright early-type galaxies. In the case of less x-ray-luminous galaxies it will have to be established first that the x-ray emission is due to a gaseous halo and not to a collection of binary x-ray sources. It will also be important to consider the effect of supernovae on the energy balance of the halo.

GALAXIES AND THE X-RAY BACKGROUND

The extragalactic x-ray background was discovered by Riccardo Giacconi, Herbert Gursky, Frank Paolini, and Bruno Rossi in 1962 in data from the same rocket flight that led to the discovery of the first extrasolar source of x-rays, Sco X-1. Since then a great deal of effort has been spent to determine if this radiation is due to the integrated contributions of different classes of discrete sources or if diffuse processes are responsible for it. The integrated emission of normal galaxies could explain ~13% of the 2-keV extragalactic x-ray background. If one includes in this estimate the contribution of low-activity nuclei present in a fraction of these galaxies, the effect of starburst activity, and even more the possibility that these types of activities were more pronounced in the past, this contribution could be significantly larger.

Additional Reading

Fabbiano, G. (1986). The x-ray properties of normal galaxies. *Pub. Astron. Soc. Pacific* **98** 525.

Fabbiano, G. (1989). X-rays from normal galaxies. *Ann. Rev. Astron. Ap.* **27** 87.

Fabian, A. C., ed. (1988). *Cooling Flows in Clusters and Galaxies.* Kluwer, Dordrecht.

Helfand, D. J. (1984). Endpoints of stellar evolution: X-ray surveys of the Local Group. *Pub. Astron. Soc. Pacific* **96** 913.

Long, K. S. and Van Speybroeck, L. P. (1983). X-ray emission from normal galaxies. In *Accretion Driven X-Ray Sources*, W. Lewin and E. P. J. van den Heuvel, eds. Cambridge University Press, New York, p. 117.

See also **Background Radiation, X-Ray; Clusters of Galaxies, X-Ray Observations; Galaxies, Starburst; X-Ray Sources, Galactic Distribution.**

Galaxy, Chemical Evolution

Curtis J. Struck-Marcell

The abundance (by number) of elements heavier than hydrogen and helium is about 2% in the Sun, and ranges from less than about 1/1000 of this value in very metal-poor stars up to about a few times this value in extremely young stars. Galactic chemical evolution is the study of the distribution and the build-up of these heavy elements, which are called metals in astronomy. In the late 1950s Geoffrey Burbidge, Margaret Burbidge, William Fowler, and Fred Hoyle and, independently, Alastair G. W. Cameron deduced that almost all of the metals could be produced in massive stars, returned to the interstellar gas, and incorporated into later generations of stars. On the other hand, only the lightest elements could be produced in any significant abundance in the high-density, high-temperature conditions in the first few minutes of the Big Bang, given standard assumptions for cosmological nucleosynthesis. These results imply the fundamental premise of this subject: Within the Galaxy the abundances of metals in stars and the interstellar gas increase with time as successive generations of stars enrich the gas, which was initially composed almost entirely of hydrogen and helium. Forty years of research on stellar populations within the Galaxy have provided broad support for this assumption. This work includes the fundamental observations of the globular star clusters located in the halo of the Galaxy. When these observations are interpreted with the aid of stellar structure and evolution theory they show that these globular clusters contain only very old stars, and that generally these stars have very low metal abundances. This contrasts with the relatively high metallicity of the young stars in the galactic disk.

INPUTS AND PROCESSES

An understanding of galactic chemical evolution can only be achieved by coherently synthesizing the results of many subfields of astronomy and astrophysics. These results provide constraints on the many rates or efficiencies that characterize the processes that must be included in a model for galactic chemical evolution. These processes include the following.

1. *Star formation.* Both the functional dependencies of the star formation rate (SFR) and the distribution of masses of the newly formed stars (initial mass function, IMF) are needed for this modeling.
2. *Nucleosynthesis.* What is the heavy element production by stars of different masses over their lifetime? This depends on nuclear reaction rates determined by laboratory measurements or theoretical calculations, and on the physical conditions in the nuclear burning zones of the stars.
3. *Internal structure and evolution.* Besides the nucleosynthesis reaction rates, many structural details, including the location of burning zones, the degree of convective mixing, and the evolution of these quantities, also determine the mass of metals that will actually be returned to the interstellar gas by the end of a star's life.
4. *Mass loss.* The flow of mass and metals back to the interstellar medium, which is a function of initial stellar mass, is a key process. A knowledge of the amount of material returned at each evolutionary stage is needed.
5. *Gas flows.* In principle, these include "small scale" flows within or between neighboring interstellar clouds, that may affect the local SFR, for example. However, in most cases it is likely that over the large length and time scales usually considered in chemical evolution models the effects of these local processes will average out. Mesoscale effects may be more important. For example, these could include radial gas flows within the galactic disk driven by spiral density waves, redistribution of gas within the disk through vertical motions in galactic fountains, or infall of gas onto the disk from some other component of the Galaxy. It also appears that large-scale processes such as the exchange of gas between interacting galaxies, the cannibalism of small galaxies, and mergers between comparably sized galaxies, may not be as rare as previously thought.

Although a variety of specialists gather the separate pieces of the jigsaw puzzle of galactic chemical evolution, the purpose of a chemical evolution model is to describe the whole picture. Usually this must be done with many of the puzzle pieces missing. Because of the wealth of data on both the stars and the gas, the most well studied example is the chemical evolution of the solar neighborhood. Many of the successes and difficulties of this subject are illustrated by this particular example.

SOLAR NEIGHBORHOOD

In chemical evolution models the solar neighborhood is frequently defined as a thick cylinder with a radius equal to the Sun's distance from the galactic center: thickness equal to about 1 kpc in the galactic disk and height equal to about 1 kpc out of the disk. This somewhat counterintuitive definition of the local neighborhood is based on the motions of nearby stars. These stars not only orbit the Galaxy on a short time scale compared to evolutionary times, but also oscillate over distances of up to a kiloparsec radially and vertically. Because there is strong shear in the orbital motions of disk stars and gas clouds, a star currently near the Sun may have been half a kiloparsec closer to the galactic center and on the opposite side of the galaxy several orbital times (about 10^9 yr) ago. This means that the solar neighborhood is likely to be well mixed and homogeneous, if not an isolated population of stars and gas clouds.

This definition of the solar neighborhood immediately suggests the first and last of the four assumptions that are the basis of a "simple" model of chemical evolution. According to Beatrice M. Tinsley these are the following:

1. The solar neighborhood can be modeled as a closed system.
2. It started as 100% metal-free gas.
3. The IMF is constant.
4. The gas is chemically homogeneous at any time.

As a realistic model for the solar neighborhood the simple model is a distinguished failure. One key aspect of its failure is the so-called G-dwarf problem, the fact that the simple model predicts far more metal-poor stars of spectral type F and G than are observed in the solar neighborhood. When it was first discovered, the evidence for the G-dwarf problem was based on rather modest samples of stars. Since that time, larger samples have confirmed the effect in F and G stars. Jeremy Mould also found the effect in low-mass M-type stars. The fact that the effect is found in stars with very different masses shows that it is not the result of surface pollution of low-metallicity stars with metals accreted from the interstellar gas, because this process would depend on stellar mass.

The G-dwarf problem is not seen as a difficulty in itself, but as an indication that the assumptions of the simple model, especially the first two assumptions, do not correctly describe the evolution of the solar neighborhood. A number of solutions to the G-dwarf problem, violating one or more of the simple model assumptions, have been studied in detail. Perhaps the most direct of these ideas is pre-enrichment, in which it is proposed that the disk gas has an initial nonzero metal abundance, either as a result of nucleosynthesis in an early generation of stars in the galactic halo before the gas settled into the disk or due to a burst of star formation very early in the life of the disk. Another family of possible solutions that exploits a halo connection is the idea of continuing gas infall. Infall could result from a slow settling of the gas into the disk or a steady feeding of the disk with gas shed from halo stars. In either case, the initial disk mass is assumed to be much lower than its current value, and so a modest amount of star formation can readily enrich this low mass of gas. Radial gas flows within the disk could also provide an effective infall into the solar neighborhood.

Both of these solutions to the G-dwarf problem are constrained by the observationally determined stellar age–metallicity relation, which indicates that the metallicity of the disk gas increased approximately linearly with age. These data together with the stellar metallicity distribution seem to imply that the current rate of star formation in the disk is not much different than its average past value. There is no evidence that star formation was generally more vigorous in the past, though there is the possibility that occasional bursts occurred.

A further difficulty with infall models is that in many cases the required rate of infall is a substantial fraction of the star formation rate. At present there is no direct observational evidence for the infalling gas either from halo sources or from radial flows. On the contrary, there is a paucity of cool gas in the galactic halo.

Two other solutions to the G-dwarf problem have been studied in some detail: metal-enhanced star formation and a variable initial mass function. The former relies on chemical inhomogeneities in the interstellar gas and the premise that stars form preferentially in high-metallicity gas. Lack of evidence for the inhomogeneities has discouraged work on such models. In the variable IMF case the idea is that most of the stars formed in low-metallicity gas are high-mass stars with high metal yields, so this is a form of prompt enrichment via enhanced yields. However, it appears that constant yield models are consistent with the observed relative abundances of oxygen (a primary element produced in massive stars) and iron (produced more slowly in low-mass stars) in the disk and halo. The observational constraints on variations in the IMF are quite weak, although the subject of the IMF in different environments is an area of active research. In particular, the possibility that the IMF is bimodal, and that the high- and low-mass stars may form by different processes has received much attention recently.

Another idea that has attracted renewed attention recently is that the galactic disk consists of two components: an old thick disk and the thin disk that contains most of the young stars. Much of the observational data relevant to the G-dwarf problem and chemical evolution in the solar neighborhood are derived from the latter component. If a distinct thick-disk component antedated the thin disk it could have pre-enriched the gas that later settled into the thin disk, as well as provided another source of infall. Alternatively, it has been suggested that the thick disk consists of stars scattered out of the central plane over the course of the disk's lifetime by encounters with giant molecular clouds, spiral density waves, and other possible density inhomogeneities. Such theories are constrained by the stellar age–velocity dispersion (i.e., random or thermal velocity) relation. Indeed, the studies of stellar kinematical and chemical evolution are becoming increasingly intertwined.

OTHER ENVIRONMENTS

Besides the solar neighborhood, other galactic environments including the galactic bulge or inner spheroidal component, the outer disk, the globular star clusters, the dwarf spheroidal satellite galaxies, and the Magellanic Clouds, have also been the subject of chemical evolution studies. In addition to providing an example of the relevant processes, the preceding discussion of the solar neighborhood also illustrates the connections between the evolution of the different morphological and dynamical components of our Galaxy. Thus, it should not be assumed a priori that these environments are isolated, although it appears that the bulge and the globular clusters have evolved more like closed systems than has the solar neighborhood. The solar neighborhood example also shows the dependence of chemical evolution processes on the details of the galaxy formation process.

Because of the interdisciplinary nature of chemical evolution studies and because astronomical research is progressing rapidly on many fronts, much progress can be expected in this area in the coming decade. The study of other galaxies is of particular interest for providing interesting comparisons with our own.

Additional Reading

Boesgaard, A. M. and Steigman, G. (1985). Big Bang nucleosynthesis: Theory and observations. *Ann. Rev. Astron. Ap.* **23** 319.

Gilmore, G. and Carswell, B. (1987). *The Galaxy*. Reidel, Dordrecht.

Scalo, J. M. (1987). The initial mass function, starbursts, and the Milky Way. In *Starbursts and Galaxy Evolution*, 22nd Recontre de Moriond, T. X. Thuan, T. Montmerle, and J. Tran Thanh Van, eds. Editions Frontieres, Paris, p. 445.

Tinsley, B. M. (1980). Evolution of the stars and gas in galaxies. *Fundamentals of Cosmic Physics* **5** 287.

Trimble, V. L. (1975). The origin and abundances of the chemical elements. *Rev. Mod. Phys.* **47** 877.

Twarog, B. A. (1985). The chemical evolution of the galaxy. In *The Milky Way Galaxy*, H. van Woerden et al., eds. Reidel, Dordrecht, p. 587.

See also **Galactic Structure, Stellar Kinematics; Galaxies, Chemical Evolution; Galaxies, Stellar Content; Solar Neighborhood; Stars, Chemical Composition; Stars, Hertzprung–Russell Diagram.**

Galaxy, Dynamical Models

Michael G. Edmunds

Modeling the dynamics of stars and gas in the Galaxy is the principal means of determining its distribution of mass. The major components of a model are the rotating *disk*, a much more slowly rotating *spheroid* of stars (in which the disk is embedded), and an invisible *corona* (which contains most of the mass). The dynamical structure is a fundamental property that any theory of the origin and evolution of the Galaxy must explain.

It is impossible in general to measure directly the accelerations of stars or interstellar matter because the time scales of significant changes in velocity are extremely large. The accelerations have to be deduced by indirect arguments from the observed velocities. Here there are problems because, except for some of the nearest stars for which "proper" motions *across* the sky are measurable, we have to rely on the Doppler effect, which gives only velocities along the line of sight to the star or gas. The one exception is a circular orbit of constant speed V_r about the center of the Galaxy because in this case the acceleration (which acts toward the center) is given by the well-known formula V_r^2/r, where r is the radius of the orbit.

The acceleration is directly related to the gravitational force experienced, which in turn depends on the amount and distribution of mass in the Galaxy. Consider first a spherically symmetric distribution of mass—that is, the mass per unit volume (density) of material can vary as a function of distance r from the center, but there is no other variation. Then a remarkable theorem first derived by Isaac Newton shows that the gravitational force experienced at radius r depends only on the mass inside that radius, so that for circular orbits,

$$V_r = \sqrt{\frac{\alpha G M(r)}{r}}$$

where $M(r)$ is the mass inside radius r, G is the gravitational constant, and α is 1. A similar result holds for ellipsoidal or spheroidal (i.e., flattened sphere) shapes, but with slightly different values of α. The speed for circular orbits in disk configurations can also be calculated, but will not be much different from the equation given here, with a value of a a little greater than 1 that may vary with r.

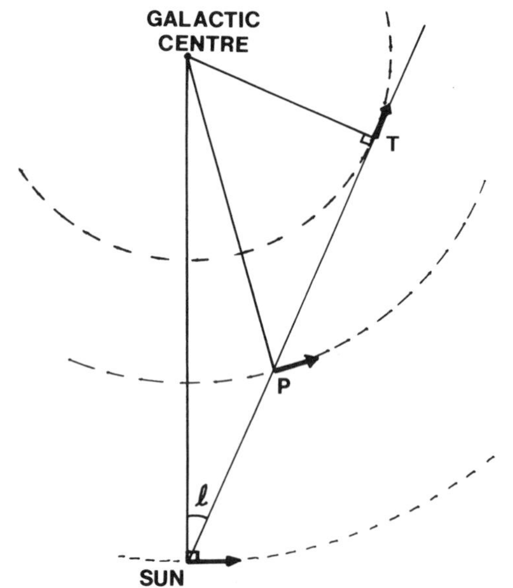

Figure 1. The geometry of the observation of galactic rotation. Observing from the Sun at a galactic longitude (the angle measured around in the disk from the line to the galactic center) emission from gas will be seen from many points along the line-of-sight. At P there will be a component of velocity along the line-of-sight that is less than the rotational velocity V_r; but V_r can be deduced if the distance to P can be estimated. The maximum line-of-sight velocity will occur at the tangent point T, where the rotational velocity is directly away from the observer.

GALACTIC ROTATION CURVE

The Galaxy has been investigated by fitting models including various shapes of mass distribution to the observed variation of circular velocity with radius. The observation of this *rotation curve* is not trivial. The Sun is located inside the disk and the interstellar obscuration (due to dust) at visible wavelengths means that optical measurements cannot reach very far in the disk, except in directions out of the disk or in the direction away from the center, known as the anticenter direction. A great deal of work has therefore been carried out at radio wavelengths, which are unaffected by this obscuration, particularly using the emission at a frequency of 1420 MHz (21-cm wavelength) from unionized interstellar hydrogen gas. This emission results from a change in the orientation of the spin of the nuclear proton of an interstellar hydrogen atom, and because the spectral line involved is very narrow it can give an accurate measure of the Doppler shift of the emitting atom.

If an observation is made through the disk at a particular galactic longitude l as shown in Fig. 1, then a complex emission profile will be found, resulting from emission all along the line of sight; but the maximum Doppler velocity will come from gas at the *tangent point T*. Features at other velocities can be identified by comparison of profiles at different longitudes l. In recent years the data have been supplemented by microwave observations of spectral lines from abundant molecules, such as carbon monoxide (CO). Outside the Sun's orbit (i.e., for $90 < l < 270°$) the determination is more difficult because there is no tangent point of maximum velocity and Doppler velocities must be interpreted with the aid of distance information, for example, using clusters of stars or using H II regions. The reliability of the resulting V_r-versus-r curve is further compromised by the known presence of some noncircular motions. The shape of the observed rotation curve is shown as the dots in Fig. 2. It is clear that the disk shows *differential rotation* and does not rotate as if it were a solid body, for which we would

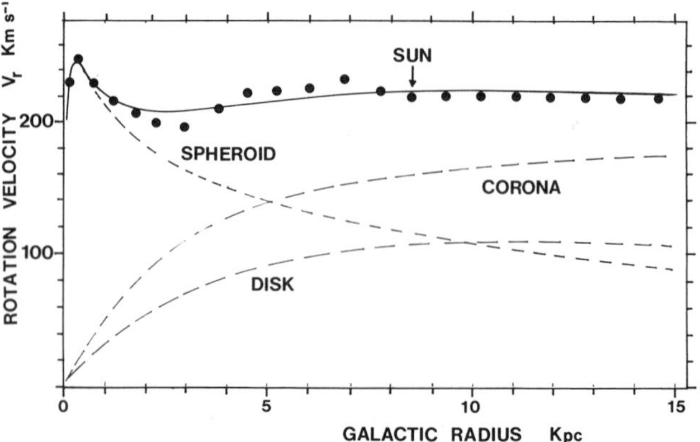

Figure 2. The rotation curve of the Galaxy. The observed curve is represented by dots, and a dynamical model is shown as the solid line. The contributions made to this model by the individual components are shown as dashed lines, which indicate what the model curve would look like with only that single component present. The dip in the observed rotation curve at radius around 3 kpc may not be real and could be an artifact introduced by noncircular motions in the gas.

expect V_r proportional to r. The existence of noncircular motions in the disk may indicate deviations from axial symmetry, in particular, that the Galaxy may have a bar-like structure in the inner few kiloparsecs (kpc). Although this will alter the detailed dynamics of stars and gas in the disk, the general conclusions from an axially symmetric model (as will be assumed) are not affected.

OBSERVABLE MASS COMPONENTS OF THE GALAXY

The components of mass used to model the Galaxy start with those that can be identified easily by direct observation. The most obvious is the flattened disk of stars and gas a couple of hundred parsecs thick, but also important is the spheroidal component, which extends many kiloparsecs above the plane and is most easily seen (both in our own and other spiral galaxies) around the central region, where it is frequently referred to as the *bulge*. The light distribution (and presumably mass) in the disk is expected (again by analogy with external galaxies) to fall off exponentially with radius from the center, whereas the spheroid, which contains only stars or stellar remnants (and virtually no gas), has a density that falls off something like r^{-3} (or perhaps not quite as steeply). It is one of the major uncertainties (not sufficiently acknowledged by some workers) that the shape, density structure, and total luminosity of this spheroid component have not been well constrained by observation. In the local solar neighborhood the actual space density of spheroid stars is poorly known because it requires the identification of the handful of spheroid stars from among many disk stars. If the overall mass of the spheroid is extrapolated from this local density using an incorrect shape and density law, the result may be seriously in error.

The identification of spheroid stars comes partly from their velocities, because it appears that the spheroid (probably a remnant from the early evolution of the Galaxy) rotates very much more slowly than the disk. The spheroid stars will not be on circular orbits but must have a considerable radial component in their motion, with their extended distribution both radially and above the plane being supported by a velocity dispersion somewhat analogous to the pressure in a gas. The spheroid stars are sometimes referred to as *high-velocity stars*, although in reality it is the disk stars like the Sun that are moving rapidly past the spheroid stars. Estimates of the total mass of the spheroid, both from star counts in different directions and from local stellar densities, give masses that vary from $\frac{1}{50}$ to over $\frac{1}{2}$ relative to the mass of the disk.

The distinction between disk and spheroid may not be completely sharp; indeed, there have been proposals that some 10% of stars may in fact belong to a *thick disk* with dynamical properties intermediate between the spheroid and the disk. Massive molecular gas clouds in the galactic plane will inevitably perturb the orbits of disk stars in such a way as to cause some vertical "puffing up" out of the plane over the age of the Galaxy, but the thick disk may well have been formed as a separate component during the early evolution of the Galaxy.

The motions of stars in the direction perpendicular to the galactic plane provide an additional source of information on the mass in the disk near the Sun. The height to which a star with a particular kinetic energy can journey above the plane is limited by the gravitational field due to the disk, and observation of the number densities and velocities of stars at different heights constrains the models of the gravitational field. One of the parameters that comes from such studies is the local mass density in the disk near the Sun, often called the *Oort limit* after the pioneering investigations of the Dutch astronomer Jan Oort. It had been thought for many years that more mass was being detected locally by this method than could be seen as either stars or interstellar material, but recent work indicates that the local disk mass density is in good agreement with the mass implied by observed stars and gas and hence that there is little, if any, hidden matter in the disk. This improved agreement does not, however solve the very fundamental problem that arises when the galactic rotation curve is fitted, as we shall now describe.

FITTING THE ROTATION CURVE

Looking again at the equation for V_r, we might expect that so long as the density of materials falls faster than r^{-2} (as would effectively be the case for an exponential disk or the observed spheroid at large radii), then $M(r)/r$ will eventually decrease as r increases and so the rotational velocity V_r will fall at large radii. In fact, as far as can be determined, the rotational velocity does not fall off and remains constant at about 220 km s^{-1} from the solar galactic radius of 8.5 kpc out to some 25 kpc or more and may, indeed, even increase slightly. This "flat" rotation-curve behaviour is not unique to the Galaxy—a similar effect is observed in many (but not all) other spirals. On the basis of the observed disk and spheroid (whatever its true mass within the limits previously quoted), the rotational velocity should decrease beyond the solar galactic radius. The inevitable conclusion must be drawn, on the basis of our current understanding of galactic physics, that there is a substantial mass component present that we cannot directly observe except by its contribution to the gravitational field. This is termed the *dark corona*, and this name is to be preferred to the term "halo" that is sometimes used, because this gives rise to confusion with the spheroid stars, which are often referred to as halo stars. The dark corona is probably fairly spherical (although there is little direct evidence to confirm this assumption), with a density distribution falling off at large radius like r^{-2}, giving $M(r)$ proportional to r, which will provide the flat (V_r = constant) rotation curve.

Figure 2 shows an example of a fit of the rotation curve including disk, spheroid, and dark corona components. It will be seen that although the disk and spheroid dominate the dynamics in the inner parts of the Galaxy, the dark corona dominates beyond the solar radius and indeed the *total* mass of the dark corona (which increases linearly with r as long as the rotation curve stays flat) is greater than the sum of the masses of the disk and spheroid. Within the solar radius (where the observable disk and spheroid dominate) the mass must be of order 1×10^{11} $M_\odot$, but if extended out to beyond 40 kpc the total could easily exceed 5×10^{11} $M_\odot$. Attempts have been made to model the rotation dynamics by increasing the mass of the disk (and spheroid) sufficiently to dispense with the necessity of invoking a dark corona. These "maximal disk" models appear to be ruled out, however, by the new limits on the local density of mass in the disk near the Sun.

Some models include an extra component of a few million solar masses at, or very close to, the galactic center. The need for such an additional component has not been unambiguously demonstrated, but even if present, it probably does not have much effect on the dynamics of the Galaxy as a whole.

OTHER DYNAMICAL EVIDENCE ON THE TOTAL MASS

There is some additional support for the high hidden corona mass implied by the rotation dynamics. A star with sufficiently high velocity in the vicinity of the Sun should be able to escape completely from the gravitational field of the Galaxy. Because we do not expect there to be a major source of stars with high velocity at the present time, then the maximum observed velocity of stars in the disk probably represents a lower limit on the escape velocity. The observed local limit of over 500 km s^{-1} does indeed imply a high total mass, perhaps four or five times that inside the solar galactic radius.

The globular cluster system of the Galaxy can be associated with the spheroidal component, and it also appears to rotate slowly compared with the disk. Although modeling the dynamics of this system must rely only on observed Doppler velocities (which cannot give the true total space motions of the clusters), nevertheless such models as have been constructed require the existence of the gravitational field of the dark corona.

The motions of other galaxies near our own are influenced by the total gravitational field of the Galaxy. Again, modeling can only be constrained by line-of-sight velocity observations, but study of the probable orbits of several small satellite galaxies also implies a total galactic mass of at least 4×10^{11} $M_{\odot}$, although with up to a 50% uncertainty in this result. The Andromeda galaxy (M 31) and our own are moving slowly toward each other; indirect arguments about the effect of their mutual gravitational field causing this turn-around from general expansion over the age of the universe lead to an upper limit on the mass of the Galaxy of about 10^{12} $M_{\odot}$.

SUMMARY

We have seen that dynamical modeling of the Galaxy leads to the conclusion that much of the mass is in a form whose nature is unknown and whose distribution is only surmised. For the regions of the Galaxy closer to the center than the Sun, a reasonable model of the gravitational field can be constructed, based on the three major components of disk, spheroid, and dark corona. The dynamics of the disk are dominated by rotation about the center, whereas the spheroid (with the associated globular cluster system) rotates much less quickly and will have quite a rich structure of stellar orbits.

Additional Reading

Binney, J. J. and Tremaine, S. D. (1987). *Galactic Dynamics.* Princeton University Press, Princeton.

Gilmore, G., Wyse, R. F. G., and Kuijken, K. (1989). Kinematics, chemistry, and structure of the Galaxy. *Ann. Rev. Astron. Ap.* **27** 555.

Gordon, M. A. and Burton, W. B. (1979). Carbon monoxide in the Galaxy. *Scientific American* **240** (No. 5) 44.

Rubin, V. C. (1983). Dark matter in spiral galaxies. *Scientific American* **248** (No. 6) 88.

Schmidt, M. (1985). Models of the mass distribution of the Galaxy. In *The Milky Way Galaxy*, IAU Symposium 106, H. van Woerden, R. J. Allen, and W. B. Burton, eds. D. Reidel Ltd., Dordrecht, p. 75.

See also **Galactic Bulge; Galactic Structure, Stellar Kinematics; Galaxy, Chemical Evolution; Interstellar Clouds, Giant Molecular; Missing Mass, Galactic; Solar Neighborhood.**

Gamma-Ray Astronomy, Space Missions

Gerald J. Fishman

Gamma-ray astronomy missions are those experiments and their associated observing platforms that are carried above the atmosphere to perform observations of the cosmos in the gamma-ray portion of the spectrum, typically at photon energies greater than 100 keV. Gamma-ray astronomy is perhaps the least developed of all of the astronomies of the electromagnetic spectrum. Even though the first observations were performed in the 1960s by balloon-borne and small satellite-borne experiments, there are still relatively few known sources in the sky and our understanding of them is rather shallow, at best. However, there is enormous potential in the study of gamma-ray astronomy because these photons are indicative of the most violent of processes in nature and are produced in regions of extremely high temperature, density, and magnetic fields. Furthermore, their high penetrating power and neutral character allow direct observations into otherwise obscured regions of space. Observations in gamma-ray astronomy embrace many diverse areas in astrophysics. These include background radiation, cosmology, cosmic rays, binary stars, supernova remnants, and solar flares.

The present status of gamma-ray astronomy is similar to that of radio astronomy in the late 1940s or that of x-ray astronomy of the late 1960s: Several dozen sources are known, as well as a general galactic distribution. Theoretical models of a cursory nature have been established but the lack of detailed observations prevents them from being tested and reformulated.

The difficulties in gamma-ray astronomy stem from the penetrating power of gamma rays, the ever-present cosmic-ray background, and the inherent low flux levels of the gamma-ray sources. The penetrating power requires that the detecting medium be dense and massive in order to contain the energy of the incoming photon. Cosmic-ray secondary radiation from the Earth's atmosphere, the spacecraft, and the detector itself provides a high background for gamma-ray observations, making it analogous to trying to observe stars during the daytime. Techniques to suppress this background further complicate experiment designs and produce additional weight. A detailed understanding of the background radiation and its systematic variations must be known in order to make reliable and accurate observations. Finally, the fluxes of gamma-ray photons from sources under study are very low, typically in the range of 10^{-4}–10^{-6} photons cm^{-2} s^{-1}. This requires that the gamma-ray astronomy experiments have large sensitive areas and long exposure times.

EARLY GAMMA-RAY ASTRONOMY EXPERIMENTS AND MISSIONS

In the low-energy gamma-ray region (100 keV to 10 MeV), the principal detector is the scintillation detector, usually a dense inorganic crystal coupled to a photomultiplier tube. Early exploratory observations were carried out on balloon flights and on detectors aboard the following spacecraft: *Ranger 3, 4,* and *5, Apollo 15* and *16,* and *Orbiting Solar Observatories* (*OSO*) *1, 2, 7, and 8.* The University of New Hampshire experiment on *OSO 7* observed the first gamma-ray lines from the sun. The leading balloon groups providing pioneering observations included the University of California at San Diego (USCD), NASA-Goddard Space Flight Center, Massachusetts Institute of Technology (MIT), Rice University, and the University of New Hampshire (UNH). These early observations not only gave us the first glimpses of the gamma-ray sky, but they also established the experimental designs and techniques that would be used on later spacecraft experiments in low-energy gamma-ray astronomy.

The first space-borne experiment in high-energy gamma-ray astronomy (energies above 10 MeV) was aboard *Explorer 11,* launched in 1961. It returned data for about six months, providing the first indication of high-energy gamma-rays from space. The same group from MIT developed a more advanced instrument for the *OSO 3* satellite. That experiment, launched on March 8, 1967, gave clear evidence of gamma rays emanating from the galactic plane, with a concentration toward the galactic center. These early spacecraft, together with balloon-flight experiments from several experimental groups in the United States, provided the technology and flux estimates that led to the first two survey missions in high-energy gamma-ray astronomy, *SAS 2* and *COS B.*

The *SAS 2* spacecraft (*Explorer 48*) was launched into an equatorial orbit on November 15, 1972 from the San Marco launch platform in Kenya. The European Space Agency (ESA) satellite *COS B* was launched into a highly elliptical orbit on August 9, 1975 from the Western Test Range in California by a Delta rocket. Together these two spacecraft provided the first look at about 20 sources of high-energy gamma rays as well as a more detailed map of the distribution of gamma-ray emission from the Galaxy.

About a dozen other gamma-ray astronomy missions were carried out in the early years of gamma-ray astronomy, some more successfully than others. Great Britain launched *UK 5* in 1974; it contained a small scintillation detector. *TD-1A*, ESRO's first satellite, which included a high-energy gamma-ray astronomy experiment, was orbited in 1972. (ESRO, the European Space Research Organization, was a forerunner of ESA.) The French *Signe 3* spacecraft, launched by a Soviet booster, also contained a gamma-ray astronomy experiment. In addition, the French and Soviets have collaborated on a number of gamma-ray astronomy experiments in the *Prognoz* series.

Military satellites played a significant role in the early years of gamma-ray astronomy. The *Vela* series of nuclear bomb detection satellites is credited with the discovery of gamma-ray bursts, first observed in the late 1960s and reported in 1973. Detectors aboard the U.S. Air Force *76-B* satellite launched in 1972 and the *P-78-1* satellite launched in 1978 performed some limited gamma-ray observations.

There have been a number of interplanetary spacecraft that have included gamma-ray astronomy experiments. The Soviets included gamma-ray burst detectors on the *Mars-5* and *Phobos* missions and on many of the *Venera* spacecraft. Some of the experiments were supplied by French collaborators. The *International Sun–Earth Explorer* (*ISEE 3*) spacecraft contained a cooled solid-state detector for high resolution observations of gamma-ray bursts. (The planned United States Mars Observer will contain a gamma-ray detector, also capable of gamma-ray burst observations.)

LARGER, RECENT GAMMA-RAY ASTRONOMY EXPERIMENTS

The *High Energy Astronomy Observatory* (*HEAO*) series of spacecraft was developed by NASA in the 1970s to explore all aspects of high-energy astronomy with experiments that were larger and more sophisticated than any that had been flown up to that time. The *HEAO 1* and *HEAO 3* spacecraft both carried gamma-ray astronomy experiments. *HEAO 1* was launched from Cape Canaveral on August 12, 1977 by an Atlas–Centaur booster into a low altitude, low inclination orbit. Of the four large experiments on board, one of them (HEAO A-4) was designed for gamma-ray observations in the energy range from 13 keV to 10 MeV. It consisted of a series of actively shielded scintillation detectors with different thicknesses and fields of view. This experiment was developed by UCSD; MIT participated in the data analysis effort. The first comprehensive survey of the sky in the low-energy gamma-ray region was performed by HEAO A-4. Selected sources were studied for their hard x-ray and gamma-ray emission, time variability, and spectral characteristics over the 17 month lifetime of the mission.

The *HEAO 3* spacecraft was launched on September 20, 1979 into a low altitude orbit with an inclination of 43.6°. It contained two cosmic-ray experiments and one gamma-ray astronomy experiment (HEAO C-1) developed by the Jet Propulsion Laboratory. This experiment was the first high resolution, solid state detector system designed for gamma-ray astronomy observations. It surveyed the entire sky and performed several concentrated scans of the galactic plane. Spectral features from the galactic center region and Cygnus X-1 were observed to change over the duration of the mission. The first known gamma-ray line from interstellar space, due to the isotope Al^{26}, was discovered by the HEAO C-1 experiment.

Although designed primarily for solar observations, the *Solar Maximum Mission* (*SMM*), launched in 1980, contained a gamma-ray experiment, that made significant contributions to gamma-ray astronomy, as well as solar physics. The gamma-ray spectrometer was a collaborative effort on the part of UNH, the Naval Research Laboratory, and the Max Planck Institute for Extraterrestrial Physics. The *SMM* experiment confirmed the earlier *HEAO 3* detection of gamma-rays from Al^{26} and detected gamma-ray line emission from Co^{56} from the supernova SN 1987A. Gamma rays up to 100 MeV were detected from several solar flares and gamma-ray bursts.

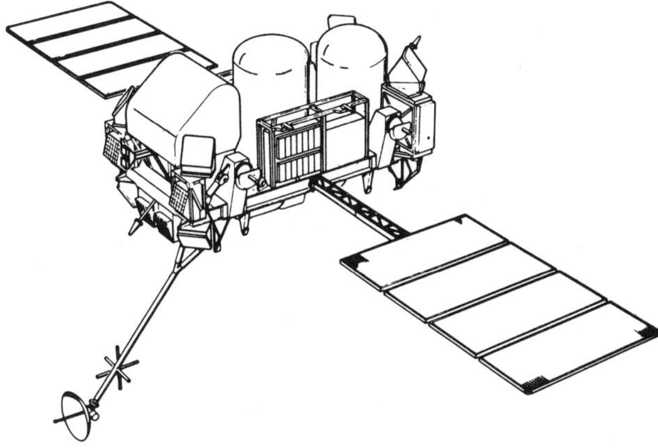

Figure 1. The *Gamma Ray Observatory* (*GRO*).

GAMMA-RAY OBSERVATORY AND GRANAT

Within the next few years, observational gamma-ray astronomy will develop into a mature science as the results from three major missions containing gamma-ray astronomy experiments become available. The 17.5 ton *Gamma Ray Observatory* [(*GRO*); see Fig. 1], developed by NASA, was successfully launched by the space shuttle *Atlantis* on April 5, 1991 into a low altitude, low inclination orbit. The spacecraft was originally designed to operate for 2 years, but it is now intended to operate for considerably longer, perhaps as long as 10–12 years. Four large gamma-ray astronomy experiments on the *GRO* can cover the entire gamma-ray spectrum above 30 keV. Two of the experiments, COMPTEL and EGRET are imaging telescopes with fields of view of about 1 sr. COMPTEL is designed to operate at energies between 1 and 30 MeV. EGRET operates at energies above 20 MeV. The OSSE experiment consists of collimated scintillation detectors that are optimized for the energy range between 30 keV and 10 MeV. The BATSE experiment is an all-sky detector array that continuously monitors the entire sky at energies above 30 keV. A comprehensive guest investigator program will allow many scientists to use data from the *GRO*.

The *GRANAT* spacecraft developed by a French and Soviet collaboration contains the Sigma experiment, the first coded aperture gamma-ray experiment to be deployed in space. The spacecraft launched in 1990, also provides significant capabilities for the study of gamma-ray bursts.

OTHER FUTURE MISSIONS IN GAMMA-RAY ASTRONOMY

The need for high spectral resolution observations of astrophysical sources has become apparent in recent years. An instrument aboard the U.S. *Wind* spacecraft in 1993 will be able to study gamma-ray bursts with high energy resolution. The highest priority future mission in gamma-ray astronomy appears to be a large, sensitive array of high resolution solid-state detectors combined with a coded aperture mask to perform imaging. Several groups in the United States have made initial balloon flights to develop this technology and to perform scientific observations of SN 1987a and the galactic center. A United States–European collaboration is studying a future mission based upon this technology. The Soviets plan to continue a program in gamma-ray astronomy with experiments on the *Mir* space station and the *Spectrum X-Gamma Observatory*. The decade of the 1990s is expected to bring several large and powerful gamma-ray astronomy missions to fruition.

Additional Reading

Fichtel, C. E. and Trombka, J. I. (1981). *Gamma-Ray Astrophysics: New Insight Into the Universe* (NASA SP-453). U.S. Government Printing Office, Washington, DC.

Gehrels, N. and Share, G. H., eds. (1988). *Nuclear Spectroscopy of Astrophysical Sources, AIP Conf. Proc. No. 170*. American Institute of Physics, New York.

Hillier, R. (1984). *Gamma Ray Astronomy*. Clarendon Press, Oxford.

Ramana Murthy, P. V. and Wolfendale, A. W. (1986). *Gamma-Ray Astronomy*. Cambridge University Press, Cambridge.

See also **Background Radiation, Gamma-Ray; Gamma-Ray Bursts, Observed Properties and Sources; Gamma-Ray Sources, Galactic Disribution; Telescopes, Detectors and Instruments, Gamma-Ray.**

Gamma-Ray Bursts, Observed Properties and Sources

Thomas L. Cline

The gamma-ray burst phenomenon has been, and may be destined to remain for some time, a curious puzzle in astronomy. Observations of these brief but intense cosmic emissions have accumulated for nearly 20 years, yet with one anomalous and controversial exception to be discussed later, no astrophysical source object has been identified. Found accidentally with satellites looking for artificial nuclear explosions hidden in space, gamma-ray bursts have been researched in the intervening decades with experiments both orbiting the Earth and traveling throughout the inner solar system: second-generation instruments will be launched in the 1990s to continue the pursuit. Their characteristics and the circumstances of their detection have inspired a variety of imaginative source concepts, from the exotica of alien warfare, antimatter annihilation, and strange modes of stellar collapse at cosmological distances to models based on various internal shifts and surface effects of relatively nearby neutron stars.

A large gamma-ray burst is significant in observational terms in that it is the instantaneously brightest flux of photons from the sky "visible" in its (100-keV) spectral region. Because the source distances not ruled out by the data could yet lie somewhere between the extrasolar and the extragalactic, the uncertainty in source energies encompasses a possible world's record for ignorance. The source volume is generally assumed (but has not been proved) to be the region containing the nearby stars. The diversity of the observed transients, however, suggests that one explanation may not be adequate. The fact that research on gamma-ray bursts continues to uncover new elements to the puzzle (which are each as unpredicted as the discovery) continues to fuel the fascination of researchers for this phenomenon.

It had been hoped that gamma-ray bursts could be associated with some other emission in a more easily researched spectral region, such as optical flashes. However, no simultaneous detection of a gamma-ray transient has yet been accomplished in any other spectral band, although an extension into the x-ray region does follow some events. Although the unique difficulties in studying these randomly occurring transients have left the source energies uncalibrated, evidence for neutron stars as source identifications has accumulated, partially by implication and partially by default. The limits to knowledge of the gamma-ray burst sources as set by the observations are reviewed here.

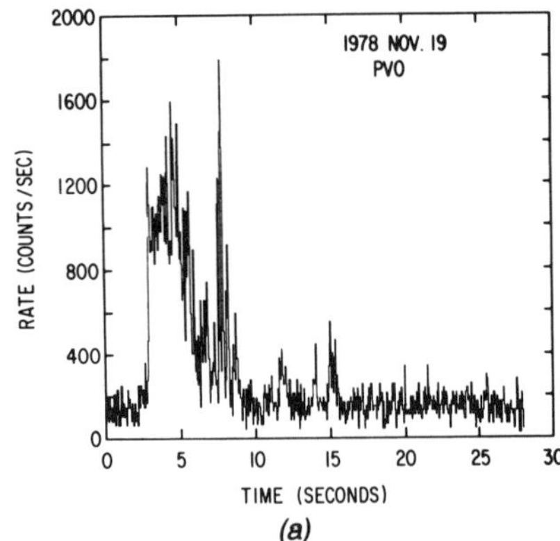

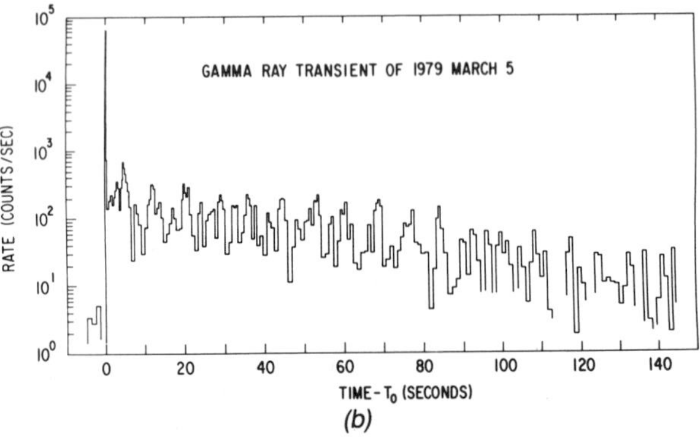

Figure 1. (*a*) The time history of the counting rate of photons of energies greater than 100 keV for a relatively typical, or classical, gamma-ray burst. (*b*) The time history of the unique 1979 March 5 event. This plot shows an 8-s periodicity on the decay but does not have the resolution for such details as the 0.15-s duration main peak, its < 0.2-ms onset time constant, or the suggestion of its 25-ms periodicity.

DETECTION OF GAMMA-RAY BURSTS

The history of this topic begins with the detection of several gamma-ray bursts per year with instruments on the *Vela* satellites designed to monitor violations of the Nuclear Test Ban Treaty. Nearly four years elapsed before Ray Klebesadel and his colleagues at the Los Alamos National Laboratory became confident that their effect was believable enough to publish, in 1973. Their orbiting detection system was capable of providing some event directional discrimination, between, e.g., a terrestrial, solar, or lunar origin. These could all be eliminated for the 14 events accumulated, as well as any source pattern in either the ecliptic or the galactic plane. No evidence for event repetition from a single source was found, implying that many source objects exist independently. The measured count rates of these transients were astonishing, competing with or exceeding those of solar gamma-ray flares, yet the distance to any source was entirely unknown, possibly even many orders of magnitude greater than that to the Sun. Also, the instrumentation was adequate to show considerable event diversity, with fluctuations down to several tens of milliseconds (ms) and durations of up to several tens of seconds. (Figure 1*a* illustrates this with a more recently obtained burst time profile.) Needless to say, the curious and chaotic nature of this powerful new natural phenomenon greatly excited the x-ray and gamma-ray astronomers, cosmic-ray physicists, and space scientists of the time.

Evidence found with a solar flare hard x-ray detector on a NASA spacecraft provided both a rapid confirmation of this discovery and the first energy spectra of the bursts. There was more consistency in peaked emission in the several-hundred-keV region, where nuclear gamma-ray lines might occur, than in the appearance of the high-energy extensions of softer, cosmic x-ray spectra. One of these confirmed events had also been observed with a collimated, or shielded, detector on another spacecraft. This fact provided the first directional confirmation, although the high galactic latitude noted for that one source provided no key to the mystery. The directional results obtained by Los Alamos (from inferring a source vector from the detection times of a burst "wavefront" at the various sensors scattered in cislunar space), as well as their other observational methods and conclusions, were all shown to be essentially correct, relieving some of the anxiety but not eliminating the mystery about this discovery.

The next step, beyond obtaining burst data accidentally was to design and fly dedicated instruments. A solar-orbiter launched in 1976 provided considerably greater precision in its source information by virtue of its great distance from the Earth: although an event timing comparison across this single baseline could determine only a ring-segment source locus, the roughly arcminute–by–tens-of-degree regions obtained were small enough in total celestial area to be found to be inconsistent with the locations of the known cosmic x-ray and other expected source objects. This result also heightened the gamma-ray burst source mystery.

The first interplanetary network of gamma-ray burst detectors was achieved in 1978 with the launches of several interplanetary probes and other spacecraft. There was a sufficient number of sensors to provide some directional overdetermination, but not to accomplish true and independent redundancy. Great efforts were made to eliminate any clock drifts or errors that could be the weak links in this long-baseline interplanetary measurement timing system. For those satellites near the Earth, at least, the interinstrument dispersion was found to be less than the event fluctuation time scale, permitting considerable confidence in the system. The results included several burst source regions determined with dimensions of less than several square arcminutes. None of these (with the exception of the 1979 March 5 event as discussed below) were consistent with the locations of known x-ray emitters or any other candidate source objects, including optical objects to exceedingly faint limits. Thus, gamma-ray burst sources and their possible companion objects, if any, remained undetectable with the existing technology of astronomy.

The properties of the bursts were also studied for clues to the source process. The brevity of the observed temporal fluctuations requires that the source regions, although not necessarily the source objects, be small. Early calculations regarding photon density considerations in the emission volume, with certain assumptions, suggested that the distances to the sources were less than several kiloparsecs. Studies of the continuum spectra, due to their curious "obliging" nature wherein a researcher can fit almost any model desired, have remained unable to dictate source limitations. However, fairly narrow spectral features seen in the 40-keV region and at 400 keV have specific implications. The more energetic feature is consistent with the 20% gravitational redshift expected of the 511-keV electron-positron annihilation line at the surface of a neutron star. This was not a necessary conclusion because a positron annihilation gamma-ray coherence effect, for example, could also fit the same observation. The low-energy feature was originally explained as due to cyclotron resonance; recently, multiple low-energy features were observed that assuredly confirm this interpretation, providing evidence for burst emission within a strong magnetic field. Thus, like the sources of x-ray bursts, the sources of these transients are very likely to be neutron stars.

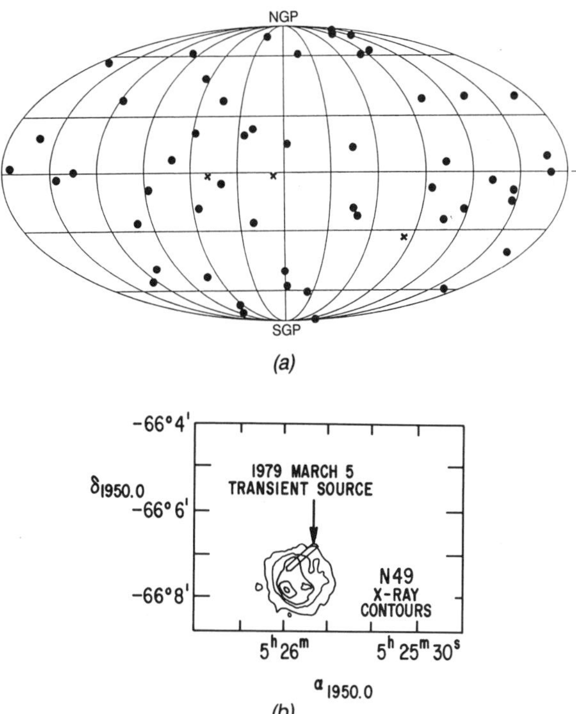

Figure 2. (*a*) A sky map of gamma-ray burst source directions, plotted in the galactic coordinate system. Also indicated, with xs, are the source directions of the three known soft repeater series. (*b*) A high-resolution plot of the source area of the 1979 March 5 event, including x-ray intensity contours of the supernova remnant N49 for reference. One of the series of soft gamma ray repeaters has rough consistency with this source.

PHENOMENOLOGY

Phenomenology became possible after a sufficiently large number of burst observations were accumulated. Although a total of several hundred events have been noted in these 20 years, only those bursts detected with one instrument in a given configuration can provide an internally consistent sample for study. Several instruments have by now provided their own all-sky source maps (e.g., as shown in Fig. 2*a*). No deviation from isotropy has been found. With no apparent clustering along the galactic disk, for example, the source volumes available to fit this result could be anything "round" or relatively featureless, including the galactic halo and the extent of the distant galaxies, as well as the nearby interstellar region.

Another property from which source knowledge can be inferred statistically is the burst "visible magnitude" or size. The number of events observed to have a given size or greater plotted as a function of size should, in principle, provide source region information, such as to distinguish a three-dimensional uniform volume (like that containing the nearby stars) from a two-dimensional slab-shaped volume (containing the distant stars in the galactic disk). This analysis, however, is subject to certain limitations, depending as it does on the unknown "luminosity function" or the equivalent distribution of sizes and types of bursts to be seen at a fixed distance, and on its variation with distance, if any. Even worse, given the white noise appearance of most bursts with temporal fluctuations at the limits of detector resolution, not to mention their extreme diversity, to assign a meaningful size to any event may very well be instrumentally or personally subjective. Nevertheless, the size spectrum has been seen to evolve over the years, with a persistent flattening at the small event end. If this is taken as noninstrumental in origin, it would be interpreted as evidence for a population transition to the galactic disk. However, an anisotropy should be exhibited in the source direction plot of the same events, providing a fundamental inconsistency that has been the subject of much concern. A recent critical result is the observation with a very large balloon-borne array of one event in four days, about the same rate as that found in deep space using much smaller and less-sensitive detectors. If all instrumental calibrations are to be believed, the conclusion could be that "that is all there are." In other words, there is no population of events less intense than those previously seen, at least in the northern hemisphere, where the balloons are flown. This could be interpreted as favoring a galactic halo source, until an extended, sensitive observation from Earth orbit (such as is expected from the Gamma-Ray Observatory, starting in 1991) can improve on or confirm it.

EVENT OF 1979 MARCH 5 AND SOFT GAMMA-RAY REPEATERS

Finally, it has become an established fact that gamma-ray burst properties appear to define distinct populations, a complication that has provided added controversy. Bursts have been grouped in terms of their behavior, with patterns that differ from each other as much as each does from the x-ray burst phenomenology. The observation of the 1979 March 5 gamma-ray transient began this process. The brightest observed at the time, it was the only burst to have a clear periodicity in its intensity-versus-time curve, a result that helped promote neutron stars as gamma-ray burst sources. Its time profile was unique in other ways, however, beginning with an exceptionally brief (under 1 ms) rise time on a single, unusually uncomplicated intensity peak, followed by the long decay with an 8-s periodicity (see Fig. 1*b*). Its spectral evolution was softer than most. Another distinctive property was the existence of a series of weaker (and also soft) sequel events that were seen on at least thirteen occasions over the following months and years, all consistent with arising in the same source direction. The different character of this event promoted its identification as a separate or unique event. Thus, conclusions drawn as to the nature of its source could be inappropriate when applied to the population of the other gamma-ray bursts that did not share its properties.

The direction of the source of the 1979 March 5 event is perhaps its most anomalous but certainly its most controversial feature. It points to the supernova remnant N49 in the Large Magellanic Cloud (see Fig. 2*b*), whereas all other sources are in the directions of blank space. The LMC distance of 55 kpc, however, caused general disbelief, even though the probability of chance fit to any of all the known supernova remnants, pulsars, and neutron stars is less than one in ten million. The problem remains: A supernova remnant should be associated with a neutron star and thus could be acceptable as a source candidate, other than for the difficulties of modeling a source at this distance. (How could the brightest gamma-ray burst come from farther away than what were assumed source distances for all the other events?) Calculations that "proved" N49 to be too far, by at least a factor of 5, have since been "disproved" with the uncovering of a basic error, keeping the controversy alive.

Part of the solution to this puzzle may be that it is only the beginning of another one. The series of repeating events that is sequel to the 1979 March 5 burst is known now to be one of three such series of so-called soft repeaters. These events contrast with the "classical" gamma-ray bursts, which are never observed to repeat from a given source, in other ways as well. The energy spectrum of the soft repeaters is characterized in the 25-keV region, in fact, geometrically in between the x-ray bursts at a few keV and the classical and spectrally harder gamma-ray bursts at several hundred keV. The spectra have not been observed to possess cyclotron resonance features. The spectral evolution seen in classical events is not found. The time histories of the repeaters are always brief and simple, with only one peak, unlike the 'white noise' effect usually exhibited by the classical events.

Finally, there may be a clue as to the sources of these soft gamma-ray repeaters, including the problem of N49, in their sky pattern. Although the number of the sources (three) is far too small to be a meaningful statistical sample, the three directions happen to be consistent individually with the galactic disk, the galactic bulge region, and N49 in the LMC (as shown in Fig. 2). Although all these three may, of course, be entirely accidental coincidences, they do hint at the possibility of a population of repeaters in our Galaxy and in our neighboring galaxy. This pattern is, at least, consistent with the fact that the typical intensities of the events having a source fitting the galactic bulge direction are brighter by the appropriate factor than those in the post–1979 March 5 series having a source direction fitting N49 in the more distant LMC, assuming that repeaters in distinct series may have similar typical intensities.

The anomaly, again, is the fact of the relative and absolute brightness of the 1979 March 5 event. That event was predicted to be in the once-in-a-lifetime category: Twenty years of data collecting have supported that guess thus far. Another, more minor, complication is that although these three series of soft repeaters have similar spectra and time profiles, the patterns of their repeating times of occurrence are not very similar. The most recently found series, from the galactic bulge direction, consists of over one hundred events seen throughout several years of observation, with the event rate clustering within several months and peaking with twenty events in one day. The sequels to the 1979 March 5 event trickle out for years and can be roughly fitted to a 164-day periodic pattern.

The diversity of all this phenomenology suggests that our view of the gamma-ray burst picture may not be complete. Continued study may or may not find another event of the 1979 March 5 category, find additional soft repeaters in the galactic disk, or even find weak classical events throughout the disk, or it may surprise us with some new twist. Given the present plans for the observations of gamma-ray transients with the next generation of instruments, we can be sure that the knowledge of their source objects and emission processes and of their relationships to other astronomical phenomena will evolve as more clues are unraveled.

Additional Reading

Higdon, J. C. and Lingenfelter, R. E. (1990). Gamma-ray bursts. *Ann. Rev. Astron. Ap.* **28** 401.

Liang, E. P. and Petrosian, V., eds. (1985). *Gamma Ray Bursts. Amer. Inst. Phys. Conf. Proc.* **141**. AIP, New York.

Lingenfelter, R. E., Hudson, H. S., and Worrall, D. M., eds. (1982). *Gamma Ray Transients and Related Astrophysical Phenomena. Amer. Inst. Phys. Conf. Proc.* **77**. AIP, New York.

Schaefer, B. E. (1985). Gamma-ray bursters. *Scientific American.* **252** (No. 2) 52.

Woosley, S. E., ed. (1984). *High Energy Transients in Astrophysics. Amer. Inst. Phys. Conf. Proc.* **115**. AIP, New York.

See also **Gamma-Ray Astronomy, Space Missions; Gamma-Ray Bursts, Optical Flashes; Stars, Neutron, Physical Properties and Models.**

Gamma-Ray Bursts, Optical Flashes

Bradley E. Schaefer

Gamma-ray bursters (GRBs) are mysterious objects that emit short intense bursts of gamma radiation. The nature of the source and the cause of the bursts are not known, although there are good theoretical and observational reasons to suspect that a neutron star is somehow involved. It is reasonable to expect that at least some small fraction of the burst energy will be emitted as optical light and will appear to astronomers as an optical flash. In 1981, three such flashes were identified on photographs at the Harvard College

Observatory. In 1987, a fourth flash was identified on a pair of photographs from Odessa, USSR. After a burst is over, the burster will appear without any disruptive glare, so that it may be possible to discern the nature of the underlying system. Much labor has been expended to find this quiescent counterpart, but to the limits of current technology at all wavelengths, no counterpart has been found.

BURSTING COUNTERPARTS

A supernova eruption emits only 1/10,000 of its energy as visible light. If a bright GRB emits a similar fraction in the visible range, then this light should appear as a bright flash. This flash may be visible to anyone standing in their backyard. The trick is to be looking at the right time and in the right direction. This is the major trouble because both the times and positions of GRB events are totally unpredictable. It is also unreasonable to pick some random position on the sky and wait for a flash, because they are so rare. For example, a region the size of the full moon has a GRB once every several centuries.

One stratagem is to observe a carefully selected direction where a GRB is already known to exist from gamma-ray data. This stratagem relies on the certainty that most GRBs burst repeatedly, perhaps as frequently as once every several years. Since 1985, the most sophisticated flash search system in the world (designed, built, and operated by amateur astronomers with the Santa Barbara Astronomy Group) has used this stratagem. However, observation time can be accumulated only at a rate of 0.1 year per year for each detector at best.

An additional stratagem is to utilize past observations with a large cumulative exposure. This is usually done by examining archival photographs of a region containing a known GRB. These studies have been performed on photographs stored in many observatories worldwide with a cumulative exposure of nearly a decade.

In this decade of monitoring time, four GRB optical flashes have been discovered. The first three were found in 1981 on photographs from 1901, 1928 (see Fig. 1), and 1944 in the Harvard collection. The fourth GRB flash was identified in 1987 on a pair of photographs from 1959 stored in Odessa, USSR.

The images found on these photographs occur at positions inside the regions on the sky known to contain GRBs. No stars (to very faint limits) appear at the flash positions. Several of the flash photographs are accompanied by other photographs taken immediately before and/or after that show nothing out of the ordinary; hence the "new" images cannot be caused by a passing comet or asteroid. The Odessa flash was recorded on two simultaneous photographs, so plate defects are not a reasonable explanation. Three of the flash photographs are slightly trailed (a condition where normal star images are elongated because of imperfect tracking of the sky motion), yet the flash images are untrailed; hence the flashes must be shorter than several minutes long. In one case, the flash image is off axis and shows a characteristic distortion (coma) due to the telescope optics, demonstrating that the flash light originated in the sky. A study was made where a control region that contained no known GRB positions was examined. The control region (with a cumulative exposure 16 times larger than the GRB search region) had no flash images, hence providing a further connection between the flashes and the GRB phenomenon.

The primary difficulty with these historic flash images is that no simultaneous gamma-ray data are available, so the relation between the optical event and the gamma-ray event is unknown. A reasonable presumption is that the optical events in 1901, 1928, 1944 and 1959 were each associated with a gamma-ray event that was comparable to the gamma-ray events later detected by satellites. In this case, the fraction of energy emerging as optical radiation is between 1/1000 and 1/10,000.

The utility in finding the four GRB optical flashes is fourfold:

1. First, the fraction of burst energy that appears in the optical can be used to deduce that the GRB system contains something much larger than a neutron star, hence demonstrating that the flash must come from some other body in the GRB system.
2. It demonstrates that the GRBs do repeat on a short time scale (one decade of observations netted four flashes), a fact that is incompatible with many theoretical explanations.
3. The optical photographs yield a position for the burster that is much more accurate than deduced from gamma-ray data, so searches for quiescent counterparts (see below) can go much deeper.
4. The mere existence of the optical flashes can (in some cases severely) constrain GRB models.

Two classes of theoretical explanations for the optical flashes have been advanced: The first class claims that the optical light is from the burst's gamma radiation being absorbed and reradiated by

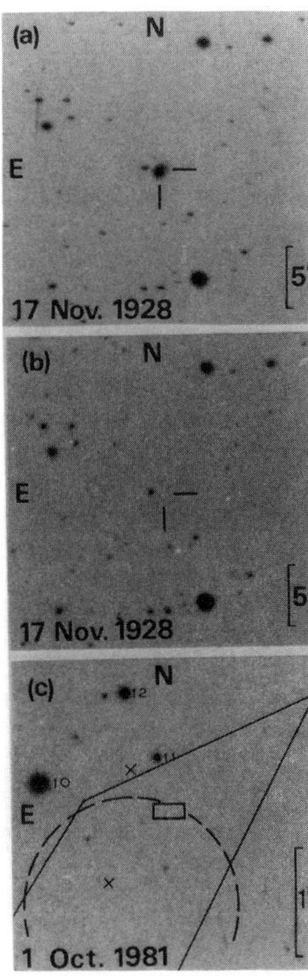

Figure 1. The 1928 optical flash. The top panel shows a blowup of a photograph taken on 17 November 1928 that shows an "extra" star image that appears in the same direction as a GRB that occurred on 19 November 1978. This photograph is fourth of a series of six 45-min exposures taken in close succession. The middle panel shows the fifth photograph, which shows nothing at the location. The star images in the top panel are all elongated in an east–west direction due to telescope trailing, but the flash image is not trailed, indicating that the flash duration is shorter than several minutes. Both normal star and flash images show the same slight deviation from circularity caused by the light passing through the telescope's optics. The bottom panel shows a blowup of the same area taken from a deep photograph (*courtesy Martha Hazen*), which shows the position of the flash (rectangle) when the GRB was quiescent. The thin solid line delineates the outer edge of the position measured for the associated GRB, crosses indicate known radio sources, and the dashed circle is the problematic x-ray source position.

a companion star or accretion disk. The second class ascribes the visible photons to cyclotron emission from plasma in a neutron star magnetosphere that was excited by the burst.

The observations merely constrain the properties of GRB flashes to be bright and of duration shorter than several minutes. We do not know if the flashes are blue or red in color or whether they last a second or a minute. Unfortunately, the night sky is full of flashes, many of which could be mistaken for GRB events. Background flash sources include fireflys, airplane strobes, head-on meteors, satellite glints, and flare stars. The rates and properties of this background are poorly known. This is illustrated by the series of flashes (the so-called Perseus flasher) seen by Canadian meteor observers from 1984 to 1985 that, only after many long investigations, were shown to be satellite glints (sunlight reflected from orbiting artificial satellites). The identification of GRB flashes is like trying to find a needle in a haystack when we do not know what the needle looks like and we have only just realized that we do not even know what the hay looks like.

QUIESCENT COUNTERPARTS

During a burst, the emitted light comes from a fireball generated by some unknown energy source. Unfortunately, the properties of this fireball are relatively insensitive to the energy source, so it is difficult to learn about the cause of the burst. After a burst, the GRB should fade back to its normal brightness, so that the underlying system can then be seen. If we could spot even one such quiescent counterpart, then classical astronomical techniques would reveal the nature of the system (e.g., a binary star, a lone neutron star, or a galaxy). This would immediately eliminate many of the proposed GRB models from consideration.

Many groups have realized that the firm identification of even one quiescent counterpart is likely to be the break point of our current ignorance. As such, great effort has been spent on determining GRB positions as accurately as possible and then searching for counterparts at any wavelength. Early on, it was realized that GRBs are not coincident with planets, bright stars, pulsars, x-ray sources, or anything else of note.

The more accurate the GRB position is known, the deeper a quiescent counterpart search can go. This is because if the GRB is only known to be within some large area on the sky, then there are likely to be many stars (some bright) within that region. The searches are usually based on the idea that the counterpart will look unusual in some way (say, it might appear very blue), but it is impractical to examine large numbers of stars for oddities. So deep searches can only be reasonably performed on small regions. The localization of GRBs on the sky is a difficult problem, and currently only seven have their position determined to within an arcminute and have had deep searches made on them.

The 5 March 1979 GRB was the brightest burst ever detected; therefore, its position could be measured with unprecedented accuracy (within 15 arcsec). This direction coincides with that of N49, a supernova remnant in the Large Magellanic Cloud, and so the GRB may be physically associated with N49. If true, this identification would imply an incredibly large burst energy ($10^{44.5}$ erg) as well as associate the burster with the neutron star presumably inside the supernova remnant. The theoretical and observational arguments for and against the association of this GRB with N49 are still raging. Even if the burst is proven to come from N49, the 5 March 1979 event is so highly unusual in so many ways that conclusions drawn from this event may not be applicable to normal bursters.

Besides this one problematic identification, no other reasonable counterparts have been identified (see Fig. 2). At radio wavelengths, over 30 h of observing time with the Very Large Array in New Mexico have revealed no counterparts. For far infrared wavelengths, the *Infrared Astronomical Satellite* has identified no counterpart for the 23 best localized bursts. Similarly, at near infrared wavelengths, several GRB locations were found to be empty even when a 4-m telescope with infrared array detectors was employed. At optical wavelengths, one GRB position was even found to be starless to a *B* and *R* magnitude of fainter than 25. At x-ray wavelengths, both the *Einstein* and *EXOSAT* satellites have spent over 40 h with only one debatable detection. In summary, current technology has been pushed to its limit and no confirmed quiescent counterpart has been found.

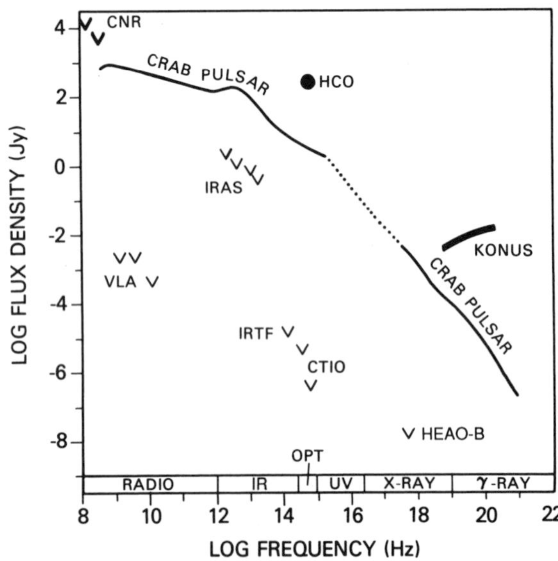

Figure 2. The spectrum of the 19 November 1978 GRB. This plots shows the spectrum for one particular GRB from radio to gamma ray. The brightness (in Janskys) along the vertical axis is given on a logarithmic scale, so that large brightness differences will be significantly compressed on the plot. The spectrum of the Crab pulsar (i.e., a lone neutron star possibly comparable to GRBs) is given for comparison. Observations made during the burst are given with thick symbols above the Crab spectrum, whereas limits on the quiescent burster appear below the Crab spectrum. The bursting GRB was detected by gamma-ray detectors (e.g., the KONUS experiment on the Soviet *Venera* spacecraft) and on archival photographs at Harvard College Observatory (see Fig. 1). A radio telescope operated by the Italian National Council of Research (CNR) detected no accompanying radio burst. The quiescent counterpart was not detected with the Very Large Array (VLA), the *Infrared Astronomical Satellite* (*IRAS*), the Infrared Telescope Facility (IRTF), the 4-m telescope at Cerro Tololo Inter-American Observatory (CTIO), nor with the *HEAO 2* satellite.

FUTURE PROSPECTS

The primary hope for advances must be based on the introduction of new technology.

For the detection of bursting counterparts, an awaited new technology is the all sky flash search system made up of the explosive transient camera (ETC) and the rapidly moving telescope (RMT). The ETC consists of two widely separated arrays of 16 charge coupled device (CCD) cameras each that will record most of the sky down to magnitude 11 at 1-s intervals. When both arrays detect a flash with no parallax (hence ensuring that the flasher is far from Earth), the data will be saved and the RMT will be notified. The RMT is a 7-in. telescope that can slew to any position on the sky within 1 s of notification. The CCD camera on the

RMT will measure the flash position to better than 1 arcsec and obtain a light curve when the flash is brighter than 15.2 magnitude. The GRB nature of these flashes will be established by time and directional coincidence with GRBs detected by the *Gamma Ray Observatory* satellite.

For the detection of quiescent counterparts, the primary new technology is the *Hubble Space Telescope*. This telescope will not only allow much deeper searches of error boxes in visible light, but will for the first time allow deep searches in the ultraviolet. The ultraviolet search capability is important because of the possibility that GRBs are lone, hot, neutron stars and hence visible only in the ultraviolet.

Other new important instruments include the burst and transient source experiment to fly on the *Gamma Ray Observatory* satellite and the *Advanced X-ray Astrophysics Facility* satellite. All these new detectors will supplement the currently existing systems (such as the Santa Barbara Astronomy Group's telescope network) and the classical astronomy techniques (such as archival photograph searches). All in all, I believe that a counterpart will be confidently identified in the next decade, with the result that the cause and nature of GRBs finally will be known.

Additional Reading

Hurley, K. (1986). Astronomical issues. In *Gamma-Ray Bursts*, E. P. Liang and V. Petrosian, eds. American Institute of Physics, New York, p. 3.

Maley, P. D. (1987). Specular satellite reflection and the 1985 March 19 optical outburst in Perseus. *Ap. J. Lett.* **317** L39.

Pedersen, H., Danziger, J., Hurley, K., Pizzichini, G., Motch, C., Ilovaisky, S., Gradmann, N., Brinkmann, W., Kanbach, G., Rieger, E., Reppin, C., Trumper, W., and Lund, N. (1984). Detection of possible optical flashes from the gamma-ray burst source GBS 0526-66. *Nature* **312** 46.

Schaefer, B. E. (1981). Probable optical counterpart of a gamma-ray burster. *Nature* **294** 722.

Schaefer, B. E. (1985). Gamma-ray bursters. *Scientific American* **252** (No. 2) 52.

Schwartz, R. E. (1986). A hunt for flashing stars. *Sky and Telescope* **72** 560.

See also **Gamma-Ray Bursts, Observed Properties and Sources; Stars, Neutron, Physical Properties and Models.**

Gamma-Ray Sources, Galactic Distribution

Giovanni F. Bignami

Astronomy with gamma rays looks at celestial objects in the light of photons emitted by high-energy processes. Practically unabsorbed by the interstellar medium, these photons are the main probe for the study of nonthermal processes. The sources visible in gamma rays are a new type of astronomical entity, not necessarily connected with objects known at other wavelengths, and, as such, frequently difficult to understand. Their distribution in the sky is interesting because it tells us about the collective properties of these objects as a population, a tool used with success in improving the understanding of other astronomical species.

Gamma-ray sources (GRS) come in two basically different types: the steady, or slowly variable, high-energy sources, detected above several tens of megaelectron volts, and the so-called gamma-ray bursters (GRB), that are characterized by a short-lived burst emission of an intense flux of lower-energy (tens of kiloelectron volts to a few megaelectron volts) photons. The sky distributions of these two classes of objects are quite different: the first is obviously galactic in nature, owing to its strong correlation with the galactic disk; the second is by-and-large uniformly distributed in the sky.

Starting with the higher-energy sources, it will be useful to briefly review the gamma-ray astronomy data that have led to the concept of GRS.

GAMMA-RAY SOURCES AND GAMMA-RAY EMISSION FROM THE GALACTIC DISK

Figure 1 shows the gamma-ray sky in galactic coordinates, as made available by the *COS-B* satellite in the energy range 100 MeV–5 GeV. The data from this European Space Agency satellite, active from 1975 to 1982, still represent the near totality of the current high-energy gamma-ray astronomy database, with over 200,000 photons detected from the sky. As is apparent, the instrument's angular resolution and point spread function (related to the point source location capability) are poor, by astronomical standards. This is due to the problem of detecting gamma-rays through their materialization process in a spark chamber, leading to an intrinsic width of the instrument's point spread function (PSF), which decreases with increasing energy, but is never below 1°.

Inspection of the figure shows that the galactic disk itself is a bright gamma-ray emitter, with a significant center-to-anticenter gradient and other large-scale characteristics similar to those of the Galaxy's spiral structure. The source of the majority of such galactic radiation is diffuse, and attributable to the interaction of both cosmic ray (CR) protons and electrons with the interstellar medium (ISM) in its various stages of clumpiness. This diffuse galactic radiation is also an all-important factor for understanding the GRS, which are seen against it. In fact, when searching for GRS, one has to treat the diffuse disk emission as a background. Operationally, then, a GRS is defined just like any other unresolved source in astronomy, that is, "an excess of photon counts which is both statistically significant with respect to the surrounding background and compatible with the instrument's PSF."

The application of an algorithm closely following this definition to the *COS-B* database has yielded a catalog of *COS-B* sources, listing over 20 objects, most of them at very low galactic latitude, with the notable exception of the bright (gamma-ray luminosity, $L_\gamma \sim 10^{46}$ erg s^{-1}), local (redshift, $z = 0.158$) quasar 3C 273.

It was clearly recognized that, because these GRS could have intrinsic angular sizes up to 1° they should not necessarily be associated with individual star-like objects, but rather could come from interstellar clouds or generally clumped material emitting gamma rays under CR irradiation. This was obviously not the case for a few bright GRSs seen somewhat away (i.e., a few degrees) from the most crowded regions of the disk, and that had already been discovered by NASA's *SAS-2* mission in 1973.

As is apparent, the process of extracting sources from the data of Fig. 1 is critically dependent on the modeling of the radiation from the disk. Under the hypothesis that it arises from CR/ISM interaction, this requires three further assumptions:

1. The knowledge of the elementary physical processes involved and of the related cross sections.
2. The distribution of the target nuclei, that is, the interstellar gas.
3. The galactic distribution of the energetic particles, that is, CR protons and electrons.

Individual cross sections are well known from accelerator work and they represent the smallest uncertainty; the galactic diffuse matter distribution in both its atomic and molecular forms is being measured with increasing precision by current radio astronomy methods. However, the galactic CR distribution and, especially, the more important proton component, cannot be measured; they must be modeled. In fact, it is in order to recall here that high-energy galactic gamma-ray astronomy was started precisely with the purpose of measuring the distribution of the CR proton component; it was recognized over 30 years ago that this would be the only observational window to CR astronomy.

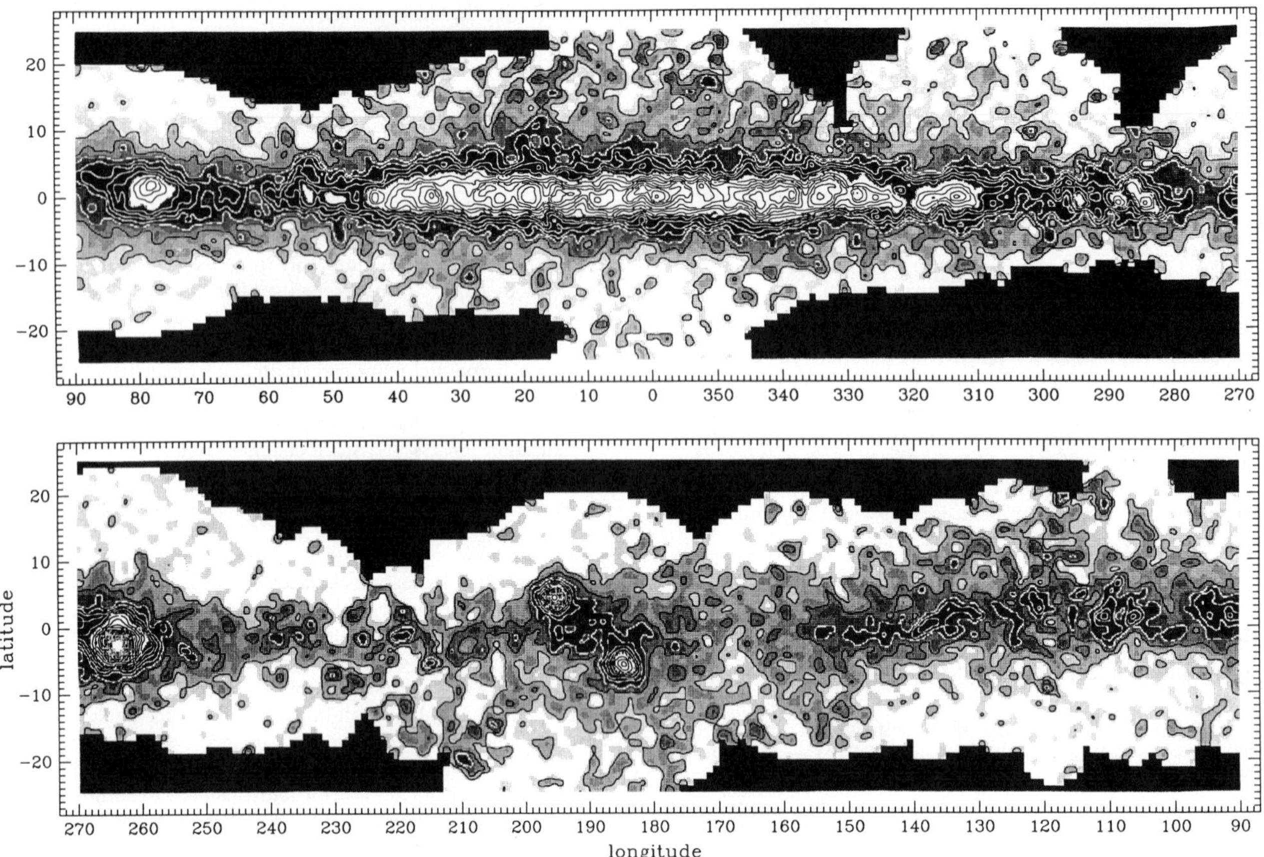

Figure 1. The galactic gamma-ray sky in galactic (l, b) coordinates. Data from the European Space Agency *COS-B* mission, including photons with energies between 100 MeV and 5 GeV. Instrumental background has been subtracted. Dark areas at the edges correspond to insufficient or lacking sky exposure. [*By permission from J. Bloemen*, Ann. Rev. Astron. Ap. *27 (1989)*.]

For the GRS search, one is then put in the uncomfortable situation of having to use the galactic gamma-ray astronomy data somewhat the wrong way and in any case, in order to do so, one has to pass the observational data through significant modeling. On the other hand, because such modeling is obviously an iterative process, it has created as a by-product the best current knowledge on the Galaxy's CR content. Moreover, diffuse gamma-ray emission has been definitely identified from such local (and diverse) interstellar cloud complexes as that in Orion and near the star ρ Oph, as well as for other, more extended local structures.

GAMMA-RAY SOURCES

Application of models accounting for the diffuse emission leads to the subtraction of the vast majority of the gamma rays from Fig. 1, leaving behind a distribution that cannot (to a given level of accuracy) be explained in terms of CR/ISM interactions. These gamma rays represent, fluxwise, a minute amount of the Galaxy's emission, but are important because they should be related mostly to compact objects, much in the same sense as, for example, in x-ray astronomy sources. (It is interesting to note how different these two modern high-energy astronomies are, with the vast majority of the celestial x-rays, at least of a given energy, being due to discrete sources and very few to diffuse emission.)

Figure 2 shows the distribution of galactic gamma rays remaining after the process just described. The figure must be looked at as somewhat model-dependent: As better data on the gas density distribution become available it will be subject to possible updating, especially on the lower-significance side. From Fig. 2 a population of true GRSs can be more easily extracted, making use of the comparison with the instrument PSF, as previously mentioned. The resulting statistically significant excesses are labeled with the acronym 2CG (2nd *COS-B* catalog of gamma-ray sources) followed by their galactic coordinate values (l, b). They are obviously a galactic population, as shown by their very narrow spread in b. However, it does not yet appear to be possible to go much beyond this general statement because there is certainly no evidence that they are a homogeneous population; rather there are suggestions to the contrary. For example, 2CG 184−05 and 2CG 263−02, the first two GRSs, discovered by *SAS-2* and confirmed by *COS-B*, are readily associated with the high-energy emission from two fast and young radio pulsars, PSR 0531+21 and PSR 0833−45, inside the Crab and Vela supernova remnants, respectively. High-energy gamma-ray emission from radio pulsars is a fundamental probe of the electrodynamics of the magnetized neutron star and its corotating magnetosphere. Its importance stems not only from the evidence for production of high-energy photons, but also from the fact that such photons carry with them a fraction of the pulsar rotational energy loss that is many orders of magnitude greater than that, for example, in radio waves. In the case of 2CG 184−05 (as opposed to 2CG 263−02), it is also remarkable that the emission is not entirely pulsed, that is, a significant gamma-ray flux may have been observed as coming from the nebula surrounding the pulsar. Both the Crab nebula and its pulsar are also sources of lower-energy radiation, whereas the nebula, at least, emits very high energy (10^{12} eV) gamma-rays as well.

Other sources in Fig. 2 (e.g., 2CG 135+01 and 2CG 78+01) are probably also related to compact objects, in particular neutron stars

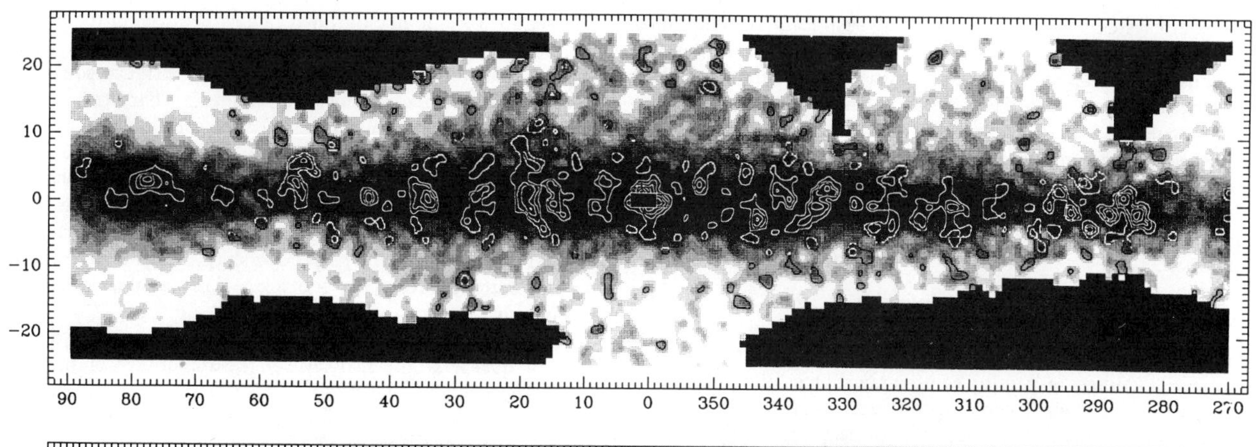

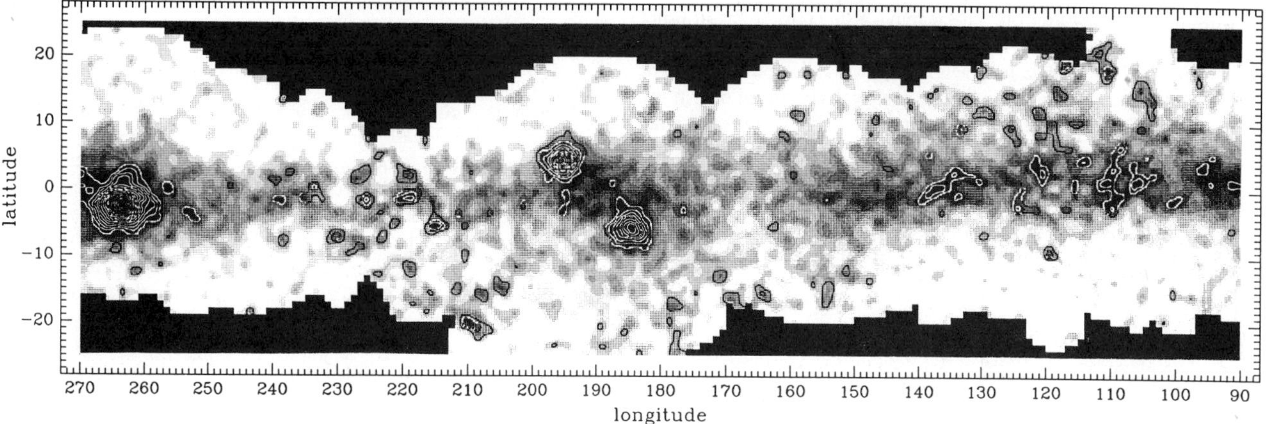

Figure 2. Galactic distribution of potential gamma-ray sources, indicated by isophotes. The map results from a subtraction of the predicted flux arising from cosmic ray/interstellar medium interactions from the data of Fig. 1 (see text). The bright sources associated with the Crab pulsar (2CG 184 − 05) and Geminga (2CG 195 + 04) are visible in the anticenter regions. [*By permission from J. Bloemen*, Ann. Rev. Astron. Ap. *27 (1989)*.]

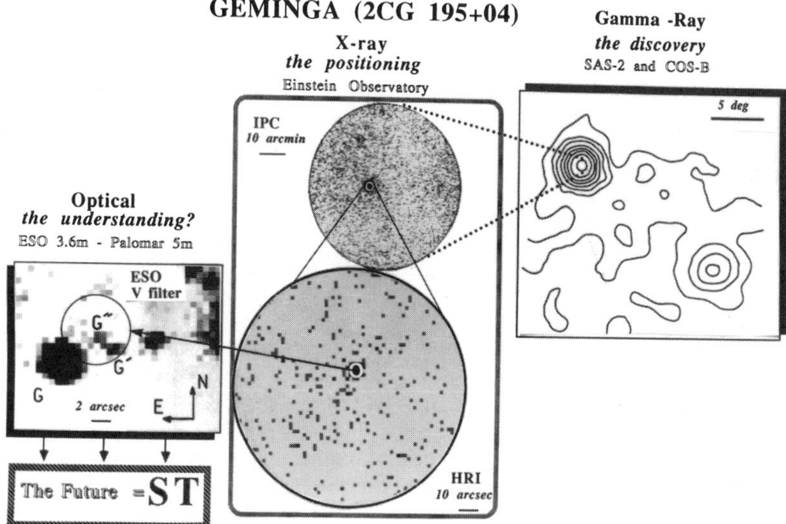

Figure 3. The Geminga counterpart search, a case for increasing angular resolution in astronomy. From right to left: anticenter gamma-ray data (see Figs. 1 and 2) showing 2CG 195 + 04 at upper left; *Einstein* Observatory x-ray data showing 1EO 630 + 178, positioned to a few arcseconds by the high resolution imager, after initial source discovery by the imaging proportional counter; optical data from large ground-based telescopes, showing the subarcsecond position of G″, the ~ 26 mag best candidate. *Hubble Space Telescope* observations, for yet better sensitivity and resolution, should be available soon.

in binary systems, that may be emitting significant x-ray luminosity through the accretion process. However, these sources, as well as those in more crowded regions near the galactic center, are yet to be unambiguously identified and, thus, physically understood. As to "population," they are certainly not radio pulsars and, on a larger scale, they are not especially well correlated to x-ray bright objects.

GEMINGA

The truly special object in the list is 2CG 195+04 (Geminga), probably the single most important discovery of high-energy gamma-ray astronomy. Again first observed by the pioneering *SAS-2* mission, Geminga [a *gemin*i *ga*mma-ray source, from its location in the Gemini (twins) constellation] was located in the sky by *COS-B* with the unprecedented level of accuracy (for gamma-ray astronomy) of ~ 24 arcmin. The search for its counterpart at other wavelengths began immediately after it became clear that it could not be related to interstellar clouds, known radio pulsars, bright x-ray binary sources, or quasars. The search that evolved over more than a decade, involving many groups and different observations, has finally succeeded in identifying Geminga both at x-ray and optical wavelengths. Figure 3 shows, in a schematic way, the search and identification process, with increasing accuracy, down to G", the proposed optical counterpart. This object is thus far unique in the sky, in that it emits 1000 times more energy in gamma rays than in x-rays, and again 1000 times more in x-rays than in visible light, where, with apparent blue magnitude $m_B = 26.5$, it is one of the faintest "stars" ever observed. Although these properties are similar to those of the Vela pulsar, Geminga does not emit in radio frequencies, at least down to the minimum flux levels of searches performed with the most sensitive available instruments. Obviously some manifestation of a compact object or system, this new astronomical entity, discovered through its high-energy gamma-ray emission, awaits more sensitive data to be fully understood.

GAMMA-RAY BURSTERS

At different energies, gamma ray bursters, the other class of GRSs, also fit quite well into this last picture. Discovered in the early 1970s as a pure-science, astronomical by-product of military-oriented research, they are sudden enhancements of hard x-ray/soft gamma-ray flux coming from the sky. Their arrival directions are difficult to pinpoint because of the very transient nature of the phenomenon. Nevertheless, simultaneous observations by distant satellites, coupled with significant ingenuity in the data analysis methods, have led to several hundreds of acceptably accurate GRB positions, which appear to be uniformly distributed in the sky.

Currently, nearly two decades after their discovery, both the spatial distribution as well as the uncertainty in their distance represent a mystery. From their sky uniformity GRBs could be either a galactic population, so local (< 1 kpc) that no nonuniformity is introduced by the disk shape, or an extragalactic population, in this case beyond 30 Mpc, that is, beyond the local supercluster. This is reminiscent of, and actually even worse than, the great debate in the 1920s over the distance of the spiral nebulae, that was finally solved by finding suitable distance indicators. No such help appears available in the case of GRBs, although useful physical indicators point to their origin in a (necessarily galactic) population of neutron stars.

Given a sky distribution of arrival directions and fluxes, one can go beyond the simple test of measuring the degree of isotropy and apply the classical astronomical tools of the $\log N$–$\log S$ and V/V_{max} tests. The first represents the number (N)–flux (S) logarithmic distribution. As for every astronomical population, it can be useful in distinguishing the case of sources uniformly filling a euclidean three-dimensional volume from that of sources filling a disk-shaped volume like our own galaxy. Because, in the first case, the available volume increases like the third power of the distance, whereas the source flux decreases with the distance squared, the source number–flux relation will follow a characteristic $S^{-3/2}$ power law. For the second case, in the basically two-dimensional volume of a disk it is intuitive that the source number will decrease as S^{-1}, thus potentially providing a suggestion of the population origin. Unfortunately, observational selection effects are important at the faint end of the $\log N$–$\log S$ distribution, precisely the one to be used for measuring the power law slope. It appears that, at least for the current GRB data, the $\log N$–$\log S$ test is substantially inconclusive.

The V/V_{max} test, a slightly more elaborate algorithm previously used in quasar work, also checks the uniformity of the source volume occupation. The result of this test appears compatible with GRBs uniformly filling a three-dimensional euclidean space, in accordance with the observed arrival directions. In summary, not much can be learned on the nature of GRBs from their properties as a population. It is more from broad energetic considerations, time variability, and spectral properties that one associates GRBs with galactic neutron stars. These must be either isolated or accompany underluminous companions, because of the difficulties of associating GRBs with objects detected at other wavelengths.

The search for counterparts of GRBs is very reminiscent of the Geminga search, with similar ratios of powers emitted in gamma, x-ray, and visible-light frequencies. It thus seems reasonable to propose that a relationship may exist between the transient and steady GRSs, and that Geminga might represent the (so far) missing link between the two.

Additional Reading

Bignami, G. F. and Hermsen, W. (1983). Galactic gamma-ray sources. *Ann. Rev. Astron. Ap.* **21** 67.

Bloemen, H. (1989). Diffuse galactic gamma-ray emission. *Ann. Rev. Astron. Ap.* **27** 469.

Dogiel, V. A. and Ginzburg, V. L. (1989). Some problems in gamma-ray astronomy. *Space Sci. Rev.* **49** 311.

Fichtel, C. E. and Trombka, J. I. (1981). *Gamma-ray astrophysics.* NASA SP-453, Washington, D.C.

Ramana-Murthy, P. V. and Wolfendale, A. W. (1986). *Gamma Ray Astronomy*. Cambridge University Press, Cambridge.

Stecker, F. W. (1971). *Cosmic Gamma Rays*. NASA SP-249, Washington, D.C.

See also **Background Radiation, Gamma-Ray; Cosmic Rays, Acceleration; Cosmic Rays, Origin; Cosmic Rays, Propagation; Gamma-Ray Bursts, Observed Properties and Sources; Gamma-Ray Bursts, Optical Flashes; Pulsars, Observed Properties; Stars, Neutron, Physical Properties and Models; Telescopes, Detectors and Instruments, Gamma-Ray.**

Gravitational Lenses

John A. Peacock

Gravitational lenses are in a sense a prediction of Albert Einstein's theory of general relativity. Given the idea that light rays will be bent by the gravitational pull of a massive body (so spectacularly confirmed during the solar eclipse of 1919), it may seem obvious that a sufficiently massive body will allow light to pass from a source to an observer by more than one path, thereby forming a lens of sorts. However, it took until 1936 before Einstein himself published such an idea, and until 1979 before the effect was first seen by astronomers—60 years after that first detection of gravitational light deflection. Since then the subject has flourished because it provides an unmatched probe of the dark matter that we

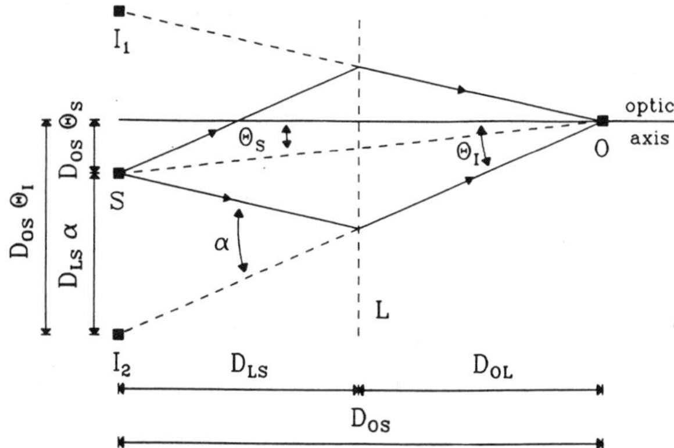

Figure 1. The lensing geometry. For a thin lens, deflection through the bend angle α may be taken as instantaneous. The angles θ_I and θ_S (two-dimensional in general) specify, respectively, the observed image position and the intrinsic source position of the object on the sky. [*Reproduced from Blandford and Kochanek (1987) by permission of the authors and publishers.*]

now believe dominates the mass content of the universe. Because the subject was purely theoretical for so long, we describe the theory first and then some of the lenses that have actually been found. Reviews containing background references for this material can be found in the additional reading at the end of this entry.

HOW LENSES WORK

Light Deflection

Light bends in response to a gravitational acceleration perpendicular to its path. The formula for this is

$$\alpha = \frac{2}{c^2} g\, dl,$$

where α is the angle of deflection, g is the acceleration, and dl is the length of path traversed. This is exactly twice the result obtained without using general relativity, and it provided a crucial test of that theory. The bending is not quite this simple in very strong gravitational fields, such as close to black holes, but this expression suffices for most purposes.

Lensing Geometry

The operation of a single gravitational lens is illustrated in Fig. 1. For simplicity, source, lens, and observer are shown lying in the same plane, although this is not generally the case. The parameters D are angular-diameter distances that may be related to redshift in a given cosmological model. The lens is *thin* when the light bending takes place within a small enough distance that the deflection can be taken as sudden. Thick lenses correspond to the combined effects of deflections at several different distances, and are harder to treat. Equating distances at the left-hand side of Fig. 1 yields the fundamental lensing equation

$$\alpha(D_L\theta_I) = \frac{D_S}{D_{LS}}(\theta_I - \theta_S),$$

where the angles θ_S, θ_I, and α are in general two-dimensional vectors on the sky. In principle, this equation can be solved to find the mapping between the object plane and the image plane [i.e., $\theta_I(\theta_S)$] and hence the positions of the images.

It is clear that gravitational lenses are not very appealing from an optician's point of view. A perfect lens would have a deflection angle that increases linearly with distance from the optic axis, whereas the opposite is often true with gravitational deflections. In the French literature, the term "gravitational mirage" is used, and this is arguably more appropriate.

Amplifications

The only other thing needed in order to find the appearance of lensed images is the fact that gravitational lensing does not alter surface brightness (rate of reception of energy per unit area of sky). This is well known to be a quantity that is independent of distance in euclidean space, but in fact is conserved generally. This comes about from the relativistic invariant I_ν/ν^3, which is proportional to the photon phase-space density and is conserved along light rays through Liouville's theorem. Hence, the amplification of image flux densities is given simply by a ratio of image areas: Larger images are brighter. In these circumstances, magnification might be a better word than amplification; however, one is often dealing with objects such as quasars where the image distortion cannot be observed; when the only observable is a change in the flux density of a point source, the term amplification is appropriate.

Image Structures

The solution of the lensing equation for a symmetrical lens may be visualized graphically as the intersection of a straight line with the bend-angle curve. Images are displaced from their intrinsic positions along radial lines; although this produces some radial distortion, the principal effect is of a transverse stretching, leading to crescent-shaped images. For cases of close alignment, the principal outer pair of images has an angular separation $(M/10^{10.49}\ M_\odot)^{1/2}([D_L D_S/D_{LS}]/\text{Gpc})^{-1/2}$ arcsec. In cases of perfect alignment, the image becomes perfectly circular, forming an Einstein ring of this diameter.

Although galaxies are not well described by point masses, in the case of circular symmetry, the size of the Einstein ring simply measures the projected mass interior to it. Thus, apart from uncertainties introduced by cosmology via the distance measures, lensing provides a clean and powerful method of estimating the masses of astronomical objects.

Odd-Number Theorem

One powerful theorem that governs the general appearance of the results of a lens event states that the number of separate images produced must be odd. The only exceptions to this occur when the lensing mass is singular (of infinite density at some point). This type of argument may, in principle, reveal the existence of a black hole or cosmic string.

Time Delays

A potentially important aspect of gravitational lensing is that the lens alters the time taken for light to reach the observer, producing a time lag between any multiple images. There are two effects at work: a geometrical delay caused by second-order increase in the path-length and a second term due to the gravitational potential (light travels more slowly in a gravitational field, partly due to gravitational time-dilation effects and partly due to spatial curvature). Sadly, although this second term is rather more model-dependent than the geometrical term, it is usually of the same order of magnitude. Where this is not the case, it would be possible to use time delays to measure the Hubble constant H_0 quite accurately. This arises because the geometry of the lensing event can usually be found from the observed images. The differential time delay between various light paths then scales with the overall size of the system. Because cosmological distances inferred from redshifts scale as H_0^{-1}, an estimate of H_0 then follows. For well-constrained lens models, one can do quite well in this respect; see the section below on observations.

Caustics and Catastrophes

Gravitational optics has an important connection with the branch of mathematics known as catastrophe theory. This provides a powerful means of analyzing general properties of lenses by concentrating on the mapping from the sky before lensing to that after lensing. The most important concept is that of critical lines or *caustics*: source positions where the amplification becomes infinite. New multiple images are created or destroyed in pairs as the source crosses a caustic. This allows important general properties to be proved: In particular, the cross section or probability of production of a high-amplification event behaves as $\sigma(>A) \propto A^{-2}$.

Statistical Lensing

Because many classes of astronomical object have a distribution of luminosities where faint objects far outnumber bright ones, the nonzero probability of high amplification alluded to previously opens the possibility that selection effects can cause lensed objects to dominate. This will tend to happen whenever the intrinsic luminosity function is as steep or steeper than the power-law tail of the lensing distribution: $\rho \propto L^{-2}$ in integral terms. It can be shown that the probability of strong lensing of a quasar at high redshift is roughly Ω, the density parameter of the universe. Because the contribution to Ω by the cores of galaxies and rich clusters is small, such events are rare. However, they may be much more common if we consider *microlensing* by objects of much lower mass. This is because lines of sight through galaxy haloes may pass many potential lenses in the form of stars; also, there is the exciting possibility that the dominant dark matter may consist of clumps. For quasars, where the intrinsic angular sizes are very small, effective lenses can be as small as 1% of a solar mass. Microlensing can thus be difficult to prove, but does leave a characteristic signature in variability as the stellar lens moves.

OBSERVATIONS OF LENSES

Observed lens events may conveniently be divided into two according to the type of object that is being lensed: ordinary galaxies or rare objects (active galaxies and quasars). Lensing in the latter is easier to recognize, not only because quasars can readily be detected to very high redshifts, but also because the typical separation of quasars on the sky is very large by comparison with the angular scale of lensing effects. A close pair of quasars then quickly raises the suspicion of lensing, whereas galaxy pairs are very common.

Multiple Quasars

In fact, the first detection of lensing was made when optical identification of a radio source yielded two stellar images 6 arcsec apart. Normally, one would be a quasar, the other a foreground galactic star. It is to the great credit of the discoverers that they took a spectrum of the second object even when the first had turned out to be "the" quasar. Similar objects have since been found at the rate of about one per year. The slow pace of discovery reflects not only the intrinsic rareness, but also the necessity for caution: True physical pairs of quasars are not completely unknown, and quite accurate spectroscopy is required to establish that the spectra are precisely identical in shape and redshift (give or take the effects of variability and differential time delays).

Probably the best example of this genre is still the first one: the double quasar 0957+561. This only displays two images, but is important because it is a bright radio source. Both images are bright enough to allow the position angle of the nuclear jets to be measured via long-baseline interferometry; this information sets much tighter constraints on the lensing potential than is available just from image positions and luminosities. Long-term monitoring has also revealed correlated variability in the two images, establishing the differential time delay between the two paths to be about 420 days. For the best lens models, this corresponds to a Hubble constant of about 60 km s^{-1} Mpc^{-1}, but the true uncertainty in this will probably not reveal itself until this estimate can be repeated using other lenses.

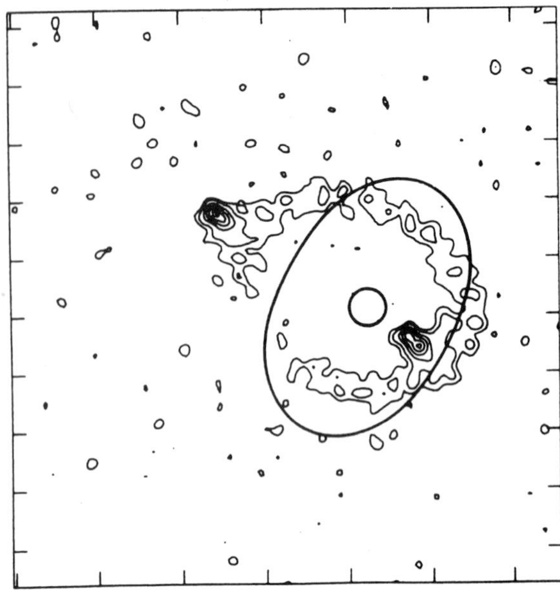

Figure 2. An example of a gravitationally lensed extragalactic object. The source 1131 + 0456 is imaged into a near-perfect ring. The 15-GHz map is 5 arcsec square and shows the critical lines for obtaining three and five images. [*Reproduced from Kochanek et al. (1989)* Mon. Not. Roy. Astron. Soc. **238** *43, by permission of the authors and publisher.*]

Einstein Rings

Searches for lensed quasars in the radiowave band have also revealed more spectacular objects. Several thousand radio sources have been examined in the hope of finding multiple point images suggestive of lensed quasars, but in many cases the radio sources possess extended lobes that improve the chance of good alignment with a lensing galaxy. Two cases have been found to date of such lobes being imaged into a near-perfect Einstein ring. One of these is illustrated in Fig. 2. Here, the proof of lensing is not redshift coincidence because the radio emission has no spectral lines, but arises in showing that the ring can be deprojected to a simple source structure.

Cluster Arcs

There are also several cases known where high-redshift galaxies have been strongly lensed by clusters of galaxies into arcs tens of arcseconds in extent. Here, the confirmation of lensing consists of showing that all the arc has the same redshift. Such lenses that probe the cluster mass distribution are very useful because clusters are the largest relaxed self-gravitating systems that exist. Deducing their masses from galaxy velocities and x-ray data on intracluster gas is of great cosmological importance; lensing adds an invaluable extra constraint.

Quasar–Galaxy Associations

Even when no multiple imaging occurs, quasars seen through the outer parts of galaxy haloes may be amplified (especially if microlensing is important). This can lead to an apparent association between quasars and low-redshift galaxies. At the time of writing, evidence is starting to emerge for the existence of such associations. Although it is hard to prove that one is seeing lensing (rather than, say, noncosmological redshifts), this type of observation has promise statistically, because weak lens events are so much more common than strong ones.

THE FUTURE

There are signs that the number of known lenses will increase more rapidly in the future. This will come not through searches for arcs in clusters, which will always be hard to recognize, but through automated surveys for quasars and radio sources using either radio maps or optical quasar searches. Present statistics suggest about one quasar in 1000 is lensed, so there must be thousands of lenses awaiting discovery. The problem is that only the bright ones are really useful for the detailed modeling that may yield the distance scale. Nevertheless, statistical studies may be just as important for understanding dark matter. In particular, lensing may be the very best way of detecting cosmic strings: relics from the early phases of the Big Bang that would manifest themselves as a line of double galaxy images. Detection of even one string would be sufficient cosmological importance to justify any amount of effort in lens searches. Come what may, then, gravitational lensing seems likely to remain an important cosmological tool for the foreseeable future.

Additional Reading

Blandford, R. D. (1990). Gravitational lenses. *Q. J. Royal Astron. Soc.* **31** 305.

Blandford, R. D. and Kochanek, C. S. (1987). *Dark Matter in the Universe*, J. Bahcall, T. Piran, and S. Weinberg, eds. World Scientific, Singapore, p. 133.

Peacock, J. A. (1983). *Quasars and Gravitational Lenses*, J.-P. Swings, ed. Universite Liège, p. 86.

Schneider, P., Ehlers, J., and Falco, E. E. (1991). *Gravitational Lenses*. Springer, Heidelberg.

Turner, E. L. (1988). Gravitational lenses. *Scientific American* **259** (No. 1) 54.

See also **Cosmology, Cosmic Strings; Gravitational Theories.**

Gravitational Radiation

Stephen P. Boughn

In the language of Albert Einstein's general theory of relativity, gravitational radiation or gravitational waves (GWs) are "ripples in the geometry of space and time." A less abstruse way to describe gravitational radiation is by drawing an analogy to the electromagnetic spectrum (light, infrared, radio, microwave, x-ray, etc.). Just as these represent forms of the free radiation or waves associated with electricity and magnetism, so GWs represent the radiation associated with the force of gravity. Einstein actually predicted their existence in 1916, the same year his paper on general relativity theory was published. He even calculated the radiation emitted from a binary star system (the strongest source known at the time) and concluded that the radiation was so weak that it had "a negligible practical effect." For the next half century gravitational radiation remained a theoretical curiosity that was of no practical astrophysical significance. In the last two decades astrophysicists have discovered several new potential sources and have come to believe that it not only may be possible to detect gravitational waves directly but also that their emission may even be the dominant process in the evolution of some astrophysical objects.

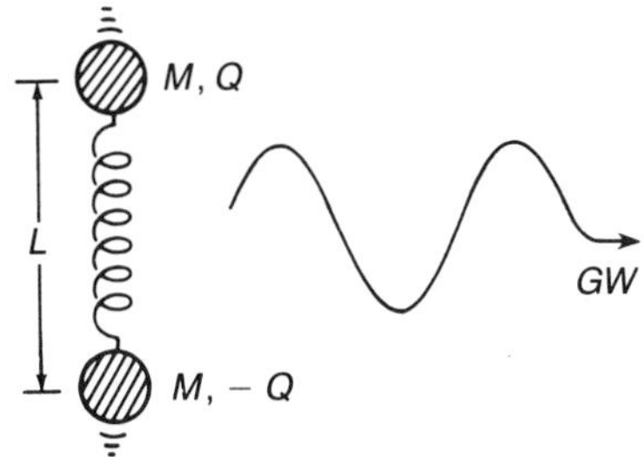

Figure 1. Oscillating masses (electric charges) are a source of gravitational (electromagnetic) waves. This same system can also absorb energy from an incident gravitational (electromagnetic) wave.

GENERATION AND DETECTION

Electromagnetism and gravity are the only two fundamental, long-range forces in nature. Just as accelerated electric charges generate electromagnetic radiation, so do accelerated "gravitational charges," that is, masses, generate gravitational radiation. Simply by analogy with electromagnetism, it is not surprising that gravitational waves are predicted by general relativity and every other viable theory of gravity. Furthermore, if gravity is to obey the laws of Einstein's special theory of relativity, then gravitational radiation must travel at the speed of light.

Because all electric charges have mass, one might expect gravitational radiation to be as abundant as electromagnetic radiation; however, this is not the case. Consider the system indicated in Fig. 1 that depicts two particles of the same mass M and opposite electric charge $\pm Q$ oscillating at opposite ends of a spring of length L. The ratio of the power P_G emitted in gravitational radiation to the power P_{EM} emitted in electromagnetic radiation by this system is

$$P_G/P_{EM} = (GM^2/Q^2)(\pi L/\lambda)^2$$

where λ is the wavelength of the radiation. If the particles are electrons (which are responsible for most of the electromagnetic radiation we observe), the first factor alone is 10^{-43}, which illustrates the incredible weakness of gravity. The second term in the preceding equation is proportional to the square of the ratio of the speed of the masses to the speed of light and is always less than 1. From this example it is clear that large, rapidly moving masses are the best sources of GWs.

Electromagnetic radiation is detected by a wide variety of instruments, all of which operate on the same principle: electromagnetic waves exert force on electric charges. Likewise, gravitational radiation can be detected by the force it exerts on masses. If a GW is incident on the spring system of Fig. 1 the masses will be driven into oscillation. The amplitude of the oscillations, however, is very small, again because of the weakness of gravity. In fact, the ratio of the energy absorbed from a gravitational wave to the energy absorbed from an electromagnetic wave of equal strength is given by exactly the same ratio as in the equation.

As is characteristic of electromagnetic waves, GWs have two possible polarizations and exert force on matter only in directions perpendicular to the direction of propagation of the wave. The energy in the wave decreases inversely with the square of the distance from the source. For a polarized GW the lines of force in the plane perpendicular to the direction of propagation are illustrated in Fig. 2. The resultant accelerations of four test masses, A,

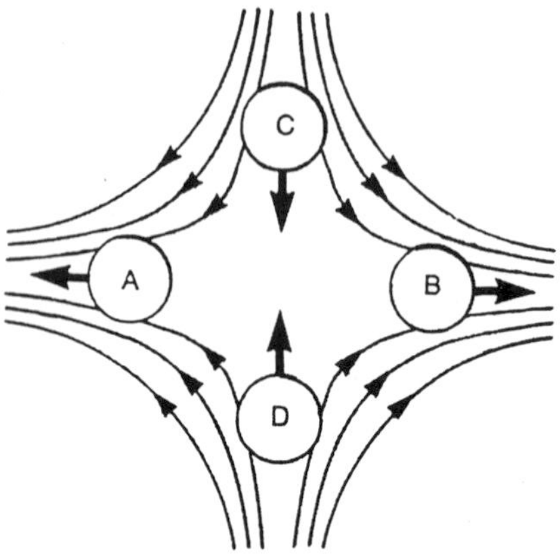

Figure 2. Lines of force of a plane, polarized gravitational wave in the plane perpendicular to the direction of propagation of the wave. The resultant accelerations of four test masses, A, B, C, and D are indicated by bold arrows.

B, C, and D, are indicated by arrows. One-half cycle later the directions of the lines of force and accelerations are reversed. A representation of the other possible polarization is obtained by rotating the figure by 45°. The primitive gravitational wave detector of Fig. 1 is obtained by simply connecting masses A and B (or C and D) by a spring.

SOURCES

Considering the inherent weakness of gravity, laboratory sources of gravitational radiation are nonexistent. For example, a 1-ton steel bar spun so rapidly that it is on the verge of being ripped apart by centrifugal force radiates less than 10^{-30} W. (This problem was considered by Einstein in 1918.) By contrast, existing detectors are only sensitive enough to detect such a source (at a distance of one wavelength) if it emits more than 10^{6} W. Current hopes for directly detecting GWs are pinned on astrophysical sources in which massive bodies undergo tremendous accelerations. Short term binary star systems emit strongly (10^{25}–10^{29} W) at frequencies from 10^{-4}–10^{-3} mHz, but there are currently no detectors capable of detecting such sources even nearby. Even though not directly detectable, the energy loss from the short term binary pulsar PSR 1913+16 due to gravitational radiation has been measured by precise timing observations of the pulsar orbit. These measurements agree with the predictions of general relativity to within 1%, a result which is an important confirmation of the existence of GWs. A much stronger source is gravitational collapse to a black hole during which a large fraction of the mass of an entire star may be accelerated to velocities approaching the speed of light. It is expected that as much as 10^{49} W of GWs will be emitted from such a source in the form of a pulse of duration 0.001 s. It has also been conjectured that massive black holes (10^{5}–10^{9} $M_{\odot}$) are located in the nuclei of many galaxies and quasars. During their formation these objects could emit as much as 10^{54} J of energy in the form of GWs at frequencies between 10^{-1} and 10^{-5} Hz. Several more exotic sources of low frequency gravitational radiation have been suggested, for example, quantum gravity fluctuations in the early universe, a quantum chromodynamic phase transition from free quarks into nucleons and oscillating cosmic strings. The period of GWs from these sources range from days to years and are out of the range of laboratory detectors.

DETECTORS

The first attempts to detect gravitational waves were made in the 1960s by Joseph Weber who acoustically suspended a massive (1400-kg) aluminum cylinder and monitored the level of excitation of its lowest vibrational mode. Resonant at 1660 Hz, this detector was designed to be sensitive to the burst of gravitational radiation predicted to accompany the formation of a solar mass black hole. His 1969 claim of detection of GWs was not confirmed and it is now the consensus of opinion within the scientific community that GWs have not yet been directly observed. Several cryogenically cooled cylinders are now in operation with a sensitivity adequate to detect the GWs emitted from gravitational collapse occurring anywhere in the Milky Way galaxy. None has been detected. Laser interferometers that detect the relative displacement of two mirrors (separated by hundreds of meters) of a Michelson interferometer are potentially more sensitive detectors that respond to a wider range of frequencies. It is hoped that several of these instruments will be in operation in the 1990s with a sensitivity sufficient to detect gravitational collapse in galaxies outside our own. At lower frequencies (10^{-3} Hz), radar ranging of spacecraft and the monitoring of solid Earth vibrations have been employed but these methods are not yet sensitive enough to detect such events as the formation of massive black holes in distant galaxies. The effect of GWs on the arrival times of pulsars has been used to place upper limits on a background of GWs with periods of a few years. These limits already have provided important constraints on some models of cosmic strings.

GRAVITATIONAL WAVE ASTRONOMY

The observations of the binary pulsar PSR 1913+16 have established the existence of gravitational radiation, but as an astrophysical source of GWs this system is the least interesting object imaginable, that is, two point particles in orbit around one another. The main motivation of researchers in the field is not simply to observe GWs directly and thereby confirm their existence, but rather to be able to use them to probe deeply into the regions of strong gravitational fields and dense matter that may block other forms of radiation. Only when GWs are detected from one or more of the spectacular astrophysical phenomena mentioned above will gravitational wave astronomy be considered a legitimate branch of astrophysics.

Additional Reading

Boughn, S. P. (1979). Detecting gravitational waves. *American Scientist* **68** 174.

Davies, P. C. W. (1980). *The Search for Gravity Waves.* Cambridge University Press, Cambridge.

Jeffries, A. D., Saulson, P. R., Spero, R. E., and Zucker, M. E. (1987). Gravitational wave observatories. *Scientific American* **256** (No. 6) 50.

Misner, C. W., Thorne, K. S., and Wheeler, J. A. (1973). *Gravitation.* W. H. Freeman, San Francisco.

Weisberg, J. M., Taylor, J. H., and Fowler, L. A. (1981). Gravitational waves from an orbiting pulsar. *Scientific American* **245** (No. 4) 74.

See also **Black Holes, Theory; Cosmology, Cosmic Strings; Gravitational Theories; Pulsars, Binary; Pulsars, Millisecond; Stars, Neutron, Physical Properties and Models.**

Gravitational Theories

Jacob D. Bekenstein

A theory of gravitation is a description of the long range forces that electrically neutral bodies exert on one another because of their matter content. Until the 1910s Sir Isaac Newton's law of universal gravitation, *two particles attract each other with a central force proportional to the product of their masses and inversely proportional to the square of the distance between them*, was accepted as the correct and complete theory of gravitation: The proportionality constant here is Newton's constant $G = 6.67 \times 10^{-8}$ dyn cm^2 g^{-2}, also called the gravitational constant. This theory is highly accurate in its predictions regarding everyday phenomena. However, high precision measurements of motions in the solar system and in binary pulsars, the structure of black holes, and the expansion of the universe can only be fully understood in terms of a relativistic theory of gravitation. Best known of these is Albert Einstein's general theory of relativity, which reduces to Newton's theory in a certain limit. Of the scores of rivals to general relativity formulated over the last half century, many have failed various experimental tests, but the verdict is not yet in on which extant relativistic gravitation theory is closest to the truth.

NEWTONIAN GRAVITATIONAL THEORY

In alternative language, newtonian gravitational theory states that the acceleration $\mathbf{a}$ (the rate of change of the velocity $\mathbf{v}$) imparted by gravitation on a test particle is determined by the gravitational potential ϕ,

$$\mathbf{a} = -\frac{d\mathbf{v}}{dt} = -\nabla\phi,$$

and the potential is determined by the surrounding mass distribution ρ by Poisson's partial differential equation

$$\nabla \cdot \nabla\phi = 4\pi G\rho.$$

This formulation is entirely equivalent to Newton's law of gravitation. Because a test particle's acceleration depends only on the potential generated by matter in the surroundings, the theory respects the weak equivalence principle: *the motion of a particle is independent of its internal structure or composition*. As the subject of Galileo Galilei's apocryphal experiment at the tower of Pisa, this principle is supported by a series of high precision experiments culminating in those directed by Baron Lorand von Eötvos in Budapest in 1922, Robert Dicke at Princeton in 1964, and Vladimir Braginsky in Moscow in 1972.

Highly successful in everyday applications, newtonian gravitation has also proved accurate in describing motions in the solar system (except for tiny relativistic effects), the internal structure of planets, the sun and other stars, orbits in binary and multiple stellar systems, the structure of molecular clouds, and, in a rough way, the structure of galaxies and clusters of galaxies (but see below).

THE GENERAL THEORY OF RELATIVITY

According to newtonian theory, gravitational effects propagate from place to place instantaneously. With the advent of Einstein's special theory of relativity in 1905, a theory uniting the concepts of space and time into that of four dimensional flat space-time (named Minkowski space-time after the mathematician Hermann Minkowski), a problem became discernible with newtonian theory. According to special relativity, which is the current guideline to the form of all physical theory, the speed of light, $c = 3 \times 10^{10}$ cm s^{-1}, is the top speed allowed to physical particles or forces: There can be no instantaneous propagation. After a decade of search for new concepts to make gravitational theory compatible with the spirit of special relativity, Einstein came up with the theory of general relativity (1915), the prototype of all modern gravitational theories. Its crucial ingredient, involving a colossal intellectual jump, is the concept of gravitation, not as a force, but as a manifestation of the curvature of space-time, an idea first mentioned in rudimentary form by the mathematician Georg Bernhard Riemann in 1854. In Einstein's hands gravitation theory was thus transformed from a theory of forces into the first dynamical theory of geometry, the geometry of four dimensional curved space-time.

Why talk of curvature? One of Einstein's first predictions was the gravitational redshift: As any wave, such as light, propagates away from a gravitating mass, all frequencies in it are reduced by an amount proportional to the change in gravitational potential experienced by the wave. This redshift has been measured in the laboratory, in solar observations, and by means of high precision clocks flown in airplanes. However, imagine for a moment that general relativity had not yet been invented, but the redshift has already been measured. According to a simple argument owing to Alfred Schild, wave propagation under stationary circumstances can display a redshift only if the usual geometric relations implicit in Minkowski space-time are violated: The space-time must be curved. The observations of the redshift thus show that space-time must be curved in the vicinity of masses, regardless of the precise form of the gravitational theory.

Einstein provided 10 equations relating the metric (a tensor with 10 independent components describing the geometry of space-time) to the material energy momentum tensor (also composed of 10 components, one of which corresponds to our previous ρ). These Einstein field equations, in which both of the previously mentioned constants G and c figure as parameters, replace Poisson's equation. Einstein also replaced the newtonian law of motion by the statement that free test particles move along geodesics, the shortest curves in the space-time geometry. The influential gravitation theorist John Archibald Wheeler has encapsulated general relativity in the aphorism "curvature tells matter how to move, and matter tells space-time how to curve." The Eötvos–Dicke–Braginsky experiments demonstrate with high precision that free test particles all travel along the same trajectories in space-time, whereas the gravitational redshift shows (with more modest precision) these universal trajectories to be identical with geodesics.

Despite the great contrast between general relativity and newtonian theory, predictions of the former approach the latter for systems in which velocities are small compared to c and gravitational potentials are weak enough that they cannot cause larger velocities. This is why we can discuss with newtonian theory the structure of the earth and planets, stars and stellar clusters, and the gross features of motions in the solar system without fear of error.

Einstein noted two other predictions of general relativity. First, light beams passing near a gravitating body must suffer a slight deflection proportional to that body's mass. First verified by observations of stellar images during the 1919 total solar eclipse, this effect also causes deflection of quasar radio images by the sun, is the likely cause of the phenomenon of "double quasars" with identical redshift and of the recently discovered giant arcs in clusters of galaxies (both probably effects of gravitational lensing), and is part and parcel of the black hole phenomenon. In a closely related effect first noted by Irwin Shapiro, radiation passing near a gravitating body is delayed in its flight in proportion to the body's mass, a time delay verified by means of radar waves deflected by the sun on their way from Earth to Mercury and back.

The second effect is the precession of the periastron of a binary system. According to newtonian gravitation, the orbit of each member of a binary is a coplanar ellipse with orientation fixed in space. General relativity predicts a slow rotation of the ellipse's major axis in the plane of the orbit (precession of the periastron). Originally verified in the motion of Mercury, the precession has of late also been detected in the orbits of binary pulsars.

All three effects mentioned depend on features of general relativity beyond the weak equivalence principle. Indeed, Einstein built into general relativity the much more encompassing "strong equivalence principle": *the local forms of all nongravitational physical laws and the numerical values of all dimensionless physical constants are the same in the presence of a gravitational field as in its absence*. In practice this implies that within any region in a gravitational field, sufficiently small that space-time curvature may be ignored, all physical laws, when expressed in terms of the space-time metric, have the same forms as required by special relativity in terms of the metric of Minkowski space-time. Thus in a small region in the neighborhood of a black hole (the source of a strong gravitational field) we would describe electromagnetism and optics with the same Maxwell equations used in earthly laboratories where the gravitational field is weak, and we would employ the laboratory values of the electrical permittivity and magnetic susceptibility of the vacuum.

SCALAR TENSOR THEORIES

The strong equivalence principle effectively forces gravitational theory to be general relativity. Less well tested than the weak version of the principle mentioned earlier, the strong version requires Newton's constant expressed in atomic units to be the same number everywhere, in strong or weak gravitational fields. Stressing that there is very little experimental evidence bearing on this assertion, Dicke and his student Carl Brans proposed in 1961 a modification of general relativity akin to a theory considered earlier by Pascual Jordan. In the Brans–Dicke theory the reciprocal of the gravitational constant is itself a one-component field, the scalar field Φ, that is generated by matter in accordance with an additional equation. Then Φ as well as matter has a say in determining the metric via a modified version of Einstein's equations. Because it involves both metric and scalar fields, the Brans–Dicke theory is dubbed scalar–tensor. Although not complying with the strong equivalence principle, the theory does respect a milder version of it, the Einstein equivalence principle, which asserts that only nongravitational laws and dimensionless constants have their special relativistic forms and values everywhere. Gravitation theorists call theories obeying the Einstein equivalence principle metric theories.

The Brans–Dicke theory also reduces to newtonian theory for systems with small velocities and weak potentials: It has a newtonian limit. In fact, Brans–Dicke theory is distinguishable from general relativity only by the value of its single dimensionless parameter ω which determines the effectiveness of matter in producing Φ. The larger ω, the closer the Brans–Dicke theory predictions are to general relativity. Both theories predict the same gravitational redshift effect, although they predict slightly different light deflection and periastron precession effects; the differences vanish in the limit of infinite ω. Measurements of Mercury's perihelion precession, radar flight time delay, and radio wave deflection by the sun indicate that ω is at least several hundred.

Initially a popular alternative to general relativity, the Brans–Dicke theory lost favor as it became clear that ω must be very large—an artificial requirement in some views. Nevertheless, the theory has remained a paradigm for the introduction of scalar fields into gravitational theory, and as such has enjoyed a renaissance in connection with theories of higher dimensional space-time.

However, constancy of ω is not conceptually required. In the generic scalar–tensor theory studied by Peter Bergmann, Robert Wagoner, and Kenneth Nordtvedt, ω is itself a general function of $\omega(\Phi)$. It remains true that in regions of space-time where $\omega(\Phi)$ is numerically large, the theory's predictions approach those of general relativity. It is even possible for $\omega(\Phi)$ to evolve systematically in the favored direction. Thus in the variable mass theory (VMT, see Table 1), a scalar–tensor theory devised to test the necessity for the strong equivalence principle, the expansion of the universe forces evolution of Φ toward a particular value at which $\omega(\Phi)$ diverges. Thus, late in the history of the universe (and today is late), localized gravitational systems are accurately described by general relativity although the assumed gravitational theory is scalar–tensor.

OTHER THEORIES

More than two score relativistic theories of gravitation have been proposed. Some have no metric; others take the metric as fixed, not dynamic. These have usually fared badly in light of experiment. Among metric theories those involving a vector field or a tensor field additional to the metric can display a preferred frame of reference or spatial anisotropy effects (phenomena that depend on direction in space). Both effects may contradict a variety of modern experiments. Table 1 gives a sample of theories of gravitation, summarizing the main ingredients of each theory and its experimental status.

Table 1. Comparison of Selected Gravitational Theories

Theory	*Metric*	*Other Fields*	*Free Elements*	*Status*
Newtonian (1687)[1]	Nonmetric	Potential	None	Nonrelativistic theory
Nordstrom (1913)[1,2]	Minkowski	Scalar	None	Fails to predict observed light deflection
Einstein's General Relativity (1915)[1,2]	Dynamic	None	None	Viable
Belinfante–Swihart (1957)[2]	Nonmetric	Tensor	*K* parameter	Contradicted by Dicke–Braginsky experiments
Brans–Dicke (1961)[1-3]	Dynamic	Scalar	ω parameter	Viable for $\omega > 500$
Generic Scalar Tensor (1970)[2]	Dynamic	Scalar	2 free functions	Viable
Ni (1970)[1,2]	Minkowski	Tensor, Vector, and scalar	One parameter 3 free functions	Predicts unobserved preferred-frame effects
Will–Nordtvedt (1972)[2]	Dynamic	Vector	None	Viable
Rosen (1973)[2]	Fixed	Tensor	None	Contradicted by binary pulsar data
Rastall (1976)[2]	Minkowski	Tensor, vector	None	Viable
VMT (1977)[2]	Dynamic	Scalar	2 parameters	Viable for a wide range of the parameters
MOND (1983)[4]	Nonmetric	Potential	Free function	Nonrelativistic theory

[1] Misner, Thorne, and Wheeler (1973).
[2] Will (1981).
[3] Dicke (1965).
[4] Milgrom (1989).

All relativistic gravitational theories mentioned so far have a newtonian limit, a tacit requirement of candidate relativistic gravitational theories until very recently. Now, if the correct gravitational theory is general relativity or any of its traditional imitations, then newtonian theory should satisfactorily describe galaxies and clusters of galaxies, astrophysical systems involving small velocities and weak potentials. But there is mounting observational evidence that this can be the case only if galaxies and clusters of galaxies are postulated to contain large amounts of dark matter. Thus far this dark matter has not been detected independently of the preceding argument.

Might not this missing mass puzzle signal instead the breakdown of the newtonian limit of gravitational theory for very large systems? In this connection several schemes alternative to Newtonian theory have been proposed. A well developed one is the modified newtonian dynamics or MOND (see Table 1), in which the relation between newtonian potential ϕ and the resulting acceleration is regarded as departing from newtonian form for gravitational fields with magnitude of $\nabla\phi$ below 10^{-8} cm s^{-2}. In galaxies and clusters of galaxies (with no dark matter assumed) the gravitational fields are weaker than this, and a breakdown of newtonian predictions having nothing to do with dark matter is expected. With its one postulated relation, MOND ties together a number of empirical relations in extragalactic astronomy. A nonrelativistic gravitational theory containing the MOND relation has been set forth, and relativistic generalizations of these ideas are currently under study.

Additional Reading

Dicke, R. (1965). *The Theoretical Significance of Experimental Relativity*. Gordon and Breach, New York.

Milgrom, M. (1989). Alternatives to dark matter. *Comments Astrophysics* **13** 215.

Misner, C. W., Thorne, K. S., and Wheeler, J. A. (1973). *Gravitation*. W. H. Freeman, San Francisco.

Will, C. (1981). *Theory and Experiment in Gravitational Physics*. Cambridge University Press, Cambridge.

Will, C. (1986). *Was Einstein Right?* Basic Books, New York.

See also **Black Holes, Stellar, Observational Evidence; Black Holes, Theory; Dark Matter, Cosmological; Gravitational Lenses; Missing Mass, Galactic; Pulsars, Binary; Stars, Neutron, Physical Properties and Models.**

H

H II Regions

Frank P. Israel

Early type (OB) stars hotter than about 25,000 K are usually surrounded by ionized regions called H II regions or Strömgren spheres. In reality, these regions contain dust, He^+, and other ionized elements in addition to H^+, are irregular in shape, and very inhomogeneous in structure. Often an H II region is excited by more than one star, further complicating the picture. Because hot ionizing stars live for a comparatively short time, H II regions are, astronomically speaking, of recent origin (at most 10 million years old). For this reason, H II regions also indicate where hot and massive stars form *at the present time*.

Diffuse nebulae such as the Orion nebula had been observed for centuries, but it was not until 1939 that Bengt Strömgren first explained their basic structure, showing how to interpret the observations quantitatively. Technological progress following World War II, especially in spectroscopy and radio astronomy, made much more detailed observations possible. The advent of aperture synthesis radio telescopes around 1970 allowed high resolution studies of H II regions to be made, unhampered by optical extinction. Since then millimeter-wave observations have shown that all but the most evolved H II regions are intimately associated with dark and massive molecular cloud complexes.

Observations of H II regions provide astronomers with information on their physical properties, evolution in time, and occurrence throughout galaxies. This can be used to find where and at what rate stars are being formed in galaxies. H II region observations also provide astronomers with the abundance (relative to hydrogen) of elements, notably He, C, N, and O as a function of galaxy type and of position in a galaxy. This in turn yields important information on galaxy evolution. Also, H II region sizes are sometimes used to estimate distances of galaxies.

H II REGION PARAMETERS

H II regions are studied by observing line emission resulting from bound–bound transitions in the hydrogen atom (optical Balmer lines, infrared Brackett lines, radio recombination lines) and in other elements, as well as by observing radio continuum emission resulting from free–free transitions. In the Galaxy, distances of H II regions are found from the properties of their ionizing stars, or the Doppler shift of their emission lines together with a galactic rotation model. Sizes are measured directly, whereas optical emission line ratios yield density ([O II] and [S II] doublets) and temperature ([O II]/[O III] and [N II]/[N III] ratios). Mean densities also follow from radio continuum observations, and temperatures follow from radio recombination lines. The observed amount of line or continuum radiation emitted by an H II region measures the rate at which the exciting stars produce ionizing ultraviolet photons, and hence yields information on the temperature (spectral type) and number of these stars. Determined with sufficiently high resolution, the observed temperature, velocity, and density distributions of hydrogen and other elements are used to construct elaborate models of the complex physical and excitation structure of H II regions. From these, the time evolution of H II regions can be reconstructed.

COMPACT H II REGIONS

Hot, massive stars form in molecular clouds, ionize their dense immediate surroundings, and heat nearby dust. The result is an ultracompact H II region less than 0.1 pc in size and of density greater than 10^4 cm^{-3}, surrounded by a dense, obscuring cocoon of warm dust. Thus, ultracompact H II regions can only be observed at radio or infrared wavelengths. They are often associated with OH or H_2O masers. Because of its high temperature and density, the ultracompact H II region expands into the equally dense, but much cooler molecular cloud (see Fig. 1): it excavates an ionized cavity. After 10^4–10^5 years, the cavity breaks through the nearest molecular cloud edge. With speeds up to 20 km s^{-1}, ionized material of high temperature and density flows through the break into the less dense surrounding medium, not unlike champagne coming out of a freshly opened bottle. At this stage, the H II region becomes optically visible as a compact H II region (size typically 0.5 pc; density 5000 cm^{-3}). The champagne concept has been successfully used to describe the flows observed in such H II regions. Because of the flow, the density in the cavity decreases, so that more ionizing photons from the exciting star can reach the ionization front that forms the boundary between the H II region and the molecular cloud. The ionization front moves slowly (typically at 1 km s^{-1}) into the molecular cloud. Neutral material is ionized and fed through the ionization front into the ionized cavity from which it streams into interstellar space. The cavity shows a steep gradient in density, highest at the ionization front, decreasing outward, and finally merging into the general interstellar medium, without a clear boundary. The Orion nebula is such a compact H II region. It also hides an ultracompact H II region, known as the Becklin–Neugebauer object (BN) at its edge.

FURTHER EVOLUTION OF H II REGIONS

As the H II region cavity further expands, it develops into a classical H II region such as the Omega nebula with typical densities of a few hundred hydrogen atoms per cubic centimeter or less and sizes of a few parsecs. Its morphology is that of a blister on the skin of a much larger molecular cloud. This (simplified) blister model is useful to describe several observed properties, such as excitation and density structures. The majority of optically visible normal emission nebulae have such a blister structure, albeit with many irregularities introduced by dense clumps, shock fronts, and additional stars. A consequence of the champagne and blister concepts is that H II regions cannot be considered as isolated entities. Their properties and evolution are closely tied to those of the (invisible) neutral atomic and molecular surroundings, and observations of these surroundings are essential to obtain the full picture.

As stars tend to form in groups, further expansion often leads to the overlap of different blisters, forming even more complex structures. This is characteristic of large H II regions (size typically 10 pc; density 50 cm^{-3}) such as the North America nebula. Finally, ultraviolet photons escaping from the blisters cause the tenuous interstellar medium to become ionized to great distances, leading to common ionized envelopes for groups of H II regions. Such giant H II regions (size 100–1000 pc; density below 10 cm^{-3}) are often seen in other gas-rich galaxies (30 Doradus in the Large Magellanic Cloud; NGC 5461 in M101) and signify the presence of as many as several thousand hot, young stars dominating a large section of a spiral arm. After about 10 million years, the exciting stars of evolved H II regions have almost completely eroded their molecular clouds, and start to die themselves; their H II regions expand ever further into the interstellar medium and fade away. Elsewhere, new hot stars and new H II regions have been formed, starting the process anew.

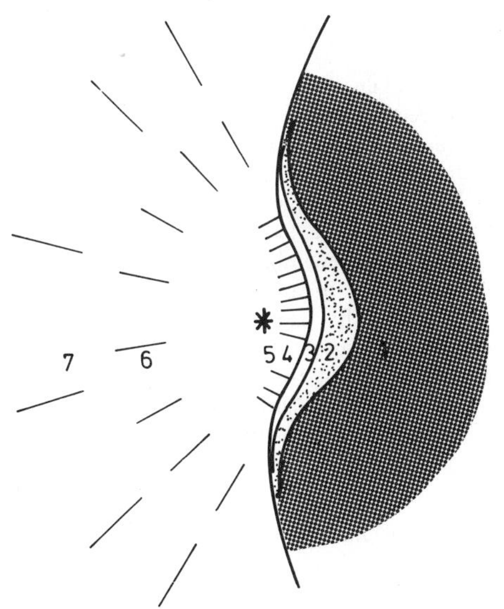

Figure 1. Schematic model of an H II region (diffuse nebula) in the blister phase. 1. Cold and dense, mostly molecular cloud. 2. Warm and dense infrared-emitting dust in the molecular cloud. 3. Very dense ionized gas and hot dust in thin ionization front on the skin of the molecular cloud. 4. Dense ionized gas and hot dust flowing away from the ionization front into the blister cavity. 5. Exciting star in the blister cavity. 6. Moderately dense ionized gas streaming out of the blister cavity and away from the molecular cloud edge. 7. Low-density ionized gas expanding into tenuous neutral atomic surroundings.

H II REGIONS IN GALAXIES

Optical telescopes can detect H II regions about a hundred times less intense than the weakest seen with radio telescopes. Most of the optical emission from stars and nebulae in the Galaxy is, however, obscured by dust clouds. Thus, radio measurements are the only way to determine the overall content and distribution of galactic H II regions. They show the Galaxy to contain several hundred large and giant H II region complexes, and many more small ones. These are concentrated between distances of 4 and 7 kpc from the galactic center, where giant molecular clouds are also concentrated (the "molecular ring"). From 4500 K at distances of 4 kpc from the galactic center, H II region (electron) temperatures increase to 10,000 K at 14 kpc. This is a consequence of the observed decrease in abundances of elements such as O and N (hence a decrease in cooling) by factors of 5–8 over the same range. Many other spiral galaxies show abundance gradients in their H II region populations. However, absolute abundances vary greatly from galaxy to galaxy, the smallest galaxies usually having the lowest abundances. Temperatures as high as 13,000 K are measured in H II regions in the low-abundance satellite galaxies of the Milky Way, the Magellanic Clouds, as well as in more distant irregular dwarf galaxies. Some of the latter are little more than a single giant H II region, such as the object II Zwicky 40.

In most other galaxies, astronomers can observe only a few dozen or less of the brightest H II regions. In spiral galaxies these occur preferentially halfway out from the center in the galactic disk; sometimes, bright H II regions are concentrated in narrow rings in the inner galaxy (e.g., NGC 1097). Sa galaxies generally have the smallest and weakest H II regions; Sc spirals and irregular galaxies have the largest and brightest. In the former, H II regions frequently nicely delineate the major spiral arms (e.g., Walter Baade's "beads on a string" in M31), whereas in the latter the distribution is more chaotic. At the edges of the nearest galaxies, large and faint ringlike H II regions have been found, of an appearance not yet seen in the Galaxy, and of unclear origin. There are many more weak than bright H II regions. Radio studies show that in a given galaxy, the number of H II regions above a certain luminosity generally decreases with the square of that luminosity. Among galaxies, the luminosity of the brightest H II region complexes varies widely, irrespective of galaxy mass or luminosity. Giant galaxies such as M101 and dwarfs such as M33 and the Large Magellanic Cloud contain H II region complexes more luminous than any found in our own Galaxy.

The resolution and sensitivity of astronomical measurements continues to increase, bringing more H II regions in more distant galaxies within reach. The observational emphasis is shifting from studies of individual H II regions to studies of H II regions as part of their atomic and molecular surroundings and to the use of H II regions as a tool to determine large-scale properties of galaxies. Thus, in only a few decades, two largely unrelated fields, the study of the (galactic) interstellar medium and the study of the evolution of galaxies, are merging into one.

Additional Reading

Gordon, M. A. (1988). H II regions and radio recombination lines. In *Galactic and Extragalactic Astronomy*, G. L. Verschuur and K. I. Kellermann, eds. Springer, Berlin, p. 37.

Habing, H. J. and Israel, F. P. (1979). Compact H II regions and OB star formation. *Ann. Rev. Astron. Ap.* **17** 345.

Osterbrock, D. E. (1989). *Astrophysics of Gaseous Nebulae and Active Galactic Nuclei*. University Science Books, Mill Valley, Calif.

Pagel, B. E. J. and Edmunds, M. G. (1981). Abundances in stellar populations and the interstellar medium in galaxies. *Ann. Rev. Astron. Ap.* **19** 77.

Shields, G. A. (1990). Extragalactic H II regions. *Ann. Rev. Astron. Ap.* **28** 525.

See also **H II Regions, Dynamics and Evolution; Interstellar Medium; Masers, Interstellar and Circumstellar.**

H II Regions, Dynamics and Evolution

Franz D. Kahn

The dominant element in interstellar space is hydrogen, present mainly in its ionized form H^+ in H II regions; the neighboring parts of space are either H I regions, containing hydrogen in its atomic form H, or molecular clouds, where the hydrogen is present as H_2. Typical temperatures in these three cases are, respectively, 10,000, 100, and 10 K; if adjacent masses of gas with different states of ionization are in hydrostatic equilibrium, say an H II region is in equilibrium next to a molecular cloud, then there must be a very large density difference between them. There is, though, a continual interchange of material between these different types of regions. A new hot and luminous star will dissociate and ionize the gas in the molecular cloud from which it was formed. The time scale on which this change takes place is typically of order 10,000 years, and the typical speeds that occur are of order 10 km s^{-1}. Evidently the conditions for hydrostatic equilibrium will often be seriously violated. A shock wave is then driven into the adjacent molecular gas, where the speed of sound is only 0.2 km s^{-1}, because the molecular gas cannot adjust smoothly to the increased pressure in the H II region that is growing next to it.

A clear distinction needs to be made between H II regions and the other parts of interstellar space where the hydrogen is also (almost) fully ionized. They include the intercloud medium (ICM), supernova remnants (SNR), and fast stellar winds, as well as the hot shocked gas that they produce. The gas in the SNRs has recently been overtaken by a violent blast wave originating in a supernova, and has been collisionally ionized and heated. The gas in the ICM is derived from SNRs; neither in the ICM nor in a SNR is the ionization maintained by radiation from a star, and although the ICM and SNRs dominate the dynamics of the interstellar medium at large, they have little to do with the properties of H II regions. The fast stellar winds, though, are emitted by the same stars that illuminate H II regions and maintain their ionization. There is an intimate connection between the dynamics of their interaction with the ambient gas and the properties of the H II regions created by the stars.

Interstellar dust can survive in H II regions, even though the surrounding gas reaches temperatures far above the melting point of all its constituents. The dust grains maintain their thermal balance largely by interaction with the radiation field, and will only be destroyed by the impact of fast moving ions, having speeds of order 100 km s^{-1} or higher, as in a stellar wind. The brightest and most important H II regions are found near molecular clouds, which themselves are very dusty. It is therefore inevitable that they are often heavily obscured and have to be observed in the infrared or by their radio emission. The dust can also play an important role in the internal dynamics of an H II region, as described below.

THERMAL BALANCE AND PRESSURE-DENSITY RELATIONS

The typical electron density ranges from 10^5 cm^{-3} [case (i)] in an H II region whose linear dimensions are of order 0.1 pc, to 10^2 cm^{-3} [case (ii)] for linear dimensions of the order of parsecs. The recombination coefficient b is typically 2×10^{-13} cm^3 s^{-1}, and therefore the characteristic rate for attaining ionization equilibrium is 2×10^{-8} s^{-1} and 2×10^{-11} s^{-1} in these two cases. At a temperature of typically 10^4 K, the thermal relaxation rate is somewhat but not much larger than the recombination rate. If either the thermal or the ionization equilibrium is disturbed, the gas will readjust in times of order 5×10^7 and 5×10^{10} s, respectively; any dynamical perturbations that are likely to occur will have much longer time scales. The ionized gas always has time to come to thermal equilibrium, and may be treated as being isothermal, with the pressure P related to the density ρ by

$$P = \rho c_i^2$$

and

$$c_i = \sqrt{kT/\overline{m}}\,,$$

where T is the fixed temperature, $\overline{m}$ is the mean mass of the particles, counting the free electrons separately, and c_i is the isothermal sound speed.

The coefficient b allows only for recombinations to excited states: Radiative recombination direct to the ground state may be ignored because it leads to the emission of a Lyman c (continuum) photon, which will cause another photoionization nearby.

Recombination to an excited state produces a cascade of further photon emission, ending in the emission of a Lyman line (i.e., resonance) photon, which is trapped in the H II region, for the following reason. The argument here is based on the properties of an H II region with $n_e \sim 10^2$ electrons cm^{-3} and radius 2.5 pc ($\sim 7.5\times10^{18}$ cm); the central ionizing star needs to have a luminosity of about $L = 10^5$ $L_\odot$ and a surface temperature $T_e \sim 40{,}000$ K. With an expansion speed of 10 km s^{-1}, such an H II region evolves on a time scale of some 250,000 years, but the recombination time, as has been shown, is only 5×10^{10} s ~ 1700 yr. Clearly only a small fraction of the Lyman c flux, 1700/250000 or one part in 150, is required to produce the new ionizations that occur when H atoms (or H_2 molecules) are incorporated into the H II region; the remaining photons go to maintain the ionization in the body of the region. The implication is that the H II region can never be fully ionized but always contains some H atoms, enough to produce an optical depth for Lyman c photons of a few (in this case about 5) between the star and the ionization front at its boundary. The optical depth is rather larger for resonance photons and all of them are trapped. The higher Lyman series photons, $L\beta$, $L\gamma$, and so on, are broken up into photons of lower energy plus a Lyman α photon during successive absorptions and re-emissions that must therefore occur. Each recombination to an excited state eventually leads to the creation of a Lyman line photon, and therefore a Lyman α photon. The radiation pressure generated in this way would become very large in the absence of interstellar dust. However, if the dust–gas mixture in the H II region has an effective opacity κ, the Lα photon can only persist, on average, for a time $(\kappa\rho c)^{-1}$, where ρ is the density, and so the radiation pressure due to Lα becomes

$$P = \frac{1}{3} b n_e^2 \chi_\alpha (\kappa\rho c)^{-1} \equiv \frac{b\chi_\alpha \rho}{3\kappa m_a^2 c}\,,$$

where $\chi_\alpha = 10.15$ eV is the energy of a Lα photon and m_a is the mass present per free electron. On comparison with the first equation it is seen that this effect will be important only if κ is not much smaller than 10 cm^2 g^{-1} and therefore not in our Galaxy, where κ is of order 100 cm^2 g^{-1}. It may possibly be important in other galaxies where metals and solid grains are significantly less abundant. The grains that are heated in this way reradiate in the infrared; this effect is easily observed.

IONIZATION FRONTS

The interface between an H II region and the neutral gas is called an ionization front, or I front; it is almost always narrow in comparison with the linear dimensions of the H II region itself. There are two ways in which the width of an I front may be defined. On the starward side the characteristic length scale is given by the recombination rate and the speed of the flow relative to the front, and is typically $\Delta_i = 5\times10^{13}$ cm [case (i)] or 5×10^{16} cm [case (ii)]. The length scale Δ_n on the dark side is determined effectively by the mean free path of a Lyman c photon in the neutral gas. In all important cases Δ_i is rather larger than Δ_n. The gas entering the front is therefore ionized and heated relatively fast, and then relaxes rather more gently to its equilibrium temperature in the H II region. The heat input into the gas derives from the excess energy $\Delta\chi \equiv \chi - \chi_L$ that the ionizing photon has above the Lyman limit χ_L. On average $\Delta\chi \sim kT_e$, where $T_e \sim 40{,}000$ K is the surface temperature of the exciting star. Each ionization therefore releases on average 2–3 eV of thermal energy.

The flux F of Lyman c photons that arrives at the I front is determined by the exciting star, by its distance from the front, and by the properties of the H II region between it and the front. There are two major classes of front: in the case of an R-type front the neutral gas ahead is *r*arefied, and the front advances into it at supersonic speed. The flow through a weak R front is supersonic throughout, so that the ionized gas streams away from it with a speed above c_i. Such a front heats the gas but passes so quickly that, seen from the rest frame of the star or the H II region, the material is left standing even though the pressure balance is severely upset. The dynamical readjustment follows much later. A necessary condition for an R-type front to exist is that

$$n < C_R F \left(\frac{m_a}{\Delta\chi}\right)^{1/2}$$

where n is the density of H atoms on the dark side and C_R is a numerical coefficient that depends on the balance between heating and cooling within the front. In simple cases there is little cooling in the zone where fresh ionizations take place, and then $C_R = 0.53$;

otherwise it is somewhat larger. Weak-R fronts typically occur early in the evolution of an H II region; strong-R fronts, which move subsonically relative to the ionized gas, seem not to be important.

In the case of a D-type front the neutral gas ahead is *d*ense and the front advances into it at subsonic speed. There are both weak-D and strong-D fronts: In the weak-D case the motion remains subsonic throughout and this condition requires that the pressure be large enough on both the dark and the illuminated sides. Again there is a critical condition, analogous to the preceding equation, that must hold for fronts to be possible: in the case of the weak-D front it is that

$$P > C_D F \left(\frac{\Delta \chi}{m_a} \right)^{1/2}$$

where P is the pressure on the dark side. $C_D = 1.14$ in the idealized case, and is somewhat smaller if the zone of new ionization overlaps with that of thermal relaxation. In a strong-D front the flow speed relative to the front is subsonic on the dark side and is supersonic on the illuminated side, where the gas streams away freely. The lack of a boundary condition on the starward side is compensated for by the stricter condition

$$P = C_D F \left(\frac{\Delta \chi}{m_a} \right)^{1/2}$$

that must be satisfied now and that demands that the flow relative to the front be exactly sonic at the point where the gas reaches its highest temperature (or strictly the highest stagnation enthalpy).

PARTIALLY IONIZED GAS CLOUDS AND HOLLOW SHELLS

According to modern ideas, an H II region has a much richer structure than was thought. The older descriptions were based on a model in which a star was located at the center of a spherically symmetrical mass of neutral gas, all smoothly distributed. At time $t = 0$ the star would suddenly begin to radiate at full strength, and an ionization front (weak-R) would race outward, later slowing and turning into a D-type front, preceded by a shock, as it moved further from the star. Real H II regions, though, show none of the regularity and order that this theoretical description implies.

Instead, as Guido Münch and Olin C. Wilson observed in their classical work on the Orion nebula, there are internal motions present that show themselves by velocity splitting along almost all lines of sight: The typical random velocity component is itself of order 10 km s^{-1}. In a sense the old model of an H II region has to be turned inside out to interpret such observations.

The Orion nebula, or any other region where massive stars are presently forming, is in the process of evolving from a giant molecular cloud, in which several (or many) subsidiary clouds have undergone gravitational collapse. In some cases new O or B stars have already formed, and are now driving a strong-D–type ionization front *into* other self-gravitating condensations. The newly ionized gas is streaming away outward from the various ionization fronts: The whole Orion nebula is a patchwork of such conflicting flows, and the individual sources from which the flows arise are the partially ionized gas clouds (PIGs).

The hollow shells arise at an earlier stage when the newly formed star breaks out of the molecular cloud in which it itself was formed. The star blows a powerful wind, with typical mass loss rate $\sim 10^{-6}\ M_\odot\ \mathrm{yr}^{-1}$ and terminal wind speed ~ 2000 km s^{-1}. The wind shocks against the surrounding gas and a bubble of hot gas forms around it, with typical temperature close to 4×10^7 K. The hot bubble itself drives a shock into the surrounding molecular gas: The H II region then forms on the inside of the shocked layer. The ionization fronts here are weak-D; such hollow shells are commonly found, but usually there is heavy obscuration by interstellar dust. The observation of hollow shells must therefore be made at radio rather than visual wavelengths: the high obscuration indicates that the ionizing star is at an earlier stage of evolution than, say, the Trapezium stars in Orion. Alternatively, the shell itself can be thought of as an evolving PIG.

Additional Reading

Dyson, J. E. (1981). The dynamical effect of hypersonic stellar winds on interstellar gas. In *Investigating the Universe*, F. D. Kahn, ed. Reidel, Dordrecht, p. 125.

Dyson, J. E. and Williams, D. A. (1980). *Physics of the Interstellar Medium*. Manchester Univ. Press, Manchester.

Habing, H. J. and Israel, F. P. (1979). Compact H II regions and OB star formation. *Ann. Rev. Astron. Ap.* **17** 345.

Osterbrock, D. E. (1989). *Astrophysics of Gaseous Nebulae and Active Galactic Nuclei*. University Science Books, Mill Valley.

Shu, F. H., Adams, F. C., and Lizano, S. (1987). Star formation in molecular clouds, observation and theory. *Ann. Rev. Astron. Ap.* **25** 23.

Yorke, H. W. (1986). The dynamical evolution of H II regions—Recent theoretical developments. *Ann. Rev. Astron. Ap.* **24** 49.

See also **H II Regions; Interstellar Medium; Interstellar Medium, Stellar Wind Effects.**

Heliosphere

W. Ian Axford

The heliosphere is the region surrounding the Sun in which the solar wind plasma has a dominant role in determining the state of the medium. This region has typical radial dimensions of the order of 100–300 AU and is bounded by an interface with the local interstellar plasma and magnetic field termed the heliopause. The interaction between the two plasma components must certainly involve stresses associated with the reconnection of magnetic fields of solar and interstellar origin that would tend to distort the heliosphere into a comet-like shape with the "tail" stretched in the direction of the local interstellar wind relative to the Sun. In addition however there is a complication associated with the fact that the neutral components of the interstellar medium (mostly H and He atoms) can penetrate throughout the entire heliosphere except in regions very close to the Sun where these atoms become ionized. The interaction of the neutral wind with the solar wind can modify the latter quite substantially and may also provide a source of nonthermal ions that can ultimately be accelerated and appear as low-energy, locally produced (anomalous) cosmic ray particles. The interaction contributes a drag and a corresponding distortion of the outer heliosphere into a comet-like shape as indicated in Fig. 1. Similar regions (astrospheres) should occur around other stars with moderate stellar winds like the Sun.

Historically, ideas concerned with the nature of the heliosphere developed piecemeal, largely because of the absence of direct observations. Indeed, the earliest ideas were mainly speculations arising from the need to understand the observed solar cycle modulation of galactic cosmic rays. The first such idea seems to have originated with Hannes Alfvén (1950), who argued that cosmic rays are in fact not cosmic at all, but instead have a local origin associated with the region of interaction between solar corpuscular streams (which we now term the solar wind) and the interstellar medium and magnetic field. In fact a brief consideration of the energy spectra and composition of the cosmic rays shows that this cannot be the case, but it is now clear that one might be able to interpret the anomalous component in such terms.

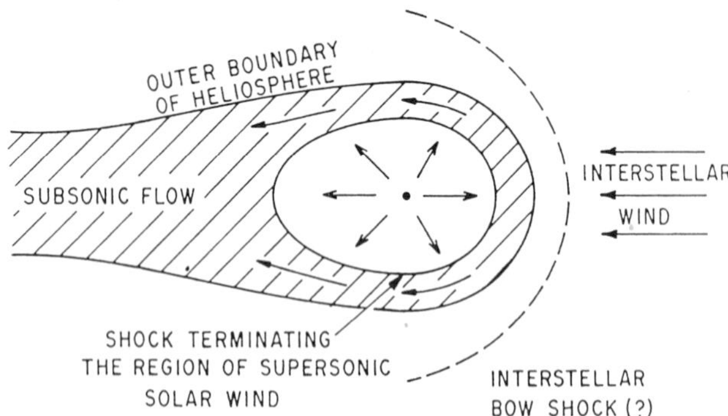

Figure 1. The general characteristics of the heliosphere. A bow shock may occur in the interstellar medium under suitable conditions but is probably not present. The interstellar and interplanetary magnetic field lines are likely to be interconnected, which has implications concerning the formation of the tail region of the heliosphere and perhaps the access of low energy cosmic rays.

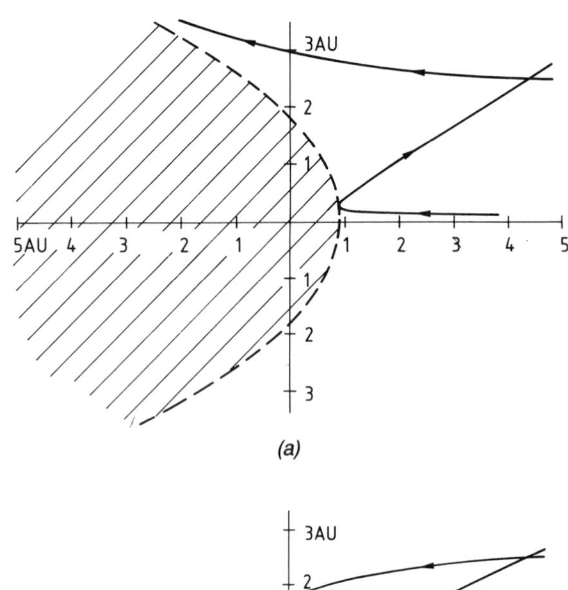

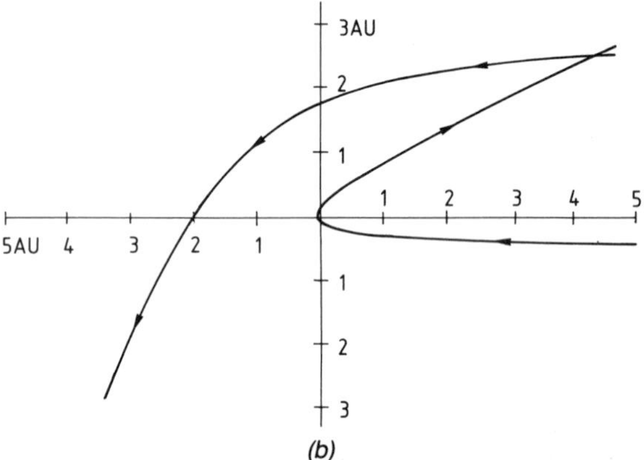

Figure 2. Trajectories of (*a*) neutral H and (*b*) He atoms originating in the very local interstellar medium (VLISM). Note that H atoms may be repelled by the Sun if the radiation pressure of solar Lyman α is sufficiently large and that the He atoms are focussed into a line on the downstream side. In general there are two trajectories through each point of space.

In 1955, Leverett Davis, Jr. proposed a model of the heliosphere with the local interstellar magnetic field distorted by solar corpuscular streams. The application of this concept to the problem of galactic cosmic ray modulation was considered by Davis, Philip Morrison, and Eugene Parker, who suggested that the modulation observed at the Earth is the result of the sweeping of cosmic ray particles out of the inner solar system by the solar wind. This implies that there should be a positive radial gradient in the cosmic ray intensity with respect to distance from the Sun. Furthermore, the observed time delay between changes in the solar wind associated with the solar cycle and changes in the modulation might be interpreted as the time taken for the changed solar wind to make itself felt throughout the heliosphere. With solar wind speeds of 400 km s^{-1} and time delays of the order of 6–12 months, this suggests that the size of the modulating region might be of the order of 40–80 AU^{-1}. In turn, the factor 2 variation in the 1–5-GeV cosmic ray intensity observed at Earth during the solar cycle would suggest that the radial gradient of the intensity must be of the order of 1–3% AU^{-1}. These estimates are roughly correct, although the nature of the solar cycle variation of the solar wind and the mechanics of the modulation phenomenon are much more complex than was originally believed.

(VERY) LOCAL INTERSTELLAR MEDIUM (VLISM)

It is easily deduced that, allowing for photoionization, collisional ionization, and ionization by charge exchange with solar wind particles, interstellar H (He) atoms can penetrate the heliosphere to distances of about 4 (0.5) AU if they have a speed of about 20 km s^{-1} relative to the Sun at large distances. In general the trajectories of the particles are hyperbolae, until they are lost, and focus onto a line behind the Sun in the downstream direction as a consequence of solar gravity. H atoms are, however, also strongly affected by the radiation pressure of solar Lyman α photons and may (especially at sunspot maximum) suffer a net repulsion so that there is an empty region downstream. This is summarized in Fig. 2.

It is possible to observe the interstellar H (He) atoms directly outside the Earth's atmosphere as a result of the resonant scattering of solar ultraviolet 1216 Å (584 Å) photons because the intensity of the scattered component is of the order of 500 (5) Rayleighs depending on direction and phase of the solar cycle. By fitting the results of model calculations with observations of the scattered light and of the corresponding solar lines one is able to deduce the properties of the neutral component of the VLISM with some certainty. The current best values for the various parameters are as follows: number density $N(\mathrm{H}) = 6.5(\pm 1) \times 10^{-2}$ cm^{-3}, $N(\mathrm{He}) = 8(\pm 4) \times 10^{-3}$ cm^{-3}; temperature $T(\mathrm{H}) = 8000 \pm 1000$ K, $T(\mathrm{He}) = 7000 \pm 2000$ K; speed relative to the Sun $V(\mathrm{H}) = 20 \pm 1$ km s^{-1}, $V(\mathrm{He}) = 21.5 \pm 2.5$ km s^{-1} and the wind points approximately in the direction 5° latitude and 253° longitude in ecliptic coordinates. At these temperatures one would expect the helium to be almost entirely neutral and the hydrogen not more than about 50% ionized; however, it is possible that the medium is not in equilibrium so that this conclusion may not be certain. The direction of the wind does not correspond to the direction of the solar apex ($\beta = 50°$, $\lambda = 269°$) nor to the direction of the wind in the LISM out to 100 parsecs.

SHAPE AND SIZE OF THE HELIOSPHERE

The solar wind is a supersonic flow of solar plasma that undergoes a shock transition to subsonic flow in the outer heliosphere but inside the heliopause. The shock is probably egg-shaped with the Sun closer to the blunt end, that is, on the face directed toward the oncoming interstellar wind (see Fig. 1). A first approximation to the minimum distance to the shock can be obtained by simply equating the ram pressure of the solar wind at a distance R AU to the total pressure of the local interstellar plasma, which includes thermal (P_t), magnetic (P_m), and ram (P_r) pressures:

$$Kn\overline{m}v^2/R = P_t + P_m + P_r,$$

where n is the ion number density, v is the solar wind speed at 1 AU, $K = 1.13$, and $\overline{m}$ is the mean ion mass in the solar wind ($\overline{m} = 2\times10^{-24}$ g). We take $P_t = 2NkT$ with Boltzmann's constant $k = 1.38\times10^{-16}$ erg per Kelvin, $P_m = B^2/8\pi$ with B the VLISM magnetic induction, and $P_r = KNmV^2$, with $m = 1.66\times10^{-24}$ g the proton mass. With $N = N(\mathrm{H})$ we find that $P_t = 1.5\times10^{-13}$ dyn cm^{-2}, $P_r = 4.9\times10^{-13}$ dyn cm^{-2}, and, with $B = 0.4$ nT, $P_m = 6.4\times10^{-13}$ dyn cm^{-2}. Accordingly, with $nv^2 = 8\times10^{15}$ cm^{-1} s^{-2}, we find that $R = 120$ AU.

Apart from increasing B substantially (for example, $B = 1$ nT would result in $R = 63$ AU), there is little prospect of reducing this estimate because the assumption $N = N(\mathrm{H})$ is probably an upper limit. It has been shown that the effects of the neutral interstellar gas and galactic cosmic rays on the ram pressure of the solar wind are small if $R < 120$ AU and it seems unlikely that there is another major contributor to the overall pressure of the interstellar medium beyond the heliopause (although very low energy galactic cosmic rays might be a possibility if they are associated with the supernova remnant on the edge of which the Sun appears to be situated).

Beyond the shock that terminates the solar wind as a supersonic flow, the heliosphere should consist of an envelope of subsonically flowing plasma and magnetic field that is forced eventually to move in the downstream direction relative to the interstellar wind (see Fig. 1). The neutral component of the VLISM penetrates the outer heliosphere freely and the occasional charge exchange between interstellar atoms and solar wind plasma ions is sufficient to cool the plasma and collapse the envelope so that the overall shape of the heliopause is perhaps better better described as tadpole-like rather than cometary.

COSMIC RAYS IN THE HELIOSPHERE

The heliospheric magnetic field, which is of solar origin, is rather weak and permits high energy cosmic rays (> 100 GeV per nucleon) to penetrate to the position of the Earth with no observable change in intensity. The highest energy particles (> 1000 GeV per nucleon) penetrate without changing their direction significantly and their directional anisotropy accordingly gives rise to a small (about 0.1%) sidereal diurnal variation of their intensity. It is interesting to note that the corresponding cosmic ray flow direction coincides neither with the solar apex nor the direction of the local interstellar wind, and presumably tells us something about the nearby sources of high energy cosmic rays and their loss from the Galaxy. The trajectories of lower energy particles are sufficiently scrambled by the heliospheric magnetic field to suppress this sidereal variation and the regular variations observed are accordingly solar-controlled.

It was shown by Parker, as a consequence of his analysis of the nature of the solar wind, that the interplanetary magnetic field lines should, on the average, have the form of Archimedes spirals on the surface of cones centered at the Sun and with their axes parallel to its axis of rotation. In addition to this average field, there is a fluctuating component, comprised to a large extent of Alfvén waves, that scatters the cosmic rays causing them to propagate diffusively as suggested originally by Morrison. Because the mean scattering time is long compared with the gyroperiod of the particles in the mean field, the diffusion is, however, anisotropic, so that a strong sense of the presence of the mean field is maintained. Thus energetic solar flare particles propagate, to a first approximation, along the spiral field from the Sun to the Earth and so appear to come at first from about 45° west of the Sun itself. Similarly, galactic cosmic rays sense the ordered mean field as well as being scattered by the fluctuations and are therefore modulated by drifts caused by the curvature of the mean magnetic field lines and the

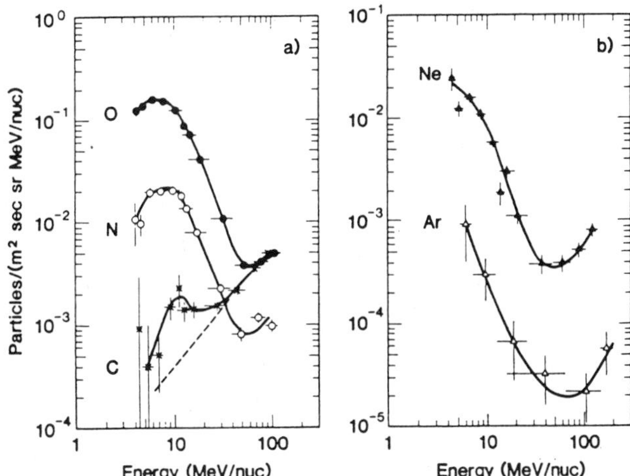

Figure 3. The spectra of cosmic ray oxygen, carbon, nitrogen, neon, and argon in 1985–6. The anomalous component is evident as a turnup in the spectra at energies < 100 MeV per nucleon. The dashed line indicates the estimated galactic spectrum of carbon. (*By permission from A. C. Cummings and E. C. Stone, in* Proceedings of the Sixth International Solar Wind Conference, *NCAR-TN 306, published by the National Center for Atmospheric Research, Boulder, Col. 1988.*)

gradients in mean field strength as emphasized by J. R. Jokipii. This greatly complicates the problem of cosmic ray modulation, which must be treated as being at least two-dimensional with particles drifting in latitude in a direction determined by the 22 year cycle of the solar magnetic field.

In 1973 it was discovered that some elements in the cosmic radiation observed in the inner solar system show enhancements in the region 10–100 MeV per nucleon relative to what appears to be a normally modulated galactic spectrum (Fig. 3). The enhancements are most prominent in the case of species that have high first ionization potentials but may be present to some extent for all species. It was pointed out by Lennard Fisk and collaborators that the origin of these particles might be the neutral VLISM that becomes photoionized in the supersonic solar wind region. There are several conceivable acceleration mechanisms, the most likely of which involves the solar wind termination shock. This is particularly interesting because shock acceleration is observed to occur in connection with solar flare particles and interplanetary shocks and is believed (in connection with supernova shocks) to be the main source of galactic cosmic rays, at least up to 10^{16} eV.

PRESENT STATUS OF EXPLORATION OF THE OUTER HELIOSPHERE

The first observations of Lyman α radiation from the VLISM were made by V. G. Kurt and his associates from a series of Soviet spacecraft. By means of parallax measurements it was shown by Jacques Blamont, J. L. Bertaux, Gary E. Thomas, and R. F. Krassa that the Lyman α is indeed local and results from the scattering of solar Lyman α by interstellar hydrogen atoms that penetrate to within the orbit of Jupiter. The helium emission was detected later, with its important enhancement in the downwind direction, but to date no other species have been found, essentially because their emissions are so weak.

Direct observations of the state of the heliosphere at large distances from the Sun have been made from the U.S. spacecraft *Pioneers 10* and *11* and *Voyagers 1* and *2*. By the year 2000 the distances achieved by *Pioneers 10* and *11* and *Voyagers 1* and *2* will be 74, 54, 76, and 61 AU, respectively, with *Pioneer 10* in the ecliptic in the downwind direction and the others in the upwind direction with the *Voyagers* at latitudes about 40° on either side of the ecliptic plane. If the solar wind termination shock is not detected by this time, the VLISM magnetic field strength will have been determined to have an upper limit of about 1 nT.

In fact the *Pioneers* may not be completely functional by the year 2000, but both *Voyagers* may be capable of performing observations and transmitting the results at least until 2010 and *Voyager 1* may therefore detect the shock if it is within 110 AU. This would be a most significant observation because it would allow us to determine the nature of cosmic ray modulation and the origin of the anomalous component with some certainty. A genuine interstellar mission may, however, have to be performed separately using a high speed trajectory, perhaps utilizing a powered solar swingby and/or a new propulsion system.

Additional Reading

Axford, W. I. (1972). The interaction of the solar wind with the interstellar medium. In *Solar Wind*. NASA SP308, Washington, D.C., p. 609.

Bertaux, J.-L. (1984). Helium and hydrogen of the local interstellar medium observed in the vicinity of the Sun. In *I.A.U. Colloquium No. 81*. NASA Conference Publication 2145, Washington, D.C., p. 3.

Grzedielski, S. and Page, D. E. (1990). *The Physics of the Outer Heliosphere*. Pergamon Press, Oxford.

Holzer, T. E. (1989). Interaction between the solar wind and the interstellar medium. *Ann. Rev. Astron. Ap.* **27** 199.

Marsden, R. G., ed. (1986). *The Sun and Heliosphere in Three Dimensions*. Reidel, Dordrecht.

Paresce, F. and Bowyer, S. (1986). The Sun and the interstellar medium (1986). *Scientific American* **255** (No. 3) 93.

Suess, S. T. (1990). The heliopause. *Rev. Geophys.* **28** 97.

Venkatesan, D. and Krimigis, S. M. (1990). Probing the heliomagnetosphere. *Eos* **71** 1755.

See also **Interplanetary Medium, Shock Waves and Traveling Magnetic Phenomena; Interplanetary Medium, Solar Wind; Interstellar Medium, Local.**

Herbig–Haro Objects and Their Exciting Stars

Richard D. Schwartz

In his observational studies of nearby galactic dark clouds carried out in the 1940s, Alfred H. Joy of Lick Observatory discovered a population of irregularly variable stars characterized by strong chromospheric-like emission lines. These stars, named after the prototypical star T Tauri, were proposed by V. A. Ambartsumian of the Byurakan Observatory in Armenia to be stars of recent birth, a proposition that was confirmed by additional observational and theoretical studies of the 1950s. The T Tauri stars are members of a broader category of objects, called young stellar objects (YSOs), which include a variety of stars (visible and infrared) in different stages of pre-main sequence evolution.

In the late 1940s, George H. Herbig of Lick Observatory and Guillermo Haro of Tonantzintla Observatory in Mexico independently discovered the presence of mysterious, semistellar emission-line nebulae in the Orion dark clouds near the Orion nebula. Under spectroscopic analysis, the objects appeared to be quite different from the T Tauri stars. Whereas the T Tauri stars exhibit strong stellar continua and emission lines (H, Ca II, Fe II) typical of higher density chromospheric conditions, the nebular objects (now dubbed Herbig–Haro objects) exhibited little continuum emission and strong nebular ("forbidden") emission lines typical of low density nebular gas. Moreover, the Herbig–Haro objects (HHs) exhibited resolved structure, consisting of multiple knots (2–3″ in diameter) embedded in diffuse nebulosity, unlike the point-like stellar images of T Tauri stars.

In the late 1950s attempts were made to understand the energy sources responsible for producing the HHs, with an eye toward understanding their origin and their connection to star forming regions. Owing to the fact that some T Tauri stars exhibit nebular ("forbidden") emission lines superposed on strong continuum and chromospheric emission, it was initially proposed that the HH knots were the sites of young stars that were veiled from direct view, but that managed to provide energy sources for the heating of the knots. It was assumed that the HHs were heated by photoionization, the process responsible for producing prominent H II regions such as the Orion nebula. This requires the presence of a hot star ($T > 20{,}000$ K) in order to provide the necessary energetic photons to ionize hydrogen. Deep photographic searches failed to reveal the presence of such stars, and spectroscopic evidence showed that hot stars could not be present in the HH knots.

OBSERVATIONAL PROPERTIES OF HERBIG–HARO OBJECTS

The prototypical HHs are characterized by semistellar knots of nebulosity, often appearing in complexes 10–60″ in extent with faint diffuse emission between knots. The spectra of HHs generally show low-excitation emission lines of H, [S II], [O I], [N I], [Fe II], [Ca II], and [O II], with higher-excitation lines of [O III] and [Ne III] appearing in some objects. A faint ultraviolet-blue continuum seen in these objects can be attributed to the nebular two-photon emission from neutral hydrogen atoms or to H_2^+ emission. The original list of 43 HHs compiled by Herbig in 1974 has now been increased to nearly 100 objects through surveys by a number of investigators. Virtually all active, nearby star-forming regions possess one or more HHs, indicating a close connection to the births of stars.

About 70% of HH objects for which radial velocities are available exhibit negative velocities. This is probably a consequence of their origin in dark clouds, which creates a selection effect in which we preferentially view objects moving toward us out of the clouds (those moving away are presumably more deeply embedded and thus obscured by the dark clouds). Also, a number of HHs have been found to show proper motions of up to 300 km s^{-1}. The prototypical objects HH 1 and HH 2 show proper motions away from one another with an apparent origin from a YSO located between the two HH complexes. Millimeter radio astronomers have found evidence for a widespread molecular outflow (detected in the millimeter lines of CO) associated with a YSO (L1551 IRS 5) in Taurus that has also produced HH 28 and HH 29. Molecular outflows that are also implicated with Herbig–Haro objects have now been discovered in association with a significant fraction of YSOs. Another recent discovery is that most HHs also exhibit near-infrared emission from molecular hydrogen (H_2) that appears to be coextensive with the low-excitation optical emission.

Most of the known HH objects are situated at distances in the range 150 pc–1 kpc from the Earth, and possess red magnitudes in the range 14–20. In one sense, HHs are remarkably feeble objects in comparison with many other astronomical objects. The total mass of gas in a typical HH object responsible for the emission probably does not exceed 10 *Earth* masses, and the total luminosity of HHs is typically less than 5% that of the Sun. Although most HHs are isolated from visible stars, a few HH nebulae are found in association with T Tauri stars. Examples include Burnham's nebula, which extends about 12″ south of T Tauri, and HL Tau and DG Tau B, which exhibit HH "jets" emanating from the stars. In the last 5 years, using sensitive imaging devices, astronomers have demonstrated the importance of supersonic jets from YSOs in the production of a number of HHs. A graphic example is that of HH 34 in which a narrow jet, consisting of at least 12 highly aligned knots, emanates from a YSO. The jet continues with very faint emission about 1.5′ to a well-formed bow shock with high-excitation nebular emission. A similar bow shock structure is seen equidistant on the other side of the YSO, suggesting the presence of a highly collimated bipolar flow from the YSO that is seen only as an infrared source.

A second example is that of the HH 46–47 system, which is located in a small globule in the Vela portion of the Gum nebula. Figure 1 displays a photograph of this system, which exhibits a jet of material flowing from a YSO embedded in the small globule. The low-excitation nebular emission lines in the jet, terminating at HH 47A, show radial velocities of -120 to -160 km s^{-1}. A low-surface brightness "bubble" appears to surround HH 47A, possibly reflecting the dynamical effects of the jet emerging from the dense dark globule and expanding freely in the lower-density Gum nebula. It has recently been shown that the jet continues at -120 km s^{-1} to a faint bow shock on the surface of the bubble along the flow axis. Also, faint HH knots (47C) are observed at $+110$ km s^{-1} emerging in projection from behind the globule, a consequence of an oppositely directed jet in the bipolar system. HH 46, close to the embedded YSO, shows a prominent and variable reflection component that evidently represents visible light from the star, which is blocked from our direct view by a circumstellar disk.

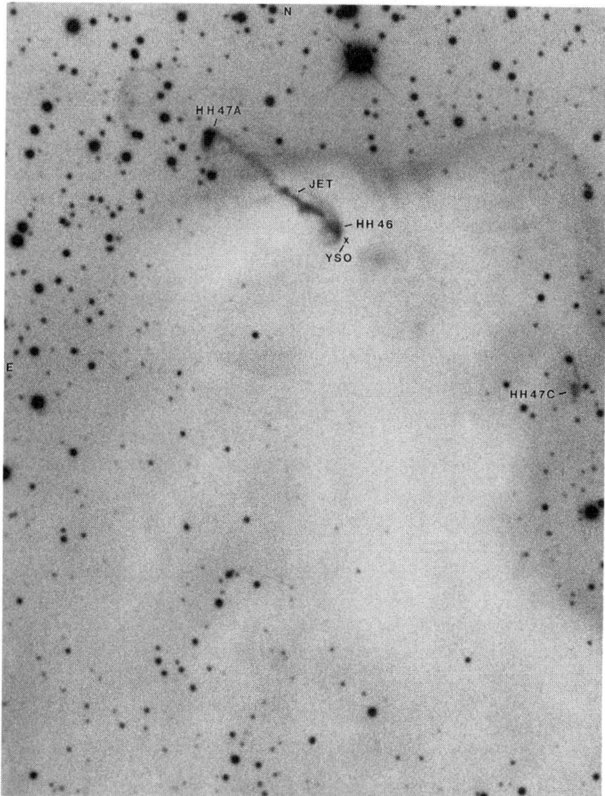

Figure 1. A young stellar object (positioned at *x*), detectable only at infrared wavelengths, is embedded in this small dark globule in the Gum nebula. Evidence suggests that the star is obscured by a thick, tilted circumstellar disk. A supersonic jet flow emerges from the polar regions of the disk, producing HH 46, the jet, and HH 47A which have been ejected out of the front side of the globule. The oppositely directed jet is not visible because of obscuration by the globule. HH 47C probably represents the end of this opposite jet emerging in projection from behind the globule. Although the distance to this system is not well determined, an estimate of 300 pc leads to a projected jet length of 0.1 pc. (*Photograph obtained by the author with the 4-m telescope at Cerro Tololo InterAmerican Observatory in Chile.*)

EXCITING STARS OF HERBIG–HARO OBJECTS

That young stars of low mass ($<2\ M_{\odot}$) are responsible for generating HH objects, seemed evident in the earlier recognition of HH nebulae close-lying to several T Tauri stars. However, the lack of visible YSOs near most HHs led astronomers in the early 1980s to carry out infrared (IR) mapping around HHs in search of probable exciting stars, which would shine only by infrared light if embedded in thick circumstellar material. Both near-IR and far-IR surveys have discovered approximately 30 candidate infrared YSOs associated with HHs. The association of an IR source and an HH object is usually inferred upon the basis of geometrical alignments between the IR source and the HH structure, or the presence of proper motion in the HH object away from the IR source. The projected distances between YSOs and their associated HHs range from a few hundred astronomical units (AU) to over a parsec. Bolometric luminosities of the infrared YSOs cluster toward those of classical T Tauri stars ($0.5 < L < 50\ L_{\odot}$), although a few higher luminosity sources ($L > 1000\ L_{\odot}$) are known to produce HHs.

In a number of instances, HH emission is found in association with reflection nebulosity. In these cases, it is evident that, although the star is not visible directly owing to obscuration by dust, nonuniformities in the circumstellar dust allow light to escape and reflect from more distant dust clouds. In most of these cases (e.g., HH 24, 46, and 100), the reflected spectrum is that of an active T Tauri star. This does not necessarily mean that the T Tauri star is responsible for producing the HH object. In the case of the star T Tauri, which appears in association with Burnham's nebulae and several other HH knots, it is known that an infrared companion star (an obscured protostar?) lies about 0.6″ (80 AU) from the visible star, and it is probable that the infrared companion is responsible for generating the HHs.

INTERPRETATION OF HERBIG–HARO ACTIVITY

A spectroscopic study by astronomers in the mid-1970s of the HH nebulosity around T Tauri led to the conclusion that the gas was heated by shock wave excitation from a supersonic stellar wind emanating from the vicinity of T Tauri. Subsequent observational studies of many HHs and theoretical shock wave modeling have firmly established the shock wave origin of the objects. Shock waves occur when an object or a gas flow penetrates an ambient gas at a speed greater than that of sound. The kinetic energy of the flow is converted to heat in the shock fronts, causing the shocked gas to radiate the emission lines seen in HHs. At the same time, other studies have demonstrated the importance of reflection components in a number of systems where the nebulosity appears to outline "walls" of bipolar flow cavities.

Several scenarios have been advanced to account for the shock wave production of HHs, and evidence suggests that no single model can account for all observed HH phenomena. In the "shocked cloudlet" model, a supersonic stellar wind from a YSO impinges upon a small dense cloud of gas, creating a higher-excitation stellar wind bow shock and a lower-excitation molecular cloud shock. Another possibility is the "interstellar bullet" model in which dense fragments of material are accelerated away from a circumstellar shell, plowing through the ambient medium at supersonic velocities with shock waves produced at the forward edges of the fragments. Still another model invokes large scale density gradients in the gas around a YSO that focus stellar winds into an ovoid cavity, creating a bipolar flow that would produce shocks as it focused the wind upon the end of the cavity. The more recent recognition of the importance of jet activity in the mass outflows suggests that focusing, in many instances, probably occurs on much smaller scales (a few stellar radii or less). Moreover, in a number of systems (e.g., HH 46–47, HH 34, HL Tau, HH 30, DG Tau B, HH 111, etc.), it is evident that the shock wave excitation is related to the hydrodynamics of supersonic jets in which shock waves can be created interior to a jet as well as at its terminus or "working surface" where the jet plows into the ambient medium and creates a bow shock structure, much as in the interstellar bullet model. Reflection shocks within a confined jet can give rise to the spatially periodic knots seen in a number of HH jets.

The bow shocks that occur at the working surfaces of jets or around dense clumps of material plowing through a medium have been the focus of considerable theoretical modeling. Primary observational signatures pointing to bow shocks are differential radial velocities and line widths between emission lines of high and low excitation, and in some cases (e.g., HH 34 and HH 111) the bow shock structure exists on a large enough scale to permit the characteristic geometrical shape of a bow shock to be seen in direct images. Theoretical bow shock modeling has shown that many of the spectroscopic features of some HHs can be understood as resulting from bow shock flows.

The recognition of the presence of highly collimated jet-like flows in a number of YSO–HH systems has resolved an earlier problem regarding the energetics of these systems. If HHs were generated by *isotropic* stellar winds at the velocities (100–200 km s^{-1}) and wind densities (~ 100 cm^{-3}) required by shock wave modeling, the kinetic energy in the flows would in many cases rival or exceed the radiative output of the YSOs, a phenomenon that would be difficult if not impossible to account for physically. However, a much smaller mass loss rate, channeled into a highly anisotropic jet flow, can easily produce the HH phenomena without excessive mechanical energy output from the YSO.

GRAND SCENARIO OF PRE-MAIN SEQUENCE EVOLUTION

Once considered to be in situ sites of star formation, Herbig–Haro objects are now understood as a by-product of star formation, representing effects of the process by which a star organizes itself into a stable configuration. The sequence of events in the formation of a low-mass star allows, in a qualitative manner, for the origin of bipolar flows and HH phenomena. Dense cloud cores in a molecular cloud complex are considered to be the progenitors of stars. The collapse of such cores, mediated either by pure self-gravitational collapse or by gravitational–magnetic effects (ambipolar diffusion), leads to a configuration conducive to a period of bipolar mass outflow. Initially, the nonhomologous collapse of a core possessing angular momentum leads to a central accreting star surrounded by a thick circumstellar disk of gas and dust. Through viscous dissipation, disk material continues to spiral inward, accreting onto the central star through a boundary layer shock that produces much of the luminosity of the system. In the earlier stages accretion occurs from all directions and the disk is both physically and optically thick, preventing visible detection of the star. The circumstellar material reprocesses the stellar and boundary shock photons, reemitting the energy in the mid and far infrared, yielding a characteristic infrared signature of a "protostar." As the central star evolves to the point of thermonuclear burning of deuterium, the star becomes fully convective and is capable of generating a stellar wind, while at the same time maintaining some degree of accretion through the disk. The disk behaves like a pressure-cooker lid. As the outward stellar wind pressure increases, the polar regions of the disk (where radial column densities are lowest) serve as pressure-release valves. Bursting through the polar valves, a supersonic bipolar flow is created. Both the predicted spatial point at which these flows emerge and their durations (10^4–10^5 years) are consistent with observations of bipolar flows and dynamical lifetimes of HHs. The supersonic flows penetrate to great distances from their parent stars whereupon their energy is released through shock wave heating.

Although the foregoing scenario provides a qualitative and semiquantitative account of the processes leading to HH formation, a number of questions remain unanswered. First, what are the details of the focusing mechanism that can produce very narrow, supersonic jets? Is it a pure hydrodynamical process, or does a strong stellar magnetic field contribute to the directed outflow? In the visible jet flows, are the shocks entirely internal to the jets, or is ambient material entrained in the flows? Are the extensive molecular (CO) outflow sources seen in a number of cases driven by the high-velocity jet flows, or do molecular outflows have a separate origin? What role is played by binary YSOs in outflow sources? Do the outflows contribute significantly to angular momentum losses, which appear to be necessary in order to transform accreting YSO–disk systems into conventional T Tauri stars? As one of the first visible signs of star formation activity, the Herbig–Haro objects will continue to yield vital information on the conditions attending the birth of stars.

Additional Reading

Cohen, M. (1988). *In Darkness Born*. Cambridge Univ. Press, New York.

Dopita, M. A. (1981). Optical observations of interstellar shock-waves. In *Investigating the Universe*, F. D. Kahn, ed. Reidel, Dordrecht, p. 405.

Lada, C. J. (1982). Energetic outflows from young stars. *Scientific American* **247** (No. 1) 82.

Mundt, R. (1988). In *Formation and Evolution of Low Mass Stars*, A. K. Dupree and M. T. V. T. Lago, eds. Kluwer, Dordrecht, p. 257.

Schwartz, R. D. (1983). Herbig–Haro objects. *Ann. Rev. Astron. Ap.* **21** 209.

Shu, F. H., Adams, F. C., and Lizano, S. (1987). Star formation in molecular clouds: Observation and theory. *Ann. Rev. Astron. Ap.* **25** 23.

See also **Shock Waves, Astrophysical; Stars, Circumstellar Disks; Stars, Pre-Main Sequence, Winds and Outflow Phenomena; Stars, Young, Jets.**

Hubble Constant

Wendy L. Freedman

The Hubble constant specifies the rate at which the universe is expanding, and it yields values for the size and age of the universe. It has been argued that the value of the Hubble constant is the most important number in cosmology. Certainly it impacts in some way on all fields of extragalactic astronomy, ranging from determining the intrinsic brightnesses and masses of stars in other nearby galaxies, to determining those same properties for more distant galaxies and galaxy clusters, determining the amount of dark (unseen) matter, the scale size of galaxy clusters, and providing tests of cosmological models. Given the importance of an accurate knowledge of the Hubble constant, it is perhaps not surprising that this field of observational cosmology has been filled with controversy, and is a very active area of current research. Moreover, it is now a particularly exciting time for this field with the recent launch of the *Hubble Space Telescope*, which is expected to provide an improved determination of the Hubble constant.

HISTORICAL PERSPECTIVE

In 1929, Edwin Hubble announced the discovery of a linear relationship between the distances to galaxies and their velocities of recession. It is the constant of proportionality in this relationship that is referred to as the Hubble constant. The implications of this relationship are profound. Largely in the preceding decade, cosmological models of an expanding universe were worked out by Albert Einstein, Willem de Sitter, Alexander Friedmann, and Georges Lemaître. Hubble's relationship could be interpreted as providing observational evidence that the universe was expanding. Einstein had speculated that the universe on a large scale was homogeneous (i.e., every region is the same as every other region) and isotropic (i.e., that the universe looks the same in every direction). Although apparently very simplistic, models based on these two assumptions have proven to be remarkably successful in describing the large scale structure of the universe to the present day. In a homogeneous, isotropic (and what has come to be termed a Friedmann) universe, the rate of change of distance is dependent only on the distance. All distances increase with a multiplicative scale factor $R(t)$ and as the universe expands, all distance ratios remain constant. The Hubble parameter H is

$$H = (dR/dt)/R(t)$$

and the current value of H is known as the Hubble constant H_0. The units of the Hubble constant are [distance/time/distance] or $[\text{time}]^{-1}$, and it is usually quoted in kilometers per second per megaparsec, where 1 Mpc = 1 million parsecs = 3.26 million ly.

In principle, the determination of the Hubble constant is very simple: Measure the distances (d) to galaxies far enough away that local gravitational perturbations on the velocity field are no longer significant, and then measure their radial velocities (V). The Hubble constant is then simply the slope of the relation between V and d. The measurement of radial velocities from the Doppler shifts of lines in spectra of galaxies is straightforward. However, accurate determinations of distances to increasingly more remote galaxies become much more difficult. The galaxies near enough to have accurate distances measured usually have their velocities perturbed by nearby density enhancements (i.e., nearby galaxies and groups of galaxies). Furthermore, the distances to all galaxies are based on a series of distance scale methods, where at every step, a more distant (and therefore more uncertain) indicator is calibrated, until distances to the freely Hubble expanding galaxies are determined.

Hubble originally obtained a value of about 500 km s^{-1} Mpc^{-1} for H_0. Although this value gave an age for the universe in agreement with those values determined by geological dating at the time, it was eventually revised downward considerably (although it was used until about 1950). There were several reasons for the change. First, there were errors in the photometric luminosity scale available to Hubble. Second, it was shown by Walter Baade that there were two different populations of the stars, known as Cepheid variables, that had been used to calibrate the relationship. These first two corrections increased the inferred distances to galaxies and thereby reduced the value of the Hubble constant to about 150 km s^{-1} Mpc^{-1}. Finally, it was shown by Allan Sandage that the objects that Hubble assumed to be resolved stars in the Virgo cluster were, in reality, H II regions and entire clusters of stars. This final major correction brought the value of the Hubble constant into the range of 50–100 km s^{-1} Mpc^{-1}. Recent estimates of the Hubble constant over the past decade have continued to fall in this range, and an uncertainty of a factor of 2 remains. A value of 50 km s^{-1} Mpc^{-1} has been obtained by Allan Sandage and Gustav Tammann. At the other extreme, Gerard de Vaucouleurs obtains a value of 100 km s^{-1} Mpc^{-1} (which implies a factor of 2 smaller age and size for the universe). Although the persistence of such a large uncertainty for this parameter is frustrating, new improvements in detector technology, specifically the availability of charge coupled devices (CCDs) and near infrared detectors, plus the launch of the *Hubble Space Telescope*, are helping and will continue to help to resolve this discrepancy.

THE DISTANCE SCALE: OBSERVATIONAL DETERMINATION OF THE HUBBLE CONSTANT

Distances to Galaxies

As a first step to obtaining the Hubble constant, the distances to nearby galaxies are needed. Here use is made of so-called primary distance indicators. The definition of a *primary* distance indicator is not fixed; some workers consider a primary indicator to be an object that can be observed within our own Galaxy, whereas others also consider objects whose calibration is based on theory. A distinction is also made between Populations I and II distance indicators. The former are found primarily in the disks of spiral galaxies where star formation is still occurring, and the latter are found in elliptical galaxies in addition to the old components of spiral galaxies. Examples of primary distance indicators are Cepheid variable stars, novae, supernovae, and RR Lyrae stars. The Cepheid variables and type II supernovae (resulting from the fatal explosion of massive stars) are Population I indicators, whereas novae, RR Lyrae stars, and the type I supernovae belong to Population II.

Many distance indicators have been tested over the years, but few have proved to be reliable. Supernovae hold promise, particularly because these extremely bright explosions can be observed to greater distances than any other primary indicator. However, there are many complications in modeling these objects theoretically, and their observational study is made difficult due to the serendipitous nature of their discovery. RR Lyrae stars are the intrinsically faintest of the primary distance indicators, and therefore cannot be observed to as great distances; thus their utility is limited. However, they do provide a useful consistency check for the distances of nearby galaxies. Problems remain in the galactic calibration of novae, and they have not been used in very many galaxies to date; thus they are not yet a well-tested indicator. Probably the physically most well-understood and empirically best-studied class of objects are the Cepheid variables, whose periods of light variation are strongly correlated with their intrinsic luminosities (the period–luminosity relation).

Calibration of the Distance Scale

Even the nearest Cepheids in our own Galaxy are sufficiently distant that direct parallax (or triangulation) measurements are not yet possible for these stars. It is at this stage that it is necessary to back up and begin the step ladder of the distance scale. Fortunately, a few of these bright, massive Cepheid variables have been found in open clusters in our Galaxy. The calibration of the cluster distances rests on the distance to one nearby cluster known as the Hyades, which is close enough and contains a sufficient number of stars that a distance measurement can be obtained by averaging over the motions of many stars (using a so-called moving cluster technique). The distances to the clusters containing Cepheids can then be obtained by comparing the more distant stars with comparable properties to those in the Hyades, and making use of the inverse square law in a technique known as main-sequence fitting. That is, the brightness of a star decreases as the square of its distance from us, and a measure of the apparent brightness of given types of stars with known intrinsic luminosities provides an estimate of the distance to the cluster. This same principle is used in applying the period–luminosity relation for Cepheids: by comparing apparent (measured) and intrinsic (calculated) brightnesses, as predicted by their observed periods, distances can be inferred for Cepheids in more distant galaxies.

A complication in determining the distances to Cepheids (or to most other types of stars, for that matter) is that dust, which is present in the interstellar regions in our own, or in other external galaxies, absorbs, scatters, and selectively reddens starlight. The effects of dust are most severe at short wavelengths where the sizes of the dust grains are comparable to the wavelengths of the radiation; thus the observations are best carried out either in the near-infrared, where the effects are less significant, or at many different wavelengths, so that the effects of obscuration can be directly measured and corrected for.

A further complication in the application of the period–luminosity relation arises from the recognition that the chemical make-up of Cepheids can vary from place to place within a galaxy and amongst galaxies. The effect of these differences on the period–luminosity relation is not yet known. Although current theoretical arguments suggest that the effect is small, no quantitative empirical data yet exist to test this. Despite these lingering uncertainties, however (which also apply with varying degrees of uncertainty to the other distance indicators), Cepheids are still the most accurate of the primary calibrators.

Cepheids are not sufficiently bright that they can be detected out to large distances where local density enhancements no longer significantly perturb the velocities of galaxies. To measure greater distances, secondary indicators are needed that can be calibrated using galaxies whose distances have been measured with primary distance indicators.

Distant Galaxies and the Value of H_0

Examples of secondary distance indicators include: the relation between the rotational velocity of a galaxy and its luminosity (known as the Tully–Fisher relation), the luminosities of planetary nebulae, supernovae (as calibrated by Cepheids, for example), the relation between velocity dispersion and luminosity (known as the Faber–Jackson relation), and the luminosities of the brightest stars in galaxies. As with the choice of a primary indicator, it is desirable to have an accurate, well-tested indicator, and one for which a physical (rather than purely empirical) basis exists. A secondary indicator, which can be observed to large distances is desirable because the contribution of deviations from a smooth Hubble flow will be less significant at greater distances.

Few of the secondary distance indicators meet all of the above criteria very well, and considerable controversy exists over the choice of, and corrections to, the various secondary indicators. The secondary indicators that can be used to large distances include the Tully–Fisher and Faber–Jackson-type relations, and supernovae. The Tully–Fisher relation has been applied to hundreds of galaxies in the field and in clusters. The excellent correlation between the width of the velocity profile in a galaxy and its near-infrared luminosity appears to provide a very accurate and powerful means of measuring distances to galaxies. This form of the Tully–Fisher relation yields a high value for the Hubble constant, $H_0 = 90$ km s^{-1} Mpc^{-1}. There is some controversy over the selection of galaxies used for this method, but the issue has not yet been resolved. This method is applicable to spiral galaxies only. Planetary nebulae may hold promise as a useful distance indicator, but they have not yet been tested in as much detail as many of the other methods, nor can they be observed as far. Faber–Jackson relations are a useful means of obtaining distances to elliptical galaxies. The elliptical galaxies cannot be directly calibrated using the Population I Cepheids, but recently the method has also been applied to the bulges of spiral galaxies and calibrated using Cepheids. The use of type Ia supernovae calibrated by primary indicators such as Cepheids may prove useful; however, there is some debate about whether these objects have too large a range of properties to be useful distance indicators. All of the different methods based on supernovae yield low values of the Hubble constant, closer to 50 km s^{-1} Mpc^{-1}. In summary, recent determinations of the Hubble constant continue to yield values in the range 40–100 km s^{-1} Mpc^{-1}.

SUMMARY AND FUTURE PROSPECTS

The largest derived values of the Hubble constant imply ages for the universe that are not consistent with independent measures of the ages of galactic globular clusters (old, spherical stellar systems containing about a million stars each). The ages of these systems are based on theories of stellar evolution. In addition, the high values are only marginally consistent with radioactive dating and isotope ages. This disagreement is often cited as a reason for preferring smaller values of the Hubble constant; however, the uncertainties that remain in each age-dating method are still large enough that it is not yet reasonable to rule out (or prefer) a particular cosmological model on the basis of the existing data.

Observations with the Hubble Space Telescope

Currently distances to galaxies based on the use of Cepheids have been determined out to roughly 3 Mpc (or 10 million ly). Plans for observing with the superior resolution of the *Hubble Space Telescope* (after its planned corrective optics are installed) include the discovery of Cepheids and obtaining distances for approximately 20 more galaxies extending out to distances about 10 times what is feasible from the ground. In addition, checks of those distances by other independent means are to be carried out, thereby considerably improving the calibration of the Hubble constant. Although the *Hubble Space Telescope* will not solve all of the problems of the extragalactic distance scale, the improvement in the accuracy of the data and the increase in the number of calibrating galaxies that can be observed from space, will considerably increase the precision to which the Hubble constant can be measured.

Additional Reading

Madore, B. F. and Tully, R. B., eds. (1986). *Galaxy Distances and Deviations from Universal Expansion*. Reidel, Dordrecht.

Rowan–Robinson, M. (1985). *The Cosmological Distance Ladder*. W. H. Freeman, New York.

Silk, J. (1989). *The Big Bang*, revised edition. W. H. Freeman, New York.

van den Bergh, S. (1989). The cosmic distance scale. *Astron. Ap. Rev.* **1** 111.

van den Bergh, S. and Pritchet, C. J., eds. (1988). *The Extragalactic Distance Scale*. Astronomical Society of the Pacific, San Francisco.

See also **Cosmology, Observational Tests; Distance Indicators, Extragalactic; Stars, Cephied Variable, Period-Luminosity, Relation and Distance Scale.**

Hubble Space Telescope

Albert Boggess

The *Hubble Space Telescope* was launched into orbit from Cape Canaveral, Florida on board Space Shuttle *Discovery* on April 24, 1990. It is designed to be the first general-purpose astronomical observatory in space. Construction of the observatory began in 1978, following seven years of careful design studies, and was completed in 1989. It consists of a telescope with a 2.4-m (94-in.) primary mirror and a suite of instruments including cameras, spectrographs, a photometer, and an astrometer. Also included is the auxiliary equipment needed to point the telescope, power it, and communicate with the ground control station. In orbit above the Earth's atmosphere it is free of the effects of atmospheric turbulence and the atmosphere's absorptions and emissions. Consequently it can have much better image quality, a wider range of wavelength coverage in both the ultraviolet and infrared, and a greater sensitivity for faint light sources than is possible with ground-based telescopes. The observatory will contribute to all aspects of optical astronomy, but one of its major objectives is to explore the structure and evolution of the universe. For this reason it has been named in honor of the American astronomer Edwin P. Hubble, who discovered, in the 1920s, that the universe expands.

TELESCOPE

The requirement for high image quality places great demands on the optics, the mechanical structure, and the pointing system of the observatory. The telescope mirrors have a Cassegrain configuration with a Ritchey–Chretien figure and should be capable of resolving stellar images separated by only 0.1 arcsec at the test

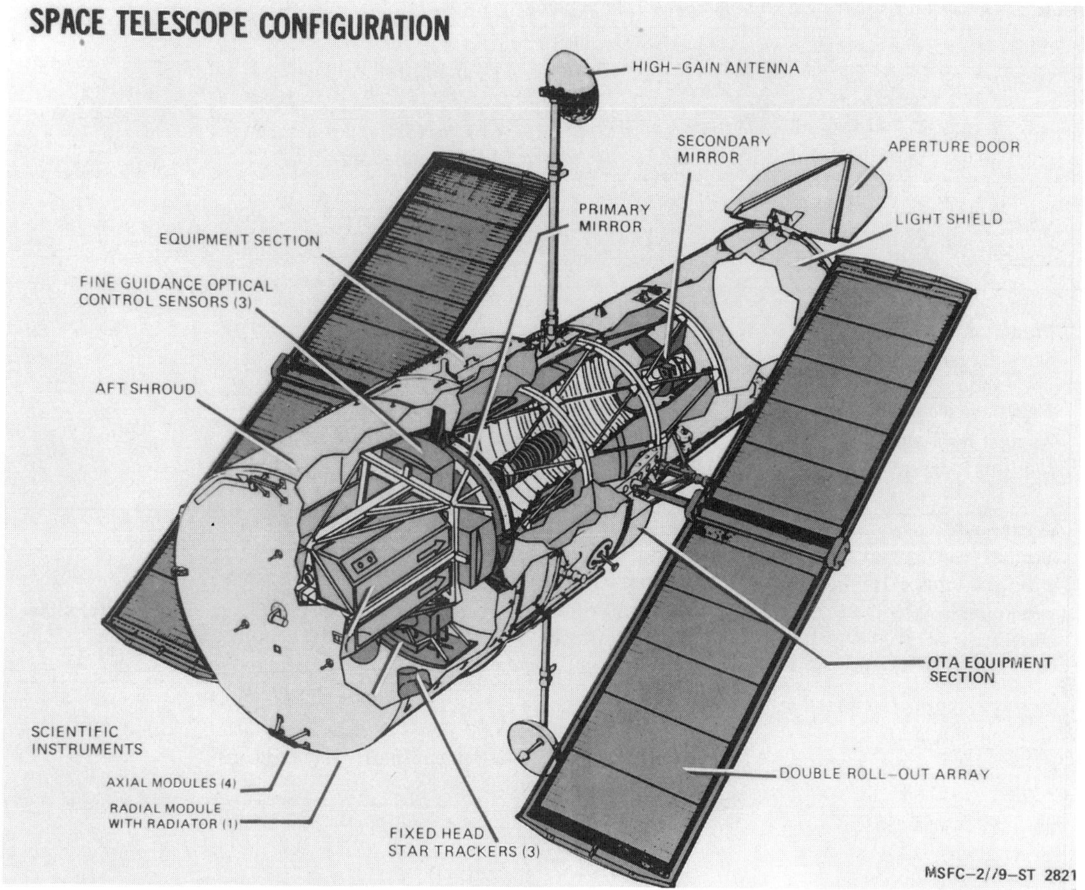

Figure 1. Cut-away drawing of the *Hubble Space Telescope*. Note the primary and secondary mirrors and internal light baffles within the telescope tube. The three fine guidance sensors and five scientific instruments are mounted in the aft section on slide-out racks to facilitate replacement on orbit. The equipment sections contain batteries, computers, and other service modules necessary for operation of the observatory. The solar arrays provide electrical power and the antennae are used for high-speed data transmission via tracking and data relay satellites.

wavelength of 6328 Å. This quality is close to diffraction-limited performance and so the images improve toward shorter wavelengths, with best performance being about 0.05 arcsec in the ultraviolet between 2000 and 3000 Å. The useful spectral range of the telescope is from about 1150 Å in the ultraviolet (limited by the optical coatings) to about 4 μm in the infrared (limited by thermal noise emitted by the 20° C telescope structure).

The pointing control system slews the telescope and points it to within 0.01 arcsec for periods up to 24 h so that jitter or drift will not degrade the image quality during the long integration times needed for very faint sources. The slewing and precise pointing is accomplished by changing the speeds of rapidly rotating reaction wheels. While the telescope is slewing from one target to the next, the reaction wheels are controlled by gyroscopes that sense the rotation and are programmed to stop the slew when the telescope arrives near the next target. Then control is passed to two star trackers, called fine guidance sensors, that are programmed to find and lock on to images of preselected guide stars located close by the target. The fine guidance sensors maneuver the guiding images until the telescope is pointed correctly and then hold those images steady for the duration of the observation.

INSTRUMENTATION

The five scientific instruments are clustered behind the telescope mirror and they simultaneously share the telescope's field of view, making it possible for two or more instruments to observe in parallel if they each have interesting targets in their shares of the field. There are two cameras with a variety of focal ratios, fields of view, and filter combinations. The widest-angle camera mode operates at f/13 with a field of 2.6 arcmin (less than one-tenth the diameter of the moon) and the size of individual image elements, or pixels, is 0.1 arcsec. It has a selection of 48 filters ranging from 1150 Å in the ultraviolet to about 1 μm in the infrared, and a mosaic of four charge-coupled devices is used to record the image. This mode is suitable for photometry of general star fields and for projects such as examining the content and structure of moderate-redshift clusters of galaxies. At the other extreme is an f/288 mode with a field of only 3.8 arcsec, which has 0.007 arcsec pixels and is able to resolve the very best images the telescope is capable of producing. Designed to operate primarily in the ultraviolet, it uses a pulse-counting television system for high sensitivity and contains an occulting finger that can block out the light from a very bright star to look for faint features nearby. Typical uses for this mode are to study very close binary stars and to search for evidence of faint galaxy structures surrounding quasars. Other camera modes are optimized for various purposes, such as imaging planets and other bodies in our solar system and recording images of the faintest possible stars. The practical faint limit for the observatory is about 29th visual magnitude.

There are also two spectrographs: one designed to obtain spectra of very faint targets; the other designed for getting very high

Table 1. Characteristics of the *Hubble Space Telescope Observatory*

Spacecraft					
Mass	11,600 kg (25,500 lb)				
Size	13.1 × 4.3 m (43 × 14 ft)				
Orbit altitude	610 km (380 mile) circular				
Orbit inclination	28.8° inclination				
Power	2400 W				
Data rate	10^6 bits per second				
Pointing stability	0.007 arcsec				
Telescope					
Figure	Ritchey–Chretien				
Primary mirror	2.4 m (94 in.)				
Secondary mirror	0.34 m (13 in.)				
Focal ratio	f/24				
Field of view for instruments	18 arcmin				
Angular resolution	0.1 arcsec at 6328 Å				
Limiting magnitude (visual)	~ 29				
Cameras					
Focal ratios	f/13	f/30	f/48	f/96	f/288
Field of view (arcsec)	158	68	22	11	4
Pixel size (arcsec)	0.100	0.043	0.043	0.022	0.007
Spectral range (Å)	1150–11,000		1200–5000		1200–3000
Detectors	CCD Arrays		Pulse counting SIT TV tube		
Spectrographs					
Resolving power ($\lambda/\Delta\lambda$)	100	200	1000	20,000	80,000
Spectral range (Å)	1150–7000			1150–3200	
Detectors	512 element linear array Digicons				
Photometer					
Field of view (arcsec)	Selectable from 0.4–10				
Spectral range (Å)	1150–8000				
Time resolution (ms)	0.01				
Detectors	Image dissecting tubes				
Astrometer					
Total field (arcmin)	4 × 18				
Instantaneous field (arcsec)	3				
Position accuracy (arcsec)	~ 0.002				
Detectors	Photomultipliers				

spectral resolution data on brighter targets. Both spectrographs use pulse-counting linear arrays, called Digicons, as detectors and so are able to record spectra of only one object at a time. (A more advanced imaging spectrograph is already being designed for on-orbit insertion into the observatory at some future date.) The faint object spectrograph is optimized for maximum sensitivity and provides a choice of eight gratings and one prism to yield spectra of low-to-moderate resolution, covering the range from 1150 to about 7000 Å. Some of the uses of the instrument are spectroscopic studies of very faint stars, chemical abundances of galaxies, and extending measurements of the redshift-distance relation to greater cosmological distances. The high dispersion spectrograph is sensitive only in the ultraviolet, from 1150 to 3200 Å. It can produce spectra of moderate-to-high resolution using a selection of five first-order gratings plus an echelle with two cross dispersers. In its highest resolution mode the instrument is able to resolve better than 0.02 Å at 1500 Å, but at such high dispersion the detector can record only 9 Å of the spectrum at a time. Consequently, the instrument will generally not be used to cover large spectral intervals, but to study specific spectral features that are critical tests for physical conditions in stars and the interstellar medium.

The fifth instrument is a photometer capable of very precise brightness measurements through 18 filter bandpasses ranging in wavelength from 1150 to 6200 Å. In addition, polarimetric measurements can be made in four ultraviolet bands. A particular feature of the instrument is its time resolution of 10 μs. Above the scintillation produced by the Earth's atmosphere, any variations that are observed in optical brightness must be intrinsic to the source, and this instrument can be used to study rapidly varying phenomena, such as light curves of cataclysmic variables and stellar occultations. When a star is occulted by a planet, the starlight passes through deeper and deeper layers in the planet's atmosphere and observations of the changing stellar brightness can serve as a sensitive probe of atmospheric conditions. When stars are occulted by the sharp edge of the moon, the shapes of the light curves can be used to calculate the diameters of individual stars or the separations of unresolved binaries.

There is in effect a sixth instrument on board the observatory. The telescope carries a total of three fine guidance sensors. Two are required to stabilize the telescope. The third unit, which is carried as a spare, can be used as an astrometer to obtain measurements of the angular positions of stars with accuracies approaching 0.002 arcsec. Data from the astrometer will improve parallax measurements for nearby stars, which should ultimately result in a more accurate distance scale and age for the entire universe. The astrometer will also be used to look for small, periodic variations in stellar positions that might be caused by the influence of dark bodies, such as planets.

OPERATIONS AND MAINTENANCE

Communication between the ground station and the observatory is accomplished via the tracking and data relay satellite system, which transmits commands to the telescope for execution and returns the scientific data to the ground. The Operations Control Center is located at NASA's Goddard Space Flight Center in Greenbelt, MD, which is responsible for the health and safety of the satellite and for its long term maintenance. The scientific data are forwarded to the Space Telescope Science Institute in Baltimore, MD, which defines the observing program and collects the data on behalf of scientists from all over the world. In order for the observatory to have a long productive life (planned for 15 years or more), NASA will send astronauts to replace subsystems as they degrade or to add new, more powerful scientific instruments. It is expected that these servicing missions will occur at about three year intervals. Already under design for future installation are a powerful new imaging spectrograph and an infrared camera-spectrograph that will extend the spectral coverage of the telescope out to about 4 μm.

PERFORMANCE IN ORBIT

The check-out and calibration of the observatory in orbit is a complex process that must be completed before the telescope can be devoted exclusively to scientific research. Most of the observatory subsystems are performing exceedingly well. However, a major problem has been discovered resulting from an error in the fabrication of the primary mirror, which produces spherical aberration at the telescope focus. The effect of this aberration is to create star images having small bright cores that preserve the very high spatial resolution of the telescope for brighter sources, surrounded by faint outer halos that limit the ability to detect very faint sources and confuse the images of tightly crowded star fields. Following the discovery of the problem two courses of action have been pursued: The short-term research program of the observatory is being revised to concentrate on observations that are not seriously affected by the aberration, and work is underway to install corrective optics during the first planned servicing mission in order to compensate for the aberration.

In the short term, observations will concentrate on spectroscopy, because the spectral data quality remains exceedingly good, and on camera images of brighter fields. Figure 2 demonstrates the image quality that is obtained with the f/13 camera. It has a resolution of 0.1 arcsec and is an image of Supernova 1987A in the Large Magellanic Cloud. The first servicing mission, which is scheduled to occur about three years after the initial *HST* launch, will carry a new camera specially modified to compensate for the error in the telescope optics. It appears to be feasible to install fore-optics in front of the remaining instruments in order to correct their images as well. At that time the full performance of the observatory should be realized and its full research program can proceed undiminished.

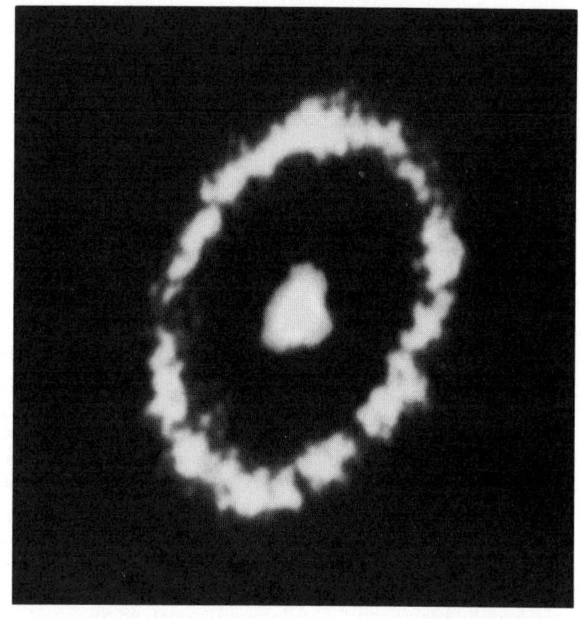

Figure 2. Image of Supernova 1987A taken with *Hubble Space Telescope*'s f/13 camera using a filter to isolate the 5007-Å radiation produced by singly ionized oxygen. The resolution of the image is 0.1 arcsec and it reveals the barely resolved expanding shell of the supernova, surrounded by an elliptical ring of gas that is a relic of the stellar wind phase from the progenitor star.

Additional Reading

Bahcall, J. N. and Spitzer, L., Jr. (1982). The Space Telescope. *Scientific American* **247** (No. 1) 40.

Beatty, J. K. (1985). HST: Astronomy's greatest gambit. *Sky and Telescope* **409**. See also related articles on pp. 295–311.

Fienberg, R. T. (1990). HST: Astronomy's discovery machine. *Sky and Telescope* **79** 366. See also related articles on pp. 373–383.

Fienberg, R. T. (1991). Hubble's agony and ecstasy. *Sky and Telescope*. **81** 14.

Hall, D. B., ed. (1982). The space telescope observatory. NASA Report No. CP-2244, Washington, D.C.

Leckrone, D. S. (1980). The space telescope scientific instruments. *Publ. Astron. Soc. Pacific* **92** 545.

Maran, S. P. (1990). The promise of the Space Telescope. *Astronomy* **18** (No. 1) 38.

See also **Telescopes, Detectors and Instruments, Ultraviolet; Ultraviolet Astronomy, Space Missions.**

I

Infrared Astronomy, Space Missions

James R. Houck

Infrared astronomy is a young science. Although pioneering observations were made by Thomas A. Edison nearly a century ago, the modern era of infrared astronomy began with the development of high-sensitivity infrared detectors using semiconductor technology. By 1964 it was recognized that great gains could be achieved by using a cold infrared telescope operating above the Earth's atmosphere. The limitations to ground-based observations arise from two phenomena: First, water vapor in the atmosphere makes it opaque over most of the infrared spectrum, 1–1000 micrometers [μm (1 μm $= 10^{-6}$ m)]. Second, the thermal emission produced by the atmosphere (even within the relatively clear atmospheric windows and the necessarily warm telescope) completely overwhelms all but the very brightest celestial sources. Indeed, for many ground-based observations, the infrared signal arising from the atmosphere and telescope is more than 100,000 times stronger than the signal from the celestial source. By cooling the telescope to near absolute zero temperature and lifting it above the atmosphere, the situation can be reversed. Under these conditions, observations that are practically impossible using a large, warm, ground-based telescope can be accomplished quickly using a small cold telescope operating above the atmosphere.

The first cryogenic telescopes were operated from sounding rockets. These telescopes made short, typically five-minute, observations before reentering the atmosphere. Despite the short length of the observations, the early instruments provided valuable scientific and technical information. The Air Force Cambridge Geophysical Laboratory conducted a highly successful series of flights, called *Hi Star*, that provided the first all-sky map of the sky at thermal infrared wavelengths: 5, 10, and 20 μm. Other sounding rocket instruments first showed the large luminosities associated with H II region complexes and the galactic center. Previous ground-based observations had been limited to small fields of view, missing much of the total flux. Interestingly enough, subsequent ground-based observations using large fields of view were in agreement with the rocket results. On the technical side, the rocket experiments demonstrated the containment of liquid helium for short periods of time as well as the usefulness of high-sensitivity photoconductive detectors over the wavelength range of 5–120 μm. By the early 1970s, it was becoming obvious that it was technologically possible to build and operate a cryogenically cooled telescope in space.

NASA issued an Announcement of Opportunity in 1974, leading eventually to the selection of the *Infrared Astronomical Satellite* (*IRAS*) and the *Cosmic Background Explorer* (*COBE*) for development and flight. *IRAS* was flown in 1983, resulting in the first high-sensitivity all-sky map for wavelengths from 12 to 120 μm. The *COBE* telescope was launched on November 18, 1989.

THE IRAS MISSION

IRAS was a joint project involving the Science and Engineering Research Council from Britain, the Netherlands Agency for Aerospace Programs (NIVR) from the Netherlands, and NASA from the United States: NASA provided the telescope and its dewar; the spacecraft and telemetry systems were supplied by the Netherlands; Britain was responsible for the ground station and operation center; the Image Processing and Analysis Center at the California Institute of Technology was responsible for the catalog preparation. In addition to the 62-detector main focal plane instrumentation provided by NASA, NIVR provided an objective prism spectrograph and a chopped photometric detector channel. The 60-cm-diameter Ritchey–Chretien beryllium optical system was surrounded by an annular volume filled at launch with 535 liters of superfluid liquid helium in the dewar. At its operational altitude of 900 km in a near-polar orbit, *IRAS* scanned a one-half-degree-wide strip of sky on each orbit. The control system directed the telescope to offset the pointing direction by one-fourth degree between scans. Therefore, there was a one-fourth-degree overlap and a source seen on one side of the focal plane would appear on the opposite side of the focal plane on the next orbit. Precession of the *IRAS* orbit insured that the orbit normal remained pointed toward the Sun throughout the mission lifetime of 300 days. Visible-light star sensors mounted adjacent to the infrared focal plane monitored the pointing direction of the telescope. Using the visible sightings and information provided by the satellite's gyros, it was possible to determine the position of an infrared source to within a few arcseconds, depending on its brightness and effective temperature. Figure 1 shows a cutaway view of the *IRAS* satellite; Fig. 2 shows a schematic view of the scan pattern.

The *IRAS* point source catalog contains nearly 250,000 sources, of which almost 20,000 are identified with extragalactic objects. One of the first and most remarkable *IRAS* discoveries was that some galaxies emit 99% or more of their luminosity in the infrared. Although it is currently believed that most of these objects derive their energy from the rapid formation of young stars (starbursts) there are also indications that nonthermal processes are at work as well. *IRAS* also discovered that our solar system is surrounded by narrow rings of dust that seem to be associated with families of asteroids sharing the same orbits. Presumably the dust rings were formed by the collisions between the asteroids. Dust rings have also been observed around several nearby stars, most notably Vega. In the case of Vega, there is growing evidence that the dust particles cannot remain in orbit for more than a small fraction of the star's current age because of Poynting–Robertson drag. Therefore, there must be some continuing source for the dust particles we now see. Presumably, asteroidal collisions similar to the ones that form the dust seen within our solar system are responsible for the dust seen in the Vega system.

COBE MISSION

COBE, an *IRAS*-sized satellite designed to measure infrared and submillimeter background radiation, successfully completed its primary mission by late 1990. *COBE* carried instruments designed to measure the spectrum and anisotropy of the cosmic microwave background, a remnant of the Big Bang. A third instrument measured the strength of the background radiation throughout the infrared. Preliminary analysis has already shown that the background radiation accurately follows the spectrum of a 2.74 K blackbody. *COBE* has also clearly revealed the motion of the Milky Way galaxy with respect to the cosmic background. These two high quality results already place severe constraints on cosmological models. Further data analysis may reveal more subtle features leading to a better understanding of the origin and evolution of our Universe.

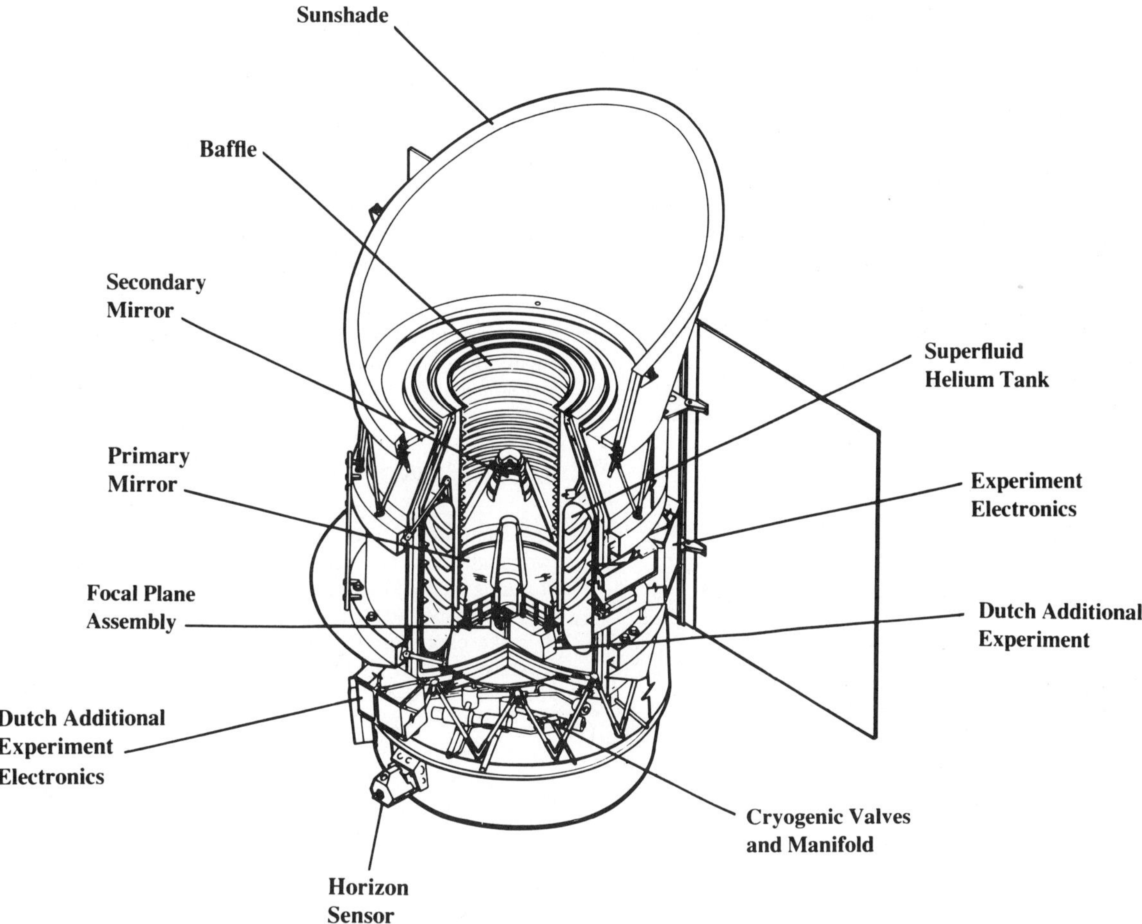

Figure 1. A cutaway view of the *IRAS* satellite showing the 60-cm telescope surrounded by a superfluid liquid helium dewar. (*NASA/JPL photo #P-25295.*)

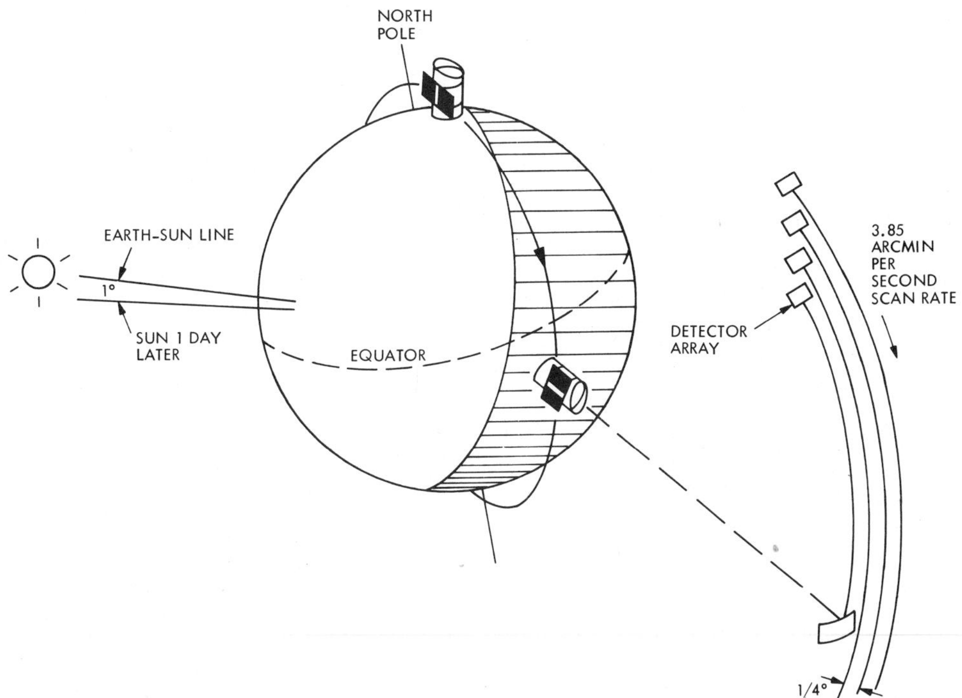

Figure 2. A schematic view of the *IRAS* scan pattern, showing the orientation of the orbit relative to the Sun. The overlap of the scan pattern from one orbit to the next is also shown. (*NASA/JPL photo #P-25786A.*)

THE SPACELAB 2 MISSION

The 15-cm Herschelian infrared telescope on the Space Shuttle's *Spacelab 2* mission during July and August 1985 was designed to complement the observations made by *IRAS*. It scanned the sky in 90°-long strips covering 60% of the sky in a single Shuttle orbit with a focal plane sensitive from 2 to 120 μm. The high infrared background from contaminants released by the Shuttle prevented observations between 8.5 and 23 μm and degraded observations at longer wavelengths. Midway through the mission, a small piece of mylar partially covered the entrance aperture, raising the background at long wavelengths. The 2–3-μm observations were unaffected and produced a high quality map of approximately 60% of the galactic plane.

CURRENT AND FUTURE MISSIONS

The Submillimeter Wavelength Astronomical Satellite (SWAS), which will employ an ambient temperature optical system to map the sky at four different wavelengths will be launched in 1993. The satellite will be sensitive to line radiation from O_2, H_2O, CO, and C II. Because the system will be sensitive over only a very narrow wavelength band, it is less affected by thermal emission from its warm optical system. However, because of its goal to study emission from gases common to the atmosphere, it is essential that it make its observations from space. The Space Infrared Telescope Facility (SIRTF) will be launched later in the decade as the final member of NASA's great observatories program. Whereas *IRAS* was designed to generate an all-sky survey, SIRTF is being designed to operate in an observatory mode in which detailed studies of the size, shape, and spectra of individual objects are possible. Plans are being developed for a *Submillimeter* mission for later in this decade.

The European Space Agency (ESA) is planning to launch the Infrared Space Observatory (ISO) in 1993 or 1994. ISO is similar in size to *IRAS* but, like SIRTF, is designed to operate in the observatory mode. ISO will contain four instruments, which can be characterized as short- and long-wavelength spectrographs and camera–photometers. This complement of instruments will enable ISO to follow up on many of the *IRAS* discoveries.

Additional Reading

Beichman, C. A. (1987). The *IRAS* view of the Galaxy and the solar system. *Ann. Rev. Astron. Ap.* **25** 521.

Gulkis, S., Lubin, P. M., Meyer, S. S., and Silverberg, R. F. (1990). The Cosmic Background Explorer. *Scientific American* **262** (No. 1) 132.

Neugebauer, G., Beichman, C. A., Soifer, B. T., Aumann, H. H., Chester, T. J., Gauttier, T. N., Gillett, F. C., Hauser, M. G., Houck, J. R., Lonsdale, C. J., Low, F. J., and Young, E. T. (1984). Early results from the *Infrared Astronomical Satellite*. *Science* **224** 14.

Rieke, G. H., Werner, M. W., Thompson, R. I., Becklin, E. E., Hoffmann, W. F., Houck, J. R., Stein, W. A., Witteborn, F. C. (1986). Infrared astronomy after *IRAS*. *Science* **231** 807.

Schwarzschild, B. (1990). *COBE* satellite finds no hint of excess in the cosmic microwave spectrum. *Physics Today* **43** (No. 3) 17.

Smith, D. H. and Page, T. L. (1986). *Spacelab 2*: Science in orbit. *Sky and Tel.* **72** 438.

Soifer, B. T., Houck, J. R., and Neugebauer, G. (1987). The *IRAS* view of the extragalactic sky. *Ann. Rev. Astron. Ap.* **25** 187.

Waldrop, M. M. (1990). *COBE* confronts cosmic conundrums. *Science* **247** 411.

See also **Background Radiation, Microwave; Galaxies, Infrared Emission; Observatories, Airborne; Telescopes, Detectors and Instruments, Infrared.**

Infrared Sources in Molecular Clouds

Ronald L. Snell

Infrared sources in molecular clouds are embedded, extremely young stars that have formed out of the gas and dust in these clouds. Interest in these objects arises because the issue of how stars form is a question of fundamental importance to astronomers. The sites of star formation are buried deep within massive interstellar clouds that are opaque to visible light. Thus, evidence that star formation is presently occurring can only be provided by observations at longer, infrared wavelengths, where light from newly formed stars can readily penetrate the dusty environment of their birth sites. From their infrared emissions, astronomers can study the nature and evolution of these new-born stars.

BACKGROUND

The study of infrared sources in molecular clouds encompasses two rapidly growing disciplines: millimeter and infrared astronomy. Stunning technological advances coupled with the construction of large telescopes over the past 20 years have permitted astronomers to make enormous improvements in sensitivity, resolution, and coverage in both these areas. To appreciate the way in which advances in these fields have allowed astronomers to study the process of star formation, it is worth providing some historical perspective.

The study of molecular clouds has emerged only recently as an important aspect of the study of the interstellar medium. In 1969 only a few interstellar molecules were known to exist and they had been detected in only a handful of sources. The importance of a molecular phase to the interstellar medium was not fully appreciated until 1970 when radio telescopes began operating at millimeter wavelengths, a region of the electromagnetic spectrum where the fundamental rotational transitions of many molecules occur. The first molecule detected at these wavelengths was carbon monoxide and subsequent surveys of its emission in the Galaxy revealed the existence of vast interstellar clouds of primarily molecular material. We now know that these molecular clouds comprise approximately one-half of the mass of the interstellar medium in our Galaxy, but more importantly, *these clouds are the sites of all star formation within our Galaxy*.

Infrared astronomy was also a fledgling discipline in the early 1960s and observations were largely concentrated at near-infrared wavelengths [1–4 micrometers (μm), 1 $\mu m = 10^{-6}$ m], targeting optically visible main-sequence and red giant stars. But by the mid- to late 1960s, a number of discoveries related to the early evolution of stars were being made. Excess near-infrared emission was detected from T Tauri stars, arising from heated dust surrounding these young, optically visible stars; the dust is a remnant of the material out of which the stars formed. Infrared sources with *no* optical counterparts were also found in star-forming regions. These were identified with young stars even more deeply embedded within the clouds than T Tauri stars and at a yet earlier stage in their development.

One of the first infrared surveys was undertaken in 1969 by the California Institute of Technology and provided a census of bright sources in the northern sky at a wavelength of 2.2 μm. However, this survey was not sensitive enough to detect many star-forming regions in the Galaxy. Although star-forming regions are much brighter and more readily detectable at longer wavelengths, the Earth's atmosphere is largely opaque to infrared radiation at wavelengths longer than about 10 μm. Therefore, unfortunately, to study this region of the electromagnetic spectrum efficiently, the use of airborne or spaceborne telescopes is necessary.

The first all-sky survey of this kind was the Air Force Geophysical Laboratory Sky Survey. Carried out through a series of rocket flights and observing the sky in the wavelength range of 4–27 μm, it was completed in 1976. The majority of the infrared sources detected by the survey were red giant stars and planetary nebulae, but roughly 20% of the sources could be identified with star-forming regions in the Galaxy. The most recent and most sensitive survey of the infrared sky was obtained with the *Infrared Astronomical Satellite* (*IRAS*), which, operating during 1983, observed at far-infrared wavelengths between 12 and 100 μm. This satellite provided the most complete census to date of star-formation activity within our Galaxy.

Over the past several years, a breakthrough in infrared detector technology has allowed the construction of two-dimensional sensor arrays containing thousands of individual detectors. Infrared cameras containing these detectors, which operate at wavelengths between 1 and 10 μm, are revolutionizing ground-based infrared astronomy and are permitting astronomers to probe the infrared sky faster and to detect much fainter sources than previously. This new technology will have an enormous impact on our understanding of star-forming regions.

SITES OF STAR FORMATION

Within the Milky Way, most (if not all) star formation occurs in dense, very cold molecular clouds. These clouds have densities that are thousands to millions of times higher than the average density of interstellar space and temperatures that are as cold as 10 K (−440°F). Such clouds form by the mutual gravitational attraction of their constituent particles and have masses that range from those of small Bok globules (10–100 times the mass of the Sun) to giant molecular clouds (with as much as 10^7 times the mass of the Sun). The constitution of these clouds is a mixture of gas-phase atoms and molecules and particulate matter called interstellar dust. Dust particles typically have sizes of 0.1 μm and are quite effective at attenuating visible light. In particular, they effectively block the light from newly born stars located in the interiors of molecular clouds, attenuating it by factors of 1,000,000 or more. In fact, the predominance of gas-phase molecules rather than atoms in these clouds is a direct result of the effectiveness of dust in screening out dissociating optical and ultraviolet radiation. This allows a complex network of chemical reactions to proceed deep within these objects, culminating in the formation of numerous molecular species.

The presence of dust produces a seemingly insurmountable barrier to the study of the star-formation process. But dust does not extinguish starlight equally at all wavelengths and is less efficient at extinguishing radiation at infrared and radio wavelengths than at optical wavelengths. Dust suspended in the Earth's atmosphere has similar properties. At sunset, not only is the Sun's brightness greatly reduced as the light traverses a longer path through the atmosphere, but the Sun also appears redder, a consequence of the gas and dust selectively scattering the blue light more effectively than the red light. Because of this selective extinction property of interstellar dust, if visible light from a young star is extinguished by a factor of one million, the near-infrared light at 2 μm is only reduced by a factor of 4. Consequently, we can study stars that are deeply embedded within their parental gas and dust by observing light at infrared wavelengths.

ANATOMY OF A STAR-FORMING CLOUD

With today's technology it is straightforward to survey the embedded populations of young stars in molecular clouds. The anatomy of a typical star-forming molecular cloud is shown in Fig. 1. The upper-left-hand picture is a contour map that represents the intensity of the carbon monoxide emission in the molecular cloud associated with the small region of ionized hydrogen known as Sharpless 228. This emission traces the extent of the molecular gas. The data indicate that the cloud has an angular extent of

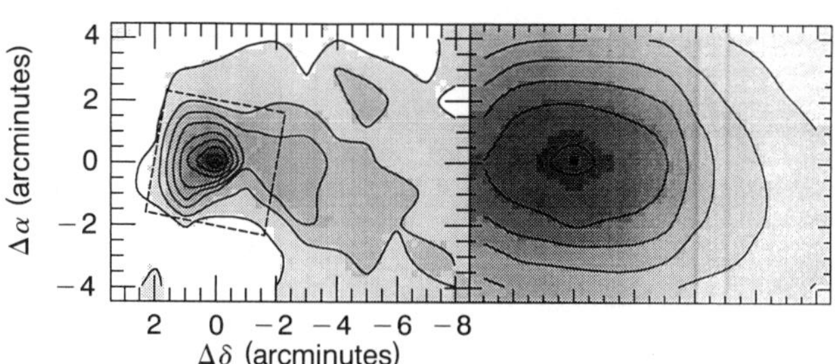

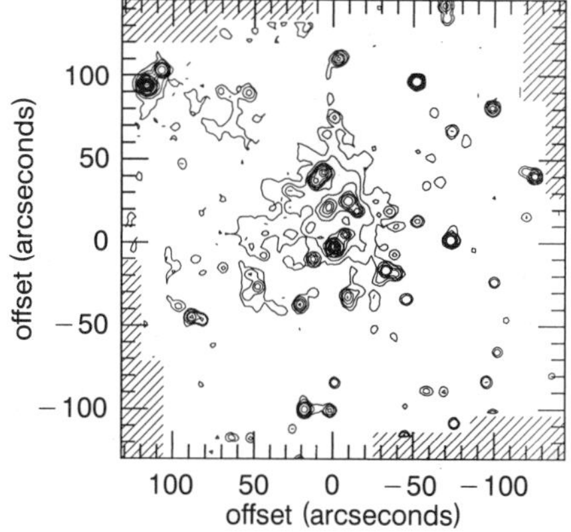

Figure 1. These three images show various aspects of the star-forming region associated with the small H II region Sharpless 228. The upper-left-hand picture shows contours of the intensity of the carbon monoxide emission obtained with the 14-m Five College Radio Astronomy Observatory telescope operated by the University of Massachusetts. The carbon monoxide emission delineates the extent of the molecular cloud. The upper-right-hand picture shows the emission from dust in the cloud emitting at a wavelength of 100 μm (data obtained with the *Infrared Astronomical Satellite*). The region within the dashed box in the upper-left-hand picture has been imaged with an infrared camera on the Wyoming Infrared Observatory 2.3-m telescope; the 1.2-μm image obtained is shown in the bottom picture. Over 60 newly born stars can be seen in this image. The data shown in this figure were obtained by John Carpenter, Ronald Snell, and Peter Schloerb of the University of Massachusetts.

roughly 6 arcminutes (6 arcmin = 0.1°). At the distance of S 228 [about 2600 parsecs (pc) or 8000 light-years (ly)] this corresponds to a diameter of 4.5 pc (about 14 ly). The mass of the molecular cloud is approximately 5000 times the mass of the Sun. The dashed square overlaying the carbon monoxide contour map indicates the part of the cloud that has been imaged with a near-infrared camera; the results of these observations at a wavelength of 1.2 μm are shown as a contour map at the bottom of Fig. 1. The image reveals over 60 individual sources of infrared emission, each of which is most likely a young star that has formed out of material in this cloud. One also can see extended infrared emission surrounding these young stars; this is infrared light reflected off the dust grains that lie near these stars.

The faintest sources detected in the infrared image are thought to be very young T Tauri-like stars that, unlike their optically visible counterparts, are still buried within the cloud. They will evolve with time into stars very similar to the Sun. By contrast, the brightest sources in the image have luminosities thousands of times greater than the Sun and will evolve to become early main sequence stars with spectral types O and B.

The light from the young stars near S 228 also has an effect on the gas and dust that surround them. The dust is heated by the starlight to temperatures that allow it to radiate at infrared wavelengths. Dust particles close to the young stars are heated to the highest temperatures (1000–1500 K) and emit at near-infrared wavelengths. Further from the stars, the dust is much colder (20–30 K) and emits at longer, far-infrared wavelengths. In fact, most of the sources have large infrared "excesses" seen at a wavelength of 2.2 μm, produced by heated dust near the stars, but the entire star-forming region is also glowing at far-infrared wavelengths (see the upper-right-hand picture in Fig. 1) due to heated dust distributed throughout the cloud.

PROPERTIES OF INFRARED SOURCES

A number of properties of infrared sources are readily observable, including location, size, and luminosity. High-resolution imaging can be used to pinpoint the location of star-formation sites within molecular clouds and to resolve individual young stars in regions where numerous stars are forming together (like that shown in Fig. 1). The size of an infrared source is largely dependent on the wavelength used for the measurement: in the far infrared, emission rises from an extended dust envelope and the source can be quite large, whereas, in the near infrared, emission arises from a much smaller region and sources are generally too small to be resolved. The total luminosity is important because it is directly related to the mass of the star that is forming: The most luminous infrared sources are newly formed stars with the greatest mass.

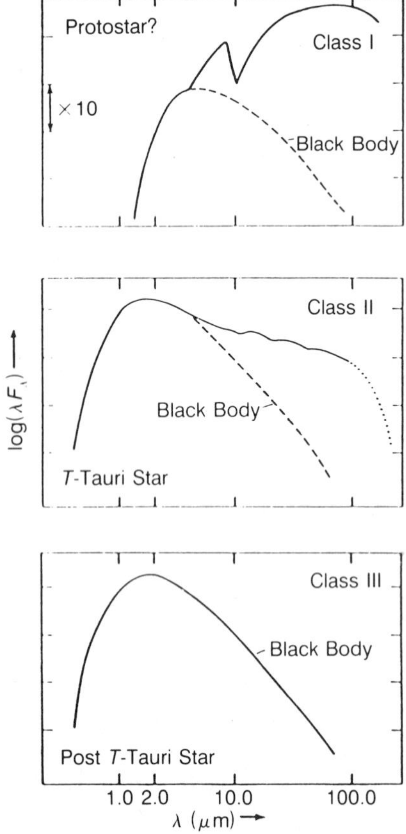

Figure 2. A schematic diagram showing the spectral energy distribution for the three broad classes of embedded, young stars. The classification is thought to be an evolutionary sequence, with class I objects the youngest and class III objects the oldest. (*This figure has been provided by Charles Lada.*)

Though location, size, and luminosity provide essential information, one of the most powerful diagnostic properties of an infrared source is its spectral energy distribution, that is, the variation in the intensity of its radiation as a function of wavelength. The spectral energy distributions found for infrared sources in molecular clouds can be divided into three broad classes, whose characteristics are shown schematically in Fig. 2. Class I sources have spectra that rise steeply to longer wavelengths and produce most of their emission at wavelengths longer than 20 μm. Class II sources (intermediate between class I and III objects) have much flatter spectra with nearly equal contributions at near- and far-infrared wavelengths. Class III sources have spectra that decline sharply beyond 2 μm and radiate most of their energy at wavelengths shorter than 5 μm.

The spectra of these three classes of infrared sources can be understood in terms of an evolutionary scenario. The class I sources are thought to be the stars in their earliest stages of development and are often called protostars. Star formation is initiated by the gravitational collapse of a portion of a molecular cloud. An embryo star is formed at the center of this collapse, surrounded by an extended opaque envelope of material that continues to fall onto it. The angular momentum present in this infalling material prevents it from falling directly onto the embryo star; instead, it forms a flattened disk surrounding the star and accretion onto the embryo star occurs through this disk. During this infall phase, radiation from the central object is absorbed by the dust in the extended envelope and is emitted primarily at far-infrared wavelengths, resulting in a class I spectrum. However, during the infall phase stars also develop powerful stellar winds. These winds eventually disrupt the extended envelope, preventing further material from replenishing the material in the accreting disk. More of the near-infrared light produced by the young star and its disk can now escape and the star now has a class II spectrum. As time goes on, material in the disk is either incorporated into planets or dispersed by the stellar winds. The infrared excess is then diminished, as most of the radiation arises from the star itself although it may still be attenuated and reddened by the dust in the surrounding molecular cloud; these stars have a class III spectrum. Thus, this broad classification scheme allows astronomers to estimate the evolutionary status of infrared sources.

There are still many questions to be answered about the star-formation process, but decades of infrared observations coupled with theoretical studies have now provided astronomers with a basic understanding of how stars form. Astronomers can locate and classify infrared sources in molecular clouds and begin to quantify the rate at which new stars are being formed in our Galaxy.

Additional Reading

Allen, D., Bailey, J., and Hyland, R. (1984). Infrared images of the Orion nebula. *Sky and Telescope* **67** 222.

Cohen, M. (1988). *In Darkness Born*: *The Story of Star Formation*. Cambridge University Press, Cambridge, England.

Gatley, I., DePoy, D., and Fowler, A. (1988). Astronomical imaging with infrared array detectors. *Science* **242** 1217.

Gehrz, R. D., Black, D. C., and Solomon, P. M. (1984). The formation of stellar systems from interstellar molecular clouds. *Science* **224** 823.

Genzel, R. and Stutzki, J. (1989). The Orion molecular cloud and star-forming region. *Ann. Rev. Astron. Ap.* **27** 41.

Habing, H. and Neugebauer, G. (1984). The infrared sky. *Scientific American* **251** (No. 5) 49.

Larson, R. B. (1987). Star formation, luminous stars, and dark matter. *American Scientist* **75** 376.

See also **Bok Globules; Galaxies, Molecular Gas in H II Regions; Interstellar Clouds, Giant Molecular; Interstellar Medium, Dust Grains; Protostars; Star Formation, Propagating; Stars, T Tauri.**

Interplanetary and Heliospheric Space Missions

James A. Van Allen

In space physics the *heliosphere* is defined to be the region (not necessarily spherical in shape) around the Sun within which there is a supersonic, radial flow of hot ionized gas, usually called the solar wind, from the solar corona. The location of the outer boundary of the heliosphere is not yet known but is estimated from a combination of observational and theoretical considerations to be at a heliocentric radial distance of about 100 astronomical units [AU (1 AU $= 150 \times 10^6$ km)]. The term *interplanetary space* means literally the region between the known planets, extending perhaps to the aphelion of Pluto's orbit, but in common parlance is more or less synonymous with the term heliosphere. The corresponding term for the region beyond the heliosphere is *nearby interstellar space* or *local interstellar medium*.

Any spacecraft whose position is outside the magnetosphere of any planet and beyond the immediate vicinity of the Sun but within the heliosphere is said to be in interplanetary space.

SCIENTIFIC OBJECTIVES OF INTERPLANETARY MISSIONS

A primary scientific objective of interplanetary missions is the in-situ observation of the physical properties of the solar wind. These properties include the bulk flow velocity, the number density and temperature of electrons and of each ionic component, the distribution of states of ionization of ions having $Z > 1$, and the vector magnetic field. Collective and large-scale phenomena are shock waves and other waves in the hot plasma, the three-dimensional structure of the interplanetary magnetic field, and the relationship of these features to solar activity, both specific and general. Ideally, measurements are made continuously by a three-dimensional array of spacecraft and over periods of time comparable to or longer than the 11- or 22-yr cycle of solar activity.

Kindred objectives are the following: the observation of the propagation of energetic electrons and ions that are emitted sporadically by the Sun (and by Jupiter); the time-varying modulation and spatial distribution of the spectral intensity of the galactic cosmic radiation and of the more recently discovered anomalous component thereof, the latter being possibly a heliospheric product of the interstellar plasma; and the local acceleration of electrons and ions in the interplanetary medium.

There are also objectives of a quite different nature. Dust (micrometeoroids) in interplanetary space can be observed and, by precise tracking of spacecraft, searches can be made for previously unknown planets and for gravitational waves. Two or more spacecraft at quite different heliographic longitudes or latitudes provide unique opportunities for stereoscopic observation of solar x-ray flares and the consequent determination of the altitude dependence of their emission in the solar atmosphere. Also, occultation of a spacecraft by the Sun provides high-quality data on the radial dependence of electron density in the solar corona.

SOME PRACTICAL CONSIDERATIONS

Of the large number of space missions to date, relatively few have had purely interplanetary objectives. Indeed most interplanetary observations have been made by spacecraft in loose orbits about the Earth and the Moon or during the cruise phases of missions to other planets along trajectories near the plane of the ecliptic, which is also near the heliographic equator. Hence, there continues to be meager knowledge of interplanetary phenomena at high heliographic latitudes and none of an in situ nature.

NOTEWORTHY INTERPLANETARY OBSERVATIONS

Early interplanetary observations were obtained on flights intended to reach the Moon. The first direct measurements of the solar wind were obtained with plasma probes on the Soviet *Luna 1, 2,* and *3* lunar missions (1959) and later on a 1961 Soviet flight toward Venus. These investigations were extended by the Earth-orbiting but short-lived *Explorer 10* in 1961. The most comprehensive and definitive of the early observations of the solar wind were the continuous ones by an electrostatic spectrometer on *Mariner 2* during four months of interplanetary flight en route to Venus (1962). Valuable cosmic-ray, solar energetic particle, and solar wind magnetic field measurements were made on the early United States lunar flights of *Pioneer 3* (1958), *Pioneer 4* (1959), and *Pioneer 5* (1960).

Solar energetic particles have been observed effectively over the polar caps of the Earth by even low-altitude spacecraft, as shown by *Explorer 7* in the period 1959–1961. The quasiinterplanetary conditions there are now understood to be the result of "open" magnetic field lines and hence geomagnetic cutoffs approximating zero energy. Also, it is feasible to investigate solar energetic particles by spacecraft in loose, low-inclination Earth orbits, either those with orbital radii of the order of 10 Earth radii or greater or those in eccentric orbits having apogee distances of that order. Especially fruitful and comprehensive interplanetary observations have been those by *Explorer 33* (1966–1971) in an Earth orbit with apogee at about 80 Earth radii, *Explorer 35* (1967–1973) in lunar orbit, and *Explorer 50* (*IMP 8*, 1973–present) in an Earth orbit ranging from 25 to 46 Earth radii.

The long duration heliocentric missions of *Pioneer 6, Pioneer 7, Pioneer 8,* and *Pioneer 9* (from 1965 onward) were purely for interplanetary purposes. They expanded knowledge of the solar wind in the radial range 0.8–1.1 AU and of the interplanetary propagation of solar energetic particles.

The interplanetary missions of the German–American spacecraft *Helios 1* (1974–1984) and *Helios 2* (1976–1981) made noteworthy advances in knowledge of cosmic rays, the solar wind, and the distribution of micrometeoroids inward to a distance of 0.29 AU from the Sun.

The *Mariner 4* (1964) and *Mariner 5* (1967) spacecraft to Mars and Venus, respectively, carried detectors for the measurement of magnetic fields, cosmic rays, solar x-rays, and solar energetic particles during the cruise phase as well as in the vicinity of the target planets. Two of the special achievements of these spacecraft were the discovery of solar electron events and, in conjunction with *Explorer 35*, the first stereoscopic observation of soft x-rays emitted by solar flares. Also, they contributed to multipoint measurements of solar energetic particles as did the *Mariner 10* mission (one flyby of Venus and three flybys of Mercury, 1973–1975).

During the period 1978–1982 the *International Sun Earth Explorer 3 (ISEE 3)* was in a halo orbit around the gravitational Lagrangian (L1) point of the Earth–Sun system at a distance of about 240 Earth radii sunward of the Earth, thus serving as an upstream monitor of solar wind characteristics that were convected past the Earth about an hour later.

The five currently active U.S. spacecraft *Pioneer Venus Orbiter* [*PVO* (1978–present)], *Pioneer 10* (1972–present), *Pioneer 11* (1973–present), *Voyager 1* (1977–present), and *Voyager 2* (1977–present) have been and continue to be the "heavy hitters" of interplanetary–heliospheric missions. *PVO* has accumulated over a decade of observations of the solar wind in the vicinity of Venus. The other four have pioneered in observations beyond the orbit of Mars (Fig. 1) and thus have provided and continue to provide an unprecedented and homogeneous body of data on the radial and temporal variations of properties of the solar wind and the intensity of the galactic cosmic radiation. *Pioneer 10* is the most remote artificial object in the universe, having reached (in late 1990) a heliocentric radial distance of 50 AU. All four of the outer helio-

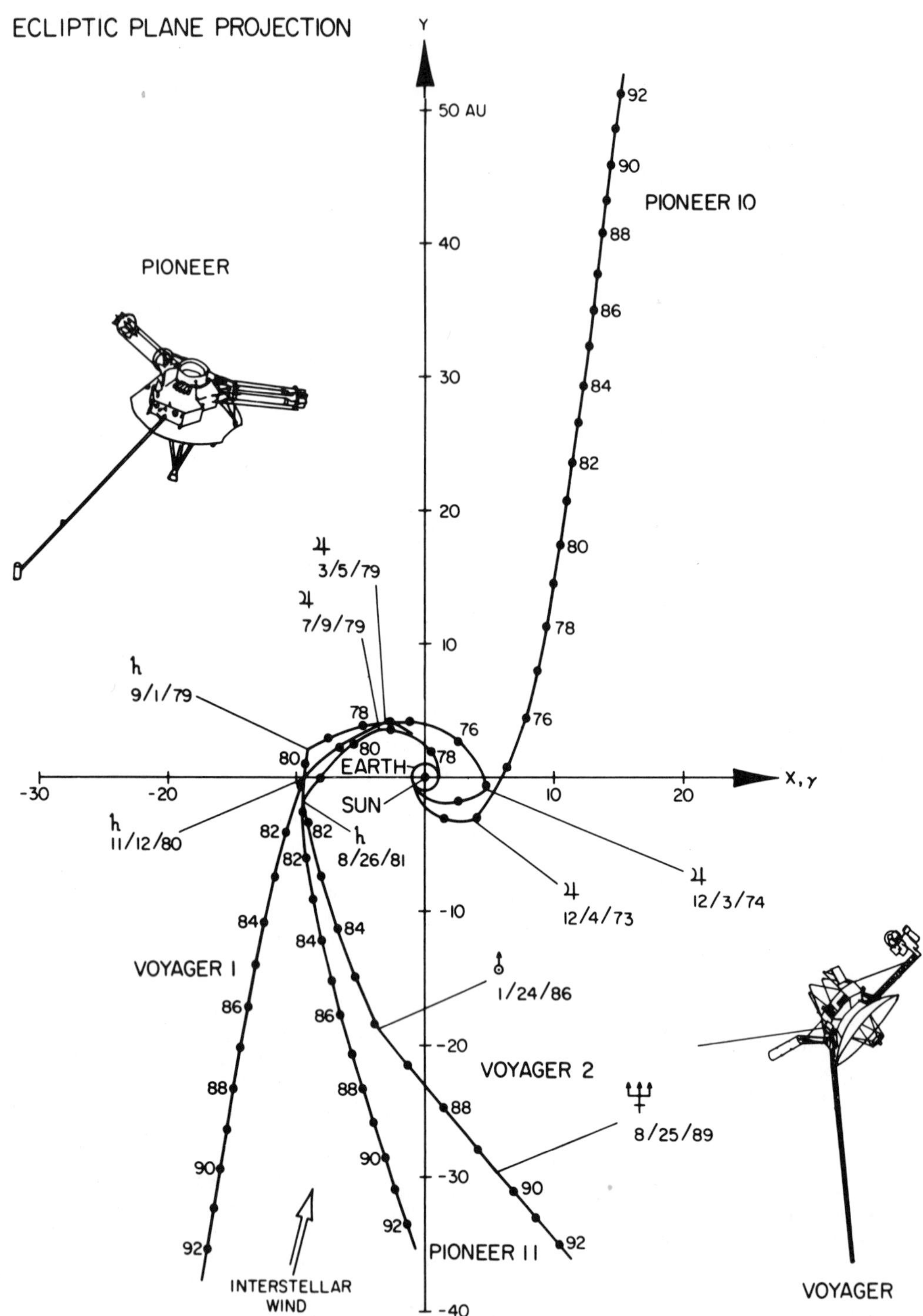

Figure 1. An ecliptic-plane projection of the trajectories of *Pioneer 10, Pioneer 11, Voyager 1*, and *Voyager 2*. The asymptotic directions of escape from the solar system are toward heliocentric longitudes λ and latitudes β as follows: 83°3, +2°90; 290°2, +12°8; 260°8, +35°5; and 310°9, −47°5, respectively.

spheric spacecraft are on hyperbolic escape trajectories from the solar system. They will reach radial distances of 100 AU during the following calendar years: *Pioneer 10*, 2010; *Pioneer 11*, 2019; *Voyager 1*, 2007; *Voyager 2*, 2013. The operational survival of one or more of them to or beyond these dates is, of course, conjectural but appears reasonable on the basis of known considerations. A special goal of their extended missions is to obtain observational data on the interstellar medium and on the intensity and other properties of the galactic cosmic radiation beyond the heliosphere.

The European Space Agency's mission *Ulysses* is the only one intended to make interplanetary observations at high heliographic latitudes. The *Ulysses* spacecraft was launched on October 6 1990. The plan is to fly it over Jupiter's north pole so that its subsequent heliocentric trajectory will pass under the south polar region of the Sun at about 1.8 AU in 1994 and over the north polar region at a similar radial distance in 1995. The results of this mission are awaited eagerly by all investigators who wish to learn the three-dimensional properties of the heliosphere.

Additional Reading

Bennett, G. L. (1990). Rendezvous with a star. *Sky and Telescope* **80** 496.

Fillius, W. (1989). Cosmic ray gradients in the heliosphere. *Adv. Space Res.* **9** (No. 4) 209.

Murray, B. (1989). *Journey into Space*. W. W. Norton, New York. R. Ramaty, T. L. Cline, and J. F. Ormes, eds. (1987). *Essays in Space Science*. NASA Conference Publ. 2464, NASA, Washington.

Simpson, J. A. (1989). Evolution of our knowledge of the heliosphere. *Adv. Space Res.* **9** (No. 4) 5.

Smith, E. J. (1989). A NASA heliospheric program. *Adv. Space Res.* **9** (No. 4) 21.

Van Allen, J. A. (1990). Magnetospheres, cosmic rays and the interplanetary medium. In *The New Solar System*, 3rd ed., J. K. Beatty and A. Chaikin, eds. Sky Publishing Corp., Cambridge, Mass. p. 29.

Wenzel, K.-P., Marsden, R. G., Page, D. E., and Smith, E. J. (1989). *Ulysses*: The first high-latitude heliospheric mission. *Adv. Space Res.* **9** (No. 4) 25.

See also **Cosmic Rays, Space Investigations; Earth Magnetosphere, Space Missions; Heliosphere; Interplanetary Magnetic Field; Interplanetary Medium, Shock Waves and Traveling Magnetic Phenomena; Interplanetary Medium, Solar Cosmic Rays; Interplanetary Medium, Solar Wind; Interplanetary Medium, Wave-Particle Interactions; Interstellar Medium, Local; Planetary Magnetospheres, Jovian Planets; Sun, Coronal Holes and Solar Wind; Sun, High Energy Particle Emissions; Zodiacal Light and Gegenschein.**

Interplanetary Dust, Collection and Analysis

Donald E. Brownlee

Interplanetary dust particles in the solar system constitute a tenuous but accessible circumstellar dust system that can be studied in situ by spacecraft and by direct collection and recovery of samples. Approximately 10,000 tons of submillimeter particles impact the Earth each year and collection from orbit, from the atmosphere, and from the Earth's surface provide an important means of obtaining detailed information on their properties. The majority of interplanetary particles have been freshly liberated from comets and asteroids and collected dust provides a unique information source on these primitive bodies. Although individual particles are small, they are large relative to early solar system grains and components contained inside them; they are also large relative to the capabilities of modern analytical instrumentation. The collected particles cover the size range that includes the bulk of the mass of extraterrestrial material that annually accretes onto the Earth (Fig. 1) and the dust samples are probably the least-biased samples of asteroids and short-period comets that are available for laboratory study. Favorable orbital dynamics and less severe atmospheric entry conditions allow small meteoroids that reach the Earth and survive atmospheric entry to be less-biased samples than are the larger meteoritic samples that are collected from the Earth's surface as conventional meteorites. Identification of generic asteroid and comet particles is important because properties of these bodies provide information on environments, processes, and materials that formed and influenced grains over a very broad range of distance in the solar nebula.

SOURCES OF INTERPLANETARY DUST

Dust in the interplanetary medium is continually depleted due to the effects of dust–dust collisions and orbital decay caused by the drag component of light pressure, the Poynting–Robertson effect (after the physicists John Henry Poynting and H. P. Robertson). The solar system loss rate of a few tons per second is replenished by fresh debris liberated from comets and asteroid collisions. By analogy to the established origin of visual meteors, it has widely been assumed that comets are the major source of dust, but the relative importance of cometary and asteroidal contributions is actually poorly known. The detection of infrared emission from dust in the asteroid belt by the *Infrared Astronomical Satellite* is evidence that asteroids are an important dust source. Dust generated by asteroid collisions spirals inward from the asteroid belt, due to Poynting–Robertson drag, and reaches 1 AU on time scales of 10,000 to over 100,000 years, depending on particle size. Comet dust is generally released on eccentric orbits that are initially Earth-crossing. Virtually all asteroids and short-period comets produce dust that reaches the Earth, a situation that is in contrast to the case of conventional meteorites where apparently no comets and only a limited number of asteroids produce samples that both reach the Earth and survive atmospheric entry.

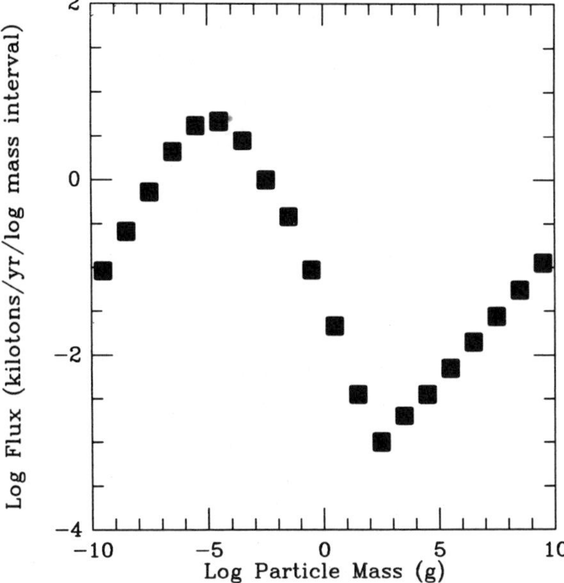

Figure 1. The mass flux of meteoroids impacting the Earth.

As a source of meteoritic material, the dust samples are distinct from meteorites because the processes that transport submillimeter objects to the Earth and that permit them to survive atmospheric entry are different and less selective than those that transport the larger bodies that become conventional meteorites. Conventional meteorites are transported to Earth-crossing orbits by rare and highly selective gravitational perturbations, and they survive atmospheric entry only if they are relatively strong rocks. In contrast, the democracy of the Poynting–Robertson effect causes all small particles to migrate toward the Sun, and once at the Earth these small particles do not have to be strong to survive atmospheric entry. They decelerate from cosmic velocity at very high altitudes where the air density and the associated maximum aerodynamic ram pressure are orders of magnitude smaller than those that must be survived by conventional meteorites.

COLLECTION

The most pristine interplanetary dust samples are collected in relatively clean stratospheric air where the slowly falling particles have decelerated and are concentrated by a factor of one million over their density in space. The stratospheric particles are the least heated and least contaminated interplanetary particles that can be collected, but they are limited to sizes below 100 μm because the flux of larger particles is too low to permit airborne collection. The infalling flux of 10-μm particles is 1 m^{-2} day^{-1} but that of 100-μm particles is less than 1 m^{-2} yr^{-1}. Using impaction plates pushed through stratospheric air by NASA U2 aircraft, it has been possible to collect nearly 1000 interplanetary particles in the 2- to 50-μm size range. Most of the particles in this size range did not melt during atmospheric entry and thus are true "micrometeorites" as defined by Fred L. Whipple. The retention of tracks caused by irradiation by solar flare particles indicates that more than half of the 10-μm particles were not heated above 600°C, a temperature at which tracks are annealed. The particles collected in the stratosphere are removed from collection surfaces and they are studied in a variety of different instruments. Generally, their distinction from contaminants is straightforward because their elemental compositions are similar to that of chondrites (stony meteorites) and quite different from terrestrial materials. In difficult cases, the extraterrestrial origin of an individual particle can be tested by the detection of either tracks caused by irradiation by solar cosmic rays or by detection of helium or neon implanted by the solar wind.

Particles larger than 100 μm are too rare to collect from the atmosphere, but they can be obtained from selected environments on the Earth's surface where they accumulate with minimal contamination by terrestrial particles of similar size. These environments include the midocean seafloor and certain polar ice deposits. Particles in the 50-μm to 1-mm size range have been collected in abundance from deep ocean sediments, from "on-ice" lakes that seasonally form near the fringes of the Greenland ice cap, and from water produced by melting tons of ultrapure Antarctic ice. In contrast to the stratospheric particles, most of the samples collected below the atmosphere are large enough that they were strongly heated during atmospheric entry and melted to form spheres. The abundances of cosmic ray produced Al^{26} and Be^{10} in the spheres indicate that typical spherules are, in fact, just melted small particles and not spray droplets released from larger meteoroids. Some particles in the 0.1- to 1-mm size range manage to survive atmospheric entry without being heated to their melting points, typically near 1400°C. These "giant" micrometeorites survive because of the combined effects of low entry velocity and low incidence angle into the atmosphere.

Interplanetary dust can also be collected from spacecraft. The great advantage of space collection is that it is possible to use electronic techniques to determine the speed and impact angle of individual particles. When this is done with a new generation of collectors proposed for the space station, it will be possible to distinguish comet and asteroid particles on the basis of their orbital parameters. In principle, this approach could also be used to identify and collect interstellar grains in transit through the solar system. The disadvantages with space collection are that the collection rate is much lower than is possible in the atmosphere or in sediments and that particles are degraded due to the high impact velocity onto collection devices. The atmosphere is an advantaged collector in this respect because particles are gently decelerated over distances of tens of kilometers whereas spacecraft experiments must stop particles on distance scales of micrometers to centimeters. Most of the analysis of meteoroids collected in space has been done by studying meteoroid residue found inside craters in metal or around perforation holes found in plastic multilayer insulation blankets. Future collections will also use low-density media, such as polymer foam and silica aerogel, that more gently capture hypervelocity particles with greatly reduced thermal and shock effects. Because particles collected in space will be generally more degraded than those collected in or below the atmosphere, one of the major goals of future dust collections will be use space techniques with orbital parameter measurements to identify the origins of major classes of extraterrestrial particles found in the atmospheric and surface collections.

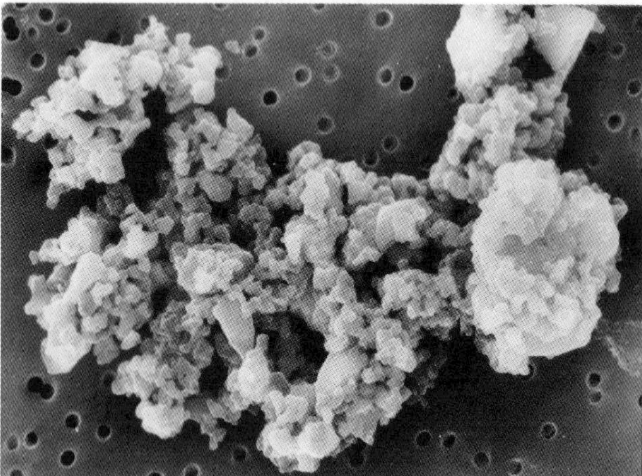

Figure 2. A scanning electron microscope image of a 10-μm porous micrometeorite collected from the stratosphere.

ANALYSIS AND PARTICLE PROPERTIES

The particles are examined by a diverse array of instruments that determine elemental, mineralogical, and isotopic composition on sample masses as small as a femtogram. Except for bulk composition, much of the analytical work has concentrated on the 10-μm stratospheric particles because they are the best preserved. Most of the stratospheric particles are black, fine-grained materials composed of large numbers of small mineral grains. Typical particles, even those only a few micrometers in diameter, have elemental abundances that closely match undifferentiated solar abundances for condensable elements. The only major common compositional difference between dust and bulk chondrites is that some dust particles are considerably more carbon-rich than the most carbon-rich meteorites. Due to their finer grain size, interplanetary particles are generally more uniform in composition at the micrometer size scale than meteorites, but on the submicrometer size scale there is considerable compositional diversity due to the preponderance of individual mineral grains of this size. The most common constituents are olivine, pyroxene, hydrated silicates, iron sulfide, glass, iron carbide, and amorphous carbon although the abundance and composition of the components varies greatly among various particle types.

There are many different types of particle but most of those with chondritic elemental composition can be grouped into two general classes. One type is dominated by hydrous minerals, such as smectite or serpentine, and the other type is dominated by anhydrous minerals, such as olivine and pyroxene. The hydrous type is usually compact without pore spaces, similar to chondrites, whereas many of the anhydrous types are very fragile and much more porous than any chondrite (Fig. 2). Some of the hydrous particles have properties that are equivalent to features in carbonaceous chondrites that have been attributed to formation by aqueous alteration occurring inside a warm and wet asteroidal parent body. These include "framboidal" clusters of magnetite, the presence of carbonates and leaching of calcium from fine-grained matrix. The similarity between the hydrous micrometeorites and carbonaceous chondrites suggests that a fraction and perhaps all of the hydrated particles are of asteroidal origin. If aqueous alteration is only possible on asteroidal parent bodies, then this would be a strong argument that the hydrous particles are asteroidal. The porous-anhydrous particles are quite unlike the hydrous micrometeorites or any meteorite type in regards to porosity, mineralogy, and structure. Because their porosity is similar to that of cometary meteoroids, it is likely that some of these particles are cometary in origin. The anhydrous and the hydrated particles could, respectively, be typical samples of short-period comets and main-belt asteroids although there is no way to prove this assertion at the present time. If, however, this is correct, then it suggests that a major difference between cometary and asteroidal materials is parent body heating. In comets, ice is lost by sublimation, whereas in some ice-bearing asteroids, internal heating melts ice and causes aqueous alteration. Such differences could be due to differences in parent body size and internal heat sources such as Al^{26} decay or inductive heating from energetic stellar winds in the early solar system. In a sense the anhydrous-porous particles might be considered to be more "primitive" because alteration of this material type inside a wet parent body could in principle produce the hydrous particle types as well as some types of carbonaceous chondrites.

It is nearly certain that the collections of interplanetary dust contain both cometary and asteroidal material but it is also possible that they contain presolar grains preserved inside particles. This is particularly the case for cometary materials that accreted in regions of the solar nebula where presolar solids were likely to have survived. The porous particles are loose aggregates of submicron grains, a structure and composition that is a reasonable result of simple accretion of interstellar grains and ice. The anhydrous nature of minerals in the most porous particles is also consistent with infrared measurements that indicate interstellar grains are also largely anhydrous. Isotopic composition provides important clues for detection of possible presolar components although it is very difficult to conduct isotopic analyses of samples the size of typical interstellar grains. The most direct evidence of a presolar component in the particles has been the discovery of micrometer-sized "nuggets" (inside particles) that have deuterium to hydrogen ratios as much as a factor of ten above the terrestrial ratio. This unprecedented level of D/H enhancement is analogous to the extreme D/H fractionation in molecular clouds caused by ion–molecule reactions and is much higher than fractionation that could have occurred by plausible reactions in the solar nebula.

Additional Reading

Bradley, J. P., Sandford, S. A., and Walker, R. M. (1988). Interplanetary dust particles. In *Meteorites and the Early Solar System*, J. F. Kerridge and M. S. Mathews, eds. University of Arizona Press, Tucson, p. 861.

Brownlee, D. (1985). Cosmic dust: Collection and research. *Ann. Rev. Earth and Planet. Sci.* **13** 147.

Mackinnon, I. D. R. and Rietmeijer, F. J. M. (1987). Mineralogy of chondritic interplanetary dust particles. *Rev. Geophys.* **25** 1527.

Sandford, S. (1987). The collection and analysis of extraterrestrial dust particles. *Fundamentals of Cosmic Physics* **12** 1.

See also **Interplanetary Dust, Dynamics; Interplanetary Dust, Remote Sensing; Meteorites, Composition, Mineralogy, and Petrology; Meteoroids, Space Investigations; Zodiacal Light and Gegenschein.**

Interplanetary Dust, Dynamics

Mark V. Sykes

During the *Apollo 15* mission, astronaut David Scott dropped a feather and a hammer at the same time. They struck the surface of the moon simultaneously. This dramatically illustrated that the laws of gravitation do not depend on the composition or mass of an object. Applied to objects in motion about the Sun, a pebble and an asteroid given the same instantaneous velocity and heliocentric distance would travel along the same orbital path. However, gravity is not the only force acting on bodies in space, and the smaller the pebble, the more its motion is affected by the solar radiation field in which it travels. In the case of interplanetary dust particles, deviations from classical Newtonian motion arise primarily from radiation pressure and Poynting–Robertson drag. To a lesser extent (except for the smallest particles) the solar wind also modifies particle orbits by "corpuscular drag." Other factors influencing the dynamical evolution of dust particles include frequent destructive collisions, which result in the redistribution of dust mass to yet smaller sizes with subsequent changes in the effects of radiation forces.

There are two principal sources of dust in the inner solar system. The most familiar is short-period and other comets, which regularly give up mass as they approach the Sun. Asteroid collisions, particularly catastrophic disruptions, are another source of dust as material is continually comminuted to smaller and smaller sizes. Studying these phenomena provides an opportunity to test theories of the dynamics of interplanetary dust particles as they evidence the effects of the various nongravitational forces described previously. In general, a particle is not fated to remain at the location at which it is created. It tends to evolve toward smaller heliocentric distances until further collisions reduce its size enough either to be ejected from the solar system or to continue its sunward spiral until vaporized. The precise orbital evolution of a given particle, however, will depend on many things, such as its initial orbital elements, its initial size, and the variation in the collisional environment it experiences over its lifetime.

RADIATION PRESSURE

Radiation pressure acts against the force of gravity as a particle absorbs or scatters solar photons, thereby picking up a component of momentum directed radially away from the Sun; this sensitivity to radiation pressure is generally described by the ratio β of radiation force to gravitational force, defined by

$$\beta = 5.7\times 10^{-5} Q_{pr}/\rho s \quad \text{(in cgs units)},$$

where ρ is the mass density, s is the radius of the particle and Q_{pr} is the radiation pressure efficiency factor averaged over the solar spectrum.

Another way of thinking of this effect is that a particle sensitive to radiation pressure perceives the Sun as having a smaller effective mass. An interesting aspect of β is that it is independent of heliocentric distance because both gravitational and radiation forces decline with the square of the distance from the Sun. Smaller particles are generally more sensitive to radiation forces than larger particles because their surface-area-to-mass ratio is greater, for a given density. However, this tendency does not continue indefinitely: Particles smaller than a particular wavelength of radiation become increasingly transparent at that wavelength; below that size, β begins to decrease with decreasing particle size. For spherical particles in orbit about the Sun, this turnaround point occurs near a radius of 0.1 μm (Fig. 1).

When $\beta = 1$, the effects of gravitation and radiation pressure cancel, and the particle escapes the solar system. However, particles might also escape with much lower values of β if their orbital velocity at the time of creation (e.g., emission from a comet) exceeds the escape velocity for the smaller effective mass of the Sun. In the case of a particle in circular orbit about the Sun (at any heliocentric distance), $\beta = 0.5$ is sufficient for ejection.

Comet dust tails are the most prominent example of the effects of radiation pressure. They consist of a stream of micrometer-sized particles (hence high values of β) extending away from the Sun. With the advent of space-based telescopes, several other dust structures have been detected that are explained by radiation forces in combination with gravitational forces. One such phenomenon is the cometary debris trail.

Debris trails were first detected by the *Infrared Astronomical Satellite* (*IRAS*) and consist of large particles ($d > 1$ mm) in the

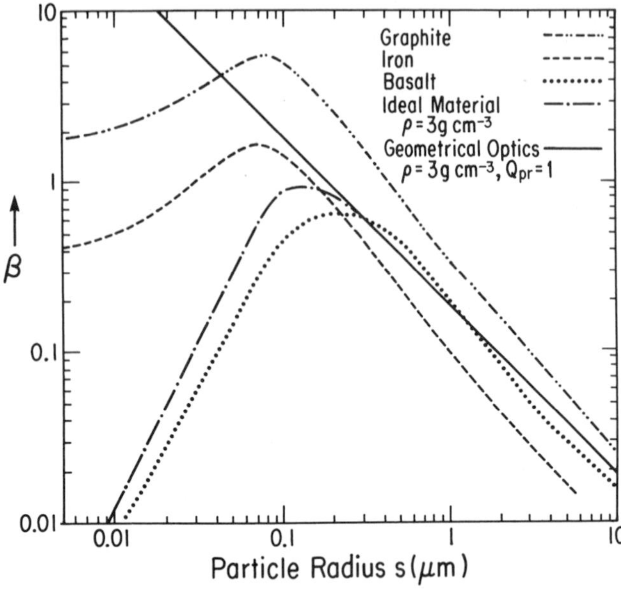

Figure 1. A log–log plot of the ratio β of radiation force to gravitational force as a function of particle size for three cosmically significant materials and two comparison standards [*from Burns, Lamy, and Soter (1979)*].

orbits of short-period comets (Fig. 2). These particles are emitted at low velocities (several meters per second) from the comet nucleus and are seen to stream tens of degrees behind the comet, looking much like an airplane contrail: long and extremely narrow. At the same time, however, they are seen to extend only a few degrees ahead of the comet. This asymmetry can be explained as an effect of radiation pressure, according to the following argument.

If the radiance of the Sun were turned off, large (low β) particles emitted isotropically from a comet at low velocity would spread evenly both ahead and behind the comet's orbital position, slowly forming a ring as they extend around the orbit. If the radiance were then turned back on, such particles would experience a sudden lessening of the Sun's gravitational grip, resulting in each particle having a slightly larger orbit, hence lower orbital velocity. In the case of most trail particles, their resultant mean velocity is slightly smaller than their parent's and they fall slowly behind the comet's orbital position. Thus, even for particles that would push ahead of the comet in its orbit when the solar radiance is turned off, when we turn the radiance of the Sun back on, the effectively reduced gravitational pull of the Sun can cause them to fall behind the comet.

POYNTING–ROBERTSON AND CORPUSCULAR DRAG

Poynting–Robertson (P–R) drag is another result of momentum transfer between solar photons and a particle in motion about the Sun. In this case, the effect is dependent upon the motion of the

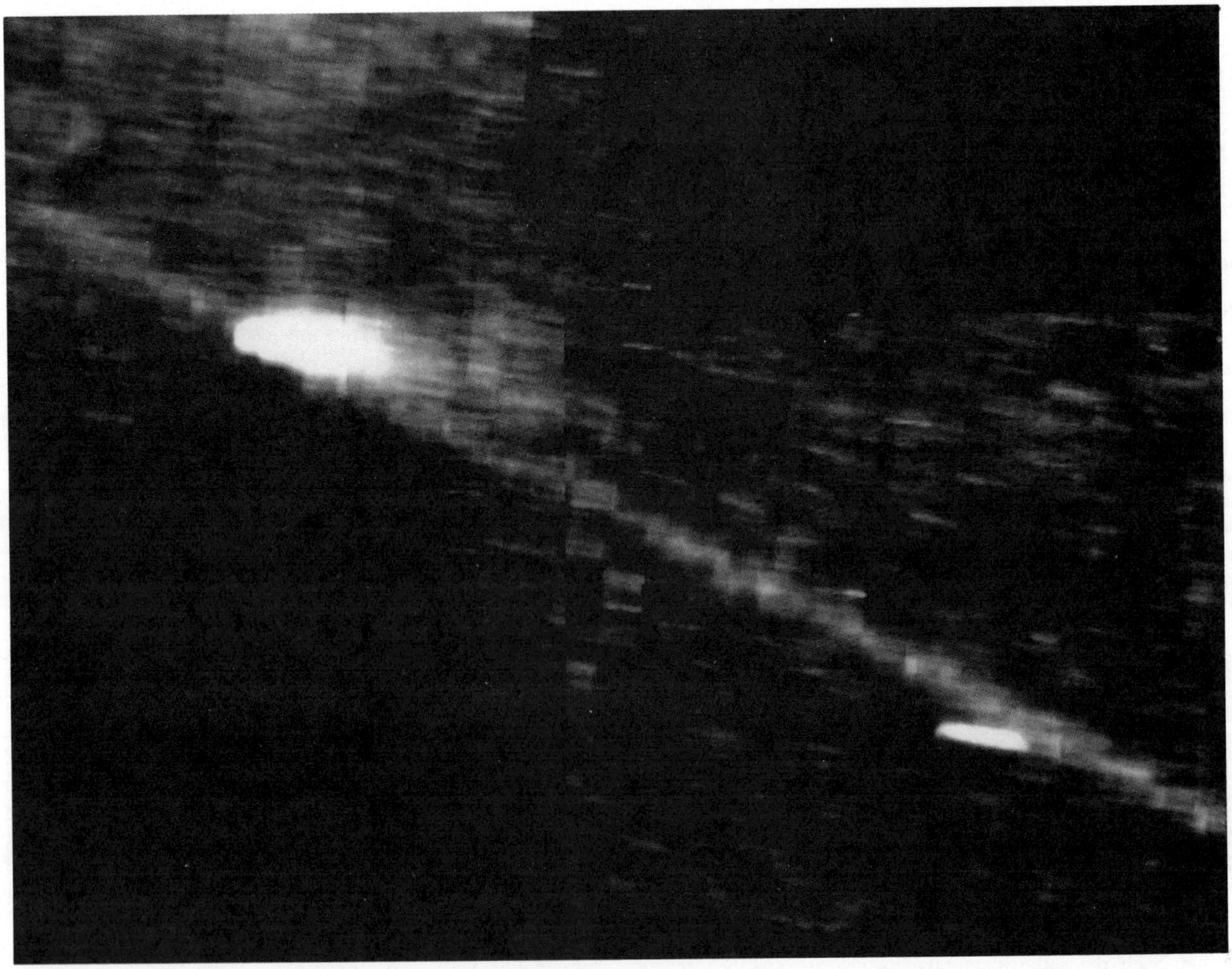

Figure 2. P/Tempel 2 and its associated dust trail as observed by *IRAS* on September 6, 1983. The composite image is constructed from scans made by the satellite at 25 μm.

particle and acts against that motion. The energy of the particle's orbit decreases, and the particle spirals in towards the Sun. As the semimajor axis of the orbit gets smaller, so does the eccentricity. Thus, as particle orbits decay, they become smaller and more circular. The time scale (in years) for P–R decay from a heliocentric distance R [in astronomical units (AU)] to the Sun is

$$t_{\mathrm{pr}} = 400R^2/\beta.$$

As in the case of radiation pressure, particles smaller than a few tenths of a micron will be less sensitive to P–R drag as a consequence of increasingly inefficient coupling to the solar radiation field. However, another force comes into play at this point as dust grains interact with the particles composing the solar wind. The new force is only one-tenth that of P–R drag until the coupling of the particle to the radiation field begins to decrease. At this point, corpuscular drag becomes dominant as the particle cross sections for this force continue to be described by geometric optics.

The effects of Poynting–Robertson and corpuscular drag on interplanetary dust are illustrated by considering the relationship between the zodiacal cloud and the major dust production regions within the asteroid belt. These regions were another discovery of *IRAS* and are referred to as the zodiacal dust bands because of their ring-like appearance (Fig. 3). The most prominent of these bands are associated with the principal Hirayama asteroid families and consist of dust being continually generated through collisions. This material remains, by and large, in the vinicity of the asteroid families until broken up into small enough fragments that their Poynting–Robertson drag lifetime is shorter than their collisional lifetime. The dust then moves from the asteroid belt toward the Earth as their orbits decay, to form the smooth background of zodiacal particles through which the dust bands are observed.

DUST AND RESONANCES

The "mobility" of dust grains experiencing orbital decay has an interesting side effect that further distinguishes their dynamics compared to large bodies: The extent to which their motions can be influenced by gravitational resonances is greatly diminished. Chaotic regions such as the 3:1 Kirkwood gap can greatly change the orbit of a main-belt asteroid on time scales of hundreds of thousands to millions of years, even shifting it into an Earth-crossing orbit. A decaying particle orbit, on the other hand, may be subject to these perturbations for only a short period of time and subsequently may be little altered.

As dust continues to spiral in toward the Sun, it may experience additional perturbations by the Earth and Venus. A fraction of these particles might be trapped in temporary (10^4–10^5 yr) resonance with the Earth, forming a toroid-like cloud in the vicinity of the Earth's orbit.

THE FATE OF DUST

Collisions play a dominant role in the orbital evolution of dust grains. At 1 AU, a grain having a mass between 1 and 10^{-4} g will survive on average $\sim 10^4$ yr—much smaller than its Poynting–Robertson lifetime. The result is that interplanetary dust is being comminuted rapidly to sizes very sensitive to radiation pressure. In fact, many (called β meteoroids) are injected into hyperbolic orbits and escape the solar system. *Pioneer 8*, *Pioneer 9*, and other spacecraft have detected β meteoroids from within the Earth's orbit throughout the asteroid belt. This process is thought to be the most efficient mechanism of dust mass loss from the solar system. However, dust mass loss at 1 AU due to P–R drag and radiation pressure is thought to be less than the amount being created through the collisional breakup of larger particles. Thus, the amount of dust near the earth's orbit may be decreasing at this time. If this is true, then large stochastic sources such as new comets or recent asteroid collisions must play a principal role in supplying the present-day zodiacal dust complex.

INTERPRETING DUST COLLECTION EXPERIMENTS

Understanding the dynamics of interplanetary dust is not only useful in explaining the dust structures found in the inner solar system, but also provides insight into the origin of dust collected at the Earth's orbit. Modern collection experiments have been conducted for decades and range from ocean bottom and polar ice core samples to stratospheric collection from aircraft. Recent compositional analysis of some of these particles [e.g., Brownlee (1987)] indicates that a majority of them have an elemental composition similar to the primitive meteorites (CI and CM chondrites), which may derive from asteroids.

Assuming an asteroidal origin, it may be possible to be even more specific as to where in the asteroid belt this dust is formed, according to the following line of argument. Radiation forces have no effect on the orbital inclinations of particles. Thus, as particles spiral in from the asteroid belt to the Earth's orbit, neither resonances nor radiation forces will greatly alter the cone along which they travel. An examination of the asteroid population having inclinations less than a few degrees shows that more than 60% of them are members of the Koronis and Themis asteroid families. These families also happen to have zodical dust bands associated with them, indicating that they are a site of significant contemporary dust production. Thus, asteroid dust collected at the Earth may derive largely from ~ 3 AU, where the Themis and Koronis asteroid families reside.

Additional Reading

Brownlee, D. E. (1987). Morphological, chemical, and mineralogical studies of cosmic dust. *Philos. Trans. Roy. Soc. London A* **323** 305.

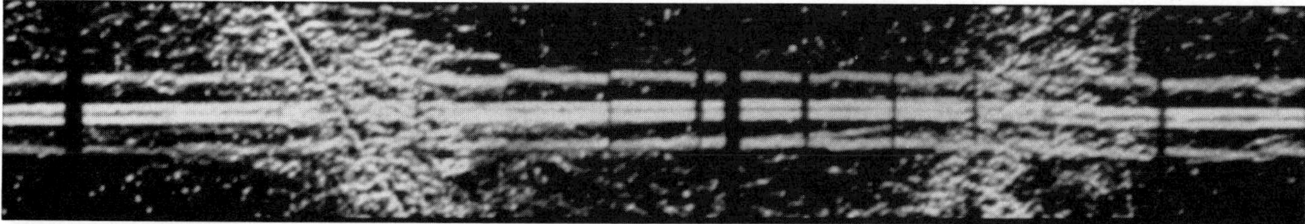

Figure 3. A map of the plane of the ecliptic as observed by *IRAS* at 25 μm. The rightmost edge corresponds to 0° ecliptic longitude, increasing to 360° at the left. Ecliptic latitude ranges between 30° (top) to −30° (bottom). The image has been high-pass filtered in ecliptic latitude to remove the background zodiacal dust. The parallel pairs of bands are associated with dust in the asteroid belt arising from collisions. Other extended linear features have been identified as cometary dust trails. The galactic plane crosses the ecliptic near 90° and 270° longitude.

Burns, J., Lamy, P., and Soter, S. (1979). Radiation forces on small particles in the solar system. *Icarus* **40** 1.

Grün, E., Zook, H., Fechtig, H., and Giese, R. (1985). Collisional balance of the meteoritic complex. *Icarus* **62** 244.

Jackson, and Zook, H. (1989). A solar system dust ring: The Earth as its shepherd. *Nature* **337** 629.

Low, F., Beintema, D., Gautier, T. N., Gillett, F., Beichman, C., Neugebauer, G., Young, E., Aumann, H., Boggess, N., Emerson, J., Habing, H., Hauser, M., Houck, J., Soifer, B., Walker, R., and Wesselius, P. (1984). Infrared cirrus: New components of the extended infrared emission. *Ap. J. Lett.* **278** L19.

Sykes, M., Greenberg, R., Dermott, S., Nicholson, P., Burns, J., and Gautier, T. N. (1989). Dust bands in the asteroid belt. In *Asteroids II*. University of Arizona Press, Tucson, p. 336.

Sykes, M., Lebofsky, L., Hunten, D., and Low, F. (1986). The discovery of dust trails in the orbits of periodic comets. *Science* **232** 1115.

See also **Asteroid and Comet Families; Comets, Dust Tails; Interplanetary Dust, Collection and Analysis; Interplanetary Dust, Remote Sensing; Meteorites, Classification; Meteoroids, Space Investigations.**

Interplanetary Dust, Remote Sensing

Anny-Chantal Levasseur-Regourd

The faint cones of light that, in the absence of the Moon and of any light pollution, are easily seen in the tropics in the west after sunset or in the east before sunrise are, together with meteors, the only visual evidence for extraterrestrial dust. Meteors are produced by dust particles crossing the Earth's atmosphere with high velocity. The cones of light detected along the ecliptic, or zodiac, are produced by sunlight scattered on dust particles spread over the whole solar system, out to 3 AU (astronomical units) from the Sun at least.

The so-called *zodiacal light* extends over the whole sky, with maximum brightness toward the Sun and the ecliptic plane. This feature suggests that the interplanetary dust cloud presents a maximum of density toward the Sun and near the ecliptic plane. The zodiacal light, which arises from two distinct physical phenomena (scattering of solar light, predominant in the visible domain, and thermal emission, predominant in the infrared), is indeed the only source of information about the integrated physical properties of the interplanetary dust.

LIGHT SCATTERED AND EMITTED BY INTERPLANETARY DUST

In the *visible domain* (and in the near ultraviolet), the zodiacal spectrum is rather solar-like, which suggests that the dust particles are larger than visible light wavelengths. Scattering on the optically thin interplanetary medium produces a partially plane-polarized light, with the electric vector perpendicular to the scattering plane (defined by the directions of illumination and observation), except for the antisolar region, where there is a polarization reversal. The measured quantities are the orthogonal components Z_1 and Z_2 (respectively, perpendicular and parallel to the scattering plane) and the orientation of the plane of polarization. The integrated degree of polarization $P=(Z_1-Z_2)/(Z_1+Z_2)$ is deduced from these quantities.

Extensive ground-based programs of observation of the zodiacal light were performed in the 1960s and 1970s, for example, from Hawaii [J. L. Weinberg] and from Tenerife [R. Dumont]. Results have also been obtained from balloons, rockets, satellites (*OSO 2*, *OSO 5*, *D2A*, *D2B*, etc.), and space stations (*Skylab*, *Salyut 6*, *Salyut 7*, etc.). The atmospheric nightglow is then straightforward to disentangle, but the observations are more fragmentary. Also, zodiacal light has been partly monitored from space probes in the inner (*Helios 1* and *Helios 2*) and outer (*Pioneer 10* and *Pioneer 11*) solar system. The observations agree quite well (see reviews listed under Additional Reading). They show that the zodiacal light is, in the visual domain, relatively smooth and stable with time; its brightness is found to decrease with heliocentric distance and to be negligible beyond the asteroid belt.

In the *near infrared* (wavelengths longward from 5 μm), the thermal emission prevails in the zodiacal light. Up to approximately 50 μm, it is the most prominent component of the sky, at least for high and medium galactic latitudes. It arises from solar radiation absorbed by interplanetary dust and predominantly reemitted at infrared wavelengths. The measured quantities are integrated thermal brightnesses B_λ, that is, brightnesses per wavelength interval. The simultaneous knowledge of B_λ for two wavelengths along the same line of sight permits the determination (with a gray-body assumption) of the integrated temperature.

Since the beginning of the 1980s, thermal emission has been observed (for various wavelengths) from balloons or rockets and from the *Infrared Astronomical Satellite* (*IRAS*) and the *Cosmic Background Explorer* (*COBE*). There are still some slight discrepancies between the various observations, but it is most likely that B_λ is, like Z, stable with time and that the zodiacal cloud is rather steady. The high resolution of the *IRAS* telescope has allowed some structures to be detected; they have been attributed to dust bands of asteroidal origin (by Stanley F. Dermott) and to narrow dust trails (by Mark V. Sykes), and have reinforced the idea (of the author and Jacques E. Blamont) that the interplanetary cloud is built from an interweaving of toroidal dust swarms.

NEED FOR AN INVERSION AND OBSERVING GEOMETRY

Both scattered light and thermal emission measurements provide *integrals along a line of sight* that extends from the observer to the outer fringe of the zodiacal cloud. Numerous efforts have been made to retrieve the elemental contributions along the line of sight in order to interpret the remote measurements in terms of local properties. Fitting the observations by a model may not provide an objective solution (as shown by R. H. Giese). Various inversion techniques have therefore been developed for the available directions of observation.

Up to now, the observations have been restricted to an observer in the ecliptic plane. They have been found (after correction for the slight inclination of the warped symmetry surface of the zodiacal cloud upon the ecliptic plane) to be *symmetric* with respect to the ecliptic and to the helioecliptic meridian plane (orthogonal to the ecliptic along the Sun–observer line). This result allows the zodiacal light integrated brightnesses to be given as functions of the heliocentric distance of the observer, of the ecliptic latitude of the line of sight, and of its helioecliptic longitude. When observations are performed in the ecliptic plane, the line of sight is only defined by its elongation (see Fig. 1).

Different *parameters* are used for the local scattering or emitting volume. Because the thermal emission is isotropic, the local thermal brightness $\mathscr{B}_\lambda$ derived from integrated measurements is only a function of the wavelength λ and of the heliocentric distance $R_\odot$ and helioecliptic latitude $\beta_\odot$ of the emitting unit volume in M. In the visual domain, the local optical brightness $\mathscr{Z}_i$ and the local polarization $\mathscr{P}$ are functions of the heliocentric distance $R_\odot$ and helioecliptic latitude $\beta_\odot$ of the scattering unit volume, and also of the scattering angle θ (or of the supplementary phase angle α).

RIGOROUS LOCAL INVERSION

The purpose of the rigorous local inversion is to derive without any assumptions the scattering and thermal properties of the dust at

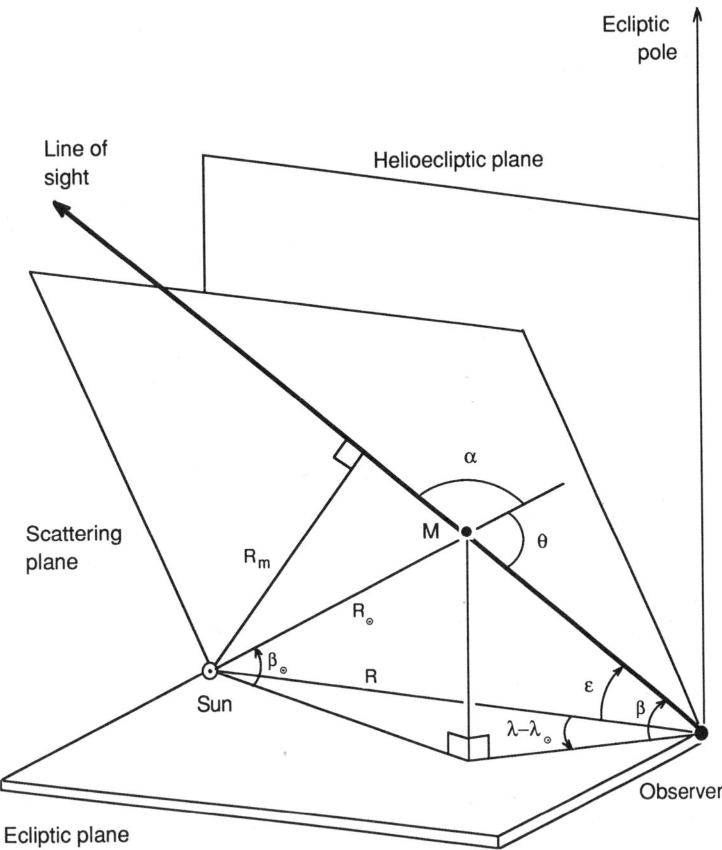

Figure 1. Observational geometry of zodiacal light observations performed along a line of sight defined by (β, $\lambda - \lambda_{\odot}$) for an observer at R AU from the Sun in the ecliptic plane. The purpose of remote sensing is to infer the optical properties of the dust located in M, defined by ($R_{\odot}$, $\beta_{\odot}$). The thermal emission in the infrared is isotropic, whereas the scattering in the optical domain depends upon the angle θ (or α).

the location of the observer. It has been demonstrated by Rene Dumont to be feasible for a line of sight that is quasitangent to the orbit of the Earth. Once derivatives with respect to ε of Z_i and B_λ are obtained at 1 AU, the local inversion provides $\mathcal{Z}_i$ ($R_{\odot}$= 1 AU, $\theta = 90°$), $\mathcal{P}$ (1 AU, 90°), and $\mathcal{B}_\lambda$ (1 AU, λ). Typical results (from work by Dumont) are that 1 km^3 of interplanetary space near the Earth's orbit scatters sunlight at a right angle from the Sun with an intensity that is 4×10^{-34} times smaller than the Sun's intensity, that the local polarization at 90° phase angle is of the order of 30%, and that the color temperature of the grains at 1 AU is about 270 K.

For directions other than the tangent, the inversion requires some derivatives of the brightness that are not yet accessible. However, rigorous local inversion would be directly applicable in the case of space observations made tangentially to the orbit of a moving probe. It has been the scientific ground for the Halley optical probe experiment (by the author and co-workers) on-board *Giotto*.

INVERSION WITH HOMOGENEITY ASSUMPTION

For sections on the line of sight other than the observer's location, arbitrary assumptions cannot be avoided. Fair assumptions are that the zodiacal cloud is stable and has a rotational symmetry axis that deviates only slightly from the Sun's ecliptic pole axis. A more disputable assumption is that the cloud is homogeneous, that is, that the size distribution and scattering properties of the dust particles are independent of their location.

Under these conditions, the local brightness is proportional to the local density and to a phase function that is the same everywhere. In the ecliptic (or, more accurately, in the symmetry plane), the density is usually assumed to decrease as $R_{\odot}^{-n}$, with n of the order of 1.3, in agreement with *Helios* results (from work by Christoph Leinert). Out of the ecliptic, various distributions have been suggested, for instance, a modified fan model (by Philippe Lamy and Jean Marie Perrin) approximately described by $R_{\odot}^{-1}\exp(-3.5 \sin \beta_{\odot})$.

The phase functions and degree of polarization have been derived by various authors (e.g., Dumont, Lamy, S. S. Hong, and D. W. Schuerman) as a function of the phase angle. Typically, the local polarization increases almost linearly from 20 to 60°, and reaches a maximum of about 30% near 100° phase angle. However, recent analysis of the observations performed from space probes in the inner and outer solar system suggests that the homogeneity assumption is highly questionable.

INVERSION WITH NODES OF LESSER UNCERTAINTY

To become free of the previous assumption, new methods have been developed recently from mathematical representations for the distribution along the line of sight of the elemental contributions. With steady state and rotational symmetry, the local brightnesses are written as functions of the directional scattering cross section C_{vis} and of the monochromatic thermal cross section C_{IR}. All the functions that characterize the distribution of C_{vis} or C_{IR} along the line of sight have to be positive (optically thin medium), rather

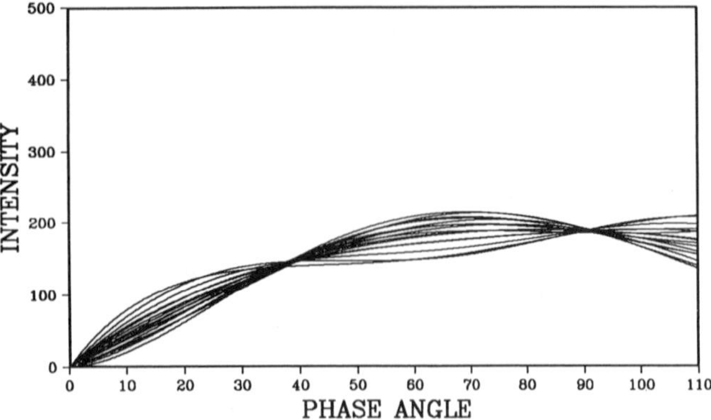

Figure 2. Typical results on local brightness near two nodes from measurements performed at $\varepsilon = 70°$ and at $\varepsilon = 110°$. The radial node is at 90° phase angle, and the martian node is near 40° phase angle.

monotonous (relative smoothness of the cloud), and must decrease asymptotically to zero with increasing solar distance (absence of interplanetary dust far away from the Sun). From the whole set of constraints, it can be demonstrated that the curves representing all the possible functions along a line of sight have to constrict in nodal regions where the local scattering cross sections are determined with less uncertainty than elsewhere.

In the symmetry plane, two integrated brightnesses are simultaneously available along the same line of sight, which intersects the Earth's orbit for a given elongation and for the supplementary angle. Two nodes are found, respectively located near the middle of the chord intersecting the Earth's orbit (radial node) and at a solar distance of about 1.5 AU (martian node; see Fig. 2).

The determination of C_{vis} ($R_\odot$, 90°) or C_{IR} ($R_\odot$, λ) at the radial node allows us to disregard the phase dependence and provides the radial dependence. Typically, the local polarization degree at 90° scattering angle is found to decrease exponentially from $\approx$ 30% at 1 AU to $\approx$ 20% at 0.5 AU, and the local temperature is found to follow a $270R^{-0.35}$ K law. These results suggest a decrease in porosity of the grains and an increase in albedo (found to be about 0.07 at 1 AU) with decreasing heliocentric distance. The determination of C_{vis} (1.5 AU, θ) and C_{IR} (1.5 AU, λ) at the martian node allows us to disregard the heliocentric dependence. The phase function and polarization function are quite reminiscent of those obtained for cometary grains.

The heterogeneity of the interplanetary dust in its symmetry plane is demonstrated by the radial dependence of polarization and albedo with gradients of power laws respectively equal to 0.8 and to −0.3 near 1 AU. The results suggest that the dark and fluffy zodiacal grains break off and evaporate while they spiral toward the Sun under the Poynting–Robertson effect; some of these grains could be cometary aggregates of silicates and organic polymers that undergo drastic alterations with increasing temperature, while other ones could be asteroidal debris.

Attempts toward an out-of-ecliptic inversion have been made recently. Polar nodal regions have been found in the plane perpendicular to the Sun–Earth line and tangent to the Earth's orbit. The local polarization (90° at 1 AU) is found to decrease as the local albedo simultaneously increases slightly with increasing height above the ecliptic or symmetry plane. The grains found above the Earth's orbit seem to be, on the average, less porous and smoother than in the ecliptic at the same solar distance. These results show that various populations of dust, the source of which could be comets, defunct comets, cometary asteroids, or asteroids, are found in the zodiacal cloud.

CONCLUSION

The remote sensing of the interplanetary dust from the zodiacal-light observations, now possible both in the visible and infrared domains, builds up a picture of a heterogeneous dust cloud. The in-situ methods of interplanetary dust collection are essential to complement this approach. However, the retrieval of local information in various regions of the solar system, some of which have never been visited by space probes, shows the contribution of the remote sensing methods. These techniques are in the process of being adapted (by Pierre Bastien and the author) to the study of accretion disks around young stellar objects, tentatively to learn more about the formation of planetary systems from the properties of cometary and interplanetary dust in our solar system.

Additional Reading

Dumont, R. and Levasseur-Regourd, A. C. (1988). Properties of interplanetary dust from infrared and optical observations. *Astron. Ap.* **191** 154.

Fechtig, H., Leinert, Ch. and Grün, E. (1981). Interplanetary dust and zodiacal light. In *Landolt-Börnstein, Neue Serie* **VI 2a**. Springer-Verlag, Berlin, p. 228.

Hauser, M. G. (1988). Models for infrared emission from zodiacal dust. In *Comets to Cosmology*, A. Laurence, ed. *Lecture Notes in Physics*. Springer-Verlag, Berlin, p. 27.

Leinert, Ch. (1975). Zodiacal light—A measure of the interplanetary environment. *Space Sci. Rev.* **18** 281.

Leinert, Ch. and Grün, E. (1990). Interplanetary dust. In *Physics of the Inner Heliosphere*, in the series *Physics and Chemistry in Space*, R. Schwenn and E. Marsch, eds. Springer, Berlin, p. 207.

Levasseur-Regourd, A.-Ch. and Hasegawa, H., eds. (1991). *Origin and Evolution of Interplanetary Dust*. Kluwer Academic Publishers, Dordrecht.

Weinberg J. L. and Sparrow, J. G. (1978). Zodiacal light as an indicator of interplanetary dust. In *Cosmic Dust*, J. A. M. McDonnell, ed. John Wiley and Sons, New York, p. 75.

See also **Interplanetary Dust, Collection and Analysis; Interplanetary Dust, Dynamics; Zodiacal Light and Gegenschein.**

Interplanetary Magnetic Field

John W. Bieber

Magnetic fields are a pervasive feature of the systems of charged particles, or plasmas, that constitute much of the matter in the universe. Embedded in plasma emanating from the Sun is the interplanetary magnetic field, one of the few examples of an astrophysical magnetic field that can be subjected to intensive scrutiny by means of in situ measurements. Interplanetary probes have now been observing the interplanetary magnetic field for more than 25 years. These measurements provide a vital observational basis for testing theoretical descriptions of space plasmas such as magnetohydrodynamics. They are also of great significance to cosmic-ray physics, because it is principally the magnetic field that controls the transport of high-energy charged particles, and to magnetospheric physics, because the magnetic field is a crucial factor governing the coupling of solar wind plasma to the plasmas surrounding planets and other solar system bodies.

The interplanetary magnetic field has been observed to vary on time scales ranging from decades to less than a second. Some of the variations can be attributed to nonstationary plasma structures such as high-speed streams and shocks, which are discussed else-

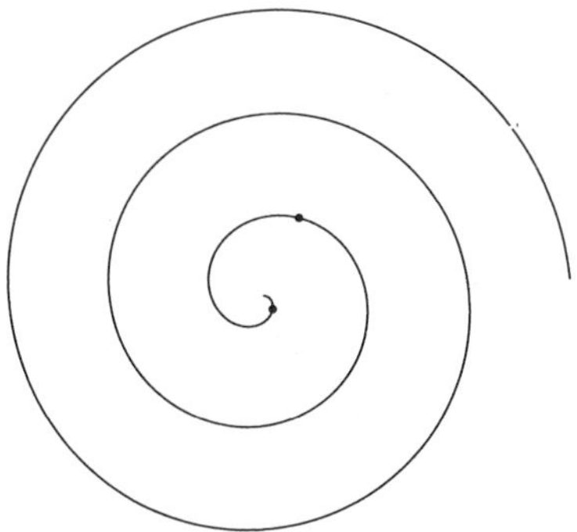

Figure 1. In Parker's model of the interplanetary magnetic field, a field line in the plane of the Sun's equator has the shape of an Archimedean spiral. In this view from the north, the Sun is at the inner terminus of the spiral, and the two dots indicate the radial distances to Earth's orbit and to Jupiter's orbit at 1 and 5.2 AU, respectively.

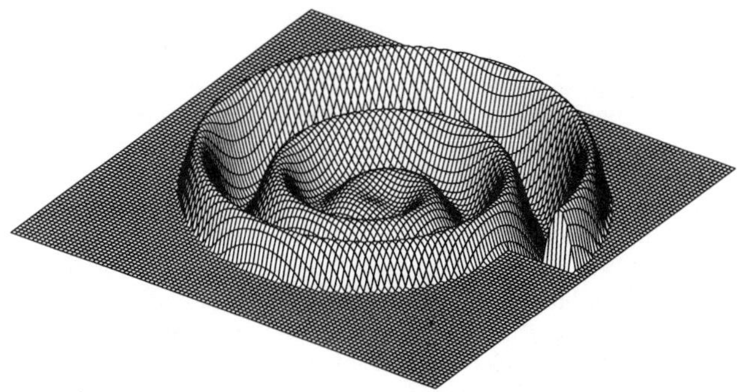

Figure 2. A wavy current sheet defines the boundary between regions of the solar wind containing opposite magnetic polarities. In this depiction, the Sun is at center and the flat surface represents a section of the Sun's equatorial plane 40 AU on a side.

where in this volume. This entry first describes the large-scale interplanetary magnetic field and its variation with the solar magnetic cycle and then summarizes the properties of interplanetary magnetic turbulence.

THE PARKER FIELD

The interplanetary magnetic field occupies a region extending from within the orbit of Mercury to beyond the orbit of Pluto. The inner boundary of this region is the so-called source surface located approximately 1 solar radius above the solar surface, where the magnetic field lines have a radial orientation. The outer boundary is thought to be the solar wind termination shock formed by the interaction of interstellar plasma with the solar wind. This boundary, which has not yet been encountered by the most distant interplanetary probes, presumably lies somewhere beyond 50 astronomical units (AU) from the Sun.

The large-scale interplanetary magnetic field is sometimes called the Parker field after Eugene N. Parker, the scientist who in 1958 correctly predicted its basic properties. In Parker's model, the footpoint of a particular magnetic field line rotates with the Sun, while at the same time the field line is drawn out from the Sun by the solar wind expansion. The field is said to be "frozen in" the plasma flow field as a result of the plasma's extremely high conductivity. Magnetic field lines in the plane of the Sun's equator thus have the shape of Archimedean spirals, as depicted in Fig. 1. Outside the equatorial plane the field lines spiral about the surfaces of cones centered on the solar rotational axis and thus have the shape of corkscrews.

Near the orbit of Earth, the interplanetary magnetic field has a strength of about 5 nT (nanotesla) on average, and it is typically inclined about 45° with respect to the radial direction. This is as expected for a solar wind speed of 430 km s^{-1}, which is also typical. The radial and azimuthal components of the Parker field vary with heliocentric distance r as r^{-2} and r^{-1}, respectively, so that the field becomes almost purely azimuthal at very large distances. In addition, the azimuthal component weakens as $\sin\theta$, where θ is the polar angle measured from the Sun's rotational axis, so that the corkscrew field lines become more loosely wound at high heliographic latitudes. The latitudinal component of the magnetic field is zero in Parker's model.

SECTOR STRUCTURE

The interplanetary magnetic field is said to have a "toward" or "away" magnetic polarity according to whether the field points inward (toward the Sun) or outward (away from the Sun) along the Parker spiral direction. The polarity is determined by the direction of the field at the source surface and thus is ultimately controlled by the solar magnetic field. Around the time of sunspot minimum, the source surface consists of two roughly hemispherical regions, one containing outward-pointing fields and the other containing inward-pointing fields. The boundary between these two regions is drawn out by the solar wind into a structure called the heliospheric current sheet, which separates regions of toward and away magnetic polarity. Because this boundary is inclined with respect to the solar rotational equator, typically by 10–15° at sunspot minimum, the current sheet has a warped shape, as shown in Fig. 2. The current sheet corotates with the Sun, so that an observer located near the equatorial plane will see the sheet sweep past twice per solar rotation. The resulting pattern of alternating toward and away magnetic polarity is termed a "two-sector" structure of the interplanetary magnetic field.

As sunspot maximum is approached, the current sheet tends to become more steeply inclined and its structure becomes more complex. A spacecraft near the equatorial plane may observe four or more sectors per solar rotation. Near sunspot maximum, the Sun's magnetic dipole reverses and the large-scale interplanetary magnetic field reverses along with it. Thus, the bipolar character of the interplanetary magnetic field varies with a period of two sunspot cycles, or approximately 22 yr. During epochs of positive solar polarity, such as occurred from 1972 to 1979, the field points predominantly away from the Sun north of the current sheet and toward the Sun south of the current sheet. During epochs of negative solar polarity, such as occurred from 1981 to 1989, the reverse pattern exists.

INTERPLANETARY MAGNETIC TURBULENCE

The subject of turbulence in space plasmas is a rich and rapidly evolving field, for which only the barest outline can be given here. In studying magnetic turbulence in the solar wind, the total magnetic field at a given instant is usually separated into mean and fluctuating parts, where the mean field is simply an average taken over a suitable time interval. At the distance of Earth's orbit, the fluctuating part of the field is typically about 30% as large as the mean field, and the fluctuations often tend to be oriented roughly perpendicular to the mean field. This suggests that the turbulence may at times be composed principally of Alfvén waves propagating nearly parallel or antiparallel to the mean field.

Magnetic turbulence in space is often characterized by its power spectrum, which measures the distribution of turbulent energy as a function of frequency. Assuming frozen-in fields, the measured power at frequency f characterizes fluctuations with length scale V_w/f, where V_w is the solar wind speed. Over several decades of frequency the power spectrum is observed to vary approximately as $f^{-5/3}$, in accord with the famous prediction of Andrei N. Kolmogoroff. This regime, called the "inertial range" of the turbulence, is bounded at large scales by the correlation length, typically 5×10^9 m, or 0.03 AU near the orbit of Earth. At scales larger than this is the "energy-containing range," where the spectrum is determined by the large-scale eddies that drive the turbulence. At the opposite extreme, the inertial range extends to scales as small as 500 km. At still smaller scales, the inertial range gives way to the "dissipation range," where the magnetic turbulence is strongly damped by dissipative interactions with the thermal plasma.

Additional Reading

Matthaeus, W. H. and Goldstein, M. L. (1982). Measurement of the rugged invariants of magnetohydrodynamic turbulence in the solar wind. *J. Geophys. Res.* **87** 6011.

Ness, N. F. (1968). Observed properties of the interplanetary plasma. *Ann. Rev. Astron. Ap.* **6** 79.

Shea, M. A. and Smith, E. J., eds. (1989). Proc. Internat. Heliospheric Study. *Adv. Space Res.* **9** (No. 4).

Svalgaard, L. and Wilcox, J. M. (1978). A view of solar magnetic fields, the solar corona, and the solar wind in three dimensions. *Ann. Rev. Astron. Ap.* **16** 429.

Wilcox, J. M., Hoeksema, J. T., and Scherrer, P. H. (1980). Origin of the warped heliospheric current sheet. *Science* **209** 603.

See also **Heliosphere; Interplanetary Medium, Shock Waves and Traveling Magnetic Phenomena; Interplanetary Medium, Solar Wind; Interplanetary Medium, Wave–Particle Interactions.**

Interplanetary Medium, Shock Waves and Traveling Magnetic Phenomena

Leonard F. Burlaga

Two types of shock wave are familiar: a *driven* shock wave (sonic boom), produced by a supersonic airplane, and a *blast wave*, produced by an explosion. A sonic boom is capable of breaking windows, because the pressure increases very abruptly by about 2 lb ft^{-2} in a few milliseconds as the shock moves past the window. The strength of a shock is the ratio of the pressure behind the shock to that ahead of the shock, which is always greater than 1. The strength of a sonic boom does not change with time if the airplane moves at a constant speed, whereas the strength of a blast wave decreases as it propagates away from the source and distributes its energy over an ever-increasing volume. A shock compresses and heats the gas through which it moves. The maximum compression is about a factor of 4, but there is no limit to the maximum possible temperature behind a shock. The speed of a shock wave relative to the ambient gas is always greater than the local speed of sound.

Shock waves also exist in the heliosphere (the region surrounding the Sun and extending far beyond the orbit of Pluto), which contains gases and magnetic fields from the Sun. A heliospheric shock is either a *propagating* shock, which moves relative to the Sun, or a *standing* shock, which does not move relative to the Sun. A propagating heliospheric shock is generally a driven shock that may be either a *transient* shock or a *corotating* shock. A transient shock is driven by an ejection (the analog of an airplane in the Earth's atmosphere) that is released impulsively from the Sun and moves radially away from the Sun. A corotating shock is driven by a long-lived stream of gas that emanates from a coronal hole on the rotating sun.

HELIOSPHERIC SHOCK WAVES

A shock wave moves faster than the characteristic speed of the medium. The medium that fills the heliosphere is a plasma, composed primarily of protons and electrons, and a magnetic field that is drawn out from the Sun; the pressure of the magnetic field is comparable to that of the plasma. Consequently, there are three characteristic speeds in the heliosphere:

1. The *sound speed*, which depends only on the plasma temperature and density.
2. The *Alfvén speed*, which depends on the strength of the magnetic field and the density.
3. The *magnetoacoustic speed*, which depends on both the sound speed and the Alfvén speed.

The sound wave is a compressive wave that moves at the speed of sound along the magnetic field direction, whose restoring force is the gradient of the plasma pressure. The magnetoacoustic wave is a compressive disturbance that moves at the magnetoacoustic speed in the direction perpendicular to the magnetic field; the restoring force is the gradient of both the plasma pressure and the magnetic field pressure. The Alfvén wave is an incompressible disturbance whose restoring force is the tension in the magnetic field.

A heliospheric shock that moves exactly along the magnetic field is strictly analogous to a gas-dynamic shock in the Earth's atmosphere. However, a heliospheric shock is usually propagating at some nonzero angle with respect to the magnetic field. In this case, two types of heliospheric shock are possible: a *fast shock*, which moves faster than the component of the magnetoacoustic speed, and a *slow shock*, which moves faster than the sound speed but slower than the Alfvén speed. The density and temperature increase across all types of shocks. The magnetic field strength increases across a fast shock, decreases across a slow shock, and remains constant across a parallel shock.

Slow shock waves are seldom seen in the heliosphere. A traveling slow shock might form ahead of an ejection near the Sun, where the magnetic field is nearly radial and the Alfvén speed is much greater than the sound speed. Theoretically, the slow shock can transform into a fast shock as the ejection moves away from the Sun. A standing slow shock might exist in the solar corona between 6 and 10 solar radii.

Transient fast shock waves occur in the solar wind near the Earth at a rate of about one per month, occurring more frequently when the Sun is active. A fast shock at the orbit of the earth is generally a transient shock driven by an ejection from either a solar flare or some other impulsive solar disturbance. Typically the speed of a fast shock near 1 astronomical unit (AU) is two or three times the Alfvén speed, and the ratio of the total pressure (magnetic pressure plus gas pressure) behind the shock to that ahead of the shock is 2 or 3.

Corotating fast shocks usually form beyond the orbit of the Earth at a distance of a few astronomical units from the Sun, although a few corotating shocks have been observed at 1 AU. Corotating shocks in the outer heliosphere often occur in pairs. A corotating shock pair consists of a "forward shock" moving away from the Sun and a "reverse shock" moving toward the Sun, relative to the ambient solar wind. Because the solar wind speed is greater than the shock speeds, the solar wind carries along the forward and reverse corotating shocks, so both shocks actually move away from the Sun.

Interactions among fast shock waves are common in the outer heliosphere. When one fast shock follows another, the trailing shock moves faster and will eventually overtake and coalesce with the leading shock. The result is a single shock that is stronger than either of the original shocks. If two ejecta move radially away from the Sun at supersonic speeds relative to the ambient solar wind, and if the second ejection moves faster than the first, then the two ejecta and their respective shocks will eventually coalesce to form a single "compound high-speed flow." Similarly, a fast corotating stream will overtake a slower corotating stream and the two corotating streams will coalesce to form a single compound stream. The two respective corotating forward shocks will coalesce to form a single corotating forward shock, and the two respective corotating reverse shocks coalesce to form a single corotating reverse shock. One can also observe a transient stream overtaking a corotating stream and vice-versa.

Frequently, there are two nearly identical corotating streams per solar rotation about 180° apart, persisting for several solar rotations. In this case the forward shock from stream B interacts with the reverse shock from the preceding stream A, and the reverse shock from stream B collides with the forward shock from the following stream C. Both the forward and reverse shocks will generally survive the collision and continue on their way, but with their strengths and speeds diminished by the collision.

TRAVELING MAGNETIC PHENOMENA

The ejecta that drive transient shocks are structures with a dimension of the order of $\frac{1}{4}$ AU when they arrive at 1 AU. The source of an ejection is generally either a solar flare or an eruptive prominence. Ejecta consist of fully ionized plasma with a negligible electrical resistivity, so the magnetic field is in effect "frozen" in the plasma. The magnetic field strength in an ejection is usually higher than the average heliospheric field strength, suggesting an origin on the Sun in the regions of closed field lines that have relatively strong magnetic fields.

The mechanism that accelerates ejecta to supersonic speeds is unknown. One possibility is that an explosion releases heat and the resulting pressure gradient propels the ejection with enough force to overcome the tension of the closed solar magnetic field lines and the Sun's gravitational field. Another possibility is that the magnetic forces associated with a magnetohydrodynamic instability propel the ejections.

The interplanetary ejecta are of two types: those in which the magnetic fields are highly variable (which we shall call complex ejecta) and those in which the magnetic field direction rotates uniformly through a large angle as the ejection moves past the spacecraft that is observing it ("magnetic clouds").

A complex ejection usually follows a shock associated with a solar flare; hence, it is often called either postshock flow or flare ejecta. However, a complex ejection can have a cause other than a solar flare and there is no reason to suppose that a complex ejection must move fast enough to drive a shock. For example, an eruptive solar prominence can produce an ejection, and an ejection can move subsonically relative to the ambient solar wind, so that it does not follow a shock. Nevertheless, most of the reports on complex ejecta concern flare-associated postshock flows.

In addition to having disturbed magnetic fields, the flare-associated postshock flows have several other distinguishing characteristics: The plasma density and temperature are variable and filamentary, and the temperature is usually lower than average. The profiles of the magnetic field and plasma parameters vary greatly from one event to another. The abundance of fully ionized helium in an ejection is often significantly higher than the average helium

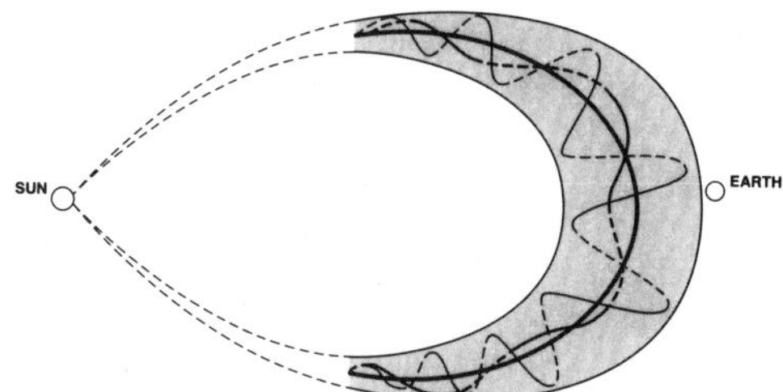

Figure 1. A sketch of the magnetic field lines in a magnetic cloud. The magnetic field line on the axis of the magnetic cloud is a simple curve. The other magnetic field lines, which are shown here in projection, are helices that wrap around the axis of the magnetic cloud. The magnetic field lines far from the axis are more tightly wound than those near the axis. It is not known whether the magnetic field lines are anchored at the Sun as shown here or are disconnected.

abundance of the solar wind. In principle, an ejection should have a distinct boundary, but no simple universal signature of a boundary has been identified.

A magnetic cloud has the following identifying characteristics: a strong magnetic field that rotates smoothly through a large angle as the magnetic cloud moves past a point (e.g., a spacecraft); a low ratio of the thermal pressure to the magnetic pressure; and a radial extent of the order of $\frac{1}{4}$ AU at 1 AU. One magnetic cloud in the solar wind at 1 AU was identified with a coronal mass ejection. It is possible that all magnetic clouds are the interplanetary manifestations of a class of coronal mass ejections. Solar flares and eruptive prominences can produce coronal mass ejections and magnetic clouds. In many cases a fast forward shock wave precedes a magnetic cloud, but this is not always the case.

The local configuration of the magnetic field lines in a magnetic cloud closely resembles that of a "force-free field," namely, a family of helices ranging from a straight line on the axis of the magnetic cloud to circles on the boundary of the magnetic cloud. It is possible that both ends of the field lines are anchored at the Sun, forming a large loop as in Fig. 1. It is also possible that the magnetic field lines are disconnected from the Sun and form some more complicated global configuration that still resembles a force-free field locally. A fast magnetic cloud can overtake and interact with a slower magnetic cloud, a classical ejection, or a corotating stream. Likewise, a slow magnetic cloud might be overtaken by a fast magnetic cloud, a fast complex ejection, or a fast corotating stream. Such interactions disturb the force-free field configuration and produce complicated flow patterns in the heliosphere.

Additional Reading

Burlaga, L. F. (1984). Magnetohydrodynamic processes in the outer heliosphere. *Space Sci. Rev.* **39** 255.

Hundhausen, A. J. (1972). *Coronal Expansion and Solar Wind.* Springer-Verlag, New York.

Hundhausen, A. J. (1985). Some macroscopic properties of shock waves in the heliosphere. In *Collisionless Shocks in the Heliosphere: A Tutorial Review*, R. G. Stone and B. T. Tsurutani, eds. Geophysical Monograph **34**, Washington, D.C., p. 34.

Parker, E. N. (1963). *Interplanetary Dynamical Processes.* Interscience, New York.

Shea, M. A. and Smith, E. J., eds. (1989). Proc. Internat. Heliospheric Study. In *Adv. Space Res.* **9** (No. 4).

See also **Heliosphere; Interplanetary Medium, Solar Wind; Interplanetary Medium, Wave–Particle Interactions; Shock Waves, Astrophysical; Solar Actvity, Coronal Mass Ejections; Solar Activity, Solar Flares; Sun, Coronal Holes and Solar Wind.**

Interplanetary Medium, Solar Cosmic Rays

Stamatios M. Krimigis

In addition to the fully ionized gas (plasma) comprising the solar wind that expands from the sun and fills the entire solar system, there are energetic nuclei and electrons, ranging from a few million electron volts (MeV) to tens of billions of electron volts that are called cosmic rays and whose origin is thought to be the Galaxy. Occasionally, however, there are transient increases in the intensity of nuclei and electrons in the interplanetary medium that can last from a few hours to several days or even weeks, which are associated with specific explosions on the sun that are called solar flares. Because the origin of these nuclei and electrons is the Sun, and because they constitute large enhancements over and above the galactic cosmic ray intensity, they are called solar cosmic rays.

Solar cosmic rays were first observed in ground-based ionization chambers by Scott Forbush in 1942 and have since been monitored with ground-based detectors and, since about 1959, by spacecraft detectors. At first, these particles were thought to be exclusively protons that were emitted at the time of the solar flare and that arrived at earth with nearly the speed of light. Following the launch of earth-orbiting spacecraft in the late 1950s and early 1960s, detailed measurements, obtained with radiation detectors, revealed that solar cosmic rays ranged in energy from a few tens of kiloelectron volts (keV) to several hundred million electron volts (i.e., speeds of about 1000 km s^{-1} to almost 300,000 km s^{-1}, the speed of light) and that they consisted of not only protons but helium nuclei, carbon, nitrogen, oxygen, and heavier nuclei all the way to iron and nickel, in effect representing a sample of solar material that was expelled into the interplanetary medium from the vicinity of the flare site on the Sun. Although radio and x-ray emissions suggested the presence of electrons at the flare site, it was thought initially that electrons could not escape magnetic confinement in solar active regions and therefore could not be observed in the interplanetary medium. More sensitive detectors, however, enabled the observation of electrons in association with solar flares by James Van Allen and the author in 1965. It is now known that electrons invariably accompany most solar flare emissions observed in the interplanetary medium. There have been more or less continuous observations of solar cosmic rays in the interplanetary medium over the past 25 years, not only in the vicinity of Earth but also by spacecraft in the inner and outer solar system, as far out as 7.5 billion kilometers from the sun (the position of the *Pioneer 10* spacecraft in 1991).

OBSERVATIONS

Intensity Profiles

Great flares produce copious quantities ($\sim 10^{34}$) of solar cosmic rays that propagate into the interplanetary medium and persist for periods ranging from several days to several weeks at a time. Such flares are typically associated with very large active regions on the Sun that are likely to appear at the ascending or descending phases of the 11-year solar activity cycle. A recent example of such unusual solar activity occurred during March 1989 when a new active region, AR 5395, was formed. This extremely large sunspot group, one of the largest ever observed, rotated past the east limb of the Sun about March 5, 1989. Flares from this region were observed during the entire passage of this group through the Earth-facing hemisphere of the Sun until it disappeared behind the west limb on or about March 20. The solar cosmic ray observations at the orbit of Earth from this series of flares are shown in Fig. 1 with the data obtained by the *Interplanetary Monitoring Platform-8* spacecraft, orbiting at an altitude of about 230,000 km. The brightest flares, designated by their hydrogen Lyman-α and x-ray magnitude, are marked at the bottom of the figure; they are all from AR 5395, with the solar longitude at the time of occurrence noted (e.g., E 69° means that the flare was observed at 69° east of central meridian, as viewed from Earth). Three of the flares produced γ-ray bursts, measured by the *Solar Maximum Mission* spacecraft.

The first high intensity flare occurred on March 6 and, as seen from the figure, resulted in large (factors of 10^3–10^4) intensity increases of electrons, low-energy protons, and high-energy protons, as well as helium, and heavier nuclei. The bottom curve shows the intensity profile of relativistic electrons, traveling at near the speed of light, that arrived at Earth about 10 min after flare onset. The next curve shows the intensity of very energetic protons whose speed is less than that of the electrons and the third curve from the bottom shows the intensity of heavier nuclei, ranging from oxygen through iron, with energies of about 10 MeV. The second curve from the top shows similar data for helium nuclei of energies of about 2 MeV and the top curve shows the intensity profile of protons at energies of $\sim$ 0.3 MeV. Note that the intensity profile of the lower-energy ($\lesssim$ 10 MeV) nuclei is quite similar but very different from those of the high-energy protons and the relativistic electrons (two lowest curves). Two more flares occurred in the same region on March 9 and 10, but, because of the already high intensities of solar cosmic rays in the interplanetary medium, the increases associated with these flares are not seen as clearly in the intensity profile. The intensity increase from the flare on March 10 is most notable in the second curve from the bottom, that is, the high-energy protons. Another intense flare occurred on March 17 that produced an evident increase in both high- and low-energy protons and in the relativistic electrons.

Propagation Characteristics

An obvious characteristic of the particle increases shown in Fig. 1 is the slow buildup of intensity following the flare of March 6, compared with the rapid onset seen in the flare of March 17. This illustrates a well-known property of solar cosmic ray events, namely that particles originating in flares that occur in the eastern solar hemisphere do not readily propagate to the location of Earth, whereas those from the western solar hemisphere arrive promptly. Evidently, solar cosmic ray propagation in the interplanetary medium is affected by the location of the observer relative to the flare site. The explanation for such effects is shown in Fig. 2, which shows lines of the solar magnetic field as it extends into the interplanetary medium in the form of Archimedian spirals, due to the rotation of the sun and the fact that the field is "frozen in" the radially flowing solar wind plasma. For an observer located at 1 AU at 0° longitude, the magnetic field line connects to $\sim$ 60° west on the Sun, so that flare particles generated in that longitude range propagate readily along the magnetic field to Earth. Conversely, particles generated in an eastern flare site on the Sun must propagate across the magnetic field to reach the observer, an inherently more difficult process.

The intensity profile of solar cosmic rays is quite complex, especially at the lower energies but also to some extent at the higher energies. In fact, the highest intensities attained prior to March 15 were not associated with direct flare injection, but occurred several days after the flare, especially the relative peak on March 8 and the maximum on March 13, both marked by the vertical, dashed lines. The latter event was associated with perhaps the largest geomagnetic storm since 1868, which caused power line outages in eastern Canada, various anomalies in many orbiting spacecraft, and auroral displays as far south as Key West, Florida. Such upheaval of terrestrial systems is caused by the arrival of a large volume of solar ejecta expelled from the vicinity of the flare into the interplanetary medium; these travel at speeds as high as 1200 km s^{-1} sweeping up the material in the normal solar wind

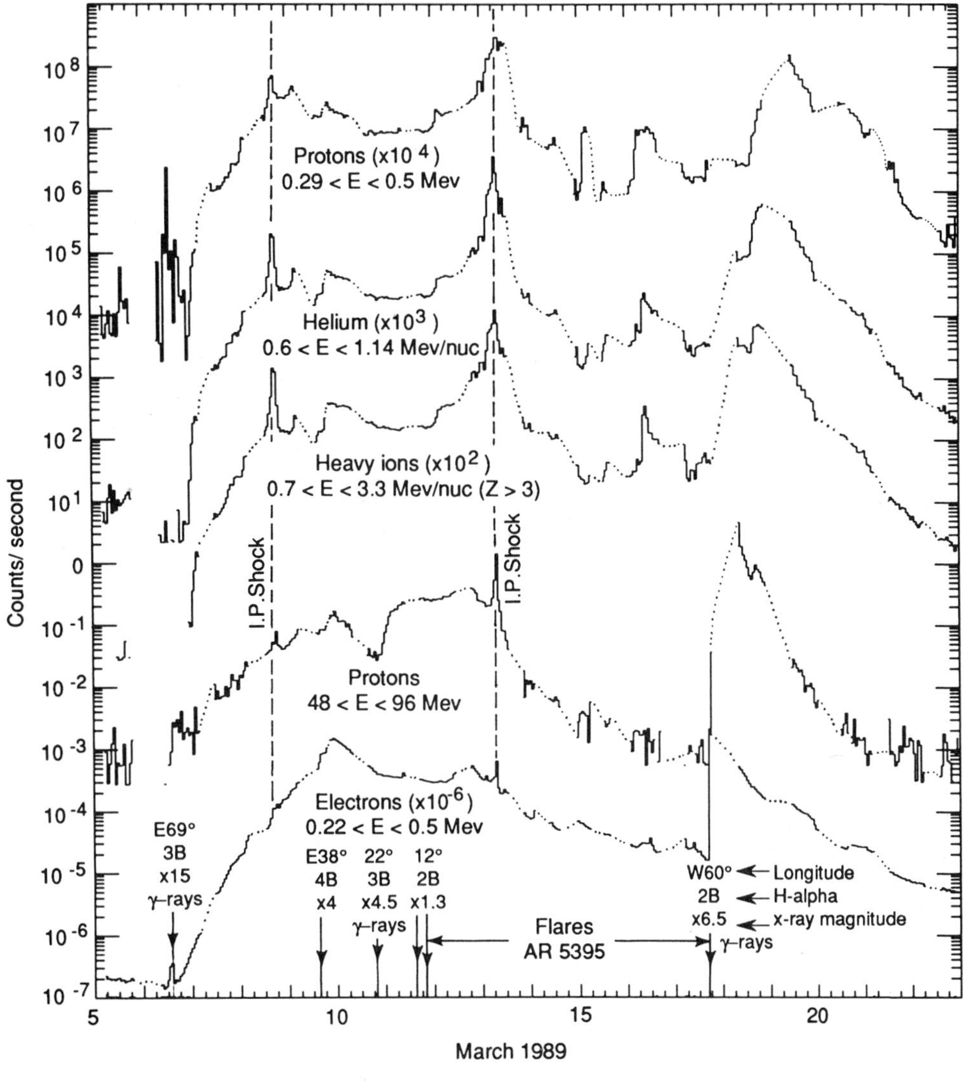

Figure 1. Intensity–time profile of solar cosmic rays for selected energies and particle species during the great flare activity in March 1989. The dotted lines connect adjacent points during data breaks. The most important flares during this period are noted at the bottom. Note that the flares on March 6 and 10, with significant x-ray brightness and γ-ray emission, were responsible for generating the shocks observed on March 8 and 13, respectively. The data were obtained with the author's experiment on the *Interplanetary Monitoring Platform-8* (*IMP-8*) spacecraft.

ahead of this blast. This sweep-up action causes a shock to be formed, and the interplanetary particle population to be accelerated to high energies, causing enhancements in solar cosmic rays over and above those that were produced by the original flare. The magnitude and intensity of these shock-associated enhancements very much depend on the location of the observer with respect to the parent flare site. The envelope sketched in Fig. 2 shows an inferred shock surface for a flare at central meridian. On the east side, the direction normal to the shock surface is nearly parallel to the interplanetary magnetic field, whereas on the west side the normal is nearly perpendicular to the field. It is now well established that it is on the west side of the shock where most of the interplanetary acceleration takes place. This is simply demonstrated in Fig. 1 where eastern flares on March 6 and 10 gave rise to large shock-associated enhancements on March 8 and 13, respectively, whereas the western flare on March 17 did not produce any shock enhancement.

Frequency of Solar Cosmic Rays

The events of March 1989, although unusually intense, are by no means unique. Typically there are 20–30 flares on the Sun every day, most of which are relatively weak and do not produce significant numbers of solar cosmic rays. Fewer flares are seen at the time of minimum solar activity, while many more occur at the time of solar maximum. As a general rule, about one flare per month produces solar cosmic ray enhancements observed in the interplanetary medium at energies greater than about 30 MeV. Generally the lower the energy, the more intensity enhancements are observed: at energies below 1 MeV there is a continuous presence of solar cosmic rays in the vicinity of Earth. Some of these intensity enhancements undoubtedly originated with smaller flares on the Sun, although many of the enhancements are typically associated with interplanetary shocks that form in the manner discussed earlier, that is, when a fast solar wind stream catches up

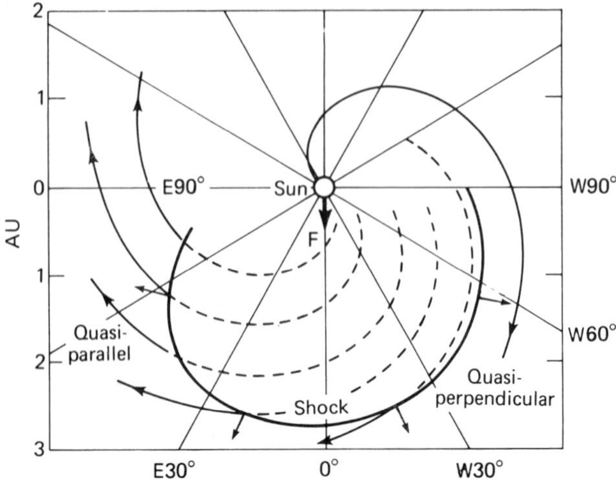

Figure 2. A schematic view of the Archimedean spiral shape of the interplanetary magnetic field and its relationship to an expanding shock front following a flare (*F*) at central meridian. Particles originating at flare sites substantially west of the Earth–Sun line propagate readily along the magnetic field to Earth; those from the eastern solar hemisphere must propagate across the magnetic field to reach Earth; hence the rise time to maximum intensity is substantially longer and can take several days. The geometry between the direction of the magnetic field and the normal to the shock surface is a crucial parameter in determining whether interplanetary acceleration at the shock takes place. When the angle is 50–90° (quasiperpendicular, west of the flare site), large enhancements are seen (peaks on March 8 and 13 in Fig. 1). When the angle is 0–40° (quasiparallel, east of the flare site), little acceleration takes place (absence of shock enhancement following the flare on March 17). [*Adapted from E. T. Sarris and S. M. Krimigis*, Solar Physics *96 413 (1985)*.]

with a slower one. The relationships between frequency of occurrence and energy threshold are illustrated in Fig. 3 for a year near the last solar maximum.

Radial Extent

Measurements by spacecraft in the outer solar system, such as *Pioneer 10* and *11* and *Voyagers 1* and *2*, have shown that solar cosmic rays from individual flares produce intensity enhancements as far out as 30–50 AU (1 AU equals the distance between the Earth and the Sun, i.e., 1.5×10^8 km). The intensities of solar cosmic rays in the outer solar system, however, are dominated by large fluxes at relatively low energies, extending at most to a few megaelectron volts. In fact, these should be called interplanetary cosmic rays because it is generally accepted that these ions, although their ultimate origin is at the Sun, have been accelerated to their observed energies by shock interactions in the interplanetary medium. In fact, the ejecta from high intensity solar flares eventually propagate to the outer solar system and cause decreases in the intensity of galactic cosmic rays in their path. These decreases are called Forbush decreases and they are associated with the interaction of cosmic rays with stronger than ambient magnetic fields that are transported to the outer solar system by these so-called plasma clouds.

Particle Composition

An important element in models of solar system evolution is the elemental and isotopic composition of solar system matter. Such composition measurements are principally derived from terrestrial (isotopic) and meteoritic (elemental) material, even though the Sun constitutes more than 99% of all solar system matter. In fact, isotopic anomalies in meteorites and interplanetary grains suggest that the solar system may be far from homogeneous, and that some nucleosynthesis may have taken place during its formation. Solar cosmic rays offer the opportunity for direct knowledge of elemental and isotopic abundances of solar system material.

It is generally found that the composition of solar cosmic rays is similar to the composition of general solar system matter, but that in certain cases there are differences between this direct sample of solar material and spectroscopically derived photospheric abundances for such key elements as carbon and iron. The electrical charge state of solar cosmic rays can also be used to determine the temperature of the region at which the acceleration of solar material took place. Typical numbers range from 10^6 to 10^7 K. Temperatures at the flare region can also be derived from observations of microwave and x-ray emissions that are characteristic of the electron population and can range up to 10^8 K.

THEORY AND MODELING

Most early theoretical work on solar cosmic rays centered on studies of propagation from the sun through the fluctuating magnetic field in the interplanetary medium to the vicinity of Earth. A Fokker–Planck equation has been used that includes terms for diffusion, convection, and adiabatic energy loss in the expanding solar wind. Although this description met with some success in fitting the data at the higher velocities (for example the flare on March 17, 1989 in Fig. 1, two lowest curves), it has had difficulty with eastern hemisphere flares and at low velocities everywhere (see top three curves in Fig. 1). It appears that the specific geometry of the flare site, together with the modification of the source spectrum due to shock acceleration effects in the interplanetary medium (i.e., intensity spikes such as those on March 8, 13 in Fig. 1) make the Fokker–Planck description a rather dubious approach for most observed cases.

More recent work attempts to take into account explicitly the state of the interplanetary magnetic field and the solar wind in order to relate specific intensity features to discontinuities in these interplanetary plasma parameters. As is evident from Fig. 1, the complexity of the profiles is such that a detailed description that includes all relevant physical parameters with some predictive capability is not likely to be attained in the foreseeable future.

Nevertheless sufficient theoretical understanding exists on the flare process itself, so that by utilizing the photon signatures (γ-rays, x-rays, microwaves) together with the particle observations in the interplanetary medium, it is possible to derive important flare site parameters: total energy release, 10^{32} erg (large flare); total energy in protons, up to 10^{30} erg; total energy in electrons, up to 10^{31} erg; acceleration times at the higher energies, less than 2 s.

STATUS

The presence of spacecraft detectors in the vicinity of Earth, in the outer solar system (*Pioneers 10* and *11*; *Voyagers 1* and *2*), and over the polar regions of the Sun (*Ulysses*), promises to bring about an unprecedented wealth of three-dimensional observations of solar cosmic rays throughout the solar system in the 1990s. It is hoped that such observations will enable the introduction of more realistic parameters into theoretical models that, in turn, should lead to an adequate description of the behavior of solar cosmic ray intensities throughout the interplanetary medium. Such a development, together with continuous observations of the Sun, should lead to accurate estimates of solar flare particle production and the prediction of possible consequences on Earth-based technological systems. It is important to remember that every few years there is

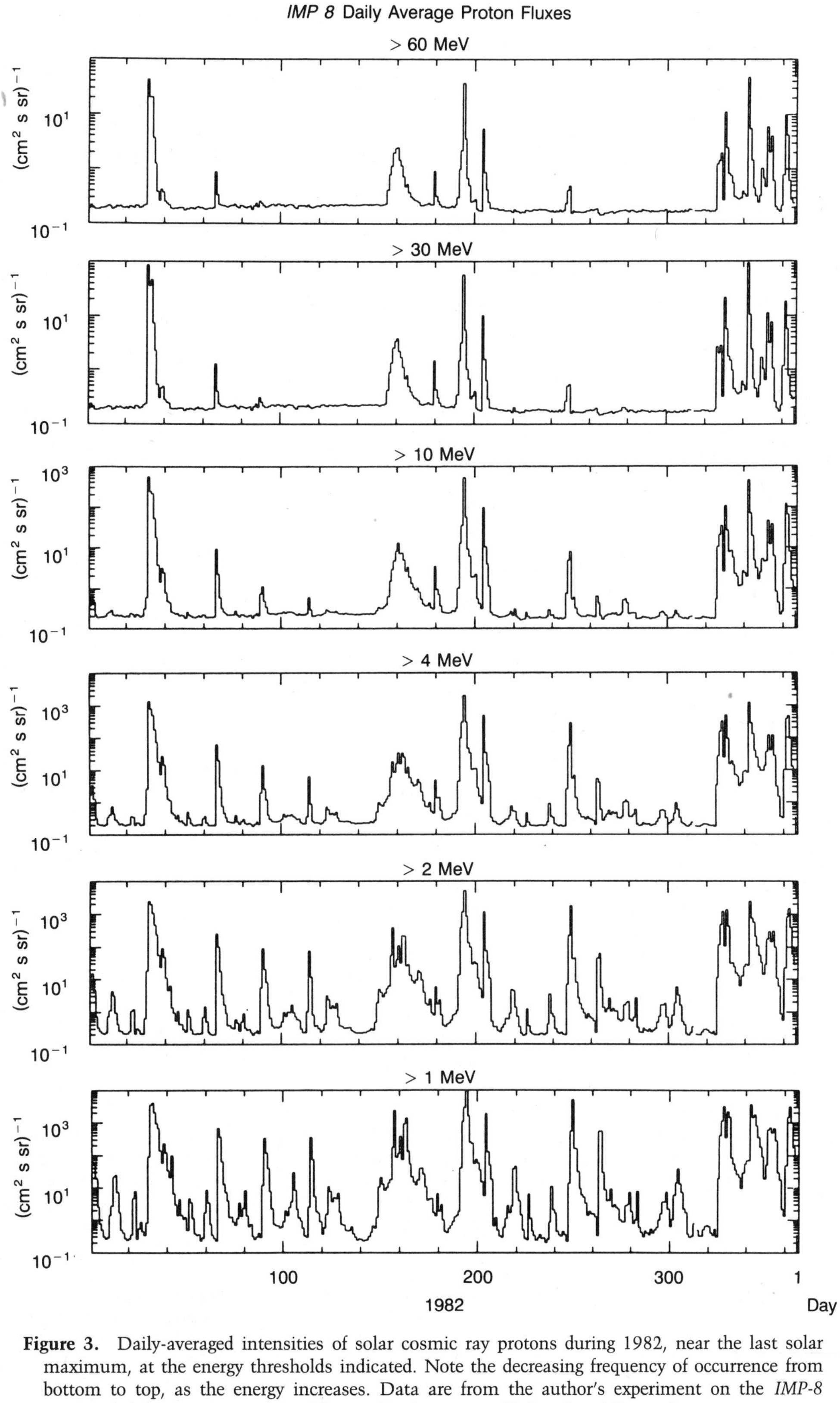

IMP 8 Daily Average Proton Fluxes

> 60 MeV

> 30 MeV

> 10 MeV

> 4 MeV

> 2 MeV

> 1 MeV

$(cm^2\ s\ sr)^{-1}$

10^{-1} 10^{1} 10^{3}

100 200 300 1

1982 Day

Figure 3. Daily-averaged intensities of solar cosmic ray protons during 1982, near the last solar maximum, at the energy thresholds indicated. Note the decreasing frequency of occurrence from bottom to top, as the energy increases. Data are from the author's experiment on the *IMP-8* spacecraft (*Plots courtesy of Dr. Thomas P. Armstrong, University of Kansas.*)

one solar cosmic ray event so intense that an astronaut exposed in interplanetary space could receive a lethal dose within a few hours.

Additional Reading

Allen, J., Frank, L., Sauer, H., and Reiff, P. (1989). Effects of the March 1989 solar activity. *EOS* **70** 1479.

Beiber, J. W., Evenson, P., and Pomerantz, M. A. (1990). A barrage of relativistic solar particle events. *EOS* **71** 1027.

Forbush, S. E. (1966). Time variations of cosmic rays. In *Handbuch der Physik*, vol. XLIX/1. Springer, Berlin, p. 159.

Kundu, M. and Woodgate, B., eds. (1986). *Energetic Phenomena on the Sun*. NASA Conference Publication 2439, NASA Scientific and Technical Information Branch, Washington, DC.

Lin, R. P. (1987). Solar particle acceleration and propagation. *Rev. Geophys*. **25** 676.

Maran, S. P. (1990). When all hell breaks out on the Sun, astronomers scramble to understand. *Smithsonian* **20** (No. 12) 32.

Smart, D. F. and Shea, M. A. (1985). Galactic cosmic radiation and solar energetic particles. In *Handbook of Geophysics and the Space Environment*, A. S. Jursa, ed. Air Force Geophysics Laboratory, Cambridge, Mass.

See also **Interplanetary and Heliospheric Space Missions; Interplanetary Magnetic Field; Interplanetary Medium Shock Waves and Traveling Magnetic Phenomena; Solar Activity, Solar Flares.**

Interplanetary Medium, Solar Wind

Thomas E. Holzer

The solar wind in interplanetary space is a highly supersonic and super-Alfvénic (moving faster than the Alfvén speed), fully ionized, magnetized plasma flowing nearly radially outward from the Sun. The plasma is composed primarily of protons, α particles, and electrons, but it is thought to contain, as minor constituents, all of the elements detected in the solar photosphere. Indeed, the abundances and ionization states of these minor constituents carry important information concerning the source region of the solar wind in the upper chromosphere and lower corona of the Sun. The large-scale spatial and temporal structure of the solar wind in interplanetary space is a direct reflection of the large-scale spatial and temporal structure of the solar corona, although the structure imposed on the wind in the corona sometimes undergoes significant dynamical evolution in interplanetary space.

It was in the eighteenth and nineteenth centuries that a number of scientists in northern Europe first began to suspect that the Sun fills interplanetary space not only with light but also with material particles. These suspicions arose from a variety of studies that revealed close relationships among the aurora, geomagnetic activity, and sunspots. A comprehensive program in solar–terrestrial research was first mounted by Kristian Birkeland around the beginning of the twentieth century, and as a result of this program, Birkeland was able to deduce (on the basis of observations from field stations around the world, laboratory experiments, and theoretical studies) that solar corpuscular radiation not only causes the aurora and geomagnetic activity, but also produces comet tails. He further suggested that virtually every star throws off mass, and that as a consequence much of the mass in the universe might be found in the interstellar medium. Birkeland's ideas were refined and expanded over the next half century by a number of workers (e.g., Carl Störmer, Julius Bartels, Sydney Chapman, V. C. A. Ferraro, Scott E. Forbush, Ludwig Biermann, and Hannes Alfvén), thus setting the stage in 1958 for Eugene N. Parker's landmark theory describing the expansion of the solar corona in a supersonic solar wind that permeates all of interplanetary space. Coming at the beginning of the space age, Parker's theory was soon confirmed by in situ observations from both Soviet and U.S. space probes, and the field of interplanetary physics was born.

Table 1. Average Bulk Properties of the Solar Wind at 1 AU

Parameter	*Units*	*Low Speed Wind*	*High Speed Wind*
n_p	cm^{-3}	10.3	3.4
v_p	$km\ s^{-1}$	330	700
T_p	K	3.4×10^4	2.3×10^5
T_e	K	1.3×10^5	1.0×10^5
T_α	K	1.1×10^5	1.4×10^6
n_α/n_p		0.05	0.05
nv	$cm^{-2}\ s^{-1}$	3.4×10^8	2.4×10^8
ρv^2	$erg\ cm^{-3}$	1.8×10^{-8}	2.8×10^{-8}
p	$erg\ cm^{-3}$	2.6×10^{-10}	1.9×10^{-10}
$B^2/8\pi$	$erg\ cm^{-3}$	1.7×10^{-10}	1.7×10^{-10}
ρu^3	$erg\ cm^{-2}\ s^{-1}$	0.59	1.9
q	$erg\ cm^{-2}\ s^{-1}$	2.7×10^{-3}	3.2×10^{-3}

BULK PROPERTIES SOLAR WIND

The solar wind in interplanetary space can generally be placed in one of three categories: (a) high speed wind, which originates in low density solar coronal holes, (b) low speed wind, which originates in or near the high density coronal region often referred to as the coronal streamer belt, and (c) transient wind, which originates in sporadic large-scale ejections of mass and magnetic field from the coronal streamer belt. Ignoring, for the moment, the transient wind and the regions of interaction of high and low speed wind, we present in Table 1 average values at the orbit of Earth (1 AU) for several parameters characterizing the bulk properties of the low and high speed flows. Consideration is restricted here to electrons, protons, and α particles (subscripts e, p, and α), because these are the only dynamically important constituents of the wind. The notation in Table 1 is standard, with the symbols n, ρ, v, T, p, q, and B standing for number density, mass density, bulk flow speed, kinetic temperature, thermal pressure, heat flux density (i.e., thermal conduction flux density), and magnetic field intensity. (Note that unsubscripted parameters, such as p, refer to the net effect of the electrons, protons, and α particles.) The observed range of variation of the listed parameters from their average values is typically about $\pm 45\%$ for low speed wind, but only about $\pm 15\%$ for high speed wind, so if there is a "quiet" state of the solar wind, it is to be found in high speed streams.

We can see from Table 1 that virtually all of the energy transported by the solar wind through interplanetary space is in the form of bulk flow energy, because $\rho v^2 \gg p$, $B^2/8\pi$, and $q/\rho v$. For this reason the nearly radial solar wind flow controls the shape of the interplanetary magnetic field. In the highly conducting solar atmosphere and solar wind, the field is "frozen" to the plasma, which requires that if elements of plasma are on the same magnetic field line at one time, then they will be on the same field line at all times. Consequently, a magnetic field line traces out the locus of plasma elements originating from a point on the solar surface, and in the case of a steady, radial, spherically symmetric flow emanating from a rotating sun, the field lines take the form of Archimedes spirals defined by $\theta = \theta_0$ and $r = (v/\Omega)(\phi - \phi_0)$. Here r, θ, and ϕ define a spherical coordinate system, $\theta = 0$ lies along the solar rotation axis, Ω is the solar angular velocity, and the spiral magnetic field is viewed in the fixed (as opposed to rotating) frame of reference. An illustration of interplanetary magnetic field lines in the $\theta = \pi/2$ plane (tilted 7° with respect to the ecliptic plane) is given in Fig. 1. Of course, near the Sun the magnetic energy density is larger than the plasma energy density, and the field channels the

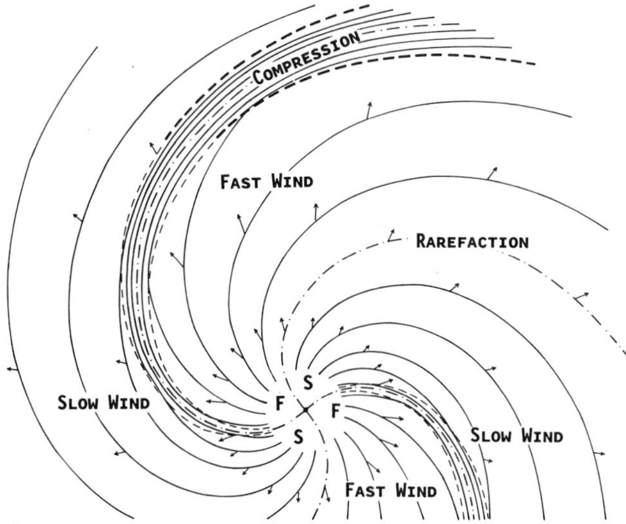

Figure 1. A view (from the north) of the solar equatorial plane to a maximum heliocentric distance of 3 AU. Fast (F) and slow (S) solar wind with average speeds at 0.3 AU of 700 km s^{-1} and 330 km s^{-1} is assumed to emanate from 90° wide longitudinal sectors. Magnetic field lines (solid lines) originating at longitudes 15° apart at the Sun are shown beyond 0.3 AU in their interplanetary configuration, and the four field lines (dash–dot lines) marking the contact surfaces separating fast and slow wind are shown from the Sun outward. Dashed lines mark the boundaries of the compression regions where fast wind overtakes slow wind, and heavy dashes indicate the region where compression fronts have steepened into shocks. Solar wind flow velocities are indicated by arrows, whose lengths indicate speed. This type of solar wind stream structure is characteristic of a solar magnetic dipole axis that is tilted significantly with respect to the solar rotation axis and that leads to magnetic fields of opposite polarity in the two high speed streams and to a polarity change (referred to as a sector boundary) near the middle of each low speed stream.

flow, but beyond some 10–20 $R_\odot$ (solar radii) the plasma dominates and the spiral structure should be a good description of the average interplanetary magnetic field.

The solar wind has been observed in situ from the orbit of Mercury (0.3 AU) to the orbit of Pluto (40 AU), but only in the vicinity of the ecliptic plane. In addition, optical coronal observations and interplanetary radio scintillation observations provide indirect information about the wind at all latitudes. Thus, we can infer how the average properties of the wind vary with heliocentric radial distance and latitude throughout the solar activity cycle. In the absence of transient wind and of interactions between high and low speed wind, the solar wind flow speed is nearly independent of radial distance (inside the shock terminating supersonic flow, which must lie somewhere beyond the orbit of Pluto) and the density declines as $\rho \sim r^{-2}$. Beyond the orbit of Earth, electron thermal conduction becomes progressively less important, and both the electron and ion temperatures would decline adiabatically (i.e., $T \sim r^{-4/3}$), were it not for the interaction of high and low speed wind streams and the presence of interstellar hydrogen atoms in interplanetary space. Charge transfer between solar wind protons and these atoms produces a population of hot protons that are not fully assimilated into the proton velocity distribution, but which nevertheless significantly modify the proton temperature, leading to an increase of T_p with increasing r beyond the orbit of Jupiter (5 AU). This effect, however, is not as large as the heating arising from the interaction of fast and slow wind, which may occur within a few tens of degrees of the ecliptic plane (see following text).

In order to understand the latitudinal structure of the supersonic solar wind we must consider the solar magnetic field, which organizes the structure of the solar corona. Near the minimum of the 11-yr solar activity cycle, the structure of the corona and the solar wind is largely organized by the dipole component of the solar magnetic field, with relatively low speed wind flowing near the dipole equator (along which the coronal streamer belt is centered) and relatively high speed wind flowing from coronal holes located at high magnetic latitudes. Throughout much of the 11-yr cycle (namely, the minimum and postminimum phases of the cycle) the dipole axis is nearly coincident with the solar rotation axis, but during the declining (postmaximum) phase of the activity cycle the dipole axis is tilted significantly (some 30°) with respect to the solar rotation axis. In this latter period the nonalignment of the rotational and magnetic axes gives rise to a strong interaction between high and low speed wind at low and middle solar latitudes. Such interactions are also produced in the ascending (postminimum) phase of the activity cycle, when the solar magnetic field exhibits a significant quadrupole (or higher order) component, which produces a significant warp of the coronal streamer belt. Finally, near the maximum of the cycle, the organizing dipole structure largely disappears, the streamer belt expands to cover virtually all of the solar surface, and relatively slow wind flows at most solar latitudes and longitudes.

The interaction between high and low speed wind streams near the ecliptic plane is illustrated in Fig. 1. The reason that wind streams from different solar longitudes interact is related to the reason that the interplanetary magnetic field has a spiral structure: the wind flows nearly radially, but because of solar rotation the wind from a given solar longitude traces out an Archimedes spiral, so wind from all longitudes lies along any given radial line. Where high speed wind overtakes low speed wind, a compression region develops, and where high speed wind runs away from low speed wind, a rarefaction region develops. Eventually (typically between 2 and 3 AU), forward and reverse shock fronts develop at the leading and trailing edges of the compression region. These fronts propagate away from the center of the compression (of course, they are both advected away from the Sun by the highly supersonic solar wind) and ultimately (near the orbit of Uranus at 20 AU for the case shown in Fig. 1) the leading shock from one solar rotation overtakes the trailing shock from the preceding rotation. At this point, all the solar wind (at the latitude under consideration) has been processed through the shocks and thus has undergone the shock heating that serves to convert relative flow energy between the high and low speed into internal energy of the plasma. This is the heating process associated with streams that was referred to previously.

The interaction of transient ejections of mass and magnetic field from the coronal streamer belt with the ambient solar wind is qualitatively similar to the interaction between high and low speed solar wind streams, although the former interaction occurs over a smaller spatial scale and does not produce the corotating structures associated with long-lived solar wind streams that dominate all of interplanetary space in the vicinity of the ecliptic plane. As in the case of stream interactions, shocks are formed and ambient wind is heated. However, in contrast to the stream interactions, for which the interplanetary magnetic field topology is not modified, the transient field structure generally exhibits a magnetic topology that is locally closed in interplanetary space (i.e., the field lines are disconnected from the Sun).

SOLAR WIND COMPOSITION

The composition of the solar wind, which refers to the ionization states and abundances of the various elements composing the wind, is of interest both because of the information it provides concerning solar and cosmic abundances and because of the inferences it allows us to draw about physical processes in the upper chromosphere and lower corona that may be relevant to the acceleration of the solar wind. It is observed that the elemental abundances (with the exception of H and He) in high speed wind are essentially the same as those of the solar photosphere, whereas in low speed wind the abundances of elements with relatively high first ionization potentials (above about 10.5 eV) are reduced by about a factor of 4. The average abundance of He is reduced by a factor of 2 in both low and high speed wind relative to other high first ionization potential (FIP) elements (the He ionization potential is 25 eV). These observations of abundance differences appear to indicate the operation of an ion–atom separation process operating at temperatures characteristic of the upper solar chromosphere, where low FIP elements are largely ionized and high FIP elements are largely neutral. Such a process likely involves the interaction of diffusive and transient mixing effects, the latter of which may play an important role in solar wind acceleration. The observation that He and the minor solar wind elements are substantially enhanced in transient wind appears to reflect the action of diffusive and transient mixing processes leading to ion–ion separation in the chromosphere–corona transition region and in the lower solar corona.

The ionization states of elements composing the solar wind are determined in the lower and middle solar corona, where the time scale over which the plasma expands due to outward flow of the wind becomes shorter than the time scales over which ionization and recombination processes operate. Beyond this point in the corona, the ionization state can be thought of as "frozen into" the flow and unable to further change. It follows that the solar wind ionization state reflects the operation of processes depending on the structure of the density, temperature, and flow in the coronal acceleration region of the solar wind. Thus, studies of the solar wind ionization state, like those of solar wind abundances, are not only of inherent interest, but also may lead to an improved understanding of the origin of the wind.

FUTURE OF INTERPLANETARY SOLAR WIND STUDIES

There are still three regions of the solar wind in interplanetary spaced that remain unexplored by in situ observations: the high latitude wind, the near-Sun wind inside the orbit of Mercury (0.3 AU), and the very distant solar wind beyond the orbit of Pluto (40 AU). *Pioneers 10* and *11* and *Voyagers 1* and *2* should penetrate into the very distant solar wind in the 1990s and the *Ulysses* mission will explore the high latitude wind. In addition, a *Solar Probe* mission, which is designed to observe the inner solar wind (outside 0.02 AU) is currently under study. Observations obtained by these spacecraft, together with the theoretical work they stimulate, should significantly improve our understanding of the solar wind over the next two decades.

Additional Reading

Birkeland, K. (1980). *The Norwegian Aurora Polaris Expedition*, **I**, 1st section (1908) and 2nd section (1913). Aschehoug Co., Christiania.

Feldman, W. C., Asbridge, J. R., Bame, S. J., and Gosling, J. T. (1977). Plasma and magnetic fields from the Sun. In *The Solar Output and Its Variation*, O. R. White, ed. Colorado Associated Universities Press, Boulder, p. 351.

Gloeckler, G. and Geiss, J. (1989). The abundances of elements and isotopes in the solar wind. In *Cosmic Abundances of Matter*, C. J. Waddington, ed. American Institute of Physics, New York, p. 49.

Hundhausen, A. J. (1972). *Coronal Expansion and Solar Wind*. Springer, New York.

Parker, E. N. (1963). *Interplanetary Dynamical Processes*. Interscience, New York.

Pizzo, V. J. (1986). The nature of the distant solar wind. *Adv. Space Res.* **6** 353.

Van Allen, J. A. (1990). Magnetospheres, cosmic rays and the interplanetary medium. In *The New Solar System*, 3rd ed., J. K. Beatty and A. Chaikin, eds. Sky Publishing Corp., Cambridge, MA and Cambridge University Press, Cambridge, U.K., p. 29.

See also **Interplanetary Medium, Shock Waves and Traveling Magnetic Phenomena; Interplanetary Medium, Wave–Particle Interactions; Sun, Coronal Holes and Solar Wind.**

Interplanetary Medium, Wave–Particle Interactions

Joseph V. Hollweg

The solar wind is the outflow of the coronal plasma into interplanetary space. It was predicted theoretically by Eugene N. Parker in 1958, and its existence was verified soon thereafter by direct observations made by spacecraft-borne particle detectors. In early solar wind models the energy source is provided by outward electron–heat conduction from the hot solar corona. These models give the essential qualitative feature of the solar wind, namely, an acceleration of the outflow from low subsonic velocities in the low corona to highly supersonic velocities far from the Sun.

However, these early models are incorrect in detail. In comparing observations with theoretical models we will concentrate on the high-speed solar wind streams, which are the most energetic features of the solar wind. The models yield flow velocities that are too slow (≈ 300 km s^{-1} at 1 AU, compared to observed velocities of ≈ 750 km s^{-1}) and proton temperatures that are much too low ($\approx 10^3$ K at 1 AU, compared to observed values of $\approx 2.5\times10^5$ K). If the electron heat conduction is described by the classical formula for a fully ionized, collision-dominated, electron–proton plasma, then the predicted electron temperatures come out several times hotter than the observed values of $\approx 10^5$ K at 1 AU; there is a qualitative difference as well, namely, the models predict that the energy flux at 1 AU is dominated by the electron heat conduction flux whereas, observationally, the energy flux is dominated by the kinetic energy of the flow.

Most of these early models concentrate on the dominant electron–proton component of the solar wind, but the so-called minor ions, the most important of which is He^{+2}, present problems of their own. The minor ions usually flow faster than the protons, with a tendency for the velocity difference to be about the Alfvén speed (see the section on magnetohydrodynamic waves). Moreover, the minor ions are hotter than the protons, with temperatures roughly in proportion to their masses.

Spacecraft plasma detectors are able to measure not only the bulk properties of the solar wind plasma, such as temperature and flow velocity, but the particle distribution functions as well. The distribution functions are observed to be organized by the interplanetary magnetic field. The protons usually show a general anisotropy, with the temperature parallel to the magnetic field somewhat greater than the perpendicular temperature. Qualitatively, this is what one expects, because particles that gyrate rapidly about a magnetic field line without collisions tend to conserve $v_\perp^2/B$, where B is the magnetic field strength and $v_\perp$ is the magnitude of the perpendicular particle velocity. Because B decreases with increasing distance from the Sun, $v_\perp^2$ (and thus the perpendicular temperature $T_\perp$) should decrease, and one expects $T_\perp \propto B$. The parallel temperature $T_\parallel$ also decreases (one expects $T_\parallel \propto \rho^2/B^2$, where ρ is the mass density) but not as rapidly as $T_\perp$ inside 1 AU, and we expect $T_\parallel > T_\perp$. The problem is that the predicted anisotropy is much greater than observed. Moreover, the lowest-energy protons in the vicinity of the peak of the distribution function usually exhibit the opposite anisotropy, that is, $T_\perp > T_\parallel$;

sometimes the proton distribution functions show a secondary peak that can be thought of crudely as a beam moving along the magnetic field faster than the general proton flow. The electron distribution functions are highly anisotropic. The higher-energy electrons are distributed in a well-collimated beam (called the strahl) that is directed along the magnetic field away from the Sun. The strahl is interpreted as escaped coronal electrons that move along the magnetic field with little interaction with waves or other particles along the way. The lower-energy electrons are much more isotropic.

During the past two decades, much effort has been put into the idea that waves in the solar wind can explain the differences between the early models and the observations. Waves exert forces on a plasma and may therefore be able to explain the high flow speeds. Waves can heat a plasma, possibly explaining the high proton and minor ion temperatures; the fact that the heaviest minor ions have higher temperatures than the solar corona probably makes wave heating indispensible. The interactions between waves and particles can depend sensitively on how the particles are moving, and thus wave–particle interactions may be responsible for the unexpected features in the observed particle distribution functions. But the converse is also possible. The non-Maxwellian distribution functions can lead to wave generation via microinstabilities, modifying the distribution functions and providing an internal source for waves in the solar wind.

The solar wind is one of the few places in the universe where magnetic fields, waves, and their interactions with plasma particles can be observed directly (planetary magnetospheres, comets, and the Earth's ionosphere are other places). However, it is generally believed that similar plasma processes may be occurring elsewhere, for instance, in stellar coronae, magnetized stellar winds, accretion disks, the magnetized plasma environments of pulsars, etc. It is expected that what we learn about the solar wind can be applied to other astrophysical objects as well.

MAGNETOHYDRODYNAMIC WAVES

Magnetohydrodynamic (MHD) waves occur in the solar wind plasma when the wave frequency is much smaller than the cyclotron frequencies of the plasma ions. These waves are well-approximated by a fluid description, in which the magnetic stresses play an essential role in the dynamics. Because there are three restoring forces available to the plasma (the fluid pressure, the magnetic pressure, and the magnetic field line tension), there are three MHD modes, usually called the fast, intermediate (or Alfvén), and slow modes, referring to their different speeds of propagation. An important feature of these modes is the "frozen-in" theorem, which states that one can think of the field lines as either being dragged with the plasma or pushing on the plasma. This theorem is strictly valid only when the electrical conductivity of the plasma is infinite, but it is a good approximation for the large conductivities in the solar corona and solar wind.

The Alfvén mode has received the most attention in solar wind studies because it has been directly observed. The solar wind contains magnetic field and velocity fluctuations that can be observed by magnetometers and plasma detectors. For time scales of several hours or smaller, these fluctuations are observed to be correlated nearly according to the predicted relation for Alfvén waves (using cgs units),

$$\delta \mathbf{V} = -\operatorname{sgn}(\mathbf{k}\cdot\mathbf{B}_0)\,\delta\mathbf{B}(4\pi\rho)^{-1/2},$$

where $\mathbf{V}$ is plasma velocity, $\mathbf{B}$ is magnetic field, the prefix δ denotes a fluctuation, $\mathbf{B}_0$ is the average magnetic field, $\mathbf{k}$ is the wave vector, and "sgn" indicates that the correct algebraic sign must be applied, depending on the angle between $\mathbf{k}$ and $\mathbf{B}_0$. The correlation between $\delta\mathbf{V}$ and $\delta\mathbf{B}$ is a consequence of the frozen-in theorem. These waves are very similar to waves on a string, with the magnetic field providing the tension. Alfvén waves have been detected almost all of the time in the solar wind, particularly in the high-speed streams, and have large amplitudes near 1 AU (i.e., $|\delta\mathbf{B}|/B_0 \approx 1$). From the algebraic sign of the correlation, it is deduced that the waves are usually propagating away from the Sun. This may mean that the Sun is the source of these waves, in which case they may provide the extra energy required to heat and accelerate the high-speed streams. The kinetic and magnetic energy of Alfvén waves propagates along $\mathbf{B}_0$ at the Alfvén speed,

$$v_A = B_0(4\pi\rho)^{-1/2}.$$

This is nearly the excess speed of the heavy ions relative to the protons, and it is likely that the waves are responsible for the excess acceleration of the heavy ions.

The most fundamental interaction of the Alfvén waves with the particles is the radiation pressure exerted by the waves on the plasma. If all ions move at the same speed, and if the wave frequency is high enough so that the Wentzel–Kramers–Brillouin approximation applies, the volume force exerted by the waves on the plasma is

$$\mathbf{F} = -\nabla\langle\delta\mathbf{B}^2\rangle/8\pi,$$

where the angle brackets denote a time average. Because $\langle\delta\mathbf{B}^2\rangle$ decreases away from the Sun, the force is outward and the flow is accelerated. The force arises from a combination of the Lorentz force on the plasma and the Reynolds stress $\rho\langle\delta\mathbf{V}\cdot\nabla\delta\mathbf{V}\rangle$.

If the ions have different flow speeds, then the various ion species acquire different accelerations from the Alfvén waves. It turns out that the waves tend to equalize the flow speeds of the various ions. This cannot explain why the heavy ions flow faster than the protons, but it helps, because without waves the heavy ions would flow slower than the protons.

To get plasma heating, the waves have to dissipate. At present, the most likely dissipation mechanism is MHD turbulence. For periods less than several hours the Alfvén wave power spectrum varies nearly as $f^{-5/3}$, where f is frequency. This is reminiscent of the inertial subrange of Kolmogorov turbulence in fluids. How this comes about in the solar wind is not understood, but the presence of an inertial subrange indicates that energy is being cascaded to high frequencies and small spatial scales via nonlinear interactions. When the energy reaches sufficiently high frequencies, cyclotron motions of the ions around the magnetic field lines become important.

In particular, the ions can absorb energy from the waves via cyclotron resonance. This occurs when the wave electric field vector rotates around the magnetic field in the same sense and at the same rate as the cyclotron motions of the ions. The ions then see a steady electric field and they can efficiently extract energy from the wave. The ions thereby increase their perpendicular temperature; the observed excess of perpendicular over parallel temperature for the lowest-energy protons is probably a signature of heating by cyclotron resonance. The ions also increase their velocities along the magnetic field as they absorb the momentum carried by the wave.

This view is qualitatively appealing with regard to the heavy ions. They have lower cyclotron frequencies than the protons, and they are the first to absorb the energy being cascaded to high frequencies by the turbulence. Thus one expects preferential heating and acceleration of the heavy ions. The mass-proportional temperatures of the heavy ions can in principle be explained by cyclotron resonance. Suppose a wave "photon" with energy proportional to angular frequency ω and momentum proportional to k is absorbed by a particle with mass m. Then

$$\Delta(\text{internal energy}) = m\,\Delta V(\omega/k - V),$$

suggesting $\Delta T \propto m$ for ions with the same velocity history, where

Δ denotes the change in the indicated quantity after absorption of the photon. This expression basically represents what happens during the cyclotron resonance interaction. The difficulty is that the heavy ions do not have quite the same velocity history as the protons. Indeed, solar wind models that incorporate a turbulent cascade and cyclotron resonances yield heavy ions that do flow faster and are hotter than the protons, but the helium–proton temperature ratio is only 2–3. The problem may be in our incomplete understanding of MHD turbulence. But the electrons may provide the answer.

ELECTRONS AND INSTABILITIES

One of the unsolved problems in the solar wind is how to treat the electron heat conduction. The classical expression, which assumes many particle collisions, is invalid in the rarefied solar wind. It is possible that the electrons can supply all the solar wind energy and that there is no requirement for MHD waves originating at the Sun. If this is true, the electron distribution functions would be highly skewed and subject to microinstabilities. The instabilities generate waves that could regulate the heat flux and interact with the protons and heavy ions. A variety of unstable waves have been suggested, for instance, ion-acoustic waves, whistlers, and low-frequency fast MHD waves. Not much is known about these possibilities, however.

Similarly, the non-Maxwellian proton and ion distribution functions, and the ion–proton drift, can generate microinstabilities that react back on the distribution functions. Little conclusive information is available with respect to this.

However, one thing is clear: By far most of the observed wave power in the solar wind resides in the low-frequency MHD waves, which are apparently of solar origin. This is why they have received more attention than the instabilities. However, it is also clear that a full understanding of the solar wind and its wave–particle interactions will not be achieved until spacecraft are sent to within a few solar radii of the Sun, where the solar wind originates.

Additional Reading

Barnes, A. (1983). Hydromagnetic waves in the interplanetary medium; Hydromagnetic turbulence in the interplanetary medium; Collisionless processes in the interplanetary medium. In *Solar–Terrestrial Physics*, R. L. Carovillano and J. M. Forbes, eds. D. Reidel Publishing Co., Dordrecht, Holland, pp. 155, 172, and 185.

Hollweg, J. V. (1981). The energy balance of the solar wind. In *The Sun as a Star*, S. Jordan, ed. National Aeronautics and Space Administration Special Publication **450**, NASA, Washington, D.C., p. 355.

Hundhausen, A. J. (1972). *Coronal Expansion and Solar Wind*. Springer-Verlag, New York.

Isenberg, P. A. (1991). The solar wind. In *Geomagnetism*, **4**, J. A. Jacobs, ed. Academic Press, New York.

Schwartz, S. J. (1980). Plasma instabilities in the solar wind: A theoretical review. *Rev. Geophys. Space Phys.* **18** 313.

See also **Magnetohydrodynamics, Astrophysical; Sun, Coronal Holes and Solar Wind.**

Interplanetary Spacecraft Dynamics

Roger E. Diehl and Louis A. D'Amario

It is desirable for planetary missions to use direct interplanetary transfers for the trajectory from Earth to the target planet or small body, such as a comet or asteroid. A direct trajectory provides the shortest flight time, which allows earlier science-data return, lower operations cost and complexity, and greater spacecraft reliability. However, as payload requirements increase, or in the absence of available launch vehicles with sufficient performance, it is necessary to use indirect trajectories with gravity-assist planetary flybys to achieve viable missions.

Several past missions have used gravity-assist flybys. *Pioneer 11*, launched in 1973, used a Jupiter gravity assist to increase spacecraft velocity sufficiently to allow a flyby of Saturn. *Mariner 10*, also launched in 1973, used a Venus gravity assist to decrease spacecraft velocity so that a flyby of Mercury could be achieved. The gravity assist from the Mercury encounter was then used to adjust the period of the orbit so that the spacecraft returned to Mercury for two additional flybys. *Voyagers 1* and *2* (1977), like *Pioneer 11*, each used a gravity-assist flyby of Jupiter to reach Saturn. *Voyager 2* continued on to Uranus and Neptune, using the gravity assist of each preceding planetary encounter to target the spacecraft to the next planet. For these early missions, the planetary gravity-assist technique was utilized to lower the launch energy requirement significantly below that for a direct non-gravity-assist transfer and to achieve multiple planetary flybys in a single mission.

Recent missions using gravity-assist transfers include the *ICE* mission to the comet Giacobini–Zinner and the Soviet *VEGA* mission to the comet Halley. The *ICE* trajectory required several close lunar flybys for gravity assist, whereas the *VEGA* mission used Venus flybys to establish the transfer trajectories to Halley for the two *VEGA* spacecraft.

Gravity-assist transfers are currently planned or underway for the *Galileo* mission to Jupiter and the *Ulysses* mission to the poles of the Sun, and will also be used in the future *Comet Rendezvous Asteroid Flyby* mission and the *Cassini* mission to Saturn. The spacecraft for these missions, except for *Ulysses*, are much more massive than either the *Pioneer* or *Voyager* spacecraft. Consequently, these missions all require the use of planetary gravity-assist flybys to reach their destinations, because in each case the available launch vehicle cannot supply sufficient energy for a direct non-gravity-assist transfer.

DIRECT TRANSFERS

Valuable insight into the behavior of interplanetary trajectories can be gained from an understanding of two-body motion, because trajectory segments can often be modeled with reasonable accuracy using a two-body approximation. The two bodies are the spacecraft and the single attracting body. The attracting body may be considered to be the Sun for an interplanetary trajectory segment or a planet for a gravity-assist flyby.

Starting from Newton's second law, the two-body equations of motion

$$\frac{d^2\bar{r}}{dt^2} = -\left(\frac{GM}{r^3}\right)\bar{r}$$

can be derived, where $\bar{r}$ is the position of a spacecraft with respect to the attracting body, t is the time, and GM is the universal gravitational constant times the mass of the attracting body. It is assumed that the mass of the spacecraft is negligible relative to the mass of the attracting body.

The solutions to the two-body equations of motion yield trajectories that are either ellipses or hyperbolas. Elliptical motion (eccentricity < 1) corresponds to bound orbits, whereas hyperbolic motion (eccentricity > 1) corresponds to escape orbits. Exactly parabolic motion (eccentricity $= 1$) or circular motion (eccentricity $= 0$) is improbable. The motion with respect to the Sun may be either elliptical or hyperbolic. Motion with respect to a planet, on the other hand, is hyperbolic in all cases addressed here: departure from Earth, a planetary gravity-assist flyby, and arrival at a destination planet.

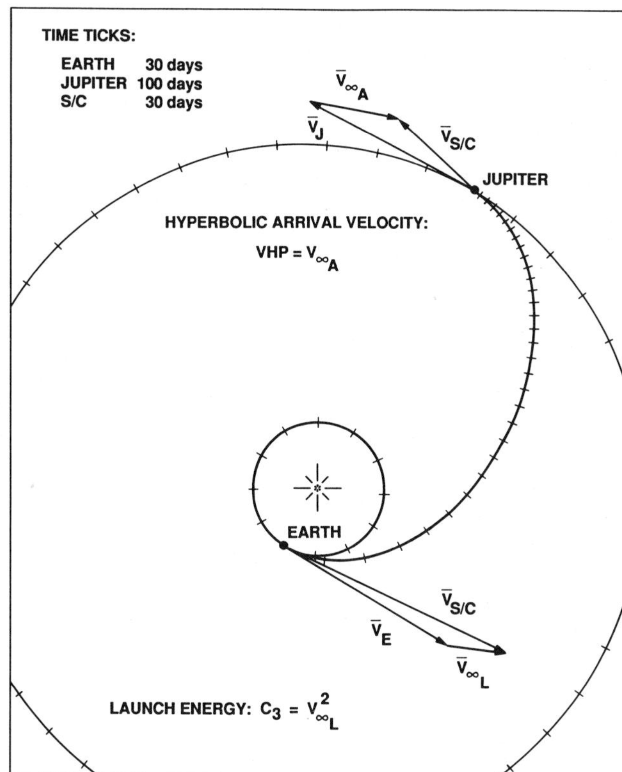

Figure 1. Direct ballistic Earth–Jupiter transfer. Velocity vector diagrams illustrate the definitions of launch energy and hyperbolic arrival velocity. The spacecraft trajectory for this transfer can be modeled with reasonable accuracy as an ellipse.

Direct interplanetary transfers are defined as transfers that do not employ planetary gravity-assist flybys. Direct transfers can be either ballistic or nonballistic. Direct nonballistic transfers have velocity changes (ΔV) between launch from Earth and arrival at the destination planet; direct ballistic transfers have no ΔV between launch and arrival and are thus purely coasting trajectories. Direct transfers may be classified according to the total transfer angle between launch from Earth and arrival at the destination planet. Type I transfers have transfer angles less than 180°; type II transfers have transfer angles between 180° and 360°.

Two quantities that are important to interplanetary trajectory design, C_3 and *VHP*, are illustrated in Fig. 1. If one models the effect of Earth's gravity at launch as an impulsive velocity change on the interplanetary trajectory, then the spacecraft's velocity with respect to the Sun ($\overline{V}_{S/C}$) is simply the Earth's velocity with respect to the Sun ($\overline{V}_E$) added to the spacecraft's asymptotic hyperbolic departure velocity with respect to the Earth ($\overline{V}_{\infty L}$). The quantity C_3 ($=V^2_{\infty L}$) is called launch energy. It is a measure of the amount of energy that the launch vehicle must add to the spacecraft to inject it onto the interplanetary trajectory. In a similar fashion, the spacecraft's arrival velocity at Jupiter (*VHP* or $\overline{V}_{\infty A}$) is the spacecraft's velocity with respect to the Sun ($\overline{V}_{S/C}$) minus Jupiter's velocity with respect to the Sun ($\overline{V}_J$). For a planetary orbiter mission, such as *Galileo*, the ΔV required to insert the spacecraft into orbit increases with increasing *VHP*. Thus, *VHP* determines the amount of on-board spacecraft propellant that must be expended to enter orbit at the destination planet.

The direction of the departure asymptote, which depends on the location of the velocity impulse that establishes the departure hyperbola, determines whether an increase or decrease in the spacecraft's velocity is achieved. For transfers to the outer planets, the net effect of the Earth departure hyperbola is to increase the spacecraft's velocity with respect to the Sun above the value corresponding to the Earth's velocity. This requires that $\overline{V}_{\infty L}$ be pointed in the same general direction as the Earth's velocity vector. For transfers to the inner planets, however, the spacecraft's velocity is decreased; that is, $\overline{V}_{\infty L}$ is opposed to the Earth's velocity.

Table 1. Characteristics of Hohmann Transfers from Earth to the Other Planets.

Planet	C_3 ($km^2\ s^{-2}$)	*VHP* ($km\ s^{-1}$)	*Flight Time*	*Synodic Period*
Mercury	57	9.6	3.5 mo	3.8 mo
Venus	6	2.7	4.8 mo	1.6 yr
Mars	9	2.6	8.5 mo	2.1 yr
Jupiter	77	5.6	2.7 yr	1.092 yr
Saturn	106	5.4	6.1 yr	1.035 yr
Uranus	127	4.7	16.0 yr	1.012 yr
Neptune	136	4.1	30.6 yr	1.006 yr
Pluto	140	3.7	45.5 yr	1.004 yr

In the idealized case of a transfer between circular coplanar orbits, there is a single unique transfer, called a Hohmann transfer (after Walter Hohmann), that has the minimum launch C_3 and arrival *VHP*. The Hohmann transfer is a cotangential 180° transfer; that is, the trajectory is tangent to both the departure and arrival planet orbits. Table 1 lists Hohmann transfer characteristics from Earth to each of the other planets.

Direct transfer opportunities repeat when the same relative planetary alignment occurs. The time interval between such opportunities is called the synodic period. The synodic period for each planet relative to Earth is also given in the table. Note that the synodic period for each of the outer planets is slightly over 1 yr. The largest synodic periods occur for the nearest planets (Venus and Mars).

If the orbits of the planets were exactly circular and coplanar, then the launch/arrival energy characteristics would be the same for each launch opportunity. However, the orbits of the planets (except Mercury and Pluto) have small eccentricities and small mutual inclinations. The effects of eccentricity (noncircular orbits) for a given launch opportunity to a planet are relatively small, because the range of achievable launch and arrival dates is a small fraction of the planet's orbit period. Comet orbits are very eccentric, and launch/arrival energy characteristics for cometary transfers are highly dependent on where the comet is encountered in its orbit. The effects of inclination (noncoplanar orbits), however, are profound regardless of the magnitude of the mutual inclination, prompting consideration of direct nonballistic transfers. For a given launch opportunity there is no longer a single minimum for C_3 or *VHP* corresponding to the Hohmann transfer. Instead, there are two minima for each parameter, one in the type I region and one in the type II region. These minima are greater than the Hohmann transfer value.

Between these minima, the transfer angle is near 180°. Such a trajectory connecting Earth's position at launch to the destination planet's position at arrival must have a very high inclination. As the transfer angle approaches 180°, the inclination of the trajectory with respect to the ecliptic (the plane of Earth's orbit) approaches 90°. As a result, the $\overline{V}_{\infty L}$ at launch must have a very large out-of-ecliptic component to establish the required inclination. For transfers between inclined orbits, the launch/arrival energy characteristics are not the same for each launch opportunity. The variation is caused by the changing position of the destination planet at arrival relative to the nodes (intersection points) of its orbit with the ecliptic. As the destination planet's position approaches the ecliptic at arrival, the type I and II minima begin to merge and their magnitudes approach the Hohmann transfer value. On the other hand, when the destination planet's position approaches a point 90° from either node, where its ecliptic latitude is at a maximum, the C_3 and *VHP* minima diverge significantly and their magnitudes differ by a considerable amount.

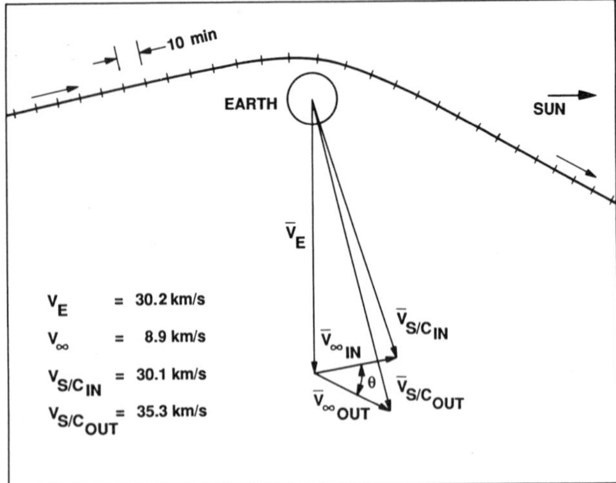

Figure 2. Gravity-assist flyby illustration. The spacecraft's V_∞ vector is rotated due to the gravitational attraction of the flyby body. This rotation has the effect of increasing the spacecraft's velocity with respect to the Sun.

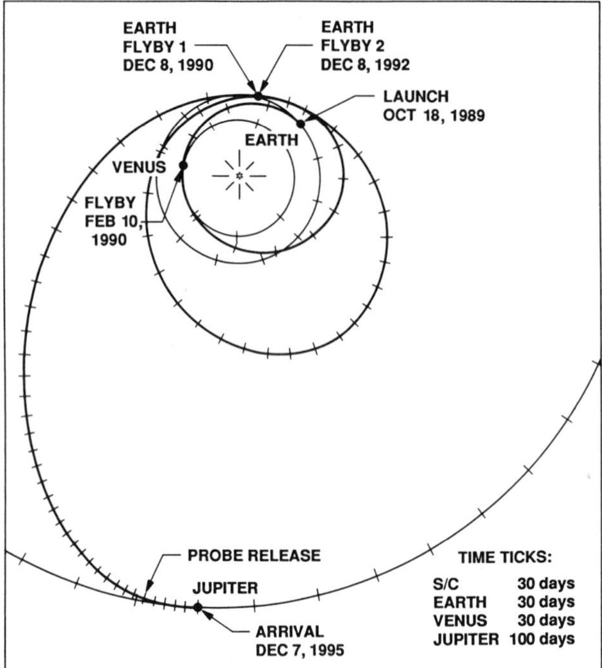

Figure 3. 1989 VEEGA trajectory to Jupiter. The shuttle/*IUS* launch vehicle could not have provided sufficient energy to launch the *Galileo* spacecraft on a direct transfer to Jupiter; it was necessary to use a VEEGA (Venus–Earth–Earth Gravity Assist) transfer, which increased the flight time from the launch on October 18,1989 to the spacecraft's arrival at Jupiter to about 6 yr.

It would appear that a significant region of the launch/arrival space corresponding to near-180° transfers must be excluded from consideration for interplanetary missions because of high launch energy requirements. By employing nonballistic transfers, however, the C_3 can be reduced considerably. These direct nonballistic transfers are referred to as broken-plane transfers. The ideal broken-plane transfer starts out in the ecliptic, and later a ΔV is performed to establish the inclination necessary to reach the destination planet. The inclination of the broken-plane transfer is minimized if the ΔV is done at a point where the remaining transfer angle to the planet is 90°. This minimum inclination is equal to the planet's ecliptic latitude at arrival. If excess launch energy is available, it can be used to perform some of the inclination change at launch, thus reducing the plane change required after launch and the associated ΔV. The design of broken-plane transfers is an optimization problem to minimize postlaunch ΔV, with postlaunch inclination and location of the plane-change maneuver as degrees of freedom.

GRAVITY-ASSIST TRANSFERS

Gravity-assist interplanetary transfers utilize one or more gravity-assist flybys to modify the trajectory and thus reduce significantly either launch energy, arrival velocity, or flight time requirements. The manner in which a gravity-assist flyby of a planet can change spacecraft heliocentric velocity can be illustrated by a velocity vector diagram. Compared with the duration of the interplanetary trajectory, the effects of the hyperbolic flyby occur practically instantaneously. Figure 2 illustrates a gravity-assist Earth flyby in the *Galileo* trajectory to Jupiter. The trajectory of the spacecraft with respect to Earth is a hyperbola; the spacecraft approaches and departs along the asymptotes of this hyperbola with a constant speed, called the V_∞. The spacecraft achieves its maximum velocity with respect to Earth at closest approach. The angle between the incoming and outgoing V_∞s is referred to as the bend angle of the flyby (θ). The velocity vector of the Earth with respect to the Sun at the time of the flyby is indicated by $\overline{V}_E$. Adding the incoming and outgoing V_∞s to the velocity of Earth yields the velocity vectors of the spacecraft with respect to the Sun before and after the flyby. The effect of the hyperbolic flyby in an Earth-centered reference frame is simply to rotate the V_∞ through an angle equal to the bend angle; there is no net energy change for the spacecraft trajectory with respect to Earth as a result of the flyby. However, the rotation of the Earth-centered V_∞ has the effect of increasing the magnitude of the velocity vector in a Sun-centered reference frame.

The physical basis of gravity assist can be seen quite easily. In Fig. 2, note that the direction of the Earth's motion is downward and the spacecraft's motion with respect to the Sun is approximately in the same direction. Because the spacecraft passes directly behind the Earth, the acceleration due to the Earth's gravity is also downward. The acceleration from the Earth is in the same direction as the spacecraft's motion and therefore increases the velocity of the spacecraft. Had the spacecraft been targeted to pass in front of the Earth, the acceleration would have been in a direction opposite to the spacecraft's motion, and the spacecraft's velocity would then have decreased. The effectiveness of this sample Earth gravity-assist flyby is impressive. The magnitude of the Sun-centered velocity has been increased by more than 5 km s^{-1}. This increase in speed results in a larger orbit and thus doubles the orbital period about the Sun from 1 to 2 yr.

An important relationship between the radius of closest approach R_P, the V_∞ magnitude, and the bend angle θ is given by

$$\sin\frac{\theta}{2} = \frac{1}{1 + R_P V_\infty^2/GM}.$$

This equation demonstrates a basic fact of gravity-assist theory; Either a smaller V_∞ or a closer flyby (smaller R_P) result in a greater rotation of the V_∞ vector and hence a larger change in the Sun-centered velocity. Atmospheric heating and navigation accuracy determine the minimum flyby altitude and thus the maximum velocity change from a gravity-assist flyby.

The VEEGA (Venus–Earth–Earth Gravity Assist) transfer for *Galileo* uses three gravity-assist flybys, one of Venus and two of the Earth, between launch and arrival at Jupiter (Fig. 3). This VEEGA transfer began with launch of the spacecraft (from the

Space Shuttle *Atlantis* on October 18, 1989) on a direct type I Earth–Venus transfer with $C_3 = 15$ km^2 s^{-2}. The Venus gravity-assist flyby altered the orbit such that the spacecraft was directed back to the Earth (type II transfer) and also added velocity to increase the orbit period to approximately 1 yr and to increase aphelion on the Venus–Earth transfer to 1.3 AU. The first Earth flyby added considerably more velocity to the spacecraft and increased the orbit period to 2 yr. On the Earth–Earth transfer the spacecraft passes through a 2.3-AU aphelion. The second Earth flyby added the final increment of velocity required for the transfer to Jupiter. After the second Earth flyby, the orbit period has been increased to 5.6 yr. The total flight time from Earth toJupiter is about 6 yr. This is significantly longer than the 2.5–3 yr required for a typical direct Earth–Jupiter transfer. On the other hand, the launch energy has been reduced from the typical C_3 value of 80 km^2 s^{-2} required for direct transfers to only 15 km^2 s^{-2}. In terms of postlaunch ΔV requirements, VEEGA transfers to Jupiter typically require less than 100 m s^{-1} compared with 200–300 m s^{-1} for a direct broken-plane transfer to Jupiter. For certain launch/arrival date combinations, the VEEGA transfer to Jupiter may be completely ballistic.

Additional Reading

D'Amario, L. A. et al. (1987). Galileo 1989 VEEGA trajectory design. AAS Paper 87-421, presented at the AAS/AIAA Astrodynamics Specialist Conference, Kalispell, Mont.

Deerwester, J. M. (1966). Jupiter swingby missions to the outer planets. *J. Spacecraft and Rockets* **3** 1564.

Flandro, G. A. (1966). Fast reconnaissance missions to the outer solar system utilizing energy derived from the gravitational field of Jupiter. *Astronautica Acta* **12** 329.

Hollenbeck, G. R. (1975). New flight techniques for outer planet missions. AAS Paper 75-087, presented at the AAS/AIAA Astrodynamics Specialist Conference, Nassau, Bahamas.

Laeser, R. P., McLaughlin, W. I., and Wolff, D. M. (1986). Engineering *Voyager 2's* encounter with Uranus. *Scientific American* **255** (No. 5) 36.

Messinger, H. F. (1970). Earth swingby—A novel approach to interplanetary missions using electric propulsion. AIAA Paper No. 70-1117, presented at the AIAA 8th Electric Propulsion Conference, Stanford, Calif.

Niehoff, J. C. (1966). Gravity-assist trajectories to solar system targets. *J. Spacecraft and Rockets* **3** 1351.

Szebehely, V. G. (1989). *Adventures in Celestial Mechanics*. University of Texas Press, Austin.

See also **Interplanetary and Heliospheric Space Missions; Outer Planets, Space Missions.**

Interstellar Clouds, Chemistry

Alexander Dalgarno

Chemistry in interstellar clouds has as its subject matter the determination of the molecular formation and destruction processes that together produce the interstellar molecules, the construction of models of the evolution of the chemical content of interstellar clouds in galaxies, and the identification of the events that have taken place in the clouds as they are reflected in the chemical composition.

A rich variety of molecular species has been discovered in the Milky Way and in external galaxies. Simple diatomic molecules are found in diffuse and translucent clouds through which the interstellar radiation fields can penetrate and in which processes of photodissociation and photoionization are significant components of the chemistry. Complex polyatomic molecules are found in more extensive dense clouds of high visual extinctions, the interiors of which are opaque to the external radiation field. The outer layers of dense clouds form molecular envelopes with a chemistry and a chemical composition similar to that of diffuse and translucent clouds.

INTERSTELLAR MOLECULES

Table 1 is a list of interstellar molecules that have been definitely identified. Many organic molecules have been found. Isotopic variants have also been detected, including several deuterated molecules. The hydrocarbon molecules are all linear with the exception of C_3H, which is present in both linear and cyclic forms, and C_3H_2. Molecular ions are contained in the list, attesting to the importance of cosmic rays to the chemistry of dense cloud interiors.

The molecular abundances relative to hydrogen range from about 10^{-4} for the ubiquitous carbon monoxide to between 10^{-7} and 10^{-10} for the complex molecules. There are substantial variations in the molecular abundances in different clouds and in different regions in a particular cloud. Of the hydrocarbons, cyclopropenylidene (C_3H_2) is found throughout the Galaxy both in dense clouds and in apparently diffuse regions. Its fractional abundance relative to hydrogen in dense clouds varies between 3×10^{-9} and 3×10^{-10}.

INTERSTELLAR MEDIUM

The interstellar gas consists of hydrogen, helium with a fractional abundance of about 0.1, and heavy elements making up a total fractional abundance of 10^{-3}. This small admixture of heavy elements provides the material for the observed interstellar molecules. The interstellar medium also contains dust grains. The high opacity of dense clouds occurs through the scattering and absorption of radiation by solid particles. The dust grains shield molecules from the destructive effects of the interstellar photons and provide a site for the formation of hydrogen molecules and of the saturated molecules H_2O, CH_4, NH_3, and H_2S.

The clouds in which molecules are found are cold. Except in localized regions, which are disturbed by interactions associated with stars, the temperatures in diffuse clouds are less than 100 K and in dense cloud interiors are only 10 K. The average densities are about 10^2 cm^{-3} in diffuse and translucent clouds and about 10^3 cm^{-3} in dense clouds. Dense clouds often contain individual clumps of gas at considerably higher densities. Nevertheless the densities are too low for three-body collisions to be effective in forming molecules and the molecules must be built by two-body reactions.

MOLECULAR HYDROGEN

The initiating reaction in which two atoms combine to form a molecule must either be a radiative process that is stabilized by the emission of a photon or it must take place on the surface of a grain. Molecular hydrogen can be formed in the gas phase by the sequence of radiative attachment.

$$e + \mathrm{H} \rightarrow \mathrm{H}^- + h\nu,$$

followed by associative detachment

$$\mathrm{H}^- + \mathrm{H} \rightarrow \mathrm{H}_2 + e,$$

and the sequence of radiative association

$$\mathrm{H}^+ + \mathrm{H} \rightarrow \mathrm{H}_2^+ + h\nu,$$

followed by charge transfer

$$\mathrm{H}_2^+ + \mathrm{H} \rightarrow \mathrm{H}_2 + \mathrm{H}^+,$$

Table 1. Interstellar Molecules

H_2	Hydrogen
CH	Methylidyne Ion
C_2	Carbon
CO	Carbon Monoxide
CS	Carbon Monosulfide
SO	Sulphur Monoxide
SiS	Silicon Sulfide
HCl	Hydrogen Chloride
H_2O	Water
HCN	Hydrogen Cyanide
HCO	Formyl
N_2H^+	Protonated Nitrogen
HNO	Nitroxyl
SO_2	Sulfur Dioxide
H_2CO	Formaldehyde
NH_3	Ammonia
HNCO	Isocyanic Acid
C_3N	Cyanoethynyl
C_3O	Tricarbon Monoxide
C_2H_2	Acetylene
CH_2CO	Ketene
HCOOH	Formic Acid
C_4H	Butadinyl
CH_2NH	Methanimine
CH_2CN	Cyanomethyl Radical
C_5H	Pentynylidyne
CH_3OH	Methyl Alcohol
HCC_2HO	Propynal
C_6H	Hexatrinyl
NH_2CHO	Formamide
CH_3NH_2	Methylamine
CH_3CHO	Acetaldehyde
CH_3C_3N	Methyl Cyanoacetylene
CH_3CH_2OH	Ethyl Alcohol
HC_9N	Cyano-octatetra-yne
CH	Methylidyne
OH	Hydroxyl
CN	Cyanogen
NO	Nitric Oxide
SiO	Silicon Monoxide
NS	Nitrogen Sulfide
PN	Phosphorus Nitride
C_2H	Ethynyl
HNC	Hydrogen Isocyanide
HCO^+	Formyl Ion
H_2S	Hydrogen Sulfide
OCS	Carbonyl Sulfide
HCS^+	Thioformyl Ion
C_2S	Dicarbon Sulfide
H_2CS	Thioformaldehyde
HCNS	Isothiocyanic Acid
HOCO	Protonated Carbon Dioxide
C_3S	Tricarbon Sulfide
C_3H_2	Cyclopropenylidene
HCOOH	Formic Acid
HC_3N	Cyanoacetylene
NH_2CH	Cyanamide
CH_3CH	Methyl Cyanide
CH_3SH	Methyl Mercaptan
CH_3C_2H	Methyl Acetylene
CH_2CHCN	Vinyl Cyanide
$HCOOCH_3$	Methyl Formate
HC_5N	Cyanodiacetylene
CH_3C_4H	Methyl Diacetylene
$(CH_3)_2O$	Dimethyl Ether
CH_3CH_2CN	Ethyl Cyanide
CH_3C_4H	Methyl Diacetylene
HC_7N	Cyanohexatriyne
$HC_{11}N$	Cyano-decapenta-yne

but the initial radiative processes are slow at low temperatures and the fractional ionization in molecular clouds is small. Hydrogen molecules, whose existence in diffuse clouds has been established by observations of interstellar ultraviolet absorption lines in the spectra of hot stars, must be formed by grain catalysis. Hydrogen molecules are destroyed by ultraviolet photons in a process of line absorption followed by spontaneous radiative dissociation. A quantitative analysis of the observed abundances of H_2 suggests an efficiency for forming H_2 of nearly unity in the collision of a hydrogen atom with a grain. In dense clouds from which the interstellar photons are excluded, the hydrogen is converted from atomic to molecular form.

OTHER MOLECULES

With the formation of H_2, other molecules can be formed by gas phase reactions driven by the ionization from cosmic rays. Neutral particle reactions usually have activation barriers and proceed slowly at cloud temperatures. Ion–molecule reactions in contrast often remain rapid at low temperatures and indeed in the case of heteronuclear molecules the rate coefficients increase with decreasing temperature.

In the chemistry driven by cosmic ray ionization, H_2^+ ions are quickly transformed to H_3^+ ions by the reaction

$$H_2^+ + H_2 \rightarrow H_3^+ + H.$$

The H_3^+ ions react with many of the neutral constituents by undergoing proton transfer. As an example, H_3^+ reacts with oxygen atoms to form OH^+:

$$H_3^+ + O \rightarrow OH^+ + H_2.$$

The proton transfer initiates an abstraction sequence

$$OH^+ + H_2 \rightarrow H_2O^+ + H,$$

$$H_2O^+ + H_2 \rightarrow H_3O^+ + H,$$

which terminates with the saturated species H_3O^+. The H_3O^+ ion then undergoes dissociative recombination

$$H_3O^+ + e \rightarrow H_2O + H$$
$$\rightarrow OH + H_2$$

to form OH and H_2O.

A similar chemistry converts neutral carbon atoms into CH_3^+ followed mostly by

$$CH_3^+ + e \rightarrow CH_2 + H$$
$$\rightarrow CH + H_2.$$

An important fraction of the CH_3^+ is converted to CH_5^+ by a radiative association with H_2:

$$CH_3^+ + H_2 \rightarrow CH_5^+ + h\nu.$$

The CH_5^+ ion reacts with carbon monoxide to form methane:

$$CH_5^+ + CO \rightarrow HCO^+ + CH_4.$$

In molecular envelopes and in diffuse regions carbon is ionized by photons and the carbon chemistry begins with the radiative association

$$C^+ + H_2 \rightarrow CH_2^+ + h\nu.$$

In dense interiors, C^+ is produced by the reaction of CO with He^+

obtained from cosmic ray ionization of helium:

$$He^+ + CO \rightarrow He + C^+ + O.$$

The destruction of CO is usually followed by its formation. Thus following the reaction of CH_5^+ to form methane, CO is produced by

$$HCO^+ + e \rightarrow H + CO$$

or by

$$HCO^+ + H_2O \rightarrow H_3O^+ + CO.$$

In dense clouds, most of the carbon is ultimately converted into carbon monoxide. However the time scale is long, of the order of 10^7 yr, and in a cloud evolving from a configuration in which the carbon is initially mostly neutral atoms, substantial abundances of hydrocarbons can be achieved in times of about 3×10^5 yr.

The important processes are insertion reactions such as

$$C^+ + CH_4 \rightarrow C_2H_3 + H,$$

which leads to acetylene through

$$C_2H_3^+ + e \rightarrow C_2H_2 + H,$$

and condensation reactions such as

$$CH_3^+ + CH_4 \rightarrow C_2H_5^+ + H_2.$$

The molecular ions react with neutral atoms to produce molecules containing oxygen, nitrogen, sulfur, silicon, and other heavy elements. Ketene is formed by

$$C_2H_4^+ + O \rightarrow C_2H_3O^+ + H,$$

$$C_2H_3O^+ + e \rightarrow CH_2CO + H.$$

The element chemistries are also mixed by radiative association, which forms complex molecules at low temperatures. Ethanol is made by the sequence

$$H_3O^+ + C_2H_4 \rightarrow CH_3CH_2OH_2^+ + h\nu,$$

$$CH_3CH_2OH_2^+ + e \rightarrow CH_3CH_2OH + H.$$

Although interstellar photons are restricted to the boundary layers of dense clouds, the electrons produced in cosmic ray ionization excite H_2, which then decays with the emission of ultraviolet photons. In particular the photodissociation of CO,

$$CO + h\nu \rightarrow C + O,$$

supplies neutral carbon atoms for the formation of hydrocarbons.

GRAIN CHEMISTRY

During the cloud evolution, the heavy atoms and molecules collide with the grains in a time scale of the order of 10^5–10^6 yr and are frozen onto their surfaces. The grain mantle material must be returned to the gas phase in a similar time scale. Desorption by cosmic ray impact and the explosive release of molecules in grain–grain collisions are possible mechanisms. Sporadic events associated with star formation destroy the mantles and return molecules to the gas phase. A further source of molecules is the outflowing gas from protostars and stars undergoing mass loss. High velocity outflows drive shocks into the interstellar medium and modify the gas and grain chemistries. Shock-heated gas may be responsible for the CH^+ ions seen in diffuse clouds. In the warm gas, the endothermic reaction

$$C^+ + H_2 \rightarrow CH^+ + H$$

can proceed.

Observations of high fractional abundances of ammonia, methanol, and water that are of varying chemical compositions in different regions of the Orion molecular cloud and other clouds suggest that molecules were evaporated off grain surfaces less than 10^4 yr ago.

In addition to the molecules in Table 1, there is some indirect evidence for a component of very small grains or very large molecules in diffuse regions. If they exist in dense clouds, the chemistry would be modified and their breakup may yield some of the interstellar molecules.

Interstellar molecules are the result of an interplay of gas and grain chemistries whose slow evolution is modified by dynamic events stemming from the formation of stars. The chemical composition is a reflection of those events and can serve as a diagnostic probe of the history of the cloud.

Additional Reading

Duley, W. W. and Williams, D. A. (1984). *Interstellar Chemistry*. Academic, London.

Hartquist, T. W., ed. (1990). *Molecular Astrophysics*. Cambridge University Press, New York.

Herbst, E. and Leung, C. M. (1989). Gas-phase production of complex hydrocarbons, cyanopolyynes, and related compounds in dense interstellar clouds. *Astrophys. J. (Suppl.)* **69** 271.

van Dishoeck, E. F. (1988). Molecular cloud chemistry. In *Millimetre and Submillimetre Astronomy*, W. B. Burton and R. D. Wolstencroft, eds. Kluwer, Dordrecht, p. 117.

See also **Interstellar Medium, Chemical Composition; Interstellar Medium, Chemistry, Laboratory Studies; Interstellar Medium, Molecules.**

Interstellar Clouds, Collapse and Fragmentation

Susan Terebey

The interstellar medium (the general term for diffuse gas and dust spread between the stars) is a repository for less than 10% of the total mass in our Galaxy. Much of the Galaxy's mass is found in stars. However, the interstellar medium has an importance belied by its mass, for we know that stars are being born by the fragmentation and collapse of these clouds of gas and dust.

According to Big Bang models of cosmology, the universe once passed through a phase where its matter was composed of cooling gas. The first generation of stars formed out of that gas. Much of the early era of galaxy and star formation is still hidden from us, but one hope of studying the birth of stars today is that it will shed light on how the very earliest stars must have formed. We know of one important difference between star formation then and now—the chemical composition of the gas has been changed through the explosions of generations of dying stars, injecting their heavier elements (built by nuclear reactions) into the interstellar medium. The composition of a gas cloud affects the ability of the gas to cool easily, which in turn affects the ability of an interstellar cloud to collapse and form stars.

Viewed on any scale, the interstellar medium is an exceedingly nonuniform environment: full of clumps, sheets, filaments, and rarefied gas, all at vastly different temperatures. The highest-density, coldest, and most massive structures, called molecular clouds, are found to be the sites of current star formation. In a general sense the study of fragmentation seeks the physics responsible for the structure of the clumpy interstellar medium. The more usual investigation focuses on how fragments that are destined to become stars, or clusters of stars, with stellar masses near one solar mass, are able to condense from molecular clouds comprising tens of thousands of solar masses. Physical processes that are known to be important in molecular clouds are gas pressure, gravitation, and magnetic fields. Energy sources include winds and radiation from young stars, whereas energy sinks include gas cooling and radiative shock waves. These effects combine to produce the power-law relationship between density, velocity dispersion, and size scale that is observed to hold over many decades in cloud size. Explaining this relationship is an important task for a successful theory of fragmentation. Given the complexity of the problem and the importance of nonlinear processes, it is perhaps not surprising that the study of fragmentation is at an early stage of development.

Massive stars ($>5\ M_\odot$, where $M_\odot$ is the mass of the Sun) are suspected to form in a significantly different fashion than low mass stars. First, they are observed to form in groups, not independently, and to form only in the densest regions of the largest molecular clouds. One suggestion for this difference involves the magnetic field. The amount of magnetic flux threading through a cloud defines a critical mass. No equilibrium between gravitational and magnetic forces is possible for a cloud whose mass is above the critical value. This suggests that clouds below the critical mass may evolve slowly, forming individual stars, whereas clouds above the critical mass evolve dynamically, fragmenting into many stars.

In nearby molecular clouds, individual low mass stars are observed to be forming in dense molecular cloud cores—barely rotating dense clumps 0.1 pc in size and perhaps 5–10 times more massive than the stars that form from them. The stars are too young and low in luminosity to modify their environments and so the physical conditions observed are thought to match closely the initial conditions present in dense cores just prior to hydrodynamic collapse.

Observations of the magnetic field indicate that it should play an important role in the formation of dense cores. Recent theoretical models show that over a period of several million years a dense core condenses quasistatically from molecular gas in a nearly balanced match between gravity, pressure, and magnetic forces. Through ambipolar diffusion the dense neutral gas is able to drift through the magnetic field lines. The consequent lowering of the magnetic pressure support relative to gravity results in higher-density gas: Eventually, the dense core becomes gravitationally unstable and collapses dynamically.

Current models of the dynamical collapse phase include the effects of gravity, gas pressure, and rotation. To the extent that the dense core is axisymmetric, the angular momentum in each gas parcel will be conserved during the collapse. The inner parts of the dense core, which may be expected to have low angular momentum, will be able to collapse all the way to the center, whereas the higher angular momentum material in the outer dense core will only reach the radius where centrifugal effects balance gravity. It is not yet known whether magnetic fields are important during the dynamical collapse phase although it is plausible that they are important in the outer regions of the dense core and, therefore, may affect the mass infall rate during the later stages of the collapse.

Dense cores are observed to be highly centrally concentrated before the collapse begins. The cloud first becomes gravitationally unstable at the center and begins to collapse from the inside outward. The interior collapses, causing the next shell out to lose pressure support from below, with the result that each shell in turn collapses, eventually involving most of the cloud. Normally, gas compression would cause the gas to heat up, generating pressure that would retard the collapse. However, at typical cloud temperatures, the mix of elements that composes the interstellar medium is able to cool very effectively. In essence the cloud collapse is isothermal. Gas velocities very quickly become supersonic and, to a high degree of precision, gas parcels follow ballistic trajectories. Fragmentation is not thought to be important during the dynamical collapse phase because small inhomogeneities are not able to grow—they get damped by the shear in the velocity field.

Material falls into the central region at a nearly constant rate that depends on the initial temperature of the dense core. The accretion rate is given approximately by a^3/G, where a is the initial sound speed in the dense core and G is the gravitational constant. Typical values for the accretion rate are on the order of $5\times10^{-6}\ M_\odot/\mathrm{yr}$. A short time after the collapse begins, a hydrostatic object (namely, a protostar) forms at the center and steadily grows in mass. Later, as material with higher and higher angular momentum falls inward, the gas is not able to fall directly onto the star and instead forms a protostellar disk. Given the low rotation rates observed for dense cores, the gas is readily able to collapse from initial radii near 0.1 pc (10^{17} cm) to form a protostar with a radius of about 10^{11} cm and a disk of about solar-system size [10 AU ($\sim10^{14}$ cm)].

The succeeding stages in the evolution of the young star and disk are not yet clear. Although we expect disks to form, their sizes and masses are very poorly known. This makes it difficult to constrain theories of how disks may evolve. It is also uncertain what determines the final mass of the young star: Do stellar winds halt the collapse of the dense core? Finally, how does the common occurrence of binary star systems fit into this picture? These questions form the basis of much active research today.

Additional Reading

Beckman, J. E. and Pagel, B. E. J., eds. (1989). *Evolutionary Phenomena in Galaxies*. Cambridge University Press, Cambridge, U. K.

Cassen, P., Shu, F. H., and Terebey, S. (1985). Protostellar disks and star formation: An overview. In *Protostars and Planets II*, D. C. Black and M. S. Matthews, eds. University of Arizona Press, Tucson, p. 448.

Shu, F. H., Adams, F. C., and Lizano, S. (1987). Star formation in molecular clouds: Observation and theory. *Ann. Rev. Astron. Ap.* **25** 23.

Scalo, J. M. (1985). Fragmentation and hierarchical structure in the interstellar medium. In *Protostars and Planets II*, D. C. Black and M. S. Matthews, eds. University of Arizona Press, Tucson, p. 201.

See also **Infrared Sources in Molecular Clouds; Interstellar Clouds, Molecular; Protostars, Theory; Star Formation, Propagating.**

Interstellar Clouds, Molecular

Neal J. Evans II

The space between stars (the interstellar medium) is, by earthly standards, practically empty, being generally much less dense than the best terrestrial vacuums. Nonetheless, stars and planets form out of this rarefied interstellar matter through processes that compress matter by about 20 powers of 10. For densities above about 100 atoms per cubic centimeter (cm^{-3}), many of the atoms combine into molecules and the resulting dense (relatively speaking!) regions are called *molecular clouds*. The largest of these are called *giant* molecular clouds.

The study of molecular clouds really began in 1968–1969, with the discovery of the first polyatomic interstellar molecules: ammonia (NH_3), water (H_2O), and formaldehyde (H_2CO). These discoveries showed that quite complex chemistry was possible in some special regions of interstellar space and encouraged many searches for other molecules. By now over 70 molecules have been detected, but only a few have been used extensively to study the physical conditions in the clouds. Most notable among these is carbon monoxide (CO), which has become the standard tracer for molecular gas.

Although we will concentrate on the properties of the molecules in these clouds, we should remember that some atoms remain unassociated into molecules (e.g., helium is the second most abundant atom in the universe, but does not form molecules) and that the clouds also contain a small fraction (about 1% by mass) of small solid particles, called dust grains. The composition of these grains is not completely certain, but there appear to be at least two main types: One type is amorphous silicate and the other is some form of carbon. The dust and gas interact in complex ways, so both will be discussed.

One of the earliest hints that some clouds could be molecular came from comparing observations of dust and atomic gas. Regions of the sky deficient in stars can be seen with the naked eye and became even more obvious with the advent of photographic astronomy; these dark clouds are the result of the dust grains blocking the light of background stars. In the more opaque clouds, there was a deficit of hydrogen atoms, relative to the dust, and it was suggested that the missing atoms may have formed molecules. This suggestion was confirmed in 1970 with the discovery of molecular hydrogen (H_2) by ultraviolet observations with a rocket-borne telescope. Although H_2 is the most abundant species in molecular clouds, it was not the first detected, nor is it generally observable because it interacts very poorly with photons. Thus, although H_2 and helium account for 99% of the mass in molecular clouds, they usually are not observed directly.

TECHNIQUES

Study of the properties of molecular clouds has proceeded primarily through spectroscopy of the rotational transitions of trace molecules. Unlike atoms, molecules in the gas phase can rotate about their center of mass (Fig. 1). Quantum mechanics requires that the angular momentum of a molecule be limited to integral multiples of $h/2\pi$, where h is Planck's constant (essentially, the molecule can only rotate at certain speeds). Thus, only a discrete set of rotational energies is possible, analogous to the energy levels of atoms. When the molecule changes to a rotational energy state that is lower by an amount ΔE, it emits a photon at the frequency $\nu = \Delta E/h$. Because we can measure the frequency of this photon, we can determine which molecule has emitted it and what energy level it was in. Furthermore, the precision of the frequency measurement (which can be as high as one part in 10^7) also allows us to detect deviations from the frequencies of the same transition measured in terrestrial laboratories. These deviations are caused by the Doppler shift due to the motion of the molecule relative to our telescope. After correction for the various motions of the telescope, we can use these shifts to study the motion of the molecular clouds. Although most of our information comes from these rotational transitions, some use is made of vibrational and electronic transitions, the last being very similar to atomic transitions.

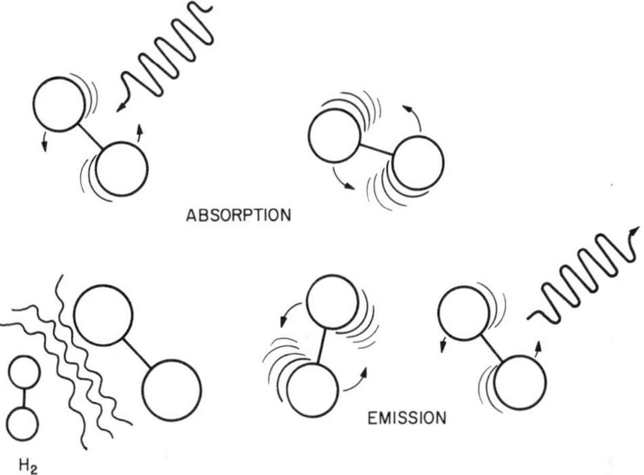

Figure 1. A molecule with two atoms can be visualized as a tiny dumbbell that can rotate about its center of mass, but only at certain speeds, allowed by quantum mechanics. For each discrete rotation speed, there is a discrete energy. If the molecule encounters a photon (represented by the wiggly line in the upper left figure) with an energy exactly equal to that required to boost the rotation to the next allowed energy state, the molecule can absorb the photon and rotate faster (upper right). Collisions with other molecules (mostly H_2 in interstellar molecular clouds) can also cause the molecule to rotate faster (lower left and center) and the molecule can emit a photon and switch to a more slowly rotating state (lower right).

GLOBAL PROPERTIES

By "global properties," we mean the properties of the cloud as a whole. These include the location and overall velocity of the cloud, which are used to study the distribution of clouds in the Galaxy. Most of the molecular gas is closer to the center of the Galaxy than the Sun, whereas the atomic gas is more evenly distributed over the Galaxy. Other global properties of the cloud include its size, shape, and total mass. These have been studied using both dust and the CO molecule as tracers for the much more abundant H_2. Clouds have sizes ranging at least from 0.3 to 300 light-years [a light-year (ly) is about 10^{18} cm]. The lower limit may be set by the difficulty of finding very small clouds, whereas the upper limit is uncertain because the definition of a cloud becomes uncertain, as the clouds blend together. Most systematic studies extend over only a restricted portion of this range, but find a continuous distribution of sizes, with more small clouds than large clouds. More specifically, the number of clouds of size d decreases as d^{-a}, with $2 < a < 3$.

Although clouds are often assumed to be spheres, many of them are elongated; the largest clouds may be elongated along the plane of the Galaxy. When observed with suitable techniques, clouds take on a very irregular appearance, reminiscent of clouds in the Earth's atmosphere. Some recent studies have suggested that cloud shapes are fractal.

The masses of clouds have been determined mostly from CO studies, which ultimately rely on the virial theorem to determine the mass of a cloud from its size and the total spread of velocities in the cloud. Masses of clouds, measured in units of the mass of the Sun ($M_\odot$), range from 10 to 10^7 $M_\odot$; the ends of this range are uncertain for the reasons discussed previously for sizes. There are more low-mass clouds than high-mass clouds [$n(M) \propto M^{-b}$, with $b \sim 1.5 \pm 0.1$, where $n(M)$ is the number of clouds of mass M], but most of the mass is in the most massive, largest clouds. There is no indication that there are fundamental differences in the internal properties between small and large clouds. In particular, very small clouds ($M \sim 10$ $M_\odot$), as well as large clouds, are engaged in forming stars. The total amount of molecular material in our Galaxy is $1-3 \times 10^9$ $M_\odot$, comparable to the total mass of interstellar atomic gas, although, as noted previously, the distributions are different.

LOCAL PROPERTIES

Local properties are those properties that are defined in each small region of the cloud and that may vary throughout the cloud. In practice, our observational techniques are limited to determining average values, with the averaging being done both along the line of sight and over our finite angular resolution. Local properties include the temperature, the density, and the local velocity variations about the average cloud velocity.

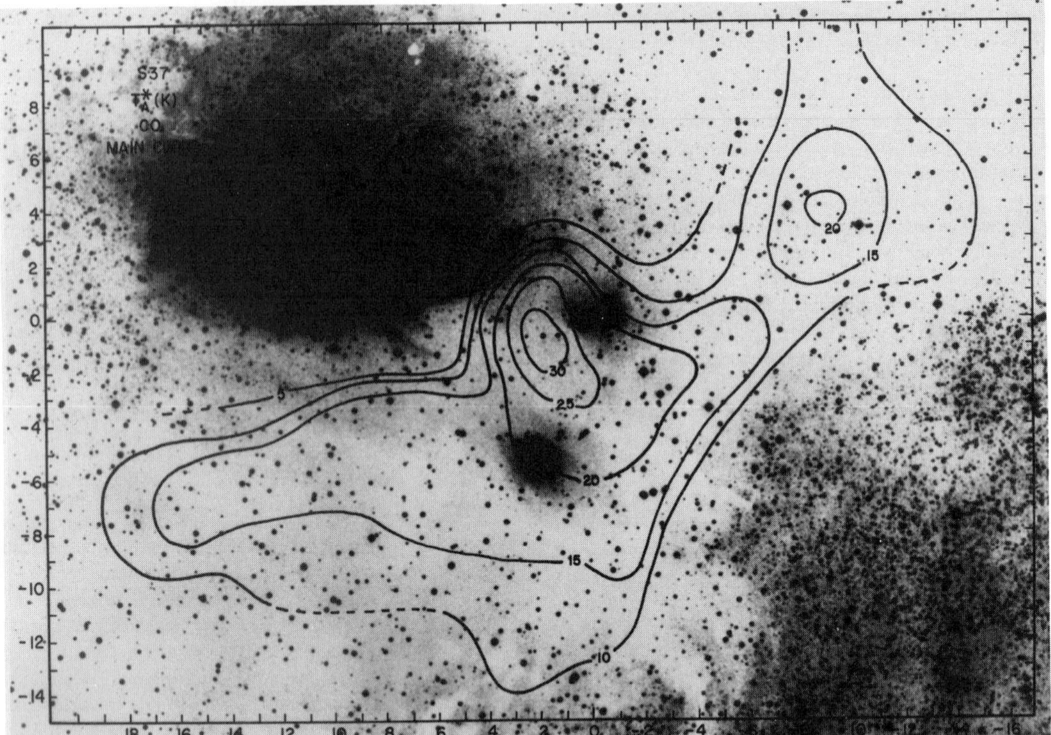

Figure 2. The photograph is a blow-up of part of the Palomar Observatory Sky Survey. Taken with a red filter, it is reproduced as a negative, so that stars are small black dots. The large black area at the upper left is glowing gas in an H II region, ionized by a recently formed star. The two smaller patches (reflection nebulae) near the center are shining by reflected light from other recently formed stars. The relatively smaller number of stars seen in a band from lower left to upper right indicates the presence of obscuring dust. The contour lines show the strength of emission of the lowest rotational transition of CO, indicating that the dust is part of a medium-sized molecular cloud (size about 30 ly and mass about $10^4 M_\odot$). The numbers on the contours are approximately the kinetic temperature, and the strong peak ($T_K = 30$ K) near the H II region suggests that the recently formed star is heating the gas. This is a typical situation in regions of star formation. [*Reprinted by permission from* Ap. J. ***253*** *115 (1982).*]

Temperature

The temperature of the gas is a measure of the amount of kinetic energy in the random motions of individual molecules; we thus refer to the kinetic temperature (T_K), which we express in kelvins (K). T_K is deduced from the relative numbers of molecules in different rotational energy states; at higher T_K, more molecules will be in higher energy states. Although, in general, the fraction in higher states depends on both T_K and the density, there are a few special cases (low energy levels of CO and certain "metastable" levels of NH_3) for which the fraction depends only on T_K. These special cases are our interstellar thermometers.

Extensive observation of CO indicates that most parts of molecular clouds are very cold ($T_K \sim 10$ K), but that warmer gas exists near stars (Fig. 2). The warmer gas has been studied recently by observations at submillimeter wavelengths, which study higher energy levels of CO; these studies find values of T_K up to about 1000 K. At temperatures of 1000–2000 K, even the normally invisible H_2 begins to emit infrared photons; such photons are observed from some regions of recent star formation. Because molecules will be dissociated into atoms if T_K is much greater than 2000 K, it appears that molecular clouds have gas at essentially all the kinetic temperatures that are possible; still, the vast majority of molecular gas is very cold.

How can we explain the range of temperatures seen in molecular clouds? T_K is set by a balance between heating and cooling processes. The main cooling process is emission of photons by CO and other molecules; these photons escape from the cloud, carrying energy away. Heating processes depend on location in the cloud. Two heat sources are always present: the microwave background radiation and cosmic rays. The former, composed of photons left over from the Big Bang, is a blackbody at about 2.7 K; this is the lowest temperature that can be reached. Cosmic rays are mostly protons that are traveling near the speed of light and that can ionize molecules or atoms; the electron produced by such an ionization moves at high speed and its subsequent collisions with molecules transfer energy to the gas, heating it. There is enough ionization by cosmic rays to heat the gas to about 10 K; thus, heating by cosmic rays can explain T_K in most parts of the cloud.

To reach higher T_K, additional heat sources are necessary. Because regions with T_K much above 10 K invariably are found near luminous stars, it is reasonable to suspect stellar photons of heating the gas. Stellar photons can heat the gas in several ways, and two of the most important heating processes in molecular clouds involve the dust. Relatively hot stars produce a substantial number of photons energetic enough to eject electrons from dust grains through the photoelectric effect. These electrons heat the gas in the same way as those ejected from molecules by cosmic rays. Near a luminous, hot star the cloud can be heated to about 1000 K, but the number of photons decreases sharply with distance into the cloud. The main question about this mechanism is whether the heating can extend far enough into the cloud to explain the amount of hot molecular gas that is seen.

Less-energetic photons can penetrate farther into the cloud; instead of ejecting electrons, they give their energy to the internal vibrations of a dust grain, thereby warming the grain. The dust temperature (T_D) is a measure of the energy in these internal vibrations and is set by a balance between the energy received by absorbing photons and the energy lost by radiating photons. Far from any star, T_D reaches about 10 K (coincidentally, the same as T_K in these regions). Near stars, T_D can be appreciably higher. Some of the dust grain's internal energy can be transferred to the gas when a molecule collides with the dust, as long as T_D exceeds T_K. In relatively dense regions, gas–dust collisions are frequent enough to make $T_K \sim T_D$. These processes can explain why T_K and T_D are higher in parts of clouds near stars. Recent evidence points to the existence of a large number of very small dust grains (or very large molecules) whose internal temperatures vary strongly with time; it is not clear what role, if any, these may play in determining T_K.

Densities

The density in molecular clouds is described by the number of molecules or atoms (mostly H_2 molecules) per unit volume and is denoted n (cm^{-3}), which is shorthand for number per cubic centimeter. Our theoretical understanding of what determines the density is less developed, and it is more difficult to determine n from observations than it is to determine T_K.

One method of obtaining n is to determine the total number of molecules along the line of sight, averaged over the resolution of our observations. Because the H_2 is normally not observed, this is done by measuring the total number of photons received from some other molecule (usually CO) and correcting for the unseen H_2. This quantity is called a column density because it is the total number of molecules in a column of cross-sectional area 1 cm^2 and depth equal to the depth of the cloud; it is denoted N (cm^{-2}). Finally, to get the average density along the line of sight, we divide by the depth of the cloud. Of course, do not really know the depth of the cloud, so it is customary to assume that the cloud is spherical and to divide by its size, as seen in projection on the sky. This complex procedure is quite uncertain but has shown that most of the cloud has a relatively low N, although some small regions may have much larger N. Further, some studies have indicated that clouds of different size have similar average N, suggesting that the average density ($\langle n \rangle = N/\text{size}$) decreases with increasing cloud size. Conversely, the small regions of high N must also have substantially higher n.

The fact that the fraction of molecules in higher energy states depends, for most molecules, on both n and T_K provides another way of measuring densities. If we have determined T_K, then measurements of the fraction in higher states can tell us the density. Molecules commonly used for this purpose include CS and H_2CO. This method almost invariably yields a higher density estimate than does the first method. The disagreement tells us that there is a range of densities along the line of sight, because the second method tends to select regions of higher n. The conclusion is then that large molecular clouds are inhomogeneous, with denser regions (clumps) in a less dense, interclump medium. Small clouds and individual clumps within large clouds may be more regular, with density decreasing smoothly with distance from the center of the region.

Measured densities in molecular clouds range from $n \sim 100$ cm^{-3} to about 10^8 cm^{-3}, with the lower values applying to the majority of the cloud. Densities above about 10^4 cm^{-3} are almost always associated with sites of star formation. In these regions, the highest densities have been found on the smallest scales, and our current upper limit is probably set by observational limitations. Because a star has an average density of about 10^{24} cm^{-3}, it is obvious that much higher densities must exist on very small scales around forming stars.

Velocities

The local velocities, relative to the cloud's center of mass, are measured by the spread of frequencies about the center frequency or, in some cases, by shifts in the center frequency with position on the cloud. The spread of frequencies is called the line width and denoted Δv [in kilometers per second (km s^{-1})]. Because T_K is a measure of the random, thermal velocities, we can predict the line width from T_K. This prediction is always much less than the observed Δv, which typically ranges from 0.5 to 5 km s^{-1}. The excess motion is attributed to turbulent motions. Except for a few small regions, the turbulence is supersonic, with the turbulent velocities exceeding the speed of sound.

In one sense, the large Δv are not surprising. If the only internal kinetic energy were the thermal energy, the pressure could not resist gravity and the clouds would all be collapsing. Because there is considerable evidence against this universal collapse, other pressures probably support the cloud, or at least slow the collapse. Turbulence is a leading candidate for supplying the needed support; this idea also underlies the method of mass determination described in the section on "Global Properties." One problem with this idea is that turbulence, especially supersonic turbulence, should dissipate fairly rapidly. Magnetic fields may slow the dissipation of the turbulence. Sufficiently strong magnetic fields may also directly provide support, at least against collapse across the field lines. The relatively constant column density from cloud to cloud has been interpreted as evidence of magnetic support, and the few measurements of actual magnetic fields in molecular clouds do indicate that they are strong enough to play a role. One suspects that both magnetic fields and turbulence are involved in cloud support, but our theoretical understanding of magnetic turbulence is rather primitive.

Although molecular clouds as a whole are not collapsing, some regions within clouds do collapse to form stars. This may occur when the magnetic field "leaks out" of the cloud or when turbulence decays. Theoretical studies indicate that a magnetically supported region may relax slowly to a centrally condensed configuration with $n \propto r^{-2}$, where r is the distance from the center, as the magnetic field weakens. At some point, the field can no longer resist gravity and collapse begins, first at the center and then spreading outward. This "inside-out" collapse produces a density distribution with $n \propto r^{-1.5}$. Observational tests of these predictions for the density distribution are just beginning, but early results are encouraging. In principle, the collapse velocities should also be detected but, in practice, it has been difficult to separate them from turbulence in the surrounding cloud. No unambiguous evidence for collapse has been found; this is one of the outstanding observational challenges at present.

Rotation of a cloud can be identified by observing the shift in the center frequency of a spectral line as one scans across the cloud. Although most clouds rotate slowly, if at all, some small clumps and small isolated clouds do show significant rotation. During collapse, any initial rotation will be amplified, producing a flattened, rotating structure. This structure may evolve into a forming star surrounded by a rotating disk, leading to the formation of a planetary system. It is thus of extreme interest to study the properties of such disks to help understand the formation of our own solar system. Recent observations have indicated that such rotating disks exist, but current observational capabilities are not sufficient to determine their detailed properties.

One complication that has made it difficult to detect collapse is that stars in the process of formation eject part of the infalling matter as a strong wind, presumably as a means of slowing the rotation. The fast wind ($v \sim 100$ km s^{-1}) sweeps up surrounding material and produces a large region of more slowly outflowing matter ($v \sim 10$–50 km s^{-1}), which may mask evidence of infall. The outflow, in turn, can contribute to turbulent velocities in the rest of the cloud, thereby preventing its overall collapse.

STATUS OF THE FIELD

Considerable progress has been made in measuring the properties of molecular clouds since their discovery about 20 years ago. One of the most important observational projects for the future will be the study of temperature and density on very small scales where stars are forming. Theoretical progress has been slower, but recent predictions of conditions in regions of star formation provide a challenge to observers. The outstanding unsolved theoretical problem is to understand the roles of turbulence and magnetic fields in cloud support.

Additional Reading

Bally, J. (1986). Interstellar molecular clouds. *Science* **232** 185.

Blitz, L. (1982). Giant molecular cloud complexes in the Galaxy. *Scientific American* **246** (No. 4) 84.

Scoville, N. Z. (1989). Molecular gas in spiral galaxies. In *Evolutionary Phenomena in Galaxies*, J. E. Beckman and B. E. J. Pagels, eds. Cambridge University Press, Cambridge, U.K.

Scoville, N. and Young, J. S. (1984). Molecular clouds, star formation, and galactic structure. *Scientific American* **250** (No. 4) 42.

Turner, B. E. (1988). Molecules as probes of the interstellar medium and of star formation. In *Galactic and Extragalactic Radio Astronomy*, G. L. Verschuur and K. I. Kellermann, eds. Springer-Verlag, New York.

Verschuur, G. L. (1987). *The Invisible Universe Revealed.* Springer-Verlag, New York, Chapter 10.

Verschuur, G. L. (1989). *Interstellar Matters.* Springer-Verlag, New York, Chapters 19–21.

See also **Infrared Sources in Molecular Clouds; Interstellar Clouds, Collapse and Fragmentation; Interstellar Medium, Dust Grains.**

Interstellar Extinction, Galactic

David A. Williams

Interstellar extinction is the attenuation of starlight caused by the absorption and scattering of the light by interstellar dust along the line of sight between the star and the observer. Extinction has been recognized for about 50 years as a general phenomenon in the Galaxy, and one that is particularly marked in the galactic plane. The effects of extinction were already evident some 200 years ago in the surveys of stars by William Herschel. The apparent absence of stars in a localized region of the sky caused Herschel to remark that this was indeed a "hole in the heavens." The application of photography to astronomy (particularly by Edward Emerson Barnard), a century later, indicated that these "holes" were in fact obscuration due to dust clouds. The fact that there is a general obscuration in the galactic disk, not merely confined to prominent isolated dust clouds, was slowly recognized. Particularly convincing arguments concerned the observations of stellar clusters whose distances could be estimated both from their angular sizes and from the luminosities of stars within them. The luminosity-derived distances are affected by extinction whereas the distances based on angular sizes are independent of it. Comparisons of these two measures established that extinction was severe and ubiquitous in the plane of the Galaxy, and confined to a relatively thin layer. It was also evident that blue light is more heavily extinguished than red, so that an apparent *reddening of starlight* is observed. This *interstellar reddening* can also be used to map the presence of dust in the Galaxy. The general trend of increasing extinction at shorter wavelengths is, with some exceptions, now known to be broadly true not only over the visual wavelengths but also from infrared to ultraviolet. The accepted origin for the bulk of the extinction, at least in the visual, is the absorption and scattering of starlight by a population of dust particles whose sizes are comparable to the wavelength of light. A variety of models consistent with the cosmic constraints on mass and chemical composition have been developed and are described elsewhere. The concept of extinction by dust particles is supported by evidence of many kinds; in particular, the serendipitous discovery of *interstellar polarization* is interpreted as the *differential extinction* caused by the partial alignment of aspherical dust particles. Radiation in the visual and ultraviolet that is *absorbed* by dust particles heats the dust and is reradiated at longer wavelengths; this longer wavelength radiation is observed directly. Radiation in the visual and ultraviolet that is *scattered* by the dust creates a diffuse background of light, the *diffuse galactic light* that is observed both in the vicinity of bright stars near dust clouds and throughout the plane of the Galaxy, similar to the zodiacal light in the solar system observed in the ecliptic plane of the sky.

INTERSTELLAR EXTINCTION CURVE

The intensity $I(\lambda)$ of radiation received at wavelength λ compared with that which would have been received, $I_0(\lambda)$, in the absence of extinction caused by an intervening slab of material, may be written

$$I(\lambda) = I_0(\lambda)e^{-\tau(\lambda)}$$

Here, $\tau(\lambda)$ is called the optical depth. More extinguishing material in the slab leads to a larger value of τ. In practice, astronomers use a logarithmic measure known as *magnitudes* to measure changes in brightness. On this scale, a difference in brightness of 100 is represented by a magnitude difference of 5. In these terms, extinction measured in magnitudes is very simply related to optical depth:

$$A(\lambda) = 1.086\tau(\lambda).$$

Differences in extinction at different wavelengths, λ and λ', cause a change in color, leading to a *color excess*

$$E(\lambda - \lambda') = A(\lambda) - A(\lambda').$$

A widely used method for studying interstellar extinction is the so-called *pair method* in which the intrinsic colors for unreddened stars are compared to the colors obtained for reddened, but otherwise similar, stars. The extinction $A(\lambda)$ measured to individual stars in this (and other) ways depends on the amount of extinguishing dust along the lines of sight to those stars. This amount depends in turn on the distances to the stars, and on the distribution and number density of the dust particles at every point along the lines of sight. To compare extinctions in a standard way it has become conventional to measure the colors $E(\lambda - \lambda')$ relative to the color at a "visual" wavelength $\lambda' = V$ (defined to be 5400 Å) and to normalize this color excess $E(\lambda - V)$ for each star so that $E(B - V) = 1$, where B represents a blue wavelength, defined to be 4200 Å. In this way, a meaningful comparison of interstellar extinctions along different lines of sight can be made. It should, however, be emphasized that this conventional and convenient normalization of extinction may be misleading if there are wide variations in the character of extinction in the region of the B and V wavelengths. The normalized extinction $E(\lambda - V)/E(B - V)$ is found to vary with wavelength. As extinction $A(\lambda)$ tends to zero as

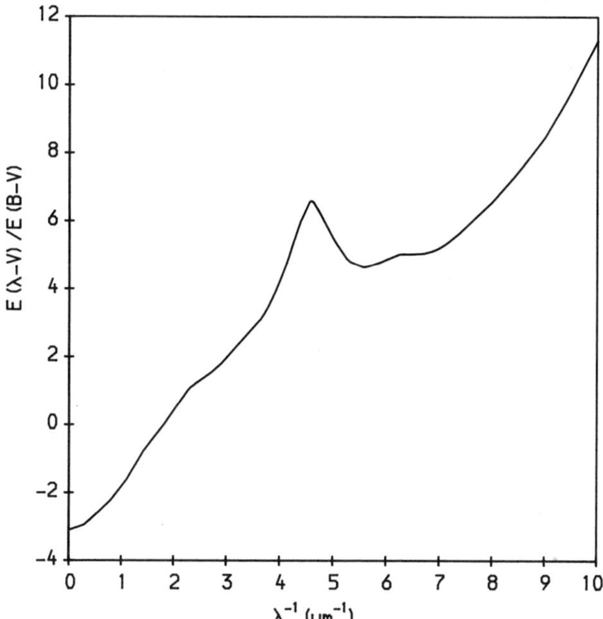

Figure 1. An average normalized interstellar extinction curve for the Galaxy, as recommended by Blair D. Savage and John S. Mathis. The normalized extinction (vertical scale), as defined in the text, is plotted as a function of the reciprocal of the wavelength (horizontal scale).

wavelength increases, the normalized extinction approaches $-R$ in this limit, where

$$R = A(V)/E(B-V)$$

is the *ratio of total to selective extinction*. Observations indicate that R has about the same value (approximately 3) along many lines of sight. This implies that dust particles along the various lines of sight tend to have similar optical properties in the visual region and may, therefore, be chemically and structurally similar throughout the Galaxy. Assuming that R is known, then the relatively straightforward measurement of the color $E(B-V)$ allows an immediate estimate of the total visual extinction A_V. An average normalized *interstellar extinction curve* for the Galaxy is shown in Fig. 1. Whereas early work on interstellar extinction curves drew attention to the near uniformity of interstellar extinction curves (apart from a few localized anomalies), it is now recognized that quite significant variations exist between different lines of sight. These variations are particularly apparent in the far ultraviolet though this is perhaps, in part, due to the normalization process. In general, however, the behavior indicated for the average in Fig. 1 represents at least qualitatively the extinction observed on particular lines of sight.

INFRARED EXTINCTION

Extinction measurements in the infrared are subject to substantial systematic errors because of the difficulty of allowing correctly for infrared emission from warm dust grains in stellar or circumstellar environments. The infrared extinction curve between about $\lambda^{-1} = 0.2$ and $0.8\ \mu m^{-1}$ ($\lambda = 5$ and $1.25\ \mu m$) has recently been reevaluated for a population of reddened early-type stars within 2–3 kpc of the Sun. The stars are widely distributed about the galactic plane and obscured predominantly by dust in atomic hydrogen clouds. The extinction is fairly accurately represented by the formula

$$A(\lambda) \propto \lambda^{-\alpha}$$

where $\alpha = 1.70 \pm 0.08$ in the range 1.6–5 μm. The data extrapolate

Table 1. Infrared Absorption Features Arising from Dust in Molecular Clouds, and their Likely Identifications. All Molecules Listed are in the Solid Phase.

λ (μm)	*Identification*
3.0	H_2O
3.3–3.6	H_2O, NH_3
3.53	CH_3OH
3.9	H_2S
4.62	OCN^-
4.67	CO
4.9	?
6.0	H_2O
6.85	?
9.7	Silicate
19.0	Silicate
42.0	H_2O

to $\lambda^{-1} = 0$ to give $R = 3.05 \pm 0.05$. There is little evidence of strong variation in infrared extinction for this sample of stars.

Deviations from this mean extinction law occur in particular *absorption bands*, which are associated with vibrational transitions in the solid matrix of the dust material. These absorptions may substantially enhance the mean extinction. Along lines of sight through darker molecular clouds (say, $A_v \gtrsim 3$) the extinction is modified by the appearance of a variety of absorption bands arising, in most cases, from molecular ices frozen onto the surfaces of dust grains. Table 1 lists the features observed and their usual identifications. The water–ice feature at 3.0 μm can be a particularly important absorption, significantly affecting total extinction in this wavelength region. The absorptions at 9.7 and 19 μm are observed even in diffuse clouds. It is accepted that these two bands arise in the silicate component of the dust—a view supported by the detection of the 9.7-μm feature in emission from hot grains near stars. The strength of the 9.7-μm feature in absorption is proportional to visual extinction according to

$$\tau(9.7\ \mu m) = (5.4 \pm 0.3) \times 10^{-2} A_v,$$

showing that it is a true interstellar extinction feature.

VISUAL EXTINCTION

Visual extinction is characterized by a monotonic rise, that occurs approximately linearly with $1/\lambda$. Differences in normalized extinction from star to star are detectable but generally minor, and the normalized extinction curves show much uniformity. The $1/\lambda$ dependence suggests that much of the visual extinction is contributed by dust particles of a size comparable with the wavelength of visual light. The ratio R of total to selective extinction is about 3.1 for many lines of sight. Recently, some departures from the average extinction have been detected at wavelengths longer than 5000 Å. The departure can be interpreted as a broad emission extending over a width of about 2000 Å and peaking near 6500 or 7000 Å. It is likely to be due to dust luminescence from particles in the vicinity of hot stars that is excited by the stellar ultraviolet radiation. In addition, certain well-known, ubiquitous but as yet unidentified absorption bands have been detected in the visual. The band at shortest wavelength, and the most prominent, is centered at 4430 Å. The origin of these bands is controversial, but may lie in the dust particles.

ULTRAVIOLET EXTINCTION

Ultraviolet extinction has been measured from the visual out to $\lambda^{-1} = 10\ \mu m^{-1}$. The characteristic shape of the extinction curve in the ultraviolet is a broad peak of extinction centered near $\lambda^{-1} = 4.6\ \mu m^{-1}$, and a rise from about $\lambda^{-1} = 6\ \mu m^{-1}$ into the far ultraviolet. The central wavelength of the peak shows little variation between the lines of sight to different stars though, when normalized to constant $E(B-V)$, the strength and width of the peak may show substantial variation. (This feature is sometimes called the "2200 Å extinction bump.") In a sample of 45 stars the mean central wavelength is 2174.4 Å with no individual star deviating more than ± 17 Å from this mean. It is interesting to note that two particular stars with extremely different far ultraviolet extinctions differ in the wavelength of the peak by only 1.4 Å. The width of the peak is substantial: The full width of the peak at half maximum strength varies from 360 to 600 Å. Broader peaks are found in dark clouds and reflection nebulae, whereas in the diffuse medium and in star formation regions the peaks are narrower. Its origin in the solid state seems certain, but remains controversial. The strength of the absorption at 4.6 μm^{-1} (as measured by the area under the peak) varies by more than a factor of 3 and correlates well with $E(B-V)$, with a few exceptions. The carrier of this feature must, therefore, either be in the same population of dust grains also responsible for visual extinction or in a population coexisting with them.

Far ultraviolet extinction is observed to show considerable variation between different lines of sight. Extreme variations occur: One star in Orion has a far ultraviolet extinction that is almost flat, that is, wavelength independent, for $\lambda^{-1} \gtrsim 6$ μm^{-1}, whereas other stars show a very steep rise. These are not exceptions to normal behavior: Statistical studies show that the strength of the far ultraviolet extinction is poorly correlated either with visual extinction or with the strength of the peak at $\lambda^{-1} = 4.6$ μm^{-1}. It appears that the dust must contain a component that contributes significantly neither to visual extinction nor to the 4.6 μm^{-1} peak, but that must be capable of substantial variation in the far ultraviolet.

Additional Reading

Allamandola, L. J. and Tielens, A. G. G. M., eds. (1989). *Interstellar Dust*. Kluwer Academic Publishers, Dordrecht.

Fitzpatrick, E. L. and Massa, D. (1986). An analysis of the shapes of ultraviolet extinction curves. I. The 2175 bump. *Ap. J.* **307** 286.

Fitzpatrick, E. L. and Massa, D. (1988). An analysis of the shapes of ultraviolet extinction curves. II. The far-UV extinction. *Ap. J.* **328** 734.

Martin, P. G. (1978). *Cosmic Dust: Its Impact on Astronomy*. Clarendon Press, Oxford.

Mathis, J. S. (1990). Interstellar dust and extinction. *Ann. Rev. Astron. Ap.* **28** 37.

Savage, B. D. and Mathis, J. S. (1979). Observed properties of interstellar dust. *Ann. Rev. Astron. Ap.* **17** 73.

Whittet, D. C. B. (1988). The observed properties of interstellar dust. In *Dust in the Universe*, M. E. Bailey and D. A. Williams, eds. Cambridge University Press, Cambridge, p. 25.

Witt, A. N. (1988). Visual and ultraviolet observations of interstellar dust. In *Dust in the Universe*, M. E. Bailey and D. A. Williams, eds. Cambridge University Press, Cambridge, p. 1.

See also **Diffuse Galactic Light; Interstellar Medium, Dust Grains; Interstellar Medium, Dust, Large Scale Galactic Properties; Stars, Circumstellar Matter.**

Interstellar Medium

Ronald J. Reynolds

The interstellar medium is a complex distribution of gas and dust particles that fills the space between the stars. It is a galaxy's atmosphere, held by the combined gravitational pull of the stars and permeated by starlight, a magnetic field, and cosmic rays. It is the material out of which new stars are created and into which old stars expel, sometimes violently, their nuclear-processed remains. The interstellar medium, therefore, plays a vital role in an ongoing cycle of stellar birth and death and galactic evolution.

Interstellar medium astronomy began about 1927 with the publication of Edward Emerson Barnard's photographic atlas of the Milky Way, which included a list of dark galactic clouds silhouetted against the background starlight, and the spectroscopic observations of John S. Plaskett and Otto Struve, which established the existence of interstellar clouds containing ionized calcium. Actually, the absorbing dust and gaseous Ca^+ in these interstellar clouds constitute only a small fraction of the total amount of interstellar matter. By number of nuclei about 90% of interstellar matter is hydrogen and 10% is helium. All of the elements heavier than helium constitute in total about 0.1% of the interstellar nuclei or about 1% of the mass. Roughly half of the heavier elements are in "interstellar dust" particles about a micrometer or less in size. Furthermore, "clouds" account for only half of the interstellar matter and occupy just 2% of the interstellar volume. The remainder of the interstellar mass and volume is associated with a more tenuous "intercloud medium" that was not identified until the mid 1960s and early 1970s when more sophisticated observational techniques were developed at x-ray, ultraviolet, visible, and radio wavelengths.

All of the major constituents of the interstellar medium now appear to be identified, at least in the portion of our Galaxy studied most extensively, a few thousand parsec (pc) radius region around the Sun. (1 pc $= 3.086 \times 10^{18}$ cm or 3.26 ly.) In this solar neighborhood a column of 1 cm^2 cross-sectional area perpendicular to the galactic disk contains on the average 1.0×10^{21} hydrogen nuclei, corresponding to a total interstellar mass surface density of 2.4×10^{-3} g cm^{-2}; this is about one-fourth the mass surface density of the stars. A little more than half of the hydrogen in the solar neighborhood is in the form of neutral atoms; one-fifth of the hydrogen is ionized and the remainder is in molecular form. Temperatures and densities range from about 10 K and 10^5 cm^{-3} or more in some of the coldest clouds to 10^6 K and 10^{-3} cm^{-3} in parts of the intercloud medium.

ATOMIC AND MOLECULAR CLOUDS

Half of the neutral atomic hydrogen and all of the molecular hydrogen in the interstellar medium is concentrated into relatively high density and low temperature regions called clouds. The properties of the atomic hydrogen (H I) clouds have been determined primarily from radio observations of the hyperfine ground state transitions of hydrogen at 1420 MHz (21 cm); interstellar absorption lines of trace elements such as Ca^+ also continue to play a very important role in the study of these clouds. For the most common clouds, from which the emitted 21-cm radiation escapes freely, the brightness of the 21-cm emission provides a direct measurement of the H I column density $N_{H\,I} = \int n_{H\,I}\,ds$, the integral of the hydrogen atom density $n_{H\,I}$ along the line of sight s through the cloud. When an H I cloud is located in front of a bright source of radio continuum emission, the decrease in the brightness of the background source at 21 cm is proportional to $\int (n_{H\,I}/T)\,ds$, where T is the temperature of the H I cloud. Thus observations of H I clouds at 21 cm in emission and absorption provide direct information about cloud temperatures and column densities. The H I clouds are found to have temperatures between 30 and 500 K and values of $N_{H\,I}$ between 10^{19} and several $\times 10^{20}$ cm^{-2}. The median cloud has $T \approx 120$ K and $N_{H\,I} \approx 0.8 \times 10^{20}$ cm^{-2}, and there are about 10 clouds along a 1000-pc line near the galactic midplane. Maps of the sky in 21-cm emission reveal that H I clouds have complex shapes resembling thin extended sheets or filaments within which are embedded small clumps. The clumps have sizes down to 1 pc, implying densities of about 25 cm^{-3}.

Throughout most of our Galaxy the clouds are confined to an average distance $\langle|z|\rangle \approx 100$ pc from the midplane, which is a third of the average $|z|$ extent of the stars.

Molecular hydrogen is confined to the interiors of the densest and most massive clouds, the dark clouds, where starlight capable of dissociating the H_2 molecules cannot penetrate. These clouds are the active star-forming component of the interstellar medium. Because H_2 has no electric dipole moment, radiative transitions of H_2 are greatly suppressed. Therefore, most of the structural information about molecular clouds in the interstellar medium is obtained through observations of the rotational transitions of the trace molecule CO at 115 GHz (2.6 mm). In addition a wide variety of other molecules, including complex hydrocarbon chains, have been detected within the H_2 clouds. The CO observations suggest that the typical molecular cloud has an average H_2 density of 200 cm^{-3}, a temperature of 15 K, and a diameter of 40 pc. The clouds have internal structure to the resolution limits of the observations (0.3 pc); some of the small condensations have densities of 10^5 cm^{-3}. The fraction of interstellar hydrogen in molecular form varies greatly with distance from the galactic center. In the inner part of the Galaxy, molecular rather than atomic hydrogen is the dominant state, whereas in the outer Galaxy there is little H_2. For the Galaxy as a whole the interstellar mass in H_2 is nearly equal to the mass in H I.

THREE COMPONENTS OF THE INTERCLOUD MEDIUM

Filling the space between the clouds is higher temperature, lower density gas that traditionally has been called the intercloud medium. In the solar neighborhood this medium contains about 50% of the interstellar matter and occupies 98% of the interstellar volume. Contrary to its name, most of this material is not located *between* the clouds but rather is in a thick disk that extends far above the cloud layer to a height $|z| \approx 1000$ pc or more. About 95% of the intercloud matter consists of regions of neutral and ionized hydrogen at a temperature of 8000 K and a midplane density near 0.2 cm^{-3}, called the *warm neutral* and *warm ionized* components of the intercloud medium, respectively. There also are regions of much lower density ionized gas at a temperature near 10^6 K, called the *hot* component of the intercloud medium. Although this hot component accounts for only 5% or less of the intercloud matter, it could occupy a relatively large fraction of the intercloud volume.

The existence of an intercloud medium was predicted in 1956 by Lyman Spitzer, who pointed out that the interstellar H I clouds would be unstable to rapid expansion unless confined by the pressure of a higher temperature, lower density ambient medium. Spitzer postulated a medium with a temperature of order 10^6 K and a density of 10^{-3} cm^{-3}. Such a gas was discovered about 17 years later with the advent of space astronomy, which made possible observations of ultraviolet absorption lines and diffuse x-ray emission from the highly ionized ions of trace elements within the hot gas. This gas is believed to be produced primarily by high velocity (100–200 km s^{-1}) shock waves that expand into the warm intercloud medium surrounding a supernova. The shock can sweep out the ambient gas creating a very hot, low density bubble, 50–100 pc in radius. If the bubbles are long lived and are created frequently enough, they may have a profound influence on the structure of the interstellar medium by interconnecting and occupying most of the interstellar volume. Furthermore, buoyant forces may carry the hot gas thousands of parsecs from the midplane. The influence of such bubbles on the structure of the interstellar medium depends in part upon whether or not their expansion is suppressed significantly by the interstellar magnetic field and the thick layer of *warm* intercloud gas. The volume filling fraction of the hot medium is currently the subject of much debate with estimates ranging from 10–70%.

The existence of the warm ionized intercloud medium was also apparent by 1973 after the discovery of pulsars at radio wavelengths and the detection in visible light of faint, diffuse hydrogen recombination line emission from the Galaxy. The progressive delay in the arrival time of a pulsar's radio pulse as the radio receiver is tuned to lower frequencies provides a direct measurement of the column density of unbound electrons between the pulsar and the receiver. Pulsars within the galactic disk plus those discovered in 1988 within globular clusters located many thousands of parsecs above the galactic midplane reveal a layer of H^+ extending to an average height $\langle|z|\rangle \approx 1000$ pc from the plane with a total column density of 2×10^{20} cm^{-2} through the disk (amounting to one-third of the H I). This intercloud H^+ contains about 90% of the ionized interstellar hydrogen (the remainder is in the hot component and in the isolated, relatively high density emission nebulae called H II regions, which surround young hot stars). The development in 1971 of large aperture Fabry–Perot spectrometers made possible the detection of faint optical line emission from recombining hydrogen and collisionally-excited metastable states of N^+ and S^+ in this warm ionized medium. These optical observations indicate that the ionized gas has a temperature near 8000 K and is clumped into regions of nearly fully ionized hydrogen that have an electron density of 0.15 cm^{-3} (and an equal proton density) and occupy approximately 20% of the intercloud volume. The source of this ionization is not yet established, although photoionization by a weak flux of ionizing radiation or a combination of photoionization and shocks appear to be likely mechanisms.

Approximately half of the interstellar H I appears to be located in the warm neutral component of the intercloud medium. This intercloud H I was first identified in 1965 as the source of the ubiquitous, relatively broad (velocity dispersion ~ 9 km s^{-1}) 21-cm

Table 1. Characteristic Properties of the Gaseous Components of the Interstellar Medium in the Solar Neighborhood

Component	*Temperature (K)*	*Midplane Density (cm^{-3})*	*Filling Fraction (%)*	*Average Height above Midplane (pc)*	*Surface Mass Density (mg cm^{-2})*
Clouds					
H_2	15	200	0.1	75	0.42*
H I	120	25	2	100	0.73
Intercloud					
Warm H I	8000	0.3†	35†	500	0.73
Warm H II	8000	0.15	20	1000	0.46
Hot H II	~ 10^6	2×10^{-3}	43†	3000†	0.04†

*The amount of H_2 increases greatly toward the inner Galaxy.
†Value is uncertain by at least a factor of 2.

emission features that had no corresponding absorption when viewed against bright background radio sources. The large velocity dispersions and the absence of absorption imply temperatures of 5000–10,000 K. Observations of the Lyman-α absorption line of atomic hydrogen toward bright stars show that this gas has a mean extent from the midplane $\langle |z| \rangle \approx 500$ pc, which is significantly larger than that of the H I clouds. If this warm H I is in pressure equilibrium with the H I clouds, then it would be clumped into regions occupying 35% of the intercloud volume with a density of 0.3 cm^{-3} at the midplane. However, the relationship of this gas to the other components is not known, and therefore its filling fraction and density are very uncertain. For example, if the warm neutral gas has the same *density* as the warm ionized gas, then its filling fraction would be about 70%.

The principal properties of these cloud and intercloud components of the interstellar medium are summarized in Table 1.

UNANSWERED QUESTIONS ABOUT THE INTERSTELLAR MEDIUM

Although the principal components of the interstellar medium have been identified and many of their properties measured, there presently is very little understanding of how they fit together into a dynamic, interacting system. What are the feedback mechanisms that determine the stellar and interstellar properties of a galaxy? It is not known, for example, what regulates the star formation rate, or the relative amounts of matter in clouds and the intercloud medium, or why the diverse gaseous components as well as the cosmic rays, magnetic field, and starlight all have roughly the same energy density of about 1 eV cm^{-3}. Another outstanding question about the interstellar medium is its topology, including the shapes, sizes, and relative positions of the various components, and the volume filling fractions of the hot and warm (neutral plus ionized) portions of the intercloud medium. Finally, there is the question of the medium's evolution. How do its properties change as it is slowly transformed into stars over the lifetime of the galaxy? Answers to some of these questions will surely come with the development of new observational methods for measuring more accurately the properties of the interstellar medium in our Galaxy, and in other galaxies, where conditions quite different from those in our Galaxy can provide valuable new perspectives on the nature of interstellar matter and processes.

Additional Reading

Cox, D. P. (1989). Structure of the diffuse interstellar medium. In *Structure and Dynamics of the Interstellar Medium*, G. Tenorio-Tagle, M. Moles, and J. Melnick, eds. Springer, New York.

Hollenback, D. J. and Thronson, H. A., Jr., eds. (1987). *Interstellar Processes*. Reidel, Dordrecht.

Kulkarni, S. and Heiles, C. (1988). H I and the diffuse interstellar medium. In *Galactic and Extragalactic Radio Astronomy*, 2nd ed., G. L. Verschuur and K. Kellerman, eds. Springer, New York, p. 95.

Spitzer, L., Jr. (1978). *Physical Processes in the Interstellar Medium*. Wiley, New York.

Spitzer, L., Jr. (1982). *Searching Between the Stars*. Yale University Press, New Haven.

Spitzer, L., Jr. (1990). Theories of the hot interstellar gas. *Ann. Rev. Astron. Ap.* **28** 71.

See also **Galactic Structure, Interstellar Clouds; Galaxies, Molecular Gas in; Interstellar Clouds, Chemistry; Interstellar Extinction, Galactic.**

Interstellar Medium, Chemical Composition

Edward B. Jenkins

Two fundamental concepts establish the framework for the chemical composition of the interstellar medium. First, we consider the relative quantities of different atomic elements that constitute whatever substances are present, regardless of their chemical form. After establishing these relative populations, we can address the second topic, namely, the concentrations of plausible chemical configurations that exist in space. Of the two considerations, in most contexts the latter shows far more diversity than the former.

GENERAL CONCLUSIONS

Distribution of Elements

The elemental compositions of the interstellar medium and relatively young (Population I) stars are closely linked by a cycle of matter: New stars form out of gravitationally contracting interstellar material, and later, as these stars evolve, most of them either continuously or impulsively cast off their outer layers, thus replenishing their environment. Some stars expel a large percentage of their mass when they perish during a supernova explosion. The interchange of material between stars and their surroundings, along with vigorous motions of gases in space, insure a general homogeneity in element concentrations. Exceptions are found in places where the matter is dominated by ejecta from a very recent supernova explosion or regions near stars that have atmospheres of unusual composition and that are losing mass rapidly. The much older Population II stars are relatively inactive in this cycle; their compositions differ appreciably from the Population I stars and the interstellar medium.

Most of the interstellar medium consists of hydrogen (H). About 8% of the atoms are helium (He), and a remaining 0.13% are elements in the Periodic Table beyond He. The elements carbon (C), nitrogen (N), oxygen (O) and neon (Ne) are a dominant fraction (93%) of this small population of heavy elements. The relative proportions of specific elements in our part of the Galaxy (the Milky Way), described previously in very broad terms, is often referred to as the "cosmic abundance scale." The composition of our Sun and nearby (normal) young stars, and also the distribution of heavier elements in meteorites and the Earth's crust, have much in common with this abundance scale. It is not truly universal for material in space, however. There are variations in remote parts of our Galaxy. Likewise, other galactic systems have different histories of element production and yield different abundances. For instance, the Magellanic Clouds (a pair of companion galaxies to our own that may be seen by the naked eye in the southern hemisphere) have heavy elements that are only about one-tenth to one-half as abundant relative to H (for the Small and Large Clouds, respectively) as they are locally

The lightest elements, H, He, deuterium (D, a stable isotope of H), and lithium (Li), were probably created in the nucleosynthesis that took place during the early development of the universe. Elements from carbon (C) to iron (Fe) were progressively built up from lighter constituents by "nuclear burning" (fusion) in stellar interiors. This steady migration to heavier elements, however, stops at Fe, the element whose nucleus has the largest binding energy per nucleon. Elements in the Periodic Table beyond Fe must have been created in brief (equilibrium and nonequilibrium) reactions where there was a plentiful supply of free neutrons, such as during a supernova explosion. Finally, the existence of beryllium (Be) and boron (B) can be traced to the splitting of heavier elements in the interstellar medium by cosmic rays.

Chemical Forms

Most of the gaseous material in the galactic disk has approximately the same (thermal) pressure. Exceptions to this balance occur in regions that are suddenly exposed to strong ultraviolet radiation from a new star, locations disturbed by supersonic motions, and the central regions of clouds that are held together by gravitation. As a consequence of Boyle's law, the near constancy of pressure insures that the product of the particle density n and temperature T is approximately the same everywhere. From one region of space to the next, however, there are enormous contrasts in temperature, with the accompanying inverse changes in density. These differing conditions have a profound effect on the physical properties of the interstellar medium and the chemical forms that dominate. Broadly speaking, we may characterize four regimes:

1. Very cool, dense clouds ($T \sim$ 10–20 K, $n > 1000$ cm^{-3}).
2. Diffuse, cool clouds ($T \sim$ 80 K, n typically 10–100 cm^{-3}).
3. Diffuse regions that are often ionized by ultraviolet photons from hot stars ($T \sim$ 8000 K, n typically 0.1–1 cm^{-3}).
4. A very hot, low-density intercloud medium (10^5 K $< T <$ 10^7 K, $n \sim 0.001$ cm^{-3}) heated by shock waves from supernovae.

High densities within the very cool clouds favor the rapid formation of free molecules and dust grains. The large concentrations of dust grains (or even ice-like particles), in turn, shield the interiors of these clouds from ultraviolet radiation from stars, thus inhibiting an important destruction process for the molecules. Except for regions near hot stars embedded within or on the edges of such dense clouds, practically all of the elements are tied up in molecules or grains, leaving relatively few free atoms (except probably the noble gases).

Molecules are minor constituents in diffuse clouds. Dust grains are still present, however, and from the amount of extinction of light they cause we know that substantial fractions of most heavier elements are condensed onto such particles (which is confirmed by direct measurements of the concentrations of such elements in free atomic form, see the following sections). Usually, ultraviolet starlight photons that could ionize hydrogen cannot penetrate a diffuse cloud if some neutral hydrogen is present, because the H atoms are very effective in blocking the radiation. Because of this shielding, the hydrogen inside a cloud is reached only by cosmic rays or x-rays, and their ionizing effect is very weak. Consequently, nearly 100% of the H atoms are neutral. It follows that elements that have a first ionization potential above that of hydrogen are likewise protected from being photoionized, because they too are shielded from photons more energetic than the Lyman series limit at 13.6 eV. Conversely, those atoms which can be ionized by photons less energetic than 13.6 eV are predominantly ionized in the diffuse clouds.

There is an abrupt transition away from neutrality when gases have an unusually low density or are close to one or more stars with surface temperatures $T \gtrsim 15{,}000$ K. When ionizing photons are plentiful enough to ionize a good fraction of the gas near the edge of a cloud, the region becomes more transparent, the radiation can penetrate more effectively, and nearly full ionization occurs up to some well-defined boundary. Thus, an interesting outcome of some hydrogen atoms shielding others from ionizing radiation is the following dichotomy: The atoms are either almost completely neutral because they are self-protected from Lyman-limit radiation or, alternatively, they are almost fully ionized because no such protection is present.

Shock waves from supernovae propagate very freely through a network of cavities and bubbles between the dense and diffuse clouds where the gas density is very low. These disturbances heat the gas to around 10^6 K, and electron impacts at such temperatures force the atoms to be multiply ionized. Dust grains probably do not survive long in this environment, so the abundances of free atoms are close to their respective cosmic abundances.

OBSERVATIONS

The centers of dense clouds are opaque to the electromagnetic spectrum from the middle infrared to hard x-rays, thus precluding this range for diagnostic purposes. The only way to probe these regions is by observing radio emission from molecules. Observations at millimeter wavelengths reveal a vast array of different, simple compounds (a few hundred have been identified so far), but sometimes quantitative measures of more abundant molecules are difficult because emission features are saturated and hard to interpret. Unfortunately, the most abundant (and also chemically important) molecule, H_2, cannot be seen in the radio domain because it has no dipole moment. The only way to detect H_2 in space is from infrared quadrupole radiation in gas excited by shocks or by absorption from electronic transitions in the far ultraviolet (seen only by orbiting telescopes above our atmosphere). For free atoms, radio recombination lines from dense clouds with strong ultraviolet illumination do give some insight on abundances.

Diffuse clouds can be studied at nearly all wavelengths. A powerful method to appraise the atomic abundances is to record stellar spectra and measure the absorption features produced by electronic transitions out of the ground states of intervening atoms. Absorption lines in the visible are not too plentiful and, more importantly, with only one exception (a group of transitions of singly ionized Ti) they indicate the abundances of atoms in stages of ionization below the favored ones. To infer total atomic abundances in such instances, one must apply an uncertain correction for the ionization equilibrium that depends on the local electron density and the intensity of ionizing radiation. In the far ultraviolet the situation is much improved because there are many useful absorption lines for a broad variety of elements in different stages of ionization, including their dominant ones.

In many fully ionized regions one may detect emission lines created by ultraviolet fluorescence or electron impacts. These emissions are also useful for abundance studies, but one must have a good understanding of the physical conditions to translate the observed brightness into relative abundances of the emitting atoms.

Important constituents of dust grains, such as water ice and silicates, can be observed in the infrared. Such observations show conclusive evidence that some grains (or the nuclei of most) are produced in the outer layers of stellar atmospheres that shed matter into their surrounding media. There are also a number of interstellar spectroscopic features in the visible and ultraviolet that provide clues to the composition of dust or very large molecules that, unfortunately, are not yet fully understood.

SPECIFIC FINDINGS

Balance between Molecules and Atomic Gas

H_2 is the most abundant molecule and is also the precursor to the formation of most of the other molecules. Ultraviolet spectra of stars show that in foreground diffuse clouds the fractional abundance of H_2 is either about 20–50% or, alternatively, very much lower (of the order of 10^{-4}), that is, there seems to be no smooth continuum of intermediate concentrations. As with ionization of H, shielding of disruptive (in this case dissociating) radiation creates this contrast. Other molecules follow this trend, because most of them are formed by reactions that start with H_2. Carbon monoxide (CO), the second most abundant molecule, is a particularly noteworthy constituent because radio astronomers use its millimeter-wave emission to study molecule-bearing clouds.

Balance between Grains and Atoms

In addition to grains being formed in stellar atmospheres (but only for cool stars, the outer atmospheres of very hot stars being too hot and moving outward too rapidly to form grains), they must undergo considerable growth in dense interstellar clouds. The condensation of free atoms onto grains results in gaseous depletions that vary from one element to another. For instance, the elements C, O, N, S, P, and Zn have moderate depletions (i.e., a factor of only 2–10), but in many cases observations are consistent with no depletion. By contrast, the abundances of Ca and Ti are sometimes down by a factor 10^4 from their cosmic abundance ratios to hydrogen. These depletion factors are driven by chemical affinities, which determine both the equilibrium concentrations where they form in stellar atmospheres and the relative ease of destruction in space by shocks. It is important to realize that for elements that have small intrinsic abundances or generally large depletions, significant modifications in relative abundances do not directly signal fundamental changes in dust composition or the amount of dust.

Theoretical calculations indicate that shocks traveling faster than about 100 km s^{-1} can destroy grains. Observations show convincingly that depletions are less severe for (1) lines of sight that have lower than average density and (2) parcels of gas that are moving rapidly. Such correlations are probably a consequence of the increased likelihood that a shock recently passed through a region and destroyed or eroded the grains.

Regions fully ionized by ultraviolet starlight show depletions similar to their un-ionized counterparts. It follows that ultraviolet radiation from stars is not important for destroying grains.

CONSEQUENCES IN OTHER AREAS

Molecules (other than H_2) and trace elements are important coolants for the interstellar medium. When they are excited by thermal collisions, they can lose energy by emitting radiation. Hence the concentrations of these chemical constituents are important for the energy balance (and equilibrium temperature) of the gas.

In regions where the gas is predominantly un-ionized, nearly all of the free electrons come from elements with ionization potentials below that of hydrogen. Of these elements, carbon is the principal contributor.

Additional Reading

Cowie, L. L. and Songaila, A. (1986). High-resolution optical and ultraviolet absorption-line studies of interstellar gas. *Ann. Rev. Astron. Ap.* **24** 499.

Spitzer, L., Jr. (1982). *Searching Between the Stars.* Yale University Press, New Haven.

Wannier, P. G. (1980). Nuclear abundances and evolution of the interstellar medium. *Ann. Rev. Astron. Ap.* **18** 399.

Winnewisser, G. and Armstrong, J. T., eds. (1989). *The Physics and Chemistry of Interstellar Molecular Clouds: mm and Sub-mm Observations in Astrophysics.* Springer, Berlin.

See also **Interstellar Clouds, Chemistry; Interstellar Medium, Chemical Composition, Galactic Distributions; Interstellar Medium, Chemistry, Laboratory Studies; Interstellar Medium Dust Grains; Interstellar Medium, Hot Phase; Interstellar Medium, Molecules; Supernova Remnants, Evidence for Nucleosynthesis.**

Interstellar Medium, Chemical Composition, Galactic Distribution

Manuel Peimbert

The distribution of the chemical composition in the interstellar medium (ISM) is due to: (i) the initial abundances with which the Galaxy was formed, (ii) the enrichment of the ISM due to mass lost by the stars during their evolution, and (iii) the existence of large scale mass flows (like infall from the halo, outflow from the disk, or radial flows across the disk of the Galaxy). Consequently a precise knowledge of the chemical composition of the ISM, as a function of position and time, strongly constrains the parameters that are responsible for the distribution of the elements in the Galaxy; the study of these parameters belongs to the realm of galactic chemical evolution. The term chemical composition will be used to describe the relative abundances of all the elements, and not the relative abundances of the molecules.

It is generally accepted that galaxies formed from gas made of hydrogen and helium and an insignificant amount of heavier elements. The pregalactic H and He abundances, about 77 and 23% by mass, respectively, seem to be the same in a large volume of space that comprises many galaxies including our own. It is thought that this chemical composition is the product of the Big Bang; the He abundance by mass is generally referred to a the primordial helium abundance. Once a galaxy is formed, mass lost by its stars during their evolution modifies the ISM, enriching it with freshly made He and heavier elements produced in their interiors by nuclear reactions. Therefore, if there are no gas flows, it is expected that the higher the fraction of gas that goes into stars, the higher the fraction of elements heavier than H in the ISM. Consequently, it is also expected that in the Galaxy the gas-to-star mass ratio should be related to the fraction of heavy elements present in the ISM.

The ISM in the disk of the Galaxy is not homogeneous in chemical composition; it shows general variations with galactocentric distances and it also shows an inhomogeneous behavior at small scales. The small scale inhomogeneities will not be discussed here. We will consider H II regions, old supernova remnants, and molecular clouds to study the present distribution of the chemical composition of the ISM. In addition, we will also discuss abundances in planetary nebulae for which excellent data based on emission lines are available.

H II REGIONS

H II regions are conglomerates of gas and dust ionized by recently formed hot stars. They are called H II regions because most of the gas is in the form of ionized hydrogen. From their emission line spectra it is found that the most abundant elements, like H, He, C, N, O, Ne, S, and Ar, are in gaseous form, which makes them excellent tracers of the chemical composition of the ISM. Alternatively, by comparing stellar abundances and H II region gaseous abundances, as well as from other considerations, it is found that Mg, Si, and Fe are mainly embedded in dust grains. In galactic H II regions about 90% of the atoms are hydrogen, about 10% are helium, and about 0.1% consist of all the other elements combined.

H II regions are confined to the plane of the Galaxy and cover a large range of galactocentric distances. Due to the extinction produced by dust particles, it is possible to observe optically only those H II regions relatively close to us or in those directions where dust obscuration is small.

Table 1. Solar Neighborhood Abundance Gradients*

	H II Regions	*Supernova Remnants*	*Planetary Nebulae*
He/H	−0.02 ± 0.01	—	−0.02 ± 0.01
C/H	−0.09 ± 0.03	—	—
N/H	−0.12 ± 0.02	−0.10 ± 0.03	—
O/H	−0.08 ± 0.02	—	−0.07 ± 0.02
S/H	−0.07 ± 0.03	−0.07 ± 0.03	−0.08 ± 0.03
Ar/H	−0.08 ± 0.03	—	−0.06 ± 0.03

*Given in $\Delta \log(X/H)/\Delta R$, with R in kiloparsecs.

Electron Temperature

From the intensity ratio of H recombination lines to the continuum in the radio spectral region, it is possible to determine the electron temperature T_e. From the radio data a clearly defined gradient in T_e is found, in the sense that the larger the galactocentric distance R, the higher the T_e. The value of this gradient is 433 ± 40 K kpc^{-1} and covers, at least, a galactocentric range from 4 to 14 kpc; for the solar vicinity T_e is about 7500 K. The observed T_e gradient implies that the main atomic coolants of the H II regions decrease with R. Since O is the most efficient coolant, it has been estimated that a decrease in the O/H ratio of about 20% per kiloparsec could explain the observed T_e gradient.

Abundance Gradients

It has been found that closer to the galactic center He, C, N, O, Ne, S, and Ar are more abundant relative to H than farther away. The gradients of the most abundant elements have been measured from optical observations of H II regions in the 6–14-kpc galactocentric range; their values are presented in Table 1. The behavior of these gradients can be divided in four groups: (i) O, Ne, S, and Ar; (ii) N; (iii) C; (iv) He.

Of the first group of abundance ratios, O/H is the best determined because O is present in H II regions as O^+ and O^{++}, and emission lines of both ions are easily observed. For Ne/H, S/H, and Ar/H it is necessary to correct for unobserved stages of ionization. Within the accuracy of the observations the O/H, Ne/H, S/H, and Ar/H ratios vary similarly with galactocentric distance; that is, Ne/O, S/O, and Ar/O ratios remain constant with R in the disk of the Galaxy, as well as in the disks of other well observed galaxies. This result implies that stars with similar masses are responsible for the production of these four elements. From stellar evolution considerations and from observations of supernova remnants it is thought that O, Ne, S, and Ar are produced by massive stars that explode as supernovae.

In the Galaxy, and in other galaxies, the radial variation of N/H is larger than that of O/H; this result probably implies that part of the N is produced from C already present in the stars when they were formed (secondary production) and part from C freshly made by the stars themselves (primary production).

Our knowledge of the C/H distribution is based on results for only three galactic H II regions. In other galaxies C/O increases with O/H, which probably means that there is a time delay in the C enrichment of the ISM due to the smaller average mass of the stars producing C than those producing O.

The He/H gradient is based on results for H II regions with a high degree of ionization, where most of the He is once ionized; unfortunately most H II regions with $R < R_\odot$ have a substantial fraction of their He in neutral form and it is not possible to derive accurate He/H values for them.

Galactic Center

It is not possible to observe the galactic center in the optical region of the spectrum due to the very large extinction. Fortunately the extinction is considerably smaller in the infrared region and the Ne and Ar abundances have been determined from the [Ne II] 12.8-μm, [Ar II] 6.99-μm, and [Ar III] 8.99-μm emission lines relative to Brackett α at 4.05 μm and to other H recombination lines. The derived Ne/H and Ar/H ratios in the galactic center H II regions are about a factor of 3 larger than in H II regions of the solar vicinity, in fair agreement with the gradients presented in Table 1.

SUPERNOVA REMNANTS

Young supernova remnants (SNR) are composed of highly enriched material from the stars that ejected them and they cannot be used to determine ISM abundances. Alternatively, old SNR can give information on the material of the ISM that has been swept up by the ejected shell; the N/H and S/H gradients derived from emission line observations of old galactic SNR are included in Table 1.

MOLECULAR CLOUDS

Molecular clouds are also concentrated in the plane of the Galaxy. From the observed molecular abundance ratios it is very difficult to establish the elemental abundances because it is necessary to know the distribution of a given element among dust grains and all the molecular species present. On the other hand, it is thought that it is possible to determine isotopic ratios of CNO atoms from observations of simple molecules like CH^+ and H_2CO. The CH^+ results come from optical data that, due to dust extinction are, restricted to molecular clouds within 2 kpc from the Sun; alternatively the H_2CO results come from radio data that include the galactic center. There are two contradictory $^{12}C/^{13}C$ recent results based on CH: One group of observers derives 43 ± 4 and another derives 77 ± 3. It is interesting to note that the average of the CH results is in agreement with the H_2CO results for the solar vicinity; see Table 2. From Table 2 it follows that the $^{12}C/^{13}C$ ratio has decreased since the solar system was formed, and that at present there is a substantial difference between the solar vicinity and the galactic center.

PLANETARY NEBULAE

Planetary nebulae (PN) are shells of gas ejected from, and expanding about, extremely hot central stars that are in the stage between red giants and white dwarfs. To study the distribution of the elements in the ISM of the galactic disk it is necessary to select PN that have nearly circular orbits and that presumably are located at galactocentric distances close to those of their birthplaces. In Table 1 we present He/H and O/H gradients for type II PN (i.e., planetary nebulae belonging to Intermediate Population I) that belong to the disk. The central stars of PN have transformed C into N during their evolution; therefore, the observed N/H gradient is

Table 2. Galactic $^{12}C/^{13}C$ Ratios

	Method		
Region	*Direct*	*H_2CO*	*CH^+*
Solar system	89	—	—
Solar vicinity	—	65 ± 10	77 ± 3 43 ± 4
Galactic center	—	25 ± 5	—

not considered to be representative of the ISM at the time the parent stars were born. The PN gradients are similar to those derived from H II regions, which indicates that since type II PN were formed, the ISM distribution of He/H and O/H has not changed significantly.

STELLAR GRADIENTS

The abundances of the various stellar populations give us useful information about the ISM that is related to the time and place where the stars were born. For the solar neighborhood, the following abundance gradients, based on stellar data, have been reported (in the same units as in Table 1):

1. Cepheid variable stars O/H = -0.08 ± 0.03.
2. Young disk stars of F and G spectral types Fe/H = -0.10 ± 0.02.
3. Old disk stars Fe/H = -0.04 ± 0.03.
4. Open clusters Fe/H = -0.08 ± 0.03.

FINAL COMMENTS

Higher accuracies in abundance determinations will permit us to separate the general trends in the composition of the ISM from the local abundance inhomogeneities due to incomplete mixing of the stellar ejecta. The variations in the chemical abundance distribution of the ISM discussed in this entry provide important constraints for models of stellar evolution and galactic chemical evolution. It is very important to find out if the gradients steepen or flatten close to the galactic center. It is also very important to determine the time and place where different stellar populations were formed, in order to enable mapping of the ISM abundances as a function of time and position.

Additional Reading

Aller, L. H. (1984). *Physics of Thermal Gaseous Nebulae* (*Physical Processes in Gaseous Nebulae*). Reidel, Dordrecht.

Osterbrock, D. E. (1989). *Astrophysics of Gaseous Nebulae and Active Galactic Nuclei*. University Science Books, Mill Valley, Calif.

Pagel, B. E. J. and Edmunds, M. G. (1981). Abundances in stellar populations and the interstellar medium in galaxies. *Ann. Rev. Astron. Ap.* **19** 77.

Peimbert, M., Serrano, A., and Torres-Peimbert, S. (1984). Interstellar matter and chemical evolution. *Science* **224** 345.

Shaver, P. A., McGee, R. X., Newton, L. M., Danks, A. C., and Pottasch, S. R. (1983). The galactic abundance gradient. *Monthly Notices of the Royal Astronomical Society* **204** 53.

Wannier, P. G. (1980). Nuclear abundances and evolution of the interstellar medium. *Ann. Rev. Astron. Ap.* **18** 399.

See also **Galaxies, Chemical Evolution; Galaxy, Chemical Evolution; Interstellar Medium, Chemical Composition; Supernnova Remnants, Evidence for Nucleosynthesis.**

Interstellar Medium, Chemistry, Laboratory Studies

J. Mayo Greenberg

Interstellar dust (or grains) are small solid particles suspended in the clouds of gas that move about in the space between the stars—a kind of galactic aerosol. Originally they were of interest to astronomers primarily because of the fact that they block the light of distant stars. Their optical properties (how they scatter and absorb light) remain an important aspect of astrophysical research and from such studies the sizes of the particles as well as their shape and morphological structure are deduced. In size they are known to span the range from tens of angstroms to tenths of micrometers with the particles of thickness of the order of 0.2 μm making up the dominant fraction of the mass of the dust. The basic constituents of these larger dust particles comprise small silicate cores that condense within the atmospheres of evolved stars and that are blown out into space, there to cool to the order of 10 K and provide nucleation surfaces as seedlings for the growth of interstellar grains. Along with the interstellar gas, the dust must play an active role in a number of processes of astrophysical importance, such as star formation. The chemistry of interstellar dust as well as the gas has now taken a place among the more active fields in astronomy.

The chemical evolution of the matter in the space between the stars involves a complex set of direct as well as indirect interactions, not only between the atoms, molecules, and ions, but critically involving the small solid particles, the ultraviolet photons, and cosmic rays as well. In the late 1960s the discovery of such polyatomic molecules as formaldehyde (H_2CO) in interstellar gas clouds provided the impetus for theoretical and laboratory investigations of atom and ion interactions that occur in the gas. However, even 20 years earlier, there had been a suggestion, based on quite reasonable assumptions as to physical processes occurring on the surface of the dust, that the grains should contain at least such molecules as H_2O, CH_4, and NH_3 as a frozen ice mixture (the dirty ice model of dust). Because the temperature of the interstellar grains is generally in the range of 10–20 K it is, in fact, rather difficult to understand why *all* of the atoms and molecules (excluding H and He) in the gas do not freeze out on their surfaces during the lifetime of the cloud. It was therefore a mystery when in 1965 a careful infrared observation of a heavily obscured star (large amount of dust) did not reveal any solid H_2O and put an upper limit on its possible abundance at less than 1/10 that expected on the basis of the "dirty ice" dust model. Could this mean that our ideas about dust acting like a vacuum cleaner in space were all wrong or did it mean that some basic physical processes were overlooked in our considerations? It turns out to have been the latter and, in fact, even before the discovery of H_2CO it had been suggested that ultraviolet photons penetrating the dirty ice could be photodissociating simple molecules like H_2O, CH_4, and NH_3, and that their molecular fragments could reconstitute themselves into new and complex molecular combinations. But how could the solid phase reactions triggered by ultraviolet photolysis be followed? How could the gas phase and solid phase reactions be combined within a coherent scheme?

Because the theory of the photoprocessing of complex ices is too difficult, a laboratory had to be created to simulate the fundamental processes of chemical evolution in space. The key elements in the simulation are the low temperature (as low as 10 K), low pressure, and ultraviolet radiation. The establishment of such a laboratory provided the basis for the Laboratory Astrophysics Department of the University of Leiden in The Netherlands in 1975, although it should be noted that efforts in this direction gave significant results as early as 1970 at the State University of New York at Albany.

LABORATORY FOR SOLID STATE PHOTOPROCESSING

The strategy that has been adopted is to simulate the photoprocessing of low temperature ices whose initial composition may be varied using different mixtures of such typical simple molecules as H_2O, CH_4, CO, NH_3, C_2H_2, N_2, and O_2. A schematic of the laboratory analog method as now practiced by various groups around the world is shown in Fig. 1. The low temperature is

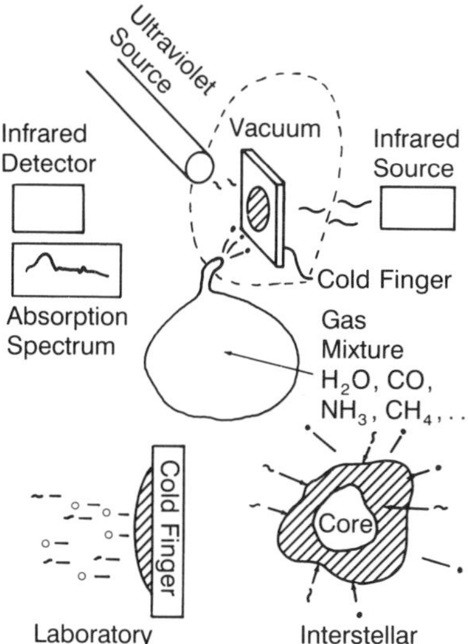

Figure 1. Schematic of the laboratory analog method for studying interstellar grain evolution. Molecules are deposited as a solid on a cold finger in a vacuum chamber and irradiated by ultraviolet photons. The infrared absorption spectrum shows the appearance and disappearance of various molecules and radicals. The cold finger may be an aluminum block (~ 3 × 3-cm block) or a glass, sapphire, or LiF window.

achieved by a cryostat (typically a closed-cycle helium cryogen system) that cools a "cold finger," which can be an aluminum block or a transparent window mounted on a storage ring. The mixture of gases enters the vacuum chamber through a narrow tube. The pressure in the chamber is as low as 10^{-8} torr, which is sufficiently low for the experiments even though the pressure in interstellar space is 10^{-19} torr, or even less. The gases condense on the cold finger, which acts like the core of an interstellar grain. Either during condensation or after (or both) the thin film of ices is irradiated using a vacuum ultraviolet lamp with emission at wavelengths as short as Ly-α ($\lambda = 121.6$ nm). The most important difference between the simulation chamber and interstellar space is the time scale of the photolysis: 1 h of laboratory irradiation may be equivalent to 1000 yr of irradiation in a low density interstellar cloud where the ultraviolet flux is about 10^8 photons per centimeter squared per second, and to a considerably longer time in the dark interior of a molecular cloud where the flux may be reduced by as much as 1000 or more. New ultraviolet lamps are under development in the Leiden Astrophysics Laboratory that will increase the equivalent time scales by as much as 100 or more and will make possible the study of photoprocessing of grains on molecular cloud time scales measured in millions of years as well as on galactic time scales measured in hundreds of millions of years.

ICES IN SPACE

In the laboratory as well as in space, infrared spectroscopy is the most important tool for following interstellar grain evolution. The wavelengths between ~ 2.5 and 25 μm are known as the fingerprint region for identifying various molecular groups from their characteristic vibration, bending, and twisting frequencies in a solid. Thus, the O—H stretch in H_2O occurs at about 3.1 μm, the C—H stretch at about 3.4 μm, the C═O stretch at about 4.6 μm, the O—H bending at about 6 μm, and similarly for other molecular groups like N—H, C═C, C═N, and so on.

In a rapidly developing field, observation and theory tend to play leapfrog. The discovery of the 3.1-μm absorption of water ice in space and its study in the laboratory is a case in point. Although the observed 3.1-μm absorption in interstellar space was almost certainly due to frozen water, it was neither at quite the right wavelength nor did it have the right shape when compared with the then "known" absorption spectrum of *pure crystalline* ice. The answer was found some years later to be in the obvious fact that not only is the H_2O in space mixed with other molecules, but it also is formed at such low temperatures that it is very amorphous —really like solid water.

THERMAL EVOLUTION

What happens after H_2O forms on interstellar grains? To answer this, a variety of molecular mixtures containing H_2O have been created in the laboratory. The infrared spectra of the mixtures have been obtained at initial temperatures from 10–165 K and the effects of annealing, warming, and recooling have been studied. Such spectra may be compared with observations of dust in cool molecular clouds, regions of star formation, and in accretion disks around young evolving stars. Changes in the shape and position of the 3.1-μm band, its width, and its wings reveal both thermal and chemical processes. Similar laboratory methods may also be used to follow the CO absorption feature—how and when CO is present in grains and how and to what degree it is evaporated by heating. Dust whose 3.1-μm ice band is characteristic of having been heated invariably has little or no CO observed in it; that is, the CO has been evaporated or never condensed.

PHOTOCHEMICAL PROCESSES

The creation of new molecules and radiation by photoprocessing of any initial mixture is detected from changes in the infrared absorption. Following the results of an experiment begun with a mixture of CO, NH_3, H_2O, and CH_4, we find almost immediately that CO_2 shows up along with formaldehyde and the formyl radical (HCO). Radicals play a very important role in the chemical processes of the solid; they are exceedingly reactive species because they have unsatisfied bonds. In fact, radicals provide the chemical basis for explosives.

A number of astrophysical objects exhibit photochemical processing of dust. An example is the dust around the protostellar source W33A, which has provided a rich source for laboratory investigations. Figure 2 shows a comparison between an early spectrum of W33A and that of an unirradiated laboratory mixture as well as that of a mixture that has been irradiated and warmed to exhibit both photoprocessing and thermal processing. The 4.6-μm feature was first identified by astronomers as simply solid CO, as in Fig. 2*b*. A higher resolution spectrum of the frequency region $2100 < \nu < 2200$ cm^{-1} ($4.76 > \lambda > 4.54$ μm) made it apparent that there was a double structure: that at 4.67 μm was indeed CO and the other feature at 4.62 μm was something else (see Fig. 2*c*). Such an extra feature always appears next to the CO absorption in experiments as a result of photoprocessing the original laboratory mixture; as the sample is warmed up the CO evaporates while the new feature remains (see Fig. 3). Considerable effort has gone into identifying the molecule responsible for the 4.62-μm absorption. With the use of isotopically labeled carbon and nitrogen in the laboratory mixtures it has been shown to be the ion OCN^-, which was initially a surprising result because it is produced by *nonionizing* radiation. Similarly the absorption in W33A (and other objects as well) at 6.86 μm has been identified as NH_4^+. The large wavelength wing on the ice band (which is so strong as to be saturated) is identified as methyl alcohol (CH_3OH), another molecule that is produced by photolysis of simple ices (containing H_2O and CH_4).

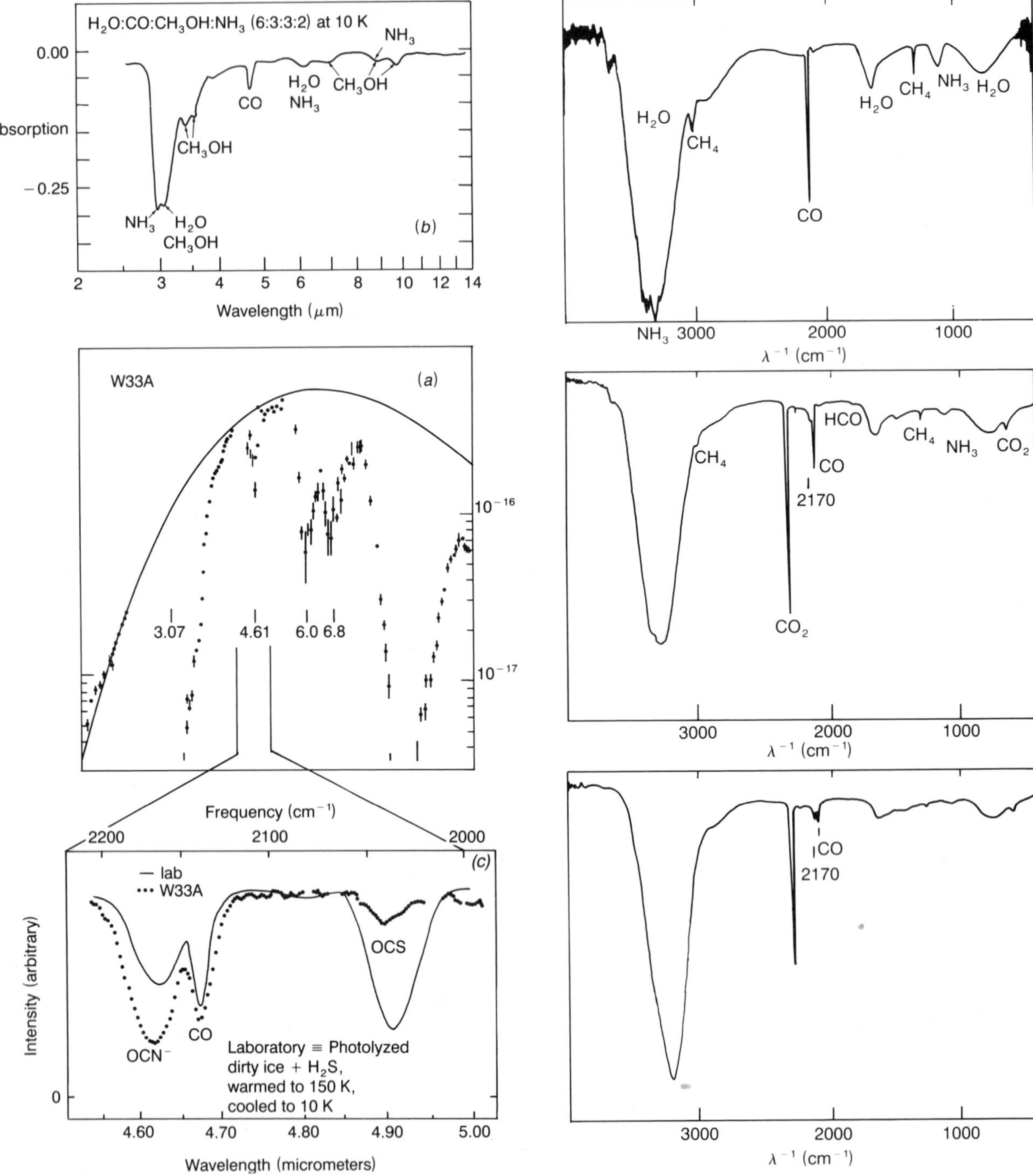

Figure 2. (*a*) Absorption spectra in an infrared source in W33A [*from Soifer, et al.*, Ap. J. Lett. ***232*** *L53 (1979)*]. (*b*) Comparison spectrum of a complex mixture in partial simulation of interstellar grain mantles (no silicates). (*c*) Laboratory spectrum of a frozen mixture of simple molecules, including H_2S, CO, and NH_3, which have been subjected to UV irradiation at 10 K and then warmed to 150 K. The spectrum of W33A is shown for comparison with its baseline flattened.

Figure 3. Spectra of a water-rich mixture (H_2O:CO:NH_3:CH_4 = 6:2:1:1). Upper spectrum, unirradiated 10 K; middle spectrum, photolyzed; lower spectrum photolyzed and warmed to 95 K. Compare the lower figure with Fig. 2*c*. Note immediate production of CO_2 by ultraviolet irradiation and the different degrees of evaporation of CH_4, CO, NH_3, and the narrowing of the H_2O ice band during warming.

COMPLEX ORGANIC MOLECULES

No H_2O ice absorption at 3.07 μm is observed in the diffuse cloud grains, where destruction conditions are harsh, even when the 9.7-μm silicate feature is quite strong. Because cosmic abundance arguments suggest that silicates (Si+Mg+Fe oxides) are insufficient to provide the observed amount of extinction, there must be *some* mantle on them even if no H_2O is present. The answer to this puzzle, which was first raised in the 1960s, is revealed in the photoprocessing of the ices. In the laboratory any initial mixture containing a carbon-bearing molecule will, after irradiation and warm-up, leave a residue (generally colored yellow) on the cold finger. Studies of these residues by such techniques as mass spectroscopy, liquid and gas chromatography, and infrared spectroscopy have shown that the residue ("yellow stuff") consists of a variety of complex organic molecules, some polymerized, in which carbon is the dominant atomic constituent in both aliphatic and aromatic combinations even when the initial mixture contains more oxygen than carbon. Mass numbers up to 1000 have been measured and the substance has been labeled organic refractory because it is, by interstellar standards, quite nonvolatile; some of it remains stable at temperatures up to 600 K and even higher. A relatively weak absorption (weak by H_2O absorption standards) that actually consists of several features appears at 3.4 μm corresponding to the C—H stretches in CH_2 and CH_3 groups. A similar absorption appears in the dust in diffuse clouds. In particular it has been well observed in the dust in the direction toward the galactic center and recently in diffuse clouds generally. Comparison of the 3.4-μm galactic center absorption with the strength of the laboratory 3.4-μm absorption leads to an estimate of the grain mantle thickness consistent with the wavelength dependence of the dust extinction characteristic of diffuse clouds. It turns out that the organic refractory dust component accounts for the major mass fraction and, in fact, about 1/2000 of the entire mass of the Galaxy ($\sim 10^8$ solar masses) is in the form of complex organics. Among the identified molecules has been an amino acid and other molecules of prebiological interest so that it is reasonable to believe that chemical evolution in interstellar space may ultimately be responsible for the origin of life.

EXPLOSIONS: SOLID AND GAS INTERCHANGE IN SPACE

The lifetime ($>10^7$ yr) of dense molecular clouds (hydrogen density $n_H = 10^4$ cm^{-3}) is generally much greater than the time it takes for a molecule to hit (and stick to) a dust grain. The mean time it takes for a gas phase molecule to encounter a dust grain, the *e*-fold accretion time scale, is $\tau = (n_d \sigma_d v_m)^{-1}$ where n_d and σ_d are the spatial number density and the cross section of the dust grains, and v_M is the molecular velocity. When $n_H = 10^4$ cm^{-3}, this time is of the order of 10^5 yr.

Why, then, are molecules most abundant in dense clouds? There must be some mechanism that desorbs them at a sufficient rate to achieve a balance with the sticking. The answer seems to be in the photoproduction and storing of free radicals frozen in the icy dust mantles that form and accrete in the molecular clouds. When an ice mixture is deposited and irradiated by ultraviolet photons, molecular bonds are broken so that, for example, $H_2O \rightarrow OH + H$ and $NH_3 \rightarrow NH_2 + H$, creating such radicals as OH, NH_2, CH_2, and so on. Some of the radicals react, but a steady state fraction of radicals remains. If the sample is heated rapidly a number of the radicals are released from their traps in the solid and many of them come into contact with each other. They then react with a burst of energy and if enough radicals react at the same time, the energy so generated heats the entire sample and releases all the remaining radicals to produce an explosive reaction that evaporates a substantial portion of the sample. In the laboratory only an extra 15 K of heating is enough to initiate the explosion. Even less is required in space *if* the triggering source is more impulsive. An inelastic collision of grains at a relative speed of > 50 m s^{-1} induced by cloud turbulence turns out to be an excellent trigger. Another one is the collision of a cosmic ray iron nucleus with a grain. Each of these types of collision occurs approximately once every 100,000 years per grain and, although this seems very infrequent, it provides sufficient molecular desorption on a sporadic basis to balance the continuous but slow accretion of molecules from the gas in a cloud.

Using the explosive desorption mechanism in combination with gas phase and dust surface chemistry leads to an explanation of one of the puzzles of gas and dust molecular abundances. It turns out that the H_2O observed as a solid molecule is produced *on* the dust and is not accreted from the gas. Thus, although H_2O is a major component observed in dust mantles (being *more* abundant than solid CO by a factor of at least 5), the gas phase H_2O is *less* abundant than CO by a factor of 10 or greater. Grain photochemistry is also needed to account for the abundances of methanol and carbon dioxide observed in the solid phase.

SOLID STATE ABSORPTION SPECTRA

The interpretation of observed infrared absorption spectra and the modeling of interstellar grains are intimately interconnected via the wavelength dependence of the grain materials as well as the size and shape of the dust. The basic data supplied by the laboratory are the wavelength dependence and strength of the absorption. From these the optical properties of the materials, that is, the real and imaginary parts of the index of refraction, may be derived from fundamental physical principles by means of the so-called Kramers–Kronig integral relations. In practice, one deals with incomplete information on one or the other, but it is generally possible to obtain excellent self-consistent values.

When the particle size (typified by a mean radius *a*) is small compared with the wavelength of an absorption feature so that $2\pi a/\lambda_{abs} \ll 1$, the laboratory absorption is a good indication of the interstellar dust absorption. However, even in this case, the *shape* of the particle may play a role in modifying the shape of the

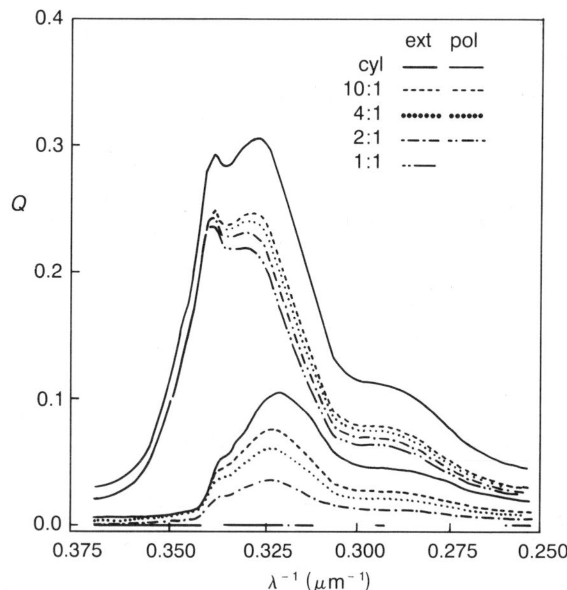

Figure 4. Extinction, polarization (where applicable), efficiencies by spheres, aligned homogeneous prolate spheroids, and circular cylinders made of H_2O:NH_3 = 3:1 ice at 12 K. Spheroids and spheres calculated in the Rayleigh (small size) approximation. Particle radii = 0.14 μm.

Table 1. Molecules Directly Observed in Interstellar Grains and / or Strongly Inferred from Laboratory Spectra and Theories of Grain Mantle Evolution.

Molecule	*Comment**		
H_2O	O		M
NH_3	O	I	M
NH_4^+	O		M
H_2S	O		M
CO	O		M
HCO		I	M
H_2CO	O	I	M
OCN^-	O		M
OCS	O	I	M
CO_2	O	I	M
CH_4		I	M
S_2		I	M
O_2		I	M
N_2		I	M
Complex Organic	O		M
"Silicate"	O		C
"Carbonaceous"	(O)		B

*O ≡ observed; M ≡ mantle; B ≡ small ball; I ≡ inferred; C ≡ core; parentheses indicate uncertain observations.

observed band. This is particularly true for the H_2O ice band at 3.07 μm, where, even though $2\pi a/\lambda$ is only ~ 0.2, there is a significant and importance difference between the bands produced by spherical and by elongated particles; both the position of the maximum absorption and the shape of the ice band depend on particle shape (see Fig. 4). This phenomenon is less important for longer wavelengths and for weaker bands.

Although the number of precisely identified molecules in dust is small compared with that for the gas, it already demonstrates a remarkably complex and interesting variety of chemical processes occurring in the solid state in space. Table 1 summarizes the current observational status of grain mantle composition. Some of the molecules shown have not yet been detected because of observational limitations imposed by the Earth's atmosphere but their presence is clearly implied by fundamental laboratory results when combined with theoretical calculations. For example CO_2 is impossible to detect from the ground because of atmospheric CO_2. But, as the laboratory spectrum shows (Fig. 2), it clearly must be made by photoprocessing of dust mantles. The reason solid H_2O can be seen is that its absorption peak is shifted by ~ 0.4 μm longward of the H_2O vapor absorption and into a relatively clear atmospheric window.

INTERSTELLAR DUST, COMETS, AND THE ORIGIN OF LIFE

Such basic problems as the origin of the solar system and the origin of life are two fields of direct application of the laboratory studies of interstellar grain chemistry. The prediction that a major fraction of comets is in the form of complex organic material was confirmed by the *Vega 1* and *2* and *Giotto* space missions to comet Halley. The photoprocessing of laboratory ices containing H_2S (which is observed in W33A) was used to show that the presence of S_2 as a parent molecule in comet IRAS-Araki-Alcock was almost certainly a consequence of the comet having been formed out of unmodified (i.e., unevaporated) interstellar dust. Based on this result we may assume that comets, in general, not only contain the most primitive constituents of the nebula out of which the solar system was born, but they also contain a direct sample of interstellar chemical evolution dating back 10 billion years, a figure obtained by adding the age of the solar system to the mean lifetime of an interstellar grain. Taken together with the impact history of planetary bodies, the very large organic fraction of comets provides a basis for suggesting that interstellar dust is the most likely source of prebiotic molecules that led to established life forms very early in the Earth's history ($\sim$ 3.8 billion years ago).

FUTURE APPLICATIONS OF SOLID STATE LABORATORY PHOTOCHOCHEMISTRY

With the advent of high resolution infrared space observations such as *SIRTF* (*Space Infrared Telescopic Facility*) and *ISO* (*Infrared Space Observatory*), the amount of spectral information on interstellar dust will increase by many orders of magnitude. Providing relevant laboratory information on photochemical and thermal modification of various ices will then become a major effort that will lead to the understanding of the evolutionary state of a host of astrophysically important regions in space.

Additional Reading

Bailey, M. E. and Williams, D. A., eds. (1988). *Dust in the Universe*. Cambridge University Press, Cambridge.

Bussoletti, E., Fusco, C., and Longo, G., eds. (1988). *Experiments on Cosmic Dust Analogues*. Kluwer, Dordrecht.

Greenberg, J. M. (1984). The structure and evolution of interstellar grains. *Scientific American* **250** (No. 6) 124.

Tielens, A. G. G. M. and Allamandola, L. J., eds. (1989). *Interstellar Dust: Contributed Papers*. Report NASA CP-3036, NASA, Washington, DC.

See also **Interstellar Clouds, Chemistry; Interstellar Medium, Dust Grains.**

Interstellar Medium, Dust Grains

John S. Mathis

Interstellar dust refers to the small solid particles that comprise about 1% of the mass of the interstellar medium, the rest being gas. Dust accounts for almost all the continuous opacity of the interstellar medium to radiation for wavelengths less than 912 nm (the ionization edge of neutral hydrogen). Dust is so opaque that visible-wavelength light is strongly attenuated along paths greater than about a kiloparsec in any particular direction in the galactic plane. Because of dust, starlight cannot penetrate the interiors of dense clouds, and therefore complex molecules can form. Grains provide the site of formation of the most abundant interstellar molecule, H_2, and possibly others as well.

Dust consists of tiny grains, distributed in sizes that range from those of single molecules up to about 300 nm. It contains most of the carbon and heavy elements associated with the interstellar medium. Its properties depend on both the compositions and size distributions of the grains.

The interstellar medium is clumped into clouds within the Galaxy, with a diffuse, low-density material between the clouds. The nature of the grains depends on their environment. We distinguish three forms of dust:

1. Diffuse dust, which lies in the diffuse interstellar medium typical of most of interstellar space.
2. Outer-cloud dust, found in the outer regions of the dark, cold molecular clouds (which are located in the plane of the Galaxy), but close enough to their edges so that there are no icy mantles on the grains.
3. Inner-cloud dust, found deep within molecular clouds and characterized by ices of water, ammonia, and other molecules in some cases.

We have few observations of inner-cloud dust, except for the absorption bands of the icy molecular mantles frozen onto the grains. Here we will discuss diffuse and outer-cloud dust, for which various observations exist.

We estimate the nature of interstellar grains by first assuming that various materials are present. Each material has a bulk index of refraction determined in the laboratory. Electromagnetic theory determines how a particle of given shape and size, with that index of refraction, will absorb and scatter light. (The size is important;

for instance, particles that are small in comparison to the wavelength do not scatter the light well, whereas larger particles will scatter efficiently.) We then assume a size distribution, calculate the absorption and scattering of that distribution, and compare their sum (called the extinction) to observations. We adjust the assumed composition and size distribution until the predictions and observations agree.

This process involves assumptions regarding both materials and size distributions, but there are constraints besides agreement with the extinction. We determine the extinction per hydrogen atom along the line of sight. The cosmic chemical composition is relatively constant throughout our local galactic neighborhood, and it provides the maximum amount of each element that we can assume is present in the dust. Our ideas should accommodate both the diffuse and outer-cloud dust in a natural way. The materials that we assume are in the grains must be produced in a plausible way by the sources of dust or else must be produced within the interstellar medium by some realistic processes.

The most important direct observations of dust are extinction measurements, both in spectral features and in the continuum, and in emission bands in the infrared. Other clues are found in the sources of dust (showing what solid materials are injected into the interstellar medium), the interstellar polarization, and the depletions of certain elements in interstellar gas (which are, presumably, in the solids).

THE OBSERVED SPECTRAL FEATURES OF SILICATES

The extinction law (which describes the amount of extinction as a function of wavelength) has broad spectral features at 9.7 and 18 μm that clearly show that silicate particles are present in the dust. These same features are sometimes seen in emission when the grains are heated by nearby stars, such as near the Trapezium stars in the Orion nebula. The profiles of the interstellar features are broader and smoother than those of terrestrial crystalline silicates, but are very similar to amorphous silicates produced in the laboratory. It is reasonable that there should be little crystalline structure in interstellar silicates, because any crystallinity originally present should be destroyed through the constant bombardment by cosmic rays in space. The silicate extinction bands are so strong that almost all of the silicon, and much of the iron and magnesium, must be contained in the silicate grains. In fact, such elements are, indeed, missing from the interstellar gas.

THE 217.5-nM "BUMP" AND INTERSTELLAR CARBON

The only other spectral feature in the extinction law of diffuse and outer-cloud dust is at 217.5 nm, usually referred to as the "bump" or as the "2200 Å extinction bump." It is so strong that only carbon, oxygen, nitrogen, silicon, iron, or magnesium, possibly in combination with hydrogen, are abundant enough to produce it. Practically every atom of the last three would be needed, even if the transition is as strong as possible (there is a fundamental limitation on the strength of an absorption feature produced by electronic transitions in a single atom). Silicates, known to be present, do not produce the bump in the laboratory. Oxygen and nitrogen can plausibly only be combined with hydrogen, but neither water nor ammonia ice has any features like the bump. In fact, both water and ammonia ice have strong absorption bands in the infrared (3.07 μm) that are not seen in diffuse and outer-cloud dust. From abundance considerations alone, we conclude that carbon, in some solid form, is responsible for the bump.

The bump must be produced by a strong resonance. The central wavelength of a resonance depends upon the material, the shape of the particle, and also the size, unless the radius of the grain is less than about 5% of the wavelength of the radiation. The central wavelength is always shifted to longer wavelengths for larger grains. Almost always the bump is observed to have its maximum extinction very close to 217.5 nm, so it must be caused by carbon particles of radii less than 10 nm, unless all lines of sight contain grains of the same size. Observations described later show that the latter is not the case.

Carbon exists in a variety of solid forms. A stable structure is a hexagonal benzene-ring arrangement, with carbon atoms at the vertices of the hexagon. Many rings can be linked together into sheets. A single such sheet of a few benzene rings, with hydrogen atoms at the edges, is called a polycyclic aromatic hydrocarbon molecule (PAH). Indeed, infrared emissions from dust (discussed subsequently) show that such molecules are almost surely present. However, small sheets of benzene rings can also be piled together in a disordered fashion, jumbled at rather random angles with respect to each other and forming a particle of comparatively large dimensions. Soot, formed in terrestrial flames, consists of such particles. The most ordered form of large solid carbon is graphite, consisting of completely ordered stacks of parallel planar sheets of benzene rings.

The optical properties of graphite and less ordered forms of carbon have been studied extensively in the laboratory. Graphite, when in particles that are small in comparison to 217.5 nm, has a resonance at the wavelength of the bump! The major doubts we have about identifying graphite as the source of the bump arise from those few stars that show the bump at other than 217.5 nm. Possibly some impurity or coating on the small graphite particles can cause the shift in these exceptional directions.

The resonance in graphite is so strong that only 10% of the carbon will produce the bump. The rest of the solid carbon is probably in a less ordered form, resembling soot, which is evidently produced easily in the carbon-rich atmospheres of certain stars. The soot helps produce the continuous extinction beside the bump.

A small amount of carbon is found as tiny (diameter about 2.6 nm) diamonds in primitive carbonaceous meteorites. It is not clear whether these diamonds are formed in the outer atmospheres of carbon stars and later ejected into the interstellar medium, later being incorporated into the meteorite, or whether the diamonds were formed in space during collisions of carbonaceous grains.

INFRARED EMISSION FROM DUST

Many dusty astronomical objects show fairly strong emission bands at 3.3, 6.2, 7.7, 8.6, and 11.3 μm. These bands are found with the appropriate strengths and wavelengths in the spectra of PAHs, which are similar to graphite except for having some attached H atoms on the edges. The sizes of the PAHs are probably distributed between values corresponding to about 20 and about 100 carbon atoms. The PAHs are mostly singly ionized in the diffuse interstellar medium. It is also possible that a slightly less ordered form of hydrocarbon is responsible for the emission bands.

There is also a red continuum emission, independent of the emission bands, in "reflection nebulae." It is produced where there is dense molecular gas that is excited by intense ultraviolet radiation. It is probably due to fluorescence of hydrogenated carbon in a form somewhat more disordered than PAHs.

DIFFERENCES IN DIFFUSE AND OUTER-CLOUD DUST; SIZE DISTRIBUTIONS

The extinction laws for diffuse and outer-cloud dust are different. Figure 1 shows the extinction (normalized to be the same at yellow light, wavelength 550 nm), plotted against the wave number, or reciprocal wavelength, for both types. Note that the curves differ systematically over the entire range of wave number. In fact, there is a family of extinction laws between the two that are plotted, and the distinction between diffuse and outer-cloud dust is rather arbitrary.

The differences in the diffuse- and outer-cloud-dust extinction laws are mainly caused by the size distributions rather than by the compositions. Small particles mainly cause extinction at small wavelengths (large wave numbers), and conversely for large grains.

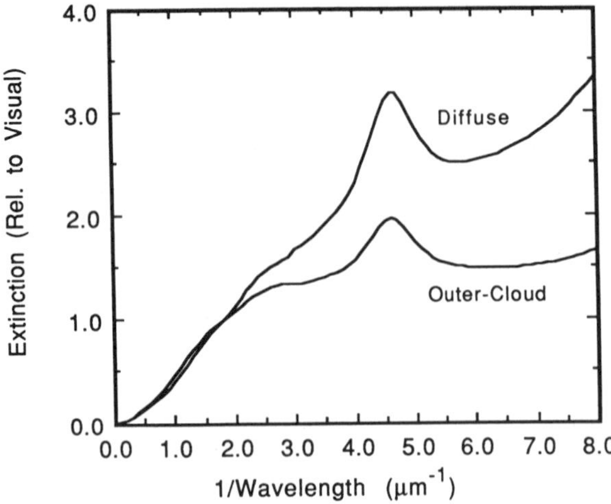

Figure 1. The interstellar extinction at various wavelengths, relative to the visual extinction (yellow light, or 550 nm), plotted against the reciprocal of the wavelength (μm^{-1}). The upper curve is appropriate to diffuse dust; the lower one to outer-cloud dust. The bump is shown in both curves at 4.6 $\mu m^{-1} = 1/(217$ nm). These are average curves, but the extinction curves determined for individual lines of sight follow them rather closely.

The figure shows that diffuse dust has more extinction at large wave numbers than does outer-cloud dust, so it must contain relatively more small particles. The bump occurs at precisely the same wavelength in both types of dust, but its strength (compared to the extinction in visible light) is smaller in outer-cloud dust. In outer-cloud dust, some of the small graphite particles responsible for the bump must be incorporated into the other grains or they are lost in some other manner.

This picture of a differing size distribution for diffuse and outer-cloud dust is corroborated by interstellar polarization. The maximum polarization occurs at larger wavelengths in outer-cloud dust, as would be expected if the average particles are larger.

The extinction is well fitted quantitatively by a power-law distribution of silicates and carbon, with the number density of particles having radii between the sizes a and $a+da$ given by $n(a)=$ constant $\times a^{-3.5}$, for sizes between a very small lower limit (a few nanometers) and about 250 nm. The size distribution for outer-cloud dust is more like $n(a)$ proportional to a^{-3}, which shifts the particle size distribution to the larger particles as required (see the preceding discussion). The same constituents (silicates and the various forms of carbon) can then explain both the diffuse- and the outer-cloud-dust extinction laws.

THE ORIGIN OF THE GRAINS

Probably the main sources of grains are giant stars whose atmospheres are so cool that solids can condense within them. These stars fall into two groups with quite different chemical compositions and types of grains, depending on whether carbon atoms are more numerous than oxygen or the reverse in the stellar atmospheres. In carbon-rich stars, the most likely solid to condense is poorly ordered elemental carbon, similar to soot. For oxygen-rich stars, silicates are observed to form. Thus, it is possible to understand how the silicate and carbon-rich grains originate. However, it is also clear that interstellar grains are also modified by the processing that occurs within the interstellar medium itself, after the grains have been ejected from the stellar atmospheres.

The picture presented here (of a power-law size distribution of amorphous silicates and carbon) fits the observations of interstellar dust well, but there is some controversy. The power-law distribution can be questioned, but there is general agreement that the size distribution is heavily weighted toward small particles. Tough mantles of organic materials, able to withstand the rigors found in the interstellar medium, are assumed by some. There is always plenty of mystery associated with any long-standing scientific question!

Additional Reading

Allamandola, L. J. and Tielens, A. G. G. N., eds. (1989). *Interstellar Dust. Proceedings of Symposium 135 of the International Astronomical Union*. D. Reidel, Dordrecht.

Bailey, M. E. and Williams, D. A., eds. (1988). *Dust in the Universe*. Cambridge University Press, Cambridge, U.K.

Martin, P. G. (1978). *Cosmic Dust*. Clarendon Press, Oxford.

Mathis, J. S. (1990). Interstellar dust and extinction. *Ann. Rev. Astron. Ap.* **28** 37.

Savage, B. D. and Mathis, J. S. (1979). Observed properties of interstellar dust. *Ann. Rev. Astron. Ap.* **17** 73.

See also **Interstellar Extinction, Galactic; Interstellar Medium; Interstellar Medium, Chemical Composition; Interstellar Medium, Chemistry, Laboratory Studies; Interstellar Medium, Dust, High Galactic Latitude; Interstellar Medium, Dust, Large Scale Galactic Properties; Molecular Clouds and Globules, Relation to Star Formation.**

Interstellar Medium, Dust, High Galactic Latitude

Loris Magnani

The interstellar medium (ISM) at high galactic latitudes ($|b| \geq 25°$) contains many low-mass gas and dust clouds that dim the light from background stars to varying degrees. This dimming, also called interstellar reddening or extinction, is produced by dust in the clouds and is measured in magnitudes. These clouds range in visual extinction from a few tenths of a magnitude for the more transparent diffuse clouds to several magnitudes for the opaque dark clouds. The clouds are identified more easily at high latitudes away from the confusing background of the interstellar matter in the galactic disk. The relative thinness of the ISM in the galactic disk ensures that clouds detected by star counts at $|b| \geq 25°$ will be local, no more than 100–200 pc from the Sun.

Before the all-sky infrared survey conducted by the *Infrared Astronomical Satellite* (*IRAS*), there were relatively few known dust clouds at high galactic latitudes. Although the more opaque objects are clearly visible on photographic plates taken during the Franklin–Adams 1911 sky survey and noted by Knut Lundmark and P. J. Melotte in 1926, it was not until the work of Beverly Lynds in the early 1960s that a complete, consistent survey of northern hemisphere dark nebulae became available. Of the 1801 dark clouds identified by her, only 19 are located at $|b| \geq 25°$. Because some of the 19 objects are related spatially, only 7 distinct groups or complexes were identified by Lynds at high galactic latitudes.

Attempts to identify less opaque dust clouds using galaxy counts and stellar reddening studies were for the most part unsuccessful because the projected distribution of sufficiently bright galaxies and stars is, in general, too low to delineate objects smaller than a few degrees. Nevertheless, in 1979 Jens Knude identified some 200 low extinction (≤ 0.15 magnitude) dust clouds in the northern galactic cap. A more successful indicator of high-latitude dust clouds is the faint, extended optical emission often seen on long exposure photographic plates. In 1965, Lynds compiled a catalog of these nebulosities, which are known as the Lynds bright nebulae. There are many more of the Lynds bright nebulae at high galactic latitudes than there are Lynds dark clouds. A correct interpretation of the optical emission from these objects was presented by Sidney van den Bergh and Allan Sandage who noted that some of these nebulosities could be dust clouds reflecting the integrated starlight of the galactic plane below them. In spite of the ability of reflection nebulae to trace the high-latitude dust, little further work on these objects was carried out. Although some optical spectra of luminous stars at high galactic latitudes showed traces of absorption lines due to intervening gas, the inability to sample more than a limited number of lines of sight prevented a comprehensive study of the clouds detected in this manner. For the most part, the ISM at high latitudes was believed to be sparsely populated by dust clouds prior to 1983.

This situation changed drastically in the fall of 1983 with the advent of the all-sky infrared survey by the *IRAS* satellite and the survey for high-latitude molecular clouds by Leo Blitz and co-workers.

IRAS CIRRUS

The *IRAS* database consists of an all-sky survey in the 12-, 25-, 60-, and 100-μm infrared bands with a resolution of ~ 2 arcmin and a sensitivity of ≤ 1 Jy (Jansky, a unit of flux density equal to 1×10^{-26} W m^{-2} Hz^{-1}). In the ISM, dust grains are heated by the absorption of ultraviolet photons from the interstellar radiation field. For a typical interstellar grain distribution, a substantial portion of this energy will be reradiated in the infrared. The 60- and 100-μm *IRAS* maps of the sky are ideal for revealing optically thin dust clouds uniformly heated by the interstellar radiation field. In 1984, the *IRAS* mission scientists announced the discovery of a large number of extended filamentary structures at high galactic latitudes especially prominent in the 60- and 100-μm bands. These objects were named "infrared cirrus" because of their morphological resemblance to terrestrial cirrus clouds.

It was soon clear that the infrared cirrus clouds were associated predominantly with atomic hydrogen clouds. Good correlations exist between the 100-μm flux of the cirrus and atomic hydrogen column density and extinction, implying that there exists little variation both in the gas-to-dust ratio of these objects and in the composition of the larger grains that produce the long-wavelength emission observed by *IRAS*. The 60/100-μm flux ratio, and thus the dust temperature, is remarkably constant over the various samples of infrared cirrus clouds that have been studied. The temperatures derived from these flux ratios, at least for the more diffuse clouds, are typically 20–30 K over the projected area of the cloud. This result implies that for these optically thin clouds there are no significant internal heating sources and the interstellar radiation field uniformly permeates and heats them. However, this situation does not hold for the denser, mostly molecular, dust clouds, where limb-brightening is seen in the 100-μm emission. This effect is probably produced by the inability of the ultraviolet field to reach and heat the dust grains in the dense, inner portions of the cloud that remain colder and less bright at 100 μm than the outer edges.

Although the above results were not surprising once the existence of a large population of diffuse dust clouds at high galactic latitudes was established, a major surprise was the significant flux of the cirrus clouds in the 12- and 25-μm *IRAS* bands. The 12/25-μm flux ratio indicates dust temperatures of hundreds of kelvins and, unlike the 60/100-μm ratio, the 12/25-μm ratio varies significantly from object to object. This result cannot be explained using standard grain models and it requires the existence of a small grain component. Graphite grains smaller than 20 Å can reach temperatures of several hundred kelvins upon absorbing a single ultraviolet photon. Grain size distribution models, which have been modified to include a significant component of small grains, can reproduce the 12- and 25-μm "warm cirrus" emission.

An alternative explanation for the 12- and 25-μm excess emission invokes the presence of line emission from molecules known as polycyclic aromatic hydrocarbons (PAHs). PAHs are large molecules composed of many rows of adjacent benzene rings with hydrogen atoms on some of the vacant sites at the edges. The idea that PAHs are responsible for the 12- and 25-μm emission is particularly attractive because it could also explain the presence of several unidentified infrared lines in the 3–12-μm wavelength range that appear in the spectra of regions such as reflection nebulae.

HIGH LATITUDE MOLECULAR CLOUDS

Independent of the discovery of the infrared cirrus, a significant addition to the galactic high-latitude gas and dust inventory was the discovery of numerous small molecular clouds at $|b| \geq 25°$. Although there were several known molecular clouds at high galactic latitudes prior to 1983, the large-scale carbon monoxide survey by Leo Blitz and co-workers revealed dozens of new objects. It soon became apparent that these objects were associated with some of the infrared cirrus, although the hydrogen gas in the great majority of the cirrus is atomic rather than molecular in form. An estimate of the fraction of the sky filled by the infrared cirrus is $\sim 20\%$, whereas the high-latitude molecular clouds cover $\sim 0.5\%$ of the sky. Despite this disparity, many interesting properties of the cirrus can be determined from the molecular clouds.

Foremost among these properties is the distance of the infrared cirrus from the Sun. Several independent methods to obtain the distance of the high-latitude molecular clouds yield estimates of a few hundred parsecs. The high-latitude molecular clouds and, consequently, the cirrus clouds are very local features of the interstellar medium, comfortably lying inside the atomic hydrogen disk of the Galaxy. We can assume that these small dust clouds are distributed uniformly throughout the galactic disk, but we can only see them locally, and then, only at latitudes far enough from the galactic plane so that the luminous emission of the interstellar matter of the Galaxy does not overwhelm them. It is important to remember that some of the cirrus features cover a large area of the sky because of their proximity to the Sun. The actual masses of typical clouds are several orders of magnitude less than those of giant molecular clouds.

The myriad of small clouds at high galactic latitudes provides us with a marvelous opportunity to better understand the interstellar medium in our Galaxy and in external galaxies. The proximity of these clouds to the Sun permits us to study *in detail* the structure, gas chemistry, energetics, and dust characteristics of small gas and dust clouds. By incorporating the small molecular clouds found at lower latitudes, a large enough sample of clouds of differing densities and opacities can be assembled and perhaps an evolutionary scheme can be developed linking the diffuse and the more dense clouds. Despite the fact that the bulk of the interstellar matter in the Galaxy is tied up in giant molecular clouds, studies of the evolution and origin of the small clouds are especially important if the large clouds are formed from agglomerations of small clouds. Understanding the infrared emission from these objects enables us to interpret the infrared energy distribution of external galaxies. Finally, an understanding of the large scale distribution of extinction and the overall transparency of the ISM at high latitudes is essential for optical studies of the distributions of external galaxies and, thus, is important in determinations of the extragalactic distance scale.

Additional Reading

Guhathakurta, P. and Tyson, J. A. (1989). Optical characteristics of galactic 100 micron cirrus. *Astrophys. J.* **346** 773.

Habing, H. J. and Neugebauer, G. (1984). The infrared sky. *Scientific American* **251** (No. 5) 48.

Hauser, M. G. (1988). Infrared cirrus: New light on the interstellar medium. *Ap. Lett. Commun.* **26** 249.

Johnson, H. M. (1968). Diffuse nebulae. In *Nebulae and Interstellar Matter*, B. M. Middlehurst and L. H. Aller, eds. University of Chicago Press, Chicago, p. 68.

Lynds, B. T. (1968). Dark nebulae. In *Nebulae and Interstellar Matter*, B. M. Middlehurst and L. H. Aller, eds. University of Chicago Press, Chicago, p. 119.

Puget, J.-L. (1987). Infrared cirrus. In *Comets to Cosmology. Lecture Notes in Physics* **297**, A. Lawrence, ed. Springer, Berlin, p. 113.

See also **Infrared Astronomy, Space Missions; Interstellar Clouds, Molecular; Interstellar Medium, Dust Grains.**

Interstellar Medium, Dust, Large Scale Galactic Properties

David Leisawitz

On a dark, moonless night, a hazy band of light interspersed with dark patches—the Milky Way—beckons our attention. The light comes from just the relatively luminous and nearby stars in our Galaxy; most of the 100 billion stars that illuminate the flat galactic disk are obscured by dusty interstellar clouds. The nearest of these clouds lie within a few hundred parsecs of the solar system and are seen as the dark patches in the Milky Way. Thus, even without optical aid, the casual observer can spot interstellar dust clouds and also unobscured regions of our Milky Way galaxy.

What is missing when we gaze at the Milky Way is the big picture. To probe the distribution and properties of interstellar dust on a *large* scale in the Galaxy requires a technique that accomplishes two seemingly conflicting objectives:

1. We must be able to *see through* the curtain of the nearby dust clouds.
2. At the same time, we must be able to *detect* the dust.

We will discuss each of these topics in the following sections.

HOW DO WE STUDY THE GALACTIC DUST?

Fortuitiously, the part of the spectrum in which interstellar dust is most radiant is a wavelength region to which the dust clouds are translucent, namely the infrared. These facts were anticipated from theoretical arguments long before infrared telescopes, which now play a vital role in our understanding of interstellar dust, were built. To estimate the infrared brightness of an interstellar dust cloud, one must know approximately how warm the dust is and, to estimate the dust temperature, in turn, one must have some idea of the size of a typical dust particle. In his classic exposition on the effects of small particles on light, the Dutch astronomer Hendrik C. van de Hulst summarized several arguments that suggest that the diameter of an interstellar dust grain is about equal to the wavelength of visible light. For example, in 1930 the American astronomer Robert Trumpler observed that the stars in distant open clusters in the Galaxy appeared redder than similar stars in nearby open clusters. This was explained as due to small dust particles that scatter and absorb blue light much more than red light, and thus make the stars seem redder than they are. The dust particles were located in interstellar space between the solar system and the star clusters. To understand the preferential scattering of blue light, consider that light characterized by a wavelength long compared to the size of a dust grain will more easily pass by the grain unattenuated than will light with a wavelength smaller than that of the grain; the short wavelength light will have a greater tendency to be diffracted (this is called Rayleigh scattering after the English physicist Lord Rayleigh [John William Strutt]) or absorbed by the grain. Thus, if the interstellar dust grains are smaller than the size of visible-wavelength light (say a few hundred nanometers), the red light (wavelength ~600–800 nm) from a star will more easily reach our telescopes unimpeded by intervening dust than will the blue light (wavelength ~200–450 nm), giving rise to the appearance that the starlight has been reddened. (Incidentally, the same principle explains why the sky is blue and sunsets are red, although those effects are due to light scattering by air molecules as well as airborne dust.) Trumpler found that the more distant star clusters, which presumably have more intervening dust, are more severely affected by reddening.

The temperature of an interstellar dust grain can be calculated by balancing the rate at which it absorbs energy against the rate at which energy is lost. A typical dust grain in the interstellar medium is heated by the absorption of short wavelength light (blue and ultraviolet). The light intensity to which a grain is exposed is a function of the grain's proximity to the light source (a star) and how much other dust there is absorbing and scattering the light along the path between the star and the grain under consideration. Other details, such as the chemical composition and shape of the grain, influence how efficiently the available energy is absorbed. Dust grains lose energy by radiating light and do so at a rate that depends most strongly on their temperature. The hotter the grains, the greater the rate at which they radiate energy. In fact, the heated grains behave in a manner analogous to the familiar glow of an iron rod in a fire: the hotter the metal, the brighter (and more yellow, or even blue) the glow. One can solve for the temperature of interstellar grains by equating the rate of energy gain to the rate of energy loss. In principle, both rates depend on the grain size, but the dependences are similar and, consequently, the dust temperature turns out to be only weakly dependent on grain size. For a typical grain not particularly close to a star but heated by the dilute radiation field of the ensemble of stars in the Galaxy, the equilibrium temperature estimated in this way is about 20 K. Grains within a few parsecs of hot, luminous stars can reach temperatures greater than 100 K. At some critical distance that depends on the temperature and luminosity of a star, the grain equilibrium temperature exceeds the melting point (which depends on the grain composition) and dust closer to a star than that distance is volatilized.

Just as an iron rod in a fire glows red, orange, or yellow depending on its temperature (which may be a few thousand kelvins), the glow of the much cooler interstellar dust grains is "redder than red," or infrared. In 1894, the German physicist Wilhelm Wien derived a formula for the wavelength λ at which a black body (an ideal, perfectly absorbing mass) of temperature T is brightest:

$$\lambda = 0.51/T,$$

where T is in kelvins and λ is in centimeters. From the estimated dust grain temperature, it is a simple matter, using Wien's law, to calculate that the interstellar grains should be brightest at a wavelength between 10 and 100 μm.

Note that the wavelength that is most strongly emitted by interstellar dust grains is much larger (nearly 1000 times) than the characteristic grain size. Therefore, most of the light *emitted* by the dust can easily penetrate and escape from the dust clouds, unlike the visible (especially blue) light that heats the grains.

In principle, it should be possible to study the distribution of dust throughout the Galaxy with an infrared telescope. But two complications arise. First, the Earth's atmosphere is a strong absorber of infrared radiation, primarily due to the atmospheric water vapor and carbon dioxide. This makes ground-based infrared observation virtually impossible from most locations except in a few narrow wavelength "windows." Second, there is a more fundamental limitation because several interstellar dust clouds may be present at different distances along the same line of sight, and there may be no way to distinguish between them. (We observe the sum of the infrared emission from all clouds in a given direction, and the warmest ones contribute the most radiation.)

The first of these two obstacles was surmounted by placing infrared telescopes on the tops of tall mountains, in balloon payloads, in jet aircraft, and finally in space. In 1983, the *Infrared Astronomical Satellite* (*IRAS*), gave astronomers their first complete view of the universe at infrared wavelengths.

Dust is but one component of the interstellar medium. In fact, about 100 times as much matter is present by mass in the form of gas (ions, atoms, and molecules) as in the form of dust, and most of the gas is hydrogen. In general, because the gas and dust are present in a fairly constant mass ratio, any observation that traces the distribution of gas in the Galaxy provides information about the dust distribution as well. Radio and millimeter-wave spectral line observations of the interstellar gas provide kinematic information that can be used, in some cases, to determine that several clouds lie in a particular direction and to estimate their distances. For example, clouds at different distances along the same line of sight may have different radial velocities with respect to the Earth that depend systematically on their position in the Galaxy.

To summarize, it is fair to think of an interstellar dust grain as a tiny device that reflects and absorbs blue light and emits infrared light. Because infrared light can penetrate an interstellar dust cloud with little attenuation, infrared telescopes can be used to study the galactic dust distribution. The infrared brightness detected in a particular direction depends not only on how much dust is present in the line of sight, but also on the dust temperature.

WHAT DO WE KNOW ABOUT THE GALACTIC DUST?

Astronomers have found that in some ways the interstellar medium resembles a living organism. Just as an organism may be composed of organs, organs consist of cells, cells are made of smaller units, and so on, galaxies contain stars and interstellar matter, the interstellar medium is composed of clouds, and the clouds have internal structure with small clumps inside bigger clumps. This kind of organization is called hierarchical structure, and the symbiotic and competitive behaviors that can be found in a biological ecosystem have astrophysical analogues: Some interstellar clouds, largely because they are dusty, are found to cool, collapse, and form stars. Stars produce the chemical elements of which dust grains are made and some stars even make dust; stars heat, sweep up, and destroy interstellar clouds. Now we consider the spatial distribution of the dust, the role played by dust in the galactic ecosystem, the composition of the dust, and the utility of dust as a probe of energy density in the interstellar medium.

Where is the Dust?

As already mentioned, dust amounts to a mere 1% of the mass of the interstellar medium in the Galaxy and is present approximately in that proportion to the dominant component: gas. So to answer the question "Where is the dust?," what one really needs to know is where the gas is.

Picture the Galaxy: Its flat disk component is a few hundred parsecs thick and about 15,000 pc in radius (about the shape of a dish), and the disk is surrounded by a spheroidal "halo" that extends at least 1000 pc in each direction away from the plane of the disk. Virtually all of the interstellar medium by mass and all the massive, young stars, are in the disk. The small amount of gas and dust in the halo is there for a short visit, either because it is falling into the disk from intergalactic space, accelerated by the Galaxy's gravity, or because it was ejected from the galactic disk by supernova explosions and the energetic winds of massive stars. In the disk, about 50% of the gas is atomic, most of the remainder is molecular, and a small fraction (less than 1%) is ionized. In terms of the fraction of the disk volume occupied by each gas component, the order is reversed: Most of the galactic disk contains tenuous ionized gas and only a small but uncertain fraction is filled by relatively dense clouds of neutral atoms and molecules. Because the dense clouds are opaque to starlight, the fraction of the disk volume that they occupy and their distribution relative to the luminous stars determines to a large degree how much of the starlight escapes from the galactic disk without being reprocessed by dust into infrared emission (see the last section, What else can the dust tell us?). The total mass of interstellar matter in the Galaxy is about 6 billion solar masses, so the total mass of the dust is about 60 million solar masses.

The interstellar clouds range in size from ~ 1 pc to ~ 100 pc and in mass from a few solar masses to more than 1 million solar masses. The most massive clouds delineate a spiral pattern in the galactic disk as do such clouds in the external spiral galaxies that are near enough to be resolved with present telescopes. Massive stars, that form in, heat, and destroy these clouds, illuminate the "spiral arms" and cause them to be prominent features of galactic structure. Small molecular and atomic clouds are found throughout the disk of the Galaxy. From our vantage point in the disk, as we look in directions away from the plane of the Galaxy, we detect some of these clouds because of the starlight that they reflect and the infrared emission from the dust that they contain; the latter signature, and their sometimes diffuse appearance, has earned these clouds the appellation "infrared cirrus." Naturally, whenever stars are present inside or near interstellar clouds, the clouds are heated and the dust that they contain glows brightly in the infrared.

What Role Does Dust Play in Interstellar Processes?

Despite its small contribution to the mass of the interstellar medium, dust plays a critical role in interstellar processes. Without dust there could be no molecular clouds; without these clouds, stars could not form; and without nucleosynthesis in stars, there could be no elements heavier than hydrogen and helium (hence no dust, and hence no life). In short, the Galaxy would be drastically different were it not for the dust.

There are two ways in which dust is influential in the formation of interstellar clouds and stars. First, dust shields the interiors of dense clouds from ultraviolet starlight, enabling the cloud centers to become quite cold (about 10 K). With correspondingly little thermal pressure for support, the collapse of the cloud in its own gravitational field, leading ultimately to star formation, is inevitable. But shutting off the heat source is not a sufficient condition for the cloud interior to cool: the temperature is controlled by a balance of energy input *and output*, not unlike the thermal balance of a dust grain. The gas in a molecular cloud cools primarily through the spectral line emission of its molecules (especially carbon monoxide), and we believe that chemical processes that occur most rapidly on the surfaces of dust grains are important in the formation of these coolants.

At the surfaces of clouds heated by nearby stars, and also in the interiors of clouds with embedded stars, the dust, warmed by absorbed starlight, may heat the gas. The dust grains collide with the molecules and give up some of their internal energy (heat), thereby increasing the average speed of the gas particles (in other words, raising the temperature of the gas).

What is the Dust Made Of?

Thus far we have said very little about the composition of the dust grains other than to mention that they are made of elements that were manufactured in stellar nucleosynthetic processes. Although the interstellar grain composition is a subject of continuing debate among astrophysicists, it is widely believed that some mixture of graphite, silicates, and amorphous carbon particles is present in space. The *IRAS* 12- and 25-μm wavelength surveys, made in 1983, led to the discovery that not all interstellar dust grains are in thermal equilibrium. A ubiquitous component of abnormally "hot" grains was found. The new component is attributed to grains that are so small that they absorb light one photon at a time and thereby are raised for an instant to a very high temperature. Then the tiny grains cool rapidly and remain very cold until struck by other photons. While hot, the small grains glow brightly at 12 μm; while cold, they produce no detectable radiation. Their composition remains undetermined, but they may well be similar to the soot particles found in industrial exhaust. Knowledge of the composition of interstellar dust could provide valuable information about interstellar chemical processes and the cosmic abundances of the elements.

What Else Can the Dust Tell Us?

Interstellar dust enables us to probe the radiation energy density throughout the Galaxy. The dust at a particular location is heated to a temperature that depends on the radiation it absorbs; then it emits infrared radiation with a spectral shape that corresponds to its temperature. By observing the dust at at least two infrared wavelengths, it is possible to derive the shape of the spectrum and hence the dust temperature and the radiation field to which the dust is exposed. From *IRAS* observations, we have learned that most of the interstellar dust, specifically the dust in diffuse atomic clouds and at molecular cloud surfaces, is heated by the interstellar radiation field to a nearly constant temperature of 24 K, but that numerous "hot spots" coincide with the ionized regions immediately surrounding massive stars, to stars inside dense molecular clouds, and to the warm surfaces of clouds heated externally by nearby stars. The energy density of the interstellar radiation field in the solar neighborhood is about 0.44 erg cm^{-3}.

The total infrared luminosity of the Galaxy is about 14 billion solar luminosities. This is approximately one-third of the power radiated by stars in the disk of the Galaxy because much of the starlight is intercepted by dust grains somewhere in the interstellar medium. The Milky Way seen in visible light constitutes the small fraction of the starlight that is not intercepted by dust.

Additional Reading

Aanestad, P. A. and Purcell, E. M. (1973). Interstellar grains. *Ann. Rev. Astron. Ap.* **11** 309.

Beichman, C. A. (1987). The *IRAS* view of the galaxy and the solar system. *Ann. Rev. Astron. Ap.* **25** 521.

Greenberg, J. M. (1978). In *Cosmic Dust*, J. A. M. McDonnell, ed. Wiley, New York, p. 187.

Greenberg, J. M. and van de Hulst, H. C., eds. (1973). *Interstellar Dust and Related Topics*. Reidel, Dordrecht.

Lynds, B. T., ed. (1971). *Dark Nebulae, Globules, and Protostars*. University of Arizona Press, Tucson.

Rowan-Robinson, M. (1989). Interstellar dust in galaxies. In *The Interstellar Medium in Galaxies*, H. A. Thronson, Jr., and J. M. Shull, eds. Kluwer Academic Publishers, Dordrecht, p. 121.

Savage, B. D. and Mathis, J. S. (1979). Observed properties of interstellar dust. *Ann. Rev. Astron. Ap.* **17** 73.

van de Hulst, H. C. (1957). *Light Scattering by Small Particles*. Wiley, New York.

See also **Interstellar Extinction, Galactic; Interstellar Medium, Dust Grains; Interstellar Medium, Dust, High Galactic Latitudes.**

Interstellar Medium, Galactic Atomic Hydrogen

William L. H. Shuter

The study of neutral atomic hydrogen, denoted H I, is important in astronomy because it is the most abundant element observed. Together with helium, the next most abundant element, it serves as a building block for creating all the heavier elements, a process that occurs in stellar interiors. In the state of lowest energy of the hydrogen atom, there is a slight excess of energy when the spins of the proton and electron that constitute it are parallel, compared to the case when they are antiparallel. Therefore, when the spins flip spontaneously, from being parallel to antiparallel, the excess energy is radiated as a photon with a frequency of 1420.405751769 MHz (corresponding to a wavelength of about 21 cm). Following the suggestion made by Hendrik C. van de Hulst in occupied Holland in 1944, this transition was first detected in the disk of the Galaxy by Harold I. Ewen and Edward M. Purcell at Harvard University in 1951. An analogous transition at a frequency of 327 MHz is known to occur in the isotope deuterium. However, despite considerable effort, this transition has not as yet been detected unambiguously by astronomers.

The detection of this first radio frequency spectral line was greeted with great enthusiasm, because it opened up a new window on the Galaxy. Previous optical studies of the stars surrounding the solar system had a horizon in the Milky Way limited to about one-third the distance from the Sun to the galactic center. This distance limit resulted from dust grains in the interstellar medium, which being of comparable size to optical wavelengths were highly efficient in scattering and absorbing starlight. However, the 21-cm line radiation, having a much longer wavelength, was essentially unaffected by the presence of interstellar dust and allowed astronomers, for the first time, to "see" right through the Galaxy. Astronomers in Leiden and Sydney collaborated to produce and interpret the first complete map of the Milky Way. These early studies yielded some fascinating results. The region of the Galaxy within a radius of 4 kpc from the galactic center was found to be highly complex and difficult to interpret. One surprising feature found was a spiral arm, at a galactocentric radius of about 3 kpc, that appeared to be expanding away from the center with a velocity of 53 km s^{-1}. Outside this radius, however, it appeared that most of the hydrogen was confined to a disk about 200 pc thick, and was rotating about the center in orbits that were very close to being circular. These studies suggested that a considerable fraction of the hydrogen in the galactic disk was concentrated into spiral arms, but locating these precisely proved to be difficult and controversial, because there was no accurate and unambiguous method of determining the distance of the emitting hydrogen. It was also found that the disk of the Galaxy was somewhat misaligned with the system of galactic coordinates in use at that time, which was based on prior optical observations. These studies led to the adoption of a new system of galactic coordinates, aligned with the plane of the disk as defined by the 21-cm line observations. The outer regions of the disk were found to be warped; although the warp was believed initially to be due to tidal effects produced by the Magellanic Clouds, its cause still has not definitely been established. In addition to these early studies regarding the large-scale structure of the Galaxy, noteworthy results of this period relating to the interstellar medium were the determination of the "average" temperature of the H I gas to be about 120 K, and the discovery that the H I column density in tenuous interstellar clouds was highly correlated with the optical extinction produced by dust grains.

It is also possible to observe the H I line in absorption. Absorption lines are detected when a cloud of cold atomic hydrogen intercepts the beam of a radio telescope directed at a more distant radio source that is emitting continuum radiation. In this case, photons from the radio source induce some of the cloud's hydrogen atoms that have their proton and electron spins antiparallel to flip to the state in which the spins are parallel, absorbing the photons in the process. The same process occurs if a cloud of cold hydrogen is in front of a hotter one, in which case the line is referred to as being self-absorbed.

After these pioneering studies, which took place 30 or more years ago, the field has evolved into five major areas of specialization. In each subfield the hydrogen line is observed in order to obtain information regarding the structure, kinematics, and physical properties (such as density, temperature, and magnetic field) in the region studied. These subfields concern the study of the following:

The large-scale structure of the Galaxy
The galactic center
The solar neighborhood
The interstellar medium
High-velocity clouds

Progress in each will be discussed briefly in the rest of this entry.

The early surveys of the large-scale structure of the Galaxy have been augmented considerably by more recent ones of the galactic plane and high galactic latitudes using larger telescopes that provide higher angular resolution, are equipped with more sensitive receivers, and have multichannel spectrometers with better frequency resolution. A particularly important result, derived from these data in 1979, was the demonstration that over most of the radial extent of the galactic disk, the rotational velocity of H I gas was essentially constant at 220 km s^{-1}, a value similar to that observed in external Sb and Sc spiral galaxies.

The central region of the Galaxy is unusual in comparison to external spiral galaxies, in that there appears to be a pronounced deficit of atomic hydrogen between the nuclear region and the "expanding" spiral arm at a radius of 3 kpc. The kinematics of this region is complex, but studies to date tend to suggest that most of the features in this region, including the "expanding" spiral arm, can be accounted for satisfactorily if it is assumed that the bulk of the atomic hydrogen is configured in the form of a bar.

Studies of the solar neighborhood have shown that there appears an inflow of hydrogen gas from the galactic poles, matched by a corresponding outflow approximately in the center–anticenter direction. After correction for this effect, the kinematics of the local hydrogen seem to be quite closely similar to those of the young O and B stars. They are consistent with circular motion about the galactic center, except for anomalies in the nearby system of bright stars known as the Gould Belt. Further, the solar system appears to be situated in a "bubble" of radius 100 pc, presumably the result of a supernova explosion, in which the hydrogen density is well below the average value in the disk of about 1 atom per cubic centimeter.

The field in which possibly the most progress has been made since the pioneering days, has been the study of the interstellar medium. It had been known that the correlation between H I column density and visual extinction, A_V, broke down at higher values of A_V. However, the relationship was subsequently shown to remain valid, down to the limits of telescope sensitivity, provided that the atomic hydrogen column density was replaced by *total* hydrogen column density, thereby taking into account the presence of atomic hydrogen in molecular form in regions of high density. This work was based on millimeter-wave observations of the carbon monoxide molecule (CO), discovered by workers at Bell Laboratories in 1970. The temperature distribution of interstellar hydrogen, observed as 21-cm line radiation, has been shown to range from about 10 to 8000 K (at which point it becomes ionized), with a definite peak at 80 K. This information has been obtained by combining data in the same region on H I emission (which to a first order gives the density only) and absorption (which gives the ratio of density to temperature). An interesting modern development has been the measurement of the magnetic field in interstellar H I clouds using the Zeeman effect on the 21-cm line of hydrogen. After 10 years of fruitless attempts by many astronomers, the first detection of this effect was accomplished at the National Radio Astronomy Observatory in 1969. It is now known that in H I clouds the magnetic fields generally lie in the range 5–20 microgauss.

The H I high-velocity clouds were discovered in 1963. They are characterized as having a line-of-sight velocity at least 80 km s^{-1} deviant from the values expected on the basis of the kinematics of the differentially rotating disk of the Galaxy. Even since the details of their distribution have been mapped and evaluated, their origin remains enigmatic, mainly because their distances are not known. One particularly striking component is the Magellanic Stream, which was detected initially at Bell Laboratories in 1972 and subsequently traced over an arc of about 100° (right up to its intersection with the Magellanic Clouds) through observations made at Parkes, Australia in 1974. The current belief is that it was produced by a tidal interaction between the Galaxy and the Magellanic Clouds.

An exciting new field in astronomy, which was spawned by the early 21-cm line observations, is SETI (the Search for ExtraTerrestrial Intelligence). In 1959 the intriguing suggestion was made that the optimum frequency for the detection of signals of extraterrestrial origin would be 1420 MHz, in the immediate vicinity of the H I line. Since that time a number of searches for such signals have been initiated.

Additional Reading

Brinks, E. (1989). The cool phase of the interstellar medium: Atomic gas. In *The Interstellar Medium in Galaxies*, H. A. Thronson, Jr. and J. M. Shull, eds. Kluwer Academic Publishers, Dordrecht.

Burton, W. B. (1988). Structure of our Galaxy from observations of H I. In *Galactic and Extragalactic Radio Astronomy*, 2nd ed., G. L. Verschuur and K. I. Kellerman, eds. Springer-Verlag, Berlin, p. 295.

Kerr, F. J. (1969). The large-scale distribution of hydrogen in the Galaxy. *Ann. Rev. Astron. Ap.* 7 39.

Kondo, Y., Bruhweiler, F. C., and Savage, B. D., eds. (1984). *Local Interstellar Medium*. NASA Conference Publication 2345.

Kulkarni, S. R. and Heiles, C. (1988). H I and the diffuse interstellar medium. In *Galactic and Extragalactic Radio Astronomy*, 2nd ed., G. L. Verschuur and K. I. Kellerman, eds. Springer-Verlag, Berlin, p. 95.

Liszt, H. S. (1988). The galactic center. In *Galactic and Extragalactic Radio Astronomy*, 2nd ed., G. L. Verschuur and K. I. Kellerman, eds. Springer-Verlag, Berlin, p. 359.

Morrison, P., Billingham, J., and Wolfe, J. (for NASA) (1979). *The Search for Extraterrestrial Intelligence*. Dover, New York.

van Woerden, H., Schwarz, U. J., and Hulsbosch, A. N. M. (1985). Highlights of high-velocity clouds. In *The Milky Way Galaxy*, R. J. Allen and W. B. Burton, eds. D. Reidel, Dordrecht, p. 387.

See also **Galactic Structure, Interstellar Clouds; Galactic Structure, Large Scale; Interstellar Medium, Galactic Molecular Hydrogen.**

Interstellar Medium, Galactic Center

Mark Morris

Within a distance of about 500 parsecs (pc) of the nucleus of the Milky Way galaxy, the interstellar medium contains an enormous reservoir of gas, mostly in molecular form. It is estimated that 10^8 solar masses of gas are present in the inner 500 pc, implying that the gas in this region is more concentrated than anywhere else in the Galaxy (Fig. 1). Thus, the gas mimics the stars, which are also centrally concentrated. However, in the vicinity of the galactic center, the interstellar medium differs markedly from that in the galactic disk; it is a violent place where high-velocity clouds are commonplace, where the clouds are unusually warm and turbulent, and where the magnetic field is far stronger than it is further out in the disk of the Galaxy. Investigating the phenomena at smaller and smaller distances from the nucleus, one finds that all of these characteristics become increasingly pronounced.

The thick column of interstellar dust lying between the Earth and the galactic center can be penetrated by only one visual photon

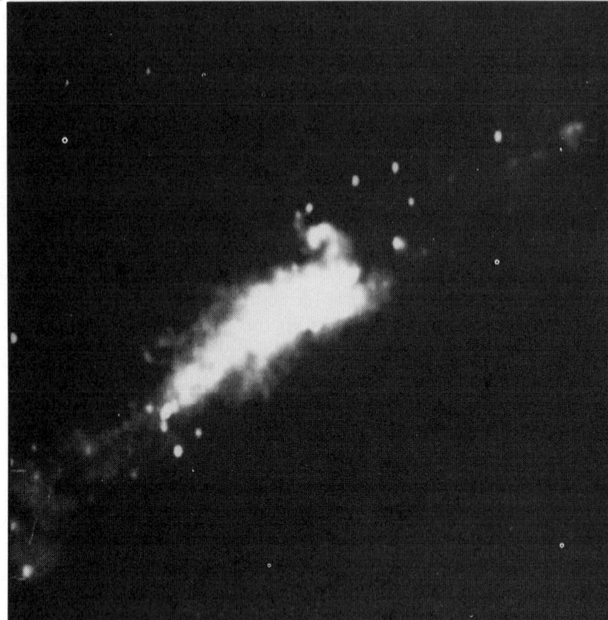

Figure 1. Image of far-infrared emission at 25 μm wavelength from dust in the galactic center cloud population. The dust emission is a good tracer of the distribution of gas. The size of the field is 6° × 6°.

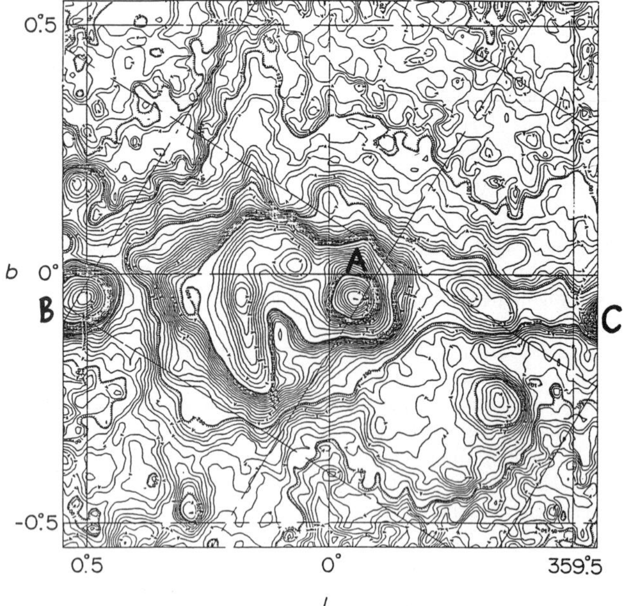

Figure 2. Contour map of radio continuum emission at 6 cm wavelength, showing the three prominent radio sources, Sagittarius A, B, and C. The map was produced by Yoshiaki Sofue and co-workers at the Nobeyama Radio Observatory.

out of every trillion. Consequently, the galactic center was not accessible to classical optical astronomy. The exploration of the galactic center region has almost all taken place since the relatively recent advent of infrared and radio astronomy, which utilize wavelength domains in which the intervening interstellar medium is essentially transparent. Fortunately, these are the domains in which the cool interstellar medium is best studied, so the picture we have of the interstellar activity near the galactic center is now becoming quite detailed.

PROMINENT RADIO SOURCES

The brightest three radio sources in the vicinity of the galactic center were identified in the 1950s and are still well known by their original names: Sagittarius A, B, and C (see Fig. 2). In the 1960s, low-resolution images made in the microwave continuum confirmed that the brightest of these, Sgr A, coincides with the dynamical center of the Galaxy. In the 1970s, radio astronomers found that Sgr A is actually comprised of two components: a thermally emitting radio source (Sgr A West), now known to be centered on the dynamical and luminosity center of the Galaxy, and a nonthermal radio source (Sgr A East), which is now known to lie behind the nucleus. Sgr A East is a hollow elliptical source that may be the remarkable juxtaposition of an unrelated supernova remnant upon our line of sight to the galactic center. It is also possible that Sgr A East resulted from an explosion at the nucleus, although its placement relative to the nucleus would not then be understood. Sgr B is also divided into two H II regions, Sgr B1 and the more famous Sgr B2. Sgr B2, located at a projected distance of 120 pc from the nucleus, is associated with a massive molecular cloud of the same name. Its mass and average density, which have been determined from measurements of centimeter- and millimeter-wavelength spectral lines of molecules, are larger than those of any other known molecular cloud in the Galaxy. Indeed, because of its very large column density of matter, the emission lines of almost every known interstellar molecule can be found in Sgr B2. The presence of compact H II regions and of H_2O and OH masers in both Sgr B2 and Sgr C indicates that star formation is occurring there. Closer to the nucleus, and particularly in the molecular cloud associated with Sgr A, there is no clear evidence for star formation. It has been suggested that the process of cloud collapse to form stars is inhibited in the innermost 50 pc of the galactic center by strong magnetic fields and by tidal forces in the presence of the central mass concentration.

CHARACTERISTICS OF THE CLOUD POPULATION

One of the most remarkable characteristics of the galactic center cloud population is that many of the clouds have velocities deviating strongly from those expected for circular orbital motion about the galactic center. Radio observations made of the spectral lines of CO, H_2CO, OH, NH_3, and CS suggest that these clouds are participating in a general expansion away from the nucleus. Following an interpretation originally published by Nicholas Scoville in 1972, it is now widely believed that the expanding system of clouds forms a ring tilted by about 20° with respect to the plane of the Galaxy, 170 pc in radius, and expanding at 150 km s^{-1}. It is difficult to escape the conclusion that this expanding molecular ring was set in motion by a very powerful explosion occurring at the nucleus about a million years ago. The hypothesis that such explosions occur from time to time in the galactic center helps to account not only for the observed variety of cloud motions, but also for the 200-pc long tongues of radio-emitting gas that rise out of the plane of the Galaxy near the nucleus. These radio features form the "galactic center lobe," which has been interpreted as the surface of a "chimney" of gas created by a rising blast from a central explosion.

Two more striking characteristics of the galactic center clouds are the large magnitude of their internal velocities and their high kinetic temperatures. Spectral lines observed in molecular clouds near the Sun typically have Doppler widths of 2 or 3 km s^{-1}, whereas in the galactic center, the line widths are about an order of magnitude larger. Such larger internal velocity dispersions may reflect an internal turbulence that happens to be violent enough to prevent these particularly dense and massive clouds from collapsing under their own self-gravity. Alternatively, the velocities corre-

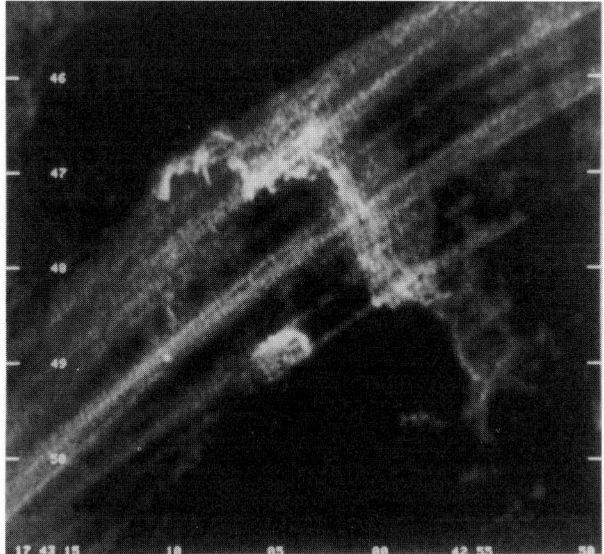

Figure 3. Radio filaments of the galactic center Arc, as observed with the Very Large Array at a wavelength of 6 cm. The sickle-shaped feature crossing the filaments marks the location where the filaments interact with gas in the galactic plane.

spond to those of magnetohydrodynamic (MHD) waves, or Alfvén waves, in a region of large magnetic field strength.

The temperature of the galactic center clouds is uniformly high, 50–70 K, as compared to a typical temperature of 10 K for clouds out in the galactic disk. The high density of stars may contribute a bit to the heating of the galactic center clouds, but it is not sufficient. The dominant heating mechanism has not yet been clearly established; it may be linked to the interaction of the clouds with the strong magnetic field in the galactic center region: As a cloud moves through the magnetic field, or as MHD waves propagate through the cloud, the energy of the cloud or of the waves is dissipated as thermal energy.

THE MAGNETIC FIELD

Much of our understanding of the interstellar medium near the galactic center depends on our knowledge of the magnetic field there. In 1984, Farhad Yusef-Zadeh and co-workers published the first of several pieces of evidence indicating that the galactic center magnetic field is strong and highly ordered. Observing a large, radio-emitting structure known as the Arc, they found that it consists of numerous, linear, magnetic filaments, all approximately parallel to each other and perpendicular to the galactic plane (Fig. 3). Subsequent observations have confirmed the magnetic character of the filaments and have revealed filamentation elsewhere in the inner 50 pc of the Galaxy. The galactic center "magnetosphere" therefore appears to have a poloidal geometry: one in which the field lines are perpendicular to the system's symmetry plane, in this case the galactic plane. The radio-emitting filaments are understood to be magnetic flux tubes that have been "illuminated" by the presence of synchrotron radiation emitted from relativistic electrons. The linearity of the filaments attests to the rigidity, or the strength, of the magnetic field; the implied field strengths of about a milligauss (mG) are 2–3 orders of magnitude larger than those inferred for the galactic disk. The implications of such a strong and extensive magnetic field have only begun to be worked out. Noting that the magnetic forces on an interstellar cloud in the inner 50 pc are comparable to the gravitational forces exerted by the Galaxy, one might well expect the magnetic field to be an essential ingredient of the description of many of the unique phenomena taking place there.

THE CIRCUMNUCLEAR DISK

Inside a radius of 5 pc, the gas is even more tumultuous than at greater radii. A circumnuclear disk of gas and dust orbits the nucleus at radii between 2 and ~8 pc. With a turbulent velocity spread exceeding 50 km s^{-1}, this warm, clumpy disk orbits a central cavity that is remarkably devoid of gas and dust, although it appears from the dynamics of the circumnuclear disk that the cavity contains about five million solar masses ($M_\odot$) of stars and perhaps nonluminous matter (a 3×10^6 $M_\odot$ black hole remains a popular possibility). Within the cavity can be found high-velocity streamers of magnetized plasma apparently falling toward the central mass concentration from the inside edge of the circumnuclear disk. What keeps the circumnuclear disk from encroaching upon the cavity is not yet clear, but disk magnetic fields of up to 10 mG may effectively constrain the gas of the disk. In any case, the long-term stability of the circumnuclear disk is in doubt: It may be a transient phenomenon, fed from the outside by matter drifting radially inward, and losing matter at its inside edge as it either accretes onto the central mass concentration or is blown away by powerful winds from the central energy source. Although the central energy source has a modest luminosity at the present time (about 10^7 solar luminosities), that luminosity might well be variable, especially if it depends on the accretion rate of matter from the circumnuclear disk. Large flare-ups would have a profound effect on the dynamics of the region and may help account for the noncircular motions and large turbulent velocities characterizing the central portions of the Galaxy.

Additional Reading

Backer, D. C., ed. (1987). *The Galactic Center. AIP Conference Proceedings No. 155.* American Institute of Physics, New York.

Brown, R. L. and Liszt, H. S. (1984). Sagittarius A and its environment. *Ann. Rev. Astron. Ap.* **22** 223.

Genzel, R. and Townes, C. H. (1987). Physical conditions, dynamics, and mass distribution in the center of the galaxy. *Ann. Rev. Astron. Ap.* **25** 377.

Güsten, R. (1989). The galactic center—a larger scale view. In *The Physics and Chemistry of Interstellar Molecules Clouds: mm and Sub-mm Observations in Astrophysics.* G. Winnewisser and J. T. Armstrong, eds. Springer, Berlin, p. 163.

Morris, M., ed. (1989). *The Center of the Galaxy, Proceedings of IAU Symposium No. 136.* Kluwer Academic Publishers, Dordrecht.

See also **Galactic Center.**

Interstellar Medium, Galactic Corona

Blair D. Savage

The gaseous galactic corona refers to the low density gaseous region extending away from the dense gas of the disk of the Milky Way galaxy into the halo for distances estimated to be at least 3000 pc. The corona is known to contain hot (~ 200,000 K), warm (~ 5000 K), and cool gas (~ 200 K). The galactic corona may be the result of violent explosive processes occurring in the Milky Way that heat and overpressurize the interstellar gas of the disk and cause it to flow away from the galactic plane into the halo, where it cools and rains down onto the galactic disk in a flow resembling that of a fountain.

GAS OF THE GALACTIC DISK

The gas in the galactic halo probably originates from the matter in the galactic disk. The astronomical observing advances brought by the space age have provided a much clearer picture of the nature of the gaseous medium found in the disk. The disk gas exists in a wide range of physical conditions. The gas temperatures range from about 30 K to 10^6 K and the densities range from about 10^4 to 10^{-3} atoms cm^{-3}. The cold gas is mostly confined to a very thin plane of the Milky Way in a region about 200 pc thick. It is in this region of the Galaxy that the process of star formation is responsible for the conversion of dense clouds of interstellar gas into stars of different masses.

The direct evidence for the existence of the hot phase of the interstellar medium in the galactic disk has come from measurements with space instrumentation of the x-rays that the gas produces and of the ultraviolet absorption toward distant stars produced by highly ionized atoms such as O^{+5} and N^{+4}. The x-rays provide direct diagnostic information on the existence of 10^6 K gas with densities of about 10^{-3} atoms cm^{-3}, whereas the ultraviolet absorption measures of highly ionized atoms reveal the existence of somewhat cooler gas. The hot gas of the galactic disk may be the fundamental substrate in which all the other gas phases are found. However, currently there is considerable uncertainty concerning the actual fraction of interstellar space filled by the hot phase.

PROPERTIES OF THE GAS OF THE GALACTIC CORONA

The existence of the hot gas in the galactic disk opens the possibility that gas may also be found at large distances away from the plane in the halo. The degree of confinement of gas to the plane of a galaxy depends on the temperature of the gas and the strength of the gravitational attraction of the matter in the disk. For a constant gravitational acceleration g toward the disk, gas at a uniform temperature T will assume an exponential density distribution away from the plane of the galaxy given by

$$n(z) = n(0)\exp[-|z|/H],$$

where n is the matter density in atoms per cubic centimeter, z is the distance from the plane of the galaxy, $H = kT/mg$ is the scale height of the gas with Boltzmann's constant $k = 1.38 \times 10^{-16}$ erg K^{-1}, and m is the average particle mass in grams. For ionized gas with a temperature $T = 10^6$ K in the solar region of the Milky Way the scale height of the gas is about 6000 pc. Therefore, gas as hot as 10^6 K in the disk of the Milky Way will tend to flow outward from the plane of the Galaxy into the halo region.

During the 1980s the evidence slowly accumulated that gas does indeed exist at large distances from the plane of the Milky Way. The evidence came from a variety of diagnostic techniques. Measures of the 21-cm radio line emission from neutral hydrogen in interstellar space revealed the existence of large expanding loop-like structures, known as supershells, with extents of up to 2000 pc. These structures probably represent the hydrogen at the swept-up boundaries of regions of gas set into motion by the collective effects of many supernova explosions that occur in groups of young massive stars known as OB associations. The large sizes of the structures imply the ejection of gas into the galactic halo.

Measures of the propagation of the pulses of radio radiation from rapidly rotating neutron stars (pulsars) revealed that the free electrons of interstellar space exist in a layer having an exponential scale height of 1000 pc from the galactic plane.

At ultraviolet wavelengths, the spectrographs aboard the *International Ultraviolet Explorer* satellite have been used to study the distribution of highly ionized gas in the interstellar medium with the result that atoms of N^{+4}, C^{+3}, and Si^{+3} were found to exist in a very thick layer with a scale height of about 3000 pc.

The gaseous region that extends away from the galactic plane to distances of 500–3000 pc or more is commonly referred to as the gaseous galactic halo or corona. The gas of the corona has a wide range of temperatures. The highest temperature so far measured is approximately 200,000 K, which is the temperature required to produce substantial quantities of the N^{+4} found in halo gas. However, gas with temperatures ranging all the way down to less than 200 K is also found and it is likely that gas with temperatures greater than 200,000 K also exist. The medium appears to be highly inhomogeneous in its conditions with there being hot low density regions and cooler clouds of higher density.

The composition of the halo gas is not well determined, but the available evidence points toward a gas phase composition similar to that found in the atmosphere of the Sun, with most of the matter being hydrogen and helium with trace amounts of the heavier elements. In halo gas most of the heavy elements appear to be in the gas phase rather than in the solid phase. This is in contrast to the composition of the gas found in the galactic disk where observations reveal a marked reduction in the gas phase abundance of many heavy elements compared to the solar composition. This reduction or depletion is interpreted as being the result of many of the heavy elements, such as silicon, iron, and magnesium, being present in the solid particles of interstellar dust. Apparently in the halo gas, the amount of dust present is greatly reduced and therefore most of a given heavy element is found in the gas phase rather than the solid phase.

Gas found at large distances from the galactic plane exhibits a complex pattern of motions. In addition to the existence of relatively quiescent regions such as found for the gaseous matter seen toward the south galactic pole, there exist disturbed regions with gas moving up away from the galactic plane and also toward the plane. Extensive rapidly moving clouds of hydrogen, known as high velocity clouds, have been found in the Galaxy. The pattern of motion exhibited by these clouds is consistent with the idea that they represent the return of matter from the halo to the disk of the Milky Way.

ORIGIN OF THE GASEOUS GALACTIC CORONA

In recent years the theoretical understanding of the new observational data on gas at large distances from the plane of the Galaxy has developed into a theory known as the galactic fountain model. In this theory the great explosions of supernovae create hot overpressurized regions of gas that can burst out of the plane of the Galaxy and allow the flow of hot highly ionized gas into the halo. The outflowing gas proceeds to cool and can then rain back down onto the galactic plane as cooler gas clouds in a flow resembling that of a fountain of water. The outflowing gas has temperatures of perhaps 10^6 K or higher. Such gas has proven difficult to detect. However, gas cooling in the fountain can explain the observed amount of N^{+4} and C^{+3} seen at large distances from the galactic plane. In the flow associated with a galactic fountain, one would expect to detect the motions of parcels of cooler gas with speeds of up to 150 km s^{-1}. Such motions might provide an explanation for the high velocity hydrogen clouds observed by radio astronomers.

In groups of massive stars known as OB associations, many supernovae may occur over time scales of about 10^7 years. Such temporally correlated explosions will create superbubbles of hot overpressurized gas that can rapidly expand and break out of the plane of the Galaxy, forming structures called "chimneys." This process is similar to that of the galactic fountain, but the ejection of gas is highly concentrated in the chimneys rather than over the entire disk of the Galaxy. Perhaps the actual situation is a combination of a widespread fountain activity with localized very active regions where chimneys are found.

Not all theorists of the interstellar medium agree with the basic ideas of the fountain model or of the chimney model. In particular, some astronomers believe that there is a more quiescent interstellar medium in which the gas found at large distances from the galactic plane is supported by the pressure of the magnetic field and cosmic rays of the Galaxy. In such a cosmic-ray-supported galactic halo, the highly ionized gas might have its origin through the process of photoionization where the ionizing radiation might come from extragalatic space or from hot stars in the halo. Although such models can explain the existence of Si^{+3} and C^{+3} found at large distances from the galactic plane, they have difficulties accounting for the observed amount of N^{+4}. Theories advocating this more quiescent picture of the interstellar medium might apply in some regions of the Galaxy, whereas the dynamic phenomena associated with fountains, superbubbles, and chimneys might apply elsewhere.

Studies of the gaseous corona of the Milky Way have so far provided information about the matter in the solar region of the Milky Way. However, there are indications that in the very central regions of the Galaxy the gas at large distances from the plane may behave differently than in the solar region. For distances from the galactic center of less than about 3 kpc there is much less hydrogen away from the galactic plane than found elsewhere. The central region of the Milky Way may be a place where the supernova activity is so great that instead of having a bound circulating corona as in the galactic fountain model, the gas leaves the Galaxy in the form of a wind. There is evidence for enhanced x-ray emission from the general direction of the central region of the Galaxy. That emission may be associated with hot outflowing gas situated above and below the center of the Milky Way.

In many respects the gaseous galactic corona bears a similarity to the hot coronae of the Sun and other stars. The solar corona contains gas at many different temperatures ranging from 5000 to 2×10^6 K. Disturbances in lower layers of the Sun spew gas into the solar corona. The ejected gas has been seen to move along lines of magnetic field, to condense into cooler matter, and to form giant prominences that then flow back to the solar surface layer. The processes that heat and thereby maintain the solar corona are believed to be various types of mechanical disturbances that arise in the turbulent surface region known as the photosphere. The disturbances propagate outward in the form of various types of mechanical waves, deposit their energy, and heat the overlying gas. Many clues about what may be happening on the galactic scale might be obtained from a study of those processes occurring in the outer atmosphere of the Sun.

GASEOUS CORONAS OF OTHER GALAXIES

Galaxies other than the Milky Way also have extended gaseous halos or coronas. The overall similarities or differences between the coronas of other galaxies and that of the Milky Way will depend critically on the frequency of occurrence of violent stellar processes that are able to create zones of hot overpressurized gas in the disks of those galaxies. A very active galaxy might be capable of converting much of its gas into hot gas and driving a galactic wind and producing a very extensive but short lived corona. In contrast, a galaxy with a low supernova rate will hardly produce any hot gas and such a galaxy would not be likely to have a dynamic gaseous corona like that proposed for the Milky Way.

The gaseous coronas of spiral galaxies do not appear to be hot enough and dense enough to be bright x-ray emitters. Data currently available regarding the possible existence of gaseous halos around external spiral galaxies mostly come from radio and optical astronomical studies. For certain galaxies viewed edge on, it has been found that neutral and ionized hydrogen sometimes occur in both confined and extended components. The extended components may represent the halo gas.

In the case of the nearby Andromeda galaxy, which is a spiral galaxy similar to the Milky Way, our viewing position is from above its disk. Radio studies of Andromeda have revealed the existence of holes in the neutral hydrogen in the disk gas with extents of 100–1000 pc. Those holes may represent the regions where multiple supernova explosions have created hot cavities of gas with the subsequent venting of gas into the halo. The phenomena occurring in the Milky Way that are responsible for the ejection of gas into the halo may be occurring in many other galaxies in the universe as well.

Additional Reading

de Boer, K. S. and Savage, B. D. (1982). The coronas of galaxies. *Scientific American* **247** (No. 2) 54.

Savage, B. D. (1988). The properties of the gaseous galactic corona. In *QSO Absorption Lines: Probing the Universe*, C. Blades, C. Norman, and D. Turnshek, eds. Cambridge University Press, Cambridge, p. 195.

Spitzer, L. (1990). Theories of the hot interstellar gas. *Ann. Rev. Astron. Ap.* **28** 71.

York, D. G. (1982). Gas in the galactic halo. *Ann. Rev. Astron. Ap.* **20** 221.

See also **Background Radiation, Soft X-Ray; Interstellar Medium; Interstellar Medium, Hot Phase; Supernova Remnants, Evolution and Interaction with the Interstellar Medium.**

Interstellar Medium, Galactic Molecular Hydrogen

Nicholas Z. Scoville

Virtually all star formation in our Galaxy takes place within dense interstellar clouds of molecular hydrogen (H_2). Because the dust in these clouds effectively absorbs the short wavelength visible and ultraviolet radiation, observational studies of molecular clouds became possible only in the last two decades with the advent of millimeter wave astronomy.

OBSERVATIONAL TECHNIQUES

Direct observations of the dominant gaseous component in these clouds (H_2) are generally precluded because the symmetry of the hydrogen molecule implies that there is no permanent dipole moment and there are therefore no strong radio-frequency transitions. The permitted electronic transitions in the ultraviolet, seen in absorption along the line of sight to nearby bright stars, cannot be studied in the denser clouds due to the very high opacity of the dust in the ultraviolet. Instead, astronomers have used radio frequency observations of trace gas constituents such as carbon monoxide (CO) to probe the interior of these clouds. The most widely used spectral line is the ground state $J=1$-0 rotational transition of CO at a wavelength of 2.6 mm (115 GHz). This transition is easily excited by collisions with molecular hydrogen even in clouds of low temperature, and hence, the observed CO photon emission rate from a particular interstellar cloud provides an excellent measurement of the density of molecular hydrogen.

One of the major developments in our understanding of these star-forming regions has been the growing recognition that the chemistry of the molecular clouds is much more advanced than previously guessed. Observations at millimeter and submillimeter wavelengths have revealed over 2000 spectroscopic transitions that are identified with 65 molecules (and their isotopic variants) with complexity ranging up to 11 atoms (HC_9N).

Most of our knowledge of the dense interstellar clouds has been derived from studies of the ubiquitous CO line. These observations yield information on the density of gas, its temperature, and the motion of gas along the line of sight. The velocity information comes from Doppler shift measurements.

The masses of molecular clouds are estimated by three independent techniques, all of which yield values generally in agreement to within a factor of 2. The first and probably most reliable method entails mapping the spatial extent and internal velocities (usually the velocity dispersion) of the cloud via the emission of a trace molecule such as CO and then assuming that the cloud has a sufficient mass to gravitationally bind the observed motions (virial theorem). The second technique involves derivation of the molecular column densities from the observed trace-gas emission intensity, and then assuming an abundance of the trace molecule relative to H_2 in order to arrive at the column density and mass of H_2. Finally, one can use observations of γ-ray emission from the clouds to deduce the column density of gas, assuming that the flux of cosmic rays giving rise to the γ rays, through their interaction with H_2 and He, is constant within the clouds. All three techniques have been applied with consistent results to clouds in both the solar neighborhood and the inner galactic plane.

GALACTIC DISTRIBUTION OF H_2

Galactic Disk

The overall distribution and properties of the molecular gas throughout the Galaxy have been determined from extensive surveys of CO emission, sampling the galactic disk. Although there do exist isolated molecular clouds in the solar neighborhood, vastly greater concentrations of molecular gas exist in the galactic center and in a ring-like distribution approximately halfway from the Sun to the center. The latter structure, termed the molecular cloud ring, contains 2 billion solar masses of gaseous molecular hydrogen within approximately 2000 giant molecular clouds (GMCs) of size greater than 40 pc (150 ly). In contrast, the distribution of atomic hydrogen (H I) is rather flat with galactic radius and, in fact, shows a deficiency in the galactic center where the molecular cloud distribution peaks. The distribution of H_2 and H I gas with radius in the Galaxy are shown in Fig. 1 together with that of giant H II regions ionized by hot, massive stars. Inasmuch as the overall galactic distribution of giant H II regions closely mimics the distribution of molecular gas, it is clear that most high-mass star formation occurs in molecular clouds rather than in atomic hydrogen clouds.

Table 1 compares the H_2 and H I masses in the galactic center region, and in the inner and outer (relative to the Sun's location) disk. Overall in the Galaxy, the H I and H_2 masses are approximately equal. On the other hand, the distributions of the two gas components in the Galaxy are entirely different; approximately 90% of the total H_2 mass is found inside the Sun's orbit whereas nearly 70% of the total H I mass is outside.

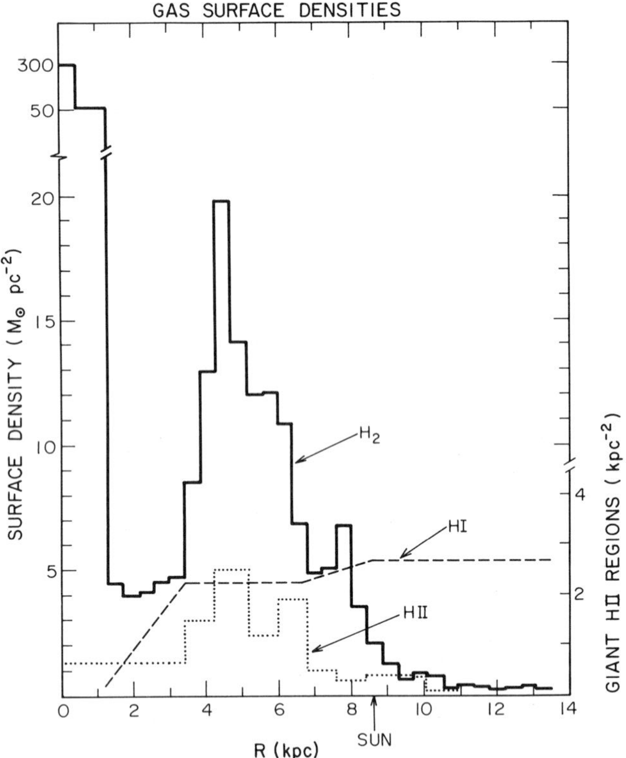

Figure 1. Comparison of gas surface densities in the Milky Way disk for molecular (H_2), atomic (H I), and ionized (H II) hydrogen. The Sun is located at a galactic radius of 8.5 kpc. The molecular gas shows a strong concentration in the galactic center and in the molecular cloud ring at 3–7 kpc as does the distribution of giant H II regions. In contrast, the atomic hydrogen shows a deficiency in the center of the Galaxy and a relatively constant density through the galactic disk out to approximately three times the distance of the Sun from the galactic center.

Galactic Center

In the center of the Galaxy, the distribution of molecular clouds is rather poorly organized. Many of the clouds in this region are apparently located in an expanding ring of radius approximately 200 pc (650 ly) from the galactic center, but there are also other clouds much closer to the central radio source Sagittarius A and some outside this ring structure. At present, it is unclear whether the expansion motion seen in the galactic nucleus ring is a result of an explosive release of energy from the central source, Sagittarius A, or is due to dynamical perturbations by a massive stellar bar such as are seen in many other spiral galaxies. The overall mass of molecular gas in the galactic center exceeds that of atomic gas by approximately a factor of 40 (see Table 1).

MOLECULAR CLOUD PROPERTIES

Nearby Molecular Clouds

The nearest molecular clouds are those in Taurus at a distance of 140 pc (450 ly), and the Orion molecular cloud at a distance of

Table 1. Total Disk Mass of H I and H_2 in the Galaxy

	*Galactic Annulus**			
Mass	*0–0.15*	*0.15–1.0*	*1.0–1.6*	*Total*
$M(H_2)\,(M_\odot)$	0.4×10^9	1.9×10^9	0.2×10^9	2.5×10^9
$M(H_2)/M_T(H_2)$	14%	77%	9%	100%
$M(\text{H I})\,(M_\odot)$	1×10^9	0.7×10^9	1.7×10^9	2.4×10^9
$M(\text{H I})/M_T(\text{H I})$	0.4%	30%	70%	100%
$M(H_2)/M(\text{H I})$	40	2.7	0.12	1.04

*Annuli in the Galaxy. Radii normalized to the Sun's orbital radius = 1.0. M_T is the total disk mass, most of which is contained in stars.

about 500 pc (1600 ly). Both can be seen as areas of high dust obscuration in optical photographs.

The Taurus clouds are often taken to be prototypical of low-mass star formation regions. The CO emission can be mapped over an extent of approximately 1° in the individual cloud components and the morphology of the gas is very filamentary. The total mass of gas in the individual Taurus clouds is in the range of 10^3–10^4 $M_\odot$ and the mean number density of H_2 molecules is approximately 10^3 cm^{-3}. In general, the embedded stars forming in these clouds are exclusively low-mass stars (< 5 $M_\odot$) and the temperature of the gas heated by these stars is relatively low, typically only 10 K. It is often speculated that much of the filamentary structure seen in these clouds is a result of asymmetric collapse of the more diffuse, presumably atomic hydrogen, parent cloud, preferentially along the magnetic field, or rotational flattening of the cloud due to increased centrifugal forces as the cloud collapses.

The Orion molecular cloud is the nearest example of what are often termed giant molecular clouds. Here the CO emissions may be traced over an extent of more than 5° corresponding to a size of 40–60 pc. The overall mass is 2×10^5 $M_\odot$, that is, a factor of 20 greater than that of the Taurus clouds. Although the mean gas density is approximately the same as in Taurus (i.e., 10^3 H_2 cm^{-3}), the gas temperatures are considerably higher (20–80 K) due to heating by the more luminous stars found in the Orion cloud. Associated with the Orion molecular cloud are two OB star clusters (NGC 2024 and the Trapezium cluster in M42—the Orion nebula) containing young stars of age approximately 1 million years with luminosities up to 10^5 times that of the Sun.

In addition to these high-mass star clusters on the front face of the Orion molecular cloud, several embedded clusters of stars have been detected over the last two decades by infrared astronomers. The best known of these, the Kleinmann–Low infrared cluster, is located approximately 1 ly behind M42 within the Orion cloud (OMC1). This cluster is totally obscured to visible wavelength observations, but it one of the strongest infrared sources in the sky with a total luminosity of approximately 10^5 times that of the Sun. Although the stars in this cluster presumably emit most of their power at optical and ultraviolet wavelengths, the radiation is absorbed within a short distance by the surrounding dust. This dust is thus heated to several hundred degrees and subsequently reemits the energy at near and far infrared wavelengths ($\lambda = 2$–200 μm).

Galactic Giant Molecular Clouds

Mapping of the individual, more distant clouds within the galactic ring indicates that most of the molecular gas is in the GMCs of mass 10^5–10^6 $M_\odot$. The properties of a typical giant molecular cloud are summarized in Table 2. Their sizes range from a few times smaller than the Orion molecular cloud to five times larger. The GMCs are the most massive gravitationally-bound units in the Galaxy. Although their orbital motion in the galactic disk is similar to that of stars at the same radius, their distribution is more tightly confined perpendicular to the galactic disk, with very few GMCs found more than 60 pc (200 ly) above or below the galactic plane.

The most massive of the galactic center molecular clouds, Sagittarius B2, contains over 2×10^6 $M_\odot$ of gas at an average density of 10^4–10^5 H_2 cm^{-3}. This single cloud harbors five clusters of high-mass star formation with luminosity greater than or equal to the Orion nebula. Sagittarius B2 is probably the most massive GMC in the Galaxy and has become the prime site for detecting the more complex organic molecules mentioned earlier.

Table 2. Properties of a "Typical" GMC

Diameter	40 pc
Total mass	4×10^5 $M_\odot$
Average H_2 density	180 cm^{-3}
Gas kinetic temperature	10 K
Internal velocities	3.8 km s^{-1}

Spiral Arms

The question of whether or not all of the observed molecular clouds in the galactic disk are confined to a regular pattern of a few spiral arms is important for our understanding of both the origin and the evolution of these clouds. In one model, they are short-lived objects that form from atomic gas as a result of compression within the spiral arm shock fronts and that are then destroyed by young stars upon leaving the arms. The spatial distribution of the clouds should then closely follow the spiral arm locations and relatively few clouds would be found between the arms. If, on the other hand, the clouds are formed by another mechanism and survive the passage of several spiral arms, they would be more widely distributed throughout the galactic disk. In this case, they will still show some concentration along the spiral arms due to the convergence of the galactic orbits at those locations.

The most recent and comprehensive analysis of the observations suggests a high degree of concentration for the largest clouds and cloud complexes along the spiral arms; the hottest molecular clouds are also found mostly in the arms. The distribution of lower-mass and cooler clouds shows them to be much more uniformly distributed between the arm and interarm regions. The latter population of clouds constitute approximately 50% of the galactic molecular gas mass.

These observational data can be reconciled in a model in which the largest clouds and cloud complexes are assembled by collisions of smaller preexisting clouds when they enter the spiral potential well. The correlation of the hotter molecular clouds with the spiral arms is probably a result of the strong heating associated with high luminosity star formation in the spiral arms. In this picture, preexisting molecular clouds arrive at the backside of the spiral arms; within the spiral arm, their orbits converge, resulting in a high rate of cloud–cloud collisions. The gas compression due to the collision of two clouds may lead to formation of high mass OB star clusters. Last, upon leaving the front side of the spiral arm, the cloud orbits diverge once again, the cloud complexes disassemble, and, within the interarm areas, most of the molecular gas resides in smaller, individual molecular clouds.

Additional Reading

Mihalas, D. and Binney, J. (1986). *Galactic Astronomy*. W. H. Freeman, San Francisco.

Scoville, N. Z. and Sanders, D. B. (1988). H_2 in the Galaxy. In *Interstellar Processes*, D. J. Hollenbach and H. A. Thronson Jr., eds. Reidel, Dordrecht, p. 21.

Scoville, N. Z. and Young, J. S. (1984). Molecular clouds, star formation and galactic structure. *Scientific American* **250** (No. 4) 42.

Thaddeus, P. (1982). Radio observations of molecules in interstellar gas. In *Molecules in Interstellar Space*, A. Carrington and D. Ramsey, eds. The Royal Society, London, p. 5.

Wilson, T. L. and Walmsley, C. M. (1989). Small-scale clumping in molecular clouds. *Astron. Ap. Rev.* **1** 141.

Young, J. (1989). Molecular clouds in spiral galaxies. In *The Interstellar Medium in Galaxies*, H. A. Thronson, Jr., and J. M. Shull, eds. Kluwer Academic Publishers, Dordrecht.

See also **Galactic Center; Galactic Structure, Interstellar Clouds; Interstellar Clouds, Chemistry; Interstellar Clouds, Molecular; Interstellar Medium, Galactic Center; Interstellar Medium, Molecules.**

Interstellar Medium, Hot Phase

Donald P. Cox

Dense interstellar clouds generally have temperatures lower than 100 K, whereas temperatures in dilute "warm" intercloud components range upward to 10,000 K. By "hot" interstellar gas one typically means that the temperature is in the range 10^5–10^7 K.

There are certainly regions of the interstellar medium [(ISM), the gas between the stars] that qualify as hot. Thermalization of high velocity gas flow is usually the cause: supernova explosions, high speed stellar winds, and high velocity clouds falling into the Galaxy all result in shock waves and high temperatures. Presently uncertain, however, are the fraction of interstellar space occupied by such hot gas and whether a hot phase has a significant role in the ecology of the interstellar system.

Three types of observation were important in bringing about the concept of a hot interstellar phase. The first, discussed by Lyman Spitzer, Jr. in 1956, is that interstellar clouds are occasionally found far from the galactic plane. Such clouds require external pressure for their confinement and he pointed out that only an intercloud gas of very high temperature could support itself at such great height. This was three decades prior to the discovery that diffuse gas with only moderate temperature is commonly found far from the plane, probably supported by the galactic magnetic field. Confinement of the clouds has not yet been restudied in this new context.

A second observation, made first from sounding rockets in the late 1960s, is that above the Earth's atmosphere the entire sky glows in low energy (soft) x-rays. The earliest assumption was that these x-rays had an extragalactic origin. (This still appears to be true for x-ray energies in excess of 1 keV.) But the extremely rapid attenuation by interstellar matter expected for the softer x-rays and the observation that such attenuation is often not found where expected have forced our interpretation of their origin ever closer to the Earth. The currently prevailing picture is that they derive from a million-degree-kelvin gas of very low density, filling a large cavity known as the Local Bubble. The solar system is inside this bubble; a typical bubble dimension perpendicular to the galactic plane is 0.1–0.2 kiloparsec (kpc), comparable to the thickness of the diffuse cloud layer. Surely, with the solar system inside such a region, the unbiased Copernican point of view is that such high temperature regions must be common in the ISM.

The third observation, made exclusively during the lifetime of the *Copernicus* (*OAO-3*) satellite (1972–1981), is that five-times-ionized oxygen (O^{+5}, or O VI) is found in absorption spectra (near 1035 Å) of hot stars. (Only hot stars provide continuum radiation this far into the ultraviolet; the absorption seems to be interstellar.) Except near extremely hot stars, where there are very energetic photons, such a high ionization state for interstellar oxygen is expected to require it to be in gas with a temperature of roughly 300,000 K. Details of the O^{+5} observations suggest that it is patchily distributed, possibly in boundary layers between hotter and cooler interstellar gas.

An impressive theoretical framework was constructed around these observations. It included a proof that warm intercloud gas of moderate density would be disrupted into a froth of clouds and very hot gas by supernova explosions, a careful analysis of the balance of energy and mass flow among the interstellar constituents, and an analysis of O^{+5}, assuming it to derive from the boundaries of interstellar clouds that are thermally evaporating in a hot environment. There were also discussions of galactic fountains and winds: high temperature gas outflowing from the galactic plane after heating by supernovae. In the heyday of such interpretations, between 1974 and 1985, the hot phase of the ISM was widely believed to occupy between 50 and 70% of interstellar space and to figure significantly in the interplay between interstellar gas, dust, cosmic rays, and their violent sources of energy and disruption, the supernovae. A key aspect was the role of the hot phase in the dispersal of new products of nucleosynthesis, recently formed heavy elements thrown into the ISM by dying stars, and its effect on the chemical evolution of the Galaxy.

As of 1989, nearly every aspect of this picture was being questioned, in part due to problems in details of things expected in this view, and in part from recent changes in perception of the ISM as a whole.

The layer of stars making up the disk of the Milky Way galaxy has very gradual boundaries but, roughly speaking, a radius of 15 kpc and a thickness of 0.6 kpc, making it flatter than a pancake. Until recently, the distribution of the interstellar medium was thought to be even thinner, with a total thickness of perhaps 0.2 kpc. More sensitive techniques, however, have identified diffuse components with quite thick distributions: Very dilute neutral material in the Lockman layer (after Felix J. Lockman) has a mean thickness of perhaps 0.8 kpc, whereas the corresponding thickness of dilute ionized material (the Reynolds layer, after Ronald J. Reynolds) is 3 kpc. The previously known ISM in the thin layer consists of condensations into denser diffuse-cloud and molecular-cloud phases. These clouds may be stabilized by the higher surrounding gas pressure at the galactic midplane (the dilute phases acting as the weighty atmosphere), or they may be stable over a broad range of heights but lack support far from the midplane.

A surprise to many has been that the total mass in the dilute phases is similar to that in the clouds. In addition, there is good evidence that the cosmic rays and galactic magnetic field pervade the thicker disk as well, carrying an appreciable energy density to high altitude.

Hot gas has not been found in the deep gravitational potential wells around normal spiral galaxies. It now seems possible that the thick layer of warm gas, magnetic field, and cosmic rays interferes with the convection of supernova-produced hot gas out of the disk, thereby suppressing fountain and wind activity.

Continued observational evidence for the existence of a pervasive warm intercloud medium may finally be beginning to be understood. This medium may arise through a stabilization brought on by the interstellar magnetic field, suppressing the frothing action of supernovae. Individual supernova remnants may be able to evolve separately, their dense shells dispersed by magnetic forces to return to the intercloud phase.

The O^{+5} ions have been reinterpreted as deriving from individual bubbles of hot gas generated in the ISM by supernovae, rather than from evaporation boundaries of clouds immersed in a pervasive hot phase. Such bubbles need occupy only 10–20% of the interstellar volume.

Certainly there are localized regions of hot gas in the interstellar medium, bubbles that are created by individual supernovae or that are reheated by subsequent supernovae. Our Local Bubble may be an example of the latter. There are also very large and violent regions surrounding the sites of OB associations, where powerful stellar winds and multiple supernovae must create large volumes of hot gas and vigorously expanding shells of denser gas. What is no longer clear is that it is appropriate to think of such active regions as being part of a phase of the interstellar medium, as opposed to being individual events in the interstellar environment. It may be that for once a Copernican view has misled us and that the local interstellar medium with its large bubble of hot gas is an unusual environment.

Additional Reading

Cox, D. P. (1989). The diffuse interstellar medium. In *The Interstellar Medium in Galaxies*. H. A. Thronson, Jr. and J. M. Shull, eds. Kluwer Academic Publishers, Dordrecht.

Cox, D. P. and Reynolds, R. J. (1987). The local interstellar medium. *Ann. Rev. Astron. Ap.* **25** 303.

Spitzer, L., Jr. (1990). Theories of the hot interstellar gas. *Ann. Rev. Astron. Ap.* **28** 71.

See also **Background Radiation, Soft X-Ray; Interstellar Medium; Interstellar Medium, Galactic Corona; Interstellar Medium, Local; Supernova Remnants, Evolution and Interaction with the Interstellar Medium.**

Interstellar Medium, Local

Frederick C. Bruhweiler

The local interstellar medium (LISM) represents the volume of space containing the interstellar gas and dust around the Sun and other stars within the immediate solar neighborhood. Although there is no universally accepted definition for the size of this region, generally it is described as a volume with a radius of 50 parsecs centered upon the Sun. [A parsec (pc) equals a distance of 3.26 light-years (ly), or 3×10^{18} cm.]

The importance of the local interstellar medium is that it can provide a basic, fundamental laboratory for probing the physical processes in the more distant interstellar gas and that it is our only means of studying what may be a "typical volume element" of the interstellar medium that prevades the disk of our Milky Way galaxy. One way of probing the interstellar medium is to study absorption by atoms or ions in the interstellar gas of the light emitted by distant stars. However, such path lengths may contain both very diffuse gaseous regions and very dense clouds. Even if one could separate the differing dense and diffuse components, there would still be a large uncertainty as to where along the line of sight each of these respective regions is located. The best means to minimize these uncertainties is to study the gas and dust morphology first over much shorter path lengths, namely, toward those objects within the local ISM.

The problem has been that, up until the early 1970s (with few exceptions), no ground-based observations were sensitive enough to detect interstellar gas or dust within 50 pc of the Sun. This seemed to present no obvious problem, because radio observations of neutral hydrogen at 21 cm over long path lengths suggested that there was a pervasive, warm (10^4 K) diffuse intercloud medium with a density of 0.1 hydrogen atom per cubic centimeter. This implied that the column density of neutral hydrogen for a 50-pc path length should be near 1.5×10^{19} cm^{-2} and thus below the detection limits of ground-based instruments. [A column density is the number of atoms or ions found in a volume with a square base (with 1-cm sides) and extending over a column (or length) that, in the case of absorption studies, is the distance to the observed star.] It was assumed that if any clouds were present within 50 pc, the neutral-hydrogen column densities would have been much larger than 1.5×10^{19} cm^{-2}. Thus, the general picture was that the LISM was free from clouds. The observations of the LISM sampled only the pervasive 0.1 cm^{-3} intercloud medium.

A NEW PICTURE EMERGES

The advent of space astronomy has changed dramatically our physical picture of the interstellar medium. The Princeton Ultraviolet Spectrometer aboard the *Copernicus* (*OAO-3*) satellite, launched in 1973, discovered the ubiquitous presence of interstellar ultraviolet absorption lines due to atomic transitions of five-times-ionized oxygen (O^{+5}). Of the many stars observed with interstellar O^{+5} in their spectra, four were at distances of 100 pc or less.

The general distribution of this ion in the interstellar medium and its high level of ionization cannot be produced by energetic photons from hot stars. It must originate in a hot coronal gas of temperature $T = 10^5$–10^6 K and a particle density of 10^{-3}–10^{-2} cm^{-3}. Assuming pressure equilibrium between this so-called coronal gas and the cooler clouds implies that the coronal gas occupies over 50% of the volume of the interstellar medium.

Repeated rocket flights clearly revealed an apparently uniform soft x-ray background over the entire sky. The uniformity of the background, especially at energies near 0.1 keV, argued that this emission is quite local to the Sun because any intervening clouds with total hydrogen column densities of 5×10^{19} cm^{-2} would severely absorb this radiation and produce pronounced "shadows" superposed upon the x-ray background. The observed characteristics of this x-ray emission were quite compatible with its being produced in a low-density (10^{-2} cm^{-3}) plasma with a temperature near 10^6 K. Furthermore, these results were quite consistent with the observations of ultraviolet absorption due to O^{+5} and argued for the local presence of a pervasive low-density, hot coronal gaseous medium near and surrounding the Sun.

Absorption line observations of interstellar neutral hydrogen at 1216 Å (the wavelength of the Lyman-alpha spectral line) toward cool stars within 5 pc of the Sun, using the *Copernicus* satellite, yielded a quite different picture. Even though these stars generally are thought to have negligible flux in the ultraviolet, they do exhibit strong intrinsic chromospheric emission lines, especially in neutral hydrogen at 1216 Å. The interstellar absorption of neutral hydrogen is seen superposed upon the corresponding, usually wider, emission of the star. These results typically showed interstellar neutral hydrogen number densities near the Sun of 0.1 cm^{-3}. This seemed to support the older view of a pervasive 10^4 K gas with density $n = 0.1$ cm^{-3}.

To resolve the dilemma posed by the two different pictures of the intercloud medium, that offered by the observed O^{+5} and soft x-ray background and that implied by observations of 1216-Å absorption within 5 pc and by the 21-cm radio observations at longer distances, a theoretical model was proposed by Christopher McKee and Jeremiah Ostriker that was a synthesis of these two intercloud medium models. In their theory the coronal 10^6-K component was the pervasive substrate, but embedded in this substrate were numerous small cloudlets largely composed of the warm 10^4-K gas with $n \sim 0.5$ cm^{-3}. The typical size of these cloudlets was on the order of 2.5 pc and the mean free path between those clouds was about 12 pc. The percentages of the volume occupied by the coronal substrate and cloudlets were about 75 and 25%, respectively.

The main result of the McKee–Ostriker model was that it could explain the observations at distances less than 5 pc and the radio observations over large distances, as well as the O^{+5} and soft x-ray data. Unfortunately, no ultraviolet absorption line studies were yet available for stars within the distance interval 5–100 pc. These stars would prove crucial for testing this important model for the interstellar medium.

THE SCARCITY OF CLOUDS

Observations of ultraviolet interstellar absorption lines in hot white dwarf stars within 50 pc of the Sun with the *International Ultraviolet Explorer* (*IUE*) satellite did provide the necessary constraints for obtaining a workable model for the LISM. Although the *IUE* (launched in 1978) could not sample the hot coronal component of the LISM, it could probe the low-ionization and neutral gas. These results indicated for the nearest white dwarfs such as Sirius B (distance from the Sun, 1.3 pc) that the average neutral hydrogen particle density along the line of sight was 0.06 cm^{-3}. Data for more distant stars showed that there is a sharp falloff in average neutral hydrogen number density correlated with distance. All directions sampled yielded very low neutral hydrogen column densities, $\sim 1 \times 10^{18}$ cm^{-2} or less. The interpretation was that all the detected neutral and low-ionization gas was located in a cloud surrounding the Sun, with no evidence of additional clouds along any of the path lengths observed. These results combined with data from the ultraviolet spectrometer aboard the *Voyager* spacecraft (which sampled interstellar neutral hydrogen continuum absorption at wavelengths shorter than 912 Å, the Lyman limit) were quite consistent and revealed no evidence for additional clouds, implying that the mean free path between clouds must be at least 100 pc, a value much larger than that suggested in the McKee–Ostriker model. In addition, *Copernicus* results for a star 200 pc away, toward the constellation Canis Major, indicated a very low neutral hydrogen column density of 1.4×10^{18} cm^{-2}.

Although the lines of sight sampled were limited, they definitely showed that a large fraction of the LISM is free of clouds.

LOCAL CLOUD

Why a cloud surrounded the Sun in a region largely devoid of clouds remained a puzzle. Further analysis of *Copernicus* and *IUE* data suggested that the cloud was not symmetrically distributed about the Sun, but that there was a density gradient that increased toward the directions of the galactic center and the constellation Scorpius.

Independent supporting evidence for this density gradient came from very sensitive ground-based interstellar polarization studies of stars within 35 pc. Because interstellar dust grains have a small charge, they will align themselves with the ambient interstellar magnetic field. The percentage of polarization of the observed starlight depends on the total amount of dust along the line of sight. Despite the sensitivity of this study, only in a small patch (with total angular extent 30–60°) centered near galactic coordinates $l = 5°$, $b = -20°$ toward Scorpius was detectable polarization found. For a typical gas-to-dust ratio, the inferred neutral hydrogen column density for this patch is $1\text{–}2 \times 10^{19}$ cm^{-2}. This column density agreed closely with that deduced from interstellar ultraviolet absorption lines for stars within 20–30 pc in the same direction.

The current picture is that the Sun is embedded in, but near the edge of, a cloud with a size less than 20 pc and a total hydrogen column density of $1\text{–}2 \times 10^{19}$ cm^{-2}. The cloud's main body and direction of highest column density lie toward the constellation Scorpius. Most directions show very low column densities, typically on the order of 10^{18} cm^{-2} or less. These directions presumably intersect only the outer skin of the local cloud and the surrounding hot, extremely low-density substrate.

THE LOCAL CLOUD AND THE LOOP I SUPERNOVA REMNANT

Analysis of additional data has revealed that the local cloud is part of a much larger nebular complex in the interstellar medium. From the Doppler shifts of the interstellar absorption, the direction of the "interstellar wind" or the motion of the cloud relative to the Sun can be deduced. The cloud is overtaking the Sun, moving at a velocity of 25–30 km s^{-1}, and coming from the general direction of the nearby young Scorpius–Centaurus (Sco–Cen) stellar association. The kinematics of the local cloud suggests that it is part of an expanding 120-pc-radius shell of filamentary clouds, known as Loop I, centered on the Sco–Cen association.

The basic structure of Loop I has been delineated at radio and visual wavelengths. This structure has an angular extent of about 90° in the sky, with the Sun lying at its outer periphery. The interpretation is that Loop I is an old supernova remnant due to an explosion of a massive star in the Sco–Cen association. Soft x-ray data at energies near 0.5 keV revealed that the interior of the Loop I shell is filled with gas at temperatures near 10^7 K, hotter than the pervasive 10^6-K region exterior to Loop I. Presumably, the 10^7-K gas was heated by the expanding shock front associated with the supernova remnant. (The region of hot gas in the solar vicinity is sometimes called the Local Bubble.)

Within a period of less than 20 years, a very coherent observational picture for the LISM has been constructed that incorporates radio, visible-wavelength, ultraviolet, and soft x-ray results. Clearly, a large fraction of the volume in the LISM is filled with the hot coronal gas. It appears that the Loop I supernova remnant is expanding into a pre-existing low-density region with $T = 10^6$ K. Whether this pre-existing region, seen in the anti-galactic-center direction and largely free of clouds, can be described as a cavity produced by some single, still older, supernova remnant or should be considered part of a more pervasive substrate cannot yet be determined. The answer to this question will help determine how representative the LISM is of a typical volume element of the general interstellar medium. These attempts to understand the physical processes occurring in the LISM are subjects of exciting, ongoing research.

Additional Reading

Bochkarev, N. G. (1987). Local interstellar medium. *Ap. Space Sci.* **138** 229.

Bruhweiler, F. C. and Cheng, K.-P. (1988). The stellar radiation field and the ionization of H and He in the local interstellar medium. *Astrophys. J.* **335** 188.

Bruhweiler, F. C. and Vidal-Madjar, A. (1987). The interstellar medium near the Sun. In *Scientific Accomplishments of the IUE*, Y. Kondo, ed. D. Reidel, Dordrecht, p. 467.

Cox, D. P. and Reynolds, J. (1987). *Ann. Rev. Astron. Ap.* **25** 303.

Kondo, Y., Bruhweiler, F., and Savage, B., eds. (1984). *The Local Interstellar Medium, International Astronomical Union Colloquium No. 81.* NASA Conference Publication 2345, NASA, Washington, DC.

See also **Heliosphere; Interstellar Medium, Hot Phase.**

Interstellar Medium, Molecules

Barry E. Turner

Although the first molecules (CH, CH^+, CN) were discovered in the interstellar medium as early as 1937–1941 by optical techniques, the study of interstellar matter was dramatically changed in 1968 with the discovery at radio wavelengths of the first complex molecules (H_2O, NH_3). Such molecules are easily destroyed by ultraviolet radiation, so their presence indicated the existence of dense, cool, opaque regions, regions now recognized as sites of star formation. Eighty-two molecular species have now been identified in these and similar regions surrounding evolved stars. They contain up to 13 atoms ($HC_{11}N$) and are dominated by organic species. The molecular component is comparable in abundance to the atomic form of interstellar gas in the Milky Way and in other galaxies. The molecular gas dominates in large regions within these galaxies. Several classical subjects in astrophysics, such as the birth and death cycles of stars and the structure and nucleosynthetic histories of galaxies, have been greatly advanced by the study of these molecules. The new field of astrochemistry has been born, and new ideas on the chemical evolution of the universe and on the origin of life on Earth have arisen. Indeed, it is believed that the formation of H_2 in the primordial universe was an essential step in the formation of primordial galaxies and subsequently of stars.

Relative to H_2, all other molecular species are just traces, ranging in abundance from CO, at about 20 parts per million (2×10^{-5}) down to $HC_{11}N$, which is just detectable at 2×10^{-12}. The very simplest species are usually the most abundant. Many complex species are also surprisingly abundant.

INTERSTELLAR MEDIUM (ISM)

The ISM is now known to comprise several different physical regimes. At one extreme is very hot (10^6 K), tenuous (10^{-2} atoms cm^{-3}) "coronal" gas that may exist only near the hottest stars or that may occupy sizable cavities swept out by supernova explosions. "Intercloud" gas is somewhat cooler (10^3–10^4 K) and denser (0.1–1 atoms cm^{-3}). Next come diffuse interstellar clouds (roughly 100 K and 0.1–100 atoms cm^{-3}). These contain a few simple molecules at the densest extreme. Finally, there are "dense"

Figure 1. Horsehead nebula in Orion (south of the star Zeta Orionis). The opaque cloud of molecules and dust (lower part of figure and beyond) is the Northern Orion molecular complex, comprising several tens of thousands of solar masses. The arc of emission nebulosity (ionized hydrogen) represents the edge of an expanding H II region progressing into the molecular cloud and producing a shock front at the edge of the Horsehead. (*From National Optical Astronomy Observatories.*)

molecular clouds that are much colder (roughly 10 K) and denser (10^2 to at least 10^{10} molecules cm^{-3}) than the other regimes.

Molecular clouds range in size from 1 pc to well over 100 pc, and in mass from perhaps 1–10^6 or 10^7 solar masses. There are countless small molecular clouds, but there are estimated to be only about 4000 molecular clouds of 10^5 solar masses or greater in the Galaxy. These objects, often called giant molecular clouds, are the most massive entities in the Galaxy (see Fig. 1). The total gaseous component of the ISM constitutes about 10% and the stars about 90% of the total mass of the Galaxy (about 10^{12} solar masses). Dust makes up 1% of the gas mass, and its particle density is only 10^{-12} that of the gas number density. However, the dust is almost entirely responsible for the extinction of starlight and it renders the molecular clouds completely opaque to visual and ultraviolet light.

TYPES OF INTERSTELLAR MOLECULES

The 82 known interstellar molecules are largely organic (see Table 1): Only 19 are inorganic; 26 are stable organic species, consisting of simple alcohols, aldehydes and ketones, acids, amides, esters and ethers, and hydrocarbons; 37 are unstable species, mostly organic, which include 16 free radicals, 8 ions, 3 simple rings, and 8 carbon-chain species reaching complexities as large as $HC_{11}N$ and CH_3C_5N. Most of the unstable species were unknown terrestrially before their interstellar identification. The most rapidly growing list is that of the ions and radicals, as a result of new laboratory techniques for their synthesis and for the determination of their microwave spectra. Carbon chemistry is dominant, and long carbon-chain species are most prominent among the organic compounds. Single, double, and triple carbon bonds are equally prevalent. Carbon–oxygen and carbon–nitrogen species are of comparable occurrence, whereas nitrogen–oxygen bonds are conspicuously absent in all species except NO itself. The molecules mostly contain H, C, O, and N, reflecting the high cosmic abundances of these elements. S and Si are seen in fewer species, although organo-sulfur compounds (e.g., C_2S, C_3S) are of increasing importance in astrophysical research. Phosphorus has recently been detected in the species PN, and chlorine has likely been seen in the form HCl. In circumstellar envelopes, the innermost parts of which are in thermochemical equilibrium, the species NaCl, AlCl, AlF, and KCl have recently been identified, as predicted for these conditions; such species are not expected in interstellar clouds. Despite cosmic abundances similar to S and Si, the refractory elements Mg and Fe are as yet unseen in interstellar molecules and are presumed confined to interstellar dust grains (although this should also apply to Si compounds).

INTERSTELLAR CHEMISTRY

Because of the very low temperatures and densities of the ISM, interstellar molecules cannot be formed from atoms by processes that normally occur on Earth. Astrochemists have focused on three different chemistries, or chemical regimes. Most important is the category of gas-phase reactions involving molecular ions. The most abundant molecule, H_2, cannot be made this way, but is believed to form by catalytic reaction of two H atoms on an interstellar dust grain. Once formed, H_2 is ionized by cosmic rays, and the resultant ions then react efficiently with other species to produce a large number of the observed molecular species. Catalytic surface reactions in the cold clouds appear able to produce only H_2, which alone is volatile enough to evaporate from the grain surface; in the warmer star-forming cores of molecular clouds, these surface reactions, as well as other grain processes that involve ultraviolet radiation, may locally produce many other molecular species, though convincing observational evidence for these processes remains lacking. A third type of interstellar chemistry is shock chemistry: massive stars produce expanding ionized regions of gas and later supernova remnants, both of which cause strong shock fronts to travel ahead of them through the ISM. These shocks heat and compress the gas for a few hundred years, producing ideal conditions for the occurrence of many high-temperature chemical reactions. Molecular mantles previously frozen onto grains are also vaporized under these conditions, releasing many additional types of molecules into the gas phase.

None of these chemistries proceeds under conditions that are even remotely near equilibrium (such as occur deep inside circumstellar envelopes), so the products depend on the particular reaction pathways that are most efficient under interstellar conditions, and on the input elemental abundances.

Ion–molecule reactions are the best understood of the three types of chemistries, and they appear to explain fairly well the simpler interstellar molecules (those containing up to 4 or 5 atoms and existing in the cooler molecular clouds that lack hot, dense, core regions). The more complex species (5–13 atoms) are observed in a few energetic star-forming clouds and have recently begun to be detected in colder clouds as well. Their formation in the latter clouds appears to proceed via simple two-body reactions (radiative association reactions) between quite complex molecular ions, which have recently been shown both theoretically and experimentally to proceed rapidly at very low temperatures.

By contrast, grain processes remain poorly understood. Catalytic reactions on grains consisting of silicates, graphite, and metal oxides have been studied. Laboratory experiments on silicates and graphites fail to produce many of the observed interstellar species and produce others that are not observed. Metal oxides appear to require ultraviolet radiation that rapidly dissociates the molecules. Ultraviolet radiation can, however, trigger chemical processes in grain mantles that eventually lead to explosive release of stored chemical energy; the products are not predictable in general. At present, grains are recognized as sites of frozen molecules, which can be released by gentle warming near star-forming regions. There is no observational evidence for the other processes.

Interstellar shock chemistry has proved difficult to simulate, especially the effect of magnetic fields. Model calculations of shock chemistry are now quite sophisticated, but results depend strongly on the type of shock assumed (e.g., dissociative or nondissociative, with or without a magnetic precursor) and cannot be said to provide a firm basis for comparison with observations as yet. Certain highly abundant species, such as OH and H_2O (in their

Table 1. Known Interstellar Molecules (February, 1989)

Inorganic Species (Stable)				
Diatomic		*Triatomic*	*4 Atom*	*5 Atom*
H_2	HCl?	H_2O	NH_3	SiH_4*
CO	PN	H_2S		
CS	NaCl*	SO_2		
NO	AlCl*	OCS		
NS	KCl*			
SiO	AlF?*			
SiS				

Organic Molecules (Stable)			
Alcohols	*Aldehydes and Ketones*	*Acids*	*Hydrocarbons*
CH_3OH (methanol)	H_2CO (formaldehyde)	HCN (hydrocyanic)	C_2H_2* (acetylene)
EtOH (ethanol)	CH_3CHO (acetaldehyde)	HCOOH (formic)	C_2H_4* (ethylene)
	H_2CCO (ketene)	HNCO (isocyanic)	CH_4 (methane)
	$(CH_3)_2CO$? (acetone)		
Amides	*Esters and Ethers*	*Organo-sulfur*	
NH_2CHO (formamide)	CH_3OCHO (methyl formate)	H_2CS (thioformaldehyde)	
NH_2CN (cyanamide)	$(CH_3)_2O$ (dimethyl ether)	HNCS (isothiocyanic acid)	
NH_2CH_3 (methylamine)		CH_3SH (methyl mercaptan)	
Paraffin Derivatives	*Acetylene Derivatives*	*Other*	
CH_3CN (methyl cyanide)	HC_3N (cyanoacetylene)	CH_2NH (methylenimine)	
EtCN (ethyl cyanide)	CH_3C_2H (methylacetylene)	CH_2CHCN (vinyl cyanide)	

Unstable Molecules					
Radicals		*Ions*	*Rings*	*Carbon Chains*	*Isomers*
CH	C_3H	CH^+	SiC_2*	C_3S	HNC
CN	C_3N	HCO^+	C_3H_2	HC_5N	CH_3NC
OH	C_3O	N_2H^+	C_3H	HC_7N	
SO	C_4H	$HOCO^+$		HC_9N	
HCO	C_5H	HCS^+		$HC_{11}N$	
C_2*	C_6H	H_3O^+?		CH_3C_3N	
C_2H	C_2S	$HCNH^+$		CH_3C_4H	
C_3*	CH_2CN	H_2D^+?		CH_3C_5N?	
C_5*	SiC				

*Seen only in circumstellar envelopes.

interstellar maser forms), and the highly refractory Si species (SiO, SiS) are almost certainly produced in shocks and are reasonably well predicted in some models. Several complex organic species are also generally predicted to have large abundances in shocks.

Interstellar chemistry is predicted to be time-dependent. Shock chemistry is an obvious example. Ion–molecule chemistry in quiescent cold clouds also predicts that the abundances of certain species, notably the complex carbon-rich compounds (HC_3N, HC_5N,...) are 100–1000 times greater at "early" times (10^5 yr) than after the gas phase chemistry reaches steady state (10^7 yr and later). Either a short lifetime for some clouds or an overabundance of carbon seems necessary to explain the surprisingly large observed abundances of these species. Thus astrochemistry offers the prospect of age-dating interstellar molecular clouds.

FUNDAMENTAL ROLE OF INTERSTELLAR MOLECULES IN ASTROPHYSICS

Interstellar molecules require the presence of dust to shield them from ultraviolet radiation. Dust grains are ejected into the ISM via mass loss from evolved stars. Such grains may continue to accrete in the ISM, but they cannot nucleate from the gas phase. To form stars, however, the interstellar gas must be cooled sufficiently, and this requires the presence of molecules. How, then, did the first stars form? It is now believed that H_2 was able to form in the early universe, shortly after the time of recombination, without the presence of dust grains to act as catalysts. The basic process is $H+H^+ \rightarrow H_2^+ + h\nu$ followed by $H_2^+ + H \rightarrow H_2 + H^+$. Alternatively, $H + e \rightarrow H^- + h\nu$ followed by $H + H^- \rightarrow H_2 + e$. The cooling properties of molecules, in this case H_2, then permitted the gravitational energy released by the initial contraction of those primordial condensations whose mass exceeded the Jeans mass to be radiated away. The collapse could then continue, leading eventually to star formation and hence supernova explosions that produced the first heavy elements, and dust grains. Thus it appears that grains play a secondary role to molecules in astrochemistry (certainly forming later in the early history of the universe); the first galaxies, which formed at a redshift of $z \sim 5$, would contain little if any dust. It appears likely, therefore, that molecular processes played a central role in galaxy formation, as they do in star formation.

Additional Reading

Andrew, B. H., ed. (1980). *Interstellar Molecules*. Reidel, Dordrecht.

Gammon, R. H. (1978). Chemistry of interstellar space. *Chem. Eng. News* **56** 20.

Turner, B. E. (1989). Recent progress in astrochemistry. *Space Science Reviews*. **51** 235.

Turner, B. E. and Ziurys, L. M. (1988). Interstellar molecules and astrochemistry. In *Galactic and Extragalactic Radio Astronomy*, G. L. Verschuur and K. I. Kellermann, eds. Springer, Berlin.

Vardya, M. S. and Tarafdar, S. P., eds. (1987). *Astrochemistry*. Reidel, Dordrecht.

See also **Interstellar Clouds, Chemistry; Interstellar Clouds, Molecular; Interstellar Medium, Chemical Composition; Interstellar Medium, Galactic Molecular Hydrogen.**

Interstellar Medium, Radio Recombination Lines

Peter A. Shaver

Radio recombination lines provide a wide-ranging and versatile observational tool in astrophysics. Above principal quantum number $n = 27$, recombination transitions between adjacent energy levels of hydrogen emit photons at wavelengths longward of one millimeter, and these radio-frequency recombination spectral lines are distributed over the entire radio spectrum. The largest bound orbit corresponds to $n \sim 1000-1500$ for typical interstellar conditions, and the lowest frequency spectral line observed to date in astronomy is a recombination line at a wavelength of 20 meters (corresponding to $n = 768$). Thus, there are a great many such lines, each one conveying slightly different information. As these "Rydberg" atoms are very large (0.1-mm diameter for $n = 1000$), they are sensitive probes of their environments. Also, as these transitions occur at radio frequencies, they can be studied across the Galaxy and in regions of high obscuration. They are therefore highly accessible and important probes of ionized gas in a variety of physical conditions, from stellar winds to the diffuse interstellar medium.

THEORETICAL CONSIDERATIONS

The physics of radio recombination lines is well understood. The relevant process is that in which a free electron combines with an ion at some high energy level. In some cases involving elements other than hydrogen, dielectronic recombination can occur, which proceeds through doubly excited states and can populate the high levels faster than radiative recombination, resulting in large overpopulations of the high quantum levels. The recombination lines arise as the recombined electron works its way down the ladder of energy levels.

The frequency of a transition between upper principal quantum level m and lower level n is given by the generalized Rydberg formula,

$$\nu = RcZ^2\left(\frac{1}{n^2} - \frac{1}{m^2}\right) \sim 2RcZ^2\frac{(m-n)}{n^3},$$

where $R = R_\infty(1 - m_e/M)$ is the Rydberg constant for an excited atom or ion of mass M and effective nuclear charge Z. The individual recombination lines are identified by the lower principal quantum number and the order of the transition ($\alpha, \beta, \gamma, \ldots$ for $m - n = 1, 2, 3, \ldots$); thus, H83α is a transition in atomic hydrogen from level 84 to level 83, and C527β is a transition in carbon from level 529 to level 527.

The energy level populations are computed by solving simultaneous equations that equate all populating and depopulating processes for each level. Although the ratio b_n of these energy level populations to the values they would have under conditions of local thermodynamic equilibrium (LTE) is very close to unity, the radio-frequency line intensities can still be affected strongly by the small remaining differences. The ratio of the true absorption coefficient κ_L to its LTE value κ_L^* is

$$\frac{\kappa_L}{\kappa_L^*} \sim b_m\left(1 - \frac{kT_e}{h\nu}\frac{b_m - b_n}{b_m}\right) \sim b_n\beta_{mn},$$

where T_e is the electron temperature. Thus, although $b_m - b_n$ is very small for these high-n transitions, so is $h\nu$, and the line intensities can be altered strongly. Usually, the result is enhancement due to stimulated emission.

The excess brightness temperature at the line center ΔT_L can be computed from the equations of radiative transfer,

$$\Delta T_L \sim \tau_L^*\left[b_m T_e - b_n\beta_{mn}\left(\frac{\tau_C}{2}T_e + T_0\right)\right]$$

where τ_L^* and τ_C are the LTE line and continuum optical depths and T_0 is the brightness temperature of the background continuum. From these equations one can get some idea of the global behavior of the recombination line intensities. At low energy levels, radiative effects (hence, downward transitions) dominate, and the corresponding high-frequency recombination lines appear in emission. At high n, however, collisions dominate and tend to thermalize the populations, and the corresponding low-frequency recombination lines will ultimately appear in absorption against a strong background continuum.

The line shapes are typically Gaussian, due to Doppler broadening characterized by the electron temperature and turbulence in the ionized gas. However, these very high energy levels are easily perturbed by collisions with free electrons, and pressure broadening therefore results. Adjacent levels are similarly affected by such collisions, so the net effect on transitions between them is reduced, but pressure broadening ultimately dominates over Doppler broadening for high-n transitions. The ratio of pressure to Doppler broadening varies approximately as n^{7-8}. Radiation broadening can also be important, and has a similarly strong dependence on n. Thus, at the lowest frequencies, the lines become broadened to the point where they merge with the continuum and each other, and are no longer detectable.

The study of radio recombination lines involves the interplay between a few basic phenomena, particularly stimulated emission (which enhances the peak line intensities) and broadening mechanisms (which diminish them). The lines convey information about electron temperatures and densities, turbulence, radial velocities, and relative ionic abundances, and so are important probes of ionized gas in a wide range of astrophysical contexts, as summarized in the following sections.

THE SUN, STELLAR WINDS, AND PLANETARY NEBULAE

Radio recombination lines from various ions are, in principle, detectable in the solar atmosphere. They could be enhanced by dielectronic recombination, and may appear either in emission or absorption. None have yet been detected, possibly due to the combined effects of pressure broadening, radiation broadening, and Zeeman splitting. The best prospects probably lie in the far-infrared and submillimeter wavelength ranges.

Some ionized stellar winds are sufficiently massive to produce significant radio emission. High-frequency radio recombination lines have recently been detected in one stellar wind source, MWC 349. These lines are almost certainly dominated by strong maser action. The profiles vary from line to line, and they also vary with time on scales of months. They provide an important new probe of the physical conditions in these objects.

The radio recombination line emission from planetary nebulae is relatively uncomplicated, as they are comparatively sharply defined objects. The derived electron temperatures and densities and helium abundances agree well with values obtained from optical studies. The optical data can be incomplete because of extinction, however, and the radio recombination line data give improved information on global properties such as the velocity structure.

H II REGIONS AND THEIR ENVIRONS

Embedded within many H II regions are compact cores with densities of up to 10^5–10^6 cm^{-3} and sizes of 0.01–0.1 pc. The sites of recent star formation, these are usually obscured optically, and difficult to detect with large single radio telescopes. Radio interferometers, however, discriminate against the more extended surrounding H II region and make it possible to study the compact cores themselves. The high densities and relative homogeneity make it possible to observe extremes in the behavior of the radio recombination lines. Pressure broadening is often observed (directly or indirectly), there can be large systematic mass motions, the line intensities can be quite anomalous, and large variations are found in the ratios of the helium to hydrogen lines (from 1 to 40%, whereas the actual abundance ratio is about 10%). These latter variations can be due to a variety of causes. A low He^+/H^+ ratio can be caused by incomplete ionization of helium as a result of excitation by a relatively cool star or can be caused by selective absorption of ionizing photons by dust. A high ratio can, in principle, be caused by an exceptionally hard ionizing radiation field or by genuine helium enrichment by stars.

These extreme effects are generally masked when the H II region is observed at lower angular resolution with a single radio telescope. The overall density distribution is inhomogeneous with strong gradients. The observed line emission is a complicated average over several components; usually, the lower densities dominate and the resulting recombination line appears to be Gaussian with approximately normal intensity. The He^+/H^+ ratio still varies significantly from one H II region to another, an effect probably due largely to variations in the ionizing radiation field.

One of the important discoveries using radio recombination lines was that the electron temperatures of most H II regions are significantly lower than the canonical 10^4 K obtained for the brightest H II regions using optical spectral lines. Some of the radio recombination lines are so narrow that the widths alone give absolute upper limits of just 3000–4000 K on the electron temperature. It is now known that the temperatures range from a few thousand degrees for diffuse, low-brightness H II regions to 10^4 K for the brightest, densest H II regions, to which the optical observations are most sensitive.

At low frequencies the outer, lowest-density components of H II regions dominate the recombination lines, as the line emission from more dense components is washed out by pressure broadening. In addition, recombination line emission from otherwise unknown diffuse H II regions along the galactic plane can be detected. The electron densities in these cases are low ($\sim$1–10 cm^{-3}) and the lines from these old, relaxed H II regions are often relatively narrow. In the presence of suitable background sources, significant stimulated emission can be observed.

At the outer boundaries of the H II regions are zones of rapidly diminishing ionization. Narrow recombination lines of hydrogen are observed that come from a cool zone where the hydrogen is only partially ionized. Still further out, in zones where hydrogen is no longer ionized, recombination lines of carbon and other heavy elements are frequently detected. Carbon dominates, as it is the most abundant element with an ionization potential lower than that of hydrogen. These lines are thought to originate in thin ($\sim$0.01 pc), cold (20 K) layers, and are often enhanced by stimulated emission on the near sides of the H II regions. Beyond this transition zone, the recombination line emission falls off and H I and molecular line emission begins to dominate.

In dark clouds, extended regions of carbon recombination line emission are sometimes seen, produced by embedded B stars.

INTERSTELLAR CLOUDS AND THE DIFFUSE INTERSTELLAR MEDIUM

Radio recombination lines are also emitted by cold interstellar H I clouds, which are partially ionized by cosmic rays, x-rays, and the ambient ultraviolet radiation field. They are most easily detected at low frequencies ($<$200–300 MHz) in the spectra of strong background sources of continuum radiation or in the general galactic nonthermal background itself. They can show up as emission features, due to stimulated emission, but at the lowest frequencies ($<$100 MHz) they must ultimately go into absorption. Very low frequency recombination lines in absorption appear to be ubiquitous along the galactic plane. Carbon again appears to dominate, as in the partially ionized zones around H II regions.

Analysis indicates electron densities of $\sim$0.1–1 cm^{-3}, low enough for the lines still to be observable at frequencies as low as 10–20 MHz. The 768α line observed at 14.7 MHz in the spectrum of Cassiopeia A is the lowest-frequency spectral line so far detected in astronomy. These very low frequency lines do, however, show clear evidence of pressure or radiation broadening and come close to the limits ($n \sim 1000$) set by ionospheric opacity and by radiation damping due to the galactic nonthermal radiation field, which ultimately causes adjacent lines to merge.

Upper limits on hydrogen recombination line emission from these clouds, combined with measurements of H I absorption, permit stringent limits to be set on the hydrogen ionization rate.

Low-frequency recombination lines can be used to set constraints on the filling factors of H II regions and partially ionized clouds, hence also on that of the hot intercloud medium. Radio recombination lines should also be emitted by the intercloud medium itself, perhaps enhanced by dielectronic recombination, but they must be very weak, and have yet to be detected.

LARGE-SCALE PROPERTIES OF THE GALAXY

Radio recombination lines play an important role in identifying the different types of discrete radio sources along the galactic plane, by discriminating between H II regions, which are strong line emitters, and nonthermal supernova remnants, which are not. The detected recombination lines can then be used, together with a model of galactic rotation, to map out the spiral structure of our galaxy. These lines, like those of H I and CO, are well suited to this task because the radio waves pass unimpeded across the Galaxy. With greater sensitivity, multiple components due to different H II complexes scattered along the line of sight are increasingly being found in individual spectra.

The global distribution of ionized gas over the Galaxy, and its relation to the radial distributions of H I, CO, and far-infrared emission, can also be studied in this way. Furthermore, it has been found that there is a galactocentric gradient in the electron temperatures of H II regions, which is almost certainly due to a gradient in elemental abundances; these gradients give information about the chemical evolution of our galaxy. Searches for a systematic gradient in the He^+/H^+ ratio with galactocentric radius have been inconclusive, due primarily to large variations in ionization.

Radio recombination lines provide a unique probe of the galactic center. Interferometric observations reveal strong velocity gradients in the central few parsec, which are well matched by a solid body rotation model and give important information about the central mass concentration. A strong gradient in electron temperature is also found, possible evidence for a central source of ionizing radiation. Limits can be placed on magnetic fields in the vicinity of the galactic center from the absence of Zeeman splitting in the recombination lines.

EXTRAGALACTIC RADIO RECOMBINATION LINES

With distance from our galaxy, spontaneous radio recombination line emission becomes increasingly difficult to detect. Such lines have been detected from a few prominent H II regions in the nearby Magellanic Clouds and from the unusual radio galaxy M 82, at a distance of 3 Mpc. This spontaneous recombination line emission is useful in determining the total mass of ionized gas in these galaxies and the star formation rate, parameters that may otherwise be difficult to obtain due to obscuration.

At much greater distances, spontaneous line emission becomes exceedingly weak, but radio recombination lines still can be detected due to stimulated emission or absorption in the spectra of distant continuum radio sources such as quasars. Stimulated emission of radio recombination lines has been detected clearly from the radio galaxies M 82 and NGC 253, establishing that such emission must in principle be detectable from much more distant sources. Promising possibilities are the narrow emission-line regions of radio-loud quasars, and the intervening galaxies and gas clouds that cause the optical absorption features in quasar spectra. Such lines could be used for a variety of purposes: obtaining redshifts of optically-unidentified radio sources, deriving absolute distances independent of redshifts, and setting limits on the variations of certain fundamental physical quantities over cosmic time scales. Still further possibilities include protogalaxies at very high redshifts and, perhaps, ultimately even the general intergalactic medium and the cosmic microwave background itself.

Additional Reading

Brown, R. L., Lockman, F. J., and Knapp, G. R. (1978). Radio recombination lines. *Ann. Rev. Astron. Ap.* **16** 445.

Dupree, A. K. and Goldberg, L. (1970). Radiofrequency recombination lines. *Ann. Rev. Astron. Ap.* **8** 231.

Gordon, M. A. (1988). H II regions and radio recombination lines. In *Galactic and Extragalactic Radio Astronomy*, G. L. Verschuur and K. I. Kellermann, eds. Springer-Verlag, Berlin, p. 37.

Gordon, M. A. and Sorochenko, R. L., eds. (1990). *Radio Recombination Lines: 25 Years of Investigation*. Kluwer Academic Publishers, Dordrecht.

Mezger, P. G. and Palmer, P. (1968). Radio recombination lines: A new observational tool in astrophysics. *Science* **160** 29.

Shaver, P. A., ed. (1980). *Radio Recombination Lines*. D. Reidel, Dordrecht.

See also **H II Regions; Nebulae, Planetary; Nebulae, Wolf–Rayet; Radio Sources, Emission Mechanisms.**

Interstellar Medium, Stellar Wind Effects

Christos D. Goudis

Stars lose mass in the form of stellar winds, that is, by continuous expansion of tenuous gas from their outer atmospheric layers. The stars lose substantial mass at various rates over extended periods of time (some of them during their whole life span). The stellar winds blowing from early type massive stars, Wolf–Rayet (WR) stars, red giants, supergiants, and young stars during their evolution to the main sequence, have important chemical and hydrodynamical effects when they impinge on the surrounding interstellar medium.

The chemical effect of such an interaction is the enrichment of the surrounding interstellar medium with elements processed in the interior of the stars and subsequently ejected in the stellar winds. Several important elements, such as He, C, O, and the Si–Fe group, are ejected by massive stars in episodes of significant mass loss.

The hydrodynamical effects are more complicated. Powerful stellar winds, along with supernova remnants, drive high velocity shocks (≥ 20 km s^{-1}) in the interstellar medium that are responsible for producing and maintaining its ubiquitous coronal component (hot tenuous gas with $T \simeq 10^6$ K). They also either play a fundamental role in or strongly contribute to the formation and evolution of certain well known structures of interstellar matter, which include interstellar bubbles, WR nebulae, giant and supergiant shells, planetary nebulae, and the various forms of molecular outflows (bipolar outflows, jets, etc.). The aim of the material that follows is to describe in some detail the interaction of stellar winds with their surrounding interstellar medium, which leads to the formation of these interstellar structures.

INTERSTELLAR BUBBLES

The presence of numerous optical H II shell structures, both in our galaxy and in the Large Magellanic Cloud, has been observationally established. Some structures are distinctly nonthermal; they are not associated with any active star and they are, beyond any doubt, defunct supernova remnants. The rest are either bubbles of ionized hydrogen (H II), with a typical radius of 10 pc surrounding single OB stars, or smaller bubbles, often ring-shaped nebulae, with radii in the range 3–10 pc surrounding a WR star. The presence of such central stars with known violent mass loss and the observational evidence of high-speed motions of the nebular material associated with them, suggests a stellar wind interaction with its interstellar surroundings that is shaping the structure of the bubbles and determining their evolution.

Stellar winds emitted spherically from these stars would sweep the interstellar material to form relatively thin and dense shells (typical thickness $\simeq 0.1$ pc and density $n \simeq 500$ cm^{-3}, respectively). The interior of a shell is occupied by hot ($T \simeq 10^6$ K) tenuous ($n \simeq 10^{-3}$ cm^{-3}) gas, whereas the shell itself is composed of ionized hydrogen (H II) of lower temperature ($T \simeq 10^4$ K) and possibly of neutral (H I) and molecular hydrogen (H_2), depending on the ionizing power of the central star.

The structure and dimensions of such an interstellar bubble resemble strongly a supernova remnant (SNR). The mechanical energy imparted to the surrounding interstellar medium by the stellar winds is also comparable to the energy of supernova shells. In a supernova explosion a star ejects 1–10 $M_\odot$ into interstellar space with terminal velocity $V_\star \simeq 10000$ km s^{-1} and kinetic energy 10^{51} erg. On the other hand, massive early type stars lose mass in powerful stellar winds with mass loss rates $\dot{M}_\star$ in the range 10^{-8}–10^{-5} $M_\odot$ yr^{-1}. The terminal velocities $V_\star$ of the winds are in the range 1000–4000 km s^{-1}. The mechanical energy slowly fed into the surrounding interstellar medium over the lifetime of such a star ($\simeq 10^6$ yr) is again of the order of 10^{50}–10^{51} erg.

To understand the structure and evolution of an interstellar bubble, some description of the complicated hydrodynamic phenomena caused by the interaction of the wind with the surrounding material is necessary. The impact of a spherical high velocity (supersonic) wind blowing constantly from an early type star on the surrounding stationary and smooth material sets up a two-shock flow pattern. In the simple case in which the star is assumed to be at rest with respect to the surrounding material, the flow pattern looks like the one shown in Fig. 1*a*. An outward-facing shock forms ahead of the leading edge of the stellar wind gas that separates the ambient interstellar gas into shocked and unshocked forms. The stellar wind gas is sharply decelerated by the shell of the shocked interstellar gas and an inward-facing shock is formed that separates the stellar wind into shocked and unshocked forms.

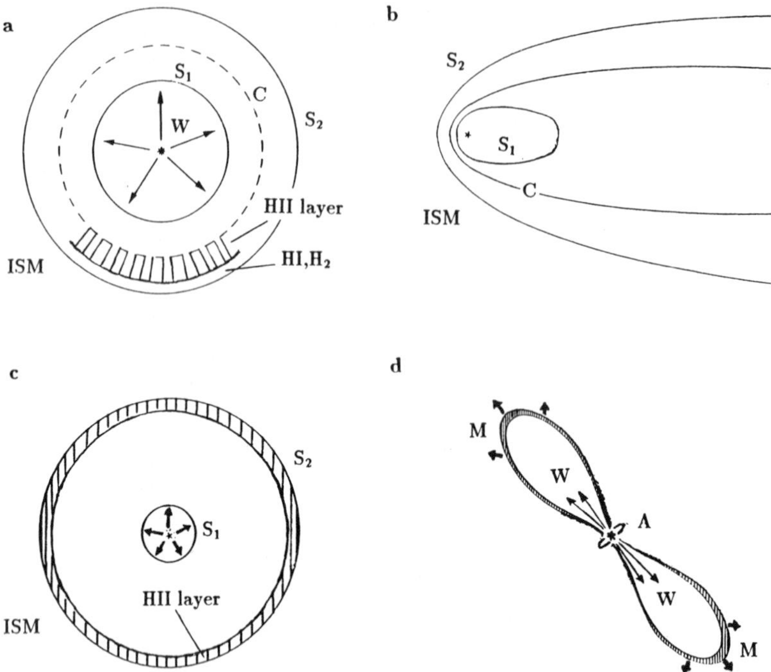

Figure 1. Interstellar structures formed by the effects of stellar winds on the interstellar medium. W is the stellar wind, S_2 the outer shock, S_1 the inner shock, C the contact discontinuity, ISM the unshocked ambient interstellar matter, M the molecular outflow, and A the accretion disk around a star embedded in a molecular cloud. (*a*) Outline of a wind-driven bubble with the star at rest with respect to the interstellar gas. In the case shown here, the shell of shocked interstellar gas is partly ionized (H II) and partly composed of neutral (H I) and molecular (H_2) hydrogen. (*b*) Outline of the bow shock (S_2) caused by the stellar wind emitted from a star moving toward the left with respect to the interstellar gas. [*Model based on T. Matsuda, Y. Fujimoto, E. Shima, K. Sawada, and T. Inaguchi (1989),* Prog. Theor. Phys. ***81*** *810–822.*] (*c*) The formation of a planetary nebula (H II layer) according to the "colliding winds model." [*Based on F. D. Kahn (1989), in* Planetary Nebulae, *S. Torres-Peimbert, ed.*] (*d*) Outline of a typical molecular bipolar flow from a young star. [*Model based on R. L. Snell, R. B. Loren, and R. L. Plambeck (1980),* Ap. J. (Lett.) ***239*** *L17.*]

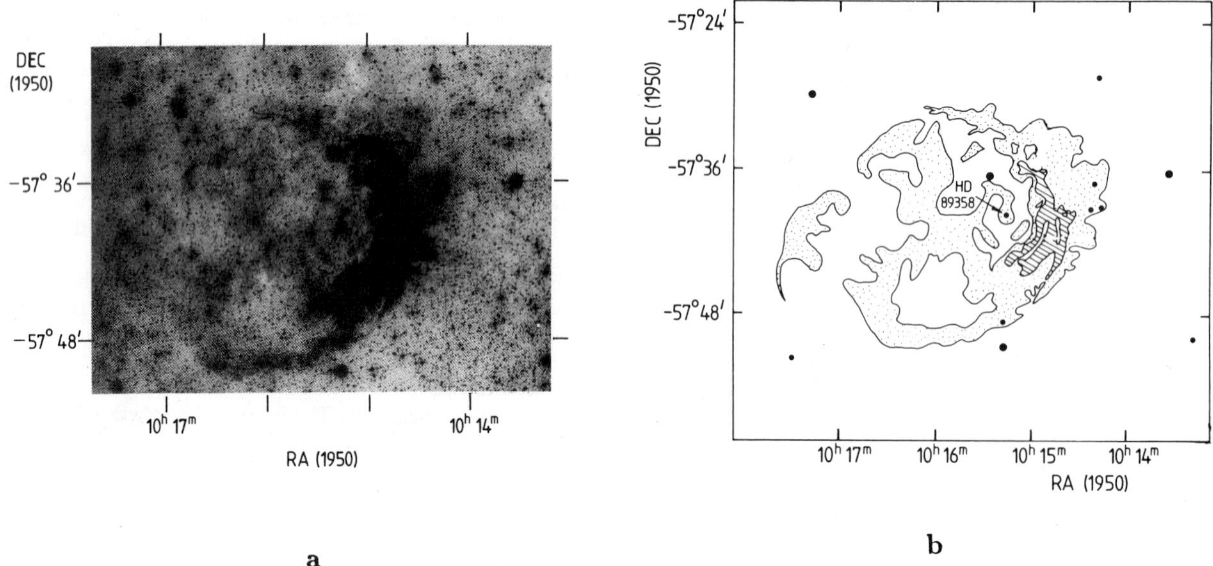

Figure 2. The WR nebula NGC 3199. (*a*) A high contrast print of NGC 3199 showing the full extent of the shell structure. (*b*) A sketch of the whole nebula including the exciting WR star HD 89358 and the observed bright arc (heavily shaded). [*From M. J. Whitehead, J. Meaburn, and C. D. Goudis (1988), The nature of the Wolf–Rayet nebula NGC 3199,* Astron. Ap. ***196*** *261.*]

The two regions of shocked gas, that is, the shocked stellar wind and the shocked (compressed) interstellar gas, are separated by a contact discontinuity. A sharp temperature difference characterizes this interface between the hot shocked stellar wind ($T \geq 10^6$ K) and the swept up compressed interstellar gas ($T \simeq 10^4$ K). The subsequent evolution of the bubble will be determined by the thermal behavior of the hot shocked stellar wind. There are two extreme cases:

1. The shocked stellar wind conserves its thermal energy. In this case, the shocked wind pressure drives the shock into the surrounding material ("energy-driven" flow).
2. The shocked wind cools very quickly by radiating its internal thermal energy away. In this case, the momentum of the wind is conserved as it impinges on the shell. Consequently the surrounding material is moved by the transferred wind momentum ("momentum-driven" flow).

Shells belonging to case 1 are expected to have more energy, larger radii, and higher velocities than those belonging to case 2.

Interstellar bubbles are distorted if the star emitting the stellar wind is moving with respect to the surrounding interstellar medium. In such a case the investigation of the interaction of a spherical supersonic wind from an object moving with respect to the surrounding interstellar material is necessary. The result of such an investigation shows that the stellar wind acts as a snowplow. The outward-facing shock has the form of a bow shock in the direction of motion, as can be seen schematically in Fig. 1*b*. The object emitting the stellar wind is surrounded by a bullet-shaped inner shock. A contact discontinuity between them separates the shocked stellar wind from the shocked interstellar matter. Consequently arc-like structures with dimensions of a few parsecs should be produced. Indeed such a structure has been observed in the case of the WR nebula NGC 3199, as shown in Fig. 2.

WR NEBULAE

From the observational point of view, the best cases to study the interaction of strong stellar winds impinging on the surrounding tenuous interstellar medium are (along with planetary nebulae) the expanding ring-like and shell nebulae around Wolf–Rayet stars (WR nebulae). These are bubbles unambiguously associated with only one plausible center of activity, a WR star. All WR stars are characterized by powerful stellar winds with a mass loss rate of the order of $\dot{M}_\star \simeq 3.10^{-5}$ $M_\odot$ yr^{-1} and terminal wind velocities $V_\star \simeq$ 2500 km s^{-1}. WR stars are massive, highly evolved objects that have lost their envelopes as a result of the very high mass loss of their progenitor Of stars, as has been established observationally. Their surface temperature is very high, in the range of 30,000–60,000 K or even higher. The formation of the shells around them, with radii in the range 3–10 pc and expansion velocities 20–80 km s^{-1}, may be due to the interaction of stellar winds not with the surrounding interstellar medium but with the debris ejected by their progenitor stars. This is supported by the fact that the masses of the majority of the shells of the WR nebulae are relatively low (5–20 $M_\odot$), which suggests a stellar origin. There are, however, some WR nebulae with distinctly larger masses (RCW 104 and NGC 3199) that are clearly of interstellar origin.

The theoretical models of these bubbles assume energy-driven flows (case 1) that, at first view, are not corroborated by estimates based on observational data that appear to indicate momentum-driven flows (case 2). According to case 1 the theoretical fraction of stellar wind energy converted into the kinetic energy of the expanding bubble is 20%, whereas the observational estimate of this conversion factor is $\simeq$ 1%. The estimate is, however, based on the amount of ionized material that is present in the shell. This is correct if the shell is fully filled with ionized matter. If, however, the ionization front is trapped inside the shell, as shown in Fig. 1*a*, the amount of neutral material in the shell is not counted and the kinetic efficiency will be underestimated.

GIANT AND SUPERGIANT SHELLS

A number of observed ionized giant and supergiant shells (radii from 20–250 pc and from 600–1400 pc, respectively), involving energy inputs of the order of 10^{52} erg, appear to be associated with clusters of from 5–50 OB stars. The formation of such gigantic structures is probably due to the combined effects of stellar winds and successive supernova explosions that can yield an energy in the range 10^{52}–10^{53} erg over a time interval of 10^7–10^8 yr.

PLANETARY NEBULAE

The formation of planetary nebulae (typical size $\simeq$ 0.1 pc and mass $\simeq$ 0.5 $M_\odot$) is also due to the action of stellar winds. The scenario starts with the evolution of a red giant, which is the progenitor of the planetary nebula and its central star. The red giant loses mass in the form of a stellar wind with $V_\star \simeq 10$ km s^{-1} and produces an extensive circumstellar, cool nebula of appreciable mass. The process continues until the hot core of the star is exposed and the fast stellar wind of the remnant star collides with the previously formed circumstellar nebula. The fast wind from the remnant star, which blows with $V_\star \simeq 1000$ km s^{-1}, sweeps into the gas of the nebula, producing a relatively dense shell of gas. The inner part of the shell, when ionized by radiation from the central star, takes the familiar form of a planetary nebula. The fast wind shocks the medium close to the central star and forms a bubble of hot gas. The hot gas cannot be cooled appreciably by contact with the surrounding ionized (H II) region and consequently expends its energy in driving the shell of slow-moving gas outward. A schematic representation of the so-called colliding winds model is shown in Fig. 1*c*.

MOLECULAR OUTFLOWS

Young stars in the first 10^5 years of their life lose mass in very powerful stellar winds, which subsequently interact with the surrounding molecular gas in which the stars are initially embedded. Shocks generated by such an interaction sweep up molecular material, producing cold molecular outflows moving with expansion velocities from 5–100 km s^{-1}, as implied by the observed velocity extent of the cold CO emission ($T \simeq$ 10–100 K). The physical size of the observed flow sources varies from 0.1–4 pc. If momentum is conserved ("momentum-driven" flows), the stellar wind velocities should be much higher because the mass of the swept-up gas is much larger than that in the stellar wind.

The observed molecular outflows often have a bipolar morphology, that is, they are collimated into two oppositely directed jets of gas. This presumably implies an initial collimation of the stellar wind itself. The winds do not always originate from the stellar surface. Accretion disks have been suggested as the place of their origin. A schematic representation of a typical bipolar outflow is shown in Fig. 1*d*. It should be stressed that the collimation and confinement of winds into jets and bipolar outflows is still a matter of intense research. It should also be kept in mind that diverse forms of outflow might reflect variations in the viewing angle or in the structure of the observed cloud.

Other noteworthy manifestations of molecular outflows driven by strong stellar winds include rapidly moving Herbig–Haro (HH) objects (semistellar emission-like nebulae), high velocity water maser sources, shock-excited molecular hydrogen emission regions, and optically visible jets originating from the immediate circumstellar environment of the young stars themselves.

Additional Reading

Dyson, J. E. and Ghanbari, J. (1989). The Wolf–Rayet nebula NGC 3199—an interstellar snow plough? *Astron. Ap.* **226** 270.

Dyson, J. E. and Smith, L. J. (1985). The energetics of nebulae associated with Wolf–Rayet stars. In *Cosmical Gas Dynamics*, F. D. Kahn, ed. VNU Science Press, Utrecht, p. 173.

Kahn, F. D. (1989). Models of planetary nebulae: Generalisation of the multiple winds model. In *Planetary Nebulae*, S. Torres-Peimbert, ed. Kluwer Academic Publishers, Dordrecht, p. 411.

Lada, C. J. (1985). Cold outflows, energetic winds, and enigmatic jets around young stellar objects. *Ann. Rev. Astron. Ap.* **23** 267.

Meaburn, J. (1983). Optical giant and supergiant interstellar shells. In *Highlights of Astronomy* **6**, R. M. West, ed. D. Reidel, Dordrecht, p. 665.

Meaburn, J. and Walsh, J. R. (1981). High velocities in the interstellar complex of M17 (NGC 6618). *Ap. Space Sci.* **74** 169.

Weaver, R., McCray, R., and Castor, J. (1977). Interstellar bubbles. II. Structure and evolution. *Ap. J.* **218** 377.

See also **Nebulae, Planetary, Origin and Evolution; Nebulae, Wolf–Rayet; Shock Waves, Astrophysical; Stars, Pre-Main Sequence, Winds and Outflow Phenomena; Stars, Winds; Stars, Wolf–Rayet Type; Sun, Coronal Holes and Solar Wind; Supernova Remnants, Evolution and Interaction with the Interstellar Medium.**

Intracluster Medium

Christopher P. O'Dea

The Virgo cluster of galaxies was detected as a source of x-rays in 1966, followed by the Perseus and Coma clusters in 1971. This was the first direct observational indication that clusters contain material other than ordinary galaxies. (The discrepancy between the cluster dynamical mass inferred from the virial theorem combined with the galaxy velocity dispersions and the observed luminous matter had suggested that there was additional unseen, or "dark," matter. However, there were no previous observations that had detected any additional material in clusters.) Since then x-ray satellites have detected many other clusters and it is now known that clusters of galaxies are very luminous sources of x-rays. The evidence is now quite strong that this emission is produced predominantly by an intracluster medium (ICM) composed of a hot, tenuous gas that fills the cluster. In principle, this medium can contain a variety of constituents (see the following sections), although the term ICM generally is used to describe the hot x-ray emitting gas.

BASIC PROPERTIES OF THE GAS

The basic properties of the hot gas are derived from the x-ray observations under the assumption that the emission is produced by thermal bremsstrahlung and line emission. (Thermal bremsstrahlung is produced by near collisions of electrons with protons. Line emission is produced as an electron bound to an ion changes from one discrete energy level to another.) Some of the best data available so far come from the *Einstein* observatory, although because of the limited instrumental capabilities (sensitivity, spectral resolution, and angular resolution) there remain many uncertainties concerning the detailed properties of the gas. Very roughly, the gas has a density of $\sim 10^{-3}$ atoms per cubic centimeter. To a very good approximation, the gas pressure is balanced by the gravitational pull of the cluster (this is called hydrostatic equilibrium). Thus, the gas density is somewhat higher in the cluster center and falls off with distance from the center. The distance at which the density has fallen to about half of its value at the center is typically a few hundred kiloparsecs. The gas typically extends much further than this, out to at least about 3 Mpc. However, the brightness of the x-ray emission is proportional to the square of the gas density and thus the emission becomes considerably fainter in the outskirts of the cluster. Thus the total extent of the gas and its total mass are hard to measure. The current rough estimates are that the mass in hot gas is about 10^{14} solar masses. This is probably about 10% of the total mass of the cluster.

Perhaps surprisingly, x-ray spectral observations detected emission lines from highly ionized ions of certain heavy elements (e.g., iron, magnesium, silicon, sulfur, and oxygen). The abundances of these elements in the ICM approach values that are half those found in the Sun. The existence of these elements provides strong support for the hypothesis that the observed x-ray emission is produced by a hot gas and not by stellar sources or by relativistic electrons. The x-ray spectral data also provide evidence for gas in many clusters with a range of temperatures from $\sim 2 \times 10^7$ to $\sim 8 \times 10^7$ K. The limited data so far indicate that the cooler temperatures are found in the cluster centers and in the atmospheres (interstellar media) of individual galaxies.

OTHER CONSTITUENTS OF THE ICM

In this entry we discuss mainly the hot x-ray emitting gas. But it is worth mentioning the other constituents of the medium between the galaxies.

In some clusters, faint diffuse light is seen that is concentrated on the cluster center around the dominant galaxy. This light is probably produced by stars that were originally bound to galaxies but that have been removed through gravitational "tidal" forces that are exerted as galaxies pass close to each other on their orbits through the cluster. The tidal forces produced by the gravitational field of the whole cluster itself may also remove stars from galaxies.

Radio-wavelength observations show that in some clusters (about half a dozen so far) there is diffuse radio emission present within the inner few hundred kiloparsecs of the cluster. These diffuse radio sources are apparently not associated with individual galaxies in the clusters but are a phenomenon of the intracluster medium and have been called cluster radio halos. By analogy with the emission from radio galaxies, the radio emission is probably produced via the synchrotron mechanism by relativistic electrons accelerated by magnetic fields. This requires two additional components of the ICM, namely, relativistic particles (or cosmic rays) and magnetic fields of strength $\sim 10^{-6}$ gauss. The origin of these particles and fields is still uncertain. One possibility is that the cosmic ray particles and magnetic fields originated in radio galaxies in the cluster. These radio galaxies have presumably turned off and have now faded away. Energy may be resupplied to the relativistic electrons in radio halos by the turbulent wakes of galaxies that occur as they pass through the core of the cluster.

Only about 20% of the total mass in clusters of galaxies presently can be accounted for by adding up the mass in galaxies and x-ray emitting gas. Thus, there must be dark matter that is bound to the *cluster* to provide the "missing" $\sim$ 80% of the total cluster mass but that cannot be bound to the individual galaxies. (If the dark matter were contained in the galaxies, these heavy galaxies would interact very strongly. This would result in the brightest and most massive galaxies being concentrated to the cluster centers, which is not observed). This suggests that there is additional material in the ICM whose nature is as yet unknown. Current speculations about the dark matter fall primarily into three groups: (1) small objects, such as very faint stars, planets, or comets; (2) remnants of massive stars, such as black holes, neutron stars, or cool white dwarfs; (3) elementary particles.

ORIGIN OF THE HOT GAS

It is possible that some of the gas that is currently in the ICM is primordial, that is, it is pristine material produced during the Big

Bang. This could be gas that was not used in the formation of the galaxies in the cluster or that fell into the cluster after its formation. This gas should be mainly hydrogen with some additional helium, deuterium, and lithium that were produced in the Big Bang. However, as discussed previously, x-ray observations have detected emission lines from heavy elements such as oxygen, silicon, and iron. Astronomers currently believe that heavy elements can only be produced though nucleosynthesis occurring in stars. This material is ejected from stars though winds or supernova explosions. If the stars that provided the "enriched" material are in the galaxies, then the galaxies must have lost large amounts of gas at some time in the past.

INTERACTIONS OF GALAXIES AND RADIO SOURCES WITH THE ICM

The interactions between galaxies and the ICM may be quite significant. As noted previously, galaxies may have lost large amounts of heavy-element-enriched gas to the ICM. The dense ICM thus created exerts a ram pressure on the galaxies as they orbit through the cluster. This ram pressure can further strip the interstellar medium from galaxies and may be responsible for the observed deficiency of atomic hydrogen in the interstellar media of spiral galaxies in clusters relative to galaxies outside of clusters.

Radio sources in clusters of galaxies are influenced by the intracluster medium in a number of ways. The ICM provides the external pressure needed to confine the radio-luminous plasma. The existence of confined radio sources at distances of ~ 3–5 Mpc from the centers of clusters indicates that an ICM sufficient to confine the source exists at those large distances. This is far beyond where the x-ray observations have the sensitivity to detect the low density gas.

Extended radio sources outside of clusters tend to have linear shapes, whereas sources inside clusters tend to be bent and twisted. In some sources, beams of radio-luminous plasma ejected in opposite directions by the galaxy have been bent together into a "tail" on one side of the galaxy. In these "tailed" radio sources the ram pressure produced by the galaxy's motion through the ICM appears to be responsible for the bending of the beams of plasma. This interaction allows the momentum flux in the radio beams to be estimated and is a powerful tool for studying the properties of radio beams.

THE PROBLEM OF COOLING FLOWS

One of the most interesting recent developments is the awareness that the denser gas in the inner ~ 100 kpc of the cluster can radiate away its energy over a time that is shorter than the lifetime of the cluster. If energy is not resupplied to the gas (i.e., if there are no sources of heat), the cooling gas will be pushed into the cluster center. The rate at which the gas in different clusters is estimated to be cooling is in the range ~ 10–1000 solar masses per year ($M_\odot$ yr^{-1}) with typical values near 100 $M_\odot$ yr^{-1}. There is evidence for cooling in about 40 clusters so far, although some astronomers believe that it is very common. In the cluster centers where the gas is dense enough that cooling is thought to be important, there is always also a very large elliptical galaxy. The inferred flow of cooling gas onto the central galaxy is called a cooling flow. It is important to realize that although there is considerable evidence for the existence of "cooling" of the gas, there is no direct evidence yet for a "flow" of this gas onto the central galaxy.

If cooling flows do exist and persist for the lifetime of a cluster (perhaps about 10 billion years), the total amount of mass accreted by the central galaxy can be quite substantial, about 10^{12} solar masses. This would have significant implications for the formation of these central galaxies. Thus, whether or not these cooling flows occur at the large inferred rates is an important (and also controversial) question. There are two main uncertainties concerning the cooling flows.

The first major uncertainty is whether heating of the cooling gas is occurring at a sufficient rate that the inferred mass accretion rate is significantly reduced. There are several ways in which the gas might be heated. The vast majority of the gas in the cluster is at a higher temperature than the central cooling gas and could, in principle, reheat the gas. However, it is not clear whether magnetic fields in the ICM could prevent the conduction of heat into the cluster center. The motions of galaxies through the cluster center might also heat the gas. The radio sources that are typically present in the central galaxies might also heat the gas through shocks as the radio source expands into the ICM or through the cosmic ray particles in the radio source. All these mechanisms are so poorly understood that it is difficult to estimate whether enough heat is produced that the cooling flow is substantially reduced.

On the other hand, if the cooling flow occurs at the estimated rates, then a very large amount of gas should be present in the central galaxy. The fate of this accreted gas is the second major uncertainty. Simply stated, there is evidence for no more than about 0.1% of the expected quantity of gas present in the central galaxies. One possibility is that the gas has formed stars. However, there are not enough young bright (massive) stars present in the central galaxies. Some astronomers believe that only very low-mass stars that are too dim to be observed are formed. This would mean that the process of star formation is very different in these central cluster galaxies than in our galaxy. Clearly, this is an exciting and evolving area of research.

FUTURE PROSPECTS

Large advances in our understanding of the ICM, as well as clusters in general, are expected when the next generation of x-ray astronomy satellites (especially *AXAF*, the *Advanced X-Ray Astrophysics Facility*) is launched. These satellites will provide much-improved imaging and spectroscopic observations that will be used to determine the density and temperature of the gas as a function of distance from the cluster center. This should allow the mass of the cluster to be determined as a function of distance from the cluster center. An accurate estimate of the mass of the cluster and the amount of dark matter will then be possible. In addition, the properties of the cooling gas in the cluster centers and the nature of the cooling flows should be greatly clarified.

Additional Reading

Fabian, A. C., ed. (1988). *Proceedings of NATO Advanced Research Workshop on Cooling Flows in Galaxies and Clusters*. D. Reidel, Dordrecht.

Forman, W. and Jones, C. (1982). X-ray imaging observations of clusters of galaxies. *Ann. Rev. Astron. Ap.* **20** 547.

O'Dea, C. P. and Uson, J. M., eds. (1986). *Proceedings of NRAO Workshop No. 16, Radio Continuum Processes in Clusters of Galaxies*. National Radio Astronomy Observatory, Green Bank.

Sarazin, C. L. (1988). *X-Ray Emission from Clusters of Galaxies*. Cambridge University Press, Cambridge.

Takano, S., et. al. (1989). Large scale extended x-ray emission from the Virgo cluster of galaxies. *Nature* **340** 289.

See also **Clusters of Galaxies; Clusters of Galaxies, Component Galaxy Characteristics; Clusters of Galaxies, Radio Observations; Clusters of Galaxies, X-Ray Observations; Dark Matter, Cosmological; Galaxies, Radio Emission.**

Io, Volcanism and Geophysics

Alfred S. McEwen

Io is the innermost of the four large Galilean satellites of Jupiter discovered by Galileo in 1610. Io's radius (1815 km) and bulk density (3.57 g cm^{-3}) are comparable with those of the Moon. However, long before the *Voyager* spacecraft encounters, it was apparent from Earth-based observations that Io is very different from the Moon: It has an unusual spectral reflectance and anomalous thermal properties, and it is surrounded by immense clouds of ions and neutral atoms. These observations were much better understood following the discovery of active volcanic plumes and hot spots on Io during the *Voyager 1* spacecraft encounter in 1979.

The bulk density of Io, its rugged topography, and models of satellite origin suggest that the bulk composition of the crust and mantle is silicate. No spectral features diagnostic of silicates have been detected, but the surface composition has been profoundly altered by volcanic resurfacing and outgassing. Sulfur dioxide (SO_2) has been positively identified both as a gas and as a surface frost. Elemental sulfur is considered a likely surface component on the basis of (1) the similarity of Io's ultraviolet through near-infrared spectral reflectance to that of powdered sulfur, (2) the detection of sulfur ions in Io's plasma torus, and (3) models for Io's volcanism and for the evolution of the satellite's volatile inventory. The detections of clouds of Na and K around Io have led to proposals that Na and K sulfides or polysulfides may be present on the surface.

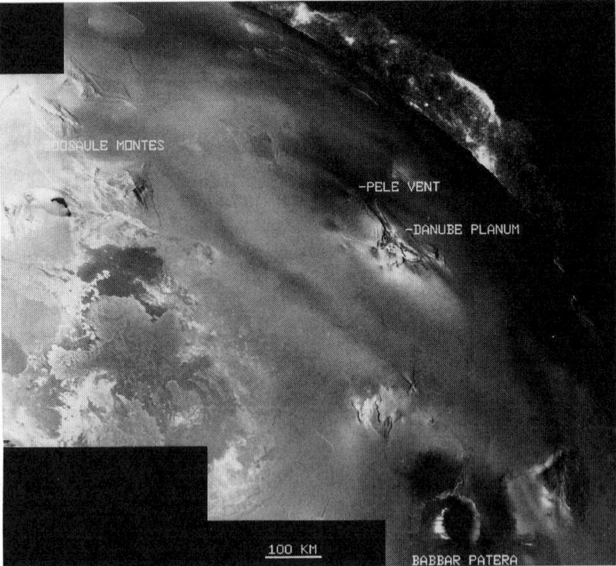

Figure 1. Mosaic showing vent, airborne plume, and fallout deposits of Pele. Vent area was a hot spot (650 K) during *Voyager 1* encounter. The plume, visible above the limb, was specially enhanced for this mosaic (it is actually very faint and barely visible). A variety of calderas (e.g., Babbar Patera), flows, and mountains (e.g., Boosaule Montes) are visible; Danube Planum is a rifted plateau.

TIDAL-HEATING MECHANISMS AND INTERNAL STRUCTURE

The importance of tidal heating in powering the enhanced heat flow and active volcanism of Io is now widely accepted. All plausible nontidal heat sources are about 2 orders of magnitude less energetic than the power output of Io (at least 4×10^{13} W). Accretional heating can supply, at most, half the gravitational potential energy of Io; the power output, averaged over 4.5 billion years, is 5×10^{11} W. However, the accretional heat was probably removed by conduction or convection very early in Io's history. Core formation can supply only a fraction of the maximum accretional energy, and it also is short-lived. Radiogenic heating (i.e., from the decay of radioactive isotopes) is a continuous process and is estimated to supply 5×10^{11} W. Heating of the surface by an electrical current running between Io and Jupiter's ionosphere provides a maximum power of 2×10^{11} W, but an ionosphere is likely to divert most of the current around Io.

The basic tidal-heating mechanism involves deformation of Io into a triaxial ellipsoid by Jupiter's gravitational field. Because Io's orbit is eccentric, the satellite undergoes a periodic deformation, which generates internal heat. The dissipation also decreases Io's orbital eccentricity, which is forced to remain nonzero by orbital resonances among Io, Europa, and Ganymede. The total rate of tidal dissipation within Io depends inversely upon its dissipation factor, which is a function of the internal structure and material properties of the satellite. However, the transfer of orbital energy from Jupiter to Io is a consequence of the bulge raised on Jupiter by Io, and the rate of such transfer depends inversely on Jupiter's dissipation factor. This value is poorly known, but adopting a lower limit for this quantity results in an upper limit to Io's energy dissipation rate (if steady state) of 9×10^{13} W.

The mechanism of heat dissipation is intimately tied to Io's internal structure. Early models, with tidal energy dissipated within a thin, elastic lithosphere and "runaway" melting of the interior, now seem unlikely. Dissipation of all the tidal energy within the lithosphere requires that it be 8–18 km thick, and such a thin lithosphere appears inconsistent with the presence of 10-km-high mountains (see below). The models may be modified by invoking a thicker lithosphere and advective heat transfer, but runaway melting might never occur because subsolidus convection would quickly remove the heat. Moreover, efficient liquid-state convection would lead to solidification of most of the satellite in 10^8 to 10^9 years, leaving a differentiated body with an iron-rich core, a solid mantle, and a partly molten asthenosphere. Therefore, most of the tidal energy is now probably dissipated by viscous or viscoelastic deformation below the lithosphere, and most of the heat is probably transported toward the surface by silicate magmas.

GEOMORPHOLOGY

The morphology and surface markings on Io are almost entirely the result of volcanic processes. Impact craters, the dominant landform on most solid planetary bodies, have not been observed on Io, despite the comet flux concentration by Jupiter's gravity. The lack of impact craters indicates that the surface is very young, which is consistent with estimates of resurfacing by volcanic activity. The intense volcanism has resulted in a variety of surface features of three general categories: calderas, mountains, and plains (Fig. 1).

Io is spotted by local dark markings; where picture resolution and illumination conditions permit discrimination of surface relief, most of these markings are seen to lie within depressions that resemble terrestrial calderas. Calderas are depressions with steep, scalloped walls, flat floors, and smooth rims; they form by collapse over shallow magma chambers following eruption of volcanic materials. Several dark markings are sources of volcanic plumes and thus are clearly the sites of volcanic vents. The calderas range from about 20 to 200 km in diameter and are as deep as ~ 2 km. Dark markings may cover all or part of a caldera floor, and they have been interpreted as sulfur lava lakes or recent basaltic lava flows.

Diffuse bright markings in and around calderas are common and may be the result of SO_2 fumarolic activity. Most calderas are not associated with an obvious edifice, but some low shields with summit calderas have been found. Several calderas were identified as hot spots (described later).

The second major class of surface feature is mountains, which typically have irregular outlines and rugged surfaces that appear to be tectonically disrupted. They may be as large as 600 km in basal diameter and 10 km high. They may be either volcanos or exposures of a silicate basement complex. Most workers agree that the mountains must be dominantly silicate rather than sulfur, because sulfur does not have sufficient strength to support the observed relief. Although superposition relations are unclear, the mountains are generally considered to be the oldest materials exposed on Io's surface, but higher resolution images are required to confirm this hypothesis.

Most of Io's surface is covered by plains with little apparent relief. They have complex albedo and color patterns, and, where relief is discernible as in the southern hemisphere, they are seen to be layered. The plains and their surface pattern have been formed by a variety of materials and processes, including lava flows, deposition from plumes, condensation of volatiles, and possible pyroclastic flows.

The relative abundance of silicates and sulfurous materials in the upper crust of Io is controversial. Most current workers agree (1) that the rise of silicate magmas through a silicate crust is the main mechanism for transporting tidal energy dissipated below the lithosphere and (2) that these magmas may mobilize more volatile species near the surface, thereby forming plumes, lakes, or eruptions of low-temperature melts such as sulfur.

VOLCANIC PLUMES

The most dramatic phenomena on the surface of Io are the active volcanic plumes. They erupt under conditions of pressure, temperature, and gravity very different from those on Earth, and they are probably driven by SO_2 and/or sulfur in which multiple phase changes can occur, so the plume dynamics and thermodynamics are unlike those of terrestrial eruptions. The low gravity and especially the very low atmospheric pressure on Io ($<10^{-7}$ bar) result in plumes up to 10^4 times larger on Io than would be produced by comparable eruptions on Earth.

Nine eruption plumes were observed during the *Voyager 1* encounter; eight of these were also observed four months later by *Voyager 2*. Many of the plumes, such as that of Prometheus (Fig. 2), were 50–100 km high. The plume of Pele (Fig. 1), active during the first encounter and having the greatest height ($\sim$ 300 km), had ceased activity sometime before the second encounter. Also, surface markings visible only in *Voyager 2* images show that two additional eruptions similar to that of Pele had occurred after the first encounter.

The dominant volatiles driving explosive volcanism on Earth, H_2O and CO_2, seem to be highly depleted on Io. Discernible H_2O or carbonate absorption bands are absent, and carbon has not been detected in Jupiter's magnetosphere near Io. These absences suggest that no significant amounts of H_2O or CO_2 gases occur in the atmosphere or in the volcanic plumes, although recent evidence for H_2S may indicate the presence of some magmatic H_2O. Volatiles such as SO_2 or sulfur probably drive the explosive volcanism.

The observed plume eruptions are of two major classes: (1) Prometheus-type eruptions are smaller (50–120 km high) and long-lived (months to years), deposit bright white materials, erupt at velocities of ~ 0.5 km s^{-1}, and are concentrated in an equatorial band around the satellite; (2) Pele-type plumes are very large ($\sim$ 300 km high) and short-lived (hours to days), deposit relatively dark red material, erupt at ~ 1.0 km s^{-1}, and occur in the region from longitude 240° to 360°. Surface patterns indicating previous Prometheus- and Pele-type eruptions have the same global distributions as the active plumes.

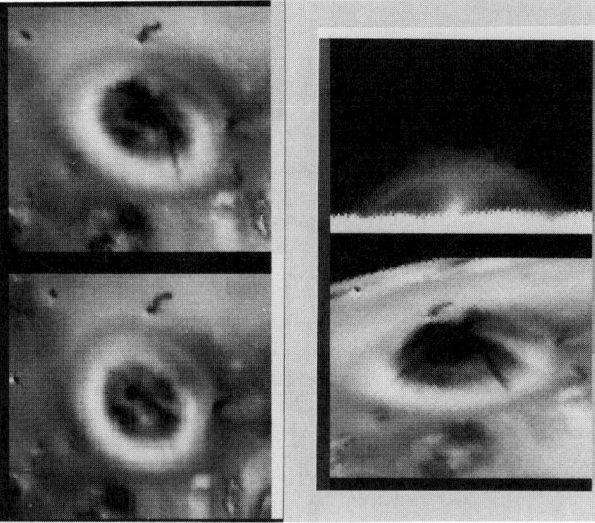

Figure 2. Four views of the active plume Prometheus from different viewing angles. Dark jets have the same albedo whether seen against a bright surface or against a space, so the jet material is optically thick. The plume's height is 75 km and the bright ring has a diameter of about 300 km.

HOT SPOTS

Evidence for hot spots or thermal anomalies on Io was acquired years before the *Voyager* encounters, but it was not understood. Thermal infrared observations had revealed two curious relations. First, when Io is eclipsed by Jupiter, its temperature falls as expected for a surface with low thermal inertia, but its minimum temperature is too high. Second, the infrared brightness temperature of Io is significantly higher at shorter wavelengths. After the *Voyager 1* encounter and the discovery of active volcanism, these observations were understood as being due to hot spots with temperatures reaching as high as ~ 650 K, whereas the surface heated only by insolation reaches temperatures no higher than ~ 140 K.

The infrared interferometric spectrometer (IRIS) carried by *Voyager 1* observed individual thermal anomalies on Io from about 4- to 55-μm wavelength. Eleven major hot spots have been identified from IRIS data; all correspond to low-albedo surface features. Most of the features are caldera-like structures; at least three appear to be associated with lava flows; three others were associated with active volcanic plumes.

Another type of thermal anomaly on Io (observed from Earth) is the "outbursts," short-lived (hours to days) enhancements at short infrared wavelengths. Most of the outbursts are consistent with ~ 600-K hot spots, but a 1986 observation indicated a temperature of at least 900 K, which suggests active silicate volcanism. Aside from the short-lived outbursts, the longitudinal pattern of heat flow from Io (dominantly at wavelengths near 8–10 μm) remained remarkably constant for at least 15 years until 1986, when some fluctuations began.

Estimates of the global hot-spot heat flow range from 1.0 to 2.0 W m^{-2}. The total power emanating from Io must also include contributions from conducted heat and kinetic energy from the active plumes. Both of these quantities are poorly known, but global averages probably do not exceed 0.5 W m^{-2}. Therefore, the total energy loss from Io may range from 1.0 to 3.0 W m^{-2}, orders of magnitude greater than the average heat loss from Earth (0.08 W m^{-2}) or from the Moon (0.02 W m^{-2}). The maximum steady-state energy flow theoretically possible from tidal heating is $\sim$2 W m^{-2}, so if Io's energy flow exceeds this value, then either episodic tidal heating or intermittent heat loss must be invoked.

Many important questions about Io remain unanswered. The *Galileo* spacecraft was launched in October 1989 and will reach Jupiter in 1995. With this mission, mapping of Io will be completed in the hemisphere poorly observed by *Voyager*, and the mission will provide observations with much better spatial and spectral resolution. Planetary scientists are eager to see how the surface of Io has changed during the 16 years between the *Voyager* and *Galileo* encounters.

Additional Reading

Belton, M. J. S., West, R. A., and Rahe, J., eds. (1989). *Time-Variable Phenomena in the Jovian System*. NASA SP-494. NASA, Washington, D.C.

Burns, J. A. and Matthews, M. S., eds. (1986). *Satellites*. University of Arizona Press, Tucson.

Johnson, T. V. (1990). The Galilean satellites. In *The New Solar System*, 3rd ed., J. K. Beatty and A. Chaikin, eds. Sky Publishing Corporation, Cambridge, MA, and Cambridge University Press, Cambridge, U.K., p. 171.

Johnson, T. V. and Soderblom, L. A. (1983). Io. *Scientific American* **249** (No. 6) 56.

Morrison, D., ed. (1982). *Satellites of Jupiter*. University of Arizona Press, Tucson.

See also **Jupiter and Saturn, Satellites; Satellites, Ices and Atmospheres.**

J

Jets, Theory of

Arieh Königl

The term "jets" was originally coined to describe the narrow, elongated features that had been discovered in optical and radio maps of active galaxies and quasistellar objects. Similar features have subsequently been found on a variety of scales in our galaxy. In particular, such narrow structures are now commonly identified in association with pre-main sequence stars embedded in dense molecular clouds. The image of a collimated gas flow that is conveyed by this term is not coincidental: It is now generally accepted that jets represent energetic outflows that emanate, often in two opposite directions, from the vicinity of compact astronomical objects. In the case of radio galaxies and quasars, the central objects are thought to be massive (10^8–10^9 $M_\odot$) black holes in the galactic nuclei, whereas in the case of newly formed stars the central masses may be less than 1 $M_\odot$. However, despite the vast differences in scale (galactic jets sometimes exceed 1 Mpc in length and can move at close to the speed of light near the origin, whereas stellar jets seldom extend beyond 1 pc and move at only a few hundred kilometers per second), the striking morphological similarities and the ubiquity of collimated outflows suggest that jets on all scales are a manifestation of a universal phenomenon that can be described in terms of certain basic dynamical principles. Explaining the origin of this apparent universality and the nature of the underlying physical processes are the main challenges of a successful jet theory.

PRODUCTION MECHANISMS

Nozzles and Funnels

Possible clues to the common occurrence of jets may be sought in the circumstances under which they are formed. Unfortunately, the formation region in any of the observed jets is not directly accessible to observations, so at this stage this question is still open. In one approach to the problem, it is postulated that the central object is the source of an isotropic gas outflow (like a stellar wind) that is channeled into two oppositely directed jets by the ambient medium. Two basic mechanisms for the formation of such channels have been proposed: the transonic nozzle, which arises from the dynamical interaction of a hot gas with a slightly flattened, confining mass distribution, and the centrifugally-supported funnel, which forms a preexisting conduit for the injected gas. The transonic flow idea is based on the well-known de Laval nozzle principle of aerodynamics (after Carl G. P. de Laval): A subsonic gas that expands into a convergent–divergent nozzle undergoes a continuous reduction in its pressure and becomes supersonic after passing through the narrowest portion of the channel. In the astrophysical application, the decreasing external pressure (rather than the nozzle geometry) is assumed to be given and it is hypothesized that the flow still undergoes a sonic transition as the channel walls adjust to maintain pressure equilibrium with the surrounding medium. The required flattening of the ambient gas could be due to rotation, in which case the jets would emerge along the rotation axis.

Rotation may also lead to the formation of funnels. If the compact object is embedded in a rotating cloud and accretes gas that conserves at least part of its angular momentum as it moves inward, then a central funnel will form by the action of the centrifugal force that excludes the accreted matter from the vicinity of the rotation axis. Alternatively, in the context of thick accretion disks around black holes, a funnel could form by the action of a purely general-relativistic effect that gives rise to an evacuated, roughly paraboloidal "zone of nonstationarity" in the innermost region of the disk. It has been suggested that the very process of funnel formation by centrifugal forces may cause some of the inflowing matter to be redirected along the funnel walls. Additional acceleration of this matter by radiation pressure forces might then result in the formation of energetic jets. Radiation pressure could similarly expel material above the photospheres of thick accretion disks in high-luminosity sources. Although the general applicability of these ideas has not yet been demonstrated, they are nevertheless attractive in that they naturally tie the formation of jets to mass accretion in compact objects.

Centrifugally Driven Outflows

Another promising mechanism that relates the existence of jets to the accretion process postulates the presence of accretion disks with embedded magnetic fields. If the field has an "open" topology and the field lines are inclined at sufficiently large angles to the disk surface, then material tied to the field will be flung out as a result of the centrifugal force, much as beads are pushed out along a rotating wire. Above the disk surface, additional acceleration is provided by magnetic pressure gradients, whereas the tension of the poloidal field component and the pinch stress of the azimuthal field component act to collimate the flow. Simple models have been constructed to illustrate the possibility that hydromagnetic outflows of this type could carry away most of the excess angular momentum and the liberated gravitational energy in the disk. The important implication of these models is that jets may not merely be an incidental by-product of accretion but that they could, instead, represent an essential ingredient in this process. In this view, the ubiquity of jets in compact astronomical objects is a manifestation of the fact that centrifugally-driven outflows from the associated accretion disks transport most of the angular momentum that needs to be removed in order for accretion to proceed. So far, the strongest support for this scenario has come from observations of pre-main sequence stars, where evidence has been accumulating for the correlated presence of circumstellar disks and energetic bipolar outflows as well as for the existence of embedded magnetic fields oriented parallel to the outflow axes. However, the same mechanism is also expected to operate in other cosmic jet sources, and, in fact, is a leading candidate for energy and angular momentum extraction from accreting black holes in active galactic nuclei.

CONFINEMENT AND COLLIMATION

Jets produced by any of the mechanisms just discussed may initially be rather poorly collimated. However, additional collimation could be achieved at large distances r from the origin by the action of a confining ambient pressure. For this to take place, the external pressure must decrease sufficiently slowly with increasing r so that the jet can remain in pressure equilibrium with its surroundings. In the case of a supersonic, narrow jet and an ambient pressure that scales as r^{-n}, this condition implies $n \leq 2$. If this condition is not satisfied, then the jet will expand freely with a constant opening angle until the pressure distribution becomes flatter, at which point it will be recollimated. This behavior has, in fact, been inferred to be present in both stellar and galactic jets. A sudden increase in the ambient pressure could, however, lead to the disruption of a supersonic jet, particularly if its Mach number M_j is comparatively low. The jet might then decelerate as a result of internal shocks and turbulent entrainment of the ambient gas, which would cause it to flare out. This could account for the observed morphologies of wide-angle-tailed radio galaxies. The gradual broadening exhibited by low-power radio jets is probably also associated with the deceleration of subsonic, turbulent flows.

Magnetic pinch stresses, already mentioned in connection with centrifugally driven outflows, provide an alternative collimation mechanism for jets. For any expanding supersonic jet that is highly conducting and not strongly sheared, it is possible to argue, on the basis of magnetic flux conservation, that the azimuthal field component should dominate the axial component sufficiently far away from the origin. In fact, this property could be responsible for the observed transition from longitudinal to transverse projected field orientation in many weak radio jets. The "hoop" stress of the toroidal field could collimate the central regions of jets that carry a net current along the axis. The return current might flow along the outer boundary of the jet, where an overall confining pressure must still be exerted by an external agent. Nevertheless, if only the innermost, high-pressure region of the jet is visible, then this mechanism could account for the apparent discrepancy between the estimated external pressure and the much higher inferred internal pressure in certain radio jets.

PROPAGATION EFFECTS

The Working Surface

The interaction of a jet with the medium through which it propagates strongly influences its structure and stability. Although some of the qualitative features of this interaction were initially deduced from general principles, only with the advent of sophisticated supercomputer simulations in the last few years has it become clear that the actual behavior could be very complex and is best studied numerically. A good example is provided by the development of our understanding of the heads of high-Mach-number jets, which were proposed as the explanation for the bright emission regions ("hot spots") observed at the outer edges of high-power radio jets. The structure of the head was originally described in terms of an effectively planar "working surface" consisting of a jet shock where the flow is decelerated and a forward bow shock where the impacted ambient gas is accelerated, with the two shocked gases being separated by a contact discontinuity. On the basis of this one-dimensional flow model, it was argued that the jet head would advance into the ambient medium with a speed $v_h \approx v_j/(1+\eta^{-1/2})$, where v_j is the jet speed and $\eta = \rho_j/\rho_a$ is the jet-to-ambient density ratio. Although the basic aspects of the predicted morphology were subsequently confirmed by two- and three-dimensional hydrodynamical simulations (see Fig. 1*a*), several important new features have emerged from the numerical work. In particular, it was found that the gas flowing near the jet boundary is not stopped in a strong perpendicular shock but is instead deflected sideways by weak oblique shocks. For "light" ($\eta \ll 1$) jets, this can lead to a substantial increase in the working surface area and to a reduction in the value of v_h. In addition, it was discovered that the shapes of the jet shock and of the contact discontinuity can be highly time variable as a result of quasiperiodic vortex shedding induced by the oblique shock configuration. The predicted emission patterns are, nevertheless, consistent with the observed radio hot spots.

In the case of high-M_j, low-η jets, the shocked jet material flows back from the head region and forms an extensive "cocoon" around the jet. However, numerical simulations of magnetized outflows have revealed that a strong toroidal magnetic field in the jet can prevent the shocked jet gas from flowing sideways and instead force it to accumulate in a narrow "nose cone" between the terminal jet shock (or "Mach disk") and the leading bow shock (Fig. 1*b*). This configuration may be relevant to the jet in the quasar 3C 273. Another modification of the basic working surface structure that

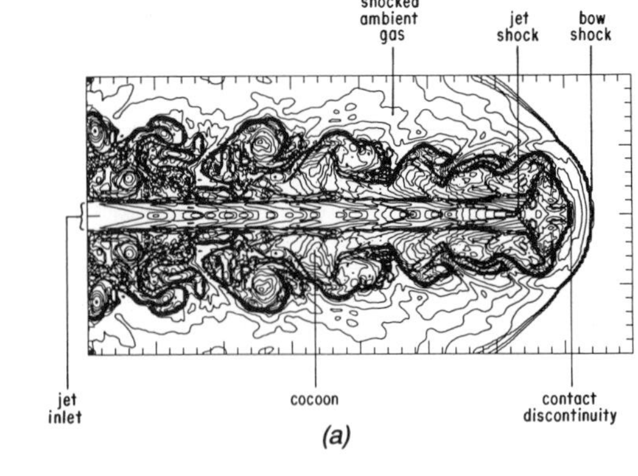

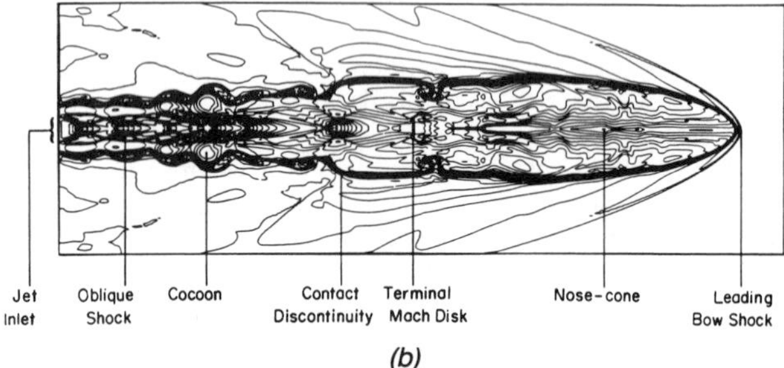

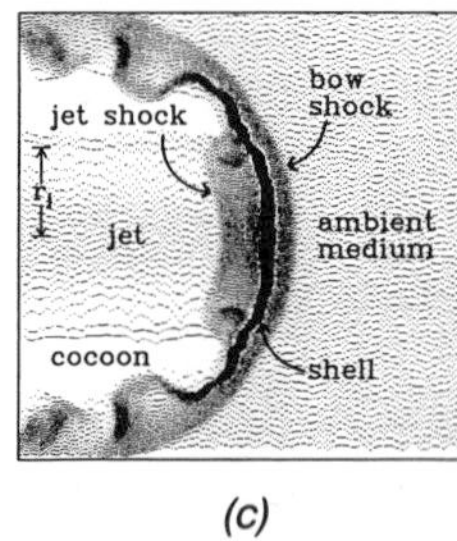

Figure 1. Supercomputer numerical simulations of supersonic jets, illustrating the density distributions in (*a*) a nonradiative jet with $M_j = 6$, $\eta = 0.1$. (*b*) A similar jet with a strong toroidal magnetic field. (*c*) The head of a radiative jet with $M_j = 20$, $\eta = 1$. [*Courtesy of John Blondin (a and c) and David Clarke (b).*]

was discovered by numerical simulations involves "radiative" jets, in which the shocked gas in the jet head can cool before it flows out of the working surface region. In this case, the cooled gas forms a dense shell between the jet shock and the bow shock (Fig. 1*c*). The shell is highly dynamically unstable and tends to fragment into individual clumps that move with different velocities. Such clumps provide a natural explanation for the high-proper-motion Herbig–Haro objects that are associated with the heads of stellar jets.

Knots and Wiggles

The relative motion between the jet and the ambient medium is expected to induce surface instabilities of the Kelvin–Helmholtz (KH) type. Such instabilities could be responsible for the common occurrence of emission knots and transverse oscillations in astrophysical jets. Specifically, these two features may correspond, respectively, to the pinch and kink modes of the KH instability. Recent analytical results on the linear development of these modes, together with numerical results on their nonlinear evolution, have provided a better understanding of the possible effects of this instability on cylindrical, supersonic jets. In the linear regime, each of these modes can be analyzed in terms of the number n of nodes in the radial direction, and a natural distinction arises between the fundamental ($n = 0$) mode and the higher-order ($n \geq 1$) reflection modes. The reflection modes, which are due to the resonant interaction between acoustic waves that propagate from one side of the jet to the other, dominate the pinch instability in jets that satisfy $M_j > 1 + \eta^{1/2}$. They develop into weak, oblique shocks that do not disrupt the jet but that could conceivably be identified with the observed quasiperiodic emission knots. A recently reported measurement of the proper motion of the knots in the stellar jet L1551 is consistent with this interpretation in view of the fact that the predicted shock pattern could, in principle, travel at a substantial fraction of the jet speed. By contrast, dense or low-Mach-number jets that do not satisfy the preceding inequality are susceptible to the fundamental pinch mode. They are expected to develop strong, planar shocks and a progressively broadening turbulent boundary layer that could eventually lead to their disruption. The formation of a turbulent boundary layer could result in the entrainment of the surrounding gas into the jet. This process was proposed as the cause of the observed star formation activity along the jet in the radio galaxy Centaurus A.

In the case of the kink instability, the fastest growing perturbation is associated with the fundamental mode that evolves into an ever-steepening pattern of zigzagging internal shocks. Recent observations of the Centaurus A jet have revealed a side-to-side limb brightening that is consistent with this pattern. However, the frequent deviation of jets from a straight trajectory may also be due to transverse pressure gradients in the ambient medium. In the extreme case of a head–tail galaxy, the two oppositely directed jets that emanate from the galactic nucleus are swept back by the ram pressure associated with the motion of the galaxy through a dense intracluster medium.

Magnetic Field Effects

Strong magnetic fields parallel to the direction of the flow could have a stabilizing effect on the development of the KH modes. However, the embedded magnetic field may have an important effect on the apparent structure of a galactic radio jet even if its overall dynamical influence is small. This is because the intensity of the radio emission, which is produced by the synchrotron mechanism, scales with the magnetic field component that is perpendicular to the line of sight. In fact, it was shown that the energetically preferred magnetic field configurations in jets with negligible internal pressure gradients could mimic the appearance of quasiperiodic knots and wiggles even when the flow remains perfectly straight and smooth. Another possible consequence of the dependence of the observed intensity on the transverse field component is that jets with a predominantly azimuthal magnetic field could be completely obscured by the radio emission from a surrounding cocoon.

RELATIVISTIC JETS

The presence of outflow speeds that approach the speed of light c has been inferred in the innermost regions of many extragalactic radio jets from apparent superluminal motions, high emission variability, and the absence of a detectable counterjet. The existence of a pair of relativistic ($0.26c$) jets has also been inferred in the galactic compact object SS 433. The origin of such high velocities is still not fully understood. According to a recent suggestion, the superluminal jets are initially accelerated to even higher speeds by an electromagnetic or a hydromagnetic process and are then decelerated to the inferred velocity range ($\lesssim 0.99c$) by the Compton scattering interaction with the ambient radiation field. The relation between the highly relativistic parsec-scale outflows and the kiloparsec-scale jets within which they are embedded is also still an open question, but in one well-studied case (the strong radio source 3C 120) it appears that the inner beam merges smoothly into the large-scale jet and that superluminal motion persists to a distance of more than 2 kpc from the nucleus. On the other hand, various dynamical arguments indicate that the speeds of the large-scale jets in weak radio sources do not exceed 10^4 km s^{-1}. Relativistic beaming of the emitted radiation and light travel time effects can strongly influence the appearance of jets, particularly in sources where the motion is not confined to one plane (as in the case of precessing jets). The relativistic focusing of the radiation in the direction of motion is, in fact, the basis for the interpretation of the most extreme members of the active galactic nuclei class in terms of relativistic jets that are observed at a small angle to the axis, for example, in the class of active galactic nuclei known as "blazars."

Additional Reading

Begelman, M. C., Blandford, R. D., and Rees, M. J. (1984). Theory of extragalactic radio sources. *Rev. Mod. Phys.* **56** 255.

Blandford, R. D., Begelman, M. C., and Rees, M. J. (1982). Cosmic jets. *Scientific American* **246** (No. 5) 124.

Henriksen, R. N., ed. (1986). Jets from stars and galaxies. *Can. J. Phys.* **64** 353.

Königl, A. (1985). The universality of the jet phenomenon. *Ann. Acad. Sci. N.Y.* **470** 88.

Lada, C. J. (1985). Cold outflows, energetic winds, and enigmatic jets around young stellar objects. *Ann. Rev. Astron. Ap.* **23** 267.

Norman, M. L. (1991). Fluid dynamics of astrophysical jets. In *Nonlinear Astrophysical Fluid Dynamics*, J. R. Buchler and S. T. Gottesman, eds. N.Y. Academy of Sciences, New York.

See also **Accretion; Active Galaxies and Quasistellar Objects, Accretion; Active Galaxies and Quasistellar Objects, Blazars; Active Galaxies and Quasistellar Objects, Jets; Active Galaxies and Quasistellar Objects, Superluminal Motion; Black Holes, Theory; Herbig–Haro Objects and Their Exciting Stars; Magnetohydrodynamics, Astrophysical; Radiation, High Energy Interaction with Matter; Radio Sources, Emission Mechanisms; Shock Waves, Astrophysical; Stars, Pre-Main Sequence, Winds and Outflow Phenomena; Stars, Winds; Stars, Young, Jets.**

Jupiter, Atmosphere

Barney J. Conrath

Jupiter, with a radius more than eleven times that of the Earth, is the largest planet in the solar system. Its atmosphere is quite deep, extending over the outer 20% of the planetary radius, and is composed primarily of hydrogen and helium. It is bounded above by an ionosphere and below by a region containing metallic hydrogen at pressures greater than 2 megabar (1 bar $= 10^5$ Pa) and temperatures in excess of 10^4 kelvins (K). Most of the direct observations available pertain to the outermost layers, where the pressure is less than 10 bar; little is known about the atmosphere at great depths.

Because markings can be seen on the disk of Jupiter with relatively modest instrumentation, an extensive history of telescopic observations exists, dating back to the seventeenth century. As Earth-based observational techniques have become more sophisticated, new information has been obtained on composition, cloud motion, and thermal structure. With the advent of the space probes *Pioneer 10, Pioneer 11, Voyager 1,* and *Voyager 2*, large advances in our understanding of the jovian atmosphere have occurred.

COMPOSITION

The jovian atmosphere consists of about 82% hydrogen (H_2) and 18% helium (He) by mass, with traces of other gases. The corre-

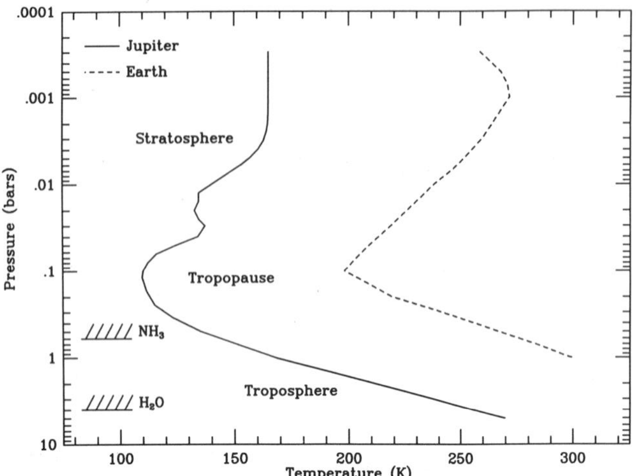

Figure 1. Atmospheric temperature on Jupiter and Earth as a function of pressure (1 bar = 10^5 Pa). The various portions of the atmosphere are indicated. Approximate levels of cloud formation in the jovian atmosphere are shown at the left for ammonia and water clouds.

sponding helium-to-hydrogen ratio appears to be slightly less than that believed to exist in the Sun. This implies that some depletion in helium has occurred in the outer envelope of the planet since the time of its formation. One possible cause is the immiscibility of helium in metallic hydrogen in the deep interior, which can result in the precipitation of helium toward the planetary center, leaving the outer envelope depleted. If this process has occurred on Jupiter, it has been much less extensive than on Saturn, where a major depletion of helium in the outer atmosphere is observed.

Many minor gaseous constituents have also been detected in Jupiter's atmosphere. Of particular interest are measurements of methane (CH_4) and ammonia (NH_3) which permit estimates of the ratios of carbon to hydrogen (C/H) and nitrogen to hydrogen (N/H). Both C/H and N/H are found to be approximately a factor of 2 larger than the corresponding values in the Sun, whereas water vapor (H_2O) appears to be depleted by a factor of almost 50 relative to that expected on the basis of the solar oxygen-to-hydrogen value. In the upper atmosphere, photochemistry resulting from the solar irradiation of CH_4 and other constituents produces additional hydrocarbons, such as acetylene (C_2H_2) and ethane (C_2H_6). Other molecules, such as germane (GeH_4) and phosphine (PH_3), are observed at levels in the troposphere where they cannot be in chemical equilibrium, suggesting that they are transported upward from deeper levels, where they form.

VERTICAL STRUCTURE AND HEAT BALANCE

Temperature as a function of atmospheric hydrostatic pressure is shown in Fig. 1. The logarithm of the pressure decreases linearly with increasing geometric height in an isothermal atmosphere and is frequently used as a convenient vertical coordinate, even though the actual temperature may vary with height. Unlike the terrestrial planets, the giant planets have no distinct solid surface to serve as a reference point for the measurement of height. However, some sense of vertical scale can be obtained by noting that the 0.1-bar level lies approximately 50 km above the 1-bar level.

Our knowledge of the thermal structure comes from various forms of remote sensing; direct measurements within the atmosphere are not yet available, but will be provided by the entry probe of the *Galileo* mission. One remote sensing approach makes use of thermal emission measurements of Jupiter in the infrared and at millimeter and centimeter wavelengths. The variation of atmospheric opacity with wavelength permits the temperature to be inferred over a range of pressure levels. The deepest penetration, to about 10 bar, occurs at centimeter wavelengths. Spatially resolved infrared measurements have been obtained from spacecraft, and infrared, millimeter, and centimeter measurements have been made using Earth-based instrumentation. Another technique, which has also contributed to our knowledge of the temperature structure, makes use of measurements of the refraction of radio signals emitted by spacecraft passing behind the planet as viewed from Earth. A combination of results from each of these techniques has been used to obtain the mean vertical temperature profile shown in Fig. 1.

The vertical temperature structure of Jupiter is qualitatively similar to that of the Earth; however, the jovian temperatures are approximately 100 K cooler than their terrestrial counterparts because of the greater distance of Jupiter from the Sun. A temperature minimum, or *tropopause*, occurs near 0.1 bar on both planets. Above this level is the *stratosphere*, where the temperature increases with increasing height (or decreasing pressure), whereas below the tropopause the temperature decreases with height in a region called the *troposphere*. Two levels in the jovian atmosphere where clouds are believed to form are indicated in Fig. 1. Ammonia condenses near the 0.6-bar level, and the cloud features that are observed on the disk of the planet are composed predominantly of ammonia ice crystals. Another cloud may exist near the 4-bar level as the result of water condensation. A third cloud layer (not shown in the figure), possibly consisting of ammonium hydrosulfide (NH_4SH), may lie between the ammonia and water clouds. The horizontal distribution of the clouds is nonuniform and is governed by dynamical processes that transport the condensates.

The warm upper stratosphere of Jupiter is a consequence of the absorption of sunlight in this region by methane gas and possibly by small particles suspended in the atmosphere. The precise nature of the particles is not well understood, but they may result in part from the condensation of hydrocarbon gases associated with the complex photochemistry occurring in the upper atmosphere. In the troposphere, the temperature decreases with increasing height (decreasing pressure) because of the existence of heat sources at deeper levels. The presence of these heat sources requires the existence of an upward heat flux in this portion of the atmosphere in order to maintain an equilibrium situation in which the temperature is neither increasing nor decreasing with time. The principal mechanisms available for vertical heat transport are radiation and convection. At pressures less than about 0.5 bar, the necessary flux can be carried by radiative transport in the infrared, and direct radiation of energy to space can occur. At higher pressures, the infrared opacity increases rapidly and the radiative exchange of energy becomes inefficient. As a consequence, convective transport of energy becomes dominant. The convective process consists of the exchange of warm parcels of atmosphere at deeper levels with cooler parcels above, resulting in a temperature–pressure relationship very close to that for an adiabat. The temperature as a function of pressure is, in fact, observed to follow very nearly an adiabat at pressures greater than approximately 0.6 bar.

The heat sources in the jovian troposphere can be attributed in part to the absorption of sunlight in this region. However, determinations of the energy balance of the planet indicate that sunlight alone is inadequate. The energy balance is defined as the ratio of the total thermally emitted power to the total absorbed solar power. Because most of the thermal emission occurs in the infrared, the total emitted power has been determined from spacecraft measurements in that portion of the spectrum. The total solar power reflected by the planet in the visible and near infrared portions of the spectrum has also been determined from spacecraft observations. By subtracting this reflected component from the known incident solar flux, the absorbed solar power can be determined. The resulting energy balance is found to be approximately 1.6. The fact that Jupiter emits about 60% more power to space than it receives from the Sun implies that an internal heat source must exist. The excess emission is believed to result from primordial heat associated with the formation of Jupiter and the continued slow contraction and cooling of the planet. There may be some additional contribution to the internal heat from the precipitation of helium in the interior as discussed previously. The helium droplets generate frictional heating as they fall to deeper levels.

Spatially resolved spacecraft measurements of tropospheric temperatures indicate that the large-scale, pole-to-equator temperature gradients in both jovian hemispheres are very small. Because more sunlight is absorbed in the troposphere at low latitudes than at high latitudes, some compensating effect must occur in order for the resulting temperature to be nearly independent of latitude. The convective interior of the planet can provide efficient redistribution of energy, resulting in a variation with latitude of the internal heat flux that just compensates for the variation in absorbed solar energy. Some redistribution of energy by dynamical processes operating above the convective region in the upper troposphere may also occur.

METEOROLOGY

Jupiter possesses a rich variety of meteorological phenomena, as is evident from the observed cloud motions. Winds have been deduced by tracking cloud features and obtaining their apparent rotation periods. This technique has been applied to relatively low-spatial-resolution observations with Earth-based telescopes, and imaging from spacecraft has recently added a large body of high-spatial-resolution data. In order to calculate wind speeds from observed rotation periods, a reference period must be adopted. Because Jupiter has no observable solid surface, a period of 9 hours and 55 minutes, obtained from measurements of the modulation of radio emissions from the planet, has been used for this purpose. This is the rotation period of the planetary magnetic field and presumably also that of the deep interior.

The winds are found to be predominantly zonal (blowing in the east–west sense). A strong eastward jet is centered on the equator, extending approximately 10 degrees of latitude into each hemisphere, with a maximum wind speed of about 100 meters per second (m s^{-1}). At higher latitudes in both hemispheres, alternating eastward and westward jets are found. The wind system is roughly symmetric about the equator, with some exceptions. For example, the maximum wind speed observed on the planet of about 150 m s^{-1} occurs in a narrow eastward jet centered at 22° North latitude, whereas no winds of this magnitude are observed anywhere in the southern hemisphere. The Earth-based historical record of the jovian winds suggests that the basic alternating jet structure has existed for at least several decades and is presumably a long-lived characteristic of the atmosphere.

The wind system is found to be strongly correlated with the horizontal temperature structure in the upper troposphere. The zonal mean (averaged over longitude) temperature varies with latitude in an oscillatory manner with a peak-to-peak amplitude of about 3 K. Temperature maxima and minima occur on the equatorward edges of eastward and westward jets, respectively, with the temperature changing most rapidly near the jet centers. On a rapidly rotating planet, the horizontal gradient in temperature can be used to determine the change in wind speed with height. The observed behavior of the temperature on Jupiter implies that the jet speed decreases with increasing height. This suggests that the jets are driven at deeper levels, with some dissipative mechanism causing them to decay in the upper troposphere and lower stratosphere.

Some correlation between the jet structure and the large-scale, zonal mean cloud albedo (reflectance) is also observed. Alternating light and dark bands occur encircling the planet parallel to latitude circles (Fig. 2). Bright "zones" are usually found on the equatorward side of an eastward jet with dark "belts" on the poleward side. Zones may be regions of upward motion where condensation of ammonia clouds is occurring, whereas the belts are associated with

Figure 2. Jupiter as imaged by the *Voyager 1* spacecraft at a distance of approximately 33 million kilometers. The limb of the planet is on the right with the sunrise terminator on the left. The Great Red Spot is seen prominently in the southern hemisphere, and one of the major white ovals lies below and to the left of it.

downward motion. This picture is consistent with the observed temperature structure, because regions of upward motion are relatively cool as a result of the adiabatic expansion of the air parcels whereas downward-moving parcels are warm because of adiabatic compression. However, it should be noted that the correlation between belts and zones and the jets exists only in a time-average sense. Substantial temporal variations in the large-scale cloud structure are observed, whereas the jet structure remains essentially invariant.

Two general classes of models have been advanced to explain the existence of the alternating jet system. In one case, the jets are the cloud-top-level manifestation of fluid motion on concentric cylinders that are parallel to the rotation axis of the planet and that extend from one hemisphere to the other. Such motion is postulated to be the result of convection in the deep layers in the presence of rapid planetary rotation. The second type of model holds that the jet system is confined to a relatively shallow "weather" layer, and that the winds are driven by mechanical forcing from the deeper convective layers, differential solar heating, latent heat effects, or some combination of these mechanisms. The observational constraints presently available do not clearly distinguish between these two types of model.

Superposed on the zonal mean jet structure and associated cloud bands are a large number of isolated meteorological features, as can be seen in Fig. 2. Chief among these is the Great Red Spot, an oval-shaped storm system extending over 20,000 km in the east–west direction and 12,000 km from north to south. The feature is nested between two counter-flowing jets and rotates in a counterclockwise sense; because it is located in the southern hemisphere, this means that its motion is anticyclonic at the observed cloud level. It is extremely long-lived, apparently having been observed as early as the seventeenth century. Three major white ovals also show anticyclonic behavior, and several brownish, elongated features are observed to be cyclonic. Many other features, including plumes and smaller spots, form and dissipate on relatively short time scales. The sources of the various colors observed in the clouds are not known with certainty, but are believed to result from the presence of small impurities in the ammonia ice crystals.

SUMMARY

We have achieved a comprehensive understanding of certain properties of Jupiter's atmosphere, including composition, thermal structure, energy balance, and winds. However, major unanswered questions remain in such areas as the formation and maintenance of the jet system, the nature of the various cloud and haze layers, and the factors responsible for the various cloud colors. The *Galileo* mission, with its orbiter and probe into the atmosphere, should provide new insight into these issues.

Additional Reading

Atreya, S. K., Pollack, J. B., and Matthews, M. S., eds. (1989). *Origin and Evolution of Planetary and Satellite Atmospheres*. University of Arizona Press, Tucson.

Beatty, J. K. and Chaikin, A. (1990). *The New Solar System*, 3rd ed., Sky Publishing Corporation, Cambridge, MA, and Cambridge University Press, Cambridge, U.K.

Ingersoll, A. P. (1981). Jupiter and Saturn. *Scientific American* **245** (No. 6) 90.

Stone, E. C. et al. (1979). *Voyager 1* encounter with the jovian system. *Science* **204** 945–1008.

Stone, E. C. et al. (1979). *Voyager 2* encounter with the jovian system. *Science* **206** 925–996.

See also **Planetary and Satellite Atmospheres; Planetary Atmospheres, Clouds and Condensates; Planetary Atmospheres, Dynamics; Planetary Atmospheres, Structure and Energy Transfer; Planetary Interiors, Jovian Planets; Planetary Magnetospheres, Jovian Planets.**

Jupiter and Saturn, Satellites

Jeffrey B. Plescia

The satellites of Jupiter and Saturn (Table 1) are a diverse and interesting group, from the cold lifeless Mimas to the seething, tumultuous, volcanically active Io. Our knowledge about Jupiter and Saturn comes largely from the *Voyager* spacecraft flybys. Most of these satellites probably accreted from the circumplanetary material remaining after the planets themselves formed and each has been bombarded by asteroidal and cometary material since its formation (4–4.5 billion years ago). Hence, the number of impact craters on a surface is indicative of its age: An ancient surface will have more craters than a young one. Variations in density of impact craters indicate that a satellite has undergone geologic activity producing the younger, less-cratered surfaces. Almost all of the large satellites rotate synchronously, that is, they keep the same side facing the planet, just as the Moon keeps the same side facing the Earth.

JUPITER

Before *Voyager* (a NASA project managed by the Jet Propulsion Laboratory), Jupiter was thought to have 5 satellites, but now is known to have 16 (Table 1). Galileo Galilei is credited with the discovery of the four large satellites (the "Galilean" satellites, Fig. 1), but there is speculation that Chinese astronomers may have seen them in the fourth century BC. Edward Emerson Barnard discovered Amalthea in 1982; the rest have been discovered in the twentieth century. Io, Europa, Ganymede, and Callisto are similar in size to the smaller terrestrial planets (e.g., Mercury); their bulk densities decrease and their sizes generally increase with increasing distance from Jupiter.

Amalthea

Amalthea is elongate, keeping its long axis pointed toward Jupiter, and may be a fragment of a larger body destroyed by impacts. It is a dark red asteroid-like body that probably has had no geologic activity. Amalthea's surface is warmer than expected, which may result from heating by electric currents traveling along Jupiter's magnetic field or from bombardment by particles from the radiation belts.

Io

This moon is perhaps the most bizarre satellite in the solar system, characterized by volcanos shooting plumes of material into space. The plumes rise to altitudes of 70–300 km and travel hundreds to thousands of kilometers across the surface. Io's volcanism changes rapidly enough that in the 18 weeks between the *Voyager 1* and *Voyager 2* observations, the appearance of the surface changed, so that new volcanos were erupting when *Voyager 2* arrived, others had ceased, and some continued.

The surface is bright orange-red (the color variations are attributed to various forms of sulfur) and mottled by large whitish patches (possibly SO_2 frost). Io's surface, covered with layers of sulfur and silicates, is dominated by volcanic landforms, typically circular depressions (calderas) with radial lava flows hundreds of kilometers long. Smooth plains between the volcanic vents are cut locally by scarps and grabens; rugged mountains are found near the poles. Virtually no impact craters are observed, indicating that Io's surface is young. This is a small satellite largely composed of silicates and it would not be expected to be currently geologically active. However, strong tides are raised by Jupiter, Europa, and Ganymede flexing Io and causing internal friction, thereby heating the satellite. That heat is released through volcanic activity.

Europa

Europa is the brightest satellite of Jupiter. Its surface is largely ice and is characterized by bright terrane and mottled dark areas. Linear dark markings tens of kilometers wide and up to 1000 km long cross these terranes. The mottled terrane is characterized by interlocking depressions and mesas a few kilometers across. Narrow cuspate ridges, about 10 km wide and 100 m high, are observed within the bright terrane. These patterns are suggestive of tectonic activity but their origin remains unclear.

As with Io, Europa's surface is very young and has few impact craters. Tidal interactions between Jupiter, Io, and Ganymede probably have caused internal heating, resulting in the geologic processes that have produced Europa's young surface. Europa is composed of about 80% rock and 20% water; it probably has a silicate core overlain by a 100-km-thick water-ice crust.

GANYMEDE

Ganymede, the largest satellite in the solar system, is composed of about half water ice and half silicates, and probably has separated into a silicate-rich core and an icy crust and mantle. The high albedo and the spectral data are consistent with a water-ice surface. The surface is composed principally of dark cratered terrane and bright grooved terrane. Bright-rayed craters are scattered across the surface and bright polar frost caps extend down to about 45° latitude.

Dark cratered terrane occurs as polygons about 1000 km across bounded by lanes of grooved terrane. A series of bright, narrow, concentric linear features (furrows 10 km wide, spaced 50 km apart) are interpreted to be rings formed as a result of a giant impact (similar to the rings around the feature on Callisto called

Table 1. Satellites of Jupiter and Saturn

Name	Orbital Radius ($\times 10^3$ km)	Mass ($\times 10^{23}$ kg)	Diameter (km)	Density (g cm^{-3})	Period (h)	Albedo
Jupiter						
Metis	128.0		40 × 40 × ?		7.08	0.05–0.1
Adrastea	129.0		120 × 20 × 15		7.16	0.05–0.1
Amalthea	181.3		270 × 170 × 155		11.95	0.06
Thebe	221.9		110 × 90 × ?		16.20	0.05–0.1
Io	421.6	0.894	3630	3.55	42.46	0.63
Europa	670.9	0.480	3138	2.97	85.22	0.64
Ganymede	1070.0	1.482	5262	1.94	171.72	0.43
Callisto	1883.0	1.077	4800	1.86	400.54	0.17
Leda	11094.0		10		5760.0	—
Himalia	11480.0		180		6024.0	0.03
Lysithea	11720.0		20		6240.0	—
Elara	11737.0		80		6240.0	0.03
Anake	21200.0		20		14808.0*	—
Carme	22600.0		30		16608.0*	—
Pasiphae	23300.0		40		17640.0*	—
Sinope	23700.0		30		18192.0*	—
Saturn						
Atlas	137.7		40 × 20		14.45	0.5
Prometheus	139.4		140 × 100 × 74		14.71	0.5
Pandora	141.7		110 × 90 × 66		15.09	0.5
Epimetheus	151.4		140 × 116 × 100		16.66	0.5
Janus	151.5		220 × 190 × 160		16.67	0.5
Mimas	185.5	0.38	394	1.17	22.62	0.77
Enceladus	238.0	0.8	502	1.24	32.89	1.04
Tethys	294.7	7.6	1048	1.26	45.31	0.8
Telesto	294.7		30 × 20 × 16		45.31	0.6
Calypso	294.7		24 × 22 × 22		45.31	0.9
Dione	377.4	10.5	1118	1.44	65.69	0.55
Helene	377.4		34 × 32 × 30		65.74	0.6
Rhea	527.0	24.9	1528	1.33	108.42	0.65
Titan	1221.9	1345.7	5150	1.88	382.69	0.2
Hyperion	1481.0		410 × 260 × 220		510.64	0.25
Iapetus	3560.8	18.8	1436	1.21	1903.9	0.04, 0.5
Phoebe	12954.0		220		13210.8*	0.06

Earth's mass = 5.976×10^{24} kg; the Moon's mass = 7.350×10^{22} kg; Jupiter's mass = 1.901×10^{27} kg; Saturn's mass = 5.68×10^{26} kg
? indicates dimension unknown; * indicates retrograde orbit.

Valhalla) that happened early in Ganymede's history. Palimpsests, circular features of relatively high albedo about 100–300 km in diameter, are widespread on the dark terrane. They are interpreted to be large impact craters whose topography was destroyed by the flow of the ice crust over time; a few still retain an interior annular topographic ring marking the original crater rim.

Grooved terrane occurs in lanes tens to hundreds of kilometers wide and is characterized by parallel grooves and ridges 5–15 km wide and perhaps a few hundred meters high. Grooves typically parallel the lane margins, but locally can form complex curvilinear patterns. The grooved terrane formed at the expense of the older cratered terrane; cratered terrane was down-dropped into graben that were filled with water ice that subsequently developed grooves and ridges. Cratered terrane dates back to a period of heavy bombardment about 3–4 billion years ago; the grooved terrane is considerably younger, but it too may be billions of years old.

Callisto

The darkest satellite of the jovian system appears to have had a relatively simple geologic history. Callisto has a dark crust saturated with impact craters. Since the crust formed, it has simply been bombarded, with little or no geologic activity. Callisto has the lowest density of the jovian satellites, suggesting that it has the highest content of water ice and a correspondingly low amount of radioactive elements. There is no evidence for widespread geologic activity driven by internal dynamics. The dearth of heat-producing elements and the absence of tidal heating condemned Callisto to remain as a cold inert ice ball.

Valhalla, a large multiring impact basin, is characterized by a circular bright patch 600 km in diameter surrounded by bright flat-topped rings spaced 50–200 km apart and extending out about 1500 km. This ring pattern, as on Ganymede, probably results from an early impact into soft icy crust.

Small Satellites

Jupiter has a total of 11 satellites smaller than Amalthea. All are very small and presumably irregularly shaped; little is known of them. The three inside Io's orbit probably represent fragments of much larger satellites that once existed in those regions. The eight outside Callisto's orbit are in orbits with very high inclinations and eccentricities and these small bodies are probably captured asteroids rather than satellites originally formed around Jupiter.

Figure 1. Jupiter and its four planet-sized moons, called the Galilean satellites, were photographed in early March 1979 by *Voyager 1*, as shown in this collage. They are not to scale but are in their relative positions. Reddish Io (upper left) is nearest Jupiter; then Europa (center), Ganymede, and Callisto (lower right). Nine other much smaller satellites circle Jupiter, one inside Io's orbit and the others millions of miles from the planet. Not visible is Jupiter's faint ring of particles, seen for the first time by *Voyager 1*.

Figure 2. This montage combines individual photos taken by *Voyager 1* and *Voyager 2* during their Saturn encounters in November 1980 and August 1981. The spacecraft photographed the planet, its rings, and all 17 known satellites. This montage includes all major satellites known before the *Voyager* launches in 1977. They are (clockwise from upper right) Titan (photo by *Voyager 1*), Iapetus (photo by *Voyager 2*), Tethys (*Voyager 2*), Mimas (*Voyager 1*), Enceladus (*Voyager 2*), Dione (*Voyager 1*), and Rhea (*Voyager 1*).

SATURN

Exploration of the saturnian system began in 1610 with Galileo Galilei's observations in which he detected the ringa but misinterpreted them as moons or "handles" to either side of Saturn. It was Christiaan Huygens who discovered the true form of the rings and also discovered Titan in 1655. Giovanni Cassini discovered Iapetus, Rhea, Dione, and Tethys in 1675. A total of 9 satellites were known by the beginning of the twentieth century (Fig. 2), the remainder, now totalling 17, have been discovered since then (Table 1). The saturnian satellites are largely composed of water ice (up to 60–70%) plus, presumably, ammonia hydrates and methane clathrates. Unlike the jovian satellites, the saturnian satellites do not display a well-defined density decrease with increasing orbital radius. Spectroscopic data indicate the presence of water ice or frost covering large fractions of their surfaces.

Saturn's satellites exhibit an interesting pairing in terms of size and mass, but the members of each pair have very different geologic histories. For example, Mimas and Enceladus are very similar in size, yet Enceladus has experienced more geologic evolution than Mimas; similarly, Dione and Tethys are similar in size, but Dione is much more evolved than Tethys. An intriguing problem is the nature of the plains-forming material found on these satellites. It is unlikely that water ice has ever melted; however, fluids of a water–ammonia eutectic composition can be produced at temperatures of only 170 K (−154°F).

Titan

Titan is the second largest satellite in the solar system and the only one to possess a significant atmosphere. Its surface, obscured by a permanent global cloud layer, remains a mystery. The diameter (5150 km) of Titan's surface was determined using the *Voyager* radio transmitter; the spacecraft's radio signals were tracked until it disappeared behind the solid surface, indicating the satellite's size. Titan is composed of a 50–50 mixture of silicates and water ice.

Titan's atmosphere has a reddish color, with a darker northern hemisphere, a bland southern hemisphere, and faint bands. The hemispheric brightness difference may be seasonal, different hemispheres being bright in alternate seasons. Surface pressure is 1.5 bar (60% more than Earth's) and the temperature is 94 K (−290°F). The atmosphere is composed principally of nitrogen, with about 10% methane and a variety of other trace compounds (acetylene, ethylene, ethane, hydrogen cyanide, and carbon monoxide), and may represent volatilization products of the early crust, rather than an atmosphere acquired during accretion.

Mimas

The innermost large satellite, Mimas is characterized by a heavily cratered surface and grooves up to 90 km long, 10 km wide and 1–2 km deep in the southern hemisphere. The most impressive feature is a giant 130-km-diameter crater, called Herschel, whose 20- to 30-km-wide central peak stands 6 km high. A larger impact might have completely disrupted the satellite and Mimas may have been shattered by large impacts and reformed repeatedly early in its history. The heavily cratered surface suggests that Mimas has evolved little since it formed.

Enceladus

The brightest satellite of Saturn, Enceladus has a varied surface, one composed of cratered terrane, cratered plains, ridged plains, and fractured craterless plains; the different surface morphologies and variable crater densities suggest that Enceladus has had a long geologic history and may still be active. Also, as with Jupiter's Io, Enceladus' young surface may be the result of tidal heating. The brightest (densest) part of Saturn's E ring lies at the same orbital distance as Enceladus and it has been suggested that the satellite is the source of material for the ring. Perhaps eruptions spew water ice into space, feeding the E ring.

Tethys

Tethys' most impressive feature is a giant canyon system (Ithaca Chasma), 100 km wide, several kilometers deep, and extending three-quarters of the way around the body. Most of Tethys' surface is heavily cratered, but younger smooth plains do occur locally. Tethys also exhibits a giant 400-km-diameter impact crater, Odysseus, which is 40% of Tethys' diameter and the largest crater in the saturnian system. Tethys was geologically active early in its history and when its liquid water froze, the moon expanded and the canyon system formed in response to the increase in volume.

Dione

Dione exhibits heavily cratered terranes and areas of smooth, relatively crater-free plains. The plains typically have narrow troughs, perhaps marking vents through which material erupted. Globally, Dione is characterized by bright wispy markings centered on the trailing hemisphere and by large variations in albedo. Viewed in high resolution, the wisps correspond to narrow troughs that may be fractures from which material erupted to form the bright deposits. These markings may have been globally distributed but later erased from the leading hemisphere by micrometeroid "gardening," a process in which the impacts of numerous small particles plow up the surface.

Rhea

From a distance, the surface of Rhea is characterized by distinct bright and dark albedo markings. The leading hemisphere is bland whereas the trailing hemisphere has complex patterns of bright markings. Like Dione, the markings may be eruptions of bright material and may have been erased from the leading hemisphere by micrometeoroid gardening.

Rhea's surface is heavily cratered, with craters (some strongly polygonal in outline) standing shoulder-to-shoulder. Local variations in the frequency of craters and a relatively smooth area near the equator suggest that some geologic processes have occurred. Overall though, the surface is ancient and Rhea may have formed and experienced almost no internally driven geologic processes.

Hyperion

This small, dark, heavily cratered irregular satellite is probably a fragment of a larger satellite disrupted by an impact. Its irregular shape is unique in that other objects of its size (e.g., Mimas) are spherical; it is one of the largest irregularly shaped objects in the solar system. Hyperion rotates nonsynchronously and may be tumbling.

Iapetus

Iapetus is notable for its great albedo contrast, an attribute known to Cassini in the seventeenth century. Half the surface is bright material whereas the other half is only one-tenth as bright. The boundary between these two terranes meanders across the surface. The bright areas, predominantly water ice, are ancient and heavily cratered. No surface detail has been seen in the dark areas and the origin of the dark material remains controversial. Its low albedo suggests that it is rich in organic material, which also is dark and commonly found in primitive meteorites. The surface is also red, consistent with a mixture of organic polymers and hydrated silicates. The dark material may have been erupted from the interior and is younger than the bright cratered terrane.

Phoebe

Phoebe is the outermost saturnian satellite, and its orbit is retrograde (and thus opposite to those of the other satellites). Its rotation is probably not synchronous. It exhibits large albedo contrasts (up to 50%) although overall it is very dark. Very little is known about its surface; its unique orbit and spectral signature suggest that it is a captured rather than an original satellite.

Small Satellites

A total of 10 objects are found between the orbit of Dione and the outer edge of Saturn's A ring. These small satellites are all bright and irregularly shaped. Some, those that share orbits with other satellites (e.g., Telesto and Calypso share Tethys' orbit) may be pieces left over after the primary satellite formed. Other small fragments may be all that remains of once larger satellites broken up by large impacts (e.g., Janus and Epimetheus, which share virtually the same orbit). Very little is known about most of these satellites. Even their sizes are uncertain and virtually nothing is known about their composition or density.

Additional Reading

Beatty, J. K. and Chaikin, A., eds. (1990). *The New Solar System*, 3rd ed. Sky Publishing Corporation, Cambridge, MA, and Cambridge University Press, Cambridge, U.K.

Burns, J. A. and Matthews, M. S., eds. (1986). *Satellites*. University of Arizona Press, Tucson.

Greeley, R. (1987). *Planetary Landscapes*. Allen and Unwin, Boston.

Hamblin, W. K. and Christiansen, E. H. (1990). *Exploring the Planets*. Macmillan Publishing Co., New York.

Morrison, D., ed. (1982). *Satellites of Jupiter*. University of Arizona Press, Tucson.

Smith, B. A. et al. (1982). *Voyager 2* encounter with the saturnian system. *Science* **215** 499.

Smith, B. A. et al. (1979). The Galilean satellites and Jupiter: *Voyager 2* imaging science results. *Science* **206** 927.

Smith, B. A. et al. (1981). Encounter with Saturn: *Voyager 1* imaging science results. *Science* **212** 163.

Smith, B. A. et al. (1979). The Jupiter system through the eyes of *Voyager 1*. *Science* **204** 951.

Soderblom, L. A. (1980). The Galilean moons of Jupiter. *Scientific American* **242** (No. 1) 88.

Soderblom, L. A. and Johnson, T. V. (1982). The moons of Saturn. *Scientific American* **246** (No. 1) 100.

See also **Io, Volcanism and Geophysics; Outer Planets, Space Missions; Planetary and Satellite Atmospheres; Planetary Rings; Satellites; Satellites, Ices and Atmospheres; Satellites, Minor; Satellites, Rotational Properties.**

M

Magellanic Clouds

John Graham

The Magellanic Clouds are the nearest of the external galaxies. Each is an independently evolving star system, actively forming stars at the present time but also containing some which are as old, about 15 billion years, as any that we know. Their importance is many-fold but two aspects especially stand out. First, they act as a mirror to our own Milky Way galaxy and provide a guide as to how it would appear if we could view it from a vantage point high above its dusty disk. Second, we can make use of them to tell us about other galaxies far too remote for any sort of detailed study. The Magellanic Clouds are fundamentally important for the calibration of extragalactic distance indicators. They represent one of the few opportunities we have to intercompare rare objects like the most luminous blue supergiant stars, variable stars, star clusters, and H II regions directly with common stars similar to the Sun, all at the same distance and all comparatively unobscured by interstellar dust. With firm calibrations in hand, we can then confidently proceed to more distant systems where only the very brightest objects may be identifiable.

POPULATIONS

Even with a small telescope trained on one or other of the two Magellanic Clouds, there is an immediate sense of looking into the heart of a galaxy. It is the young population which is immediately the most striking. These are the massive stars which expend energy so profusely that their nuclear fuel is used up after only a few million years. They tend to clump in close groups of associations and often illuminate the surrounding gas to form bright H II regions. These are the classic markers of Population I as defined by Walter Baade in the middle of this century. The older Population II is much less conspicuous, contributing a faint substratum of stars which have long ago arranged themselves in extended, rather uniform distributions.

Population I and Population II are very much extreme categories and we have learned, after Baade's fundamental work, that there is a continuous transition between them. Stars and clusters of all ages covering a wide range of chemical composition are found in both the Magellanic Clouds and the Galaxy. As well as discussing the two extreme groups, it is appropriate to give special mention to this intermediate population as it is comparatively so prominent in the Magellanic Clouds.

Population I

Among the representatives of Population I, the brightest stars provide a unique opportunity to study the evolution of massive stars and the upper limit to the mass that a star can have and remain stable. Stars like this are sufficiently rare and widespread that this is an impossible job to do within the Milky Way. For most of this century, it was thought that a mass of 70–80 times that of the Sun was the maximum that a star could have and remain vibrationally stable. In the last decade, largely through Magellanic Cloud research, it has been shown that stellar winds dampen incipient instabilities very effectively and that stellar masses 100–200 times that of the Sun are not only possible but probable.

All massive stars end as supernovae and we were incredibly privileged in 1987 to witness in the Large Magellanic Cloud the brightest supernova since the invention of the telescope nearly 400 years ago. Among many other things, this event founded a whole new science of extrasolar neutrino astronomy and provided indisputable observational evidence that nucleosynthesis actually occurs inside stars. SN 1987A was formerly a normal, undistinguished blue supergiant star in one of the rich Population I regions of the Large Magellanic Cloud.

Also included among Population I are the Cepheid variable stars. Cepheids have become one of the standard distance indicators for galactic and extragalactic research through their period-luminosity relation and its validity from galaxy to galaxy. Fortunately, encouraging progress is being made in removing this uncertainty by observing many of the Cepheids in the Magellanic Clouds. With new instrumentation, the accuracy of the brightness and color measurements is being refined and observations in the infrared are proving especially useful.

Neutral and molecular hydrogen gas has a close association with Population I. 21-cm radio surveys have shown that each Magellanic Cloud has an abundant supply remaining for future generations of stars. Molecular hydrogen is harder to detect as we rely mainly on measurements at mm wavelengths of carbon monoxide, a tracer molecule. Both are present in regions where luminous stars are now forming in the Magellanic Clouds.

Population II

When we come to study the older populations of the Magellanic Clouds, we look past the brilliant associations with their blue supergiants and H II regions, past the Cepheid variables, and the numerous open star clusters until we see in each Cloud only the faint amorphous background which is made up of stars and planetary nebulae a billion years or more old. Relieving the general uniformity, old globular star clusters similar in appearance to those of our own galaxy are seen but, except for the occasional nova, every star in this old population is faint. Yet it is this component which forms the structural backbone of each Cloud, accounts for most of the mass and thereby determines the internal dynamics. Of the oldest objects, those which tell us that the Magellanic Clouds have existed for as long as our galaxy, the short period RR Lyrae stars are perhaps the easiest Population II objects to discover. Most information about the chemical composition of Population II comes from observing red giants in the oldest globular clusters. As in our galaxy, there is a good correlation between age and heavy element abundance although, in neither Cloud does the heavy element enrichment reach the level that is found in the youngest galactic stars.

Study of the old populations is important for investigating the origin of the Magellanic Clouds and the differences between them and the Galaxy at the earliest epochs. The distribution of faint red stars on photographs taken with wide field Schmidt telescopes is a guide to the mass distribution within each body. Magellanic Cloud novae occur two or three times a year and, as the list lengthens, these too are giving us a better idea about where most of the matter in each Cloud is located.

Intermediate Populations

Both Magellanic Clouds have been found to have a major component of intermediate age which spans the two extremes of Population I and II. It is much more developed than the analogous age group in the Milky Way. One example is the large number of rich populous star clusters. In our own galaxy, such clusters are invariably old with ages in excess of 10 billion years. In the Magellanic

Clouds, similarly aged clusters exist but they are outnumbered by a strong representation of populous clusters in the 1–10 billion year age range. Partly through their dynamical history, but more directly through their star forming history, the Magellanic Clouds have been able to create and to maintain massive clusters like this at all times. In the Milky Way, populous, globular clusters were only made at the earliest epoch. Similar intermediate age clusters either never formed or have long since been destroyed.

The distribution of the more numerous open clusters that appear at all times in all three systems gives us some hints. Recent work comparing age distributions of complete samples show that the Cloud clusters can survive to much longer lifetimes as they are evidently not subject to disrupting forces as strong as those that exist in the Galaxy.

Independent of such effects, we find that even in the general field, the representation of intermediate age stars is proportionally much larger than in the Milky Way. Apparently bursts of star formation have been occurring throughout the lifetime of each Magellanic Cloud which are much greater than anything we find in the Milky Way. It is from the study of intermediate populations more than from any other that we can observe the long-term effects of differing star-forming histories and apply the knowledge gained to more distant stellar systems which cannot be resolved into individual stars.

DYNAMICS

A fundamental difference between the Magellanic Clouds and our own galaxy is that neither Cloud has a strong central concentration of stars which can maintain dynamical order in the rest of the system. Both Magellanic Clouds are very vulnerable, not only to gravitational interaction with the Galaxy but also to gravitational interaction with each other.

The Large Magellanic Cloud rotates in a fairly regular manner. The best measurements come from young supergiants and from planetary nebulae as well as from neutral hydrogen. To reconcile the observed rotation (line of nodes) with the distribution (major axis orientation) of the old populations, a transverse motion of the Large Cloud amounting to about 300 km s^{-1} is required. This is sufficiently large to be directly measureable now from the proper motions of Cloud stars with reference to background galaxies. However, detailed interpretation of the rotation curve is hampered by the mutual gravitational interactions mentioned above.

Observation constrains the two Clouds to form a bound system which is currently orbiting our galaxy. At their last approach, approximately 200 million years ago, substantial damage was done to the Small Magellanic Cloud which is still apparent in its very disordered structure today. A long stream of gaseous material was drawn out of the Small Cloud by the interaction which is observed in neutral hydrogen radiation and has been called the Magellanic Stream. It has no associated stellar component. This close approach and others which have taken place at earlier epochs may have given rise to the bursts of star formation in both clouds which we now observe as the intermediate age population. The calculation for this intercloud orbit is remarkably explicit and it tells us that there have been several such close encounters in the last 10 billion years. However, there is still some doubt as to whether the orbit of the Small Cloud around the Galaxy is bound or unbound. The fact that the Magellanic Clouds have survived for the last 10 billion years indicates that there can have been no approaches much closer than the one we are experiencing now.

EVOLUTION

As independently evolving galaxies, the Large and the Small Magellanic Clouds are being gradually enriched with heavy elements. A finite step in this direction is taking place before our eyes as we observe the remnant of the 1987 supernova dispersing into interstellar space. Dynamically, the Small Cloud may be breaking up as a result of its last encounter with the Large Magellanic Cloud. Never held together very tightly, it is now strung out mostly along the line of sight over a distance of about 20 kpc. This is of the same order as the current distance between it and the Large Cloud. The Large Magellanic Cloud remains stable as shown by its well-defined rotation curve. Up until the present, the evolution of both Clouds has differed greatly from that of our galaxy. As low-mass galaxies from the beginning, neither underwent a major rapid collapse with a concurrent burst of star formation. This is shown by the lack of a strong central concentration of RR Lyrae stars, novae, and planetary nebulae to the degree that we see in the Galaxy. Instead star formation, and consequently heavy element enrichment, has proceeded at a much more gradual pace, with star bursts irregularly occurring every 10 million years or so wherever and whenever there is enough raw material. These are punctuated by system-wide star forming events whenever the two Clouds approach closely to each other. Through the work done over the past decade, the history of the Magellanic Clouds has become a lot clearer and we have been able to see how they relate to more massive star systems like our own galaxy.

Additional Reading

Bok, B. J. (1966). Magellanic Clouds. *Ann. Rev. Astron. Ap.* **4** 95.

Feast, M. W. (1988). The Magellanic Clouds and the extragalactic distance scale. In *The Extragalactic Distance Scale*. S. van den Bergh and C. J. Pritchet, eds. Astronomical Society of the Pacific, San Francisco, p. 9.

Mathewson, D. (1984). The Clouds of Magellan. *Scientific American* **252** (No. 4) 106.

Muller, A. B., ed. (1971). *The Magellanic Clouds*. Kluwer Academic Press, Dordrecht.

Murdin, P. (1989). The structure of the Magellanic Clouds. *Astronomy Now* **3** (No. 10) 16.

van den Bergh, S., and de Boer, K. S., eds. (1984). *Structure and Evolution of the Magellanic Clouds*. Kluwer Academic Press, Dordrecht.

See also **Galactic Structure, Stellar Populations; Galaxies, Local Group; Stars, Cepheid Variable, Period-Luminosity Relation and Distance Scale.**

Magnetohydrodynamics, Astrophysical

Max Kuperus

Early in this century George Ellery Hale discovered that sunspots have strong magnetic fields, a discovery that resulted in Horace W. Babcock's assertion of the 22-year magnetic cycle of solar activity. Meanwhile, it was found that interstellar polarization could be understood by the alignment of interstellar dust particles in a magnetic field, whereas Hannes Alfvén, Karl Kiepenheuer, and Vitaly L. Ginzburg suggested that the nonthermal radio emission of the Galaxy should be interpreted as synchrotron radiation of relativistic electrons gyrating in a magnetic field. In the late 1930s, Bernard Lyot and Bengt Edlén discovered that the solar corona is a million degree hot plasma. Soon it became clear that the major part of the observable universe consists of plasma frequently dominated by strong magnetic fields. Stimulated by these discoveries, magnetohydrodynamics developed into a discipline of great importance to astrophysics.

Magnetohydrodynamics (MHD) is the study of the dynamics of an electrically conducting fluid in a magnetic field. By its very nature, MHD is macroscopic in the sense that on large scales the effects of the individual charged particles on the dynamics of the fluid as a whole are negligible.

COSMICAL MAGNETOHYDRODYNAMICS

The strength of cosmical magnetohydrodynamics is that on cosmic scales the diffusion of plasma with respect to the magnetic field is negligible. The magnetic Reynolds number is extremely large so that the field is effectively frozen to the plasma. This has far reaching consequences for the evolution of magnetic fields in astrophysical plasma. Any compressive motion causes a field amplification because of flux conservation and any shearing motion will amplify the magnetic field. Hence the state of motion of an astrophysical plasma is of paramount importance for the evolution of magnetic fields. Magnetic fields decay because of ohmic dissipation, but because the dimensions of cosmical plasmas are so large this decay is usually negligible. However, there are a number of magnetohydrodynamical and plasma instabilities occurring when the stressing of the magnetic field by motions surpasses certain specific thresholds. Some of these instabilities do create small scale magnetic structures, whereas others cause a drastic increase in the resistivity. The combination of magnetic fine structure and anomalous plasma resistivity results in a much faster field decay, creating an explosive energy release as is observed in solar flares. Characteristic for cosmic plasmas is the strong effect of gravity as compared with laboratory plasmas. Magnetic fields in a gravitating medium such as a stellar atmosphere are lighter than the ambient plasma. This effect, called magnetic buoyancy, is due to the fact that in pressure equilibrium between a magnetic flux tube and its surroundings, the internal gas pressure is smaller than the ambient gas pressure because the magnetic pressure has to be added. Magnetic buoyancy is one of the major mechanisms that limit the amplification of magnetic fields and that causes magnetic flux to be expelled from gravitating objects, such as stars and accretion disks, thus forming extended, tenuous, magnetically-dominated outer atmospheres, such as the solar corona. I will now discuss a few selected areas in astrophysics wherein magnetohydrodynamics made a vital contribution.

MAGNETOSPHERE

The Earth's magnetosphere is the best cosmic plasma laboratory because many of the processes that are assumed to take place in distant and frequently poorly resolved astrophysical plasmas actually are observed to take place in our magnetosphere, as has been confirmed by numerous space probe experiments. The in situ measurements of magnetohydrodynamic and plasma processes have provided a much better understanding of the physics of particle acceleration, collisionless shock waves, magnetic reconnection, the generation of electric fields parallel to the magnetic field, and the more general problem of storage and release of magnetic energy. The magnetosphere originates from the interaction of the solar wind with the Earth's magnetic field. The result is an elongated deformed dipole field separate from the heliosphere by a bow shock and a boundary layer called the magnetopause. The plasma tail contains a neutral sheet current system where current-driven instabilities provide dissipation of stored energy, partly by the production of strong electric double layers generating electron beams that cause the aurorae at polar latitudes as well as causing geomagnetic storms.

SOLAR CYCLE

The solar surface is covered with magnetic concentrations, for example, the sunspots, which have been known for centuries. The bipolar magnetic active regions are concentrated in belts around the solar equator and show the well known 11-year cycle. Both hemispheres have opposing magnetic polarity that changes sign every 11 years. The dynamo mechanism underlying the solar cycle is based on the interaction of the differential rotation and cyclonic convective and turbulent motions in the convective solar envelope with the solar magnetic field. These cyclonic motions originate from the Coriolis effect due to the solar rotation. The screw-like nature of the convective and turbulent motions in a rotating star is essential for the transformation of toroidal fields into poloidal fields, which then can be stretched again by the differential rotation into two concentrated antiparallel strands of field situated at the sunspot latitudes. This model of the solar dynamo satisfactorily explains the observed global properties of solar magnetic fields but leaves many unanswered questions regarding the small scale fields that are found all over the solar surface and that are likely the result of the interaction of the cellular convective motions with individual magnetic flux tubes that are concentrated on the edges of the cells, thus forming a magnetically structured network. Similar dynamo mechanisms are operating in many other late type stars.

SOLAR CORONA

The formation and the heating of the solar corona has been a challenging puzzle in astrophysical magnetohydrodynamics for about half a century. From early eclipse observations as well as from radio observations it was already clear that the corona was a highly structured plasma strongly associated with the underlying active regions. Space observations in the ultraviolet and x-rays show that the corona consists of closed magnetic loops with their footpoints in photospheric magnetic field concentrations and large open regions, called the coronal holes, which are associated with unipolar magnetic regions. The x-ray emission originates primarily from the loops, whereas the coronal holes are the seats of the high-speed solar wind streaming. Because the corona is a magnetically dominated plasma, the heating of the corona is presumably electrodynamic. Two mechanisms seem to be appropriate: First, shear Alfvén waves generated by the subphotospheric oscillations can be trapped in loops where they can viscously dissipate by cascading their energy to small scales through a process of phase mixing, which essentially occurs because of the inhomogeneous structure of loops. Second, random footpoint motions of the coronal field lines give rise to a magnetic energy buildup in the corona that leads to the formation of extended thin current sheets where the energy is dissipated. The first mechanism seems particularly appropriate for loops and the second for coronal holes. In the open magnetic regions the large thermal conductivity transports the dissipated heat outwardly. The temperature distribution in the corona does not allow the atmosphere to be in hydrostatic equilibrium. It is therefore the expanding outer solar atmosphere that gives rise to the solar wind.

PROMINENCES AND FLARES

Solar prominences are manifestations of magnetic activity in the solar corona that are invariably found above the division line between magnetic polarities in the photosphere. They are cool filaments persisting for long times because they are efficiently shielded by the magnetic field against thermal conduction from the ambient hot corona. Moreover, the magnetic field offers several possibilities for magnetohydrostatic support. Oscillations of prominences demonstrate that the large scale field structure supporting prominences is stable to relatively large perturbations, but sometimes this stability is lost and the prominence suddenly rises, resulting in a coronal mass ejection. Solar flares are frequently related to prominences in the sense that a prominence eruption precedes a flare explosion. During a solar flare a large amount of magnetic energy (10^{30}–10^{32} erg), partitioned over heating, bulk motions, and particle acceleration, is released in about half an hour. The fast particles cause nonthermal emission, which is observed in radio and x-ray wavelengths during the short impulsive phase marking the onset of a flare. Flares often give rise to large coronal mass ejections.

STAR FORMATION

Stars form from the contraction of interstellar gas clouds due to self-gravitation as soon as the cloud mass surpasses the so-called critical Jeans mass. Star formation is strongly facilitated in regions of enhanced density, such as in the shock waves associated with the spiral arms and supernova remnants. Magnetic fields play a dominant role in this process because the contraction perpendicular to the field is inhibited. This in turn gives rise to modifications of the Jeans mass that lead to fragmentation of the contracting protostar. Another effect of the magnetic field is the braking of the rotation of the collapsing star and the transport of angular momentum via the field, which actually is a necessity for the contraction to proceed. A natural consequence of the removal of angular momentum from the contracting protostar is the formation of a protoplanetary disk.

GALACTIC DYNAMO

Most of the plasma in a spiral galaxy is concentrated in the disk and arms. It is nonuniformly distributed and strongly turbulent ($V \approx 10$ km s^{-1}). The high level of turbulence is probably induced by supernova explosions. From radio observations it is known that the gas follows the general rotation of our galaxy as determined from the stellar motions. The galactic magnetic field is of the order of a few microgauss and extends over scales of a few kiloparsec. This means that any diffusion of galactic magnetic fields must be due to the turbulence. Although the field is not very strong, it is able to confine cosmic ray particles with energies up to 10^{18} eV. The galactic field is generated in the disk by the differential rotation. This basically azimuthal field is transformed into a meridional component by the turbulent motions that have a screw-like nature due to the Coriolis effect. The galactic rotation then stretches the meridional field and thereby regenerates the azimuthal field. The characteristic time scale for this cycle is 5×10^{8} yr, which is relatively large given the age of the Galaxy.

ACCRETION DISKS

Accretion disks in close binary systems originate when mass overflow occurs from the primary star onto the secondary star. When the secondary star is a neutron star or a black hole, the inner parts of the thin disk extend to the Alfvén radius for neutron stars or a few times the Schwarzschild radius for a black hole. In the keplerian rotating highly turbulent inner parts of the accretion disk magnetic fields are strongly amplified and expelled from the disk, thus leading to the formation of a magnetically structured accretion disk corona, sandwiching the disk to which it is electrodynamically coupled. The magnetic energy supplied to the corona is radiated by inverse Compton scattering of soft x-ray photons produced in the disk by the viscous heating of the accreting matter. This may explain why certain x-ray sources show a very large fluctuating hard x-ray component. The interaction of the inner parts of an accretion disk with the magnetosphere around a neutron star leads to channeled accretion onto the magnetic poles, resulting in the phenomenon of x-ray pulsars with the associated spin variations due to angular momentum transfer. The interaction of disk structures with the relatively weak magnetic fields of old fast-spinning neutron stars leads to quasiperiodic phenomena around the so-called beat frequency, which is the difference between the neutron star rotation frequency and the Kepler frequency at the site of the interacting structures.

PULSAR MAGNETOSPHERES

Pulsars are rapidly spinning neutron stars (milliseconds to seconds) with extremely strong magnetic fields (10^{12} G). In the magnetosphere of a neutron star very strong electric fields are induced because of the fast spinning. The magnetosphere is thus filled with charged particles that are accelerated to highly relativistic energies. The basic mechanism for understanding the radio pulsar is that the ultrarelativistic particles radiate gamma-ray quanta in those regions where the magnetic field lines are open but still have a sufficiently large curvature (the so-called hollow cone polar cap model). These gamma quanta produce electron–positron pairs that are again accelerated to such high energies that their emission of gamma quanta results in new pair production. Such pair production avalanches cause sparks of radio emission that are typical of the microstructure of radio pulses. The pulsar emission mechanism requires a coherent radiation mechanism because of the extremely high radio brightness temperature.

Additional Reading

Alfvén, H. (1981). *Cosmic Plasmas*. D. Reidel, Dordrecht, Holland.

Low, B. C. (1990). Equilibrium and dynamics of coronal magnetic fields. *Ann. Rev. Astron. Ap.* **28** 491.

Parker, E. N. (1979). *Cosmical Magnetic Fields*. Clarendon Press, Oxford.

Priest, E. R. (1982). *Solar Magnetohydrodynamics*. D. Reidel, Dordrecht, Holland.

Zeldovich, Ya. B., Ruzmaikin, A. A., and Sokoloff, D. D. (1983). *Magnetic Fields in Astrophysics*. Gordon and Breach Science Publishers, New York.

See also **Accretion; Cosmic Rays, Acceleration; Earth, Magnetosphere and Magnetotail; Galactic Structure, Magnetic Fields; Plasma Transport, Astrophysical; Solar Activity, Coronal Mass Ejections; Solar Activity, Solar Flares, Theory; Stars, Magnetism, Theory.**

Magnitude Scales and Photometric Systems

Michael S. Bessell

More than any other aspects of astronomy, the subjects of magnitude scales and photometric systems are encumbered by history. The *intensity of light* from stars and other cosmic objects is usually expressed in *magnitudes*, an inverse logarithmic scale that frustrates physicists who work with SI units, but that is practical for astronomers. The *apparent* magnitude of an object is a measure of the intensity of radiation received at the Earth. The *absolute* magnitude is the magnitude that the object would have if it were situated at a distance of 10 parsecs [(pc), about 32.6 light-years (ly)] from the Sun. The relation between apparent magnitude m and absolute magnitude M is

$$m - M = 5\log d - 5,$$

where the distance d is in parsecs.

Early astronomers compared star with star, a procedure that still retains great benefits. The surface temperatures of common stars range from 30,000 K down to 3000 K, and their apparent brightnesses cover a range of almost a factor of 10^{10}, from the sky background upwards (this range does not include the Sun). The majority of stars are constant in total light output and in temperature. In groundbased astronomy, they must all be observed through the Earth's atmosphere. No laboratory sources of light have energy distributions similar to those observed in stars and it is natural that astronomers seek to use the standard candles in the sky rather than inferior and more expensive ones in the laboratory. *Photometric systems* represent attempts to define standard bandpasses and sets of standard sources, measured with these bandpasses, that are well-distributed about the whole sky. Different photometric systems measure different wavelength bands. All photometric systems enable the measurement of relative fluxes, from which can be inferred particular properties (such as temperature) of the emitting object, but different systems are claimed to do it more precisely, more quickly, or more easily compared with other systems. Some photometric systems are suited for hot stars, others for cool stars. Most of the systems were developed and modified by different astronomers over many years and the literature contains confusing versions and calibrations. Some people have despaired that it is too confusing and have suggested that we should start again with a well-defined ultimate system, but recent analysis has shown that modern versions of the existing photometric systems can be placed on a firm quantitive basis and that more care with passband matching will ensure that precise and astrophysically valid data can be derived from existing, though imperfect, systems.

THE MAGNITUDE SCALE OF HIPPARCHUS; THE INTENSITY SCALE OF POGSON

In the earliest recorded star catalog, Hipparchus (second century BC) divided the stars in the sky into six groups. Twenty of the brightest stars that could be seen were called first-magnitude stars and those at the limit of visibility were called sixth-magnitude stars. Intermediate-brightness stars were put in intermediate magnitude classes. In the eighteenth century, astronomers were using telescopes and had begun to measure the light intensities of stars by closing down the telescope aperture until the image of the star under study just disappeared (the disappearance aperture). By taking the ratio of the squares of the disappearance apertures of two different stars, the relative intensity of the stars' light could be calculated. This was the beginning of astronomical visual photometry. Norman R. Pogson (in 1856) at the Radcliffe Observatory compared his measurements of stellar brightness with stellar magnitudesgiven in contemporary star catalogs (such as those of Stephen Groombridge and the zone observations of Friedrich Argelander and Friedrich Wilhelm Bessel) and suggested the simple relationship $m = 5 \log a + 9.2$ to relate the magnitude m of a star and the disappearance aperture a (in inches). This relation implies a coefficient of 2.5 for the relation between magnitude and the logarithm of the intensity as $I \sim a^2$. Around that time, Gustav Theodor Fechner and Wilhelm Edward Weber (1859) were investigating the response of the eye to light and proposed the following psychophysical law: $m - m_0 \doteq s \log I/I_0$, where m is a perceived brightness and the constant s defines the scale. Pogson's work implied a scale of -2.50 for astronomical visual (eye) photometry, and we thereby have the basis for the inverse logarithmic scale. There was continuing disagreement concerning the adoption of this exact scale of 2.5 and it was not until almost 30 years later, after the Harvard photometry was published in 1884, that adoption was assured. The constant m_0, which defines the zero point, has undergone much refinement since Pogson's estimate and is officially set by the specified visual magnitudes of stars in the "north polar sequence." The early photometry catalogs are based on this sequence but the magnitude scale today is established by the contemporary whole-sky photometric standard star catalogs.

NEW MAGNITUDES FROM NEW DETECTORS

Technological advances over the last 100 years have provided a series of light detectors to supplement the eye. These detectors, in general, respond differently to light of different wavelengths than does the eye; that is, they are more sensitive to blue light or to red light than is the eye. The advent of photography in the late nineteenth century revolutionized astronomy, as did the introduction of photomultiplier tubes with their light-sensitive photocathodes in the mid-twentieth century and sensors such as silicon charge-coupled devices (CCDs) and infrared detectors over the last 10 years. Light intensities, or magnitudes, measured with these new detectors naturally differ from the visual magnitudes and depend on the color of the star. Initially, there was only the difference between visual magnitudes and "blue" photographic magnitudes to be considered, but several factors resulted in a proliferation of different passbands and photometric systems: the extension of photographic and photocathode sensitivities to a wider wavelength range, and the use of colored glass filters and interference filters to sample the starlight in narrower bands within the total wavelength sensitivity range of the detectors. The total light, integrated over all wavelengths, received at the Earth from an object is also expressed as a magnitude, the *bolometric magnitude*. The difference between the bolometric magnitude m_{bol} and the magnitude m_λ at wavelength λ is called the *bolometric correction* BC_λ. BC without a qualifier normally refers to the correction to the visual magnitude; BC, or BC_V, is usually defined to be zero for stars with temperatures similar to the Sun.

RATIONALE FOR MULTICOLOR PHOTOMETRY

Photometry of astronomical objects is carried out in order to measure the apparent total brightness of objects and their relative brightnesses at different wavelengths, that is, their energy distributions. It is possible to characterize the temperatures of most objects from the overall shapes of their energy distributions. It is also possible to infer the metal content (in astrophysics, "metal" refers to any element heavier than He) of stars from depressions in their energy distributions at particular wavelengths. These depressions (absorption lines) are due to the absorption of flux, principally by Fe and Ti (which have very rich line spectra), Ca, Mg, and the molecules CN and CH (which have very strong lines in the blue-violet region of cool stars). There are many other molecular absorption bands (such as TiO, CO, and H_2O) that depress the continuum in very cool stars; such molecular features are also used to provide information on the temperature, chemical composition, and luminosity. The energy distributions of galaxies and star clusters can be analyzed to extract the relative numbers of different kinds of stars making up these composite objects. Redshifts of very distant quasistellar objects also can be measured from the positions of depressions or peaks in their energy distributions.

Multicolor photometry is best thought of as *very-low-dispersion spectroscopy*. The entire high-resolution spectrum of a star or other cosmic object contains a large amount of information, but when dealing with extremely faint objects or with large numbers of objects, it is a great advantage to measure a small number of wavelength bands in as short a time as possible. Such a minimal technique is invaluable if it enables the derivation of many of the same parameters obtainable from a complete (and very redundant) description of the spectrum. A great deal of effort therefore has gone into accurately measuring and calibrating colors and depressions in terms of temperatures, metal abundances, and other parameters, and investigating which of competing minimal descriptions of a star's spectrum is the most accurate or most practical.

PHOTOMETRIC SYSTEMS: NATURAL AND STANDARD

A light detector, a telescope, a set of colored filters, and a method of correcting for atmospheric extinction make up a *natural* photometric system. Each observer therefore has their own natural system. The *standard* system is indirectly defined by a list of standard magnitudes and colors that have been measured for a set of typical stars, using the natural system of the originator. These are often called the *primary standards*. Later lists comprising more stars and fainter stars but based on the primary standards are called *secondary standards*. However, in the case of all photometric systems, recently published secondary standards effectively redefine the standard system because they tend to be more accurately measured than the primary lists and to represent contemporary detectors, filters, and practice.

The term "color" is an abbreviation for *color index*, which is the difference between the apparent magnitudes in two different spectral regions. Photometry is generally published as a series of colors and a single magnitude. The zero points of many color systems are set so that α Lyrae (Vega) has zero colors. In the southern hemisphere (where Vega is inaccessible) and often also in the north, the zero point is set by requiring that an ensemble of unreddened A0 stars have zero colors.

The most influential of the early works of photoelectric photometry were the so-called broadband Johnson *UBVRI* and Kron *RI* systems, which covered the wavelength region between 310 and 900 nm (3100 and 9000 Å). The natural systems of Harold L. Johnson and of Gerald E. Kron and co-workers served as "standard" systems for many other users who attempted with varying success (due to differences in detectors, filters, telescopes, and techniques) to duplicate the originators' natural systems. That is, using their own detectors and filters, astronomers measured stars from the Johnson and Kron lists and linearly transformed their natural magnitudes and colors to be the same as the Johnson and Kron colors and magnitudes. They then applied those same linear coefficients to transform the colors and magnitudes of unknown stars onto the Johnson or Kron system.

The original blue and yellow filters were chosen by Johnson from readily available glasses so that when used with the 1P21 photomultiplier tube they approximated the ordinary blue (B) photographic response (~ 436 nm) and the visual (V) response (~ 545 nm). A more violet magnitude U (~ 367 nm), which is useful for very hot stars, was obtained by using a common violet glass. In retrospect, these choices should have been based more on astrophysics and less on glass availability, but so much work has been done in this *UBV* system that the weight of history assures its continuation. Intercomparison of much of the published broadband photometry (in particular, photometry taken more than 15 years ago) often shows scatter of more than 0.03 magnitudes, but more recent photometry obtained using better equipment, better matched natural systems, and better secondary standard stars agrees to better than 0.01 magnitude, or 1%.

The 1P21 phototube was a remarkable invention and its high blue sensitivity dominated the development of photometric systems for over 30 years. There were red-sensitive devices available but observations were made only for bright stars because for many years these devices were much less sensitive, noisier, and less reliable than the 1P21. In the mid-1970s new detector materials became available; in particular, the gallium-arsenide and multialkali phototubes, which provided high (> 15% quantum efficiency) sensitivity between 300 and 860 nm, and the infrared-sensitive InSb (indium antimonide) photodiodes together with low-noise preamplifiers, which revolutionized photometry between 1000 and 4000 nm. Both developments enabled photometry to be done on faint stars that had hitherto been the sole province of the blue-sensitive detectors.

Photometry done with the new red-sensitive tubes was placed on either the Kron or the Johnson standard system, again with mixed success, and it has only been in the last few years that A. W. J. Cousins' *RI* "near-natural" standard system (based on the Kron system) has gained widespread acceptance. It has also been very useful that the Cousins system's R (~ 638 nm) and I (~ 797 nm) bands are similar to the contemporary photographic R and I bands.

Johnson also introduced the infrared alphabetic JKLMN (approximately 1.22, 2.19, 3.45, 4.75, and 10.4 μm) system in the mid-1960s, using PbS (lead sulfide) detectors and bolometers. The water vapor in the Earth's atmosphere defines a series of wavelength bands (windows) through which observations from the ground can be made. Johnson used interference filters (and, unfortunately, the atmospheric H_2O absorption bands) to define what he called the J, K, L, M, and N bands. Ian S. Glass, in his early observations with an InSb detector, used the additional band H (~ 1.63 μm) between J and K, and in his choice of filters attempted to match the other Johnson bands. All infrared observers have proceeded in a similar fashion and have concentrated mainly on copying the Johnson K magnitude scale. Identical detectors have been used but a range of slightly different filters and observatory altitudes have produced subtly different systems. The publication of sufficient numbers of stars in common from the different natural systems has helped delineate the differences, and transformations between the systems are now quite reliable.

Passbands or Sensitivity Functions

The most important specifications of a photometric system are the passbands or sensitivity functions of its magnitudes. For a variety

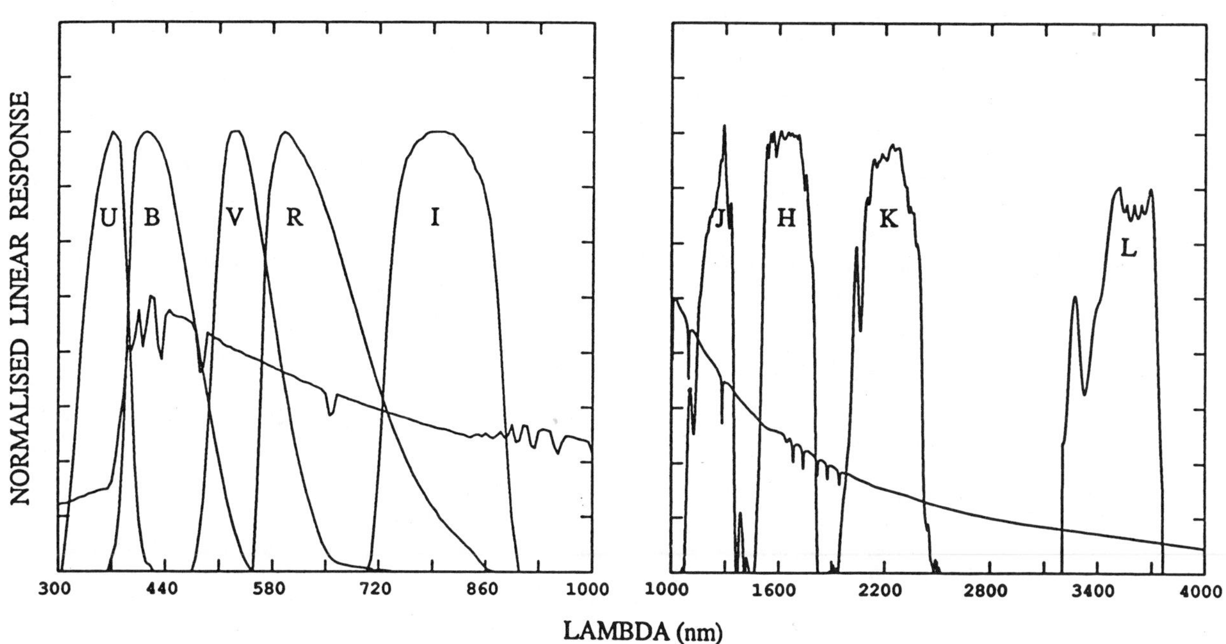

Figure 1. The passbands of the *UBVRIJHKL* system, plotted as functions of the wavelength in nm.

of reasons, technical and historical, the passbands of the broadband photometric systems have not been known with certainty and this has inhibited close matching of natural systems and has prevented computation of accurate synthetic colors from theoretical spectra. The recent availability of spectrophotometry for many stars combined with the increased precision of second-generation photometric catalogs has, however, enabled the passbands to be derived indirectly by computing synthetic colors from spectrophotometry of stars with well-defined standard colors and adjusting the passbands until the computed and standard catalog colors agree. This technique has enabled the passbands of the major systems to be well-defined, which in turn has permitted filters to be designed that still will result in good passband matches with a variety of detectors. In addition, when it is not possible to match passbands exactly with some detectors, such as photographic plates, it is possible to predict accurately the differences between photographic and photoelectric magnitudes by computing the synthetic magnitudes using the different passbands.

In Fig. 1 the linear normalized passbands of the Johnson–Cousins–Glass *UBVRIJHKL* system are shown. The F_ν (flux per unit frequency interval) spectrum of an A0 star is shown for orientation. Table 1 lists the effective wavelengths λ_{eff}, the approximate bandwidth $\Delta\lambda$, which is the full width at half maximum of the passband, and the absolute calibration of this system, based on the flux of Vega, for a zero-magnitude A0 star. Note that the effective wavelengths of the broad bands change with the color of the objects. The effective wavelengths listed are for an A0 star.

OTHER PHOTOMETRIC SYSTEMS

Real or perceived drawbacks in existing photometric systems (the *UBV* system in particular) stimulated the design of other photometric systems better suited for measuring temperatures, metal-line blanketing, effective gravity, and interstellar reddening. Some of these systems used broad bands comparable to the *UBVRI* system, whereas others used narrower bands defined by different mixes of glass filters or interference filters. Effective wavelengths and other specifications of some of the better-known systems are given in Table 2 and are discussed in the rest of this entry.

Geneva and Walraven Systems

Difficulties with matching natural systems have been eliminated by the strategy employed by proponents of the Geneva ($UBB_1B_2VV_1G$) and Walraven (*VBLUW*) systems. The latter takes its name from Th. and J. H. Walraven. These multiband photometric systems are supervised by small groups who control the instrumentation and supervise the data reduction and calibration. The colors have been well-calibrated in terms of gravity, temperature, and abundance. Such closed systems have excellent precision, but not necessarily greater than that possible from the open Cousins *UBVRI* system with careful bandpass matching.

Washington System

This CMT_1T_2 system was devised to use the wideband sensitivity of the extended-red detectors, to improve the sensitivity of blue-violet colors to metallicity and gather more violet light in cool stars, and to try to separate the effects of CN from other metal lines. We have found that the violet *C* band is a very useful metallicity indicator for faint K giants but that the *M* band contains no better information than does *V*; T_1 and T_2 have no advantages over *R* and *I*. We find a minimal *CVI* system very useful for metal-weak K stars.

Strömgren Four-Color System

The *uvby* system was devised by Bengt Strömgren to measure better the Balmer discontinuity, the metallicity, and the temperature of A, B, and F stars. The bands are essentially separate unlike the *UBV* bands, which overlap. The *u* band is completely below the Balmer jump; *v* measures the flux near 400 nm, a region with much absorption due to metal lines; *b* is centered near 460 nm and is affected much less than *B* by metal-line blanketing; y is essentially a narrower *V* band. The *u* filter is colored glass, the others are interference filters. Two special indices are derived: $m_1=(v-b)-(b-y)$, which measures metallicity, and $c_1=(u-v)-(v-y)$, which measures the Balmer discontinuity. The index $(b-y)$, like $B-V$, is used primarily as a temperature indica-

Table 1. Johnson–Cousins–Glass *UBVRIJHKLM* System

	U	*B*	*V*	*R*	*I*	*J*	*H*	*K*	*L*	*M*
λ_{eff} (nm)	367	436	545	638	797	1220	1630	2190	3450	4750
$\Delta\lambda$ (nm)	66	94	88	138	149	213	307	390	472	460
F_ν ($V=0$) (10^{-30} W cm^{-2} Hz^{-1})	1780	4000	3600	3060	2420	1570	1020	636	281	154

Table 2. Effective Wavelengths (nm) and FWHM Bandpasses (nm) for Selected Systems

	λ_{eff}	$\Delta\lambda$		λ_{eff}	$\Delta\lambda$		λ_{eff}	$\Delta\lambda$
Geneva *U*	350	47	Walraven *W*	323.3	15.4	Washington *C*	391	110
B	424	76	*U*	361.6	22.8	*M*	509	105
B_1	402	38	*L*	383.5	21.9	T_1	633	80
B_2	448	41	*B*	427.7	49.0	T_2	805	150
V	551	67	*V*	540.6	70.3			
V_1	541	44						
G	578	47						
Strömgren *u*	349	30	DDO 35	349.0	38.3	Thuan–Gunn *u*	353	40
v	411	19 (12)	38	381.5	33.0	*v*	398	40
b	467	18	41	416.6	8.3	*g*	493	70
y	547	23	42	425.7	7.3	*r*	655	90
β_w	489	15	45	451.7	7.6			
β_n	486	3	48	488.6	18.6			

tor. The system is capable of very high precision but, unfortunately, errors in the width of v filters manufactured some years ago resulted in nonstandard filters being supplied to many users. Since then, published photometry has exhibited some systematic differences in c_1 and m_1 and there are difficulties in synthesizing c_1 and m_1 from theoretical spectra, particularly for cool stars. Recent standard catalogs of new and more homogeneous observations are of high precision and internal consistency, and it should now be possible to define better the v band. Two additional interference filters (15 and 3 nm wide) centered on the Hβ line are often used together with the four colors. The Hβ index is used to derive luminosities in B stars and reddening in F and G stars. The Strömgren system was the first photometric system devised to measure specific stellar features. Because of the short wavelength range of its four color filters, 1% photometry at least is required to utilize the system's advantages over the *UBVRI* system.

DDO (35, 38, 41, 42, 45, 48) System

This system (also built around the sensitivity of the 1P21 photomultiplier) was designed for the analysis of G and K dwarfs and giants. The 35 filter is the u filter of the four-color system; the 38 filter is also a glass filter and better measures metal blanketing than the v filter, being further to the violet and wider; 41 measures the CN band; 42, 45, and 48 are continuum filters. The color $35-38$ (the 3538 index) measures the Balmer jump, 3842 measures the metallicity, and 4245 and 4548 are used for gravity and temperature measurements. By restricting the measurements to the blue spectral region, complicated corrections for spectral line blanketing are necessary to derive temperatures and gravities. Good results, especially for faint K dwarfs, can be obtained by using $V-I$ or $R-I$ as the temperature indicator. Because of the narrow bandwidth of some of the filters, the DDO (David Dunlap Observatory) system has been mainly restricted to relatively bright stars.

Thuan–Gunn System

The *uvgr* system of Trinh Xuan Thuan and James E. Gunn was devised in the mid-1970s from the *UBVR* system for use with an S20 photocathode detector and in order to avoid the strong mercury emission lines from city lights and [O I] lines in the night sky. The g and r bands are of similar width to the V and R bands whereas the u and v bands are about half the width of the U and B bands. The $g-r$ color has a longer baseline than $V-R$ but they transform well.

Johnson 13-Color System

Johnson measured many stars with a 13-color filter system. These data are very useful but the use of this system has not spread.

Photographic Systems

Originally photographic emulsions were only sensitive to light blueward of 490 nm. These were the O emulsions. Different chemical sensitizing shifted the red sensitivity cutoff to longer wavelengths: G 580 nm; D 650 nm; F 700 nm; N 880 nm, approximately. By using blue-cutoff glass filters and the red cutoff of the emulsions, various photographic passbands were made. Photographic U used a violet filter for both blue and red cutoffs. Attempts were made to convert the photographic colors onto the photoelectric *UBVR* system but these were not often very accurate because of limitations in iris photometry and poor matches of the bandpasses. In recent years, astronomical photography has undergone a renaissance caused, first, by the development of new fine-grain emulsions (Kodak IIIaJ and IIIaF) and the utilization of methods of greatly increasing the sensitivities of the J and F emulsions (using hydrogen gas) and of the N emulsions (using silver nitrate solution) and, second, by the use of new scanning microdensitometers and better methods of intensity calibration. Averaging of several wide-field Schmidt camera plates or higher scale prime-focus plates can now produce photometry to a few percent to very faint limits. Theoretical investigation of bandpasses enables better filter design for bandpass matching or predicts the relevant transformations and systematic differences between photoelectric and photographic photometry. Photographic photometry these days is usually restricted to attempted matches to the Johnson U and B or the Thuan–Gunn g systems using IIIaJ plates, to Cousins R or Thuan–Gunn r using IIIaF plates, and to Cousins I using IVN plates. Direct photographic calibration from step-wedges is usually supplemented by direct magnitude measurements of stars in each field using a CCD array.

CCD Photometric Systems

The high quantum efficiency of CCDs and their inherent linearity have made them the detectors of choice in recent years for most areas of photometry, especially for color–magnitude diagrams of clusters. Unfortunately, the advantages of the CCDs were not fully attained because many users paid insufficient care to defining their passbands and to standardizing their photometry. This resulted in internally precise results but an inability to relate these results with much confidence to the photoelectric system data or to theoretically derived magnitudes and colors. Most astronomers now realize the importance of matching the CCD passbands to the photoelectric passbands or at least of measuring their instrumental passbands.

Additional Reading

Bessell, M. S. (1983). *VRI* photometry: An addendum. *Publ. Astron. Soc. Pacific* **95** 480.

Bessell, M. S. (1986). On the Johnson *U* passband. *Publ. Astron. Soc. Pacific* **98** 354.

Bessell, M. S. (1986). Photographic and CCD *R* and *I* bands and the Kron–Cousins *RI* system. *Publ. Astron. Soc. Pacific* **98** 1303.

Bessell, M. S. and Brett, J. M. (1988). *JHKLM* photometry: Standard systems, passbands, and intrinsic colors. *Publ. Astron. Soc. Pacific* **100** 1134.

Davis Philip, A. G., ed. (1979). *Problems of Calibration of Multicolor Photometric Systems*. Dudley Observatory Report No. 14.

Hermann, D. B. (1977). N. R. Pogson and the definition of the astrophotometric scale. *J. British Astron. Assn.* **87** 146.

Lamla, E. (1982). Magnitudes and colors. In *Landolt–Börnstein New Series* **1 / 2b**, *Astronomy and Astrophysics*, K. Schaifers and H. H. Vogt, eds. Springer-Verlag, Heidelberg, p. 35.

Rufener, F. and Nicolet, B. (1988). A new determination of the Geneva photometric passbands and their absolute calibration. *Astron. Ap.* **206** 357.

See also **Star Catalogs and Surveys; Star Catalogs, Historic; Stars, Masses, Luminosities and Radii.**

Mars, Atmosphere

Mikhail Marov

Mars, the fourth planet from the Sun, possesses a substantial and rather rarefied atmosphere. The pressure at the surface is less by a factor of more than 2 orders of magnitude than the corresponding quantity in the atmosphere of the Earth. The martian atmosphere consists mainly of carbon dioxide and it undergoes profound seasonal and diurnal variations, with temperature changes exceeding 100 K. The seasonal variations incorporate the interaction of the atmosphere with the surface through phase transfer (solid to gas and vice versa) of principal constituents and, thus, partly control the atmospheric pressure. Strong winds that result both from complicated processes of global circulation and from mesoscale dynamics blow in the martian atmosphere. Dynamics is responsible for such peculiar phenomena as the global dust storms and dust devils and, hence, for the dramatic changes of the atmospheric transparency and visibility of the surface of the planet.

These introductory remarks briefly characterize contemporary knowledge of the martian atmosphere and reflect the greatly improved information that space exploration has provided. In the last century it was recognized that Mars had an atmosphere. The idea was supported by observations of white clouds over some regions of the planet at certain seasons, the existence of polar caps that waxed and waned with the seasons, and the occasional development of giant dust storms that obscured details on the planetary disk. More difficulties were encountered, however, in attempts to determine the thickness and gas content of the atmosphere. For a long time ground-based spectrometry, photometry, and polarimetry techniques were the only source for developing concepts about atmospheric parameters, structures, chemical composition, and dynamics. These techniques initially suggested a rather dense atmosphere with a surface pressure of about 0.1 bar (one-tenth that of the Earth) with nitrogen as the dominant constituent, even though spectroscopic findings indicated the presence of carbon dioxide. Only in the early 1960s, as spectroscopic methods improved, were these first estimates of the pressure and composition reduced to indicate a much thinner atmosphere of CO_2, which was soon dramatically confirmed by the first measurements of *Mariner 4* during its flyby of Mars in 1964. This close-up probing of the martian atmosphere showed that the surface pressure was only 0.0065 bar and that the thin gas envelope was at least 95% carbon dioxide.

The first in situ measurements of the height profiles of atmospheric temperature and pressure were carried out in 1974 by the *Mars 6* space probe. The most accurate direct atmospheric measurements were obtained along two descent trajectories in 1976 by *Viking 1* and *2*; the data included detailed measurements of atmospheric composition as well as measurements of the weather at the two landing sites (in the Chryse and Utopia regions) over two martian years.

Table 1. Principal Parameters of the Atmosphere of Mars

Chemical Composition	
Main Constituents (fraction percent)*	
Carbon Dioxide (CO_2)	*95 (0.034)*
Nitrogen (N_2)	*2.7 (78.1)*
Argon-40 (^{40}Ar)	*1.6 (0.93)*
Oxygen (O_2)	*0.13 (20.9)*
Carbon Monoxide (CO)	*0.07 (5×10^{-6}–2×10^{-5})*
Water Vapor (H_2O)	*0.03† (0.1–3.0)*
Trace Constituents (parts per million)*	
Argon-36 (^{36}Ar)	5 (30)
Neon (Ne)	2.5 (18)
Krypton (Kr)	0.3 (1)
Xenon (Xe)	0.08 (0.09)
Ozone (O_3)	0.04–0.2 (0.01–0.4)
*Isotopic Ratios**	
Carbon: $^{12}C/^{13}C$	90 ± 5 (89)
Oxygen: $^{16}O/^{18}O$	500 ± 25 (499)
Nitrogen: $^{14}N/^{15}N$	165 ± 15 (277)
Argon: $^{40}Ar/^{36}Ar$	3000 ± 500 (292)
Xenon: $^{129}Xe/^{132}Xe$	2.5^{+2}_{-1} (0.97)
Mean Molecular Mass*	43.4 (28.97)
Surface Temperature (K) (middle latitudes)*	
T_{max}	270 (310)
T_{min}	200 (240)
Mean Surface Pressure* (bar)	6×10^{-3} (1)
Mean Surface Density* (g cm^{-3})	1.2×10^{-5} (1.27×10^{-3})
Nominal Optical Depth, τ‡	0.4 (0.6; 0.2)
Nominal Single-Scattering Albedo ω‡	0.860 (0.975; 0.750)

*The respective values for the Earth are given in parentheses.
†Abundance varies with season and location.
‡The surface albedo is assumed to be $A = 0.25$. Within the parentheses are the maximum and minimum values of τ and ω that correspond to maximum and minimum solar irradiance with $A = 0.30$ and 0.15, respectively.

COMPOSITION, TEMPERATURE, AND PRESSURE

The data on chemical composition and other parameters of the martian atmosphere are summarized in Table 1. The major constituents are CO_2 (95%), N_2 (2.7%), and Ar (1.6%), whereas the abundances of O_2, CO, and H_2O are small; water vapor exhibits strong seasonal and locational variations. The distribution of noble gas abundances and isotopic ratios of the elements composing the main volatiles is of special interest because it serves as a clue to the early history of Mars and its atmosphere; in particular, it seems that Mars has outgassed more Ar^{40} (the isotope produced by the decay of radioactive K^{40}) relative to Ar^{36} (the primordial isotope). Nitrogen on Mars is highly enriched in its heavier isotopes; assuming the escape process, this allows one to calculate the original nitrogen abundance and to conclude that Mars must have started out with ~ 10 times the nitrogen we now find in its atmosphere. The same conclusion was reached from the evidence for the relative amounts of primordial noble gases, normalized to the mass of the planet (in other words, referred to grams of its matter), which are about the same on Mars and Earth suggesting that other volatiles, such as carbon and nitrogen, should also be released in similar relative abundances on the two planets. The ratio of total nitrogen and carbon dioxide to atmospheric neon on Earth together with the amount of neon on Mars argue for a much denser atmosphere in Mar's early history, with a comparable or even higher sea-level pressure than on Earth. This could explain the presence of liquid water and the remnants of ancient river beds on the martian surface as a result of the greenhouse effect with a temperature about 0°C.

There are quite reasonable scenarios to explain the climatic change on Mars: depleted radionuclides in its interior and the resultant predominance of buried dissolved atmospheric CO_2 through rock weathering and deposition of carbonates; a gradual decrease in the greenhouse mechanism and declining atmospheric temperature and pressure; finally the escape of nitrogen into space and partial loss of water. This led to the contemporary atmosphere of Mars as shown schematically in Fig. 1. The temperature profiles within the troposphere (corresponding to the first 10–30 km, depending on time of day) and the stratosphere–mesosphere region up to about 100 km are based on the results of direct measurements. The solid curve corresponds to the model based on data available, according to the *COSPAR Reference Atmosphere*. Because the surface pressure varies depending on planetary relief and the seasons, a pressure of 6 mbar can be adopted as an average value attributed to thesurface of a standard reference ellipsoid or a mean figure for summer in the martian northern hemisphere. For northern winter the mean pressure is approximately 1 mbar higher. Depending on local elevations the pressure can increase up to 10 mbar in depressions and drop to as low as 0.5 mbar at the summits of volcanic domes. Following the large-amplitude seasonal oscillations the pressure varies by more than 3 mbar with the lowest value in late summer in the northern hemisphere before the autumnal equinox (southern winter) and the highest value in early

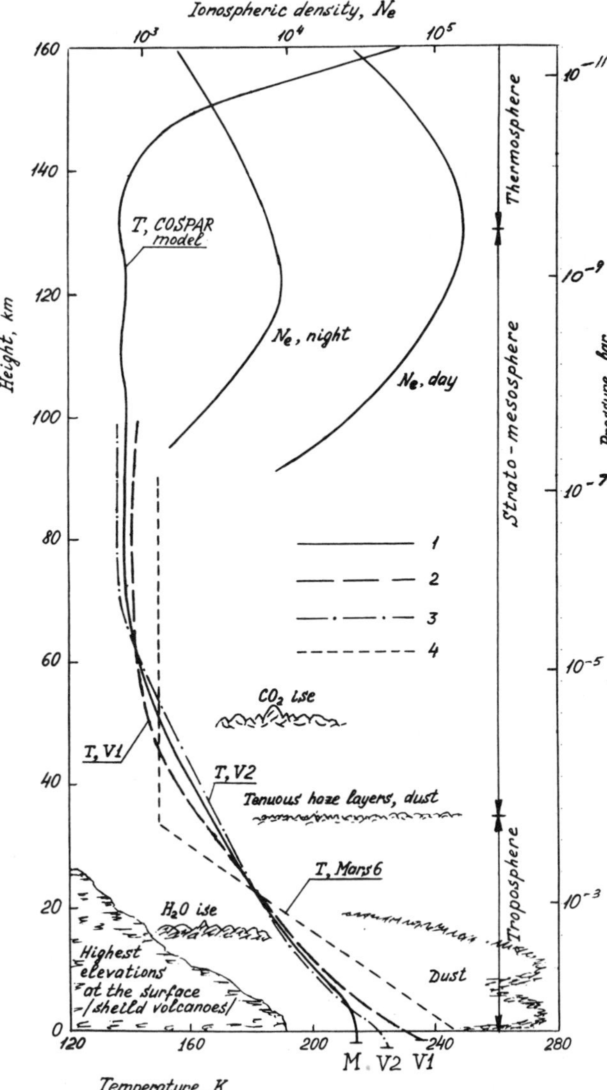

Figure 1. Schematic representation of the atmosphere of Mars. Temperature height profiles are based on the *COSPAR Reference Atmosphere* and the available direct *in situ* measurements. Regions in the atmosphere are designated as follows: 1—*COSPAR* model, nominal midlatitude summer mean temperature; 2—*Viking 1*, $\lambda = 96°$, $\theta = 22.3°$N, 4 P.M. MLT (Mars Local Time); 3—*Viking 2*, $\lambda = 116°$, $\theta = 47.6°$N, 10 A.M. MLT; 4—*Mars 6*, $\lambda = 19°$, $\theta = 24°$S, 3:30 P.M. MLT. The surface reference level corresponds to a pressure of 6 mbar; lower levels for the different landing sites are indicated. Usual heights for clouds and haze locations are shown as well as electron density profiles in the martian ionosphere for both day and night conditions.

winter in the northern hemisphere before the winter solstice (southern summer). Around the vernal equinox in the northern hemisphere the pressure is about 2 mbar higher. The seasonal variations based on the *Viking* meteorological measurements on the surface of Mars are shown in Fig. 2. The reason for seasonal oscillations is the variation in atmospheric mass because CO_2 condenses out on the winter polar caps. Thus, the deepest pressure minimum (near the areocentric longitude of the Sun $L_s = 150$) corresponds to maximum CO_2 accumulation in the south polar cap and the secondary minimum (near $L_s = 350$) corresponds to maximum accumulation in the northern cap. As a result the average atmospheric surface pressure changes by about 20% with the seasons.

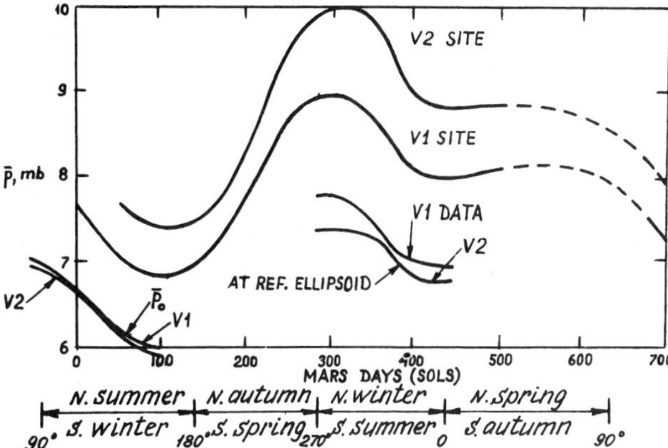

Figure 2. Seasonal variation of surface pressure on Mars according to *COSPAR*. The lower curves are *Viking* data referenced to a reference ellipsoid of Mars.

The temperature of the atmosphere of Mars near the troposphere experiences not only seasonal but also wide daily fluctuations. The reason is that the atmosphere is thin and does not contain moderating quantities of water; also the low thermal conductivity of the soil and the absence of water reservoirs on the surface prevent the smoothing out of diurnal thermal contrasts as the insolation varies. The daily average temperature at the base of the atmosphere can be adopted as 200–220 K at middle latitudes for the late summer of the northern hemisphere; the temperature of the atmosphere is 10–15 K lower than that of the surface. There is a near-surface thermal boundary layer (estimated to be ~ 4–6.5 km thick) within which the atmosphere is likewise subject to large diurnal temperature variations. Air temperatures at the two *Viking* sites reached summer daytime highs of about 244 K, and the ground itself only occasionally exceeded the freezing temperature of water, even in the strong afternoon sun.

These conditions provoke convection within the boundary layer of several kilometers thickness where there is a near-adiabatic temperature lapse rate (overlying a turbulent superadiabatic surface boundary layer whose depth is typically a few meters) and in the subadiabatic region beyond it. The convective thermal driver on Mars is ~ 10 times as effective as the diurnal forcing on the Earth (per unit mass). In contrast, night-time air temperature dropped to 188 K at the *Viking* sites and the models suggest even lower ground temperatures with strong inversion of the air temperature profile within the boundary layer. This means a highly stable atmosphere with damped convection (again in contrast to the Earth's troposphere that has much smaller day/night variations because convection rarely ceases completely). Even lower temperatures are attained at the poles in winter: the nominal value is assumed to be 147 K, which just corresponds to the sublimation of CO_2. 50 km above the surface, the temperature averages about 140 K and a lapse rate close to zero (isothermy) is attained, but there are large-amplitude, large-scale irregularities superimposed on the height profile. The explanation for this is found in the atmospheric dynamics.

WIND, CLOUDS, AND DUST

Unfortunately very few direct observations of winds are available. From the motions and morphology of the clouds, some predominant seasonally dependent wind directions were deduced and the *Viking* landers measured diurnally variable surface winds that had small daily mean values. Radio occultation and infrared sounding techniques provided temperature cross sections, which, using the thermal wind equation, also yielded zonal wind profiles.

The planetary circulation on Mars is very different from that on Earth. A general circulation model (GCM) that approximates geostrophic balance predicts distributions of surface winds as well as winds aloft and it shows eastward surface winds in winter high latitudes and in the summer subtropics, and generally westward winds elsewhere. Because the thermal wind equation is not applicable to the evaluation of mean meridional winds, the latter are less certain. One may assume both variable insolation and the exchange of CO_2 between the atmosphere and the polar caps as the main seasonal drivers. The GCM predicts a strong thermally driven circulation at the solstices and the formation of a pattern known as a Hadley cell, which exhibits flow toward the summer pole at low levels, rising in the summer subtropics, return flow toward the winter pole aloft, and descending flow in winter midlatitudes.

There is some evidence that wind patterns are strongly influenced by the global scale and local topography; thus topographically forced as well diurnally varying winds associated with diurnal heating and cooling on regional slopes appear to further complicate the picture that follows from the GCM. Topographic variations also force quasistationary horizontal long waves as well as internal gravity waves propagating upward and observed near the edges of the winter polar caps. These waves are responsible for the previously mentioned irregularities superimposed on the temperature height profile in the strato–mesosphere. In turn, the patterns of gravity waves may be indicative of vertical wind profiles. When discussing atmospheric dynamics on Mars the transient planetary scale waves arising from baroclinic instability should also be taken into account; interestingly, these waves (more regular than baroclinic waves in the Earth's atmosphere) were generated in GCM calculations and they resemble the regular wave regime of rotating differentially heated tank experiments.

Finally, an important dynamical phenomenon is thermally generated diurnal and semidiurnal tides, which are likely involved in a peculiar positive feedback with suspended dust that ensures strong internal heating of the martian atmosphere and thus enhances the tides (the enhancement varying with the dust load). The high dust content suspended in the atmosphere is among the most remarkable features of the planet and because of it, the nominal optical depth usually remains above 0.4 (see Table 1) even under quiet conditions.

Dramatic changes in the atmosphere of Mars are experienced during planetwide dust storms when aerosols are distributed nearly globally (with the exception of lower content in the polar regions) and uniformly by height up to 50 km, without signs of layering. Changes in the planetary circulation and wind patterns during a dust storm should be brought about by a combination of a shift in the heating to high levels and high latitudes (due to increasing opacity corresponding to $\tau \gtrsim 5$) with a heating decrease near the surface because dust prevents solar radiation from penetrating deep into the atmosphere. Although many different mechanisms have been proposed to explain dust storm genesis, the most plausible process involves a coupling between dust heating, dust raising, and global scale winds (tides and mean meridional circulation). Because global dust storm generation is favored only when the optical depth is near unity and the circulation is suppressed by very large tropospheric opacity, the reasonable explanation of storm decay can be drawn from the idea of negative feedback, in other words, the commencement of storm decay immediately after the storm is triggered. The decay phase can take several months and damping the progressive development of a dust storm can localize it to rather small areas (so-called dust devils). The most striking feature of the circulation in the atmosphere of Mars is the powerful role of heating by atmospheric dust in generating an intense global scale wind system.

There is another category of aerosols of condensate nature: ice crystals of water and carbon dioxide. Localized regions of Mars are heavily clouded during particular seasons and larger areas are obscured by haze for long periods. Local dust storms in certain areas contribute to the atmospheric hazes. The two regions of greatest condensate cloud activity are the north polar hood and the elevated Tharsis and Valles Marineris region, which is particularly active in spring and summer when the northern hemisphere is moist. Extensive clouds are regularly found around the high shield volcanoes and morning fogs occur in low-lying areas.

Discrete condensate clouds can be seen over other areas of the planet but are never as persistent as those in the two regions mentioned above. Certain types of clouds give useful information about various properties of the martian atmosphere, such as wind speeds and direction, vertical wind shear, atmospheric stability, wave processes, and so on.

UPPER ATMOSPHERE

Above about 130–140 km the temperature of the martian atmosphere increases rapidly due to direct absorption of solar short-wave (both extreme ultraviolet and soft x-ray) radiation (see Fig. 1). From here (or slightly lower) the thermosphere overlying the strato–mesosphere begins. This region is partly ionized and thus is also referred to as the ionosphere.

The composition and temperature of both neutral and ionized gas of the upper atmosphere were measured using different techniques (mass spectrometry, air-glow emissions, radio occultation). The temperature of the thermosphere of Mars rises from ~ 140 K at 130 km to ~ 300 K at 200 km. The latter value is referred to as the exospheric neutral temperature T_∞ because it remains nearly constant higher up in the exosphere. In contrast to the Earth's thermosphere it depends weakly on the solar activity cycle. The ion temperature exhibits rather close thermal contact with the neutral gas below 175 km and appears to exceed T_∞ in the exosphere.

The neutral composition above the turbopause (homopause) at about 130 km (the level where molecular diffusion begins to prevail over turbulent mixing) is represented by the progressively dominant species (CO_2 in this case) produced due to photodissociation of the main components. Densities of O, CO, and H along with CO_2 and O_2 are best fitted to a model-assumed eddy diffusion coefficient at turbopause level $K = 10^7 – 10^8$ cm^2 s^{-1}.

The martian ionosphere is weaker than the Earth's. The mean ionosphere has a peak density of about 2×10^5 cm^{-3} at the subsolar point and a topside scale height of 36 km. It is located at an altitude of 135–140 km. A second maximum was also found at about ~ 110 km with a lower peak density, 7×10^4 cm^{-3}. The dominant ion near the ionization peak is O_2^+; CO_2^+ is $\sim 13\%$. The O_2^+ ion is mainly produced due to charge-exchange processes of CO_2^+ with neutral atomic oxygen. At higher altitudes ionic O^+ also contributes to the ionospheric composition. Of great importance and interest are the complicated processes of the interaction of solar wind plasma with the gas envelope of Mars—a planet with seemingly no intrinsic magnetic field.

Additional Reading

Kliore, A., ed. (1978). *The Mars Reference Atmosphere* (*COSPAR*). Jet Propulsion Laboratory, California Institute of Technology, Pasadena, Calif.

Leovy, C. B. (1979). Martian meteorology. *Ann. Rev. Astron. Astrophys.* **17** 387.

Marov, M. Ya. (1987). *Planets of the Solar System*, 2nd ed. Nauka, Moscow.

Moroz, N. I. (1978). *Physics of the Planet Mars*. Nauka, Moscow.

Morrison, D. and Owen, T. (1987). *The Planetary System*. Addison-Wesley, Reading, Mass.

Pollack, J. B. (1990). Atmospheres of the terrestrial planets. In *The New Solar System*, J. K. Beatty and A. Chaikin, eds. 3rd edn. Sky Publishing Corp., Cambridge, MA, and Cambridge University Press, Cambridge, U.K., p. 91.

See also **Mars, Space Missions; Planetary and Satellite Atmospheres; Planetary Atmospheres, Clouds and Condensates; Planetary Atmospheres, Dynamics; Planetary Atmospheres, Escape Processes; Planetary Atmospheres, Structure and Energy Transfer.**

Mars, Space Missions

Joseph M. Boyce and Sandra L. Dueck

To date, 24 spacecraft have been launched to Mars, with 16 of those reaching the vicinity of the planet (Table 1). The first attempts to explore Mars began in 1960 with a series of unsuccessful missions launched by the Soviet Union. It was not until July 14, 1965, that *Mariner 4* became the first spacecraft to reach Mars.

Mariner 4 was one of two identical spacecraft launched by the United States to Mars in November 1964 and was one of a series of similar spacecraft sent to Mars by the United States; all were three-axes-stabilized vehicles consisting of an octagonal framework, or "bus," with four attached solar panels. *Mariner 3* was launched but it failed en route because its solar panels did not open. During *Mariner 4*'s successful flyby of Mars, it acquired 22 close-up photographs that showed a cratered surface resembling the lunar highlands. Other instruments aboard *Mariner 4* confirmed that Mars has a thin carbon dioxide (CO_2) atmosphere (5–10-mbar range) and a small intrinsic magnetic field.

The exploration of Mars continued in 1969 with the launch of another pair of Mariner spacecraft (*Mariner 6* and *Mariner 7*), which were much more sophisticated than *Mariner 4*. *Mariner 6* flew by the equatorial region of Mars on July 30, 1969, and *Mariner 7* flew by the south polar region on August 4. Together they made measurements of surface temperature, the molecular composition, temperature, and pressure of the atmosphere, and took more than 200 pictures of Mars.

In May 1971, five more missions were launched to Mars: two by the United States (*Mariner 8* and *Mariner 9*), and three by the Soviet Union (*Kosmos 419*, *Mars 2*, and *Mars 3*). The launch vehicle for *Mariner 8* failed, and *Kosmos 419* failed before transit to Mars. *Mars 2* was a lander–orbiter combination, launched on May 19, 1971, that weighed 4650 kg (10,230 lb) and was nearly identical to *Mars 3*, which was launched on May 28, 1971. Both made measurements of solar plasma and cosmic rays en route to Mars. On November 27, 1971, the lander separated from *Mars 2* on approach to Mars but crash-landed when its braking rockets failed. Despite this failure, *Mars 2* was the first object to "land" on Mars.

Mars 3 arrived at Mars on December 2, 1971, releasing its lander for the first "successful" landing on Mars; however, after relaying 20 seconds of video imaging (to the *Mars 3* orbiter), the signal failed. A dust storm raging on the surface may have initiated a failure on the lander, or the orbiter may have had a malfunction. Although the lander failed, the orbiter made remote measurements of the surface temperature, photometric properties, and composition of the martian atmosphere.

Mariner 9 was placed into orbit about Mars on November 24, 1971, becoming the first American spacecraft to orbit a planet beyond the Moon. The *Mariner 9* spacecraft was a heavier [546 kg (1204 lb)], more sophisticated version of the previous *Mariner*s. It carried a large rocket motor and fuel tanks, which were used to place it into orbit, and a larger payload of scientific instruments. Most of its scientific observations of the planet had to be postponed for several months until the global dust storm that obscured the surface had subsided. Before *Mariner 9* successfully completed the mapping of Mars, it had provided the first high-resolution look at the martian moons, Phobos and Deimos, and had discovered rivers and channels, which renewed speculation about life and showed that Mars was not the dead, moon-like planet that the earlier *Mariners* had suggested.

The Soviets continued their quest for a successful Mars mission in 1973 at the next favorable opposition. *Mars 4* and *5* were launched in July 1973 and *Mars 6* and *7* in August. *Mars 4* and *5*

Table 1. Launch and Arrival Data for Soviet and American Mars Missions

Mission	*Launch Date*	*Fate*
Mars 1960A	October 10, 1960	Failed to reach Earth orbit
Mars 1960B	October 14, 1960	Failed to reach Earth orbit
Mars 1962A	October 24, 1962	Failed to leave Earth orbit
Mars 1	November 1, 1962	Communications failed
Mars 1962C	November 4, 1962	Failed to leave Earth orbit
Mariner 3	November 5, 1964	Stuck shroud prevented flyby
Mariner 4	November 28, 1964	July 14, 1965 at 9920 km
Zond 2	November 30, 1964	Communications failed
Mars 1969A	March 27, 1969	Failed to reach Earth orbit
Mariner 6	February 24, 1969	July 1, 1969 at 3437 km
Mariner 7	March 27, 1969	August 5, 1969 at 3551 km
Mariner 8	May 8, 1971	Failed to reach Earth orbit
Kosmos 419	May 10, 1971	Failed to leave Earth orbit
Mars 2	May 19, 1971	Lander crashed
Mars 3	May 28, 1971	Lander transmitted 20 s
Mariner 9	May 30, 1971	November 3, 1971
Mars 4	July 21, 1973	Failed to enter Mars orbit
Mars 5	July 25, 1973	February 12, 1974
Mars 6	August 5, 1973	Lander descent data only
Mars 7	August 9, 1973	Lander missed planet
Viking 1	August 20, 1975	June 19/July 20, 1976
Viking 2	September 9, 1975	August 7/September 3, 1976
Phobos 1	July 1988	Failed
Phobos 2	July 1988	January 1989 acquired data; March 27, 1989 lost contact

were orbiters and *Mars 6* and *7* were orbiter–lander combinations; *Mars 4* and *5* were planned to be used for scientific and landing-site studies and as communication relays for the *Mars 6* and *7* landers. *Mars 4* passed within 2200 km (1375 miles) of Mars but failed to enter an orbit around the planet when its braking engine malfunctioned. *Mars 5* was successfully inserted into Mars orbit on February 12, 1974, to await the arrival of *Mars 6* and *7* and to acquire imaging data. On March 9, *Mars 7* passed the planet instead of entering Mars orbit, apparently because of a premature release of the lander. On March 12, *Mars 6* successfully launched its lander after entering orbit. The lander transmitted atmospheric data, but failed as it descended.

The United States was ready for the next favorable opportunity to launch to Mars with *Viking*—the most ambitious planetary mission to date. It consisted of two identical orbiter–lander combination spacecraft. Each orbiter was of *Mariner* design, weighing 900 kg (1984 lb), each with an attached 600-kg (1328 lb) heat-sterilized lander within its capsule. Each lander was a horizontal platform structure, 1.5 m (4.9 ft) across and 0.5 m (1.6 ft) thick, mounted on three legs.

In addition to instruments to investigate the geological and geophysical nature of Mars, both landers carried instruments specifically designed to search for life (biology, molecular analysis). These instruments found no unambiguous evidence of life on Mars. However, *Viking* provided the first close-up look at the surface of Mars and the first direct measurements of soils and atmospheric properties.

Both orbiters were deactivated once they had depleted their attitude control gas: *Viking Orbiter 1* in August, 1980 and *Viking Orbiter 2* in July, 1978. *Viking Lander 2* was using *Viking Orbiter 2* as a communications relay to Earth and as a result had to be deactivated at the same time as the orbiter. *Viking Lander 1* continued to operate until October, 1982.

Most recently the Soviet Union sent the *Phobos 1* and *2* spacecraft to investigate Phobos, the largest martian moon. The missions consisted of two orbiter–lander combinations. One orbiter carried a small lander; the second orbiter carried a similar lander together with another lander capable of hopping across the surface. Both missions were launched in July 1988. *Phobos 1* failed en route to Mars. *Phobos 2* was inserted into Mars orbit in January 1989. Its orbit was synchronized with Phobos and it moved to within 800 km of Phobos before it failed.

Both the United States and the Soviet Union plan to continue the exploration of Mars. Currently, the United States is building the Mars Observer spacecraft, scheduled for launch in 1992. This mission is designed to make remote sensing and atmospheric measurements of Mars to characterize the surface composition and the climatology of the planet. Both the United States and the Soviet Union are considering future robotic missions during the 1990s and early 2000s, with human exploration following in the early- to mid-2000s.

Additional Reading

Chicarro, A. F., Scoon, G. E. N., and Coradina, M. (1989). Mission to Mars: Report of the Mars exploration study team. Report ESA SP-1117, European Space Agency, Paris.

Collins, M. (1990). *Mission to Mars*. Grove Weidenfeld, New York.

Goldman, S. J. (1990). The legacy of *Phobos 2*. *Sky and Telescope* **79** 156.

Hartmann, W. K. and Raper, O. (1974). *The New Mars; The Discoveries of* Mariner 9. NASA SP-337, NASA, Washington, DC.

Horowitz, N. H. (1977). The search for life on Mars. *Scientific American* **237** (No. 5) 52.

Mariner 9 Television Team and Planetology Program Principal Investigators (1974). *Mars as Viewed by* Mariner 9. NASA SP-329, NASA, Washington, DC.

Reiber, D. B., ed. (1989). *The NASA Mars Conference*. Univelt, Inc., San Diego.

See also **Mars, Atmosphere; Mars, Surface Features and Geology; Planetary Volcanism and Surface Features.**

Mars, Surface Features and Geology

Michael H. Carr

Geology is the study of near-surface rocks in order to reconstruct the formation and history of a planet. Mars has been observed with telescopes for centuries, but it was not until 1965, when the first spacecraft arrived at Mars, that the study of martian geology could realistically begin. Even now, geologic interpretation is based largely on remote sensing data, such as imaging and infrared observations, rather than on more definitive measurements made directly on rock materials at the surface. Remote sensing data have been acquired primarily from two orbiter missions, from *Mariner 9*, which arrived at Mars in 1971, and from the two *Viking* orbiters that observed the planet continuously between 1976 and 1980. From these missions we now have photographic coverage (see Fig. 1) of the entire planet at a resolution of about 500 m and of small fractions of the planet at resolutions as low as 20 m. The *Viking* mission also included landers that made a variety of observations at the surface at two locations.

With a diameter of 6788 km, Mars is intermediate in size between the Earth (12,756 km) and the Moon (3476 km). Mars also has geologic characteristics that are intermediate between these two bodies. It resembles the Moon in that a large part of the surface dates back to 3.8 billion years ago when meteorite impact rates were very high. It resembles the Earth in having experienced sustained volcanism and tectonism and in having a surface that has been sculpted by wind, water, and ice. Despite these similari-

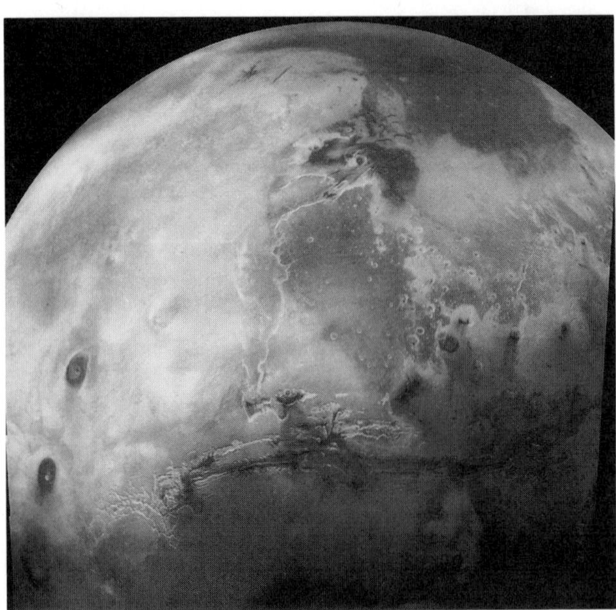

Figure 1. Synoptic view of Mars. The three large Tharsis volcanoes appear as dark spots on the left side of the image. The 4000-km long equatorial canyon system extends roughly east–west across the bottom center. Chryse Planitia, on which the *Viking 1* spacecraft landed and on which many large flood channels converge, is in the middle of the upper right quadrant. (*Courtesy of U.S. Geological Survey, Flagstaff, AZ.*)

ties with the Earth, Mars' geology is very different from the Earth's. One reason is that Mars lacks plate tectonics, which affect almost every geologic process here on Earth. A second reason is that climatic conditions throughout much of Mars' history, caused liquid water to be unstable at the surface. At present, with a 6-mbar atmosphere of CO_2 and mean daily temperatures that range from 150 K at the poles to 220 K at the equator, the atmosphere can contain only minute amounts of water vapor; liquid water will freeze everywhere, water ice is stable at the surface only at the poles, and ground (subsurface) ice is stable only at latitudes higher than 40°. As we shall see later, however, conditions were more permissive of liquid water early in the planet's history.

Mars is markedly asymmetric in the distribution of its surface features. Most of the southern hemisphere stands at 1–3 km above the elevation datum and is heavily cratered like the lunar highlands. In contrast, much of the northern hemisphere is only sparsely cratered, with elevations 1–2 km below the datum. Why the heavily cratered surface survives primarily in the south is not clear. A possible explanation is that, very early in the planet's history, a gigantic impact excavated a huge basin in the northern hemisphere. The heavily cratered terrain has survived only outside this large impact basin. Superimposed on the north–south asymmetry are two bulges. The first, about 6000 km across and 10 km high, is centered on the equator at 110°W in a region called Tharsis. The second, much smaller bulge, is centered in the region of Elysium at 25°N, 210°W. Both bulges have been sites of sustained volcanic activity. They may have formed directly, or indirectly, as a result of the convective upwelling of the warm mantle below the two regions.

CRATERED HIGHLANDS

The cratered highlands cover almost two-thirds of the planet and are characterized by several large impact basins and numerous large craters standing almost shoulder to shoulder. Two large, relatively well preserved impact basins are the 2000-km diameter Hellas basin at 43°S, 291°W and the 1200-km diameter Argyre basin at 50°S, 42°W. By analogy with the Moon, the rates of impacts by meteorites are believed to have been high early in Mars' history, prior to 3.8 billion years ago. They then declined substantially so that the present low rates prevailed for much of the planet's history. Heavily cratered surfaces like the southern highlands are thus thought to date back to at least 3.8 billion years ago. The heavily cratered highlands of Mars differ from those on the Moon in several respects, but most distinctively in the appearance of impact ejecta around craters and in the presence of numerous branching valleys. The ejecta around lunar impact craters are generally disordered and hummocky near the rim crest and grade outward into radial ridges and strings of secondary craters. In contrast, the ejecta around most martian craters are commonly disposed in discrete petal-like lobes, each outlined by a low ridge. The favored explanation for the distinctive martian patterns is that the surface materials contain large amounts of water or ice. The ejecta were thus saturated with water, giving them a mud-like consistency that caused them to flow along the ground after ejection from the crater.

The presence of branching valleys throughout the cratered highlands is one of the main reasons for believing that the highlands were water-rich and that climatic conditions on early Mars were different from climatic conditions that prevailed for most of the rest of the planet's history. The valleys are generally several kilometers across and may be up to several hundred kilometers in length. Most have tributaries and increase in size downstream. They usually end in a local basin. Their characteristics strongly suggest that they formed as a result of slow erosion by running water, and that groundwater rather than rainfall was the source. The valleys are found everywhere that the heavily cratered terrain is exposed, except at high latitudes, where they may have been destroyed by ice-abetted creep of the near-surface materials. They are rarely found on younger terrains. The valleys appear to have formed, not by massive floods, but by streams of modest discharges that would freeze quickly under present climatic conditions. The valleys appear to indicate, therefore, that conditions on early Mars were more conducive to the presence of liquid water than are present conditions. The scarcity of valleys on younger surfaces suggests that climatic conditions subsequently changed so that, for most of the planet's history, small streams could rarely survive long enough to do any significant erosion. A likely possibility is that Mars had an initially thick CO_2 atmosphere that provided sufficient greenhouse warming that surface temperatures were close to 273 K. Subsequently, most of the CO_2 was irreversibly lost from the atmosphere by reacting with the surface materials to form carbonates.

VOLCANISM

Between the large craters in the cratered highlands are plains on which lava flows are clearly visible. It appears that while the cratered terrain was forming, high rates of volcanism were occurring simultaneously. However, the most spectacular manifestations of volcanism are the huge shield volcanoes in Tharsis and Elysium. Shield is a term used to describe volcanoes, like those in Hawaii, that are built almost entirely of very fluid lava. Three enormous shield volcanoes form a NE–SW line along the crest of the Tharsis bulge. Each is about 400 km across and has a large summit caldera that stands over 15 km above the surrounding terrain. Several hundred kilometers to the northwest of these volcanoes is the tallest volcano on Mars, Olympus Mons. It is 600 km across and has a 90-km diameter summit caldera that is 27 km above the surrounding terrain. Elsewhere throughout Tharsis are many other smaller volcanoes. Parts of the volcanoes are devoid of impact craters, indicating very young surfaces and suggesting that some of the volcanoes may still be active. On most of the shield volcano flanks are numerous lava flows, superimposed one on another, many with leveed lava channels, as is the case with terrestrial shield volcanoes. Indeed the martian shield volcanoes closely resemble their terrestrial equivalents in everything except size. The huge size of the martian volcanoes is probably due in large part to the lack of plate tectonics. Terrestrial volcanoes are relatively short-lived because the plate on which they stand moves slowly and carries them away from the source of magma below the plate. On Mars, however, a volcano sits stably over the magma source and will continue to grow as long as magma is available and can get to the surface.

The shield volcanoes in Elysium are smaller than those in Tharsis, but one peculiarity deserves special mention. Many large channels start close to the volcanoes in Elysium and extend several hundred kilometers to the northwest, down the regional slope. The most plausible explanation is that volcanic heat melted ground ice in the region, thereby leading to periodic release of meltwater, which carved the channels. Shield volcanoes, which are built with relatively fluid lava, are not the only type of volcano present on Mars. Some volcanoes surrounded by deeply eroded deposits are most likely built of ash. In addition many of the plains, particularly in Tharsis and Elysium, are covered with lava flows, which appear to be the results of fissure eruptions rather than of eruptions from central volcanic vents.

The Tharsis bulge, in addition to being a focus of volcanic activity, is the cause of most of the deformational features observed on the planet. The bulge is at the center of a vast system of radial fractures that affect about a third of the planet. The fractures generally occur in pairs with a down-faulted block between. Also surrounding the bulge are circumferential ridges indicating compression. Both the fractures and the ridges are probably the result of stresses created in the crust by the bulge.

CANYONS AND FLOOD FEATURES

On the east flank of the Tharsis bulge is a vast system of interconnected canyons and dry valleys that appears to have formed by a combination of faulting and water erosion. The largest canyons, the Valles Marineris, extend from close to the summit of the Tharsis bulge, eastward for almost 4000 km. Most of the canyons are over 3 km deep and over 100 km across. In the central section, three parallel canyons merge to form a depression over 7 km deep and 600 km wide. The canyons are mostly flat floored with steep, gullied walls. Many contain thick, partly eroded, layered sediments. The origin of the canyons is probably complex. Straight walls parallel to the radial fractures around Tharsis suggest that faulting has been important in creating the relief, but deeply gullied walls demonstrate that significant erosion has also occurred. The layered sediments indicate that the canyons may at one time have contained lakes, and streamlined, sculpted landforms at the lower, eastern ends of the canyons suggest that the lake waters may ultimately have drained eastward down the canyons.

The canyons merge eastward into what has been termed "chaotic terrain." In these areas, which cover thousands of square kilometers, the ground has seemingly collapsed to form a jumbled array of blocks at a lower elevation than the surrounding terrain. Several enormous channels emerge from the chaotic terrain east of the canyons and extend northward, down the regional slope for over 2000 km, causing extensive erosion of the volcanic plains of Chryse Planitia. Other large channels start north of the canyons and also converge on Chryse. The channels ultimately disappear in the low-lying northern plains. Large channels occur elsewhere on the planet. Those in Elysium have already been mentioned. Others drain into the large impact basin Hellas.

The channels are enormous by terrestrial standards. Many are several tens of kilometers wide. Kasei Vallis, one of the largest, is in places over 200 km across. All start full-size and have few, if any, tributaries. They contain numerous tear-drop shaped islands, the walls are streamlined, and the channel floors generally have a distinct longitudinal scour. When the channels were first observed, there was scepticism that features this large could have formed by floods, but the close resemblance to large terrestrial flood features and other supporting evidence for abundant water on the planet have led to general acceptance of the features as water worn. They appear to have formed by catastrophic release of groundwater under large pressure. Massive excavation of the groundwater was followed by collapse to form the chaotic terrain.

HIGH LATITUDES

The high latitudes differ from the rest of the planet in several respects. In some areas repeated deposition and removal of eolian (wind-deposited) debris has given the surface a finely etched appearance. Most heavily cratered areas poleward of about latitude 30° have a muted appearance compared with similar terrains at low latitudes. This has been attributed to slow downslope movement of the near surface materials as a result of the presence of ground ice. At high latitudes, wherever escarpments are found, aprons of debris commonly extend 20–30 km outward from the foot of the escarpment. Such debris flows are not found at low latitudes. The simplest explanation is that at high latitudes ice lubricates talus, allowing it to flow away from steep slopes. At lower latitudes where ground ice is unstable, similar debris flows do not form.

Surrounding both poles, just outside the 80° latitude circle, are numerous dune fields. Those in the north merge to form an almost continuous collar of dunes around the pole. At the poles themselves, and extending out to about latitude 80°, are layered deposits several kilometers thick. The layering is clearly visible on the walls of valleys deeply incised into the sediments. The deposits are believed to be composed of dust and ice, the layering being the result of varying proportions of the two components. The almost total lack of impact craters superimposed on the deposits suggests that they are relatively young. The layering has been attributed to cyclical erosion and deposition in response to small climatic changes induced by periodic variations in the planet's orbital and rotational motions.

SUMMARY

Mars, like the Earth, has had a long history of volcanic and tectonic activity, and its surface has been modified by wind, water, and ice. However, the two planets differ greatly in geologic style. Materials that form at the surface of the Earth are continually being cycled deep within the planet as a result of plate tectonics, and are continually being redistributed across the surface as a result of the vigorous action of liquid water. On Mars, no deep recycling occurs; volcanic materials brought to the surface simply remain in huge volcanic piles. Moreover, although some water erosion has occurred, the cumulative effect over the life of the planet has been trivial. Once relief is created, it largely remains. The result is a planet on which a record of sustained geologic activity is beautifully preserved in huge volcanoes, enormous canyons, extensive fracture systems, and giant flood channels.

Additional Reading

Baker, V. R. (1982). *The Channels of Mars*. University of Texas Press, Austin.

Carr, M. H. (1981). *The Surface of Mars*. Yale University Press, New Haven.

Carr, M. H. (1990). Mars. In *The New Solar System*, 3rd ed., J. K. Beatty and A. Chaikin, eds. Sky Publishing Corporation, Cambridge, MA and Cambridge University Press, Cambridge, U.K., p. 53.

Hamblin, W. K. and Christiansen, E. H. (1990). *Exploring the Planets*. Macmillan Publishing Co., New York.

Morrison, D. and Owen, T. (1988). *The Planetary System*. Addison-Wesley, New York.

Murray, B., Malin, M. C., and Greeley, R. (1981). *Earthlike Planets*. W. H. Freeman, San Francisco.

See also **Mars, Atmosphere; Mars, Space Missions; Planetary Tectonic Processes, Terrestrial Planets; Planetary Volcanism and Surface Features.**

Masers, Interstellar and Circumstellar

Shuji Deguchi

Maser emission from astronomical objects, discovered in 1965, serves as a powerful probe for the physical and chemical conditions of the dense gas around protostars and late-type stars. The term "maser" stands for microwave amplification of stimulated emission of radiation and is applicable to these objects because millimeter and submillimeter wavelengths are involved. No optical or infrared "maser" has yet been observed in the sky; hence, use of the term laser (light amplification of stimulated emission of radiation) is less common in the astronomical community. Molecules in the interstellar gas normally radiate thermal line emission with an intensity that is comparable to or less than the intensity expected from a blackbody having the temperature of the gas. However, spectral line emission due to some molecules is observed to be much stronger than the intensity expected from the gas, assuming blackbody radiation. (For example, the ground state Λ-doublet transition of OH at 1667 MHz from the W 49 molecular cloud is stronger than expected by approximately a factor of 10^{10}.) It is now well-established that this phenomenon is a result of the

Table 1. Molecules exhibiting maser action

Molecule	Frequency (GHz)	Typical Transition	Characteristics†
OH	1.612	$^2\Pi_{3/2}, J = \frac{3}{2}, F = 1\text{–}2$	M
	1.665	$^2\Pi_{3/2}, J = \frac{3}{2}, F = 1\text{–}1$	O, M
	1.667	$^2\Pi_{3/2}, J = \frac{3}{2}, F = 2\text{–}2$	O, M
	1.720	$^2\Pi_{3/2}, J = \frac{3}{2}, F = 2\text{–}1$	O
H_2CO	4.829	$1_{10}\text{–}1_{11}$	O
CH_3OH	12.178	$2_0\text{–}3_{-1}\ E$	O
SiS	18.155	1–0	C
H_2O	22.235	$6_{16}\text{–}5_{23}$	O, M
NH_3	23.870	3,3–3,3	O
SiO	43.122	$v = 1, J = 1\text{–}0$	M, S, O
	86.243	$v = 1, J = 2\text{–}1$	M, S
HCN	89.087	$v = 2, J = 1\text{–}0$	C

†O means that the maser emission is frequently found in star-forming regions; M, in M stars; S, in S stars; C, in carbon stars.

amplification of the microwave radiation by stimulated emission, which is caused by the inversion of level populations of interstellar and circumstellar molecules.

Maser emission in astronomical objects is characterized by the following properties: Within a cloud the maser emitting region is confined to a region of size $\leq 10^{16}$ cm; the spectral-line profiles of masers consist of sharp peaks having widths comparable to the thermal width of the gas; the intensity of a line can vary on a time scale of several months to a year; occasionally, the radiation is linearly or circularly polarized. In general, masers are associated with protostellar objects in star-forming regions and with late-type stars that lose their mass into circumstellar space. In detail, the maser characteristics are somewhat different in the various objects and molecular transitions. Table 1 summarizes the molecules that exhibit maser action in celestial objects.

The power of astronomical masers is expressed in terms of the number of maser photons per second emitted from the source. For the usual H_2O masers in star-forming regions (e.g., the Orion molecular cloud, W 3, W 51, etc.), this is about $10^{46}\ s^{-1}$, and for the H_2O masers in W 49, about $10^{49}\ s^{-1}$ (if isotropic emission is assumed); W 49 is the strongest maser source in the Galaxy. OH and H_2O masers have also been found in the nuclei of active galaxies, such as NGC 3079 and NGC 1068. The powers of these extragalactic masers are stronger [by about 10^6 for OH masers ("megamasers") and by 10^3 for H_2O masers] than the known maser sources in our own galaxy.

The intensity of maser emission can vary dramatically on a time scale of a year. One of the most spectacular examples is the H_2O outburst in Orion that occurred in late 1979 and which lasted for 8 years. The intensity of this one maser component in Orion suddenly increased by a factor of 1000 making it the brightest H_2O maser source in the sky. The position of this flare was near IRc 4 in the Orion molecular cloud and the maser spot size was measured to be about 10^{13} cm.

The distributions of OH and H_2O emission are studied using very long baseline interferometry (VLBI): Radio signals are recorded on magnetic tapes at several different telescopes located on different continents; these data are then correlated at a later time. By this technique, maser emission in star-forming regions is found to come from clusters of maser spots in molecular clouds. The size of an individual H_2O maser spot is about 10^{13} cm, and the size of the cluster is about 10^{16} cm. OH masers tend to be larger than H_2O masers by a factor of about 10. It is not well-understood whether these spots are individual protostellar objects or just fragments formed by a protostellar outflow. Proper motions of individual maser spots have been measured by VLBI; these motions seem to be random except for the masers in Orion, where a systematic outflow is found. From the random or systematic motions of maser spots, the distance to a molecular cloud can be obtained by assuming that the average radial velocity is comparable to the mean motion tangential to the line of sight. For example, a distance of 7.1 ± 1.5 kpc is obtained for the Sgr B2 molecular cloud.

OH maser emission in star-forming regions is strongly circularly polarized due to the weak magnetic field in molecular clouds. This is partly due to Zeeman splitting and partly due to maser amplification. However, the Zeeman patterns (two circular components and one linear component) are not very clearly observed in the spectral line profiles, even with the spatial resolution of VLBI. The magnetic field strength deduced from the Zeeman splitting is several milligauss (mG). Linear polarization of a few percent up to 50% is observed occasionally in the spectra of H_2O masers. However, the H_2O masers in the W 49 molecular cloud (the most powerful H_2O masers in our galaxy) exhibit negligible polarization.

Some transitions of the molecules NH_3, H_2CO, and CH_3OH, are also found to produce maser emission; these are relatively weak and have not been observed as extensively as the OH or H_2O masers. There are numerous transitions of NH_3 at frequencies around 23 GHz associated with different energy levels; some of the transitions exhibit maser characteristics and have been determined to be masers by measurments of the size of the emitting region. Maser emission from the $1_{10}\text{–}1_{11}$ line of H_2CO at 4.8 GHz is found in the star-forming region NGC 7538. This transition of H_2CO has been found previously in absorption against the cosmic microwave background in many dark clouds. The size of the H_2CO maser cloud in NGC 7538 is known to be less than 10^{17} cm. The molecule CH_3OH has many rotational transitions at microwave frequencies and some of them are found as weak masers. Maser emission from CH_3OH is usually extended ($\sim 10^{18}$ cm), but emission from the $2_0\text{–}3_{-1}\ E$ transition at 12.2 GHz is found to be quite strong and to come from a compact cloud ($\leq 10^{15}$ cm).

In spite of large numbers of observations, no simple picture has been established for the maser clouds in star-forming regions.

Maser emission has been also found in the circumstellar envelopes of late-type stars with optical spectral types of M, S, and C. These stars contain molecules in their atmospheres and expel them into circumstellar space at a rate of $10^{-7}\text{–}10^{-5}\ M_\odot\ yr^{-1}$ due to radiation pressure on grains and due to the pulsation of the stars. Masers from the OH, H_2O, SiO, SiS, and HCN molecules have been found. The species of masers found in circumstellar space are well-correlated with the spectral type of the associated stars. In M stars, where oxygen is more abundant than carbon, masers from oxygen-bearing molecules (such as OH, H_2O, and SiO) are found. It is well-established by interferometry that the OH 1612-MHz masers occur at the outer side of the stellar envelope (r about 10^{16} cm), where interstellar ultraviolet radiation dissociates H_2O and forms OH. The line profile of OH 1612-MHz masers consists of characteristic double peaks that are interpreted to be emissions from the approaching and receding parts of the expanding shell of the star (Fig. 1).

H_2O and SiO masers occur closer to the star. From VLBI, the radius of an H_2O emitting region is known to be about 10^{15} cm and that of SiO about 10^{14} cm. Intensities of these two types of maser vary on a time scale of a month to a year. The line profiles are also very complex and it is difficult to interpret them with a simple model for an expanding shell. H_2O and SiO circumstellar masers emit typically 10^{43} photons per second.

In carbon stars, where carbon is more abundant than oxygen, masers from the carbon-bearing molecule HCN and the sulfur-containing molecule SiS have been found. Strong masers from the vibrationally excited state of HCN recently have been found in carbon stars such as IRC +10216 and CIT 6. HCN masers are used as a probe of the inner region of the envelope in carbon stars, whereas the SiS maser is a probe of the outer envelope. In S stars,

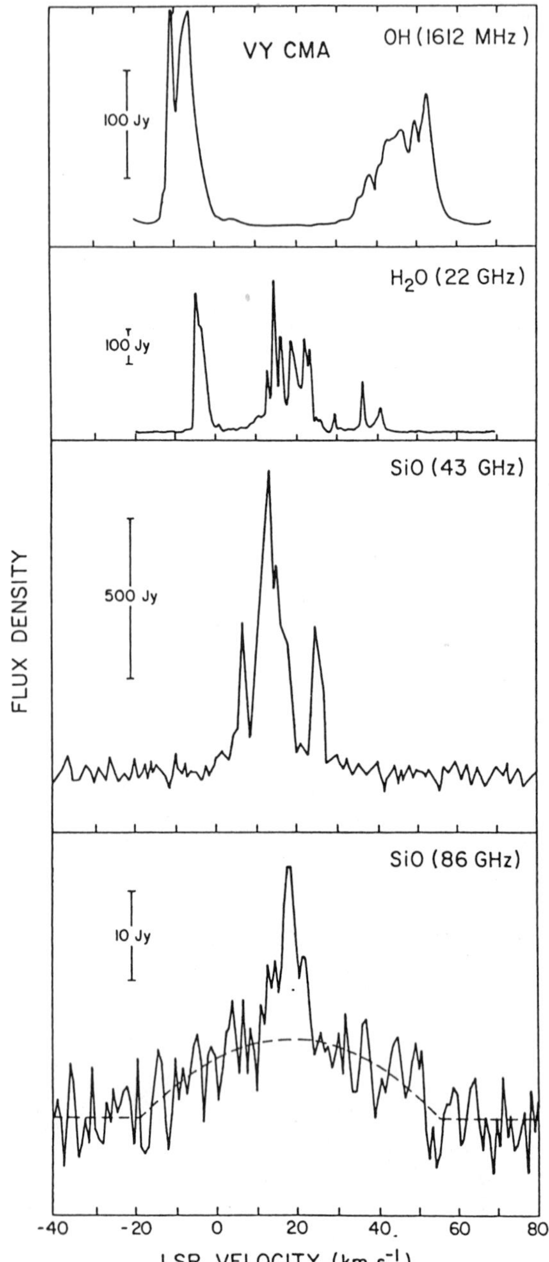

Figure 1. Line profiles of circumstellar masers in VY CMa. Flux density is shown as a function of velocity with respect to the Local Standard of Rest (LSR). [*Reid and Moran (1981)*, Annual Review of Astronomy and Astrophysics.]

where oxygen is as abundant as carbon, SiO masers frequently are found.

SiO masers have been found in late-type stars that are in a late stage of evolution. The only exception being the SiO masers found in Orion, a well-known star-forming region. It had been suspected that the SiO source in Orion might be an evolved object. However, SiO masers recently have been found in other star-forming regions (the W 51 and Sgr B2 molecular clouds) so that SiO masers are now recognized to be associated with young objects as well.

Masers due to molecules containing isotopically replaced atoms, ^{29}SiO and H^{13}CN, are also found in late-type stars. It is expected that these emissions will provide a clue to understanding the pump mechanism of astronomical masers.

Additional Reading

Cohen, R. J. (1989). Compact maser sources. *Rep. Prog. Phys.* **52** 881.

Moran, J. M. and Ho, P. T. P., eds. (1988). *Interstellar Matter*. Gordon and Breach, New York.

Reid, M. J. and Moran, J. M. (1982). Masers. *Ann. Rev. Astron. Ap.* **19** 231.

See also **Interstellar Medium, Molecules; Masers, Interstellar and Circumstellar, Theory; Stars, Evolved, Circumstellar Masers; Stars, Long Period Variables.**

Masers, Interstellar and Circumstellar, Theory

Moshe Elitzur

The radio radiation detected in some lines of certain astronomical molecules is attributed to the natural occurrence of the maser phenomenon (microwave amplification by stimulated emission of radiation), the same as that produced by artificial means in laser devices. Astronomical maser radiation is produced by population inversion of the pertinent transitions and is usually identified as such when at least one of the following properties is observed:

1. Enormous radiation intensity, as measured by its equivalent temperature.
2. Very narrow linewidths.
3. Abnormal line ratios, indicating deviations from equilibrium.

In at least one astronomical maser source there is also direct evidence for population inversion and amplification of radiation. The mechanism that leads to inversion, which is caused by the cycling of molecules through energy levels, is called the *pump*. The pump process is initiated by excitations from the ground state, due to either external radiation or collisions. The inversion results from a combination of various factors, specific to the particular maser molecule. The strong maser emission detected in our galaxy occurs around late-type stars, where it is called *circumstellar*, and in the cores of dense molecular clouds that are regions of active star formation, which is then termed *interstellar*. Maser radiation probes small-scale structure in these sources and is now used to measure distances by kinematic means, the equivalent of the classic *moving cluster method*. In recent years, maser emission has also been detected in many nearby galaxies.

MASER EFFECT

Population exchange between the two levels of any transition is governed by both collisional and radiative processes. The latter include spontaneous decays from upper to lower level and absorption of external radiation, accompanied by an excitation from lower to upper level. The frequency of the absorbed photon must match the energy separation of the transition. The inverse process, *induced* or *stimulated emission*, is the essence of the maser phenomenon. In it, a downward transition is induced by an incoming photon with a matched frequency. To conserve energy and momentum, the transition is accompanied by the emission of another photon whose properties are identical to those of the initial parent photon. The process effectively acts as negative absorption—increasing, instead of decreasing, the number of photons in the radiation field. If for any reason the population density of the upper level is larger than that of the lower level, the rate of stimulated emission exceeds absorption and the medium amplifies the propagating radiation rather than attenuating it. When the contribution of stimulated emission is included in the absorption coefficient, the latter becomes negative. The same applies to the optical depth τ and the standard attenuation term $\exp(-\tau)$ becomes an amplification factor $\exp|\tau|$. The absolute value of the optical depth is then referred to as the maser *gain*. A gain of more than 20 leads to amplification in excess of 10^8 and could explain in part the observed exceptional intensities. Additional intensity enhancement results from the focusing of the radiation into a narrow beam, caused by the fact that amplification is proportional to incoming intensity, so stronger rays are amplified more strongly. Observed brightness is frequently expressed in terms of *brightness temperature*, the temperature of an equivalent blackbody that would be needed to produce the same intensity. Maser brightness temperatures can be as high as $\sim 10^{15}$ K.

The exponential growth of the intensity cannot continue indefinitely because it would eventually lead to infinite energy density for sufficiently long masers. A self-limiting process is built right into the maser effect itself: Induced emission removes particles from the upper level, thus reducing the inversion. Because of the excess of induced emissions over absorptions, the inversion decreases once the population exchange between the maser levels is dominated by the interaction with the maser radiation. The intensity then approaches a limit and the maser *saturates*. The brightness is highest during saturated operation because every pumping event produces a maser photon with the maximal intrinsic efficiency allowed by the pump.

ENVIRONMENTAL CONSTRAINTS

The prevalence of astronomical masers shows that the interstellar medium is apparently capable of producing the maser effect relatively easily, although special efforts are required to achieve the same end in the laboratory. This is a result of the great differences in densities and geometrical dimensions between the two environments. In thermodynamic equilibrium, the populations per substate n of the two levels of a transition separated by energy gap ΔE obey

$$\frac{n_{\text{upper}}}{n_{\text{lower}}} = \exp\left(-\frac{\Delta E}{kT}\right)$$

where k is the Boltzmann constant and T is the temperature. This equilibrium distribution is established when population exchange is dominated by collisions, which is the case for sufficiently high densities. For typical parameters of molecular rotational transitions, the required densities are in excess of $\sim 10^4$ cm^{-3}. This is an extremely low density in the laboratory, achieved only with state of the art vacuum techniques (pressures of $\sim 10^{-12}$–10^{-13} torr), but it is quite high by interstellar standards. Equilibrium populations are therefore the rule in terrestrial circumstances but are the exception in interstellar space due to the large difference in relevant densities. Once thermal equilibrium is violated, population inversion is a priori almost as likely as its reverse.

An appreciable maser effect requires large gain, which in turn implies a substantial number of molecules along the line-of-sight. This conflicts with the necessary deviation from thermal equilibrium, which requires that the densities be small, and the only way to reconcile these opposite demands is with large dimensions. At a density of $\sim 10^4$ cm^{-3}, an appreciable gain is achieved only for linear dimensions in excess of $\sim 10^{10}$ cm—almost as much as the radius of the Sun. One of the methods to overcome this difficulty in the laboratory is to increase the radiation path length by bouncing the laser light between mirrors, effectively increasing the linear dimensions of the system by the many passes of the laser beam in the resonant cavity. This technique is possible only in systems that can maintain a high degree of phase coherence. On the other hand, typical lengths of astronomical masers are at least $\sim 10^{13}$ cm (the same as the radius of the Earth's orbit around the Sun), and the required gains are obtained during simple photon propagation as in a single-pass laser. Interstellar space is therefore a natural environment for maser operation. Using laboratory terminology, astronomical masers are single-pass, lossless, gaseous lasers without feedback.

The observed profiles of interstellar lines usually indicate the presence of highly supersonic motions in the emitting regions. Because maser amplification is achieved by induced emission, the maser photons must seek paths that maintain good coherence in the component of the velocity along the line-of-sight; otherwise the transition frequencies of molecules encountered by the maser photons would be shifted by the Doppler effect, making amplification impossible. Observations show indeed that maser sources are comprised of many emission spots, each with its own well defined velocity. These single features are often shaped like elongated tubes, or cylinders, which need not be well-defined physical entities, but rather directions that developed the required velocity coherence by chance. Because of their high brightness, the spots can be studied individually using techniques of high-resolution interferometry. This provides the opportunity for probing small scale structure in the host environments.

SOURCES AND PUMPING

Maser emission is associated with both the early and late stages in the life of a star. This is a fortunate coincidence because these are generally regarded as the most interesting phases of stellar evolution. Late-type stars display strong maser radiation in transitions of all three "classical" maser molecules—OH, H_2O, and SiO. The masers occur in distinct regions located at different distances from the central star. The SiO masers involve rotational transitions inside excited vibration states, which lie high above the ground state. These levels can maintain substantial populations only close to the star where the excitation rates are high, and that is where SiO masers are located. The H_2O and OH masers, on the other hand, emanate from transitions in the ground vibration states of these molecules and do not require such extreme conditions for pumping. Both are located in shells that are part of an expanding wind that blows away from the star. The H_2O masers require higher temperatures than OH masers and occur at distances of up to $\sim 10^{15}$ cm from the central star (by comparison, the radius of Pluto's orbit around the Sun is about 6×10^{14} cm). The OH maser shell extends further out, to a radius larger by another order of magnitude at least.

Regions of active star formation, located at the cores of many molecular clouds, display the most powerful and spectacular maser emission observed in the Galaxy in both OH and H_2O, and in at least one case also in SiO. The OH masers appear to be surface phenomena on the edges of very compact H II regions, ionized spheres around young and very hot stars. The H_2O masers usually trace high velocity flows (velocities of ~ 200 km s^{-1}) from some centers of activity that presumably erupt at a certain stage of the star formation process.

Pump analysis involves the construction of models capable of producing the observed maser output with parameters that can be checked against other observations. The pumping mechanism is termed *collisional* or *radiative* according to the nature of the process that dominates the molecular excitation from the ground state. In general, both types of pumping can produce inversion under the right conditions (although in certain, specific circumstances only one may be capable of inversion) and the nature of the pump in any particular source is determined by the relative strengths of both processes. Radiative pump models can be more easily confronted with observations because they relate two directly observed quantities—the number of photons observed in the maser transition and in the pump bandwidth. Indeed, the masers whose detailed modeling has been most successful are the OH masers in late-type stars that are pumped by infrared radiation resulting from the reemission of the stellar radiation by the dust particles that permeate the stellar wind. Detailed models of the H_2O masers in these sources show that pumping is controlled by collisions. The model predicts correctly the location of the H_2O maser region and its variation with the stellar mass loss rate, but these observational tests are not as direct. Current models of masers in star-forming regions are not as detailed yet, reflecting perhaps the somewhat poorer overall understanding of these sources.

Additional Reading

Cohen, R. J. (1989). Compact maser sources. *Rep. Prog. Phys.* **52** 881.

Elitzur, M. (1982). Physical characteristics of astronomical masers. *Rev. Mod. Phys.* **54** 1225.

Genzel, R. (1986). Strong interstellar masers. In *Masers, Molecules and Mass Outflows in Star Forming Regions*, A. D. Haschick, ed. Haystack Observatory, p. 233.

Herman, J. and Habing, H. J. (1985). OH/IR stars. *Phys. Rep.* **124** 255.

Reid, M. J. and Moran, J. M. (1981). Masers. *Ann. Rev. Astron. Ap.* **19** 231.

See also **Masers, Interstellar and Circumstellar.**

Mercury and the Moon, Atmospheres

Wing-Huen Ip

Most of the bodies in the solar system (such as Mercury and the Moon, all of the asteroids, most of the icy satellites orbiting the jovian planets, and the two martian satellites Phobos and Deimos) have only very tenuous atmospheres if any at all. The atmospheric densities are so low that the gases are essentially collisionless and the exobases are simply the planetary and satellite surfaces. This means that the motion and transport of the gas particles can be described in terms of ballistic trajectories instead of with a hydrodynamical description. Consequently, the main physical ingredients are the atmospheric sources and sinks and the gas–surface interaction process. The major sources are solar wind accretion (which contributes H, He, ^{20}Ne, ^{36}Ar, and other rare gases), radiogenic outgassing from the interior, meteoroidal impact, and thermal evaporation from the surface regolith. The lack of an intrinsic magnetic field on the Moon permits the direct access of solar wind particles to the lunar surface. In addition, as the Moon crosses the geomagnetic tail, its surface is exposed to the Earth's magnetospheric plasma environment. At Mercury, on the other hand, with the presence of a dipole magnetic field capable of standing off the solar wind, the surface would be shielded effectively from the solar wind plasma, with only about 1% of the charged particles entering the mercurian magnetosphere. The subsequent magnetospheric acceleration of the solar wind particles and of ions released from the planetary surface would provide a population of energetic ions and electrons constantly irradiating the surface. Photoionization is the main mechanism of atmospheric loss for the heavy gases, whereas Jeans escape is the predominant route for light atoms and molecules such as H, H_2, and He. As will be discussed later, the gas–surface interaction involves thermal accommodation between the gas particles and the surface material as well as involving ballistic random walk over regions of different surface temperatures such as the dayside and nightside hemispheres.

As a result of previous spacecraft observations and more recent ground-based work, a large body of data has been accumulated for Mercury and the Moon, providing very interesting comparisons and

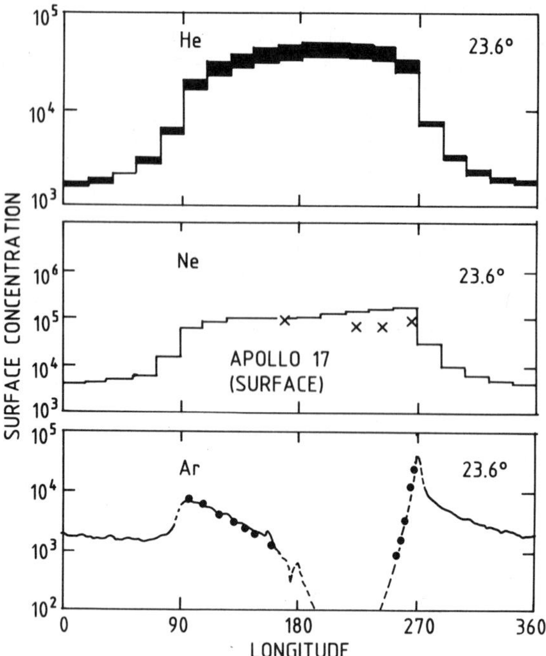

Figure 1. Preliminary data from the *Apollo 17* mass spectrometer superimposed on a theoretical curve for the diurnal variations of He, Ne, and Ar. Note that more detailed considerations suggested that incomplete compensation for $H_2{}^{18}O^+$ and H_2DO^+ in the mass spectra could have resulted in significant overestimates of the neon abundance in the lunar atmosphere (R. R. Hodges, Jr., private communication, 1989). [*From Hodges, Hoffman, and Johnson (1974).*]

tests for theoretical models that have been developed to explain the behavior of the mercurian and lunar atmospheres. It should be stressed that, purely scientific issues aside, there are very good practical reasons why this problem should be investigated in detail. For example, future exploration of the Moon with the eventual establishment of lunar bases would require a close monitoring of the surface atmospheric environment.

THE MOON

The lunar atmospheric environment has been examined by an array of instruments during the *Apollo* missions, including mass-spectrometer experiments both on the orbiter vehicles and on the lunar surface, the ultraviolet spectrometer on the *Apollo 17* orbiter, and, of course, by studying the returned lunar samples. Figure 1 shows data on the helium concentration as a function of longitude as determined from the *Apollo 17* mass spectrometer. The diurnal variation can be understood in terms of the dependence of the step size Δs of the ballistic random walk on the surface temperature T. Taking the ballistic time t_b as $\sqrt{2}(V/g)$, where V is the launch velocity at an inclination of 45° to the surface and g is the surface gravity, we have

$$t_b = (2/g)(kT/m)^{1/2},$$

with k as the Boltzmann constant and m as the mass of the hopping particle. The scale height is given as $H = kT/mg$; with $\Delta s \approx H$, the average number of random-walk steps across the lunar surface can be expressed as $n_{rw} = (r/\Delta s)^2 \approx (r/H)^2$, where r is the radius of the Moon. The characteristic time of ballistic transport is thus

$$t_m = t_b n_{rw} \propto T^{-3/2}.$$

At equilibrium the upward flux of particles must be nearly the same everywhere on the surface of the Moon, that is, $n_0 H/t_m =$ constant so that the surface density $n_0 \propto T^{-5/2}$. Because the day-to-night temperature ratio is about 4, the expected density ratio is 32. The *Apollo 17* mass-spectrometer observations indicated a factor of 15. This difference may be accounted for if effects such as Jeans escape and uncertainties in the surface thermal accommodation are invoked.

If no loss process other than Jeans escape and photoionization is involved, the solar wind accretion should be able to maintain a surface density of about 600 cm^{-3} for atomic hydrogen on the dayside. That the *Apollo 17* orbital ultraviolet spectrometer indicated a dayside concentration of less than 10 hydrogen atoms per cubic centimeter suggests that a very efficient loss mechanism must be in operation. One such possibility is that the hydrogen atoms are converted to H_2 molecules via surface recombination.

The solar wind is also an efficient source of ^{20}Ne. The total influx can be estimated to be $\approx 2 \times 10^{21}$ Ne ion s^{-1}. The release of the neutralized atoms from the surface leads to the buildup of a layer of Ne atoms with a nightside concentration as high as 10^5 cm^{-3}. In the theoretical calculations (see Fig. 1*b*) it was assumed that there was no adsorption of the Ne atoms and that all solar wind Ne ions were converted to Ne atoms. On the other hand, in order to fit the observed density profile of Ar (Fig. 1*c*), a certain degree of surface absorption probability (up to 0.05) on the nightside must be invoked. Although ^{36}Ar is of solar wind origin, ^{40}Ar comes from the radiogenic decay of ^{40}K in lunar rock. The total rate of ^{40}Ar is of the order of 6×10^5 atom cm^{-2} s^{-1}, which corresponds to the release of the entire radiogenic argon production from the first 3 km of the lunar crust. The actual venting of ^{40}Ar and other gases of lunar origin (i.e., ^{222}Rn) may be quite inhomogeneous and sporadic. In this context it should be noted that a transient gas event was detected by the mass-spectrometer experiment on the *Apollo 15* orbiter as the spacecraft passed by Mare Orientale, thereby demonstrating that mass-spectrometer observations are a powerful tool for monitoring subsurface venting from the lunar interior.

Recent ground-based observations have detected sodium emission from the lunar atmosphere. The surface density on the dayside disk is of the order of 50–100 cm^{-3}. In principle, such a sodium exosphere can be maintained by meteoroidal impact. With a scale height of about 100 km and a photoionization time scale of 10^5 s, the total loss rate of sodium is about 5×10^{20} atom s^{-1}. The mass influx onto the lunar surface from interplanetary meteoroids can be calculated to be ~ 25 g s^{-1}. With a mixing ratio of 10^{-3} for chondritic composition, the sodium influx is then of the order of 6×10^{20} atom s^{-1}. Hence the photoionization loss rate is balanced by such an external source. It is interesting to note that a significant amount of H_2O could be brought to the lunar surface in the same manner. With a 10% mixing ratio, the H_2O influx could amount to 8×10^{22} molecules s^{-1}, that is even higher than the solar wind accretion rate of ^{20}Ne. Photodissociation of the water molecules would lead to the formation of O atoms and O_2 molecules, which subsequently might interact with the lunar soil. Trace quantities of CO, CO_2, and NO may form in this manner.

MERCURY

The *Mariner 10* encounters with Mercury in 1974 and 1975 provided the first opportunities to survey the atmospheric environment of this innermost planet. Lyman alpha emission of atomic hydrogen, He I emission at 584 Å and O I emission at 1304 Å were detected off the limb. A summary of the observed gas concentrations is given in Table 1. It is interesting to observe the relatively large abundance of oxygen on Mercury ($n_0 \approx 7 \times 10^3$ cm^{-3}) in comparison with the lunar case. The difference of a factor of 200–400 in the surface density of sodium is also striking. Meteoroidal impact might not be the main supplier of the sodium

Table 1. Atmospheric Species Detected on Mercury and the Moon*

Species	*Wavelength (Å)*	*Mercury n_0 (cm^{-3})*	*Moon n_0 (cm^{-3}) Day*	*Night*
H	1216	8.0 (hot), 82 (cold)	< 10	—
He	584	4.5×10^3	2×10^3	4×10^4
O	1304	7.1×10^3	—	—
Na	5890, 5896	$1.7–3.8 \times 10^4$	50–100†	—
K	7664, 7699	5×10^2	—	—
Ar	869	$< 8.7 \times 10^5$	1.6×10^3	4×10^4

*From Hunten et al. (1988).

†From Tyler et al. (1988) and Morgan and Potter (1988).

atmosphere on Mercury. An important additional source could be thermal evaporation from the regolith layer, which may be enriched in alkalis. Potassium, as with sodium was also detected via ground-based observations, but with a much smaller brightness.

The sodium emissions at 5895.92 and 5889.95 Å show that the average column density is of the order of $1–2 \times 10^{11}$ atom cm^{-2} and the temperature of the gas is about 500 K. The sodium atmosphere has an optical depth ~ 1 and is thus optically thick. Two additional factors affect the random walk of the sodium atoms on the planetary surface: First, because the photoionization time of $\sim 10^4$ s is comparable to the transit time, a significant fraction of the sodium atoms would be ionized and lost as they move from the dayside hemisphere to the nightside; thus, the equilibrium density distribution would not follow the temperature dependence (i.e., $n_0 \sim T^{-5/2}$) as discussed previously. Second, the radiation pressure acceleration (A_{pr}) could be as large as 200 cm s^{-2}; consequently, there is a systematic acceleration of the sodium atoms from the dayside to the nightside. Because Mercury's orbit has a large eccentricity, the Doppler shift effect in the radial direction can be very significant. The value of A_{pr} varies a great deal (20–200 cm s^{-2}) according to the orbital phase of Mercury. The radiation-driven transport of the sodium atoms thus may be diagnosed by monitoring the brightness variation of the sodium D-lines as a function of planetary orbital phase. Some indication of such a correlation has indeed been found. The bulk of the sodium atoms should have relatively low velocity (~ 0.5 km s^{-1}); their motion, although perturbed by solar radiation pressure, is firmly under the control of Mercury's gravitational field. On the other hand, a small component of hot sodium vapor can be generated by meteoroid impact and the fast atoms with velocities exceeding 2 km s^{-1} would have a good chance of being accelerated away from the planet, forming a tenuous gaseous tail. However, such an intriguing structure has not been detected.

Once ionized, the exospheric ions will be subject to the acceleration of the $-\boldsymbol{V} \times \boldsymbol{B}$ electric field in the planetary magnetosphere (or, in the solar wind in the case of the Moon). Although ions created on one hemisphere will be accelerated toward the planetary surface and hence recycled, those created on the opposite hemisphere will be ejected away and hence lost. The total loss rate of the sodium atoms as a result of photoionization and such a plasma pickup process amounts to about 1.3×10^{24} atom s^{-1}. Meteoroidal impact could supply no more than 1.3×10^{22} atom s^{-1}. Only under the extreme condition that the whole planetary surface is shielded from the magnetospheric electric field would the interplanetary meteoroids be an adequate source. Otherwise an additional supplier such as the regolith layer is required. One important consequence of the existence of the sodium (and potassium) exosphere is that the mercurian magnetosphere must be enriched with such heavy ions. In addition, surface sputtering by energetic charged particles should provide a component of metallic (e.g., Fe^+, Mg^+, etc.) and even compound (such as $Fe_2O_3^+$) ions in the magnetosphere of Mercury.

Solar wind accretion is the main source of H and He atoms. Figure 2 illustrates the He I emission at 584 Å as scanned across the disk of Mercury by the ultraviolet spectrometer (UVS) on *Mariner 10*. The spatial distribution of the He atoms, with a day-to-night ratio of about 50, in principle, can be modeled by considering their random-walk motion, Jeans escape, and surface interactions. In the case of hydrogen atoms, the *Mariner 10* UVS observations of the Lyman alpha emission on the subsolar side suggest that two components of H atoms with very different temperatures (i.e., 420 K versus 100–150 K) may exist. However, there is no satisfactory explanation for such a bimodal velocity distribution. One major uncertainty in theoretical modeling concerns the thermal accommodation effieiency (α) as the gas particles interact with the surface grains. Besides its dependence on the chemical composition of the surface soil and the porosity of the regolith layer, the surface temperature is also a relevant parameter. The present understanding is that $\alpha < 0.1$ for light gases (i.e., little energy exchange) and about 0.5 for heavier ones like Na. Furthermore, surface recombination such as $H + H \rightarrow H_2$ may be important in explaining the deficiency of H atoms at both Mercury and the Moon (see Table 1).

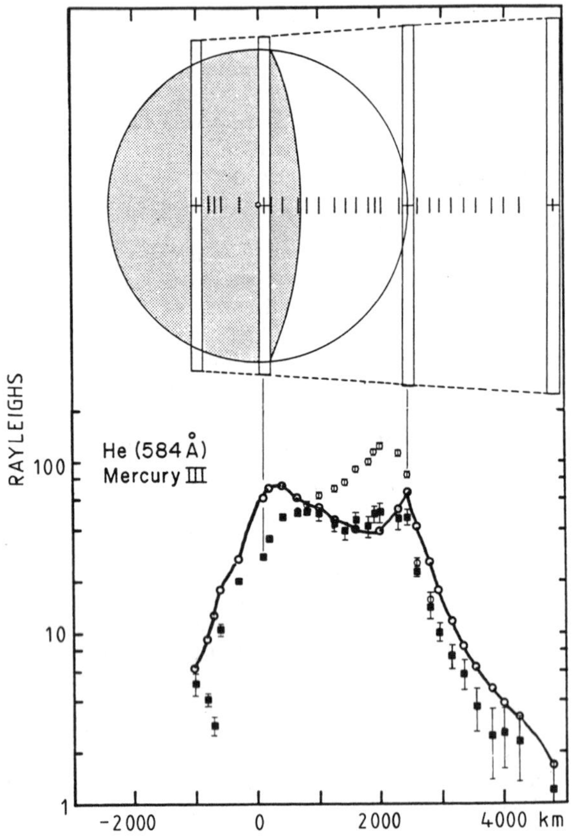

Figure 2. Helium emission from the *Mariner 10* ultraviolet spectrometer data. The light squares represent data obtained during a drift across Mercury at an average range of 85,000 km on the incoming pass (evening side). The filled squares represent observational data corrected for planet surface albedo. The error bars represent statistical variations at that point. The circles joined by a curve represent a model by Smith et al. (1978), which starts with a Maxwell–Boltzmann flux distribution at the surface and includes losses by Jeans escape and photoionization. [*Reproduced from Smith, Shemansky, Broadfoot, and Wallace,* Journal of Geophysical Research **83** *3783 (1978), copyright by the American Geophysical Union.*]

OTHER OBJECTS

Even though the atmospheres of both Mercury and the Moon are extremely tenuous, their origin and evolution are by no means simple. In fact there are still a number of important issues to be investigated before we can fully understand the past spacecraft measurements and ground-based observations. As exemplified by Mercury, many of these factors are linked to interactions of the planetary surface with the external plasma (and meteoroidal) environment. Such magnetospheric effects would be even more evident in the case of the icy satellites of the jovian planets. Except for Io, which has a very active volcanism, the other Galilean satellites should have very dilute atmospheres maintained essentially by charged particle sputtering. Again, apart from Titan, a similar situation exists for the icy satellites and rings in the saturnian system. In the near future both the *Galileo* and *Cassini* missions will provide a wealth of new information concerning the atmospheric origins and formations of these bodies.

Additional Reading

Goldstein, B. E., Suess, S. T., and Walker, R. J. (1981). Mercury: Magnetospheric processes and the atmospheric supply and loss rates. *J. Geophys. Res.* **86** 5485.

Hodges, R. R., Jr., Hoffman, J. H., and Johnson, F. S. (1974). The lunar atmosphere. *Icarus* **21** 415.

Hunten, D. M., Morgan, T. M., and Shemansky, D. E. (1988). The Mercury atmosphere. In *Mercury*, F. Vilas, C. R. Chapman, and M. S. Matthews, eds. University of Arizona Press, Tucson, p. 562.

Potter, A. E. and Morgan, T. H. (1985). Discovery of sodium in the atmosphere of Mercury. *Science* **229** 651.

Potter, A. E. and Morgan, T. H. (1988). Discovery of sodium and potassium vapor in the atmosphere of the Moon. *Science* **247** 675.

Potter, A. E. and Morgan, T. H. (1990). Evidence for magnetospheric effects on the sodium atmosphere of Mercury. *Science* **248** 835.

Smith, G. R., Shemansky, D. E., Broadfoot, A. L., and Wallace, L. (1978). Monte Carlo modeling of exospheric bodies: Mercury. *J. Geophys. Res.* **83** 3783.

Tyler, A. L., Kozlowski, R. W. H., and Hunten, D. M. (1988). Observations of sodium in the tenuous lunar atmosphere. *Geophys. Res. Lett.* **15** 1141.

See also **Planetary and Satellite Atmospheres; Planetary Atmospheres, Escape Processes.**

Mercury and Venus, Space Missions

Stephen Cole

Mercury and Venus represent extremes in our endeavors to investigate the solar system with spacecraft (the successful missions are summarized in Table 1). Venus, the first planet to be encountered by a spacecraft, has been visited by 20 missions that have photographed its veil of clouds, analyzed its dense atmosphere, sampled its soil, and mapped its surface with a wide array of instruments and vehicles. Mercury, on the other hand, has been visited by a single spacecraft (*Mariner 10*). Despite the fact that this solitary mission flew by the planet on three separate occasions, only half of Mercury's surface was photographed; the other half remains unexplored. Despite that fact, no return missions to Mercury are currently planned.

FIRST CONTACT AT VENUS (1962–1972)

Beginning in 1961, the Soviet Union made the first attempts to send a spacecraft to Venus, but they did not achieve a successful

Table 1. Successful Missions to Venus and Mercury

Mission	*Launch*	*Arrival*	*Mission Type*
Mariner 2	Aug. 27, 1962	Dec. 14, 1962	Venus flyby
Venera 4	June 12, 1967	Oct. 18, 1967	Venus atmospheric probe
Mariner 5	June 14, 1967	Oct. 19, 1967	Venus flyby
Venera 5	Jan. 5, 1969	May 16, 1969	Venus atmospheric probe
Venera 6	Jan. 10, 1969	May 17, 1969	Venus atmospheric probe
Venera 7	Aug. 17, 1970	Dec. 15, 1970	Venus atmospheric/surface probe
Venera 8	Mar. 27, 1972	July 22, 1972	Venus atmospheric/surface probe
Mariner 10	Nov. 3, 1973	Feb. 5, 1974	Venus flyby
		March 29, 1974	Mercury flyby
		Sept. 21, 1974	Mercury flyby
		March 16, 1975	Mercury flyby
Venera 9	June 8, 1975	Oct. 22, 1975	Venus orbiter/lander
Venera 10	June 14, 1975	Oct. 25, 1975	Venus orbiter/lander
Pioneer Venus 1	May 20, 1978	Dec. 4, 1978	Venus orbiter
Pioneer Venus 2	Aug. 8, 1978	Dec. 9, 1978	Venus atmospheric probes (4)
Venera 11	Sept. 9, 1978	Dec. 25, 1978	Venus flyby/lander
Venera 12	Sept.14, 1978	Dec. 21, 1978	Venus flyby/lander
Venera 13	Oct. 30, 1981	March 1, 1982	Venus flyby/lander
Venera 14	Nov. 4, 1981	March 5, 1982	Venus flyby/lander
Venera 15	June 2, 1983	Oct. 10, 1983	Venus orbiter
Venera 16	June 7, 1983	Oct. 14, 1983	Venus orbiter
Vega 1	Dec. 15, 1984	June 11, 1985	Venus lander, balloon
Vega 2	Dec. 21, 1984	June 15, 1985	Venus lander, balloon
Magellan	May 4, 1989	Aug. 10, 1990	Venus orbiter

mission until 1967 (*Venera 4*). The first spacecraft to fly by the planet was the United States' *Mariner 2*, a modified lunar *Ranger* craft. Although it carried a modest set of scientific instruments that scanned the planet for less than an hour, its flyby of Venus at a distance of 34,800 km produced some fundamental data about the planet. The radiometers revealed a surface temperature of 425°C with little variation between the night and day side of the planet. *Mariner 2* found no strong magnetic field nor radiation belt at Venus.

The *Mariner 5* mission to Venus in 1967 flew much closer to the planet, passing just 3900 km above the surface. The ultraviolet photometer searched for atomic hydrogen and oxygen emissions in the upper atmosphere and found a hydrogen corona. The magnetometer placed an upper limit to the strength of Venus' magnetic field at 1% that of Earth's. The radiation team, led by James van Allen, found no trapped particle radiation around the planet. Despite the lack of a major magnetic field, the solar wind does not penetrate to the surface but flows around Venus. Carbon dioxide was found to be the major component of the atmosphere (85–99%).

The Soviet *Venera 4* mission, which arrived at Venus 1 day before *Mariner 5*, placed the first probe directly into the venusian atmosphere. Carrying a radio altimeter, temperature and pressure sensors, a gas analyzer, and accelerometers, the spherical probe found no nitrogen in the atmosphere and 90–95% carbon dioxide. Although the probe did not survive to the surface, its temperature and pressure readings indicated a surface temperature of 500°C and a surface pressure of 75 bar.

The Soviets repeated the mission in 1969 with the twin *Venera 5* and *Venera 6* spacecraft. Atmospheric data from these missions were consistent with 93–97% carbon dioxide, 2–5% nitrogen and inert gases, and less than 4% oxygen. Contact with both probes was lost before they reached the surface. With the aid of an internal cooling system, the *Venera 7* probe survived to transmit data from the surface in 1970. A surface temperature of 475°C and an atmospheric pressure of 90 bar were recorded. *Venera 8*, the last of the first-generation *Venera* probes, arrived in 1972. Its anemometer measured dramatic variation in wind speeds at different altitudes: 100 m s^{-1} above 48 km, 40–47 m s^{-1} at 48–42 km, and 1 m s^{-1} below 10 km.

ONE MISSION, TWO PLANETS (1974–1975)

The first imaging system flown to Venus was carried on the *Mariner 10* spacecraft. This was a dual-planet mission—the first of its kind. Following a gravity-assisted flyby of Venus, *Mariner 10* was sent in an orbit to encounter Mercury. With an orbital period exactly twice that of Mercury's, *Mariner 10* rendezvoused with the planet two more times providing our first—and only—images of the planet in its three flybys.

Using a twin television camera system equipped with ultraviolet filters, *Mariner 10* took over 4000 images of Venus. These images recorded a 4-day circulation of the upper atmosphere. It was unclear at the time whether the rotation of cloud markings was the result of wave motions in the atmosphere or of the movement of air masses. The infrared radiometer showed the temperature at the cloud tops to be −23°C on both the day and night sides. The occultation of the spacecraft's radio signal as it disappeared behind the planet revealed at least two different cloud layers in the atmosphere.

Mariner 10's first flyby of Mercury brought the spacecraft to within 270 km of the planet. During this 11-day encounter over 2000 television pictures were returned, revealing a heavily cratered, Moon-like surface. The three encounters produced over 10,000 pictures covering 57% of the planet. The first flyby detected a weak magnetic field that was confirmed by the third flyby and measured to be less than 1% the strength of Earth's. Surface temperatures varied from −183°C on the night side to 187°C on the day side. Ultraviolet spectrometers failed to detect the mercurian atmosphere. Radio tracking of the spacecraft yielded a precise radius for the planet that showed Mercury to be a more perfect sphere than either the Earth or Mars. A search for mercurian moons proved futile.

PROBING THE ATMOSPHERE AND SURFACE OF VENUS (1975–1985)

A new generation of *Venera* probes began to plumb the depths of the venusian atmosphere and investigate the surface beginning in

1975. The *Venera 9* and *10* missions consisted of orbiters and landers equipped with imaging systems to return the first photographs from the surface of another planet. In their dozen or so revolutions around the planet, the orbiters photographed the cloud cover and probed the upper atmosphere. Three cloud layers were found at 70–57, 57–52 km, and 52–49 km altitudes. Each lander transmitted for about an hour from the surface. The black-and-white images with a 40×180° field of view showed sharp-edged, flat rocks at the *Venera 9* landing site and a more eroded terrain at the *Venera 10* site. Gamma-ray spectrometer and radiation densitometer measurements indicated a basaltic terrain.

The *Venera 11* and *12* missions in 1978 consisted of a flyby spacecraft bus for data relay and an improved lander. Although the imaging systems presumably failed because no photos were released, *Venera 12* recorded frequent electrical discharges, possibly produced by lightning.

The *Venera 13* and *14* missions conducted the first soil analysis in 1982. The landers drilled under the surface and returned a 1-cm^3 sample of soil to a chamber where an x-ray fluorescence spectrometer analyzed it. *Venera 13*'s sample was a type of basalt rare on Earth (leucite basalt) that was undergoing chemical weathering. *Venera 14*'s sample (tholeiitic basalt) was similar to that found in terrestrial mid-ocean ridges. Two cameras on both landers produced several high-resolution panoramic surface photos, some of which were in color. Acoustic sensors confirmed the low wind velocity on the surface.

In 1978 the United States launched its most ambitious Venus mission: *Pioneer Venus*, a dual mission carrying four separate atmospheric probes on one spacecraft and an orbiter that would produce the first radar map of the surface. The four probes consisted of a large probe (carrying a neutral mass spectrometer, gas chromatograph, and cloud particle size spectrometer) and three identical smaller probes named for the regions they would investigate (north, day, and night). The probes revealed a fine haze layer at 70–90 km, little atmospheric convection between 10–50 km, and a clear atmosphere below 30 km. The deuterium/hydrogen ratio was discovered to be 100 times higher than Earth's, suggesting that at one time in its past Venus may have been partially covered by oceans.

The *Pioneer Venus* orbiter is the longest-lived spacecraft to visit Venus. Operating continuously since 1978 (it is expected to survive until September 1992 when it plunges into the venusian atmosphere), the orbiter sampled in situ the upper atmosphere, ionosphere, and solar wind interaction. The orbiter detected no intrinsic magnetic field, not even a very weak one. The electron field detector recorded radio bursts that could be the result of lightning (although this interpretation is controversial). One interesting result from a decade of measurements is a decrease in the sulfur dioxide in the atmosphere since 1978. Only 10% of the 1978 level was recorded in 1988. Some investigators hypothesize that this could be due to a large volcanic eruption shortly before *Pioneer Venus* arrived and the subsequent decline in sulfur dioxide particles over a decade. This, however, along with the relationship between the apparent lightning signals that supposedly are correlated with areas of volcanic terrain on Venus, remains controversial.

Atmospheric probing continued in a new and unique form with the Soviet *Vega 1* and *2* missions in 1985. On their way to a flyby of comet Halley, the two spacecraft dropped *Venera*-style landers at Venus, each of which also carried a balloon experiment to investigate Venus' middle cloud layer. The two balloons floated in the atmosphere for nearly 48 h and traveled for 11,000 km across Venus as they were tracked by a network of antennas on Earth that determined their positions using very long baseline interferometry. Cruising at an altitude of 54 km, the balloons encountered numerous downward gusts averaging 1 m s^{-1} and revealed wind velocities of 240 km h^{-1}. Neither balloon detected lightning. Although the *Vega 1* lander soil experiment failed, *Vega 2* sampled anorthosite-troctolite, which is rare on Earth but is found in the lunar highlands.

GLOBAL MAPPING OF VENUS (1978–1992)

The Soviet Union and the United States have both sent spacecraft to penetrate the constant cloud cover of Venus and to map the surface using radar. The *Pioneer Venus* orbiter conducted the first survey using a radar altimeter. This produced a map of 93% of the surface, from 74°N to 63°S, revealing a primarily smooth planet with extremes represented by the 10.8-km high Maxwell Montes and the 2.9-km deep Diana Chasma. Because the resolution of the altimeter was limited (75 km), only very large surface features were detected.

The *Venera 15* and *16* missions placed a higher-resolution radar imaging system into orbit around Venus in 1983. This produced images with a resolution of 1–2 km and revealed craters, hills, major fractures, and mountain ranges. These missions produced a map of the northern hemisphere from 30°N to the pole. Infrared spectrometers produced a thermal map of the northern hemisphere and revealed several "hot spots" where temperatures were as great as 700°C, which is 200°C above the average surface temperature. These areas may be sites of volcanic activity.

Radar mapping is the primary agenda of the most recent mission to Venus: the U.S. *Magellan* mission launched in 1989. Using a synthetic-aperture radar, *Magellan* will map 90% of the planet with a resolution of 300 m during its primary mission, a substantial improvement over the resolution of previous missions. *Magellan* arrived at the planet in August 1990.

Additional Reading

Beatty, J. K. (1985). A Soviet space odyssey. *Sky and Telescope* **70** 310.

Beatty, J. K. and Chaikin, A., eds. (1990). *The New Solar System*, 3rd ed. Sky Publishing Corporation, Cambridge, MA and Cambridge University Press, Cambridge, U.K.

Schubert, G. and Covey, C. (1983). The atmosphere of Venus. In *The Planets*, introduction by B. Murray. Scientific American, Inc., p. 16.

Strom, R. G. (1987). *Mercury: The Elusive Planet*. Smithsonian Institution Press, Washington, D.C.

Wilson, A. (1987). *The Solar System Log*. Jane's Publishing Company Ltd., London.

See also **Mercury and the Moon, Atmospheres; Mercury, Geology and Geophysics; Mercury, Magnetosphere; Venus, Atmosphere; Venus, Geology and Geophysics; Venus, Magnetic Fields.**

Mercury and Venus, Transits

LeRoy E. Doggett

A planet is said to transit the Sun when it passes between the Sun and the Earth, casting a shadow on the Earth. Because Mercury and Venus are the only major planets inside the Earth's orbit, only they can transit the Sun. If the orbital planes of Mercury and Venus coincided with the plane of the Earth's orbit, the ecliptic, there would be a transit at every inferior conjunction. Usually, however, Mercury and Venus pass north or south of the Sun. A transit occurs when the Earth and one of these two planets arrive nearly simultaneously at a node of the planet's orbit (the line AB in Fig. 1). The required degree of simultaneity depends on the distance and velocity of the planet.

Because the lines of nodes remain nearly stationary in space, the Earth passes a node about the same time each year. The Earth passes the nodes of Mercury's orbit in May and November. Passage through the nodes of Venus occurs in June and December. Thus transits can only occur in these months.

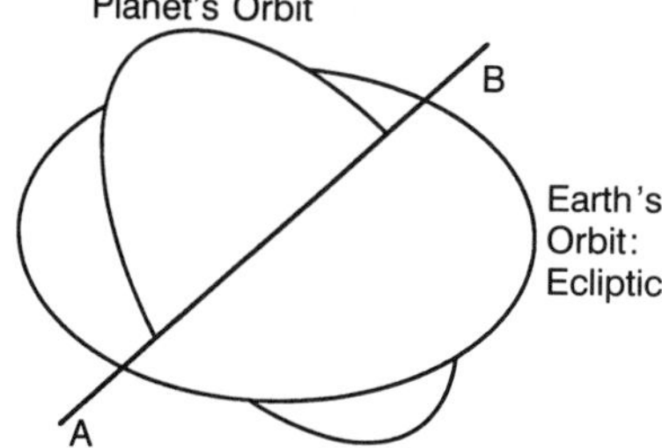

Figure 1. Nodes of the orbit of an inner planet.

TRANSITS OF MERCURY

The synodic period of Mercury (i.e., the average interval between inferior conjunctions) is 115.88 days, whereas the Earth passes a particular node of Mercury's orbit once every year. Because 22 synodic periods take 1 week less than 7 yr, there is occasionally a pair of transits occurring at one of the nodes, 7 yr apart. After another 7 yr, however, conjunction occurs too early to allow a transit. But because 41 synodic periods is only 3 days longer than 13 yr, this forms a useful cycle. More accurate cycles consist of 46 yr and 217 yr.

November transits occur about twice as often as May transits. This is because Mercury's orbit is quite eccentric and perihelion lies close to the November node. When Mercury is moving rapidly near perihelion, it is more likely to catch the Earth near the node. May transits are normally separated by 13 or 33 yr, with an occasional interval of 20 yr. Between a pair of May transits, there will be 1–4 transits in November. November transits are normally separated by 13 yr, but occasionally by 7 yr and in rare instances by 6 yr.

In the nineteenth century Urbain LeVerrier discovered from observations of transits that the motion of Mercury's perihelion could not entirely be explained by newtonian gravitation. Eventually this was explained with Einstein's general theory of relativity. In our own century transit observations helped reveal fluctuations in the rate of rotation of the Earth.

A transit of Mercury may last for more than 3 h if the planet crosses the center of the Sun's disk. Future transits of Mercury through the twenty-first century are listed in Table 1.

TRANSITS OF VENUS

The synodic period of Venus is 583.92 days. Because 5 synodic periods is only 2 days short of 8 yr, a transit may be followed 8 yr later by a transit at the same node. After 8 more years, however,

Table 1. 1993–2098 Transits of Mercury

	6 Nov 1993
	15 Nov 1999
7 May 2003	
	8 Nov 2006
9 May 2016	
	11 Nov 2019
	13 Nov 2032
	7 Nov 2039
7 May 2049	
	9 Nov 2052
10 May 2062	
	11 Nov 2065
	14 Nov 2078
	Nov 2085
8 May 2095	
	10 Nov 2098

conjunction will occur 4 days early, with no transit resulting. However, because 152 synodic periods is only 2 days longer than 243 yr, this forms a stable cycle.

Because the orbit of Venus is nearly circular, June and December transits occur with nearly equal frequency. Normally a pair of transits, separated by 8 yr, occurs at one of the nodes. Following a June pair, a period of 105.5 yr elapses before a December pair occurs. Then 121.5 yr elapse before the next June pair. In future centuries the pairs will dissolve into single transits at each node, but the 243-yr cycle will remain. The following list of transits covers the eighteenth through twenty-first centuries: 6 June 1761, 3 June 1769, 9 December 1874, 6 December 1882, 8 June 2004, and 6 June 2012.

A transit of Venus may last as long as 8 h if the planet passes across the center of the Sun's disk.

FIRST TRANSIT OBSERVATIONS

Although both Claudius Ptolemy and Nicholas Copernicus discussed the possibility of Mercury and Venus transiting the Sun, Johannes Kepler was the first to predict particular transits. Using his new elliptical models of planetary motion he predicted a transit of Mercury on 7 November 1631 and a transit of Venus on 7 December 1631. Although the Venus transit was not observable in Europe, three astronomers saw the transit of Mercury. The planet proved to be astonishingly small.

A transit of Venus in December, 1639 was the next to be observed. Kepler predicted that Venus would pass below the Sun, without transiting. However, after correcting Kepler's tables, the English astronomer Jeremiah Horrox successfully predicted and observed the transit.

DETERMINING THE ASTRONOMICAL UNIT

Transits of Mercury and Venus are now observed merely as curiosities. However, from the seventeenth through the nineteenth centuries, they provided useful information about the orbits of the inner planets and the scale of the solar system.

Prior to the seventeenth century, the mean distance from the Earth to the Sun (the astronomical unit) was generally believed to be about 1200 Earth radii (ER), which was based on the estimate of Ptolemy in the second century AD. As a result of the first transit observations, however, a few authorities in the mid-seventeenth century concluded that the Earth might be 7000–15,000 ER from the Sun. From observations of Mars in 1672, Gian Dominico Cassini deduced a value of 22,000 ER. An accurate determination of the astronomical unit became an urgent goal of astronomy.

The Scottish mathematician James Gregory proposed using transits of Mercury and Venus. The basic idea is shown in Fig. 2, where S, V, and E denote the Sun, Venus, and Earth, respectively. Two widely separated observers are indicated by e and e'. It is assumed that the distance between the observers is known. As seen by observer e, Venus crosses the Sun along the chord v, whereas observer e' sees Venus cross along chord v'. After determining the angular distance between the two chords, giving the angles eVe' and vVv', the distance EV is found from simple geometry. Applying Kepler's laws of motion then gives VS and hence ES. The idea was further developed by Edmond Halley, who showed that Venus would provide more accurate data than Mercury.

Astronomers traveled around the world to remote observing sites for the Venus transits of 1761 and 1769. Unfortunately, observations of the 1761 transit did not provide the accuracy promised by Halley. Determinations of the astronomical unit ranged from 19,500–24,000 ER. The 1769 expeditions provided somewhat better but still inconsistent results. Clearly, however, the small universe of Ptolemy and Copernicus had been destroyed.

The Venus transit of 1874 inspired an even greater international effort, with more astronomers participating and a new method of

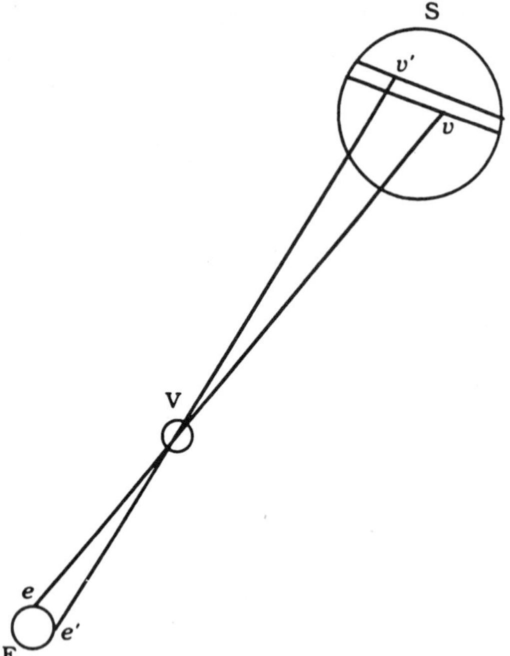

Figure 2. Projected path of Venus on the disk of the Sun during a transit, where E is the Earth.

recording the phenomenon—photography. Determinations of the astronomical unit were again disappointing: 23,200–23,600 ER. Although groups traveled to observe the 1882 transit, much of the enthusiasm had died and astronomers turned to other methods for determining the astronomical unit. In retrospect the transits of Venus expeditions are most notable as early efforts at international cooperation in astronomy.

Today we know the astronomical unit is 149,597,870 km or 23,454.8 ER. For this accurate determination, however, astronomy had to await radar and laser ranging and interplanetary space missions.

Additional Reading

Chapman, A. (1990). Jeremiah Horrocks, the transit of Venus, and the "New Astronomy" in early seventeenth-century England. *Q. J. Royal Astron. Soc.* **31** 333.

Janiczek, P. M. and Houchins, L. (1974). Transits of Venus and the American expedition of 1874. *Sky and Telescope* **48** 366.

Meeus, J. (1982). *Astronomical Tables of the Sun, Moon and Planets*. Willmann-Bell, Inc., Richmond, Va.

Nunis, D. B. (1982). *The 1769 Transit of Venus*. Natural History Museum of Los Angeles County, Los Angeles, Calif.

Van Helden, A. (1985). *Measuring the Universe*. University of Chicago Press, Chicago, Ill.

Woolf, H. (1959). *The Transits of Venus*. Princeton University Press, Princeton, N.J.

See also **Coordinates and Reference Systems; Moon, Eclipses, Librations, and Phases; Sun, Eclipses.**

Mercury, Geology and Geophysics

Robert G. Strom

Mercury is the closest planet to the Sun and has the most eccentric (eccentricity, 0.205) and inclined (7°) orbit of any planet except Pluto. Mercury's average distance from the Sun is 0.39 AU (57.9×10^6 km), but this varies from 0.31 AU (46×10^6 km) at perihelion to 0.47 AU (70×10^6 km) at aphelion because of the large eccentricity. The rotation period is 58.646 days and the orbital period is 87.969 days. Therefore, Mercury has a unique 3:2 resonant relationship between its rotational and orbital periods: It makes exactly three rotations around its axis for every two orbits around the Sun. As a consequence, a solar day (sunrise to sunrise) lasts two mercurian years (176 Earth days). Mercury's surface temperature ranges from a maximum of 427°C at perihelion on the equator to about −183°C on the nightside; the greatest temperature range of any planet or satellite in the solar system.

Mercury is smaller than any other planet except Pluto. Its diameter is 4876 km and its mass is 3.3×10^{23} kg, or 0.055 times the mass of Earth. Consequently, the mean density is 5.44 g cm^{-3}, which is larger than any other planet except Earth (5.52 g cm^{-3}). This high density poses a severe problem for the origin of the planet, to be discussed later.

Mercury represents an end member in solar system origin and evolution because it formed closer to the Sun than any other planet and, therefore, in the hottest part of the solar nebula from which the entire solar system was formed. Most of our current knowledge of Mercury is derived from data returned by the *Mariner 10* spacecraft, which flew by the planet three times in 1974–1975, and from theoretical models of its interior and thermal evolution. *Mariner 10* imaged only about 45% of the surface at an average resolution of about 1 km and less than 1% at resolutions between 100 and 500 m. This coverage and resolution are somewhat comparable to Earth-based telescopic coverage and resolution of the Moon before the advent of spaceflight. Consequently, there are still many uncertainties and questions concerning this unusual planet.

MAGNETIC FIELD

Aside from Earth, Mercury is the only terrestrial planet with a magnetic field. It has an intrinsic dipole field with a dipole moment equal to about 0.004 that of the Earth. Although weak compared to that of the Earth, the field has sufficient strength to hold off the solar wind, creating a bow shock and accelerating charged particles from the solar wind. Like Earth's, the magnetic axis of Mercury is inclined about 11° from the rotation axis, but Mercury occupies a much larger fraction of the volume of its magnetosphere than does the Earth. At times of highest solar activity the solar wind actually reaches the surface. Because of the small size of Mercury's magnetosphere, magnetic events happen more quickly and repeat more often than in Earth's magnetosphere. The maintenance of planetary magnetic fields is thought to require an electrically conducting fluid outer core. Therefore, Mercury's intrinsic dipole magnetic field is taken as strong evidence that the planet currently has a fluid outer core of unknown thickness.

INTERIOR

Mercury's internal constitution is unique in the solar system. Because Mercury's internal pressures are so low, its uncompressed density is still a high 5.3 g cm^{-3} compared to Earth's uncompressed density of 4.4 g cm^{-3}. Therefore, Mercury contains a much larger fraction of iron than any other planet or satellite. This iron is probably concentrated in a metallic core that occupies about 75% of the planet's diameter (42% of the volume) and is surrounded by a silicate mantle and crust only about 600 km thick. Because the presence of a magnetic field indicates that the outer core is currently fluid, there must be a light alloying element in the core to lower the melting point and retain a partly molten core over geologic history (4.5 billion years), as otherwise the core would have solidified long ago. Although oxygen is such an element, it is not sufficiently soluble in iron at Mercury's low internal pressures and, therefore, sulfur is the most likely candidate. The present thickness of the outer fluid core depends on the abundance of sulfur in the core. If the sulfur abundance is less than 0.2% then the entire core should be solidified at present. If it is 7% then the core should be entirely fluid at present. Therefore, Mercury's core probably contains somewhere between 0.2 and 7% sulfur. This large iron core and its sulfur content raise severe problems concerning the origin of Mercury and, consequently, the other terrestrial planets.

ORIGIN

In 1972, John S. Lewis proposed that the densities of the planets could be explained by the condensation, under conditions of chemical equilibrium, of gases and solids of solar composition from a solar nebula whose temperature decreased with increasing distance from the proto-Sun. Although this chemical equilibrium condensation model works very well for most planets, it cannot explain Mercury's very high fraction of iron. For Mercury's position in the solar nebula the model predicts an uncompressed density of only about 4 g cm^{-3}, rather than the observed 5.3 g cm^{-3}, and it also predicts the complete absence of sulfur.

Three very different hypotheses have been proposed to explain the large discrepancy between the iron abundance indicated by Mercury's high density and that predicted by the chemical equilibrium condensation model: One hypothesis (selective accretion) proposes a mechanical accretion process for concentration of the required iron at Mercury's position in the solar nebula, whereas the other two hypotheses (postaccretion vaporization and giant impact) account for Mercury's high density by removing a large fraction of Mercury's silicate mantle early in its history. In the selective accretion model, a differential response of iron and silicates to impact fragmentation and aerodynamic sorting leads to iron enrichment owing to higher gas densities and shorter dynamical time scales in the innermost part of the solar nebula. The postaccretion vaporization model proposes that intense bombardment by solar electromagnetic and corpuscular radiation in the earliest phases of the Sun's evolution (T Tauri phase) vaporized and drove off much of the silicate mantle. In the giant impact hypothesis, a collision of a planet-sized object with Mercury ejects much of the planet's silicate mantle, which was later swept up by the other planets or ejected from the solar system. Each hypothesis predicts a significantly different composition for Mercury's mantle that could be tested by a future space mission to Mercury.

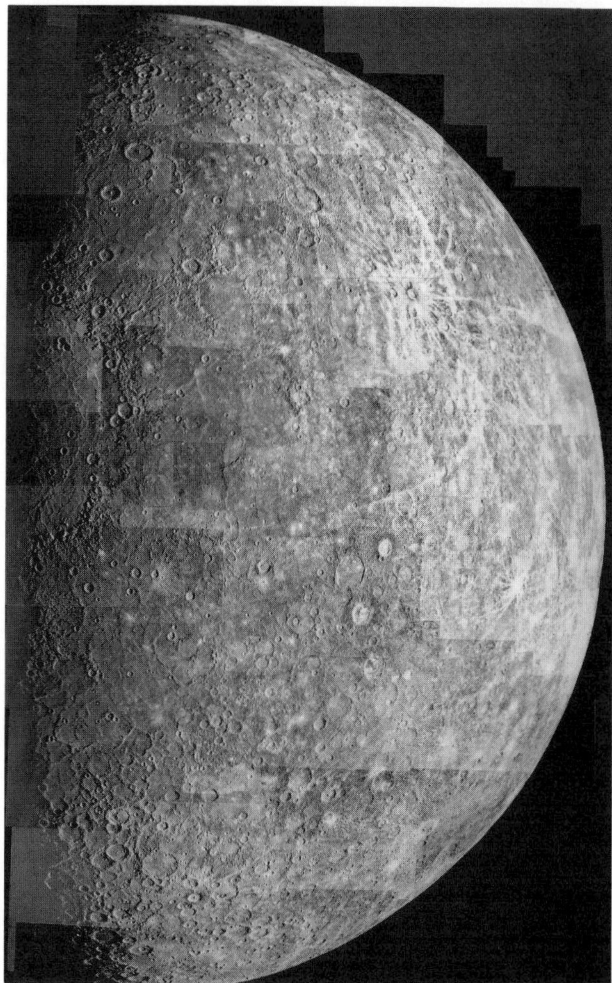

Figure 1. Photomosaic of Mercury as seen by the outgoing *Mariner 10* spacecraft in March 1974. At the left on the terminator is the Caloris basin surrounded and filled by younger smooth plains. (*From the Jet Propulsion Laboratory.*)

GENERAL SURFACE CHARACTERISTICS

Although Mercury superficially resembles the Moon, there are significant and geologically important differences. Like the Moon, it has heavily cratered upland regions and large areas of smooth plains that surround and fill impact basins (Fig. 1). Infrared temperature measurements from *Mariner 10* indicate that the surface is a good thermal insulator and therefore consists of a porous cover of fine-grained soil like the lunar regolith. Unlike the Moon, however, Mercury's heavily cratered terrain contains large regions of gently rolling intercrater plains, the major terrain type on the planet. Mercury also displays a unique tectonic framework consisting of compressive thrust faults called lobate scarps. The largest structure photographed by *Mariner 10* is the Caloris impact basin some 1300 km in diameter (Fig. 2). Antipodal to this basin on the opposite side of Mercury is a region of broken-up terrain, the "hilly and lineated terrain," probably caused by focused seismic waves from the Caloris impact (Fig. 3).

The albedo of surface features is significantly higher on Mercury than for comparable features on the Moon, suggesting a different composition. Color differences between the Moon and Mercury have been inferred to mean that the surface of Mercury is depleted in iron and titanium compared to the Moon. In 1985–1986, Earth-based telescopic observations discovered a tenuous atmosphere of sodium and potassium surrounding Mercury. These elements are probably derived from the surface, possibly by solar wind sputtering or micrometeoroid impact vaporization of surface material, but their surface abundance is unknown.

GEOLOGIC SURFACE UNITS

Because of the limited photographic surface coverage and resolution, the interpretation of some of the major terrains and the inferred geologic history of Mercury are somewhat uncertain. The heavily cratered terrain certainly records the period of intense meteoroid bombardment that ended about 3.8 billion years ago on the Moon and, presumably, at about the same time on Mercury. This period of heavy bombardment occurred throughout the inner solar system and is also recorded by the heavily cratered regions on the Moon and Mars. The origin of the two plains units is somewhat uncertain. They have been attributed to either impact basin ejecta deposits or volcanism. The older intercrater plains were emplaced over a range of ages contemporaneous with the period of heavy bombardment. They are generally believed to be lava flows that erupted early in mercurian history, but this interpretation is less certain than for the smooth plains. The smooth plains primarily fill and surround the Caloris basin and occupy a large circular area in the north polar region. Their youth relative to the basins they occupy, their great areal extent, and other stratigraphic relationships suggest that they are volcanic deposits that erupted rela-

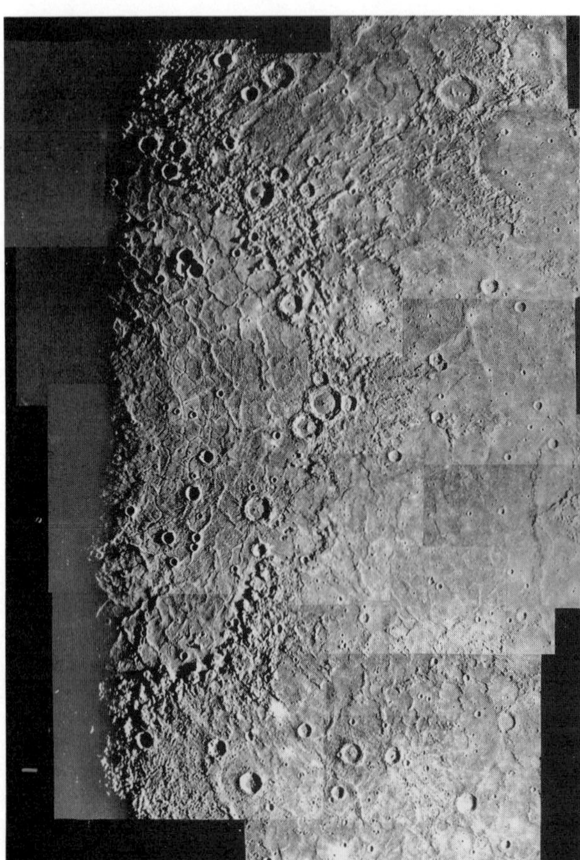

Figure 2. Photomosaic of the 1300-km-diameter Caloris basin, the largest impact structure observed by *Mariner 10*. (*From the Jet Propulsion Laboratory*.)

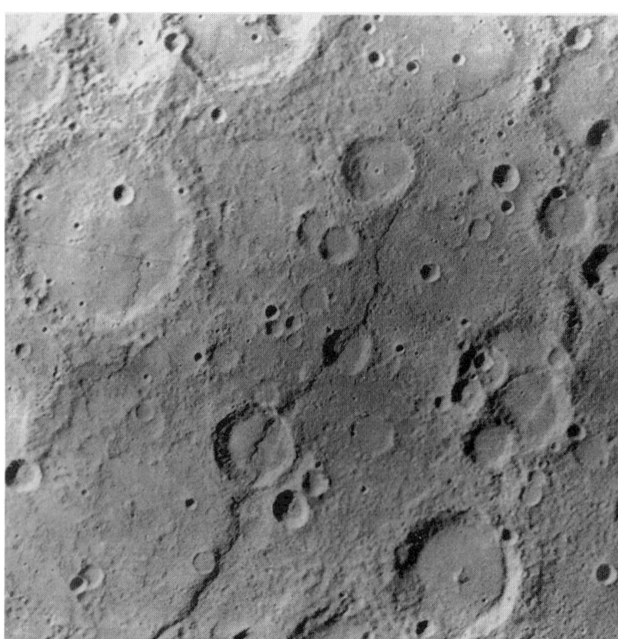

Figure 4. A close view of Mercury from *Mariner 10*, showing a heavily cratered region interspersed with intercrater plains. Running diagonally across the center is Discovery Scarp, a major compression thrust fault about 500 km long and up to 3 km high. (*From the Jet Propulsion Laboratory*.)

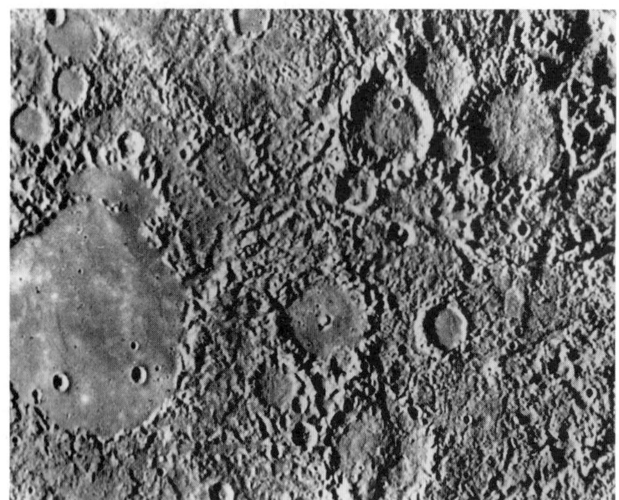

Figure 3. This view of the hilly and lineated terrain was taken by *Mariner 10* in March 1974. The terrain is antipodal to the Caloris basin and was formed by focused seismic waves from the Caloris impact. (*From the Jet Propulsion Laboratory*.)

tively late in mercurian history. Based on the shape of the size frequency distribution of superimposed impact craters, the smooth plains appear to have formed near the end of heavy bombardment and may average about 3.8 billion years old. If so, then they are, in general, older than the lava deposits that constitute the lunar maria.

The Caloris impact basin has a floor structure that is unique. It consists of closely spaced ridges and fractures arranged in both a concentric and radial pattern (Fig. 2). This pattern was probably caused by subsidence and subsequent uplift of the basin floor. Directly opposite the Carolis basin, on the other side of Mercury (the antipodal point of Caloris), is the hilly and lineated terrain, which disrupts preexisting landforms (Fig. 3). This terrain is thought to be the result of seismic waves focused at the antipodal region and caused by the Caloris basin impact.

TECTONIC FRAMEWORK

Mercury has a system of compressive thrust faults that occur everywhere on the imaged half of the planet and that presumably have a global distribution (Fig. 4). Transection and stratigraphic relationships indicate that they post-date the formation of the intercrater plains, suggesting that they formed relatively late in mercurian history. This tectonic framework was caused by crustal shortening resulting from a decrease in the planet radius due to cooling of the mantle and core. The amount of radius decrease estimated from the number and length of the thrust faults is about 2 km.

GEOLOGIC HISTORY

The general picture of Mercury's geologic history that has emerged from analyses of *Mariner 10* data is that soon after Mercury formed it became almost completely melted by heating from the decay of radioactive elements, tidal dissipation, and the inward migration of the large amount of iron metal that formed its enormous core. This led to expansion of the planet and tensional fracturing of a thin solid lithosphere that provided egress for lavas to reach the surface and form the intercrater plains during the period of heavy bombardment. As the core and mantle began to cool, Mercury's radius decreased by about two or more kilometers, and the crust was subjected to compressive stresses that resulted in the unique tectonic framework. At about this time, the Caloris basin was formed by a gigantic impact that caused the hilly and lineated terrain, from seismic waves focused at the antipodal region. Further eruption of lava within and surrounding the Caloris and other large impact basins formed the smooth plains. Volcanism finally ceased when compressive stresses in the lithosphere became strong enough to close off magma sources. All these events probably took place very early, perhaps during the first 700 or 800 million years, in Mercury's history. Since that time only occasional impacts of comets and asteroids have occurred.

This geological history of Mercury is based only on rather limited data that pertain to only half of the planet. Future missions to Mercury that image the unseen half will surely resolve many of the ambiguities and provide a better understanding of this unusual planet.

Additional Reading

Head, J. W., III. (1990). Surfaces of the terrestrial planets. In *The New Solar System*, 3rd ed., J. K. Beatty and A. Chaikin, eds. Sky Publishing Corp., Cambridge, MA, and Cambridge University Press, Cambridge, U.K. p. 77.

Strom, R. G. (1987). *Mercury: The Elusive Planet*. Smithsonian Institution Press, Washington, DC.

Strom, R. G. (1990). Mercury: The forgotten planet. *Sky and Telescope* **80** 256.

Various papers reporting *Mariner 10* results: *Science* **185** 141–180, July 12, 1974; *J. Geophys. Res.* **80** 2341–2514, June 10, 1975.

Vilas, F., Chapman, C. R., and Matthews, M. S., eds. (1988). *Mercury*. University of Arizona Press, Tucson.

See also **Mercury and the Moon, Atmospheres; Mercury and Venus, Space Missions; Mercury, Magnetosphere; Planetary Interiors, Terrestrial Planets; Planetary Rotational Properties; Planetary Tectonic Processes, Terrestrial Planets; Planetary Volcanism and Surface Features.**

Mercury, Magnetosphere

Stephen P. Christon

The discovery of and all direct information about Mercury's intrinsic magnetic field and magnetosphere result from measurements made in the years 1974 and 1975 by magnetic fields and charged particle instrumentation on board a single NASA spacecraft, *Mariner 10* (also known as *Mariner Venus–Mercury*). After a gravity-assisted trajectory maneuver (the first ever) at the planet Venus, *Mariner 10* attained a heliocentric orbit that allowed it repeated encounters with the planet Mercury at approximately 6-month intervals. Three successful encounters were completed before the spacecraft attitude control gas was exhausted: Mercury 1, 2, and 3, for which the closest approach distances were 707, ~50,000, and 327 km, respectively (or 1.29, ~20.5, and 1.13 R_M, where 1 $R_M = 2439$ km).

Critical information about the structure and dynamics of Mercury's magnetosphere were gathered during the Mercury 1 equatorial and Mercury 3 high-latitude passes, with durations ~17 and ~13 minutes, respectively (see Fig. 1). Magnetic field and plasma measurements allowed unambiguous identification of the bow shock and magnetopause at both encounters. The Mercury 3 trajectory was chosen to maximize intrinsic magnetic field observations because variations resulting from dynamic magnetospheric current systems dominated observations during a portion of Mercury 1. The only relevant magnetospheric measurements at Mercury 2 closest approach, far upstream in the solar wind, were a few possible plasma electron flux enhancements similar to those observed upstream of the Mercury 1 encounter. In addition to dynamic magnetic field and plasma variations, our knowledge of and/or inferences drawn about magnetospheric dynamics at Mercury depend on the measured sequence of associated electron flux enhancements at Mercury 1 [the A, B (including B′ and B_{tail}), C, and D events] and the Mercury 3 electron burst.

OVERVIEW

Mercury's magnetosphere, that region of space in which Mercury's intrinsic planetary magnetic field dominates and stands off the solar wind, is populated by plasmas whose electron properties are distinctly different from those of the nearby interplanetary solar wind plasma. Mercury has an apparently Earth-like magnetotail, drawn out in the anti-sunward direction by the solar wind and composed of two oppositely directed magnetic hemispheres, which are separated by a cross-tail current sheet imbedded in a plasma sheet. Indication of the north tail lobe comes from the Mercury 3 polar low plasma flux region, the expected high-latitude, low-altitude lobe signature. Although the structure of Mercury's magnetotail appears similar to Earth's, direct measurements were only made close to the planet, $<1.65\ R_M$ downstream of the planet's center. Much of Mercury's near-planet magnetosphere, a factor of ~20 smaller than Earth's in absolute dimensions, would fit inside an Earth-sized sphere (1 $R_E = 6378$ km). Mercury's effective intrinsic magnetic dipole moment ($\mu_M \sim 6 \pm 2 \times 10^{22}$ G cm^3) is much smaller than Earth's ($\mu_E = 8 \times 10^{25}$ G cm^3). Its polarity orientation, in the same sense as Earth's with respect to the interplanetary magnetic field (IMF), results in solar–magnetospheric energy coupling similar to Earth's, with energy transfer controlled by IMF BZ, the IMF component along the planet's dipole axis. Mercury occupies a volume relative to its magnetosphere ~125–500 times larger than does Earth relative to its magnetosphere, thus most likely precluding the existence of trapped particles and a plasmasphere. Mercury possesses no measured ionosphere or auroras, distinct components of Earth's magnetospheric current system. Nevertheless, morphologically and dynamically, Mercury's magnetosphere appears to be a scaled down (~5–8 times smaller relative to planetary dimensions) and speed enhanced (~30 times faster in absolute units) version of Earth's. Table 1 lists structural and dynamic phenomena that, when considered together, functionally define the terrestrial magnetosphere. Mercury 1 and 3 observations are compared separately to Earth observations.

Mercury's plasma sheet electron temperatures are cooler than Earth's by a factor of ~4 during quiet times, but may be as hot as Earth's during disturbed periods. The shapes of the quiet time plasma sheet electron spectra of Mercury and Earth are identical, suggesting similar plasma processes in the two systems. The plasma energy spectrum was not fully measured during dynamic disturbances at Mercury. However, spectral changes with respect to the quiet time shape in the measured range appear to be similar to those observed during dynamic substorm injections at Earth. Energy is transferred to the magnetosphere from the solar wind and magnetically stored temporarily in the magnetotail during periods of southward IMF BZ. Magnetospheric disturbances known as substorms are a combined set of phenomena representing intense, rapid dissipation of this stored energy through particle heating, which allows the magnetic field to reconfigure and relax to a less stressed state. The dynamic disturbances at Mercury, with field and particle characteristics strikingly similar to substorms in Earth's

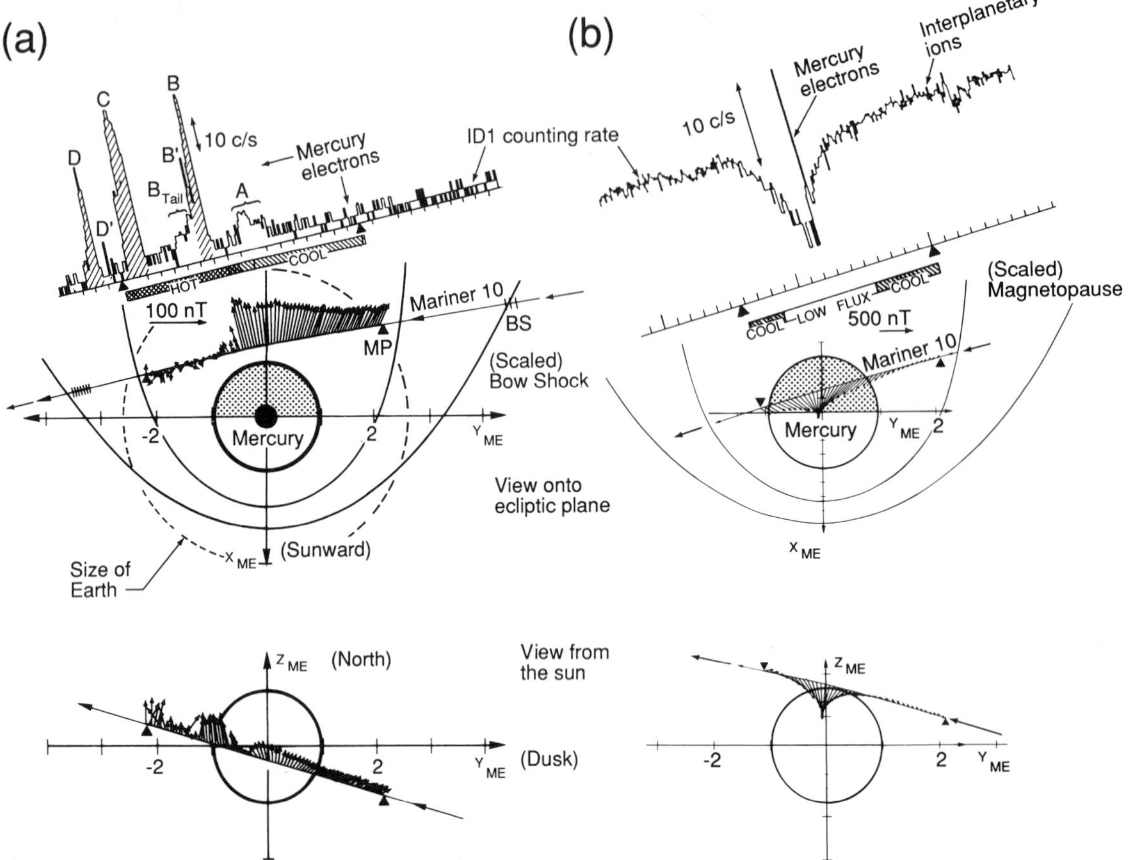

Figure 1. Schematic drawings of the *Mariner 10* trajectory and Mercury magnetic field, plasma, and high-energy electron measurements highlight important features at the (*a*) Mercury 1 and (*b*) Mercury 3 encounters [*after figures in Connerny and Ness (1988), Eraker and Simpson (1986), and Christon (1987)*]. Bow shock (BS) and magnetopause (MP or ▲) crossings are shown. Mercury's nightside face is shaded. The true size of the planet Earth (dashed circle in panel *a*) is shown for reference, whereas the small darkened circle at the origin shows Earth scaled to magnetospheric dimensions. Bow shock and magnetopause curves, scaled from Earth, are shown. Cool, hot, and low-flux plasma regions are indicated. The "ID1" (meaning the wide angle [half-cone ≈ 47°] operational mode of the *Mariner 10* main cosmic-ray telescope) counting rate timeline is offset from the *Mariner 10* trajectory. Shaded portions of the ID1 counting rate (with nominal response to both > 175-keV electrons and/or > 620-keV protons) indicate times of probable > 35-keV electron response. Observed magnetic field vectors are projected onto *X-Y* (top) and *Y-Z* (bottom) planes in the Mercury ecliptic coordinate system, where *X-Y* is a plane parallel to the ecliptic plane and the positive *Z* axis points north. (*By permission from* Mercury, *the University of Arizona Press, 1988 and* The Journal of Geophysical Research, *1986; 1989.*)

near-planet (~5–15 R_E) magnetosphere, argue for an extended, but proportionately scaled down, magnetotail at Mercury.

MARINER 10 MEASUREMENTS AT MERCURY

As the result of an unexplained problem, the *Mariner 10* plasma ion instrument never operated and all plasma information is based upon measurements from a hemispherical electrostatic analyzer that sequentially measured the electron energy spectrum from 13–688 eV in fifteen 0.4-s energy steps over 6 s. The *Mariner 10* main (MT) and low-energy (LET) solid state cosmic-ray telescopes were designed, respectively, to measure >175-keV electrons and >620-keV ions (MT) or >530-keV ions only (LET) at 0.6-s resolution. Individual 0.4- and 0.6-s electron measurements, as well as the high-resolution 0.04-s magnetic field measurements, have been essential in understanding rapid magnetospheric plasma variations.

Just as charged particle instrument saturation provided the first observations of Earth's trapped radiation zones, instrument saturation of the MT at Mercury 1, known as electron pulse pileup, provided unexpected and unplanned observations of Mercury's magnetospheric energetic (>35 keV) electron population. Initially reported ~300-keV electron response during the B, B', C, and D events was later found to result most probably from intense >35-keV electron flux. Possible electron pulse pileup precluded unambiguous LET identification of >530-keV ions.

Prior to the Mercury 1 encounter, IMF BZ was northward, antiparallel to Mercury's intrinsic field. *Mariner 10* entered the magnetosphere at low latitude, observing a cool plasma region similar to Earth's quiet outer plasma sheet until near closest approach, in the nightside equatorial region, after which episodes of intense plasma heating and magnetic field disruption occurred. These variations closely resemble substorms at Earth. Southward IMF BZ was measured immediately outside the outbound magne-

Table 1. Comparison of the Magnetospheres of Mercury and Earth†

	Mercury		
	1	*3*	*Earth*
Structure			
Bow shock	c	c	c
Upstream waves	c	c	c
Upstream plasma electron spikes	c	—	c
Magnetosheath (between bow shock and magnetopause)	c	c	c
Magnetosheath flux transfer events	—	c	c
Magnetopause	c	c	c
Outer magnetosphere			
Magnetospheric boundary layer (= LLBL, mantle, entry layer)	s	—	c
Magnetospheric flux transfer events	c	c	c
Plasma sheet (electrons*)	c	c	c
Plasma sheet boundary layer (PSBL)	—	—	c
Cross-tail current sheet	c	—	c
Neutral lines			
Near-planet	—	—	—
Distant tail	—	—	c
Tail lobes	—	s	c
Field-aligned currents (steady-state FAC)	—	—	c
Inner magnetosphere			
Trapped radiation zones (ring current)	—	—	c
Plasmasphere	—	—	c
Polar cap low plasma flux region (: tail lobes)	—	c	c
Ionosphere	n	—	c
Auroral oval (cusp/cleft: precipitating sheath/LLBL_PSBL)	—	—	c
Auroral electrojets	—	—	c
Auroral radiation (AKR)	—	—	c
Atmosphere/exosphere	c	c	c
Intrinsic magnetic field	c	c	c
Dynamics in Near-Planet Region (Solar–Magnetosphere Energy Coupling)			
Storms	—	—	c
Substorms	c	s	c
Storage (stretching of quiet magnetic field)	c	s	c
Dissipation (energy release and field reorganization)	c	s	c
Magnetic field dipolarization (⇒ transient FAC)	c	—	c
Plasma sheet electron* heating	c	—	c
Electromagnetic/electrostatic plasma wave activity	s	—	c
Dynamic plasma injections			
Plasma electron* flux decrease	c	—	c
Energetic electron* flux increase	c	—	c
Multiple onsets or injections	s	—	c
Disturbance triggered by IMF BZ south-to-north turning	s	—	c
Relativistic electron* flux increase	c	c	c

†An asterisk indicates no unambiguous ion measurement at Mercury; c, measured and confirmed; s, suggested by the measurements; n, negative or upper limit measurement; —, no measurement; 1 and 3, first and third *Mariner 10* Mercury encounters; respectively.

topause, suggesting that the observed disruptions had been produced by dynamic release of energy transferred from the solar wind, just as occurs at Earth. It has been suggested that the escaping D-event energetic electrons, measured in the outbound magnetosheath, result from another substorm, which was triggered by a south-to-northward IMF BZ reorientation (as is the case at Earth).

At Mercury 3, cool plasma was again observed upon entering the magnetosphere. Close to the planet, plasma flux decreased in a polar region expected to be magnetically connected to Mercury's magnetotail lobes. Measurement of Mercury's intense polar magnetic field confirmed its intrinsic nature. A single burst of 175–360-keV magnetospheric electrons, superposed on interplanetary ion flux pervading Mercury's magnetosphere, was measured before closest approach.

Enhanced fluxes of relativistic (> 175–600 keV) electrons were measured during the A and B_{tail} events at Mercury 1 and the electron burst at Mercury 3. The dynamically energized A and B_{tail} relativistic electrons have energy spectra similar to, and may be related to, the lower-intensity jovian electron fluxes present in the nearby interplanetary region. (The Sun and Jupiter are the strongest electron sources in the solar system.) Cosmic-ray telescope measurements strongly suggest that > 175-keV electrons were also measured during the most intense portions of the B and C events. The > 175-keV electron energies observed during the A, B, and C events and the Mercury 3 burst argue strongly that dynamic processes at Mercury accelerate electrons (and, implicitly, ions) to high energies. Mercury's magnetosphere may be as efficient as Earth's in the context of astrophysical particle acceleration.

The magnetospheres of Earth and Mercury, unlike those of Jupiter, Saturn, and Uranus, are not significantly affected by corotation. Venus and Mars apparently do not possess strong intrinsic magnetic fields or terrestrial magnetospheres. Mercury's magnetosphere, a nearby astrophysical entity to which we can directly compare Earth's, is an important scientific discovery, but its physics will remain tantalizing and enigmatic until explored further. Its apparent detailed similarity to Earth's magnetosphere, both morphologically and dynamically, was not anticipated and is not presently understood.

Additional Reading

Christon, S. P. (1987). A comparison of the Mercury and Earth magnetospheres: Electron measurements and substorm time scales. *Icarus* **71** 448.

Christon, S. P. (1989). Plasma and energetic electron flux variations in the Mercury 1 C event: Evidence for a magnetospheric boundary layer. *J. Geophys. Res.* **94** 6481.

Connerny, J. E. P. and Ness, N. F. (1988). Mercury's magnetic field and interior. In *Mercury*, F. Vilas, C. R. Chapman, and M. S. Matthews, eds. University of Arizona Press, Tucson, p. 494.

Eraker, J. H. and Simpson, J. A. (1986). Acceleration of charged particles in Mercury's magnetosphere. *J. Geophys. Res.* **91** 9973.

Ogilvie, K. W., Scudder, J. D., Vasyliunas, V. M., Hartle, R. E., and Siscoe, G. L. (1977). Observations at the planet Mercury by the plasma electron experiment: *Mariner 10*. *J. Geophys. Res.* **82** 1807.

Russell, C. T., Baker, D. N., and Slavin, J. A. (1988). The magnetosphere of Mercury. In *Mercury*, F. Vilas, C. R. Chapman, and M. S. Matthews, eds. University of Arizona Press, Tucson, p. 514.

See also **Earth, Magnetosphere and Magnetotail; Magnetohydrodynamics, Astrophysical; Mercury and Venus, Space Missions.**

Meteorites, Classification

Harry Y. McSween, Jr.

Meteorites are samples of other solar system bodies, mostly of asteroids but in a few cases of the Moon and possibly Mars. Classification provides a means of sorting meteorites into groups with similar properties, so that their origins can be better understood. Modern classifications attempt to group meteorites that formed on the same parent body together, but this is not always possible.

The simplest classification of meteorites is a subdivision into stones, irons, and stony irons, based on their relative proportions of stony (mostly silicate) material and metal. The great majority of meteorites seen to fall are stones; irons and stony irons taken together comprise less than 5% of falls. Stony meteorites probably sample as few as 15 different parent bodies, and several types may be derived from one object. Although few in number, iron metorites are very diverse and represent samples of many different parent bodies. They are apparently more resistant to collisional disruption in space and thus have longer orbital lifetimes than stones.

CHONDRITES

The most common meteorites are chondrites (pronounced "kondrites"), stony meteorites that take their name from the tiny spherules (chondrules) that they contain. Radiometric dating of chondrites indicates that they are 4.55 billion years old, which we believe is the approximate age of the solar system. They also have chemical compositions that are similar to that of the Sun, except that chondrites are depleted relative to solar values in volatile gases like hydrogen and helium. Because the Sun contains more than 99% of the mass of the solar system, this chemical equivalence means that chondrites can be viewed as having the composition of the average solar system. The planets also have nearly chondritic compositions and may have been formed by the accretion of chondrite-like bodies. All chondrites are thought to have been derived from asteroidal-sized planetesimals.

Several types of chondrites are recognized. Carbonaceous chondrites have the highest proportions of volatile elements and are the most oxidized. Enstatite chondrites contain the most refractory elements and are reduced. The ordinary chondrites, so called because they are the most common type, occupy an intermediate position with respect to volatile element abundances and oxidation state. Some meteoriticists have suggested that these properties correlate with formation location: the enstatite chondrites formed in the inner solar system, ordinary chondrites in the inner asteroid belt, and carbonaceous chondrites at even greater solar distances. Each of these chondrite classes can be further subdivided into smaller groups that have distinct properties that presumably represent different parent objects. For example, the ordinary chondrites can be divided into three groups, called H, L, and LL, based on minor differences in chemistry and oxidation state.

Many chondrites have been affected by secondary processes on their parent bodies, and these processes have changed their properties. Ordinary and enstatite chondrite parent bodies were heated soon after their formation, and parts of these objects reached temperatures approaching 1000°C. Meteorites from these regions have been metamorphosed and recrystallized so that their original textures are nearly unrecognizable. Chondrite metamorphism occurred under nearly dry conditions, so that no minerals containing water were produced. Carbonaceous chondrites were probably formed within bodies that originally contained ice. Heating of such objects caused the ice to melt, but temperatures were buffered to low values. These meteorites have suffered aqueous alteration by cold water, producing a variety of hydrous minerals. Other chondrites are breccias, formed from broken bits and pieces of material that were pulverized by impacts onto the parent body surface. Modern classification schemes also provide information on secondary processing of chondrites.

ACHONDRITES

Achondrites are igneous rocks, formed by crystallization of melts on or within meteorite parent bodies. These stony meteorites have distinctive textures and mineralogies that are indicative of igneous processes. Five major groups have been recognized. The HED (howardite, eucrite, diogenite) group consists of basaltic and ultramafic rocks from an asteroidal body that was probably originally chondritic in composition. They have radiometric ages around 4.5 billion years and represent the earliest known igneous activity. The SNC (shergottite, nakhlite, chassignite) group are much younger, about 1.3 billion years, and are commonly thought to be martian samples. These range in composition from basaltic to ultramafic and are similar in many respects to terrestrial igneous rocks. The ureilites consist primarily of silicates and carbon compounds and may have formed by melting of a carbonaceous chondrite-like asteroid. The aubrites consist almost entirely of enstatite, a magnesium silicate, and apparently formed through melting of an enstatite chondrite-like body. A small number of lunar achondrites have been recovered in Antarctica during the last few years. They all are regolith breccias similar to those returned from the Moon during the Apollo program. Mesosiderites are stony irons that are brecciated mixtures of HED-like materials and metal, and these meteorites probably relate in some way to the HED achondrite group.

IRONS AND PALLASITES

Another type of melting behavior within meteorite parent bodies is demonstrated by the occurrence of iron meteorites. These meteorites formed when molten metal segregated from less dense silicate material and cooled.

Table 1. Meteorite Classification*

Stony meteorites
Chondrites
Carbonaceous (four types: CI, CM, CV, CO)
Enstatite (two types: EH, EL)
Achondrites
HED group (howardite, eucrite, diogenite)
SNC group (shergottite, nakhlite, chassignite)
Aubrites
Ureilites
Stony iron meteorites
Pallasites (two types: main group and Eagle Station group)
Mesosiderites
Iron meteorites (thirteen types: IAB, IC, IIAB, IIC, IID, IIE, IIF, IIIAB, IIICD, IIIE, IIIF, IVA, IVB)

*All of the major groups listed contain anomalous members that do not fit into this classification scheme and presumably represent samples of distinct parent bodies.

Iron meteorites consist essentially of iron–nickel alloys with minor amounts of carbon, sulfur, and phosphorus. The alloy commonly segregates into two minerals, one low in nickel (kamacite) and the other with higher nickel concentration (taenite). Irons composed only of kamacite are called hexahedrites; those appearing only to contain taenite (they actually contain microscopic grains of kamacite as well) are ataxites; those with both phases are octahedrites. Slabs of octahedrite etched with acid display a characteristic geometric structure of kamacite plates in taenite, called a Widmanstätten pattern, after Aloys J. B. von Widmanstätten. The structural appearance of irons depends partly on their compositions and partly on their rates of cooling. Some irons also contain inclusions of silicate material. Radiometric dating of these inclusions indicates that iron meteorites have ancient ages, similar to those of chondrites and most achondrites.

Classification of iron meteorites by their structure does not produce well defined groups, so modern classification schemes are based primarily on their chemical compositions. In addition to their nickel content, other important chemical distinctions between irons occur in their content of germanium and gallium, two trace elements that substitute in metal structures. Initially four groups (I–IV) were distinguished on this basis, but data obtained subsequently forced some of these groups to be subdivided; for example, group IV has been now separated into groups IVA and IVB, which have nothing to do with each other. To make matters more confusing, a few subdivided groups have now been recombined after transitional members were discovered, giving rise to names like IIAB. The resulting array of names is a confusing but otherwise very servicable classification system. Thirteen important groups of irons have now been recognized.

Pallasites are stony iron meteorites composed of abundant grains of olivine, a magnesium silicate mineral, enclosed in metal. There are two recognized groups. Most pallasites have metal with similar chemical composition to group IIIAB irons, and it is thought that these meteorites formed at the core–mantle boundary of the IIIAB parent body as olivine was submerged into the liquid metal core. A few pallasites contain metal that is apparently unrelated to any recognized iron group.

SUMMARY

Table 1 presents a summary of the classes of meteorites just discussed. This table has not changed a great deal over the last century of meteorite studies, but our understanding of how different meteorite groups relate to each other has improved dramatically. The fundamental distinction in meteoritics is between chondrites, which are more nearly pristine samples of early solar system matter, and other meteorite types, which have been geologically processed in one way or another. Even chondrites, however, have in many cases experienced thermal metamorphism or aqueous alteration that have modified their properties somewhat. Achondrites, irons, and stony irons have extended our knowledge of the applicability of igneous processes learned from studies of the Earth and Moon to other bodies with different compositions and sizes. Future meteorite discoveries may permit us to gain additional information on chemical diversity in the early solar system and the geologic processes that have affected meteorite parent bodies.

Additional Reading

Burke, J. G. (1986). *Cosmic Debris: Meteorites in History*. University of California Press, Los Angeles.

Dodd, R. T. (1986). *Thunderstones and Shooting Stars: The Meaning of Meteorites*. Harvard University Press, Cambridge, Mass.

Hutchison, R. (1983). *The Search for Our Beginning: An Enquiry Based on Meteorite Research into the Origin of Our Planet and of Life*. Oxford University Press, London.

Kerridge, J. F. and Matthews, M. S., eds. (1988). *Meteorites and the Early Solar System*. University of Arizona Press, Tucson.

McSween, H. Y., Jr. (1987). *Meteorites and Their Parent Planets*. Cambridge University Press, New York.

Wasson, J. T. (1985). *Meteorites: Their Record of Early Solar-System History*. Freeman, San Francisco.

See also **Interplanetary Dust, Collection and Analysis; Meteorites, Composition, Mineralogy, and Petrology; Meteorites, Origin and Analysis.**

Meteorites, Composition, Mineralogy, and Petrology

James L. Gooding

DEFINITION

The three broadest groupings of meteorites (stony, stony iron, and iron) are defined by the material compositions of the respective meteorites. Indeed, measurement of compositional properties has become the root of all other studies in meteoritics, including the classification of new specimens and the development of models to explain the nature of meteorites and the planetary bodies from which they come.

Meteorite "composition" is understood to include three attributes:

Chemical composition: concentrations of chemical elements; compounds other than minerals.

Mineralogical composition: types and abundances of minerals and related natural glasses.

Isotopic composition: ratios of isotopes of a given chemical element.

Meteorite *petrology* is the scientific specialty that seeks to integrate information about chemical and mineralogical compositions into interpretational models that allow the origins and histories of meteorites to be understood. Isotopic compositions contribute absolute ages of the meteorites and place powerful limits on the viability of alternative petrological models for meteorite origin.

HISTORY AND METHODS OF ANALYSIS

Meteorites have been seriously studied as rocks since the 1860s, following invention of the petrographic microscope, which allowed thin slices (about 25 μm thick) of rocks to be studied with polarized light. Polished sections of meteorites viewed microscopically either in reflected light (e.g., opaque slabs of iron and stony iron meteorites) or in transmitted light (e.g., thin sections of stony meteorites) reveal the identities and physical histories of the minerals. Even though sophisticated analytical instruments now dominate petrological studies, petrographic examination of polished sections remains the most important first step in the study of each new meteorite specimen. Since the middle 1960s, mineral identification has relied heavily on chemical compositions of individual grains as obtained by electron-probe microanalysis of polished sections. The electron microprobe bombards a mineral grain with a tightly focussed beam of electrons and measures the characteristic x-rays that are emitted by the target; identities and relative abundances of the x-rays form the basis of the quantitative analysis. Quantitative analysis of grains as small as about 10 μm can be routinely made for all of the important major and minor elements (atomic numbers of 11 or higher) in minerals; with special x-ray detectors, lighter elements (e.g., C) can also be analyzed. In the 1980s, ion-probe analyses also became common for a more restricted set of elements in meteoritic minerals. The ion probe bombards a polished mineral surface with a focussed beam of ions to sputter secondary ions from the mineral into a mass spectrometer where the sample elements are identified and their abundances quantitatively measured. Ion probes excel at trace-element analyses of small spots and, therefore, complement electron microprobe analyses. Physical details of mineral grains are studied by electron microscopy, including transmission electron microscopy that can identify minerals through their electron-diffraction patterns from grains as small as a few tens of angstroms. Such methods, along with microtome procedures adapted from biological studies, have made possible electron micropetrography of ultrathin slices of selected meteorites.

Reliable bulk chemical analyses have been performed on meteorites since the early 1900s, beginning with classical wet-chemical procedures and ending with instrumental methods that sometimes are completely nondestructive of the specimen. Chemical studies of meteorites in the 1930s and 1940s played a crucial role in the development of the science of *geochemistry*, particularly in the geochemical classification of the elements according to their behavior in nature. Definition of the lithophile, chalcophile, and siderophile groups of elements was strongly influenced by observations of which chemical elements were concentrated in meteoritic silicate (or oxide), sulfide, and metal minerals, respectively. In 1953, a critical and comprehensive screening of early wet-chemical analyses led to recognition of possible groupings of chondritic meteorites according to their content and chemical forms of iron. The so-called ordinary (noncarbonaceous) chondrites were thereby divided into the H (high iron), L (low iron), and LL (low iron and low metal) groups. A subsequent electron microprobe study in 1964 revealed that the same groups could be recognized by the iron contents of their silicate minerals.

The most commonly used methods of bulk chemical analysis now are x-ray fluorescence spectrometry and neutron activation analysis. In the x-ray fluorescence method, an homogenized powder (usually a few grams) of the bulk meteorite is irradiated with x-rays of selected energies to stimulate emission of characteristic x-rays; thereafter, quantitative analysis proceeds much the same as in the electron microprobe method. In contrast, neutron activation stimulates characteristic γ radiation from a sample through bombardment with neutrons; quantitative analysis is based on identities and relative abundances of the emitted γ rays. Neutron activation requires comparatively little sample preparation (unless radiochemical separations of different elements are attempted), can be performed on very small samples (a few milligrams in some cases), and is sensitive to many elements in ultratrace concentrations (a few parts per million or billion). Most current knowledge about trace-element compositions of meteorites has been obtained by neutron-activation methods.

Since the early 1970s, with the development of ultrasensitive isotope-ratio mass spectrometers, studies of isotopic compositions have blossomed into a separate subdiscipline that has produced some of the most rapid and exciting advancements of knowledge. Measurement of daughter/parent ratios for radioactive nuclides point not only toward formational ages but can sometimes reveal different events in the development of especially complex meteorites. Important radiometric chronometers include K-Ar, Rb-Sr, Sm-Nd, and U-Pb. Measured ratios of stable isotopes can significantly constrain possible genetic interrelationships among different meteorite types and lead to recognition of exotic materials, possibly including presolar grains, hidden among more ordinary solar system matter. The more important stable-isotope ratios are $^{2}H/H$, $^{13}C/^{12}C$, $^{15}N/^{14}N$, $^{18}O/^{16}O$, and $^{17}O/^{16}O$. Other key measurements include the isotopic ratios of the individual noble gases (He, Ne, Ar, Kr, and Xe) that occur at trace concentrations in nearly all meteorites.

MINERALOGICAL COMPOSITIONS

Most of the minerals in meteorites (see Table 1) are also known as minerals on Earth although a few trace minerals either have no documented terrestrial counterparts or were recognized in terrestrial rocks only after their discovery in meteorites. Brian Mason's 1979 compilation included 26 minerals (mostly rare oxides, sulfides, nitrides, or phosphides) that have been found in meteorites but not in terrestrial rocks.

Chondrites

Chondrites are stony meteorites composed of iron- and magnesium-bearing silicate minerals with minor amounts of sulfide and metal minerals. Chondrites show a wide variety of compositions, ranging from anhydrous, nearly pure enstatite rocks to specimens composed almost entirely of hydrated minerals such as serpentine. Most chondrites are distinguished by their chondrules, which are millimeter-sized frozen droplets of olivine, enstatite, pigeonite, or glass, that testify to high-temperature melting of free-floating grain aggregates. Carbonaceous chondrites of the CM (C2) and CV (C3) varieties also contain irregularly shaped millimeter- to centimeter-sized, light-colored inclusions composed of melilite, diopside, perovskite, and other minerals that form only at very high temperatures. Those so-called calcium–aluminum-rich inclusions (CAIs), or refractory inclusions, have been intensely studied because of their exotic chemical and isotopic compositions that represent extremities among meteoritic materials.

Ordinary chondrites are close-packed masses of olivine–pigeonite chondrules with traces of Na-rich plagioclase (or glass of equivalent chemical composition) as well as accessory kamacite, taenite, and troilite. Matrix areas between chondrules are composed of essentially the same minerals. Equilibrated ordinary chondrites are so named because the Fe/Mg ratios in their olivine and pigeonite grains cluster around fixed values that seem to divide the meteorites into three different (H, L, LL; defined above) groups. In addition, individual chondrules are indistinct, having apparently been integrated into a continuous fabric after the meteoritic material was assembled. Unequilibrated ordinary chondrites are distinguished by wide-ranging Fe/Mg ratios and by trace minerals, such as smectite and magnetite, that indicate a lower temperature origin involving water and other volatile compounds. Furthermore, individual chondrules are well preserved and commonly are set in a fine-grained dark matrix, dominated by Fe-rich olivine, that is not found in equilibrated counterparts. Carbonaceous chondrites, which

Table 1. Principal Minerals in Meteorites*

Name	Chemical Formula
	Minerals of Widespread Occurrence
Kamacite	α-(Fe, Ni) with 4–7% Ni
Taenite	γ-(Fe, Ni) with 30–60% Ni
Troilite	FeS
Olivine	$(Mg, Fe)_2SiO_4$
Pigeonite	$(Mg, Fe, Ca)SiO_3$
Diopside	$Ca(Mg, Fe)Si_2O_6$
Plagioclase	$(Na, Ca)(Al, Si)AlSi_2O_8$
	Minerals of Important but Limited Occurrence
Enstatite	$MgSiO_3$
Melilite	$Ca_2(Mg, Al)(Si, Al)_2O_7$
Perovskite	$CaTiO_3$
Serpentine	$(Mg, Fe)_3Si_2O_5(OH)_4$
Smectite	$(Ca_{0.5}, Na)_{0.3}(Mg, Fe, Al)_3(Al, Si)_4O_{10}(OH)_2 \cdot nH_2O$
Magnetite	Fe_3O_4
Epsomite	$MgSO_4 \cdot 7H_2O$
Dolomite	$Ca(Mg, Fe, Mn)(CO_3)_2$

*Not a comprehensive list; refer to Dodd (1983), Mason (1979), and Kerridge and Matthews (1988) for more complete inventories.

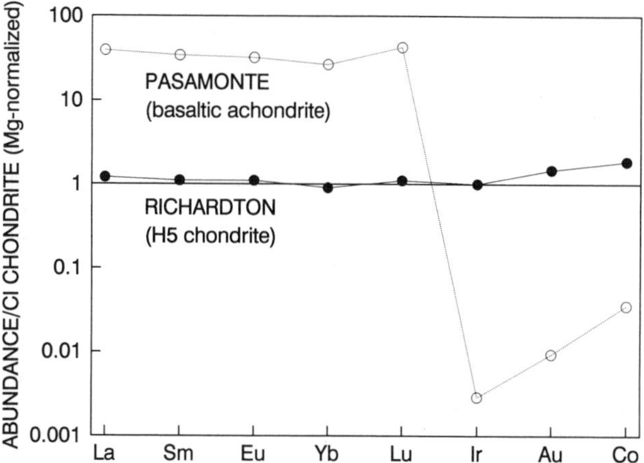

Figure 1. Bulk elemental compositions of the Richardton and Pasamonte meteorites normalized to the composition of the Orgueil (CI) meteorite for selected lithophile (La–Lu) and siderophile (Ir, Au, and Co) elements. Richardton shows "chondritic" abundances of the elements (i.e., enrichment factor of approximately 1 for each element) that approximate those in average, condensible matter in the solar system. In contrast, Pasamonte, a basaltic achondrite, represents a more highly processed planetary material for which melting of the source material fractionally separated lithophiles from siderophiles. Accordingly, chondrites are "primitive" meteorites whereas achondrites are "differentiated" meteorites.

are named for their bulk carbon contents of about 0.1–3 wt%, are also dominated by olivine and pigeonite, with admixed CAIs, but contain substantial portions of low-temperature minerals. CI (or C1) chondrites consist almost entirely of serpentine, smectite, and related hydrous silicate minerals with accessory salt minerals such as epsomite and dolomite.

Achondrites

Achondrites are stony meteorites that resemble ordinary chondrites in terms of major minerals but that have a greater abundance of (Ca-rich) plagioclase and diopside and a much lesser abundance of metal and sulfide minerals. Unlike the chondrites, most of which are mechanical aggregates of various particles, achondrites seem to be products of more broadly based melting and recrystallization. In fact, many achondrites resemble volcanic rocks and are interpreted as products of large-scale melting on their parent bodies. Therefore, achondrites are viewed as differentiated or reprocessed matter, whereas ordinary chondrites, and some carbonaceous chondrites, are taken as more primitive samples of early solar system materials.

Irons and Stony Irons

Irons are composed mostly of kamacite, taenite, and troilite but with a host of unusual trace minerals. The dense, integrated fabrics of iron meteorites imply that they formed by slow cooling and crystallization of molten metal laced with impurities. The coarseness of crystallinity and the distribution of Ni between its major host minerals have been used to infer differences in cooling rates among individual meteorites. Stony irons can be viewed as iron meteorites with mechanically admixed inclusions of silicate minerals that resemble those in various achondrites or ordinary chondrites.

Both chondrites and achondrites often occur as *breccias*; that is, rocks that have been assembled from broken pieces of preexisting rocks. For example, inclusions of chondrites of one type within chondrites of other types is fairly common. Consequently, clues about the earliest histories of meteorites must be filtered from overprints of later events that apparently included energetic collisions, partial melting, and mechanical mixing of projectile and target materials.

CHEMICAL COMPOSITIONS

Although their mineral compositions clearly indicate a complex history, the bulk elemental compositions of CI (or C1) carbonaceous chondrites have been accepted as the best indications of the average composition of matter (other than H_2 and He) in the early solar system. That conclusion developed from comparison of elemental abundances in CI chondrites with elemental abundances in the Sun, as derived by spectroscopy. Accordingly, it is common practice to display compositional data for a given meteorite as an element-by-element normalization to the average composition of CI chondrites. Furthermore, because the absolute concentrations of elements in CIs are influenced by the somewhat variable hydration states of the CI minerals, an additional normalization to a nonvolatile element (e.g., Mg or Si) is commonly included.

Figure 1 shows how CI-normalized elemental abundances can reveal meteorite histories. In this illustration, selected data are shown for some of the rare-earth elements (La–Lu), which are classified geochemically as lithophiles, and for three siderophiles (Ir, Au, and Co). During natural melting and crystallization of geologic materials, lithophiles tend to concentrate in silicate and oxide minerals whereas siderophiles concentrate in metallic minerals (e.g., kamacite and taenite). When their concentrations are normalized to those in CI chondrites, it is clear that the Richardton chondrite has not been affected by any process that separated lithophile from siderophile elements whereas, in contrast, the Pasamonte basaltic achondrite has been significantly affected. The nearly flat line with an abundance ratio of 1 shows that the elemental composition of Richardton has not been *fractionated* relative to average solar system matter. Pasamonte has been greatly fractionatedrelative to primitive compositions, however; it has been enriched in lithophiles and depleted in siderophiles as would be expected on a planetary body that experienced melting and metal-core formation. Although this illustration is quite simple, the same technique is used to compare a wide variety of meteorites using dozens of different chemical elements. CAIs represent intriguing cases where not only have lithophiles been fractionated from siderophiles but lithophiles, as a group, have been significantly fractionated relative to each other on the basis of volatility. (See Table 1 in the entry on Meteorites, Origin and Evolution, for further information on the composition of meteorites of various types.)

Carbonaceous chondrites are further distinguished from other meteorites by traces of organic compounds of extraterrestrial origin. Documented compounds include saturated and unsaturated hydrocarbons (both chain- and ring-structured), fatty acids, purines, pyrimidines, and amino acids, as well as solvent-insoluble, high molecular weight material known as kerogen. Despite the surprising variety of compounds seen, the organic components can be understood as products of reactions that did not require biological participation. Instead, they probably formed through inorganic pathways involving heat, water, and catalytically active minerals that were available during formation of the carbonaceous chondrites.

ISOTOPIC COMPOSITIONS

Oxygen isotopic compositions are particularly noteworthy because there exist fundamental differences in the oxygen three-isotope ratios among groups of stony meteorites as well as among individual mineral grains within the same meteorite. In fact, a plot of $^{17}O/^{16}O$ versus $^{18}O/^{16}O$ for stony meteorites becomes equivalent to a fingerprint map with data points for each meteorite group clustering together in particular places on the diagram. Intermeteorite differences in bulk oxygen–isotopic composition indicate different meteorite parent bodies whereas intrameteorite variations point toward the materials and processes that affected an individual meteorite. Under favorable circumstances, oxygen-isotopic ratios in coexisting minerals in a given meteorite can be used to deduce the temperature of formation of the minerals or, at least, the lowest temperature at which the minerals participated in oxygen-exchange reactions. Extreme enrichment of CAIs in ^{16}O, relative to other meteoritic materials, has been interpreted to mean that CAIs contain grains that originated outside the solar system.

Other diagnostic ratios of importance include $^{87}Sr/^{86}Sr$ and $^{143}Nd/^{144}Nd$, which serve as indices of the amount of geochemical processing that a meteorite has experienced. The "initial" values of the Sr- and Nd-isotopic ratios, which are found as the *y* intercepts in Rb–Sr and Sm–Nd isochron diagrams (see Fig. 1 in the entry for Meteorites, Isotopic Analysis), are lowest when the meteorite has been least processed relative to the original solar system matter. Most meteorites have lower Sr- and Nd-isotopic initial ratios than do Earth rocks or Moons rocks, thereby corroborating the geochemical antiquity of meteorites.

COMPOSITIONS AS PARENT-BODY SIGNATURES

A major goal of meteoritics is to determine the origins and histories of the meteorite parent bodies. With only one group as an exception, however, the sources of specific meteorites remain unproven. Several achondrites recovered from Antarctica since 1981 have been conclusively shown to be of lunar origin, based on their compositional matches (including oxygen three-isotope ratios) to rocks returned from the Moon by the Apollo missions in 1969–72. A separate group of eight achondrites (the so-called shergottite, nakhlite, and chassignite subgroups or "SNC meteorites") is strongly suspected to have come from Mars, based on trapped atmospheric gases (in shock-melted minerals) that resemble the composition of the martian atmosphere as measured by the *Viking* lander spacecraft in 1976. On the belief that the SNC meteorites clan represent martian rocks, fairly sophisticated geochemical models of Mars have been proposed.

All other groups of meteorites are presumed to have originated on asteroids. Matches between meteorite and asteroid types have been made with various degrees of success by correlating laboratory visible and infrared reflectance spectra (controlled by minerals, rather than by elements or isotopes) of meteorites with similar data obtained telescopically for asteroids. C-type asteroids seem to correlate with carbonaceous chondrites whereas S-type asteroids resemble certain achondrites. Finding asteroid equivalents of other known meteorite types continues to be an active area of research.

Additional Reading

Burke, J. G. (1986). *Cosmic Debris—Meteorites in History*. University of California Press, Berkeley.

Chyba, C. F. (1990). Extraterrestrial amino acids and terrestrial life. *Nature* **348** 113.

Dodd, R. T. (1983). *Meteorites: A Chemical-Petrologic Synthesis*. Cambridge University Press, New York.

Kerridge, J. F. and Matthews, M. S., eds. (1988). *Meteorites and the Early Solar System*. University of Arizona Press, Tucson.

Mason, B. (1979). Data of geochemistry. In *Cosmochemistry, Part 1. Meteorites*, 6th ed. U.S. Geological Survey Professional Paper No. 440-B-1, Washington, D.C., Chapter B.

Nagy, B. (1975). *Carbonaceous Meteorites*. Elsevier, New York.

See also **Asteroids; Interplanetary Dust, Collection and Analysis; Meteorites, Classification; Meteorites, Isotopic Analyses; Meteorites, Origin and Evolution.**

Meteorites, Isotopic Analyses

Robert M. Walker

Isotopic studies on meteorites have determined certain of the conditions under which the solar system originated and evolved—the initial composition, age of formation, the time sequence and duration of various events, and the nature of different physical processes, such as condensation, accretion, melting, and cooling. Information about energetic particles from the Sun and Galaxy has also been obtained. Recently, crystals having unique isotopic records have been isolated from meteorites. These grains are probably interstellar in origin, and their study may provide new insights into stellar evolution and nucleosynthesis.

CHRONOLOGY OF THE EARLY SOLAR SYSTEM

Conventional Chronometers

The time when a solid object cooled to the point where isotopic exchange ceased is determined by measuring correlations between radioactive parent nuclei and stable daughter products. For example, ^{40}K (half-life, $t_{1/2}=1.2\times10^9$ yr) decays to ^{40}Ar, and the amount of ^{40}Ar relative to ^{40}K measures the time since samples cooled to the point where gas retention occurred. Complementary isotopic systems include U-Th-Pb, ^{87}Rb-^{87}Sr, and ^{147}Sm-^{143}Nd.

Information about the period *before* final isotopic closure can also be obtained. This can be seen by referring to Fig. 1, which shows a Rb-Sr diagram for a metamorphosed stony meteorite. Subsamples containing different amounts of Rb define a straight line "isochron" whose slope is a measure of the age. Extrapolation of this line to zero Rb defines the *initial* isotopic ratio of $^{87}Sr/^{86}Sr$. The lowest (least evolved) ratio has been measured in the Allende carbonaceous chrondite. Variations in initial ratios are due to the growth of ^{87}Sr in Rb-containing reservoirs prior to rock formation. These can be converted to time intervals by assuming Rb/Sr values for these reservoirs. If a solar value is taken, the precision of the measurements translates into a time resolution of $\approx 1\times10^6$ yr.

Some meteorites, such as irons and basaltic achondrites, appear to have been produced by melting and differentiation of larger

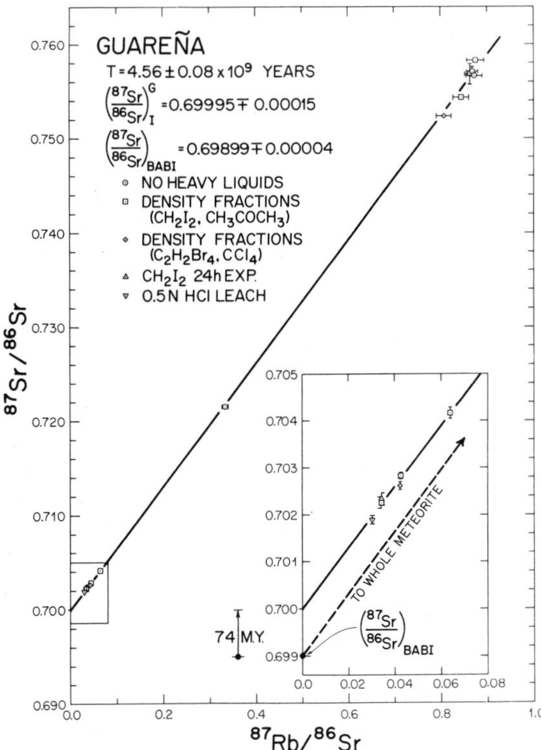

Figure 1. This shows an internal Rb-Sr evolution diagram measured from separated components of the metamorphosed stony meteorite Guareña. The initial $^{87}Sr/^{86}Sr$ ratio is higher than the value 0.6987 measured in Allende because of a period of metamorphism. [*From G. J. Wasserburg, D. A. Papanastassiou, and H. G. Sanz*, Earth and Planet. Sci. Lett. *7 33 (1969)*.] The BABI referred to in the figure is the initial value determined from basaltic achrondites that showed the least evolved ratio that was known at the time the diagram was prepared.

parent bodies. Others, such as carbonaceous chondrites, are undifferentiated, primitive assemblages. Application of the various dating methods shows that both primitive and differentiated objects are all $\approx 4.55 \times 10^9$ yr old. There is evidence for a period of heating for the latter objects, but this typically is $< 100 \times 10^6$ yr. The heat sources responsible for melting the parent bodies of differentiated objects must therefore have been present at the beginning; the parent bodies must also have been small enough (≤ 100 km), or the samples close enough to the surface, to permit rapid cooling.

Extinct Isotopes

The observation that short-lived isotopes synthesized in galactic processes were present in meteorites proves that the 4.55×10^9 yr age coincides closely with the formation of the solar system. The first such isotope discovered was ^{129}I ($t_{1/2} = 16 \times 10^6$ yr), which decays into ^{129}Xe. An excess of ^{129}Xe was found in Richardton chondrite, and the association of the ^{129}Xe with I was demonstrated by irradiating samples with thermal neutrons to produce a built-in tracer via the reaction $^{127}I(n,\gamma\beta^-)^{128}Xe$. The release of excess ^{129}Xe was found to parallel the release of ^{128}Xe for a range of temperatures.

If it is assumed that ^{129}I was originally uniformly mixed with the stable isotope ^{127}I, then differences in the amounts of ^{129}Xe relative to neutron-induced ^{128}Xe can be used to determine small differences in the times at which different objects cooled to the temperature of gas retention. Measurements show that both primitive and evolved meteorites apparently formed within a total time span of $\leq 50 \times 10^6$ yr. However, the results do not present a simple pattern; some of the most processed meteorites appear to predate some of the more unequilibrated assemblages. Although the time inferences may be correct, the assumption of a universal initial value for the ratio $^{129}I/^{127}I$ may not be strictly valid.

Another extinct isotope is ^{244}Pu ($t_{1/2} = 82 \times 10^6$ yr), which spontaneously fissions, giving rise to a characteristic pattern of heavy Xe isotopes. After a new fission-like component was identified from Xe measurements, it was shown that crystals rich in this component contained an excess of fossil fission tracks. Proof that ^{244}Pu was responsible for the fission effects was provided when the spontaneous fission yields were measured in a sample of manmade ^{244}Pu and found to agree with the pattern observed in meteorites. Unlike iodine, there is no stable isotope of plutonium that can be used as a reference to correct for variations in initial concentrations in different mineral phases; thus Pu is of limited use in obtaining fine scale chronological information.

However, both ^{129}I and ^{244}Pu can be used to estimate the time interval ΔT between isolation of the solar system from material exchange with the Galaxy and the formation of solid objects. Estimates necessarily depend on models of galactic nucleosynthesis, but are consistent with $\Delta T \leq 100–200 \times 10^6$ yr.

Evidence for extinct ^{26}Al has also been found in a number of refractory inclusions found in primitive meteorites. Because of the short half-life of ^{26}Al ($t_{1/2} = 0.72 \times 10^6$ yr), this observation was originally interpreted as requiring a late injection of freshly synthesized material into the solar system, possibly causally related, via a supernova shock wave, to the gravitational collapse of the nebula; however, recent γ-ray measurements have shown that there is a higher abundance of ^{26}Al in the general interstellar medium than originally calculated. Moreover, it is not yet known whether the extinct ^{26}Al was uniformly distributed in the nebula; unlike ^{244}Pu, it is possible that ^{26}Al was locally produced by energetic particle reactions. The amount of ^{26}Al seen in some inclusions, if uniformly distributed, would be sufficient to provide the needed heat source for the melting of small bodies. It should also be noted that the existence of live ^{26}Al in the solar system is not universally accepted; alternative models based on "chemical memory" effects in interstellar grains have also been proposed.

MORE RECENT EVENTS: METEORITES FROM THE MOON AND (MAYBE) MARS

A unique group of meteorites of igneous origin, the SNCs (after the type meteorites Shergotty, Nahkla, and Chassigny), give evidence of recent formation. For example, Nakhla has ^{40}K-^{40}Ar and internal Rb-Sr isochron ages of $\sim 1.3 \times 10^9$ yr. Mars represents a reasonable source of such young rocks and, in certain temperature release ranges, the isotopic compositions of both noble gases and nitrogen in the SNC meteorites are similar to those measured for the atmosphere of Mars by *Viking*.

Although it is difficult to prove that SNC meteorites come from Mars, the discovery of several samples of unquestioned lunar origin among meteorites collected in Antarctica shows that samples sufficiently intact to permit detailed isotopic and mineralogical studies can be blasted off planetary objects and transported to the Earth.

COMPOSITION AND HISTORY OF ENERGETIC PARTICLES FROM THE SUN AND THE GALAXY

Nuclear reactions induced in meteorites by galactic cosmic rays and their cascade secondaries produce isotopes that are used to determine the times since meteorites were exposed as small ($\sim$ 1-m) objects in space. Exposure ages for stony meteorites range from $< 1–80 \times 10^6$ yr, and those for irons from 100 to $\leq 1000 \times 10^6$ yr (presumably due to their greater resistance to destruction by collisions).

Cosmic rays with atomic number $Z \geq 20$ produce tracks as they pass through minerals, and measurement of the distribution of track lengths gave the first determination of the abundances of cosmic rays with $Z > 32$. Because heavy cosmic rays are absorbed more rapidly than cosmic ray protons or their secondaries, measurements of both tracks and isotopes can be used to determine the preatmospheric sizes of meteorites.

Solar wind ions and solar flare particles produce effects only in surface samples. Although the record of recent exposure has been lost by ablation of the meteorite surface during atmospheric entry, a number of meteoritic breccias preserve a record of solar radiations in grains removed from their interiors. These grains are rich in solar wind gases, contain large densities of solar flare tracks, and have concentrations of spallation isotopes in excess of those found in the bulk meteorite. All these effects must have been produced *before* the grains were incorporated into the meteorite, probably close to the beginning of the solar system.

Measurements of cosmic-ray–produced isotopes with different half-lives are used to determine secular variations in cosmic-ray intensities. For example, comparison of the activity of cosmic-ray–induced ^{40}K with shorter-lived isotopes in iron meteorites indicates that the average flux of cosmic rays over the last 1×10^9 yr may have been 50% lower than now. However, this is an upper limit for the variation because the samples may have been more shielded and the production rates correspondingly lower early in their exposure. Although the record is still being deciphered, neither isotopic nor fossil track studies have shown unambiguous evidence for significant time variations in the average fluxes or energy spectra of solar or galactic radiation.

STABLE ISOTOPE ANOMALIES: INTERSTELLAR GRAINS IN METEORITES

Numerous stable isotope anomalies have been discovered in recent years. Some, such as large enrichments in $^2H/H$ found in meteorites and interplanetary dust, were probably produced by ion–molecule reactions similar to those believed to be responsible for $^2H/H$ enrichments observed in interstellar dust clouds. Anomalies in other elements are certainly nucleosynthetic in origin; still others appear to result from a combination of nuclear and physical effects.

A seminal discovery was the observation that a plot of $^{17}O/^{16}O$ versus $^{18}O/^{16}O$ in mineral separates from refractory inclusions in meteorites gives a line with a slope of 1. Physical fractionation usually gives a line with a slope of $\frac{1}{2}$, and the result was interpreted as having been caused by the addition of a nucleosynthetic component rich in ^{16}O. The observation of a 5% isotopic effect in a major element quickly stimulated additional work on inclusions. Anomalies in other elements were soon found, with the magnitudes of the effects becoming larger as smaller parts of the inclusions were studied.

Laboratory experiments have since shown that chemical reactions can yield oxygen isotopic fractionations of the type found in the refractory inclusions; there is thus no longer the need to invoke a nucleosynthetic ^{16}O component. However, some of the anomalies in other elements are nucleosynthetic. For example, the neutron-rich isotopes ^{48}Ca and ^{50}Ti show correlated variations in different grains, and it is difficult to envisage a chemical origin for this effect. Surprisingly, the largest ^{50}Ti and ^{48}Ca effects are found in grains with normal Mg whereas those grains with the largest ^{26}Mg anomalies (indicative of the prior presence of ^{26}Al) generally contain normal Ti and Ca. The available evidence suggests that the inclusions were formed in the solar nebula, and it is not yet understood how the nucleosynthetic effects were incorporated.

In contrast, other studies of primitive meteorites have led to the identification of isolated phases with isotopic properties so unusual that they appear *not* to have been formed in the solar nebula. Noble gas studies have played a key role in locating these phases.

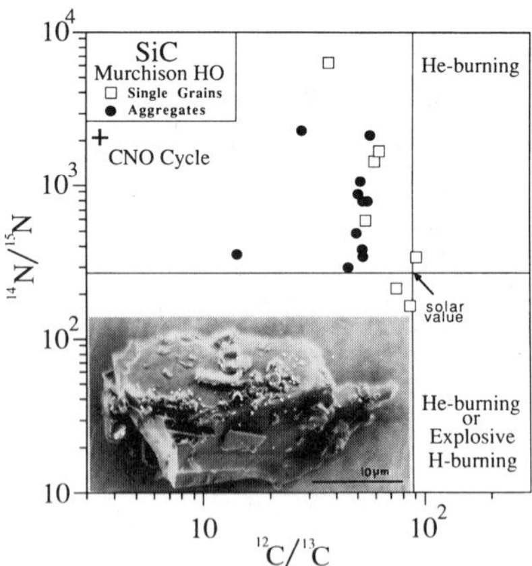

Figure 2. C and N isotope ratios of meteoritic SiC. Most points lie in a region where nuclear H-burning by the CNO cycle is dominant. The insert photo is a scanning electron microscope photograph of one of the largest SiC grains found. Points identified as aggregates consist of ensembles of submicron grains. [*Figure adapted from T. Ming, E. Anders, P. Hoppe, and E. Zinner*, Nature ***139*** *351 (1989)*.]

Monitoring the isotopic composition of gases that are released at different temperatures in a stepwise heating (or combustion) experiment provides a method of finding isotopically distinct components that are associated with specific phases. Three of the most interesting noble gas components are Ne-E, S-Xe, and Xe-HL. The first is almost pure ^{22}Ne and occurs in at least two different carriers, Ne-E(L) and Ne-E(H) that release Ne-E at low and high temperatures, respectively. S-Xe is released at high temperatures and the pattern of Xe isotopes matches that previously calculated for the specific nucleosynthesis process of slow neutron addition (s process). Xe-HL is enriched in both heavy and light isotopes when normalized to ^{130}Xe.

Once a component has been identified, it can be followed as a bulk sample is separated into constituent parts, giving mineral separates that contain ever-larger proportions of the gas-rich phases. It is remarkable that acid residues of primitive meteorites, containing only 1×10^{-4} of the starting material, contain almost all of the unique noble gases found in the starting bulk samples.

Residues rich in Xe-HL consist primarily of clusters of small diamond grains, each containing only several hundred atoms. Surprisingly, the isotopic composition of the carbon is normal. In contrast, residues rich in Ne-E(H) and S-Xe consist largely of SiC grains that are themselves isotopically anomalous (see Fig. 2). The $^{12}C/^{13}C$ ratios vary from 3–180, compared to the normal value of 89. Although smaller in magnitude, anomalies are also found in the Si isotopes; anomalous N is present as well. The magnitude of the effects, as well as the association of anomalies in different elements, gives strong evidence that the grains are preserved interstellar materials.

CONCLUDING REMARKS

Experimental progress in this field has resulted from a continuing interplay between the availability of different types of extraterrestrial samples and the development of improved methods of experimental analysis. Early work depended on the development of high precision mass spectrometry, whereas more recent advances have resulted from the ability to make isotopic measurements on a small spatial scale. This trend should continue for some time to come until the fundamental limit imposed by the number of atoms in a sample is reached.

Although work in the near term will rely on the continuing study of available materials—meteorites (particularly, the growing

Antarctic collections), micrometeorites found in ocean and polar sediments, and interplanetary dust collected in the stratosphere—in the long term, it will be important to undertake sample return missions to primitive bodies such as comets and asteroids.

Additional Reading

Clayton, D. D. (1982). Cosmic chemical memory: A new astronomy. *Q. J. Roy. Astron. Soc.* **23** 174. The paper outlines a different interpretation of some isotopic anomalies, particularly ^{26}Al, and discusses other possible relationships with interstellar dust.

Fleischer, R. L., Price, P. B., and Walker, R. M. (1975). *Nuclear Tracks in Solids*. University of California Press, Berkeley. A complete description of tracks is given, including applications to extraterrestrial materials.

Kerridge, J. F. and Matthews, M. S. (1988). *Meteorites and the Early Solar System*. University of Arizona Press, Tucson. This book consists of a series of review articles that treat the topics discussed in this brief entry in the detail that they deserve.

Lewis, R. S., Tang, M., Wacker, J. F., Anders, E., and Steel, E. (1987). Interstellar diamonds in meteorites. *Nature* **326** 160.

Wetherill, G. W. (1975). Radiometric chronology of the early solar system. *Ann. Rev. Nucl. Sci.* **25** 283. This review includes a further discussion of the physical implications of the chronological results.

Zinner, E., Tang, M., and Anders, E. (1987). Large isotopic anomalies of Si, C, N and noble gases in interstellar silicon carbide from the Murray meteorite. *Nature* **330** 730.

See also **Interplanetary Dust, Collection and Analysis; Meteorites, Classification; Meteorites, Composition, Mineralogy, and Petrology; Meteorites, Origin and Evolution.**

Meteorites, Origin and Evolution

Dieter Heymann

ORIGIN

The topic of the origin of meteorites has three major aspects: the origin of the atoms in them, the origin of their mineral grains, chondrules, and inclusions, and the origin of the meteorites themselves. (Table 1 gives a simplified classification of meteorites.) The overwhelming majority of the atoms in the meteorites originate from stellar nucleosynthesis; comparatively few atoms are products of in situ radioactive decay of elements such as ^{238}U. Still others were formed by high-energy nuclear reactions when the meteorites were exposed to cosmic rays while they were relatively small stray bodies in the solar system.

With regards to the origin of the major structural elements of meteorites, that is, their mineral grains, chondrules, and inclusions, three distinct environments have been widely considered: the circumstellar environment; the solar nebula (the parent cloud of the Sun and planets) in various stages of development; and the parent bodies proper of the meteorites. There has long been doubt that presolar, circumstellar grains occur in meteorites. However, the discovery of mineral grains with clear-cut isotopic anomalies, such as diamond and silicon carbide, have removed such doubts. The most recent summary includes alpha-carbon, poorly crystalline graphite, SiC, diamond, theta-carbon (amorphous), and aleph-, kappa-, and lambda-carbon. It is widely accepted that these grains formed in the environments of carbon-rich stars. The diamonds are enigmatic because they contain an isotopically anomalous xenon that seems to have formed in supernova explosions. The formation of solid grains in the expanding matter of a supernova, called SUNOCONS, has been postulated, but has never been directly observed.

When the hypothesis was accepted that solid grains had condensed directly from a gas phase in the solar nebula and had eventually been incorporated into meteorites, the "traditional" mineralogical–petrological–geochemical theory for the origin of most grains in chondritic meteorites came into its own. It was postulated that a significant portion of the inner solar nebula was heated to at least 2000 K and was a perfectly homogenized gas from which grains condensed upon cooling. The grains reacted chemically with the gas phase and participated in the formation of planetesimals, planets, and satellites, including the parent bodies of meteorites. The "early" condensates must have been rich in "refractories," that is, calcium-aluminum-rich phases, metals, and mafic silicate minerals, but relatively poor in "volatiles" such as sulfides, clay minerals, and organics.

The condensation theory can account for the origin of certain refractory inclusions in stony meteorites, but not for the origins of chondrules, (the nearly spherical objects in chondrites, generally less than 1 mm) nor for the so-called xenolithic inclusions such as "pieces" of, say, a carbonaceous chondrite embedded in, say, an ordinary chondrite. It is assumed that these xenolithic inclusions were formed by "impact gardening" on surfaces of parent bodies.

Chondrules are thought to have formed by the rapid congealing of liquid droplets; therefore any theory about their origin is constrained by their compositions, by the nature of the energy source for the melting of the precursor matter, and by the nature of the cooling process. Numerous distinct theories for chondrule origin

Table 1. A Simplified Meteorite Classification

		Percentage (by weight)									
	Class	*SiO_2*	*MgO*	*FeO*	*Al_2O_3*	*CaO*	*C*	*H_2O*	*Metal*	*Fe*	*Ni*
Stony	Ordinary chondrites*	38–40	23–25	9–14	3	2	Small		8–19	—	—
	Carbonaceous chondrites	23–34	16–23	10–24	2–3	2	0.2–5	1–20	0–4	—	—
	Calcium-rich achondrites	44–49	8–12	8–19	2–13	8–24			—	—	—
	Calcium-poor achrondrites	39–54	26–36	1–16	1	1–2			1–8	—	—
Iron	Octahedrites	—	—	—	—	—			100	89	6–12
	Hexahedrites	—	—	—	—	—			100	93	5–6
	Ataxites	—	—	—	—	—			100	80–92	7–19
Stony–iron	Pallasites	17	—	7	—	—			54	—	—
	Mesosiderites	20	—	6	4	2			51	—	—

*Chondrites are meteorites that contain chondrules (spherical silicate objects, usually smaller than 1 mm).

have been advanced, some of which place their formation in the gas–dust environment of the solar nebula, while others place it on parent bodies of chondrites. The following are among the current models: direct condensation in a cooling, chemically unfractionated nebula; condensation in a fractionated nebula; condensation in the atmosphere of some protoplanet; melting of solid grains by transient events in the solar nebula (even lightening has been proposed); melting of solids by impacts on the surface of a large parent body; impacts in already molten matter on the surface of a large parent body; melting by impacts between small bodies in the solar nebula; and melting of interstellar dust during the accretion of the solar nebula, for example, by grain–grain collisions. This last hypothesis seems to be gaining broad acceptance today.

On Earth, new minerals, or else new crystals of the same mineral, form by igneous processes (melting and congealing), metamorphic processes, and the evaporation of aqueous solutions (evaporites). Evaporite-like minerals are rare in meteorites, but they do occur; for example, the epsomite (magnesium sulfate) in carbonaceous chondrites, which almost certainly formed on the parent body of these meteorites. There is also general agreement that the mineral grains of the achondritic and the iron meteorites formed by crystallization of melts and by exsolution from the solid metallic state, respectively, on parent bodies proper.

EVOLUTION AND PARENT BODIES

Evolution of a meteorite, in the broadest sense, means its complete history from nucleosynthesis until its recovery on the Earth. Attempts have been made to reconstruct the history of matter during its sojourn in interstellar space, especially in the context of the now-extinct radioactivities ^{26}Al and ^{129}I, but meteorites do not seem to have preserved many records from that sojourn. Moreover, the radiometric formation ages of whole meteorites and of component minerals are not greater than the accepted age of the solar system (4.53 Gyr).

It is perhaps useful to present the evolution of meteorites in "reverse time," that is, from well-understood to quite poorly understood history. By the time a meteorite is found it may have lain on the Earth from a few hours to apparently more than one million years. During this time it is exposed to weathering, as are all terrestrial rocks, hence acquiring weathering products, mainly rust. Prior to its terrestrial era, a meteorite is in a Sun-bound, hence keplerian, orbit with perihelion distance inside the Earth's orbit and aphelion distance (based on known last orbits of meteorites) between Mars and Jupiter. One of the best-known orbits is that of the Pribram meteorite, which fell in 1949 in Czechoslovakia: semimajor axis = 2.42 AU; eccentricity = 0.674; perihelion = 0.790 AU; aphelion = 4.05 AU; inclination = 10.4°. Such orbital elements seem to place Pribram's parent body among the main-belt asteroids, the "traditional" location for meteorite parent bodies. Theoretical treatments show that the orbit may have changed substantially several times by the meteoroid passing close to Mars, Earth, Venus, or even Jupiter. Significant mineralogical changes during this era can have occurred only if the meteoroid came close to the Sun, that is, within the orbit of Mercury. The principal alteration during this era was the formation of new radioactive and stable atomic nuclei by cosmic-ray-induced nuclear reactions. Times spent in orbits do not seem to have been longer than 1 Gyr for iron meteorites and only a few stony meteorites have "orbital" ages in excess of 100 Myr.

Meteorites contain ample evidence for changes having occurred on comparatively large parent bodies from which they were eventually removed (presumably broken off by impacts, to be placed in the first of many revolutions). One class of such changes is composed of those changes clearly located at the outer surface of the parent body where impact gardening is the principal process. Impact gardening is thought to have formed the many brecciated stony meteorites. Impacts are also recorded as shock-metamorphism e.g., the formation of glass, veins, blackening, and, by reheating, the resetting of certain radiometric "clocks". Metamorphism proper has long been a controversial topic in meteoritics, but has now become widely accepted as the mechanism for the changes, principally recrystallization, seen in the petrologic sequences of the ordinary chondrites. The upper temperature limit seems to have been near 1100 K.

It is widely accepted that the "basaltic" stony meteorites and the iron meteorites were formed by melting in planet-like bodies. By analogy with the Earth, it has long been assumed that the iron meteorites formed in the cores of one or several planetlets and that the basaltic meteorites are extrusive rocks, also from several parent bodies. Although the analogy of Earth and parent bodies cannot be perfect, it seems to be a fairly good working hypothesis for most of these meteorites.

There remains the issue of whether we can identify or else rule out specific celestial objects as parent bodies for either all meteorites, a given class of meteorites, or a given meteorite proper. We can state with firm certainty that one parent body for all meteorites will not do. For this conclusion there are several arguments; one of the strongest is the occurrence of several groups of meteorites with distinct oxygen isotopic compositions.

The Earth and Venus can be ruled out for obvious reasons. The Moon seems to have been the source of a very few meteorites; the orbital ages of most meteorites are much too long for a lunar origin. Mars has long been in the running as the parent body of stony meteorites. Currently, it is assumed that the so-called SNC meteorites come from Mars on the grounds of their high contents of volatiles, relatively short igneous ages, and severe shock-metamorphism. Still, there remains the question of how any solid object could have been lifted off of Mars intact.

One currently very active field of investigation involves the comparison of albedos and reflectance spectra of meteorites and asteroids. A large number of asteroid classes has been established on the basis of their optical properties, which have been compared with laboratory-based data on meteorites. It is in this context that the relatively large asteroid Vesta has been suggested as parent body for the eucrites. The dynamical problem of bringing eucrites from Vesta to the Earth seems to have been alleviated with the recent discovery of at least three smaller Vesta-like asteroids whose orbits come close to that of the Earth. In the same context it has been suggested that the surfaces of the S asteroids are very metal-rich. It is still debated, however, whether these are the parent bodies of the ordinary chondrites. The low-albedo C-type asteroids have been proposed as parent bodies of carbonaceous chondrites, but their spectra seem to be a much better match for the heavily shocked black (ordinary) chondrites. The optical studies have, however, raised fresh problems such as the observation that the C-types in the Trojan region appear not to contain any hydrated minerals, whereas those in the inner belt, closer to the Sun, do contain such minerals. Comets have long been considered potential parent bodies, especially for the carbonaceous meteorites. For various reasons, they cannot be accepted as parent bodies for any but perhaps a few of the known meteorites.

Additional Reading

Anders, E. (1964). Origin, age and composition of meteorites. *Space Sci. Rev.* **3** 583.

Chapman, C. R. (1976). Asteroids as meteorite parent bodies: The astronomical perspective. *Geochim. Cosmochim. Acta* **40** 701.

Grossman, L. and Larimer, J. W. (1974). Early chemical history of the solar system. *Rev. Geophys. Space Phys.* **12** 71.

Kerridge, J. F. and Matthews, M. S., eds. (1988). *Meteorites and the Early Solar System*. The University of Arizona Press, Tucson.

King, E. A., ed. (1983). *Chondrules and their Origins*. The Lunar and Planetary Institute, Houston, Texas.

Lewis, R. S., Ming, T., Wacker, J. F., Anders, E., and Steel, J. (1987). Interstellar diamonds in meteorites. *Nature* **326** 160.

See also **Asteroids; Asteroids, Earth-Crossing; Meteorites, Classification; Meteorites, Composition, Mineralogy, and Petrology; Meteorites, Isotopic Analyses.**

Meteoroid Streams, Dynamical Evolution

David W. Hughes

The year 1867 saw the first calculation of a meteoroid stream orbit, this being the stream responsible for the November Leonids. It was realized in the same year that this orbit was similar to that of comet P/Tempel–Tuttle (1866 I). (The P stands for "periodic," and I indicates that this comet was the first to pass perihelion in the year 1866.) Since that time comets and meteoroid streams have been closely associated, and the influence of cometary decay and orbital evolution on the dynamics of meteoroid streams has been studied in detail.

The meteoroid streams that are detected as meteor showers in the Earth's atmosphere are produced by the decay of comets with periods less than 200 yr. 10, 20, and 30% of this group of comets have present-day nuclear masses greater than 1.1×10^{16}, 1.2×10^{15}, and 3.6×10^{14} g, respectively. The assumption can be made that the average observed short-period comet is about half way through its inner solar system life, so the diameter of its nucleus has decayed to about half of what it was at the time when a series of close approaches to Jupiter changed the initial long-period orbit into a short-period inner solar system orbit. When the diameter has decreased by 50%, the comet will have lost about 87% of its initial mass and will have passed perihelion around 1000 times (taking about 7000 yr). These estimates are in keeping with the recent findings that the total masses of the meteoroids in the streams responsible for the Quadrantid, Perseid, Orionid/Eta Aquarid, and Geminid meteor showers are 1.3×10^{15}, 3.1×10^{17}, 3.3×10^{16}, and 1.6×10^{16} g, respectively. These four streams are heavyweights among the population that intersects the Earth's orbit. They are well developed and middle aged. Observations of meteor showers using radar, telescopic, television, and visual techniques have been used to measure the flux of meteoroids over a large mass range. The results indicate that each 10^{15} g of meteoroid mass would contain on the order of 10^{21}–10^{22} particles with individual masses in the range 10^{-10}–10^{5} g. It is the dynamical evolution of these 10^{22} particles that we are going to consider. Remember, however, that 70% of the stream mass is in the form of 0.001–10-g meteoroids.

STREAM FORMATION

The vast majority of cometary decay, and thus meteoroid stream production, occurs near the Sun. In the inner solar system, at heliocentric distances r, where the nucleus of the comet has a temperature that is above the sublimation temperature of water snow, the rate of mass loss is proportional to r^{-n}, where n is in the range 2–3. Here, nearly all the absorbed solar radiation is used to sublimate the cometary snow. Loosely bound dust particles on the surface of the nucleus are detached and pushed away by momentum transfer from the outflowing gas, against the influence of the extremely weak gravitational field. This meteoroid dust is given a velocity V_m with respect to the nucleus, which is proportional to, among other things, $(r_m\rho_m)^{-0.5}$, where r_m is the radius of the meteoroid and ρ_m is its density. The heliocentric velocity of the meteoroid is obtained by vectorially adding its velocity to that of the cometary nucleus. The resulting orbit is only slightly different from that of the parent comet. For example, the difference in

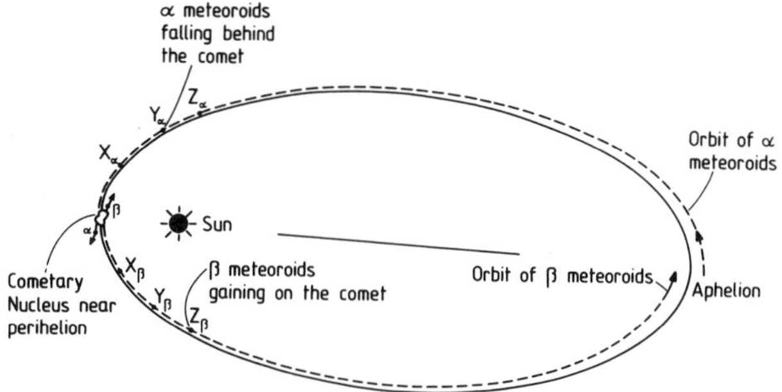

Figure 1. The formation of a meteoroid stream by cometary decay. The two test meteoroids, α and β are emitted at the comet's perihelion such that α has a velocity vector in the direction of the comet's motion ($\phi = 0°$) and β moves in the opposite direction ($\phi = 180°$). When the comet returns to perihelion, one orbit after the emission of the two meteoroids, they now have positions X_α and X_β. Two orbits later they are at Y_α and Y_β. Slowly the stream develops.

period ΔP between the two orbits is given by

$$\Delta P_m = 3.4\times10^{-3} a_c V_c V_m P_c \cos\phi \text{ yr},$$

where a_c, V_c, and P_c are the comet's semimajor axis (in astronomical units) heliocentric velocity at the time of meteoroid emission (in kilometers per second), and period (in years), respectively. Angle ϕ is the emission direction of the meteoroid measured with respect to the comet's direction of motion. The result is illustrated in Fig. 1. Two meteoroids emitted at the comet's perihelion, in opposite directions (i.e., $\phi = 0$ and 180°), will pass each other at aphelion after a time T given by

$$T = \frac{P_c^2}{2\,\Delta P_m} - \frac{P_c}{2}.$$

This is the minimum time required to form a complete loop of meteoroids around the orbit and (as the first term in the preceding expression is dominant) is roughly proportional to $m^{-1/6}$, where m is the mass of the meteoroid. The orbital end-product is illustrated in Fig. 2. In this exercise, the Quadrantids have been assumed to be a typical meteoroid stream. If we consider the meteoroids responsible for the formation of visual meteors (these meteoroids having a mass of around 0.1 g), the time taken to form a loop will be on the order of 1500 yr. This time will decrease to 500 yr for the 2×10^{-4}-g meteoroids that typically produce radar meteors. The general cross-sectional profile of the stream is also mass dependent, the width being proportional to the velocity with

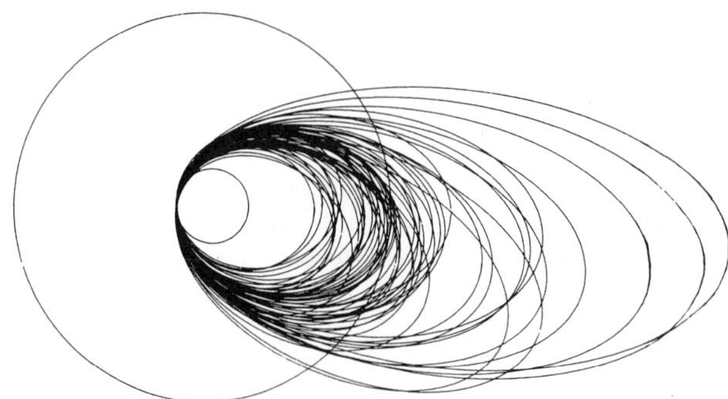

Figure 2. The orbits of 57 radar Quadrantid meteoroids, obtained using the Radio Meteor Project's six station radar network at Havana, Illinois in January 1962–1964. Differences in inclination have been ignored. The small and large circles have radii of 1 and 5.2 AU, respectively.

which the meteoroids are emitted from the cometary nucleus. Initially the "visual" stream will have a cross section that is about one-third that of the "radar" stream shown in Fig. 2.

DYNAMIC EVOLUTION

The dynamic evolution of meteoroid streams can be divided into four stages.

1. Stream formation starts after the first perihelion passage of a new short-period comet, that is, one that has just been transferred into a short-period orbit by an encounter with Jupiter. The meteoroids in this new stream will only extend around a fraction of the orbit of the parent comet. The symmetry of the stream with respect to the position of the comet in the orbit will depend on the distribution of mass emission with respect to the angle ϕ (see Fig. 1). A comet with a direct spin mode will emit most of its dust in the "afternoon" sector where ϕ is between say 40 and 80°. The majority of the dust will thus trail behind the comet and the stream will take more than twice as long to form a complete loop in comparison to the symmetrical case shown in Fig. 1. The narrower the ϕ interval, the lower the velocity dispersion and the longer it takes for loop formation. Note that just such a strong directionality of dust emission was exhibited by P/Halley.

The resultant new meteor shower, as seen from Earth, will be of short duration (a few hours). Because of the incomplete loop of meteoroids there will also be a distinct periodicity in the observed annual rates, corresponding to the orbital period of the parent comet. The Leonids are a good example; meteor storms occurred in 1799, 1833, 1866, and so on.

2. In the second stage the stream meteoroids have fallen behind and/or gained on the parent comet to such an extent that their orbits overlap at aphelion. At this stage the stream produces a shower every year and, as the overlap increases, the annual shower flux becomes more and more constant. The stream is, however, still narrow in cross section and the Earth observer sees a shower of short duration. The radiants (the points on the sky from which the meteors appear to emanate) are concentrated in a small area. The Quadrantids, Lyrids, and Ursids are typical examples.

3. The third stage sees the stream becoming broader and broader due to interparticle collisions and the cumulative action of the Poynting–Robertson effect and planetary gravitational perturbation. The radiant distribution on the celestial sphere becomes broader and the spatial density of the meteoroids in the solar system decreases. The Taurids, Orionids, Eta, and Delta Aquarids and Geminids are examples.

4. Finally the meteoroid spatial density becomes so low that the shower is too weak and diffuse to be detected. It becomes indistinguishable from the sporadic meteoroid background, which is itself simply the combined debris from decayed meteoroid streams.

POYNTING–ROBERTSON EFFECT

The Poynting–Robertson effect is the name given to the orbital changes that occur simply because a meteoroid can lose energy when it absorbs and then reemits radiation. Radiation is absorbed from the solar direction but, due to the meteoroid's transverse motion, it is reflected and diffracted and emitted over a range of directions and with asymmetrical wavelength shifts. Radiation leaving the particle in the particle's direction of motion is blue-shifted and radiation leaving in the opposite direction is red-shifted. The net result is the imposition of a drag force, which leads to a decrease in the eccentricity of the meteoroid's orbit and subsequently to it spiraling into the Sun on a near circular orbit of ever decreasing radius. The time t taken for this to happen is given by

$$t = 7 \times 10^6 r_m \rho_m \nu q^2 E(e) \text{ yr},$$

where q (AU) and e are the perihelion distance and eccentricity of the initial orbit, ν is the fraction of radiation that the meteoroid absorbs, and $E(e)$ is a function that has values of 1.0, 2.73, 5.28, and 10.4 for eccentricities of 0, 0.5, 0.75, and 0.9, respectively. A perfectly absorbing meteoroid of radius 0.4 cm and density 0.3 g cm^{-3} (i.e. one responsible for a typical visual meteor) will take 800,000 yr to spiral into the solar photosphere from a circular orbit of radius 1 AU. A "radar" meteoroid of radius 0.04 cm and density 1 g cm^{-3} would take 300,000 yr. So small meteoroids are preferentially leached from the stream. If we assume that a meteoroid has been lost from the stream when it has had its perihelion changed by 0.05 AU by the Poynting–Robertson (PR) effect, the time taken to do this is 10% of the times given above. The PR effect makes the smaller meteoroids migrate to the sunward side of the stream and this characteristic has often been observed.

METEOROID COLLISIONS

Collisions between stream meteoroids and the sporadic interplanetary meteoroids are more important than the less frequent and lower-relative-velocity collisions that take place between meteoroids in the same stream. The mass distribution of the interplanetary dust cloud is such that the number of particles increases drastically as the mass under consideration goes down. Two types of sporadic-stream collisions are important. In the first case, the stream meteoroids are catastrophically fragmented by smaller impacting bodies. The fragments then leave the collision site with very low relative velocities. Little mass is lost from the stream—it merely widens slightly—and the number of meteoroids in the stream goes up and the mean meteoroid size goes down. In the second case, the colliding particle is very much smaller than the meteoroid and this leads to the erosive cratering of its surface. For a meteoroid of radius 0.4 cm both the catastrophic and the erosive collisions lead to a lifetime of about 100,000 yr. For the 0.04-cm meteoroid the fragmentation lifetime is around 90,000 yr, whereas the erosion lifetime is 300,000 yr.

GRAVITATIONAL PERTURBATION

Nearly half the meteoroids responsible for bright photographic meteors have orbits that coincide with the identifiable streams. Also, those meteoroids with aphelia within the orbit of Jupiter show no semimajor axis alignment with that of Jupiter. This indicates that the streams have lifetimes less than about 10^4–10^5 yr.

Meteoroid streams are severely perturbed by the gravitational fields of the major planets. Due to the similarity between the stream orbits and those of the decaying short-period comets, the major culprit is Jupiter. The orbits of many of the meteoroids in the Quadrantid stream straddle that of Jupiter, so the observed perturbation of these meteoroids provides a telling example of the process. This process may be modeled by replacing the myriad of stream meteoroids by a small set of test particles placed around the mean orbit. The equations of motion of these particles can then be solved using, for example, a Runge–Kutta technique with self-adjusting step lengths. A small computer can easily integrate the orbits of a hundred test particles for a few millenia. As is to be expected, the only parameter to remain reasonably constant is the semimajor axis. At the present time the Quadrantids have an inclination i of 72.5° and a perihelion distance q of 0.977 AU. A mere 1500 yr ago the majority of the meteoroids in this stream had inclinations of 12° and perihelia that were on average only 0.09 AU from the Sun.

Over shorter time intervals the variability is quantified by $d\Omega/dt = 0.017°\ \text{yr}^{-1}$, $di/dt = 0.0049°\ \text{yr}^{-1}$, and $dq/dt = 0.0012$ AU yr^{-1}. All these theoretical values have been confirmed by observation. What is even more important to the Earth-based observer is the fact that the closest distance of approach between the median stream orbit and the orbit of Earth varies as a function of time. Our ability to see the Quadrantid stream is short lived. Gravitational perturbation slowly changes and moves the orbit. The stream first started to intersect the Earth's orbit some 110±50 yr ago. In 210±100 yr time the stream will have been pulled past the Earth's orbit and will be seen no more. These modeling predictions agree well with the fact that the Quadrantids were first recorded in 1835. The Geminids suffer in a similar way and were first seen in 1862. Some streams are well away from the major planets and have been intersected by Earth for many hundreds of years. The Perseids, for example, were first recorded in 36 AD and the Leonids in 902 AD.

The orbital perturbation of the parent comet can considerably change the position and density of the associated meteoroid stream. The orbit of P/Halley seems to rock about its major axis through an angle of about 25° and this produces a stream with a fat, ribbon-like cross section. The steady decay of a comet might be interspersed by periods when large fragments break away from the nucleus. This would produce a large increase in the stream mass at a single point in time. The parent comet could also suffer from a close encounter with a major planet and thus be completely removed from its original orbit and its meteoroid stream. It could then start to produce another meteoroid stream in another place.

Additional Reading

Carusi, A. and Valsecchi, G. B. (1985). *Dynamics of Comets: Their Origins and Evolution. IAU Colloquium 83*. D. Reidel, Dordrecht.

Chebotarev, G. A., Kazimirchak-Polonskaya, E. I., and Marsden, B. G. (1972). *The Motion, Evolution of Orbits and Origin of Comets. IAU Symposium 45*. D. Reidel, Dordrecht.

Dohnanyi, J. S. (1972). Interplanetary objects in review: Statistics of their masses and dynamics. *Icarus* **17** 1.

Hughes, D. W. (1986). The relationships between comets and meteoroid streams. In *Asteroids, Comets, Meteors II*, C.-I. Lagerkvist, B. A. Lindblad, H. Lundstedt, and H. Rickmann, eds. Uppsala Universitet Reprocentralen HSC, Uppsala, Sweden.

McDonnell, J. A. M., ed. (1978). *Cosmic Dust*. Wiley, New York. See especially Chapter 3 (Meteors) by D. W. Hughes and Chapter 8 (Particle Dynamics) by J. S. Dohnanyi.

Whipple, F. L. (1967). On maintaining the meteoric complex. In *The Zodiacal Light and the Interplanetary Medium*, J. L. Weinberg, ed. NASA SP-150, p. 409.

Williams, I. P., Murray, C. D., and Hughes, D. W. (1979). The long-term orbital evolution of the Quadrantid meteor stream. *Monthly Notices Roy. Astron. Soc.* **189** 483; see also **195** 625 and **199** 313.

Yeomans, D. K. (1991). *Comets—A Chronological History of Observation, Science, Myth, and Folklore*. Wiley, New York. See the section on comets and meteors, p.188.

See also **Comets, Dust Tails; Interplanetary Dust, Dynamics; Meteors, Shower and Sporadic.**

Meteoroids, Space Investigations

Herbert A. Zook

The word "meteoroid" derives from meteor, the "shooting star" phenomenon that one may observe at night when a meteoroid passes rapidly through the upper atmosphere. It is loosely defined to be any object in interplanetary space that is too small to be labeled an asteroid or a comet; at the small size end it is often called a micrometeoroid or a cosmic dust grain, a category that includes any interstellar grain that should happen to penetrate into our solar system.

HISTORY

As soon as it became possible to do so, investigators attempted to measure the flux of meteoroids directly in space. The first attempts were made on suborbital rocket flights with only short exposures above the sensible atmosphere. As early as 1949, an instrumented piezoelectric sensor was bonded to the skin on the nose of a V-2 rocket. Such pressure-sensitive, or "acoustic," sensors had been shown to be quite sensitive to the oscillatory waves induced when something strikes the plate to which they are bonded; it was expected that they would sense all but the smallest meteoroids that impact the plate. Laboratory calibrations indicated that they should easily record impacts by meteoroids as small as 10^{-10} g.

Both in the United States and in the Soviet Union, meteoroid investigators with experiments on the early orbital and suborbital flights utilized, almost exclusively, acoustic sensors attached to impact plates. Some of the early meteoroid experiments on U.S. orbital spacecraft included those flown on (with year of launch) *Explorer 1* (1958), *Pioneer 1* (1958), *Vanguard 3* (1959), *Midas 2* (1960), and *Explorer 8* (1960). Similarly, for the Soviet Union, there were acoustic meteoroid impact experiments on *Sputnik 3* (1958), *Luna 1* (1959), *Luna 2* (1959), *Luna 3* (1959), and others. These early space flight investigations resulted in impact rates (presumably due to meteoroids in the 10^{-10}–10^{-7}-g range) that were so high near the Earth that, by 1961, at least three papers had appeared that suggested that a concentrated cloud of meteoroids must be orbiting the Earth. The decreasing impact rate with increasing orbital altitude added weight to such an idea.

By 1965, however, meteoroid experiments on the near-Earth orbiting *Explorer 16* satellite (launched Dec. 16, 1962) were already casting doubt on the validity of the high impact rates obtained by the acoustic detectors. Five types of meteoroid experiments were flown on this satellite:

1. Acoustic detectors;
2. Cd-S cells covered by aluminized mylar that were designed to sense the increase in sunlight penetrating the mylar as the aluminum coating was abraded away due to meteoroid impacts;
3. Conducting wires arrayed in fine grids that sensed when a meteoroid impact broke a wire and opened a circuit;
4. A similar card of fine wire grids covered by a thin sheet of metal that must be penetrated before a wire was broken and a circuit was broken;
5. Hollow cells of different wall thicknesses pressurized with helium gas where a large decrease in gas pressure indicated wall perforation.

The largest exposed area, about 1.6 m^2, was devoted to the pressurized cell experiment, with most of that devoted to cells with 25-μm-thick Be-Cu alloy walls. The *Explorer 16* wire grid and the pressurized cell experiments both gave meteoroid impact rates, over the satellite's 7 month lifetime, that were 2–4 orders of magnitude less than those deduced from the previously flown acoustic sensors. By 1967 it had been further shown that the acoustic sensors were very probably responding much more to acoustic and thermal noise generated as the spacecraft to which they were bonded orbited into and out of sunlight, than they were to meteoroid impacts. Also, additional meteoroid penetration data had, by this time, been collected by other spacecraft (see Table 1) that confirmed the *Explorer 16* penetration data. Surfaces that had been exposed to space and returned to Earth gave dramatic visual evidence of the very low flux of meteoroids. Microscopic examination of 14 windows (each of area 550 cm^2) from eight separate Gemini spacecraft revealed only a single 30-μm-diameter impact pit (centered within a 110-μm-diameter spallation zone) of probable meteoritic origin. There is now little belief in the proposition of a high flux of Earth-orbiting meteoroids—although a low flux is still possible.

Table 1. Important Spacecraft Sources of Meteoroid Data

Spacecraft	*Yr Launched*	*Sensing Method*	*Key Results*[1]
Explorer 16	1962	Cell penetration; gas loss	A
Mercury Atlas 9[2]	1963	Impact crater observation	A
Explorer 23	1964	Cell penetration; gas loss	A
Ariel 2	1964	Foil penetration; detect sunlight	A
Pegasus 1,2,& 3	1965	Capacitor sheet penetration	A
Gemini 4–12[3]	1965–1966	Impact crater observations	A
Lunar Orbiter 1–5	1966–1967	Cell penetration; gas loss	A, B
Pioneer 8 & 9	1967–1968	Impact generated ion detection	A, C, E, G
Apollo 7–17[4]	1968–1972	Impact crater observations	A
HEOS 2	1972	Impact generated ion detection	A, C, D
Explorer 46[5]	1972	Cell penetration; gas loss	A
Pioneer 10, 11	1972, 1973	Cell penetration; gas loss	A, E, F
Skylab 2–4	1973	Impact crater observations	A, H
Helios 1, 2	1974, 1976	Impact generated ion detection	A, C, E, I
Solar Maxium Mission	1980	Impact feature observations	A, H, I
Space Shuttle STS-3	1982	Impact feature observations	A

[1]A: meteoroid flux versus mass determination; B: Earth-to-Moon flux decrease; C: β meteoroid sensing; D: possible observations of lunar ejecta, intermittent high rate near Earth; E: heliocentric radial distribution; F: enhanced meteoroid flux near Jupiter and Saturn; G: trajectory measurement; H: detection of impacts by Earth-orbital debris; I: meteoroid composition measured.
[2]Spacecraft hull looked at on two other Mercury flights, but no impacts observed.
[3]*Gemini 8* surfaces not examined; windows plus S-10 and S-12 surfaces examined for impacts.
[4]*Apollo 11* surfaces not examined; *Surveyor 3* spacecraft parts brought back on *Apollo 12* and examined for impact pits; lunar ejecta meteoroid (LEAM) experiment left on lunar surface during *Apollo 17*.
[5]This is a double-walled structure.

The United States has flown over 50 spacecraft in geocentric or in heliocentric orbits from which efforts to collect meteoroid impact data have been made. Those 43 spacecraft from which the most reliable meteoroid impact data have been obtained are listed in Table 1, along with the year that the spacecraft were launched, the sensing technique used, and the key results from each spacecraft. Included are experiments on the European spacecraft *Ariel 2*, *HEOS 2* (*Highly Eccentric Orbiting Satellite 2*), and *Helios 1* and 2. No results from the Soviet Union are included because their early meteoroid data were derived from the unreliable acoustic sensors, and little of their data on other types of meteoroid sensing have appeared in the western press. It should be noted, however, that the Soviets appear to have obtained reliable results from acoustic sensors flown on the *Kosmos 135* (1966) and *Kosmos 163* (1967) spacecraft. They lowered the sensitivity of their acoustic sensors and made other careful efforts to avoid thermal noise problems, so it is perhaps not quite fair to leave these data out of the table. It is also possible that the acoustic detector data on the *Mariner 2* (1962) Venus flyby and the *Mariner 4* (1964) spacecraft to Mars are valid but, because of reliability concerns, they are not included in the table.

The four meteoroid sensing techniques accepted as quite reliable and included in the table are:

1. Cells pressurized with gas;
2. Impact craters on surfaces and holes in thin foils observed in the laboratory on surfaces and foils exposed to, and returned from, space;
3. Metal sheet penetration followed by transient shorting of a penetrated capacitor by the ionized plasma;
4. Direct sensing of the ionized plasma generated by meteoroid impact.

The table includes experimental results where only a few, or even no, impacts were registered (e.g., *MA-9*, *Ariel 2*, *Gemini* experiments, etc.) which are, therefore, of low statistical weight, but they are included because they add to the reliability of the combined data set (especially in the earlier years). Because of their singular purpose, the excellent dust impact and compositional data obtained in 1986 by the European *Giotto* and the two Soviet *Vega* spacecraft as they flew by comet Halley are left out of the table. Note must also be made of the copious and rich meteoroid impact pit and crater data that have been observed on the returned samples of lunar rocks. Although it is difficult to independently establish space exposure periods, and therefore to establish absolute meteoroid fluxes from the lunar rock data, these data are nevertheless very useful for determining the shape of the meteoroid abundance versus meteoroid mass curve. Finally, it is noted that the plasma wave experiment on the *Voyager* spacecraft, which carried no dedicated meteoroid experiment, sensed multiple dust impacts as this spacecraft flew through the ring planes of Saturn, Uranus, and Neptune.

HIGHLIGHTS

We have learned the following from the meteoroid experiments so far flown in space:

1. The near-Earth flux of meteoroids versus meteoroid mass is now fairly well established and is presented in log-log form in Fig. 1. Valid spacecraft data have been obtained for meteoroid masses between about 10^{-15} and 10^{-5} g. For meteoroid masses larger than about 10^{-5} g, values are established from radar and photographic meteor data. There are no reliable data for meteoroids smaller than about 10^{-15} g, and the line shown to 10^{-18} g in the figure is merely an extrapolation from larger mass meteoroids. The uncertainty in establishing the mass of the impacting meteoroid could be as much as a factor of 10 at 10^{-15} g and at 1 g, and is perhaps a factor of 4 or 5 for masses between 10^{-12} and 10^{-6} g.

2. The penetration rate for meteoroid sensors on spacecraft in near-lunar orbits is about a factor of 2 less than that for identical sensors in near-Earth orbit. This result is primarily ascribed to the gravitational focusing of meteoroids passing near the Earth with a mean near-Earth velocity of about 17 km s^{-1} relative to the Earth.

3. The *Pioneer 8* and *9* spacecraft showed that most meteoroids smaller than about 10^{-13} g in mass are leaving the solar system on hyperbolic orbits to become interstellar grains. These meteoroids

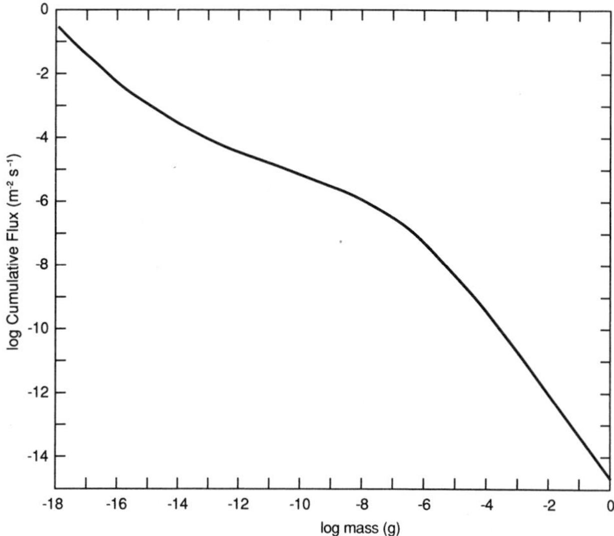

Figure 1. Logarithm of the cumulative flux of meteoroids that are larger than a given mass versus the logarithm of the meteoroid mass for meteoroids near the Earth. [*Adopted from Grün et al.* Icarus ***62*** *244 (1985).*]

are commonly called β meteoroids and appear to produce a loss rate of about 10 ton s^{-1} of meteoritic material from the solar system. These observations have been confirmed by data from the *HEOS 2* and the *Helios 1* and *2* spacecraft.

4. The *Pioneer 10* and *11* spacecraft detected strong increases in meteoroid penetration rates near Jupiter and Saturn. These increases are doubtless due, primarily, to particles in orbit about these planets. Also *Pioneers 8, 9, 10, 11*, and *Helios 1* and *2* have measured the variation in meteoritic impact rate as a function of heliocentric distance. Although the data generally appear to be quite reliable, it is not yet clear how they should be reconciled with the somewhat different radial distributions obtained by zodiacal light photometers on the *Pioneer 10* and *11*, and on the *Helios 1* and *2* spacecraft.

5. The meteoroid size distribution obtained at 1 AU as well as the heliocentric radial data obtained have made it possible to do some rather detailed, if preliminary, analyses of the dynamics of meteoroids in the inner solar system. It has been found that meteoroids more massive than about 10^{-5} g are usually destroyed via collisions with other meteoroids; collision lifetimes at 1 AU range from a minimum of a few thousand years for 10^{-2}-g meteoroids to about 10^5 yr for meteoroids of about 10^{-5} g in mass. Below about 10^{-5} g, it appears that meteoroids are usually destroyed by vaporization, rather than by collisions, as they drift close to the Sun under Poynting–Robertson (PR) and solar wind drag. The PR and solar wind drag lifetime is proportional to particle radius as well as to the original heliocentric distance squared. The drag lifetime is about 10^5 yr (equal to the collision lifetime) for a 10^{-5}-g meteoroid originally in circular orbit at 1 AU.

6. Earth-orbiting spacecraft debris is becoming a significant factor in interpreting spacecraft measurements in near-Earth orbit. Either trajectory information or compositional information on the impacting particle is required to securely establish particle origin.

FUTURE

Important new meteoroid experiments are currently in progress on the *Galileo* and *Ulysses* spacecraft launched in October, 1989 and October, 1990, respectively. The meteoroid detectors on these two spacecraft are identical to each other and similar in design to the already proven *HEOS 2* impact plasma detection design, except that they are about 10 times larger in area, with about 0.1 m^2 of sensing area each. They are designed to sense meteoroids down to 10^{-16} g in mass. *Galileo*, after a 6-yr tour past Venus and Earth (the latter twice for gravity assists) and three passes through much of the asteroid belt, will go into orbit about Jupiter; there it will collect detailed data on meteoroids either in orbit about Jupiter, or that are gravitationally concentrated toward it. *Ulysses* also will go past Jupiter where it will be flung back inward and will orbit nearly over the poles of the Sun at a distance of a little over 2 AU. Good results on meteoroid fluxes versus heliocentric latitude should be obtained for the first time. In addition to their near-jovian and heliocentric latitude studies, these two spacecraft should provide many more details than hitherto obtained on the heliocentric radial distribution of meteoroids as a function of meteoroid size. Early meteoroid data from these two spacecraft appear to be of excellent quality.

On January 12, 1990, the *Long Duration Exposure Facility* (or LDEF) was recovered by the Space Shuttle *Columbia* after nearly 6 yr in space. With about 130 m^2 of surface area exposed to space, the area–time space exposure product of this spacecraft was nearly 2 orders of magnitude greater than the total of all previously space-exposed surfaces that have been returned to Earth and examined for meteoroid impact. Because LDEF was stabilized in both local vertical via gravity gradient and in roll about the vertical axis, important meteoroid impact directionality studies can be done. Many meteoroid investigations of LDEF surfaces are currently underway and should result in a significantly more accurate meteoroid flux–mass curve, as well as new information on Earth-orbiting debris. Meteoroid impact data have also been obtained from the Soviet *MIR* space station and from the United States space shuttle flights; these data could not be included in Table 1 because they were so recently obtained.

Among other meteoroid experiments being planned, the Cosmic Dust Collection Facility (CDCF), proposed for deployment on the Freedom Space Station in the late 1990s, should particularly be mentioned. It is planned, on this facility (with a proposed 10 m^2 of sensing area), to both measure meteoroid trajectories and to capture and return meteoroids for detailed analyses in terrestrial laboratories. It is expected that the orbits calculated for the captured meteoroids will make it possible, in many cases, to relate them to the individual comet or asteroid from which they derive. The compositional, isotopic, mineralogical, and other analyses of each of these samples will, then, provide detailed information on the corresponding individual comet or asteroid; and this will be done without even visiting the larger body.

Additional Reading

Fechtig, H., Grün, E., and Kissel, J. (1978). Laboratory simulation. In *Cosmic Dust*, J. A. M. McDonell, ed. Wiley, New York, p. 607.

Grün, E., Zook, H. A., Fechtig, H., and Giese, R. H. (1985). Collisional balance of the meteoritic complex. *Icarus* **62** 244.

Hörz, F., Morrison, D. A., Gault, D. E., Oberbeck, V. R., Quaide, W. L., Vedder, J. F., Brownlee, D. E., and Hartung, J. B. (1977). The micrometeoroid complex and evolution of the lunar regolith. In *The Soviet–American Conference on the Cosmochemistry of the Moon and Planets*. NASA SP-370, Part 2, U.S. government Printing Office, Washington, D.C., p. 605.

Humes, D. H. (1980). Results of *Pioneer 10* and *11* meteoroid experiments: Interplanetary and near-Saturn. *J. Geophys. Res.* **85** 5841.

Laurance, M. R. and Brownlee, D. E. (1986). The flux of meteoroids and orbital space debris striking satellites in low Earth orbit. *Nature* **323** 136.

McDonnell, J. A. M. (1978). Microparticle studies by space instrumentation. In *Cosmic Dust*, J. A. M. McDonnell, ed. Wiley, New York, p. 337.

McDonnell, J. A. M., Stevenson, T. J., Tulloch, S. M., and Green, S. F. (1989). In-situ studies of cosmic dust and primordial bodies: The plan ahead. *J. Br. Interplanet. Soc.* **42** 317.

See also **Interplanetary and Heliospheric Space Missions; Interplanetary Dust, Collection and Analysis; Meteoroid Streams, Dynamical Evolution; Meteors, Shower and Sporadic.**

Meteors, Shower and Sporadic

Robert L. Hawkes

The term *meteor* applies to the streak of light (and related phenomena such as ionization) produced when an interplanetary dust particle (a *meteoroid*) enters the Earth's upper atmosphere (typical heights of ablation are 80–110 km above the Earth's surface). The meteoroid complex can be divided conveniently into two parts: a *stream* component made up of particles in highly correlated orbits and a more or less random *sporadic* component. It is believed that most meteor streams are formed by the decay of a cometary nucleus, and hence stream meteoroids will be spread around the original cometary orbit (subject to differential ejection velocities from the nucleus and to subsequent planetary and radiative perturbations). When the Earth's orbit intersects the meteor stream, an enhancement of the meteor rate (a *meteor shower*) is observed. A particularly intense meteor shower is termed a *meteor storm*. The sporadic meteor component is believed to be derived from the shower component by gradual loss of orbital coherence due to collisions and radiative effects (subsequently enhanced by differential gravitational perturbations). Hence the division between shower and sporadic meteors is not precise, and there continues to be debate regarding the validity of a large number of minor meteor showers that may in fact be simply random statistical enhancements of the sporadic background.

A perspective effect causes the approximately parallel paths of shower meteors to appear to converge to a distant point when projected backward (similar to the apparent convergence of straight railway tracks). The constellation in which this point of convergence, or *radiant*, seems to lie is used in naming the meteor shower; for example, the Perseid meteor shower has a radiant within Perseus. When two or more meteor showers have radiants in the same constellation, a nearby bright star is used in the designation (e.g., the Eta Aquarids). There are occasional exceptions to the rule for naming meteor showers: Several showers are commonly named after the parent comet (e.g., the Giacobinids), whereas sometimes a month is used to differentiate between several showers with radiants in the same constellation (e.g., the June Boötids).

A typical meteor shower will be active over a number of days, and the apparent radiant will move regularly during this period. This daily motion of the radiant can be found in the references cited as additional reading.

The activity of a meteor shower is usually expressed by giving the zenithal hourly rate (ZHR). This is the number of meteors a single experienced observer would see (from a shower with a radiant directly overhead) in a dark sky location under ideal conditions where 6.5-magnitude stars are visible. The number of meteors that can be observed is a strong function of limiting sensitivity. For example, if one were observing a shower with a ZHR of 100 from a location with a limiting magnitude of 5.0, the number of meteors per hour observed would be reduced to about 25, whereas a limiting magnitude of 4.0 would reduce the number to about 10. Sporadic and shower meteors have different size distributions so that although the number of bright meteors observed during a shower may be many times the sporadic meteor rate on that night, the enhancement of the number of faint meteors is far less strong. The observed meteor rate is highest when the radiant is near the zenith (because this corresponds to perpendicular projection of the stream cross section on the Earth's surface).

HISTORICAL SURVEY

Even though the first recorded meteor observation dates to 1809 BC and the first recorded shower observation (the Lyrids) to 687 BC, the extraterrestrial nature of meteoroids was only confirmed in 1798, when triangulation was used to deduce the height and speed of a number of meteors. On the night of Nov. 12–13, 1833, the Leonid meteor shower produced a spectacular display over North America, with tens of thousands of meteors per hour visible to the unaided eye. Impressive Leonid displays also occurred in 1799 and 1866. This periodicity, as well as similarities in orbital elements, quickly suggested a link between the meteors in this shower and Comet Tempel–Tuttle. A number of other meteor showers were soon associated with parent comets.

With the widespread use of radar techniques in meteor science, starting in the late 1940s, it became possible to expand the investigation of meteor showers to those that peaked during daylight and to much smaller meteoroids. One of the most intense meteor showers, the Arietids in June, is confined to daylight hours. In the past two decades intensified video techniques have been used to extend coverage to faint meteors, free of some of the biases inherent in the radio investigations. As noted earlier, the size distribution of shower versus sporadic meteors results in the majority of faint meteors detected by these techniques being sporadic rather than members of the major meteor showers.

MAJOR METEOR SHOWERS

Table 1 provides a listing of the major meteor showers. Several of the meteor showers (designated with a D for daytime shower) have their radiants so close to the Sun that the meteors are only observable by using radar techniques. The peak of the Eta Aquarid shower occurs during the day although some of the meteors are observable during the predawn period. The Quadrantid shower is named after the obsolete constellation Quadrans Muralis. For each shower the time of maximum activity is given both in terms of calendar date and solar longitude. Some showers (like the Quandrantids) are highly concentrated, with strong displays lasting for only a few hours, whereas others are spread over weeks (like the Taurid complex). Concentrated showers are more likely to be of recent formation and there are no records of the Quadrantid shower further back than the nineteenth century, although many of the other showers have been observed for many hundreds of years. The duration of each shower (defined as the period during which the activity is one-half maximum or more) is given in the table. The radiant position at the peak of the shower is specified in degrees of right ascension and declination. Typically, radiants are diffuse (about a degree) due to variations in orbits between different members of a given stream. The geocentric speed of the meteors in each shower is given in kilometers per second. The higher the speed of the meteors, then generally the greater the height above the Earth's surface where ablation occurs. Where orbital similarity permits a link between a meteor shower and a parent object, that circumstance is indicated in the table. All of the early associations were with comets until 1983 when a clear link between the Geminid shower and the asteroid Phaethon was established (although some astronomers consider Phaethon itself to be a "dead" cometary nucleus). A link between the Zeta Perseids and the asteroid Icarus has been suggested, but is not definitive.

IMPORTANCE OF METEOR STUDIES

Meteor showers provide a means of studying cometary dust. Such characteristics as the bulk density of the dust particles can be inferred from the apparent deceleration of the meteors. It is impor-

Table 1. Major Meteor Showers

Shower	*Maximum Date*	*Activity (Solar Longitude)*	*Duration (days)*	*Activity (ZHR)*	*Radiant*		*Geocentric Velocity ($km\ s^{-1}$)*	*Parent Object*
					RA (degrees)	*Decl. (degrees)*		
Quadrantid	Jan. 3	282.7	0.4	80	230	+49	42	
Lyrid	Apr. 22	31	1	15	272	+33	48	Comet Thatcher
Eta Aquarid	May 4	43	6	60	336	0	66	Comet Halley
Arietid (D)	Jun. 7	76	12	100	45	+23	39	Asteroid Icarus?
Zeta Perseid (D)	Jun. 9	78	8	80	62	+24	29	
Beta Taurid (D)	Jun. 29	96	5	40	86	+19	30	
Southern Delta Aquarid	Jul. 29	125	8	30	331	−16	41	
Northern Delta Aquarid	Aug. 12	139	8	20	339	−5	41	
Perseid	Aug. 12	139	3	100	46	57	60	Comet Swift–Tuttle
Orionid	Oct. 21	208	2	30	95	+16	66	Comet Halley
Southern Taurid	Nov. 3	220	30	15	53	+12	29	Comet Encke
Northern Taurid	Nov. 5	222	30	15	54	+21	30	Comet Encke
Leonid	Nov. 16	234	2	20	152	+22	72	Comet Tempel–Tuttle
Geminid	Dec. 13	261	3	90	112	+32	36	Asteroid Phaethon
Ursid	Dec. 22	270	1	20	217	+76	34	Comet Tuttle

tant to realize that most of the shower meteors currently detected were released from the parent comet hundreds or thousands of years ago. Hence the study of shower meteors provides a means of estimating the active ages of comets, and the uniformity in cometary activity from one perihelion passage to the next. The meteor size distribution provides an indication of the mass distribution of cometary grains, whereas the orbital distribution of shower meteors provides indirect evidence for differential ejection velocities from the parent comet.

Additional Reading

Halliday, I. (1988). Geminid fireballs and the peculiar asteroid 3200 Phaethon. *Icarus* **76** 279.

Kronk, G. W. (1988). *Meteor Showers: A Descriptive Catalog.* Enslow Publishers, Hillside, NJ.

MacRobert, A. M. (1988). Meteor observing I; Meteor observing II. *Sky and Telescope* **76** 131; 363.

McKinley, D. W. R. (1961). *Meteor Science and Engineering.* McGraw-Hill, New York.

Roggemans, P. (1987). *Handbook of Visual Meteor Observations.* Sky Publishing Corp., Cambridge, MA.

Yeomans, D. K. (1991). *Comets—A Chronological History of Observation, Science, Myth, and Folklore.* Wiley, New York. See the section on comets and meteors, p. 188.

See also **Meteoroid Streams, Dynamical Evolution; Meteoroids, Space Investigations.**

Missing Mass, Galactic

Jack G. Hills

About 90% of the mass of the Galaxy is in dark matter whose existence is not in serious doubt but whose nature is almost a complete mystery. The missing mass or dark matter is detected from the effect of its gravitational field on the motions and spatial distributions of stars and gas in the Galaxy. The gravitational effect of the dark material was detected more than half a century ago, but until recently there were no viable theories for the dark matter, so investigators either ignored it or questioned the observations or their interpretation.

There are two components of dark matter in the Galaxy: The first is the Oort dark matter, which is strongly concentrated to the galactic disk and is a small fraction of the galactic mass. The second is the Zwicky dark matter, which forms a spherical halo around the Galaxy and that may extend out as far as 100 kpc and comprise 90% of the galactic mass. We shall discuss the observational evidence for each of the two types of dark matter and the constraints that can be placed on their nature.

OORT DARK MATTER IN THE GALACTIC PLANE

The visible Milky Way, a spiral galaxy, is a highly flattened disk. At the Sun, which lies very nearly in its plane at a distance R_0 about 8 kpc from its center, the characteristic half-thickness of the galactic disk is about 100 pc or about 0.01 R_0 for interstellar gas and 300 pc or about 0.04 R_0 for stars. This flattening is the result of most of the stars and gas in the galactic disk being in nearly coplanar, circular orbits around the center of the Galaxy. The material in the disk must have undergone considerable energy dissipation before its stars formed, which allowed it to settle to the minimum energy state determined by its angular momentum.

An object with a given velocity away from the galactic plane will only go a fixed distance from it before the gravitational attraction of the disk draws it back. As was recognized by Jan H. Oort, the mean (scale) height of a given class of stars above the galactic plane depends on their distribution of velocities perpendicular to the disk and on the total mass density in the galactic plane. By determining the velocity dispersion of a group of stars (usually giants) and their scale heights, it is possible to find the mass density in the galactic plane. After subtracting out the mass density of the stars and gas, the mass density of the Oort dark matter in the galactic plane is about 0.1 $M_\odot\ pc^{-3}$ (solar masses per cubic parsec), or more than half the total density in the plane.

The thickness of the Oort dark-matter layer is much less than that of the stars and may even be less than that of the interstellar gas. If its scale height is less than that of the interstellar medium, the dark matter may comprise less than 10% of the stellar mass in the galactic disk, even though it dominates the density in the galactic plane. The small scale height of the dark matter implies that it has undergone considerable dissipation, which suggests that it is made up of baryons.

The most fashionable guess for the nature of the Oort dark matter is that it is in brown dwarfs, substellar objects less massive than 0.08 $M_\odot$, the minimum mass needed for them to tap the energy of hydrogen fusion. However, these dwarfs still release some energy by gravitational contraction. Even Jupiter and Saturn, with masses of about 0.001 and 0.0003 $M_\odot$, respectively, radiate several times more energy than they receive from the Sun. Infrared searches have failed to detect brown dwarfs with certainty, although they cannot yet be ruled out as the source of the dark matter.

If the dark matter is a small fraction of the stellar mass in the galactic disk, it could conceivably be a product of stellar evolution such as dust or dust aggregates (e.g., comets). However, only a small fraction of the mass of the dark matter can be in low-mass baryonic objects without coming in conflict with observations: If the dark matter were in dust grains, the extinction of starlight in the galactic disk would be up to 2 orders of magnitude larger than observed. Dark-matter grain clumps with masses ranging from that of grains of sand to several tons would dominate the meteor flux on the Earth, but interstellar meteors make up less than 1% of the total, if they exist at all. Still more massive objects would produce several times the number of craters found on the Earth and other planets. If the dark matter is in comets, their average masses would have to exceed 10^{22} g to be infrequent enough to have avoided detection to date.

ZWICKY DARK MATTER IN THE GALACTIC HALO

More than half a century ago, Fritz Zwicky noted that applying the virial theorem to the observed radial velocity dispersion of galaxies in clusters gave cluster masses that are 1–2 orders of magnitude larger than the sum of the masses of their individual galaxies. Masses of individual galaxies were obtained from rotation curves extending out to their visible edges. The Zwicky dark mass must either lie outside the visible parts of galaxies or lie between galaxies in the clusters unless the law of gravitation takes on unexpected behavior on scales larger than the radii of galaxies.

Neither the possibility of dark halos nor new physics on large scales gained favor in the astronomical community, so the Zwicky results were largely ignored (although nearly universally known in the community). *N*-body simulations were made to test the validity of the virial theorem, but the results just strengthened the case for dark matter in clusters.

The intellectual stagnation in this field was broken by a theoretical study by Jeremiah P. Ostriker and P. J. E. (James) Peebles. They found that the visible, highly flattened galactic disk was violently unstable on a dynamical time scale (period of orbital revolution of about 2×10^{8} yr), which is much less than the age of the Galaxy. The only reasonable way they found to stabilize the galactic disk was by adding a spherical halo of dark matter that constituted at least half the total mass *within* the Sun's orbit around the galactic center.

The halo needed for stabilizing the galactic disk does not have to extend much beyond the solar distance. However, the objects in the Ostriker–Peebles halo must have a velocity dispersion comparable to the orbital velocity in the galactic disk or about 200 km s^{-1}, which gives that halo a scale height comparable to the local radius of the Galaxy. This large scale height produces a relatively slow radial decrease in the halo density, which is consistent with a massive halo extending far beyond the luminous Galaxy. Subsequent observational work indicated that the density of dark matter in galactic halos falls off as the inverse square of the radius, so the total mass interior to a point is proportional to its radius. If the radius of the galactic halo extends out to 100 kpc, as it does in many other spiral galaxies, the mass of the Galaxy exceeds 10^{12} $M_\odot$, or about 10 times the mass of the Galaxy inside the solar orbit. The local density of the galactic halo dark material is about 0.008 $M_\odot$ pc^{-3}, or an order of magnitude less than the local density of Oort dark matter.

As was the case for the Oort dark matter, the Zwicky dark matter cannot be in the form of baryonic objects less massive than about 10^{22} g. Baryonic objects with masses between 10^{22} g and the minimum mass required for hydrogen fusion, about 0.08 $M_\odot$, cannot be ruled out by current observations, but there is no theory that suggests that such objects should form in galactic halos. Observations also do not rule out the galactic halo dark matter being in the form of black holes with masses up to 10^{6} $M_\odot$.

Is the Galaxy Held Together by Its Weakest Members?

Baryons (protons, neutrons, and their antiparticles) may not be the only stable massive particles to form in the Big Bang. Baryons are conspicuous because they can condense into optically opaque or even solid objects. All baryons interact with each other directly by the strong force as determined by nuclear cross sections. If they are charged, they interact by the electromagnetic force with its much larger atomic cross sections. Weakly interacting massive particles (WIMPs) are an interesting class of hypothetical nonbaryons that may have formed in the Big Bang. They are electrically neutral. They interact with each other and with other particles only through the weak force. WIMPs are a good candidate for the dark matter in the halos of galaxies and in clusters of galaxies. As is the case for their low-mass cousin, the neutrino, they can pass through an object such as the Earth without difficulty, so they can coexist with baryonic objects, including ourselves, in large numbers without much effect except by their gravitational attraction. The weak interaction does not allow significant collisional dissipation of energy, so the WIMPs cannot form subgalactic objects such as stars. Axions and photinos are examples of proposed WIMPs.

The local density of halo dark matter is about 0.008 $M_\odot$ pc^{-3}, or about 0.1 that of the Oort dark matter. If the halo dark matter is in WIMPs with a rest mass of 1 GeV, which is comparable to that of a proton, there would be 0.32 WIMPs cm^{-3} throughout all local space. A person weighing 165 lb (75 kg) has a volume of about 8×10^{4} cm^{3}, so would have about 26,000 WIMPs passing through their body at any given time. The WIMPs go through the body at 200 km s^{-1} or 2×10^{7} cm s^{-1}, so if the average travel distance through the body is about 20 cm, each WIMP spends only about 1 μs in the body and there is a flux of about 2×10^{10} WIMPs s^{-1} through the person.

The weak interaction makes direct detection of WIMPs difficult despite their large flux through any laboratory apparatus, but they may soon be detected by using highly cooled silicon crystals. A WIMP passing through the crystal has a small, but finite chance of exciting a quantized sound wave, a phonon, which can be detected if the crystal is at a sufficiently low temperature. The rise in temperature of the crystal may also be detected.

Additional Reading

Audouze, J. and Tran Tranh Van, J., eds. (1988). *Dark Matter*. Editions Frontières, Gif-sur-Yvette.

Bahcall, J. N. (1984). Self-consistent determinations of the total amount of matter near the Sun. *Ap. J.* **276** 169.

Davis Phillip, A. G. and Lu, P. K., eds. (1989). *The Gravitational Force Perpendicular to the Galactic Plane*. L. Davis Press, Schenectady, N.Y.

Hills, J. G. (1986). Limitations on the masses of objects constituting the missing mass in the galactic disk and the galactic halo. *Astron. J.* **92** 595.

Ostriker, J. P. and Peebles, P. J. E. (1973). A numerical study of the stability of flattened galaxies: Or, can cold galaxies survive? *Ap. J.* **186** 467.

Spitzer, L., Jr. (1978). *Physical Processes in the Interstellar Medium*. Wiley, New York.

Trimble, V. (1987). Existence and nature of dark matter in the Universe. *Ann. Rev. Astron. Ap.* **25** 425.

See also **Dark Matter, Cosmological; Galactic Structure, Stellar Kinematics; Galaxy, Dynamical Models.**

Molecular Clouds and Globules, Relation to Star Formation

Philip C. Myers

Understanding the process of star formation is a fundamental problem in astrophysics. In the last few decades, great progress has been made in defining this problem: through optical and infrared observations, which identify young stars; through centimeter- and millimeter-wavelength observations, which identify the clouds where stars are forming; and through theoretical models, which account for observed relationships between cloud and star properties.

Current estimates indicate that in the Milky Way, several thousand observable stars have not yet reached "adult" status, and some ten new stars are "born" each year. Thus stars in a wide range of developmental stages exist simultaneously and can be studied systematically. On the other hand, the youngest stars are also the most heavily obscured, and the circumstellar structure and motions most relevant to the formation of a star occur on size scales smaller than those detectable with present-day resolution and sensitivity. Thus the study of star formation is at once fundamental, practicable, and challenging.

This entry presents a brief history of our knowledge of young stars, and of molecular clouds. It reviews links between young stars and molecular clouds, and it presents current ideas about the physical mechanisms at work. It assumes a modest knowledge of physical concepts, but requires no specialized knowledge of astronomy.

YOUNG STARS

A "low-mass" star like the Sun shines with its present brightness for some 10 billion years, whereas a "massive" star like those in the sword of Orion has about 20 times the mass of the Sun and shines for several million years. These times represent the adult lifetime of a star, when the star steadily burns its nuclear fuel and maintains a nearly constant luminosity. In contrast, a star is "young" for a much shorter period than its adult lifetime: about 100,000 years for a massive star to about 10 million years for a low-mass star. During the youth of a star, its luminosity varies with time, its nuclear fires are just starting, and its interaction with its parent gas cloud is evident.

This interaction was first seen in the diffuse emission nebulae that surround young massive stars. They shine not by reflection, but by emission, from the circumstellar gas ionized and heated by the energetic photons from the massive star. The best known nebula, in Orion, was first noted by Nicolas C. F. de Peiresc in 1611, shortly after Galileo's first use of the telescope. The first drawing of the nebula was published by Christiaan Huygens in 1656, and the first photograph was made by Henry Draper in 1880.

The nature of this nebulosity was poorly understood for decades. Support for the idea that such nebulosity is an illuminated remnant of the cloud that formed the star came in 1939, when Bengt Strömgren showed that the hydrogen around a massive star is nearly fully ionized inside a sharply bounded volume and is neutral beyond the boundary. This explanation accounted for the size of the nebula, its proximity to a luminous star, and the radiation temperature of the nebular gas.

But the association of a luminous star with an ionized emission nebula is only a circumstantial indication that the star is young, because the star could, in principle, travel away from its birthplace, lighting up any cloud it might encounter. Also, some stars of similar luminosity have no associated nebula. More conclusive indications of the youth of certain stars came in the 1940s, when Alfred H. Joy noted three distinct characteristics of low-mass "T Tauri" stars, named for the most prominent example [historically, the third variable star designated in the direction of the Taurus constellation (R, S, T, . . .)]:

1. Spectral lines of hydrogen, calcium, and other atoms in emission, in contrast to the absorption of these lines seen against the continuous stellar emission for most stars.
2. Significant variation in time of the intensity of the stellar emission, usually irregularly.
3. Location only in regions of dark or opaque nebulosity, indicating their association with clouds of interstellar matter thick enough to absorb the light from background stars.

These properties of T Tauri stars led to the current understanding that they are young, because the properties are specific enough to allow clear identification and because the T Tauri stars are numerous enough to allow statistical studies (today nearly 1000 are known). In the 1950s, George H. Herbig and others showed that the number of T Tauri stars projected on an interstellar cloud is too great to arise from the chance alignment of the cloud with stars of normal space density. Thus, T Tauri stars are spatially concentrated, and it follows that they are physically associated with the clouds on which they are projected. Also, the typical speed of a T Tauri star with respect to its cloud is too small to allow its travel to the cloud from any birthplace other than within the same cloud. Thus, T Tauri stars are born in the clouds where they are seen. Herbig also showed that T Tauri stars are typically brighter than normal stars of the same temperature; a fact that can be explained if the T Tauri stars are radiating some energy derived from their gravitational contraction. This contraction is a sign that the stars are very young.

The idea that stars form by gravitational contraction dates to Isaac Newton, as early as 1692. Such contraction was also proposed by Hermann von Helmholtz and Lord Kelvin in the nineteenth century as the source of the Sun's heat, that is, of the luminosity of an adult star. However, the geological and fossil records showed that the Earth is some 200 times older than the expected lifetime of the Sun, if the Sun were radiating due to gravitational contraction alone. Instead, the luminosity of adult stars was proposed and shown by Arthur Stanley Eddington, Subrahmanyan Chandrasekhar, William Fowler, and others to arise from transmutation of hydrogen into helium, in and around the 1930s. But gravitational contraction is still thought to power substantial radiation from young stars, first described in terms of temperature and luminosity by Henry Norris Russell in 1913.

In 1983 a substantial population of stars as luminous as the T Tauri stars, but probably even younger than them, was found by the *Infrared Astronomical Satellite* (*IRAS*), which surveyed nearly the entire sky at infrared wavelengths from 12 to 100 micrometers (μm). These optically invisible stars are surrounded by extremely thick blankets of dust and are thought to be "protostars," or young stars still in the process of assembling their ultimate adult mass. We discuss these candidate protostars further in the section on molecular clouds and star formation.

MOLECULAR CLOUDS

Evidence for clouds of interstellar matter was known as early as 1784, when William Herschel noted that luminous nebulae, such as that in Orion, are frequently associated with a marked decrease in the projected density of nearby stars. This extinction of distant stars by intervening matter was first quantified by Friedrich Georg Wilhelm von Struve in 1847. By the 1920s it was recognized by Edwin P. Hubble, Russell, and others that solid particles ("dust grains") less than 0.1 mm in diameter are responsible for the observed extinction. In the same decade, absorption line studies of bright stars showed that lines of ionized calcium arise from material extending far from the stars. These absorption line observations led to the current concept of an interstellar medium, consisting of patchy clouds and complexes of gas and dust, extending in some cases for hundreds of light-years (ly), and containing enough mass to form hundreds or even thousands of stars.

Understanding the temperature, ionization, and energy balance of clouds and of the interstellar medium was placed in the framework of gas kinematics and dynamics by Eddington in the 1920s and was extensively developed by Lyman Spitzer, Jr. in the 1940s and 1950s.

In 1945 Hendrik C. van de Hulst predicted that interstellar atoms of neutral hydrogen would radiate in a spectral line at 21-cm wavelength, corresponding to the minute change in atomic energy between the state where the electron and proton spins are aligned and the state where they are antialigned. In 1951 Harold I. Ewen and Edward M. Purcell detected such radiation. In the 1950s extensive maps were made of atomic hydrogen clouds throughout the Milky Way, and it became possible to measure the size, density, and temperature of interstellar gas clouds. But, as seen in the 21-cm line, interstellar clouds are relatively tenuous and rarefied: They blend together and lack the specific definition needed to allow their clear identification with regions of star formation.

Interstellar molecules of CH, CH^+, and CN were known from the optical absorption line studies of Walter S. Adams in the later 1930s and early 1940s. But in the 1950s the conventional view held that interstellar space was too bright with ultraviolet light to maintain a significant density of molecules: The energetic ultraviolet photons from massive stars would dissociate most molecules into their constituent atoms. Nonetheless, radio astronomers pursued and discovered spectral lines from simple interstellar molecules, starting with OH in 1963 (by Alan H. Barrett and colleagues), followed by ammonia and formaldehyde in 1968, and by more than 90 other molecules emitting at centimeter, millimeter, and submillimeter wavelengths.

Among these interstellar molecules the brightest and most widely detectable line is the fundamental rotational line of carbon monoxide (CO) at 2.6-mm wavelength, first observed by Robert W. Wilson, K. B. Jefferts, and Arno A. Penzias in 1970. In the last 20 years this tracer has been used to map clouds throughout the Galaxy. Unlike the clouds seen in atomic hydrogen, clouds seen in CO are closely associated with emission nebulae and young OB stars, optically invisible infrared stars, and with the extensive dark clouds and complexes seen in optical photographs. In some cases hydrogen and CO clouds coincide, but generally the CO clouds are distinct.

The CO clouds are now understood to be concentrations of dust and molecules, with relatively little atomic gas. Atomic gas is the main component of lower-density regions that occasionally surround CO clouds. The main ingredient of molecular clouds is diatomic molecular hydrogen; CO is a trace constituent, with about one CO molecule for every 10,000 H_2 molecules. Molecular clouds are cold (with temperature 10 K), clumpy (with complex internal structure), and massive (with total mass ranging from a few to a few million times the mass of the Sun). The largest of these are called "giant molecular clouds." In the 1970s and 1980s Patrick Thaddeus, Philip Solomon, and others showed that such clouds contain most of the mass of the interstellar medium and that they contain virtually all known young stars: Stars form in molecular clouds.

In the late 1970s and early 1980s many radio astronomers showed that a cloud map in one molecular spectral line has a consistently different size from a map in another line. The relation of a spectral line to the map size in that line is usually consistent with the idea that a line that needs higher gas density for its excitation is observed over a smaller region than a line, like that of CO, that requires relatively low density for its excitation. Thus lines of NH_3, CS, and other molecules that trace molecular hydrogen densities of some 10,000 molecules per cubic centimeter have map sizes of about 0.3 ly, whereas the same region mapped in the CO line would appear some 10 times more extensive and would contain several distinct "dense cores" revealed by the high-density probes.

MOLECULAR CLOUDS AND STAR FORMATION

In the late 1970s and early 1980s the paths of research into young, optically invisible stars and into dense cores in molecular clouds began to join. Emission in high-density-tracer lines from molecules such as ammonia was found to be brightest toward or near the positions of optically invisible stars. At first, sensitivity limitations restricted this conclusion to the relatively rare massive stars, most of which occur in close groups and clusters. For these stars, the complexity of the observed line emission suggested a somewhat distorted record of the gas properties needed to form a star. The complexity could be understood, in that the extremely energetic winds and radiation from the stars should ionize and dissociate some molecules and should dramatically increase the thermal and turbulent motions of others. But the initial conditions and physical processes of star formation remained uncertain.

As the sensitivity of centimeter- and millimeter-wavelength receivers improved, it became possible to measure the properties of dense cores in regions of low-mass star formation. These cores, like low-mass stars, are dimmer but far more numerous than their more massive cousins. More than 100 low-mass cores have been detected in the 1.3 cm inversion line of ammonia and (more than half of these have been mapped) by Priscilla J. Benson and the author in the mid-1980s. Their typical size (0.3 ly), temperature (10 K), and density (10,000 hydrogen molecules per cubic centimeter), and their relatively small component of nonthermal internal motions, are well-established.

In 1983 *IRAS* found single, point-like infrared sources associated with about half of the known dense cores in dark clouds. In turn, about half of these sources are optically visible T Tauri stars; the other half, previously unknown, have too much circumstellar dust to be seen in visible light. As a group, the optically invisible sources are presumed to be younger than the T Tauri stars and are the best candidates to be protostars. Their association with dense cores indicates that such cores are regions in molecular clouds where low-mass stars form. The candidate protostars probably become visible T Tauri stars, because they are found near T Tauri stars and have luminosities similar to those of T Tauri stars.

Analysis of these results by Charles A. Beichman, the author, and others showed that the dense cores with associated stars have somewhat greater internal motions than do starless cores, but that they are otherwise similar in size, density, and temperature: Evidently, low-mass stars disturb their birthplaces much more slowly and gently than do massive stars. Furthermore, the properties of starless cores can be considered as initial conditions for the formation of low-mass stars.

The essential idea of star formation from a dense core is that of gravitational collapse: A region initially in balance between its internal pressure and self-gravity is disturbed and develops a runaway instability, where matter falls toward the center of gravity faster than pressure can develop to oppose it. This process continues on the small scale until the protostellar temperature and density increase enough to halt the collapse. It continues on the large scale until the supply of gas available to collapse is exhausted or is deflected by other forces, such as those from stellar winds. The dependence of time scale on density for this instability was first described by Sir James Jeans in 1928. For the typical core, this time is of the order of a few hundred thousand years.

In 1969 Richard B. Larson made the first detailed calculation of the gravitational collapse of a sphere to form a low-mass star, using a computer. This work showed that the collapse is highly nonuniform, with a small part of the central mass achieving stellar density, when the outer part of the cloud has moved only slightly. In 1977 Frank H. Shu showed that a spherically symmetric core with sufficient initial central concentration will collapse from the "inside out," increasing the mass of the central protostar at a constant rate. If a core rotates as it collapses, it is expected to form a circumstellar accretion disk, first described by Carl F. von Weizsäcker in 1948 and elaborated by Alastair G. W. Cameron,

Chia-Chiao Lin, and others in the 1970s and 1980s. In the mid-1980s Fred C. Adams and Shu showed that the spectrum of infrared emission observed from optically invisible stars in dense cores can be matched by a model protostar, disk, and core, which arise from the inside-out collapse of a rotating core with observed properties.

PROSPECTS

Many questions remain unanswered about the process of star formation in molecular clouds, including the following: the role of magnetic fields in cloud support and dense core formation; the structure, motions, and time scales of disks and other circumstellar matter during collapse; the transition from optically invisible protostar to visible T Tauri star; the formation of binary stars and multiple star groups; and the similarity of the processes of massive and low-mass star formation. Yet the prospects for progress seem bright, for several reasons. Many new tracers of very young stars are now available, especially the high-velocity, bipolar winds evident in the wings and maps of the CO line profile from nearly all young stars. New instruments with substantially improved resolution and sensitivity are in the process of becoming available, including the *Hubble Space Telescope*, the *Space Infrared Telescope* facility, and several interferometric arrays at millimeter and submillimeter wavelengths. Most important, the field of star formation continues to attract scientists with imagination and skill.

Additional Reading

Herbig, G. H. (1967). The youngest stars. *Scientific American* **217** (No. 2) 30.

Myers, P. C. (1985). Molecular cloud cores. In *Protostars and Planets II*, D. C. Black and M. S. Matthews, eds. University of Arizona Press, Tucson, p. 81.

Shu, F. H., Adams, F. C., and Lizano, S. (1987). Star formation in molecular clouds. *Ann. Rev. Astron. Ap.* **25** 23.

Winnewisser, G. and Armstrong, J. T., eds. (1989). *The Physics and Chemistry of Interstellar Molecular Clouds: mm and Sub-mm Observations in Astrophysics*. Springer, Berlin.

See also **Bok Globules; Interstellar Clouds, Collapse and Fragmentation; Interstellar Clouds, Molecular; Star Formation, Propagating; Stars, T Tauri.**

Moon, Eclipses, Librations, and Phases

Ewen A. Whitaker

MOON'S PHASES

The changing shape of the Moon (Greek *phasis*—appearance, shape) from night to night, as it circles the heavens in its monthly course, has intrigued and puzzled mankind for untold centuries. Although the correct explanation had been reached at least two millenia ago, misconceptions are common even today. The Moon is, of course, a solid opaque sphere that emits no light of its own. It is illuminated by the Sun and thus always has a "day" side and a "night" side. The invisible far side of the Moon is often incorrectly called the dark side of the Moon. Because the Sun is not a point source, but subtends an angle of about 0°.5, very slightly more than half the Moon (about 180°.5) receives sunlight.

The plane of the Earth's orbit around the Sun is known as the *ecliptic* (not to be confused with elliptic) because solar and lunar eclipses can take place only in or very near this plane. The Moon's orbit around the Earth does not coincide with the ecliptic, but is tipped at an angle of just over 5° to it. Thus the Moon usually passes above or below the Sun–Earth line. When the angle between the Sun and the Moon reaches its minimum value, as seen from Earth (strictly, from the center of the Earth), it is the astronomical new moon. Under these conditions, the Moon presents its unilluminated (i.e., night) side toward us and is invisible in the bright daytime sky.

As the Moon moves eastward away from the Sun at its average rate of about 12° per day, an increasing amount of illuminated lunar surface is visible from Earth, and the familiar shape of the thin crescent (Latin *crescere*—to increase) may be seen in the western sky some time after sunset. The first sighting of the crescent Moon, an important event in the Islamic religion, is popularly regarded as marking the time of new moon, even though it lags 1–3 or even more days behind the astronomical new phase. It is affected considerably by the season, viewing location, sky clarity, and so forth.

The average length of the month (new moon to new moon) is about $29\frac{1}{2}$ days. The phases are new, first quarter, full, last quarter, and back to new (see Fig. 1); they occur on the average at intervals about 7.4 days apart. The first quarter is defined as that point at which the longitude of the Moon, measured along the ecliptic, is 90° greater than that of the Sun. At this phase, the week-old Moon appears to present a half-disk to us, and it lies on or fairly near the north–south meridian at 6 PM local time. This marks the end of the crescent phase and the beginning of the gibbous phase, in which over 50% of the Moon's total disk is illuminated. This continues until the full phase, at age about 14.8 days, when the disk appears, by naked eye, to be truly circular. This half of the lunation is usually referred to as the waxing (i.e., growing) Moon.

As in the case of astronomical new moon, the full moon usually passes above or below the Sun–Earth line, thus missing the Earth's shadow. Astronomical full moon occurs when the Moon's longitude on the ecliptic is 180° greater than that of the Sun. To the naked eye, however, the disk appears to be complete for a couple of days each side of that event. The full moon rises at about sunset, peaks near midnight, and sets at about sunrise.

The phases now repeat in the opposite sequence; this second half of the lunation being known as the waning moon. The gibbous phase, with the defect of illumination now on the west side of the disk, progresses to last quarter, where the Moon's longitude on the ecliptic is 270° greater than that of the Sun and we see a half-disk again. The last quarter peaks at about 6 AM. This is followed by the crescent shape of the old moon in the dawn sky, and thus back to new moon. Strictly, this should be named the "decrescent" phase; the term "waning crescent" is a contradiction in terms.

Because the boundary between the illuminated and dark portions of the Moon (the terminator) is a circle, it presents an elliptical shape when viewed from various angles. Thus the shape of the sunlit portion is always defined by the intersection of a semicircle —the edge ("limb") of the Moon—and half an ellipse of varying ellipticity. The cusps or horns always point directly away from the Sun, along a segment of a great circle on the sky. This is even the case at full moon, when the terminator may be observed (with a telescope) to lie along either the north or south half of the lunar disk.

Because the Moon never strays far from the ecliptic, the optimum chances for viewing a very young phase occur when the ecliptic is most nearly vertical soon after sunset. For Earth's northern hemisphere, this is during early March; the Moon's horns are then pointing steeply upward, and the Moon is said to be "holding water". The autumnal new moon is far less noticeable, lying much closer to the horizon. For related reasons, the winter full moon rides high in the sky, whereas that at midsummer rides low from southeast to southwest. At the crescent phase, the whole lunar disk can usually be seen to be faintly illuminated—"the old moon in the new moon's arms"—caused by sunlight reflected from a gibbous (as seen from the Moon) Earth. The decrescent Moon displays the same phenomenon.

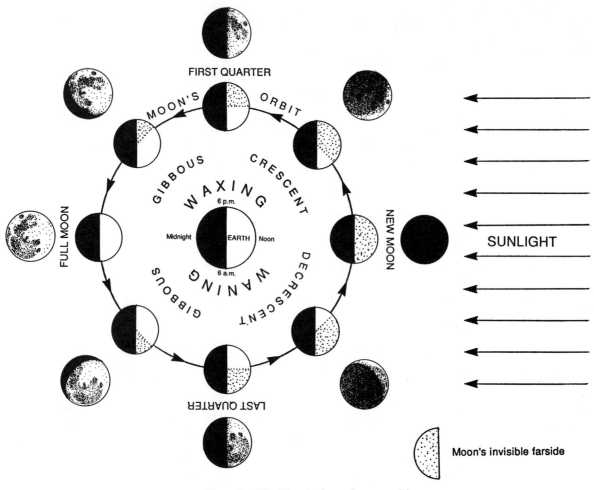

Figure 1. The Moon's phases (not to scale).

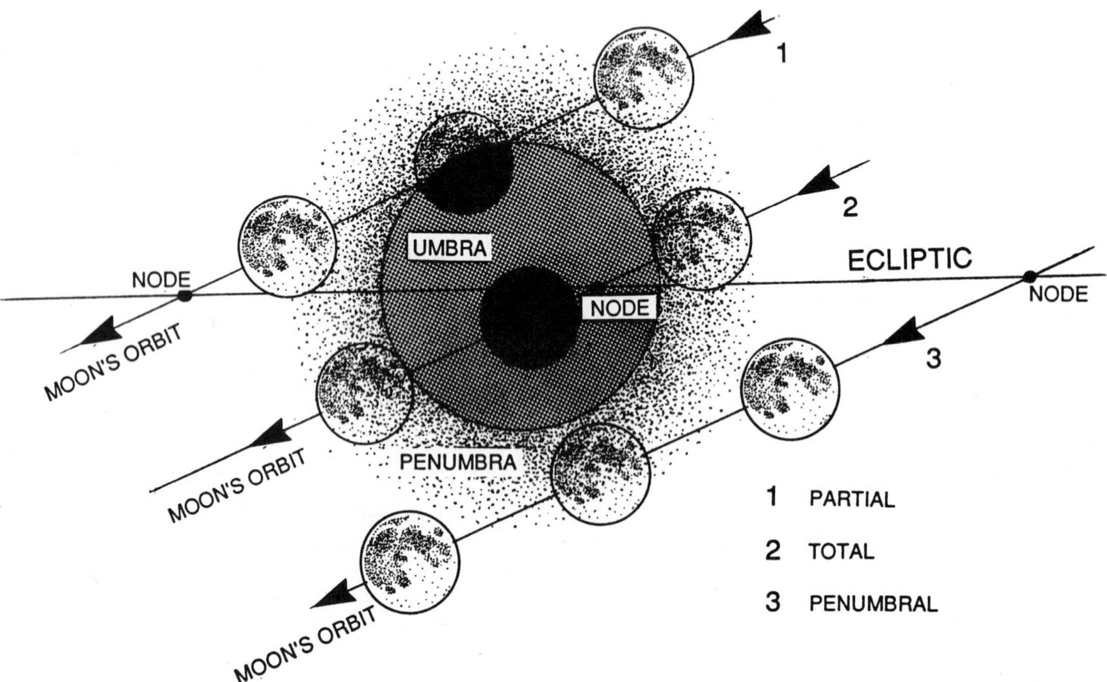

Figure 2. Typical lunar eclipses. The umbra is shown gray for clarity.

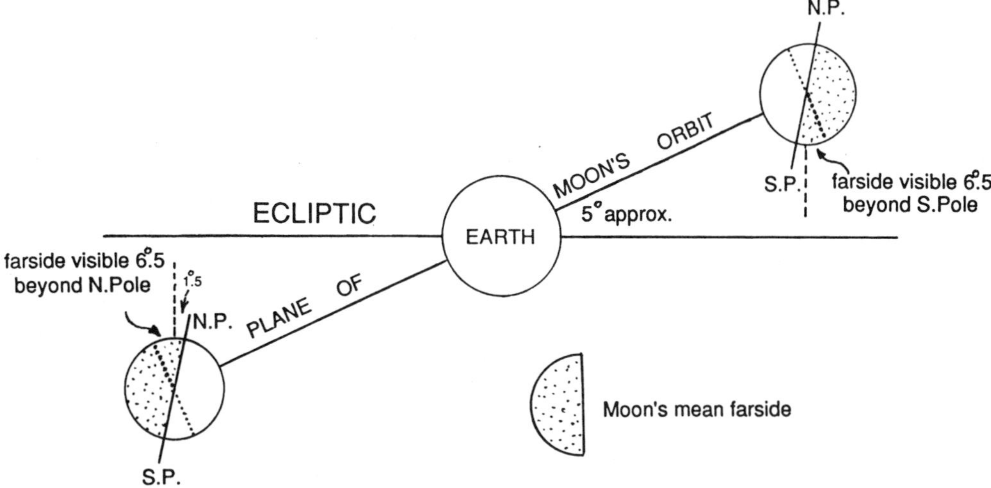

Figure 3. Libration in latitude. All angles are exaggerated for clarity.

LUNAR ECLIPSES

As already noted, the Moon's orbit is tilted to the ecliptic by just over 5°. Thus although the Moon usually passes above or below the Earth's shadow at full, on occasion it will pass through this shadow (see Fig. 2). The points where the lunar orbit crosses the ecliptic are called the nodes. If the full moon is at or near a node when these are aligned with the Sun, then a lunar eclipse can occur. Because the Sun is not a point source and is larger than the Earth, the Earth's shadow is a long, fuzzy-edged cone stretching over 800,000 miles into space in the antisolar direction. If Earth had no atmosphere, this shadow, if cast on a huge screen at the Moon's distance, would appear somewhat as in the diagram. The solid shadow, in which no sunlight falls, is named the umbra. The annular area surrounding this is the penumbra (Latin *paene*—almost), which grades smoothly from the total shadow of the umbra to complete sunlight at its outer edge.

The Moon's path can cross this shadow at any position. If the umbra never falls on the Moon, it is a penumbral eclipse; if the umbra covers part but not all of the Moon at the middle of the eclipse, then it is a partial eclipse. When the Moon passes completely into the umbra, it becomes a total eclipse. The magnitude of the eclipse is reckoned as the fraction of the lunar diameter covered by the umbra at mid-eclipse. Thus magnitude 1.0 means that the Moon is just completely covered; magnitude 1.5 signifies that the edge of the umbra lies half a lunar diameter beyond the edge of the Moon. Lunar eclipses are visible from anywhere that the Moon is above the horizon.

If the umbra were perfectly dark, the Moon would completely vanish during each total eclipse. A few eclipses are so dark that the Moon is barely discernible, but much more frequently it remains visible, glowing with a reddish or coppery colored light. The illumination is usually uneven, with a lighter, whiter segment most often in the direction of the nearest part of the penumbra. This is caused by sunlight passing horizontally through the Earth's atmosphere and being refracted into the umbra. This light consists largely of the reddish sunset hues, with some whiter light refracted by the higher levels of our atmosphere. Cloudy or dusty conditions along the sunrise–sunset line on Earth influence the general darkness of the eclipse and the location of any lighter areas.

LUNAR LIBRATIONS

By casual observation, the Moon appears to present exactly the same face toward us at all times. By the middle of the seventeenth century, telescopic observations showed that this was not the case. The Moon was seen to exhibit both a north–south nodding ("yes") motion and an east–west ("no") motion, the period for each being roughly a month. It was soon realized that these librations (Latin *libra*—a balance) were not actual oscillatory motions of the Moon, but apparent motions arising from Earth's changing viewpoint, a consequence of certain properties of the Moon's orbit and axial direction. They are termed the optical librations.

Libration in Latitude

As already noted, the plane of the Moon's orbit is tilted to the ecliptic by an angle of just over 5°, the actual mean value being 5°8′.7. The Moon's rotational axis, however, is tilted at only 1°32′.1 from the ecliptic pole. The pole of the Moon's orbit is not a fixed direction in space, but precesses westward around the ecliptic pole in a period of $18\frac{2}{3}$ yr. In the late 1600s, Giovanni Domenico Cassini discovered that the Moon's axis also precessed around the ecliptic pole in the same direction at the same rate. Furthermore, this axis always lay on a line from the Moon's orbital pole continued past the ecliptic pole. Referring to Fig. 3, this means that the Moon's north and south poles are alternately tilted toward Earth each month (approximately) by a value of 6°40′.8. Actual values vary slightly from this figure due to various perturbations of the lunar orbit. The mean period for this latitudinal libration is 27.2 days, 2.3 days shorter than the lunation. Thus the lunar phase at which maximum libration in latitude takes place is somewhat different each successive lunation.

Libration in Longitude

The Moon's orbit around the Earth is not a circle with Earth at the center. Rather it is an ellipse of somewhat variable ellipticity, the Earth being situated eccentrically at one of its two foci. According to Kepler's second law, orbiting bodies sweep out equal areas in equal times; thus the Moon travels faster when closer to the Earth, and vice versa. However, the Moon's axial rotation proceeds at a constant rate, apart from very minor perturbations, and the two motions get out of step as shown in Fig. 4. The greatest discrepancies occur halfway, in time, between perigee and apogee. The mean value of these longitudinal librations is 6°9′ each side of the mean position, but orbital perturbations can raise this to over 8°. Their mean period is about 27.6 days, somewhat different from the latitudinal libration period. Thus over a period of a few years, the mean central point of the Moon's disk, if it were always visible, would appear to exhibit a motion resembling Lissajous' figures, the resultant of two mutually perpendicular simple harmonic motions

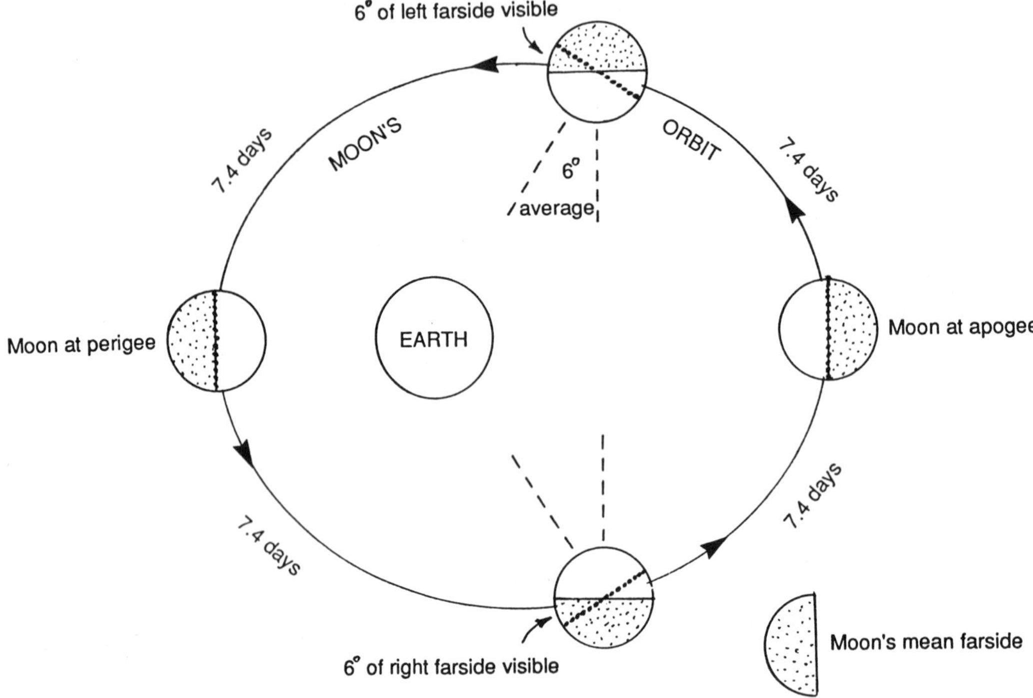

Figure 4. Libration in longitude. All angles and the orbital ellipticity are exaggerated for clarity.

with different periods. When the two librations reach a maximum simultaneously, the displacement of the mean center of the lunar disk amounts to over 10°.5.

Diurnal Libration

The preceding librations refer to a viewpoint from Earth's center of gravity. As observers on the Earth's surface, we view the Moon from a changing viewpoint from moonrise to moonset. At these two times we can see about 1° farther over the Moon's upper limb; this amount decreases with increasing lunar altitude, varying as the cosine of that angle.

Physical Librations

An unsymmetrical distribution of mass in the Moon has allowed Earth's gravity to keep one lunar hemisphere permanently turned earthward. If the optical librations in latitude and longitude were true simple harmonic motions of constant amplitude and period, the induced torques on the Moon would balance out, leaving no preponderance in any one direction. However, perturbations in the Moon's orbit due to the Sun's gravity, the eccentricity of the Earth's orbit, and so forth allow cyclical preponderances of torque in both east–west and north–south directions. These are sufficient to produce very small but real librations (maximum about 0°.04) about the Moon's center of mass.

Additional Reading

Espanak, F. (1989). *Fifty Year Canon of Lunar Eclipses: 1986–2035*. NASA Reference Publication 1216. NASA, Washington, D. C.

Meeus, J. and Mucke, H. (1979). *Canon of Lunar Eclipses, −2002 to +2526*. Astronomisches Buro, Vienna.

Spencer Jones, H. (1934). The moon and eclipses and occultations. In *General Astronomy*. Edward Arnold and Co., London, Chaps. 6 and 8.

See also **Calendars; Mercury and Venus, Transits; Sun, Eclipses.**

Moon, Geology

Paul D. Spudis

The task of lunar geology is to reconstruct the events and processes that have shaped the Moon's composition and surface appearance throughout its history. Although the Moon is a relatively primitive body compared to some of the other terrestrial planets, it has undergone a long and protracted geologic evolution. The Moon is a differentiated object, meaning that its surface composition is not representative of its bulk composition, and its long and complex magmatic record is complemented by an even longer impact bombardment history, which produced most of the visible landforms on the Moon's surface.

SURFACE FEATURES AND STRATIGRAPHY

The Moon's surface consist of two major types of terrain: the relatively bright, heavily cratered highlands (Latin *terrae*) and the darker, smoother lowland plains (Latin *maria*). (Figure 1.) The terrae consist of an apparently endless sequence of overlapping craters, ranging in size from small craters at the limit of resolution on even the best photographs to large multiringed basins (some of which are over 1000 km in diameter). More than two decades of study has demonstrated that nearly all of these craters and all of the basins formed by meteoritic impact. Thus, the more heavily cratered highlands are older than the maria, having accumulated more craters. The large number of impact craters in the lunar highlands further indicates that at least the top few kilometers of the crust has been repeatedly mixed by impacts. Such a bombardment creates brecciated rocks, some of which contain many fragments of older rocks that may be genetically unrelated.

The dark, relatively lightly cratered maria cover about 16% of the lunar surface. During the pre-*Apollo* geological mapping of the Moon, geologists recognized that a substantial amount of time had elapsed between the heavy bombardment of the highlands and the final emplacement of the visible maria. Small flow lobes seen in some mare regions in the highest resolution Earth-based pho-

Figure 1. View of the Moon obtained from the homeward-bound *Apollo 16* spacecraft. The heavily cratered lunar highlands are evident, as are the darker, smoother maria. Most of this scene is of the lunar farside; the large circular mare at the upper left is Mare Crisium, which is on the eastern limb of the lunar nearside.

tographs led to the idea that the lunar maria consist of volcanic lava flows. Better photographs taken from lunar orbit show numerous landforms attributable to volcanic activity, including lava channels (sinuous rilles), domes, cones, and collapse pits.

The maria are concentrated on the lunar nearside and are almost everywhere contained within an impact basin. Thus, it is important to distinguish between such terms as the *Imbrium basin*, the large and ancient multiring impact structure, and *Mare Imbrium*, which refers to the younger, dark volcanic fill of the Imbrium basin. Remote sensing data show that over 30 types of basalt occur within the maria, at least two thirds of which are dissimilar to any kind of basalt present in the *Apollo* or *Luna* samples.

According to the geological law of superposition, younger units overlie, embay, or intrude older units. This simple but powerful methodology has allowed us to make geological maps of the entire Moon and to produce a formal stratigraphic classification to help unravel its history (Table 1). The youngest stratigraphic system is the Copernican System; it encompasses the freshest lunar craters, most of which have preserved rays. The Eratosthenian System includes those craters that are slightly older, as evidenced by degraded form and no visible rays; this system also includes most of the youngest mare deposits. The Imbrian System, the base of which is defined by the deposits of the Imbrium basin, also includes the spectacular Orientale basin on the western lunar limb, most of the visible mare deposits, and numerous large impact craters. The base of the Nectarian System is defined by the deposits of the Nectaris basin, another large multiring basin on the lunar nearside, and includes over four times as many large craters and basins as does the Imbrian System. The Nectarian System may also contain some volcanic deposits, although most volcanic surfaces of this age have been destroyed by impacts. The oldest system, the pre-Nectarian, includes crater and basin deposits and any other units formed before the Nectaris basin impact; this period began with the origin of the Moon and witnessed the formation of its crust and the creation of its most heavily cratered surfaces.

The stratigraphic analysis of the Moon can indicate the relative ages of surface units, but it cannot by itself determine the absolute ages of these units. Our understanding of emplacement times, as well as sample compositions and rock types, had to await the return of samples from the Moon. These tasks were begun by the six manned U.S. *Apollo* missions and the three unmanned Soviet *Luna* missions, which returned pieces of the Moon to Earth. Geologic studies are still continuing as we try to place these samples into their geological context and to make inferences about regional geologic events from these few spot samples.

TYPES OF LUNAR ROCK

Maria

The return of rock samples from the lunar maria resolved any lingering doubts as to their volcanic origin. The mare rocks are basalt, which is also a widespread terrestrial volcanic rock. Different mare flows were erupted over a substantial interval of time. Basalts returned from the mare plains range in age from 3.8 to 3.1 billion (10^9) years. Continuing study of small fragments of mare basalt found in highland breccias has revealed evidence of even more ancient basalts, as old as 4.3 billion years. Because the landing sites selected for the *Apollo* missions do not include very young maria, we do not know the ages of the youngest mare basalts on the Moon. However, stratigraphic evidence from high-resolution photographs suggests that some mare flows actually embay (and therefore postdate) some rayed (young) craters. These lava flows could be as young as 1 billion years; thus, mare volcanism may have been active for as long as 3.3 billion years. However, study of the regional distribution of basalt units and their ages suggests that most of the visible mare deposits were emplaced between 3.8 and about 3.0 billion years ago.

In addition to the mare basalts, varied homogeneous, mafic glasses (of volcanic, not impact origin) were found in the soils at virtually all the landing sites. Some of these glasses are similar (but not identical) in chemical composition to the mare basalts and apparently were formed within about the same time interval. The surfaces of the glasses have been studied in great detail and are found to be coated by amorphous mounds of volatile elements, including zinc, lead, sulfur, and chlorine. These glasses were formed during volcanic eruptions that were at least partly contemporaneous with the local mare basalt eruptions. They are similar to small airborne particles produced during the eruption of Hawaiian volcanoes in the so-called fire fountains and so they are the lunar equivalents of terrestrial ash formed during basaltic eruptions. These glasses appear to have been erupted from deep within the lunar mantle, with little or no chemical changes during their ascent to the surface.

Terrae

The lunar highlands are composed of rocks having high contents of plagioclase feldspar (a mineral rich in calcium and aluminum). The dominant rock type collected from the terrae is a breccia, that is, a rock consisting of clasts (including a wide variety of rock types) assembled into a single rock by impact processes. Most terra breccias are composed of still older breccia fragments, attesting to a long and protracted bombardment history. The highland samples also include several fine-grained crystalline rocks. Although it was initially thought that some of these were fragments of early highland volcanic rocks, it is now recognized that they are impact melts, formed by shock melting due to the very high pressures that accompany an impact event.

Virtually all of the highland breccias and impact melts formed in a restricted interval from about 4.0 to 3.8 billion years ago. This surprising relation suggested to some workers that the Moon experienced a pronounced increase in the meteorite bombardment rate at that time, the so-called terminal cataclysm. Alternatively, it is possible that this restricted age range merely reflects the tail end of a long history of intense bombardment that began 4.6 billion years

Table 1. The Lunar Time-Stratigraphic System

System	*Age (10^9 yr)*	*Remarks*
Pre-Nectarian	Began: 4.6 Ended: 3.92	Includes crater and basin deposits and many other units formed before the Nectaris basin impact; includes formation of lunar crust and its most heavily cratered surfaces.
Nectarian	Began: 3.92 Ended: 3.85	Defined by deposits of the Nectaris basin (a large multiring basin on the lunar nearside); includes almost four times as many large craters and basins as the Imbrian system; may also contain some volcanic deposits.
Imbrian	Began: 3.85 Ended: 3.15	Defined by deposits of the Imbrium basin; includes the striking Orientale basin on the Moon's extreme western limb, most visible mare deposits, and numerous large impact craters.
Eratosthenian	Began: 3.15 Ended: about 1.0	Includes those craters that are slightly more degraded and have lost visible rays; also includes most of the youngest mare deposits.
Copernican	Began: about 1.0 Ended: (to present)	Youngest segment in the Moon's stratigraphic hierarchy; it includes the freshest lunar craters, most of which have preserved rays.

ago, the estimated time of lunar origin. The resolution of this question requires resumed lunar exploration and sampling of carefully selected geological sites.

Some of the clasts in lunar breccias are plutonic igneous rocks, that is, rocks that cooled slowly from internally generated magmas. At least two distinct magmas were involved in the production of the plutonic rocks of the lunar highlands. Rocks composed almost completely of plagioclase feldspar are called anorthosites and form the ferroan anorthosite suite, so-called because the few mafic minerals in these rocks (pyroxene and olivine) are relatively rich in iron (Fe). The other plutonic rock suite of the lunar highlands is also rich in plagioclase feldspar, but it contains substantial amounts of the mafic minerals olivine and orthopyroxene. It is referred to as the Mg-suite, so-called because its rocks are rich in magnesium (Mg). Petrological and geochemical studies have demonstrated that the Mg-suite and the ferroan anorthosites could not have crystallized from the same parental magma; thus, at least two, and probably more, internally generated magmas contributed to the formation of the early lunar crust.

During the early study of the *Apollo* samples, an unusual chemical component was identified that is enriched in incompatible trace elements, that is, those elements that do not fit well into atomic sites within the common lunar minerals (plagioclase, pyroxene, and olivine) during crystallization. This element group includes potassium (K), the rare-earth elements (REE, e.g., samarium), and phosphorous (P); because of its enrichment by these elements, this component was given the name KREEP. KREEP has been found within many highland soils, breccias, and impact melts; it appears to result from the impact admixture of one or more now-obliterated rock types with more common lunar rocks.

HISTORY AND EVOLUTION OF THE MOON

About 4.6 billion years ago, a large disk of debris orbited the Earth, possibly a result of a giant impact. This material was depleted in iron and volatile elements and enriched in refractory elements with respect to other terrestrial protoplanetary material. On the order of a few tens of millions of years later, this disk of debris coalesced to form the Moon. The coalescence on such a short time scale released a large amount of thermal energy, enough to melt almost completely at least the outer few hundred kilometers of the Moon. The detailed nature of this melted material, called the lunar magma ocean, is only dimly perceived at present. In this global, anhydrous magma system, low-density plagioclase feldspar tended to float to the top, whereas the higher-density mafic minerals, olivine, and pyroxene, sank toward the bottom of the molten shell. Thus, the protocrust of the Moon consisted of floated plagioclase cumulates (ferroan anorthosites) tens of kilometers thick, and the protomantle consisted of ultramafic olivine–pyroxene cumulates, the future source regions of mare basalts. This magma ocean may not have been everywhere completely molten, as the production of other magmas (the Mg-suite) and their intrusion into the crust may have been contemporaneous with this stage in lunar history.

This intense, global magmatic activity was accompanied by a very high meteoritic bombardment rate; the impacts fragmented and mixed the uppermost portions of the crust. The early global magmatism was largely completed by about 4.3 billion years ago, when the last incompatible residual phases of the original magma system crystallized as the KREEP source region. This was not the end of regional magmatism, however, and the emplacement of numerous smaller intrusive bodies and the eruption of volcanic lavas began at this time. Both nonmare, KREEP volcanics and mare basalts were erupted, but most of their surfaces were destroyed by the heavy bombardment. As the impact rates began to decline, the oldest surfaces were preserved. Numerous large impacts thoroughly mixed the upper crustal materials into polymict breccias, largely destroying the original igneous contacts within the primordial crust. Some of the larger multiring basins penetrated this impact-processed debris layer, producing basin impact melts and bringing up pristine rock types from depth for our collection and inspection 3.9 billion years later.

The Imbrium and Orientale impact basins were the last major basins formed on the Moon. We now estimate the age of the Imbrium impact as about 3.85 billion years; the Orientale impact probably occurred within a few tens of millions of years later. Because the cratering rate was declining very rapidly at about this time, more volcanic flows began to be preserved. The visible maria, which cover about 16% of the Moon's surface, probably should be considered to represent the declining stages of lunar mare volcanism; although we are not certain of the extent of mare volcanism prior to the end of the heavy bombardment, it may well have been more extensive before the Imbrium basin was formed. Numerous large impact craters continued to form from 3.8 to 3 billion years ago, after which the cratering rate appears to have been relatively constant.

The Imbrian–Eratosthenian boundary is placed between the mare deposits of the *Apollo 15* site (uppermost Imbrian, 3.3 billion years old) and the mare flows of the *Apollo 12* site (lowermost Eratosthenian, 3.15 billion years old). Mare basalt eruptions continued to flood the surface, although in rapidly decreasing volumes after 3 billion years ago. On the order of 1.5–1 billion years ago, the large crater Copernicus formed and the current phase of lunar history, the Copernican period, began. It is possible that some very small amounts of mare basalt were extruded at the beginning of the Copernican, but the dominant geological activity on the Moon since that time has been the occasional large impact and the continued formation of the regolith. For all practical purposes, the Moon is now geologically "dead."

This general history of an intensively active early Moon and a quiescent current Moon gives us a picture of a planetary body fossilized in time. The period of active geological processes on the Moon, from 4.6 to about 3 billion years ago, perfectly complements the observable geological record of the Earth, for which rocks older than 3 billion years have been almost completely destroyed by active processes. Thus, the Moon holds secrets of planetary processes that we could barely imagine before the Space Age.

Additional Reading

Heiken, G., Vaniman, D., and French, B. M., eds. (1991). *The Lunar Sourcebook*. Cambridge University Press, New York.

Masursky, H., et al., eds. (1978). *Apollo over the Moon: A view from orbit*. NASA **SP-362**.

Mutch T. A. (1972). *The Geology of the Moon: A Stratigraphic Approach*. Princeton University Press, Princeton, N.J.

Spudis, P. D. (1990). The Moon. In *The New Solar System*, 3rd ed., J. K. Beatty and A. Chaikin, eds. Sky Publishing Corp., Cambridge, MA, and Cambridge University Press, Cambridge, U.K., p. 41.

Wilhelms, D. E. (1987). *The Geologic History of the Moon*. U.S. Geological Survey Professional Paper **1348**.

See also **Mercury and the Moon, Atmospheres; Moon, Origin and Evolution; Moon, Rock and Soil; Moon, Seismic Properties; Planetary Tectonic Processes, Terrestrial Planets; Planetary Volcanism and Surface Features.**

Moon, Lunar Bases

Paul D. Lowman, Jr.

Neil Armstrong's electrifying report "Tranquillity Base here—the Eagle has landed" marked a giant step toward the dream of many decades, a base on the Moon, although the *Apollo 11* crew's brief visit could hardly be considered colonization. Lunar bases and colonies had been a science fiction staple for years, but serious technical studies did not begin until after World War II when the V-2 ballistic missile showed that spaceflight was technically possible in the near future. A number of essentially modern descriptions of possible lunar bases were published as early as 1946, in parallel with serious and technically sound treatments of spaceflight in general.

When the Apollo Program began in 1962, it was assumed by NASA that the Apollo missions would be only the beginning of manned lunar exploration and occupation. A series of detailed engineering studies was begun by NASA and its contractors for modular lunar bases that could be established using modifications of the Saturn V launch vehicle and the lunar module (LM). The base concepts ranged from long Apollo missions, with stay times on the Moon of up to two weeks, to permanent modular bases designed for habitation by as many as 18 people. One major study was the Apollo Logistics Support System (ALSS), in which an improved LM ("LEM Truck") would be used to land a 7000-lb payload consisting of either a shelter or a two-person roving vehicle, for a 14-day stay. Although ALSS never became an approved program under this name, the *Apollo 15*, *16*, and *17* missions had capabilities that came surprisingly close: lunar rovers with a range of several kilometers and up to 3 days on the Moon.

The most ambitious of the NASA lunar base plans during the 1960s was termed Lunar Exploration Systems for Apollo (LESA), in which the Saturn V, with a new upper stage, would be used to land 25,000-lb shelter modules permitting, as a beginning, 90-day occupations by 3 people. An extensive series of detailed engineering studies of all aspects of a program based on LESA was carried out. They are of largely historical interest now if for no other reason than that the Saturn V is no longer available. However, in light of the successful Apollo missions, based on similar technology and assumed environmental conditions, it is safe to say that the United States could have established a lunar base by the late 1970s had a decision to do so been made.

A still-relevant lunar base concept was developed in 1971 under the title "Lunar Base Synthesis Study," funded by NASA. Although not carried out, this study was notable in at least two aspects. First, being completed after the first Apollo missions, it could embody realistic and well-grounded environmental models. Second, it was the first base concept designed for modules that could be carried into Earth orbit by the Space Shuttle, then recently approved.

With the revival of the American manned space flight program in 1981 came a comparable revival of interest in lunar bases. Despite the hiatus since the *Apollo 17* mission, considerable low-level work had been carried on by various organizations. A conference on "Lunar Bases and Space Activities in the 21st Century" was held in 1984 at Washington, at which one of the opening talks was given by George A. Keyworth, President Reagan's science advisor, a good indication of administration interest. Loss of the *Challenger* in 1986, of course, derailed American space efforts in general, and the ambitious program of solar system colonization proposed by the National Commission on Space, issued shortly after the *Challenger* disaster, was largely ignored. Nevertheless, planning continued. An important development, in 1987, was establishment of human expansion into the solar system as an official NASA goal.

A major study, clearly marking the beginning of a new lunar program, including a base, was completed in 1989 under the title "Lunar Outpost." Although recognizably derived from the Apollo hardware, the system of orbital transfer vehicles, lunar landers, and shelters was different in many fundamental ways. Transportation between Earth and Moon, for example, would use an Earth-orbiting space station in conjunction with aerobraking, unlike the Apollo system.

The Lunar Outpost concept also differed from Apollo-based systems in using large-volume inflatable shelters (Fig. 1), insulated and protected by lunar soil. It was also designed for pilot plant extraction of oxygen from lunar soil, a process studied in detail since the end of the Apollo Program. Such activities would require large amounts of power, and the study considered photovoltaic, solar dynamic, and nuclear systems. Solar power systems suffer from the problem of intermittent energy supply, except for polar locations, and would need supplemental energy storage devices such as fuel cells.

The Lunar Outpost study recommended a site in the Orientale Basin at about 13°S on the west limb. Such a site would have astronomical advantages, such as intermittent occultation of the Earth and thus of terrestrial radio interference, and would permit exploration of a geologically important area. An independent study

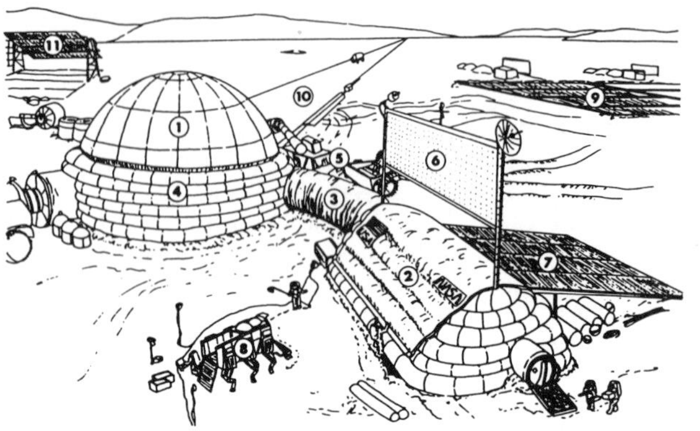

Figure 1. Lunar Outpost concept (1989): (1) the inflatable habitat; (2) the construction shack; (3) connecting tunnel; (4) continuous, coiled regolith bags for radiation protection; (5) regolith bagging machine, coiling bags around the habitat while bulldozer scrapes loose regolith into its path; (6) thermal radiator for shack; (7) solar panel for shack; (8) experimental six-legged walker; (9) solar power system for the outpost; (10) road to landing pad; (11) solar power system for the lunar oxygen pilot plant.

Figure 2. Characteristics of NE Orientale basin site: (1) *Latitude* (equator)—gives access to entire celestial sphere over one month; launch at any time to equatorial lunar transportation node. (2) *Longitude* (80°W)—continuous line of sight to Earth for communications and observations; close to farside for radio astronomy instrument emplacement and maintenance. (3) *Terrain*—generally trafficable and workable, subject to detailed orbital reconaissance, Earth-based study, possibly remotely controlled rover traverse. (4) *Geologic interest*—Orientale Basin, Oceanus Procellarum volcanic features, Aristarchus Plateau, LO 5 crater, Flamsteed Ring, Reiner Gamma, *Luna 8*, *Luna 9*, *Surveyor 1*. (5) *Volcanic features*—nearby (mare basalt, Kopff, Marius Hills) and favorable for resource exploration.

of lunar "platform science" also recommended a site in this region, specifically centered on the equator at 80°W longitude, that is, on the northeast flank of the basin. Such a location would permit, over a month's time, astronomical observation of essentially the entire celestial sphere while retaining continuous line-of-sight contact with Earth, for ease of communications and data relay. This area and its characteristics are shown in Fig. 2. The author suggested in 1991 that the central peak of Riccioli crater, on the northeast flank of Orientale Basin, at 3°S, 73°W, would be an excellent candidate site for a lunar observatory.

Looking toward the future, it appears certain that a lunar base will be established by one or a combination of countries within the next 20 years. Establishment and maintenance of such a base, after the initial "outpost" stage, would permit detailed exploration of the Moon and assessment of its resources. Astronomical and space physics observations from the Moon would perhaps furnish the main scientific justification for a permanent base, because a lunar observatory site would have inherent advantages over observatories in low Earth orbit, such as reduced terrestrial interference and extremely long data collection or exposure times. The regular Earth–Moon flights necessary to maintain a base would build up technical and operational experience for interplanetary missions. Similarly, lunar surface operations would help establish the infrastructure for bases on Mars, presumably the next target of human exploration after the Moon.

Taking the very long view, a lunar base would represent the first step toward dispersal of the human species beyond the Earth, a dispersal long recognized as absolutely essential if humanity is to escape destruction by inevitable and uncontrollable natural catastrophes such as major impacts, glaciation, nearby supernovae, and the like. Such catastrophes are now recognized as probable causes of many extinctions in the geological record. As long as our species is confined to the Earth, it is vulnerable; colonization of the Moon and eventually other bodies in the solar system may help insure long-term survival.

Additional Reading

Burgess, E. (1957). *Satellites and Spaceflight*. Macmillan, New York.

Burns, J. O., Duric, N., Taylor, G. J., and Johnson, S. W. (1990). Observatories on the Moon. *Scientific American* **262** (No. 3) 42.

Clarke, A. C. (1951). *The Exploration of Space*. Harper, New York.

Mendell, W. W. (1985). *Lunar Bases and Space Activities of the 21st Century*. Lunar and Planetary Institute, Houston.

Mumma, M. J. and Smith, H. J., eds. (1990). *Astrophysics from the Moon*. American Institute of Physics, New York.

National Commission on Space (1986). *Pioneering the Space Frontier*. Bantam Books, New York.

See also **Moon, Rock and Soil; Moon, Space Missions.**

Moon, Origin and Evolution

H. Jay Melosh

Planetary scientists' ideas on the origin of the Moon have changed drastically since 1984 when a seminal conference took place in Kona, Hawaii. The giant impact theory that emerged in the aftermath of that conference is now widely accepted, although many important questions remain to be answered. The need for a new theory of the Moon's origin grew out of the confrontation between the three "classical" theories (capture, fission, and coaccretion) and the flood of new data on the Moon provided by the U.S. *Apollo* missions from 1969 to 1972. Analysis of the lunar rocks returned by these missions, along with geophysical measurements and photographs made both on the surface and from orbit yielded a picture of the Moon that is incompatible with any of the classical theories and provided the foundation for an entirely new approach to the Moon's origin.

MOON'S PRESENT STATE

It has been known for over a century that the Moon is, on average, much less dense than the Earth. The modern value of the Moon's mean density is 3.344 g cm^{-3}. Although this is suggestively close to the density of the Earth's mantle, it is less than 0.6 of the Earth's mean density and is taken to indicate that the Moon lacks a significant metallic iron core. The absence of a core is consistent with the Moon's principal moment of inertia ratio $C/MR^2 = 0.3904$, which is nearly the same as that of a uniform sphere (0.4000). (C is the principal moment of inertia, M is the mass of the moon, and R is the radius of the moon.) If the Moon does have a metallic iron core (this has not yet been proved), the core is less than 450 km in radius and comprises less than 4% of the Moon's total mass. Although the Moon has no intrinsic magnetic field at present, remnant magnetism has been found in ancient lunar crustal rocks, suggesting the activity of a dynamo early in its history.

The Moon has a crust composed predominantly of the plagioclase-rich rocks anorthosite, norite, and troctolite, which appear to have crystallized from melts. Seismic and gravity investigations show that the average thickness of this crust is less than about 60 km. The crust is generally thicker on the far side of the Moon than on the near side, where it locally thins to 20 km beneath the lava-filled mare basins. The lunar mantle extends to a depth of about 1000 km and is composed mainly of the minerals olivine and pyroxene. The mantle and crust appear to have separated chemically about 4.5×10^9 yr ago. The rare earth element europium, which has an affinity for plagioclase, is enriched in the crust but depleted in the mantle, a complementary pattern suggesting that the two separated during an era of widespread melting in the ancient Moon. Below about 1000 km depth seismic waves are strongly attenuated, indicating rocks below this depth are partially molten. Very small "moonquakes" occur regularly between 700 and 1100 km depth in response to tidal stresses.

The estimated bulk chemistry of the Moon shows both striking similarities and differences with the Earth's mantle. The oxygen isotopic signatures of the Earth and Moon are basically identical, indicating a close relation between the two bodies. The major elements Si and Mg are generally similar in both bodies. Iron oxide was believed to be more abundant in the Moon than in the Earth's mantle, but recent work suggests that the FeO content of the Earth's lower mantle may be much higher than previously thought, bringing the Moon and Earth into agreement on this element as well. On the other hand, the refractory elements Al, Ca, Cr, Ti, and U may be enriched in the Moon compared to the Earth, whereas the volatile elements Na, K, and Rb are strongly depleted. The highly volatile molecule H_2O has not been found at all in any lunar sample. Elements with an affinity for metallic iron (siderophiles) such as Ni, Ir, Mo, and Ge are much more strongly depleted in the Moon than in the Earth's mantle.

The ratio between the Moon's mass and the Earth's mass is 0.0123, larger than any other planet–satellite pair in the solar system except for Pluto–Charon. The Moon's mean distance from the Earth is currently 60.3904 Earth radii and the Moon is receding from the Earth at a rate of 3.74 cm yr^{-1} due to tidal friction. An important fact is that the Moon is not now, and almost certainly never has been, in the Earth's equatorial plane, to which its inclination varies between 18.4 and 28.6° as its orbit precesses. The angular momentum of the Earth–Moon system is anomalously high compared to the other planets (except, again, for the Pluto–Charon system). If all of the mass and angular momentum of the Earth–Moon system were put into the Earth it would rotate with a period of only about 4 h.

CLASSICAL THEORIES OF ORIGIN

There are three basic models for lunar origin that were under study at the time of the *Apollo* landings. Although there are many variants of these models and even some hybrids, three pure "end-members" can be described under the headings of fission, intact capture, and coaccretion.

Fission

One of the best-studied theories of the Moon's origin is derived from the high angular momentum of the Earth–Moon system. In 1879, George H. Darwin suggested that if the Earth ever did spin with a period of 4 h, it would be subject to rotational instabilities and a chunk (or a debris ring in more modern theories) would spin off at the equator and could eventually clump together outside the Roche limit to form the Moon. This theory neatly predicts the Moon's depletion in metallic iron and is consistent with the similarity of oxygen isotopes and major elements in the Earth and Moon. However, it does not account for the differences in refractory or volatile elements, incorrectly predicts that the Moon was once in Earth's equatorial plane, and altogether sidesteps the question of how the Earth began in a state of such anomalously rapid rotation.

Intact Capture

The most direct way to explain the differences between the Moon and the Earth is to suppose that they were formed in different parts of the solar system under different conditions and that the Earth subsequently captured the Moon. This theory also has a long history of investigation, none of which has succeeded in overcoming the problem that capture of a Moon-size body is a very low-probability process. Capture is plausible only if the relative velocity of the Earth and Moon is very low, and for the velocity to be low the two bodies must have formed at nearly the same radial distance from the Sun, making it hard to understand how they could differ chemically. Although capture could account for some of the differences between the Earth and Moon, their similarities are fortuitous in this theory.

Coaccretion

If the Moon neither broke off the Earth nor came from elsewhere in the solar system, then perhaps it grew along with it. A group of theories has developed recently that posits that the Moon accumulated from debris in orbit around the growing Earth. These theories run immediately into the problem of explaining how the Earth happened to acquire a metallic iron core totaling about 32% of its mass whereas the Moon remained nearly iron-free. Although a variety of "compositional filters" have been proposed, most of which rely on the differences in mechanical strength between iron and silicate planetesimals, none has gained general acceptance as a plausible explanation of the density difference between the Moon and the Earth. Coaccretion theories also have difficulty explaining the other chemical differences, predict that the Moon grew in the Earth's equatorial plane, and, for somewhat more subtle reasons, fail to account for the angular momentum of the Earth–Moon system. Nevertheless, theories of this kind enjoyed widespread support at the time of the *Apollo* landings.

GIANT IMPACT HYPOTHESIS

The idea that the Moon might have been formed in an unusually large impact with the proto-Earth (see Fig. 1) was suggested inde-

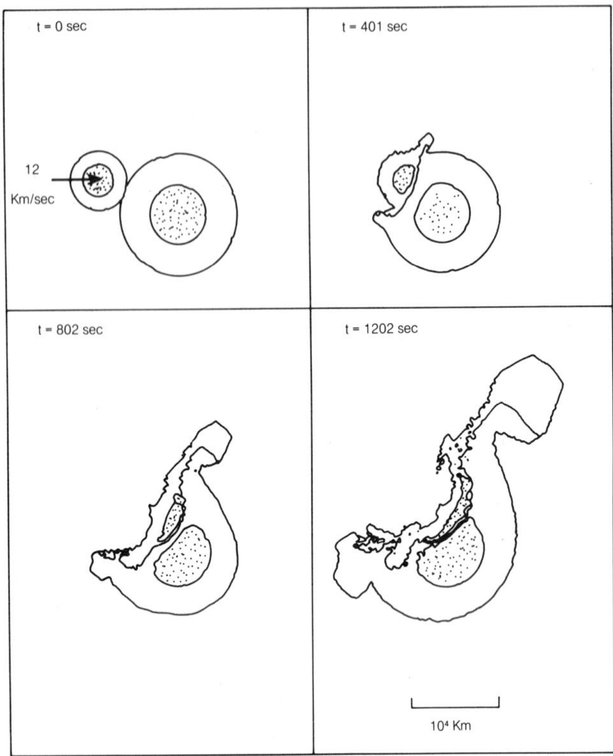

Figure 1. Four stages in the collision of a Mars-size protoplanet with the proto-Earth at an impact velocity of 12 km s^{-1} and impact parameter equal to 0.59 of the Earth's radius. A hot jet of vaporized mantle material forms during the collision and expands away from the impact site. This jet contains material from both the projectiles and target planet's mantles, but no core material. [*Reproduced, with permission, from H. J. Melosh (1988),* Impact Cratering: A Geologic Process, *Oxford University Press.*]

pendently by Alastair G. W. Cameron and William R. Ward and by William K. Hartmann and Donald R. Davis in 1975. Earlier, in 1970 and 1972, Alfred E. Ringwood had argued that the chemistry of the Moon indicates that its material was derived from the Earth's mantle by impacts, but he was unable to propose a plausible dynamical scenario in which this occurred. Cameron and Ward noted that the angular momentum problem would be solved if the Earth were struck glancingly by a Mars-size protoplanet at a speed roughly comparable with the Earth's escape velocity, 11.2 km s^{-1}. The angular momentum of the Earth–Moon system does not provide enough information to determine separately the angle of impact, mass, and velocity of the projectile, but given plausible values of these parameters a mass ranging from about $\frac{1}{2}$–3 times the mass of Mars is probable. Hartmann and Davis approached the problem by examining the distribution of protoplanetary sizes in the solar nebula near the end of planetary accretion. They concluded that, in addition to a few large protoplanets, there were probably several times that number that were half as large, still larger numbers that were one-fourth as large, and so on down to the smallest sizes. In their view the last stages of planetary accretion in which these objects were swept up must have been characterized by extraordinarily violent collisions among these planetary-size objects. This view was also discussed in detail by George W. Wetherill in his 1985 numerical simulations of the final stages of planetary growth.

An apparent problem with the giant impact idea is that ordinary solid ejecta from an impact follow keplerian orbits, which are either open hyperbolae, so that the ejecta escapes the Earth, or closed ellipses that the ejecta follow back to their starting point—the Earth's surface. In neither case do any ejecta achieve orbit. However, in a high speed impact much of the projectile and an equal or larger quantity of the target vaporize. The expansion of this vapor plume is driven by the non-keplerian pressure gradients and, in a sufficiently large impact, may achieve altitudes of several Earth radii before the plume condenses and the solid condensates take up keplerian orbits that, in this case, do not intersect the Earth. Alternatively, because of the large mass of the projectile, noncentral gravitational forces may act powerfully enough among a portion of the ejecta to modify its trajectory into a stable orbit. Detailed numerical computer codes that have been constructed to study the early phases of a planetary-scale impact demonstrate that both factors are important. The immediate result of most of these impact scenarios is a ring of debris in orbit about the Earth. Interactions within this debris ring circularize the orbits of the ejecta particles, reduce their orbital inclinations, and permit them to clump together into the body we now recognize as the Moon. Because of the short orbital period, most of this accretion takes place only a few hundred years after the impact event.

On the basis of this model the similarities between the Earth and Moon stem from the incorporation of a fraction of the Earth's mantle in the ejecta plume. The size of this fraction has not yet been resolved, however. In some models that successfully orbit a lunar mass, the Moon forms entirely from the projectile, whereas in others the Moon incorporates about 50% of Earth mantle material. In all simulations the Moon receives little metal if both the proto-Earth and protoplanet projectile had cores. The metal in the cores remains separate from the major activity of the collision and is unlikely to vaporize. In some simulations of very glancing collisions the massive core of the projectile separates from its mantle and follows a short ellipse after the impact, gravitationally slinging its silicate mantle into a stable orbit while the core's perigee is lowered, after which it reimpacts the Earth.

The chemical differences between the Earth and Moon in this model are either inherited from the projectile or arise from conditions in the impact itself. The Moon's depletion in volatiles and enrichment in refractory elements relative to the Earth's mantle could have developed in Earth mantle material as a result of vaporization. Mantle material vaporized by the impact expands rapidly and cools in a vacuum so that the late-condensing volatiles become physically separated from the more refractory elements. The near identity of the oxygen isotopic signatures of the Earth and Moon is hard to understand if the Moon were mostly composed of projectile material, barring a fortuitous coincidence between oxygen isotopic signatures of the proto-Earth and the projectile. Vaporization and recondensation of Earth mantle material, on the other hand, does not affect the oxygen signature.

Although most discussions of the giant impact hypothesis give prominence to only a single impact, it is clear that the Earth or Moon suffered at least two such impacts and almost certainly was struck many more times by objects smaller than Mars. More than one impact is necessary because the Moon is not in the Earth's equatorial plane. A single impact on a nonrotating initial Earth is bilaterally symmetrical and cannot lead to a Moon inclined to the Earth's equator, so the Earth must have either been struck by another large projectile before the Moon-forming impact to give it an initial spin, or the Earth or Moon was struck afterward to tilt the equator away from the plane of the Moon's orbit.

EVOLUTION OF THE MOON AFTER ITS FORMATION

Isotopic dating of lunar samples shows that the Moon originated about 4.5×10^9 yr ago, a date that is currently not resolvable from the age of the Earth. With current accuracies, the two bodies must have originated within 50 Myr (million years) of each other. The earliest chemical event in the Moon's geologic history was the separation of crust from the mantle in what is believed to have been a moonwide "magma ocean." This event is also indistinguishable from the age of the Moon itself, with an uncertainty of 50 Myr. The Moon's surface is scarred with thousands of impact craters ranging in size from the Imbrium basin with a diameter of 1340 km down to the smallest micrometeoroid craters that are only a few micrometers across. The flux of meteoroid impacts has not been constant through time, however. During the era of "heavy bombardment" between the Moon's origin and about 3.8×10^9 yr ago the impact flux was at least several thousand times the present flux. The great basins that form the basis for lunar stratigraphy have ages that date from near the end of this period. Lava rose from the Moon's mantle and flowed out over its surface, partly filling the low-lying nearside mare basins for nearly 10^9 yr after their formation. The bulk of lunar volcanism then ceased about 3×10^9 yr ago. The Moon is a one-plate planet: plate tectonics never occurred on the Moon because its small size prevents the vigorous convection necessary to break cold surficial rocks into discrete plates. The Moon's few geologic faults all seem to be related to impact basins or to sagging of the crust beneath the load of extruded lavas, although an ancient "tectonic grid" or fabric of faults and fractures in the crust may have developed as it was flexed by tidal forces during the Moon's recession from the Earth. Subsequent to its violent birth and active youth the Moon has lapsed into a state of geologic senescence marked only by occasional impacts and the regular, tidally-triggered moonquakes.

Additional Reading

Burns, J. A. (1986). The evolution of satellite orbits. In *Satellites*, J. A. Burns and M. S. Matthews, eds. University of Arizona Press, Tucson, p. 117.

Drakes, M. J. (1990). Experiment confronts theory. *Nature* **347** 128.

Hartmann, W. K., Phillips, R. J., and Taylor, G. J., eds. (1986). *Origin of the Moon*. Lunar and Planetary Institute, Houston.

Kaula, W. M., Drake, M. J., and Head, J. W. (1986). The moon. In *Satellites*, J. A. Burns and M. S. Matthews, eds. University of Arizona Press, Tucson, p. 581.

Stevenson, D. J. (1987). Origin of the moon–The collision hypothesis. *Ann. Rev. Earth Planet. Sci.* **15** 271.

Taylor, S. R. (1982). *Planetary Science: A Lunar Perspective*. Lunar and Planetary Institute, Houston.

Wilhelms, D. E. (1987). *The Geologic History of the Moon*. U.S. Geological Survey Professional Paper 1348.

See also **Moon, Geology; Moon, Seismic Properties; Solar System, Origin.**

Moon, Rock and Soil

Steven B. Simon

Over the past two decades, our understanding of the formation, petrology, and geology of the Moon has improved greatly. In the minds of many scientists the Moon has evolved from a distant object of curiosity to a natural laboratory for planetary geology. Due to the lack of continental drift and chemical weathering on the Moon, rocks more than 4 billion years (4×10^9 yr) old still exist there. Such rocks yield information about the early history of the solar system that is unavailable on Earth. Furthermore, understanding of the Moon's formation and early evolution helps us learn about the Earth, because the two bodies appear to have formed close together in space and time.

Only recently have we approached a consensus on the origin of the Moon. Compositional similarities indicate that the Earth and Moon are somehow related and argue against random capture of the Moon by the Earth. However, compositional differences, such as the Moon's depletion in volatile elements and iron and its enrichment in refractory elements, eliminate formation of the Moon directly from the Earth as a possibility. Through improved understanding of the dynamics of the early solar system and the growth of planets as well as advances in computational capabilities, the "collisional ejection" hypothesis has emerged as the favored theory. Simply put, the model involves impact of a Mars-sized body into the Earth, producing a disk of material from which the Moon forms, containing components of both the impactor and the Earth. The model, though not proven, may be able to account for the compositional similarities and differences of the Earth and Moon and also for the high angular momentum of the Earth–Moon system.

From the time of formation until about 3.8×10^9 yr ago, the Moon and most of the other planets and moons in the solar system experienced a period of abundant meteorite impacts. With planets sweeping up debris, cratering rates 4.6×10^9 yr ago approached 10^{10} times the present rate, declining to about 100 times the present rate by 3.8×10^9 yr ago. Evidence of this early epoch in planetary formation can be seen in the densely cratered highland surfaces of the Moon, as well as Mercury and many of the moons of Mars, Jupiter, and Saturn. Certainly the Earth experienced intense cratering as well, but Earth's active geological processes have erased that part of the record. The large multiringed basins on the Moon testify to the size and force of some of the projectiles.

This period of intense bombardment is something of a mixed blessing for geologists. The impacts thoroughly mixed many of the original rocks of the ancient crust, melted, buried, or obliterated many others, and reset isotopic clocks. On the positive side, the impacts excavated deep rocks that would otherwise be inaccessible and distributed rock fragments laterally away from their places of origin.

This last point is very important with respect to lunar sample studies. Although only nine localities on the Moon have been sampled directly, we have samples of many different rock types due to transport of "exotic" materials by meteoritic impacts. Thus, a single sample of lunar soil, itself actually a powder of impact-derived debris, may contain a variety of rock types from different localities.

Another important aspect of the basin-forming events is that they thinned and cracked the underlying crust, allowing molten basalt from the interior to reach the surface. Although they occupy impact basins, lunar basalts crystallized from internal melts, not impact melts. They are too Fe- and Ti-rich to have formed from the crust, and their clast-free textures are also inconsistent with an impact origin. The basalt flows have relatively few craters because their eruption postdates the time of high meteorite flux. Their dark, smooth surfaces looked like seas to the early European astronomers, who called the basins *maria*, which is the Latin word for seas; hence the terms mare basalt, Mare Tranquillitatis, etc.

The light-colored, feldspar-rich rocks of the highlands and the mare basalts are the two major lunar rock types. A third, less-abundant rock type is called KREEP [basalt rich in potassium (K), rare earth elements (REE), and phosphorus (P)], some samples of which may have formed by volcanism in the highlands. These three rock types and the lunar soil will be considered in this entry.

LUNAR HIGHLAND ROCKS

The returned lunar samples have provided much evidence supporting the existence of a moonwide magma ocean or "magmasphere" soon after the formation of the Moon. The arguments will be summarized here.

The lunar crust averages 75% plagioclase feldspar, a light-colored, Ca- and Al-rich silicate, and we know from sample studies that the rocks are cumulates. The most plausible way to account for such a large volume of plagioclase is not by the generation of a melt of this composition, but by concentration of crystals by flotation of plagioclase on an anhydrous basaltic melt.

Ages of pristine (having no meteoritic contamination) highland rocks are in the $4\text{–}4.5 \times 10^9$-yr range. These ages indicate crystallization within a few hundred million years of the formation of the Moon, not enough time for large-scale internal heating.

Anorthosite

The anorthosites (rocks containing $>90\%$ plagioclase) exhibit a wide range of Fe/Mg ratios in mafic (ferromagnesian) silicates but there is no corresponding Na enrichment in plagioclase, which would be expected as crystallization proceeded. The large amount of Ca-rich plagioclase relative to the mafics may have buffered the plagioclase composition. Alternatively, the anorthosites may simply be reflecting very low Na_2O contents in the parent liquid.

The crystals in most lunar anorthosites have been cracked and broken by impacts, the degree of crystallization ranging from none to complete. Original plagioclase grain size was probably about 5 mm, Mg/(Mg+Fe+Ca) of low-Ca pyroxene is generally in the range 0.55–0.65, and Mg/(Mg+Fe) of olivine is in the range 0.40–0.66 and most commonly is around 0.60. Augite and orthopyroxene are present in many samples, and in a suite of *Apollo 16* anorthosites indicate equilibration to below 800°C. Exsolution lamellae, however, are rare. Minor minerals are ilmenite, Fe metal, chromite, and silica.

Chondrite-normalized abundances of the rare earth (lanthanide) elements in lunar anorthosites are low ($<1 \times$ chondrite) except for Eu, which exhibits a strong positive Eu anomaly. This is attributed to the reduction of Eu^{3+} to Eu^{2+}, which can substitute for Ca^{2+} in plagioclase.

Mg-Suite

Another suite of lunar plutonic rocks includes a variety of lithologies with a wide range of mineral compositions. Pyroxene and olivine compositions extend to higher values of Mg/(Mg+Fe), and these rocks are known collectively as the Mg-suite. This suite includes a range of rocks from dunite (almost pure olivine) through gabbroic, noritic, and troctolitic anorthosites (10–25% clinopyroxene, orthopyroxene, and olivine, respectively), and many modally intermediate types. Norites (plagioclase+orthopyroxene) are probably the most abundant among the returned highland rock samples. New rock types are still being discovered: Mg-anorthosite and Mg-gabbronorites are just two examples.

The bulk compositions of these rocks reflect their lower plagioclase and higher mafic mineral contents relative to anorthosites. Trace elements contents are higher, and typical REE abundances are approximately $10 \times$ chondrite.

Petrogenesis

The variety of endogenous (internally generated) highland lithologies argues for more than one magma.

Although a magma ocean appears unable to account for all the observed lithologies, it is not ruled out; rather, a second episode of magmatism is invoked. According to one popular model, partial melting began in the primitive mantle after crystallization of the magma ocean was complete except for a discontinuous KREEPy layer (discussed below) between the mafic cumulates and the anorthosite. Thus some of the rising magmas acquired a KREEP component and others did not. After intrusion into the anorthositic crust, the magmas formed layered plutons. These plutons were then exposed, brecciated, granulated, and then melted by meteoritic bombardment. The variable KREEP assimilation and the mixing of lithologies from layered plutons could account for the observed compositional diversity. Most workers now agree that the Mg-suite rocks formed from intrusions that postdate the anorthositic crust, but the source of the melts is uncertain.

MARE BASALTS

Although mare basalts compose only about 1% of the lunar crust and about 17% of the surface, they provide much information that is necessary for an understanding of lunar thermal history, geochemistry and the nature of the lunar interior. The basalt flows provide smooth, relatively sparsely cratered terrain on which to land. The *Apollo 11* (*A-11*), *A-12*, *Luna 16* (*L-16*) and *Luna 24* (the last two being Soviet unmanned missions) landing sites are all on basalt, and *Apollo 15* and *Apollo 17* landed on basalt near mare–highland interfaces.

The chemistry, mineralogy, and petrology of mare basalts have been thoroughly studied. As with the Mg-suite highland rocks, in spite of a large amount of data, the sources and depths of origin of the mare basalts remain uncertain.

Different types of basalts were discovered at each site. They are generally divided into three groups on the basis of TiO_2 content. High-Ti basalts occur at the *Apollo 11* and *Apollo 17* sites, low-Ti at *Apollo 12* and *Apollo 15* sites, and very low-Ti at *Apollo 17* and *Luna 24* sites. The wide gap between the low- and high-Ti groups is probably a function of sampling. It is likely that a continuum of TiO_2 contents exists, and remote sensing studies indicate that there are unsampled mare basalt varieties.

Textures cover the full range for basalts, from vitrophyric (glassy, indicating rapid cooling) to coarse intergrowths of plagioclase, pyroxene, and olivine, with crystals up to several centimeters in length. Bulk compositions range from olivine-normative to quartz-normative.

Pyroxene is more abundant than feldspar, and opaques (mostly ilmenite) increase with increasing TiO_2. Most minerals are chemically zoned, reflecting the relatively rapid cooling experienced by these rocks.

Mare basalts do not contain the Mg-rich orthopyroxene (opx) found in Mg-suite plutonic rocks. The basalts have a wide range of clinopyroxene (cpx) compositions, extending to extremely Fe-rich compositions. In addition to the major pyroxene components (Fe, Mg, and Ca), systematic variations of Al, Ti, and Cr abundances make pyroxenes effective recorders of basalt crystallization histories.

The Mg/(Mg+Fe) of basaltic olivines ranges from 0 to 0.8. Mare basalts are known to have up to 35 vol.% olivine. In samples in which it is present, it is one of the first minerals to crystallize.

Mare basalt feldspar compositions have a much narrower range than olivine and extend to higher Ca contents than terrestrial basalts. For example, in ocean floor basalt, plagioclase Ca/(Ca+Na) ranges from 0.25 to 0.9, whereas in mare basalts, it ranges from 0.7 to 1.0.

A variety of opaque phases is found in mare basalts. Ilmenite is the most common and occurs in all basalt types. In most samples it is almost pure $FeTiO_3$; MgO contents range up to 6 wt.% and probably reflect the bulk Fe/Mg of the rock. Ilmenite will also be affected by whether it is an early phase (high-Ti basalts) or a late one (low-Ti basalts). Unlike the terrestrial variety, lunar ilmenite has no Fe^{3+}. Spinel, troilite, armalcolite, and native iron also occur in mare basalts.

Volcanic Glasses

In addition to the basalts described previously, lunar mare volcanism is also manifested in the form of volcanic glass. Such glass forms when molten material is erupted and cooled very quickly, before crystals can grow. Concentrations of glass beads were found at the *Apollo 15* and *Apollo 17* sites, and through study of soil and soil breccia samples from all the sampling sites, 25 varieties of volcanic glass have been identified. They provide important clues to lunar basalt generation because they are more primitive than crystalline basalts and have generally undergone less chemical change and crustal contamination than basalts; thus they are probably more representative of their original liquid compositions. Also, glass beads exhibit surficial enrichments in many volatile elements, indicating that eruption was associated with a vapor phase. The glasses are thought to have been erupted in fire-fountaining events.

Ages of Mare Basalts

Ages of mare basalts not only are important for determining the geochronology of lunar basaltic volcanism, but they allow us to obtain minimum ages for the large basins that contain the flows and to calibrate relative ages obtained from crater counts.

The ages of the major mare basalt units that were sampled range from $3.86 \pm 0.7 \times 10^9$ to $3.08 \pm 0.5 \times 10^9$ yr with some high-Al basalts from *Apollo 14* at about 4×10^9 yr. Dating of basalt clasts extracted from *Apollo 14* breccias has extended the range of basalt ages to older than 4.2×10^9 yr. It is not known how extensive basaltic volcanism was at that time, however. These basalts formed before the great bombardment ended, so the basins that these flows probably occupied have been obscured by later basins and ejecta. Although there are older samples and younger (unsampled) units, if the returned samples are an accurate indication, the main epoch of mare basalt volcanism was from 3.0 to 3.9 billion years ago.

No simple model for mare basalt petrogenesis explains all the data. Most likely, a heterogeneous cumulate source was melted at depths between 150 and 500 km. No mare basalts have been identified as having been derived from a primitive, unfractionated source. All bear the trace element imprint (e.g., negative Eu anomaly) of the early lunar differentiation event, including a 4.2×10^9-yr clast. The presence of a negative Eu anomaly in such an ancient rock supports the theory that mare basalts were derived from Eu-depleted cumulate sources and indicates that mare volcanism began soon after these sources formed.

KREEP BASALTS

Simply put, models for the lunar magma ocean suggest three major products: anorthositic crust, mafic cumulates, and a highly fractionated residuum between the two. This incompatible-element–rich residuum is best known by the acronym KREEP. Although it has not been recognized as a rock type, it is a chemical component in many lunar rocks and soils. It is thought to have originated in the magma ocean, based on its Rb–Sr and Sm–Nd model age of $\sim 4.4 \times 10^9$ yr and its relatively uniform composition.

Most KREEPy samples are impact-produced mixtures of several rock types, but at most of the landing sites igneous samples known as KREEP basalts have been recognized.

Mineralogy

In contrast with mare basalts, KREEP basalts have Mg-rich orthopyroxene as the dominant pyroxene. In *A-15* samples the opx is zoned to pigeonite, which is followed by augite later in the crystallization sequence.

Plagioclase generally occurs as narrow crystals with $Ca/(Ca+Na)$ in the range 0.70–0.92. Most grains are between 50 and 500 μm, averaging about 200 μm. Despite the relatively K-rich bulk compositions, the plagioclase does not exhibit a significant K component, except for the most Na-rich grains. Instead, in most samples, the K and other incompatible elements tend to concentrate in the Si-rich groundmass. Fe-rich pyroxene, Ba–K feldspars, and REE-rich phosphates (whitlockite and apatite) also occur in the groundmass.

Petrogenesis

Although KREEP gives a model age of 4.4×10^9 yr, the basalts give crystallization ages of about 3.9×10^9 yr and have been interpreted as impact melts, possibly related to the event that formed the Imbrium basin. However, lack of meteoritic contamination in the *A-15* KREEP basalts, and recent petrographic observations indicating a two-stage cooling history, make a strong case for a volcanic origin for the *A-15* KREEP basalts.

Another problem in understanding their origin is the contradictory degrees of partial melting required to produced the *A-15* KREEP basalts. This stems from their relatively unfractionated mineral compositions (Mg-rich pryroxene and Ca-rich plagioclase), which indicate high degrees of partial melting, combined with their incompatible-element-rich bulk compositions, which suggest small degrees of partial melting.

REGOLITH

The lunar regolith, a several-meter-thick layer of unconsolidated debris, forms the interface between the Moon and its space environment. It virtually covers the entire lunar surface, with the possible exception of steep crater and valley walls.

The regolith forms as a result of the breakdown of bedrock into rock and mineral fragments by repeated meteorite impacts. Also present in the regolith are volcanic glasses, impact glasses, lithified soil fragments known as regolith breccias, and particles that form by micrometeorite impacts into the regolith; these impacts form new particles by creating small amounts of impact melt that bond soil grains together. The resulting constructional particles are agglutinates. They form only at the surface and their abundance increases with increasing surface residence time. They are therefore useful as recorders of surface exposure and are helpful in the interpretation of the depositional histories of soil core samples.

Because the Moon has no atmosphere and no water, the components of lunar soils do not weather chemically as they would on Earth. In general, the composition of the regolith at a given location is strongly dependent upon the local geology and rock types. However, the regolith does evolve with time, as gain or loss of material by meteorite impacts may change the soil's composition and, with continued exposure at the surface, rock fragments are broken down into mineral fragments, and agglutinates and impact glass increase in abundance.

Additional Reading

Hartmann, W. K., Phillips, R. J., and Taylor, G. J., eds. (1986). *Origin of the Moon*. Lunar and Planetary Institute, Houston.

Heiken, G., Vaniman, D., and French, B. M., eds. (1991). *The Lunar Sourcebook*. Cambridge University Press, New York.

Norman, M. D. and Ryder, G. (1979). A summary of the petrology and geochemistry of pristine highlands rocks. In *Proceedings of the Tenth Lunar and Planetary Science Conference*. Pergamon Press, New York, p. 531.

Papike, J. J., Hodges, F. N., Bence, A. E., Cameron, M., and Rhodes, J. M. (1976). Mare basalts: Crystal chemistry, mineralogy and petrology. *Rev. Geophys. Space Phys.* **14** 475.

Papike, J. J., Simon, S. B., and Laul, J. C. (1982). The lunar regolith: Chemistry, mineralogy and petrology. *Rev. Geophys. Space Phys.* **20** 761.

Spudis, P. D. (1990). The Moon. In *The New Solar System*, 3rd ed. J. K. Beatty and A. Chaikin, eds. Sky Publishing Corp., Cambridge, MA, and Cambridge University Press, Cambridge, p. 41.

Taylor, S. R. (1982). *Planetary Science: A Lunar Perspective*. Lunar and Planetary Institute, Houston.

See also **Moon, Geology; Moon, Origin and Evolution; Moon, Space Missions.**

Moon, Seismic Properties

Anton M. Dainty and M. Nafi Toksöz

The seismic properties of the Moon are its velocity structure, seismicity, noise environment, and the nature of seismic wave transmission through it. Because determination of these properties requires seismometers placed on the Moon, all of our information has been derived from the scientific work carried out as part of the Apollo project, which landed astronauts on the Moon. By the time the project ended in 1972, four long term seismic stations had been installed on the front (Earth-facing) side of the Moon at the landing sites of the *Apollo 12* in Oceanus Procellarum, *Apollo 14* in the Fra Mauro highlands nearby, *Apollo 15* in the Hadley–Appenine region, and *Apollo 16* in the Descartes region of the Central Highlands. These four stations were powered by small nuclear reactors and returned data for up to 6 yr before being shut down in 1977; a seismic station powered by solar cells was put in at the *Apollo 11* landing site but did not survive the first lunar night. The seismic instruments at the stations consisted of a long-period instrument recording the vertical component of ground motion and two orthogonal horizontal components, and a short-period vertical component instrument. The response of the short-period instrument peaked at 8 Hz and the long-period instrument could be operated either in a peaked mode with maximum sensitivity at 0.45 Hz or in a flat mode with constant gain between 0.1 and 1 Hz. Both instruments were operated at very high gain, with the smallest detectable signals corresponding to about 1 Å of ground motion.

The high gain was made possible by the lack of seismic noise due to fluids, such as wind, atmospheric pressure fluctuations, and ocean waves, present on Earth. The main sources of seismic noise on the Moon are due to thermal stresses associated with the heating of the lunar surface starting at sunrise and the cooling around sunset. The lunar module descent stage, which was left on the Moon when the astronauts ascended to lunar orbit, and other equipment left on the Moon, were strongly affected by thermal stress around sunrise, and small micromoonquakes that recur at specific time offsets relative to lunar sunrise also contribute to the noise. In addition, micrometeorite impacts close to the stations cause a continuous noise background.

The first seismogram (of the crash of the *Apollo 12* lunar module ascent stage) recorded by the network was completely unlike seismograms seen on Earth. The first compressional wave arrivals were extremely small and were followed by an extremely long reverberating wave train that built up to a maximum after several minutes. Detectable seismic energy was present 1 h after the crash; on Earth a seismogram at the same range (73 km) would have a duration of about 1 min. An example of a lunar seismogram is shown in Fig. 1. This difference in seismic expression is primarily due to a combination of intense scattering and low anelastic attenuation (attenuation due to conversion of seismic energy to heat) in the near-surface material of the Moon. The scattering, which is caused by the intense fracturing of the lunar surface under continuous meteorite bombardment, diminishes the amplitude of seismic waves traveling in direct paths through the material. The energy scattered out of these direct phases, however, reappears after a delay due to travel in a circuitous path after many scatterings, producing the reverberating wave train following the small first arrivals previously described. On a gross scale the seismic energy

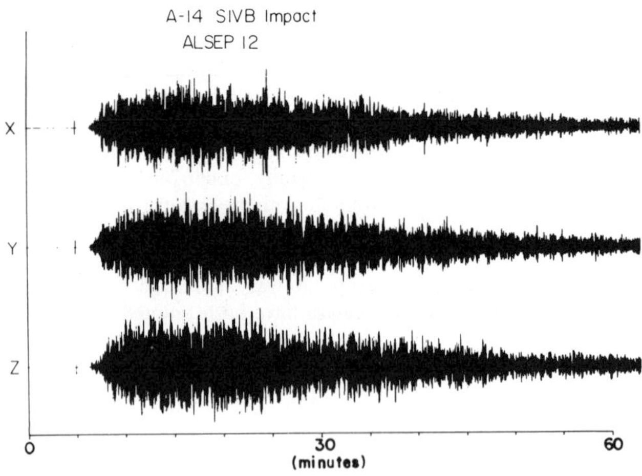

Figure 1. Seismogram recorded by the *Apollo 12* long period seismometer from the crash of the *Apollo 14* third stage booster rocket 172 km away. Three components of ground displacement are shown: X is south–north (S + *ve*), Y is west–east, and Z is vertical. The tick marks on the seismic traces indicate the crash time. The dominant signal frequencies are between 0.5 and 1.0 Hz. Note that the time scale is in minutes; such long reverberating seismograms are recorded for all types of lunar events. [*From, Dainty et al. (1974), Seismic scattering and shallow structure of the Moon in Oceanus Procellarum*, The Moon ***9** ii*, *copyright D. Reidel*.]

diffuses through the surficial fractured layer. It has not been possible to determine the thickness of the scattering zone unambiguously, but indirect evidence suggests it is between 1 and 20 km thick. The low attenuation plays an even more critical role because it allows the scattered energy to persist instead of being converted to heat. The standard measure of seismic attenuation is the seismic quality factor

$$Q = 2\pi \frac{E}{\Delta E},$$

where E is the peak strain energy stored in an elastic cycle and ΔE is the energy lost per cycle due to anelasticity; note that high Q means low attenuation. On the Moon the Q of the near surface is at least 3000, whereas on Earth it is about 100.

SEISMIC STRUCTURE

Seismic data are the main source of information about the interior of the Moon, as they are for the interior of the Earth. The information is obtained by measuring the travel time and variation of amplitude of seismic waves as a function of distance between source and receiver. Spherically symmetric models of the internal velocity structure may then be constructed to match the measurements. The sparsity of the measurements does not permit more complicated models, but no data contradict the assumption of spherical symmetry. For the purposes of structure determination, sources may be classified as artificial and natural. The artificial sources were the deliberate crashes of the third stage booster rocket for each mission prior to landing and of the lunar ascent module after return to lunar orbit. These crashes occurred in accurately known locations and were sources of two standard sizes, thus providing excellent data for the study of structure. However, they were not strong enough to provide useful information at source–receiver distances greater than 400 km, and because the depth of penetration of observed seismic waves depends on range, it was only possible to determine structure to about 70 km depth using these impacts. The natural sources consisted of large meteorite impacts, near-surface moonquakes, and deep focus moonquakes.

The analysis of the artificial impacts demonstrated that the Moon has a crust about 60 km thick in the center of the near side. Like the Earth's crust it is chemically distinct from the underlying mantle and is apparently derived from it by igneous processes. This is a first order datum of considerable importance because it indicates that the Moon has been through one or more major episodes of differentiation. If this crust were uniform over the Moon it would constitute about 10% of the Moon's material by volume (the equivalent figure for the Earth is less than 1%). The materials and thus the origin of the lunar crust can be surmised by comparing the compressional wave velocity with that of samples collected from the lunar surface. Starting at the surface, the first 1 km has very low velocities (0.1–1 km s^{-1}) due to the intense fracturing; that is, the physical state of the material (fractured) rather than the composition controls the velocity. From 1–20 km depth, velocity increases steadily from 4–6 km s^{-1}. This matches the expected increase of velocity with depth due to pressure for mare basalt samples. In this case, both the composition (basalt) and the physical state (somewhat fractured) are important, and other compositions could also fit the velocities. If the mare basalt identification is correct, this layer was formed during the last episode of extensive volcanism on the Moon from 3–4 billion years before present. The main body of the crust, from 20–60 km, has a velocity close to 6.8 km s^{-1}, which corresponds well with unfractured rocks having the mean composition of the lunar highland samples, which are rich in the plagioclase mineral anorthosite. These rocks were formed during an initial differentiation of lunar crust coincident with or shortly after the formation of the Moon at around 4.5 billion years before present. Knowledge of the seismic structure of the lunar crust has been an important datum in deriving quantitative models of lunar igneous processes and lunar evolution.

Below the crust is the lunar mantle. Information about this part of the lunar interior is less extensive than for the crust because most of the data consist of the arrival times from natural events whose position and time of occurrence are not known a priori. Thus some of the available information must be used to determine these parameters. Nonetheless, the mantle compressional wave velocity is close to 8 km s^{-1} and no velocities deviating by more than 0.5 km s^{-1} have been detected to at least a depth of 1100 km. The ratio of compressional to shear velocities is in the range 1.7–1.8 typically observed for terrestrial rocks. Seismic Q is over 1000 above 1100 km, higher (~ 3000) in the upper 500 km, and appears to decline substantially below 1100 km, perhaps due to high temperatures. Structure within the lunar mantle has been the subject of disagreement among investigators: two "best models" exist, but large variations in velocity are not found in either model. The velocities suggest that the composition of the lunar mantle could show similarities to the upper mantle of the Earth with olivine and pyroxene as important minerals, and this is confirmed by detailed petrological models of the early lunar crust differentiation and the later basalts. In both cases an olivine–pyroxene-rich mantle is predicted either as a residuum after crustal differentiation or as a source for the mare basalts. An equally important negative result is the failure to detect an iron-rich core up to a depth of 1100 km; there is very little information below this depth and such data as do exist are too uncertain for meaningful statements. Thus any such core has a maximum possible radius of 680 km, or is at most 5% of the Moon's volume. This is a first order datum because the corresponding figure for the Earth is 15%; that is, the Moon is deficient in iron metal relative to the Earth. This result has been a major constraint on theories of the Moon's origin.

SEISMICITY

The seismicity of a planet is a reflection of the tectonic and other forces operating on it, with the rheology as the reflecting agent: Material must be relatively cool and brittle to fail in a moonquake. The tectonic styles of the Earth and the Moon appear to be very different, with plate tectonics producing volcanism and large horizontal motions on Earth throughout much, if not all, of its history up to and including the present, whereas plate tectonics is not evident on the Moon and the last reliable indication of volcanism is 3 billion years ago. This difference in visible tectonic style is reflected in a markedly lower rate of seismicity on the Moon: The rate of seismic energy release per volume in the Moon is smaller than that in the Earth by a factor of around 10^6. Most of this energy is released by occasional moonquakes occurring in the upper mantle between 50 and 300 km below the surface; the largest of these moonquakes was recorded on 12 March 1973 and had a body wave magnitude of about 5. The epicenters of such events appear to be randomly scattered over the moon, although there is a faint suggestion of association with the edges of maria. Likewise the times of occurrence seem to be random, although it has been suggested that there is some correlation with the tidal cycle. About five events per year of body wave magnitude greater than 2 were observed during the 8 yr of operation of the lunar seismic network. The most likely cause for these events is gradual shrinking or expansion of the entire Moon due to cooling of the surface layers and radioactive heating of the interior.

Although these shallow moonquakes represent the major seismic energy release in the Moon, they are not the commonest class of internal seismic event. A most remarkable seismic phenomenon on the Moon is the regular occurrence of tidal moonquakes. These events were discovered through the observation of repeating identical seismograms, within a scale factor. Because the seismograms are (nearly) identical, the locations of the events that produced them must be separated by substantially less than a wavelength of the observed seismic waves of 1-Hz frequency; it is believed that these repeating events occur at sites that are within 1 km of each other. During the 6-yr operation of the lunar seismic network about 80 separate hypocenters where such moonquakes occur were identified (only the seismograms from a specific hypocenter are identical). In contrast to the shallow moonquakes, those hypocenters that could be located were all found to lie at depths of 800–1000 km. With one exception, the located hypocenters are on the near side of the Moon in two belts trending east–west and north–south. The absence of hypocenters on the far side is suspected to be due to the inability of recording these low amplitude events at such great distance rather than due to an absence of events. The events are substantially smaller than the shallow moonquakes, with the largest event having a body wave magnitude of 3. Thus although about 500 such moonquakes with a magnitude greater than 1.5 were detected each year by the lunar network, only about 1/2000 of the energy of the shallow moonquakes was released.

The association of these events with the tides of the Moon caused by the Earth is shown by their time history. Events at a given hypocenter occur with great regularity near a fixed point in the tidal cycle of 27.5 days (an anomalistic month, the time from apogee to apogee; because the Moon always presents the same face to the Earth, tidal variations are due to such effects as the variation of the Earth–Moon distance over the orbital cycle and have this periodicity). Furthermore, the number of such events per month shows regular variations of 206 days, corresponding to solar perturbation of the Moon's orbit, and also shows partial evidence of a 6-yr variation corresponding to the beat period between the anomalistic month and the nodical month (time to return to a node, i.e., a point in the orbit where the orbit crosses the plane of the ecliptic). Indeed, within the 6-yr period of observation, events at a few hypocenters have been observed to reverse polarity as a presumed response to this 6-yr cycle. For example, events at one hypocenter recorded in mid-1973, when compared to events recorded earlier or later, had seismograms that were identical to the earlier or later events except that they were exactly reversed in sign, indicating the same hypocenter but an opposite sense of movement.

In spite of this close association with the tidal cycle, the processes causing these moonquakes remain enigmatic. The fidelity with which the events track the tidal cycle might suggest that they are solely a response to tidal stresses. This idea receives support from the depth of the moonquakes and from estimates of the stress released in each event. The depth, around 900 km, is just above the point (1100 km) at which seismic attenuation appears to increase substantially, as discussed in the previous section. If this is interpreted as indicating a temperature increase with depth, then it is likely that the lunar material is substantially weakened; that is, there is a relatively rapid change in elastic compliance at this depth. Numerical calculations indicate that tidal stress would be enhanced at such a point. Also, the stress released in a deep moonquake is about 0.1 bar, whereas the tidal stress amplitude is about 1 bar. However, there are two major arguments against the idea that these moonquakes are merely a response to tidal forces. One is that the location in geographical coordinates of the events does not bear any discernible relationship to the geographical variation of tidal stress. The second is that if only tidal forces were involved, events at a given hypocenter should occur twice a month, not once as they do, and should reverse polarity every half month. Again this is not observed; although some reversals occur, as already discussed, they occur at long intervals; for long stretches of time events retain the same polarity, indicating the same direction of movement. These arguments suggest a combination of tidal stress and stress due to some other, as yet unknown, cause.

To close this section, meteorite statistics as determined by the recordings of impacts by the lunar seismometers will be presented. Strictly speaking meteorite impacts are not lunar events in the sense that they are not of intrinsic lunar origin, but ~ 300 per year were recorded by the lunar seismic network. Historically they had to be distinguished from other signals and they were used extensively to determine lunar mantle structure. Meteorite impact statistics may be expressed as the cumulative number $N(m)$ of meteorites of mass m (grams) or greater impacting per square kilometer per year. This number determined for the Moon should be directly comparable with similar numbers for the Earth determined by observations from satellites, radar, falls and finds, and camera observations. It has been found that such observations may be summarized by the relation

$$\log N(m) = B + \gamma \log m,$$

where logarithms are to the base 10 and B and γ are constants. This relation is believed to reflect the influence of collisions in the origin of meteorites. For the Earth-based data, satellite and radar observations give $B = -0.25$ and $\gamma = -1.05$ in the mass range 10^{-4}–10 g; falls and finds give $B = 0.7$ and $\gamma = -1.0$ in the mass range 1 g to 1000 kg; and camera observations give $B = -1.6$ and $\gamma = -0.6$ in the mass range 100 g to 1000 kg. (A "fall" is a meteorite found after it was observed to fall; "finds" are all other recovered meteorites.) Although the constants are somewhat different, these relations lead to similar values of $N(m)$ in the mass ranges where they overlap. For the Moon, analysis of data from the short-period seismometer at the *Apollo 14* station gave $B = -1.8$ and $\gamma = -0.7$ in the mass range 1–100 g; data from the long-period seismometers of the network gave $B = -1.1$ and $\gamma = -1.1$ in the mass range 50 g to 50 kg. If these two estimates are compared where they overlap they give similar values of $N(m)$ to each other, but the lunar values are an order of magnitude lower than the Earth values. The reason for this discrepancy is not known.

SUMMARY

The seismic properties of the Moon show a planet like the Earth in some respects but unlike it in others. Figure 2 shows an equatorial section through the Moon illustrating one version of the structure and the projected hypocenters of the deep focus moonquakes. The quietness of the lunar seismic environment is due to the lack of an atmosphere and oceans and was forseen before the first Apollo landings. The unique high scattering–low attenuation near-surface layer is due to the combined effects of extreme outgassing of volatiles and meteorite bombardment. The seismic structure shows a layered planet, indicating differentiation by igneous (i.e., melting) processes, but there is either no iron-rich core or only a small one, indicating a substantially lower metallic iron content than the Earth. The rocks making up the crust and mantle of the Moon appear to be composed of minerals familiar to Earth geologists. The seismic determination of lunar structure has greatly influenced theories of the formation and evolution of the Moon. The seismicity of the Moon shows a planet with only residual tectonic activity due to cooling and tidal forcing.

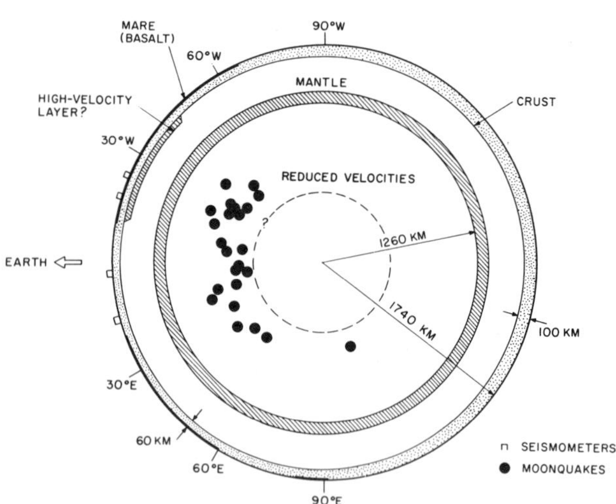

Figure 2. Equatorial section through the moon showing structural units and deep moonquake foci locations projected onto the plane. [*From Goins et al. (1981), Seismic energy release of the Moon*, J. Geophys. Rev. **86** *378, copyright by American Geophysical Union*.]

Additional Reading

Dainty, A. M. and Toksöz, M. N. (1977). Elastic wave propagation in a highly scattering medium—A diffusion approach. *J. Geophys.* **43** 375.

Goins, N. R., Dainty, A. M., and Toksöz, M. N. (1981). Lunar seismology: The internal structure of the Moon. *J. Geophys. Res.* **86** 5061.

Goins, N. R., Dainty, A. M., and Toksöz, M. N. (1981). Seismic energy release of the Moon. *J. Geophys. Res.* **86** 378.

Hamblin, W. K. and Christiansen, E. H. (1990). *Exploring the Planets.* Macmillan Publishing Co., New York. See "The internal structure of the Moon," p. 86.

Lammlein, D. R., Latham, G. V., Dorman, J., Nakamura, Y., and Ewing, M. (1974). Lunar seismicity, structure and tectonics. *Rev. Geophys. Space Phys.* **12** 1.

Latham, G. V., Ewing, M., Press, F., Dorman, J., Nakamura, Y., Toksöz, N., Lammlein, D., Duennebier, F., and Dainty, A. (1973). Passive seismic experiment. *Apollo 11 Preliminary Scientific Reports.* NASA SP-330, Sec. 11.

Nakamura, Y. (1983). Seismic velocity of the lunar mantle. *J. Geophys. Res.* **88** 677.

Toksöz, M. N., Dainty, A. M., Solomon, S. C., and Anderson, K. R. (1973). Velocity structure and evolution of the Moon. *Proc. Fourth Lunar Sci. Conf.* **3**. Pergamon, New York, p. 2529.

See also **Moon, Geology; Moon, Origin and Evolution; Planetary Interiors, Terrestrial Planets; Planetary Tectonic Processes, Terrestrial Planets.**

Moon, Space Missions

Paul D. Lowman, Jr.

The first spacecraft from Earth, *Luna 2*, landed on the Moon (more precisely, it crashed in the Imbrium basin) in 1959, thus opening a golden age of lunar exploration, culminating in the six Apollo expeditions during which men lived on the Moon for as long as three days. There has been a prolonged pause in lunar exploration since the last mission (also Soviet, the 1976 *Luna 24* landing in Mare Crisium). This pause may soon be ending, but this is a convenient time to summarize the opening period.

The total number of missions *attempted* during this period may surprise some: a comprehensive compilation by Patrick Moore lists 73, including many failures. However, we will concentrate on those missions that were at least partly successful, the failures being of interest chiefly to engineers.

UNMANNED MISSIONS

Unmanned missions (Table 1) will be summarized in the following section. Because the Soviet Union was consistently first in lunar achievements until the Apollo landings, Soviet missions will be covered first.

The *Luna* (or *Lunik*) missions resulted in a wide range of accomplishments, from the uninformative though historic impact of *Luna 2* in 1959 to the extraordinarily informative sample return of *Luna 24*—a type of robotic mission not yet even attempted by the United States. The first real accomplishment was that of *Luna 3*, which photographed the previously unseen far side of the Moon. These photographs revealed a geologic anomaly not yet explained: the near-absence of maria on the farside. Later Apollo geophysical investigations showed that this asymmetry is reflected in crustal thickness, that of the farside being about twice as thick as that of the nearside. *Luna 9*, the first soft lander, simply by surviving and returning pictures of the mare surface, dispelled lingering fears about an unstable deep dust layer. Later Luna missions further demonstrated Soviet capabilities, with the Lunokhod vehicles that slowly crawled scores of kilometers and the unmanned sample returns by *Lunas 16*, *20*, and *24*. These sample returns fueled controversy in the West over manned versus unmanned missions, but it was generally realized that the few hundred grams of lunar soil they returned were not as valuable as the hundreds of *kilograms* brought back by the Apollo astronauts.

The Zond missions, involving circumlunar flights and return to Earth, represent another category never attempted by the United States with the notable exception of the *Apollo 8* flight. The *Zonds* were generally considered Soviet preparation for manned missions, involving as they did small animal payloads—the first earthlings, unfortunately inarticulate, to visit the Moon's vicinity. It has only recently been admitted that the U.S.S.R. was preparing to land people on the Moon, but their many unmanned missions showed that the Soviets were well aware of its importance for future space exploration and colonization.

American efforts to reach the Moon with unmanned spacecraft opened with a dismal string of failures between 1958 and 1964 that was finally broken with *Rangers 7*, *8*, and *9*, which provided superb close-up television pictures of the impact sites. Although these pictures did not produce new fundamental discoveries, and are rarely seen in the scientific literature later than 1970, they were absolutely invaluable at the time as the first high-resolution look at the lunar topography. Analysis of the *Ranger* pictures provided experience in the interpretation of lunar geology that would be applied within 2 yr to the much higher quality pictures from Lunar Orbiters.

Surveyor 1 made the first American soft landing in 1966, opening a spectacular series of 11 successful lunar missions in less than 2 yr: 5 Surveyors, 5 Lunar Orbiters, and 1 Explorer, a pace that from the viewpoint of the 1980s appears unreal, especially considering the 10 manned Gemini missions and many satellites launched in about the same period. The increase in fundamental knowledge of the Moon from these missions was of course enormous. The Surveyor and Lunar Orbiter data were complementary, Surveyors producing detailed information such as chemical composition on their landing sites and the Orbiters returning global imagery that eventually covered almost the entire surface of the Moon. Tracking of the Orbiters, furthermore, permitted mapping of the lunar gravity field, including the positive gravity anomalies over the circular maria termed mascons by the discovers. An important demonstration of the Lunar Orbiter missions was the superiority of orbital missions to brief flybys (e.g., *Mariner 4* to Mars) or impactors (e.g., the Ranger series). *Mariner 4*, for example, gave us the misleading impression that Mars was a heavily cratered and by implication unevolved planet, simply because of the limited areal coverage of its images. The Lunar Orbiter photographs, in contrast, provided nearly complete coverage of the Moon, and are still in use years after acquisition. *Mariner 9*, a Mars orbiter, provided comparably valuable global coverage of Mars in 1971.

As a result of the various unmanned lunar missions, we had by the time of the first *Apollo* landing in July 1969, a general understanding of the Moon's physiography and at least the main features of its evolution. The circular maria, in particular, were found by Surveyor and Lunar Orbiter to be flood basalts, occupying large impact basins and blanketed by a regolith formed by innumerable smaller impacts over a long though indeterminate time. Craters of the Tycho variety and smaller analogues were interpreted as essentially the result of impact, though many of these craters were probably the site of later internal activity or basalt eruptions. The main first-order questions that were completely unanswered by the unmanned missions up to 1969 were the

Table 1. Unmanned Lunar Missions*

Series Name, Number	*Dates*	*Objectives*
		Soviet
Luna 1–3, 9–14, 16, 17, 19, 20–24	1959–1976	Impact Moon; photograph lunar terrain from orbit; soft landing with photography; land and deploy Lunokhod; obtain samples and return to Earth
Zond 3–8	1965–1970	Lunar flyby with photography; circumlunar flight with return to Earth; some carried biological payloads
		American
Pioneer 1, 4	1958–1959	Fields and particles studies; neither impacted Moon
Ranger 7–9	1964–1965	Hard-landing with TV of impact sites
Surveyor 1, 3, 5–7	1966–1968	Soft-landing; TV, soil mechanics, chemical composition, magnetic properties, Earth and astronomical TV
Lunar Orbiter 1–5	1966–1967	Orbital photography of *Apollo* landing sites and other areas; micrometeorite flux; radiation measurements
Explorer 35	1967	Fields and particles measurements in lunar orbit
Explorer 49	1973	*Radio Astronomy Explorer*; very-low-frequency radio astronomy from lunar orbit

*Totally unsuccessful missions not included

Table 2. Manned Lunar Missions

Number	*Date*	*Landing Site*	*Accomplishments*
Apollo 8	Dec. 1968	Circumlunar	First manned mission to Moon; photography; visual observations
Apollo 10	May 1969	Circumlunar	Manned mission with lunar module separation; 50,000-ft perilune without landing; photography
Apollo 11	July 1969	Mare Tranquillitatis	First manned landing on Moon; sample collection; EASEP emplacement; photography (surface and orbital)
Apollo 12	Nov. 1969	Oceanus Procellarum	Second manned landing; sample collection; ALSEP emplacement; photography (surface and orbital)
Apollo 13	April 1970	Lunar flyby	Service module explosion forced landing abort; emergency return to Earth; far side photography accomplished
Apollo 14	Jan. 1971	Near Fra Mauro crater	Investigated Imbrium ejecta blanket; ALSEP emplacement; photography (surface and orbital); active seismic investigations
Apollo 15	July 1971	At base of Apennines	First use of lunar rover; sample collection; ALSEP emplacement; orbital remote sensing; subsatellite launch
Apollo 16	April 1972	Near Descartes, Central Highlands	First landing in true highland terrain; sample collection; ALSEP emplacement; orbital remote sensing; subsatellite launch
Apollo 17	Dec. 1972	Taurus–Littrow Valley	Last *Apollo* landing; sample collection; ALSEP emplacement; orbital remote sensing; first geologist on Moon

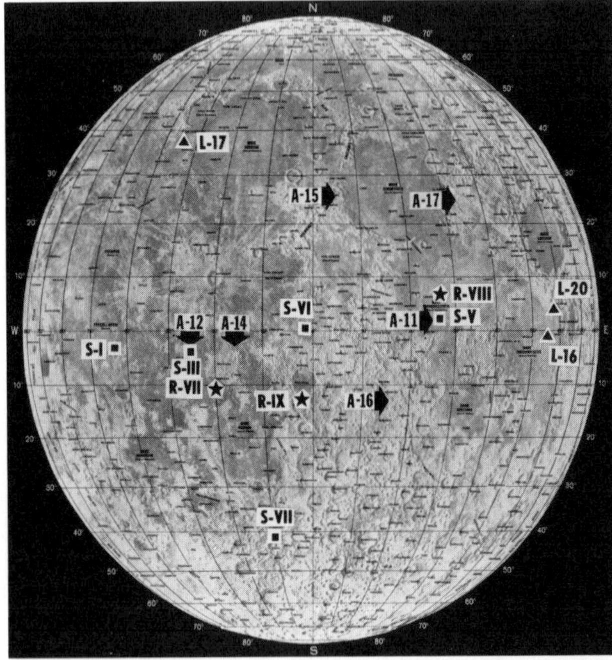

Figure 1. Map of the lunar nearside showing landing sites of *Apollo* (A), *Luna* (L), *Ranger* (R), and *Surveyor* (S) spacecraft. (*Courtesy of NASA.*)

absolute time scale of lunar geology, the composition of the lunar highland crust, the deep structure of the Moon, and the origin of the Moon. On January 24, 1990, Japan demostrated its ability to participate in lunar missions with the launch of the *Muses A* satellite to the Moon.

MANNED LUNAR MISSIONS: APOLLO

The Apollo program was generally viewed as the greatest technological achievement in history, a judgement that two decades later seems sound. As shown in Table 2, Apollo included six landings on the Moon (see Fig. 1), one flight around the Moon, and two Earth-orbital missions. The Apollo launch vehicles and spacecraft were later used for the first American space station, *Skylab*, and for the first joint American–Soviet space mission. To summarize a program that already fills many library shelves is difficult, but the high points of the Apollo lunar missions include the following.

Perhaps the most surprising of these missions was the very first, *Apollo 8*, which did not land. It was "surprising" because, unlike the first landing mission, it was planned and executed within a few months in response to several factors, in particular Lunar Module (LM) readiness and the apparent imminence of a Soviet circumlunar flight. *Apollo 8* was the first manned flight of the *Saturn V* and the first to send astronauts to escape velocity. The launch vehicle did not carry a Lunar Module, a slight additional risk because the LM provided a backup propulsion system for returning to Earth should the Command and Service Module (CSM) engine fail. The *Apollo 8* mission was a complete success, though its scientific results were minor, and it opened the door for later missions.

After *Apollo 9*, an Earth-orbital test of the complete "stack" including the Lunar Module, the *Apollo 10* was a test of all elements of a lunar landing mission except the landing itself. The Lunar Module descended to within 50,000 ft of the surface, then underwent stage separation and rendezvous with the CSM in lunar orbit. (The LM could not, incidentally, have landed on the Moon, though it carried a full load of fuel, because the crew did not have the computer software needed for landing.) Apart from a large number of excellent 70-mm photographs of the lunar terrain, including the far side, there was again relatively little scientific gain from the mission.

The first landing on the Moon, *Apollo 11*, has been described, discussed, and debated for 20 years. Scientifically, it was enormously successfully, settling almost immediately several major questions about the Moon through the 48 lb of rock and soil collected. Most important was the establishment of a chronologic benchmark by radiometric dating of the mare basalts from Tranquillity Base: 3.7 billion years. Because *relative* dates of major lunar features, such as ray craters, maria, and highlands had been established through telescopic and photographic study by stratigraphic relationships, the approximate time scale of the Moon's history could immediately be established. The maria—argued by some as late as 1969 to be water-laid deposits—were found to be heavily-impacted basaltic lava flows. The nature of the highlands was inferred, somewhat speculatively, from small fragments of feldspar-rich rock (anorthosites and related types), although initial descriptions of the highland crust as anorthosite proved inaccurate. The formation age of the highlands was put at well over 4 billion years. Even these early and somewhat misleading results demonstrated clearly that, unlike pre-Apollo concepts of a cold, undifferentiated Moon, the Moon had undergone early global differentiation following a high temperature primordial phase.

Succeeding Apollo missions did much to fill out the picture evolving from the *Apollo 11* results. *Apollo 12*, on Oceanus Procellarum, was the only mission to visit the site of a previous landing, that of *Surveyor 3*, study of which provided invaluable data on the effects of the lunar environment. (Equipment from *Surveyor 3*, including a television camera, was retrieved by the *Apollo 12* crew and returned to Earth for study.) The *Apollo 12* astronauts also emplaced the first of five nuclear-powered Apollo lunar surface experiment packages (ALSEPs), networks of geophysical instruments that operated successfully for several years until turned off for budgetary reasons.

The *Apollo 13* mission was successful only in that the crew survived; an explosion in the Service Module component of the CSM during translunar coast destroyed both power and propulsion systems. Fortunately, the Lunar Module had not yet been separated, and was able to serve as a lifeboat, returning the CSM and crew to Earth where a normal Command Module reentry on battery power was made. The *Apollo 13* crew did manage to carry out 70-mm photography during their swing around the Moon, a remarkable achievement under the circumstances. But their main achievement was coming back alive.

The concept of circular mare basins as huge impact craters was largely confirmed by the *Apollo 14* mission, which landed on the Fra Mauro formation, interpreted as part of the ejecta blanket from the Imbrium basin to the north. This was the first mission to land on anything like highland crust, and analysis of the *Apollo 14* samples—chiefly breccias and impact melts—began to narrow constraints on highland composition and age.

The "J series"—*Apollos 15, 16,* and *17*—represented a major increase in capability; these flights could be termed expeditions rather than simply missions. Each carried a Lunar Roving Vehicle (LRV), permitting surface travel of several kilometers from the landing site. Landing site restrictions were considerably relaxed from the low-latitude, level-terrain requirements of the first three missions. *Apollo 15* landed at the foot of the Apennine Mountains, *Apollo 16* in the central highlands near the crater Descartes, and *Apollo 17* in the Taurus–Littrow Valley on the rim of Mare Serenitatis. The crews stayed on the Moon as long as three days, making long surface excursions and collecting large amounts of rock and soil, in addition to emplacing the ALSEPs. Of almost equal importance, the J series Command Modules were equipped with complex remote sensing packages, capable of carrying out composition mapping, radar sounding, and topographic profiling along the flight path. *Apollos 15* and *16* even launched subsatellites carrying magnetometers. Tracking of these subsatellites, and of the various Apollo spacecraft themselves, permitted increasingly refined maps of the lunar gravity field.

The scientific return from the Apollo missions is still being exploited long after the last astronaut left the surface. Most recent work has focussed on the 845 lb of rock and soil collected. As demonstrated by meteorite studies, it is important not to consume all of a sample at once, precluding studies by more advanced techniques when they become available. This philosophy is being followed with the Apollo samples, and they will undoubtedly still be under analysis many decades from now.

What has been learned scientifically from the Apollo missions and their automated counterparts? The first-order answers seem to be the following.

First, it is now clear that the Moon is a globally differentiated, evolved body, not the primitive chondritic object expected by Harold Urey and others in the 1960s. This differentiation, furthermore, happened very early in the Moon's history, probably 4.4 billion years or more ago. It was followed by a second differentiation, in which basaltic magmas were generated and erupted over hundreds of millions of years.

The dominant role of impact in shaping the Moon's surface now seems confirmed. Apart from the direct evidence of shock metamorphism found in lunar samples, the impact origin of most large craters is implied by the fact that erupting extraterrestrial volcanos have been found on the jovian satellite Io. They *look* like volcanos, not like Tycho and its many relatives. But balancing the importance of impact in lunar geomorphic evolution is the fact that except for breccias (fragmental rocks), almost all lunar rocks are igneous rocks. Furthermore, there are many landforms (not counting the maria) that are almost certainly of volcanic origin, such as the Marius Hills and the sinuous rilles such as Schröter's Valley. Radon emission from the latter area was in fact detected on the *Apollo 15* and *16* missions, supporting telescopic observations of transient reddish glows seen in 1963. Seismic data from the ALSEP seismometers, in addition, show attenuation of shear waves in the deep interior of the Moon, implying a partly liquid and by implication hot center.

An important and very general result of the manned and unmanned lunar missions is a reasonably firm knowledge of the lunar surface environment. This environment is harsh by terrestrial standards: an unprotected person would be unconscious in 15 s and dead in 5 min. But a winter night in northern Ontario could kill an unprotected person almost as rapidly. Fortunately, the lunar hazards are apparently predictable, and therefore can be allowed for. Instruments left behind on the Moon operated for years until terrestrial "hazards" in the form of congressional budget cuts forced a shutdown, but the completely passive laser retroreflectors continue to reflect normally after 20 yr exposure on the surface.

In the long run, the most important scientific results of the lunar missions are not answers but questions. The origin of the Moon is still not agreed upon: the currently popular giant impact theory has, to some, an uncomfortably ad hoc flavor. The source of the bodies that excavated the mare basins is still unknown: fragments from the giant impactor, from the nearly demolished Earth, or from outside the Earth–Moon system? The origin of the highland crust is not agreed upon; the magma ocean concept is still popular, but petrologic considerations support partial melting of the primordial Moon. Can the timing of highland crust formation be consistent with origin of the Moon by a "giant impact"? Answers to these questions may have to await the renaissance of lunar exploration, a renaissance that will owe much to lunar missions accomplished during the dawn of space exploration.

Additional Reading

Aldrin, E. E. and McConnell, M. (1989). *Men from Earth*. Bantam Books, New York.

Collins, M. (1988). *Liftoff*. Grove Press, New York.

French, B. M. (1977). *The Moon Book*. Penguin Books, New York.

Lang, K. R. and Whitney, C. A. (1991). *Wanderers in Space*. Cambridge University Press, New York.

Lowman, P. D., Jr. (1972). The geologic evolution of the moon. *J. Geology* **80** 125.

Moore, P. (1981). *The Moon*. Rand McNally, New York.

Rycroft, M., ed. (1991). *The Cambridge Encyclopedia of Space*. Cambridge University Press, New York.

See also **Moon, Geology; Moon, Lunar Bases; Moon, Rock and Soil.**

N-Body Problem

Myron Lecar

Since Isaac Newton formulated the laws of motion in 1687, astronomers and mathematicians have focused on the problem of the motion of the planets around the Sun; this study is called celestial mechanics. Elegant approximation techniques have been developed that rely on the fact that the Sun is 1000 times more massive than the planets. In 1950, this field was revolutionized by the advent of fast, digital computers. It was transformed again by the introduction of KAM theory (named for its developers, Andrei N. Kolmogorov, Vladimir I. Arnol'd, and J. Moser) and the theory of "chaos." This oldest branch of physics is now vibrant.

Another form of the *N*-body problem arises when we study the interactions of bodies of comparable mass; for example, the dynamical evolution of a globular cluster of stars. This subject started with computer studies; the first numerical experiment was published in 1960 by Sebastian von Hoerner.

CELESTIAL MECHANICS

Newton solved the two-body problem and determined that the orbit of a planet around the Sun described an ellipse as deduced by Johannes Kepler. Because even Jupiter, the most massive planet, has a mass less than 1/1000 that of the Sun, a reasonable first approximation neglects the forces of the planets on each other and assumes that the planets describe ellipses around the Sun. The next approximation calculates the forces of each planet on the others assuming that their positions are given by the first approximation. The changes to their orbits are called perturbations. According to Dirk Brouwer, the first calculation of the mutual perturbations of Jupiter and Saturn was given by Leonhard Euler in 1744.

A question that occupied early celestial mechanicians, and that is still of interest, is whether the solar system is stable. The method of approximation is such that the time dependence of the planetary orbits is expressed as a sum of periodic terms (sines and cosines). The primary question is whether, in the expression for the semimajor axes of the planets, there also exist terms that are proportional to the time itself; these are called secular terms. If such terms exist, they indicate that the distances of the planets from the Sun increase (or decrease) without bound. Joseph Louis Lagrange, in 1776, showed that there were no such terms to the first order in the ratios of the planetary masses to the mass of the Sun. Siméon-Denis Poisson, in 1809, showed that there were no pure secular terms in the second order. The situation is not clear for the next order.

In any case, numerical integration of the equations of motion has cast doubt on the accuracy of the theory in higher orders. In 1951, Wallace Eckert, Brouwer, and G. M. Clemence integrated the motion of the outer planets for 400 years. Charles J. Cohen and E. C. Hubbard, in 1964, using the NORC computer, extended the calculation to 120,000 years. In 1973, using the IBM 7030, the same authors joined by Claus Oesterwinter extended the integration to one million years. This was the longest integration until the "Digital Orrery" (Fig. 1), a special-purpose computer designed just to integrate the motion of the planets, was developed. In 1986, the Orrery was used to integrate the motions of the outer planets for 210 million years. When compared with the perturbation theory of P. Bretagnon, it was discovered that higher order terms, not included in the theory, were larger than many of the terms considered (third order in the eccentricities and inclinations, second order in the masses). One conclusion of this study was that the orbit of Pluto was stable for the duration of the integration. However, in 1988, the Digital Orrery produced an even long integration of 845 million years. Gerald J. Sussman and Jack Wisdom reported the results of this study in a paper entitled "Numerical evidence that the motion of Pluto is chaotic."

This notation of "chaos" is the most recent revolution in celestial mechanics. In 1905, Jules Henri Poincaré introduced a powerful technique for studying the topology of the solutions of differential equations. Imagine a particle, driven by gravitational interactions with other bodies, whose motion is confined to a plane. This system is said to have two degrees of freedom, and a specification of four quantities [say its two coordinates of position (X and Y) and its two components of velocity (V_x and V_y)] completely determines the orbit. Although it is impossible to visualize, it is easy to construct mathematically a four-dimensional space with the axes X, Y, V_x, and V_y. Such a space is called the phase space and, as time progresses, the phase point of the particle moves around in that space. It is often the case that a function of the four coordinates, for example, the energy, remains constant. In that case, the phase point cannot go anywhere but is confined to a surface in the four-dimensional space. Now a surface in four-dimensional space is a volume in (ordinary) three-dimensional space. Such a volume is something that we can visualize: a ball, for example. Say we used the energy constant to eliminate one coordinate, V_x. Now, we integrate the orbit, and every time $V_y = 0$, we plot a point on the X-Y plane. That is an example of a Poincaré return map. In 1964, M. Hénon and C. Heiles made effective use of the Poincaré map to study the orbits of stars in our galaxy. That now classic paper revived the use of Poincaré maps.

In our example, these points would fill some area; the intersection of our volume with a plane. This would be expected to be the case if there were no other functions of the four quantities that remained constant. Until about 1960, we thought we knew all such quantities. They were called integrals of the motion and resulted from symmetries built into the equations of motion. For example, because the equations of motion are unconcerned with what time the experiment is conducted (i.e., in technical language, they are invariant to a translation in time), energy is conserved. Similarly, invariance to a translation in coordinates (it does not matter where the experiment is conducted) results in the linear momentum integral, and invariance to rotation of the coordinates results in the conservation of angular momentum. And that was it.

About 1960, Kolmogorov, Arnol'd (a student of Kolmogorov), and Moser discovered that there were other quantities that were constant for all time. When such quantities exist, the trajectory on the Poincaré return map no longer fills an area; instead, it lies on a curve. It has been found that orbits whose trajectories lay on a curve are quasiperiodic. These orbits are stable. If one analyzes the angular variables of a stable orbit, one finds periodic terms with a finite, discrete set of frequencies. In contrast, for the orbits that produce an area on the Poincaré map, the frequencies seem to blend into each other. Those orbits are called chaotic. Wisdom has been foremost in studying chaotic orbits in the solar system and has explained, for example, how pieces chipped from asteroids can later impact the Earth as meteorites.

For chaotic orbits, if the initial conditions are changed slightly, the distance between the orbits increases exponentially. The e-folding time (the time for the distance to increase by a factor of $e = 2.718\ldots$) is called the Lyapunov exponent. In their integration of Pluto's orbit, Sussman and Wisdom found a Lyapunov time of

Figure 1. The Digital Orrery, a special-purpose parallel-processing computer built to simulate orbital evolution in the solar system. (*Photo courtesy of Gerald Sussman.*)

20 million years. It should be emphasized that their integration extended for 40 times longer than the Lyapunov time and that no catastrophe occurred; Pluto did not hit another planet or escape from the solar system. In March 1989, J. Laskar found that the Lyapunov time for the solar system was only 5 million years. Again, no catastrophe occurred in his integration and, indeed, we have every reason to believe that the planets have had reasonably regular orbits for 4.5 billion years.

However, the small Lyapunov time restricts the predictability horizon of the equations of motion of the solar system. One might imagine starting with very precise initial conditions so that the initial error was quite small; say, one millionth of a degree in angular position. That error would grow to 485° in a mere 100 million years. We have yet to untangle the consequences of these new results.

GLOBULAR CLUSTERS

The dynamics of the solar system is dominated by the Sun, which provides a high degree of stability to the motions of the planets; in fact, we believe the planetary orbits have been relatively unchanged for 4.5 billion years and the main aim of celestial mechanics is to rationalize that stability.

In contrast, globular clusters (which are collections of about 100,000 stars) are evolving, and the aim of stellar dynamics is to follow that evolution. The evolution is driven by encounters (flybys) between stars, which exchange energy and momentum and serve to randomize the dynamical parameters, much like collisions between molecules in a gas. However, globular clusters cannot reach an equilibrium state, but continue to evolve indefinitely. The main feature of that evolution is that the inner portion (i.e., the "core") of the cluster contracts, at first slowly and later more rapidly. The central density increases and binary stars are formed; when one of the partners is a neutron star, these binary stars become visible as x-ray sources.

In 1960, von Hoerner integrated the equations of motion of 16 stars, and that started the field of "experimental stellar dynamics." The main practitioner of this trade for the past 25 years has been Sverre J. Aarseth. The difficulties of performing the numerical integration are due to the fact that for N stars, there are $N(N-1)/2$ pairs of forces to be calculated. Another difficulty is due to the enormous spread in times scales in the stellar motions; a close binary may have an orbital period one million times shorter than the period of an average star orbiting about the cluster.

The main tricks of the trade are as follows:

1. *Individual time steps*. Each star is integrated with a time step appropriate to its immediate environment. The time step is short if the star is in close interaction with a neighbor; otherwise it is longer.
2. *Regularization*. The force between a pair of stars is proportional to the inverse square of their distance. As the distance becomes very small, the force increases dramatically. Regularization is a method of recasting the equations of motion in new variables that eliminate this singularity in the force (at the expense of contracting the time variable).
3. *Treating neighbors and distant stars differently*. The first practical algorithm for accomplishing this separation was introduced by A. Ahmad and L. Cohen and resulted in a reduction of the time to compute a crossing time from being proportional to N^2 to approximately $N^{3/2}$

Using all these tricks and the fastest computers allows simulations for about 3000 bodies. These investigations have provided many insights into the evolution of globular clusters but the limitation to relatively small values of N has distorted one important feature of the evolution: the formation and importance of binary stars. In the numerical integrations, a binary star forms early and soon acquires a substantial fraction of the energy of the entire cluster. For real clusters, this process takes much longer and the binary never acquires more than about 1% of the energy of the cluster.

Statistical methods have had to be used to model the evolution of realistic clusters. These are well-described in a recent book by Lyman Spitzer, Jr. one of the leaders in this field.

Very recently, new algorithms have been introduced for ordering the force calculation that reduce the N-dependence of the computing time to $N \ln(N)$ ("tree codes" introduced by A. Appel and by J. G. Jernigan) and to N (by L. Greengard and V. Rohklin and by F. Zao). These new algorithms implemented on fast, massively parallel processors may enable realistic simulations. We can expect to gain a firmer understanding of the evolution of globular clusters from numerical experiments on the 100,000-body problem.

Additional Reading: Celestial Mechanics

Arnol'd, V. I. and Avez, A. (1968). *Ergodic Problems of Classical Mechanics*. W. A. Benjamin, Inc., New York.

Brouwer, D. and Clemence, G. M. (1961). *Methods of Celestial Mechanics*. Academic Press, New York.

Cajori, F. (1934). *Sir Isaac Newton's Mathematical Principles of Natural Philosophy and his System of the World*. University of California Press, Berkeley.

Cohen, C. J., Hubbard, E. C., and Oesterwinter, C. (1973). Elements of the outer planets for one million years. *Astronomical Papers of the American Ephemeris* **XXII** II.

Hénon, M. and Heiles, C. (1964). The applicability of the third integral of motion: Some numerical experiments. *Astron. J.* **69** 73.

Laskar, J. (1989). A numerical experiment on the chaotic behavior of the solar system. *Nature* **338** 237.

Poincaré, H. (1967). *New Methods of Celestial Mechanics* **I**, **II**, and **III**. Dover Publications, New York, and NASA **TT F-450**, **451**, and **452**. (A translation of *Les Méthodes Nouvelles de la Mécanique Céleste*.)

Sussman, G. J. and Wisdom, J. (1988). Numerical evidence that the motion of Pluto is chaotic. *Science* **241**, 433.

Wisdom, J. (1987). Urey prize lecture: Chaotic dynamics in the solar system. *Icarus* **72** 241.

Additional Reading: Globular Clusters

Aarseth, S. J. (1985). Direct methods for N-body simulations. In *Multiple Time Scales*, J. U. Brackbill and B. I. Cohen, eds. Academic Press, New York.

Aarseth, S. J. and Lecar, M. (1975). Computer simulations of stellar systems. *Ann. Rev. Astron. Ap.* **13** 1.

Goodman, J. and Hut, P., eds. (1984). *Dynamics of Star Clusters*. D. Reidel, Dordrecht.

Henon, M. (1973). Collisional dynamics in spherical stellar systems. In *Dynamical Structure and Evolution of Stellar Systems*, L. Martinet and M. Mayor, eds. Geneva Observatory, Sauverny.

Lecar, M., ed. (1970). *Gravitational N-Body Problem*. D. Reidel, Dordrecht.

Spitzer, L. (1987). *Dynamical Evolution of Globular Clusters*. Princeton University Press, Princeton, N.J.

von Hoerner, S. (1960). Die numerische Integration des n-Körper-Problemes für Sternhaufern I. *Z. Astrophys.* **50** 184.

See also **Star Clusters, Globular, Formation and Evolution; Star Clusters, Globular, Gravothermal Instability; Star Clusters, Globular, Mass Segregation.**

Nebulae, Cometary and Bipolar

Luis F. Rodríguez

Most emission and reflection nebulae can be described as roughly spherical regions. However, an important fraction of them exhibit remarkably well-defined morphologies that clearly depart from the spherical case. The most important classes of these nonspherical nebulae are the cometary and the bipolar nebulae. The cometary nebulae receive this name because they resemble comets, with a bright "head" and a fainter "tail" extending in one direction. The exciting (if the nebula has an emission nature) or illuminating (if the nebula has a reflection nature) star of the object is located at the head of the nebula. As it happens with many names in astronomy, there is no relation between comets and cometary nebulae, but the nomenclature remains. The bipolar nebulae are formed by two relatively symmetrical bright lobes with a star between them. Until about two decades ago, all known bipolar and cometary nebulae had been detected by their radiation in the visible region of the electromagnetic spectrum. Now we know of many radio and infrared cases that, as a result of heavy extinction by dust, are not detectable in the optical.

MORPHOLOGIES

The morphologies of the bipolar and cometary nebulae can be attributed to an anisotropy of the exciting or illuminating radiation, to an anisotropy of the gaseous medium surrounding the stars, or to a combination of both. At present it is believed that the main reason for the morphologies of these nebulae is that there is an anisotropic density distribution of gas around the associated star. Basically two distributions are considered to be relevant. If the star is located in a medium with a large-scale density gradient (see Fig. 1), a cometary nebula will result. If the star is located in the midplane of a flattened distribution of gas (see Fig. 1), a bipolar nebula will result. Under this interpretation, the cometary nebulae can also be considered as monopolar (one lobe) nebulae. These conclusions have been corroborated by detailed theoretical models that take into account the radiation and wind of the star as well as the distribution of the gas, and reasonable agreement has been obtained for the morphology of some prototypical objects. In most models a wind from the star clears a cavity in the gaseous environment. The shape of the cavity will be dictated by the density gradients of the surrounding medium. The light from the star then shines on the walls of the cavity, producing the nebula. If the radiation comes from a hot star ($T_* \geq 30{,}000$ K), there will be ionization of the walls and, consequently, an emission (H II region) nebula. If the star is cooler, the nebula will mainly have a reflection nature. If the wind clearing the cavity is powerful enough, significant shock excitation may occur, giving rise to Herbig–Haro-like emission nebulosities.

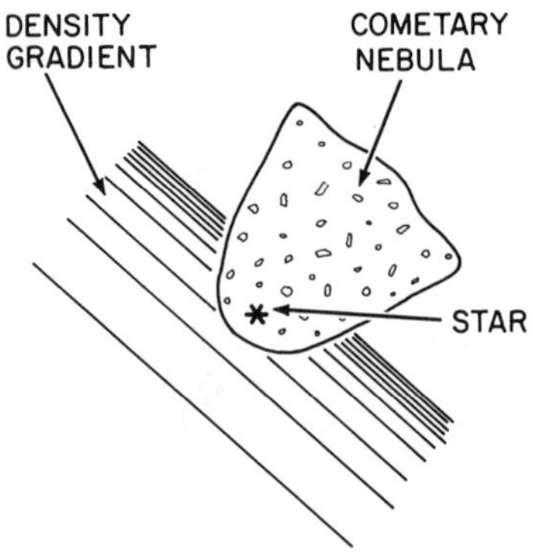

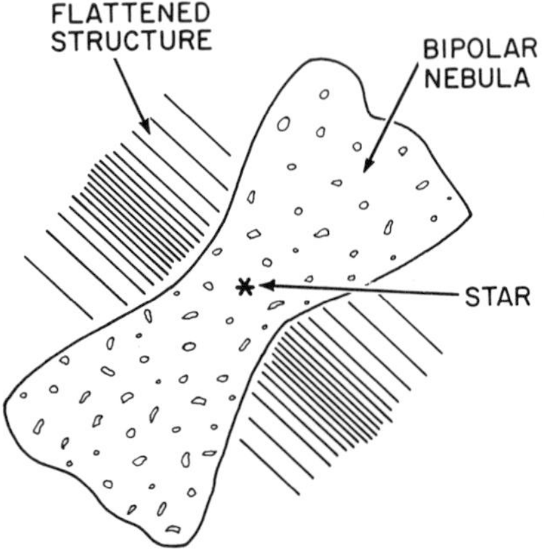

Figure 1. The morphologies of cometary (top) and bipolar nebulae (bottom) are believed to result from the interaction of the radiation and wind of a star with an anisotropic medium. In the case of cometary nebulae, the star is located in a gaseous distribution with a large-scale density gradient. In the case of bipolar nebulae, the star is located in the midplane of a flattened gas distribution.

NEBULAE ASSOCIATED WITH YOUNG STARS

It is known that cometary and bipolar nebulae appear preferentially in association with either very young of very evolved stars (see Table 1, where the luminosities of the associated stars are given in units of the luminosity of the Sun). This apparently puzzling conclusion is understood, at least qualitatively. We discuss first the case of young stars. Stars are believed to form from the gravitational contraction of the gas in a molecular cloud. Star formation processes are not 100% efficient and not all the gas involved in the contraction ends up in the star. Hence, the young star initiates its life surrounded by dense, *placental* gas. This envelope can be more or less spherical but, if it has significant angular momentum or magnetic flux, it will more likely have contracted to a flattened geometry. Models of the contraction of a rotating or magnetized cloud support this notion. This could account for the bipolar nebulae observed in association with young stars. In other cases, the large scale density gradient of the cloud may dominate the

Table 1. Examples of Cometary and Bipolar Nebulae

Object	*Morphology*	*Nature of Nebula*	*Luminosity of Star* ($L_\odot$)	*Evolutionary Stage of Star*
NGC 2261	Cometary*	Reflection†	800	Young
S106	Bipolar	Emission	25,000	Young
G34.3+0.2	Cometary‡	Emission	150,000	Young
NGC 6334(A)	Bipolar‡	Emission	250,000	Young
NGC 6302	Bipolar	Emission	13,000	Evolved
OH 231.8+4.2	Bipolar	Emission†	~10,000	Evolved
MWC 349	Bipolar‡	Emission	~10,000	Evolved
CRL 2688	Bipolar	Reflection	40,000	Evolved

*Radio studies suggest that it may actually be a bipolar nebula with one obscured lobe.
†Also contains Herbig–Haro objects.
‡Detected at infrared or radio wavelengths.

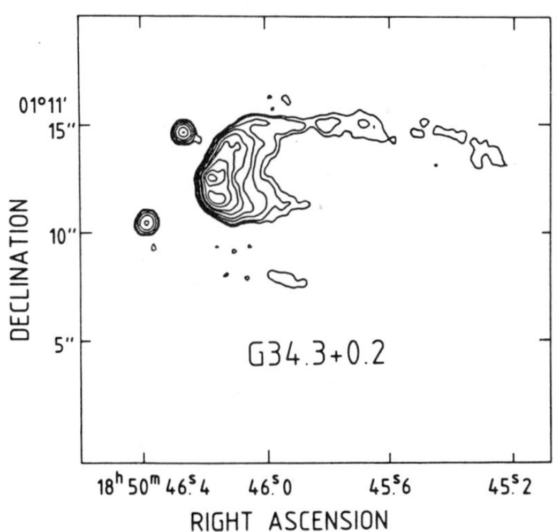

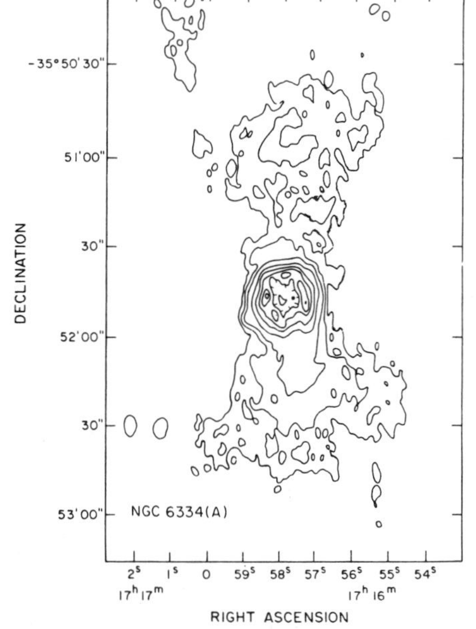

Figure 2. The radio H II regions G34.3 + 0.2 (top) and NGC 6334(A) (bottom) are examples of cometary and bipolar nebulae, respectively. The contour maps shown here trace free–free (bremsstrahlung) radio continuum emission from ionized gas and they were obtained with the Very Large Array radio interferometer in New Mexico.

situation and a cometary nebula will form (Fig. 2). Indeed, in the case of compact (having dimensions smaller than 0.1 pc) radio H II regions, a recent study of 75 such sources indicates that cometary morphologies are evident in 15 (20%) of them with no obvious bipolar cases. Of course, this does not mean that there are no bipolar H II regions: two well-known cases are S106 and NGC 6334(A) (Fig. 2).

Another interesting aspect of this topic has been the discovery of bipolar nebulae that appear to be cometary (that is, with one lobe) in the optical. This situation can be understood by looking at the bipolar model in Fig. 1; if the flattened gas structure has sufficient dust, one of the lobes will be obscured. Two of the objects usually considered as prototypes of the class of cometary nebulae, R Monocerotis and PV Cephei, have turned out to be actually bipolar nebulae with one obscured lobe.

NEBULAE ASSOCIATED WITH EVOLVED STARS

Although both cometary and bipolar nebulae are found in association with young stars, only bipolar nebulae are usually found in association with evolved stars. The reason for this is that, near the end of its evolution, solar-type stars create a gaseous envelope with the help of a massive (mass loss rate of up to 10^{-5} solar masses per year), slow (velocity of 10–20 km s^{-1}) stellar wind. Then, both very young and very evolved stars are embedded in dense gaseous envelopes, whereas stars in the main sequence are relatively "naked." In the case of the evolved stars this wind is, at least in some cases, anisotropic, creating an envelope with larger densities at its equator. It is not understood how this anisotropic wind is created, but stellar rotation or the presence of a companion have been proposed as possible mechanisms. In any case, it is clear that these circumstances are propitious for the formation of a bipolar nebula and not for that of a cometary nebula, which requires the gas distribution to be asymmetrical with respect to the star. Once the expanding, flattened envelope is created around the evolved star, the stellar core (on its way to becoming a white dwarf) contracts and heats considerably, producing sufficient ultraviolet radiation to ionize the envelope. At the same time, a fast (velocity of a few 1000 km s^{-1}), relatively tenuous (mass loss rate of 10^{-8} solar masses per year) stellar wind replaces the previously present massive and slow wind. This fast wind may clear a bipolar cavity similar to that found in bipolar nebulae associated with young stars. A bipolar planetary nebula is then formed. A recent study of the morphologies of planetary nebulae concludes that as much as 20–30% of them can be classified as bipolar. In contrast, there is no clearly established case of a cometary morphology in a planetary nebula.

RELATIONSHIP BETWEEN BIPOLAR NEBULAE AND BIPOLAR OUTFLOWS

By definition, the cometary and bipolar nebulae have lobes that emit radiation that in principle is detectable in the visible spectral region (although sometimes extinction by dust obscures them). The bipolar outflows, a very important phenomenon detected and studied in the last decade, are related to the cometary nebulae. The bipolar outflows are commonly present in association with young stars and their formation mechanism is, quite probably, very similar to that discussed previously for bipolar nebulae associated with young stars. The main difference seems to be that the stars powering bipolar outflows have a large mechanical (wind) to radiation luminosity ratio, $L_{mech}/L_{rad} \simeq 0.01–1$, whereas the stars powering bipolar nebulae have $L_{mech}/L_{rad} \leq 0.01$. In both types of bipolar objects a cavity is created by a wind, but in the case of the bipolar outflows the radiation is insufficient to ionize or illuminate significantly the walls of the cavity. Also, because the accelerating effect of the wind is more important in bipolar outflows than in bipolar nebulae, we expect gas motions to be more important in the former. We can then argue that bipolar outflows are, in some sense, bipolar nebulae with "dark" lobes and larger motions than typical. The dark lobes of the bipolar outflows can be easily detected and studied in the radio emission of several molecules and at present their study is a very active field of research.

Additional Reading

Armstrong, J. T. (1989). Molecular outflows. In *The Physics and Chemistry of Interstellar Molecular Clouds: mm and Sub-mm Observations in Astrophysics*, G. Winnewisser and J. T. Armstrong, eds. Springer, Berlin, p. 143.

Barral, J. F. and Cantó, J. (1981). A stellar wind model for bipolar nebulae. *Rev. Mexicana Astron. Astrof.* **5** 101.

Calvet, N. and Cohen, M. (1978). Studies of bipolar nebulae. V. The general phenomenon. *Monthly Not. Roy. Acad. Sci.* **182** 687.

Morris, M. (1981). Models for the structure and origin of bipolar nebulae. *Ap. J.* **249** 572.

Parsamian, E. S. and Petrosian, V. M. (1979). Catalogue of cometary nebulae and related objects ($-42° < \delta < +66°$). Publication No. 51, Byurakan Observatory, USSR.

See also **H II Regions; Herbig-Haro Objects and Their Exciting Stars; Nebulae, Planetary; Nebulae, Reflection; Stars, Pre-Main Sequence, Winds and Outflow Phenomena.**

Nebulae, Planetary

Yervant Terzian

The theory of late stellar evolution indicates that stars with masses between ~1 and ~6–8 $M_\odot$ (or somewhat higher, see entry on Nebulae, Planetary, Origin and Evolution—Ed.) will undergo core collapses reaching densities of about 10^7 g cm^{-3} and radii on the order of 7000 km. The formation of these well known white dwarf stars is accompanied by a mild ejection of the outer parts of the evolved stars. The ejected material forms an envelope that is known as a planetary nebula. These nebulae have low expansion velocities of ~20–30 km s^{-1} and have a lifetime of about 30,000 yr before they disperse into the diffuse interstellar medium. Initially, the ejected material is cold and neutral and contains molecular species such as H_2, CO, and OH and also dust particles. Eventually the surface of the remnant stellar core becomes hot enough and dissociates and ionizes part of the expanding envelope, which in turn emits strongly in the ultraviolet, visible, and radio spectral regions.

Although planetary nebulae are not very luminous, they are easily recognizable and can be observed within 10 or more kiloparsecs from the Sun. At the present time, some 1500 planetary nebulae have been identified in our galaxy, and several hundred other such nebulae have been detected in other systems, especially in the Andromeda galaxy and the Magellanic Clouds.

Planetary nebulae can be found by examining photographic plates of the sky and distinguishing them by their shell-type or disk shapes. Such a method is limited to nebulae that have angular sizes of at least 10 arcsec, most of which are within a few kiloparsecs from the Sun. Almost all other discoveries of planetary nebulae are made by detecting the nebular emission line spectra, which show strong Balmer emission lines and the [O III] lines at about wavelength (λ) 5000 Å.

The space distribution of planetary nebulae shows a concentration along the galactic plane, and a strong concentration of the smaller nebulae in the direction of the galactic center. This distribution, in galactic coordinates, is shown in Fig. 1. A histogram as a function of galactic latitude, shown in Fig. 2, also indicates this

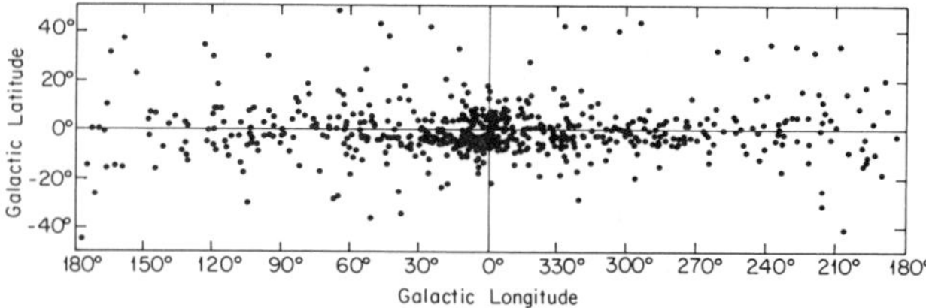

Figure 1. The galactic distribution of planetary nebulae, showing a concentration to the galactic plane and a larger density in the direction of the galactic center.

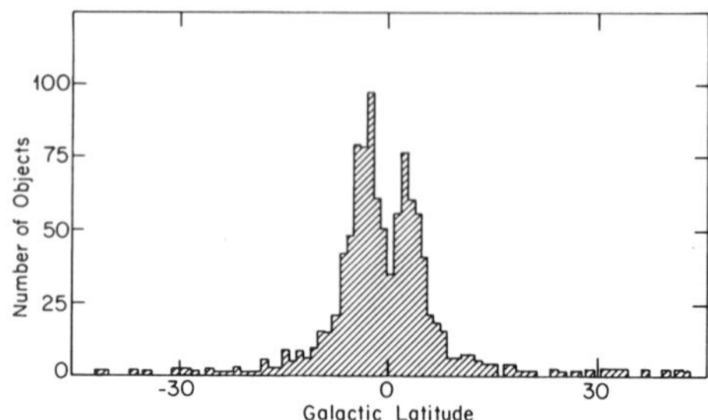

Figure 2. A histogram of planetary nebulae as a function of galactic latitude. The central deficiency of nebulae is probably due to the extreme interstellar extinction by dust particles preventing the detection of the nebulae in that direction.

distribution and in addition, reveals a deficiency of nebulae at galactic latitudes near zero. This probably reflects the strong effects of interstellar extinction, particularly in the direction of the galactic center.

Even though the population of planetary nebulae in the Galaxy is large and their apparent distribution is well defined, details about their real space distribution remain uncertain because of difficulties in determining their individual distances. The trigonometric parallax method can only be applied to a very few planetary nebulae, and the spectroscopic parallax method is not applicable to the central stars of these nebulae. Hence other methods must be used to estimate the distances to individual objects. One method uses the spectroscopically observed radial velocities of the nebular expansion together with the observed angular expansion rate to derive the distance. Another method uses estimates of the amount of interstellar extinction, due to interstellar dust, in the direction of a nebula to estimate the distance. Still another method uses the λ21-cm neutral hydrogen absorption line measurements in the line of sight to a planetary nebula to estimate the distance. Very few planetary nebulae are known members of stellar clusters whose distances are known by other independent methods. One such object is K648 (Ps-1) in the globular cluster M15 at a distance of 10 kpc. These methods apply only to a few dozen nebulae, and the majority have distances derived from approximate statistical methods. In general, the available distance estimates of planetary nebulae are very uncertain and individual distances may be in error by factors of 2 or even 3.

The uncertainty in the distance scale of planetary nebulae unfortunately propagates into many vital characteristics of these objects, such as the nebular mass, the absolute luminosity, and the total number of planetary nebulae in the Galaxy. However, from the available information, it appears that the ionized nebular mass in planetary nebulae has a range from ~0.01–1 $M_\odot$. There is also a correlation indicating that the larger the mass is, the smaller is the electron density of the hot gas. The ionized mass can be a small fraction of the total mass.

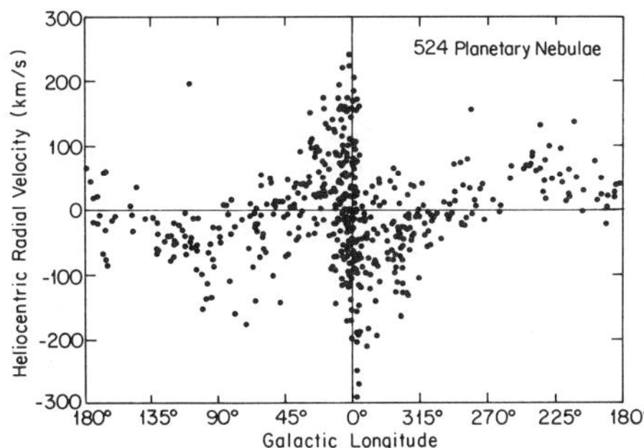

Figure 3. The distribution of the observed radial velocities of planetary nebulae as a function of galactic longitude.

The space within 1 kpc from the Sun contains at least 50 planetary nebulae and perhaps as many as 100, given the distance uncertainties of these objects. The scale height of their distribution above and below the galactic plane is about 200 pc, and the total number of planetary nebulae in the Galaxy can be as low as 10,000 and as high as 100,000 depending on the adopted space density. Statistics of planetary nebulae in other nearby galaxies suggest that there are $1-4 \times 10^{-7}$ nebulae per solar mass. This implies that in our galaxy there must be 20,000–80,000 nebulae if the Galaxy has a total mass of 2×10^{11} $M_{\odot}$. These numbers, though uncertain, indicate that the birthrate of planetary nebulae in the Galaxy is ~ 1 yr^{-1}, which agrees well with the rate of formation of white dwarf stars.

The kinematics of planetary nebulae can be investigated by measuring their radial velocities. More than 500 nebulae have such measured velocities and Fig. 3 shows their galactic distribution. It is seen that a large velocity dispersion exists near the galactic center of ~ 140 km s^{-1} and a total range of about ± 250 km s^{-1}. The kinematics of the planetary nebulae outside the galactic center regions are in agreement with the expected radial velocities assuming circular orbits at various distances from the galactic center. The galactic center objects do not appear to follow the regular differential circular galactic motion, and in this regard they are similar to the motions of Population II objects, such as RR Lyrae stars, globular clusters, and OH/IR stars. The kinematical data indicate that planetary nebulae near the galactic center essentially are members of an older population similar to globular clusters. It is, however, also true that planetary nebulae that are confined to the galactic plane, outside the galactic central regions, are of a younger Population I type.

It is also important to consider the influence that planetary nebulae have on the chemical evolution of the interstellar matter. Most of the matter returned to the interstellar medium comes from mass ejection from cool giant stars such as Miras and OH/IR stars, which are considered possible progenitors of planetary nebulae. If the mean mass in a planetary nebula is 0.2 $M_{\odot}$, and if one such object is formed per year, then 20 $M_{\odot}$ of processed material are returned to the interstellar medium per century from the formation of these nebulae. This mass is at least comparable to the mass ejected by novae and supernovae, so that the formation of planetary nebulae seems to play a central role in the chemical evolution of the Galaxy.

Additional Reading

Maciel, W. J. (1989). Galactic distribution, radial velocities and masses of PN. In *Planetary Nebulae*, S. Torres-Peimbert, ed. Kluwer Academic, Dordrecht, p. 73.

Pottasch, S. R. (1984). *Planetary Nebulae*. D. Reidel, Dordrecht.

Schneider, S. E., Terzian, Y., Purgathofer, A., and Perinotto, M. (1983). *Ap. J. Suppl.* **52** 399.

Terzian, Y. (1980). Planetary nebulae. *Quart. J. Roy. Astronom. Soc.* **21** 82.

See also **Interstellar Medium, Chemical Composition, Galactic Distribution; Nebulae, Planetary, Energetics and Radiation; Nebulae, Planetary, Extragalactic; Nebulae, Planetary, Origin and Evolution.**

Nebulae, Planetary, Energetics and Radiation

James B. Kaler

A planetary nebula is a cloud of gas that surrounds an old, hot star that is in a particular phase of its development (see Fig. 1). It represents a transition stage that takes a star from the giant into the white dwarf state, and is produced when a distended asymptotic-branch giant sloughs off its envelope—the nascent nebula—to reveal its hot, nuclear-burning core. The nebula then expands away, lit by the hot, diminished star at its center until it disappears into interstellar space. The planetary nebulae, of which the order of a thousand are known, were named and revealed as a class by Sir William Herschel about 200 yr ago, as the apparent disks reminded him of the disks of planets. He was well aware that they bore no relation to any of the bodies of our solar system, and was the first to suggest that the glowing substance did not consist of unresolved stars but of a "shining fluid," which we now know to be an ionized gas.

The metamorphosis of a giant star the dimension of the martian orbit into a white dwarf the size of Earth is a dramatic event, and the stars and nebulae involved consequently possess a wide range of properties. The illuminating stars have temperatures that range from $\sim$ 25,000 K to well over 200,000 K, and luminosities from 10 to over 10,000 solar luminosities ($L_{\odot}$). The youngest nebulae are only a few hundredths of a parsec in radius, with densities of 10^4 atoms cm^{-3} or more, whereas the oldest approach 1 pc in size, spanning the distances between stars, and have densities of perhaps 10 cm^{-3} or under, not much above that of the interstellar medium.

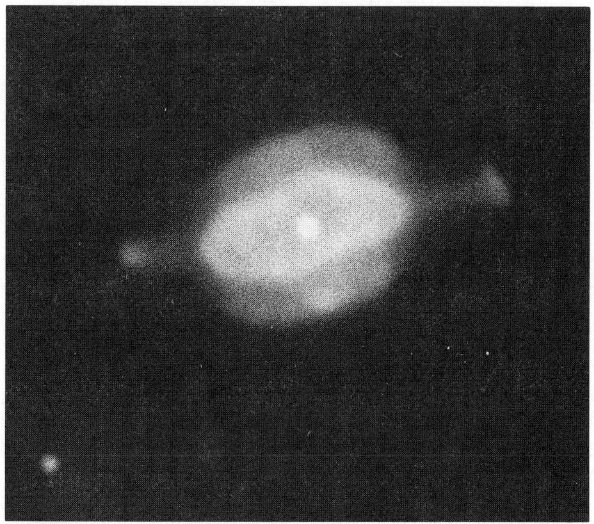

Figure 1. A typical planetary nebula, NGC 7009. The central star, with a temperature somewhat under 100,000 K, illuminates the surrounding nebula, which has a radius of ~ 0.1 pc. The nebula is part of the old red giant envelope, and the central star is in the process of becoming a white dwarf. (*National Optical Astronomy Observatories.*)

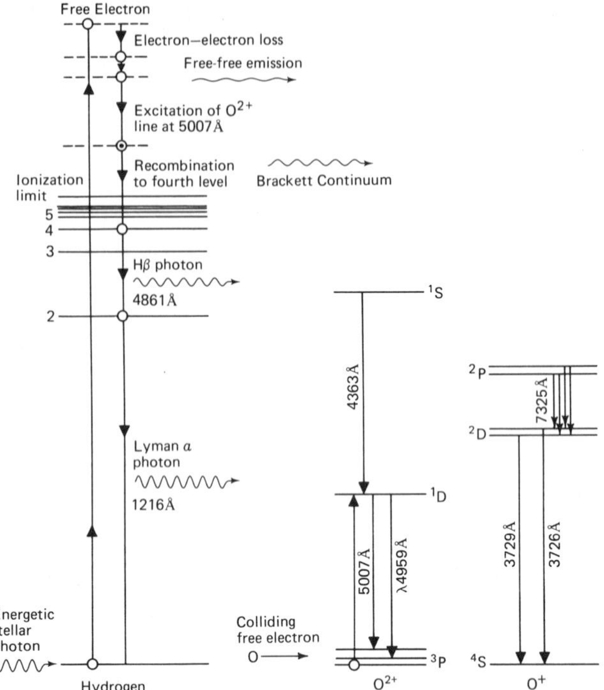

Figure 2. The formation of the nebular spectrum. The hydrogen electron is ionized by ultraviolet starlight. It then loses energy by radiationless collisions with other electrons (thermalizing it), loses some to a passing proton creating free–free emission seen in the radio spectrum, then loses more by colliding with an electron in the ground state of O^{2+}. Eventually it recombines, in this case to the fourth level with the production of a Brackett continuum photon, followed by Hβ and Lyman α photons. The Lyman α photon is trapped and may be broken down into two quanta, producing more continuum emission. The O^{2+} electron is knocked up to a metastable state, and decays backs to the ground state by the creation of a forbidden quantum. Other forbidden lines of O^{2+} and O^+ are also shown. The scale of the oxygen energy-level diagrams is doubled relative to that of hydrogen for clarity. The separations between the substates of the levels called ^{3}P, ^{2}D, and ^{2}P are greatly exaggerated, as is the size of the free–free transition. [*Adapted from* Stars and Their Spectra: An Introduction to the Spectral Sequence, *J. B. Kaler, Cambridge University Press, Cambridge (1989).*]

IONIZATION AND RECOMBINATION

The mechanism of the illumination of planetary nebulae was worked out 60 yr ago by Hermann Zanstra. For the sake of simplicity, consider a pure hydrogen nebula and follow the action on Fig. 2. The atom will be ionized by stellar radiation at wavelengths shortward of the Lyman limit at λ912 Å. In order for the nebula to be in equilibrium, which it obviously is from observation, each photoionization must ultimately be followed by recapture of the electron by some proton. The electron can go to the ground state, with the production of a Lyman continuum photon, or it can land on any of the upper energy levels. If it goes directly to $n=1$, the stellar photon is (in number, not in energy) simply re-created, resulting in another photoionization, and it is as if the recapture had never taken place. Eventually, the electron must land in an upper level with the production of a photon in a subordinate continuum; here $n=4$ (the Brackett continuum) is chosen as an example. Given the short lifetimes of the excited states ($\simeq 10^{-7}$ s), they will contain an insignificant number of electrons at any one time; consequently the Brackett photon cannot be absorbed and will escape the nebula.

From $n=4$, the electron will jump very quickly down to a lower level. If it goes to $n=1$, a Lyman line photon (here Lyman γ; not shown), will be produced. If the nebula is opaque (or optically thick) enough to capture Lyman continuum photons, it must be *very* thick in the Lyman lines. Consequently, the Lyman γ photon will immediately be reabsorbed, placing the electron back into level 4. The Lyman γ photon may be scattered like this several times until finally by chance the electron takes an alternative route and goes to, say $n=2$ with the creation of an Hβ quantum, which must also escape. If the transition is from $n=4$ to $n=3$ (Paschen α), then we would see the same scenario, with the scattering of a Lyman β photon ($3\rightarrow 1$) until the $3\rightarrow 2$, Hα photon is created. Every recapture must then eventually result in one photon of the Balmer series, each of which escapes, and each Balmer photon must be followed by a Lyman α quantum. Thus we have the relation of the *Zanstra Mechanism*:

$$\begin{aligned} N(\text{stellar UV photons}) &= N(\text{photoionizations}) \\ &= N(\text{recombinations}) \\ &= N(\text{Balmer photons}) \\ &= N(\text{Lyman } \alpha \text{ photons}), \end{aligned}$$

where N is the total number of processes or photons produced per second.

The nebula acts as an ultraviolet photon counter. By counting the number of Balmer photons, we know the number of ultraviolet photons radiated from the star, which yields $\int_{\nu_1}^{\infty}(B_\nu/h\nu)\,d\nu$, where B_ν is the Planck function and ν_1 is the Lyman limit. A brightness measurement of the star in the optical provides B_ν (λ5500). The ratio of the integral to the monochromatic optical flux is exquisitely sensitive to temperature, which can then easily be derived, and which is termed the *Zanstra temperature* T_z(H). The theory of the relative strengths of the Balmer lines is well known, so that N(Balmer) is easily related to N(Hβ). T_z(H) is then simply found from the difference between the nebular Hβ flux and the stellar magnitude.

The nebula is also a mixture of other atoms, notably helium. If the star is sufficiently hot, He^+ will absorb all the photons with frequencies higher than ν_3, which corresponds to wavelength λ228 Å. We can similarly use the optical He II lines to determine $\int_{\nu_3}^{\infty}(B_\nu/h\nu)\,d\nu$, resulting in the He II *Zanstra temperature*. If the star is a blackbody and if all the stellar radiation is absorbed (i.e., the nebula is optically thick), then $T_z(\text{H}) = T_z(\text{He II})$. Normally, $T_z(\text{He II}) \geq T_z(\text{H})$, the inequality implying that either the stellar ultraviolet spectrum deviates from a blackbody law or that the nebula is optically thin to the hydrogen Lyman continuum; that is, it is not counting all the lower-energy photons. The problem is still being debated.

The ionization balance for a hydrogen nebula is given by

$$N_1 \int_{\nu_1}^{\infty} \frac{F_\nu^\star}{h\nu} a_\nu \, d\nu = N_i N_e \alpha_B,$$

where N_1 is the number of atoms in the ground state (in cubic centimeters), $F_\nu^\star$ is the dilute stellar flux at a point in the nebula, a_ν the photoionization cross section, N_i and N_e the number of ions and electrons (in cubic centimeters), and α_B the total recapture coefficient. Upper levels need not be included because of their short lifetimes, and consequently $N_1 = N_\text{H}$, the number of neutral H atoms cm^{-3}. Given the size and density of a typical nebula, and the luminosity and temperature of a typical star, the lifetime in the ground state is about 3 weeks, whereas that of a free electron is about a century. The N_H/N_i ratio from the preceding equation is then $\approx 10^{-3}$–10^{-4}, and the nebula is almost a perfect plasma, with just enough neutral atoms in the ground state to absorb the ultraviolet starlight.

FORBIDDEN LINES

As a consequence of the foregoing mechanism, the spectrum of a planetary nebula is filled with recombination lines of hydrogen and a variety of other atoms and ions, notably those of helium, oxygen, carbon, and nitrogen. The strongest features, however, the so-called nebulium lines at λ5007 and λ4959 Å, are not directly related to recombination and they remained chemically unidentified until 1928, when Ira S. Bowen established them as *forbidden lines* of doubly ionized oxygen. The forbidden lines are produced by downward transitions from metastable states with long lifetimes. The transition probabilities are so low that the lines cannot be seen in laboratory spectra, requiring a huge mass of low density gas to produce sufficient flux.

Return to the energy-level diagram in Fig. 2 and follow in detail the electron that has been freed from the hydrogen atom. Upon ionization it will immediately suffer electron–electron interactions that will lower its energy, and that, taken over the whole nebula, will thermalize the electron gas and create a Maxwellian velocity distribution. The electron may also experience non-recombining (free–free) interactions with protons, which are described below. The separation between the ^{3}P ground state of O^{2+} and the ^{1}D metastable state is about 2.5 eV. If the free electron has a kinetic energy greater than this value it can, upon colliding with the oxygen ion, loft the bound electron upward into ^{1}D; the free electron then leaves the scene with 2.5 eV less energy. Eventually the ^{1}D electron drops back down to one of the ^{3}P levels, with the production of the forbidden line. If the free electron has at least 5.3 eV of kinetic energy it can even excite the O^{2+} ^{1}S state, which can decay to ^{1}D, producing a weaker forbidden line at λ4363 Å. The forbidden features are denoted by square brackets; the three lines of the figure belong to the [O III] spectrum. A large number of ions possess these metastable states: another, showing forbidden lines of O^+ (called [O II]) is also shown in the figure. The [Ne III], [N II], [S II], [S III], [Ar III], and [Ar IV] lines, as well as several others in the ultraviolet, optical, and infrared, are also prominent. Because the production of the forbidden lines removes energy from the electron gas, they are the principal coolant of the nebula. The electron temperature (that appropriate to the Maxwell distribution) is determined by the balance between the stellar heating rate and the cooling rate, and usually establishes itself at about 10,000 K. It can go considerably higher, up to $\sim$ 20,000 K, if oxygen is deficient.

The forbidden lines provide a superb set of diagnostics for the nebula. The ratio of the strength of [O III] λ4363 to λ5007 depends upon the ratio of the number of electrons with energies $>$ 5.3 eV to the number $>$ 2.5 eV, and is consequently an accurate indicator of electron temperature. The upward collision and downward transition probabilities involved in the production of [O II] λ3726 and λ3729 are such that their intensity ratio gives the nebular density. A variety of line ratios of other ions is also available in the planetary nebula spectrum. Once the temperature and density are known, the strengths of the forbidden lines are then used to calculate the ionic abundances with considerable accuracy.

FLUORESCENCE

A third, although minor, mechanism for nebular illumination was also found by Bowen. The Lyman α line of He^+, at 303 Å, is very close to an ultraviolet line that arises from the ground state of O^{2+} and can selectively excite an upper state of that ion. Decay downward will produce enhanced permitted lines of O III, principally at λλ3444, 3429, 3133, and λ3760. Then, oddly, the decay back to the ground state is coincident with an ultraviolet transition of N^{2+}, which similarly selectively excites the group of N III lines at λ4640, plus λλ4103, 4097. When He II is present in the nebular spectrum, these lines will all be anomalously strong.

CONTINUUM

Not only does the nebula possess a rich line spectrum, it also exhibits a complex continuum (see Fig. 2). First, electrons are subject to free–free interactions with protons that can produce photons. Because the assembly of electrons has a continuous energy distribution, so must the assembly of emitted free–free quanta. This radiation dominates the radio spectrum, making the planetary nebulae quite radio bright. The mechanism produces lesser amounts at shorter wavelengths, but is still significant even in the optical.

The optical and ultraviolet spectra of planetary nebulae are dominated first by free-bound processes that take the recombining free electrons into various bound states. The set of Brackett transitions illustrated here produces a continuous array of photons shortward of λ14585 Å; free-bound transitions to $n=3$ and 2 produce the Paschen and Balmer continua shortward of λ8204 and λ3646 Å, respectively, with their characteristic sharp edges, and slow declines toward shorter wavelengths. Second, it is possible that in the decay of a hydrogenic electron from $n=2$ to $n=1$ (Lyman α), two electrons rather than one can be produced. It is as if the atom conjures up an intermediate orbit between the first and second. Because there are no restrictions on the energy of this quasiorbit, there are none on the wavelengths of the two photons except that their energies must sum to that of Lyman α: thus a continuum is produced with a shortward limit of λ1216 Å. The intrinsic probability of two photons being released rather than one is low, but Lyman α is scattered so many times on its course through the nebula that the two-photon continuum builds up considerable strength and is of great importance in the optical and ultraviolet.

Finally, the infrared spectrum of a planetary nebula is dominated by a thermal continuum from dust that is both within and around the nebula. The dust, made of silicates or of carbon grains depending on the composition of the nebula, is heated to a few hundred (up to a thousand) kelvins by absorption of resonance photons such as Lyman α and of the stellar continuum. Reradiation from regions at various temperatures then produces a continuum of overlapping blackbody curves.

In conclusion, the planetary nebulae present us with excellent examples of a variety of radiative processes that are amenable to accurate theoretical analysis and that provide a means to understand clearly a critical phase of stellar evolution.

Additional Reading

Aller, L. H. (1984). *Physics of Thermal Gaseous Nebulae. Astrophysics and Space Science Library* **112**. Reidel, Dordrecht.

Bowen, I. S. (1928). The origin of the nebular lines and the structure of the planetary nebulae. *Ap. J.* **67** 1.

Kaler, J. B. (1985). Planetary nebulae and their central stars. *Ann. Rev. Astron. Ap.* **23** 89.

Pottasch, S. R. (1984). *Planetary Nebulae: A Study of Late Stages of Stellar Evolution*. Reidel, Dordrecht.

Torres-Peimbert, S., ed. (1989). *Planetary Nebulae*. Kluwer Academic Publishers, Dordrecht.

Zanstra, H. (1927). An application of quantum theory to the luminosity of diffuse nebulae. *Ap. J.* **65** 50.

See also **Nebulae, Planetary; Nebulae, Planetary, Extragalactic; Nebulae, Planetary, Origin and Evolution.**

Nebulae, Planetary, Extragalactic

Stephen J. Meatheringham

Planetary nebulae (PN) represent a final stage in the life of a star like our Sun (i.e., stars with masses between approximately 1–8 $M_{\odot}$). Their importance to astronomers lies in the information that they provide on their role in the chemical enrichment of interstellar space as well as in providing the link between stars and white dwarfs. Many are known in our galaxy, but those in other galaxies are especially important. A fact that has hampered progress in the study of galactic PN is that the distances of many are only poorly known. This means that the estimated luminosities of the nebulae and the properties inferred about their central stars are inaccurate, and hence details of the evolution are similarly uncertain. Extragalactic PN may be placed into two distinct groups: those in the Magellanic Clouds and those in more distant galaxies. PN in the Small and Large Magellanic Clouds (SMC and LMC) are bright enough to permit detailed study, with the assurance that all the nebulae are at the same known distance. Most other galaxies in which PN have been detected (e.g., M 31, NGC 5128, M 87) are at such great distances that the nebulae are extremely faint.

In the study of PN extensive use is made of the fact that most of the light comes from nebular emission lines. Spectroscopic measurements of the lines yield velocity information and quantities such as temperature, density, and chemical composition. Radial velocity measurements of PN are used to probe the structure of a galaxy by comparing their motion (velocity dispersion) with that of other distinct populations of objects of different ages. For example, globular clusters form an old group of objects, whereas neutral hydrogen (H I) gas is dynamically young. Velocity dispersion increases with age and hence the velocity dispersion of these nebulae gives information on the age and mass distribution of stars giving rise to PN. Comparison of velocity dispersions and light distributions permits testing of dynamical models of galaxies.

MAGELLANIC CLOUDS

Nebulae in the SMC and LMC were first classified as PN in the 1950s. Early studies were mainly concerned with the distribution and motion of the PN populations, a fact partly due to the difficulty of obtaining calibrated, high-quality photographic spectra to measure the fainter nebular spectral lines. This situation improved greatly with the advent of high-quantum-efficiency detectors, which now permit detailed spectroscopy of the faintest lines. Approximately 200 PN have been detected in the Magellanic Clouds, and estimates of the total number range from 10,000 to 50,000, depending on the adopted model of their formation.

Both the SMC and LMC are of considerable interest from a kinematical viewpoint. Their close proximity to our galaxy and to each other appears to have generated significant tidal interactions in the recent past. The SMC, in particular, seems to have been considerably disrupted by a close passage of the LMC. H I radio data indicate that the SMC is comprised of two distinct gaseous components. PN velocities indicate they are mainly associated with one component, suggesting that the other component is young. Velocity observations of PN in the LMC show that the PN population forms a flattened disk with almost the same spatial distribution and rotation pattern as the H I gas and with a typical age of $2–4 \times 10^9$ yr.

Chemical composition of PN in the Magellanic Clouds yields vital information regarding the role of these objects in enriching the interstellar medium. It is through nuclear fusion that significant carbon, nitrogen, oxygen, and other heavier elements are synthesized, and PN in turn eject this material back into space. Similar abundances are found for the SMC and the LMC, with differences attributed to the higher gas content of the SMC. A single value of the helium abundance is consistent for both PN and H II regions in the Magellanic Clouds and our galaxy, although a class of objects with high helium abundance exists in all three galaxies. Nitrogen is underabundant by about five times relative to galactic PN, but greatly enhanced (three to eight times) compared with local H II regions. This enrichment of nitrogen indicates processing of carbon to nitrogen in the stars prior to PN formation. Neon and argon are the most easily observed elements and the least affected by nuclear processing, and their abundances are lower by factors of 4 and 2 for the SMC and LMC, respectively, compared to galactic PN.

The overall picture from velocity studies and chemical abundances is one of lower metal abundance in the Magellanic Clouds than in our galaxy, with the Clouds having been chemically enriched only in the past few billion years. This is a result consistent with a recent burst of star formation having occurred in them.

OTHER GALAXIES

The first galaxies more distant than the Magellanic Clouds to have PN identified in them were the Andromeda galaxy (M 31) and its companions (M 32 and NGC 205). To find PN in these and other galaxies, a technique is used that makes use of the spectral characteristics of the nebulae. The dominant emission lines are the forbidden [O III] pair at 4959 and 5007 Å. A pair of images [either photographic or using a charge-coupled device (CCD)] are taken through narrow-band filters, on and off the [O III] lines. Only those objects having substantial line emission and no continuum are detected by comparing these images. When these detectors are screened with the requirement that the objects appear stellar (extended objects are H II regions) a large number of PN are found with relative ease.

Well over 300 PN are known in M 31. They are seen out to 34 kpc (well into the galactic halo) and to within a few parsecs of its center. Velocity measurements have been used to separate the PN into separate disk and halo populations. Those PN in the outer disk are rotating slightly slower than the gas, in a manner similar to galactic PN, whereas the halo nebulae form a slowly rotating group similar to the globular clusters. The close similarity between M 31 and our galaxy means that studies of the spatial distribution and luminosity function (the number of nebulae of different luminosities) of its PN will lead to a better understanding of those in our galaxy.

PN AS DISTANCE INDICATORS

PN are potentially one of the best distance indicators to galaxies. They are the only easily seen old component of galaxies at large distances and are found in all types of galaxies. More important, stellar evolution works to ensure a relative constancy in the PN luminosity function across galaxies with a wide range in mass, composition, and age. As well, a strong dependence on evolution rate with stellar mass ensures observation of PN with a very small range in central star masses. These points suggest PN may form good standard candles and provide an accurate method to measure distances out to the Virgo cluster of galaxies.

Additional Reading

Barlow, M. J. (1989). Planetary nebulae in the Magellanic Clouds. In *IAU Symposium 131, Planetary Nebulae*, S. Torres-Peimbert, ed. Reidel, Dordrecht, p. 318.

Ford, H. C. (1978). Planetary nebulae in the Andromeda galaxy and its companions. In *IAU Symposium 76, Planetary Nebulae*, Y. Terzian, ed. Reidel, Dordrecht, p. 19.

Ford, H. C. (1983). Planetary nebulae in Local Group galaxies. In *IAU Symposium 103, Planetary Nebulae*, D. R. Flower, ed. Reidel, Dordrecht, p. 443.

Ford, H. C., Ciardullo, R., Jacoby, G. H., and Hui, X. (1989). Planetary nebulae in galaxies beyond the Local Group. In *IAU Symposium 131, Planetary Nebulae*, S. Torres-Peimbert, ed. Reidel, Dordrecht, p. 335.

Henry, R. B. C. (1990). Abundance patterns in planetary nebulae. *Astrophys. J.* **356** 229.

Jacoby, G. H. (1983). Planetary nebulae in the Magellanic Clouds. In *IAU Symposium 103, Planetary Nebulae*, D. R. Flower, ed. Reidel, Dordrecht, p. 427.

See also **Distance Indicators, Extragalactic; Magellanic Clouds; Nebulae, Planetary; Nebulae, Planetary, Energetics and Radiation.**

Nebulae, Planetary, Origin and Evolution

Icko Iben, Jr., and James B. Kaler

A planetary nebula is an outwardly expanding gaseous shell of highly ionized matter that has been ejected by a star of modest mass (1–10 times that of the Sun [estimates of the upper mass limit range as low as 6 solar masses—see other entries on planetary nebulae—Ed.]) during its last major phase of nuclear burning as a red giant prior to becoming a white dwarf. The high state of ionization is maintained by the absorption of energetic ultraviolet photons emitted by the remnant central star as it contracts and develops a surface of high temperature (higher than 25,000 K). A significant portion of the energy absorbed by the nebular material is reemitted at longer, visible wavelengths in a series of cascade transitions, principally those that lead to Balmer line radiation from recombined hydrogen atoms and to collisionally excited forbidden lines of oxygen and other atoms. The number of hydrogen and helium line photons leaving the nebula provide determinations (by the Zanstra method) of the effective temperature and luminosity of the star, which for a typical bright object are, respectively, 50,000–100,000 K and several thousand solar luminosities. The central star is thought to maintain itself at this high luminosity by continued nuclear burning in a narrow shell near the surface. As nuclear fuel is progressively depleted, the central star contracts and at first develops progressively higher surface temperatures. As the system evolves, the flux of ionizing photons from the central star then increases and the nebula becomes brighter in the visible. Observational evidence and theoretical arguments suggest that the nuclear-burning lifetime of the central star is typically 10,000 yr and that, once the nuclear fuel is effectively exhausted, the remnant shrinks to become a cooling and dimming white dwarf; the nebular shell, which is no longer irradiated by such a high flux of ionizing radiation, dims and finally becomes invisible.

This picture of the origin and evolution of planetary nebulae has been developed over the past four decades through the efforts of observers (including George O. Abell, Lawrence H. Aller, C. Robert O'Dell, and others) and of theoreticians (including Iosif S. Shklovskii, Michael J. Seaton, Bohdan Paczynski, Detlef Schönberner, and Alvio Renzini) and is summarized in Figs. 1 and 2. The immediate stellar precursor of a planetary nebula system is thought to have been an asymptotic giant branch (AGB) star. All stars less massive than about 10 solar masses ($M_\odot$) evolve into this phase after they have exhausted hydrogen and helium at their centers. An AGB star is a very large red giant, with a radius between 100 and 1000 solar radii, and it contains a very hot white dwarf at its center that is composed of carbon and oxygen, the products of complete hydrogen burning followed by complete helium burning. Just above the white dwarf core is a thin shell, containing helium and carbon, that undergoes recurring helium-burning thermonuclear runaways (helium shell flashes). Between the helium-burning phases, which last for about 10% of the duration of the entire AGB phase, hydrogen burns quiescently in a very thin shell above the intermittently flashing helium–carbon layer. At the start of the helium-flashing phase, the mass of the white-dwarf core varies from about one-half of the Sun's mass (for a star initially of solar mass) to about 1.4 solar masses (for stars of initial mass near the upper limit of about 10 $M_\odot$). Thus, most of the mass of an AGB star initially resides in the distended red giant envelope at a composition that is essentially that of the matter out of which the star first condensed, that is, mostly unburned hydrogen and helium. From an analysis of the number–magnitude distributions of AGB stars in individual Magellanic Cloud globular clusters, one discovers that a real star must eject most of its hydrogen-rich envelope before its white-dwarf core has grown in mass much beyond its mass at the beginning of the helium-flashing phase. Specifically, solar-mass stars eject all but about six-tenths of a solar mass (0.6 $M_\odot$) of material and 10-solar-mass stars eject all but about 1.4 $M_\odot$ of material. Part of the ejected material becomes the fluorescing planetary nebula, and the compact remnant becomes the exciting central star. It is believed that the objects known as OH/IR masers contain AGB stars in the act of ejection, at least in the case of relatively massive progenitors.

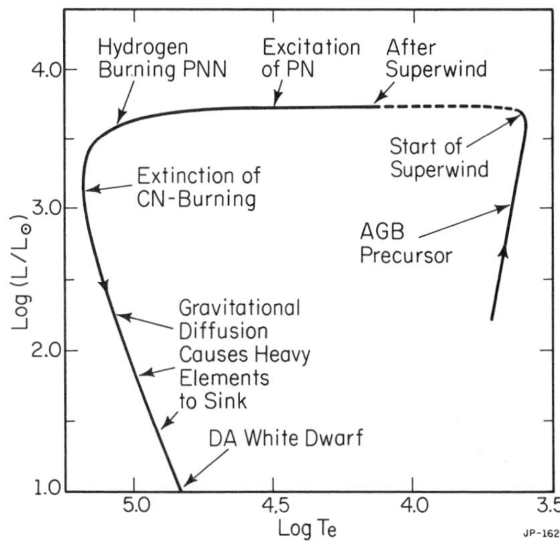

Figure 1. The evolutionary track in the Hertzsprung–Russell diagram of a model star that ejects nebular material during the quiescent hydrogen-burning phase on the asymptotic giant branch (AGB). It has been assumed that ejection occurs via a stellar superwind at a rate of 10^{-5} $M_\odot$ yr^{-1}. The star evolves along the dashed portion of the curve. (PN = planetary nebula; PNN = planetary nebula nucleus [central star].) Following the cessation of mass loss, the model then evolves for about 10^4 yr to the point where hydrogen burning by CN-cycle reactions ceases, whereupon the luminosity of the model drops by a factor of 10 in 10^3 yr. From then on dimming occurs at an ever decreasing rate. Gravity at the stellar surface is large enough to cause element segregation: Helium and other heavier elements sink into the interior while hydrogen floats to the top. The surface composition becomes essentially pure hydrogen. In short, the model has become a DA white dwarf.

The ejection mechanism is not well-understood theoretically. In particular, precisely how much hydrogen-rich material remains in the outer layers of the remnant star at the end of the ejection episode cannot be determined from first principles. Nor can it be stated definitively when in the flashing cycle the final ejection occurs. We can, however, place limits on the amount of hydrogen-rich material that remains and assess the consequences of two possible alternatives. If the final ejection occurs during the quiescent hydrogen-burning phase, then enough hydrogen must remain

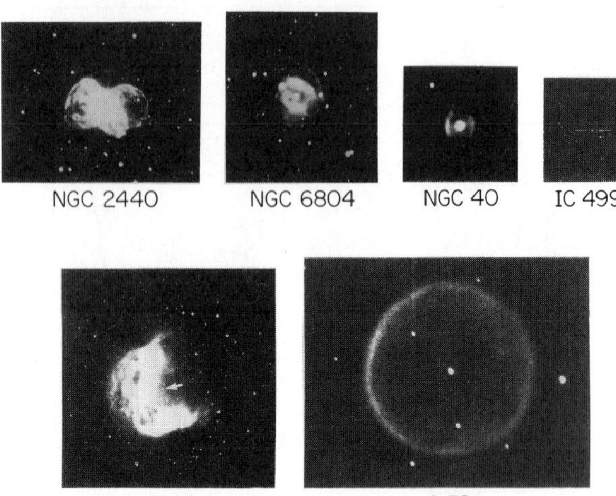

Figure 2. The evolution of planetary nebulae and their nuclei. All photos are reduced to the same linear scale. The four on top show the development (right to left) along the horizontal portion of the evolutionary track of Fig. 1. At first the nebula is nearly stellar. Then we see a bright star in a small planetary nebula and at far left the star fades visually into the nebular background. The two bottom pictures show development (left to right) on the descending track. A dimming star is now revealed in a large low-surface-brightness nebula. A 39 is one parsec across and will soon become invisible. (*Photo of IC 4997 courtesy of Lick Observatory, from Heber D. Curtis 1918; NGC 40 from the University of Illinois Prairie Observatory; NGC 2440 from Palomar Observatory; NGC 6804, YM 29, and A 39 from Kitt Peak National Observatory, the first courtesy of Allan G. Millikan, the latter two courtesy of George H. Jacoby.*)

at the surface to permit the remnant central star to continue to burn hydrogen for the known typical lifetimes of planetary nebulae. For a central star of 0.6 $M_\odot$, the lower limit is about one ten-thousandth of a solar mass (10^{-4} $M_\odot$). An upper limit comes from the fact that the surface temperature of the remnant must exceed 25,000 K if the ejected nebula is to be excited into fluorescence, and from the fact that the size of a model remnant increases in proportion with the mass of the hydrogen-rich layer near the surface of the model. The upper limit is about 10^{-3} $M_\odot$ for a 0.6-$M_\odot$ remnant. If the final ejection occurs during a helium shell flash or during the subsequent quiescent helium-burning phase, the observational constraints just employed can be met if the mass of hydrogen-rich matter that remains is much less than 10^{-4} $M_\odot$. This is because the luminosity and nuclear-burning lifetime of a helium-burning model remnant are nearly the same as those of a hydrogen-burning model remnant.

There are potentially several ways of deciding from the observations which of the two alternative energy sources is more likely to power the central star of a planetary nebula (or, conceivably, which is the more frequently tapped source). One method relies on the fact that hydrogen-burning models cease nuclear burning abruptly and dim by a factor of 10 bolometrically in only 1000 yr, whereas helium-burning models dim at a much slower rate after the cessation of nuclear burning. Thus, with enough examples of central stars with determinable luminosities to construct a meaningful number–luminosity distribution, one eventually may be able to make a choice between the two possibilities. Schönberner believes he has already demonstrated in this way that most central stars are hydrogen burners. Another method relies on the fact that central stars ultimately evolve into white dwarfs and on the fact that the vast majority of white dwarfs can be classified into one of two groups: DA white dwarfs, whose surface composition is nearly pure hydrogen, and non-DA white dwarfs, whose surface composition is almost pure helium. The distributions in number versus luminosity of the two types are very puzzling. The brightest and hottest white dwarfs appear all to be non-DA white dwarfs. There is a (lower) luminosity range where no non-DA white dwarfs are found, and at the very lowest luminosities, the ratio of non-DAs to DAs increases with decreasing luminosity. These facts are very difficult to explain if one assumes that all white dwarfs have a similar ancestry (with, for example, all central star precursors burning helium and not hydrogen) and it seems almost obvious that some degree of mass loss during post-AGB phases must be invoked to understand the increase in the non-DA-to-DA ratio with decreasing luminosity at low luminosity. Nevertheless, a very complex "common-origin" scenario has been proposed by Gilles Fontaine and Francois Wesemael. This scheme requires that the amount of hydrogen remaining in surface layers be of the order of one ten-thousand-billionths of a solar mass (10^{-13} $M_\odot$). If true, this means that all bright central stars must be helium burners and that the last phase of nebular ejection occurs during a helium shell flash or during the subsequent quiescent helium-burning phase.

The nebula evolves along with the star, beginning as a cold, neutral cloud of ejected matter. It is still not clear exactly how the planetary-nebula-to-be is formed. Prior to nebular ejection, the AGB precursor loses mass in a relatively quiet wind. The nascent nebula may be produced by a period of very rapid mass loss (a *superwind*) that terminates AGB life, and the ejected material may be shaped by a high-speed wind from the central star that compresses the inner part of the mass lost during the terminal AGB phase. In any case, it is clear that the bright planetary nebula usually will be surrounded by a much larger low-density, neutral, molecule-rich cloud.

A planetary nebula typically expands at a speed of about 20 km s^{-1}, and becomes ionized, hence visible, when the contracting and heating central star first reaches about 25,000 K and the nebula is a few hundredths of a parsec across. At this time all of the stellar ultraviolet light is absorbed, only the inner portion of the dense cloud is ionized, and the nebula is said to be *optically thick* or *ionization-bounded*. The central star may exhibit its own emission lines and even display Wolf–Rayet characteristics. It is too cool to doubly ionize nebular helium.

As the surface of the central star heats, the planetary nebula grows larger and, in response to an increased ultraviolet flux and lower density, the ionization front expands at an even faster rate. At about 60,000 K, when the nebular radius is of the order of a tenth of a parsec, the front reaches the mass boundary of the nebula, ionizing radiation can escape, and the object is said to be *optically thin* or *mass-bounded*. The escaping radiation can now illuminate outer shells, or even a good portion of the low-density surrounding cloud lost by the AGB star before planetary nebula formation. The Balmer lines of hydrogen no longer count the number of ionizing photons, and the Zanstra method gives a lower limit to temperature and an upper limit to luminosity. However, now helium becomes doubly ionized. The nebula is optically thick to the He^+ Lyman limit and the Zanstra method can be applied to the He II lines to derive stellar parameters. Only in extreme circumstances will the planetary nebula even be optically thin to this higher-energy radiation. The higher-temperature central stars display a variety of spectral characteristics: Some show a mixture of absorption and emission. The more luminous exhibit evidence for continuing mass loss at rates of up to 10^{-8} $M_\odot$ yr^{-1} or so, which is a factor in their evolutionary rate. At their most extreme, the stars exhibit powerful O VI lines.

Eventually the central star reaches its maximum surface temperature, which is predetermined by its mass. The mass distribution of planetary nebula nuclei (central stars) is uncertain, but certainly covers the range 0.55 to 1.2 $M_\odot$, which produces stars with maxima between about 100,000 and 400,000 K. The hottest detected central star has a temperature of about 250,000 K. As the central star heats and more of its radiation is emitted in the ultraviolet, it dims in the visible. Above 150,000 K or so it may be nearly lost in the nebular haze.

After the central star "turns the knee" (Fig. 1) of its evolutionary track and begins to cool and dim, the wind rapidly weakens and the emission spectrum is replaced by a nearly pure absorption spectrum. The planetary nebula then grows to very large size, the surface brightness dropping dramatically as its nucleus fades. After some 30,000 yr, when the nebula has grown to perhaps a parsec in radius (the exact numbers depending upon stellar and nebular mass), the nebula fades to invisibility, leaving behind a hot white dwarf of some 70,000 K or so with a total luminosity approaching one solar luminosity (1 $L_{\odot}$). The expanding nebula then merges with the general interstellar medium, carrying with it any by-products of thermonuclear fusion that the precursor AGB star may have dredged to its surface before nebular ejection.

Additional Reading

Iben, I., Jr. (1989). Peculiar red giants—What kind of white dwarfs do they become? In *Evolution of Peculiar Red Giant Stars*, H. R. Johnson and B. Zuckerman, eds. Cambridge University Press, Cambridge, p. 205.

Kaler, J. B. (1989). *Stars and Their Spectra: An Introduction to the Spectral Sequence.* Cambridge University Press, Cambridge, Chapters 11 and 12.

Kwok, S. (1987). *Physics Reports* **156** 111.

Renzini, A. (1989). Thermal pulses and the formation of planetary nebula shells. In *Planetary Nebulae*, S. Torres-Peimbert, ed. Kluwer, Dordrecht, p. 391.

Schönberner, D. (1989). Evolutionary tracks for central stars of planetary nebulae. In *Planetary Nebulae*, S. Torres-Peimbert, ed. Kluwer, Dordrecht, p. 463.

See also **Interstellar Medium, Stellar Wind Effects; Stars, White Dwarf, Structure and Evolution; Stellar Evolution, Intermediate Mass Stars; Stellar Evolution, Low Mass Stars.**

Nebulae, Reflection

Kristen Sellgren

Reflection nebulae are regions of interstellar space where radiation from a source is reflected by interstellar dust grains, resulting in a diffuse emission called reflection nebulosity. Many different wavelengths of light may be reflected by dust, including ultraviolet, visible, and infrared light. Reflection nebulosity can arise in many situations: The zodiacal light in our solar system is due to sunlight reflected from interplanetary dust particles; reflection nebulosity is seen when dust at high galactic latitudes reflects the light of our galaxy; in some external galaxies, the emission of an obscured active nucleus is believed to be reflected from surrounding dust clouds. In this entry the focus will be on reflection nebulae where a central star too cool to ionize hydrogen illuminates dust in a surrounding or nearby cloud.

Classical visual reflection nebulae are illuminated by visible stars, most often B stars. Very young stars or protostars, as well as very evolved stars, also illuminate reflection nebulae; these nebulae often have a bipolar geometry. The most famous reflection nebula is the Pleiades reflection nebulosity, illuminated by the brightest stars in the Pleiades star cluster. (This star cluster, in the constellation Taurus, is also known as the Seven Sisters and is easily visible to the naked eye.)

SCATTERING PROPERTIES OF INTERSTELLAR DUST GRAINS

Reflection nebulae provide the only opportunity for determining how starlight is reflected by interstellar dust grains. The scattering properties of a grain are determined by the albedo, which is the ratio of the scattering efficiency to the total extinction efficiency of a dust grain, and by the phase function, which describes the angular dependence of scattering. The phase function is often characterized by the phase asymmetry factor g, defined as the average of the cosine of the scattering angle; g is 0 for isotropic scattering and 1 for complete forward scattering (i.e., no change in the direction the photon is traveling). The scattering properties are important not only for understanding the way in which grains will scatter starlight, but also in determining the size and composition of dust grains. The value of g is most sensitive to the ratio of the grain size to the wavelength of scattered light, with $g = 1$ for large grains and $g = 0$ for small grains. The albedo depends on both the grain size and the composition.

Classic work by Edwin P. Hubble in 1922 derived the albedo of dust from the diameters of reflection nebulae and the apparent brightnesses of their central stars. In more recent work, observations of the surface brightness distributions in reflection nebulae are compared to detailed computer models to derive the albedo and g. These comparisons are complicated by the nonuniqueness of models. At visual wavelengths, the albedo is around 0.6, whereas g is thought to be about 0.7. The values of the albedo and g at ultraviolet wavelengths are controversial. The albedo and g are unknown at infrared wavelengths.

The process of scattering by an interstellar dust grain polarizes the reflected starlight. The resultant electric vector is usually perpendicular to the line between the dust grain and the illuminating star. The percentage of polarization depends on the angle through which the light is scattered and is maximum when the light is scattered through 90°.

REFLECTION NEBULAE AND STAR FORMATION

Surveys show that reflection nebulae are strongly concentrated in the galactic plane. Reflection nebulae are also often found near associations of recently formed low-mass stars called T Tauri stars. Reflection nebulae, in many cases, clearly are associated with recent star formation; many of the brightest reflection nebulae, such as M 78 and NGC 7023, arise because a young B star illuminates the dust in the dense cloud in which it has recently formed. The bright Pleiades reflection nebulosity, however, is thought to be due to a chance encounter between the Pleiades star cluster and an unrelated dust cloud. Reflection nebulae associated with young B stars have typical gas densities around 10^4 hydrogen atoms per cubic centimeter and diameters of 0.1–1 parsec (pc). The dust that scatters the star's light accounts for only 1% of the total mass of gas and dust. These nebulae are usually theoretically modeled as spheres of uniform density, although observations show filamentary and clumpy structure.

Although visual reflection nebulae are often found in regions of star formation and in some cases are illuminated by young stars that are not yet on the main sequence, infrared reflection nebulae represent an even earlier stage of star formation. Visual reflection nebulae are illuminated by visible stars that are relatively unobscured by dust, and thus the central star has evolved enough to dissipate the dense clump of dust and gas from which it formed. At earlier stages of star formation, however, the forming star is surrounded by a very dense cloud (with a density of 10^6 hydrogen atoms per cubic centimeter and a diameter of 0.01 pc) that completely obscures the star at visual wavelengths. Dust extinction is 10 times less at infrared wavelengths than at optical wavelengths, however, so that it is far easier to detect the embedded young star in the infrared. Recent observations of these young stars have shown that their circumstellar material is often in the form of a disk, so that the emission of the star suffers very high extinction in the plane of the disk but much less extinction perpendicular to the disk, resulting in a bipolar infrared reflection nebula. This reflected starlight is an important probe of the conditions under which the earliest phases of star formation occur. In some regions, the central star is completely obscured even at infrared wavelengths; in this case, the star can be located only by observing the polarization position angles in the reflection nebula.

REFLECTION NEBULAE AROUND EVOLVED STARS

Reflection nebulosity is also studied to probe the later stages of stellar evolution, when a star is undergoing rapid mass loss. Dust forms in the outflow of gas from the star, and this dust can then reflect light from the central star. In some cases, such as when a star becomes a protoplanetary nebula, the mass loss is highly asymmetric, forming a disk around the star and resulting in a bipolar reflection nebula with a size around 0.1 pc. In some stars the disk reaches gas densities of 10^5 hydrogen atoms per cubic centimeter, completely obscuring the central star, even at infrared wavelengths. In these cases the star can only be studied in light reflected from the lower density poles of the bipolar nebula.

REFLECTION NEBULAE AND STUDY OF THE INTERSTELLAR MEDIUM

Much reflection nebula research in recent years has focused not on the reflection nebulosity itself, but on the reflection nebula as a laboratory for the study of the interstellar medium. The presence of a dense cloud of gas and dust near a bright star provides a perfect opportunity for making detailed observations of the interaction of starlight with interstellar gas and dust.

Because, by definition, a reflection nebula is illuminated by a star too cool to ionize hydrogen, the hydrogen gas in a reflection nebula is either in a neutral or molecular form. Millimeter-wavelength observations of the carbon monoxide molecule show that most reflection nebulae are embedded in or near molecular clouds. Molecular clouds are composed largely of the hydrogen molecule (H_2), yet radio- and millimeter-wavelength observations are unable to detect H_2 directly, because it has no spectroscopic lines at these wavelengths. At infrared wavelengths, however, H_2 has many emission lines that can be excited either by very energetic collisions of the H_2 or by absorption of ultraviolet radiation. The infrared fluorescent emission of H_2 excited by ultraviolet radiation was first detected in 1985, in the reflection nebula NGC 2023. Reflection nebulae are ideal for studying fluorescent H_2 because of the presence of plentiful molecular gas illuminated by intense ultraviolet emission from the hot central star of the reflection nebula.

Reflection nebulae are also ideal for studying the absorption and emission properties of interstellar dust grains, because the physical relationship between the central star and the surrounding dust is well-understood. The same grains that reflect some of the stellar photons to create the visual reflection nebulosity also absorb many of the remaining stellar photons. This absorbed starlight heats the dust grains (to temperatures of 40–60 K) and the absorbed energy is re-radiated by the dust as thermal emission at far-infrared wavelengths of 50–100 μm. The temperature that the dust reaches depends on the ratio of the absorption efficiency at the ultraviolet and visual wavelengths where stellar radiation is absorbed to the emission efficiency in the far infrared where the dust re-radiates most of its energy. This ratio, which has values between 100 and 5000, is an important constraint on models of the composition of interstellar grains. Grains are believed to be composed primarily of silicates and carbon in the form of graphite or amorphous carbon, although grain constituents such as silicon carbide and water ice have also been identified.

A new and unexpected component of the interstellar medium was first identified in near-infrared observations of reflection nebulae. These observations discovered continuum emission at 1–25 μm that was too bright to be attributed to reflected starlight or to thermal emission from dust. The continuum emission was explained by emission from very small grains, small enough that a single stellar photon can heat the grain to very high temperatures (1000 K) for a short time. Grain sizes of 0.001 μm were required to explain the near-infrared emission, 100 times smaller than the sizes of grains responsible for absorption and reflection at visual wavelengths and for thermal emission at far-infrared wavelengths.

Superposed on this near-infrared continuum emission in reflection nebulae are strong emission features at 3.3, 3.4, 6.2, 7.7, 8.6, and 11.3 μm. These emission features had been previously observed in planetary nebulae and H II (ionized hydrogen) regions, but the composition of the emitting material was unknown. The wavelengths of the features are typical of aromatic hydrocarbons, and this, together with the small sizes required for the continuum emission, suggests the 0.001-μm-sized particles perhaps exist in the form of large aromatic molecules rather than very small grains. This identification is still being debated, but it is clear that near-infrared observations of reflection nebulae have uncovered evidence of the existence of very small particles, whose size is intermediate between previously known interstellar grains and interstellar molecules.

The discovery of near-infrared emission from very small grains in reflection nebulae inspired a search for related phenomena at visual wavelengths. Observations of reflection nebulae have discovered continuum emission at 0.5–1 μm in excess over the expected reflected light. An emission feature near 0.67 μm has also been discovered. The relationship between the excess visual emission and the near-infrared emission is under study.

SUMMARY

Reflection nebulae have proven to be valuable tools in understanding interstellar dust and they provide a fundamental technique for determining the scattering properties of dust. Reflection nebuale are associated with regions of star formation, often arising when a young star illuminates the dark cloud from which it has formed. Infrared reflection nebulae provide a glimpse into the earliest stages of star formation, when the young star is still obscured by circumstellar dust from our direct view. Reflection nebulae are also observed in later stages of stellar evolution, particularly when stars rapidly lose mass on the way to becoming a planetary nebula. Reflection nebulae are uniquely suited to the study of the interaction of starlight and the interstellar medium. Reflection nebulae observations have proven crucial to the study of fluorescent H_2 and to the discovery of a new component of the interstellar medium with sizes intermediate between interstellar molecules and previously known interstellar grains.

Additional Reading

Savage, B. D. and Mathis, J. S. (1979). Observed properties of interstellar dust. *Ann. Rev. Astron. Ap.* **17** 73.

Sellgren, K. (1989). Infrared emission from reflection nebulae. In *IAU Symposium 135, Interstellar Dust*, L. J. Allamandola and A. G. G. M. Tielens, eds. Kluwer Academic Publishers, Dordrecht, p. 103.

Smith, D. H. (1985). Reflection nebulae: Celestial veils. *Sky and Telescope* **70** 207.

van den Bergh, S. (1966). A study of reflection nebulae. *Astron. J.* **71** 990.

Vanysek, V. (1969). Reflection nebuale and the nature of interstellar grains. In *Vistas in Astronomy* **11**. Pergamon Press, Oxford, p. 189.

Witt, A. N. (1989). Visible/UV scattering by interstellar dust. In *IAU Symposium 135, Interstellar Dust*, L. J. Allamandola and A. G. G. M. Tielens, eds. Kluwer Academic Publishers, Dordrecht, p. 87.

See also **Diffuse Galactic Light; Interstellar Medium, Dust Grains; Interstellar Medium, Dust, High Galactic Latitude; Nebulae, Cometary and Bipolar.**

Nebulae, Wolf–Rayet

You-Hua Chu

The bright ring-shaped nebulae NGC 2359 and NGC 6888 (Fig. 1*a*) were once considered candidates for supernova remnants, but each of them appears to be radiatively ionized by a central Wolf–Rayet (WR) star. The link between the ring morphology of the nebula and the WR spectral type of the central star was not evident until the third case was found in the ring nebula S 308 (Fig. 1*b*) around the WR star HD 50896. "Ring nebulae around WR stars" thus emerged as a new class of astronomical object. Systematic searches using photographic surveys have turned up nearly 20 cases in our galaxy, 11 in the Large Magellanic Cloud (LMC), and about 10–15 in M 33. These ring nebulae around WR stars are often called WR rings or WR nebulae for short.

The galactic WR rings have sizes ranging from a few parsecs to a few tens of parsecs (pc), with the majority being between 5 and 20 pc. The WR rings in the LMC appear much larger; half of them are 100–200 pc in diameter and the other half 20–50 pc. The WR rings in M 33 are similarly large, if not larger. This apparent

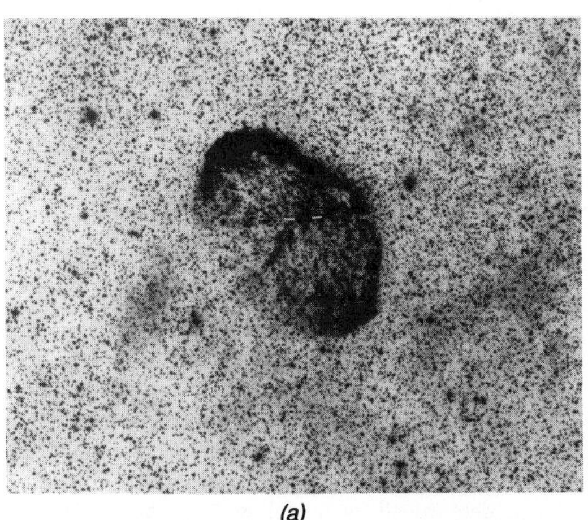

(*a*)

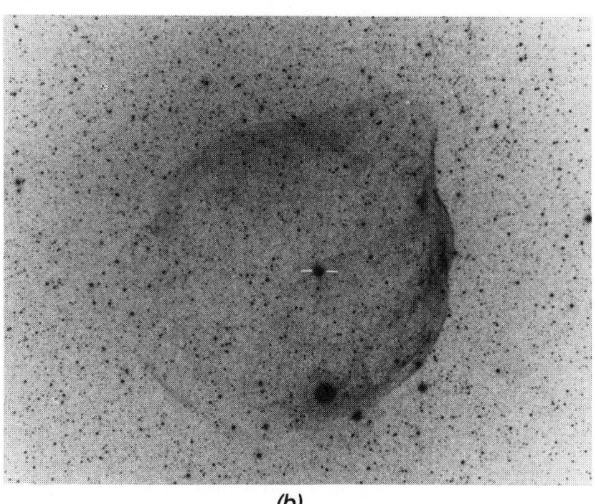

(*b*)

Figure 1. (*a*) NGC 6888; reproduced from an E-plate of the Palomar Sky Survey. (*©1960 National Geographic Society–Palomar Sky Survey. Reproduced by permission of the California Institute of Technology.*) (*b*) S 308; photographed with the Curtis Schmidt telescope at Cerro Tololo Inter-American Observatory. A filter centered at the [O III]λ5007 line was used to take this picture. (*c*) RCW 58; photographed with the 3.9-m Anglo-Australian Telescope by David Malin. (*Photograph courtesy of Anglo-Australian Observatory.*) (*d*) DEM 39 in the LMC; photographed with an Hα filter and an image-tube camera on the Yale 1-m telescope at Cerro Tololo Inter-American Observatory.

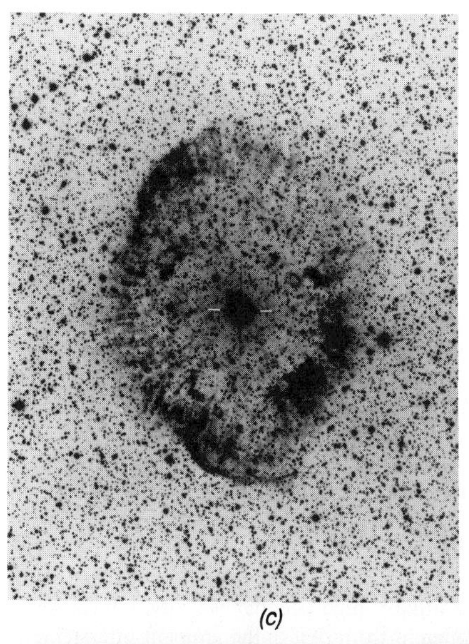

(*c*)

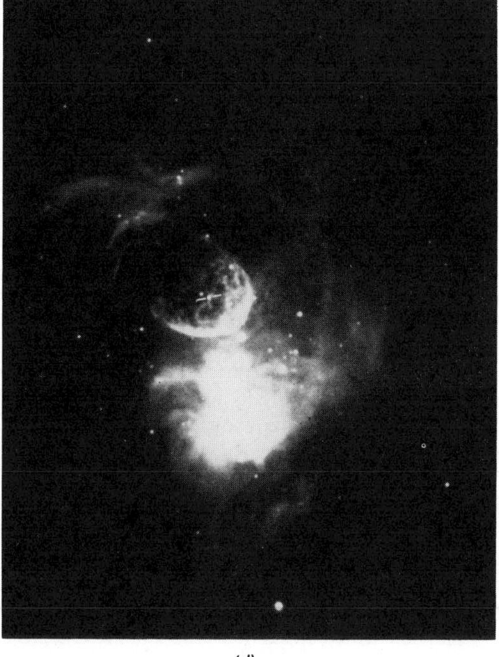

(*d*)

Figure 1. (*continued*)

discrepancy in size between the galactic and the extragalactic objects is an observational effect. Smaller WR rings, with a diameter of 5 pc or less, can be overlooked easily in the LMC and they are usually unresolved in M 33 for the ground-based observations. The largest extragalactic WR rings may contain uncatalogued OB stars or evolved OB associations, in which case they would be superbubbles instead of simple WR rings.

PHYSICAL NATURE OF WOLF–RAYET RING NEBULAE

An important clue to the nature of WR rings comes from the nebular mass derived from radio flux measurements. The mass in a WR ring, a few to a few hundred solar masses, is usually too high for the ring to contain purely stellar material. At least part, if not most, of the nebular mass must be of interstellar origin. Because WR stars are known for their powerful stellar winds, it has been suggested that WR rings are formed by stellar winds sweeping up the ambient interstellar material as in "wind-blown bubbles." However, when morphological and kinematic observations become available, it will be discovered that not every WR ring was formed by the wind from its WR star.

Some ring nebulae were present well before the WR stars, because their dynamical ages, computed from their observed sizes and expansion velocities, are much larger than the lifetime of a WR phase. For these nebulae, the wind from the central WR star could not have contributed much to the formation of the ring; the ring must have been shaped by the progenitor of the WR star and/or other stars via stellar winds and/or supernovae. About one-third of the galactic WR rings and one-half of the LMC WR rings are of this nature, and their expansion velocities are typically 10–15 km s^{-1}.

For the WR rings that do have dynamical ages smaller than the lifetime of a WR phase, the WR stars may have taken an active role in forming the rings. At least three WR nebulae, M 1–67, NGC 6888, and RCW 58 (Fig. 1*c*), show clear evidence that they consist mostly of material ejected by the central star. These bits of stellar material have been called stellar ejecta to denote a different type of mass loss as opposed to the hot tenuous stellar winds. The central WR star in M 1–67 has a very high peculiar velocity, and the nebula has a similarly high peculiar velocity, which is more than 150 km s^{-1} different from that of the ambient interstellar medium. NGC 6888 and RCW 58 each has a bright clumpy Hα ring enveloped in a faint smooth ring that is best shown in the high-excitation forbidden optical line of doubly ionized oxygen, [O III]λ5007. The morphology, internal motion, and excitation of these two rings can only be explained if the inner Hα ring is a fragmented shell of stellar ejecta, and the stellar wind flows through the gaps to sweep up both stellar and interstellar material to form a bubble, the [O III] ring. These three shells of ejecta are characterized by uniform expansion; their expansion velocities are 40–110 km s^{-1} and their densities are a few hundred to a few thousand hydrogen atoms per cubic centimeter.

The other WR rings are probably indeed shaped by the stellar winds of the WR stars. Nevertheless, the situation is more complex than a classical bubble model in which stellar winds plow into the interstellar medium, because a WR star is always surrounded by the hydrogen envelope lost by its progenitor. The physical conditions and dynamics of this circumstellar material depend on the initial mass of the progenitor, whether the star is single or in a close binary system, the ejection process itself, and the elapsed time since ejection. The stellar wind from a WR star must interact with this circumstellar material before reaching the interstellar material. Because the subtypes of WR stars may have evolved differently and even stars of the same subtype may have different masses and evolutionary histories, it is conceivable that WR rings are very individualistic objects. RCW 58 and NGC 6888 represent an early stage of interaction between the stellar wind and the ejected stellar envelope; the nitrogen and helium abundances in their ejecta are distinctly higher than those in their ambient interstellar medium. S 308 may be a bubble blown mostly out of the ejected stellar material, because it is far away from the galactic plane in a gas-poor environment and because its chemical abundance anomaly is similar to that of the known stellar ejecta. In a late stage of evolution, the mass in a WR ring is dominated by the interstellar material and the signature of stellar ejecta would have been diluted below detection. Abundance anomalies are not seen in the older, slowly expanding (expansion velocity ≤ 20 km s^{-1}) nebulae, such as NGC 2359 and NGC 3199.

ASTRONOMICAL APPLICATIONS OF WOLF–RAYET NEBULAE

Studies of WR rings find important applications in several areas. First, the elemental abundances in WR rings give direct evidence of heavy element enrichment of the interstellar medium by massive stars. The nitrogen and helium abundances of several WR rings (e.g., NGC 6888, RCW 58, and S 308) have been shown to be above the local level, whereas the oxygen abundance appears to be lower than the local value. These abundance anomalies provide clues to the stellar nucleosynthesis processes.

It is certain that the progenitor of a WR star has to lose its hydrogen envelope, but it is not known how and when the mass loss takes place. The formation mechanism of a WR ring implied by the physical structure of the ring may place constraints on the mass loss process of the central star. At least two kinds of ejection processes can be inferred from the ejecta in WR rings. In RCW 58 and NGC 6888, the clumpy morphology of the ejecta shell implies that the ejection was instantaneous and the ejected shell has since fragmented due to Rayleigh–Taylor instability. In RCW 104, nitrogen-enriched, high-velocity knots are observed near the bright filamentary ridges; the kinematic properties of these knots indicate an anisotropic ejection different from the aforementioned shell ejection.

Because WR rings are photoionized by their central stars, the ionization and excitation levels of the nebulae should reflect the shape of the stellar ionization spectrum. Stellar effective temperatures can be determined by modeling the nebulae and fitting the observed nebular line strengths. A cool temperature of 35,000 K was determined for the WN 8 central star of RCW 78, whereas higher temperatures of 60,000–90,000 K are derived for the WN 5 central stars of NGC 3199 and S 308.

WR rings have been used to test wind-blown-bubble models; however, they may not be ideal for this purpose because the invisible stellar ejecta can confuse the bubble dynamics. The initial kinetic energy at ejection affects the nebular evolution but is an unknown parameter. The stellar ejecta nature of NGC 6888 explains why its observed x-ray emission is much lower than that expected from an energy-conserving wind-blown-bubble model.

Statistical properties of WR rings have been used to speculate on stellar evolution, but small-number statistics and the incompleteness in the survey should be considered carefully. It was once suggested that WC stars are older than WN stars because no WC stars were surrounded by ring nebulae, implying that ring nebulae had already dissipated. However, this argument became invalid when DEM 39 (Fig. 1*d*) was found around a WC star in the LMC.

Massive stars explode as supernovae at the end of their evolution. The supernova ejecta are likely to encounter the local environment created by stellar winds and stellar ejecta before reaching the unperturbed ambient interstellar medium. The physical structures of WR ring nebulae provide the most realistic boundary and initial conditions for theoretical calculations of supernova remnant evolution.

RECENT DISCOVERIES AND FUTURE WORK

It was shown recently that shock excitation may be important in some WR nebulae. These shocks are best manifested in the

[O III]-to-Hβ line ratio, which reaches 15–20 in the outer rims of NGC 2359 and NGC 6888; the derived electron temperatures may reach well over 20,000 K, as opposed to the normal electron temperatures of ~10,000 K for photoionization.

The most exciting recent discovery is the He II λ4686 emission in some WR nebulae. The presence of the nebular He II emission is indicative of a very high stellar effective temperature, well in excess of 75,000 K. The central stars of He II–emitting nebulae are usually of the earliest type of WO or WN.

Ring nebulae are traditionally identified using optical surveys, which are appropriate for emission lines of ionized gas at 10^4 K. For neutral hydrogen it is necessary to use 21-cm-line observations. Recently, several WR stars have been found to be in neutral hydrogen holes or shells. Future 21-cm-line surveys in the radio should reveal interesting ring structures in the neutral material around WR stars. A wind-blown bubble with ionization front trapped in the cold shell is expected to possess a neutral shell outside. This neutral shell is often the cause for the discrepancy between the optically observed bubble dynamics and theoretical predictions.

The recently available infrared survey by the *Infrared Astronomical Satellite* (*IRAS*) provides another promising way to search for ring nebulae around WR stars. S 308 is a faint incomplete ring in the optical photographs, but appears bright and complete in the infrared images. The *IRAS* survey may yield many more WR rings.

Additional Reading

Chu, Y.-H. (1981). Galactic ring nebulae associated with Wolf–Rayet stars. I. Introduction and classification. *Ap. J.* **249** 195.

Chu, Y.-H. and Lasker, B. M. (1980). Ring nebulae associated with Wolf–Rayet stars in the Large Magellanic Cloud. *Publ. Astron. Soc. Pacific* **92** 730.

Chu, Y.-H., Treffers, R. R., and Kwitter, K. B. (1983). Galactic ring nebulae associated with Wolf–Rayet stars. VIII. Summary and atlas. *Ap. J. (Suppl.)* **53** 937.

Esteban, C., Vilchez, J. M., Manchado, A., and Edmunds, M. G. (1989). Chemical abundances of the WR-ring nebulae NGC 2359 and RCW 78. *Ap. Space Sci.* **157** 3.

Johnson, H. M. and Hogg, D. E. (1965). NGC 2359, NGC 6888, and Wolf–Rayet stars. *Ap. J.* **142** 1033.

Kwitter, K. B. (1984). Nitrogen and helium enrichment in four Wolf–Rayet ring nebulae. *Ap. J.* **287** 840.

Smith, L. F. and Batchelor, R. A. (1970). 11 cm observations of small nebulae associated with Wolf–Rayet stars. *Australian J. Phys.* **23** 203.

See also **Interstellar Medium, Stellar Wind Effects; Stars, Wolf–Rayet Type.**

Neutrino Observatories

Maurice M. Shapiro

The nascent science of neutrino astronomy is based observationally upon the rare interactions of neutrinos with a suitable target and the detection of their charged secondaries. For this purpose, massive targets and arrays of particle sensors (or radiochemical methods of detection) are employed. Neutrinos are well-recognized members of the basic elementary-particle families. They are the most elusive of the established fundamental particles because of their extremely weak interaction with matter. Yet it is this same attribute—their very low cross section—that makes neutrinos (the most penetrating of all known particles) unique tracers of phenomena occurring in "hidden" regions otherwise inaccessible to observation. A famous example is the interior of the Sun, which is powered by nuclear interactions that give rise to neutrinos.

The salient properties of astrophysical neutrinos are treated elsewhere in this volume. Here we shall outline the main conditions that must be fulfilled in order to detect cosmic neutrinos. Then we shall briefly describe the principal neutrino "telescopes", that is, the arrays of sensors that have been designed to observe the various types of cosmic neutrinos expected from the more likely celestial sources. Neutrino observatories are nearly always located deep underground or deep undersea in order to be shielded from atmospherically generated charged-particle secondaries of the cosmic rays.

NEUTRINO ASTRONOMY AND ASTROPHYSICS

Neutrino detectors are designed to investigate a variety of astrophysical sites and processes by means of the neutrinos that they emit. Among these are the following:

1. Celestial regions and events in which high-energy phenomena are prominent, for example, stellar collapse events, supernova remnants (including neutron stars), compact binary stars, and active galactic nuclei.
2. The nuclear processes that power the Sun.
3. The acceleration and interactions of cosmic rays in the interstellar medium, in which the particles propagate.
4. The nature of dark matter in the universe.
5. The mesons, muons, tau particles, and neutrinos generated by cosmic rays in the Earth's atmosphere.
6. Solar flares.

Some underground installations originally designed mainly for other purposes have turned out to be useful neutrino observatories. Conversely, such observatories can be used to investigate the properties of various elementary particles (known or suspected to exist), for example, to probe the possible masses of neutrinos.

SOLAR NEUTRINOS AND COSMIC NEUTRINOS: PIONEERING EFFORTS

Attempts to observe neutrinos from the core of the Sun and to observe cosmic neutrinos (defined as extrasolar in origin), began several years after the existence of electron-antineutrinos ($\bar{\nu}_e$) had been firmly established by Frederick Reines and Clyde L. Cowan, Jr. Seeking to test the theory of nuclear energy production in the Sun's interior, Raymond Davis, Jr., employed a radiochemical technique for the detection of MeV neutrinos. He collected the few argon atoms produced via the absorption of electron-neutrinos (ν_e) by atoms of chlorine in a huge tank placed in a mine deep underground (see Fig. 1). His experiment engendered the puzzling "solar neutrino problem" (an apparent deficit of these several-MeV particles).

Soon thereafter, Reines and his collaborators in South Africa, and M. G. K. Menon, S. Miyake, and their colleagues in the Kolar Gold Fields of India undertook to detect cosmic neutrinos. After several years and more than 200 registered muon-neutrino (ν_μ) events, no clear evidence had emerged for extraterrestrial neutrinos in these searches. All of the observed interactions were attributable to neutrinos generated in the Earth's atmosphere by cosmic-ray collisions.

Indeed, the first definitive observations of *cosmic* neutrinos came from underground detectors set up for a different purpose: to explore the question of nucleon stability. In February 1987, two groups [the Irvine–Michigan–Brookhaven (IMB) collaboration in the United States and the Kamiokande consortium in Japan] detected events due to neutrinos emitted from the stellar core of Supernova 1987A in the Large Magellanic Cloud. This marked the birth of extrasolar neutrino astronomy as an observational science, and it was a dramatic example of serendipity in scientific discovery.

TYPES OF NEUTRINO OBSERVATORIES

Neutrino observatories may be classified according to the following characteristics and principal aims: the types of neutrinos to be detected [ν_e, $\bar{\nu}_e$, ν_μ, $\bar{\nu}_\mu$, ν_τ, $\bar{\nu}_\tau$ (electron, muon, and tau neutrinos and antineutrinos)]; the energies involved; the "size" of the target (its mass, sensitive area, or sensitive volume); the techniques employed for detection; whether they are situated underground or deep underwater; the instruments' angular resolution. Of the observatories listed in Table 1, some are operational, some are in the prototype mode for feasibility studies, whereas others are under construction.

Usually the observatories are situated in subterranean caverns, for example, in mines such as Homestake, or in deep excavations like Gran Sasso. Some caverns, for example, Gran Sasso, Baksan, and Homestake, are sufficiently commodious to house several large neutrino detectors.

NEUTRINO-ENERGY DOMAINS

The wide range of detectable neutrino energies may be subdivided into various experimental ranges according to principal aims of the investigators:

< 40 MeV: especially for solar neutrinos, and for neutrinos emitted within seconds of stellar collapse.
$> 10^8$ eV: solar flare neutrinos.
> 1 GeV: cosmic-ray-generated neutrinos from the Earth's atmosphere (above 1 GeV, the designation "HEν", for high-energy neutrinos, is often used).
> 1 TeV: HEν from a variety of celestial sources, notably sites of cosmic-ray production.
> 10 TeV: diffuse neutrino emission from the interstellar medium, generated by cosmic rays.

Table 1. Existing Neutrino Observatories

Detector	*Main Aims**	*"Size" of Target*	*Depth** (mwe)*	*Sensors,† Detection Techniques*	*Remarks*
Baksan, N. Caucusus	SN, n, HEν	330 ton 250 m^2	~ 1000	LS	One of the oldest underground neutrino observatories
Homestake Mine, S. Dakota	HEν, n	140 ton	4200	LS	
Artyomovsk, Ukraine	SN	100 ton		LS	
NUSEX, Mt. Blanc	n, SN	150 ton	5000	Plastic tubes in limited streamer mode	
LSD, Mt. Blanc	n, SN	90 ton	5000	LS	
Frejus France	n, SN	912 ton	4800	Flash chambers, Geiger tubes	
MACRO Gran Sasso, Italy	SN, HEν	3240 m^2	3600	LS, streamer tubes, track-etch detectors	
LVD, Gran Sasso, Italy	SN, HEν	1800 ton		LS, streamer tubes	
DUMAND Hawaii	HEν	2×10^4 m^2	4500	Čerenkov	Undersea; later stage, 10^5 m^2, 3×10^7 m^3
NT-200, Lake Baikal, Siberia	HEν	5000 m^2	2400	Čerenkov	"NT", stands for neutrino telescope
Soudan, Minn.	n, HEν	1100 ton		Honeycomb drift chamber	Iron calorimeter
Kolar Gold Fields (2), India	n, HEν	140 ton	7200	Proportional counters, calorimeter	Earlier KGF(1) was devoted to HEν alone, as was the South African experiment of Frederick Reines, *et al.*
Kamiokande, Japan	n, SN, HEν	130 m^2	2400	Čerenkov	Detected ν_e from SN 1987A
IMB, Ohio	n, SN, HEν	3300 ton (fiducial vol.)	1580	Čerenkov	Detected ν_e from SN 1987A
Homestake Mine,‡ (solar)	sol	600 ton	4200	Radiochemical (perchloroethylene)	$^{37}Cl + \nu_e \rightarrow {}^{37}Ar + e^-$ Detects ^{8}B neutrinos
Baksan (solar, Ga)	sol	60 ton Ga	~ 1000	Radiochemical	$^{71}Ga + \nu_e \rightarrow {}^{71}Ge + e^-$ Detects *p-p* neutrinos
GALLEX Gran Sasso, Italy	sol	30 ton Ga	4000	Radiochemical	Detects *p-p* neutrinos
Sudbury, Canada	sol, SN	1000 ton D_2O, 5000 ton H_2O	5900	Čerenkov	$\nu_e + d \rightarrow p + p + e^-$ $\bar{\nu}_e + d \rightarrow n + n + e^+$ $\nu_x + e \rightarrow \nu_x + e$
SunLab, Australia	sol	100 ton H_2O			$\nu_x + e \rightarrow \nu_x + e$

*SN, supernova bursts; n, nucleon decay; HEν, high-energy neutrinos; sol, solar neutrinos.
**mwe, meters water-equivalent.
†"Sensors" means detectors of neutrino secondaries, for example, muons; LS, liquid scintillators; Čerenkov light from charged secondaries activates photomultipliers, which are used in nearly all the observatories.
‡A ^{37}Cl detector tank, similar to that of Raymond Davis, Jr., is also being completed in the Baksan mine.

For observations of HEν_μ a working threshold of ~1 TeV is favored for several reasons. Muons created in the underlying (or surrounding) medium, whether it be rock or water, have a range that increases with energy. This enhances the effective target volume at higher energies. The neutrino cross section also increases with energy, and cosmic neutrinos have a "hard" spectrum with an index close to 2. Considerations of atmospheric background also favor energies greater than 1 TeV for this work.

Detection of muons coming from below (i.e., of upward-moving muons) is often important in neutrino observations, because these muons can be attributed to ν_μ events.

DISTINCTIVE FEATURES AND CAPABILITIES OF NEUTRINO OBSERVATORIES

Nearly all neutrino observatories are characterized by the following features:

1. Massive shielding.
2. Massive targets.
3. "Events" (interactions) can occur quite far from the sensors (and can be detected by means of penetrating secondaries).
4. Sensitivity to particles coming from the other side of the Earth (i.e., to very-long-range upward-moving neutrinos).
5. Very wide aperture, typically several steradians, approaching omnidirectionality.
6. The detector can view many sources at the same time and can observe them continually, night and day.
7. Temporal variations in the emissivity of a source can be observed over a wide range of time scales.
8. The detector can be used to investigate neutrino physics independently of celestial sources, for example, with terrestrial accelerators.

Table 1 lists the existing neutrino observatories (as of 1990), with their main purposes and some of their salient characteristics. Figures 1–3 illustrate several of these. Additional observatories are being designed and proposed, one of which is GRANDE, a multipurpose detector to be sited in an artificial "lake" in Arkansas. It would investigate both extensive air showers (EAS) and neutrino sources. Another is the Large Muon Detector (LMD), with an area of 624 m^2, that works as a calorimeter. Several of the existing underground detectors also have EAS arrays at the surface, connected for operation in coincidence with the neutrino detectors far below. These could contribute significantly to the study of ultra-high-energy cosmic rays.

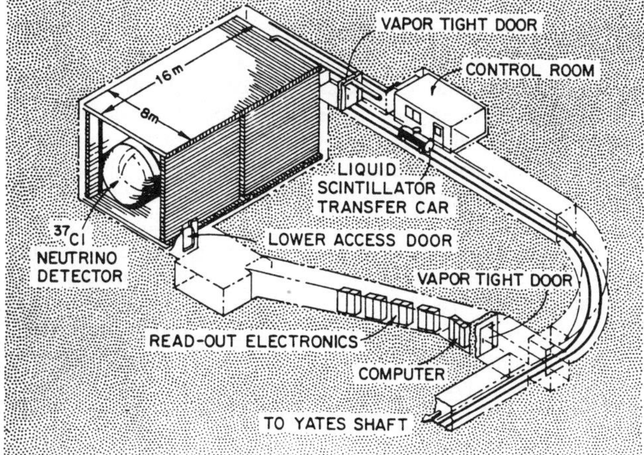

Figure 1. The original Homestake Observatory was the cylindrical tank at the left, employed by Raymond Davis, Jr., for detection of solar neutrinos. Subsequently, Kenneth Lande added the surrounding detector array to search for cosmic neutrinos.

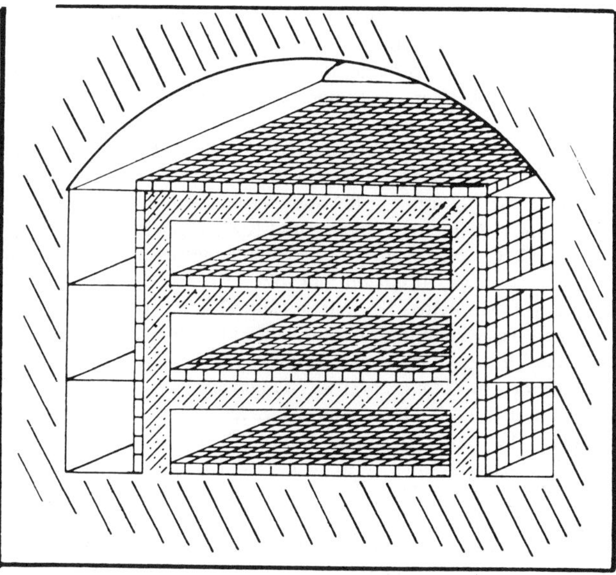

Figure 2. A. E. Chudakov's large scintillation detector for cosmic neutrinos in the Baksan Observatory, near Mt. Elbrus.

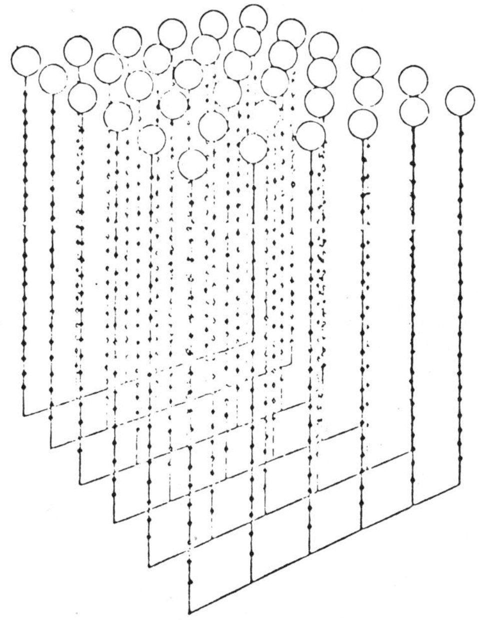

Figure 3. Basic configuration of the Deep Underwater Muon and Neutrino Detector (DUMAND). The large circles represent flotation spheres. The dots on the vertical strings represent photomultipliers for detecting the Čerenkov light emitted by neutrino secondaries, mainly muons. In this version of DUMAND, the instrumented volume of water is 30 million cubic meters (30 × 10^6 m^3).

Early in the twenty-first century, or even sooner, we may expect further significant astrophysical discoveries and insights from the young science of neutrino astronomy.

Additional Reading

Bahcall, J. N. (1989). *Neutrino Astrophysics*. Cambridge University Press, Cambridge.

Berezinsky, V. S. and Zatsepin, G. T. (1984). High-energy neutrino astrophysics (10^2–10^7 GeV) with emphasis on Cyg X-3. In *Proceedings XI International Conference Neutrino Phys. and Astrophys.*, K. Kleinknecht and E. A. Paschos, eds. World Scientific, Singapore, p. 589.

Cline, D., ed. (1989). *Observational Neutrino Astronomy*. World Scientific, Singapore.

Davis, R., Jr., Mann, A. K., and Wolfenstein, L. (1989). Solar neutrinos. *Ann. Rev. Nuclear and Particle Science* **39**.

Learned, J. G. and Eichler, D. (1981). A deep-sea neutrino telescope. *Scientific American* **244** (No. 2) 104.

Lo Secco, J. M., Reines, F., and Sinclair, D. (1985). The search for proton decay. *Scientific American* **252** (No. 6) 42.

Peterson, V. Z. (1982). Deep underwater muon and neutrino detection. In *Composition and Origin of Cosmic Rays*, M. M. Shapiro, ed. D. Reidel, Dordrecht, p. 251.

Shapiro, M. M. and Silberberg, R. (1979). Neutrinos, supernovae, and related astrophysics. In *Relativity, Quanta, and Cosmology* **2**, M. Pantaleo and F. de Finis, eds. Johnson Reprint Corp., New York, p. 745.

See also **Neutrinos, Cosmic; Neutrinos, Solar; Neutrinos, Supernova; Radiation, High-Energy Interaction with Matter.**

Neutrinos, Cosmic

David Eichler

The neutrino is by definition a weakly interacting, spin $\frac{1}{2}$ particle that has no charge and no strong interactions. Its existence was predicted in 1933 by Wolfgang Pauli, who reasoned that another spin $\frac{1}{2}$ particle must accompany the electron and proton that emerge from β decay. A description of the neutrino's weak interaction with matter was provided by Enrico Fermi in the same year. The parameter that expresses the strength of the interaction was determined by the decay rate of the neutron. The neutrino (actually the antineutrino) was finally detected directly more than two decades later by Clyde L. Cowan, Jr. and Frederick Reines, who put a scintillation detector near a nuclear reactor and measured the emergent positron in the reaction $\bar{\nu} + p \rightarrow e^+ + n$ (here $\bar{\nu}$ is an antineutrino, p is a proton, e^+ is a positron, and n is a neutron). Not long afterwards, evidence was found that the weak interactions governing β decay do not conserve parity. That is, only the neutrinos with left-handed helicity participate in these interactions. The right-handed neutrino, if it exists, is constrained experimentally to interact even more weakly.

Neutrinos are manufactured in nature by a variety of astrophysical processes. They are produced in thermonuclear reactions at the centers of stars, by thermal processes in supernovae and the Big Bang, and in high-energy collisions between cosmic rays and ordinary nuclei. Thus far, neutrino physicists have detected neutrinos from cosmic ray collisions with the Earth's atmosphere, neutrinos from Supernova 1987A, and (probably) neutrinos from the Sun's core. It is also presumed that a fossil background of neutrinos should be leftover from the Big Bang. All of these neutrino-generating processes will be reviewed following a brief discussion of the basic physics that governs neutrino interactions.

BASIC NEUTRINO PHYSICS

The weak force field is actually about as strong as the electromagnetic field sufficiently close to a source particle, but the weak force extends only over a very short range, about 1/100 of the proton radius. Hence the cross section resulting from a purely weak interaction is extremely small and the interaction appears weak. There can be "weak force" radiation just as there is electromagnetic radiation. For example, suppose a weakly interacting particle is accelerated in a collision so suddenly that the surrounding weak force field cannot immediately respond. Then the force field is perturbed and the perturbation propagates to infinity in the form of radiation. Because the range of the weak force field is so short, however, the collision must be very close range if radiation is to be released. By the Heisenberg uncertainty principle, a close encounter requires extremely high momentum and energy. Not until the 1980s were collisions of such high energy possible in laboratory accelerators, and only then was radiation via the weak force directly detected.

Quanta of weak force radiation are called intermediate vector bosons, just as quanta of electromagnetic radiation are called photons. Unlike photons, however, these bosons can each carry a charge of $+1$ or -1 (these are called the W^+ and W^- particles, respectively) or they can be neutral, in which case they are called Z particles. This implies that the weak force, in addition to transmitting momentum from one particle to another, can also transmit charge. If, for example, a neutrino emits a W^+, it must take on a negative charge by charge balance. Thus, in the modern theory of electroweak interactions, every neutrino has a charged counterpart. Neutrinos that emerge from nuclear processes have electrons as their charged counterparts, and are called electron neutrinos. Similarly, there are muon neutrinos and tau neutrinos, where muons and tau particles are just like electrons except that they are more massive. A muon neutrino is never observed to transmute into, say, an electron by the emission of a W^+. Rather, the neutrinos appear to remain faithful to their respective charged partners. Physicists suspect that this rule may be broken at a minute level, but such an effect has not yet been observed in a convincing manner.

There are at least two distinct categories of techniques for detecting individual neutrinos. The *chemical* technique involves isolating atoms that have been transmuted into one element from another via a neutrino interaction with their nucleus. The *electronic* technique involves detection of electromagnetic radiation from an energetic electron or positron that emerges from a neutrino interaction. Since the de Broglie wavelengths of most astrophysical neutrinos are typically longer than the range of the weak interaction, their interaction cross sections increase with energy (i.e., increase as the wavelength decreases). This is somewhat analogous to the fact that long wavelength light is less likely to be scattered by dust than shorter wavelength light.

ASTROPHYSICAL PRODUCTION SITES

Thermal Neutrino Emission

An electron can become a neutrino by "sending" a W^- particle to a proton. In absorbing the W^-, the proton becomes a neutron. As the rest energy of the neutron exceeds that of the proton by 1.29 MeV, the process requires at least this much energy from the electron. If an astrophysical plasma can be heated to a temperature of order 10^9 K, then a significant number of thermal electrons will have enough energy to become neutrinos via this process. The neutrons can revert back to protons via decay or collisions, generating antineutrinos in the process. The neutrinos and antineutrinos so generated rapidly escape the system, carrying off energy.

Similarly, in a hot plasma, thermal fluctuations in the electron charge density (called plasmons) can create neutrino–antineutrino pairs via an interesting quantum mechanical process. This process also becomes important at temperatures of 10^9 K and above. It produces neutrinos of all known types.

Given the preceding processes, thermal emission of neutrino–antineutrino pairs is generally an important cooling mechanism above 10^9 K. This is especially true in very dense environments, where standard radiative cooling is suppressed by the very efficient trapping of photons.

The interiors of compact objects, particularly neutron stars, are examples of astrophysical environments where neutrino emission can be the main cooling mechanism. For example, when a stellar core collapses, over 10^{53} erg is released. Roughly 99% of this energy is released as thermal neutrinos during the formation of the neutron star; only about 1% of the energy goes into the supernova explosion. Most of this is thermal emission from the hot, freshly formed neutron star, and it is expected that it is distributed fairly evenly among the six types of neutrinos and antineutrinos. The energies of the individual neutrinos are of the order of 10 MeV. The detection by two experimental groups of electron antineutrinos from Supernova 1987A confirmed this picture.

Because neutrinos have such long mean free paths, they provide the most effective means of outward energy transport during stellar collapse, and it may well be that this energy transport is what powers the supernova explosion.

There is a related neutrino emission mechanism that does not rely on thermal energy: If material is compressed to a sufficiently high density that the zero-point energy of the electrons exceeds the neutron–proton mass difference, then it becomes energetically favorable for the electron–proton pair to convert into a neutron plus a neutrino. If the compression is due to gravity, as in stellar core collapse, then it is ultimately the release of gravitational energy that powers the neutronization of protons and the associated neutrino production.

The total cosmic energy density in the form of supernova neutrinos is probably comparable to that in starlight. This follows from the fact that, although only about 10% of collapsing stars are thought to produce neutron stars, those that do put about 10 times as much energy into neutrinos during the collapse as was put into starlight during the entire history of the star. It is therefore probable that all supernova neutrinos are stable over cosmological time scales to decay into lighter leptons plus gamma rays. If they could undergo such decay, the universe would be inundated with gamma rays far more than is observed. However, physicists have speculated on more exotic decay modes.

A large starquake in a neutron star would put most of its energy into neutrinos if the matter were heated well above 10^9 K. This is unfortunate because the neutrinos would not be produced in observable quantities, and the small remainder of energy that could plausibly escape as photons would be hard to detect.

Finally, the Big Bang was very hot and very dense. A thermal, blackbody background of each of the six known types of neutrinos and antineutrinos should have been leftover just as there is a blackbody background of photons. These neutrinos would have cooled down to below 2 K with the adiabatic expansion of the Universe, and therefore would be difficult to detect directly. However, comparison of present day abundances of the light elements with calculations of nucleosynthesis in the early Universe provides evidence that the thermal neutrino background did exist at the time of primordial nucleosynthesis, and that there could not have been much more present in not-yet-discovered types of neutrinos. If there are more types of neutrinos, they could not have contributed significantly to the mass density of the Universe at the time of nucleosynthesis. In the standard Big Bang picture this implies that they must either interact far more weakly than the known types or they must be extremely massive.

Assuming neutrinos are stable over cosmological time scales, then the background still exists, and our knowledge of the total mass density of the Universe limits the rest energy of all known neutrino types to less than about 30 eV.

It is possible that there is a heretofore undiscovered neutrino with a rest energy lying between several gigaelectronvolts to several teraelectronvolts. Within this range, production of heavy neutrinos in the early Universe would have ceased while the Universe was still dense enough that all but an acceptably low number would have annihilated with their antiparticles. Serious efforts are currently underway to detect such neutrinos, as well as other hypothetical particles, with very advanced detection techniques. Thus far, none has been detected and experimental constraints on their nature have recently become tighter.

Neutrinos from Nuclear Burning

Even when the temperature is well below 10^9 K, neutrinos can be emitted via nuclear processes. Here the "strong" (or "nuclear") interactions, by pulling nucleons together, release the energy needed to convert a proton into a neutron, positron, and neutrino. For example, when hydrogen is burned to helium in a main sequence star, two protons are converted to neutrons for every helium nucleus that is produced; hence, two neutrinos must be produced as well. The total rate of neutrino emission can be directly related to the star's luminosity. However, the nuclear burning takes place via a variety of possible reaction chains, whose relative contributions to the neutrino detection rate depend very sensitively on temperature.

The only star that is presently burning nuclear fuel from which we could hope to detect the resulting neutrinos is our own Sun. A positive signal has been reported from two experiments—one a chemical detection technique and the other electronic—but the levels reported appear to be lower than theoretical expectations. This discrepancy, known as the solar neutrino problem, is one of the outstanding puzzles in modern astrophysics. It has motivated speculation that, after being emitted in the Sun's core, electron neutrinos oscillate to other types and thus evade detection. Other speculations include questioning basic assumptions that go into solar structure calculations.

Ultrahigh Energy Neutrinos

When relativistic protons collide with ambient nuclei in a particular astrophysical environment, mesons, including charged ones, are generated. Charged mesons decay primarily into muons and muonic neutrinos and their associated antiparticles.

Cosmic rays, the chief component of which is relativistic protons, are believed to be produced in astrophysical shock waves such as those produced by expanding supernova shells in the interstellar medium. It is known that about 10% of galactic cosmic rays collide with ambient protons in the interstellar medium; hence, diffuse emission of ultrahigh energy neutrinos is expected from the galactic disk. In addition, shock waves could occur near accreting compact objects, such as neutron stars in binary star systems and such as quasars. Because there is a significant thickness of gas near the object, any relativistic protons that are produced there have a good chance of hitting an ambient gas nucleus, thereby producing ultrahigh energy neutrinos. The neutrino luminosities that are conceivable from this process could be detectable with very large detectors.

This same production mechanism operates when cosmic rays strike the Earth's atmosphere. Atmospheric neutrinos have been detected by several groups.

Production of neutrinos by high-energy proton collisions has the property that more neutrinos than antineutrinos are produced. This follows from the fact that, because the proton has positive charge, more positively charged mesons are produced than negatively charged ones. If neutrinos could be detected from other galaxies, they would, in principle, provide a direct test of whether the galaxies are made of matter or antimatter.

SUMMARY

Neutrinos play several important roles in astrophysics. They shape the dynamics of collapsing stars, even the Universe itself, and invariably bear valuable information about the systems that emitted them. Neutrino astrophysics depends on the microscopic physics of neutrinos; hence, these particles constitute an important interface between elementary particle physics and astronomy.

Additional Reading

Bahcall, J. N. (1986). Solar neutrinos. In *Highlights of Modern Astronomy*, S. L. Shapiro and S. A. Teukolsky, eds. Wiley, New York, p. 85.

Bahcall, J. N. (1989). *Neutrino Astrophysics*. Cambridge University Press, Cambridge.

Learned, J. and Eichler, D. (1981). A deep sea neutrino telescope. *Scientific American* **244** (No. 2) 138.

Mohapatra, R. N. and Pal, P. B. (1991). *Massive Neutrinos in Physics and Astrophysics*. World Scientific, Singapore.

Reines, F. (1978). Reminiscences: The early days of experimental neutrino physics. In *Proceedings of the Clyde Cowan Memorial Symposium*, A. W. Saenz and H. Uberall, eds. Catholic University, Washington DC, p. 1.

Shapiro, S. L. and Teukolsky, S. A. (1985). *Black Holes, White Dwarfs and Neutron Stars*. Wiley, New York.

See also **Antimatter in Astrophysics; Cosmology, Big Bang Theory; Neutrino Observatories; Neutrinos, Solar; Neutrinos, Supernova.**

Neutrinos, Solar

Lincoln Wolfenstein

Thermonuclear reactions taking place near the center are generally believed to be the major source of energy for stars on the main sequence. Most of this energy goes into heat, which is transferred to the stellar surface and emitted as electromagnetic radiation. A small fraction of the energy is emitted in the form of neutrinos, which can travel directly from the stellar interior to the Earth without significant absorption. Over the last 20 years efforts have been made to detect neutrinos emerging from the Sun. It now seems clear that such neutrinos have been observed, providing direct evidence for the nuclear reactions in the solar interior.

The neutrino is an elementary particle with no electrical charge and little or no mass. The existence of the neutrino was suggested by Wolfgang Pauli in 1930 to explain the apparent missing energy in nuclear β decay. The basic β decay processes are

$$n \rightarrow p + e^- + \bar{\nu}_e,$$

$$p \rightarrow n + e^+ + \nu_e.$$

Note we distinguish the neutrino ν_e emitted together with a positron (e^+) and the antineutrino $\bar{\nu}_e$ emitted together with an electron (e^-); n and p are the neutron and proton, respectively. The $\bar{\nu}_e$ emitted from a nuclear reactor was first detected by Clyde L. Cowan, Jr. and Frederick Reines in 1953 by the inverse of the β-decay process

$$\bar{\nu}_e + p \rightarrow e^+ + n.$$

The corresponding detection of ν_e requires a nuclear target containing neutrons:

$$\nu_e + n \rightarrow e^- + p$$

Because neutrinos interact only weakly, they can penetrate large amounts of matter without being absorbed and also they are very difficult to detect.

The main sources of energy in the Sun are the proton–proton interaction chains beginning with the weak reaction

$$p + p \rightarrow {}^2\mathrm{H} + e^+ + \nu_e. \tag{1}$$

Table 1. Solar Neutrino Fluxes

		Energy (MeV)	*Flux at Earth ($cm^{-2}\ s^{-1}$)*	*SNU**	
				^{37}Cl	^{71}Ga
(1)	pp	< 0.42	6.0×10^{10}		71
(2)	^{7}Be	0.86(90%)	5×10^9	1.1	34
(3)	^{8}B	< 15	6×10^6	6.1	14
CNO	^{13}N	< 1.2	6×10^8	0.1	4
CNO	^{15}O	< 1.7	5×10^8	0.3	6
Total				7.9 ± 2.6	132 ± 20

*Detection rate in SNU for ^{37}Cl and ^{71}Ga detectors from indicated sources. Total SNU differs somewhat from the sum due to small omitted sources. (1 SNU ≡ 10^{36} interactions per atom per second.) 1, 2, 3 refer to the numbered reactions in the text.

The resultant deuterium (^{2}H) reacts quickly through the reaction $p + {}^2\mathrm{H} \rightarrow {}^3\mathrm{He} + \gamma$. The main chain ends when two ^{3}He combine to form ^{4}He plus two protons. About 15% of the time, however, calculations show that ^{3}He combines with ^{4}He present in the Sun to form ^{7}Be. Most of the ^{7}Be is converted to ^{7}Li by electron capture,

$$e^- + {}^7\mathrm{Be} \rightarrow {}^7\mathrm{Li} + \nu_e, \tag{2}$$

whereas about one in a thousand ^{7}Be react with a proton to form ^{8}B, which then β decays

$$^8\mathrm{B} \rightarrow {}^8\mathrm{Be} + e^+ + \nu_e. \tag{3}$$

Reactions (1)–(3) represent the three main neutrino sources in the Sun; in addition, some neutrinos come from β decays of ^{13}N and ^{15}O in the CNO cycle.

The energies and calculated fluxes of these main sources are shown in Table 1. The largest flux and the one that can be calculated to an accuracy of a few percent comes from the primary reaction (1). Because these neutrinos have a low energy they are difficult to detect. In practice the experiments detecting neutrinos up through 1989 have been sensitive primarily to the higher energy neutrinos from ^{8}B decay that have a flux 10^{-4} that of the *pp* neutrinos. Furthermore, it is much more difficult to calculate this flux both because it is sensitive to the rates of competing nuclear reactions and because it is very sensitive to the central temperature, T, of the Sun (approximately varying as T^{18}). The results shown come from a detailed calculation of the standard solar model (SSM) made by John N. Bahcall and Roger K. Ulrich. By varying the input parameters (reaction rates, composition, opacity) in accordance with measured or estimated uncertainties, they determined a "3-standard-deviation error" in the ^{8}B flux of ±37%. The uncertainty associated with the simplifying assumptions of the SSM is difficult to estimate.

For a long time the only attempt to detect solar neutrinos was the radiochemical experiment of Raymond Davis, Jr. and collaborators based on the reaction

$$\nu_e + {}^{37}\mathrm{Cl} \rightarrow e^- + {}^{37}\mathrm{Ar}.$$

The detector consists of 615 metric tons of perchloroethylene (C_2Cl_4) located in a mine at a depth of 4850 ft. The argon is extracted about once every two months and the amount (about 0.5 atom per day) is measured by observing ^{37}Ar decays. The expected rate shown in Table 1 is 7.9±2.6 SNU (1 SNU equals 10^{36} interactions per atom per second), whereas the average value observed over the period 1970–1988 was 2.33±0.25 SNU. The difference between the calculation and the experiment is often referred to as the "solar neutrino problem." The data from the Davis experiment from the 1987–1988 run (about 18 months) give a value of 4.2 ± 0.7 SNU, much closer to the expected rate. No reason is known for the higher values for this period although it has been noted that it is consistent with an apparent anticorrelation of the neutrino flux with solar activity.

Recently there has been a second observation of solar neutrinos using a large H_2O Čerenkov detector in the Kamioka mine in Japan. The detection is based on the elastic scattering of ν_e by electrons. An important feature is that the scattered electrons are expected to be closely aligned with the incident neutrino direction. As a result this experiment definitely shows that the neutrinos are coming from the Sun. The experiment is only sensitive to ^{8}B neutrinos with energies above 9.3 MeV. The result in 1987–1988 is 0.46 ± 0.13(stat)± 0.08(syst) times the SSM prediction (where "stat" and "syst" identify the statistical and systematic errors, respectively). This is in very good agreement with the Davis result in the same period.

It is possible that the ν_e flux reaching the Earth may be less than the original flux produced in the solar interior as a consequence of the effects of neutrino mass. The mass of ν_e is known to be less than 20 eV from experiments on ^{3}H β decay as well as from the arrival times of $\bar{\nu}_e$ from SN 1987A, but many theories suggest that neutrinos are not exactly massless. If neutrinos have a mass, it is necessary to consider not only ν_e but also two other types of neutrinos ν_μ and ν_τ, the muon neutrino and tau neutrino, respectively, which interact differently than ν_e. (For example, ν_μ is detected by the reaction $\nu_\mu + n \rightarrow p + \mu^-$, which has a threshold of 110 MeV, the mass of the muon.)

The neutrino mass eigenstates are expected to be mixtures of ν_e, ν_μ, and ν_τ. As a result there exists a quantum-mechanical oscillation effect in which the ν_e may be transformed into ν_μ and ν_τ. This transformation may be enhanced by the material medium as the neutrinos propagate from the solar interior to the surface. This effect of the medium is called the Mikhaeyev–Smirnov–Wolfenstein (MSW) effect. If such a transformation took place, the resultant ν_μ (or ν_τ) could not be detected by the Davis experiment and would produce a very small signal in the Kamioka experiment. One reason for the interest in solar neutrinos is that they provide a possibility for studying neutrino properties that cannot practically be explored using terrestrial sources.

Two experiments designed to observe the *pp* neutrinos are now in progress in the Soviet Union and Italy. These are radiochemical experiments based on the reaction

$$\nu_e + {}^{71}\text{Ga} \rightarrow e^- + {}^{71}\text{Ge},$$

which has a threshold of 0.23 MeV. The expected rate is shown in Table 1; slightly more than half the rate would be due to *pp* neutrinos. A number of other experiments are now in the planning stage. One of the goals is to measure the spectrum of the ^{8}B neutrinos that could be distorted if oscillations took place. Another goal is to measure the total neutrino flux including ν_μ and ν_τ by using weak interactions that are the same for all types of neutrinos; namely, those due to neutral currents in which a nucleus is excited or disintegrated without a change of charge. Only a varied program of solar neutrino observations can elucidate possible problems with the standard solar model or possible unknown properties of neutrinos.

Additional Reading

Bahcall, J. N. (1989). *Neutrino Astrophysics*. Cambridge University Press, New York.

Bahcall, J. N. (1990). The solar-neutrino problem. *Scientific American* **262** (No. 5) 54.

Bahcall, J. N. and Ulrich, R. K. (1988). Solar models, neutrino experiments, and helioseismology. *Rev. Mod. Phys.* **60** 297.

Beier, G. and Wolfenstein, L. (1989). Neutrino oscillations of solar neutrinos. *Physics Today* **42** (No. 7) 28.

Bilenky, S. M. and Petcov, S. T. (1987). Massive neutrinos and neutrino oscillations. *Rev. Mod. Phys.* **59** 671.

Davis, R., Mann, A. K., and Wolfenstein, L. (1989). Solar neutrinos. *Ann. Rev. Nucl. Particle Sci.* **39** 467.

Schwarzschild, B. (1990). Solar neutrino update: Three detectors tell three stories. *Physics Today* **43** (No. 10) 17.

See also **Neutrino Observatories; Neutrinos, Cosmic; Sun, Interior and Evolution.**

Neutrinos, Supernova

Lawrence M. Krauss

Supernovae, exploding stars, increase in luminosity by many orders of magnitude within a space of hours and can shine for months with a luminosity almost as large as that of an entire galaxy containing billions of normal stars. Nevertheless, in a type II supernova, this visual extravaganza is merely a sideshow to the main event. In this case the explosion results from the shock wave induced by the collapse and subsequent bounce of the core of a large star. This collapse reduces the volume of the core [containing over 1 solar mass ($M_\odot$) of material] by a factor of a million within less than a second, to form a dense object tens of kilometers across, which cools over a period of seconds to become a neutron star. This catastrophic event involves an energy release that is more than two orders of magnitude larger than that involved in the visual explosion and takes place almost entirely by the emission of elementary particles called neutrinos. These very weakly interacting, perhaps massless particles were postulated in the 1930s in order to explain the observed features associated with the process of beta decay in radioactive materials. Because their interactions with normal matter are so weak, they are very difficult to observe, with the first direct detection of neutrinos taking place in 1956. Since that time, large underground detectors have been developed that would be sensitive to the neutrino burst from a nearby supernova in our galaxy. On February 23, 1987, two such detectors, located on opposite sides of the Earth, detected neutrino events associated with the observed supernova SN 1987A in the Large Magellanic Cloud. These observations allowed the theory of stellar collapse to at last be tested directly, confirming its general features, and allowed several new tests of particle physics and general relativity.

NEUTRINO BURST FOLLOWING STELLAR COLLAPSE

The iron or oxygen–neon–magnesium cores of stars whose masses are greater than about 8 $M_\odot$ build up and eventually reach the Chandrasekhar mass ($M_c = 1.2$–$1.8\ M_\odot$), named after the astrophysicist Subrahmanyan Chandrasekhar. At this point the electrons in the core become relativistic, that is, they have kinetic energies that are larger than the energy associated with their mass. This occurs because the electrons have become so squeezed in the core that the Pauli exclusion principle (named after the Austrian physicist Wolfgang Pauli), which states that no two electrons can occupy the same quantum state at the same time, forces some of the electrons to occupy high energy states. Once the electrons become relativistic, the pressure they exert can no longer grow sufficiently fast as the core contracts to compensate against the inward-directed gravitational force. This leads to a catastrophic collapse of the core, with a dynamical free-fall time of less than one second. This compression of the core further forces the electrons to occupy yet higher energy levels, eventually pushing the most energetic electrons above the threshold to be captured by protons in the atomic nuclei present in the core. This process,

$$\text{proton} + \text{electron} \rightarrow \text{neutron} + \text{electron-neutrino},$$

leads to copious electron-neutrino production. These neutrinos could escape without interacting from a region the size of the core (initially about 10^8 cm across) if it was at a normal stellar core density, thus providing a mechanism in principle both for dissipating the gravitational potential energy that is lost as the core collapses and for converting the nuclei of a normal stellar core into the amalgamation of neutrons that constitutes a neutron star.

If this were the case, all of the gravitational potential energy released in the stellar collapse would come out in electron-neutrinos (the particles emitted in β decay). However, as the core collapses, two other things happen. First, as the density approaches 10^{12} grams per cubic centimeter, which is 10–100 times the initial density in the outer part of the core, neutrinos become trapped, that is, the probability of interacting before escaping from the core approaches unity. Once this occurs, the capture of electrons by protons does not continue unabated, because the neutrinos that are emitted in this reaction themselves come into thermal equilibrium with the rest of the matter in the core. Once their equilibrium number density is achieved, the rate of the reaction that is shown in the previous paragraph becomes equal in both directions, and no further net neutrino production results. Because the electron number density is no longer being reduced by capture, the electrons in the core once again become degenerate, that is, the quantum mechanical pressure due to the Pauli exclusion principle becomes important, and the most energetic electrons are driven to energies of around 200 MeV.

Once the core density reaches nuclear density, of about 10^{14} g cm^{-3}, the equation of state of this "nuclear matter" stiffens, that is, the pressure gradient increases, as the protons and neutrons are subjected to the same quantum mechanical pressure due to the Pauli exclusion principle that began to govern the behavior of electrons at much lower densities. This causes the collapsing core to rebound and to drive a shock wave into the outer mantle of the star. Whether this shock survives its traversal through the stellar core and mantle then determines whether we witness a type II supernova. Independent of this, the core remains very hot and barely bound. The total electron-neutrino losses during the collapse stage amount to only about 10^{51} erg, roughly equal to the total kinetic energy imparted to the outgoing supernova material. However, about 100 times this energy, about 10^{53} erg, equivalent to about 0.1–0.2 $M_\odot$, must be radiated to account for the gravitational potential energy lost by the collapsing core.

Essentially all of this energy is emitted in neutrinos, but not via the same mechanism by which the initial electron-neutrino burst is generated. Rather, neutrinos and their antiparticles are emitted thermally near the surface of the proto-neutron star. At the high temperatures and densities present in the interior, neutrinos are emitted copiously, as are photons. However, neutrinos can escape on a time scale that is significantly shorter than that of the escape of photons because the neutrino mean free path is much larger. Thus, neutrinos carry out the excess thermal energy of the collapsed core on a time scale determined by the neutrino opacity in the dense nuclear-matter core.

Because the neutrinos that are emitted during this cooling phase of the proto–neutron star are emitted via thermal processes, including pair annihilation of charged particles, plasmon decay, and bremsstrahlung, all species of neutrinos and antineutrinos are produced in roughly equal abundance. Thus the dominant neutrino burst no longer consists of merely electron-neutrinos. At the "neutrinosphere," that is, the surface that is located one mean free path for neutrinos above the surface of the star, the thermal emission of these neutrinos will carry away the neutron star binding energy in a period of a few seconds. Because electron-neutrinos may scatter by charged current as well as neutral current interactions with matter, the mean free path for electron-neutrinos in the core will be slightly smaller than for the other species. Because neutrino scattering reactions are proportional to the neutrino energy (or energy squared) electron-neutrinos are emitted from the neutrinosphere with a slightly lower average temperature (and energy) than muon and tau neutrinos. It has been estimated that the electron-neutrino spectrum in this cooling phase is thermal, with a temperature of about $kT \approx 3$–5 MeV, whereas the muon neutrino and tau neutrino distribution has $kT \approx 7$–10 MeV.

To summarize, the ≈ 1.5-$M_\odot$ core of a star whose main sequence mass exceeds about 8 $M_\odot$, with a main sequence lifetime of about 10^7 yr, collapses during a type II supernova within a second to a proto–neutron star. The cooling of this object to a neutron star, with the release of about 10^{53} erg associated with its gravitational binding energy, takes on the order of seconds, by the thermal emission of neutrinos and antineutrinos of all types in roughly equal abundance. The shock wave induced by a bounce in the core travels to the surface of the star in a period of hours, and leads to the observed supernova explosion that lasts on the order of months. The energy radiated in light is a mere 10^{49} erg, and the kinetic energy of the explosion ejecta is about 10^{51} erg. The neutrino burst is therefore by far the dominant "signal" associated with the roughly simultaneous death of a massive star, the birth of a neutron star, and the visual supernova explosion.

OBSERVATIONS OF SUPERNOVA NEUTRINOS

The preceding theoretical picture, which was developed over a period of 50 years, had no direct empirical confirmation until 1987. Supernovae are predicted to occur in our galaxy at a rate of about 1 every 10–50 yr, and only since about 1980 have there existed detectors that, in principle, might have been sensitive in real time to a supernova in the region of our galaxy. The largest such detectors consist of huge water tanks located deep underground, surrounded by sensitive phototubes. These can detect the Čerenkov light emitted when sufficiently energetic charged particles travel in the water. When supernova neutrinos traverse the detector, they may scatter elastically off of electrons and, in so doing, impart energy to these charged particles, which might then be detectable via their Čerenkov radiation. Alternatively, electron-antineutrinos ($\bar{\nu}_e$) may interact with the protons making up the nuclei of the hydrogen atoms in the water, inducing the transmutation

$$\bar{\nu}_e + p \rightarrow n + e^+.$$

The positrons (e^+, antiparticles of electrons) created in association with the neutrons (n) also produce Čerenkov light as they travel in the water. Finally, sufficiently energetic electron-neutrinos and antineutrinos may induce transmutations of oxygen nuclei in the water into fluorine or nitrogen in the process producing electrons or positrons, respectively.

The proton transmutation reaction has a much higher probability of being induced than any of the other reactions in the energy range of supernovae neutrinos, so that water detectors are most sensitive to electron-antineutrinos coming from supernovae. In fact, even though these presumably make up only about one-sixth of all neutrinos emitted in a type II supernova, the signal induced by electron antineutrinos is predicted to be almost 15 times larger than that due to all other neutrino types combined. In this reaction the positrons are produced almost isotropically and so provide no directional information about the incident neutrino burst.

Even though the interaction rate of antineutrinos with protons in the water is larger than the rates for other reactions, it is still minuscule. In a cubic water detector that is 10 m on a side (i.e., ≈ 1000 ton), a given supernova neutrino traversing the detector has a probability of less than about one part in 10^{15} of interacting in the detector volume.

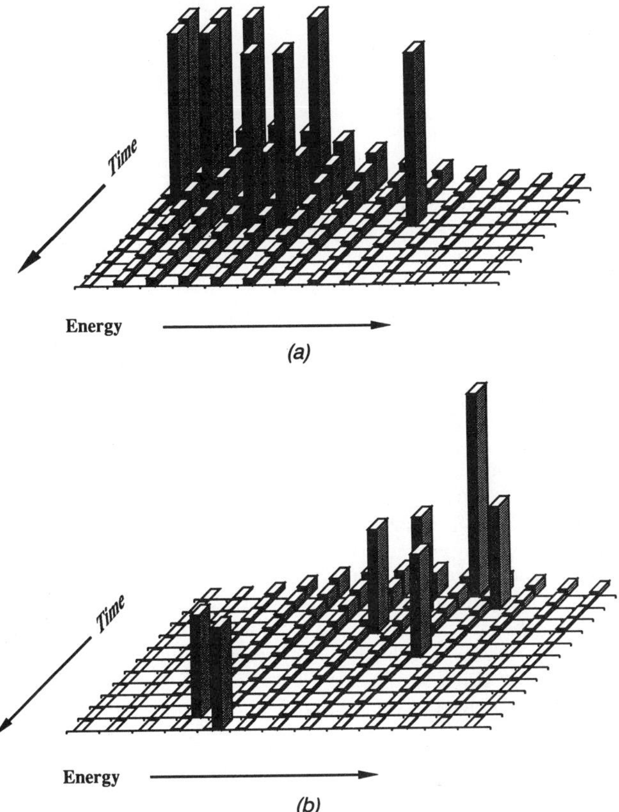

Figure 1. (*a*) The first 4.5 s of the observed Kamiokande neutrino burst from SN 1987A compared to the predicted distribution of events for one choice of parameters: a thermal pulse with $kT = 5$ MeV, total luminosity of approximately 2×10^{53} erg, and an exponential falloff with time constant of 2 s. In this figure both the data and predictions have been binned into time and energy bins: Each time bin is 0.5 s, and each energy interval is approximately 4 MeV, so that the range covered is 5.7–54.3 MeV. (*b*) The first 6 s of the observed IMB neutrino burst from SN 1987A compared to the predicted distribution for the same parameter choices and the same binning procedure. [*Adapted from Krauss (1987).*]

Nevertheless, on February 23, 1987, at 23.316 Universal Time, in a 12-s period, 2 large underground water detectors detected 19 neutrino-induced events, about 6 h before the first visual sighting of supernova SN 1987A in the Large Magellanic Cloud. One detector, in Japan (built by the Kamiokande II collaboration), had a fiducial volume of about 2100 ton and a threshold for detecting Čerenkov radiation deposits of energy above about 7 MeV, whereas the other, in Ohio [built by a collaboration of the University of California at Irvine, Brookhaven National Laboratory, and the University of Michigan (IMB)], had a fiducial volume of almost 7000 ton of water and an energy threshold closer to 15–20 MeV. Neither detector, in fact, had been built with the purpose of detecting supernova neutrinos, but instead each had been designed to search for the possible decay of protons, predicted in certain particle physics models.

Shortly after these observations were reported (along with other reported neutrino events in underground detectors in Baksan in the Soviet Union, and under Mont Blanc at the border between France and Italy) scores of researchers began to examine the data (see Fig. 1) both for consistency among the detectors and for consistency with theoretical expectations.

Although 19 events are not enough to pin down definitively the detailed parameters of stellar collapse, it is remarkable that the Kamiokande and IMB events both support the general features of the theoretical description given previously. Allowing for the statistical and systematic uncertainties in the data as well as in astrophysical parameters such as the distance to the LMC, and even folding in possible backgrounds, the antineutrino signal measured by the two water detectors provided a striking verification of current theory. The only feature of the events that was in mild disagreement with expectations was the fact that several of the events seemed to point back to the direction of the supernova. Statistically this is improbable if all of the events resulted from antineutrino–proton scattering. This suggests that perhaps 1–2 of the events resulted from neutrino–electron scattering, which produces predominantly electrons traveling in the same direction as the incident neutrinos. In any case, both sets of data were consistent with a type II supernova in the LMC at a distance of about 55 kiloparsecs (kpc), involving an energy emitted in neutrinos in the range $1.5\text{–}5 \times 10^{53}$ erg, with a mean electron-antineutrino emission temperature kT of 3–6 MeV and a cooling time in the range of 2–10 s.

Aside from confirming the general theory of stellar collapse and neutron star formation, *a phenomenon that was never before directly observed*, the neutrino signal from SN 1987A provided physicists with important new information to probe various phenomena in particle physics and general relativity. The observed time span of the 12-s neutrino signal allowed an upper limit to be placed on the mass of the electron-antineutrino, due to the dispersion in arrival time expected if this particle has nonzero mass. This was competitive with limits obtained after over a decade of dedicated experiments in terrestrial laboratories. The most recent detailed analysis suggests that at a 95% confidence level the electron-antineutrino mass must be less than 23 eV, or at least 20,000 times smaller than the electron mass. In addition, the general agreement of the signal with theoretical expectations put new limits on the existence and couplings of other possible weakly interacting particles that might be emitted in a type II supernova explosion, as well as putting new limits on possible new neutrino types and interactions. Finally, the fact that the neutrino signal was observed within hours of the photon signal allowed the first test of the equivalence principle for the motion of relativistic elementary particles in a background gravitational field.

New detectors for supernova neutrinos are being developed that should supplement the existing detectors in the event of another nearby supernova. Should a supernova occur anywhere in our own galaxy, up to 10,000 events may be expected in any of these detectors. This would allow one to probe the detailed time structure of the signal as well as the detailed temperature dependence of the signal as the proto–neutron star cools. In addition, upper limits on the electron-neutrino mass of the order of 3–5 eV may be expected, as well as new mass limits on muon and tau neutrinos. The emerging field of neutrino astronomy has already resulted in a major advance in astronomy and astrophysics, and it promises, via the observation of supernova neutrinos, many exciting results in years to come.

Additional Reading

Arnett, W. D., Bahcall, J. N., Kirshner, R. P., and Woosley, S. E. (1989). Supernova 1987A. *Ann. Rev. Astron. Ap.* **27** 629.

Bethe, H. A. and Brown, G. (1985). How a supernova explodes. *Scientific American* **252** (No. 5) 60.

Burrows, A. (1987). The birth of neutron stars and black holes. *Physics Today* **40** (No. 9) 28.

Kafatos, M., ed. (1987). *Supernova 1987A in the Large Magellanic Cloud, Proceedings of the George Mason Workshop*. Cambridge University Press, Cambridge.

Krauss, L. M. (1987). Neutrino spectroscopy of Supernova 1987A. *Nature* **329** 689.

See also **Neutrino Observatories; Neutrinos, Cosmic; Supernovae, Type II, Theory and Observation.**

Observatories, Airborne

Frank J. Low

Since the mid-1960s, astronomers have made increasing use of jet aircraft to carry telescopes and their instruments above most of the Earth's atmosphere. Their greatest use has been in observing at infrared wavelengths where telluric water vapor and carbon dioxide attenuate cosmic radiation before it reaches the Earth's surface. At the operational altitudes of modern jet transports, 40,000–50,000 ft, most of the infrared and submillimeter spectrum, between 1 and 1000 μm, becomes accessible, whereas, from the ground only certain narrow atmospheric windows are sufficiently transparent. Quite large infrared telescopes can be operated "open-port," that is, without intervening windows, and, because such airborne observatories can fly to many locations around the world, the entire celestial sphere is observable with negligible interference from weather systems. This permits critical observations of time-dependent events, such as occultations and eclipses, that might otherwise be lost from ground-based observatories. Another important characteristic of airborne astronomy is that observers, engineers, and technicians accompany their instruments and can make adjustments and improvements in near real time. Many different types of astronomical instruments can be flown on relatively short notice. Thus, airborne observatories share many advantages of both space and ground-based observatories.

Gerard P. Kuiper, from the University of Arizona, was one of the first astronomers to realize the great potential of modern jet transports as platforms for high altitude observations. As a visiting scientist he initiated a series of flights at the NASA Ames Research Center (ARC) where a variety of aeronautical research projects were underway and research aircraft of several types were in use. Near infrared spectra of the planet Venus were obtained in May 1967 with a small telescope observing through a quartz window. The cabin of the Convair 990 jet transport was maintained fully pressurized, providing a "shirt sleeve" environment for the telescope and observers. Accurate guiding and pointing was accomplished with a gyrostabilized heliostat developed by ARC, and the Venus spectra, nearly free of telluric carbon dioxide and water vapor contamination, clearly showed the presence of water vapor in the upper atmosphere of Earth's sister planet.

Following these pioneering observations, the CV 990 research aircraft was equipped to serve a larger number of observers who made complementary observations with separate instruments. James R. Houck, from Cornell University, obtained spectroscopic observations of Mars and found water of hydration in the soil. Tragedy struck the program in April 1973 when the CV 990 and a Navy P-3 collided on approach to Moffett Field, California.

NASA LEARJET OBSERVATORY

In a series of high altitude flights begun in September 1966 and completed in March 1967, the author and Carl Gillespie, from Rice University and the University of Arizona, used a liquid-helium–cooled bolometer and a small, 1-cm telescope to measure the absolute brightness temperature of the Sun at a wavelength of 1000 μm. This was the first measurement of the solar temperature at these wavelengths. To overcome atmospheric attenuation they mounted their hand-pointed instrument in the sextant port of a Douglas A3-B naval aircraft, operated at the China Lake, Naval Ordinance Test Station, California. Tragically, the aircraft and two pilots were lost on a transcontinental flight, just after the solar observations were completed.

The scientific and technical success of this experiment led to a proposal to design and build a 30-cm diameter infrared telescope that would be operated open-port in a small high-performance aircraft shared with other research projects at NASA Ames. The objective was to operate the telescope in an environment that would allow observations at wavelengths inaccessible to the highest ground-based observatories but permit development of the techniques necessary to improve and understand the measurements that would be made. Among NASA's research aircraft, the twin engine Learjet model 23 was selected for its excellent performance and economy. Work on the telescope began in April 1967 and the first astronomy flights took place at Moffett Field, California, where NASA Ames is located, in October 1968. The feasibility and safety of open-port operation was established by Glen Stinnett, one of the NASA test pilots at ARC.

During the late 1960s infrared observations from the ground were still carried out with great difficulty. In the thermal infrared, wavelengths longer than about 2 μm, blackbody radiation from the telescope mirrors and baffles combine with fluctuating emission from the atmosphere to create "sky noise" at levels much higher than the sensitivity of the best infrared detectors. The first Learjet flights used the same techniques developed on ground-based telescopes where observations are confined to the relatively narrow atmospheric windows between 1 and 25 μm. Because of the wider bandwidths and much larger thermal gradients associated with the airborne telescope, excess noise problems were greatly amplified and only the brightest astronomical sources could be observed.

For this reason the first experiment chosen was a measurement of the total infrared emission of Jupiter. Utilizing the broad spectral response of a superfluid-helium–cooled bolometer, Jupiter's thermal emission was compared with the emission of a very bright calibration star. It was shown that the giant planet emits significantly more energy than it receives from the Sun. Similar results were obtained for Saturn, proving that these bodies radiate heat generated internally by their slow but never-ending gravitational contraction. In order to nearly eliminate the effects of atmospheric attenuation between 2 and 100 μm, the band over which the measurements were acquired, the Learjet was operated at its rated ceiling of 50,000 ft. Stress on the telescope was reduced by pressurizing the cabin to only 30,000 ft and the two pilots and the observer breathed pure oxygen.

In order to overcome the effects of noise generated by nonrandom fluctuations in the thermal background from the telescope optics and the residual atmosphere, caused by fluctuations in temperature or opacity, it was necessary to devise a method to modulate infrared signals from weak celestial sources without modulating the much more intense infrared background. Although the solution to this seemingly complex problem is simple, it was found only as the result of many unsuccessful experiments on the ground and in the air. By rapidly scanning the image of a source on and off the detector using a rocking motion of the small cassegrain secondary mirror of the telescope it was possible to achieve the desired result. With this technique the Learjet telescope was then able to observe a wide variety of far-infrared sources for the first time. This same technique, the modulating secondary, is now widely used on ground-based and airborne infrared telescopes.

KUIPER AIRBORNE OBSERVATORY

Even before the open-port performance of the Learjet telescope was demonstrated, Michel Bader and Robert M. Cameron realized that

Figure 1. The Kuiper Airborne Observatory, showing the viewing port (dark rectangle) for the 90-cm infrared telescope located just ahead of the wings of the Lockheed C-141 jet aircraft.

a much larger instrument would be needed to fully exploit the relatively unexplored region of the spectrum from roughly 1 to 1000 μm. In April 1967 a proposal was written by Bader's group at ARC to design and build a "36-inch aperture airborne infrared telescope." However, it was not until January 1974 that the newly dedicated Kuiper Airborne Observatory (KAO) made its first flights from Moffett Field.

Whereas the 30-cm Learjet telescope was mounted in the existing escape hatch of the aircraft, the much larger Lockheed C-141 was modified extensively to receive the permanent installation of the 90-cm telescope. This required construction of an isolated compartment with large pressure bulkheads located immediately ahead of the wings and a large slot in the fuselage, which is sealed at low altitude by a sliding door. Support for the telescope and its attached instruments is accomplished by means of a single large air bearing, which also serves to isolate the pressurized cabin from the telescope chamber: The bearing is hollow so that the bent cassegrain focus can be located inside the cabin where it is easily accessed during flight. Figure 1 shows the location of the telescope and the size of the slot through which the telescope views the sky. Unlike the Learjet, which was limited in its access to the sky, the KAO can reach a large part of the celestial sphere for extended periods of time.

In addition to the bent cassegrain focus where most instruments are located, there is a direct cassegrain focus that eliminates one mirror. Computers are used to aid in pointing and guiding the telescope, which is of good optical quality. Sensitive electronic cameras are available to monitor stars optically, and the guiding is accurate to about 2 arcsec for extended periods of time. As of 1991 the KAO is the world's largest airborne observatory dedicated to astronomical observations.

MAJOR SCIENTIFIC RESULTS

As mentioned, the discovery of internal energy production on Jupiter and Saturn was a major accomplishment for airborne astronomy and it proved that reliable infrared observations are possible from jet aircraft flying above most of the Earth's atmospheric absorption and emission. The Learjet project led to a number of important firsts for far-infrared astronomy, such as studies of the galactic center, which emits 10 million times the power of a single star such as the Sun, and of bright H II regions emitting comparable luminosities at wavelengths from 25–300 μm.

Once the KAO became operational, the number of observing programs greatly expanded. It was possible to study the far-infrared emission of galaxies other than our own and to measure infrared luminosities in such objects at levels found only in quasars. Returning to the solar system, the technique of stellar occultations was used to discover the presence of a system of rings around the planet Uranus. High resolution spectra of comets Halley and Wilson showed the presence of water vapor in those bodies formed early in the evolution of the Earth's planetary system.

One of the most fertile areas of research explored by infrared observations is that of star formation. This follows from the fact that most of the emission of very young stars and pre-stellar clouds of gas and dust is confined to infrared wavelengths. Stellar nurseries in our galaxy, the giant molecular clouds, are best observed through their cool emission, which occurs as thermal emission from cosmic dust particles or as spectral line emission or absorption from the numerous species of atoms and molecules comprising their gaseous components.

Most recently, the sudden appearance of the supernova SN 1987A in the Large Magellanic Cloud created an opportunity for the KAO to fly into the southern hemisphere with its powerful instruments. Among the discoveries that have been reported are the detection of iron, nickel, cobalt, and carbon monoxide formed in the supernova. This ongoing study is contributing directly to our knowledge of how elements are formed in these cataclysmic events and how they are ejected into interstellar space. It is from this newly formed cosmic material that future generations of stars and planets will form.

The number of astronomical papers published as the result of observations made by users of the Learjet Observatory and the Kuiper Airborne Observatory number more than 500 and this number is still increasing. This research, covering a wide range of subdisciplines, has been the training ground for many graduate students and young scientists. Thus, research projects of this nature help rebuild the ranks of scientists needed for future space research while adding to our store of astronomical knowledge and technical expertise.

SOFIA

Because NASA's airborne astronomy program has been so successful and because there is a great need for a larger instrument that can more fully exploit the potential for astronomy at infrared and submillimeter wavelengths a detailed study has been made to determine the largest feasible telescope of this type based on existing technology. The proposed Stratospheric Observatory for Infrared Astronomy (SOFIA) is the result of a joint project under U.S. leadership with the Federal Republic of Germany supplying a major part of the telescope system and participating in the flight program. During its proposed 20-yr lifetime, this 3-m class telescope mounted in a Boeing 747 will serve a wide variety of scientific applications. Its design is based heavily on that of the KAO, but it involves innovative solutions to a number of challenging engineering problems, such as an exceedingly light weight 3-m mirror. If this large and powerful telescope is built, it will ensure the continued growth of this field of astronomy.

Additional Reading

Ewald, R., Himmes, A., and Dahl, A. F. (1989). The stratospheric observatory for infrared astronomy—a 3 m class airborne telescope. In *Physics and Chemistry of Interstellar Molecular Clouds: mm and Sub-mm Observations in Astrophysics*, G. Winnewisser and J. T. Armstrong, eds. Springer, Berlin, p. 421.

Maran, S. P. (1987). Seeing red. *Air and Space/Smithsonian* **2** (No. 5) 88.

Soifer, B. T. and Pipher, J. L. (1978). Instrumentation for infrared astronomy. *Ann. Rev. Astron. Ap.* **16** 335.

See also **Infrared Astronomy, Space Missions; Telescopes, Detectors and Instruments, Infrared.**

Observatories, Balloon

Jacques Blamont

Among the several suggestions made for vehicles that could explore a planet, balloons appear as the most simple and efficient because they require no propulsive engine or fuel. Significant advances in the design and use of free balloons flying in the Earth's atmosphere have been brought about in the last 30 yr, primarily by the development of extruded plastic fibers. This is why balloons are the only machines that have ever moved on a planet (the Vega balloons in 1985) and are also today part of an officially approved program of martian exploration.

A word of caution: Such a simple device seems to be thoroughly understood after 200 yr of use. However the balloon is a complicated thermodynamic and hydrodynamic system, which makes it difficult for the designer to find an optimal design. It is difficult to take into account the specific constraints imposed by planetary ballistics and atmospheric conditions. It also seems that balloon literature is generally ignored by engineers, who try to "reinvent the wheel" when thinking about balloons.

AEROSTATIC ANALYSIS

Assuming a balloon is at an altitude z, in an atmosphere of density ρ_a and contains a lifting gas of density ρ_g, with g being the local acceleration of gravity, the gas-supplied lifting force per unit volume is f_L:

$$f_L = g(\rho_a - \rho_g)\ \mathrm{N/m^3}.$$

Assuming the atmosphere and lifting gases behave as ideal gases, density is given by

$$\rho = \frac{MP}{RT},$$

where M is the gas molecular weight, R is the universal gas constant, and P and T are the pressure and absolute temperature. Consequently, the lifting force per unit volume can be written:

$$f_L = \rho_g g\left(\frac{\rho_a}{\rho_g} - 1\right) = \rho_g g\left(\frac{M_a}{M_g}\frac{P_a}{P_g}\frac{T_g}{T_a} - 1\right).$$

The temperature and pressure of the gas filling the balloon depend on conductive, convective, and radiative heat exchanges with the outside atmosphere. These exchanges create differences with the ambient temperature and pressure ΔT and ΔP.

If we assume the balloon temperature and pressure are given by

$$T_g = T_a + \Delta T,$$

$$P_g = P_a + \Delta P,$$

the lifting force equation can be written

$$f_L = \rho_g g\left(\frac{M_a}{M_g}\frac{1 + (\Delta T / T_a)}{1 + (\Delta P / P_a)} - 1\right).$$

Computation of ΔT and ΔP, which are the essential parameters defining the behavior of a balloon system, requires a complete knowledge of the thermooptical properties of the balloon (gas and envelope) and a detailed analysis of the atmospheric environment.

Balloons are unstable: Once they have lifted off, they will ascend all the way to their ceiling. These statements hold when the vertical temperature gradient keeps ΔT positive, as is the general case in an unstable atmosphere. If the atmospheric temperature gradient becomes positive, as for instance in an inversion layer, ΔT can become negative and the buoyant force can vanish. Conversely, once balloons start falling, they will descend to ground level without stopping; once again, the velocity will depend on the sign and magnitude of the vertical temperature gradient.

We will define *floated mass* as the sum of the masses of the fabric, the buoyant gas, and the gondola.

BALLOON TYPES

Balloons may be classified into four categories: extensible, superpressure, equal pressure, and hot air. Limitations and advantages of each type differ greatly. All types with the exception of the last, use helium or hydrogen as filling gas.

Extensible Balloon The extensible balloon is inflated with a given mass of gas, sealed, and allowed to rise until it bursts. At no altitude is it in buoyant equilibrium. Used for vertical sounding on Earth, its limited lifetime makes it unattractive for planetary transport.

Superpressure Balloon The superpressure balloon, made of nonextensible fabric, is designed to float stably with an internal pressure greater than that of the ambient atmosphere. It is only partly inflated before it is launched. It rises until the gas inside expands to fill the balloon's fixed volume, and soon afterward reaches an equilibrium altitude. From there on, it will stay at an isodensity level, moving nearly horizontally with the wind. During the day, the temperature of the buoyant gas increases with the solar heating and reaches a maximum. It is essential that the fabric strength withstands the maximum corresponding pressure. At sundown, the buoyant gas temperature drops rapidly, and the internal pressure varies directly with the temperature, eventually reaching a minimum. For the balloon to stay aloft, the minimum internal pressure has to be superior to the local atmospheric pressure. If the skin's material is strong enough to resist the superpressure, the lifetime is limited by the leak rate of the buoyant gas.

Because the skin tension per unit area increases with the radius of curvature, the balloon's diameter is limited to a few meters. Being small, the balloon can only carry a very limited payload in addition to its own mass. The thickness of the skin, imposed by the strength needed to withstand the maximum overpressure, therefore sets an ultimate maximum to the mass of the payload. This limit is the main drawback of the superpressure balloon. Another drawback stems from the inferior reliability of suitable fabrics. Mylar is strong but presents highly stressed points and holes. Polyethylene is not strong enough for superpressurized balloons.

Equal Pressure Balloon The equal pressure balloon has an internal pressure that cannot exceed atmospheric pressure. During ascent, it behaves as do superpressure balloons. When the internal pressure reaches the atmospheric pressure, the gas is valved out. The balloon rises to an equilibrium floating altitude where the buoyant force becomes zero and pressure is equalized in and out.

If the internal temperature increases, more gas is vented but the balloon remains fully inflated. Should the temperature now decrease, the gas within the balloon would contract, the balloon would lose buoyancy, and fall. To stay aloft, it must drop ballast. Ballast is needed for the balloon to survive more than one day-to-night transition; but at noon, the following day, the ceiling will be reached at a higher altitude because the floated mass is smaller, and the situation will require more ballast dropping.

The Hot Air Balloon The hot air balloon or montgolfiere relies for lift entirely on the temperature differential between the atmospheric gases inside the balloon and the air outside. The air inside is heated and thus made less dense than the air at the same pressure outside. The balloon rises and floats at the altitude where it is neutrally buoyant. In the case of a solar heated machine, the temperature differential is entirely due to a strong absorption coefficient of the fabric in the visible coupled with a small emissivity in the infrared.

The hot air balloon is far less efficient than the buoyant gas balloon because the difference between the densities of the gas inside and outside is very small; therefore it must be much larger in order to carry the same payload. The larger size increases storage and inflation problems.

USE OF BALLOONS IN PLANETARY EXPLORATION

Balloons can only be used on celestial objects possessing a heavy atmosphere: Venus, Earth, Mars, and Titan. The atmosphere is CO_2 on Venus and Mars; N_2 on Titan.

The constraints imposed by the planetary nature of the mission are very severe:

Mass limitation. The balloon is inherently an inefficient system as far as mass is concerned. The following table (in kilograms) gives the characteristics of the Vega Venus balloons and of the Mars-94 balloons: The gondola mass is a small fraction of the floated mass; the mass of the tanks is about 10 times the mass of the lifting gas; deployment and filling subsystems are also very heavy.

Characteristic	*Vega*	*Mars-94*
Suspended mass	6.9	28.5
Floated mass	21	66
Inflation deployment and structure	100	115
Parachute	30	67
Total	151	238

Structures and housekeeping have to be added to the total. A mass of 200–300 kg (not counting aeroshells and reentry mechanisms) has to be deorbited for an instrument mass of 3–7 kg.

Bit rate limitation. The largest energy consumer on board is the transmitter, whose emitted power has to reach 1–5 W in usual cases. The total number of transmitted bits depends on the energy available on board, which is nonrenewable. Therefore, the factor defining the mission is the number of coulombs available at balloon launch, which itself is limited by the mass that can be devoted to batteries. Because the total mass is limited to a few kilograms, the mass of batteries cannot exceed 1–3 kg and the stored energy cannot exceed 300–1000 Wh. This corresponds in the case of the Mars-94 balloon to a total of about 1 gigabit for the whole mission.

Lifetime limitation. The balloons use nonrenewable resources which have to be stored at the beginning of the mission, for buoyancy and energy. It is difficult to envision a lifetime longer than a few days, since the mass of ballast is limited and the leak rate through the skin is finite.

VENUS BALLOONS

The Venus atmosphere is characterized by its superrotation: a zonal motion present everywhere, peaking at 100 m s^{-1} near the altitude of 60 km. A balloon launched in a location visible from Earth will circle the planet in 4 days, become very warm at the subsolar meridian, cool off quickly, and encounter a day-to-night transition terminator in < 4 days. Active control (value and ballast) are needed both for superpressure and equal pressure balloons to cross the terminator and stay aloft. Ballistic constraints demand that the balloon be launched when the terminator is situated nearly at the sub-Earth point. A maximum duration of 2 days can be achieved by a passive balloon launched at midnight, which will move downwind to the subsolar point.

The pressure and temperature of the atmosphere increase to 90 bar and 735 K on the ground. Balloons were proposed to the Soviet Union by the author in 1967 for the exploration of Venus. Various types of balloons, adjusted to the conditions characterizing different altitude levels, were studied by the joint Soviet and French team during the 1970s. A superpressurized balloon of 9-m diameter, carrying a gondola of 220 kg was shown to be feasible; however, it was finally decided to use the much smaller Vega balloons.

Vega Balloons

The Vega mission consisted of two identical and independent space probes that were launched from Baikonour, U.S.S.R., on 15 and 21 December 1984. On 11 and 15 June 1985 the probes reached Venus, where each separated into two modules. One of the modules entered the venusian atmosphere and the other swung by the planet on a trajectory that led it to a flyby of comet Halley in March 1986.

Each entry module separated into two parts. One of them was a descent probe similar to the previous Venera landers, and the other was a canister containing a balloon. During the descent, the balloon was inflated to a pressure slightly greater than that expected at the float altitude (superpressurized) with a predetermined quantity of helium from tanks that were then jettisoned at an altitude of about 50 km.

Each balloon had a nominal diameter of 3.4 m and supported a floated mass of 21 kg, including a 6.9-kg gondola. The balloons floated in a region of the clouds identified as convective at an initial equilibrium float altitude of 53.6 km, corresponding to a pressure of 535 mbar and a temperature of 305 K. This altitude was chosen partially because of the relatively benign temperature and pressure conditions.

Both balloons were inserted near Venus midnight at 2:00 h universal time (UT) on their respective encounter dates and at a Venus longitude of about 180°. The *Vega-1* balloon entered 7° north of the equator, and the *Vega-2* balloon entered 7° south of the equator. They drifted westward with the predominant zonal wind and were expected to follow closely a parallel of latitude. Each balloon was tracked over a distance of over 11,000 km, encountering dawn about 33 h (8000 km) after injection and then penetrating far into the daylit hemisphere. It is assumed that the float time was longer than the 46 h of transmission for each balloon, which was limited by battery lifetime. The capacity of the batteries restricted the effective earthward radiated power to 2–4 W, depending on the position of the balloon. As a result, it was necessary to use exceptionally sensitive receiving stations in the terrestrial tracking network (Figs. 1–3).

The primary objective of the balloon mission was to obtain information about the large- and small-scale motions, structure, and cloud properties of the venusian atmosphere at the float altitude. Two types of measurements were planned: in situ measurements transmitted by telemetry, and ground-based determination of balloon motion by differential very long baseline interferometry (VLBI) between each balloon and its associated flyby spacecraft. Both types of measurements were provided by signals transmitted directly from the balloons to Earth at 1667 MHz. Because of Earth's rotation, continuous reception of telemetry data required the use of antennas widely distributed in longitude. In addition, the VLBI measurements required a high density of antennas widely spaced in both longitude and latitude. Therefore, the tracking network consisted of an array of 20 antennas distributed over the globe (Fig. 4).

The Vega balloons made in situ measurements of pressure, temperature, vertical wind velocity, ambient light, frequency of lightning, and cloud particle backscatter. Both balloons encountered highly variable atmospheric conditions, with periods of intense vertical winds occurring sporadically throughout their flights. Downward winds as strong as 3.5 m s^{-1} occasionally forced the balloons to descend as much as 2.5 km below their equilibrium float altitudes. Large variations in pressure, temperature, ambient light level, and cloud particle backscatter (*Vega-1* only) correlated well during these excursions, indicating that these properties were strong functions of altitude in those parts of the middle cloud layer sampled by the balloons.

The VLBI data showed that the two balloons followed similar trajectories with, respectively for the *Vega-1* and *Vega-2* balloons, a zonal velocity (meters per second) of 69.8 ± 2.3 and 66.7 ± 1.9 and a meridional velocity (meters per second) of -0.1 ± 1.3 and $+2.6 \pm 1.4$ (+means toward north).

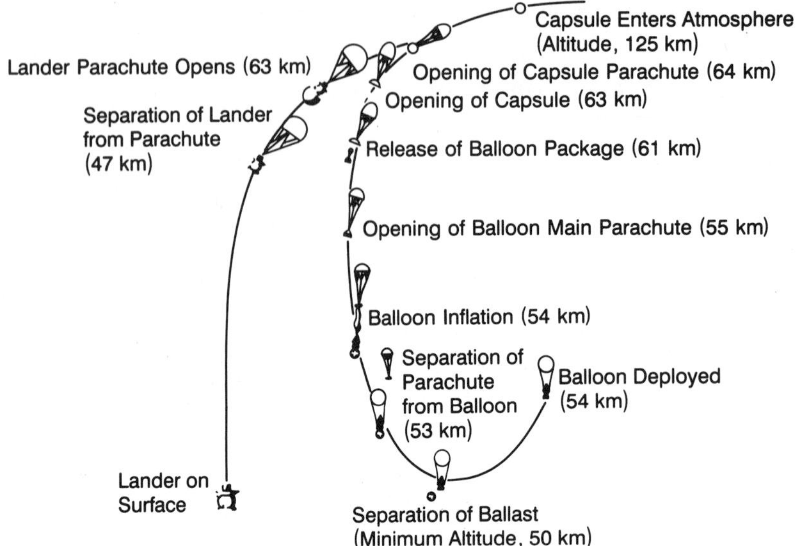

Figure 1. Balloon entry sequence, Vega mission.

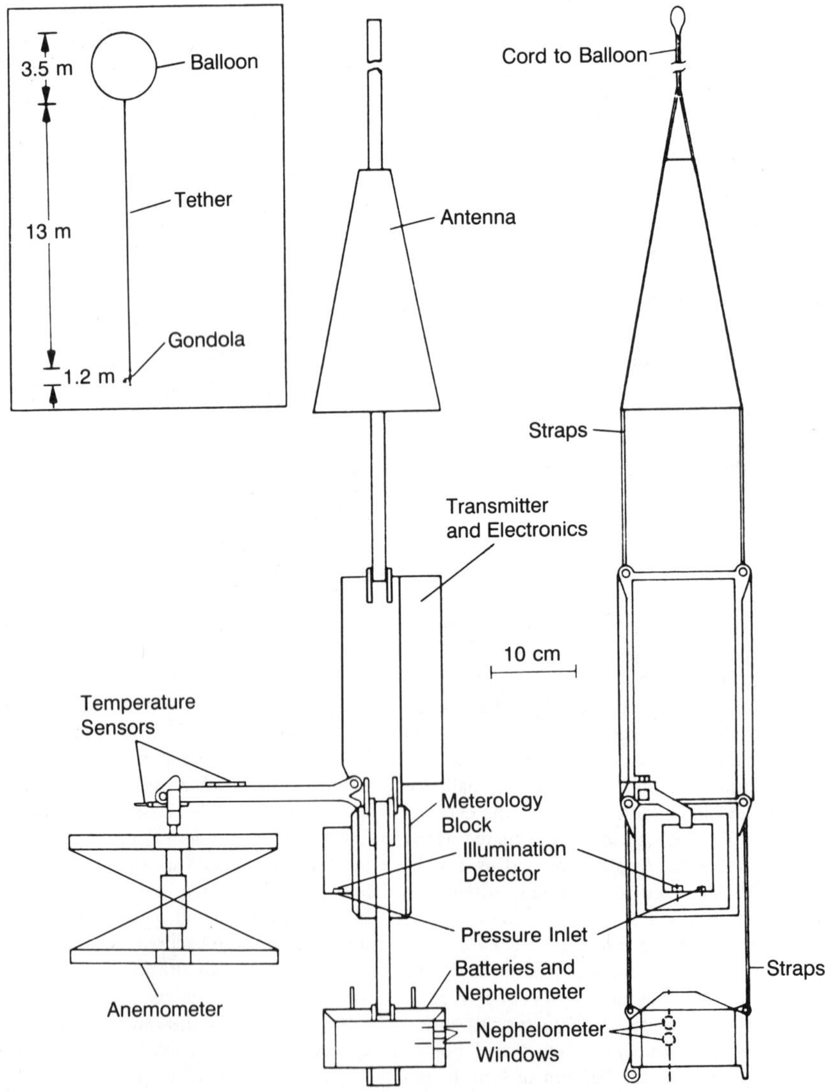

Figure 2. Schematic drawing of the Vega gondola.

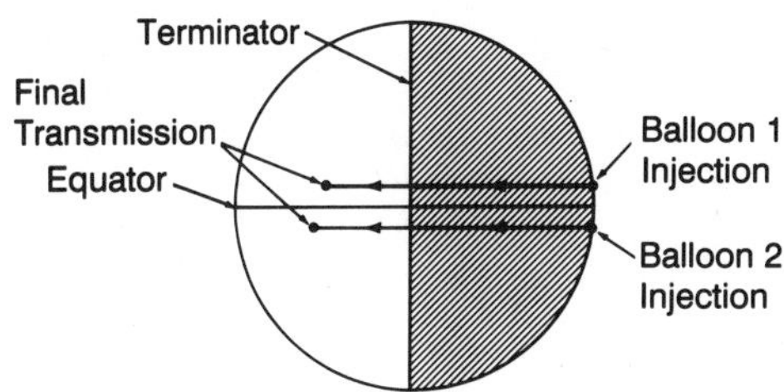

Figure 3. Track of Vega balloons across the face of Venus as viewed from Earth.

Thermal structure measurements showed the venusian middle cloud layer to be generally adiabatic. Temperatures measured by the two balloons at locations roughly symmetrical about the equator differed by about 6.5 K at a given pressure. The *Vega-2* temperatures were about 2.5 K cooler and those of *Vega-1* about 4 K warmer than temperatures measured by the *Pioneer Venus* large probe at these levels. Data taken by the *Vega-2* lander as it passed through the middle cloud agreed with those of the *Vega-2* balloon. Study of individual frames of the balloon data suggests the presence of multiple discrete air masses that are internally adiabatic but lie on slightly different adiabats. These adiabats, for a given balloon, can differ in temperature by as much as 1 K at a given pressure.

Both balloons encountered vertical winds with typical velocities of 1–2 m s^{-1}. These values are consistent with those estimated from the mixing length theory of thermal convection. However, small-scale temperature fluctuations for each balloon were sometimes larger than predicted. The approximate 6.5-K difference in temperature consistently seen between *Vega-1* and *Vega-2* is probably due to synoptic or planetary-scale nonaxisymmetric disturbances that propagate westward with respect to the planet. Surface topography may influence atmospheric motions experienced by the *Vega-2* balloon, which seems to have encountered a large disturbance in the wake of the air flow behind Aphrodite Terra, a 6-km altitude mountain.

The result of the experiment was new insight gained into such phenomena as turbulence, eddy motions, waves, meridional flow, and heat and momentum transfer.

Future Balloon Missions on Venus

The Venus atmosphere is ideally suited for exploration by balloons, but it is so complex that its understanding will require a number of missions, none of which is currently planned by any space agency.

Table 1 provides a slate of options that cover the complete range of altitudes. To this list can be added a more exotic system, which uses two gases in one (or two) closed balloon(s). The condensation (or evaporation) of one of the gases at a certain temperature produces a vertical oscillation around an altitude level that can be fixed by adjusting the relative masses of the two gases. The vertical amplitude of this oscillation can reach 10–15 km and the period can be hours. The H_2O–toluene couple will provide oscillations around the altitude 30 km; the H_2O–paraxylene system will oscillate around an altitude of 20 km.

The main constraints of the Venus balloon missions in the atmosphere below the clouds (50-km altitude) are due to the high temperature, which severely limits the use of energy services and electronic components. Commercial electronics and batteries operate up to 200°C, but developments are needed for higher temperature (below 35 km of altitude). Below this level, the use of vacuum tubes for electronic components may not be impossible, but in the foreseeable future batteries exist only for temperatures below 350 K and between 600 and 650 K (Fig. 5).

Scenario for Venus Sample Return

R. M. Jones, the author, and K. T. Nock have presented a possible scenario that shows the Venus sample return mission to be similar in challenge, magnitude, and cost to recent Mars sample return mission concepts.

The mission would begin in 1999 and require 2 yr for a sample to be returned to Earth orbit. The concept uses small, simple samplers that, after release from an orbiter, are parachuted to the surface and acquire samples. These "grab samples" are then carried in 4 h to an intermediate altitude (50 km) by ammonia-filled balloons. At this level the samples are autonomously retrieved by small airplanes and transferred to a 3-ton ascent rocket, carried by a large montgolfiere that has been deployed from a second orbiter. The ascent rocket carries the samples (1 kg) to Venus orbit, where a rendezvous is performed and the samples are transported back to Earth orbit by a return vehicle.

In this concept, no complex, high technology vehicles are placed into the hostile environment of the Venus surface; the two types of balloons that are being used (small ammonia balloons and large montgolfiere) appear feasible. The planes and their guidance controls are similar to drones already operational for military purposes in a number of countries; their engines could be derived from already-tested hydrazine motors.

The mission would require only a reusable 20-ton orbital transfer vehicle and one transfer orbit stage. The whole package would be assembled in Earth orbit with the help of three space shuttle launches.

MARS BALLOONS

The presence of boulders, stones, and dust on the surface of Mars is a major impediment to the motion of an automatic rover.

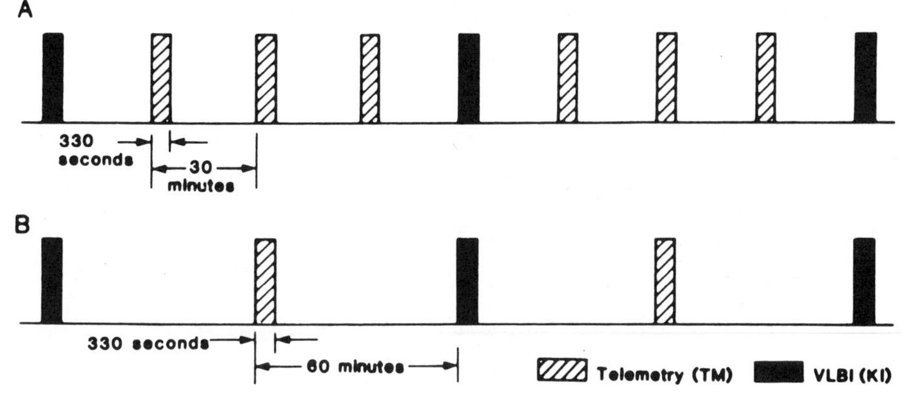

Figure 4. Telemetry and VLBI modulation sequences, Vega balloons. (*A*) 0–12 h and 24–36 h (no transmission 0–2 h); (*B*) 12–24 h.

Table 1. Future Venus Balloon Systems

Altitude (km)	*Temperature (average) (K)*	*Balloon Type*	*Volume (m^3)*	*Diameter (m)*	*Balloon Mass (kg)*	*Gas Mass (kg)*	*Gondola Mass (kg)*	*Floated Mass (kg)*	*Inflating System Mass (kg)*	*Total Mass (kg)*	*Mission*
65–70	232–243	0 pressure; 25-kg ballast	500–1700	10–15	30–65	9–14 Helium	30	94–134	185–275	280–410	More than 1 complete turn. Injection at midnight, vertical excursion (2–3 km)
65–70	235–250	Solar Montgolfiere	1000	12	90	Air	30	120	60	180	Day only; constant level. Injection at subsolar point
55	298	Superpressurized plus valve; 25-kg ballast	250	8	75	24 Helium	90	214	450	665	Could release 12 drop sondes, over 36° longitude
40–48	420	0 pressure plus 12-kg ballast	20	3.5	8	3.2 Helium	10	33	62	95	Half a turn; injection at midnight. Vertical excursion (3–8 km)
20–25	586	0 pressure; 20-kg ballast	4	2	6	23 H_2O	10	59	30	89	Depends on unknown winds

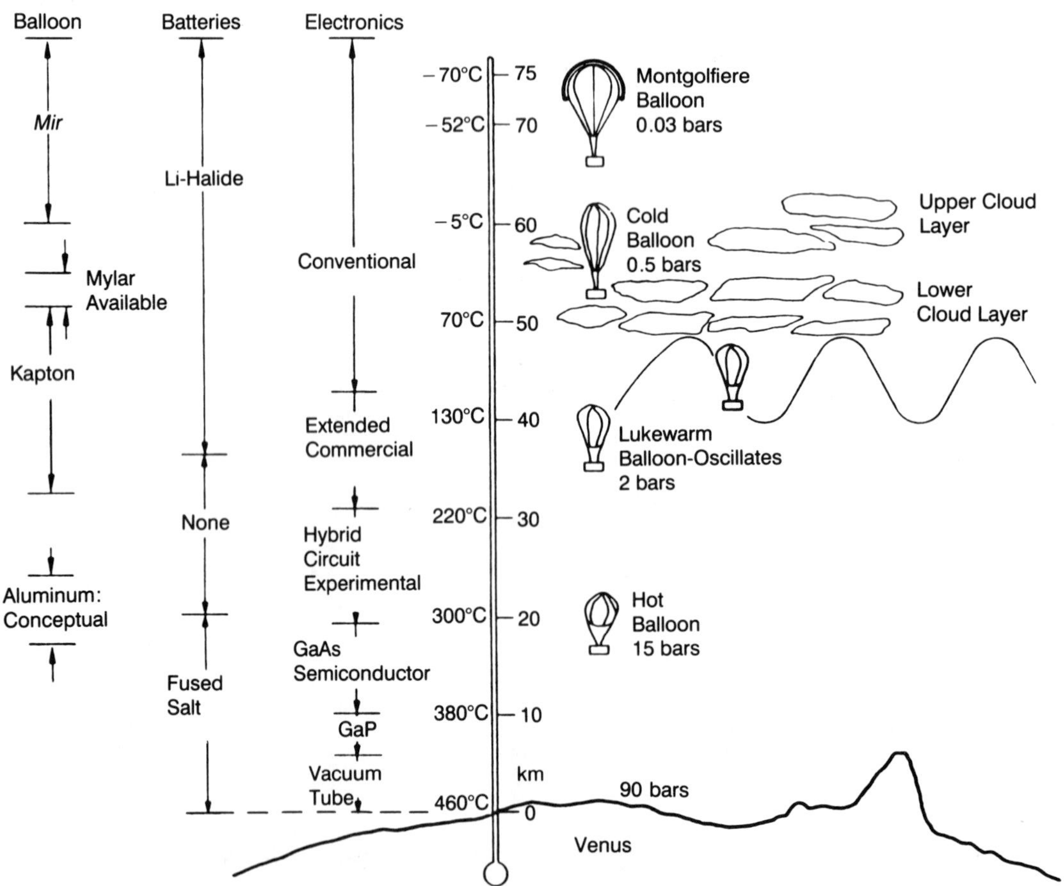

Figure 5. Materials and technologies applicable to balloon missions at various altitudes in the Venus atmosphere.

However, the existence of an atmosphere provides the possibility of easy mobility. It is very hard to walk on Mars, but it could be easy to fly there.

Three modes can be envisioned for the use of balloons:

Data collection from instruments placed on board a long duration flight.
Network deployment by the release of small packages.
Rock collection.

Mars Balloon Concept

The concept that forms the basis of the accepted Mars-94 mission idea presented and established on a quantitative basis by the author in 1987 is the use of solar energy, which provides a variable buoyancy to the balloon.

Near the ground, the atmosphere of Mars has a pressure around 6 mbar and a highly variable temperature (280–160 K). Because the atmospheric density is very low, martian balloons have to be very large: the buoyancy of 1 m^3 of gas is only a few grams and therefore the fabric has to be ultralight (less than 20 g m^{-2}, including attachments).

A major characteristic of the martian atmosphere is its very low thermal inertia. It follows that *any martian balloon is a thermal balloon*, whose design depends critically on the large excursion of the radiation field with local time, atmospheric dust load, altitude, and season. The solar energy source is practically negligible in this radiation field, compared to the ground source. Therefore, the albedo and the thermal inertia of the surface, which present huge differences from one site to another, introduce large local variations in the buoyancy force.

The driving factor in the design of a balloon is the difference between day and night conditions:

During the night, the lifting gas cools to a temperature much lower than the temperature of the ambient atmosphere (ΔT = 20–30°C). The quantity of gas has to be computed with the constraint of keeping the balloon afloat in this condition.
During the day, the internal gas temperature increases to values much larger than the ambient temperature (ΔT = 30–60°C). In the case of a zero pressure balloon, a large amount of gas has to be vented and the mass of ballast needed to keep the balloon flying is impossibly large. In the case of a pressurized balloon, the pressure increases and the fabric has to be extremely strong, therefore heavy, and the payload becomes vanishingly small.

Therefore, the simplest solution, a zero pressure open balloon with ballast, does not provide an acceptable lifetime. The other solution, a pressurized balloon, does not provide an acceptable payload.

The only solution compatible with present film technology is a balloon that flies only part of the time, that is, only during the day. The following concepts have been investigated by the French Space Agency (Centre National d'Études Spatiales, CNES) for a joint mission with the Soviet Union to Mars, approved for launch in 1994.

Double Balloon (Montgolfiere + Gas Balloon, or Canniballoon) Such a system would be made of two separate balloons: one inflated with helium and closed; the other inflated by martian CO_2 and open. The helium balloon provides buoyancy during day and night. The open balloon is the equivalent of a montgolfiere heated by the Sun, which provides the hot air lift effect only during the day, which adds to the buoyancy of helium. This vehicle is on the ground during the night and in flight during the day.

The canniballoon is similar, but the two parts are coupled together and form one single balloon with two separate compartments.

The two options of the double balloon concept were unfortunately found to reach large superpressure during flight. Moreover, the hot air balloon, nearly empty at night, showed a very high sensitivity to transverse wind when on the ground (spinnaker effect), which could be destructive. The manufacturing process also proved to be extremely difficult. For these reasons the concept was not retained for further development.

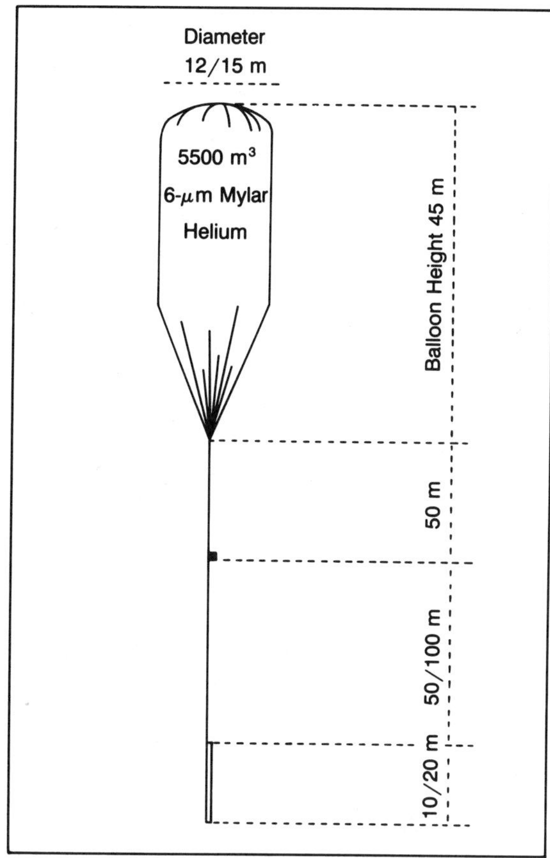

Figure 6. Nominal balloon configuration, daily landing single balloon concept for Mass exploration.

Daily Landing Single Balloon One single balloon is filled with a quantity of helium that provides buoyancy only for half the total mass of the system. During the day, absorption of solar energy by helium adds a new buoyancy, and liftoff of the total payload is obtained. The balloon is closed in order to imprison helium, and the volume of the balloon is chosen in order to keep the overpressure at a manageable level.

This balloon is simpler to build than the double balloon. It is made with a fabric as transparent as possible in order to minimize radiation cooling or heating; even with these effects being kept small, radiative exchanges still dominate the physics of the system.

A part of the suspended mass is on the ground at night when the gas is cold. In the morning the buoyancy increases and the balloon ascends to a cruise altitude; it lands in the evening. This concept allows the maximum superpressure to decrease inside the envelope, but it needs to incorporate a ground interface such as a guide rope. The gondola is divided in two parts: the atmospheric gondola, which during the night stays suspended at 100-m altitude, and the guide rope, which is partly suspended and partly on the ground during the night (Fig. 6).

Description of the Mars-94 Balloon

The daily landing single balloon has been selected for the Mars-94 mission.

Balloon Material and Shape The material of the balloon envelope should have high resistance/mass ratio, low permeability to helium, adaptability to manufacturing, and wide temperature range operating capability. A polyester film similar to Mylar was selected to meet these requirements. The film is transparent to minimize the day/night gas temperature variations.

Spherical and cylindrical envelope designs were both investigated. The cylindrical shape was selected in order to enable use of very thin film material (6–8 μm thick). The volume of the balloon when fully inflated has been chosen to be 5500 m^3.

Lifting Gas Helium and hydrogen are widely used as aerostatic gas. Hydrogen is more effective for a given volume and floated mass but unfortunately causes brittleness of titanium tanks, thereby requiring heavier steel tanks. A tradeoff analysis led to the selection of helium.

Suspension Geometry The suspension lengths determine the behavior of the gondola, the spatial motions during flight, and altitude above ground at landing, at night, and at liftoff. They have to be greater than to 100 m.

Mass Breakdown (in kilograms)

Gondola	15
Balloon	25.6
Gas	6.3
Guide rope	13.5
Margins	5
Floated mass	65.4

Mission Profile

The Injection, Deployment, and Inflation of the Balloon The mission profile starts with the injection of the probe into interplanetary transfer orbit. After injection into a martian polar orbit, this probe will send a descent module into the Mars atmosphere.

The orbital velocity is reduced by two aero-braking mechanisms:

At the first step, the braking of the descent module is achieved with the help of a conical aeroshell (3–3.5 m diameter).

The second step provides braking by parachute.

After the parachute is stabilized, the balloon is deployed. A few seconds later, the on-board sequencer initiates the inflation. This operation is made in two steps:

First step: The envelope is unfolded and the inflation started at a low regime.

Second step: The total required quantity of gas is now injected into the balloon in 200 s.

After inflation, the balloon is released from the parachute system. A ballast attached under the balloon provides different descent velocities to the two mobiles for separation. The ballast is removed at ground level to release the balloon in the mission configuration.

Potential landing sites were selected taking into account the scientific objectives, orbitography limitations, and balloon criteria. Northern latitudes are more suitable for relief and terrain characteristics. In the selected areas, the ground altitude is in the −3 to −1 km range (6.1 mbar zero reference). Three regions were investigated in more detail: Utopia Planitia, Arcadia Planitia, and Acidalia Planitia.

The Daily Profile In the morning when the Sun heats the lifting gas, the balloon buoyancy increases until it lifts the entire guide rope off the ground. The takeoff occurs around 8–9 a.m. and the balloon ascends slowly to the ceiling in about 1 h. The balloon stays there during the day, moving horizontally with the prevailing winds. In the afternoon the gas cools when the solar flux decreases, so that the balloon lands smoothly on its guide rope. During the night the balloon drifts horizontally until the next morning.

Figure 7 shows a typical daily altitude profile and the corresponding overpressure, which should reach 50 P after 2 h. The fabric is built to sustain overpressures up to 100 P. With a target lifetime of 10 days, it is assumed that the balloon system will travel a distance of the order of 2000 km, reaching during the day an altitude of 2–4 km above the 0-km reference altitude.

Payload

Main Scientific Objectives The drift of the balloon opens the opportunity for direct measurements of both atmosphere and soil parameters in various separated areas within a single balloon mission. Based on the scientific objectives and the balloon capabilities, the following investigations are contemplated:

High and very high resolution imaging of the Mars surface from various altitudes.

Day and night in situ measurements of the atmospheric parameters defining the physical state of the boundary layer: pressure, temperature, humidity, optical thickness, aerosols.

Studies of the wind by balloon tracking.

Studies of the magnetic field and magnetic anomalies.

Characterization of rocks by their infrared reflectance spectrum.

In situ measurements of the soil parameters: chemical composition of the surface by γ spectrometry analysis and characterization of permafrost down to a depth of 3 km by electromagnetic sounding.

Gondola Subsystems

Data handling is carried out on board the gondola: housekeeping, data management and storage, and picture compression.

The radio communication system transmits the data to the orbiters (the Soviet Mars-94 orbiters and the American Mars Observer). The balloon is planned to have about 2 h daily total radio link sessions. The radio communication system contains a receiver and a transmitter operating in the 400 MHz band, with an output power of 5 W. The radio link bit rate depends on the distance between the balloon and the orbiters. For the foreseen trajectories the bit rate is in the range of 10 kbits s^{-1} for the Soviet orbiter and 128 kbits s^{-1} for the American orbiter. The total number of bits collected, constrained by network considerations, is 150 Mbits per day. The data collected by instruments placed in the guide rope are transmitted to the gondola and then relayed to the satellites.

Electrical power. It is assumed that electrical power will be provided by lithium batteries. The thermal protection of these components against the cold temperatures is of prime importance. It is planned to use thermal heating provided by plutonium oxide generators.

Guide Rope Subsystem

Operating Mode of the Guide Rope The balloon descends in the late afternoon, with a vertical velocity determined by the balance between the negative lift of the cooling balloon and the aerodynamic drag of the envelope (Fig. 8). Upon nearing the surface, the balloon's descent is decelerated by the deposition of the linear guide rope on the ground, and the resultant increase in balloon buoyancy. The vertical descent velocity must be fully stopped without allowing a long part of the tether line to contact the ground; such contact could result in irreversible snagging of the system.

Once the vertical speed has been dissipated, the balloon is a captive of the ground and enters its nocturnal operation phase. The balloon will move at a horizontal speed, which is a function of the balance between the motive force (the wind moving past the balloon) and the resistive force (the dynamic friction of the portion of the guide rope on the ground). This achieves a good balance of forces without allowing excess wind effects to damage the balloon envelope, and prevents excessive leaning of the balloon toward the ground. The drift velocity during the night is less than the wind velocity only by 2–3 m s^{-1}. Because winds at night should be as strong as during the day, and possibly stronger (8–12 m s^{-1}), the velocity on the ground could reach 15–18 km h^{-1} and the distance

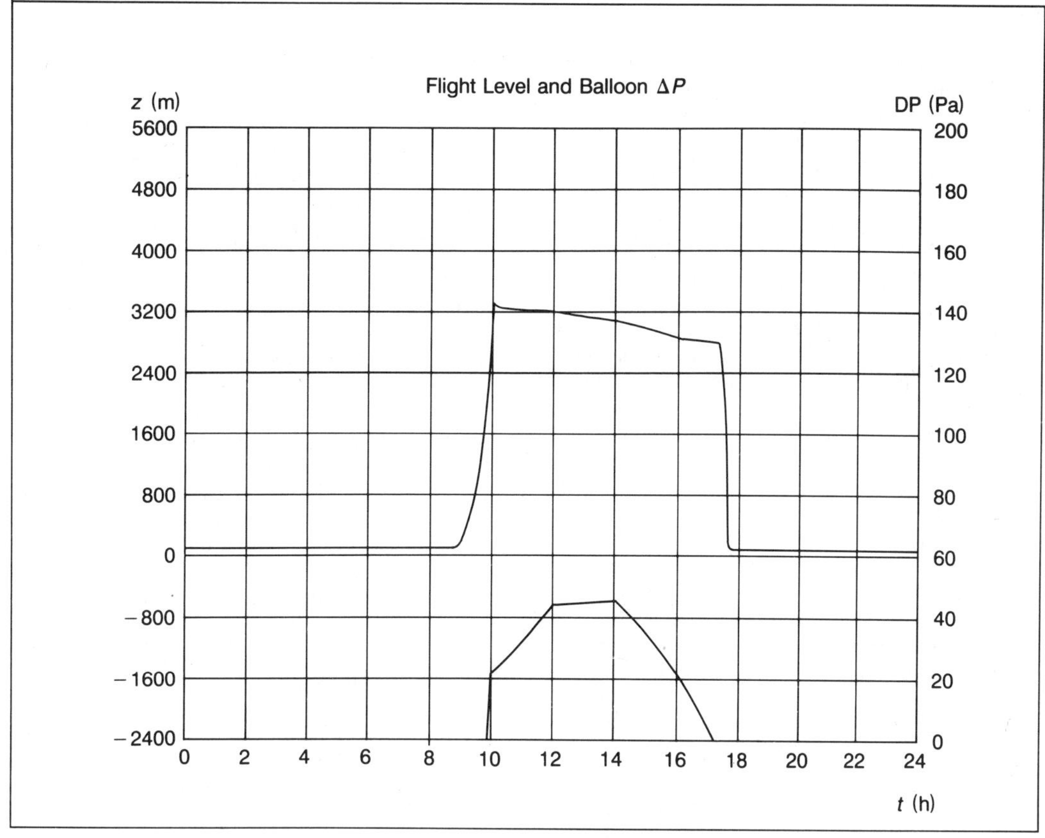

Figure 7. Flight simulation profile (daily), Mars-94 mission.

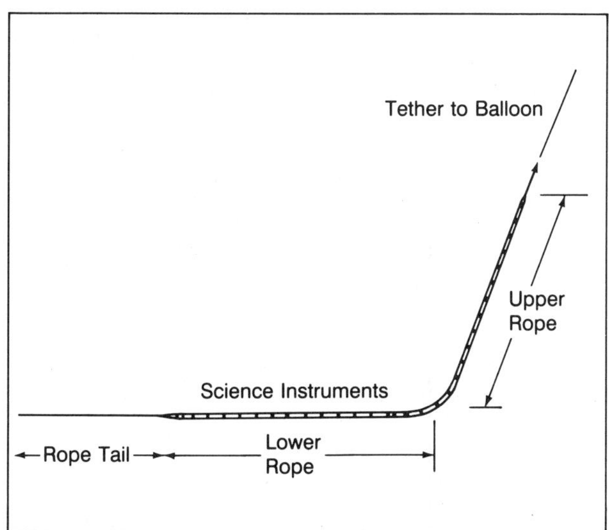

Figure 8. Baseline guide rope configuration, Mars-94.

covered during the night could be greater than the distance covered during the day.

In the morning the balloon's buoyancy increases with the rising of the Sun. The guide rope must be smoothly lifted from the ground. Near the end of this process, the guide rope is moving horizontally with the ground wind velocity.

Guide Rope Concept The current guide rope design is 5 cm in diameter and 10 m long. The structural mass is evaluated to be more than 60% of the total mass in the current design.

The guide rope is split into three distinct sections: the upper part, which is suspended above the surface, the lower part, which contacts the ground, and the tail, which also contacts the ground.

The upper part must contain a mass reserve sufficient to properly ballast the balloon above the ground and to minimize the tether angles such that the balloon does not contact the ground. Furthermore, the nose must remain well above the surface at all times during operation to minimize the risk of snagging on surface features. For safe operation, no more than 50% of the guide rope mass should be allowed to rest on the ground.

The lower part has a rough life dragging along the surface during the nighttime traverse. Here, it will experience many impacts with surface rocks at speeds up to 10 m s^{-1}, superimposed on a 1000-km-long abrasion path of sand and rock.

The tail, reduced to a titanium cable, is necessary for stabilizing the motion of the lower part, adding length to the overall guide rope system and dissipating the whipping energy of the end.

Scientific Payload of the Mars-94 Balloon

Location	*Instrument*	*Mass (kg)*
Gondola	Cameras	2.1
	Meteorological package	0.5
	Infrared photometer	0.5
	Magnetometer	0.5
	Altimeter	0.3
	Solar sensor	0.2
		4.1
Guide rope	Gamma ray spectrometer	1.8
	Sounding radar	1.8
	Shock characterization	0.2
		3.8

Future Mars Balloons

A mission with a reasonable number of balloons built on the preceding principles would provide a map of the permafrost, a map of the chemical composition of the surface, and the values of a large number of other parameters. Balloons appear, therefore, as powerful adjuncts to the survey of the planet by orbiters and by a permanent network of small fixed stations. Note that the presence of an orbiter is essential for retrieving the data collected by the balloon, even if it is interesting and important to receive balloon transmission at Earth, for instance for VLBI tracking and wind determination.

Balloons should be incorporated in rover missions. Rovers are considered by space agencies essentially as a means of providing mobility over 500 km in support of a sample return. It is difficult for them to move in an unknown environment, and tethered balloons appear to provide a great help for the locomotion. A zero pressure closed blimp with a volume of 900 m^3 would be 95% inflated during the day and could lift a payload of 1 kg to an altitude of 100 m, on all martian locations and conditions. During the night, the balloon would only be 70% inflated and would lift no weight. Then a "mouse" carrying a camera, easily accommodated with batteries and transmitter inside a mass of 1 kg, could ascend the tether during the day and descend during the afternoon. The floated mass would be 8 kg with 0.9 kg of helium, and the total mass needed for the system including tanks, structure, inflation devices, and containers would be less than 25 kg.

A second, and potentially much more important, use of balloons in a rover sample return mission would provide it with a global dimension. Suppose that the rover plus balloon systems are placed in martian orbit. Then the balloon only is deorbited and deployed. The gondola carries essentially a grab system, which would function at each landing, collecting material each time, and documenting the sample by pictures and chemical analysis of the environment. Each sample would be placed in a sealed box. After many such landings and samplings (at least 10), the balloon would be released and a cairn of sample boxes would be left on the ground with a beacon. The rover would then be deorbited in the near vicinity of the cairn. It could carry out a detailed analysis of samples inside its radius of action, and also retrieve less documented samples taken from very distant points and collected in the cairn.

There is a real possibility of navigating the balloons in a predetermined direction and forcing them to land in a predetermined region. Following a procedure that is currently employed in balloon races on Earth, the vertical profile of the wind would be measured, for instance, by correlating pairs of pictures taken by the onboard camera in real time. Attempts to perform real time velocity measurements with the correlation technique are planned during the Mars-94 mission. It is known that the wind hodograph varies with altitude. The altitude at which the wind flows in the predetermined direction would be selected by constantly adjusting the buoyancy of the aerostat, for instance by the use of a canniballoon whose hot air compartment would be kept empty or full by means of a valve. This exotic navigation is not essential to the concept of sample collecting by balloons.

Powered Mars Aerostats

Powered aerostats offer the opportunity of controlled, extended duration missions. By flying a powered aerostat at an altitude of 10 km, approximately 95% of the planet is accessible. However, such systems with any speed and altitude capability will be significantly larger than the free-flying aerostat systems.

The U.S. Navy HASPA aerostat was designed for operation in very similar atmospheric conditions to those near the surface of Mars. The aerodynamic shape of the Mars powered aerostat could be assumed geometrically similar to HASPA.

HASPA Specifications

Hull length	101.5 m
Maximum diameter	20.42 m
Volume	22,653 m^3
Hull surface area	5,314 m^2
Fin surface area	0.223 m^2
Total surface area	5,537 m^2
Required thrust	
At 12.9 m s^{-1}	204 N
At 7.7 m s^{-1}	80 N

For martian conditions, a total system mass of 2400 kg would be needed in a similar design providing 120 kg for payload; this payload includes power and propulsion. The flight would take place at an altitude of 10 km.

The following figures are given as an example: the volume of the aerostat would be 135,000 m^3, its surface 17,470 m^2, the length 184 m; 5 hp would be required to deliver 204-N thrust at 12.9 m s^{-1}. A propeller weighing 12 kg hp^{-1} was used on HASPA. The mass needed for propulsion including controls for fins would be 25 kg; solar cells and batteries providing 6 h of flight at 10 m s^{-1} per day would require 50 kg. Therefore, the aerostat could transport a payload of about 50 kg. Clearly the modest payload and large mass appear to disqualify such a system for near term missions.

TITAN BALLOON

The atmosphere of Titan (Fig. 9) is the most favorable of all the planetary atmospheres for balloon flight because it is rather dense, being essentially nitrogen with a zero altitude pressure of about 1.4 bar and very cold (less than 100 K in the 0–70-km altitude range). Because there is no foreseeable opportunity for such a mission, the description of a Titan balloon will be succinct.

The minimal option could consist of a very small helium-filled zero pressure open balloon. Vinyl polyfluoride (30 g m^{-2}) is an adequate material, resistant at low temperatures. The balloon container would be launched with a descent probe, kept warm on the ground by the thermal losses of an RTG (radioisotope thermoelectric generator), and reawakened during a following return of the orbiter. The balloon would then be inflated and launched. The diameter of this balloon would be 6 m, with a volume on the

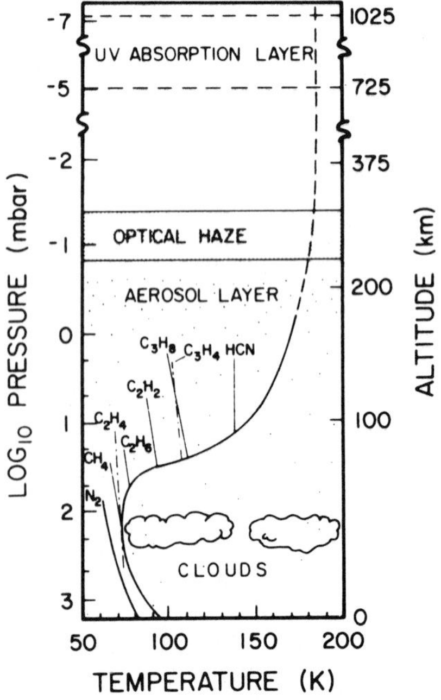

Figure 9. Atmosphere of Titan.

Table 2. Buoyant Station Probe Mass Breakdown

Station Subsystem	*50 km Small Balloon (kg)*	*5 km Large Balloon (kg)*	*5 km Large Blimp (kg)*
Science	20	80	80
Structure & devices	28	72	85
Thermal control & cabling	12	20	25
Telecommunications	11	11	11
Command & data	14	23	23
Power & heat source	15	74	102
Propulsion	—	—	14
Tether system	—	—	20
Δ for dropped package	—	20	—
Total station payload	100	300	360
H_2O buoyant gas	10	27	32
Reserve gas & tank	5	11	15
Balloon fabric	12	6	7
Total floated mass	127	344	414
Gas transport	90	243	288
Sounding rocket & payload	—	— (60)	—
Total entry payload	217	587 (647)	702
Aerodeceleration module	79	178 (219)	219
Total entry probe mass	296	765 (866)	921

ground of 2 m^3 and a volume at ceiling of 100 m^3. In 1000 min it would ascend to 90 km. Such a balloon would fly at some level between 50 and 100 km altitude where stability is provided by the positive temperature gradient. The diffusion losses would be compensated by small releases of ballast. The lifetime would be 1 month; with a long (10-m) venting tube, the system would survive the day-to-night transition. Below 50 km, pressurized balloons are needed; it is not clear that a fabric can withstand the overpressure at very low temperatures. Total payload would be 5 kg including 1 kg for science, 1 kg for transmitter, and 3 kg for RTG. This simple system would require a mass allocation of 40 kg on the descent probe.

More ambitious missions are described in Table 2. They would study upper and lower atmosphere chemistry, including haze at high altitude and precipitation and clouds at lower altitude, together with atmospheric structure and global circulation. They would also study the surface features and morphology, including oceans and continents. Finally, a network deployment mode could open the way to seismometry and tectonics.

Additional Reading

Blamont, J. (1981). Balloons on other planets. *Adv. Space Res.* **1** 63.

Blamont, J. (1987). Balloons for Mars missions. *Acta Astronautica* **15** 523.

IKI (1988). Mars-94 mission: Proposals for the Mars exploration programme. Preprint 351, U.S.S.R. AS, IKI.

Jones, R. M., Blamont, J., and Nock, K. T. (1986). A concept for a Venus sample return mission. AIAA paper 86-2013.

Kremnev, R. S., Karyagin, V. P., Balyberdin, V. V., and Klevtsov, A. A. (1985). *Balloons in the Venusian Atmosphere*. Naukova Dumka, Kiev.

Kremnev, R. S., Pichkhadze, K. M., Linkin, V. M., Kerzhanovich, V. V., Blamont, J., Tarrieu, C., Heinsheimer, T. F., Cantrell, J., Murray, B., Friedman, L., and Schurmeier, B. (1989). Alternate concepts for a Mars balloon exploration vehicle. AIAA/JPL International Conference, Pasadena, 22–24 August.

Kremnev, R. S., Rogovsky, B. N., Pichkhadze, K. M., Martynov, B. N., and Vorontsov, V. A. (1989). Design of scientific and technical experiment in Mars-94 mission. Technical report, Babakin Center, Moscow.

Kremnev, R. S., Rogovsky, G. N., Pichkhadze, K. M., Martynov, B. N., and Vorontsov, V. A. (1989). Basic data and technical specifications in development of the scientific equipment to be installed on the landing module and released probes. Mars-94 mission, Part 2. Report, Babakin Center, Moscow.

Petrone, F. J. and Wessel, P. R. (1975). HASPA design and flight objectives. AIAA paper 75-924.

Tarrieu, C., Mauroy, P., Sablé, C., Sirmain, C., and Vecten, A. (1988). Martian aerostat preliminary feasibility. CNES report BA/AM/53.

Tarrieu, C. and Sirmain, C. (1988). Mars balloon: A French contribution to the Mars-94 Soviet mission. IAF report 88-396.

Vega Balloon Team (1986). (Seven papers on the Vega balloons.) *Science* **231** 1341–1480.

See also **Mercury and Venus, Space Missions; Telescopes, Detectors and Instruments, Gamma Ray; Venus, Atmosphere.**

Outer Planets, Space Missions

J. Kelly Beatty

Spacecraft destined for the outer solar system differ from other spacecraft in that they (1) need a source of power other than sunlight-illuminated photovoltaic cells; (2) must operate reliably for years with a high degree of self-sufficiency; and (3) must have electronic systems able to endure traverses through the magnetic fields and populations of energetic charged particles surrounding their target planets. During the 1970s, the National Aeronautics and Space Administration (NASA) conducted two such outer-planet missions, named *Pioneer* and *Voyager*. Their success paved the way for the *Galileo* orbiter-probe combination, which was launched toward Jupiter in 1989. Two more spacecraft under construction are to revisit Saturn and to fly past an asteroid en route to a comet.

PIONEERS 10 AND 11

Formal plans to explore the outer solar system began in 1967, when NASA's Lunar and Planetary Missions Board endorsed sending a pair of relatively simple spacecraft to Jupiter. The agency received funding for such a mission in 1969. Earth's first Jupiter-bound spacecraft, *Pioneer 10*, was lofted by an *Atlas-Centaur* rocket from Cape Canaveral, Florida, in March 1972. The launch of its twin *Pioneer 11* followed 13 months later (see Table 1).

Each 250-kg spacecraft is built around a six-sided equipment compartment and a paraboloidal antenna 2.7 m in diameter. Mounted to the box are 10 of the scientific instruments and a 6-m-long boom with the eleventh, a magnetometer, at its end (*Pioneer 11* has two magnetometers). Two shorter booms carry the spacecraft's source of power; radioisotope thermoelectric generators, or RTGs, make electricity from heat produced during the fission decay of plutonium-238. To maintain orientation in space and keep its large dish antenna pointed toward Earth, each spacecraft spins 4.8 times per minute.

After crossing the asteroid belt without incident, *Pioneer 10* reached Jupiter in December 1973 and passed 132,250 km from the planet's cloud tops. Throughout the encounter, *Pioneer 10* relayed more than 500 images of Jupiter and its major satellites. But the probe's true legacy is the huge volume of data collected on the planet's magnetic field, trapped charged particles, and interaction with the solar wind. Fortunately, *Pioneer 10* suffered little radiation damage as it crossed the vast jovian magnetosphere, and by surviving it paved the way for future spacecraft.

One year later, *Pioneer 11* came much nearer to Jupiter—only 42,900 km at its closest. Mission managers sent the spacecraft past the planet along a steep south-to-north trajectory that minimized the time spent in the dangerous charged-particle zone in the

Table 1. Outer-Planet Missions: Key Events

Spacecraft	Launch Date (UT)	Flyby Target	Flyby Date (UT)	Distance to Target's Center (km)
Pioneer 10	3 Mar 1972, 1:49	Jupiter	24 Dec 1973, 2:25	204,200
Pioneer 11	6 Apr 1973, 2:11	Jupiter	3 Dec 1974, 5:22	114,400
		Saturn	1 Sep 1979, 16:31	80,400
Voyager 1	5 Sep 1977, 12:56	Jupiter	5 Mar 1979, 12:05	348,900
		Saturn	12 Nov 1980, 23:46	185,200
Voyager 2	20 Aug 1977, 14:29	Jupiter	9 Jul 1979, 22:29	721,700
		Saturn	26 Aug 1981, 3:24	162,300
		Uranus	24 Jan 1986, 17:59	107,000
		Neptune	25 Aug 1989, 3:56	29,200
Galileo	18 Oct 1989, 16:54	Venus	10 Feb 1990, 5:59	22,200
		951 Gaspra	29 Oct 1991	1,100
		243 Ida	28 Aug 1993	1,100
		Jupiter	7 Dec 1995	*
Comet Rendezvous and Asteroid Flyby	22 Aug 1995 (planned)	449 Hamburga	22 Jan 1998	3,400
		Comet Kopff	14 Aug 2000	†
Cassini	8 Apr 1996 (planned)	66 Maja	29 Mar 1997	3,400
		Jupiter	6 Feb 2000	3,855,000
		Saturn	6 Dec 2002	‡

*Main spacecraft enters variable jovian orbit; probe enters jovian atmosphere.
†Long-term rendezvous with variable comet–spacecraft separation.
‡Main spacecraft enters variable saturnian orbit; probe reaches Titan on 17 Mar 2003.

plane of Jupiter's magnetic equator. This flyby permitted much better imaging of the planet and allowed detectors on board to make comprehensive measurements of the intense magnetic field and charged-particle environment.

In coming so close to Jupiter, *Pioneer 11* was accelerated to an unprecedented velocity of 48 km s^{-1}. More important, the planet's gravity swung the spacecraft onto a much different path headed back toward the inner solar system and inclined 15°.6 above the ecliptic plane. The new course would bring it to Saturn's vicinity after a cruise of nearly five years.

Deciding how close to go past Saturn was not easy. A flyby deep through the magnetosphere—about halfway between the inner C ring and the planet's cloud tops—would be most productive scientifically. But passing well outside the known rings would test the path that *Voyager 2* would eventually use en route to Uranus. Ultimately, *Pioneer 11* was commanded to take the latter, pathfinding route.

By the time it reached Saturn, *Pioneer 11* had traveled 3.2×10^9 km, yet the health of its electronic systems and instruments remained excellent. On September 1, 1979, the probe passed 3500 km from the outer edge of Saturn's A ring, then slipped underneath the ring system to a point 20,930 km from the planet's clouds. During its brief visit, *Pioneer 11* discovered a faint ring and several small satellites, while confirming scientists' expectations that Saturn has an intrinsic magnetic field and well-developed magnetosphere.

VOYAGERS 1 AND 2

About every 176 years, the outer planets line up in a way that allows a spacecraft to pass Jupiter and then, accelerated by its gravity, to reach Saturn, Uranus, Neptune, and Pluto in much less time than is possible using simple, ballistic trajectories from Earth. Flight dynamicists realized that such an opportunity would occur during the late 1970s. Consequently, NASA attempted, but failed, to obtain funding to send spacecraft on a "grand tour" of the outer solar system. Instead, Congress agreed in 1972 to let the agency pursue a more modest mission—later named *Voyager*—to Jupiter and Saturn only.

Despite the scaled-down objectives, NASA endowed the twin *Voyagers* with many of the capabilities that would have been used on the grand tour spacecraft, such as redundant computers and communication systems. They also drew heavily on a long heritage of successful *Mariner* designs. The electronics are housed in a 10-sided "bus," and the main communication antenna is a dish 3.7 m across. Two outward-extending booms provide mounts for a magnetometer and the RTG powerplants; attached to a third are more experiment packages and a separately-articulated platform for two telescopic cameras and two spectrometers. Each spacecraft weighs 815 kg, and rather than spinning, it is stabilized in a fixed reference system based on the Sun and certain bright stars.

Voyager 1's launch in the autumn of 1977 actually took place 16 days after that of *Voyager 2*. The latter craft utilized a slower trajectory that took four months longer to reach Jupiter but retained the option of flying on to Uranus and even Neptune after reaching Saturn. Thus, in theory most of the grand tour's objectives could still be accomplished. An interesting footnote to this mission involves the *Titan III/Centaur* rockets used to loft the spacecraft. To reach all four planets, *Voyager 2* demanded a perfect, full-thrust launch, which it got. But *Voyager 1*'s *Titan III* shut down prematurely; had it been used with *Voyager 2*, the encounters with Uranus and Neptune would have been impossible. By sheer chance, *Voyager 2* got the better rocket.

However, it proved to be the less fortunate spacecraft, for soon after launch, problems developed with *Voyager 2*'s attitude-control system. Eight months later, the primary radio receiver failed, and its backup proved to be "tone deaf"—unable to lock onto a particular frequency as planned. *Voyager 2* could still receive instructions beamed from Earth, but only in a very narrow and changeable bandwidth. Thus, for each communication session mission, engineers had to determine carefully which frequency to use.

Voyager 1 reached Jupiter after an interplanetary cruise of 19 months and came no closer to the planet than 280,000 km. However, the spacecraft passed much closer to Jupiter's large satellites Io (22,000 km), Ganymede (115,000 km), and Callisto (126,000 km). None of its many discoveries was more dazzling or unexpected than the realization that numerous volcanos were erupting on Io as the spacecraft swept by. Twenty months later, *Voyager 1* arrived at Saturn. Actually, its first encounter there was with the large satellite Titan, passing just 4000 km from its haze-covered surface on November 12, 1980. About 18 hours later, the spacecraft swept past Saturn at a distance of 124,200 km.

Although hampered by electronic difficulties, *Voyager 2* also proved remarkably successful during its visits to Jupiter in July 1979 and to Saturn in August 1981. Thanks to the successful close-range observations made by *Voyager 1* during its brush with Titan, *Voyager 2* did not have to pass near the satellite and could continue on its course to Uranus.

Misfortune struck the spacecraft again as it crossed the plane of Saturn's rings. The gear mechanism in the instrument-pointing platform seized, causing a good deal of data to be lost. Although the platform was returned to partial service in the days thereafter, many months passed before engineers back on Earth understood the problem fully and coaxed the gearing back into dependable—if somewhat restricted—operation. *Voyager 2*'s 53-month journey between Saturn and Uranus also allowed engineers to make major improvements in how the spacecraft obtains and relays its data.

These changes proved essential once the spacecraft reached Uranus, where sunlight is some 400 times weaker than at the Earth and the exposure times for images were consequently much longer than they were at Jupiter and Saturn. The Uranian system proved even more challenging because of the high obliquity of its equatorial plane and satellite orbits. As *Voyager 2* approached in early 1986 the system had a bull's-eye aspect. The desire to continue to Neptune dictated a unique transit point through this target, which fortunately was very close to the satellite Miranda.

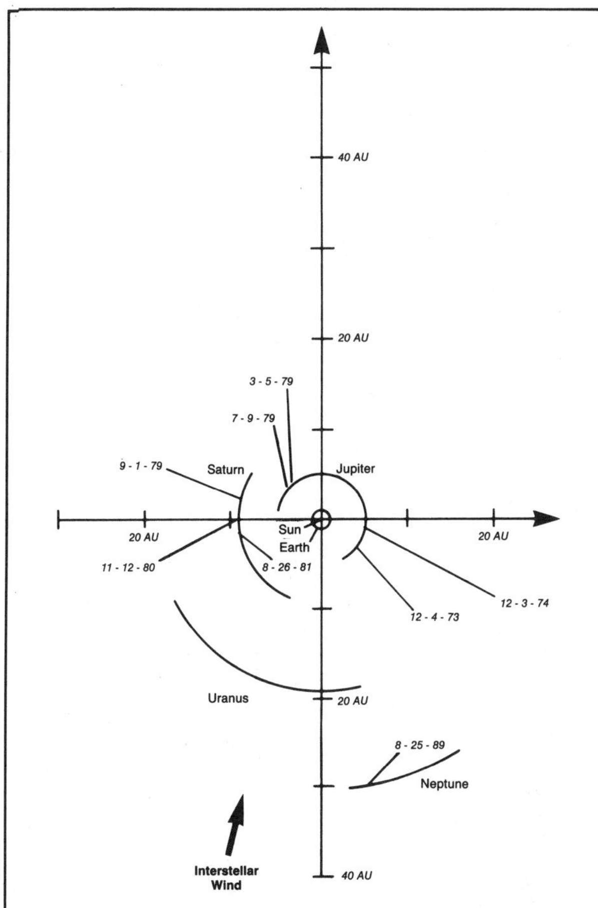

Figure 1. Four spacecraft are now headed out of the solar system, as indicated by this plot of their trajectories (projected on the ecliptic plane). Dates of planetary encounters are indicated, and the tick marks are spaced at one-year intervals. (*Reproduced with permission from* The New Solar System, *Beatty and Chaikin, eds.*)

As the spacecraft cruised toward Neptune, NASA engineers continued to modify its on-board software and to upgrade the receiving stations in the worldwide Deep Space Network. By the time *Voyager 2* reached its last planetary destination in August 1989, it was a much more capable spacecraft than when launched, as attested to by the exquisite data it returned about the planet and its magnetosphere, rings, and huge satellite Triton.

All four *Pioneers* and *Voyagers* are traveling fast enough to leave the solar system (see Fig. 1). They should eventually cross the heliopause, the boundary some $9\text{–}12\times10^9$ km distant that marks the limit of the solar wind's dominance and the beginning of interstellar space. *Pioneer 11* and the two *Voyagers*, now roughly 5×10^9 km from the Sun, could well reach the heliopause before the end of their electronic lifetimes. *Pioneer 10* is already much farther away, more than 7×10^9 km, but it is moving down the "tail" created as the heliosphere plows through the interstellar medium and may thus take longer to reach the boundary.

FUTURE MISSIONS

The Soviet Union has yet to attempt missions to the outer planets and has no firm plans to initiate them through the early years of the twenty-first century. However, NASA launched the *Galileo* mission in October 1989 to extend our knowledge of the jovian system. The spacecraft has two components: an orbiter that will spend several years surveying Jupiter, its magnetosphere, and major satellites; and a probe that will plunge into the atmosphere and return data to a depth of perhaps 20 bars.

NASA has two more deep-space missions under construction. The Comet Rendezvous and Asteroid Flyby (CRAF) spacecraft is to accompany the periodic comet Kopff over much of its orbit, including a pass through perihelion, beginning in the year 2000. The Cassini mission will reach the Saturnian system in 2002. NASA's contribution to Cassini is a long-life Saturn orbiter; the European Space Agency is providing an instrumented probe, called Huygens, that will descend through the dense, hazy atmosphere of Titan and land on its surface. Both CRAF and Cassini employ a new basic spacecraft design, the Mariner *Mark 2*.

Additional Reading

Beatty, J. K. (1990). Getting to know Neptune. *Sky and Telescope* **79** 123.

Beatty, J. K., and Chaikin, A., eds. (1990). *The New Solar System*, 3rd ed. Sky Publishing Corp., Cambridge, MA and Cambridge University Press, Cambridge, U. K.

Fimmel, R. O., Van Allen, J., and Burgess, E. (1980). *Pioneer: First to Jupiter, Saturn, and Beyond*. National Aeronautics and Space Administration, Washington, D.C.

McLaughlin, W. I. (1989). *Voyager's* decade of wonder. *Sky and Telescope* **78** 16.

Morrison, D. (1982). *Voyager to Saturn*. National Aeronautics and Space Administration, Washington, D.C.

Morrison, D., and Samz, J. (1980). *Voyager to Jupiter*. National Aeronautics and Space Administration, Washington, D.C.

See also **Interplanetary Spacecraft Dynamics; Io, Volcanism and Geophysics; Jupiter, Atmosphere; Jupiter and Saturn, Satellites; Planetary Magnetospheres, Jovian Planets; Saturn, Atmosphere; Uranus and Neptune, Atmospheres; Uranus and Neptune, Satellites.**

P

Planetary and Satellite Atmospheres

John J. Caldwell

A planetary atmosphere is an outer, gas-phase layer overlying the solid or liquid surface of a planet. In the four giant planets, the atmospheres comprise a major fraction of the total planet, whereas in the terrestrial planets, the fraction of the mass involved is minute. Within each broad category, the diversity is so great that it is difficult to generalize properties in a meaningful way.

Hydrogen (H) and helium (He) constitute the overwhelming majority of the known mass of the universe. These two light species have been retained to a significant extent in all of the giant planets. In Jupiter and Saturn, the amounts of H_2 and He are close to what would be expected in a cosmic mixture of the elements, whereas in Uranus and Neptune, the fractional abundances of the elements heavier than H_2 and He are significantly enhanced, by about an order of magnitude. The atmospheres of the giant planets therefore, not surprisingly, are composed primarily of these gases, with admixtures of chemically reduced substances (both as gases and as particulates).

To illustrate the difficulty of generalization, it should be noted that helium is relatively less abundant in the atmosphere of Saturn than it is in Jupiter, probably as a result of gravitational differentiation. This is not important chemically, as He is inert, but the cumulative effect of the heavier He atoms settling toward the center of Saturn is to enhance significantly the amount of heat generated in the planetary interior, with important consequences for the thermal structure of the atmosphere.

The terrestrial planets and the satellites of the giant planets have lost essentially all of their original H_2 and He. A small amount of hydrogen is retained chemically in heavier molecules, such as water (H_2O) and sulfuric acid (H_2SO_4). Helium is so scarce on Earth that it was first discovered in the spectrum of the Sun.

The atmospheres of the terrestrial planets therefore are not indicative of the original material from which the whole planets formed. They are the result of outgassing and volcanism in the interior, possibly bombardment by volatile-rich comets, and, in the case of one planet, biological activity.

Of the numerous planetary satellites in the solar system, three have atmospheres sufficiently important that they must be included in this review: Io, Titan, and Triton. The most extensive satellite atmosphere is Titan, and its chemistry is exotic. The atmosphere of Io is transient, with rapid losses to space and replenishment by the most active volcanic regime in the solar system. The atmosphere on Triton is very thin, but shows interesting dynamical and chemical features. Pluto does not fit well with any category of object, but it may be physically similar to Triton. It is also known to have a thin atmosphere. Together, these small objects are as diverse as the atmospheres of the major planets.

TERRESTRIAL PLANETS

The volume of space immediately above the surface of Mercury contains enough gas-phase molecules to be detectable from the Earth, but these are transient, being derived from the interaction of the solar wind with the surface. It may be compared with the exospheres of other atmospheres (regions where the molecular density is so low that individual particles are on ballistic trajectories), and it is beyond the scope of this review.

Venus, Earth, and Mars form an interesting trio, with instructive comparisons between the individual planets. Table 1 summarizes their atmospheric compositions. It should be noted that many of the so-called minor gases show great variations in concentration with altitude and with climatological conditions. Of major importance is the difference, by two orders of magnitude, between the atmospheric masses of Venus and Earth.

From Table 1, it may be noted that the absolute abundances of N_2 and Ar in the atmospheres of Venus and Earth are similar, suggesting that the great difference in CO_2 between the two planets is anomalous. The presence of O_2 on Earth is, of course, not comparable to any other planet because of the uniqueness of life. Furthermore, Mars is clearly just different.

To begin to understand some of these differences, it is first necessary to define the concept of equilibrium temperature T_e: the temperature that the outermost radiative surface of a planet (cloud or solid) would have if the sunlight absorbed by the planet were distributed uniformly around the surface and radiated to space. Although Venus is closer to the Sun than is Earth, the clouds covering the former reflect such a large fraction of the incident sunlight that the equilibrium temperature of Venus is actually lower than the equilibrium temperature of Earth. (The clouds on Venus are composed of droplets of H_2SO_4.)

However, some sunlight does penetrate beneath the clouds of Venus, where it is trapped, greatly heating the lower atmosphere and surface. This trapping occurs because of the great opacity of the atmosphere (primarily of CO_2) at infrared wavelengths and is generally referred to as the "greenhouse effect." For Venus T_e = 231 K, which occurs at the cloud tops, whereas the surface temperature T_s = 731 K. One implication of this extremely high T_s is that there is no liquid water on or near the surface of Venus.

For Earth T_e = 246 K, but a more modest greenhouse effect raises the global mean surface temperature above 273 K so that liquid H_2O may exist at most places. The existence of H_2O oceans at the surface of Earth (sufficient to cover the entire surface to a depth of 3 km in the absence of topographic features) moderates the entire surface environment and is also responsible for the great difference in the atmospheric composition and thermal structure between the two atmospheres.

On Earth, CO_2 dissolves in the oceans and precipitates as carbonates, such as limestone, on the ocean floor. It is thus permanently removed from the atmospheric system. In fact, the inventory of CO_2 bound in rocks on Earth is very nearly the same as the atmospheric CO_2 on Venus. If the oceans had not removed Earth's atmospheric CO_2, the greenhouse effect here would be much greater that it is. It is probable that the surface environment would closely resemble that of Venus and that life would not have evolved!

It is a great concern of contemporary Earth scientists that anthropogenically induced changes in the Earth's atmospheric CO_2, caused by burning fossil fuels, may produce irreversible changes in the atmospheric opacity, leading to an uncontrollable greenhouse warming of the surface, with catastrophic results.

On Mars, the major atmospheric component, CO_2, partially condenses at the winter pole, producing unique seasonal interhemispheric flows. The second most abundant species, N_2, can escape from Mars through exothermic photodissociation. These differences from the Venus–Earth class of planet preclude simple comparisons. However, it is of great interest to note that Mars also has huge, permanent polar caps, probably consisting of H_2O ice. Furthermore, although liquid H_2O cannot exist on the surface of Mars today, a variety of geological forms on its surface are explica-

Table 1. Atmospheric Compositions for Venus, Earth, and Mars

	Venus	*Earth*	*Mars*
Surface pressure (atm)	92	1	0.006
Atmospheric mass (kg cm^{-2})	107	1.03	0.018
Volume mixing ratios: (major gases only):			
CO_2	0.965	320 ppm	0.953
N_2	0.035	0.781	0.027
O_2	—	0.209	—
Ar	70 ppm	9340 ppm	0.016
Additional minor gases	CO, H_2O, SO_2 He, Ne, Kr, H_2S, HCl, HF, C_2H_6, O_2	Ne, He, CH_4, Kr, CO, SO_2, H_2, N_2O, O_3, Xe, NO_2, Rn, NO	CO, H_2O, Ne, Kr, Xe, O_3, NO

Table 2. Properties of the Giant Planets

	Jupiter	*Saturn*	*Uranus*	*Neptune*
Solar distance (AU)	5.2	9.5	19	30
Effective temperature (K)*	125	97	58	58
Enhancement of heavy elements†	2.3	4	25	50

*A measure of the total thermal flux emitted by the planet, including components due to absorbed sunlight and internal sources.

†With respect to solar abundances, as determined from CH_4 measurements in the planetary atmospheres. Uncertainties in this estimate increase sharply from Jupiter to Neptune.

ble only by erosion by significant amounts of flowing water at some earlier time. The implication is that the climate of Mars has been much different in the past than it is now. Whether the change is cyclical or secular is open to question.

GIANT PLANETS

Although the giant planets' atmospheres are all composed mainly of H_2 and He (which are colorless), their visual appearances are quite diverse, in part because of their differing distances from the Sun and in part because of their own internal properties.

The equilibrium temperature, as defined for terrestrial planets, is equally applicable to the giant planets. However, infrared measurements show that three of the four giant planets, Jupiter, Saturn, and Neptune, emit more total radiation than expected for their respective T_es, by a factor of about 2, implying internal energy sources, whereas Uranus has no detectable internal source. The resulting effective temperatures are shown in Table 2. Also shown there is an estimate of the enhancement of elements heavier than hydrogen and helium, with respect to their abundance in the Sun, as represented by measurements of carbon species.

Jupiter and Saturn have grossly similar compositions, but Saturn is colder. On Jupiter, ammonia (NH_3) is partially frozen, producing the prominent bright zones that are present in images of Jupiter. The organization of these zones parallel to the equator is a consequence of the strong coriolis forces due to the rapid rotation of the planet. In other latitude belts, clear of condensed NH_3, one observes to deeper levels, where other clouds, containing coloring agents such as sulfur, are present. The result is the familiar alternating bright/dark, zone/belt structure. The situation is similar to the Earth, where clouds of H_2O particles or droplets cover part but not all of the surface. However on the colder planet Saturn, the NH_3 is more completely frozen, with no clear belts, resulting in much-reduced contrast of zonal features. The same coloring agents are undoubtedly present on Saturn, but they are not visible from space.

Uranus and Neptune, which have similar temperatures and compositions, have similar visual appearances. They are sufficiently cold that methane (CH_4) condenses in both planets' atmospheres, unlike Jupiter and Saturn. This produces an additional thick cloud on these planets, overlying the other cloud layers and further reducing gross contrast. However, there is enough residual gaseous CH_4 above the clouds that the prominent absorption of CH_4 at red wavelengths results in a perceptible bluish color on both planets. The situation here is again analogous to the Earth, where the abundance of H_2O vapor at altitudes above cumulus clouds is not zero.

There are additional detailed features on some of the giant planets. On Jupiter, the most prominent of these is the Great Red Spot, a long-lived atmospheric storm that is larger in diameter than the Earth. Many smaller features, generally with different colors, are also seen. Jupiter also exhibits aurorae and lightning in its atmosphere. At very high altitudes, nonequilibrium chemistry is driven by photons at the equator and by magnetospheric particle precipitation at the poles, producing such species as C_2H_2, C_2H_4, C_2H_6, C_3H_4, and C_6H_6 plus smog-like particles. On Neptune, high-altitude clouds of CH_4 particles, analogous to cirrus clouds on the Earth, were seen in the 1989 *Voyager 2* encounter images. Also seen there was a feature comparable in scale to the jovian Great Red Spot, but without its characteristic color, called the Great Dark Spot on Neptune. Other spots at latitudes different from the Great Dark Spot were observed to rotate around Neptune at distinctly different rates, implying significant differences in zonal and/or altitude structure of wind fields.

SATELLITES AND PLUTO

The extensive volcanism of Io was discovered by the *Voyager 1* spacecraft. The images showed plumes of material ejected from volcanos, and infrared spectroscopy revealed that gaseous SO_2 was present near one of the "hot spots." The pressure of this SO_2 was 10^{-7} times the atmospheric pressure on Earth. The particulates that make the plumes visible are probably condensed SO_2; SO_2 frost is also known to be present on part of the surface, and numerous dissociation fragments of SO_2, such as SO^{2+}, S III, S IV, and O III, have been observed in a torus around Jupiter (the "Io torus") coincident with Io's orbit. These fragments are stripped from Io by the co-rotating jovian magnetic field as soon as they are changed from neutral SO_2 into a charged state. The totality of these observations of Io is consistent with SO_2 being the driving fluid behind the volcanic eruptions, producing a very thin, possibly transient atmosphere, in which the individual molecules survive only for a very short time before condensing or escaping.

Saturn's large satellite, Titan, has the most extensive atmosphere of any satellite in the solar system. In fact, the column abundance (the number of molecules above a square centimeter on the surface) there is 10 times greater than on Earth and, despite the lower gravity, the atmospheric pressure is 1.5 times Earth's. The bulk constituent of the atmosphere is N_2, and the next most abundant species, CH_4, is variable with altitude and condenses to form thick clouds. It is likely that it rains on Titan. Argon may also be present. Photochemistry produces very dark aerosols at extremely high altitudes and also produces a large number of complex hydrocarbons and nitriles, including possible precursors to amino acids. The atmosphere is sufficiently optically thick that the surface can only be observed from space at radio wavelengths. Trace amounts of CO and CO_2 have been detected on Titan. The oxygen required to make them is probably of external origin, coming either from the small icy satellites or the rings, as a result of sputtering of magnetospheric particles.

One of the most important research questions in all of atmospheric science is the degree to which the atmosphere of Titan is primordial or has evolved from something else. If the atmosphere is largely primordial, Titan may hold important clues to the origin of the solar system itself.

Neptune's large moon Triton was discovered to have an atmosphere during the 1989 *Voyager 2* encounter and studies of it are not far advanced. It is known, however, to have a very thin atmosphere, with high-altitude haze layers, which may be similar chemically to the high-altitude haze on Titan. One instance of an active volcanic geyser was observed, ejecting a stream of material vertically to an altitude of several kilometers, where the flow was deflected horizontally by a very rapid jet stream. Triton itself clearly has a very interesting geological history, with condensed N_2 and/or CH_4 on the surface, and the thin atmosphere is an important boundary condition on the surface environment.

Pluto is known to have CH_4 in a thin atmosphere and condensed on its surface. The surface pressure, as measured at a time when Pluto occulted a star, is about 10^{-5} times the surface pressure on Earth. It is possible that another gas, more difficult to detect spectroscopically than CH_4 may also be present.

Additional Reading

Atreya, S. K., Pollack, J. B., and Matthews, M. S., eds. (1989). *Origin and Evolution of Planetary Atmospheres*. University of Arizona Press, Tucson.

Beatty, T. K. and Chaikin, A., eds. (1990). *The New Solar System*, 3rd edn. Sky Publishing Co., Cambridge, MA and Cambridge University Press, Cambridge, U.K.

Hunter, D. M., Colin, L., Donahue, T. M., and Moroz, V. I., eds. (1983). *Venus*. University of Arizona Press, Tucson.

Gehrels, T. and Matthews, M. S., eds. (1984). *Saturn*. University of Arizona Press, Tucson.

See also **Earth, Atmosphere; Io, Volcanism and Geophysics; Jupiter, Atmosphere; Mars, Atmosphere; Mercury and the Moon, Atmospheres; Pluto and Its Moon; Satellites, Ices and Atmospheres; Saturn, Atmosphere; Uranus and Neptune, Atmospheres; Venus, Atmosphere.**

Planetary Atmospheres, Clouds and Condensates

Robert E. Samuelson

Clouds and aerosols in planetary atmospheres arise from many sources. Photochemistry and charged particle bombardment in the upper atmosphere create ions and radicals from the ambient gas. These recombine into more complex molecules, which in turn polymerize into submicrometer-sized particles to form stratospheric hazes or smog. Lower in the atmosphere (usually the troposphere), regions may exist where certain gases become supersaturated, forming condensation clouds. The shapes and properties of these clouds depend on composition, temperature, and atmospheric dynamics. Near the surface, high winds can give rise to dust clouds over dry, desert-like terrain. Fog may form over moist or wet surfaces under calm conditions. Volcanoes can eject large quantities of subsurface material into the atmosphere, sometimes to quite high altitudes.

Particulates play an active role in atmospheric dynamics. Sunlight can either be reflected, leading to a cooling at lower levels, or absorbed, resulting in direct atmospheric heating. The thermal structure thus induced will be modified by the subsequent emission and absorption of thermal radiation, and particulates are important sources of opacity affecting this redistribution of radiant energy. Ultimately, the dynamics of the atmosphere depend on the magnitudes and directions of the thermal gradients established by these radiative processes.

Seven planets and one satellite have substantial atmospheres, and all contain important quantities of particulate matter: Venus, Earth, Mars, Jupiter, Saturn, Uranus, Neptune, and Titan, the largest satellite of Saturn. We will consider these objects in order with the exception of the Earth, the study of which is better undertaken elsewhere. Most of our information comes from such spacecraft as the Pioneer orbiter–probe and Venera descent–lander missions to Venus, the *Mariner* flyby and Viking orbiter–lander missions to Mars, and the Voyager flyby missions to the outer planets, although extremely useful complementary information (especially in connection with long time-base studies) has been obtained from numerous ground-based observations.

VENUS

Venus is essentially 100% cloud-covered. Features are absent at visible wavelengths, but are readily discernible in the ultraviolet: Streaks, bows, and bands, mark the cloudy atmosphere giving way to broken, mottled areas, especially near the subsolar point. Cell sizes range between 100 and 1000 km across. Lifetimes range from 1–2 h for the smaller features to in excess of 12 h for the larger. Bright rings often mark the polar regions, whereas a dark, horizontal Y-shaped feature frequently extends across the disk at low and middle latitudes. Even when no "Y" is present (and in spite of the short-lived aspect of the smaller features), there is a subtle 4–5-day quasiperiodicity in the distribution of these features across the disk.

Vertically, the clouds are quite stratified and separate roughly into five regions: An upper haze layer, consisting of submicrometer-sized particles, extends from 70 to 90 km above the surface. A lower counterpart lies between 31 and 47.5 km. Sandwiched between these two layers lies the main cloud deck, consisting of three levels with sharp transition boundaries between. The upper level (between 56.5 and 70 km) contains droplets of concentrated sulfuric acid with mean diameters of 2–3 μm. The middle and lower levels (between 50.5–56.5 km and 47.5–50.5 km, respectively) contain crystals of unknown composition with effective mean linear dimensions of about 7–8 μm, in addition to the sulfuric acid droplets found in the upper cloud. The ubiquitous haze pervades all three levels.

The clouds are in a region of the atmosphere that is stable against strong convection but that supports a strong vertical wind shear. As a result, the clouds are highly stratified; local deviations in cloud density are quickly spread out horizontally, whereas vertical propagation is very slow. Particle growth is governed by diffusion. Consequently, growth rates are small and particle lifetimes can span months and even years.

Formation of the clouds is incompletely understood. Sulfur dioxide gas diffuses upward from the lower atmosphere to levels above 60 km, where it is oxidized photochemically to form SO_3. Homogeneous reactions between SO_3 and H_2O form H_2SO_4 vapor. The atmosphere quickly saturates, and concentrated H_2SO_4 droplets are formed on condensation nuclei already present. Mist or drizzle eventually reaches the 48-km altitude where evaporation and thermal decomposition of H_2SO_4 take place. Because precipitation cannot reach the surface, the atmosphere is never cleansed of condensation nuclei, and they are recycled to higher levels again. Their composition is unknown (sulfur has been suggested), but some property allows them to absorb ultraviolet radiation, resulting in cloud contrast at these wavelengths. The composition of the crystals found in the middle and lower cloud levels is also unknown, but sulfates, chlorides, or even hydrates are possibilities.

MARS

Water-ice clouds on Mars are thin and sparse by terrestrial standards. Low humidity tends to inhibit the formation of thick clouds in late autumn and winter in both northern and southern hemispheres at high latitudes. Thick clouds are also rare in early winter at southern midlatitudes. Humidity increases during spring and summer, and condensation clouds become more abundant at midlatitudes, more so in the north than in the south, where dust clouds become influential. Low latitudes tend to have fewer clouds than midlatitudes, and generally there are fewer clouds in the south than in the north.

Several types of condensation clouds occur. Lee waves are periodic ridges, often chevron-shaped, that appear on the downwind side of obstacles such as mountains or craters. They are probably caused by trapped, forced internal gravity waves. Wave clouds are also found and consist of rows of clouds not associated with any obstacle. Cloud streets exhibit a double periodicity, resembling linear arrays of marble-like convective cells. Sometimes clouds are streaky without sharp edges, at other times they resemble hazes or fogs at the surface, especially in low-lying areas such as crater basins or over the polar caps.

Dust on Mars is pervasive. Local dust storms may occur at any season, and some dust is probably always in the atmosphere. When Mars is nearest the Sun during northern fall and winter, however, the great dust storms appear, engulfing large areas of the planet and occasionally reaching global scale. One or two of these storms may occur each martian year.

A combination of circumstances appears necessary to generate these great storms. Local terrain features give rise to dynamical phenomena such as slope winds and turbulent wakes that raise dust locally. Sunlight heats the dust, and the resulting diabatic forcing raises still more dust. Enhanced solar irradiation near martian perihelion intensifies the general circulation, including atmospheric tides, especially in the subtropical regions. Planetary-scale winds are thus induced, and these provide the necessary background to maintain local storms driven by the diurnal cycle, enabling them eventually to spread across the planet.

JUPITER

Jupiter presents a banded structure between 45° north and south latitude, with the bands running parallel to the equator. Bright *zones* and darker *belts* give way to a more mottled appearance at higher latitudes. These belts and zones are fairly permanent, although they can fluctuate in width with time, and occasionally certain ones will reverse their appearance (e.g., transform from zone to belt).

The largest local feature is the Great Red Spot (GRS), an anticyclonic vortex centered at about 24° south latitude and measuring approximately 11,000 by 22,000 km. Dynamically similar white ovals, about one-half to one-tenth the linear size of the GRS, are dotted around the planet at southern midlatitudes. Dark brown spots ("barges") can be found in northern low latitudes. Sometimes white cloud filaments extend over these barges, implying the dark material lies deeper.

Strong vertical mixing appears to occur just north of the equator. Small, bright cloud nuclei appear to convect upward where they are sheared out to the west, forming plumes with spreading tails, becoming fully formed after 5–10 jovian revolutions. Other disturbed regions include "wake-like" trains westward of the GRS and other anticyclonic features. North–south mixing over thousands of kilometers can take place at certain zone–belt interfaces.

A stratospheric haze, thickest at high latitudes, is present at altitudes below the 10 mbar pressure level. Effective particle radii are estimated to range between 0.2 and 0.5 μm, although their shapes appear to be highly aspherical. Their composition is unknown, although hydrazine laced with polyacetylenes or sulfur is one possibility.

A cloud of variable thickness exists in the upper troposphere. The top of this region is at about 200 mbar at the equator, lowering to 400 mbar at about 45° north and south latitudes. The bottom is between 600 and 800 mbar. Its composition is thought to be a mixture of hydrazine and ammonia ice, probably contaminated with chromophores (groups of atoms or molecules that produce visible color). Hydrazine particles should be micrometer-sized and mixed throughout. Ammonia particles can range between 1 and 100 μm in size and are probably restricted to altitudes below 500 mbar and densest in the zones.

There also appears to be an inhomogeneous, time-variable cloud near the 2-bar level, possibly composed of a mixture of NH_4SH and chromophores, provided H_2S is present in solar abundance. A still deeper H_2O–NH_3 cloud may be present at 5–6 bar if oxygen is present in solar abundance, a controversial issue at this time.

SATURN

Like Jupiter, Saturn presents a banded appearance, though contrast is much more subdued and there are relatively few conspicuous features. Similarities include several long-lived symmetric ovals, both brown and white, associated with anticyclonic shear zones that rival Jupiter's in size. Occasional eruptions of convective cells with subsequent shearing, analogous to Jupiter's equatorial plumes, also exist. Bright, irregular, short-lived convective clouds are associated with a westward jet at 39° north latitude, but not elsewhere. A unique cloud ribbon, dark and wavy, encircles the globe at 46° north latitude. Anticyclonic and cyclonic vortices, each about 5000 km across, fill the crests and troughs of the ribbon, respectively.

A haze of aerosols with particles approximately 0.1 μm in radius appears to exist between 30 and 70 mbar at low latitudes and above 20 mbar at latitudes greater than 60°. Between 100 and 300 mbar the particles are larger, with number densities in the equatorial region exceeding those at temperate latitudes by about a factor of 2. Still larger particles are found between 300 and 500 mbar, with single-scattering phase functions resembling those of ammonia crystals several micrometers across, as measured in the laboratory. The upper haze is likely to be formed by photochemistry and charged-particle bombardment. Lower down, ammonia crystals may be intermixed, and condensed P_2H_4, analogous to hydrazine on Jupiter, is also possible.

The base of an ammonia cloud is expected at 1.4 bar. There is some evidence from reflection spectra for a relatively clear region between roughly 0.5 and 1 bar. If true, this suggests that the clouds above 500 mbar do not contain ammonia.

URANUS AND NEPTUNE

Uranus is the most subdued of all the outer planets, showing only a very few, relatively small local features at low latitudes, greatly distended in east–west directions. Neptune reveals many more features, both bright and dark, with the largest being the Great Dark Spot in the southern tropics. Many of Neptune's smaller features alternately appear and disappear, suggesting condensation and evaporation on rising and falling wave crests. Both planets exhibit a banded structure and general haze, variable with latitude, although Neptune's atmosphere is sufficiently clear at some level above the main cloud deck to show cloud shadows over tens of kilometers at certain locations.

Submicrometer-sized haze particles have been inferred for Uranus at levels corresponding to a few tens of millibars, as well as a general cloud deck between 0.9 and 1.3 bar, the latter presumably consisting of methane-ice crystals. A comparable methane cloud is expected for Neptune.

TITAN

The appearance of Titan is extremely diffuse and bland. A photochemical aerosol between 40 and 250 km was very faintly banded and slightly brighter in the southern hemisphere at the time of the *Voyager* encounters in 1980–1981. This north–south brightness asymmetry is thought to alternate with the seasons. Particle sizes tend to increase with depth, being of submicrometer size near the top and a few micrometers in diameter near the bottom. There is some evidence for an additional thin layer of highly aspherical particles, several micrometers in length, near the top. A detached haze exists between 300 and 350 km, although the clearing between it and the aerosol below may fill in occasionally at certain locations.

Many hydrocarbons and nitriles, formed in the vicinity of the mesosphere from photochemistry and energetic particle bombardment of CH_4 and N_2, condense in the lower stratosphere. An increased global opacity appears to occur between 50 and 80 km, the most likely sources being condensed C_2H_2, C_2H_6, and C_3H_8. Direct spectroscopic evidence exists for condensed HC_3N and C_4N_2 (and possibly C_3H_4 and H_2O) at these levels near the north polar hood. Other likely condensed organics include C_4H_2 and C_2N_2. Probably advection rather than diffusion is responsible for the layered structure.

Methane appears to condense in the troposphere between 15 and 35 km. The clouds are probably quite inhomogeneous, both horizontally and vertically, with particle sizes ranging from 1 to 1000 μm. Precipitation is almost inevitable, and, unlike Venus, the troposphere appears to be largely cleansed of condensation nuclei.

Additional Reading

French, R. G., Gierasch, P. J., Popp, B. D., and Yerdon, R. J. (1981). Global patterns in cloud forms on Mars. *Icarus* **45** 468.

Hunten, D. M., Tomasko, M. G., Flasar, F. M., Samuelson, R. E., Strobel, D. F., and Stevenson, D. J. (1984). Titan. In *Saturn*, T. Gehrels and M. S. Matthews, eds. University of Arizona Press, Tucson, p. 671.

Kahn, R. (1984). The spatial and seasonal distribution of martian clouds and some meteorological implications. *J. Geophys. Res.* **89** 6671.

Knollenberg, R. G. and Hunten, D. M. (1980). The microphysics of the clouds of Venus: Results of the *Pioneer* Venus particle size spectrometer experiment. *J. Geophys. Res.* **85** 8039.

Knollenberg, R. G. et al. (1980). The clouds of Venus: A synthesis report. *J. Geophys. Res.* **85** 8059.

Rossow, W. B., Del Genio, A. D., Limaye, S. S., and Travis, L. D. (1980). Cloud morphology and motions from *Pioneer* Venus images. *J. Geophys. Res.* **85** 8107.

Samuelson, R. E. (1985). Clouds and aerosols of Titan's atmosphere. In *The Atmospheres of Saturn and Titan, ESA SP-241*, p. 99.

Smith, B. A. et al. (1979). The Jupiter system through the eyes of *Voyager 1*. *Science* **204** 951.

Smith, B. A. et al. (1979). The Galilean satellites and Jupiter: *Voyager 2* imaging science results. *Science* **206** 927.

Smith, B. A. et al. (1981). Encounter with Saturn: *Voyager 1* imaging science results. *Science* **212** 163.

Smith, B. A. et al. (1982). A new look at the Saturn system: The *Voyager 2* images. *Science* **215** 504.

Smith, B. A. et al. (1986). *Voyager 2* in the uranian system: Imaging science results. *Science* **233** 43.

Smith, B. A. et al. (1989). *Voyager 2* at Neptune: Imaging science results. *Science* **246** 1422.

Tomasko, M. G. and Doose, L. R. (1985). Clouds and aerosols on Saturn. In *The Atmospheres of Saturn and Titan, ESA SP-241*, p. 53.

Tomasko, M. G., West, R. A., Orton, G. S., and Tejful, V. G. (1984). Clouds and aerosols in Saturn's atmosphere. In *Saturn*, T. Gehrels and M. S. Matthews, eds. University of Arizona Press, Tucson, p. 150.

West, R. A., Strobel, D. F., and Tomasko, M. G. (1986). Clouds, aerosols, and photochemistry in the jovian atmosphere. *Icarus* **65** 161.

Zurek, R. W. (1982). Martian great dust storms: An update. *Icarus* **50** 288.

See also **Jupiter, Atmosphere; Mars, Atmosphere; Satellite Ices and Atmospheres; Saturn, Atmosphere; Uranus and Neptune, Atmospheres; Venus, Atmosphere.**

Planetary Atmospheres, Dynamics

Michael Allison

Planetary atmospheres are the circulating fluids of world-sized thermodynamic engines. Their observed wind and wave motions are the dynamic response to differential heating imposed by their absorbed solar radiation and, in the case of the giant outer planets, to the internal energy generated by their formative gravitational contraction. Although partly derived from the observational and analytical tools of astronomy, the comparative study of dynamic atmospheres has emerged as an extension of terrestrial meteorology to the rich variety of size, rotation, mass, temperature, and composition exhibited by the other planets in the solar system. The subject has begun to mature only over the past two decades, largely as a result of the preliminary reconnaissance of the planets by spacecraft. Although based on well-understood laws of mechanics, thermodynamics, chemistry, and radiation, the aggregate behavior of these macroscopic systems has proven outstandingly difficult to deduce and predict. There is as yet no general theory of atmospheric dynamics that can comprehensively account for the observed features of extraterrestrial wind patterns. Some progress has been made, however, in diagnosing the basic physical balances and kinematics. Aside from the challenge to solve individual problems of fascinating complexity, the comparative investigation of motions in planetary atmospheres proceeds with the expectation of eventually attaining a fundamental improvement in the unified understanding of geophysical fluid dynamics.

PARAMETERS AND SCALING RELATIONSHIPS

The dynamics of planetary atmospheres is characterized by the combined effects of rotation, stratification, and dissipation. Although model representations typically require sophisticated mathematics and computation, important insights may be derived from the consideration of simple measures of the relevant parameters and scaling estimates of the force balances. One of the most fundamental constraints on the large-scale motions of a rapidly rotating atmosphere is the approximate balance between horizontal gradients in pressure and the Coriolis acceleration. As discovered in the nineteenth century by Gaspard Gustave de Coriolis, any moving object, such as a cannon ball or a puff of air, traversing a certain horizontal distance will experience a deflection at right angles to its course, as measured with respect to the surface of the planet turning beneath it. The effect may be witnessed as the slow rightward turning, in the northern hemisphere, of the swinging plane of a Foucault pendulum, with a complete period equal to one-half the planetary rotation period, divided by the trigonometric sine of the local latitude, or $\tau_{pend} = \tau_{rot}/2\sin\lambda$. The corresponding angular rotation frequency $f = 2\pi/\tau_{pend}$, is called the Coriolis parameter or planetary vorticity.

The time required for an atmospheric parcel moving with a speed U to traverse a horizontal distance L is just L/U. Planetary rotation will significantly affect the motion if the corresponding inertial acceleration U^2/L is small compared to the Coriolis acceleration, given as $f \cdot U$. The ratio of inertial to Coriolis acceleration is measured by the Rossby number,

$$\text{Ro} \equiv U/fL,$$

after the pioneering twentieth-century meteorologist Carl-Gustaf Rossby. For large-scale motions at mid-latitudes in the Earth's atmosphere, for example, $\text{Ro} \approx 0.1$. Wherever the Rossby number is small (and viscous effects can be neglected), horizontal gradients in pressure are in approximate *geostrophic* balance with the Coriolis acceleration of the flow. As a result, low-pressure centers are observed as cyclonic weather patterns, locally rotating in the same direction as the planet, whereas high-pressure centers are anticyclonic. Where the Rossby number is large, the Coriolis acceleration is negligible and steady motions represent a balance between horizontal pressure gradients and the centripetal acceleration associated with the flow curvature. Examples of this so-called *cyclostrophic* regime include tornados and the winds of Venus.

The thinness of planetary-scale dynamics, as measured by the small ratio of the characteristic vertical and horizontal scales of the motion, $D/L \ll 1$, generally insures that the vertical pressure gradient in the atmosphere is "hydrostatically" balanced by the weight per unit volume at a given level. (Violent updrafts in localized storm systems and perhaps also deep convective motions in the atmospheres of the giant planets are exceptions for which the vertical acceleration must also be taken into account.) When combined with the equation of state relating the atmospheric pressure to the product of the density and temperature, hydrostatic balance yields the barometric law for the vertical (exponential) drop of pressure with altitude. The associated pressure scale height is given as $H = RT/g$, where R is the gas constant for the mixture, itself fixed in inverse proportion to the mean molecular weight, T is the temperature, and g the gravitational acceleration.

Under special circumstances the atmospheric structure may be regarded as *barotropic*, with constant-density (or isopycnic) surfaces parallel to the constant-pressure (or isobaric) surfaces, for which the horizontal motion is constant with altitude and effectively is decoupled from the thermodynamics. More generally, however, planetary atmospheres exhibit *baroclinic* structure, with isopycnals inclined to the isobars, for which the motion is vertically sheared. In this case the hydrostatic pressure balance, together with the equation of state and the geostrophic (or cyclostrophic) balance of horizontal gradients, prescribe the thermal wind shear. The corresponding scaling relation for the horizontal velocity is

$$U \sim \begin{cases} -(gD/fL)(\Delta T/T), & \text{for Ro} \ll 1, \\ \pm\sqrt{(gD\Delta T/T)}, & \text{for Ro} \gg 1, \end{cases}$$

where D denotes the depth of the vertically sheared motion, typically at least as large as the scale height, ΔT is the horizontal contrast in temperature, and again g is the gravitational acceleration, f the Coriolis parameter, L the horizontal scale of the motion, and T the absolute temperature. In the geostrophic ($\text{Ro} \ll 1$) case, the correct sign for the thermal contrast requires that for a positive Coriolis parameter, the velocity U be directed 90° to the right of the direction of decreasing temperature. In the cyclostrophic ($\text{Ro} \gg 1$) case, the sign of the flow velocity is undetermined, except by the further consideration of its forcing. In both cases, the application of the thermal wind balance to the zonal motion breaks down at the equator, where the Coriolis and (zonal) centripetal accelerations vanish. Although extremely useful for the diagnostic inference of horizontal motions from remotely sensed temperature gradients, the thermal wind equation is by itself insufficient to describe the causal establishment or time-evolution of the circulation.

In addition to the horizontal gradients in temperature, the dynamics of an unevenly heated atmosphere inevitably entails the consideration of its vertical stratification. Most of the incoming sunlight is absorbed either at the surface or, in the case of a thick atmosphere, within a deep cloud layer. The resulting excess heat is reradiated at infrared wavelengths, for which the atmosphere is largely opaque, and must be carried by rising motions up to the emission level, where it can be cooled efficiently to space. Rising parcels cool as they expand in equilibration with the decreasing pressure of their ambient environment. For neutrally stable conditions, the drop in temperature with altitude, the so-called lapse rate, is just sufficient to match the cooling of an adiabatically rising parcel (similar to that prescribed by the Schwarzschild criterion for the onset of convection in stars). The static stability is measured as the difference between the adiabatic lapse rate, given as the ratio of the gravitational acceleration to the specific heat at constant pressure ($\Gamma_{ad} = g/c_p$), and the ambient lapse rate Γ. (In the presence of moist convection or other phase changes, the adiabat must also take account of latent heat and variations in molecular weight.) At the top of the convective troposphere, the stability is very nearly the same as Γ_{ad} and increases aloft. Within the troposphere, the weaker but still generally stable stratification, with respect to large-scale motions, depends upon a complicated interplay of radiation, chemistry, and dynamical transports. Although the internal heating within the deep atmospheres of Jupiter and Saturn might support a negative static stability, it is presumed that convective transports will efficiently adjust this to a value nearly indistinguishable from zero. For rapid, nearly adiabatic motion within a statically stable atmosphere, the conservation of heat demands that temperature fluctuations δT balance the associated vertical displacement δz in proportion to the stability, so that

$$\delta T \sim -\delta z \cdot (\Gamma_{ad} - \Gamma),$$

again in the sense that rising motions are cooled.

For geostrophic motion there is a further constraint on the evolution of the flow, given by the conservation of vorticity. Assuming negligible dissipation, this specifies that the change in the rotation or vorticity of the motion is balanced by the vertical stretching of the fluid. Estimating the characteristic scale of the vorticity as the incremental change in velocity δU over a horizontal length L, and expressing the stretching in terms of the vertical displacement, the corresponding scaling relationship is

$$\delta U/L \sim f(\delta z/D).$$

Assuming that the geostrophic ($\text{Ro} \ll 1$), vorticity-conserving motion is coupled to the stratification according to the heat equation, with $\delta U \sim U$ and $\delta T \sim \Delta T$, these may be eliminated from the three balance relations reviewed previously to infer the associated horizontal "deformation" scale, given as

$$L^2 \sim L_D^2 \equiv [g(\Gamma_{ad} - \Gamma)/T] \cdot D^2/f^2.$$

L_D is called the Rossby radius of deformation and represents the characteristic horizontal scale for geostrophic baroclinic motion. For the midlatitude troposphere of the Earth, for example, L_D is approximately 1000 km.

The size and form of dissipation in planetary atmospheres are more difficult to quantify than the inertial and thermodynamic balances. On the time scales appropriate to observed large-scale weather patterns ($\sim L/U$), the motions are effectively inviscid and generally adiabatic. Nevertheless, dissipation plays an essential role

Table 1. Dynamical and Meteorological Parameters for Planetary Atmospheres

	Earth	*Mars*	*Venus*	*Titan*	*Jupiter*	*Saturn*	*Uranus*	*Neptune*
Equatorial radius (km)	6,378	3397	6051	2575	71,490	60,270	25,560	24,760
Rotation period (h)	23.93	24.62	5832 (243 days)	382.7 (15.9 days)	9.925	10.66	17.23	16.12
Axial obliquity (°)	23	25	177	—	3	27	98	30
Emission temperature (K)	255	210	229	85	124	95	59	59
Surface (or cloud) temperature (K)	288	214	731	94	(NH_3) 152	(NH_3) 156	(CH_4) 80	($CH_4 + H_2S$) 80–120
Pressure (bar)	1.013	0.007	92	1.5	0.7	1.4	1.2	1–3
Gravity (m s^{-2})	9.8	3.7	8.9	1.4	24	10	8.8	11
Scale height (km)	8	11	16	20	20	40	30	30
Adiabatic lapse rate (K km^{-1})	9.8	4.4	10	1.3	2.0	0.7	0.8	0.8
Radiative cooling time (s)	$\sim 10^7$	$\sim 10^5$	$\sim 10^7 \times (p/\text{bar})^{1.3}$	$\sim 10^9$	$\sim 10^8$	$\sim 10^9$	$\sim 10^{10}$	$\sim 10^{10}$
Horizontal motion scale (km)	1000	600	6000	3000?	2,000	3,000	8,000?	7,000?
Jet speed (m s^{-1})								
midlatitude	+15	+30	+90	+80?	±50	±100	+200	+200?
equatorial	−4	?	+100	+?	+120	+500	−100	−400

in the steady (or time-averaged) equilibration of the flow in response to the continual input of solar energy. Near the lower solid surface of terrestrial-type atmospheres, the dissipation of momentum by small-scale eddies appears to act analogously to friction, and supports a boundary layer of turbulent motions and strong vertical shear. Thermal damping occurs over a characteristic radiative "cooling time," given in proportion to the pressure p and temperature T at a given level as

$$\tau_{\text{rad}} \sim pT/\Gamma_{\text{ad}}\sigma T_e^4,$$

where T_e is the atmospheric temperature at the infrared emission level, and $\sigma = 5.67\times10^{-5}$ mW m^{-2} K^{-4} is the Stefan–Boltzmann constant. Some dissipation of both momentum and heat may occur at high stratospheric levels, as a result of mixing by vertical wave propagation. In addition, significant horizontal mixing may be imposed by unstable eddies in the zonal flow, but probably cannot be described accurately as viscous dissipation and so must be either parametrized or explicitly calculated as nonlinear, small-scale motions.

OBSERVATIONS AND THEORIES

Table 1 summarizes the most essential external parameters for the dynamic planetary atmospheres in the solar system including Saturn's large satellite Titan, along with observational estimates of the characteristic length and velocity scales for their global motions. Although each dynamic atmosphere is unique, certain pairs exhibit sufficient similarity for comparative discussion.

Earth and Mars: Cyclone Engines

The atmospheric dynamics of the Earth is obviously the best studied of all the planets and constitutes a fundamental paradigm for general circulation studies. Its zonal motion is characterized by a "westerly" midlatitude jet in each hemisphere, strongest at upper tropospheric levels, and at low latitudes the weak easterly "trade winds," as discovered by the early transoceanic explorers. (Terrestrial meteorologists customarily label the winds by the direction from which they flow, so that an "easterly" wind blows westward.) The characteristic speeds for the Earth's zonal winds given in the table refer to annual averages at the 500-mbar level, as measured by airborne radiosondes. As early as 1686 Edmond Halley identified the Sun as the ultimate cause of atmospheric motions. In 1735 George Hadley proposed that the preferential heating of low latitudes would drive a slow meridional circulation, rising at the equator and sinking toward the poles, and that returning low-level flow would be rotationally deflected to the west as observed, with an associated surface torque necessarily compensated by eastward flow at higher latitudes.

Although Hadley's proposal proved to be a qualitatively correct picture for the tropical circulation (which bears his name), it did not account for the observed latitudinal gradients in pressure and is now known to be mediated by a reverse meridional circulation at midlatitudes, as suggested by William Ferrel in the nineteenth century. Ferrel demonstrated that the midlatitude winds were geostrophically balanced by the pressure field, with a surface maximum at the descending branch of the Hadley cell, near 30° latitude, and a minimum near the pole. The associated equator-to-pole temperature contrast is about 40 K, as required by thermal wind balance for the midlatitude jet, over a vertical scale of about one scale height. This balance is, however, unstable to longitudinal eddy fluctuations and, as originally proposed by Vilhelm Bjerknes in 1937, gives rise to the pattern of traveling cyclones displayed on synoptic weather maps. Instability theory, as developed by Jule Charney and E. T. Eady in the late 1940s, suggests that the characteristic horizontal scale of the unstable eddies is about four times the deformation scale, or $4L_D \approx 4000$ km. This is observed as the characteristic horizontal spacing of pressure centers, vividly apparent in photographs of the Earth from space as the typically "wave-6" pattern of cyclonic (low-pressure) cloud swirls, encircling the planet at midlatitudes. In the 1920s, even before the unstable origin of the cyclones was understood, Albert Defant had proposed that they transfer heat poleward and restrict the upper-level wind strength, whereas Harold Jeffreys suggested that they are also responsible for the poleward transport of angular momentum required to maintain the surface westerlies against viscous dissipation. Modern computer simulations demonstrate that the general circulation represents an elaborate nonlinear balance of meridional flow and eddy transports of both heat and angular momentum against dissipation, complicated by the radiative and latent heat effects of clouds.

Mars is about half the size of the Earth, but has nearly the same rotation period and axial obliquity. The surface pressure of its thin carbon dioxide atmosphere is less than 1% that of the Earth's but is nevertheless sufficient to support a rich variety of weather phenomena, as revealed by orbiter and lander spacecraft. The radiative cooling time is only about three days and, as a consequence, diurnal variations in temperature are extreme, varying by as much as 60 K. The alignment of the elliptical orbit of the planet with its seasonal solstices produces a hemispheric asymmetry in the evaporation and sublimation of its polar caps, attended by a regular but extreme variation in the surface pressure from about 7 mbar in the northern summer to 9 mbar in the winter. The seasonal meridional mass flow produces a cross-equatorial Hadley circulation, rising in the summer subtropical latitudes and sinking to the other side of the equator. The summer season at the locations of the Viking landers has a monotonously repetitive weather pattern, with upslope winds during the afternoon and downslope winds during the early morning. The spacecraft observations for the northern hemisphere winter, however, suggest a peculiarly bimodal behavior: In some years, this season is characterized by a strong westerly jet, with intense but regular traveling weather fronts. In other years, planet-encircling dust storms develop, extending up to 50 km altitude, accompanied by an intensified Hadley circulation. The first of these appears to be analogous to the terrestrial baroclinic cyclone regime. The midlatitude deformation scale may be estimated from the parameters in Table 1 as $L_D \approx (g\Gamma_{ad}/2T)^{1/2}H\tau_{rot}/2\pi \approx 960$ km. Then, assuming the wavelength of the eddy cyclone pattern is again $4L_D$, and assuming a midlatitude circumference of $\sqrt{2}\pi a$ (where a is the planetary radius), the zonal wave number for baroclinic disturbances may be estimated as $\sqrt{2}\pi a/4L_D \approx 4$, in good agreement with that inferred from the analysis of pressure and wind data at the Viking landers. The global dust storm regime, occurring in occasional northern winters, near the time of perihelion, is perhaps the most remarkable feature of the martian meteorology. This undoubtedly involves some feedback between the radiative heating of the airborne dust, intensified horizontal thermal gradients and winds, and the resulting injection of more dust within the increasingly turbulent surface layer. This is an important problem for further investigation by modeling and spacecraft observations, and may serve as a safe and natural laboratory for the study of "nuclear winter" conditions.

Venus and Titan: Windy Overdrives

Venus is nearly the same size and density as the Earth and, although it intercepts about twice as much sunlight in its closer orbit to the Sun, the high reflectivity of its globally pervasive, high-altitude cloud deck gives it a slightly lower emission temperature. The Venus clouds are composed of sulfuric acid, condensed at an altitude of 65 km, and overlie a thick carbon dioxide atmosphere with a surface pressure of 92 bar and a corresponding "greenhouse" temperature of 730 K. Perhaps the most extraordinary feature of the atmosphere is its zonal superrotation, as measured by the tracking of ultraviolet cloud features in the orbiter imaging data and corroborated by the Doppler radio tracking of several descent probes. The motions amount to about 100 m s^{-1}, with global cloud patterns traveling completely around the planet in 4–5 days, in the same direction but some 50 times as fast as its own (243-day) sidereal rotation. The large Rossby number (~ 50) implies that the motions are in global cyclostrophic balance with the (remotely sensed) temperature field, which indicates a gradual reduction in the wind speed with altitude above the cloud deck. The superrotation represents an excess of angular momentum, as compared with that conserved by meridional motions in the absence of forcing. The excess must be supplied either externally by solar thermal tides or somehow internally by the viscous torque of the planet itself. Although the radiative cooling time near the surface is about 100 yr, at the cloud deck, near the 50-mbar level, it is comparable to the atmospheric rotation period and therefore plausibly supports the efficient diurnal pumping of Sun-following waves, which might provide the requisite momentum driving. According to the celebrated proposal by Peter Gierasch, the superrotation might alternatively be maintained by a global Hadley circulation, which vertically redistributes the viscous torque imposed on its lower branch by the planet's surface, assuming a strong eddy-diffusive transfer of angular momentum aloft from high to low latitudes. Although the Gierasch mechanism assumes an anisotropic mixing of momentum and a negligible mixing of heat, these requirements might be satisfied by "barotropic" eddies arising from horizontal shear instabilities in the global motion field. This idea is consistent with the estimated inefficiency of baroclinic eddy motions, given the largeness of the baroclinic deformation scale at the stable cloud-top levels, in comparison with the planetary radius, due to the slow planetary rotation. The details of the angular momentum mixing are not as yet understood, however, and might also involve some transfer by vertically propagating gravity waves.

Titan is the large satellite of Saturn, about $1\frac{1}{2}$ times the size of the Earth's moon. It has a nitrogen atmosphere, with a surface pressure of 1.5 bar, enshrouded with a thick hydrocarbon haze. The satellite is presumed to be tidally locked to Saturn, with a rotation period equal to its approximately 16-day orbital period. Although the absence of discrete cloud features has prevented the direct tracking of winds on Titan, some tentative information about the motions has been inferred from Voyager infrared measurements of latitudinal variations in the atmospheric brightness temperatures. These imply an equaltor-to-pole contrast of about 20 K at stratospheric levels (~ 1 mbar) and of about 2 K in the lower troposphere (around the 1-bar level). Assuming cyclostropic balance, as appropriate for the slow planetary rotation, and a characteristic flow depth of one scale height, the thermal wind relation implies a zonal velocity of $U \sim \sqrt{(gH\Delta T/T)} \approx 80$ m s^{-1} at upper levels and about 20 m s^{-1} below. Both estimates are consistent with a large Rossby number, as required for cyclostrophic balance, and suggest that Titan may represent a superrotational regime analogous to Venus. If this inference is borne out by further spacecraft observations, it will support a view that atmospheric superrotation is a robust feature of slowly rotating, differentially heated planets.

Jupiter and Saturn: Banded Giants

Jupiter and Saturn are each about 10 times the size of the Earth and rotate over twice as rapidly. Both emit about twice as much energy as they absorb, implying internally heated and deeply convective interiors. Although they differ in their axial tilt, gravitational potential, emission temperature, and visual contrast, the two hydrogen–helium gas giants appear to have qualitatively similar flow regimes. Both planets have high-velocity superrotating equatorial currents and at high latitudes an alternating, axisymmetric pattern of counter-flowing jet streams, as revealed by the longitudinal drift rate of their ammonia clouds. On Saturn, however, the currents are stronger, with equatorial and midlatitude velocities of about 500 and 100 m s^{-1}, respectively, as compared with 120 and 50 m s^{-1} at low and midlatitudes on Jupiter. (The frame of reference for wind velocities on the giant planets is provided by the radio measurement of the rotation of their magnetic fields, presumably tied to their deep interiors.) The predominantly axisymmetric character of the motions on both planets is emphasized by the visually prominent banding of their cloud features. On Jupiter the bright so-called zones generally correlate with regions of anticyclonic shear on the equatorial sides of the prograde jets, flanked by the darker cyclonic "belts." On Saturn the correlation of cloud brightness variations with the motions is more obscure. On both planets, however, infrared measurements indicate that the anticyclonic regions are relatively cool over their cloud tops, plausibly as a result of rising motions, whereas the cyclonic regions are relatively warm. The associated thermal wind shear implies a reduction of the cloud-top winds with altitude. In addition to the zonal cloud bands, the giant planets also show localized regions of swirling motion, or vorticity, such as Jupiter's Great Red Spot.

Unlike the bright synoptic-scale cloud swirls in the Earth's atmosphere, however, most of these tend to be anticyclonic against the planetary rotation.

Although the large-scale motions of Jupiter and Saturn are characterized by small Rossby numbers, comparable to that for midlatitude winds on the Earth, the multiplicity of the jets and the superrotational equators clearly represent a radically different dynamical regime. The differences are probably related in some way to the deep stratification. Unlike the terrestrial planets, the jovian atmospheres have no rigid lower boundaries to support strong horizontal pressure contrasts and are heated from below over depths of several scale heights. The depth of the zonal motion itself is a fundamental uncertainty. According to one view, the observed cloud-top winds extend throughout the molecular hydrogen envelope on convective cylinders of motion, concentric with the planetary spin axis. Alternatively, the motions might be confined to a shallow thermal wind layer, supported perhaps by thermal gradients associated with a latitudinally-variable water cloud, some 2–5 scale heights below the ammonia deck. Whatever the depth, the observed velocities appear to relate to the spacing of the jets according to

$$U \sim \beta \cdot L^2,$$

where $\beta \equiv 4\pi(\cos\lambda)a\tau_{\rm rot}$ is the so-called planetary vorticity gradient, with a denoting the planetary radius and $\cos\lambda$ the trigonometric cosine of the latitude. A second relation between velocity and the horizontal motion scale is needed to complete the specification of the dynamics. It is tempting to consider the possibility that the length scale is set by a baroclinic deformation radius L_D, but is difficult to estimate because the static stability at deep levels is unknown, but presumably small, whereas the depth scale D might be as large as several scale heights. It is interesting to note that at upper tropospheric levels on both Jupiter and Saturn the deformation radius may be estimated as $L_D \approx (g\Gamma_{\rm ad}/2T)^{1/2}H\tau_{\rm rot}/2\pi \sim$ 2000 km, comparable to the spacing of the jets. This may be only a coincidence, however, and at any rate is only diagnostic.

Uranus and Neptune: Topsy-Turvy Mysteries

Uranus and Neptune are roughly one-third the size of Jupiter and Saturn but are similarly constituted from predominantly hydrogen–helium mixtures, laced with significant traces of heavier elements. Both are sufficiently cold to effect the condensation of methane, which forms the top of their visual cloud decks. Unlike the other giant planets, Uranus appears to have a negligible internal heat source. Its rotation axis is peculiarly almost coincident with the plane of its orbit, so that on average its polar regions receive more sunlight than its equator, unlike any other planet in the solar system. Elementary reasoning, based on experience with the other planets, anticipates that this reverse solar forcing might induce a reverse meridional circulation, rising at high latitudes and sinking at the equator, along with a retrograde geostrophic wind. The *Voyager 2* observations tentatively indicate that there is indeed a retrograde (approximately 100 m s^{-1}) wind at the uranian equator but, surprisingly, also indicate an even stronger prograde flow (in excess of 200 m s^{-1}) at high latitudes. Neptune is only slightly more massive than Uranus but has a sufficient internal heat source to support almost the same emission temperature, despite its much greater distance from the Sun. Neptune has an upright 30° obliquity so that, as for Jupiter and Saturn, its weak solar forcing must be greatest at the equator. As if to confound further the systematic understanding of the outer planet atmospheres, however, the neptunian winds are if anything more similar to those on Uranus. As revealed by cloud-tracked wind observations during the 1989 *Voyager 2* reconnaissance, the equatorial velocity is 300 m s^{-1} retrograde, gradually diminishing toward the pole, with some prograde flow near 70° latitude, as measured with respect to its radio rotation rate. The dynamical maintenance of the zonal circulation of both planets eludes the grasp of any presently available theory.

FUTURE PROSPECTS

Despite the enormous progress of recent years in the study of planetary atmospheres, it is not yet possible to answer even the most fundamental questions regarding the maintenance of their dynamics and general circulation. The prospects for further observational constraints from the 1993–1995 *Mars Observer*, 1995–1997 *Galileo* (Jupiter), and 2002–2006 *Cassini* (Saturn and Titan) missions are encouraging, however. These will involve extended orbital coverage over several years and, in the case of Jupiter and Titan, the first in-situ measurements by descent probes. As the planetary arrivals of these new spacecraft are patiently awaited, it will be equally important to develop theories and computer models of increasing maturity and imagination.

Additional Reading

Allison, M., Beebe, R. F., Conrath, B. J., Hinson, D. P., and Ingersoll, A. P. (1990). Uranus atmospheric dynamics and circulation. In *Uranus*, J. Bergstralh and E. D. Miner, eds. University of Arizona press, Tucson, p. 251.

Hammel, H. B., Beebe, R. F., De Jong, E. M., Hansen, C. J., Howell, C. D., Ingersoll, A. P., Johnson, T. V., Limaye, S. S., Magalhaes, J. A., Pollack, J. B., Sromovsky, L. A., Suomi, V. E., and Swift, C. E. (1989). Neptune's winds speeds obtained by tracking cloud in *Voyager* images. *Science* **245** 1367.

Hunten, D. M., Tomasko, M. G., Flasar, F. M., Samuelson, R. E., Strobel, D. F., and Stevenson, D. J. (1984). Titan. In *Saturn*, T. Gehrels and M. S. Matthews, eds. University of Arizona Press, Tucson, p. 671.

Ingersoll, A. P. (1981). Jupiter and Saturn. *Scientific American* **245** (No. 6) 90.

Leovy, C. B. (1977). The atmosphere of Mars. *Scientific American* **237** (No. 1) 34.

Schubert, G. and Covey, C. (1981). The atmosphere of Venus. *Scientific American* **245** (No. 1) 66.

See also **Jupiter, Atmosphere; Mars, Atmosphere; Planetary and Satellite Atmospheres; Saturn Atmosphere; Uranus and Neptune, Atmospheres; Venus, Atmosphere.**

Planetary Atmospheres, Escape Processes

Siegfried J. Bauer

Planetary atmospheres can lose mass by several escape processes provided atmospheric particles are able to gain outward velocities greater than the escape velocity from the planet and suffer no further collisions. Such a situation prevails in the *exosphere*, the region of the atmosphere characterized by the requirement that the mean free path become greater than the characteristic scale length of exponential decay of density or pressure with height, the so-called scale height. This condition defines the columnar content of the exosphere $N_{\rm ex} \approx 3\times10^{14}$ cm^{-2}. (For comparison, the total number of particles in a column of unit cross section of the Earth's atmosphere is $N_0 \approx 3\times10^{25}$ cm^{-2}.) The critical level above which escape begins, the base of the exosphere, can thus be specified for any planet: For Earth the "exobase" lies at ~ 500 km, for Mars and Venus slightly below 200 km, and for Saturn's moon Titan at about 1400 km, whereas for Mercury and the Moon exospheric conditions prevail already at the surface. There are two classes of processes leading to atmospheric escape: thermal and nonthermal, according to the source of energy of the escaping particles.

THERMAL ESCAPE

Although the idea that atmospheric particles could escape by virtue of their thermal motion seems to have originated in the middle of the last century (attributed to J. L. Waterston), the thermal escape process was formulated mathematically about 75 yr later by James Jeans. According to his concept, the existence of particles having appropriate velocity components for hyperbolic orbits leads to an escape flux (now called the Jeans flux) that depends on the number density of the escaping constituent at the exobase and a so-called effusion velocity, which in turn is a measure for the portion of particles of a Maxwell–Boltzmann distribution with velocities necessary to escape the planet. The "efficiency" of escape depends on a parameter usually defined by the ratio of the square of the escape velocity to the most probable velocity of the Maxwell–Boltzmann distribution; it is thus directly proportional to the mass of the escaping constituent and the acceleration of gravity and inversely proportional to the *exospheric* temperature (the characteristic temperature of the upper atmosphere of a planet). Whereas an escape parameter (for hydrogen) of greater than 10 implies negligible escape, values of 4–8 typical of the terrestrial planets indicate that a fraction of up to a few percent of the Maxwell–Boltzmann distribution have escape velocities; the escaping constituent (usually hydrogen) is resupplied to the exosphere by diffusion from a reservoir below. Whereas the concept of thermal escape originated by Jeans is based on a kinetic theory approach treating individual particles, there is a critical value of the escape parameter of 1.5 (corresponding to the mean thermal energy of the escaping constituent being equal to its escape energy) when there is a transition to *hydrodynamic* escape. This critical situation was first recognized by Ernst J. Öpik, who termed this form of escape blow-off because of its high efficiency. Hydrodynamic escape is somewhat analogous to the solar wind, depending on the level where the gas motion becomes supersonic. This type of escape is sometimes invoked in the explanation of extensive mass loss from early atmospheres (e.g., from a possibly H_2O-dominated early atmosphere of Venus, implied by the measured high deuterium-to-hydrogen ratio). Hydrodynamic escape, however, is of no consequence for present day planetary atmospheres (with the possible exception of Titan), because typical exospheric temperatures lie well below the critical temperature required for dynamic escape (Table 1). On Titan such efficient escape of hydrogen, originating from the secondary hydrocarbon constituents, populates the huge hydrogen torus surrounding Titan's orbit around Saturn.

Thermal (Jeans) escape with escape fluxes smaller than 10^8 cm^{-2} s^{-1} has been found to be not the most important process for the mass loss from planetary atmospheres; a number of nonthermal processes lead to much more effective escape, even of atmospheric constituents heavier than hydrogen.

NONTHERMAL ESCAPE PROCESSES

Atmospheric particles can also gain escape energies by chemical processes, especially involving ionized constituents. An escape process that has been recognized as the most important one for the loss of hydrogen from the Earth is resonant charge transfer between energetic (hot) protons and (cold) hydrogen atoms. A hot proton having an energy greater than that required for escape from the gravitational field, as a charged particle, is confined to the Earth's magnetic field, but can escape upon exchanging its charge with a cold hydrogen atom. This charge exchange process leads to an escape flux 3–4 times larger than that due to Jeans escape. In the exosphere of Venus an ion–molecule reaction involving atomic oxygen ions and molecular hydrogen is the major source of hot hydrogen atoms.

Table 1. Escape Characteristics of Venus, Earth, Mars, and Titan*

	V_{esc} (km/s)	T_{ex} (K)	T_{crit} (K)
Venus	10.4	~ 300	~ 4000
Earth	11	~ 1000	~ 5000
Mars	5	~ 350	~ 1000
Titan	2.4	~ 190	~ 200

*V_{esc} is the escape velocity, T_{ex} the exospheric temperature, and T_{crit} the critical temperature for dynamic escape of hydrogen.

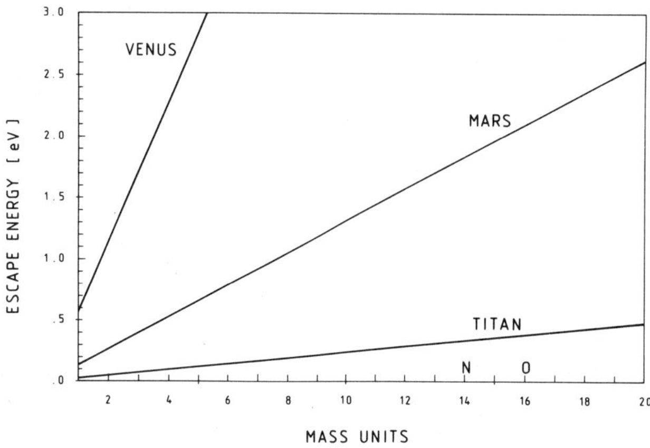

Figure 1. Required energy for atmospheric constituents (in atomic mass units) to escape from Venus, Mars, and Titan.

Additional nonthermal escape processes involving atoms even heavier than hydrogen are exothermic chemical reactions occurring in planetary ionospheres near the exobase level. Dissociative recombination of molecular ions can lead to atomic species reaching escape energy. (Typically 2–3-eV excess energy is made available to the resulting atomic products.) This process is of particular importance for Mars where the principal ionospheric constituent is O_2^+, whose dissociative recombination produces "hot" oxygen atoms, some of which are able to escape. Similarly, the dissociative recombination of the minor ion N_2^+ can also lead to the escape of nitrogen atoms. The anomalously high isotopic ratio of $^{15}N/^{14}N$ measured in the Mars atmosphere can be explained in this fashion, because escape is mass dependent.

In Titan's molecular nitrogen atmosphere, impact ionization by energetic electrons of Saturn's magnetosphere (Titan usually lies within Saturn's magnetosphere) produces energetic nitrogen atoms; the dissociative recombination of molecular nitrogen and nitrile molecule ions present in Titan's ionosphere also leads to hot nitrogen atoms capable of escape. Figure 1 shows the required energies for escape of atmospheric constituents from Venus, Mars, and Titan. Nonthermal chemical escape processes are responsible for atmospheric mass loss rates of the order of hundreds of grams per second (Earth and Venus) to kilogram per second (Mars and Titan). This nonthermal escape, if sustained over the age of the solar system, implies for Mars an atmospheric mass loss corresponding to several times the present mass of the martian atmosphere.

For essentially nonmagnetic planets (Venus and Mars), the solar wind interacting with the planetary ionosphere can also lead to mass loss by ion pickup. This solar wind "scavenging" is limited by the possible mass loading of the solar wind—usually about one-third of its mass flux (mass density times wind velocity)—and the solar wind/ionosphere interaction volume. For Venus, mass loss via solar wind ion pickup amounts to about 1 kg s^{-1}; for Mars this process seems to be about comparable to the loss via hot neutrals produced by exothermic chemical reactions. When Titan lies within the magnetosphere of Saturn its corotating plasma (a subsonic magnetospheric wind) can also remove ions from Titan's ionosphere, which in this case acts as a supplier of "heavy" ions found in Saturn's magnetosphere. At times when Titan is outside Saturn's magnetosphere, solar wind scavenging will contribute to its atmospheric mass loss.

Although energetic ions are confined to the Earth's magnetosphere (which is produced by the interaction of the solar wind with the geomagnetic field), above the polar caps where magnetic field lines extend into an open tail, such ions, particularly H^+ and He^+, can escape. Because of the small cross section of the polar caps, this escape (called polar wind) is not very important for mass loss but may play a role in the isotope fractionation of hydrogen and helium.

On Venus and on Mars heavier ions can also escape in the form of plasma clouds that become detached from the ionosphere, possibly due to accelerating processes, such as plasma instabilities. Such mass loss has been experimentally confirmed from observations by the U.S. *Pioneer Venus* orbiter and the recent U.S.S.R. *Phobos* mission to Mars.

Whereas thermal (Jeans) escape once was considered to be the principal process by which atmospheric mass loss—particularly over extended periods such as the age of the solar system—could occur, it is now recognized that nonthermal escape processes play the dominant role in the mass loss from planetary atmospheres.

Additional Reading

Bauer, S. J. (1973). *Physics of Planetary Ionospheres*. Springer, New York.

Chamberlain, J. W. and Hunten, D. M. (1987). *Theory of Planetary Atmospheres*, 2nd ed. Academic, Orlando, Fla.

Cheng, A. F. and Johnson, R. E. (1989). Effects of magnetosphere interactions on origin and evolution of atmospheres. In *Origin and Evolution of Planetary and Satellite Atmospheres*, S. K. Atreya, J. B. Pollack, and M. S. Matthews, eds. University of Arizona Press, Tucson, p. 682.

Hunten, D. M. (1982). Thermal and nonthermal escape mechanisms for terrestrial bodies. *Planet. Space Sci*. **30** 773.

Hunten, D. M., Donahue, T. M., Walker, J. C. G., and Kasting, J. F. (1989). Escape of atmospheres and loss of water. In *Origin and Evolution of Planetary and Satellite Atmospheres*, S. K. Atreya, J. B. Pollack, and M. S. Matthews, eds. University of Arizona Press, Tucson, p. 386.

Walker, J. C. G. (1977). *Evolution of the Atmosphere*. Macmillan, New York.

See also **Planetary Atmospheres, Ionospheres; Planetary Magnetospheres, Jovian Planets.**

Planetary Atmospheres, Ionospheres

Andrew F. Nagy

The ionosphere is considered to be that region of an atmosphere where significant numbers of free electrons and ions are present. All bodies in our solar system that have a surrounding neutral-gas envelope, due either to gravitational attraction (e.g., planets) or some other process such as sublimation (e.g., comets), have an ionosphere.

The first suggestions for the presence of charged particles in the terrestrial atmosphere were made more than 150 years ago. Karl Friedrich Gauss, Lord Kelvin, and Balfour Stewart hypothesized the existence of electric currents in the atmosphere to explain the observed variations of the magnetic field at the surface of the earth. In 1901, Guglielmo Marconi succeeded in sending radio signals across the Atlantic, which implied that radio waves were deflected around the earth in a manner not immediately understood. The following year, working independently, Oliver Heaviside in England and Arthur E. Kennelly in the United States proposed that a layer of free electrons and ions is responsible for the reflection of these radio waves. During the next 40 years, radio reflection techniques were used to remotely study these ionized, reflecting layers. The term ionosphere was coined by Robert A. Watson-Watt in 1926. The introduction of sounding rockets soon after World War II and satellites about a decade later led to tremendous advances in our understanding of the chemical and physical processes controlling our terrestrial ionosphere.

The first measurement of an ionosphere other than the terrestrial one was made by *Mariner 5*, which reached Venus on October 19, 1967 (*Venera 4* actually reached Venus a day earlier but did not make ionospheric measurements). Since that time, we have also obtained some information on the ionospheres of Mars, Jupiter, Saturn, Uranus, Neptune, the moons Titan and Triton, and comets Halley and Giacobini–Zinner. We also know that Mercury and Io have significant electron/ion populations; it is only a question of semantics whether to call these ionospheres. Of all the ionospheres in our solar system, that of Venus is the one that we know the most about other than the terrestrial one, mainly because the satellite *Pioneer Venus* has been observing it since 1978. The very basic processes of ionization, chemical transformation, and diffusive as well as convective transport are analogous in all ionospheres; the major differences are the results of factors such as differing background gas composition, intrinsic magnetic field strength, and distance from the Sun. In this entry, we will use the ionosphere of Venus as a representative example for describing the basic chemical and physical processes that control the behavior of all ionospheres. We will also give a brief summary of the ionospheres of Saturn and comet Halley as further examples of ionospheres in our solar system. The three ionospheres to be discussed are representative of those associated with (1) planets having no intrinsic magnetic fields, (2) those having strong dynamo fields, and (3) bodies in the solar system that have negligible gravitational attraction.

VENUS

Photochemistry

The major source of ionization at Venus is the solar extreme ultraviolet (EUV) radiation, which photoionizes the neutral gas constituents in the upper atmosphere. This photoionization process creates free electrons and ions, which can then undergo chemical reactions or be transported to some other regions of the ionosphere. At Venus the photoionization rate peaks at around 140 km above the surface of the planet. At this altitude, the major neutral atmospheric constituent is CO_2 along with about 10–20% of atomic oxygen. This predominance of CO_2 at the ionization peak led to early predictions that the main ion in the Venus ionosphere is CO_2^+; however, it was realized even before direct measurements could confirm it that chemical reactions quickly transform CO_2^+ to O_2^+. The main chemical reactions affecting the major ion species at these altitudes ($\sim$120–200 km) are:

$$CO_2 + h\nu \rightarrow CO_2^+ + e,$$

$$CO_2^+ + O \rightarrow O_2^+ + CO$$

$$\rightarrow O^+ + CO_2,$$

$$O^+ + CO_2 \rightarrow O_2^+ + CO,$$

$$O_2^+ + e \rightarrow O + O.$$

The last of these chemical reactions indicated above, called dissociative recombination, is the major loss process for ions. Figure 1 shows modeled and measured ion densities for the dayside ionosphere, indicating that the peak total ion (and electron) density is

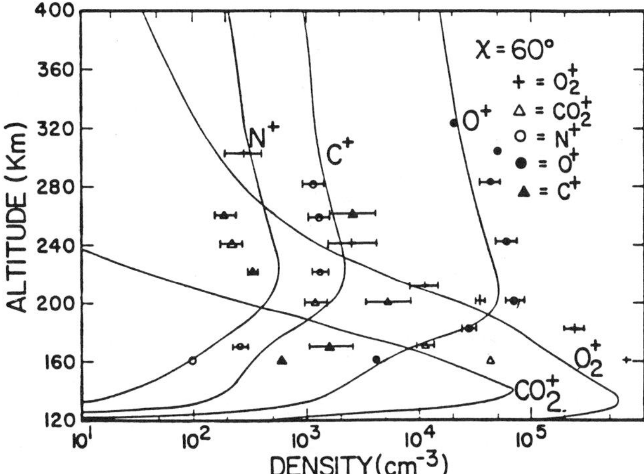

Figure 1. Calculated (solid lines) and measured ion densities in the Venus ionosphere for a solar zenith angle of 60°. [*By permission of the author; from Nagy et al.,* J. Geophys. Res. ***85*** *7795 (1980).*]

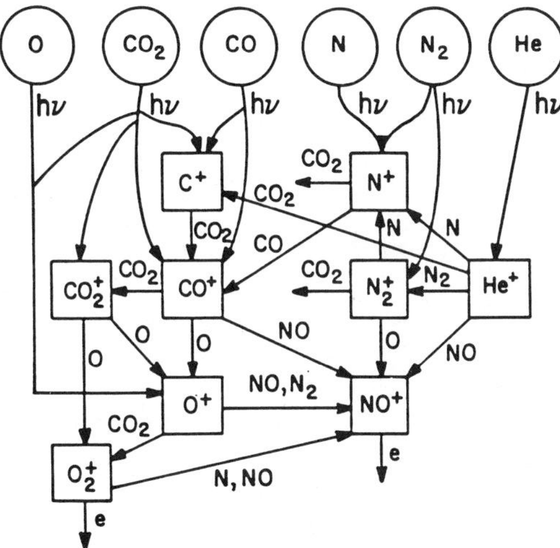

Figure 2. Block diagram of the major ion chemistry scheme for the Venus ionosphere. [*By permission of the author; from Nagy et al.,* in Venus, *ed. D. M. Hunten et al., University of Arizona Press (1983), p. 841.*]

near 140 km, that the major ion is O_2^+ and that CO_2^+ is truly a minor ion. That figure also shows that there are many other ion species present, which are the result of a large variety of photochemical processes, some of which involve metastable species. A block diagram of the main ion chemistry of Venus is shown in Fig. 2.

In the altitude region where photochemical processes dominate, the ion/electron densities can be calculated by simply equating the production and loss rates:

$$P_i = L_i.$$

As mentioned before, the dominant loss process is the dissociative recombination of O_2^+, and this combined with the fact that O_2^+ is the major ion below about 180 km, allows us to approximate the total ion/electron loss rate as

$$L_i = \alpha n_i^2,$$

where α is the dissociative recombination rate and n_i is the total ion density. Hence for the altitude region below about 180 km, where chemical processes dominate, we get the expression for the total ion/electron densities

$$n_i = \sqrt{P_i/\alpha}.$$

Topside Ionosphere

The lifetime for chemical interactions increases with decreasing densities and thus with increasing altitude. The chemical lifetime becomes long enough above about 200 km to allow transport processes, due to diffusion or bulk plasma drifts, to dominate. Venus has no significant intrinsic magnetic field, although at times of high solar wind dynamic pressure, a significant (~ 100 nT) horizontal induced magnetic field is present in the ionosphere. Therefore, except when the induced field is significant, plasma can move freely both vertically and horizontally. The vertical distribution of the ion/electron density near the subsolar region is believed to be controlled mainly by vertical diffusion, while horizontal plasma flows become dominant at larger zenith angles. Calculations, involving the solutions of the coupled continuity, momentum, and energy equations, have been carried out both in one and two dimensions to study these transport processes. One-dimensional magnetohydrodynamical calculations of the magnetized ionospheres have also been successful in reproducing the observed ion densities and magnetic fields. Ion velocity measurements have indicated that the horizontal plasma velocities increase with altitude and solar zenith angles, reaching a few kilometers per second at the terminator. The model calculations indicate that these measured velocities are driven by day-to-night pressure gradients.

There is a sharp break in the topside ionosphere at an altitude where the thermal plasma pressure is approximately equal to the magnetic pressure and their sum is equal to the dynamic pressure of the unperturbed solar wind outside the bowshock. This very sharp gradient in the ionospheric thermal plasma density is called the ionopause; at this altitude there is a transition from the plasma pressure to the magnetic pressure dominated region in an altitude increment, which at most times, is only a few tens of kilometers.

Nightside Ionosphere

The effective night on Venus lasts about 58 Earth days, during which time the ionosphere would be expected to disappear, because no new photoions and electrons are created to replace the ones lost by recombination. Therefore it was very surprising, at first, when *Mariner 5* found a significant nightside ionosphere at Venus. More recent and extensive measurements have confirmed the presence of a significant, but highly variable nightside ionosphere, with a peak electron density of about 2×10^4 cm^{-3}. Plasma flows from the dayside along with impact ionization caused by precipitating electrons are responsible for the observed nighttime densities.

SATURN

Our knowledge of the ionosphere of Saturn is much more limited than that of Venus. The flybys of Saturn by *Pioneer 11* and *Voyagers 1* and *2* provided us with a number of electron density profiles of its ionosphere; however, much remains still unknown, as the brief summary below will indicate.

The major neutral constituent in Saturn's upper atmosphere is H_2, therefore the major primary ion which is formed by either photoionization or particle impact ionization is H_2^+. The relative importance of photoionization is still not clear at this time. For the sake of brevity, we will only discuss photoionization in this section, because particle ionization leads to products which are similar to those created by photoionization. Photodissociation and ionization of the main neutral constituent, H_2, in the upper atmosphere lead to

$$H_2 + h\nu \rightarrow H + H$$

$$\rightarrow H_2^+ + e$$

$$\rightarrow H^+ + H + e.$$

The resulting neutral atomic hydrogen can also be ionized:

$$H + h\nu \rightarrow H^+ + e.$$

H^+ can only recombine directly via radiative recombination, which is a very slow process ($\sim 10^{-12}$ cm^{-3} s^{-1}). H_2^+ is very rapidly transformed to H_3^+, which then undergoes dissociative recombination:

$$H_2^+ + H_2 \rightarrow H_3^+ + H,$$

$$H_3^+ + e \rightarrow H_2 + H.$$

The dissociative recombination rate of H_3^+ is believed to depend very strongly on its vibrational excitation. As we do not know the vibrational distribution of H_3^+, this is an issue which needs further attention.

The ionospheric models which are based on the above chemical processes, predict an ionosphere, which consists mainly of H^+, because of its long lifetime ($\sim 10^6$ s). In these models, H^+ is removed by downward diffusion to the vicinity of the homopause (~ 1100 km), where it undergoes charge exchange with heavier neutral gas molecules, mostly hydrocarbons such as methane, followed by rapid dissociative recombination. The trouble with these "hydrogen only" upper atmosphere models is that:

1. The predicted electron density at the apparent main peak is about an order of magnitude larger than the observed value.
2. The altitude of the calculated ionospheric main peak is much lower than the observed one.
3. The predicted long lifetime of H^+ is inconsistent with the inferred diurnal variations of the electron density.

To overcome these difficulties, scientists have looked at the role of vibrationally excited neutral molecular hydrogen to reduce the calculated electron densities. H_2 in a vibrationally excited state greater than $v = 4$ is energetically capable of charge exchanging with H^+. The resulting H_2^+ can then rapidly recombine dissociatively. However this scenario requires significant vibrational excitation of the molecular hydrogen in the upper atmosphere, which does not appear to be the case. The most recent ionospheric models are based on the assumption that water molecules and water cluster ions "raining down" on the ionosphere from the rings around Saturn are causing the smaller densities. The reactions leading to the rapid loss of ions are:

$$H^+ + H_2O \rightarrow H_2O^+ + H,$$

$$H_2O^+ + H_2 \rightarrow H_3O^+ + H,$$

$$H_3O^+ + e \rightarrow H_2O + H.$$

Figure 3 shows the results of model calculations which include the above-mentioned processes along with electron density profiles obtained with the *Voyager 2* spacecraft, using the radio occultation technique. A block diagram of the photochemistry of Saturn is shown in Fig. 4 (it is also useful for understanding the main H_2O ion chemistry at comets).

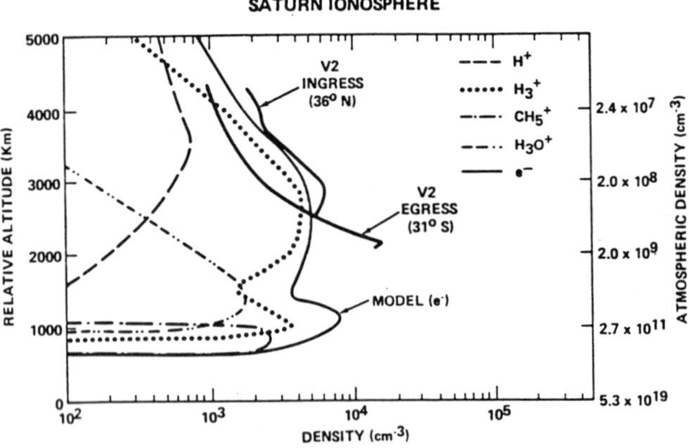

Figure 3. Calculated and measured (solid lines) ion densities for the ionosphere of Saturn. [*By permission of the author; from Waite and Cravens (1987).*]

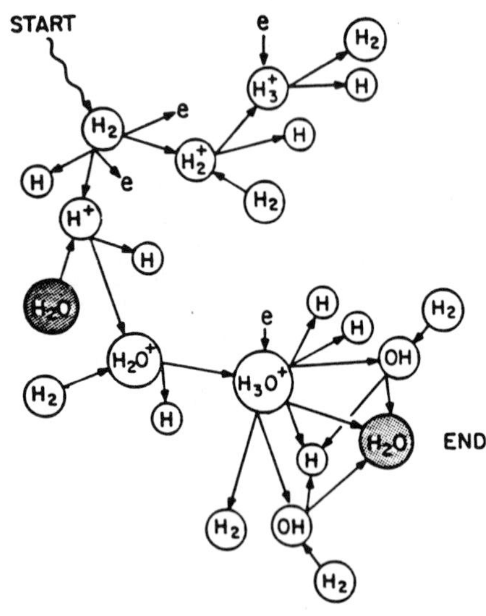

Figure 4. Block diagram of the major ion chemistry scheme, involving water, for the ionosphere of Saturn. [*By permission of the author; from Waite and Cravens (1987).*]

COMET HALLEY

The atmospheres of comets, commonly referred to as comas, are different from conventional planetary atmospheres in a number of important ways. The most important distinguishing chacteristics of comas are:

1. The lack of any significant gravitational force.
2. Relatively fast radial outflow velocities (~ 1 km s^{-1}).
3. The rapidly varying, time-dependent nature of their physical properties.

A direct consequence of the first two of these characteristics is the presence of a very extended neutral atmosphere around active comets such as Halley.

The neutral mass spectrometer instrument carried by the *Giotto* spacecraft, which came within about 600 km of the nucleus of comet Halley, established that water vapor was the major gaseous constituent of the atmosphere. More specifically it was found that water vapor accounts for about 80% of the gases escaping from the nucleus, with NH_3, CH_4, and CO_2, making up about a further 10%, 7%, and 3.5%, respectively (the first two values are upper limits). This instrument also obtained an estimate of about 900 m s^{-1} for the neutral gas expansion velocity.

The predominance of water vapor in the atmosphere of comet Halley means that the following photochemical processes are believed to control the behavior of the ionosphere:

$$H_2O + h\nu \rightarrow H_2O^+ + e$$
$$\rightarrow H^+ + OH + e$$
$$\rightarrow OH^+ + H + e,$$
$$H_2O^+ + H_2O \rightarrow H_3O^+ + OH,$$
$$H_3O^+ + e \rightarrow OH + H_2$$
$$\rightarrow OH + H + H$$
$$\rightarrow H_2O + H.$$

The very rapid rate at which H_2O^+ transforms to H_3O^+ means that in comets which have water-dominated atmospheres, such as Halley, H_3O^+ is the dominant ionospheric constituent. Model calculations have shown that the electron density varies roughly as $1/r$, where r is the radial distance from the nucleus, under both photochemical and transport-controlled conditions, as long as the transport velocity is constant.

The *Giotto* spacecraft carried two spectrometers which were capable of measuring the ion composition in Halley's ionosphere. The neutral spectrometer, operating in its ion mode, found that the H_3O^+ to H_2O^+ ratio increases with decreasing distance from the nucleus and it exceeds unity at distances less than about 20,000 km. The unperturbed ionosphere proper is located inside the contact surface, which was found to be located at about 4700 km, at the time of the *Giotto* encounter. The detailed variations of the different ion densities measured by the ion mass spectrometer, which are shown in Fig. 5, are in qualitative agreement with model calculations of the ion composition and structure. Our understanding of the energetics and dynamics of cometary ionospheres is more limited; model studies are continuing to advance our understanding of these processes, but more data are necessary to quantify the controlling processes.

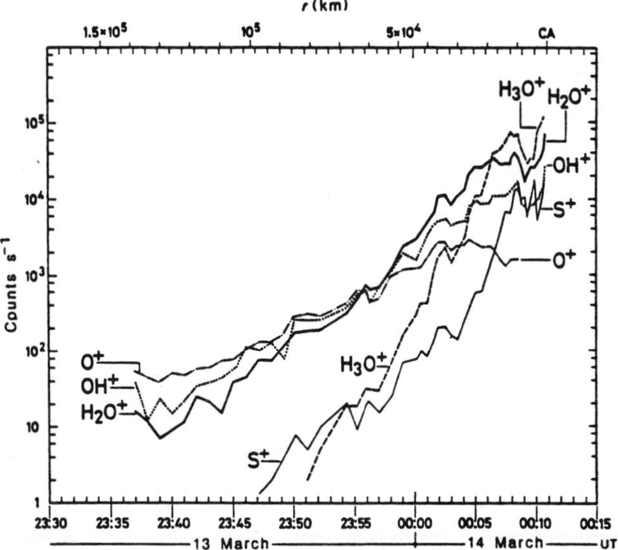

Figure 5. The measured ion densities at Comet Halley. [*By permission of the author; from Balsiger et al.,* Nature **321** 330 *(1986).*]

SUMMARY

As indicated at the beginning of this entry, the three planetary ionospheres selected for discussion were intended to demonstrate some of the basic similarities (e.g., role of photochemistry and diffusion) of planetary ionospheres, while also emphasizing some major differences (e.g., the specifics of the photochemical processes, magnetic field effects). The best explored planetary ionosphere, besides the terrestrial one, is that of Venus and even in the case of Venus many of the basic processes are still not well understood. Significantly more information needs to be obtained in the years ahead to assure that we have a good understanding of the physics and chemistry of planetary ionospheres.

Additional Reading

Barth, C. A., Bauer, S., Bougher, S. W., Hunten, D. M., Nagy, A. F., and Stewart, A. I. F. (1991). *Aeronomy, Mars.* University of Arizona Press, Tucson.

Brace, L. H. and Kliore, A. J. (1991). The structure of the Venus ionosphere. *Space Sci. Rev.*

Nagy, A. F. (1987). Photochemistry of planetary ionospheres. *Adv. Space Res.* 7 89.

Schunk, R. W. and Nagy, A. F. (1980). Ionospheres of the terrestrial planets. *Rev. Geo. Space Sci.* **18** 813.

Waite, J. H., Jr. and Cravens, T. E. (1987). Current review of the Jupiter, Saturn and Uranus ionospheres. *Adv. Space Res.* 7 (No. 12) 119.

See also **Comets, Atmospheres; Earth, Atmosphere; Saturn, Atmosphere; Venus, Atmosphere.**

*Planetary Atmospheres, Solar Activity Effects**

Alan E. Hedin

The search for possible changes in planetary atmospheres that are related to the sunspot cycle has been of interest for more than a century. In recent years, artificial satellites have provided detailed knowledge on solar emissions, their variations, and the parts of an atmosphere that are the most responsive. Only the atmospheres of Earth and Venus have been examined with sufficient detail to reveal solar activity variations and the clearest correlations with solar activity are found in the upper atmospheres.

The Sun is the primary external source of energy determining the structure and dynamics of planetary atmospheres. Most of the solar energy arriving at the planets is electromagnetic radiation in the visible, near-infrared, and near-ultraviolet wavelengths. Part of this energy is reflected back to space and most of the rest is absorbed in the lower atmosphere (troposphere) or at the surface. The variability of the solar radiation at these wavelengths is very small and thus related variations in an atmosphere would be difficult to distinguish from natural meteorological variations.

The solar radiation at short wavelengths [ultraviolet (UV), extreme ultraviolet (EUV), and x-ray] and long wavelengths (radio), however, varies with both the 11-yr sunspot cycle and with the rotation period of the Sun (approximately 27 days). UV and EUV radiation are absorbed in the upper atmospheres of the planets, the stratosphere, mesosphere, and particularly in a region called the thermosphere (above 85 km on Earth). Although the energy available in this wavelength region is very small compared to the energy at visible wavelengths, the effect of variations on the thin upper atmosphere can be quite dramatic. A planetary thermosphere is the region where absorbed radiation is balanced by thermal heat conduction to the lower atmosphere and this region is very sensitive to any change in solar emission at the absorbed wavelengths.

The absorption of EUV and x-ray radiation in the thermosphere produces ions and electrons (the ionosphere) that are responsible for reflecting radio signals. The degree of ionization increases from sunspot minimum to maximum with significant implications for radio communication. With the radio occultation technique, the electron density and scale height of a planetary atmosphere can be determined by analyzing the effects of radio signals from planetary probes as they pass behind a planet.

Solar radio emissions, although they have no direct effect on planetary atmospheres, are frequently used as diagnostic indices of UV and EUV radiation because they can be measured routinely from the ground and have been shown to be highly correlated with variations in the UV and EUV wavelength regions of the solar spectrum which can only be observed from a satellite. The emission at 10.7 cm is frequently used for this purpose because it is emitted from some of the same outer layers of the Sun that emit UV and EUV radiation. Although it is not a perfect index for any particular wavelength, it has been found to be a good index of the overall effects of UV and EUV light in the atmosphere.

In addition to electromagnetic emissions from the Sun, there are high-energy particle emissions associated with solar flares, and solar wind particles and magnetic fields which are highly variable and interact with the thermosphere, ionosphere, and magnetosphere of a planet. This interaction often results in a magnetic storm with associated aurora and other phenomena. Although solar activity in general encompasses all the various emissions related to sunspots and active regions on the Sun, this term is also used in a more restrictive sense to refer only to variations in the UV and EUV emissions, excluding magnetic storm effects. The storm effects in the upper atmosphere, characterized by special indices, are distinctly different from those caused by variations in UV radiation.

EARTH

Astronomers studying the orbits of the first artificial Earth satellites, which traversed the thermosphere and the even more distant exosphere, soon noticed systematic changes in orbital periods resulting from atmospheric drag on the satellites, even though the density there is very low (roughly one hundred thousand millionth of sea-level density near 300 km). Subsequently, Luigi Jacchia of the Smithsonian Astrophysical Observatory and Wolfgang Priester of the University of Bonn found that the atmospheric drag varied with an approximate 27-day period and they correlated this variation with variations in the solar radio flux.

In addition to studies of atmospheric drag on satellites, ground-based studies of the thermosphere by incoherent scatter radar have shown that the ion and neutral temperatures are correlated with the 10.7-cm flux. Satellite-borne accelerometers and mass spectrometers and Fabry–Perot interferometers from both ground and space have made detailed measurements of the density, composition, temperature, and wind changes in the thermosphere related to solar activity including magnetic storms.

Changes in the thermosphere correlated with short-period (27-day) variations in the 10.7-cm solar flux are found to be two to three times smaller than thermospheric variations associated with a long-term (solar cycle) variation in the 10.7-cm flux of similar magnitude. Direct measurements of solar EUV from the Atmospheric Explorer satellites showed that this same relation held between the EUV and the radio flux. The processes on the Sun that are responsible for the 27-day variations and the long-term solar cycle variations are different, thus producing different mixes of EUV and radio flux. To reflect these differences, long- and short-term variations are usually separated by using the running average of the solar flux over three (or more) 27-day periods as indicative of the long-term variations. The difference between the daily flux and the average flux is used to represent the short-term variations.

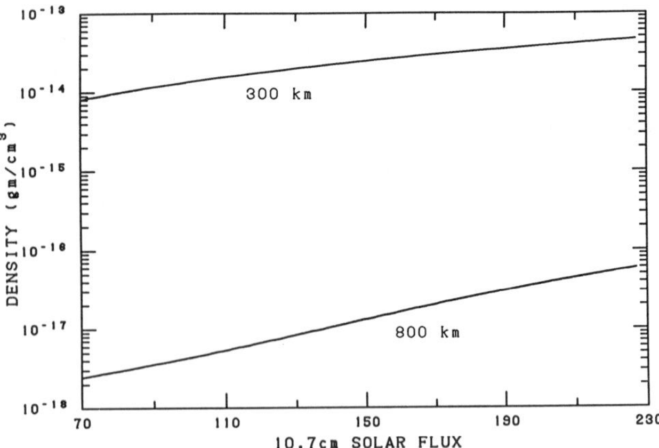

Figure 1. Global and time-averaged total mass density (g cm^{-3}) as a function of the 81-day averaged solar 10.7-cm flux (10^{-22} W m^{-2} Hz^{-1}) at 300 and 800 km, calculated from the COSPAR International Reference Atmosphere 1986.

Figure 1 shows examples of the global average density variation at 300 and 800 km from near sunspot minimum (10.7-cm flux index of 70) to a rather high sunspot maximum (10.7-cm flux index of 230). The increase in total mass density in the thermosphere, for a given increase in solar activity, is much greater at higher altitudes. The thermospheric temperature increases on average from 740 K at sunspot minimum to 1250 K at sunspot maximum. In the ionosphere, daytime peak electron densities increase around a factor of 5. The increased neutral density in the thermosphere at solar maximum significantly reduces the lifetime of artificial satellites in low Earth orbit and the lack of a good quantitative predictive capability for future sunspot cycles is a significant complication for satellite orbit selection.

Magnetic storms are a manifestation of the interaction of the solar wind with the Earth's magnetosphere. During a storm there are large increases in energetic particle fluxes entering the thermosphere from the magnetosphere and corresponding increases in currents flowing in the ionosphere and magnetosphere, both of which act to heat the thermosphere preferentially at high latitudes. Global wind systems redistribute this energy so that the density of the thermosphere increases over the whole globe. Magnetic storm effects are second only to solar cycle variations in their effects on density and temperature in the thermosphere. The lack of a good predictive capability for storm occurrence, coupled with their sudden onset and localized effects in the thermosphere, make magnetic storm effects as important as the general sunspot cycle for satellite operations.

The composition of the thermosphere changes in a complex pattern from sunspot minimum to sunspot maximum and during magnetic storms. The changes depend strongly on altitude. In general, heavier gases such as molecular nitrogen and oxygen increase relative to the lighter gases such as atomic oxygen and helium during sunspot maximum and during magnetic storms. Hydrogen, however, decreases from sunspot minimum to sunspot maximum in the thermosphere.

Systematic winds blow in the thermosphere from the day to night side of the Earth with speeds ranging from 100 m s^{-1} near the equator to several hundred meters per second at high latitudes. The effects of the solar cycle on these winds is fairly complicated. The increase in the horizontal pressure gradients at times of higher EUV flux should lead to increased wind speeds, but the winds in the thermosphere are slowed by friction with ions of the ionosphere whose density has also increased. The net effect on the winds is small with a complicated dependence on season and location. The winds at high latitudes are also driven by a complex pattern of ion drifts that are produced by the interaction of the solar wind with the magnetosphere. These processes are more effective during high solar activity. Changes in the direction of the interplanetary magnetic field associated with the solar wind produce changes in the ion convection pattern and, in turn, the neutral wind pattern at high latitude. The relatively localized heating of magnetic storms can also produce rather large winds (observed up to 1000 m s^{-1}).

Solar activity effects in the thermosphere are clearly present and have been quantitatively related to variations in solar emissions using theoretical models. At altitudes below the thermosphere, solar activity variations are more controversial. Their amplitude is apparently small and more easily masked by other natural variations and quantitative physical mechanisms have not been clearly established. Correlations of weather with sunspots have been claimed, but have often not withstood the test of new data. A 22-yr cycle in severe midwest droughts is an example of a possibly real solar cycle influence. An extreme example of sunspot influence may be a persistent period of abnormally low temperatures in Europe coincident with a near absence of sunspots in the Maunder Minimum (1645–1715).

Satellite measurements in the stratosphere and mesosphere indicate a probable correlation of ozone with 27-day variations in solar activity. High-energy particles from the Sun (solar proton events) have also been shown to disturb the chemistry of the mesosphere. Nitric oxide, produced in the thermosphere at high latitudes during magnetic storms, is likely to be carried by circulation into the mesosphere where it will affect the chemistry there.

Recently, it has been discovered that solar cycle effects in the stratosphere and troposphere are more clearly seen if data are separated according to the phase of a 2-yr change in average winds in the stratosphere (the quasi-biennial oscillation). The physical mechanism for this connection is unknown, but it is one of the more promising and exciting results to date.

VENUS

The only other thermosphere whose density, temperature, and composition have been measured extensively is that of Venus. Mass spectrometers carried on the *Pioneer Venus Orbiter*, as well as analysis of the tracking data from this satellite for atmospheric drag, demonstrated that temperature and density correlate with the 10.7-cm index in the same sense as they do on Earth. The variation in temperature for a given change in 10.7-cm flux is only 10% of the variation in the Earth's thermosphere, because of different heat conduction and infrared emission properties for the different mix of gases. Unlike Earth, solar cycle effects have only been detected on the day side of Venus. The night-side thermosphere is very cold and highly variable for reasons not completely understood. The peak daytime electron density changes by about 50% from sunspot minimum to maximum.

Venus has no detectable intrinsic magnetic field and, as a result, the solar wind has a more direct effect on the ionosphere than on Earth. Increases in solar wind pressure result in compressing the altitude of the ionopause (where electron and ion densities drop rapidly) closer to the planet. The solar wind also apparently strips ionospheric plasma from the top of the sunlit ionosphere and carries this plasma to the night side and into interplanetary space.

Additional Reading

Hedin, A. E. and Mayr, H. G. (1987). Solar EUV induced variations in the thermosphere. *J. Geophys. Res.* **92** 869.

Herman, J. R. and Goldberg, R. A. (1978). *Sun, Weather, and Climate*. National Aeronautics and Space Administration, Washington, DC.

Hunten, D. M., Colin, L., Donahue, T. M., and Moroz, V. I., eds. (1983). *Venus*. University of Arizona Press, Tucson.

Jacchia, L. G. (1975). The Earth's upper atmosphere. II. *Sky and Telescope* **49** 229.

Jastrow, R. (1959). Artificial satellites and the Earth's atmosphere. *Scientific American* **201** (No. 2) 37.

Mahajan, K. K., Mayr, H. G., Brace, L. H., and Cloutier, P. A. (1989). On the lower altitude limit of the venusian ionopause. *Geophys. Res. Lett.* **16** 759.

See also **Earth, Atmosphere; Planetary Atmospheres, Ionospheres; Planetary Atmospheres, Structure and Energy Transfer; Solar Activity; Sun, Coronal Holes and Solar Wind; Venus, Atmosphere.**

Planetary Atmospheres, Structure and Energy Transfer

Laurence M. Trafton

In a general sense, atmospheric structure refers to the spatial variation of physical variables in a planetary atmosphere. These include the variation of temperature, pressure, density, and mean molecular weight with altitude at arbitrary locations on the planet. When winds are present or turbulence exists, the wind field and the boundaries separating convective and radiative zones may be thought of as describing aspects of atmospheric structure even though they are primarily related to the energy transfer. More generally, the dynamical state of the atmosphere influences the atmospheric structure. Hence global circulation patterns, static stability, and eddy diffusion help to characterize the structure.

Energy transfer pertains to the processes that occur as a result of solar and planetary heating of the atmosphere (including its haze and aerosol particles) and the radiation of heat to space by the atmosphere and planetary surface. Other sources of heating are not important on a global scale for most atmospheres. However, galactic cosmic rays contribute as much as solar extreme ultraviolet radiation to the ionospheres of Uranus and Neptune because of the great distance of the Sun. Moreover, Jupiter's polar regions exhibit hot spots from the local dumping of auroral particles into the atmosphere. Volcanic effluence on Jupiter's innermost Galilean satellite Io may be the source of an SO_2 atmosphere that is localized to the vicinity of the active volcanos because of the tendency of the gas to freeze out at the ambient temperature. Energy transfer processes include radiation, conduction, phase changes such as evaporation, cloud formation, and precipitation, and dynamical effects such as winds, turbulence, and convection. Most processes occur in local thermodynamic equilibrium but nonequilibrium phenomena are found to occur at very high altitudes. These include diffusive separation, vibrational relaxation, and atmospheric escape. Diffusive separation of the lighter and heavier gases occurs at the homopause, where the diffusion coefficient for the atmosphere equals the atmospheric eddy diffusion coefficient. Vibrational relaxation occurs where collisions are inadequate to repopulate vibrationally deexcited energy levels of molecules. Atmospheric escape causes a truncation of the high-velocity tail of the molecule velocity distribution near the exobase and leads to a deviation of the density from that predicted by the barometric equation.

HYDROSTATIC EQUILIBRIUM

Most planetary atmospheres are very close to being in hydrostatic equilibrium where inertial and dynamical forces are small enough that the pressure difference between two nearby altitudes equals the weight per unit area of the enclosed gas column. A possible exception is the atmosphere of Io which is thought to be localized near active volcanos that forcefully replenish the rapidly freezing and escaping atmosphere. As a consequence of hydrostatic equilibrium, the thicknesses of the visible portions of most planetary atmospheres are small compared to the apparent planetary radius and the atmospheric structure does not vary appreciably over a local region. Under these circumstances, the atmosphere may be approximated by horizontal plane parallel layers where variations occur only with height. The scale height in such an atmosphere is defined by

$$H = kT/\mu mg,$$

where T is the local temperature, k is Boltzmann's constant, μ is the mean molecular weight, m is the proton mass, and g is the local acceleration due to gravity. For an isothermal atmosphere, this is the altitude range over which the pressure falls by the factor e. This result also holds for an extended, spherically symmetric atmosphere if g is computed for the appropriate distance from the planet. Scale heights can be determined observationally from the atmospheric differential refraction of starlight which takes place during a stellar occultation (or the differential refraction of a flyby spacecraft's radio signal) or else measured directly by a spacecraft atmospheric probe. An example is Pluto which has the smallest radius, 1145 km, of any moon or planet known to possess an atmosphere. Pluto's atmospheric scale height has been determined by stellar occultation to be 60 km at the occultation radius, 1214 km. The scale height of the other planets are of this order of magnitude but their radii, mass, and gravity are larger so that the plane–parallel atmosphere approximation is more valid for them. From the observed scale heights, the T/μ ratio can be determined near the occultation level. Determination of the temperature then yields the mean molecular weight for the atmosphere.

TEMPERATURE STRUCTURE

In the scientific literature, atmospheric structure is frequently presented as the variation of temperature with pressure. From this, the mean molecular weight, and the equation of state, the variation of density with pressure can be found. The altitude dependence of pressure is determined from the equation of hydrostatic equilibrium. In the absence of direct measurement of the atmosphere by a spacecraft probe, determination of the temperature variation with pressure depends on knowing the thermal opacity of the gas; that is, the absorption coefficient at thermal wavelengths. This requires knowledge of the atmospheric composition and radiative transfer. Composition has been largely determined from Earth-based and spacecraft-based spectroscopy. Remote sounding in conjunction with radiative transfer models using the wings of strong molecular absorption bands, such as the 8-μm CH_4 band, has yielded the temperature variation over an extended pressure range for most atmospheres. This works because thermal radiation at a series of wavelengths selected where the opacity in an absorption band becomes progressively weaker originates from (and therefore probes) progressively deeper pressure levels.

Thermal opacities heat a planetary atmosphere by a process similar to the greenhouse effect. Like a greenhouse, planetary atmospheres are fairly transparent at visible wavelengths where the solar radiation is strong. The absorbed sunlight is converted to heat and reradiated deep within these atmospheres where the opacity at thermal (infrared) wavelengths is high, as for the glass in a greenhouse. Because the radiation of heat to space is restricted by the thermal opacity, the local energy density builds up until the heat leaking to space balances the absorbed sunlight. The increased energy density is manifested as a temperature increase which becomes greater with depth owing to the greater difficulty for heat to radiate to space directly from deeper regions of the atmosphere. The temperature increases with depth until the maximum depth of sunlight penetration is reached. In the absence of any source of heat interior to the planet, such atmospheres would then become isothermal at the elevated temperature level.

For Venus, CO_2 provides the major thermal opacity. Although the atmosphere is translucent owing to thick clouds, the scattering of sunlight in the clouds results in sufficient heating of the deep atmosphere to provide a strong greenhouse effect and to produce high temperatures on the surface. For Mars, CO_2 is also the dominant thermal opacity (in the absence of dust storms) but the martian atmosphere is thin because the CO_2 is thermally buffered in the polar caps. A relatively weak greenhouse effect therefore results for Mars. In the case of the deep atmospheres of the major planets (Jupiter, Saturn, Uranus and Neptune), the dominant thermal opacity is the pressure-induced absorption of H_2 and its enhancement by He. Hydrogen dominates the composition of these atmospheres and He is the second most abundant gas with mass fractions ranging from 6–26%. The major planets also contain significant amounts of CH_4 and in some cases NH_3. These minor constituents are ineffective sources of thermal opacities because their strong absorption bands lie on the fringes of the Planck function at the temperatures of these atmospheres, 80–135 K, for which radiative equilibrium dominates and for which these gases are not frozen out. Jupiter, Saturn, and Neptune are known to have internal heat sources with strengths roughly comparable to their solar heating. Consequently, their temperatures must keep increasing with depth down to the depths of the internal heat source. Such a source of heat has not been detected for Uranus; its fluid interior may be considerably cooler than the interiors of the other major planets.

In real planetary atmospheres, sunlight is absorbed at various altitudes owing to absorption by various gases. Absorption by the near-infrared CH_4 bands in the atmospheres of the major planets causes their stratospheric temperatures to increase with altitude rather than to decrease as predicted by the greenhouse model. This is because the pressure-induced thermal opacity, which is proportional to the square of the pressure, becomes ineffective for radiating heat at high altitudes (poor absorbers make poor emitters). Most of the heat is radiated by the 8-μm CH_4 band which is located on the outer fringe of the thermal spectrum. This stratospheric trapping of solar energy causes the temperature to build up with increasing altitude.

RADIATIVE, CONVECTIVE, AND DYNAMICAL EQUILIBRIUM

At those upper tropospheric levels where the absorption of sunlight is negligible, the diurnally averaged atmospheric structure is approximated by radiative equilibrium. Within these regions, the thermal flux is constant. Model atmospheres employing radiative equilibrium provide a reasonably accurate description of the variation of temperature with pressure at those atmospheric levels for which radiative equilibrium is a good approximation, that is, below the stratospheric temperature inversion and between any convectively unstable regions. The accelerating increase of temperature with pressure fostered by the thermal opacity in the atmospheres of the major planets soon causes the deeper regions of these atmospheres to become convectively unstable. This occurs when the temperature of a rising parcel of gas cools adiabatically by a lesser amount than the decrease in the ambient air temperature. The resulting decrease in density relative to the surrounding air causes the parcel to rise further leading to positive feedback and convection. When the mean molecular weight is changing with depth, as in the case of Uranus owing to methane's being frozen out of the upper atmosphere, this argument is still valid provided that all references to temperature are replaced by references to potential temperature. This effect causes the top of Uranus' convective zone to lie much lower than it would otherwise lie.

Unlike the situation in stellar atmospheres, the radiative portion of the atmosphere has little influence on the convective gradient, $d\ln T/d\ln P$, which is set essentially by the thermal properties of the gas. Hence convection carries essentially all of the thermal flux when it occurs. Hybrid radiative–convective models terminate the radiative $T(P)$ solution at the top of the convective boundary and continue it with the convective solution. These have been shown to give accurate representations of $T(P)$ in the atmospheres of all the major planets except possibly Uranus (which has greater stability against convection) when modified to include the physics of the temperature inversion and the effect of the increasing mean molecular weight with depth in the Uranian atmosphere. This agreement occurs in spite of the much longer radiative time constants (years) in the radiative portions of these atmospheres than dynamical time constants (hours). Of course, the time constant for phase changes can be quite short because no temperature change is required. The reason for the agreement pertains to the low static stability for these planets, a consequence of their internal heat sources and large planetary scale. Therefore, the mean temperature and horizontal temperature gradients should change slowly during which dynamical quantities rapidly relax over the dynamical time scale and then slowly relax to the final equilibrium state over the radiative time scale. Hence radiative–convective models with seasonally varying parameters should represent the structure of most major planets adequately.

The Richardson number is the most important parameter governing the dynamical regime:

$$\mathrm{Ri} = \frac{g\,\partial\theta/\partial z}{T(\partial u/\partial z)^2}.$$

Here, θ is the potential temperature, z is the altitude relative to an arbitrary reference level, and u is the zonal velocity. Large negative values imply vigorous small-scale convection and large positive values imply the most stable possible regime. For $\mathrm{Ri} > 1$, the dominant kind of instability is geostrophic baroclinic instability such as occurs on Earth and Mars. This kind of regime can occur only at high latitudes on Jupiter; its lower-tropospheric latitudes are fundamentally unstable.

CRYOGENICALLY BUFFERED ATMOSPHERES

A special case arises with the atmospheres that are cryogenically buffered, that is, those of Mars, Io, Triton, and Pluto. Here, seasonal variations can control the atmospheric mass because it is supported by the vapor pressure of volatile ices on the surface. The vapor pressure is controlled by the temperature of these ices which in turn is controlled by the seasonal insolation. Only a small temperature change is required to produce a substantial change in the atmospheric bulk of the cooler bodies. When the atmosphere is thick enough to envelop the planet uniformly, which is the case for all of these bodies except Io, then the surface temperature of the volatile ices is regulated to a globally uniform value. This is accomplished by sublimation from the subsolar or high-insolation regions and deposition at polar or low-insolation regions. Sublimation cools the subsolar region through the latent heat of sublimation and heats the night or dimly lit regions through the latent heat released during freezing. This process is most dramatic for those bodies having large obliquities, Pluto and Triton. In the case of Triton, whose orbit precesses about Neptune, the process causes a complicated seasonal pattern consisting of a series of summers and winters with gradually increasing severity followed by solstices of gradually decreasing severity. Triton should be approaching a major summer early in the twenty-first century.

Additional Reading

Atreya, S. K. (1986). *Atmospheres and Ionospheres of the Outer Planets and Their Satellites*. Springer-Verlag, New York.

Barbato, J. P. and Ayer, E. A. (1981). *Atmospheres*. Pergamon, New York.

Chamberlain, J. W. and Hunten, D. M. (1987). *Theory of Planetary Atmospheres: An Introduction to Their Physics and Chemistry*. Academic, Orlando.

Conrath, B. J., Hanel, R. A., and Samuelson, R. E. (1989). Thermal structure and heat balance of the outer planets. In *Origin and Evolution of Planetary and Satellite Atmospheres*, S. K. Atreya, J. B. Pollack, and M. S. Matthews, eds. University of Arizona Press, Tucson, p. 513.

Marov, M. Y. (1978). Results of Venus missions. *Ann. Rev. Astron. Ap.* **16** 141.

Trafton, L. M. (1981). The atmospheres of the outer planets and satellites. *Rev. Geophys. Space Phys.* **19** 43.

See also **Planetary and Satellite Atmospheres; Planetary Atmospheres, Dynamics; Satellites, Ices and Atmospheres; Uranus and Neptune, Atmospheres.**

Planetary Formation, Gaseous Protoplanet Theories

Peter Bodenheimer

The giant planets Jupiter, Saturn, Uranus, and Neptune, which have a substantial gaseous component, formed by accumulation and condensation of matter from the solar nebula, a disk-like, rotating structure of gas and dust that emerged in the same process by which the Sun itself was formed. Two general theories for their origin have been extensively discussed. The first involves gravitational instability of the gas in the solar nebula. Under the proper conditions of density and temperature, elements of gas with mass comparable to those of the giant planets become gravitationally bound and can form subcondensations in the nebula. A condensation evolves through a contraction by a factor of several thousand to its present size. The second theory involves the gradual accretion of small particles to form a solid core, which later gravitationally attracts a gaseous envelope from the nebula. Upon the termination of accretion, the object evolves at constant mass to its present state. Although both theories have strengths and weaknesses, and neither is entirely satisfactory in explaining all observed properties of the present-day giant planets, the second theory is now regarded as more appropriate.

IMPORTANT QUESTIONS

The formation of the giant planets is of interest for a variety of reasons. For example, with regard to star formation, one might ask whether the process of formation of a planet like Jupiter was similar to that of a binary stellar companion, only with an exceptionally large mass ratio, or whether the giant planets formed by a completely different process. Furthermore, the processes that resulted in the formation of a giant planet accompanied by its regular satellite system may provide clues regarding the formation of the solar system itself. Did the giant planets each form with a surrounding disk composed of material from the solar nebula, or was the material that was to form the satellites originally included as part of the contracting interior of the planet and later left behind in the form of a disk as a consequence of angular momentum transport? Of observational interest is the fact that giant planets evolve through an early stage when their energy output (primarily in the infrared) is six orders of magnitude larger than the present values, suggesting the opportunity for the detection of planetary systems around young stars. Can the radiation from a forming planet be distinguished from that of the still-present nebular disk? Are the disks now deduced to exist around young stars actually suitable locations for the formation of giant planets? Further important questions include the following: What was the formation time scale of the giant planets compared with that of the terrestrial planets and with the lifetime of the solar nebula? How can the similarity of the masses of the rock–ice components among the giant planets and the wide variation of the hydrogen–helium components be explained? How are the differing chemical compositions in the atmospheres of the giant planets explained? How can we explain the extreme inclination (98°) of the rotation axis of Uranus relative to its orbital plane?

OBSERVATIONAL CONSTRAINTS

The principal observational information available to check theories of the origin of the giant planets is that determined at the present time, 4.7×10^9 yr after formation. The data include masses, radii, luminosities, chemical compositions, gravitational moments, and the orbits and physical characteristics of the major satellites. The observational data, in conjunction with theoretical models of the present giant planets, show that they all have an excess of elements heavier than helium, with respect to solar abundances. Furthermore, in the atmospheres of the giant planets the ratio of carbon to hydrogen, relative to solar values, increases with increasing distance from the Sun, from about a ratio of 2 at Jupiter to 40 at Neptune. Current planetary models, therefore, are composed of two components: a solid or liquid core composed basically of ice and rock; and a fluid envelope composed primarily of hydrogen and helium, roughly in solar proportions, and with a heavy-element abundance somewhat in excess of solar. Generally, for each planet, several possible models are consistent with observations, each having a different fraction of the excess heavy elements in the core. Recent models give core masses of 8–14 $M_{\oplus}$ (Earth masses) for Jupiter, 10–20 $M_{\oplus}$ for Saturn, 11–12 $M_{\oplus}$ for Uranus, and approximately 16 $M_{\oplus}$ for Neptune. The range of values for each planet corresponds not only to observational uncertainties but also to theoretical uncertainties, particularly in the equation of state in regions where hydrogen is in liquid metallic form, and to nonuniqueness of the models. Nevertheless, it is striking that the core masses are so similar, in spite of the fact that the masses of the gaseous envelopes range from about 2 $M_{\oplus}$ in the case of Uranus to more than 300 $M_{\oplus}$ in the case of Jupiter. Another observational constraint involves the astronomically deduced maximum lifetimes of disks around young stars, now estimated at 10^7 yr. Apparently, formation times must be less than this value; otherwise the supply of gas becomes insufficient to account for the present amount of hydrogen and helium in the giant planets. Furthermore, it is generally accepted that Jupiter was fully formed before accretion in the asteroid belt was completed. Gravitational scattering by this massive planet may have prevented the accumulation of a terrestrial-sized object there and may also have limited the mass of Mars.

FORMATION THEORIES

The first of the formation theories assumes that the process proceeded through an early stage involving a giant gaseous protoplanet. The general requirement is that the nebula be gravitationally unstable, which means that its mass during its early history must have been considerably greater than that deduced from the present masses of the planets. Then, for typical conditions in the region of formation of the giant planets (density 10^{-10} g cm^{-3}, temperature 100 K), a mass about that of Jupiter's can become gravitationally bound and can contract as a subcondensation in the nebula. The condensation forms rapidly and then evolves either at constant mass or with decreasing gas mass as a function of time. During the evolution, solid particles or planetesimals are captured from the nebula; some of these sink to the center to form the core. Alternatively, some of the condensible materials in the gas precipitate to the center. The planet forms with a radius about 5000 times its present size, contracts rapidly to a few times the present size over a time of about 10^5–10^6 yr, heats to internal temperatures of 10,000–30,000 K, and then contracts more slowly and cools, reaching its present size after a time of several billion years. The principal advantage of this theory is that the short formation time is consistent with the lack of a planet in the asteroid belt and the small mass of Mars, and there are no problems regarding the relatively short lifetime of the nebula. The main disadvantage has to do with the formation of the core. In general, pressure–temperature conditions are not favorable for precipitation of heavy elements, and small planetesimals accreting from the outside would dissolve in the gas rather than settle down to the core. The hypothesis does not explain in a natural way the fact that the rock–ice masses are similar for all giant planets. Also, it is difficult to explain the origin of Neptune and Uranus by this process, because the amount of hydrogen and helium is so small compared with the amount of heavy material; much of the gas would have had to escape, which is difficult to explain.

The presence of the core has led to the second hypothesis, in which the buildup of planets occurs according to the following steps.

1. Dust particles in the nebula gradually accumulate to form a solid core of approximately 1 $M_{\oplus}$.
2. Gravitational capture of surrounding gas by the growing core results in the development of a bound envelope in hydrostatic equilibrium. Until the core reaches a few Earth masses, the envelope mass is negligible compared with that of the core, but the radius of the envelope is many times larger. The energy supplied by the accretion of the solid particles keeps the gaseous envelope in equilibrium. As the core mass grows, the envelope slowly contracts, and more gas is gradually added at the outer edge.
3. As the core mass approaches the so-called "critical" stage, the envelope mass becomes comparable to the core mass and the evolutionary time scale rapidly changes from that of accretion of solid material by the core to that of rapid gravitational contraction of the envelope. After this time the envelope accretes much faster than the core.
4. The accretion of gas onto the protoplanet stops; for Jupiter the reason probably is tidal truncation of the nebula by the gravitational effects of the protoplanet itself. Once the proto-Jupiter has grown to about its present mass, it opens up a gap in the nebula, resulting in a shutoff of the gas supply. In the case of the other giant planets, termination of gas accretion may occur simply because of dissipation of the nebula.
5. The gaseous envelope then contracts by a factor of several hundred in radius in a time of about 10^5 yr, radiating at a relatively high rate (10^{-3}–10^{-4} $L_{\odot}$ in the case of Jupiter and Saturn).
6. When the radius has decreased to a value comparable to the present size of the planet, the further evolution involves slow contraction, internal cooling, and a steady decrease in luminosity on a time scale of 4.5×10^9 yr.

Detailed numerical calculations show that the value of the critical core mass depends on the rate of accretion of solid particles onto the core and on the opacity and mean molecular weight in the gaseous envelope. For standard values of these latter quantities, the critical core mass is about 11, 17, and 29 $M_{\oplus}$ for assumed constant accretion rates of 10^{-7}, 10^{-6}, and 10^{-5} $M_{\oplus}$ yr^{-1}, respectively. The calculations also show that these values are independent of the position of the planet in the nebula. Therefore, the main point supporting this hypothesis is that it predicts that the core masses of Jupiter and Saturn should be similar, which is consistent with observations. The calculated values for the critical core mass are similar to those deduced for the giant planets. The model can also be modified to take into account the fact that some of the solid accreted material will not actually reach the core but will be dissolved in the gaseous envelope. In this way the observed carbon enhancement in the giant planets over solar values can be explained.

There are also some problems with the models. The suggested evolutionary time for the formation of Jupiter and Saturn, based entirely on the values of their core masses and the accretion model, is 10^7–10^8 yr. The times to accumulate a core of about 10 $M_{\oplus}$, according to direct calculations of the accretion process, fall in the same range. Yet nebular evolution times are presumed to be shorter.

A second problem is that the structure of Uranus and Neptune cannot really be explained by the critical core mass argument, because their envelope masses are much smaller than their core masses, whereas in the theoretical models the critical core mass is not reached until the envelope mass is comparable to the core mass. Accretion of gas must have stopped in these cases simply because the nebular gas was dissipated. Furthermore, the direct calculations of the accretion of planetary cores give extremely long times, up to 10^{10}–10^{11} yr, for the accumulation of the solid material in Uranus and Neptune. Even if the time scale is on the order of 10^9 yr, consistent with the age of the solar system, the 1 $M_\oplus$ or so of hydrogen-rich material in their atmospheres is difficult to explain.

THE PROBLEM OF FORMATION TIMES

Current work on protoplanets is focused on the above-mentioned problem of time scale, particularly for Jupiter. A number of suggestions are currently being investigated.

1. It is possible that the value of the critical core mass was actually only about 0.2 $M_\oplus$. This value would be appropriate if the grains in the envelope had accumulated into planetesimals, thereby sharply reducing the grain opacity. Rapid accumulation of gas onto the core coupled with possible collisions of different cores could then result in a short time scale for planet formation.

2. In the standard calculations of accretion times, the density of solid material in the solar nebula near the orbit of Jupiter is usually assumed to be consistent with the heavy-element mass in Jupiter at present. It is possible that the actual density at formation was about 5–10 times higher than this standard value. Under these conditions, the core of Jupiter can undergo runaway growth up to its present value on a time scale of 10^6 yr. The excess mass is presumably scattered out of the solar system by the gravitational influence of Jupiter. The nebular surface density would have had to be nearly constant interior to Jupiter's position, so that the amount of mass (in heavy elements) in the nebula in the regions of Earth and Venus would have been nearly equal to their present mass; ejection of much excess mass from the inner solar system is not possible.

3. Another way to increase the surface density at Jupiter's position and thereby to accelerate the core-formation process is by diffusive transport of water vapor from the inner solar nebula to the position of Jupiter. The vapor would condense at a temperature of 170 K and thereby enhance the surface density of solids at that point, which is presumably where Jupiter formed. These last two possibilities form consistent scenarios for the formation of Jupiter, but they run into difficulties with the formation of Saturn.

4. One might also speculate that the gas remains in a disk-like configuration around at least some stars for times longer than those deduced from the disappearance of significant infrared radiation (around 10^7 yr) in young stellar objects. Once the dust has settled and coagulated, the infrared radiation is sharply reduced, but the gas, which is much more difficult to detect in these sources, could still be present.

5. It is still possible that the gravitational instability model, which gives formation times on the order of only a few years, is correct after all.

FURTHER PROBLEMS OF GIANT PLANET FORMATION

Numerous interesting questions remain to be answered. *What caused the high inclination of the axis of Uranus*? An impact of a single object of around 1–2 $M_\oplus$ late in the accretion process could have provided the required angular momentum if it hit the core off-center; furthermore, such an impact could have left the gaseous envelope of the object intact.

How did the regular satellites form? There are three possibilities.

1. The satellite accretion disk was formed from material that came directly from the solar nebula during late stages of accretion.
2. Satellite material was first incorporated into the protoplanet. During contraction of that object with conservation of angular momentum, or with transport of angular momentum outwards, some material was left behind in the form of a disk.
3. A giant impact ejected some material into orbit, which later condensed to form satellites.

For an evaluation of the first two of these processes, one must consider the effect of angular momentum. The rate of transfer of mass and angular momentum from the solar nebula into the region of gravitational influence of the protoplanet must be calculated; this calculation must be combined with that of the evolution of the protoplanet including the effects of rotation.

How are the variations in the helium abundances and the heavy-element abundances in the giant planets to be explained? In connection with evolutionary models, the amount and distribution of solid material dissolved in the envelope, as well as the possible insolubility of helium in the deeper layers, must be taken into account. Calculations of the preferential capture of icy versus rocky material could help to explain the deduced ice-to-rock ratios in the planets. Because of compositional changes with time in the planet, the chemical composition of the satellite disk will depend on its mode of origin; thus the deduced compositions of the planetary envelopes and of the satellites could help to solve the problem regarding how the disk originated. Spacecraft measurements will play a key role in future improvements in our knowledge of the formation processes of the outer planets.

Additional Reading

Bodenheimer, P. (1989). Structure and evolution of gaseous protoplanets. *Quart. J. Roy. Astron. Soc.* **30** 169.

Cameron, A. G. W. (1988). Origin of the solar system. *Ann. Rev. Astron. Ap.* **26** 441.

Hayashi, C., Nakazawa, K., and Nakagawa, Y. (1985). Formation of the solar system. In *Protostars and Planets. II*, D. C. Black and M. S. Matthews, eds. University of Arizona Press, Tucson, p. 1100.

Pollack, J. B. (1984). Origin and history of the outer planets: theoretical models and observational constraints. *Ann. Rev. Astron. Ap.* **22** 389.

Pollack, J. B. and Bodenheimer, P. (1989). Theories of the origin and evolution of the giant planets. In *Origin and Evolution of Planetary and Satellite Atmospheres*, S. K. Atreya, J. B. Pollack, and M. S. Matthews, eds. University of Arizona Press, Tucson, p. 564.

Stevenson, D. J. (1982). Formation of the giant planets. *Planetary Space Sci.* **30** 755.

See also **Planetary Interiors, Jovian Planets; Planetary Systems, Formation, Observation, Evidence; Solar System, Origin; Satellites.**

Planetary Interiors, Jovian Planets

William B. Hubbard

The jovian planets (Jupiter, Saturn, Uranus, and Neptune) comprise 99.6% of the total mass of the planetary system, or some 445 $M_\oplus$ (Earth masses) of material. These planets differ in fundamental ways from the four terrestrial planets (Mercury, Venus, Earth, and Mars), and comparisons of the structure of jovian and terrestrial planets provide important clues to the processes that lead to

Table 1. Physical Properties of the Jovian Planets

Planet	Mass ($M_\oplus$)	a (km)	e	Rotation Period (h)	q
Jupiter	318	71,500	0.067	9.9	0.089
Saturn	95	60,200	0.101	10.7	0.153
Uranus	15	25,600	0.023	17.2	0.030
Neptune	17	24,800	0.017	16.1	0.026

formation of planets and low-mass stars. Recent observational and theoretical studies of the jovian planets have indicated that these bodies are in many respects more similar to stars than to the solid terrestrial planets.

Jovian planets (Table 1), are in the outer parts of the solar system whereas terrestrial planets are closer to the Sun. This spatial dichotomy along with the great difference in masses of the two types of planets implies that the jovian planets formed of material that was much more abundant in the primordial nebula than terrestrial-planet material and that was largely excluded from planet-forming processes in the inner regions of the nebula.

The last jovian planet (Neptune) was discovered in 1846. As all of the jovian planets have at least one satellite that can be readily observed from Earth, their large masses were soon determined from Kepler's third law. But their compositions remained essentially unknown because the prime constituents of their atmospheres were not at first accessible to available spectroscopic techniques. Neglected by geophysicists and astronomers alike (with a few notable exceptions such as Harold Jeffreys and Rupert Wildt), the jovian planets gave few clues as to their bulk composition and possible origin. However, in the 1920s, Jeffreys developed an essential technique for studying the jovian planets. He noted that their rapid rotation rates (Table 1) lead to significant distortion from spherical symmetry. This distortion is measured by two parameters, the oblateness

$$e = (a - b)/a$$

and the rotation parameter

$$q = (\omega^2 a^3)/GM,$$

where a is the planet's equatorial radius, b is its polar radius, ω is the angular rotation rate, G is the gravitational constant, and M is the mass. These numbers are of comparable magnitude for the jovian planets, and their ratio for a given planet gives a measure of the planet's interior structure via the equation of hydrostatic equilibrium. Thus the value of e/q takes its minimum value of $1/2$ for a planet with all its mass concentrated in the center (i.e., with zero moment of inertia), and its maximum value (except for unphysical models) of $5/4$ for a uniform-density object. The results for Jupiter and Saturn showed that these planets were far more centrally condensed than the terrestrial planets, implying that they were made up of much more compressible material. But the results also showed that they could not be gaseous either, because they were less centrally condensed than highly compressible gaseous stars. This analysis could not be reliably carried out for Uranus and Neptune because of lack of information.

An important article by E. Wigner and H. B. Huntington in 1935 led to great improvements in the understanding of Jupiter and Saturn. This article showed that solid hydrogen would, under great pressure, become a simple alkali metal whose properties were amenable to theoretical calculation. The pressure of metallization seemed to lie in the vicinity of a few million bars, substantially lower than the central pressure of Jupiter. In 1938, Wildt and D. S. Kothari found that the theoretical hydrogen compression curve was compatible with possible interior structures for Jupiter and Saturn. For the first time it became apparent that Jupiter and Saturn might be rather similar to the Sun in interior composition.

MODERN DEVELOPMENTS

The advent of modern detectors for infrared bolometry and spectroscopy in the late 1960s and the precise measurement of various planetary properties from the Earth and from flyby spacecraft led to great improvement in our understanding of the jovian planets. Other factors were new laboratory measurements of the compressibility of hydrogen and other relevant light elements, theoretical and observational work that yielded precise determinations of the external gravitational potentials of jovian planets, and advances in the theory of matter at high pressure.

Frank J. Low made pioneering measurements of the thermal radiation of Jupiter and Saturn in the infrared and found that both of these bodies radiate into space more energy than they receive from the Sun. The average surface heat flux from Jupiter was subsequently accurately measured by spacecraft experiments and found to be about 5400 erg cm^{-2} s^{-1}; the corresponding value for Saturn is 2000 erg cm^{-2} s^{-1}. For comparison, the average surface heat flux from the Earth's interior is only about 60 erg cm^{-2} s^{-1} and is swamped by the heat flux due to thermalized sunlight.

These results caused a revolution in the interior models of Jupiter and Saturn. Wildt and his student Wendell C. DeMarcus had constructed interior models that were composed of solid hydrogen and helium, but Wildt presciently remarked that the adiabatic lapse rate in gaseous hydrogen is so large that red heat (1000 K) would be reached at a depth of about 500 km on Jupiter and about 1000 km on Saturn, even if the atmospheres were at a temperature of only 150 K at 1 bar. Later studies of heat transport mechanisms in dense hydrogen then revealed that the measured surface heat fluxes could not be sustained by any conceivable thermal conduction or radiation process; the effective opacities were far too high.

These results implied that the observed heat flux must be transported to the surface by convection in the deep interiors of Jupiter and Saturn, and that for efficient convection, the lapse rate should be essentially identical to the adiabatic lapse rate calculated by Wildt.

A self-consistent model for Jupiter's thermal structure is shown in Fig. 1, which is a modern representation of the high-pressure hydrogen phase diagram. The dashed curve marked "Jupiter" shows an adiabatic temperature profile in Jupiter, terminating at the central pressure of about 40 Mbar and a central temperature of

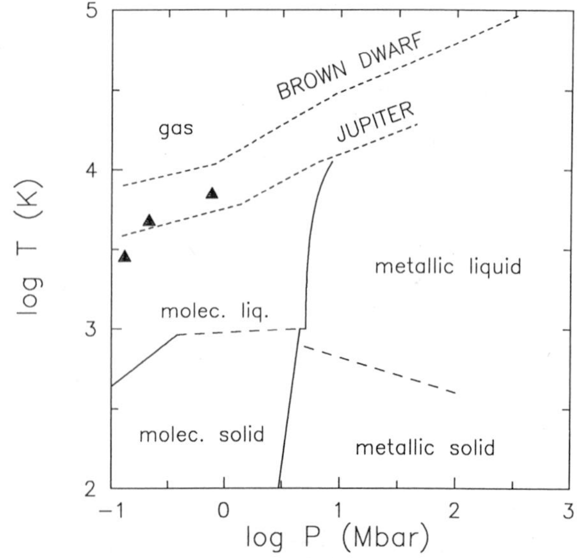

Figure 1. A phase diagram for hydrogen (dashed phase boundaries are particularly uncertain). Triangles show shock data points from laboratory experiments.

about 20,000 K. For comparison, a thermal profile of a similar object, a so-called brown dwarf, is also shown. A brown dwarf is a hydrogen-rich object with mass approximately $50 M_J$ (Jupiter masses), which is too small to become a star. Various phases of dense hydrogen are shown in Fig. 1. The transition from liquid to gas is gradual and does not have a discrete change of phase in the jovian temperature range. Thus there is no well-defined surface in Jupiter's hydrogen-rich layers, and the state of the hydrogen gradually changes with increasing depth from a cold molecular gas to a red-hot molecular liquid, and then to a blue-hot metallic liquid. It is not known whether the molecular–metallic transition in Jupiter is supercritical, as is sketched in the figure.

The high-temperature thermal model of Jupiter and Saturn provides a natural explanation of the heat flow from these planets. They are slowly radiating into space heat that was stored from gravitational accumulation of hydrogen and helium when these planets formed. The interior temperatures are so high that the characteristic time for radiating the heat into space (the so-called kelvin time) is comparable to or exceeds the age of the solar system. A more detailed assessment of the energy balance shows that although this model works fairly well for Jupiter, there may be a shortfall for Saturn. In Saturn, additional energy for radiation into space may be released from gradual formation of a core from a dense component.

COMPOSITION AND STRUCTURE OF JUPITER AND SATURN

Modern interior models of Jupiter and Saturn are constructed in a manner analogous to Jeffreys' approach, but take into account not just the oblateness but also smaller-scale modulations of the planetary figure produced by its response to rotation. These modulations, which are called gravitational harmonics, depend upon the pressure–density relation in the planet's outer layers and thus are sensitive to composition. Accurate determinations of the atmospheric compositions of Jupiter and Saturn are now available and can be compared with compositions of the deep interior deduced from modeling.

Both Jupiter and Saturn resemble the Sun in overall composition, but differ in detail. Standard solar composition is represented by the mass fractions X, Y, and Z, respectively the mass fractions of hydrogen, helium, and everything else, with

$$X + Y + Z = 1.$$

In the Sun, $Y \simeq 0.24$ and $Z \simeq 0.02$. In Jupiter's atmosphere, $Y = 0.18 \pm 0.04$, whereas in Saturn's atmosphere, $Y = 0.06 \pm 0.05$. Interior models confirm the near-solar value of Y in Jupiter, but not necessarily the subsolar value in Saturn. Theorists now understand that, unlike the Sun, the composition of Jupiter and Saturn's atmospheres need not closely resemble the composition of their deep interiors. Although large-scale convection would tend to homogenize the composition in these bodies, phase transitions could cause substantial partitioning of elements between the observable atmosphere and the deep interior. The phase transition between liquid–molecular hydrogen and liquid–metallic hydrogen shown in Fig. 1 could be one such transition. Another could be a postulated transition in the metallic phase only, leading to the partitioning of helium into a separate helium-rich phase, which then sinks toward the center of the planet, depleting the helium in the atmosphere.

Models of Jupiter and Saturn show that both planets have substantially more of the Z component than would be expected for solar abundances. This shows up both in atmospheric abundances (both planets have excess carbon in their atmospheres with respect to the Sun) and in interior models. Much of the excess Z component appears to be separated out in the form of a dense core in each of these planets, with a total mass of about 10–15 $M_\oplus$.

COMPOSITION AND STRUCTURE OF URANUS AND NEPTUNE

In contrast to Jupiter and Saturn, the fraction of hydrogen and helium in Uranus and Neptune is quite small. This is known in an essentially model-independent fashion, because at pressures prevailing in these bodies (~ 10 Mbar), hydrogen has a much larger volume than any other material. Thus the relatively small size of Uranus and Neptune compared with Jupiter and Saturn puts strict limits on their hydrogen abundance. This limit is assumed to apply to the entire hydrogen and helium budget because it is considered unlikely that these elements could have separated when the jovian planets were first forming. The amount of hydrogen/helium in Uranus thus appears to be only about 3 $M_\oplus$ in Uranus, and perhaps half that in Neptune.

The composition of the major component in Uranus and Neptune is poorly constrained, because there is no uniqueness in the pressure–density relations of many compounds or mixtures of compounds that could conceivably be in these planets. But from the point of view of nebular theory, the most abundant condensate at the time that Uranus and Neptune formed was probably water ice, and in fact the masses, radii, and gravitational harmonics of Uranus and Neptune are compatible with the compression curve of water, with a surface layer of hydrogen and helium. From cosmic abundances, the oxygen should be accompanied by a substantial amount of carbon and nitrogen, although the chemical state of these elements in Uranus' and Neptune's deep interior is poorly understood. Heat flow data for Neptune show a surface flux of about 300 erg cm^{-2} s^{-1}, suggesting that this planet may also have a hot interior, and that the water and other molecules may be liquid. Curiously, no intrinsic heat flow has yet been detected for Uranus.

CONSTRAINTS ON FORMATION MECHANISMS

Its seems suggestive that the masses of Uranus and Neptune are comparable to the masses deduced for the cores of Jupiter and Saturn. Current theories for the formation of the jovian planets propose the initial formation of large solid cores, with masses approximately 15 $M_\oplus$. These cores were mainly composed of icy molecules such as H_2O, NH_3, and CH_4 or CO, which could condense in the outer reaches of the solar nebula. An instability triggered by these cores then led to accretion of the surrounding nebular gas onto the cores, forming the planets. The smaller amount of gas captured by Uranus and Neptune is explained as a consequence of these bodies' large distances from the Sun, where the solid condensates were more dispersed, took longer to aggregate into cores, and may have been large enough to accumulate gas only when the nebular gas had largely dispersed.

Additional Reading

Hubbard, W. B. (1984). *Planetary Interiors*. Van Nostrand Reinhold Company, New York.

Hubbard, W. B. (1990). Interiors of the giant planets. In *The New Solar System*, 3rd ed., J. K. Beatty and A. Chaikin, eds. Sky Publishing Corp., Cambridge, MA, and Cambridge University Press, Cambridge, U.K., p. 131.

Hubbard, W. B. and Stevenson, D. J. (1984). Interior structure of Saturn. In *Saturn*, T. Gehrels and M. S. Matthews, eds. University of Arizona Press, Tucson, p. 47.

Stevenson, D. J. (1982). Interiors of the giant planets. *Ann. Rev. Earth Planetary Sci.* **10** 257.

Zharkov, V. N. and Trubitsyn, V. P. (1978). *Physics of Planetary Interiors*, W. B. Hubbard, ed. Pachart, Tucson.

See also **Jupiter, Atmosphere; Planetary Formation, Gaseous Protoplanet Theories; Saturn, Atmosphere; Uranus and Neptune, Atmospheres.**

Planetary Interiors, Terrestrial Planets

Lon L. Hood

The bulk compositions and internal structures of the inner planets represent basic constraints on the manner in which the solar system formed and on how planets similar to the Earth have evolved. In addition, studies of the internal structures and thermal histories of the terrestrial planets are required to provide a context for the interpretation of surface expressions of internal activity such as tectonic deformation and volcanism. The terrestrial planets are generally distinguished from the jovian planets by their larger mean densities (averaging about 5 g cm^{-3} versus about 1 g cm^{-3} for Jupiter and Saturn), smaller sizes, and location in the inner solar system within 1.6 AU of the Sun. Their larger densities immediately suggest formation in a hot central nebular environment where cosmically more abundant (and less dense) volatile substances could not have easily condensed into solid phases (the equilibrium condensation model of solar system origin). The increasing uncompressed mean densities of the terrestrial planets with decreasing distance from the Sun (Table 1) also suggest an increasing concentration of refractory (high temperature) condensates, including metallic iron, in protoplanetary materials with decreasing radial distance in the solar nebula. On the other hand, terrestrial planets do contain some volatile substances (such as crustal carbonates and surface ices) that would suggest a heterogeneous accretion model in which more refractory (early condensing) substances accreted first followed by the addition of some less refractory and volatile (late condensing) substances. In addition, the accretion of the terrestrial planets generally resulted in the release of heat sufficient to melt and differentiate these bodies; this may have resulted in modifications of bulk composition (losses of volatiles) that would not be predicted by an equilibrium nebular condensation model. Figure 1 illustrates the probable structures of the terrestrial planets.

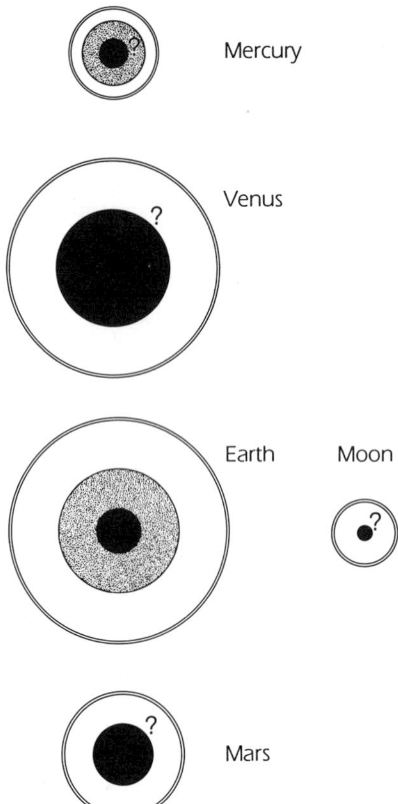

Figure 1. Schematic illustration of the probable structures of the terrestrial planets including the Moon. Mercury and the Earth are shown with solid inner cores (dark shading) and molten outer cores (lighter shading). Each body is also shown differentiated into a mafic mantle and an aluminous crust.

EARTH

The detailed structure of the Earth's interior has been deduced from a combination of indirect geophysical measurements and analyses of meteorites and of igneous rocks originating in the mantle. From these data, it has become apparent that the Earth is differentiated; that is, it was originally heated and melted allowing it to be segregated according to the densities of major solid phases. Original melting occurred as a result of the dissipation of energy during accretion but long-term heating is a consequence of the existence of radioactive elements such as uranium and thorium in the interior. The present-day globally averaged heat flow is about 80 mW m^{-2}. The primary compositional divisions of core, mantle, and crust, inferred from meteorite compositions and determined using seismic data, have been known for nearly 80 years. The crust is composed of aluminum-rich silicates that would have crystallized from the near-surface melt of a self-gravitating magma composed of minerals similar to those found in chondritic meteorites (meteorites with a nearly solar composition minus lost volatiles). The core is composed of (mainly iron) metal that would have migrated to the bottom of such a system. In between is the mantle composed predominantly of mafic (magnesium- and iron-rich) silicates with an intermediate density. The terrestrial crust is generally thicker beneath the continents (30–40 km) than beneath the oceans (5–7 km). The thin oceanic crust is relatively young and is continually generated by the eruption of basaltic lavas along mid-ocean ridges. The thickness of the older continental crust increases with increasing surface elevation (up to 100 km beneath the Himalayas) due to isostatic compensation of the increased elevation by low-density roots (analogous to buoys floating on the mantle). In the uppermost part of the mantle, to a depth of 50–70 km, seismic waves have slightly higher velocities and experience little attenuation implying a relatively rigid zone. This part of the mantle together with the crust is known as the lithosphere. Beneath the lithosphere, to a depth of 200–300 km, seismic velocities are somewhat lower and seismic waves experience greater attenuation, suggesting the existence of partial melting. This zone, termed the asthenosphere, represents a relatively deformable layer over which rigid lithospheric plates can slide. At greater depths in the mantle, seismic velocities show several discontinuous increases

Table 1. Gross Properties of the Terrestrial Planets

	Equatorial Radius (km)	*Mean Density ($g\ cm^{-3}$)*	*Uncompressed Density ($g\ cm^{-3}$)*	*Moment of Inertia Factor*	*Surface Magnetic Field (gamma)*
Mercury	2,439	5.42	5.3–5.4	?	300
Venus	6,050	5.25	4.0–4.6	?	~ 0
Earth	6,378	5.52	4.2–4.6	0.3308	31,000
Moon	1,738	3.34	3.3	0.391 ± 0.002	~ 0
Mars	3,398	3.94	3.6–3.8	$0.345 - 0.365$	< 60

that mark changes to denser mineralogical phases (e.g., olivine to spinel) but do not necessarily imply changes in chemical composition. Analyses of inclusions in magmas originating in the upper mantle (depths less than 200 km) indicate that the dominant minerals are all ultramafic: olivine, orthopyroxene, and small amounts of clinopyroxene and garnet.

The moment of inertia about a given axis in a planet is defined by

$$I = \int \rho r^2 \, dv,$$

where ρ is the mass density of a volume element dv and r is the perpendicular distance from the axis. For a uniformly dense sphere, $I = 0.4MR^2$, where M is the mass and R is the radius; the existence of a central density concentration leads to smaller values of the constant factor. The Earth's principal moment of inertia (about the rotation axis) is accurately determined as $0.3308\,Ma^2$, where M is the Earth's mass and a is the mean equatorial radius. To obtain this value, the astronomically measured rate of precession of the Earth on its axis (produced by torques exerted on the internal mass distribution by the gravitational attractions of the Sun and Moon) first yielded an accurate estimate of the difference between the moments of inertia about the rotation axis and about an axis in the equatorial plane. Second, measurements of the departure of the global gravity field from that of a central point mass (due mainly to the Earth's "flattening" or equatorial bulge) yielded an independent precise estimate of this difference. Combining the two produced numerical values of the moments of inertia. Unfortunately, the radial-density profile for a given moment of inertia value cannot be uniquely determined although the range of allowed models is limited. Also, uncertainties in the deep temperature profile and in the behavior of mineral phases at large pressures essentially preclude absolute compositional inferences at large mantle depths.

A discontinuous decrease in seismic pressure (P) wave velocity from 13.6–8 km s^{-1} occurs at a depth of 2920 km in the Earth marking the base of the mantle and the top of the outer metallic core. Transverse or shear (S) waves are not observed to propagate at depths below this discontinuity implying that this part of the core is molten. The existence of an additional inner core at a depth of about 5100 km was inferred in the 1930s from the fact that the shadow zone expected for a continuous low-velocity core was not complete, indicating refraction of rays through an inner higher-velocity core. The inner core is solid as deduced from its transmission of S waves but may be near the melting point. The composition of the core is primarily iron with added nickel. However, the inferred density is somewhat less than expected for a purely metallic composition, indicating admixture of lighter elements, probably sulfur or possibly oxygen.

MOON

The gross structure of the interior of the Moon has been determined from Apollo-era geophysical measurements although many details (including the structure of the lower mantle and the existence of a metallic core) remain unresolved. The initial return of Apollo samples showed that lunar surface rocks are enriched in aluminous silicate phases indicating that the Moon has a differentiated crust analogous to that of the Earth. The depth of the crust was determined by active and passive seismic methods beneath one of the Apollo mare sites as 60 ± 5 km. Beneath the lunar highlands, the thickness is likely to be significantly greater although this was not definitely determined. The existence of such a thick aluminous crust implies that a large volume of the lunar interior (to more than 500 km depth) must have been melted at or shortly after the Moon formed. Analyses of seismic event data from the limited (four-station) Apollo seismic network showed that upper mantle seismic velocities decrease slightly with increasing depth. This is believed to be mainly due to the rapidly rising temperature in this zone. At a depth of approximately 500 km, however, an increase in velocity was inferred that may represent a composition change or a phase change (e.g., from spinel to garnet for aluminous silicates). In general, the accuracy of lunar seismic velocity determinations at depths greater than about 500 km is low due mainly to the limited areal distribution of the network. The outermost 800 km of the Moon is characterized by very low seismic wave attenuation, consistent with a very rigid volatile-free lithosphere. It is therefore very unlikely that solid-state convection occurs in the lunar interior at depths less than 800 km. At depths greater than 800 km, seismic wave attenuation was observed to increase rapidly and S waves were not observed to be transmitted, implying the existence of a partially molten asthenosphere. Electromagnetic sounding of the lunar interior using Apollo surface magnetometers showed that the mantle electrical conductivity increases rapidly with increasing depth, consistent with that expected for nominal temperature models and mantle compositions. The two Apollo surface heat flow experiments yielded final estimates of 21 and 16 mW m^{-2}. Global extrapolations of these values are necessarily difficult.

Direct physical measurements relating to the existence of a lunar metallic core are suggestive but not definitive. In order to determine the moment of inertia, Apollo laser ranging data first refined earlier astronomical measurements of lunar librations (slight departures from the Moon's Cassini alignment caused by external torques on its internal mass distribution) to yield estimates for the differences between two of the three moments. Later analyses of Apollo spacecraft orbit data yielded accurate values for a third difference between the three moments from measurements of the gravity field. Combining these difference estimates, a value was obtained for the principal moment factor, I/Ma^2, of 0.391 ± 0.002. Modeling of the lunar mantle density structure to be consistent with mean density, moment of inertia, and the limited seismic velocity data suggested that the presence of a small dense core with radius greater than 300 km was likely. Electromagnetic sounding data provided an upper limit of 450–500 km for the radius of a highly conducting metallic core. Seismic data for one large farside meteoroid impact suggested the presence of a small low-velocity core; however, no confirming measurement for other events was obtained during the lifetime of the Apollo network. Thus the available data indicate that an approximately 400-km core (representing as much as 4% of the lunar mass) may be present but a definite determination of this must await new measurements in the course of future lunar exploration. In general, the small size of the lunar core, as indicated by its anomalously low mean density (Table 1), suggests that the Moon may have experienced an exceptional origin that would not be predicted by solar nebula and accretion models per se. For example, one recently popular model supposes that a giant impact on the Earth during the late stages of accretion ejected a small quantity of both the impactor and the terrestrial mantle that could have formed a Moon relatively depleted in metallic iron.

MERCURY

The internal structures of the remaining terrestrial planets, Mercury, Venus, and Mars, are essentially undetermined observationally but some inferences are possible. In the case of Mercury, the higher uncompressed density relative to the Earth (Table 1) implies that 60–70% of the planet consists of metal phases (as compared to 32% for the Earth). Because Mercury is large enough to have been heated to melting temperatures by accretion alone, the interior is probably differentiated. A metallic core with a radius of about 1800 km (75% of the planetary radius) is therefore to be expected and is probably surrounded by a segregated mantle and crust. Mercury also has a weak global magnetic field suggesting that the core is at least partly molten in order to generate the field by hydromagnetic dynamo action. Finally, the identification of large compressive thrust faults of the surface (unlike the Moon) in spacecraft pictures indicates that the planet experienced a global cooling and contraction epoch a few hundred million years after formation. This may have been accelerated by solid-state convection in the mantle subsequent to core formation.

VENUS

The uncompressed mean density of Venus is not significantly different from that of the Earth. With the reasonable assumption that the interior is differentiated, it follows that Venus most probably has a core, mantle, and crust with composition and density not very different from that of the Earth. Solid-state convection in the mantle is also likely. However, the higher venusian surface temperature and the correspondingly more buoyant lithosphere may preclude subduction of crustal plates as occurs on the Earth. The absence of an intrinsic venusian magnetic field could imply that the core is not molten although the slow planetary rotation rate might be expected to inhibit dynamo action even if the core is molten.

MARS

In the case of Mars, the uncompressed mean density is significantly larger than that of the Moon indicating a substantial (10–30%) mass fraction of metallic phases. Mars is sufficiently large that it was almost certainly melted and differentiated at the time of its formation. Consequently, it is expected that the interior is segregated into a core, mantle, and crust. The martian principal moment of inertia is only weakly constrained observationally. The gravity field is accurately known, providing two relationships between the three principal moments. However, the axial precession rate remains undetermined so it is not possible to uniquely calculate any of the moments. Instead, estimates of the moment values require assumptions to be made about the nature of the axially asymmetric mass distribution in the planet. Different assumptions yield moment factors in the range 0.345–0.365 although the higher value may be favored. Interior models consistent with this range of the moment of inertia may be constructed for possible core and mantle compositions with core radii ranging from 1400–2000 km. The martian magnetic field is either very weak or nonexistent, suggesting the absence of a molten core.

Additional Reading

Anderson, D. L. (1990). Planet Earth. In *The New Solar System*, 3rd ed., J. K. Beatty and A. Chaikin, eds. Sky Publishing Corp., Cambridge, MA, and Cambridge University Press, Cambridge, U.K., p. 65.

Carr, M. H., Saunders, R. S., Strom, R. G., and Wilhelms, D. E. (1984). *The Geology of the Terrestrial Planets, NASA SP 469*, M. H. Carr, ed. NASA Scientific and Technical Information Branch, Washington, DC.

Hood, L. L. (1986). Geophysical constraints on the lunar interior. In *The Origin of the Moon*, W. K. Hartmann, R. J. Phillips, and G. J. Taylor, eds. Lunar and Planetary Institute, Houston, p. 361.

Phillips, R. J. and Malin, M. C. (1983). The interior of Venus and tectonic implications. In *Venus*, D. M. Hunten, L. Colin, T. M. Donahue, and V. I. Moroz, eds. University of Arizona Press, Tucson, p. 159

Wood, J. A. (1979). *The Solar System*. Prentice-Hall, Englewood Cliffs, NJ.

See also **Earth, Figure and Rotation; Moon, Seismic Properties; Planetary Magnetism, Origin; Planetary Rotational Properties.**

Planetary Magnetism, Origin

Palmer Dyal

Magnetism associated with the planets represents one of four basic forces—electromagnetic, gravitation, strong nuclear, and weak—that are used to describe all interactions in the universe. Both magnetic and gravitational forces act over large distances compared to the strong nuclear and weak forces and are the most prevalent physical force fields encountered near planetary bodies.

William Gilbert in 1600 wrote a textbook titled *De Magnete* in which he asserted that "the earth itself is a great magnet." During the subsequent 400 years, magnetic fields have been measured as intrinsic properties of most of the planets, the Sun, and many objects observed throughout the universe. These measurements and studies have concentrated on the Earth's field and as a result, the magnetic field models for the other planets rely heavily upon models developed from the geomagnetic measurements of the Earth's crust, mantle, and core. The time dependence of the geomagnetic field has also yielded significant information on the origin and dynamics of the Earth's interior. For many centuries the strength and direction of the Earth's magnetic field has been measured by a compass needle and utilized for navigational purposes. Chinese legends attribute the use of a compass to the emperor Huang-ti in 2634 BC and the first recorded use was also in China and appears in a work entitled *Ping-chou-K'o-t'an* at the end of the eleventh century AD. Subsequent measurements have shown that the Earth's magnetism originates from two sources: (1) electrical currents in the core and (2) magnetized crustal material. The compass needle used by ancient and modern people points along north–south arcs which indicate that the magnetic force is in general alignment with the Earth's rotation axis. Variations of the strength and direction of the Earth's magnetic field over decade periods indicate that the source of the main field is due to electrical currents in the core rather than to remanent magnetism in the crust. The crustal material is static over periods of decades and time-dependent measurements show that it cannot be the source of the Earth's main field.

Seismic and heat flow measurements indicate that the material below the crust is hotter than the Curie temperature, at which magnetized material loses its magnetic properties. These seismic measurements also indicate that the core is molten and could be the moving electrical conducting matter that generates electrical currents and their associated main magnetic field. The crustal material is cool enough, however, to retain its magnetism and has recorded the Earth's main field over time as lava and other sources of molten rocks cooled below the Curie temperature. These paleomagnetic rock measurements indicate that the Earth's field changed polarity approximately twice every million years during the last 165 million years. Rock measurements also indicate that the core was generating a field for the last 3.5 billion years. This shows that the field is not due to electrical currents initiated at the time of Earth's formation because the electrical conductivity of the core material is too high and these ohmic currents would have dissipated in less than a few times 10,000 years. It is now generally accepted by the geophysical community that the Earth's magnetic field is maintained against dissipative ohmic decay by a self-excited dynamo in the outer core. Inversion of seismic measurements through the core and the knowledge of the density and inferred compositional model of the Earth all point to a molten iron core. The movement of this electrically conducting fluid in a magnetic field that generates electric current, which in turn induces a magnetic field, describes the self-sustaining dynamo. A recent development in the study of geomagnetic field variations over time periods much shorter than core ohmic decay times and longer than mantle decay times has permitted the depth to the core–mantle boundary to be calculated. Accurate measurements of the Earth's field from 1900 to the present have been used to calculate a core–mantle boundary at a radius of 3484 ± 48 km which is in close agreement with the seismic-determined core radius of 3485 ± 3 km.

During the last 20 years space-borne magnetometers have measured significant magnetic fields associated with the planets Mercury, Earth, Jupiter, Saturn, Uranus, and Neptune. Future measurements of these planetary fields over time, along with a more complete theoretical development of the planetary dynamo model, will yield a much better understanding of the origin of planetary magnetism.

PLANETARY CORE DYNAMOS AND CRUSTAL REMANENT MAGNETISM

The two main sources of magnetic fields associated with planetary-sized bodies in the solar system are dynamo-driven currents in the core and remanent magnetism frozen in the crustal rocks of terrestrial planets and moons. The remanent and induced magnetism in the crust of the Earth is only about 0.05% of the dynamo field; however, it is one of the most intensely studied areas of geophysics because these studied provide direct information on the past history of the core. The time history of the dynamo field has been recorded in the oceanic crust that is spreading away from the mid-oceanic ridges and cooling below the Curie temperature, thereby becoming permanently magnetized. These measurements not only proved sea-floor spreading, and inferred the existence of continental drift and plate tectonics, but also showed that the Earth's magnetic field reversed polarity at an average rate of two reversals per million years during the last 50 million years. The most recent reversal occurred 730,000 years ago and the frequency of polarity reversals has been extended back in time for 170 million years. These records show that the polar transition periods span about 10,000 years and that the main dipolar field is stable for most of the intervening time. Paleomagnetic records also indicate that the dynamo field underwent significant changes of intensity at times of polar transitions.

In addition to magnetic remanent studies of Earth, there have been crustal magnetic fields measured on the lunar surface with magnetometers operated by the Apollo astronauts and remanent magnetism was measured in the rock samples returned to Earth. These measurements indicated that the original magnetizing field was more intense than presently exists; however, no large scale dipolar form was inferred. Lunar surface and orbiting satellite magnetic field and electron reflection measurements showed that the present crustal field has a scale size that is small with respect to lunar dimensions and is randomly oriented.

As previously mentioned, the main planetary magnetic field is probably due to a self-exciting dynamo in the core of the planet. This model was developed by Walter M. Elsasser and Edward C. Bullard in the 1940s and 1950s and its basic principle is the generation of electromagnetic energy from mechanical energy. The motion of an electrical conductor through a magnetic field induces an electromotive force that causes a current to flow at right angles to the direction of motion and to the magnetic field. By assuming a suitable geometry for both the fluid flow and electrical current flow in the molten core of the Earth, it has been demonstrated that the magnetic field observed at the surface can be generated. Because the magnetic energy in the dynamo is dissipated as ohmic heat, there is a need to continually resupply the core material with energy to keep it moving. One plausible source is radioactive uranium, thorium, and potassium that is combined with the metallic iron of the core. Friedrich H. Busse has developed a model of the core fluid flow based upon thermal convection constrained by gravitational buoyancy and the Coriolis force. The gravitational buoyancy causes the less dense material to move upward and the Coriolis force accelerates this upward moving material in a direction both perpendicular to the upward motion and to the direction of the Earth's rotation. Measurements of magnetic fields associated with other planets, our Sun, and other stars show a strong correlation between the alignment of the magnetic dipole and the axis of rotation of the body. These measurements along with the amount of uranium, thorium, and potassium in the Earth's mantle and crust further support the dynamo model. Busse showed that the components of gravity and the temperature gradient perpendicular to the axis of rotation are the most important for determining the scale size and modes of the fluid flow in the outer core that are required for a dynamo.

Another direct piece of evidence that the fluid outer core is in motion relative to the Earth's rotation is the slow variation of the Earth's field with time. Measurements obtained during the last two centuries show many anomalies in the dipolar magnetic field. The anomalies are approximately 5000 km in size, have an amplitude of about 10% of the main field, and drift slowly westward at a rate of 10 minutes of arc per year. This drift implies that the surface of the fluid core rotates at a slower rate than the overlying mantle and crust.

One striking feature of the Earth's main magnetic field that has not been adequately modeled is the observed polar reversals that appear in the paleomagnetic records and the steady decrease in the intensity of the field. Future developments in dynamo theory should permit these field properties to be calculated from the state of rotation, the physical and chemical state of interior, and the energy source that produce the fluid convection.

SOLAR SYSTEM PLANETARY MAGNETIC FIELDS

During the last two decades, space-borne magnetometers have measured strong magnetic fields associated with the planets Mercury, Earth, Jupiter, Saturn, Uranus, and Neptune. Mars and Venus do not possess a measurable dipolar field and no spacecraft has encountered the planet Pluto to determine its magnetic field properties. Most of the planetary magnetic fields are centered dipoles and almost all are aligned with the planet's spin axis. (Uranus and Neptune, with dipoles inclined 60 and 50° to the rotation axis, are the major exceptions to the planets and most stars.) Because these magnetic fields are generated by fluid flows, it has become apparent that one of the best methods to study the dynamics and history of the planetary core is to study the dynamo-induced field.

One of the most important aspects of the dynamo theory during the last two decades has been its utility in developing scaling laws to predict magnetic fields for the unexplored planets in our solar system. Because of the extreme physical conditions in the planetary interior and because there has never been a direct observation of fluid motions in the core, the ability to calculate a planet's magnetic field from external physical observables has been very important for experiment designers of planetary spacecraft missions.

I. A. Eltayeb and P. H. Roberts extended the theoretical development of the dynamo theory by examining the effect of the magnetic field upon the convective fluid flow. They developed a scaling law for planetary dynamo magnetic field strength based upon a balance between the fluid's Coriolis force and the magnetic field's Lorentz force. Steven A. Curtis and Norman F. Ness reformulated this scaling law to include only terms of externally observable physical parameters. They obtained the relationship

$$B_p \sim r_p^{-3} p^{1/3} M_p^{1.54} W^{1/2} E^{1/6},$$

where B_p is the equatorial magnetic field strength at the terrestrial planetary surface or at the 1-bar level for gaseous planets, r_p is the planetary radius, M_p is the planetary mass, W is the planetary rotation rate, and E is the observed planetary heat flux. They assumed that the mean planetary density is proportional to the core density, that the core rotation rate is proportional to the planetary rotation rate, and that the observed heat flux is proportional to the energy flux associated with the core convection veloc-

Table 1. Planetary Physical Properties

Planet	*Mass (Earth = 1)*	*Density ($g\ cm^{-3}$)*	*Spin Frequency (Day^{-1})*	*Radius (km)*	*Estimated Core Radius (km)*	*Heat Flux ($W\ M^{-2}$)*	*Surface Magnetic Field (Calculated) (G)*	*Surface Magnetic Field (Measured) (G)*
Mercury	0.0555	5.46	0.017	2,425	1,800			2.8×10^{-3}
Venus	0.82	5.23	0.0041	6,070	3,000			$< 5 \times 10^{-5}$
Mars	0.11	3.92	0.976	3,395	1,700			$< 5 \times 10^{-4}$
Moon	0.011	3.35	0.037	1,738		0.017	1.90×10^{-3}	$\sim 2 \times 10^{-3}$
Earth	1	5.52	1	6,378	3,486	0.062	0.31	0.31
Jupiter	317.8	1.31	2.45	71,300	52,000	5.40	3.23	4.20
Saturn	95.2	0.70	2.35	60,100	28,000	2.00	0.57	0.20
Uranus	14.5	1.19	1.41	25,500	14,500	< 0.18	0.25	0.23
Neptune	17.2	1.66	1.50	24,800	16,000	0.29	0.45	0.13
Pluto		2.07		1,123				

ity. Using the scaling law and these assumptions, they calculated Neptune's equatorial magnetic field at the 1-bar level in 1986 to be between 0.4–0.5 G, in good agreement with their measured value of 0.2 G in 1989.

Observations of the magnetic fields of the inner eight planets and the Moon as shown in Table 1 indicate that the planets with measurable fields are in reasonable agreement with this scaling law; however, it is a scalar and not a vector equation and does not yield information about the direction of the magnetic dipole axis in relation to the planetary spin axis. Venus, Mars, and the Moon have no measurable dipolar fields, so that the field strengths are much lower than that predicted from the scaling law, and they apparently do not possess convecting fluid cores. The failure of the scaling law in these three cases is probably due to the nonlinear nature of the problem. The magnetic field may have an effect upon the state of the interactions between the Coriolis and Lorentz forces.

Future developments in the study of the general planetary dynamo and more extensive measurements of the planetary magnetic fields will significantly increase our knowledge of the interior fluid properties of the planets. Opportunities to measure the vector magnetic field as a function of time for periods of months to decades with planetary orbiting spacecraft have the potential to make dramatic progress in elucidating the dynamic properties of the conducting planetary cores. Even paleomagnetic studies of those bodies lacking a dipolar field will yield critical information on the state of the planetary interior before dynamo action ceased or show if it had ever existed. These developments will also improve our understanding of other large, electrically conducting, rotating fluid bodies in the Galaxy.

Additional Reading

Busse, F. H. (1978). Magnetohydrodynamics of the Earth's dynamo. *Ann. Rev. Fluid Mechanics* **10** 435.

Carrigan, C. R. and Gubbins, D. (1979). The source of the Earth's magnetic field. *Scientific American* **240** (No. 2) 118.

Parker, E. N. (1983). Magnetic fields in the cosmos. *Scientific American* **249** (No. 2) 44.

Russell, C. T. (1980). Planetary magnetism. *Rev. Geophys. Space Phys.* **18** 77.

Stevenson, D. J. (1983). Planetary magnetic fields. *Rept. Prog. Phys.* **46** 555.

Van Allen, J. A. (1990). Magnetospheres, cosmic rays, and the interplanetary medium. In *The New Solar System*, 3rd ed., J. K. Beatty and A. Chaikin, eds. Sky Publishing Corp., Cambridge, MA, and Cambridge University Press, Cambridge, U.K., p. 29.

See also **Earth, Magnetosphere and Magnetotail; Planetary Magnetospheres, Jovian Planets.**

Planetary Magnetospheres, Jovian Planets

Christoph K. Goertz

The four known magnetospheres of the giant, gaseous planets Jupiter, Saturn, Uranus, and Neptune are huge regions with diameters ranging from several tens to hundreds of planetary radii, in which the dynamics of plasma is controlled by the planetary magnetic fields. They are characterized by significant internal plasma sources and rapid rotation rates. In the 1950s, it was discovered that Jupiter does not only emit thermal radiation but also nonthermal radiation in two frequency bands: the decimetric radiation (DIM) between 1000 and 3000 MHz and the decametric radiation (DAM) between 5 and 40 MHz. Radiation below 5 MHz is absorbed by the Earth's ionosphere and cannot be observed on the ground. It was very quickly realized that DIM is due to synchrotron radiation from energetic electrons trapped in the inner region of a jovian magnetosphere. Thus the "discovery" of the jovian radiation belts actually preceded the discovery of the Earth's radiation (Van Allen) belts in 1957. The origin of the powerful DAM radiation remained a mystery for almost two decades. The most intriguing aspect of DAM is the control the innermost Galilean satellite, Io, has over its occurrence. DAM occurs when Io is either at the extreme east or near the extreme west elongation with respect to Jupiter and if it happens to be, at the same time, above the longitudes toward which Jupiter's northern and southern magnetic poles are pointing. No equivalent radiation was observed from Saturn, Uranus, and Neptune, although we now know that all three are emitters of radio waves which, however, are too weak to be detected at the Earth. These waves were discovered by experiments aboard the two U.S. spacecraft, *Voyager 1* and *2*, as they approached these planets in the 1980s.

THE INNER MAGNETOSPHERE

Due to the fact that the jovian magnetosphere was traversed by four spacecraft (*Pioneer 10* in 1973, *Pioneer 11* in 1974, and *Voyager 1* and *2* in 1979), whereas only three visited Saturn and only one the planets Uranus and Neptune, our knowledge of the jovian magnetosphere is the most advanced, although far less than that of the Earth's magnetosphere. Here, we will deal mostly with the jovian magnetosphere. The inner magnetosphere extends from Jupiter's surface to a distance of 6 R_J (jovian radii) and to about $4R_S$ from Saturn's surface. In this region the magnetic field is mainly due to the currents in Jupiter's interior. It is dipolar, although the magnetic dipole is tilted by about 10° with respect to the planet's rotation axis. The equatorial magnetic field at the surface of Jupiter is 4 G, that is, about 20 times stronger than that of the Earth. (At Saturn it is 0.2 G and at Uranus it is also 0.2 G. At Saturn the dipole is almost exactly parallel to the rotational axis.

At Uranus it is tilted by 60° with respect to its rotational axis which lies in the ecliptic plane.) Electrons and ions with energies in excess of several megaelectron volts have been found here. At Jupiter their intensity peaks at a radial distance of about $2R_J$ which is precisely the region from which DIM is emitted. The early explanations of DIM as being generated by such electrons gyrating about the planetary magnetic field have been fully confirmed by detailed calculations based on the observed intensities of electrons.

It is believed that the energetic electrons and ions are transported from the outer magnetosphere toward the planet by diffusion. The time scale for transporting the particles from $6R_J$ to the radiation belts is about one year. The diffusion coefficient increases with radial distance in a manner that strongly suggests that the diffusion is due to random fluctuations of the electric field in the inner magnetosphere which are created by turbulent winds in the jovian ionosphere. Particles convected inwards by the electric field gain energy, those that are driven outwards lose energy. Because the source of these particles is in the outer magnetosphere, there are always more particles being driven inwards than outwards. As in any diffusion process the direction of net particle transport is from a source toward a sink or opposite a density gradient. Theoretical expectations that the energy of a particle is proportional to the magnetic field, based on the conservation of the magnetic moment of the gyrating particle, have been confirmed by the observations in the inner magnetosphere (but not in the outer magnetosphere). Thus the energy of a particle transported from $6–2R_J$ increases by a factor of 27, because a dipolar magnetic field varies as r^{-3}, where r is the radial distance from the planet's center measured in the equatorial plane. This increase of energy as the particles get closer to the planet is responsible for the increase of synchrotron radiation power. Inside $2R_J$ the energy loss due to the emission of DIM becomes greater than the energy gained by inward motion and the particle fluxes at constant energy decrease inside of about $2R_J$. It is clear from this discussion why Saturn does not emit much synchrotron radiation. On the one hand, the magnetic field is smaller. In addition, the A ring absorbs the inward diffusing particles before they reach large enough energies to generate appreciable fluxes of synchrotron radiation. The number of energetic electrons in the uranian magnetosphere is also much smaller than at Jupiter. This and the weaker magnetic field of Uranus provides a simple explanation for the absence of detectable synchrotron radiation from Uranus.

PLASMA TORUS

The moon, Io, orbits Jupiter at a radial distance of $6R_J$. This moon is the most active volcanic body in the solar system. Even before *Voyager 1* discovered active volcanos on Io, it had been predicted that the tidal compression and expansion due to the second Galilean moon (Europa) and Jupiter would heat the interior of Io. Almost all of Io was predicted to be in a hot, liquid state with a very thin crust which should break occasionally and allow for volcanic eruptions. Even though it is still not entirely clear how, we know that a significant amount of gas (sulfur, oxygen, sodium, potassium) escapes the gravitational field of Io and populates the magnetosphere. Best estimates suggest that nearly 1 ton of material is injected every second. The gas is ionized by solar ultraviolet radiation and the hot magnetospheric electrons. It is then trapped by the magnetic field of Jupiter and forced to corotate with the planet. Because this Io-produced plasma cannot easily move in radial distance, it remains at $6R_J$ and forms the great Io torus, a dough-nut-shaped region of dense plasma (see Fig. 1). This torus has been mapped by the Voyager ultraviolet instrument and has also been observed from the Earth. The torus is quite variable, presumably related to the sporadic volcanic activity which ultimately is responsible for maintaining it. The presence of such a strong internal source of matter and hence plasma makes the jovian magnetosphere unique. Our own Moon is inactive and therefore does not

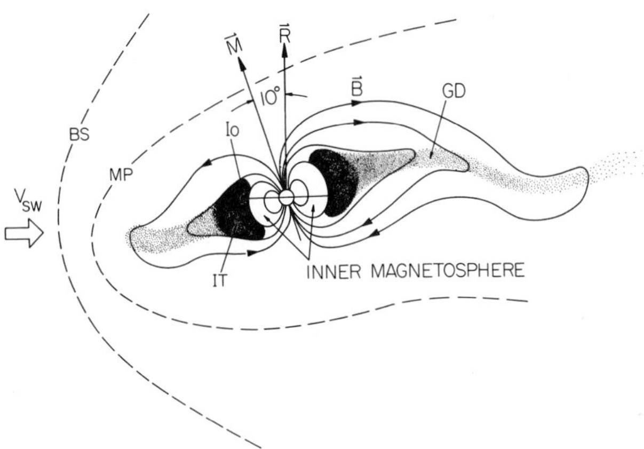

Figure 1. A cross section through the noon–midnight meridional plane in Jupiter's magnetosphere. The solar wind comes from the left and is slowed down and heated in the bow shock (BS). The magnetopause (MP) position is highly variable. The Io torus (IT) circulates the planet between $6R_J$ and about $10R_J$. The magnetic field lines are severely stretched in the outer magnetosphere where the plasma is confined to a thin Gledhill disk (GD).

influence the Earth's magnetosphere significantly. Although Saturn's magnetosphere contains many moons, none is such a strong source of plasma. There are indications that Enceladus and Dione are responsible for maintaining a similar torus which is, however, much less dense. In addition, it is not clear whether this torus is produced by these moons or by erosion of the dust particles in the tenuous E ring that coincides with it. Whether some of the uranian moons are active plasma sources or not is not clear. Certainly, none rivals Io. In the dense torus, electromagnetic waves are generated which derive their energy from energetic electrons and ions diffusing inwards from the outer magnetosphere. These waves have frequencies in the kilohertz range and scatter the energetic electrons into the jovian atmosphere where they produce aurora-like emissions in the ultraviolet and visible range. Lower-frequency waves are also believed to exist there and they scatter the energetic ions. However, their frequency is below the lowest frequency detectable by the Voyager wave experiments. The rate at which particle energy is deposited into the jovian atmosphere has been estimated to be as large as 10^{14} W. The torus thus acts as a barrier to the inward diffusing energetic particles. And, indeed, all spacecraft observed a significant decrease of the number of energetic particles as they flew across the torus.

The Io torus is symmetric about the magnetic equator which is inclined by 10° with respect to the rotational equator in which Io orbits Jupiter. Io thus moves up and down relative to the magnetic equatorial plane. As it does so the plasma density in its vicinity varies and the nature of its interaction with the plasma changes. Like any object moving through a fluid, Io generates perturbations in the plasma which propagate away from it. Because the plasma is a conducting magnetized fluid, three types of perturbations are produced: a slow acoustic mode, an intermediate Alfvén mode, and a fast mode. The most important one in this case is the Alfvén mode which involves not only a perturbation of the flow but also of the magnetic field. The energy of this mode is guided along the magnetic field and can thus propagate a long distance without geometrical attenuation. It has been estimated that the power in this wave generated by Io is as large as 10^{11} W. This power is transmitted along the magnetic field connecting Io with the ionosphere where through some as yet not well understood mechanism part of it is transformed into kinetic energy of electrons which in turn radiate energy away in the form of electromagnetic waves with a frequency roughly equal to the gyrofrequency of the electrons.

The power of DAM is highly fluctuating and can reach maximum values of about 10^{10} W. Near the ionosphere the gyrofrequency has a maximum value of 40 MHz which is precisely the upper frequency limit of DAM. Because the plasma density changes along Io's orbit, the power in the Alfvén waves also changes, which may explain why DAM is observed mainly when Io is above the longitudes toward which the northern and southern poles of Jupiter's magnetic dipole point. The details of this model for the Io control of DAM are rather complex and not completely understood.

THE INTERMEDIATE AND OUTER MAGNETOSPHERE

Even though the torus is dense (several thousand ions per cubic centimeter) by magnetospheric standards, it is not as dense as one might expect if the only loss of plasma were recombination. Of course, the torus plasma particles are subject to the same fluctuating electric fields mentioned previously as the energetic particles. Hence the torus particles will diffuse away, both inwards and outwards. However, this is too slow. Theory predicts that the outer edge of the torus cannot be stable. This can be understood by realizing that the torus is subject to an outward centrifugal force which at $6R_J$ is much larger than the inward gravitational force of Jupiter. Thus the torus is like a dense fluid "on top" of a less dense one further out. Such an interface between a dense fluid and a less dense one is unstable as everyone knows who has ever observed ink on top of water. The interface breaks up into small ripples and the two fluids quickly mix. The inner edge of the torus which is "below" the tenuous plasma of the inner magnetosphere is stable and is thus very sharp. In the outer region of the torus, the density decreases and merges with the plasma in the intermediate magnetosphere. Because of the turbulent mixing in the outer region of the torus, most of the torus plasma is transported outwards. Because the magnetic field decreases with distance from Jupiter, it cannot enforce rigid corotation any more and the plasma in the intermediate magnetosphere lags behind corotation. The plasma is dense and hot enough to cause a significant distortion of the magnetic field from its dipolar form. The field lines become stretched as shown in Fig. 1. This stretching is even more pronounced in the outer magnetosphere.

The presence of MeV electrons and ions in the intermediate and outer magnetospheres came as a total surprise. If these particles come from Io, they should be less energetic because outward moving plasma cools. If they come from the solar wind, an additional energization mechanism must be active in these regions. Several mechanisms have been discussed in the literature but none can explain all aspects of the observations. The most straightforward mechanism involves neutrals ejected from the torus at high speeds. If a corotating ion recombines with an electron in the torus, the resulting neutral will have a velocity much larger than the keplerian velocity. It will thus move on an elliptical orbit with an apogee in the outer magnetosphere or even in the solar wind. Some of these neutrals are reionized in the outer and intermediate magnetospheres. They will be picked up by the magnetic field. This represents a local source of plasma. Particles diffusing inward gain energy and could have large energies in the intermediate magnetosphere. This mechanism cannot, however, account for MeV electrons and ions in the outer magnetosphere. A second model involves the recirculation of particles through cycles of inward diffusion (energy gain) and outward transport conserving energy. This recirculation model can also only account for energetic particles in the intermediate magnetosphere. A third model invokes sporadic energization events due to induced electric fields created by rapid changes of magnetic field (perhaps magnetic reconnection events). The most detailed mechanism invokes the day–night asymmetry of the jovian magnetosphere which is shown in Fig. 1. Even though the plasma does not strictly corotate with the planet, it still makes complete orbits around it. As the plasma moves from the day side to the night side, it expands, because the solar wind compresses the front side but not the night side magnetosphere, and cools adiabatically. The reverse process, compression and adiabatic heating, occurs as the plasma moves back from the night to the front side. Thus the plasma is periodically compressed and expanded as it circulates around Jupiter once every 10 hours. If nonadiabatic processes such as scattering by waves occur, as they invariably will, the temperature of the plasma will increase very much like the air in a bicycle pump that is being worked through several cycles of compression and expansion. It has been shown that both electrons and ions can gain up to several megaelectron volts of energy in each orbit, that is, every 10 hours. This mechanism is called magnetic pumping. Ultimately, the energy is derived from the rotational energy of Jupiter. Of course, the resulting spindown of Jupiter is unmeasurably small. At the present time it appears that all mechanisms contribute to the observed energization of the plasma. The plasma in the outer magnetosphere is therefore very hot with a temperature of almost 0.5 billion degrees. Some of the energized electrons escape from the magnetosphere and can be detected in interplanetary space, even as far away as the orbit of Mercury and inside the Earth's magnetosphere.

In the intermediate magnetosphere the composition is similar to that of the torus (rich in heavy ions such as S^+). In the outer magnetosphere the energetic ions have a more solar-wind-type composition (rich in He^{++} ions). In the outer magnetosphere the particles are concentrated in the equatorial plane, whereas in the intermediate magnetosphere they are more evenly distributed along the magnetic field. The hot plasma in the outer magnetosphere forms a plasma disk which had been predicted by John A. Gledhill in 1967. This Gledhill disk is almost parallel to the magnetic equator and thus moves up and down as the planet rotates. Because the outer parts lag behind (the plasma does not strictly corotate), a snapshot of the disk would reveal a curvature as shown in Fig. 1, which resembles the skirts of a pirouetting ballerina. A similar disk exists in the saturnian magnetosphere, although it is much smaller. It seems to be a direct consequence of the internal plasma source and the rapid rotation of the planet which causes the plasma to move outwards and hence stretch the field lines into a disk-like configuration, very much like the solar wind produces such a magnetic field configuration in interplanetary space.

Magnetospheres are embedded in the solar wind which compresses the field on the front side and extends it into a long tail on the night side. We know from the Earth's magnetosphere that the solar wind induces a convection pattern in the magnetosphere. The outer layer of the magnetosphere is dragged along with the solar wind in the antisolar direction. In the interior of the magnetosphere a return flow toward the Sun is set up. The net convection of plasma is the superposition of this solar-wind-induced flow and corotation. At the Earth the solar-wind-induced convection dominates the magnetosphere and corotation is confined to the inner regions. At Jupiter and Saturn corotation dominates almost the entire magnetospheres. At Uranus the two convection patterns are orthogonal to each other because the rotation axis of Uranus lies in the ecliptic. Thus solar-wind-induced convection is important even in the inner regions of the uranian magnetosphere. The solar-wind-induced convection causes a mixing of solar wind and magnetosphere plasma. The fact that solar wind (He^{++}) ions are observed in the jovian magnetosphere confirms that this mixing occurs. However, the size and configuration of this mixing zone is not known. It appears that hot magnetosphere plasma is ultimately lost into the solar wind somewhere in the night-side magnetosphere. It has been suggested that the magnetic field is too weak there to maintain even approximate corotation and that the plasma moves freely away from the planet. This requires the escape of plasma from magnetic flux tubes. Because our knowledge of the night-side magnetosphere of the jovian planets is very limited, the physical processes that are responsible for this are not known. After the plasma has escaped the flux tubes, they return to the front side with a greatly reduced plasma density and one expects that there exists a region inside the front-side magnetopause where the plasma density is very small. This outer layer has been clearly identified in the saturnian magnetosphere but at Jupiter the evidence is less clear, probably because the flux tubes are more rapidly refilled by diffusion. In both magnetospheres this transition region is characterized by a more dipolar field orientation, with large fluctuations of magnetic field and plasma density.

The outer boundary of any planetary magnetosphere is the magnetopause which separates solar wind plasma from magnetospheric plasma. At the magnetopause the magnetic field changes from its magnetospheric value to the, usually, smaller magnetosheath value. The stronger magnetospheric magnetic pressure is balanced by the pressure of the magnetosheath plasma which is solar wind plasma heated in the bow shock that exists in front of the magnetopause (see Fig. 1). Because of the internal plasma sources, the magnetospheres of the giant planets are inflated and the magnetopauses are usually much further away from the planets than at the Earth. At Jupiter the magnetopause has been observed as far out as $120R_J$. Thus the jovian magnetosphere can be 10 times larger than the Sun. If it could be seen with the naked eye, it would appear bigger than the Sun. On the other hand, these magnetospheres can be more easily compressed by solar wind pressure increases and the jovian magnetopause has been observed as close in as $50R_J$. At the Earth such large variations of the magnetopause position are not observed.

These giant magnetospheres display a remarkably complex topology and dynamics. They generate radio waves, synchrotron radiation, and particles with energies in excess of several megaelectron volts which populate the entire solar system. As such they resemble stellar magnetospheres more than the Earth's magnetosphere. It is a triumph of technology that we have made in situ observations of these objects; it is an equal triumph of space plasma physics that we understand at least the fundamental processes that shape them.

Additional Reading

Dessler, A. J., ed. (1983). *Physics of the Jovian Magnetosphere*. Cambridge University Press, New York.

Gehrels, T., ed. (1976). *Jupiter*. University of Arizona Press, Tucson.

Gehrels, T. and Shapley, M., eds. (1984). *Saturn*. University of Arizona Press, Tucson.

Goertz, C. K. (1986). Jovian magnetospheric processes. *Magnetospheric Phenomena in Astrophysics*, Vol. 144, AIP Conference Proceedings, 208.

Lanzerotti, L. J. and Uberoi, C. (1989). The planets' magnetic environments. *Sky and Telescope* 77 149.

Van Allen, J. A. (1990). Magnetospheres, cosmic rays, and the interplanetary medium. In *The New Solar System*, 3rd ed., J. K. Beatty and A. Chaikin, eds. Sky Publishing Corp., Cambridge, MA, and Cambridge University Press, Cambridge, U.K., p. 29.

See also **Interplanetary and Heliospheric Space Missions; Outer Planets, Space Missions; Planetary Magnetism, Origin; Planetary Radio Emissions.**

Planetary Radio Emissions

Donald A. Gurnett

A wide variety of planetary radio emissions are known to be generated in our solar system. For our purposes, the term *planetary radio emission* is defined to be any naturally occurring electromagnetic emission that can propagate freely away from the planet. This definition excludes various types of waves known as plasma waves that are generated in planetary ionospheres and magnetospheres, but which cannot escape because of propagation constraints. It also excludes radio emissions produced by radio transmitters and other human activities.

Planetary radio emission can be conveniently classified into two types: thermal and nonthermal. Thermal radio emissions are simply part of the thermal radiation spectrum emitted by the planet. Thermal radiation is usually strongest at infrared frequencies and is very weak and difficult to detect at radio frequencies. Nonthermal radio emissions are caused by nonequilibrium charged-particle distributions in the atmosphere, ionosphere, and magnetosphere of the planet. Nonthermal radio emissions are classified into two types, incoherent and coherent, depending on whether the charged-particle motions are uncorrelated or correlated. For incoherent radiation, the electric fields of the individual particles add randomly. The total radiated power is then the sum of the powers radiated by the individual particles (i.e., proportional to N, the number of radiating particles). For coherent radiation, the electric fields of the individual particles add in phase. The total radiated power is then proportional to N^2, which increases very rapidly as N increases. Because of the N^2 dependence, coherent radiation is generally much more intense than incoherent radiation. However, some mechanism is required to produce correlated motions. For planetary radio emissions, this mechanism usually involves a plasma instability or an impulsive phenomenon such as lightning.

THERMAL RADIATION

Thermal radiation has been detected from all of the planets. At radio wavelengths the spectrum follows the Rayleigh–Jeans law,

$$B = \frac{2kT}{\lambda^2}, \tag{1}$$

which gives the brightness of the source (in W m^{-2} Hz^{-1}), as a function of the temperature T and wavelength λ. The quantity k is Boltzmann's constant. Although the thermal radiation is stronger at infrared wavelengths, infrared radiation is absorbed by clouds and sometimes does not give a reliable measurement of the surface temperature. Because planetary atmospheres are generally transparent at radio wavelengths, measurements in the radio part of the spectrum provide the best method for determining the surface temperature of a planet. The best known example of the use of radio measurements to determine the surface temperature of a planet occurred at Venus, when Cornell H. Mayer and collaborators, using microwave measurements, estimated the surface temperature to be nearly twice as hot as on Earth. These very high temperatures were later confirmed by the *Venera 7* lander, which measured a temperature of 740 K on the surface of Venus. If the planet does not have a solid surface, as is the case at Jupiter, Saturn, Uranus, and Neptune, then the spectrum gives the temperature deep in the interior of the atmosphere. An example of the thermal spectrum of Jupiter is shown in Fig. 1. The thermal spectrum can be seen on the right-hand side of the plot, at frequencies greater than about 10 GHz. The straight dashed line labeled "thermal" shows the $1/\lambda^2$ variation predicted by the Rayleigh–Jeans law. The effective disk temperature of Jupiter at these frequencies ranges from about 200–320 K.

INCOHERENT NONTHERMAL RADIATION

Although several types of incoherent nonthermal radio emissions can occur, the only planetary radio emission of this type that has been observed is synchrotron radiation. Synchrotron radiation is caused by the accelerated motion of charged particles moving in a magnetic field. The radiated power increases rapidly as the particle speed approaches the speed of light. Because of their smaller mass, electrons are much more effective radiators than ions. If the motion is circular, with no motion along the magnetic field, then the radiation occurs at harmonics of the cyclotron frequency, and extends up to a critical harmonic number, above which the inten-

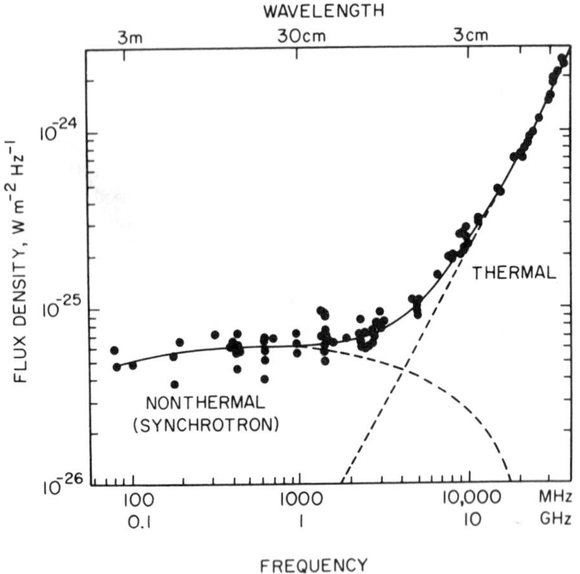

Figure 1. The thermal radiation and synchrotron emission spectrum of Jupiter. [*From Carr et al. (1983).*]

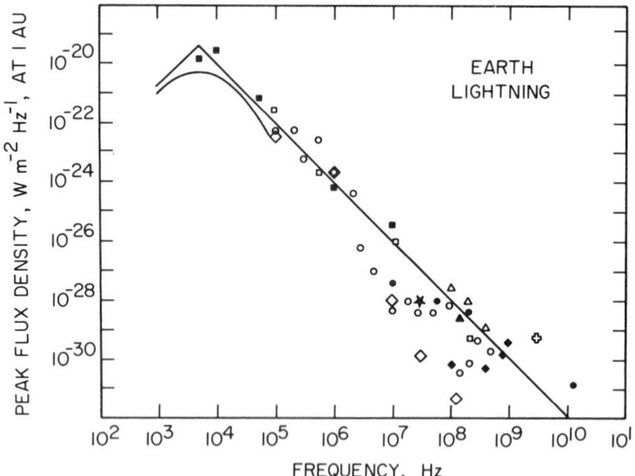

Figure 2. The spectrum of terrestrial lightning, adjusted to a distance of 1 AU. This radiation can only escape through the ionosphere at frequencies above about 10^7 Hz. [*From Pierce (1977).*]

sity decreases rapidly with increasing frequency. The critical harmonic number varies approximately as E^3, where E is the particle energy. The total power radiated is proportional to E^2B^2, where B is the magnetic field strength. Thus both a strong magnetic field and high energies are required for efficient generation of synchrotron radiation. When a distribution of velocities exists along the magnetic field, as is normally the case, the discrete harmonic spectrum is converted to a continuum. This process is called Doppler broadening. For nonrelativistic energies most of the radiation is concentrated near the first few harmonics. This type of radiation is usually called gyrosynchrotron radiation.

The only known example of planetary synchrotron radiation occurs at Jupiter. This radiation was first discovered by Russel M. Sloanaker in 1959. A spectrum of the jovian synchrotron radiation is shown on the left-hand side of Fig. 1. As can be seen, the synchrotron radiation is stronger than the thermal spectrum at frequencies below about 3 GHz, and has a broad peak centered on about 1 GHz. Because the highest intensities occur at wavelengths in the decimeter range, this radiation is often called jovian decimetric radiation.

Jupiter is an intense synchrotron source because the planet has a strong magnetic field and a very intense, energetic radiation belt. Measurements by the *Pioneer 10* and *11* spacecraft showed that the surface magnetic field of Jupiter is about 10 G and that the inner region of the magnetosphere is populated with intense fluxes of electrons with energies up to 40 MeV.

COHERENT NONTHERMAL RADIATION

Several types of coherent nonthermal planetary radio emissions have been identified. Compared to the thermal and incoherent nonthermal emissions described in the previous sections, coherent emissions are more highly structured and vary considerably from planet to planet. The basic mechanisms involved in the generation of coherent radio emissions are also more complex and poorly understood. At present, three distinct types of coherent nonthermal radio emissions are known to exist: lightning, cyclotron maser radiation, and mode conversion from electrostatic waves.

LIGHTNING

Radio signals from lightning have now been detected at six planets: Earth, Venus, Jupiter, Saturn, Uranus, and Neptune. At Earth, lightning discharges can be heard on a simple AM radio receiver. At Venus, lightning was first detected by a low-frequency (10–80 kHz) receiver on the *Venera 11* probe which entered the atmosphere of Venus on December 21, 1978. At Jupiter, impulsive very low frequency (1–10 kHz) signals called whistlers, which are known to be produced by lightning, were detected by the plasma wave instrument on the *Voyager 1* spacecraft which flew by Jupiter on March 5, 1979. Whistlers were also observed by the *Voyager 2* spacecraft as it flew by Neptune on August 25, 1989. At Saturn and Uranus, impulsive high-frequency radio signals, believed to be caused by lightning, were detected by the radio astronomy instruments on *Voyager*.

Lightning generates coherent radio emissions because the electrical breakdown initiates an impulsive, highly correlated motion of electrons along the discharge path. The coherence is very high at low frequencies (~ 10 kHz) and decreases at higher frequencies. On Earth, the discharge produces a spectrum that has a peak at about 5 kHz, decreasing approximately as f^{-2} with increasing frequency. A spectrum of terrestrial lightning is shown in Fig. 2. The energy emitted per discharge varies over a wide range. The high-energy terrestrial discharges are often referred to as "superbolts." Because of the limited amount of data available at other planets, it is difficult to carry out quantitative comparisons with terrestrial lightning. The comparisons that have been made suggest that lightning at the outer planets is considerably more energetic than terrestrial lightning, possibly comparable to terrestrial superbolts.

Cyclotron Maser Radiation

Five planets, Earth, Jupiter, Saturn, Uranus, and Neptune, are known to produce powerful radio emissions via a mechanism known as the cyclotron maser instability. This radiation was first observed from Jupiter by Bernard F. Burke and Kenneth L. Franklin in 1955 at a frequency of 22 MHz using a ground-based radio receiver. Further studies showed that the jovian radio emissions occurred over a broad range of frequencies extending up to about 40 MHz. Because the maximum intensities occurred at decameter wavelengths, this radio emission is called jovian decametric radiation (DAM). About 10 years after the discovery of the jovian decametric radiation, a similar type of radio emission was detected from the Earth's magnetosphere by the *Electron-2* satellite. Subsequent studies using a variety of Earth-orbiting spacecraft showed that the terrestrial radio emission is very intense (total radiated power $\sim 10^7$–10^9 W) and is generated at high altitudes over the

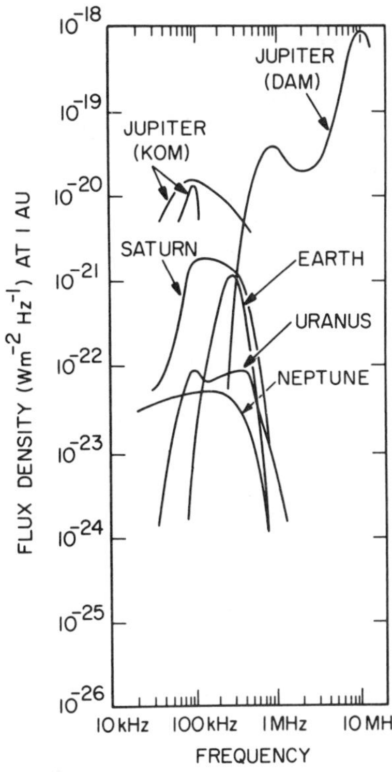

Figure 3. A comparison of the radio emission spectrums of coherent radiation from Earth, Jupiter, Saturn, and Uranus, adjusted to a distance of 1 AU. [*Provided by Michael L. Kaiser (1990).*]

evening auroral zone in association with discrete auroral arcs. The frequency range of the intense terrestrial radio emission typically extends from about 100–500 kHz. Because the maximum intensities occurred at kilometer wavelengths, the radiation is called terrestrial kilometric radiation, or auroral kilometric radiation. Similar types of radio emissions were also discovered from Saturn, Uranus, and Neptune by the *Voyager 2* spacecraft, which flew by these planets in the mid and late 1980s. A comparison of the radio emission spectra of Jupiter, Earth, Saturn, Uranus, and Neptune is shown in Fig. 3.

After several years of study, it is now clear that the intense radio emissions from these five planets have several common characteristics. First, in all cases the radiation is associated with a strong planetary magnetic field. Second, the emission occurs at a frequency very close to the electron cyclotron frequency, $f_c = (1/2\pi)eB/m_e$, where e is the electronic charge, B is the magnetic field strength, and m_e is the electron mass. Third, the polarization of the emitted radiation is primarily right-hand polarized with respect to the magnetic field in the source region. Fourth, the emission usually occurs in regions of very low plasma density. These characteristics are all consistent with a plasma instability called the cyclotron maser instability. For this instability an electromagnetic wave acts to organize the phase of the cyclotron motion in such a way that the electrons radiate in phase with the original wave. In this respect, the mechanism is similar to stimulated emissions from an optical laser. However, the free energy is not of atomic origin, as in a laser, but rather from anisotropies in the electron velocity distributions. At Earth, it is estimated that as much as 0.1–1.0% of the energy of the auroral electron precipitation is converted to radio emission via this process. Such high conversion efficiencies can only be obtained from a coherent process.

Mode Conversion From Electrostatic Waves

In addition to the cyclotron maser radiation, a wide variety of much weaker radio emissions are observed that are believed to be generated by mode conversion from electrostatic waves. These radio emissions were first discovered at Earth by the *IMP-6* spacecraft, and then were later discovered at Jupiter, Saturn, Uranus, and Neptune by the *Voyager* spacecraft. At Jupiter, this radiation occurs at kilometer wavelengths and is called jovian kilometric radiation (KOM). A typical spectrum of KOM is shown in Fig. 3. This type of radiation often consists of many narrowband emission lines that sometimes merge into an essentially continuous spectrum that has been called by various names, including "continuum radiation," "myriametric radiation," "narrowband electromagnetic emissions," and "narrowband kilometric radiation." In specific cases it can be shown that the radiation originates from localized electrostatic waves generated by intense fluxes of low-energy (few kiloelectron volts) electrons. The association with electrostatic waves strongly indicates that the radiation is produced by a mode conversion process that transfers some of the electrostatic energy to escaping electromagnetic radiation. Usually the electrostatic waves occur near half-integral harmonics of the electron cyclotron frequency, $(n+1/2)f_c$. These waves become particularly strong when the emission frequency is near a characteristic frequency known as the upper hybrid resonance, f_{UHR}. Although many details are poorly understood, the mode conversion process appears to be most efficient in regions with steep density gradients.

Additional Reading

Burke, B. F. and Franklin, K. L. (1955). Observations of a variable radio source associated with the planet Jupiter. *J. Geophys. Res.* **60** 213.

Carr, T. D., Desch, M. D., and Alexander, J. K. (1983). Phenomenology of magnetospheric radio emissions. In *Physics of the Jovian Magnetosphere*, A. J. Dessler, ed. Cambridge University Press, Cambridge, p. 226.

Farrell, W. M., Desch, M. D., and Kaiser, M. L. (1990). Field-independent source localization of Neptune's radiobursts. *J. Geophys. Res.* **95** 19,143.

Gurnett, D. A. (1974). The Earth as a radio source: terrestrial kilometric radiation. *J. Geophys. Res.* **79** 4227.

Gurnett, D. A., Shaw, R. R., Anderson, R. R., and Kurth, W. S. (1979). Whistlers observed by *Voyager 1*: detection of lightning on Jupiter. *Geophys. Res. Lett.* **6** 511.

Jackson, J. D. (1962). *Classical Electrodynamics*. Wiley, New York, p. 485.

Kaiser, M. L. (1989). Observation of non-thermal radiation from planets. In *Plasma Waves and Instabilities at Comets and in Magnetospheres*, B. T. Tsurutani and H. Oya, eds. Geophysical Monograph 53, American Geophysical Union, Washington, DC, p. 221.

Mayer, C. H., McCullough, T. P., and Sloanaker, R. M. (1958). Observations of Venus at 10.2 cm wavelength. *Ap. J.* **127** 1.

Pierce, E. T. (1977). *Lightning*. Academic, New York, p. 356.

Sloanaker, R. M. (1959). Apparent temperature of Jupiter at a wavelength of 10 cm. *Astron. J.* **64** 346.

See also **Earth, Aurora; Interplanetary and Heliospheric Space Missions; Mercury and Venus, Space Missions; Outer Planets, Space Missions; Planetary Magnetospheres, Jovian Planets.**

Planetary Rings

Laurance R. Doyle

When Galileo first spotted the rings of Saturn in 1610, he thought they might be "handles" or large moons on either side of the planet. He was perplexed, however, years later when he viewed the planet again and found that the rings had disappeared (were edge-on). In 1655 (based on the vortex theory of René Descartes),

Christiaan Huygens proposed that Saturn was surrounded by a *solid* ring, explaining many of the intervening observations reporting Saturn's strange shape and brightness variations. However, his solid ring had difficulties in remaining stable (the edge-on disappearance requiring it to be quite thin). Then, in 1675, Giovanni Cassini discovered a division or gap in the rings which, in 1789, led William Herschel to suggest that Saturn must be surrounded by two solid rings (these were later called the A ring—later found to be azimuthally asymmetric—and the B or bright ring extending radially inward from the Cassini division). For dynamical reasons, Pierre de Laplace argued in 1787 that the system must be composed of a large number of narrow solid rings. However, when the innermost C (or crepe) ring was discovered in 1848 and it was seen that the planet was visible through it, the idea of solid rings was discredited. The work of Edouard Roche further showed that the rings were inside the limit where tidal forces from a planet would break them up, lending credibility to the notion that rings were composed of orbiting small particles. The classic work of James Clerk Maxwell on the stability of the rings in 1857 finalized this conclusion, being observationally confirmed by James E. Keeler and William Wallace Campbell's spectroscopic detection of a keplerian radial-velocity profile for the rings in 1895. Observations of the rings in the next three-quarters of a century included possible "cloud-like" radial features in the B and A rings, an opposition effect (abrupt increase in ring brightness when viewed with the Sun almost directly behind the observer), and the discovery, by Walter A. Feibelman in 1967, of a large but faint ring outside Saturn's A ring, now known as the E ring; an unconfirmed ring inside the C ring was meanwhile dubbed the D ring. The latter half of the twentieth century, however, was really to be the renaissance of planetary ring observations.

In March 1977, the uranian rings were discovered serendipitously by James L. Elliot and collaborators using the *Kuiper Airborne Observatory* and Robert L. Millis at Lowell Observatory by observing the occultation by Uranus of the star SAO 158687. Ring structure around the planet Jupiter was discovered as well in March 1979, by the *Voyager 1* spacecraft on its grand tour of the solar system. During its sweep through the Saturn system in 1980, it was to discover that the rings of Saturn were really a diverse collection of thousands of ringlets while the *Voyager 2* spacecraft, during its encounter with the uranian system in January 1986, discovered two more rings there. Earth-based observations of stellar occultations by Neptune had also indicated that ring "arcs" existed around the planet, and *Voyager 2* observations in August 1989 revealed a ring system there. We will now take a look at each of these unique ring systems in more detail.

JUPITER'S RING SYSTEM

The jovian ring system consists of three components, the so-called bright ring, the vertically extended halo ring inside the bright ring, and an even sparser ring outside the bright ring called the "gossamer" ring. The bright ring extends from a rough inner boundary of about 122,000–129,130 km from the center of Jupiter, with a vertical extent of about 30 km. It consists of slightly reddish (probably silicate), micrometer-sized grains with an optical depth of about $\tau = 1\text{–}6 \times 10^{-6}$, very thin compared to other rings systems, as we shall see. (As light penetrates the rings, it decreases exponentially with depth. The diminishment can be expressed in units of factors of the constant e. This parameter is called the optical depth.) Mutual collisions between particles are thus rare, and the particles making up this ring are expected to be dominated (after gravity) by electromagnetic effects and radiation drag forces (plasma drag dominates). Small particles should also be short lived in the high-radiation environment of Jupiter. The bright ring also shows three evenly spaced enhancements (of about 10%) that may be caused by the moons Metis (embedded in the ring) and Adrastea (just inside the inner gradual boundary of the bright ring).

The halo ring is a toroidal cloud with a denser central core running from the bright ring's inner boundary to about halfway to the jovian cloud tops (90,000 km) extending vertically about 15,000 km. It is composed of particles a bit larger than the bright ring, which are thought to be slightly perturbed by the jovian magnetic field. The inner portion of this ring may be defined by a nearby Lorentz resonance (the grains' orbital periods here are commensurate with one of the vertical forcing periods of the magnetic field).

The gossamer ring extends from the bright ring outward to about 210,000 km and is about 5% the brightness of the bright ring (the optical depth is likely around $\tau \approx 10^{-7}$). It shows a 20% brightness enhancement at synchronous orbit (the location where a particle in keplerian motion would coorbit with the period of the planet). It is probably no thicker than about 4000 km, and is likely made up of very small particles. The moons Amalthea and Thebe are likely contributors to the particles in this ring, and Jupiter itself may also contribute. Overall, the jovian rings are interesting examples of short-lived small particles and their interaction with a very strong planetary magnetic field.

SATURN'S RING SYSTEM

Saturn has, by far, the most extensive ring system in the solar system, and it has been the most extensively observed—from radio wavelengths through the ultraviolet. It is from using these different wavelength "probes" that the ring particles' sizes and, to some extent, composition have largely been determined (Saturn's rings are mostly water ice). Moving radially outward from Saturn (in units of Saturn radii $R_S = 60{,}330$ km), the general ring areas are the D ring (from 1.11–1.225 R_S), the C ring (1.235–1.525), the B ring (1.525–1.949), the Cassini division (1.949–2.025), and the A ring (2.025–2.267). In between the outer edge of the A ring and the F ring are two moons, Atlas and Prometheus. The F ring is next (2.324–2.329), then the moons Pandora, Epimetheus, and Janus. The G ring is next (2.82–2.87), followed by the larger moon Mimas, the E ring (extending from 3.0–8.0 R_S), and finally the large moon Enceladus. We will first discuss the tenuous D, G, and E rings.

The innermost or D ring has about 1/100th the surface brightness of its neighbor, the C ring and, contrary to some previous reports, cannot be seen from Earth. The G ring, significantly outside the major rings, is the most tenuous of Saturn's rings ($\tau \approx 10^{-5}$, similar to the jovian halo ring), is several thousand kilometers in width, and is about 100 km in vertical extent. The *Voyager 2* spacecraft passed quite close to the outer portion of this ring recording a significant increase in small particle impacts there. The E ring, about the same optical depth as the G ring, extends from the outer edge of Mimas's orbit with its density peaking at the orbit of Enceladus, which is hypothesized to be the source for the E ring material (similar to the Titan torus produced further out). The E ring extends vertically about 2000 km and, together with the G ring, contributes significantly to the edge-on brightness of the rings altogether.

The F ring, just outside the major rings, is fairly unique in the Saturn system (ringlets in the Encke gap of the A ring as well as the neptunian rings show some similarities). The F ring clumps into bright regions and also twists on apparent time scales from hours to months. It is "shepherded" on either side by the moons Prometheus and Pandora, a possible cause for some of its transient morphology. Comparing radio observations with visible wavelengths, it appears to have a central core of larger particles ($\geq$ millimeters) while being composed of about 90% small particles ($\approx$ microns). At present, its dynamical behavior is not fully understood.

The least massive of the major ring regions is the C ring, the innermost of the three major rings. A small gap known as the Maxwell gap can be seen in its outer regions. With an optical depth of about $\tau \approx 0.1$, the C ring contains a fairly uniform distribution of particles that are less red than either the A or B ring particles. Radio wavelength observations indicate mostly large particles in the C ring ($\geq$ centimeters), whereas thermal emission (for a monolayer ring model) indicates that both hemispheres of the particles are heated, indicative of rapid spin rates for these particles.

Next in mass is the outermost of the major rings, the A ring. When seen edge-on, it is less than 200 m in vertical extent. Overestimates of the ring thickness had likely come from the bending waves (like warps in a plastic phonograph record left in the Sun) that propagate there. Spiral density waves (like the ones suggested to account for the arms in spiral galaxies) can also be found in the A ring. These are density increases resulting from the superposition of many elliptical ring particle orbits of increasing and slightly rotated semimajor axes. A great deal of the structure in the A ring can be accounted for by these phenomena—the density waves being caused by orbital resonances with various moons, and the bending waves likely caused by the slightly inclined orbit of Mimas. What may cause the large-scale asymmetry is still being debated. In the outer portion of the A ring is the Keeler gap, whereas still farther in is the larger Encke gap. On either side of this latter gap, a "wake" of ring material can be seen to propagate, being the result of a recently discovered moon orbiting there. Kinky ringlets have also been imaged propagating across this gap.

Radially inward of the A ring is the Cassini division, which is actually not empty, but a lower-density area (about the density of the C ring), cleared by the Mimas 2:1 resonance (particles here orbit Saturn twice every orbital revolution of Mimas). It contains about five additional gaps, the inner one being dubbed the Huygens gap. (This phenomenon is similar to the Kirkwood gaps created by Jupiter in the asteroid belt.)

Radially interior to the Cassini division is the B ring of Saturn, containing probably about three-quarters of the total ring mass. When the *Voyager* spacecraft recorded the attenuation of the star δ Sco as it passed behind the rings (showing that ring structure extended down to the 100-m scale), the star was completely obscured by the outer B ring implying $\tau \geq 3$ (compared to the A and inner B rings with an average $\tau = 0.4$ and 0.7, respectively). The major constituent particles of the B ring extend to slightly larger sizes than the A ring (centimeters to a hundred meters) and were shown from infrared spectra and microwave observations to be composed almost exclusively of water ice (albedos of about 0.55). Both rings show a deficiency in freely floating micrometer-sized particles. The exception is the so-called "spokes" that form in the outer B ring. These are radial dark clouds about 8000 km long and 2000 km wide that develop in a matter of minutes, evolve (widen from the nearly corotating radial origin line), and then shift to keplerian rotation, broadening to a triangular form (base innermost), and disappearing in a matter of several hours. The spoke particles are enhanced when a certain region of Saturn's magnetic field (whose lines pass through the rings) is rotating past. Spokes are dark when backlit (Sun behind the observer) but become brighter (by contrast with the underlying B ring) when forwardlit, an indication that they are micrometer-sized particles. (Wavelength-sized particles diffract light forward instead of reflect it back. An example is the brightening of the dust on a car windshield as it turns toward the Sun.) One theory for the origin of spokes is that they are the result of meteoroid impacts into the rings producing a radially propagating plasma (Lorentz forces from Saturn's magnetic field force the plasma into a radial direction). The plasma then electrostatically levitates small ring particles off the regolith of the larger particles as it rushes outward (or inward if the origin is inside corotation).

One may ask the question here whether the rings are left over from the condensation of Saturn, or were created later as the result of, for example, a moon passing within the tidal Roche limit. (The rings' mass is about the same as the mass of the moon Mimas.) Studies of the rings presently seem to favor the latter theory when meteoroid erosion of the particles, angular momentum transport of the material outward, and the amount of meteoritic material in the rings (how long have the rings been catching meteoroids if they started out as pure ice?) are considered.

URANUS' RING SYSTEM

The uranian rings, in order radially outward (in 10^3 km), are UR2 (37–39.5), 6 (41.85), 5 (42.24), 4 (42.58), α (44.73), β (45.67), η (47.18), γ (47.63), δ (48.31), λ (50.04), and ε (51.16). In addition, there are broader regions of material in between these major ring features. Each major ring besides UR2 and λ (previously UR1) were discovered by Earth-based infrared stellar occultation observations. The uranian rings are, in general, sharp-edged, narrow, and have (visual) optical depths of $\tau \geq 0.3$ (the γ and ε ring optical depths are closer to ≈ 2). Rings 6, 5, 4, α, and β are inclined up to 46 km above the ring plane, whereas all rings but the η ring are eccentric (the largest, the ε ring, being the most eccentric at ≈ 0.0079). The η ring, as narrow as rings 6, 5, and 4 (about 3.5 km), also has a 60 km wide "shoulder" of optical depth $\tau \leq 0.1$. All the rings show variations in azimuth, the ε ring varying in width from 20 km at periapse to 98 km at apoapse (the α and β rings vary from 5–12 km). These narrow ringlets (also very thin, about 150 km on edge) first suggested the idea of shepherding satellites. However, only two such satellites, Cordelia and Ophelia, respectively, were detected orbiting just inside the λ ring and immediately outside of the ε ring by *Voyager 2*. This may be because the ring particles are very dark (albedo ≈ 0.15, similar to carbonaceous chondritic meteorites). Detection of any moons smaller than about 10 km of this albedo is additionally difficult because the Sun is so far away. The particles composing the uranian rings turned out also to be centimeter-sized or larger and very rough-surfaced, with very few micrometer-sized particles ($\leq 0.1\%$). Dust could have been removed rapidly, however, due to gas drag from the extended uranian exosphere. A very tenuous distribution of micrometer-sized dust between the rings was, however, detected in extreme forward scatter. Because the lifetime of micrometer-size particles is short in this ring environment, a source for the renewal of the dust in the uranian rings (meteoritic impacts or moonlets, for example) remains an interesting question. (See Table 2, page 939.)

NEPTUNE'S RING SYSTEM

Ground-based stellar occultations of Neptune revealed that ring-like material exists around the planet. However, dips in the stars' brightness going into occultation were not accompanied by a dip on the opposite side of the planet. This led to the idea of ring "arcs" around Neptune. The *Voyager 2* spacecraft flew past the Neptune system in August 1989, and detected six rings. They are all prograde, equatorial, and circular. Two narrow rings (named 1989N1R and 1989N2R) lie at 62,900 km and 53,200 km from Neptune, respectively (well within the Roche limit). 1989N1R has three arcs of 4°, 4°, and 10° separated in order by 14° and 12° of fainter ring material. Both narrow rings lie just 1000 km outside the moons Despoina and Galatea. The inter-arc material of 1989N1R has an optical depth of about $\tau \approx 0.01$–0.02, whereas the arcs here have $\tau \approx 0.04$–0.09. The small-particle fraction of ring 1989N1R is also about 30%, whereas that of the other rings appears to be twice as high. From the brightening of the rings in forward lighting, the arcs likely have a dust content as high as 70%, similar to Saturn's E and F rings. The first broad ring (1989N3R) lies at 41,900 km with diffuse material extending outward to about 1700 km. Extending from 1989N2R to about 59,000 km is 1989N4R, dubbed the plateau. It has a brighter edge at 57,500 km now called 1989N5R. In addition, there is also an extended sheet of material that may fill the inner neptunian system. The two broad rings have optical depths of about $\tau = 10^{-4}$. However, this indicates that the total dust content of the ring system is about two orders of magnitude greater than that of, for example, the jovian system. Because the dust in this system is also a short-lived phenomenon, the meteoritic flux at Neptune may have to be significantly greater to replenish the rings. Also of interest, is that a comparison of Earth-based data over several years with Voyager data shows that some of the arcs have existed for at least five years. What makes this arc-like structure is presently a new and open question in planetary ring studies.

In conclusion then, planetary rings are one of the best-studied representatives of the "other shape" (besides spheres) in the universe. In addition to planetary rings, the primitive solar nebula was likely a disk (as the planar planetary orbits indicate), and similar disks can be seen forming around certain other stars. Also, the spiral galaxies are ring disks of sorts, and many of the mechanisms active in them (density waves, for example) apply on all scales. In the final consideration, however, planetary rings are not only interesting scientifically, they are also certainly some of the most beautiful objects in the universe.

Additional Reading

Cuzzi, J. N. and Esposito, L. W. (1987). The rings of Uranus. *Scientific American* **257** (No. 1) 52.

Doyle, L. R., Dones, L., and Cuzzi, J. N. (1989). Radiative transfer modeling of Saturn's outer B ring. *Icarus* **80** 104.

Gehrels, T. and Matthews, M. S., eds. (1984). *Saturn*. University of Arizona Press, Tucson.

Greenberg, R. and Brahic, A., eds. (1984). *Planetary Rings*. University of Arizona Press, Tucson.

Ockert, M. E., Cuzzi, J. N., Porco, C. C., and Johnson, T. V. (1987). Uranian ring photometry: Results from *Voyager 2*. *J. Geophys. Res.* **92** 14969.

Pollack, J. B. and Cuzzi, J. N. (1981). Rings in the solar system. *Scientific American* **245** (No. 5) 104.

Showalter, M. R., Burns, J. A., Cuzzi, J. N., and Pollack, J. B. (1987). Jupiter's ring system: New results on structure and particle properties. *Icarus* **69** 458.

Smith, B. A. et al. (1979). The Jupiter system through the eyes of *Voyager 1*. *Science* **204** 951.

Smith, B. A. et al. (1979). The Galilean satellites and Jupiter: *Voyager 2* imaging science results. *Science* **206** 927.

Smith, B. A. et al. (1981). Encounter with Saturn: *Voyager 1* imaging science results. *Science* **212** 163.

Smith, B. A. et al. (1982). A new look at the Saturn system: The *Voyager 2* images. *Science* **215** 504.

Smith, B. A. et al. (1986). *Voyager 2* in the uranian system: Imaging science results. *Science* **233** 43.

Smith, B. A. et al. (1989). *Voyager 2* at Neptune: Imaging science results. *Science* **246** 1422.

See also **Jupiter and Saturn, Satellites; Outer Planets, Space Missions; Satellites, Minor; Uranus and Neptune, Satellites.**

Planetary Rotational Properties

Alan W. Harris

Rotational motion is ubiquitous among solid bodies, from atoms to planets. The rates and orientations of planetary rotations can tell us something of the way the bodies were formed and of their subsequent evolution. Several planets, the Earth included, have experienced substantial evolution of their spin rates and/or orientations since their formation. The distortions of the physical shape and gravity field of a fluid planet can be used to deduce the radial density profile of the planet.

Table 1 is a summary of the present rotational characteristics of the planets. The rotation periods listed for Jupiter, Saturn, Uranus, and Neptune are the "internal" rotation rates, determined from radio emissions modulated by the rotating magnetic fields. The cloud tops of the first three of those planets rotate at faster rates, corresponding to wind speeds of hundreds of meters per second. The cloud tops of Neptune rotate slower than the deep interior, with similar wind speeds. Detailed measurements for each of these planets were made by the Voyager spacecraft.

Table 1. Rotational Properties of the Planets

Planet	*Rotation Period*	*Obliquity*	*Precession Period (yr)*
Mercury	$58\overset{d}{.}646225$	0°	(2,000)
Venus	$243\overset{d}{.}0250$	$177\overset{\circ}{.}33$	21,500
Earth	$23\overset{h}{.}934471$	$23\overset{\circ}{.}45$	26,000
Mars	$24\overset{h}{.}622962$	$25\overset{\circ}{.}19$	173,000
Jupiter	$9\overset{h}{.}925$	$3\overset{\circ}{.}45$	450,000
Saturn	$10\overset{h}{.}656$	$27\overset{\circ}{.}34$	1.8×10^6
Uranus	$17\overset{h}{.}24$	$97\overset{\circ}{.}02$	5.7×10^8
Neptune	$16\overset{h}{.}11$	$27\overset{\circ}{.}67$	1.3×10^8
Pluto	$6\overset{d}{.}38723$	$115\overset{\circ}{.}42$	7.3×10^6

ORIGIN OF ROTATIONAL MOTION

The origin of planetary rotational motion is related to the formation process itself. As the planets accumulated from dust and gas in the "solar nebula," their own gravitational fields distorted the trajectories of incoming material slightly from isotropic infall, to produce a slight prograde vorticity to the incoming flux of matter. It has been shown that streamlines infalling from perfectly circular heliocentric orbits result in a slight retrograde vorticity; if the heliocentric orbits are slightly eccentric, a net prograde vorticity results, but diminishes with increasing eccentricity. To match the observed spin rates of the planets, matter forming them must have arrived from orbits of very low eccentricity, so low in fact that the radial excursions implied are insufficient to deliver matter from between the planets to the adjacent neighbors. This is an unresolved issue concerning the origin of the planets.

A second source of rotational motion is the stochastic impulses received by a growing planet from collisions with very large secondary bodies. Unlike the previous effect, this effect leads to randomly oriented spin impulses. The obliquities of the planets have been interpreted as a measure of this random component of spin, to deduce the size of the largest planetesimals falling on the planets in the formation process. One must be cautious of the statistics of small numbers, in that at best, we have only nine cases, and of them, several are highly evolved spin states. Thus it cannot be ruled out (particularly for the terrestrial planets and Pluto) that the spin orientations are totally random and completely dominated by the stochastic component.

EVOLUTION OF ROTATIONAL MOTION

A spinning planet experiences a torque couple between its equatorial bulge and the Sun, which causes the spin axis to precess. Satellites close to the planet, such as the Galilean satellites of Jupiter, are tightly coupled by the planet's equatorial bulge and coprecess with it. Thus they both increase the effective moment of inertia of the system, which tends to lengthen the precession period, and increase the strength of the torque couple with the Sun, which tends to decrease the precession period. The combined effect, for all of the outer planet satellite systems, is to decrease the precession periods.

A more distant satellite, such as the Moon, is dominated by solar perturbations and does not follow the planet's equator. The Moon contributes only to the torque on the Earth and not to the effective moment of inertia, thus increasing the precession rate of the Earth. In fact, the Moon contributes about twice as much torque as the Sun and decreases the period of precession by about a factor of 3 from what it would be without the Moon. The precession rates of the planet spin vectors are listed in the last column of Table 1.

The orbits of the planets also precess about the "invariable plane" of the solar system (nearly the plane of Jupiter's orbit). If the rate of precession of the spin axis is fast compared to the orbital precession rate, then the spin axis precesses at a nearly constant angle with respect to the orbit plane, and the motion tracks the moving orbit plane. This is the case for the inner three planets, and the obliquities listed in Table 1 are those angles between the spin axes and the respective orbital planes.

If the spin axis precession is slow compared to the precession of the orbital plane, then the spin axis precesses at a nearly constant angle to the mean plane of the orbital precession, the invariable plane of the solar system. This is the case for the outer five planets, and the obliquities listed are with respect to the invariable plane.

Mars is an intermediate case: Its rate of precession is currently very close to one of the eigenfrequencies of its orbital precession, and the resultant precessional motion is very complex, so that Mars' obliquity with respect to its orbit plane can vary from 15–35°, on a time scale of hundreds of thousands of years. Furthermore, the precession rate is significantly influenced by the Tharsis Ridge of volcanos, which are geologically young. It is likely that as this ridge grew, Mars' precession rate passed through the resonance, with resultant excursions in obliquity of 9–46°. This wide range of obliquities has probably had a profound effect on the past climate of Mars. The obliquity listed in Table 1 is the current value with respect to its orbital plane.

In addition to the resonant variation of obliquity, mentioned previously, several spin states are highly evolved as a result of tidal friction. It has long been known that the Earth's rotation is slowing down due to the energy dissipation in the ocean tides. The other half of this torque couple is causing the Moon to recede from an earlier orbit closer to the Earth. Exactly how close the Moon once was is not certain. Whereas the present rate of recession is very accurately measured, it is generally realized that the rate of dissipation is dominated by the arrangements of the continents and shallow seas, which have changed considerably over geologic time. Thus the rate of recession has undoubtedly been different at different times. If the Moon were once very near the Earth (say, less than 10 $R_\oplus$ away), which seems likely, then the "primordial" spin period of the Earth would have been 6–8 h, and the obliquity would have been about 10^0.

Mercury and Venus have also apparently had their spins dramatically altered by tidal friction, in these cases by the Sun. Mercury is presently locked in a 3:2 resonance between its rotational and orbital rates. Its spin axis is also locked in what is called a "Cassini state," such that the spin axis coprecesses with the orbital plane. The Moon is also in a Cassini state, such that its spin axis is inclined about 1°.5 to the orbital plane and coprecesses with it with an 18-yr period. The lunar case was first described by Giovanni D. Cassini about 300 years ago, but only recently has it been shown that this is a natural outcome of tidal evolution. In the case of Mercury, the obliquity corresponding to the Cassini state is immeasurably close to 0. Venus' rotation rate is curiously close to a resonance with the synodic period between the Earth's and Venus' orbital motions (the time between conjunctions of the two planets), but it is distinctly different, so the present or past existence of a resonant lock is highly unlikely. Nevertheless, Venus' spin is undoubtedly highly changed from its primordial rate. It seems very likely that the initial spin was retrograde, like the present state, because it is difficult to devise a mechanism that would reverse the direction of the spin, but otherwise, it is unknown what the initial rate or obliquity might have been. Pluto has also evolved from its initial spin state. Like Mercury it has reached a tidal end state, in this case having reached synchronism with its satellite, Charon. In this state, both bodies are spinning synchronously with their orbital motion, so that the two behave as a single "rigid body." All relative motions between the two bodies have been damped by tidal friction.

ROTATION DYNAMICS AS A PROBE OF INTERNAL STRUCTURE

The spin of a planet induces a distortion in the physical shape of the body, the equatorial bulge, and a corresponding distortion in the gravitational field, characterized to lowest order by the quadrupole moment of the gravitational field. For a homogeneous fluid, these distortions can be calculated exactly knowing only the spin frequency and mean density, or alternatively, if one knows a body is homogeneous and fluid, the mean density can be found from the spin period and physical or gravitational distortion. If the body is fluid but radially stratified (i.e., increasing density toward the center), then the equilibrium state is more complex. The planet can be characterized by four parameters:

1. The dimensionless quantity $G\rho/\omega^2$, where ρ is the mean density and ω is the spin frequency (essentially the ratio of the gravitational acceleration to the centrifugal acceleration at the equator).
2. The degree of central condensation, characterized by I/Mr^2, where I is the equatorial moment of inertia and M and r are the mass and equatorial radius, respectively ($I/Mr^2 = 0.4$ for a homogeneous sphere, and less for a centrally condensed sphere).
3. The optical flattening, which is defined as the difference between the equatorial and polar radii divided by the equatorial radius.
4. The quadrupole moment of the gravitational field, characterized by the difference between the equatorial and polar moments of inertia.

It can be shown that if any two of the previous four parameters are known, the other two are uniquely determined. In most cases, the spin period can be observed, and the quadrupole moment of the gravitational field can be determined from the precession of satellite orbits or by accurate tracking of spacecraft near the planet. The optical flattening can also be observed, serving as a consistency check on the assumption of fluid equilibrium. These observations thus lead to a measure of the central condensation of the planet, of fundamental importance in understanding the structure and composition of a planet. In the case of the terrestrial planets, the deviation of the figure from the fluid equilibrium shape can be taken as a measure of the planet's rigidity or plasticity. For example, the Earth is somewhat more flattened than the equilibrium amount, indicating a lag in relaxing to the equilibrium figure as the spin rate slows down due to tidal friction. Higher-order gravitational harmonics of the terrestrial planets have been used to infer crustal strength and thickness.

Additional Reading

Davies, M. E. et al. (1989). Report of the IAU/IAG/COSPAR working group on cartographic coordinates and rotational elements of the planets and satellites: 1988. *Celestial Mechanics and Dynamical Astronomy* **46** 187.

Harris, A. W. (1977). An analytical theory of planetary rotation rates. *Icarus* **31** 168.

Harris, A. W. and Ward, W. R. (1982). Dynamical constraints on the formation and evolution of planetary bodies. *Ann. Rev. Earth Planetary Sci.* **10** 61.

Kaula, W. M. (1968). *An Introduction to Planetary Physics: The Terrestrial Planets.* Wiley, New York.

Safronov, V. S. (1969). *Evolution of the Protoplanetary Cloud and Formation of the Earth and Planets* [Translated from Russian (1972)], NASA TT F-677. Israel Program for Scientific Translations, Jerusalem.

Stevenson, D. J. (1982). Interiors of the giant planets. *Ann. Rev. Earth Planetary Sci.* **10** 257.

See also **Earth, Figure and Rotation; Satellites, Rotational Properties; Solar Systems, Origin.**

Planetary Systems, Formation, Observational Evidence

David C. Black

A picture is emerging as to how planetary systems like the solar system form. It is believed that planetary systems form as a consequence of the process of star formation that takes place in a rotating dissipative disk formed through the gravitational collapse of high-density cores within molecular clouds. Although this theoretical picture has reached the status of a paradigm, its ultimate test will come through observations. These observations are generally of two kinds: those relating to the process of star formation (i.e., planetary systems in formation), and those relating to the existence of other planetary systems (i.e., the end product of the process).

The paradigm that has emerged in recent years is one in which formation begins with the gravitational collapse of molecular cloud cores, objects that are nearly in equilibrium under the combined effects of gravity, gas pressure, magnetic fields, and rotation. As collapse takes place, rotation plays an increasingly prominent role relative to gravity, leading to the formation of a flattened disk. This disk is thought to be highly dissipative. An inescapable consequence of this dissipation is that angular momentum is transferred outward in the disk with a resulting flow inward of mass to form the young star. This same disk is thought to be the nursery of planets that form in association with the star. Although this brief summary omits many problems and details, it is adequate to set out the key ingredients in a program to test this paradigm through observations.

EVIDENCE OF DISKS AND THEIR EVOLUTION

The last decade has witnessed a significant increase in the ability of astronomers both to detect and to study disks in regions of star formation. The diffuse, low temperature, and often heavily obscured nature of disks associated with forming stars makes them difficult to study observationally; it also leads to an emphasis on observations in the radio and infrared part of the spectrum.

The discovery of energetic and, in many cases, highly collimated bipolar outflows or winds from young stellar objects, was the first convincing evidence that astronomers had for disk structures associated with young stars. Although these discoveries provided only circumstantial evidence for the presence of disks associated with stars during their formation, much of that evidence has been supported by more recent and direct studies.

Imaging of young stellar objects, notably HL Tau, R Mon, and L1551/IRS 5, by both direct and speckle observations in the near-infrared part of the electromagnetic spectrum gives compelling evidence of the presence of dusty disk structures. The observations sense light from the star that is scattered by submicrometer-size and larger dust grains in the circumstellar disks around those stars. Evidence for gas disks has been provided by radio observations of continuum radiation at millimeter wavelengths as well as emission from CO.

A powerful method of observation of disks associated with pre–main-sequence stars such as Herbig emission stars and T Tauri stars involves the study of relatively broad infrared spectral features arising from thermal emission by dust grains in disks around these stars. The dust, which is heated by radiation from the star, appears to be located as close to the star as a few tens of astronomical units (AU) in some cases, and as far from the star as a few hundred astronomical units in others. The dust that is close to the star is characterized by temperatures near 1000 K, whereas the dust that resides in the outer reaches of a disk is typically at temperatures of a few to several tens of degrees kelvin.

Estimates of the amount of dust required to account for the radiation range from 10^{-5}–10^{-4} $M_{\odot}$. The ratio (by mass) of gas to dust in the interstellar medium is roughly 100:1. If that ratio applies to these disks as well, and it may not, then the amounts of dust would imply a total mass in these disks ranging from 0.001–0.01 $M_{\odot}$.

An intriguing set of studies has been conducted recently that uses ground-based optical and infrared observations in conjunction with data from the *Infrared Astronomical Satellite* (*IRAS*) that was launched in 1983 and operated for nearly a year. These observations are intended to address the issue of whether pre–main-sequence stars that will end up being similar to the Sun show evidence for disks and whether there is any indication of how those disks evolve with time.

That study revealed a number of interesting results. One can define an index that is a measure of an excess of infrared flux, relative to a stellar standard, and that index can be used to indicate both the presence of a dust disk and to a lesser degree the amount of dust in the disk. Nearly 60% of the stars with ages less than 3×10^{6} yr that were studied in the survey showed significant excess. This represents a lower limit to the fraction of such stars that might have had disks at some time early in their history. The survey also showed that only 10% of the stars with ages greater than 10^{7} yr showed significant excess.

If it is assumed that all stars of this type have disks, then the evidence would indicate that the disks evolve significantly on time scales ranging from much less than 3×10^{6} yr up to around 10^{7} yr. This study provides the first observational constraint on possible time scales for disk evolution and, therefore, for building of giant planets similar to Jupiter and Saturn.

A schematic summary of results from this recent study is shown in Fig. 1. The dotted curves in the two panels in Fig. 1 indicate the spectrum that one would expect from the star if it had no disk. The solid lines in the two panels are intended to indicate, in a *generic* way only, the type of spectra that were observed in the study by Karen M. Strom and collaborators. An intriguing aspect of the difference in spectra between the upper and lower panels is that they may be providing evidence for evolution within disks. Disks of the type represented in the upper panel appear to be intact; there is hot dust as well as cooler dust contributing to the spectrum indicated by the solid line. In contrast, there is a paucity of short wavelength infrared (i.e., hot dust) emission from the disks represented in the lower panel, but emission from the cooler dust is present. This may result from a clearing out of dust from the inner regions of disks, either due to loss processes or because the process of planet building has taken place in these disks and has removed the dust. Much more work needs to be done in this area of observational research, but the promise for exciting results is clear.

EVIDENCE FOR OTHER PLANETARY SYSTEMS

Planets are small and dim objects compared to most denizens of the astronomical zoo. This, coupled with the fact that they are in close physical association with a relatively massive and bright object (their central star), makes detection of other planetary systems difficult.

One can identify three basic objectives for a search effort. In order of increasing observational difficulty they are: Detect another planetary system, gather statistical information concerning planetary systems (e.g., their frequency of occurrence as a function of

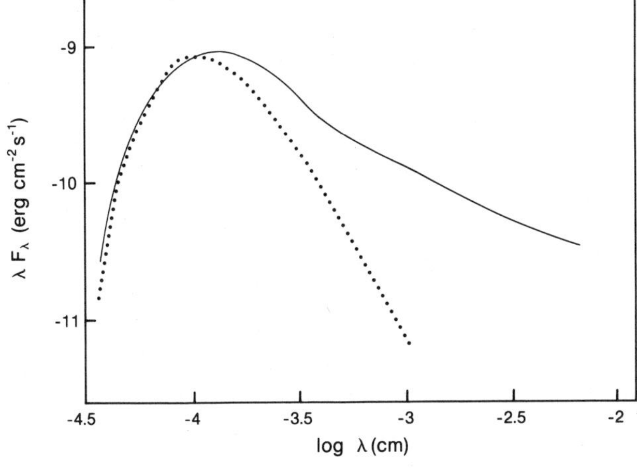

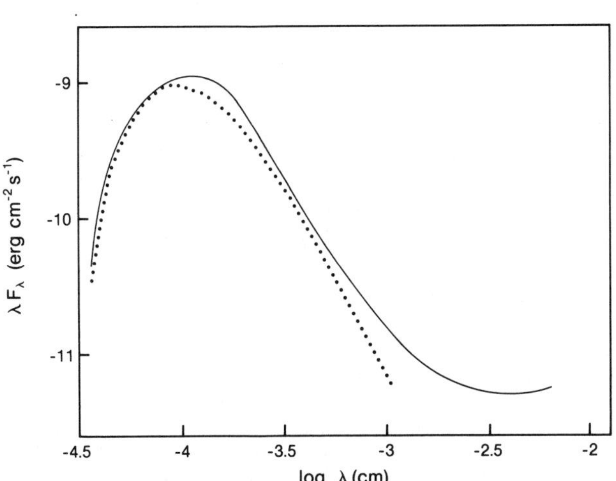

Figure 1. Plot of the spectral energy distribution for standard stars (dotted curves) and of a schematic representation of data from young stellar objects (solid curves). The upper figure is typical of stars that show evidence for disks with a full and continuous range of temperatures ranging from nearly 1000 K down to a few hundred degrees kelvin. The upper figure is typical of stars that show some evidence for a lack of high-temperature dust in the disk (i.e., dust close to the star) suggesting that those disks may be evolving through planet building or some other process that removes dust.

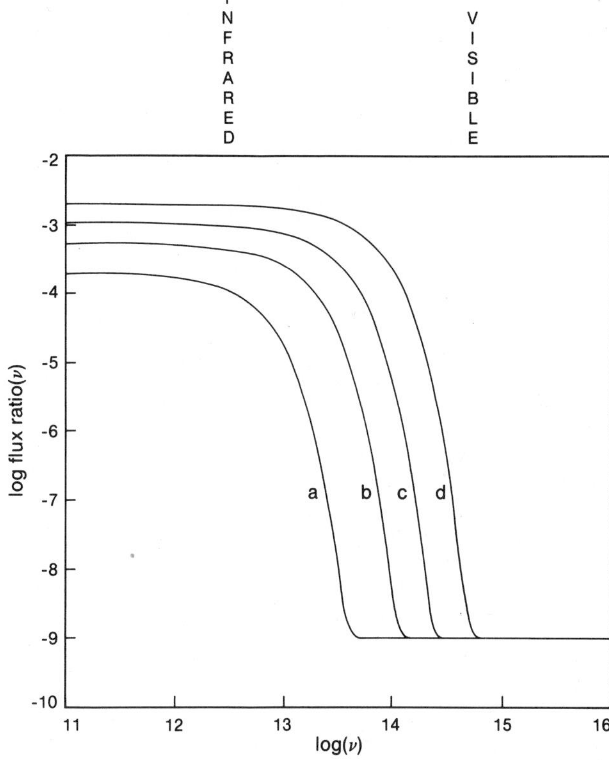

Figure 2. Plot of the ratio of power emitted as a function of frequency by Jupiter to that emitted by the Sun. Curves *a*, *b*, *c*, and *d* correspond, respectively, to the output from present-day Jupiter, and to Jupiter at ages of 10^8, 10^7, and 10^6 yr. The Sun is nearly a billion times brighter than is Jupiter at visible wavelengths, but is only several thousand times brighter at infrared wavelengths. A young, hot Jupiter-like planet is much brighter in the infrared than is a mature Jupiter and may therefore be easier to detect.

their star's spectral type), and gather detailed information about specific planetary systems. It is worthwhile stressing that although the discovery of another planetary system would be exciting, significant scientific return begins with the more difficult task of gathering statistical information. Only then will we be able to view the formation of our own planetary system in context.

Techniques to search for other planetary systems can be placed into two general categories: *direct* and *indirect*. Direct techniques are those by which an observer detects radiation (e.g., reflected visible light) emanating from a planetary companion to another star, whereas indirect techniques are those by which an observer infers the presence of companions to another star through observation of some effect those companions have on that star (e.g., a perturbation in the star's motion). Each of these types of techniques have strengths and weaknesses which will be discussed.

A pictorial representation of the nature of the direct detection problem is shown in Fig. 2. What is shown is the ratio of power emitted by the Sun at a given frequency to that emitted by Jupiter (or a Jupiter-like planet) at various stages in its history. The curve labeled *a* corresponds to present-day Jupiter, and the curves labeled *b*, *c*, and *d* correspond, respectively, to what Jupiter was like at ages of 10^8, 10^7, and 10^6 yr. Note that the Sun is much brighter, nearly a factor of a billion, at visible wavelengths. The ratio is much more favorable at infrared wavelengths, particularly for the progressively younger and therefore warmer Jupiter. This suggests that looking for Jupiter-like planets around young stars may be a promising approach.

As promising as the infrared appears, detection of planets around other stars through studies in the visible part of the spectrum is difficult. Imagine the problem of directly detecting Jupiter if we were to view the solar system from a distance of 10 pc. Jupiter's orbit of 5 AU would subtend a maximum angle of 0.5 arcseconds. Imagine further that we were to use a telescope comparable in optical performance to the *Hubble Space Telescope* (when corrected for its spherical aberration). The task of detecting even a planet as bright as Jupiter is very difficult, indeed beyond the capability of the *Hubble Space Telescope* for all but possibly the very nearest stars. There are optical systems currently under study that are designed to search for other planetary systems, and they may be available in the early part of the next century.

The two most frequently mentioned methods of indirect detection of other planetary systems are astrometry and spectroscopy. A common physical effect underlies both of these methods. If a star has a companion, or companions, the center of mass of the system (often referred to as the barycenter) is not at the center of mass of the star; that is, the mere presence of companions gravitationally bound to a star causes a shift in the center of mass of the system. This in turn means that the star undergoes orbital motion about the barycenter. Astrometry seeks to detect and quantify the amplitude of the star's offset motion about the barycenter, whereas spectroscopy is used to detect evidence of a line-of-sight component of this orbital motion relative to the observer. Both methods also detect the period of the orbital motion. A drawback to the spectroscopic method is that it is sensitive to the angle between the line of sight to a star and the orientation of the orbital plane of that star's possible planetary system; it is much harder to detect planetary systems seen more face-on than edge-on. Also, stars give rise to a variety of intrinsic phenomena that could mimic the type of spectroscopic behavior due to the presence of planetary companions. Astrometry is less sensitive to these possible sources of confusion or ambiguity as it measures the apparent displacement of a star from the barycenter due to the presence of companions.

A recently completed study that involved a combination of spectroscopic and astrometric observations indicates that planetary systems, or at least systems with relatively large companions, may be relatively rare. Of the 70 stars that were studied, *none* showed evidence for low-mass (i.e., nonstellar) companions down to a limit of a few tens of Jupiter masses. Indeed, the nondetections indicate that the likelihood of planets with masses greater than about 10 times the mass of Jupiter is *less than* 2%.

The next few decades will be ones in which the observational underpinning of our model of how our planetary system and others like it were formed will be put on firm footing. We are witnessing the completion of the revolution started by Copernicus more than three centuries ago.

Additional Reading

Black, D. C. (1980). In search of other planetary systems. *Space Sci. Rev.* **25** 35.

Black, D. C. (1991). Worlds around other stars. *Scientific American* **264** (No. 1) 76.

Black, D. C. and Matthews, M. S., eds. (1985). *Protostars and Planets. II.* University of Arizona Press, Tucson.

Marcy, G. W. and Benitz, K. J. (1989). A search for substellar companions to low-mass stars. *Astrophys. J.* **344** 441.

Strom, K. M., Strom, S. E., Edwards, S., Cabrit, S., and Skrutskie, M. F. (1989). Circumstellar material associated with solar-type pre–main-sequence stars: A possible constraint on the timescale for planet building. *Astron. J.* **97** 1451.

See also **Solar System, Origin; Stars, Low Mass and Planetary Companions; Stars, Pre–Main-Sequence, Winds and Outflow Phenomena.**

Planetary Tectonic Processes, Terrestrial Planets

Matthew P. Golombek

Tectonics is the field of geology that studies the formation, evolution, and origin of major structural or deformational features present on planetary surfaces. Structural features form when sufficient stress and strain causes fracture and deformation of all or part of the lithosphere (the rigid outermost part of a planet). A complete understanding of tectonics typically involves study of the geometry, kinematics (deformation through time), and dynamics (stresses that led to failure) of major structural features. Because tectonic features form in the lithosphere, whose thickness and mechanical properties are a function of the thermal gradient, composition, strain rate, and thickness of the crust (the outermost chemical layer of a planet), their formation can be used to understand the thermomechanical evolution of a planet.

Our solar system has five terrestrial (predominantly rock or silicate) planets: Mercury, Venus, Earth, Moon (although a satellite, its large size has led most planetary scientists to consider it as a planet), and Mars. Tectonic activity correlates with size; Earth, the largest, is the most active, followed by Venus, Mars, and Mercury, and the Moon, the smallest, is the least active. This relation between size and tectonic activity is understandable considering that most geologic and tectonic processes are driven by the heat engine of a planet, and large planets take more time to cool and probably have had greater contributions of heating from accretion, differentiation, and radioactivity than small planets. Here, we will describe current understanding of the tectonic processes active on each of the terrestrial planets, starting with the Earth, the most familiar, as a reference.

EARTH

The Earth is unique among the terrestrial planets in having great oceans of liquid water, the development of life, and plate tectonics. Plate tectonics is the motion of about a dozen large (up to thousands of kilometers across), thin (up to 100 km thick), fairly rigid plates on the Earth's surface at rates of up to about 15 cm yr^{-1}. The movement of the plates results from convection in the mantle, in which oceanic crust and lithosphere are created at mid-ocean ridges, cools by contact with seawater, thickens and subsides with age, ultimately to be subducted (recycled) back into the mantle (a layer beneath the crust and above the core) at a trench; the oldest oceanic crust is thus only 2×10^8 yr. old. Continental crust, which is created by volcanism and tectonism at subduction zones, does not subduct due to its low density and is not recycled back into the mantle; as a result, the cores of some continents are as old as 3.8×10^9 yr old. Large earthquakes, volcanos, and mountain belts result from interactions at the edges of the plates.

Continental rifts and grabens are linear troughs, bounded by inward dipping normal faults, about 100 km wide and a few kilometers deep where the lithosphere is extending apart. Rifts are the first phase of opening of an ocean, although not all succeed in becoming oceans (e.g., African rift system, western U.S. Basin and Range, Rio Grande rift, Rhinegraben, and Baikal rift); rifts that have just opened small oceans include the Gulf of California and the Red Sea. Mountain belts and volcanic arcs occur at subduction zones, where oceans close. Enormous mountains form when two continental plates collide, such as the Himalayas between India and the Tibetan plateau. Mountain belts form under compression by folding and imbricate (overlapping) thrust faulting that thickens the crust. Great strike-slip faults known as transforms occur where two plates move side by side, such as the San Andreas fault between the Pacific and North American plates in California.

VENUS

Venus is a virtual twin of the Earth in size and density, yet it is covered by a thick, cloudy, water-free atmosphere whose surface temperature is about 475°C. Our knowledge of the geology of Venus is limited by its opaque atmosphere and will not approach our understanding of the other terrestrial planets until the *Magellan* mission completes its mapping of the surface with radar. Nevertheless, data from a number of missions and ground-based radar reveal geologic and tectonic features that are similar to those on Earth.

Like the Earth, Venus has a number of continent-sized highlands that stand several kilometers high, although most of the planet is composed of fairly low-relief rolling plains. Ishtar Terra, the highest region, is about the size of the continental United States and is composed of a large plateau, Lakshmi Planum, surrounded by mountain ranges that resemble fold and thrust belts on the Earth. Another highland, Beta Regio, is split by a linear trough with large volcano-shaped constructs that may be analogous to continental rifts and associated volcanos on the Earth. There are many other disrupted zones that may also be locations where the venusian lithosphere has ruptured in extension. Within the rolling plains are regions of banded terrain that have been suggested to be compressional folds. Some parts of the venusian surface may contain enough craters to be a billion years old; other crater-free areas are substantially younger.

Substantial debate has focused on whether the array of tectonic features found on the surface of Venus could be indicative of plate tectonics that occurs on the Earth. A variety of arguments such as the extreme buoyancy of the hot, thin lithosphere and the lack of oceans to cool the lithosphere suggest that plate tectonics as we known it on Earth, cannot be occurring on Venus. Nevertheless, Venus appears to be substantially more active than the Moon, Mercury, or Mars. It is even possible that some form of plate mobility is, or has occurred on Venus; resolution of this question awaits the analysis of *Magellan* data.

MARS

The tectonic and geologic evolution of Mars lies between the protracted histories and great activity of Venus and the Earth and the abbreviated histories and low activity of the Moon and Mercury. The surface of Mars is characterized by heavily cratered terrain in the southern hemisphere that dates from heavy bombardment, lightly cratered northern plains that date since heavy bombardment, and large volcanic provinces that have been active over a few billion years. Mars is dominated by a large, high-standing province with a large positive gravity anomaly (indicating excess mass), known as Tharsis. Tharsis covers about one-quarter of the planet, rises to about 10 km above the datum, and is unique in scale in the solar system. At its center are the four largest shield volcanos (analogous to, but three times larger than Mauna Loa of Hawaii) in the solar system. An enormous radiating graben system and a sweeping concentric system of wrinkle ridges (compressional structures discussed later) cover an entire hemisphere of the planet. Tharsis probably formed by a combination of volcanic and tectonic processes involving large-scale uplift, volcanic construction, and lithospheric loading, all of which may be ultimately tied to convection deep within the planet. Elysium is a smaller region on Mars that shares many of the same characteristics as Tharsis. Even though the martian lithosphere has undergone hemisphere-wide faulting, Mars, like the Moon and Mercury, is most likely a single-plate planet with a thick to moderately thick lithosphere.

Grabens are the most common tectonic feature on Mars. The grabens are fairly simple normal fault-bounded valleys a few kilometers wide, a few hundred meters deep, and up to hundreds of kilometers long (Fig. 1). A few, however, are larger and more complex structures up to 100 km wide with multiple border faults and deeper (up to a few kilometers deep), multiply faulted floors, similar to rifts on the Earth. Valles Marineris is a large system of canyons, a few thousand kilometers long, up to hundreds of kilometers wide, and up to 10 kilometers deep, that probably represent large rifts where the entire martian lithosphere has ruptured in extension. Wrinkle ridges (discussed more fully later) and other ridges, interpreted to result from compressional stresses, have been mapped over the entire globe of Mars, perhaps due to cooling and shrinkage of the planet. Cases of strike-slip faulting have also recently been tentatively identified on Mars.

MERCURY

The half of Mercury that was imaged by the *Mariner 10* flyby shows a single plate that had a short tectonic and geological history ending about 3×10^9 yr ago. Its surface is composed of ancient heavily cratered terrain that dates to the period of heavy bombardment, slightly younger, smoother intercrater plains, and smooth

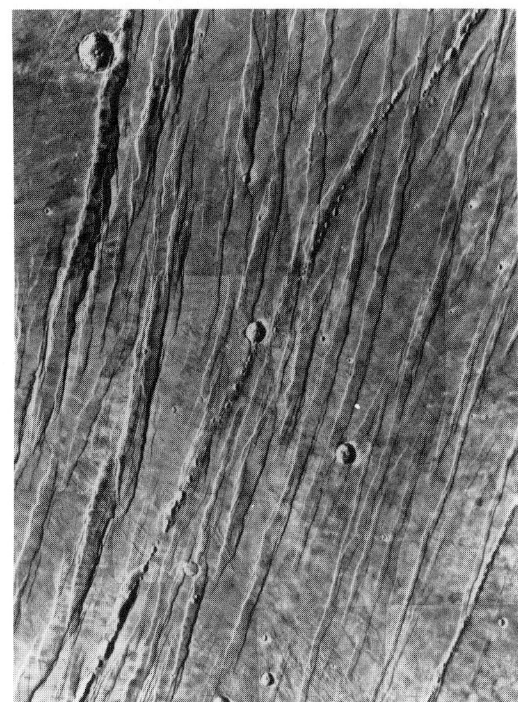

Figure 1. Example of extensional deformation on the terrestrial planets. The image shows a swarm of grabens on Mars. The negative relief and symmetry of the structures imply that each trough is bounded by at least two inward-dipping normal faults. Note the fairly simple grabens bounded by two scarps and the more complicated structures bounded by many faults. Pits along two grabens suggest extensional voids at depths to accommodate subsurface drainage of material. Mosaic is 1:500,000 scale, covering 5° of latitude and longitude (about 225 km wide). North is up.

plains in and around the large Caloris basin. The principle tectonic features on Mercury are lobate scarps that break the lithosphere. They are large (hundreds of kilometers long and a few kilometers high) scarps that cut all three major geological units. Craters cut by the scarps show marked foreshortening (Fig. 2), indicating that the scarps are most likely thrust faults that may have resulted from a few kilometers of planet-wide contraction, due to planetary cooling.

An ancient putative grid of orthogonal lineaments has been proposed based on a limited number of images with uniform lighting. This grid may have resulted from strike–slip faulting related to stresses likely to have accompanied despinning. Mercury now has a very slow spin rate that is in resonance with its orbit around the Sun. It probably formed with a spin roughly comparable to that of the Earth or Mars. Because Mercury is so close to the Sun, large solar tides would have despun the planet within a few billion years. Despinning would result in relaxation of a centrifugal equatorial bulge, resulting in equatorial thrust faulting, mid-latitude strike-slip faulting, and polar normal faulting.

Smaller tectonic features related to the Caloris basin have also been identified and are roughly analogous to the grabens and wrinkle ridges associated with large basins on the Moon (discussed next). Their distribution in and around Caloris argues for a thick lithosphere that was loaded and flexed.

THE MOON

The Moon has had the simplest and most abbreviated geological and tectonic history of the terrestrial planets. The surface is dominated by ancient heavily cratered terrain and younger (a few billion

Figure 2. Example of compressional deformation on the terrestrial planets. The image shows Discovery scarp on Mercury. North is up. Note upper crater is foreshortened, indicating that the scarp is a thrust fault with the west side overthrusting the east side (fault plane dips to the west). Scarp is about 2 km high and about 500 km long. (Also see Figure 4, page 426.)

years old), smoother, and darker maria ("seas") that fill in giant impact basins that date to a period of heavy bombardment early in solar system history. The small size, sparse number, and location of lunar tectonic features indicate they are relatively minor structures with small strains that are related to regional scale subsidence of mare basins and perhaps mild global contraction. No large-scale tectonic features involving protracted and extensive deformation have been found. The lithosphere of the planet is thus likely to be a single thick plate that has not experienced any widespread, throughgoing lithospheric failure of endogenic origin.

Two types of tectonic features have been identified on the lunar surface—grabens and wrinkle ridges. Grabens are ancient linear to arcuate fault-bounded valleys that resulted from extension of the shallow crust. Normal faults bounding grabens on the Moon dip inward with the interior floor of the graben having been downdropped relative to the sides. Lunar grabens are typically a few kilometers wide, up to a few hundred meters deep, and have experienced hundreds of meters of extension. Virtually all lunar grabens are found around large lunar basins that have large positive gravity anomalies suggesting that they formed due to extensional bending stresses at the periphery of basins whose centers have subsided.

Wrinkle ridges (also called mare ridges) are linear to arcuate broad, low positive-relief features that are found in virtually all lunar maria. They are up to tens of kilometers wide, a few hundred meters high, and are composed of three components: a broad rise or arch, a superposed hill or ridge (the "ridge" of wrinkle ridges), and a smaller high-relief crenulation (the "wrinkle" of wrinkle ridges). Mare wrinkle ridges also extend into highland regions. The origin of wrinkle ridges has been attributed to the intrusion and extrusion of thick viscous lavas due to compressional faulting and folding. The identification and study of analogous structures on Earth has argued for an origin due to thrust faulting and folding under compressional stresses. Their location within mare basins is consistent with the proposed compressional bending stresses within the interiors of subsiding basins. The occurrence of ridges and highland thrust faults outside basins has been cited as evidence for a period of global contraction due to planetary cooling, consistent with the generally younger ages of wrinkle ridges.

In conclusion, the terrestrial planets reveal a wide variety of tectonic features and histories. The Moon, Mercury, and Mars are single-plate planets having thick, rigid lithospheres dating back to the period of heavy bombardment. Tectonic features formed in response to local, regional, and global processes that produced extensional grabens, rifts and compressional scarps, and wrinkle ridges. The Earth is dominated by lateral motion of large, thin lithospheric plates, some of which recycle back into the mantle. Venus is intermediate, with many Earth-like tectonic features, such as folded and faulted mountain belts, yet it probably lacks the plate tectonics that occur on the Earth.

Additional Reading

Carr, M. H. (1981). *The Surface of Mars.* Yale University Press, New Haven.

Carr, M. H., Saunders, R. S., Strom, R. G., and Wilhelms, D. E. (1984). *The Geology of the Terrestrial Planets*, SP-479. NASA, Washington, DC.

Head, J. W. (1990). Surfaces of the terrestrial planets. In *The New Solar System*, 3rd ed. J. K. Beatty and A. Chaikin, eds. Sky Publishing Corp., Cambridge, MA, and Cambridge University Press, Cambridge, U.K., p. 77.

Head, J. W. and Solomon, S. C. (1981). Tectonic evolution of the terrestrial planets. *Science* **213** 62.

Head, J. W., Wood, C. A., and Mutch, T. A. (1977). Geological evolution of the terrestrial planets. *American Scientist* **65** 21.

Hunten, D. M., Colin, L., Donahue, T. M., and Moroz, V., eds. (1983). *Venus*. University of Arizona Press, Tucson.

Vilas, F., Chapman, C. R., and Matthews, M. S., eds. (1988). *Mercury*. University of Arizona Press, Tucson.

Wilhelms, D. E. (1987). *The Geologic History of the Moon*. Professional Paper 1348. U. S. Geological Survey, Washington, DC.

See also **Mars, Surface Features and Geology; Mercury, Geology and Geophysics; Moon, Geology; Moon, Seismic Properties; Venus, Geology and Geophysics.**

Planetary Volcanism and Surface Features

Kenneth L. Tanaka

Volcanism is the geological process in which molten material (produced by interior heating) and associated gases rise through a planetary crust and are extruded over the surface. On Earth, volcanism continues to produce new crust, demonstrating that our planet is a dynamic body with sources of energy from deep within. Particularly since the 1960s, volcanologists have made great progress in determining the relations among volcanic processes and landforms on Earth. In the recent explosion of planetary data, this work has assisted geologists in the interpretation of volcanic landforms in spacecraft images of the Moon, the terrestrial planets (Earth, Mars, Venus, and Mercury), and the outer-planet satellites. Remote-sensing, lander, and returned-sample data are also crucial in identifying and interpreting volcanic materials on other planets. The spacecraft data show that, of all resurfacing processes in the solar system that originate within planetary bodies, volcanism is probably the most widespread, and it is generally the inferred mechanism for resurfacing on planets devoid of atmospheres, even where diagnostic volcanic landforms have not yet been identified.

Volcanic landforms have varied morphologies, because the emplacement of volcanic materials is affected by such diverse yet interrelated factors as crust and mantle structure; composition, flow properties, and volatile content of the magma; gravity; atmospheric conditions; and discharge rate and volume of the eruptions. Recognition of volcanic landforms in spacecraft images depends on the size and topographic expression of the feature and the quality and resolution of available images. Relatively quiescent eruptions produce flows and fill low areas; this eruption style volumetrically dominates over more spectacular pyroclastic eruptions that yield a variety of landforms. We can deduce the general volcanic history of a planet—as well as its thermal history—by mapping variations in relative ages of materials over the body.

VOLCANISM AND RESURFACING ON PLANETARY BODIES

Earth

According to the theory of plate tectonics, now generally accepted, the oceanic crust of the Earth has been recycling for at least the past 2×10^8 yr and probably since Archean time ($>2.5\times10^9$ yr ago). Volcanism has been most active along plate boundaries, where heat flow is greatest. Ocean-floor basalts, the most abundant rock type in the Earth's crust, are erupted mainly at mid-ocean ridges where new oceanic crust is being formed. Over plate-subduction zones and related island arcs and basins, dominantly andesitic rocks have erupted in violent, pyroclastic explosions. Complex suites of volcanic rocks are associated with continental rifting. Kilometer-thick sequences of voluminous, low-viscosity, flood or plateau basalts have erupted on continental crust through fissures in areas undergoing extensional tectonics. Other intraplate volcanism includes fixed-mantle "hot spots," that produce chains of seamounts and islands in the oceans (e.g., the Hawaiian Islands) or fields of cinder cones and lavas on the continents.

Early volcanism on Earth was probably widespread. Volcanic rocks dominate Archean rock sequences (many of which are > 50% volcanic). A minor constituent of these ancient volcanic rocks (but more common than in younger rocks) is an ultramafic rock known as komatiite, which indicates that the early Earth had a hotter mantle and greater magma production than it does now.

Moon

The Lunar Orbiter and Apollo missions have provided a wealth of photographic, remote-sensing, and sample data for the study of the materials that make up the lunar surface. Volcanism on the Moon is expressed by the dark, smooth maria that cover about one-sixth of the lunar surface. The maria embay densely cratered highland rocks that formed prior to the termination of heavy bombardment in the solar system nearly 4×10^9 yr ago. The cratering therefore has largely obscured and reworked highland lavas. Most material of the maria is composed of low-viscosity basalts largely erupted between 3.9×10^9 and 3.1×10^9 yr ago from fissures commonly controlled by impact-basin ring structures. Volcanic intensity then waned rapidly; minor activity continued until about 1×10^9 yr ago. Flows of the maria appear to have been voluminous and fluid (lava-flow scarps are rare) and they commonly display sinuous rilles. The rilles are probably lava channels or collapsed lava tubes produced during channelized outpourings of lava. In places, constructional features such as domes, shields, and cones (attributed to centralized volcanism) range from hundreds of meters to kilometers across, and some occur as volcanic fields. Minor pyroclastic volcanism has resulted in dark-mantle deposits; possible pyroclastic vents include cones and dark-haloed craters that commonly lie along fractures.

Mercury

Our understanding of the geology of Mercury comes from the three flybys of *Mariner 10*, which photographed nearly half of the planet's surface. Resurfacing of Mercury was short lived, resulting in widespread intercrater plains material emplaced at the end of heavy bombardment and in younger smooth plains material largely emplaced near and within large craters and basins. The regional expanses of smooth plains material that clearly postdate local major impacts appear to be made up of fluid basalts similar to the lunar maria; however, diagnostic volcanic landforms and remote-sensing data are lacking. Isolated patches of plains materials may have either volcanic or impact-ejecta origins.

The intercrater plains material may be volcanic rocks erupted during probable mantle heating and global expansion associated with early differentiation. However, later global compression, resulting in ridge formation, would have closed conduits to the surface, generally shutting off volcanism. The smooth plains material, which was emplaced during ridge formation, may have been erupted from local areas of extensional stress associated with impact basins.

Venus

The venusian surface has recently been imaged by radar (which can penetrate Venus' cloud cover) from Earth-based observatories and the *Pioneer Venus*, Venera, and *Magellan* spacecraft. Radar images, having resolutions as high as 100 m, demonstrate that the surface of Venus is mainly made up of intensely faulted and folded tectonic zones and vast volcanic plains. Both types of terrain reveal a plethora of volcanic flows and edifices.

The venusian plains are marked by tens of thousands of volcanic domes and shields only a few kilometers across. Larger shields (as much as several hundred kilometers across) are common in association with deformed uplands, elliptical volcanotectonic structures, and zones of linear fractures and ridges. These shields have low relief (a few kilometers at most) because of the thin lithosphere, low viscosities and high eruption rates of magma, and efficent dispersal of lava through flank vents and lava tubes. Also, emanating from some tectonic zones are huge rilles; the rilles are associated with linear depressions and apparently channeled lavas for hundreds of kilometers. These styles of volcanism, as well as geochemical data from four Venera landers, indicate a dominantly basaltic composition for volcanic rocks on Venus. However, the morphology of some domes and data from the *Venera 8* lander suggest that intermediate to silicic rocks may also be present.

The modest density of impact craters on Venus may indicate that the average surface age of the planet is only a few hundred million years. If so, we can expect Venus to remain volcanically active for many eons to come.

Mars

The red planet, owing its color to oxidized iron compounds, was shown by the Viking landers to have surface materials that were probably derived from the weathering of basalts. The striking volcanic constructs on Mars viewed by the *Mariner 9* and Viking orbiters (e.g., Olympus Mons; see Fig. 1) can be attributed to long-lived, extensive basaltic volcanism on a one-plate planet. The extent of early volcanic products emplaced during and shortly following heavy bombardment is unknown because of the probable low viscosity of the lavas (resulting in relatively featureless flows) and subsequent resurfacing and gradation aided by the appreciable atmosphere and water and ice in the crust. In intermediate stages

Figure 1. Olympus Mons, perhaps the largest shield volcano in the solar system. The volcano covers an area equivalent to that of the state of Arizona and reaches an elevation of nearly 90,000 ft (over 26 km) above Mars datum—the highest known mountain on any planet. Olympus Mons may have been active for hundreds of millions of years as a result of hot-spot volcanism in a stationary lithosphere.

of martian geologic history, ridged plains material interpreted to be broad lava flows covered at least 10% of the planet (much more may be buried in the northern lowlands). These flows may have been mostly fed by fissures; some ridged plains are cut by circular calderas. Several low, dissected volcanos known as highland paterae occur within some of the ridged plains near impact basins; these volcanos may be the result of explosive eruptions caused by interaction between magma and water or ice.

Following ridged plains formation, complex fields of lobate lava flows were erupted at centralized vents as well as fissures, particularly in the Tharsis and Elysium regions. Although no active volcanism on Mars has been observed, some of the most recent flows are so lightly cratered that volcanism cannot be regarded as finished. The large volcanic shields of Mars, several of which attain heights of over 20 km and diameters of 500 km (see Fig. 1), may have been active for hundreds of millions of years. Volcanic domes also formed, which may be made up of more viscous lavas or combinations of lava flows and pyroclastic materials. Flows extend from the volcanic centers as far as 1500 km. Many other materials, some of considerable volume and extent, have been attributed to volcanism, but more data and research are required to determine their origin with confidence.

Io

Scientists of the Voyager mission discovered spectacular, explosive sulfuric volcanic eruptions in progress on Io. This dominantly silicate body maintains a molten or partly molten upper mantle that feeds eruptions through a thin lithosphere. Volcanism may account for a surface burial rate of 100 m per million years, which helps to account for the absence of impact craters on Io's surface.

The eight active volcanic plumes observed by *Voyager 1* and *2* were apparently fed by molten SO_2 that, when reaching the surface, produces a spray of rapidly expanding vapor that drives plumes as high as 300 km and deposits bright SO_2 frost below them. More than 300 vents have been recognized; most form calderas tens of kilometers across (the largest is 250 km in diameter). Volcanic structures range from simple shields and disk-shaped constructs to complex calderas with radiating flow fields. Calderas are commonly irregular and have dark floors that may contain molten sulfur. Some of the high, steep caldera walls, as well as kilometer-high mountains and scarps in many other places on Io, indicate material stronger than sulfur compounds; they are probably made up of silicate volcanic rocks. Flows and flow units extend as much as hundreds of kilometers from the vents. The yellow hues of the flows indicate that they were emplaced as highly fluid molten sulfur (or sulfur-rich silicate material). Plains materials between vents probably are made up of various volcanic products including airfall deposits, flows, and fumarolic materials.

Icy Satellites of Jupiter, Saturn, Uranus, and Neptune

Most of the larger, icy, outer-planet satellites show histories of resurfacing since intense bombardment. The low surface temperatures of these bodies require volcanism of materials that are mobile at low temperatures, such as water or ammonia–water mixtures (in liquid form or as ices that flow viscously at temperatures close to the melting point) and methane, which can be incorporated in ice as a clathrate. The icy satellites are thus capable of solid-state "cryovolcanic" resurfacing in addition to the eruption of liquid flows and explosive, "cryoclastic" activity.

On the Galilean satellites Europa and Ganymede, plentiful fractures and grooves occur that contrast in albedo with older terrains; some smooth plains are associated with the fractures. Europa is light colored, nearly devoid of impact craters, and criss-crossed by dark fracture zones tens of kilometers wide. The fracture zones may have been injected with water and ice slushes propelled by gas effervescence. On Ganymede, fracture zones are light colored and produce grooved terrains cut into a darker crust. These terrains may have been produced by the slush-injection mechanism proposed for the fracture zones on Europa or by the flow of warm ice over the surface, perhaps released by impacts. The outermost satellite, Callisto, shows only minor resurfacing associated with the Valhalla and Asgard impact features.

In the saturnian system, Enceladus displays long ridge systems and smooth plains probably caused by tectonism and volcanism. The orbit of Enceladus is coincident with Saturn's E ring, suggesting to some researchers that the ring comprises volcanic ejecta from the satellite. The outer satellites Tethys, Dione, and Rhea also show resurfaced areas as well as pit chains suggestive of volcanism (Rhea's apparently resurfaced region may also be an artifact of photometric geometry). In addition, Dione and Rhea have bright, wispy streaks that may be frost-like material deposited by explosive fissure eruptions occurring when ammonia–water magmas heat methane clathrate until it flashes into ice and pressurized methane gas. The inner satellite Mimas shows no signs of volcanism, and the surface of Titan, the largest saturnian satellite, is veiled by an atmosphere.

Several of the uranian satellites also show evidence of resurfacing. In particular, the large satellite Titania and the smaller, innermost satellites Ariel and Miranda show complex resurfaced and fractured terrains that include possible ice flows (whose mobility may be enhanced by interstitial liquid methane or nitrogen). Albedo patterns on Umbriel suggest possible early resurfacing in conjunction with tectonism, whereas Oberon apparently was relatively inactive.

The observed surface of Triton, the large moon of Neptune, is made up of various smooth materials, a global system of criss-crossing sets of ridges, and a dimpled and ridged terrain resembling the surface texture of cantaloupe rind; heavily cratered terrain is absent. The smooth materials, which are layered and marked by pits (many appear structurally controlled), may include solidified ammonia-water ice flows and lakes (in topographic basins), as well as cryoclastic deposits generated by the eruption of mixtures containing highly volatile compounds such as methane and nitrogen. The ridge sets appear to be made up of grabens along which extrusion of ices formed ridges. The enigmatic "cantaloupe terrain"

Table 1. Volcanism and Resurfacing of Selected Planetary Bodies

Planet / Satellite	*Dominant Type(s) of Volcanism*	*Volcanic / Resurfacing History*
Earth	Silicate, particularly basaltic; along lithospheric plate margins and at intraplate hot spots	Active recycling of plates from Archean to present
Moon	Basaltic flows effused along impact structures; flows fill mare plains	Waning from heavy bombardment; ended about 1×10^9 yr ago
Venus	Basaltic plains; small domes and low shields in tectonic regions	Probable active planetwide crustal recycling
Mars	Basaltic; flows from impact structures, central volcanos, and fissure systems	Early widespread flows; later centralization and development of constructs
Mercury	Possible basaltic plains near impact structures	Cessation soon after heavy bombardment
Io	Explosive sulfur dioxide eruptions and sulfur and silicate flows from central vents	Continuous eruptions, mainly in equatorial band, from tidal heating
Europa	Injection of water–ice slush along fractures	Recent activity obscuring ancient impact cratering
Ganymede	Injection of water–ice slush along fractures or flow of warm ice from impacts	Active during waning stages of heavy bombardment
Callisto	Water–ice flows erupted through fissures in floors of and surrounding impacts	Minor; flows emplaced shortly after impact events
Mimas	None apparent	
Enceladus	Possible fissure-fed flows and explosive eruptions of volatiles	Long history of activity; may still be active
Tethys, Dione, Rhea(?)	Possible plains flows and explosive eruptions of water-volatile mixtures from fissures	Early regional burial of large impact features
Iapetus	Dark crater-filling material?	Early(?) activity causing albedo irregularities
Miranda	Flows from banded terrain	Intermediate age
Ariel	Flows from fractures	Widespread at end of heavy bombardment
Umbriel	Possible early plains material	Before end of heavy bombardment
Titania	Possible flows from fractures	Widespread early and local intermediate activity
Oberon	Minor dark flows in crater floors	Probably at end of heavy bombardment
Triton	Flows of ammonia-water ice, fissure-fed extrusives, and explosive eruptions of volatile mixtures from circular vents	Intermediate and recent activity obscuring ancient impact cratering; may still be active

makes up Trition's oldest observed surface and may be volcanic or tectonic in origin.

VOLCANIC HEAT SOURCES

Volcanic histories of the planets are related to the histories and mechanisms of heating. The larger the body, the more likely that internal core and mantle differentiation has occurred, providing longer-lived heat release and localized concentrations of radiogenic elements. Bodies smaller than several hundred kilometers in diameter have not differentiated and thus have been geologically inactive. Differentiation of intermediate-sized bodies such as the Moon, Mercury, Mars, and the larger outer-planet satellites releases gravitational energy in the form of heat and causes global expansion. This process may produce widespread volcanic resurfacing during or after heavy bombardment by providing the heat to generate magmas and the fractures to serve as eruptive conduits. On the Earth and Venus, energy released from the interior has been sufficient to induce recycling of lithospheric plates, along whose margins volcanism is concentrated. Where the crust has not recycled, major impact scars (e.g., those on the Moon and Mercury) or fractures produced by extension (e.g., those on Europa and Ganymede) are the focus of volcanism, which, if centralized for long periods, may result in major volcanos (as on Mars). Long-lived volcanism on smaller bodies such as Io and Enceladus is generated by tidal heating due to gravitational interactions with their parent planets.

STATUS OF PLANETARY VOLCANISM

We are still in the discovery and preliminary interpretive stages regarding volcanism and resurfacing on the rocky and icy planets and satellites beyond the Earth–Moon system, where spacecraft data are necessary to document volcanic activity. (See Table 1 for a summary of what is known.) Some of the outer planetary bodies (such as Pluto and Titan) remain to be observed up close by spacecraft, whereas others have been incompletely or inadequately imaged, let alone sampled or remotely sensed by sophisticated instruments. Recent work, however, has revealed the remarkable extent and variety of known or inferred volcanic surfaces and landforms in the solar system. These discoveries continue to stimulate research into the composition, histories, and origins of volcanic materials in diverse planetary settings, as well as our anticipation of further discoveries.

Additional Reading

Basaltic Volcanism Study Project (1981). *Basaltic Volcanism on the Terrestrial Planets*. Pergamon, New York.

Cas, R. A. F. and Wright, J. V. (1987). *Volcanic Successions, Modern and Ancient*. Allen and Unwin, London.

Greeley, R. (1987). *Planetary Landscapes*, revised ed. Allen and Unwin, Boston.

Hamblin, W. K. and Christiansen, E. H. (1990). *Exploring the Planets*. Macmillan Publishing Co., New York.

Soderblom, L. A. (1980). The Galilean moons of Jupiter. *Scientific American* **242** (No. 1) 88.

See also **Io, Volcanism and Geophysics; Mars, Surface Features and Geology; Mercury, Geology and Geophysics; Moon, Geology; Venus, Geology and Geophysics.**

Plasma Transport, Astrophysical

Daniel S. Spicer

Plasma transport in astrophysics is the study of transport processes that control the flow of mass, momentum, and energy in the astrophysical environment. Historically, the term "transport processes" has been used to identify plasma properties associated with collisional effects, such as thermal conductivity, viscosity, and electrical resistivity in the context of the fluid approximation. However, the term "transport processes" has broadened to include transport processes that occur in collisionless plasmas. Because the collisionality of a typical astrophysical plasma within a given volume can vary from fully collisional to fully collisionless, transport "coefficients" within the volume can likewise change, often dramatically. The atmosphere of a typical star with a hot wind, such as the Sun, is an example. Complicating matters are ubiquitous magnetic fields, which alter transport in a variety of ways.

TRANSPORT COEFFICIENTS AND THE FLUID APPROXIMATION

Astrophysicists have used fluid models to model phenomena involving both collisional and collisionless processes. Phenomena of note are solar and stellar flares, supernova remnants and associated shocks, stellar winds, accretion disks, and jets. Collisional transport theory uses assumptions that can also be valid when collective effects control transport but are not necessarily valid for all applications. In general, classical transport is valid when studying changes occurring on time scales much greater than an "effective" collision time, if the "effective" mean free path (λ_{mfp}) of the particles is much less than the scale (δL) over which the physical parameters vary, and the distribution function of the fluid is close to a Maxwellian. The first-order correction to the Maxwellian is then proportional to the effects that disturb the Maxwellian (e.g., electric fields, etc.). As a Maxwellian distribution function and its derivatives are uniquely determined by the density, flow velocity, and temperature, and their derivatives, these quantities are used to parameterize the correction, which then permits parameterization of the viscosity tensor, the heat flux, and the momentum exchange vector. These latter quantities are proportional to the effects that produce the deviation from a thermal equilibrium. The corresponding coefficients of proportionality are the transport coefficients (i.e., electrical resistivity, etc.). It is the basic goal of classical kinetic theory to compute these coefficients, which is accomplished using the expansion procedure due to Sydney Chapman and David Enskog. This procedure uses the Knudsen parameter, ε, which is the ratio of λ_{mfp} to δL parallel to any attendant magnetic field and is assumed to be no larger than the ratio of the gyroradius to the characteristic length perpendicular to any attendant magnetic field for a given particle species. When $\varepsilon \ll 1$, collisions are dominant and classical transport theory is an excellent approximation. However, experience has shown that when $\varepsilon \gtrsim 10^{-2}$ classical transport theory breaks down. When this occurs the plasma is said to be semicollisional or collisionless depending on the degree to which ε exceeds $\approx 10^{-2}$.

The fluid equations are obtained from moments of the Boltzmann equation. The set of equations obtained, expressed in terms of the *ensemble average* velocities and densities, contain parameters determined by higher-moment equations and therefore are not closed. How closure of the moment equations is made can eliminate various properties of a plasma. A common closure technique is an expansion about a local Maxwellian. A limitation to the moment approach is loss of information about the distribution.

CLASSICAL AND ANOMALOUS TRANSPORT COEFFICIENTS

Collisional transport is usually referred to as "classical" transport, while collisionless transport, controlled by plasma microinstabilities, is referred to as "anomalous" transport ("anomalous" transport has also been used to denote the effect macroscopic instabilities have when they relax the gradients in configuration space that drive the plasma configuration unstable; see the following discussion). However, the term "anomalous" is unfortunate because it is the general mechanism of transport in high-temperature, low-density plasmas. Nevertheless, a unifying picture of both transport regimes exists. This picture is obtained by considering the scattering of test particles in the stochastic electric fields of a plasma. Stochastic electric field fluctuations with wavelengths $\lambda < \lambda_{\text{De}}$ originate in the incoherent motions of single particles, whereas stochastic fields with $\lambda > \lambda_{\text{De}}$ result from collective effects that are treated as coherent waves, where λ_{De} is the Debye length. For thermal plasmas, the ratio of the electric field energy density from the coherent fluctuations, to the electric field energy density from incoherent fluctuations is small and so collective effects are negligible. However, nonthermal plasmas are more the rule than the exception. The free energy that exists in a nonthermal plasma can drive the coherent fluctuations of a plasma to large amplitudes that easily exceed those of thermal fluctuations, thus modifying transport coefficients by orders of magnitude and thus producing "anomalous" transport coefficients. Plasma instabilities are the mechanisms by which coherent fluctuations of a plasma usually grow and modify the transport coefficients of a plasma. (Note that a nonthermal distribution that is stable still produces enhanced levels of coherent fluctuations, albeit smaller than from microinstabilities.) The transport coefficient's change depends on the amplitude and spectral properties of the coherent electric fields excited by the instability.

Plasma instabilities are categorized as macroinstabilities (derivable from a fluid description) and microinstabilities (derivable only from a kinetic description). Macroinstabilities driven by free energy residing in configuration space (e.g., the tearing mode, Rayleigh–Taylor, and Kelvin–Helmholtz instabilities) do not modify the microscopic transport coefficients, although an effective transport coefficient to model the effect of these instabilities can be computed. Macroinstabilities cause large-scale and chaotic behavior in the plasma, but from afar macroinstabilities appear as enhanced plasma transport. Macroinstabilities are used to explain matter transport and anomalous shear viscosity within accretion disks. On the other hand, microinstabilities are short wavelength, high frequency, and grow on time scales short compared to a typical fluid time scale. Microinstabilities are sometimes further subdivided into *fluid* and *kinetic*. The use of the term "fluid" implies the distribution function's shape is unimportant for instability, whereas "kinetic" implies the contrary. Examples of microscopic fluid instabilities are the ion–ion two-stream instability and the Buneman instability. The ion–ion two-stream instability is robust and leads to strong momentum and energy exchange between the streams. This kind of instability is important during the development of supernova remnants. An example of a kinetic instability is the *bump-on-tail* instability. Microscopic instabilities grow rapidly into the nonlinear regime and have small amounts of free energy available to them, so their growth is limited. Thus they cause fast, small-scale, but relatively ordered transport.

The primary sources of free energy that drive plasma microinstabilities are: (1) velocity–space anisotropy and inhomogeneity, specifically counterstreaming cold beams, distortions in the distribution function associated with currents or heat fluxes, and differences in distribution functions for particles trapped and untrapped in a magnetic field or differences in the shape of the distribution function in the direction parallel to and perpendicular to any attendant magnetic field; (2) plasma expansion energy that resides in gradients in density, temperature, and so forth.

Microinstabilities tend to drive an unstable plasma toward a thermal distribution but only collisions make the distribution a true Maxwellian. Furthermore, microinstabilities are localized in "boundary layers" where parameters such as density, velocity, temperature, and magnetic fields very rapidly. The resulting anomalous transport coefficients reestablish, within these layers, the local relationship that already exists in collisionally dominated plasmas between effects that disturb the plasma equilibrium and quantities that parameterize the distribution correction. For example, Ohm's law, a local relationship between electric field and current density, breaks down in a collisionless plasma unless microinstabilities produce an effective resistivity sufficient to reestablish the local relationship.

Stable collisionless plasmas are more difficult to treat than those with microinstabilities. This is because the local definition of coefficients that classical and anomalous transport permits is not usually possible in a stable collisionless plasma. This is understood by examining the evolution of an electron temperature profile. A temperature profile is controlled by three effects: (1) the local "heating of the electrons"; (2) the transport of these electrons in the presence of scattering; and (3) electron energy loss to a "cold" plasma. The effect of (2) is to spatially separate velocity diffusion from friction. If the spatial separation introduced by (2) is small, that is, $\varepsilon \ll 1$, (1) and (3) balance one another to form a local Maxwellian distribution, to lowest order, and the Chapman–Enskog approximation (CEA) is valid. However, when $\varepsilon \gtrsim 10^{-2}$ the separation introduced by (2) prohibits a local equilibrium and the CEA is invalidated. This is because a relaxation time and relaxation length are essentially equivalent in the CEA, but not equivalent when $\varepsilon \gtrsim 10^{-2}$. Hence the relaxation length becomes the important quantity and transport becomes nonlocal. Detailed Fokker–Planck studies of heat transport in the semicollisionless regime have confirmed this model, which has been used in solar flare modeling.

MARGINAL STABILITY

To develop a self-consistent anomalous transport model in the context of the fluid approximation is not easy. The conventional approach is to have a thorough understanding of the linear and nonlinear properties of the relevant plasma instabilities. Next phenomenological transport coefficients are computed, parameterized in terms of bulk fluid parameters (i.e., density, drift velocity, and temperature) as are threshold conditions for the relevant instabilities. The use of these transport coefficients in a fluid model is predicated on the assumption that the instability producing the anomalous transport grows and approaches marginal stability in a short time compared to a fluid time scale. When finite difference numerical methods are used to solve the fluid equations, grid resolution sufficient to resolve the gradient scale lengths is required. The parameterized threshold conditions are recalculated within every grid cell and at every time step. If the threshold conditions are satisfied within a cell the appropriate phenomenological transport coefficient is used. This approach was used in modeling solar flare plasmas and supernova explosions. An alternative technique exists that is powerful when steady-state conditions exist and a fluid model is used. This approach is best illustrated with an example. Suppose a temperature gradient drives an instability and thermal conduction is the only transport process. The heat balance equation is given by

$$\frac{\partial}{\partial s}\kappa\frac{\partial T}{\partial s} + S = 0, \tag{1}$$

whereas the marginal stability condition is

$$\alpha\frac{\partial T}{\partial s} = D, \tag{2}$$

where κ is the thermal conduction coefficient, S represents energy sources and sinks, D represents damping, and $\alpha(\partial T/\partial s)$ is the growth rate as determined by the temperature gradient. The technique is just the reverse of the conventional approach. Within the spatial domain where instability occurs, use Eq. (2) as the temperature equation and Eq. (1) as the transport coefficient equation. This approach is particularly powerful because it does not rely strongly on detailed knowledge of the nonlinear properties of instability, just its linear properties.

MULTISTREAMING

Multistreaming of a collisionless plasma results from the natural tendency of velocity gradients to steepen as plasma flow evolves. One common form of multistreaming is the reflected ions observed in collisionless shocks. A single-fluid model cannot treat multistreaming nor a multifluid model unless algorithms are used to exchange fluid from one stream to another. Whereas plasma instabilities can prevent multistreaming, this is the exception and not the rule. For example, a plasma composed of charge-neutralizing electrons and two counterstreaming ion beams cannot be prevented from multistreaming if the electron thermal energy is less than the energy of the two ion streams. Standard fluid numerical techniques using either artificial viscosity or some form of "shock capturing" scheme suppress a collisionless fluid from multistreaming by forming a strong shock. A supernova remnant (SNR) is an example that demonstrates this point. Because the thermal speed of ions that form an SNR is much less than their streaming energy and the λ_{mfp} of the SNR ions is of order parsecs during the earliest phase of an SNR, the SNR interaction with the surrounding circumstellar material is collisionless and involves the multistreaming of SNR plasma and ambient plasma. A single-fluid model of this interaction is inadequate and leads to the incorrect prediction of stagnation flow.

Magnetic Fields

Classical transport is unaffected parallel to any attendant magnetic field **B**, but can be sharply reduced perpendicular to **B**. When **B** is turbulent and the field fluctuations, $\delta \mathbf{B}$, vary slowly compared to a Larmor radius there are three known mechanisms by which $\delta \mathbf{B}$ can enhance cross-field transport: (1) resonant scattering of particles; (2) field line wandering of open field lines, which is unlike resonant scattering because the particles remain tied to the field lines; and (3) neoclassical, resulting from the inhomogeneity of **B** forming localized magnetic wells. Both (1) and (2) are believed to be important in cosmic ray diffusion and propagation. A dramatic form of enhanced cross-field transport can occur if field lines are topologically closed. In this case field lines can ergodically fill the volume enclosing the closed field lines. This possibility is similar to field line wandering except that a particle on a closed field line can make a larger effective diffusion step without a $\delta \mathbf{B}$.

Additional Reading

Benney, D. J., Shu, F. H., and Chi, Y., eds. (1988). *Applied Mathematics, Fluid Mechanics, Astrophysics*. World Scientific, Singapore.

Braginskii, S. I. (1963). Transport processes in plasma. *Rev. Plasma Phys.* **1** 205.

Davidson, R. C. and Krall, N. A. (1977). Anomalous transport in high-temperature plasmas with application to solenoidal fusion systems. *Nuclear Fusion* **17** 1313.

Galeev, A. A. and Sagdeev, R. Z. (1983). Current instabilities and anomalous transport. *Handbook of Plasma Phys.* **2** 271.

Hinton, F. L. (1983). Collisional transport in plasma. *Handbook of Plasma Phys.* **1** 148.

Manheimer, W. M. (1979). Anomalous transport from plasma waves. *J. Phys.* **40** C7.

Montgomery, D. C. and Tidman, D. A. (1964). *Plasma Kinetic Theory*. McGraw-Hill, New York.

See also **Comets, Solar Wind Interactions; Interplanetary Medium, Wave-Particle Interactions; Magnetohydrodynamics, Astrophysical; Planetary Radio Emissions; Radio Sources, Emission Mechanisms; Shock Waves, Collisionless, and Particle Acceleration.**

Pluto and Its Moon

Marc W. Buie

At the outer limits of our solar system, Pluto and its moon, Charon, endure a cold and lonely existence. Their combined orbit around the Sun takes them from 29 AU inside the orbit of Neptune out to 49 AU. Pluto's location gives it the distinction of being the coldest planet in our solar system. A full understanding of Pluto and its place in the scheme of the solar system is only now starting to fall into place.

EARLY UNDERSTANDING

Clyde W. Tombaugh discovered Pluto in 1930 shortly after beginning his search for a ninth planet. The justification for his endeavor was to find the planet that was causing irregularities in the motion of Neptune. These theoretical predictions did not aid in the actual discovery of Pluto. Instead, the success of Tombaugh's search lies with the meticulous methods he applied to searching for any object that might be present. Therefore, it is no surprise that the mass attributed to Pluto from its presumed effect on Neptune is completely wrong. This erroneous mass together with early attempts to measure the size of Pluto in 1950 combined to disguise the true properties of Pluto until the late 1970s.

Progress in the study of Pluto before 1978 was slow. The first photometric observations revealed a large-amplitude light curve with a period of 6.4 days. Then more sets of observations in 1964 and 1973 showed the mean brightness of Pluto was decreasing. These light curve observations provided the first clue about the orientation of the polar axis of Pluto. Similar to Uranus, Pluto's rotational axis is nearly in the plane of its orbit. The dimming of Pluto occurred as our viewpoint shifted from nearly pole-on in 1954 to nearly equator-on in 1973.

The first significant advance in our understanding of Pluto occurred in 1976 with the first infrared observations of Pluto. An observation specially designed to take advantage of the spectroscopic signatures of different ices revealed the unmistakable presence of methane. From this single observation came the first realistic picture of Pluto: a small, cold, yet highly reflective planet.

RENAISSANCE

The second advance, perhaps the most significant of all, came in 1978 with James W. Christy's serendipitous discovery of Pluto's satellite, Charon. Without this important discovery, we might not know any more about Pluto now than we do about most other small bodies in the solar system. The period of the orbit (which matches the light curve period to very high accuracy) and the distance between Pluto and Charon determine the total mass of the system. This new mass for the system is much less than the previous estimates from assumed perturbations on Neptune. A second surprise came with the realization that for once Nature had been kind; the sub-Earth latitude at the time of discovery was approaching the orbital plane of Charon and would cross the plane sometime in the 1980s. The time we are near the orbit plane crossing is important because Pluto and Charon will eclipse each other then.

Starting in the 1980s, a growing number of astronomers began to take interest in Pluto. New high-precision photometry of Pluto's light curve provided a much improved period. Spectroscopic observations aided by a new generation of electronic detectors confirmed the presence of methane but suggested that the methane might not be present in the form of frost on the surface. The new data suggested that the methane might be present in the form of a tenuous atmosphere. Such an atmosphere would have a surface pressure a million times lower than the Earth's. Despite its low pressure, a methane atmosphere has a very profound effect on the climate. The constant cycle of sublimation of methane from the sunlight side and its condensation on the night side can efficiently smooth out temperature variations across the surface of Pluto.

The annual cycle of seasons on Pluto is unique. During 1989, Pluto reaches perihelion. Coincidentally, the Sun crosses the equator of Pluto at nearly the same time as it is closest to the Sun (perihelion). The equator crossing is the same as the time that we refer to as the vernal equinox. Unlike the Earth, the equinoxes have little effect on Pluto's weather. True summer (defined to be the warmest time of the year regardless of where the Sun is in the sky) occurs at perihelion. This time happens to coincide with the beginning of solar spring (as defined by the location of the Sun) for the northern hemisphere and the end of solar autumn for the southern hemisphere. Pluto's winter occurs 125 years later at aphelion which coincides with the autumnal equinox. At aphelion, the atmosphere will completely freeze out causing all weather to cease until the next perihelion passage. In this way, Pluto's year bears a striking resemblance to that of a comet. After all, the coma of a comet is an atmosphere that occurs only near perihelion.

The first crude maps of the surface of Pluto were attempted in the mid-1970s based solely on the observed brightness variations of

Pluto as it rotates. Efforts to improve on these maps continue to this day as an active research topic. The more recent maps of Pluto show that the poles are systematically brighter than the rest of the planet.

Current studies have also been aided by space-based observatories. The *Infrared Astronomical Satellite* (*IRAS*) made the first measurements of the thermal emission from the Pluto–Charon system. These observations are consistent with the very cold conditions assumed for a planet at 30 AU from the Sun. Detailed analyses of the *IRAS* measurements reveal that the surfaces of Pluto and Charon are perhaps quite different. The observed flux is not consistent with both Pluto and Charon having normal asteroid-like surfaces. Only Charon can have a normal surface. Pluto, on the other hand, must have a very unusual surface; consistent with the presence of the methane atmosphere and the polar caps.

MUTUAL ECLIPSES

Twice during a single plutonian year the Sun crosses the orbital plane of Charon. For a short time before and after the plane crossing, Pluto and Charon alternate in casting shadows on each other. In addition to the shadowing, their apparent disks will also overlap as seen from Earth. These phenomena cause the combined brightness of the system to decrease in proportion to the fractional area either covered or shadowed.

By carefully measuring brightness as a function of time, it is possible to determine a number of important physical quantities. Of the full set of Charon's orbital parameters, the mutual eclipses provide accurate values for all except the semimajor axis. One important number deduced from the orbital and physical parameters is the density of the combined Pluto–Charon system. Despite the presence of CH_4 (a low-density ice), the bulk density is 2.1 g cm^{-3}, more than twice that of water. The density is much lower than for the terrestrial planets but more than that of the icy satellites of Saturn. Silicates (rock) must contribute nearly 75% of the volume of Pluto, considerably more than previous guesses. This density also implies that Pluto condensed out of a CO-dominated (H_2O-poor) nebula and that the CH_4 we now observe is only a small fraction of the total composition of Pluto.

In addition to simple photometric observations, some spectroscopic observations have been obtained during the mutual events. The power of this technique cannot be overemphasized. Under the most favorable conditions, that is, perihelion, the separation between Pluto and Charon is at most 0.9 arcseconds. This tiny angular separation is the reason Charon remained undiscovered for so many years; most of the time Pluto and Charon appear as a single unresolved point of light. Therefore, any observation of Pluto also includes the light from Charon. A measurement taken during a total eclipse of Charon includes only the light from Pluto. Thus subtracting the Pluto measurement from a measurement of both reveals the properties of Charon. By applying this technique, the individual spectra of Pluto and Charon are now known. These spectra show that CH_4 frost covered the surface Pluto and water frost covers the surface of Charon (see Fig. 1). Considering its size, albedo, and composition, Charon is most similar to the satellites of Uranus. Pluto, on the other hand, is quite different from everything else in the solar system with the possible exception of Neptune's principal satellite, Triton.

LONG-AWAITED EVENT

As if the onset of the mutual events was not enough to keep researchers busy, another long sought after celestial configuration materialized during the summer of 1987. Pluto occulted a bright star. Stellar occultations are very useful for measuring atmospheric properties and sizes of most of the planets as well as the sizes and shapes of a number of asteroids. In fact, short of visiting an object with a spacecraft, there is no more sensitive method for measuring

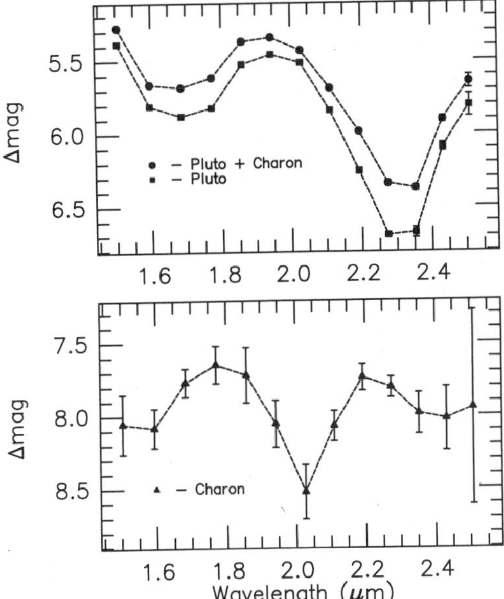

Figure 1. Individual infrared spectra of Pluto and Charon. These curves indicate the brightness as a function of wavelength relative to a solar-type star. The topmost curve is the spectrum of the combined system just prior to an eclipse. The bottom curve in the top panel is the spectrum of Pluto alone, taken while Charon was completely eclipsed. The bottom panel displays the result of subtracting the top two spectra leaving the spectral signature of Charon. The spectra are very different. The absorptions seen at 1.7 and 2.3 μm in the Pluto spectrum are due to CH_4. The absorptions seen at 1.6 and 2.0 μm are due to water frost.

molecular composition of the atmosphere and the absolute size of the system.

Astronomers have observed a number of close passes between Pluto and other stars in the past with no results. However, shortly after Charon's discovery, Charon occulted a star. The Charon occultation provided a single chord that was almost perfectly central but revealed no trace of any atmosphere. On June 9, 1988, astronomers were watching from all available telescopes hoping to see Pluto pass in front of a star. Fortune shined on a few of the groups that night and they obtained the first direct confirmation of Pluto's atmosphere. Current analyses of the occultation data show that the principal constituent of Pluto's atmosphere is CH_4. Though somewhat dependent on the assumed model atmosphere, the size of Pluto measured from the occultation is very close to the numbers derived from mutual event observations and speckle interferometry measurements.

FUTURE DIRECTIONS

At long last astronomers are beginning to agree on the basic properties of Pluto and Charon. Pluto is now being drawn fully into its proper place in the solar system rather than being considered as a distant interloper. We now know the basic properties of this planet (mass, size, density, albedo, and atmospheric composition to name a few; see Table 1) and can now begin to concentrate on some of the more fundamental questions that relate Pluto to other objects in the solar system.

The last year of mutual events remains as of this writing. Though these data will have little effect on the measured sizes of Pluto and Charon, the final events will be very important for completing the albedo maps of the Charon-facing hemisphere of Pluto. The last regions to be traversed are the south polar regions

Table 1. Current Orbital and Physical Parameters of the Pluto–Charon System

Semimajor axis	19,640 ± 320 km
Eccentricity	0.00009 ± 0.00038
Inclination (1950.0)	98.3 ± 1.3°
Ascending node (1950.0)	222.37 ± 0.07°
Argument of periapsis	290 ± 180°
Mean anomaly (from ascending node)	259.90 ± 0.15°
Epoch	JDE 2,446,600.5 = June 19, 1986
Period	6.387230 ± 0.000021 days
Pluto radius	1142 ± 9 km
Charon radius	596 ± 17 km
Pluto blue geometric albedo	0.43–0.60
Charon blue geometric albedo	0.375 ± 0.018
Mean density	2.065 ± 0.047 g cm^{-3}
Total mass of system	(1.47 ± 0.07) × 10^{25} g
Surface gravity of Pluto	6% of Earth's gravity

and the albedo found will help establish important limits on the atmospheric transport of CH_4 frost from pole to pole.

High-precision photometry of the Pluto–Charon system will continue to be important in the future. We still do not know if the current surface of Pluto is static (like the surface of the Moon). It is possible that the surface is slowly changing in response to the variation in surface temperature throughout Pluto's year.

Some questions remain that the *Hubble Space Telescope* can address. With *HST*'s improved spatial resolution, obtaining separate images and spectra of Pluto and Charon has become possible at any time, not just at the time of the mutual events. *HST* will also be able to make accurate measurements of the relative densities of Pluto and Charon, thus addressing the question of similar or different origins.

One final hope for the future rests with the continued development of the world's space program. With a little hope, perhaps one day we will all see a spacecraft make its way out to Pluto. This double planet will certainly have surprises for us that will equal any we have uncovered during our exploration of the other planets in our solar system.

Additional Reading

Beatty, J. K. (1987). Pluto and Charon: the dance goes on. *Sky and Telescope* **74** 248.

Beatty, J. K. (1988). Discovering Pluto's atmosphere. *Sky and Telescope* **76** 624.

Binzel, R. P. (1990). Pluto. *Scientific American* **262** (No. 6) 50.

Hoyt, W. G. (1980). *Planets X and Pluto*. University of Arizona Press, Tucson.

Tombaugh, C. W. and Moore, P. (1980). *Out of the Darkness: The Planet Pluto*. Stackpole Books, Harrisburg, PA.

See also **Planetary and Satellite Atmospheres; Planetary Atmospheres, Structure and Energy Transfer; Satellites, Ices and Atmospheres.**

Pre–Main-Sequence Objects: Ae Stars and Related Objects

Istanvan Jankovics

The Ae stars and related objects are young stellar objects (YSOs) of intermediate mass, which have not yet evolved to the main sequence in the Hertzsprung–Russell diagram. They have extended atmospheric envelopes, like other YSOs, and are related to the well-known T Tauri stars, which are low-mass YSOs.

The early stages of stellar evolution have been intensively studied since the mid-1950s, when much observational evidence had already proven that recently formed stars exist. It was also clear by the mid-1950s that groups of YSOs still exist in the regions where they formed, because there has not been enough time for the stars to escape from the vicinity where they were born. Also, it was known that these young stellar groups are associated with molecular clouds and complexes of nebulosity. A great many peculiar stellar objects, both with and without emission lines in their visible-light spectra, were found in the vicinity of dark and bright nebulae in the dust-obscured regions in the plane of the Milky Way.

Among the ensemble of presumed young objects, which included stars of all luminosities (e.g., stars of the so-called Orion Population), the T Tauri type stars (TTSs) were first classified as a distinct class of YSOs. Further investigations located them on the Hertzsprung–Russell (HR) diagram above the main sequence (MS), and they were thought to be in a particular evolutionary stage, probably still in the phase of gravitational contraction evolving toward the MS.

Over three decades ago, it became evident, and astronomers had no doubt whatsoever, that the irregular emission line variable stars of the T Tauri type that were discovered in very young star clusters and associations are young, low-mass pre–main-sequence (PMS) objects.

YOUNG STELLAR OBJECTS WITH INTERMEDIATE MASSES

In 1960, on the basis of earlier experience with TTS investigations and after statistical calculations using the expected PMS lifetime and the observed population of the MS from the spectral type of A0 toward the higher masses, George H. Herbig thought that early-type stellar objects could be identified that had an evolutionary status similar to the TTS group. Herbig's selection criteria for the early-type counterparts of TTSs were: "the spectral type is A or earlier, with emission lines, the star lies in an obscured region; the star illuminates fairly bright reflection nebulosity in its vicinity."

Observations obtained during the past two decades on a group of Ae- and Be-type stars associated with nebulosity (called Herbig emission stars or HES) and selected in the previously mentioned way demonstrate a close relationship between the two classes of stars, namely the TTS and HES. It is now generally established that HES, like other YSOs, possess extended envelopes containing matter in different physical states.

RECENT OBSERVATIONAL RESULTS ON HERBIG EMISSION STARS

Emission Lines and Their Line Profiles

High-resolution spectroscopic studies in the visual and ultraviolet wavelength ranges, following Herbig's work, established that the Balmer emission features in the HES spectra are dominated by the strong and broad Hα emission. Line equivalent widths at Hα usually range from 10 Å $< W_\lambda(\mathrm{H}\alpha) <$ 100 Å, where the typical line widths at half maximum (FWHM) exceed 200 km s^{-1}. (This average FWHM value illustrates the characteristic velocity scale in an HES envelope; see also Fig. 1.)

Apart from the complex hydrogen lines and the other emission lines that occasionally exist in this spectral range (e.g., Na D, He I, and forbidden lines), the spectra of HES exhibit no significant difference from those of MS stars of the same spectral type.

Depending on the shape of their Hα profiles, HES are classified into three subclasses: from a sample containing 57 HES: HES with single-peak emission (25%); HES with double-peak emission (50%); and those showing a P Cygni profile (20%).

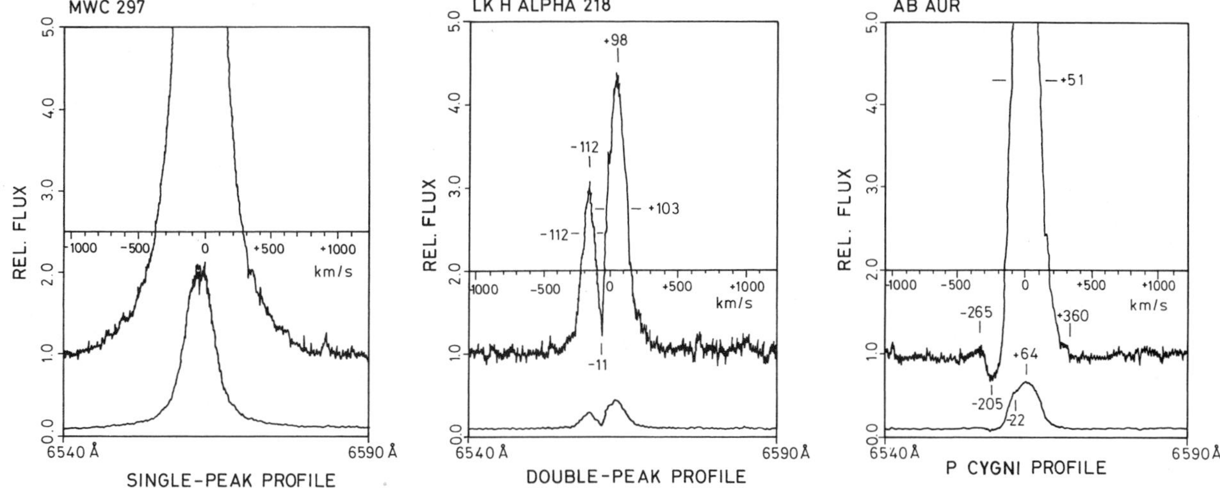

Figure 1. High-resolution Hα line profiles of three Herbig emission stars illustrate each type of profile found in these stars. The velocity scale, wavelength scale, and characteristic emission and absorption features are indicated. The values at the labels are the corresponding heliocentric radial velocities (in kilometers per second).

Figure 1 shows the observed Hα line profiles to illustrate each type of profile. The velocity scale indicates the velocity range of the emitting volume. The characteristic emission and absorption features are also labeled with values of the corresponding heliocentric radial velocities (in kilometers per second).

The type of profile of the higher Balmer lines Hβ, Hγ, and so on, for a given HES does not change within the series; it is the same as that classified at Hα. However, the flux of the emission components rapidly decreases with increasing Balmer numbers, and in the higher Balmer lines the photospheric absorption feature becomes more dominant.

Because the strong Hα emission originates in the extended stellar atmospheres, the considerable differences in the line profiles from star to star may indicate differences in the geometry of the extended gaseous envelopes of HES. Alternatively or additionally, the line profile differences could reflect different envelope–velocity–field, density, or temperature conditions in the HES.

Direct evidence for stellar winds was obtained initially only in the "P Cygni" subclass of HES. Detailed studies made during the late 1980s focused on the basic similarities within a given subclass, with the object of establishing the basic differences between the subclasses. In addition to the common Hα profile, definite similarities have also been found in the shapes of other line profiles for many "P Cygni" type HES. Because certain lines are formed in different regions in the HES envelopes (e.g., the Ca II lines originate close to the photosphere, the Na I D line is formed in a zone further out in the atmosphere, and the Hα and Mg II resonance lines arise in a very extended volume), the observed phenomenon indicates a significant structural similarity: HES of the "P Cygni" subclass have the same characteristic structure.

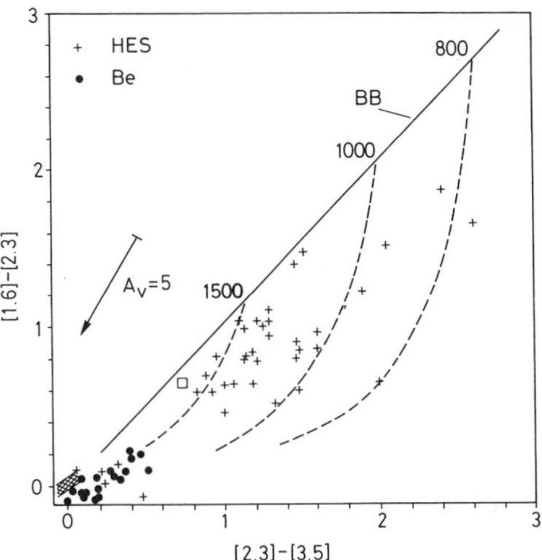

Figure 2. Near-infrared color–color diagram of Herbig emission stars (crosses). Filled circles represent a number of classical Be stars and at the lower left corner the hatched area corresponds to the main sequence stars between the spectral types of B0 and F0. The corresponding blackbody radiation is shown as a straight line, labeled BB, and the locus of the hydrogen gas emission (free–free, free–bound, and bound–bound) at $T_{eff} = 10^4$ K is marked by an open square. The reddening law is indicated by the vector $A_V = 5$. The three dashed lines, labeled with values 800, 1000, and 1500 K, represent the photospheric radiation of an A0-type star overlaid with the thermal emission of hot circumstellar dust grains.

Infrared Properties

Using photometric measurements in visible light, HES are found to occur close to the MS on the HR diagram (M_V versus $B-V$). This picture, however, drastically changes when the flux coming from the near-infrared wavelength ranges (H[1.6], K[2.3], L[3.5], M[5.0], where the number in brackets indicates the wavelength, in micrometers, at which the magnitude is measured) is taken into account. When the flux of all stellar radiation is calculated, the bolometric luminosities locate the HES above the MS on the HR diagram (log T_{eff} versus log L).

All of the HES show a large infrared excess: The typical values are 3–5 magnitudes at wavelength 3.5 μm. In the [1.6]–[2.3] versus [2.3]–[3.5] infrared two-color diagram (Fig. 2), HES show infrared colors significantly different from those of the classical Be stars. The two groups of stars are well separated in this diagram. The infrared properties thus provide a physical criterion for the distinction between the HES and classical Be stars as separate classes of objects, namely the presence of substantial amounts of warm dust emission in HES. In Fig. 2 the filled circles indicate classical Be stars and the crosses refer to HES. In the lower left corner a small hatched area represents the MS between the spectral types of B0 and F0.

Figure 2 also demonstrates the interpretation, favored by many astronomers, that the infrared excess of HES and related YSOs could be due to thermal emission from hot, circumstellar dust grains. The straight line in Fig. 2, labeled BB, represents blackbody radiation and the open square indicates the locus of the hydrogen gas emission (free–free, free–bound, and bound–bound) at effective temperature $T_{eff} = 10^4$ K. The vector labeled $A_v = 5$ shows the reddening law. Furthermore, the dashed lines refer to the photospheric radiation of an A0-type star overlaid by the thermal emission of dust grains with temperatures of 800, 1000, and 1500 K.

The origin of this circumstellar matter, which surrounds the stellar objects in the form of shells and extended disks and which emits radiation over the range from the near infrared to the far infrared, and its role in the further evolution of pre–main-sequence objects is of crucial interest.

Association of Herbig Emission Stars with Molecular Clouds

The observed radial velocities indicate no systematic motion of HES with respect to nearby molecular clouds. A study of the radial velocities of the interstellar absorption components in the optical HES spectra with the molecular cloud velocities determined from radio measurements places the HES in or behind the molecular clouds. The HES are physically associated with the molecular clouds and are embedded in or intermixed with clouds or cloud complexes.

Photometric and Spectroscopic Variability of Herbig Emission Stars

A number of stars now classified as HES have long been known to be variable. The typical values of the amplitudes in the photographic wavelength range are 2 magnitudes (in some cases up to 4 magnitudes).

No systematic investigation has yet been made of the role of the photometric variability of HES. However, on the basis of measurements available in the literature, it has been established that almost all strongly variable HES have spectral types later than B8, whereas all stars with no significant variation (< 0.05 magnitude) have spectral types earlier than A0. The typical amplitudes and behavior of variations of the latter-type HES are similar to those of T Tauri stars.

In many cases there are earlier observations of spectroscopic variability in stars later classified as HES. Considerable variations in line profiles may occur on time scales ranging from months to years, or even on time scales of about one day.

Various observations carried out at different dates have clearly shown the variable nature of HES, but there is as yet no detailed study that gives a characteristic picture of all the spectroscopic variables within this class of stars. Nevertheless, investigations of selected HES have provided important specific results. For example, a recent study of short-term spectroscopic variability showed that several lines are modulated by the rotation of the star and of its envelope.

Evolutionary Aspects: A Semiempirical Model

The location of HES above the MS on the effective temperature versus surface gravity diagram or the effective temperature versus bolometric luminosity diagram, the anomalous infrared properties of these stars, which are due to the presence of circumstellar dust envelopes and opaque flattened disks, and the unambiguous physical association of HES with nearby molecular clouds all support the conclusion that HES are young, pre–main-sequence objects. These pieces of evidence (among others) prove that according to theoretical evolutionary sequences, HES are the very young, intermediate mass representatives of upper–main-sequence stars.

No comprehensive theoretical model exists to explain the HES phenomenon itself. However, in the past few years, a semiempirical approach has been taken to understand the wind properties. Also, strong efforts have been made to form a picture of the structure, temperature, and velocity laws of the HES envelope.

As the brightest of the HES in the northern hemisphere, AB Aurigae (A0ep, $m_v = 7.2$) has turned out to be a good representative of its class. The interpretation of several observed lines in its spectra (e.g., Hα, and the Mg II, and C IV resonance lines) is characteristic of the whole class.

In the course of investigations made over the past few years, the semiempirical model created about five years ago was repeatedly updated. According to the basic assumption in this model, the central star with a classical stellar photosphere (observations support a model of a photosphere in hydrostatic and radiative equilibrium, with $T_{eff} = 10^4$ K and $\log g = 4$) possesses an extended and expanding atmosphere.

In this framework the average structure of AB Aurigae's wind is the following:

Temperature law: Very close to the star, above the photosphere, the temperature drops to 7100 K, then increases again and reaches its maximum value between 16,000 and 18,000 K in the chromosphere. The extent of the chromosphere is smaller than 2.5 stellar radii. Further out, in the postchromosphere region, the temperature decreases again.

Velocity law: At the bottom of the chromosphere the velocity gradient is low, then the wind accelerates so that the wind velocity reaches about 150 km s^{-1} in the outer layer of the chromosphere. Beyond the chromosphere the flow keeps on accelerating outward until its maximum velocity is about 300 km s^{-1}. Farther away from the star, in the postchromospheric region where the stellar wind is strongly cooled, the flow also decelerates to a terminal wind velocity of about 150 km s^{-1}. The location of the cooling and deceleration region is unknown.

Mass-loss rate: The estimated mass-loss rate of AB Aurigae is between 10^{-8} and 1.8×10^{-8} $M_{\odot}$ yr^{-1}.

SYNOPSIS

In our present understanding, HES are the higher-mass counterparts of the T Tauri stars. These two classes of stars represent PMS objects with ages less than 3×10^6 yr and cover a continuous mass spectrum from low to intermediate masses between about 0.1–6 $M_{\odot}$ (solar masses).

Current research on HES is addressing the problem of the phases of pre–main sequence evolution in the context of the mass and age, the driving mechanism of the wind in HES envelopes and particularly the building of a comprehensive hydrodynamical model of the HES phenomenon.

The causal connection of the YSOs to the surrounding disk-like circumstellar structures has become almost obvious, and in many recent articles the mass outflows from very young stars have also been stressed as a common characteristic of their PMS nature.

HES, like other YSOs, are engaged in very complex interactions with their environment. The opaque circumstellar disks and the collimated outflows associated with them have become a topic of investigation as well.

Additional Reading

Catala, C. (1984). The envelopes of the Herbig Ae/Be stars. *Lecture Notes in Phys.* **237** 198.

Catala, C. (1989). Herbig Ae and Be stars. In *Low Mass Star Formation and Pre-Main Sequence Objects*, B. Reipurth, ed. European Southern Observatory, Garching, p. 471.

Catala, C. and Kunasz, P. B. (1987), Line formation in the winds of Herbig Ae/Be stars. *Astron. Ap.* **174** 158.

Finkenzeller, U. and Jankovics, I. (1984). Line profiles and radial velocities of Herbig Ae/Be stars. *Astron. Ap. Suppl.* **57** 285.

Finkenzeller, U. and Mundt, R. (1984). The Herbig Ae/Be stars associated with nebulosity. *Astron. Ap. Suppl.* **55** 109.

Herbig, G. H. (1960). The spectra of Be- and Ae-type stars associated with nebulosity. *Ap. J. Suppl.* **4** 337.

See also **Herbig–Haro Objects and Their Exciting Stars; Stars, Pre-Main-Sequence, Winds and Outflow Phenomena; Stars, T Tauri.**

Protogalaxies

Gregory D. Bothun

Over the last 20 years, rather compelling evidence has been gathered which suggests that there is more mass contained in galaxies than can be accounted for by the amount of light that they emit. This blurs our concept of what actually constitutes a galaxy and suggests that the luminous stars that form from the gas in galaxies may simply be tracers of some large concentration of nonluminous matter, whose nature is unknown. This complicates our concept of galaxy formation because it is not clear whether it fundamentally refers to the problem of the formation of dark matter potentials or to the problem of the condensation of baryonic material within these dark potentials. The formation of the luminous component of galaxies, will be discussed here. In this context, the term *protogalaxy* refers to a galaxy that is experiencing its first generation of star formation, having consisted totally of hydrogen gas trapped in some dark matter potential prior to this point.

Although nearby galaxies presently exist in a wide variety of shapes and forms, there are three general features of the galaxy population that must be explained by any theory:

1. Galaxies come in two basic kinds, namely, elliptical galaxies that are supported by the internal velocity dispersion of the stars (isotropic orbits) and spiral galaxies that are flattened disk systems in which the stellar orbits are circular.
2. Most spiral galaxies contain a mini-elliptical galaxy at their center (i.e., a bulge).
3. Galaxies are not randomly distributed in space but are highly clustered.

Galaxies may be conveniently parameterized by their bulge-to-disk (B/D) ratio, where the bulge component is a spheroidal distribution of stars with isotropic orbits and the disk component is a highly flattened distribution of stars and gas with circular orbits. Interestingly, the B/D ratio of a given galaxy seems to be dependent on its environment in that galaxies located in the cores of rich clusters are preferentially bulge dominated, whereas those located in the lower-density regions of the universe are preferentially disk dominated. This indicates that protogalaxies rarely form in isolation and hence interactions with other protogalaxies at the time of formation may greatly determine the evolutionary course of a given galaxy.

BUILDING BLOCKS OR FRAGMENTATION

The formation of large-scale structure (i.e., clusters of galaxies, superclusters of clusters) is intimately related to the formation of individual galaxies. Currently, there are two competing scenarios for the formation of structure in the universe and current observations are incapable of distinguishing between them. Clearly, galaxies represent density enhancements in the universe which makes their existence somewhat difficult to understand in terms of the hot big bang theory for the origin of the universe. In particular, it is known from observations of the cosmic microwave background (CMB) that the early universe was very homogeneous on large scales. Furthermore, when the universe was less than approximately 300,000 years old, the energy density contained in this radiation field far exceeded that which was contained in matter. In this physical situation, gravity, in effect, was nonexistent and the distribution of matter was governed by the distribution of radiation, which we now measure to be quite homogeneous. Hence it is quite paradoxical that any inhomogeneities would form and so our understanding of the existence of galaxies is challenged at a very fundamental level.

Because of this paradox, it is common practice to use the present-day distribution of galaxies as a fossilized imprint of what the original spectrum of density perturbations must have been after matter and radiation decoupled and gravity became important. This leads to two possible scenarios for the formation of galaxies and clusters of galaxies. In the fragmentation picture, only very massive perturbations (mass up to 10^{16} $M_{\odot}$ (solar masses); a typical galaxy has a mass of $2\text{–}5\times10^{11}$ $M_{\odot}$) survived the radiation-dominated era. These large-mass perturbations are identified today as the largest known superclusters of galaxies. Subsequent cooling and fragmentation within the overall density perturbation then produces smaller-scale clusters of galaxies (of mass $10^{14}\text{–}10^{15}$ $M_{\odot}$). Further fragmentation within those individual clusters then produced individual galaxies. In this scenario, virtually all galaxies that formed should be members of clusters and/or superclusters. To a large degree, this seems to be verified by current observations. However, this scenario also suggests that, because galaxies are forming via the process of fragmentation and collapse within a much larger cloud of gas, protogalaxies originally started out as objects that were about 10 times larger than galaxies today. Because protogalaxies are already clustered by this point, their environment fosters strong interactions between them. Moreover, because the disk component forms over a significantly longer time scale than the bulge component, these interactions do not favor the production of disk galaxies due to the tidal disruption of the gas destined to collapse into a disk. Hence, whereas this scenario does qualitatively predict the correlation between B/D and local density, it is not at all clear that any disk-dominated galaxies should have formed.

The alternative to the cooling and fragmentation picture is the idea of hierarchical clustering of small-mass units. In this picture, the first objects to form in the universe were $10^5\text{–}10^6$ $M_{\odot}$. objects that gravitationally coalesced into larger-mass units (i.e., galaxies). The process of gravitational coalescence then continues with galaxies forming clusters of galaxies and then with clusters forming superclusters (and this process may still be continuing). Like the earlier scenario, the gravitational clustering idea also predicts that virtually all galaxies should be members of a cluster or a supercluster. However, in this picture, galaxies are allowed to originally form in isolation before becoming part of the cluster environment and this would tend to favor the production of disk galaxies. Moreover, this scenario also qualitatively predicts the correlation between B/D ratio and local density in the sense that bulge-dominated systems originally formed in denser areas than disk-dominated systems. Hence gravitational clustering around bulge-dominated galaxies will occur more quickly and disk-dominated systems will then infall at some later time into these newly formed clusters. This process of disk galaxy infall is something that is observed in the case of the Virgo cluster.

THE EPOCH OF GALAXY / CLUSTER FORMATION

Although the two scenarios discussed previously can both give rise to galaxies that are distributed in a highly clustered manner, each makes different predictions regarding the redshift at which galaxy and cluster formation should occur. In the cooling and fragmentation picture, it is possible that virialized clusters were present at redshifts as large as 2 (80% of the age of the universe). The best signature of a virialized cluster is the emission of x-rays from the hot intracluster medium. The strength of the x-ray emission is directly proportional to the mass of the cluster because the gas is heated by the cluster gravitational potential. Current x-ray surveys are only sensitive to x-ray emitting clusters out to a redshift of $z \approx 0.7$ (55% of the age of the universe) and hence can place no stringent limits on the epoch of cluster virialization. ROSAT and future x-ray satellites such as AXAF may prove invaluable in this regard.

One of the main difficulties with the hierarchical clustering idea is the amount of time it takes for the process to begin. Specifically, gravitational amalgamation of subunits can only occur efficiently when the subunits themselves are reasonably close to one another. Because the universe is expanding, then, at a given mass density, there will be an epoch past which this process can never get started because the average spacing between subunits is too great. At present, there is at least an order of magnitude range in our estimate of the mass density of the universe. Various theoretical arguments favor a mass density that actually closes the universe. Observations of large-scale deviations from Hubble flow, however, indicate that the mass density is between 10 and 20% of the closure density. In such a low-density universe, the process of hierarchical clustering must start at redshifts of 50–100 (0.1% of the age of the universe) and the physical details regarding the merger of subunits at such an early time remain obscure.

DETECTION OF PROTOGALAXIES

Of course, speculation regarding the formation of protogalaxies would cease upon their discovery. In fact, one can make a simple argument that suggests that protogalaxies might easily be detectable. A typical elliptical galaxy has $10^{11}\ M_\odot$ of stars. Characteristics of the stellar population in present-day ellipticals suggest that the bulk of this star formation occurred very early on, perhaps in the phase of protogalactic collapse. The collapse or free-fall time of any cloud is given by $\sqrt{G\rho}^{\,-1}$, which is 4×10^8 yr for a typical elliptical with an initial radius of 50 kpc. (G is the gravitational constant and ρ is the density.) In this case the predicted star formation rate would be approximately 250 $M_\odot$ yr^{-1}, which is a factor of 50 larger than the current star formation rate in nearby spiral galaxies. A star formation rate this high produces a great deal of luminosity in hot young stars as well as a supernova rate of about 1 per year! Most of this luminosity comes out at 1000 Å in the rest frame of the galaxy and so the redshift of this hypothesized protogalaxy would have to be $3 \geq z \leq 10$ in order to be accessible to ground-based observations.

Before discussing possible reasons why this population of galaxies has yet to be detected, it is worth pointing out that protogalaxies with initial densities that are rather low will take a correspondingly longer time to form. In general, their formation process will be interrupted by the tidal shearing forces of neighboring galaxies. However, if these objects are in relative isolation, then some may be collapsing now, particularly in the case of flattened disk galaxies. Interestingly, a number of very low surface mass density disk galaxies have been discovered over the last two years. Although they are not protogalaxies by our adopted definition, their discovery in nevertheless important because these galaxies represent examples of galaxy formation that has been quiescent and that extended over a considerable period of time.

So the task of discussing protogalaxies now becomes one of suggesting reasons why they have escaped detection. We close with the following four possibilities. The most obvious possibility is that the first phase of vigorous star formation was essentially complete prior to $z = 10$ (which, coincidentally, is when the universe is approximately 1 galactic free-fall time old). However, in this case the radiation from the hot young stars is redshifted into the near-infrared part of the electromagnetic spectrum (e.g., 1–5 μm). Searches for this population with newly developed near-infrared imaging arrays are presently underway. Second, because the ultraviolet radiation from young stars is strongly absorbed by dust, it is possible that the early universe is not transparent to this radiation. This provides a rather effective screen that can potentially hide the process of galaxy formation from us. In this case, the absorbed ultraviolet photons should heat the dust to a temperature of 50–100 K. The radiation from this heated dust will be redshifted to the submillimeter portion of the electromagnetic spectrum. The energy density of these hypothetical sources is thought to be detectable with the SIRTF mission to be launched sometime around the year 2000. Third, it is well known that the peak in the redshift distribution of quasars is at $z \approx 3$. Because quasars are now known to be in the nuclei of galaxies and the energy source is thought to be the infall of gas onto a massive black hole, it is possible that this infall process is facilitated by protogalactic collapse. In that case the radiation from the quasar effectively overwhelms that from the rest of the galaxy, imposing a sort of cosmic censorship against observing galaxies in the act of formation. Finally, our expectation that protogalaxies should be observable stems from a relatively naive physical argument. Perhaps, galaxy formation is a far more gentle and quiescent process than we have assumed and there simply is no ultraluminous phase which marks the birth of a galaxy. Future advances in telescopes and instrumentation will hopefully yield firm detections of protogalaxies, from which we can finally study their formation in detail.

Additional Reading

Hensler, G. and Burkert, A. (1989). The initial conditions of protogalaxies—constraints from galactic evolution. In *Progress Report on Cosmology and Gravitational Lensing*, G. Börner, T. Buchert, and P. Schneider, eds. Max Planck Institute for Physics and Astrophysics, Garching, p. 203.

Meier, D. L. and Sunyaev, R. A. (1979). Primeval galaxies. *Scientific American* **241** (No. 5) 130.

Peebles, J. (1984) Origin of galaxies and clusters of galaxies. *Science* **224** 1385.

Rees, M. J. and Silk, J. (1970). The origin of galaxies. *Scientific American* **222** (No. 6) 26.

Silk, J., Szalay, A. S., and Zeld'ovich, Y. B. (1983). The large-scale structure of the universe. *Scientific American* **249** (No. 4) 72.

See also **Cosmology, Galaxy Formation; Galaxies, Formation; Galaxies, High Redshift.**

Protostars

Claude Bertout

Protostars are primitive stars that have just been formed. These stellar objects have been so elusive that C. G. Wynn-Williams started his 1982 review of the topic by stating that "Protostars are the Holy Grail of infrared astronomy." Whether a protostar has ever been seen is a question even today because what theoretical models tell us to expect a protostar to look like does not match actual observations of young stars.

THEORETICAL EXPECTATIONS

Stars are believed to form from the gravitational contraction of dense molecular cores with density $\approx 10^{-19}$ g cm^{-3} and dimensions of ≈ 0.1 pc that are seen in the giant interstellar clouds of our galaxy. How these dense cores become gravitationally unstable is not well understood yet. A current model assumes that dense cores are held in dynamical equilibrium by the magnetic field and that gravitational instability is reached after the field diffuses slowly out of the densest cloud parts. One then expects the density profile of the forming core to be $\rho \propto 1/r^2$ (where r is distance from the center) at the onset of the gravitational collapse. Other models hypothesize instead that the collapse of individual cloud fragments is triggered by an external compression that is itself perhaps caused by an exploding supernova or by the winds from nearby young massive stars. Hydrodynamical collapse simulations then show that even if the density of the molecular core is constant at the beginning of the collapse, it becomes proportional to $1/r^2$ after a time comparable to the core's initial free-fall time τ_{ff}, that is, the time necessary for the core to collapse onto itself under the action of gravity alone. It is given by

$$\frac{\tau_{ff}}{2.1\times10^5 \text{ yr}} = \left[\frac{\rho_0}{10^{-19} \text{ g cm}^{-3}}\right]^{-1/2}$$

Thus τ_{ff} depends only on the initial density ρ_0 of the fragment, and denser parts of the core collapse on shorter time scales than the more diffuse parts. Whatever the initial conditions of the collapse, then, a runaway density increase is likely to occur at the core center where collapse proceeds on faster and faster time scales.

As the fragment's center becomes denser, it also becomes more opaque to its own thermal radiation. The temperature and pressure of this central region then increase until finally the pressure is strong enough to counteract gravity. A stable protostellar core thus forms at the center of the collapsing fragment and is surrounded by an extended, diffuse envelope that is still in a state of free-fall. The radius of the protostellar core at this stage is only a small fraction (typically 1/100,000th) of the initial fragment's radius.

This picture is more complicated if the observed rotation of dense molecular cores is taken into account in the simulations. Whenever every mass element conserves its angular momentum, then the collapse usually creates a rotating ring rather than a stellar core. There are, however, several possible angular momentum transport mechanisms, such as turbulent friction and magnetic braking, that can help separate mass and angular momentum during the collapse. A circumstellar disk containing most of the angular momentum is then formed together with the protostellar core. The expected disk diameter will be about 100 AU, which is comparable to the diameter of the solar system.

Once a protostellar core has formed, it grows hydrostatically by accreting matter from the infalling envelope and from the circumstellar disk. Because gravitational energy is the main energy source at this stage, the relevant time scale for evolution of the protostar is the so-called Kelvin–Helmholtz contraction time τ_{KH} given by

$$\frac{\tau_{KH}}{3\times10^7 \text{ yr}} = \left[\frac{M_\star}{M_\odot}\right]^2\left[\frac{R_\star}{R_\odot}\right]^{-1}\left[\frac{L_\star}{L_\odot}\right]^{-1},$$

where $M_\star$ is the mass, $R_\star$ the radius, and $L_\star$ the luminosity of the protostellar core. $M_\odot$, $R_\odot$, $L_\odot$ denote the mass, radius, and luminosity of the Sun.

Subsequent evolution depends on the protostar's mass. The Kelvin–Helmholtz time scale of a massive protostar is shorter than the infall time scale, so that it starts burning hydrogen while still accreting mass. For these massive objects, radiation pressure from the luminous core is then expected to reverse the matter infall, and one can calculate that this mechanism should prohibit the formation of stars with masses larger than about 7 $M_\odot$. Because young OB stars are definitely more massive than this, one suspects that the simple picture of radial infall summarized previously is incorrect and that a highly nonspherical collapse leads instead to infall of matter onto an equatorial disk at a large distance from the star, where infalling matter is shielded from direct stellar radiation. The star then grows more massive by accreting disk matter.

The Kelvin–Helmholtz contraction of low-mass, solar-type protostars is longer than the infall time scale, so that accretion ends before the star reaches the main sequence. In current numerical simulations of protostellar collapse, infall terminates "naturally" when envelope matter has entirely rained down onto the protostar. But although these simulations self-consistently solve the basic hydrodynamic equations, they also ignore such complexities as the role of magnetic fields in driving stellar winds, as well as the role of winds in removing angular momentum from the protostar. That nature does not confine itself to the simple physical equations that astronomers are able to set up and solve is once again demonstrated by observations of young stellar objects. The extensive infalling regions models predict will surround forming stars have not been observed yet, in spite of considerable effort by observers. Studies of star-forming regions reveal instead extensive outflows from young stellar objects.

THE OBSERVATIONS

Star-forming interstellar clouds are often called dark clouds because they contain dust that blocks light from the stars behind them and appear as the dark patches in the Milky Way. Dust is opaque to radiation emitted at all visible and shorter wavelengths (i.e., ultraviolet and x-ray ranges). Thus, as long as astronomers were limited to studying the visible light emitted by stars, they had no way of knowing what was going on within these dark clouds; they could only study the cloud outskirts. However, dust becomes transparent for radiation with wavelengths greater than the dust grain size: infrared and radio. In the last 20 years or so, astronomers were able to build efficient infrared, millimeter-wave, and radio-wave detectors that now allow them to "see" through interstellar clouds. The two closest and best-studied interstellar clouds in which stars are believed to form today are the Taurus–Auriga and the ρ Ophiuchi regions, both located in the vicinity of the galactic plane at a distance of about 150 pc from the Sun, but in different directions.

An Example: The ρ Ophiuchi Star-Forming Region

The population of young stellar objects in the core of the ρ Ophiuchi dark cloud provides an illustration of these new techniques' power. Prior to the 1970s, only a dozen or so optically visible young stars—named T Tauri stars after their prototype—had been discovered during Hα-emission surveys; they are located at the periphery of the cloud where dust-caused extinction is lowest. More recent observations using the *Einstein* x-ray observatory uncovered another population of optically visible young stars with little activity (besides their x-ray emission) located predominantly at the outskirts of the cloud. Observations in the near- and mid-infrared revealed a large number of stellar objects deeply embedded in the cloud, many of them with bolometric luminosities in the 0.1–25 $L_\odot$ range. A recent radio continuum survey using the Very Large Array, a powerful radio interferometer, detected yet another, albeit smaller, class of objects within the cloud boundaries. Figure 1 displays spatial relationships between all these objects, which are predominantly low-luminosity and thus low-mass young stars. Because of the difference in their distance from the cloud's center, the various stellar populations seen in these surveys are believed to represent several evolutionary stages of low-mass stars. Stars that are more concentrated around the center of the molecular cloud have had less time to diffuse away from their place of birth, which means that they are younger than stars farther out from the cloud center. Implicit in this idea is the hypothesis that gravitational collapse leading to star formation is more likely to initiate in the densest cloud parts. The ρ Ophiuchi infrared sources can be separated into three distinct morphological classes based on the shape of their spectral energy distributions in the $\log\lambda F_\lambda$ versus $\log\lambda$ diagram, where λ is the wavelength and F_λ the flux per unit wavelength interval; it is now clear that these classes correspond to different stages in an evolutionary sequence from protostellar object to pre–main-sequence star.

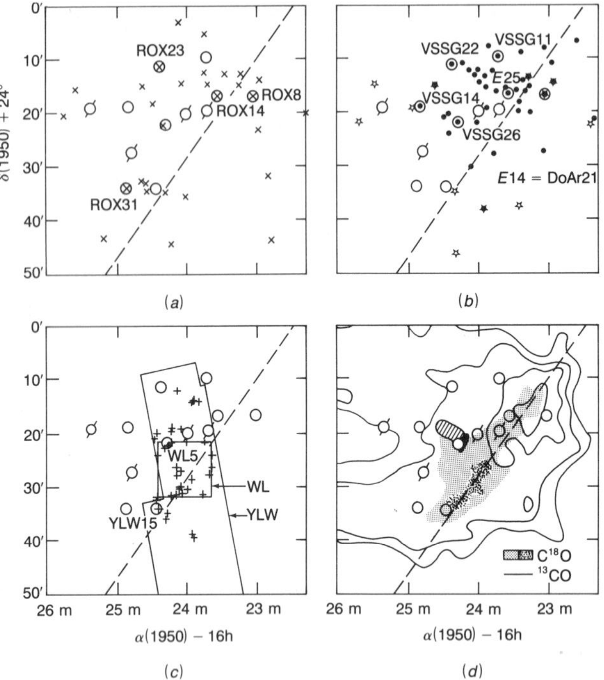

Figure 1. The various populations of young stars seen at different wavelengths in the central part of the ρ Ophiuchi star formation region. Open circles denote the radio continuum sources, their counterparts at various wavelengths are named according to the survey that led to their discovery, and the dashed line shows the orientation of the high-density gas. (*a*) Crosses are definite x-ray sources. (*b*) Dark circles are bright infrared sources, and star symbols are optical young stellar objects of various masses. (*c*) Plus symbols are faint embedded infrared sources. Two different survey regions (denoted WL and YLW) are delineated. (*d*) Contours show ^{13}CO emission and increasingly shaded areas represent high density C^{18}O. The hatched and black areas are dense regions seen respectively in DCO$^+$ and H_2CO. (*Courtesy of Dr. Thierry Montmerle.*)

In this scheme, Class I sources are deeply embedded, very young stellar objects that are seen only in the infrared and millimeter spectral ranges. Their infrared spectral energy distributions are either flat or rising toward long wavelengths. One model that includes a star and its circumstellar disk both surrounded by a dusty envelope successfully reproduces at least some of the observed spectral energy distributions. Class II objects have energy distributions typical of T Tauri stars; they can be either visible or infrared sources, and are probably somewhat more evolved than Class I objects. Their spectra have been shown to originate from the star and a surrounding accretion disk. Class III sources display reddened blackbody-like energy distributions; they are often found at the cloud periphery and correspond presumably to the more evolved cloud stars. Objects detected primarily in the x-ray range often belong to Class III, and their x-ray emission is presumably caused by magnetic activity similar to that of the Sun, albeit stronger.

It thus appears that the youngest stellar objects known so far are the Class I sources. Although they have some similarities with "theoretical" protostars—they look like stellar cores surrounded by disks and dusty remnants of the cloud from which they were born—they also display a range of unexpected, albeit exciting, properties.

Properties of Young Stellar Objects

Observations of molecular lines in the millimeter range, which are used to study physical conditions in the vicinity of newly formed stars still embedded in their parental molecular cloud, reveal that young stellar objects are violently interacting with their surroundings. Extensive surveys of star-forming regions in the rotational lines of CO have detected cold, high-velocity molecular gas ($T \sim$ 10–20 K, $V > 5$ km s^{-1}) in the direction of about 100 embedded young stellar objects with luminosities ranging from $10^5\ L_\odot$ down to a few $L_\odot$. That the protostellar phase is accompanied by strong, collimated mass loss originating from the forming star is supported by the following arguments:

1. Mass inside the high-velocity emission region is too small for velocities to be due to gravitationally-bound rotation or collapse.
2. In most cases, the high-velocity emission region is bipolar with outflowing gas symmetrically distributed in two lobes about the central source, which rules out turbulence or shear between colliding clouds as the origin of the supersonic motions.
3. In several cases, direct evidence for outflow is provided by the presence in the high-velocity knots of fast-moving nebulosity called Herbig–Haro objects, the proper motion vectors of which point away from the stellar source (see the following discussion).

CO data allow astronomers to derive mass, velocity, spatial extent, and age of the flow, and hence to indirectly measure the mass-loss rate from the central object averaged over the flow lifetime. In most cases, the mass of the high-velocity molecular gas is a large fraction of the mass of the central object, so that the flow is believed to consist mostly of ambient cloud material entrained by the wind from the young stellar source. Derived mass-loss rates span the range 10^{-7}–$10^{-5}\ M_\odot$ yr^{-1}, and velocities of the CO gas range from a few kilometers per second to more than 100 km s^{-1}. It is now widely believed that most stars go through a phase of energetic bipolar outflow during the early phase of their formation. Whether this outflow is necessary to the formation of the star—its role in carrying away excess angular momentum could conceivably be crucial—is, however, unknown.

Luminous molecular outflow sources ($L_{bol} > 10^3\ L_\odot$) often possess detectable radio centimetric continuum emission attributed to thermal free–free radiation from an ionized envelope. The spectral index of the radio flux is usually close to the value expected for a constant-velocity, fully ionized, spherical wind, although sources with moderate luminosity tend to show spectra that suggest optically thin emission which could be produced either in an infalling envelope or in a collimated wind whose opening angle increases with distance from the star. High-resolution maps reveal collimated radio jets in a few young stellar objects.

Optical jets and Herbig–Haro (HH) objects offer the most obvious and spectacular examples of strong outflows from young stars. Jets are highly collimated gaseous outflows that can be seen in emission lines such as the Hα hydrogen line and in forbidden lines from ionized sulfur (denoted by S II in the following). They can have projected lengths up to 0.5 pc, and typically have length-to-width ratios of 5–20. HH objects are small nebular emission knots that emit lines normally thought to arise in shocked interstellar material. They are sometimes found in a linear chain that points back to a known pre–main-sequence star or embedded infrared object. Proper motions can be found for a number of HH objects, and their space motions show that they are moving at several hundred kilometers per second away from the exciting star.

Perhaps the best example of a collimated outflow from a young stellar object is HH 34. Figure 2 shows the environment of this object in the light of S II. Here a straight jet that is partially broken into a series of knots extends for more than 30″ out of a point source. The jet then disappears; but directly along its extension another knot or two can be seen, followed by a large conical nebula (HH 34S) with its apex pointing away from the star. HH 34S looks like a bow shock that marks the end of the flow as the jet bores its way into the interstellar medium. A similar but fainter nebula

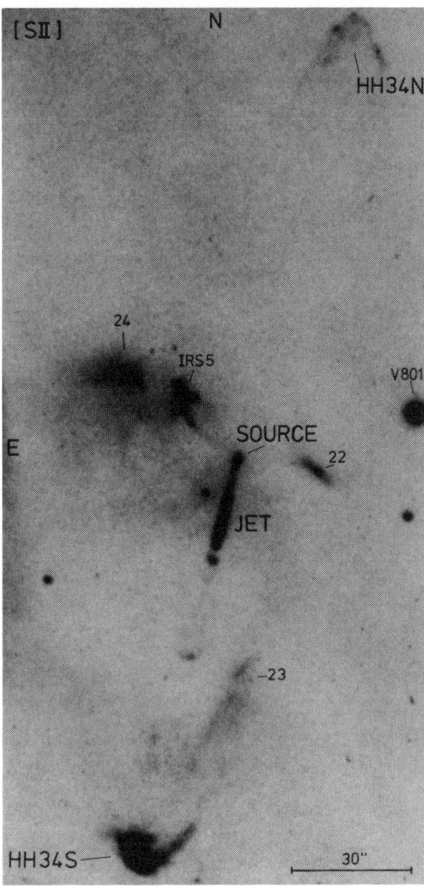

Figure 2. Image of the HH 34 region in the light of S II. A knotty jet is pointing away from the protostar (denoted "source") toward HH 34S. Both this Herbig–Haro object and the fainter one (HH 34N) on the opposite side of the source are shaped like bow shocks. HH 34S and the jet are moving toward us, whereas HH 34N is moving away from us. (*Courtesy of Dr. Reinhard Mundt.*)

with the symmetry reversed (HH 34N) can be seen emanating in the opposite direction from the star. The jet toward HH 34N is not seen because it is directed away from us and into the dense, opaque molecular cloud.

Jets have a typical velocity of several hundred kilometers per second and electron densities on the order of 10^3 cm^{-3}. Properties of their exciting sources resemble those of solar-type young stars. Rates of mass loss in the jet are estimated to span the range 10^{-10}–10^{-8} $M_\odot$ yr^{-1}, which means that they are appreciably smaller than mass-loss rates derived from the molecular lines. It is therefore unlikely that jets drive their associated large-scale molecular outflows. The existence of an additional atomic wind component has therefore been hypothesized and confirmed by the detection of massive atomic hydrogen outflows associated with some of the best-studied bipolar molecular flows.

The driving mechanisms of the various outflow manifestations displayed by young stellar objects are the topic of much work and speculation. Several proposals involve the action of magnetic fields in extracting and collimating the protostellar wind. Because young low-mass stars do not appear energetic enough to drive extensive outflows, the possible role of circumstellar disks in driving winds is currently being investigated by several groups.

Evidence for Circumstellar Disks

Although we have mentioned a number of theoretical arguments favoring the presence of circumstellar disks around protostars, we know very little today about their properties. Current resolution limits of millimetric interferometers prohibit detection of disks smaller than several thousand astronomical units. Even in the case of a detection, lack of resolving power makes it quite difficult to decide whether a flattened molecular condensation is a disk rather than an envelope. One can try to distinguish between these two possibilities by studying the velocity distribution of the molecular condensation; for example, it was concluded from such a study that the flattened molecular condensation with diameter 4000 AU that surrounds the young star HL Tau has a velocity distribution compatible with that expected from a rotating disk.

Subarcsecond resolution can be attained in the near-infrared through direct imaging, lunar occultations, and speckle interferometry. Although both the resolution and sensitivity of these techniques are still too poor to detect solar-system-sized disks, they are expected to dramatically improve over the next decade. Meanwhile, one must be content with indirect evidence of cold disks comparable in size to the solar system (less than 100 AU) around many young stars, including not only protostellar Class I objects but also T Tauri stars. That indirect evidence consists of

1. Their spectral energy distributions, which are best explained if a disk contributes heavily to the overall emitted flux.
2. Maps of polarization vectors in the vicinity of young stellar objects, which can be reproduced if the stellar light is multiply scattered in a geometry involving a disk and bipolar lobes.
3. The presence of blue-displaced forbidden lines in the spectra of T Tauri stars, which indicates that an opaque disk is hiding the receding part of the ionized wind from the star.

Observations of the millimeter continuum of T Tauri stars suggest that relatively massive circumstellar disks (mass ~ 0.01–1 $M_\odot$) may exist for as long as 10^7 yr around young stars. Whether they ultimately form planets is unknown.

CONCLUSIONS

Whereas theoretical models have led astronomers to expect that star formation would be accompanied by extensive infall of matter onto the forming stars, observations have unexpectedly revealed that extensive mass loss from the forming star takes place instead during the first observable stages of star formation. This may indicate either that the protostellar stage cannot be observed because it takes place on a much shorter time scale than envisioned so far, or, more likely, that the necessary spatial and spectral resolution powers necessary to detect the earliest phase of collapse are not attainable with current technology. In any case, current theoretical models of star formation do not take into account all the relevant physical processes. For example, the exact role of outflows in the process of star formation remains mysterious, although strong mass loss appears linked to the presence of an opaque disk around the young stellar object. Although this recent finding suggests that the protostellar wind is powered by gravitational energy stored in the disk, the exact mechanism by which this is achieved remains unknown today.

Additional Reading

Lada, C. J. (1985). Cold outflows, energetic winds, and enigmatic jets around young stellar objects. *Ann. Rev. Astron. Ap.* **23** 267.

Mundt, R. (1988). Flows and jets from young stars. *Formation and Evolution of Low-Mass Stars*, A. K. Dupree, ed. Reidel, Dordrecht, p. 257.

Sellwood, J. A., ed. (1990). *Dynamics of Astrophysical Disks.* Cambridge University Press, Cambridge, U.K.

Shu, F. H., Adams, F. C., and Lizano, S. (1987). Star formation in molecular clouds: Observation and theory. *Ann. Rev. Astron. Ap.* **25** 23.

Wynn-Williams, C. G. (1982). The search for infrared protostars. *Ann. Rev. Astron. Ap.* **20** 587.

See also **Herbig–Haro Objects; Protostars, Theory; Stars, Pre-Main-Sequence, Winds and Outflow Phenomena; Stars, Pre-Main-Sequence, X-Ray Emission; Stars, T Tauri.**

Protostars, Theory

Frank H. Shu

Without stars, the natural world would hold far less interest than it does. If the raw material of the cosmos had remained with the original composition that emerged from the Big Bang, the Universe would today contain virtually nothing except hydrogen and helium. From these two elements, nature can make only one molecule, H_2, and except for unlikely dense regions that drop to temperatures near absolute zero, there would then exist no states of matter other than structureless gases.

The emergence of stars changed everything. Stars constitute the great transformers of matter in the Universe, converting by nuclear reactions simple elements to more complex ones. When the first stars died, they spewed their internally processed elements back into interstellar space. From the enriched material condensed vast interstellar clouds of molecular gas and dust particles. These dusty molecular clouds became the sites for the formation of new generations of stars, perhaps encircled by objects capable of sustaining solid surfaces and liquid oceans. On one such object—the planet Earth—the diversity of structures accessible through carbon chemistry made possible living organisms, a species of which evolved sufficient curiosity and intelligence to ask questions about its own origins and those of the natural world.

Why does the Universe ubiquitously form luminous self-gravitating balls of gas with just the right mass range—between approximately one-tenth to one hundred times the mass of our Sun—to transform the primary products of the Big Bang into the versatile set of chemical elements that occupy the 92 natural entries of the periodic table? At its root, this fundamental problem confronts any attempt to understand star formation on a scientific basis. In the quest to answer the question of the origin of stellar masses, observational and theoretical discoveries made during the 1970s and 1980s have revealed a number of astonishing surprises. In particular, astronomers have learned that the basic entities that give birth to stars—giant molecular clouds—have themselves masses hundreds of thousands times larger than that of a typical star. Moreover, sunlike stars do not arise, for example, because a $10^5\ M_\odot$ cloud yields 10^5 stars, each of 1 $M_\odot$ (solar mass). Giant molecular clouds generally have quite a low efficiency for forming stars; there exists in principle much more raw material than actually goes into the final products.

Astronomers now believe that the formation of relatively low-mass stars like our own Sun proceeds in four stages. In the first stage (Fig. 1), cold dark pockets of gas and dust slowly contract from the background of a much larger cloud. Such pockets are called *molecular cloud cores* by the astronomers who observe such objects at radio wavelengths. The separation of a cloud into cores and envelopes occurs by a process called ambipolar diffusion or plasma drift, because giant molecular clouds are supported against their considerable self-gravity in large part by interstellar magnetic fields. However, the magnetic fields directly affect only the motion of the electrically-charged component of the medium, and the matter of a molecular cloud is only lightly ionized, with the fractional ionization decreasing as the overall density increases. Thus the electrically-neutral part of the matter content continuously slips with respect to the ionized plasma and the magnetic field, with dense pockets getting ever denser under the influence of their own gravity.

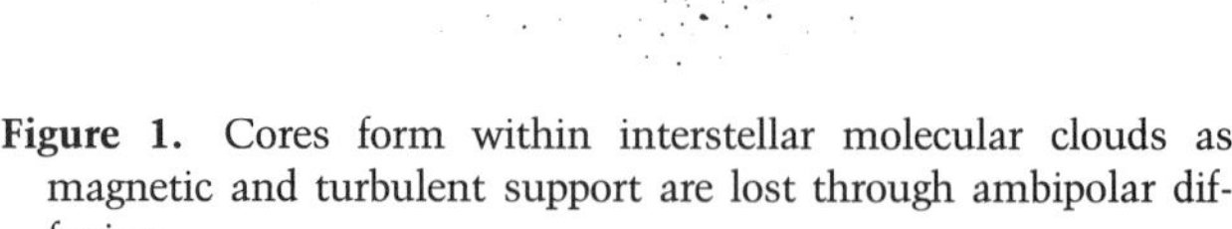

Figure 1. Cores form within interstellar molecular clouds as magnetic and turbulent support are lost through ambipolar diffusion.

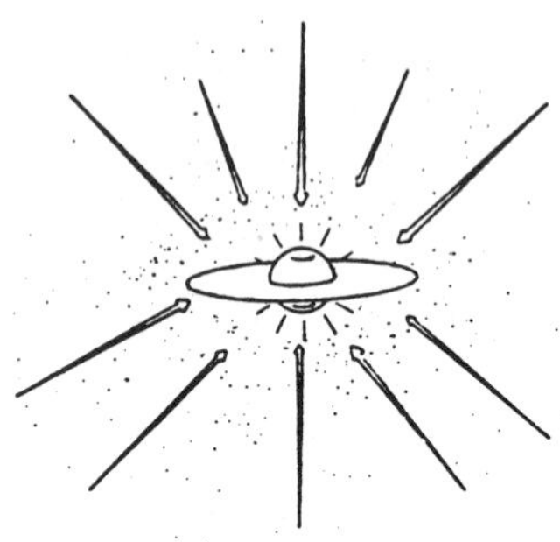

Figure 2. A protostar with a surrounding nebular disk forms at the center of a molecular cloud core collapsing from the inside-out.

At some point, the growing concentration becomes unstable to a runaway increase of density in the middle portions, and the molecular cloud core collapses gravitationally from inside out, forming at the center a protostar whose gases contain sufficient opacity to trap much of the internal heat and to stabilize the central configuration. For about a hundred thousand years, gas and dust continues to rain down from the molecular cloud core, effectively obscuring the system from external view at all but the longest infrared wavelengths. Rotation of the cloud core causes much of the matter to fall into a swirling disk (Fig. 2), whose hotter portions spiral inward to add to the growing mass of the central protostar, and whose cooler regions spread outward to become the raw material for the formation of a planetary system, or, perhaps, a companion star.

Astronomers lack detailed knowledge concerning the physical mechanism responsible for the basic feature of astrophysical "accretion disks," the inward transport of mass accompanied by the outward transport of angular momentum. Two promising mechanisms have been identified in the case of the rotating disks that surround young stellar objects: the turbulent friction associated with the onset of thermal convection in the disk, and the gravitational torques associated with the development of nonaxisymmetric instabilities similar to those believed to account for the spiral structure of some galaxies. The first process may dominate in disks with relatively low masses; the second, in disks with masses comparable to that of the parent star.

Some astronomers have theorized that when the central temperature of the protostar has increased enough to ignite fusion reactions of deuterium (or "heavy hydrogen"), the gases of the protostellar envelope will begin to boil from the release of internal heat, and strong magnetic fields generated by dynamo action will rise to the surface of the protostar. When combined with the rapid rotation, such magnetic fields act like whirling rotary blades to drive the surface gases into a protostellar wind that blasts back out through the infalling envelope of gas and dust. Such protostellar winds have been measured to contain a hundred million times more power than the present solar wind.

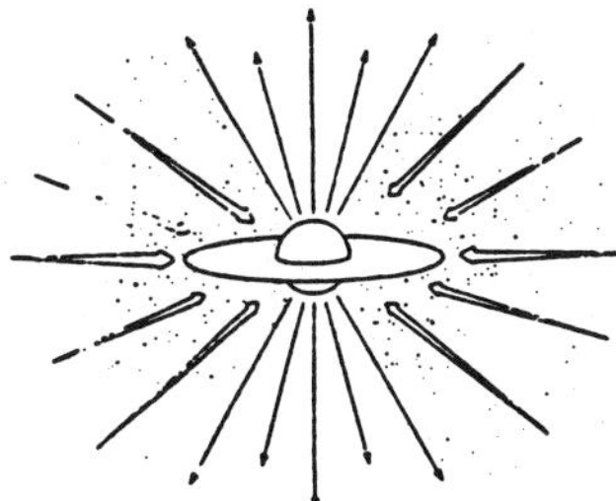

Figure 3. A stellar wind breaks out along the rotational axis of the protostar system, creating a bipolar flow.

In the preceding picture, an outflow inevitably develops because the central star cannot continue to accept matter from a rapidly rotating disk without flinging a significant fraction of it back out. Other astronomers believe protostellar outflows to originate in the body of the nebular disk that surrounds the protostar, or, alternatively, in the boundary layer where the disk attaches to the star. The latter scenarios do not resolve the basic difficulty of continuous spin-up of the star by the disk. In any case, for incompletely understood reasons, the outflow process is sufficient at first to reverse the inflow only over the polar regions. The system, which possesses simultaneous inflow from the equatorial regions and outflow over the polar regions, corresponds to the *bipolar flow* phase of protostellar evolution (Fig. 3).

The outflowing stellar wind in a bipolar flow source sweeps up the gas in the ambient molecular cloud into thin shells. These shells can frequently be detected by radio astronomers in the emission of the carbon monoxide molecule as two oppositely moving lobes of gas. Accompanying this unusual phenomenon are often fantastic sinuous jets, studded by bright knots called *Herbig – Haro objects*, which represent gas shocked to optical radiance by interactions at high speeds. Using special cameras sensitive to the radiation from excited hydrogen molecules, infrared astronomers have obtained images that show these jets in a number of cases to extend several light-years on both sides of the embedded protostars along the lengths of the bipolar flow axes.

The cone of outflow over the rotational poles of a particular source probably widens with time until eventually the pattern of infall has been reversed from the poles to the equator. Were it not for this reversal, the system would continue to accumulate matter from the overlying molecular cloud core to become much more massive than any ordinary star. The timing of the shut-off in infall thus fixes the total amount of molecular cloud material that ultimately gets incorporated in the system. In this picture, the forming star helps to determine its own mass; objects capable of thermonuclear fusion then naturally result because the ignition of such reactions triggered the outflow that shuts off the build-up in mass of the final object.

Because the infalling blanket of gas and dust does ultimately get removed, the central star and disk become revealed even at visible wavelengths (Fig. 4). For systems of relatively low total mass, such objects are known to optical astronomers as *T Tauri stars*. T Tauri stars have larger sizes and cooler surface temperatures than normal sunlike stars; these manifestations occur because T Tauri stars are too young to have yet contracted to the so-called *main-sequence* phase of stellar evolution, where stars burn at their centers ordinary hydrogen, rather than heavy hydrogen. The relatively large amounts of the element lithium found in their atmospheres also hint at the youth of T Tauri stars; in mature stars of this type, convection drags the surface lithium to layers deep and hot enough to lead to its rapid destruction by nuclear reactions.

Figure 4. The infall of cloud material terminates, revealing a newly formed star with a circumstellar disk.

T Tauri stars also have large excesses of infrared radiation in comparison with ordinary stars. This excess probably arises from the thermal emission of warm dust embedded in the surrounding nebular disk. By examining the properties of the infrared radiation, astronomers have deduced that the disks have sizes equal typically to a few hundred times the distance from the Earth to the Sun, and masses spanning from a few percent of the mass of the Sun to an amount comparable to 1 $M_{\odot}$. These properties lie within the range that theorists have deduced on other grounds to be necessary for the formation of planetary systems like our own, or for the origin of the majority of binary stars.

For more than a million years, a typical T Tauri star will contract toward its main-sequence state, all the while accreting mass from its remnant nebular disk. The friction generated by the mismatch in angular speeds between the rapidly rotating inner edge of the disk and the slowly spinning star heats up the boundary layer and causes it to emit considerable ultraviolet radiation. The amount of ultraviolet light in T Tauri stars in excess above normal stars with otherwise similar surfaces therefore constitutes a diagnostic for the level of accretion taking place at relatively advanced stages of a young star's early evolution. An interesting result from such studies has been the deduction that systems possessing higher levels of disk accretion also exhibit more powerful stellar winds. This conclusion supports the basic idea that disk accretion provides the underlying driver for the winds that emanate from young stars.

In the interim, the bulk of the disk cools by emitting infrared radiation to space. If the disk contains sufficient mass, it may fragment gravitationally to form another star. (Binary star formation may also occur during earlier phases of evolution when infall from the molecular cloud core brings matter into the disk faster than it can be emptied into the central protostar.) On the other hand, disks that contain too little mass to be gravitationally unstable may constitute excellent candidates for the birth of a planetary system. Through mutual collisions and chemical or physical adhesion, the solids (initially in the form of tiny dust grains) in a low-mass disk may coagulate to form larger bodies, eventually forming the rocky and icy cores of terrestrial and giant planets. This phase of evolution ends when the stellar wind, or some other agent, succeeds in clearing the disk of remaining gas and small solid particles.

The formation of stars with masses appreciably larger than the Sun probably occurs by processes similar in general outline to that given previously for sunlike stars, but there exist notable differences. First, unlike low-mass stars that frequently condense from well-separated individual cores, high-mass stars usually form in tight groups from large dense regions containing appreciable substructure and undergoing overall gravitational collapse. Second, powerful stellar winds also appear to blow clear the surrounding cocoons of gas and dust from which high-mass stars are born, but the trigger for this wind remains more mysterious than for low-mass stars. (For example, the trigger may not correspond to the onset of deuterium burning, which may not release enough energy to make the envelope of a high-mass star unstable to the onset of convection.) Third, because the interior evolution of high-mass stars occurs much more rapidly than that of low-mass stars, the former reach the main sequence while still heavily embedded in obscuring gas and dust. Thus, when high-mass stars become optically revealed, they do not possess the pre–main-sequence characteristics that help astronomers to date the ages of their T Tauri counterparts. Finally, the radiation field and powerful outflows that emanate from young bright stars of high mass so obliterate their surroundings that it becomes difficult for astronomers to reconstruct prior events. Hence mysteries still remain in our understanding of the processes by which stars are born in the cosmos.

Additional Reading

Cohen, M. (1984). The T Tauri stars, *Phys. Rep.* **116** (No. 4) 173.

Lada, C. J. (1985). Cold outflows, energetic winds, and enigmatic jets around young stellar objects. *Ann. Rev. Astron. Ap.* **23** 267.

Lin, D. N. C. and Papaloizou, J. (1985). On the dynamical origin of the solar system. In *Protostars and Planets. II*, D. C. Black and M. S. Matthews, eds. University of Arizona Press, Tucson, p. 981.

Shu, F. H., Adams, F. C., and Lizano, S. (1987). Star formation in molecular clouds: Observation and theory. *Ann. Rev. Astron. Ap.* **25** 23.

Strom, S. E., Strom, K. M., and Edwards, S. (1988). Energetic winds and circumstellar disks associated with low mass young stellar objects. In *Galactic and Extragalactic Star Formation*, R. Pudritz and M. Fich, eds. Kluwer Academic Publishers, Dordrecht, p. 53.

Weaver, H. A. and Danly, L., eds. (1989). *The Formation and Evolution of Planetary Systems*. Cambridge University Press, Cambridge, U.K.

See also **Herbig–Haro Objects and Their Exciting Stars; Interstellar Clouds; Collapse and Fragmentation; Protostars; Stars, Pre–Main Sequence, Winds and Outflow Phenomena.**

Pulsars, Binary

Daniel R. Stinebring

The majority of the 23 known binary radio pulsars were discovered in globular clusters during 1988 and 1989 (see Table 1). They have taught us much about the origin and evolution of neutron stars and have been key probes of gravitational physics. Binary radio pulsars are star systems in which one of the members is a radio pulsar. The unseen companion is either another neutron star or a white dwarf. Orbital periods of these systems range from 32 min to 3.4 yr and the orbits are either highly circular or have a substantial ellipticity. In half of these binary systems, the radio pulsar has a rotational period of about 10 ms or less and is known as a millisecond pulsar. Because almost half of the known millisecond pulsars are in binary systems, whereas only 2% of the normal pulsar population are in binaries, this linkage between binary and millisecond pulsars is not accidental, as we will discuss later. The first binary pulsar discovered, PSR 1913+16, has been a rich laboratory for studying the dynamical consequences of the theory of gravitation (Fig. 1). Observations made over the 15 years since its discovery in 1974 have conclusively demonstrated that this binary system is losing orbital energy (presumably by emitting gravitational radiation) at the rate predicted by general relativity, thereby providing the only evidence that gravitational radiation exists. Several other binary radio pulsars provide some useful tests of gravitation theory, including one system that may coalesce into a single object in less than 10^6 yr. Another unusual binary radio pulsar, PSR 1957+20, is eclipsed by an extremely low-mass companion every 9 h. The eclipse is caused by absorption of the radio signal in a wind that is being driven off the surface of the white dwarf companion, eventually causing the companion to evaporate.

BASIC PROPERTIES

Radio pulsars are highly magnetized, rotating neutron stars. By recording the pulses due to the spinning star as its radio beam sweeps periodically past the Earth, the rotational period can be determined. All but one of the 500 known radio pulsars have periods, P, that are *increasing* at a regular rate that is in the range $10^{-13} < \dot{P} < 10^{-20}$ s s^{-1}. If the energy loss due to magnetic dipole radiation is set equal to the rotational energy-loss rate, an estimate

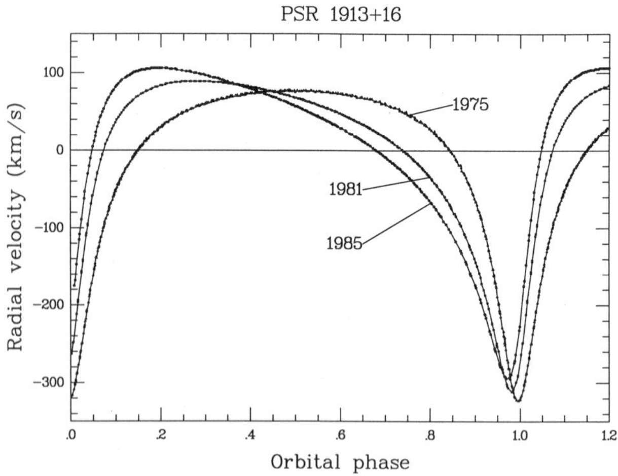

Figure 1. The radial-velocity curve for PSR 1913 + 16 is plotted versus the orbital phase for three epochs. The orbital period of this pulsar is 7 h 45 min and its eccentricity is 0.62. The precession of periastron (a purely general relativistic effect in this system) is 4.22° yr^{-1}, which causes the dramatic change in the shape of the velocity curve over the 10-yr period, as plotted here. Phase-coherent timing observations of this pulsar have been conducted since soon after its discovery in 1974. These allow a complete solution of the orbital parameters of the system within the framework of general relativity. The orbital decay due to the emission of gravitational radiation has been measured for this system. It agrees to within 1% of that predicted by general relativity, providing compelling evidence for the reality of gravitational wave production. (*Courtesy of Joseph H. Taylor.*)

of the pulsar magnetic field can be made ($B_0 \approx 3.2 \times 10^{19}[P\dot{P}]^{1/2}$ G). Radio pulsars in binary systems generally have magnetic fields that are small compared to those of single pulsars, typically about 10^8–10^9 G as opposed to 10^{11}–10^{12} G. In addition, they are spinning more rapidly on average. The median rotational period for the pulsars in binary systems is 12 ms compared to 650 ms for that of single pulsars.

Binary radio pulsars can be divided naturally into two groups. Most have nearly circular orbits (eccentricity, $e < 0.01$) and a low-mass companion. The others have relatively massive companions and large eccentricities ($e > 0.1$). The former systems are thought to consist of a white dwarf companion and the latter cases consist of two neutron stars in orbit, with only one observable as a radio pulsar.

GLOBULAR CLUSTERS, MILLISECOND PULSARS, AND THE MISSING LINK

The field of binary radio pulsars has been transformed in the last few years by the discovery of many interesting binary systems in globular clusters. Fourteen of the binary pulsars are in globular clusters. The presumption is that most of these systems result from tidal captures in the crowded environs of cluster cores, but there is too high a percentage of close binaries for this to be the full picture. Unlike the disk of the Galaxy, cluster cores are dense enough that tidal captures occur between a neutron star and a main-sequence star with some regularity. Radio pulsars are active for only about 10^7 yr after their birth (before the accelerating electric potential at their surface, proportional to B/P^2, drops below a critical value). Tidal captures give the dead pulsars another chance at life because the evolution of the main-sequence star may include an era of mass transfer onto the neutron star, "recycling" the neutron star as a rapidly rotating, weak field (10^9 G) pulsar. Millisecond pulsars ($P < 10$ ms) are the extreme cases of this

Table 1. Binary Radio Pulsars

PSR	Period (ms)	log $\dot{P}$ ($s\,s^{-1}$)	P_{orb} (day)	log Mass Function ($M_\odot$)	Eccentricity	Association	Discovered
0021 − 72A	4.5	< −18	0.02	−7.80	0.33	47 Tucanae	1988
0021 − 72E	3.5	—	~ 2	—	—	47 Tucanae	1989
0021 − 72H	3.2	—	—	—	—	47 Tucanae	1989
0021 − 72I	3.5	—	—	—	—	47 Tucanae	1989
0021 − 72J	2.1	—	0.12	—	—	47 Tucanae	1989
0021 − 72K	1.8	—	—	—	—	47 Tucanae	1989
0655 + 64	195.7	−18.20	1.03	−1.15	< 0.00005	Field	1982
0820 + 02	864.9	−16.0	1232.47	−2.52	0.012	Field	1980
1310 + 18A	33.2	—	255.84	—	< 0.01	M53	1989
1516 + 02B	7.9	—	6.85	−3.19	0.13	M5	1989
1534 + 12	37.9	—	0.42	−0.49	0.27	Field	1990
1620 − 26A	11.1	−18.1	191.44	−2.10	0.025	M4	1988
1639 + 36B	30.0	—	1.25	—	—	M13	1990
1744 − 24A	11.6	−19.3*	0.07	−3.49	< 0.003	Terzan 5	1990
1802 − 07A	23.1	—	2.62	−2.01	0.22	NGC 6539	1990
1820 − 11	279.8	−14.9	357.76	−1.17	0.79	Field	1989
1831 − 00	520.9	−16.8	1.81	−3.92	0.0001	Field	1987
1855 + 09	5.4	−19.8	12.33	−2.25	0.00002	Field	1986
1913 + 16	59.0	−17.1	0.32	−0.88	0.62	Field	1975
1953 + 29	6.1	−19.5	117.35	−2.62	0.00033	2CG 095	1983
1957 + 20	1.6	−19.8	0.38	−5.28	< 0.00004	Field	1988
2127 + 11C	30.5	−17.3	0.34	−0.82	0.68	M15	1989
2303 + 46	1066.4	−15.2	12.34	−0.61	0.66	Field	1985

*$\dot{P}$ for 1744 − 24A is a negative quantity.

scenario, and 12 of the 26 known millisecond pulsars are in binary systems.

There is a missing link in the resurrection/spin-up scenario, however. In addition to the millisecond pulsars in binary systems, there are a nearly equal number without companions. This absence of a companion had been difficult to explain, but nature provided a probable answer in the form of a binary system discovered in 1988. This system is the only binary pulsar system in which the pulsar is eclipsed by its companion. The companion blocks out the radio signal for 45 min of the 9.2-hr binary period. The proximity of the pulsar to the low-mass (0.02 $M_\odot$) companion (they are separated by 2 $R_\odot$; $M_\odot$ and $R_\odot$ are the mass and radius of the Sun, respectively) and the presence of a substantial time delay in the pulsar signal as the signal emerges from behind the companion indicate that there is significant mass loss from the companion system, probably driven by the high-energy flux of particles and radiation directed at it from the pulsar. In a few hundred million years or so the pulsar will have completely disintegrated the companion that gave it a new life in the first place, leaving behind only the spun-up, solitary millisecond pulsar.

TIMING A BINARY PULSAR

The most fundamental timing measurement of a pulsar is an observation of its rotational period. Once a single pulsar has been discovered, it is usually easy to determine its rotation rate and its spin-down rate. The job is not so easy for a binary pulsar because the orbital motion of the pulsar causes a Doppler shift of the pulsar period. If the pulsar period is measured over the entire orbital period it traces out the "velocity curve" of the orbit. Figure 1 shows three such velocity curves (in this case for the same pulsar at three different epochs).

Five classical parameters can be extracted from an analysis of this velocity curve, in analogy to the analysis of a single-line spectroscopic binary star system: the binary orbital period, P_b; orbital size a; and eccentricity e can be determined from the shape and amplitude of the velocity curve; additionally, the longitude of periastron (measured relative to the line of the nodes, where the orbital plane and the plane of the sky intersect) and the epoch of periastron passage can also be determined. Once the basic form of the velocity curve has been determined, it becomes possible to predict the rotational period of the pulsar well enough that the arrival phase of the pulse can be measured. This phase-coherent timing can be extended across the gaps between the observing sessions, resulting in an unambiguous numbering of the pulses observed in each session. This dramatically improves the accuracy of the timing results and allows further refinement of the orbital model parameters.

From this information alone it is impossible to determine the individual masses of the two stars or to determine the inclination of the orbital plane with respect to the plane of the sky. It is possible, however, to combine the orbital parameters to obtain the "mass function" of the system:

$$f(m_1, m_2) = \frac{4\pi^2}{G}\frac{(a\sin i)^3}{P_b^2} = \frac{(m_2 \sin i)^3}{(m_1+m_2)^2},$$

where m_1 is the pulsar mass, m_2 is the mass of the companion, G is the gravitational constant, and i is the inclination of the orbital plane with respect to the plane of the sky. The mass of the companion can then be estimated by using either supplementary information (the behavior of interstellar scintillation parameters in one case or the use of relativistic effects in two other cases) or by making plausible assumptions about the inclination angle and the mass of the observed pulsar. Only four of the binary systems have companions that are massive enough to be neutron stars. The other companions are probably white dwarfs because the highly circular orbits indicate that the companion star has evolved off of the main sequence.

GRAVITATIONAL PHYSICS

The compact size, high velocities, and compact nature of the companions—as well as the phase-coherent timing that can be achieved—make binary pulsars powerful probes of gravitational theory. The original binary pulsar, PSR 1913+16, discovered by Russell A. Hulse and Joseph H. Taylor in 1974 at the Arecibo Observatory, is still the leading system for studying the dynamic effects of gravitation. The first "post-Newtonian" effect to manifest itself is the precession of the longitude of periastron (also known as the apsidal advance). For two point masses in general relativity, the rate of this advance, $\dot{\omega}$, is given by

$$\dot{\omega} = \frac{3G^{2/3}}{c^2}\left(\frac{2\pi}{P_b}\right)^{5/3}\frac{(m_1+m_2)^{2/3}}{1-e^2}$$

where c is the velocity of light. The effect of periastron advance can be seen vividly in Fig. 1. There, the velocity curve for this pulsar is shown at three different epochs: 1974, 1981, and 1984. There are substantial differences in the shape of the velocity curve, caused by the precession of the orbit during this interval. An accurate fit to this precession yields a uniform rate of 4.2° yr^{-1} for this pulsar. The rate is about 30,000 times larger than the general relativistic contribution to the precession of Mercury's elliptical orbit in the gravitational potential of the Sun. The apsidal rate has been measured for several other binary pulsars.

Two other relativistic effects are detectable with a compact, high-velocity system such as PSR 1913+16. A combination of gravitational redshift and special relativistic time dilation causes an "Einstein time delay" as the pulsar moves in and out of the gravitational potential of the other neutron star. A second effect, known as the Shapiro time delay, is introduced by the bending of light as it crosses the curved space–time in the vicinity of the companion mass. Both of these effects have been measured for PSR 1913+16, which allows the separate determinations of the stellar masses and the inclination of the orbit.

According to general relativity, dynamical systems with a time-varying "quadrupole moment," such as an elliptical binary pulsar system, produce gravitational radiation. Although the gravitational radiation produced by a binary pulsar system is too weak to be observed directly by the current generation of gravity wave detectors, this is an energy loss from the orbital kinetic energy of the system. The net effect of this energy loss is to shrink the orbit and speed up the motion of the stars. Because this will, in turn, increase the energy-loss rate due to gravitational radiation, the orbit decays at an accelerating rate until the stars finally coalesce in a burst of high-frequency gravitational radiation. Using the independently determined masses for the Hulse–Taylor system (derived from the relativistic effects), this orbital decay rate can be very accurately predicted. Observations of the orbital decay over the last 15 years show spectacular agreement with the predicted value (to within 1%)! This is the most far-reaching and thorough test of general relativity ever performed, because agreement depends on the dynamical details of the theory and not just static tests of gravitation such as the classical bending of light measurements. This agreement, and further work continuing on this pulsar, has ruled out essentially all other proposed theories of gravitation (or constrained their parameters to the point that they are synonymous with general relativity).

We can expect much activity in the study of binary radio pulsars in the upcoming years. The discovery rate of these systems is on the rise, and they continue to increase our understanding of the birth and evolution of neutron stars, particularly in globular clusters. Based on past experience, we can be sure that newly discovered binary radio pulsars will continue to surprise us with their variety and utility.

Additional Reading

Backer, D. C. and Hellings, R. W. (1986). Pulsar timing and general relativity. *Ann. Rev. Astron. Ap.* **24** 537.

Lyne, A. G. and Graham-Smith, F. (1990). *Pulsar Astronomy*. Cambridge University Press, Cambridge, U.K.

Manchester, R. N. and Taylor, J. H. (1977). *Pulsars*. W. H. Freeman, San Francisco.

Shaham, J. (1987). The oldest pulsars in the universe. *Scientific American* **256** (No. 2) 50.

Shapiro, S. L. and Teukolsky, S. A. (1983). *Black Holes, White Dwarfs, and Neutron Stars*. Wiley, New York.

Taylor, J. H. and Stinebring, D. R. (1986). Recent progress in the understanding of pulsars *Ann. Rev. Astron. Ap.* **24** 285.

See also **Binary Stars, Spectroscopic; Binary Stars, X-Ray, Formation and Evolution; Gravitational Radiation; Pulsars, Millisecond; Pulsars, Observed Properties.**

Pulsars, Millisecond

Donald C. Backer

In 1982, a pulsar, PSR 1937+21, was discovered at the Arecibo Observatory with a rotation period of 1.557 ms. The slow rate of decay of its period on a time scale of 250 million years indicates that this object has a weak magnetic field 4×10^8 G and is ancient in comparison with the more slowly rotating pulsars. The weak field is inferred on the basis of the standard model of pulsars as magnetized neutron stars that slow down by radiating magnetic dipole radiation. Most of the other millisecond-period pulsars that have been discovered since 1982 ($\sim$ 24) are members of low-mass binary systems with orbital periods between 10 and 100 days. Ongoing search efforts, which require intensive supercomputer analysis of large data sets, are likely to increase the numbers of millisecond pulsars substantially in the coming decade.

ORIGIN AND EVOLUTION

A scenario for the origin and evolution of millisecond pulsars is emerging which proposes that these objects descend from low-mass x-ray binaries. The fundamental dynamical process in this model is mass and angular momentum transfer from a rapidly evolving companion in the late stages of stellar evolution onto an old neutron star. The old neutron star is initially rotating slowly and has a weak magnetic field. Spin-up of the neutron star from the angular momentum transfer is limited to a period established by the keplerian velocity at a radius where the stellar magnetic field pressure balances the gas pressure of infalling material. One consequence of this model was that it led to a successful prediction of an abundant population of millisecond pulsars in globular clusters that are rich in low-mass x-ray binaries, which has been verified in recent discoveries.

The synthesis of models for rotation-powered neutron stars that emit at radio wavelengths and accretion-powered neutron stars that emit at x-ray wavelengths can be extended to include all binary pulsars independent of rotation period. The millisecond-pulsar designation is then an artificial categorization that highlights the extreme of the process. The minimum period of a solar mass neutron star is near 1 ms and is established by gravitational radiation from rotation-driven distortions.

The origin of a single, millisecond-period pulsar such as PSR 1937+21 remains unexplained. The discovery in 1988 of an eclipsing, binary millisecond pulsar, PSR 1957+20, at the Arecibo Observatory suggests a possible model. The eclipses are explained by a wind that is being ablated from the surface of the pulsar's companion by high-energy particle emission from the pulsar. The subsequent mass loss could lead to total removal of a companion. Alternatively, single millisecond pulsars could be formed from a single main-sequence star if in rare cases the initial magnetic field strength is suppressed many orders of magnitude below the common value of 10^{12} G. The coalescence of a neutron star–neutron star binary, which is driven by gravitational radiation of its orbital energy, could also produce a single millisecond pulsar.

BEAMING

The widely accepted model for pulsar beaming is based on a dipole magnetic field oriented at an angle a with respect to the rotation axis. The closed magnetosphere that rotates with the star is bounded by the dipole field lines that loop out to the light cylinder radius $R_c = cP/2\pi$, P is the rotation period. In this model the radio emission is beamed tangentially to the field lines in the open magnetosphere at an altitude R near the stellar surface and has an angular width proportional to $(R/R_c)^{0.5}$. Pulse widths are then expected to grow with $P^{-0.5}$. Whereas the broad widths and multiplicity of components of the millisecond pulsars support this model, the complex emission patterns are difficult to explain in detail. Assessment of the two-dimensional beaming pattern is essential to the determination of the total population of neutron stars with millisecond rotation periods. For slowly-rotating pulsars, analysis of beaming effects suggest that we see only 20% of the population. A simple extrapolation to millisecond periods suggests that we see nearly 100% of these objects.

CELESTIAL CLOCKS

Comparison of the pulse arrival times of millisecond pulsars with atomic clocks with precisions better than 1 ms has shown that the rotations of these stars are extremely stable in comparison to the more slowly rotating pulsars. The stability is associated with the slow evolution of the period which leads to a slow, and evidently uniform, rate of adjustment of the neutron star's equilibrium figure. The more rapidly rotating stars adjust to their equilibrium figure unevenly and sometimes abruptly.

Precise timing observations require precise pulsar positions to correct for the motion of the Earth. Variations of these positions due to proper motion and trigonometric parallax can be measured at the level of 0.0004". Motion of the pulsar in a binary orbit as well as the acceleration of a pulsar in the core of a globular cluster can also be determined.

Observations at the Arecibo Observatory of PSR 1937+21 indicate that its stability exceeds that of the best international atomic time scale on intervals exceeding one year: We can count rotations of this distant star with more precision than cycles associated with a quantum transition in cesium atoms in an Earth laboratory clock. Further tests of pulsar stability will require measuring one object relative to another. In addition, pulsar timing requires precise knowledge of the orbit of the Earth in an inertial reference frame. Precise timing of an array of millisecond pulsars can provide the data for improved models of solar system dynamics. Millisecond-pulsar timing array experiments are underway at radio observatories in the United States, the United Kingdom, France, Australia, and Japan.

A number of chaotic processes in the early universe will contribute to a stochastic background of long-wavelength gravitational radiation, for example, dissipation of large-scale cosmic strings that some models require for galaxy formation. This background parallels the more familiar microwave background. Gravitational radiation passing through the solar system will produce apparent Doppler shifts in apparent pulsar rotation rates whose signature for an array of pulsars distributed across the sky is distinct from the effects of an atomic clock error or an Earth location error. The magnitude of the shift is equal to the dimensionless strain amplitude; levels of 10^{-14} are detectable at the present time. The aforementioned PSR 1937+21 observations have placed a limit on the logarithmic spectrum of gravitational radiation with periods near one year that is a factor of 10^{-6} of the energy density required to close the universe. This limit is close to the level of radiation predicted by cosmological models that use cosmic strings as seeds for large-scale structure in the distribution of luminous galaxies.

Additional Reading

Lyne, A. G. and Graham-Smith, F. (1990). *Pulsar Astronomy*. Cambridge University Press, Cambridge, U.K.

Manchester, R. N. and Taylor, J. H. (1977). *Pulsars*. W. H. Freeman, San Francisco.

Ögelman, H. and van den Heuvel, E. P. J. (1989). *Timing Neutron Stars*, NATO ASI Series C, vol. 262. Kluwer Academic Publishers, Dordrecht.

Thorne, K. S. (1991). *Gravitational Radiation*. Cambridge University Press, Cambridge, U.K.

See also **Binary Stars, X-Ray, Formation and Evolution; Pulsars, Binary; Pulsars, Observed Properties; X-Ray Sources, Galactic Distribution.**

Pulsars, Observed Properties

Joanna M. Rankin

The characteristics of radio pulsars were a complete surprise. Four sources of precisely timed, radio frequency (RF) pulses, with periods of between 0.25 and 1.3 s, were discovered in late 1967 by graduate student Jocelyn Bell and her professor Antony Hewish at Cambridge University. Although the strength of the pulses fluctuated strongly over a few minutes and from day to day, Bell was able to show that these sources were very distant, because the periods of their pulses appeared to be compeletly constant apart from slight changes produced by the Doppler shift of the Earth's motion about the Sun. Because of their rapid and precise periodicity, these "pulsars" were soon associated with rotating neutron stars. The existence of these neutron stars—having as much mass as the Sun, but only about 10 km in radius—had long been predicted. This identification was thus quite satisfying, but their copious radio frequency emission was completely unanticipated.

POPULATION AND DISTRIBUTION

Some 450 pulsars are now known out of a galactic population which may number as many as 10^5. These stars have rotation periods ranging between 0.0015 s and just over 4 s, though most fall between 0.5 and 1 s. The few pulsars with periods shorter than about 0.01 s are known as "millisecond pulsars" and apparently constitute a distinct group, many for instance occurring in binary systems. Apart from these, most radio pulsars are single; only five are known to have companions.

Detailed observations of pulse timing have revealed that all pulsars gradually spin ever more slowly, typically slowing down about 1 part in 10^{15} each rotation. Simple physical models then indicate that pulsars have magnetic fields of 10^{12}–10^{13} G at their surface, and that, overall, pulsars gradually convert their stored rotational energy into radiation, some of which we directly observe at radio wavelengths.

Most pulsars are found close to the disk of our galaxy at typical distances of a few kiloparsecs from the Earth. Overall, the population has a mean height above (or below) the galactic plane of some 300 pc. Pulsar distances can be estimated by measuring the effects of interstellar dispersion, that is, the amount to which free electrons in the interstellar medium delay a pulsar's low-frequency pulses relative to those at higher frequencies. The closest pulsars then lie at distances of only a few hundred parsecs, whereas the most distant lie beyond the galactic center some 20–25 kpc away (one pulsar, PSR 0042-73, has been discovered in the Small Magellanic Cloud).

If pulsars are produced by supernovae, we might expect to find them in close proximity to known supernova remnants, but only a very few of the youngest (most notably the pulsars in the Crab nebula and Vela-X remnants) are so associated. Scintillation studies as well as proper motion measurements have shown, however, that pulsars have peculiar velocities of between 30 and 300 km s^{-1} and thus may move away from the supernova remnant during its lifetime of some 30,000 yr or so.

RADIATION AND BEAMING

Most pulsars have one region of emission (the "main" pulse) whose total duration is about 10% of the rotation period, but some dozen pulsars also have a second region (an "interpulse"), spaced by about half a period. A few pulsars also have broader profiles with durations up to about half a period.

Interpulses seem to indicate that the beaming of pulsar pulses is closely associated with the polar regions of a predominantly dipolar magnetic field. Differences in profile width then result simply from variations in the orientation of their magnetic axes relative to their rotation axes and to the direction to the Earth. Close alignment between the rotation and magnetic axes would produce a wide region of emission, and were the magnetic axis perpendicular to the spin axis, emission might be received from both polar regions—very much as in the case of a lighthouse beacon.

Most pulsars are observable over a broad band of radio frequencies, and only a very few pulsars (such as the Crab nebula and Vela-X pulsars) have been detected at optical or x-ray wavelengths. Their radio emission is strongest in the band between about 100 and 1000 MHz, although a few pulsars have been detected at frequencies as low as 15 MHz or as high as some 25 GHz. This bright radio frequency emission suggests that pulsars must radiate through coherent processes, that is, through the joint action of many relativistic particles.

AVERAGE PROFILES

The emission characteristics of pulsars are often studied using averages, computed over many rotation cycles, of the emission from each phase (or "longitude" of the spinning star) of the main pulse or interpulse region. Many weak pulsars can only be detected via such average "profiles." The shapes of these profiles vary markedly from pulsar to pulsar and from low to high frequency. Moreover, it is possible to observe not only the intensity of this average radiation pattern, but also its polarization, both linear and circular.

Many pulsars have symmetrical, Gaussian-shaped profiles, which are referred to as having a "single" emission "component." Others have two such components with a bridge of emission connecting them ("double" profiles), and still others have either three or five components ("triple" or "multiple" profiles). Pulsars with four, or more than five, components have not been observed. Figure 1 gives an example of both a double and a triple profile.

These profiles are often highly polarized: Linear polarization is typically about 50%, but ranges from near 0 to essentially complete. Circular polarization is less frequent, but 10–25% circular polarization is not unusual, and several instances of more than 50% have been observed. Typically, the direction of linear polarization rotates smoothly over the duration of the profile by an angle of up to about 180°.

INDIVIDUAL-PULSE PHENOMENA

The trains of individual pulses that constitute a pulsar's profile exhibit a number of characteristic phenomena. The individual pulses of some pulsars differ little from one to another in shape, amplitude, polarization, or position, whereas others vary markedly. A few of the most common effects are as follows.

Subpulses and Micropulses

Individual pulses often appear to be constituted of "subpulses," which have typical scales of somewhat less than 1% of the period. In some pulsars, these subpulses tend to occur at longitudes corresponding to the profile components and differ little from them in shape, position, and polarization. More frequently, however, profile components are a complicated average of subpulses with different characteristics. At higher resolution the subpulses of some pulsars exhibit structure on even finer scales. Such "micropulses" can be extremely intense and have durations of about 10^{-4} of a period. They appear to represent temporal fluctuations in the coherent emission process. A series of individual pulses showing both subpulses and micropulses is given in Fig. 1*a*.

Subpulse Modulation

Subpulse sequences often exhibit some form of periodic modulation. In certain pulsars "drifting subpulses" are observed, wherein a given subpulse is emitted progressively earlier or later in successive periods. Groups of such subpulses in successive rotations then form "drift bands," typically spaced by 2–15 periods. In other pulsars, fluctuations in the intensity of subpulse modulation are observed at particular longitudes.

Nulling

Many pulsars turn off and on abruptly, thus skipping from one to many thousands of pulses. (A number of null pulses are visible in Fig. 1*a*.) Sensitive observations have failed to detect any residual emission during these "null" pulses. This nulling appears to represent a cessation of the fundamental processes of pulsar emission. Interestingly, in pulsars that have drifting subpulses, the position of the last subpulse before a null can sometimes be correlated with the first one after, indicating some continuing "memory" throughout the null.

Profile Mode Changing

A number of pulsars exhibit several different types of individual-pulse trains, which can be identified by differences in their resulting average profiles. A wide range of different characteristic times are observed for changes in "mode," ranging from many thousands of pulses down to a very few. Distinct subpulse modulation and polarization patterns are often associated with these mode changes.

Polarization Mode Changing

At a given longitude, most pulsars emit radiation in two distinct polarization states which are approximately orthogonal (i.e., with linear polarization directions at about right angles and opposite senses of circular polarization). The stronger of these two polarization states, or "modes," determines the polarization angle. If one is dominant throughout the profile, the overall rotation of the position angle can be little altered, but shifts in the dominant mode produce abrupt changes in the polarization angle of about 90°. The interaction of the two modes represents the principal means by which average profiles become depolarized, and this modal interaction may depolarize pulsar radiation on shorter times scales as well.

EMISSION GEOMETRY

In 1969, V. Radhakrishnan and D. J. Cooke noted that single profiles had small rotations of the linear polarization angle across the pulse, whereas rotations of up to 180° were observed in double profiles. They identified the direction of the linear polarization with the direction of the projected magnetic field in the polar-cap emission region and suggested that pulsars have emission beams in the shape of hollow cones. "Conal" single profiles then result when our sight line cuts through this beam obliquely, whereas double profiles are produced by a central cut.

Two groups, Peter Goldreich and William H. Julian, and James E. Gunn and Jeremiah P. Ostriker, also in 1969, outlined theoreti-

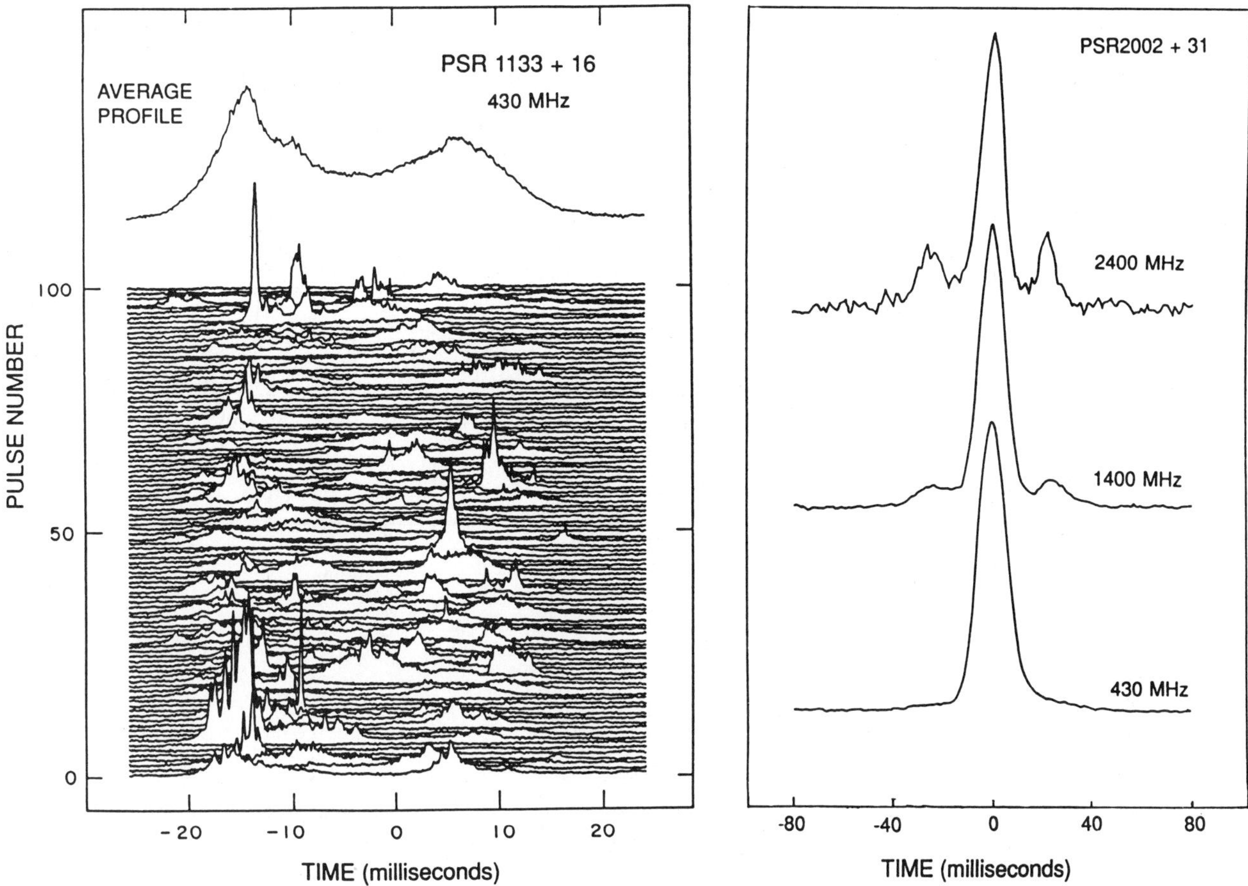

Figure 1. (*a*) The lower curves give a sequence of individual pulses from pulsar 1133 + 16 which has a period of 1.19 s. Only the narrow interval of emission is shown. Note the broad subpulses and narrow micropulses as well as the null pulses. The upper curve is an average of the 100 individual pulses shown. This pulsar has two principal "components" and is thus said to have a "double" profile. (*b*) The three curves show the profile of pulsar 2002 + 31 (period 2.11 s) at 431, 1400, and 2400 MHz, respectively. Again, only a narrow region around the profile is plotted. Note the evolution of the profile: The outriding conal components are barely discernible at 430 MHz, but are quite prominent at 2400 MHz.

cal models that supported this general picture. The exceedingly strong, approximately dipolar, magnetic fields of pulsars form a "magnetosphere" around the star out to a distance where any material (corotating with the star through the action of the magnetic field) would reach the speed of light. Only a narrow bundle of field lines around each magnetic pole, however, can carry particle currents across this "velocity-of-light cylinder" to distances far from the pulsar (see Fig. 2).

These bundles of "open" field lines form a "polar-cap" region around each magnetic pole. Charged particles are then emitted in the polar-cap region through the action of powerful ($\gtrsim 10^{12}$ V) electric potentials induced by the spinning, magnetized, highly conducting star—and travel along the open field lines like "beads on a string." "Curvature" radiation, in the form of a hollow-conical beam, is apparently produced by the transverse acceleration which bunches of these particles experience as they follow the curved magnetic field lines some distance from the pulsar.

Pulsars with triple profiles, however, imply a more complicated picture. Their outer components ("outriders") have properties similar to those of the double profiles, suggesting that a "core" beam is responsible for the central component. Indeed, these core components frequently exhibit circular polarization that symmetrically changes sense and, once identified, are also found singly in core-single profiles (as well as in the central components in five-component profiles).

The angular widths of core components are very nearly equal to the angular extent of the polar-cap region at the pulsar surface, and thus appear to be emitted close to the surface. Thus there are two types of emission and two radiation mechanisms: conal emission, which is emitted at heights of some 5–50 stellar radii in a hollow-conical beam, and core emission, which is emitted near the surface in a pencil beam within the hollow cone.

TYPES OF PULSARS

Understanding that there are two distinct classes of single profile, the core and the conal, suggests a system for classifying the entire range of pulsar phenomena according to profile type.

Pulsars with "conal" single and double profiles are closely related, differing only by whether our sight line cuts the hollow-conical emission pattern tangentially or centrally. These profiles become progressively broader (and the conal single profiles tend to bifurcate) at lower frequencies, because the lower-frequency radiation is emitted at somewhat greater heights in the flaring hollow-conical emission zone of the star (see Fig. 1*b*). Both have polarization angle traverses that follow the projected field direction as well as prominent polarization-modal depolarization on the edges of their profiles.

In contrast, the "core" single pulsars are similar to the central components of triple profiles, and indeed many add pairs of adja-

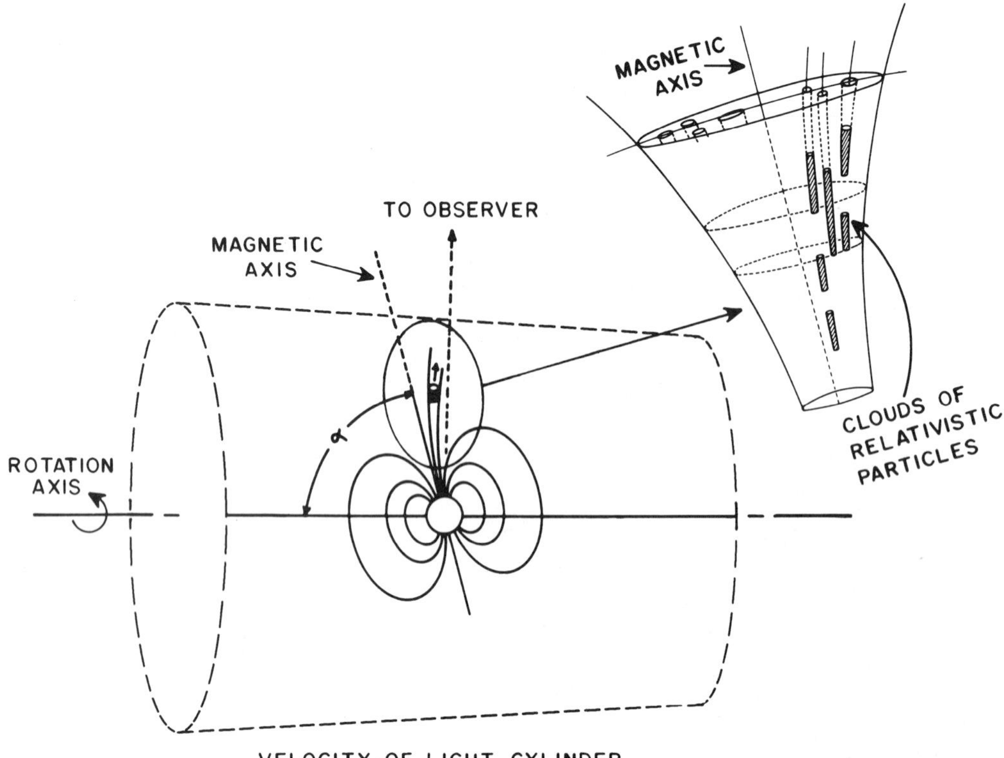

Figure 2. Schematic diagram showing the geometry of pulsar emission according to the polar-cap model. The magnetic axis of the spinning pulsar is not generally aligned with the rotation axis. Potentials induced across the conducting star accelerate charges along the "open," polar field lines, that is, those that do not close within the velocity of light cylinder. The observed characteristics of pulsars depend critically on the direction to the observer relative to the magnetic and rotation axes of the pulsar.

cent conal outriders at high frequency. (Figure 1*b* shows the frequency evolution of a pulsar with a triple profile.) These core components exhibit strong circular polarization, sometimes of symmetrically alternating sense. The leading and trailing components of the triple profiles have properties very similar to the double species as would be expected from a "core" beam within the hollow cone of emission. Finally, "multiple" or five-component profiles are observed which exhibit a central core component and a double set of conal outriders.

Some 60% of the observed pulsar population have core-dominated profiles. Core emission, and especially the triple profile, emerge as most generally prototypical of pulsar emission.

Core-single profiles do not null, nor show evidence of any sort of mode changing; neither do they exhibit much ordered modulation, although some display low-frequency periodicities (15–50 periods per cycle) with which no orderly drift is apparently associated. Similarly, the core components in triple, double (the "bridge" or "saddle"), and multiple profiles also show these longitude-stationary, low-frequency fluctuations.

Drifting subpulses are then an exclusively conical phenomenon. Stationary subpulse modulation (2–15 periods per cycle) is associated with the conal components of stars with double, triple, and five-component profiles, but progressive, orderly drifting is observed only in conal single stars, that is, pulsars where the sight line has a nearly tangential trajectory. This periodic subpulse modulation can then be pictured as a pattern of excitation which circulates around the polar-cap emission region.

Mode changing and nulling are associated with both core and conal emission, and thereby provide important clues to the relationship between them. Mode changing is most readily identified in stars with triple and five-component profiles and manifests itself as a reorganization of the conal emission components about the profile's central core component. By their joint effect on both main-pulse and interpulse emission, mode changing and nulling suggest magnetospheric changes of global extent.

We then have the following evolutionary picture: the core single stars are by far the youngest, both in terms of spin-down age and galactic scale height. Triple stars are of intermediate age, and the remaining species (conal single, double, and five-component) are all relatively old. These young core-single pulsars emit a bright, steady core beam. As they age, their conal emission becomes competitive at ever lower frequencies (triple stars), and the emission of the older species is primarily conal throughout their spectrum.

Additional Reading

Goldreich, P. and Julian, W. H. (1969). Pulsar electrodynamics. *Ap. J.* **157** 869.

Lyne, A. G. and Graham-Smith, F. (1990). *Pulsar Astronomy*. Cambridge University Press, Cambridge.

Manchester, R. N. and Taylor, J. H. (1977). *Pulsars*. W. H. Freeman, San Francisco.

Radhakrishnan, V. and Cooke, D. J. (1969). Magnetic poles and the polarization structure of pulsar radiation. *Ap. Lett.* **3** 225.

Rankin, J. M. (1990). Toward an empirical theory of pulsar emission IV. Geometry of the core emission region. *Ap. J.* **352** 247.

Srinivasan, G. (1989). Pulsars: their origin and evolution. *Astron. Ap. Rev.* **1** 209.

Taylor, J. H. and Stinebring, D. R. (1986). Recent progress in the understanding of pulsars. *Ann. Rev. Astron. Ap.* **24** 285.

See also **Pulsars, Binary; Pulsars, Millisecond; Radio Sources, Emission Mechanisms; Stars, Neutron, Physical Properties and Models.**

Quasistellar Objects, Absorption Lines

Bradley M. Peterson

Because of their high luminosities, quasars are detectable at very large distances. Except for a handful of extraordinarily bright emission line galaxies, they are the only luminous objects that have been detected at redshifts $z > 1/2$. Quasars thus provide one of the few direct probes of the history of the universe. Light from distant quasars that is received at the Earth now was emitted when the universe was only a fraction of its current age, and it has been traveling through the universe ever since.

As light from these distant sources travels to us, there is a reasonable chance that it may be intercepted by an intervening galaxy or intergalactic gas cloud along the way. The signature of such an occurrence will be an absorption spectrum, produced by the intervening gas, superposed on the quasar spectrum (continuum plus broad emission lines). The absorption lines will be at a lower redshift than the emission lines of the quasar itself because the intervening object is closer to us and has a lower cosmological recession velocity. Quasars can thus be used simply as background sources against which we can observe less luminous objects which themselves may be at very large distances or "lookback" times. A fundamental difficulty in using absorption lines to study cosmological evolution is that one does not know precisely what kind of object is producing the absorption lines. Indeed, one cannot always clearly distinguish between truly intervening gas (in which the physical properties of the gas have nothing to do with the background quasar) and absorbing gas in the immediate vicinity of the quasar. This point historically has been a source of great controversy, although most researchers now agree that only a minority of the observed absorption features can be attributed to material associated with the quasars themselves.

The strongest absorption lines seen in quasars are generally hydrogen Ly-α λ1216, C IV $\lambda\lambda$1548, 1550, and Mg II $\lambda\lambda$2795, 2802. Whether or not a specific line is detected in any absorption system depends on the redshift of the absorber, which determines whether the redshifted line will lie in the observable spectral window of a given detector, as well as on the physical parameters of the absorbing cloud (column density, elemental abundances, state of ionization, and velocity dispersion).

The absorption lines seen in quasar spectra generally can be placed into three categories: (1) heavy-element systems, (2) Ly-α forest systems, and (3) broad absorption line (BAL) systems. In the first two types, the absorption lines are so narrow that they are unresolved in optical spectra. The heavy-element systems are characterized by lines of ionized and neutral metals (meaning elements heavier than helium), and, of course, Ly-α when it is accessible. Neutral hydrogen column densities in these systems are typically 10^{17}–10^{21} cm^{-2}, and metals are probably slightly underabundant relative to solar values. The velocity dispersions determined from curve-of-growth analyses are typically tens of kilometers per second for these systems, which is much too large to be attributed to purely thermal motions in gas that is not highly ionized. A high-redshift quasar may have up to a few such systems at different absorption redshifts in the observed spectral window. Important subclasses of such systems are the damped Ly-α systems and the Lyman-limit systems, both of which are characterized by very high column densities. This is shown in the former case by the damping wings on the Ly-α absorption line, and in the latter case by the high optical depth in the Lyman continuum, which results in an abrupt disappearance of the quasar continuum at wavelengths shortward of redshifted 912 Å. In a few cases, H I 21-cm absorption has been detected, but molecular absorption (H_2) has been detected in only one case. There is no clear evidence for dust absorption in any of these systems.

At wavelengths shortward of the Ly-α emission line in quasars, the quasar spectrum is riddled with strong, narrow absorption features, nearly all of which are attributable to Ly-α absorption by clouds at lower redshift. The name "Ly-α forest" has been applied to this spectral region on account of the high density of high-contrast absorption features. These systems have low neutral hydrogen column densities, 10^{13}–10^{16} cm^{-2}, and velocity dispersions of a few tens of kilometers per second. These systems are apparently devoid of corresponding metal features, although metal lines should be detectable only in the larger column density systems. In a few cases, it has been shown that the metal abundances in the Ly-α clouds must be less than 0.001 the solar value.

In a minority of quasars ($\sim 5\%$), broad absorption features are seen in the short-wavelength wings of the ultraviolet resonance lines. These are almost certainly due to gas flowing outward from the quasars at velocities up to $\sim 30{,}000$ km s^{-1}. It is not known whether all quasars have BAL regions that cover only a fraction of the sky as seen from the quasar or only a small fraction of quasars have BALs when seen from any direction. It has been claimed that there are some differences between the emission spectra of quasars with BALs and other quasars. The true situation is very unclear, however, because the very presence of BALs severely alters the appearance of a QSO spectrum.

It has long been supposed that the narrow quasar absorption lines arise in the disks or extended halos of galaxies. The required galaxy cross sections for producing absorption lines stronger than a given equivalent width can be computed by using the Holmberg radius–luminosity relation for galaxies ($r \propto L^{5/12}$), the Schechter luminosity function, and the measured incidence of absorption systems per unit comoving path length. It is found that the sizes of galaxies must exceed their Holmberg radii by a factor of 2–5 to account for all observed narrow absorption lines with rest equivalent widths larger than ~ 300 mÅ. Furthermore, the absence of absorption in the spectra of many quasars where our line of sight falls close to a known galaxy indicates that the gas would have to be distributed in a very patchy fashion, and thus extend even farther out than a few times the optical radius of the galaxy. The absorbing clouds are thus probably small and far more numerous than galaxies. However, they must somehow be associated with galaxies, because deep searches of the fields of quasars with relatively low-redshift absorption lines have turned up galaxies near the absorption redshift in a few cases. Moreover, the very fact that heavy elements are observed argues for some association with galaxies, at least in the case of the heavy-element systems.

Sizes of the absorbers can be estimated from the incidence of identical absorption features in the spectra of pairs of quasars that appear close together on the sky, or in the different images of a gravitationally lensed quasar. Typical sizes derived this way are greater than or of order 10 kpc, though it is clear that there must be a considerable range in sizes. The patchiness of the absorbers supports the idea that the absorption occurs in small clouds within larger structures. The multicomponent structure of many of the heavy-element systems as well as their large velocity dispersions is consistent with such a picture.

If the clouds producing the narrow absorption lines are distributed uniformly per unit comoving volume and their cross sections for producing lines larger than a given equivalent width are constant, the number of detected absorption systems per unit redshift will be

$$\frac{dN}{dz}=\frac{n_0\sigma_0 c}{H_0}(1+z)(1+2q_0 z)^{-1/2},$$

where n_0 and σ_0 are the space density and cross section, respectively, of clouds at the current epoch. This expression is often approximated as $dN/dz \propto (1+z)^{\gamma}$, where $\gamma=1$ for $q_0=0$ or $\gamma=1/2$ for $q_0=1/2$. Thus, in any plausible cosmology, $\gamma>1$ would imply evolution in that the number of absorbers or their cross sections were larger in the past than they are today. Analysis of known absorption systems containing Mg II lines (typically at $z<1$) gives $\gamma \approx 1.5 \pm 0.5$ (although values exceeding 2 have also been reported), and the data on the Lyman-limit systems ($z>2.7$) give $\gamma \approx 0.7 \pm 0.5$. However, the Ly-$\alpha$ forest lines show stronger evidence for evolution, with $\gamma>2$. In any given quasar, however, the density of Ly-α forest lines *decreases* close to the quasar redshift, an effect known as the "inverse" or "proximity" effect. This is likely due to the absence of neutral gas near the quasar, which is capable of photoionizing all of the diffuse gas within several megaparsecs. The spectral density of C IV absorption lines (observed at $z>1$) by contrast seems to show an *overall* decrease with redshift. It has been speculated that this is due to low cosmic metal abundance at high redshift. There appears to be, however, an enhancement of C IV absorbers near the emission line redshift in steep-spectrum, radio-loud quasars, though not in other types of quasar.

Additional Reading

Bergeron, J. (1988). Metal-rich absorption-line systems. In *The Post-Recombination Universe*, N. Kaiser and A. N. Lasenby, eds. Kluwer Academic Publishers, Dordrecht, p. 202.

Blades, J. C., Turnshek, D. A., and Norman, C. A., eds. (1988). *QSO Absorption Lines: Probing the Universe*. Cambridge University Press, Cambridge.

Peterson, B. A. (1986). QSO absorption lines: heavy elements and Lyman-α clouds. In *Quasars*, G. Swarup and V. K. Kapahi, eds. Kluwer Academic Press, Dordrecht, p. 555.

Turnshek, D. A. (1986). Broad absorption line QSOs. In *Quasars*, G. Swarup and V. K. Kapahi, eds. Kluwer Academic Publishers, Dordrecht, p. 317.

Weymann, R. J., Carswell, R. F., and Smith, M. G. (1981). Absorption lines in the spectra of quasi-stellar objects. *Ann. Rev. Astron. Ap.* **19** 41.

See also **Quasistellar Objects, Spectroscopic and Photometric Properties.**

Quasistellar Objects, Host Galaxies

Alan Stockton

The *host galaxy* of a QSO is the galaxy in which the QSO is embedded and of which it is, presumably, the nucleus. The original definition of QSOs specified that they should have essentially stellar images. At the time (i.e., the mid- to late-1960s), this statement meant that little or no difference from the images of stars was apparent on deep photographic exposures obtained with large telescopes. Although it was widely believed that QSOs were extremely luminous galactic nuclei, only a few were known to show faint extended luminous material that might be taken as evidence for a surrounding galaxy, the most notable of these being 3C 48. An important early study by Jerome Kristian showed that this general lack of visible extended material was nevertheless consistent with QSOs being in giant elliptical galaxies: Scattering of the QSO light in the atmosphere, telescope optics, and photographic emulsion was sufficient to obliterate any evidence of a normal surrounding galaxy. In fact, somewhat ironically, it appeared that if the faint material seen around 3C 48 was to be interpreted as a galaxy, it would have to be an abnormally large and luminous galaxy.

More recently, advances in detector technology and the availability of large telescopes at sites with consistently good atmospheric image quality (i.e., what astronomers call "seeing") have resulted in the resolution of extended luminous material around virtually all QSOs with redshifts $z<0.5$ and around many at higher redshifts. The focus of studies of this material has shifted from simple detection to elucidating its nature and origin.

VARIETIES OF EXTENDED LUMINOUS MATERIAL

Applying the term "host galaxy" to the extended luminous material around a QSO presupposes that the extended luminosity is due to a distribution of stars. That such was the case was tacitly assumed in most of the early work; it therefore came as a surprise to find a few objects whose extranuclear luminosity in bandpasses of a thousand angstroms or more was dominated by a single strong emission line from extended ionized gas surrounding the QSO. Although such objects are in the minority, they are by no means uncommon, so it is dangerous to draw conclusions regarding the nature of the host galaxy from images whose bandpass includes one or more of the emission lines known to be strong in gas at low densities and moderately high ionization. The principal relevant lines are Ly-α, [O II]λ3727, [O III]$\lambda\lambda$4959, 5007, and Hα. In most cases, at least among the low-redshift ($z<0.5$) QSOs, the distributions of the extended continuum sources and of the ionized gas are quite different, so a misidentification of an emission feature as indicating the structure of the stellar component of a host galaxy could lead to serious errors in interpretation.

Even bona-fide continuum features must be viewed with some caution, although problems of interpretation here are less likely. *Synchrotron jets* producing strong optical radiation occur infrequently on the sorts of scales likely to cause confusion for ground-based studies of QSO host galaxies, with that associated with 3C 273 being a notable exception. Such jets will presumably be found more frequently in images obtained with the *Hubble Space Telescope*: They can be distinguished from thermal sources such as stars by their high polarization, their power law spectra, and their coincidence with radio features. For some source geometries and viewing angles, *scattered light* from the nuclear continuum source may be important, particularly in cases for which direct radiation from the nucleus does not reach the observer. One would generally expect in such cases that the scattered light would show the spectrum of the QSO's broad-line region as well as the nuclear continuum, and that both the continuum and the broad lines would show significant polarization. Scattered nuclear light with these properties has been found associated with the Seyfert galaxy NGC 1068.

The surest positive sign that an observed extended distribution of luminosity around a QSO is indeed due to stars is the spectroscopic detection of stellar absorption lines. Because of the faintness of the material and the problem of scattered light from the adjacent bright nucleus, convincing detections exist for only a few QSOs. These few, however, are sufficient to establish the existence of bona-fide QSO host galaxies and make plausible the assumption that most of the extended continuum emission seen in deep images of QSOs is due to stars, even in the absence of spectroscopic confirmation.

PROPERTIES OF QSO HOST GALAXIES

Assuming that the worries mentioned previously are taken care of, and we are reasonably confident that we are actually dealing with a stellar system, what can we say about the host galaxies of QSOs?

As may be expected, it has proved to be quite difficult to say anything very definite about many of the most luminous QSOs, for which the strong nuclear component tends to overwhelm the extended stellar distribution. The usual criteria for distinguishing different classes of galaxies include: (1) radial luminosity profiles; (2) structural features, such as spiral arms; and (3) colors. Of these, the last may be unreliable for QSO host galaxies, for reasons to be discussed shortly; the first two are difficult to apply in practice for ground-based observations of normal galaxies with $z > 0.3$, and the presence of a luminous nucleus only aggravates the situation. Considering the large amount of telescope time that has been spent on observations of QSO host galaxies, the results in terms of definitely established properties are relatively meager. It is widely expected that observations with the *Hubble Space Telescope* will settle many of these remaining questions.

Because of these difficulties, much of the ground-based work has concentrated on less-luminous objects and on extrapolations from what we know about related forms of nuclear activity in well-resolved, low-redshift active galaxies. One example is the widely held view that radio-quiet QSOs are simply the high-luminosity tail of the Seyfert galaxy population, which are largely spirals, and that radio-loud QSOs bear a similar relation to the broad-lined radio galaxies, which appear always to be ellipticals (though often peculiar in some way). What evidence there is from QSO imaging surveys tends to support this division, although to date the evidence that many radio-quiet QSOs are in spirals appears to be firmer than that for radio-loud QSOs being in ellipticals.

The few cases for which good spectroscopy of the host galaxies have been obtained show stellar spectra in the optical region ranging from a dominantly A-type spectrum (for 3C 48) to those typical of a smaller admixture of young stars in a predominantly older population. This spectroscopic evidence is in general agreement with what color information exists: Those host galaxies that morphologically look most like ellipticals nevertheless have bluer colors than elliptical galaxies generally, indicating recent star formation.

EVIDENCE FOR INTERACTIONS

Essentially all of the large surveys have resulted in the conclusion that a large fraction of QSO host galaxies have distorted morphologies or are otherwise peculiar. An unusual number seem to have close companions. Even though the statistical bases for these claims have not been thoroughly established, this evidence, together with the color information indicating recent widespread star formation is not unusual in QSO host galaxies, have led to the common supposition that galaxy interactions have played a major role in the QSO phenomenon.

Even before there was any observational evidence to support such a conjecture, some of the components of a link between galaxy interactions and nuclear activity were being discussed. In a classic paper on interactions, Alar and Juri Toomre included a speculative section (entitled "Stoking the Furnace"), in which they suggested that interactions might often bring a fresh supply of gas deep into the centers of galaxies. Although the Toomres spoke only in terms of enhanced star formation in galaxy nuclei, this gas could also presumably fuel a black hole (it was about this time that an effective consensus had been reached that QSOs were very likely powered by energy released as matter fell onto a compact supermassive object, presumably a black hole). These thoughts were made explicit in a paper ("Feeding the Monster") by James E. Gunn, which primarily dealt with the problem of removing angular momentum from the gas so that it can come within reach of the central black hole. This issue remains a thorny one to this day.

The most obvious signatures of tidal interactions are large-scale distortions in the luminosity distributions of galaxies, particularly the tidal bridges and tails that formed the main subject of the Toomres' paper. Such features are likely to be really prominent only for disk galaxies. In the absence of these or other morphological clues, the presence of a close companion galaxy may suggest an interaction, but cannot be conclusive in any individual case because of the possibility of projection effects. However, a statistical excess of close companions can be strong circumstantial evidence in favor of the importance of interactions. A final piece of evidence is enhanced star formation in QSO host galaxies, because it is well known that nearby interacting galaxies often show galaxy-wide enhanced star formation.

In some cases, apparent tidal tails occur in situations where only one galaxy is visible. These can be explained as cases where the two interacting galaxies have merged or where one of the galaxies is hidden in the region obscured by the brilliant QSO nucleus.

From the evidence available so far, it appears that a significant fraction of QSOs are in interacting systems. Because of the observational difficulties involved and the subtlety of some kinds of interaction signatures, it is remarkable that there is as much evidence as there is. Whether essentially *all* QSOs are due to interactions will, once again, be a program for the *Hubble Space Telescope*.

ULTRALUMINOUS IRAS GALAXIES

One of the major discoveries of the *Infrared Astronomical Satellite* (*IRAS*) was the detection of a population of galaxies emitting most of their energy in the far-infrared region of the spectrum. The most luminous of these emit as much power as do many QSOs, and the most plausible interpretation for some of them, at least, is that an active QSO-like nucleus is present at their centers but hidden by massive amounts of dust. The dust absorbs the optical and ultraviolet radiation from the nuclear continuum source and reradiates it in the far infrared. These ultraluminous IRAS galaxies are virtually all members of strongly interacting or merging disk systems, and the dust is a result of the unusually vigorous, galaxy-wide star formation that has been induced by the interaction. A speculative scenario is that some fraction of such galaxies, once the star-formation rate moderates and the dust is reduced, will be visible as normal QSOs (not all can be, because the number density of the ultraluminous IRAS galaxies is larger than that of QSOs, and it seems unlikely that the dust enshrouded stage would last significantly longer than the "normal" QSO stage). If this connection can be demonstrated, it would provide another link between interactions and QSO activity.

QSOs IN GROUPS AND CLUSTERS

The host galaxies of QSOs at low redshifts are usually found in small groups, but almost never in rich clusters. At larger redshifts, however, QSOs are found in richer environments and possibly in more luminous galaxies. Most of the conclusions regarding QSO host galaxies mentioned previously are based on samples of QSOs strongly biased toward low redshifts and may not apply to those at higher redshifts. In particular, the effect of interactions is likely to be more important in small groups, where relative velocities of galaxies are low, than in rich, relaxed clusters. On the other hand, it has been suggested that nuclear activity may sometimes be fueled by gas from cooling flows resulting from thermal instabilities in the hot intracluster medium, and this process may be more important for QSO host galaxies that are central galaxies in rich clusters.

FUTURE PROSPECTS

When the *Hubble Space Telescope* is optically corrected to perform as planned, many of the present uncertainties regarding QSO host galaxies should be cleared up. The fivefold increase in spatial resolution over the best ground-based images, the ability to obtain high-quality images in the ultraviolet, and the absence of the strong airglow background in the near infrared—all of these should make a substantial difference in our ability to interpret these elusive objects.

Additional Reading

Balick, B. and Heckman, T. M. (1980). Extranuclear clues to the origin and evolution of activity in galaxies. *Ann. Rev. Astron. Ap.* **20** 431.

Courvoisier, T. and Mayor, M., eds. (1991). *Active Galactic Nuclei*. Springer, Berlin.

Heckman, T. M. (1990). Galaxy interactions and the stimulation of nuclear activity. In *Paired and Interacting Galaxies, IAU Colloquium 124*, J. W. Sulentic, W. C. Keel, and C. M. Telesco, eds. NASA CP-3098, Washington D.C., p. 359.

Soifer, B. T., Beichman, C. A., and Sanders, D. B. (1989). An infrared view of the universe. *American Scientist* 77 46.

Stockton, A. (1986). The environments of QSOs. *Ap. Space Sci.* **118** 487.

See also **Active Galaxies and Quasistellar Objects, Interrelations of Various Types; Galaxies, Binary and Multiple, Interactions; Galaxies, Infrared Emission; Hubble Space Telescope.**

Quasistellar Objects, in Galaxy Clusters and Superclusters

Patrick S. Osmer

Quasars often occur in groups of galaxies but rarely are found in clusters of galaxies at redshifts less than 0.5. At larger redshifts certain types of quasars do occur in galaxy clusters, and groups of quasars the size of galaxy clusters and superclusters are being found. Quasars are the only available tracers of the large-scale structure of the universe at redshifts larger than 1 because normal galaxies are too faint to study. The relation of quasars to neighboring galaxies is important to the study of the formation and evolution of both quasars and galaxies.

BACKGROUND

Although quasars were originally defined as star-like objects of large redshift, application of higher-resolution imaging showed that extended emission occurred around quasars with redshifts less than 0.5. Since then, the generally accepted picture of quasars being the extremely luminous nuclei of galaxies has been developed. With the realization that quasars occur in galaxies came a number of questions: Do quasars occur in groups or clusters of galaxies? What can they tell us about the conditions in such groups and clusters? What can be learned about the formation and evolution of quasars by studying their environments? Does the spatial distribution of quasars provide clues to the large-scale structure of the universe and its evolution? There has been considerable progress on some of these questions in recent years. Others are just now being studied, and substantial progress can be expected in the future.

This entry first gives a brief history of the development of the topic and then considers the subjects of quasars in groups and clusters of galaxies, quasars in superclusters, and, finally, the distribution in space of quasars themselves.

The first observations that low-redshift quasars have resolvable structures around their nuclei prompted a number of studies on the nature of the surrounding "fuzz." During the course of that work, which provided evidence for the cosmological nature of the redshifts, it was noted that low-redshift quasars were frequently accompanied by nearby galaxies, some of which appeared to be interacting with the quasar. Although such quasars were not found in rich clusters of galaxies, they often did occur in groups of galaxies. Subsequent, extensive studies of the environment of quasars at successively larger redshifts form the basis of much of our knowledge about quasars in groups and clusters of galaxies. Over the same interval much work was being done on the nature and evolution of galaxies, galaxy clusters, and the large-scale structure of the universe, and it has become clear that the mechanisms that trigger the ignition of quasars are closely related to fundamental events in galaxy groups and clusters themselves.

Table 1

Redshift z	*Distance** *(Mpc)*	*Lookback Time*†
0.5	2201	0.13
1.0	3512	0.65
2.0	5068	0.81

*The distance from us at the present time, in megaparsecs, where 1 Mpc = 3.26×10^6 ly.
†Expressed as a fraction of the time since the Big Bang.

As work continued on quasar surveys, the possibility arose of using quasars themselves as tracers of the structure of the universe at distances significantly beyond those attainable with normal galaxies. The light travel time to such a quasar covers more than three-quarters of the age of the universe, therby offering a chance to study the evolution of structure over an interval that could be decisive for choosing among different theories. At the same time the first groups of quasars at redshift 2 were found, and their possible relation to superclusters was noted. Subsequently the first detection of quasar clustering at small scales was made, a topic which is currently very lively and not completely settled. The addition of large, systematic surveys for quasars has made it possible to address the question in detail.

Observationally, work on galaxies in the vicinity of quasars is made difficult by the great disparity of their luminosities and forms: Quasars have bright, stellar-appearing nuclei that can easily be 100 times brighter than galaxies, which are spatially extended and often faint compared to the brightness of the night sky. The development of modern detectors such as CCDs (charge coupled devices), which have high quantum efficiency, excellent stability, and wide dynamic range, has been key to advances in the field. Even so, with the present generation of telescopes it is difficult to pursue the problem at redshifts above 0.6. Therefore, in view of the strong evolution of quasars toward higher redshifts, it must be recognized that the available sample for study may not show all the phenomena that occur.

Throughout this entry redshifts are used as a measure of distance and cosmic epoch. Distance increases with redshift, as does lookback time, which is defined as the travel time for light from an object at a given redshift divided by the elapsed time since the Big Bang (for a Friedmann cosmology). For reference, if the values of $H_0 = 50$ km s^{-1} Mpc^{-1} and $q_0 = 0.5$ are adopted, then the values in Table 1 are obtained.

QUASARS IN GROUPS AND CLUSTERS OF GALAXIES

We now know that both quasars which are sources of radio emission (radio-loud quasars) and those which are not (radio-quiet quasars) are not found in rich clusters of galaxies for redshifts less than 0.5. However, they do occur in areas where the density of galaxies is enhanced by a factor of 2–3 over the density of the background, consistent with earlier remarks on the tendency of such quasars to have companions or occur in groups. For redshifts less than 0.5, no difference in the environment of radio-loud and radio-quiet quasars is apparent.

However, a striking result is that radio-loud quasars with $0.5 < z < 0.65$ occur in regions where galaxy density is eight times larger than average, with half of them occurring in clusters of galaxies of Abell richness 1 or more (clusters with 50–79 members within 2 magnitudes of the third brightest member). The implication is that radio quasars in clusters of galaxies evolve strongly between redshifts 0.4 and 0.6 and have dimmed by several magnitudes by the epoch corresponding to redshift 0.4. Also, the galaxies themselves are brighter than normal in the clusters containing the radio-loud quasars, an indication that they have also evolved. However, radio-loud quasars in less dense environments have not evolved and therefore dominate the population at redshifts less than 0.5.

The result implies that the environment has a strong effect on quasars and that conditions in rich clusters have changed significantly between redshifts of 0.6 and 0.4. It subsequently was found that some faint, radio-loud quasars occur in Abell 1 clusters, consistent with the idea. It appears that the rate of decay is 1 magnitude (a factor of 2.5) per 0.5 Gyr. Subsequently, it has been found that low-level activity in radio galaxies of low redshift can occur in rich clusters; this is consistent with a rapid fading of quasar activity with time.

The addition of new data on faint quasars at low redshifts shows that radio-loud quasars do occur in regions of enhanced galaxy density at all accessible redshifts. This is consistent with the hypothesis that radio-loud quasars occur in elliptical galaxies and radio-quiet quasars in spirals. An unresolved possibility at present is that radio-quiet quasars may occur on the outskirts of rich clusters.

QUASARS IN SUPERCLUSTERS; GROUPS OF QUASARS

At redshifts larger than 0.6 the space density of quasars increases significantly but the detection of accompanying galaxies becomes very difficult. Therefore, the topic of quasars in superclusters generally covers the finding of groups of quasars with similar redshifts that have the dimensions of superclusters. During the course of large surveys for quasars, several such groups have been found. They have dimensions ranging from the size of galaxy clusters, 10 Mpc or less, to greater than 100 Mpc.

The interest in quasar groups is at least twofold:

1. The spatial distribution of quasars and their relation to galaxies yields information on the processes and conditions leading to the formation of quasars. Quasars can be observed at high enough redshift to cover the evolution of such effects over a substantial part of the age of the universe.
2. To the extent that quasars trace the distribution of matter in the universe, they give us the only available information on how the large-scale structure of the universe itself has evolved. This information is of critical importance to cosmology and to theories of galaxy formation.

DISTRIBUTION OF QUASARS IN SPACE

Quasar surveys have reached the point that analyses of their own space distribution at high redshift can now be done in analogy with the early studies of the galaxy distribution. These programs are a generalization of searches for groups of quasars and are intended to probe the basic question of how the population of quasars as a whole is distributed in space.

The first analyses of quasar surveys showed no evidence for quasar clustering and were consistent with quasars being distributed uniformly at random in space. Recognized shortcomings of the first work, however, were that the samples contained only a few hundred objects; they were based on surveys whose selection effects were not necessarily well known; and the surveys did not reach quite faint enough to attain the density of quasars needed to probe structure on small scales. Therefore, they did not place very stringent limits on the presence or absence of clustering. However, the surveys all tended to contain intriguing groups of the type mentioned in the preceding section.

The first positive detection of quasar clustering came from an ingenious test applied to the inhomogeneous but large sample of all known quasars. It provided evidence for quasars being clustered on a scale of 10 Mpc. Subsequent analyses have continued to show clustering on this scale, with some indication that it is present at $z < 1.5$ but not at larger redshifts. The amplitude is intermediate between that of galaxies and clusters of galaxies. Some researchers do not agree that the result is generally applicable to the quasar population as a whole, although they recognize that there are sizable individual groups of quasars.

Attempts have also been made to analyze the distribution of quasars on the very large scale by considering their occurrence over the entire sky. Some intriging nonuniformities have been noted, but it is difficult to establish their reality, for lack of a sufficiently well-calibrated and homogeneous survey for quasars over the whole sky.

SUMMARY

The subject of the environments of quasars and the three-dimensional distribution of quasars in space is new enough for few results to be definite. At the same time, the results to date show that quasar environment are of great interest to determining how quasars evolve and under what conditions and also promise to tell us about galaxy evolution. At large redshifts, the clustering properties of quasars provide our only information on the distribution of matter. Large, systematic surveys for quasars are crucial to future progress in the field. (Preliminary x-ray observations from the ROSAT satellite provided suggestions of quasar clustering, reported in 1991—Ed.)

Additional Reading

Hutchings, J. B. (1983). QSOs: Recent clues to their nature. *Publ. Astron. Soc. Pacific* **95** 799.

Osmer, P. S. (1981). The three-dimensional distribution of quasars in the CTIO surveys. *Ap. J.* **247** 762.

Powell, C. S. (1991). X-ray riddle. *Scientific American* **264** (No. 3) 26.

Shaver, P. A. (1988). Quasar clustering and the evolution of structure. In *Large Scale Structures of the Universe*, IAU Symposium 130, J. Audouze et al., eds., Kluwer, Dordrecht, p. 359.

Yee, H. K. C. (1990). The evolution of galaxies and galaxy clusters associated with quasars. In *The Evolution of the Universe of Galaxies: The Edwin Hubble Symposium*, R. G. Kron, ed. Astronomical Society of the Pacific Conference Series, vol. 10, San Francisco.

See also **Active Galaxies and Quasistellar Objects, Central Engine; Cosmology, Clustering and Superclustering; Quasistellar Objects, Host Galaxies.**

Quasistellar Objects, Spectroscopic and Photometric Properties

Beverley J. Wills

Historically, QSOs (quasistellar objects) were defined as point-like optical counterparts of radio sources having broad, redshifted emission lines, strong ultraviolet emission (UV excess), and strong time variability of the optical light. Although first discovered by their strong radio emission, more than 90% are only weak radio sources [flux density $F_\nu <$ a few mJy, where 1 Jy (jansky) = 10^{-26} W m^{-2} Hz^{-1}]. "Radio-loud" QSOs are generally called "quasars" (from "quasistellar radio sources"). We adopt this convention here.

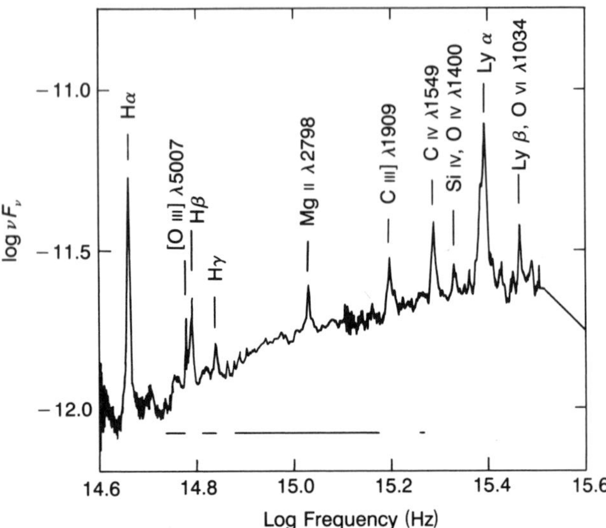

Figure 1. The optical-ultraviolet spectrum of a typical QSO, formed from a composite of several. The ordinate is in arbitrary units and the abscissa is frequency in the rest frame. Some discrete emission lines are marked and regions of the strongest Fe II blends are indicated by horizontal lines.

However, this distinction between QSOs and quasars is often confused, and "quasars" rather than "QSOs" is sometimes used to refer to the class as a whole, whereas "QSOs" refer to those that are selected only by optical properties.

The redshifts ($z = \lambda/\lambda_0 - 1$, where λ and λ_0 are the observed and rest frame wavelengths) almost certainly indicate cosmological distances. Of thousands known, only about 20 QSOs have $z > 4$, and the highest redshift (early 1991) is 4.73, corresponding to a look-back time of 93% of the age of the universe (deceleration parameter $q_0 = 0$). Ground-based observations can reach a wavelength of about 600 Å in the rest frame, satellite observations about 200 Å.

Optical flux densities range from approximately 38 mJy (apparent visual magnitude $V \sim 12.5$) to the limit of photographic sky surveys (~ 6 μJy or $V \sim 22$) and fainter. As expected for cosmological redshifts, there is a clear trend between V and redshift; however, the distribution of flux density at a given redshift (at any wavelength) is very broad.

The implied luminosities of the QSOs are as high as $\nu L_\nu \sim$ a few $\times 10^{47}$ erg s^{-1} (for a Hubble constant, $H_0 = 100$ km s^{-1} Mpc^{-1}, and $q_0 = 0$). The lower-luminosity limit is often arbitrarily set at $M_V = -21.5$ (10^{44} erg s^{-1}). This is the approximate luminosity at absolute visual magnitude which the nucleus dominates light from the host galaxy. Thus there is no clear distinction between the QSOs and other active nuclei of galaxies, especially at low redshifts ($z \sim 0.1$).

EMISSION LINES

There are two kinds of emission lines: relatively narrow lines from both permitted and forbidden transitions, with Doppler widths (FWHM, or full-width at half maximum intensity) from 300–1500 km s^{-1}, typically 600 km s^{-1}; and broad permitted lines with FWHM from 1000–10,000 km s^{-1}, typically 4000 km s^{-1}. The full-width at zero intensity for broad lines can approach 50,000 km s^{-1}. The broad-line region and most of the narrow-line region has never been spatially resolved. In some cases, especially at low redshift, some of the narrow-line emission is observed out to many kiloparsecs from the nucleus. Table 1 lists typical relative intensities for broad lines. For some strong lines we also give typical equivalent widths (rest frame), which are a measure of the line strength relative to the local continuum. Unlike other discrete lines the Fe II emission consists of thousands of blended broad lines forming a pseudocontinuum, mostly between 2000 and 5500 Å. In general, the relative strengths among narrow lines, those among broad lines, and the broad-line equivalent widths are similar from one QSO to another, and there are very few for which there is clear evidence for reddening by dust. This may be a result of observational bias as QSOs are often selected by UV excess and for their luminous quasistellar nuclei. The strengths of the narrow lines can be dominant to very weak compared with the broad lines; for example, the ratio [O III] λ5007/Hβ is inversely correlated with both optical continuum and x-ray luminosity.

Table 1. Relative Strengths of QSO Broad Emission Lines

Line	*Wavelength (Å)*	*Relative Intensity*	*Equivalent Width (Å, rest frame)*
Ly β	1,028	8	
O VI	1,034	25	
Ly α	1,216	100	60
N V	1,240	15–40	
O I	1,304	5:	
C II]	1,335	1.5	
Si IV	1,397	4	
O IV	1,402	9	
N IV]	1,486	6	
C IV	1,549	40	40
He II	1,640	5	
O III]	1,663	4	
N III]	1,750	4	
C III]	1,909	16	20
C II]	2,326	1.6	
Mg II	2,798	12–30	
Hδ	4,101	2.0	
Hγ	4,340	3.9	
He II	4,686	< 0.6–1.5	
Hβ	4,861	11	110
He I	5,876	6	
Hα	6,563	50	
Pα	18,751	5	
Fe II UV	2,000–3,000	81	
Fe II	3,000–3,500	11	
Fe II optical	3,500–6,000	< 8–16	
Fe II (total)		116	
Fe I	3,710, 3,860	1	
H7-Ba Limit	3,970–3,646	12	
Balmer continuum	< 3,646	89	

The profiles of the narrow lines, where measurable, are generally smooth and symmetric, although there are often small red or blueward asymmetries. The broad-line profiles are also generally quite smooth and symmetric, suggesting many emission clouds rather than a few that dominate the profile. The wings follow an approximately logarithmic intensity distribution. The overall shape may be peaked or stubby. The profiles are generally very similar from one line to another within the same QSO. However, in detail, C IV λ1549 often shows a stronger wing toward short wavelengths; the high ionization lines and Ly α tend to be broader and blueshifted by about 1000 km s^{-1} with respect to the lower ionization lines, which appear to be at rest with respect to the narrow lines, and probably with respect to the systemic velocity as indicated by stellar absorption features. In determining profiles, blending with other weak lines can be a serious problem; for example, C III] λ1909 can be blended with Al III λ1858, Si III] λ1892, and Fe III UV34, and weak Fe II blends can seriously affect many discrete lines.

The broad emission lines in the rare class of broad absorption line (BAL) QSOs are different in having larger Al III/C III] intensity ratios, probably stronger Fe II and, compared with NV $\lambda 1240$ and C III] $\lambda 1909$, unusually weak C IV $\lambda 1549$ emission lines whose profiles do not agree well with those of lower ionization lines.

CONTINUUM

An interesting approximation to the continuum spectral energy distribution is that the power per decade of frequency is roughly constant from radio to x-ray frequencies, except at radio wavelengths for radio-quiet QSOs, where the power may be down by a factor of 10,000 or more. This suggests an underlying relation between energy production processes at different frequencies. However, the same mechanism is not responsible for the whole continuum. After subtracting the contributions from discrete and blended lines and Balmer continuum, there are at least three broad spectral components, even in the near infrared through ultraviolet—the Big Blue Bump, the 3-μm Bump, and a "power law" component. The first two of these seem to be associated with the broad emission line region.

Big Blue Bump

In most QSOs the optical-ultraviolet region is dominated by the Big Blue Bump, with $\nu F_\nu \sim \nu^{0.5}$ in the optical and near ultraviolet. An observed minimum in νF_ν near approximately 1 μm suggests that this component declines in the near infrared. Statistically, this power law continues to Ly α in the higher-redshift QSOs, but falls slightly for lower-redshift QSOs. This may be intrinsic to the QSO continua but the results are also consistent with dust "reddening" in lower-luminosity QSOs. (Observational selection in a flux-limited sample results in the least luminous objects being detectable only at small redshifts, introducing a spurious correlation between redshift and luminosity.) Beyond Ly α ($\lambda < 1216$ Å), for the continua of intermediate- and high-redshift QSOs the flux density falls markedly, and increasingly with increasing redshift—as steep as $\nu F_\nu \sim \nu^{-1.5}$ for $\lambda < 912$ Å, the Lyman limit, sometimes steeper. This is attributed to increasing absorption by intervening neutral hydrogen at lower redshifts, with increasing distance to the QSO. Recent extreme ultraviolet observations of one luminous quasar, HS 1700+6416, suggest an intrinsic continuum that is a simple extrapolation of the optical-ultraviolet continuum to 300 Å, although Lyman line and continuum absorption severely depresses this continuum between 850 and 450 Å. This same Big Blue Bump component may be responsible for the soft x-ray excess above an extrapolation of the hard x-ray spectrum.

Power Law Component

It is often assumed that there is a power law component ($\nu F_\nu \sim$ constant) dominating in the near infrared that can be extrapolated into the optical-ultraviolet and even to the x-ray region. Quasi-power-law continua do indeed dominate in QSO-like objects called blazars, and appear to be a smooth continuation of their high radio frequency continuum. Actually, this component in the blazars is only approximately power law, typically varying with wavelength from $\nu F_\nu \sim$ constant in the near infrared, generally steepening to as much as $\nu F_\nu \sim \nu^{-2}$ in the ultraviolet. Flux densities vary rapidly, often by factors of 2, occasionally by factors of approximately 100. The time scales are hours to weeks for smaller variations, and weeks for larger amplitude changes. This component also has high linear polarization, often extremely variable in degree and position angle. This component, attributed to electron synchrotron radiation, is strongly correlated with the presence of a luminous, compact radio core. It has been shown recently that any QSOs with luminous compact cores, not just the traditional blazars, are likely to have such a power law component whose strength probably depends on the luminosity of the radio core. Thus a power law component has not yet been demonstrated in QSOs with weak radio cores, or in the radio-quiet QSOs.

TIME VARIABILITY

For the Big Blue Bump component (contributing the entire optical-ultraviolet continua of radio-quiet and weak-radio-core QSOs), the amplitudes of variability range from less than 10% to factors of 2 or more, over time scales of months to years. Variability seems to be of larger amplitude toward shorter wavelengths. For the power law component, the amplitude and time scale of variability of both flux density and linear polarization appears to be similar to those observed for the blazars.

Broad emission line variability has been claimed in a few QSOs, with a lag behind the continuum variations of less than a few weeks, suggesting light crossing times for the line-emitting regions much smaller than predicted by standard photoionization models.

LINEAR POLARIZATION

The high and variable polarization of the synchrotron-emitting core-dominant quasars contrasts with that in other QSOs, where the mechanism may be dust or electron scattering, or transmission by aligned grains.

In the weak-core, radio-loud quasars, there is weak linear polarization ($p \le 1$–2%) that tends to be aligned with the radio structure. In a few well-observed cases it is the continuum, not the emission lines, that is polarized. One exceptional weak-core, radio-loud quasar, OJ 287 (0752+258), shows $p \sim 8\%$, independent of wavelength and aligned in the direction of the radio lobes.

The rare broad absorption line QSOs and infrared selected QSOs (i.e., detected by the *IRAS* satellite) are the only radio-quiet QSOs to show high polarization (p up to 20%, and steady), and a large fraction have measurable polarization ($p > 1\%$), suggesting a possible relation between these classes.

RELATIONS BETWEEN OBSERVED QUANTITIES

The most important observed relationship is probably the strong, almost linear relation among the line and continuum luminosities. When combined with AGN of lower luminosity, this relation extends over a factor of 10^5 in luminosity. This is most readily explained in terms of photoionization of the emission line regions by the QSO ultraviolet continuum. The next most important is the small but definite departure of the previous relation from a linear one—the "Baldwin effect." The equivalent widths of the broad C IV $\lambda 1549$ emission line become smaller with increasing continuum luminosity. Some other lines show slightly different dependences on luminosity.

There is a possible trend for higher-luminosity QSOs to have broader C IV lines (FWHM). The trend may be most marked, however, for the full width of Hβ, measured at zero line intensity, and has been used to deduce a mass–luminosity relation, consistent with accretion near and below Eddington luminosities.

Emission Lines and Radio Properties

An important radio parameter is R, the ratio of the radio core luminosity to the luminosity in the extended radio emission (usually double lobes). In the relativistic beaming hypothesis, high values of R imply strong beaming toward the observer. There is an anticorrelation between widths of broad lines (at least Hβ and C IV $\lambda 1549$) and R, the dominance of the radio core—suggesting dominant motions of emission line gas in a plane perpendicular to the direction of radio beaming. There are anticorrelations between R and the ratios of emission line strengths to local or x-ray continuum intensities, supporting beaming of these continua.

Also relevant to the geometry and dynamics of the emission line and continuum emission regions are positive correlations between broad line widths and equivalent widths, although these are not well understood.

The only clear dependence on radio properties is in the strength of the optical power law continuum mentioned previously. It has been suggested that the blended optical Fe II lines are stronger in core-dominant (high R) than in lobe-dominant quasars, and in radio-quiet compared with radio-loud QSOs. The greater strength of [O III] emission in lobe-dominant quasars is also controversial. It remains to be seen whether these are real differences, or a dependence on some other property, such as intrinsic luminosity.

Additional Reading

Burbidge, G. and Burbidge, M. (1967). *Quasi-Stellar Objects*. W. H. Freeman, San Francisco.

Courvoisier, T. and Mayor, M. eds. (1991). *Active Galactic Nuclei*. Springer, Berlin.

Elvis, M. (1989). The ultraviolet continua of active galactic nuclei. *Comm. Ap.* **14** 177.

Gondhalekar, P. M., ed. (1987). *Emission Lines in Active Galactic Nuclei*. Seventh Workshop on Astronomy and Astrophysics, RAL-87-109. Rutherford Appleton Laboratory, Chilton.

Osterbrock, D. E. and Miller, J. S., eds. (1989). *Active Galactic Nuclei. IAU Symp.* Kluwer Academic Publishers, Dordrecht, 134.

Strittmatter, P. A. and Williams, R. E. (1976). The line spectra of quasi-stellar objects. *Ann. Rev. Astron. Ap.* **14** 307.

Weedman, D. W. (1986). *Quasar Astronomy*. Cambridge University Press, Cambridge.

See also **Active Galaxies and Quasistellar Objects, Emission Line Regions; Quasistellar Objects, Absorption Lines.**

Quasistellar Objects, Statistics and Distribution

Malcolm G. Smith

The discovery of very high redshift ($z > 3$) galaxies and quasars offers some hope that we shall be able to make direct tests of models for the formation and evolution of both classes of object. Complete samples of radio galaxies have been studied at optical and infrared wavelengths, and it is found that the faintest radio sources in these samples are so readily detected in the infrared that very few of these galaxies could lie at redshifts $z > 2.5$. The idea of a formation epoch for all massive galaxies (some of which may host quasars) suggests itself; furthermore, in a recent sample of radio galaxies, all the galaxies with $z > 0.8$ were found to show colors consistent with star formation activity in excess of that expected from an old, passively evolving, elliptical galaxy. It is, however, still difficult to obtain redshifts for the bulk of the galaxy population beyond $z \sim 0.5$.

Quasars, which are much brighter than most galaxies, offer the opportunity to look back over 80–90% of the time since the Big Bang. Their relevance to galaxy formation models depends on how much can be deduced about the formation and space density of galaxies in general by observing quasars that have a space density about two orders of magnitude lower. In this context it is important to try to deduce the masses of quasar host systems at high redshift. There is a growing body of evidence that typical masses of central objects in nearby normal and active galaxies are less than $10^9\ M_\odot$. If these objects are derived from luminous quasars, and the luminous quasars were powered by gravitational accretion of matter without significant mass loss, then the masses of the central objects in nearby galaxies will be greater than the masses of the high-redshift progenitors. This in turn will constrain the total amount of accreted matter and radiated energy over the lifetime of the quasar. The luminosity function (defined in the next section), coupled with suitable models, also provides, in principle, a constraint on the integrated energy and hence the (possibly episodic) active lifetime of the quasar. The duty cycle of activity may be a guide to the fraction of all galaxies that have harbored quasars. Physical models such as this can be used to compute the evolving luminosity function. Unfortunately, it turns out that the reverse process of deriving an unambiguous physical model from the observed data on fluxes and redshifts has proved to be much less secure.

In this entry, it is assumed that redshift is a measure of distance in a standard Big Bang cosmology, as described elsewhere in this encyclopedia. This assumption has yet to be rigorously proven.

LUMINOSITY FUNCTIONS AND THEIR EVOLUTION AT $z < 2.2$

Surveys of large samples of quasistellar objects have, as one aim, provision of a statistical basis for deductions about the distribution and time evolution of these objects. These observational deductions can then be compared with a series of alternative concepts, or models, to gain a better understanding of the underlying astrophysics. The statistics used to describe the distribution of quasars in space depend on the concepts of space density and luminosity function.

Consider first a volume of space containing a number of quasars. The volume must be specified with respect to the redshift or epoch of the universe at which the volume is seen, because redshift is a measure of the scale factor of the universe at the time the radiation was emitted. In conventional Big Bang models, the galaxies and quasars are roughly stationary in space; space itself is expanding, stretching the wavelengths of radiated photons passing through it. Comoving volumes scale with this expansion of the universe. The concepts of comoving volume and comoving space density provide corrections for the cosmological changes associated with the expansion of the universe. An unchanging population of objects has a comoving space density that is constant with time and is therefore independent of redshift. We find, however, that the number of quasars observed in a given comoving volume begins to increase rapidly as we look back to earlier times—corresponding to greater redshifts. The comoving space density of quasars appears to reach a maximum somewhere between $2 < z < 3$ and then may decrease again.

The quasars in any given volume will, like stars, usually differ from each other in their intrinsic luminosity. At great distances, many are too faint to be detectable in surveys that are necessarily flux limited. Any description of the three-dimensional distribution of astronomical objects such as stars and quasars must therefore be restricted to a finite range of intrinsic luminosities. This range is set by the flux limit of the survey and by relations between flux, luminosity, and distance. The luminosity function at a particular redshift is the variation of the comoving space density per unit luminosity as a function of luminosity $\rho = \rho(L, z)$. Luminosity functions are also often quoted per unit magnitude, for which the symbol $\phi = \phi(M, z)$ is used. Once this function is determined for all redshifts and luminosities, one has a complete evolution function for the given object population. Figure 1 presents model fits to the optical luminosity functions of nearby galaxies, $z \sim 0$, and for quasars at $z \sim 0.5$ and $z \sim 2$. High-redshift quasars are clearly much more luminous than the population of galaxies as a whole. On the other hand, the space density of galaxies today ($z \sim 0$) is several orders of magnitude greater than the space density of quasars at any redshift (as little significant increase in quasar space density has been found beyond $z \sim 2$).

We see that Fig. 1 is apparently consistent with a model in which the population of quasars gradually dims with time. The luminosity function for quasars with $z = 2$ can be fitted over the one for Seyfert galaxies near $z = 0$ (assuming the centers of nearby Seyfert galaxies are intrinsically faint quasars) by sliding the function over from the right toward the left of the diagram. The comoving space density of the entire population does not appear to change apprecia-

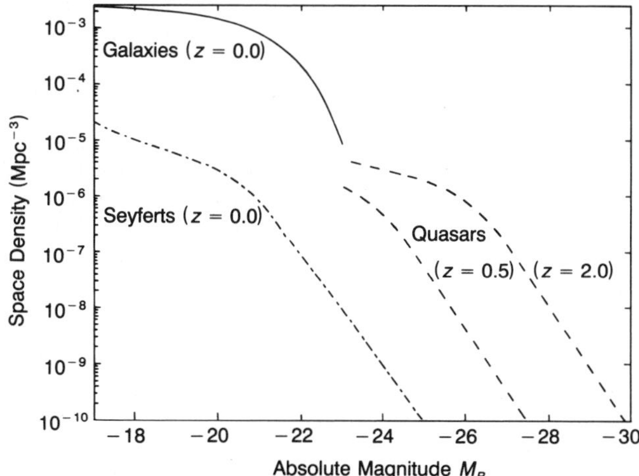

Figure 1. Model fits to the space density, ϕ, per cubic megaparsec, as a function of absolute blue magnitude, M_B, of normal galaxies ($z = 0$), Seyfert galaxies ($z = 0$), and quasars ($z = 0.5$ and $z = 2.0$). (*Courtesy of Paul C. Hewett and Steven J. Warren.*)

bly with time, whereas the population as a whole has dimmed by two orders of magnitude (5 magnitudes in M_B, the absolute magnitude in blue light). Such horizontal evolution of the luminosity function is referred to as luminosity evolution. An alternative scenario in which the luminosity of a feature in the luminosity function (e.g., the break between the flat and steep sections) remains constant, but the entire function shifts vertically in the diagram as redshift increases, is referred to as density evolution. A diagonal movement of the function across the $M-z$ diagram can be caused by a mixture of luminosity and density evolution. The data for quasars with $z < 2.2$ are consistent with pure luminosity evolution.

The evolution of the luminosity function is also important in determining the contribution of QSOs to the x-ray background and to the ultraviolet ionizing field at high redshift.

FINDING QUASARS: RECENT SURVEY WORK

It is unfortunately necessary to have a very clear understanding of the nature of the observational samples before one can attempt to work out what can be concluded from them. Traps abound. In the early days of work on statistics of quasars, for example, it was widely assumed that all quasars are strong radio sources. Most quasars known today have not been detected at all at radio wavelengths. Some optical selection methods work well only over a restricted range of wavelengths. Surveys at x-ray and infrared wavelengths are still too restricted in sensitivity to provide much information on the global properties of quasars beyond $z \sim 0.5$. One therefore does one's best to work on well-defined subsets of the total population of quasars.

Our knowledge of the evolution of the majority of quasars at $z < 2.2$ (summarized in Fig. 1) is drawn from surveys based on sets of direct photographic plate pairs taken in ultraviolet and blue light. The quasars on these plates were separated from most of the stars and very distant normal galaxies on account of their stronger ultraviolet emission. They were separated from nearby galaxies on account of their star-like appearance. These ultraviolet-excess (UVX) surveys appear to be nearly complete for objects with redshifts $z < 2.2$; surveys using other techniques find very few additional quasars up to this redshift.

At redshifts $z > 2.2$, quasars normally lose their ultraviolet excess with respect to many galactic stars, which makes them much harder to separate out for further study. At $z = 2.2$, the strong Lyman-α emission line of hydrogen (rest wavelength 121.6 nm) has been redshifted to 389.1 nm, and moves out of the passband of the ultraviolet filter used in most photographic surveys. A variety of techniques therefore has had to be used to locate objects at higher redshifts. Until recently, different techniques used in the same area of sky often produced rather different lists of objects, which led to concerns that each list was unlikely to be representative of the total population.

The largest sample of quasars with $z > 3$ selected in a homogeneous manner consists of about 50 objects. The technique of original candidate selection is based on automated classification of some 200,000 objects, with apparent red magnitudes $16 < m_R < 20$, measured with a high-speed automated plate measuring machine (APM) on UK Schmidt telescope plates; sets of plates were taken through five different colored filters with the telescope centered in turn on two fields giving an effective survey area of 50 square degrees. Again the technique of measuring ratios of fluxes through filters has proved the most effective, but it is necessary to use more than two filters to separate out the high-redshift quasars from the bulk of normal stars and high-redshift galaxies; the method is based on a measure of the degree of isolation of each image in a four-dimensional flux-ratio diagram. Only quasars whose flux ratios (colors) are similar to those of galactic stars escape detection by this technique. The selection function of the survey—a measure of the probability of detecting different known classes of objects having distinct spectral energy distributions—can be determined because the candidate selection procedure is automated.

Dispersive elements such as prisms and transmission-grating/prism (grism) or grating/lens (grens) combinations when placed ahead of the focal plane of a telescope can produce spectra of a large number of objects at once. Automated detection of emission lines in such spectra, recorded with charged-coupled device (CCD) detectors, is producing well-defined, gradually increasing subsamples of large-redshift quasars—including a recently discovered object with $z = 4.37$.

LUMINOSITY FUNCTIONS AND THEIR EVOLUTION AT REDSHIFTS $z > 2.2$

The highest measured galaxy redshift is 3.8. The highest redshifts known for quasars are approaching 5. Most of the highest-redshift quasars are bright—rather few faint quasars have been found at high redshifts, even though the techniques used appear to be sufficiently sensitive to detect them if they exist. The luminosity function therefore appears to steepen rapidly at fainter flux levels. The result is that the decline in comoving space density appears to be sharper for the fainter objects. It may also set in at lower redshifts than for brighter objects. Only few very bright objects can be found, even over a search area of 50 square degrees, so that one cannot yet be certain whether, for the very brightest classes of quasar, there is any turn-down in the comoving space density. The situation at present is that only the simplest models for the evolution of the luminosity function can be tested. Constraints on the detailed shape parameters for the luminosity function are still very weak. Extensive survey work currently underway using CCD detectors on large telescopes is very costly in large telescope time but currently seems the most likely way to improve the situation.

CLUSTERING OF QUASARS

We are still a long way from an astrophysical understanding of the observed evolution of the quasar luminosity function. If the black hole mass is increasing, why is the characteristic luminosity of the quasar population decreasing? One of the many uncertainties is the effect on the quasar of its environment. Assuming that gravitational interaction of the quasar with close companion galaxies is responsible for triggering quasar activity, evolution of the quasar luminosity function would be interpreted in terms of a change in the interaction rate with cosmic time. Observational tests of this idea are difficult; for example, low-power sources may not be seen in dense environments because they cannot supply enough ram pressure to overcome the static pressure of the intergalactic medium; the more luminous quasars and galaxies would have no such difficulties. Therefore, one has to check each sample of quasars carefully to ensure that any apparent correlation with redshift is not in fact produced by a stronger correlation with intrinsic luminosity.

The extent to which quasars cluster among themselves is important if only because the standard cosmological models assume a homogeneous and isotropic universe, whereas the distribution of galaxies and even clusters of galaxies is patently not homogeneous. Current studies of quasar association provide no evidence for quasar–quasar clustering on the large scale; however, evidence for quasar–quasar clustering on scales of order 10 Mpc or less has recently been reported.

Additional Reading

Begelman, M. C., Blandford, R. D., and Rees, M. J. (1984). Theory of extragalactic radio sources. *Rev. Mod. Phys.* **56** 255.

Boyle, B. J., Shanks, T., and Peterson, B. A. (1988). The evolution of optically selected QSOs—II. *Monthly Notices Roy. Astron. Soc.* **235** 935.

Osmer, P. S. (1982). Quasars as probes of the distant and early universe. *Scientific American*, **246** (No. 2) 96.

Peacock, J. A. and Miller, L. (1988). Radio quasars and radio galaxies—A comparison of their evolution, environments and clustering properties. In *Proc. Workshop on Optical Surveys for Quasars*, P. S. Osmer et al., eds. *Astron. Soc. Pacific Conf. Ser.* **2** 194.

Schmidt, M., Schneider, D. P., and Gunn, J. E. (1988). Spectroscopic CCD surveys for quasars at large redshift. In *Proc. Workshop on Optical Surveys for Quasars*, P. S. Osmer et al., eds. *Astron. Soc. Pacific Conf. Ser.* **2** 87. (See also the many other relevant articles in this volume.)

Warren, S. J. and Hewett, P. C. (1990). The detection of high-redshift quasars. *Reports on Progress in Physics.*

Weedman, D. W. (1986). *Quasar Astronomy*. Cambridge University Press, Cambridge, U.K.

See also **Cosmology, Observational Tests; Galaxies, High Redshift; Quasistellar Objects, in Galaxy Clusters and Superclusters.**

Radiation, High-Energy Interaction with Matter

Demosthenes Kazanas

Our knowledge and understanding of astrophysical phenomena is attained by observation and deepened by modeling and interpretation of the observational data. Led by observation, astronomers have come to the realization that certain astrophysical phenomena and objects involve the interaction among particles and photons at high energies. These high-energy particles have either been measured directly (e.g., in cosmic rays and solar flares) or their presence has been inferred from their observed photon emission (as in quasars, extragalactic jets, accreting black holes, and neutron stars in our galaxy). Therefore, the detailed knowledge of the high-energy interactions between photons and matter is an indispensable tool in order to successfully model these sources.

The qualification of an interaction as "high energy" requires a qualifier (a scale) with respect to which the interactions will be considered as "high energy." This scale is determined by the fundamentals of each interaction. Thus for the interactions between electrons and photons (the electromagnetic interactions), this scale is the energy corresponding to the electron rest mass, $m_e c^2$, where c is the velocity of light. For the interactions between protons or protons and photons (hadronic interactions), this scale is the rest energy corresponding to the mass of the pion ($m_\pi c^2$), the particle responsible for transmitting the nuclear force. Interactions are therefore termed "high energy" when the energy of at least one of the participating particles is larger than its characteristic energy scale. In practice, however, interactions involving particles (or photons) of energies greater than $m_e c^2$ are termed "high-energy" ones.

The great depth and breadth of the subject do not allow a detailed, comprehensive study of all these processes in the limited space of this entry. However, their general characteristics (cross sections, range of validity, etc.) can be obtained in a qualitative fashion by general considerations of their kinematics, classical electrodynamics, and quantum mechanics. Similar general considerations can also be applied to obtain the energy loss rates for particles associated with each such process; these loss rates are of great importance in modeling and are given in a separate section.

The division of the interactions into electromagnetic and hadronic (strong) is a natural one and the two sets are discussed separately. Also, because electromagnetic interactions are better understood and more prevalent, they are discussed in greater detail than hadronic ones.

ELECTROMAGNETIC INTERACTIONS

The electromagnetic interactions are by far the best understood interactions in physics to date. Given the fact that there exists a quantum theory of the electromagnetic field in the form of a perturbation theory which is convergent and finite (renormalizable), one can in fact calculate all the relevant quantities of the interaction (i.e., cross section, energy loss, etc.) from first principles; the agreement of the theory with the experimental findings has so far been excellent. We will discuss individually the electron–photon, electron–magnetic field, photon–photon, and bremsstrahlung interactions. The case of photon–electron scattering is used to introduce the general notions involved and is therefore presented in greater detail.

Electron–Photon Interactions

One of the most common processes in high-energy astrophysics is the collision of a relativistic electron of Lorentz factor γ (i.e., of energy $\gamma m_e c^2$) with a photon of energy ε, more commonly known as inverse Compton (IC) scattering. In the center of momentum (CM) frame (this is the electron rest frame if $\gamma\varepsilon \ll m_e c^2$), which moves with Lorentz factor γ relative to the lab, the photon energy $h\nu_{\text{CM}}$ (where h is Planck's constant and ν_{CM} is the frequency) is γ times larger than seen in the lab (i.e., $\gamma\varepsilon$); in addition, the collision in this frame is almost elastic (the electron preserves its energy, $\gamma\varepsilon$, during the collision). Transformed to the lab frame, the photon energy after the collision appears larger by yet another factor γ; that is, the scattered photon energy in the lab is $\gamma^2\varepsilon$. This energy gain is hence the result of the two successive Lorentz transformations and the elastic scattering in the center of momentum. This energy is much smaller than the total energy of the electron $\gamma m_e c^2$, so that an electron has to suffer a large number of collisions before it loses a sizable fraction of its energy. The cross section for this interaction can be calculated classically by considering the electron as a free point-like charge that responds to an external electric field according to the laws of classical electrodynamics. This cross section is roughly the square of the classical electron radius $r_0^2 = (e^2/m_e c^2)^2$ (the only length scale in the classical theory; e is the charge on the electron); it is the so-called Thomson cross section, and its precise value is $\sigma_T = (8\pi/3)r_0^2 \simeq 6.65\times 10^{-25}$ cm^2.

As the energy of the photon or the electron increases, a breakdown of the preceding arguments is expected. Kinematic modifications are needed when $\gamma^2\varepsilon \simeq \gamma m_e c^2$, or when the photon energy in the center of momentum frame $h\nu_{\text{CM}} = \gamma\varepsilon$ approaches $m_e c^2$. The CM frame is no longer the electron rest frame and the collision is not elastic in this frame. The electron recoil is significant and the photons lose energy in these collisions. As seen in the lab frame, the electrons lose a large fraction of their energy in a single collision. In addition, quantum mechanical effects become important when the wavelength of the photon in the CM frame becomes comparable to the Compton wavelength of the electron $\lambda_c = \hbar/m_e c$. The electron can no longer behave as a single, structureless, point particle in interactions with radiation; according to quantum theory there exists a "cloud" of virtual particles around the electron of size $\simeq \lambda_c$, whose constituents are "felt" by the photon when its wavelength becomes of order λ_c. The response of the electron to the electric field of the incident radiation is no longer coherent; the different parts of the electron interfere destructively with each other, causing a decrease of the cross section. The corresponding photon energy is $h\nu_{\text{CM}} = \gamma\varepsilon = hc/\lambda_c = m_e c^2$; that is, the quantum mechanical effects become important at the same energy as the electron recoil effects. This regime in the energetics of the collisions, in which both quantum and electron recoil effects become important, is the so-called Klein–Nishina regime and the cross section for interaction drops substantially from that of the Thomson value.

Electron–Magnetic Field Interactions

Bearing in mind that a magnetic field of strength B can be considered as a collection of virtual photons of energy $h\nu_B$ ($\nu_B = eB/m_e c$ is the Larmor frequency of the electron), the interactions of relativistic electrons with the magnetic field (synchrotron radiation) are qualitatively very similar to their interactions with real photons as described above (IC scattering). All the arguments of the collision kinematics and the onset of the Klein–Nishina modifications in the cross section are qualitatively applicable to this process, though the specific details are quite different. There is, however, an important distinction: Synchrotron radiation converts "virtual" photons into "real" ones and hence manifests itself as a photon-producing process, whereas IC involves the scattering of "real" photons and preserves their number.

Photon–Photon Interactions

This is a purely quantum mechanical process in which a collision between two photons of energies ε_1 and ε_2 yields an electron–positron pair. This process does not exist within the framework of classical electrodynamics, because this theory, being linear, does not allow for interactions between photons. However, quantum mechanics (and the uncertainty principle) allow for a "cloud" of virtual e^+-e^- (electron-positron) pairs around each photon; it thus becomes possible for the other photon to "knock off" such a virtual pair, thus allowing the interaction between photons. In the simplest case both photons disappear producing an electron–positron pair. Because of energy conservation, the energies of the two photons in the CM frame $E_{\rm CM} \simeq (\varepsilon_1/\varepsilon_2)^{1/2}\varepsilon_2 = (\varepsilon_2/\varepsilon_1)^{1/2}\varepsilon_1 = (\varepsilon_1\varepsilon_2)^{1/2}$ must be at least equal to $m_e c^2$ for the process to occur (threshold condition). At threshold, the cross section is roughly that of electron–photon scattering, that is, the Thomson cross section, whereas as $E_{\rm CM} \simeq (\varepsilon_1\varepsilon_2)^{1/2} \gg m_e c^2$ the process moves into the Klein–Nishina regime and the cross section drops accordingly. This process, which allows for the absorption of a $\simeq 10$ MeV gamma ray by a 30–50 keV x-ray, is the main source of opacity for gamma rays in accreting compact objects, as discussed below.

Bremsstrahlung

This process consists of the emission of a photon in the Coulomb scattering of two charges (in most astrophysical applications this occurs between an electron of charge e and a nucleus of charge Ze). In analogy with synchrotron radiation, bremsstrahlung can be considered as the scattering of a high-energy electron by the virtual photons that make up the electrostatic field of the nucleus. As with synchrotron radiation, because this scattering converts a virtual photon into a real one, bremsstrahlung is also a photon-producing process. The relevant cross section is therefore the Thomson one, σ_T, multiplied by the probability dN of emitting a photon of energy ω within an interval $d\omega$ (or the virtual photon spectrum). This probability is $dN = (2\alpha Z^2/\pi)(d\omega/\omega)$, where $\alpha = e^2/\hbar c \simeq 1/137$ is the fine-structure constant and $\hbar = h/2\pi$. The effective cross section is therefore α times smaller than σ_T, and despite the apparent $1/\omega$ divergence in the number of photons the total emission is finite, because the energies of the emitted photons correspondingly decrease. In a large number of astrophysical applications, one usually considers bremsstrahlung emission from a thermal electron distribution. The resulting differential photon spectrum diverges at low energies, as argued previously and cuts off exponentially for energies much larger than the gas temperature, reflecting the cutoff in the electron distribution.

STRONG (HADRONIC) INTERACTIONS

Our lack of a quantum theory of the strong interactions as complete and successful as that of quantum electrodynamics does not allow the a priori calculation of interactions between strongly interacting particles. However, the fact that the nuclear force is mediated by the exchange of pions does provide a scale to the theory, namely the Compton wavelength of the pion, $\lambda_\pi = \hbar/m_\pi c \simeq 1.4\times10^{-13}$ cm. An estimate for the cross section of order $\sigma \simeq \lambda_\pi^2$ is in good agreement with the experimental value of 3×10^{-26} cm^2. For strong interactions, high-energy collisions are hence those with CM energies larger than the rest energy corresponding to the mass of the pion, $m_\pi c^2$. The result of such collisions is the copious production of pions (π^+, π^-, π^0), which, however, are very short lived and upon their decay produce relativistic electrons and neutrinos (from the decay of π^+, π^-) or photons (from the decay of π^0). Because the resulting electrons and photons interact electromagnetically, it is very difficult to trace their origin to hadrons. In the absence of detectors capable of observing these neutrinos from pion decay, the presence of strong interactions in astrophysics can be unequivocally inferred only in rare occasions (e.g., interactions of high-energy cosmic rays in the atmosphere). Interactions of photons (of energy ε) with high-energy protons (of Lorentz factor γ_p), resulting in pion production, are also possible when allowed kinematically, that is, $\gamma_p\varepsilon \geq m_\pi c^2 \simeq 140$ MeV. However, these require protons of extremely high energies [$E_p \geq (m_\pi c^2/\varepsilon) m_p c^2 \simeq 10^{14}$–$10^{15}$ eV for $\varepsilon \simeq 1$ keV], whose presence has been indicated in only a small number of sources.

ENERGY LOSS RATES—ASTROPHYSICAL APPLICATIONS

The apparently nonthermal character of observed radiation from most high-energy sources precludes the use of thermal particle distributions in modeling their photon emission. The determination of the emitting particle distributions requires, in general, the solution of the kinetic equation. In its simplest form this is the continuity equation in energy space

$$\frac{\partial f}{\partial t} = \frac{\partial}{\partial \gamma}\left(\dot{\gamma}\frac{\partial f}{\partial \gamma}\right) + Q(\gamma, t),$$

where f is the unknown distribution, $Q(\gamma, t)$ is the high-energy particle injection rate which is determined by the dynamics and the acceleration mechanism, and $\dot{\gamma}$ is the total particle energy loss rate. In most cases one is interested in the particle distribution f under steady-state conditions ($\partial f/\partial t = 0$), for a given injection $Q(\gamma)$. Under these conditions, the energy loss rate $\dot{\gamma}$ is the single most important quantity in determining f and hence the emitted photon spectrum.

The loss rate can be estimated by multiplying the energy emitted per collision with the number of collisions a particle suffers per unit time, $n_{\rm sc}\sigma c$ ($n_{\rm sc}$ is the number density of scatterers, σ is the cross section for the particular interaction, and c is the speed of light).

For inverse Compton scattering $n_{\rm sc}$ is the number density of photons $n(\varepsilon)\,d\varepsilon$ in an energy interval $d\varepsilon$, σ is the Thomson cross section, and the energy loss per scattering is $\gamma^2\varepsilon$. The total energy loss rate can be estimated by integrating over the photon distribution (assuming that Klein–Nishina effects do not become important), that is,

$$m_e c^2\left(\frac{d\gamma}{dt}\right)_{\rm IC} = \sigma_T c\int n(\varepsilon)\gamma^2\varepsilon\, d\varepsilon = \sigma_T c \rho_{\rm rad}\gamma^2.$$

The integral $\int n(\varepsilon)\varepsilon\, d\varepsilon$ is equal to the total energy density in radiation $\rho_{\rm rad}$, which for a source of luminosity L and size R is equal to $L/\pi R^2 c$. Owing to the analogy of synchrotron radiation to inverse Compton scattering, the energy loss for synchrotron radiation has precisely the same form with the energy density of the radiation $\rho_{\rm rad}$ replaced by the energy density in the magnetic field $\rho_B = B^2/8\pi$.

For bremsstrahlung the number density of scatterers is equal to the number density of the ambient particles n, and the cross section is roughly σ_T, whereas the energy per scattering is equal to the energy of the emerging photon ω, integrated over virtual photon spectrum $dN/d\omega$, that is,

$$m_e c^2 \left(\frac{d\gamma}{dt}\right)_{\rm br} = nc\sigma_T \int \frac{dN}{d\omega} \omega \, d\omega$$

$$\simeq \alpha \sigma_T c Z^2 n \int d\omega \simeq \alpha \sigma_T c Z^2 n \gamma m_e c^2.$$

The energy loss rate is proportional to the energy of the electron, because the energy of the emitted photons ω extends up to the energy of the electron $\gamma m_e c^2$.

The need to use the kinetic equation for calculating the function f, rather than simply using the injection $Q(\gamma)$, can be assessed by comparing the electron loss time scales ($\tau_{\rm loss} = \gamma/\dot{\gamma}$) with the fastest possible dynamical time, namely the light-crossing time $\tau_d \simeq R/c$ across the source. If $\tau_{\rm loss} \ll \tau_d$, the source cannot be considered to be in a quasistatic state and solution of the kinetic equation is in order. (In cases where more than one loss process is involved, one should consider the dominant process, i.e., the one with the shortest loss time scale.)

The dominant loss process depends on the conditions encountered in the particular astrophysical source. As is apparent from the previous expressions, bremsstrahlung dominates in sources of high particle densities, whereas inverse Compton and synchrotron dominate in sources of high photon densities and magnetic fields. However, even in the same source, due to their different energy dependence, alternate processes may dominate in different energy regimes. Inverse Compton and synchrotron losses dominate in sources such as active galaxies, quasars, and extragalactic jets, whereas bremsstrahlung is important in high-energy events such as solar flares, where the particle densities are high relative to those of photons or magnetic fields.

The physical parameters of the sources cannot in general be deduced from observation, and thus it is not possible to determine the importance of the various radiation loss mechanisms. This is possible, however, for IC losses because the $\rho_{\rm rad}$ can be estimated from the observed luminosity L and the size of the source R, which can be inferred from variability measurements. The requirement that the light-crossing time be longer than the IC loss time scale ($R/c \geq \tau_{\rm IC}$) provides a relation ($L/R \geq \pi m_e c^3/\sigma_T \gamma \simeq 1.15 \times 10^{29}$ erg s^{-1} cm^{-1}) which can determine the dynamical importance of IC losses. So when the ratio L/R (called the compactness of the source) is greater than $\simeq 1.15 \times 10^{29}$ erg s^{-1} cm^{-1}, IC loss effects are important even for subrelativistic ($\gamma \simeq 1$) electrons.

Another important process that can be estimated easily in terms of measurable source parameters is the photon opacity to pair production, $\tau_{\gamma\gamma}$. If the source luminosity in gamma rays is L, the number of photons of energy $\varepsilon_{\rm ph} \simeq m_e c^2$ is $n_\gamma \simeq L/\pi R^2 c \varepsilon_{\rm ph}$ and the $\gamma-\gamma$ opacity is

$$\tau_{\gamma\gamma} \simeq n_{\rm ph} \sigma_T R \simeq \frac{L}{R} \frac{\sigma_T}{\pi c \varepsilon_{\rm ph}} = \frac{L}{R} \frac{\sigma_T}{\pi m_e c^3}.$$

This opacity, like the ratio $\tau_d/\tau_{\rm IC}$ of light crossing to IC loss times, depends only on the ratio L/R, that is, the compactness of the source. It is roughly unity when $L/R \simeq 1.15 \times 10^{29}$ erg s^{-1} cm^{-1} (i.e., the value at which $\tau \simeq \tau_{\rm IC}$ for subrelativistic electrons) and in most sources it increases with the energy of the gamma ray (more precisely it depends on the spectrum of photons at $\varepsilon \leq m_e c^2$). Therefore, a source with $L/R \geq 1.15 \times 10^{29}$ erg s^{-1} cm^{-1} will convert all its high-energy radiation into $e^+ - e^-$ pairs which will also have to be incorporated into the kinetic equation when modeling these sources.

The compactness parameter is of paramount importance for modeling high-energy sources in which IC scattering is the dominant loss mechanism, such as accreting compact sources (black holes and neutron stars). For these sources, in fact, L/R turns out to be roughly independent of the mass of the accreting object, because both their luminosity and size scale linearly with its mass M. The luminosity is believed to be a fraction, $F \simeq 0.01-1.0$, of the Eddington luminosity [$L_{\rm Edd} \simeq 1.3 \times 10^{38}$ $(M/M_\odot)$ erg s^{-1}], whereas the radius is a multiple, $x \simeq 10$, of the Schwarzschild radius [$R_s = 3 \times 10^5$ $(M/M_\odot)$ cm], so that $L/R \simeq 4 \times 10^{32} (F/x)$ $\simeq 4 \times 10^{29} - 4 \times 10^{31}$ erg s^{-1} cm^{-1}, independent of the mass. Moreover, these values for the compactness are larger than the critical one for which $\tau_{\gamma\gamma} \geq 1$. It is therefore expected that these sources will be pair dominated and will have suppressed gamma ray emission. It is somehow ironic that in the most luminous high-energy sources, the high-energy ($\geq m_e c^2$) radiation remains hidden, precisely because of their high luminosity!

Additional Reading

Blumenthal, G. R. and Gould, R. J. (1970). Bremsstrahlung, synchrotron radiation, and Compton scattering of high-energy electrons traversing dilute gases. *Rev. Mod. Phys.* **42** (No. 2) 237.

Gaisser, T. K. (1991). *Cosmic Rays and Particle Physics*. Cambridge University Press, Cambridge, U.K.

Jackson, J. D. (1962). *Classical Electrodynamics*. Wiley, New York.

Jauch, J. M. and Rohlich, F. (1980). *The Theory of Photons and Electrons*. Springer-Verlag, New York.

See also **Cosmic Rays, Origin; Neutrinos, Cosmic; Radio Sources, Emission Mechanisms; Sun, Radio Emissions.**

Radiation, Scattering and Polarization

Peter Mészáros

A beam of electromagnetic radiation can be characterized by its direction of propagation $\hat{k}$, its frequency ω, and its state of polarization. These remain constant as the radiation propagates in empty space, unless the beam interacts with matter that scatters it. In this case a fraction of the incident radiation changes its direction and its polarization, and in some situations also its frequency, in a manner that depends on the characteristics of the incident radiation and on the nature of the scattering centers. The scattering centers can be electrons, atoms, molecules, or solid dielectric grains or aggregates made up from these.

POLARIZED RADIATION

Polarized radiation can be defined as that for which the plane of oscillation of the electric and magnetic field vectors, **E** and **B**, remains constant during propagation, or else varies in a predictable way. The polarization vector $\hat{\varepsilon}$ at a particular point along the propagation trajectory is defined as a unit vector along **E** at that point. The polarization can be either linear, if the tip of the electric vector **E** always oscillates in the same plane, circular if the tip of **E** moves in a circle in the plane perpendicular to the direction of motion, and elliptical when $|\mathbf{E}|$ is not constant during the rotation of the plane of polarization. In the latter two cases, the plane of polarization rotates at an angular frequency equal to that of the wave. One can define the Stokes parameters I, Q, U, and V, where I gives the total intensity of the radiation, Q and U give the orientation of the ellipse of polarization, and V gives the circularity, or the ratio of principal axes of the polarization ellipse. Monochromatic light is always 100% polarized, and one of the four Stokes parameters is redundant, because they satisfy the equality $I^2 = Q^2 + U^2 + V^2$. Realistic radiation sources usually have a frequency spread $\Delta\omega$, and in general such light is not 100% polarized. Measurements of the polarization over periods of time $\Delta t \gg 1/\Delta\omega$ would detect chaotic variations of the polarization state. However, for quasimonochromatic light with $\Delta\omega \ll \omega$, it is possible to define appropriately time-averaged values of the Stokes parameters, valid

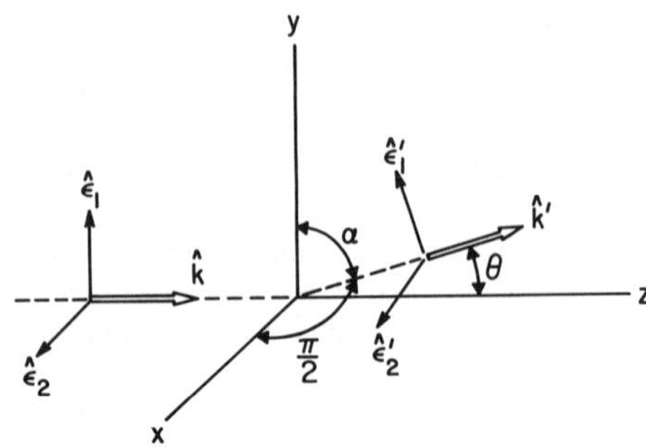

Figure 1. Radiation propagating along $\hat{k}$ with polarization vectors $\hat{\varepsilon}_1$ and $\hat{\varepsilon}_2$ being scattered in the direction $\hat{k}'$ making an angle θ with respect to the original direction.

for measurement times less than the so-called coherence time of the radiation, $\Delta t \sim 1/\Delta\omega$.The meaning of the Stokes parameters is the same as before, the only difference being that in this case there is the possibility of having partially polarized light. This can be described as a mixture of 100% polarized light and completely unpolarized light. All four Stokes parameters are needed, because now $I^2 \geq Q^2 + U^2 + V^2$, and the fourth number describes the ratio of polarized to total intensity. The equality holds for purely polarized light, whereas for purely unpolarized light one has $Q = U = V = 0$.

SINGLE SCATTERING AND POLARIZATION

One of the earliest and best-known studies of radiation scattering is that of Lord Rayleigh, who investigated this phenomenon in connection with the scattering of sunlight in the Earth's atmosphere. The purpose was to find out why the color of the sky is blue, and such a study also provided an explanation of why scattered light becomes polarized. Although this was done originally for small spherical dielectric particles whose dimension was small compared to the wavelength of the light, it is of wider applicability, being representative also of the scattering by free electrons and by nonresonant electrons bound in atoms.

The change in the polarization caused by a scattering event is most easily illustrated for linearly polarized radiation that is scattered by cold electrons (Thomson scattering), in which case the frequency ω remains unchanged to a first approximation. Consider an electromagnetic wave, or beam of radiation traveling initially in the direction $\hat{k}$ along the z axis (Fig. 1), which is completely linearly polarized in the y, z plane; that is, its electric vector $\hat{\varepsilon}_1$ oscillates in the y direction (because in vacuum, the **E** and **B** vectors of an electromagnetic wave must be perpendicular to $\hat{k}$). The radiation scattered by a single electron into a direction $\hat{k}'$ making an angle $\theta = \pi/2 - \alpha$ with the original direction will have an intensity characterized by the differential cross section for this state of polarization. This is just the ratio of the flux scattered in this direction to the incident flux, which in terms of the angle α between the original polarization vector $\hat{\varepsilon}_1$ and the scattered direction $\hat{k}'$ is given by $(d\sigma(\alpha)/d\Omega)_p = r_0^2 \sin^2 \alpha$, where $r_0 = e^2/m_e c^2 = 2.82 \times 10^{-13}$ cm is the classical radius of the electron, and $d\Omega$ is the differential of solid angle along the $\hat{k}'$ direction. The scattered radiation will also be linearly polarized in the x, y plane, with a polarization vector perpendicular to $\hat{k}'$. A similar expression will be valid for radiation incident along $\hat{k}$ but which is linearly polarized perpendicular to the plane of scattering, that is, with $\hat{\varepsilon}_2$ along the x axis. In this case, the differential cross section is the same as the previous one, but with $\alpha = \pi/2$, and the radiation scattered into the direction $\hat{k}'$ will be linearly polarized parallel to the original polarization $\hat{\varepsilon}_2$.

To see how polarized light can be produced in a scattering medium, it is useful to consider what happens to initially completely unpolarized light that suffers scattering. Because unpolarized light can be represented as an equal mixture of light polarized along two mutually orthogonal directions, such as $\hat{\varepsilon}_1$ and $\hat{\varepsilon}_2$, the differential scattering cross section for unpolarized light will be given by an average of the polarized cross sections for $\hat{\varepsilon}_1$ with α and $\hat{\varepsilon}_2$ with $\alpha = \pi/2$, that is,

$$\left(\frac{d\sigma}{d\Omega}\right) = \frac{1}{2} r_0^2 (1 + \sin^2 \alpha) = \frac{1}{2} r_0^2 (1 + \cos^2 \theta).$$

The total scattering cross section is obtained from this by integrating over the differential of solid angle $d\Omega$ for all possible final directions, giving the well-known Thomson value, $\sigma_T = (8\pi/3) r_0^2 = 6.65 \times 10^{-25}$ cm^2. In the preceding equation, the two terms give the relative proportion of scattered intensities polarized along the two perpendicular directions, the first one having $\hat{\varepsilon}'_1$ in the y, z plane of scattering, and the second having $\hat{\varepsilon}'_2$ along x, perpendicular to the plane of scattering, for a particular angle of scattering θ. The two intensities are in the ratio 1:$\cos^2 \theta$. Thus unpolarized light acquires after a single scattering a degree of linear polarization given by $p = [1 - \cos^2 \theta]/[1 + \cos^2 \theta]$. This degree of linear polarization depends on the angle of scattering, ranging from a value of 0 (no net polarization) for viewing along the incident direction $\theta = 0$ (because by symmetry all directions about $\hat{k}$ are equivalent), to a value of unity (100% linearly polarized) for viewing perpendicular to the incident wave, $\theta = \pi/2$ (because the electron motion is in a plane which contains the viewing direction).

The Thomson scattering law described previously is an example of the more general Rayleigh scattering function, applicable whenever the scattering particle radiates as a simple dipole. This occurs for scattering by atoms at frequencies away from any resonances and for spherical dielectric particles. The phase function, that is, the angular and polarization dependence, remains the same as before, but the numerical coefficient in front becomes dependent on the wavelength of the radiation and the characteristics of the atoms or the particles. In the case of scattering by anisotropic particles, such as elongated grains or molecules, the scattering does not follow a simple dipole pattern, but is characterized by a differing polarizability along the major axes of the scattering particles. The phase function is consequently more complicated, and in general it is no longer true that radiation scattered through an angle of $\pi/2$ is 100% polarized.

POLARIZATION BY MULTIPLE SCATTERINGS

The examples discussed previously considered the polarization caused by a single scattering event. When radiation is scattered many times, such as when radiation escapes from a medium that has characteristic dimensions equivalent to many scattering mean free paths, the cumulative effect of the many individual scatterings does not lead to a simple coherent addition of the polarization caused by one scattering, because the successive angles of scattering are given by a random walk process. As a consequence, a significant fraction of the polarization achieved along one polarization direction $\hat{\varepsilon}_1$ in one scattering can be undone in the next scattering event, by reapportioning the radiation along a different polarization direction $\hat{\varepsilon}_2$. Knowing, however, the behavior for one scattering event allows one to calculate the degree of polarization of the radiation escaping from a scattering atmosphere of arbitrary dimensions, after a varying number of scatterings. In the general case of arbitrary elliptical polarization, one can define an intensity vector $\mathbf{I} = (I_r, I_l, U, V)$ made up of the four Stokes parameters, which are a function of the angles θ and ϕ of propagation, and of the position r in the atmosphere. Here I_l and I_r are the intensities polarized in the plane of scattering and perpendicular to it, which can be used instead of the usual I and Q. The scattering is represented through a 4×4 scattering matrix corresponding to redistribution between the four Stokes parameters, $S(\theta, \phi; \theta', \phi')$, which is a function of the angles after and before scattering. Defining an optical depth $d\tau = -n_s\sigma_s\,dr$ as in unpolarized radiative transfer (where n_s and σ_s are the space density of scatterers and their angle-integrated total scattering cross section, respectively), the general transfer equation for polarized radiation is

$$\frac{d\mathbf{I}(\theta,\phi,\tau)}{d\tau} = \mathbf{I}(\theta,\phi,\tau) - \frac{1}{4\pi}\int_0^{\pi}\int_0^{2\pi} \mathrm{S}(\theta,\phi;\theta',\phi')\mathbf{I}(\theta',\phi',\tau)\sin\theta'\,d\theta'\,d\phi'.$$

For a plane parallel atmosphere, one finds from the preceding equation that the components $U = V = 0$. This is to be expected, because in this case there can be no ϕ dependence, and there is no reason for circular polarization to be present. The previous system of four equations reduces to a simpler system of two equations with two unknowns:

$$\mu\frac{d}{d\tau}\begin{pmatrix} I_l(\tau,\mu) \\ I_r(\tau,\mu)\end{pmatrix} = \begin{pmatrix} I_l(\tau,\mu) \\ I_r(\tau,\mu)\end{pmatrix} - \frac{3}{8}\int_{-1}^{+1}\begin{pmatrix} 2(1-\mu^2)(1-\mu'^2)+\mu^2\mu'^2 & \mu^2 \\ \mu'^2 & 1\end{pmatrix} \times \begin{pmatrix} I_l(\tau,\mu') \\ I_r(\tau,\mu')\end{pmatrix} d\mu',$$

where $\mu = \cos\theta$ is the cosine of the angle made by the radiation with the normal to the atmosphere surface. Solutions of this equation show that the degree of linear polarization p varies as a function of optical depth and angle of viewing. For infinite optical depth atmospheres, the linear polarization degree has a maximum value of 11.7% at $\mu = 1$, or grazing viewing angles, with the electric vector parallel to the surface. The linear polarization decreases to 0 at $\mu = 0$ (normal), because there one has no preferred direction. For lower optical depths, the direction of polarization remains initially the same, but the degree of polarization decreases, whereas for scattering optical depths of order less than a few, the direction of polarization switches to an electric vector perpendicular to the surface, and the degree of polarization reaches again maximum values of order 10%. An example for radiation sources uniformly distributed in depth through a finite plane parallel atmosphere of scattering depth 2τ is shown in Fig. 2.

Notice that in the cold electron scattering approximation discussed previously, the intensity is grey, that is, independent of frequency. The polarization degree is obtained from considering *all* the escaping photons. However, the number of scatterings suffered by individual photons varies significantly about a mean value, given by τ^2. It is possible to expand the transfer equation discussed previously, expressing the intensity as a sum of intensities made up of photons that have suffered none, one, two,... scatterings. If this is done, one finds that for low optical depth atmospheres the polarization of the photons that have been scattered much more than the average number of times can reach rather higher values than those mentioned previously. This is of interest when Comptonization is taken into account, that is, in the cumulative effects of the changes of frequency in individual scatterings, which are important in a hot plasma. In this case, the scattering matrix for nonrelativistic thermal electrons at temperature T is given by

$$\mathrm{S}(\omega',\Omega',\hat{\varepsilon}';\omega,\Omega,\hat{\varepsilon}) = \frac{3}{2}\frac{\omega'}{\omega}\frac{m_e c}{|\Delta\mathbf{k}|}\frac{|\hat{\varepsilon}'\cdot\hat{\varepsilon}|^2}{(2\pi m_e kT)^{1/2}}\exp\left[-\frac{(m_e\Delta\omega + \hbar/2\,\Delta k^2)^2}{\Delta k^2 2m_e kT}\right].$$

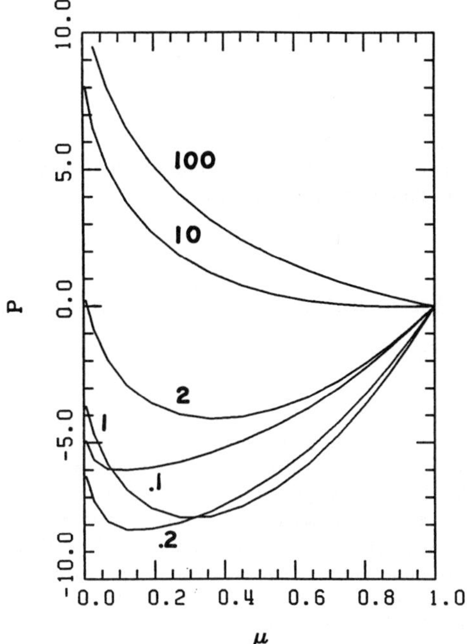

Figure 2. Degree of linear polarization (P) in percent as a function of $\mu = \cos\theta$, where θ is the angle of the escaping radiation with respect to the normal, for a plane parallel Thomson scattering slab. The curves are labeled by the value of τ, the scattering optical depth of the half thickness of the slab. [*From Phillips and Mészáros, (1986); reprinted with permission from the* Astrophysical Journal.]

For other scatterers, such as atoms or spherical particles, the frequency dependence is also important. In the scattering by atoms (actually, by electrons bound in atoms with a characteristic binding frequency ω_0), the total scattering cross section is given by $\sigma_T(\omega/\omega_0)^4$, if the incident light has frequencies $\omega \ll \omega_0$. This is the classical Rayleigh scattering regime, which was used to explain why the sky is blue (higher frequencies are scattered more efficiently), when not looking directly at the Sun, and why the Sun is red at sunset or sunrise. The degree of polarization is dependent on both the frequency and the angular distance from the Sun. Other, more complicated examples involve scattering by dielectric spheres (Mie scattering) and by grains of arbitrary shapes, which find applications in the scattering of light by interplanetary and interstellar dust grains. The latter case leads similarly to polarization of light because the anisotropic grains generally align themselves in response to the interstellar magnetic field.

Additional Reading

Chandrasekhar, S. (1960). *Radiative Transfer*. Dover, New York.

Collins, G. W., II (1989). *The Fundamentals of Stellar Astrophysics*. W. H. Freeman, New York.

Mészáros, P. and Bussard, R. W. (1986). The angle-dependent Compton redistribution function in x-ray sources. *Ap. J.* **306** 238.

Phillips, K. C. and Mészáros, P. (1986). Polarization and beaming of accretion disk radiation. *Ap. J.* **310** 284.

Rayleigh, Lord (1899). *Scientific Papers*, pp. 87, 104, 518. Cambridge University Press, Cambridge, U.K.

Rybicki, G. B. and Lightman, A. P. (1979). *Radiative Processes in Astrophysics*. Wiley, New York.

Sobolev, V. V. (1963). *A Treatise on Radiative Transfer*. Van Nostrand, New York.

Sunyaev, R. A. and Titarchuk, L. G. (1985). Comptonization of low-frequency radiation in accretion disks: Angular distribution and polarization of hard radiation. *Astron. Ap.* **143** 374.

van de Hulst, H. C. (1981). *Light Scattering by Small Particles.* Dover, New York.

See also **Interstellar Medium, Dust, Large Scale Galactic Properties; Pulsars, Observed Properties; Quasistellar Objects, Spectroscopic and Photometric Properties.**

Radio Astronomy, Receivers and Spectrometers

Lewis E. Snyder

The wavelength range for radio astronomy is not rigidly defined but in general it extends from approximately 350 μm to about 150 m. This range is heavily influenced by ground-based observations, where the short-wavelength limit is dictated by atmospheric transparency and the long-wavelength limit is determined by ionospheric transparency. Radio astronomers study both continuum and spectral-line radiation from celestial sources over this broad range of wavelengths. Celestial energy is collected by a radio telescope, detected and amplified over a narrow-frequency range by a radio receiver mounted on the telescope, and recorded in a "backend" system as either a continuum measurement of flux density or as one or more spectral lines superimposed on an underlying continuum background. The power level of a typical radio signal can be very low, perhaps 10^{-16} W for a continuum source and as small as 10^{-21} W for a weak spectral line. Receiver systems designed exclusively for continuum measurements are sometimes called radiometers to distinguish them from systems for spectral-line measurements. A spectral-line system must have higher frequency stability than an ordinary radiometer because of the small frequency width of astronomical spectral lines and it must have a more complicated backend system, a spectrometer, for recording spectral-line data.

MODERN RECEIVERS FOR RADIO ASTRONOMY

At infrared and optical wavelengths, astronomers obtain optimum sensitivity by using direct detectors such as bolometers, charge-coupled devices, and photoelectric detectors, but at radio wavelengths the superheterodyne receiver is usually best. The crossover point between direct and heterodyne detection techniques lies in the far-infrared region; its precise location depends on wavelength, resolution, bandwidth, and background emission of the experiment. This discussion will be confined to heterodyne detection techniques, which employ the process of mixing a weak radio frequency (RF) signal with a local oscillator signal to produce an intermediate frequency (IF)—the sum or the difference frequency of the two signals—which is more readily amplified than the initial RF signal. Superheterodyne receivers used in radio astronomy have a predetection or high-frequency section, which includes all components before the detector (e.g., RF and IF amplifiers, mixer, and local oscillator), and a postdetection or low-frequency section, which includes everything else. One postdetection section using a standard frequency will accommodate several predetection sections, each of which can be tuned to different radio frequencies of observational interest. Many different types of superheterodyne receivers are used at radio observatories. Further details have been published by M. E. Tiuri and A. V. Räisänen who discussed nine astronomical receivers and by R. M. Price who compared the sensitivities of 35 systems.

System Temperature

The sensitivity of a radio astronomical receiver is measured by a figure of merit called the system temperature, T_{sys}, which is the temperature that would be required for an ideal resistor to produce the same power as that produced by an ideal receiver if the resistor were placed across the input to the receiver. T_{sys} is composed of temperature contributions from the antenna, the transmission line, and from the receiver itself. Its importance stems from the fact that the minimum detectable radio signal is proportional to $T_{sys}/(Bt)^{1/2}$, where B is the postdetection bandwidth and t is the integration time. The values for T_{sys} range from 10 or 15 K to a few thousand degrees kelvin, depending on the observing frequency, antenna, and receiver. Bandwidths vary from a few kilohertz for spectral-line measurements to a few gigahertz for wideband continuum observations at short wavelengths. Effective integration times might vary from fractions of a second for strong, time-varying pulsars to a few days for weak spectral lines.

Predetection Section

The predetection section of the receiver accepts an incoming celestial emission signal (which can be visualized as a superposition of many independent waveforms with random polarizations) and converts it to a low-frequency input to the postdetection section. Radio observers often refer to the frequency of the center of the bandwidth of the celestial RF signal as the sky frequency, ν_s. A typical block diagram for the predetection section of a modern astronomical superheterodyne receiver is shown in Fig. 1. The sky signal with center frequency ν_s enters the receiver via a feed horn on the left-hand side. A long-wavelength receiver often has an RF amplifier as the first stage followed by a mixer, but receivers constructed to observe wavelengths shorter than about 7 mm usually begin with the mixer because appropriate amplifiers are not readily available. The first-stage RF amplifier at long wavelengths may be a bipolar transistor (for $\nu_s \leq 1$ GHz), a FET [the field effect transistor category includes the HEMT (high-electron mobility transistor)], a parametric amplifier, or a maser.

In the next stage, the incoming sky frequency ν_s is mixed with the local oscillator frequency ν_0 to produce a fixed intermediate frequency ν_{IF}. This process of converting ν_s to ν_{IF} may appear complicated because the mixer is a nonlinear device that produces a current containing sum, difference, and overtone combinations of sky and local oscillator frequencies. An astronomical observer usually never detects all of these frequencies, however, because a

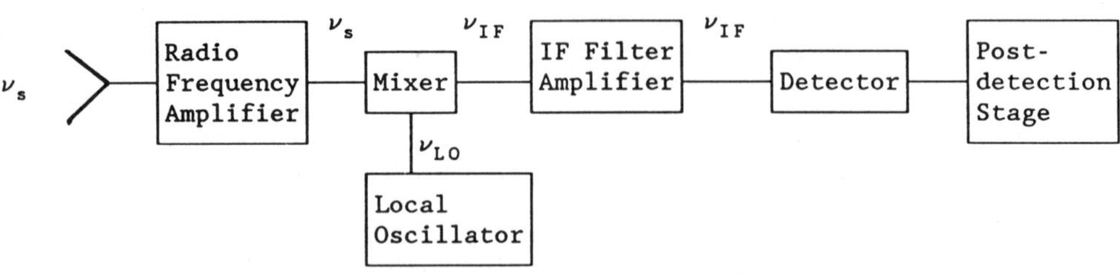

Figure 1. A typical block diagram for the predetection section of a modern astronomical superheterodyne receiver.

filter following the mixer and centered on the intermediate-frequency band passes only bands centered on the sum frequency, $\nu_{LO} + \nu_{IF}$, and difference frequency, $\nu_{LO} - \nu_{IF}$. Because two bands of sky frequencies can be detected, this is called a double-sideband system. By convention, the sum denotes the center frequency of the upper sideband, $\nu_s(\text{usb}) = \nu_{LO} + \nu_{IF}$, and the difference the center frequency of the lower sideband, $\nu_s(\text{lsb}) = \nu_{LO} - \nu_{IF}$. Double-sideband systems are usually used in the millimeter- and submillimeter-wavelength ranges. Some improvement in image rejection may be gained by using a dual conversion receiver, where a second mixing stage is introduced after the first IF amplifier, but in effect two sidebands are still present. At longer wavelengths, single-sideband systems are more common because they can be created by removing one sideband with appropriate filtering or enhancing one sideband by narrowband amplification.

Diodes are used as mixer elements in the centimeter wavelength region. Schottky diode mixers have dominated the millimeter-wavelength region in the past because of their high reliability and acceptable system noise temperatures, whereas both Schottky diode and InSb bolometer mixers have been used with some success in the submillimeter region. Today these mixers are being replaced by superconductor–insulator–superconductor quasiparticle mixers, commonly called SIS junction mixers, in both wavelength regions. A major advantage of the SIS junction mixer is a low noise temperature which potentially can reach the limit established by the Heisenberg uncertainty principle, but a major disadvantage is that current SIS junctions require complicated cryogenic systems because they have to be cooled to 4 K in order to operate.

Solid-state local oscillators are used in receivers built for observations at wavelengths around 10 cm and longer. In the centimeter-wavelength region, Gunn oscillators are common but in the millimeter-wavelength region, both the reflex klystron and the Gunn oscillator with a frequency multiplier are used as local oscillators. Research at the Radio Astronomy Laboratory at the University of California, Berkeley, has demonstrated that the Gunn oscillator arrangement is particularly well suited for millimeter array applications for the frequency range from 65–115 GHz. Carcinotrons, molecular lasers, or klystron-driven or Gunn oscillator-driven harmonic generators are commonly used at submillimeter wavelengths. Spectral-line and continuum receivers used for interferometry and spectral-line receivers (including pulsar receivers) used on single-element telescopes require frequency stable (or phase-locked) local oscillators, but this is unnecessary for ordinary continuum radiometers used on single-element telescopes.

SPECTROMETERS FOR RADIO ASTRONOMY

Postdetection Section for Single-Element Radio Telescope Systems

In a single-element telescope a typical postdetection section receives the IF signal, a voltage which varies in time, and passes it first through a detector with an output voltage directly proportional to the input power (a square law detector), and then through an integrator. In a continuum radiometer, the data are averaged over a time period which is long compared to the reciprocal of the effective observing bandwidth of the system, so each continuum data point is an average across a bandwidth. Because continuum measurements are used to determine the flux density of a source, usually they are recorded as a function of both time and sky position. In a spectral-line receiver, the data are sampled in times much less than the reciprocal effective bandwidth, "binned," and recorded in narrow, adjoining frequency bands. Modern spectrometer systems used in radio astronomy are filter banks, digital autocorrelators, or acoustooptical systems.

The filter bank spectrometer divides the broadband IF signal into discrete, contiguous, narrow channels or filters. Each filter has its own square law detector and integrator. The spectrometer bandwidth is the collective output of these adjoining filters, usually viewed as channels in frequency space, and it gives a display of the amplitude of a spectral line versus its frequency. Almost all early spectral-line work in radio astronomy was done with some type of filter spectrometer because the operating principle is straightforward. The autocorrelation spectrometer, or autocorrelator, is based on the fact that the power spectrum of the time-varying IF signal is the Fourier transform of the signal's autocorrelation function. Digital sampling is used to divide the IF signal into quantized time channels which are delayed and multiplied to ultimately form an autocorrelation function. The autocorrelation function of the time series of samples is accumulated and then a discrete Fourier transform is used to obtain the power spectrum. Thus the autocorrelator is a small, special-purpose computer. A major disadvantage of a filter bank spectrometer can be its fixed bandwidth, but this is not a problem with an autocorrelator because its bandwidth is changed by changing the sampling rate. An example of a spectrum taken with an autocorrelator is shown in Fig. 2. The acoustooptical spectrometer (AOS) has become important at millimeter wavelengths where the wide bandwidths required make both autocorrelators and filter banks complicated and expensive. In the AOS, the IF signal is fed into a piezoelectric transducer attached to a Bragg cell where it is converted into acoustical waves. Laser light—fixed in wavelength, amplitude, and direction—is beamed into the cell, where the acoustical waves have altered the index of refraction. Some of the emerging laser light exhibits a diffraction pattern which is the power spectrum of the IF signal. Photodiodes or other optical sensors may be used to digitize this spectrum for computer analysis.

Postdetection Section for Radio Telescope Arrays

The postdetection section for an array of radio telescopes is more complicated than for a single-element telescope because arrays produce high spatial resolution maps or images of the radio brightness of a source by using Fourier synthesis. This is based on the principle that the radio brightness distribution of a source and the response of an interferometer to that source are connected through a Fourier transform relationship. As a practical point, a synthesis array is often treated as an ensemble of two-element interferometers. Most interferometers obtain optimum sensitivity by using correlation receivers; each pair of receivers in the array measures one Fourier component of the brightness distribution over the part of the sky contained within the envelope pattern of the array. Just as in a single-element telescope, each interferometer element produces an amplified IF signal which is given an appropriate time delay and combined in a correlator with the IF signal from another element. A correlator consists of a multiplier and an integrator and the output from the integrator is the integrated product of two voltages—which approximates a cross-correlation function. The Fourier transform of the source sky brightness distribution is systematically mapped by sampling all pairs of receivers in the array or by changing the interferometer spacing. The sky brightness distribution is calculated later by using the Fourier transform of the cross-correlation function produced by each pair of elements, which gives the cross power spectrum of the signal for each map point. For spectral-line measurements, the correlator output can be viewed as a number of small adjoining continuum bandwidths (usually treated as channels in frequency space) which give a display of the amplitude of a spectral line versus its frequency. The plot of amplitude versus source position may be published as a spectral-line map, as shown in Fig. 2. An example of a spectrometer for interferometry which combines the advantages of both filter and correlation spectrometers is the hybrid analog–digital spectrometer, which has been discussed by W. L. Urry, D. D. Thornton, and J. A. Hudson. This important technique allows simultaneous observations of multiple spectral lines in windows defined by analog filters with the spectral decomposition in each window provided by correlation.

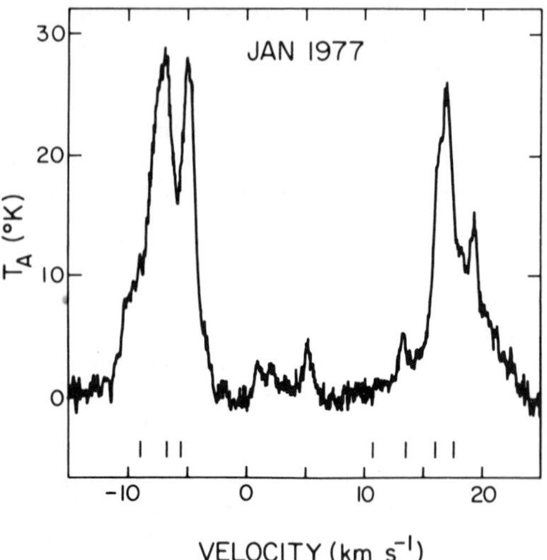

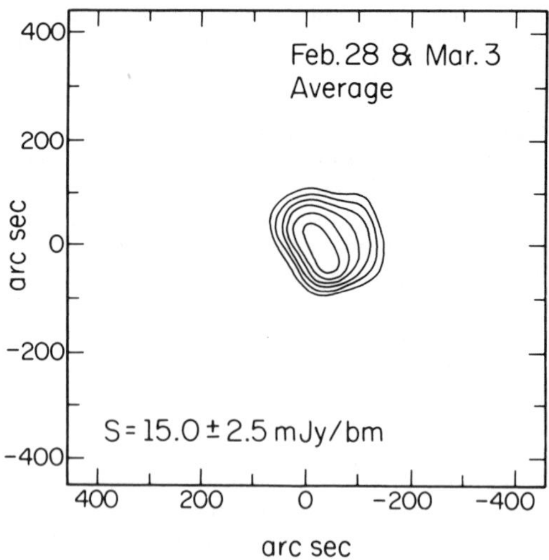

Figure 2. Example of spectra. Left-hand side: An example of a spectrum taken with an autocorrelator at the NEROC Haystack Observatory. This is the emission spectrum of the $V = 1$, $J = 1\text{–}0$ SiO maser emission from the source Orion A. Ordinate: antenna temperature T_A; abscissa: radial velocity with respect to the local standard of rest calculated for a rest frequency of 43,122.027 MHz. The frequency resolution is 16.28 kHz (0.1132 km s^{-1}). [*Reproduced with permission from L. E. Snyder, D. F. Dickinson, L. W. Brown, and D. Buhl*, Ap. J. ***224*** *512 (1978).*] Right-hand side: An example of a contour map made with the NRAO Very Large Array. This is a spectral line map which shows the average intensity of the maser emission from the $^2\Pi_{3/2}$ $J = 3/2$, $F = 2\text{–}2$ transition of OH at 1667.3590 MHz observed in Comet Wilson (1986l) on February 28, 1987, and March 3, 1987. The contours are plotted in steps of 0.5σ, beginning with 3σ. The peak flux and σ are given at the bottom of the panel. The coordinates are relative to the predicted position of the nucleus. [*Reproduced with permission from P. Palmer, I. de Pater, and L. E. Snyder*, Astron. J. ***97*** *1791 (1989).*]

SOME CONCLUDING THOUGHTS

It is clear that the past progress in the science of radio astronomy was coupled to progress in the technologies of certain applied fields such as radar and communications. Future progress in radio receivers and spectrometers probably will owe much to developments in solid-state physics and to faster, more powerful computers. For example, it is believed that the 4 K cooling requirement for the SIS junction may be improved to a more favorable temperature through research on new superconductors with high critical temperatures. More powerful computers will allow a more sophisticated correlation treatment of spectral data, particularly for arrays, which will be invaluable for advanced spectral data processing tasks such as retrieving weak spectral lines from beyond the current limit of approximately 10^{-21} W.

Additional Reading

Ball, J. A. (1975). Computations in radio-frequency spectroscopy. In *Methods in Computational Physics*, Vol. 14, B. Alder, S. Fernbach, and M. Rotenberg, eds. Academic Press, New York, p. 177.

Penzias, A. A. and Burris, C. A. (1973). Millimeter-wavelength radio-astronomy techniques. *Ann. Rev. Astron. Ap.* **11** 51.

Phillips, T. G. (1988). Submillimeter and far-infrared detectors. In *Interstellar Matter*, J. M. Moran and P. T. P. Ho, eds. Gordon and Breach, New York, p. 141.

Phillips, T. G. and Woody, D. P. (1982). Millimeter and submillimeter-wave receivers. *Ann. Rev. Astron. Ap.* **20** 285.

Price, R. M. (1976). Radiometers. In *Methods of Experimental Physics*, Vol. 12B, M. L. Meeks, ed. Academic Press, New York, p. 201.

Thompson, A. R., Moran, J. M., and Swenson, G. W., Jr. (1986). *Interferometry and Synthesis in Radio Astronomy*. Wiley, New York.

Tiuri, E. and Räisänen, A. V. (1986). Radio-telescope receivers. In *Radio Astronomy*, 2nd ed., J. D. Krause, ed. Cygnus-Quasar Books, Powell, OH, p. 7-0.

Urry, W. L., Thornton, D. D., and Hudson, J. A. (1985). The Hat Creek millimeter-wave hybrid spectrometer for interferometry. *Publ. Astron. Soc. Pacific* **97** 745.

Welch, W. J. (1988). Techniques and results of millimeter interferometry. In *Millimetre and Submillimetre Astronomy*, R. D. Wolstencroft and W. B. Burton, eds. Kluwer Academic Publishers, Dordrecht, p. 95.

See also **Radio Astronomy, Space Missions; Radio Telescopes and Radio Observatories; Radio Telescopes, Interferometers and Aperture Synthesis.**

Radio Astronomy, Space Missions

Robert G. Stone

Radio waves reach the ground-based observer over a wavelength range from approximately 1 mm to 20 m. Limits are imposed at short wavelengths by atmospheric water vapor and oxygen absorption and at long wavelengths by terrestrial ionosphere shielding. With the launching of the first artificial satellite in 1957, astronomers began to plan for the expansion of this window. However, there were (and still remain) severe technological problems for space-borne observations, ranging from inadequate system sensitivity and antenna pointing accuracy at submillimeter wavelengths to difficulties of deployment and stabilization of physically large structures required for even modest resolution at long wavelengths. In fact, the *first* submillimeter astronomical mission has only recently been approved for *development*. Therefore, we will focus solely on long-wavelength space-borne observations. Even though directive arrays were not developed for space flight, exploratory

Table 1. Missions of Particular Historical or Scientific Significance for Radio Astronomy

Mission (Year)	*Orbit*	*Antenna*	*Frequency*	*Key Historical or Scientific Result(s)*
Eta 1 (1960)	640 × 1040 km	Loop antennas	3.8 MHz	First satellite observations at long wavelengths
Canadian *Alouette* Top-side sounder (1964)	1000-km circular 80.5° inclination	23- and 46-m dipoles	0.5–12 MHz	First routine observations of solar radio bursts and the cosmic backgroun
Soviet *Elektron II* (1965)	? × 70,000 km	3.75-m dipole	0.725 and 1.525 MHz	Cosmic noise background Sporadic solar system noise
OGO-5 Science payload (1968)	292 × 147,000 km	9.12-m monopole	0.050–3.5 MHz	Detailed study of Type III solar burst properties
Radio Astronomy Explorer RAE-1 First mission devoted to long-wavelength observations (1968)	6000-km circular	Two oppositely directed 223-m V traveling wave antennas (~ 60° beamwidth) and a ~ 100-m dipole	0.25–10 MHz	Cosmic noise maps Study of solar Type III and Type III storms Detection of interplanetary shocks from Type II solar radio bursts Study distribution of terrestrial noise sources
IMP-6 Science payload (1971)	354 × 206,000 km	92-m spinning dipole	0.03–4.9 MHz	Cosmic noise spectrum to 200 kHz Tracking solar radio bursts using spinning dipole direction finding Detect radio emission from Jupiter and Saturn
RAE-2 Long-wavelength observations (1973)	1100-km circular lunar orbit	Same as *RAE-1*	0.02–13 MHz	Cosmic background, and solar observations Terrestrial auroral kilometric radiation Showed lunar shielding from terrestrial noise
Helios I and *II* Solar science payload (1974 and 1976)	Solar orbit ~ 0.3 AU and ~ 1 AU	10-m spinning dipole	0.03–3 MHz	Association of Type III radio emission with Langmuir waves and solar electrons Two-spacecraft triangulation of solar burst positions
ISEE-3 Science payload (1978)	Lagrange point sunward of Earth at ~ 250 $R_{\oplus}$	Dipole in spin plane and one along spin axis allowing tracking in azimuth and elevation	0.03–2 MHz	Three-dimensional tracking of solar radio bursts Detailed solar Type III storm properties Detect bursts behind the limb Type III interactions with shock waves (Type II)

low-frequency observations could be carried out with physically short (~10 m) dipole antennas deployed from spacecraft. The dipole, having minimal resolution, observes the spatially averaged cosmic noise background. Dipoles are also routinely used to observe dynamic spectra of solar system radio sources. The *Radio Astronomy Explorer* (*RAE*), providing modest resolution, was a logical second-generation mission. Future progress must await higher-resolution arrays either in space or preferably on the far side of the Moon. Table 1 summarizes key space missions.

COSMIC RADIO WAVES

Discrete Radio Sources

A major unachieved goal has been the observation of spectra of a large number (~100) of galactic and extragalactic radio sources at frequencies around 1 MHz. These observations, requiring a resolving power of the order of 1 square degree, would be of major astrophysical significance by revealing the effects of synchrotron self-absorption and plasma processes in the source regions themselves.

Cosmic Noise Background

Detailed mapping of the cosmic noise background at frequencies in the range 0.5–10 MHz with a resolution of the order of 1 square degree would add significantly to our knowledge of galactic structure and in particular to the distribution of cosmic rays and free electrons (H II regions). Thus far, however, observations have been limited to those obtained with short dipoles and with the modest directivity provided by the *RAE* satellite as will be discussed.

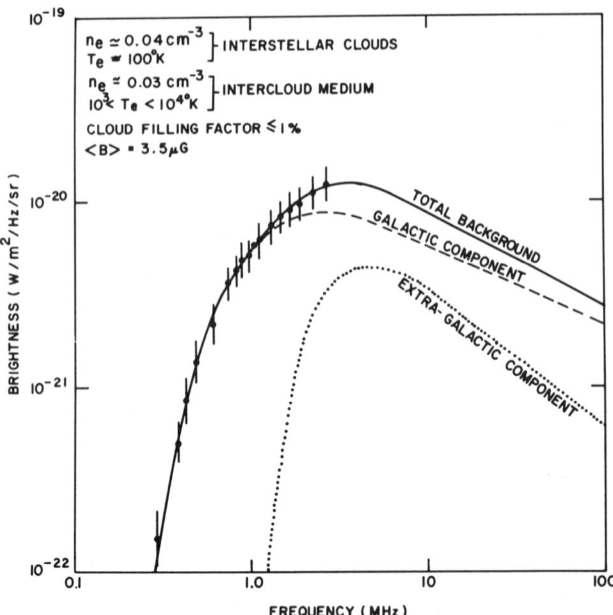

Figure 1. The integrated cosmic radio noise spectrum, observed below 3 MHz with space-borne observations and by Earth-based observations above 10 MHz. A combination of free–free absorption and Razin-effect suppression of the radiation accounts for the decrease of intensity at low frequencies for an interstellar medium whose properties are summarized in the figure.

Integrated Cosmic Noise Background. The first observations of the cosmic noise background from above the ionosphere were obtained in 1960 by Canadian investigators. A number of exploratory sounding rocket investigations followed in 1964 and 1965. Thereafter, observations of the cosmic background and solar bursts were obtained from a number of satellite experiments. Many of these are listed in Table 1. Fred T. Haddock's Space Radio Project at the University of Michigan should be credited with developing many of the new techniques used in this early period. Figure 1 shows the cosmic noise spectrum observed at low frequencies. Also illustrated is a fit to a simple model composed of a galactic disk of thickness L and an isotropic extragalactic component. Above 10 MHz, the intensity I_ν, due to synchrotron emission, varies with frequency ν as $I_\nu \sim \nu^{-0.6}$. The observed intensity is given by

$$I_\nu = I_z e^{-\tau_0} + \langle j_\nu L \rangle (1 - e^{-\tau_0})/\tau_0,$$

in which I_z is the isotropic component, τ_0 is the optical depth, and j_ν is the volume emissivity for synchrotron emission. The intensity decrease below 5 MHz is due in large part to free–free absorption in ionized hydrogen. The optical depth, which varies inversely with frequency squared, is a function of electron density and temperature and path length in the medium. These parameters have been adjusted to fit the observations and are listed in Fig. 1. Below 1 MHz, the intensity decrease results from free–free absorption and from suppression of radiation due to the presence of plasma in the source region (Razin effect). This leads to a sharp cutoff for frequencies of order $20N/B_0$, where B_0 is the magnetic field intensity in gauss and N is the electron density (number of electrons per cubic centimeter).

A Low-Resolution Cosmic Noise Background Survey. The *Radio Astronomy Explorer* (*RAE-1*), launched in 1968, was the first successful attempt to obtain more spatial resolution. Two oppositely directed, 229-m, traveling wave V antennas were deployed to form a large gravity-gradient-stabilized X configuration in a 6000-km circular Earth orbit. Observations were obtained in the frequency range between 1.31 and 9.18 MHz both in the direction of the celestial sphere (upper V) and simultaneously in the direction of the Earth (lower V). Even with antennas of this physical size, the resolution of the radio maps is low (~ 60°). The *RAE-1* observations demonstrated that because of the intensity and duration of terrestrial emissions, observations of cosmic sources from Earth orbit would be extremely difficult at best. The *RAE-2* was placed into an 1100-km circular *lunar* orbit in 1972. It provided data on the cosmic noise distribution with far less interference from terrestrial sources and also demonstrated the much quieter environment offered by the lunar farside.

DYNAMIC RADIO PROCESSES IN THE SOLAR SYSTEM

Solar Radio Bursts

Solar electron beams and shock waves produce intense radio emission which provides significant information on the ejecta itself and on the electron density distribution and large-scale magnetic field configuration in the solar wind from about 10–215 $R_\odot$ from the Sun. These observations bridge the gap between ground-based radio observations and in-situ measurements of solar ejecta at 1 AU. The spinning dipole direction-finding technique developed by Joseph Fainberg has made it possible to track solar radio bursts between about 10 $R_\odot$ and 1 AU from the Sun. Donald A. Gurnett and Robert P. Lin, using data from the FRG/U.S. *Helios* space probe, demonstrated that Type III bursts were, as theory predicted, generated by Langmuir waves excited by solar electron streams. A French/U.S. experiment aboard *ISEE-3*, launched in 1978, was instrumented to determine both source azimuth and elevation. Using ISEE data, Jean-Louis Bougeret conducted comprehensive studies of the properties of storms of Type III bursts in which he made use of these long-duration phenomena to determine electron exciter trajectories, electron densities along the storm trajectories, and the solar wind acceleration as a function of radial distance.

Planetary Nonthermal Radio Emissions

Orbiters detected radio emission from Earth, Jupiter, and Saturn. However, it has required the close flyby of the Voyager spacecraft to investigate in detail the radio properties of the planets. Nonthermal radio waves are observed from Earth, Jupiter, Saturn, Uranus, and Neptune. This radio emission is generated at frequencies near the electron gyrofrequency which is proportional to the magnetic field intensity. Thus remote radio observations can be related to the magnitude and orientation of the planet's magnetic field as well as to the magnetospheric region where strong electron beams interact with the ambient plasma medium to generate emission.

FUTURE AND PLANNED MISSIONS

Higher-Resolution Telescopes

A directivity of 1° at 1 MHz requires deploying and stabilizing an antenna of the order of 20 km in extent! Additionally, intense terrestrial noise from human activities, atmospherics, and natural emission (auroral kilometric radiation) can reach a radio telescope in Earth orbit. Successful surveys of galactic and extragalactic sources may require placing a radio telescope on the far side of the Moon, shielded from terrestrial radio sources.

Submillimeter-Wave Astronomy Satellite

SWAS, with a 55-cm telescope, will survey submillimeter emission from cold extended galactic molecular clouds in H_2O, O_2, and ^{13}CO. The mission will obtain a complete survey of water and O_2 in our galaxy which can only be performed outside of the Earth's atmosphere.

Additional Reading

Alexander, J. K. and Novaco, J. C. (1974). A survey of the galactic background radiation at 3.93 and 6.55 MHz. *Astron. J.* **79** 777.

Bougeret, J. L., Fainberg, J., and Stone, R. G. (1983). Determining the solar wind speed above active regions using remote radio-wave observations. *Science* **222** 506.

Fainberg, J. (1974). Solar radio bursts at low frequencies. *Coronal Disturbances*, G. Newkirk, ed. Reidel, Dordrecht, p. 183.

Stone, G. (1973). Radio physics of the outer solar system. *Space Sci. Rev.* **14** 534.

Weber, R. R., Alexander, J. K., and Stone, R. G. (1971). The radio astronomy explorer satellite, a low frequency observatory. *Radio Sci.* **6** (No. 12) 1085.

See also **Interplanetary Medium, Shock Waves and Traveling Magnetic Phenomena; Planetary Radio Emissions; Sun, Radio Emissions.**

Radio Sources, Cosmology

Rogier A. Windhorst

When the radio window first opened up for astronomical observations, it soon became clear that studies of the cosmos at radio wavelengths would have far-reaching consequences for our models of the universe. In the 1930s, Karl G. Jansky and Grote Reber discovered a celestial signal at long radio wavelengths. This radio hiss has a variable component that appeared to go around the Earth in one solar day and is caused by the Sun, as well as a steady component with a period of one *sidereal* day which was soon identified with the center and plane of our own galaxy.

The postwar refinement of radio antennae and interferometers in the 1950s provided ever-increasing resolution and sensitivity to map this steady signal into two main components. One turned out to be a rather smooth continuum associated with the galactic interstellar medium and is caused by synchrotron emission from cosmic ray electrons circling in interstellar magnetic fields. Superimposed on this is 21-cm-line radiation from neutral hydrogen (or H I) gas in interstellar clouds. The other component consists of discrete or individual radio sources, which remained largely unresolved at the then-available resolution. A subset of these are concentrated toward the galactic plane and are primarily associated with stars and some H II regions. The majority of discrete sources, however, are *uniformly distributed* across the entire sky. These sources were also referred to as "radio stars" until the late 1950s, when Sir Martin Ryle discovered that their distribution in space was far from uniform. Their number appeared to increase much more rapidly toward weaker radio levels than could be explained by any nonevolving population in any reasonable world model (cosmology). This result ruled out the Euclidean or steady-state models and implied that the number of radio sources per unit volume must increase strongly toward larger distances, even in any one of the relativistic world models.

In the early 1960s, the increased positional accuracy of radio interferometers led to more reliable optical identifications of radio sources at high galactic latitudes. Rudolph L. Minkowski identified some of the *strongest* radio sources in the sky with *optically faint* galaxies at redshifts of up to 0.5. This was a great mystery, because *if* these "radio galaxies" had been at redshifts well beyond 1, they would have been completely invisible to even the largest *optical* telescopes, while they still would have been among the *brightest* radio sources in the sky! Not long thereafter, Allan R. Sandage identified some bright radio sources with mysterious stellar objects, whose spectra did not look like normal stars. These were referred to as quasistellar radio sources, or briefly "quasars." Maarten Schmidt identified the unusually strong emission lines in the spectra of several quasars with highly redshifted hydrogen lines, reaching redshifts of up to $z = 2.0$.

It soon became clear that quasars might indeed be the very high redshift counterparts of the more nearby radio galaxies, and that these two classes of radio sources *together* might cause the excessive increase of sources toward weaker radio levels. This process is commonly referred to as "the cosmological evolution of radio sources" and implies that the radio source population must have had larger radio luminosities or space densities in the past, or a combination thereof. In addition to the Hubble expansion, the microwave background radiation and the observed helium abundance, the discovery of the cosmological evolution of radio sources (and the mild evolution in the optical spectra of galaxies discovered in the 1980s) constitute our best evidence that the universe has not always been the same, but once originated through a hot big bang.

We now know that radio galaxies and quasars evolve in most of their radio properties on time scales much longer than the human, but smaller than the Hubble time. Hence we can only observe the evolution of an *entire population* with cosmic epoch, not that of individual objects. When faint objects are selected from deep radio (or optical) images, they generally occupy a handful of independent pixels, so that the studied properties are usually limited to: (1) their brightness, (2) their density in space, (3) their morphology or characteristic size, and/or (4) their colors or spectra. Each of these parameters may appear to evolve with cosmic epoch, either because of poorly understood systematic observational errors and selection effects, or because of good physical reasons, or both. In the following discussion, we will address how the radio source population evolves with cosmic time in each of these individual properties, and study its impact on cosmology. It must be emphasized that *any* evolution of these properties dominates the minute observational differences expected from world geometry [H_0, q_0, Λ (Hubble constant, deceleration parameter, cosmological constant), etc.]. Hence the impact of radio astronomy on cosmology is indirect: It follows only after a fairly complete picture has been formed of the evolutionary history of the radio source population.

RADIO SOURCE COUNTS AND COSMOLOGICAL EVOLUTION

The intensity of a radio source received by a telescope on Earth is called "flux density," or briefly "flux." It is measured in units of 1 Jy (1 jansky), equivalent to 10^{-26} W Hz^{-1} m^{-2} (so 1 mJy = 10^{-3} Jy; 1 μJy = 10^{-6} Jy). One could sum up the integrated flux in all components of an extragalactic radio source (i.e., the radio core and the two opposing radio jets and lobes) and do this for all radio sources in the sky. The resulting number of sources as a function of flux is called the "radio source count," which can be directly compared with the predictions for a given world model and the radio luminosity function (or RLF, which describes their space density ρ as a function of radio power P).

To obtain a reliable source count, one should first estimate the fraction of weak sources with angular sizes much larger than the instrumental beam size, because such sources would be resolved out and missed in a complete sample. In the next section, we describe the radio source angular size distribution, from which one can determine a correction for the fraction of large sources missed. For this reason, radio surveys have been made during the last three decades with instrumental beam sizes similar to the *expected* angular size at every flux level. In the 1960s, primarily single dish telescopes were used to survey areas of steradian size, typically down to levels of 100 mJy. Interferometers were used in the 1970s to survey areas of many square degrees typically down to a few mJy, and in the 1980s to make ultradeep surveys currently reaching noise values as weak as a few μJy. Important contributions were made by the UK Cambridge One Mile Telescope, the Dutch Westerbork Synthesis Radio Telescope (WSRT), and the U.S. Very Large Array (VLA).

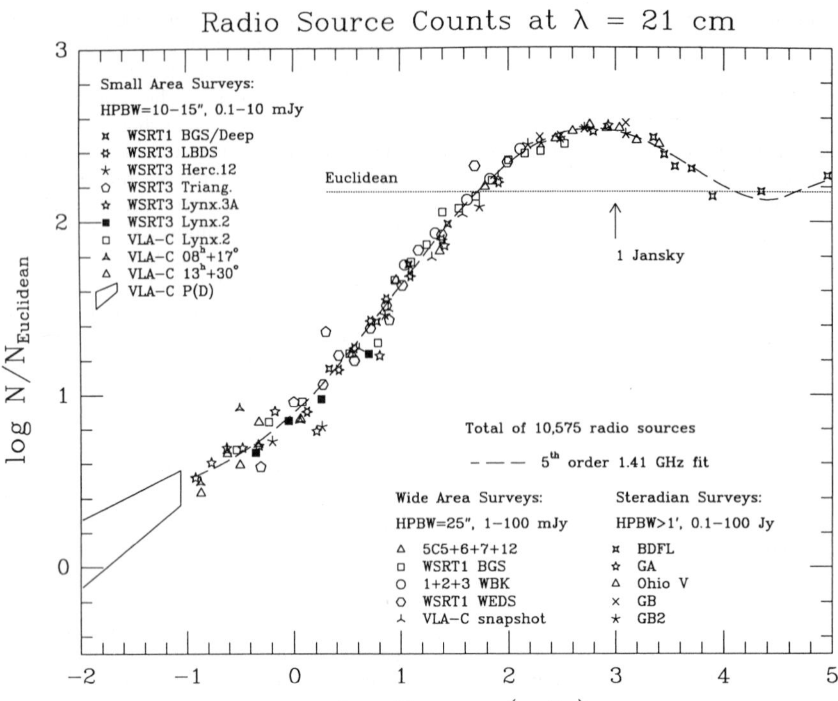

Figure 1. The radio source counts at 1.41 GHz. The counts are differential, so that the data points are all independent and normalized to the expected prediction in a nonevolving Euclidean world model ($N_0 = S_{1.4}^{-2.5}$, drawn as the dotted horizontal line). Plotted are 108 independent data bins with 10,575 radio sources from all 24 available surveys (as of 1989). The skewed box is the likely range of the weakest counts from a noise fluctuation analysis. The maximum excess of radio sources with respect to the Euclidean prediction is caused by the cosmological evolution of giant elliptical radio galaxies and quasars; the upturn in the weakest source counts is caused by the other population of spiral and starburst radio galaxies.

Figure 1 shows the differential source counts at 1.41 GHz (or 21 cm), normalized to those expected in a homogeneous, nonevolving Euclidean universe. In Euclidean space, a population of radio sources of *constant* radio power P (in W Hz^{-1}) and *constant* space density ρ (numbers Gpc^{-3}) would result in observed numbers N that increase with distance proportional to R^3, whereas their observed flux S decreases with distance proportional to R^{-2}. So when we count the integral number of sources N brighter than a certain flux S, we expect $N(>S) \propto S^{-3/2}$ in Euclidean space, or $dN/dS \propto S^{-5/2}$ in differential form. Hence the differential counts in a Euclidean universe would be a straight horizontal line in Fig. 1. The observed counts are clearly not Euclidean.

Let us consider the source counts from bright to weak radio fluxes (from right to left in Fig. 1). Only at the brightest levels (between 100 and 10 Jy) are the counts approximately Euclidean, because these sources are indeed radio galaxies at local redshifts where space curvature and evolutionary effects are unimportant. Between about 10 and 1 Jy, the source counts show an enormous excess over Euclidean, which discovery won Ryle the Nobel Prize. (The observed counts are even more in excess of the prediction in a nonevolving relativistic world model, not shown in Fig. 1.) This excess of sources is caused by the cosmological evolution of the radio source population. For some reason, galaxies and quasars produced much more powerful radio sources, or much more frequently were radio sources, in the past. In the preceding example, this means that their radio power P and space density ρ are *not constant*, but must have been much larger in the past, causing the excess of sources at those bright flux levels.

At flux levels between 1 Jy and a few mJy, the differential source counts decline steadily, diving well below the Euclidean prediction. Because the counts shown are differential, the integral number of sources still does increase toward weaker radio fluxes, but not as quickly as at brighter flux levels. Optical identification work has shown that radio sources down to fluxes of 10 mJy consist of *one* fairly homogeneous population (primarily giant elliptical galaxies and quasars). The fact that the integral counts from this dominant source population converge so strongly down to 10 mJy is our best evidence that space cannot be Euclidean. One expects exactly this decline in a relativistic world model for a population with a given (epoch-dependent) RLF. The counts converge, because at weaker radio fluxes one has seen increasingly more of the whole source population, and the available volume per unit redshift is much less than in the Euclidean case, certainly for $\Omega \sim 1$. The peak in the differential source counts around 1 Jy must then be caused by those sources with statistically the largest redshifts, which was only confirmed by observations in the late 1980s.

At fluxes below a few milli-jansky, the 1.4-GHz counts slowly turn up again, although they never quite attain the (horizontal) Euclidean slope. This gradual change in slope was measured in the mid-1980s by independent surveys with different radio telescopes at different frequencies of the same and different areas of sky. Hence the upturn in the weak source counts is *not* due to instrumental effects, but is a property of the universe. It starts below 10 mJy and is visible in all survey fields that contribute significantly below a few milli-Jansky. This upturn is significant, because it cannot easily be attributed to the canonical giant elliptical radio galaxies or quasars, which dominate the counts only above 10 mJy. Neither can it be explained by normal spiral or Seyfert galaxies, unless these objects have undergone significant cosmological evolution in the recent past. Various models have been developed to explain the steeper differential slope of the weak source counts, such as local, nonevolving, low-luminosity radio galaxies, or evolving Seyfert, spiral, or starburst galaxies. The most likely explanation is that spiral or starburst galaxies underwent cosmological evolution similar to that of giant ellipticals and quasars, which would bring the number of evolving populations to at least two. If these two evolving populations do not evolve into each other, one can extrapolate their epoch-dependent RLFs beyond the redshift to which it was measured, and reach 100% of the source counts for fairly low redshift values. At a maximum redshift of 2 ± 0.5, about 95% of the sources would have been seen. Some researchers have identified this redshift as the formation epoch of luminous galaxies, where they formed the bulk of their stars and first turned on as radio sources.

The cosmological evolution of the radio source population has been rather well established over the last two decades by studying the evolution of their RLFs, but this process is still poorly understood physically. It is now generally agreed upon that the most powerful radio sources, giant ellipticals and quasars, have undergone strong cosmological evolution in their RLFs. They were at least 100 times more powerful and/or more numerous in the past. At radio powers below the characteristic break in the RLF (or $P < P^*$), radio galaxies have probably undergone cosmological evolution as well, but at a more modest level that still has to be established by measuring the redshift distribution of weak radio galaxies. According to the Schmidt model, the cosmological evolution of radio sources can be globally described as:

$$P(z) \quad \text{or} \quad \rho(z) \propto (1+z)^p, \tag{1}$$

where $p \sim 5$ for giant ellipticals and quasars, and possibly $p \sim 3$ for spiral or starburst galaxies. The redshift dependency of the radio luminosity P is referred to as "luminosity evolution," that of space density ρ as "density evolution." Because the epoch-dependent radio luminosity function, $\rho(z, P)$ has not been measured yet for the full range in radio power and redshift, we cannot currently distinguish which one of these two physical evolution mechanisms causes the process of cosmological evolution. In other words, we do not know whether radio galaxies and quasars were on average more luminous in the past, or more numerous (per unit volume), or perhaps a combination of both. As for its physical cause, the cosmological evolution of radio galaxies may be a *direct consequence* of the process of galaxy formation itself. Galaxies evolve, often through episodic starbursts that are triggered by dynamic disturbances (such as mergers), which also indirectly trigger the central engine to produce a radio source. Therefore, the physical process that causes the cosmological evolution of radio sources is probably intricately related to the process of galaxy formation and evolution itself.

RADIO SOURCE ANGULAR SIZES AND LINEAR SIZE EVOLUTION

With the advent of large synthesis radio telescopes, especially the VLA, the measurement of radio source structure and angular sizes has become feasible at resolutions often even inaccessible to optical telescopes. Figure 2 gives the typical (or median) angular of radio sources versus their 1.4-GHz flux for several radio source samples. In all samples, the median angular size Θ_{med} decreases monotonically from approximately 200″ around 100 Jy to 2″ below 1 mJy. This is not likely due to selection effects, because each survey was done with an instrumental beam size similar to Θ_{med}, so that not many large, low surface brightness sources have been missed.

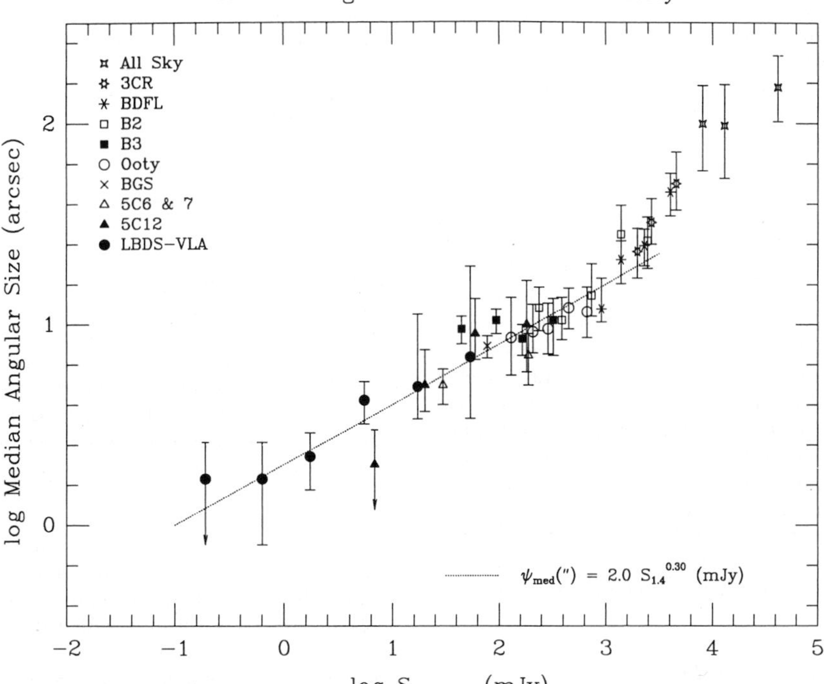

Figure 2. The median angular size–flux density relation of radio sources at 1.4 GHz. The median angular size declines rapidly from 100–1 Jy because the median redshift of sources in this flux range increases strongly (from $z_{med} \sim 0.05$–1.5). Below 1 Jy, the median redshift actually decreases to 0.7 at weaker fluxes, but the median angular size continues to decline because radio sources in giant elliptical galaxies and quasars undergo reverse linear size evolution (as indicated by the dotted line), and because at the weakest fluxes the radio source population is dominated by compact sources in starburst and spiral galaxies.

The angular size, Θ, of a rigid rod decreases with distance roughly as $\Theta \propto z^{-1}$ (with noticeable departures at $z \gtrsim 0.5$ due to world geometry). So let us consider what causes Θ_{med} of radio sources to decline toward lower radio fluxes (moving from right to left in Fig. 2, just like the counts in Fig. 1). In the flux range between 100 and 1 Jy, angular sizes decrease steadily with flux, because the median redshift increases strongly from $z_{med} \lesssim 0.05$ around 100 Jy to $z \sim 1.2$–1.5 around 1 Jy. This is a direct consequence of the cosmological evolution of giant ellipticals and quasars, which are the dominant radio source population at bright flux levels. Below 1 Jy, the median redshift starts to decrease again toward weaker fluxes, until it slowly reaches $z_{med} \sim 0.7$ around 1 mJy. This decline in z_{med} occurs because at weaker radio fluxes we have actually seen *all* of the powerful radio sources in the whole universe, and the other sources that show up are much less powerful and somewhat less distant. As a consequence, one expects Θ_{med} to slowly *increase* again below 1 Jy.

However, Fig. 2 shows clearly that the median angular sizes continue to decline toward weaker fluxes (as indicated by the dotted line). This occurs because the intrinsic or linear size D (in kiloparsecs) between the two outermost hot spots in the opposite radio beams of giant elliptical radio galaxies and quasars evolves strongly with redshift and is also weakly dependent on the intrinsic radio power ($P_{1.4}$). Combination of a large number of surveys shows that:

$$D(z, P_{1.4}) \propto P_{1.4}^{0.3}(1+z)^{-2.7} \quad \text{for } H_0 = 50,\ q_0 = 0. \qquad (2)$$

That is, radio sources of a given power are intrinsically much smaller at higher redshifts, and at any given redshift the two jets will expand further from the parent galaxy for the more powerful radio sources. The strong redshift dependence of radio source linear sizes is best explained by models in which the pressure balance between the hot intergalactic medium and the cooler galaxy halo determines the maximum distance out to which the radio source can expand. This pressure balance defines a transition surface whose distance from the galaxy is redshift dependent, resulting in a predicted redshift dependence of the linear size approximately proportional to $(1+z)^{-3}$, in good agreement with the observations. Hence radio lobes cannot escape as far from the parent galaxy at high redshift.

This (reverse) linear size evolution outweighs any geometrical effects ($\Theta \propto z^{-1}$) by a large margin, and the net effect is that the observed Θ_{med} *continues to decline* toward the weakest observable flux levels, even though z_{med} starts to decrease again below 1 Jy. We have to be very grateful that nature is like this, because otherwise we could never have made such deep radio surveys. The weakest sources (of which there are many more than the brighter ones!) would have blended into each other if they did not have much smaller angular sizes than the brighter ones, which would make deep surveys impossible. When this article went to press in early 1991, the deepest available VLA survey had in fact shown that the median angular size has leveled off to $\Theta \simeq 1''.5$ at the 30 μJy flux level. It appears that ultradeep radio surveys are now picking up spiral disks at large redshifts (which are indeed a few arcseconds across), and will from here on out always be limited by this "natural confusion."

EVOLUTION OF RADIO SPECTRA WITH COSMIC TIME?

The radio emission in the jets, lobes, and hot spots of extended extragalactic radio sources is caused by synchrotron emission of relativistic electrons circling in interstellar/intergalactic magnetic fields. This is a nonthermal radiation process with a power law spectral index of order $\alpha \sim 0.75$ (α follows from $S_\nu \propto \nu^{-\alpha}$). Most radio spectra measured between three or more frequency points are straight (with $\alpha \sim 0.75$) at frequencies $\nu \sim 1$ GHz, although many sources show flatter spectra ($\alpha < 0.5$ or even < 0) at lower frequencies due to synchrotron self-absorption, and steeper spectra ($\alpha > 1$) at frequencies much greater than 1 GHz because the most energetic synchrotron electrons (which radiate mostly at the highest frequencies) lose their energy more quickly. This results in convex (downward-turning) spectra. A typical synchrotron source ages on time scales of 10^7–10^8 yr, yielding spectra that steepen with time.

In addition to these physical considerations, the whole radio source population might show spectra whose slope changes on average with redshift due to cosmological effects. At frequencies around 1 GHz, one samples primarily the steep spectrum class at any flux level. The *observed* median spectral index α_{med} measured in many surveys (between ~ 0.5 and 5 GHz) shows a behavior with flux density much like Fig. 1. At the brightest flux levels, we find $\alpha_{med} \sim 0.8$, steepening to $\alpha_{med} \sim 0.9$ around 1 Jy where the differential source counts reach their maximum excess (with respect to the Euclidean model), whereas at lower fluxes α_{med} flattens again to approximately 0.75 or less. The most straightforward explanation is that the redshift distribution of radio sources causes the observed α_{med} to steepen with redshift, just like it causes the excessive number of radio sources around 1 Jy. Remember that the median redshift follows a similar trend as a function of radio flux, reaching a maximum of $z_{med} \sim 1.2$–1.5 around 1 Jy (see the previous section).

Is this steepening of α_{med} with redshift necessarily caused by evolution in the spectral index, or does it just indicate that one samples different parts of the source spectra at different redshifts, because many spectra are not a straight power law, but convex? This would result in a nonnegligible radio K correction for sources at the highest redshifts, much like the well-known K correction for optical galaxies. Especially for surveys around 1 Jy, we expect a substantial radio K correction, because they have the largest median redshift so that spectral curvature at high frequencies will affect the observed value of α the most. A combination of many surveys (with radio spectra measured between frequencies that correspond as much as possible to the same rest frame frequencies) suggests that the median spectral index is *apparently* a strong function of redshift and only weakly dependent on radio power, roughly as following:

$$a_{med}(z, P_{1.4}) \sim 0.66 + 0.5\log(1+z) + 0.03\log(P_{1.4}/P^*) \quad (3)$$

(with $P_{1.4}$ in W Hz^{-1} for $H_0 = 50$, $q_0 = 0$ and $\log P^* = 25.0$ W Hz^{-1}). Because the radio spectrum of a typical extended synchrotron source maximally steepens by approximately 0.5 (from 0.75–1.25) between frequencies of approximately 1 and several GHz, the corresponding spectral curvature can account for most of the second term in Eq. (3) through the radio K correction. Hence radio sources have not necessarily undergone any evolution in their radio spectra at all. However, the third or $\log P$ term itself may have a hidden dependency on redshift if the cosmological evolution of radio sources is caused by pure luminosity evolution, so that $P(z) = P(0)(1+z)^p$, with $p \sim 4$ as in Eq. (1). In that case, the second term of Eq. (3) has an additional term of $+0.12\log(1+z)$, so that the expected K correction does not explain all of the apparent spectral steepening with redshift. The apparent dependence of α_{med} on redshift is then in a sense an artifact of the cosmological evolution of the radio source population, and the true physics lies in the relations between α_{med} and $\log P_{1.4}$, plus $\log P_{1.4}$ and redshift. The former may be explained by stronger synchrotron losses for more powerful sources (because they radiate their energy more quickly), the latter by an epoch-dependent fueling that was somehow more efficient in the past. Equation (3) also explains why astronomers in the late 1980s were so successful in finding the highest-redshift galaxies among strong radio sources with the *steepest* spectra.

Additional Reading

Condon, J. J. (1988). Radio sources and cosmology. In *Galactic and Extragalactic Radio Astronomy*, 2nd ed., G. L. Verschuur and K. I. Kellermann, eds. Springer-Verlag, New York, p. 641.

Longair, M. S. (1978). Radio astronomy and cosmology. In *Observational Cosmology*, Eighth Advanced Course of the Swiss Society of Astronomy and Astrophysics, A. Maeder, L. Martinet, and G. A. Tammann, eds. Geneva Observatory, Geneva, p. 127.

Minkowski, R. (1975). The identification of radio sources. In *Stars and Stellar Systems: Galaxies and the Universe*, A. Sandage, M. Sandage, and J. Kristian, eds. University of Chicago Press, Chicago, p. 177.

Ryle, M. (1957). The spatial distribution of radio stars. In *Radio Astronomy*, IAU Symp. 4, H. C. van der Hulst, ed. Cambridge University Press, Cambridge, p. 221.

Scheuer, P. A. G. (1975). Radio astronomy and cosmology. In *Stars and Stellar Systems: Galaxies and the Universe*, A. Sandage, M. Sandage, and J. Kristian, eds. University of Chicago Press, Chicago, p. 725.

Schmidt, M. (1972). Statistical studies of the evolution of extragalactic radio sources. III. Interpretation of source counts and discussion. *Ap. J.* **176** 303.

Windhorst, R. A., Mathis, D. F., and Neuschaefer, L. W. (1989). The evolution of weak radio galaxies at radio and optical wavelengths. In *The Evolution of the Universe of Galaxies, Edwin Hubble Centennial Symposium*, R. G. Kron, ed. ASP Conference Series. Book Crafters, Inc., Provo, UT, p. 389.

See also **Cosmology, Big Bang Theory; Cosmology, Observational Tests; Galaxies, High Redshift.**

Radio Sources, Emission Mechanisms

Stephen L. O'Dell

Until the mid-twentieth century, astronomical observations had been confined to visible light. The 1888 discovery by Heinrich Hertz of radio transmission and reception, followed by the development of radio technology for communications, provided the capability to extend our vision of the universe to the radio portion of the electromagnetic spectrum. However, the 1932 serendipitous detection of cosmic radio emission by Karl G. Jansky did not immediately initiate the science of radio astronomy. Although Jansky determined that this radio emission was concentrated in the direction of the center of the Galaxy, he was unable to investigate further the phenomenon he had discovered. During the decade following Jansky's discovery, Grote Reber was essentially the world's sole radio astronomer. Incredibly, Reber personally financed and built a large (10-m diameter) paraboloidal reflecting telescope, that is, a radio "dish," and radio receivers to search systematically for cosmic radio emission! Reber's mapping of the celestial distribution of radio emission not only confirmed Jansky's discovery of diffuse emission from the Milky Way peaking near the galactic center (in the constellation Sagittarius), but also indicated two additional concentrations of radio emission (one in the constellation Cassiopeia and the other in Cygnus).

In 1940, Louis G. Henyey and Philip C. Keenan concluded that known emission mechanisms could not account for the intensity and spectrum (frequency distribution) of the cosmic radio emission measured by Jansky and Reber. Until then, astrophysicists had been able to interpret all observed cosmic radiation—which had been exclusively visible light from stars and emission nebulae (clouds of hot ionized gas)—in terms of thermal emission (radiation of the heat contained in matter in thermal equilibrium at a nonzero temperature). Cosmic sources do emit some thermal radiation at radio frequencies; however, in the powerful cosmic radio sources, the dominant mechanism of continuum emission (radiation whose intensity varies smoothly over a broad range of frequencies) is a nonthermal process. It would be another decade before theoretical physicists recognized the relevance and prevalence of the *synchrotron* process [radiation from electrons moving relativistically (nearly at the speed of light) in a magnetic field]. In the late 1940s, the observation of light from synchrotron machines (large devices for accelerating charged particles to relativistic speeds, while confining them to orbits in a magnetic field) renewed interest in this radiation process, which G. A. Schott had previously (in 1912) investigated theoretically. Because physicists had already detected cosmic rays (high-energy particles from extraterrestrial sources), Hannes Alfvén and N. Herlofson (1950) suggested that the synchrotron process produces the radio emission from strong discrete sources; Karl O. Kiepenheuer (1950) proposed that it produces the diffuse nonthermal radio emission from the Milky Way.

Meanwhile (beginning in 1944), Jan H. Oort emphasized the importance of radio spectral lines (radiation whose intensity peaks sharply at a well-defined frequency) in studying the structure and motion of interstellar gas, utilizing the Doppler effect (the shift in observed frequency due to the relative motion of a source and an observer). Soon thereafter, Hendrick C. Van de Hulst predicted the astrophysical relevance of the 21-cm line of atomic hydrogen—a prediction which Harold I. Ewen and Edward M. Purcell later confirmed in 1951.

In the late 1940s, following Reber's pioneering research and spurred by wartime advances in radar technology and by radio detections of nonthermal activity in the Sun and of thermal emission from the quiet Sun and from the Moon, radio astronomy began its development into a major subfield of astronomy and astrophysics. Within a few years, radio astronomers had detected dozens of cosmic sources of radio emission; however, no cosmic radio source—other than the Sun, Moon, and Milky Way—was identified until 1949.

Crucial to the growth of radio astronomy have been improvements in sensitivity and in angular resolution: Enhanced sensitivity, achieved using low-noise receivers and large telescopes, permits detection of a weak source and measurement of its strength; improved angular resolution, accomplished through interferometry or aperture synthesis (a technique that combines the signals from two or more telescopes, in order to mimic parts of a very large telescope), allows accurate determination of a source's position and its angular structure. Establishing the precise position of a radio source is critical to identifying it with a visible object.

COSMIC RADIO SOURCES

Identifications (beginning in 1949) of discrete cosmic radio sources showed that the radio sky is very different from the (optically) visible sky: The radio brightest sources are not the optically bright, nearby stars; instead, they are identified with optically faint objects, for example, active galaxies (galaxies exhibiting phenomena not associated with normal stellar processes) and supernova remnants (debris of exploded stars) in the Galaxy. In the decades following these early discoveries, radio astronomers have investigated radio continuum and spectral lines, with continually improving sensitivity and angular resolution. Some investigations have examined known or expected phenomena (e.g., radio extrapolations of optically visible thermal emission), whereas others have led to remarkable discoveries. Indeed, during the 1960s, radio astronomers contributed substantially to three of their generations' most significant astrophysical discoveries—the cosmic microwave background (relic radiation of the Big Bang), quasars (extremely active galactic nuclei), and pulsars (rapidly rotating, collapsed stars). This section briefly describes these and other major categories of cosmic radio sources.

Cosmic Microwave Background

The cosmic microwave background is diffuse, isotropic thermal emission, with a spectrum like that of a blackbody at a temperature of about 3 K. This background radiation seems to be remnant thermal emission from the hot, early universe—the Big Bang—when radiation and matter were in thermal equilibrium. In 1963, Arno A. Penzias and Robert W. Wilson detected the cosmic microwave (3°) background, while accounting for sources of noise in a very sensitive (low-noise) system, being developed for communications satellites.

Active Galactic Nuclei

An active galactic nucleus (AGN) is the central region of a galaxy exhibiting activity not seen in normal galaxies. The radio emission originates via the synchrotron process, indicating an abundance of relativistic electrons. Many AGNs are extremely powerful, nonthermal radio sources, exhibiting radio-emitting compact cores, often connected by long, narrow jets to gigantic radio lobes up to a million light-years from the core. In contrast to these radio-loud AGNs, other AGNs are radio quiet. Astronomers have grouped AGNs into several categories, based primarily on their optical properties.

A quasar is an optically point-like object with strong, broad emission lines in its optical spectrum. The redshift of these spectral lines indicates that quasars are very distant and very luminous. In 1963, Maarten Schmidt first identified a quasar, using a precise radio position determined by Cyril Hazard, M. B. Mackey, and A. J. Shimmins during the previous year (during lunar occultations of the radio source). Although the discovery of the quasar phenomenon relied upon radio positions, only about 10% of the quasars are radio loud. About 10% of the radio-loud quasars exhibit highly polarized optical emission, indicating optical synchrotron emission.

A blazar is similar to a highly polarized, radio-loud quasar, except that it lacks strong emission lines. For some blazars, the point-like optical emission is sufficiently weak, that an underlying elliptical galaxy is visible. Some astronomers view blazars as a link between radio-loud quasars and radio galaxies.

A radio galaxy is an elliptical galaxy with strong, nonthermal radio emission in the core, jets, and lobes. Their radio properties are similar to those of the radio-loud quasars; however, their optical properties are those of an elliptical galaxy with a point-like source in its nucleus. Many of the early (beginning in 1949) radio-source identifications were radio galaxies.

A Seyfert galaxy is (usually) a spiral galaxy with an optically point-like nucleus with strong, broad emission lines. Most astronomers consider the nuclei of many Seyfert galaxies to be low-luminosity, radio-quiet quasars.

Galactic Radio Sources

The Galaxy contains many types of radio sources, both discrete and diffuse. Because the Galaxy is a rather "normal" spiral galaxy, other normal galaxies exhibit similar radio properties. Of course, because of their proximity, galactic radio sources can be observed in more detail in the Galaxy than in other normal galaxies.

A pulsar is a highly magnetized, rapidly rotating neutron star [a collapsed star with a mass density close to that of atomic nuclei—about a hundred million million (10^{14}) times denser than the Earth]. Pulsars emit radio pulses with well-defined periods, ranging from milliseconds to seconds. In 1967, Jocelyn Bell and Antony Hewish discovered pulsars, while searching for rapidly scintillating ("twinkling") radio sources. Pulsar radio emission is decidedly nonthermal; however, it is not synchrotron radiation.

A supernova remnant (SNR) is a large cloud of expanding gas, comprised of the material from an exploded star. The core of the exploded star may survive as a neutron star or as a black hole in the center of the supernova remnant. The synchrotron process generates the radio emission either in a shell (if the relativistic electrons are shock-accelerated as the expanding remnant slams into the interstellar medium) or throughout the SNR (if the relativistic electrons are accelerated by a central object).

The interstellar medium of the Galaxy is a diffuse source of radio emission. The nonthermal radio continuum is synchrotron radiation from relativistic electrons (originating primarily in supernova remnants), trapped in the large-scale interstellar magnetic field. In addition, the 21-cm line of atomic hydrogen is an invaluable tool in mapping the structure of the Galaxy.

An H II region is a volume of hydrogen photoionized (into protons and electrons) by the ultraviolet radiation from a central, hot object. In the Galaxy, the young, massive stars photoionize the interstellar medium near them, thus producing an H II region. H II regions emit radio-through-ultraviolet thermal continuum and line radiation, produced as an electron passes close to a proton or when it recombines with a proton to form atomic hydrogen.

A planetary nebula is essentially an H II region. However, for a planetary nebula, a hot, central white dwarf (a dead star, about a million times denser than the Earth) photoionizes the star's outer, expanding envelope, which was shed as the star left the red giant phase of its evolution.

A molecular cloud is a huge volume of interstellar matter, in which gas and dust are sufficiently dense that atoms can combine to form molecules. Such conditions are conducive to star formation; thus molecular clouds are often associated with peripheral or encapsulated H II regions produced by bright, recently formed stars. In addition to the radio emission associated with the H II regions, the molecules emit radio spectral lines. Often the radio molecular lines are much more intense than expected for thermal equilibrium of the molecular states. A radio source of this intense molecular radiation is called a maser (microwave amplification by stimulated emission of radiation) in analogy to the similar phenomenon for (visible) light—namely, a laser (light amplification by stimulated emission of radiation).

Many other galactic objects produce observable radio emission. In scaled-up versions of the radio emissions from the Sun, some stars produce detectable radio emission from giant flares or from a corona (rarefied, very hot plasma surrounding a star) or wind. More exotic phenomena may occur in a binary system (a pair of stars orbiting their common center of mass), especially if one of the stars is compact (black hole, neutron star, or white dwarf). Matter transferred to one star from its companion may initiate explosions detectable at radio frequencies. Furthermore, the transferring matter may settle into an accretion disk, as the continually heated matter spirals toward the compact star. Such an accretion disk is often associated with a jet (a collimated flow) of material streaming away from the accretion disk. Astrophysicists believe that this disk–jet geometry also occurs in the active galactic nuclei—but on a much larger scale.

Solar System Radio Sources

The Sun and planets are intrinsically rather weak radio sources. Their radio emission would be undetectable from the next nearest star. Nevertheless, because of their proximity, radio astronomers have been able to study the Sun (particularly) and the planets in great detail.

The quiet Sun generates radio emission from thermal electrons in its lower corona. In addition, the Sun exhibits several types of radio bursts, often originating above active regions near the solar surface. Whereas many of these radio bursts emit via the synchrotron (or related) mechanism, other radio bursts emit via other nonthermal mechanisms—related to oscillations or instabilities in the plasma.

The planets (and other inert solar system bodies) not only reflect sunlight but are heated by the Sun. In turn, the planets reradiate this heat as thermal emission. Because planetary temperatures are a few hundred degrees kelvin, planetary blackbody spectra peak in the infrared and extend with decreasing intensity into the radio. Some planets (e.g., Jupiter and Earth) have strong magnetic fields that trap electrons and protons in a large circumplanetary volume, called the magnetosphere. Electrons in the radiation belts (magnetospheric regions containing a concentration of high-energy electrons and protons) radiate synchrotron (or related) radio emission. In addition to these now-well-understood phenomena, a few planets (most prominent among them, Jupiter) emit low-frequency nonthermal radio bursts, probably related to plasma oscillations and instabilities.

BASIC PRINCIPLES

The radio portion of the electromagnetic spectrum comprises wavelengths longer than about 1 mm (i.e., frequencies lower than about 300 GHz shorter than which normal electronic techniques for detection are impractical). Earth-based radio observations are possible for frequencies higher than about 10 MHz (below which ionospheric electrons reflect incident radio waves) and lower than about 300 GHz (above which atmospheric molecular absorption—most seriously, by water vapor—becomes quite severe). Because radio emission is electromagnetic radiation, the basic principles are those of electromagnetism.

Waves and Photons

For any (sinusoidal or harmonic) wave, the wavelength λ (in m), frequency ν (in Hz, equal to one cycle per second, honoring Heinrich R. Hertz, the discoverer of radio transmission and reception), and wave speed c (in m s^{-1}) are related through

$$c = \lambda\nu. \tag{1}$$

The wave speed of electromagnetic radiation (in a vacuum) is the speed of light (300,000 km s^{-1}). Electromagnetic radiation is a transverse wave, with oscillating electric E and magnetic B fields, orthogonal (perpendicular) to each other and to the direction of propagation. The direction of the electric field defines the polarization of the wave. Because the energy carried by a wave is proportional to the square of the wave's amplitude, the power transmitted by a simple electromagnetic wave goes as $|E|^2$ (or as $|B|^2$).

Radiation, as does matter, exhibits wave–particle duality; that is, it behaves both as a wave and as a particle. The quantum of electromagnetic radiation is the photon, a massless particle with an energy related to the frequency through Planck's equation:

$$E = h\nu, \tag{2}$$

where h is Planck's constant (6.6×10^{-34} J Hz^{-1}). The energy of a single radio photon is very small, generally much less than thermal energies kT, where k is Boltzmann's constant (1.4×10^{-23} J K^{-1}) and T is the temperature (in K). Consequently, radio astronomers (in contrast with x-ray astronomers) cannot detect individual photons.

Measures of Source Strength

Radio astronomers usually express the strength of a source, at a frequency ν, in terms of the spectral flux or flux density S_ν, which measures the received power (in W) per unit area (in m^2) per unit frequency. Because cosmic sources are so weak, a practical unit of flux density S_ν is the jansky (Jy), equal to 10^{-26} W m^{-2} Hz^{-1}, honoring the discoverer of extraterrestrial radio emission.

If the angular structure of a source is resolved, then the spectral intensity I_ν, that is, the flux density S_ν per unit solid angle Ω, of the source may be mapped. A very informative measure of the spectral intensity I_ν is the brightness temperature T_b:

$$2kT_b \equiv \lambda^2 I_\nu = \lambda^2\, dS_\nu / d\Omega. \tag{3}$$

[The factor of 2 (multiplying the kT_b) occurs because electromagnetic radiation has two independent polarizations.]

Emission and Absorption

Classically, when a single, electrically charged particle accelerates, the distortion in its electric field propagates through space as electromagnetic radiation. The polarization (the direction of the electric field of the radiation) is aligned with the direction of the acceleration; the radiated power (energy radiated per unit time) is proportional to the square of the electric field (and hence to the square of the acceleration). Thus, from Newton's third law of motion ($F = ma$), the lightest charged particle—the electron—is the most effective radiator for a given force. The next lightest stable particle—the proton—is 1836 times more massive: It is over three million (1836^2) times less effective as an emitter and thus radiates insignificantly.

Quantum mechanically, a transition from one quantum state to another, with a lower energy (and different electric or magnetic configuration), may result in the spontaneous emission of a photon, that is, radiation. (This corresponds to the classical emission of an electromagnetic wave.) The inverse process, absorption, may also occur, whereby an incident photon is absorbed, raising the energy of the quantum-mechanical system. An incident photon may also trigger the emission of a second photon (with the same energy and direction), thus lowering the energy of the quantum-mechanical system. This process, stimulated emission, differs from spontaneous emission, in that it depends upon the intensity of photons (with the appropriate energy), as well as on the density of emitters (systems in the higher-energy state). Absorption depends upon the intensity of photons (with the appropriate energy) and the density of absorbers (systems in the lower-energy state). Because radio photons are usually much less energetic than the radiators (e.g., $h\nu \ll kT$ in most cases), stimulated emission is nearly as important as absorption.

Under most circumstances, the lower-energy state is more populated than the higher. Consequently, absorption usually exceeds stimulated emission: The net absorption coefficient is then positive, resulting in attenuation of the intensity. This is always the case for thermal (equilibrium) distributions (and for most nonthermal distributions as well). However, if the populations are inverted, that is, if the upper state is more populated than the lower, the net absorption coefficient may be negative, resulting in amplification of the intensity and hence the laser or maser phenomenon.

Allowed and Forbidden Transitions

Classically, the emission of radiation by a single charge is calculated by computing the electromagnetic radiation field produced as a point charge accelerates. When, however, the radiating charge is not a single point charge, physicists often describe the distribution of charges (or currents) in terms of a multipole expansion or moments of the charge (or current) distribution. The lowest-order radiating moments are dipole, quadrupole, and so on, because the monopole moment, having no orientation, cannot radiate.

In atomic and molecular systems, the rates for electric-quadrupole and magnetic-dipole transitions are each much lower than those for electric-dipole transitions. (This follows from the fact that, for atoms and molecules, the ratios of characteristic diameters to relevant radiation wavelengths are small; rates for successively higher multipoles scale as the square of this ratio.) Consequently, electric-dipole transitions are called "allowed" (or permitted); electric-quadrupole and magnetic-dipole transitions are termed "forbidden," although, in fact, they are merely less likely.

Incoherence and Coherence

When radiators operate independently of one another, the phases of the generated waves are random. The emitted power from an ensemble of radiators is then simply the sum of the emissions of the independent radiators: This is incoherent emission. If the radiators are not independent, then the phases of the waves may be nonrandom. Under such conditions, the radiation is coherent, so that the waves may interfere constructively or destructively. For constructive interference, the amplitudes for the waves add; hence the emitted power (proportional to the square of the amplitude) from an ensemble of radiators may greatly exceed the sum for independent radiators (at least in certain directions).

Radio and television stations transmit coherent radio emission: Voltages applied to the transmitting antenna force electrons in the antenna to oscillate together. On the other hand, most (but not all) astrophysically relevant emission mechanisms are incoherent. For incoherent radiation, the maximum brightness temperature T_b of the source does not differ substantially from the temperature T of the radiators. Thus an extremely high brightness temperature constitutes excellent evidence for coherent emission.

RADIO EMISSION MECHANISMS

There are several astrophysically relevant mechanisms for radio continuum emission and for radio spectral-line emission. Observational discriminators among the various continuum mechanisms include the maximum brightness temperature, the spectral shape (Fig. 1), the polarization, and the variability time scales. Laboratory or theoretical knowledge of the line spectra of various atomic and molecular species facilitates the identification of radio spectral lines.

Magnetobremsstrahlung

Magnetobremsstrahlung (magnetic braking radiation), the incoherent emission of electrons spiraling in a magnetic field, is the most prevalent emission mechanism in cosmic radio sources. It occurs in the stronger radio sources (active galaxies and supernova remnants) and in the weaker (the Sun and planetary magnetospheres).

A charged particle in a uniform magnetic field executes a circular orbit (more generally, a spiral) about the magnetic-field direction. For an electron, the orbital repetition frequency (number of orbits per unit time) is

$$\nu_o = \nu_e/\gamma = (2.8\ \text{MHz})B/\gamma, \tag{4}$$

where ν_e is the electron cyclotron frequency and B is the magnetic field (in G, a unit comparable to the magnetic field at the Earth's surface). The Lorentz factor γ is a consequence of special relativity and is defined by

$$\gamma \equiv 1/\sqrt{1-(v/c)^2}, \tag{5}$$

with v the electron's speed and c the speed of light. (It becomes large for ultrarelativistic electrons and is nearly unity for nonrelativistic electrons.)

The orbiting electron emits radiation at the orbital repetition frequency and at higher harmonics (multiples) of this frequency. At nonrelativistic electron speeds ($v^2 \ll c^2$), only radiation at the fundamental (first harmonic) frequency is important: The magnetobremsstrahlung in this case is cyclotron or gyroresonance radiation, which is intrinsically circularly polarized. At mildly relativistic electron speeds, several to many harmonics are important: The magnetobremsstrahlung here is gyrosynchrotron radiation. At ultrarelativistic electron speeds ($\gamma \gg 1$), higher harmonics (up to harmonic numbers of order γ^3) dominate: The magnetobremsstrahlung is synchrotron radiation, which is intrinsically linearly polarized (for the higher harmonics, in which most of the power is radiated).

Because typical synchrotron-emitting electrons in cosmic radio sources have Lorentz factors of order 10^3 or more, the synchrotron spectrum of a single relativistic electron spans a very large frequency range (a factor of about 10^9 or more). However, the single-particle emission spectrum peaks near a characteristic synchrotron frequency:

$$\nu_s \approx \nu_e\gamma^2 = (2.8\ \text{MHz})B\gamma^2. \tag{6}$$

Because astrophysical mechanisms for accelerating particles to relativistic speeds tend to produce power law distributions of electron energies ($E_e = \gamma m_e c^2$, with m_e the electron's rest mass) and because the characteristic frequency ν_s and radiated power for synchrotron emission also depend upon powers of the Lorentz factor γ, the incoherent synchrotron spectra of cosmic radio sources tend to be power laws ($S_\nu \propto \nu^{-\alpha}$, where α is the spectral index, usually between 0.3 and 1.0). At sufficiently high frequencies (typically in the infrared), synchrotron radiative losses steepen the spectrum.

At sufficiently low frequencies, an incoherent synchrotron source becomes self-absorbed such that the maximum brightness temperature remains less than 10^{12} K. For a homogeneous, self-absorbed synchrotron source, the flux density $S_\nu \propto \nu^{-5/2}$. However, so simple a spectrum (as in Fig. 1) is seldom seen because sources are generally inhomogeneous, with the lower-frequency radio emission originating in larger, less dense structures.

In addition to the random ultrarelativistic motion of the synchrotron-radiating electrons, the relativistic plasma itself may move at speeds approaching the speed of light. Such bulk relativistic motion (observed or inferred for many strong, compact radio sources) produces several curious special-relativistic effects, for example, superluminal (apparent faster-than-light) motion, enhanced brightness temperature, Lorentz beaming (relativistic aberration, concentrating emission in the direction of motion), Doppler-shifted frequencies, and dilation of time scales.

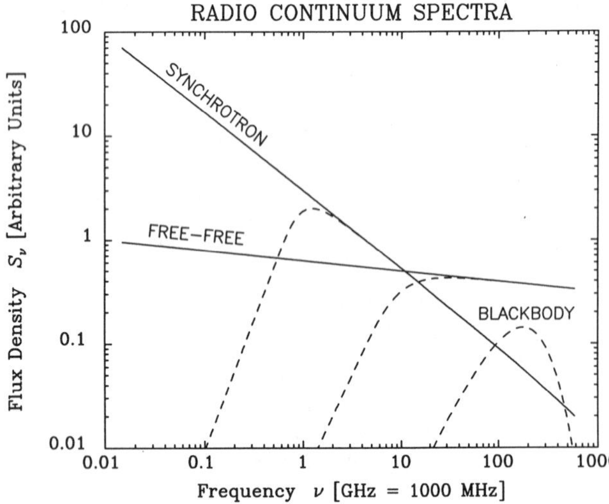

Figure 1. Indicative spectra for the principal continuum emission mechanisms in cosmic radio sources. Solid curves indicate spectra at frequencies for which the emitting component is transparent (negligible self-absorption); dashed curves, opaque (substantial self-absorption). Spectra of actual cosmic sources are often more complex, owing to multiple components or other inhomogeneous structure.

Blackbody Radiation

When radiation is in thermal equilibrium with matter, that is, when emission and absorption are in balance at all relevant frequencies, it is incoherent and unpolarized and its spectrum is blackbody, described mathematically by the Planck distribution. The brightness temperature T_b of a blackbody is rather simply related to the thermal temperature T of the radiators:

$$kT_b(\nu) = h\nu / (e^{h\nu/kT} - 1). \tag{7}$$

For $\nu \ll kT/h = (21\ \text{GHz})T$, $T_b \to T$. In fact, this low-frequency limit of the Planck distribution, which is called Rayleigh–Jeans behavior, is the physical basis for defining the brightness temperature.

In terms of the flux density S_ν, the spectrum of a blackbody goes as ν^2 at low frequencies (Rayleigh–Jeans limit) and falls nearly exponentially for $\nu > kT/h$ (basically because the photon energy $h\nu$ cannot exceed the typical energy kT of the radiator). The peak flux density of a blackbody occurs at

$$\nu_{\text{bb}} = 2.8kT/h = (59\ \text{GHz})T. \tag{8}$$

The cosmic microwave background (see Fig. 1) has a blackbody temperature of about 3 K; the planets and Moon, a few 100 K. Because astrophysical sources are warmer than 3 K, radio blackbody emission is invariably in the Rayleigh–Jeans regime.

Thermal Bremsstrahlung

In an ionized gas, electrons passing near an ion accelerate in the electric field of that ion. If the velocities of the electrons are thermally distributed (Maxwell–Boltzmann distribution), the incoherent, unpolarized radiation resulting from this acceleration is called thermal bremsstrahlung or free–free emission [because the electron is free (unbound) both before and after passing the ion]. The thermal-bremsstrahlung spectrum is flat (S_ν nearly independent of ν) up to

$$\nu_t = kT/h = (21\ \text{GHz})T, \tag{9}$$

and then falls exponentially. Because the characteristic temperatures of ionized hydrogen are about 10,000 K or greater, ν_t typically falls in the ultraviolet. Consequently, thermal-bremsstrahlung (free–free) spectra (see Fig. 1) are nearly flat at all radio frequencies for which absorption is unimportant. Such radio spectra are indicative of H II regions and similar clouds of ionized hydrogen.

At some low frequency, free–free self-absorption becomes important because the brightness temperature T_b of an incoherent thermal source cannot exceed the kinetic temperature of the emitting electrons. The spectrum of a self-absorbed isothermal source goes as the frequency squared ($S_\nu \propto \nu^2$), characteristic of Rayleigh–Jeans behavior. However, if the temperature of the bremsstrahlung-emitting plasma depends upon depth into the source, then other types of spectral behavior may occur.

Atomic Transitions

As an electron passes near an ion, it may recombine with the ion to form an atom (or, more generally, to form a lower ionization state). In order to drop into a bound atomic level (orbit) from a free state, the electron must shed some energy. If this shed energy is in the form of a photon, the resulting emission is recombination continuum or free–bound continuum [because the electron is free (unbound) before and bound after].

The most probable recombinations are those to low-lying atomic levels: Such recombinations produce optical or ultraviolet continua (or, in heavier elements, x-ray continua), but *no* radio emission. Nevertheless, some—albeit a very small fraction—of the recombinations are to the high-lying atomic levels. In principle, such recombinations may produce radio recombination continuum; however, it is impractical to distinguish this very weak radio continuum from the accompanying thermal bremsstrahlung. On the other hand, as the electron in a high-lying atomic level cascades downward to low-lying atomic levels, in a series of discrete steps (because the atomic energy levels are quantized), the atom emits spectral lines, recombination lines from bound–bound transitions. For hydrogen, the recombination lines occur at

$$\nu_{qq'} = (3.3 \times 10^{15}\ \text{Hz})\left(\frac{1}{q^2} - \frac{1}{q'^2}\right), \tag{10}$$

where q and q' are principal quantum numbers (integers), describing the atomic level of the electron after and before the (allowed) transition. Typical radio recombination lines of hydrogen are for $q' = q + 1$, with q of order a hundred. Radio astronomers observe such recombination lines from H II regions.

Outside H II regions, interstellar hydrogen may be atomic (H I), that is, neutral or unionized. Furthermore, temperatures are sufficiently low that most of the atomic hydrogen is in the ground (lowest-lying) principal state. However, the ground state of hydrogen is split into two slightly different energy levels: This hyperfine structure results from the very small energy difference between the case of parallel proton and electron magnetic moments and that of antiparallel magnetic moments. The hyperfine transition, which occurs as the electron's magnetic moment flips from parallel to antiparallel to the proton's magnetic moment, emits a low-energy photon, corresponding (through Planck's equation) to a frequency of 1.4 GHz or a wavelength of 21 cm. This is the 21-cm line of atomic hydrogen, which has proven so useful in deducing the structure and dynamics of the Galaxy and of other galaxies. Although this is a magnetic-dipole (forbidden) transition and hence occurs at considerably slower rates than electric-dipole (permitted) transitions, atomic hydrogen is so abundant, that this hyperfine-structure line is relatively strong.

Molecular Transitions

Molecules consist of two or more atomic nuclei, surrounded by their associated cloud of electrons, which are responsible for binding the molecule. In addition to the electronic levels and the fine-structure and hyperfine-structure splitting (analogous to the atomic transitions), molecules exhibit other types of energy levels resulting from the presence of more than one nucleus. Vibrational spectra, resulting from spring-like oscillations in the separations between nuclei, occur primarily in the infrared; rotational spectra, resulting from the spinning of the molecule about an axis, occur primarily in the radio to far infrared.

Molecular spectral lines occurring in the radio wavelength region result primarily from transitions between the quantized rotational levels or between the two levels of Λ doublets. (For nonrotating molecules, the quantum-mechanical energy of the molecule depends upon the square of the projected electronic orbital angular momentum Λ and thus is independent of the sign of Λ. Rotation of the molecule removes this degeneracy, causing Λ-doublet splitting.) Because both types of emission are electric-dipole (allowed) transitions, they are considerably stronger than hyperfine lines [which are magnetic-dipole (forbidden) transitions].

Although radio astronomers have detected dozens of molecular species, they have not detected the most abundant—viz., molecular hydrogen H_2. (Ultraviolet transitions of molecular hydrogen have, however, been identified.) Molecular hydrogen, like other symmetric molecules, has no electric-dipole moment (hence no allowed transitions) because the center of charge is at the center of mass.

Many astronomical molecular sources are masers. Physical conditions required for astronomical masers are high gas density (about a million times denser than the average interstellar medium) and a luminous source (10,000 times brighter than the Sun). Under such conditions, molecules may be pumped, overpopulating the upper molecular level. This population inversion causes the stimulated emission to exceed the absorption, resulting in amplification of the radiation and consequent very high brightness temperatures (up to 10^{15} K) in the lines, as well as anomolous line ratios and polarization. When the maser's brightness temperature grows too large, the pumping may not be sufficient to maintain the population inversion (which is diminished by the stimulated emission). If this is the case, the maser is saturated and amplification along the ray path ceases, thus limiting the brightness temperature.

Coherent Processes

Coherent emission can produce much higher brightness temperatures than incoherent processes, because the radiators act collectively. A maser emits coherently because the stimulated emission has the same direction and phase as the incident radiation. In addition to molecular-line masers in dense clumps of the interstellar medium, electron-cyclotron (gyroresonance) masers may occur in the magnetospheres of planets and in the lower corona of the Sun (and other stars). For the cyclotron maser, the loss of electrons through scattering or the redistribution of electrons through plasma instabilities may invert the population of electrons, resulting in maser activity at the fundamental (or other low harmonic) of the electron-cyclotron frequency ν_e. Cyclotron masers produce circularly polarized emission at high brightness temperatures (up to about 10^{13} K, inferred for solar microwave spike bursts).

Another category of coherent emission mechanisms is plasma radiation resulting from coupling electromagnetic radiation to plasma collective modes (waves or oscillations in the plasma). Although theories for this process are complex and the models not thoroughly confirmed, they usually predict coherent radiation at the plasma frequency (or its second harmonic):

$$\nu_p = (9.0 \text{ kHz}) n_e^{1/2}, \tag{11}$$

where n_e is the number density of electrons (in cm^{-3}). In a medium containing free electrons, the plasma frequency is the natural frequency for oscillation of electrons about their equilibrium positions, where the restoring force is the electric field resulting from the displacement of the electrons from the positively charged ions. Electromagnetic waves of frequency $\nu \leq \nu_p$ cannot propagate through the medium (essentially because the plasma can respond sufficiently rapidly to cancel the electric field of the electromagnetic wave). This plasma cutoff accounts for the reflection of low-frequency radio waves by the Earth's ionosphere and for the reflectivity of metals at optical frequencies. Indeed, one of the problems faced by models for plasma radiation (and for cyclotron radiation) is getting the generated electromagnetic wave out of the region in which it is produced (without substantial reflection or absorption). Plasma radiation from solar radio bursts is circularly polarized (to a degree depending upon the ambient magnetic field) and has high brightness temperatures (up to 10^{15} K at the lowest radio frequencies).

Other means for producing coherent emission rely upon geometrical bunching or clumping. If like charges clump or bunch together and experience the same acceleration, then, for wavelengths much larger than the dimensions of the clumping, they effectively emit as one (possibly quite large) point charge. Hence the coherent radiation from a clump of N charges may be as much as N times as powerful as the incoherent radiation would have been, that is, N^2 times the single-charge radiated power. Such a situation may occur in radio pulsars, resulting in coherent curvature radiation. Plasma instabilities apparently lead to bunching of electrons (or positrons), as they stream relativistically, confined to follow the very strong magnetic field (10^{10}–10^{12} G) in the vicinity of a neutron star. Because the charges must accelerate to follow the curving magnetic field, they radiate; because they are bunched together, the radiation is coherent (over a suitable range of wavelengths). Except for coherence, this curvature radiation is quite similar to synchrotron radiation, with the repetition frequency replaced by

$$\nu_o = \frac{v_{\text{bunch}}}{2\pi R_c} \rightarrow \frac{c}{2\pi R_c} < 5 \text{ kHz}, \tag{12}$$

where R_c is the radius of curvature ($R_c > 10$ km for a neutron star) and v_{bunch} is the streaming speed of bunched charges ($v_{\text{bunch}} \rightarrow c$ for relativistic streaming of bunches). Observations require that the emission mechanism in radio pulsars produce radiation with extremely high maximum brightness temperatures (up to 10^{30} K!), high linear and circular polarization, and steep spectral indices ($1 < \alpha < 3$).

Additional Reading

Dulk, G. A. (1981). Radio emission from the Sun and stars. *Ann. Rev. Astron. Ap.* **23** 169.

Hey, J. S. (1973). *The Evolution of Radio Astronomy*. Neale Watson Academic Publications, New York.

Pacholczyk, A. G. (1970). *Radio Astrophysics*. W. H. Freeman, San Francisco.

Reid, M. J. and Moran, J. M. (1981). Masers. *Ann. Rev. Astron. Ap.* **19** 231.

Rybicki, G. B. and Lightman, A. P. (1979). *Radiative Processes in Astrophysics*. Wiley, New York.

Verschuur, G. L. (1987). *The Invisible Universe Revealed*. Springer, New York.

Verschuur, G. L. and Kellermann, K. I., eds. (1988). *Galactic and Extragalactic Radio Astronomy*, 2nd ed. Springer, Berlin.

See also **Galaxies, Radio Emission; Interstellar Medium, Radio Recombination Lines; Pulsars, Observed Properties; Radiation, High-Energy Interaction with Matter; Radiation, Scattering and Polarization; Sun, Radio Emissions.**

Radio Telescopes and Radio Observatories

George W. Swenson, Jr.

A radio telescope is an antenna designed to collect electromagnetic radiation in the wavelength range between, say, 1 mm and 30 m, plus the necessary electronic appurtenances to amplify, select, classify, or otherwise sort the incoming data for presentation to the astronomer. The short wave antenna, used in 1932 at the Bell Telephone Laboratories in New Jersey by Karl G. Jansky to investigate the background noise that interfered with transatlantic radiotelephone signals, is generally recognized as the first radio telescope. Jansky's demonstration that the noise originated in the Milky Way initiated the science of radio astronomy. Grote Reber, a young radio engineer, established the basic principles of radio telescope design by a perceptive analysis and in 1937, as a hobby project, built the first antenna and receiver specifically designed for radio astronomy. This was a 9.6-m-diameter parabolic reflector, the prototype for several generations of larger instruments; it has been reconstructed on the grounds of the National Radio Astronomy Observatory in Green Bank, West Virginia.

The desirable attributes of a given radio telescope depend upon the type of cosmic source to be observed and which of its characteristics are to be determined. The principal observable attributes of a source are total power per unit of bandwidth, overall angular size, degree of angular detail, spectra of the overall source and of its local elements, positions on the celestial sphere of any features, and temporal variations of any of these attributes. Telescope properties that are under the control of the designer include collecting area, angular resolution (width of the antenna "beam"), usable wavelengths and bandwidths, and spectrographic, power measurement, and precision position measurement capabilities. The match of these telescope qualities to the observational requirements determines its suitability for the application. Because no single instrument can have appropriate parameters for every application, every radio telescope is specialized for certain types of observations.

There are two major categories of radio telescope, which differ in the way in which electromagnetic rays reaching different parts of the antenna are processed. In the first category, the different components are added together with careful attention to the phases of the components. This category includes reflector antennas of the type so familiar as spacecraft tracking or satellite communication facilities and the so-called phased arrays of small antennas connected together by waveguides. The second category includes interferometers and synthesis arrays (see below).

REFLECTOR ANTENNAS

Grote Reber's pioneering radio telescope was of the reflector type, actually a paraboloid of revolution with a collector antenna (usually called the feed antenna) located at the focus of the paraboloid. The radio wave from a very distant source of small angular extent is approximately a plane wave when it impinges on the telescope. A plane wave approaching in a direction parallel with the axis of revolution of the paraboloid is transformed by reflection into a converging wavefront that arrives at the focus with all components in phase. Most reflector antennas for radio astronomy are paraboloids of revolution, varying in size from 1–100 m in diameter. Most are mounted on two-axis mobility systems, which permit the paraboloidal axes to be pointed to desired points on the celestial sphere.

ANTENNA MOUNTING SYSTEMS

Two pointing systems are in general use with reflector antennas. Telescopes built between the late 1940s and the middle 1960s were generally of the "equatorial" type, in which one axis of motion, the hour–angle axis, is parallel with the Earth's axis, and the other axis of motion, the declination axis, is normal to the first. This system, long used in optical astronomy, permits the telescope to follow a constant declination source across the sky by driving the hour–angle axis with a constant-speed motor. Figure 1 shows a large, equatorially mounted, paraboloidal radio telescope. The equatorial mount has two advantages: It requires no computer to convert motion along a circle of constant declination into the component motions about two orthogonal axes, and the plane of polarization of the antenna and its feed system is constant with respect to the celestial sphere as the antenna tracks a source of constant declination. Its disadvantages are that it possesses inherently limited coverage of the sky and is generally more structurally complicated and more expensive to build than is the principal alternative type.

Figure 1. An equatorially mounted, paraboloidal radio telescope: the 37.5-m antenna of the Vermillion River Observatory, University of Illinois.

Figure 2. An interferometric synthesis array: the Very Large Array of the National Radio Astronomy Observatory in New Mexico. The antennas are az-el-mounted and are movable along railroad tracks to permit observations with different array configurations.

The second common type of telescope mounting (see Fig. 2) is the so-called azimuth-elevation (az-el) mount [sometimes called altitude-azimuth (alt-az)]. Here the entire structure can rotate about a vertical axis, to give motion in azimuth. Orthogonal to that axis is a horizontal axis to permit the antenna's pointing direction to rotate in a vertical plane. Motion about each axis is provided by a variable speed motor. A computer provides drive signals to the motors after a real-time conversion from celestial coordinates to az-el coordinates. The advantages of the az-el system over the equatorial mount are: easy achievement of all-sky coverage, ability to track a circumpolar source for 24 h, and more economical construction. The main disadvantages are: the requirement for real-time coordinate conversion, inability to track a source through the zenith, and rotation of the plane of polarization as a source is tracked. The polarization rotation can be corrected by an appropriate mechanical (or electronic) rotation of the feed antenna at the focus; the necessary real-time computations can be performed in the same inexpensive computer used for the celestial-to-az-el coordinate converter. The problem of tracking a source exactly through the zenith is insoluble; it would require infinite angular velocity about the azimuth axis. However, this limitation can usually be tolerated because the excluded portion of the sky can be made as small as necessary. The availability of inexpensive, fast, general-purpose computers has given a clear cost advantage to the az-el system and most large reflector telescopes built since the late 1960s have been of this type.

SIZE, SHAPE, AND PRECISION OF REFLECTOR ANTENNAS

The angular resolution of an antenna determines the fineness of detail in an image reconstructed from observations with that antenna. It can be defined as the minimum angular separation of two point sources at which the antenna can just distinguish between the sources. The angular resolution approximately equals the width of the antenna "beam," that is, the angle within which radiation can be received by that antenna. The beamwidth thus determines the fineness of detail that can be mapped by the antenna. It is approximately 60° divided by the diameter of the reflector in wavelengths. Thus, an antenna 10 m in diameter, observing at a wavelength of 1 m, could determine that two sources 6° apart on the sky were actually two sources. Two sources 2° apart would appear as one slightly broadened source.

The wavelength at which a reflector antenna is effective is set by the accuracy with which the reflecting surface is constructed and with which the feed antenna is positioned and maintained. In general, to be reasonably efficient the errors must be very small compared with a wavelength. In statistical terms, if the surface has mean random errors in surface shape of one-sixteenth wavelength (root-mean-squared), the power received from an incident radio wave will be reduced by one-half from that received by a perfect reflector of the same design parameters. Antennas for very short wavelengths (very high frequencies) must be built with very great precision. For example, an antenna operating at 1.0-cm wavelength must have a surface precision of 0.6 mm or better in order to achieve 50% of the effective collecting area of a perfect reflector.

The third of the principal antenna attributes is collecting area, that is, the amount of power the antenna can extract in a given bandwidth from an impinging radio wave with unit intensity in the same bandwidth. The main determinant is the size of the reflector; the larger its geometrical area, the greater its collecting area. Thus, as size affects both angular resolution and collecting area, radio telescopes tend to be very large in order to optimize these parameters. Neither parameter is uniquely determined by size alone: both collecting area and angular resolution can be affected somewhat by the characteristics of the feed antenna at the focus and by intentional or unintentional distortions of the reflecting surface. In fact, though traditionally the general rule was that a well-constructed antenna system could achieve 50% efficiency (could extract half the power from the incident wave), in recent years efficiencies as high as 85% have been achieved with reflectors representing departures from true paraboloids, ranging from small perturbations of paraboloids up to spherical reflectors. As only the paraboloid has a point focus, all of the so-called shaped-reflector antennas incorporate elaborate feed antennas to compensate for the arrival of the reflected rays on an extended surface or line rather than at a point. Most large, modern reflector antennas employ shaped-reflector technology to some degree in order to achieve high efficiency. The spherical reflector antenna has an additional advantage that its beam can be steered without mechanical movement of the reflector. Limited beam steering is achieved by causing the feed antenna to revolve about the center of the sphere.

The largest fully steerable paraboloidal antenna in current use is the 100-m-diameter, az-el-mounted radio telescope of the Max Planck Institut für Radioastronomie near Bonn, West Germany. It achieves an angular resolution of 0.7° and an effective area of 4000 m^2 at a wavelength of 11 cm. A new 100-m class az-el mounted Green Bank Telescope is planned for the National Radio Astronomy Observatory in West Virginia. Preliminary plans call for an operating wavelength as short as 3.0 mm and for use of an "offset" feed system. Here the paraboloidal reflector sector actually constructed is offset from the paraboloidal axis so that the aperture is not even partially obstructed by the feed antenna. The largest shaped-reflector antenna is the 305-m spherical reflector radio telescope of the National Astronomy and Ionosphere Center at Arecibo, Puerto Rico. With upgrading of the surface precision and of the feed system now under way, it is expected that its short-wavelength limit will be 3.7 cm, at which an approximately 230-m diameter portion of the spherical surface is included in the reception "beam" of the feed antenna to obtain an effective collecting area of 30,000 m^2. The beamwidth at this wavelength will be approximately $\frac{1}{2}$°.

Many variations of the paraboloidal and shaped-reflector principles have been employed in radio telescopes for special purposes. These include meridian transit instruments, in which motion is provided only in the declination axis so that a given source can be observed only during a very short interval each day. Other specialized instruments have been built for monitoring a single source for temporal changes in its characteristics.

PHASED ARRAYS

Phased arrays are used extensively in radar applications and at one time were used in a number of radio astronomy applications at the longer wavelengths. An arrangement of closely spaced small (in wavelength) antennas, each feeding its intercepted power to a common detector with close control of phase relationships, can have properties equivalent to those of a reflector antenna of similar projected area. The main advantage of the phased array is its low cost relative to an equivalent reflector antenna. Very large collecting area and very narrow beamwidth can be achieved quite economically and beam shapes can be designed for specific applications. The disadvantages include substantial power losses in the waveguides or transmission lines and phase-control devices, and restricted wavelength flexibility. Two-dimensional beam pointing is possible by controlling the relative phase delays between various antennas and the common detector. Phased arrays having collecting areas of many hectares, far greater than could conceivably be achieved with a reflector antenna for comparable cost, have been constructed for special classes of observations at wavelengths longer than 1 m. Another application is the use of a phased array as a feed antenna for a parabolic cylindrical reflector.

A long, narrow, phased array of small antennas, either by itself or as a feed antenna for a long, narrow parabolic cylindrical reflector, can be used to produce a fan-shaped reception beam useful for certain specialized observations. Two such arrangements at right angles can produce intersecting fan beams whose electrical outputs can be electronically combined to produce a pencil-shaped beam somewhat equivalent to that of a paraboloid of revolution. This arrangement is an example of a class of radio telescopes known as interferometers.

INTERFEROMETERS AND SYNTHESIS ARRAYS

Two antennas looking at the same point of the sky, whose electrical outputs are multiplied together, constitute a radio interferometer. The antennas may be separated by substantial distances; interferometers are routinely operated with separations of continental or intercontinental scale (so-called Very Long Baseline Interferometry). A given interferometer, in a given orientation with respect to a cosmic source, is sensitive only to features smaller than a specific angular size. If the source contains no features of this size or less, the interferometer is "blind" to that source. The characteristic size is uniquely determined by the separation of the two antennas, the "baseline," as projected on the celestial sphere at the source. A single interferometer can provide only very limited though possibly valuable information about a source, but an ensemble of interferometers, with various baselines can yield enough data for a detailed map of a source. Interferometric synthesis has proved to be an extremely effective means of observation and much of the recent development in radio astronomical instrumentation has involved this technique.

RADIO OBSERVATORIES

An observatory for radio astronomy must satisfy different environmental criteria from those of an optical observatory. Even within the radio category the requirements differ depending upon the type of observations to be made and the characteristics of the instruments involved. The Earth's atmosphere is one of the environmental components that most seriously affects the observations. At long wavelengths, say 10 cm and greater, the atmosphere near the Earth is quite well behaved; that is, it does not distort the downcoming radio waves very seriously. At these wavelengths, humidity, cloud cover, temperature variations, or turbulence have only minor perturbing influences on the observations. Telescopes for these wavelengths can be sited with minor concern for these weather factors. On the other hand, high winds may be a safety hazard to a telescope made up of a large, elevated, movable structure, and also may seriously affect the precision of its surface and its movement. An instrument intended for operation at the longer wavelengths may well be sited in a mountain valley where it is sheltered from high winds. Shorter waves may be distorted by varying moisture content and other atmospheric parameters to the extent that observations are seriously in error. These effects are increasingly troublesome as wavelengths become shorter than, say, 3.0 cm. Telescopes for wavelengths shorter than about 1.5 cm are preferably sited at high altitudes, on mountains or plateaus. The deleterious effects of the atmosphere are mitigated to some extent when the wave travels a shorter distance through the air. These geographical arrangements are not sure proof against meteorological problems, but represent sensible precautions against loss of valuable observing time due to bad weather.

Aside from the atmosphere, the other most important environmental influence on radio astronomy observations is the pollution of the electromagnetic spectrum. Humans have found radio waves to be extremely useful for communications, entertainment, remote sensing, industrial processing, and other applications. In addition, many human activities generate useless radio noise that is unintentional; examples are automobile ignitions, computers, lighting, motors, and many others. Thus, the radio spectrum is jammed throughout with radio signals and useless noise, often at levels thousands or millions of times stronger than the faint cosmic radio "signals" the radio astronomer is attempting to study. Many ingenious technical efforts have been employed to circumvent this jamming of the cosmic radio waves, but the only really effective means is geographical isolation. The mountaintop location preferred for very short wavelength observations is particularly vulnerable to interference, whereas a remote mountain valley affords some shielding from signals of terrestrial origin. A few "reservations" in the frequency spectrum have been designated by international treaty to protect radio astronomy and other passive (listening only) users of radio waves; unfortunately, these reserved bands are too few, too narrow, and too loosely protected from unintentional noise and from transmitters nearby in the spectrum to be an adequate solution to the problem. Geographical isolation from concentrations of noise and transmitters is also essential.

The various field stations of the National Radio Astronomy Observatory (U.S.) represent examples of the siting of radio telescopes to mitigate atmospheric and interference problems. The original observatory location is at Green Bank, West Virginia, in a deep valley of the Allegheny Mountains. By government legislation a "radio quiet zone" was established surrounding the observatory, with provisions for limiting and coordinating transmitters within the zone. There is, of course, no protection against transmitters on aircraft or spacecraft, which represent increasing problems to radio astronomers. This location is relatively free from high winds but is subject to the other meteorological disadvantages of a relatively humid, low altitude site. Because the region is sparsely peopled, the level of unintentional noise is not high, but it is necessary to restrict access to the extensive observatory grounds to certain types of vehicles and to maintain constant watch for other sources of noise. When a specialized telescope was acquired for observations at wavelengths shorter than 1 cm, it was located on a mountaintop near Tucson, Arizona, at the Kitt Peak National Observatory. Here the atmosphere is less dense and less humid than at Green Bank, and observations are possible during most of the year except for the summer thunderstorm season. At millimeter wavelengths the levels of unintentional noise and the density of transmitters are relatively low, so the mountaintop location is not a severe disadvantage from those standpoints. An extensive array of interferometers was designed to operate at wavelengths in the centimeter range. This required a site on which a large number of 25-m-diameter paraboloidal antennas could be moved about relatively freely, over an area of hundreds of square kilometers. Clearly a large, flat area was needed. Because the instrument was intended to produce extremely high resolution, atmospheric perturbations would have to be minimized even though they are not extreme at such wavelengths. Again, geographical isolation from terrestrial sources of interference was needed. This instrument, the Very Large Array (Fig. 2), was built on the Plains of San Agustin, a 2126-m-high, mountain-shielded, desert plateau in central New Mexico. The observatory is now constructing a 12-antenna Very Long Baseline Array (VLBA) extending from Hawaii to the Virgin Islands. In selecting the 12 sites the preceding criteria had to be considered, of course, but the dominating consideration had to be the geometrical relationships of the various antennas to one another. These are constrained by mathematical requirements related to the quality of the images of cosmic sources to be produced by the VLBA.

Additional Reading

Christiansen, W. N. and Hogbom, J. A. (1985). *Radio Telescopes*, 2nd ed. Cambridge University Press, London.

Committee on Radio Frequencies (1989). *Radio Astronomy Observatories*. National Academy Press, Washington, DC.

Kraus, J. D. (1986). *Radio Astronomy*, 2nd ed. Cygnus–Quasar Books, Powell, Ohio.

Swenson, G. W., Jr. (1977). Radio telescopes. In *Yearbook of Science and the Future*. Encyclopedia Britannica, Chicago, Ill.

Swenson, G. W., Jr. (1980). *An Amateur Radio Telescope*. Pachart Publishing Co., Tucson, Ariz.

See also **Radio Astronomy, Receivers and Spectrometers; Radio Astronomy, Space Missions; Radio Telescopes, Interferometers and Aperture Synthesis.**

Radio Telescopes, Interferometers and Aperture Synthesis

Barry G. Clark

THEORY OF APERTURE SYNTHESIS: FOURIER OPTICS

The capacity of an optical instrument to separate the images of two closely spaced objects is determined by the instrument's aperture divided by the radiation wavelength. That is, the angular separation of two objects that can just be resolved cannot be much smaller than the wavelength of operation divided by the size of the instrument. This fundamental principle of optics has special implications for the study of astronomical objects at radio wavelengths. Radio waves may have wavelengths of a few centimeters, whereas the wavelength of visible light is about $\frac{1}{2}$ micrometer (μm). Therefore, to achieve equal resolving power, radio instruments must be about 100,000 times larger than their optical counterparts.

This is a fundamental physical principle and cannot be avoided. The required large size of radioastronomical instruments has strong implications for the cost of constructing them. Although this principle dictates a restriction on the overall size of the instrument, it does not require that the entire aperture of the instrument be occupied by optical surfaces. In fact, radio instruments with high resolving power are constructed as interferometers or interferometer arrays, synthesizing instruments of many kilometers in size, with only a tiny fraction of that area occupied by receiving surfaces.

The science upon which this form of instrument is based is called Fourier optics or, occasionally, modern optics (classical optics deals with the properties and aberrations of lens systems). A fundamental principle of the science is that there is a Fourier transform relationship between the electromagnetic field in the aperture plane of the instrument and that in the focal plane:

$$E_f(x,y) = \iint E_a(u,v)e^{2\pi i(ux+vy)}\,du\,dv.$$

In this equation, E is a complex function of spatial location, giving the amplitude and phase of the electromagnetic field, where the coordinates (x, y) define the location of a point in the focal plane and (u,v) give the location of a point in the aperture plane.

The image is the detected intensity, or absolute square of the electromagnetic field in the focal plane. The product of the field at two different locations is called the *correlation*, and by a theorem in Fourier transform theory, squaring a quantity in the focal plane corresponds to correlation in the aperture plane:

$$I(x,y) = |E(x,y)|^2$$

$$= \iint E_a(u_0,v_0)E_a^*(u_0+u,v_0+v)e^{2\pi i(ux+vy)}\,du\,dv,$$

where $I(x, y)$ denotes the image intensity as a function of location and the asterisk indicates the operation of complex conjugation.

An interferometer is a device that performs correlation. A probe, called an interferometer element, is constructed that picks up and amplifies the electromagnetic field at a point (u_0, v_0) in the aperture plane. The field from this element is then multiplied by that from a second aperture-plane point $(u_0 + u, v_0 + v)$, and the product is averaged. This correlation, whose usual symbol in radio interferometry is V (for visibility), is

$$V(u,v) = \langle E(u_0,v_0)E(u_0+u,v_0+v)\rangle,$$

where the angle brackets denote time averaging. [It can be shown that this correlation is independent of the choice of (u_0, v_0).] The expression for the image given previously can be written as

$$I(x,y) = \iint V(u,v)e^{2\pi i(ux+vy)}\,du\,dv.$$

If correlation $V(u,v)$ is measured at a large number of element separations (u,v), these measured data can be used to evaluate the preceding integral numerically, producing a numerical description of the image $I(x, y)$. This procedure is known as the technique of *aperture synthesis*.

EARTH ROTATION SYNTHESIS AND SYNTHESIS WITH ARRAYS

In principle, the image can be reconstructed by measurements made with a single pair of elements, moving one with respect to the other to all separations (u,v). In practice, this would be exceedingly tedious, and two techniques are employed to lessen the effort. The first relies not on physically moving the elements, but on utilizing the fact that the rotating Earth is already moving the elements. This is most readily visualized by considering the appearance of the interferometer from the very distant object being observed. From that extreme distance, the distance from the center of the Earth to the interferometer elements may be neglected entirely, and it may be conceived that the whole Earth is revolving about one of the elements, carrying the other element with it. The path traced by the second element is clearly a circular ring, lying in a plane parallel to that of the Earth's equator. Viewed from the perpendicular direction, that of the north celestial pole, the path of one element relative to the other is a circle; all loci (u,v) along this circle are visited by the interferometer element and $V(u,v)$ may be measured at those points. If the object of interest does not lie at the north celestial pole, this ring is, clearly, viewed somewhat from one side and is thus forshortened into an ellipse. In the extreme case, for an object lying in the plane of the Earth's equator, the ellipse degenerates into a straight line.

The technique of observing objects for substantial portions of a day, to exploit this movement of the interferometer element, is a variant of aperture synthesis called *Earth rotation synthesis*.

In practice, even Earth rotation synthesis becomes impracticably tedious for a two-element interferometer for the case of large and complex images. Such images require that, over a large area of the (u,v) plane, almost all points should be close to one of the elliptical tracks of an interferometer. Very large savings in time can be accomplished by the use of an *interferometer array*, that is, an array of more than two interferometer elements. Because correlations may be done between every pair of elements, the number of (u,v) points measured simultaneously rises quadratically with the number of elements. That is, an array with two elements measures a single (u,v) point, one with three elements measures three (u,v) points, one with four elements measures six (u,v) points, and one with five elements measures 10 (u,v) points.

HISTORY OF RADIO INTERFEROMETERS

The prospect of making astronomical observations at radio wavelengths was inaugurated by the serendipitous discovery of radio waves from the galactic center by the physicist Karl G. Jansky at Bell Telephone Laboratories in 1933. That discovery in turn grew out of the increasing use of radio for communication since the beginning of the century and an investigation into the ultimate physical limitations of its use for that purpose. Within a few years, the advancing technology of radio made it, in the form of radar, into one of the most formidable weapons of World War II. Because of its defense importance, vast technical strides were made in the use of radio waves in a few years. With the return of peace in 1945, there were many new techniques to be exploited for astronomical purposes, and many brilliant radio engineers who had worked in the field for military reasons who remained in the field to develop it for basic science.

Early in the postwar years, the frustrating lack of resolution intrinsic to radio instruments of modest size was felt by many practitioners, who soon turned to interferometry and aperture synthesis, among other techniques. Soon after the war, several investigations (most notably by the Australians, John G. Bolton and Bernard Y. Mills in particular) provided finding lists of these discrete sources. They were known to be smaller than the beamwidth of the antennas with which they were discovered, a few degrees, and very little else was known about them. It was conceived at the time that these objects might be "ordinary" (or rather extraordinary) stars, and the conundrum of how to demonstrate this (by showing that they, like stars, were tiny points) aroused interest among several investigators. A hint was given by the experiment in which diameters of stars had been measured with a large optical interferometer (by Albert A. Michaelson and Francis G. Pease early in the century). Finally, measurements of the angular sizes of radio sources were made interferometrically by Roger C. Jennison and co-workers in England.

These early efforts were taken up by a capable and productive group at Cambridge University in England, led by Sir Martin Ryle, that developed the techniques of aperture synthesis and Earth rotation synthesis and that has remained among the leading practitioners for more than three decades. The first applications of aperture synthesis instruments at Cambridge were for survey instruments, to locate a large number of discrete sources, and to provide finding lists of a large number of these objects. These synthesis instruments operated at the relatively low frequencies of 85, 159, and 178 MHz, producing the radio source catalogs called the 2C, 3C, and 4C (second, third, and fourth Cambridge) catalogs.

These instruments were followed, in the mid-1960s, by the first great Earth rotation synthesis instrument, the One Mile Telescope, which operated at 408 and 1400 MHz. Meanwhile, other groups had recognized the value of the technique, and aperture synthesis radio telescopes appeared at several great observatories around the world. The following are especially noteworthy: Westerbork Synthesis Radio Telescope (an array of 10 25-m elements in the Netherlands), the Very Large Array [(VLA), an array of 27 25-m elements located near Socorro, New Mexico], the Multi-Element Radio Linked Interferometer Network [(MERLIN), a number of hetrogeneous elements in northern England, operated by the Nuffield Radio Astronomy Observatory associated with the University of Manchester], and the Australia Telescope [(AT), an array of eight 25-m elements near Molongolo, NSW, also used with various other Australian elements to form a long baseline array].

In a further quest for higher resolution, several very large instruments are being operated, with a varying number of elements and less than full-time operation. The largest separations used approach the diameter of the Earth, which, at centimeter wavelengths, gives a resolving power better than 0.001 arcsec (roughly 1000 times better than achieved with conventional, ground-based, optical photography). The principal arrays are the European Very Long Baseline Interferometry Network (EVN), the U.S.-based VLBI Network Consortium, the U.S. Very Long Baseline Array (VLBA), currently under construction, and the AT array mentioned previously.

IMAGE PROCESSING FOR INTERFEROMETER ARRAYS

Although interferometer arrays provide a large number of measurements of $V(u,v)$, they do not provide complete coverage of the whole (u,v) plane. For one thing, the measurements do not extend beyond the length of the longest baseline. Therefore, the image produced is not the true brightness distribution of the radio source, but is the true brightness convolved with an instrumental response of finite size. This is both inevitable and well-understood. A more tractable, though more vexing, problem is the fact that the coverage of the (u,v) plane is not complete within the radius of that longest baseline. A straightforward execution of the Fourier transform, ignoring the locations where measurements were not made, results in a very much degraded image, colloquially known to radio astronomers as the *dirty image*. The procedure used in going from this dirty image to what is believed to be a physically better image of the radio sky is known as deconvolution. By adding surprisingly small amounts of knowledge of things other than the $V(u,v)$, substantial improvements can be made. There are currently two popular procedures. One is called the variational method (VM) or maximum entropy method (MEM), which uses the fact that the image must be positive everywhere and uses iterative variational methods to maximize some measure of image goodness. There are various measures of image goodness in use (integrals over the logarithm of image intensity, which enforces the nonnegativity of the image). There are analogies between these goodness measures and entropy in thermodynamics, although the physical interpretation of the "entropy" so defined is unclear.

The other popular method of deconvolution is called Clean, which is based on the fact that, in practice, the brightness of a radio source is rather spiky, that is, the source looks like a collection of points or near points. The Clean method is based on iteratively subtracting from the input $V(u,v)$ data the response of the instrument expected from a point source located at the brightest point of the image, and remaking the image.

As one might expect from the preceding qualitative descriptions, the variational method tends to work better than Clean on images that have large, smooth areas of distributed brightness. Clean tends to work somewhat better and, especially, to converge faster for images with large areas of zero emission and a generally spiky appearance.

There is a deep connection between the two methods, in that Clean can be regarded as a way of implementing the variational method with a particular choice of the goodness measure, which in this case is *not* an integral over the logarithm of the image.

There is another class of errors due to propagation effects between the radio source and the interferometer elements, in much the same way that optical images are degraded by atmospheric fluctuations called seeing. Optical seeing is caused by density fluctuations due to turbulent heat transport in the atmosphere. At radio wavelengths, the propagation irregularities are due to fluctuations in either the water-vapor distribution function (water vapor has a large index of refraction at radio wavelengths because of the high dipole moment of the water molecule) or the electron content of the ionosphere. At most wavelengths, the structure in the water vapor is not so strong on small spatial scales that the gain of an interferometer element is affected, but the signal from that element is delayed by a slowly varying amount.

For simple images, a large interferometer array produces much more information than is needed to reconstruct the image. Even for complex images, there is redundant information in the collected data because, for one thing, the same (u,v) locus may be visited by various interferometer pairs. For another, it is possible to construct invariant sums of measured phases that are independent of propagation effects, as described previously. These invariants and redundancies can be used (in good cases with high signal-to-noise ratios) to solve for the propagation variations and thus to remove the time-variant effects of the atmosphere. The technique is called *self-calibration* or, occasionally, *hybrid mapping*.

Although the preceding techniques have been pioneered in the relatively conceptually simple environment of radio-interferometer data, they can be applied to image data from other sources as well: Very similar procedures are used in the optical image processing techniques known as speckle interferometry.

Additional Reading

Born, M. and Wolf, E. (1975). *Principles of Optics: Electromagnetic Theory of Propagation, Interference, and Diffraction of Light*, 5th ed. Pergamon Press, Elmsford, N.Y.

Christiansen, W. N. and Hogbom, J. A. (1985). *Radiotelescopes*, 2nd ed. Cambridge University Press, New York.

Downes, D. (1989). Radio telescopes: Basic concepts. In *Diffraction-Limited Imaging with Very Large Telescopes*, D. M. Alloin and J. M. Mariotti, eds. Kluwer Academic Publishers, Dordrecht, p. 53.

Perley, R. A., Schwab, F. R., and Bridle, A. H., eds. (1989). *Synthesis Imaging*. *Publications of the Astronomical Society of the Pacific: Conference Proceedings*.

Thompson, A. R., Moran, J. M., and Swenson, G. W. (1986). *Interferometry and Synthesis in Radio Astronomy*. Wiley-Interscience, New York.

See also **Radio Astronomy, Receivers and Spectrometers; Radio Telescopes and Radio Observatories.**

S

Satellites

Paul J. Thomas

Our knowledge of the satellites of the solar system has expanded enormously with the advent of space exploration. Before 1957, detailed information was available only for the Moon; the remaining known satellites appeared as points of light through the most powerful telescopes. Currently, more than fifty satellites have been observed, over twenty of these by close spacecraft encounters. In addition to the enormous flood of data from these observations, progress has been made in our theoretical understanding of the origin and evolution of these bodies.

The satellites of the solar system consist of several distinct groups of objects (see Table 1). The Moon, as discussed below, is an anomalous object in many ways, and does not fit into any category suitable for the other satellites. The major satellites of Jupiter, Saturn, and Uranus appear to have regular (equatorial, low-inclination) prograde orbits. The regularity of these systems suggests a common formation mechanism, perhaps similar in some ways to that of the solar system (although there are also significant differences: see below). In addition, however, there are numbers of satellites of the giant planets that have irregular or retrograde orbits. These, including the outer jovian satellites and Phoebe, Triton, and Nereid, may have been gravitationally captured during close encounters in the past. The small satellites of the outer planets, including the ring particles also represent a distinct class (one that is discussed elsewhere); these probably arise from satellite fragmentation during or after the establishment of a satellite system.

The satellites can also be categorized compositionally (Table 2). The majority of the known satellites occur in the outer solar system, where the temperatures during formation were sufficiently low for water ice to be incorporated in the satellites. Where this is not observed (for example, in Io), early heating by accretional energy released by the formation of Jupiter is probably responsible. As we shall see, the presence of ice has an important influence (because of its high mobility and low melting point) on the subsequent thermal and structural evolution of the icy satellites. These bodies exhibit extremely active geological histories, despite their small size and low ambient temperature. In the saturnian and uranian systems, where formation temperatures were low enough to permit the incorporation of volatiles such as ammonia and methane into the ice mix, melting and mobilization of the satellites' interiors can occur at temperatures much lower than that necessary to mobilize pure ice.

Finally, several satellites (Phobos, Deimos, Phoebe, and many of the satellites of Uranus and Neptune) have surface reflectivities similar to those of carbonacous chondrites. Study of these objects (as a by-product of studies of Mars and Saturn) may thus reveal important information about the most primitive bodies of the solar system.

FORMATION

Three formation mechanisms for satellites have been extensively discussed by scientists:

1. Formation from a disk of gas and dust orbiting a planet.
2. Formation elsewhere in the solar system and subsequent capture by a planet.
3. Formation from the debris of a major impact on the primary planet.

The third mechanism has recently found favor with scientists as a possible explantion for the origin of the Moon and will be discussed separately below in that context.

Quantitative discussion of these ideas began in the late eighteenth century. Pierre Simon de Laplace proposed that the Moon condensed from a ring spun off from a rotating gaseous proto-Earth. This idea, essentially an adaptation of his model for the formation of the solar system, still survives today as a viable model for the formation of the satellite systems of the giant planets, as we shall see below. An alternative model, proposed by Sir George Darwin in 1878, was that the Moon was formed by fission from the Earth during an early period when the Earth supposedly was molten.

By contrast to the satellite systems of the outer planets, progress toward an accepted theory for the formation of the Moon was much more difficult. Part of the problem may well have been the vast amount of information known about the Earth–Moon system: This provided many constraints for theoretical models. The fission theory of Darwin was eventually abandoned due to problems understanding how the newly-separated Moon could resist tidal disruption or even how oscillations could successfully cause fission to begin with. More contemporary models are discussed below.

PROTOSATELLITE DISKS

Models involving satellite formation from a protosatellite disk, proposed for the regular satellite systems of Jupiter, Saturn, Uranus, and Neptune, resemble those for the solar nebula (from which the solar system is believed to have formed). In these models, satellite formation is accepted as a natural by-product of planetary formation. While these systems superficially resemble the solar system, there are important differences: Most of the angular momentum of the solar system is found in the planets, whereas most of the angular momentum of the satellite systems is found in the primary planets. In the solar system, a compositional gradient is observed, with rocky material more common closer to the Sun and volatiles more common in the outer solar system. A similar gradient is seen in the Jupiter system, with more refractory satellites closer to Jupiter. This is certainly indicative of accretion in an environment with a strong thermal gradient. However, this is not observed in the other regular satellite systems. In the Saturn system, for example, no such compositional gradient is observed. Furthermore, Saturn's satellite Titan is clearly anomalously large, compared to the rest of the system (it also contains most of the angular momentum of the satellite system).

Formation of a protosatellite disk around a planet may occur as a result of several distinct processes: A disk of gas may be captured from the surrounding solar nebula by the accreting nucleus of a giant planet, the material may have arisen by disruption from the equator of the primary during a period of rapid rotation, by disruption of a planetesimal during a close passage, by collisions between existing satellites, or by a collision with the primary planet.

These models are appropriate for both terrestrial planets and the giant planets. However, in the latter case the accreting giant planet probably was able to capture a large amount of gas from the surrounding solar nebula. In this gas-rich case, drag forces effectively decay the orbits of small particles and remove energy and angular momentum: Viscous drag in the disk ensures that material condensing from the nebula will assume circular, low eccentricity,

Table 1. Satellites of the Solar System

Planet	*Satellite*	*Distance from Planet (1000 km)*	*Radius (km)*	*Density (10^3 kg m^{-3})*
Earth	Moon	384	1738	3.34
Mars	Phobos	9.4	9–13	2.2
	Deimos	23.5	5–8	1.7
Jupiter	Metis	128		
	Adrastea	129		
	Amalthea	181	67–131	
	Thebe	222	45–55	
	Io	422	1815	3.57
	Europa	671	1569	2.97
	Ganymede	1,070	2631	1.94
	Callisto	1,883	2400	1.86
	Leda†	11,094	~ 8	
	Himalia†	11,480	90	
	Lysithea†	11,720	~ 20	
	Elara†	11,737	40	
	Ananke*†	21,200	~ 15	
	Carme*†	22,600	~ 22	
	Pasiphae*†	23,500	~ 35	
	Sinope*†	23,700	~ 20	
Saturn	Atlas	138	14–19	
	Prometheus	139	34–74	
	Pandora	142	31–55	
	Epimetheus‡	151	55–69	
	Janus‡	151	77–97	
	Mimas	199	197	1.17
	Enceladus	238	251	1.24
	Tethys	295	524	1.26
	Telesto§	295	11–12	
	Calypso§	295	8–15	
	Dione	377	559	1.44
	Helene¶	377	15–18	
	Rhea	527	764	1.33
	Titan	1,222	2757	1.881
	Hyperion	1,481	100–175	
	Iapetus	3,561	718	1.21
	Phoebe*†	12,952	105–115	
Uranus	Cordelia	50	~ 25	
	Ophelia	54	~ 25	
	Bianca	59	~ 25	
	Cressida	62	~ 40	
	Desdemona	63	~ 30	
	Juliet	65	~ 40	
	Portia	66	~ 40	
	Rosalind	70	~ 30	
	Belinda	75	~ 30	
	Puck	86	85	
	Miranda	130	242	1.26
	Ariel	191	580	1.65
	Umbriel	266	595	1.44
	Titania	436	800	1.59
	Oberon	583	775	1.50
Neptune	Naiad	48	~ 27	
	Thalassa	50	~ 40	
	Despoina	53	75	
	Galatea	62	~ 90	
	Larissa	74	95	
	Proteus	118	200	
	Triton*†	354	1350	2.08
	Nereid†	551	170	
Pluto	Charon	20	1230	< 2.4

*Retrograde.
†Irregular.
‡Coorbital satellites.
§Coorbital with Tethys.
¶Coorbital with Dione.

prograde orbits. This is the mechanism from which the regularity of the satellite systems of the giant planets arises.

Models of gas accretion from the solar nebula predict that a disk of gas approximately 20 times the current radius of the giant planets may be formed. This is roughly the size of the regular satellite systems of the outer planets.

ACCRETION OF SATELLITES

Once material has settled down to the equatorial plane of the protosatellite disk, accretion of individual satellites begins. The timescales for accretion of the satellites are extremely short ($\sim 10^4$ yr) compared to those for the planets ($> 10^7$ yr). During the accretion process, forming satellites may well be shattered by high velocity collisions with accreting fragments. Support for this conclusion comes from the fact that craters are observed on Phobos and Mimas that are remnants of impacts almost large enough to have disrupted the satellites. Disruption and subsequent reaccretion may have occurred several times during the formation process for the inner satellites of Jupiter and Saturn. The rings may be the remnant of disrupted satellites within the Roche limit, where reaccretion would have been prevented by tidal forces. It should be noted that for satellites of large bodies, the debris will remain in the same orbit and may reaccrete but for small bodies, it may be lost completely. This may be why asteroids are not observed to have satellites.

Belief that the satellites of Mars are captured asteroids is based principally on their densities and dark surface appearances, both characteristic of carbonaceous chondritic asteriods. An alternate suggestion is that they formed in orbit around Mars from accretional debris left over from the formation of the planet. In either case they appear to be extremely primitive bodies: Their low densities and albedos indicate that they have not been melted and differentiated.

The satellites of the Uranus system are all prograde and equatorial, despite the high obliquity of the planet. This strongly implies that they formed from a protoplanetary disk that was effectively coupled to the planet during the period in which the obliquity change occurred, perhaps as the result of a massive impact. Alternatively, a succession of mutual gravitational interactions may have brought the satellites into the equatorial plane of Uranus.

The Neptune system is far more difficult to understand in terms of formation by a protosatellite disk. While the six small satellites newly discovered by *Voyager 2* all have regular, prograde orbits, Triton, a satellite comparable in size to the Galilean satellites, has a retrograde orbit with a large inclination. Nereid, a much smaller satellite, orbits at a much greater distance in a highly eccentric orbit. Triton and Neptune were probably formed in the solar nebula and captured by Neptune.

CAPTURE OF SATELLITES

For a satellite to be successfully captured into a bound orbit around a planet, it must lose a significant fraction of its kinetic energy. In general, it is difficult to propose plausible mechanisms for this process, with the exception of gas drag within a protoplanetary nebula.

The outer satellites of the outer planets, with highly inclined or retrograde orbits, were very likely captured by this process, perhaps during the terminal stages of planetary formation.

ORBITAL EVOLUTION

Tidal perturbations by an orbiting satellite produce varying gravitational forces throughout the planet. However, the cohesion of the planet causes internal strains to develop that prevent the planet from moving freely in response to the perturbations. Due to their mutual tidal interactions, the planet and the satellite develop

Table 2. Known Satellite Compositions

Moon	Rocky, iron, and volatile poor
Phobos, Deimos	Carbonaceous chondritic?
Io	Rocky, sulfur-rich crust
Europa	Rocky, ~ 100 km icy surface layer
Ganymede	60% rock, 40% ice
Callisto	60% rock, 40% ice
Mimas, Enceladus, Tethys, Dione, Rhea	~ 40% rock, 60% ice ($NH_4 \cdot H_2O$ ice?)
Titan	60% rock, 40% ice, N_2, CH_4
Iapetus	40% rock, 60% ice
Phoebe	Carbonaceous chondritic?
Miranda, Ariel, Umbriel, Titania, Oberon	40% rock, 60% ice ($CH_4 \cdot 7H_2O$ ice?)

elongated shapes, such that (in the absence of dissipation) their longest axes lie along the line connecting their two centers. In practice, however, internal dissipation of the tidal energy within the planet and satellite will prevent such an alignment from occurring. Tidal bulges will be formed on the planet either rotating behind the satellite (if the planet rotates more slowly than the satellite's mean motion) or ahead of the satellite (if the planet rotates more rapidly than the satellite's mean motion).

The gravitational attraction between the satellite and the asymmetric bulge causes the planet to eventually rotate with the same period as the satellite revolves (this has apparently occurred in the Pluto–Charon system). A similar tidal attraction also tends to slow the rotation of the satellite such that it always presents one face to the primary: This has occurred for every regular satellite in the solar system. To conserve angular momentum in the system, the excess energy is added to the satellite's orbit, causing it to expand. It is this process that has caused the Moon's orbit to expand significantly over geological history.

If several satellites orbiting the same planet have mean motions that have ratios near to those of small integers, a resonance results. Under these conditions, mutual perturbing forces may be reinforced, producing significant changes in orbital parameters and substantial tidal heating. We find this situation, for example, in the satellites Io, Europa, and Ganymede in the Jupiter system and Enceladus and Dione in the Saturn system.

Tidal forces also arise due to solar perturbations. These prevent the existence of satellites around Mercury and Venus: The orbits of such bodies would have decayed over timescales shorter than the history of the solar system.

The internal friction, or anelasticity, of the planet and satellite determine the rate at which tidal energy will be dissipated. This parameter is a function of the frequency of the perturbing force and of the internal constitution of the body. Most geological materials behave in both a viscous and an elastic manner at tidal frequencies; a good application of this behavior to planetary interiors is only just emerging as a result of numerical models.

INTERNAL EVOLUTION

Satellites with radii less than ~100 km evolve principally by impact. However, within larger satellites interior heat sources may substantially alter the thermal and mechanical structure. The heat sources available to a satellite include the energy released during accretion and differentiation of the core, heat produced by the decay of radioactive elements within the rock fraction of the satellite and heat liberated by internal friction due to tides.

The dominant heat source in the early history of the satellite is accretional energy, from the gravitational potential energy of the accreted material. While it is clear that larger satellites will be heated to a greater extent by this source, the exact variation of heating for different satellites is highly model-dependent. It is believed that accretional heating was sufficient to melt and differentiate the interiors of the Galilean satellites, with the possible exception of Callisto. Callisto, unlike Ganymede, appears to have an ancient unevolved surface. Because the two satellites are so similar in size and density, this is hard to understand. It may be that Ganymede, with a slightly greater fraction of rock, may have just been able to undergo differentiation, while Callisto remained unmelted. This question is currently unresolved and awaits more data from the *Galileo* mission to the Jupiter system.

Satellites composed largely of ice may have sufficiently great pressures in their interiors for various high pressure phases of ice to occur. The physical properties of these phases, which differ significantly from ice I (the familiar form of ice) have only recently become better understood as the result of extensive laboratory measurements. All of these phases are denser than water, unlike ordinary ice: Their differences in thermal conductivities, viscosities, and other material parameters complicate models of the thermal and structural evolution of the icy satellites.

Over much of the history of the satellites, radiogenic heating by the decay of U, Th, and K is the major heat source. The volumetric heating rate is proportional to the fraction of the mass of the satellite composed of silicates. Larger satellites generally would undergo greater heating, because radiogenic heating is proportional to volume and cooling is proportional to surface area. However, many other factors need to be taken into account, among them the silicate mass fraction of the satellite. For example, a 500-km radius satellite composed entirely of ice will not experience radiogenic heating. A 500-km radius satellite composed entirely of rock will heat up substantially due to radiogenic heating but will undergo little geologic activity, as the temperatures reached will be insufficient to liquefy rock. A satellite with a mixture of rock and ice will undergo the most geologic activity, as the heat released by the silicates will mobilize the ice, allowing surface modification by tectonism and resurfacing by ice flows.

Trace components in the ice may also substantially affect the geologic evolution of ice-rich satellites. Ammonia can form a hydrate with ice that has a melting point of 173 K, allowing substantial melting, even in the intermediate-size satellites of Saturn. Such melting has been proposed as a mechanism for the resurfacing of Enceladus, perhaps following an episode of tidal heating. The incorporation of methane with ice in the form of a clathrate may permit solid-state ice flows at temperatures below 100 K: This process may be responsible for producing viscous flow features observed on the uranian satellites Ariel and Miranda.

Triton exhibits signs of an active geologic history, including smooth lake-like plains and a complex fractured surface known as "cantaloupe terrain." These features are almost certainly the result of viscous flow of ammonia- or methane-rich ice. Active volcanic eruptions, probably of solar-heated nitrogen from frozen subsurface deposits, were also observed. These eruptions may be the source of the thin (10^{-5}-bar) atmosphere.

Tidal heating may be a very important heat source for certain satellites at particular times. The rate of tidal heating is substantially enhanced where melting can produce an inviscid layer, allowing the core and elastic lithosphere to flex separately. This appears to have occurred within Io, and perhaps Europa. In the latter satellite, tidal heating may maintain a liquid ocean underneath the icy surface. Tidal heating may have also been responsible for resurfacing observed on Enceladus and Miranda.

The dominant mechanisms for removing heat from a satellite's interior are conduction and subsolidus convection. The latter is strongly dependent on the rheology of the satellite's interior. If solid material within the satellite can undergo sufficient creep deformation for it to flow as a fluid over geological timescales, then the thermal state of the interior will self-regulate because the effective viscosity of the material is strongly temperature-dependent: An increase in internal temperature will produce more vigorous convection, and vice versa. Although a substantial body of data describing the rheology of various rock types has been available for some time, our knowledge of the rheology of ice at the low temperatures appropriate to the icy satellites is considerably more recent and less complete.

The tectonic past of the icy satellites appears to be dominated by a history of extension. Given our incomplete knowledge of the evolution of the icy satellites, such extension may have arisen as a result of thermal expansion, differentiation, phase transitions of the ice within the satellite (due to temperature changes), or some combination of these processes.

THE MOON

Although any introductory discussion of the satellites will strive for generality, it is clear that the Earth–Moon system is unique. If one ignores the Pluto–Charon system because of its own anomalies, the Moon's mass as a fraction of the Earth's ($\frac{1}{81}$) is the largest in the solar system by an order of magnitude. The Moon is also unusually distant from the Earth: some 60.4 planetary radii. In addition, the mean density of the Moon is significantly lower than that of the other inner planets. Rocks returned from its surface show chemical anomalies: water is virtually absent, compared to terrestrial rocks. However, oxygen isotope ratios are similar to those of basaltic rocks on Earth (but very different from those of meteorites).

Of all of these properties, the large Earth–Moon distance may be clearly understood from our knowledge of the Earth. Tidal dissipation in the oceans of the Earth has caused the Moon's orbit to evolve outwards from perhaps one-sixth of its current distance over 4.5 Gy (the exact amount is dependent on assumptions concerning the ancient lunar orbit and the extent of tidal dissipation in the oceans).

Understanding other aspects of lunar composition and history, however, is more difficult. Although the Moon's orbit lies in the ecliptic, it is unlikely that it was gravitationally captured during a close encounter. On the other hand, an origin via fission from the Earth, as was suggested by Darwin, is dynamically implausible. If the Moon accreted in orbit around the Earth, the protosatellite disk must have been much more massive, relative to the Earth, than was the case in the outer solar system. Various chemical similarities and differences between Earth and Moon rocks, as briefly described above, only complicate the problem further: The Moon's crust resembles the Earth's isotopically, but not in terms of volatile content.

One recent model that is currently popular in the scientific community that attempts to address these issues is the impact-trigger hypothesis, which proposes that, during the final stages of accretion, a Mars-sized body impacted the Earth at a glancing angle. The vapor erupted from this impact would contain shocked and heated material from the primordial crust and upper mantle from which volatiles would have been depleted. The subsequent evolution of the heated cloud of material can be numerically modeled using large hydrodynamic codes: It appears that suitable impact parameters can be found that will permit the formation of a Moon-sized body. At this time, only the dynamical environment of the collision may be calculated with any confidence: We are able to say very little about the thermal and chemical states of the post-impact cloud. A more complete understanding of this recent, but important, model requires a great deal of progress in these areas.

Additional Reading

Beatty, J. K. and Chaikin, A., eds. (1990). *The New Solar System*, 3rd ed. Sky Publishing Corp., Cambridge, MA, and Cambridge University Press, Cambridge, U.K.

Burns, J. A. and Shapley, M. S., eds. (1986). *Planetary Satellites*. University of Arizona Press, Tucson.

Hartmann, W. K., Phillips, R. J., and Taylor, G. J., eds. (1986). *Origin of the Moon*. University of Arizona Press, Tucson.

Morrison, D., ed. (1982). *Satellites of Jupiter*. University of Arizona Press, Tucson.

See also **Io, Volcanism and Geophysics; Jupiter and Saturn, Satellites; Moon, Origin and Evolution; Planetary and Satellite Atmospheres; Pluto and Its Moon; Satellites, Ices and Atmospheres; Satellites, Minor; Uranus and Neptune, Satellites.**

Satellites, Ices and Atmospheres

Jonathan I. Lunine

The moons, or satellites, of the outer solar system span a wide range of sizes, compositions, and physical properties. Because of the low ambient temperature beyond the asteroid belt, both at present and at the time of formation, many of the satellites are made up largely of water ice. In addition, a number of molecular species that exist in the gaseous phase on Earth are inferred to be present in or on these bodies as solids or liquids. Three of the moons, Io, Titan, and Triton, have tangible atmospheres. This combination of exotic ices and, in some cases, atmospheres is responsible for a variety of fascinating and occasionally bizarre features on these distant worlds. The information on satellite ices and atmospheres has been gathered from a combination of powerful ground-based techniques, such as spectroscopy, and flybys of the outer planets by the United States unmanned probes *Pioneers 10* and *11* and *Voyagers 1* and *2*.

The first direct confirmation of an atmosphere around a moon of the outer planets was made by the astronomer Gerard P. Kuiper, who, in 1944, detected the spectral absorption lines of methane while observing Saturn's satellite Titan. The Nobel Prize winning chemist Harold C. Urey developed a theoretical model for the chemical makeup of the outer solar system, predicting that with increasing distance from the Sun, progressively more volatile materials would be found in planetary satellites. (Volatility refers to the propensity of a chemical species to exist in the vapor phase, rather than condensed state, over a given range of temperatures.) Spectroscopic observations in the 1970s indicated the presence of a thick haze around Titan, and the occurrence of methane on the surfaces or in the atmospheres of Neptune's moon Triton and the outermost, moon-sized planet Pluto. It remained for the spacecraft flybys of the 1970s and 1980s to reveal the full nature of the atmospheres of Titan and Triton, to detect and measure a tenuous atmosphere around Jupiter's volcanically active moon Io, and to reveal a variety of unusual geologic features on many of the icy satellites of the outer solar system.

ICES IN THE OUTER SOLAR SYSTEM

The planets and their moons are thought to have formed from a disk of gas spun out during the formation of the Sun 4.5 billion years ago. This disk, called the solar nebula, probably had an elemental composition very similar to that of the Sun. Observations of the Sun, meteorites, and interstellar clouds indicate that the most abundant elements in the solar neighborhood, after hydrogen and helium, are oxygen, carbon, and nitrogen (we exclude the noble gases from this discussion). These three elements form a variety of molecular species, but chemical models predict that water (H_2O), methane (CH_4), carbon monoxide (CO), carbon dioxide (CO_2), ammonia (NH_3), and molecular nitrogen (N_2) are the most stable molecular forms likely to have been present. Some of the oxygen, along with silicon, magnesium, and other elements, combine to form silicates, or "rock." Together with iron and nickel, rock is the most abundant planet-forming material in the inner solar system. Beyond the asteroid belt, however, temperatures in the nebula were low enough that condensed water was an important moon-building material.

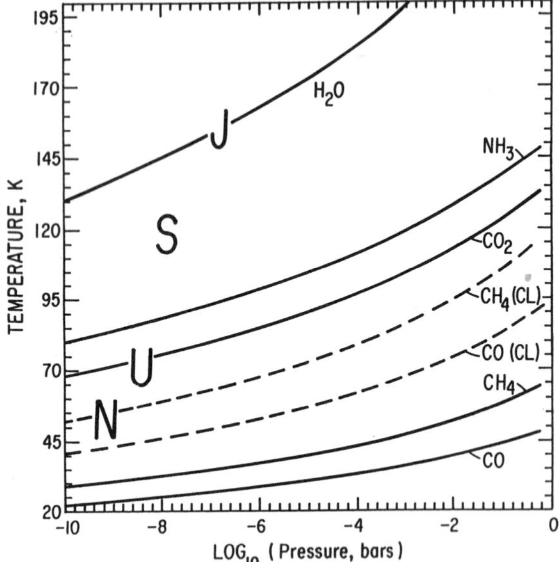

Figure 1. Low-temperature condensation in the outer solar system for a gas whose elemental composition is that of the Sun. The plot is condensation temperature versus total pressure in the nebula. The approximate positions of the forming giant planets Jupiter, Saturn, Uranus, and Neptune (indicated by J, S, U, N) are shown based on solar nebula models. The dashed lines show the temperatures at which water ice would convert to clathrate in the presence of methane and carbon monoxide.

Figure 1 shows a "condensation sequence" for materials in the outer solar system. At any given distance from the early Sun, and at a particular time, the solar nebula had a fixed temperature and gas pressure (of mostly hydrogen along with much smaller amounts of other molecular species). The condensation sequence is a guide to what molecules could have condensed from the nebula in the outer solar system. The approximate locations of the giant planets are indicated. Any molecular species with a condensation line above a particular planet on the figure was likely to be present in condensed form. Rock condenses out at such a high temperature that it is not shown on the figure. The diagram demonstrates that water ice, and perhaps ammonia, were ubiquitous condensates in the early outer solar system, and were incorporated in most of the solid bodies there. The condensation of methane and the other species is more problematic: Temperatures may not have been low enough to allow this to happen even at Neptune. Physical chemists have proposed that these species may instead have been trapped in water ice in an arrangement known as a clathrate hydrate. The temperatures at which such trapping occurs are shown by the dashed lines. Although such clathrates could have contained a great deal of volatile species, such as methane and carbon monoxide, the physical properties, such as density, of the clathrates are very similar to normal water ice.

A complication to the preceding story is that the abundances of the molecular species in the gas phase varied from place to place in the solar nebula. Close to the giant planets Jupiter and Saturn, gas densities were high, and CH_4 would have been more abundant than CO, and formation of NH_3 favored over that of N_2. Also, because of the high gas densities, species would have condensed out at higher temperatures than shown on Figure 1. Far from the giant planets (and possibly in regions near Uranus and Neptune), gas densities were lower, and CO and N_2 were the dominant carbon and nitrogen species. Thus it is thought that the satellites of Saturn, such as Titan, incorporated large amounts of methane and ammonia during formation, whereas Pluto received more carbon monoxide than methane, and perhaps very little ammonia.

By abundance, water ice and rock were the most ubiquitous condensates in the outer solar system. Therefore, the present-day physical properties of the satellites should be determined by these materials. Most of the satellites of the outer solar system (as well as Pluto) have densities between 1 and 2 g cm^{-3}, consistent with a mix of rock and water ice. The other, more volatile substances have perhaps played a role in shaping the tectonic features seen on some of the satellites, and in a few cases are present in satellite atmospheres, surface frosts, and possibly liquids.

ICES ON SATELLITE SURFACES

Satellites of Jupiter

We consider here the classical galilean satellites. Ganymede and Callisto, the two outermost satellites, have densities consistent with mixtures of rock and water ice and planet-sized radii (roughly 2600 km for Ganymede). Both bodies are very heavily cratered, but Ganymede contains large areas of resurfaced terrain containing features resembling grooves in the Voyager images. These areas are thought to represent tectonic activity largely generated in the icy component of this largest moon of the solar system. Moving inward, Europa has a density close to 3 g cm^{-3}, indicative of a largely rocky composition, but a surface of water ice, largely devoid of craters. Apparently Europa suffered significant heating early in its history, which brought most of the water to the near-surface and perhaps resulted in substantial loss of water to space. Current speculation centers on whether there is a liquid water ocean beneath the icy surface, or a solid ice mantle instead. The innermost galilean satellite, Io, was revealed by the *Voyager 1* spacecraft to be very active volcanically. The source of the satellite's impressive activity is tidal heating incurred as Io is gravitationally tugged by Jupiter and by the other galilean satellites. The strong heating has removed any vestiges of the water ice that may have been present early on, and Io now has a density consistent with pure rock. The volcanically active regions contain sulfur compounds in addition to silicates. The Pioneer and Voyager spacecraft found evidence of a tenuous atmosphere, perhaps associated with some of the volcanic vents; sulfur dioxide is a part of this atmosphere at a maximum pressure of one-millionth of an atmosphere near the Loki volcanic area (the pressure at the Earth's surface is 1 atm). The global distribution of this atmosphere and its full composition remain unknown. The United States space probe *Galileo* will arrive at Jupiter in the mid-1990s to investigate in much greater detail these diverse, planet-sized moons.

Satellites of Saturn

With the exception of Titan, the moons of Saturn are substantially smaller (radii 200–800 km) and contain proportionately more ice than do the galilean satellites. The most unusual of the Saturn satellites, Iapetus, exhibits a heavily cratered, relatively bright hemisphere, whereas the other side is covered with a very dark material. Iapetus is a mixture of rock and water ice, based on the density, but neither material is likely responsible for the dark component. One suggestion is that the material had its origin in methane that was incorporated in the satellite, then outgassed and was transformed by radiation into a high molecular weight, organic substance (crudely akin to tar). Alternatively, the dark material may be an organic dust, similar to that found in meteorites, blasted off the moon Phoebe and splattered across Iapetus. The remaining icy saturnian satellites display a range of tectonic features and heavily cratered terrains. Most noteworthy is Enceladus, which contains broad areas resurfaced by flows and globally coated with bright material. The physical chemistry of liquid ammonia and water mixtures makes them strong candidates for the flow features, and ammonia ejected volcanically from the interior could be responsible for the bright coating as well as for Saturn's E-ring,

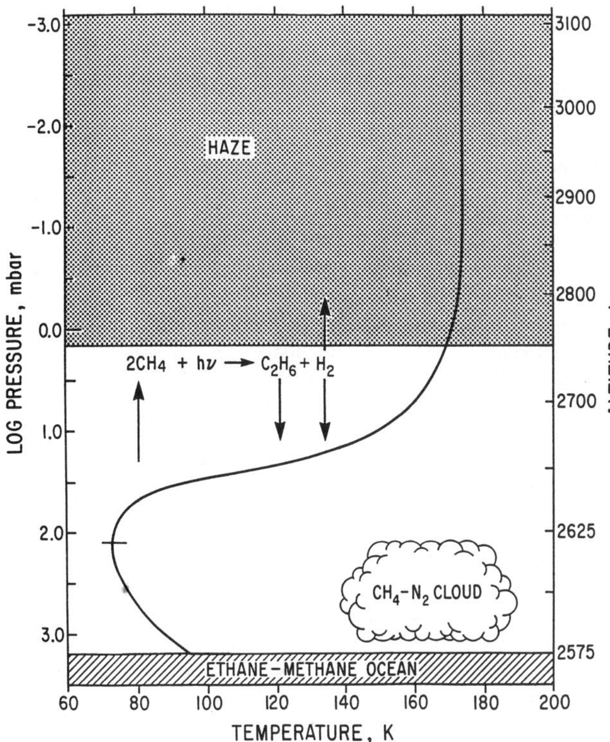

Figure 2. The atmosphere and surface of Titan based on Voyager data. The curved line plots temperature versus atmospheric pressure (left axis) and distance from the center of Titan (right axis). The haze is actually broken up into discrete layers. The methane photochemical cycle is indicated schematically.

within which the orbit of Enceladus is imbedded. No positive identification of ammonia has been made on any of the icy satellite surfaces, however.

Saturn's giant satellite Titan distinguishes this system. Almost as large as Ganymede, and with the same density, this moon is encased in a thick, hazy atmosphere that completely hides the surface. After decades of spectroscopic study from Earth, the atmosphere was investigated in detail during the flyby of *Voyager 1* in 1980. The surface pressure is 1.5 atm, with a composition of over 90% molecular nitrogen and between 1 and 10% methane at the surface. The noble gas argon may contribute upward of 10% to the composition as well, although this cannot be confirmed from *Voyager 1* data. The surface temperature is 95 K, decreasing to 70 K several tens of kilometers above the surface, and increasing again to perhaps 170–200 K high in the atmosphere, as shown in Fig. 2. The way the temperature of Titan's atmosphere changes with altitude bears a remarkable resemblance to that of the Earth: The main differences are that Titan's atmosphere is much colder and more extended than that of the Earth.

Based on detailed study of the *Voyager 1* data, scientists believe that methane plays a role closely analogous to that of water in the Earth's atmosphere: It forms clouds in the lower atmosphere, likely produces rain, and may form a substantial liquid reservoir on Titan's surface (akin to our water oceans). However, methane on Titan participates in a chemical cycle more complex and elaborate than that of water on Earth. Methane molecules rise into Titan's upper atmosphere, are broken apart by sunlight into fragments, and recombine to form heavier molecules with loss of hydrogen to space. This "photochemical cycle" produces molecules of acetylene (C_2H_2), ethane (C_2H_6), and heavier species, that condense and rain out onto the surface. The acetylene forms a solid at Titan's surface, but the ethane remains liquid and so contributes to the surface ocean. A third major component of the ocean is molecular nitrogen, dissolved from the atmosphere, so that the surface of Titan may have a chemically-complex liquid layer of ethane, methane, and nitrogen, with a "sediment" of acetylene. Many other hydrocarbons are formed in the upper atmosphere and contribute to the ocean and sediment.

Models of the long-term evolution of Titan's complex surface and atmosphere suggest a wealth of chemical and meteorological processes that control the climate of this satellite, and that may erode surface features in the bedrock, such as ancient craters. Equally intriguing is the origin of Titan's atmosphere, which contains both molecular nitrogen and methane, which are expected to have been abundant in different parts of the outer solar nebula. An outstanding question is whether ammonia exists, locked in the bedrock of Titan. A look at this intricate world is planned by the United States and Europe after the turn of the century, when the *Cassini* spacecraft maps the Titan surface by radar and sends a probe into Titan's atmosphere.

Satellites of Uranus

Similar in size to the icy saturnian satellites, the major moons of Uranus are darker, and slightly denser, than their saturnian counterparts. No spectroscopic evidence for material other than rock and water ice is present on their surfaces, which exhibit a variety of tectonic features. Most bizarre is Miranda, the smallest of the five major uranian satellites. Its surface is dominated by regional, oval-shaped assemblages of ridges and canyons called coronae. Such geologic activity indicates that more than just water ice must be present on this body, because so far from the Sun water ice itself is difficult to mobilize. Compounds of water, ammonia, and possibly methane or carbon monoxide are implicated.

Satellites of Neptune

Neptune's major satellite, Triton, orbits retrograde and has an atmosphere. Most of what we know of its physical properties is derived from the *Voyager 2* flyby in August, 1989. Triton's radius is approximately 1350 km; it has a density of 2 g cm^{-3}, indicating a mix of rock and ice. Spectroscopic observations from the Earth and *Voyager 2* indicate that both molecular nitrogen and methane are present on the surface and in the atmosphere. Being the more volatile gas, nitrogen dominates the atmospheric composition. Other gases, such as carbon monoxide and argon, could be present in trace amounts. The surface pressure and temperature were determined by *Voyager 2* to be 15 millionths of an atmosphere and 38 K, respectively. This makes Triton the coldest body yet explored in the solar system. Because Triton orbits Neptune inclined to the planet's equator, and the planet itself is tipped relative to the Sun, Triton undergoes bizarre, complex seasons, varying from tens to hundreds of years in duration, with the Sun (as seen from Triton) swinging up over 50° latitude at times. Nitrogen in the form of surface frosts is sublimed into the atmosphere by sunlight, and moves globally in response to the changing seasonal Sun angle. Plumes of nitrogen seen by *Voyager 2* in Triton's summer hemisphere may be caused by solar heating of nitrogen ice, initiating weak geysering and hence jetting of material into the atmosphere.

Pluto and Charon

The outermost planet is smaller (radius about 1100 km), than Triton and has a moon about half its own size. Ground-based spectroscopy revealed methane on Pluto's surface and/or in its atmosphere, whereas an infrared Earth-orbiting satellite (*IRAS*) measured a surface temperature of roughly 50–55 K. An occultation of Pluto by a star, observed in 1988, revealed that the atmosphere is tenuous indeed: only several millionths of an atmosphere pressure. Data from the occultation also point strongly toward an atmospheric composition of more than just pure methane: carbon monoxide and/or molecular nitrogen may be present. Because of Pluto's low gravity, the atmosphere is escaping over time, and must be resupplied, perhaps from polar caps several kilometers in thickness. Ground-based data indicate that Pluto's moon Charon has no atmosphere, but may tidally perturb the gas escaping from Pluto.

Additional Reading

Burns, J. A. and Matthews, M. S., eds. (1986). *Satellites*. University of Arizona Press, Tucson.

Encrenaz, T. and Bibring, B. (1990). *The Solar System*. Springer, Berlin.

Johnson, T. V. and Matson, D. L. (1989). Io's tenuous atmosphere. In *Origin and Evolution of Planetary and Satellite Atmospheres*, S. K. Atreya, J. B. Pollack, and M. S. Matthews, eds. University of Arizona Press, Tucson, p. 666.

Lunine, J. I., Atreya, S. K., and Pollack, J. B. (1989). Present state and chemical evolution of the atmospheres of Titan, Triton, and Pluto. In *Origin and Evolution of Planetary and Satellite Atmospheres*, S. K. Atreya, J. B. Pollack, and M. S. Matthews, eds. University of Arizona Press, Tucson, p. 605.

See also **Io, Volcanism and Geophysics; Jupiter and Saturn, Satellites; Planetary and Satellite Atmospheres; Pluto and Its Moon; Uranus and Neptune, Atmospheres.**

Satellites, Minor

Peter C. Thomas

Satellites of the planets fall into two size and morphologic classes: large, ellipsoidal (radius $r > 200$ km) and small, irregularly shaped ($r < 200$ km). The small, or "minor," satellites are difficult to study, but because they are considered to be analogs of many asteroids and comet nuclei, they have attracted considerable research effort. Prior to exploration of the planets by spacecraft, knowledge of these objects was largely restricted to their orbital elements and to spectra of some of the outer jovian satellites. Spacecraft data, largely consisting of images, have shown the small satellites to have a variety of surface features and to pose special problems in interpreting their histories. These data have come from the Mariner and Viking missions to Mars (1971–1972; 1976–1980), and from the Voyager flybys of Jupiter (1979), Saturn (1980, 1981), and Uranus (1986). Resolution of features on the martian satellites is generally a few meters; on the other satellites it is a few kilometers.

ORBITS AND ROTATION

Table 1 lists the known small satellites. They occur in three main circumstances: in close, nearly circular and independent orbits of the primary; in libration with much larger satellites sharing the same orbit (saturnian satellites Telesto, Calypso, and Helene); and in inclined, even retrograde, distant orbits. Many of those in close circular orbits are associated with planetary rings, and impacts on these satellites may supply the particles that compose the rings. The satellites themselves gravitationally confine some of the ring systems. Hyperion is an exception to this simple orbital division in that it orbits Saturn between the large satellites Titan and Iapetus.

The small satellites that orbit close to primaries are all in synchronous rotation: Their long axes point toward the planet. Hyperion exhibits chaotic rotation because of torques applied to its irregular shape by Titan. The very distant small satellites have rotation periods independent of their orbital periods because of the minimal tidal influence at those distances from their respective planets.

COMPOSITIONS

Ground-based spectral data for Phobos and Deimos show they have compositions similar to carbonaceous meteorites or to C-type asteroids. Their densities, about 2 g cm^{-3}, are consistent with such a composition. Yet the uncertainty is great enough that there is still speculation that ice might be a significant component of the martian moons.

The composition of the inner jovian satellites is unknown, but cosmogonic considerations suggest that they are rocky. The color of the surface of Amalthea implies contamination by sulfur from volcanoes on Io, and thus its bulk composition may be hidden from remote sensing measurements. The spectra of the outer jovian satellites appear similar to those of C-type asteroids in the outer asteroid belt, and these satellites are thus classified as rocky.

The high albedos of most of the saturnian small satellites indicate they are composed of ice, or are at least covered by it. The spectroscopic detection of water ice on the larger saturnian satellites, Hyperion, and the rings, combined with the densities of the saturnian satellites, suggests that the small ones are also largely made of ice. Masses of two of the small satellites, Janus and Epimetheus, can be approximately calculated from orbital resonances, and the results give densities consistent with porous ice. Phoebe is dark (visual albedo of 0.06 with patches having albedos of up to 0.08) and could be rocky or icy with a dark surface layer. Hyperion appears to be slightly dirty ice from its spectrum and albedo.

The small uranian satellites appear to be dark (only two have measured albedos, both about 7%) and may either be composed entirely of one of the dark materials found in outer solar system objects, or may be icy with dark surface residues. They appear to be slightly brighter than the material that makes up the uranian rings.

SIZES AND SHAPES

The minor satellites range from 6–150 km in mean radius. All are irregularly shaped, that is, their shapes are not well fitted by spheres or ellipsoids and they can present distinctly asymmetric appearances (Fig. 1). Those that are well measured have long dimensions that average about 1.7 times their shortest dimension. The greatest ratio is 2.0. Local topography can be difficult to define on such objects, but some crater walls are 15 km high on Amalthea.

SURFACE FEATURES

Impact craters are the major surface features of the small satellites. In good images they appear quite similar to those on the Moon and other airless objects, with some minor variations. Ejecta blocks as large as 150 m across are present on Phobos and Deimos.

The upper layer of the Moon was found to consist of fragmental debris from impacts (the regolith), and both Phobos and Deimos are covered with particulate debris, evidently derived from impact ejecta. Layers exposed in craters suggest that the depth of regolith on Phobos may be well over 100 m. On Deimos, loose material fills in some craters to depths of tens of meters. Some regolith creeps downslope (see Fig. 1) and may have filled low areas to depths greater than 200 m. It is sometimes regarded as surprising that gravity less than 1/1000 that on Earth would be effective at moving material, but thermal expansion and contraction and shaking by impacts may assist in allowing the gravity to direct the resultant motion of regolith particles.

Phobos' surface is crossed by dozens of linear troughs, usually termed grooves. They are mostly a few meters deep, tens of meters wide, and several kilometers in length. Most occur in sets of parallel members, a pattern that suggests a relationship to fractures within the body of Phobos. Theories of their origin focus on fracture mechanisms (impact or tidal forces) and the means by

Table 1. Small Satellites (For Neptune, also see tables on pp. 607, 616.)

Primary Satellite	*Semimajor Axis of Orbit ($\times 10^3$ km)*	*Satellite Radii a, b, c (km)*	*Albedo*	*Composition*
Mars				
Phobos	9.4	13.4, 11.2, 9.2	0.06	Carbonaceous?
Deimos	23.5	7.5, 6.1, 5.2	0.07	Carbonaceous?
Jupiter				
Metis	128.0	20, —, 20	0.05–0.1	Rock?
Adrastea	129.0	12, 10, 8	0.05–0.1	Rock?
Amalthea	181.3	131, 73, 67	0.06	Rock?
Thebe	221.9	55, —, 45	0.05–0.1	Rock?
Leda	11,110	5	—	Rock?
Himalia	11,470	90	0.03?	Rock?
Lysithea	11,710	10	—	Rock?
Elara	11,740	40	0.03?	Rock?
Ananke	20,700	10	—	Rock?
Carme	22,350	15	—	Rock?
Pasiphae	23,300	20	—	Rock?
Sinope	23,700	15	—	Rock?
Saturn				
Atlas	137.7	18, —, 14	0.5	Ice?
Prometheus	139.4	74, 50, 34	0.5	Ice?
Pandora	141.7	55, 44, 31	0.5	Ice?
Janus	151.4	97, 95, 77	0.5	Ice?
Epimetheus	151.5	69, 55, 55	0.5	Ice?
Telesto	294.7	—, 12, 11	0.6	Ice?
Calypso	294.7	15, 8, 8	0.9	Ice?
Helene	378.1	18, —, 14	0.6	Ice?
Hyperion	3,560	170, 120, 100	0.25	Ice?
Phoebe	13,210	115, 110, 105	0.06	Rock?
Uranus				
Cordelia	49.8	13	0.065	Rock?
Ophelia	53.8	16	0.07	Rock?
Bianca	59.2	22	0.07?	Rock?
Cressida	61.8	33	0.07?	Rock?
Desdemona	62.7	29	0.07?	Rock?
Juliet	64.4	42	0.07?	Rock?
Portia	66.1	55	0.07?	Rock?
Rosalind	69.9	29	0.07?	Rock?
Belinda	75.2	34	0.07?	Rock?
Puck	86.0	77	0.074	Rock?
Neptune				
Nereid	551.5	170		????

which the regolith was disturbed along the fractures (drainage into fractures or ejection from area above fracture). Grooves have not been observed on any other satellite, but only for Deimos do the data allow a firm negative finding.

Albedo markings showing downslope movement also occur on Amalthea. Some of the icy satellites have fuzzy albedo markings, but distinct downslope movement or crater ejecta blankets are not visible on the small icy satellites. Some of the albedo features on Phoebe may be more icy material exposed by impact craters.

The small satellites lack any signs of tectonics driven by internal heat sources. This is hardly a surprise given the very small amount of heating possible in most such objects.

IMPACT CRATERS AND THE SHAPING OF SMALL SATELLITES

The diameter of the largest well defined craters averages about 0.7 times the body's mean radius. There are large indentations seen in the limbs of most small satellites that are not immediately identifiable as craters, but which appear to be fragmentation or spallation scars; they are commonly over 1 satellite radius across. A large indentation in the southern hemisphere of Deimos has a width of 1.7 times the mean radius.

Many of the small satellites orbiting close to planets are expected to have more large craters than are in fact seen. The expected number of craters can be inferred from the numbers of craters (as a function of size) seen on larger satellites orbiting farther from the planet. More and larger craters are formed closer to the planet because of three effects of a satellite being located deeper in the planet's gravity well: impacting particles have higher velocities, satellites orbit faster, and the number of impactors is increased. This scaling suggests that the inner satellites of Jupiter, Saturn, and Uranus should have experienced impacts sufficiently energetic to form craters with diameters equal to or greater than the satellites' diameters. Thus, there is the strong expectation that these objects may have been completely fragmented, perhaps many times, during their histories. Complete fragmentation by impacts would form a temporary cloud of debris in the satellite's orbit, but most of the debris would reaccumulate by its self-gravity. Such might not be the case for asteroids or objects in very long period orbits. Indeed, the outer satellites of Jupiter appear to be fragments that have not been gravitationally reassembled. Many of the asteroids

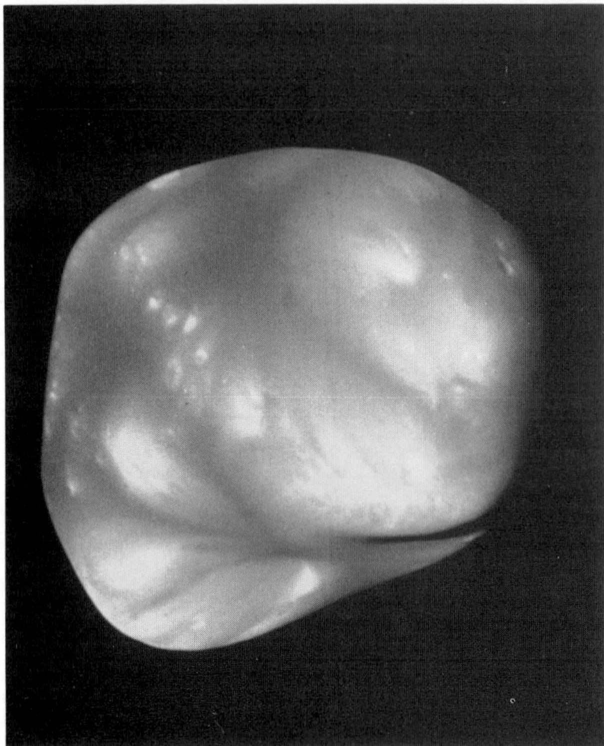

Figure 1. The martian satellite Deimos. In this view the satellite is 12 km across. Bright streamers show that surface material creeps downslope. The smooth appearance of the satellite shows the cumulative effects of such movement. The other martian moon, Phobos, is much more lunar in appearance. A major mystery is why these two satellites, similar in composition, have such different surfaces.

have almost certainly suffered complete fragmentation, which is one reason small satellites are considered likely analogs, at least in some characteristics, of asteroids.

Why are the small satellites irregular and the larger ones ellipsoidal? Two major factors are involved: The smaller objects can be fragmented, or spalled, by impacts, and on the larger ones gravity can force relaxation of topography. Gravitational binding energy varies as the fifth power of the radius; binding due to material strength varies as the cube of the radius. Thus the relative effect of gravity, tending to make objects ellipsoidal, to strength, which can support irregular shapes, varies as the square of the radius. This implies there should be a rapid change with increasing radius from shapes determined by strength to those determined by gravity. Relaxation of topography on icy satellites appears to become effective over a narrow range of mean radii because several factors, all dependent on radius, contribute to viscous relaxation. Gravity varies linearly with radius and density. The larger the satellite the more internal heat it will retain, and the more likely that tidal heating will be effective. Because the viscosity of ice is very strongly dependent on temperature there should be a point where a small increase in radius will mean a very great reduction of the resistance to relaxation. For rocky objects the story may not be so simple, and some asteroid data suggest that the transition from irregular to smooth objects may occur over a broader diameter range than for icy bodies.

ORIGINS AND ORBITAL EVOLUTIONS

The orbital positions of the small satellites probably reflect three major modes of origin. The close satellites in regular orbits probably formed in place from the circumplanetary material that condensed to form the large satellites. Many of these satellites, however, may be fragments from much larger precursor objects that condensed close to the planet. Satellites in trojan orbits with larger ones may represent debris from very large impacts on the large satellites. The distant satellites of Jupiter, with their highly inclined orbits—one set are prograde, the other set are retrograde—are almost certainly captured asteroids. The two sets of satellites may represent two asteroids, captured and fragmented. Phoebe, likewise, would appear to be a captured object.

Orbits of satellites can evolve by tidal effects and orbital resonances that add or remove energy. Although some small satellites may not have orbitally evolved very much, a few may have moved considerably closer to their primaries. Phobos is a particularly interesting case, because estimates of the rate of its evolution inward toward Mars suggest a remaining lifetime of order 10^8 yr, a very small fraction of the age of the solar system. There may have been other small martian satellites that already evolved inward and impacted the surface of Mars.

FUTURE OF RESEARCH ON SMALL SATELLITES

Fundamental properties of many small satellites remain unknown (see Table 1). Some of the mysteries of the martian moons may be investigated by spacecraft in the 1990s, and some limited data on the small jovian satellites should come from the *Galileo* orbiter in the mid-1990s. Further spacecraft exploration of the saturnian satellites may happen only after the end of the century. Comparison studies of asteroids and comet nuclei are likely to provide some of the stimulus for more research on the small satellites.

Additional Reading

Burns, J. A. (1986). The evolution of satellite orbits. In *Satellites*, J. A. Burns and M. S. Matthews, eds. Univ. Arizona Press, Tucson, p. 117.

Hamblin, W. K. and Christiansen, E. H. (1990). *Exploring the Planets*. Macmillan Publishing Co., New York.

Housen, K. R. and Wilkening, L. L. (1982). Regoliths on small bodies in the solar system. *Ann. Rev. Earth and Planetary Sci.* **10** 355.

Thomas, P., Veverka, J., and Dermott, S. (1986). Small satellites. In *Satellites*, J. A. Burns and M. S. Matthews, eds. Univ. Arizona Press, Tucson, p. 492.

Veverka, J. and Burns, J. (1980). The moons of Mars. *Ann. Rev. Earth and Planetary Sci.* **8** 527.

See also **Jupiter and Saturn, Satellites; Planetary Rings; Satellites; Satellites, Rotational Properties; Uranus and Neptune, Satellites.**

Satellites, Rotational Properties

Stanton J. Peale

Properties of satellite rotation that are observable in principle include the rotation period, the orientation of the spin axis relative to the orbit plane, precession of the spin axis due to gravitational torques, non–principal-axis rotation (wobble), and deviations from uniform principal-axis rotation (libration). Considerable order is observed in current satellite rotation states, and it is of interest to ascertain how this order came about and why some satellites do not conform to the norm. There is a strong coupling between the spin and orbital motions that is primarily responsible for maintaining the ordered rotation states in most cases, but this coupling is equally responsible for destroying any chance of orderly rotation for Saturn's satellite Hyperion. Understanding the processes that constrain current rotation states, as well as processes of an evolutionary nature that could have brought the individual satellites to their observed rotation and orbit states, allows us sometimes to infer interior properties of some satellite or even of its primary planet, although attempts to deduce primordial rotation states are usually frustrated. This entry summarizes the observed rotational properties of the planetary satellites, outlines our understanding of the processes maintaining and of those leading to the observed states, and indicates some of the inferences that can be drawn about intrinsic properties of the bodies themselves.

Other than our own Moon, on which we can easily see many distinct features, a satellite's rotation period usually has been determined by photometric observations of sunlight scattered from the surface. Such scattering varies periodically both because of variations in surface albedo around the satellite and, for the small satellites, because of variations in the observed cross section of a nonspherical shape. Spacecraft observations have determined rotation periods of several smaller satellites. All but two of the observed satellite rotations have been found to be synchronous with their orbital motions, which means that rotational and orbital periods are the same, with one satellite hemisphere always facing the planet. Our own Moon rotates synchronously with its orbital motion and, before spacecraft observations became possible, we had seen only one hemisphere.

The Moon's spin axis maintains a constant inclination of 1°32.′5 relative to a vector that is perpendicular to the plane of the Earth's orbit about the Sun. (A vector that is perpendicular to a plane is called a normal to the plane, and the Earth's orbital plane is called the ecliptic plane.) Moreover, the Moon's spin axis, its own orbit normal, and the ecliptic plane normal remain coplanar as the lunar orbit (inclined by a little over 5° with respect to the ecliptic plane) precesses due to gravitational torques, mainly from the Sun. The Moon's spin axis maintains an inclination of 6°41′ with respect to its own orbit normal in this coplanar configuration. These properties of the lunar rotation are called Cassini's laws, and they are generalizable to those other satellites and planets in the solar system whose orbits precess about the normal of some nearly invariable plane while maintaining a nearly constant inclination to that plane. For each satellite, this invariable plane lies between the equatorial and orbital planes of the planet. It is close to the former for satellites near a planet and close to the latter for distant satellites.

The Moon's rotation rate and that of Mars' satellite Phobos also are observed to be not quite constant. Periodic variations lead to observable angular deviations (called librations) from what would be a mean orientation at a particular point in the orbit. Another irregularity occurs when a spin axis does not coincide with the axis of maximum moment of inertia of a solid satellite (e.g., the shortest axis of a uniformly dense ellipsoid). In this case the axis precesses in the frame of reference fixed in the satellite, although such "wobble" amplitudes are currently too small to be observable. In Table 1 the names and properties of the known satellites are grouped under their respective primary planets in the order of increasing orbit semimajor axis. The Roman numerals associated with the satellites designate roughly the order of discovery.

Satellites, like the terrestrial planets, are thought to have formed by the gravitational accretion of small, solid pieces. As such a process leads to a dispersion of rotation periods and spin axis orientations, the current order must be the result of an evolutionary process, which is identified as the consequence of the tidal distortion of a satellite in the field of its primary. All coastal dwellers are very much aware of the twice-daily rise and fall of the ocean level that we call the tide; almost none realize that the solid body of the Earth also rises and falls twice daily from the same gravitational effects of the Moon and Sun but with a considerably smaller amplitude. A similar tide is raised on a solid, but not absolutely rigid, satellite by the gravitational field of its primary planet. The tidal distortion results because the gravitational force of attraction by the primary decreases as $1/r^2$ across the finite dimension of the body, where r is the distance to the center of mass of the planet. A formerly nearly spherical body assumes a slight football shape with the long axis tending to line up with the tide-raising body. On the Earth this accounts for the two high and two low ocean tides per day as the Earth rotates under its tidal bulge, whose dominant progenitor is the Moon.

An exaggerated tidal bulge on a satellite still rotating relative to its planet in the same sense as its orbital motion is shown schematically in Fig. 1. Dissipation of energy due to the tidal flexing of the satellite allows the relative rotation to carry the tidal bulge away from the direction toward the planet. The gravitational attractions of the displaced bulges are not equal, as indicated by the force vectors $\mathbf{F}_1 > \mathbf{F}_2$ in Fig. 1. The resulting force couple, or torque, on a circularly orbiting satellite retards the spin until synchronous rotation is attained. For the at least slightly eccentric orbits of the satellites, the tidal torque averaged around the orbit and, likewise, the change in the spin rate, vanish at a rotation period somewhat smaller than the synchronous value. Nevertheless, synchronous rotation is attained and maintained by gravitational torques due to the permanent deviations from axial symmetry, which are far larger than the tidal distortions. The tides need only bring the satellite close to synchronous rotation, where these far larger torques take over. These latter torques tend to align the long axis with the direction toward the planet when the satellite is at the periapse, the point of its orbit nearest the planet. The satellite will oscillate about synchronous rotation with its long axis swinging to either side of the planet direction at periapse. These so-called free librations eventually damp down due to tidal dissipation, but a small "forced" or "physical" libration persists and has been observed for the Moon and for Mars' satellite Phobos. This latter libration results from the periodically reversed torque on the permanent deformation as the satellite rotates nearly uniformly, but traverses its eccentric orbit in a nonuniform manner. The amplitude of libration can be used to determine the ratio $(B-A)/C$ for the satellite, where $A < B < C$ are the principal moments of inertia.

If the satellite spin axis is inclined relative to its orbit normal (the angle between the two vectors is called the obliquity), the tidal bulge also will be carried out of the orbit plane, leading to a component of the torque perpendicular to the spin axis. Thus both the magnitude and direction of the spin vector evolve from their primordial states from tidal interaction, with rotational angular momentum being transferred to the orbital motion.

The time required for the tides to change the spin angular velocity $\dot{\psi}$ by $\Delta\dot{\psi}$ is

$$T = 2.4\times 10^{10}\frac{P_o^4\,\Delta\dot{\psi}\,Q}{R^2}\ \text{yr},$$

where P_o is the orbit period in days, $\Delta\dot{\psi}$ is measured in radians per second, R is the satellite radius in kilometers, and $1/Q$ is the specific dissipation function typically used to parameterize dissipation in oscillating systems. We infer values of Q near 100 for most solid satellites from observations of Q in the solid Earth, of Mars, and of rocks in the laboratory. It is clear that the satellites closer to their primaries are retarded much more rapidly than those further away. Maximum times of slowing each satellite from an initial period of 2.3 h (the period at which a fluid, spherical body would be rotationally unstable) to a rotation rate that is synchronous with its current mean orbital motion are given in the last column of Table 1. These times are short compared with the 4.6×10^9-yr age of the solar system for plausible values of Q near 100 for all of the synchronously rotating satellites except Iapetus, but we can accommodate the observed synchronous rotation of Iapetus by assuming it must have had a much less extreme initial rotation rate. Consistently, the two satellites with observed nonsynchronous rotation (Himalia and Phoebe) could not have had their spins changed significantly by the tides. Tides raised on a planet with a single satellite retard the planet's spin toward synchronous rotation as

Table 1. Satellite Data

Planet	*Satellite*	*Orbital Semimajor Axis (10^3 km)*	*Orbital Period (days)*	*Rotation Period (days)*	*Radius (km)*	*Mass (10^{20} kg)*	*τQ^{-1} (10^3 yr)*	*TQ^{-1} (10^4 yr)*
Earth	Moon	384.4	27.3217	s	1738	734.9	110	330
Mars	MI Phobos	9.378	0.319	s	13.5 × 10.7 × 9.6	1.26×10^{-4}	3.5	0.12
	MII Deimos	23.459	1.263	s	7.5 × 6.0 × 5.5	1.8×10^{-5}	870	120
Jupiter	JXVI Metis	127.96	0.2948		? × 20 × 20		1.0	0.023
	JXV Adrastea	128.98	0.2983		12.5 × 10 × 7.5		4.1	0.096
	JV Amalthea	181.3	0.4981	s	135 × 82 × 75		0.19	0.012
	JXIV Thebe	221.90	0.6745		? × 55 × 45		1.9	0.013
	JI Io	421.6	1.769	s	1815	894	0.026	0.005
	JII Europa	670.9	3.551	s	1569	480	0.28	0.11
	JIII Ganymede	1,070	7.155	s	2631	1482.3	0.83	0.67
	JIV Callisto	1,883	16.689	s	2400	1076.6	13	24.1
	JXIII Leda	11,094	238.72		~ 8		16	9.1×10^{10}
	JVI Himalia	11,480	250.57	0.4	90		0.12	8.7×10^{8}
	JX Lysithea	11,720	259.22		~ 20		2.5	2.0×10^{10}
	JVII Elara	11,737	259.65		40		0.62	5.1×10^{9}
	JXII Ananke	21,200	631 R		~ 15		4.4	1.2×10^{12}
	JXI Carme	22,600	692 R		~ 22		2.1	8.5×10^{11}
	JVIII Pasiphae	23,500	735 R		~ 35		0.82	4.2×10^{11}
	JIX Sinope	23,700	758 R		~ 20		2.5	1.5×10^{12}
Saturn	SXV Atlas	137.64	0.602		19 × ? × 14		13	0.77
	SXVI Prometheus	139.35	0.613		70 × 50 × 37		1.0	0.059
	SXVII Pandora	141.70	0.629		55 × 43 × 33		1.6	0.095
	SXI Epimetheus	151.422	0.694	s	70 × 58 × 50		1.3	0.088
	SX Janus	151.472	0.695	s	110 × 95 × 80		0.52	0.036
	SI Mimas	185.52	0.942	s	197	0.38	0.34	0.033
	SII Enceladus	238.02	1.370	s	251	~ 0.8	0.64	0.094
	SIII Tethys	294.66	1.888	s	524	7.6	0.38	0.079
	SXIII Telesto	294.66	1.888		? × 12 × 11		730	150
	SXIV Calypso	294.66	1.888		15 × 13 × 8		535	110
	SIV Dione	377.40	2.737	s	559	10.5	1.0	0.31
	SXII Helene	377.40	2.737		18 × ? × (< 15)		1400	430
	SV Rhea	527.04	4.518	s	764	24.9	2.5	1.3
	SVI Titan	1221.85	15.945		2575	1345.7	9.5	17
	SVII Hyperion	1481.1	21.277	Chaotic	175 × 120 × 100		6,700	16,000
	SVIII Iapetus	3561.3	79.331	s	718	18.8	15,000	140,000
	SIX Phoebe	12,952	550.48R	0.4	115 × 110 × 105		0.083	1.4×10^{10}
Uranus	Cordelia	49.75	0.336		~ 25		0.95	0.026
	Ophelia	53.77	0.377		~ 25		1.3	0.043
	Bianca	59.16	0.435		~ 25		2.1	0.080
	Cressida	61.77	0.465		~ 30		1.7	0.074
	Desdemona	62.65	0.476		~ 30		1.9	0.082
	Juliet	64.63	0.494		~ 40		1.2	0.059
	Portia	66.10	0.515		~ 40		1.3	0.064
	Rosalind	69.63	0.560		~ 30		3.0	0.16
	Belinda	75.25	0.724		~ 30		6.6	0.48
	Puck	86.00	0.764		85		0.96	0.074
	UV Miranda	129.8	1.413	s	242	0.71	0.75	0.11
	UI Ariel	191.2	2.520	s	580	14.4	0.74	0.21
	UII Umbriel	266.0	4.144	s	595	11.8	3.1	1.5
	UIII Titania	435.8	8.706	s	800	34.3	16	16
	UIV Oberon	582.6	13.463	s	775	28.7	63	98
Neptune	Naiad	48.0	0.296		~ 27		0.56	0.011
	Thalassa	50.0	0.313		~ 40		0.30	0.0076
	Despoina	52.5	0.333		~ 90		0.071	0.0020
	Galatea	62.0	0.428		~ 75		0.22	0.0084
	Larissa	73.6	0.554		~ 95		0.29	0.016
	Proteus	117.6	1.121		~ 200		0.55	0.066
	NI Triton	354.8	5.877R	s	1352	214.2	1.8	1.2
	NII Nereid	5513.4	360.16		~ 170		0.035	1.1×10^{9}
Pluto	PI Charon	19.1	6.387		596	18.3	11.4	8.3

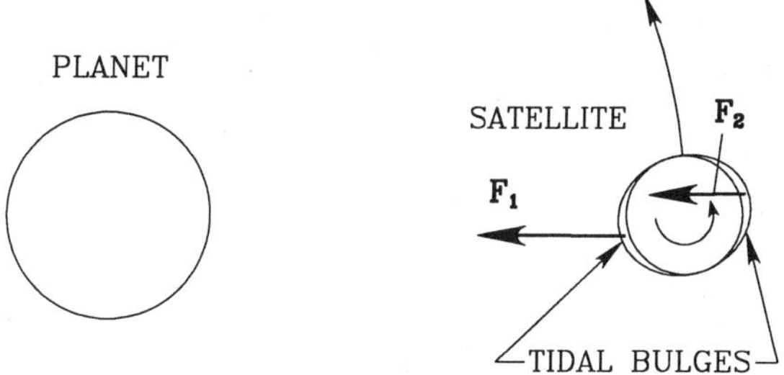

Figure 1. Geometry of the retardation of satellite rotation by gravitational tides.

well, but only the Pluto–Charon system has reached the final evolutionary state of dual synchronous rotation. The almost complete arbitrariness in initial conditions that ultimately result in synchronous rotation precludes knowledge of the primordial rotation of any tidally evolved satellite and thereby of any rotation-dependent constraints on their origin.

Throughout most of their tidal evolution, the satellites have rotated nearly about axes of maximum moment of inertia even though the accretion process would have left them initially with a significant wobble. This follows from the short time constants for wobble decay given by

$$\tau = \frac{38}{5}\frac{\mu Q}{\rho R^2 \dot{\psi}^3}$$

and evaluated for each satellite in the penultimate column of Table 1. Here $\mu = 5 \times 10^{11}$ dyn cm^{-2} is a rather high rigidity assumed for all of the satellites, $\rho = 2$ g cm^{-3} is the density assumed for all, R is a mean radius for each satellite, and the rotational angular velocity $\dot{\psi}$ is assumed to be the synchronous value for all satellites with short tidal decay times and to be the value corresponding to a period of 0.4 days otherwise. These times represent upper bounds because of the high rigidity assumed and because synchronous satellites were rotating faster when their wobbles decayed. The physical process leading to wobble decay is the dissipation of energy while the satellite flexes as the position of the instantaneous equator varies. The rotating object evolves secularly toward the minimum energy configuration for conserved angular momentum, which is rotation about the axis of maximum moment of inertia. Given the short time constants in Table 1, any current measurable wobble amplitude would indicate a recent disturbance such as a large meteorite impact.

The evolution of the satellite obliquities can be understood qualitatively. If the orbit plane of a satellite is fixed in inertial space, the spin axis of a satellite with a nonzero obliquity will precess about the normal to the orbit plane due to the gravitational torques on the equatorial bulge. A unit vector along the spin angular momentum thus traces out a cone or, equivalently, a line of latitude on a unit sphere centered on the satellite center of mass. The spin vector remains stationary in the orbit frame of reference only when it is parallel or antiparallel to the orbital angular momentum. For spin angular velocities larger than $2n$, where n is the mean orbital angular velocity, tides tend to drive the spin toward an obliquity between 0 and 90°. The value of the obliquity where the tidally induced change vanishes is closer to 90° the more the spin angular velocity exceeds $2n$, and it approaches 0° as the spin approaches $2n$. The evolution toward large obliquities for high spin rates is understood in terms of the relative rates at which the components of the spin angular momentum parallel and perpendicular to the orbit plane are reduced. If a satellite had its spin vector lying in its orbit plane (much like the planet Uranus), then twice during each orbit the spin axis would be pointing toward the primary and at those instants there would be no tidal reduction in the spin angular velocity. Thus, averaged around the orbit, the tidal reduction of a satellite spin with an obliquity of 90° would be less than a similarly averaged reduction for a satellite whose obliquity was 0°. With a spin vector represented by its components in and perpendicular to the orbit plane, the perpendicular component is reduced almost twice as fast as the parallel component for equal values of the two components, and the obliquity is increased. The change in the obliquity vanishes when the relative rates of reduction of the two components of the spin just maintain their constant ratio. The time scale for changing the obliquity is thus comparable with that for retarding the spin magnitude. The shrinkage of the obliquity for which there is no tidal change toward zero obliquity as the spin is reduced means that tides, if acting alone, always bring a satellite in a *fixed orbit* to zero obliquity provided the endpoint of the tidal evolution corresponds to $\dot{\psi} \lesssim 2n$. As $\dot{\psi} = n$ corresponds to synchronous rotation, the synchronously rotating satellites should have nearly zero obliquities.

The orderly evolution of a satellite's rotational properties toward synchronous rotation with an obliquity near zero, a state in which we seem to find so many of the satellites, can be frustrated if the satellite's orbit is somewhat more eccentric or inclined to the invariable plane defined previously. The permanent deviation from axial symmetry offers the possibility that a tidally evolved satellite in a highly eccentric orbit could have been captured into a state where the rotational angular velocity is a half-integer multiple of the orbital mean motion. For example, the planet Mercury rotates at a rate that is precisely $1.5n$. Synchronous rotation is simply a special case of the more general spin–orbit resonances that are stabilized by the torque on the permanent deformation tending to keep the long axis of the satellite aligned with the primary–satellite line when the satellite is at its orbit periapse. The stability of a spin–orbit resonant rotation state decreases as the half-integer characterizing the state increases and, for a given state, the stability decreases as the eccentricity decreases. No satellite has been observed in any spin–orbit resonance other than the synchronous one. Those satellite orbits with substantial eccentricities are not tidally evolved, and those that are tidally evolved generally have orbits that are too circular for any but the synchronous rotation state to be likely. The exception is Saturn's satellite Hyperion with a mean orbital eccentricity of 0.1. However, Hyperion is not in *any* spin–orbit resonance for reasons described later.

We have noted previously that tides eventually bring a satellite in a fixed orbit to zero obliquity (spin axis perpendicular to its orbit) if its final spin–orbit state has $\dot{\psi} \leq 2n$. But an orbit that is inclined relative to a local nearly invariable plane will precess about the normal to that plane with no secular change in its inclination. The orbit normal toward which the tides are tending to drive the spin axis vector is no longer fixed in inertial space. As a consequence the final spin state is not one of zero obliquity, but the spin vector will eventually occupy one of two Cassini states, which are named after Cassini's laws for the Moon. In a Cassini state, the spin vector is again fixed in the orbital frame of reference but at an obliquity that is determined by the deviation of the satellite from spherical symmetry. The latter deviations are characterized by the ratios $(C-A)/C$ and $(B-A)/C$. The position of the Cassini state is in the plane defined by the orbit normal and the normal to the invariable plane. These two normals and the spin vector remain coplanar as the orbit precesses about the invariable plane normal. Cassini states 1 and 2 lie, respectively, on the side of the orbit normal opposite the invariable plane normal and on the side of the invariable plane normal opposite the orbit normal. Tides eventually drive the spin to either state 1 or state 2, depending on the initial conditions or on the conditions at the time of capture into a spin–orbit resonance with $\dot{\psi} < 2n$. For the Moon, state 1 does not exist and the tides have selected state 2 as the only possible end point of the dissipative evolution. Most of the other tidally evolved satellites should occupy state 1 with very small obliquities, where the latter follow from the very small orbital inclinations. Hence, for most of the tidally evolved satellites, accounting for the precession of the orbits of small inclination only slightly modifies the trend toward the zero obliquity final state. None of these small obliquities have been measured accurately. The obliquity in the Cassini state is a measure of the response of the satellite to an external gravitational torque and its measurement yields a value of $(C-A)/C$.

Saturn's satellite Hyperion is a very special case: It occupies an orbital resonance with Titan, where the ratio of their orbital mean motions is nearly $\frac{4}{3}$, and this resonance maintains Hyperion's orbital eccentricity near 0.1. In addition, images from the *Voyager* spacecraft showed Hyperion to be shaped like a thick hamburger with half-axes of approximately 100, 125, and 175 km. The combination of high orbital eccentricity and very aspherical shape prevents Hyperion from rotating in any spin–orbit resonance state. Hyperion's "attempt" to librate simultaneously about two adjacent spin–orbit resonances when the spin is normal to the orbit and the instability of the orientation of its spin axis for spin angular velocities near the synchronous value force it to tumble chaotically with large changes in angular velocity, wobble amplitude, and spin orientation on time scales comparable with the orbital period.

Tidal dissipation does not cease when the satellite reaches synchronous rotation *if* the orbit is eccentric. This follows from the periodic motion of the subplanet point and thereby the tidal maximum on the surface of the satellite and the periodic variation in the amplitude of the tidal bulge as the satellite traverses near and far points of its eccentric orbit. Normally this dissipation leads to torques that reduce the eccentricity toward zero, where the dissipation vanishes. However, if the satellite occupies an *orbital* resonance with a nearby satellite, where the orbital mean motions of the satellites are in the ratio of two small integers, the eccentricity can be forced to a nonzero value and the dissipation thereby maintained. This mechanism has heated Jupiter's satellite Io so drastically that it is the most volcanically active body in the solar system.

Laser ranging to corner-cube retroreflectors placed on the Moon by *Apollo* astronauts has allowed us to monitor the Moon's orientation to the unprecedented accuracy of a few centimeters. There has been one inference of a recent large meteorite impact from possibly observed free librations of less than 2 arcsec and another inference of a dissipative liquid core to account for a 0.23-arcsec displacement of the spin axis from the precise location of the Cassini state. Although both inferences are tentative, they add to our growing list of what can be learned of recent history and interior properties from rotational properties of the satellites.

Additional Reading

Davies, M. E. et al. (1989). Report of the IAU/IAG/COSPAR Working Group on cartographic coordinates and rotational elements of the planets and satellites: 1988. *Celest. Mech. Dyn. Astron.* **46** 187.

Goldreich, P. and Peale, S. J. (1968). Dynamics of planetary rotations. *Ann. Rev. Astron. Ap.* **6** 287.

Mulholland, J. D. (1980). Scientific achievements from ten years of lunar laser ranging. *Rev. Geophys. Space Physics* **18** 549.

Peale, S. J. (1973). Rotation of solid bodies in the solar system. *Rev. Geophys. Space Physics* **11** 767.

Peale, S. J. (1977). Rotation histories of the natural satellites. In *Planetary Satellites*, J. A. Burns, ed. University of Arizona Press, Tucson, p. 87.

Wisdom, J., Peale, S. J., and Mignard, F. (1984). The chaotic rotation of Hyperion. *Icarus* **58** 137.

Yoder, C. F. (1981). Free librations of a dissipative moon. *Philos. Trans. Roy. Soc. London A* **303** 327.

See also **Earth, Figure and Rotation; Moon, Eclipses, Librations, and Phases; Planetary Rotational Properties.**

Saturn, Atmosphere

F. Michael Flasar

Saturn is the second largest planet in the solar system, with a radius approximately 10 times the Earth's (cf. Table 1). Like Jupiter, it is composed primarily of hydrogen (H_2) and helium (He), and in this it is more similar to the Sun than to the Earth. The base of its atmosphere is not particularly well-defined, because there are no solid or liquid surfaces. The profile of temperature and pressure with depth is thought to be so warm that there is no distinct phase transition of the H_2–He gas to a solid or liquid state. Instead, the gaseous atmosphere gradually changes with depth into a degenerate fluid, for which intermolecular forces are important. Current theoretical models indicate that the fluid does undergo a phase transition from molecular to metallic hydrogen at approximately one-half Saturn's radius, where the pressure is 2×10^6 bar (1 bar $=10^5$ Pa) and the temperature is 8000 degrees kelvin (K). However, little is known about the deep atmosphere; most Earth-based and spacecraft observations have only probed altitudes for which the pressure is less than 10 bars, comprising the outermost "skin" of the planet. At high altitudes, about 1000 km above the 1-bar level, where pressures are only 10^{-8} bar, Saturn's neutral atmosphere is bounded by an ionosphere, where ions and free electrons are produced by the absorption of sunlight of ultraviolet wavelengths.

Vertical Structure

Figure 1 illustrates the vertical variation of atmospheric temperatures and the locations of various clouds and hazes on Saturn and Earth. To facilitate comparison of the two planetary atmospheres, the common vertical axis is labeled in terms of atmospheric pressure, instead of altitude. However, altitude and pressure are related, because the pressure at a specified altitude is just the weight of the overlying atmosphere; altitude scales are included as insets. Any specified interval of atmospheric pressure on Saturn encompasses a larger range of altitude than on Earth, because the molecular weight of Saturn's atmosphere is much lower.

After hydrogen and helium, the next most abundant species that has been identified unambiguously is methane (CH_4). Photolysis of CH_4 by sunlight at ultraviolet wavelengths is thought to initiate the production of hydrocarbon polymers that might account for the tenuous hazes that have been inferred to exist at pressures less than 70 mbars (cf. Fig. 1). Saturn's atmosphere is too warm for CH_4 to condense, but ammonia (NH_3) does condense and presumably accounts for the whitish clouds that have been observed. The NH_3 cloud base should lie between 1 and 2 bar, but a tenuous

Table 1. The Atmospheres of Saturn and Earth

	Saturn	*Earth*
Length of day (h)	10.7	24.0
Orbital period (y)	29.5	1.0
Radius (km) at 1 bar	60,000	6378
"Surface" gravity (m s^{-2})	9.1	9.8
Distance from Sun (AU)	9.5	1.0
Inclination of equator to orbital plane (degrees)	26.7	23.5
Major constituents	Dry gas*:	Dry air*:
	H_2 (96.3%)	N_2 (78.1%)
	He (3.3%)	O_2 (21.9%)
	CH_4 (0.4%)	Ar (0.9%)
	Condensibles:	Condensibles:
	H_2O, NH_3	H_2O ($\leq$ 2.2%)†

*Composition (by volume) ignores presence of condensibles.
†Spatially and temporally variable.

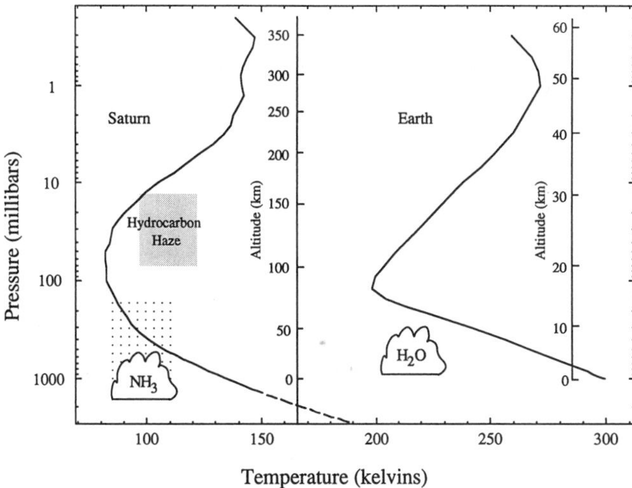

Figure 1. Vertical profiles of temperature on Saturn and Earth as a function of atmospheric pressure (1000 mb = 1 bar = 0.987 atmospheres). The saturnian temperatures were obtained from the radio occultation experiment on the Voyager spacecraft. The dashed curve is an extrapolation along a dry adiabat. Clouds and hazes are indicated schematically.

haze of NH_3-ice crystals probably extends to nearly the 100-millibar (mbar) level. Much lower in the atmosphere (pressure approximately 10 bar), water (H_2O) clouds may form; indeed at these levels H_2O may be more abundant than CH_4, but it has yet to be detected.

The temperatures in Saturn's atmosphere have been determined by remote sensing. In the range of pressures depicted in Fig. 1, one method consists of measuring the atmospheric refraction of radio waves transmitted by a spacecraft as it is occulted by the planet. The amount of refraction depends on the temperature–pressure profile of the atmosphere. This method was used in the flybys of the *Pioneer 11* spacecraft in 1979 and of the *Voyager 1* and *2* spacecraft in 1980 and 1981, respectively. Another method is to observe Saturn's electromagnetic radiation at infrared, submillimeter, and radio wavelengths (micrometers through centimeters). The opacity of the atmosphere varies with wavelength, and radiation observed at different wavelengths originates at different altitudes. Because the intensity of radiation emitted depends on the temperature of the atmosphere at the emitting level, one can, in principle, retrieve atmospheric temperature as a function of altitude. In practice, this is only possible if the abundance of the opacity source is known. Over much of the pressure range depicted in Fig. 1, the dominant species, hydrogen, is the principal source of opacity. At deeper levels, where pressures are greater than 1 bar, condensible gases and condensates, such as ammonia (NH_3) and water (H_2O), are thought to comprise much of the opacity. Insofar as these are horizontally and vertically distributed nonuniformly, it is not possible to unambiguously retrieve temperatures at these levels.

The atmospheric temperatures on Saturn are colder than those on Earth, because Saturn is much farther from the Sun (Table 1). At atmospheric pressures less than 100 mbar on Saturn, the temperatures in Fig. 1 result from absorption of sunlight by gaseous methane and hydrocarbon hazes and from radiation back to space at infrared wavelengths. This accounts for the increase of temperature with altitude. Lower in the atmosphere, where temperatures decrease with altitude, another process, which is dynamical and linked to Saturn's energy balance, determines the thermal structure.

Saturn, like Jupiter, radiates nearly a factor of 2 more energy at infrared wavelengths than it absorbs from sunlight. The excess energy must come from its interior. One possible source of energy is the thermal reservoir that was created when these planets formed from the solar nebula and contracted to their present size. While the protoplanets were sufficiently diffuse that the hydrogen–helium mix behaved as an ideal gas, much of the compressional work under the influence of self-gravitation went into increasing the internal temperatures. As the planets became more compressed, the interiors behaved less like an ideal gas, intermolecular forces become more important, and as a result less of the compressional work went into increasing internal temperatures. At present, the interiors are more like a liquid than a gas, and further contraction, albeit small, can be achieved only by cooling of the interior. This process can account for Jupiter: Current models of the evolution of its interior indicate that its thermal reservoir is large enough to supply most if not all of its excess luminosity over its lifetime of 4.5 billion years. On the other hand, these models cannot account for Saturn: They predict that Saturn would have cooled down 2 billion years ago and would not have enough thermal energy today to maintain its observed luminosity. Another source of energy is needed. Giants that they are, both Jupiter and Saturn are still not massive enough to produce the high internal temperatures required for nuclear fusion reactions, the source of the Sun's luminosity. However, at the temperatures believed to characterize the interiors of Jupiter and, particularly, Saturn, hydrogen and helium are thought to be only partially soluble in one another. Separation of a heavier, helium-rich phase from a lighter, hydrogen-rich phase, with the heavier phase sinking, would release gravitational energy that would ultimately be converted to heat. This could well provide the needed energy reservoir for Saturn. Evidence supporting the occurrence of extensive gravitational differentiation within Saturn came from the determination of its atmospheric helium abundance from measurements by the infrared spectroscopy experiment on *Voyager*. The ratio of helium to hydrogen was found to be substantially depleted in Saturn's atmosphere relative to that determined on Jupiter and also relative to that on the Sun, which presumably characterizes the gas from which Saturn formed.

The observed increase of temperature with depth at pressures greater than 100 mbar on Saturn is a manifestation of the need to transport energy from the interior to the 400-mbar level, where infrared radiation can escape directly to space. This upward transport from the interior must be dynamical, because thermal conduction and radiation are much too inefficient. Sufficiently deep in the atmosphere, the dominant mode of vertical transport is probably by convection, in which hotter fluid at deeper levels rises and mixes with the cooler ambient environment higher up. The convective motions are thought to be highly efficient and to establish a vertical temperature profile that is close to adiabatic, that is, nearly identical to that experienced by a parcel of gas displaced upward without exchanging heat or mass with its environment. For the transport of heat to be upward, temperatures must decrease with altitude slightly faster than the adiabat. The observed temperatures depicted in Fig. 1 indicate that the vertical profile is already close to adiabatic at pressures as low as 700 mbar.

At altitudes above the 1-mbar level, there are no measurements from which temperatures can be retrieved until one reaches 10^{-5} mbar, approximately 700 km higher. At these high altitudes, spectrometers on the *Voyager* spacecraft tracked the attenuation of ultraviolet radiation from the Sun and from several stars as they were occulted by Saturn's atmosphere. At these wavelengths, absorption of radiation by hydrogen is more important than refraction. The vertical resolution afforded by these measurements is crude, but they indicate that temperatures are approximately 140 K at 10^{-5} mbar, and increase to nearly 800 K at 3×10^{-6} mbar. This increase may be produced by dissipation of electrical currents in the ionosphere.

HORIZONTAL STRUCTURE AND WINDS

With an axial tilt of nearly 27°, Saturn's atmosphere might be expected to exhibit large meridional contrasts and seasonal varia-

Figure 2. Photograph of Saturn, obtained with the Hubble Space Telescope. The scalloped white formation, stretching nearly across the whole facing hemisphere, is the "great white spot" that was first seen on September 25, 1990, and that grew rapidly in longitudinal extent. (*NASA photo.*)

tions in temperature. However, the temperatures are so low that Saturn's atmosphere responds sluggishly to seasonally varying sunlight, and its meridional thermal structure is more a product of the annual mean solar heating. Infrared observations by Voyager did detect some asymmetry in the temperature field about the equator at pressures of 100 mbar and less, and this was interpreted as a seasonal variation, but this asymmetry decreased with depth. In addition, the observed difference in temperature between the equator and poles was quite small, despite the fact that more than twice as much sunlight is absorbed at low latitudes as at high latitudes. The atmosphere is evidently quite efficient in meridionally redistributing heat by some dynamical process. Possibly, the thermal convection in Saturn's interior just described effects the redistribution. Preferential deposition of solar energy at low latitudes will lessen the vertical temperature gradient there relative to that at high latitudes. Because the vertical energy transport by convection is a sensitive function of the small difference between the vertical temperature gradient and the adiabat, small adjustments in the former will effect large changes in the convective heat flux. As a result, convection will transport more heat from the interior to the atmosphere at the poles than at the equator. This can act to compensate the latitude-dependent absorption of solar radiation and reduce the meridional gradient in atmospheric temperature that otherwise would exist.

The most direct method of determining motions in Saturn's atmosphere has been to track distinct cloud features (Fig. 2). The altitudes of these features are not well-known, but they are probably near the NH_3 condensation level, around 1 bar. With respect to inertial space, the atmosphere rotates from west to east, but the rotation is not uniform: There are winds. The measured winds are predominantly zonal, that is, they lie along latitude circles. Unlike Earth, Saturn does not have a visible surface to which one can refer the wind speeds. Instead, the (presumably uniform) rotation of its interior is used as a reference. The internal rotation rates of Saturn and the other outer planets (Jupiter, Uranus, and Neptune) have been inferred from the periodic modulation of radio signals observed at each body. The radio emissions are thought to be from charged particles that are tightly bound to the magnetic field, which itself is rigidly coupled to the planetary interior where it is generated. Jupiter's period thusly derived has remained unchanged during over 30 years of ground-based and spacecraft observations. Saturn's periodic radio emission was only first detected during the *Voyager 1* and *2* flybys, which suggested a period of 10 hours 39 minutes. However, unlike the other outer planets, Saturn's magnetic field, as determined by the *Pioneer 11* and the two Voyager spacecraft, is axisymmetric and it is not clear how it can have any periodically modulated radio emission.

Despite possible reservations concerning the reliability of the radio modulation as an indicator of the internal rotation rate, the cloud tracking studies in any case show that Saturn's winds vary markedly with latitude, forming a pattern of eastward and westward currents. The structure is reminiscent of Jupiter's winds, but the latitudinal extent of the currents is broader and the peak-to-peak variation in the winds is far greater. For example, Saturn has a very strong broad eastward current at low latitude. The winds near the equator are approximately 500 m s^{-1} relative to the westward blowing winds at $\pm 40°$ latitude. In contrast, the eastward equatorial current in Jupiter's atmosphere is only about 200 m s^{-1} relative to its westward currents at $\pm 20°$ latitude. Saturn's winds are also large compared to those on Earth, where the strongest mean zonal winds are high in the atmosphere near 60 km altitude and exhibit a latitude contrast of 150 m s^{-1} or less. If the interior rotation period is as long as 10.7 hours, Saturn's winds are predominantly eastward at all latitudes; the westward currents are typically less than 25 m s^{-1}. Just how Saturn is able to maintain such a high degree of superrotation, particularly at low latitudes, remains a mystery.

Saturn's atmosphere, like Jupiter's, exhibits an alternating sequence of brighter and darker zonal bands. The width of these is comparable to the width of the zonal currents. However, in contrast to Jupiter, there is little correlation between the albedo of the bands and the winds. Instead, the winds are correlated with small undulations of temperature with latitude, typically 5 K or less. The temperature maxima and minima occur on the poleward edges of the eastward and westward currents, respectively. From terrestrial meteorology it is well-known that horizontal temperature gradients imply vertical variations in the horizontal winds. On Saturn the temperature undulations indicate that the zonal winds decay with altitude above the 700-mbar level. Thus, Saturn's strong wind system appears to be driven at deeper levels, and some unknown dissipative process acts to brake them at higher altitudes.

Although much has been learned about Saturn's atmosphere, particularly in recent years with the advent of space probes, several fundamental questions concerning its meterology remain. Comparative studies of Saturn and the remaining outer planets may provide the best means of unlocking its secrets.

Additional Reading

Gehrels, T. and Matthews, M. S., eds. (1984). *Saturn*. University of Arizona Press, Tucson.

Ingersoll, A. P. (1981). Jupiter and Saturn. *Scientific American* **245** (No. 6) 90.

Ingersoll, A. P. (1990). Atmospheres of the giant planets. In *The New Solar System*, 3rd ed., J. K. Beatty and A. Chaikin, eds. Sky Publishing Corp., Cambridge, MA, and Cambridge University Press, Cambridge, U.K., p. 139.

Morrison, D. (1982). Voyages to Saturn. NASA Special Publication No. 451, U.S. Government Printing Office, Washington, D.C.

Stone, E. C., et al. (1981). *Voyager 1* encounter with the saturnian system. *Science* **212** 159. *Voyager 2* encounter with the saturnian system. *Science* **215** 499.

Stone, E. C., et al. (1983). The *Voyager* mission: Encounters with Saturn. *J. Geophys. Res.* **88** 8639.

See also **Jupiter, Atmosphere; Planetary and Satellite Atmospheres; Planetary Atmospheres, Cluds and Condensates; Planetary Atmospheres, Ionospheres; Uranus and Neptune, Atmospheres.**

Scientific Spacecraft and Missions

Roger M. Bonnet

Our understanding of the Universe depends in an unavoidable way on the instrumentation used to carry out the observations. Every time a new technology or a new technique is used, we witness immediate progress in our knowledge. This certainly was the case when Galileo used his telescope for the first time, but also when radioastronomy, big computers, CCDs, and obviously space techniques were introduced in the field of astronomy.

When Victor F. Hess made his historic balloon flight from Aussig in Austria to Berlin in August 1912, he discovered a source of unknown radiation coming from outer space, which was later baptized cosmic radiation by Robert A. Millikan. Hess' discovery gave the first evidence that the atmosphere of the Earth is screening our view of the Universe and that special techniques are necessary to unveil it. Thirty-four years later, Richard Tousey, an American scientist from the Naval Research Laboratory, using a German V2 rocket made the first observation of the Sun's ultraviolet spectrum. In 1957, the historic and successful launch of *Sputnik 1*, the first artificial satellite of the Earth, opened a completely new era in humanity's quest for knowledge and created a genuine revolution in our approach to observing the Universe.

Table 1 presents a list of the major past and future space astronomy events. Evidently, without space-borne instrumentation our knowledge of the Universe would indeed be very fragmentary: Access to space is now indispensable for any further progress in astronomy.

ADVANTAGES OF SPACE

Astronomical observatories operating in space do present considerable advantages with respect to those that operate from the ground.

Elimination of Atmospheric Perturbations

The atmosphere absorbs a substantial part of the electromagnetic spectrum with the exception of a few "windows" located in the visible as shown in Fig. 1, and only from space can we get full access to it from γ rays to the far infrared, through the x-ray, ultraviolet, and infrared domains. Yet, two physical limits affect observations from space. Very high energy photons ($\lambda < 3\times10^{-19}$ cm) collide with the far infrared cosmic background radiation producing electron–positron pairs and, hence, cannot be observed. Photons of wavelengths larger than 10 km are absorbed by the interstellar medium.

On the ground, motions in the atmosphere limit the quality of the seeing to at best a fraction of an arcsec, and astronomical telescopes are never actually used at their intrinsic diffraction limits. Speckle interferometry partly overcomes this effect for a few bright objects, but only from space can we get rid of this detrimental effect.

The gases of the Earth's atmosphere (e.g., O I, N I, H I, N_2, N^+, O_2, and OH) also emit light in discrete visible and infrared bands. These emissions, which can reach thousands of kilorayleighs, make astronomical observations very difficult in the corresponding spectral domains.

Elimination of Gravity-Related Perturbations

The disappearance of gravitational effects makes it possible to use large structures without major mechanical distortions and thereby improve the performance of high precision optical systems. In space it is possible to observe the whole celestial sphere with the same instrument. This is particularly important in the case of astrometry.

Access to Long Uninterrupted Observing Sequences

From special orbits or from a lunar station we get access to long sequences of full darkness or full sunlight (for solar observations) that increase our ability to detect faint or highly variable objects, as shown by *EXOSAT*, which discovered the so-called quasiperiodic oscillators, a class of x-ray sources. Uninterrupted observations of a single star, or the Sun, offer unique conditions for astroseismology and helioseismology. Only from space can we eliminate from the Fourier spectra of the source oscillations, any spurious frequencies due to Earth-connected periodic interruptions in observations.

Very Long Baseline Interferometry

Ground-based very long baseline interferometry is limited by the Earth's diameter, a limitation that does not exist in space. Likewise optical interferometry is limited to baselines of a few tens of meters maximum, due to atmospheric turbulence. Space techniques not only get rid of this problem but they allow nearly the whole electromagnetic spectrum to be explorable by this last technique.

In Situ Observations

Only with space vehicles can we explore in situ the interplanetary and the interstellar medium, the solar system, its planets and the comets, and even to land on solar system bodies, explore their surfaces, and bring samples of their soil back to Earth.

CHARACTERISTICS OF VARIOUS OBSERVING SITES

Table 2 lists a few characteristic orbits and gives some examples of the related missions.

At first, space astronomy missions were conducted from low altitude ($\simeq$ 500-km) orbits that were easy to reach and that provided the main results in solar physics (with the *Orbiting Solar Observatories* and *Skylab*). Typically a 350–450-km, 28° inclination orbit with a period of 95 min offers solar visibility for nearly 60 min and a nighttime of 35 min when the satellite's orbit crosses the shadow of the Earth. This is a typical orbit for the U.S. Shuttle and consequently for the *Hubble Space Telescope*. Later missions, for example, the U.S.–Dutch–U.K. *IRAS*, ESA's *EXOSAT*, and its future *Infrared Space Observatory* (*ISO*), use highly eccentric orbits that offer longer observing time. Such orbits are also frequently and naturally used for in situ magnetospheric measurements.

The geostationary circular orbit at 36,000 km above the Earth offers interesting advantages. With a period of exactly 24 h, the satellite is fixed in the Earth rotating frame of reference and can be in permanent view from the ground: The *International Ultraviolet Explorer* is used by astronomers alternatively from two stations, one at NASA Goddard Space Flight Center and one an ESA station near Madrid (Spain). The *Hipparcos* satellite was designed to be operated in a geostationary orbit from the Odenwald ESA station in Germany.

Very interesting prospects are offered by orbits centered on the two Lagrangian points L1 and L2. They offer either full sunlight uninterrupted observations or complete solar eclipses, respectively.

The ESA–NASA SOHO mission and the Soviet *Relict* missions are designed to take full advantage of these two possibilities.

The exploration of the solar system, involving planetary and cometary flybys, is based on a precise adjustment of the orbits to the aim of the mission and is making use of powerful launchers like the Shuttle, *Titan* (U.S.), *Ariane* (Europe), and *Proton* (U.S.S.R.) boosters, with additional motors to kick the spacecraft out into interplanetary space. Sometimes gravity assistance by the Earth, Venus, or Jupiter is necessary to change the orbit or to communicate extra energy to the spacecraft. Typical examples in this category are *ICE*, *Galileo*, *Ulysses*, and the future Cassini.

MAJOR SPACE AGENCIES

The chief contributions to space astronomy are the result of four main space agencies.

NASA in the United States was set up in 1958 for the peaceful exploration of space. It has pioneered in many areas of space science (Table 1). Its budget is one order of magnitude larger than that of any other space agency in the world. It reached $12 billion in 1990, 20% of which is devoted to space science exclusively.

In the Soviet Union, the Soviet Academy of Science plays the major role in the definition and management of missions through specialized research institutes, like the prestigious Institute for Space Research (IKI) in Moscow, which conducted several of the pioneering planetary missions including *VEGA 1* and *2* to Halley's comet. The Academy has also set up a special council for international cooperation, which coordinates space activities with the eastern countries and several important space-faring nations in western Europe and with the United States.

In Europe, the European Space Agency represents the main organization for the peaceful exploration of space. It is in essence

Table 1. Major Past and Future Space Astronomy Events / Missions

Name of Event or Mission	*Year of Launch*	*Main Emphasis*
Balloon flight (Europe)	1912	Discovery of cosmic rays
NRL V-2 rocket (U.S.)	1946	First observation of Sun's UV spectrum
NRL V-2 rocket (U.S.)	1949	First observation of solar x rays
Sputnik-1 (U.S.S.R.)	1957	First artificial satellite
Explorer III (U.S.)	1958	Discovery of Earth's radiation belts
Luna-1 (U.S.S.R.)	1959	Discovery of the solar wind
Luna-2 (U.S.S.R.)	1959	First image of the Moon's hidden side
Aerobee Rocket (U.S.)	1962	First observation of an x-ray star
Luna-9 (U.S.S.R.)	1966	First picture from the lunar surface
OAO-2 (U.S.)	1968	First orbiting astronomical observatory
Apollo-11 (U.S.)	1969	First human on the Moon
Copernicus (U.S.)	1970	First far ultraviolet observatory
Skylab (U.S.)	1973	High resolution images of solar corona in x rays
Mariner 10 (U.S.)	1973–74	First detailed picture of Mercury
COS-B (ESA)	1975	First map of Galaxy in γ rays
Venera 9 (U.S.S.R.)	1975	First picture of venusian surface
Viking 1, 2 (U.S.)	1976	First pictures taken on the martian surface
Voyager 1, 2 (U.S.)	1977–79	First images of Jupiter, Saturn, Uranus, and Neptune satellite and ring systems
Einstein (U.S.)	1978	First observatory devoted to x-rays
IUE (U.S.–U.K.–ESA)	1978	First international space observatory
Relict-1 (U.S.S.R.)	1983	First large scale measurement of cosmic background anisotropy
IRAS (U.S.)	1983	First large infrared survey from space
Vega 1 (U.S.S.R.)	1986	First close encounter with Halley's comet
Giotto (ESA)	1986	First high resolution image of Halley's nucleus
TDRSS (U.S.)	1987	First VLBI measurements from space
Hipparcos (ESA)	1989	First space astrometry mission
Granat (U.S.S.R)	1989	X- and gamma ray observatory
Galileo (U.S.)	1989	First orbiter and probe to Jupiter
COBE (U.S.)	1989	Cosmic background explorer
Hubble Space Telescope (U.S.–ESA)	1990	First large space observatory
ROSAT (F.R.G.–U.K.–ESA)	1990	X-ray astronomy mission
GRO (U.S.)	1991	Gamma ray observatory
Ulysses (ESA–U.S.)	1990–95	First flight above solar poles
EUVE (NASA)	1991	Extreme UV astronomy explorer
ISO (ESA)	1993	Infrared space observatory
AXAF (NASA)	1996	Advanced x-ray astronomy facility
Cassini/Huygens (NASA–ESA)	1995–2002	Saturn orbiter and probe of Titan's atmosphere
XMM (ESA)	1998	High throughput x-ray spectroscopy mission

Table 2. Classification of Missions by their Orbit

Orbit Type	*Usage*	*Example of Mission*
Near Earth, low to moderate inclination (300–800 km)	Solar observatory, stellar observatory, earth science	*OSO*s, *Skylab*, *Hubble Space Telescope*, space station
Polar (800 km to several Earth radii) Sun synchronous	Auroral zones, magnetosphere, solar observatory	*Spot*, *Polar*, EXOS-D Cluster
Lagrangian point L1	Solar observatory	*ISEE*, Wind, SOHO
Lagrangian point L2	Astronomy, earth science	*Relict*
High solar latitude	Solar wind, interplanetary medium	*Ulysses*
Geostationary	Astronomy, magnetosphere	*IUE*, *GEOS*
Highly eccentric	Magnetosphere, astronomy	*GEOTAIL*, *IRAS*, *EXOSAT*, XMM, ISO
Low perihelion	Solar physics	Solar probe
Circumplanetary	Planetary	*Viking*, *Venera*, *Mariner 10*, *Magellan*, *Galileo*, Cassini
Interplanetary	Planetary, cometary	*Voyager 1* and *2*, *ICE*, *Vega*, *Giotto*, *Suisei*, *Sakigake*
Landers	Planetary	*Apollo*, *Viking*, *Venera*, Cassini/Huygens, Rosetta

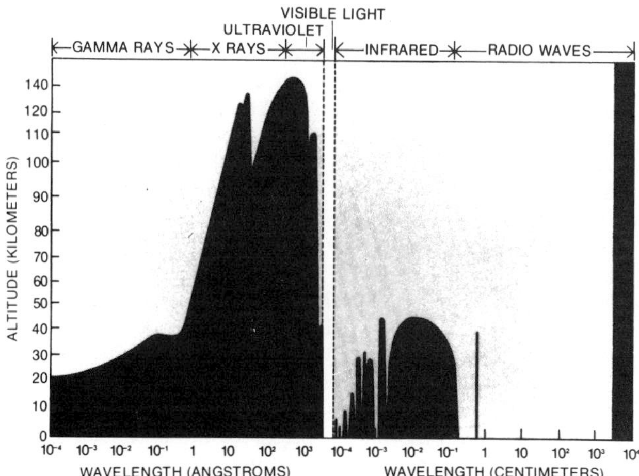

Figure 1. Altitude at which the Earth's atmosphere is 50% transparent to incoming electromagnetic radiation as a function of wavelength. [*From Ultraviolet Astronomy, Leo Goldberg,* Scientific American *(1969)*.]

an international organization composed of 13 western European states, with Canada as an associate member. ESA has several pioneering missions in its programs, like *Giotto*, *Hipparcos* and *ISO*, and was responsible for the development of the *Ariane* launcher. It devotes 13% of its budget to space science (excluding earth sciences and research in microgravity), representing the equivalent of $200 million in 1990.

In addition several European countries have their own space agencies. The most important is CNES in France, a country that has shown a constant determination toward space and has been involved in many prestigious projects in cooperation with the U.S.S.R. and the U.S., in addition to purely national activities based essentially on the use of balloons. More modest space agencies also exist in Italy, Germany, and the United Kingdom.

In Japan, space science is placed under the responsibility of ISAS, the Institute of Space and Aeronautical Science, which depends on the Ministry of Education. ISAS develops its own launchers, the *MU3S* and the *M5*. It launches one satellite regularly every year and contributes very actively to the progress in the area of x-ray astronomy, solar physics, comet science, and plasma physics. Its annual budget is greater than $150 million.

Since 1982, the four main space agencies have coordinated their efforts in the framework of the Inter Agency Consultative Group (IACG), first for the exploration of Halley's comet and presently in the area of the solar terrestrial science program.

In the rest of the world, the Indian Space Research Organization (ISRO), Brazil, and, to a lesser extent, China, are also involved in space science.

Additional Reading

Allen, C. W. (1983). *Astrophysical Quantities*. Athlone Press, London.

Harwit, M. (1988). Space science and fundamental physics. ESA Report SP-283, p. 205.

Perryman, M. A. C. (1985). Ad Astra Hipparcos. ESA Report 24.

Rycroft, M., ed. (1991). *The Cambridge Encyclopedia of Space*. Cambridge University Press, Cambridge, U.K.

Van der Klis, M. and Jansen, F. A. (1986). Discovery of quasi-periodic oscillations in accreting neutron stars. In *Proc. 4th Marcel Grossman Meeting on General Relativity*, R. Ruffini, ed. North Holland, Amsterdam, p. 847.

See also **Earth, Magnetosphere, Space Missions; Gamma Ray Astronomy, Space Missions; Hubble Space Telescope; Infrared Astronomy, Space Missions; Interplanetary and Heliospheric Space Missions; Interplanetary Spacecraft Dynamics; Mars, Space Missions; Mercury and Venus, Space Missions; Moon, Lunar Bases; Moon, Space Missions; Observatories, Airborne; Observatories, Balloon; Outer Planets, Space Missions; Radio Astronomy, Space Missions; Solar Physics, Space Missions; Sounding Rocket Experiments, Astronomical; Ultraviolet Astronomy, Space Missions; X-Ray Astronomy, Space Missions.**

Shock Waves, Astrophysical

David J. Hollenbach and Christopher F. McKee

Astrophysical shock waves are like the sonic booms produced by supersonic aircraft. Because shock waves move faster than the speed of sound, the medium ahead of the shock cannot respond until the shock strikes: The shock is a "hydrodynamic surprise." Astrophysical shocks compress, heat, and accelerate the gas and dust that they encounter. They range in size from a millionth of a centimeter for the collision of two dust grains to millions of light years for shocks in the intergalactic medium. Shock waves are ubiquitous in astrophysics because there are many mechanisms that generate velocities in excess of the local sound speed, which ranges from 0.3 km s^{-1} in cold ($T = 10$ K) molecular clouds, to 100 km s^{-1} in the hot ($T \simeq 10^6$ K) component of the interstellar medium (ISM), to more than 1000 km s^{-1} in the cores of evolved stars or in the hot gas in clusters of galaxies. For example, gas accreting onto a neutron star goes through a shock at a velocity of order 10^5 km s^{-1}; jets from radio galaxies produce shocks in the ambient gas with velocities that may be many thousands of kilometers per second; supernova explosions can drive shocks with velocities of up to 2×10^4 km s^{-1} in both the stellar envelope and the surrounding ISM; stellar winds have velocities of up to about 3000 km s^{-1}; galaxies collide at velocities of hundreds of kilometers per second and ionization fronts and cloud–cloud collisions in the ISM produce velocities of about 10 km s^{-1}. Shocks with velocities close to the speed of light have been conjectured to exist near the pulsar in the Crab nebula and in quasars. Astrophysical shocks also occur closer to home, in the interplanetary medium and at planetary magnetospheres. These shocks can be observed directly by spacecraft, and are the subject of a separate article.

Most of the information we have about shocks outside the solar system comes from the study of interstellar shocks. Until the 1970s, astronomers generally pictured the ISM as quiescent, with interstellar clouds of gas and dust floating about, gradually cooling and gravitationally contracting to form stars and planetary systems. However, since the mid-1970s, it has become increasingly clear that the ISM is violent, rent with supersonic motions and shock waves. These interstellar shock waves play a dominant role in determining the structure of the ISM because they transmit energy from the stars to the gas, thereby fixing the gas pressure. They can compress the gas to the point that it becomes gravitationally unstable, inducing star formation. Because shocks are effective at destroying interstellar dust, they play a key role in determining the gas phase abundances of the elements in the ISM. Furthermore, they are likely to be responsible for the acceleration of cosmic rays. From the point of view of astronomers, however, shocks are important primarily because they cause the gas to radiate, providing an invaluable diagnostic for violent events in the ISM.

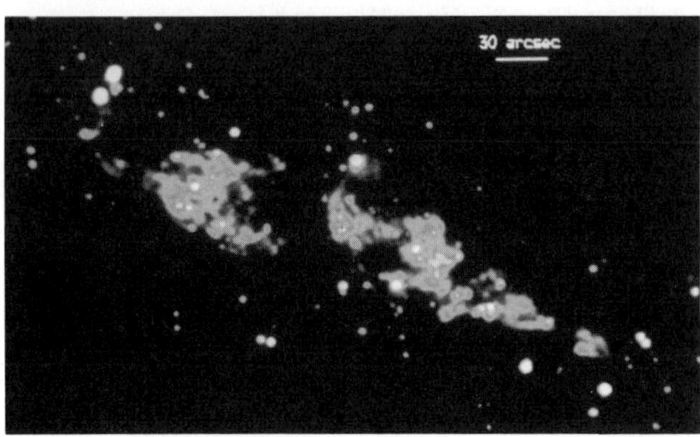

Figure 2. A molecular outflow in the star forming region DR 21, which is at a distance of about 3 kpc. The photograph shows the emission from shocked H_2 extending over a distance of 5 pc. The shocks are driven by a protostellar wind from a newly forming star near the center of the outflow. The bipolar shape of the outflow is characteristic of many protostellar outflows. (*Photograph provided courtesy of Ron Gardner, University of California, Irvine.*)

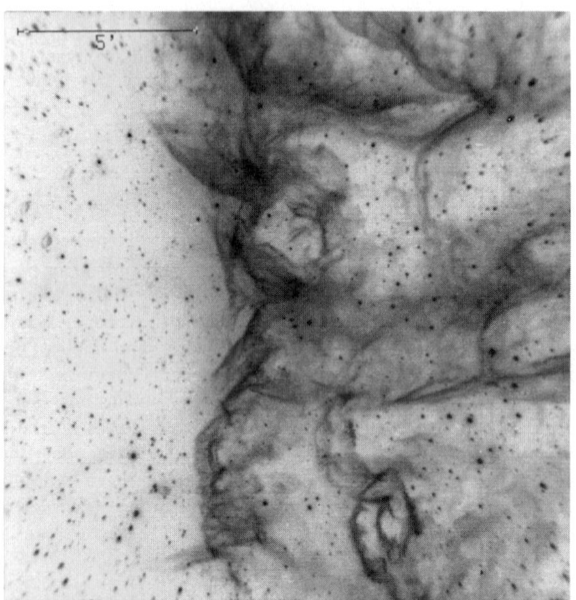

Figure 1. Radiative shocks in the eastern limb of the Cygnus Loop supernova remnant (sometimes called the Veil nebula). This photograph was made in the light of the 5007-Å line of the O^{++} ion, and shows the complex pattern of shocks established by the expanding remnant of a supernova. The entire remnant is 3° across; the region portrayed is 16′ × 16′ in angular extent, or about 3 pc × 3 pc in linear extent. (*Photograph provided courtesy of Jeff Hester, Palomar Observatory.*)

The first identification of the emission from interstellar shocks was made by the Soviet astrophysicist S. B. Pikel'ner, who in 1954, showed that the filaments in the Cygnus Loop, a 10,000-yr-old supernova remnant, are due to shocks propagating in the diffuse ISM at about 100 km s^{-1} (see Fig. 1). Subsequent observations of the Cygnus Loop at x-ray wavelengths have revealed gas that was shocked to millions of degrees by faster shocks earlier in its history. In the mid-1970s, a comparison of the optical spectra of Herbig–Haro objects, bright nebulosities associated with regions of star formation, with theoretical spectra led to the conclusion that the emission from these objects was also due to shock waves; in this case, the shocks are caused by the interaction of protostellar winds with the ambient medium. In 1976, the first infrared emission from shock waves was detected at a wavelength of about 2 μm and identified with a vibrational transition of molecular hydrogen (H_2). This emission arises from the dark, star-forming, molecular cloud behind the Orion nebula, and is believed to have been triggered by strong mass outflow from a newly forming star hidden (at visible wavelengths) in the dusty molecular cloud in which it was born. This observation marked the beginning of a new era of discovery of interstellar shock waves at infrared wavelengths, which can penetrate the obscuring dust that hides from view the centers of some galaxies (including the Milky Way) and regions of star and planet formation (see Fig. 2). At the present time, interstellar shock waves are being observed over most of the wavelength range accessible to astronomers, ranging from radio to x-rays.

STRUCTURE OF ASTROPHYSICAL SHOCK WAVES

The structure of a shock wave can be subdivided into four zones:

1. The *precursor*, in front of the oncoming shock front, where radiation and/or fast particles heat and possibly dissociate and ionize the gas.
2. The *shock front*, in which the gas is accelerated, heated, and compressed by interactions with already-shocked particles.

3. The *relaxation layer*, in which the shocked gas can change its chemical state (for example, it can be ionized by collisions just behind the shock front or it can recombine farther downstream), and in which it can cool and compress due to the emission of radiation.
4. The *thermalization layer* in which the radiation from the relaxation layer is absorbed and reradiated as quasi-blackbody radiation.

Not all shocks have all four zones. Fast shocks in tenuous gas often do not have enough time to radiate a significant amount of energy; such shocks have only a part of the relaxation layer and no thermalization layer, and are termed nonradiative. Fast shock waves in young supernova remnants are often nonradiative, and are generally detected by their x-ray emission. The majority of interstellar shocks are able to radiate away most of their internal energy behind the shock front, and are said to be radiative. On the other hand, the density in a typical interstellar shock is too low to thermalize the radiation, so the thermalization layer is absent. Shocks in dense interstellar gas and shocks in stars generally have all four zones.

In most astrophysical shocks there is a clear separation between the shock front and the relaxation layer. Such shocks are termed J shocks because the properties of the gas jump from their preshock values to their postshock values in a short distance. In a neutral gas, or in a dense plasma, this jump is effected by collisions between the particles, and the distance is of the order of a collisional mean free path. Outside the shock front itself, the collisions maintain equal temperatures among all the species of particles in the gas. In a tenuous plasma, such as the solar wind or the ISM, however, it is possible for the jump to occur as a result of plasma instabilities between the upstream and downstream plasmas, and the shock is termed collisionless. The thickness of a collisionless shock can be orders of magnitude less than a mean free path; for example, the Earth's bow shock is often only several hundred kilometers thick, although the mean free path is about ~ 1 AU (the distance from the Earth to the Sun). Because collisions are unimportant, there is no guarantee that the temperatures of the electrons and ions will be the same behind the shock; the determination of the electron/ion temperature ratio behind collisionless shocks is a major unsolved problem with important consequences for the x-ray spectra of these shocks.

Some interstellar shocks, however, have a structure in which the shock front and the relaxation layer are essentially coincident. This structure occurs in slow shocks in weakly ionized, magnetized plasmas. The ions, electrons, and charged dust particles couple with the compressing magnetic field and are accelerated. The friction of the trace amount of charged particles moving through the neutrals slowly heats and accelerates the neutrals. Because the heating is slow, the neutrals are able to radiate the heat as fast as it is generated. As a result, the properties of the gas vary continuously through the front, and the shock is called a C shock. If the shock is weak (i.e., its velocity is not much greater than the sound speed), then it is possible for the shock to be C type even in the absence of radiation. The thickness of a C shock is of order the mean free path for a neutral particle to hit a charged particle; because there are very few charged particles, this distance is large compared to the thickness of a J shock.

All steady shocks, whether J type or C type, satisfy conservation of mass and momentum. Imagine moving with the shock front, with the gas ahead of the shock, which has a density ρ_1, flowing toward you at a velocity v_1. After the gas has passed through the shock front it has a density ρ_2 and flows away at a velocity v_2. Because mass is conserved, the rate at which mass enters the shock $\rho_1 v_1$ must equal the rate at which it leaves, $\rho_2 v_2$:

$$\rho_1 v_1 = \rho_2 v_2.$$

Similarly, conservation of momentum leads to

$$p_1 + \rho_1 v_1^2 = p_2 + \rho_2 v_2^2,$$

where the two terms on each side of the equation represent the momentum flux in thermal motions and in directed motion, in order, respectively. The effects of magnetic fields (which are essential for C shocks) can be readily included in the momentum equation. For J shocks, energy conservation across the shock front gives

$$\frac{5}{2}\frac{p_1}{\rho_1} + \frac{1}{2}v_1^2 = \frac{5}{2}\frac{p_2}{\rho_2} + \frac{1}{2}v_2^2,$$

for an ideal monatomic gas. These three relations for mass, momentum, and energy conservation are called the Rankine–Hugoniot relations after the physicists who discovered them, or the jump conditions for short.

The solution of these equations depends on the *Mach number* of the shock M, which is the ratio of the shock velocity v_1 to the speed of sound in the unshocked gas. More precisely, the sound speed is proportional to the thermal velocity of the particles in the gas, and the Mach number is $M = (3\rho_1 v_1^2/5p_1)^{1/2}$. A shock exists only if M is greater than 1. Sonic booms generally have M slightly greater than 1, corresponding to a weak shock. Astrophysical shocks can be strong, with M much greater than 1. For strong shocks, the solution of the jump conditions is very simple: The density jumps by a factor of 4 and the pressure and temperature jump by a factor proportional to M^2. For example, for an ionized gas, the temperature behind the shock is

$$T_2 = 1.4 \times 10^5\ \mathrm{K}\left(\frac{v_1}{100\ \mathrm{km\ s^{-1}}}\right)^2$$

under the assumption that the electrons and ions are at the same temperature. In a young supernova remnant, this temperature can be quite high: In Cassiopeia A, for example, the shock velocity is estimated to be about 6000 km s^{-1}, so that the shock temperature is about 5×10^8 K (500 million degrees kelvin).

The mass and momentum jump conditions remain valid in the relaxation layer of the shock as well as in the shock front, but the energy condition is violated as the internal energy of the shocked gas is radiated away. In a radiative shock, most of the internal energy is radiated, and the flux of radiation from the shock is about $\rho_1 v_1^3/4$ in each direction. This can be simply understood by noting that in the frame of the shock wave, the preshock gas flows into the shock front at v_1 so that the flux of particles into the shock is $n_0 v_1$ and the flow energy per particle is proportional to $v_1^2/2$. After the gas is shocked, essentially all this energy is radiated away as the gas comes to rest in the postshock gas; half the energy is radiated upstream and half downstream. In shocks with a thermalization layer, the temperature of the emitted radiation is determined by equating the blackbody flux σT^4 to the incident energy flux $\rho_1 v_1^3/2$. At very high temperatures and densities, such as occur in the interior of a supernova explosion, this energy flux is carried by neutrinos rather than by photons.

Magnetic fields can play an important role in radiative shocks because they are very resistant to compression. If an atomic gas is compressed adiabatically (i.e., in the absence of radiative losses), then the pressure goes up as $\rho^{5/3}$; if radiation is allowed to escape freely so that the temperature remains constant, then the pressure increases only as ρ. Magnetic fields have a pressure $B^2/8\pi$, where B is the field strength in gauss. In most astrophysical plasmas, the field is "frozen" to the plasma, so that the field strength varies linearly with the density ($B \propto \rho$) in a compression perpendicular to the field, and as a result the magnetic pressure varies as ρ^2. If the density increases by a large factor, as it often does in a radiative shock, then the magnetic pressure eventually dominates the gas pressure and the final density at the back of the relaxation layer is determined by the magnetic pressure (mathematically, the final field strength B_f is given by $B_f^2/8\pi = \rho_1 v_1^2$). For typical interstellar magnetic fields, the final compression ratio of the gas ρ_f/ρ_1 is about 100 (v_1/100 km s^{-1}).

SPECTRA OF INTERSTELLAR SHOCK WAVES

A great deal of both observational and theoretical effort has gone into understanding the emission lines produced by interstellar shocks. The emission spectrum is strongly affected by whether the shock is J type or C type. Strong shocks in atomic or ionized gas are generally J shocks; if the shock is advancing into a molecular gas it will be J type if its velocity is greater than about 30–50 km s^{-1}. On the other hand, C shocks are generally slow shocks ($v_1 \lesssim 30$–50 km s^{-1}) in molecular clouds. Their peak temperatures rarely exceed about 3000 K because at higher temperatures, molecules collisionally dissociate and the reduced cooling leads to a runaway situation of increasing temperature and collisional ionization, producing a J shock. C shocks are therefore nondissociative shocks.

In J shocks, where the shock heat is added impulsively and the cooling occurs in a relaxation layer behind the shock front, the cooling is often partially offset by heating due to the absorption of ultraviolet photons from hotter gas upstream and by the heat of (re)formation of H_2 molecules dissociated in the upstream gas. The cooling of gas in J shocks is dominated by atomic and ionic electronic transitions at high temperatures $T > 5000$ K, and by atomic and ionic fine structure transitions [especially O I (63 μm)] and molecular rotational transitions (especially those of H_2, CO, OH, and H_2O) at low temperatures. The chemistry in the relaxation zone of a J shock is marked by collisional dissociation and ionization at high temperatures $T \gg 10^4$ K, recombination of ions and electrons at $T \sim 10^4$ K, and the reformation of molecules at $T \lesssim 3000$ K. The reformation of molecular hydrogen initiates the formation of all other molecules, and this process occurs slowly on the surfaces of dust grains. Often, the gas has cooled to much less than 100 K before the formation of the molecules is complete. However, for dense J shocks, the heat of formation of H_2 maintains the gas temperature at about 400 K while the molecule formation occurs. Chemical reactions with moderate activation energies proceed rapidly in neutral gas with temperatures 300–3000 K, and these reactions lead to large quantities of CO, OH, and H_2O in warm postshock molecular gas.

The gas in a strong C shock never gets too hot, because that would dissociate the molecules that provide the cooling. As a result, the emission spectrum is dominated by the rotational and vibrational transitions in the molecules in the shock. The drift of the charged particles through the neutral gas heats it to 300–3000 K and produces large quantities of molecules such as CO, OH, and H_2O. A notable feature of C shocks is that the trace ions are not tightly coupled with the neutrals, so that their temperature can far exceed the neutral temperature. These hot ions can reproduce themselves by collisionally ionizing the neutrals; as already noted, if the shock velocity is too high, this process runs away and the shock switches to J type.

CURRENT OBSERVATIONS AND APPLICATIONS

Shocks play an important role in inducing the formation of new stars. The compression of clouds in the shocks driven by spiral density waves accelerates star formation there and may explain the long spiral pattern of massive star formation seen in spiral galaxies. Sequential star formation has also been proposed, in which the shock waves driven by expanding H II regions or supernovae compress the ambient gas and trigger further star formation, although the details of this process remain obscure. Thus, either the birth or death of massive stars can trigger further star formation, and it is possible that a chain reaction can be set up in which star formation propagates through huge molecular clouds in a galaxy, and perhaps even from cloud to neighboring cloud. Colliding galaxies have recently been observed to be copious emitters of infrared radiation, and it has been suggested that this emission originates directly from the shock waves associated with the collision, or, alternatively, that it is produced by embedded stars whose formation was triggered by the shock waves.

The formation of a star is a violent event that has recently been observed to be accompanied by strong interstellar shocks. Newly forming stars lose an appreciable fraction of their mass by ejecting supersonic outflows ("protostellar winds") for a period of 10^4–10^6 yr during their births. These outflows drive shocks into the ambient molecular gas, where they can be detected by observation of the high velocity molecules (particularly CO) accelerated by the shocks. The infrared emission of shocked H_2, CO, OH, and atomic oxygen has also been used to detect, map, and analyze these outflows. The morphology of the outflows is often bipolar, as in the case portrayed in Fig. 2. In addition, H_2O masers (extremely intense bursts of radio emission from a rotational transition of H_2O) and Herbig–Haro objects (bright visible nebulae seen in and near molecular clouds) are now believed to be shock wave emission produced by the interaction of protostellar winds with the ambient gas.

Some stars die in titanic explosions and are observed as supernovae. It has been suggested that the interstellar shocks driven by supernovae may determine the structure of the ISM. A three phase model of the ISM has been developed in which most of the volume of the interstellar medium is hot ($T \approx 10^6$ K) gas that has been shocked by supernova blast waves. Most of the mass of the ISM is in cold ($T \lesssim 100$ K) clouds. These clouds are surrounded by warm envelopes that are heated to $T \approx 8000$ K and are partially ionized by ionizing radiation. The cold, warm, and hot phases are all at about the same pressure. Inside nonradiative blast waves, evaporation of the clouds and their envelopes injects material into the hot gas phase; after the blast wave becomes radiative, the hot gas cools and returns to the warm and cold phases.

The elements heavier than lithium are made in the hot interiors of stars by nuclear reactions. Most of these reactions occur during the course of the evolution of the star, but some are made in the shock wave that accompanies a supernova explosion. The atoms of these newly minted elements are injected into the ISM at high velocities, and by processes that are not well understood, many of them, such as silicon and iron atoms, become constituents of tiny interstellar dust grains. Interstellar shocks can then act to destroy these grains by sputtering (collisions with high velocity ions) and by collisions with other dust particles. The charged grains spiral in the postshock magnetic field and are actually accelerated as the postshock gas and field are compressed (a process called betatron acceleration). Refractory grain materials such as graphite and silicates are destroyed in shocks of velocity $\gtrsim 100$ km s^{-1}. Shock waves dominate the destruction and shattering of interstellar dust, and therefore play a key role in determining the abundance and size distribution of this important component of the ISM, as well as the gas phase abundances of the condensible elements.

Shocks driven by supernovae may also account for cosmic rays, which are energetic particles that fill the Galaxy and can be observed on Earth and, more easily, in Earth orbit. High energy particles are observed in association with solar flares, interplanetary shocks, and supernova remnants, and this has led to the hypothesis of ion acceleration in shocks. The primary mechanism thought responsible is called the first order Fermi acceleration mechanism: Suprathermal particles in the shock front are scattered into the preshock gas by turbulence. Because these charged particles are streaming through unshocked gas, they generate Alfven turbulence, which scatters them back across the shock. Thus, the shocked and unshocked fluids act like converging magnetic mirrors that efficiently accelerate particles. Particles can be accelerated to velocities very near the speed of light by this mechanism.

Finally, theorists have suggested that the compression and cooling behind shock waves may trigger the formation of galaxies as well as of stars. In the early universe, in the era of galaxy formation, the propagation of huge shock waves driven by the collective effect of supernovae in "seed" galaxies, or possibly by superconducting loops of cosmic string, may have swept up and compressed the hydrogen gas to galactic proportions, and triggered the collapse of protogalactic gas clouds. In addition, the shock waves can produce H_2, enabling the nearly pure hydrogen gas to cool significantly below 10^4 K and thereby form stars.

Additional Reading

Flower, D. R. (1989). Atomic and molecular processes in interstellar shocks. *J. Phys B* **22** 2319.

McCray, R. and Snow, T. P. (1979). The violent interstellar medium. *Ann. Rev. Astron. Ap.* **17** 213.

McKee, C. F. and Hollenbach, D. J. (1980). Interstellar shock waves. *Ann. Rev. Astron. Ap.* **18** 219.

Shull, J. M. and Draine, B. T. (1987). The physics of interstellar shock waves. In *Interstellar Processes*, D. J. Hollenbach and H. A. Thronson, eds. Reidel, Dordrecht, p. 283.

Zel'dovich, Ya. B. and Raizer, Yu. P. (1966). *Physics of Shock Waves and High Temperature Phenomena* **1** and **2**. Academic, New York.

See also **Herbig–Haro Objects and Their Exciting Stars; Interplanetary Medium, Shock Waves and Traveling Magnetic Phenomena; Interstellar Medium, Stellar Wind Effects; Shock Waves, Collisionless, and Particle Acceleration; Supernova Remnants, Evolution and Interaction with the Interstellar Medium.**

Shock Waves, Collisionless, and Particle Acceleration

Martin A. Lee

A shock wave is a transition in a fluid (or solid) across which there is a mass flux and an increase in density and entropy per unit mass. Collisionless shocks are shock waves in which the dissipation or irreversibility is provided not by collisions between individual particles but by collective electromagnetic interactions between ions and electrons in a plasma. The solar corona and solar wind are totally ionized; direct collisions in these plasmas are negligible over spatial and temporal scales of interest for shocks, and the solar wind is supersonic. Thus a variety of collisionless shocks occur throughout the heliosphere: Solar flares drive shock waves which propagate deep into interplanetary space, planets and comets produce bow shocks in the solar wind, fast and slow solar wind streams from neighboring regions on the Sun with similar latitude interact due to solar rotation to generate shock pairs beyond a few AU (1 AU $\approx 1.5\times10^8$ km), and finally the solar wind itself decelerates to subsonic flow at the solar wind termination shock to merge with the local ionized interstellar medium. A distinguishing characteristic of heliospheric collisionless shocks is their ability to accelerate ions (and, less often, electrons) to energies well beyond thermal plasma energies of ~ 100 eV/nucleon, corresponding to temperatures of 10^6 K. These particles are either energetic particles (e.g., solar flare particles) reaccelerated by the shock or thermal particles extracted from the solar wind plasma by the shock itself.

HELIOSPHERIC SHOCKS AND ASSOCIATED ENERGETIC PARTICLES

Collisionless shocks are responsible for many (perhaps most) of the energetic particle populations throughout the heliosphere. The energy spectra, anisotropies, composition, and spatial distributions of these populations have been studied intensively over the last 15 years based on measurements by a fleet of spacecraft in the solar wind including *Helios* probes that penetrated to within 0.3 AU of the Sun, *ISEE* and *IMP* satellites near Earth, and the *Voyager* and *Pioneer* probes in the outer heliosphere.

Associated with interplanetary traveling shocks produced by flares or coronal mass ejections on the Sun are energetic storm particles (ESP), named for their temporal correlation at Earth with shock-induced magnetic storms. The bulk of these particles are ions of solar wind origin with energies of up to ~ 500 keV/nucleon near 1 AU. In addition, ambient solar flare energetic ions, often originating in the flare which produced the shock, may be reaccelerated up to ~ 1 GeV/nucleon. ESP events peak near the shock, decay with energy-dependent scalelengths in the range 10^5–10^7 km, and exhibit power-law energy spectra at lower energies near the shock.

During periods of solar minimum, solar coronal holes, which are the origin of fast solar wind streams, often migrate to low latitudes. As the Sun rotates, the streams overtake slow solar wind and an interaction region characterized by high magnetic field and pressure is formed in the solar wind beyond 3–5 AU. The configuration of this region is stationary in a frame which corotates with the Sun and it is bounded by a forward and a reverse shock. The corotating ion events exhibit peaks in intensity at both shocks. They have an ion composition which is approximately that of the solar wind or solar flare particles, spatial gradients of 100%/AU, and energy spectra which are exponential in ion speed with a characteristic energy of ~ 1 MeV/nucleon.

At Earth's bow shock, energetic ions with approximately the composition of the solar wind are observed in the energy range 5–150 keV/nucleon. Either they stream away from the shock along the solar wind magnetic field as field-aligned beams or reflected ions, or they form a diffuse distribution that is nearly isotropic in the frame of the shock, or they form a distribution intermediate between the two. The diffuse ions have energy spectra which are exponential in energy per unit charge, are spatially uniform in the magnetosheath, and decrease in intensity upstream of the shock with scalelength $\sim 4\times10^4$ km at 30 keV per unit charge. Electrons in the energy range 1–20 keV are also observed to stream upstream away from the shock. Similar energetic particle populations have been observed at Jupiter.

Other associations of energetic particles with heliospheric shocks are less direct. Some solar cosmic rays may be accelerated at coronal shocks. The anomalous cosmic ray component, which consists primarily of energetic He^+, N^+, O^+, and Ne^+ in the outer heliosphere in the energy range 1–40 MeV/nucleon, originates from interstellar neutral gas which is ionized in the solar wind, convected to the outer heliosphere, and probably accelerated at the solar wind termination shock. Finally, galactic cosmic rays may be accelerated at interstellar (probably supernova) shocks. Indeed a major motivation for studying collisionless shocks and shock acceleration in the heliosphere, where theory is guided by in situ observations, is to help understand astrophysical shocks, and the origins of cosmic rays and the energetic electrons which generate nonthermal radio emission from galactic and extragalactic sources.

COLLISIONLESS SHOCKS

A shock can be viewed as a transition region between upstream and downstream states of a fluid through which there is a net mass flux and across which mass, momentum, and energy flux are conserved but entropy per unit mass and mass density increase. A shock may form through the steepening of a large-amplitude compressive wave or through the deceleration and/or deflection of a supersonic flow, as at a planetary bow shock.

As implied by the former origin, a shock corresponds to a particular (compressive) wave mode with phase speed C. The shock propagates into the upstream fluid with speed V_s where the Mach number $M = V_s/C_u > 1$. Let the subscript or superscript u (d) denote upstream (downstream). In hydrodynamics there is an acoustic shock for which C_u is independent of the direction of the shock normal, **n**. In magnetohydrodynamics (MHD), there are fast and slow shocks corresponding to the two compressive modes, for which the phase speeds depend on the angle θ^u between **n** and the upstream ambient magnetic field, $\mathbf{B}_u$. For $\theta^u = 90°$, $C_{\text{fast}}^2 = C_s^2 + C_A^2$, where C_A is the Alfvén speed and C_s is the sound speed. Although Alfvén shocks, corresponding to the noncompressive MHD mode, are not allowed in a stationary configuration, they have been identified in numerical simulations which allow them to evolve with time and in which a dissipative process (e.g., electrical resistivity) is included; their possible role in space is still unclear.

Because the heliospheric shocks mentioned in the last section arise from large scale disturbances with characteristic frequencies much smaller than the gyrofrequencies of the thermal ions, they correspond to MHD shocks. Actually they are virtually all fast shocks for which $B_u < B_d$. A few slow shocks ($B_u > B_d$), limited to $C_{\text{slow}}^u < V_s < C_A^u \cos(\theta^u)$, have been identified with some certainty in the solar wind within 1 AU; slow shocks are also expected to occur at sites of magnetic reconnection (i.e., in solar flares, at Earth's dayside magnetopause, and in Earth's magnetotail).

The mechanism by which the upstream flow is decelerated and heated depends on the type of shock. Binary collisions provide the heating in a gas shock. Historically it was first questioned whether a shock could exist in a tenuous gas such as the solar wind in which collisions are negligible. However, the sudden commencement of geomagnetic storms at Earth approximately two days after a flare on the Sun was first interpreted by Thomas Gold (and later supported by direct observations) to be due to the increased mass flux behind an interplanetary shock impinging on Earth's magnetosphere. In an ionized gas the electric and magnetic field fluctuations supported by the collective motions of the plasma play the role of collisions in scattering and heating the individual ions and electrons.

A key parameter is θ^u. If $\theta^u \geq 45°$, the shock is quasiperpendicular. In this case for $M < M_c$, where M_c is a critical Mach number which depends on θ^u but is ~ 2, the dissipation is provided by resistive heating via electron-current-driven plasma instabilities. For $M > M_c$, it is provided by ion deflection and reflection resulting from the large scale electric and magnetic fields within the shock transition layer. If $\theta^u \leq 45°$, the shock is quasiparallel. In this case **B** does not inhibit counterstreaming of the upstream and downstream plasmas. Based on recent numerical simulations, dissipation appears to be dominated by the action of large-amplitude transverse waves which are excited by, and scatter and heat, the counterstreaming ions. Particularly for quasiparallel shocks an important additional channel of dissipation is the acceleration of particles to suprathermal energies.

Major insights into collisionless shock structure have arisen from the detailed observations of Earth's bow shock by the *ISEE* (International Sun–Earth Explorer) satellites.

PARTICLE ACCELERATION

It was recognized more than 20 years ago that particles could gain energy by interacting with the large-scale electric and magnetic fields (**E** and **B**) at a fast shock. In the frame of reference in which the upstream plasma flow, $\mathbf{V}_u$, is parallel to $\mathbf{B}_u$ and the shock is at rest (the deHoffman–Teller frame), an upstream energetic particle approaching the shock with sufficiently large pitch angle will be mirrored by the increased field strength at the shock and reflected back upstream elastically. Viewed in the frame of the upstream fluid, or in a frame in which the shock is stationary and $\mathbf{V}_u$ is more nearly parallel to the shock normal, **n**, the reflected particle gains energy by curvature/gradient drifting along the shock front parallel to the electric field ($\mathbf{E} = -c^{-1}\mathbf{V}\times\mathbf{B}$). Energy gains by this mechanism are limited to at most an order of magnitude and are largest at quasiperpendicular shocks. The mechanism, often called shock drift acceleration, accounts for electron acceleration at Earth's bow shock and the reacceleration of solar cosmic rays at quasiperpendicular interplanetary traveling shocks.

In 1977–1978 it was recognized independently by G. F. Krymskij, by W. Ian Axford, Egil Leer, and George Skadron, by A. R. Bell, and by Roger D. Blandford and Jeremiah P. Ostriker that shocks can accelerate particles by many orders of magnitude in energy and that the process can be very efficient. The first condition for acceleration is that there be seed particles sufficiently energetic that they can pass from the downstream to the upstream plasma against the oblique orientation of **B**. If the shocked downstream plasma provides the only seed particles, then the mechanism is limited to quasiparallel shocks. The second condition is that there be magnetic fluctuations in the upstream and downstream plasma which can scatter the particles in pitch angle and reverse their guiding-center motion along **B**.

If these two conditions are satisfied, particles can traverse the shock many times. Viewed in the shock frame, particles returning to the shock from upstream have been scattered by approaching irregularities and have gained energy, whereas those returning from downstream have been scattered by receding irregularities and have lost energy. Because a shock is compressive, $\mathbf{V}_u\cdot\mathbf{n} > \mathbf{V}_d\cdot\mathbf{n}$ and the particles experience a net energy gain during a complete traversal of the shock. Because the essence of this acceleration mechanism is repeated head-on collisions with each being elastic in the frame of the scatterer, it is often described as a first-order Fermi process. However, the particles do still undergo gradient/curvature drift parallel to **E** while traversing an oblique ($\theta^u \neq 0$) shock so that a (frame-dependent) portion of the energy gain is due to the shock-drift mechanism.

The statistics of the process, which determines actual energy spectra and spatial distributions, is often calculated in the limit of nearly isotropic particle distributions and large particle speeds ($\gg \mathbf{V}_u, \mathbf{V}_d$) using a transport equation which balances convection with the plasma, curvature, and gradient drift, spatial diffusion due to scattering, and compressive energy gain at the shock. In this limit the process is called diffusive shock acceleration. For stationary acceleration at an infinite planar shock and seed particle injection at low energy, the theory predicts a differential density at the shock, $N(p)$ ($\int_0^\infty dp\,N(p) =$ particles cm^{-3}), which is a power law in momentum, $N(p) \propto p^{-\alpha}$, where $\alpha = (r+2)/(r-1)$ and r is the shock compression ratio $[=(\mathbf{V}_u\cdot\mathbf{n})/(\mathbf{V}_d\cdot\mathbf{n})]$. If particle escape, finite acceleration time, or energy loss is included, softer energy spectra are derived. The theory accounts quantitatively for many features of the shock-associated particle distributions described above.

WAVE EXCITATION

The presence of accelerated particles either streaming into the upstream plasma or as a stationary upstream distribution implies at a fast shock that the particles are streaming along $\mathbf{B}_u$ with a bulk speed greater than C_A. If these particles are ions, they are scattered effectively by MHD waves which are cyclotron-resonant with the ions, that is, with a wavelength along $\mathbf{B}_u$ equal to the distance the ion transverses in one gyroperiod. If the waves propagate in the same direction along $\mathbf{B}_u$ as the bulk flow of the ions, the ions lose some energy in the plasma frame as they scatter elastically in the wave frame. That energy is transferred to the waves, which are amplified as they are convected toward the shock. The waves are further compressed at the shock transition layer to contribute to the downstream turbulence. Thus, the ions themselves can excite the waves required to confine them near the shock for acceleration in the first place.

For ions in the energy range 10–200 keV/nucleon, the resonant wave period in the spacecraft frame is about 10–50 s at 1 AU. Large-amplitude wave enhancements in this frequency range have been observed in association with intermediate and diffuse ion distributions at Earth's bow shock and ESP events at quasiparallel interplanetary traveling shocks. Enhancements have also been observed upstream of Jupiter's bow shock and the shocks bounding corotating interaction regions.

The region upstream of a collisionless shock containing the accelerated ions and associated MHD waves is often called the foreshock, a term coined by E. W. Greenstadt. It is interesting to note that the coupling of the upstream plasma and energetic ions by MHD waves excited by their counterstreaming is analogous to the coupling of the upstream and downstream thermal plasmas by a large-amplitude transverse wave at the shock transition layer of a quasiparallel shock.

INJECTION, MODIFICATION, AND ELECTRONS

Ion shock acceleration out of the thermal plasma is one channel of dissipation for a collisionless shock. Thus the scalelength of the foreshock is one measure of the shock transition width. Collisionless shocks may be broad structures with a separate scalelength for each dissipation mechanism. Plasma heating generally occurs on a smaller scale, denoted here as the transition layer, and involves other plasma waves, for example, ion acoustic waves.

How many ions are able to escape from the shock-processed downstream plasma back upstream to initiate shock acceleration depends sensitively on shock structure (e.g., resonant wave spectra) at the scale of plasma ion thermalization. This injection rate decreases as θ^d increases, depends on energetic ion rigidity, and probably determines the composition of the energetic ions. The theory of diffusive shock acceleration cannot describe the more complex process of injection. Research in this area focuses on detailed observations at Earth's bow shock and numerical simulations.

Although much of the theoretical work on shock acceleration treats the energetic particles as test particles, in principle their pressure gradients can substantially alter shock structure. Such a shock is called a cosmic-ray-modified or mediated shock. Observed energetic particle intensities at virtually all heliospheric shocks are too small to modify shock structure. Nevertheless, interstellar shocks are probably strongly modified by galactic cosmic rays, and the solar wind termination shock may be modified by anomalous cosmic ray hydrogen and helium. Strongly modified shocks with large M have further complexities: The hydromagnetic waves are excited by the energetic ions to amplitudes greater than $\mathbf{B}_u$, which implies nonlinear saturation and preheating of the upstream plasma. Furthermore, the energy spectrum produced at a stationary planar shock with large M is strongly divergent, which may imply that shock acceleration cannot be stationary at the highest energies. Also, as pointed out by David Eichler, shock modification by the energetic ions controls injection by decelerating and heating the foreshock plasma which decreases the Mach number M of the thermal plasma subshock, or transition layer, and hence the injection rate.

Although diffusive electron acceleration can in principle occur at collisionless shocks, diffuse energetic electron distributions are not observed directly at heliospheric shocks. There is only indirect evidence that solar flare electrons in the energy range 1–10 MeV are accelerated at coronal shocks. Because electrons resonate with whistler waves, which are right-hand circularly polarized for propagation approximately parallel to $\mathbf{B}$, unstable whistler waves do not easily provide both senses of polarization which are required to scatter electrons back to the shock. Only at MeV energies can electrons scatter on proton-excited turbulence. Thus, if the ambient intensity of whistler waves is small, shock acceleration could be very inefficient for electrons.

Additional Reading

Blandford, R. and Eichler, D. (1987). Particle acceleration at astrophysical shocks: A theory of cosmic ray origin. *Physics Reports* **154** 2.

Drury, L. O'C. (1983). An introduction to the theory of diffusive shock acceleration of energetic particles in tenuous plasmas. *Reports on Progress in Physics* **46** 973.

Forman, M. A. and Webb, G. M. (1985). Acceleration of energetic particles. In *Collisionless Shocks in the Heliosphere: A Tutorial Review*, R. G. Stone and B. T. Tsurutani, eds. American Geophysical Union, Washington, D.C., p. 91.

Kennel, C. F., Edmiston, J. P., and Hada, T. (1985). A quarter century of collisionless shock research. In *Collisionless Shocks in the Heliosphere: A Tutorial Review*, R. G. Stone and B. T. Tsurutani, eds. American Geophysical Union, Washington, D.C., p. 1.

Scholer, M. (1985). Diffusive acceleration. In *Collisionless Shocks in the Heliosphere: Reviews of Current Research*, B. T. Tsurutani, and R. G. Stone, eds. American Geophysical Union, Washington, D.C., p. 287.

Thomsen, M. F. (1985). Upstream suprathermal ions. In *Collisionless Shocks in the Heliosphere: Reviews of Current Research*, B. T. Tsurutani and R. G. Stone, eds. American Geophysical Union, Washington, D.C., p. 253.

See also **Heliosphere; Interplanetary Medium, Shock Waves and Traveling Magnetic Phenomena; Magnetohydrodynamics, Astrophysical; Supernova Remnants, Evolution and Interaction with the Interstellar Medium.**

Solar Activity

Frank Q. Orrall

Solar activity is a generic term used to designate a number of different transient phenomena, with a wide range in size, lifetime, energy, and complexity, that occur in the atmosphere of the Sun. Sunspots, faculae, active regions, flares, radio and x-ray bursts, corpuscular emission, prominences, coronal streamers, coronal mass ejections, and coronal holes, are some specific examples of these phenomena. This activity is in some ways analogous to meteorological phenomena (e.g., clouds, rain, frontal depressions, and lightning) that occur in the Earth's lower atmosphere. However, a major difference is that the Sun and its atmosphere are composed of ionized, electrically conducting gas (plasma), so that the Sun's complex magnetic field is coupled to the plasma and is strongly implicated in all solar activity.

Although some active solar phenomena occasionally can be seen with the unaided eye, systematic study of solar activity had to await the invention of the telescope in the first decade of the seventeenth century. (Large sunspots can sometimes be seen with the naked eye when the Sun is close to the horizon or dimmed by haze. Solar prominences and coronal streamers can be seen with the naked eye during a total solar eclipse.) The first telescopic observations of sunspots were made and reported independently by David Fabricius, Galileo Galilei, Thomas Harriot, and Christopher Scheiner between 1610 and 1612 and led to the discovery of the Sun's rotation with a period of about 27 days as seen from Earth. These early observations established that large spots have a discernable structure, consisting of a dark *umbra* and surrounding *penumbra*. They also revealed the tendency of spots to appear in groups, the existence of bright photospheric *faculae*, and gave the first tentative hint that the Sun does not rotate as a solid body.

CYCLES OF SOLAR ACTIVITY AND MAGNETISM

In 1843 Heinrich S. Schwabe presented the first evidence that the frequency of sunspot occurrence varies with a period of roughly 10 years. Rudolf Wolf, in 1848 introduced a relative daily sunspot number or index that is proportional to $(10g + s)$ where g and s, respectively, are the total numbers of spot groups and individual spots counted on the visible disk of the Sun. He reconstructed this number back to 1610, using all available historical data, and found the mean length of the *sunspot cycle* to be 11.1 yr. This cycle is also referred to as the *activity* cycle because almost all solar active phenomena in some way follow it. In 1852 Edward Sabine, Wolf, and others independently recognized that the frequency and magnitude of disturbances in the Earth's magnetic field follow the sunspot cycle. This was the first clear evidence of a direct link between solar and terrestrial activity. The Wolf (or Zürich) number is still the reference index of solar activity although other indices (e.g., daily measurements of the 2800-MHz solar radio flux) are of increasing usefulness.

Most sunspot groups have lifetimes much shorter than a single solar rotation although a few may persist for several rotations. Most are found at solar latitudes between 5 and 35° from the equator. Between 1853 and 1873 Richard C. Carrington, and later Gustav Spörer, made careful studies of the positions of sunspots and established two important effects: First, they found that the Sun rotates, not as a solid body, but differentially, with the equator rotating faster than the poles. The synodic period in days is given approximately by $26.75 + 5.7 \sin^2 \phi$, where ϕ is the solar latitude. Second, they showed that the mean latitude at which sunspots appear decreases systematically during the cycle. The first spots born in a new cycle have an average latitude of about 28° in each hemisphere, whereas by the last year of the cycle this average latitude has decreased systematically to about 7°.

In 1908 George E. Hale established by polarimetric measurement that the splitting of certain spectral lines observed within sunspots and their surroundings was due to the *Zeeman effect* (which had been discovered in the laboratory in the previous decade). This implied strong magnetic fields within sunspots ranging from a few hundred to a few thousand gauss. Between 1908 and 1924 Hale and Seth B. Nicholson made systematic measurements of the polarity and strength of magnetic fields within sunspot groups. They found that the magnetic field in a group is usually *bipolar*, that is, both magnetic polarities are present and tend to be separated from each other. The magnetic polarity (positive or negative) on the western side of the group they called *preceding* and that on the eastern side, *following*. Most spot groups in a given hemisphere and sunspot cycle have the same preceding polarity, whereas groups in the opposite hemisphere have preceding polarity of the opposite sign. However, in the next sunspot cycle the sign of the preceding polarity reverses in both hemispheres, corresponding to a general reversal of the Sun's magnetic field. Consistent with this, the weak magnetic fields at the poles change sign a few years after the start of the new cycle. Thus the Sun has a *magnetic* cycle that is 22 yr in duration, which gives rise to two 11-yr *activity* cycles. Successive sunspot cycles overlap in that the first high-latitude spot groups of the new cycle appear while the last low-latitude groups of the old cycle are still present, each having the preceding polarity appropriate to their own cycle.

The global ordering of sunspot groups and of their magnetism implied by the preceding classical observations suggest that below the solar surface there exist strongly ordered toroidal magnetic fields that from time to time rise to the surface and produce sunspot groups with their magnetic fields and attendant activity. An additional observational fact is the obvious irregularity of the activity cycle. Cycles have been observed to vary in length (between minima) from about 9 to 14 years, and to vary widely in amplitude as indicated by sunspots and other indicators of activity. Moreover, between 1645 and 1715 very few sunspots were observed at all (the "Maunder minimum"), suggesting that the magnetic cycle may very nearly shut down from time to time and then restart. This conjecture is supported by proxy evidence of prehistoric solar activity, such as that contained in measurements of radioactive carbon (^{14}C) found in dated tree rings.

The basic mechanism that maintains this magnetic cycle and thereby drives solar activity is by no means well understood and constitutes an important problem of solar and stellar astrophysics. Most recent theories have evoked the idea of a *hydromagnetic dynamo* that is reversing and self-regenerating and that is driven by the interplay between convection and differential rotation. This dynamo effect is thought to take place in and above the convection zone, with the transition shell between the convective zone and the deep interior playing an important role. Crucial questions concern the depth of the convective zone, how the Sun rotates internally, and how angular momentum is transported. Studies of solar oscillations are beginning to provide such information. Studies of solar-like magnetic activity in other stars provides insight concerning how dynamo action depends on stellar mass, rotation, and stage of evolution.

SOLAR ACTIVE REGIONS

The term *active region* (AR) is used to designate the whole evolving structure that comes into being in the Sun's photosphere, chromosphere, and corona with the birth and development of a sunspot group. The more fundamental event is, however, the emergence of magnetic flux tubes from below the photospheric surface, which gives rise to the sunspots and the other active phenomena. These *emerging flux regions* (EFRs) and the process by which sunspots and other magnetic structures develop from them have been well-observed. Most ARs are small and short-lived, have a simple bipolar structure, and are produced by only one or two EFRs. Long-lived ARs require the periodic emergence of new flux. (About 50% of all sunspot groups have lifetimes of less than 2 days, and 90% less than 11 days; however, some large groups may last for 100 days or more.)

The AR gives rise to regions of bright chromospheric emission called *plage*, which are most easily seen in the cores of strong Fraunhofer lines (absorption lines in the solar spectrum) such as the Hα line of hydrogen or the K line of ionized calcium. The overlying transition region and corona are also enhanced in brightness by the formation of hot dense plasma-filled magnetic loops that extend into the corona. ARs give rise to a major portion of the Sun's extreme ultraviolet and soft x-ray emission, which determine the ionization and structure of the Earth's upper atmosphere.

As ARs disperse due to convective motions in the atmosphere, their magnetic fields diffuse poleward to produce large-scale *unipolar magnetic regions*. These are responsible for reversing the polar fields a few years after the start of a new cycle, impose large-scale structure on the corona, and determine the magnetic fields in interplanetary space.

PROMINENCES

Prominence is a term used to designate several quite different types of structure, all of which have in common that they are relatively cool (mean temperature $\overline{T} = 5\text{–}10 \times 10^3$ K) and dense, yet exist surrounded by the hot ($T = 10^6$ K) tenuous corona. All have a spectrum superficially similar to that of the chromosphere. Prominence material is nonbuoyant in the corona, being about 100-fold more dense than its surroundings. Hence static prominences must be supported against gravity (by the magnetic field) or have their material continuously resupplied. Prominences are most easily observed in the Hα line in emission above the Sun's limb or in absorption when seen against the disk. They are associated in some way with most forms of solar activity.

The most commonly observed and best studied type of prominence appears as a dark elongated structure when seen on the disk, where it is called a *filament*. When the same structure is seen above the limb, it appears bright and reveals a complex internal structure and is called a *hedgerow* or *quiescent prominence*. Filaments are found along the neutral lines that separate regions of opposite magnetic polarity. They are found both inside and outside of active regions. Some types of prominences are found only within active regions: *Surges* are jets that rise into the corona and fall back along the same path. *Sprays* consist of material thrown explosively into the corona during major flares, sometimes at a velocity exceeding the escape velocity. Systems of *postflare loops* develop in the aftermath of some very energetic flares.

SOLAR FLARES

Flares are the most complex and energetic manifestation of solar activity. They occur within active regions, often near the magnetic neutral line. They range in duration from a few minutes to hours, but their effect on the interplanetary medium may last for days. They are believed to be due to the sudden release of magnetic energy stored in stressed, current-carrying magnetic fields. Consistent with this, the active regions most productive of energetic flares have a complex magnetic structure, usually produced by the intrusion of new magnetic flux. The first reported flare (in 1859) was seen in white light and was followed by a strong disturbance of the Earth's magnetic field: the first indication that flares have important effects in the Earth's magnetosphere and upper atmosphere. This was a large and unusually energetic event. Most flares are not easily detected in white light, but are most commonly recorded in the Hα line, in which they appear as a transient brightening of the plage. The most commonly used measure of flare importance is based on the measured area of this region of enhanced chromospheric emission, together with an estimate of flare brightness. Even for the largest flares this area is less than 1% of the area of the disk. Another commonly used index is based on the enhancement in the Sun's soft x-ray flux due to the flare, which is a measure of the hot coronal plasma produced by the event. No simple index can describe adequately the complexity and variety of individual flares. However, great energetic flares do often show three well-defined phases: (1) A *precursor* or *preflare* stage of gradual brightening; (2) a *flash* or *impulsive* phase lasting seconds to minutes and marked by bursts of hard x-rays, centimetric and metric radio waves, γ-rays, and the onset of a shock wave; (3) a *main* or *extended* phase characterized by soft x-ray emission, the development of the optical flare, and complex radio emission (the aftermath of the impulsive phase).

Additional Reading

Eddy, J. A. (1977). The case of the missing sunspots. *Scientific American* **236** (No. 5) 80.

Foukal, P. V. (1990). The variable Sun. *Scientific American* **262** (No. 2) 34.

Parker, E. N. (1975). The Sun. *Scientific American* **233** (No. 3) 43.

Sturrock, P. A., Holzer, T. E., Mihalas, D. M., and Ulrich, R. K., eds. (1986). *Physics of the Sun* **1**, **2**, and **3**. Reidel, Dordrecht.

Zirin, H. (1988). *Astrophysics of the Sun*. Cambridge University Press, Cambridge.

Zwaan, C. (1987). Elements and patterns in the solar magnetic field. *Ann. Rev. Astron. Ap.* **25** 83.

See also **Interplanetary Medium, Solar Wind; Planetary Atmospheres, Solar Activity Effects; Solar Activity, Solar Flares; Solar Activity, Sunspots and Active Regions, Observed Properties; Stars, Activity and Starspots; Sun, Atmosphere, Corona; Sun, High Energy Particle Emissions; Sun, Magnetic Field; Sun, Radio Emissions.**

Solar Activity, Coronal Mass Ejections

Stephen W. Kahler

The expulsion of solar coronal material into the interplanetary medium occurs in two ways. The primary way is through the quasisteady streaming of coronal gas along open magnetic field lines. The second takes place when regions of closed magnetic field expand outward to form transient eruptions known as coronal mass ejections (CMEs).

Observations of CMEs are made by space-borne coronagraphs that observe photospheric white light scattered by the free electrons in the CME. The coronagraph occults the bright solar disk to allow observations of the corona at distances of approximately 2–10 solar radii from Sun center. Unfortunately, this excludes the lower region of the corona where CMEs apparently form. Another limitation is that although CMEs originating near the solar limb are easily observed (with a coronagraph) against the background corona, those originating near disk center are difficult to detect.

Observations made by the *Skylab*, *P78-1*, and *Solar Maximum Mission* spacecraft spanning more than an 11-yr solar activity cycle have provided the basis for statistical studies of CMEs. A given CME can be characterized by its speed (measured at its leading edge), its mass, its angular width (measured from Sun center), and its position (in solar latitude). CME speeds range from less than 50 to 2000 km s^{-1}, masses range from 10^{15} to 10^{17} g, and the angular widths range from less than 10 to more than 90°. CMEs occur where the polarity inversion lines of solar magnetic fields are found. During high solar activity, the inversion lines extend beyond the latitudes of the active regions, and the CMEs occur over a broad range of latitudes at an observed rate of 1–2 CME per day. Near solar activity minimum the inversion line is considerably simplified and lies close to the equator. Accordingly, the CMEs at that time are confined to lower latitudes and occur at a rate of only 0.1–0.3 CME per day. In addition, the average CME speeds decline from about 400 km s^{-1} at solar maximum to about 200 km s^{-1} at solar minimum.

BASIC STRUCTURE AND ASSOCIATIONS

Coronagraph observations show a variety of CMEs loosely described as loops, spikes, fans, clouds, streamer brightenings, and so on. The simple structure of loop CMEs led early investigators to suppose that they were planar loops arising from the expansion of coronal loop structures. Later studies suggested that a bubble-like or cone geometry is more compatible with loop and other CMEs. The loop appearance results from viewing an optically thin bubble.

One approach to understanding the dynamics of CMEs is to seek other transient solar phenomena associated with the CMEs. Studies of CMEs observed by *Skylab* and *Solar Maximum Mission* revealed that CMEs were more frequently associated with erupting prominences than with flares. Erupting prominences are now recognized as one of the three basic components of many CMEs. These components are (1) a bright outer loop enclosing (2) a region of depleted density with (3) a bright core, as shown in Fig. 1. They are associated with the preevent (1) ambient corona, (2) region of depleted particle density known as the coronal cavity, and (3) cool prominence, respectively.

Fast-mode magnetohydrodynamic shock waves are expected from the many CMEs that exceed the coronal fast-mode wave speed (about 600 km s^{-1}). The fastest CMEs do show a good (but not perfect) correlation with reported metric type II bursts, the slow-drift bursts attributed to coronal plasma oscillations at an expanding shock front. These fast CMEs generally show a bright frontal structure expected from the compression of coronal plasma. Slow CMEs, on the other hand, show a smooth or diffuse front resulting from a gradual compression of the plasma and are not associated with type II bursts. There is some evidence that slow-mode magnetohydrodynamic shocks may be driven by CMEs with speeds between the coronal sound speed (200 km s^{-1}) and the coronal Alfvén speed (600 km s^{-1}).

Early models of CMEs assumed that associated flares provided the thermal energy to drive a pressure pulse in the lower corona that produced the CME. However, most CMEs occur without accompanying flares and, when a flare does occur, it usually follows the CME onset and often lies near one leg of the CME rather than directly under the CME. The angular size of a CME is also

Figure 1. A loop CME occurring over the northeast limb of the Sun, observed on 14 April 1980 with the coronagraph on the *Solar Maximum Mission* spacecraft. The radius of the occulting disk is 1.6 solar radii. The bright inner loop is the eruptive prominence. (*Courtesy of A. Hundhausen.*)

about 2–5 times that of an associated flare. Furthermore, CME accelerations are sometimes seen up to 10 solar radii. These observations show that the flare is a secondary and nonessential feature of the eruptive process.

Recent modeling of CME development involves closed coronal magnetic fields that evolve out of equilibrium states. Then the coronal plasma is no longer bound by gravity and magnetic tension and begins to rise. As the CME evolves, a magnetic neutral sheet develops at which magnetic reconnection occurs. This process results in large loop arcade structures, known as postflare loops, which develop at the previous site of the prominence. These loops can be observed in soft x-ray and Hα wavelengths as a characteristic signature lasting for hours after the CME.

INTERPLANETARY EFFECTS

The average mass contribution of CMEs to the solar wind is estimated to be only 2–10%. However, CMEs are the most energetic of all transient solar phenomena and are therefore of interest for the effects they produce in the interplanetary medium. Nearly all large solar energetic (>1 MeV) particle events observed at the Earth are associated with CMEs. In addition, the fastest interplanetary shocks are associated with large, bright, and fast CMEs. Conversely, nearly all large, bright CMEs lying in the ecliptic plane are associated with interplanetary shocks, for which they are assumed to be the drivers. These shocks and the associated driver gas cause geomagnetic disturbances when the plasma interacts with the magnetosphere.

Several distinct signatures of CMEs are expected in the solar wind near the Earth, based on statistical studies of known CMEs or on their predicted properties. These include the following:

1. An enhanced helium–hydrogen abundance ratio.
2. An enhanced abundance of singly ionized helium resulting from the cool solar prominence material.
3. Low proton and electron temperatures.
4. Enhanced magnetic fields.
5. Rotations of enhanced magnetic fields, known as magnetic clouds, resulting from the CME loop structure.
6. Bidirectional particle fluxes resulting from reflections in closed magnetic loops.

Convincing cases of all these signatures have been found, but the agreement among the various signatures for individual cases is often poor. Thus the evolution of CMEs in the solar wind is not well-understood.

Additional Reading

Dryer, M. (1982). Coronal transient phenomena. *Space Sci. Rev.* **33** 233.

Gosling, J. T., Hildner, E., MacQueen, R. M., Munro, R. H., Poland, A. I., and Ross, C. L. (1974). Mass ejections from the Sun: A view from *Skylab*. *J. Geophys. Res.* **79** 4581.

Howard, R. A., Sheeley, N. R., Jr., Koomen, M. J., and Michels, D. J. (1985). Coronal mass ejections: 1979–1981. *J. Geophys. Res.* **90** 8173.

Hundhausen, A. J. (1988). The origin and propagation of coronal mass ejections. In *Proceedings of the Sixth International Solar Wind Conference*, **1**, V. J. Pizzo, T. E. Holzer, and D. G. Sime, eds. National Center for Atmospheric Research, Boulder, Colorado, p. 181.

Kahler, S. (1987). Coronal mass ejections. *Rev. Geophys.* **25** 663.

Schwenn, R. and Marsch, E., eds. (1990). *Physics of the Inner Heliosphere. I. Large-Scale Phenomena*. Springer, Berlin.

Wagner, W. (1984). Coronal mass ejections. *Ann. Rev. Astron. Ap.* **22** 267.

Wentzel, D. G. (1989). *The Restless Sun*. Smithsonian Institution Press, Washington, D.C.

See also **Interplanetary Medium, Shock Waves and Traveling Magnetic Phenomena; Solar Activity, Flares; Sun, Atmosphere, Corona; Sun, Radio Emissions.**

Solar Activity, Solar Flares

George A. Doschek

Solar flares are explosive outbursts that occur in the Sun's atmosphere. An average flare releases between 10^{30}–10^{31} erg in tens of minutes and larger flares release more than 10^{32} erg. This energy is in the forms of electromagnetic radiation and the kinetic energy of accelerated particles. A flare can affect all regions of the Sun's atmosphere, notably the chromosphere, transition region, and corona; sometimes brightenings in the photosphere may be seen.

The radiation and particle emissions from flares also produce important effects in the Earth's atmosphere. The particle emissions cause aurora, geomagnetic storms, and disruptions of communications.

Solar flares were discovered by Richard C. Carrington and R. Hodgson, on September 1, 1859. They observed a so-called white light flare, that occurs when part of the energy released by a flare penetrates into and perturbs the photosphere.

Flares are intimately related to the Sun's magnetic field. Flares occur in the active regions of the solar atmosphere that overlay sunspots. Consequently, the number of flares and their locations on the Sun essentially mirror the similar quantities for sunspots. The number of flares waxes and wanes with the 11-yr sunspot cycle, and most flares are confined to the latitude regions in which sunspots are found, usually between 5–30° in both hemispheres. Recently, data on flare positions on the Sun indicate that there may exist preferred solar longitudes for the occurrence of flares. These preferred longitudes may persist over several solar cycles.

Because of the close association of flares with magnetic fields, flares are believed to be produced by the conversion of stored magnetic energy to kinetic energy and radiation. Theories on how the magnetic energy is converted are discussed in the next entry in this volume.

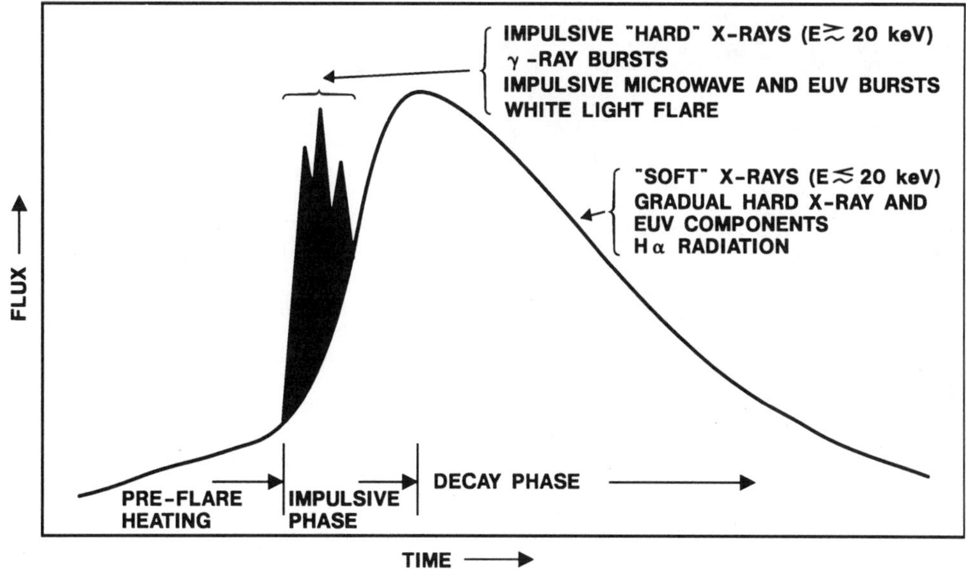

Figure 1. Typical solar flare radiative light curves.

Flares emit radiation throughout the electromagnetic spectrum, from decameter radio waves to gamma rays. The reason is that different flare regions have different temperatures. Radiation is also produced under nonthermal conditions by accelerated particles. A summary of important radiative emissions from flares is shown in Fig. 1.

The spatial size of flares varies considerably. They are small compared to the active regions. Compact flares have a soft x-ray emitting spatial scale of about 10 arcsec. (1 arcsec, as measured from Earth, corresponds to about 725 km at the Sun's surface.) However, some flares can be considerably larger, with characteristic lengths of about 2 arcmin. Regardless of size, flares can be extremely bright in energetic radiation such as x-rays, completely outshining the million degree K corona.

The shapes of flares depend on how they are observed, that is, what radiation they are detected in. Different regions of the flare have much different temperatures and morphologies as described below.

CHROMOSPHERIC FLARE

Most ground-based observations of flares are made in the light of the Hα Balmer hydrogen line at 6563 Å. Flare plasma emitting this radiation is at a temperature of about 10,000–20,000 K. Until the advent of space research, almost all flare observations were made in Hα light and there is, therefore, a very large data base for chromospheric flares.

A flare first appears in Hα as a group of small brightenings in an active region. Reference to photospheric magnetograms shows that the brightenings form along opposite sides of a magnetic neutral line, that is, the boundary between magnetic fields of opposite polarity. In many flares the brightenings quickly connect with each other to form two ribbons of emission. These ribbons usually move away from one another at speeds of a few kilometers per second. Frequently, a filament that had previously existed over the neutral line is observed to disrupt during the rise (also called impulsive) phase, or phase of increasing Hα flare brightening.

Flares tend to occur more frequently in active regions with complicated magnetic field configurations. They also tend to occur in regions of very large magnetic field gradients, that is, regions where the strength or direction of the field changes rapidly with position.

Inspection of the Hα features in flares, which are believed to outline the magnetic field, suggest that flares begin in regions of highly sheared magnetic fields, that is, those where the lines of magnetic force are aligned more along the neutral line rather than at right angles to it. As the flare progresses, and after the filament erupts, the field configuration becomes less sheared.

Observations of large and intense flares on the limb show that during the flare decay phase (time of decreasing radiation) bright loops develop in the flare region. The loops cross the neutral line and the ribbons previously mentioned are believed to be the footpoints of the loops. These loops appear to rise at speeds of a few kilometers per second as the flare decays, but some observers feel that the motion is only apparent and that what is observed is the successive heating of plasma trapped in higher magnetic loops.

THERMAL SOFT X-RAY FLARE

Hot solar flares emit copious x-ray emission. This emission arises in coronal plasma at multimillion degree temperatures. The x-ray emission at wavelengths corresponding to photon energies of a few kilovolts ("soft x-rays") rises rapidly to a maximum flux and usually decays away on longer time scales. The soft x-ray emission light curves are roughly similar to the chromospheric Hα light curves. Flare rise times can vary from about 15 s to as long as about 40 min. Flare decay times can be as short as a minute and as long as many hours. There appears to be a particular class of flare with very long decay times of many hours. These flares are frequently associated with large ejections of mass in the corona [called coronal mass ejections (CMEs).

The soft x-ray emission is produced primarily by spectral line radiation from highly ionized atoms, and from continuum bremsstrahlung and free-bound radiation. Bremsstrahlung, from the German for "braking radiation," refers to photons emitted when a free electron moves past a positively charged ion, and loses speed due to the electrical attraction of the ion. Hydrogen- and helium-like ions (ions with only one or two electrons, respectively) are the most abundant ion species in the soft x-ray emitting flare plasma.

Observations obtained from extreme ultraviolet and x-ray telescopes on the *Skylab* manned space station in 1973 show that the soft x-ray emission in flares is confined to magnetic loop-like structures with field strengths of several hundred gauss. The x-ray emission occurs between the Hα ribbons, and, therefore, the hot

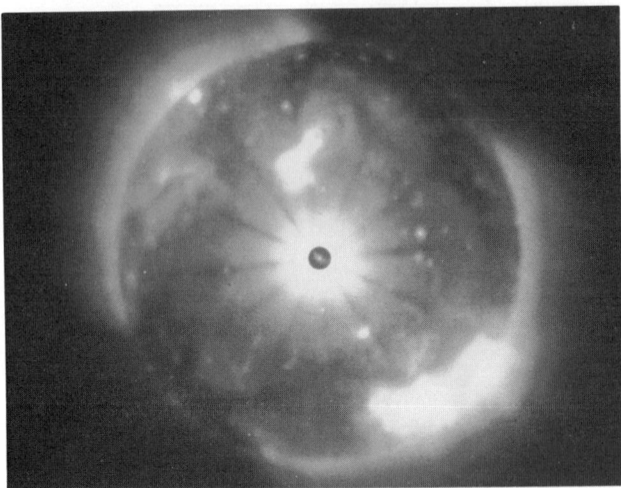

Figure 2. An x-ray photograph of a solar flare recorded by the American Science and Engineering x-ray telescope on the NASA *Skylab* manned space station. The symmetric halo surrounding the flare near Sun center is the overexposed image of the flare. A shorter exposure image has been overlaid onto the longer exposed image.

loops probably have footpoints either in or near the ribbons. However, not every point in the Hα ribbon is the footpoint of a hot coronal loop. An x-ray photograph of a flare in progress, compared to the x-ray emission of the entire solar disk, is shown in Fig. 2.

The absolute soft x-ray flux and the observed volumes of flare loops have been used to obtain electron density estimates of the multimillion degree plasma. Typical densities are $10^{10}-10^{12}$ cm^{-3}, several orders of magnitude higher than the density of the quiet solar corona.

Observations obtained with Bragg crystal x-ray spectrometers reveal a highly dynamic multimillion degree plasma in flares. During the rise phase of soft x-ray flux, plasma is observed to move upward with average speeds of about 300 km s^{-1}, but a smaller amount of plasma can move upward with higher speeds that can reach 1000 km s^{-1}. All of this plasma is probably chromospheric in origin and is ablated from the chromosphere due to intense heating. Some workers feel that the bulk of the hot coronal flare plasma has its origin in this ablated or evaporated plasma. Others feel that this plasma represents a surge-like eruption not necessarily connected to the hot loops that confine the nonmoving or very slowly moving x-ray emitting plasma.

The x-ray spectral lines also show considerable excess broadening in flares, much beyond the so-called Doppler broadening caused by the ordinary thermal motions of individual atoms. The excess broadening is interpreted as due to a nonthermal mass motion, with speeds that typically vary between about 60 and 160 km s^{-1}.

HARD X-RAY FLARE

Hard x-ray (10 to hundreds of kilovolts) bursts occur during the soft x-ray rise phase of most flares. The character of these bursts is entirely different from the soft x-rays. The soft x-ray light curves are smooth with either no, or very small, brightness fluctuations that occur on a scale of minutes. On the contrary, hard x-ray emission is extremely impulsive, with bursts that range from 1 min to milliseconds in duration. These impulsive bursts occur after the onset of a small amount of soft x-ray emission; the soft x-ray emission indicates preheating of the flaring loops. The main impulsive bursts begin when the soft x-ray emission begins to increase rapidly and they last for several minutes. In some flares there is also a gradual hard x-ray component that is much longer lived and can persist beyond the impulsive phase.

The energy of the hard x-ray emission extends to a few hundred kilovolts, and the spectral distribution can be fitted with either a multithermal source with a peak temperature of several hundred million degrees (a flaring region with zones at different, high temperatures), or by a nonthermal power law distribution of energies with a spectral index that is usually -2 or steeper, with a typical value of about -5. The low energy cutoff of the hard x-ray spectrum is not known, because it merges with the soft x-ray thermal spectrum. The energy in the electrons producing the hard x-ray bursts is comparable to the energy of the thermal multimillion degree plasma.

The hard x-rays are produced by bremsstrahlung between accelerated electrons and ambient ions. It is believed that the particles are accelerated and travel down flaring loops into the dense chromosphere. Bremsstrahlung can be produced all along the loop; in the coronal loop it is termed thin-target emission and in the chromosphere it is termed thick-target emission because of the much greater density in the chromosphere.

It is technically difficult to obtain flare images in hard x-rays. At present the best images are only of 8 arcsec quality. However, it would be highly desirable to obtain images of about 1 arcsec resolution, in order to determine the physical relationship of the hard x-ray producing regions to flaring regions emitting in other wavelength bands.

The hard x-ray bursts are accompanied by microwave and ultraviolet bursts that mimic very closely the time profiles of the hard x-ray bursts, suggesting that the same acceleration mechanism is responsible for all three emissions. The microwave emission is produced by gyrosynchrotron emission as the accelerated electrons spiral around the magnetic fields of the flare loops. The ultraviolet bursts are believed to be produced by chromospheric plasma heated by the accelerated electrons to temperatures of several hundred thousand degrees.

In large and intense flares small (about 2000–4000-km) knots of white light emission appear simultaneously with the hard x-ray bursts. These may also appear in less intense events; the observational coverage is poor. The knots represent an excess of white light continuum emission in the blue region of the visible spectrum, the origin of which is presently unclear. They later may form ribbons that are co-spatial with the Hα ribbons and persist beyond the impulsive phase (but not into the decay phase).

GAMMA RAY FLARE

Some electrons are accelerated to relativistic energies in the megaelectronvolt region by flares. Ions are simultaneously accelerated as well, and because of the high energies, nuclear reactions occur producing nuclear line emission. Strong lines due to carbon and oxygen are produced at 4.44 and 6.13 MeV, respectively, as well as the electron–positron annihilation line at 0.511 MeV. Another strong gamma-ray line from flares, at 2.223 MeV, is produced when neutrons from the disintegration of ^{4}He and heavier nuclei are captured by hydrogen nuclei. In addition to neutrons, the highly energetic particles produce positrons, pi mesons, and radioactive nuclei. The neutrons are captured in the photosphere, but some escape the Sun entirely and have been detected at Earth by orbiting detectors.

The gamma-ray spectrum is a probe of the flare acceleration mechanism and can also be used to determine chromospheric abundances. (Abundances are determined in the ultraviolet through x-ray regions as well using high resolution spectroscopy and plasma diagnostics.)

Continuum gamma-ray emission is predicted to be anisotropic. Observationally, the locations of gamma-ray flares are biased toward the solar limb, consistent with anisotropic emission. That is, solar flare gamma rays appear to be emitted more strongly in directions parallel to the solar surface. In principle, the center-to-limb variations of gamma-ray flare number and intensity can be used to obtain information on the physics of the particle propagation mechanism and energy transport from the acceleration site into the photosphere.

Additional Reading

Bray, R. J., Cram, L. E., Durrant, C., and Loughhead, R. E. (1990). *Plasma Loops in the Solar Corona*. Cambridge University Press, Cambridge, U.K.

Emslie, A. G. (1987). Explosions in the solar atmosphere. *Astronomy* **15** 18.

Sturrock, P. A., ed. (1980). *Solar Flares (A Monograph from Skylab Workshop II)*. Colorado Associated University Press, Boulder.

Svestka, Z. (1976). *Solar Flares*. Reidel, Dordrecht.

Tandberg-Hanssen, E. and Emslie, A. G. (1988). *The Physics of Solar Flares*. Cambridge University Press, London.

Wentzel, D. G. (1989). *The Restless Sun*. Smithsonian Institution Press, Washington, D.C.

Zirin, H. (1988). *Astrophysics of the Sun*. Cambridge University Press, London.

See also **Interplanetary Medium, Solar Cosmic Rays; Solar Activity, Coronal Mass Ejections; Solar Activity, Solar Flares, Theories; Sun, High Energy Particle Emissions; Sun, Radio Emissions.**

Solar Activity, Solar Flares, Theories

John T. Mariska

A moderately sized solar flare releases about 10^{31} ergs of energy over a time interval of a few hundred to a thousand seconds. This is enough energy to provide for the electrical needs of the United States for more than 100,000 years. To understand what we see when a flare takes place, we must identify the source of this energy, how and where it is stored and released, and how it is converted to the radiation and particles that we see as a flare evolves. It is now generally believed that the source of this energy is the Sun's magnetic field. Energy is stored in the field before the flare, and then is released suddenly during the impulsive phase and more slowly during the gradual phase that follows. Here we describe our current picture of the theoretical mechanisms that are at work as a flare progresses. Recent detailed observations from space and with ground-based instruments have allowed us to piece together much of what is going on when a flare takes place. A unified picture is beginning to emerge. Much of it, however, is still controversial.

PREFLARE PHASE

Because of processes taking place below the surface, the Sun's magnetic field is continuously changing. The changes are quasiperiodic with roughly an 11-yr period. Near the maximum of this cycle, large areas of enhanced magnetic field strength break through the visible surface of the Sun. As seen from space at ultraviolet and x-ray wavelengths, these active regions are characterized by loop-like structures that outline the magnetic field. Loop sizes can vary from a few thousand kilometers up to 100,000 km. It is in these active regions that flares are born.

The feet of active-region loops are anchored beneath the visible surface of the Sun, where the solar convection is continuously stirring the gas. These convective motions also move the magnetic field around beneath the surface and over a period of time build up stresses. Through the constant buffeting of the magnetic field lines beneath the surface, energy is taken from the motions of the gas and stored as increased stress in the magnetic field lines. It is much like the process of gradually winding up the rubber band on a toy

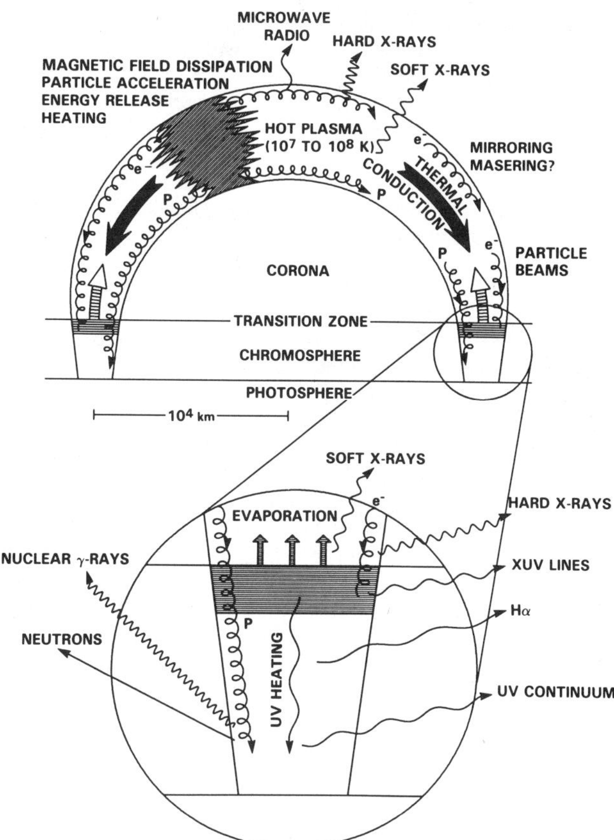

Figure 1. An idealized flaring loop, showing where all of the radiation and particles come from as the flare evolves. (*Courtesy NASA Solar Maximum Mission Project Scientist.*)

airplane. The kinetic energy used to wind up the rubber band is stored in the twisted rubber band until it can be released. Electric currents associated with the magnetic loops maintain this stressed magnetic field. As the stresses build up in an active region, the magnetic field structure can become quite complex. Often during this preflare phase there will be small releases of energy. Once the stresses have built up sufficiently, however, something triggers the sudden release of all of the energy, and the impulsive phase begins.

IMPULSIVE PHASE

Although the impulsive phase only lasts a few minutes, many complex events are taking place: Gamma rays, hard and soft x-rays, and ultraviolet, visible, and microwave radiation are all produced during this phase, along with neutrons. Current theories assume that all of the events take place within one or more of the magnetic loops in an active region. Figure 1 shows an idealized flaring loop, indicating where all of the radiation and particles come from in the flare.

At the beginning of the impulsive phase, the trigger event starts a process in which the stored magnetic energy is converted back into other forms of energy. In this process, which is usually called magnetic reconnection, the electric currents that maintain the stressed magnetic field are rapidly dissipated and the magnetic field geometry simplifies to a more relaxed form. The release of energy associated with this magnetic reconnection must either directly or indirectly produce all of the radiation and particles associated with the impulsive phase. For the theorist, getting this reconnection to proceed rapidly is not easy in the solar corona. The plasma there has almost no electrical resistance. Studies of the physics of the reconnection process show that the currents can be dissipated rapidly if the dissipation regions are small, well below the roughly 500-km size scale that is observable with current instruments. Even with small regions of dissipation, there may be a need for enhanced resistivity. This "anomalous" resistivity is thought to be due to instabilities in the plasma located in the reconnection region.

The most widely accepted theoretical model for how the impulsive phase proceeds is referred to as the thick-target model. In this model the primary outcome of the magnetic reconnection is the production of large quantities of energetic electrons and protons. These charged particles stream away from the reconnection site along the magnetic field lines. They move rapidly away from the corona, where the reconnection is thought to take place, toward the denser lower portions of the atmosphere. As they encounter the denser plasma in the chromosphere, they are slowed and eventually stopped. The dense lower atmosphere is the "thick target" in the model name.

A charged particle that is accelerated will emit radiation as long as the acceleration takes place. Thus the charged particles from the magnetic reconnection site emit radiation as they interact with other particles on their trip down the legs of the loop. For the streaming electrons, this radiation is called electron–ion bremsstrahlung (German for "braking radiation"). It is thought to produce the hard x-ray and gamma-ray emission seen in the impulsive phase.

Because of the Lorentz force, the energetic particles move along the magnetic field lines in a spiral pattern. Thus, along with being decelerated by Coulomb interactions with other charged particles, they also experience centripetal acceleration as they gyrate around the field lines. The resulting gyrosynchrotron radiation is emitted in the microwave part of the spectrum and produces the microwave burst associated with the impulsive phase.

The most energetic of the particles in the beam are powerful enough to cause nuclear reactions. They interact with the ions in the loop atmosphere knocking neutrons out of the target nuclei. These neutrons are detected at the Earth after a large flare. They also participate in additional nuclear reactions that produce gamma rays, which are observed by spacecraft experiments.

Along with accelerating large numbers of charged particles, the magnetic reconnection event also heats the plasma in the loop at the reconnection site. Just what fraction of the energy stored in the magnetic field goes into energetic particles and what fraction results in local heating is controversial. In fact, in one theory for the impulsive phase, the thermal model, all of the energy goes into raising the temperature of the plasma at the reconnection site to very high levels (100×10^6 K or more). At such high temperatures, some of the electrons will be very energetic and will stream down to the lower portions of the loop, much like the particle beams of the thick-target model. In fact, they will emit much the same radiation that is a signature of the electron beams of the thick-target model. The majority of the electrons will, however, remain confined near the initial energy release site. This results in a somewhat different evolution of the plasma in the loop, because the energetic particle beams of the thick-target model move substantial amounts of energy to the lower portions of the atmosphere much faster than occurs in the thermal model.

GRADUAL PHASE

As the charged particles encounter the denser chromospheric portions of the loop, collisions with the gas there remove the kinetic energy they were given when they were accelerated by the reconnection event: The beam thermalizes. Considerable energy is involved, and the chromospheric plasma temperature is raised suddenly from a few thousand degrees to tens of millions of degrees. The heated plasma immediately begins to emit large quantities of ultraviolet and soft x-ray radiation. This radiation carries away some of the energy that has been deposited. Part of the ultraviolet radiation that is directed downward is then absorbed in the denser lower portions of the chromosphere and in the upper photosphere, resulting in still more heating and the emission of more radiation, this time at visible wavelengths. The onset of this thermal radiation is usually referred to as the gradual phase.

Although all of this radiation does carry away some of the energy initially deposited when the beams thermalize, it cannot fully cool the plasma. The increased temperature at the site where the beams thermalize causes the gas pressure there to rise dramatically. This region of high temperature and pressure can no longer be confined by the lower-pressure material above and below it, so it expands. Because the overlying coronal plasma has relatively low pressure, there is little to resist the expansion and very high upward velocities are produced as the high-pressure and high-density region literally explodes into the corona. This explosive evaporation carries large quantities of high-temperature plasma up into the magnetic flux tube, where it is observable because of the soft x-ray emission it produces.

At some point after the loop has filled with hot plasma, all of the heating ends. It is not clear yet how long after the initial impulsive burst this takes place. In fact, there are some flares in which no impulsive phenomena are observed, suggesting that heating unrelated to the formation of particle beams is taking place.

Once the heating begins to decrease, the plasma cools by radiating away its energy. Some of this radiation comes from the hot plasma in the upper part of the loop. A high-temperature gas of solar composition is, however, a poor radiator. Hydrogen and helium, the most abundant atoms, have lost all of their electrons and so cannot emit photons in their primary spectral lines. They thus do not participate significantly in the cooling of the hot plasma. Instead, it takes place through emission from highly ionized trace atoms, such as iron. Much of the energy put into the corona by the flare is gradually moved by thermal conduction (heat flowing from a hot region to a cool one) to the cooler chromospheric regions of the atmosphere, where it can be radiated away in the strong emission lines of hydrogen and helium.

As the hot plasma cools, it reaches a point where all of the material that was lifted into the upper portions of the loop when the flare began can no longer be supported against the pull of the Sun's gravity. The plasma then slowly flows back down the legs of the loop, and the top portion returns to the low-density preflare state. These flows are often seen in cooling postflare loops. Finally, after several hours, the loop system in which the flare took place returns to the preflare state.

This picture of the mechanisms taking place as a flare progresses explains many of the observed features of flares. It is still, however, lacking in detail. For example, there is still no clear understanding of the trigger that initiates the impulsive phase. The best current idea is that new magnetic field structures emerge from beneath the visible solar surface and collide with the stressed magnetic field system, triggering a sudden reconfiguration of the magnetic field. The details of the conversion process that takes energy from the magnetic field and accelerates large numbers of particles once the flare has started are also uncertain. Our overall picture of how a flare proceeds seems to be sound, but the details must await better observations from new balloon, rocket, and satellite instruments.

Additional Reading

Emslie, A. G. (1987). Explosions in the solar atmosphere. *Astronomy* **15** (No. 11) 18.

Haisch, B. M. and Rodonò, M., eds. (1989). *Solar and Stellar Flares*. Kluwer Academic Publishers, Dordrecht.

Noyes, R. W. (1982). *The Sun, Our Star*. Harvard University Press, Cambridge, Mass.

Tandberg-Hanssen, E. and Emslie, A. G. (1988). *The Physics of Solar Flares*. Cambridge University Press, London.

Wentzel, D. G. (1989). The *Restless Sun*. Smithsonian Institution Press, Washington, D.C.

Zirin, H. (1988). *Astrophysics of the Sun*. Cambridge University Press, London.

See also **Solar Activity, Coronal Mass Ejections; Solar Activity, Solar Flares; Stars, Red Dwarfs and Flare Stars.**

Solar Activity, Sunspots and Active Regions, Observed Properties

Ronald L. Moore

Active regions are islands of enhanced magnetic field in the solar atmosphere, large enough and strong enough to stand out from the magnetically weaker and quieter background atmosphere in images of the whole Sun. They are formed by the emergence of magnetic flux loops from below the photosphere, which makes them bubbles of closed, bipolar magnetic field. Within the strong-field bubble of an active region, the magnetic field makes striking alterations in all levels of the atmosphere from low in the photosphere to high in the corona. At most times, the most obvious of these signatures are dark sunspots in the photosphere, bright areas (called plage) and dark filaments in the chromosphere, and bright loops in the corona and corona–chromosphere transition region. The magnetic structure of an active region gradually changes over its life, which is as short as a day for the smallest active regions and as long as a few months for the largest. In addition, at places within active regions where the magnetic field is more rapidly evolving, highly contorted, and unstable, there are sporadic bursts of magnetic energy release (called flares) that drive explosive motion and catastrophic heating of the chromosphere and corona within the detonated magnetic field. The continually evolving magnetic structure and restless flaring give active regions their name.

Active regions are key to understanding both the cause of the Sun's magnetic field (i.e., the field-generating dynamo process driven by the Sun's rotation and convection below the photosphere) and the effects of the field throughout the solar atmosphere. In turn, the Sun's magnetic phenomena, because they can be observed in great detail, are key to understanding similar magnetic phenomena in other astrophysical objects, including planetary magnetospheres, stellar flares, stellar coronas, stellar winds, magnetospheres of collapsed stars, and plasma ejections from galactic nuclei. Active regions and all larger-scale components of the Sun's magnetic field, which are built from remnants of the fields that emerge in active regions, are basic facets of the dynamo. The entire solar atmosphere above the photosphere is permeated, heated, and controlled by these fields together with finer-scale fields that are also injected through the photosphere. The stronger the magnetic field, the greater its effect on the Sun's atmosphere. The fields are strongest in active regions, and the strongest fields within active regions are rooted in the largest sunspots. The greatest flares are produced by those regions in which both polarities of such strong fields are crammed together in large complex sunspots. These flares are the greatest explosions in the solar system, blasting out through the solar wind and disrupting the magnetospheres and ionospheres of the planets. So, we study sunspots and active regions because they are the seats of the strongest magnetic fields and strongest magnetic action on the Sun, and because the origin and effects of these magnetic fields are of general importance in astrophysics and space physics.

LARGE-SCALE PROPERTIES

Figure 1 shows several examples of typical active regions as seen in images spanning the Sun. The gross magnetic form of individual active regions and the spatial order in the collection of active regions are plainly evident. The magnetogram and the coronal image together show that each active region is a bipolar magnetic bubble arching into the corona. The global arrangement of these active-region bipoles illustrates four attributes of the solar cycle. One is that active regions large enough to have sunspots are restricted to two low-latitude belts, one north and one south of the equator. Another is that almost all such large bipoles are aligned roughly parallel to the equator. A third attribute of the cycle is that in each of the two belts, with few exceptions, these bipoles all point in the same direction, so that the leading end (i.e., leading in the direction of the Sun's rotation) of each active region has the same polarity. Finally, the direction of the bipoles in each belt is the reverse of that in the other belt.

Measured by the span of the area that is largely filled by strong (>100 G) magnetic field, the two largest active regions in Fig. 1 are each about a tenth the size of the Sun in diameter. These are fairly large active regions; 9 out of 10 active regions with sunspots are smaller. Only a hundred or so active regions this large or larger are produced per 11-yr cycle. The very largest active regions have about twice the span of the largest ones in Fig. 1; only a few are produced per cycle. The smallest active regions with sunspots are about the size of a supergranule (30,000 km across), or about five times smaller in girth than the largest active regions in Fig. 1.

The life of an active region begins when the bundle of magnetic flux tubes that forms it begins to emerge through the photosphere. In the first few hours of emergence, the region of new magnetic flux is smaller than a supergranule and has no sunspots. At this stage, the magnetic field typically has an overall bipolar structure, but the bipole may be oriented far from east–west. If flux emergence continues for many hours, dark pores the size of granules (about 1000 km across) start to form in the photosphere at the places of strongest field in the opposite ends of the bipole. With further flux emergence, the pores coalesce and grow into full-fledged sunspots while (usually) the overall bipole becomes aligned nearly east–west. To produce active regions as big as the larger ones in Fig. 1, flux emergence continues for several days, causing the opposite polarity domains and the sunspots in them to continue to grow and spread apart. Although the overall form of the emerged field is bipolar, the flux emergence is usually rather chaotic on smaller scales within the overall bipole. This produces a jumble of both polarities in the interior of the growing active region, so that the magnetic complexity and its accompanying flaring and heating are usually greatest while the active region is growing. After emergence stops, the concentration of strong magnetic field and its effects (sunspots, flaring, and enhanced heating) gradually diminish and disappear over the course of weeks to months. Most of the magnetic flux disappears within the active region, probably by reconnecting into short loops that submerge. The rest of the active region's magnetic flux, a small fraction, diffuses away to become incorporated into the global-scale components of the Sun's magnetic field.

Comparison of the photospheric image in Fig. 1 with the magnetogram shows that the sunspots are located in the strongest magnetic fields. Each of the larger sunspots has a roundish dark center, the umbra, surrounded by a less dark penumbra. The strength of the magnetic field is ~ 1000 G in the penumbra and increases to as much as a few thousand gauss in the center of the umbra. At these strengths, the magnetic field greatly inhibits the convective flow of heat to the surface, which results in a cooler and hence darker photosphere than in areas of weaker magnetic field.

Figure 1 shows that sunspots are also dark in the Hα chromosphere. The bright chromospheric plage areas are mostly in other parts of the active regions where the field is still quite strong (a few hundred gauss) but not as strong as in the sunspots. From careful comparison of the chromospheric image and the magnetogram in Fig. 1, it can be seen that the plages are brightest at places where strong fields of opposite polarity are in close contact [so that the dividing line (called the polarity inversion line) between them is sharply defined]. The enhanced brightness results from enhanced heating of the solar atmosphere at these sites. In contrast to the more limited extent of plage within active regions, the enhanced

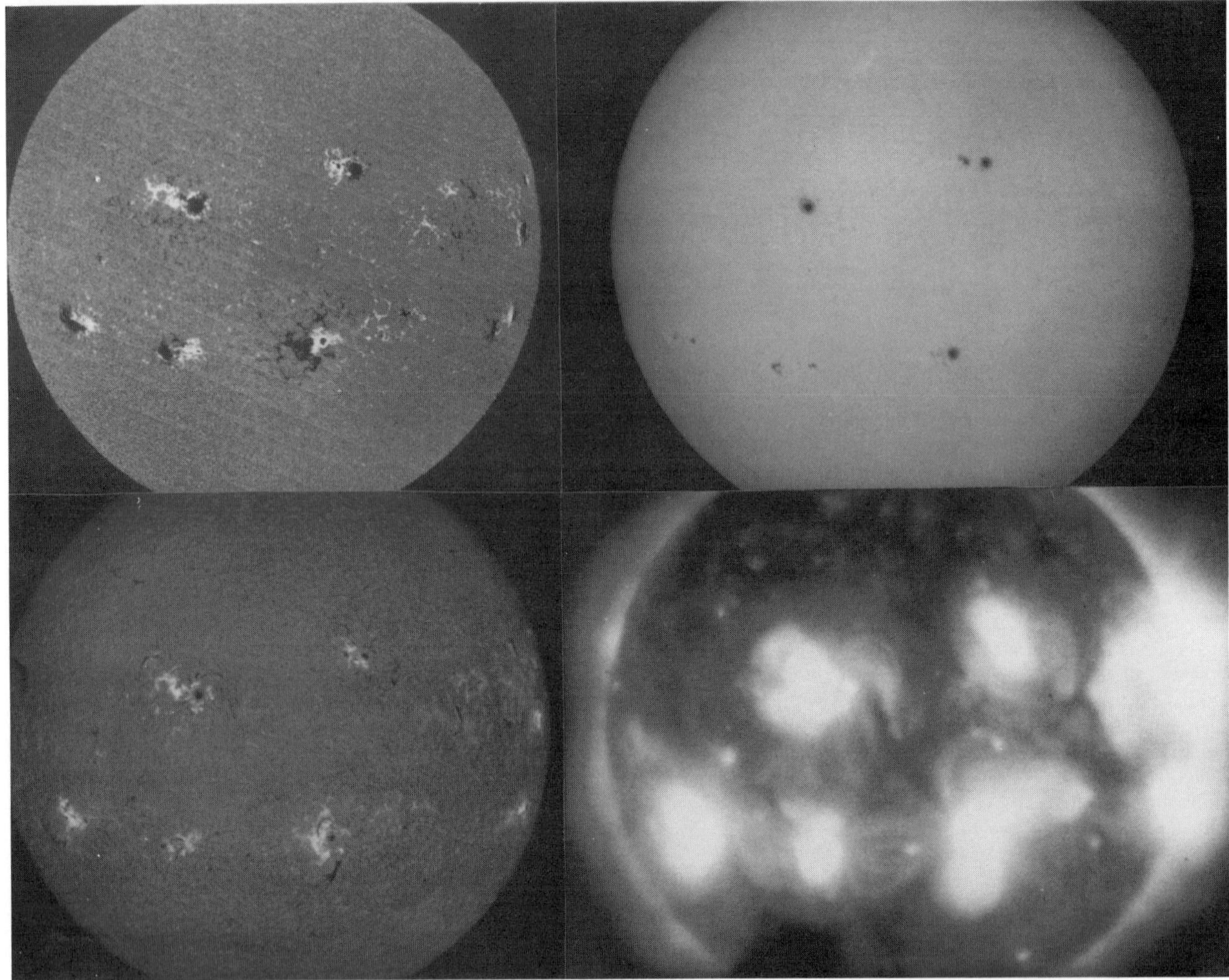

Figure 1. Typical active regions distributed across the Sun. These four images were taken nearly simultaneously; each shows the same active regions observed in a different way. The upper left panel is a magnetogram, a map of the polarity and strength of the photospheric roots of the magnetic field. The upper right panel shows the photosphere photographed in ordinary visible light. The lower left panel shows the chromosphere photographed in Hα, the red spectral line of the hydrogen atom. The lower right panel shows the corona photographed in soft x-rays. The temperature of the plasma seen here in the active regions is about 6000 K in the photosphere (decreasing to about 4000 K in the umbras of sunspots), about 10,000 K in the chromosphere, and about 2,000,000 K in the corona. (*Pictures from the Marshall Space Flight Center archive of the* Skylab *missions.*)

brightness and hence enhanced heating shown by the coronal image is more widespread, covering the whole of each active region. Chromospheric and coronal images such as these demonstrate that solar atmospheric heating increases markedly with field strength for field strengths in the range of 1 to several hundred gauss. The concentration of plage brightness on sharp inversion lines is an indication that the heating depends on the structure of the field as well.

FINE-SCALE PROPERTIES

The smallest resolved solar features in the pictures in Fig. 1 are a few thousand kilometers across, roughly a tenth the diameter of the larger sunspots present. At this resolution, an essential property of active regions remains hidden. In Fig. 2, a 10-fold increase in resolution reveals a wilderness of chromospheric fibril structure full of striations down to the resolution limit of a few hundred kilometers. This fine-scale structure is apparently imposed by the magnetic field, and hence indicates that up through the reach of the chromosphere (several thousand kilometers above the photosphere), the magnetic field, or at least its activity, is strongly structured on scales of a thousand kilometers and less. In the photosphere outside of sunspots, it is observed that the roots of the magnetic field are bunched and separated on these scales by the action of the granular convection; this is a likely cause of the fibril structure in the overlying chromosphere.

Although the sweep and pattern of aggregates of fibrils evolve slowly along with the large-scale distribution of the magnetic field in an active region, the individual fibrils are much more transient, changing appreciably in length, width, or darkness in tens of minutes. These changes result from in situ changes in brightness

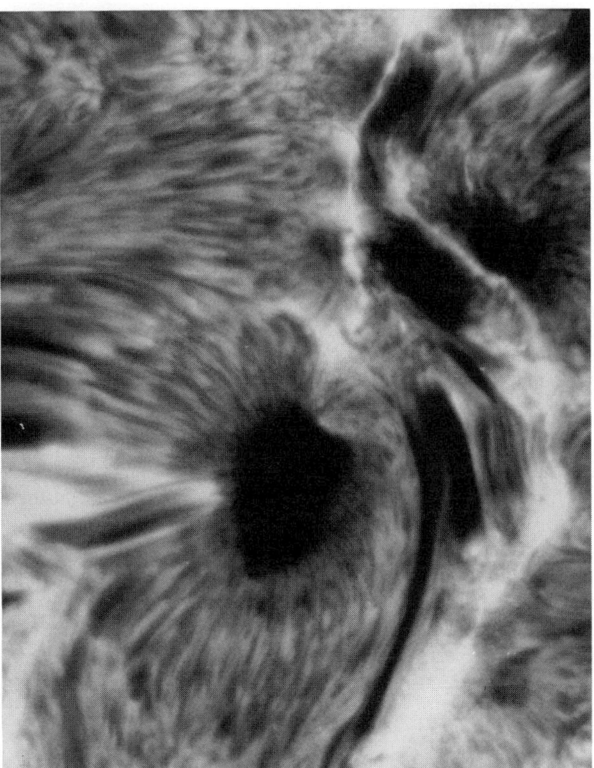

Figure 2. Typical active region viewed in the chromospheric Hα line at high spatial resolution. The narrowest discernible fibrils are a few hundred kilometers wide. The photospheric polarity inversion line snakes between the two sunspots; it runs down the center of the dark chromospheric filament seen here. At this resolution, this filament is seen to be a low-lying magnetic arcade that is strongly sheared across the inversion line, that is, the opposite polarity footpoints of the magnetic loops are displaced from each other along the inversion line much farther than they are offset from the inversion line. (*Photograph from Big Bear Solar Observatory.*)

in combination with flows of plasma along the magnetic field lines at speeds of order 10 km s^{-1}. The widths and transient behavior of chromospheric fibrils in active regions are similar to those of the chromospheric spicules in the magnetic network in quiet regions; the main differences are that active-region fibrils are longer and appear to be more nearly horizontal than spicules on average.

In the largest sunspot in Fig. 2, in contrast to the stark umbra, the penumbra shows only faintly because it is obscured by the many radial fibrils that are rooted in the penumbra and outer umbra and that arch beyond the outer edge of the penumbra. The wreath of these chromospheric fibrils, called the superpenumbra, is a normal feature of mature sunspots. It graphically demonstrates that the sheaf of magnetic field rooted in a sunspot splays out rapidly with height. Consistent with this magnetic fountain picture, the fibril material in the superpenumbra flows inward and downward. Similar filamentary downflow has been observed in the chromosphere–corona transition region over umbras. Fibril structure similar to that of the chromospheric superpenumbra persists down into the photospheric levels of the penumbra. At spatial resolution as good as that in Fig. 2, the photospheric penumbra is seen to be full of interlaced bright and dark radial striations only a few hundred kilometers wide. Many of the dark strands are low-lying versions of the elevated fibrils in the superpenumbra in that they extend out over the granulation at the edge of the penumbra.

In addition to the dark magnetic fibrils, the widths of which may well be set by the scale of granular convection in and below the photospheric penumbra, the photospheric penumbra also shows more direct effects of convection. Foremost, in contrast to the inflow in the superpenumbra, in the photospheric penumbra there is horizontal outflow (called the Evershed flow) of a few kilometers per second. Further, some of the dark component of the photospheric penumbra is apparently the magnetically modified counterpart of the dark intergranular lanes in the normal photosphere. Finally, the other component of the photospheric penumbra, the bright component, consists of elongated grains that fill somewhat less than half the area of the penumbra. A typical grain is about a thousand kilometers long, a few hundred kilometers wide, nearly as bright as the normal photosphere, and lives for about an hour, during which it migrates toward the umbra by about its own length. In the photospheric umbra, there are bright dots that are the counterparts of the penumbral grains, having roughly the same filling factor, diameter, brightness, and lifetime. The umbral dots and penumbral grains are probably magnetically modified counterparts of the granule convection cells in the normal photosphere. In any case, the presence of the dots and grains suggests that the photospheric roots of the magnetic field inside sunspots are similar to those outside in being highly nonuniform on scales of a few hundred kilometers.

Vertical velocity oscillations with periods of 2–3 min are present in the interiors of most umbras. The amplitude increases with height to as much as 5–10 km s^{-1} in the upper chromosphere. When the oscillations are this strong, especially in the core of the violet K line of singly ionized calcium, a flash of emission is produced with each upward pulse, perhaps because the wave front steepens into a shock wave. The horizontal span, or wave-front size, of each oscillation is several thousand kilometers; a large umbra has room for a few oscillations but not many. In addition to the 2–3-min oscillations well inside the umbra, along the edge of the umbra there are sometimes 4–5-min vertical oscillations that launch waves that propagate out across the penumbra at speeds of 10–20 km s^{-1}. The periods and front sizes of both the umbral oscillations and the running penumbral waves fall in the range of the Sun's global p-mode oscillations. This suggests that the oscillations and waves in sunspots are excited by the bobbing and shaking of the whole sunspot flux tube by the global oscillations in the Sun's surface and subsurface layers on which the sunspots float like lily pads.

MAGNETIC STRUCTURE AT SITES OF BIG FLARES

The main inversion line in the bipole of an active region is oftentimes traced by a dark chromospheric filament. The inversion line in the large southern active region near the central meridian in Fig. 1 is typical in that the filament is very narrow (and faint because it is unresolved) whereas the inversion line is sharp and the opposite-polarity fields that it divides are strong. In Fig. 2, the magnetic structure of such an inversion line and low-lying filament is seen at high resolution. The essence of the structure is that the magnetic field is strongly sheared across the inversion line. Maps of the magnetic field vector in the photosphere under such filaments confirm the magnetic shear by showing that the horizontal direction of the field on and near the inversion line is nearly parallel to the inversion line. In Fig. 2, note that the sheared chromospheric loops are rooted in a ribbon of bright plage on each side of the inversion line. This illustrates in more detail than in Fig. 1 that enhanced heating of the solar atmosphere depends on the structure of the magnetic field in addition to its strength.

The presence of magnetic shear across the inversion line means that this field is charged with stored energy in the form of the magnetic stress of the shear. If the field had no stored energy, the field would be relaxed to its so-called potential state in which there is no shear; the magnetic field would arch orthogonally across the inversion line. Flares often occur in sheared magnetic fields like that in Fig. 2. Field configurations of this form are especially typical of the largest and most powerful flares. The flares are apparently releases of the energy stored in the sheared field. The greater the degree of shear and the longer the interval of strong shear along the inversion line, the more likely a flare will occur. The greater the field strength in the area of strong shear, the greater the stored energy and the more powerful the flare. The chromospheric filament on the sheared inversion line typically erupts in the midst of the flare, in step with the onset and impulsive peak in flare intensity. It appears that the filament eruption and flare are consequences of a loss of equilibrium of the whole sheared field configuration, and that the filament motion and flare energy release are driven by the explosion of the sheared field, the core of which carries the filament.

NEED FOR LARGE SOLAR TELESCOPES IN SPACE

Observations to date have clearly established that the magnetic field in an active region is the basic cause of the structure, heating, and flaring on all scales, that the field is highly filamentary on scales of a few hundred kilometers, and that most of the field eventually disappears within the area of the active region. But we have little observational knowledge of how the magnetic field does these things. To see how sunspots come and go, how magnetic shear builds up, how flares are triggered and how their energy is released from the field, how the solar atmosphere is heated, and how the magnetic field disappears, we need to clearly see and follow the elements of the magnetic field, the individual magnetic structures that are only a few hundred kilometers wide. This is yet to be achieved either from the ground or in space, although recent imaging of the photosphere and chromosphere has come within a factor of 2 or 3 of the needed sustained resolution of 100 km. The corona, at least on the face of the Sun as in Fig. 1, has yet to be imaged at 1000-km resolution.

To see the whole extent of magnetic loops and the activity within them, simultaneous observations must be made of all levels from the photosphere to the corona, and hence of their radiations from the visible to x-rays, with 100-km resolution throughout. Adequate coverage of this broad spectral range requires at least three separate telescopes, one for visible and near ultraviolet light (longward of about 2000 Å), at least one for x-rays (shortward of about 100 Å), and at least one for the 100–2000-Å range. Sustaining 100-km resolution hour after hour in the visible requires a telescope of large aperture (1 m or more in diameter) and removal of blurring by the Earth's atmosphere. At wavelengths shorter than about 3000 Å, observations can be obtained only from above the Earth's atmosphere, and apertures of tens of centimeters are needed to collect enough photons to adequately see and spectrally analyze small faint features. Finally, the whole range of simultaneous high-resolution observations needs to be continuous over the lifetime of active regions. Hence, to greatly advance our observational knowledge and understanding of solar magnetic activity, we need a complement of large co-observing solar telescopes in space.

Additional Reading

Bray, R. J. and Loughhead, R. E. (1964). *Sunspots*. Chapman and Hall, London.

Bray, R. J., Cram, L. E., Durrant, C., and Loughhead, R. E. (1990). *Plasma Loops in the Solar Corona*. Cambridge University Press, Cambridge, U.K.

Bruzek, A. and Durrant, C. J. (1977). *Illustrated Glossary for Solar and Solar-Terrestrial Physics*. Reidel, Boston.

Moore, R. L. (1981). Dynamic phenomena in the visible layers of sunspots. *Space Sci. Rev.* **28** 387.

Moore, R. and Rabin, D. (1985). Sunspots. *Ann. Rev. Astron. Ap.* **23** 239.

Zirin, H. (1988). *Astrophysics of the Sun*. Cambridge University Press, London.

Zwaan, C. (1987). Elements and patterns in the solar magnetic field. *Ann. Rev. Astron. Ap.* **25** 83.

See also **Solar Activity; Solar Activity, Sunspots and Active Regions, Theories; Sun, Magnetic Field.**

Solar Activity, Sunspots and Active Regions, Theories

Marcos E. Machado

The Sun has been studied for millenia by Chinese astronomers, so that sunspots are known to have been present on the Sun for at least 2000 years. Yet most of the recent reviews on sunspot theory state the fact that, in spite of being the longest observed phenomenon, the sunspot remains as one of the most challenging theoretical problems in solar physics. This is not surprising, because the sunspot problem is intimately linked to a large variety of physically complex issues spreading over topics such as the large- and fine-scale structure of magnetic fields, the solar cycle, and the global structure of the convection zone. Moreover, sunspot groups form active regions, where we see the effects of magnetic energy storage and its dissipation over time scales of a few seconds to several days. The panoply of active-region-related phenomena extends in the solar atmosphere from the invisible layers below the photosphere to the optically thin regions in coronal loops, where most atomic species are highly ionized, resistivity is low, and thermal conductivity is high. Furthermore, between these two regimes, all physical parameters and transport properties of the solar plasma vary by many orders of magnitude.

It is in this complex scheme where theories must give answers to the following questions:

Why do sunspot form and why do they disappear?
What makes sunspots cool and cohesive?
How and where is the heat flux from below blocked, and where and when does it eventually reappear?
What is the structure of the magnetic field and how does it interact with the velocity field?

The first two questions address the fundamental problem of stability in a plasma–magnetic-field configuration. In the solar photosphere, the magnetic flux is found to be concentrated in a great spatial range of structures, from small elements of a few hundred kilometers to the large sunspots. It has been recognized for a long time that an interchange, or flute instability, is likely to occur in a flux tube and cause its disruption. In the words of Eugene N. Parker, more than two decades ago: "Sunspots are too unstable to form and, if once formed, should immediately break apart." However, F. Meyer, H. U. Schmidt, and N. O. Weiss demonstrated that in a gravitationally stratified atmosphere concave fields with a flaring angle in excess of a given critical value can be stable. The most accepted view is then that sunspots are the result of the merging of many individual small-scale flux tubes that, in fact, would not clump together under normal circumstances. Peter R. Wilson showed, however, that the temperature of the tubes relative to their surroundings is a critical factor; if the conditions are such that an interruption or decrease in the energy supply leads to cooling of the region, then clustering of tubes into a lower energy state of a single tube (or a packed collection of small tubes) is possible. This same "mechanism" that leads to the clustering of tubes into full-fledged sunspots also provides a means for their disruption. It would simply suffice to heat up the region, and the tendency to clump would immediately disappear.

Another possible way of explaining why sunspots are stable would be to assume that they are formed by helically twisted flux ropes originating within the convection zone. This type of configuration also prevents the occurrence of fluting, and the twist is an essential ingredient of their equilibrium and stability. The decay of a sunspot would then relate to the unwinding of the flux rope. However, weighing the observational evidence for and against the implications of such an alternative, one finds that the twisted flux-rope model does not provide a firm alternative to the clumping model.

Next in the sequence of our interconnected questions is, Why do sunspots have low temperature? This, together with their field strength, is one of the most distinctive properties of sunspots, and it is rather intriguing to note that sunspots are cool (3500–4000 K) compared to the surrounding photosphere, but not cold. When one accepts the classical view that spots are dark because of the magnetic field inhibition of convective energy transport, it is by no means straightforward to explain why the umbral radiative energy flux is 20% instead of, say, 1% or even less.

The classical explanation (due to Ludwig Biermann) of the sunspot energy deficit is that the strong magnetic fields inhibit the convective energy transport. Yet, it is by no means clear why sunspots are not cooler than observed, and it is not clear what happens to the missing flux. In a model proposed by E. N. Parker, the sunspot's magnetic field separates into a bundle of discrete flux tubes at a certain depth below the surface of the umbra. This separation depth should be rather small, of the order of 1000 km or less, and independent of the radius of the spot. Below this depth, energy can be transported upward by convection in the field-free region between discrete small (or slender) flux tubes, explaining why the energy flux in the umbra is as large as 20% of that in the photosphere and why it is independent of the spot's radius. Parker proposes that the clustering is held by the buoyancy of the (observed) Wilson depression at the visible surface and by a (postulated) downdraft beneath the spot. The downdraft contributes to the cooling, which in this model is also effected by the convective generation of Alfvén waves emitted preferentially downward.

In their model, Meyer and coworkers directly relate the formation and evolution of sunspots to supergranular convection. They show that in the region from 2000 to 10,000 km below the umbra, the energy flux may be carried laterally from ambient supergranules by small-scale motions and then transported upward by elongated convection eddies at a lower efficiency than in the normal photosphere. Thus, even though they do not explicitly invoke a separation of the field into discrete flux tubes, the eddies transporting the (reduced) energy flux must separate the field in a way similar to that proposed by Parker.

These are just two examples, but it should be noted that whatever the reason, observational or theoretical, it is quite clear that spots are structured and cannot be formed of a single isolated flux rope. Although models may differ, there is an almost general consensus that this should be the case. In the most general, model-independent terms we can also safely state that because the clumping is in opposition to the mutual repulsion of the fields as seen in the visible surface of the Sun, it is also obvious that it must be driven by forces below this layer. Unseen hydrodynamic forces must then be able to build up a total flux of 10^{22} Maxwells (Mx), characteristic of a large spot from, say, the equivalent of 10^4 individual elements of 10^{18} Mx each.

The question of the ultimate fate of the missing flux is currently a nagging problem for spot models and modelers. Solar irradiance measurements have shown, with high accuracy, that the presence of large spot groups on the solar disk causes dips in the solar "constant," with amplitudes ranging from 0.1% up to 0.25%. This reduction in the overall irradiance can be due to the presence of spots (!) or to a temporary storage of the missing energy over a time scale longer than the spot's lifetime. Additional aspects of this correlation, which may have profound physical importance when studied over a long-term basis, are that the largest observed decreases seem to be associated either with newly formed active regions that have large spots or with rapidly evolving spot groups of high complexity.

This and other unsolved issues, like the longitudinal distribution of active complexes as well as their periodicities, are not just linked to the physics of sunspots but also to that of the solar cycle, which is reviewed elsewhere in this volume.

On the other hand, there is more to active region phenomenology than just sunspots. When these form, they are accompanied by bright continuum emission knots within the otherwise undisturbed atmosphere. A large conglomerate of these elements or knots, of sizes of the order of 1000 km or less, form what are known as faculae. Each facular element is associated with a magnetic flux tube of similar dimension. Owing to the fact that the magnetic field contributes to their pressure balance, it turns out that the facular knots are less dense, at equal geometrical height, than the surrounding atmosphere. According to currently accepted models, the lower density causes a depression of the continuum optical depth unity level (where most of the observed emission originates) of the order of 200–300 km. This affords a straightforward explanation of the increased brightness of facular elements, because radiation from hotter subphotospheric layers can leak into and through the facular walls and this results in an atmosphere that appears hotter than its surroundings. This effect is only important in the case of small flux tubes, and does not play any significant role in the larger and equally depressed cool sunspots.

As seen from the ground, sunspots and faculae are the most striking phenomena in active regions. However, an active region extends through all levels of the solar atmosphere, including the chromosphere and corona. In particular, spacecraft observations obtained with ultraviolet, extreme ultraviolet, and x-ray telescopes, capable of observing high temperature radiation, show that the coronal structure in active regions consists of many closed magnetic flux tubes or loops, filled with plasma at temperatures ranging from 10^5 to a few 10^6 K. These loop structures are the basic building blocks of the corona in active regions, indicating a cause–effect relationship between the magnetic field and the enhanced temperature. A number of possible mechanisms for depositing heat in coronal loops have been studied, but none has been able to provide a definite answer. Among these are the models that attribute the heating to the propagation and dissipation of magnetohydrodynamic (MHD) waves. These have observational support from the presence of Alfvén waves measured in the solar wind, which heat and accelerate protons and ions. The problem with wave theories as a loop heating mechanism is that the waves have a tendency to be reflected by the very steep Alfvén speed gradient in the chromosphere and transition region, so that they do not reach coronal heights (but may efficiently heat the active region chromosphere, as needed). However, it can be estimated that sufficient energy can be provided to the loops if global resonances can be excited, which seems to be easy for short (<10 km) loops but not as obvious for intermediate scale structures (10^4–5×10^4 km).

A second class of models, currently gaining widespread acceptance, invokes the gradual buildup of magnetic free energy in coronal loops, by random walk of photospheric flux tubes. This energy is subsequently released impulsively, by the formation of current sheets and reconnection. The models, whose details are still being worked out, are supported by observations of weak high energy events in active regions. These so-called microflares are much more frequent than the larger full-fledged flare events and still involve the same type of impulsive energy release processes. A problem with these models is that they are unable to provide sufficient chromospheric heating. Therefore, unless we see in the future a major unexpected theoretical breakthrough, the solution to the overall active region heating problem may reside in a hybrid model where various processes can contribute. There is no reason why this could not be the case.

Additional Reading

Bray, R. J. and Loughhead, R. E. (1964). *Sunspots*. Chapman and Hall, London.

Cram, L. E. and Thomas, J. H., eds. (1981). *The Physics of Sunspots*. Sacramento Peak Observatory, Sunspot, New Mexico.

Eddy, J. A. (1979). A new Sun, the solar results from Skylab. NASA report SP-402.

Foukal, P. V. (1990). *Solar Astrophysics*. Wiley Interscience, New York.

Spruit, H. C. (1981). Magnetohydrodynamics of sunspots. *Space Sci. Rev.* **28** 435–448.

See also **Solar Activity, Sunspots and Active Regions; Sun, Solar Constant.**

Solar Magnetographs

William C. Livingston

It was Galileo who discovered that the Sun departed from the prevailing concept of a uniform sphere of light and that it was contaminated by dark spots. Today we recognize not only sunspots, but many other indicators of surface activity such as faculae, prominences, and flares. These phenomena are all a direct consequence of solar magnetic fields, and what we know about these fields is derived almost entirely from the observation of the spectroscopic Zeeman effect with instruments called solar magnetographs.

ZEEMAN EFFECT

When a spectral line of wavelength λ is formed in the presence of a magnetic field it will, in general, split into two "sigma" components whose separation $\Delta\lambda$ is proportional to the field strength H. If the direction of the field is parallel to the line of sight (a longitudinal field), the two components are circularly polarized in opposite directions. If the field is perpendicular (transverse), there are three components: the two sigma components, which are plane-polarized in the same sense, and an undisplaced "pi" component that is plane-polarized at right angles to the outer pair. This is the Zeeman effect, where $\Delta\lambda = 9.34(10^{-13})g\lambda^2 H$. The Landé parameter g is a quantum-mechanical parameter that ranges from 0 to about 3; most commonly $g = 1$.

HISTORY OF SOLAR MAGNETIC MEASUREMENTS

In 1908 George Ellery Hale found that spectral lines in sunspots were polarized and split. Assuming the Zeeman effect, he deduced field strengths of up to 3000 gauss (G). His observations were both visual and photographic. He noticed that the lines were broadened immediately outside sunspots, but any splitting was only a fraction of the line width and not readily measurable. With Albert E. Whitford, Hale made some initial attempts at using a photocell to study nonsunspot magnetism, but the available equipment was not up to the task. It was not until after World War II that photoelectronic techniques became sufficiently advanced for the problem to be attacked.

DEVELOPMENT OF THE MAGNETOGRAPH

In the case of the longitudinal Zeeman effect, because the components are of opposite circular polarization, a spectral line will shift a small fraction of its width when viewed alternately through right- and left-circular polarizers. In a first attempt to detect what seemed like weak nonspot fields, Karl O. Kiepenheuer placed a motor-driven circular polarizer in front of his spectrograph. As the favorable Fe 5250.2-Å ($g = 3$) line shifted minutely back and forth by $\Delta\lambda$ an AC intensity signal was created by a photomultiplier in one of its wings. This signal was then synchronously detected by a lock-in amplifier. Although the scheme worked fairly well, false modulation arose because of a slight polarization introduced by the telescope mirrors and by imperfections in the rotating polarizer. Shortly thereafter, Horace W. Babcock solved both problems. Instrumental polarization was eliminated by using *two* photomultipliers, one positioned in each line wing. By taking the difference signal, again synchronously detected, instrumental polarization was virtually eliminated because its signal was common to both wings. The rotating polarizer was replaced by an equivalent electrooptical crystal, removing the need for any mechanical motion in the optical train. Babcock's magnetograph was found to be limited in sensitivity only by photon noise and he was able to measure fields as weak as 1 G. Following the development of equipment at the Hale Laboratory in Pasadena (circa 1952), Babcock installed an improved version at the 150-ft tower telescope on Mt. Wilson. He initiated a program of producing daily full-disk magnetic maps of the Sun: "magnetograms," created by making a raster scan of the solar disk with a 12-arcsec spectrograph aperture. These have proved to be of great value for the prediction of flares and geomagnetic storms, and for understanding the physics of the Sun itself. Begun during the 1957 International Geophysical Year (IGY), the program continues today and is reported in *Solar Geophysical Data*.

Figure 1. Full-disk Kitt Peak magnetogram taken near a time of maximum solar activity (12 February 1989). White represents N polarity, black represents S polarity.

IMPROVED-RESOLUTION MAGNETOGRAPHS

Scanning the solar image with a single-channel detector is inefficient, and a number of modern magnetograph schemes have evolved that realize better spatial resolution. At the Kitt Peak Vacuum Telescope, for example, a pair of 512-element diode arrays is effectively caused to cross the solar image in four swaths, producing a seeing-limited full-disk magnetogram (see Figs. 1 and 2) having 1-arcsec pixels. Observations are obtained at a daily pace and these also are published in *Solar Geophysical Data*. At Big Bear Solar Observatory a narrow-band optical filter precedes a Vidicon television-type transducer to produce "video" magnetograms. Like Kiepenheuer's arrangement, a video magnetograph is susceptible to

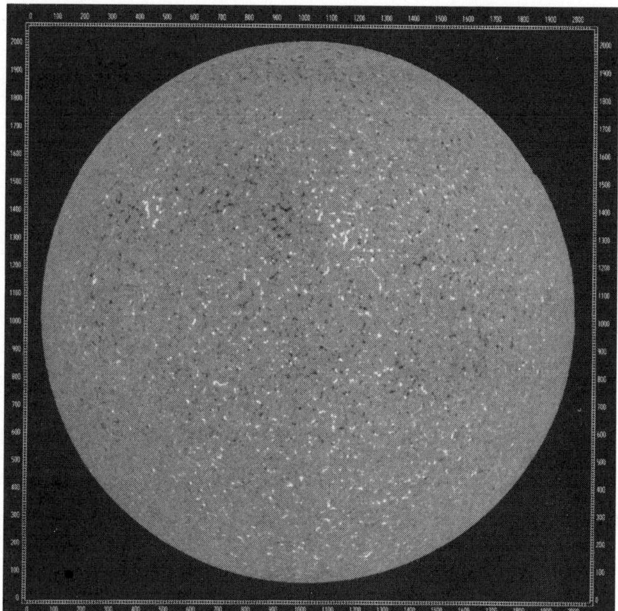

Figure 2. Full-disk magnetogram taken near solar minimum (10 January 1986).

instrumental polarization, but because the resolution is much better the fields are stronger (see the next section) and residual polarization bias becomes less important.

FIELD STRENGTH VERSUS FLUX AMBIGUITY

It was noticed that as resolution on the solar disk improved, the deduced field strength went up. Mt. Wilson low-resolution magnetograms typically display fields of a few tens of gauss. Kitt Peak maps show fields of 100 G or more. Magnetographs do not actually measure field *intensity* (gauss) but rather field intensity averaged over the effective aperture, that is, magnetic *flux*. This leads to an ambiguity in how we interpret magnetograph data. For example, suppose that under conditions of fairly good seeing (2 arcsec) a magnetograph indicates that a region has a field of 10 G. This could mean that there is indeed a uniform field of 10 G on the Sun in the area being sampled. Alternatively, it could mean that there is a 1000-G field element there that occupies only 0.2×0.2 arcsec ($1000 \times 0.04/4 = 10$ G). The strength of nonsunspot fields has been a subject of continuing uncertainty in solar physics for many years. Jan O. Stenflo has argued that, based on simultaneous data from spectrum lines of differing g values, most of the fields on the Sun must be concentrated into subarcsecond and kilogauss elements. But does this mean that truly weak fields are nonexistent?

INFRARED MAGNETOGRAPHS

In order to measure field intensity rather than flux, particularly where the magnetic elements are spatially unresolved, the Zeeman splitting must be greater than the Doppler width of the line. For Fe 5250 Å, 1000 G leads to a splitting of 0.077 Å, but the Doppler width is 0.1 Å and the blend cannot be disassembled. By observing at infrared wavelengths the situation improves. Recall that $\Delta\lambda = 9.34(10^{-13})g\lambda^2 H$. Taking into account that the Doppler line width itself increases with λ, there remains a net gain in splitting that is proportional to λ. At 1.5 μm the splitting due to kilogauss fields begins to separate from the Doppler cores. At 12.2 μm this separation is complete. Several emission lines due to magnesium are present at this wavelength and are suitable for magnetic measurements. Future magnetographs can be expected to take advantage of operation in the infrared.

TRANSVERSE FIELD MAGNETOGRAPHS

Transverse fields are much more difficult to measure because their Zeeman sigma components have the same polarization state and the Babcock differencing method does not work. This means that any instrumental polarization must be compensated for, and the splitting must be strong enough to disengage at least partially the sigma and pi components. Another problem is that both the intensity–flux ambiguity and an indeterminate azimuthal direction complicates any deductions about the three-dimensional structure of the field. Nevertheless, several "vector field" magnetographs are in routine use. At the Marshall Space Flight Center, Huntsville, AL, for example, transverse field measurements in and about sunspots are proving to be of help in the prediction of flares.

Because the interpretation of both longitudinal and transverse magnetic data can be expected to simplify when spatial resolution becomes equal to the size of magnetic elements, the ultimate magnetograph must be space-borne. Magnetograph experiments thus make up an important part of future space missions such as the Orbiting Solar Laboratory and SOHO.

Additional Reading

Canfield, R. C. and Mickey, D. L. (1989). An imaging vector magnetograph for the next solar maximum. In *Solar System Plasma Physics*, J. H. Waite, Jr., J. L. Burch, and R. L. Moore, eds. American Geophysical Union, Washington, D.C., p. 37.

Solar Geophysical Data. SEL, NOAA, Boulder, Colo.

Stenflo, J. O. (1989). Small-scale magnetic structures on the Sun. *Astron. Ap. Rev.* **1** 3.

See also **Stars, Magnetism, Observed Properties; Sun, Magnetic Field; Telescopes and Observatories, Solar.**

Solar Neighborhood

Gerard F. Gilmore

The solar neighborhood may conveniently be defined as that volume of space near the Sun in which we can determine reasonably complete and reliable information describing what is really present. The volume of space resulting from this definition is perhaps surprisingly small. Our census of the population of normal stars is thought to be complete only to about 5 pc from the Sun ($1\ \text{pc} = 3.086 \times 10^{18}\ \text{cm} = 3.2616$ ly), though even within that region no useful information is available on entire classes of objects. Important examples of known objects that are not found near the Sun, and hence that cannot be studied in great detail, include molecular clouds in the interstellar medium and regions of current star formation. Thus knowledge of the interstellar medium is not available in such detail as is possible for common types of stars. Similarly, some classes of objects, such as interstellar planets, quite possibly exist in large numbers near the Sun but could not be detected by present methods.

Our concept of the solar neighborhood depends strongly on the type of information used to study it. Only about one in three normal stars that must exist within 25 pc of the Sun have as yet been identified. For comparison, the most distant objects thought to be part of the Milky Way are roughly 100,000 pc from its center. Thus the solar neighborhood includes only one part in 1000 billion of the observable volume of the Milky Way. It should not then be surprising that only the most common types of object that exist in the Galaxy are identified near the Sun, though conversely it is only very near the Sun that the most common types of stars can be studied in detail, as such stars are of too low luminosity to be detectable at large distances.

There are four basic features of the solar neighborhood that are amenable to detailed study. These are the stars, the interstellar material, the light, and the mass. Each requires different methods of study, and relates to other aspects of astronomy in many different ways. For the remainder of this article we will discuss each of these briefly in turn.

OUR STELLAR NEIGHBORHOOD

Looking at the night sky by eye provides almost no information about the solar neighborhood, and such information as it does provide is quite unrepresentative. Of the 20 visually brightest stars in the sky, only 6 (Alpha Centauri at 4.4 ly, Sirius at 8.6 ly, Procyon at 11.4 ly, Altair at 16 ly, Fomalhaut at 23 ly, and Vega at 26 ly) are within 30 ly of the Sun. The others range in distance up to 1600 ly (Deneb). A total of about 1900 stars and stellar systems (two or more stars orbiting each other sufficiently closely that they are bound into a single system by their mutual gravity) have been identified within 25 pc of the Sun. However, examination of their distribution in space makes clear that only for the intrinsically brightest stars—those brighter than absolute visual magnitude $M_V = +9$, or equivalently of luminosity greater than about 2% that of the Sun or of mass about one-half that of the Sun—is the census effectively complete.

Nearby stars must of course be recognized as such in some way, and the form of this recognition is an important factor in understanding the limitations of the stellar census in the solar neighborhood. The determination of stellar distances can be done in only two ways: either directly, using trigonometric parallax or indirectly, by measuring both its apparent luminosity and some parameter of a star that is correlated with its intrinsic luminosity. The inverse square law then allows a distance estimate from the difference between the intrinsic and the apparent luminosity of the star. In practice the color of a star is the most easily measured reliable measure of intrinsic luminosity (though the spectral type is also useful, especially for hotter stars) so that modern surveys measure colors (or spectral types) for very large samples of stars, and select from samples of many millions of faint stars those very few that are intrinsically very faint and nearby, rather than intrinsically luminous but more distant. Automated techniques now allow this to be carried out over useful areas of sky with acceptable ease. Available studies of this type can identify reliably all stars sufficiently massive to burn hydrogen, and thus be considered as bona fide stars, within about 100 pc of the Sun. As yet, however, only a small fraction of the entire sky has been studied in this way.

Trigonometric parallax measurements are more precise distance measures, but are very time consuming. Hence one requires an efficient method to select stars with a high probability of really being nearby before a parallax study is initiated. Prior to the automated color measurements previously noted, surveys tended to isolate stars whose apparent motions on the sky—proper motions—are unusually large. Large angular motion can correspond to a star with a large speed across the sky, or to a star with a normal speed but a large apparent speed due to its proximity to us. Proper motion studies of reasonably bright stars covering the whole sky have been available for some years, and have provided the starting list for those trigonometric parallax measures that have provided the complete census of nearby bright stars.

In addition, the parallax studies have highlighted another difficulty in identifying the true distribution of stars near the Sun: Stars tend to occur in pairs, with the companion often being difficult or impossible to see in the glare of the brighter star. This situation is illustrated for possibly the most famous case in Fig. 1. This figure shows the path on the sky of Sirius, the brightest star to the eye. The motion shows a pronounced wiggle (discovered in 1844 by Friedrich W. Bessel), which indicates that the visible star is being wobbled in its motion through the Galaxy by an unseen companion. Thus Sirius is two stars, the companion now being known to be a white dwarf. Another important example is the Alpha Centauri system. The bright star is in fact a double, with the brighter being about 10% more massive than the Sun, and the fainter being about 10% less massive. These two stars are the second and third closest to the Sun. The closest star to the Sun is Proxima Centauri, 4.22 ly distant, and also in orbit as part of the Alpha Centauri system, which thus contains three stars. Proxima was discovered by searching the sky for stars with apparent motions similar to those of known nearby stars. Proxima Centauri is a low luminosity (1/25,000 that of the Sun) and low mass (1/10 that of the Sun) star with a rather faint apparent brightness in spite of its proximity (sic) to the Sun. It is 2°.2 or about 450 times the radius of the orbit of Neptune, from Alpha Centauri. Many similar stars that do not happen to have brighter companions or that have smaller speeds across the sky undoubtedly remain to be discovered.

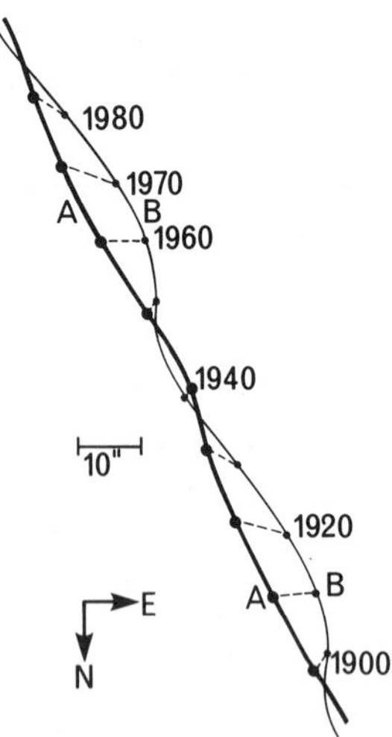

Figure 1. A representation of the path of Sirius on the sky, showing the wobble that first provided evidence for a companion star. The star marked A is the apparently brightest star in the sky, whereas the companion (B) is a white dwarf. The two stars are bound together by their mutual gravity, and orbit about their common center of mass, in the same way as the Sun and planets. About one-half of all stars have companions; triple systems (Alpha Centauri, the nearest star system to the Sun being an example) are also common.

The distribution in space of the nearest stars to the Sun is illustrated in Fig. 2. This figure is adapted from a model of the nearby stars kept at the Sproul Observatory, Pennsylvania, where much of the most important work in mapping the stellar population in the immediate solar neighborhood has been carried out.

From detailed study both of small areas of sky and of the volume of space within a few light years of the Sun, using the methods previously outlined, it is well established that the number of stars as a function of absolute magnitude (the luminosity function) continues to rise to a maximum near absolute visual magnitude $M_V = +12$, which is equivalent to a luminosity of about 1/10,000 that of the Sun, and a mass about 1/4 that of the Sun. At even lower luminosities the number of stars declines, with the lowest luminosity single star near the Sun being of absolute visual magnitude about $M_V = +19$, or about 2 parts in one million of the solar luminosity. Such low luminosities do not correspond to extremely low masses, but to a mass of a little below 1/10 that of the Sun (the precise value being problematic).

At even lower masses stars are of too low mass to burn hydrogen, and so do not shine brightly or for very long. Thus extremely low mass objects (planets or "brown dwarfs") would be intrinsically very faint and difficult to detect. Some rather weak limits exist

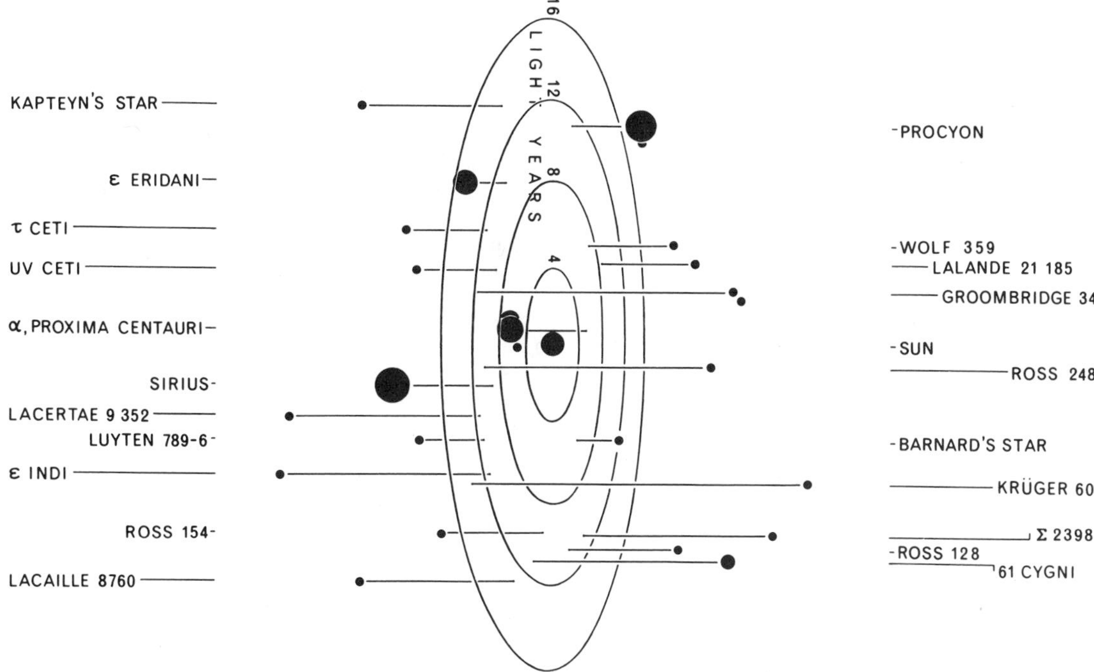

Figure 2. The distribution in space of the nearest stars to the Sun. Because most stars are very much less luminous than is the Sun, fewer than one-half of these stars can be seen by the unaided eye in spite of their proximity. Conversely, it is only in the immediate solar neighborhood that the most common types of star in the Universe can be detected and studied.

from the lack of such sources in sky surveys by infrared satellites, which are sensitive to their heat output. Substantial numbers of objects whose mass is too low to burn hydrogen must exist, unless the physical processes involved in star formation have conspired not to produce objects we would have difficulty detecting. Because we live on an object we could certainly not detect were it much further away, this seems extremely unlikely.

However, even though very low mass "stars" must exist in large numbers, the total mass in such objects is only a small fraction of that in more massive stars. Smooth extrapolation of the distribution of the number of known luminous stars as a function of their mass to planetary masses does not suggest a significant mass can be hidden there, with only about 10% of the total mass in stars likely to be contained in brown dwarfs and interstellar planets. Thus brown dwarfs, at least insofar as their properties and numbers can be determined in the immediate solar neighborhood, do not seem to provide a solution to the problem of the missing mass.

OUR INTERSTELLAR NEIGHBORHOOD

The interstellar medium is a complex mix of hot bubbles, cold clouds, and structures in between, and is composed mostly of hydrogen (both atomic and molecular) and helium, with some contribution by other elements and molecules. Thus, any sample on a scale of a few light years will be unlikely to be representative of the larger scale properties. Insofar as the solar neighborhood in the interstellar medium is concerned, the most important result is that we live in a low density "hole," with a size of a few hundred light years, further complicating the situation. A useful general result however is that the average amount of interstellar material in a column through the Milky Way at the same distance from the galactic center as the Sun (and perpendicular to the galactic plane) is about 13 $M_\odot$ pc^{-2}. That means that if all the interstellar medium in a column through the Milky Way near the Sun were compressed to a density equal to that of water (1 cm^{-3}) the interstellar medium would be all contained in a layer 0.025 mm thick.

OUR LUMINOUS NEIGHBORHOOD

Although most stars are intrinsically faint, most luminosity comes from the few bright stars. Thus, within 5 pc of the Sun half the total luminosity is contributed by one star: Sirius. More distant samples, which are reliable for bright stars, show that the mean luminosity in the solar neighborhood is close to 0.06 $L_\odot$ pc^{-3}. This rather low value is an indication of the rarity of stars as bright as the Sun.

Were we to observe the solar neighborhood from outside the Galaxy, the surface brightness seen could be calculated from star counts, giving about 1.3 $L_\odot$ pc^{-2} sr^{-1}. That is, the total luminosity of the entire thickness of the galactic disk near the Sun is not much brighter than that due to one star like the Sun every parsec in a flat distribution. Again, this indicates the rarity of luminous stars in the solar neighborhood.

OUR MASSIVE NEIGHBORHOOD

The total amount of mass near the Sun is probably the most fundamental property of the solar neighborhood of relevance to other aspects of astronomy. This can be deduced from measurements of both the thickness of the disk of the Milky Way and the velocity of the stars in the disk. The physical principle is the same as spinning a weight on a piece of string. As the weight is spun faster one feels a harder pull on the string. Similarly, if stars move quickly there must be a lot of mass below them to prevent them spreading out more than they have, whereas if they move slowly less mass is required. Applications of this principle have been fraught with technical difficulties, though the most comprehensive recent studies show that there is no unidentified mass associated with the galactic disk.

The total mass in the solar neighborhood is about 0.11 $M_\odot$ pc^{-3}, or the mass of one star of the lowest possible mass to burn hydrogen per cubic parsec. This total mass is made up of 0.05 $M_\odot$

pc^{-3} in normal luminous stars, 0.005 $M_\odot$ pc^{-3} in stellar remnants (white dwarfs and neutron stars), an estimated 0.006 $M_\odot$ pc^{-3} in objects of too low mass to burn hydrogen (brown dwarfs, planets), 0.001 $M_\odot$ pc^{-3} in stars associated with the galactic halo, and 0.01 $M_\odot$ pc^{-3} of mass of an unknown type ("missing mass") that dominates the outer parts of galaxies, and provides the dark halos deduced from the kinematics of stars and gas in the very outer parts of galaxies.

Additional Reading

Frisch, P. and York, D. (1986). Interstellar clouds near the Sun. In *The Galaxy and the Solar System*, R. Smoluchowski, J. Bahcall, and M. Matthews, eds. University of Arizona Press, Tucson, p. 83.

Gliese, W., Jahreiss, H., and Upgren, A. (1986). Stars within 25 parsecs of the Sun. In *The Galaxy and the Solar System*, R. Smoluchowski, J. Bahcall, and M. Matthews, eds. University of Arizona Press, Tucson, p. 13.

Kuijken, K. and Gilmore, G. (1989). The mass distribution in the galactic disk. I, II, III. Mon. Notices Royal Astron. Soc. **239** 571, 605, 651.

van de Kamp, P. (1971). The nearby stars. *Ann. Rev. Astron. Ap.* **9** 103.

van de Kamp, P. (1975). Unseen astrometric companions to stars. *Ann. Rev. Astron. Ap.* **13** 295.

See also **Astrometry; Interstellar Medium, Local; Missing Mass, Galactic; Stars, Distances and Parallaxes; Stars, Proper Motions, Radial Velocities and Space Motions; Stars, White Dwarf, Observed Papers.**

Solar Physics, Space Missions

Guenter E. Brueckner

EARLY YEARS: DISCOVERIES FROM SOUNDING ROCKETS

Solar physics from space in particular and space science in general began on October 10, 1946, when Richard Tousey photographed the ultraviolet spectrum of the Sun below the atmospheric cutoff at 3000 Å from a captured German V2 rocket. Successive exposures taken at different altitude showed the continuation of the photospheric spectrum down to 2200 Å and its absorption by ozone in the Earth's atmosphere. Using thermoluminescent phosphor plates, in 1948 Tousey established from sounding rockets the existence of the chromospheric emission line of hydrogen Lyα at 1216 Å. In September 1949, Herbert Friedman and his colleagues from the Naval Research Laboratory (NRL) quantitatively measured for the first time x-rays (1–8 Å), Lyα (1216 Å), and Schumann continuum radiation (1425–1600 Å). A photographic spectrum of hydrogen Lyα was obtained in 1953 by William A. Rense. A profile of Lyα obtained in 1960 by J. D. Purcell and Tousey showed the existence of the extended hydrogen geocorona. Hans E. Hinteregger in 1960 recorded for the first time the solar emission line spectrum below Lyα. At that time, it was demonstrated that the integrated intensity of all emission lines below Lyα is less than the intensity of Lyα itself. In 1963 B. C. Fawcett and coworkers demonstrated that the XUV emission lines are emitted from the hot (1.5–2$\times 10^6$ K) coronal plasma. Alan H. Gabriel, Fawcett, and Carole Jordan classified the high ionization Fe lines in 1965. During the same year, Leo Goldberg, William H. Parkinson, and Edmond M. Reeves identified CO molecular bands as the dominant absorbers in the extreme ultraviolet spectrum between 1600 and 1800 Å.

In 1956, Friedman observed for the first time the strong enhancement of solar x-ray emission during a flare and established the connection between solar x-rays and SIDs (sudden ionospheric disturbances). Firing a series of sounding rockets during successive phases of the total solar eclipse of October 12, 1958, Friedman demonstrated that solar x-rays originated in the hot corona and are particularly intense above active regions. Highly resolved spectroheliograms in Lyα were recorded by Tousey in 1959. The first image of the Sun in x-rays was photographed by Friedman in 1960 using a pinhole camera mounted on a sounding rocket. This picture showed a clear correlation between enhanced Lyα emission and Ca II plages.

Using an externally occulted coronagraph invented by John W. Evans, Tousey in 1963 created an artificial solar eclipse from a sounding rocket above the atmosphere. He found a highly structured corona out to 10 solar radii dominated by narrow coronal streamers.

EARLY SATELLITES: VANGUARDS, SOLRADS, AND OSOS

Vanguard-3, launched in September 1958, was the first artificial satellite to monitor solar x-ray and Lyα radiation. It failed because the detectors were swamped with high energy particles from the Van Allen belt, which was unknown when the small six-inch satellite was designed. However, a group at NRL had set up a very precise radio tracking system to determine with high precision the satellite's orbit. From orbital variations, the diurnal expansion and contraction of the ionosphere caused by solar ultraviolet and x-ray radiation as well as the longterm ionospheric density changes connected with the solar activity cycle were found.

The beginning of solar research from satellites can be identified with the launch of the NRL solar radiation (*Solrad 1*) satellite on June 20, 1960. This satellite measured Lyα and x-rays (1–8 Å) for about five months. The measurements established beyond doubt that sudden ionospheric disturbances were caused by enhanced solar x-ray emission but not by increased Lyα radiation. This confirmed earlier models proposed by Friedman and Talbot A. Chubb in 1954. *Solrad 1* also demonstrated a high degree of coronal variability in x-rays (several orders of magnitude) compared with rather low variability of the Lyα line, which is of chromospheric origin.

The *Solrad* series of satellites continued through 1975 producing a continuous record of solar UV and x-ray radiation over more than a solar cycle. During the eclipse of 1966 it was found from *Solrad* measurements that soft x-rays (8–20 Å) were concentrated in areas less than 50 arcsec in diameter and that hard x-rays originated from areas less than 30 arcsec in diameter.

John C. Lindsay's initiative resulted in a series of NASA-launched orbiting solar observatories (OSOs) which for the first time provided three-axis-stabilized platforms. Imaging of the sun over long time periods was now possible. The first OSO was launched in 1962. The pointing capability of the OSO satellites was far better than the imaging quality of early ultraviolet and x-ray instruments; they were matched in 1975 when Roger M. Bonnet flew a 1 arcsec resolution Lyα telescope on *OSO-8*. However, even coarse imaging of the early OSOs (35 arcsec on *OSO-6* by a Harvard extreme-ultraviolet spectrometer) revealed that the magnetic field at the boundaries of supergranular cells must be responsible for ultraviolet emissions in the chromosphere and transition zone. It could also be seen that this picture changes in the corona, which does not show a supergranular structure. Quantitative measurements of solar radiation from the transition zone and corona led to a one-dimensional picture of the outer solar atmosphere in which the chromospheric temperature rise is followed by a steep temperature gradient in a thin ($\sim$ 50 km) transition zone. Heat conduction from the overlaying corona supplies the energy to maintain the transition zone.

Figure 1. The *Skylab* satellite (1973–1974). The Apollo Telescope Mount (ATM), in the center of the large solar cell array, carried five large instruments for solar observations in the x-ray and ultraviolet wavelength regimes and one externally occulted coronagraph. (*Photograph by Ed Gibson*, Skylab *astronaut.*)

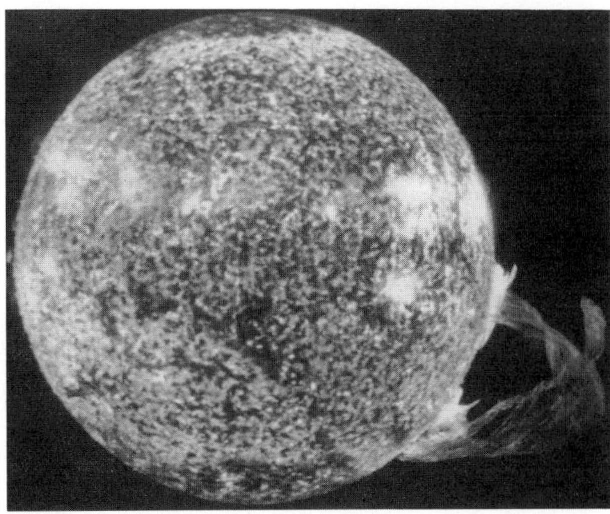

Figure 2. A large erupting prominence (December 19, 1973) photographed in the 304 Å He II resonance line by the Naval Research Laboratory's spectroheliograph on board *Skylab* (NRL and NASA).

A white light coronagraph flown on *OSO-7* in 1972 for the first time detected coronal transients, material being ejected from the solar surface with large velocities $\simeq 1200$ km s^{-1}, and correlated them with type II radio bursts. Gamma rays from solar flares were detected from *OSO-7* by Edward L. Chupp in 1972.

SKYLAB

Skylab provided solar physicists with enormous advantages: For the first time, film could be retrieved from a satellite and instruments could be much bigger than those on OSOs (see Fig. 1). X-ray (American Science and Engineering Corporation) as well as XUV (NRL) telescopes reached spatial resolution of 2 arcsec. The high resolution pictures showed that magnetic fields are governing the detailed structure of the chromospheric transition zone and coronal plasma (see Fig. 2). In closed magnetic areas all coronal plasma is organized as loops. Magnetic loops were identified as the location of flares, sometimes exploding during flares, sometimes containing the flaring plasma. The high temperature component ($\sim 20 \times 10^6$ K) of flares was made visible as well as flare-related high speed ejecta. The twisted structure of coronal loops was seen for the first time. Most importantly, *Skylab* opened the door for an entirely new interpretation of the coronal heating mechanism: Changes in evolving magnetic field are now supplying the energy to maintain the hot corona replacing earlier theories of sound wave heating. Models of coronal holes based upon *Skylab* measurements clearly defeated the long established theory that the solar wind is a result of coronal thermal expansion.

SOLAR MAXIMUM MISSION AND SOLWIND

The *Solar Maximum Mission* satellite was launched in 1980 to be in orbit during the solar maximum of cycle 21 (1981). It was equipped mostly with high energy instruments, which did not fly on *Skylab*. Repaired in 1984, it reentered the Earth's atmosphere on December 3, 1989. Because of its long time in orbit, *Solar Max* collected many observations of phenomena which are dependent on the solar cycle. For the first time, Richard C. Willson demonstrated beyond doubt variation of the total solar luminosity not only with the disk passage of sunspots ($\leq 0.25\%$) but also with the solar cycle ($\sim 0.08\%$). This was confirmed by the Earth Radiation Budget (ERB) experiment on board *Nimbus 7*. A thus far unexplained period of 155 days in the occurrence of high energetic gamma ray flares was detected. Flare observations indicated that the extremely energetic radiation (hard x-rays and gamma rays) is caused by the dissipation of energy of high energy charged particles which are accelerated in magnetic loops. Blue shifts found in x-ray spectra are interpreted as upward moving material from the region of high energy deposition. However, flare models remain controversial because exact timing of flare emissions at different energies indicates a highly filamentary structure of the flare plasma and constrains the region where all these processes occur to very small sizes of only a few hundred kilometers. An externally occulted coronagraph (*Solwind*) flown on *P78-1* accumulated a wealth of coronal images over an eight-year period from 1978–1985. One of its most significant results was the correlation between coronal mass ejections and shock waves in the interplanetary medium.

MODERN SOUNDING ROCKETS AND SPACELAB

A sub-arcsecond solar pointing system for sounding rockets (SPARCS) was introduced in 1968. Using it, a group at American Science and Engineering Corporation for the first time obtained high resolution (~ 10 arcsec) x-ray images with grazing incidence optics. The x-ray bright points which anticorrelate in numbers with the solar cycle were detected. Coronal holes became visible on the solar disk. Their correlation with high-speed solar wind streams was established by Allen S. Krieger, Adrienne F. Timothy, and Edmond C. Roelof in 1970. The author and O. Kenneth Moe during the same year photographed for the first time an ultraviolet spectrum in which most of the chromospheric and transition zone lines were spectrally resolved. They demonstrated the existence of strong nonthermal motions in the transition zone. A very high resolution telescope and spectrograph for ultraviolet work was used by the author for the first time in 1975. From the *Spacelab 2* flight of this instrument in 1985 it was demonstrated that the transition zone turbulence is caused by newly emerging flux and its reconnection with old flux. This added additional strong evidence that the corona is heated by changes in the local magnetic field of the Sun. Alan M. Title showed from high resolution white light pictures obtained also from *Spacelab 2* that proper motions of granules lead to the supergranular structure of the Sun's magnetic field.

FUTURE: SOLAR-A, SOHO, AND OSL

A Japanese satellite called *Solar A* will be launched in 1991. It will be equipped with x-ray and gamma ray imaging experiments for flare work. The European Space Agency together with NASA will launch the Solar Heliospheric Observatory (SOHO) in 1995. This satellite will carry a large number of very powerful instruments which will cover with spectroscopic diagnostics the whole outer solar atmosphere from the chromosphere to the outer solar corona at 30 $R_{\odot}$. For the first time, the newly emerging technology of high resolution imaging with charge coupled devices (CCDs) will be used extensively throughout this satellite. NASA will launch the Orbiting Solar Laboratory (OSL) as early as 1996. This satellite covers wavelengths from the x-rays to visible light with very high spatial and spectral resolution (0.1 arcsec in the photosphere, 0.5 arcsec in the outer solar atmosphere). This will be the first solar space mission that carries instruments for simultaneous observations from the photosphere into the corona.

Additional Reading

Eddy, J. A. (1979). *A New Sun, the Solar Results from* Skylab. NASA SP-402. National Aeronautics and Space Administration, Washington, D.C.

Friedman, H. (1974). Solar ionizing radiation. *J. Atmos. Terrest. Phys.* **36** 2245.

Friedman, H. (1987). Origins of high-altitude research in the Navy. The Charles H. Davis Lecture Series. National Academy of Sciences, Washington, D.C.

Friedman, H. (1990). *The Astronomer's Universe.* W. W. Norton and Co., New York.

Poland, A. and Domingo, V., eds. (1988). *The* SOHO *Mission, Scientific and Technical Aspects of the Instruments.* ESA AP-1104, European Space Agency, Paris.

The SMM Principal Investigator Teams (1987). *NASA's Solar Maximum Mission: A Look at a New Sun*, J. B. Gurman, ed. National Aeronautics and Space Administration, Goddard Space Flight Center.

Tousey, R. (1967). Some results of twenty years of extreme ultraviolet solar research. *Ap. J.* **149** 239.

See also **Coronagraphs, Solar; Scientific Spacecraft and Missions; Sounding Rocket Experiments, Astronomical.**

Solar System, Origin

Alan P. Boss

One of the most profound questions in astronomy is to understand how the Sun and planets of our solar system formed, because the answer to this question will not only satisfy our curiosity about the origin of our own planet and ultimately our species, but will also have strong implications for the likelihood of similar planetary systems occurring around other stars. Although astronomers will never know with absolute scientific certainty how our solar system formed because of the irreproducible nature of that grand experiment, we can hope to assemble a reasonable scenario through the development of theoretical models constrained by the clues provided by observations of the bodies in our solar system. The central theoretical concept of solar system origin, attributed to Immanuel Kant (1755) and Pierre Simon de Laplace (1796), is that formation of the planets and the Sun occurred in a rotating, flattened, gaseous cloud termed the *solar nebula*. This hypothesis ensures that the resulting planets will indeed form with the gross orbital properties of our solar system namely, revolution of the planets around the Sun in nearly circular orbits, in the same direction, and largely confined to a single plane. The challenge to cosmogonists for the last 200 years has been to discover how the solar nebula formed and evolved into our solar system.

SOLAR NEBULA FORMATION

Our solar system formed about 4.56 billion years (the oldest meteorite age) ago, after a dense, interstellar cloud of gas and dust began to collapse inward because of the force of its own self-gravity. Assuming that our solar system was formed in much the same way as solar-type stars are forming in the Galaxy today, we can use observations of such star formation to provide critical information about the properties of the interstellar cloud that collapsed to form the solar nebula; this information can be used to infer the structure of the solar nebula.

Interstellar clouds contain debris ejected from earlier generations of stars, particularly the elements heavier than hydrogen (H) and helium (He), produced through stellar nucleosynthesis. The heavier elements that constitute the terrestrial planets (and human beings) resided primarily in dust grains in the solar nebula. Dust grains contain only about 2% of the mass of interstellar clouds, with most of the matter being gaseous H (about 77%) and He (about 21%).

The densest interstellar clouds (number density about 10^4–10^6 molecules per cubic centimeter) are quite cold (about 10 K), because dust grains are efficient at radiating away the thermal energy of the clouds. Such a cloud is unstable: It tends to collapse because thermal pressure in a cold cloud is unable to resist the inward pull of gravity. Collapse involves rapid (less than a million years) contraction of the cloud at supersonic velocities; collapsing clouds that form stars are termed *protostars*. Astronomical observations of protostars are hampered by the short time period involved and by the presence of obscuration by the interstellar cloud itself, so much of our understanding of the collapse phase comes from theoretical modeling.

Theoretical models show that collapse of the presolar nebula continued until the center of the cloud became so dense that heat produced by cloud compression was trapped, increasing the thermal pressure enough to halt collapse. A quasistatic core, the proto-Sun, formed, surrounded by an infalling envelope. As more mass accreted onto the proto-Sun and after thermonuclear reactions began, the central temperature increased and reached the temperature of the present solar interior ($\sim 10^7$ K); the center of the proto-Sun was about 10^{20} times denser than the initial interstellar cloud.

Because of conservation of angular momentum, much of the infalling cloud formed a flattened disk (the solar nebula) in the equatorial plane of the proto-Sun. The existence of our planets is due to the angular momentum of the initial cloud: Without angular momentum, the interstellar cloud would have collapsed to form a single star without a nebula. On the other hand, collapse of a cloud with too much angular momentum leads to binary star formation; evidently, our solar system resulted from the collapse of a cloud with an intermediate amount of angular momentum.

SOLAR NEBULA EVOLUTION

The Sun contains 99.9% of the mass of the solar system, but only 2% of the angular momentum. Because the Sun and planets formed from the same interstellar cloud, one might expect the Sun and the planetary system to contain roughly the same amount of angular momentum per unit mass. This depletion of the Sun's angular momentum can be accounted for partially by magnetic braking by the solar wind and by preferential formation of the Sun from the lowest-angular-momentum material in the initial cloud. However, considerable depletion must have been caused by processes that transported nebula mass inward to form the Sun, and angular momentum outward to the preplanetary region.

Three separate processes for such transport have been proposed, involving respectively, viscous stresses, gravitational torques, and magnetic fields. Although each process individually may have been able to transport enough angular momentum to form the Sun and planets out of the solar nebula, it also may have been that all three processes were involved to some extent. Understanding these three processes is currently a central focus of solar nebula research.

1. *Viscous stresses* require the presence of turbulence to generate an effective viscosity, because ordinary molecular viscosity is too small to be important. Convective instability (caused by the "boiling" upward of hot matter from the midplane of the solar nebula to its surface), is the most likely means of exciting turbulent motions. The viscous stresses result in the outward transport of angular momentum through the friction between adjacent fluid parcels trying to move past each other with the different speeds caused by nearly Keplerian rotation in the nebula.

2. *Gravitational torques* arise from the gravitational forces between segments of asymmetric mass distributions in the solar nebula, such as between a prolate proto-Sun and a bar-like solar nebula or between the inner and outer regions of trailing spiral arms in the nebula. Gravitational torques also result in significant outward transport of angular momentum, providing that a source of asymmetry exists; asymmetry can arise from rotational instability of the nebula or proto-Sun or from instability of the matter accreting onto the nebula, or the asymmetry could be a residue of the initial interstellar cloud.

3. *Magnetic fields*, produced either by amplification of the magnetic field of the initial cloud during collapse or by a dynamo mechanism in the proto-Sun or solar nebula, can also transport angular momentum outward through resistance of the field lines to the winding caused by Keplerian rotation.

The evolution of the solar nebula thus involved inward transport of mass onto the central proto-Sun and, possibly, expansion of the preplanetary regions through the addition of angular momentum. Most of the mass of the nebula fell onto the proto-Sun; the remainder formed the planets or else was swept away by the early solar wind. Two distinct processes exist for planet formation in the solar nebula: formation by accumulation or through gaseous disk instability.

PLANET FORMATION BY ACCUMULATION

Interstellar dust grains have mean sizes of about 0.1 micrometer (μm), sizes that are about 10^{14} times smaller than the Earth. The favored means of explaining the formation of the terrestrial planets and the central cores of the giant planets is through collisions and sticking together (*coagulation*) of dust grains and the subsequent *accumulation* of much larger, solid bodies into the planets.

Compared to the situation in interstellar clouds, growth of dust grains through coagulation was rapid within the solar nebula, where relatively high densities and relative motions among dust grains (caused by drag forces, between the grains and the gas, that depend on grain size) ensured frequent collisions. Even while growing, the dust grains "sedimented" down to form a thin disk in the midplane of the solar nebula, because the gaseous portion of the nebula was supported by thermal pressure and hence was distributed over a much thicker disk. Sedimentation may have had to await the cessation of turbulence in the nebula, because vigorous turbulent motions would have kept the grains well mixed with the gas. Once the nebula became only weakly turbulent, however, dust grain sedimentation occurred rapidly, within about 1000 yr. The grains coagulated to sizes of about 1 cm to 1 m during sedimentation.

The next phase of accumulation is generally thought to have involved a collective gravitational instability of the dust disk. That is, when the thin dust disk became massive enough (through ongoing sedimentation), its self-gravity rapidly broke it up into a large number of "planetesimals," solid bodies about 1 km in size. The planetesimals in the terrestrial planet region alone constituted a swarm of about 10^{12} bodies on nearly circular orbits about the Sun.

Subsequent planetesimal growth occurred through accumulation following random collisions among these self-gravitating bodies. Two distinct phases can be isolated. The first phase (I) involved accumulation of nearby (in orbital radius) planetesimals and ended when there were no more nearby planetesimals. Accumulation in this phase was completed within about 10^4 years in the terrestrial planet region and produced planetesimals ("planetary embryos") about 500 km in size. The second phase (II) involved collisions between the initially widely separated planetary embryos produced during phase I and resulted in the final planetary system. Accumulation in this phase required large changes in the orbital radius of the planetary embryos (caused by the random effects of gravitational forces between them), especially during close encounters. Phase II required about 10^7–10^8 yr to produce the terrestrial planets.

Two possible extremes exist for the outcome of either phase of accumulation: Runaway accretion may occur, where one body grows by accreting all the others, because its cross section for collision is enhanced by its own gravity as the runaway body becomes larger. Alternatively, growth can be much more uniform, with most of the mass residing in a number of planetesimals with roughly equal mass. Deciding whether runaway accretion or uniform growth dominated these two phases of accumulation is another active area of current cosmogonical research; present indications are that runaway growth dominated phase I of accumulation but not phase II. This means that the final phases of planetary accumulation were likely characterized by violent collisions between relatively large bodies of roughly equal mass. A giant impact, between a Mars-sized protoplanet and the Earth, may have resulted in the formation of the Moon.

Perhaps the most outstanding problem in the accumulation theory of planetary formation is explaining the rapid formation of the giant planets. Jupiter and Saturn must have formed prior to the removal of the gaseous portion of the solar nebula, if they were to accrete their H- and He-rich envelopes from the solar nebula. Astronomical observations imply that solar-type young stars have dispersed their nebulae within 10^5–10^7 yr after their formation; these times set an upper limit for giant planet formation. The problem is that most estimates of the accumulation time for the giant planets are 10^8 years or longer (in some cases, longer than the age of the solar system!) when the older theories are used. Rapid formation only appears to be possible if the solar nebula was substantially more massive in the giant planet region than is usually assumed and if runaway accretion proceeded all the way to bodies 10 times as massive as Earth within about 10^6 yr; bodies this massive could then have accreted gas quickly from the nebula and acquired the H- and He-rich envelopes of the giant planets.

PLANETARY FORMATION BY GASEOUS DISK INSTABILITY

The alternative to accumulation for explaining planetary formation involves a gravitational instability of the gaseous portion of the solar nebula. If the nebula was massive enough, the instability would have led to fragmentation of the gaseous nebula and the formation of *giant gaseous protoplanets* on a time scale of about 10 yr. Considering that this process rapidly produces giant planets in a single step, there is no time scale problem for giant planet formation by this scheme. However, a number of even more serious problems have led cosmogonists to doubt that gaseous disk instability led to planet formation in our solar system.

The most fundamental problem with giant gaseous protoplanet formation is that the instability requires a nebula much more massive than our planetary system; it is unlikely that the residual mass could be removed without also removing the protoplanets that have already formed. In order to form the rocky cores inferred for the giant planets, rocky matter in the giant gaseous protoplanet envelope must be able to sediment to the center of the protoplanet, but these materials are thought to be miscible with the hydrogen–helium envelope and so cannot sediment to the core. Even if rocky cores could have formed by this means, the entire gaseous envelope must be stripped away if a terrestrial planet is to result from a giant gaseous protoplanet. Also, if an efficient stripping process did exist then it must have operated only in the inner and outer solar system, but *not* at Jupiter and Saturn, in order to account for the dominantly rock and ice compositions of the terrestrial planets and of Uranus and Neptune, respectively. Both of the likely stripping processes, tidal forces and thermal evaporation by the Sun, are inconsistent with this requirement.

OUTLOOK

Although many of the details of the theory of solar system origin will undoubtedly change with time, the fundamental concept of solar system formation appears to be irrefutable: The Sun and planets formed about 4.56 billion years ago out of a solar nebula produced by the collapse of a rotating interstellar cloud of gas and dust. A plausible outline now exists for understanding the formation of the terrestrial and giant planets through collisional accumulation of planetesimals, and it is through further development of this outline that we expect to sharpen our understanding of solar system origin.

Additional Reading

Black, D. C. and Matthews, M. S., eds. (1985). *Protostars and Planets II*, University of Arizona Press, Tucson.

Boss, A. P. (1985), Collapse and formation of stars. *Scientific American* **252** (No. 1) 40.

Cameron, A. G. W. (1988). Origin of the solar system. *Ann. Rev. Astron. Ap.* **26** 441.

Encrenaz, T. and Bibring, J. B. (1990). *The Solar System*. Springer, Berlin.

Safronov, V. S. (1969), *Evolution of the Protoplanetary Nebula and Formation of the Earth and the Planets*, Nauka, Moscow. Translation, NASA TTF-677 (1972).

Weaver, H. A. and Danly, L., eds. (1989). *The Formation and Evolution of Planetary Systems*. Cambridge University Press, Cambridge, U.K.

Wetherill, G. W. (1981) The formation of the Earth from planetesimals. *Scientific American* **244** (No. 6) 162.

See also **Binary and Multiple Stars, Origin; Interstellar Clouds, Collapse and Fragmentation; Meteorites, Origin and Evolution; Moon, Origin and Evolution; Planetary Formation, Gaseous Protoplanet Theories; Planetary Systems, Formation, Obsevational Evidence; Protostars; Protostars, Theory.**

Sounding Rocket Experiments, Astronomical

George R. Carruthers

Sounding rockets were the first vehicles used for space astronomy, that is, astronomical observations from above the Earth's atmosphere. Space-based observations are important in astronomy, largely because they allow measurements of radiation from celestial objects that fall in wavelength ranges inaccessible to ground-based telescopes, because of absorption in the atmosphere. These wavelength ranges include the entire ultraviolet, x-ray, and γ-ray portions of the spectrum shortward of 3000 Å (300 nm), as well as much of the infrared and submillimeter spectrum. Sounding rockets provided the first astronomical observations in the ground-inaccessible ultraviolet and x-ray wavelength ranges.

Sounding rockets are defined as vehicles that are launched on more or less vertical trajectories, to altitudes of typically 100–500 km, followed by immediate return to Earth (i.e., they do not go into orbit). In its application to astronomy, a figure of merit for a sounding rocket is the amount of time it provides above that portion of the Earth's atmosphere that interferes with astronomical observations (typically, above 150 km). Also important are the size and weight of payload that can be carried to a given altitude, and the availability and performance capabilities of attitude control systems for pointing the instruments at the selected targets.

Sounding rockets typically remain above the atmosphere for only a few minutes before falling back to Earth; hence the observing time available is much less than that achievable with orbital vehicles. However, sounding rockets were the only means available for obtaining astronomical observations from space from 1946 until the early 1960s. Even today, sounding rockets are still used for space astronomy despite the existence of, and capability for, long-term space observatories launched into Earth orbit by unmanned or manned satellite vehicles.

The continued usefulness of sounding rockets stems from their relatively low cost and short mission preparation time, compared to those of satellite vehicles carrying comparable instrumentation. Also, instrumentation planned for use in a short-duration, recoverable sounding rocket payload can be much less expensive, and made ready for flight in a much shorter time, then can similar instrumentation used in orbital mission payloads.

HISTORICAL BACKGROUND

The first astronomical observations from above the atmosphere were obtained in 1946, using captured German V-2 rockets. These were used to obtain the first ultraviolet and x-ray observations of the Sun. In the late 1940s and early 1950s, the V-2s were replaced by U.S.-developed rockets specifically intended for space research. These included the *Viking* rocket, similar to the V-2 in size and performance capability, and the smaller *Aerobee* rocket, which was still in use as late as 1980.

The liquid-propellant *Aerobee* rocket, in various versions, was the most widely used sounding rocket for space astronomy from 1950 until 1980. In recent years, however, it has been replaced for most applications by the solid-fueled *Black Brant* vehicle. Figure 1 illustrates a version of the *Black Brant* used for recent ultraviolet measurements of comet Halley. Other vehicles of note include the *Aries* rocket, based on the second stage of the *Minuteman I* ICBM, which can carry larger and heavier payloads than the *Aerobee* or *Black Brant*.

The following sections describe some of the specific "firsts" and other significant results in space astronomy that were obtained using instrumentation flown on sounding rockets.

Solar Rocket Astronomy

The very first observations of the solar ultraviolet (UV) spectrum at wavelengths below the ground-based limit of 3000 Å were obtained by the Naval Research Laboratory from a V-2 rocket launched in October, 1946. The spectra revealed the middle-UV (2000–3000 Å) range of solar radiation as the rocket ascended through the ozone layer (30–60 km) in the Earth's stratosphere. Spectra taken in later flights extended the coverage through the far-UV (1000–2000 Å) and extreme-UV (below 1000 Å) ranges. The spectra showed that the solar spectrum changes from an absorption-line (Fraunhofer) type to an emission-line type in the far-UV. Important features revealed were the atomic hydrogen (Lyman α) emission line at 1216 Å and the emission lines of neutral and ionized helium at 584 and 304 Å, respectively. Measurements of the far- and extreme-UV emission features allowed much more

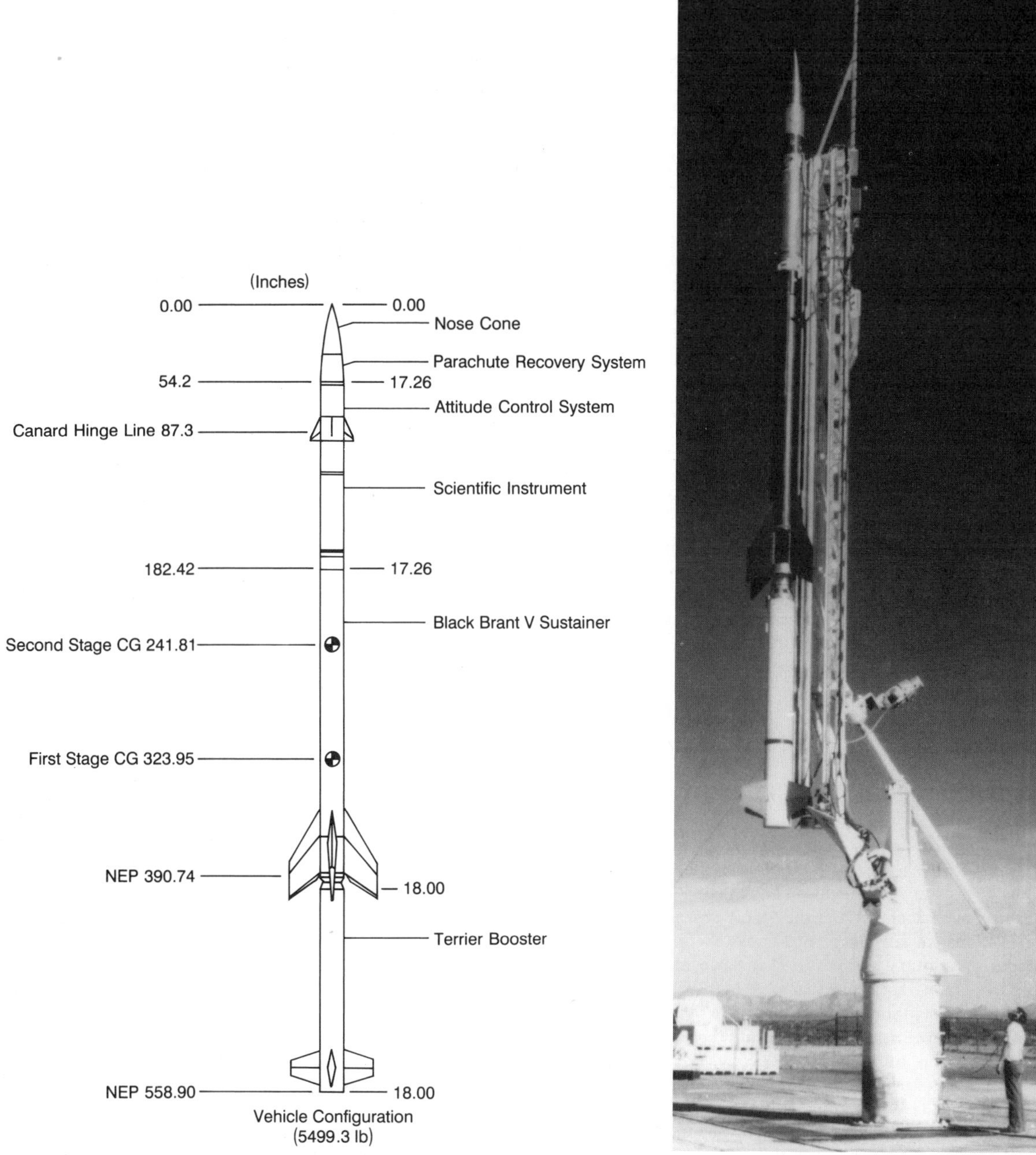

Figure 1. Left is a diagram and right is a photograph (*courtesy U.S. Army*) of a *Terrier*-boosted *Black Brant* rocket. This is typical of sounding rockets presently used for space astronomy. This particular vehicle, provided and supported by NASA Wallops Flight Facility, carried a UV camera and spectrograph payload developed by the Naval Research Laboratory for observations of comet Halley in February 1986. Dimensions shown on the diagram are length from nose tip (left) and diameter (right).

accurate determinations of the very high temperatures associated with solar flares, solar active regions, and the solar corona.

V-2 rockets also were used for the first detection of x-rays from the Sun, using photographic films behind various metal filters. Sounding rockets launched from a ship during a solar eclipse in 1958 showed for the first time that much of the solar x-ray emission was associated with solar active regions. Later flights using *Aerobee* rockets provided the first x-ray images of the Sun (in 1960, using a pinhole camera) and the first x-ray spectra (using Bragg crystal spectrometers).

Over the years, improvements in instrumentation along with concurrent improvements in rocket attitude control systems and increased rocket performance, resulted in steady improvements in the spectral range, spectral resolution, and photometric quality of the UV and x-ray solar measurements. In the 1960s, long-duration satellites began to replace sounding rockets for such programs as long-term monitoring of the solar UV and x-ray intensities, but sounding rockets were still used for the initial tests of new instruments, such as high-resolution spectrographs and coronagraphs later used in orbital missions. They were also used for special studies, such as of solar flares or solar eclipses. Sounding rocket flights of well-calibrated instruments were used to calibrate similar instruments on board orbiting spacecraft (such as *Skylab* or the *Solar Maximum Mission* satellite). Sounding rockets were used to obtain the first x-ray images of the Sun using focusing x-ray optics (by American Science and Engineering, Inc. in 1970) and, more recently (in 1987) using normal-incidence multilayer-interference-coated optics (groups at Stanford University, Lawrence Berkeley Laboratory, and Lockheed Missiles and Space Company).

Planetary and Cometary Rocket Astronomy

Sounding rockets were used for the first UV spectroscopic measurements of the planets, particularly Venus and Jupiter, and have also been a primary tool for cometary UV studies. The objectives of such measurements include determination of the compositions and physical processes in the outer atmospheres of the planets, and of the production rates of gases in the comae of comets. In particular, the hydrogen Lyman α line (at 1216 Å) provides a very sensitive means for detecting and measuring the concentration of atomic hydrogen. Likewise, the resonance lines of atomic oxygen (1304 Å), carbon, and other common light elements, and simple molecules such as H_2, N_2, and CO, can be used for sensitive measurements of these species.

UV cameras and spectrographs have been flown on sounding rockets for observations of the recent bright comets Kohoutek (1973), West (1975), and Halley (1986). A particular advantage of rockets for such observations is that experiments can be prepared and flown with relatively short (a few months) notice. The apparitions of most bright comets cannot be predicted in advance, and hence satellite observations are usually restricted to the use of instruments already in orbit, which were designed primarily for other purposes.

Celestial Rocket Astronomy

The first UV measurements of stars other than the Sun were obtained from *Aerobee* rockets beginning in the late 1950s and early 1960s. Groups involved included NASA Goddard Space Flight Center, Princeton University, and the Naval Research Laboratory. The investigations initially involved simple photometers, looking out the sides of spinning rockets to scan large areas of the sky. Later, objective-grating spectrometers were used in a similar manner to obtain low-resolution stellar UV spectra. With the advent of rocket attitude control systems, it became possible to obtain higher-resolution and more-sensitive spectral measurements by pointing objective grating spectrographs at preselected regions of the sky. The first stellar UV spectra with high enough resolution to reveal absorption and emission lines were obtained in the mid-1960s. With further improvements in the pointing systems, including the use of startrackers for fine pointing, it became possible to use larger telescopes with focal-plane slit spectrographs for detailed studies of single stars (as is typically done in ground-based observatories).

The *Orbiting Astronomical Observatories* (launched in 1968 and 1972), the *International Ultraviolet Explorer* (*IUE*) satellite (launched in 1978), and other astronomical satellites supplanted sounding rockets for most survey work and for monitoring of individual objects over long time periods. However, rockets were still useful for initial tests of advanced instrumentation, such as wide-field electronic imaging cameras and imaging spectrographs, high-resolution spectrographs, and measurements in the low-wavelength end of the far-UV (i.e., below 1200 Å) where most of the satellite instruments were insensitive.

As an example, Princeton University has developed an Interstellar Medium Absorption Profile Spectrometer (IMAPS) that covers the wavelength rage 950–1150 Å with a spectral resolution $\lambda/\Delta\lambda = 2\times10^5$. This is the highest-resolution spectrograph ever flown for stellar UV measurements, and in addition covers an important wavelength range to which *IUE* and the *Hubble Space Telescope* are insensitive. Figure 2 shows results from a 1985 rocket flight in which the star π Scorpii was observed.

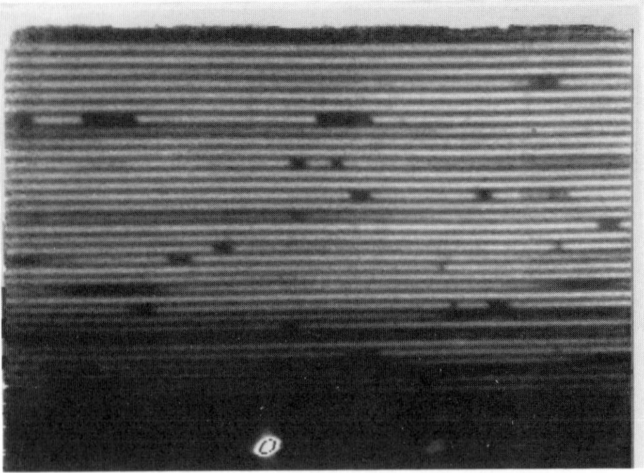

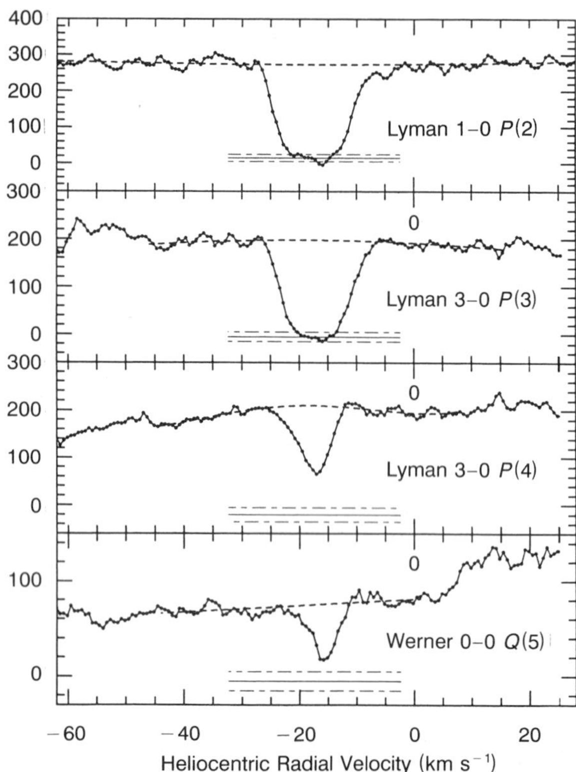

Figure 2. Top, a portion of the 1003–1172 Å far-UV spectrum of the star π Scorpii, obtained in an April, 1985 sounding rocket flight of Princeton University's Interstellar Medium Absorption Profile Spectrograph. This spectrum has a resolution of better than 0.01 Å. Several orders of the echelle-mode spectrogram are shown, and dark interstellar absorption features are apparent. At bottom is shown intensity versus radial velocity plots for some interstellar absorption lines of molecular hydrogen. (Radial velocity is related to wavelength by the Doppler relation $\Delta\lambda/\lambda = \Delta v/c$; hence $\Delta v = 10$ km s^{-1} corresponds to $\Delta\lambda = 0.033$ Å.) (*Courtesy of Edward B. Jenkins, Princeton University Observatory.*)

Sounding rockets also were used for the first detection of x-rays from sources outside the solar system. The measurements were made by groups at American Science and Engineering, Inc., the Naval Research Laboratory, and the Massachusetts Institute of Technology beginning in the early 1960s. The first nonsolar x-ray source detected, Scorpius X-1, at first could not be identified with a known visible star. Only later did ground-based astronomers identify the visible counterpart of this x-ray source. Another strong x-ray source was identified as the Crab nebula, the remnant of a supernova explosion that occurred in the year 1054. A sounding rocket flight that observed the Crab as it was being eclipsed by the moon showed that much of the x-ray emission was produced in the nebula, rather than in a central star-like source. Later rocket flights also revealed pulsed x-ray emission from the central neutron star (pulsar) that corresponded to previously observed pulsed radio and visible radiation. Rockets also produced crude maps of the x-ray sky, revealing a large number of additional point sources and measurements of the diffuse x-ray background, prior to the launches of the first x-ray astronomy satellites.

Infrared space astronomy began with rocket observations by groups at the Naval Research Laboratory, Cornell University, and what is now called the Air Force Geophysics Laboratory in the mid-1960s. The advantages of rockets (and satellites) for infrared measurements stem not only from being outside the infrared-absorbing (and emitting) atmosphere, but also from the ability to cool (in a vacuum environment) the telescope and detectors to cryogenic temperatures (such as that of liquid helium, 4 K). This greatly improves the sensitivity and photometric accuracy of infrared measurements, especially in the case of diffuse sources.

The rocket observations surveyed large areas of the sky to map both point and diffuse infrared sources, and provided the first space-based far-infrared observations of objects outside the solar system. The *Infrared Astronomical Satellite* (*IRAS*) was launched in 1983. *IRAS*'s 11-month operational period provided much more sensitive and complete mapping of the infrared sky than was possible with the previous sounding rocket investigations. However, we have been (since November 1983) without an orbital infrared astronomy facility. Hence, special-purpose infrared rocket observations may still be useful.

CURRENT STATUS AND FUTURE PROSPECTS OF ROCKET ASTRONOMY

Sounding rockets are still useful in several areas of space astronomy, despite the availability of the space shuttle and unmanned orbital launch vehicles. They are especially adaptable to one-of-a-kind, short-notice observations, such as of a new comet or a supernova, because payloads can be prepared and flown more quickly and at much lower cost than payloads for orbital missions. Also, rocket flights provide opportunities for testing new instrumentation concepts in the space environment, thereby reducing the risk involved in their later use in more costly long-duration space missions.

However, for more routine types of observations, such as sky surveys and other investigations requiring observations of a large number of objects (or long observations of one or a few objects), sounding rockets have diminished in usefulness in recent years. This is not due so much to increased availability of orbital flight opportunities, as to the fact that in most fields of space astronomy, all of the "easy" measurements have been done; that is, in most cases, the observing time available in a rocket flight is no longer sufficient for obtaining significant new scientific data. Exceptions include the IMAPS investigation previously mentioned, which explores a combination of resolution and wavelength range not covered by any existing or currently planned orbital mission, and recent rocket observations of bright comets and of Supernova 1987A.

For the foreseeable future, it will remain necessary to maintain a viable rocket astronomy program, to provide the capability for short-notice observations, for special-purpose observations requiring instrumentation not available on satellite payloads, and for testing new instrumentation concepts in the space environment prior to use in longer-duration missions.

Additional Reading

There are no comprehensive reviews of sounding rocket astronomy specifically, but most reviews of space astronomy mention results obtained with sounding rockets. The following are recent introductory to intermediate level books. The history of sounding rockets is reported by Frank H. Winter.

Cornell, J. and Gorenstein, P., eds. (1983). *Astronomy from Space: Sputnik to Space Telescope*. M.I.T. Press, Cambridge.

Field, G. B. and Chaisson, E. J. (1985). *The Invisible Universe*. Birkhäuser, Boston.

French, B. M. and Maran, S. P., eds. (1981). *A Meeting with the Universe*. NASA EP-177, U.S. Government Printing Office, Washington, D.C.

Friedman, H. (1990). *The Astronomer's Universe*. W. W. Norton and Co., New York.

Hanle, P. A. and Chamberlain, V. D., eds. (1981). *Space Science Comes of Age*. Smithsonian Institution Press, Washington, D.C.

Henbest, N. and Marten, M. (1983). *The New Astronomy*. Cambridge University Press, London.

Lawton, A. T. (1979). *A Window in the Sky*. Pergamon Press, New York.

Winter, F. H. (1990). *Rockets into Space*. Harvard University Press, Cambridge.

See also **Infrared Astronomy, Space Missions; Radio Astronomy, Space Missions; Solar Physics, Space Missions; Ultraviolet Astronomy, Space Missions; X-ray Astronomy, Space Missions.**

Spectrographs, Astronomical

Alex W. Rodgers

HISTORY

The application of spectroscopy to the study of the light of celestial objects marked the beginning of the subject of astrophysics. This occurred in the late nineteenth century.

Radiation from celestial objects can be parameterized as depending on frequency (or wavelength), polarization, the spatial coordinates of the objects, and their variations with time. Radiation detectors in optical astronomy, ranging from the eye through the photographic plate to the more recent panoramic electronic detectors, record the intensity of stellar radiation across two spatial coordinates in the plane of the detector. Thus most astronomical observations using these detectors must select a limited range of the total number of parameters describing the radiation.

In the most general case one would wish that it were possible to register on a detector the total angular field of a telescope with an arbitrary number of monochromatic images indicating their spatial structure, as well as the variation of intensity in each of the sources with wavelength. However, the normal form of the astronomical spectrograph that has developed over the last 100 years attempts to image as much as possible the total spatial distribution of stars in one coordinate of the telescope focal plane with the maximum of wavelength resolution.

REQUIREMENTS FOR A STANDARD ASTRONOMICAL SPECTROGRAPH

The standard astronomical spectrograph simultaneously records the variation of intensity from a star over a range of wavelengths, used as one spatial dimension of the detector. This is achieved by using a spectrograph slit that lies in the focal plane of the telescope, followed by an optical system containing a dispersing element that spatially separates adjacent images of the slit corresponding to different wavelengths of the light received. The design of such an optical system then becomes similar to that of a standard reimaging system, the *redacteur focale*, in which the focal plane of the telescope is reimaged through an optical system onto a camera, which generally has a smaller focal ratio than is used in the main telescope optics. Because the optical elements that produce the dispersion of the light in the spectrograph must operate in a collimated beam, the first optical element of the standard spectrograph is a collimator, either lens or mirror, in which the light of different wavelengths is dispersed by different angular amounts and then reimaged by the system camera. In Fig. 1*a* we show the way in which a focal plane of a telescope is transferred to the detector in a typical redacteur focale. In Fig. 1*b* we show that a dispersing element in the collimated beam produces a specific angular dispersion that is recorded by the camera as a displacement in the image of the telescope focal plane depending on the wavelength of the light. Figure 1*b* shows the essential configuration of most stellar spectrographs. For the last 50 years, the optical element of the spectrograph collimator, the dispersing element, and the camera have usually been made of reflecting components. In this way chromatic effects within the spectrograph are minimized and allow major simplifications in the optical design leading to wide wavelength coverage. Let the focal ratio of the telescope focus be F; let the focal ratio of the camera be f. If the linear resolution of the detector is d, the projected slit width corresponding to the detector resolution in the focal plane of the telescope is Fd/f. If the entrance slit width Fd/f is smaller than the typical size of the seeing disk of the star in the focal plane of the telescope, light will be lost at the edges of the spectrograph slit. Therefore, a fundamental design requirement of spectrographs is to arrange the ratio F/f such that most of the starlight in the focal plane of the telescope is transmitted through the entrance slit. It is then important, in order to achieve the required wavelength resolution for the spectrograph, to have high angular dispersion in the dispersive element in the collimated beam of the spectrograph. Over the last 50 years, high angular dispersion coupled with high transmissivity has been achieved through the use of reflective diffraction gratings and echelles. The designer is offered a wide choice of resolutions that can be incorporated in a single spectroscopic instrument.

A major advance in spectrograph practice was the adoption of the Schmidt camera for spectrograph cameras. This was pioneered by Theodore Dunham at Mount Wilson Observatory in the 1930s. A Schmidt camera permits a wide angular field together with a large spectral coverage and high spectral resolution. Additionally, the Schmidt camera could be designed and made to operate at low focal ratios f, thereby allowing entrance slits for the spectrograph that approach the size of the typical seeing disk.

Another technique for astronomical spectroscopy that has been extraordinarily influential in studies of stellar astrophysics and galactic structure in the first 50 years, arose from the use of dispersing elements (prisms) placed in front of the objectives of wide field photographic telescopes. Each stellar image in the field gave rise, on the photographic emulsion, to a one-dimensional spectrum. By this means catalogs, such as the famous Henry Draper survey, were assembled and they gave the possibility of spectral classification of all the stars in the sky brighter than 10th magnitude. The price to be paid for having multiplexed spectral and spatial information is that, while the stellar light is dispersed, light of all wavelengths produces background sky fog on the photographic plate.

Even with this limitation, wide field spectroscopic surveys made with objective prisms, or combinations of prisms and transmission

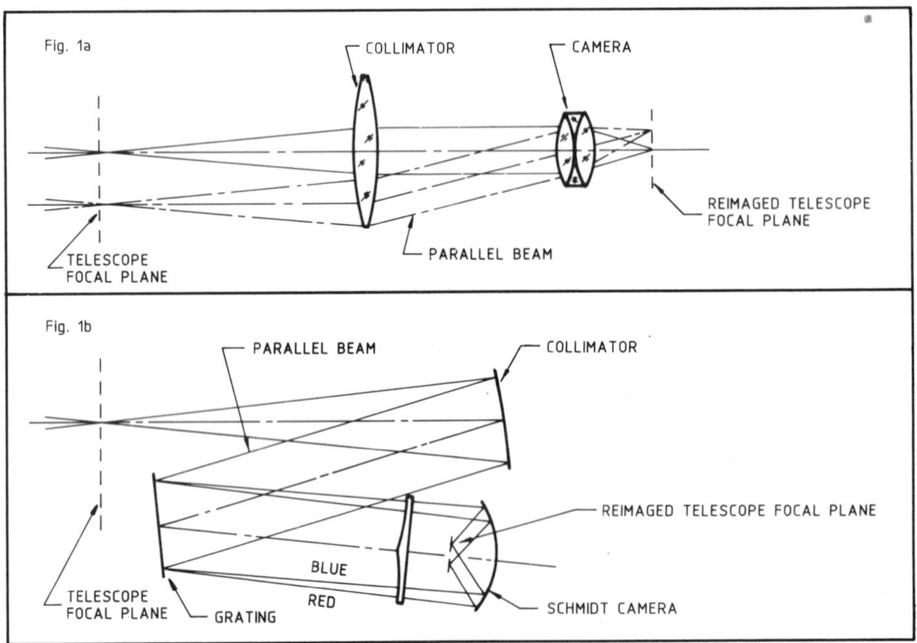

Figure 1. (*a*) A schematic arrangement whereby the focal plane of the telescope is reimaged through a collimator/camera system at the focal plane of the camera. In the event that the focal ratio of the collimator (hence the telescope) is F and the camera focal ratio is f, the linear displacement at the camera focal plane d corresponds to a displacement of Fxd/f in the focal plane of the telescope. (*b*) The schematic layout of a conventional astronomical spectrograph that introduces a dispersing element (in this case a reflection grating) into the collimated beam of the system shown in (*a*). The reimaged telescope focal plane is then displaced along one dimension of the detector for different wavelengths of light passing through the entrance aperture of the telescope focal plane. In (*b*) the more commonly used reflective optics of astronomical spectrographs are illustrated.

diffraction gratings, have provided an enormous data base that has determined much of the course of stellar astronomy in this century.

CLASSICAL SPECTROGRAPH DESIGN

The designer of the spectrograph must achieve goals of high wavelength resolution, as much spatial information as is available, and ensure that the spectrograph has sufficient versatility in wavelength coverage and ease of operation as is possible. The precept of high throughput leads to the aphorism that in astronomy, and in astronomical spectrographs in particular, "the least optics is the best optics."

Design considerations for optimization of the throughput of astronomical spectrographs were summarized by Ira S. Bowen when he described the spectroscopic equipment being built for the Hale 5-m reflector. This discussion covered a large range of conditions that are encountered in classical spectrograph design. The single most important parameter was the relative size of the spectrograph collimator beam compared to the telescope aperture. The spectrograph input slit width is determined by the resolution of the detector (i.e., pixel size) and the ratio of the telescope and spectrograph camera focal lengths. In the case of high resolution spectroscopy, Bowen showed that

$$\text{speed} \propto rd^2\Delta^3/as,$$

where r is pixel size, d is spectrograph beam diameter, Δ is linear dispersion of the spectrum, a is angular dispersion of the prism or grating, and s is the size of the stellar seeing disk.

In the case where the slit width is equal to or larger than the seeing disk,

$$\text{speed} \propto \Delta D^2,$$

where D is the telescope aperture diameter.

Bowen commented on the disconcerting result that the spectrograph efficiency was independent of telescope size when, in the high resolution case, the seeing disk was larger than the entrance slit width of the spectrograph. The aim of the designer must be to ensure Bowen's second equation is the domain in which the spectrograph operates. A method of doing this is to increase the angular dispersion of the grating or the beam size of the collimator so that the conditions of Bowen's second equation more nearly hold. It is these precepts that will have to be considered most carefully in the design of efficient spectrographs for the large reflectors ($D > 8$ m) that are projected for construction.

MULTIOBJECT SPECTROSCOPY

In recent years, the technology of image transfer from the focal plane of a telescope to the input slit of a spectrograph, through the use of multiple fiber optic conduits, has been developed at the Anglo–Australian Telescope and is now in use and under construction at many telescopes. The input end of the individual fiber defines the entrance slit width and the focal ratio of the exit beam is set by the fiber optical properties. Generally, because of reflections at the fiber ends and because of fiber losses, some 40% of the light is lost from each image. Nevertheless, typical installations with approximately 60–100 fibers allow the simultaneous acquisition of data on up to 100 stars in the focal plane of the telescope. To date, many of the spectrographs that are fed by fibers have not been optimized in terms of collimator focal length or off-axis performance. It is evident now that such optimization can be achieved and, additionally, several spectrographs can be used simultaneously, each fed by a bundle of fibers illuminated by different parts of the telescope field.

By the placement of aperture plates in the focal plane of the telescope with the apertures corresponding to the positions of the desired objects, spectra of many stars can be obtained without the losses associated with the use of fibers feeding the spectrograph. The advantage of multiaperture spectroscopy over fiber spectroscopy is only realized in the faintest objects where, if detectors with a finite readout noise are used, it is crucial to maximize the flux through the spectrograph/detector system.

MULTIPLE BEAM SPECTROGRAPHS

A crucial advance in the techniques of increasing the efficiency of spectrographs was made by J. Beverly Oke, who, in the construction of a multichannel spectrophotometer and again in a double beam spectrograph attached to the Hale Reflector, divided the spectrograph beam into red and blue wavelength regions by means of a dichroic reflective mirror. In this way he was able to effectively build a dual spectrograph fed by light passing through a common entrance slit. Large advantages in spectrograph efficiency follow from the use of this technique. The dichroic mirrors have a transition wavelength at around 550 nm. The blue side of the spectrograph can then use reflective coatings, gratings, and detectors that are optimized for efficiency in that wavelength region. Similarly, high reflectance coatings for the red and near infrared region, such as silver, and high efficiency detectors can be used separately in the red arm of the spectrograph. There is the additional gain that the spectral coverage of the dual spectrograph exceeds the possibilities available in conventional spectrograph design.

When spectrographs are designed for the large reflectors currently envisaged, the challenges to the designer to maintain the high efficiency of which many current spectroscopic instruments are capable, will require application of all the techniques that have been described and continued ingenuity in optical and mechanical practice.

Additional Reading

Bowen, I. (1952). The spectroscopic equipment of the Hale Telescope. *Ap. J.* **116** 1.

Dunham, T. (1956). Methods in stellar spectroscopy. *Vistas in Astronomy* **2** 1223.

Oke, J. B. and Gunn, J. B. (1982). An efficient low and moderate-resolution spectrograph for the Hale Telescope. *Publ. Astron. Soc. Pacific* **94** 586.

Schmidt, G. D., Weymann, R. J., and Foltz, C. B. (1989). A moderate-resolution, high-throughput CCD channel for the Multiple Mirror Telescope. *Publ. Astron. Soc. Pacific* **101** 713.

See also **Telescopes, Large Optical; Telescopes, Wide Field.**

Star Catalogs and Surveys

Wayne H. Warren Jr.

Astronomy is an observational science dealing, to a great extent, with the study of individual objects. The objects may be, for example, stars, emission and reflection nebulae, novae and supernovae, galaxies (composed of all of the previously mentioned), or clusters of galaxies. Because objects must be identified, designated for future reference, and located by position in the sky, and because the characteristics of objects must be preserved for future work, it is necessary to catalog them. Thus an astronomical catalog can be defined in general terms as a collection of data compiled from observations. This definition is opposed to that, for example, in computer science and other disciplines, where a catalog is generally thought of as a list of items such as data sets on a computer, or in the commercial market where a catalog is a book that lists merchandise. Even within astronomy there are various types of catalogs with important differences that should be clearly understood by their users. The different kinds of astronomical catalogs are defined and described in an article by Carlos Jaschek.

The earliest catalogs, which are now mainly of historical interest only were compilations of bright (naked eye) stars visible to the ancients. These "lists" of stars contained, for each object, a designation (star name), a position in some coordinate system, and an estimated brightness. This basic principle still holds today, although modern catalogs contain a much wider variety of information for many different kinds of astronomical objects.

Before proceeding with a more detailed discussion of modern star catalogs, it is necessary to distinguish among the various types of catalogs and to break them down into individual categories based upon the kinds of data they contain. Following Jaschek, we define the basic types of catalogs:

1. *Observational:* Lists of observations by a single author, or specific data gathered as a result of a particular project.
2. *Bibliographical compilation:* Lists of observations of a particular type, compiled from many different sources. The purpose of the compilation catalog is to gather all observations into one source where it is convenient for users to locate the information. It is essential, however, that the original source for each observation be referenced so that users can refer back to it if necessary.
3. *Critical compilation:* Lists of observations of a particular type of data as before, but made by an expert in the field who evaluates the quality of the data from the various sources. The quality of each observation may be indicated by assigning weights or quality codes. The evaluation usually results in the reporting of a single "best" datum for each object. This has the advantage for users of having the opinion of an expert as to which of a variety of measurements is best, but it has the disadvantage of not including all observations available; thus the user must be entirely confident in the opinion of the compiler.
4. *General compilation:* A collection of a wide variety of observational data for a list of objects, usually of a particular type. Original sources may or may not be given and the compiler has usually had to choose among multiple observations of each type.

All of the preceding types of catalogs are represented in the discussion of star catalogs to follow. A distinction must also be made between primary and secondary data, at least in the first three types of catalog. Each catalog is usually assembled for the purpose of reporting specific types of observations. In order to make a catalog that will be most useful to a wide audience of astronomers, however, secondary data may be included. For example, in a catalog of star positions, the equatorial positions and their errors are the only primary data, but the catalog may contain cross identifications to various designations, magnitudes, and spectral types as secondary data. Another catalog of spectral types may contain secondary positional data so that the objects can be located in the sky. Although it was useful to include secondary data in printed catalogs that were used without the aid of computers, it is less important to do this in the case of machine-readable catalogs that can be merged in a straightforward way by computer. However, in the latter case, it is extremely important that objects be designated unambiguously, completely, and uniformly enough to allow the cross identification of each object by the computer.

MODERN STAR CATALOGS

Modern star catalogs include a wide variety of information and can be divided among several categories depending upon the type of primary data they contain. Brief mention of a few of the major catalogs in each category will be given to alert the reader as to the different kinds of catalogs available and to the primary data included. A discussion of surveys, which may or may not become available in catalog form, is deferred to a later section. Additional information about the catalogs and surveys that are mentioned is given in Table 1. The list is not meant to be complete, but provides a representative sample of important catalogs and surveys that are currently available.

Astrometric and Positional Data

Astrometry (the study of star positions and motions) has naturally produced a large number of catalogs because of the need to remeasure accurate positions at regular intervals. This is necessary because of inaccuracies in stellar proper motions and in the motion of the reference frame. Astrometric catalogs can be divided into basic positional (usually including proper motions) and fundamental catalogs, the latter being assembled for purposes of defining a reference frame. Also included in this category are catalogs of stellar proper motions and parallaxes, because the determination of these quantities involves the accurate measurement of positions.

Examples of positional catalogs are those of the *Astronomische Gesellschaft* (*AG*), which have a long history dating back to the nineteenth century. The most recent catalog of this series is the *AGK3*, a catalog of positions and proper motions for 183,145 stars, as measured on photographic plates. A reference catalog of higher-accuracy positions for a fewer number of *AGK3* stars is known as the *AGK3R*. Another example is the *Smithsonian Astrophysical Observatory Star Catalog*, a compilation of positions and proper motions for 258,997 stars (there are some duplicate entries) assembled from a wide variety of earlier catalogs that were reduced to a common system. The *SAO* was assembled for satellite-tracking purposes and contains a uniform number of stars per sky area; thus there are cases where stars brighter than its nominal faint magnitude limit of $9^{\mathrm{m}}.0$ were not included. A similar positional catalog has recently been completed at the Astronomisches Rechen-Institut (ARI) in Heidelberg and is based on multiple positional epochs rather than on the usual two as in the *SAO* catalog. The largest by far of the positional catalogs is known as the *Astrographic Catalogue*, a group of catalogs for ranges of declination zones, resulting from an international collaborative survey project known as the *Carte du Ciel*. Unfortunately, not all of the plates were reduced and the plates for certain zones have now been lost or destroyed, so, even though there is an effort at present to rereduce all of the astrographic zones to a common system, many of the original plates cannot be remeasured using modern techniques.

There are also several regional astrometric catalogs covering small areas of the sky that are of special interest, an example being the Pleiades catalog of Heinrich K. Eichhorn and collaborators.

Eventually emerging from the large collection of positional and proper-motion data was a series of German fundamental catalogs, including the *Fundamentalkatalog* (1879, 1883), the *Neuer Fundamentalkatalog* (1907), the *FK3* (1937, 1938), and the *FK4* and its supplement (1963). The successor to the *FK4*, the *FK5*, has now just been published by the ARI. The basic catalog contains the same number of stars as the *FK4*, but a supplement is planned that will include several thousand additional stars as faint as magnitude 8.

Before leaving fundamental catalogs, it is appropriate to briefly mention the American series, which began with Simon Newcomb's (1899) compilation of 1257 stars and continued with the *General Catalogue of 33,342 Stars for the Epoch 1950* (*GC*) and the *N30* catalog.

As mentioned earlier, catalogs of stellar proper motions and parallaxes, even though they may sometimes not contain primary positional data, are also included with positional catalogs. Examples of proper-motion catalogs include a series by Willem J. Luyten and his collaborators that resulted from a survey for stars exhibiting large proper motions (NLTT) and those of the Lowell Observa-

Table 1. List of Representative Catalogs and Surveys

Catalog	Description	Reference(s)
Astrometric and Positional Data		
AGK3	Positions and proper motions 183,145 stars	Dieckvoss (1975). Hamburg-Bergedorf
AGK3R	Positions and proper motions 20,194 *reference stars*	Corbin (1978). *IAU Colloquium 48, Modern Astrometry*, p. 505
FK5 Basic	Positions and proper motions 1535 stars	Fricke, Schwan, and Lederle (1988). *Veröff. ARI Heidelberg* (No. 32)
GC	Positions and proper motions 33,342 stars	Boss (1937). Carnegie Inst. Wash. Publ. 468
Lowell survey	Proper motions 8989 (N), 2758 (S) stars	Giclas, Burnham, and Thomas (1971, 1978). Lowell Observatory
N30	Positions and proper motions 5268 stars	Morgan (1952). *Astron. Papers Amer. Ephemeris* **13** part III
NLTT	Proper motions 58,855 stars	Luyten (1979–80). University of Minnesota
Pleiades positional	Astrometric positions 502 stars	Eichhorn et al. (1970). *Mem. Roy. Astron. Soc.* **73** 125
PPM	Positions and proper motions 181,731 stars	Röser and Bastian (1988). *Astron. Ap. Suppl.* **74** 449
Yale parallax	Trigonometric parallaxes	Jenkins (1952, 1963). Yale University
Photometric Data		
HST Guide Star Photometric	BV sequences 1477 stars	Lasker et al. (1988). *Ap. J. Suppl.* **68** 1
IRAS Point Sources	About 246,000 infrared sources	Infrared Processing and Analysis Center (1986)
UBV Photoelectric	Colors and magnitudes on *UBV* system (about 87,000 stars)	Mermilliod (1987). *Astron. Ap. Suppl.* **71** 413
UBVRI Photoelectric	About 20,000 stars	Lanz (1986). *Astron. Ap. Suppl.* **65** 195
uvbyβ Photoelectric	About 40,000 stars	Hauck and Mermilliod (1985). *Astron. Ap. Suppl.* **60** 61
Spectroscopic Data		
MK spectral types	MK types and bibliography	Jaschek et al. (1964). *Publ. La Plata Obs.* **28** (No. 2) Morris-Kennedy (1978). Mount Stromlo Observatory
Michigan MK/HD stars	MK types for all HD stars	Houk et al. (1975–88). University of Michigan
MK selected	"Best" types for classified stars	Jaschek (1978). *Bull. Inf. CDS* (No. 15) 121
Radial velocities	Bibliography and supplement	Abt and Biggs (1972). Kitt Peak National Observatory; Barbier-Brossat and Petit (1987). Observatoire de Marseille
Radial velocities	Quality assessment	Wilson (1953). Carnegie Inst. Wash. Publ. 601; Evans (1967). IAU Symp. 30; Barbier-Brossat (1989). *Astron. Ap. Suppl.* **80** 67
Rotational velocities		Boyarchuk and Kopylov (1964). *Publ. Crimean Ap. Obs.* **31** 44; Bernacca and Perinotto (1970–73). *Asiago Ap. Obs.*; Uesugi and Fukuda (1982). Kyoto University
Cross Identifications		
HD, HDE, DM Open Clusters		Mermilliod (1986). *Astron. Ap. Suppl.* **33** 293
SAO-HD-GC-DM	All stars in *SAO* catalog	NASA Astronomical Data Center (1983)
WDS-DM-HD-ADS	Double and multiple stars	Roman (1987). NASA Astronomical Data Center
Combined and Derived Data		
Bright stars	9096 stars to magnitude 6.5	Hoffleit and Warren (1991). NASA Astronomical Data Center
	1628 stars to magnitude 5.0	Ochsenbein and Halbwachs (1987). CDS Strasbourg
Nearby stars	Stars closer than 22 pc	Gliese (1969). *Veröff. ARI Heidelberg* (No. 22)
	Stars within 25 pc of Sun	Woolley et al. (1970). *Roy. Obs. Ann.* **5**
Spectroscopic binaries	Spectroscopic orbits	Batten et al. (1989). *Dominion Ap. Obs.* **15** 121
Visual binaries	928 orbits for 847 systems	Worley and Heintz (1983). *Publ. U.S. Nav. Obs.* 2nd Ser. **24** Part VII
Other Catalogs		
Spectroscopic bibliography	Miscellaneous information	Parsons et al. (1980). *Bull. Inf. CDS* (No. 18) 46
Variable stars	Bibliography	Huth and Wenzel (1986). Zentralinstitut für Astrophysik
RR Lyrae variables	Bibliography	Heck (1988). *Bull. Inf. CDS* (No. 34) 137

Table 1. *(continued)*

Catalog	*Description*	*Reference(s)*
Other Catalogs		
Visual binaries	*General catalog*	*Worley and Douglass* (1984). *U.S. Naval Observatory*
	Finding list	*Wood et al.* (1980). *University of Florida*
	Interferometric measurements	*McAlister and Hartkopf* (1988). *Georgia State University*
Am and Ap stars		*Bertaud and Floquet* (1974). Astron. Ap. Suppl. **16** 71; *Bidelman and MacConnell* (1973). Astron. J. **78** 687
Am stars	*Known spectral types*	*Hauck* (1986). Astron. Ap. Suppl. **64** 121
Carbon stars		*Stephenson* (1989). Publ. Warner & Swasey Obs. **3** (*No.* 2)
S-type stars		*Stephenson* (1976). Publ. Warner & Swasey Obs. **2** (*No.* 2)
White dwarfs		*Luyten* (1970). *University of Minnesota*; *McCook and Sion* (1987). Ap. J. Suppl **65** 603
B emission stars		*Page* (1984). *Mt. Tamborine Observatory*
O stars		*Cruz-González et al.* (1974). Rev. Mex. Astron. Astrof. **1** 211; *Garmany et al.* (1982). Ap. J. **263** 777; *Goy* (1980). Astron. Ap. Suppl. **42** 91
Wolf–Rayet stars		*van der Hucht et al.* (1981). Space Sci. Rev. **28** 227
Subdwarfs		*Kilkenny et al.* (1988). SAAO Circ. (*No.* 12)
Surveys		
Durchmusterungen	*Bonner, Southern*	*Argelander* (1859–62). *Schönfeld* (1886). *Bonn Observatory*
	Córdoba	*Thome* (1892–1932). *Córdoba Observatory*
	Cape Photographic	*Gill and Kapteyn* (1895–1900). Ann. Cape Obs. **3–5**
Feige	*Faint blue stars*	*Feige* (1958). Ap. J. **128** 267
Palomar–Green	*Faint blue stars*	*Green et al.* (1986). Ap. J. Suppl. **61** 305
Downes	*Hot white dwarfs and subdwarfs*	*Downes* (1986). Ap. J. Suppl. **61** 569
Steward	*Near infrared*	*Craine et al.* (1979). *Steward Observatory*
Caltech	*Infrared*	*Neugebauer and Leighton* (1969). *NASA SP*-3047
AFGL	*Infrared*	*Price and Murdock* (1983). *Air Force Geophysics Laboratory*
Henry Draper	*Spectroscopic*	*Cannon and Pickering* (1918–36). Ann. Astron. Obs. Harvard College **91–100**
OB stars		*Stephenson and Sanduleak* (1971). Publ. Warner & Swasey Obs. **1** (*No.* 1); *Hardorp et al.* (1959–65). *Hamburg-Bergedorf*

tory. The major catalogs of parallaxes are compilations of trigonometrically determined parallaxes (the only highly accurate method at present) and include the Yale parallax catalogs. A successor to the earlier catalogs is presently nearing completion by William F. van Altena (also at Yale University) and will be a greatly expanded compilation.

Photometric Data

The primary data of photometric catalogs are the magnitudes and colors of astronomical objects. Any other data included in a photometric catalog are of a secondary (or ancillary) nature unless the catalog contains another kind of original observations. This is generally not the case because all of the major photometric catalogs are compilations of observations culled from the literature. Such compilations are made by specialists in a particular discipline and are the only practicable way of producing a comprehensive catalog, because one observer could never hope to create such a large catalog from his or her own observations.

Because there are many different photometric systems, there is a rather large collection of photometric catalogs; in fact, in the archives of machine-readable catalogs held by the astronomical data centers, the photometric category contains the largest number of active catalogs (presently 122). Also included with photometric catalogs, because they contain related primary data, are compilations of polarization data and of variable stars (whose variability classification is based on the existence of varying brightness).

Among photometric compilations, we can only mention a few catalogs based on the most widely used photometric systems. These include the *UBV* compilations of Jean-Claude Mermilliod, the *UBVRI* of Thierry Lanz, and the *uvbyβ* of Bernard Hauck and Mermilliod. In a slightly different category are the *IRAS Catalog of Point Sources*, based on the *Infrared Astronomical Satellite*'s survey of infrared sources, and the *Hubble Space Telescope Guide Star Photometric Catalog*, prepared expressly for calibrating the photographic magnitudes and colors of the much larger *Guide Star Catalog*. Somewhat different again is the *Catalog of Infrared Observations*, which is a compilation and bibliographical source reference for all types of observations made in the infrared.

For the compilation catalogs, one can also distinguish between compilations of actual observations (including reference sources) and catalogs of the best available values based on some statistical method of accurately combining all of the observations of each object. The latter data are what the user actually wants, but the weighting procedure is complex and many users do not have confidence in mean values computed by someone else. (That is why it is of extreme importance to make the individual data and their sources available.) Catalogs of weighted (so-called "homogeneous") means are often produced with the observations compilation, but in the case of the *UBV* system, a separate catalog was published.

Spectroscopic Data

Catalogs of spectroscopic data contain collections of spectral types or primary data determined by spectral analysis, such as radial and rotational velocities, equivalent widths, and listings of emission stars. The first large compilation of spectral types on the Morgan–Keenan (MK) system appeared in 1964 and has been supplemented in several editions by Pamela Morris-Kennedy. The work has been extended by William Buscombe, but Buscombe's catalog does not contain source references for the individual spectral types and is, therefore, much less useful than the others. The largest uniform collection of MK spectral types is being assembled by Nancy Houk of the University of Michigan, who is reclassifying, on objective-prism plates, all stars contained in *The Henry Draper Catalogue*. Four volumes of this catalog have been published to date. The Michigan work has the advantage of being a very large collection of homogeneous MK types, because the spectra have been essentially classified by one person, but the objective-prism spectra are of lower dispersion than that normally used to obtain slit spectra, and thus the spectral types are of lower accuracy. A critical catalog of MK types was compiled by Mercedes Jaschek, who reviewed all available types in the Jaschek compilation and an earlier version of the Morris–Kennedy catalog and selected a single type that she thought was best, based on classifier, dispersion, and other factors. A newer version of this catalog is desirable, because it is very difficult to choose among the multitude of MK types often available for many stars. An index-card catalog of such "selected" types is maintained by Robert F. Garrison, who is in the process of computerizing the data.

Catalogs of radial velocities include the bibliographical type, which contain individual measurements and their references, and the critical evaluation, where observations are examined and assigned quality codes. Compilations of stellar rotational velocities and atmospheric abundances are also associated with spectroscopic catalogs.

Cross Identifications

Compilations of cross identifications for astronomical objects are much more important than they might at first seem. When an astronomer is working on a particular collection of objects and wants to gather existing information about them, he or she must consult a large number of catalogs with differing object identifications. It is not a simple matter to identify the same object in many different catalogs, especially for binary stars and objects that are close together in crowded fields. For this reason, and to enable the construction of combined catalogs and databases by computer, it is important to have a variety of cross-index catalogs available. As a precursor to SIMBAD (Set of Identifications, Measurements and Bibliography for Astronomical Data), which is now the largest on-line astronomical database in the world, the *Catalog of Stellar Identifications* (CSI) was constructed almost 20 years ago. Although it took nearly a decade to build and refine the CSI, the construction of SIMBAD by the merging of computerized data would not have been possible without the cross index. Several other important cross-identification catalogs have been prepared to allow astronomers to easily identify entries for the same object in multiple catalogs.

Combined and Derived Data

As mentioned throughout, most catalogs are compiled for specific astronomical applications and consist of primary and secondary data. For reference purposes, it is most efficient to build collections of primary data in combined catalogs for particular classes of objects. These catalogs allow researchers to find a wide variety of information by consulting only a few sources rather than having to find the objects in dozens of individual catalogs. Examples of such collections include catalogs for nearby stars, high-velocity stars, bright stars, visual binaries, and spectroscopic binaries. There are many other collections of combined data that cannot be itemized here, but are available through the astronomical data centers.

Other Catalogs

Even the broad range of classifications mentioned previously cannot encompass all of the different types of star catalogs available. Only a few examples of other types of catalogs are cited in Table 1; interested readers should obtain a list of available catalogs from one of the data centers.

These catalogs provide a wealth of information useful to present and future astronomers. Fortunately, it was realized in 1970 that the great collections of data should be permanently archived and made available to astronomers, especially because many catalogs were, at the time, becoming available in machine-readable form. At its Brighton General Assembly in 1970, the International Astronomical Union took action that resulted in the creation of the Centre de Données Astronomiques de Strasbourg (CDS). Along with cooperating data centers in Potsdam, Moscow, Tokyo, the United States (at NASA/Goddard Space Flight Center), and several other locations, the CDS is active in acquiring data from scientists, creating new catalogs, and distributing them to astronomers throughout the world.

ASTRONOMICAL SURVEYS

A survey might be defined as an examination for purposes of ascertaining the quality or quantity of something. In astronomy this means, depending on the nature of the survey, looking at a large number of objects in an attempt to discover those meeting specific criteria. A survey frequently, but not always, results in a catalog of objects that meet the defined criteria. The important point here is that the selection criteria be well defined and that only objects meeting the specifications be selected for further study.

Stellar surveys are only applicable to the first three categories of catalogs discussed earlier, namely positional, photometric, and spectroscopic data. We discuss a few examples of surveys in each of these categories.

The surveys of positional data are the *Durchmusterungen* (*DM*), a German term first used by F. W. A. Argelander. The criterion used was very broad and consisted only of listing all stars visible to a certain limiting magnitude. DM surveys were done in both the northern and southern hemispheres and resulted in the publication of four large catalogs. The northern surveys were realized at the Bonn Observatory and resulted in the *Bonner Durchmusterung* (*BD*) and its "southern" extension, the *Southern Durchmusterung* (*SD*). The *SD* actually covers zones in the southern hemisphere ($-02°$ to $-23°$), but is considered a northern survey because it was done from a northern-hemisphere observatory as an extension to the *BD*. The southern-hemisphere *DM* surveys were done from Córdoba, Argentina (a visual survey, as are the northern *DM*) and from the Cape of Good Hope (a photographic survey), the resulting catalogs being known as the *Córdoba Durchmusterung* and the *Cape Photographic Durchmusterung*. The *DM* surveys were monumental projects resulting in catalogs of 325,037, 134,834, 613,959, and 454,877 stars, respectively. Computerized versions of all four catalogs have recently been completed by the data centers in the United States and France following a 15-year effort. The *Carte du Ciel* was another large photographic survey that resulted in the series of astrographic catalogs discussed previously.

Also placed under the first category are the proper-motion surveys, where the criterion is a lower limit on stellar proper motion. These surveys look exclusively for stars of large proper motion, an indication of nearness to the Sun. Examples of large proper-motion surveys are those of Luyten (*Proper Motion Survey with the Forty-Eight Inch Schmidt Telescope*) that resulted in, among others, the NLTT catalogs mentioned earlier, and the survey of the Lowell Observatory that resulted in the catalogs by Henry L. Giclas and collaborators. Proper-motion surveys were reviewed by Luyten.

Photometric surveys generally consist of finding objects meeting certain color criteria (blue, red, etc.) because objects of extreme color are almost always deserving of further study. (Although not discussed here, the quasistellar objects, or quasars, were discovered in this manner.) Among surveys for faint blue objects are ones by Luyten, resulting in 50 publications in the years 1953–1969; a small survey by J. Feige; the Palomar–Green survey; and a galactic plane survey by Ronald Downes designed to detect hot white dwarf and subdwarf stars. Surveys for red objects include the Dearborn Observatory survey (visual) for faint red stars, a *Steward Observatory Near Infrared Photographic Sky Survey*, and infrared surveys from Caltech, and the Air Force Geophysics Laboratory. The *Infrared Astronomical Satellite* (*IRAS*) carried out an infrared survey of the whole sky that resulted in the *IRAS Catalog of Point Sources* cited previously.

Spectroscopic surveys are similar to photometric surveys except that the criteria are based upon particular features displayed in stellar spectra. The earliest large survey of stellar spectra resulted in *The Henry Draper Catalogue*. Another early survey done in specific "selected area" fields at the Hamburg Observatory resulted in the *Bergedorfer Spektral-Durchmusterung* published in the 1930s in numerous volumes. These were general surveys wherein all stars to certain magnitude limits were classified, so they were more along the lines of the *DM* positional surveys. Spectroscopic surveys for particular types of objects are generally carried out with objective prisms because the spectra of many stars can be recorded on a single photographic plate. Numerous objective-prism surveys have been carried out for early-type (OB) stars and for interesting stars in particular regions of the sky. Surveys for late-type (red) stars have been reviewed by A. N. Vyssotsky.

FUTURE NEEDS

Catalogs and surveys will continue to be important tools for astronomical research, because astronomy deals with large numbers of objects that must be categorized in some orderly fashion to allow future work to be built upon past measurements and discoveries. As accumulated data become more accurate and new detectors and observations from space allow fainter objects to be observed and cataloged, astronomical data management must keep pace. In astronomy, perhaps more than in other physical sciences, future work depends on the utilization of older data; for example, in the determination of accurate parallaxes and proper motions. Therefore, data collected in the past, mainly in the form of catalogs, must be permanently archived and maintained so that they are always available to future workers.

The advent of future space observatories, such as the *Hubble Space Telescope* (*HST*) and *HIPPARCOS*, poses new challenges for quality data management. For example, the *HST Guide Star Selection Catalog* recently completed at the Space Telescope Science Institute contains approximately 18 million stars (> 500 megabytes of data). The storage and distribution of such large catalogs on the commonly used media of today are not practical, but new storage technologies, such as optical disks, CD-ROM, and high-density tapes, promise to alleviate the problem for now. For future missions, even those currently being contemplated by the space agencies, the new media will not suffice and even more advanced storage technologies will be needed.

The problem of data access to information contained in many catalogs and archives has a long history. Fortunately, this difficulty is now being solved by modern data banks like SIMBAD, which allows an on-line user to easily access a vast collection of data and bibliographical information on an object-oriented basis. Another system, called STARCAT (Space Telescope Archive and Catalog), developed at the Space Telescope–European Coordinating Facility (ST–ECF) and the European Southern Observatory (ESO), both in Garching, FRG, allows a user to access and manipulate data from a large number of complete astronomical catalogs and databases. These systems have already revolutionized astronomical data retrieval and, as they continue to expand and become more sophisticated, the astronomer's most difficult problem, that of locating previously known data and assembling complete information about individual objects, becomes manageable at last.

In preparation for what NASA calls the era of the great observatories, a highly advanced Astrophysics Data System (ADS) is now being constructed. This system will allow astronomers to locate, retrieve, and analyze data stored at a large number of distant facilities by the use of high-speed computer networks. A similar effort is under way in Europe at the European Space Agency facility in Frascati; hence the future promises to allow astronomers to access and analyze data across international boundaries from the comfort of their own homes and offices.

ASTRONOMICAL DATA CENTERS

We give the complete addresses of the principal members of the international network of astronomical data centers where machine-readable astronomical catalogs are archived and distributed. These centers have lists of all the catalogs available. Electronic addresses, if they exist, are given, followed by the principal contact for information and data requests. The centers are listed alphabetically by name.

Astronomical Data Center
National Space Science Data Center
Code 933
NASA/Goddard Space Flight Center
Greenbelt, MD 20771
USA
BITnet: TEADC@SCFVM or W3WHW@SCFVM
SPAN: NSSDCA::ADCREQUEST
Internet: TEADC@SCFVM.GSFC.NASA.GOV or
W3WHW@SCFVM.GSFC.NASA.GOV
Dr. W. H. Warren, Jr.

Astronomical Data Center of the German Federal Republic
Zentralinstitut für Astrophysik
R-Luxemburg Strasse 17a
0-1591 Potsdam
Federal Republic of Germany
Dr. Elena Schilbach

Centre de Données Astronomiques de Strasbourg
11, rue de l'Université
F-67000 Strasbourg
France
BITnet: U01117@FRCCSC21
Dr. M. Crézé

Japanese Astronomical Data Center
National Astronomical Observatory
Mitaka, Tokyo 181
Japan
BITnet: A32404@JPNKUDPC
Dr. Shiro Nishimura

Soviet Center for Astronomical Data
Astronomical Council
Academy of Sciences of the USSR
48 Pjatnitskaya Street
Moscow 109017
USSR
Dr. O. B. Dluzhnevskaya

Additional Reading

Davis Philip, A. G. and Upgren, A. R., eds. (1989). *Star Catalogues: A Centennial Tribute to A. N. Vyssotsky*. L. Davis Press, Schenectady.

Eichhorn, H. (1974). *Astronomy of Star Positions*. Frederick Unger, New York.

Jaschek, C. (1984). *Quart. J. Roy. Astron. Soc.* **25** 259.
Luyten, W. J. (1963). In *Basic Astronomical Data*, K. Aa. Strand, ed. University of Chicago Press, Chicago, p. 46.
Villard, R. (1989). The world's biggest star catalogue. *Sky and Telescope* **78** 583.
Vyssotsky, A. N. (1963). In *Basic Astronomical Data*, K. Aa. Strand, ed. University of Chicago Press, Chicago, p. 192.
See also **Astrometry; Star Catalogs, Historic.**

Star Catalogs, Historic

Heinrich Eichhorn

"Star catalog" as a technical term is much more restrictive than the naive interpretation of its meaning would lead one to believe: It is a list of estimates of star positions or "places" (i.e., angular polar coordinates), referred to an unambiguously defined coordinate system which is very little (ideally not at all) accelerated with respect to an inertial system. A complete catalog implies, in addition, the transformation that keeps the system inertial and contains the estimates of the time derivatives of the angular polar coordinates (the proper-motion components) with respect to this system. It is further understood that the precision of the position estimates listed in a catalog (in contrast to those in a mere finding list) is of the order of the best achievable at the time the positions were estimated (observed). In most contemporary investigations, the positions (polar angles) are referred to the fixed mean system Q of the equator, as oriented at a given epoch (almost invariably J2000). In this system, right ascension and declination function as longitude angle and latitude angle, respectively.

It is not clear what first prompted someone to organize the perceived celestial sphere by a set of coordinates as a standard of reference and to record the positions of the stars. We only know from Aratos' astronomical poem that Eudoxos of Cnidus (409 B.C.–356 B.C.) had a star catalog of 25 principal stars whose positions were given in ecliptic (E-system) coordinates (i.e., ecliptic longitudes and latitudes), for the orientation of the E-system at an epoch around 360 B.C. Unfortunately, this has not been preserved, although Tycho Brahe extracted the declinations of 47 stars from Aratos' poem. Aristyllos and Timocharis most likely also had—around 300 B.C.—compiled star catalogs or perhaps more modestly, lists of positions.

There is a record that the Chinese astronomers Han Hun and Shih Shen had made a catalog, giving the coordinates of about 800 stars with respect to the E-system as oriented about 360 B.C. A Babylonian tablet at the British Museum dating from about the seventh century B.C. contains fragments of star lists. Another one gives the differences between the culmination times (as measured by a water clock) of 26 zenith stars.

Eratosthenes (276–195 B.C.) compiled a list of probably about 860 stars in the various constellations, but only the names of the constellations and the number of stars in each have been preserved.

Even though the ancient Greeks regarded the equator as the principal great circle on the celestial sphere—not surprisingly, because the peculiarities of the Earth's kinematics make it an easily observable great circle—the reference coordinate system for the stars was then universally the E-system, with the vernal equinox being the zero mark for the longitudes. This makes it plausible that the star positions were mostly used as markers with respect to which the apparent paths of the planets could be traced, and they were mostly required for the practice of astrology, a superstition which did much, however, to generate support for astronomy and astronomers in the past. (Frederick II of Denmark supported Tycho Brahe so he could compute horoscopes for the royal family.) The vernal equinox, that place in the sky where equator and ecliptic intersect when the Sun goes from the southern to the northern hemisphere is, in principle, observable even with primitive means and therefore suggests itself as the zero point for reckoning the longitude angle (ecliptic longitude). This would have been a perfect marker if Newtonian mechanics and later observations had not revealed that the great circle of the ecliptic changes—albeit very slowly—its situation with respect to the background of stars. In the long run, it became evident that the vagaries of the perturbations of the Earth's orbit make the ecliptic an extremely poor—and difficult—choice as a fundamental plane; none of this could, however, have been foreseen by the ancient astronomers who first utilized the ecliptic. Had Newtonian mechanics been known then, there is little doubt that the physically well-defined invariable plane of the planetary system would have been adopted as one of the fundamental planes, even though there are certain difficulties associated with observing it.

Hipparchos (ca. 190 to ca. 125 B.C.), the greatest astronomer of antiquity, apparently decided to observe a list of the stars' positions and magnitudes after the appearance of a nova in Scorpio. When he compared his measured positions with those that Aristyllos and Timocharis (and perhaps also Eudoxos) had obtained for the same stars, he noted that all longitudes had increased by the same amount whereas the latitudes had remained unchanged. This was probably the discovery of precession. Unfortunately, we know of Hipparchos' catalog only through references in the writings of other ancient authors. The coordinates were most likely referred to the E-system as oriented at about 128 B.C. [Plinius ("Pliny"), who reported on Hipparchos' undertaking, criticized it as blasphemous.]

The oldest catalog that has come down to the present in its original form is that of Klaudios Ptolemaios ("Ptolemy") who was active around 130 A.D. in Alexandria. His catalog is a part of his *Megale Syntaxis* (the *Great Compilation*, commonly known as the *Almagest*). Even though its 1025 stars are not exactly the same as those whose positions were listed in Hipparchos' catalog, it had been suggested that Ptolemaios' catalog is simply Hipparchos' catalog transferred to the E-system as oriented at the epoch 128 A.D. by using the constant of general precession derived by Hipparchos, namely 1°/100 yr. (In the meantime, this suspicion has been laid to rest, and Ptolemaios' catalog is now regarded as based at least in part on his own observations.) This constant of precession deviates considerably from the correct value of about $1\overset{\circ}{.}4$/100 yr so that the continued use of Hipparchos' constant of precession had led to gross discrepancies between calculated and observed longitudes over the centuries, so that toward the end of the ninth century A.D. Thabit ibn Qurra contrived a "trepidation" that would contribute to producing a variable rate of precession—one of the earliest of the many examples in which erroneous observations were readily "explained" by a quickly postulated, and in the long run often tenacious, theoretical model. (Georg Peuerbach relates Thabit's theory of trepidation in his *Theoricae novae planetarum* published in the fifteenth century and only Tycho Brahe showed that a trepidation did, in fact, not exist.)

Even though Ptolemaios' catalog was frequently precessed and reissued, no records of new observations of the positions of the stars in Ptolemaios' catalog had been reported to exist until the Mongol Prince Ulugh Begh (also spelled Beg, Beigh, or Bey) (1394–1449), who was Tamerlane's grandson, observed at Samarkand positions for a new star catalog around 1420–1437. It seems that Ulugh Begh observed azimuths and altitudes and calculated ecliptic longitudes and latitudes from them by a procedure of which we know no details. None of the many Arab astronomers in the early middle ages seem to have done more than precess the coordinate system of Ptolemaios' catalog to contemporary epochs. Abd-el Rahman Al-Sufi, however, observed his own magnitudes. This may sound strange by contemporary standards of conducting science, but the ancient view was that the stars were physically attached to the "celestial sphere," that is, their positions relative to each other were unchangeable; once they had been measured, there was therefore no point in repeating this laborious exercise. The discovery in 1718 by Edmond Halley (of comet fame) that the stars have proper motions still lay far in the future.

Ptolemaios claims to have included in his catalog all stars that are visible to the naked eye. Although his catalog does, in fact, contain several stars of about 6^m, it is complete only down to about $4\overset{m}{.}5$.

The Landgrave Wilhelm IV (The Wise) of Hesse (1532–1592) observed with his assistant Christoph Rothmann the positions of 1004 stars in the coordinate system as oriented at the epoch 1594. The technique he used (most likely at the suggestion of Tycho Brahe) for the measurements of the longitudes in this catalog is a landmark insofar as he established the angular distance between the Sun and a star not by using the Moon, but by using Venus as an intermediary, which leads altogether to more accurate as well as more precise results. Ecliptic longitudes and latitudes, even though they are the principal and final data in his catalog, were not observed directly but computed from meridian altitudes and right ascension differences that had been inferred from measuring differences of meridian transit times. This appears to be the first recorded instance that clocks were used for measuring right ascensions. His second assistant Justus Bürgi (who was one of the inventors of logarithmic calculations) was reputed to have also been a master clockmaker.

Landgrave Wilhelm the Wise was much impressed with Tycho Brahe, accepted his model of the planetary system and followed his example in considering refraction in the reduction of his meridian altitude measurements.

The precision and accuracy of star positions observed without a telescope reached a high plateau with the work of the Danish nobleman Tyge (Tycho) Brahe (1546–1601), who derived at his private observatory "Uraniborg" on the island Hven the positions of 1005 stars from observations during the years 1578–1597. The declinations were determined on his large mural quadrant of 9-feet radius (or with an armillary sphere), whereas he measured the (angular) distances between the stars with a sextant. (The vernal equinox was determined by observing Venus with respect to the Sun and relating α Arietis in turn to Venus.) The position of α Arietis was in this way determined with an error of only 15 arcsec. The positions of nine stars forming a fundamental reference frame that was referred to α Arietis were accurate to 25 arcsec. Brahe used a vernier-like device on the large mural quadrant whose division marks were accurate to 10 arcsec, better than the precision of the observations that allowed one to derive positions with standard errors of the order of 1 minute of arc. Brahe embarked on the work on his catalog because of the supernova of 1572, an event similar to the one that had made Hipparchos start his catalog. Only 777 star positions are of high precision; the precision of the remaining 228, which were observed and added later, is significantly lower.

It was Johann Hewelke (or Höwelcke)—"Hevelius"—(1611–1687) who reached the highest precision in a star catalog observed without the use of a telescope. Although Hevelius is mostly remembered for his chart of the Moon and physical observations of the planets for which he put telescopes to good use, he produced his catalog of 1553 star positions—the most precise ever observed up to his time—without a telescope. Halley, who traveled with his telescopic equipment to Danzig to visit Hevelius, found the precision of his own telescopic positions to be poorer than that of Hevelius who observed 1553 stars at Danzig with the naked eye. Hevelius' *Prodromus Astronomiae* (1690) contained two catalogs which also utilized the positions derived by his predecessors: The *Catalogus Maior* contains 1563 star positions; the *Catalogus Minor* contains 1540 of them. Hevelius also incorporated into this catalog the positions of 335 southern stars which Halley (1656–1742) had observed with an instrument that used a telescope for setting. Halley had traveled to St. Helena (latitude −15°) for the purpose of making these observations to create a foundation for navigating in southern waters.

These catalogs of Hevelius are the last ones observed without a telescope (Hevelius had excellent eyes and could see stars of the seventh magnitude) and the first ones to give, besides ecliptic coordinates, also the Q-system coordinates, namely right ascension and declination. Hevelius also introduced the "new" constellations—Sextans, Scutum, Lacerta, Vulpecula, Canes Venatici, Leo Minor, and Lynx.

The precision of Hevelius' catalogs might have been even better if astronomers had then had a better understanding of the nature of observing errors. Observers as recent as Flamsteed did not realize that there were unavoidable accidental observing errors and that the arithmetic mean has a higher precision than the individual observations have by themselves. All observers sought to identify which of their like measurements (e.g., of the position of the same star) were the "best" or the most trustworthy, and the published data were then based on these exclusively, the other observations having been simply discarded. This may appear quaint and even ignorant by the standards of modern practice, until one realizes that the same mistake is even nowadays frequently repeated in situations where certain parameters (e.g., the constant of precession) can be estimated by several independent methods. One still finds scientists arguing that the results of only one of the methods—that which they consider the one most likely to be free of systematic errors and thus the "most reliable"—should be adopted.

Halley also determined accurate positions of, among others, Sirius, Procyon and Arcturus and discovered the phenomenon of proper motion (in 1718) by comparing the positions he had determined with those in Ptolemaios' catalog. Thus came to an end the myth of the "fixed" stars; the term lingered on, however, for at least 200 years longer.

The use of a telescope as the sight in the instruments designed for measuring star positions led to an explosion in the number of their observations, which was started by Halley's somewhat older contemporary and predecessor in and first holder of the office of Astronomer Royal, the Rev. John Flamsteed (1646–1719). In 1675, Flamsteed started observing star positions with a sextant. He continued after 1684 on a mural quadrant with the aid of a clock. After 1704, he measured declinations on a transit circle (*rota meridiana*) designed by Ole Rømer, and he measured right ascensions on a transit instrument. He observed the positions with respect to a set of 36 stars whose absolute positions (i.e., with respect to the Sun) had been measured in 1690. Because Halley discovered the existence of proper motions only one year before Flamsteed died, Flamsteed knew of no good reason why he should have worried about a wide spread in observing epochs. In 1901, F. Ristenpart called Flamsteed's *Historia Coelestis Britannica*, which contains position estimates of 3310 stars down to 8^m, the oldest star catalog "still of use to the contemporary astronomer," a judgment that few will subscribe to now, almost a century later.

The definitive edition of the *Historia Coelestis Britannica* was prepared by the indefatigable Francis Baily, who reedited almost all historical star catalogs. It appeared (referred to the Q-system 1690) in the year 1835. Baily corrected many of Flamsteed's errors and added 458 stars whose positions Flamsteed had measured, but had not published in his original catalog. The standard deviation of Flamsteed's positions is about 10 arcsec (according to F. W. Bessel), quite inadequate by contemporary standards but a huge improvement over the positions that his predecessors had measured without the aid of a telescope. Flamsteed also observed the Sun and the Moon in the course of the work on the *Historia Coelestis Britannica*. His catalog was reedited several times, also outside of England.

Modern precision astrometry may be said to have had its beginning with the *vir incomparabilis*, as the great Bessel himself called him, the Rev. James Bradley (1692–1762), as successor to Flamsteed and Halley the third Astronomer Royal. Introduced to astronomical observing by Samuel Molyneux (1689–1728), Bradley announced in 1728 the discovery of annual stellar aberration (this, rather than the first successful measurement of an annual parallax, was the first direct proof of the Earth's motion) and in 1748 that of nutation, both from observations of γ Draconis on a zenith sector by Graham. Bradley never compiled and reduced his observations into a catalog; this was undertaken first by Bessel and later anew by Artur Auwers, who found the standard errors of the positions derived from Bradley's observations to be $0\overset{s}{.}23$ and $1\overset{''}{.}5$, respectively, in right ascension and in declination. The catalog, referred to the Q-system 1755, contains the positions of 3268 stars down to 8^m. Bradley was the first astrometrist to use on a transit instrument the eye–ear method for observing right ascensions, in which the position of a star between two crosshairs in a reticle is estimated at the moment the observer hears the tick of the clock.

The self-taught Johann Tobias Mayer (1723–1762), who at the time of his death was professor of mathematics at Göttingen University, observed from 1756–1760 a catalog of 1027 zodiacal stars on a transit circle, and developed the formula for the reduction of meridian observations that bears his name. At about the same time, the Abbot Christian Mayer (1719–1783) observed during 1776–1778 at Mannheim "trabants" of "fixed" stars and in 1779 published a catalog of 409 multiple and double stars, thus ushering in the observations of double stars.

Bradley's successor in the office of Astronomer Royal was Nathaniel Bliss, whose very brief tenure was followed by that of Nevil Maskelyne (1732–1811) who observed, mostly between 1765–1772, 1779–1785, and 1803–1807, the positions and proper motions of 36 fundamental stars, which became established as the backbone of absolute position determinations and the skeleton for the later fundamental catalogs. A rereduction (1865) of Maskelyne's observations was one of the earliest scientific endeavors of the later-celebrated astrophysicist Ejnar Hertzsprung.

The Abbé Nicolas Louis de La Caille (Lacaille), who lived from 1713–1762, undertook the first systematic large-scale cataloging of the stars in the southern sky. During 1751–1752, he observed on a transit circle at the Cape of Good Hope the positions of 9766 stars south of $-23°$ declination relative to 398 absolutely observed principal stars. The definitive edition, sponsored by the British Association for the Advancement of Science, was published in 1847. Lacaille died over the observation (1760–1761) of 515 additional stars which were worked into and published as a catalog by Baily.

Work on the—up to then by far most ambitious—catalog was undertaken by Joseph Jerome le François de Lalande (or Delalande), 1732–1807. His aim was the cataloging of all stars down to 9^m, for which he used a transit instrument and a mural quadrant by John Bird; he later (1748) used a transit circle. Because Lalande himself utilized only a part of his observations for his *Histoire Celeste Française*, Baily obtained the support of the British Association for the Advancement of Science to reduce all of Lalande's observations and in 1837 published a catalog of 47,390 star positions, which are on the average not very precise, because Lalande observed most stars only once. Perhaps for this reason he missed recognizing several minor planets that he had observed, as well as the planet Neptune. The definitive reduction (*Catalogue de l'Observatoire de Paris*) first appeared in 1928.

The most important catalog up to his time was constructed by the Theatine priest Giuseppe Piazzi (1746–1826), mostly remembered as the discoverer in 1801 of Ceres, the minor planet whose existence had been suspected long before because of the "gap" in Bode's sequence. He soon exchanged his vocation as a parish priest with astronomy, founded the Palermo (Sicily) observatory in 1786, and became its first director. The first to use microscopes on his instruments to read the circles, he started observations in 1791 on several instruments and soon recognized the gain in precision by observing fewer stars, but observing those rather frequently. He measured the positions of 220 stars absolutely (except for the relation to the vernal equinox) and relative to them observed the positions of 7646 stars, aided by his assistant Niccolò Cacciatore. His catalog was, in the judgment of Simon Newcomb (1906), "vastly superior to any that preceded it." A definitive reduction is, however, not yet available, but F. Porro did homogenize the observation material, a Herculean task without a computer.

Friedrich Wilhelm Bessel (1784–1846) was the most accomplished astrometrist of the nineteenth century and introduced the principles and formulas that, until only a few years ago, were the standard for reducing observed star positions. He was the first astrometrist to measure right ascensions and declinations simultaneously on the same transit circle. The accomplishments of his relatively short life could have distinguished—by their quality as well as by their bulk—the careers of several individuals. He reduced and edited Bradley's observations (and published a catalog based on them) in a fashion that became the paradigm for such work for over a century and used his own about 75,000 observations (made between 1821 and 1833 in zones, which he introduced) for the computation of zone catalogs, which later inspired Argelander to embark on the work toward the *Bonner Durchmusterung*. It is to Bessel that science owes the first reliably measured parallax, Bessel functions, and the discovery of the variable proper motions of Sirius and Procyon that eventually led to the discoveries of their white dwarf companions. By computing proper motions from his own and Bradley's earlier positions, he derived proper motions and thus created the first complete (and also) fundamental star catalog in the modern sense.

Additional Reading

Eichhorn, H. (1974). *Astronomy of Star Positions*. Frederick Ungar, New York.

Knobel, E. B. (1877). The chronology of star catalogues. *Mem. Roy. Astron. Soc.* **43** 1.

Newcomb, S. (1906) (reprinted in 1960). *A Compendium of Spherical Astronomy*. Dover, New York, p. 380

Proverbio, E. (1988). The third reduction of Giuseppe Piazzi's star catalogue. *Mapping the Sky, Proc. 133rd Symp. IAU*, S. Debarbat, J. Eddy, H. Eichhorn, and A. Upgren, eds. Kluwer Academic Publishers, Dordrecht, p. 75.

Ristenpart, F. (1901). Sterncataloge und -Karten. *Handwörterbuch d. Astronomie*, W. Valentiner, ed. Trewendt, Breslau, p. 455.

Ristenpart, F. (1909). Fehlerverzeichniss zu den Sterncatalogen dcs 18. und 19. Jahrhunderts. *Astron. Abhandlungen als Ergänzungshefte zu den Astron. Nachr.* (No. 16).

Ševarlić, B. M. (1978). Fundamental astrometry—A look through the past. *Epitome Fundamentorum Astronomiae. I. Catalogues of Star Positions*, B. Ševarlić and G. Teleki, eds., University of Beograd, Beograd.

Wagman, M. (1991). Who numbered Flamsteed's stars? *Sky and Telescope* **81** 380.

Zinner, E. (1934). *Die Geschichte der Sternkunde*. Springer-Verlag, Berlin.

See also **Constellations and Star Maps; Coordinates and Reference Systems; Stars, Catalogs and Surveys; Telescopes, Historical.**

Star Clusters, Globular

James E. Hesser and Michael J. Bolte

The most luminous remnants of the formation epoch of our Milky Way galaxy are some 154 clusters of stars known as the galactic globular clusters (GGCs), which represent $\sim\frac{1}{100}$ of the known mass in stars in the galactic halo. These clusters span considerable ranges in total mass and luminosity, central concentration, chemical composition, and orbits in the Galaxy, yet seem to exhibit a relatively small range in age. A GGC traditionally has been viewed as a self-gravitating collection of extremely old, low-mass stars deficient (compared to the Sun) in chemical elements with atomic numbers (Z) greater than 2. For the known clusters, the combination of mass and orbit in the Galaxy have made them stable against disruption by interaction with other components of the Galaxy. (It remains a matter of conjecture how many clusters may have been formed and subsequently destroyed, perhaps thereby dispersing their stars throughout the halo.) The exact definition of what constitutes a globular cluster grows murkier in other galaxies, where objects with the same physical appearance as GGCs, but composed of much younger stars, are found. Recently, even in our Milky Way some of the traditional criteria for classifying objects as GGCs have been relaxed as careful searches have found low-luminosity, relatively loose clusters at large galactocentric radii, and it has been determined that some GGCs have chemical compositions nearer to solar values than heretofore suspected.

GGCs play a central role in several areas of astronomy. They provide one of the few probes of conditions that existed during the earliest stages of the Milky Way's formation, act as bright tracers of the structure of the galactic halo, and present unparalleled opportunities to compare stars in different evolutionary stages but at the same distance (hence, accurate relative luminosities and temperatures are easy to obtain). They also represent the simplest systems for the study of stellar dynamical processes. Traditionally, much effort has gone into trying to understand the structure of the "turnoff" from the main sequence, and of the red giant and horizontal branches in the cluster color–magnitude diagrams (CMDs). Much investigation has been focused on the understanding of the behavior of the RR Lyrae variables, which are important distance indicators. The resultant comparisons provide invaluable constraints on theories of stellar evolution, which have been remarkably successful in explaining the observations with relatively few parameters. More recently, the discovery of x-ray sources and millisecond pulsars in the very cores of GGCs have opened exciting new insights into evolutionary and dynamical effects previously hidden from view in the optical region and have added to the growing evidence for binary stars within GGCs.

GGCs span a wide range of integrated- or composite-light colors, $0.56 \lesssim (B-V)_0 \lesssim 0.84$, and integrated spectral types, F2 to G5 (more extreme examples are found in the GC systems of other galaxies), arising primarily from the effects of differing abundances of the chemical elements in their stars. State-of-the-art (1991) broadband observations of GGC stars extend to $V \gtrsim 26$, which corresponds to $M_V \gtrsim 13.5$ and $\gtrsim 4$ in the nearest and most distant clusters, respectively. Intermediate resolution (~ 3 Å) spectra can be obtained of stars as faint as $V \sim 18$, which allows metal-abundance estimates of stars in the main-sequence turnoff region of the CMD for the nearest clusters; the high-resolution spectral data required for fine analyses are limited to the brighter evolved stars.

The "globular" clusters are well-named in that their mean axial ratio is $a/b = 0.93$ (reported by Raymond E. White and Stephen J. Shawl). Detailed stellar velocity measurements as a function of radial distance from the center for ω Centauri, a cluster with one of the larger deviations from circular symmetry ($a/b = 0.83$), suggest that its flattening is due to a mild rotation of the cluster. In the Magellanic Clouds and other galaxies some globular-like clusters are much more elliptical than any GGC, possibly indicating that they acquired or retained more angular momentum during their formation.

A number of basic characteristics of each GGC were tabulated by Ronald F. Webbink and recent reviews and references were published by Jonathan E. Grindlay and A. G. Davis Philip and by J. E. Hesser.

SPATIAL DISTRIBUTION AND KINEMATICS

The GGCs have been the classical tracers of the galactic halo. Early in this century, Harlow Shapley correctly felt that the center of a system as massive as the GGCs would also be the center of the Galaxy. He used the distribution of clusters on the sky to locate the galactic center and to demonstrate for the first time that the Sun was located well away from the hub of our galaxy. Subsequently, our picture of the GGC system has evolved from one in which the clusters all are members of a single population tracing a filled, spherical halo centered on the Galaxy, to one that contains at least two (and possibly more) distinct subsystems (demonstrated by Robert Zinn in 1985).

Table 1. Disk and Halo Globular Cluster Properties

	Halo GGCs	*Disk GGCs*	*Thin-Disk Stars*
$\overline{[Fe/H]}$	−1.5	−0.5	0.0
$V_{rotation}$ (km s^{-1})	45	185	220
$V_{dispersion}$ (km s^{-1})	115	60	15

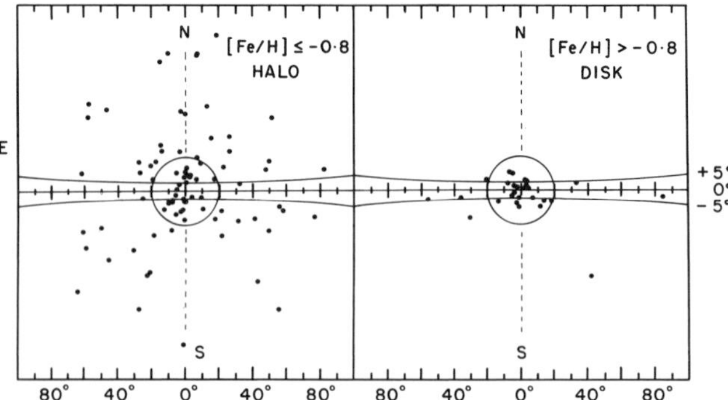

Figure 1. The distributions of the halo and disk GGCs in a polar coordinate system centered on the galactic center [*adapted with permission from Robert Zinn* (Astrophysical Journal ***293*** *1985 p. 430)*].

From analyses of radial velocities, spatial distributions, and chemical compositions, Zinn found that approximately three-fourths of the known GGCs belong to the "halo" subsystem. The remaining clusters belong to what is being called the disk subsystem, which may be similar or related to the so-called thick-disk population of the Galaxy. Table 1 summarizes the metal abundances, the rotational velocity around the Galaxy, and the line-of-sight velocity dispersion of these two systems and the much younger "thin-disk" population of stars. Figure 1 illustrates the differences in spatial distributions of the two subsystems.

It is possible that there are additional distinct groups in the halo. The distribution of clusters in the halo drops off smoothly (space density $\rho \propto R^{-3.5}$) to a galactocentric radius R of ~ 30 kpc. Beyond this there is a gap out to ~ 60 kpc, and then there is what may be a separate population of nine clusters that are found from 60 kpc out to 120 kpc.

A typical halo GGC is thought to pass through the galactic plane every $\sim 3 \times 10^8$ yr or so, during which the ejecta (primarily stellar winds from the evolving stars) that have accumulated in the central potential well will be swept out. One of the exciting prospects for the 1990s is measurements of all three components of the space motion for many GGCs, thereby revealing the true nature of their orbits for the first time.

LUMINOSITY FUNCTIONS

The distribution of total or integrated luminosity for every cluster in the system as a whole (the luminosity function) contains information concerning the range in formation conditions convolved with the processes by which clusters are modified or destroyed by the galactic tidal field, encounters with giant molecular clouds, and so forth. The GGCs span a large range in total visual luminosity

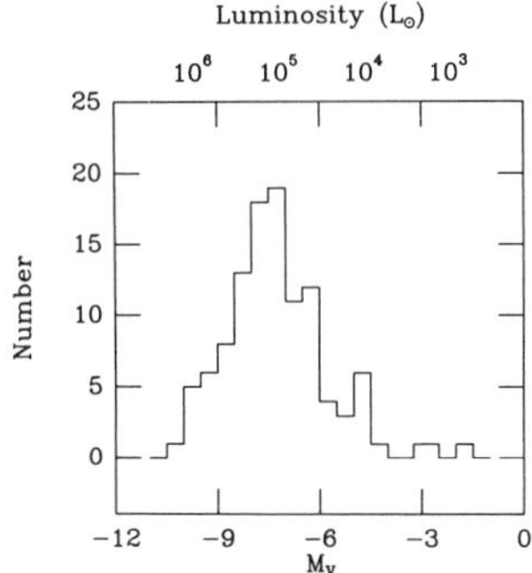

Figure 2. The distribution of cluster integrated absolute luminosity given in absolute visual magnitude (bottom scale) and in solar luminosity units (top scale). These data, and those shown in Fig. 3, are taken from Ronald F. Webbink (1985).

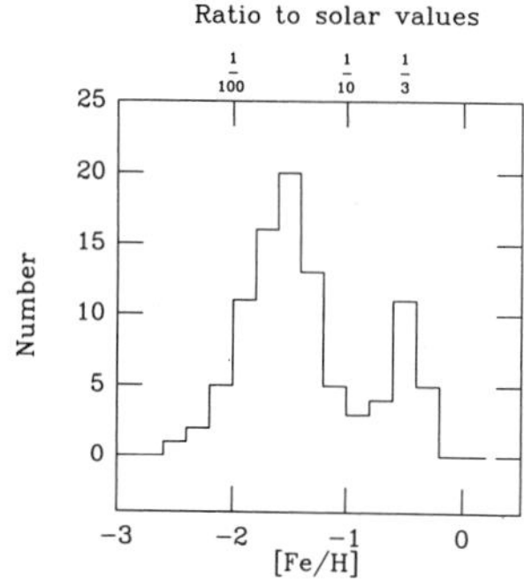

Figure 3. The distribution of the abundance of "metals" for the GGC system. The bottom scale is logarithmic, with $[\mathrm{Fe/H}] \equiv \log[n(\mathrm{Fe})/n(\mathrm{H})] - \log[n(\mathrm{Fe})/n(\mathrm{H})]_\odot$. The top scale gives the ratio of the abundance of $Z > 2$ elements in the cluster stars relative to the Sun. The vertical scale indicates the number of GGCs having the corresponding abundances.

from the extremes of $L_V \sim 1.2 \times 10^6\ L_\odot$ for ω Centauri to $L_V \sim 400\ L_\odot$ for the very sparse AM 4. Empirically, the distribution is found to be a Gaussian distribution with its peak at $M_V \sim -7.4$ and a dispersion of ~ 1.3 magnitudes, as seen in Fig. 2; there is at present no theoretical explanation for the Gaussian distribution. The luminosity functions of the disk and halo subsystems are similar. Indeed, available evidence suggests that the luminosity functions of extragalactic GC systems are remarkably similar to that of the GGC system, which has led to the possibility they might be used as "standard candles" for extragalactic distance determinations.

The luminosity function of the stars in individual clusters is a topic of importance whose potential is only now beginning to be tapped after the introduction of new detectors in the mid-1980s. One puzzling early result, due to Robert D. McClure and coworkers, was the discovery of an apparent correlation between the slope of the luminosity function for main-sequence stars and the chemical composition of the clusters. This correlation exhibits considerable scatter and it is unclear whether it is telling us primarily about a composition dependence in the original mass distribution of stars within the cluster, or whether it reflects some equally poorly understood difference between the way clusters formed in the halo and in the disk subsystems.

CHEMICAL COMPOSITIONS

The abundances of elements with $Z > 2$ in globular cluster stars are extremely uniform within individual clusters, yet vary by a factor of several hundred from cluster to cluster. These facts, combined with the small age range between clusters provide key clues to the chemical evolution of the Galaxy. From the viewpoint of cosmology, accurate knowledge of the chemical composition of globular star clusters is an essential ingredient for estimating their ages, which in turn provide a lower limit to the epoch of galaxy formation.

Traditionally, the overall chemical composition of a cluster has been characterized by the abundance of the heavy elements in the iron (Fe) peak relative to hydrogen; solar values are used as the reference. (The astronomical use of the term "metals" to refer to all elements with $Z > 2$ derives from the role of these elements as donors of the electrons essential to the generation of opacity within a star and from the fact that absorption lines due to Fe-peak elements dominate spectra of cool stars like the Sun in the visual region.) The GGC system shows a distinctly bimodal distribution in iron abundance: Halo clusters range in Fe abundance from $\sim \frac{1}{300}$ to $\sim \frac{1}{10}$ of solar, with a median value near $\frac{1}{40}$. The inner disk clusters are significantly more Fe-rich, with values ranging from $\sim \frac{1}{10}$ solar to near solar. The distinction between the subsystems and the total range can be readily appreciated in Figs. 1 and 3.

However, it has become clear in the past few years that within GGC stars the pattern of element abundances relative to Fe may be quite different from that of the Sun. Indeed, specific abundances of individual elements provide important insights into the chemical enrichment process, although much remains to be deciphered. Although there is a three-orders-of-magnitude range in abundance of any particular element among the clusters, star-to-star abundance differences within any one cluster are generally small except among carbon (C), nitrogen (N), and, probably, oxygen [(O) but O is *much* harder to measure accurately]. Only two clusters (ω Centauri and M 22) are known whose stars generally show measurable differences of elements heavier than C, N, and O [although stars with strong CN bands have been shown in some clusters to have enhanced aluminum (Al) and magnesium (Mg) spectral lines].

Patterns of chemical inhomogeneities among the CNO elements appear to separate into at least two regimes of stars, evolved and unevolved, as well as into at least two metallicity groupings. In many clusters, star-to-star differences are seen in the abundances of C and N, as well as variations in the ratio of C isotopes, as stars evolve up the giant branch. There is a tendency for the strengths of the CN bands to be bimodally distributed among evolved stars within a single cluster (in all but the most "metal"-deficient clusters), and for CN and CH band strengths to be anticorrelated. Such variations could arise from episodes when the convection zone reaches deep into the stellar interior, where the ratios of C, N, and O are being modified by one or more nuclear reaction cycles. The subsequent convective mixing of nuclear-processed material then alters the surface abundances. However, in spite of significant effort over the past decade and a half, it has not been established unequivocally whether all the observed inhomogeneity patterns are due to mixing events. An alternate possibility is that they arose from abundance differences present in the material from which the stars formed. More and better spectroscopic abundance determinations are required, as are observations of much fainter stars, near or on the main sequence. It should be noted that the extremely sharp loci determined for the upper main sequence and turnoff regions in the broadband color–magnitude diagrams for several GGCs tightly constrain the range of star-to-star composition differences within a cluster.

The overall chemical composition controls the color and position of the principal sequences in each cluster's color–magnitude diagram, with clusters of increasing heavy element content having main-sequence turnoffs, and giant and horizontal branches, lying progressively redwards. This effect can be seen clearly in Fig. 4.

However, there are a number of clusters with apparently similar chemical composition whose features, particularly the distribution of stars along their horizontal branches, mimic clusters of very different abundances. This phenomenon, referred to as the "second

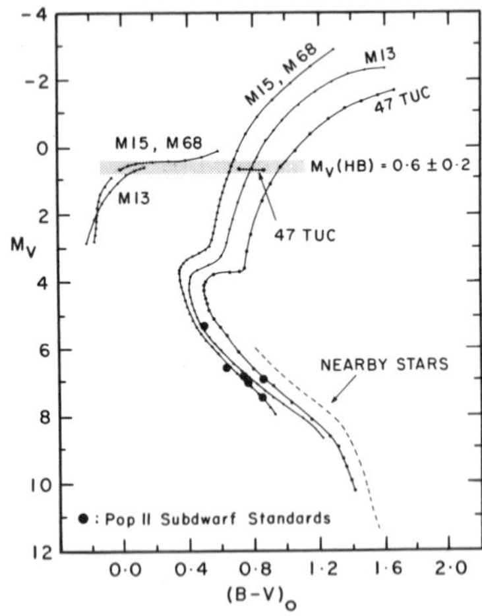

Figure 4. A schematic representation of the color–magnitude diagrams of four nearby GGCs spanning a wide range in chemical composition (*from J. E. Hesser et al.,* Pub. Astron. Soc. Pacific ***99*** *1987 p. 739, courtesy of the* Publications of the Astronomical Society of the Pacific). The diagrams for each cluster have been shifted to account for the distance to and the interstellar reddening toward the cluster. The [Fe/H] ratios (see Fig. 3) for the clusters are as follows: M 15 and M 68, −2.1; M 13, −1.6; 47 Tuc, −0.7. The metal deficiency of the GCs causes them to lie blueward of the locus of the solar-composition nearby stars. Also shown (as filled circles) are the magnitudes and colors of the nearby metal-deficient field halo stars.

parameter problem," has received much attention in the past two decades. Although the phenomenon manifests itself at all galactocentric radii, it appears to be more commonly seen in clusters lying far from the galactic center. Stellar models reveal that the distribution of stars on the horizontal branch is very sensitive to many physical parameters (e.g., helium abundance, specific heavy-element ratios, the ratio of the stellar envelope-to-core mass, mass loss, and age) at levels below current measurement thresholds. Although evidence favoring identification of the second parameter with age differences (of up to a few billion years) between the clusters has been mounting recently, it is fair to say that whether age is *the* second parameter or only a dominant component thereof remains unresolved.

CLUSTER DYNAMICS

Comparing photometric and radial velocity observations of GGC stars with cluster models provides tests of dynamical processes in clusters and a means of probing those cluster populations that are too faint to be observed directly. The dynamical processes that may lead to a runaway collapse of cluster cores, mass segregation, and a tidal truncation of clusters are described in other entries.

The total mass of directly observed stars in a cluster can be estimated by integrating the observed luminosity function and using a mass–luminosity relation from stellar models. A comparison of this value to the total mass inferred from radial velocities of individual stars and an application of the virial theorem gives the amount of "dark" matter contained in a cluster. This includes main–sequence stars below the faint limit of the observations and stellar remnants from stars more massive than the present-day turnoff (e.g., white dwarfs, neutron stars, and black holes). Because reasonable assumptions concerning their numbers have been sufficient to explain the dynamics of GGCs, there has been no need to appeal to the mysterious "dark" matter that seems to be required to explain the dynamics of galaxies and groups of galaxies. As the limits on the population of lower main-sequence stars improve through observations of even fainter objects, the constraints on the mass and numbers of stellar remnants in clusters will tighten. In turn this will provide clues to the original mass spectrum of stars [called the initial mass function (IMF)] in the clusters. Knowledge of the IMF, in combination with detailed abundance studies, will give insight into the chemical enrichment process of individual clusters and the Galaxy as a whole.

ABSOLUTE CLUSTER AGES

Star clusters are among the few objects in the Galaxy for which relatively precise ages can be measured. Because of their extreme age, GGCs provide a unique probe into the history of the formation of the Galaxy and give what has proven to be a very useful lower limit to the age of the universe.

The measurement of *absolute* (i.e., accurate) stellar ages free from systematic errors has proven to be a difficult task. To measure a cluster's age to an accuracy of 2 or 3 billion years requires, in addition to very accurate observations of the cluster-star magnitudes and colors, determinations of the cluster's distance, element ratios, and amount of interstellar reddening to a higher accuracy than has been achieved to date in all but a few clusters. For those few clusters having the best studies, it appears that the remaining hurdle is an accurate determination of the abundance of oxygen. Oxygen is a key element because it is a catalyst for the energy generation of stars in the age-sensitive main-sequence turnoff region. It is also, unfortunately, one of the most difficult of the elements to measure in stars. Even given the improvement in observations (which have been dramatic following the introduction of charge-coupled devices and other innovations in the latter half of the 1980s), there remain possible systematic errors in the stellar models to which the observations must be compared to derive ages. For example, recent models that contain a more realistic mixture of elements for GGC stars (metal-poor stars do not have a scaled solar composition) have led to a 20% reduction in cluster age estimates. Incorporation within the models of heretofore neglected physical phenomena, such as helium diffusion, may introduce comparable shifts.

GALAXY FORMATION AND RELATIVE CLUSTER AGES

As relics of an early epoch in the formation of the Galaxy, GGCs provide a unique probe into the process of galaxy collapse and formation. At this time any understanding that we possess of how GCs formed is extremely rudimentary. Nonetheless, they *did* form, and the examples we know best, the GGCs, show remarkable internal chemical homogeneity, suggesting the material from which they formed was extremely well-mixed. An important issue has centered on whether a heavy-element-deficient, collapsing cloud of globular cluster mass (e.g., $\sim 10^6\ M_\odot$) could retain and thoroughly mix the heavy-element-enriched ejecta from violent supernovae prior to the formation of the low-mass stars seen in GGCs today. The concern that it could not was one factor that led to the suggestion of Leonard Searle and Robert Zinn (in 1978) that perhaps the GGCs were formed in systems with mass $\sim 10^8\ M_\odot$ that then merged to form the galactic halo.

One important detail from which any theory of galaxy formation must begin is the time scale for the collapse of the galaxy halo. The width of the distribution of cluster ages gives this time scale directly. Because of its importance, the question of age differences among clusters is the topic of much current debate and research. Although far from settled in 1991, it does appear that in at least a few cases, age differences of more than 2 billion years exist between clusters, and it is possible that the age extremes of the cluster system differ by as much as 5 billion years—a significant fraction of the ~15-Gyr age of the Galaxy. As the age distribution of the GGCs becomes better known, correlations between age and kinematics, abundances, and position in the Galaxy will provide the basis for the next generation of models of galaxy formation and the process of chemical enrichment of galaxies. Perhaps then we will know if our galaxy formed from the inside out or from the outside in.

Additional Reading

Grindlay, J. E. and Philip, A. G. D., eds. (1988). *Globular Cluster Systems in Galaxies, IAU Symposium 126*. Kluwer Academic, Dordrecht.

Hesser, J. E. (1988). Globular clusters in the Galaxy and beyond. In *Progress and Opportunities in Southern Hemisphere Optical Astronomy*, V. M. Blanco and M. M. Phillips, eds. *Astronomical Society of the Pacific Conference Proceedings Series*. Astronomical Society of the Pacific, San Francisco, p. 161.

Hesser, J. E., Harris, W. E., VandenBerg, D. A., Allwright, J. W. B., Shott, P., and Stetson, P. B. (1987). A CCD color-magnitude study of 47 Tucanae. *Pub. Astron. Soc. Pacific* **99** 739.

McClure, R. D., VandenBerg, D. A., Smith, G. H., Fahlman, G. G., Richer, H. B., Hesser, J. E., Harris, W. E., Stetson, P. B., and Bell, R. A. (1986). Mass functions for globular cluster main sequences based on CCD photometry and stellar models. *Ap. J. (Lett.)* **307** L49.

Searle, L. and Zinn, R. (1978). Compositions of halo clusters and the formation of the galactic halo. *Ap. J.* **225** 357.

Webbink, R. F. (1985). Structure parameters of galactic globular clusters. In *Dynamics of Star Clusters, IAU Symposium No. 113*, J. Goodman and P. Hut, eds. D. Reidel, Dordrecht, p. 541.

White, R. E. and Shawl, S. J. (1987). Axial ratios and orientations for 100 galactic globular star clusters. *Ap. J.* **317** 246.

Zinn, R. (1985). The globular cluster system of the Galaxy. IV. The halo and disk subsystems. *Ap. J.* **293** 424.

See also **Galaxy, Chemical Evolution; Pulsars, Binary; Star Clusters, Globular (all entries); Star Clusters, Mass and Luminosity Functions; Star Clusters, Stellar Evolution; Stars, Blue Stragglers; Stellar Associations and Open Clusters, Extragalactic.**

Star Clusters, Globular, Binary Stars

Carlton Pryor

The identification and study of binary stars in globular clusters is challenging because even the closest clusters are about 1000 times more distant than the nearest stars. This distance causes even widely separated binaries to appear as one rather than two points of light. It also makes cluster stars faint and thus difficult to observe. The most successful search strategies have either relied on the unusual properties of some binaries, such as the emission of x-rays, or looked for orbital motion among the brightest cluster stars. Starting in the mid-1970s, observations at wavelengths from the radio to x-rays have discovered a wide variety of binary stars in globular clusters. Our ultimate goals are to determine the numbers of binaries, their location within clusters, and such properties of each binary as the shape and period of the orbit and the kinds of the stars in the system. Such information is important because globular clusters contain some of the oldest stars in our galaxy and are also one of the densest stellar environments known. More than half of the stars in the neighborhood of the Sun belong to binary systems, a consequence of the poorly understood star-formation process. Is the same true for stars in globular clusters, which were formed at the very beginning of the history of our galaxy? Globular cluster dynamics relates the motions of stars in the cluster to the cluster's spatial structure and also studies how the motions and structure change with time. The high density of stars in globular clusters causes the stars to approach each other closely much more often than elsewhere in the Galaxy. Over billions of years this can affect the numbers and kinds of binaries present. It can also allow the binaries to affect the structure of the cluster itself.

X-RAY BINARIES: EVIDENCE FOR THE IMPORTANCE OF DYNAMICS

In the mid- and late 1970s a succession of x-ray astronomy satellites, beginning with *Uhuru* and culminating with the *Einstein* observatory, discovered eight luminous x-ray sources in globular clusters. Their luminosities and other properties strongly suggested that these systems were close binaries in which mass was falling onto a neutron star from its companion. Here a "close" binary means a system where the stars are separated by a few times their combined radii or less. In 1986 and 1987 it was discovered that the x-ray sources in M 15 and NGC 6624 (as well as the M 15 source's optical counterpart) fluctuated in brightness with periods of 11.4 min and 8.54 h, respectively. Both times are probably orbital periods, strengthening the case that these sources are short-period binaries with small separations. Soon after the discovery of the x-ray sources in globular clusters, it was realized that they are about 100 times more common in the clusters than elsewhere. The high density of stars in the clusters leads to the creation of close binaries by the mechanism of tidal capture. When two stars approach each other to within a few times their radii, the tidal bulges raised on their surfaces transfer energy from their relative motion into oscillations of the surfaces. These oscillations are similar to waves on the surface of the ocean, though with a much larger distance between wave crests. If enough energy is transferred, the stars will become bound to each other and thus form a binary. The initial orbit is very elliptical, but more energy is transferred each time the stars come close and the two stars end up in circular orbits with a separation of about twice their initial closest approach. This process produces only very close binaries, because an approach to within about three stellar radii is necessary to transfer enough energy into surface oscillations. Such close approaches are much more likely in the high density environment of a globular cluster and greatly supplement the original population of close binaries.

The location of the cluster x-ray sources generally fits well with this picture. Seven of the eight are near the centers of globular clusters that have very high central densities. The eighth is in a cluster that is not especially dense and remains a puzzle.

ROLE OF BINARIES IN THE DYNAMICAL EVOLUTION OF GLOBULAR CLUSTERS

The spatial structure of globular clusters is expected to evolve on time scales of billions of years. One of the most important changes is that the cluster "core," the central region where the density reaches a plateau, contracts to become increasingly dense, a development called core collapse. One of the best candidates for a process to halt core collapse is the interaction of binaries and other stars in the core. The binaries can either be "primordial" systems created when the stars were originally formed or ones created by tidal capture. When a third star approaches to within a few times the separation of the components of a close binary, the gravitational interaction of the three stars on average transfers energy from the orbital motion of the binary to the motion of the third star. Another possible outcome is the collision of two of the stars and their coalescence, but probably most of the time the net effect of the interaction is a more tightly bound, hence closer, binary and a third star flying away with increased energy. The resulting release of energy to the stars in the core of the cluster halts the collapse.

All binaries, not just those in the core, will exchange energy with stars that approach closely. How do these exchanges affect the primordial population of binaries in the cluster? Analytical estimates and numerical experiments have determined that the average outcome depends on the properties of both the binary and its environment. If the orbital velocities of the stars in the binary are slower than the typical velocity of stars in the cluster, then the binary on average gains energy in the interactions and grows less bound and hence wider. If the orbital velocities are larger, the binary on average loses energy and becomes closer. The dividing line corresponds to a binary with a separation of about 14 AU and a period of 40 yr in a typical globular cluster where the stars have a root-mean-square velocity of 5 km s^{-1} (assuming a binary of two 0.8-$M_{\odot}$ stars). Such a binary changes its separation by about 14 AU in 100 million years when immersed in an environment with a density of ten thousand stars per cubic parsec, which is present at the centers of many clusters. This process could drive the two stars of a binary so far apart that the system is disrupted or so close together that the two stars coalesce. Thus primordial binaries are expected to be destroyed over time in globular clusters.

OBSERVATIONS OF PRIMORDIAL BINARIES

Two kinds of observations that have discovered primordial binaries in globular clusters are searches for stars with variable radial velocities and searches for stars that are unusually bright for their color. The component of a star's velocity along the line-of-sight, its radial velocity, is measured from the Doppler shift of the absorption lines in the spectrum of the star. The velocities expected from orbital motion are about 10 km s^{-1} or less, so the measurements must be accurate to about 1 km s^{-1}. This precision was impossible for even the brightest globular cluster stars before the introduction of high quantum-efficiency detectors in the 1970s and generally still requires the largest telescopes. There are probably over 750 globular cluster stars with at least two velocity measurements. However, many of these data have been taken so recently that analysis has just begun. Only two of the stars with variable radial velocities have a sufficient number of measurements to determine periods. About 10% of the stars in globular clusters appear to be in binaries, based on the analysis of a sample of slightly less than 400 stars. The uncertainty in this fraction is at least a factor of 2. This fraction is significantly less than that for stars in the solar neighborhood, but whether this is due to the destruction of primordial binaries or to fewer being formed to begin with is the subject of current research. Studies of low density globular clusters, where destruction should have been less important, do appear to find a higher fraction of binaries, suggesting that destruction is the cause of the difference.

When the brightnesses and colors of the stars in a globular cluster are plotted against each other in a color-magnitude diagram, the stars fall along definite sequences. Most fall along the main sequence, which represents the line along which stars of different masses spend most of their lives. The combined light of the two stars in a binary will cause it to appear as a star above the normal sequence. Many stars at once can be searched for binaries in this way by taking pictures with two different color filters and then measuring the brightnesses of all the stars. The two biggest stumbling blocks are that the main sequence stars for which the technique works best are very faint and that the crowding of stellar images produces chance superpositions of stars that look like binaries. The crowding is manageable in the outer parts of clusters; however, we are uncertain whether such studies miss a large population of binaries in the cluster centers. This question will be answered by observations with the *Hubble Space Telescope*, whose high resolution will make this technique feasible even in the centers of clusters. Three clusters appear to have a significant population of stars above the main sequence (Pal 5, E 3, and NGC 288), but others do not. Pal 5, E 3, and NGC 288 have much lower densities than average, allowing the observations to be made unusually close to the cluster centers.

MILLISECOND PULSARS

A recent exciting development has been the discovery since 1987 of 11 pulsars with periods between 3 and 110 ms in the clusters M 4, M 5 (two pulsars), M 13, M 15 (three pulsars), M 28, M 53, NGC 6440, and 47 Tucanae. Pulsars with such short periods are probably the result of mass transfer in close binaries, so their presence in globular clusters is not surprising. The very accurate "clock" provided by the radio pulses makes it possible to measure the motion of the pulsar very precisely using the Doppler effect. Four of the eleven show orbital motion and two have had periods determined (8 h and 191 days). The other periods are uncertain because of the small amount of data yet available. The pulsars without companions may have lost them in interactions with the other cluster stars. The pulsar clocks are so accurate that, with a few years of observations, it will be possible to measure the changes in the accelerations of the pulsars as they move through the complex gravitational field of the globular cluster core. The accelerations themselves cannot be measured because they cause only a change in the pulsar period with time and the intrinsic rate of change is not known. However, the period of a pulsar in M 15 is decreasing, which is extremely unusual. The acceleration needed to produce a spurious apparent decrease is somewhat larger than is expected in the core of M 15, but the pulsar may just currently be unusually close to another cluster star.

FUTURE

Though this field is still in its infancy, the rapid progress of the last 15 years should continue. The existence of both primordial and created binary stars in globular clusters has been demonstrated: We now need to determine their precise numbers, properties, and histories.

Additional Reading

Cohn, H., Hut, P., and Wise, M. (1989). Gravothermal oscillations after core collapse in globular clusters. *Ap. J.* **342** 814.

Elson, R., Hut, P., and Inagaki, S. (1987). Dynamical evolution of globular clusters. *Ann. Rev. Astron. Ap.* **25** 565.

King, I. R. (1985). Globular clusters. *Scientific American* **252** (No. 6) 78.

Ostriker, J. P. (1985). Some summary remarks. In *Dynamics of Star Clusters, Proceedings of I.A.U. Symposium 113*, J. Goodman and P. Hut, eds. D. Reidel, Boston, p. 511.

Spitzer, L., Jr. (1987). *Dynamical Evolution of Globular Clusters.* Princeton University Press, Princeton.

See also **Binary Stars, X-Ray, Formation and Evolution; Pulsars, Binary; Pulsars, Millisecond; Star Clusters, Globular, Mass Segregation.**

Star Clusters, Globular, Chemical Composition

John E. Norris

Globular clusters comprise hydrogen ($X \sim 0.75$) and helium ($Y \sim 0.23$), with a trace of the heavier elements ($Z = 0.0001$–0.02), where X, Y, and Z are fractions by mass and where, by definition, $X + Y + Z = 1$. Because many of these systems are believed to have been among the first objects to form, some 15 Gyr ago, their chemical abundance patterns contain essential clues for an understanding of the origin of the elements, of the clusters themselves, of the Galaxy, and, indeed, of the universe.

A major complicating factor in understanding the observed abundance patterns in these systems is that most of our knowledge is gleaned from spectroscopic analysis of the outermost layers of red giants and horizontal-branch stars, which in some cases have been polluted by the mixing up from their interiors of material that has

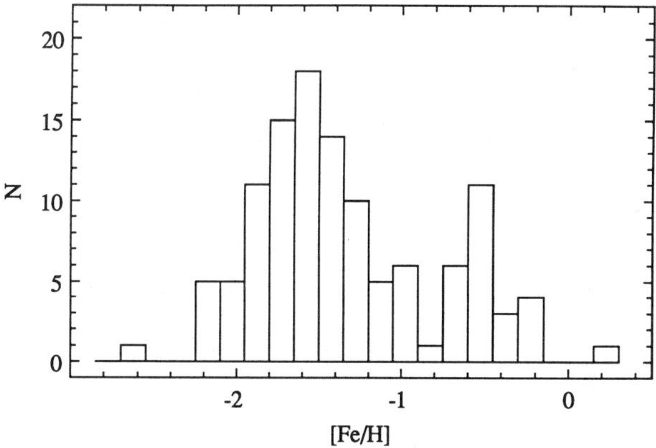

Figure 1. The abundance distribution of 119 galactic globular clusters, based on the work of Robert Zinn and Taft Armandroff [see Astronomical Journal ***89*** 92 (1988)]. Note the bimodality of the distribution: The low-abundance subsystem corresponds to objects that occupy a spheroidal distribution some 80 kpc in diameter, whereas the more metal-rich one comprises objects that delineate a disk-like configuration.

Figure 2. Globular cluster ω Centauri, photographed at wavelength 160 nm with the Ultraviolet Imaging Telescope on the *ASTRO-1* mission of the Space Shuttle *Columbia* on December 8, 1990. (*Courtesy Theodore P. Stecher, NASA/Goddard Space Flight Center.*)

undergone processing by nuclear reactions, principally in the carbon–nitrogen–oxygen (CNO) bi-cycle. Care must be taken to separate those factors that pertain to the conditions that existed when the clusters formed from those that are connected with stellar evolutionary effects. For convenience, we discuss first the elements heavier than oxygen, of which iron is representative and the most readily measured, then helium, and finally CNO—deferring to the end the elements for which the effects of stellar evolution are most important.

HEAVY ELEMENTS

With the exception of ω Centauri (the most massive of the galactic globular clusters) and M 22, individual systems appear to be chemically homogeneous with respect to iron to within an observational limit of 40%. This points to the cluster stars having been formed from material that was very well mixed and places strong constraints on the amount of self-enrichment in heavy elements by a series of stellar generations within the cluster. Because of this homogeneity, each cluster may be characterized by the abundance parameter $[Fe/H] = \log(Fe/H)_{cluster} - \log(Fe/H)_{Sun}$, which for our galaxy is found to span the range from -2.6 (iron underabundant by a factor of 400 relative to solar) to ~ 0.0 (solar abundance). The data for the galactic globular cluster system are shown in Fig. 1.

The bimodality of the abundance distribution is clear. The low-abundance peak corresponds to clusters that occupy a spheroidal volume of galactocentric radius of some 40 kpc (the halo subsystem), whereas the more metal-rich group comprises clusters that occupy a disk-like configuration confined near the plane of the Galaxy and having an exponential scale height of approximately 1 kpc (disk subsystem). The simplest explanation of the halo abundance distribution is provided by a one-zone model in which gas was progressively enriched chemically by being processed through massive stars *and* in which material was expelled from the system at a rate proportional to that of star formation. The latter condition is driven by the requirement that because the peak of the distribution occurs at $[Fe/H] = -1.5$, gas exhaustion should have occurred before the abundance reached higher values. The fact that there appears to be a well-defined break between the halo and disk peaks in Fig. 1 shows that the halo phase was distinct from that of the disk and is indicative of there having been two formation phases for the cluster system. The interrelationship of these subsystems is the center of considerable current investigation.

An extremely important fact about the halo subsystem is that there appears to be no gradient of cluster heavy element abundance as a function of galactocentric distance, in contradistinction to what might have been expected from earlier ideas about the formation of the Galaxy involving a rapid monolithic collapse in which the proto-Galaxy enriched itself as it contracted. The lack of a gradient is suggestive of a more chaotic and clumpy formation of the halo.

As implied previously, the case of ω Cen is of particular interest. Here there are well-established star-to-star variations of most of the elements. Iron varies by a factor of at least 4. The simplest explanation is that the system experienced self-enrichment over several generations of stars. Because objects of different mass synthesize various elements to different extents, the quantity [Metal/Fe] as a function of [Fe/H] has the potential to place strong constraints on our understanding of both the sites of production of various elements and the mass function of the earliest stellar generations. Most of the interesting results of this type have come from studies of Population II field stars, because of the greater ease with which the very high resolution spectra necessary for the endeavor can be obtained. ω Cen presents an unparalleled opportunity, as yet not fully realized, to examine this general question of how a system chemically enriches itself (see Fig. 2).

HELIUM

One of the basic predictions of the simplest Big Bang cosmological models is that the abundances of helium was initially $Y = 0.20$–0.30 (depending on initial parameters). Globular clusters, as the oldest well-dated objects in the universe, provide an important constraint on this question. The interpretation of helium lines in the spectra of hot horizontal-branch stars in clusters provides a relatively direct determination of this quantity, and it was exciting in the 1960s to find that most observations favored a value some 10 times lower than that found in the solar neighborhood. The issue was resolved by the discovery that the field star counterparts of these stars are peculiar in the sense that the abundances of other elements, such as phosphorus, appeared anomalously high (presumably as the result of surface diffusion effects as are believed to be responsible for the peculiar A star phenomenon), which led to the conclusion that horizontal-branch B stars are unreliable indicators of a cluster's helium abundance.

Indirect methods are therefore necessary. In the 1960s it was also argued that the effective temperature of the blue edge of the RR Lyrae instability strip was a good indicator of helium abundance, and values were determined that were quite consistent with the predictions of the standard Big Bang model. The method, however, was shown to be very sensitive to the treatment of convection in the model calculations and appears to have fallen into disrepute. Fortunately, stellar evolution calculations show that the ratio of the number of horizontal-branch stars to that of red giants is very sensitive to Y. The most recent comparisons of observations of a number of clusters with theory suggest that $Y=0.23$ with dispersion 0.02, providing an important limit to any suggested range in helium in the globular cluster system.

CARBON, NITROGEN, AND OXYGEN

The abundance patterns of CNO are extremely complicated. No cluster has been found to be chemically homogeneous with respect to these elements. Although the situation for oxygen is not yet clear, intracluster variations of carbon and nitrogen of factors of 2–10 are well-established and common. The simplest explanation is that stellar evolution, via some mixing phenomenon, has modified the abundances of the elements in the outer layers of the cluster stars. Such large variations are not produced by current standard stellar evolution theory, which predicts variations of only tens of percent for these elements. Clearly the data point to a need for a revision of the theory.

Of particular importance is the fact that the behavior of carbon and nitrogen differs systematically in passing from the more metal-poor clusters to the more metal-rich ones. In the very metal-poor M 92 ([Fe/H] $= -2.2$), for example, there is evidence that the carbon abundance steadily decreases as stars evolve from subgiant to giant to asymptotic giant branch. In contrast, in the more metal-rich system 47 Tucanae ([Fe/H] $= -0.7$), although there appear to be anticorrelated bimodal distributions of carbon and of nitrogen, there is no compelling reason to believe that the degree of abundance anomaly increases with advancing evolutionary status. This basic difference is suggestive of a need for a mixing mechanism that operates more readily at lower heavy element abundance; indeed, standard stellar evolution calculations show that in red giants the shell in which CNO processing is occurring is more readily accessible for mixing to the surface in stars of lower abundance. Thus, although little of such mixing occurs in the standard theory there is a growing suspicion that more realistic calculations, including the effects of phenomena such as rotation, may indeed come closer to reproducing the observations.

For completeness it should be noted that evolutionary mixing is probably not the full solution to the problem. Variations in the cyanogen bands in the spectra of stars at the main-sequence turnoff of 47 Tuc, together with star-to-star variations in the apparent abundance of sodium and aluminum that correlate with nitrogen variations in several clusters, although well-documented, are little understood and may require an explanation involving effects primordial to the cluster, rather than related to evolutionary phenomena.

The determination of the oxygen abundance of globular clusters is technically very difficult because it is based on analysis of two very weak forbidden O I lines. Further, the measurements to date pertain of necessity to the bright cluster giants, and it has been argued in some cases at least that the oxygen abundances show the signature of CNO processing, leading to the suspicion that giant-based oxygen abundances may not reflect a cluster's initial abundance. There are important repercussions for the age dating of globular clusters in this problem, because the value of [O/Fe] is one of the factors influencing the position of the main-sequence turnoff in the color–magnitude diagram. One is faced with two choices: Either one accepts current globular cluster determinations, which yield [O/Fe] ~ 0.3 with large scatter or, if one believes that the cluster giant results are invalid, one assumes that the Population II field dwarfs, which yield [O/Fe] ~ 0.6–1.2, give a better estimate of the primordial cluster value. This difference in [O/Fe] leads to an uncertainty in globular cluster age determinations of ~ 1–3 Gyr.

Additional Reading

Cayrel de Strobel, G., Spite, M., and Lloyd Evans, T., eds. (1989). *The Abundance Spread within Globular Clusters: Spectroscopy of Individual Stars*. Observatoire de Paris, Paris.

Freeman, K. C. and Norris, J. (1981). The chemical composition, structure, and dynamics of globular clusters. *Ann. Rev. Astron. Ap.* **19** 319.

Kraft, R. P. (1979). On the nonhomogeneity of metal abundances in stars of globular clusters and satellite subsystems of the Galaxy. *Ann. Rev. Astron. Ap.* **17** 309.

Shaver, P., Kunth, D., and Kjär, K., eds. (1983). *Proceedings of ESO Workshop on Primordial Helium*. European Southern Observatory, Garching.

Smith, G. H. (1987). The chemical inhomogeneity of globular clusters. *Pub. Astron. Soc. Pacific* **99** 67.

See also **Galaxy, Chemical Evolution; Star Clusters, Globular, Formation and Evolution; Star Clusters, Globular, Stellar Populations; Star Clusters, Globular, Variable Stars.**

Star Clusters, Globular, Extragalactic

William E. Harris

The extremely old star clusters that inhabit the *halo* or spheroidal region of a galaxy are rather commonly found in galaxies other than our own Milky Way. A *globular cluster system* (GCS) is the ensemble of all such clusters belonging to one galaxy, and can be treated as a dynamically distinct subsystem of the halo (see Fig. 1). Typically, the GCS comprises only about 1% of the total halo light (or mass in the form of visible stars), and so it must represent a special subset of all the matter found in the halo: Either the globular clusters we see now are merely the ones that have survived a long series of erosive processes since their epoch of formation in the early Galaxy, or else they were formed under initial conditions unlike those for the vast majority of halo stars. Because the globular clusters are among the oldest visible objects anywhere, comparing the GCSs in different galaxies is a way to find out how different (or how similar) the first processes of star formation were in different protogalactic environments.

The intrinsic luminosity of an average globular cluster is about $10^5\ L_{\odot}$ (solar luminosity units) and the largest ones attain almost $2\times10^6\ L_{\odot}$. They can therefore be detected in relatively distant galaxies (with CCD imaging on large ground-based telescopes, the brightest globular clusters are visible around large galaxies as remote as ~ 100 Mpc), but the detail in which they can be studied falls dramatically with distance. For galaxies more distant than $d\sim 2$ Mpc, the individual stars in the globular clusters are no longer directly resolvable; for $d \gtrsim 5$ Mpc, the clusters themselves have angular sizes $\lesssim 1$ arcsec and so appear star-like on photographs or digital images (these distance limits should be multiplied by ~ 5 for the higher angular resolutions that are possible by imaging from space). Thus for most galaxies, the GCS is detectable primarily as an excess population of star-like images concentrated around the central spheroid and halo.

The existence and structure of the GCS in our own Milky Way was first recognized in the early part of this century by Harlow Shapley; in his classic work of 1918 he used the space distribution of the globular clusters to derive the distance of the Sun from our galactic center for the first time. In the early 1930s, Edwin P. Hubble discovered the GCS around the nearest large galaxy, M 31 (the "Andromeda nebula"); in the 1950s and early 1960s, especially with the work of William A. Baum, Allan R. Sandage, and

Figure 1. The giant elliptical galaxy Messier 87, at the center of the Virgo cluster. The hundreds of faint star-like images concentrated around its halo are its brighter globular clusters; the entire M 87 globular cluster system extends detectably far beyond the borders of this illustration. Photograph taken by Malcolm G. Smith with the 4-m telescope at Cerro Tololo Interamerican Observatory *(AURA Inc.)*.

Rene Racine, GCSs were found around many of the large elliptical galaxies in Virgo. In M 31, which contains roughly twice as many globular clusters as does the Milky Way, it is possible to study the integrated spectral properties of the clusters in much detail, and even to measure color-magnitude diagrams for many individual clusters. These measurements reveal certain differences from the Milky Way globular clusters that may be due to slightly different ages or abundance ratios, but in general they demonstrate that the globular clusters in these Local Group galaxies are basically the same objects. For clusters in much more distant systems, low-dispersion spectra and photometric indices must be used as more approximate indicators of composition, but the pattern of first-order similarity is maintained to a remarkable degree. When other global characteristics of the CGSs are added (see below), the existing evidence suggests that the clusters we see in the halos of spiral galaxies, disk galaxies, dwarf ellipticals, and giant ellipticals are all generically similar; that is, they are old, populous star clusters with heavy-element abundance ratios ranging from near-solar down to about 1/100 solar. The presence of GCS therefore appears to represent a common theme in the very earliest evolutionary stages of large galaxies.

Obtaining accurate observations for globular clusters in galaxies much beyond the Local Group was a formidable task when photographic emulsions were the principal detectors for astronomical imaging and spectroscopy. The advent of CCD detectors in the mid-1980s, with their enormously higher quantum efficiency and linearity, began to turn this field into a mature observational subject and brought a far wider range of galaxies into reach. Nevertheless, the large-scale properties of a GCS must, in most cases, be characterized by just a few simple quantities:

The *total number* of clusters present in the system.
Their *spatial distribution* around the parent galaxy.
The internal *dynamics* of the system (the average motions of the clusters in the halo).
The number of clusters at any given brightness (i.e., their *luminosity function* or LF).
The spectroscopic properties of the clusters or (more crudely) their photometric colors, indicative of their *chemical composition*.

GCSs have now been observed in several dozen galaxies. In general, bigger galaxies have more clusters, the number increasing roughly in direct proportion to the total spheroid luminosity (i.e., the amount of old stellar population present) of the parent galaxy. For elliptical and S0 galaxies, the "spheroid" luminosity in this sense means essentially that of the entire galaxy. For disk galaxies of type Sa to Sb, the halo is relatively less prominent, but virtually all the star clusters in the halo and central spheroid region still resemble the classic old globular type. For galaxies still further along the Hubble sequence (types Sc, Sd, and irregular) the stellar component is almost completely dominated by the younger Population I, the visible halo is nearly negligible, and very few star clusters can be found that are plainly classifiable as old globular. The prototypes of this category are the Magellanic Clouds, in which only a handful of star clusters have the unambiguous defining marks of extreme age (halo-type space motions, low metallicity, and color-magnitude diagrams with well developed red giant and horizontal branches or RR Lyrae variables).

The ratio of total cluster population to *spheroid* (halo) luminosity is called the *specific frequency* S of the GCS. This ratio is a convenient index of comparison for galaxies of different types and sizes. Elliptical galaxies generally have 2–3 times higher S-values (that is, relatively more globulars) than disk or spiral galaxies. The surrounding environment of the parent galaxy may, however, have an equally important effect: E galaxies in rich groups such as the Virgo cluster tend to have higher specific frequencies than those in sparse groups, and some giant E galaxies that are sitting at the centers of rich clusters have enormously larger globular cluster populations, by about 3 times the normal numbers. Because it is a *ratio*, S is relatively unaffected by interactions between galaxies (except for the most extreme total mergers), because encounters between galaxies exchange or eject both halo stars and clusters in similar relative numbers. Thus differences in specific frequency from galaxy to galaxy suggest that the formation rate of globular clusters in protogalaxies varied depending on the surrounding density and type of other protogalaxies.

The true spatial extent of a GCS can be enormous (reaching to galactocentric radii larger than 100 kpc in the biggest giant ellipticals). An important feature of their spatial distribution in large ellipticals is that the GCS is often *less centrally concentrated* than the halo of the galaxy itself. If their space distribution (number per unit volume) is approximated by a power law $\rho \sim r^{-n}$, then within the same galaxy the exponent n(GCS) may be smaller than n(halo) by $\gtrsim 0.5$. However, in other galaxies, such as smaller ellipticals or disk systems, the GCS space distribution seems to follow the spheroid light distribution more closely. A common phenomenon for large galaxies is that the innermost "core" ($r \lesssim 2$ kpc) of the GCS has a flat, near-uniform density; either the clusters did not form there as easily, or (perhaps more likely) they were depopulated by the destructive processes that become extremely effective very close the nucleus of the galaxy, such as dynamical friction and tidal shocking.

The internal dynamics of a GCS may be studied by measuring the radial velocities of the individual clusters. Data of this type have been obtained for only a few large galaxies (such as the Andromeda galaxy M 31, the large nearby elliptical NGC 5128, and the Virgo ellipticals M 87 and M 49). However, these data confirm the overall pattern seen in the Milky Way halo clusters, that is, that the GCS is a dynamically "hot" system in which the clusters follow a wide range of randomly oriented, elliptical orbits with little or no overall rotation around the galactic center. At any galactocentric radius, the *velocity dispersion* (the scatter of the individual cluster velocities around the mean) can be employed as a direct estimate of the mass of the galaxy $M(r)$ contained within radius r, through the generalized viral expression

$$\alpha GM(r) = \langle r \cdot v_r^2 \rangle,$$

where v_r is the measured radial velocity of each cluster relative to the mean of the entire system and α is a constant determined by the mean orbital characteristics of the clusters ($\alpha \sim 0.1$ for an isotropic velocity distribution). Results of this type have been used for selected galaxies (notably the giant ellipticals M 87 and NGC 5128, as well as the Milky Way) as important confirmation that large amounts of "dark matter' dominate the mass distribution of galaxies in their outer regions [the velocity dispersion v_r is found to be approximately the same at any radius r, so $M(r)$ must increase in direct proportion to r].

The luminosity function (LF) for globular clusters is likely to be largely determined by their original mass spectrum of formation, because most clusters lie in the low-density halo regions of their parent galaxy where tidal disruption and other destructive effects are minor. Thus another important clue to the apparent near-universality of their formation process is that the LF has a characteristic shape that seems to be reproduced from one galaxy to another: Clusters with sizes near $10^5\ L_\odot$ are the most common, with ones at either higher or lower luminosity being less frequent. The number of clusters at a given *magnitude* (i.e., the logarithmic luminosity) is empirically well matched by a simple gaussian function. Thus to first order, the LF is described by only three parameters: the total sample population N, the magnitude m_0 of the peak frequency, and the dispersion $\sigma(m)$ of the distribution. Existing data for several galaxies of a wide range of types and sizes indicate that $\sigma(m)$ is consistently in the range 1.3 ± 0.2 mag, and that m_0 is roughly constant (to within ~ 0.3 mag or $\pm 30\%$), but not enough is yet known about the LF parameters to establish in detail their systematic variation with galaxy type. If these features can be fully calibrated and better understood theoretically, they will become attractive "standard candles" for calibrating the extragalactic distance scale, because they are such luminous objects and can be found readily in quite distant galaxies.

Additional Reading

Grindlay, J. E. and Philip, A. G. D., eds. (1988). *Globular Cluster Systems in Galaxies. IAU Symposium* **126**. Kluwer Academic Publishers, Dordrecht.

Harris, W. E. (1991). Globular clusters in distant galaxies. *Sky and Telescope* **81** 148.

Harris, W. E. and Racine, R. (1979). Globular clusters in galaxies. *Ann. Rev. Astron. Ap.* **17** 241.

See also **Andromeda Galaxy; Magellanic Clouds; Star Clusters, Globular; Star Clusters, Globular, Stellar Populations.**

Star Clusters, Globular, Formation and Evolution

Richard B. Larson

The globular star clusters are notable as the largest and oldest clusters in our galaxy. There is no sharp dividing line between the globular clusters and the smaller open clusters; a convenient definition is that globular clusters are systems with masses greater than 10^4 solar masses ($M_\odot$). Few, if any, of the young or intermediate-age clusters in our galaxy are this massive; all such massive clusters appear to be very old, with ages exceeding 10^{10} yr. Therefore they are relics of the early history of our galaxy; indeed, understanding the origin of these massive clusters may even hold the key to understanding how the galaxy was formed. Most of the globular clusters in our galaxy are relatively metal-poor and are distributed in an extended spherical halo, but the most metal-rich ones appear to belong to the galactic disk and are distributed like the oldest disk stars. The globular clusters may thus trace the early development of both the halo and the disk components of our galaxy.

Because of their larger sizes, globular clusters have longer lifetimes than open clusters, but all star clusters eventually disintegrate. It is of interest to understand how they are destroyed as well as how they are formed, because the presently observed clusters may be just the survivors of a once much larger population; their observed properties will then depend on both formation and destruction processes. Also, many field stars in both the halo and the disk of our galaxy may have originated in now-disrupted clusters; if so, fundamental stellar properties such as the stellar mass spectrum and the incidence of binaries may be determined in part by processes that occur in clusters.

FORMATION

Because no well-studied young clusters are as massive as a typical globular cluster, we have no direct observational information about the formation of globular clusters. However, the apparent continuity in properties between the globular and the open clusters suggests that their formation processes may share some basic similarities. Infrared studies of nearby regions of star formation have revealed that the birthplaces of open clusters are the dense and heavily obscured cores of giant molecular clouds. Embedded in some of these cores are extremely compact clusters of young stars; a well-known, partly visible example is the Trapezium cluster in the Orion molecular cloud. The efficiency of cluster formation is always very small: typically only about 2×10^{-3} of the mass in a star-forming cloud complex goes into a probable bound open cluster. The efficiency of cluster formation is thus only about 10% of the overall efficiency of star formation.

Among known systems, the closest analog to a young globular cluster may be the massive young cluster embedded in the 30 Doradus nebula of the Large Magellanic Cloud. This system may have a bound mass of the order of $10^4\ M_\odot$, marginally qualifying as a globular cluster. Like the Trapezium cluster and several other young clusters, the 30 Doradus cluster is highly centrally condensed and is dominated by a central multiple star system containing some of the most massive stars in the cluster. Also, like the others, the 30 Doradus cluster contains only about 10^{-3} of the total mass of the associated star-forming region.

Evidently the formation of a typical globular cluster requires more extreme conditions than are found locally. It has been speculated that such conditions may exist in starburst galaxies, where exceptionally vigorous star formation is taking place, perhaps caused by the accumulation of gas into particularly large and dense clouds. If the formation of globular clusters is as inefficient as that of open clusters, very massive star-forming complexes may be required that have the masses of small galaxies. A current hypothesis is that most of the globular clusters in galaxies were formed in protogalactic subsystems that evolved for a time as independent small galaxies before being merged to make galactic halos. If the globular clusters in our galaxy have an age spread of several billion years, as some of the present evidence suggests, this would support the idea that the globular clusters formed in separate subsystems that later merged.

EVOLUTION

After their formation, star clusters expand considerably as a result of mass loss, which tends to unbind them. First the residual gas is blown away, and some massive stars are soon ejected by the disintegration of compact subsystems like the Trapezium. On a longer time scale, mass loss associated with stellar evolution causes continuing gradual cluster expansion; in globular clusters, whose time scales for dynamical evolution are relatively long, mass loss may remain the dominant effect driving cluster evolution for several billion years. Most star clusters probably do not survive this initial period of expansion and thus are soon disrupted.

Eventually, after most of the stars more massive than 1 $M_{\odot}$ have already evolved, mass loss becomes relatively unimportant and stellar-dynamical effects begin to dominate. The primary dynamical effect is two-body gravitational encounters, which tend to cause "relaxation" of the stellar velocity distribution toward the Maxwellian form characteristic of perfect thermodynamic equilibrium. The rate of dynamical evolution is governed by the relaxation time at the radius containing half of the cluster mass, which numerically is about 0.01 times the number of stars in the cluster multiplied by a typical stellar crossing time.

Because the stars in the tail of a Maxwellian velocity distribution are not bound to the cluster, the effect of two-body encounters is to cause a steady evaporation of stars from the cluster. The binding energy of the cluster is then shared among progressively fewer stars, so that the remaining cluster, or part of it, must contract. Many calculations have shown that the cluster in fact becomes increasingly centrally condensed and that, although the half-mass radius changes little, the central density increases indefinitely and approaches infinity after a finite time, a phenomenon known as core collapse. For a cluster of equal-mass objects without mass loss, core collapse occurs after 15 half-mass relaxation times. A spread in stellar masses can accelerate core collapse, because the most massive stars tend to sink rapidly toward the center; however, this acceleration is almost cancelled by the effects of stellar mass loss, and in the most realistic models core collapse occurs after about 12 relaxation times. By this time, the cluster has already lost about three-quarters of its original mass, if subjected to a realistic galactic tidal force field, and it is well on its way to complete evaporation; this is predicted to occur after about 18 relaxation times.

The estimated relaxation times of globular clusters range from about 10^8 yr to 3×10^{10} yr, with a median of 1.4×10^9 yr. Thus, for a typical cluster, core collapse is expected to occur after about 17×10^9 yr and complete disruption after perhaps 25×10^9 yr. At least one-quarter of the observed globular clusters in our galaxy should already have undergone core collapse; in fact, about one-fifth of them exhibit central cusps in their light profiles and thus appear to contain collapsed cores. The observed cusps are less steep than the predicted surface density profiles of collapsed cores, but this is probably because the cores are dominated by dark stellar remnants that are more massive and more centrally concentrated than the visible stars, so that the visible stars do not closely trace the mass distribution.

Core collapse is eventually halted when the central region contains only a few tens of stars and interactions involving three or more stars begin to produce binaries. These binaries become progressively more tightly bound and transfer energy to the rest of the core, reversing core contraction and even ejecting stars that come too close. Eventually the binaries themselves are ejected by very violent encounters, and core contraction resumes and generates more binaries. Some calculations predict large oscillations in the core density during this poorly understood "postcollapse" phase of evolution. Binaries present from the time of formation of the cluster and binaries formed by tidal capture during very close encounters could also play important roles during the late stages of core evolution.

Another possibility is that cluster cores are dominated by heavy stellar remnants almost from their time of formation, when the central regions are dominated by massive stars. The two most massive remnants, possibly stellar-mass black holes, might then form a relatively stable and long-lived central binary. Runaway coalescence of stars and/or stellar remnants conceivably could even lead to the formation of a single massive black hole at the center. Such an object would then halt core collapse by swallowing any stars that come too close.

Whatever may happen in detail at the center, the bulk of the cluster remains relatively unaffected and continues to evaporate at a nearly constant rate driven by two-body relaxation. During the final stages of cluster evolution, the visible stars are preferentially ejected and the system becomes increasingly dominated by stellar remnants. The relaxation time continues to decrease as the mass decreases, and finally the cluster disintegrates completely in a short time, presumably leaving only a single binary system or black hole.

DESTRUCTION

For most globular clusters, the evaporation process just described is by far the most important destruction mechanism. For a typical cluster, complete evaporation is predicted to occur after about 25 billion years, a little longer than the present age of the universe. This coincidence is probably not an accident, because it is likely that many clusters initially had lifetimes shorter than the present age of the universe and that we now see only those that happened to have been formed with longer lifetimes.

Because the evaporation time of a cluster is proportional to its mass and inversely proportional to the square root of its mean density, smaller clusters generally have shorter lifetimes; thus, most clusters with masses much smaller than the present typical mass would not have survived to the present time. Also, because galactic tidal forces limit the radii of clusters and constrain their mean densities to be higher in the inner regions of galaxies, clusters in these regions will generally have shorter lifetimes. Some support for this prediction is provided by the fact that the inner part of our galaxy contains a relatively high proportion of clusters with cusped light profiles, which are believed to be the most evolved clusters.

The destruction of some clusters may be hastened by effects associated with their orbital motions. Massive clusters in the innermost regions of a galaxy may spiral into the center as a result of the "dynamical friction" caused by gravitational interaction with the background field stars. Also, clusters whose orbits take them through the central bulge or through the inner disk of a galaxy are subjected to "tidal shocks" that tend to disrupt or strip away the outer parts of the cluster. This accelerates the eventual dispersal of the cluster and also preferentially removes the low-mass stars, which are concentrated in the outer parts of the cluster.

All of the destruction mechanisms that have been mentioned operate most rapidly and effectively in the inner parts of galaxies, so they should preferentially deplete the globular cluster populations in these regions; this may help to explain why globular cluster systems are less centrally concentrated than the background stars in the galaxies. In fact, it is only the relative weakness of tidal and other destructive effects that has allowed some fragile clusters, such as the relatively sparse and remote Palomar clusters, to survive until the present in the outer reaches of our own galaxy.

Additional Reading

Chernoff, D. F. and Weinberg, M. D. (1990). Evolution of globular clusters in the Galaxy. *Ap. J.* **351** 121.

Elson, R., Hut, P., and Inagaki, S. (1987). Dynamical evolution of globular clusters. *Ann. Rev. Astron. Ap.* **25** 565.

Goodman, J. and Hut, P., eds. (1985). *IAU Symposium No. 113, Dynamics of Star Clusters.* D. Reidel, Dordrecht.

Grindlay, J. E. and Philip, A. G. D., eds. (1988). *IAU Symposium No. 126, Globular Cluster Systems in Galaxies.* Kluwer Academic, Dordrecht.

Merritt, D., ed. (1989). *Dynamics of Dense Stellar Systems.* Cambridge University Press, Cambridge, U.K.

Spitzer, L., Jr. (1987). *Dynamical Evolution of Globular Clusters.* Princeton University Press, Princeton.

See also **Galactic Structure, Globular Clusters; N-Body Problem; Star Clusters, Globular, Galactic Tidal Interactions; Star Clusters, Globular, Gravothermal Instability; Star Clusters, Globular, Mass Loss; Star Clusters, Globular, Mass Segregation.**

Star Clusters, Globular, Galactic Tidal Interactions

Luis A. Aguilar

Globular clusters are not isolated systems, but live within the gravitational fields of their parent galaxies. The effect of the external potential is manifested as a tidal force that can be split into two components: a constant term and a time-dependent part. The constant term imposes a cutoff to the spatial extent of the cluster, the tidal radius r_t, which is the place where a star is no longer bound to the cluster but to the galaxy. The second term, being time-dependent, does work on the cluster and thus produces a secular evolution of the cluster. These theoretical arguments are in agreement with observations: Globular clusters in our galaxy have a surface brightness profile that approaches zero at a finite distance from the cluster center, and, furthermore, the observed tidal radii vary as expected with position in the Galaxy. The clusters closer to the galactic center are the smaller ones. This last trend is only approximately due to the eccentricity of the clusters' orbits, because their tidal radii are imposed at the time of closest approach to the galactic center, which is not, in general, the place where they are likely to be at the present time. Because the boundary of the cluster relaxes on a time scale comparable to its orbital period around the Galaxy, the cluster does not have time to expand to a larger tidal radius before it passes close to the galactic center again.

TIME-INDEPENDENT EFFECTS

The effect of the constant part of the tidal force is the easier to model. In this case we consider the motion of a star that moves subject to the gravitational attraction of the cluster and the galaxy, the latter two moving in circular orbit around each other. The energy of the star, measured in the rotating frame of reference fixed with respect to the cluster and the galaxy, plus a term that can be thought of as the "potential energy" of the centrifugal force, is a constant (called the Jacobi energy). It is clear that a star whose Jacobi energy is less than that of the first equipotential that no longer encloses the cluster alone, cannot escape from the cluster. Traditionally, the intersection of this equipotential (the Roche lobe) with the line that joins the cluster and the galactic center, is taken as the theoretical definition of tidal radius. This argument, however, neglects the influence of the Coriolis force which reinforces the cluster's attraction for stars in retrograde orbits (orbiting the cluster in the opposite sense from that in which the cluster moves around the galaxy), whereas it opposes it for prograde orbits. Stars in, or near, retrograde orbits can thus extend beyond the Roche lobe. This effect makes it difficult to define a tidal radius because whether a particular star escapes or not depends on factors other than just its energy. This ambiguity, however, is not very important because the observed tidal radii are very difficult to measure and because the cluster's orbits are, in general, not circular, so that we have to contend with the time-fluctuating part of the tidal force.

TIME-DEPENDENT EFFECTS

The amplitude of the time-dependent part of the tidal force upon a star depends on the ratio of the time scale for change in the tidal force to the orbital period of the star around the cluster center. If this ratio is much larger than 1 (the so-called adiabatic regime), the external force "sees" the star as a distribution of mass smeared over the stellar orbit and it affects the overall position of the orbit, but not its shape. This means that the star does not acquire energy from the tidal force because it is protected by adiabatic invariants. When the ratio of time scales is much less than 1 (the so-called impulsive regime), the star experiences an "instantaneous" encounter with the galaxy that injects energy into it and makes it less bound to the cluster. This phenomenon is called a tidal shock. The pioneering work on this subject was that of Lyman Spitzer, Jr., in 1958, who was interested in the destruction of galactic clusters due to tidal shocks. Spitzer modeled this phenomenon by making an analogy with a forced harmonic oscillator: He assumed that the cluster's potential is harmonic and then computed the energy transferred to a star by an external tidal force. He found that for nonimpulsive encounters, the efficiency of energy transfer to the cluster diminishes exponentially with the ratio of time scales we referred to above. He then used the orbital period of a star in circular orbit at the half-mass radius to characterize the effect of the shock on a cluster. Because in a globular cluster the mass density, and therefore the orbital period of stars, varies over several decades from the center to the tidal radius, the efficiency of a tidal shock changes dramatically as well. The author, Piet Hut, and Jeremiah P. Ostriker have most recently extended this model by evaluating the efficiency of the shock at each radial position within the cluster and then integrating over its density profile. They found that the net efficiency of a tidal shock depends not only on the half-mass radius but also on the concentration of the density profile, being more important for clusters with the most extended envelopes. Two clusters of the same mass and half-mass radius that move on the same orbit can thus be affected very differently.

TIDAL SHOCKS

The physical phenomena responsible for the tidal shocking of the globular clusters in our galaxy are: encounters with the central bulge of our galaxy, crossing of the galactic disk, and encounters with large inhomogeneities in the mass distribution of the Galaxy (e.g., molecular clouds or any putative massive black holes in the galactic halo).

Encounters with the central bulge of our galaxy have been considered recently by the author, Hut, and Ostriker, who found that this destruction mechanism can be very effective provided that the cluster's orbit has a small perigalacticon and a large eccentricity. The small perigalacticon is necessary to have a large tidal force; the large eccentricity is required to make the encounter impulsive. These authors find that if we extrapolate the effect of these encounters for the age of the Galaxy, the selective destruction of clusters in radial orbits can significantly alter the observed kinematics of the surviving population of globular clusters and can account for the observed difference between the velocity distributions of globular clusters and other tracers of the halo population. This phenomenon could perhaps also explain why the present-day spatial distribution of globular clusters is less centrally peaked than that of the spheroidal population.

Another source of tidal shocks is the crossing of the galactic disk. Because the scale height of the galactic disk is larger than the tidal radii of globular clusters, the clusters are completely immersed in the disk at some point in their orbits, and, because this crossing lasts only a small fraction of the cluster's orbital period, the crossing is impulsive to the stars close to the tidal radius and can lead to the cluster's destruction. David F. Chernoff and Stuart L. Shapiro have made a very detailed study of this phenomenon. They follow the evolution of a globular cluster under the combined effect of tidal shocks due to galactic disk crossings and giant molecular clouds under the assumption that the cluster evolves through a series of King models. They find that disk heating dominates and can strongly affect the evolution of clusters within 5 kpc of the galactic center. Clusters within 3 kpc of the center evolve very rapidly and attain a state of infinite central density (the so-called core collapse). Clusters within 3 and 5 kpc are either forced to undergo core collapse or are dissolved, depending on their initial concentration. This trend may have been observed already; Stanislav Djorgovski recently reported that the most concentrated clusters, as well as the clusters that present a power-law density profile in their central region and thus are candidates to be post–core-collapse clusters, tend to be closer to the center of the Galaxy. The effect of giant molecular clouds, on the other hand, is completely negligible according to Chernoff and co-workers.

Various authors have postulated the existence of massive (10^6 $M_\odot$) black holes as the major constituents of the galactic dark halo responsible for the flat rotation curve of the Galaxy. Roland Wielen has studied the effect of such a population of massive components on the population of globular clusters. He finds that the effect is strongly dependent on the mean density of the clusters, and that for clusters of median mean density, the vast majority may have already been destroyed.

SUMMARY

The effect of the galactic tidal field upon the system of globular clusters of our galaxy is quite important and has to be taken into account when we study or draw conclusions from it. It affects present day globular clusters and may have destroyed the vast majority of an initial population of clusters. In this sense, the present day characteristics of globular clusters may be largely dependent upon these external forces that shape them. The effect of the tidal field of the parent galaxy may also explain, at least in part, the observed differences in number of globular clusters per unit luminosity between spiral and elliptical galaxies, because the latter do not have a disk that can disrupt their clusters.

Additional Reading

Aguilar, L., Hut, P., and Ostriker, J. P. (1988). On the evolution of globular cluster systems I. Present characteristics and rate of destruction in our Galaxy. *Ap. J.* **335** 720.

Chernoff, D. and Shapiro, S. L. (1988). Tidal heating of globular clusters. In *Globular Cluster Systems in Galaxies. IAU Symposium* **126**, J. Grindlay and A. G. Davis Philip, eds. Kluwer Academic, Dordrecht, p. 283.

Chernoff, D. F. and Weinberg, M. D. (1990). Evolution of globular clusters in the Galaxy. *Ap. J.* **351** 121.

Djorgovski, S. (1988). Surface photometry of globular clusters. In *Globular Cluster Systems in Galaxies. IAU Symposium* **126**, J. Grindlay and A. G. Davis Philip, eds. Kluwer Academic, Dordrecht, p. 333.

Wielen, R. (1988). Dissolution of star clusters in galaxies. In *Globular Cluster Systems in Galaxies. IAU Symposium* **126**, J. Grindlay and A. G. Davis Philip, eds. Kluwer Academic, Dordrecht, p. 393.

See also **Galactic Structure, Globular Clusters; Star Clusters, Globular, Formation and Evolution; Star Clusters, Globular, Mass Loss.**

Star Clusters, Globular, Gravothermal Instability

Piet Hut

Globular clusters do not live forever. Just as a glass of water, left by itself, slowly loses its contents by evaporation, so do star clusters lose stars, in a process that resembles evaporation. In a glass of water, interactions between molecules occasionally give an unusually high velocity to an individual molecule, which can then escape from the water surface. In a star cluster, gravitational interactions between stars occasionally give a star a velocity that exceeds the escape velocity, causing the star to leave the cluster. This process of random exchange of energy between stars, as a cumulative effect of near and far hyperbolic encounters, is called two-body relaxation. As a consequence of the resulting escape of stars, the inner parts of a star cluster lose energy and therefore contract, a process that leads to a higher central density. In turn, this increase in density leads to a higher rate of two-body relaxation, which leads to a higher rate of escapes. Not all the stars that evaporate from the central region will actually leave the system; to sustain the central contraction, it is enough that they are removed from the core. The vicious circle of increased density leading to an increased rate of contraction leading to an even more increased density is a runaway process: Two-body relaxation predicts the occurrence of an infinite central density in a finite time.

Before discussing the solution to this paradox of predicting infinite increase in observational quantities, let us look at a few other self-gravitating systems that show similar behavior. The birth of a star, for example, is the end result of the contraction of a gas cloud: The cloud loses energy by radiation from its surface and therefore tends to grow colder, which lowers the internal pressure that was holding up the cloud against its own gravity. As a result, gravity takes the upper hand and the cloud contracts, paradoxically being heated (by an amount larger than the amount of heat lost by radiation at the surface) in the process. This is a consequence of the virial theorem, which states that the internal kinetic energy is proportional to the potential energy (half as large, in fact, and of opposite sign).

This phenomenon of *increasing* temperatures as a result of a *loss* of heat runs counter to our normal intuition, based on laboratory experiments, where the exact opposite holds true. It seems that the heat capacity of a self-gravitating object has a *negative* sign. In fact, this is a very basic aspect of gravity and is related to the fact that we cannot apply gravity to an infinite static and homogeneous medium: The Jeans catastrophe will break up such a medium into clumps with masses of the order of the Jeans mass, within a time scale corresponding to a crossing time through such a clump. In other words, we cannot apply the usual framework of thermodynamics to self-gravitating systems, partly because there is no meaningful thermodynamic limit. Another way of seeing this is by realizing that the energy of a self-gravitating system grows with the square of its mass, which implies that the energy grows faster than an extensive quantity. This poses a much more serious problem than some other laboratory phenomena, such as the surface tension of a soap bubble, where the associated energy grows at a speed in between that of extensive and intensive quantities.

Another, even simpler, example of the negative heat capacity associated with gravity can be found closer to home, in the form of a satellite orbiting the Earth: When such a satellite encounters some friction from the upper atmosphere, it tends to lose energy and, thereby, altitude. As a consequence, however, its orbital speed is actually *increased*, even though it has *lost* energy. Returning to our discussion of the dynamics of a globular cluster, we see that the accelerated collapse of the inner regions is a consequence of the inherent instability of gravity, which can be characterized by assigning a negative heat capacity to a self-gravitating system. This runaway process was first discovered, in the simpler but analogous case of a self-gravitating gas confined within a rigid sphere, by V. A. Antonov in 1962. It was later analyzed in more detail by Donald Lynden-Bell and Roger Wood who termed it a "gravothermal catastrophe."

GRAVOTHERMAL CATASTROPHE

Antonov found that isothermal equilibrium configurations can be stable only when the density contrast (the ratio between the density near the confining sphere and the central density) is less than a critical value, which he determined to be ≈ 709. He showed that for larger values, a small increase in central temperature leads to an accelerated contraction of the core. Conversely, a small decrease in the central temperature leads to a runaway expansion of the core. Direct confirmation of the validity of such gas models for stellar dynamical applications was provided by Shogo Inagaki and Haldan Cohn in 1980. Inagaki performed a linear stability analysis of a stellar dynamical system enclosed in a sphere, using the Fokker–Planck approximation to describe the interactions between the particles, and recovered Antonov's instability criterion for stellar dynamics. Cohn solved the Fokker–Planck equations for diffusion in energy space numerically, and closely reproduced the results of self-similar collapse calculations by Lynden-Bell and Peter P. Eggleton, based on gas dynamics. For more details we refer the reader to the monograph on globular cluster dynamics by Lyman Spitzer, Jr.

With the existence of the gravothermal catastrophe being firmly established, we have to investigate the fate of a globular cluster after core collapse. For guidance, let us again look at the corresponding situation of a gas cloud. The contraction of a protostar is halted when the core of the star starts burning nuclear fuel at a rate that balances the energy loss at the surface. Similarly, gravothermal collapse of the core of a globular cluster will be halted when a central source generates more energy than is effectively conducted out of the core by two-body relaxation. This notion of a central energy source was first discussed by Michel Hénon, who suggested that a small subsystem of the core could shrink enough to generate the energy required. For example, even the formation of a single close binary, with an orbital velocity about an order of magnitude larger than the velocity dispersion in the cluster, would release an amount of energy comparable to that of a hundred single stars.

Hénon's ideas were confirmed several years later by *N*-body calculations. For all values of N explored (10^1–10^3) a contraction of the central region produced at least one tight binary star with a binding energy comparable to the total binding energy of the system. Formation and hardening of such a binary then fueled the expansion and evaporation of the *N*-body system. (A review of this topic was given by Sverre J. Aarseth and Myron Lecar in 1975.)

However, a significant difference between the *N*-body calculations performed to date and globular clusters is that for clusters $N \sim 10^5$–10^6, whereas for the calculations $N \leq 10^{3.5}$. The large number of stars in a globular cluster implies that a single hard binary can absorb at most $\sim 1\%$ of the binding energy of the cluster: Attempts to form harder binaries lead either to a merging of the stars or to an escape of the binary from the cluster by the recoil momentum gained in an encounter with a third star. Thus, although in simulations each binary plays a dominant role, in real globular clusters only the cumulative effect of many binaries can change the local energy budget of the cluster significantly.

The first attempt to study the effects of binaries in systems with large N was made by Hénon in 1975 using a Monte Carlo Fokker–Planck code. He introduced an artificial energy source in the innermost part of his cluster model, which was tuned to give off just the amount of energy necessary to avoid collapse locally at the inner shell. He found that the cluster reached a maximum central density, after which the collapse was reversed into an overall expansion.

A decade later, several authors confirmed Hénon's results by following the evolution of much more detailed models. Around that time, observations by Stanislav G. Djorgovski and Ivan R. King of central brightness excess of the cores of several globular clusters suggested that a significant fraction of all globular clusters may already have undergone core collapse. A review of both the theoretical and observational situation was given by Rebecca Elson and co-workers in 1987.

POSTCOLLAPSE EVOLUTION

The evolution of a globular cluster after core collapse has only recently been studied intensively, and many aspects of our understanding of it remain uncertain and may change in the years following this review. The *mean* behavior of the cluster after core collapse, however, is firmly established: The half-mass radius of an isolated cluster expands according to $r_h(t) \propto t^{2/3}$, where t is the time since core bounce, whereas the velocity dispersion drops according to $v \propto t^{-1/3}$. This relation may be derived from general principles, without any knowledge of the mechanism of energy generation in the core, as was done by Hénon in 1965 and 1975, in a manner analogous to Arthur S. Eddington's prediction in 1926 of the mass–luminosity relation for stars, which requires no precise knowledge of the nature of their internal energy generation. The derivation goes as follows: (1) The half-mass relaxation time t_{hr} in a self-similar solution scales as $t_{hr} \propto t$, the time since core bounce; (2) $t_{hr} \propto N t_{hc}$, where N is the number of stars in the cluster, t_{hc} is the crossing time at the half-mass radius, and we have neglected a factor log N; (3) if we neglect the slow change in mass and particle number due to escape, the virial theorem gives $t_{hc} \propto r_h^{3/2}$; (4) combining these gives $t \propto r_h^{3/2}$, which leads to the results quoted previously. In contrast, the rate of expansion of the *core* does depend on the details of the central engine. This was illustrated by detailed calculations, performed by Cohn, by Ostriker, and by Statler and coworkers.

There are, however, strong indications that the evolution of globular clusters after core collapse is quite a bit more complicated than the simple picture described here. Several years ago, Erich Bettwieser and Daiichiro Sugimoto followed the evolution of gas sphere models, and found large oscillations in the size of the core radius, which they interpreted as a new physical phenomenon: gravothermal oscillations. They interpreted these gravothermal oscillations as yet another consequence of the negative heat capacity of gravity. In the previous section we saw how the gravothermal instability can lead to core collapse: If the central temperature is slightly too high, the core will lose more heat than it gains, and this will lead to a contraction and therefore a density increase, which in turn will produce a higher central temperature. After core collapse is reversed into core reexpansion, the opposite may occur: The expansion may lower the central temperature, leading to an energy flow into the core and, in turn, to a lowering of both core density and temperature. The result is a runaway expansion that proceeds much faster than dictated by the boundary conditions at the surface of the cluster. The expanding region in the cluster center will grow radially until it reaches a region of radially decreasing temperature. At this point the expansion halts and the central region starts to collapse again.

The gravothermal character of core oscillations in gas sphere models was confirmed explicitly by Jeremy Goodman, in a linear stability analysis of a new regular self-similar model for postcollapse evolution, which he constructed in the same paper. Also for a more realistic stellar dynamics model, based on Fokker–Planck approximations for the two-body relaxation, Cohn, the author, and Michael Wise confirmed the existence and gravothermal nature of core oscillations. However, it is not yet clear to what extent these models apply to real globular clusters, as opposed to gas-spheres and equal-mass point-particle models. The fact that the core of a cluster around the time of core collapse typically contains fewer than a hundred stars makes the use of statistical models questionable. Ultimately, full *N*-body calculations are needed to provide the answer to the question of the nature of gravothermal oscillations, as was shown in a detailed analysis by Hut and coworkers in 1988.

SUMMARY AND OUTLOOK

In many respects, models of globular cluster evolution have reached a point comparable to the state of stellar evolution theory in the 1950s, when nuclear reaction rates describing the physics of energy generation became available, and the computers necessary for the construction of detailed models were just being developed. The analogous two- and three-body gravitational reactions between stars were studied in detail by the author in 1985. These reactions turn on after the initial core collapse and heat the central regions of the cluster both directly, through the effects of energetic reaction products, and indirectly, via mass loss. They power the "main-sequence" phase of a globular cluster, and drive a slow but steady loss of stars by evaporation. Computationally, the ongoing development of new supercomputers and parallel computers promises, within a few years, speeds orders of magnitude greater than those available today.

The progress in our understanding of globular cluster dynamics during the last three decades has been impressive indeed. We now have a consistent standard picture of precollapse evolution, initiated during the 1960s and developed in detail during the 1970s. Some of the main ingredients of postcollapse evolution have emerged during the 1980s, filling in parts of what was still a blank spot on the map only a few years ago. However, major questions about the postcollapse phase remain unanswered. Our insight into the further stages of evolution, during and after core bounce, is much less complete. Fundamental questions, such as whether gravothermal oscillations will occur in realistic cluster models, are largely unsettled. We are still far from the point where we can construct models that can be compared directly with observations.

Additional Reading

Aarseth, S. J. and Lecar, M. (1975). Computer simulations of stellar systems. *Ann. Rev. Astron. Ap.* **13** 1.

Antonov, V. A. (1962). Die wahrscheinlichste phasenverteilung in sphärischen sternsystemen und ihre existenzbedingungen. *Vestnik Leningrad Univ.* **7** 135.

Bettwieser, E. and Sugimoto, D. (1984). Post-collapse evolution and gravothermal oscillation of globular clusters. *Monthly Notices Roy. Astron. Soc.* **208** 493.

Cohn, H. (1980). Late core collapse in star clusters and the gravothermal instability. *Ap. J.* **242** 765.

Cohn, H., Hut, P., and Wise, M. (1989). Gravothermal oscillations after core collapse in globular cluster evolution. *Ap. J.* **342** 814.

Djorgovski, S. G. and King, I. R. (1986). A preliminary survey of collapsed cores in globular clusters. *Ap. J. (Lett.)* **305** L61.

Eddington, A. S. (1926). *The Internal Constitution of the Stars.* Cambridge Univ. Press, London.

Elson, R., Hut, P., and Inagaki, S. (1987). Dynamical evolution of globular clusters. *Ann. Rev. Astron. Ap.* **25** 565.

Goodman, J. (1987). On gravothermal oscillations. *Ap. J.* **313** 576.

Hénon, M. (1961). Sur l'évolution dynamique des amas globulaires. *Ann. d'Astrophys.* **24** 369.

Hénon, M. (1965). Sur l'évolution dynamique des amas globulaires. II. Amas isolé. *Ann. d'Astrophys.* **28** 62.

Hénon, M. (1975). Two recent developments concerning the Monte Carlo method. In *Dynamics of Stellar Systems, IAU Symposium No. 69*, A. Hayli, ed. D. Reidel, Dordrecht, p. 133.

Hut, P. (1985). Binary formation and interactions with field stars. In *Dynamics of Star Clusters, IAU Symposium No. 113*, J. Goodman and P. Hut, eds. D. Reidel, Dordrecht, p. 231.

Hut, P., Makino, J., and McMillan, S. (1988). Modelling the evolution of globular star clusters. *Nature* **336** 31.

Inagaki, S. (1980). The gravothermal catastrophe of stellar systems. *Publ. Astron. Soc. Japan* **32** 213.

Lynden-Bell, D. and Eggleton, P. P. (1980). On the consequences of the gravothermal catastrophe. *Monthly Notices Roy. Astron. Soc.* **191** 483.

Lynden-Bell, D. and Wood, R. (1968). The gravo-thermal catastrophe in isothermal spheres and the onset of red-giant structure for stellar systems. *Monthly Notices Roy. Astron. Soc.* **138** 495.

Ostriker, J. P. (1985). Physical interactions between stars. In *Dynamics of Star Clusters, IAU Symposium No. 113*, J. Goodman and P. Hut, eds. D. Reidel, Dordecht, p. 347.

Spitzer, L., Jr. (1987). *Dynamical Evolution of Globular Clusters.* Princeton University Press, Princeton, N.J.

Statler, T. S., Ostriker, J. P., and Cohn, H. (1987). Evolution of N-body systems with tidally captured binaries through the core collapse phase. *Ap. J.* **316** 326.

See also **N-Body Problem; Star Clusters, Globular, Formation and Evolution; Star Clusters, Globular, Mass Loss; Star Clusters, Globular, Mass Segregation.**

Star Clusters, Globular, Mass Loss

Donald J. Faulkner

Current ideas about the late stages of the evolution of low mass stars, such as those presently ending their lives in the globular clusters of our galaxy, suggest that they lose an appreciable fraction of their mass before collapsing to become white dwarfs. In spite of this, observational searches for interstellar material within clusters have largely proved negative and have set upper mass limits for the cluster gas and dust content that are well below the amounts that would be present, in many clusters, were they to retain all the gas lost from stars. Many attempts at resolving this discrepancy have been made during the last 20 years, but no proposed explanation has, as yet, been conclusively verified.

STELLAR MASS LOSS IN CLUSTERS

Evidence that mass loss occurs from stars within globular clusters is as follows:

1. Studies of field white dwarf stars indicate that their masses are typically 0.5–0.6 times that of the Sun. However, a star that began life with such a mass would take considerably longer than the age of the Universe to evolve. In a 0.6 $M_\odot$ star, the hydrogen-burning phase takes well over 25 Gyr for any reasonable stellar composition. Thus observed white dwarfs must have lost mass at some earlier stage of their evolution. In globular clusters, the stars that are presently finishing the long-lived, hydrogen-burning, main sequence phase of their lives are ~ 0.8 $M_\odot$, so that we can expect them to lose $\sim 25\%$ of this mass before becoming white dwarfs.

2. After a low mass star has exhausted its central hydrogen and has swollen up to become a red giant for the first time, there follows a phase in which it burns both helium at its center and hydrogen in a surrounding shell. In the case of globular clusters, such stars occupy a distinctive horizontal branch in the cluster color-magnitude diagram. An acceptable match between this observational feature and the theoretical models for the stars that correspond to it can be obtained only if the mass assumed for them is ~ 0.2 $M_\odot$ less than that of stars presently leaving the main sequence. Thus an appreciable fraction of the anticipated pre-white-dwarf mass loss seems to occur as early as during the first excursion into the red giant domain.

3. A number of spectroscopic studies of the brighter giant stars in globular clusters show that many of them display asymmetric Hα emission signatures indicating possible mass loss. Loss rates reported are 10^{-9}–10^{-7} $M_\odot$ yr^{-1}, which could account for a total loss of ~ 0.2 $M_\odot$ over the lifetime on the brighter part of the giant branch, fitting in well with the horizontal branch mass requirements. Furthermore, infrared observations of long period variables (LPVs) in globulars have shown 3.5-, 10-, 12-, and 25-μm excesses and strong H_2O absorption, indicating that circumstellar dust shells and extended atmospheres are present. Loss rates have been estimated at $\sim 5\times10^{-6}$ $M_\odot$ yr^{-1}. The LPVs are *asymptotic* giant branch stars, so mass loss seems to occur during the second excursion into the giant domain as well.

4. It is known that low mass stars in the galactic field can experience still further mass loss immediately prior to the white dwarf collapse, through the ejection of a shell of surface material in the form of a planetary nebula. The shell expands and disperses on a time scale of about 30,000 yr. If the stars evolving in globular clusters also participate in this form of mass loss, then, on statistical grounds, one would expect one or two planetary nebulae to be observable in the globulars of the Galaxy. One has indeed been observed, K 648 in M 15, confirming the existence of planetary nebula gas ejection in globulars.

Thus there is strong evidence that all the usual processes of mass loss known to occur in the late stages of the evolution of low mass stars are active in the stars of globular clusters, and we need to consider the fate of the ejected material in the cluster systems.

Globular clusters have orbits in the Galaxy that carry them far out of the galactic plane, and most of their time is spent in the galactic halo. Periodically, however, a cluster's orbit crosses the plane (typically every few hundred million years), and, during this crossing, the cluster will be swept clean of any accumulated gas by the interstellar medium in the galactic disk.

Thus, as a starting point in estimating the typical gaseous content of a globular cluster, we should integrate the total rate of gas release due to the mechanisms already mentioned, over the crossing interval for a typical cluster orbit. The simplest way of doing this is to combine observational estimates of the number of horizontal branch stars in well observed globulars with the theoretical horizontal branch lifetime to obtain an estimate for the rate at which stars are finishing their lives in each cluster. Then, assuming that each dying star loses 0.2 $M_\odot$ between the main sequence and its collapse to the white dwarf domain, the total rate of stellar mass loss follows. It is found that, if all such ejected material is retained within the cluster, 100–1000 $M_\odot$ of intracluster gas should accumulate between passages through the galactic plane.

SEARCHES FOR INTRACLUSTER MATERIAL

Over the last two decades, there have been numerous searches for gas, dust, and molecules arising from the mass loss processes in globulars. They have spanned the whole wavelength domain from the x-ray to the radio. Almost all surveys have failed to detect any intracluster material. A recent, comprehensive bibliography of these searches has been given by Morton S. Roberts. Altogether, over 80 globulars have been investigated in some manner or other—roughly half of the total known in the Galaxy. Some of the results of these investigations are as follows:

Over 30 clusters have been investigated for the presence of neutral hydrogen using radio searches in the H I 21-cm line. No detections have been made, and the upper limits for the amount of neutral hydrogen that can be present in individual clusters have been brought down from several hundred solar masses in the earliest studies to about 0.1 $M_\odot$ in the more recent ones. Whether any intracluster hydrogen present would be neutral or would be ionized by ultraviolet radiation from the horizontal branch stars depends upon the temperature of these stars in any particular cluster. Some clusters have very hot, blue-horizontal-branch stars that would cause photoionization.

In view of this possibility, searches for *ionized* hydrogen have also been carried out, both optically (in the Hα line) and at radio wavelengths (in the free–free emission continuum). Most of the optical searches have proved negative, the upper limits for H^+ mass in particular clusters being just a few times 0.1 $M_\odot$. The only exception is a reported detection in the cores of a small number of clusters, notably NGC 5824 and NGC 6624, by means of filter photometry. The corresponding detected H^+ mass is very small, however; less than 0.1 $M_\odot$. Similarly, radio searches for free–free emission from ionized hydrogen have proved negative, although the limits in this case are not as stringent; typically 20 $M_\odot$.

Some globular clusters display another feature that may be due to intracluster matter: small irregularities of surface brightness that appear to be too sharply defined to be attributable simply to statistical effects in the distribution of the cluster stars. Such dark patches have been noted for over a century, and it has frequently been speculated that they are due to small, dense clouds of dust (and molecules). Dust grains are expected to form in the stellar winds of cluster stars losing mass, and to be ejected into the intracluster medium. Most workers who have investigated the statistics and extinction properties of the cluster "dust patches" believe that they are real and located in the globulars. Nevertheless, it would be satisfying to confirm this by direct observations indicating that they have properties analogous to those of the dark clouds in the galactic disk. There are two ways in which one can seek to do this.

The first is to search in the infrared for thermal emission from the dust grains, which will absorb cluster starlight and reradiate it primarily at wavelengths about 100 μm (having been heated to temperatures around 40 K). A recent study using data from the *Infrared Astronomical Satellite* has reported such a 100-μm excess in the central regions of 47 Tucanae. The flux is very small, however, corresponding to only 0.0003 $M_\odot$ of silicate dust. This is 100–1000 times less than the amount that one would expect to have accumulated within 47 Tuc since it last passed through the galactic plane, which indicates that dust is being removed from the central regions of the cluster on a time scale of 10^4–10^5 yr.

The second method of seeking to establish the presence of dense clouds in globulars is to search for the 2.6-mm $J = 1 \rightarrow 0$ rotational transition of the CO molecule, a very prominent signature in the case of galactic clouds. Such searches have now been attempted in about a dozen clusters, but no detections have been made. For several clusters the detection limits are, as yet, only of about the same order as the signals to be expected were the "dark patches" real and of similar properties to clouds in the galactic disk. In a few cases, however, more stringent tests have been possible, resulting in upper limits for the mass of H_2 present in individual clusters of between 0.2 and 8 $M_\odot$ (provided the clouds have excitation temperatures less than 100 K, as is the case for clouds in the galactic disk).

POSSIBLE EXPLANATIONS FOR THE LACK OF DETECTED MATERIAL

The discrepancy between the expected intracluster gas content in globulars and the upper limits obtained in observational searches has prompted a good deal of model calculations to try to account for the removal of gas from the clusters.

One obvious possibility is that the gas injected into the intracluster medium has (or subsequently acquires) an energy per unit mass in excess of the escape energy at cluster center, and that it simply leaves the cluster in a steady-state flow. Several different energy inputs have been suggested: the original ejection energy of the gas from the atmospheres of mass-losing cluster stars; photoionization energy associated with the absorption of ultraviolet photons from hot, blue-horizontal-branch stars; energy injected by novae; and so forth.

Quite detailed model calculations have been made to investigate the properties of the gas systems that will result from these mechanisms. Some studies have produced models of steady-state cluster winds that are applicable to the type of gas flows that would be expected when the gas energy exceeds that required for escape. Others have produced time-dependent solutions simulating the gradual accumulation of gas in a cluster after its passage through the galactic plane; time-dependent models can simulate both those situations where a steady-state flow is eventually established and those where the gas cannot escape and a continuous buildup in the cluster results. The outcome of these studies is best understood by reference to Fig. 1, which plots the masses and radii of globulars. Also shown are loci of equal central escape velocity, which, for many clusters, is quite low, < 20 km s^{-1}. Gas flow calculations have shown that, for these clusters, the gas has no problem in leaving the cluster in a steady flow; the stellar wind velocities reported for mass loss from giant stars are 20–40 km s^{-1}, corresponding to gas ejection energies that alone are sufficient to ensure outflow. However, for the most tightly bound clusters (those at the top left of Fig. 1) a problem certainly exists. These would require stellar wind velocities in excess of 100 km s^{-1} for outflow to occur (allowing for radiative cooling in the gas flows), and even then there would be enough gas in the steady-state flow itself to violate some of the observational upper limits.

Following the discovery that about 15 globular clusters have compact, moderately centrally-located x-ray sources (several of them bursters), there was speculation that such sources might be

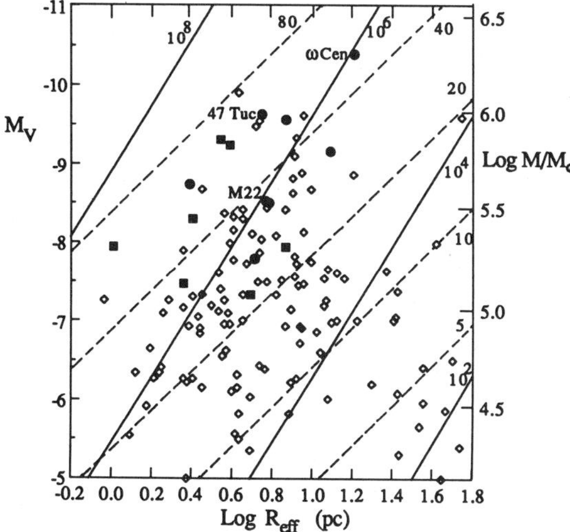

Figure 1. Plot of absolute visual magnitude against effective radius (the geometric mean of the core and tidal radii) for galactic globular clusters. The data are from the tabulation of Ronald F. Webbink [Dynamics of Star Clusters, IAU Symposium 113, J. Goodman and P. Hut, eds. Reidel, Dordrecht, p. 541 (1985)]. The figure shows points for 133 clusters in Webbink's table; a further 15 are fainter than $M_V = -5$. The right-hand axis indicates cluster mass, calculated from the M_V value using Garth Illingworth's calibration, $\langle M/L_V \rangle = 1.6(M/L_V)_\odot$. Clusters known to contain x-ray sources are shown with full points—dots for the lower x-ray luminosity class sources; squares for the higher x-ray luminosity sources. Also shown are loci of constant central escape velocity, in units of kilometers per second (broken lines), and loci of constant central surface density × escape velocity, in units of solar mass per square parsec per kilometer per second (full lines). Both these sets of loci are approximate, depending somewhat upon the cluster concentration parameter.

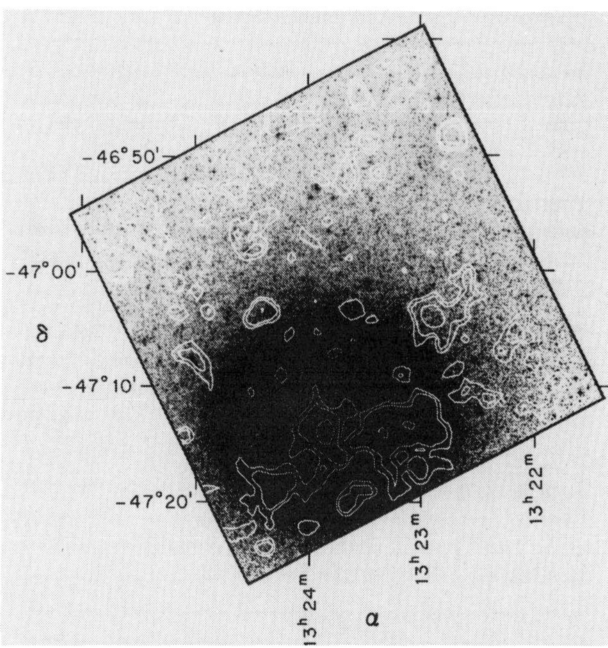

Figure 2. The extended soft x-ray source in the globular cluster ω Centauri (NGC 5139). This source may be due to a bow shock occurring between the intracluster gas in ω Cen and the galactic halo medium. *(Reprinted courtesy of F. D. A. Hartwick, A. P. Cowley, J. E. Grindlay and* The Astrophysical Journal ***254** L11, published by the University of Chicago Press. ©1982, The American Astronomical Society.)*

due to the accretion of intracluster gas by a massive black hole at cluster center. If that were the case, this accretion might account for an appreciable loss of gas, and the observed x-ray properties might provide information about the gas content of clusters, and hence the stellar mass loss processes. This hope has now evaporated, however, because there is good evidence that the compact sources are due to accretion processes within binary systems formed by stellar "collisions" in the central regions of those clusters with the highest central densities (also those at the top left of Fig. 1). The x-ray luminosities of these compact sources have a distinct bimodal distribution, the fainter being thought to be accreting white dwarf systems, and the brighter binary systems involving neutron stars. Thus the compact x-ray sources convey no information about the gas content of the cluster at large.

The x-ray sources might still be relevant to the problem at hand, however, because models have indicated that the presence of such a source in a globular can result in a considerable energy input into the cluster gas system (by photon absorption), and, consequently, an enhanced gas outflow can occur. Figure 1 indicates that the x-ray sources occur predominantly in the same clusters for which the gas content problem is most severe. This might help explain the observed paucity of intracluster gas, at least for some clusters.

Another possible explanation for the observed gas deficiency in globulars is that gas might be removed by the ram pressure of a hot, gaseous galactic halo through which the clusters are moving. The cluster parameter that describes its resistance to such stripping, is the product of the surface density of stars at cluster center and the central escape velocity. Loci of constant values for this quantity are also shown in Fig. 1; once again it is the clusters at the top left that are hardest to strip of gas.

An x-ray observation that seems to support the ram pressure possibility, is the *Einstein* satellite detection of soft (0.5–3.5-KeV) x-ray emission in 47 Tuc, ω Cen, and M 22. In all three cases there is significant extended emission offset from the cluster center (see Fig. 2). Such a feature could be produced by high temperature intracluster gas interacting with the hot galactic halo gas to form a bow shock. If this explanation is confirmed, it will be quite important because these three clusters are among those from which it is most difficult to remove gas (see Fig. 1). There is one additional piece of evidence in favor of this interpretation: the direction of motion of ω Cen through the halo is known (300 km s^{-1} toward the south) and it agrees with the direction in which the extended x-ray emission is offset from cluster center. This interpretation of the extended x-ray emitting region in ω Cen implies that its gas has a number density of about 0.1 cm^{-3} and has been shocked to a temperature of 10^5–10^6 K. In order to obtain agreement with this bow shock model, there would, however, have to be some quantitative modifications to the flow model predictions for the intracluster gas content of ω Cen. In particular, the distance of the observed shock from cluster center implies a higher density of the intracluster gas than can easily be explained.

FUTURE OUTLOOK

Even after two decades of work, the fate of the material ejected by the cluster giants in the most tightly bound globulars is still an open question. Some promising explanations have been suggested, but one will have confidence that the right gas removal mechanism has been found only when (i) detailed models are calculated for the mechanism under consideration, that provide quantitative predictions (in the case of specific clusters) for the values of the various resulting gas diagnostics and their distribution, and (ii) observational searches yeild positive detections in those clusters that correspond with these model predictions. At present, we seem to have some way to go in both these respects.

Additional Reading

Dupree, A. K. (1986). Mass loss from cool stars. *Ann. Rev. Astron. Ap.* **24** 377.

Faulkner, D. J. and Freeman, K. C. (1977). Gas in globular clusters. I. Time-independent flow models. *Ap. J.* **211** 77.

Forte, J. C. and Méndez, M. (1989). Dust clouds within globular clusters: polarization. *Ap. J.* **345** 222.

Grindlay, J. E. (1985). X-raying the dynamics of globular clusters. *Dynamics of Star Clusters. IAU Symposium 113,* J. Goodman and P. Hut, eds. Reidel, Dordrecht, p. 43.

Roberts, M. S. (1988). Interstellar matter in globular clusters. *The Harlow-Shapley Symposium on Globular Cluster Systems in Galaxies. IAU Symposium 126,* J. E. Grindlay and A. G. Davis Philip, eds. Kluwer Academic Publishers, Dordrecht, p. 107.

VandenBerg, D. A. and Faulkner, D. J. (1977). Gas in globular clusters. II. Time-dependent flow models. *Ap. J.* **218** 415.

VandenBerg, D. A. (1978). Gas in globular clusters. III. Time-independent outflow models including photoionization. *Ap. J.* **224** 394.

See also **Star Clusters, Globular, Stellar Populations; Star Clusters, Globular, X-Ray Sources; Stars, Winds.**

Star Clusters, Globular, Mass Segregation

Stanislav G. Djorgovski

Mass segregation is naturally expected to occur in all self-gravitating stellar systems, such as the globular star clusters. It is a consequence of the tendency towards thermal equilibrium and equipartition of energy: heavier or binary stars would tend to sink closer to the cluster center. This process may be coupled with the radial instabilities which cause core collapse in globular clusters. The observable consequences of mass segregation may be color or population radial gradients, radial variations in the stellar luminosity and mass functions, and the details of the cluster surface brightness and velocity dispersion profiles. Some evidence exists for all of these phenomena, but none of it is very firm or completely unambiguous at this point.

DYNAMICAL BACKGROUND

A self-gravitating cluster of stars can be considered as a gas sphere with stars as molecules, interacting gravitationally. The stars would exchange energy in binary encounters and by interacting with the collective gravitational potential of the cluster as a whole. The net long-term effect of the encounters is a tendency towards the equipartition of energy. Generally, a spectrum of stellar masses is present. The tendency towards thermal equilibrium (i.e., equipartition) would cause the more massive stars to acquire on the average smaller velocities. The balance of the kinetic and potential energy would then require that the slower stars are found closer to the center of the potential well. The complete thermal equilibrium is never reached in any real cluster, mainly because star clusters are open systems, liable to mass loss by escaping stars, and subject to tidal shocks, and so on. However, a considerable mass segregation may be reached in a finite time, with important dynamical consequences, and with, at least in principle, observable effects. A cluster in a quasistatic thermal equilibrium, such as the King–Michie models describe, would then consist of stratified mass groups, with the most massive stars (e.g., neutron stars, white dwarfs, binaries) most concentrated towards the cluster center.

Sometimes even the quasistatic equilibrium may be difficult to establish. Consider, for example, a simplified case of a cluster composed of two stellar groups, with individual stellar masses $m_1 \ll m_2$, and the total masses in the two groups M_1 and M_2. Energy exchange between the stars of unequal masses is on the average more efficient than the exchange between stars of equal mass. Thus, the lighter stars would form a more efficient thermal bath for the heavier stars than would the stars in the same mass group, resulting in an accelerated pace of energy exchange and mass segregation. The heavier stars lose energy to the lighter stars, sink closer to the cluster center, and so on. If the inequality

$$\left(\frac{M_2}{M_1}\right)\left(\frac{m_1}{m_2}\right)^{3/2} > 0.16$$

is satisfied, the process becomes runaway. This is the Spitzer mass segregation instability. It enhances and speeds up the Antonov–Hénon–Lynden-Bell gravothermal catastrophe, which is a consequence of the negative specific heat of self-gravitating systems, and can occur even in a cluster composed of single-mass stars. The two instabilities may act together and lead to a faster core collapse. In the case of galactic nuclei, this joint instability may lead to the formation of a massive central object, a core of an incipient active nucleus.

Numerical simulations of core collapse in globular clusters always predict that a strong mass segregation should set in. The shape of the resulting surface brightness profile, for example, the power-law slope of the central cusp, would depend on the mass spectrum which is present, and may change with the radius. Because about 20% of all known galactic globulars have the characteristic post-core-collapse morphology, the effects of mass segregation in post-collapse cores may be studied statistically.

High central densities may be reached during the core collapse, and the rates of binary star formation via tidal capture, or even stellar mergers, may be enhanced as a result. Both products are by definition heavier than the average ambient stars, and would participate in the mass segregation. This may be an important factor for the formation of low-mass x-ray binaries and millisecond pulsars in globular clusters.

The escape of stars through the tidal cutoff is more efficient for the low-mass stars, because on the average they have higher velocities. Evaporation of stars from cluster envelopes thus causes an inverse mass segregation near the tidal cutoff, but the data available so far are insufficient to check for this effect.

OBSERVATIONAL EVIDENCE

Mass segregation can lead to several observable consequences. There is much ongoing work in this field, and several interesting results. Whereas it would be perhaps premature to claim that mass segregation in globular clusters has been unambiguously detected (as of late 1989), the sum of the available evidence is at least suggestive.

If a population of binaries or of neutron stars can be identified in some way (e.g., in the form of x-ray sources, possibly blue stragglers, pulsars, etc.), it is expected that their distribution would be more concentrated than that of the red giants, which provide most of the light. There are still too few x-ray sources known in globular clusters, but the preliminary result is that they are indeed more concentrated towards the cluster centers than the red giants. Discoveries of millisecond pulsars in globular clusters are rapidly growing in number, and may soon become one of the most interesting and powerful probes of globular cluster dynamics. The nature of blue stragglers is still controversial, but it is quite possible that they are either binaries, or stellar merger products. Star counts in at least two clusters examined so far show that blue stragglers are more centrally concentrated than the red giants, which can be understood if either of the two hypotheses about their nature is correct, and mass segregation is present.

Dynamical modeling of some well-observed clusters (e.g., 47 Tucanae and ω Centauri) using multimass King–Michie models with velocity anisotropy and rotation, and a range of initial mass function slopes, essentially requires that mass segregation be present, in the manner predicted by the King–Michie (after Ivan R. King and Richard Michie) models. Ad hoc models without mass segregation can be fitted to the data, but would require a very special adjustment of the parameters.

Luminosity and mass functions determined directly from the star counts as a function of radius suggest that mass segregation is present in M 30, M 71, and possibly a few other clusters as well. Star counts are subject to difficult completeness and crowding corrections, and the evidence presented so far is suggestive, but not yet clearly compelling. Still, this method is very promising. Mass segregation effects must also be taken into account when the global mass function shape is derived from local measurements in any limited radial interval.

Color and population gradients have been detected in several post-core-collapse clusters, most notably M 15 and M 30. These should not be confused with the possible metallicity gradients in clusters like 47 Tucanae or ω Centauri. The new results are based on the data from linear imaging detectors (CCDs), and avoid the systematic errors which plagued the earlier attempts to measure color gradients in globular clusters, using single-channel photomultipliers. The color gradients are always in the sense that the clusters are bluer towards the center, which is the opposite from what may be expected from possible systematic errors caused by the seeing and crowding effects. The origin of the gradients is not clear. Direct star counts in at least some of the affected clusters indicate that the color gradients are caused by the differences in distributions of red giants and blue horizontal branch stars, but the interpretation of the stellar types is not yet fully secure. If this is indeed the case, the effect would imply an *inverse* mass segregation (since the blue horizontal branch stars are lighter), which would be hard to understand dynamically, unless some heretofore unknown dynamical mechanism related to the core collapse is operating. Other experiments indicate that there is a diffuse blue component present near the center, composed of fainter stars of as yet undetermined nature; cataclysmic binaries are a possibility, and would imply a mass segregation in the normal sense. The issue is still open, and a subject of an active research.

Finally, a central cusp in the velocity dispersion has been detected in M 15, a classical post-core-collapse cluster. The cause of the cusp is still unknown, but it is possible that a sharp *density* cusp composed of dark stellar remnants is responsible. Efforts are under way to determine whether similar velocity dispersion cusps exist in other post-core-collapse clusters, and to model them dynamically.

FUTURE PROSPECTS

The most interesting near-future observations should be with the *Hubble Space Telescope*. With it, star counts can be carried deep into the cores of globular clusters, and compared with the outer regions which can be studied from the ground. Detection and good positional measurements of large numbers of x-ray binaries can be expected from *ROSAT* and the future *AXAF* mission.

Perhaps the most promising new development from the ground are the searches for millisecond pulsars in globular clusters. Pulsars can be an excellent probe of the distribution of neutron stars, which are otherwise detectable only through the x-ray binaries, and which may be an important mass component. Neutron stars are also good test particles for the cluster potential as a whole, since their masses are reasonably well known (viz., 1.4 $M_{\odot}$).

Further progress can also be expected from the CCD star counts and measurements of color gradients, especially if they are coupled with seeing-compensating techniques.

Additional Reading

Chernoff, D. F. and Weinberg, M. D. (1990). Evolution of globular clusters in the Galaxy. *Ap. J.* **351** 121.

Elson, R., Hut, P., and Inagaki, S. (1987). Dynamical evolution of globular clusters. *Ann. Rev. Astron. Ap.* **25** 565.

Goodman, J. and Hut, P., eds. (1985). *Dynamics of Star Clusters. IAU Symposium 113*. Reidel, Dordrecht.

Grindlay, J. and Philip, A. G. D., eds. (1988). *Globular Cluster Systems in Galaxies. IAU Symposium 126*. Kluwer, Dordrecht.

Lightman, A. and Shapiro, S. (1978). The dynamical evolution of globular clusters. *Rev. Mod. Phys.* **50** 437.

Merritt, D., ed. (1989). *Dynamics of Dense Stellar Systems*. Cambridge University Press, Cambridge, U.K.

Spitzer, L., Jr. (1987). *Dynamical Evolution of Globular Clusters*. Princeton University Press, Princeton.

See also **N-Body Problem; Pulsars, Binary; Star Clusters, Globular, Gravothermal Instability; Stars, Blue Stragglers.**

Star Clusters, Globular, Stellar Populations

Gary S. Da Costa

Globular clusters have long been objects of study by astronomers, primarily because a globular cluster can be regarded as a homogeneous entity: All the stars it contains have the same age and, in most cases, the same chemical composition. In this sense the stellar population of a globular cluster can be defined as the relative numbers of cluster stars at different brightnesses and colors. The first studies of the stellar population of globular clusters, such as that of Harlow Shapley in the early part of this century, revealed that the brightest cluster stars were also the reddest, with the fainter stars being bluer. This property was opposite to that exhibited by the stars in the vicinity of the Sun, where the bluer stars are generally the brighter. This difference led Walter Baade to suggest in the mid-1940s that the stellar content of entire galaxies could be viewed as being made up of varying amounts of two distinct populations: Populations I and II. Population I was exemplified by the stars of the solar neighborhood and of open clusters, whereas the stellar population of globular clusters formed the prototype for Population II.

The modern subject of the stellar population of globular clusters is best discussed with reference to a color–magnitude, or C-M, diagram. In such a diagram, a measure of the brightness of the cluster stars, usually the visual (V) magnitude, is plotted against any quantity related to the color or temperature of the stars, such as the color index $B-V$. The diagram is oriented so that brightness increases upward, with redder colors or cooler temperatures falling to the right. Because all the stars in a cluster are effectively at the same distance from the Sun, the C-M diagram accurately represents the true relative brightnesses and colors of the cluster stars. A typical C-M diagram for a globular cluster is shown in Fig. 1; all but a small number of faint red stars are likely to be cluster members. When viewing this diagram, however, two things should be kept in mind. First, the apparent cutoff in the number of cluster stars at faint V magnitudes represents the limit of the observations, not a limit on the brightness of faint cluster stars. Second, most of the principal sequences outlined by the cluster stars are broadened by observational errors in the individual stellar colors, especially at fainter magnitudes. In the absence of observational errors the main sequence, for example, would appear as an extremely narrow band rather than the fairly wide span of color shown in the figure. It is clear however, from this C-M diagram and those of other globular clusters that the vast majority of cluster stars are confined to well-defined sequences. These are labeled on the figure with their usual names.

The *main sequence*, as the name implies, is the relation between brightness and color that is followed by the majority of cluster stars. Because this phase of evolution is the longest in the

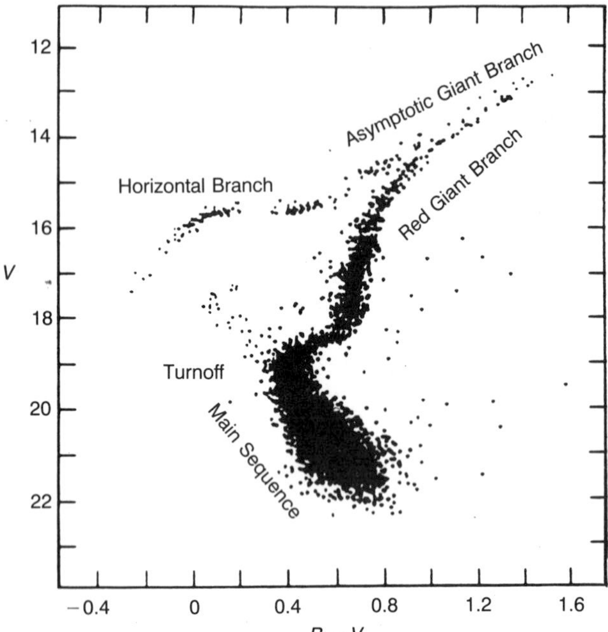

Figure 1. A color–magnitude diagram for the globular cluster M 3 (NGC 5272). [*Reproduced by permission of Kluwer Academic Publishers, Dordrecht, from R. Buonanno et al. (1988), High precision photometry of 10,000 stars in M 3. In* The Harlow Shapley Symposium on Globular Cluster Systems in Galaxies, IAU Symposium 126, *J. E. Grindlay and A. G. D. Philip, eds., p. 621.*]

life of a star, there are more main-sequence stars in a globular cluster than stars of any other type. The masses of individual main-sequence stars decrease with decreasing brightness but in most clusters the number of such stars increases with decreasing brightness. Thus the main sequence stars make up most of the mass of a cluster though, because of their relative faintness, they do not contribute significantly to the total light output. The relative number of main-sequence stars at different masses is called the present-day mass function (PDMF) of the cluster and it may, or may not, differ from the mass function over the same interval of stellar mass at the time the cluster formed, the so-called initial mass function (IMF). The PDMFs of globular clusters and their relation to the corresponding IMFs are a subject of active research at the present time.

The main sequence terminates at the *turnoff point* where the main-sequence stars begin to evolve toward the red. The absolute brightness of the turnoff in a globular cluster, or turnoff luminosity, is an important quantity because from it the cluster's age can be determined. In order to fix the turnoff luminosity, however, the distance to the cluster must be known. Unfortunately, all globular clusters are too distant for direct measurement of their distances and indirect methods are not as precise as one would like. However, it is generally agreed that the most metal-poor globular clusters have turnoff luminosities that correspond to ages in the range of 14–18 billion years. Such clusters are then the oldest objects in our galaxy for which age determinations are available.

The *red giants*, that is, the stars that lie on the red giant branch sequence in the C-M diagram, are the stars that dominate the visual light output of a globular cluster. During this phase of evolution many physical processes occur, some of which are better understood than others. For example, in some metal-poor globular clusters, there is observational evidence for the mixing of reaction products from the interior of the star into the surface layers. This mixing is much more extensive than theoretical models predict and the origin of this discrepancy is a subject of much current work.

The color (or, equivalently, the surface temperature or radius) of a red giant is sensitive to the metal abundance, being redder for higher abundances. This property can be put to use in a number of ways. First, once calibrated with clusters of known abundance, the color of the giant branch can be used to determine a cluster's metal abundance. Second, the intrinsic color width of the giant branch, that is, the width that remains after observational errors are removed, is a measure of the metal abundance homogeneity of the cluster giants. In all globular clusters studied, except the cluster ω Centauri (NGC 5139), this intrinsic width is immeasurably small, indicating that the giants have very closely similar metal abundances. This is not the case in ω Centauri for reasons that are not yet understood. Third, because the visual light ouput of a globular cluster is dominated by these stars, the metal abundance sensitivity is reflected in the integrated spectral type and integrated color of a globular cluster. More metal-rich clusters have later spectral types and redder intrinsic colors than the more metal-poor clusters.

Because the time scale for evolution along the giant branch is much shorter than the main-sequence lifetime, the number of red giants in a globular cluster is only a small fraction of the number of main-sequence stars. For the same reason, the number of red giants decreases with increasing brightness; the evolutionary time scale becomes progressively shorter. Evolution up the giant branch comes to a halt when the temperature in the central helium core becomes hot enough for helium to ignite, forming carbon. Because it happens under explosive conditions, this ignition is referred to as the helium flash and the corresponding brightness is called the red giant branch termination point or red giant branch tip. It occurs at an absolute visual magnitude of approximately $M_V = -2.5$ in a globular cluster of typical composition. Once equilibrium is restored in the star, its brightness and color are such that it is now found on the horizontal branch sequence in the C-M diagram.

The *horizontal branch* (HB) stars have approximately constant brightness (hence the term "horizontal") and lie in the C-M diagram to the blue of the giant branch some 3.5 magnitudes above the turnoff and 2.5–3 magnitudes fainter than the red giant branch tip. Although HB stars account for only 15% or so of the total cluster light output at visual wavelengths, they are bluer than the red giants and thus their contribution becomes more significant at shorter wavelengths. Indeed, for wavelengths shorter than about 0.4 nm they dominate the cluster light, especially if there are large numbers of blue HB stars. The number of HB stars (N_{HB}) is observed to exceed the number of red giants of equal or greater brightness (N_{RGB}) by about 40%. This ratio, $R = N_{HB}/N_{RGB}$, is important because it is sensitive primarily to the helium abundance of the cluster stars, a vital quantity that is difficult to determine precisely. A value of $R = 1.4$, when interpreted with current HB and RGB theory, suggests a helium abundance Y of 0.23, which is in good agreement with determinations in other old and/or metal-poor systems.

In Fig. 1 the horizontal branch is evenly populated with both blue and red HB stars present. The apparent gap at $B - V \cong 0.3$ is an artifact; the HB stars that fall in this color range vary in brightness and as a result they are not plotted on the figure. Horizontal branch stars that vary periodically in brightness, with typical periods of approximately $\frac{1}{2}$ day, are known as RR Lyrae variables; such stars are relatively common in globular clusters and are often used in estimating cluster distances. The distribution of HB stars with color, the so-called HB morphology, varies considerably from cluster to cluster, with some clusters having only blue HB stars whereas others have horizontal branches that lie entirely to the red of the variable star gap. Theoretical calculations have shown that the HB morphology is sensitive to a number of quantities. The most important of these is metal abundance, with the higher-abundance clusters having generally redder horizontal branches. The existence, however, of different HB morphologies between clusters of similar metal abundance indicates the need for (at least) a second parameter. Candidates for this second parameter include age, variations in the abundance (relative to iron) of the elements carbon, nitrogen, and oxygen, and internal rotation. At the present time age appears to be the most promising candidate but the issue is far from completely settled. If age is the second parameter, then differences of 2–4 billion years are required to explain the observed differences in HB morphology at constant abundance.

Although stellar evolution calculations can describe successfully the HB morphologies of globular clusters, the models require stellar masses that are approximately 0.2 solar masses less than the turnoff mass, which is typically about 0.85 solar masses, in order to reproduce the HB observed luminosities and colors, especially for the bluest stars. Thus it appears that globular cluster stars must lose mass at some point during the red giant phase of evolution. Although by no means definitively established, most of this mass loss is believed to occur when the star is near the red giant branch tip. Alternatively, it is possible that the mass is lost during the rapid structural readjustment that occurs after the helium flash when an HB star forms from a red giant.

After the HB stars exhaust the supply of helium in their central cores, they again evolve towards higher luminosity and redder colors. The extent of this evolution however, is determined by the envelope mass, the mass above the hydrogen-burning shell in the star. Stars with small envelope masses evolve almost vertically in the C-M diagram for a time before eventually becoming white dwarfs. Those HB stars with larger envelope masses, however, evolve back toward the red giant branch and outline the *asymptotic giant branch* (AGB) in the C-M diagram. In a typical globular cluster there are about one-third as many AGB stars as RGB stars at the same brightness. The evolution of a star up the AGB is terminated when the envelope mass becomes very small. This can occur as the result of gradual mass loss during the AGB evolution or as a result of a short period of catastrophic mass loss at the AGB tip that culminates in the formation of a planetary nebula. A number of variable stars commonly found in globular clusters are found on or blueward of the AGB. These include Population II Cepheids, and Mira-type, semi-regular and irregular red variables. Interestingly, in most globular clusters the AGB and the RGB terminate at very similar luminosities, although in younger clusters AGB stars can evolve to luminosities above the red giant branch tip because they have larger envelope masses.

The C-M diagram of Fig. 1 also shows a small number of stars that lie brighter and bluer than the cluster turnoff: stars that are apparently younger than the rest of the population. These stars are known as *blue stragglers* and in recent years it has been recognized that they are found in many globular clusters. Although it is possible to exclude a younger age for blue stragglers in globular clusters, there is no generally accepted explanation for their origin; the most plausible suggests that they result from mass transfer in a close binary star system.

The preceding paragraphs dealt with the stars that make up a significant fraction of the mass of and that contribute all of the light from a globular cluster. There are, however, a number of other objects that need to be included for a complete description of the stellar population of a globular cluster. These additional objects can be split into two groups: those that are too faint to be observed with current instrumentation (i.e., an "unseen" component) and those that occur so rarely that their numbers within individual globular clusters are extremely small.

The principal unseen component in globular cluster stellar populations are the *remnants* that result from the completed evolution of the stars more massive than the current turnoff mass. These remnants are likely to be mostly white dwarfs but the existence of luminous x-ray sources in some globular clusters testifies to the presence also of neutron stars. The number of remnants and their contribution to the total mass of the cluster is not easily determined. The standard approach is to take the PDMF of a cluster and extrapolate it to higher masses to estimate the number of remnants; their mean mass and mass distribution must be assumed. The validity of these assumptions and estimates can be investigated to some extent by comparing theoretical cluster models with observations of the velocity dispersion of cluster stars. The models and the velocity dispersion measurements yield the total mass; comparison with that expected from the (mostly visible) nonremnant stars then reveals the contribution of the remnants. This generally lies in the range of 5–30% of the total mass. Unfortunately, the accuracy of this method depends critically on the validity of the models; these may or may not be an adequate description of the dynamics of real globular clusters. With the advent of the Hubble Space Telescope, however, it may be possible to observe directly the white dwarf populations of some nearby globular clusters. In this context it is worth noting, however, that globular clusters do not appear to contain any "dark matter" in the sense of that required to explain, for example, the flat rotation curves of spiral galaxies.

Among the exotic or rare populations seen in globular clusters, we may include planetary nebulae, so-called UV-bright (ultraviolet) stars, and x-ray sources. The UV-bright stars lie in the C-M diagram to the left of the giant branch above the horizontal branch. They are stars that are either on short-lived blueward excursions from the AGB or stars that are evolving across the top of the C-M diagram on the way to becoming white dwarfs. The x-ray sources, which are found in many globular clusters, are of two types: The higher-luminosity sources are thought to be binary systems in which a main-sequence star is losing mass to a neutron star companion whereas the lower-luminosity systems have a white dwarf in place of the neutron star. The binaries are thought to be formed by dynamical capture processes in the dense central regions of globular clusters.

However, perhaps the most bizarre objects in the stellar populations of globular clusters are the recently discovered *millisecond pulsars*. These are rapidly spinning neutron stars but their occurrence in old stellar systems such as globular clusters came as a complete surprise because pulsars are normally considered very young objects (cf. the Crab pulsar). The most probable explanation is that we are observing the end product of the evolution of a neutron-star–main-sequence-star binary. The angular momentum of the mass accreted by the neutron star from the main-sequence companion causes the neutron star to "spin up"; by the time the main-sequence star has been completely destroyed, the neutron star is (once more) spinning rapidly.

The discovery of these objects shows once again that the total stellar population of a globular cluster is far from completely categorized; undoubtedly, more exciting discoveries lie ahead.

Additional Reading

Freeman, K. C. and Norris, J. (1981). The chemical composition, structure and dynamics of globular clusters. *Ann. Rev. Astron. Ap.* **19** 319.

Hesser, J. E. (1988). Globular clusters in the Galaxy and beyond. In *Progress and Opportunities in Southern Hemisphere Optical Astronomy*, V. M. Blanco and M. M. Phillips, eds. *Astronomical Society of the Pacific Conference Series* **1**, p. 161.

Iben, I., Jr. (1970). Globular cluster stars. *Scientific American* **223**, (No. 7) 27.

Lee, Y.-W., Demarque, P., and Zinn, R. (1990). The horizontal-branch stars in globular clusters. I. The period-shift effect, the luminosity of the horizontal branch, and the age-metallicity relation. *Ap. J.* **350** 155.

Renzini, A. and Fusi Pecci F. (1988). Tests of evolutionary sequences using color-magnitude diagrams of globular clusters. *Ann. Rev. Astron. Ap.* **26** 199.

See also **Pulsars, Binary; Pulsars, Millisecond; Star Clusters, Globular, Mass Loss; Star Clusters, Globular, X-Ray Sources; Star Clusters, Mass and Luminosity Functions; Stars, Blue Stragglers.**

Star Clusters, Globular, Variable Stars

Carla Cacciari

The discovery of the first variable star in a globular cluster was announced by Edward C. Pickering in 1890. Extensive search for and study of cluster variables began in the early decades of this century, and by the 1940s a good deal of information had been collected, as one can see in *A Catalogue of Variable Stars in Globular Star Clusters* compiled by Helen B. Sawyer-Hogg in 1939, which contained 1116 variables in 60 galactic globular clusters. The most recent edition of her catalog, which appeared in 1973, contained 2119 entries in 108 clusters.

Galactic globular clusters are among the best examples of chemically homogeneous, old (i.e., Population II) stellar systems. Their variables are evolved stars and can be placed on the Hertzsprung–Russell (HR or color–magnitude) diagram of a typical globular cluster according to their evolutionary stage. Figure 1 shows a sketch of the color–magnitude diagram of M 3 and the main areas that can be populated by variable stars. These areas are defined by the physical conditions of the stellar atmospheric surface, that is, by the values of chemical composition, gravity, and temperature that, in appropriate combinations, induce the star to pulsate:

1. *The instability strip*: RR Lyrae stars, type-II Cepheids (BL Her, W Vir, and RV Tau stars), and anomalous Cepheids. The instability strip covers a narrow region on the horizontal branch, and up to ~ 3.5 magnitudes brighter. The strip is ~ 1300 K wide in effective temperature at any given luminosity; because it is slightly tilted in the luminosity–temperature plane, it covers a temperature range between 5000 and 7500 K, the higher luminosities going with the lower temperatures.

2. *The cool, evolved stars with unstable envelopes*: Mira variables, yellow semiregular variables, and red semiregular and irregular variables; they appear near the tip of the red giant and asymptotic giant branches.

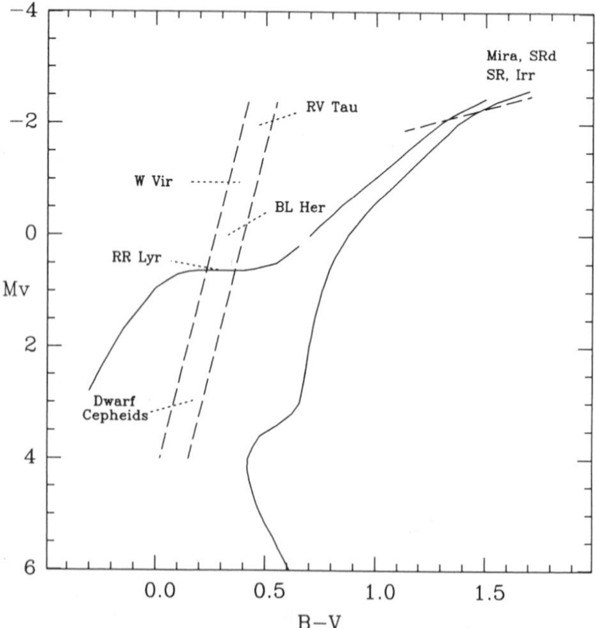

Figure 1. The color–magnitude diagram for Population II variables. The solid lines represent the globular cluster M 3 (*from the data by R. Buonanno et al. (1986), in* Mem. S. A. It. **57** *391; with permission from the authors)*; the dashed lines indicate the main areas that can be populated by variable stars in globular clusters.

3. *The instability strip near the main sequence*: dwarf Cepheid stars; blue stragglers can fall within this area, probably as a result of mass transfer in binaries.

In addition, it is likely that other types of variable stars are present in globular clusters.

Along their cooling sequence, white dwarfs can enter the instability strip for their chemical composition and become ZZ Ceti or DB variables; these may be quite common among Population II white dwarfs, but at present we have little information on such variables in the field and no information on them in globular clusters (where, if present, they elude detection by ground-based instruments because these stars are so faint).

All of the preceding variables pulsate. According to the theory of stellar pulsation, a well-defined relation ($P\sqrt{\rho} \sim$ constant) exists between the average density ρ and pulsation period P, which can be used to derive the physical characteristics of these variables.

Other nonpulsating variables are also found in galactic globular clusters but are extremely rare mainly because they represent intrinsically rare events in a globular cluster stellar population (e.g., flare stars, novae, cataclysmic variables, and x-ray variables) but also because of selection effects in the observing methods (e.g., eclipsing variables): Stars with very long or very short periods and/or small brightness variations are difficult to detect.

Only the most common types of variable star will be reviewed in this entry.

RR LYRAE VARIABLES

RR Lyrae stars are the most frequent (~ 90%) type of variable star in globular clusters. According to the stellar pulsation theory, they can pulsate in the fundamental mode (ab-type variable star) or in the first overtone (c type), with a period that is about three-fourths of the fundamental one. They have periods less than one day, and lie on the horizontal branch with surface effective temperature between ~ 6000 and 7500 K.

In the early 1940s P. Th. Oosterhoff found that globular clusters could be divided into two groups according to the mean periods of their ab-type variables: group I clusters with ab-type RR Lyrae stars of mean period ~ 0.55 day, are mostly clusters of intermediate metal abundance (i.e., ~ 3–4% of the solar metal abundance); group II, with mean period ~ 0.65 day, are mostly metal-poor clusters (i.e., ~ 2% of the solar metallicity, or less). A small but significant number of clusters have very few or no RR Lyrae stars: They are either among the most metal-rich clusters (i.e., $\gtrsim$ 15% of the solar value) and have a very red horizontal branch, or have intermediate metallicity and have a very blue horizontal branch. In both cases the horizontal branch does not extend far enough to enter the instability strip. The reason some intermediate-metallicity clusters have plenty of RR Lyrae stars (e.g., M 3) and others of comparable chemical abundance have almost none (e.g., M 13) is not clear yet: This obviously suggests the existence of some still unidentified parameter(s), causing the nonmonotonic horizontal branch location with metal abundance.

Many attempts have been made to explain the "Oosterhoff dichotomy" between group I and group II clusters in terms of differences in luminosity, mass, temperature, or evolution across the horizontal branch. The most recent one dates from the early 1980s, when Allan R. Sandage noticed that, at a given temperature, a period–metallicity relation exists not only among the clusters (using the average periods of their variables), but also among the individual stars. Recall that the basic period–density relation for pulsating stars corresponds to a relation between period, mass, luminosity, and temperature, because the density can be expressed in terms of mass and radius, and the radius is related to the luminosity and temperature of the star via the Stefan–Boltzmann law. This relation, which is widely used to derive the physical parameters of the RR Lyrae stars from their pulsation properties, is

$$\log P = 11.497 - 0.68 \log M + 0.84 \log L - 3.48 \log T_e,$$

where P is the fundamental period, M and L are, respectively, the mass and luminosity of the star in solar units, and T_e is the effective temperature. Therefore under "appropriate" assumptions for the stellar mass, the period–metallicity relation found by Sandage was accounted for by a luminosity–metallicity relation, in the sense that the metal-poor RR Lyrae stars would be significantly brighter than the metal-rich ones. The slope of this relation is $\Delta M_{bol}/\Delta[Fe/H] \sim 0.35$, resulting in a difference between the horizontal-branch luminosity of the most metal-rich and most metal-poor globular clusters of ~ 0.5 mag. [The notation [Fe/H] stands for $\log(Fe/H)_{star} - \log(Fe/H)_{Sun}$, where (Fe/H) indicates the stellar abundance of iron with respect to hydrogen and the iron is taken to represent all elements heavier than helium (which are called "metals"); the relation is normalized to the Sun, which has (Fe/H) = 0.017. According to this notation, the globular clusters have metal abundances ranging from $[Fe/H] = -0.6$ to -2.2.

The "Sandage effect" has been the subject of considerable debate: First, because the derived relation between luminosity and metallicity is contradicted by some independent determinations, both empirical and theoretical, of the RR Lyrae absolute luminosity. Second, because it introduces a significant factor of variation, namely, the metallicity, into the absolute luminosity of the RR Lyrae stars (which traditionally had been considered "constant," hence making these stars ideal distance indicators). As a matter of fact, RR Lyrae stars are relatively bright among globular cluster stars, and their apparent magnitude only varies by ± 0.15 mag around the average in any given cluster. It is this observation that first suggested their use as distance indicators, and a number of independent determinations, both empirical and theoretical, suggested a value for the absolute visual magnitude of RR Lyrae stars of $M_V(RR) \sim 0.6$ until recently.

If, however, $M_V(RR)$ varies significantly with metallicity, then the entire galactic and extragalactic distance scale based on the RR Lyrae luminosity has to be adjusted accordingly, and RR Lyrae stars can be used as distance indicators only when their chemical composition is known also. A number of different independent methods have been applied to RR Lyrae stars both in globular clusters and in the field in order to estimate their absolute luminosity. The average of these estimates is the relation

$$M_V(RR) = 0.22[Fe/H] + 0.94,$$

where the slope (0.22 in the above relation) can vary from 0.0 to 0.40 depending on the method, and where the zero-point (0.94) is uncertain by at least 0.2 mag.

The absolute magnitude of the RR Lyrae stars in a globular cluster allows the derivation of the distance of the cluster and hence the absolute magnitude of the main-sequence turnoff point $[M_{bol}(TO)]$, which is directly related to the age of the cluster via stellar evolution theory. For a value of the helium abundance $Y = 0.24$ and assuming that the chemical abundances of all elements (oxygen in particular) scale with iron as in the Sun, then

$$\log t \text{ (yr)} = 8.319 + 0.41 M_{bol}(TO) - 0.15[Fe/H].$$

This leads to an average value for the age of the globular clusters of $t \sim 16 \times 10^9$ yr, with an uncertainty of $\pm 3 \times 10^9$ yr. The observational errors and the theoretical assumptions and approximations do not yet allow a more precise determination of a globular cluster's age using its RR Lyrae stars as distance indicators. Therefore one cannot say whether the above spread in age is (at least in part) real, in which case the globular cluster system formed over a span of a few billion years during a relatively slow collapse of the Galaxy, or whether it is entirely a statistical error, in which case the globular clusters formed simultaneously with the rapidly collapsing Galaxy.

POPULATION II CEPHEIDS

Approximately 50 variable stars in globular clusters are presently known to be Population II Cepheids. They fall within the instability strip above the RR Lyrae region and can be subdivided into three types, in order of increasing luminosity and period and decreasing effective temperature:

BL Her stars have periods between 1 and 10 days and are evolving through the instability strip from the blue part of the horizontal branch or are entering the instability strip from the red side, during the "blueward hooks" that may occur on the lower asymptotic giant branch. Period variations have been detected that are in agreement with the values expected from the evolution rates of post-horizontal-branch stellar models.

W Vir stars have periods between 10 and 20 days and are crossing the instability strip on blue loops from the asymptotic giant branch in response to helium shell flashes or are making their final transition toward the state of hot white dwarf.

RV Tau stars have periods longer than 20 days and are in a similar evolutionary stage as the W Vir stars. Their very long periods indicate, according to the pulsation theory, that they have large radii and high luminosities, because they have the lowest temperatures among the Population II Cepheids. In fact, they can be as cool as 5000 K and as bright as $M_V \sim -3.5$. Their basic characteristic is to show alternate deep and shallow minima in the light curve. These have been attributed to either a modulation of the convective flux of energy by pulsation or the presence of a very extended envelope around the star. This latter hypothesis is supported by the detection of an excess of infrared emission in a significant fraction of W Vir stars and in most RV Tau stars, which suggests the presence of circumstellar dust shells (probably due to recent mass loss).

Anomalous Cepheids have periods between about 0.5 and 3.0 or more days and are found in a region of the HR diagram brighter than the RR Lyrae stars or the Population II Cepheids of corresponding periods. The only galactic globular clusters that are known to contain anomalous Cepheids are NGC 5466 and ω Cen, but they are quite common in extragalactic systems such as the Small Magellanic Cloud and the dwarf spheroidal galaxies Draco, Leo, Sculptor, and Ursa Minor. From the period–mean-density relation, their anomalously high luminosity translates into an anomalously large mass, which in fact has been estimated around 1.5 $M_\odot$ (solar masses) on the assumption that they pulsate in the first overtone mode (the mass would be $\sim 57\%$ larger were they fundamental pulsators). Although in the extragalactic systems mentioned previously a small fraction of metal-poor intermediate age (i.e., larger mass) stars cannot be ruled out, such a "young" stellar component is highly implausible in galactic globular clusters. However, NGC 5466 also contains numerous blue stragglers, and the similarity of masses derived for anomalous Cepheids (from their pulsation properties) and for the blue stragglers (from their dynamical segregation in the cluster) suggests a common origin. The current explanation for these variables is therefore mass transfer in binaries or coalescence of two or more stars, where the stars involved are in a later stage of evolution (e.g., red giants) than the main-sequence stars that produce blue stragglers.

Population II Cepheids have been found only in metal-poor globular clusters with blue horizontal branches, but the occurrence of these variables seems to be more closely related to the presence of very hot horizontal branch stars than with the metallicity content. Because the other property (besides metallicity) that can account for a horizontal branch star being blue is a small total mass, this may indicate a substantial mass loss during the previous life of the star on the red giant branch.

Cepheid stars are bright, easy to recognize, and follow a well-defined period–luminosity (P-L) relation; therefore they are excellent distance indicators and provide important information on our galaxy's halo as well as on the outer regions of the nearer external galaxies. Their P-L relation is different from the analogous relation for Population I Cepheids, which are generally used in extragalactic distance determinations because they are brighter and more frequent than their Population II counterparts. Therefore, the appropriate Cepheid P-L relation must be used in distance determinations, in order to avoid a historical mistake that lasted over 30 years: In 1918 Harlow Shapley attempted the first calibration of the

Cepheid P-L relation using 11 Population I Cepheids for which the proper motions and statistical parallaxes were available. He neglected, however, the contribution of interstellar absorption and the effects of galactic rotation on the proper motions, which were poorly known at that time, and the zero-point of his P-L relation was ~1.4 magnitudes too faint. By a fortuitous coincidence, this was just the right amount to bring the globular cluster Population II Cepheids into the same P-L relation, because they are intrinsically ~1.5 mag fainter and, being mostly at high galactic latitude, are much less affected by interstellar absorption. Ever since Edwin P. Hubble discovered Cepheid variables in the Galaxy NGC 6822 in 1925, Shapley's P-L relation had been used to determine the distance to nearby galaxies. In the late 1940s, the distance to the Andromeda galaxy was thought to be ~300 kpc, based on the preceding calibration, in spite of the growing evidence of the existence of interstellar absorption and of two separate classes of Cepheids in the galaxy, with quite different spectroscopic and kinematic properties. It was only in 1952, however, after the first observations of Andromeda with the new 5-m (200 in.) Palomar telescope, that Walter Baade realized the mistake and suggested the reevaluation of the distance scale with the correct P-L calibration for Population I Cepheids; this "simple" recognition doubled the size and the age of the universe. The present calibration of the P-L relation for Population II Cepheids is based on RR Lyrae-derived distances to globular clusters.

RED VARIABLES

Approximately 135 red variable stars [also called long-period variables (LPVs)] are currently known in globular clusters. They can be subdivided into three types, according to our present understanding of stellar evolution:

1. Mira variables, with periods between ~100 and ~300 days and large pulsation amplitudes ($\Delta m \geq 2.5$ mag); those in globulars occur only in relatively metal-rich clusters and are believed to be very luminous asymptotic giant branch stars.
2. Yellow semiregular variables (SRd), with periods between 75 and 120 days; they appear to be lower-luminosity asymptotic giant branch stars, probably undergoing blueward loops in the HR diagram in response to helium shell flashes, or on their final blueward track toward the white dwarf region. They are found preferentially in metal-poor globular clusters with blue horizontal branches.
3. Red semiregular and irregular variables (SR, Irr); they may be either asymptotic giant branch or red giant branch stars, and they fall at or near the top of their globular cluster giant branches. They occur equally in metal-rich or metal-poor clusters.

Potentially the most interesting among red variables are the Mira variables, for several reasons: (a) They are the brightest stars in globular clusters and can be used for the determination of distances to remote stellar systems, once their P-L relation is defined and calibrated. (b) Their luminosity is related via the stellar evolution theory to the age of the parent cluster. (c) They represent the last stage of the nuclear burning life of the star; therefore their mass provides important information on the subsequent stellar evolution to the planetary nebula and white dwarf stages. (d) They are losing a significant amount of mass that has been enriched by various nuclear burning products and, therefore, can play a role in the chemical enrichment of the galactic interstellar medium.

The pulsation modes, the fundamental periods, and the exact P-L relation for these variables are still being debated, mainly for two reasons: First, the very extended atmospheres of these stars make it difficult to define quantities such as radius and effective temperature and to calculate accurate model atmospheres. Second, the dominance of convective energy transport in the interiors, combined with the lack of an adequate theory of convective transport, hampers accurate quantitative studies of the pulsation of the envelopes. It appears, however, that the red edge of the Cepheid instability strip tends to become horizontal at approximately $M_{bol} = -2.5$ and that red stars brighter than this value pulsate in the fundamental mode if they are hotter than ~3700 K (generally the metal-poor stars) and in the first overtone mode if they are cooler than 3700 K (the metal-rich ones). On the assumption that this is correct, these stars then seem to follow a well-defined P-L relation (where P is the fundamental period and L is the bolometric luminosity, mostly determined from infrared measurements), whose exact slope depends on the value adopted for the absolute magnitude of the RR Lyrae variables used as distance calibrators. The empirical P-L relation determined for red variables seems to join with the analogous relation for Population II Cepheids, thus suggesting that there may be a unique relationship between fundamental period and bolometric magnitude for cluster variables with periods between 1 and 300 days. More work, however, is still needed in order to understand and define in detail this relationship and the properties of the red variables.

DWARF CEPHEIDS

The Population II dwarf Cepheids are so called because they pulsate with very short periods, between 0.04 and 0.14 day, and with very small light variation amplitudes, of only a few tenths of a magnitude. Some of these variables had been distinguished from their Population I counterparts in the field about 10 years ago, thanks to their metal deficiencies and high velocities. Around 1984 three of them were found in the globular cluster ω Cen, but only very recently a systematic search and study of these variables was undertaken in globular clusters, focusing on low-concentration clusters containing a good number of blue stragglers. These variables, in fact, are located toward the cluster centers and in the same region of the HR diagram as the blue stragglers. In addition, their short periods and small pulsating amplitudes also suggest a higher mass (~1.5 $M_{\odot}$) than for a single globular cluster star.

The current explanation, therefore, is for mass transfer occurring in a binary system, which pushes a main-sequence star up along the main sequence to a brighter and hotter location, as appropriate to the newly acquired larger mass. If this blue straggler star happens to have the right mass to enter the instability strip for its chemical composition, then it pulsates as a dwarf Cepheid.

As in the case of the other variable stars that form in binary or multiple stellar systems (e.g., anomalous Cepheids, novae, and eclipsing variables), these stars can provide important information on the dynamical structure and evolution of the parent cluster and on the conditions of mass density and angular momentum that allow the formation of binary or multiple stellar systems in a crowded stellar environment.

Additional Reading

Harris, H. C. (1986). Population II variables. In *Stellar Pulsation*, A. N. Cox, W. H. Sparks, and S. G. Starrfield, eds. Springer-Verlag, Berlin, p. 274.

Rosino, L. (1978). Problems of variable stars in globular clusters. *Vistas in Astronomy* **22** 41.

Sandage, A. (1990). The Oosterhoff period effect: Luminosities of globular cluster zero-age horizontal branches and field RR Lyrae stars as a function of metallicity. *Ap. J.* **350** 631.

Sawyer-Hogg, H. B. (1973). A third catalogue of variable stars in globular clusters. *Publ. David Dunlap Observatory, University of Toronto* **3** (No. 6).

Wallerstein, G. and Cox, A. N. (1984). The Population II Cepheids. *Publ. Astron. Soc. Pacific* **96** 677.

Wood, P. R. (1986). Long-period variables. In *Stellar Pulsation*, A. N. Cox, W. H. Sparks, and S. G. Starrfield, eds. Springer-Verlag, Berlin, p. 250.

See also **Star Clusters, Globular, Binary Stars; Star Clusters, Globular, X-Ray Sources; Stars, Blue Stragglers; Stars, Chemical Composition; Stars RR Lyrae Type.**

Star Clusters, Globular, X-Ray Sources

Paul L. Hertz

The first survey of the sky in x-ray wavelengths was conducted by the *UHURU* satellite in the early 1970s. It was noted that the 100 or so bright cosmic x-ray sources discovered in the Milky Way included at least 5 located in globular clusters. This meant that, relative to the number of stars present, bright x-ray sources were more than 100 times more likely to be present in a globular cluster than in the disk or central bulge of the Galaxy. In the late 1970s, the relationship between x-ray sources and globular clusters was strengthened when two previously unknown globular clusters, heavily obscured by intervening interstellar dust, were discovered at the positions of bright x-ray sources. Astronomers believe that these bright globular cluster x-ray sources are similar to the bright x-ray binaries in the galactic plane and consist of a neutron star accreting matter from a normal stellar companion. When the imaging x-ray telescope on board the *Einstein* observatory satellite was turned on globular clusters between 1978 and 1981, more than 10% of the clusters observed were found to have x-ray sources in them, which were 1000 to one million times fainter than those first discovered a decade earlier. These low-luminosity sources may be systems similar to cataclysmic binaries, with a white dwarf taking the place of the neutron star, or they may be similar to the bright systems, but with their accretion turned off.

BRIGHT GLOBULAR CLUSTER X-RAY SOURCES

The nature of the galactic low-mass x-ray binaries is well established. They are close binary star systems. One member is a main-sequence star, with a mass typically between 0.5 and 1 $M_\odot$ (solar mass) and of spectral type F, G, K, or M. The other is a neutron star—the collapsed remnant of a star that has completed its evolution. Neutron stars have masses slightly larger than the Sun, but are only 10 km in radius. The binary orbit is so small (ranging between 0.01 and 0.10 the size of Mercury's orbit) that gas flows over the gravitational saddle at the inner Lagrangian point between the two stars and onto the neutron star. The gas is decelerated abruptly at the surface of the neutron star, and the kinetic energy is converted into heat and radiation—primarily x-rays. The x-ray properties of bright globular cluster x-ray sources resemble those of galactic x-ray binaries in every particular, and it is assumed that they are similar in nature.

Globular cluster x-ray sources are of special interest for several reasons. First, x-ray sources are more than 100 times more populous, relative to the number of field stars, in globular clusters than in the galactic disk and central bulge. This means that the environment within a globular cluster must favor the formation of an x-ray binary. Once enough bright globular cluster x-ray sources were known (there are approximately 10 known today; see Table 1), it was noted that they tend to be found in the densest globular clusters. Imaging observations with *Einstein* revealed that the sources are located near the cluster centers where the stellar density is highest. These x-ray sources probably formed through the process of tidal capture. If a neutron star passes within 3 stellar radii of a normal star, it raises tides in the atmosphere of the normal star. These tides then undergo damped oscillations, and in doing so they dissipate enough energy that the neutron star can no longer escape from the tidally distorted star—the two stars are now gravitationally bound. Because this process only takes place during very close encounters, the resulting binary will undergo a phase of mass transfer and x-ray emission in the aftermath of capture. Tidal capture does not occur in the galactic plane because stellar densities are low and close encounters are too rare—the exact formation mechanism for x-ray binaries outside of globular clusters is not known.

A second reason for interest in globular cluster x-ray sources is that their distances, and hence their intrinsic brightnesses, are known. X-ray sources in globular clusters are known to be at the distance of their host cluster, whereas the distances to x-ray sources in the galactic plane are uncertain by factors of 2–5. The measurable luminosities of globular cluster x-ray sources fall between 10^{36} and 10^{38} erg s^{-1}. (The solar luminosity is 3.8×10^{33} erg s^{-1}.) They are limited on the bright end by the Eddington luminosity, that luminosity above which radiation pressure would exceed gravitational attraction. An accreting source cannot exceed the Eddington luminosity because the outgoing radiation would blow away the incoming matter, preventing accretion onto the neutron star. The lower limit on the x-ray luminosity is less well understood, but it is related to the physical processes that force the stars closer together and drive the accretion flow.

Finally, some of the most interesting x-ray binaries known are located in globular clusters. The shortest period for any known binary is the 11.5-min binary period of the x-ray source 4U1820−30 in the globular cluster NGC 6624. Recent studies of this binary with *EXOSAT*, the *European X-Ray Observatory Satellite*, in 1984 and 1985 revealed that, in order to have such a short orbital period, the companion star must be a helium degenerate dwarf star with a

Table 1. Globular Cluster X-Ray Sources

Globular Cluster	*X-Ray Source*	*Luminosity*	*Comments*
NGC 104 (47 Tuc)	1E0021.8 − 7721	Low	
NGC 1851	4U0513 − 40	High	Burster
NGC 1904 (M 79)	1E0522.1 − 2433	Low	
NGC 5139 (ω Cen)	Five sources	Low	
NGC 5272 (M 3)	1E1339.8 + 2837	Low	
NGC 5824	1E1500.7 − 3251	Low	
Terzan 2	4U1722 − 30	High	Burster
Grindlay 1	4U1728 − 34	High	Slow Burster
Liller 1	MXB 1730 − 335	High	Rapid Burster
Terzan 1	XB 1733 − 30	High	Burster
Terzan 5	XB 1745 − 25	High	Burster
NGC 6440	MX 1746 − 30	High and low	Transient
NGC 6441	4U1746 − 37	High	Burster
NGC 6541	1E1804.4 − 4343	Low	
NGC 6624	4U1820 − 30	High	11.5-min binary
NGC 6656 (M 22)	Three sources	Low	
NGC 6712	4U1850 − 08	High	Burster
NGC 7078 (M 15)	4U2127 + 12	High	8.5-h binary

binary separation distance less than half the distance between the Earth and the Moon.

A second x-ray source of interest is the x-ray binary 4U2129+12 in the globular cluster M 15. It is the only globular cluster x-ray source for which an optical counterpart has been identified—the ultraviolet-excess star AC 211. Observations of AC 211 with ground-based telescopes found that the orbital period is 8.5 h and that the star is expelling gas in excess of 170 km s^{-1}. Analysis of archival x-ray data obtained with the *HEAO-1* satellite in 1977 showed that the x-rays are modulated at the 8.5-h orbital period, most likely by scattering from inhomogeneities in the hot gas surrounding the neutron star.

A unique x-ray source is MXB 1730−335, the Rapid Burster. This source is located in the obscured globular cluster Liller 1, which was discovered during a search for the optical counterpart of the Rapid Burster. The Rapid Burster has periods of activity during which it produces x-ray bursts of 10^{38} erg s^{-1} in quick succession; intervals as short as 10 s have been observed. These bursts are probably due to accreted material being "gated" from a holding area above the surface of the neutron star. The holding area may be caused by the interaction of the neutron star's magnetic field and the magnetized material in orbit around it.

LOW-LUMINOSITY GLOBULAR CLUSTER X-RAY SOURCES

The *Einstein* satellite carried the first imaging x-ray telescope to be used for obtaining x-ray pictures of cosmic sources. Among the thousands of targets observed with *Einstein* were 71 globular clusters, a majority of all the known galactic globular clusters. These observations were made during the late 1970s in order to measure the exact position of the bright globular cluster x-ray sources and to search for new x-ray sources fainter than 10^{36} erg s^{-1}. A total of 14 low-luminosity sources were discovered in 8 globular clusters (see Table 1). Surprisingly, the brightest of these sources had a luminosity of 10^{34} erg s^{-1}, and no x-ray sources were observed in the "x-ray luminosity gap" between the brightest low-luminosity source and the faintest high-luminosity source. They represent a new class of objects, because a single class would be expected to have a continuum of possible luminosities. Sources as faint as the detection threshold were observed, so no lower limit on luminosity can be set from the observations.

Three theories have been advanced concerning the nature of the low-luminosity globular cluster x-ray sources. One theory predicts that the low-luminosity sources are close binaries containing a white dwarf and a low-mass star, similar to cataclysmic binaries. Cataclysmic binaries are known to emit x-rays with luminosities up to 10^{32} erg s^{-1}. According to this theory, cataclysmic binaries in globular clusters formed by tidal capture, and the higher luminosity sources may represent the brighter "tail" of the distribution of the large number of cataclysmic binaries that may be present in globular clusters.

A second theory suggests that these sources are a class of x-ray binaries known as transient x-ray sources. Transients undergo brief periods of high x-ray luminosity, but spend much of their time at low luminosities. A recent calculation of the dynamics in the companion star's atmosphere near the inner Lagrangian point indicates that accretion flow may be unstable. The only allowed luminosities may be above 10^{36} erg s^{-1} and below 10^{34} erg s^{-1}, thus explaining the luminosity gap.

In order to determine which of these theories is correct, these x-ray sources need to be detected optically. The first theory predicts that these sources will resemble cataclysmic binaries optically, with blue colors and rapid flickering. The second theory predicts that only the normal stellar companion will be visible when the source is quiescent, but with clear variability due to its binary nature. Extensive searches for blue and/or variable objects in the large (1-arcmin radius) x-ray error circles have been undertaken, with negative results to date. This has led to a third theory—that the low-luminosity sources are either foreground stars or background galaxies and quasars, and are not associated with the globular clusters.

X-RAY SOURCES AS PROBES OF GLOBULAR CLUSTERS

Globular cluster x-ray sources can be used to probe the nature of both the globular clusters in which they are found and the x-ray sources themselves. The bright globular cluster x-ray sources are the only nonpulsing x-ray binaries for which a mass can be determined. Globular clusters undergo a process called mass segregation. As stars pass each other in the dense cluster environment, they are able to exchange energy. If two stars of unequal mass pass, it is energetically easier to accelerate the light star and decelerate the heavy star. In this way more massive stars tend to lose kinetic energy and sink toward the center of the cluster, whereas less massive stars may be found further out in the cluster halo. The position of the star is thus an indication of its mass.

The positions of the bright globular cluster x-ray sources were measured with arcsecond accuracy using *Einstein*. From their positions in their respective globular clusters, it was determined that the typical x-ray binary has twice the mass of a typical star in a globular cluster. This is consistent with the x-ray binary containing a 1.4-$M_{\odot}$ neutron star and a 0.5-$M_{\odot}$ companion star. The positions of the low-luminosity x-ray sources are not so well known, but they tend to be located further from the center of the cluster. This favors the cataclysmic binary theory, because white dwarfs are less massive than neutron stars.

The relative number of high- and low-luminosity sources is a probe of the efficiency of tidal capture and of the relative number of isolated neutron stars and white dwarfs available for tidal capture in globular clusters. Detailed models of these processes, including the effects of mass segregation, indicate that white dwarfs are not tidally captured as efficiently as neutron stars, and predict a relatively low number of cataclysmic binaries. This prediction argues against the cataclysmic binary theory of low-luminosity globular cluster x-ray sources.

The number of globular cluster x-ray sources also tests our understanding of the formation and evolution of millisecond radio pulsars. These rapidly spinning (up to 1000 rotations per second) neutron stars are believed to have evolved from x-ray binaries. Accretion onto the neutron star speeds up the neutron star's spin, in the same way that ice skaters spin faster when their arms are pulled in closer to their bodies. Increasing numbers of millisecond pulsars are being found in globular clusters by current searches. Current theories of their evolution and of the number of bright x-ray sources observed today in globular clusters imply that a great many millisecond pulsars may be present, but undetected, in globular clusters.

FUTURE RESEARCH

Understanding the nature of low-luminosity globular cluster x-ray sources and studying the formation mechanisms and evolutionary scenarios of all globular cluster x-ray sources require further observations. X-ray observations, which must be made from above the atmosphere because x-rays are absorbed by the atmosphere, depend on current and future x-ray observatories. These include the German satellite *ROSAT*, launched in 1990, and the NASA x-ray observatory *AXAF*, planned for the mid-1990s. Optical observations of these sources in the crowded regions at the centers of globular clusters are being facilitated by the *Hubble Space Telescope*, which was launched in April 1990.

Additional Reading

Bailyn, C. D., Grindlay, J. E., and Garcia, M. R. (1990). Does tidal capture produce cataclysmic variables? *Ap. J.* (*Letters*) **357** L35.

Grindlay, J. E. (1988). X-ray binaries in globular clusters. In *The Harlow–Shapley Symposium on Globular Cluster Systems in Galaxies*, J. E. Grindlay and A. G. Davis Phillip, eds. Kluwer Academic Publishers, Dordrecht, p. 347.

Ilovaisky, S. A. et al. (1987). CCD photometry of AC211/X2127 + 119: The 8.5 h period of the x-ray binary in the M 15 globular cluster. *Astron. Ap.* **179** L1.

See also **Binary Stars, Cataclysmic; Binary Stars, X-Ray, Formation and Evolution; Pulsars, Binary; Pulsars, Millisecond; Stars, Neutron, Physical Properties and Models; X-Ray Bursters.**

Star Clusters, Mass and Luminosity Functions

Graeme H. Smith

The luminosity function, together with its counterpart the mass function, specify the distribution of intrinsic luminosities and masses among a collection of stars. They provide fundamental information that must be accounted for by successful theories of stellar formation and evolution. Because observations suggest that the stars within a given open or globular cluster were born at similar times and with similar abundances, many of the parameters upon which the luminosity function depends may be considered constant within a cluster. As a result of this simplification, considerable observational and theoretical effort has been devoted to the study of cluster luminosity functions.

DEFINITIONS

The luminosity function of a sample of stars, $\phi(L)$, is defined such that $\phi(L)\,dL$ is the number of stars per unit volume having luminosities in the range L to $L+dL$. Stellar luminosities are typically measured on a magnitude scale, so that the luminosity function can be expressed as $\phi(M_{\text{bol}})$, the number of stars per unit bolometric magnitude interval. Because bolometric magnitudes are not directly observable, the luminosity function is generally determined with respect to magnitudes measured through a well-defined filter system, such as the Johnson *V*-magnitude system. The stellar mass function, $\xi(\log m)$, as defined by Edwin E. Salpeter, is the number of stars per unit volume per unit logarithmic (base 10) mass interval. This function can be related to a mass spectrum, $n(m)$, by the expression $\xi(\log m) = (\ln 10) m n(m)$, where $dN = n(m)\,dm$ is the number of stars per unit volume in the mass interval $(m, m+dm)$. In the astronomical literature, $n(m)$ is also often referred to as the mass function.

THE FORM OF THE GLOBULAR CLUSTER LUMINOSITY FUNCTION

The luminosity of a star is a function of its mass, age, and chemical composition. A globular cluster provides a relatively large sample of approximately 10^5–10^6 stars, which, to good approximation, have identical ages and abundances. The differences in luminosity among the stars in any given cluster can therefore be related directly to the differences in their mass (except over small regions of the color-magnitude diagram in which stellar evolutionary tracks are nonmonotonic in luminosity). Consequently, the luminosity function of a cluster can be related to its mass function by the formula

$$\phi(M_{\text{bol}}) = \left|\frac{dN}{dM_{\text{bol}}}\right| = \frac{dN}{d\log m}\left|\frac{d\log m}{dM_{\text{bol}}}\right| = \xi(\log m)\left|\frac{d\log m}{dM_{\text{bol}}}\right|.$$

In this expression $d\log m/dM_{\text{bol}}$ is the slope of the mass–bolometric magnitude relation, which is a function of the cluster age and metal abundance, and is determined by the physics of stellar interior structure. The mass function which enters into the preceding equation is determined initially by the processes of star formation, plus subsequent modifications brought about by dynamical evolution within the cluster and the interaction of the cluster with its environment. It is common to represent the mass function by a power law expression of the form

$$\xi(\log m) = \frac{dN}{d\log m} \propto m^{-x}.$$

The logarithm of the luminosity function, $\log \phi(M_V)$, of the relatively massive cluster 47 Tucanae has been studied in considerable detail and is shown in Fig. 1. Three main regions can be identified: the unevolved main sequence ($M_V \gtrsim +5$), the main-sequence turnoff region and evolved main sequence ($+2.8 \lesssim M_V \lesssim +5$), and the giant branch ($M_V \lesssim +2.8$). The exact magnitude ranges of these three regions depend on the metal abundance and age of a cluster. We discuss them in turn.

On the unevolved main sequence, stars are in a long-lived phase of evolution, burning hydrogen in their cores. It is possible to relate the luminosity of such stars to their mass by an expression of the form $L \propto m^a$, where the exponent a depends on the structure of the star, and can be considered constant over limited magnitude intervals. Such a mass–luminosity relation, combined with the preceding power law expression for $\xi(\log m)$, produces a luminosity function of the form

$$\log \phi(M_{\text{bol}}) = \frac{x}{2.5a} M_{\text{bol}} + \text{constant}.$$

This expression serves to illustrate the basic features of the observations and the more realistic theoretical luminosity functions that are plotted in Fig. 1, such as the increase in ϕ with increasing magnitude and the dependence on the power law index x for $M_V \gtrsim +5$. The slope of the luminosity function depends upon both x and a when fainter than this magnitude. This second dependence explains the relatively abrupt change in $\phi(M_V)$ near $M_V = +9.6$. This local increase in the luminosity function slope is produced by a decrease in the value of a, which is connected with the formation of the H_2 molecule near the surfaces of stars of mass less than approximately 0.45 $M_\odot$.

The main-sequence luminosity function therefore contains information about the stellar mass function and the mass–luminosity relation. However, only with the advent of charge-coupled device detectors has it become possible to measure $\phi(M_V)$ for globular clusters to sufficiently faint magnitudes. Observations indicate that the value of x varies from cluster to cluster. In 47 Tuc, the example shown in Fig. 1, the observed main sequence is characterized by a flat mass function with $x \sim 0.2$. Other clusters, such as M 15 and NGC 6752, have been found to show much steeper mass functions, with $x \sim 1$–2.

Main-sequence luminosity functions that are determined for only a small region within a globular cluster are likely to be sensitive to the presence of mass segregation within these systems. Dynamical relaxation due to stellar encounters drives the stars within a cluster toward equipartition of energy. As a consequence, lower-mass stars will, on average, develop greater velocities and so fill a larger volume than higher-mass stars, with the result that luminosity functions will become increasingly dominated by low-

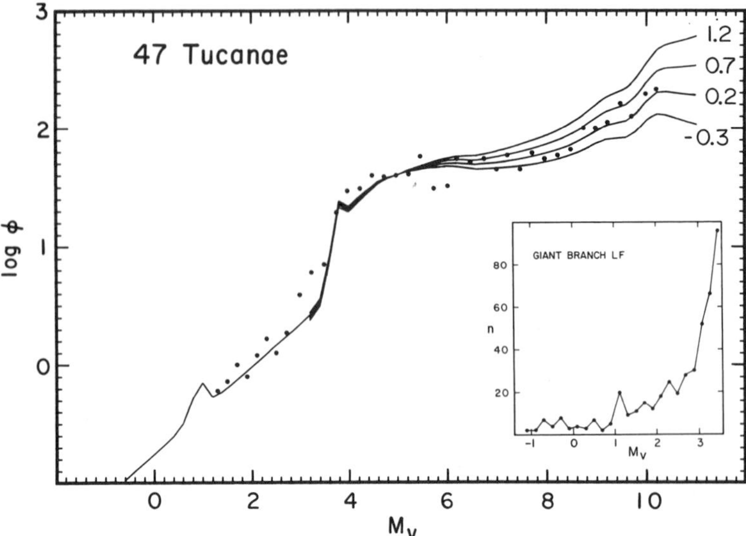

Figure 1. A plot of the luminosity function of the globular cluster 47 Tucanae. The main diagram shows the observed luminosity function for the subgiant branch and main sequence (dots) for absolute visual magnitudes in the range $+1 < M_V < +10$. Also shown are theoretical luminosity functions (solid lines) computed for a cluster age of 13×10^9 yr. The models employ a power law representation for the mass function and are labeled by the values of the power law index x. The inset in the lower right-hand corner of the figure shows the observed luminosity function of the 47 Tuc giant branch. Of particular interest is the local peak near an absolute magnitude of $M_V \sim +1.1$, which is thought to be due to the passage of the hydrogen-burning shell through a chemical composition discontinuity within the 47 Tuc stars. [*This diagram has been adapted from two figures from J. E. Hesser, W. E. Harris, D. A. VandenBerg, J. W. B. Allwright, P. Shott, and P. B. Stetson (1987).* Publ. Astron. Soc. Pacific **99** *739. Reproduced with permission; ©1987 by the Astronomical Society of the Pacific.*]

mass stars with increasing distance from the cluster center. Evidence that main-sequence luminosity functions have been altered by mass segregation has been obtained for a small number of globular clusters, notably M 71 and M 30, although again the relevant observations are difficult to make. The mass function of a cluster can be altered in other ways. During passage through the galactic plane, a globular cluster is subjected to the disruptive effects of tidal shocks, which produce internal heating and an enhancement in the rate of evaporation of the outermost stars. This process will preferentially remove low-mass stars from clusters in which mass segregation has been established.

The main-sequence turnoff region of the luminosity function is populated by stars that are exhausting hydrogen in their cores. The pronounced break in the luminosity function of 47 Tuc near $M_V \sim +3.4$ marks the transition from the main-sequence turnoff to the base of the giant branch and is due to the change from a relatively slow rate of luminosity evolution for stars on the main sequence to a much faster rate among red giants. As can be seen from the models in Fig. 1, $\phi(M_V)$ near the turnoff is relatively insensitive to the form of the mass function because there is little difference in mass among the turnoff stars. In principle, the morphology of the luminosity function in the turnoff region provides a means of measuring the abundance of a cluster. Theoretical calculations show that the slope of the break in $\phi(M_{\text{bol}})$ over the magnitude range $+2.8 < M_{\text{bol}} < +4$ increases with a decrease in the helium abundance and an increase in the heavy element metallicity. So far, however, observations have not been sufficiently sensitive to utilize this property for precise abundance measurements. Theoretically, this region of the luminosity function can also provide information about the age of a cluster, but in practice this is again difficult. At a fixed metallicity the location of the break in ϕ moves to fainter magnitudes as the cluster age increases; however, the shape of the break remains relatively unaltered. Consequently, the age of a cluster cannot be derived from observations of the break without an accurate knowledge of the cluster distance modulus.

On the red giant branch the relatively rapid rate of stellar evolution becomes the dominant factor determining the morphology of the luminosity function. Consider two stars of luminosity L and $L + dL$ with masses m and $m + dm$ on the giant branch. The difference in the time since these stars evolved past the main-sequence turnoff is dt. Because these cluster stars are assumed to be of the same age, dt is just equal to the difference in their main-sequence lifetimes, and in the absence of mass loss dm is equal to the difference in their main-sequence masses. Hence the luminosity function on the giant branch can be written as

$$\frac{dN}{dM_{\text{bol}}} = \frac{dN}{dm}\frac{dm}{dt}\frac{dt}{dM_{\text{bol}}}.$$

One can think of $(dN/dm)(dm/dt) = dN/dt$ as the rate of flow of stars onto the giant branch from the main sequence, and dt/dM_{bol} as the inverse of the rate of luminosity evolution on the giant branch. The second factor in the preceding equation is determined by the main-sequence mass–lifetime relation, and for a given helium abundance and metallicity is a function only of the stellar mass. Within a globular cluster the turnoff mass differs negligibly among the giants, so that the first two factors on the right-hand side of the preceding equation are essentially constant on the giant branch, yielding the result that $\phi(M_{\text{bol}}) \propto (dt/dM_{\text{bol}})$. Consequently, the luminosity function is determined by the inverse of the rate of luminosity evolution up the giant branch; the more rapid the evolution, the less densely will a luminosity interval be populated.

To a reasonable approximation, the rate of evolution of a globular cluster red giant having a constant mass can be described by the relation $dt/d\log L \propto L^{-b}$, which produces a luminosity function

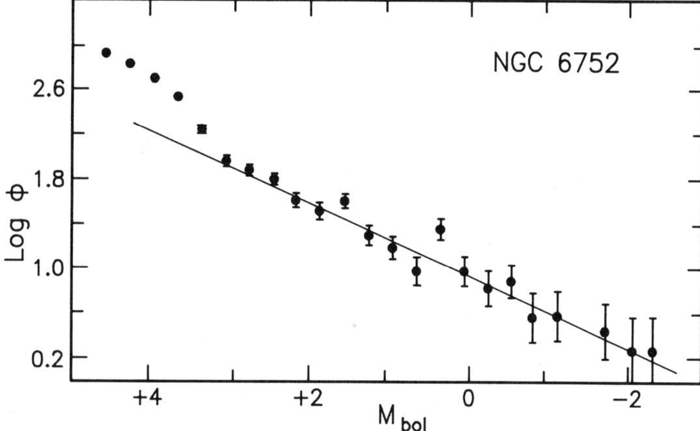

Figure 2. The observed giant branch luminosity function (dots) of the globular cluster NGC 6752 as measured from an extensive program of photographic photometry. The straight line has a slope of $\Delta \log \phi / \Delta M_{\text{bol}} = 0.32$ and represents the theoretical luminosity function derived in the text. This locus provides a good match to the observations. [*From R. Buonanno, V. Caloi, V. Castellani, C. Corsi, F. Fusi Pecci, and R. Gratton (1986).* Astron. Ap. Suppl. Ser. **66** *79. Reproduced with permission; ©1986 by the European Southern Observatory.*]

of the form

$$\log \phi(M_{\text{bol}}) = \frac{b}{2.5} M_{\text{bol}} + \text{constant}.$$

The giant branch luminosity function is thus predicted to be relatively featureless, with $\log \phi(M_{\text{bol}})$ varying linearly with magnitude. In addition, stellar models indicate that there is little change in the form of $\phi(M_{\text{bol}})$ over the metallicity and age ranges applicable to globular clusters. A value of $b = 0.8$ provides a good approximation to the results of detailed models, so that the slope of the luminosity function in the $(\log \phi(M_{\text{bol}}), M_{\text{bol}})$ diagram is predicted to be approximately 0.32. A plot of the giant branch luminosity function for the globular cluster NGC 6752 is shown in Fig. 2. The straight line shows the theoretical luminosity function given by the preceding equation, and provides a good fit to the observations.

Within a cluster red giant the hydrogen-burning shell is very thin, and the rate of evolution is sensitive to the hydrogen abundance at the position of the shell. As a star evolves up the giant branch the shell moves outward, and any changes in abundance that it encounters will be reflected in the morphology of the luminosity function. This leads to the one feature that standard stellar evolution theory predicts in the giant branch luminosity function of a globular cluster. During evolution away from the main sequence, the convective envelope of a model low-mass star moves inward, ultimately reaching a maximum depth near the base of the giant branch before retreating again throughout the upper giant branch evolution. A discontinuity in the hydrogen abundance is left at the point of maximum inward extent of the convection zone. As the red giant evolves to higher luminosities the hydrogen-burning shell moves outward and eventually reaches this discontinuity, at which point the rate of evolution up the giant branch is slowed. The resultant increase in dt/dL, which occurs over a narrow range in L, produces a local maximum in the luminosity function. A peak close to the predicted luminosity has been observed in a small number of globular clusters, notably 47 Tuc and NGC 362, and can be identified at $M_V = +1.1$ in the giant branch luminosity function of 47 Tuc, which is shown as an inset to Fig. 1. Features not predicted by standard stellar evolution theory have also been identified in the luminosity functions of some globular clusters. Among these, a gap near the base of the subgiant branch in the color-magnitude diagrams of NGC 288, ω Cen, M 68, and NGC 6752, seems to be the best documented, and has yet to be explained. In principle, such gaps and peaks may provide important information about the evolution of low-mass stars, but in practice their reality remains to be demonstrated. Even for clusters as populous as the globulars, statistical fluctuations in star counts may conspire to produce such features.

OPEN CLUSTER LUMINOSITY FUNCTIONS

Open clusters, being of much lower mass, are not as well suited to the study of luminosity functions as are globular clusters, and most investigations of these systems have concentrated on the main sequence. No systematic trends have emerged from observational studies to date. The main-sequence luminosity function above 1 $M_{\odot}$ for clusters such as the Hyades, Praesepe, NGC 6811, M 35, and M 11 is similar to that of the field stars. Studies of composite luminosity functions compiled from large numbers of clusters indicate that the average open cluster luminosity function is similar in slope to that of the field stars for stellar masses in the range 1 $M_{\odot} < m < 10$ $M_{\odot}$ (solar masses), a value of $x \sim 1.7$ being typical for the power law index of the mass function. Differences of $\Delta x \sim 0.5$ or more may exist between individual clusters, however. Several objects, such as NGC 2420, exhibit steeper luminosity functions than do the field dwarfs. By contrast, a turnover is observed near $m = 1$ $M_{\odot}$ in the luminosity function of other clusters, such as M 67 and NGC 2506, with the number of stars per unit magnitude decreasing toward fainter magnitudes rather than increasing. Interpretation of this turnover is complex. Although it may be a reflection of the star formation process within these clusters, it could also be produced by disruptive physical processes associated with the motion of a cluster through the galactic disk. Of particular relevance to the origin of a turnover is the presence of mass segregation, which can produce a depletion of low-mass stars in the inner parts of an open cluster, where observations are typically made. Radial mass segregation has been observed in several open clusters including the Pleiades, M 11, and M 67. In the Pleiades, for example, it has been found that the luminosity function changes with distance from the cluster center, the outer regions of the cluster exhibiting a greater proportion of faint dwarf stars than the inner regions. It is apparent that progress in the study of the global mass functions of open clusters will require much time-consuming effort in identifying faint cluster members at large distances from the cluster cores.

Additional Reading

Buonanno, R., Caloi, V., Castellani, V., Corsi, C., Fusi Pecci, F., and Gratton, R. (1986). The giant, asymptotic and horizontal branches of globular clusters. III. Photographic photometry of NGC 6752. *Astron. Ap. Suppl. Ser.* **66** 79.

Demarque, P. (1988). Globular cluster luminosity functions. In *The Harlow–Shapley Symposium on Globular Cluster Systems in Galaxies, IAU Symposium 126*, J. E. Grindlay and A. G. Davis Philip, eds. Kluwer Academic Publishers, Dordrecht, p. 121.

Hesser, J. E., Harris, W. E., VandenBerg, D. A., Allwright, J. W. B., Shott, P., and Stetson, P. B. (1987). A CCD color-magnitude study of 47 Tucanae. *Publ. Astron. Soc. Pacific* **99** 739.

Ratcliff, S. J. (1987). Theoretical stellar luminosity functions and the ages and compositions of globular clusters. *Ap. J.* **318** 196.

Richer, H. B., Fahlman, G. G., Buonanno, R., and Fusi Pecci, F. (1990). Low-luminosity stellar mass functions in globular clusters. *Ap. J. (Letters)* **359** L11.

Scalo, J. M. (1986). The stellar initial mass function. *Fundamentals of Cosmic Physics* **11** 1.

van den Bergh, S. and Sher, D. (1960). The luminosity functions of galactic star clusters. *Publ. David Dunlap Observatory* **2** (No. 7) 203.

See also **Star Clusters, Globular, Mass Segregation; Star Clusters, Globular, Stellar Populations; Star Clusters, Stellar Evolution.**

Star Clusters, Open

Kenneth A. Janes

The open star clusters are perhaps the most recognizable of stellar systems; the well-known Pleiades star cluster, easily visible to the naked eye, is a thoroughly typical open cluster and a dozen or so open clusters can be seen without a telescope. Many other open clusters are observable in small telescopes; Galileo was able to telescopically resolve Praesape and the Perseus double cluster, visible as fuzzy patches to the unaided eye, into individual stars. An open cluster is a system of stars containing as few as a dozen or as many as several thousand stars; the designation "open" refers to their appearance in the sky—even when viewed through a modest-sized telescope the individual stars can be clearly distinguished. In contrast, stars in the other major category of star cluster, the globular clusters, are often so numerous that near the center of a globular cluster, the images of individual stars are so crowded that one cannot see through the cluster. At the other extreme are the stellar associations, large, very loose groups of generally very young and luminous stars. In practice the distinction among these categories is not a precise one and the categorization of a particular cluster as open or globular is based on additional information such as the age of the cluster, its location in the Galaxy, or its composition. Many stellar associations actually include one or more open clusters within their boundaries.

Astronomers have long recognized the importance of the open clusters for our understanding of stellar evolution and for galactic structure and evolution. Because a star cluster forms a physical system, more or less bound gravitationally, its member stars are all essentially at the same distance from us, and they were presumably also formed approximately at the same time, out of material with a uniform composition. These properties make it possible to measure the distance, age, and composition of a cluster with some confidence. In contrast, it is very difficult to estimate, in general, the age of a single star in the general field of the Galaxy. For these reasons, the open clusters constitute powerful test cases for theories of the evolution of stars, and they provide a fundamental step in determining the distance scale of the universe. In fact, to a considerable degree, the distance scale depends on a precise determination of the distance to a single cluster, the Hyades, which, at a distance of 45 pc, is the nearest of the major star clusters.

CLASSIFICATION OF CLUSTERS

There is a great variety in the appearance of open clusters, so that some sort of classification scheme is useful. Several factors enter into the visual appearance of a cluster: its angular size (which in turn depends of course on its linear size and distance), the degree of central concentration of cluster stars, the range in the brightness of cluster members, and an estimate of the number of stars in the cluster. In 1930, Robert J. Trumpler in his major work on clusters, derived a classification scheme based on these ideas. He defined four concentration classes, designated I through IV in order of decreasing central concentration and degree of contrast with the background field of stars. He further specified three categories of range in brightness from 1, where the cluster stars are mostly the same brightness, to 3, where there are both bright stars and faint ones. Finally, he defined three richness classes: p, referring to clusters with fewer than 50 stars; m, consisting of clusters with 50–100 stars; and r, for clusters with more than 100 members. Thus a Trumpler class I3r cluster would be a rich, centrally concentrated cluster with a large range in magnitude, whereas a type IV1p would be a rather small, undistinguished group of stars. The Trumpler class is a useful way to catalog the visual character of a cluster, and is often listed in cluster catalogs, but it is only roughly correlated with physical properties such as age, linear size, and mass.

OPEN CLUSTER CATALOGS

Open clusters are well represented in the classic catalogs of nonstellar objects: The Messier Catalogue contains 30 open clusters and the New General Catalogue (NGC) of John L. E. Dreyer lists 347 clusters. In recent years, a large number of additional clusters have been found, both accidentally (e.g., Clyde Tombaugh found several clusters while searching for Pluto) and in systematic searches. There are now just over 1200 known open clusters. The principal modern catalogs are the *Catalogue of Star Clusters and Associations* by G. Alter, Jaroslav Ruprecht, and Vladimir Vanysek which lists references to each cluster between 1901 and 1967, and the Lund Observatory (Sweden) *Catalogue of Open Cluster Data* which is a compilation of the basic information for each cluster—its location, size, age, mass, composition, and other properties.

The known open clusters represent a small fraction of the total number in our galaxy. The principal factor limiting the number of known clusters is the obscuration along the galactic plane, because more distant clusters are much more likely to be located behind dust clouds than the nearby clusters. There is also some difficulty in identifying a distant cluster against the rich background of galactic stars; in many cases the contrast between the cluster and the surrounding field is so low that the cluster may not be noticed. Although it is difficult to quantify these selection effects, nearly all clusters within perhaps 2000 pc have been identified, but relatively few of them beyond that distance are known. The entire Galaxy probably contains between 50,000 and 100,000 open clusters.

PHYSICAL PROPERTIES: SIZE, MASS, AGE, AND COMPOSITION

As judged from visual examination of photographs, the open clusters range in size from about 2 pc to more than 10 pc. The median diameter is close to 4 pc. Most of the larger clusters are very young systems located near the centers of associations or star-forming regions and are probably not gravitationally bound. The relative uniformity in the sizes of the open clusters was used by Trumpler in 1930 as a method to determine the distances to the open clusters. When he compared the distances as inferred from their diameters to the distances as inferred from their luminosities, he realized that the more distant clusters were systematically too faint. This led him to the major discovery of interstellar absorption, caused by the presence of dust clouds distributed along the galactic plane.

The number of apparent cluster members, as inferred from their appearance on photographs, ranges from fewer than a dozen to several thousand stars. The total mass of a typical open cluster is not at all well known, however. The mass of a stellar system can be estimated either from dynamical measurements or from direct counts of all stars at each mass. Because the velocities of stars in open clusters tend to be very small, one has to rely on the latter method in general. Unfortunately, few clusters have been examined sufficiently closely to permit a definitive measurement of their masses. It has been estimated that the typical mass for an open cluster is roughly 300 $M_{\odot}$ (solar masses), but this is almost certainly far too low. All of the nearby clusters are more massive than that and the nearest clusters are rather modest compared to the more distant ones. It is probable that the typical mass for an open cluster is in excess of 1000 $M_{\odot}$.

Very few open clusters are more than a few hundred million years in age. A diagram of the numbers of clusters as a function of age shows an approximately exponential decline in their numbers with a characteristic time scale of 100 million years. Because there is no reason to believe that the rate of formation of open clusters has changed dramatically in the last few billion years, the implication is that the open clusters dissolve in a relatively short period of time. Although most open clusters are bound gravitationally, they are fragile structures, easily destroyed by interactions with other structures in the Galaxy, such as giant molecular clouds. Some of the youngest open clusters are probably not actually gravitationally bound at all and their appearance as clusters is a consequence of their recent formation from individual giant molecular clouds; they simply have not yet had time to disperse. Nevertheless, for many purposes the distinction is not important: Even such systems consist of stars formed at the same time from a single cloud of material.

A few open clusters survive for very much longer periods of time, perhaps as long as the galactic disk itself. Although there is some dispute as to their ages, the oldest open clusters such as NGC 6791 or NGC 188 are of the order of 10 billion years in age. Their longevity is apparently due to two factors: They are much more massive than the usual open cluster (NGC 6791 presently contains at least 10,000 stars), and they have orbits that carry them far from the plane of the Galaxy. Because they spend most of their time out of the disk, they are not often affected by encounters with giant molecular clouds.

The vast majority of stars in the disk of the Galaxy near the Sun, including stars in open clusters, have chemical compositions close to that of the Sun. There is, however, a distinct gradient in their compositions as a function of distance from the galactic center. The further from the galactic center a cluster is, the more metal-poor it is. There is evidence that the same relation holds true for stars not in clusters, but because it is possible to estimate the distances of clusters better than those of individual field stars, the cluster data are more definite. It is less clear whether there is any correlation of age with composition among the open clusters. Modern theories of galactic evolution would indicate that the oldest stars in the solar vicinity should be metal-poor relative to the youngest stars, but the cluster data are somewhat ambiguous. Only two clusters known to be more than half the age of the Galaxy have known chemical compositions (NGC 6791 and NGC 188), and both of them are similar in composition to the Sun. The implication is that the composition of the interstellar medium has evolved little in the past 10 billion years, in apparent contradiction of the theory.

DISTRIBUTION OF OPEN CLUSTERS

The open clusters are strongly concentrated to the fundamental plane of the Galaxy, and, in fact, the proximity to the galactic plane is one of the defining characteristics of the open clusters. Only a few of the known clusters lie more than 100 pc from the galactic plane. If the open clusters are assumed to be distributed in an exponential fashion about the galactic plane, then the characteristic scale height is about 65 pc. The youngest clusters are just emerging from the dark clouds in which they formed; their distribution in space traces the pattern of spiral arms in the Galaxy. In principle, it should be possible to use this fact to deduce the character of the spiral arms near the Sun. Unfortunately, because interstellar dust clouds are intimately associated with very young clusters, most of the clusters that have gone undiscovered are precisely those that would define the spiral pattern of the Galaxy. Thus the observed distribution is likely to be severely affected. Although there are distinct concentrations of young clusters in some regions near the Sun suggestive of spiral arms, there is little evidence among the open clusters for large-scale systematic spiral arms in spite of frequent claims to the contrary and in spite of a great deal of effort to uncover such evidence. If the observed cluster distribution has not been hopelessly compromised by observational selection effects, then the galactic spiral arms in our galaxy resemble the spiral arms found in the "flocculent" type of spiral galaxy.

OPEN CLUSTERS AS PROBES OF STELLAR EVOLUTION

In a diagram of the luminosities of stars in an open cluster versus their temperatures (a Hertzsprung–Russell or HR diagram), the stars are found to fall along a narrow path in the diagram, the shape of which depends on the cluster age. Because all the cluster stars are at the same distance, the same age, and the same composition, the only physical property that distinguishes one star from another is its mass. The HR diagram of a cluster, then, represents the surface properties of stars of various masses as they appear at a fixed age. By comparing one cluster with another of a different age, it is possible to see how stars of various masses change as they age. Over the past 30 years, theoretical work on the evolution of stars has led to detailed models of how stars should evolve; comparisons of the theoretical HR diagrams with those of observed clusters can be used first to date individual clusters and second to test the predictions of the theory. One of the major triumphs of twentieth century astronomy is the fact that the theoretical models of stars predict rather well the large range in the shapes of the HR diagrams of star clusters.

Additional Reading

Goodman, J. and Hut, P., eds. (1985). *Dynamics of Star Clusters*. D. Reidel, Dordrecht.

Hesser, J. E., ed. (1980). *Star Clusters*. D. Reidel, Dordrecht.

Janes, K., ed. (1991). *The Formation and Evolution of Star Clusters*. Astronomical Society of the Pacific, San Francisco.

Janes, K. A., Tilley, C., and Lyngå, G. (1988). Properties of the open cluster system. *Astron. J.* **95** 771.

Lyngå, G. (1985). Computer-based catalogue of open-cluster data. In *The Milky Way Galaxy*, H. van Woerden, R. J. Allen, and W. B. Burton, eds. D. Reidel, Dordrecht, p. 143.

Ruprecht, J., Balasz, B., and White, R. E. (1981). *Catalogue of Star Clusters and Associations*, Supplement 1. Akademiai Kiado, Budapest.

See also **Star Clusters, Mass and Luminosity Functions; Star Clusters, Stellar Evolution; Stellar Associations and Open Clusters, Extragalactic.**

Star Clusters, Stellar Evolution

Pierre Demarque

Star clusters teach us about stellar structure and evolution in a variety of ways. One of the by-products of stellar evolution theory is the determination of the ages of star clusters. Stellar evolution enables us to study the changes with time of color–magnitude-diagram morphology and of luminosity function, and therefore also of the spectral distribution and intensity of integrated light from the whole cluster. This knowledge permits study of stellar populations in distant galaxies, where individual stars cannot be resolved, and where one must decipher the evolutionary status and age of the system on the basis of the integrated light of many stars.

Because all the stars in a star cluster have a common origin and can be regarded as coeval and formed out of a chemically homogeneous gas cloud, they play a central role in the study of stellar evolution. Their color–magnitude (C–M) diagrams then become snapshots of the distribution of stars of different masses at a given age.

The first C–M diagrams were obtained by photographic techniques. Images of individual stars on photographic plates obtained in selected handpasses were measured, and from the intensity of these images, apparent magnitudes could be derived. The development of photoelectric photometry greatly improved the accuracy of such measurements, but because each star had to be measured separately, it proved very demanding in terms of observing time, particularly when faint stars in globular clusters were observed. The recent advent of two-dimensional charge-coupled device detectors enormously increased the efficiency of these measurements because large numbers of images can be stored simultaneously on a single frame. These detectors have proved ideal for globular cluster research: They are more sensitive than previous array detectors, thus enabling astronomers to reach fainter magnitudes; at the same time, like photoelectric cells, they have a linear response, which increases greatly the accuracy of magnitude measurements.

These improvements, together with the increased power of computers, which has facilitated much more elaborate theoretical calculations of stellar physics and stellar evolution than previously possible, have led to major advances in this field during the last few years.

We will emphasize the older star clusters, in which stars with masses not too different from that of the Sun are found. These clusters are relatively well populated and have been excellent testing grounds for stellar evolution theory. They have been extensively used in stellar chronology. Younger star clusters differ in that they still have in them a population of massive stars that evolve rapidly. Massive stars play an important role in galactic nucleosynthesis and are major contributors to the luminosities of the stellar systems to which they belong. However, massive-star evolution is at this point not as well understood as low-mass star evolution, and will not be considered here because of the relatively less decisive role played by star clusters in its study.

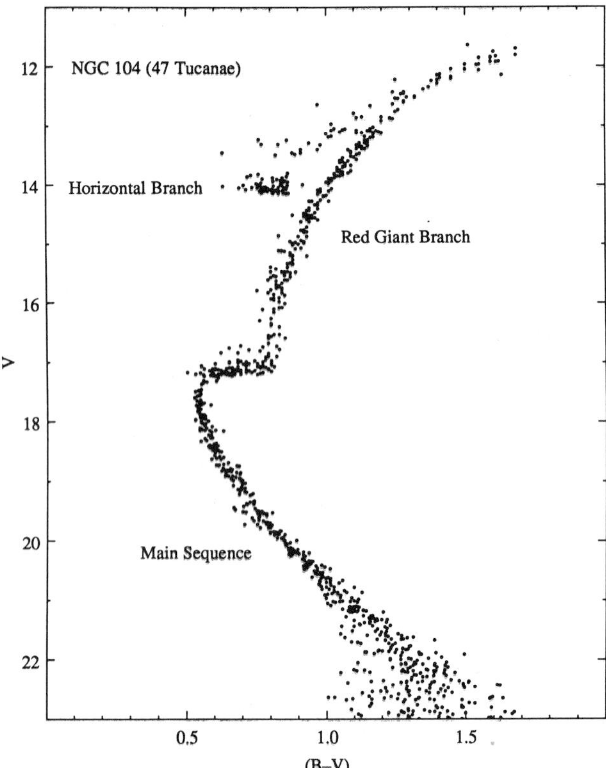

Figure 1. Color-magnitude diagram for 47 Tucanae.

STAR CLUSTERS AND THE THEORY OF STELLAR EVOLUTION

Historically, the theory of stellar evolution was primarily developed in the context of the interpretation of C–M diagrams of star clusters. The principal tool has been the construction of theoretical isochrones and luminosity functions derived from theoretical evolutionary tracks. The standard theory has been extremely successful in describing the main features of the C–M diagrams of old star clusters. Figure 1 shows a typical globular cluster C–M diagram. The main phases of post–main-sequence evolution are represented: the hydrogen burning phases on the main sequence, turnoff, subgiant and giant branches, and the subsequent phases during which stars derive their energy from both hydrogen and helium burning. Given an initial chemical composition and mass, the theory predicts the evolutionary path of a star in the Hertzsprung–Russell (HR) diagram as a function of time. Combining a set of such evolutionary tracks and assuming a uniform chemical composition from star to star, a theoretical isochrone can then be constructed in the theoretical HR diagram. An additional step in the comparison between theory and observation entails the conversion from the theoretical plane (i.e., effective temperatures and luminosities) to an observational plane [e.g., $(B-V)$ colors and absolute visual magnitude]. This is the color isochrone. Similar steps are involved in the construction of the luminosity function which, based on the relative rates of evolution at each magnitude interval, gives the relative number of stars to be found on the isochrone in this magnitude interval.

The final step in the comparison of the color isochrones with the C–M diagram for a given star cluster requires a determination of the amount of interstellar reddening in front of the cluster, and of its distance.

There are many other ways in which stars in star clusters can be used to test stellar evolution theory. One important class of tests has to do with the properties of variable stars. For example, variable stars that are members of a star cluster offer an opportunity to test the sensitivity of stellar pulsational properties and luminosities to mass and chemical composition. Evolutionary lifetimes also depend sensitively on chemical composition. This is because of the controlling effect of opacities and of the equation of state in determining the mass–luminosity relation of individual stars, given the radial composition profile in their interiors. In addition, the nuclear processes that drive the evolutionary process are also dependent on composition. Star clusters of different chemical composition thus provide necessary tests of the sensitivity of stellar evolution calculations to chemical composition.

In this context, we note the special opportunity provided by the star clusters in the Large and Small Magellanic Clouds, our nearest neighbor galaxies. The Magellanic Clouds are distant enough from us that to a first approximation, all clusters within the same Magellanic Cloud are at nearly the same distance, and thus can be compared more easily. In addition, the Magellanic Clouds contain clusters with combinations of ages and chemical compositions not found in our galaxy. Finally, the stellar population of the Magellanic Cloud clusters can conveniently be observed in two different modes: star by star (for the construction of a C–M diagram) or in integrated light (in the same way as distant extragalactic clusters). They should thus enable us to calibrate the ages of star clusters in very distant galaxies, which can only be observed in integrated light.

We have already mentioned the remarkable success of the standard theory of stellar evolution in explaining the major features of the color–magnitude diagrams of star clusters. However, an increasing number of observations cannot be explained by the standard theory (e.g., surface abundance peculiarities among the light elements lithium, beryllium, and boron, or elements processed by the CNO cycle). The standard theory assumes spherical symmetry in the models and therefore ignores the effects of rotation and magnetic fields, and in general the effects of internal dynamics and particle transport other than convection (e.g., rotationally induced instabilities or diffusion). A great deal of effort is currently being expended to understand nonstandard effects in stellar evolution, both observationally and theoretically. In this research also, observations of stars in clusters are crucial because they provide a snapshot of a stellar population with a given age and composition. In general, star cluster observations also permit a more accurate estimate of luminosity and evolution status for each cluster member than observations of field stars can provide.

STAR CLUSTERS IN THE GALACTIC DISK

Young Disk Star Clusters

Dating the young disk star clusters is done by fitting the unevolved main sequence in the C–M diagram to a theoretical isochrone for the same chemical composition. A by-product of this procedure is the distance modulus of the cluster. The main uncertainties on the observational side are in the cluster chemical composition, and on the theoretical side are in the calculation of the radii of stellar models. These combined uncertainties may amount to a factor of 2 in the ages.

The special case of the Hyades open cluster, for which it is possible to derive a distance using the moving cluster method (as well as trigonometric parallaxes of individual stars), provides a strong test of stellar structure theory. Because there are binary star systems with known periods in the Hyades, there are, for each adopted distance modulus, corresponding values for the masses of the components of the binary systems. Combining together the absolute magnitudes implied by the adopted distance with the astrometric masses yields a mass–luminosity relation that must be consistent with that given by the stellar models.

Old Open Clusters

The ages of the old open clusters, such as M 67, NGC 188, and NGC 6791, give us information on the chronology of the old disk population. These clusters contain stars that are very similar in mass and chemical composition to the Sun, and that probably have ages of the same order as the Sun (4.5 Gyr). Recent calculations that make use of the solar calibration for these clusters' helium abundance and convective efficiency in the surface convection zone yield ages in the vicinity of 4, 6, and 8 Gyr for M 67, NGC 188, and NGC 6791, respectively. Because of uncertainties in the reddening, the ages of M 67 and NGC 188 could be underestimated by as much as 2 Gyr.

AGES OF THE GLOBULAR CLUSTERS

There is currently a great deal of interest and controversy relating to the way the halo of our galaxy formed and evolved. In the context of star clusters, two questions are being asked: (1) Are the globular clusters coeval (within the free-fall time scale of the galactic halo, less than 1 Gyr, as in the ELS theory, after Olin J. Eggen, Donald Lynden-Bell, and Allan R. Sandage), or did the halo take a long time to form, of the order of several gigayears, due to more chaotic early stages, and perhaps mergers and the accretion of fragments, as in the scenario offered by Leonard Searle and Robert J. Zim? (2) Is there among halo star clusters, in the mean, a correlation between galactocentric distance, age, and metallicity?

There is growing evidence that the answer to question 1 favors a long time scale of halo formation, and that the globular clusters in the galactic halo formed over a period of several gigayears. This is suggested by two separate lines of evidence: (1) Age differences offer the simplest interpretation of the "second parameter" of horizontal-branch morphology in globular clusters. (2) The luminosity dependence of RR Lyrae variables in globular clusters on metallicity predicts, in the mean, larger ages for the most metal-poor clusters in the inner parts of the Galaxy. Although these results are still controversial, they represent the most straightforward interpretation of the data compatible with current theory. Observations with the *Hubble Space Telescope* should greatly help clarify these questions about the globular clusters in our galaxy.

It is important to note that the absolute chronology of globular clusters is more difficult to establish than relative ages and galactic time scales. A definitive answer about absolute ages may have to await advances in stellar evolution theory relating to the effects of internal rotation, magnetic fields, and diffusion processes in the stellar envelope. At the same time, we need better determinations of the chemical abundances of stars [in particular, the (O/Fe) ratio which affects sensitively the nuclear energy generation by the CNO cycle, and therefore the luminosity of the main-sequence turnoff]. For all these reasons, although changes in the age estimates of the oldest stars by more than 50% seem unlikely, expansions or contractions of the age scale, or of some of its parts, cannot be ruled out.

One cannot at this point but be impressed by the progress made in the last quarter century. However, the nagging question still remains whether the remarkable agreement (now within a factor of 2) of the nuclear age of the oldest star clusters (about 18 Gyr), and of the expansion time scale of the universe (Hubble time [about 10 Gyr]) will continue to improve. It is hoped that the next few years will tell if the current stubborn discrepancy between the two time scales will vanish, or if it is real, will make clear what profound implications for cosmology this discrepancy may have.

INTEGRATED LIGHT OF STAR CLUSTERS

We conclude on another cosmological note. The study of stellar evolution in star clusters will also play an increasingly important role in understanding stellar populations in external galaxies, which are so distant that individual stars cannot be resolved, and where only the integrated light of whole star clusters, or other subpopulations in a galaxy, can be observed. It is not a simple matter to extend our knowledge of star cluster color–magnitude diagrams to integrated properties, but we have already seen how the Magellanic Cloud clusters could play a special role in these studies. In this way, here again, our understanding of stellar evolution in star clusters can contribute a key piece of information in deciphering the cosmological puzzle.

Additional Reading

Blanco, V. M. and Phillips, M. M., eds. (1988). *Progress in Southern Hemisphere Optical Astronomy*. Astronomical Society of the Pacific, San Francisco.

Grindlay, J. E. and Philip, A. G. D., eds. (1988). *Globular Cluster Systems in Galaxies, IAU Symp. 126*. Kluwer Academic Publishers, Dordrecht.

Hesser, J. E., ed. (1980). *Star Clusters, IAU Symp. 85*. D. Reidel, Dordrecht.

Lee, Y.-W. and Demarque, P. (1990). The evolution of horizontal branch stars: Theoretical sequences. *Ap. J. (Suppl.)* **73** 709.

Norman, C. A., Renzini, A., and Tosi, M., eds. (1986). *Stellar Populations, Space Telescope Science Institute Symp. Ser., No. 1*. Cambridge University Press, Cambridge.

Philip, A. G. D., ed. (1988). *Calibration of Stellar Ages*. L. Davis Press, Schenectady, NY.

See also **Magellanic Clouds; Star Clusters, Globular, Chemical Composition; Star Clusters, Globular, Extragalactic; Star Clusters, Mass and Luminosity Functions; Star Clusters, Open.**

Star Formation, Propagating

Guillermo Tenorio-Tagle

Propagating star formation is a theory based on the idea of continuous reproduction which defines the principal agent(s) or physical processes required to duplicate the conditions appropriate for the formation of stars. The theory ignores the actual detailed physics of star formation and it is not aimed at specifying a criterion for the onset of the final collapse of a star or a group of stars in either of the two suggested modes of propagation. On a small scale, it is restricted to a particular molecular cloud and has been applied to both massive and low-mass stars. In both cases the main assumption is that stellar formation occurs in the densest parts of a molecular cloud, in the cloud cores, where the particle density exceeds values of 10^4 cm^{-3}. Upon star formation, the deposition of stellar energy locally enhances the pressure, leading to the compression of the remaining cloud matter, and with it to the formation of new cloud cores in which another generation of stars will form. The long-range propagating star formation is, on the other hand, applied to the whole volume of the galactic disk. The key issue here is how to form giant molecular cloud complexes ($M_{cloud} \geq 10^5$–10^6 solar masses [$M_\odot$]), places where the microphysics of star formation should take over, triggering the birth of new stellar clusters.

The various proposed scenarios have led us to realize that the propagation of star formation is in fact induced by the stars themselves (self-propagation). Also, from the observed efficiency of star formation, it has been suggested that it could be a self-limiting process. Furthermore, propagating star formation under certain circumstances, as when applied to the whole galactic system, may achieve self-regulation. A definite proof of these properties of propagating star formation is still lacking, although several observed examples have been cited.

SMALL-SCALE PROPAGATING STAR FORMATION

The first documentation of propagating star formation was given in 1964 by Adriaan Blaauw who showed that some well-spaced subgroups of OB associations present a linear sequence of stellar ages. The observational evidence has grown rapidly in recent years, largely stimulated by the ideas proposed by Bruce G. Elmegreen and Charles J. Lada. In their view, a sudden large-pressure enhancement, caused by the ionization of the gas following the formation of massive stars, leads to the compression of the surrounding cloud material into a thin layer of shocked gas. As a function of time t, the layer collects a column density of matter $\sigma = t(P\rho)^{0.5}$, where P is the driving pressure and ρ the density of the cloud, until it eventually becomes gravitationally unstable and collapses. The collapse causes a further compression and leads to the buildup of dense cloud cores in which another generation of massive stars will rapidly form. The difference in age between neighboring subgroups is, in this scenario, given by the time required for the onset of the gravitational instability. This problem has been thoroughly studied under a wide variety of boundary conditions, including either a thermal pressure or an ionization front at the trailing side of the layer and a shock at the leading edge. The latter accounts for the transverse flow behind the shock, which may erode a growing perturbation, as the layer slows down and sections of the shock, pushed by the condensation, become oblique to the incoming gas. The calculations are consistent with earlier results indicating that an isothermal layer of shocked gas becomes gravitationally unstable when its thickness L ($=\sigma c^2/P$, where c is its sound speed) becomes comparable to the Jeans length in the layer [$L_{Jeans} = c^2/(GP)^{0.5}$], or when $\sigma = (P/G)^{0.5}$. This occurs at a time $t = 1/(G\rho)^{0.5}$, which is also the gravitational collapse time in the preshocked gas. (G is the gravitational constant.) At first glance, therefore, it may seem that the shock is doing nothing to stimulate the collapse. It is, however, well known that molecular clouds are not collapsing on a free-fall time scale because of their inefficiency in dissipating their internal turbulent and magnetic energy. A large fraction of this energy is provided by the winds from low-mass stars, thought to have infested the whole cloud. To keep star formation going, some authors have argued for self-regulated propagation of low-mass stellar formation. Others have postulated an external regulating agent, for example, the far-ultraviolet interstellar radiation field, which by suppressing ambipolar diffusion inhibits star formation. Thus only parts of the cloud shielded from this radiation collapse to form new low-mass stars. In either case, however, the wind energy has been found sufficient to keep a typical molecular cloud from collapsing. Thus the role played by the shock passage, as proposed by Elmegreen and Lada, is to enhance dissipation of the internal energy of the cloud. Note that displacement and compression shields the gas from the local energy sources, allowing it to collapse on its free-fall time scale.

LARGE-SCALE PROPAGATING STAR FORMATION AND THE PHYSICS OF SELF-REGULATION

Large-scale propagating star formation was at first idealized as a simple extension of its small-scale counterpart. The large amount of energy deposited by a collection of massive stars was thought to result in a large-scale remnant that would eventually become unstable and collapse to form a new generation of stars. The model has been applied to giant shells (radius $\geq$ 400 pc), with H II regions at their inner edges, found in our galaxy and in other galaxies such as the Large Magellanic Cloud.

Large-scale propagating star formation has evolved into a theory that implies a close relationship between star formation, stellar evolution, interstellar matter (ISM), and galactic dynamics that establishes a global star-forming cycle. Massive stars are born in groups or associations in molecular clouds; during their lifetime they produce large amounts of energy in the form of photoionizing radiation, stellar winds, and supernova explosions. This energy causes the rapid disruption of the parent cloud and can also affect the surrounding medium on scales of several hundred parsecs. Therefore, the structure of the ISM strongly depends on the rate of stellar formation and thus globally the coexistence of the various ISM phases is not accidental; it is a consequence of the deposition of energy generated mainly by the massive ($M \geq 8\ M_\odot$) stars. On

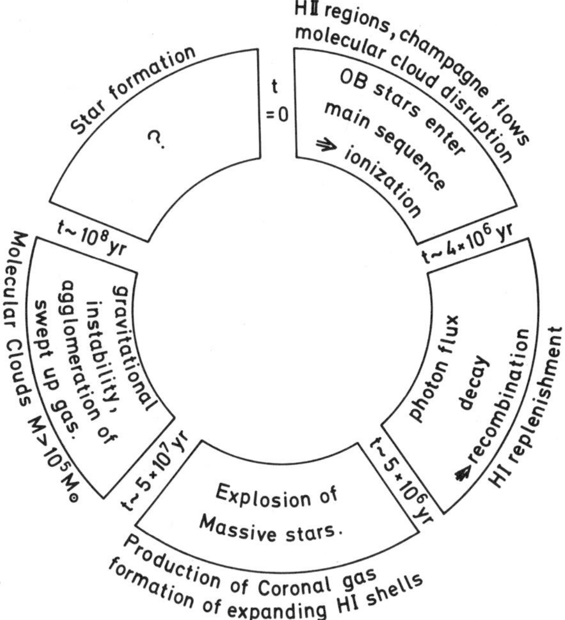

Figure 1. The star formation cycle. The possible tight relationships between star formation, stellar evolution, and interstellar matter are illustrated. In the boxes are indicated the "fundamental" events or processes that lead to phase transitions in the interstellar medium. The times indicate the possible starting points for each process under the assumption that all stars in an association form at the same time. However, note that two or more processes may be at work at any given time. For example, the first supernova may occur while ionization is still in the process of disrupting the molecular cloud. Outside the boxes are noted some of the consequences of the processes. The question mark indicates the still-unknown physical process that causes interstellar matter to collapse into stars. [*Adapted from Tenorio-Tagle and Bodenheimer (1988); reproduced with permission from* Ann. Rev. Astron. Ap. ***26*** *145*.]

the other hand, star formation itself seems to be kept under control within a given galactic system and, at least in our galaxy, it seems to have proceeded at an almost constant rate for the last few billion years. This reasoning implies an "order," a direct control, or a feedback mechanism on star formation that concurrently defines an almost steady-state structure of the ISM. A theory, which in general terms can lead to the observed "order," has to point out the factors whose dominance, or relative absence, play a crucial role in defining the star-forming cycle. These factors should also be identified as the "fundamental" physical processes that result in the various ISM phases. Figure 1 highlights the various links thought to compose the star-forming cycle. These are:

1. Massive star formation brings star formation to an end in a molecular cloud because these stars ionize the remaining cloud matter. Even if ionization-driven shocks compress some of the leftover cloud gas and cause secondary star formation, much more gas is ionized than is converted into stars by this process. Photoionization also leads to the disruption of the parent cloud by causing a supersonic expansion of the ionized matter. This begins when the H II region reaches the edge of the cloud and enters its "champagne" phase, as described by the author in 1979.

2. Once the most massive stars begin to evolve away from the main sequence, the production of ionizing radiation decays, leading to recombination of the dispersed material. Ionization is identified as a fundamental factor in the star-forming cycle, because it limits star formation, or defines the efficiency of the process, and because it leads to molecular cloud disruption. Recombination is a fundamental process because it replenishes H I gas in the system.

3. The explosion of massive stars is also regarded as a fundamental process because it leads to the production of "coronal" (hot) gas. It also causes the formation of large-scale remnants (radius ≥ 500 pc) by sweeping the surrounding disk matter into massive ($M \geq 10^5\ M_\odot$) giant shells. Typical remnants are likely to exceed the dimensions of the galactic disk and to discharge their overpressure into the halo, long before the last supernova explosion occurs. Such events reheat the galactic halo and in some cases may even cause a galactic wind, implying that the energetics from star-forming cycles affect in fact the whole galactic volume.

4. Up to this point in the cycle, stellar evolution has defined the physical processes that lead to phase transitions. To complete the cycle, one should identify the mechanism responsible for the formation of molecular clouds. Several possibilities have been proposed, ranging from large-scale gravitational instabilities to magnetic Rayleigh–Taylor instabilities, density wave compression, swing amplification, or cloud coagulation. All of these imply that an independent agent closes the cycle. True self-regulation has been studied in various ways. The computer models of "stochastic star formation" assume that further star formation is promoted by the energy deposition from the existing stars, with little further physical justification. The steady-state models of galactic disks involve self-sustained cycles that act to fill the whole volume with a collection of remnants resulting from "typical" OB associations. The collection of remnants yields values of star formation and supernova rates in agreement with the observations, suggesting that propagating star formation and the steady state are reasonable assumptions. More detailed calculations of remnants caused by evolved OB association in the presence of differential galactic rotation have shown an organized agglomeration of the swept-up disk matter, leading to the formation of well-spaced giant molecular clouds. In this case, the formation of molecular clouds, seeds of future generations of OB associations, is also promoted by the rate of formation of massive stars, and thus the cycle is self-regulated. Calculations with such a scheme have shown that the star-forming cycle runs at different rates depending on the environmental conditions, causing a well-defined spiral structure and an almost constant star formation rate once it propagates and achieves self-regulation on a galactic scale.

Clearly, structural changes in the galactic environment may affect the cycle in a variety of ways, raising the problem of stability. For example, the star formation efficiency may be strongly affected in clouds where the density differs from the values found in galactic molecular clouds. The reason is that the number of photons, and thus of massive stars, required to ionize a given mass is, as a first approximation, proportional to the cloud number density. In the case of a burst of star formation, the only requirement then is a large cloud compression. The amount of gas involved may be the same as in the case of galactic molecular clouds, but the efficiency of the process could be enhanced. This would also bring about a greater than average number of supernovae, leading to strong heating and/or to the likely dispersal of the remaining matter. In either case, further star formation would have been inhibited and basically only one generation would be present, a situation that may have occurred in elliptical galaxies and/or globular clusters. Star formation in a galactic system will also be affected by the physical interaction of neighboring star-forming centers. However, regardless of the mechanism(s) that led to cloud formation, after the formation of massive stars, the physics of self-regulation will begin to operate. A sequence of changes will occur in search of the establishment and stabilization of the star-forming cycle. Locally, therefore, the selective evolution will proceed in an almost "Darwinistic" manner. This process should also affect the global evolution and ultimately the appearance of the galactic system, whose structure depends only on the ability to form stars and thus on the global stability of the star-forming cycle, or the success of propagating star formation.

Additional Reading

Blaauw, A. (1964). The O associations in the solar neighborhood. *Ann. Rev. Astron. Ap.* **2** 213.

Elmegreen, B. G. (1985). Primary and secondary mechanisms of giant cloud formation. In *Birth and Infancy of Stars*, R. Lucas, A. Omont, and R. Stora, eds. North-Holland, Amsterdam, p. 215.

Elmegreen, B. G. (1989). On the gravitational collapse of decelerating shocked layers in OB associations. *Ap. J.* **340** 786.

Elmegreen, B. G. and Lada, C. J. (1977). Sequential formation of subgroups in OB associations. *Ap. J.* **214** 725.

Franco, J. and Cox, D. P. (1983). Self-regulated star formation in the Galaxy. *Ap. J.* **273** 243.

Franco, J. and Shore, S. N. (1984). The Galaxy as a self-regulated star-forming system: The case of the OB associations. *Ap. J.* **285** 813.

Gerola, H. and Seiden, P. (1978). Stochastic star formation and spiral structure of galaxies. *Ap. J.* **223** 129.

Larson, R. (1987). Star formation rates and starbursts. In *Starbursts and Galaxy Evolution*, T. X. Thuan, T. Montmerle, and J. Tran Thanh Van, eds. Editions Frontières, Gif-sur-Yvette, p. 467.

McKee, C. F. (1989). Photoionization-regulated star formation and the structure of molecular clouds. *Ap. J.* **345** 782.

Mueller, M. W. and Arnett, W. D. (1976). Propagating star formation and irregular structure in spiral galaxies. *Ap. J.* **210** 670.

Palouš, J. (1990). In *Structure and Dynamics of the Interstellar Medium, IAU Colloquium 120*, G. Tenorio-Tagle, M. Moles, and J. Melnick, eds. Springer-Verlag, New York.

Tenorio-Tagle, G. (1979). The gas dynamics of H II regions. I. The champagne model. *Astron. Ap.* **71** 59.

Tenorio-Tagle, G. and Bodenheimer, P. (1988). Large-scale expanding superstructures in galaxies. *Ann. Rev. Astron. Ap.* **26** 145.

See also **Interstellar Clouds, Collapse and Fragmentation; Interstellar Medium, Stellar Wind Effects; Supernova Remnants, Evolution and Interaction with the Interstellar Medium.**

Stars, Activity and Starspots

Sallie L. Baliunas

Stellar activity comprises starspots and other time-variable and localized magnetic phenomena on stars. It is the extrapolation of solar activity to other stars. Magnetism in the outer layers of the Sun, for which we have direct observational evidence, powers and dominates the structure of the solar atmosphere. Spatially distinct features on the Sun, for example, sunspots, active regions, plages, faculae, coronal holes, and the solar wind, are ultimately caused and determined by the strength and configuration of magnetic fields. Furthermore, solar magnetic fields evolve and thereby create the time-dependent phenomena such as solar flares and the 11-year sunspot cycle.

No detailed understanding of the origin and evolution of solar (and stellar) magnetism is yet available. The basis of our knowledge of stellar magnetism is primarily empirical and began with Heinrich Schwabe's detection in 1843 of the 11-year periodicity in the number of sunspots and George Ellery Hale's discovery in 1908 of magnetic fields in sunspots. The most popular notion of solar magnetism and its perplexing periodicity is the magnetohydrodynamic dynamo. In the dynamo model, an internal, poloidal magnetic field of the Sun is converted to a toroidal one which, in turn, refreshes the poloidal field approximately every 11 years by the subsurface action of convection and differential rotation. Unfortunately, the keys to our insight, namely, the physics of the magnetic processes and internal solar conditions, remain elusive.

Hence the impetus for studies of stellar magnetic activity and its variability arises in part to help determine the physical principles responsible for solar magnetism and its periodicities. By turning to the stars, we are offered not only clues to help understand the mysterious principles of solar magnetism but also otherwise unpredictable information on the past and future history of solar magnetism.

DETECTING STELLAR ACTIVITY

The phrase "stellar activity" has two related and oft-times confusing connotations: first, the *presence* of spatially restricted magnetic features similar to solar phenomena such as sunspots; second, the *variability* of proxies of magnetic phenomena which helps to reveal their existence. Confusion may arise from the nature of the measurements that are used to infer the properties of stellar magnetic features: Because the stellar surface features are spatially unresolvable, the time variations of magnetic proxies are important in detecting the *presence* of spatially confined magnetic regions. Thus the two important aspects of "stellar magnetic activity," namely, its time-averaged value and its fluctuations, are revealed through *cross-sectional* and *time-serial* observations and are therefore empirically intertwined. Moreover, the *physics* of the gross level of stellar activity and of its variability are also intermingled!

Proxies of stellar magnetic activity are available from either space-based or ground-based observatories. For example, chromospheric and coronal emission lines occur predominantly in the ultraviolet and x-ray spectrum regions and are prominent markers of stellar magnetic activity in the form of bright areas similar to solar active regions. The ultraviolet and x-ray emission lines must, however, be measured from satellite experiments, which are difficult to muster for extended, dense time coverage of a large group of stars. Therefore, extensive observations historically have been derived from ground-based measurements of photospheric light and the chromospheric Ca II H and K emission cores, near 400 nm. In the Sun, the 11-year cycle can be tracked by an easily detected 20–40% variation in the flux of the Ca II H and K emission lines. The importance of the Ca II emission lines for detailing magnetic activity in cool dwarf stars has been widely recognized and systematically studied since the 1950s.

To detect sunspots on other stars, that is, "starspots," ground-based photometry is employed. In the case of the Sun, spots darken photospheric light at an almost imperceptible level for photometric measurements—a fraction of 1%. As will be described, however, the small, photometric variability produced by sunspots is sometimes outmatched by the effects of spots on other stars. Moreover, highly precise photometry has recently succeeded in revealing starspots on stars with modest levels of magnetic activity, nearly comparable to those of the Sun. In conjunction with the chromospheric measurements, the photometric observations are providing an unprecedented view into the time scale and nature of solar and stellar magnetic variability.

STELLAR MAGNETIC ACTIVITY

The bulk of our knowledge of stellar magnetic activity derives from lower main-sequence, or dwarf, stars, which are in a long-lasting evolutionary state similar to that of the Sun. Magnetic activity is apparently ubiquitous among cool dwarf stars, that is, main-sequence stars ranging in mass from 1.2 times greater than that of the Sun down to several tenths of its mass. (The question of the existence of a lower limit to the stellar mass that can support magnetic activity remains unanswered.)

Average Level of Magnetic Activity

The dynamo model is tailored to explain the universality of cool-star magnetism as well as the dependence of the global average of magnetic activity on stellar macroscopic parameters. This dependence derives from empirical results of cross-sectional studies, which detailed that the average level of activity is a function of lower main-sequence mass and rotation (or, equivalently, age). For example, the young (a few percent of the age of the Sun), solar-mass stars in the Pleiades star cluster show the highest average value of chromospheric emission and the swiftest rotation, compared to those in the Hyades, at an age of about 10% of the Sun, with intermediate activity levels and rotation rates, in turn compared to the Sun and other, several-billion-year-old stars in the field, which are magnetically weakest and rotationally slowest.

A dynamo explanation of these data asserts that the Sun appeared on the main sequence with strong magnetic activity and rapid rotation. The pervasive magnetic field in active stars forces angular momentum loss through a persistent, magnetic wind which results in spin-down. As angular momentum is lost, the efficacy of the dynamo process in renewing magnetic activity is inexorably weakened. Having undergone an erosion spanning several billion years, the Sun now is weak in magnetic activity and slow in rotation.

MAGNETIC VARIABILITY

The Solar Cycle

One of the fundamental attributes of solar magnetism is its dominant, 11-year magnetic periodicity exemplified by the waxing and waning of the number of sunspots. (The fundamental period is doubled if the orientation of sunspot magnetic polarity is considered along with the variation of sunspot number.) The period of 11 years is, however, an average period. Additional nearby and closely spaced frequencies as well as longer periods (such as the 90-year Gleissberg cycle) may also be present in order to explain the variation in amplitude and periodicity in the time series of sunspot numbers. Finally, epochs of suppressed or nonexistent cycles, for example, the Maunder Minimum which began in the seventeenth century and lasted about 70 years, may have occurred during as much as one-third of the Sun's recent history. Any explanation of solar magnetism must successfully produce an average, 11-year periodicity, its accompanying frequencies, and interregna of inactivity. The imprecise clock of solar magnetism is a challenge to magnetic theory.

STELLAR MAGNETIC VARIABILITY

Recent technological achievements allow stellar magnetic inhomogeneities to be studied in solar-type stars, enabling us to refine magnetic theories. Hence *time-serial* studies of magnetic activity are a powerful technique that reveal the complex, time-dependent behavior of stellar magnetism.

Starspots on RS CVn and BY Dra Variable Stars

Historically, photometric measurements of stellar brightness variations with amplitudes of 10–20% or more first signaled pronounced, dark inhomogeneities, or starspots, on the BY Draconis and RS Canum Venaticorum variable stars. Temporal changes in stellar color and spectrum signatures imply that starspots are cooler and are darker in visible light than the surrounding, quiescent photosphere. Such substantial photometric darkenings suggest a correspondingly large coverage by starspots which can occupy tens of percent of the visible stellar surface in contrast to the spot coverage on the Sun of less than 1%. In addition to extensive surface coverage by visibly dark areas, the BY Dra and RS CVn variables exhibit contemporaneous, bright chromospheric and coronal emission lines in the ultraviolet and x-ray spectrum regions. Remarkably, the ultraviolet and x-ray surface fluxes integrated over the visible stellar hemisphere of such a star are as bright (in surface flux) as those of the individual active regions that often accompany sunspots on the Sun. Thus the magnetic activity on the BY Dra and RS CVn stars is elevated far above disk-averaged solar levels. Those variable stars are, however, not typical dwarf stars but main-sequence or slightly-evolved stars in which the rotation rate is extremely rapid as a consequence of youth or tidal interaction with a close, binary star companion.

Activity Cycles of Lower Main-Sequence Stars

More pertinent to the Sun are the studies of single, cool dwarf stars. For them, chromospheric fluctuations signaling magnetic activity from the stellar counterparts to solar active regions are straightforward to detect through variations of the chromospheric Ca II H and K emission fluxes with amplitudes of tens of percent that are similar in the disk average to sunspot cyclic fluctuations. The subtle photometric variations of accompanying starspots with brightness changes of less than 1% are difficult but not impossible to document.

One unique and extensive program, the HK Project, has been investigating the Ca II H and K chromospheric levels and its variations on cool dwarf stars for over two decades. Begun by Olin C. Wilson in 1966 at the Mount Wilson Observatory with the 100-inch Hooker telescope, the project has continued to monitor, with the 60-inch telescope, the fluxes in the chromospheric Ca II H and K emission lines relative to the nearby, steady stellar photospheric flux. The observations are both cross sectional and time serial so quantitative information about the level of magnetic activity and its variability are revealed.

From time-series observations spanning a decade of a sample of almost 100 young and old dwarf stars in a range of masses, Wilson discovered that long-term magnetic variability falls into three general classes: smooth, long-term variations that appear cyclic with a period near a decade and similar to the Sun's behavior; erratic variations with no clear period; and essentially constant Ca II emission.

Wilson's program has been continued and expanded to produce longer and denser time series for determining accurate periodicities and to provide a broader sample of stars that may help to understand the parameters influencing magnetic variations. After two decades, the diversity of long-term chromospheric fluctuations is evident (see Fig. 1). The periods, however, either do not obey a simple dependence upon stellar properties such as mass, rotation, and age, or the periods have not been sufficiently accurately measured for any conclusion to be obvious.

In the case of the Sun, a 20-year interval barely covers two 11-year cycles. Longer time series are desired to determine accurate periods in other stars, especially if multiple frequencies are present. Several firm and positive results are, however, evident in just the 20-year time span:

1. All the active (i.e., young) stars in the sample vary over long time scales.
2. The *majority* of dwarf stars display fluctuations that are apparently or possibly cyclic. The periods range from 2.5 years to as long as 20 years.
3. No clear relation exists between the apparent cycle period and stellar properties such as rotation (or age), mass or activity level, or combinations of them.
4. Cycle periods shorter than five years, which have been well defined by the two-decade time span, are seen only on stars more massive than the Sun. Such short, nonsolar periods are never seen on the Sun; the stars more massive than the Sun have both shorter rotation periods and presumably shallower subsurface convection zones than the Sun.

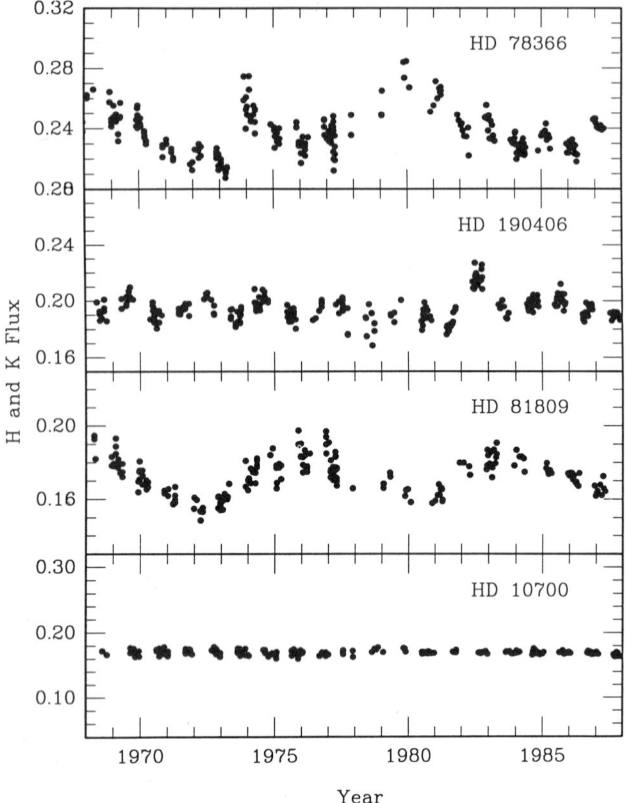

Figure 1. The Ca II H and K fluxes of four solar-type stars, revealing long-term chromospheric activity variations, observed from the Mount Wilson Observatory, are plotted as a function of time from 1966 to the present. The top two panels show young, magnetically active stars with a higher mean value of the H and K flux ratio compared to the stars in the bottom two panels which are old, magnetically weak stars. Of the two young, active stars, HD 78366 varies significantly but not with one particular cycle period; the other young star, HD 190406, has a cycle period of 2.6 years. The period of the old star HD 81809 is nearly 8 years. The star HD 10700 is essentially constant to within the uncertainties of the measurements and may resemble the Sun as it was during the Maunder Minimum in the seventeenth century.

5. The observations of stars of solar mass suggest that the young Sun, at an age of about one billion years, would have persistently shown long-term variability. In contrast, about one-third of the stars at an age of several billion years, comparable to the age of the present Sun, are essentially constant. This agrees with the time-series analysis of isotopic abundances in tree rings, which indicates that the Sun spends about one-third of its current life in a state of little or no cycle variability, exemplified by the Maunder Minimum.

Activity and Brightness Variations of Lower Main-Sequence Stars

Recent photometric measurements of solar-type stars and recent satellite experiments affirm a previously suspected, direct effect of the Sun's magnetic activity variations on the terrestrial climate. Over the previous sunspot cycle, satellite experiments have recorded irradiance fluctuations, that is, changes in the total energy delivered to the terrestrial atmosphere, amounting to nearly 0.1%, where the Sun is brightest at sunspot maximum, and vice versa. Models of the irradiance changes predict that the bright faculae ultimately prevail over dark sunspots in governing the irradiance changes during the sunspot cycle. This small change in solar irradiance over the cycle does not last long enough to overcome the thermal inertial of the Earth's atmosphere and noticeably alter the climate. During the seventeenth century, however, the Maunder Minimum in solar activity could have produced a dimming in the solar radiative output of several tenths of 1% that was *prolonged over decades*. In simple climate models, such a small but sustained irradiance change produces a cooling of the global temperature by nearly 1°C compared to today. Thus the well-documented "Little Ice Age" of seventeenth-century Europe can be accounted for by a slightly but persistently faint Sun at that time.

Measurement of subtle brightness changes in solar-type stars is difficult but has been achieved at the Lowell Observatory in a program of high-precision photometry conducted in tandem with the Mount Wilson Observatory's HK Project. The early results from the dual time series reveal information about the past and future of solar magnetism and irradiance variations:

1. For old stars, representing the Sun at its present age, the irradiance and magnetic activity variations are correlated over the course of starspot cycles; that is, the stars are brightest at starspot maximum and dimmest at starspot minimum, *as is true for the Sun*.

2. For young stars, representing the Sun several billion years ago, the irradiance and magnetic activity fluctuations are *inversely* correlated during starspot cycles. Surface coverage by dark sunspots in the young Sun would probably have been at least an order of magnitude greater than for the current Sun. Large sunspot areas would have governed the irradiance variations of the young Sun, contrary to the dominance by the bright facular areas of today's Sun. Hence the young Sun would have behaved opposite to the current Sun. Furthermore, the changes in the young Sun's irradiance delivered to the early terrestrial environment could have been as large as a few percent over time scales of the sunspot cycle. Although they would have occurred over time scales of only years, short-lived irradiance changes of several percent may have noticeably impacted the young Earth's climate.

Additional Reading

Baliunas, S. L. and Jastrow, R. (1990). Evidence for long-term brightness changes of solar-type stars. *Nature* **348** 520.

Baliunas, S. L. and Vaughan, A. H. (1985). Stellar activity cycles. *Ann. Rev. Astron. Ap.* **23** 379.

Eddy, J. A. (1977). The case of the missing sunspots. *Scientific American* **236** (No. 5) 80.

Lean, J. and Foukal, P. (1988). A model of solar luminosity modulation by magnetic activity between 1954 and 1984. *Science* **240** 906.

Parker, E. N. (1979). *Cosmical Magnetic Fields*. Clarendon Press, Oxford.

Radick, R. R., Lockwood, G. W., and Baliunas, S. L. (1990). Stellar activity and brightness variations: A glimpse at the Sun's history. *Science* **247** 39.

Wallerstein, G., ed. (1990). *Cool Stars, Stellar Systems, and the Sun*. Astronomical Society of the Pacific, San Francisco.

Wilson, O. C., Vaughan, A. H., and Mihalas, D. (1981). The activity cycles of stars. *Scientific American* **244** (No. 2) 104.

See also **Binary Stars, RS Canum Venaticorum Type; Magnetohydrodynamics, Astrophysical; Solar Activity; Stars, Atmospheres, X-Ray Emission; Stars, Magnetism, Observed Properties; Stars, Red Dwarfs and Flare Stars; Stars, Rotation, Observed Properties; Sun, Solar Constant.**

Stars, Atmospheres

Steven N. Shore

Stellar atmosphere theory deals with two separate but complementary problems: the calculation of the structure of the outer layers of a star knowing the radiation flux, radius, and surface gravity (the direct problem); and the determination of the abundances and physical stellar properties of a star from an observed spectrum (the inverse problem).

THE EQUATION OF RADIATIVE TRANSFER

The passage of radiation through a medium is governed by the equation of transfer. This equation is a very generalized one, treating processes schematically by rate coefficients. The intensity, $I_\nu(z,\mu)$, of a monochromatic pencil beam of radiation of frequency ν emergent from a layer at depth z and with directional cosine μ is altered by absorption and scattering, as well as by the intrinsic emissivity of the layer. For a plane parallel medium, which is a good approximation for a thin atmosphere such as that encountered in most main-sequence and giant stars (although this may fail for the supergiants):

$$\mu \frac{dI_\nu}{dz} = -\kappa_\nu \rho I_\nu + j_\nu \rho, \tag{1}$$

where κ_ν and j_ν are the mass absorption and emission coefficients and ρ is the mass density. The source function, defined as $S_\nu = j_\nu/\kappa_\nu$, is the term that describes the radiation field. If the medium is in local thermodynamic equilibrium (or LTE), the source function is uniquely given by the Planck function which, by Kirchhoff's law, is a unique function only of the temperature and independent of any other property of the local medium (like pressure, composition, density, etc.):

$$B_\nu(T) = \frac{2h\nu^3}{c^2}(e^{h\nu/kT}-1)^{-1}, \tag{2}$$

where h and k are Planck's and Boltzmann's constants, respectively, c is the speed of light, and T is the local gas temperature. In general, it is impossible to specify from observation the physical depth from which the photon we observe has emerged because of the strong frequency dependence of the opacity κ_ν. We can, however, define an optical depth:

$$d\tau_\nu = -\kappa_\nu \rho\, dz, \tag{3}$$

which is a dimensionless variable, characteristic of each frequency, which is measured inward from the depth of vanishing density and/or opacity. A layer is called optically thick if τ_ν is greater than or of order unity. For a plane parallel atmosphere, τ_ν and the direction cosine to the surface normal μ, which is independent of τ_ν for each ray, completely specify the path of each photon. Then the formal solution to the equation of transfer is

$$I_\nu(\tau_\nu,\mu) = \int_{\tau_\nu}^{\infty} S_\nu(t_\nu,\mu) \exp\left[-\frac{t_\nu - \tau_\nu}{\mu}\right]\frac{dt_\nu}{\mu}. \tag{4}$$

Given the conditions at the base of the atmosphere, that is, for some depth $\tau_\nu \to \infty$ having the flux and surface gravity specified and having the run of temperature and density through the atmosphere along with the chemical composition, it is possible to compute the emergent intensity at each frequency and for each angle of viewing.

Although the transfer equation has a compact formal solution, its numerical solution involves severe computational difficulties. First, we require a detailed knowledge of the opacity at all depths in the atmosphere and at all frequencies at which we wish to sample the emergent spectrum. That is, we need a grid of absorption coefficients of the constituent mixture of elements, in all of their ionization stages and excitations, for a wide range of densities and temperatures and in a fine enough set to make interpolation accurate and quick. This requires an enormous catalog of atomic energy levels (and even molecular properties should the atmosphere be cool enough), ionization potentials, transition probabilities, and collision cross sections. Then we need to solve for the populations of every atomic level contributing to the opacity at each depth in order to properly calculate the optical depth for a specific line of sight. And most important of all, we need the temperature and density of the atmosphere as a function of some standard optical depth scale to compute any of these physical quantities.

The task facing a model atmosphere theorist is therefore fundamentally iterative. The temperature and density structure of the atmosphere are required to compute the emergent spectrum. But because this is not perfectly known in advance, it is necessary to first guess a structure, calculate the properties, solve the equation of transfer, compare the answer with the requirements of radiative equilibrium, and correct for the deviations resulting from a less-than-perfect guess. Armed with a grid of such atmospheric models, one attempts to reproduce the observed properties of stars of known physical properties such as mass, temperature and radius.

THE DIRECT PROBLEM

Currently, the computation of a stellar atmosphere from first principles requires that three conditions be met. First, the atmosphere is assumed to be in radiative equilibrium. That is, we assume that the bolometric flux in a plane parallel atmosphere is constant. Then we add that the medium through which the radiation passes is mechanically stable. Finally, we assume chemical homogeneity, or that the elements are completely mixed throughout the atmosphere. Each of these conditions can, in principle, be dropped.

To illustrate the principles of the direct problem, assume that we know some approximate solution to the equation of radiative transfer. For the case of a frequency-independent, or gray, opacity and an isotropic source function, we can assume that $T(\tau)$ is known (the Rosseland mean opacity is often used as the depth scale). However, τ is clearly fictitious in the complete problem, because we know that the absorbance of the atmosphere will depend sensitively on frequency. We then solve the equation of hydrostatic equilibrium, written in the optical depth scale rather than in terms of physical depth:

$$\frac{dp}{d\tau} = \frac{g}{\kappa} - \frac{dp_{\text{rad}}}{d\tau}, \tag{5}$$

where p_{rad} is the radiation pressure and p is the gas pressure. With a first solution of p and T in hand, we compute the populations of the ionization states using the constraint that the atmosphere remains electrically neutral in the face of photoionization and collisional ionization, and that the rate of ionizations is balanced by the rate of recombinations (ionization equilibrium). In local thermodynamic equilibrium, the ionization is governed by the Saha equation:

$$\frac{N_{r+1}n_e}{N_r} = \left(\frac{2\pi mkT}{h^2}\right)^{3/2} \exp\left(-\frac{\chi_r}{kT}\right)\left(\frac{2U_{r+1}}{U_r}\right), \tag{6}$$

where U_r is the ion partition function (the total statistical population of all of the accessible atomic levels at a temperature T) in ionization state r, χ_r is the ionization potential, N_r is the ion fraction, and n_e is the electron density; n_e can be computed once the composition is specified by computing all of the ionization fractions and enforcing the condition of charge neutrality.

The populations of individual atomic energy levels can be simply computed only in the case of LTE. Here, the levels have the same relative populations as those expected for a gas in strict Maxwell–Boltzmann statistical equilibrium:

$$\left(\frac{n_j}{n_i}\right)_{\mathrm{LTE}} = \frac{g_j}{g_i} e^{-(E_j - E_i)/kT}, \tag{7}$$

where g_j is the statistical weight of the atomic state j with energy E_j relative to the ground state. The partition function is then

$$U_r(T) = \sum_{j=0}^{\infty} g_j e^{-E_j/kT}, \tag{8}$$

so that in a more complete treatment of the ionization equilibrium, one must carefully calculate the atomic populations as well. If the atmosphere is *not* strictly in LTE, for example, if it is heated from without and is very optically thin, or if it is very distended, or if it is not precisely stationary, the solution of the full equations of statistical equilibrium is required. This means assuming that all collisional plus radiative processes that lead to atomic excitation are balanced, statistically and completely, by deexcitations. To compute this requires detailed knowledge of collision cross sections and radiative lifetimes for each of the relevant energy levels. Because in LTE collisions are assumed to dominate over the radiative rates for all levels, it is not required to know the atomic parameters as precisely. Once an LTE model has been successfully calculated, it can serve as the basis for a test of self-consistency by relaxing, for individual ions and levels, the LTE population assumption.

The frequency-dependent opacity is defined by

$$\kappa_\nu = (n_i B_{ij} - n_j B_{ji})\phi_\nu, \tag{9}$$

where $i < j$, B_{ij} is the Einstein stimulated transition probability, and ϕ_ν is the frequency-dependent line profile. The broadening of a spectral line depends both on atomic radiative time scales and on the rate of collisional perturbations by background molecules, atoms, and electrons. The primary environmental broadening agents are the Stark effect (due to the fluctuation of the local ionic electrical microfield) and collisional broadening by electrons. Because the atoms are also in thermal and turbulent motion, the lines are broadened by the Doppler effect. Thus the shape of the line reflects the atmosphere that forms it: The core is produced mainly by the Doppler profile which is a Gaussian function (reflecting the velocity distribution of the atoms); the wings are due to the larger perturbations by the local ions and the occasional strong field produced by close electron encounters with the absorbing atom. The profile is important as well for the opacity calculation, because the frequency domain over which a single atomic line has influence is a function of the environmental conditions, the phenomenon of line blanketing: A large enough density of lines or broadening simulates an opacity continuum and greatly adds to the local opacity, thus steepening the temperature gradient.

If the opacity is large enough, the medium will become convectively unstable and will preferentially transfer energy by turbulent motion until the depth is reached at which the atmosphere is again optically thin enough to radiate the excess energy. Lacking a fundamental theory of convection, especially in stellar environments, phenomenological prescriptions must be employed for those parts of the atmosphere that are convectively unstable. These are the layers where the temperature gradient required to transport energy by radiation exceeds the adiabatic gradient.

Finally, the equation of transfer is integrated at each frequency and averaged over each angle to provide the total flux and the integrated intensity. The resultant flux is compared with the input value as a function of depth, and the process is repeated until the two agree to some pre-chosen tolerance.

THE INVERSE PROBLEM: SPECTRUM SYNTHESIS

The process of computing large grids of model atmospheres is very labor intensive and has been successfully executed by only a few groups. These grids are the raw material for the inverse problem. Model plane-parallel atmospheres are characterized by the surface gravity, g, and the effective temperature, T_{eff} (defined from the bolometric flux). The detailed temperature distribution can be determined only for the solar atmosphere, as manifested by the limb darkening and angle-dependent variations in line strengths, so that most stellar atmosphere spectrum synthesis assumes some limb-darkening law for angular averages.

The problem of synthesis is especially difficult for rapidly rotating stars, where the surface may be distorted from a sphere and the temperature and density profiles become latitude-dependent, an effect known as gravity darkening. Angle averaging requires using a grid of local atmospheres, consistent with the expected temperature and gravitational acceleration, to be patched together and summed in order to calculate the emergent spectrum.

The effective temperature can be guessed using the bolometric flux for the star, and the surface gravity can be approximately chosen on the basis of the spectral type. If the radius is known from interferometric measurements, or if the star is in an eclipsing binary, or if its parallax is known, these first tries can be greatly improved. Differential coarse analyses start with a star whose properties are well known and appear to differ only slightly from those of the unknown. Generally, the aim of such analyses is more limited in scope, chiefly concerned with abundance determinations.

Whereas calculating the basic properties of a stellar atmosphere in the direct problem does not require a complete list of *all* absorption lines (because, in general, average opacities in some frequency band suffice for most applications), spectrum synthesis makes far greater demands on the laboratory data. To determine the abundance of an element, or set of elements, from a portion of spectrum, we require precise knowledge of all contributors to the opacity at any frequency. Usually, the combined effects of instrumental resolution and intrinsic broadening conspire to blend features, rendering the absorption at any frequency composite. To deconvolve these effects and determine abundances and check atmospheric properties, one computes the emergent spectrum for a grid of atmospheres chosen to closely bracket the effective temperature and surface gravity of the star. The atmospheric structure is checked using profiles that sample great depth ranges, specifically the hydrogen and helium lines for which the Stark-broadened wings provide a sensitive measure of the temperature and density as a function of optical depth. The size of the Balmer jump, the opacity discontinuity of the $n = 2$ level of neutral hydrogen at 3647 Å, is also a good indicator of surface gravity.

The process of synthesis is much the same as the first step of the direct problem. Level populations are calculated, although often the LTE assumption is relaxed for synthesis (because the atmosphere will not have to be iterated—it is an initial-value problem rather than a two-point boundary-value problem), and the resultant line profiles are compared with the observed profiles. A grid of abundances, T_{eff}, and $\log g$ is employed. Although the abundances should be exactly the same as those used for the calculation of the input atmosphere, most studies allow for small departures from these values for trace elements.

Additional Reading

Collins, G. W. (1989). *The Fundamentals of Stellar Astrophysics.* W. H. Freeman, New York.

Conti, P. S. and Underhill, A. B., eds. (1988). *O Stars and Wolf–Rayet Stars*, NASA SP-497.

Gray, D. (1976). *The Observation of Stellar Photospheres.* Wiley, New York.

Johnson, H. R. and Querci, F. R., eds. (1986). *The M-Type Stars*, NASA SP-492.

Jordan, S., ed. (1981). *The Sun as a Star*, NASA SP-450.

Mihalas, D. (1978). *Stellar Atmospheres*, 2nd ed. W. H. Freeman, San Francisco.

Mihalas, D. and Mihalas, B. M. (1986). *Foundations of Radiation Hydrodynamics*. Oxford University Press, New York.

Osterbrock, D. (1989). *Astrophysics of Gaseous Nebulae and Active Galactic Nuclei*. University Books, Mill Valley, CA.

Underhill, A. B. and Doazan, V., eds. (1982). *B Stars with and without Emission Lines*, NASA SP-456.

Wolff, S. C. (1983). *The A-Type Stars: Problems and Perspectives*, NASA SP-463.

See also **Stars, Atmospheres, Radiative Transfer; Stars, Atmospheres, Turbulence and Convection; Stars, Temperatures and Energy Distributions; Sun, Atmosphere, all entries.**

Stars, Atmospheres, Radiative Transfer

David G. Hummer

Radiative transfer in stellar atmospheres refers to the quantitative treatment of the flow of radiant energy through the outer layers of stars and the interaction of radiation with the gas in these regions. In addition to radiant energy, in certain types of stars a small amount of energy is transported through the atmosphere by the large-scale motion of the gas itself; both forms, of course, arise ultimately from thermonuclear processes deep inside the star. The division between these two modes is not fixed, as radiation can transfer momentum to the gas, causing outflows known as stellar winds, and flow energy can be transformed into radiation by shocks.

Radiation undergoes repeated interactions with the gas, which collectively determine the rate at which it can escape from the atmosphere. Because the rate at which energy is produced deep in the star is essentially fixed, these interactions establish the equilibrium distribution of temperature and density in the atmosphere. In addition to this constructive role, radiative transfer also plays a diagnostic role, for the spectrum of the escaping radiation carries with it all that we can possibly learn about the internal physical conditions of the atmosphere and the underlying layers.

PHYSICAL MECHANISMS

Radiation interacts with matter via a number of mechanisms, which are here best discussed in the photon picture. Photons are created from electron energy by radiative decay of collisionally excited bound states of atoms or molecules, by radiative recombination following a collisional ionization, or by free–free emission, in which a free electron in the vicinity of an ion or atom loses kinetic energy (bremsstrahlung). The inverse processes, collisional deexcitation of radiatively excited states, three-body recombination, and free–free absorption, convert photon to electron energy. Thus collisional processes couple the electron and photon gases. When photons are removed from the radiation field by excitation, ionization, or free–free processes, their energy goes into the thermal pool before reappearing in the radiation field, so that there is no correlation between the frequencies of absorbed and emitted photons. Such processes are known as true absorption and emission.

Photons can also interact with the gas so as to change their direction with little or no change of frequency by the excitation and immediate deexcitation of an atomic state to the original state; if the excited state is a bound state of the atom the process is resonance scattering, otherwise it is Rayleigh scattering. Thomson scattering from free electrons can also be significant. The relatively small frequency change is important only if a photon scatters repeatedly in the same transition. Such processes are called noncoherent, whereas in stellar atmospheres all other scattering processes are normally regarded as coherent, that is, occurring with no frequency change. Except in the vicinity of strong spectral lines, Thomson scattering is regarded as coherent, although in many types of hotter objects, the exchange of energy between photon and electron—Compton scattering—is crucial. The combined processes—absorption and scattering—which remove photons from the radiation field are referred to as extinction.

The correlation among the frequencies and directions of initial and final photons in noncoherent resonance scattering arises from a number of processes, which are also responsible for the broadening of spectral lines. Because excited atomic levels are not perfectly sharp by virtue of natural and collisional broadening, changes in photon energy on the order of the line widths induced by these processes are to be expected. Moreover, because the incoming and outgoing photons have different directions relative to the direction of the atom's motion, the amount of Doppler shift will also differ. These correlations are described by redistribution functions. For most purposes, it is sufficient to assume complete redistribution, in which no correlation exists and the probabilities of photons being removed and reemitted in a line are proportional to the same function of frequency, the profile function. Only for the first resonance line, or similar lines with a metastable state as the lower level, must a detailed redistribution function be used.

In certain cases, the polarization of the radiation must be considered along with the mechanisms by which it is created and destroyed; the formalism for treating this phenomenon is similar to that for noncoherent scattering.

Macroscopic factors are often important in determining the radiation field in stellar atmospheres. Thermal and turbulent velocity fields influence primarily the shapes of spectral lines. Even more drastic are large-scale flows, such as the predominantly radial flows in stellar winds, which by radically enhancing the escape of line photons can change the degree of excitation and ionization of the gas. Also important is the degree of geometrical extension of the atmosphere, which influences the escape of radiation.

TRANSFER EQUATION

The radiation field in transfer theory is described in terms of the specific intensity, or simply intensity, defined as follows. The intensity $I(\nu, \mathbf{r}, \mathbf{n}, t)$ of radiation with frequency ν at point $\mathbf{r}$ flowing in direction $\mathbf{n}$ at time t is defined as the amount of energy per unit frequency passing through a unit area normal to $\mathbf{n}$ into a unit solid angle about $\mathbf{n}$ per unit time. The transfer equation is a phenomenological relation, based on empirical evidence (Beer's law), that expresses the change in intensity $dI(\nu)$ over a distance dl along a ray with direction $\mathbf{n}$ in terms of extinction and emission coefficients per unit length k and j, respectively:

$$dI(\nu,\mathbf{r},\mathbf{n})/dl = j(\nu,\mathbf{r},\mathbf{n}) - k(\nu,\mathbf{r},\mathbf{n})I(\nu,\mathbf{r},\mathbf{n}).$$

If the properties of the medium are changing very rapidly, it may be necessary to add the term $dI/c\,dt$ to the left-hand side. In view of the importance of this equation, its phenomenological nature and the lack of detailed experimental verification were disturbing until John Cooper and P. Zoller derived it from Maxwell's equations under conditions appropriate for astrophysics (previous attempts were not satisfactory).

The extinction coefficient is composed of a finite number of terms proportional to $N_i A_i(\nu)$, where N_i is the number density of the ith species including specification of charge and excitation state if appropriate, and $A_i(\nu)$ is the cross section for the radiative process in question; if stimulated emission is important, additional terms appear. The emission coefficient contains the corresponding terms proportional to the upper state populations, or for coherent scattering processes, to $N_i \int d\Omega\, I(\nu, \mathbf{r}, \mathbf{n})$. The ratio of the emission to the extinction coefficients is known as the source function

$S(\nu, \mathbf{r}, \mathbf{n})$ and is a very useful quantity in practice. As the coefficients j and k depend on atomic level populations, which in general are determined by the radiation field, we see immediately that radiative transfer is a nonlinear process! It is thus incompletely specified until the relation of the atomic level populations to the radiation is given.

The statistical equilibrium equations at each point in the gas equate the rate at which each state is populated and depopulated and have the form

$$N_i \sum_{j \neq i} (R_{ij} + C_{ij}) = \sum_{j \neq i} N_j (R_{ji} + C_{ji}), \qquad i = 1, 2, \ldots,$$

where R_{ij} and C_{ij} represent the radiative and collisional transition probabilities between all states of interest; again possible time dependence and the transport of particles into and out of the unit volume in question have been ignored. The collisional rates are simply proportional to the densities of the collision partner and the temperature-dependent rate coefficients, whereas the radiative rates are proportional to the intensities. Thus these equations are linear in both level populations and radiation field, and therefore bilinear in the unknown quantities of the problem. As this system of equations has one redundant member, it must be supplemented by a condition of conservation of particle number.

In much of the early work in this field, the level populations were assumed to be given by the Saha and Boltzmann distributions at the local electron temperature. This assumption of local thermodynamic equilibrium (LTE) is now known to be grossly incorrect for hotter objects and for the outer regions of cooler stars.

ATMOSPHERIC EQUATIONS

In a stellar atmosphere the conservation conditions for energy and momentum together with an equation of state, such as the ideal gas law, make it possible to determine the distribution of temperature and density with depth. If all of the energy is transported by radiation, then the condition of radiative equilibrium, which requires that the energy absorbed in each volume element must equal that emitted, determines the temperature at each point. For many purposes it is useful to define an effective temperature T_{eff} related to the luminosity L or total radiative power by $L = 4\pi R^2 \sigma T_{\text{eff}}^4$, where R is the stellar radius and σ is the Stefan–Boltzmann constant. One of the goals of the theory is to relate T_{eff} to the electron temperature in the outer regions of the stellar atmosphere where the observable spectral features are formed.

If the force of gravity is sufficient to overcome the pressure due to gas, radiation, and turbulence, the atmosphere is stationary and the balance between these competing forces gives the equation of hydrostatic equilibrium. This condition makes it possible to determine from the stellar spectrum the surface gravity $g = GM/R^2$, where M is the stellar mass and G is the gravitational constant. When radiation and turbulent pressure are relatively small, the thickness of the region where the observable visual spectrum is formed—the photosphere—is very small relative to the radius of the star. Therefore, it can be represented as a slab extending to infinity in the two dimensions perpendicular to the radial direction, in which the properties of the gas depend on only the vertical coordinate. This is known as the plane parallel approximation and is usually made. If the radiation and gas pressures approach closely the force of gravity, the photosphere becomes extended and the more general case of spherical symmetry must be used.

Finally, many types of objects are sufficiently luminous so that the radiation and/or gas pressure overcome gravity. In this case a radial outflow of mass results, which is the stellar wind, and the density, flow speed, and temperature are found from the gas dynamical equations representing the conservation of mass flux, momentum, and energy.

TECHNIQUES

The modern age of stellar atmosphere theory began in the late 1960s when Dmitri M. Mihalas and Lawrence H. Auer developed the technique of complete linearization for the solution of the strongly nonlinear set of equations discussed previously. This procedure is a multidimensional generalization of the Newton–Raphson method. The variables are the temperature, gas density, level populations, and radiation density at a large number of discrete frequencies, all at a number of discrete depth points. This technique, with various modifications, has been the standard tool since then. The textbook by Mihalas summarizes this work through the mid-1970s.

Whereas complete linearization has been very effective, the enormous computer resources required limits in practice the number of elements that can be included. Recently, Lawrence S. Anderson has developed a powerful method for solving this problem. Another potent general technique has evolved in several hands, known as approximate (or accelerated) lambda iteration (ALI). The obvious method of iterating between the equations of radiative transfer and statistical equilibrium can be shown to converge, but in practice it is much too slow to be useful. In the early 1970s, George B. Rybicki identified the cause of the problem and developed the core saturation method which led to useful convergence. This was the first application in radiative transfer of a general technique in numerical mathematics, known as operator splitting, in which the operator acting on the level populations to give the radiation field, traditionally called Λ, is replaced by an approximate operator with a simple inverse and the difference between the two is treated iteratively. Subsequent work on this problem has been largely concerned with the choice of the approximate operator. Stellar atmosphere modeling codes based on ALI are already showing significant advantages over complete linearization.

Approximate methods based on estimates of the escape probability of a photon continue to be useful. The Russian astrophysicist V. V. Sobolev showed that when the flow speed exceeds the thermal velocities of the ions, escape probability methods become nearly exact and provide a dramatic simplification of the line formation problem.

The collection edited by Wolfgang Kalkofen contains reviews on polarization, on ALI methods and applications to stellar modeling, and on Anderson's method. The application of the ideas and techniques sketched previously to the stellar atmospheres and winds has been very fruitful in recent years. Work on hot stars and the appropriate modeling techniques have been recently reviewed by Rolf-Peter Kudritzki and the author, and work on cool stars has been reviewed by Jeffrey L. Linsky and Robert E. Stencel.

Additional Reading

Cooper, J. and Zoller, P. (1984). Radiative transfer equations in broad-band, time-varying fields. *Ap. J.* **277** 813.

Hubeny, I. (1985). Redistribution and radiation transfer in astrophysical situations. In *Spectral Line Shapes*, Vol. 3, F. Rostas, ed., Walter de Gruyter, Berlin, p. 501.

Kalkofen, W., ed. (1987). *Numerical Radiative Transfer*. Cambridge University Press, Cambridge.

Kudritzki, R. P. (1988). The atmospheres of hot stars: Modern theory and observation. In *Radiation in Moving Gaseous Media*. Geneva Observatory, Sauverny-Versoix, p. 3.

Kudritzki, R. P. and Hummer, D. G. (1990). Quantitative spectroscopy of hot stars. *Ann. Rev. Astron. Ap.* **28** 303.

Linsky, J. L. and Stencel, R. E., eds. (1987). *Cool Stars, Stellar Systems, and the Sun*. Springer-Verlag, Berlin.

Mihalas, D. (1978). *Stellar Atmospheres*, 2nd ed. W. H. Freeman, San Francisco.

Rybicki, G. B. (1984). Escape probability methods. In *Methods in Radiative Transfer*, W. Kalkofen, ed. Cambridge University Press, Cambridge, p. 21.

See also **Radiation, Scattering and Polarization; Stars, Atmospheres; Stars, Temperatures and Energy Distributions; Stars, Winds.**

Stars, Atmospheres, Turbulence and Convection

David F. Gray

Movement of the gases in a stellar atmosphere is referred to as turbulence. It typically amounts to a few kilometers per second. Most motion results from convection, the upward streaming of hotter gas and the downward return streaming of cooler gas. Other possible motions include sound waves, magnetic waves, and various types of pulsations, but in normal stars these are much smaller than the convective motions. Convection in stellar atmospheres is also called granulation, a name stemming from the grainy appearance of the convection streams as we see them on the surface of the Sun. Some of the motion at the smallest geometrical scales may actually be turbulence in the sense of fluid-flow mechanics, that is, highly chaotic motion, but most of the motion is more highly organized. In particular, the convection has hotter material rising and cooler material falling. All stars have some atmospheric turbulence. In past years, large-scale motion was called macroturbulence, whereas small-scale motion was called microturbulence.

Studies of turbulence and convection are important in understanding:

1. Convection as a heat-flow phenomenon: Convection zones in cool stars carry most of the internal heat to the surface, but the physics is poorly understood. The tops of such convection zones penetrate into the atmosphere, allowing us to detect the motions and structure they induce.
2. Convection as a major ingredient in the generation of stellar magnetic fields: Convection zones interact with the stellar rotation in a process called a dynamo to generate a magnetic field. In the deep atmosphere and below, the magnetic field is dragged about by the gas as it moves and twists and flows. The gas controls the magnetic field lines when its density is high. In areas of starspots, the tables are turned, and the magnetic field is so strong that it inhibits convection, less energy is lifted to that part of the surface, and the spot appears dark compared to its surroundings. In the outer atmosphere, the gas density is low and the magnetic fields are in control, giving structure to the outer layers (the chromosphere and corona). Magnetic fields cause enhanced dissipation of angular momentum, resulting in braking of stellar rotation, they control and channel mass escaping from stars, and they may modulate the total power output by a small amount as the magnetic field varies over time scales of years, as with the solar activity cycle.
3. A source of power for the super-heated chromosphere and corona: Some of the magnetic field extends through the deeper atmosphere (photosphere) into the outer atmosphere. Wigglings and oscillations of the magnetic field by gas motions in the photosphere may travel up the field lines to heat the outer atmosphere to temperatures of millions of degrees. Twisting of the magnetic field by swirling motion of the convection may lead to direct dissipation of magnetic energy in the outer atmosphere, again heating it. Acoustic waves are also generated by the turbulence. Some of this mechanical power may be lifted to the chromosphere where it dissipates as shock waves.

Gas motions stir the stellar atmosphere, ensuring homogeneous chemical composition.

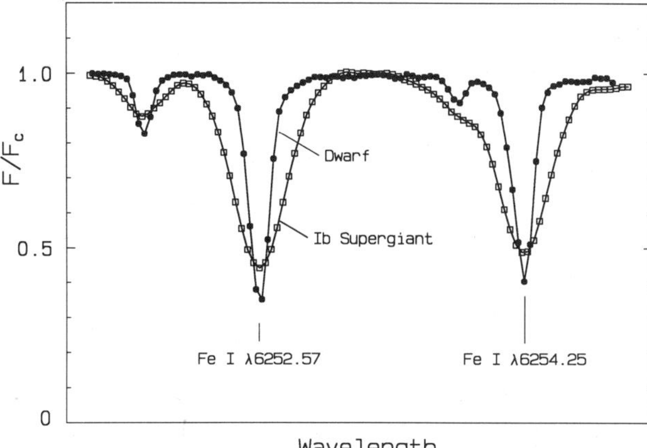

Figure 1. Differences in line profile broadening can be seen in this comparison between a dwarf star and a supergiant of comparable temperature. [*From Gray (1988) with permission of The Publisher*.]

DETECTION AND MEASUREMENT

The Doppler effect is the mechanism of detection. Atmospheric motions of any origin produce wavelength shifts proportional to the velocity along our line of sight. For example, rising material near the central portion of the apparent stellar disk produces a blue shift (shift to shorter wavelength), whereas rising material at the edge of the disk moves across our line of sight and causes no Doppler shift. Falling gas behaves in a similar way except that any Doppler shifts are redward. Most stars are too far away for us to see their surfaces, and we are forced to decipher the integrated effect of the different Doppler shifts coming from all parts of the unobservable disks.

Because the velocities of turbulence are only a few kilometers per second, the Doppler shifts can only be detected by using sharp features in the stellar spectra, namely the spectral absorption lines. But it is not the shift of the whole spectral line (that gives the velocity of the whole star relative to us) but rather its *broadening* that arises from turbulence (see Fig. 1). In effect, the spectral line as we observe it is a composite of many overlapping spectral lines coming from all parts of the stellar surface, each with its own Doppler shift. The resolution of the spectrograph must be high enough to clearly resolve the spectral lines in order to be useful for turbulence measurements.

Doppler shifts arising from rotation of the star also broaden the spectral lines. In cooler stars ($T \lesssim 6000$ K), rotation is usually slightly smaller than turbulence; but in hotter stars, rotation can be much larger. Careful measurement of the detailed line shape is needed to separate the two broadeners.

Asymmetries of the line broadening are used to explicitly identify stellar granulation. In a typical cool star, the rising gas (composed of granules) contributes more strongly to the star's light because it is hotter and brighter and/or covers more of the stellar surface than the falling material (termed dark lanes). About 80% of the starlight comes from hot, rising material, but techniques of analysis have not progressed to the point where the 80% can be separated into the two relevant physical factors of surface area and temperature excess. For the Sun where the individual granules can be seen, these numbers are approximately 60% and 100–200 K. In the stellar case, dominance of the light from rising material heavily weights the distribution of Doppler shifts toward the blue, resulting in asymmetric spectral lines. Expressed in net Doppler shift, the asymmetries produce differential effects of only a few hundred *meters* per second and less, depending on the star. The actual convective velocities are approximately 2 km s^{-1} for rising material and 4 km s^{-1} for falling material, and in general the velocities of fall are found to be about twice as large as the velocities of rise.

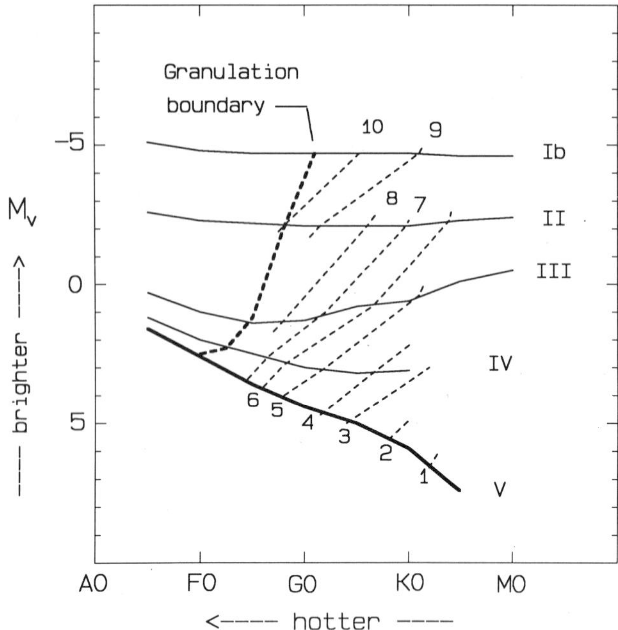

Figure 2. This Hertzsprung–Russell diagram shows contours of constant macroturbulence dispersion labeled with their values in kilometers per second. The ordinate, M_V, is the absolute magnitude, a measure of intrinsic stellar brightness. The spectral types on the abscissa are a logarithmic measure of temperature (F0 ≈ 7200 K; K0 ≈ 5200 K, but exact temperatures depend on M_V). Luminosity classes are labeled with Roman numerals. The main sequence is class V, subgiants IV, giants III, bright giants II, and supergiants Ib. The Sun is a G2 V dwarf. Asymmetries of spectral lines are reversed on opposite sides of the granulation boundary.

These are mean values; there is a wide range of velocities about these means from point to point on the stellar surface. Granulation velocities are expected to decline with height in a stellar atmosphere. The observations give qualitative support to this idea, although the details cannot yet be mapped out with confidence even for the Sun. Signal-to-noise ratios of several hundred and resolving power, $\lambda/\Delta\lambda \geq 10^5$, where λ is the wavelength and $\Delta\lambda$ is the spectral resolution, are needed to study stellar granulation.

The amount of line asymmetry is observed to go hand in hand with the line broadening, and it seems likely that most of the atmospheric motion previously lumped under the generic name turbulence is physically granulation which in turn is the top of a deeper convection zone.

Classical microturbulence is detected through the strengthening of intermediate-strength lines, and it acts like an enhanced thermal broadening. Values of 0–1 km s^{-1} are typical for dwarf stars; values as high as 2.5 km s^{-1} are seen in supergiants.

VARIATIONS WITH SURFACE GRAVITY AND TEMPERATURE

Figure 2 is a Hertzsprung–Russell (HR) diagram (absolute magnitudes of stars plotted as a function of their spectral types) illustrating how atmospheric motions vary systematically with the temperature and surface gravity of stars. Loci of constant macroturbulence dispersion (dashed lines) are labeled with the dispersion values in kilometers per second. The Roman numerals label curves of constant stellar luminosity class, and to an order of magnitude, these numerals equal the logarithm of the surface gravity in cm s^{-1}. In general, both the line broadening and the line asymmetry are observed to be larger in hotter stars and larger in stars having lower surface gravity. Stars with low surface gravity are the giants and supergiants. The amplitudes of the line asymmetries are observed to run from approximately 600 m s^{-1} at spectral type F5 ($T \sim 6500$ K) to approximately 50 m s^{-1} at K5 ($T \sim 4300$ K) for dwarf stars. As indicated earlier, the actual convective velocities are 10–15 times larger than the asymmetries they produce. Convection theory is in qualitative agreement with these observed variations across the HR diagram, but the theory is not well enough developed, nor is the extraction of velocities from the asymmetries sufficiently accurate for quantitative comparison. Numerical integration of the equations of hydrodynamics is now being done with some success. Atmospheric motions in stars hotter and cooler than the range of Fig. 2 have not been adequately studied.

Figure 2 also shows a recently discovered granulation boundary. Stars on the hot side of the boundary have line profiles of opposite asymmetry compared to those on the cool side. The difference in asymmetry signals differences in the atmospheric motions, with stars on the hot side having an order of magnitude higher velocities (~20 km s^{-1}), but with only a small fraction (~10%) of the surface showing such velocities. Hot stars are thought to have only thin convection zones, whereas cool stars have deep convection zones that actively carry heat out of the star. The granulation boundary apparently separates these two domains.

Additional Reading

Dravins, D. (1982). Photospheric spectrum line asymmetrics and wavelength shifts. *Ann. Rev. Astron. Ap.* **20** 61.

Gray, D. F. (1978). Turbulence in stellar atmospheres. *Solar Phys.* **59** 193.

Gray, D. F. (1988). *Lectures on Spectral-Line Analysis: F, G, and K Stars.* The Publisher, Arva, Ontario.

Rutten, R. J. and Severino, G., eds. (1989). *Solar and Stellar Granulation.* Kluwer, Dordrecht.

Spruit, H. C., Nordlund, Å., and Title, A. M. (1990). Solar convection. *Ann. Rev. Astron. Ap.* **28** 263.

See also **Stellar Atmospheres; Sun, Atmosphere, all entries.**

Stars, Atmospheres, X-Ray Emission

Robert A. Stern

A tenuous corona of ionized gas (plasma), heated to temperatures of 10^6–10^7 K or more, forms the outermost atmosphere of our Sun and of many stars. Soft x-ray emission at photon energies of approximately 0.1–10 keV (wavelengths ~120–1.2 Å) is a characteristic signature of coronal plasma. However, these x-rays cannot be detected by ground-based astronomical telescopes because they are strongly absorbed in the Earth's atmosphere. Coronae are also difficult to see optically: Until the dawn of space astronomy, the Sun was the only star with a detectable corona, briefly glimpsed as the faint "crown" of scattered visible light surrounding a totally eclipsed Sun. The solar corona's temperature was first determined —not by x-ray observations—but through the identification of the puzzling "coronium" lines seen during eclipses in the visible solar spectrum. Grotrian and Edlén identified these lines in the 1940s as atomic transitions in highly ionized iron and calcium at million-degree temperatures.

The detection of solar x-rays was one of the first results of space-age astronomy. Sounding rocket experiments in the 1960s and most notably the *Skylab* x-ray telescopes in 1973–1974 saw for the first time the complex and ever-changing nature of the solar corona. In 1975, the first corona of another star was detected in soft x-rays from Capella (α Aurigae) through rocket and satellite observations. The late 1970s and early 1980s saw a burst of new stellar x-ray observations, most notably from NASA's *HEAO-1* and the *Einstein* x-ray observatory, and the ESA *EXOSAT* mission. With this rapidly increasing observational database, astronomers have begun to piece together the nature of coronal x-ray emission.

Classical models of stellar atmospheres show a steady decrease in temperature from the star's core to its surface. In the Sun, for example, the temperature declines from about 15 million K at the center to approximately 5800 K at the photosphere. What, then, causes the formation of a million-degree or hotter corona *above* the photospheric layer? From a combination of solar and stellar x-ray observations and the application of some theoretical modeling, the answer appears to lie in a complex coupling of plasma motions in the stellar convective zone (the turbulent, "boiling" region lying beneath cool star photospheres) with stellar differential rotation (differential rotation of the solar atmosphere can be seen in the faster rotation rates for sunspots near the equator compared to those at higher solar latitudes). This coupling induces a dynamo effect, generating magnetic fields below the photosphere which extend to the corona in the form of loop-shaped structures. Although the magnetic field is not directly visible, magnetic loop structures can be identified on x-ray photos of the Sun via their enhanced x-ray brightness. It is in these loop structures that the coronal plasma is confined and heated: The coronal magnetic field is strong enough, and the gas pressure is low enough, that coronal plasma is trapped in the loops. The corona can thus be thought of as an ensemble of loop structures with varying temperatures and pressures. Astronomers have not yet identified a universally accepted explanation for coronal heating; however, heating probably occurs via the dissipation of energy accumulated through slow twists of magnetic field lines or by wave motion built up from the vigorous shaking of field lines at their anchor points below the stellar photosphere. The temperature distribution within a coronal loop is then determined by the energy balance between this heat input and the cooling mechanisms of electromagnetic radiation (seen as x-rays) and heat conduction to the lower stellar atmosphere.

Which types of stars possess coronae? As x-ray satellites have become increasingly sensitive, we are discovering that, in fact, *most* types of normal stars emit x-rays. The stellar coronae detected with these satellites are, in general, far more luminous in x-rays than the solar corona. The Sun's x-ray luminosity (L_x) ranges from about 10^{26}–10^{27} erg s^{-1} during the solar cycle, reaching up to 10^{28} erg s^{-1} for brief periods during solar flares. Stellar coronae detected so far exhibit $L_x \sim 10^{26}$–10^{31} erg s^{-1}, in some cases reaching higher levels during large stellar flares. Coronal temperatures in some of the most x-ray luminous stellar coronae may reach 50 million K or more, also increasing during stellar flares.

O and B Stars O and B main-sequence stars with surface temperatures of 15–30,000 K or more exhibit x-ray emission at a level of approximately 10^{-7} of their overall (bolometric) luminosities. The O and B stars are perhaps the least well understood of the "normal" stellar x-ray emitters, because they should not undergo the dynamo action seen in later-type (F and cooler) stars with outer convective zones. Though they may possess a corona at the base of their outer atmosphere, their x-ray emission is more likely to come from shock-heated material in a strong stellar wind.

A Stars Stars of spectral type approximately B7–A5 do not, in general, emit x-rays at a detectable level. The few exceptions to this rule have almost all been shown to be binary systems with faint cooler stars as the likely x-ray emitters. The general lack of x-ray emission is in accord with expectations of no dynamo action in stars lacking convective zones.

F–M Stars These "cool" main-sequence stars show the largest range of coronal x-ray emission, with $L_x \sim 10^{26}$–10^{31} erg s^{-1} as seen in the *Einstein* stellar surveys. Observations have shown that the level of x-ray emission varies by orders of magnitude within main-sequence stars of a given spectral type, that the average level of x-ray emission also varies as a function of spectral type, and, most importantly, that stellar rotation plays a key role in determining the level of coronal x-ray emission.

Pre–Main-Sequence Stars The detection of x-ray emission from pre–main-sequence (PMS) stars in star-forming regions such as the ρ Ophiuchi Dark Cloud and the Taurus-Aurigae region were important discoveries of *Einstein*. PMS x-ray emission is highly variable, often exhibiting flare activity. The most likely explanation for such strong and variable x-ray emission is some form of stellar corona. Yet the relationship between PMS coronae and those of solar-like stars is poorly understood at present.

Giant Stars X-ray emission has been observed from the hotter (approximately K0 and earlier) giants and supergiants. However, the coolest giants do not seem to possess coronae.

White Dwarfs Searches for white dwarf coronae have thus far proved unsuccessful. However, a number of very hot (30–60,000 K) white dwarfs such as HZ 43 and Feige 24 emit ultrasoft x-rays (~ 0.1 keV energy) from their photospheres.

Stellar Binarity and Rotation A number of key observational facts confirm the importance of stellar rotation for coronal heating.

1. The most active coronal x-ray emitters are the RS CVn binary systems, which consist of two cool (F–K) main-sequence, subgiant, or giant stars (luminosity classes V, IV, or III, respectively) in synchronous rotation about a common center of mass with periods of typically days to weeks. Because of this enforced tidal synchronism, one or both stars rotate substantially faster than comparable single stars of the same age and spectral type. This enhanced rotation is probably the major factor in their enhanced x-ray activity.

2. Studies of main-sequence stars show that solar-type (F–G main sequence) stars in clusters such as the Pleiades and the Hyades (see Fig. 1) have average x-ray luminosities over 100 times that of the Sun. Because these main-sequence stars are more than 10 times younger than the Sun, they are, on the average, rotating considerably faster: G stars in the Hyades, for example, rotate about five times faster than the Sun. This increased rotation seems to produce increased dynamo action, hence increased magnetic field generation and coronal heating.

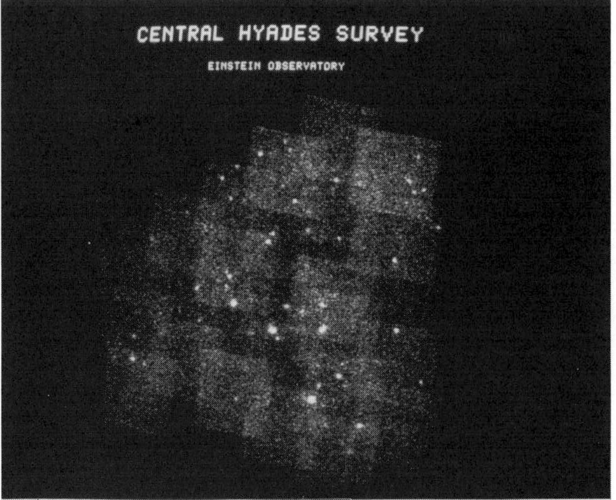

Figure 1. Composite *Einstein* satellite x-ray image of the central Hyades cluster. The photograph is a mosaic of 27 overlapping *Einstein* Imaging Proportional Counter exposures, each approximately 1 degree square. The bright "blobs" are individual x-ray sources, most of them normal main-sequence stars in the cluster. The speckled background seen on each exposure is a combination of celestial diffuse x-ray background emission and charged particle background in the x-ray detector. (*Photo supplied by the author.*)

3. Finally, studies of field (noncluster) stars have shown a clear correlation of x-ray emission with rotation. Although this rotation–activity connection is well established, the exact mathematical form of the correlation is still controversial.

X-Ray Flaring and Variability Stellar x-ray emission is often highly variable. The classic K–M dwarf "flare stars" exhibit x-ray as well as optical flares. On the Sun, we see x-ray flares that can occasionally increase the solar x-ray luminosity by an order of magnitude or more for periods of perhaps a few minutes to a few hours. Flares are thought to come from a rapid rearrangement of magnetic field geometry in the corona, resulting in the release of stored magnetic energy in the form of energetic particles, high-energy radiation, and heat. Even so, solar flares rarely have a peak $L_x > 10^{28}$ erg s^{-1} and involve only a small fraction of the x-ray emitting loops that produce the bulk of the solar x-ray emission. By comparison, some stellar flares last as long as a day, with a peak $L_x > 10^{32}$ erg s^{-1}. The largest stellar flares occur in RS CVn binaries and involve a huge volume of flare material, up to 10^6 times greater than in solar flares. Flare temperatures can approach 60–80 million K, several times hotter than in most solar flares. Because astronomers lack a detailed physical description of the flare process, we can only speculate that the enhanced magnetic field generation in highly active stars accounts for large stellar flares.

Spatial Structure of Coronae Except in the case of the Sun, we have very limited knowledge of coronal spatial structure. For a few RS CVn binaries that are eclipsing pairs, we may derive some information concerning the size scale of each star's corona by watching the variation of x-ray emission as the star is eclipsed by its companion. Such observations have been performed with the *Einstein*, *EXOSAT*, and *Ginga* satellite observatories for the systems AR Lac, TY Pyx, and Algol (β Per), the famous "demon" star. The case of Algol is perhaps the most intriguing: It consists of a relatively hot (B8) main-sequence star and a cooler and slightly larger (K0) subgiant in a 2.87-day orbit. In visible light, Algol dims by about 2 magnitudes when the cooler K star passes in front of the B star. In x-rays, neither star appears to eclipse the other by an appreciable amount, even, in one case, during a flare! The lack of x-ray dimming when the B star is optically eclipsed seems understandable in light of the fact that few stars of late-B or early-A spectral type show much x-ray emission. Yet the lack of a strong eclipse for the K star is less understandable, unless the corona of the K star is located far above the stellar surface. Algol may thus hold further surprises in store for astronomers.

Thermal Structure of Coronae Astronomers also have only rudimentary facts regarding the temperatures of stellar coronae. In the case of the RS CVn systems, for which we have the most spectral information, the x-ray spectrum cannot be modeled by a single-temperature plasma. At least two temperature components or loop ensembles are required. Typically, one component has a temperature of several $\times 10^6$ K (somewhat hotter than solar active regions), whereas the second has a temperature of several $\times 10^7$ K (hotter than most solar flares). Although such two-component models can adequately explain the current observations, the detailed distribution of coronal plasma as a function of temperature for the RS CVn systems remains uncertain, primarily because of the lack of high-resolution spectroscopic data in the x-ray range. For other main-sequence stars, astronomers are limited by the number of x-ray photons detected: In general, fainter stellar coronae can be characterized by single-temperature models with the higher luminosity coronae possessing higher temperatures.

The Coronal Dividing Line Another outstanding question concerns the coronal "dividing line." X-ray emission commonly occurs in the hotter giant and supergiant stars (roughly spectral type K and earlier, varying with luminosity class). However, the coolest M-type giants and supergiants do not show evidence of x-ray emission. There is no generally accepted explanation for this "coronal dividing line" on a Hertzsprung–Russell diagram. Some researchers believe that the magnetic field lines which would normally confine a corona in the cool giants do not form closed loop structures, but instead produce a geometrically "open" type field configuration, allowing gas to escape in the form of strong stellar winds. Another group of researchers believes it is the lower force of gravity on the surface of these stars that only allows loop structures with relatively low temperature ($< 10^5$ K) gas to form, too cool to produce measurable x-ray emission. The answer to the puzzle of the corona-less cool giants will have to await the next generation of x-ray satellites.

The exciting new observational data on stellar x-ray emission has provided as many new questions as answers. With the recently launched German/U.S. *Röengtensatellit* (*ROSAT*), the anticipated launches of the Japanese/U.S. *Astro-D X-Ray Telescope* in 1993, and the future NASA *AXAF* (*Advanced X-Ray Astrophysics Facility*) and *ESA XMM* (*X-Ray Multi-Mirror Mission*) in the late 1990s, stellar x-ray astronomers should make great progress in answering these unsolved problems.

Additional Reading

Pallavicini, R. (1989). X-ray emission from stellar coronae. *Astron. Ap. Rev.* **1** 177.

Rosner, R., Golub, L., and Vaiana, G. S. (1985). On stellar x-ray emission. *Ann. Rev. Astron. Ap.* **23** 413.

Stern, R. A. (1983). *Einstein* observations of cool stars. *Adv. in Space Res.* **2** (No. 9) 39.

Stern, R. A. (1984). Stellar coronas, x-rays, and *Einstein*. *Sky and Telescope* **68** 24.

Wallerstein, G., ed. (1990). *Cool Stars, Stellar Systems, and the Sun*. Astronomical Society of the Pacific, San Francisco.

See also **Binary Stars, RS Canum Venaticorum Type; Magnetohydrodynamics, Astrophysical; Solar Activity, Solar Flares; Stars, Activity and Starspots; Stars, Pre–Main-Sequence, X-Ray Emission; Stars, Red Dwarfs and Flare Stars; Sun, Atmosphere, Corona.**

Stars, Beta Cephei Pulsations

Myron A. Smith

Many massive stars of spectral classes O and B are observed to pulsate. Astronomers observe the brightness, color, and radial-velocity variations caused by these pulsations and use the measurements to learn about the internal structure of stars, how they rotate, why they pulsate, and how they evolve. The important classes of chemically normal, pulsating massive stars include the β Cephei stars (which should not be confused with the "classical Cepheid" variable stars of which δ Cephei is the prototype). Other luminous O- and B-type stars, known as S Doradus, Hubble–Sandage variables, and "slow variable" stars are not known to be regular pulsators and are not discussed here. The study of periodic variable B-type stars is one of the most exciting frontier areas in the field of stellar pulsations. The subject is still not sufficiently well understood that it can be summarized with the same confidence as the study of other classes of periodic variables. The nonradially pulsating B stars are described in another entry.

OBSERVATIONAL AND PULSATIONAL PARAMETERS

Since β Cephei was discovered to be variable by Edwin B. Frost in 1902, the "β Cephei variables" have come to be defined as a group of spectral type B1–2 giants with periodic variations in their light, color, and radial-velocity curves. The pulsation frequencies of these variables range from 4–7 cycles day^{-1}. On the Hertzsprung–Russell (HR) diagram they are located along an "instability strip" that is roughly parallel to but just above the main sequence.

Table 1. Typical Parameters for β Cephei Stars

Star	*Spectral Type*	M_V	*Frequency (cycles day^{-1})*	*Double Amplitude (mag)*	*Pulsation Mode Type*	*Source*
β Cephei	B2 III	−3.8	5.249575	0.0365	Radial	Lesh and Aizenman; Smith
β Canis Majoris	B1 II	−4.4	3.979331	0.0210	Radial?	Lesh and Aizenman; Smith
			3.9995	0.0044	Nonradial?	
			4.1834	0.003	Nonradial	
12 Lac	B2 III	−3.6	5.06632	0.029	Radial	Lesh and Aizenman; Smith
			5.17915	0.082	$l = 2, m = 0$	
			5.33510	0.012	$l = 2, m = 0$	
			5.49104	0.026	$l = 2, m = 0$	

Present estimates of their physical parameters include surface temperatures of 20,000–25,000 K, masses of 8–16 $M_{\odot}$, radii of 7–12 $R_{\odot}$, and luminosities of about 10^4 $L_{\odot}$, where the subscript symbol denotes solar units. Their chemical compositions are thought to be solar-like. Some relevant parameters are listed in Table 1 for three typical β Cephei stars. About half of the β Cephei stars exhibit cycle-to-cycle variations that reveal the excitation of one to three additional small amplitude pulsation modes near the primary mode frequency. For example, in the star 12 Lac, there are three weak pulsations that cluster around a primary frequency of 5.2 cycles day^{-1}. It is difficult to detect the weaker pulsations with only slightly differing frequencies because observations must be obtained over a long interval, and daily and seasonal interruptions seriously complicate the frequency analysis. The close spacing of primary and weaker frequencies led Pierre Ledoux to propose in 1951 that "nonradial oscillations" are excited in β Cephei stars. Nonradial pulsations had long been known to occur in the solid Earth and oceans after an earthquake, but they were new to the interpretation of stellar phenomena.

In a star a radial pulsation is a process in which the entire surface moves in and out together; the star's radius alternately expands and contracts. Nonradial pulsations (NRP), on the other hand, are a more general type of oscillation in which waves move around the star within fixed latitude zones. The modes of NRP fall between two extremes: at one extreme, some latitude belts on the star move in and out in unison, whereas other belts move in exactly opposite phase, contracting when the first belts expand, and vice versa. At the other extreme, a single pattern of waves travels around the star's equator.

Ledoux's suggestion that nonradial pulsations occur in some β Cephei stars was widely accepted because the earlier radial pulsation theory had predicted much larger frequency spacings between fundamental and overtone radial pulsation frequencies than were observed in any β Cephei star. Recent work has led many astronomers to believe, however, that β Cephei stars are actually hot B giants or subgiants in which radial pulsations always occur and which in some cases are accompanied by NRP. The radial mode is usually the oscillation with the largest amplitude, whereas the weaker frequencies, if present, arise from nonradial pulsation. When this occurs, about 50% of the time, one of the nonradial modes generally is observed to have almost the same frequency as the radial mode, suggesting that a "resonance" occurs between the radial and nonradial modes.

The conclusion that all β Cephei stars pulsate in a single radial mode is not universally accepted. Still, it is supported by three observations:

1. The velocity and light amplitudes (typically 10–30%), which are large relative to those of B stars that pulsate nonradially;
2. The large increase in light amplitude at shorter wavelengths, chiefly because of temperature variations during the pulsation cycle;
3. The manner in which spectral absorption lines change during the pulsation cycle (the line core shifts in wavelength, the wings more so, and the line width is preserved).

To investigate the nature of pulsations in β Cephei stars, we need to understand whether they pulsate in a fundamental or overtone mode. In many respects this is not too important at the surface because matter there merely follows the "piston driver" from underneath. However, in the interior the mode's overtone number (the number of times the radial wave amplitude becomes 0 and changes sign along a radius through the star) gives a clue about how the pulsations are excited. The frequencies of the first overtone in B stars are expected to be 26–30% larger than the fundamental frequencies in the same stars. This is a large enough difference to discriminate between the fundamental mode and various overtones for a given star, despite errors in estimating its absolute luminosity and radius. Unfortunately, efforts to identify the excited mode from its frequency alone have not always been successful. For example, the studies of nearby "field" β Cephei stars tend to imply that the modes are usually fundamental, whereas the studies of β Cephei stars in clusters imply a tendency for modes to be first and even second overtones. It may be that a β Cephei star can pulsate radially in either the fundamental or first overtone mode; no star seems to be pulsating in both of these modes. This conclusion is consistent with the fact that searches for a period–luminosity relation in β Cephei stars have not been altogether successful. It may be that collections of data on periods and luminosities of these stars lump together members of the fundamental and overtone subgroups of pulsators so that the period–luminosity relations of each subgroup contaminate those of the other and cause scatter in the period-luminosity diagrams. An additional complication is that the distances of individual β Cephei stars are so uncertain that their luminosities are in error by amounts nearly as large as the total range in luminosity among all members of the class.

EVOLUTIONARY STATE AND CAUSES OF PULSATIONS

Among the B stars, β Cephei variables exhibit the most stable pulsations both in amplitude and frequency. Still, over a few decades a typical β Cephei star can exhibit changes in the fifth or sixth significant figure of its frequency. For example, from about 1955 to 1980 the frequency of the β Cephei star δ Ceti decreased by 2.7 parts in 100,000. This decrease is several times the change in frequency expected from the slow change in the star's radius over most of its main-sequence lifetime. Some astronomers suspect that frequency changes in β Cephei stars occur discontinuously rather than smoothly as theory predicts. Whether or not this is the case, there may be nonevolutionary processes that cause the observed minute changes in frequency.

Luis A. Balona and Christian Engelbrecht and Robert R. Shobbrook have recently shown that there is an instability strip in the HR diagram which is defined by 17 β Cephei stars in the open clusters NGC 3293 and NGC 6231. This fact suggests that β Cephei variability is a natural stage of evolution, through which most or all stars of about 8–16 $M_{\odot}$ pass briefly, late in their core-hydrogen-burning stage. However, what causes this pulsating is unclear. Current observations rule out most or all of the physical mechanisms that seem to be responsible for pulsation in other variable stars. For example, if the β Cephei stars have actually evolved enough to have exhausted the hydrogen fuel in their cores, then an unusual pulsation excitation mechanism could be at work. However, theoretical investigations have shown that several proposed mechanisms, such as those depending upon chemical composition gradients or intermittent turbulence across narrow hydrogen-burning shells, are too weak to overcome damping in the stellar envelope. Likewise, only an abnormal composition of certain elements in the envelope, such as C, N, and O could permit the "kappa mechanism" in the star's envelope, which accounts for radial pulsations in other types of stars, to overcome damping in the core. Possibly, it takes two mechanisms to make a star unstable to radial pulsations. However, the existence of β Cephei pulsations is so common to B1–2 giant stars that most astronomers believe that a single physical mechanism (or two that are interrelated) is responsible for β Cephei pulsations. Unfortunately, the nature of the actual pulsation mechanism in any massive star is unknown. As we have seen, the β Cephei stars extend over a small range of stellar masses, surface temperatures, ages, and chemical composition. The small size of the β Cephei strip in the HR diagram is one of the main reasons that the pulsations of these stars has defied explanation. The actual mechanism must be so effective as to come into play in nearly all stars with just the right parameters and yet be almost completely ineffective when the conditions are slightly different. It remains to be seen whether these stars have an altogether different or merely a more specific reason for pulsating than the much larger group of B stars that pulsate only nonradially.

At a 1990 meeting of the American Astronomical Society, A. Cox announced that new theoretical opacities of iron ions (which were calculated at Livermore) at 150,000 K are large enough to "drive" pulsations of β Cephei stars after all. A year later R. Stellingwerf made the same announcement. We anticipate that the correctness of these new opacities of iron will be tested by the "Opacity Project," led by Dmitri Mihalas. If these new sets of iron opacities are in agreement, the long sought pulsation mechanism for these stars will have been resolved, as due to a "kappa mechanism" arising from iron opacities.

Additional Reading

Cox, A. N. (1987). Pulsations of B stars. In *Stellar Pulsations*, A. N. Cox, ed. Springer-Verlag, New York. p. 36.

Cox, A. N. and Morgan, S. M. (1989). An opacity mechanism for the pulsations of B stars. *Bull. American Astron. Soc.* **21** 1095.

Lesh, J. R. and Aizenman, M. L. (1978). The observational status of beta Cephei stars. *Ann. Rev. Astron. Ap.* **16** 215.

Percy, J. R. (1981). The evolutionary state of beta Cephei stars. In *Workshop on Pulsating B Stars*, C. Sterken, ed. Nice Observatory Publication. p. 119.

See also **Stars, Nonradial Pulsations in B-Type; Stars, Pulsating, Overview; Stars, Pulsating, Theory.**

Stars, Be-Type

Arne Slettebak

Be stars are defined as main-sequence or somewhat higher luminosity B-type stars whose spectra have, or had at one time, one or more of the Balmer lines of hydrogen in emission. They are, as a class, characterized by rapid axial rotation and are variable in both light and spectrum on time scales of hours to years.

The first Be star to be detected was γ Cassiopeiae, which was seen to have Hβ in emission in 1866 during visual spectroscopic observations by the Vatican astronomer Angelo Secchi. The subsequent introduction of photography in stellar spectroscopy resulted in the discovery of many more Be stars, particularly in the objective-prism surveys undertaken at Harvard (1886–1912) and at the Mount Wilson and Lamont–Hussey Observatories (1919–1951). Recent catalogs of early-type stars with emission lines list several thousand such objects.

Statistical studies show that Be stars are not exotic objects: They represent up to 20% of the B-star population per volume of space. The relative number of Be to B stars is largest in the spectral type range B3–4, with lower frequencies for the early- and late-B types. The Be phenomenon (defined by the presence of circumstellar gas) actually extends beyond the B-star range to stars as early as O7 on

Figure 1. Pictorial representation of a Be star.

the hot side and into the A- and F-star range on the cool side, though the relative numbers are quite small.

The first physical model of a Be star was presented by Otto Struve at the Yerkes Observatory in 1931, who suggested that the rapid rotation that characterizes the class is responsible for the ejection of an equatorial envelope of gas from which the Balmer emission arises. This picture (see Fig. 1) is still the basis for later models, although more recent observations, especially from space, have led to modifications.

THE UNDERLYING STARS

The spectra of the underlying stars are often distorted by both emission and absorption contributions from the circumstellar envelopes, making it difficult to determine the stellar physical parameters. Be stars are variable stars, however, and at those times when the envelopes become optically thin or disappear altogether, the underlying stars look spectroscopically like normal (although usually with greatly broadened lines, due to their rapid rotation) B stars. At such times, the spectrum is also that of a main-sequence or somewhat higher luminosity star (subgiant or giant, but never supergiant), a result which is confirmed by many studies of Be stars in open clusters and double-star systems.

Rotational velocities of Be stars, derived from measurements of line broadening, have average projected values ($v \sin i$) in the neighborhood of 250 km s^{-1}, significantly higher than are found for absorption-line B-type stars of corresponding spectral types. (The inclination, i, is the angle between the line of sight and the axis of rotation.) The largest measured rotational velocities for individual Be stars are about 400 km s^{-1}, which are, however, smaller than the critical velocities at which centrifugal force balances gravitational attraction at the Be star's equator. Thus, whereas rotation must play a crucial role in the Be phenomenon, an additional trigger is required for the formation of the shell.

PHYSICAL CHARACTERISTICS OF THE ENVELOPES

Studies by Struve and others during the 1940s and early 1950s led to a better physical understanding of the envelopes surrounding Be stars. Ultraviolet radiation from the hot underlying star ionizes hydrogen atoms in the envelope and recombination produces the observed Balmer emission. Although all Be stars could be called "shell stars," the term is reserved for that subclass whose spectra show sharp hydrogen absorption cores plus narrow absorption lines of ionized metals, in addition to the Balmer emission and broad, rotationally widened absorption lines of neutral helium. The Struve model suggests that shell stars are Be stars oriented such that we view their equatorial envelopes edge-on. The sharp hydrogen absorption cores and narrow metallic absorption lines can then be interpreted as arising from that part of the envelope (which has very small line-of-sight Doppler broadening) that is projected against the photosphere of the rapidly rotating underlying star.

Spectroscopic studies of shell stars show that the shells are analogous to the atmospheres of supergiant stars: cooler and less dense than the photospheres of the underlying B-type stars. These and studies of Balmer emission-line profiles also suggest a lenticular model for the emitting regions, with radius several times the stellar radius, and predominantly rotational motions.

Strong support for the disk-like nature of the circumstellar shells also comes from polarization measurements. The fact that the optical radiation from Be stars is observed to be intrinsically linearly polarized in itself argues against spherical symmetry and in favor of a disk geometry. Models based on polarization data suggest disks with electron temperatures of $\simeq 10{,}000$ K, electron densities of $\simeq 10^{12}$ cm^{-3}, and extent 3–10 times the radius of the star. These numbers are consistent with similar values obtained from optical spectroscopic observations.

Infrared observations also contribute to our understanding of the envelopes. Observed infrared excesses for Be stars were shown to be due to emission from hot gas rather than dust, and correlations of infrared radiation with Hα emission and intrinsic polarization of Be-star radiation are consistent with the infrared excesses being produced in a flattened envelope or disk.

Whereas all of the aforementioned observations support the Struve model of an equatorial disk of gas around the Be star, ultraviolet observations from space beginning in the 1970s revealed the existence of another component of the Be-star circumstellar envelope: a low-density, expanding envelope of hot gas which is more nearly spherically symmetrical than the flattened, cooler disk which gives rise to the Balmer emission, polarized radiation, and infrared emission. This hot component is observed to contain highly ionized atomic species moving outward at high velocities (up to 1000 km s^{-1} or more), often in the form of discrete blobs or shells, leading to mass loss of the order of 10^{-11}–10^{-9} $M_{\odot}$ yr^{-1}. Obviously, Be-star envelopes are more complex than was suggested by the Struve model.

VARIABILITY

Optical spectroscopic, photometric, polarimetric, ultraviolet, and x-ray observations of Be stars have shown them to be variable on various time scales. The causes of the variations, which may be periodic or quasiperiodic but are more often irregular, are generally uncertain or unknown.

On time scales of decades, Be stars may develop optically thick envelopes (i.e., pass through a "shell phase", as defined previously) or may lose their emission shells altogether, to look like ordinary B-type stars. Well-known examples are the bright Be stars γ Cassiopeiae and 28 Tauri (Pleione).

Another type of variation is found within the Balmer lines themselves. These are usually split into violet (V) and red (R) emission components, which vary both in intensity and wavelength. This V/R variation is often quasiperiodic on time scales of years. Explanations in terms of the rotation of an elliptical emitting shell, or gaseous flows in a close binary system, have been suggested but none satisfy all of the observations.

All Be stars are probably variable on short time scales. Changes in the strengths and shapes of the Balmer emission components occur from month to month, whereas photometric and/or spectroscopic variations also occur on time scales of a few days to a few hours. Two hypotheses have been suggested to explain the rapid variations: nonradial pulsations and rotation. The former is supported by spectroscopic observations of absorption line profile variability of the underlying star (transient bumps and wiggles in the line profiles) in some Be stars (see Fig. 2), whereas the latter (rotational modulation of bright or dark regions of the photosphere and/or circumstellar envelope) is suggested by analysis of the light curves. There is no agreement as to the correct explanation at this time.

The aforementioned variability refers generally to the cool, equatorial disk component of the Be circumstellar envelope. The hot, expanding component is also variable, as evidenced by the appearance and disappearance of shortward-shifted, discrete absorption components of highly ionized atoms such as C IV, Si IV, and N V. The stellar wind from Be stars is evidently not homogeneous but consists of discrete blobs or shells.

MODELS OF Be STARS

Various explanations for the presence and behavior of circumstellar matter around Be stars have been proposed in the years since the original Struve model. These include rotationally enhanced stellar wind models, nonradial pulsation models, interacting binary models, and magnetic loop models. Although each has attractive features, no single model seems able to explain all of the observations. Be stars are obviously complex objects and it seems more likely

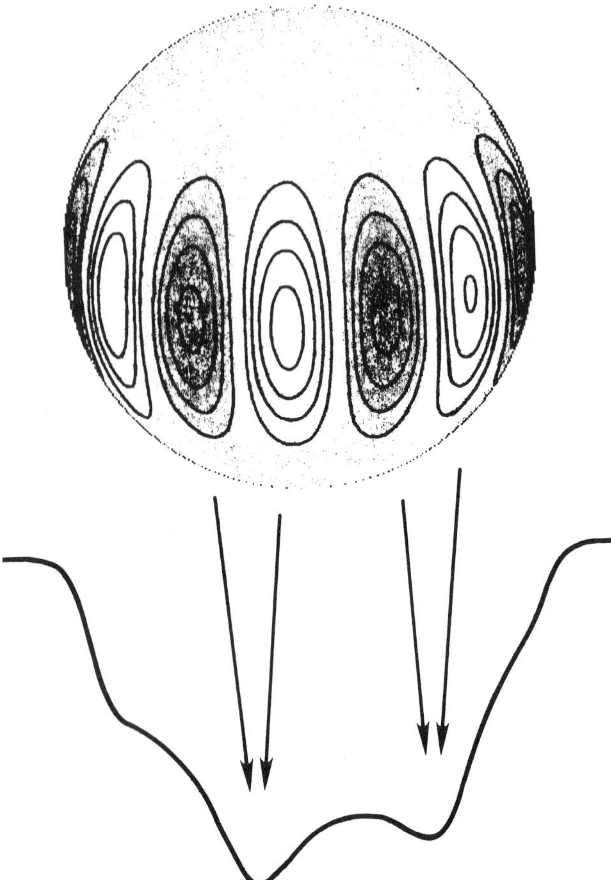

Figure 2. Illustration of how bumps are formed in the line profile of a rapidly rotating, nonradially pulsating star. This is a velocity map of the star with the resultant line profile shown below. The darkest shaded regions correspond to material moving away from us; the lightest regions correspond to material moving toward us. [*Courtesy of Steven S. Vogt and G. Donald Penrod (1983).*]

that their behavior is due to a combination of the preceding, and perhaps other, mechanisms, and that these will vary from star to star.

There seems to be no question that rotation must play a major role in the Be phenomenon. Statistical studies of line broadening in Be stars lead to the conclusion that *all* Be stars rotate rapidly, the observed differences in line widths being due primarily to differences in the inclination of the rotation axes. Rotation is therefore a necessary but not sufficient (because observed rotational velocities seem to be below the critical velocities required for instability, and a trigger is needed in any case) condition for the Be phenomenon.

It was recognized many years ago that in a hot star, radiation pressure forces acting selectively on ions may result in an outward atmospheric momentum transfer and the production of a stellar wind. Such winds will be enhanced by the presence of rotation, and the equatorial wind will be much denser than the polar wind, in agreement with observations supporting disk models. Magnetic fields would produce a further enhancement, although such fields have not been observed in Be stars. On the other hand, highly disordered fields (analogous to those on the solar surface) would produce only very small mean field components when averaged over the entire stellar surface and would, therefore, be very difficult to detect.

None of the aforementioned mechanisms explains the episodic nature of the Be phenomenon: Why are the envelopes variable, and why are they thrown off quasiperiodically every decade or so? Nonradial pulsation may play a role here. The amplitudes of the pulsations, as determined from line-profile variability studies, were found to be correlated with the occurrence of emission episodes in some Be stars. The changing nature of the nonradial pulsations might add matter to the circumstellar envelope in a time-varying way, accounting for the variable emission. The time scale between shell episodes might then reflect the relaxation oscillation cycle of the pulsations. This explanation is still conjectural, however, until the existence of nonradial pulsations in Be stars and their relationship to variations in emission is more firmly established.

EVOLUTIONARY STATUS OF Be STARS

The relatively high frequency of Be stars compared with the frequency of normal absorption-line B-type stars suggests that the Be phenomenon might represent a stage in the normal evolution of a B star. Color-magnitude and Hertzsprung–Russell diagrams of open clusters containing Be stars show, however, that whereas Be stars are located one-half to one magnitude above the main sequence on average, they may be found anywhere between the zero-age main sequence and the giant region. This suggests that Be stars may exist in all evolutionary states from essentially unevolved to the giant branch. Other effects may cause Be stars to appear off the main sequence in such diagrams, however, including intrinsic envelope reddening and rotation-induced gravity darkening of the underlying star.

The best that can be said about the evolutionary status of Be stars at this time is that although some, at least, are relatively unevolved, the majority are located somewhat off the main sequence. This is probably largely due to evolution but may be due in part to circumstellar reddening and, probably to a lesser degree, to gravity darkening of the underlying, rapidly rotating star.

Additional Reading

Doazan, V. (1982). Be stars. In *B Stars with and without emission lines*, A. Underhill and V. Doazan, eds. NASA SP-456, p. 277.

Jaschek, M. and Groth, H.-G., eds. (1982). *Be Stars*. Proc. IAU Symp. 98. D. Reidel, Dordrecht.

Slettebak, A., ed. (1976). *Be and Shell Stars*. Proc. IAU Symp. 70. D. Reidel, Dordrecht.

Slettebak, A. (1979). The Be stars. *Space Sci. Rev.* **23** 541.

Slettebak, A. (1988). The Be stars. *Publ. Astron. Soc. Pacific* **100** 770.

Slettebak, A. and Snow, T. P., eds. (1987). *Physics of Be Stars*. Proc. IAU Colloquium 92. Cambridge University Press, Cambridge, U. K.

See also **Stars, Nonradial Pulsation in B-Type; Stars, Rotation, Observed Properties; Stars, Winds.**

Stars, BL Herculis, W Virginis, and RV Tauri Types

Hugh C. Harris

The stars classified as BL Herculis, W. Virginis, and RV Tauri stars are all pulsating stars of low mass but high luminosity. Together, they comprise the old (and often metal-poor, Population II) stars whose post–main-sequence evolution has carried them into the Cepheid instability strip. However, they differ from the RR Lyrae stars (a similar class of stars that also lie in the Cepheid instability strip) and other core-helium-burning stars in their more advanced stage of evolution, so that they have higher luminosities and longer pulsation periods than the RR Lyrae stars. They have luminosities ranging from about 10^2 to 3×10^3 times greater than the luminosity of the Sun, pulsation periods ranging from about one day to more than one month, and surface temperatures typically 4000–6500 K. Their luminosities are closely related to their periods, with the long-period stars being the most luminous. Therefore, the observed period of a star can be used to estimate the distance of the star, which is so helpful in studying these stars as to make them useful for studying stellar evolution and for probing the Galaxy. The most luminous of these stars are probably close to ejecting their envelopes, perhaps forming planetary nebulae, and evolving quickly to become white dwarf stars.

The names for these classes of stars, as for other classes of variable stars, come from a prototype star for each class. The classes differ from each other by increasing period and increasing luminosity from the BL Her stars to the RV Tau stars. For example, the star BL Her itself has a period of 1.3 days, W Vir has a period of 17.3 days, and RV Tau has a period of 39.4 days. All together, these stars are often called Type II Cepheids or Population II Cepheids. The anomalous Cepheids, which are anomalously bright for their periods, are related, but are more massive than the Type II Cepheids and are likely to have a different origin in binary star systems. In globular star clusters, we now know of 24 BL Her stars, 17 W Vir stars, and 6 RV Tau stars. Among field stars outside of clusters, we know of about 100 BL Her stars, about 100 W Vir stars, and about 100 RV Tau stars. These numbers are approximate, because it is often difficult to distinguish reliably these Type II variable stars from the younger and more massive Population I variable stars. Our discovery of these stars is nearly complete only in the solar neighborhood, and there must be hundreds more of each type, scattered throughout our galaxy.

EVOLUTIONARY STATUS

These variable stars are in the late stages of stellar evolution, and they have a structure similar to stars on the asymptotic giant branch. In their interiors, there are an outer shell where hydrogen fusion is occurring and an inner shell where helium fusion is under way. The general evolution of these stars as they consume their nuclear fuels can change their surface parameters relatively quickly, as do instabilities in the rates of shell burning. When a star reaches a radius such that its surface temperature is within certain limits (the instability strip), its atmosphere becomes unstable to pulsation. It then appears as one of these kinds of Type II Cepheid variable stars and will remain so for typically 10^5 yr or less. Computer models show that a metal-poor star is most likely to be carried into the instability strip. Models also show that a low-mass star with a metallicity similar to that of the Sun is likely to evolve through the instability strip if it loses an unusually large fraction of its envelope by mass-loss processes. In fact, all of the Type II Cepheids known in globular clusters are metal-poor, with metallicities approximately one-tenth to one-hundredth that of the Sun. However, among field stars outside of clusters, some are metal-poor whereas others are not. Apparently, many field variables are stars from our galaxy's old-disk population, with only moderate metal deficiencies and with intermediate ages, whereas others are true Population II stars of the galactic halo.

CURRENT OBSERVATIONAL RESEARCH

There are ongoing programs to discover and classify these stars and to collect basic data defining their light curves and measuring their temperatures, velocities, and so forth. These data give improved knowledge of the distances, masses, metallicities, and space motions of individual stars. A few stars have been found to have binary star companions, providing additional information about the stars.

A defining characteristic of RV Tau stars is that their light curves do not always repeat from one cycle to the next, instead sometimes exhibiting alternating deep and shallow minima. Models of luminous W Vir stars begin to show this behavior. The models indicate that the period between adjacent minima should be taken as the more physically meaningful period, rather than the longer "formal period" between primary minima. Some stars at times have such small secondary amplitudes, however, that this conclusion becomes doubtful. Possible explanations for the alternating pulsation include the excitation of more than one mode of pulsation, the resonance between different modes, the influence of shock waves in the atmosphere, and a transition into chaos.

The field RV Tau stars, in particular, show a confusing variety of pulsation properties, abundance peculiarities, and circumstellar dust shells. They are often subclassified by their spectroscopic appearance that, in turn, results from their initial mass and metal abundance, the current relative abundances of carbon and oxygen atoms in their atmospheres, their amount of circumstellar material, and their surface gravity. The carbon/oxygen ratio sometimes exceeds unity in the more luminous field stars (mostly in RV Tau stars, rarely in W Vir stars), turning them into carbon stars. The well-studied stars often show apparent deficiencies of elements produced by *s*-process nucleosynthesis which might be explained by either mass loss or by mixing nuclear-processed material into the star's atmosphere. Most RV Tau stars have at least some excess infrared emission indicating that, for many of these stars, significant mass loss has recently occurred.

A few field variable stars found at high galactic latitudes appear so peculiar that they have sometimes been given a separate classification as UU Her stars. These are probably low-mass RV Tau stars with luminosities as high as, or perhaps somewhat higher than, the most luminous RV Tau variables found in globular clusters. Recent far-infrared observations of a few of them show significant infrared emission indicating that moderate to large mass loss has occurred. The metal-poor, semiregular variables sometimes called SRd stars are generally cooler stars more closely related to the metal-rich Mira variables than to the RV Tau stars. Finally, some other high-latitude supergiant stars are known with characteristics similar to RV Tau stars except for their lack of pronounced variability: They are probably in the same evolutionary state as the RV Tau stars, but have temperatures sufficiently different that they are not unstable to pulsation, or they may be variable with small amplitudes, perhaps in nonradial modes, but have not yet been sufficiently studied to classify them reliably.

COMPUTER MODELS

Although the pulsation of these stars is generally understood, many details about their atmospheric structure throughout the pulsation cycle are unknown. Research is being conducted on the importance of shock waves that propagate through the atmosphere of most (perhaps all) of these stars each cycle, the effect of convection in damping the pulsation, the degree to which the atmospheres are extended by the pulsation and lead to occasional or ongoing mass loss, the analysis of the elemental abundances in their atmospheres, the evidence for dredge-up from the interior of material that has undergone nuclear processing, and a variety of other processes. Generally this research involves the construction of models of the stellar atmosphere and/or interior and the matching of model parameters to those of observed stars.

The most luminous of these stars are extremely difficult to model accurately, however, for several reasons. They have very low surface gravities and they have amplitudes of pulsation at their surfaces that are comparable to their radii. Their pulsation contributes to their having extended atmospheres, where the simplifying assumptions of plane-parallel geometry and local thermodynamic equilibrium break down. The passage of shock waves through their atmospheres alters the atmospheric structure and the radiative energy transfer. Convective energy transport is important in their outer envelopes, but we lack an adequate description of convection as it varies through the pulsation cycle. Circumstellar material, with dust forming in the outer atmospheres, has an uncertain but possibly significant influence on the atmospheres of some stars. Today's increasingly sophisticated computer models help enormously in addressing these problems, and our understanding will undoubtedly increase in the coming years.

Additional Reading

Cox, J. P. (1980). *Theory of Stellar Pulsation*. Princeton University Press, Princeton.

Fernie, J. D., ed. (1972). *Variable Stars in Globular Clusters and in Related Systems*. D. Reidel, Dordrecht.

Harris, H. C. (1985). Population II Cepheids. In *Cepheids: Theory and Observations*, B. F. Madore, ed. Cambridge University Press, Cambridge, p. 232.

Nemec, J. M. (1990). Anomalous Cepheids and Population II blue stragglers. In *The Use of Pulsating Stars in Fundamental Problems of Astronomy*, E. Schmidt, ed. Cambridge University Press, Cambridge, p. 215.

Rosino, L. (1978). Problems of variable stars in globular clusters. *Vistas in Astron*. **22** 39.

Wallerstein, G. and Cox, A. N. (1984). The Population II Cepheids. *Publ. Astron. Soc. Pacific* **96** 677.

See also **Star Clusters, Globular, Variable Stars; Stars, Cepheid Variable; Stars, Pulsating, Overview; Stars, RR Lyrae Type; Stellar Evolution, Pulsations.**

Stars, Blue Stragglers

Linda L. Stryker

Blue straggler stars appear in the color–magnitude diagrams of many star clusters. These stars lie above the turnoff region in these diagrams, where, if the stragglers had been normal stars, they should already have evolved away from the main sequence. (See Figure 1 in the entry on Star Clusters, Globular, Stellar Populations —Ed.) These enigmatic stars lie *blueward* of the turnoff and appear to linger or *straggle* in their evolutionary process, hence the name "blue stragglers."

Why do blue stragglers remain behind in their evolution? Evidence is accumulating that there may be more than one possible mechanism for creating a straggler, because they exist in all stellar populations. These mechanisms generally fall into two broad categories: delayed formation (DF), where stars form in a more recent burst of starbirth; or delayed evolution (DE), where stars gain an extended lifetime through the ability to mix more fuel into the stellar core, or through the transfer of mass from a close binary companion.

Knowledge of the nature of blue stragglers is crucial, not only for gaining a complete understanding of how stars evolve, but also for learning about the stellar populations found in galaxies, because the presence of numerous stragglers can provide a significant hot component in the integrated light of galaxies.

Stragglers are found abundantly in open clusters of all ages (Population I: young disk, intermediate-age disk, old disk), in globular clusters (Population II: halo), and in dwarf galaxies.

SOME OBSERVED PROPERTIES

Very Young to Young Disk ($< 2 \times 10^8$ yr old)

Example: Pleiades. Numerous blue stragglers are found in these clusters as well as in intermediate-age open clusters. Many show abnormal spectra, such as peculiar A (Ap) stars, metallic A (Am) stars, and emission-line early-type stars (Be, Of). These stragglers sometimes fall in the extended instability strip where ultra-short-period ($P \lesssim 0.275$ days) Cepheid variables lie. For clusters with ages less than 3×10^7 yr, exotic Ap stars (showing lines of strontium, chromium, and europium) are absent.

Many early (O6–B2) blue stragglers show rapid rotation and some are nitrogen rich (OBN); at least one OBN star is known to be a binary. Enhanced nitrogen on the surface may occur from binary mass transfer. Core mixing caused by rotational support might be the cause of lengthened lifetimes for these massive (40 $M_\odot$ [solar masses]) stars, but early carbon-rich stars (OBC) have the same rotational velocities and are not observed to be stragglers or binaries.

Intermediate Age (3×10^8–10^9 yr)

Examples: Hyades and Praesepe. Over 60% of the blue stragglers are Ap stars, belonging, if not to the cluster, to its surrounding "supercluster." These stars are strong in silicon, chromium, europium, titanium, and strontium, have small light variations, and some are magnetic variables. Field Ap stars are generally slow rotators and deficient in short-period binaries; Am stars are often binaries. Stragglers have later spectral types, B3–A2, and slower rotations, but have strong magnetic fields of about 10^3 G, at least 10 times greater than the field strength of the Sun.

Old Disk ($> 10^9$ yr)

Examples: M 67, NGC 188, NGC 7789, NGC 2506, and NGC 6791. Numerous blue stragglers are found and provide about 10% of the V (visual) light of their clusters. Theoretical models using the observed blue straggler mass function, binary mass transfer process, and fractions, periods, and mass ratios of close binaries in old clusters indicate that binary orbital periods may be long (100–1000 days) with consequent low radial velocity amplitudes ($\lesssim 10$ km s^{-1}).

Halo (17×10^9 yr)

Examples: M 3, M 71, NGC 5466, NGC 5053, NGC 6352, and ω Cen. Blue stragglers are found in several globular clusters, especially in ones known to have very low central densities (NGC 5466 and NGC 5053); equally important, they are absent in others (or at least their outer parts). Of the stragglers in ω Cen, three are known dwarf Cepheids and one is an eclipsing binary (period less than 3 hr) with small light variations. Visual spectra show no unusual chemical characteristics.

Field stars show Population II metallicity and kinematics so they are old stars. Field Ap (Si) stars show a deficiency of binaries, especially short-period ones, which may indicate that Ap (Si) blue stragglers are also unlikely to have high binary frequency. No chemical abundance anomalies were found in a recent high-resolution study of three nonvariable field stragglers, but finding anomalies does not distinguish between theories.

Dwarf Galaxies

The presence of many variable stars, stragglers in the mass range 1.3–1.6 $M_\odot$ (solar masses), and anomalous Cepheids in dwarf galaxies allows a test of the binary mass transfer hypothesis by using the ratio of stragglers to anomalous Cepheids. The ratio is expected to be from 1 to 10 if coalesced-binary blue stragglers are precursors of anomalous Cepheids. The Draco dwarf galaxy has a ratio of about 14 and Ursa Minor has a ratio of about 25, which is not far from prediction. Draco shows a spread in metallicity along the giant branch, and a red horizontal branch; these may indicate more than one era of star formation (as in the DF theory), although lower helium or higher CNO abundance could be the cause. Sculptor may be a "young" galaxy with (DF) stragglers about 5×10^9 yr old.

THEORIES AND DISCUSSION

Several theories to explain the existence of blue stragglers have emerged, beginning with Fred Hoyle and William H. McCrea in the 1960s. Some theories are more readily testable than others, but none are without ambiguity. The basic question, still unanswered, *is whether stragglers are single or double stars.*

1. *Are blue stragglers simply field stars and not actually members of the cluster?* No. We now have good proper motions confirming membership for stragglers in a number of clusters. Radial velocity and polarization studies are less conclusive, but generally support the cluster membership of blue stragglers. Memberships for the very brightest stragglers (ones crucial to distinguishing between theories because their masses are at or above the limit allowed by binary mass transfer theory) remain ambiguous.

2. *Are blue stragglers in a post-main-sequence phase (i.e., evolved, perhaps horizontal branch stars), and only coincidentally appearing near the main sequence as they evolve toward later phases?* No. Evolved stars have lower surface gravities and lower masses than main-sequence stars and they would be distributed less centrally than more massive stars. Studies by Karen M. and Stephen E. Strom and by Olin J. Eggen in the 1970s showed, in fact, that the surface gravities and effective temperatures of blue stragglers are those of unevolved main-sequence stars.

Blue stragglers evidently have normal masses for their positions, that is, masses larger than those at turnoff. Studies of the spatial distribution of stragglers within a cluster show that there is a significant concentration similar to that of known binaries toward the central regions. This is consistent with the mass segregation process and implies that blue stragglers are either binaries or single stars with masses of at least 1.5 $M_{\odot}$.

An evolved star undergoing full mixing at the helium flash may end up blueward of the normal zero age main sequence (ZAMS), but blue stragglers lie on it or to the red. Partial mixing could place an evolved star on or near the ZAMS, but in any case, the mass derived would be less than that for turnoff stars.

3. *Are blue stragglers normal main-sequence stars born in a later formation era than the cluster turnoff stars (DF)?* Maybe. However, other evidence for more recent star formation in the cluster such as dark nebulae, T Tauri stars, 21-cm excess showing neutral hydrogen, or signs of clumpy dust is not observed. Abundance differences might be expected to be seen among stars of the two (or more) epochs. Globular clusters move through the galactic disk so gas that might have remained to form stars will be swept out.

Recently, Eggen and Icko Iben have discussed the idea of successive modes of star formation for the very young disk population: mode A for low-mass stars (0.3–5 $M_{\odot}$), where formation proceeds quietly in the gas-and-dust nest; and mode B, where higher mass stars ($> 5\ M_{\odot}$) explode and sweep out the star-building materials, thus canceling any further mode A activity. According to these models, mode B stars are younger (these models also include very large convective overshooting, which gives a star twice the normal lifetime). Mode B stars are the ones that appear as DF blue stragglers; other very bright stragglers, such as θ Car (an Ap and known binary star) combine DE and DF.

4. *Have blue stragglers gained additional mass from a binary companion (or merged with their companion) to increase their lifetimes (DE)?* Maybe. For old stragglers, this model says that the binary system's component stars formed with similar masses; the original primary overflowed its Roche lobe and transferred mass to the secondary which now appears as the more massive straggler. The combined mass is constrained by this model to be no more than twice the turnoff mass. A full 97% of the stragglers obey this constraint, with about four notable exceptions and their memberships are still in question. If blue stragglers are binaries, then variations in their radial velocities should be observable and allow orbital parameters to be obtained. The spatial distribution of stragglers in old clusters is similar to that of known binaries. There is debate on whether *all* blue stragglers are members of close binary systems, because velocity variations are not detected in most cases —but the amplitudes may be too low. Also, velocity variations could be caused by pulsation; few orbits have been determined. There is convincing observational evidence that at least some blue stragglers are binaries: for example, θ Car; NJL 5, an eclipsing binary in ω Cen; and F190 in M67. Stragglers are found in low-density environments where collisions to disrupt binaries would be minimized. If periods are long, 100–1000 days, then perhaps the semiamplitudes of velocity variation will be small, less than 10 km s^{-1}, and hard to detect.

The companion may be a white dwarf, which should be detectable in the ultraviolet (< 1500 Å) for stragglers later than A0, but a recent *IUE* study does not detect any excess ultraviolet light. Or it could be a red dwarf or infrared subgiant, perhaps with an eccentric orbit, again causing hard-to-detect, low-amplitude velocities.

High-mass binaries could become x-ray sources. If a close binary has coalesced and merged, fast rotations would be expected. There may also exist an "excretion" disk, as is exhibited in FK Comae, to carry off excess angular momentum. Lifetimes would be increased, but early in the process, the surface gravity of a coalesced star would be that of a "bloated" star, a giant—not the main-sequence gravities as seen. An old coalesced star might more easily resemble a blue straggler. The constraint on the combined mass would now be lifted. Although the probability of coalescence in the not-very-dense environment of open clusters and some globular clusters must be low for single stars, if the primordial cluster binary population is of order 10%, then coalescence remains a viable mechanism. If coalescence follows as a result of contact, as in W Ursae Majoris contact binaries, then we should ascertain whether the fraction of such systems in a cluster is sufficiently large to produce the number of observed stragglers.

5. *Might blue stragglers be explained as having increased lifetimes due to some nonthermal pressure support (larger rotations or magnetic support) (DE)?* Maybe. Stars undergoing quasihomogeneous evolution in which substantial mixing supplies the core with hydrogen beyond the normal time of exhaustion can have increased lifetimes. Likely there is a nonthermal source of support, provided by rotation or magnetic pressure, which lowers the temperature and luminosity, thus lengthening the lifetime of core hydrogen burning. Younger clusters show evidence of these sources of pressure support.

6. *Might stragglers be the result of stellar collisions and coalescence (DE)?* Maybe. Some blue stragglers in the high-density cores of globular clusters must be products of collisions and mergers. As a test of this, there should be more stragglers in clusters with short relaxation times, small core radii, or x-ray sources; there may also be correlations between chemical inhomogeneities and dynamical properties. Coalesced blue stragglers should have a higher helium abundance and live longer due to the mixing of additional hydrogen into the core.

7. *Could there be mass loss for some main-sequence stars that evolve downward, leaving the "stragglers" behind?* Unknown. A recent idea suggests that certain main-sequence stars with masses from 1–3 $M_{\odot}$ and lying in the instability strip might experience mass loss and evolve downward on the main sequence. Clusters could then have turnoffs which indicate an age older than they really have; this would bring globular cluster ages more in line with the somewhat younger age being recently proposed for the universe. "Stragglers" would then be the more slowly rotating stars left behind.

ADDITIONAL QUESTIONS

Do main-sequence stars show any bimodal characteristics, such as increased rotations or presence of a companion? Is there evidence of more than one epoch of star production for stars on the main sequence? Do blue stragglers show different characteristics (abundances, rotation, magnetism, etc.) than those of main-sequence stars in the same cluster? Are stragglers the progenitors of anomalous Cepheids? Are stragglers the progenitors of Ba II, CH, subgiant CH, or carbon stars? What fraction of blue stragglers are ultrashort-period dwarf Cepheids?

PROSPECTS

The next steps are to: (1) increase the number of observations of blue stragglers to improve statistics; (2) determine cluster membership, stellar masses, velocity variations, light variations, rotations, chemical anomalies, and magnetic strengths; (3) model the binary-system mass transfer, coalescence and mergers, and the nonthermal, pressure-supported mixing; (4) check catalogs for nearness of x-ray sources' positions to stragglers' positions. Further observational tests of the binary hypothesis are needed, such as infrared searches for late-type, under-massive, semidetached companions.

STATUS

There are several conflicting results that have not yet been resolved. Stellar crowding in cluster cores and the faintness of many objects make observations difficult. Lively debate continues on all fronts. That this is still the case after almost 25 years strongly suggests that several mechanisms must be responsible for the blue straggler phenomenon.

Additional Reading

Eggen, O. J. and Iben, I., Jr. (1988). Starbursts, blue stragglers, and binary stars in local superclusters and groups. I. The very young disk and young disk populations. *Astron. J.* **96** 635.

Eggen, O. J. and Iben, I., Jr. (1989). Starbursts, blue stragglers, and binary stars in local superclusters and groups. II. The old disk and halo populations. *Astron. J.* **97** 431.

Leonard, P. J. T. (1989). Stellar collisions in globular clusters and the blue straggler problem. *Astron. J.* **98** 217.

McCrea, W. H. (1964). Extended main-sequence of some stellar clusters. *Monthly Notices Roy. Astron. Soc.* **128** 147.

Renzini, A., Mengel, J., and Sweigart, A. (1977). The anomalous Cepheids in dwarf spheroidal galaxies as binary systems. *Astron. Ap.* **56** 369.

Saio, H. and Wheeler, J. C. (1980). The evolution of mixed long-lived stars. *Ap. J.* **242** 1176.

Wheeler, J. C. (1979). Blue stragglers as long-lived stars. *Ap. J.* **234** 569.

See also **Star Clusters, Globular, Binary Stars; Star Clusters, Globular, Mass Loss; Star Clusters, Globular, Mass Segregation; Star Clusters, Globular, Stellar Populations; Star Clusters, Stellar Evolution.**

Stars, Carbon

I. Juliana Sackmann and Arnold I. Boothroyd

The only way to create carbon from the lighter elements is the triple-α reaction, in which three helium nuclei fuse to produce one carbon nucleus, as first proposed by Hans Bethe in 1939. But it was only in the 1950s that this nuclear reaction was really understood, due to work by Edwin E. Salpeter, Fred Hoyle, William A. Fowler, Ward Whaling, and collaborators at the Kellogg Radiation Laboratory. The Big Bang at the universe's birth produced almost no carbon (nor heavier elements). Practically all the carbon that we see in the universe today has been created in the interior of stars. Because life on Earth is built largely from carbon compounds, we are in fact "a little bit of stardust" (William A. Fowler). Carbon stars occur relatively frequently among red giants. Some stand out as superluminous stars, visible over great distances. Their spectra are among the richest stellar spectra known. A carbon star can be defined as one that has $C/O > 1$ in the surface layers, where "C/O" is the number ratio of carbon atoms to oxygen atoms. In cool stars, as much carbon as possible will combine with oxygen to form the stable CO molecule. It is only if there are more carbon than oxygen atoms (i.e., $C/O > 1$) that other carbon compounds can be formed from the leftover carbon, and their molecular bands observed in the stars' spectra.

R STARS (EARLY C STARS)

The R stars are giant stars, but are relatively hot (roughly 4000–5000 K) with a low luminosity [below about 2000 $L_\odot$ (solar luminosities)] compared to most other carbon stars. They are enriched in the isotopes ^{13}C and ^{14}N, with $^{12}C/^{13}C$ in the range 4–15, and with ^{14}N about four times as abundant as in the Sun. (Note that the solar value of $^{12}C/^{13}C$ is 90.) The R stars differ from most other carbon stars in showing *no* enhancement of the *s*-process elements, which are heavy elements that can be produced by the slow absorption of neutrons upon iron nuclei. The R stars tend to be old stars, situated in the disk of our galaxy.

The most probable explanation for the enrichment of carbon in R stars is the *helium core flash*, which is encountered by stars of initial masses ranging from slightly less than 1 $M_\odot$ up to slightly over 2 $M_\odot$. All stars begin their life on the *main sequence*, burning hydrogen into helium in their interiors. As a "side-effect," CNO-cycle burning of hydrogen quickly converts enough ^{12}C to ^{13}C to yield a ratio of $^{12}C/^{13}C \approx 3.5$; thereafter, both ^{12}C and ^{13}C are depleted in concert, yielding ^{14}N. After the central hydrogen fuel is exhausted, the star expands to become a cool and luminous red giant, and some CNO-processed material from the interior is dredged up to the surface, where it is observed (the *first dredge-up*). The surface is thus somewhat depleted in ^{12}C and enriched in ^{13}C and ^{14}N. The stellar core contracts and heats, until the center reaches a temperature near 100 million K and the helium at the center ignites. Because of the high central density, this ignition is a *very* violent event, called the *helium core flash*. Calculations of this violent event are still only approximate, but it seems that the ashes of this violent helium burning are usually *not* mixed up to the surface. However, under *some* conditions (such as perhaps a rapidly rotating core), they might be mixed outwards far enough to be picked up by the outer convective zone and then brought up to the surface. Because this flash helium burning produces primarily ^{12}C, the surface would be enriched in carbon. *If* the convection from the core flash managed to mix any hydrogen downwards, it would burn very quickly, converting some of the ^{12}C into ^{13}C and ^{14}N, and resulting in further surface enrichment of these isotopes. After the helium core flash is over, the star settles down to quiet core helium burning, decreasing a good deal in luminosity and growing somewhat hotter and bluer at its surface. The R stars are observed in this stage.

N STARS (LATE C STARS OR "CLASSICAL CARBON STARS")

The N stars are giant stars, very cool (around 3000 K) and very luminous (roughly from 2000–20,000 $L_\odot$). They display strong molecular bands of C_2, CH, CN, and C_3 in their spectra—they are the "classical" carbon stars, first discovered in 1868 by Angelo Secchi using visual spectroscopy. Most of these carbon stars also display enrichment in the *s*-process elements; on the average, they show moderate ^{13}C and ^{17}O enrichment, and moderate ^{18}O and ^{14}N depletion. They are observed to be losing mass rapidly, from 10^{-7}–10^{-4} $M_\odot$ yr^{-1}. They populate the disk of our galaxy and are bright enough to be seen in the Magellanic Clouds and other nearby galaxies.

These stars are much further evolved than the R stars. They start out with initial masses that range from slightly less than 1 $M_{\odot}$ (solar mass) up to (very roughly) 3 $M_{\odot}$. They evolve through the same stages as those described previously for the progenitors of R stars, except that the more massive stars do not undergo a helium core flash but rather ignite helium quietly. They evolve through the core helium burning stage and up to the double shell burning stage, becoming cool and luminous stars on the asymptotic giant branch. At this stage, the helium burning shell is squeezed sufficiently thin (geometrically) that an instability is triggered. The result is violent, runaway nuclear burning, called a *helium shell flash* (or a *thermal pulse*). These flashes occur at regular intervals (tens of thousands to hundreds of thousands of years) and produce a ^{12}C-rich pocket reaching from the helium shell almost to the hydrogen shell. Flash-driven expansion can cause the convective envelope to reach far enough down to dredge up part of the carbon pocket (this is called *third dredge-up*). In this way a carbon star can be formed. *If* a little hydrogen gets mixed into the carbon pocket, it will burn some ^{12}C into ^{13}C; the latter in turn can burn with helium, releasing neutrons and thus producing *s*-process elements, which would be dredged up just like the carbon. Some CNO-processed material is dredged up, but not much; more work needs to be done to explain the CNO isotopic ratios in detail. It is possible that in some stars the base of the convective envelope may be hot enough that CNO burning in the quiescent interflash periods could affect the surface CNO abundances.

If not enough ^{12}C has been dredged up to form a carbon star, there may still be *s*-process enhancement; such stars are called S stars, and are observed to have C/O somewhat below unity. Some have C/O within 1% of unity; these are called SC stars. S and SC stars are observed to occur at slightly lower luminosity than the classical carbon stars: As time goes by, they become more luminous and dredge up more carbon, finally becoming carbon stars.

From observations of the luminosities of carbon stars in the Magellanic Clouds, one deduces that they are stars of relatively modest initial mass, less than about 3 $M_{\odot}$. However, the theoretical models predict that more massive (and more luminous) stars, up to 6 or 8 $M_{\odot}$, should also become carbon stars; such stars are not seen. The most probable reason is that mass loss in these more massive stars is so severe that they never reach the shell flash stage, and thus cannot become carbon stars. Recent radio observations do indeed indicate high mass-loss rates for stars in this stage, up to 10^{-4} $M_{\odot}$ yr^{-1}.

R CrB AND HdC STARS

The R Coronae Borealis (R CrB) and HdC stars are hydrogen-deficient supergiant carbon stars. Their surface temperatures range from those of stars on the giant branch (about 3000 K) to much hotter temperatures (up to 15,000 K), with corresponding bluer spectral types, and their luminosities are similar to those of the classical carbon stars. The R CrB stars are variable in their optical brightness, whereas their infrared luminosity remains constant; it appears that they are obscured intermittently by blotchy circumstellar clouds. HdC stars look much like R CrB stars, but are not variable. R CrB stars and HdC stars are best understood as stars slightly less massive than the Sun which have shed all (instead of merely most) of their hydrogen envelopes at the end of the previously described asymptotic giant branch stage. This would expose the carbon pocket produced by the shell flashes. Another possible explanation is that a star cooling toward the white dwarf stage could encounter a final helium shell flash, causing it briefly to reach the luminosity and effective temperature region occupied by the R CrB stars. Such stars would have extremely small hydrogen envelopes, which might more easily be ejected or mixed downwards and burned, resulting in an R CrB type surface composition.

CH STARS, sgCH STARS, AND BARIUM STARS

CH stars are carbon stars showing especially strong CH, CN, and C_2 bands in their spectra, as well as enhanced bands due to the *s*-process elements. They are metal-poor stars found in the halo of our galaxy; some of them are found in globular clusters and are on the giant branch. The barium stars are very similar to the CH stars, except that they are more metal-rich, and they probably do not have quite enough carbon enhancement to be carbon stars. They are found in the disk of the Galaxy, rather than in the halo. The subgiant CH stars (sgCH stars) comprise a less luminous and somewhat hotter extension to both of these groups; some (such as G 77-61 and HD 88446) appear to be on or near the main sequence. If one considers these three groups of stars as a whole, they range from 4000–6000 K and from 1–2000 $L_{\odot}$.

None of these stars are bright enough to be asymptotic giant branch stars, and therefore they *cannot* have undergone helium shell flashes with their subsequent carbon and *s*-process enrichment. The key to understanding the CH stars, sgCH stars, and barium stars is that they all appear to be members of *binary star systems*, with white dwarf companions. The enrichment in their surfaces seems to have resulted from mass transfer. The companion, originally more massive and thus evolving faster, has already passed through the helium shell flash stage, becoming a carbon star (with *s*-process enhancement) at that point. This massive companion has suffered severe mass loss on the way to becoming a white dwarf; at least some of this mass has been deposited on the less massive, less evolved companion, which now displays these carbon and *s*-process enrichments.

WOLF–RAYET STARS: THE WC STARS

These stars are very luminous (somewhere between 30,000–500,000 $L_{\odot}$) and very hot (somewhere between 25,000–100,000 K) with no evidence of hydrogen. Roughly half of the Wolf–Rayet stars show strong nitrogen lines in their spectra and are called WN stars, whereas the other half show strong carbon lines (the WC stars, which are carbon stars). Many are observed to be hotter and bluer than normal main-sequence stars of corresponding luminosity. They are observed to be losing mass at a very high rate, up to 10^{-4} $M_{\odot}$ yr^{-1}.

Wolf–Rayet stars can be understood in terms of very massive stars, with initial masses from about 20 $M_{\odot}$ up to perhaps as much as 60 $M_{\odot}$. The mass loss has exposed the hydrogen-exhausted core. First, one sees the ashes of hydrogen burning, namely helium and some nitrogen, accounting for the WN stars. Further mass loss exposes the ashes of partial helium burning, primarily ^{12}C in the outer core, accounting for WC stars. Further in, where burning had proceeded nearer to completion, there is more ^{16}O than ^{12}C; if this is exposed, one gets an oxygen-rich WO star.

IMPACT OF CARBON STARS ON THE UNIVERSE

The carbon stars all undergo mass loss—in some cases so strongly that the star is obscured by its cocoon of ejected mass and can only be detected at infrared or radio wavelengths. The *Infrared Astronomical Satellite* was a gold mine for the discovery of such stars. Carbon stars enrich the interstellar medium with considerable carbon, some nitrogen and oxygen, and *s*-process elements. In fact, the classical carbon stars produce most of the carbon in the universe.

Additional Reading

Epchtein, N., Le Bertre, T., and Lépine, J. R. D. (1990). Carbon star envelopes: near-IR photometry, mass loss and evolutionary status of a sample of *IRAS* stars. *Astron. Astrophys.* **227** 82.

Fowler, W. A. (1984). Experimental and theoretical nuclear astrophysics: the quest for the origin of the elements. *Rev. Mod. Phys.* **56** 149.

Jaschek, M. and Keenan, P. C., eds. (1985). *Cool Stars with Excesses of Heavy Elements*. D. Reidel, Dordrecht.

Kwok, S. and Pottasch, S. R., eds. (1987). *Late Stages of Stellar Evolution*. D. Reidel, Dordrecht.

Vangioni-Flam, E., Audouze, J., Cassé, M., Chieze, J.-P., and Tran Thanh Van, J., eds. (1986). *Advances in Nuclear Astrophysics*. Editions Frontières, Gif-sur-Yvette.

See also **Stars, R Coronae Borealis; Stars, Wolf–Rayet Type; Stellar Evolution, Intermediate Mass Stars; Stellar Evolution, Low Mass Stars; Stellar Evolution, Massive Stars.**

Stars, Cepheid Variable

T. Lloyd Evans

Cepheid variable stars are supergiants with luminosities 500–30,000 times greater than that of the Sun, although their surface temperatures are similar to the Sun's temperature. They undergo regular radial pulsations (i.e., the star expands and contracts), with periods mainly in the range 1–50 days, and can be distinguished at great distances. More than 400 Cepheids are known in the Galaxy and about 1000 Cepheids have been found in each of the two nearest galaxies, the Magellanic Clouds, as well as substantial numbers in other nearby galaxies. The close relationship between period and luminosity which was found by Henrietta S. Leavitt in 1912 has given Cepheids a unique role in establishing the distances of the nearer galaxies and hence the distance scale of the universe.

The regularity of the light curve of a Cepheid variable star is matched by that of the radial-velocity curve, which is almost a mirror image of the light curve with minimum radial velocity (i.e., maximum velocity of approach) at light maximum. The light amplitude is typically between 0.5 and 2 magnitudes in visual light and the velocity amplitude usually lies in the range 30–60 km s^{-1}. The first Cepheid velocity curves were measured toward the end of the nineteenth century and were interpreted as the results of orbital motion. It was only after orbits had been computed for a substantial number of Cepheids that it was realized that these orbits were physically implausible.

The pulsation hypothesis gained increasing acceptance after 1910, especially because the surface temperature changes over the cycle. Sir Arthur Eddington's theoretical work from 1917 onwards showed that Cepheids are single stars that undergo radial pulsations because they function as a heat engine. Later work by S. A. Zhevakin, J. P. Cox, Robert F. Christy, and others has provided a deeper understanding of the mechanism. Energy is stored in the form of the second ionization of helium during the compression stage of the cycle and then released as the helium recombines during the expansion stage. The restriction of Cepheid pulsations to stars in a limited temperature range follows from the requirement that the second helium ionization zone lies near the transition from the nearly adiabatic interior, where any driving is almost canceled by an equal amount of damping, to the nonadiabatic exterior where the thin outer layers lack the heat capacity to modulate the outward flow of radiation. The pulsation is a property of the stellar envelope and is independent of the nuclear-energy-generating core.

CLASSICAL AND TYPE II CEPHEIDS

Classical Cepheids are comparatively young stars with masses of several times the solar mass. This follows from their strong concentration toward the plane of the Milky Way and their low space velocities. Their presence in star clusters allows their ages to be estimated as up to about 10^8 yr. Observations of the Cepheids in the Magellanic Clouds show that the classical Cepheids are confined to a narrow strip in the period–luminosity diagram, whereas the less common Type II Cepheids are fainter than them at a given period. The presence of Type II Cepheids in globular clusters and in the galactic halo population allows their age to be estimated as up to 15×10^9 yr. so that they must be much less massive than the classical Cepheids. The Type II Cepheids can also be distinguished from the classical Cepheids by the shape of the light curves and by spectroscopic peculiarities.

CEPHEID LIGHT CURVES

The light curves of most classical Cepheids are asymmetrical, with a rapid rise to maximum light and a slower fall. The form of the light curve changes with period in a systematic way known as the Hertzsprung progression. A bump appears on the descending branch of the light curve of stars with periods of about a week and is found at earlier phases in stars of successively longer periods so that the bump is near maximum light in stars of 10-day period which may show a double maximum. The bump falls on the rising branch in stars of longer period. Stars of the shortest or longest periods have smooth light curves. The amplitude of the pulsation increases slowly with period up to about 10 days, where there is a drop in amplitude; it then increases more rapidly to longer periods. The bumps may represent an echo of the surface pulsation from the deep interior; an alternative explanation is that they result from a resonance when the second overtone period is about one-half of the fundamental period.

Some Cepheids of short period have nearly sinusoidal light curves with amplitudes of only about 0.5 magnitude. These stars are uncommon in the Galaxy but account for about 10% of the Cepheids known in the Magellanic Clouds. A monumental study by Cecilia Payne-Gaposchkin and Sergei Gaposchkin showed that they are systematically brighter than the period–luminosity relation defined by the stars with asymmetrical light curves. The ratio in periods at a given luminosity is 0.6, which in view of the observational uncertainties is probably identical to the well-defined ratio $P_1/P_0 = 0.71$ found for double-mode Cepheids in the Galaxy. The latter are Cepheids, mostly in the period range $P_0 = 2.0$–4.3 days, whose light curves may be represented as the sum of simultaneous pulsations in the fundamental and first overtone periods of P_0 and P_1 days, respectively. An analogous situation is found in the RR Lyrae stars in globular clusters.

THEORETICAL EVOLUTIONARY TRACKS OF CEPHEIDS

Evolutionary tracks covering the mass range and evolutionary states in which Cepheids occur have been calculated by Icko Iben, Jr. and others. The mass range 3–9 $M_\odot$ (solar masses) covers all but stars of very long period. The most important point is that a star may cross the Cepheid instability strip on the Hertzsprung–Russell (HR) diagram more than once during its evolution (see Fig. 1). The star leaves the main sequence after the exhaustion of H in the core and then expands to become a red giant while burning H in a shell surrounding the temporarily inert He core. It crosses the instability strip rapidly on a Kelvin–Helmholtz or thermal time scale. It climbs the red giant branch to the red giant tip and after the ignition of He burning in the core it may make a loop to higher temperature in the HR diagram. This loop may extend to sufficiently high temperature (or blue color) to intersect the instability strip, in which case two more crossings will occur. Core helium burning is a relatively long-lived evolutionary stage and the star may remain in the instability strip for much longer, by perhaps a factor of 50, than it did in the first crossing. The exact location of the blue loops is a function of mass and of chemical composition,

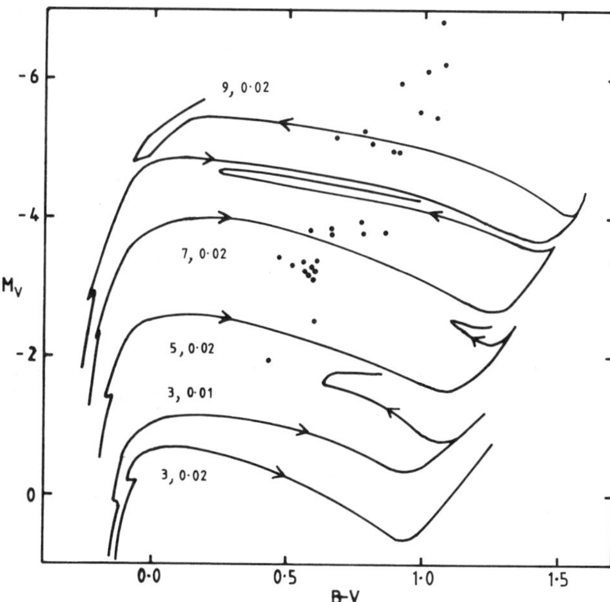

Figure 1. Color–magnitude diagram (a convenient form of the Hertzsprung–Russell diagram) comparing theoretical evolutionary tracks with the absolute magnitudes and B–V colors of Cepheid variable stars of known distance. The numbers given by the evolutionary tracks, which all assume a helium content by mass fraction of 0.28, are the mass in solar masses (here ranging from 3 to 9) and the content of heavy elements, Z, by mass fraction (0.01 or 0.02).

so that above and below the most favorable mass, the number of Cepheids will decline quickly. The more massive stars evolve more rapidly in any case, so that the peak residence time will correspond to a relatively low mass with a decline in the numbers of longer-period stars which is accentuated by the relative rarity of massive stars. All the Cepheids at the extremes of the mass (period) distribution must be on their first crossing of the instability strip. It has been estimated that they account for about 10% of all Cepheids.

CEPHEID PERIOD DISTRIBUTIONS

The distribution of periods has been found for Cepheids in the Galaxy, the Magellanic Clouds, and several other nearby galaxies. There is inevitably selection in favor of the brighter long-period Cepheids and those of larger amplitude at a given period, but differences between the period distributions in the Galaxy and the two Magellanic Clouds are well established. The distributions in these three systems may be characterized by the shortest periods, the periods that are most common, and the longest periods for fundamental mode pulsators:

Galaxy	2.3 days	5.4 days	67 days
LMC	1.6	3.4	135
SMC	1.1	2.9:	210

(The second, third, and fourth columns above list the shortest, maximum frequency peak, and longest periods, respectively.) The maximum frequency peak period in the SMC, about 2.9 days, is uncertain because there is a wide double maximum in the period distribution. There is some doubt as to whether the Cepheid-like stars of the very longest periods are genuine Cepheids or are some other type of variable star. Even so, the overall distributions are very wide and are difficult to account for theoretically unless there is a spread of metal content (abundances of elements heavier than helium) within each galaxy.

The principal difference between the Cepheids in the three galaxies is the successively smaller value of the shortest and maximum frequency peak periods in the order Galaxy–LMC–SMC, which is in order from high to low metal abundance. This is readily explained by the differences in the evolutionary tracks resulting from the differences in composition. The blue loops in the core helium burning stage extend to higher temperature the lower the metal abundance, so that the maximum frequency of Cepheids occurs at lower mass and luminosity and shorter period. Stars of lower mass are relatively more numerous, so that a star system of lower metal abundance will contain more Cepheids, other factors being equal. The more metal-deficient Small Magellanic Cloud contains roughly as many Cepheids as the Large Magellanic Cloud whose total mass is four times greater.

CEPHEIDS IN STAR CLUSTERS

The approximately 400 known classical Cepheids in the Galaxy include only 17 which are well-established members of star clusters and as these are divided between 14 clusters, there are too few stars in any one cluster to study the distribution of stars in the Hertzsprung–Russell diagram. The Magellanic Clouds contain much richer clusters and NGC 1866 in the Large Magellanic Cloud contains at least seven Cepheids with periods in the range 2.6–3.5 days. Comparison of the observed and theoretical HR diagrams leads to an estimated metal (elements heavier than helium) abundance $Z = 0.016$, a Cepheid mass of about 4.9 $M_\odot$, and an age of 86×10^6 yr. There are still differences in detail in the numbers and positions, especially of the red giant stars, between the observed cluster and that which is calculated theoretically.

The main value of the Cepheids in the small clusters in the Galaxy is that they may be used to establish the zero point of the period–luminosity relation. The 17 stars noted previously and another 8 of longer period which belong to the loose stellar groups known as associations have been used to establish the zero point to an accuracy of ± 0.1 magnitude, excluding uncertainties in the distance scale for star clusters.

CEPHEIDS IN BINARY SYSTEMS

Our knowledge of the masses of stars is obtained from binary systems. Spectroscopic orbits need to be obtained for both components of a binary and the inclination of the orbital plane to the line of sight must be found or the mass of one component has to be established independently. The orbital periods of Cepheid binaries are generally long: The shortest known is the 507-day orbital period of the 9.7-day variable star S Muscae. This means that the velocity amplitudes are quite small and that eclipses that would establish the orbital inclination are unlikely to occur. The mass of the companion must be deduced from its spectrum. The light of the Cepheid is always dominant at visible wavelengths so the spectrum of the companion, which is usually a much hotter B star, is only readily distinguishable in the far ultraviolet. This means that observations must be made from a satellite. A few estimates of around 5–6 $M_\odot$ for Cepheids of periods of 4–10 days are available so far, in broad agreement with the mass estimates from evolutionary tracks. The latter have generally been found to give larger masses than methods based on the pulsation properties of the stars and more precise dynamical mass estimates are needed to clarify the situation.

The limiting orbital period below which a Cepheid on its first crossing of the instability strip would suffer disturbance to its evolution is about 20 days but a Cepheid crossing the instability strip for the second time on a blue loop has previously expanded to a much larger radius at the red giant tip. The minimum period to avoid overflowing the Roche lobe with consequent severe mass loss is several hundred days in this case. The longer periods of all the Cepheid binaries studied to date are in accordance with the theoretical estimate that most are on their second or a subsequent crossing of the instability strip.

Additional Reading

Christy, R. F. (1966). Pulsation theory. *Ann. Rev. Astron. Ap.* **4** 353.

Cox, A. N. (1980). The masses of Cepheids. *Ann. Rev. Astron. Ap.* **18** 15.

Cox, J. P. (1980). *Theory of Stellar Pulsation.* Princeton University Press, Princeton.

Fernie, J. D. (1990). The structure of the Cepheid instability strip. *Astrophys. J.* **354** 295.

Payne-Gaposchkin, C. and Gaposchkin, S. (1966). Relation of light curve to period for stars in the Small Magellanic Cloud. *Vistas in Astronomy* **8** 191.

Pel, J. W. (1985). Fundamental parameters of Cepheids. In *Cepheids: Theory and Observations*, B. F. Madore, ed. Cambridge University Press, Cambridge, p. 1.

Zhevakin, S. A. (1963). Physical basis of the pulsation theory of variable stars. *Ann. Rev. Astron. Ap.* **1** 367.

See also **Star Clusters, Globular, Variable Stars; Stars, BL Herculis, W Virginis, and RV Tauri Types; Stars, Cepheid Variable, Dwarf; Stars, Cepheid Variable, Period-Luminosity Relation and Distance Scale; Stars, Pulsating, Overview; Stars, Pulsating, Theory.**

Stars, Cepheid Variable, Dwarf

D. Harold McNamara

In the 1950s it became apparent that there was a class of pulsating variable stars with periods from approximately one hour to one-quarter of a day with visual light amplitudes greater than or equal to 0.3 magnitude that range in spectral type between late A and early F. Their light curves are unsymmetrical (faster rise time from light minimum to light maximum than the decline from light maximum to light minimum), in the same sense as Cepheids and RR Lyrae variables. Because the periods suggested that they were less luminous than either of these classes of longer-period variables but yet they exhibited similarly shaped light curves, H. J. Smith coined the term dwarf Cepheids to describe them. They have also been called RRs, AI Velorum, ultra-short-period, large-amplitude δ Scuti, and SX Phoenicis variables. As to whether these stars form a distinct class, or are a large-amplitude version of the δ Scuti stars, has never been completely satisfactorily answered. We shall come back to this issue after we have discussed the properties of the stars.

The light (V magnitude) and color variations (B–V) versus phase of CY Aquarii, a typical dwarf Cepheid, are displayed in Fig. 1*a*. The light amplitude in V is $0\overset{\mathrm{m}}{.}7$ and the variations in (B–V) correspond to a change of approximately 1000 K. The radial velocity varies in the same period, and is nearly a mirror image of the light curve, with minimum velocity occurring at phase approximately 0.05. Some of the stars in this group exhibit two oscillations of period ratios $P_1/P_0 \sim 0.77$, where P_0 is the fundamental period and P_1 is the first overtone. They are designated with an asterisk in Table 1.

One of the interesting properties of these variables is the decrease in their surface gravities g and mean effective temperatures $\langle T_{\mathrm{eff}} \rangle$ with increasing period. This is evident in the third and fourth columns of Table 1. The variation of log g with log P is shown in Fig. 1*b*. The changes are a consequence of the $P\sqrt{\langle\rho\rangle} = Q$ (P is the period, $\langle\rho\rangle$ is the mean density, Q is a constant) relation and the period–luminosity relation of these variables (compare the second and sixth columns of Table 1). As the stars become more luminous, their radii increase and, consequently, their gravities decrease. Although an increase in mass accompanies the increase in period, it is not sufficient to overcome the increase in radius so that a decrease in gravity results.

Another interesting property of the dwarf Cepheids is the tendency of the metal-poor stars to be restricted to the shorter-period stars. In the fifth column of Table 1 are listed the [Fe/H] values of these stars ([Fe/H] = $\log[\mathrm{Fe/H}]_{\mathrm{star}} - \log[\mathrm{Fe/H}]_{\mathrm{Sun}}$). It is apparent that short-period variables such as BL Camelopardi and KZ Hydrae are extremely metal-poor with [Fe/H] = −2.4. At the other extreme are long-period variables such as VZ Cancri [Fe/H] = 0.2 and DY Herculis [Fe/H] = 0.3. A plot of the [Fe/H] values versus log P is exhibited in Fig. 1*c*. There is clearly a continuous decrease in [Fe/H] with decreasing period. The so-called SX Phoenicis variables are clearly a metal-poor version of the metal-strong variables. The SX Phoenicis category needs to be dropped from the *General Catalog of Variable Stars* (GCVS).

One of the most exciting developments in the study of these stars is the discovery of dwarf Cepheids in globular star clusters. NGL 220 and NGL 79 are two examples. They provide a check on the metal abundances but, more importantly, the absolute magnitudes of the variable stars because the distances to the clusters are quite well known. The absolute magnitudes inferred from membership in globular star clusters are consistent with those inferred from a period–luminosity relation based on surface gravities and Wesselink radii determined for a few of the well-observed stars.

What is perhaps the most significant aspect of the discovery of these variable stars in globular clusters is their blue straggler status. Blue stragglers, first recognized by Allan R. Sandage, are stars that do not fall on either the cluster main sequence or the horizontal branch. Instead, the blue stragglers are found to the blue of the main sequence on the Hertzsprung–Russell (HR) diagram. They appear to be main-sequence stars more massive than the most massive stars near the "turn-off" of the cluster main sequence (but see below).

The ages of the dwarf Cepheids, calculated on the basis of the pulsation equation $P\sqrt{\langle\rho\rangle} = Q$ and their position on the HR diagram, are given in the eighth column of Table 1. The ages of the metal-poor stars are much younger than one would anticipate. For example, NGL 79 which appears to be a member of the star cluster ω Centauri, a cluster with an age in excess of 10 billion years, has an apparent age of 5 billion years. Even more striking is XX Cygni with a [Fe/H] = −1.0. On the basis of classical evolutionary theory, we would assign the star to the Population II halo population with an age of approximately 10 billion years—yet its present position in the HR diagram suggests an age of about 1 billion years, fully one-tenth the age of typical stars in the halo. Thus we are faced with the following dilemma: Either the metal-poor dwarf Cepheids are younger than the star aggregates they are typically identified with or some mechanism must be responsible for delaying their evolution. It has been suggested that blue stragglers are a consequence of mass transfer in a binary system or possibly stars that have undergone complete chemical mixing. The majority of the variables exhibit little if any direct evidence for current duplicity. The fact that the variables have masses, luminosities, radii, and so on, consistent with evolutionary theory argues against the mixing and consumption of more fuel than other main-sequence stars. Thus we are left with the "born late" or "born again through stellar mergers" hypotheses or some yet unknown mechanism to explain their relative youth.

Are the dwarf Cepheids a distinct class of objects or simply δ Scuti stars of large light amplitude? In a number of respects such as distribution of periods, compositions, space motions, and masses they are similar to δ Scuti variables. In other respects they differ markedly. A case in point is their rotational velocities. The ninth column of Table 1 lists their $v \sin i$ values. Note that they are *all* very small—no evidence of rotational line broadening—and have

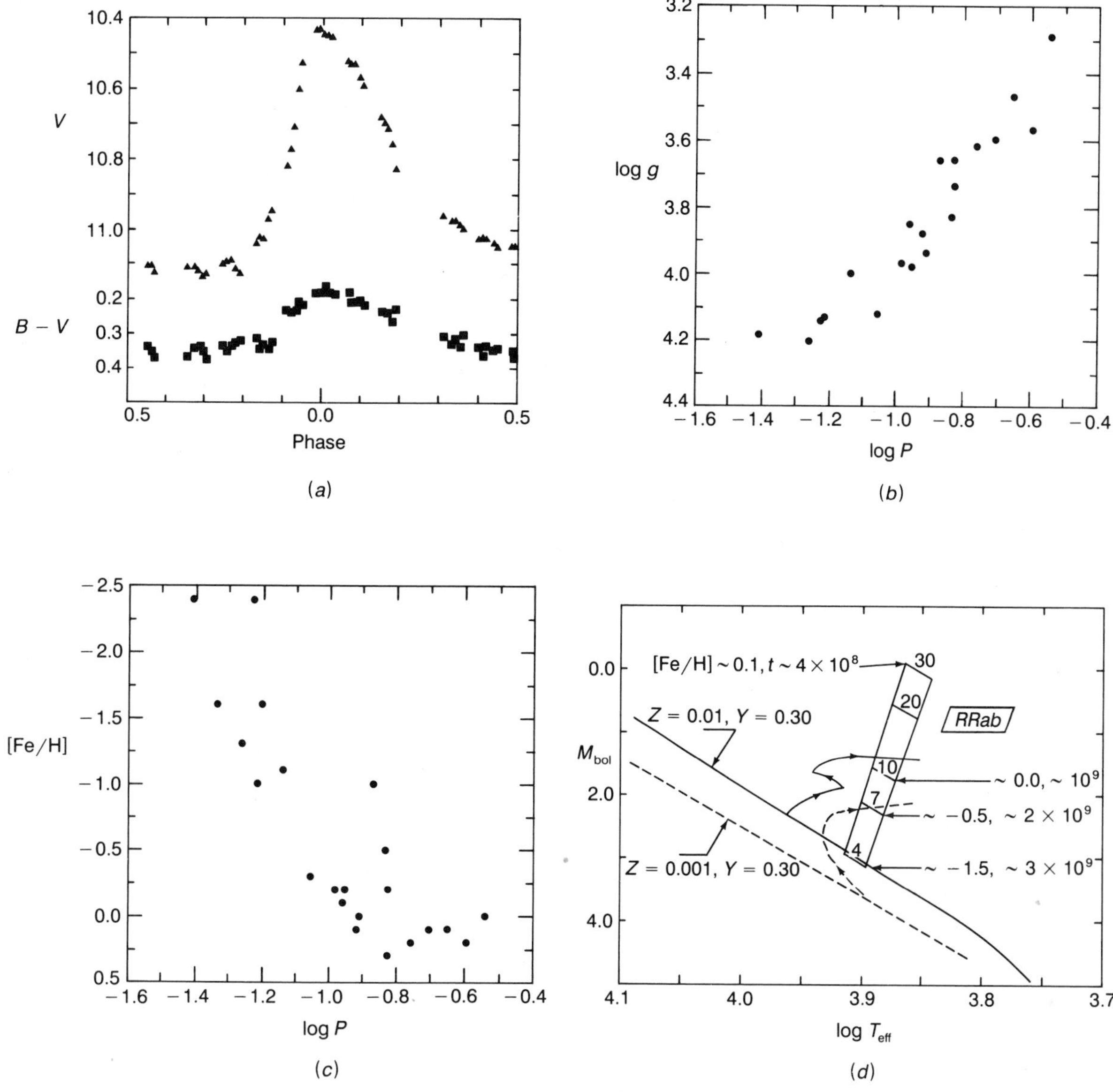

Figure 1. (*a*) The light and color curve of CY Aqr. (*b*) A plot of the log *g* values against log *P* for dwarf Cepheids. (*c*) The [Fe/H] values of the dwarf Cepheids plotted against log *P*. (*d*) The instability strip of the dwarf Cepheids. Lines of constant period (hundredths of a day) are shown for five periods. The [Fe/H] values and ages (yr) of variables at four periods are listed next to the arrows. Note that the metal-poor stars are confined primarily to the shorter periods. Evolutionary tracks for two stars are shown (see final paragraph of text).

much smaller $v \sin i$ values (where v is the rotational velocity and i is the inclination of the rotational axis to the line of sight) than the δ Scuti variables where $\langle v \sin i \rangle \sim 100$ km s^{-1}. In addition, we note that the dwarf Cepheids are restricted to a much narrower domain in the HR diagram than δ Scuti variables, are all post–main-sequence stars, and exhibit only radial pulsations in addition to their larger light and velocity amplitudes. The differences are sufficient that at least they deserve to be treated as a separate class in the GCVS.

Finally, we refer the reader to Fig. 1*d*. Two zero-age main sequences are represented for two heavy-element abundances $Z = 0.01$ and $Z = 0.001$. The evolutionary tracks of two stars—a 1.55-mass star with $Z = 0.01$ and a 1.1-mass star with $Z = 0.001$—into the dwarf Cepheid instability strip are shown. Lines of constant period in hundredths of a day, typical [Fe/H] values, and ages are also shown. The tracks suggest that the variables are in the early phases of the hydrogen-shell-burning stage of evolution and the tracks also account successfully for the restriction of the metal-poor variables to the shorter periods.

Additional Reading

Andreason, G. K. (1983). Delta Scuti variables. *Astron. Ap.* **121** 250.

McNamara, D. H. (1985). The rotational velocities of the dwarf Cepheids and related stars. *Publ. Astron. Soc. Pacific* **97** 715.

McNamara, D. H. and Feltz, K. A., Jr. (1978). GD 428 and the nature of the dwarf Cepheids. *Publ. Astron. Soc. Pacific* **90** 275.

Nemec, J. M. (1989). Anomalous Cepheids and Population II blue stragglers. In *The Use of Pulsating Stars in Fundamental Problems of Astronomy*, E. G. Schmidt, ed. Cambridge University Press, Cambridge, p. 215.

See also **Star Clusters, Globular, Variable Stars; Stars, Blue Stragglers; Stars, δ Scuti and Related Types; Stars, RR Lyrae Type.**

Table 1. Properties of Dwarf Cepheids

Star	*log P*	$\langle T_{eff}\rangle$	$\langle \log g\rangle$	*[Fe / H]*	M_{bol}	$M / M_\odot$	*Age* (10^9 *yr*)	*v* sin *i* (*km* s^{-1})
BL Cam	−1.408	8050	4.18	−2.4	3.1	0.96	6.3	≤ 18
NGL 220†	−1.334	7100:		−1.6	3.4:	0.80:	6.0:	
SX Phe*	−1.260	7850	4.20	−1.3	2.8	1.2	1.6	≤ 18
KZ Hya	−1.225	7650	4.14	−2.4	2.6	0.9	6.5	≤ 45
CY Aqr	−1.214	7930	4.13	−1.0	2.4	1.2	2.3	≤ 18
NGL 79†	−1.201	7550		−1.6	2.6	0.9	5.3	
DY Peg	−1.137	7800	4.00	−1.1	2.0	1.2	2.1	≤ 16
EH Lib	−1.054	7930	4.12	−0.3	1.6	1.8	0.8	≤ 16
YZ Boo	−0.983	7650	3.97	−0.2	1.4	1.8	1.0	≤ 16
BP Peg*	−0.960	7470	3.85	−0.1	1.4	1.7	1.3	≤ 18
AI Vel*	−0.952	7620	3.98	−0.2	1.3	1.8	1.0	≤ 18
SZ Lyn	−0.919	7540	3.88	+0.1	1.3	1.8	1.0	≤ 40
AD CMi	−0.910	7580	3.94	0.0	1.2	1.8	0.9	≤ 20
XX Cyg	−0.870	7530	3.66	−1.0	1.3	1.5	1.1	≤ 18
RS Gru	−0.833	7600	3.83	−0.5	0.9	2.0	0.7	≤ 40
DY Her	−0.828	7130	3.66	+0.3	1.2	1.8	0.7	≤ 20
V567 Oph*	−0.825	7450	3.74	−0.2	1.1	2.1	1.0	≤ 18
VZ Cnc*	−0.759	7100	3.62	+0.2	1.0	2.0	0.8	≤ 10
BS Aqr	−0.704	7200	3.60	+0.1	0.6	2.1	0.6	≤ 5
VX Hya*	−0.652	6980	3.47	+0.1	0.6	2.2	0.6	
DE Lac	−0.596	6960	3.57	+0.2	0.4	2.3	0.5	≤ 20
SS Psc	−0.541	7300	3.29	0.0	−0.1	2.6	0.4	≤ 18

*Double mode variables.
†NGL is the Niss, Jorgensen, and Laustsen star number in the globular cluster ω Centauri. [See Niss, B., Jorgensen, H. E., and Laustsen, S. (1978). *Astron. Ap.* **32** 387.]
: uncertain.

Stars, Cepheid Variable, Period–Luminosity Relation and Distance Scale

Barry F. Madore and Wendy L. Freedman

In the early decades of this century, Henrietta S. Leavitt recognized that there is a statistical relationship between the average brightness of a Cepheid variable star and the period of pulsation. The periods of these stars range from a few days to a few hundred days, and their luminosities (all of which are intrinsically much greater than that of the Sun) span factors of several hundred in brightness. In general, the longer the period of a Cepheid, the brighter is its average *intrinsic* luminosity.

Because the period of pulsation can be measured independently of distance and because the *apparent* luminosity of a Cepheid depends on the square of the distance, one can use the intrinsic luminosity, as predicted by the period, in combination with the apparent luminosity to derive a distance. As tools in the extragalactic distance scale, Cepheid variables provide one of the most accurate means of determining distances to the nearby spiral and irregular-type galaxies of the Local Group and somewhat beyond.

THE BASIC PHYSICS

One very appealing attribute of Cepheid variables as a class of distance indicators is that the physics for the Cepheid period–luminosity relation is well understood. Put simply, all self-luminous objects, including Cepheids, give off light in proportion to both their area and the surface brightness over that area. For spherical bodies of radius R and surface temperature T, the total luminosity L (integrated over all wavelengths) is found using Stefan's law where $L = 4\pi R^2 \sigma T^4$ and σ is the Stefan–Boltzmann constant. Furthermore, the fundamental period of oscillation P of any mechanical system depends only on one thing, the mean density ρ. Low-density systems have longer periods of oscillation than high-density systems, with the equation relating these two quantities having the following form, $P\rho^{1/2} = Q$, where Q is a constant. Combining Stefan's law, the $P\rho^{1/2}$ law, and an assumed mass–luminosity relation indicates that the largest stars will have the highest luminosities and the longest periods. And that is exactly what is observed.

Of course, with a more rigorous (and more complex) application of physics, we can relax these simplifying constraints and redo the calculations for stellar models having a reasonable range of mass and with a variety of surface brightnesses. But these are refinements, certainly necessary for detailed calculations, but not absolutely necessary for an understanding of the underlying process leading to a period–luminosity relation itself. The refinements do tell us that the relation between period and luminosity will have scatter, and that the scatter will have a physical origin due to differences in surface temperature between stars of the same density (i.e., same mass and radius), for it is temperature that determines the surface brightness of self-luminous objects. Ultimately, Cepheids are well described by a PLC (period–luminosity–color) relation.

It is of interest to note here that the physics that applies to the ensemble of Cepheid variables (that is to the period–luminosity relation as a whole) also applies to the individual stars as they each cycle through their oscillations. Radius variations in a single star give rise to changes in area, which result in changes in luminosity; surface temperature variations around the cycle drive changes in the surface brightness which also affect the total luminosity. There are again two parameters to the problem: The area is a geometrical property of the star and therefore has the same effect on the luminosity nearly independent of wavelength, whereas the surface brightness, radiation theory tells us, is very sensitive to where in the spectrum one observes the star. In the infrared, temperature variations make only a slight contribution to luminosity differences, whereas in the blue and ultraviolet, the same temperature variations can dominate the luminosity variations. These changing effects with wavelength can be seen in Fig. 1 where the cyclical

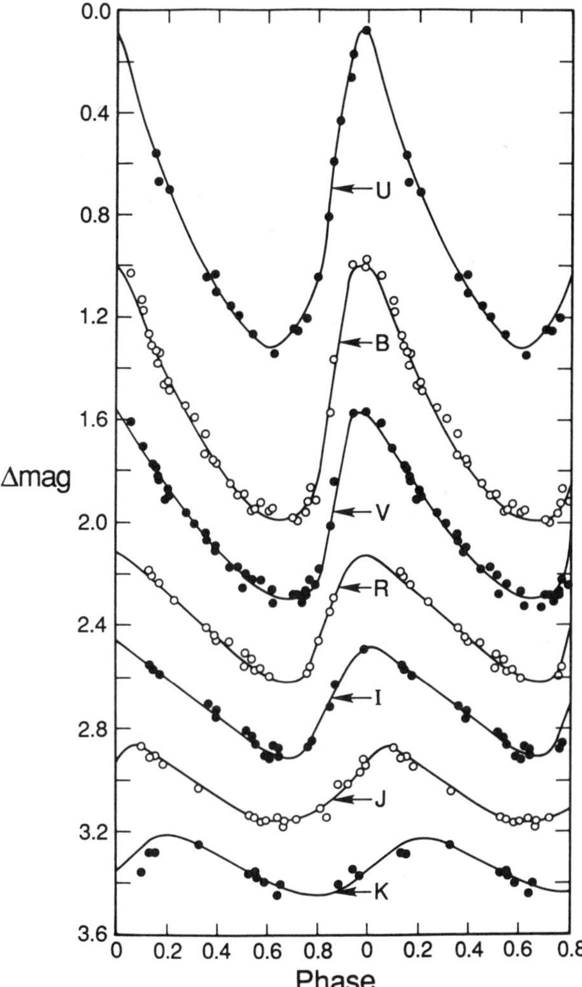

Figure 1. Typical light variations as a function of phase for a Cepheid variable as observed at wavelengths ranging from the ultraviolet (top light curve, labeled U) through the blue, visual and red parts of the spectrum (labeled B, V, and R, respectively) out to the near infrared (ending with the bottom light curve at $K = 2.2\ \mu m$). Note the decreasing amplitude of the light variation, as well as the change in shape and the shift in the phase of maximum brightness as longer and longer wavelengths are examined.

luminosity variation for a Cepheid is shown at different wavelengths. In the near ultraviolet the amplitudes are large and driven primarily by surface temperature variations. In the far infrared almost all of the luminosity change is a reflection of the radius variation which is known to be out of phase and distinctly different in shape with respect to the temperature variations.

The absolute calibration of the Cepheid period–luminosity relation is based on a small number of Cepheids found in galactic star clusters. These clusters have independent distances, obtained from main-sequence fitting techniques. Additionally, the same main-sequence stars can be used to independently estimate the amount of interstellar dust obscuring and reddening the light of the Cepheids. Unfortunately, the statistics are poor and the intrinsic luminosities and colors of many of the *cluster* Cepheids are still uncertain. Moreover, most *field* Cepheids are too far away to have direct parallax measurements made with the present technology. However, an independent calibration will be forthcoming when the refurbished *Hubble Space Telescope* provides improved direct determinations of the distance to the Large and Small Magellanic Clouds in which hundreds of Cepheids, all essentially at the same distance, will enter the calibration.

MODERN APPLICATIONS

Optical (blue/visual) images are essential for the discovery of Cepheid variables and for the definition of their light curves and periods. Observed in the blue, Cepheids can be picked out easily as they periodically rise and fall in luminosity against the sea of constant stars surrounding them. By contrast, when observed in the infrared, Cepheids are hard to detect as variables at all.

Practical advantages arise from a careful inspection and interpretation of Fig. 1. If in the course of observing a known Cepheid variable, one wishes to know the average brightness, then choosing to observe the star at wavelengths farther and farther into the infrared means that the star will be seen to vary less and less (its observed amplitude is a decreasing function of increasing wavelength). Accordingly, long-wavelength determinations of mean magnitudes will be less dependent upon complete sampling of the light curve. Indeed, for some applications a single infrared observation of a Cepheid (at random phase) can be as useful in distance determinations as are many dozen such observations made in the blue. This is so because, not only do the amplitudes of the Cepheids themselves decrease with increasing wavelength, but also the width of the observed period–luminosity relation itself decreases with increased wavelength, thereby providing a better and more precise estimator of the distance.

INTERSTELLAR OBSCURATION

Unfortunately, the effects of interstellar reddening due to intervening dust both in our own galaxy and in the parent spiral galaxies of extragalactic Cepheids can be quite severe and are very much stronger in the blue as compared to the red and infrared. A multicolored approach to the application of Cepheids to determining the distance scale is therefore required. Using modern electronic detectors (such as charge-coupled devices) and a variety of filters running from the blue to the very near infrared, this latest, hybrid approach allows for both the distance and the total reddening to individual external galaxies to be simultaneously solved for by a careful application of the period–luminosity relation, given an a priori knowledge of the interstellar extinction law. Without a determination of the extinction, all other distance estimates are upper limits, overestimating the true distance. Examples of period–luminosity relations constructed by this technique can be seen in Fig. 2 which consists of a montage of optical wavelength observations of Cepheids in the Large and Small Magellanic Clouds, followed by the dramatically narrower period–luminosity relations found at infrared wavelengths. These latter relations are so well defined and narrow that they have allowed astronomers to calculate not only the distances to the Magellanic Clouds but also the three-dimensional shape and orientation of these galaxies in space.

THE LOCAL (CEPHEID) DISTANCE SCALE

With the application of modern techniques to the Cepheid period–luminosity relation, it is now widely agreed that the extragalactic distance scale is relatively secure for galaxies within, and slightly beyond, the Local Group. For galaxies with Cepheids identified and observed, the distance estimates, out to approximately 2 Mpc, are now agreed upon at the 10% level. However, a factor of 10 further away at 20 Mpc, where Cepheids have not yet been discovered, the uncertainty, as judged by rival factions, rises to a factor of 2 difference in opinion. Much of this uncertainty is expected to disappear with the optically corrected imaging phase of the *Hubble Space Telescope* expected to begin in late 1993 or 1994. One of the

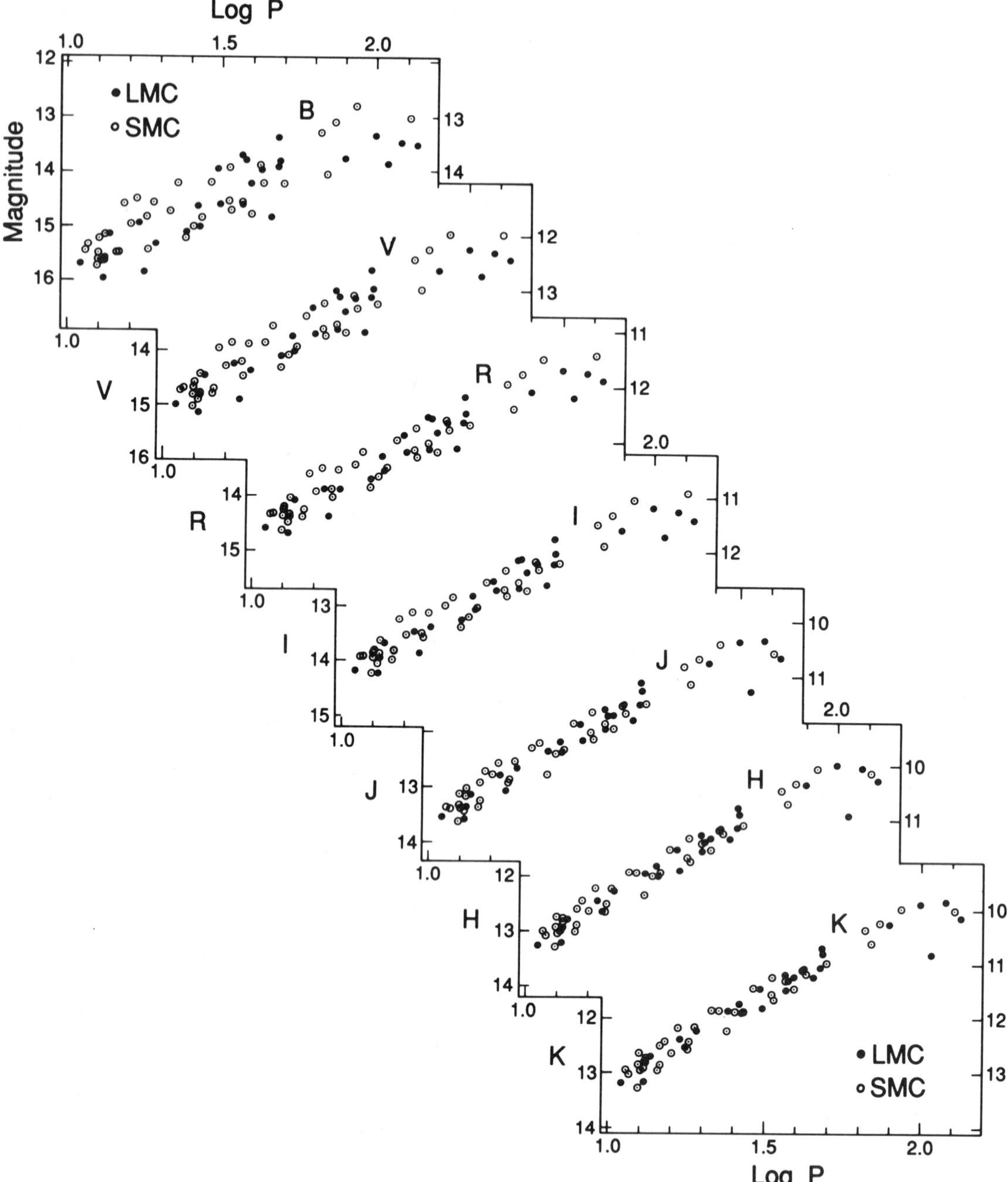

Figure 2. Multiwavelength period-luminosity relations for Cepheids observed in the Small Magellanic Cloud (open circles) shifted into magnitude registry with data for Cepheids in the Large Magellanic Cloud. Note the decreased scatter as one goes from the blue wavelength data (at the top left of the figure) to the near-infrared data (at the bottom right). The individual PL relations are shifted vertically in magnitude and horizontally in period for display purposes only.

major missions that *HST* is committed to working on is the extragalactic distance scale. *HST* will be capable of discovering Cepheids in galaxies that sample a much larger volume of space than can ever be imaged from the ground (because of atmospheric turbulence). A Cepheid-based distance to one or two galaxies at 20 Mpc (for instance in the Virgo Cluster, which is probably the practical limit even for *HST*) will not end the controversy about the size and age of the universe; but such observations will certainly go a long way toward narrowing the divergence of opinion. On the other hand, with many more Cepheid-based distances to galaxies inside that volume limit, secondary distance indicators can be calibrated with some statistical certainty, and these alternate techniques can then be pushed deep into the extragalactic space where the pure Hubble flow is expected to be revealed.

But the history of science teaches us that even our most modest expectations are not always met as we venture into new regimes. To be sure, questions of importance, such as those concerned with the age, size, and structure of the universe, will never completely go away; given time they will only become more interesting. Without Cepheid variables to lead the way in the extragalactic distance scale, our view of the universe would certainly be far less secure than we hope it is today.

Additional Reading

Freedman, W. L. (1988). Cepheid distances to nearby galaxies. In *The Extragalactic Distance Scale*, S. van den Bergh and C. Pritchet, eds. Astronomical Society of the Pacific, San Francisco, p. 24.

Madore, B. F., ed. (1985). *Cepheids: Theory and Observations*. Cambridge University Press, Cambridge.

Madore, B. F. and Tully, R. B., eds. (1986). *Galaxy Distances and Deviations from Universal Expansion*. Reidel, Dordrecht.

Sandage, A. R., Sandage, M., and Kristian, J., eds. (1975). *Galaxies and the Universe*. *Stars and Stellar Systems*, Vol. 9, University of Chicago Press, Chicago.

See also **Cosmology, Observational Tests; Distance Indicators, Extragalactic; Stars, Cepheid Variable; Stars, Pulsating Theory.**

Stars, Chemical Composition

R. Earle Luck

In simplest terms, the chemical composition of the stars can be summarized in these words: hydrogen and helium. The remaining 90 natural elements are found only in trace amounts in the stars, and in fact, in the universe as a whole. The general composition of stars is by mass 70% hydrogen, 28% helium, with the remaining elements comprising 2%. By number of atoms the composition is even more remarkable: The fraction of hydrogen is 90.8%, helium 9.1%, and the other elements 0.1%. This result, so contrary to our experience on the Earth, is one of the cornerstones of astronomy.

The realization of the dominance of hydrogen and helium in stellar composition was due to the work of Cecilia Payne-Gaposchkin, first set forth in 1925 in her doctoral thesis. This work was confirmed in 1929 by Henry Norris Russell in the first rigorous analysis of the composition of the Sun. In the period 1930–1950, studies of individual stars were undertaken in an effort to gain an understanding of the total range of stellar compositions. Most stars were found to have compositions similar to that of the Sun. However, there were small numbers of objects found with apparently different compositions. These objects included stars that appeared metal-poor; that is, they have a lesser amount of the heavy elements (elements heavier than helium, called "metals" by astronomers) in them than does the Sun. Also detected were objects in which strong features due to diatomic molecules appeared. Sometimes these appeared to be a result of the star being cooler than the Sun and thus favoring the formation of molecules (i.e., the K and M stars), but in other cases the composition appeared to be truly anomalous (the carbon stars). The difficulty in understanding these results was that there was no detailed framework in which to place them; the fields of stellar and galactic chemical evolution were only in their beginning phases.

The application to astronomy of high-speed digital computing in the period around 1960 led to a revolution in the study of stellar composition. Not only do computers allow more rapid computation of individual stellar abundances, they enable the utilization of more realistic physics and encourage astronomers to investigate larger samples of objects. Studies of large numbers of related objects allow exploration of the interrelations that exist among the various chemical elements. Also possible are studies of the variations in chemical composition that exist among the stars themselves.

Besides providing more tools for the astronomer interested in chemical abundances, the computer indirectly provided another enormous benefit: a theory of stellar and galactic chemical evolution. Stellar and galactic chemical evolution theory provides the framework for understanding stellar chemical abundances. Since the early 1970s, very few abundance studies have been designed without direct reference to furthering our understanding of stellar or galactic chemical evolution.

The basic method used for stellar abundance determinations is that of spectroscopic analysis. Secondary methods exist (e.g., photometric systems) but their calibration depends on the results obtained from spectroscopic analysis.

PHYSICAL DATA AND METHODS

For any type of stellar abundance analysis, three basic sets of data are needed. First, one needs the stellar spectrum. High-resolution data ($\lambda/\Delta\lambda = 30{,}000-100{,}000$) are used in order that the spectral features may be accurately measured. The wavelength ranges observed depend strongly on the type of object and the goals of the analysis. For example, if one were interested in determining the iron content of a G giant star, one would generally observe in the 5000–7000 Å region where there are numerous easily measurable iron lines.

Next, one needs atomic (and molecular) data about the elements to be investigated. These data include (but are not limited to) wavelengths of spectral lines, energy levels in the atom (or molecule), and transition probabilities. The last item is the probability of an electron moving between any two energy states in the atom and in doing so absorbing (or emitting) a photon. Transition probabilities can be determined from laboratory experiment or calculated from first principles.

Finally, one needs a model of the stellar atmosphere. A model atmosphere defines the structure of the photosphere of a star. The photosphere is the part of the star where the observed light and spectral features originate. The quantities of interest are the temperature and pressure within the atmosphere as a function of height above (or below) some predefined reference level.

With these physical quantities one still needs a set of assumptions under which to do the analysis. The assumptions generally used are: (1) plane-parallel geometry, (2) hydrodynamic equilibrium, and (3) local thermodynamic equilibrium. The first assumption means that the atmosphere is considered to be a vertically stacked series of infinitely thin planes without curvature. This approximation is valid for the Sun where the photosphere has a thickness of 500 km compared with a diameter of 700,000 km, but may fail for more extended atmospheres where the photosphere may extend to one-third of the radius. The assumption of hydrodynamic equilibrium is equivalent to stating that the various parts of the atmosphere are not moving with respect to one another. Finally, local thermodynamic equilibrium implies that the energy state populations are determined by collisions among the particles and thus that the physics of statistical mechanics can be used to describe those populations.

The type of stellar abundance analysis in common current use is the "fine" analysis. This technique operates under the previous physics using an iterative algorithm to determine abundances. Using a first guess at the abundance, the analysis code (computer program) determines the strength of the spectral feature in question, compares it to the observed strength, computes an abundance correction if necessary, and then redoes the calculation until the observed and calculated strengths match. This technique is useful for spectral features that can be separated from adjacent and overlapping features; however, this is often not feasible. In such cases a variation known as spectrum synthesis is used. In a synthesis all transitions (from all atomic species) contributing to the spectral feature are combined to predict the strength of the observed feature. It is of obvious importance to have extensive data on many species so that an accurate prediction can be made for the spectral feature of interest.

It is known that the basic assumptions of the "fine" analysis can fail. The assumption of local thermodynamic equilibrium is the

Table 1. The Most Abundant Elements in the Sun

Element	*Atomic Number*	*Abundance**	*Mass Fraction*
H	1	12.00	0.7012796
He	2	11.00	0.2787047
C	6	8.69	0.0040962
N	7	7.99	0.0009531
O	8	8.91	0.0090554
Ne	10	8.0	0.0014051
Mg	12	7.58	0.0006434
Si	14	7.55	0.0006939
S	16	7.21	0.0003620
Fe	26	7.67	0.0018189

*Logarithmic scale.

most prone to failure. The cause of the failure is that the radiation field interacts with the atoms, causing transitions among the energy states. Thus the populations are no longer predictable by the equations of statistical mechanics. The populations are still determinable, but at the expense of requiring much more detailed information about atomic structure (i.e., cross sections for collisions and photon interactions), and with a very significant increase in necessary computer resources.

THE SUN

In discussions of stellar compositions, the Sun has a central role as it serves as the standard reference point. There is a combination of reasons. First, it is possible to acquire excellent data (signal-to-noise, wavelength coverage, spectral resolution, and spatial resolution) for the Sun. These data allow construction of a model atmosphere for the Sun based on empirical evidence, not based on theoretical constructs as is necessary for other stars. Next, experience has shown us that the Sun has chemical abundances not significantly different from those of most stars. Thus we can base our results on the firmest physics available, have excellent data, and have an object not vastly different in composition than a large number of other objects: a perfect standard.

Many analyses have sought to determine the composition of the Sun. Recent results are given in Table 1 for the 10 most abundant elements in the Sun. The system used to quote abundances is logarithmic with the abundance of hydrogen set equal to 12. It is also a relative system; for example, if in a given volume of solar material there were 10^{12} hydrogen nuclei, then $10^{7.67}$ iron nuclei would also be present. Eighty of the ninety-two naturally occurring elements have been detected on the Sun. The solar photospheric abundances generally agree with abundances determined from meteorites. Meteoritic abundances can be accurately determined in terrestrial laboratories and thus serve as a comparison for solar photospheric abundances. The agreement between the two abundance sets points to a common origin and chemical history for the material of the solar system.

THE STARS

In a discussion of chemical abundances stars can be divided into three groups: (1) the oldest stars, (2) stars of the galactic disk, and (3) peculiar stars. Note that the number of analyses the discussion is based upon is very limited: The number of "fine" analyses is about 2000, encompassing some 1000 stars.

The Oldest Stars

If a star were composed of truly primordial material, its composition would consist of hydrogen and helium with a trace of lithium. Such a star has yet to be found. The most metal-deficient stars known have heavy-element abundances roughly 0.0001 that of the Sun. These objects are not globular cluster stars, but are a component of the galactic halo interspersed among the general field of stars. Globular cluster stars have abundances that range from 1/1000 to 1/4 of the abundances of the Sun. This range of abundance is regarded as evidence of a difference in age, the oldest clusters being the most metal-poor. Among these metal-poor objects, the elemental abundance ratios can vary systematically with the amount of iron (e.g., the O/Fe ratio increases from star to star with decreasing iron, whereas the Ba/Fe ratio decreases). These systematic variations are thought to reflect changes in the mass of supernovae as a function of time.

The Galactic Disk

Most stars in the galactic disk have chemical compositions much like that of the Sun. The approximate range in metal abundances in the disk is from 0.25 to 1.5 times that of the Sun. This range roughly parallels time in that the older stars are generally thought of as being more metal-poor. For nonevolved stars in the disk, the elemental abundance ratios are usually solar. For evolved stars (K giants and late-type supergiants for example), the abundances of carbon, nitrogen, and oxygen have been modified by stellar nucleosynthesis and mixing events so that carbon is decreased from the solar value by a factor of 2 with a corresponding increase in the nitrogen abundance. For heavy elements, normal evolved stars have solar abundance ratios.

Peculiar Stars

Among the stars of the disk and halo, there exist groups of stars with peculiar abundances. These stars are explained by either nonstandard evolutionary scenarios or by invocation of transitory phases of stellar evolution. We shall not go into the evolutionary arguments, but merely enumerate the more important classes of these objects and their composition anomalies.

1. The carbon stars: These are cool stars with excess carbon relative to their oxygen content. For the Sun the C/O ratio (the ratio of the abundances of these elements by numbers of atoms) is 0.6, whereas in the carbon stars the ratio is greater than 1.
2. The hydrogen-deficient stars: Helium is the dominant element in these stars; instead of a solar He/H ratio of 0.1 these stars can have ratios of 100–1000.
3. The barium stars: G giants with apparently normal abundances except for the *s*-process elements (e.g., Sr and Ba) which may be overabundant by a factor of 10 relative to the Sun.

STATUS OF CHEMICAL COMPOSITION STUDIES

The determination of stellar chemical abundances continues to be an important and vibrant area of astronomical research. Each new analysis builds on the previous work using better physics, better transition probabilities, and better spectra with the aim of producing more and better stellar abundance data. These studies are today inextricably linked to questions of stellar and galactic chemical evolution as well as the physics of stellar atmospheres.

Additional Reading

Grevesse, N. (1984). Accurate atomic data and solar photospheric spectroscopy. *Physica Scripta* **T8** 49.

Lambert, D. L. (1990). The chemical composition of stars. In *Astrophysics—Recent Progress and Future Possibilities*, B. Gustafsson and P. E. Nissen, eds. Munksgaard, Copenhagen, p. 75.

Mihalas, D. (1978). *Stellar Atmospheres*. W. H. Freeman, San Francisco.

Struve, O. and Zebergs V. (1962). *Astronomy of the 20th Century*. Macmillan, New York.

Wheeler, J. C., Sneden, C., and Truran, J. W., Jr. (1989). Abundance ratios as a function of metallicity. *Ann. Rev. Astron. Ap.* **27** 279.

See also **Star Clusters, Globular, Chemical Composition; Stars, Atmospheres; Stars, Carbon; Stars, Magnetic and Chemically Peculiar.**

Stars, Circumstellar Disks

Steven V. W. Beckwith

A circumstellar disk is a flattened cloud of gas or small particles in approximately circular motion about a star, in which the material velocity is determined primarily by the balance of gravity and centrifugal force. The dimension of the cloud normal to its motion, its thickness, is small compared to its radial extent. A disk may have a sharply defined inner radius or may extend into the surface of the star; the outer radius may also be truncated, or the material density may fall off slowly with increasing radius. The density of matter above and below the plane of the disk is small compared to the density of material in the interior of the disk. Saturn's rings are a well-known example of a particle disk that has the general shape of a circumstellar disk.

Circumstellar disks occur around stars of all ages including exotic objects such as neutron stars and black holes. The disks are generally less massive than the stars they encircle, but usually much larger in size. Whenever large quantities of gas or particulate matter orbit a star, a disk will form in the plane perpendicular to the aggregate angular momentum vector of the material.

Very young stars sometimes have disks as remnants of the star formation process. The stars are born from the collapse of large, tenuous clouds of gas and dust. Slight rotational motions of the cloud material prior to collapse result in the flattened disks around the young stars. Gas with high initial angular momentum (rotational speed) orbits the centrally condensed star at large radii; vertical motions of the gas out of the disk plane are quickly eliminated by collisions with material in the disk. If there is little damping of the orbital motion, a disk may persist long after the star is formed. Eventually, coagulation of the disk material into larger bodies is thought to produce planetary systems. The solar system presumably formed from such a disk. Remnant materials from pre-planetary disks is seen around some main-sequence stars.

Old stars and exotic objects often occur in binary systems. A disk can develop around one of the stars if its companion loses mass, either through the action of a stellar wind or because its atmosphere becomes very extended and spills onto the primary star. The balance of forces near the orbiting stars forces the gas into a circumstellar disk around the primary, and the matter slowly spirals in toward the star. These accretion disks, so named because gas accretes from the disk onto the star, provide a mechanism whereby the kinetic and gravitational energy of the orbiting material can be released as radiation. This energy is thought to be observed as the outbursts from recurring novae, cataclysmic variables, and the emission from some black holes. Some disks emit as much as 10,000 times the power of the Sun entirely in hard x-rays, an enormous amount of energy even by astronomical standards.

Isolated stars may also accrete material from the interstellar medium into small disks. The strong emission from some galactic nuclei and quasistellar objects is believed to result from emission by large accretion disks around massive black holes. The disks are supplied from the interstellar medium and from stars that are torn apart by the strong gravitational forces near the massive objects. Quasistellar objects (QSOs) can have luminosities more than one trillion times that of the Sun; if they are powered by accretion disks around black holes, these accretion disks are the most luminous objects in the universe.

DISK STRUCTURE

The velocity of the disk material at small radii is the Keplerian velocity determined by the star's gravitational field; it decreases as the inverse square root of the distance from the star. For stars without nearby companions, this behavior usually describes the velocity at all radii. The Keplerian velocity v at distance r from a star of mass $M_\star$ is

$$v = \sqrt{\frac{GM_\star}{r}},$$

where G is the universal constant of gravitation. This behavior is modified at large distances when a companion star is present or when the mass of material in the disk is comparable to the mass of the star.

Usually, the disk temperature also decreases away from the star. When the inner radius extends to the stellar surface, a thin boundary layer exists at the star/disk interface, through which material slows abruptly from the orbital velocity to the (much slower) velocity of the stellar atmosphere, releasing energy in the form of heat. The boundary layer is responsible for approximately half the radiation of the disk itself and is the hottest part of the disk. Material at larger radii may be heated by the loss of energy as gas spirals slowly toward the star, by radiation from the star and boundary layer, by the dissipation of magnetic fields, by stellar winds, and by matter falling directly onto the disk. In the special cases where only accretion energy and/or stellar radiation heat the disk, the disk temperature falls as the $-3/4$ power of the radius.

Because the orbital velocity of material decreases with increasing distance from the star, there is a continual shearing or friction in which the material is accelerated by matter at smaller radii which is in turn decelerated in reaction. In an accretion disk, matter flows inward toward smaller radii, and angular momentum flows outward through the action of viscosity. Energy is lost, allowing the material to slowly spiral toward the star. By this means, accretion disks provide a steady flow of material to the star itself, fueling strong radiation from the boundary layer and causing the stellar mass to increase, albeit slowly in most cases. This inevitable transfer of material limits the lifetime of most disks, unless matter is replaced in the disk. In a binary system, disk material is constantly replenished from the atmosphere of the companion. It is presumed that active galactic nuclei and QSOs eventually run out of material and cease to be powerful sources of luminosity.

The material density almost always decreases with distance from the star, the exact variation depending on local physical variables such as viscosity and temperature. Theoretically, the density falls exponentially in the direction orthogonal to the plane of the disk. The exponential scale height, h, the distance at which the density has decreased by a factor of e (the base of the natural logarithms, about 2.7), is approximately equal to the radius times the speed of sound divided by the orbital angular velocity.

OBSERVATIONAL SIGNATURES

Direct images of disks are rare; most disks are smaller than the resolution of the most powerful telescopes. In a few cases, radio interferometric images of the gas show the disks directly. Occasionally, light from above and below the disk is reflected by residual matter falling onto the disk plane, revealing the disks in shadow and confirming the confinement of the densest matter to a plane surrounding the star. Small disks are sometimes revealed during lunar occultations, when the effective angular resolution of the stellar observations is enormously enhanced.

Because the disk material is confined to a plane, the appearance of the star depends strongly on the viewing angle. Viewed from above the plane ("face-on" disk), the star will suffer little or no obscuration, yet the disk may still be seen from its radiated light. Many isolated stars are known to have substantial amounts of circumstellar material, but their light is unattenuated, indicating a disk-like distribution of circumstellar matter that is viewed face-on. Occasionally, a star will be very heavily obscured, yet with a small amount of circumstellar matter, suggesting a strong concentration of the matter into a plane with the star viewed within that plane ("edge-on" disk). Although a disk-like distribution does not uniquely explain these observations, it is sometimes the only plausible means to account for a large number of observations of similar stars without invoking a separate distribution for each star.

Circumstellar disks may give rise to singular spectral energy distributions, patterns of Doppler-shifted emission lines, and time variability of light emission. Most disks are studied indirectly through their signatures on the energy spectrum. If the temperature in the disk is a function only of radius, the radiative power is usually a unique function of the emission wavelength. When the temperature has a power-law radial dependence, for example, the flux density is a power law in wavelength (generally with a different, but well-known, power-law index). Similarly, spectral lines formed in different parts of a disk will normally have Doppler shifts relative to the star which are determined by the orbital velocity. Doppler shifts in a series of spectral lines sometimes vary in a manner that could only come about from gas in orbit, thus making a strong circumstantial case for the existence of a disk. Theories of accretion disks used to explain the x-rays from binary stars and galactic nuclei, the time variations of cataclysmic variables, and the spectra of pre–main-sequence stars gain acceptance through their predictions of these observations.

Additional Reading

Gursky, H. and van den Heuvel, E. P. J. (1975). X-ray-emitting double stars. *Scientific American* **232** (No. 3) 24.

Lynden-Bell, D. and Pringle, J. E. (1974). The evolution of viscous discs and the origin of the nebular variables. *Monthly Notices Roy. Astron. Soc.* **168** 603.

Padman, R., Lasenby, A. N., and Green, D. A. (1991). Jets in the Galaxy. In *Beams and Jets in Astrophysics*, P. A. Hughes, ed. Cambridge University Press, Cambridge, p. 484.

Pringle, J. E. (1981). Accretion disks in astrophysics. *Ann. Rev. Astron. Ap.* **19** 137.

Sellwood, J. A., ed. (1990). *Dynamics of Astrophysical Disks.* Cambridge University Press, Cambridge.

Shu, F. H. (1982). *The Physical Universe.* University Science Books, Mill Valley, CA, pp. 193, 320 and 475.

Shu, F. H., Adams, F. C., and Lizano, S. (1987). Star formation in molecular clouds: observations and theory. *Ann. Rev. Astron. Ap.* **25** 23.

See also **Accretion; Active Galaxies and Quasistellar Objects, Accretion; Binary Stars, Cataclysmic; Planetary Systems, Formation, Observational Evidence; Protostars; Protostars, Theory.**

Stars, Circumstellar Matter

Robert E. Stencel

Circumstellar matter refers to material wholly, or at least partly, within the gravitational sphere of influence of a star. Virtually all stars have a mantle of circumstellar matter left over from the star's birth or expelled during the life of the star. The density of circumstellar matter differs greatly from star to star, depending on its origins. Such matter can be in the proces of accreting, orbiting, or being driven away. The physical and chemical state of circumstellar matter can include ionized or neutral atoms, molecules, and/or solid-state ensembles—such as amorphous particles—called circumstellar dust. Material closest to the star may be part of the stellar outer atmosphere; that furthest away mingles with the local interstellar medium.

All stars have some circumstellar matter, but certain stars can be completely enshrouded by circumstellar matter at times during their evolution. Young protostars and old post–main-sequence stars are examples of this latter situation. Thus, the current state of the circumstellar matter can provide an indication of the star's evolutionary history. The detectability of circumstellar matter varies, but it can sometimes be observed as a reflection nebula, or spectroscopically as additional spectral features (beyond those of a normal star with minimal circumstellar matter). It can also, under certain conditions, be detected by radio telescopes as molecular maser emission.

An important measure of matter in a circumstellar environment is the *crossing time*. The crossing time is defined by the ratio of characteristic size of the circumstellar environment to the characteristic velocity of the circumstellar matter. Sizes can be determined either by images (direct or interferometric), or by inference using spectroscopic data. For example, H_2O masers in red giant stars might be observed to be hundreds of stellar radii above the stellar surface (i.e., at a distance of $\approx 10^{17}$ cm), with an expansion velocity of ~ 10 km s^{-1} (10^6 cm s^{-1}). Assuming linear expansion, the crossing time for material leaving the star and reaching the observed maser radius is 10^{11} s, or nearly 10^4 yr. This number compares well with estimates of the "dwell time" for a star in this phase of its evolution, the duration of the phase. The outer boundary of such shells is often limited by penetration of ionizing radiation from the interstellar medium, which dissociates the molecules and switches off the diagnostic masers.

Another example of a crossing time can be estimated for a hot Wolf–Rayet star, with a 1000 km s^{-1} (10^8 cm s^{-1}) outflow, as measured spectroscopically. In this case, the spectroscopic line profile can be synthesized with a computer model using a stellar-wind acceleration law, wherein the terminal or plateau velocity for the wind is reached within a certain distance (say, at 5 $R_\odot$ or $\sim 10^{13}$ cm). Thus, the crossing time is ~ 1 d. This time scale is consistent with rapid variations seen in the light and spectra of such stars.

Finally, stars intermediate in mass and surface temperature, like the Sun, may possess planetary systems, disks of solid debris and/or cometary clouds—all "leftovers" of the stellar formation epoch. For orbiting circumstellar matter, crossing times are less meaningful, so additional measures such as the total mass and opacity can be estimated from infrared excesses above the star's normal spectral distribution. The first all-sky, far-infrared survey by the *Infrared Astronomical Satellite* (*IRAS*), which was launched in 1983, suggests that at least 10% of main-sequence stars possess substantial post-formation debris disks. Knowledge of the distribution of debris masses may ultimately be correlated with the likelihood of forming companion stars, planets, and/or comets. An interesting question concerns the fate of such disks and cometary clouds once the star evolves away from the main sequence and swells into a red giant star.

A variety of physical and chemical states of matter are possible in circumstellar environments. The surface temperatures of most stars are so high that matter near the surface exists as partially or fully ionized atoms. However, much of interstellar space is extremely cold, so temperatures must eventually decline with increasing altitude above the star. Depending on the output energy from the star, and whether the heating is in the form of radiative heating, acoustic waves, or even magnetic Alfvèn waves, the material injected into the circumstellar environment can remain as an ionized plasma, or in the process of cooling, become molecular and even particulate.

In hot stars, the outflows remain at coronal temperatures (~ 10^6 K). The solar wind also retains its coronal character for huge distances. Still cooler stars, which have too little energy input to develop coronal temperatures, generally exhibit cold circumstellar outflows instead. Once the plasma cools sufficiently, efficient cooling by a few molecules helps create conditions conducive to the formation of more molecules, further enhancing the overall cooling and helping to form still more molecules, in nearly a runaway process. This so-called "condensation instability" is literally a chemical phase change, much like steam condensing into liquid water.

Such condensations occurring within the circumstellar environment also are chemically active, permitting cluster bonding of molecules into amorphous silicates [like the mineral olivine—$(Mg, Fe)_2SiO_4$] in oxygen-rich stars (those where the ratio of carbon to oxygen exceeds unity). These particles coagulate and grow wispy, fractal-like structures in the decreasing density of circumstellar space, and become important constituents of cool-star circumstellar matter. They are accelerated by starlight and, through collisions with the gas, can enhance stellar mass loss by several orders of magnitude. Enhanced mass loss can hasten the demise of a highly evolved star. This dust is also the source for solids in the interstellar medium. Such solids play an important role in subsequent star and planet formation by selectively enriching the interstellar medium with refractory elements, which are tied up in the dust particles themselves. The study of meteorites has yielded particles with isotopic compositions that are unlike any solar system material, but that match the signatures observed in red giant stars and supernova ejecta. Somehow, such aggregates survived the long trip from their stellar origin, through the heat, cold, and shock fronts of the interstellar medium, and ultimately found their way to the terrestrial laboratory.

As previously mentioned, measuring the circumstellar matter can reveal the history of the star, and by extrapolation, something about its future. A dense circumstellar environment suggests the star is undergoing a rapid mass loss phase. Some stars have estimated rates of 10^{-4} $M_\odot$ yr^{-1} where $M_\odot$ is the mass of the Sun, which suggests they cannot continue for more than a few tens of thousands of years without completely evaporating! Such objects are thought to be immediate precursors of so-called planetary nebulae and of supernovae (depending on the original total mass of the star). The analysis of light from the vicinity of Supernova 1987A in the Large Magellanic Cloud suggest the presence of circumstellar matter at a radius of 4.5 pc, as the result of a presupernova red supergiant star wind. The crossing time for a typical 15 km s^{-1} red supergiant wind is about 10^4 yr in this case, suggesting the supernova precursor had recently been a red star. The Hubble Space Telescope should detect the evolution of this (see Figure 2 of the entry on Hubble Space Telescope) and other circumstellar matter with its imaging instruments.

Circumstellar matter is the interface between the star and the rest of the Galaxy. Through that medium, the life story of the star is communicated, for the enrichment and replenishment of the interstellar medium.

Additional Reading

Appenzeller, I. and Jordan, C., eds. (1987). *Circumstellar Matter* (*Proceedings of IAU Symposium 120*). Reidel, Dordrecht.

Chevalier, R. and Emmering, R. (1989). Illuminating a red supergiant wind around SN 1987A. *Astrophys. J. (Letters)* **342**, L75.

Delache, P., Laloë, S., Magnan, C., and Tran Thanh Van, J., eds. (1989). *Modelling the Stellar Environment: How and Why?* Editions Frontières, Gif-sur-Yvette.

Ming, T., Anders, E., Hoppe, P., and Zinner, E. (1989). Meteoritic silicon carbide and its stellar sources; implications for galactic chemical evolution. *Nature* **339** 351.

Nuth, J. and Stencel, R., eds. (1985). *Interrelationships Among Circumstellar, Interstellar and Interplanetary Dust*, NASA Conference Publication No. 2403. U.S. Government Printing Office, Washington, DC.

Tinsley, B. (1980). Evolution of the stars and gas in galaxies. *Fundam. Cosmic Phys.* **5** 287.

Wallerstein, G. (1988). Mixing in stars. *Science* **240** 1743.

See also **Interstellar Medium, Stellar Wind Effects; Planetary Systems, Formation, Observational Evidence; Stars, Circumstellar Disks; Stars, Evolved, Circumstellar Masers; Stars, Pre-Main Sequence, Winds and Outflow Phenomena; Stars, Red Supergiant; Stars, Winds; Stars, Wolf–Rayet.**

Stars, δ Scuti and Related Types

Michel Breger

The extension of the Cepheid instability strip on the Hertzsprung–Russell diagram crosses the main sequence among the A- and F-type stars. Not surprisingly, pulsational variability can also be found in this region. δ Scuti variable stars are short-period pulsators situated on and near the main sequence inside the instability strip with periods from about 30 min up to 8 h (see Fig. 1). The amplitudes of pulsation are usually small, with visual amplitudes near 0.01 mag, but can be as large as 0.8 mag. More than 200 such variable stars have been discovered. The rapidly oscillating Ap stars (which have strong magnetic fields) are also found on the main sequence and are probably related to the δ Scuti variables. These classes of pulsators can be considered to represent a transition between the relatively simple Cepheid-like radial pulsation and the complex nonradial pulsation found in other parts of the Hertzsprung–Russell diagram. For most stars the combination of radial as well as different nonradial pulsation modes leads to complex multiperiodicity. The identification and study of these modes makes these variables ideal probes for the effects of chemical composition, magnetic fields, evolutionary changes, and rotation on stellar structure.

A subgroup of the δ Scuti variables may be astrophysically different from the normal Population I δ Scuti variables: the SX Phe stars (Population II). Attempts have also been made to subdi-

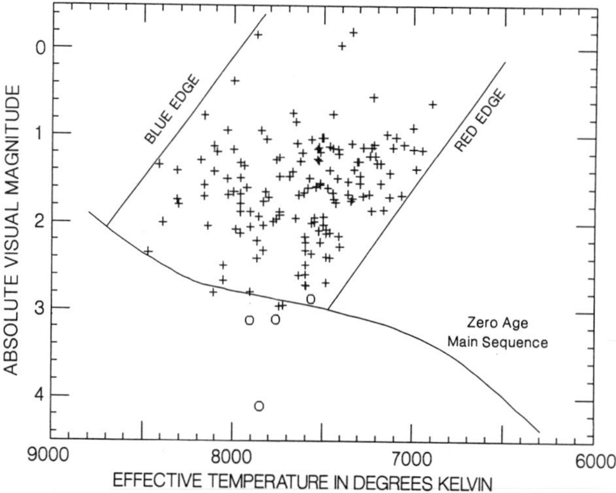

Figure 1. Position of δ Scuti variables in the Hertzsprung–Russell diagram. The open circles denote the Population II SX Phe stars with known distances.

vide the δ Scuti stars according to amplitude. The names "dwarf Cepheid" and "AI Vel stars" refer to the δ Scuti stars with visual amplitudes in excess of 0.3 mag. The name "δ Scuti star c" ("c" for cluster) is used by the *General Catalog of Variable Stars* for the members of the δ Scuti class with amplitudes smaller than 0.1 mag. These designations may not be astrophysically meaningful, as the amplitudes of many of the stars are variable and it has been shown that the amplitudes are not an indicator of evolutionary history.

The pulsation of δ Scuti stars is driven by the κ mechanism operating in the He II and hydrogen ionization zones. For the rapidly oscillating Ap stars, magnetic overstability may be an important excitation mechanism.

PERIODS AND PULSATION MODES

The observations show that the normal Population I δ Scuti stars obey a period-luminosity-color relation of the form

$$M_v = -3.05 \log P + 8.46(b-y)_0 - 3.12,$$

where M_v is the absolute visual magnitude, P is the period in days, and $(b-y)$ is a color index in magnitudes on the Strömgren four-color system, as defined in Table 2, page 406 of the entry, Magnitude Scales and Photometric Systems.

Because most variables of this type actually have multiple periods, the observed relation shows considerable scatter and refers to an "average" period. The observed coefficients agree with predictions from the theoretical period-density relation ($P^2\rho/\rho_0 = Q^2$), where the value of the pulsation constant Q depends on the pulsation mode, for example, for the radial fundamental mode $Q = 0.033$ d.

Because of multiperiodicity the frequencies of pulsation are observationally difficult to determine. Lengthy data sets, often obtained in coordinated campaigns on different continents, are required. The excited radial and nonradial pulsation modes are inferred from the length and ratio of the frequencies, amplitude ratios and phase differences for different wavelengths and radial velocity changes, as well as the study of line profiles. The present state of our knowledge can be summarized as follows:

(i) Some δ Scuti variables pulsate with one or more purely radial modes. These stars are usually luminous, rotate slowly, and have large amplitudes (e.g., AD CMi). Radial pulsators with small amplitudes also exist (e.g., HR 6434).

(ii) The majority of the stars show a complex mix of different nonradial p (pressure) and radial modes. Only a few of these stars, such as 1 Mon and θ^2 Tau, can be considered as observationally "solved." The modes are described by radial, angular, and azimuthal quantum numbers (k, ℓ, m, respectively). Values of k from 0–3, of ℓ from 0–2, and m values from $-\ell$ to $+\ell$ have been found. The study of line profiles has led to suggestions that p modes with much higher ℓ values may also be present. A problem is the presently unknown nature of the "filter," which selectively excites certain modes. The search for this filter requires the observationally difficult determination of the multiple frequencies of more stars in order to study the dependence on stellar properties such as luminosity, temperature, rotation, binary nature, and chemical abundances.

Reports in the literature of frequencies varying over a few years are probably caused by overinterpreting insufficient observations or not considering variable amplitudes. However, among the more luminous δ Scuti variables, slow evolutionary frequency changes on the order of 10^{-11} d d^{-1} are expected. Their observation provides an excellent method to test the calculated evolutionary time scales.

AMPLITUDES

The large range of pulsation amplitudes for different stars provides important clues about the effects of the secondary stellar properties on the structure and pulsation of stars. In fact, about half of the stars in the lower instability strip show no detectable light variations. Statistical analysis of the observations shows that stellar rotation and metallicism inhibit large amplitudes and may even lead to stability. Large amplitudes in excess of 0.1 mag are generally limited to subgiant or giant stars with slow rotation ($v \sin i \leq 30$ km s^{-1} where v is the rotational velocity and i is the inclination of the rotation axis to the line of sight). Among the dwarfs, on the other hand, the slowly rotating classical metallic-line A stars are generally stable or show only very small amplitudes of a few thousandths of a magnitude. The latter effect is usually attributed to diffusion. According to this picture, in a stable atmosphere helium will tend to sink due to gravity. The helium content in the He II ionization zone, which is the primary driving mechanism for pulsation, may be reduced sufficiently to prevent pulsation. A disagreement between the large size of the calculated amplitudes and the small observed amplitudes (called the "main sequence catastrophe") might be resolved if rotation were included in the models.

The large majority of the δ Scuti variables have stable amplitudes associated with each excited frequency. A few stars show amplitude changes on a time scale of years; for example, the nonradial amplitudes of 4 CVn vary by more than a factor of 2 within 15 yr.

SX Phoenisis STARS

A number of δ Scuti variables show the weak metal content and high space velocity typical of Population II stars. The period ratio of the two radial periods of the prototype SX Phe is 0.778, which is significantly higher than the radial period ratio of 0.772 observed for radially pulsating Population I δ Scuti stars. The difference has been successfully explained by the different chemical composition. SX Phe stars have also been detected in the globular clusters ω Centauri and NGC 5053. A common characteristic of metal-poor variable stars is their large amplitude, in excess of 0.3 mag. The period-density relation shows that the mass of SX Phe stars is one solar mass or less, in contrast to the 1.5 to 2.5 solar masses of the Population I δ Scuti variables.

The position of the SX Phe stars in the stellar evolution scheme is still unclear. They are considerably hotter than the main-sequence turnoff points on the Hertzsprung–Russell diagrams of the globular clusters and are too faint to lie on the horizontal branch. Two possible explanations have been proposed: (i) The SX Phe stars are Population II blue stragglers situated in the Hertzsprung–Russell diagram on an extension toward higher temperatures of the observed globular cluster main sequence, or (ii) they are evolved stars evolving towards the white dwarf stage with luminosities much lower than expected for normal horizontal branch stars. At present, the mass determinations of SX Phe stars are not accurate enough to distinguish between the blue straggler and evolutionary hypotheses.

RAPIDLY OSCILLATING Ap STARS

Small-amplitude pulsation has been detected in 14 peculiar A and F stars with high magnetic fields. They pulsate with high-overtone nonradial modes showing extremely short periods between 4 and 15 min. These periods are about a factor of 5 shorter than those of pulsators with normal spectra of the same luminosity and temperature. In most rapidly oscillating Ap stars the amplitude of pulsation is modulated with the rotational period of the star in the sense that the times of maximum pulsation coincide with the times of extrema of the measured magnetic field. The work of Donald W.

Kurtz has shown that the observed amplitude variations can be described by nonradial *p* (pressure) modes with the axis of oscillation aligned with the magnetic axis of the star (oblique pulsator model).

Additional Reading

Breger, M. (1979). Delta Scuti and related stars. *Publ. Astron. Soc. Pac.* **91** 5.

Cacciari, C. and Clementini, G., eds. (1990). *Confrontation Between Stellar Pulsation and Evolution*. Astronomical Society of the Pacific, San Francisco.

Kurtz, D. W. (1988). Multiple-mode and non-linear pulsation in rapidly oscillating Ap stars. In *Multimode Stellar Pulsations*, G. Kovács, L. Szabados, and B. Szeidl, eds. Konkoly Observatory, Budapest, p. 107.

Shibahashi, H. (1987). Rapidly oscillating Ap stars and Delta Scuti variables. In *Stellar Pulsation*, A. N. Cox, W. M. Sparks, and S. G. Starrfield, eds. Springer-Verlag, Berlin, p. 112.

Wolff, S. C. (1983). *A-Stars: Problems and Perspectives*. NASA Spec. Publ. 463, Chapter 6, The δ Sct stars, NASA, Washington, DC, p. 93.

See also **Magnitude Scales and Photometric Systems; Stars, Blue Stragglers; Stars, Cepheid Variable, Dwarf; Stars, Magnetic and Chemically Peculiar; Stars, Nonradial Pulsation in B-Type; Stars, Pulsating, Theory.**

Stars, Distances and Parallaxes

Conard C. Dahn

Observations of remote astronomical objects can usually only be compared with theoretical predictions and thus properly interpreted if the distances to these objects can be deduced. Consequently, astronomers have expended considerable effort to establish reliable distance scales—first within the solar system itself, then extending to the nearest stellar neighbors of our Sun, then on to more remote objects within our Milky Way galaxy, and finally to extragalactic objects reaching to cosmological distances. The dimensions of the solar system are now accurately established from a combination of radar-ranging measures of the inner planets (Mercury, Venus, and Mars) and laser-ranging measures of the Moon, which are then interpreted by a consistent dynamical theory. The fundamental unit of distance for the solar system is the astronomical unit (AU) —formally defined as the radius of a circular orbit in which a body of negligible mass, free from perturbations, would revolve around the Sun in $2\pi/k$ d, where k is the Gaussian gravitational constant. Effectively, the AU is the mean distance between the Earth and the Sun, now measured to be 1.49597870×10^8 km with a formal uncertainty of only 1 or 2 km. This quantity provides the basis for extending our distance scale to the nearby stars via the technique of trigonometric parallaxes.

DIRECT DISTANCE DETERMINATIONS FOR INDIVIDUAL STARS: TRIGONOMETRIC PARALLAXES

Distances to individual stars in the immediate vicinity of the Sun can be measured geometrically by making use of the Earth's annual revolution around the Sun to form a triangle whose two long sides are lines of sight to a star. Observations of the position of a nearby star against the background of more distant stars enables one to solve for the small angular shift caused by the varying perspective. The derived angle π, known as the heliocentric relative parallax and expressed in arcseconds, yields the distance to the star in AU of $206{,}265/\pi$. The technique, although extremely straightforward in principle, is made difficult due to the remoteness of even the nearest stars. The first successful measure of stellar parallax is attributed to Friedrich W. Bessel, who in 1838 reported a value of $\pi = 0''.33$ (compared with the modern value of $0''.289$) for the binary star 61 Cygni. (Because even the nearest stars are very remote, it is no longer convenient to express distances in kilometers or even astronomical units, but to quote parallaxes themselves or, alternatively, to employ units of parsecs—the distance at which a star would have a parallax of 1 arcsecond. Hence, a star's distance in parsecs is simply the reciprocal of its parallax in arcseconds.)

Nearby stars generally exhibit a measurable linear motion on the plane of the sky due to projection of their motion through space relative to the Sun. This annual "proper motion" often exceeds the annual parallactic shift and necessitates that observations carried out for a parallax determination cover at least 1.5 yr to separate the two effects. Typically a parallax determination will make use of 30–50 observations extending over 4–7 yr. Today, making use of modern fine-grain photographic emulsions measured on laser-encoded microdensitometers, electronic detectors such as charge-coupled devices (CCDs), or Ronchi-ruling–modulated photometric systems, relative stellar parallaxes good to between $\pm 0''.002$ and $\pm 0''.001$ (standard error) can readily be measured. However, the accuracy realized for any particular star is heavily dependent upon the availability of suitably located reference stars of appropriate brightness and color in the field of the parallax star. The best results obtained in the U.S. Naval Observatory CCD program to date yield relative parallaxes with formal internal standard errors of approximately $\pm 0''.0006$.

Distances so determined are merely relative to random field stars, which are themselves often located at distances perhaps less than an order of magnitude more distant than the parallax star. Hence, a correction must be applied to transform these relative parallaxes to true or "absolute parallaxes." These transformations are generally carried out using statistical corrections derived from observed star counts modeled to the Milky Way galaxy. Hence, the magnitude of the correction depends both on the brightness of the reference stars and on the direction of the parallax star in space. At apparent visual magnitude $V \approx 10$ these additive corrections range from approximately $0''.0047$ at the galactic pole to about $0''.0023$ at the galactic equator, while at $V \approx 15$ the corresponding values are $0''.0015$ and $0''.0007$. Because they have only statistical validity, application of them to reference frames containing small numbers of stars (typically four or five) involves considerable uncertainty, which then must be reflected by an appropriate increase in the estimated error of the absolute parallax over that formally derived for the relative parallax.

A new compilation combining published trigonometrically determined parallaxes from all sources with appropriate allowance for systematic errors between the contributing observatories is in preparation at the Yale University Observatory. This new catalogue presents results for roughly 7500 stars and will serve as the primary compilation of such distances throughout the 1990s. An inventory made from a prepublication copy of that catalogue (kindly made available by Dr. William F. van Altena) reveals some interesting statistics about the present status of the fundamental distance-scale database. First, an overall perusal reveals that the vast majority of the parallax measures are photographic determinations made before the development of measuring machines capable of automatic, impersonal image centroiding. Consequently, the formal errors of these relative parallax determinations are large compared with those carried out in the past decade or two. A mere 44 of the catalogued stars (counting binaries and multiple stars as single objects) have absolute parallaxes with formal errors $\leq 0''.0020$ and the number with formal errors $< 0''.0030$ totals fewer than 300. For the catalogue as a whole the median error of the absolute parallax is $\pm 0''.0090$, indicating that the database is dominated by determinations that are clearly inferior by today's standards.

A more relevant measure of a parallax's usefulness in applications is the ratio of the uncertainty in the parallax to the parallax itself (ε_π/π). For example, the uncertainty in a star's absolute magnitude due solely to the parallax determination employed and its associated uncertainty is equal to 2.17 (ε_π/π). Likewise, the uncertainty in a tangential velocity (V_t, in kilometers per second) due to the parallax uncertainty alone will be $(4.74\ \mu/\pi)(\varepsilon_\pi/\pi)$, where μ is the total proper motion expressed in arcseconds per year. A survey of the new Yale catalogue reveals only 24 stars (or systems; amounting to a total of 33 confirmed components) with $\varepsilon_\pi/\pi \leq 0.01$.

The number of stars (systems) with $\varepsilon_\pi/\pi \leq 0.03$ still totals only 155 (with roughly 200 confirmed components). Due to the stellar luminosity function in the solar neighborhood, the 3% sample is heavily dominated by lower-luminosity main-sequence stars. Fully 144 (73%) of the presently identified components are dwarf M stars with $M_V > 8.4$. If we include the 13 dwarf K stars, lower main-sequence K and M stars constitute nearly 80% of the $\varepsilon_\pi/\pi \leq 0.03$ sample available today. The remaining 20% of this sample is made up by (a) 11 white dwarfs (most relatively cool), (b) 8 main-sequence stars with spectral types A, F, and G (including such well-known bright stars as α CMa, α Gem, α Cen AB, and α Aql), and (c) two subgiants (α CMi and ζ Her). Noticeably absent are all of the higher luminosity stars—OB stars, supergiants (especially the Cepheid variables), KM giants, carbon stars, and RR Lyr variables—important for extending the distance scale to more remote regions of the Galaxy and beyond.

At the 10% quality level, the corresponding uncertainty in absolute magnitude is ± 0.22 mag, making a parallax determination for an individual star of questionable value for setting distance-scale zero points. Such lower-quality determinations could, in principle, provide a useful statistical zero point if sufficient numbers of appropriate objects were available. That is, unfortunately, seldom the case. The two most intrinsically luminous stars among the 1000 individual stars (or system components) with 10% or better trigonometric parallax are γ Gem (absolute visual magnitude $M_V = -0.38$; A0 IV) and β Per ($M_V = -0.18$; B8 V). Since the most luminous stars—the all-important Cepheids and the other supergiants—are not represented in the near-solar neighborhood, alternatives to trigonometric parallaxes must be utilized.

DIRECT DISTANCE DETERMINATIONS FOR STAR GROUPS: MOVING-CLUSTER PARALLAXES

The moving-cluster parallax method provides a geometrical technique for determining the distance to a nearby stellar cluster. Since high-luminosity stars including Cepheids exist in several remote open clusters and associations, the distance scale can be extrapolated through a series of cluster main-sequence fittings. The application of the moving cluster method to the Hyades open cluster provides the keystone for this extension.

The basis for the method is the assumption that individual members of a physical cluster of stars moving through space have essentially parallel space-velocity vectors—that overall expansions, contractions, or rotations are small, as are the random motions of the stars with respect to the center of mass of the cluster. The measured proper motions of such a cluster moving through the solar neighborhood will appear either to converge or diverge, depending upon whether the cluster possesses a radial velocity component away from or toward the Sun. The angular separation λ on the sky between each individual cluster member and the convergent (or divergent) point is simply related to the star's parallax, proper motion, and radial velocity (V_r, in kilometers per second) via the relation $\pi = 4.74\ \mu\ (V_r \tan\lambda)^{-1}$.

An alternate analysis of the same proper motion and radial velocity data is possible without solving explicitly for the cluster convergent point. Here one equates (1) the fractional time rate of change of the cluster's angular diameter caused by the cluster's radial component of motion relative to the Sun, with (2) the negative of the fractional time rate of change of the cluster's distance. The former can be derived independently from either of the proper-motion gradients across the cluster (i.e., from either $d\mu_\alpha/d\alpha$ or $d\mu_\delta/d\delta$ where α and δ are right ascension and declination, respectively), and the latter is simply the mean radial velocity of the cluster divided by the distance. The independent solutions provided by the motion components in both celestial coordinates yield a further check on possible systematic errors in the measured proper motions.

Modern applications of the above two analyses yield a weighted mean parallax for the Hyades cluster of $0''.0218 \pm 0''.0006$. Combined with direct trigonometric parallax determinations for cluster members made primarily at the Lick and Van Vleck Observatories, one arrives at a weighted, geometrically determined distance corresponding to $\pi = 0''.0219 \pm 0''.0005$ for the Hyades.

Successful application of the moving cluster method is predicated upon having a well-populated cluster sufficiently nearby that the members cover a significant solid angle on the sky. Unfortunately, only the Hyades cluster adequately satisfies these requirements. Attempts at applying the method to other clusters (e.g., the UMa group and the Sco-Cen group) have been only moderately successful.

Extensions of the distance scale by main-sequence fitting of the Hyades to more remote open clusters and associations are discussed in the entries on Stars, Hertzsprung–Russell Diagram and Galactic Structure, Optical Tracers.

INDIRECT DISTANCE DETERMINATIONS: PHOTOMETRIC AND SPECTROSCOPIC PARALLAXES

The trigonometric parallax and moving cluster methods have been designated as direct determinations, because they are based almost entirely on geometrical principles and do not require assumptions about the physical nature of the stars being measured. Indirect methods—ultimately calibrated by the results obtained from the direct determinations—often represent the only available distance estimates.

After trigonometric parallaxes, calibrated relations between intrinsic luminosity (M_V, M_{bol}, $\log L/L_\odot$, etc.) and various photometric colors or spectroscopic classes provide the most useful and reliable distance estimators for individual stars. Well-calibrated photometric relations and accurate photometry can provide absolute magnitudes good to ± 0.2–± 0.4 mag for main-sequence or degenerate-sequence objects. Avoidance of misinterpretations due to unresolved duplicity, chemical abundance differences, or interstellar absorption (and reddening) can often be avoided by the use of multiple colors spread over as wide a range of wavelength as possible and by the use of proper-motion data (or limits) for implied tangential velocity constraints. Spectroscopic parallaxes rely on the assignment of stars to certain discrete classes based on atomic or molecular features. High signal-to-noise spectra offer an advantage over broadband photometric techniques for distant high-luminosity stars and, with appropriate resolution, avoid the pitfalls due to unresolved duplicity and interstellar reddening. The discrete nature of established spectroscopic classes makes photometric parallaxes preferable for lower main-sequence and cooler degenerate stars. Obviously, for very faint objects the scarcity of photons often dictates the use of a photometric distance rather than a spectroscopic one.

INDIRECT DISTANCE DETERMINATIONS: STATISTICAL METHODS

A variety of methods has been applied to classes of single field stars to derive mean absolute luminosities for the class as a whole. Included are the methods of "secular" and "statistical" parallaxes, the details of which are beyond the scope of this entry. Basically, both make use of the Sun's motion relative to the group of stars under consideration to effectively extend the baseline of 2 AU provided by the Earth's orbit. The solar motion with respect to relatively nearby members of these groups amounts to about 20 km s^{-1}, corresponding to 4.2 AU yr^{-1}. Hence, proper motions measured over extended time intervals (e.g., 20 yr and longer) can effectively utilize a greatly extended baseline to provide a mean distance to the group members. Although these methods have been successfully applied to such high-luminosity objects as long-period variables, supergiants, carbon stars, and RR Lyr variables, rapidly improving direct trigonometric techniques combined with improved extension of cluster main-sequence fitting are relegating such methods to secondary importance.

SPACE ASTROMETRY AND THE FUTURE

The failure of *HIPPARCOS* (the European Space Agency's astrometric satellite) to achieve its intended geostationary orbit following launch on August 9, 1989 represented a severe blow to the field of direct trigonometric parallax determinations. The planned 2.5-yr lifetime of the mission was projected to yield absolute trigonometric distances for some 120,000 stars, including all stars brighter than $V = 7.3$ and reaching as faint as $V = 12.4$ for selected objects. The estimated accuracy was to have been $\pm 0''.002$ for stars down to $B \approx 9$—similar to the best ground-based photographic determinations. However, preliminary simulations indicate that parallaxes obtained from *HIPPARCOS* in its elliptical orbit will be degraded to the $\pm 0''.004$–$\pm 0''.006$ range, depending on the lifetime of the satellite.

Shortly following launch in April 1990, the primary mirror of the *Hubble Space Telescope* (*HST*) was discovered to have been figured with serious spherical aberration. Furthermore, telescope oscillations induced by thermal effects during terminator crossings severly degrade the capability to maintain precise pointing. Although *HST* is not primarily an astrometric mission, it possessed the design capabilities for measuring trigonometric parallaxes of stars with $4 < V < 17$ to a projected accuracy of $\pm 0''.0005$. Among the limited parallax determinations scheduled were a selection of subdwarfs of spectral types F and G covering a range of metallicity. Such objects would provide important calibration standards for measuring the distances to globular clusters. However, it now appears that these *HST* parallax observations must be deferred until the optical aberrations and pointing instabilities have been corrected.

As the accuracy of relative trigonometric parallax determinations —both from space and from the ground—press below the milliarcsecond level, the contribution to the uncertainty in distance from the correction to absolute parallax becomes an increasing limitation for the technique. At some point it will become important to determine the distances to individual reference stars via spectrophotometric observations rather than relying on statistical estimates.

Additional Reading

Gondhalekar, P. M., ed. (1989). *Astrometry: Into the 21st Century*. Report RAL-89-117. Rutherford Appleton Laboratory, Chilton.

Hanson, R. B. (1980). The Hyades cluster distance. In *Star Clusters, IAU Symposium 85*, J. E. Hesser, ed. D. Reidel, Dordrecht, p. 71.

Mihalas, D. and Binney, J. (1981). *Galactic Astronomy*. W. H. Freeman, San Francisco.

van Altena, W. F. (1983). Astrometry. *Ann. Rev. Astron. Ap.* **21** 131.

van de Kamp, P. (1981). *Stellar Paths*. (Astrophysics and Space Science Library, Vol. 85). D. Reidel, Dordrecht.

See also **Astrometry; Astrometry, Techniques and Telescopes; Coordinates and Reference Systems; Distance Indicators, Extragalactic; Galactic Structure, Optical Tracers; Hubble Space Telescope; Star Clusters, Stellar Evolution; Stars, Cepheid Variable, Period-Luminosity Relation and Distance Scale; Stars, Hertzsprung–Russell Diagram; Stars, Proper Motions, Radial Velocities, and Space Motions.**

Stars, Evolved, Circumstellar Masers

Harm J. Habing

Cosmic masers are small sources in the sky that emit very intense radiation in an isolated spectral line. By "small" we mean that the diameter is less than 1 arcsec. "Very intense" implies that the amount of energy received within the maser line is more than that received from a very hot body of similar size, namely, one with a temperature of 10^{12} K. Such temperatures are expected only in the earliest seconds of the universe, whereas the cosmic masers are definitely present-day objects. The conclusion is that the cosmic sources cannot produce this line emission by normal or thermal equilibrium processes and that nonthermal processes are required. An important consequence is that the radiation may be (and sometimes is) polarized; in turn, when polarization is measured this is often taken as support for a nonthermal origin of the radiation. The most likely radiation processes are similar to those in laboratory masers and hence the name: cosmic masers. Yet, although the fundamental processes in cosmic and laboratory masers are similar, the more detailed conditions differ so much that one cannot apply the theories about laboratory masers in any detail to those of cosmic masers; nature has its own rules. The first discovered maser lines belong to hydroxyl (OH); at about 18-cm wavelength OH has four different lines that are very close together. Sometimes one and sometimes three OH lines were observed in a most irregular fashion. Other maser molecules that were later found in cosmic sources are H_2O, SiO, CH_3OH, and recently HCN. (See Table 1, page 415, in the entry on Masers, Interstellar and Circumstellar.) This list is certainly not yet complete; new discoveries are to be expected. All of the known maser lines are at centimeter and millimeter wavelengths.

There are three main categories of cosmic masers: (i) masers occurring in the immediate surroundings of highly evolved, dying stars (the so-called "circumstellar masers"), (ii) masers in the immediate surroundings of recently formed stars (usually, but less precisely, called "interstellar masers"), and (iii) "megamasers" discovered in distant, active galaxies. This entry concentrates on circumstellar masers.

In spite of the extremely high intensity of the radiation from a circumstellar maser, there is relatively little energy contained in the lines: The intensity occurs only over a very small bandwidth. In circumstellar shells, maser lines carry less than one billionth of the energy of the source; most of the source energy is usually emitted in the infrared.

WHY STUDY COSMIC MASERS?

Maser sources are studied for a variety of reasons: (i) The basic processes are complex and, except in broad outline, unknown. This is a challenge by itself: How does nature do it? (ii) The radiation is so strong that masers can be found at large distances; they are easily detected signposts of interesting objects such as young stars or very old stars. (iii) Because of the great intensity, the sources can be studied with all the advanced techniques that radio spectroscopy and radio interferometry offer: Maps have been obtained showing details of a few milliarcseconds in size. An example is shown in Fig. 1.

Because spectral lines are involved, Doppler shifts, and thus radial velocities, are measured. The combination of detailed maps and velocity information sometimes provides such a detailed pic-

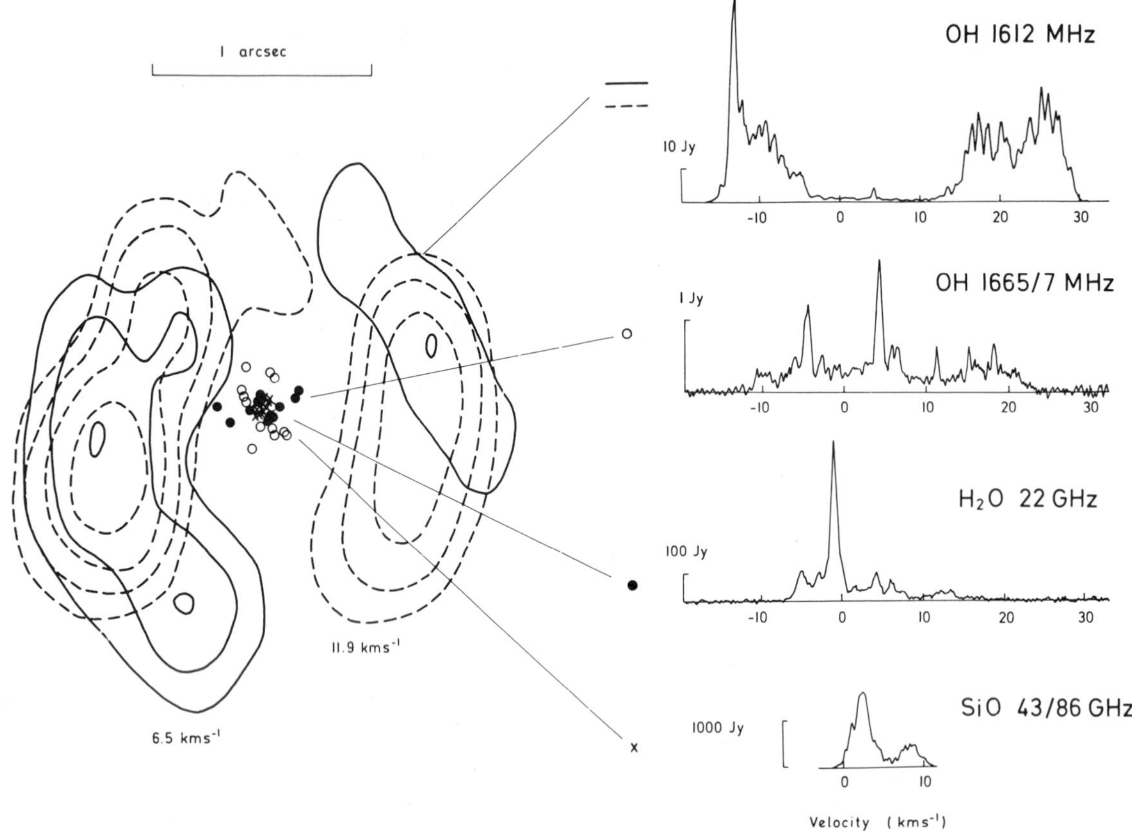

Figure 1. Left: A detailed map of the distribution of maser emission around the red supergiant star VX Sagittarii. Right: The line profiles of maser emissions of OH, H_2O, and SiO. The symbols on the map represent the spots where the maser emission from SiO (×), H_2O (black dots), and the so-called "main lines" of OH (1665–1667 MHz [small circles]) are observed. The strongest emission is from the 1612 MHz "satellite" line of OH; this emission has a wide distribution (shown by isophotes). The position of the star VX Sgr on the map is poorly known; presumably it is at the center. Notice at the top of the map the scale that corresponds to 1 arcsec. [*From Chapman, J. M. and Cohen, R. J. (1986),* Mon. Not. R. Astron. Soc. ***220*** *513*.]

ture of the object that one can derive the distance to the object by purely geometric arguments. Geometric distance determinations are rare in astronomy and very precious, more so when the distance measured is larger than a few hundred light-years. Results already obtained on circumstellar OH masers and on interstellar H_2O masers promise that in the next 10 or 20 years we may measure geometric distances to the center of our Galaxy and even to other galaxies.

HOW MASERS WORK

In a simplified way masers work as follows. Molecules and atoms can absorb energy, and release it at a later time. Quantum physics decrees that the amount of energy is fixed in discrete parcels (quanta). For simplicity, assume a molecule that can absorb and emit only one amount of energy, E; it then contains either zero or E energy. Let there be n_E molecules with energy E and n_0 molecules with zero energy in each cubic centimeter. In a gas there are many processes at work on a microscopic scale, even when the gas looks quiet on a macroscopic scale. Continuously, molecules absorb quanta of energy E and get rid of it. On average, however, n_E and n_0 will be constant. Assume that the gas is at a uniform temperature and has a uniform density; it is then said that the gas is in thermal equilibrium. By very general statistical means one can calculate n_0 and n_E and find that whatever the temperature may be, it is always the case that $n_0 > n_E$. Now consider what happens when a beam of light, that is, a beam of photons with energy E, enters this gas. Once a photon collides with a molecule one of two events can happen: If the molecule has no energy, it will absorb the photon; if the molecule has an energy E, the incoming photon triggers the molecule to eject another photon with energy E. This photon ejection is called stimulated emission; it thus leads to the creation of a new photon (and is sometimes called negative absorption). In a thermal gas $n_0 > n_E$, and thus there are always more atoms that can absorb than can emit. The net result is absorption. A beam of photons passing through a thermal gas will lose photons on its way and its intensity will decline exponentially.

Next consider a gas that is not thermal; an example is a cool gas subjected to radiation from a much warmer background source. Then it may happen, although rarely, that for some molecules and for some energy levels $n_E > n_0$. This is called a population inversion. When a photon enters such a gas, stimulated emission dominates over absorption and every incoming photon will set free many others. Therefore, a beam of photons entering the gas will gain in number and will grow exponentially in intensity. The intensity can reach values not attainable under thermal conditions. This is the maser situation. Some reflection on this situation will show that the energy added to the beam comes from the process that inverts the population; this process is called "the pump." A maser is thus a contraption or a process that converts other kinds of energy (the pump) efficiently into energy of one specific wavelength. In cosmic masers the pumps have not yet been identified with certainty, although suggestions have been made. Definitive conclusions probably will depend on future detailed spectroscopic studies in the infrared.

OBSERVATIONS OF COSMIC MASERS

The first masers were discovered around 1965 in regions of star formation and were thus interstellar: A maser in the Orion nebula serves as a good example. A major step forward came in 1967 when the position of the Orion maser was found to coincide very precisely with that of a newly discovered infrared star, the so-called Becklin–Neugebauer object or BN. The infrared star discovery was the product of the first survey of the sky at the wavelength of 2.2 μm. This coincidence between infrared point source and maser turned out to be typical: All maser sources coincide with strong infrared point sources. BN is a very young star, still embedded in the gas from which it originated. In the catalogue of the 2.2-μm survey, BN is an exception, because the large majority are very old stars.

The first circumstellar maser was discovered in 1968 when William J. Wilson and Alan H. Barrett started to search for OH maser emission in some of these old stars. Subsequently many searches have been made. These include successful, recent searches for maser emission that were based on the Point Source Catalogue obtained with the *Infrared Astronomical Satellite* (*IRAS*), the U.S.A./Netherlands/U.K. infrared survey spacecraft that operated in 1983. As a result we now know of more than 1500 circumstellar maser sources.

MIRA VARIABLES AND OH/IR STARS

There are three major groups of evolved stars with circumstellar shells: those in which the chemistry is dominated by reactions starting with oxygen and hydrogen (oxygen-rich stars, such as Mira variables), those where the reactions start with carbon and hydrogen (carbon stars), and a small intermediate group in which oxygen and carbon play an equal role (SC stars). OH, H_2O, and SiO masers occur only in oxygen-rich stars. Recently, HCN masers have been discovered in carbon stars. The relation between the three groups of stars is not yet understood. The most likely explanation is that all stars initially are oxygen-rich, but that they may become carbon stars during some moment in their evolution when carbon, which has been produced by nuclear fusion in the center of the star, is mixed through the rest of the star during a short event known as a thermal pulse.

The various searches for circumstellar masers have revealed several general trends. Circumstellar masers are mostly associated with variable stars of the longest periods, those exceeding 300 d. Such stars are actually a subclass of the so-called Mira variables. Stellar matter flows away from the star at a modest velocity (15 km s^{-1}) and at a high mass loss rate that nevertheless varies from star to star (from one solar mass in ten million years to one solar mass in ten thousand years). Because the star and the outflowing gas are cool, a small fraction of the gas condenses into small solid particles, called dust. This dust absorbs some of the stellar light and radiates it away in the infrared. As a result, all stars with maser emission have a certain amount of "extra" radiation in the infrared. This infrared excess is proportional to the total amount of dust and gas that recently has moved away from the star. The amount determines rather precisely what maser line is present and what the relative strengths of the various maser lines are (of SiO, of H_2O, and of OH).

The search for masers in redder and redder stars has led to a surprise, namely, the discovery of circumstellar masers without an optically visible counterpart. Ultimately a counterpart to the maser was always found, but it was totally invisible to the eye, a pure infrared source. Figure 2 shows the spectrum of such a star and compares it to that of Betelgeuse (Alpha Orionis), one of the reddest stars visible to the naked eye. Such infrared objects, called OH/IR stars, still contain a long-period variable inside their thick dust, as repeated observations of many of these objects have shown. Their total radiation varies with a large amplitude (a factor of 5) and with periods between 500 and 2000 d, similar to, but even greater than, the periods of the Mira variables. (The fact that the periods are about a factor of 2 longer than those of the Mira stars may point to a fundamental change in the structure of the underlying long-period variable star.) These OH/IR stars lose matter at the highest rates mentioned above: one solar mass in 10,000–100,000 yr. It is obvious that such an outflow will have dramatic consequences for a star when it continues for several thousands of years. How long the process continues can be estimated from the number of such stars existing at any time in our Galaxy. The number of candidate objects in the *IRAS* catalogue suggests that there are several tens of thousands, and thus it is concluded that all stars (except the most massive ones) lose practically all their mass in this (now final) outflow. This discovery—that there is a critical phase in the evolution of a star, a phase that has escaped us until recently because the star is "invisible" during this period—solves some long-standing problems in astronomy. The first is that only a few stars can become supernovae at the end of their lives: If all stars did this, there would be many more supernovae than we observe. The existence of the OH/IR stars proves that only the stars of the highest masses can become supernovae, and there are only a few of those.

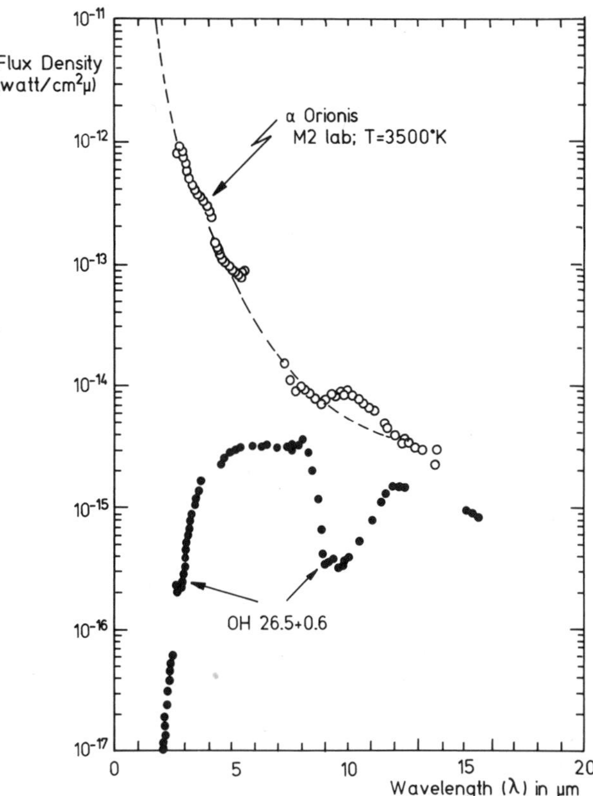

Figure 2. The energy distributions of two evolved giant stars. Betelgeuse (Alpha Orionis) is the brightest star in the constellation Orion. It is a red supergiant of spectral type M2. The dashed line corresponds to the spectrum that the star would have without extra infrared radiation from the circumstellar dust shell; notice the observed excess radiation around wavelength 10 μm. The other object, OH26.5 + 0.6, is an OH/IR star discovered independently as an OH maser and as a strong infrared point source. In its spectrum there is a local minimum at 10 μm; this is caused by absorption by the same type of material that produces the infrared excess in Betelgeuse.

The second problem needs a bit of introduction. It is clear what will remain behind after the star has bled away its outer layers: a core consisting of carbon and oxygen. Actually, this core is the "ash heap" from the nuclear fusion processes that kept the star alive previously. After the star has lost all matter outside of the core, this ash heap will emerge, glow for some time, and then fade away. When we observe a white dwarf star we are examining such an ash heap. The problem mentioned above is that the mass measured for white dwarfs is much smaller than that of the dying red giant stars; white dwarfs measure on average 0.6 solar mass, and the dying stars have masses of at least 1.0 solar mass. Where did the difference go? The discovery of the OH/IR stars has solved that riddle: The difference was slowly ejected during the hidden phase.

A third point of consideration is that planetary nebulae are now explained. When the mass loss stops, the stellar core emerges. At first the core is so hot that it will ionize the little bit of the outflowing gas that still lingers around the core. Thus a planetary nebula is born. When the core becomes too cool to ionize the surrounding gas (it cannot produce new energy), the planetary nebula fades from sight and the white dwarf remains.

A NEW KIND OF CIRCUMSTELLAR MASER SOURCE

The large majority (say, 90%) of circumstellar masers are thought to reside in a spherical, expanding shell around a pulsating red giant star. A few percent may be associated with supergiants. In recent years a new category of circumstellar masers has been found that again probably contains not more than a few percent of all sources. In this category the outflow is not spherical, but rather bipolar—similar to the flows often seen around very young stars. Yet this newly discovered category definitely consists of very old, evolved stars. The objects are presently under study; at the moment they are thought to be transition objects between asymptotic giant branch stars (such as the long-period variables) and planetary nebulae. Such bipolar transition objects are expected to exist: The morphology of planetary nebulae makes it clear that most have not been formed in a spherical fashion, but rather aspherically.

Additional Reading

Cohen, R. J. (1989). Compact maser sources. *Rep. Prog. Phys.* **52** 881.

Habing, H. J. and Neugebauer, G. (1984). The infrared sky. *Scientific American* **251** (No. 5) 48.

Reid, M. K. and Moran, J. M. (1988). Astronomical masers. In *Galactic and Extragalactic Radio Astronomy*, G. L. Verschuur and K. I. Kellermann, eds. Springer-Verlag, Berlin, p. 255.

Shepherd, M. C., Cohen, R. J., Gaylard, M. J., and West, M. E. (1990). OH-IR stars as precursors to protoplanetary nebulae. *Nature* **344** 522.

See also **Masers, Interstellar and Circumstellar; Masers, Interstellar and Circumstellar, Theory; Nebulae, Planetary, Origin and Evolution; Stars, Carbon; Stars, Long-Period Variable; Stellar Evolution, Intermediate Mass Stars.**

Stars, Hertzsprung–Russell Diagram

Robert F. Garrison

Ejnar Hertzsprung in 1911 used a diagram to illustrate the differences between giant stars and dwarf stars in clusters. Henry Norris Russell in 1913 independently presented a diagram illustrating the relationship between absolute magnitude (ordinate) and spectral type (abscissa) for all stars with parallax measurements. Similar diagrams are referred to by astronomers as Hertzsprung–Russell or HR diagrams. An example of a standard HR diagram, for the nearby Hyades open cluster, is shown in Fig. 1.

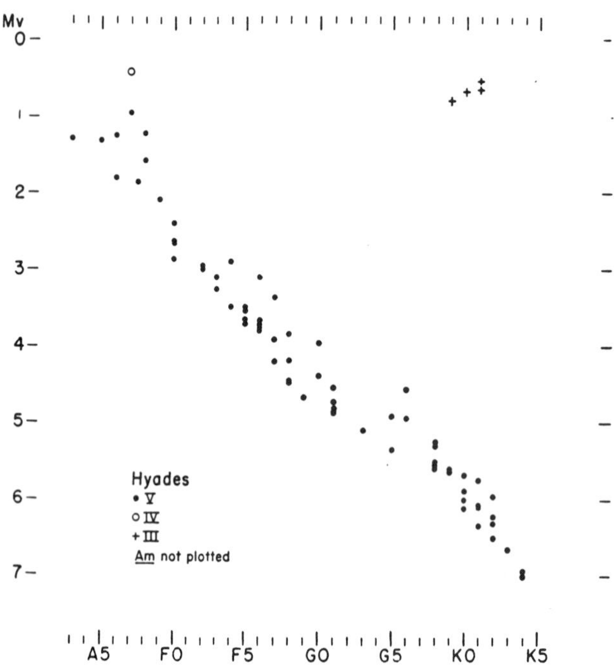

Figure 1. The HR diagram for the Hyades star cluster. [*From an article by W. W. Morgan and W. A. Hiltner (copyright 1965),* Astrophysical Journal.]

Astronomers have found the HR diagram very useful for illustrating most of the important aspects of stellar astronomy. The main reason for its significance is that the absolute magnitude (luminosity, surface gravity) and spectral type (temperature, color) are observationally determined parameters that are closely related to the fundamental parameters of mass, age, and chemical composition, which in turn determine the structure and evolution of a star. Just by looking at an HR diagram, an astronomer can estimate the mass and evolutionary state of a star, as well as many of its other characteristics.

DESCRIPTION OF THE HR DIAGRAM

The absolute magnitude is a measure of the intrinsic brightness (luminosity) of a star; it is equivalent to the apparent brightness referred to a standard distance of 10 parsecs (pc) [32.6 light years (ly)]. In the HR diagram, the luminosity increases from bottom to top (as the numerical value of the absolute magnitude decreases).

Spectral types originally were assigned letters in alphabetical order according to the decreasing strength of the hydrogen absorption lines; they are most prominent in A-type stars and are the only lines that persist throughout the entire sample of stars. It soon became apparent, however, that the highest stellar temperatures did not correspond to the stars with the strongest hydrogen lines. Thus, the spectral sequence in the HR diagram was reordered according to temperature, resulting in a nonalphabetical order with temperature decreasing from left to right. The spectral sequence, in order of decreasing temperature, is now OBAFGKM, with subdivisions 0–9; the maximum strength of the hydrogen lines occurs at A0. Examples of A stars with strong hydrogen lines are Sirius and Vega, the two brightest stars in the sky. Spectra of various stellar types are illustrated in the entry on Stars, Spectral Classification.

Because luminosity increases from bottom to top and temperature increases from right to left, stars in the upper left of the HR diagram are very hot and large, but quite uncommon. Those in the upper right are cool, very large (up to a few thousand times the radius of our Sun), and extremely rare. Stars in the lower right are cool and small, about a tenth the radius of our Sun, and very common. Located in the extreme lower left (often beyond the limits of a particular HR diagram) are the white dwarfs, which are quite hot and about as small as the Earth itself. Black holes and neutron stars are tiny (between 0 and 20 km in diameter) and emit either no radiation or very peculiar radiation unrelated to temperature in the normal way; so they are not represented on HR diagrams.

The distribution of stars in the HR diagram is not random; nor are the stars distributed uniformly. The majority of stars lie on the main sequence, a diagonal line extending from the large, hot stars in the upper left to the small, cool stars in the lower right. This locus of points corresponds to the hydrogen-fusion phase of the life of each star. All normal stars spend the major portion (about 90%) of their lives near the main sequence; thus the main sequence is the most populated region of the HR diagram.

The stars in the upper-right part of the diagram (red supergiants) are very rare, but because they are so luminous, they are visible from a larger distance than others. Several of them are among the very brightest stars in the night sky (e.g., the apparently brightest star in each of Orion, Scorpius, and Hercules is a cool supergiant). Conversely, the stars in the lower right are the most numerous of all, but they are so intrinsically faint that not one of them is visible among the brightest stars.

STELLAR CONTENT OF GALAXIES AND THE HR DIAGRAM

This leads to an interesting puzzle for galaxy observers, who would like to know the mass-to-light ratios for different kinds of galaxies. The small stars contribute much mass but little light; the opposite is true for the large stars. The HR diagram for nearby stars (a volume-limited sample) is quite different from that for the brightest stars (a magnitude-limited sample). The former diagram is very heavily populated in the lower-right portion, whereas the latter contains very few, if any, of the small, cool stars (red dwarfs) in the lower right. It is not a simple matter to decide whether the luminous, but very rare, red giants and red supergiants or the faint, but numerous, red dwarfs dominate in a galaxy, where hundreds of billions of stars contribute to the integrated visible light.

In order to determine the masses and mass-to-luminosity ratios for galaxies, it is important to know or to be able to deduce which of the various stars in different parts of the HR diagram contribute the most light. From a study of the spectrum of the integrated light, William W. Morgan was able to show in the late 1950s that, although the red dwarfs are most numerous, the GK giants (giant stars of spectral types G and K) contribute the most light in the blue-violet region of the spectrum. The study of the stellar content and chemical composition of galaxies remains a topic of great interest in astronomy today.

OTHER USEFUL FORMS OF THE BASIC HR DIAGRAM

There are several modifications of the basic HR diagram. The most important are the color-magnitude diagram and the temperature-luminosity diagram. The former is a plot of absolute magnitude (or visual magnitude for a cluster of stars, the members of which are at the same distance) against the color, which, like the spectral type, is related to the temperature. (Subtracting the visual magnitude from the blue magnitude, as in $B - V$, gives an indication of color of a star; cool, red stars have very different $B - V$ from hot, blue stars. A quantity like $B - V$ is called a color index.)

The temperature-luminosity diagram is also referred to as the "theoretical HR diagram," because the parameters are not directly observed, but are deduced from observational data using theoretical calibrations. It is on this diagram that theoretical models of stellar evolution are usually plotted, allowing a very clear illustration of the evolutionary paths of stars. A variation of this diagram, first suggested by Morgan, is a plot of the logarithm of the temperature against the logarithm of the surface gravity.

STELLAR DISTANCES AND THE HR DIAGRAM

Distances must be known before stars can be plotted on an HR diagram, because actual luminosities or absolute magnitudes are required. The most direct method of distance determination is trigonometric parallax. However, trigonometric parallaxes can only be used for the nearest stars, because the accuracies of the derived absolute magnitudes are strongly distance dependent.

Other methods of distance determination have been developed, but by far the most powerful one is the method of spectroscopic parallaxes. This technique involves the careful comparison of the spectrum of a star, the distance of which is required, with the spectrum of a standard star of known distance. It is assumed that two stars with identical spectra will have identical size and temperature, and therefore identical luminosity. The development by Morgan and Philip C. Keenan of the MK system of spectral classification enabled the evolution of more sophisticated modern spectroscopic parallax methods that allow the determination of distances with relatively high precision. The MK system is virtually the only system of spectroscopic parallax determination in use today.

STAR CLUSTERS AND THE HR DIAGRAM

HR diagrams for star clusters have several advantages over those for stars in the general field; for example, the stars in a cluster are more or less equidistant, coeval, and chemically homogeneous, at least initially. Provided that foreground and background stars have been eliminated, and that the absorption by interstellar dust is either uniform or has been corrected for, the cluster members will populate a portion of the HR diagram that corresponds to a given age for the entire cluster. It is usually assumed that all the members of a cluster formed at the same time and have the same chemical composition, so differences in the positions of stars in the HR diagram of a particular cluster can be ascribed primarily to differences in mass and angular momentum.

The HR diagrams for clusters like the Hyades (Fig. 1) and 47 Tucanae (Fig. 2) are quite dissimilar. The differences are due to age and initial total mass; the Hyades is a relatively young open cluster (less than a billion years old) and contains only a few hundred stars, whereas the globular cluster 47 Tuc is very old (more than ten billion years) with many tens of thousands of member stars. The initial chemical composition also affects the appearance of the HR diagram, though less dramatically.

CLUSTER-FITTING METHOD OF DISTANCE DETERMINATION

Because the variables of distance, age, and chemical composition are constant for a given cluster, the HR diagram provides another powerful technique for determining the distances to other clusters. One can determine the distance to uncalibrated clusters by comparing their main sequences with that of a standard cluster of known distance and shifting the standard main sequence up or down to fit (assuming that stars of the same spectral class, color, or temperature have the same intrinsic luminosity; Fig. 3).

Not only is this technique essential for determining distances to remote clusters, but some types of stars are so peculiar or uncommon that distances cannot be determined in the usual ways. If they are located in a cluster, and if the distance to the cluster can be determined by cluster fitting, the distance to the peculiar or uncommon cluster star is then known. As none of the hot, bright, young stars (even those on the upper left of the main sequence) are nearby, cluster fitting has allowed astronomers to derive distances for them. Such bootstrapping techniques are sometimes the only ways in which astronomers can determine the basic parameters of stars.

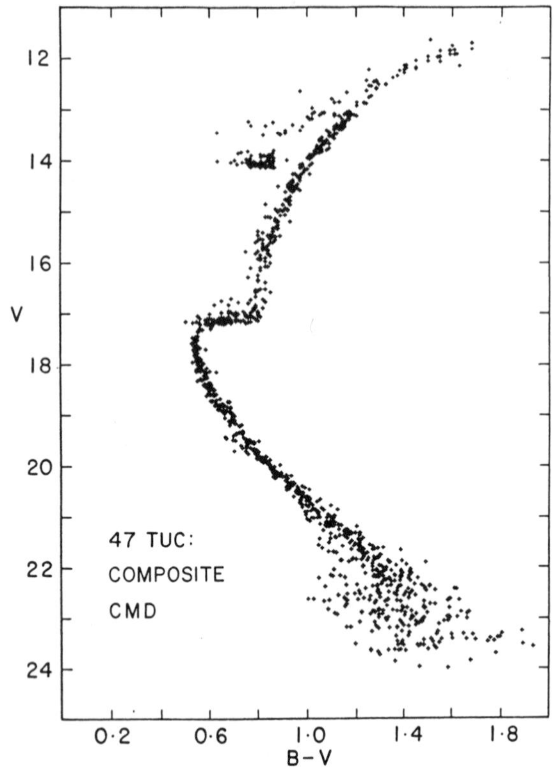

Figure 2. The HR diagram for globular cluster 47 Tucanae. [*From an article by J. E. Hesser et al. (copyright 1987), Astronomical Society of the Pacific.*]

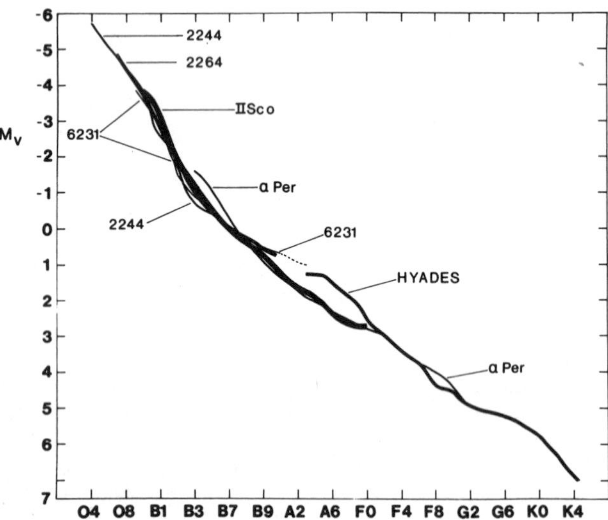

Figure 3. The HR diagrams for six open clusters combined in a cluster-fitting diagram. [*From an article by R. F. Garrison (copyright 1978), International Astronomical Union and D. Reidel Publishers.*]

FUTURE DIRECTIONS IN HR DIAGRAM RESEARCH

With new digital detectors, astronomers are able to plot HR diagrams that include much fainter stars, and with much higher precision than is possible with photographic techniques. Such improvements will inevitably lead to better theoretical models and new observational limits.

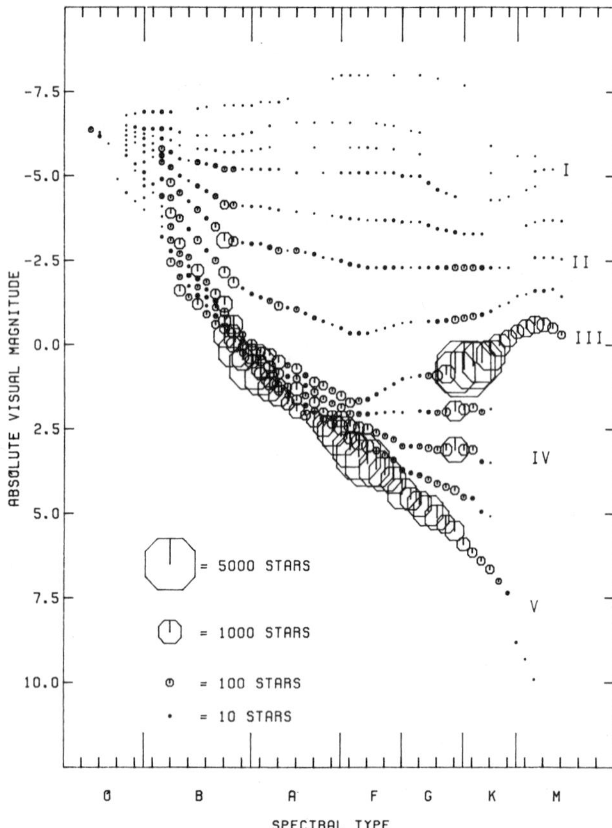

Figure 4. The HR diagram for 93,000 stars. [*From the* Michigan Spectral Catalogue *by N. Houk (copyright 1989), University of Michigan.*]

Observations have already been extended to individual stars in external galaxies, an exciting new area of research. The exploration of new kinds of stars and different population characteristics will depend on and extend the established methodology involving the HR diagram.

Nancy Houk, at the University of Michigan, is reclassifying the brightest quarter of a million stars according to the MK system. The impressive HR diagram illustrating some of these results schematically is shown in Fig. 4. With this remarkable sample of relatively homogeneous types for a magnitude-limited sample, meaningful statistical studies can be carried out. Because this task is taking most of her lifetime, it is unlikely that such a project will be repeated by a human; machines will be used.

The next tens of millions of stars will be classified automatically, using digital data. The difficulty is in retaining or improving the precision that has been achieved with current techniques. Pattern-recognition techniques for digital computers are being developed for the task, and some astronomers are looking forward to the development of neural computers as the ultimate answer.

The HR diagram undoubtedly will continue to be a useful medium for communicating the results of stellar astronomy. Modern and future versions may have to be three dimensional, with chemical composition as the third dimension. Some attempts in this direction have already been made, especially with color-magnitude diagrams.

Additional Reading

Garrison, R. F. (1988). Comments on the method of spectroscopic parallaxes. *Publ. Astron. Soc. Pac.* **100** 1036.

Hearnshaw, J. B. (1986). *The Analysis of Starlight*. Cambridge University Press, Cambridge, especially pp. 208–228 and 283–293.

Kaler, J. B. (1989). *Stars and Their Spectra: An Introduction to the Spectral Sequence*. Cambridge University Press, Cambridge.

Philip, A. G. D. and Hayes, D. S., eds. (1978). *The HR Diagram*. D. Reidel, Dordrecht.

Struve, O. and Zebergs, V. (1962). *Astronomy of the 20th Century*. Macmillan, New York, especially pp. 198–203 and 258–283.

See also **Magnitude Scales and Photometric Systems.**

Stars, High-Energy Photon and Cosmic Ray Sources

Robert Rosner

The Sun has been long known as a copious source of high-energy particles and photons. Before the era of space-borne instruments, this knowledge was largely inferential, as, for example, the association between solar flares and the high-energy particles responsible for terrestrial auroras; but with the placement of particle and photon detectors above the Earth's atmosphere (which shields the Earth's surface from virtually all photons in the ultraviolet and at shorter wavelengths), it became possible to directly detect these high-energy emissions and to study their production. These studies showed that the Sun was indeed responsible for a significant fraction of the high-energy particles—cosmic rays—and photons in the near-Earth environment, but that by no means was the Sun the only such source. A wide variety of both galactic and extragalactic objects are now known to be sources of both types of emissions; and a central question of modern astrophysics is: What is the fractional contribution of these various candidate sources, such as stars, supernovae and supernova remnants, accreting compact binary star systems, active galactic nuclei, and quasars to the high-energy radiation? Here we discuss the role of stars as contributors to the high-energy photon and particle environment in space.

SOLAR HIGH-ENERGY PHOTON EMISSION

High-energy photons emitted from the Sun—ranging from the extreme ultraviolet and x-rays to gamma rays—are totally absorbed by the Earth's atmosphere and can therefore only be detected by space-based instruments. The first such measurements were carried out in the late 1940s. At that time, American scientists used film images, obtained with pinhole cameras mounted on captured German V2 rockets, to show that the Sun was a source of x-ray emission, and that the particular region of emission could be identified with the outer layers of the Sun, the solar chromosphere and corona. This was an important discovery, because until the 1940s, it was not universally agreed that these layers were associated with gas temperatures some 2–3 orders of magnitude hotter than the temperature of the photosphere, the visible surface of the Sun. The detection of x-rays from these layers, together with the earlier definitive identification by Bengt Edlén in 1940 of the so-called coronium lines with radiation from highly ionized species of iron (including Fe X and Fe XIV) and calcium (Ca XV), settled this temperature dispute, and led to our present understanding of the solar corona and chromosphere: The Sun's outer layers increase in temperature, from roughly 6000 to several million degrees kelvin, within several thousand kilometers of the visible surface. The thin interstitial layer (the chromosphere and transition region) separating the corona from the photosphere is the source of most of the ultraviolet emission, while the far more extensive, and much hotter, corona is the source of the x-ray emission. For photon energies up to a few kiloelectronvolts, these emissions are due virtually entirely to radiation from bound-bound and free-bound transitions in the "metals," including calcium, carbon, iron, neon, nitrogen, oxygen, silicon, sodium, sulfur, and so forth; the contributions from the principal constituents of solar gas, namely, hydrogen and helium, are relatively negligible at these temperatures. There is, however, a significant contribution from free-free transitions, that is, from electron bremsstrahlung emission, which becomes more and more dominant for gas temperatures above 10^7 degrees, temperatures that are, in fact, reached in solar flares. These flares are the result of the impulsive release of a large amount of energy in the solar corona, and are responsible for essentially all of the Sun's photon emission at photon energies above a few kiloelectronvolts, and ranging up to gamma ray energies. The gamma rays, the most energetic, solar emissions are mostly the result of nuclear processes—such as neutron capture or excitation of nuclear levels—as fast protons and neutrons produced during a flare in the solar corona slow down upon encountering the higher-density lower layers of the solar atmosphere. For example, the strongest gamma-ray line, at 2.223 MeV, is the result of neutron capture by photospheric protons, leading to the production of deuterium and the associated emission of a gamma ray. However, other processes also contribute, the most prominent of which is electron-positron annihilation, which leads to the well-observed 0.511-MeV line.

Why are these layers so hot when compared to the visible solar surface? The reasons for the presence of this hot material are entwined with the problem of solar activity, discussed elsewhere in this volume; suffice it to say here that the heating is tied to the presence of magnetic fields in the outer layers of the Sun, and that these magnetic fields also play a central role in the physical confinement of these hot plasmas. Furthermore, there is considerable evidence that the heating processes are highly transient in nature; this transient behavior reaches its extreme in the case of the heating that must take place during solar flares. In this case, it appears quite certain that some form of magnetic field reconnection process must take place, leading to the dissipation of energy stored in the coronal magnetic fields by the surface motions at the photospheric level and below; detailed calculations show that such reconnection processes can have the requisite short time scale for energy dissipation (which may be on the order of 1 s or less), as well as being able to provide the observed amount of energy. Recent observations of very-small-amplitude hard x-ray (>20 keV) transient emissions, which have been associated with so-called "microflares," suggest that transient heating may well occur in the quiescent corona as well. Future, more sensitive, observations should settle this question.

SOLAR COSMIC RAY PRODUCTION

The solar flares that lead to such spectacular levels of photon emission are also prolific sources of highly energetic particles, ranging from electrons with energies of up to 10 MeV to nucleons with energies up to hundreds of megaelectronvolts. Most of these particles remain confined by the Sun's magnetic field, and end up thermalizing in the lower reaches of the solar outer atmosphere (thereby leading to copious emission of photons as they are slowed). However, a certain fraction of the more energetic flares leads to the escape of a substantial fast-particle population (the so-called solar energetic particles, or SEP) into the interplanetary medium, where they have been directly detected. These observations, when taken together with measurements of the associated high-energy photon emission, have proved to be a powerful tool for exploring the processes of particle acceleration. For example, relative timing measurements of gamma ray line emission and continuum hard x-ray emission arriving at Earth have shown that protons and electrons ought to be virtually simultaneously accelerated within time scales of 1 s. These observations cast some doubt on the existence of a "two-stage" particle-acceleration process, and instead suggest that theoretical explanations ought to be focusing on acceleration processes which are single stage, that is, that can accelerate both electrons and protons on comparable time scales.

An often neglected, but very crucial, aspect of cosmic ray studies is the light that cosmic ray composition can shed on the elemental abundances in stars, and on the physical conditions in the regions in which they are accelerated. In general, solar cosmic ray abundances agree with coronal abundances; but some very significant exceptions do exist. The most striking is the He^3 enrichment in certain, relatively small flares; in these events, the He^3/He^4 ratio changes from normal values of 10^{-2} or less to values in excess of 0.2.

HIGH-ENERGY PHOTON EMISSIONS FROM OTHER STARS

Until the mid-1970s, the Sun stood as an isolated example of an "active" star. Radio observations of transient emissions associated with low-mass main-sequence stars (stellar flares) and optical observations of periodic low-amplitude stellar bolometric light modulations (starspots) and white light transients (stellar flares) were not universally agreed upon as signatures of solar-like surface activity. This situation changed drastically with the launch of the *International Ultraviolet Explorer* (*IUE*) and those of the *High Energy Astronomy* satellites *HEAO-1* and *HEAO-2* (also known as the *Einstein* observatory), which were designed to peer at the sky at x-ray wavelengths. These telescopes were vastly more sensitive than those flown before; and *Einstein* combined this sensitivity with an imaging capability almost comparable to that of large ground-based optical telescopes. As a result, it proved possible to survey virtually all types of stars in the Hertzsprung-Russell diagram, from young massive OB stars to the low-mass dwarf M stars, in the photon-energy range of 0.1–4.5 keV.

These observations revealed two distinct types of stellar x-ray emission. Early-type stars, which are relatively young and massive, showed themselves to be vigorous sources of x-ray emission, with x-ray output levels typically 10^{-7} of their bolometric output and with characteristic temperatures of the x-ray emitting gas in the range of 10^7 degrees kelvin. Attempts to correlate these emission levels with other stellar attributes showed that the only correlation was with the total luminosity of the underlying star, or with other attributes that were already known to be correlated with the stellar luminosity. There is as yet no definite explanation for these results. Most theoretical models produced prior to these observations envisaged a type of stellar corona that would lie at the stellar surface, and thus at the base of the massive winds that are known to emanate from these stars. The x-ray observations are not easily reconciled with these models, principally because the expected degree of absorption of x-ray photons below roughly 0.2 keV by the stellar wind itself is not universally seen (some OB stars do show such extensive absorption, but others do not). More recent theoretical calculations have suggested that the winds are themselves unstable, and that shocks produced in the winds by these instabilities may lead to the observed x-ray–emitting plasma.

In contrast, late-type stars, which are roughly of one solar mass or less, show highly variable levels of x-ray emission behavior. Their x-ray emission levels can range from roughly 10^{-3}–10^{-7} or less of their bolometric luminosity and the x-ray emissions show relatively little correlation with the spectral type of the underlying star; this latter fact speaks powerfully against the idea that acoustic waves generated in stellar photospheres could be responsible for the necessary plasma heating (because the level of acoustic wave generation is very sensitive to the temperature of the stellar photosphere, and hence to the spectral type of the underlying star). Only a relatively small fraction of the observed range of emission levels can be attributed to variability associated with the stellar counterpart of the solar activity cycle; instead, the x-ray emission levels of these stars correlate best with their rotation rate, and show a systematic decline in level with stellar age, in concert with the well-established spin down of stars with time. The temperatures of the plasmas that emit the x-rays are typically in the range 2–10×10^6 degrees kelvin, with total x-ray luminosities in the range 10^{26}–10^{30} erg s^{-1}; the x-ray emitting plasmas are thus rather similar to the coronal plasma we see directly on the Sun. Indeed, our understanding of stellar x-ray emission is today entirely based on the solar analogy, that is, on a relatively straightforward model for stellar activity that uses the Sun as a touchstone, and is based on the idea that all late-type stars that both rotate and convect in their outer layers will produce magnetic fields in their interiors via a magnetic dynamo. The emergence of these magnetic fields to the stellar surfaces, as they couple to the ambient turbulent motions in these photospheres, then leads to both plasma heating and confinement of the resulting hot plasma. Because the rate of magnetic field generation is a function of the stellar rotation rate, this model naturally accounts for the x-ray luminosity–stellar rotation rate correlation in a qualitative way. Furthermore, the level of activity affects the rate of stellar mass loss via winds (presumably, the more vigorous the activity level, the larger the mass loss rate), as well as the strength of the large-scale stellar magnetic field. It is this mass loss that leads to the spin down of stars, as mediated by the large-scale magnetic field (which, by forcing the outflowing gas into corotation in the immediate stellar vicinity, acts to increase the effective lever arm for despinning the star by the outflow). This model thus also predicts rapid spin down during stellar youth: When stars rotate relatively rapidly, the mass loss rates and large-scale magnetic fields are relatively large, and hence the spin down efficiency of the outflow is also relatively great; furthermore, one then expects a gradual decrease in the spin down rate as the star ages. These effects again qualitatively agree with observations.

Finally, we note that because of their large numbers in our galaxy, stars can in principle contribute significantly to the diffuse galactic soft x-ray background. The *Einstein* stellar surveys have been used to construct the x-ray luminosity functions for late-type stars and, by using a model for the distribution of such stars in the Galaxy, to compute the stellar contribution to the galactic x-ray background. Results obtained to date suggest that stars are relatively minor contributors to the overall background, at levels most likely well below 20% of the galactic component of the total diffuse background in the energy range 0.1–1.0 keV. This result, in concert with the stellar x-ray luminosity functions, is generally consistent with more recent observations of stars in this energy range with the *EXOSAT* satellite; but observations carried out with the *ROSAT* satellite, which was launched in 1990, are expected to definitively resolve the remaining uncertainties via a full-sky imaging survey in this approximate energy range.

COSMIC RAYS FROM OTHER STARS

Straightforward extrapolation of the observed solar cosmic ray flux to other stars in our galaxy shows immediately that these stars could not possibly be responsible for the observed extra-solar cosmic rays. Clearly, the acceleration of these energetic particles must take place elsewhere than in the outer atmospheres of stars. However, one must clearly distinguish between the nature of the acceleration process and the process of extracting the energetic particles from the thermal background of particles. In particular, it was suggested in the 1970s that certain classes of stars, principally the so-called flare stars, could act as "injectors" of moderately energetic particles to the interstellar medium (these particles would be the counterpart of the solar-flare–produced solar cosmic rays). However, this suggestion remained largely speculative until much more recently, when it became possible to study in considerable detail the composition of galactic cosmic rays near the Earth, principally with facilities such as *HEAO-3*.

The idea that ordinary solar-like stars play a role in galactic cosmic ray injection solves two problems in standard models for cosmic ray particle acceleration: First, it resolves the "injection problem" itself, namely, the problem of what exactly is the seed population of particles for the acceleration process; that is, a number of particle-acceleration processes for producing extremely high-energy cosmic ray particles are relatively inefficient in accelerating ambient thermal particles to such energies. The resolution provided by stellar cosmic ray injection is that this first step of starting with ambient thermal particles to produce particles of moderate energy (i.e., a few megaelectronvolts per nucleon) is accomplished by the stellar particle-acceleration process. The interstellar medium acceleration processes are then called upon to only further accelerate these particles to gigaelectronvolt per particle energies. Second, stellar cosmic ray injection solves the abundance problem. This is, extensive studies of cosmic ray composition show that the seed material for these energetic particles must be mainly matter that emerged from the atmospheres of normal stars (and thus does not resemble the bulk yield of, for example, the ejecta of Type II supernovae). It thus remains possible that the galactic cosmic rays originate mainly from interstellar medium material; this possibility is, however, severely constrained by the fact that galactic cosmic rays do not at all show the depletion of refractory elements (including carbon and silicon), which are condensed in grains in the interstellar medium, and are hence highly depleted in the gas phase. Mechanisms for selectively accelerating grain destruction products at, for example, supernova remnant shocks, have been proposed to avoid this basic problem; but it remains the case that the simplest explanation for the observed composition is that the seed material for galactic cosmic rays is simply ordinary stellar surface matter from stars of spectral types F to M. Finally, we note that compositional studies also indicate that a small fraction (of the order of a few percent) of galactic cosmic rays ought to originate in material that has undergone nuclear burning of helium; one possible source of these particles are Wolf–Rayet stars.

Additional Reading

Forman, M. A., Ramaty, R., and Zweibel, E. G. (1986). The acceleration and propagation of solar flare energetic particles. In *Physics of the Sun*, P. Sturrock, ed. D. Reidel, Dordrecht, p. 249.

Haisch, B. M. and Rodonò, M., eds. (1989). *Solar and Stellar Flares*. Kluwer Academic Publ., Dordrecht.

Rosner, R., Golub, L., and Vaiana, G.S. (1985). On stellar x-ray emission. *Ann. Rev. Astron. Ap.* **23** 413.

See also **Cosmic Rays, Acceleration; Cosmic Rays, Origin; Solar Activity, Solar Flares; Sounding Rocket Experiments Astronomical; Stars, Magnetism, Observed Properties; Stars, Red Dwarfs and Flare Stars; Sun, High-Energy Particle Emissions; Stars, Activity and Starspots.**

Stars, High Luminosity

Roberta M. Humphreys

The most luminous and therefore most massive stars have always intrigued astronomers. As the brightest stars, they are our first probes of stellar populations and stellar evolution in other galaxies as well as at large distances in our own Milky Way. They evolve very quickly, with lifetimes of less than 10 million years, and influence their environments and future generations of stars via strong stellar winds and other forms of mass loss, and eventually as supernovae.

During the past decade or so, both observational and theoretical studies of luminous stars and their evolution have progressed very rapidly and in new directions. Observationally, this development was spurred by (1) the recognition of the importance of mass loss in stars with masses greater than 20 $M_{\odot}$ (solar masses), thanks largely to results from the *International Ultraviolet Explorer* and from infrared observations; and (2) modern analyses of the population of luminous stars in our Milky Way galaxy and the Magellanic Clouds, plus observations of the brightest and most massive stars in other nearby galaxies. In theoretical work, progress came from the inclusion of mass loss and internal mixing in the models for stellar structure and evolution of massive stars, which made them physically more realistic and gave better agreement with observations.

OB SUPERGIANT STARS

The most massive, most luminous stars are also the hottest. These luminous stars or supergiants are designated *O- and B-type supergiants* on the basis of the appearance of their spectra. The spectra of these supergiants are dominated by lines of H, He I, and He II, and their temperatures range up to 50,000 K. In a diagram of luminosity versus temperature, known as the Hertzsprung–Russell (HR) diagram, the OB supergiants occupy the upper-left corner (see Fig. 1). The physical parameters of luminous stars of different temperatures are summarized in Table 1. In addition to their high temperatures and luminosities, OB supergiants are distinguished by their mass loss and stellar winds, which are driven by radiation pressure.

To a first approximation the mass loss rates of the luminous, hot stars depend on the stellar luminosity, although there is considerable scatter in this relation, which suggests that other factors such as temperature, radius, chemical composition, and evolutionary state play a role. The measured mass loss rates of the OB supergiants are typically between 10^{-7} and a few times 10^{-6} solar masses per year.

When plotted on the HR diagram, the most luminous OB supergiants in the Galaxy and the Magellanic Clouds reveal an upper luminosity boundary in the diagram, an envelope of declining luminosity with decreasing temperature, which for stars with temperatures less than 8000–10,000 K becomes an upper boundary of essentially constant luminosity. There are no cooler counterparts to the most luminous OB supergiants. The observed upper region of the HR diagram and the characteristics of some of its most luminous stars provide the empirical evidence for an upper luminosity boundary that is most likely due to the instability of the photospheres of the evolved, most massive, most luminous stars. The luminosity/stability limit is shown in Fig. 1.

LUMINOUS BLUE VARIABLES

The *luminous blue variables* (LBVs) are a small group of high-luminosity, unstable, hot supergiant stars whose behavior provides important insight into understanding the luminosity/stability limit in the HR diagram. The LBVs include such well-known stars as η Carinae and P Cygni in the Galaxy, S Doradus in the Large Magellanic Cloud, and stars known as the Hubble–Sandage variables in the spiral galaxies M31 and M33. Their most distinguishing characteristic is the occurrence of irregular eruptions or ejections that result in a greatly enhanced mass outflow (10^{-5}–10^{-4} $M_{\odot}$ yr^{-1}), which leads to the formation of a pseudophotosphere when a star is at maximum light at visual wavelengths. At this stage, the slowly expanding (100–200 km s^{-1}) envelope is cool (8000–9000 K) and dense (number density $N \simeq 10^{11}$ cm^{-3}), and the star resembles a very luminous A-type supergiant. At its minimum light (in the visual), or the quiescent stage, the LBV has a much higher photospheric temperature (> 20,000 K–25,000 K), and resembles an OB supergiant, but with prominent emission lines of hydrogen, He I, and permitted and forbidden ionized iron in its spectrum. At visual minimum the mass loss rate of an LBV is lower by a factor of 10–100. During the variations in visual light, the total luminosity of the LBV remains essentially constant. The visual light variations are caused by the apparent shift in the star's energy distribution driven by its instability. The schematic HR diagram in Fig. 1 shows the location of the well-studied LBVs at visual minimum and maximum. Notice that at visual maximum (marked by the dots at the right ends of the dashed horizontal lines) they all have essentially the same temperature, near 8000 K.

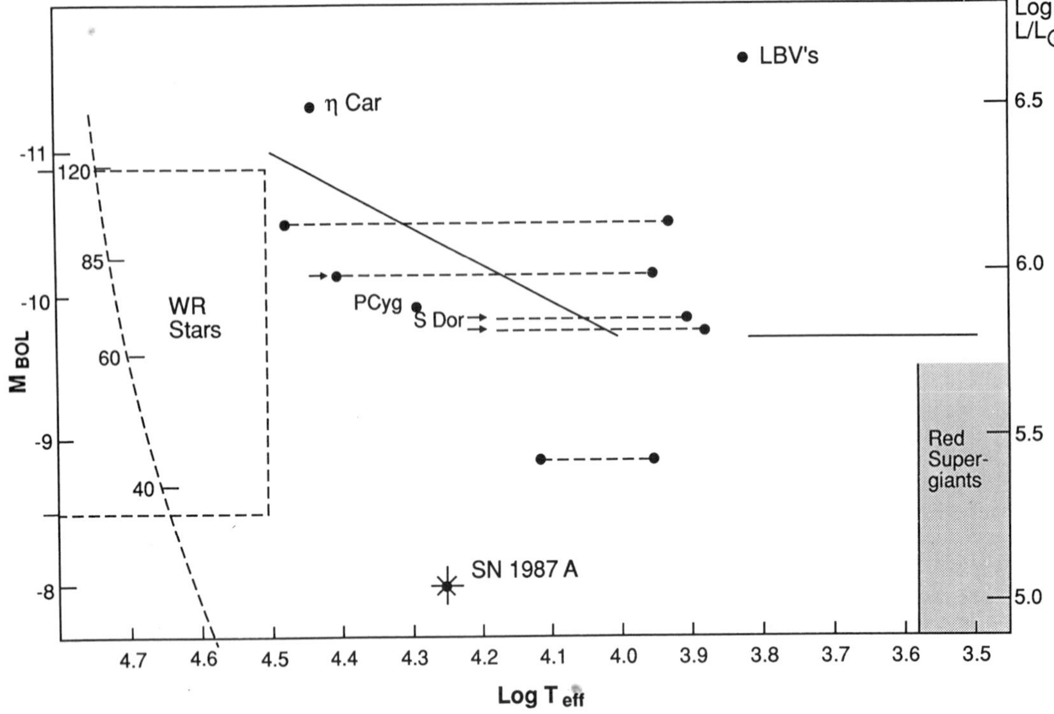

Figure 1. A schematic HR diagram (M_{bol} or log $L/L_{\odot}$ versus log T_{eff}) showing the locations of the well-studied LBVs at minimum and maximum visual light. The dashed lines represent their transition between these two states. The region of the red supergiants, the location of the Wolf–Rayet stars and the position of Supernova 1987A are shown. The zero-age main sequence, marked at the positions of stars of initial masses 40, 60, 85, and 120 $M_{\odot}$, is at the left. The solid straight lines at center and at the right show the locations of the observed upper luminosity boundary for the hot and cool stars, respectively.

In addition to their light variations, many of the LBVs are surrounded by ejected material from previous eruptions. Analysis of this circumstellar material shows that it contains processed material (elements that have undergone nuclear reactions) such as nitrogen and helium from the star's interior. This shows that the LBVs are evolved, post-hydrogen-burning stars.

With these remarkable properties, it is not surprising that the LBVs have come to play a major role in our current thinking about the evolution of the most luminous stars. Based on their luminosities and their location on the HR diagram, most LBVs have evolved from stars with initial masses $\geq 40\ M_{\odot}$. With their high luminosities, enriched ejecta, and high mass loss rates, which are similar to those of the Wolf–Rayet stars, LBVs are commonly considered evolved, massive stars in transition to Wolf–Rayet stars. Thus the LBVs may be a relatively short-lived ($\approx 10^4$ yr), highly unstable, but important stage in the evolution of the most luminous stars.

The most likely cause of the instability in the LBVs and of the observed upper luminosity boundary is radiation pressure. However, other processes, such as interior evolution, or hydrodynamic effects, such as atmospheric turbulence, may drive the star to the boundary between radiation pressure and gravity, the so-called Eddington limit. The Eddington limit of a star is that luminosity at which the inward acceleration due to gravity (g_{grav}) is balanced by the outward acceleration (g_{rad}) due to radiation pressure (thus, the ratio of these accelerations, $\Gamma = g_{rad}/g_{grav} \rightarrow 1$). But the classical Eddington limit due to electron scattering is independent of temperature, unlike the observed luminosity boundary for hot stars. However, as the temperature of the stellar photosphere decreases, the opacity increases, reducing the Eddington luminosity. Thus an instability related to this modified Eddington limit may be responsible for the upper luminosity boundary.

The typical LBV has eruptions of approximately 2 mag in visual brightness on time scales of decades, but a few LBVs have been observed in much more violent outbursts of more than 3 mag in visual brightness. The best example is η Carinae's famous outburst between 1837 and 1860, when it became the second brightest

Table 1. Summary of the Physical Properties of the Luminous Stars of Different Temperatures

Parameter	*Hot Stars*	*Intermediate Temperature Stars*	*Cool Stars*
Spectral types	O, B, A	F, G, K,	M
Surface temperature (K)	50,000–10,000	10,000–4000	< 4000
Color	Blue	Yellow	Red
Luminosity range ($L/L_{\odot}$)	10^4–~ 5×10^6?	10^4–8×10^5	10^4–5×10^5
Mass range ($M/M_{\odot}$)	20–200?	20–50	20–50
Size range ($R/R_{\odot}$)	10–100	30–1000	300–2000
Mass loss ($M_{\odot}$ yr^{-1})	10^{-8}–10^{-5}	10^{-7}–10^{-4}	10^{-7}–10^{-4}

$M_{\odot} = 2 \times 10^{33}$ g, $L_{\odot} = 4 \times 10^{33}$ erg s^{-1}, $R_{\odot} = 7 \times 10^5$ km, and the subscript $\odot$ denotes the Sun.

star in the sky. Other possible examples are P Cygni's behavior in the 1600s, Var 12 in the spiral galaxy NGC 2403, and possibly Supernova 1961V in NGC 1058, which may have been an outburst of a very massive star instead of the stellar collapse involved in an actual supernova.

THE FAMOUS CASE OF η CARINAE

η Carinae is the most extreme member of the LBVs. It is the most luminous LBV, with $\log L/L_{\odot} \simeq 6.5$ (meaning that it is about 3 million times more luminous than the Sun), and presumably had the highest initial mass. After its 1837–1860 outburst, η Car rapidly faded to below naked-eye visibility due both to a subsiding of the eruption and to the formation of dust. The extensive dust shell reradiates the star's ultraviolet and visual energy in the infrared, permitting a very accurate measurement of its current luminosity. At a wavelength of 10 μm η Car is the brightest extra–solar-system object in the sky.

Today η Car is distinguished by its associated nebulosity, the "homunculus," which is the material ejected during the 1840s outburst. The emission lines in the spectrum of these ejecta show an enhanced abundance of nitrogen over carbon and oxygen; helium is also overabundant. This composition is consistent with material that has been processed by nucleosynthesis in the stellar core, brought to the surface by mixing, and ejected. Thus there is no doubt that η Car is an evolved, very massive star.

During its great eruption or outburst, the total absolute luminosity of η Car exceeded its current absolute luminosity by more than 1 mag and at its actual visual maximum in 1843, by more than 2 mag, unlike the constant total luminosity maintained by other LBVs during their more modest eruptions. This excess energy must have been the thermal and radiative energy trapped in the layers ejected by the star. To liberate enough radiated energy, more than 1 and probably 2 or 3 $M_{\odot}$ (a remarkable amount) must have been ejected by η Car during its famous eruption, corresponding to a mass loss rate of about 10^{-1} $M_{\odot}$ yr^{-1}. Its current mass loss rate is estimated at 10^{-3} $M_{\odot}$ yr^{-1}. With these mass loss rates, η Car must be near the end of its life.

The violent behavior of η Car may seem extreme even for the unstable LBVs, but the existence of circumstellar material around other LBVs suggests that they also may have experienced more violent outbursts in the past. We suspect that most of the mass shed by an LBV may occur in these giant eruptions, but the frequency of such events is not known. The dynamical age of these circumstellar shells is about 10^4 yr. Since a brief brightening in 1889, η Car has been relatively stable. It has been slowly brightening since 1940 due to a decrease in circumstellar extinction as the dusty ejecta expand. P Cygni has also not shown any significant variations since its seventeenth century eruption.

Based on its luminosity and location on the HR diagram, η Car probably began its life as a 200-$M_{\odot}$ star. But recent speckle interferometry observations show that η Car may be multiple, in which case the primary component would have had an initial mass of close to 100 $M_{\odot}$. The three other components would be hotter, less luminous (about 60 $M_{\odot}$), and more stable. The components would be too far apart to have affected each other's evolution, so the nature of the primary star is unchanged.

R136a AND THE POSSIBILITY OF SUPERMASSIVE STARS

Any discussion of the most luminous and most massive stars always raises the question of what is the most massive star, whether observed or theoretical. For many years the theoretical upper limit to stellar masses was believed to be near 60 $M_{\odot}$, a value due to vibrational instabilities in the interiors of the stars. Recent studies have shown that the upper mass for vibrational stability on the main sequence is nearer to 130 $M_{\odot}$. We know of many stars in the Galaxy and the Magellanic Clouds with initial masses likely to be greater than 60 $M_{\odot}$ and several stars with masses near 100 $M_{\odot}$. Suggestions that the object R136a near the center of the 30 Doradus complex (the Tarantula nebula) in the Large Magellanic Cloud might be a supermassive star of 1000–3000 $M_{\odot}$ created renewed interest in the theoretical and observed upper limit to stellar masses.

It was assumed that R136a provided most of the ionizing radiation for the Tarantula nebula; then the very high stellar temperature inferred from its ultraviolet spectrum implied a mass of 3000 $M_{\odot}$. But this conclusion hinged on the assumption that R136a was *not* a compact cluster of several hot very massive stars. Several years later, photographic speckle interferometry resolved R136a into eight separate components within 1 arsec or 0.3 pc. Thus R136a is not a single *supermassive* star, but a group of massive stars. The most luminous member of R136a is probably no more massive than 250 $M_{\odot}$, but the entire 30 Doradus region may contain 15–20 stars with initial masses of 100–200 $M_{\odot}$. This is indeed a remarkable region.

The *super*massive-star idea has fallen to closer scrutiny, but it has resulted in recognition of a relatively large number of *very* massive stars which exist in some very special regions, including 30 Doradus in the Large Cloud, the great Carina nebula, and the cluster and nebula NGC 3603, both in our Galaxy, for example. Questions about the formation of these very massive stars, and the reasons they form in large groups, are outstanding problems for the future.

Additional Reading

Davidson, K., Moffat, A. F. J., and Lamers, H. J. G. L. M., eds. (1989). *The Physics of Luminous Blue Variables*. Kluwer Academic Publ., Dordrecht.

Humphreys, R. M. and Davidson, K. (1984). The most luminous stars. *Science* **223** 243.

Humphreys, R. M. and Davidson, K. (1986). Sizing up the superstars. *New Scientist* **112** (No. 1538) 38.

de Jager, C. (1980). *The Brightest Stars*. D. Reidel, Dordrecht.

Lamers, H. J. G. L. M. and de Loore, C. W. H., eds. (1980). *Instabilities in Luminous Early-Type Stars*. D. Reidel, Dordrecht.

See also **Stars, Red Supergiant; Stars, Wolf–Rayet Type; Stellar Evolution, Massive Stars.**

Stars, Horizontal Branch

Klaas S. de Boer

Horizontal branch (HB) stars form a special group among the stars in the later phases of evolution. They represent a rather stable phase in which the star has, after its red giant phase, again a rather condensed atmosphere and is thus bluish in color. HB stars derive their name from the location of their observed properties in the color-magnitude diagram (Hertzsprung-Russell diagram), where they occupy a horizontal (equal brightness) strip to the left of the area where the red giant stars are located. In color-magnitude diagrams of globular clusters, HB stars stand out very clearly because of their blueness and because they are among the brighter stars in the clusters (see Fig. 1). HB stars are also present in the field of the Galaxy and are then called field horizontal branch (FHB) stars, but they are more difficult to find.

The HB stars can be divided into a few categories, based on their appearance in globular cluster color-magnitude diagrams, as follows. The blue HB (BHB) stars are to the left and the red HB (RHB) stars are to the right of a gap in the HB. This gap is sometimes marked by the appearance of cluster RR-Lyr–type variable stars. The BHB stars superficially appear like stars of spectral types B8 to

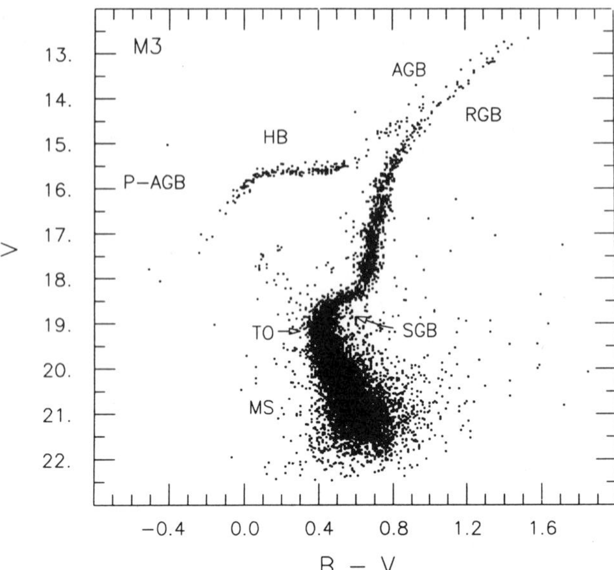

Figure 1. The color-magnitude diagram (visual brightness V versus color index $B-V$) of M3 shows the large variety of stars present in a globular star cluster. In this case the density of data points in the diagram is roughly proportional to the true number of the corresponding kinds of stars, and therefore indicates how long stars exist in the respective phases of evolution. Stars spend most of their time on the main sequence (MS), burning hydrogen in their cores. The turn-off point (TO) marks the end of the MS phase and thus is an indication of the age of the cluster. The TO defines the start of the subgiant branch (SGB) phase with shell hydrogen burning, in which the star increases in brightness and evolves to the red giant branch (RGB). Here the star loses mass and eventually becomes a horizontal branch (HB) star. The final evolution of the star may lead up to the asymptotic giant branch (AGB) and into the very hot post-AGB phase. *(From A. Renzini and F. Fusi Pecci. Reproduced, with permission, from* Annual Review of Astronomy and Astrophysics ***26***, *©1988 by Annual Reviews Inc.)*

A3. Further to the blue side on the HB, other gaps may appear, and stars found there are generally hotter and fainter. The horizontal branches of some globular clusters extend to quite faint magnitudes, and stars there are called extended HB (EHB) stars. Finally, in some clusters, stars are found with parameters such that they appear "above" the HB in color-magnitude diagrams. These stars are sometimes called supra HB (SHB) stars.

The distribution of the stars on the HB is neither homogeneous in the color-magnitude diagram, nor the same for all globular clusters. Therefore, one uses the "shape" of the HB as a parameter to characterize globular clusters. Some clusters have mostly BHB stars, and some others have a rather red horizontal branch. The proportion between the two kinds of stars is commonly quantified by the fraction

$$f = n(\mathrm{BHB})/(n(\mathrm{BHB}) + n(\mathrm{RHB})),$$

with $n(X)$ the number of a given kind of star as counted in an observed color-magnitude diagram. The values for f cover the full range from 1–0 indeed. There seems to be a relation between the blueness f of the HB and the metal content of the cluster stars in the sense that the more metal rich the stars are, the redder the HB is.

When, in a main-sequence star, the hydrogen in the core is exhausted and the star begins burning hydrogen in a shell around the core, the structure of the star changes. It develops an extended envelope and becomes a red giant star. In this stage, a star will also lose material through stellar wind. After such a star has lost some of its outer shell during the red giant phase and after ignition of helium in the core, the stellar structure readjusts so that it more closely resembles that of a normal main-sequence star. For massive stars, evolution calculations indicate that further burning stages will occur. But for stars with masses well below 10 $M_\odot$ (solar masses), the substantial mass loss which took place previously leaves a rather modest stellar object. Based on stellar models, we know that HB stars form a sequence by mass along the horizontal branch, reaching from 0.45 $M_\odot$ at the blue end to about 0.85 $M_\odot$ at the red end, which intersects with the red giant branch. The RR-Lyr-star gap of pulsational instability spans the small mass range from approximately 0.60 to 0.62 $M_\odot$, with surface temperatures near 7000 K.

The mass retained in the HB star from the initial mass of the main-sequence star is determined by many parameters. These include (1) the stellar structure at the onset of H-shell burning, (2) the amount of mass lost due to the wind in the red giant phase, and (3) the moment of the core helium ignition. The structure of any star depends critically on the abundance of the metals (elements heavier than helium), because metallicity governs the opacity in general. In addition, the strength of the stellar wind (and thereby the mass loss) specifically is a function of metallicity. Thus, knowledge of the metal content is an essential parameter in the investigations of HB stars.

The fact that HB stars are found in globular clusters indicates that they are old stars. The age of HB stars in a given cluster should be the same as that of the cluster as a whole (for practical purposes, all stars of the cluster formed at the same time). Age determinations of globular clusters are based on the readily-estimated ages of the most massive main-sequence stars still present in the clusters. These lifetimes are derived from calculations of stellar models with evolution. In general, the large ages found for globular clusters are consistent with the low metal content of their stars, indicating that the clusters were formed in the early phases of the formation of the Galaxy, but there is not a one-to-one correlation of cluster age and metallicity.

It is relatively difficult to recognize HB stars in the field of the Milky Way, as compared to the ease with which they are found in globular clusters. However, based on our general knowledge of globular cluster stars it is possible to define parameters to discriminate HB stars in the field from main-sequence or other Population I stars. The most important difference between normal field stars and field HB stars is the smaller metal content of the latter. One simple way to select metal-poor stars from a sample is through observations in the Strömgren photometric system, defined in Table 2 of the entry on Magnitude Scales and Photometric Systems, page 406.

With the four measurements normally made in the Strömgren system, one can determine for three intervals in the observed-wavelength range the slopes of the spectral intensity distribution. These spectral slopes can be analyzed to calculate indexes for temperature and metallicity, solving at the same time for interstellar reddening effects. In particular, the c index reaches high values for stars near 10^4 K if they have low metal content. This index describes the steepness of the Balmer jump, the spectral discontinuity near 365 nm (3650 Å). This discontinuity is on the one hand a function of hydrogen excitation and ionization (and thus of temperature), and on the other hand a function of the continuum slope at the long-wavelength side of the Balmer jump, a slope that, in the Strömgren system, is governed by the density in the spectrum of atomic absorption lines (and thus by the metal abundance). Many FHB stars have been found by photometry in searches that started over 20 years ago.

Another way to identify FHB stars is to use spectroscopy and to look for stars with a paucity of stellar absorption lines compared to Population I stars. However, this method requires a large amount of observing time. Therefore, spectroscopy is only used to further

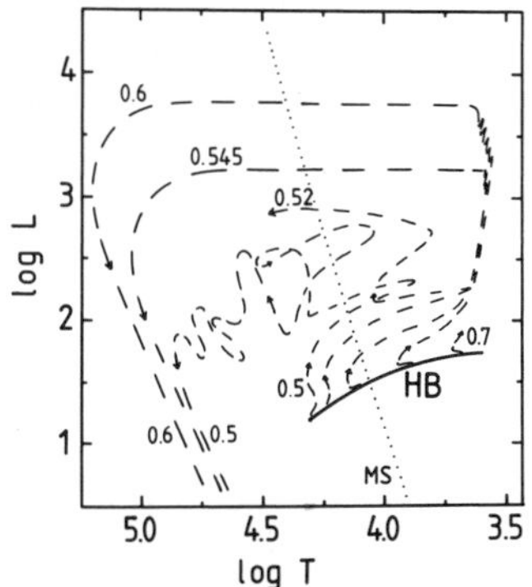

Figure 2. The final evolutionary tracks of stars beyond the horizontal branch are shown schematically on a diagram of total luminosity versus surface temperature. HB stars with more than about 0.54 solar masses evolve into the luminous asymptotic giant branch (AGB) stars, which in turn become post-AGB (PAGB) stars with very high surface temperatures. They may excite the gases lost in previous evolutionary phases, which may then become luminous as planetary nebulae. The less massive HB stars evolve away from the HB along shorter and partly erratic loops in the diagram, entering the final white dwarf state. The tracks are labeled according to the stellar mass, in units of the solar mass. The HB phase lasts roughly 10^7 yr, and subsequent evolution through the PAGB phase ranges from approximately 10^4 yr for the more massive stars to 10^8 yr for the lower mass stars. The least massive HB stars evolve in less than 10^8 yr to the white dwarf cooling tracks. The location of the main sequence (MS) is indicated by the dotted line.

investigate the stars whose FHB nature has been suspected from other observational programs. From spectroscopy of the brighter FHB stars it was learned that most FHBs have a metal content of about $\frac{1}{10}$ the solar value.

A pronounced effect of the lower metallicity of HB stars is found in their ultraviolet spectra. The lower metallicity results, in addition to the obvious weakness of spectral lines, in a lower overall opacity. This means that a HB star can radiate away more ultraviolet photons than a Population I star of normal metal content, but with a slightly fainter absolute visual magnitude in consequence. In fact, among stars with effective temperatures near 8500 K, it was demonstrated that the FHB stars were brighter by a factor of 2 at 150 nm than Population I stars.

After the HB phase there are two channels for the final evolution of stars (Fig. 2). The HB stars with masses of less than 0.54 $M_\odot$ will brighten a little and appear in cluster color-magnitude diagrams as the SHB stars mentioned earlier. They have started He-shell burning but cannot sustain it. Such a star, after "an attempt" to become a giant star, readjusts and shrinks, thereby having an enhanced luminosity for just several hundred years. It then starts its final contraction phase to become a white dwarf, cooling and dimming to the ultimate stage of an extinct stellar remnant.

The more massive HB stars (with masses of more than 0.54 $M_\odot$) evolve from core He-burning stars into stars burning He in an inner shell and H in an outer shell. Such a star therefore readjusts its structure to that of a giant star and it reaches the asymptotic giant branch (AGB), becoming red and substantially more luminous for 10^4–10^5 yr. As a giant, the star again loses mass, perhaps to the extent that the H-burning shell may disappear. The AGB phase is followed by a final contraction in the post-AGB stage, in which the star continues with its enhanced luminosity. This implies that the stellar surface must, while shrinking, radiate away the same energy as before. Therefore, the atmosphere becomes hotter. These stars are among the bluest in globular clusters and they may, because of their large luminosity, contribute substantially to the total photon-energy production of a globular cluster. Ultimately, such a star will contract and the He-burning shell will be extinguished, leaving the star on its final cooling track to become a white dwarf.

During the red giant phases (the RG and the AGB phases), mass is lost and it may accumulate as gaseous shells around the star. Such shells may be readily dispersed from the vicinity of stars in a globular cluster due to interactions with other stars. For stars in the field, however, the shells may remain around the star. When the finally contracting star becomes blue and hot, it produces large amounts of ionizing radiation. These photons will excite any gas remaining in the windblown shell, resulting in a period of several tens of thousand of years of strong and brilliant nebular radiation from the ionized gas shell, a so-called planetary nebula.

Additional Reading

de Boer, K. S. (1985). UV-bright stars in galactic globular clusters, their UV spectra and their contribution to the globular cluster luminosity. *Astron. Ap.* **142** 321.

Iben, Jr., I. and Renzini A. (1983). Asymptotic giant branch evolution and beyond. *Ann. Rev. Astron. Ap.* **21** 271.

King, I. R. (1985). Globular clusters. *Scientific American* **252** (No. 6) 66.

Philip, A. G. D. (1987). Four-color observations of field horizontal-branch stars. In *The Second Conference on Faint Blue Stars, IAU Colloquium 95*, A. G. D. Philip, D. S. Hayes, and J. W. Liebert, eds. L. Davis Press, Schenectady, NY, p. 67.

Renzini, A. and Fusi Pecci, F. (1988). Tests of evolutionary sequences. *Ann. Rev. Astron. Ap.* **26** 199.

Sweigart, A. V. (1987). Theoretical horizontal-branch evolution. In *The Second Conference on Faint Blue Stars, IAU Colloquium 95*, A. G. D. Philip, D. S. Hayes, and J. W. Liebert, eds. L. Davis Press, Schenectady, NY, p. 57.

See also **Nebulae, Planetary, Origin and Evolution; Star Clusters, Globular, Stellar Populations; Star Clusters, Globular, Variable Stars; Stars, RR Lyrae Type; Stellar Evolution, Low Mass Stars.**

Stars, Interiors, Radiative Transfer

T. Richard Carson

The concept of radiative equilibrium in stars, that is, the thermal balance supported by the radiative transfer of energy, was first introduced by Ralph A. Sampson in 1894, and adopted by Arthur Schuster in 1903 and by Karl Schwarzschild in 1906 in the context of stellar atmospheres. However, it was only in 1916 that Arthur S. Eddington firmly established radiative equilibrium as a fundamental feature of stellar interiors. S. Rosseland in 1924 established the relation between the radiative energy flux and the radiative absorption coefficient. The full development of the theory had to await a satisfactory treatment of the interaction of radiation with matter as provided by quantum mechanics, which superseded the earlier semiclassical theory of H. A. Kramers. The first application of the quantum mechanical results to stellar structure calculations was made in 1932 by Bengt Strömgren.

STELLAR STRUCTURE

A star is formed when a sufficiently large mass contracts under self-gravitation to become a compact spherical body, with a heat content derived from the release of gravitational potential energy. Contraction ceases when the compression of the material under gravity increases the density and pressure until a hydrostatic equilibrium is reached between the pressure gradient and the local gravity at each point. Even if the temperature were uniform initially, the loss of energy by radiation from the surface will establish a thermal profile with temperature also decreasing from the center outwards, supporting an energy flow from the interior. When the pressure is temperature dependent, hydrostatic equilibrium can only be maintained if the energy loss of each element of mass is balanced by differential energy flow, together with energy generation or release, thus establishing the condition of thermal equilibrium.

ENERGY TRANSPORT IN STARS

The actual transport of energy in the interior of a star is to a large extent controlled by the temperature gradient. In a typical star like the Sun, with central temperature $T_c = 14 \times 10^6$ K and radius $R_\odot = 7 \times 10^{10}$ cm, the mean temperature gradient is only 2×10^{-4} K cm^{-1}. Other factors affecting the energy transport are the number density and average velocity of the energy carriers and their mean free path, that is, the average distance they can travel before losing their excess energy. Which of the mechanisms of energy transport, conduction, radiation, or convection will be most effective depends upon the competition between them as determined by local physical conditions. In the case of conduction, more properly thermal or heat conduction, the energy is carried by individual atomic particles, electrons, and ions. Because of their smaller mass and therefore greater speed the electrons are usually more mobile than the ions and more effective in transporting energy than the more massive particles. However, under conditions similar to those at the center of the Sun, where the density is 100 g cm^{-3}, the mean free path of an electron is only about 10^{-9} cm. Only at temperatures and densities where the electrons become degenerate, so that interactions are inhibited by the Pauli exclusion principle, does the electron mean free path become sufficiently large for electron thermal conduction to become really effective. Photons, on the other hand, under similar conditions have a mean free path of about 10^{-2} cm, considerably larger than that for electrons. Also, because of their high speed, photons become efficient transporters of energy. For convection, in which the energy is carried by bulk mass motions driven by buoyancy forces, the mean free path or "mixing length" can be an appreciable fraction of the radius of the star. However, the buoyancy forces are only operative when the temperature gradient set by other processes exceeds the adiabatic temperature gradient of the material. Therefore, convection tends to be only effective as the means of energy transport when the other processes, conduction and radiation, require large temperature gradients to carry the flux of energy demanded by thermal equilibrium. This occurs typically in stellar cores, where highly temperature-dependent energy production is taking place, and in the outer layers of cool stars. Thus throughout the main bulk of the interiors of main-sequence and other stars, radiative transfer is the main process by which energy is transported.

RADIATIVE TRANSFER

The basic equation of radiative transfer takes the form of a differential equation for the change of radiation intensity with distance:

$$dI(\nu)/ds = -\rho k(\nu) I(\nu) + \rho j(\nu),$$

where $I(\nu)$ is the specific intensity (flux of energy per unit area per unit time per unit frequency interval per unit solid angle), ν being the radiation frequency, $k(\nu)$ is the mass absorption coefficient, $j(\nu)$ is the specific mass emissivity, ρ is the density of the material, and s is the distance measured along the direction of transfer. The mass absorption coefficient is the total absorption cross-section per unit mass, that is, the cross-section or target area per absorber times the number of absorbers per unit mass. The quantity $\mu(\nu) = \rho k(\nu)$ is therefore the total absorption cross-section per unit volume and is usually known as the volume absorption coefficient, and its reciprocal $\lambda(\nu) = 1/\mu(\nu)$ is the mean free path. The mass emissivity is the total radiation energy emitted per unit time per unit frequency interval per unit solid angle per unit mass, and the quantity $\varepsilon(\nu) = \rho j(\nu)$ is therefore the volume emissivity. It should be noted that $k(\nu)$ is really only an effective absorption coefficient because, as well as terms representing pure absorption, it also contains terms representing stimulated emission (as negative absorption) and scattering (as positive absorption), both of which are proportional to the incident intensity. Likewise, $j(\nu)$ is only an effective emissivity because, as well as terms representing pure emission, it also contains terms representing scattering. The transfer equation may be conceived of as a transport equation for the radiation or photon "fluid" traversing a space in which there are distributed sources and sinks.

The absorption (and emission) of radiation by atoms is determined by quantum mechanical laws which describe the radiative processes in terms of the initial and final states of the atom and of the radiation field. Thus the spectrum depends on the initial state or, in the case of an assembly of atoms, on the distribution of the atoms among their possible initial states. In addition, for a given initial state, an atom may interact with a radiation field in a variety of ways. It is usual, and convenient, to distinguish between free-free, bound-free, and bound-bound transitions according to whether the initial and final states belong to the free (continuous) or bound (discrete) spectrum of states. Whereas the free-free spectrum is characterized by a continuous variation with frequency, the bound-free spectrum shows the sharp discontinuities or edges at the threshold frequency, and the bound-bound spectrum consists only of sharp lines, indicating that transitions are only possible at or near certain frequencies, so the $k(\nu)$ and $j(\nu)$ are exceedingly complex functions of frequency. Fortunately, two circumstances serve to alleviate the problem of handling the frequency-dependent radiative transfer equation. First, it may be assumed that in the stellar interior thermodynamic equilibrium between matter and radiation obtains to a high degree locally (the assumption of local thermodynamic equilibrium). Justification of this assumption rests on noting that the mean free path of a photon is small and, associated with this, the temperature gradient is also small. The net flux of radiative energy then only arises from the extremely small departure of the radiation field from isotropy in the direction of the temperature gradient. Thus locally the radiation intensity may be taken, except in calculating the flux, to be given by the Planck function for the local temperature T, and the distribution of atomic states may be taken as given by the statistical thermodynamic laws of excitation (Boltzmann equation), ionization (Saha equation), and dissociation at the local temperature T and density ρ. Second, the overall structure of a star is conditioned by the total rate of energy transfer integrated over all frequencies rather than by the energy at particular frequencies. Rosseland showed that the total energy flux carried by radiation is then given by

$$F = -(4acT^3/3\rho k)\, dT/dr = -K\, dT/dr,$$

where a is the radiation constant, c is the velocity of light, and dT/dr is the radial temperature gradient in the star. The quantity k is known as the Rosseland mean absorption coefficient or opacity, defined by

$$1/k = \int [1/k(\nu)][dB(\nu,T)/dT]\, d\nu \Big/ \int [dB(\nu,T)/dT]\, d\nu,$$

where $B(\nu, T)$ is the Planck specific intensity for blackbody radiation at temperature T. The quantity K may be regarded as the effective radiative conductivity. In other words, the Rosseland mean absorption coefficient k is a harmonic mean of the frequency-dependent absorption coefficient $k(\nu)$ weighted with the temperature derivative of the Planck function. That it is a harmonic mean is due to the fact that while conductivity combines additively, its reciprocal, resistivity (here represented by opacity) combines harmonically.

OPACITY CALCULATIONS

The calculation of the opacity, at a given temperature and density for a specified chemical composition, therefore involves as a first step the solution of the equations of thermodynamic equilibrium to obtain the state of dissociation, ionization, and excitation of the atoms. The second step is to calculate the mass absorption coefficient at each of a number of frequencies sufficient to perform the final frequency averaging. A knowledge of the cross-section for absorption by all the more important classes of absorbers is thus required. For some cases, such as the scattering of radiation by free electrons or the bound-free and bound-bound absorption by hydrogen-like ions, there exist relatively simple analytical formulas derived from quantum mechanics. In most cases, however, particularly for atoms with more than one electron, appeal has to be made to the results of detailed numerical calculations and/or to laboratory measurements. Additional factors, such as the perturbations of atomic levels and the broadening of spectral lines by interactions with neighboring particles, mean that corrections have to be applied to the results of calculation or measurement for isolated atoms.

At the lowest temperatures found in stars ($T < 5000$ K) the main sources of opacity are band absorption by molecules, Rayleigh scattering by neutral molecules and atoms, and bound-free and free-free absorption by negative ions, of which H^- is particularly effective in solar-type stars. At higher temperatures ($T > 5000$ K) the opacity is chiefly due to bound-bound and bound-free absorption by atoms and positive ions. At still higher temperatures (typically around 10^6 K) free-free absorption by electrons in the fields of positive ions becomes the major contribution to opacity. Finally, at the highest temperatures the only important opacity source is the (Thomson or Compton) scattering of photons by free electrons. Thus at a given density, as the temperature increases, the radiative opacity rises sharply from a low value to a maximum around the ionization temperature of the most abundant elements. Thereafter the opacity decreases to the scattering limit $k = 0.2(1 + X)$ cm^2 g^{-1}, independent of temperature and density, where X is the mass fraction of hydrogen in the stellar material. At the highest densities, which are associated with the highest temperatures in the cores of stars, particulary in advanced stages of evolution, for example, red giants and white dwarfs, the mean free path of electrons becomes sufficiently long for electron thermal conductivity to compete with radiative conductivity in the transport of energy.

Additional Reading

Carson, T. R. (1972). Stellar opacity. In *Stellar Evolution*, H.-Y. Chiu and A. Muriel, eds. The MIT Press, Cambridge, p. 427.

Chandrasekhar, S. (1939). *An Introduction to the Study of Stellar Structure*. University of Chicago Press, Chicago.

Cox, J. P. and Giuli, R. T. (1968). *Principles of Stellar Structure*. Gordon and Breach, New York.

Kippenhahn, R. and Weigert, A. (1990). *Stellar Structure and Evolution*. Springer-Verlag, Berlin.

Schwarzschild, M. (1958). *Structure and Evolution of the Stars*. Princeton Univeristy Press, Princeton.

See also **Radiation, Scattering and Polarization; Stars, Atmospheres, Radiative Transfer; Stars, White Dwarf, Structure and Evolution; Stellar Evolution, all entries; Sun, Interior and Evolution.**

Stars, Long Period Variable

Mahendra S. Vardya

Long period variable stars (LPV) are pulsating red giant and supergiant stars with visual amplitudes from about 2.5–9 mag (i.e., they are 10–4000 times brighter at maximum visible light than at minimum) and periods between 80 and 1000 days. Near maximum light, their spectra show bright hydrogen emission lines. Most of the stars are of spectral type Me, though some are of types Se and Ce (or R and N) also. A typical representative is Mira Ceti (o Ceti).

The first star to be found to vary periodically in brightness was omicron (o) in the constellation Cetus, the Whale. It was discovered as a previously unlisted star on 13 August 1596 by the amateur astronomer David Fabricius. It was the Dutch astronomer Phocylides Holwarda who found its variable nature in 1638, but it was not until 1662 that the star was shown to vary from second to tenth magnitude in a period of 330 d. At that time, this being the only known variable star, it was named Mira, the wonderful (or magical). Chi (χ) Cygni was discovered to be a Mira-type variable by Gottfried Kirch in 1681, R Hydrae by G. F. Maraldi in 1704 and R Leonis by Koch in 1782.

Long period variable stars show considerable irregularity not only in their light amplitude and the shape of light curve (i.e., variation of brightness with time), but in the length of a cycle, which may vary by as much as 10% (Fig. 1). For example, the visual light maximum of Mira can vary from first to fifth magnitude; the minima can also vary, but not to the same extent.

LIGHT CURVES

The fact that period, amplitude, and shape of the light curve varies from cycle to cycle, requires regular observations of long-period variables. In this task, amateur astronomers have rendered immense help to the professional astronomers by observing these stars. The American Association of Variable Star Observers (AAVSO) in Cambridge, MA is the central organization that collates observations from different observers and societies throughout the world and makes them available to users.

The shapes of the light curve, though different for different stars, and even for the same star from one cycle to the other, can be classified into three broad groups: (a) α type: the ascending or rising branch is noticeably steeper than the descending branch, and the minimum is always broader than the maximum; (b) β type: the light curve is basically symmetric; (c) γ type: the ascending branch has step(s) or hump(s), or the light curve has a double maximum. The first two types are subdivided into four subclasses each and the γ type into two. Most of the M-spectral-class long period variables

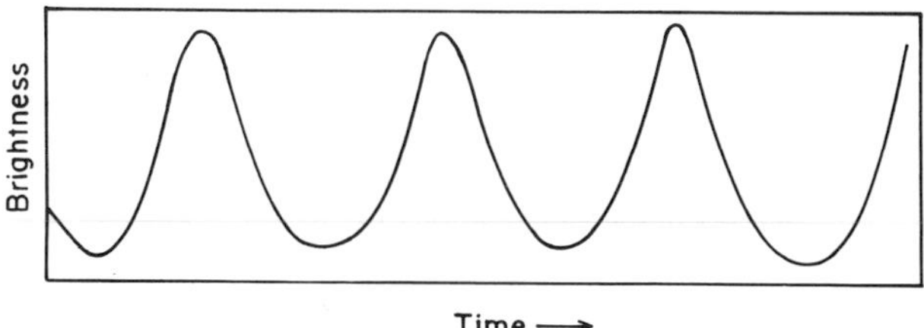

Figure 1. Visual light curve of a long period variable star.

belong to the α type, the C class are mostly of the β and γ types, whereas the S class are equally distributed among the types.

The asymmetry of the light curve is also defined by the factor f, the fraction of time spent from the minimum of light curve to the next maximum relative to the period. If f is one-half, the light curve is symmetric and if it is less than one-half, then the ascending branch is faster than the descending branch.

PERIOD AND AMPLITUDE

Long period and Mira-type variable stars are synonymous, though a true Mira has been arbitrarily defined as having a visual light amplitude of at least 2 or 2.5 mag. This implies a lower limit to the period of a Mira variable of 80 days. In practice this does not matter, as stars between 50 and 80 days are, in general, irregular and do not show characteristic hydrogen emission lines. The distribution of the numbers of stars of given period shows broad peaks around 300 days for M-spectral-type stars, 360 days for S stars, and at 400 days for C stars.

Though the mean period of a long period variable remains more or less constant, some stars show (a) decrease or increase in mean period with time, including an abrupt change, and (b) change in the light-curve shape between two or several consecutive periods. For example, the period of R Hydrae was 507 days in the early eighteenth century, but has now come down in several steps to 389 days.

Though the amplitudes of long period variable stars are large in visual light, they are only a few magnitudes in the infrared or in the total light output (bolometric magnitude). This is because these stars radiate mostly in the infrared and the dark molecular oxide bands in the visual greatly suppress the intensity at light minimum. The mean visual amplitude is 5.4 mag for M-spectral-type stars, 6.4 mag for S stars, and 4.6 mag for C stars. The average visual amplitude increases as the period increases. A representative light curve is shown in Fig. 1.

Type II OH/IR objects are strong maser sources with the 1612-MHz satellite line of OH amplified, and are associated with infrared point sources. Most of them are not observable in the visible, as they are surrounded by thick circumstellar dust shells. Their infrared light curves show properties akin to those of long period variables, and hence the objects can be considered as the extension of long period variable stars to periods as long as 2000 days or even more.

SPECTRA

M-spectral-class stars are characterized by broad, dark (absorption) bands of titanium and vanadium oxides (TiO, VO), S stars are characterized by zirconium oxide (ZrO), and C stars by molecular carbon, cyanogen, and silicon carbide (C_2, CN, SiC_2) bands. Stars of longer period have later spectral types, show stronger absorption bands, and have cooler atmospheres. When emission lines are present, *e* is added to the spectral class, as in Me. In M and S stars, the elemental abundance ratio of oxygen to carbon is greater than one; though for S stars it is very close to unity, whereas for C stars it is less than one. As carbon monoxide (CO) is a very stable compound, almost all carbon is tied up as CO in M and S stars and the excess oxygen is available for the formation of molecular oxides, whereas in C stars, all oxygen is tied up as CO and excess carbon is available for formation of carbon compounds. The lines of hydrogen are normally absent except when they appear in emission around the time of light maximum. M stars show bands of yttrium oxide, lanthanum oxide, and silicon hydride; some of the stars show broad silicate grain emission in the infrared at 9.7 and 18 μm, and maser emission lines of OH, H_2O, and SiO in the millimeter and centimeter wavelength regions of the radio spectrum.

Many variable S stars show absorption lines of technetium, and some M and C stars do so as well. This was an amazing discovery by Paul W. Merrill in 1951, as Tc is a very unstable radioactive element with a half-life of only 200,000 years for the most likely isotope.

C stars exhibit in the visible bands of CN, C_2, SiC_2, and CH, and in the infrared exhibit bands of CN, C_2, HCN, C_2H_2, and some of them show a silicon carbide (SiC) grain-emission feature at 11.3 μm. In the radio region, emission lines due to CO, CN, CS, C_2H, HNC, HCN, C_3N, and other complicated molecules have been detected. In the remarkable carbon star, CW Leo, better known as IRC+10°216, a large number of molecules have been detected in the infrared and radio region, including C_3, C_5, OH, SiO, several molecules of the cyanopolyyne family ($HC_{2n+1}N$), and the ring molecule C_3H_2.

LONG PERIOD VARIABLES IN CLUSTERS AND EXTRAGALACTIC SYSTEMS

Long period variable stars have been found in open clusters, globular clusters, and in our two neighboring irregular galaxies—the Large Magellanic Cloud (LMC) at a distance of 46 kpc (1 kpc equals approximately 3×10^{16} km or 3.3 thousand light years), and the Small Magellanic Cloud (SMC) at 53 kpc. In globular cluster 47 Tucanae, six long period variables have been found; these have high space velocity with short periods, less than 250 days, and belong to Population II, that is, they are old stars. Population I stars, that is, young stars, have generally longer periods with low space velocities. In the Magellanic Clouds, a very large number of long period variables have been studied, giving valuable information about the nature of these stars. These are asymptotic giant branch stars, with degenerate carbon/oxygen cores in the center and hydrogen- and helium-burning shells around the core, going up the giant branch for the second time in the Hertzsprung–Russell diagram. Their maximum luminosity is 5×10^4 times the solar luminosity (solar luminosity $= L_\odot = 3.9\times10^{33}$ erg s^{-1} $= 3.9\times10^{26}$ J s^{-1} $= 3.9\times10^{26}$ W), and their masses lie between 0.7 and 7 solar masses (solar mass $= M_\odot = 2\times10^{30}$ kg). Asymptotic giant branch stars increase their luminosity with time and, correspondingly, increase their period. These stars are pulsating radially and are driven mainly by the ionization of hydrogen. Most of them are pulsating in the fundamental mode. The galactic Type II OH/IR stars with a period of 1000 days or more may be in a transition phase between long period variables and planetary nebulae.

PHYSICAL PROPERTIES

Masses of long period variable stars are not known accurately. Stellar evolution calculations indicate that the masses lie between 0.7 to 1.5 $M_\odot$ for stars with period less than 500 days. The masses may be larger for longer-period stars.

The maximum radius of long period variable stars ranges from about 1.5–2.7×10^{11} m (or 215–390 $R_\odot$, where $R_\odot$= radius of the Sun $= 6.96\times10^8$ m) at minimum brightness. Radius is minimum at the time of light maximum and the radius variation is about 20%. The effective temperature of these stars lies between 3700 and 2500 K, though the temperatures are not yet known accurately.

The visual absolute magnitude of Miras lies between 0 to -3 mag (80–1300 times the solar luminosity). Whereas visual absolute luminosity decreases with increase in period for stars with periods greater than 200 days, the total or bolometric luminosity increases with increase in period. It is found that the period–luminosity relation for periods greater than 200 days is satisfied by Type II OH/IR sources also.

Long period variable stars have stellar winds with velocities of 10–20 km s^{-1}, through which they are subject to great mass loss. The rate of mass loss varies between 10^{-7}–10^{-5} solar masses per year. The mass loss from OH/IR objects is even larger. M and C stars have similar mass loss rates. Mass loss can be continuous or sporadic or both.

Mass loss creates a circumstellar shell or envelope around the star that contains solid grains (dust) that have condensed from the gas. The grains absorb visual light and reradiate it as excess infrared radiation. Sometimes, the envelope adds its direct signature in the form of emission by silicates at wavelengths of 9.7 and 18 μm and by silicon carbide at 11.3 μm. The dust grains thermalize the radiation received from the star and then emit the absorbed radiation at a temperature from about a few hundred to 1000 K.

Long period variables are pulsating radially. This causes shock waves in the outer layers of these stars. Velocities associated with the shock wave front are rather small, between 10 and 15 km s^{-1}, but this is three–four times the speed of sound in these cool tenuous layers. Asymmetry of light curves, variability of emission lines, doubling of absorption lines in the near infrared, and even condensation are manifestations of shock waves.

EVOLUTIONARY STATE

Stellar evolutionary calculations indicate that long period variable stars are objects of about one solar mass and are in the asymptotic giant branch phase of evolution as described earlier. These stars are pulsating and losing mass. The rate of mass loss increases as the period increases. These stars are not only losing mass, but material dredged up from the deep interior is also brought to the surface during the course of evolution. It is thought that the evolution proceeds from M stars to S stars and to C stars, but recent evidence suggests that the intermediate S phase may occasionally be skipped. These stars may end up as very condensed white dwarf stars or as planetary nebulae, which have degenerate central condensed nuclei of about half a solar mass at a very high temperature, each surrounded by an ionized tenuous envelope.

Additional Reading

Campbell, L. and Jacchia, L. (1946). *The Story of Variable Stars*. The Blakiston Company, Philadelphia.

Glasby, J. S. (1968). *Variable Stars*. Constable and Company, London.

Gorbatskii, V. G. (1969). *Physics of Stars and Stellar Systems*. Israel Program of Scientific Translations, Jerusalem, Chap. 8.

Hoffmeister, C., Richter, G., and Wenzel, W. (1985). *Variable Stars*. Springer-Verlag, Berlin.

Johnson, H. R. and Querci, F. R., eds. (1986). *The M-Type Stars*. NASA SP-492, National Aeronautics and Space Administration, Washington, DC.

Kholopov, P. N. (1985). *General Catalogue of Variable Stars*. Nauka Publishing House, Moscow, Vols. 1–3.

See also **Stars, Carbon; Stars, Circumstellar Masers; Stars, Pulsating; Stellar Evolution, Low Mass Stars.**

Stars, Low Mass and Planetary Companions

James Liebert

One of the most intriguing observational problems in astronomy is the search for planets around stars other than our Sun. Substellar bodies could encompass a much wider range of mass than is represented in our solar system, as the hydrogen-burning mass limit is some 80 times the mass of Jupiter (M_J) or 0.08 $M_\odot$ (solar masses). Because cool, low-mass main-sequence stars are traditionally called red dwarfs or M dwarfs, the term brown dwarfs has been adopted in recent years to refer to these generally dimmer, substellar analogues that might be found either as stellar companions or existing alone in space. Near the time of formation, substellar objects emit substantial rates of primarily infrared radiation due to the release of gravitational energy. Young brown dwarfs may thus resemble the lowest-mass hydrogen-burning stars in radius, luminosity, temperature, and observable spectrum, as we shall see. Moreover, theoretical calculations indicate that transition objects with masses just below 0.08 $M_\odot$ may burn hydrogen in the core for a time $\geq 10^9$ yr. This energy release is insufficient to stop completely the gravitational contraction, however, so that these objects become electron degenerate and begin cooling. As a result of these ambiguities, it is likely that some objects announced as being brown dwarfs in recent years are more likely to be stable, hydrogen-burning stars. It has been extremely difficult in the flurry of recent searches for low-luminosity companions to establish even one object of certifiably substellar mass.

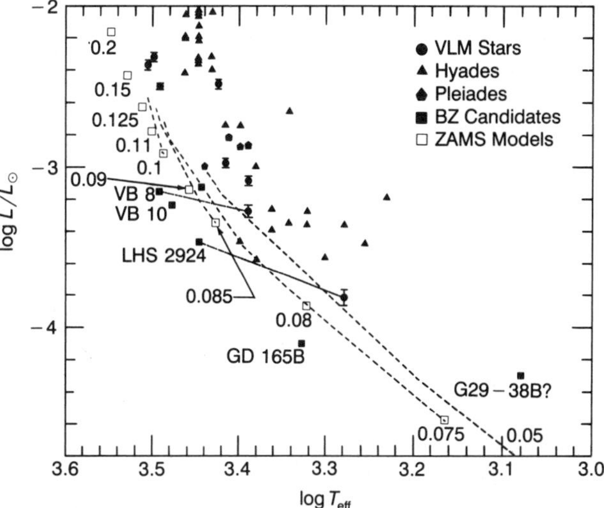

Figure 1. The Hertzsprung–Russell diagram of M dwarfs and brown dwarf candidates (solid circles), plus several of the recently discovered companions discussed in this entry with best estimates for log L and T_e taken from the literature. Positions of very low mass (VLM) stars are taken from Berriman and Reid (1987). Hyades and Pleiades candidates are taken from Leggett and Hawkins and Stauffer et al. (1989), respectively. The BZ candidates are from Becklin and Zuckerman (1989). Two published positions for the stars vB8 and LHS2924 are connected by lines. Model tracks (dashed lines) labeled by masses in units of the solar mass are courtesy of Dr. A. Burrows.

For stars not in fairly close binary systems, the mass is one of the most difficult of the fundamental stellar parameters to estimate. This is a two-step process. First, one needs to determine from fits to observations the basic parameters luminosity L and effective temperature T_e that specify where a stellar object lies, on or near the low end of the main sequence in the Hertzsprung–Russell diagram. Then we can compare this position with stellar interior evolutionary tracks in order to estimate its mass, and in particular, to determine whether it is substellar. However, as can be seen from Fig. 1, the $L - T_e$ slopes predicted for the very low-mass main-sequence and for the brown dwarf contraction-cooling sequence are remarkably similar and nearly overlapping at radii near $\sim 0.1\ R_\odot$. The precision with which L and especially T_e can be determined is far less than that needed to resolve the two sequences. Also shown in Fig. 1 are data points representing best estimates for objects believed to be among the intrinsically faintest stars and brown dwarf candidates; these include both field stars and objects found as companions. The observations follow at least approximately the predicted slope, but the radii appear to be larger than predicted for stable, main-sequence stars of lowest mass. However, the derived T_e values currently depend on fits to the observed energy distributions using blackbodies, that is, without the benefit of synthetic spectra predicted from realistic model atmospheres. The luminosities are on somewhat less shaky ground, because accurate trigonometric parallaxes are available for many objects and the bolometric correction has only a modest dependence on the uncertain T_e values. It is little wonder that the agreement between theory and observation in Fig. 1 is poor.

Another parameter might in principle be used to resolve this ambiguity: For the substellar object to be emitting substantial luminosity, it must still be in a phase of gravitational contraction and/or limited nuclear burning, and therefore generally should be younger than a main-sequence star of similar luminosity. However, the age of a field star—one not a member of a cluster or moving group—may be estimated only crudely. The kinematics (space motions) and the degree of chromospheric or coronal activity may be used to estimate ages for statistical samples of stars, although it is precarious to apply such an approach to an individual case. Moreover, the decline in chromospheric-coronal activity with age is not well calibrated for solar-type stars; it is not well understood for very low-mass configurations having completely convective interiors.

For stellar objects in close binary systems, the masses may be determinable directly by analysis of the binary orbits. The observational techniques may include direct photography of visual binaries near enough to the Sun (or more distant ones that can be imaged from space), astrometric perturbation analyses, and speckle interferometry for barely resolvable pairs and/or spectroscopic measurements of radial velocities for unresolved and close pairs. Such studies of binary systems offer the best hope for unambiguous identification of a substellar companion to a nearby star. Results from these and other techniques applicable to more distant companions are discussed below.

DISCOVERY AND ANALYSIS OF LOW-MASS COMPANIONS

Faint Companions with Common Proper Motion

Photography of the fields around nearby stars has been employed for many decades to discover wide, common proper motion companions of much lower luminosity. Actually these discoveries are often just a consequence of the series of photographic plates required to determine the trigonometric parallax of the brighter binary component. The very low-luminosity benchmark main-sequence stars van Biesbroeck 10 and 8 were discovered in this manner; with an estimated $\log L/L_\odot = -3.3$, vB10 for a long time was the least-luminous known M dwarf star. Single objects of such low luminosities have been found in follow-up photometric and spectroscopic observations of the faintest stars of large motion in proper-motion surveys; LHS2924 from the Luyten Half Second Survey (which lists stars with proper motions of greater than 0.5 arcseconds per year) is believed to be significantly dimmer than vB10. It is generally believed that these may have masses at or slightly above the hydrogen-burning limit. Neither has a companion close enough for an astrometric mass determination.

Photographic Image Perturbations and Speckle Interferometry

Astrometric analyses of time series of photographic plates have revealed closer companions with masses near or below the hydrogen-burning mass limit. Barnard's Star, the Sun's nearest neighbor after the Alpha Centauri system, has long been suspected to have one or two planets with masses comparable to Jupiter's. Wolf 424 may itself be a pair of substellar mass objects, with masses near 0.05 and 0.06 $M_\odot$ (50 and 60 M_J), circling each other with an orbital period near 16 yr; this result is based on 50 years of observations at Sproul Observatory, although it should be noted that the derived masses may depend on the precision with which actual separations may be determined from somewhat blended images. Additional candidates have been identified in particular from the parallax plate series (and now from images obtained with charge-coupled devices) of the U.S. Naval Observatory.

Speckle interferometry at infrared wavelengths is an optimal tool for the discovery of dim, red, close companions. A systematic study of 27 known stars out to 5 pc from the Sun found a few new low-mass stellar companions, but no convincing cases of substellar companions. For this sample, the technique is sensitive to projected separations of 0.2″–5″, corresponding to physical separations of a few to 25 a.u., and for magnitudes $M_K = +11.5$ or brighter. Note that LHS2924 has $M_K = +10.5$ and GD165B (see below) has $M_K = +11.8$; only young brown dwarfs or objects with long, transitory nuclear-burning phases would be detectable. Infrared speckle interferometry has also provided many more data points for the comparison of the relation of L and mass, although the ages of the objects are generally unknown. For several cases the derived error bars for the mass straddle the stellar lower limit.

Searches for Velocity Variations

A complementary technique for uncovering unresolved companions to nearby stars is the search for radial-velocity variations due to orbital motion around the center of mass with unseen companions. A combination of precision and high spectral resolution is required, but the stars surveyed do not necessarily need to be solar neighbors. The most systematic search to date of some 70 low-mass stars, sensitive generally to orbital separations of up to several astronomical units, uncovered only one possibly substellar companion, associated with the star Gliese 623. However, follow-up studies using techniques discussed previously suggest that the mass of the companion is very close to the 0.08-$M_\odot$ limit. In another survey, British Columbian researchers at the Canada-France-Hawaii Telescope employed a unique spectrograph able to measure radial velocities of bright stars to a relative accuracy of 13 m s^{-1}. However, multiyear observations of 16 more luminous, F–K dwarfs and subgiants yielded no companions with masses in the 10–80-M_J range for separations <10 a.u. A few stars showed significant but small long-term variations indicative of the possibility of planetary companions with masses below 10 M_J.

In a somewhat circumstantial discovery, workers at the Smithsonian Center for Astrophysics discovered that the G dwarf HD 114672 shows modest, periodic velocity variations indicative of a companion which could have a mass as low as 11 M_J. The "catch" is that, as in any noneclipsing single-lined spectroscopic binary solution, the inclination of the orbit is unknown; if the system were by chance viewed close to pole-on, the companion to HD 114672 might exceed the stellar lower mass limit.

Imaging with Solid State Array Detectors

The development of sensitive, two-dimensional astronomical detectors—both optical charge-coupled devices and several types of infrared arrays—may finally help astronomers establish whether brown dwarf companions exist in large numbers, or whether they are rare. Imaging of the fields of nearby stars at 2.2 μm is able to test for separated companions of much lower luminosity than the use of optical wavelengths. Note, however, that if the brown dwarf were fairly old, it would not emit enough infrared flux for detection. Systematic surveys covering several dozen solar neighbors have so far failed to yield any definitive brown dwarfs.

Ironically, the best candidate to be found in this manner is a companion to the white dwarf GD165, spatially resolved in infrared images and having an indicated luminosity $\log L/L_\odot \leq -4$. Depending on the manner in which the T_e is crudely estimated, and the theoretical tracks adopted, even this object might have a mass as high as 80 M_J. Earlier, another white dwarf (G29–38) showed an unresolved infrared excess indicative of a possible companion that could be even dimmer and cooler at $\log L/L_{\odot} \sim -4.3$ and ~1200 K; however, the origin of this excess is currently in dispute.

The most promising use of the detector arrays is in the search for low-mass objects in young clusters and stellar associations, in which any substellar object would still be in a stage of gravitational contraction and emitting significant infrared flux. Studies of the Hyades and Pleiades clusters, in particular, have uncovered numerous candidates whose nature must be established by careful follow-up observations.

STATUS OF THE SEARCH

The preceding is a brief summary of the approaches currently being utilized in the numerous attempts to find stellar companions of very low mass and to identify objects of substellar mass. The mass function (distribution function of stellar masses) of stars in the local galactic disk is dominated by M dwarf stars, and a fair fraction of the well-studied nearby stars have low-mass companions including some very near the $0.08\text{-}M_{\odot}$ limit. However, there is currently no established, unambiguous case of a brown dwarf companion to a star other than our Sun. The current working hypothesis, resulting especially from the systematic infrared speckle and radial-velocity surveys, is that substellar companions between 80 and perhaps as low as 10 M_J are relatively rare. Substellar objects most likely to have been missed are old, low-mass, and/or very distant companions. It is also possible that some objects with estimated masses near 80 M_J are in fact substellar in mass, including objects which may be undergoing a transition phase of thermonuclear burning. Improved astrometric analyses using infrared arrays and/or space-based imaging should improve the accuracy of the mass determinations. The current flurry of activity also promises more interesting discoveries and analyses in the relatively near future.

Additional Reading

Becklin, E. E. and Zuckerman, B. (1988). A low-temperature companion to a white dwarf star. *Nature* **336** 656.

Berriman, G. and Reid, N. (1987). Observations of M dwarfs beyond 2.2 μm. *Mon. Not. Roy. Astron. Soc.* **227** 315.

Campbell, B., Walker, G. A., and Yang, S. (1988). A search for substellar companions to solar-type stars. *Astrophys. J.* **331** 902.

D'Antona, F. and Mazzitelli, I. (1985). Evolution of the very low mass stars and brown dwarfs. I. The minimum main-sequence mass and luminosity. *Astrophys. J.* **296** 502.

Heintz, W. D. (1989). The substellar masses of Wolf 424. *Astron. Ap.* **217** 145.

Henry, T. J. and McCarthy, D. W., Jr. (1990). A systematic search for brown dwarfs orbiting nearby stars. *Astrophys. J.* **350** 334.

Jameson, R. F., and Skillen, I. (1989). A search for low-mass stars and brown dwarfs in the Pleiades. *Mon. Not. Roy. Astron. Soc.* **239** 247.

Kafatos, M. C., Harrington, R. S., and Maran, S. P., eds. (1986). *Astrophysics of Brown Dwarfs*. Cambridge University Press, Cambridge, U.K.

Latham, D. W., Mazeh, T., Stefanik, R. P., Mayor, M., and Burki, G. (1989). The unseen companion of HD114762: A probable brown dwarf. *Nature* **339** 38.

Leggett, S. K. and Hawkins, M. (1989). *Mon. Not. Roy. Astron. Soc.* **238** 145.

Liebert, J. and Probst, R. (1987). Very low mass stars. *Ann. Rev. Astron. Ap.* **25** 473.

Marcy, G. W. and Benitz, K. J. (1989). A search for substellar companions to low-mass stars. *Astrophys. J.* **344** 441.

Probst, R. and Liebert, J. (1983). LHS 2924: A uniquely cool low-luminosity star with a peculiar energy distribution. *Astrophys. J.* **274** 245.

Stauffer, J., Hamilton, D., Probst, R., Rieke, G., and Mateo, M. (1989). Possible Pleiades members with $\mathscr{M} \simeq 0.07\ M_{\odot}$: Identification of brown dwarf candidates of known age, distance, and metallicity. *Astrophys J. (Lett.)* **344** L21.

Stevenson, D. J. (1991). *Ann. Rev. Astron. Ap.* **29**.

van Biesbroeck, G. (1961). The star of lowest known luminosity. *Astron. J.* **51** 61.

van de Kamp, P. (1971). The nearby stars. *Ann. Rev. Astron. Ap.* **9** 103.

Zuckerman, B. and Becklin, E. E. (1987). Excess infrared radiation from a white dwarf—an orbiting brown dwarf? *Nature* **330** 138.

See also **Planetary Systems, Formation, Observational Evidence; Stellar Evolution, Low Mass Stars.**

Stars, Magnetic and Chemically Peculiar

Sidney C. Wolff

It was nearly a century ago that astronomers identified a distinct subclass of peculiar stars that distinguished themselves from the more common normal stars by having abnormally strong lines of particular elements. Much more recently, observations have shown that these so-called magnetic and chemically peculiar stars also have strong magnetic fields, and that these strong fields hold the key to accounting for the many unusual characteristics of the peculiar stars.

SPECTRAL PECULIARITIES

The prototypical member of this class of stars is α^2 CVn, which was first classified as peculiar in 1897, primarily because it had anomalously strong lines of Si II. We now recognize that there is a sequence of peculiar stars ranging in temperature from slightly more than 20,000 to 8,000 K, or, equivalently, from spectral types early B to about F0. Because many of these stars have temperatures equivalent to stars of spectral type A, and others have unusually weak helium lines for their temperature and so mimic spectral type A0, the entire group is often referred to as Ap ("peculiar A") stars.

Although Ap stars of a given temperature can differ substantially in their spectroscopic characteristics, there is a tendency overall for the most obvious peculiarities to depend systematically on temperature. Various subgroups of peculiar stars have been identified and named according to the dominant peculiarity observable at low spectral resolution. The hottest subgroup is characterized by weak lines of He. At slightly lower temperature, unusually strong lines of Si II become the defining characteristic. At temperatures of about 10,000 K, lines of Cr and Eu are strongly enhanced, and at still cooler temperatures Sr lines are overly strong.

Coexisting in temperature with the magnetic peculiar stars is a second class of peculiar stars without magnetic fields and with somewhat different spectral anomalies. Included in this group are the Mn stars, which have temperatures equivalent to normal stars of late-B spectral types, and the metallic-line stars, which are late-A-type stars. All of these peculiar stars, with or without magnetic fields, are main-sequence stars.

MAGNETIC FIELDS

The Ap stars were the first stars other than the Sun in which magnetic fields were directly measured. In principle, it is possible to measure either transverse or longitudinal magnetic fields in stars. In practice, for stars it has proven much easier to measure longitudinal fields, which produce spectral line shifts, than transverse fields, which produce line broadening. Accordingly, for most

Ap stars, we have data only on the longitudinal component of the magnetic field.

The fields in the Ap stars are found to be strong and well ordered, and both of these characteristics are essential in making it possible to detect magnetic fields in the integrated light of a star. In the case of the Sun, the fields are strong, often reaching several thousand gauss. However, the strong solar fields are characteristic only of local regions. Integrated over the solar surface, there are approximately as many regions of positive as negative field, and so the net longitudinal component of the solar field, that is, the component along the line of sight, is nearly zero.

In magnetic Ap stars, observations show that to a good first approximation the magnetic fields are approximately dipolar. The net longitudinal fields are typically in the range of a few hundred to 1000–2000 G, and the fields at the poles in exceptional cases exceed 10,000 G. Approximately 5–10% of all main-sequence stars in the spectral range early B to F0 either have magnetic fields or, if their spectral lines are too broad to permit measurement of the field directly, have spectral peculiarities that are apparently found only in stars with strong magnetic fields.

The magnetic fields hold the key to explaining many of the other properties of the Ap stars. For example, the Ap stars as a class rotate more slowly than normal stars of similar temperature. This slow rotation is attributed to magnetic braking, which can be achieved through mass loss. It appears likely that most of this magnetic braking occurs before the Ap stars reach the main sequence. Ap stars with the lowest masses also have the slowest rotations, presumably because they evolve more slowly than more massive Ap stars and the braking mechanism has a longer time to slow the rotation.

RADIATIVE DIFFUSION

The magnetic field combined with slow rotation also plays an essential role in accounting for the chemical peculiarities of the Ap stars. The basic premise of the theory is that, in a stable atmosphere, those elements that experience an excess force caused by radiation pressure, transferred either through bound-bound or bound-free atomic transitions, will be driven upward in the atmosphere and concentrated in the line-forming regions. Those elements that have few transitions in the wavelength region where the stellar flux is at its maximum will tend to sink. This process is called radiative diffusion. It will occur only in slowly rotating stars where meridional circulation is too weak to lead to significant mixing.

Magnetic fields can influence the diffusive separation of elements by suppressing macroscopic motions and by influencing the motions of ionized particles. If densities are low and collisions infrequent, then ions will tend to spiral along magnetic field lines. As an ion diffuses upward in a stellar atmosphere into regions of decreasing density, it will follow the field lines more and more closely. Diffusion will stop where the field lines are horizontal. The distribution of elements over the stellar surface will then reflect the magnetic geometry. Depending on the details of the magnetic geometry, diffusion may cause the elements to be distributed in patches or rings.

RIGID ROTATOR MODEL

The radiative diffusion model offers an explanation of one of the other properties of Ap stars—namely, that they are spectrum variables. The strengths of the spectral lines of certain elements vary in a regular cycle, and observations show that the period of variation is equal to the period of rotation of the star. Long before calculations of the effects of diffusion were carried out, observations of the spectrum and magnetic variability of the Ap stars had led to the rigid rotator or oblique-dipole rotator model. The basic premise of the model is that, in analogy with the Sun and Earth, the axis of the stellar magnetic field, which is taken to be approximately dipolar, is inclined at an angle with respect to the rotation axis. The field is assumed to be locally constant, that is, "frozen in" to the surface of the star, and to corotate with it. To a distant observer, the magnetic field will, therefore, appear to vary in intensity as the star rotates. The spectrum variations can be explained by the assumption that the variable elements are concentrated in patches on the stellar surface. Detailed diffusion calculations for specific elements yield in many cases surface distributions for elements with respect to the magnetic geometry that are consistent with observations.

The nonuniform distribution of chemical abundances over the stellar surface also accounts in part for the fact that, in addition to being spectrum and magnetic variables, the Ap stars vary in brightness, typically by a few percent. The light variations are in large part a direct consequence of the spectrum variations. Changes in the opacity in the ultraviolet region of the spectrum will cause a redistribution of flux in the visible and, hence, will produce changes in brightness. An anticorrelation between brightness variations in the ultraviolet and optical regions of the spectrum is frequently observed. Various elements, including Si, Fe-peak elements, and the rare earths, have all been suggested as being the primary source of variable opacity, and probably they all contribute.

In addition to varying in brightness in a period equal to the rotation period, many Ap stars are pulsationally unstable and vary on time scales of minutes with typical amplitudes of 1%. The observations can be interpreted on the assumption that the star pulsates about the magnetic axis rather than the rotation axis. Maximum radial displacements occur at the magnetic poles, where the motion is along field lines. No radial motion occurs at the equator, where the field lines are parallel to the stellar surface.

ORIGIN OF THE MAGNETIC FIELD

There are two fundamental questions that must be resolved by theoretical treatments of magnetism in A-type stars: What is the origin of the observed fields? Why are fields observed in some, but not all, A stars?

There are two possibilities for accounting for magnetic fields in stars. In the case of solar-type stars, the magnetic field is generated and maintained by a dynamo mechanism. Although this process cannot be ruled out for Ap stars, it is thought to be unlikely. For example, in Ap stars there is no tendency for magnetic fields to increase in strength with increasing rotational velocity, as is true for solar-type stars.

The alternative theory postulates that the magnetic fields are remnants of the field originally present in the interstellar medium, compressed and amplified by the star formation process. Some magnetic flux must be dissipated in the process, because conservation of flux during the protostellar collapse would yield a configuration in which the magnetic energy would exceed the gravitational energy, and clearly under those circumstances, collapse could not occur. The ohmic decay time for A-type stars with dipolar fields is about 10^9–10^{10} yr, and so a fossil field can, in principle, survive long enough to be observed in main-sequence stars.

The second question—why magnetic fields are observed in only some A-type stars—remains unanswered. A part of the explanation probably lies in the competition between meridional circulation and magnetic fields. If meridional circulation is dominant, then the circulation currents will tend to pull the magnetic lines of force beneath the surface of the star and the magnetic field will be unobservable.

There are, of course, many slowly rotating, nonmagnetic stars. The absence of a detectable field may reflect a condition at the time of star formation. One can imagine that stars form with a range of values of angular momenta and magnetic fluxes, and that only those with the right balance of magnetic field and rotational velocity become magnetic stars. Because the factors that shape star formation and the interaction of rotation and magnetic fields early in the history of a star are not well understood, any speculation on such issues must remain no more than that—speculation.

Additional Reading

Borra, E. F., Landstreet, J. D., and Mestel, L. (1982). Magnetic stars. *Ann. Rev. Astron. Ap.* **20** 191.

Cowley, C. R., Dworetsky, M. M., and Mégessier, C., eds. (1985). *Upper Main Sequence Stars with Anomalous Abundances.* D. Reidel, Dordrecht.

Vauclair, S. and Vauclair, G. (1982). Element segregation in stellar outer layers. *Ann. Rev. Astron. Ap.* **20** 37.

Wolff, S. C. (1983). *The A-Type Stars: Problems and Perspectives.* NASA Spec. Publ. **463**. National Aeronautics and Space Administration, Washington, DC.

See also **Stars, δ Scuti and Related Types; Stars, Magnetism, Theory.**

Stars, Magnetism, Observed Properties

Mark S. Giampapa

Magnetic fields play a central role in virtually all aspects of astronomy and especially in the area of stellar astrophysics. Among the most distinctive features that characterize the surface of the Sun and, by implication, the late-type stars, are the atmospheric inhomogeneities that are the sites of magnetic fields. Magnetic field structures are now recognized as a fundamental property of stellar atmospheres.

The magnetic fields on solar-like stars appear in various structures that are consistent with an interpretation within the framework of stellar analogs of sunspots, plages, prominences, and coronal loops. A variety of compelling circumstantial evidence indicates that the origin of stellar chromospheres and coronae is intimately related to interactions between the footpoints of magnetic structures and the turbulent motions of the photospheric gas. Both solar observations and indirect evidence in the case of stars suggests that transient activity, which sometimes appears in the dramatic form of explosive flare events, is also associated with magnetic fields. A key discovery of the past decade was the observation of small ($\sim 0.1\%$) but measurable changes in the solar irradiance. These brightness changes are correlated with magnetic activity on the Sun, such as the disk passage of sunspots as the Sun rotates. Parallel studies of solar-type stars also revealed fluctuations in their luminosity that coincide with the presence of spots and plage-like regions on the stellar surface.

The emergent magnetic flux itself arises from the interaction between convective gas motions, rotation, and magnetic fields through dynamo processes that are believed to occur deep within the stellar interior. Stellar cycles, analogous to the solar cycle, are a manifestation of a fundamental property of the interior dynamo mechanism that is responsible for the generation of stellar magnetic fields. During the 1960s, Robert P. Kraft discovered that solar-type stars rotate more slowly with increasing age. Evry Schatzman had suggested somewhat earlier that the spin down of a star could occur through the application of a kind of magnetic torque by a process referred to as "magnetic braking." The torque is exerted when the magnetic field, through magnetic stresses along curved field lines, forces the matter that composes the wind to rotate with the star out to large distances, where it then carries away significant angular momentum per unit mass. Thus, the interaction between magnetic fields and winds in a rotating, late-type star determines the evolution of the rotation rate of at least the outer convective layers.

Realizing that magnetic field generation and rotation are linked through the dynamo led to the hypothesis that magnetic-field-related activity must also decrease with time. This conjecture was confirmed when Olin C. Wilson noted that the mean level of chromospheric emission in the resonance lines of singly ionized calcium, as observed in stars that are members of open clusters of known ages, exhibited a similar decline with increasing cluster age.

In all these phenomena, an understanding of the generation, emergence, and evolution of magnetic flux becomes crucial because magnetic fields modulate the radiative outputs of stars and affect their angular-momentum evolution. Despite its enormous importance, the theory of stellar magnetism is not well developed, mainly because of the lack of measurements of the properties of magnetic fields on stars.

Methods for the direct measurement of stellar magnetic fields will therefore be discussed, with particular emphasis on measurements in solar-type stars. Other stellar types, such as pulsars, the magnetic white dwarfs, and the metallic-line Am stars, exhibit striking and well-studied magnetic characteristics. They also contribute to the investigation of stellar magnetism and are discussed in other entries in this volume. The techniques that are used to study the magnetic properties of late-type stars, along with the results that have thus far been obtained, is the subject of this entry.

METHODS

The quantitative analysis of stellar magnetism relies on the detection of the *Zeeman effect* in stellar spectra. This effect, which is named after the Dutch physicist Pieter Zeeman, who discovered the phenomenon in 1896, occurs when atoms radiate in the presence of an external magnetic field. Radiation that is normally emitted at a single wavelength is instead split into emissions at two or more slightly different wavelengths. In quantum-mechanical terms, the presence of the external field removes the degeneracy that is associated with the quantized orbital angular momentum of the electron about the nucleus. The quantum states, denoted by m_l, have the property that for transitions corresponding to $\Delta m_l = 0$ in a simple triplet, the emitted radiation is unshifted and linearly polarized. The unshifted feature is referred to as the π component. The π component is seen most strongly when observing with the line of sight perpendicular to the field. It does not appear at all when viewed in a direction along the field lines.

The shifted or σ components that correspond to transitions with $\Delta m_l = \pm 1$ are circularly polarized in opposite senses when the star is viewed along a line of sight that is parallel to the field, although they appear linearly polarized when the star is viewed perpendicular to the field direction. The σ components of a simple triplet transition are each shifted from the line center by an amount

$$\Delta\lambda(\text{Å}) = 4.67\times 10^{-13} g_{\text{eff}} \lambda^2 B, \qquad (1)$$

where λ is the central wavelength of the line in angstroms, B is the magnetic field strength in gauss, and g_{eff} is the effective Landé g factor. This factor is an atomic parameter that essentially is a measure of the magnetic sensitivities of the σ components. A value of $g_{\text{eff}} = 2.5$ would represent a spectral line that is highly sensitive to magnetic splitting. The magnitude of the Zeeman splitting can be estimated from (1) using typical values that are encountered in observations of solar active regions, that is, $B \sim 1000$ G, $\lambda = 6000$ Å, and $g_{\text{eff}} = 2.5$. Substituting into (1) yields $\Delta\lambda = 0.04$ Å.

The degree of splitting can be compared to other effects that broaden spectral lines. In particular, thermal or Doppler broadening is given by

$$\Delta\lambda_D = \frac{\lambda}{c}\left[\frac{2kT}{M} + \xi^2\right]^{1/2}, \qquad (2)$$

where T is the temperature of the gas, M is the mass of the emitting atom, ξ is the microturbulence velocity, k is the Boltzmann constant, and c is the speed of light. Substituting values appropriate for the solar photosphere in (2) and assuming an iron atom yields $\Delta\lambda_D = 0.06$ Å. Thus, Doppler broadening and magnetic splitting are comparable in this example. But there are additional nonmagnetic line-broadening mechanisms, including opacity effects, macroturbulence due to convective motions, and rotation, that contribute to the widths of spectral features. In fact, lines in the visible spectrum of the Sun are typically characterized by full widths at half maximum of about 0.15 Å. Hence, the contribution of magnetic splitting or broadening to the total line width is relatively small in the visible portion of the spectrum, even in spectral features that are highly magnetically sensitive.

Fortunately, the opposite polarizations of the σ components enable the measurement of the subtle signatures of magnetic splitting, even when the lines are strongly broadened by additional nonmagnetic effects. This principle, the detection of polarized radiation, is the basis for magnetographs. The magnetograph is an instrument that has been successfully utilized since the early 1950s to obtain observations of localized magnetic fields on the solar surface. The line centers of the polarized components of the total line profile are measured and the result inserted in the relation given in (1) to derive the value of the magnetic field strength B. In practice, the inferred field strength is a lower limit to the true field strength, because nonmagnetic regions that are usually present in the observing aperture "dilute" the magnetic signal.

The observational approach just outlined provides extensive data on the magnetic fields on the solar surface. The field strength in the umbrae of sunspots is generally in the range of 1500–3000 G. The majority of spots first occur in bipolar groups where the preceding spot (with respect to the direction of the solar rotation) is characterized by a polarity opposite to that of the following spot. In addition, the polarities of the preceding (following) spots in one hemisphere are opposed to the polarities of the preceding (following) spots of the other hemisphere. The polarities then reverse with each new 11-yr cycle.

A further fundamental magnetic structure on the Sun is the flux tube. Magnetic flux tubes are found when the Sun is observed with magnetograph devices at the highest possible spatial resolutions. The flux tubes, which are only about 100 km in diameter, are the sites of emergent magnetic flux with field strengths of approximately 1500 G. Active (plage) regions are composed of aggregates of flux tubes, and the combination of active regions and spots occupies ~0.1–1% of the solar surface, a percentage that is a function of the solar cycle. Although localized sites of intense magnetic field are present on the solar surface, the global average field of the Sun is relatively weak, with a strength in the range of about 1–15 G.

STELLAR OBSERVATIONS

The techniques that have proven so successful in the study of magnetic fields on our nearest star, the Sun, are virtually useless in the investigation of stellar magnetism. Standard polarization methods are inappropriate for measurements in the integrated light of solar-type stars. The sites of significant magnetic flux on their surfaces are, as in the case of the Sun, characterized by tangled field topologies and opposing polarities that, in the integrated light of the unresolved stellar disk, cancel to yield no net polarization. Positive detections of stellar-surface magnetic fields using polarization techniques therefore require fortuitous geometrical circumstances involving the angle between the line of sight and the field direction, or the presence of large-scale, coherent fields such as those that characterize the magnetic A stars and white dwarfs. The null results of past investigations to detect to a significant degree any net linear or circular polarization demonstrates that large-scale, coherent longitudinal fields are not present on the surfaces of late-type stars.

In view of the difficulties introduced by random vector magnetic field orientations, it has become necessary to develop techniques to directly detect Zeeman splitting in stellar spectra through unpolarized, or so-called "white light," observations in order to determine the field strength and fractional area coverage, or "filling factor," of magnetic fields on stellar surfaces. The new methods rely on the interpretation of the detailed shapes of magnetically sensitive photospheric lines in order to extract the Zeeman-splitting pattern. The line profile is expected to contain contributions from magnetic regions, such as analogs of solar plages and cool spots, and a central, unsplit component from the nonmagnetic photosphere. The relative importance of the contributions of these regions to the line profile depends on their respective filling factors and contrasts (i.e., brightness relative to the surrounding photosphere) and the magnitude of the magnetic field. The multicomponent nature of the Zeeman-broadened flux profile in a stellar spectrum can be approximately represented by the expression

$$F_\lambda = \sum_i \alpha_i(\lambda) f_i F_{i_\lambda}(B) + \left(1 - \sum_i f_i\right) F_\lambda(0), \tag{3}$$

where F_λ is the observed profile, f_i is the filling factor of magnetic region i, and $\alpha_i(\lambda)$ is the region contrast. The latter is a function of wavelength, particularly in the case of cool starspots, which are relatively brighter in the infrared than in the visible portion of the spectrum. In (3) the unsplit profile in the quiet photosphere where $B \approx 0$ is denoted by $F_\lambda(0)$. The Zeeman-split profile that appears in active region i with a mean magnetic field strength B is described by $F_{i_\lambda}(B)$.

The actual technique entails comparing two nearby spectral lines of similar formation properties but differing in their magnetic sensitivities as indicated by the Landé g value for each transition. Features that arise from the same multiplet in an atom and that are nearby in wavelength are ideal for this purpose. Any excess broadening in the "magnetic line" relative to the nonmagnetic, or "insensitive line," is attributed to the effects of significant magnetic flux present on the stellar hemisphere facing the observer. The magnetically insensitive reference line represents $F_\lambda(0)$ in (3) in actual observation. The ratio of the mathematically transformed line profiles in the Fourier frequency domain is derived and the values of the magnetic field strength and the filling factor are inferred with the aid of a mathematical model which describes the Zeeman-broadened profile.

This approach was first introduced and applied by Richard D. Robinson, Simon P. Worden, and John W. Harvey. They inferred the presence of kilogauss-level fields extending over significant fractions of the visible surfaces of the two chromospherically active, solar-type stars, ξ Bootis A (G8 V) and 70 Ophiuchi A (K0 V). These investigators analyzed the profiles of the Fe I lines at 6842.7 Å ($g_{\text{eff}} = 2.5$) and 6810.27 Å ($g_{\text{eff}} = 1.17$) using spectra with a resolution of 0.06 Å and a signal-to-noise ratio of approximately 50 (Fig. 1).

Subsequent research extended this technique to the infrared, which offers several advantages. Magnetic sensitivity, and therefore detectability, are enhanced in the infrared, following the wavelength dependence of Zeeman splitting as given in Eq. (1). Although the relation suggests this advantage is proportional to the square of the wavelength, the fact that Doppler linewidths scale linearly with wavelength [Eq. (2)] implies that magnetic sensitivity is only directly proportional to wavelength for a given field strength and effective Landé g factor (i.e., $\Delta\lambda/\Delta\lambda_D \propto \lambda$). Thus, resolution requirements are correspondingly reduced. Additional advantages include the fact that those magnetically active stars that are cooler than solar-type stars, such as the dMe flare stars, are brighter in the near infrared. The infrared brightness of spots relative to the quiet photosphere on these red dwarfs and solar-type stars is larger as well. The lower number density of lines in the infrared further reduces the contamination of candidate spectral features by weak

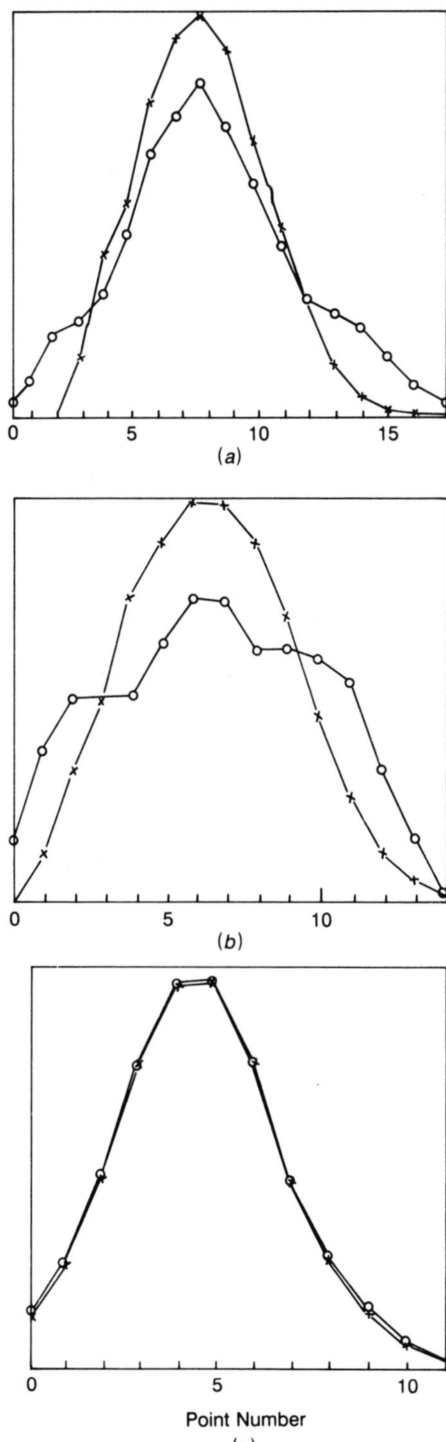

Figure 1. Magnetically sensitive (open circles) and insensitive (crosses) line profiles for (*a*) the magnetically active G8 dwarf star ξ Bootis A, (*b*) a sunspot umbra, and (*c*) the nonmagnetic photosphere of the quiet Sun. The absorption-line profiles have been inverted in order to more clearly illustrate the effects of Zeeman line broadening. Zeeman line broadening due to the presence of split σ components is clearly evident in the line wings in both ξ Boo A and the sunspot in (*a*) and (*b*), respectively. The two profiles are virtually indistinguishable in the quiet solar photosphere in (*c*). The wavelength separation between each point is 0.0279 Å. (*Reproduced from Robinson, Worden, and Harvey 1980. Reproduced from* The Astrophysical Journal, ***236***. *© 1980 by the American Astronomical Society.*)

blends. Finally, infrared observations obtained with a Fourier transform spectrometer can yield numerous lines simultaneously. The availability of many spectral lines for the Zeeman-broadening analysis increases the accuracy of the results.

Although current techniques emphasize the analysis of magnetically sensitive and insensitive features within a single stellar spectrum, a novel variation of this approach involves observation of a magnetically sensitive line in both an active star and a quiet comparison star of the same or similar spectral type. The line profile in the quiet comparison star is assumed to have a negligible contribution to its width from Zeeman splitting. The profile is adjusted for differences in nonmagnetic broadening between the two stars, and then the set of profiles is utilized in the Fourier analysis procedure to infer the field strength and filling factor on the active star. A further promising approach involves the use of model atmospheres that explicitly take into account radiative transfer effects in the spectral diagnostics, the presence of weak line blends, and the exact Zeeman-splitting patterns. The combination of these methods is proving successful, though there are limitations in their applicability. In particular, they can only really be applied to stars with projected rotational velocities of $v \sin i \lesssim 10$ km s^{-1}, where i is the inclination of the rotation axis with respect to the line of sight. Rotation in excess of this amount leads to profiles that are so dominated by rotational broadening that the subtle signature of Zeeman broadening cannot be extracted. The ubiquitous occurrence of weak blends and the limited availability of suitable magnetically sensitive and insensitive line pairs with well-known atomic parameters pose further obstacles. Despite these limitations, important results are emerging that could not be obtained in any other way.

RESULTS

Selected magnetic field measurements for a small sample of late-type stars are given in Table 1. While the data set is limited, significant trends are present. The inferred stellar magnetic field strengths B do not exceed an upper limit determined by the equipartition magnetic field strength B_{eq}, such that $B \leq B_{\text{eq}}$. This quantity is determined by an equilibrium between the ambient photospheric gas pressure P_{gas} and the magnetic pressure so that

$$B_{\text{eq}} \equiv \sqrt{8\pi P_{\text{gas}}}\,, \tag{4}$$

where P_{gas} is evaluated at continuum optical depth unity in stellar photospheric models. This result suggests that the external photospheric gas pressure confines magnetic flux tubes and thereby limits the highest value that the surface magnetic field strength can attain.

An important preliminary result is that the observed magnetic flux, as represented by the product $f \times B$, decays with stellar age on the main sequence and also declines with lower rotation rates. Further analysis indicates that it is the filling factor f rather than the field strength B that actually changes in time. The value of the field strength appears to be independent of age or rotation rate; rather, it is determined purely by photospheric properties. The limited data available at this time suggest that the filling factor increases with angular rotational velocity Ω according to $f \propto \Omega^{1.1}$, for the conditions where $f \lesssim 0.80$ and $\Omega \lesssim 0.25$ day^{-1}. As a star ages on the main sequence, the mean fractional area coverage of magnetic fields declines approximately according to $f \propto t^{-0.65}$. On shorter time scales that are less than a rotation period, variability in the strength and area coverage of magnetic active regions is observed. The variability is attributed to either the effects of rotational modulation of active complexes near the limb of the star or to intrinsic changes in the surface active regions themselves.

Direct detections of magnetic fields on post–main-sequence stars of the kind that the Sun will eventually become, namely, yellow and red giants, have yet to be obtained. In general, it appears that

Table 1. Magnetic Field Parameters for Late-Type Stars

Star	Spectral Type	B (G)	f (%)	Source
χ^1 Ori	G0 V	1000	60	1
HD 190406	G1 V	1800	10	1
HD 1835	G2 V	1400	32	1
ξ UMa B	G5 V	1970	32	2
κ Cet	G5 V	1500	35	1
HD 28099	G6 V	1700	30	1
ξ Boo A	G8 V	1800	37	1
ξ Boo A	G8 V	1600	30	3
ξ Boo A	G8 V	1200	40	4
HD 152391	G8 V	1700	18	1
λ And	G8 III–IV	1290	48	5
λ And	G8 III–IV	600	30	6
VY Ari	G9–K0 IVe	2000	66	7
70 Oph A	K0 V	1200	18	3
σ Dra	K0 V	1900	30	8
36 Oph	K1 V	1500	13	3
ε Eri	K2 V	1000	30	3
ε Eri	K2 V	1000	35	4
HR 222	K2 V	1600	12	3
HR 6806	K2 V	1500	15	4
61 Cyg A	K5 V	1200	24	3
EQ Vir	K5 Ve	2500	80	9
BY Dra	K7–M0 Ve	2800	60	9
Gliese 229	M1 V	2500	20	9
AU Mic	M1.6 Ve	4000	90	9
AD Leo	M3.5 Ve	4300	70	9
EV Lac	M4.5 Ve	5200	90	9

References:

(1) S. H. Saar (1987); unpublished Ph.D. thesis, University of Colorado, Boulder. The results are preliminary and subject to revision.

(2) S. H. Saar (1988).

(3) G. Basri and G. W. Marcy (1989). Physical realism in the analysis of stellar magnetic fields. II. K dwarfs. *Ap. J.* **345** 480.

(4) G. W. Marcy and G. Basri (1988). Physical realism in the analysis of stellar magnetic fields. *Ap. J.* **330** 274.

(5) M. S. Giampapa, L. Golub, and S. P. Worden (1983). The magnetic field on the RS Canum Venaticorum star Lambda Andromedae. *Ap. J. (Letters)* **268** L121.

(6) Ph. Gondoin, M. S. Giampapa, and J. A. Bookbinder (1985). Stellar magnetic field measurements utilizing infrared spectral lines. *Ap. J.* **297** 710.

(7) B. W. Bopp, S. H. Saar, C. Ambruster, P. Feldman, R. Dempsey, M. Allen, and S. P. Barden (1989). The active chromosphere binary HD 17433 (VY Arietis). *Ap. J.* **339** 1059.

(8) D. F. Gray (1984). Measurements of Zeeman broadening in F, G, and K dwarfs. *Ap. J.* **277** 640.

(9) Saar, Linsky, and Giampapa (1987).

field strengths are lower than in dwarf stars, a result that is consistent with the lower photospheric pressures that characterize the tenuous atmospheres of giants. However, some detections have been reported for highly active subgiants, particularly those that are members of RS CVn systems.

The pre–main-sequence stars that are precursors to solar-type stars exhibit strongly enhanced counterparts of solar magnetic activity. Nevertheless, surface magnetic fields have not yet been directly detected. The null results are due to their generally faint visual magnitudes and to the high rotational velocities of the few relatively bright stars available.

SUMMARY

The results thus far obtained imply the following tentative conclusions for stars on the main sequence. The degree of magnetic activity depends on the fractional area coverage, or filling factor, of magnetic fields present on the stellar surface at the time of observation. The actual field strengths that can be attained are determined solely by the properties of the photosphere. The filling factor itself is a function of surface rotational velocity up to a saturation limit that likely corresponds to a maximum surface density of magnetic flux tubes. The observed decline in stellar activity with rotation and age is due to a decrease in the filling factor of magnetic active regions with time. The decline of the filling factor with time is a consequence of a reduction in the action of the rotation-dependent dynamo, which is responsible for the generation of magnetic flux. This is, in turn, due to a gradual reduction in the rotation rate as a result of the magnetic braking that occurs during evolution on the main sequence.

The exact role of the convection zone in dynamo processes is not yet fully understood. The dynamo in the Sun is believed to operate in an interface region just below the base of the convection zone and outside the radiative core. In this region, the field is amplified in strength through a mechanism referred to as the "shell" dynamo. Interestingly, fully convective dwarf stars, including late-M dwarf flare stars, do not have such an interface region (by the nature of their interior structure), yet they have very strong fields covering a significant fraction of their surfaces. A so-called "distributed" dynamo model is evoked to describe magnetic field generation in these stars. Neither the details nor the applicability of either the shell or the distributed dynamo models are well established.

The scenario just outlined still requires further examination based on a significantly enlarged data base of stellar magnetic field measurements. The study of stellar magnetism is an active field of research where the techniques are still being developed and refined, and the quantitative results adjusted in the light of new knowledge. The advent of large ground-based telescopes and new instrumentation will enable astronomers to obtain spectra of unparalleled quality for fainter objects, particularly in the crucial infrared region, where Zeeman splitting is enhanced. By so doing, our understanding of the nature and role of magnetism in stars from the pre–main-sequence through the post–main-sequence stages of evolution will be advanced.

Additional Reading

Giampapa, M. S. (1984). Direct and indirect methods of measurement of stellar magnetic fields. In *Space Research Prospects in Stellar Activity and Variability*, A. Mangeney and F. Praderie, eds. Observatoire de Paris, Meudon, p. 309.

Hartmann, L. (1987). Stellar magnetic fields: Optical observations and analysis. In *Cool Stars, Stellar Systems, and the Sun*, J. Linsky and R. E. Stencel, eds. Springer-Verlag, New York, p. 1.

Linsky, J. L. and Saar, S. H. (1987). Measurements of stellar magnetic fields: Empirical constraints on stellar dynamo and rotational evolution theories. In *Cool Stars, Stellar Systems, and the Sun*, J. Linsky and R. E. Stencel, eds. Springer-Verlag, New York, p. 44.

Saar, S. H. (1987). The photospheric magnetic fields of cool stars: Recent results of survey and time-variability programs. In *Cool Stars, Stellar Systems, and the Sun*, J. Linsky and R. E. Stencel, eds. Springer-Verlag, New York, p. 10.

Saar, S. H. (1988). The magnetic fields on cool stars and their correlation with chromospheric and coronal emission. In *Hot*

Thin Plasmas in Astrophysics, R. Pallavicini, ed. Kluwer Academic Publ., Dordrecht, p. 139.

Saar, S. H., Linsky, J. L., and Giampapa, M. S. (1987). Four meter FTS observations of photospheric magnetic fields on M dwarfs. In *Proceedings of the 27th Liège International Astrophysical Colloquium on Observational Astrophysics with High Precision Data*, L. Delbouille and A. Monfils, eds. Universite de Liège, Liège, p. 103.

Wallerstein, G., ed. (1990). *Cool Stars, Stellar Systems, and the Sun*. Astronomical Society of the Pacific, San Francisco.

See also **Binary Stars, RS Canum Venaticorum Type; Magnetohydrodynamics, Astrophysical; Solar Activity; Solar Magnetographs; Stars, Activity and Starspots; Stars, High-Energy Photon and Cosmic Ray Sources; Stars, Magnetism, Theory; Stars, Pre–Main Sequence, X-Ray Emission; Stars, Red Dwarfs and Flare Stars.**

Stars, Magnetism, Theory

David Moss

Magnetic fields are observed to be present at the surfaces of several types of stars. The major groups of such magnetic stars are the lower main-sequence stars, the chemically peculiar (CP) stars, and degenerate objects such as white dwarfs and pulsars. Observations gives (rather limited) information about the mean properties of the magnetic fields in the outermost, optically thin, stellar layers, typically comprising a fraction of a percent of the stellar mass. Only theoretical studies can deduce properties of the fields inside stars from those observed. Such studies involve both magnetohydrodynamics (the study of the motion of electrically conducting fluids containing magnetic fields) and stellar structure and evolution.

One of the major problems is to explain the origin of the observed fields. The currently favored theories fall into two major groups. Fields may be inherited from an earlier epoch or they may be generated (perhaps from a very small "seed" field) within the star itself while it is in its present evolutionary phase. Theories of the first type are known as "fossil field" theories. The latter group includes the dynamo, battery, and thermomagnetic instability theories. It appears unlikely that any one mechanism can explain the complete range of stellar magnetism, which encompasses a very wide range of stellar objects in very different phases of evolution. This entry describes theories of the origin of magnetic fields and outlines their application to various types of magnetic stars.

FOSSIL THEORY

The interstellar medium in which stars form is electrically conducting and contains a large-scale magnetic field (typically of order 3×10^{-6} G). If a spherical region of gas of density 10^{-24} g cm^{-3} were to collapse isotropically to form a star of several solar masses with mean density about 1 g cm^{-3} while retaining all the magnetic flux initially pervading the gas, then the final field strength would be about 10^{10} G. In fact, this field would be much too strong to allow the star to form, and this crude discussion massively overestimates the field strength. In its simplest form the fossil theory claims that enough of the interstellar field is able to survive the process of star formation and pre–main-sequence evolution to produce the fields that are observed in main-sequence stars of a few solar masses.

A fundamental test of any theory of field origin for the CP stars is to explain how apparently similar stars have very different field strengths. If the external fields of the CP stars are not very much larger than the surface fields, then it seems plausible that the distribution of field strengths in this group of stars could result from either the precise amount of flux loss during star formation or the initial, primordial, magnetic flux varying from star to star, so that only stars that arrive on the main sequence with larger than average fluxes appear magnetic. This theory is subject to all of the uncertainties of the star formation process. In particular, there may be a phase of evolution for stars of a few solar masses before the main sequence (the "Hayashi" track), where the bulk of the stellar material undergoes large-scale convective motions. It is possible, although not inevitable, that these motions could dissipate or expel the surviving primordial field from the star. If this happens, or even if not, it might be that a dynamo then operates in the convective envelope, generating a magnetic field that persists after the convection dies away. The observed fields would then be fossil relics from a pre–main-sequence dynamo rather than from the interstellar medium.

White dwarfs are stars with mean densities of about 10^6 g cm^{-3}, compared with about 1 g cm^{-3} for a typical main-sequence stars. Neutron stars have mean densities estimated as of order 10^{14} g cm^{-3}. If a CP star with mean field about 3000 G were to contract to the radius of a typical white dwarf without loss of flux, its field would be amplified to about 10^8 G. If the collapse continued to neutron star densities then the field would be compressed to about 10^{13} G. These considerations have prompted the suggestion that the fields of white dwarfs and pulsars are fossils from their main-sequence precursors (irrespective of the origin of the main-sequence fields). There is some evidence that the magnetic white dwarfs originated from somewhat more massive main-sequence stars, and the proposal is consistent with the observed numbers of magnetic CP stars and degenerate objects with strong fields.

DYNAMO THEORY

A magnetohydrodynamic dynamo has the property that an arbitrarily small initial field immersed in a conducting fluid can be amplified by the fluid motions, so that, when suitably averaged, the strength of the large-scale fluid is much larger than that of the "seed." The field may either be steady, or it may oscillate regularly or irregularly.

For a dynamo to operate, generally it is necessary for a magnetic Reynolds number, the product of the electrical conductivity, the fluid velocity, and the dimension of the region occupied by the motions, to be large enough. This condition is far from sufficient—for example, Thomas G. Cowling's "anti-dynamo" theorem states that axisymmetric motions cannot drive a dynamo to generate an axisymmetric field. However, the combination of differential rotation and convection in spherical geometry plausibly can give a working dynamo. In simple terms the mechanism can be envisaged as follows.

Consider a magnetic field line that lies initially in a plane containing the rotation axis. Suppose now that part of the field line rotates more rapidly than the remainder. The field line will become distorted and will gain a component parallel to the azimuthal direction (Fig. 1). Thus the fluid motions feed energy into the azimuthal part of the field, but this mechanism cannot similarly maintain the meridional part of the field, lying in planes through the rotation axis. Now consider an element of fluid that rises through a region containing an azimuthal field. As it rises it expands, and the fluid velocity is approximately as shown by the single-headed arrows in Fig. 2*a*. Coriolis forces then rotate the element, and the embedded magnetic field line, as shown by the double-headed arrow. If the rotation is roughly 90° the azimuthal field generates a component lying in meridian planes, approximately as shown in Fig. 2*a*. If the process occurs more or less simultaneously through the convecting region, then resistive reconnection can join the rotated field loops (Fig. 2*b*), giving a large-scale meridional field. In this very simplified picture, the energy of the differential rotation maintains the azimuthal field against decay, and energy from the convective motions maintains the meridional

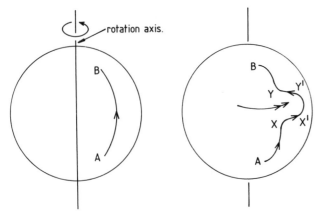

Figure 1. (*a*) A meridional field line *AB* is (*b*) distorted by differential rotation (double-headed arrow) and azimuthal components (*XX′*, *YY′*) appear.

field. Similar ideas lie behind what is known as the $\alpha-\omega$ mechanism of mean field electrodynamics, where rotation imparts a "helicity" (handedness) to the turbulence. In any dynamo there is inevitably a continuing loss of magnetic energy, for example by conversion to heat by the finite resistivity (which must be present for reconnection to operate), and the ultimate energy source is that which drives the fluid motions.

A dynamo mechanism is widely (but not universally) believed to be responsible for the solar magnetic field and, by inference, for the generation of analogous fields in lower–main-sequence and other stars with substantial subsurface convective regions. (Fossil theory, as outlined above, requires a relatively quiescent envelope, and so is not thought to be applicable to these stars.) The proposal that the fields of the magnetic chemically peculiar stars are also of dynamo origin is considerably more controversial. It is plausible that a dynamo mechanism does operate in the convective cores of these stars, which extend over 15–20% of the stellar radius. The question is whether these fields can penetrate the overlying envelope to give fields at the stellar surfaces that are of the strength and geometry observed.

Still more speculatively, it has been suggested that a dynamo might operate in the convective cores of certain stars as they evolve away from the main sequence, and that the strong fields observed in some white dwarfs are dynamo relics from this epoch.

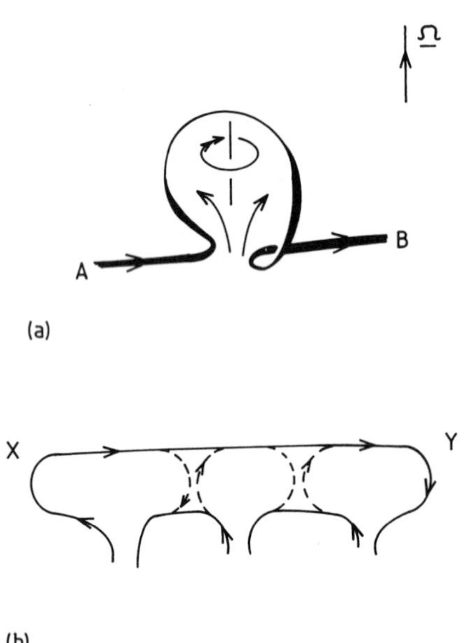

Figure 2. (*a*) A rising element of fluid distorts an initially azimuthal field line, *AB*, and the resulting field loop is rotated by Coriolis forces (double-headed arrow). (*b*) Three neighboring meridional field loops (dashed lines) reconnect to form a large-scale meridional field *XY*. The final configuration is shown by the solid lines.

BATTERY THEORY

Under stellar conditions atoms are ionized so that free electrons are present. The gravitational force experienced by the less massive electrons is negligible compared with that felt by the positive ions. In a nonmagnetic, uniformly rotating, chemically homogeneous star this difference in gravitational force results in a slight charge separation, and so an electric field, of a magnitude determined by the condition that the electric and pressure gradient forces on the electrons are in approximate balance. This electric field is radial with zero curl and so, by the Maxwell equation

$$\frac{\partial \mathbf{B}}{\partial t} = -\operatorname{curl} \mathbf{E}, \tag{1}$$

there is no associated magnetic field. If the stellar material is chemically inhomogeneous with molecular weight not a function of radius only, then the electric field of the separated charges is not, in general, curl-free and a magnetic field is generated according to Eq. (1). (There are other forms of "battery" mechanism, but this seems the most efficient.) For realistic parameters it seems difficult to generate fields of more than a few hundred gauss and, in the simplest cases, these fields are predominantly azimuthal. Moreover the time taken for such fields to grow is likely to be a significant fraction of the main-sequence lifetime of a CP star. Such considerations make it improbable that a battery mechanism can be the sole source of the magnetic fields of the CP stars. However, it may play a role in determining the overall geometry and stability of a predominantly fossil field. It might also provide a seed field for a dynamo to operate. A mechanism of this type has also been invoked to explain the fields of neutron stars.

THERMOMAGNETIC INSTABILITIES

In a white dwarf both heat and electric current are almost entirely transported by free electrons. Consider such a star with a purely radial temperature gradient, associated with an outward flow of heat. In the presence of a small, nonradial, magnetic field $\mathbf{B}$, the slightly hotter electrons, moving with mean velocity $\mathbf{v}$ directed radially outwards will be deflected by the Lorentz force $((-e)\mathbf{v}\times\mathbf{B})$ in a direction perpendicular to both the field and the temperature gradient. This deflection will be slightly greater than, and in the opposite direction to, that of the cooler-than-average electrons, which have a mean inwards motion. The net effect is to generate a (nonradial) perturbation to the heat flux, parallel to the deflection of the hotter electrons, together with associated temperature and electron pressure gradients. In equilibrium, this extra pressure gradient per particle is balanced by a (nonradial) electric field with curl $\mathbf{E}\neq\mathbf{0}$. By the Maxwell equation (1), an additional magnetic field is generated that, if parallel to the original field $\mathbf{B}$, will amplify it. If the amplification rate exceeds the ohmic dissipation rate, the field will grow. Suitable conditions for field growth may exist in white dwarfs and neutron stars. This linear argument cannot predict the magnitude or large-scale structure of the field generated, but strong fields might be generated on relatively short time scales. The relation of such fields to those seen at the surfaces of some white dwarfs and inferred to be present in some neutron stars is, however, uncertain.

OTHER EFFECTS

Magnetic fields are thought to play a role in many stellar phenomena not discussed here. For example, the evolution of the AM Her systems ("polars") may be partly controlled by magnetically controlled accretion. Abundance anomalies on CP stars probably are intimately connected with their magnetic fields. The recently discovered short-period oscillations in some cool CP stars seem to be channelled along the field axis. Magnetic fields certainly play an important role in star formation.

SUMMARY

At the moment there is no unanimous agreement about the origin of magnetic fields in any of the major classes of magnetic stars. In the Sun (and similar stars) there is a strong consensus that the field is dynamo-generated, although the details are not fully understood. An active debate is in progress between proponents of dynamo and fossil theories for the origin of the fields of the CP stars. Battery, dynamo, fossil, and thermomagnetic instability theories have all been invoked to explain the strong fields both of the white dwarfs and of the neutron stars at the seat of the pulsar phenomenon. Some of these uncertainties may be resolved by more precise and plentiful observational data, but theoretical advances are also needed.

Additional Reading

Borra, E. F., Landstreet, J. D., and Mestel, L. (1982). Magnetic stars. *Ann. Rev. Astron. Ap.* **20** 191.

Cowling, T. G. (1981). The present status of dynamo theory. *Ann. Rev. Astron. Ap.* **19** 115.

Dolginov, A. Z. (1988). Magnetic field generation in celestial bodies. *Phys. Rep.* **162** 338.

Moss, D. (1986). Magnetic fields in stars. *Phys. Rep.* **140** 1.

Parker, E. N. (1983). Magnetic fields in the cosmos. *Scientific American* **249** (No. 2) 36.

See also **Binary Stars, Polars (AM Herculis Type); Magnetohydrodynamics, Astrophysical; Pulsars, Observed Properties; Stars, Magnetic and Chemically Peculiar; Stars, Magnetism, Observed Properties; Stars, Neutron, Observed Properties and Models; Stars, White Dwarf, Observed Properties; Sun, Magnetic Field.**

Stars, Masses, Luminosities, and Radii

Wulff D. Heintz

The characteristics of stars which are the subject of this entry are interrelated, first, by the Stefan–Boltzmann law

$$L \sim R^2 T^4, \tag{1}$$

where L = luminosity, R = radius, and T = effective temperature of the radiating "surface" layer (photosphere), determined from color photometry or from spectral-line classification. Second, stars are for most of the time (except in brief transition phases) in hydrostatic and radiative equilibrium. That is to say, pressure balances gravity everywhere in the stellar interior, and the accumulation of energy is balanced by its outward flow. These are reasons why, for stars of similar chemical makeup (in particular, the main-sequence stars, which are those not yet showing substantial effects of aging), the mass alone determines the other characteristics (the Vogt–Russell theorem); there is then especially a relation between mass and luminosity. The color-luminosity (Hertzsprung–Russell) diagram of stars therefore implicitly describes their radii (increasing from lower left to upper right), and—for stars on and near the main sequence—also their masses (increasing from lower right to upper left).

Masses are expressed in units of the solar mass (2×10^{30} kg), and stellar radii are often stated in units of the solar radius (7×10^8 m). Luminosities can also be given in terms of the solar luminosity ($L_\odot = 3.8\times10^{26}$ W); more convenient is the so-called absolute magnitude M as a logarithmic scale:

$$M = 4.8 - 2.5 \log L/L_\odot. \tag{2}$$

Absolute magnitudes M are defined by the formula

$$M = m - 5 \log(D/10), \tag{3}$$

where m is the directly observed or apparent magnitude, and D the distance in parsecs. With respect to luminosities and magnitudes, distinction is made as to whether they represent the radiation in a limited-wavelength range of, say, visible light as observed (subscript v = visual), or whether they represent the total radiation output (subscript bol = bolometric); the difference between these depends on the temperature T.

Knowledge of luminosities thus requires distance determinations: usually by trigonometric measurement of parallaxes, or for some star clusters by analyzing the motions or the color-magnitude relations of their member stars. In all other cases formula (3) is used the opposite way, that is, a star whose M and D are known is compared with a more distant one of the same type (and hence presumably of the same M), so that the distance of the latter is inferred. If this M calibration is extended stepwise to more luminous and more distant object types, the distance scale is constructed from (3) up to extragalactic distances.

Masses are determined by measuring the gravity interaction between stars in binary or multiple systems, that is, from orbits through Kepler's third law:

$$(\text{mass})_1 + (\text{mass})_2 = a^3/P^2, \tag{4}$$

where the semimajor axis a is expressed in astronomical units (1.50×10^{11} m), and the period P in years (3.16×10^7 s). If SI units are used instead, the left side of (4) carries the factor $G/4\pi^2 = 1.69\times10^{-12}$, which includes the Newtonian or universal constant of gravitation, G.

Some data additional to the orbital elements are needed to get the orbit scale a in linear measure (and free from foreshortening due to perspective), and to separate the mass sum in formula (4) into the individual masses of the two stars:

(a) For eclipsing binaries the scale has to be determined by spectroscopic Doppler-shift measures of the spectral lines of both components. (This method is especially applicable to aging, expanded stars because they are more likely to eclipse; on the other hand, the increasing luminosity of the aged star makes a companion harder to detect.)

(b) For positionally separated pairs (visual and interferometric binaries), the orbit scale has to be converted from the observed angular separation into linear measure, usually by direct distance determination (parallax) together with an analysis for the mass ratio of the components. This method is limited to stars nearer than about 25 pc, where the parallax is large enough to be determined with a small percentage error. Otherwise spectroscopic measures may sometimes substitute for (or strengthen) the parallax result.

As several high-precision data need to be combined, well-determined masses are known for at most 100 binary stars, mostly main-sequence and subgiant components.

Table 1. Typical values of stellar parameters.

	Main Sequence			*Giants and Supergiants*		
M_v*	*Mass*†	*Radius*‡	*Type*	*Mass*	*Radius*	*Type*
−6	> 20	> 10	O			
−4	20	9	B0 V	> 5	> 100	M I, II
−2	8	5	B3 V	4:	50	K II
0	3	2	B8 V	3	15	K III
2	1.9	1.5	A5 V	2	5	G III
4	1.2	1.1	F8 V			
6	0.8	0.8	K0 V			
8	0.4	0.5	M0 V			
11	0.2	0.2	M4 V			
14	0.1	0.1	M7 V			

*Absolute visual magnitude.
†In units of the solar mass.
‡In units of the solar radius.

Differences in the energy generation and absorption mechanisms are probable causes why the mass-luminosity relation of main-sequence stars is not a single formula:

$$\log L = 2.6 \log \text{mass} - 0.3, \quad \text{for masses under } 0.5\ M_\odot,$$

$$\log L = 3.8 \log \text{mass}, \quad \text{for middling masses}, \qquad (5)$$

$$\log L = 2.6 \log \text{mass} + 0.5, \quad \text{for masses over } 2.5\ M_\odot.$$

If the components of a binary are known to be main-sequence stars, the orbit data and the apparent magnitudes can be used in conjunction with one of formulas (5) to find masses and luminosities simultaneously (method of dynamical parallaxes).

Radii have been measured by interferometer for large and for some nearby stars, and in some cases by observing occultations by the Moon when the diffraction pattern at the lunar limb indicates a finite extension of the stellar light source. These angular data need conversion into true size by distance determinations. Most results come from eclipsing binary stars, where the durations of eclipses determine the component diameters, once the orbit size has been found from the Doppler-shift measures. For other stars the radii are computed from the luminosities by formula (1). Some typical values of these parameters are shown in Table 1.

Although stellar luminosities and radii vary enormously in the course of evolutionary time, the masses are sensibly constant for most of the stellar life—except for mass transfer in close interacting binary stars, and for some modest loss due to stellar wind. Only at late stages does the mass loss become severe.

There are few good mass determinations for giant stars, and none for supergiants. Therefore, masses of these stars are inferred from theoretical "tracks of evolution," that is, by computer models showing how stars of given masses will evolve in such a way as to match the observed colors and luminosities. It is concluded that masses about in the range 3–8 $M_\odot$ will have only a brief lifetime as giants or supergiants before entering phases of significant mass loss, and that stars of still higher masses will not reach the supergiant phase at all, because the supernova collapse intervenes.

The stars with the highest masses believed to be well determined (despite the presence of circumstellar matter, which interferes with the spectroscopic measures) are around 40 solar masses and are very rare. Theory has reasons for an upper limit of stellar masses somewhere between 60 and 100 $M_\odot$: The rapidly increasing radiation pressure counteracts the condensation to a star, and the very high angular momentum associated with such masses will foster their splitting into two or more bodies. (Larger masses are sometimes suggested; see the entry on Stars, High Luminosity.—Ed.)

Owing to more frequent formation and to longer lifetimes, the numbers of stars increase rapidly toward lower masses and luminosities. The large range of stellar luminosities (from 10^{-5}–10^{5} in solar units) together with this asymmetric mass frequency distribution causes the well-known disparity in star counts: The bright naked-eye stars are practically all of types more luminous than the Sun. A true census of the star population shows these types to constitute only 10% by number; there are about 5% degenerates (white dwarfs), and the remaining 85% are red dwarfs on the lower main sequence, less massive and much less luminous than the Sun.

Because the mass is roughly proportional to the stellar "fuel" available, and the luminosity measures its consumption rate, the ratio of mass to luminosity expresses the life expectancy of stars. By comparison with the main-sequence lifespan of a one-solar-mass star, about 9×10^9 yr, it is found that highly luminous O-type stars do not reach an age even of a million years, whereas the life of red dwarfs will outlast the entire current age of the universe by a factor of 10 or more.

Most common are the red dwarf stars with masses around one-fifth of the solar mass (visual luminosity about 1/1000 of the solar luminosity) and type M3 V. Below that mass, the abundance decreases again, as observed and as theoretically expected from a lesser efficiency of the formation process. The condition that a certain temperature must be reached in the core, in order to generate at least a small amount of nuclear energy, imposes a lower limit to masses of stars, that is, according to present models about 0.08 solar masses. There is now good evidence supporting the expectation that substellar masses below 0.08 (sometimes called brown dwarfs) are rare, as the lack of nuclear resources makes their lifetime much shorter than that of red dwarf stars. (Brown and red dwarfs cannot be distinguished by radiative properties such as luminosity, but only by mass.) At the time of this writing, one binary with masses in the substellar range is known and apparently qualifies for a brown dwarf pair (Wolf 424), and there is evidence for an unseen brown dwarf companion of the star DT Vir. Most of the 90 stars known to be within 6 pc of the Sun have been well studied by now, and none of them shows definite evidence for an invisible companion. The discovery chance diminishes with the mass, but some low-mass objects would have been found if they existed in larger numbers.

Degenerate stars (class D, also known as white dwarfs) remaining after exhaustion of their energy sources and after mass loss have an inverse mass–radius relation: The higher the mass, the more will gravity outpower the internal electron pressure, and thereby compress the star to a smaller volume. In rare cases of binaries containing a white dwarf, when its mass and its amount of relativistic (nonorbital) redshift are known from comparative measures of the other (nondegenerate) component, the radius of the white dwarf is found directly. White dwarfs have no mass–luminosity relation, as their remnant radiation fades with age. The better-known cases have radii around 1/100 of the solar radius, and luminosities of 10^{-2}–10^{-4} in solar units; smaller and fainter objects are difficult to find. Few masses are known: One white dwarf (40 Eri B) has a mass of only 0.4 $M_\odot$, but generally—in order to have reached that advanced stage of evolution by now—masses of 0.7 $M_\odot$ and over are expected. The upper mass limit is near 1.4 solar masses; condensing stars which retain more than that mass are not expected to stabilize but to collapse further to become neutron stars.

Documentation Machine readable (computerized) catalogues in the astronomical data centers, some also appearing in print—are primarily concerned with more or less directly observed stellar data (apparent magnitudes and colors, also binary star positions and orbits, etc.). The parameters discussed here are less directly obtained, and the quality selection and combination of basic data depend on judgment and on the purpose at hand. Only for eclipsing stars are the radii catalogued as elements of the light-curve interpretation.

Additional Reading

Allen, C. W. (1985). *Astrophysical Quantities*. Athlone Press, London.

Hayes, D. S. and Pasinetti, L. E., eds. (1985). *Calibration of Fundamental Stellar Quantities*. D. Reidel, Dordrecht.

Heintz, W. D. (1978). *Double Stars*. D. Reidel, Dordrecht.

Liebert, J. and Probst, R. G. (1987). Very low mass stars. *Ann. Rev. Astron. Ap.* **25** 473.

Popper, D. M. (1980). Stellar masses. *Ann. Rev. Astron. Ap.* **18** 115.

Weidemann, V. (1990). Masses and evolutionary status of white dwarfs and their progenitors. *Ann. Rev. Astron. Ap.* **28** 103.

Zeilik, M. and Gaustad, J. E. (1990). *Astronomy: The Cosmic Perspective*. Wiley, New York, Chaps. 17, 18, and 21.

See also **Binary Stars, Astrometric and Visual; Binary Stars, Eclipsing, Determination of Stellar Parameters; Star Catalogs and Surveys; Stars, Distances and Parallaxes; Stars, High Luminosity; Stars, Low Mass and Planetary Companions; Stars, White Dwarf, Observed Properties.**

Stars, Neutron, Physical Properties and Models

Jonathan I. Katz

Neutron stars are the densest stars known. Their interiors are composed primarily of free neutrons, with smaller numbers of electrons, protons, and exotic nuclei. Their central densities are of order 10^{15} g cm^{-3}, several times that of ordinary atomic nuclei.

Neutron stars are important in many fields of modern astrophysics. Radio pulsars are rotating, magnetized, neutron stars. Many compact binary x-ray sources are neutron stars that are accreting matter from their stellar companions. Gamma rays of extremely high (10^{12} and 10^{15} eV) energy have been reported from the environs of some of these accreting neutron stars. Supernovae involve the collapse of matter to neutron star densities, and may leave neutron stars behind as compact remnants. Some naturally occurring elements and isotopes may be synthesized during this process of collapse and reexplosion. The most sensitive tests of general relativity depend on the binary pulsar PSR 1913+16, which is believed to consist of two neutron stars in orbit around a common center of mass. Widely accepted arguments indicate that gamma ray bursts are produced just above the surfaces of neutron stars.

Although neutron stars are essential to many astronomical phenomena, and have been the subject of extensive research, comparatively little is known empirically about their properties. Most of the observed phenomena depend only on the qualitative fact that they are very dense and compact.

MASSES

The most fundamental property of neutron stars is the relation $M(R)$ between their masses M and radii R. Because their interior temperatures are much less than their characteristic Fermi (degeneracy) temperature of 10^{12} K, their interior equation of state [the relation $p(\rho)$ between pressure p and density ρ], is very close to its zero-temperature value. This is also the case for white (degenerate) dwarf stars, but not for ordinary stars, whose richness and variety of structures are determined by the temperature dependence of their equation of state. Neutron stars are expected to be much simpler, with a unique $M(R)$ relation.

Unfortunately, the equation of state $p(\rho)$ of nuclear matter depends on incompletely understood physics, and is controversial. Calculations using popular assumptions indicate that neutron stars have radii of about 10–12 km, nearly independent of mass. Possible masses of stable neutron stars are calculated to lie in the range from a few tenths of a solar mass ($M_\odot$) to 1.5–2.5 $M_\odot$, with the exact limits depending on the equation of state chosen. It is believed unlikely that neutron stars can be formed with masses much less than 1.40 $M_\odot$, the Chandrasekhar limiting mass of white dwarfs. The binary pulsar PSR 1913+16 has a mass, measured from its orbit and the theory of general relativity, of 1.44 $M_\odot$. Its companion, probably also a neutron star, has a mass of 1.39 $M_\odot$. No other neutron stars' masses are accurately known, though rougher measures are all consistent with masses of 1.4 $M_\odot$.

RADII

Measurement of a neutron star's radius, especially if its mass is known, will be important to nuclear physics by constraining the equation of state $p(\rho)$ of nuclear matter. Unfortunately, no accurate measurements of neutron star radii exist. In principle, their radii could be inferred from gravitational redshifts (which depend on the ratio M/R) of spectral lines emitted from their surfaces. The best candidates for such measurements are the 7-keV x-ray line of 25-times ionized iron and the 511-keV positron annihilation line. The 7-keV line has been observed in x-ray sources, but its gravitational redshift has not been measured. If the radiation does not originate at the neutron star's surface, as may be the case, then its redshift cannot be used to determine the stellar radius. A 511-keV line has been reported from some gamma ray bursts, and theoretical arguments indicate that its source should be positrons stopped at the neutron star's surface. A redshift roughly consistent with expected neutron star surface values has been claimed, but the data are imprecise. Cyclotron lines have been reported in x-ray sources and gamma ray bursters but cannot be used to determine gravitational redshifts because their unshifted wavelengths are not independently known.

ROTATION

The presence of neutron stars was first inferred by Thomas Gold in 1968, when he explained the periods of radio pulsars as the rotational periods of magnetized neutron stars. The rotational periods are therefore measured very accurately. They are in the range 0.0015–4 s. Accreting neutron star x-ray sources have rotation periods in the range 0.1–1000 s. The gamma ray burst of March 5, 1979 showed a period of 8 s, which is attributed to a neutron star's rotation. The periods of radio pulsars are generally very stable, but slowly lengthen. In some cases the time required for the period to double is longer than the age of the universe (the expected life of the neutron star as an observable pulsar is less than 10^7 yr). Other pulsars slow their rotation more rapidly; the Crab nebula pulsar, born in the year 1054, has a period-doubling time of about 2000 yr. This spin down is explained as the result of the loss of rotational energy, which accelerates energetic particles.

The spin periods of accreting neutron stars in binary systems vary irregularly. In some cases both increases and decreases are observed, reflecting a complicated process of exchange of angular momentum among the neutron star, a surrounding gas flow, and the orbital motion of the companion star. The mean trend is usually a decreasing spin period, as would be produced by the accretion of rapidly orbiting matter. The period-halving times range from about 100 yr for the slowest neutron stars, with the least angular momentum of their own, to of order 10^6 yr for those rotating more rapidly.

MAGNETIC FIELDS

Most observations of neutron stars depend on their magnetic fields. These fields are essential to the radiation of radio pulsars; nonmagnetic neutron stars cannot emit pulses. Magnetic, accreting neutron stars produce radiation which is periodically modulated by the rotation of their magnetic fields; nonmagnetic accreting neutron stars produce aperiodic radiation. A number of accreting neutron stars and gamma ray bursters show lines in their spectra at energies between 10 and 100 keV, which are believed to result from the cyclotron resonance of electrons in the magnetic field.

The magnetic dipole moments of pulsars may be estimated from the rates at which their rotation is slowing. The implied surface magnetic fields are in the range 10^8–10^{13} G. There may be higher fields confined to the neutron stars' interiors or in disordered surface patches, which do not contribute to the external dipole fields. There is a strong anticorrelation between pulsars' spin rates and their magnetic fields. This is not surprising, because a rapidly spinning neutron star with a high field will rapidly lose its rotational energy, and will be observable for a significant period of time only as a slowly spinning pulsar; neutron stars with small fields many spin rapidly for a very long time. The measured magnetic fields of accreting neutron stars and of those which have gamma ray bursts are in the range 10^{12}–10^{13} G; this may be explained as an observational selection effect, for if the field were much outside this range its cyclotron line would be hard to observe.

Magnetic fields of 10^{12} G or more change the properties of matter at the relatively low densities ($\ll 10^6$ g cm^{-3}) near a neutron star's surface. The magnetic forces on electrons in fields this strong exceed the electrostatic forces. The thermal conductivity is strongly anisotropic, with a very high conductivity parallel to the field, and a much smaller conductivity perpendicular to it. The rates of absorption and emission of radiation are strongly dependent on its frequency, angle, and polarization, and on the magnetic field, so the observable radiation from a magnetic neutron star may be quite different from that of a nonmagnetic neutron star.

TEMPERATURE

Neutron stars' interiors are hotter than their surfaces, from which they lose energy by radiation. Heat is conducted from the interior to the surface by degenerate electrons, just as in an ordinary metal. This transport process is efficient, and neutron star interiors are believed to be nearly isothermal, resembling a hot copper ball wrapped in asbestos. Central temperatures are expected to be roughly 100 times the surface temperatures. Neutron stars are born hot, with interior temperatures as high as 10^{11} K. They cool rapidly at first, and then more slowly, so that a million-year-old pulsar is expected to have an interior temperature of order 10^7 K.

The very hot interiors of newly formed neutron stars radiate neutrinos. Because an enormous amount of energy (of order 10^{53} erg) was radiated in a few seconds, the neutrinos produced by Supernova 1987A were observed at Earth, despite the very small cross-sections for neutrinos to interact with detectors. This has provided direct confirmation of the predicted high temperatures in the core of a supernova. Their connection with neutron stars remains controversial, because although some supernovae (such as the Crab nebula supernova of 1054) undoubtedly leave neutron star remnants, it has not been proved that Supernova 1987A did so. The surfaces of neutron stars have been predicted to be observable sources of x-rays. Several candidates for such sources have been proposed, but are not universally accepted. In some cases observational upper bounds to the x-ray emission significantly constrain theories of neutron star cooling.

INTERIOR STRUCTURE

Calculations of the interior structure of neutron stars are very elaborate. Most, but not all, of the qualitative results are generally accepted, but quantitative results [such as the maximum mass and the $M(R)$ relation] depend on the assumed equation of state. In addition, subtler properties such as the dynamic coupling among various interior components remain controversial.

The outermost layer of a neutron star is believed to be a solid crust a few hundred meters thick, consisting of atomic nuclei and electrons. In the course of their formation these nuclei have captured many electrons, and have many more neutrons (and proportionately fewer protons) than nuclei found on Earth. The nuclei in neutron stars would rapidly decay by emission of electrons and antineutrinos if they were produced in our laboratories, but because they are immersed in a dense gas of degenerate electrons, the Pauli exclusion principle prevents such decays. Similarly, our familiar nuclei, if placed in a neutron star's interior, would rapidly capture electrons (inverse beta decay). Coulomb forces between the positively charged nuclei arrange them in a stiff crystal lattice or glassy state that has great (by terrestrial standards) mechanical strength, but which may break or flow when placed under significant stress.

Underneath this outer crust is an inner crust in which the nuclei are so neutron-rich that they cannot hold any further neutrons. The total crustal thickness is 1–2 km. The nuclei in the inner crust are accompanied by free neutrons. These neutrons are believed to be superfluid, like the atoms in superfluid helium. Because neutrons have spin $\frac{1}{2}$ they more closely resemble the rare isotope helium-3 (also spin $\frac{1}{2}$) than the abundant isotope helium-4 (spin 0). Superfluids can flow without any viscous friction, even through very fine openings. Neutron superfluidity is important because it permits persistent differential rotation of the neutron superfluid with respect to the stiff solid crust, even though the superfluid flows between the closely packed nuclei of the crust. However, mechanisms exist by which angular momentum may be slowly coupled between the crust and the superfluid neutrons.

Below the crust is a region, probably comprising the bulk of the neutron star, in which there are no nuclei. Most of the matter is superfluid neutrons, accompanied by smaller numbers of electrons and superconducting protons.

In the innermost parts of the neutron star the density exceeds severalfold that of ordinary atomic nuclei. A number of exotic suggestions have been made for the state of matter in this region, including free quark matter, strange matter, hyperon matter, a neutron solid, and a pion condensate (a high density of pions, which are ordinarily unstable). Unfortunately, these intriguing speculations lead to few unambiguous predictions.

Empirical information about the interior structure of neutron stars comes chiefly from the study of "glitches" in pulsar spin periods. Accreting neutron stars are less informative because their rotation is affected by external torques. A glitch is a sudden decrease in a pulsar's spin period, briefly interrupting its usual steady lengthening. Some glitches have decreased the spin period by about a part in 10^6, while others did so only by a part in 10^9. Some pulsars show "timing noise" consisting of very slight variations about their mean rate of spin down. This may be the effect of many very small "microglitches." One possible explanation of glitches is a breaking of the rigid neutron star crust, whose equilibrium shape changes as its rotation slows, reducing its moment of inertia. Another explanation is sudden coupling of the crust to an internal reservoir of angular momentum in a differentially rotating component of the neutron superfluid. Glitches are followed by a period of more rapid slowing, as the dependence of spin period on time gradually returns partway to its original path. Because the variation of pulse period following a glitch is measured in detail and with great accuracy, it may be possible to determine quantitatively the physical conditions inside a neutron star. This will require a generally accepted model for glitch and postglitch dynamics, and the ability to calculate in detail the physical properties of neutron star interiors.

Additional Reading

Anonymous (1986). Internal dynamics of neutron stars. *Los Alamos Science* Spring (No. 13) 29.

Baym, G. and Pethick, C. (1979). Physics of neutron stars. *Ann. Rev. Astron. Ap.* **17** 415.

Harding, A. K. (1991). Physics in strong magnetic fields near neutron stars. *Science* **251** 1033.

Helfand, D. J. and Huang, J. H., eds. (1987). *The Origin and Evolution of Neutron Stars*. Reidel, Dordrecht.

Ruderman, M. A. (1972). Pulsars: Structure and dynamics. *Ann. Rev. Astron. Ap.* **10** 427.

Shapiro, S. L. and Teukolsky, S. A. (1983). *Black Holes, White Dwarfs, and Neutron Stars*. Wiley-Interscience, New York.

See also **Neutrinos, Supernova; Pulsars, all entries; Stars, Magnetism, Theory.**

Stars, Nonradial Pulsation in B-Type

Myron A. Smith

The close spacing of frequencies excited in some β Cephei variable stars strongly suggests that nonradial pulsations (NRP) are excited in them. Such pulsations are now thought to occur among a much larger range of massive B-type stars than is represented by the β Cephei stars on the Hertzsprung–Russell (HR) diagram. It is important to understand that NRPs are not merely partially successful or "out of phase" radial pulsations. On the contrary, they are a large family of pulsations, of which radial pulsations are themselves only a part, that are subject to a complex set of physical coupling interactions in three dimensions. For example, NRPs can transfer angular momentum from one region of a star to another. In addition, nonradial pulsations describe a complex pattern of "hills and valleys" over a star's surface that do not occur in radial pulsations. These surface distortions vary with time, for example, with the hills becoming valleys. The mathematics that describe this behavior employ the same spherical harmonic functions, Y_l^m, used in the quantum mechanics of the atom. According to this description, l (the NRP degree or "family") corresponds to the number of zero-amplitude lines on a star's surface, and m (the order) equals the number of zero-amplitude lines running through the poles. As in quantum mechanics, m runs from $-l$ to $+l$, so that there are $(2m+1)$ orders for each degree, or family, l. However, unlike in the atom, it is stellar rotation, and not magnetism, that ordinarily splits the frequencies of the m orders. Frequency splitting of the orders can be observed and used to determine the mass-averaged stellar rotation rate. Altogether, NRP modes can be excited over a great many frequencies because an entire sequence of mode degrees occurs for each overtone, and then another sequence of mode orders occurs for each degree. In the Sun, for example, some 10 million modes may be excited at any one time.

The radial velocity and light curves produced by different NRP orders of a given family have the same shape, and conceivably even the same amplitude, if the star is viewed from a particular inclination angle relative to its rotation axis. (That favored angle is different for each m order.) Because this inclination angle is in general indeterminate for a given star, one cannot tell what the mode order is from a light curve alone. Fortunately, a technique exists that permits us to determine which m mode has been excited. That technique is the analysis of how spectral absorption line shapes change with time. When we observe an absorption line in the integrated light from the disk of a rotating star, its profile represents the addition of a complex set of contributions from many regions on the star with different components of rotational and pulsation velocities along the line of sight. These velocities map differently for each m mode onto a typical spectral absorption line profile in the integrated-disk light. Consider, for example, the $m=0$ mode in which the star's poles oscillate in and out together, and various latitude ranges oscillate in or out of phase with them. The spectroscopic result is that the line mainly moves back and forth in wavelength (radial velocity) and changes its shape rather little. In this case the variations in a line profile during a pulsation cycle mimic those from radial pulsation, especially when the star is viewed looking down on a rotation pole. An imaginary observer located above the star's equator would see that modes with $m=\pm l$ (equatorial traveling waves) would cause the absorption line to move back and forth only a little in velocity. At the same time, the shape of the absorption line would develop a pattern of adjacent absorption and emission "wiggles" that would move across the line profile from bluer to redder. These traveling wiggles would correspond to the physical movement of traveling waves sweeping around the star's equator. Donald Penrod has shown that, contrary to intuition, it is easier to detect the traveling wiggles from high m modes in rapidly rotating stars than in slowly rotating ones. This is because rapid rotators distribute more absorption across a wider range of wavelengths from the Doppler effect, and this line broadening makes it easier to resolve the wiggles spectrally. Our ability to detect high m modes is limited, however, because the wavelength separation between adjacent wiggles goes roughly as $1/m$, and eventually becomes smaller than the resolution of the spectrograph. The sign of m can be determined by comparing the wave velocity with the star's rotational velocity as obtained from the line profile width. This comparison shows whether the waves move east to west or vice-versa relative to an imaginary observer on the rotating star.

Since the discovery of NRP in 1977, astronomers have suspected that they occur in certain B stars that are not β Cephei variables. The basis for this suspicion is the presence in their absorption lines of traveling wiggles like those described above. When a single set of wiggles is present, it suggests that a single mode is operating strongly at the surface. This mode is described by $m=-l$ or $+l$, or by a traveling set of waves around the star's equator. Although multiple modes may be observed, generally the secondary oscillations are much weaker, making their detection difficult. In contrast to the β Cephei stars, each additional mode arises from a different l family. High-resolution spectroscopy is almost essential to the diagnosis of NRP, and the disentangling of all excited frequencies requires painstaking observations over a long, well-sampled time.

In practice, this disentangling is difficult due to other demands on the relatively large telescopes required for high-resolution spectroscopy. The practical solution is to search for light variations with a small telescope in a star already suspected from spectral variations of having NRP. As with β Cephei stars, one searches for regular periods in the resulting light curves. However, unlike those of the β Cephei stars, the periods of NRP stars fall within a wide range, so they cannot be estimated a priori. In addition, because the temperature changes of hills and valleys tend to cancel across the surface, temporal light variations are small in NRP stars. For these reasons the systematic study of NRP in B stars has advanced slowly. Nonetheless, years of photometric studies, especially by Christoffel Waelkens in Leuven, Belgium, has demonstrated that as many as four modes can be excited. Table 1 shows typical parameters for three nonradially pulsating stars from a combination of spectroscopic and photometric studies.

The occurrence of equally spaced, close frequencies, as in β Cephei stars, is generally taken as establishing the existence of NRP beyond doubt. However, when this circumstance occurs at only one position in the frequency spectrum, the preceding conclusion can sometimes be incorrect. For example, line profile variations in a rapidly rotating B star's spectrum can sometimes arise without the presence of pulsators: a large dark star spot carried across the surface of a rotating spot can mimic a wave on the line profile. Since starspots are known to exist on some B stars, it is conceivable that NRP and starspots can be confused in the most rapid rotators.

NRP seems to be common, perhaps universal, among chemically normal, types O9–B5 main-sequence stars. There are strong indications that it occasionally occurs in spectral types as hot as O4 and as cool as B7–8. OB supergiants show an erratic light variability. Such behavior may arise from the beating of a variety of low-amplitude NRP modes. The evidence is circumstantial in that

Table 1. Typical Parameters for Nonradially Pulsating β Stars

Star	*Spectral Type*	M_V	*Frequency (cycles / day)*	*Double Ampl. (mag)*	*Pulsation Mode Type*	*Source*
53 Persei	B4 IV	−3.0:	0.430	0.04 (var?)	$l = 2$ (uncertain)	Smith, Balona (Capetown)
			0.595	0.03 (var?)		
			0.29 (?)	—		
HD 160124	B4 IV	−2.7:	0.52073	0.052	unknown	Waelkens (Leuven)
			0.52007	0.048		
			0.52161	0.022		
			0.70624	0.015		
ε Persei	B0.7 III	−4.8	5.38	—	$-m = l = 3$	Smith, Gies, and Kullawajara
			6.25	0.02	$-m = l = 4$	
			7.87	—	$-m = l = 5$	
			10.6	(0.006)	$-m = l = 6$	
λ Eri	B2 III–IVe	−3.3	1.4252	0.01 to 0.06	$+m = l = 2$	Percy, Penrod, Bolton, and Smith
			3.0–3.4	—	$+m = l = 6$	

the variability time scales fall between the fundamental radial periods and the rotation periods. NRP periods among B stars show the largest range of any on the HR diagram, a factor of 50. The best known subgroups of NRP B stars are those with the largest observable variations, that is, those with a low-degree mode. Among the slow and fast rotators these are the 53 Persei stars and a subset of the so-called B-emission (Be) stars, respectively. (The latter display intermittent and often sudden episodes of emission in the Balmer alpha line. The origin of this activity is still unknown after a century of study.) Both the 53 Persei and Be stars exhibit periods between a few hours and a few days. It is not yet known whether these are different aspects of the same pulsation phenomenon. One fascinating observation is that the traveling waves in 53 Persei stars run faster than rotation, whereas they run slower than rotation in B-emission stars that exhibit NRP. In addition, one or more high-degree modes are excited in many moderately and rapidly rotating stars. Whenever this occurs the modes all run one way. This characteristic distinguishes them from NRP in the Sun, in which all m orders ($2l+1$ of them) are observed for each l degree. The degree of the observed modes in 53 Persei and Be stars is usually even. Typical values of high-degree modes are 6, 8, or 10, although values up through 16 have been found. The amplitudes of the modes are often observed to vary over a month or less. Occasionally, mode amplitudes can be variable in time, so that a new "primary mode" begins to dominate an old one. Typical amplitudes of NRP modes detected from line profile variations are a few kilometers per second. The typical depth of a valley or a hill is about 1% of a star's radius. Amplitudes as high as 20 km s^{-1} (the atmospheric sound speed) are observed. The nonlinearities in such pulsations can mimic other periodicities, adding great complexity to the analysis of modes. It is possible that the techniques of chaos theory may serve to distinguish between apparently nonperiodic variability and complicated "beating" patterns of many strictly periodic pulsation modes.

THE UNCERTAIN ORIGINS OF NRP

The observed range of NRP frequencies in B stars is so large that either gravity or hydrostatic pressure could operate as the restoring force. At first glance, the excitation of individual modes with frequencies of 0.3–0.5 cycles/day, as in 53 Persei itself or in α Virginis A, seems to imply that isolated modes of very high overtone (25 or even 100) can be excited without exciting the neighboring overtones. High-overtone modes are closely spaced in frequency according to gravity-mode theory. However, the natural selection of isolated high-overtone modes is unusual in any vibrating system, and the existence of very low frequencies calls into question whether these pulsations are high-overtone modes at all. Perhaps they are fundamental modes of another type of nonradial pulsation? A dual challenge now exists to find a theory that can explain both how nonradial pulsations originate and why only certain modes are excited among many of nearly the same frequency.

Another theoretical problem is that no pulsational excitation mechanism has been found that can drive oscillations strongly enough to overcome damping, the tendency of the pulsations to be dissipated by the smoothly varying absorption of radiation through the rest of the star. Some specialists believe that the large rotational energy of B stars, about 1% of their gravitational binding energy, could permit novel mechanisms to operate that could not operate in other stars. Others believe that if rotation does not actually excite NRP modes, it may help *select* the modes that are excited. For example, the fact that equatorial traveling-wave modes are most commonly excited suggests a preferred geometrical axis; the NRP axis is indeed defined by rotation. Also, the direction of motion of *traveling waves* in the prograde (with rotation) or retrograde direction seems to be decided by the star's rotational velocity. A rotational velocity of 175 km s^{-1} seems to be a rough demarcation: Those rotating slower have prograde waves and those rotating faster almost always have retrograde waves. Theoretical excitation mechanisms involving rotational energy have been suggested by Yoji Osaki, Hiroyasu Ando, Hideyuki Saio, and U. Lee at Tokyo. This group has studied whether an instability deep within the star, for example, associated with differential rotation or with interactions between rotation and convection, can excite pulsations. In this case, internal oscillations would pump mechanical energy and angular momentum into the pulsations of the whole star. Recently, Lee and Saio concluded that oscillatory convection in the stellar core can overcome dissipation. If these ideas are correct, the oscillation frequencies observed at the surface are controlled more by the star's internal rotation rate than by its most natural vibrational models. However, these mechanisms are unproven, and in some cases the proposed driving agents may not even exist.

The presence of variable amplitudes of NRP modes in B stars could imply that the energy source that drives pulsations is located near the surface. However, other causes that are considered ad hoc today, such as subsurface magnetic structures, may prove important. Because nonradial pulsations in massive stars are very common, it appears that the discovery of their cause will give us insights into fundamental physical processes inside massive stars that we currently cannot even envision.

Additional Reading

Gies, D. R., and Kullavanijaya, A. (1988). The line profile variations of epsilon Persei. I. Evidence for multimode nonradial pulsations. *Ap. J.* **326** 813.

Osaki, Y. (1986). Nonradial modes in line profile variable stars. *Seismology of the Sun and Distant Stars*, D. Gough, ed. Reidel, Dordrecht, p. 453.

Smith, M. A. (1986). Observations of nonradial pulsations in OB stars. *Hydrodynamic and Magnetohydrodynamic Problems in the Sun and Stars*, Y. Osaki, ed. University of Tokyo, Tokyo, p. 145.

See also **Stars, Beta Cephei Pulsations; Stars, Be-Type; Stars, δ Scuti and Related Types; Stars, Pulsating, Theory; Sun, Oscillations.**

Stars, Population III

Steven W. Stahler

It is now generally accepted that all of the hydrogen and most of the helium in the universe were produced cosmologically, during the first few minutes after the Big Bang. The elements heavier than helium were created by nuclear fusion in the interiors of stars and were subsequently dispersed into space through stellar winds and supernovae. An important consequence of this picture is that the earliest stars contained negligible amounts of these heavy elements, known collectively as "metals." Such primordial, metal-free stars are said to belong to Population III.

Spectroscopic studies of Population I stars, which comprise the disk of the Milky Way and other spiral galaxies, show them to have a metallicity (mass fraction in heavy elements) of $Z \approx 10^{-2}$. The metallicity is lower by a factor of 10–100 in Population II stars, which are the chief constituents of elliptical galaxies and are found in the halos of spirals. Stellar evolution theory shows that any star, regardless of metallicity, should have a lifetime longer than the age of the Galaxy if its mass is less than about 0.8 $M_\odot$ (solar masses). To date, however, no truly metal-free stars have been found, although a few stars with Z less than 10^{-5} have been identified. Thus, if these findings in the solar neighborhood are representative, it may be that Population III stars exist only in distant, younger galaxies. Alternatively, the first stars could have been so massive that they exploded long before the epoch of galaxy formation. Cosmological theory is not sufficiently developed to choose which of these alternatives, if either, is correct.

Despite this uncertainty concerning their origin and present existence, Population III stars have proved an attractive concept to astronomers in a number of different contexts. Massive primordial stars have been invoked to provide rapid heavy-element enrichment of the interstellar gas and to distort the cosmic microwave background, while their low-mass counterparts have been suggested to constitute the dark halos in our own and other spiral galaxies. Such issues will not be reviewed here; we will concentrate on the results concerning the evolution of individual metal-free stars that can be obtained from star formation and stellar structure theory.

CLOUD CHEMISTRY AND FRAGMENTATION

At the present epoch, stars form from the collapse of high-density knots of gas, known as "cold cores," which are embedded within much larger interstellar clouds. These cores have masses not much greater than typical stellar values, a fact that is a direct consequence of their extremely low temperatures. In any gas cloud, the critical mass for which self-gravity just balances internal pressure is given by

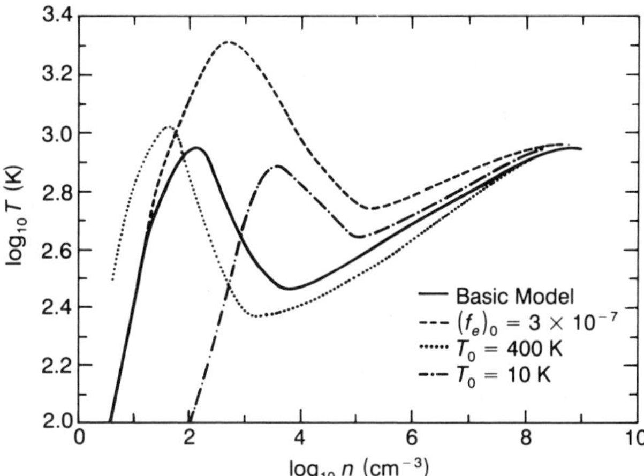

Figure 1. Evolution of the temperature in a collapsing primordial cloud. The "basic model" used an initial temperature $T_0 = 100$ K, a number fraction of electrons $(f_e)_0 = 2.8 \times 10^{-5}$, and a total number density $n_0 = 2.9$ cm^{-3}. Notice how a variety of initial states all converge by the stage at which a density of 10^8 cm^{-3} is reached. (*From Stahler, 1986. Reproduced, with permission, from the* Publications of the Astronomical Society of the Pacific ***98***.)

$$M_{\text{crit}} = 30T^{3/2}n^{-1/2},$$

where T is the cloud temperature in degrees kelvin, n is the number of gas atoms or molecules per cubic centimeter, and M_{crit} is measured in solar masses. In present-day cold cores, T is only about 10 K, allowing M_{crit} to be several solar masses at the typical n of 10^5 cm^{-3}.

The low temperatures in the cores are a result of efficient cooling by radio emission from gas-phase molecules, such as CO. In primordial clouds, such coolants were absent. In the past, this fact led many astronomers to conclude that Population III stars must have been much more massive than stars today. However, this conclusion only follows if primordial clouds fragmented into stars at densities similar to those of the present cold cores. If the clouds could have reached higher densities, they would have been unstable to collapse at lower masses so that lower-mass stars might have formed.

To examine this situation more carefully, let us consider the idealized example of a primordial cloud undergoing free-fall collapse after separating out from the large-scale cosmological expansion (Fig. 1). Cosmological theory tells us plausible initial values of density and temperature for such a cloud. We find that the hydrogen is almost entirely in atomic form, but that a very small fraction of it is ionized. As the collapse begins, the temperature in the cloud indeed rises very quickly. At higher densities, however, the temperature reaches a maximum, falls temporarily, and then resumes a much slower climb. By the time n has reached about 10^8 cm^{-3}, T is 1000 K, regardless of the precise initial conditions adopted for the cloud. The subsequent temperature rise is so gentle that T only reaches 2000 K if n is allowed to increase to 10^{14} cm^{-3}.

The temperature ceases its rapid rise at high densities because the cloud is able to radiate away much of the heat generated by gravitational compression. Although the original atomic hydrogen is a very poor radiator, chemical reactions gradually convert the hydrogen to molecular form as the density increases. The first reactions of importance are the coupled pair

$$H + e^- \rightarrow H^- + \gamma,$$

$$H^- + H \rightarrow H_2 + e^-,$$

where the electrons are those which were already present at the start of the collapse. Only about 10^{-3} of the hydrogen atoms become molecular as a result of these reactions, but their cooling is sufficient to overturn the temperature. At densities exceeding 10^8 cm^{-3}, the three-body reactions,

$$H + H + H \rightarrow H_2 + H,$$

$$H + H + H_2 \rightarrow 2H_2,$$

quickly convert *all* the hydrogen to molecular form. Because the resulting temperature rise is slow, M_{crit} falls quickly with rising density, reaching 1 $M_\odot$ at a density of 10^{14} cm^{-3}.

The large complexes that fragmented into star-forming clouds were undoubtedly *not* freely collapsing prior to fragmentation. However, our example does show that the primordial analogs to cold cores need not have been extremely massive. If we believe that they were the masses of ordinary stars, then we must also accept that the clouds were somehow able to contract to very high densities prior to forming stars. It is currently thought that the repulsion of embedded magnetic fields plays a role in limiting the densities of present-day cold cores. If primordial magnetic fields were smaller, then much higher cloud densities would indeed have been possible.

PRIMORDIAL PROTOSTARS

The collapse of cold cores to form present-day stars is thought to proceed in an inside-out fashion. The deep interior of the cloud quickly forms a primitive star, or protostar, which is supported against further collapse by internal thermal pressure. Meanwhile, the diffuse outer region gradually falls onto this central object. The protostar is highly luminous, since there is considerable heat generated at the shock front where the in-falling cloud material impacts the protostellar surface. The photons from the shock become degraded into the infrared region of the spectrum as they diffuse outward through the dusty cloud envelope. Thus protostars forming now are not optically visible.

If we accept the same picture of inside-out collapse for the formation of Population III stars, then it is possible to describe in detail the structure of primodial protostars (Fig. 2). A basic difference from the Population I case is that the time scale for collapse of the diffuse cloud is considerably shorter. This fact follows from the higher density, and therefore shorter free-fall time, of primordial star-forming clouds. Using a typical cloud temperature of 1500 K, it can be shown that a protostar of 1 $M_\odot$ builds up in about 10^3 yr, compared with 10^5–10^6 yr for its present-day counterpart.

This shorter evolutionary time scale leads to other differences between primordial and Population I protostars. First, because the luminosity generated at the shock depends on the rate at which in-falling matter crosses the stellar surface, a Population III protostar of a given mass would have been more luminous. For a mass of 1 $M_\odot$, a typical luminosity was 2000 $L_\odot$, about a factor of 100 greater than for the Population I case. Second, the rapid accretion rate means that any gas which has just hit the stellar surface is unable to cool very much before it is buried beneath the next settling layer. Consequently, primordial protostars were thermally distended objects. The radius of a 1-$M_\odot$ protostar, for example, was 50 $R_\odot$ (solar radii), 10 times larger than in the Population I case.

Because primordial clouds lacked the dust that impedes the radiation emanating from present-day protostars, *Population III protostars are optically visible objects*. The shock luminosity still does not directly stream out of the cloud, however, hydrogen in the

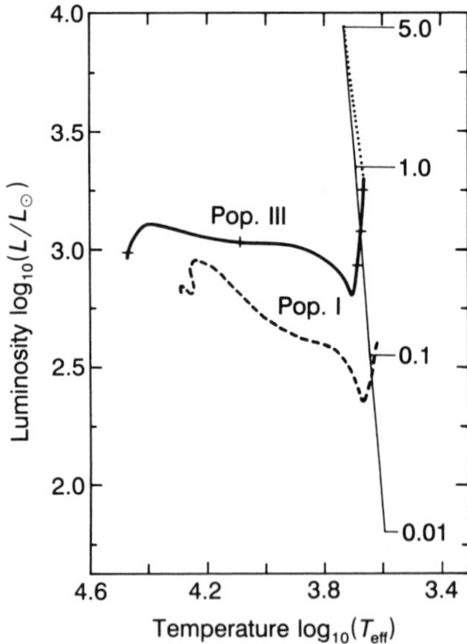

Figure 2. Primordial stellar evolution in the Hertzsprung–Russell diagram. The light solid curve is the photosphere of the collapsing protostar, with the hydrostatic core masses (in solar masses) indicated. The heavy solid curve is a 5-$M_\odot$ primordial pre–main-sequence star, with the tick marks indicating times of 10^2, 10^3, 10^4, 10^5, and 10^6 yr after the end of protostellar accretion. The dotted line connecting the light and heavy curves represents schematically the end of the accretion process. Finally, the dashed curve is the Population I 5-$M_\odot$ star of Iben (1965, *Ap. J.* **141**, 993). (*From Stahler, 1986. Reproduced, with permission, from the* Publications of the Astronomical Society of the Pacific ***98***.)

in-falling gas becomes collisionally ionized before striking the stellar surface, and the photons from the shock front are again degraded in frequency as they are absorbed and reradiated by this ionized gas. The photons break free from the gas at a temperature of about 5000 K. In summary, Population III protostars look quite similar to ordinary red giants. If they exist in distant galaxies, they should in principle be detectable, but their extremely short lifetimes make observation difficult.

SUBSEQUENT EVOLUTION

The precise manner by which Population III stars ended their accretion phase is not known, but we may speculate, in analogy to stars today, that a strong stellar wind blew back the in-falling parent cloud. Regardless of the mechanism involved, the young star, thereby divested of its in-falling cloud envelope, began to contract gravitationally. The star entered the pre–main-sequence phase, in which the contraction rate is controlled by cooling from the surface layers. (See Stellar Evolution, Pre–Main Sequence.)

The lack of metals in Population III stars had a steadily diminishing effect on successive stages of their evolution. Thus, the initial radii of these stars were extremely large, but the stars rapidly contracted to more familiar sizes once they could radiate freely into space. The ignition of hydrogen eventually halted their gravitational contraction. In present-day stars more massive than the Sun, carbon, nitrogen, and oxygen participate in a network of nuclear reactions (the CNO cycle) that mediates the final conversion of hydrogen into helium. In contrast, Population III stars of any mass could only use reactions involving hydrogen or helium (the PP chains). Since the PP chains are less temperature sensitive than the CNO cycle, main-sequence stars of primordial composition had higher central temperatures and lower radii (by about a factor of 2) than their Population I counterparts of equal mass. The lower opacity in these stars also made them more luminous and bluer.

In the near future, it should be possible to confirm these unusual characteristics in nearby stars of sufficiently low metallicity. In practice, a Z of 10^{-6}, just below the current lower limit, is small enough for all the effects described here to be manifested. Observation of more distant primordial stars could also be possible. As mentioned above, the dark halos of spiral galaxies may consist of primordial stars with masses too low to ignite hydrogen. If this is the case, then the luminosity generated during the prolonged gravitational contraction of these objects could be seen, if the galaxies are detected at a sufficiently young age.

Additional Reading

Fujimoto, M. Y., Iben, Jr., I., and Hollowell, D. (1990). Helium flashes and hydrogen mixing in low-mass Population III stars. *Astrophys. J.* **349** 580.

Jones, J. E. (1985). Early galactic evolution and the nature of the first stars. *Publ. Astron. Soc. Pac.* **97** 593.

Palla, F. (1988). Primordial star formation. In *Galactic and Extragalactic Star Formation*, R. E. Pudritz and M. Fich, eds. Kluwer Academic Publ., Dordrecht, p. 519.

Silk, J. (1987). Galaxy formation. In *Dark Matter in the Universe*, J. Kormendy and G. R. Knapp, eds. D. Reidel, Dordrecht, p. 335.

Stahler, S. W. (1986). The formation of primordial stars. *Publ. Astron. Soc. Pac.* **98** 1081.

See also **Cosmology, Population III; Interstellar Clouds, Collapse and Fragmentation; Stars, Chemical Composition.**

Stars, Pre-Main Sequence, Winds and Outflow Phenomena

Suzan Edwards

Matter is outwardly expelled into the surrounding interstellar medium by stars during most stages of their evolution; however, the amount of energy and momentum carried by these winds, the rate at which material is expelled, and the physical mechanism responsible for the expulsion, varies widely with stellar mass and age. Those winds which emerge from newly formed stars are among the most powerful a star will produce during the course of its evolution. For example, the mass-loss rate in a wind from a young (10^6 yr) solar mass pre-main-sequence star can exceed that of the present-day solar wind by a factor of one million; when a comparison is made between the mechanical energy in the wind and the stellar radiant energy, the numbers are equally staggering: For the Sun, the ratio of wind mechanical luminosity to stellar luminosity is about 10^{-5}%, and for a young solar mass pre-main-sequence star this ratio can be between 1 and 10%.

Energetic winds from young stars affect both the star and its surroundings by (1) providing a means of shedding the angular momentum contained in the molecular cloud core out of which the star formed, (2) limiting the mass of the accreting star, (3) dispersing the surrounding star-forming material, and (4) providing pressure support to the surrounding molecular cloud. Energetic winds are also widely invoked, at least in the case of our own youthful Sun, as a means of sweeping the early solar system clear of debris left over from the era of planet formation.

Energetic winds are seen to emerge from young stars in two different phases of their evolution:

Embedded infrared source: In this earliest era the star is hidden from direct view, lying deep inside a dusty molecular cloud "core" out of which it formed, which in turn usually lies within a larger molecular cloud complex. The emergent luminosity is entirely in the infrared and radio spectral region, because any optical light will be absorbed by dust grains along the line of sight, and then reemitted at longer wavelengths. The distribution of the infrared spectral energy from embedded young stellar objects (YSOs) has been interpreted to arise in a composite system characterized by a central star, an opaque circumstellar disk, and an in-falling envelope of gas and dust.

Optically visible, pre-main-sequence stars: In this later stage, the stars have sufficiently accreted and dispersed their dense cores that the stellar photospheres are visible. [Only lower-mass stars, with masses comparable to our Sun, are observed in this pre-main-sequence (PMS) phase because their accretion/dispersal time scales are short compared to the time scale for the star to reach the zero-age main sequence]. Many of these T Tauri stars are characterized by considerable excess infrared emission, which is believed to arise in highly flattened circumstellar disks.

In this entry, wind diagnostics will be described and our current understanding of the ubiquity, the morphology, and the wind energetics of both the embedded infrared sources and the optically visible (low mass) pre-main-sequence stars will be summarized. Although the origin and evolution of energetic winds from young stars is not well understood, the possibility will be explored that these early winds derive their energy not from the stars themselves, but from their circumstellar disks, which, in turn, may be precursors to planetary systems.

EMBEDDED INFRARED SOURCES

Systematic surveys of infrared sources in the vicinity of dense cores within molecular clouds suggest that most, if not all, embedded young stars undergo a phase of energetic mass loss. The most common observational diagnostic of this phase is a "molecular outflow"—a spatially extended, often bipolar, stream of cold molecular gas expanding supersonically away from an embedded infrared source. These outflows are traced via Doppler-shifted millimeter-wave line emission from abundant molecular species such as CO. As drawn in Fig. 1, the observations are best explained as swept-up shells of ambient molecular gas which have been accelerated to speeds of 10–30 km s^{-1} by a high speed (50–200 km s^{-1}) atomic/molecular wind that emerges from near-stellar regions. In at least some well-studied examples, shock-excited ionized gas, located in both elongated, highly collimated "jets" and in bow shocks formed at the working surfaces of these jets [Herbig–Haro (HH) objects], is also found to be moving away from the YSO with speeds of up to several hundred kilometers per second.

The majority of the molecular outflows from embedded YSOs are channeled into two oppositely directed, collimated streams. Although the degree of collimation of the high-velocity molecular gas, which extends 0.5–1 pc from the outflow source, is modest (length/width ratios < 5), the ionized jets, when observable, extend only for 1000s of astronomical units from the outflow source and show remarkable collimation (length/width ratios of 5–20). There are several clear examples, notably L1551 IRS 5 and HL Tau, where favorable inclinations and luminosities allow one to see that the outflow axes are perpendicular to the plane of the YSO's circumstellar disk. The collimation mechanisms for either the small-scale or the large-scale structures have yet to be identified, but the alignment of multiple adjacent outflows with the direction of the magnetic field through the host molecular cloud suggests that the cloud magnetic field can play a role, at least in the initial orientation of collapsing cores.

The energetics of the winds associated with embedded YSOs can be derived from the observable properties of an outflow: its size, velocity field, and mass. Most notably, there is a strong correlation between the total radiant luminosity of the central YSO, and both the mechanical luminosity of the flow and the force or thrust

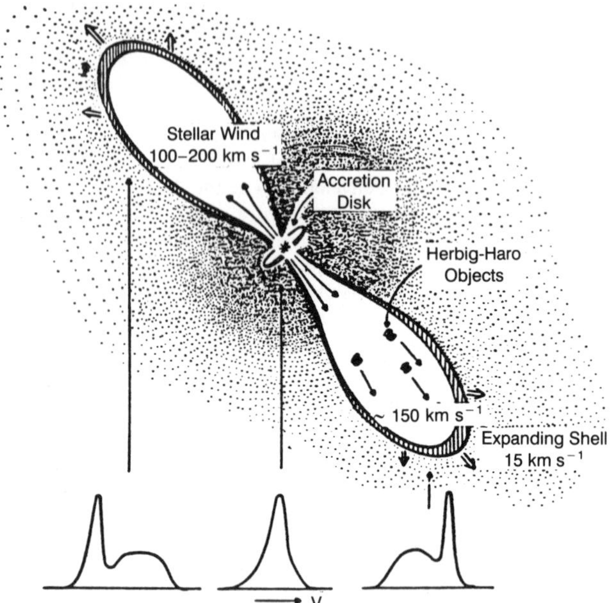

Figure 1. An artist's conception of a bipolar molecular outflow from an embedded infrared source. The young star is surrounded by a disk of solar system dimensions, and a collimated wind emerges perpendicular to the disk plane. The wind sweeps up ambient molecular cloud material into two expanding bipolar lobes of cold molecular gas. [*From Snell, Loren and Plambeck, 1980,* Ap. J. (Lett.) ***239****, L17.*]

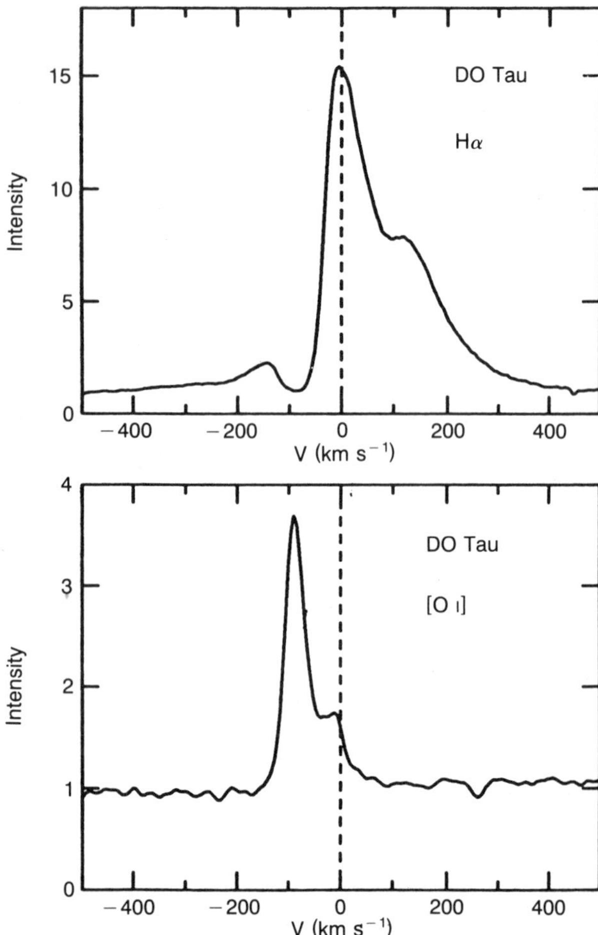

Figure 2. Representative T Tauri star emission line profiles revealing the presence of an energetic wind in the star DO Tau. In both panels, the brightness at a given frequency is plotted against the corresponding radial (Doppler) velocity measured with respect to the stellar photospheric velocity. The upper panel shows the Hα line, formed in the dense inner wind, with broad, symmetric wings cut by a *blueshifted* reversal. The lower panel shows the forbidden line [O I] λ6300 Å, formed in the low-density outer wind, with broad and almost entirely *blueshifted* emission. The absence of redshifted gas in the forbidden lines is attributed to occultation of the receding hemisphere of the outer wind by an opaque circumstellar disk of solar system dimension.

needed to drive the flow. These correlations hold over 5–6 orders of magnitude and suggest that (1) a common physical mechanism drives the flows in all sources and (2) the energetics of the outflow are determined in some way by the luminosity of the YSO, which is comprised of unknown relative contributions between stellar and accretion luminosity.

OPTICALLY VISIBLE, LOW-MASS PRE-MAIN-SEQUENCE STARS: T TAURI STARS

Very few PMS stars show the *spatially extended* signatures characteristic of the embedded YSOs, such as molecular outflows and collimated ionized jets; those that do are the most similar to the embedded sources in terms of their excess infrared emission, suggesting that they are transition objects between the embedded infrared source phase and the optically visible PMS star phase. However, at high spectral resolution, many show velocity structure in strong emission lines indicative of mass outflows, most notably at Hα and in low-excitation metallic forbidden lines such as [O I] or [S II].

Representative T Tauri star line profiles for Hα and for [O I] λ6300 Å are shown for DO Tau in Fig. 2. The Hα line, which is very optically thick and formed close to the star ($r < 10\ R_\star$), has broad, symmetric wings and is cut by a blueshifted reversal. Such P Cygni-like profiles, formed in a dense wind, are found at Hα in many T Tauri stars. The forbidden lines, which are optically thin and formed in low-density, ionized outflowing gas far from the star ($r \gg 1$ AU), are also broad, but the observed emission is almost entirely blueshifted. The lack of redshifted emission is characteristic of nearly all T Tauri star forbidden lines, and is attributed to occultation of the receding hemisphere of the outflow by a highly flattened, opaque circumstellar disk.

The morphology of the energetic winds from T Tauri stars is not yet established; it is clear, however, that the observed velocity structure of the forbidden lines, which is often double peaked, cannot arise in a spherically symmetric outflow. A variety of possible T Tauri wind morphologies have been suggested, but the final answer awaits high-spatial-resolution imaging, which will be available to astronomers in the coming decade.

Our understanding of the energetics of T Tauri winds is based on theoretical interpretations of the wind-sensitive line profiles, which is not as definitive as interpreting the spatially resolved outflows from the embedded YSO population. One theoretical expectation is that the luminosity in both the Hα and the forbidden lines will increase in proportion to the wind mass-loss rate, although for the temperature-sensitive Hα line, this would occur only if all T Tauri stars have similar wind ionization and velocity structure. In a recent study of T Tauri stars of age $< 3\times10^6$ yr, it was found that the luminosities in the forbidden lines, formed in the outer wind, and the luminosities in the Hα lines, formed in the inner wind, range proportionately over nearly 20 orders of magnitude. This observed correlation can thus be interpreted as indicating that a large range in wind energetics and mass-loss rates characterizes PMS stars of similar mass and age. There is reasonable agreement in the range of mass-loss rates found among the PMS stars as estimated from these two lines, from a high of about 10^{-7} $M_\odot$ yr^{-1} to a low of about 10^{-9} $M_\odot$ yr^{-1}.

ORIGIN OF ENERGETIC WINDS?

The origin and evolution of energetic winds from young stars is still unknown. A crucial question to be answered in the coming years is whether or not the energetic winds from the embedded YSOs and from the optically visible PMS stars share a similar origin. Another important question is whether the energy source for the wind is the young star itself, its surrounding circumstellar disk, or some interaction between the two. The possibility that the energetic winds from both types of young stars are similar in origin and ultimately derive from gravitational potential energy associated with the accretion of material through their disks and released in near-stellar regions is discussed in the remainder of this section.

The stellar versus disk origin for the winds is best addressed by looking at the optically visible T Tauri stars. In contrast to the embedded YSO population, the underlying stellar photospheres can be observed and compared for stars of differing wind strengths. What is striking is that the *stellar* properties, such as photospheric luminosity, mass, age, rotational velocities, and presence of dark star spots and the properties of those spots, and the x-ray flux arising from enhanced solar-like magnetic activity, appear to be quite similar among stars with both large and small wind mass loss rates. This implies that the energy source driving the winds from T Tauri stars is *external* to the star.

A link between the strength of the wind and the infrared brightness of the circumstellar disk among the T Tauri stars is suggested by an observed correlation of the luminosities in the wind-sensitive emission lines (Hα and the forbidden lines) with the luminosity of the excess infrared emission, which is believed to come from the disks. The interpretation of this correlation requires an understanding of the origin of the excess infrared emission from the disk. There are basically two alternatives: (1) the excess infrared luminosity is due solely to a passive "reprocessing" disk, which simply reradiates energy in the infrared that was absorbed from optical and ultraviolet stellar photons or (2) the excess luminosity arises from a combination of reprocessing and an additional intrinsic disk luminosity generated by accretion of material through the disk. If internal-disk heating from accretion is an important contributor to the infrared excess, then the observed correlation would imply a proportionality between the mass loss rate in the T Tauri star winds and the mass accretion rate through the T Tauri star disks.

In the embedded YSOs the inability to observe the stellar photospheres makes it difficult to address the issue of a stellar versus a disk origin for the energetic winds. The observed wind/disk morphology in these sources, however, argues in favor of a disk origin for the energetic winds. If the total luminosity of the embedded star plus disk systems is largely due to a disk accretion luminosity, then the observed correlation of wind mechanical energy and total system luminosity in these sources would also be explained as a proportionality between the wind mass-loss rate and the disk mass-accretion rate.

At this writing, the link between energetic winds and circumstellar disks in young stars is not fully understood, but the possibility that the winds may serve as a probe of disk accretion is an exciting one. The means by which the gravitational potential energy associated with disk accretion would be transferred outward into a wind in the near-stellar regions is still mysterious, but not implausible. If this scenario proves to be correct, then the termination of the phase of energetic winds and active disk accretion in young stars may be linked to disk clearing and the formation of planetary systems. These issues will be resolved as we advance theoretically in our understanding of the physics of accretion disks and observationally by routinely achieving a factor of 10 increase in angular resolution with optical telescopes in space and with actively controlled optical imaging systems on the ground.

SUMMARY

Energetic winds are seen to emerge from young stars in two evolutionary stages: In the earliest stage, the stars are still embedded in the molecular cloud cores out of which they formed (embedded YSO), and in the second stage the stars have sufficiently accreted and dispersed their molecular cores that the stellar photospheres have become optically visible (T Tauri stars). In the former case, spatially extended structures show that the winds are highly collimated and bipolar in nature, emerging perpendicular to the plane of the solar-system-sized disks around these forming stars. In the latter case, although the wind/disk geometry is not yet known, the strength of the energetic winds from the T Tauri stars is correlated with the brightness of their solar-system-sized circumstellar disks. We speculate that it is the disks which provide the energy reservoir for the energetic winds from both groups of young stars. If this scenario proves to be correct, then the termination of the phase of energetic winds and active disk accretion in young stars may be linked to disk clearing and the formation of planetary systems.

Additional Reading

Bertout, C. (1989). T Tauri stars: Wild as dust. *Ann. Rev. Astron. Ap.* **27** 351.

Bieging, J. H. and Cohen, M. (1985). Multifrequency radio images of L1551 IRS 5. *Astrophys. J. (Letters)* **289** L5.

Cabrit, S., Edwards, S., Strom, S. E., and Strom, K. M. (1990). Forbidden line emission and infrared excesses in T Tauri stars: Evidence for accretion-driven mass loss? *Astrophys. J.* **354** 687.

Hartmann, L. (1986). Theories of mass loss from T Tauri stars. *Fundam. Cosmic Phys.* **11** 279.

Lada, C. J. (1985). Cold outflows, energetic winds, and enigmatic jets around young stellar objects. *Ann. Rev. Astron. Ap.* **23** 267.

Lada, C. J. and Shu, F. H. (1990). The formation of sunlike stars. *Science* **248** 564.

Mundt, R. (1988). Flows and jets from young stars. In *Formation and Evolution of Low Mass Stars*, A. K. Depree and M. T. V. T. Lago, eds. Kluwer Academic Publ., Dordrecht, p. 257.

Pringle, J. E. (1989). A boundary layer origin for bipolar flows. *Mon. Not. R. Astron. Soc.* **236** 107.

Sargent, A. and Mundy, L. G. (1988). High resolution observations with the Owens Valley Millimeter Wave Interferometer. In *Galactic and Extragalactic Star Formation*, R. E. Pudritz and M. Fich, eds. Kluwer Academic Publ., Dordrecht, p. 261.

Schwartz, R. D. (1983). Herbig–Haro objects. *Ann. Rev. Astron. Ap.* **21** 209.

Shu, F. H., Adams, F. C., and Lizano, S. (1987). Star formation in molecular clouds: Observations and theory. *Ann. Rev. Astron. Ap.* **25** 23.

Snell, R. L. (1989). Molecular outflows. In *Structure and Dynamics of the Interstellar Medium*, G. Tenorio-Tagle, M. Moles, and J. Melnick, eds. Springer-Verlag, Berlin, p. 231.

Strom, S. E., Strom, K. M., and Edwards, S. (1988). Energetic winds and circumstellar disks associated with young stellar objects. In *Galactic and Extragalactic Star Formation*, R. E. Pudritz and M. Fich, eds. Kluwer Academic Publ., Dordrecht, p. 53.

Strom, S. E., Edwards, S., and Strom, K. M. (1989). Constraints on the properties and environments of primitive stellar nebulae from the astrophysical record provided by young stellar objects. In *The Formation and Evolution of Planetary Systems*, H. A. Weaver and L. Danly, eds. Cambridge University Press, Cambridge, p. 91.

See also **Herbig–Haro Objects and Their Exciting Stars; Planetary Systems, Formation, Observational Evidence; Stars, T Tauri; Stars, Young, Continuum Radio Observations; Stars, Young, Jets.**

Stars, Pre-Main Sequence, X-Ray Emission

Eric D. Feigelson

Prior to the 1980s, pre-main sequence stars were not considered to be likely sources of significant x-ray emission. A star passes through its pre-main sequence phase of evolution after it forms by gravitational collapse of an interstellar molecular cloud, and before hydrogen-fusion nuclear reactions power the long-lived main sequence phase. Pre-main sequence stars have relatively cool surface temperatures (typically 3000–4000 K) compared to main-sequence stars, are often straddled by disks of dust and gas with temperatures below 500 K, and are located in or near gas clouds with temperatures below 20 K. X-rays, on the other hand, are produced by gas with temperatures of 10^6–10^7 K, or by particles accelerated to high speeds in magnetic fields. A few researchers speculated that x-rays might be produced in shocks associated with the gas ejected by some young stars, but the energy available in these outflows would be sufficient only to produce temperatures up to 1×10^6 K.

It was therefore quite a surprise when satellite-borne x-ray telescopes found hundreds of x-ray–emitting pre-main sequence stars. The first hint of these sources emerged in the 1970s from observations with satellites that had x-ray detectors but that could not produce two-dimensional x-ray images. They showed that the Orion nebula, one of the largest nearby regions of star formation, is an extended x-ray source. The cause of this x-ray emission was not clear; it was much fainter than x-ray sources involving accreting compact objects, but much more luminous than a typical main-sequence star like the Sun. The enigma was explained when the *Einstein* observatory, the first satellite equipped to focus x-rays into an image, resolved the Orion nebula source into dozens of individual pre-main sequence stars. *Einstein* images of other nearby star-forming regions—such as the ρ Ophiuchi cloud, Chamaeleon cloud, and especially the large Taurus–Auriga cloud complex—showed that x-ray–emitting pre-main sequence stars are very common.

The x-ray emission of these stars was found to have little relationship with the ejecta (winds, bipolar flow, or Herbig–Haro objects) or dusty disks that characterize protostars and "classical" T Tauri stars. Some of these x-ray stars are classical T Tauri stars, but most are not. Labeled "weak" T Tauri stars, these x-ray–emitting, pre-main sequence stars do not have the strong broad optical emission lines, infrared and ultraviolet excesses, and large-amplitude variabilities of classical T Tauri stars. X-ray surveys reveal that the population of weak T Tauri stars (sometimes called "naked" T Tauri stars) is at least comparable to, and probably considerably larger than, the population of classical T Tauri stars. The *Einstein* x-ray images thus greatly increased the number of known pre-main sequence stars, and highlighted a new topic for investigation: Why do weak T Tauri stars produce x-rays at levels thousands of times stronger than those typical of main-sequence stars?

Recent observational study of these stars at bands across the electromagnetic spectrum—radio, optical, ultraviolet, and x-ray—has shed light on this question. Radio observations of weak T Tauri stars, made with the Very Large Array and very long baseline interferometry, show that some emit radio continuum radiation at levels 10^5–10^6 that of the Sun. These radio sources are spatially compact, temporally variable, and have spectral indices like those seen on the Sun during its occasional powerful flares. Optical photometry shows that weak T Tauri stars often have cool starspots, similar to but much larger than the dark sunspots seen on our Sun. The presence of huge starspots is inferred from

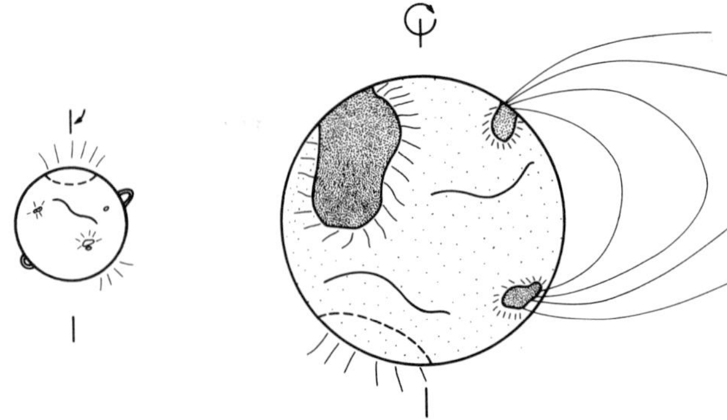

Figure 1. Sketches comparing the appearance of the Sun today (left) with its appearance as a 1×10^6-yr-old weak T Tauri star (right). Active regions, rotation rates, and magnetic loops are shown.

periodic brightness variations; these stars rotate on their axes every few days, and appear fainter and cooler when the side facing us has a large starspot. Optical and ultraviolet spectra of weak T Tauri stars show emission lines characteristic of an enhanced solar-type chromosphere. In one such star, a powerful flare lasting about 20 min was seen in blue light. The x-ray luminosities change by factors of 2–50; all sources vary on time scales of days, and a few are seen to change x-ray brightness in hours or minutes.

All of these properties are familiar to solar researchers. The Sun has sunspots, chromosphere, and occasional powerful flares that cause sudden huge increases in its radio and x-ray emission. A diagram comparing the young and contemporary Sun is shown in Fig. 1. The weak T Tauri stars, however, show these characteristics at levels 10^3–10^5 times stronger than those seen in the contemporary Sun. This finding is consistent with, and extends, trends in solar-type activity seen in stars of intermediate ages, such as those in the Pleiades open cluster. Empirical relationships indicate that younger stars generally rotate faster and have larger star spots, more prominent chromospheres, and more powerful flares. This is all qualitatively understood in terms of the generation and dissipation of magnetic fields in stars. The discovery of strong x-ray emission in pre-main sequence stars thus suggests that low-mass stars exhibit their highest levels of magnetic activity during their pre-main sequence phase of evolution.

The study of x-ray emission from young stars is only a decade old. During the 1990s, several new satellites with x-ray telescopes will gather much more observational data. For example, the advent of x-ray imaging at shorter wavelengths, where absorption by foreground gas is reduced, may permit x-rays from deeply embedded protostars to be detected. Complementary studies at ultraviolet, optical, infrared, and radio wavelengths will deepen our understanding of the x-ray–emitting stars as well. Many intriguing issues are still not resolved. It is not clear whether x-rays from classical T Tauri stars are also due to magnetic activity, or rather to their circumstellar disks and outflows. The ionizing radiation or particles produced by magnetically active pre-main sequence stars may affect the surrounding molecular cloud or disk. An intriguing relationship has been suggested between energetic particles formed in the giant flares on the pre-main sequence Sun and unusual properties of certain meteorites. There may thus be subtle links between x-rays emitted by young stars and the processes by which stars and planetary systems form.

Additional Reading

Bertout, C. (1989). T Tauri stars: Wild as dust. *Ann. Rev. Astron. Ap.* **27** 351.

Cohen, M. (1988). *In Darkness Born: The Story of Star Formation.* Cambridge University Press, Cambridge.

Feigelson, E. D., Giampapa, M. S., and Vrba, F. J. (1991). Magnetic activity in pre-main sequence stars. In *The Sun in Time*, C. P. Sonett and M. S. Giampapa, eds. Arizona University Press, Tucson.

Rosner, R., Golub, L., and Vaiana, G. S. (1985). On stellar x-ray emission. *Ann. Rev. Astron. Ap.* **23** 413.

See also **Herbig–Haro Objects and Their Exciting Stars; Solar Activity; Stars, Activity and Starspots; Stars, Atmospheres, X-Ray Emission; Stars, Magnetism, Theory; Stars, Pre-Main Sequence, Winds and Outflow Phenomena; Stars, T Tauri; Stars, Young, Jets.**

Stars, Proper Motions, Radial Velocities, and Space Motions

Arthur R. Upgren

The determination of the total motion of a star in space is simple in concept, but quite difficult in practice. One of the problems encountered is the orientation of that space motion to a fundamental or dynamically meaningful reference frame. Another is the determination and removal of the motion of the observer. But the most significant and involved problem is the calculation and combination of the radial and transverse components of that motion, which are obtained from two entirely different kinds of observations. The radial component is measured directly, but the transverse (or tangential) component must be derived from the apparent angular motion or proper motion of the object and its distance. An understanding of proper motion and radial velocity is essential to understanding the calculation of the space motion and its level of precision, and thus the grouping of the three subjects in the title is a natural one. The equation relating the three necessary parameters and the space motion is

$$S=[R^2+T^2]^{1/2}=\left[R^2+(4.74\mu/\pi)^2\right]^{1/2},$$

where S, R, and T are the total space velocity and its radial and transverse components, respectively, and are customarily given in units of kilometers per second (km s^{-1}). The angular quantities μ and π are the annual proper motion and heliocentric parallax, respectively, and 4.74 is the factor necessary for the conversion into kilometers per second from astronomical units per year, the natural units of the ratio μ/π. This ratio is sometimes replaced by the term μd, because the distance d, in parsecs, is simply the reciprocal of the parallax in seconds of arc.

PROPER MOTIONS

The individual motions of stars have been known to exist since 1718, when Edmond Halley noticed that a few of the brightest stars had shifted their positions within their constellations from those given by Hipparchus in his catalog almost 2000 years earlier, even after the motion attributable to precession had been removed. This phenomenon is called proper motion, and is usually given in units of seconds of arc per year. It can only be interpreted as the motion perpendicular to the line of sight and, like any angular quantity, requires knowledge of a distance to be converted into linear units. The proper motion is often also defined as the component in the annual variation of position of a star that is not due to the effects of precession and nutation.

The proper motions of bright stars have commonly been calculated from meridian-circle observations. Those of fainter stars are most easily found from photography. Some of the proper motions of highest precision have been found in the course of the determination of the heliocentric or trigonometric parallax, because both parameters are found from solutions of the same equations of condition. In all cases, proper motions are derived from changes in position and their precision increases with the time span between first and last observations. The precision also increases with decreasing distances of stars, because distant objects appear to move more slowly, and a longer time span between the epochs of observation is required to achieve a given level of precision for them. Because the measurement of precise proper motions of even the nearest stars requires years or even decades, the introduction of new methods of light collection will not totally replace the photographic plate for some time to come.

Proper motion is, along with position and apparent magnitude, the most frequently known parameter among the common stars. Among the reasons for this is the existence of a number of surveys that locate stars of significant proper motion and distinguish them from the many other background stars, the proper motions of which are negligibly small. This can be accomplished for many thousands of stars by means of the examination of two well-aligned photographic plates taken years apart. The stars with detectable proper motion will appear to jump back and forth in the field of a lens alternating between the two plates, whereas the background stars appear stationary. Many hundreds of thousands of stars have been examined in this way, and those that are intrinsically faint and nearby become identified as such. Proper-motion surveys have been of critical importance in determining the stellar luminosity function, the number of stars per unit volume as a function of intrinsic luminosity. In cases where parallaxes and radial velocities are impossible or impractical to obtain, proper motions are useful for the calibration of luminosities of stars by means of methods of statistical and secular parallax.

RADIAL VELOCITIES

The radial velocity of an object is the component of its total velocity along the line of sight between it and the observer. It is a consequence of the principle of the dependence of the observed wavelength of light on the relative motion of the light source and the observer; the dependence is attributed to Christian J. Doppler, who is credited with first describing this so-called Doppler shift. In the case of a star, the shift is the displacement in wavelength of the spectral lines away from their wavelength at rest as determined in the laboratory. Stellar radial velocities are denoted as positive when directed away from the observer and negative when approaching, and are typically in the range of tens of kilometers per second. The velocity can be found from the ratio between the shift in the wavelength and the wavelength itself, which can be assumed to be proportional to the ratio between the velocity of a star and that of light. Radial velocities of all sources outside the solar system are referred to the Sun, or more properly, to the barycenter of the solar system. This calls for the evaluation and removal of the component of the orbital motion of the Earth about the Sun along the direction to the object and in the most precise work, of the Earth's rotational velocity along the same direction as well.

Stellar radial velocities were accumulated steadily throughout the first half of the twentieth century. The observations were made using slit spectrographs. Great care had to be taken in order to obtain reliable data because of the smallness of the Doppler displacements and because of the faintness of the light sources. In most cases, comparison spectra from local light sources were photographed alongside the stellar spectrum and were exposed simultaneously on the same spectrogram in order to calibrate the observed wavelengths of the spectral lines. The mechanical and optical tolerances of the spectrographs had to be severely restricted in order that systematic errors were held to a minimum. The collimation of the optical components, the rigidity of the spectrographs and their supporting systems attached to the telescopes, and the surrounding temperatures were among the features that needed to be controlled carefully, lest unacceptably large systematic errors arise.

Radial velocities obtained in this manner have errors typically of about 3–5 km s^{-1}. With few exceptions the stars observed are bright; only a handful are fainter than the tenth magnitude and most are brighter than the seventh magnitude. The results of this body of observations were compiled and printed in the *General Catalogue of Stellar Radial Velocities*, last published in 1953. It contains the radial velocities of 15,107 stars and is a compilation only, with all of the problems inherent in any combination of heterogeneous data. But it gives a good picture of the state of the art of the time. A more recent (1972) source for published radial velocities is the *Bibliography of Stellar Radial Velocities*.

In the last decade, the field of radial velocities has been revolutionized through the use of new techniques. The uncertainty in the velocity of a single star has been reduced by nearly an order of magnitude from a few kilometers per second to a few hundreds or even to tens of meters per second in the best cases. Furthermore, large numbers of stars extending to faint limiting magnitudes can now be routinely observed. In addition to the great improvement in the understanding of problems investigated by studies of radial velocities in the past, other problems may now be attacked that were beyond the reach of earlier work. Among these are the detection of stellar oscillations and the potential for better insight into the internal constitutions of stars, modes of pulsation for low-amplitude pulsating variable stars, the possible existence of fine structure in the motions of members of nearby star streams such as the Ursa Major group, and the detection of companions of substellar mass around nearby stars.

A number of new radial-velocity measurement methods have been developed. Most have some major features in common. Chief among them is the superimposition of the wavelength reference upon the starlight. In the conventional design, the stellar and comparison beams could not be made to follow the same path through the optics of the spectrograph. But with light beams from both sources made to follow the same route, a number of systematic errors are avoided or greatly reduced.

SPACE MOTIONS

The combination of radial velocities and proper motions with parallaxes (or other secondary distance indicators) to form total space motions was briefly described at the beginning of this entry. The resultant motions will naturally be oriented in a coordinate system aligned with the plane of the sky and the axis that is normal to it and directed towards the observer. But the motions of stars have the greatest physical meaning when related to the plane of the Galaxy, and a simple rotation of the coordinate axes accomplishes this conversion. The axes are most commonly labeled U, V, and W in velocity space, with the U axis lying along the direction towards the galactic center, the V axis lying in the galactic plane and positive in the direction of circular galactic rotation about the center and the W axis normal to the plane and positive toward the north galactic pole. Note that the positive direction along the U axis appears in the literature as either directed toward or directed away from the galactic center, and care must be taken not to confuse the two definitions.

One of the difficulties involved in the use of space motions to study the kinematical and dynamical properties of stars in general and of the galactic system as a whole is due to the paucity of stars for which precise motions are known. Until recently, radial velocities were known for only the brighter stars with very few exceptions. Individual proper motions, on the other hand, are usually only accurate for nearby stars, those within a few tens of parsecs. It is well known that the overlap between the brightest and the nearest stars is very small. As an example of this mutual exclusion, all but 1 of the 300 stars appearing brightest in our night skies are intrinsically brighter than the Sun and most of them are many times as luminous. Yet among the 300 stars closest to us, only about 10% are intrinsically brighter than the Sun, and most are much fainter. Coverage of two such disparate samples with radial velocities, parallaxes, and proper motions, all to the high precision necessary for determining a reliable space motion has not been made for many stars until the last decade.

Space motions naturally incorporate the Sun's own orbital motion about the galactic center, as well as those of the stars in question. However, a reference frame tied to the solar system is not practical for the investigation of stellar dynamics. A dynamical local standard of rest or LSR, orbiting the galactic center at the distance of the Sun, makes much more sense, but is not easy to determine. In lieu of this, the mean group space motions of stars with well-defined properties are often used as approximations. These kinematical LSRs are also assumed to reflect the individual motion of the Sun relative to the stars forming the groups. The solar motion can be and has often been derived from either radial velocities or proper motions alone, but the total space motions are much more efficient and precise for the purpose. Results of recent years show that there exists no single basic or standard solar motion, but instead the motion components along both the U and V axes vary with the average ages of the stars to which the Sun's motion is referred. Since the true motion of the Sun is unique, we can conclude that the average motions of young and old stars are different along both of these axes. One of the major results of the accumulation of space motions is that old stars move toward the galactic center and more slowly with respect to young stars (although not necessarily with respect to the dynamical LSR).

Parallaxes and proper motions, as well as radial velocities, are now calculated much more precisely than they were just a few years ago. As a result we can look forward to many more total space motions of higher accuracy, and the solution of any number of vexing problems of our Milky Way galaxy.

Additional Reading

Abt, H. A. and Biggs, E. S. (1972). *Bibliography of Stellar Radial Velocities.* Latham Process Corp., New York.

Delhaye, J. (1965). Solar motion and velocity distribution of common stars. In *Stars and Stellar Systems, Vol. III, Galactic Structure*, A. Blaauw and M. Schmidt, eds. University of Chicago Press, Chicago, p. 61.

Griffin, R. (1989). The radial-velocity revolution. *Sky and Telescope* **78** 263.

Hayes, D. S., Pasinetti, L. E., and Philip, A. G. D., eds. (1985). *Calibration of Fundamental Stellar Quantities, IAU Symposium 111.* D. Reidel, Dordrecht.

Mihalas, D. and Binney, J. (1981). *Galactic Astronomy, Structure and Dynamics*, 2nd ed. W. H. Freeman and Co., San Francisco.

Philip, A. G. D. and Latham, D. W., eds. (1985). *Stellar Radial Velocities, IAU Colloquium 88.* L. Davis Press, Schenectady, N.Y.

Wilson, R. E. (1963). *General Catalogue of Stellar Radial Velocities.* Publication 601, Carnegie Institution of Washington, Washington, D.C.

See also **Astrometry; Astrometry, Techniques and Telescopes; Coordinates and Reference Systems; Stars, Proper Motions, Radial Velocities, and Space Motions.**

Stars, Pulsating, Overview

Michael W. Feast

Stars, like the strings of a musical instrument (or indeed, like any object), have various possible modes of oscillation or tone. Depending on the star's structure and the way energy is transported and stored within it, these oscillations may, if excited, be rapidly damped out or be self-sustaining. As with a violin string, more than one frequency of oscillation may be present (i.e., fundamental, first overtone, etc.), leading to complex patterns of oscillation. (The terms stellar pulsation and stellar oscillation are frequently used interchangeably.)

That radial pulsation, the regular, spherically symmetrical expansion and contraction of a star, might be responsible for the observed light variability of some stars was suggested by the German physicist August Ritter in 1879. Later (1914) Harlow Shapley showed that radial pulsation was the only tenable hypothesis for the variability of the star δ Cephei and similar objects (the classical Cepheids). The light variations are due to changes in the star's surface area and to the accompanying changes in its surface temperature. In addition, a regular inward and outward motion of the star's atmosphere is revealed by the varying (Doppler) displacement of the star's spectral lines. In more recent times it has been found that some stars (including the Sun) are nonradial pulsators; these stars change shape rather than volume, becoming nonspherical. A combination of radial and nonradial motions is also possible.

Pulsating variable stars are of importance to astronomy for the following main reasons:

1. Whilst stars in general have a wide range of characteristics (mass, age, luminosity, surface temperature, etc.), pulsators may generally be classified by their light variations into groups that have a limited range of physical characteristics. It thus becomes possible to discuss, with some success, the ages, masses, and evolutionary states within each group.
2. Several groups (Cepheids, Miras, RR Lyrae stars) have well-defined brightnesses (which may depend on the star's period, temperature, and chemical composition). Because of this, these groups are of fundamental importance in establishing galactic and extragalactic distance scales.
3. The pulsational characteristics of a star depend on its mass, radius, and internal structure, and therefore, in favorable cases, pulsation studies provide information on these quantities that cannot be obtained in any other way.
4. The changes in luminosity and surface temperature of a star as it evolves (i.e., as it ages) are generally far too small to be detected directly over the period of time during which measurements of these quantities have been made. (Measurements of stellar brightnesses and colors with an accuracy of a few percent have only been possible for about 100 yr, though rough estimates for some stars exist from a few thousand years ago.) Thus, in constructing theories of stellar evolution we are in a similar position to the extraterrestrial who tries to establish human relationships and ages from a single, unmarked, snapshot showing several families together. However, the periods of some pulsating stars can be found with high accuracy and the very small period changes expected, due to the slow change in the star's radius as it ages, can be detected from observations over periods of 50 yr or even less. Thus one pulsating pre-white dwarf with a period of 8.6 min was found, from 96 h of observing time spread over 4.4 yr, to have an evolutionary time scale (the period divided by the rate of change of the period) of 1.4×10^6 yr, close to the predicted value.

NOMENCLATURE OF VARIABLE STARS

The system of nomenclature for variable stars (including pulsators) uses the division of the sky into constellations (Andromeda, Antilia, Apus, etc.). This division has, nowadays, no significance other than one of convenience. Within each constellation the variables are named, in order of discovery, by the letters R, S,...,Z, RR, RS,...,RZ, SS, ST,...,SZ, etc., AA, AB,...,AZ, etc., and after QZ by V335, V336, etc. The letter J is omitted throughout. An example is UW Cen (three-letter abbreviations for constellation names are used). Exceptions to this scheme are certain bright stars, for example, *o* (Mira) Ceti; variables in star clusters; and variables in galaxies other than our own. But there are even exceptions to these exceptions, and it is likely that the whole scheme will eventually be replaced by a numerical one based on the measured positions of the variable stars in a specified coordinate system. The International Astronomical Union delegates the naming of newly discovered variables to the Moscow Variable Star Bureau (a unit of the Soviet Academy of Science), which publishes name lists and also the *General Catalogue of Variable Stars*, the latest edition of which contains data on 28,450 variable stars. Older data (to 1954) on variable stars are summarized in the various volumes of the *Geschichte und Literatur des Lichtwechsels der Veränderlichen Sterne* (prepared by R. Prager and H. Schneller, in Berlin). References to recent work are most easily obtained through computer access to one of the astronomical data bases such as SIMBAD (Strasbourg Astronomical Data Centre, France).

TYPES OF PULSATING STARS

The following is a brief summary of the main types of pulsating stars. In most cases a comprehensive discussion is given in individual entries in this volume.

Radial Pulsators

1. Classical Cepheid variable stars or Type I Cepheids, or simply, Cepheids, are named after the type star (δ Cep). They are intermediate-mass stars (ranging from about 3–10 times the solar mass) with surface temperatures similar to that of the Sun (about 6000 K) but that are 300–40,000 times more luminous (i.e., they are supergiants). Their periods range from about 1 to about 100 days. They are distributed with young stars in the disk and spiral arms of our galaxy. Most Cepheids are pulsating in the fundamental mode, but some are in the first (or rarely, second) overtone, and a few show both fundamental and first-overtone pulsations simultaneously.
2. W Vir stars (or Type II Cepheids), BL Her stars and RV Tau stars are named in each case after a typical star of the group. These variables have some resemblance to classical Cepheids, but are evolved, low-mass (about one solar mass) objects. They occur in the old disc and halo populations of our galaxy (including the membership of halo globular clusters). Also in this mass range are the R Coronae Borealis or RCB stars (type star, R CrB) some at least of which pulsate. Both the RCB and RV Tau stars are examples of objects that eject shells or blobs of gas and dust, very probably due to the effects of their pulsation.
3. RR Lyrae variable stars are also called cluster variables because they occur in globular clusters (typically 10 to 100 in one cluster). A principal characteristic is the short period (less than 1 day). The abundance of elements heavier than helium (the "metals") varies from one RR Lyrae star to another. The total range is from near solar abundance to about 100 times less. Metal-poor RR Lyrae stars are the most commonly used tracers of the halo population in our galaxy. RR Lyrae variables are also found in large numbers in the central bulge region of our galaxy. They are of importance for studies of the distance, stellar population, and space density in the galactic bulge. The forms of their light curves divide RR Lyrae variables into two groups, type ab and type c. The former are fundamental pulsators; the latter pulsate in the first overtone.
4. Mira variables (after Mira Ceti, the type star) are cool (surface temperature typically 2500 K) and luminous variables (bolometric luminosities of typically 3000 times that of the Sun) with long periods (100–500 days). Their visual light amplitudes are large (a brightness change from maximum to minimum light of a factor of at least 10 and sometimes as great as 1000). Miras are sometimes called long period variables, al-

though this term includes stars of smaller light amplitude and less regular period. Because of their low surface temperatures most of their energy is radiated in the infrared (mainly between 1- and 2-μm wavelength). They are believed to be the last evolutionary state of stars of about one to perhaps one and a half solar (initial) masses before the ejection of a shell of matter and the formation of a planetary nebula. In fact, considerable mass is being lost in the Mira phase (typically 10^{-6} solar masses per year). At the long-period end (500–2000 days) the Mira sequence is continued by the OH/IR Miras. These OH/IR stars are losing mass at a much greater rate (up to about 10^{-4} solar masses a year) and are surrounded by dense circumstellar dust shells, which absorb much of the radiation from the star itself and reradiate it as low-temperature dust emission, much of it at (infrared) wavelengths of 10 μm or more. Mira variables are found in some of the less-metal-poor globular clusters and they are an important component of the bulge and old disc populations of our galaxy. Although it is generally accepted that they are radial pulsators there has been uncertainty as to whether they are in the fundamental or first-overtone modes.

Nonradial Pulsators

The prime example of a nonradial pulsator is the Sun itself, which undergoes complex oscillations, including a principal one with a period of about 5 min. The solar oscillations are of very small amplitude, both in light and radial velocity, and require special techniques to detect. Their complexity is due to the large number of modes (frequencies) that are simultaneously excited. A full understanding of these oscillations will allow us to study in some detail the internal structure of the Sun. This rapidly developing subject is called solar seismology. Attempts are in progress to detect similar phenomena in other stars (asteroseismology). Other important groups of high-overtone, nonradial pulsators are the rapidly oscillating Ap stars (stars with peculiar surface chemical compositions that vary in light by perhaps 1% with periods of about 10 min) and pulsating compact objects (white dwarfs and pre-white dwarfs including ZZ Cet and GW Vir stars, with light amplitudes of typically a few percent and periods of about 10 min).

General

The β Cep variables (visual luminosities of typically 3000 times that of the Sun, surface temperatures of about 30,000 K and principal periods in the range 1–100 days) and the δ Sct variables (somewhat cooler, less luminous objects with principal periods of about 0.1 day) are probably examples of objects which show both radial and nonradial pulsations.

Only the major classes of pulsators have been mentioned, and, especially for the smaller classes, different workers adopt different groupings and names. This reflects the uncertainty still surrounding the current interpretation of the observational phenomena in these cases. It should be particularly noted that pulsars (q.v.) are not pulsators.

Most surveys for variable stars have been carried out photographically. Certain areas of the sky, for example, globular clusters, some small areas of the galactic bulge, the Magellanic Clouds, and other small regions have been rather thoroughly investigated. In other regions the discoveries are quite incomplete, even among stars bright enough to be visible with ordinary binoculars. Although much modern work on pulsating variables requires sophisticated equipment for observation and analysis, there are still areas in which amateur astronomers with small telescopes can make valuable contributions to astrophysics. For instance, the Mira variable R Hya with a very large light amplitude (a range in brightness of a factor of nearly 1000) is of naked-eye brightness for 25% of its cycle. Extensive observations (many by amateurs) show that its period has decreased from 500 days in the year 1700 to 389 days at the present time. Such rapid changes are expected if the star happens to be undergoing the type of energy generation known as helium shell flashing, and the work on R Hya is taken as rather direct observational evidence that this process actually occurs.

Additional Reading

Christensen-Dalsgaard, J. and Frandsen, S., eds. (1988). *Advances in Helio- and Asteroseismology, IAU Symposium 125*. D. Reidel, Dordrecht.

Cox, J. P. (1980). *The Theory of Stellar Pulsation*. Princeton University Press.

Kholopov, P. N., ed. (1985, 1987, 1990, and in press). *General Catalogue of Variable Stars*, five volumes. Nauka, Moscow.

Schmidt, E. G., ed. (1989). *The Use of Pulsating Stars in Fundamental Problems of Astronomy, IAU Colloquium 111*. D. Reidel, Dordrecht.

Reports on Variable Stars (Commission 27), published triennially in *Trans. IAU*. See, for example, *Trans. IAU* **20A** 257 (1988).

Unno, W., Osaki, Y., Ando, H., and Shibahashi, H. (1979). *Non-Radial Oscillations of Stars*. Tokyo University Press, Tokyo.

See also **Star Clusters, Globular, Variable Stars; Stars, Beta Cephei Pulsations; Stars, BL Herculis, W Virginis, and RV Tauri Types; Stars, Cepheid Variable, all entries; Stars, δ Scuti and Related Types; Stars, Long Period Variable; Stars, Nonradial Pulsations in B-Type; Stars, Pulsating, Theory; Stars, R Coronae Borealis; Stars, RR Lyrae Type; Stars, White Dwarf, Observed Properties; Stellar Evolution, Pulsations; Sun, Oscillations.**

Stars, Pulsating, Theory

Arthur N. Cox

Stars vary in light because of intrinsic and extrinsic reasons. This discussion will focus on predictions for stars that are intrinsically unstable against pulsations because of their structure and internal material properties. Most stars evolve during their lifetime to a state where they naturally pulsate, and, for some stars, they enter and leave configurations that are pulsationally unstable several times. We observe these stars as varying both in light and in radial velocity. The theory of pulsating stars is able to account successfully for these observations in most, but not all, cases.

The classical variable stars, such as the Cepheids and the RR Lyrae variables, were observed in spherical (radial) modes originally by noting that the total output of light varied in time. However, spectra showed long ago that the absorption lines oscillate between negative and positive velocities around the mean velocity of the star. For stars near or on the main sequence, pulsations are also observed to occur in nonradial modes, so that the shape of the surface is nonspherical. These modes can be detected by variations of the absorption line shape with time using Doppler imaging. The Sun is unique in that spatial resolution of the surface can be obtained, giving a direct measure of how each part of the surface moves and changes its output intensity with time.

BASIC THEORY

Stellar astrophysics can predict many properties of observed stars by considering the basic equations of fluids. These are the conservation of mass, momentum, and energy. Most stars are in thermal equilibrium, meaning that the energy produced by various internal processes, such as condensing into a smaller volume, cooling, or thermonuclear reactions, is radiated away at the same rate as it is produced. The stars are also in hydrostatic equilibrium, meaning that the attraction of gravity due to the internal mass at any level in the star is balanced by a pressure gradient. The very slow evolution that all stars undergo is due to the slow change of the deep composition as the hydrogen fuel is converted to helium and heavier elements. These changes are so slow that the star can always be considered to be in thermal and hydrostatic equilibrium.

One in a million stars is found in a state that can allow a perturbation in these equilibria to grow with time. Then the star is a variable star, often with the variations being quite significant. There is another class of variable stars that has been considered lately that is not exactly intrinsically unstable against pulsations, as are the classical variables. The only confirmed member of this class is the Sun, but surely many other solar-like stars must exist that also have very small pulsational amplitudes. The pulsations of these latter stars are excited by coupling with the acoustic noise generated by the violent turbulent convection in the surface layers.

There are several available procedures for calculating the stability of stars. They are divided into linear and nonlinear, radial and nonradial, adiabatic and nonadiabatic, and Lagrangian and Eulerian methods. All of these approaches assume that a stellar model is available with detailed data for the temperature, density, pressure, luminosity, energy source, opacity, and thermodynamic properties as a function of both mass and radius position in the unperturbed model.

The linear theory takes the three basic equations for the stellar structure and substitutes in them for each quantity its equilibrium value plus a small perturbation. The general procedure, then, is to note that the equilibrium quantities can be grouped together to produce terms that collectively all cancel, because the model is in equilibrium. Then the perturbation terms in the three basic equations relate to one other by an eigensolution that contains the three independent-variable perturbations and an eigenvalue representing the mode frequency. The eigensolution is really an eigensolution of the matrix of the coefficients for the variable perturbations, with the proper boundary conditions for the basic equations, appropriately, as the top and bottom rows of the matrix.

In nonlinear theory, the basic equations are not developed with small perturbations, but are taken with the proper time-dependent terms and followed in time. These time-dependent terms arise in the momentum equation because the pressure gradient is no longer balanced by the gravity. Accelerations and velocities are nonzero, and evolve into periodic changes in the configuration. These lead also to a nonlinear energy flow.

Nonlinear theory has not been developed to follow nonspherical pulsations, because computers with sufficient storage are not yet available. For the radial pulsations, however, extensive calculations have been made since the mid-1960s.

For both linear and nonlinear theories, the solutions give the spatial and temporal behavior of all the main variable properties and those that can be calculated from them. Thus, although the observations can only give simple data for the remote stars, theory can interpret an observed period for a star of approximately known mass, radius, luminosity, and composition in terms of detailed internal variations. The theory can set limits on masses, radii, etc., because only certain of these global quantities are allowed in order for the solutions to match the observed periods and the fact that the star is indeed pulsating.

The theory of stellar pulsation comprises the procedures required to make detailed predictions and also the physical reasons for intrinsic stellar pulsation. This more interesting second aspect has been developed only enough to explain some of the classes of variable stars. Causes for the instabilities to occur and for the pulsation amplitudes to grow in time are understood for most of the yellow and red giant stars and the white dwarfs. Exactly how the oscillations of the Sun and solar-like stars are driven is currently being debated. It seems, however, that coupling with the convective motions that have the same spatial and temporal scales as the trapped normal oscillation modes is responsible. The mechanisms of pulsations in the B stars, the hottest R CrB variables, and the Hubble–Sandage variables seem to be known now, but all details have not been accepted. The proposed mechanisms for pulsations in planetary nebula central stars are under study.

For more technical details about all aspects of stellar pulsations, one can refer to two books. *Theory of Stellar Pulsation* discusses both the radial and nonradial modes, complementing the earlier work *Nonradial Oscillations of Stars*, which concentrates almost entirely on nonradial modes. Both books point out that rotation and magnetic fields may be important, but many aspects of the theory for these two important effects have not been developed yet.

TYPES OF PULSATION MODES

The solutions for the variations in stellar models are often divided into the p, f, and g modes. For the first category the restoring force is the gas and radiation pressure. Radial modes seen for all the yellow giant pulsators are merely p modes with no node lines on the surface. For the g modes the restoring force is gravity or buoyancy force that produces pressure gradients and mostly horizontal motions. The intermediate f or fundamental mode has no nodes in the radial direction, but has, as the others do, node lines on the stellar model surface, and they extend radially all the way to the center.

The p modes are those seen at very small amplitude on the Sun. Possibly a few p modes are selected by δ Scuti variables, because they are observed among their radial modes. The B stars and the very luminous Hubble–Sandage variables may have many g modes as well as some p modes. The white dwarf stars seem to show only high-radial-order g modes.

LINEAR, RADIAL, NONADIABATIC METHODS

The most efficient method for the calculation of radial pulsations in the linear nonadiabatic approximation is that presented by John I. Castor. The mass equation is immediately satisfied by use of a Lagrangian mesh for the solution with shells that have a mass fixed in time. The linearized momentum equation and the linearized energy equation are written for each mass shell with the independent variables being δr and $T\delta S$. Here the r is the radius of the mass shell and T and S are, respectively, the temperature and specific entropy. The usual linear theory assumption is that all quantities vary sinusoidally in time (t) with the form:

$$\delta r / r = (\delta r / r)_0 \exp(i\omega t),$$

where the δr is the Lagrangian variation of r, i is the square root of minus one, and the eigenvalue ω is complex when there is an energy equation allowing energy to flow from one mass shell to another.

Note that there are n shells with known conditions, each with $2n$ equations for the $2n$ independent variables that vary over the pulsation cycle. However, there is a need to solve for the ω also. Thus one of the independent variables needs to be defined separately and removed from the list of variables. This normalization is almost always that $\delta r / r = 1.0$ at the surface.

The two-variables-per-zone procedure can be made into a one-variable method by assuming that there is no need for an energy equation. Then the calculation is adiabatic, because there is no equation specifying how energy flows between shells. In this case the eigenvalue ω is purely real, and it represents the angular oscillation frequency. With nonadiabatic effects allowed, the imaginary part of the eigenvalue indicates the growth or decay rate of the mode from its arbitrarily small amplitude.

The adiabatic eigensolutions can be used in an application of a variational principle that can be used to improve the accuracy of the eigenvalue. A discussion of this principle for both the radial and nonradial adiabatic cases appears in the book *Theory of Stellar Pulsation*.

The physical mechanisms that cause a perturbation to grow in the linear (and nonlinear) nonadiabatic radial mode solutions are caused by the stellar material thermodynamic properties: the opacity and its variation with temperature and density; the time-dependent effects of convection; the nuclear energy production and its variations with respect to temperature and density; and the slow contraction of the central cores of stars as they evolve. To achieve pulsational instability, the common damping from the normal rapid radiation flow to smooth out fluctuations needs to be overcome.

All the yellow and red variable stars are driven into radial pulsation by the cyclical ionization of hydrogen and helium. This periodic blocking of the outflowing luminosity gives a lag in the luminosity such that it increases somewhat during expansion to reinforce the outward motions. Also during compression, the luminosity is lower than its mean, so that contraction can proceed more easily. These two main microscopic mechanisms are called the κ and γ effects. The first, the kappa mechanism, is an opacity increase that causes the radiation blocking, and the second is an ionization increase that causes the luminosity to be periodically hidden during the compression part of the pulsation cycle. The hotter yellow giants derive their driving mostly from helium ionization, whereas hydrogen ionization powers pulsations in the cooler stars.

The only confirmed mechanism involving the modulation of the nuclear energy generation (the ε mechanism) seems to be applicable to the very massive, high-luminosity main-sequence stars. These stars have considerable radial pulsation motions for the fundamental mode right at the center, and the temperature and density variations are enough to cyclically increase and decrease the energy production.

LINEAR, NONRADIAL, NONADIABATIC METHODS

A method to include nonadiabatic effects in linear nonadiabatic calculations was presented by Hideyuki Saio and John P. Cox. This method is based on an Eulerian mesh. Centering of the equations in space has proved to be difficult to do to maintain calculation stability. A better Lagrangian method has been developed by W. Dean Pesnell, where the concepts are much like those for the Castor method. Other methods, which use only the adiabatic approximation, have been described by Wojciech Dziembowski and by Jorgen Christensen-Dalsgaard.

All the methods use the spherical harmonic functions $Y_{l,m}$ which are well known from the theory of the hydrogen atom. The momentum equation, when displacements are represented by

$$\delta\boldsymbol{\xi} = \delta r Y_{l,m}\mathbf{r} + \delta h \nabla Y_{l,m},$$

is satisfied by an equation that does not involve the $Y_{l,m}$ at all. This is because horizontal terms can be represented with factors $l(l+1)Y_{l,m}$.

The usual nomenclature is that the l value is called the degree of the spherical harmonic function, and is an integer that represents the number of radial nodes on any surface at any depth into the star. The m integer is the longitudinal degree indicating how many of these node lines pass through the poles, and most often it is just called m.

The Pesnell method uses four equations. One is the radial-momentum equation that calculates the radial direction acceleration, and by simple development, the radial displacements. A second equation is the momentum equation for the horizontal direction. The structure of the horizontal displacement of any Lagrangian element is given by the spherical harmonic function, with δh, the amplitude of this displacement, part of the eigensolution. The third equation is the linearized Poisson equation that expresses the gravitational potential, which does not have to be purely radial (central) in nonradial problems. Finally, the linearized energy equation is included, now with the possibility that energy can flow both radially and horizontally. The four variables that go with these four equations are δr, δh, the perturbation of the gravitational potential, γ, and the Castor variable $T\delta S$ that is discussed above.

There are many published results using the Cowling approximation that there is no perturbation of the gravitational potential, and therefore the Poisson equation and its perturbations do not have to be included in the eigenvalue problem. For large l values, the average of the hills and valleys of the eigenmode is near zero for the gravitational perturbation, and these solutions are frequently reasonably accurate with significantly less computing cost.

It now seems that the cause of the radial and nonradial B star pulsations and hot R CrB star pulsations is again a κ effect, but in this case is due to the rapid increase of the opacity from bound-bound absorption lines of iron. The g-mode pulsations of the central stars of planetary nebulae and of the pre-white dwarfs seem also to be a κ effect, but due to carbon and oxygen ionizations rather than those of hydrogen and helium. The blocking effects of luminosity at the bottom of the convection zone in the hydrogen and helium surface composition pulsating white dwarfs apparently produces the g-mode driving in them.

NONLINEAR RADIAL, NONADIABATIC METHODS

Initial value integrations of the momentum and energy equations through time for stellar models can produce detailed descriptions of how pulsations can grow or decay, as well as how the steady state behaves. Light and velocity curves can then be predicted at the real amplitudes at which stars are observed to pulsate. Many results are available for the classical Cepheids and the RR Lyrae variable stars, mostly because they have growth rates for the modes of interest and decay rates for contaminant modes that are rapid enough. Integrations for even thousands of pulsation cycles have sometimes been made, to assure that the solution is truly steady state.

It was shown soon after the first nonlinear calculations of Cepheids that the observed phase lag between the minimum radius and the maximum light output was an intrinsic thermodynamic property of the stellar material. Later, the linear eigensolutions were refined to show that much of the phase lag is largely due to linear, not nonlinear, effects.

An important ingredient for the nonlinear (and even the linear) calculations is the behavior of the convection. In the cyclically changing configuration, the convection "tries" always to adapt and transport the luminosity according to the current stellar structure. But the convection in pulsation-driving regions frequently cannot change rapidly enough, and this lagging needs to be considered in the calculation. Therefore, usually the effects of time-dependent convection are ignored, and the solutions for light curves are not strictly comparable to observations of real stars.

Robert F. Stellingwerf developed a strictly periodic method for finding nonlinear pulsation solutions. A trial cycle at a trial period is followed with the usual initial value integration through time. At the end of the trial period, the closure of the variations in three variables, T, r, and the velocity $\dot{r}$, is calculated. During the time integration a matrix is generated that has as elements the partial derivatives of every one of the three variables in every one of the mass shells with respect to all others. With the closure residuals and the matrix, a correction list for all the variables in all the zones can be calculated, so that for the next trial period the closure will be smaller, or even close to zero. This is a Newton–Raphson method in many dimensions. Without the complications of the time-dependent convection, it works well to obtain, with only a few iterations, a strictly periodic, full amplitude, pulsating star solution.

An interesting application for both linear and nonlinear studies has been the one-zone models. They enable us to understand the essential physics even though they have no spatial resolution, as discussed in *Theory of Stellar Pulsations*. It was in the application of such rudimentary models that many of the pulsation mechanisms were discovered by actually inspecting the terms of the simple equations.

SPECIAL PROBLEMS FOR VARIABLE STAR CLASSES

Theoretical pulsation solutions in both linear and nonlinear theory for the nonadiabatic radial pulsations of Cepheid variable stars have been available since the mid-1960s, but more work is needed. Problems have arisen in studying the dozen or so double-mode Cepheids. Observations of these stars show that the ratio of the two periods (for fundamental and first-overtone modes) is not appropriate for their typical masses of about 7 $M_{\odot}$ (solar masses). The ratio seems to indicate a stellar mass of only perhaps 3 $M_{\odot}$. The explanation seems to be that there is a large opacity increase due to the many previously neglected iron lines that occur in gas at temperatures between 100,000 and 500,000 K. The internal temperature gradient is therefore increased, the density gradient decreased, and the star appears to the fundamental and first-overtone modes to be less concentrated. This increases the periods of both modes, and decreases the first-overtone to fundamental mode ratio in accord with observations.

There is a parallel problem for the Population II stars with low abundances of elements heavier than helium. These RR Lyrae variable stars, seen in globular clusters and in the galactic field, are sometimes also found with two pulsation modes. The period ratio for the same two modes as for the Cepheids does indicate the correct masses for the RR Lyrae variables, but still we do not know why the pulsation driving in both the Cepheids and the RR Lyrae variable stars is such that it allows the two modes to exist together. Minor discrepancies between observation and theory also exist in the prediction of the light and velocity curves.

A very important question needs to be answered for the Mira variable stars. In which mode are these stars pulsating? Periods and growth rates have been calculated assuming that the masses are near the expected 3 $M_{\odot}$. However, the periods for a fixed mass and radius for the fundamental and first-overtone modes differ by a factor of 3. Nevertheless, observations are not able to determine the radius accurately enough to indicate in which mode the star is observed to pulsate. Currently most workers prefer to think that these stars also are in the radial fundamental mode, but this concept has not been completely accepted by all experts. The mode switch that may occur to produce a great increase in the rate of mass loss or just the cause for a great mass-loss increase without any pulsation-mode switch would be interesting to know for these dying stars.

An even more interesting situation exists for the Population II red variable stars as they evolve up the asymptotic giant branch. If their luminosity becomes very large for their mass, they may leave the regime of regular pulsations and become chaotic. There are difficulties in observationally determining their periods, a problem that can possibly be attributed to this theoretically predicted chaos.

It now appears that even after evolving along the asymptotic giant branch with considerable mass loss, the stars still can pulsate as their surfaces become hotter during blueward evolution. These are the UU Herculis and RV Tauri variables, with again hydrogen and helium κ and γ effects driving. Their pulsations are frequently irregular.

Most yellow giant and red giant stars can pulsate only in radial modes because nonradial motions deep in the star are strongly damped. The δ Scuti variable stars, however, are less evolved and less centrally concentrated, so that this deep damping is not so large for, at least, the low-radial-order and low-angular-degree nonradial modes. Some nonradial modes are seen in these stars, which have masses of a few solar masses each, but how can one identify the observed modes among the many possible ones that are predicted?

Solar oscillations need much more study, not only to refine knowledge of the solar structure and the internal rotation of the Sun, but also to discover the mechanisms that cause the p modes to exist in a star that is intrinsically stable according to theory. Forcing of the modes by coupling to convection needs to be better understood.

A prediction that has not been verified by observation is that stars cooler or slightly brighter than the Sun on the main sequence should be forced by convection to display oscillations. These oscillations are hard to observe in integrated (white) light, perhaps explaining the lack of detection.

An important question concerns the pulsations of the several classes of B stars. These stars consistently have been shown to be theoretically pulsationally stable with standard opacities, and yet such stars do pulsate, with sometimes several periods near a few hours. These stars are rich with nonradial modes, as seen in the spectral line variations over their periods. Can the recently proposed opacity increase due to iron lines destabilize these stars?

For the white dwarf stars, many nonradial modes are predicted to occur by the interaction of convection and radiation at the bottom of the convection zone. Dozens of modes are sometimes predicted for specific models, and yet typically only a few periods can be seen in a given star. The pre-white dwarf (GW Vir) variables, also with g modes, give a unique opportunity to watch how the cooling evolution proceeds. However, they show period decreases, whereas theory presently predicts the opposite.

How are the very high-radial-order g modes found in the planetary-nebula central stars excited? It now appears that these stars are excited by the ionization of carbon and oxygen, as opposed to hydrogen and helium, in the classical Cepheids.

The R CrB helium stars are seen to pulsate at very high surface temperatures. The current best idea for them and even for the extremely luminous Hubble–Sandage variables is that the sudden increase in the opacity in the temperature range between 100,000 and 500,000 K from iron lines can give periodic κ-effect driving to overcome the normal radiative damping.

Additional Reading

Castor, J. I. (1971). On the calculation of linear nonadiabatic pulsations of stellar models. *Astrophys. J.* **166** 109.

Cox, J. P. (1980). *Theory of Stellar Pulsations*. Princeton University Press, Princeton.

Dziembowski, W. (1971). Nonradial oscillations of evolved stars. I. Quasiadiabatic approximation. *Acta Astron.* **21** 289.

Kippenhahn, R. and Weigert, A. (1990). *Stellar Structure and Evolution*. Springer-Verlag, Berlin.

Saio, H. and Cox, J. P. (1980). Linear nonadiabatic analysis of nonradial oscillations of massive near main sequence stars. *Astrophys. J.* **236** 549.

Stellingwerf, R. F. (1974). The calculation of periodic pulsations of stellar models. *Astrophys. J.* **192** 139.

Unno, W., Osaki, Y., Ando, H., and Shibahashi, H. (1979). *Nonradial Oscillations of Stars*. University of Tokyo Press, Tokyo.

See also **Stars, Beta Cephei Pulsations; Stars, BL Herculis, W Virginis, and RV Tauri Types; Stars, Cepheid Variable; Stars, δ Scuti and Related Types; Stars, High Luminosity; Stars, Long Period Variable; Stars, Nonradial Pulsation in B-Type; Stars, Pulsating, Overview; Stars, R Coronae Borealis; Stars, RR Lyrae Type; Stars, White Dwarf, Observed Properties; Stellar Evolution, Pulsations; Sun, Oscillations.**

Stars, R Coronae Borealis

Detlef Schönberner

The R Coronae Borealis (RCB) stars form a very spectacular subgroup of hydrogen-deficient stars of which only about 30 objects are known. Their spectra show strong carbon features, except for CH, which is always very weak. Similarly, the Balmer lines of hydrogen are rather weak or even absent. Named after the prototype R CrB, the most striking property of these variable stars is the occurrence of deep light minima, which occur at random times and may involve a decrease of 8 mag in the visual. The average time between successive light drops is about 1000 d, and they may last for a year or sometimes even more. For more than 50 years these light minima have been attributed to obscuration by soot, that is, by some sort of amorphous carbon grains. This idea is supported by the apparent hydrogen deficiency and carbon richness of the atmospheres of RCB stars. Despite the fact that these very peculiar stars pose a challenge to our understanding of stellar structure and evolution, our knowledge about them is rather limited and comes mainly from studies of the three brightest ones. Even a spectroscopic verification of the atmospheric hydrogen deficit does not exist for many RCB stars.

STELLAR PARAMETERS OF RCB STARS

The luminosities of RCB stars are only poorly known because reliable distance indicators do not exist. Fortunately, three RCB stars were detected in the Large Magellanic Cloud, and their absolute magnitudes range between -4 and -5. These results indicate that the RCB stars are giants with luminosities close to 10^4 $L_\odot$ (solar luminosities).

Similarly, their effective temperatures are not well known. Only for the three brightest members, R CrB, RY Sgr, and XX Cam, are detailed spectroscopic analyses based on model atmospheres available. All three of these stars have nearly the same effective temperature of $T_{\text{eff}} = 7000$ K. The surface gravities are low, with $\log g \approx 1-3$, where g is in cm s^{-2}, thus indicating high luminosities also for the galactic RCB stars. Estimates for other RCB stars give effective temperatures between 4000 and 7000 K, with only three exceptions: DY Cen (10,000 K), MV Sgr (16,000 K), and V 348 Sgr (20,000 K). The surface gravity of MV Sgr was estimated from the wings of the helium lines to be about 300 cm s^{-2}. Thus, all the available information from spectroscopic analyses indicate very large luminosity–mass ratios $L/M \sim T_{\text{eff}}^4\, g^{-1}$ of $\approx 10^4$ $L_\odot/M_\odot$, as is typical for supergiants.

Concerning the surface composition of RCB stars, it appears evident from spectrograms that the hydrogen content is extremely low. Again, only a few objects have been analyzed so far, and the main results are summarized in Table 1. The tabulated numbers are averages of the results of different authors. Despite the fact that all the listed abundances are quite uncertain by at least ± 0.4 dex, it is evident that the abundance pattern of the lighter elements is very unusual. Hydrogen is reduced by more than a factor of 10^4, and although the spectra of the cooler objects are dominated by carbon lines, the main constituent of the surface layers is always helium. The rather low carbon content of the hot RCB star MV Sgr is somewhat at variance with the carbon abundance of the other analyzed RCB stars, for which a high carbon content appears to be typical. From the absence of $^{13}C_2$ bands it can be argued that the carbon is entirely ^{12}C, that is, the product of the triple-α process. Estimates give $^{12}C/^{13}C > 40$ for R CrB and > 50 for RY Sgr, respectively. The small sample and the rather low accuracy of the analyses do not permit us, however, to draw any further conclusions.

Table 1. Surface Abundances of RCB Stars

	Logarithm of the Elemental Abundance*						
Star	*C/He*	*H*	*He*	*C*	*N*	*O*	*Fe*
R CrB	0.01	7.3	11.5	9.7	8.5	9.3	7.0
XX Cam	0.01	< 4.2	11.5	9.5	8.7	9.2	7.1
RY Sgr	0.03	7.4	11.5	10.0	8.8	8.5	8.3
MV Sgr	0.0002	—	11.6	7.8	8.0	—	—
Sun	0.005	12.0	11.0	8.7	8.0	8.9	7.7

*Normalized to 12.15, taken as the total abundance of all elements.

The important question of the masses of RCB stars is also difficult to answer. From their space distribution and radial velocities it has been concluded that they belong to a rather old population, and that hence their mass could not be much greater than 1 $M_\odot$. However, one must be cautioned against such reasoning because it makes implicit assumptions about the evolutionary history of RCB stars that may not be justified (see the discussion below about the origin and evolution of RCB stars).

Cepheid-like pulsations, with more-or-less regular periods, are quite common for RCB stars. The best-known cases are RY Sgr and RS Tel, with periods of 38.6 and 45.8 d, respectively. There exists a loose correlation between pulsational periods and effective temperatures, in the sense that at 7000 K periods around 40 d dominate, and at 5000 K, periods between 45 and 60 d dominate. The coolest RCB stars, with temperatures of about 4000 K, have periods of about 100 d. Obviously, the coolest RCB stars are also the biggest ones, indicating a roughly constant luminosity for all of these objects. Application of the pulsation theory gives mass estimates of ≈ 0.9 $M_\odot$.

THE OBSCURATION MODEL

The now widely accepted model for the deep light minima of RCB stars involves the ejection of soot in random directions. An ejection in the line of sight then causes obscuration. Broad emission lines (i.e., the sodium D lines, the H and K lines of singly ionized calcium, and the 3888-Å line of helium), visible when the stellar disk is being obscured, indicate a stellar envelope in which gas is streaming radially outwards from the star with velocities of about 200 km s^{-1}. Of course, the carbon grains will not form in the vicinity of the stellar surface, because there the temperature highly exceeds the grain condensation temperature. Farther out, however, grain formation is possible. The exact position depends on the stellar surface temperature and on the mass-loss rate. Computations show that carbon-grain formation occurs at a distance from the surface that varies between about 5 stellar radii for $T_{\text{eff}} = 4000$ K and about 30 stellar radii for $T_{\text{eff}} = 7000$ K.

One has to assume that density fluctuations lead to the rapid formation of an opaque cloud of soot that obscures an increasing part of the stellar disk. As the cloud moves outwards it expands and becomes more transparent, and the star gradually regains its normal brightness. This model explains the asymmetry of the light curves: The luminosity drop caused by the opaque cloud is fast, with practically no color change, whereas the recovery to normal light occurs more gradually with substantial color changes according to the optical properties of the newly formed circumstellar material. This new matter replenishes the dust shells, which are present in all RCB stars. Infrared observations indicate dust temperatures between 400 and 900 K and dust-shell radii between 10 and 90 stellar radii. In particular, the corresponding values for R CrB are 650 K and 50 stellar radii. The observations also show that the infrared radiation from the shell does not change during a light minimum. Obviously, the mass ejected in one puff is small compared to the total dust-shell mass.

The cause of the triggering of the irregular behavior of RCB stars is still rather unclear. Studies of the RCB star RY Sgr indicate a tendency for the light drops to occur within a narrow range of the pulsation phases. The stochastic distribution of the puffs over the stellar surface may be connected with the surface convection zone. The total mass-loss rate is difficult to estimate, but a value on the order of $10^{-6}\ M_{\odot}\ yr^{-1}$ appears to be realistic.

The light fadings of RCB stars offer the possibility to study in situ the extinction properties of newly formed dusty material. It turns out that the extinction of this dust differs from that of the mean interstellar dust: the new circumstellar matter has its ultraviolet absorption peak at about wavelength 2500 instead of near 2200 Å. Laboratory measurements and theoretical calculations (Mie theory) confirm that such an extinction property is due to small glassy or amorphous carbon grains (i.e., soot) formed in a hydrogen-poor environment. More graphitic grains have their extinction bump close to wavelength 2200 Å.

STRUCTURE AND EVOLUTION OF RCB STARS

The existence of hydrogen-depleted stars is still an unsolved problem of the theory of stellar evolution. The only way to create a virtually hydrogen-free star is mass exchange in a binary system, unless one wishes to invoke unrealistic mass-loss and/or mixing processes in later evolutionary phases. Indeed, four binaries are known in which the primaries are helium supergiants, but none of these behave like an RCB star. Furthermore, none of the RCB stars shows any evidence for binarity. Thus, whereas a formation of helium giants by mass exchange is possible, such a scenario does not seem to create RCB stars.

Before taking up this point again, it appears useful to discuss the structure and present evolution of RCB stars by comparing the observations with stellar models. Despite our ignorance about the origin of these stars, it is possible to construct stellar models which encompass the observed luminosity and effective temperature ranges. These models are inhomogeneous in the sense that they consist of

1. An inert, electron-degenerate, carbon–oxygen core, growing due to helium shell burning, which supplies the stellar luminosity.
2. An extended, mostly convective, helium envelope, which may contain a substantial fraction of the total stellar mass.

The useful range in total mass is restricted to between 0.75 and, probably, much less than 2 $M_{\odot}$. The properties of these models depend critically upon their internal structure: For given mass and effective temperature (always assumed to be $< 20{,}000$ K), there exists a low-luminosity branch with a "thick" (masswise) envelope, and a high-luminosity branch with a "thin" envelope. A model with a thick envelope will evolve toward the Hayashi limit at $T_{eff} \approx 4000$ K with roughly constant luminosity. During the evolution upwards along the Hayashi line (at $T_{eff} \approx 4000$ K) the envelope is fully convective, and when the envelope becomes too thin by nuclear burning and/or mass loss, the model evolves bluewards, again with constant luminosity. (The Hayashi line or Hayashi limit is a theoretical line on the HR diagram of luminosity versus temperature. It represents the minimum temperatures for which self-consistent models can be formulated for stars of given luminosities and composition.)

A thin envelope on the high-luminosity branch contains less than 0.1 $M_{\odot}$ (total stellar mass $\approx 0.9\ M_{\odot}$), in contrast to an envelope mass of 0.3 $M_{\odot}$ for the same model on the low-luminosity branch. The luminosity–mass ratios are also quite different in both evolutionary phases: A 0.9 $M_{\odot}$ model has $\approx 10^{3.8}\ L_{\odot}\ M_{\odot}^{-1}$ on the low-luminosity branch, but $\approx 10^{4.5}\ L_{\odot}\ M_{\odot}^{-1}$ on the high-luminosity branch. The rather limited accuracy of the known spectroscopic analyses does not yet allow us to distinguish between both evolutionary stages.

An important additional piece of information is provided by the previously mentioned pulsational properties of RCB stars. For the best-known cases, R CrB and RY Sgr, the calculations show that only on the upper, that is, the high-luminosity, branch does the blue edge of the instability region extend far enough to higher effective temperatures to explain the pulsation of these two objects. The case of RY Sgr is particularly interesting, because its period is the best known, $P = 38.6$ d, but decreases slowly according to $\dot{P}/P = -3 \times 10^{-4}\ yr^{-1}$. Interpreting this as an evolutionary effect, that is, as shrinking of the thin envelope with constant luminosity, one has to place RY Sgr onto the high-luminosity branch. Indeed, a stellar model of 0.9 $M_{\odot}$ and 18,000 $L_{\odot}$ has $P = 37$ d and this theoretical period is changing at the rate of $\dot{P}/P = -5 \times 10^{-4}\ yr^{-1}$, very close to the observed value. We conclude that RY Sgr evolves as predicted by inhomogeneous helium-star models, namely, towards higher effective temperatures with constant luminosity while burning helium in a shell. Typical evolutionary rates would then be $\lesssim 1\ K\ yr^{-1}$. As for R CrB itself, a very cool (≈ 30 K) and extended (radius ≈ 0.7 pc) "fossil" dust shell was detected by the *Infrared Astronomical Satellite*. This shell indicates that R CrB was in the past much closer to the Hayashi line; that is, it was cooler and bigger, and that the mass-loss rate was greater than the present one. Despite the fact that the observational evidence concerning the present evolution of RCB stars is sparse, we will assume in the following that only the high-luminosity branch occurs in nature.

The known periods of RCB stars range from 38.6 (RY Sgr) up to ≈ 135 d (S Aps). The coolest models, at about 4000 K, predict fundamental periods of about 400 d, far greater than 135 d, which corresponds to $T_{eff} \approx 5000$ K. Thus one may conclude that either very cool RCB stars with, say, T_{eff} below 5000 K do not exist, or that they are hidden behind optically thick dust shells. Support for the latter idea comes from nonadiabatic pulsational calculations, which indicated that for $M < 1.6\ M_{\odot}$ and $T_{eff} < 6000$ K the pulsational amplitudes grow without bound. In reality one must therefore expect substantial mass ejections in all these cases. However, a better knowledge of effective temperatures and pulsational periods of RCB stars is important to further investigate the relation between periods and stellar temperatures.

Further evolution towards the white dwarf stage will carry the RCB stars through the B-type spectral region, and indeed two RCB stars are known to be there: MV Sgr and V 348 Sgr (cf. preceding discussion). They resemble, except for the emission lines, another group of peculiar stars: the extreme helium stars. These objects show strong helium and singly ionized carbon lines, but hydrogen is weak or even totally absent. Their surface abundances are very similar to the ones listed in Table 1, but RCB-like light variations are not known in the extreme helium stars. Thus their genetic relation to the RCB stars appears questionable. The extreme helium stars may, however, be related to the so-called hydrogen-deficient carbon stars, of which only five are known. Their spectral appearance resembles that of RCB stars, but again, RCB-like light variations are not known to occur, and infrared excesses have also not been found. Obviously these objects have, for some unknown reason, substantially lower mass-loss rates. They may represent an extension of the RCB group to lower masses, and hence also to lower luminosity–mass ratios.

It is clear from the above discussion that the genetic relationships between these three groups of extremely hydrogen-deficient stars are still only poorly understood, and that any statements concerning their origin and further evolution must remain highly uncertain. Originally, it was thought that RCB stars are descendants from the asymptotic giant branch (AGB), and that some sort of deep-mixing and mass-loss processes are responsible for the disappearance of hydrogen. This hypothesis, however, has difficulty in explaining the rather large helium-envelope masses ($\approx 0.1\ M_{\odot}$) of the RCB models discussed previously. An AGB star with a carbon–oxygen core of, say 0.8 $M_{\odot}$ has a helium intershell mass of only about 0.01 $M_\odot$. It is impossible to construct a cool RCB model with such a low-mass helium envelope.

At present, it appears that a likely scenario for the creation of RCB and related stars is the merging of two close white dwarfs, driven by the energy loss of gravitional radiation. Both white dwarfs would be the remains of successive mass exchanges within a close binary system that had rather massive components ($M_1 \lesssim 10\ M_\odot$). One of the white dwarfs consists of a carbon-oxygen core, and the other is a helium core. The still-hypothetical merging process is expected to lead to the formation of a star of about 1 $M_\odot$ with a carbon–oxygen core and a helium envelope, the further evolution of which should then proceed in the same manner as that of the models discussed previously. The envelope of the newborn giant star is also expected to display an abundance pattern similar to that of Table 1: a mixture of CN-processed and triple-α-processed matter with virtually no hydrogen.

Additional Reading

Cottrell, P. L., Lawson, W. A., and Buchhorn, M. (1990). The 1988 decline of R Coronae Borealis. *Mon. Not. R. Astron. Soc.* **244** 149.

Drilling, J. S. (1986). Basic data on hydrogen-deficient stars. In *Hydrogen Deficient Stars and Related Objects*. K. Hunger, D. Schönberner, and N. K. Rao, eds. D. Reidel, Dordrecht, p. 9.

Fadeyev, Yu. A. (1986). Theory of dust formation in R Coronae Borealis stars. In *Hydrogen Deficient Stars and Related Objects*, K. Hunger, D. Schönberner, and N. K. Rao, eds. D. Reidel, Dordrecht, p. 441.

Feast, M. W. (1986). The RCB stars and their cirumstellar material. In *Hydrogen Deficient Stars and Related Objects*, K. Hunger, D. Schönberner, and N. K. Rao, eds. D. Reidel, Dordrecht, p. 151.

Iben, I., Jr. and Tutukov, A. V. (1985). On the evolution of close binaries with components of initial mass between 3 $M_\odot$ and 12 $M_\odot$. *Astrophys. J. (Suppl.)* **58** 661.

Lambert, D. L. (1986). The chemical composition of cool stars: The hydrogen-deficient stars. In *Hydrogen Deficient Stars and Related Objects*, K. Hunger, D. Schönberner, and N. K. Rao, eds. D. Reidel, Dordrecht, p. 127.

Saio, H. (1986). Pulsations of hydrogen-deficient stars. In *Hydrogen Deficient Stars and Related Objects*, K. Hunger, D. Schönberner, and N. K. Rao, eds. D. Reidel, Dordrecht, p. 425.

Schönberner, D. (1986). Evolutionary status and origin of extremely hydrogen-deficient stars. In *Hydrogen Deficient Stars and Related Objects*, K. Hunger, D. Schönberner, and N. K. Rao, eds. D. Reidel, Dordrecht, p. 471.

Warner, B. (1967). The hydrogen-deficient carbon stars. *Mon. Not. R. Astron. Soc.*, **137** 119.

See also **Stars, Carbon; Stars, Pulsating, Overview; Stars, Pulsating, Theory.**

Stars, Radio Emission

Robert M. Hjellming

Radio stars are single stars, binary systems, or multiple star systems where atmospheric or intrastellar outflows generate plasmas capable of emitting radio waves that can escape and be observed with Earth-based radio telescopes. In practice this means observations of either hot plasmas in very large volumes or relativistic electron plasmas in a wide range of volumes.

Before 1970 there were two major "false alarms," where objects were prematurely identified as "radio stars." In the years before the earliest known strong radio sources were identified with nonstellar objects most were described as radio stars because of the presumption that astronomical radiation was dominated by stars. This misassociation of names became rare as the strong radio sources were found to be distant galaxies, nonstellar objects in our galaxy, or quasars. In the late 1950s and early 1960s nearby flare stars were extensively observed with radio telescopes, with many reports of radio flares. Unfortunately, many of these reports were probably unknowing detections of man-made interference. Since that time nearby flare stars have become one of the most important categories of stellar radio emission, but they are now known to be weaker radio sources, with significantly different properties from those reported before 1970.

SIZE AND SURFACE BRIGHTNESS

The observable strengths of all radio sources are determined by the combination of their surface brightness and the size of their emitting regions. However, this is more critical in the case of radio stars, because the small angular size of stars does set the size scale of associated radio emission in one way or another. Because the Planck radiation function reduces to the Rayleigh–Jeans formula at radio wavelengths, one can use this formula multiplied by the apparent solid angle (Ω) subtended by the emitting regions to determine a total observed flux density (S_ν), at a frequency ν (and wavelength λ), in terms of an average brightness temperature T_B. The result is $S_\nu = (2kT_B/\lambda^2)\Omega$, where k is the Boltzmann constant. Using an equivalent angular radius θ (in units of arcseconds) and a minimum detectable flux density S_{min} (in units of Janskys where 1 Jy $= 10^{-26}$ W m^{-2} Hz$^{-1} = 10^{-23}$ erg s^{-1} cm^{-2} Hz^{-1}), the requirement for stellar radio emission to be observable is

$$T_B \cdot \theta^2 \geq 1970\lambda^2\, S_{\text{min}},$$

where λ is in centimeters and T_B is in kelvins. Typically S_{min} for modern radio telescopes is in the range 10^{-4}–10^{-3} Jy. Since angular diameters of stars are typically much less than 0.05 arcseconds (for the nearest red supergiant, Betelgeuse) one can see that for typical radio wavelengths from 2–90 cm, brightness temperatures considerably higher than the effective temperatures ($\leq 50{,}000$ K) of stars, and/or angular sizes much larger than diameters of the stars, are required before there will be detectable radio emission.

Radio emission processes can be described as thermal or nonthermal. For thermal emission the maximum possible brightness temperature is the kinetic temperature of the emitting regions. Nonthermal processes are predominantly emission from relativistic electrons interacting with magnetic fields, producing cyclotron, gyrosynchrotron, or synchrotron radiation, depending upon whether the electrons are subrelativistic, relativistic, or very relativistic. The brightness temperature of these nonthermal processes can have any value, but there is an upper limit of $\approx 10^{12}$ K, because at that level the radiation fields produced by synchrotron emission reach the point where the radiating electrons lose their energy due to inverse Compton interactions with the radio photons they produce. Plasma radio emission processes and various types of masers play major roles in solar radio emission and in a few types of stellar radio emission. These are the only processes that allow unlimited brightness temperatures, but they are inherently phenomena occurring only in small volumes with very short time scales.

SOLAR EXTRAPOLATION

Radio emission from the Sun has been known since the 1940s. A simple extrapolation of solar radio emission processes to other stars would indicate that the thermal emission from the general solar surface would be seen out to 0.2 pc, and the strongest solar radio bursts could be seen out to roughly 5–10 pc. The existence of stellar radio emission from stellar systems at distances of hundreds and thousands of parsecs shows that stellar radio sources are typically 10^5–10^6 stronger than the strongest radio events on the Sun, indicating some combination of higher brightness temperatures or larger emitting regions. In practice they are mostly the result of much larger regions of emission at brightness temperatures not that different than those found on the Sun.

RED DWARF FLARE STARS

Nearby red dwarf flare stars, like UV Ceti, AD Leo, YZ CMi, etc., have long been known to exhibit optical flares; hence they were obvious candidates for solar-like radio flares. In Fig. 1 we show an example of 30 min of flaring at 1400 MHz in AD Leo. The top panel shows "typical" flare star behavior with highly circularly polarized bursts (LCP and RCP are left and right circular polarization). The bulk of the radio emission is due to the nonthermal emission process called gyrosynchrotron emission. The lower panel in Fig. 1 shows 90 s of an LCP event that has unresolved "spikes" (labeled 1, 2, and 3), which may be due to either cyclotron maser events or plasma radiation from Langmuir waves.

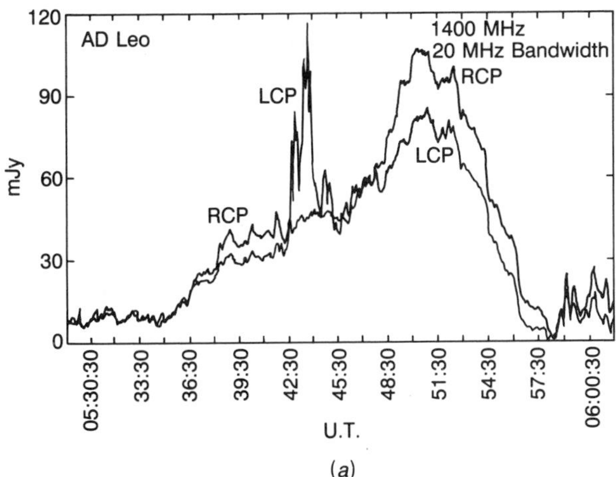

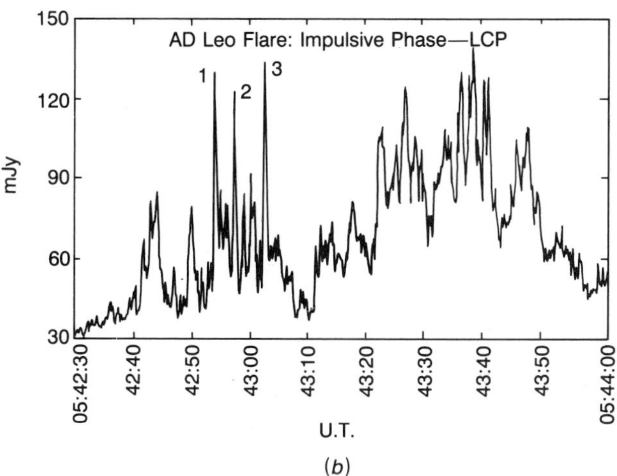

Figure 1. A 1400-MHz radio event in the flare star AD Leo. The top portion shows a plot of flux density versus time for a period of 30 min, in which there is a complete flaring "event" with moderate circular polarization (LCP and RCP are left- and right-hand circular polarization, respectively), and the bottom portion shows a "spiky" LCP event with a time resolution of 200 ms, in which the spikes labeled 1, 2, and 3 are not resolved in time. [*From K. R. Lang, J. Bookbinder, L. Golub, and M. M. Davis (1983), Astrophys. J. (Letters)* **272** *L15.*]

"ACTIVE" BINARIES

The active binaries are double star systems with intense flaring activity seen at x-ray, optical, ultraviolet, and radio wavelengths. Two types of systems that commonly show such activity are the RS Canis Venaticorum (RS CVn) and Algol binaries. These systems exhibit much stronger radio emission that is found in flare stars, but their time scales of variations are usually tens of minutes to hours. Active binary radio emission is probably due to gyrosynchrotron radiation from mildly relativistic electrons interacting with magnetic fields. The radiating mixtures of relativistic electrons and magnetic fields are thought to be associated with either large regions of stellar surface activity or with the intrastellar regions between two stars with interacting magnetospheres. In both cases large starspots on a cool, usually K-type star, are the basic sources of activity, but for one case all emission takes place in and around magnetic field loops over these starspots, and for the other case magnetic field configurations have intrastellar dimensions and magnetic field lines connect the two stars.

The leading working hypothesis for why the solar-like activity in active binaries is $\approx 10^5$ that of the Sun is that the active star or stars are "spun up" through interaction with their companion, resulting in higher stellar rotation rates. These higher rotation rates then increase the effectiveness of "dynamo" phenomena to generate extensive magnetic field configurations and associated starspot regions. The empirical correlation of binaries of the RS CVn-type with strong radio emission (typically up to a few tenths of a Jansky) is an association with objects with large fractions of their surface area covered by starspots. Unfortunately, aside from this broad association, it has been hard to establish correlations between specific radio source parameters and stellar properties.

Active binaries are the only stellar radio sources with solar-like phenomena where there is the possibility of directly imaging the radio emission. Some RS CVn and Algol systems have angular sizes of the order of milliarcseconds, so very long baseline interferometry (VLBI) techniques involving trans- and intercontinental linkages of radio telescopes can resolve or image many of these systems. Algol and the RS CVn binary UX Ari have been shown to have both unresolved radio components and components with the dimensions of stellar separations. Increases of instrumental sensitivity and the growth of worldwide VLBI antennas will make imaging of radio-emitting binaries one of the most important areas of stellar radio astronomy in the 1990s and beyond.

STELLAR WINDS AND EJECTA

As discussed above, stellar environments with plasmas with temperatures on the order of 10^4 K can be detectable radio sources if their size scales are much larger than the stars themselves. A large variety of stellar radio sources are now known to be variations on this particular theme. The ones that are most predictable from known stellar characteristics are the radio-emitting stellar winds. Hot ionized winds have apparent radio "photospheres" with diameters of the order of ≈ 0.1 arcsec at radio wavelengths of about 2 cm. Radio observations of these winds have been used to determine mass-loss rates and electron temperatures for the cases where the emission processes are thermal. Unfortunately, in many cases nonthermal processes occur in these winds that produce other, time-variable components of radio emission. Even the cool winds of a few red giants and supergiants have detectable thermal radio emission from their weakly ionized inner regions. The red supergiant Antares and a number of other binaries (VV Cephei-type) with B-stars orbiting inside cool supergiant winds have two types of thermal radio emission. One type is emission from the weakly ionized, inner portions of the wind, and the other is emission from small ionized regions where the B-star "lights up" the cool wind in its immediate vicinity. Antares, with its B-star companion offset by 2.9 arcseconds, has been imaged at radio wavelengths showing both the unresolved thermal emission from the inner part of the cool wind and the ionized subregion of the wind around the B star.

Much more common, but much more complicated, are the stellar systems with two interacting winds. Symbiotic stars are the most common objects with radio emission of this type. In addition to normal wind and ionized subregion phenomena, the shocks involved in such regions produce dense, hot components and in some cases jet-like outflows. CH Cygni was the first symbiotic star shown to have thermally emitting ejected material exhibiting proper motions of the order of 1 arcsecond per year.

The shell ejecta produced by the nova and recurrent nova phenomena provide another type of large scale, thermal radio source. The classical novae HR Del 1967, FH Ser 1970, V1500 Cyg 1975, V1370 Aql 1982, PW Vul 1984, QU Vul 1984, V1819 Cyg 1986, and V827 Her 1987 have had transient radio sources, lasting from 2–6 yr, that behave like expanding shells of gas with velocity gradients across the shells. The 1985 outburst of RS Oph was the first recurrent nova for which a radio source was observed, and in this case the interaction of the ejected material with the wind of the red giant companion star produced strong interactions and the type of radio source decay previously seen only in radio supernovae. The thermal emission of novae and recurrent novae usually reaches angular sizes where the shells can be either imaged or resolved by high-resolution radio arrays.

RADIO EMITTING X-RAY BINARIES

The strongest and most spectacular stellar radio emission occurs in binary systems that are also strong x-ray sources, because one star is transferring matter to an accretion disk around a neutron star or black hole. Figure 2 shows a sequence of radio flares in Cyg X-3 during the period in 1972 when it was first found that these objects produced strong radio flares due to the production and rapid expansion of large volumes of highly relativistic electrons mixed with magnetic fields. Theories originally used to interpret radio flares in quasars apply very well to these types of synchrotron radiation events. When Cyg X-3 is not exhibiting strong flares it shows fluctuating radio emission with a dominant component with a periodicity of 4.8 h, close to the period seen in both x-ray and infrared emissions of this source. Flares of the type best known from Cyg X-3 occur occasionally in weaker form in other x-ray binaries, and have also been produced by transient x-ray sources like A0620−00, Cen X-4, and GS2000+25.

The x-ray binary SS433 shows synchrotron emission in radio images of corkscrew-like twin jets, with apparent proper motions of 3 arcseconds per year, in ballistic trajectories reflecting the same kinematics seen in optical and x-ray emission. The jets are apparently the result of axial outflows at 78,000 km s^{-1} (0.26c, where c is the speed of light) from a precessing accretion disk. SS433 ejects relativistic plasma continuously, with only a modest amount of flaring. Cyg X-3 has occasionally been shown to have transient extended structures produced by ejection velocities of the order of 0.1c–0.3c.

Amongst the x-ray binaries with strong radio emission are Sco X-1, which is usually the strongest x-ray source in the sky, and Cir X-1. Both have circumstellar flaring and both are surrounded by extended structures. Sco X-1 has an apparent double radio source, whereas Cir X-1 has a large, complicated nebulosity. Cir X-1 and another x-ray binary, LSI+61°303, are cases where radio flaring seems to be related to particular epochs in the period of binary systems.

Extensive, simultaneous, multiwavelength observations of some low-mass x-ray binary systems have revealed a common type of correlation between their radio, x-ray, optical, and ultraviolet emission. Sco X-1, Cyg X-2, and GX17+2 are three of the systems with clearly identifiable x-ray states that are correlated the state of the object at other wavelengths. Observations with the *Ginga* x-ray satellite in several independent energy ranges between 1.4 and 20 keV have been used to show that x-ray "color-color" plots are the key to identifying the states of these systems. An x-ray color, usually called hardness, is computed from the ratio of the measured fluxes in two energy ranges. Two independent hardness parameters plotted against each other exhibit a Z-shaped diagram with the lower part of the Z called the flaring branch, the upper part of the Z called the horizontal branch, and the connecting section of the Z called the normal branch. These branches probably reflect differences in the accretion rate onto the compact object, presumably a magnetized neutron star, with the accretion rate increasing from the horizontal branch to the normal branch, and then to the flaring branch. The empirical radio–x-ray emission correlation shown for Sco X-1, Cyg X-2, and GX17+2 is that the strongest, most variable, radio emission occurs when the x-ray source is on

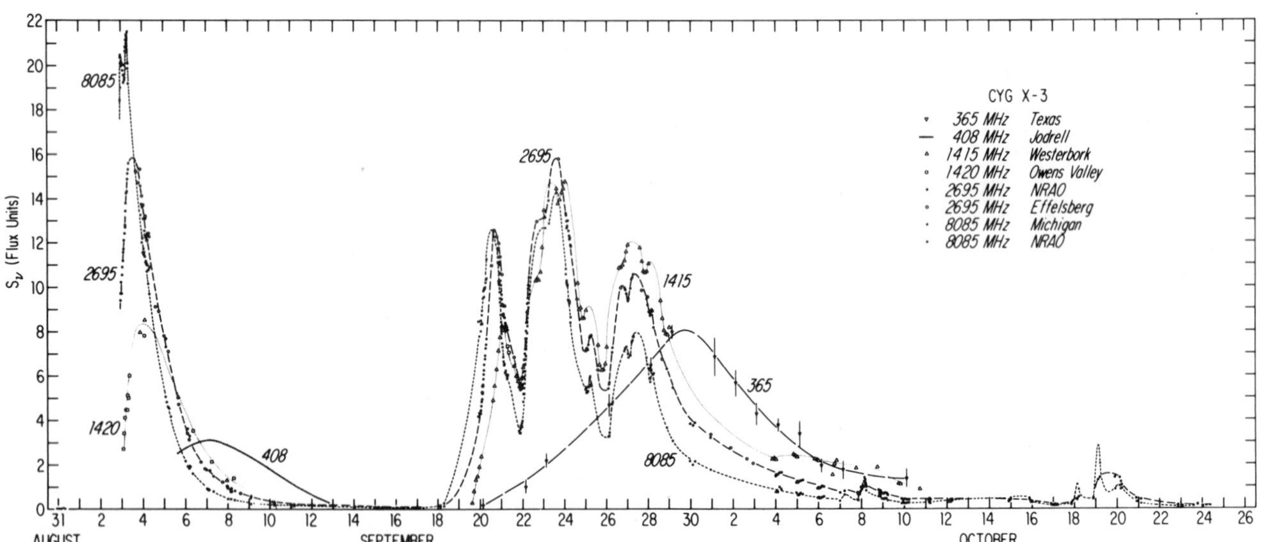

Figure 2. Multifrequency plots of radio flux as a function of time from data on the x-ray binary Cyg X-3 taken during August–October in 1972 when it was first found to have strong synchrotron-flaring events. Light curves are marked with the frequency in MHz.

the horizontal branch, and the weakest occurs while it is on the flaring branch. This means the radio emission is strongest when the compact object and a small accretion disk dominate the x-ray emission, and becomes weaker and possibly more steady as the accretion environment increases in size. In addition, the coupling between state changes in the x-ray emission and state changes in the radio emission can be on time scales of tens of minutes. Because the size scales of the radio and x-ray emitting environments are roughly 10^{13}–10^{14} and 10^{7}–10^{8} cm, respectively, this implies effective propagation speeds $\geq 0.1c$. The implication is clear that there is either direct ejection of relativistic plasmoids from the magnetosphere of the compact object or there are jet-like gas flows, perpendicular to the axis of the accretion disk, with associated relativistic electrons and magnetic fields.

Future studies of stellar radio emission will greatly depend upon observing with very large, single radio telescopes or large arrays of radio telescopes, providing increased sensitivity, better angular resolution, and in some cases, higher time resolution.

Additional Reading

Dulk, G. A. (1985). Radio emission of the Sun and stars. *Ann. Rev. Astron. Ap.* **23** 169.

Hjellming, R. M. (1988). Radio stars. In *Galactic and Extra-Galactic Radio Astronomy*, 2nd ed., G. L. Verschuur and K. I. Kellermann, eds. Springer-Verlag, Berlin.

Hjellming, R. M. and Gibson, D. M., eds. (1985). *Radio Stars*. D. Reidel, Dordrecht, p. 381.

Kuijpers, J. (1989). Radio emission from stellar flares. *Solar Phys.* **121** 163.

See also **Binary Stars, Cataclysmic; Binary Stars, RS Canum Venaticorum Type; Binary Stars, Semi-Detached; Binary Stars, X-Ray, Formation and Evolution; Stars, Red Dwarfs and Flare Stars; Sun, Radio Emissions.**

Stars, Red Dwarfs and Flare Stars

P. Brendan Byrne

Red dwarf stars are among the least massive stellar objects in the universe. They range in mass from ≈ 0.7 to $\approx 0.1\ M_{\odot}$ (we use the symbol $M_{\odot}$ to indicate the mass of the Sun, $L_{\odot}$ its luminosity, and $R_{\odot}$ its radius), in luminosity from ≈ 5 to $\approx 0.01\%\ L_{\odot}$ and in radius from ≈ 0.8 to $\approx 0.25\ R_{\odot}$. In spite of their relative insignificance, however, red dwarf stars are probably the most common stars in the universe. Therefore, an understanding of their properties is of the most fundamental importance to astronomy.

Like the Sun, red dwarf stars generate their energy by nuclear burning of hydrogen in their cores. Also, as in the Sun, an important means of transporting this energy to the surface of the star is convection. These convective motions, in combination with rotation, provide a stellar dynamo that generates a magnetic field. This magnetic field is, in turn, locally concentrated by the convective motions themselves. The result is a range of magnetically initiated phenomena, similar to those occurring on the Sun, namely, active regions, spots, flares, hot chromospheres, and coronae. Because the convective motions in red dwarfs extend over a much more significant fraction of the stellar radius, however, many of these phenomena are proportionately more energetic than in the Sun.

There exists a subset of these red dwarfs that display all of the above signatures of magnetic activity, but at a very much greater scale than average. Their coronae are hotter, their flares more energetic and frequent, and their spots larger. These are the flare stars.

STELLAR FLARES

Stellar flares were first observed more than 50 years ago as sudden, impulsive brightenings in the visible light of certain red dwarf stars. Typically a stellar flare will rise to maximum light in a time between 1 s and 1 min and decay rather more slowly, taking between 1 min and 1 hr to fade (see Fig. 1). Depending on total energy content, flares may occur as frequently as 3–4 per hour to 1 per day for the most energetic. The total energy content of the largest flares in optical light alone may be as much as 10^{28}–10^{29} J.

It was realized very early on in flare star research that the overall color of flare light was very different from that of the underlying star. A red dwarf, as its name suggests, has a deep red color, reflecting its relatively cool surface temperature (≈ 2500–4000 K). By way of contrast, the visible light output of a typical stellar flare peaks in the near ultraviolet, indicating an effective temperature of the emitting gas of nearer 10,000–20,000 K. Thus, in the course of a stellar flare, gas is heated to temperatures high above that of the underlying stellar photosphere and has a density high enough to emit continuum radiation.

Continuum radiation is not the only optical signature of a stellar flare, however. Even the earliest stellar flare spectra showed clearly that there is very strong hydrogen Balmer emission-line enhancement during a flare (Fig. 2). The development of the emission-line flare proceeds at a slower pace than the continuum flare. As a result, early in the flare (the so-called flash phase) continuum radiation dominates; later, during the decay (the slow phase), the radiation in the emission lines increasingly takes over. Indeed, it is the transition from rapidly decaying continuum emission to the slower decay of the emission lines that gives the light curve of the later stage of the optical flare its characteristic quasiexponential form. The behavior of the emission lines mimics closely their behavior in solar flares, but until recently the continuum behavior was considered to be unique to the red dwarf flare stars. Recent results on the Sun have confirmed that solar continuum flares or, as they are more generally known, "white light" flares do occur, but they are relatively rare.

Recent high-speed spectroscopy has revealed that not only are emission lines greatly enhanced during flares, but they are also broadened. Some of this broadening is undoubtedly caused by the greatly increased gas density during the flare, but the overall level of broadening is too great, and often asymmetric, to be fully explained in this way (Fig. 3). The interpretation of these asymmetric, rapidly varying broadenings is that the flare produces rapid motions in the atmosphere of the star with velocities of up to 1000 km s^{-1}. These are presumably the counterpart of solar flare surges and disrupting filaments. Such velocities are greater than the escape velocity from a typical red dwarf, and so the matter ejected in flares may well be lost from the parent star.

WHY ARE SOME STARS ACTIVE?

Because rapid rotation underlies the activity phenomenon in late-type stars, we are naturally led to the question of why some late-type stars are in a highly active state and others, apparently similar in all other respects, are not. The generally accepted theory of star formation requires a cloud of interstellar hydrogen gas to collapse under its self-gravitation to the point where core energy generation stabilizes it against further collapse. Conservation of angular momentum during this collapse means that stars arrive on the hydrogen-burning main sequence with a high rotation rate. Subsequent early stages of the evolution of the star lead to a shedding of this excess angular momentum. Indeed, the T Tauri phenomenon is thought to be closely associated with this process.

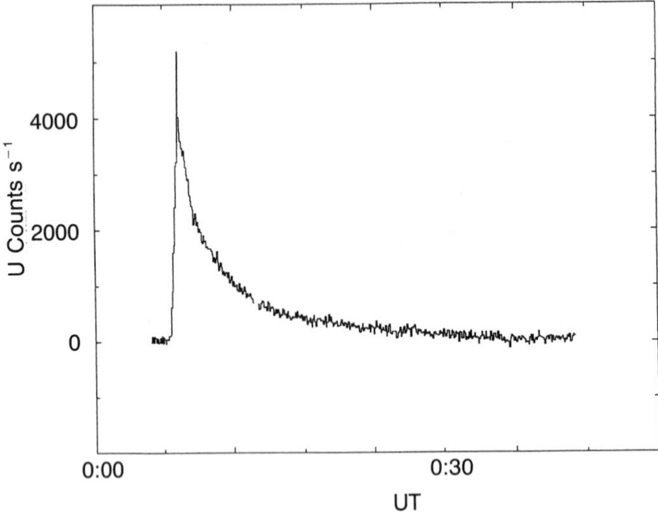

Figure 1. The light curve of a typical stellar flare from Gliese 867A (FK Aqr) in broadband near-ultraviolet light. The brightness in the U band is plotted versus universal time. *(Reprinted from Byrne and Doyle, 1987.)*

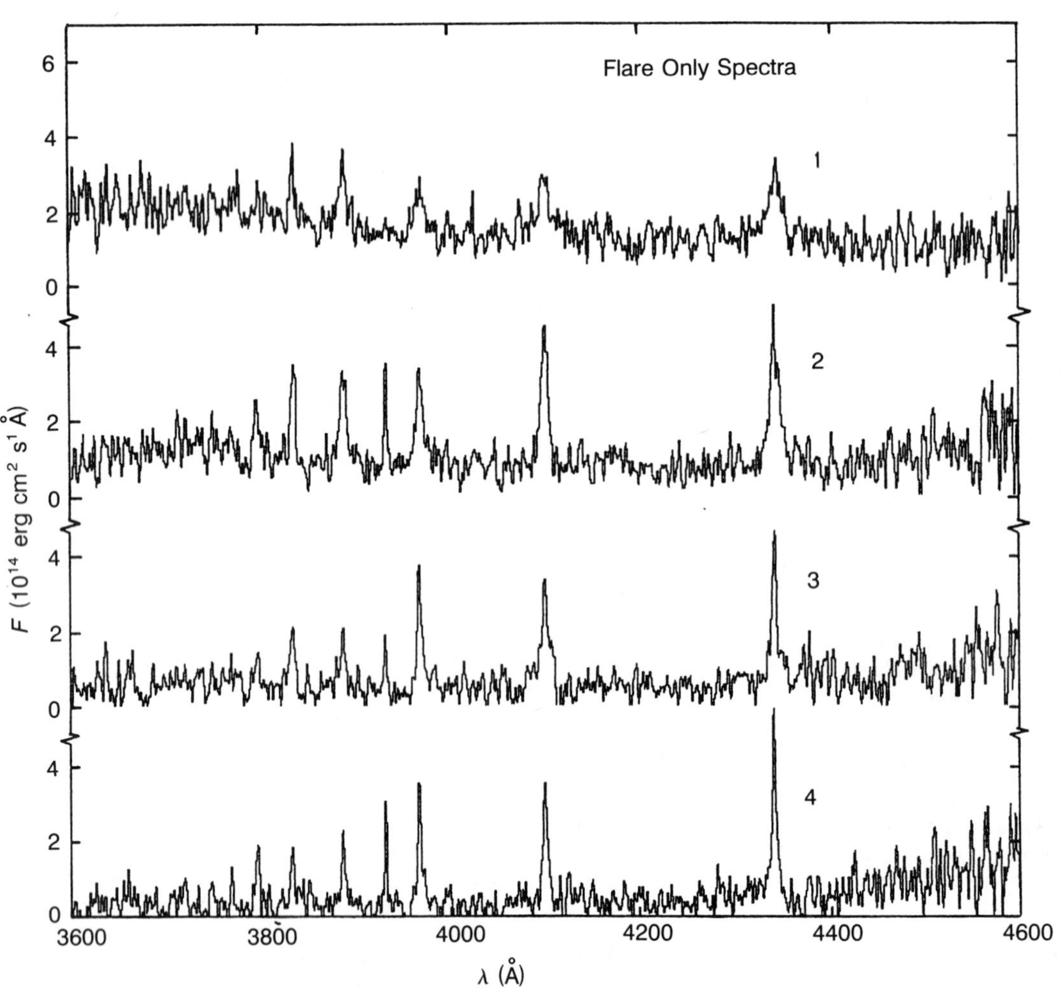

Figure 2. Spectrum of optical stellar flare light, showing the higher members of the hydrogen Balmer series. *(Reprinted from Doyle et al., 1988.)*

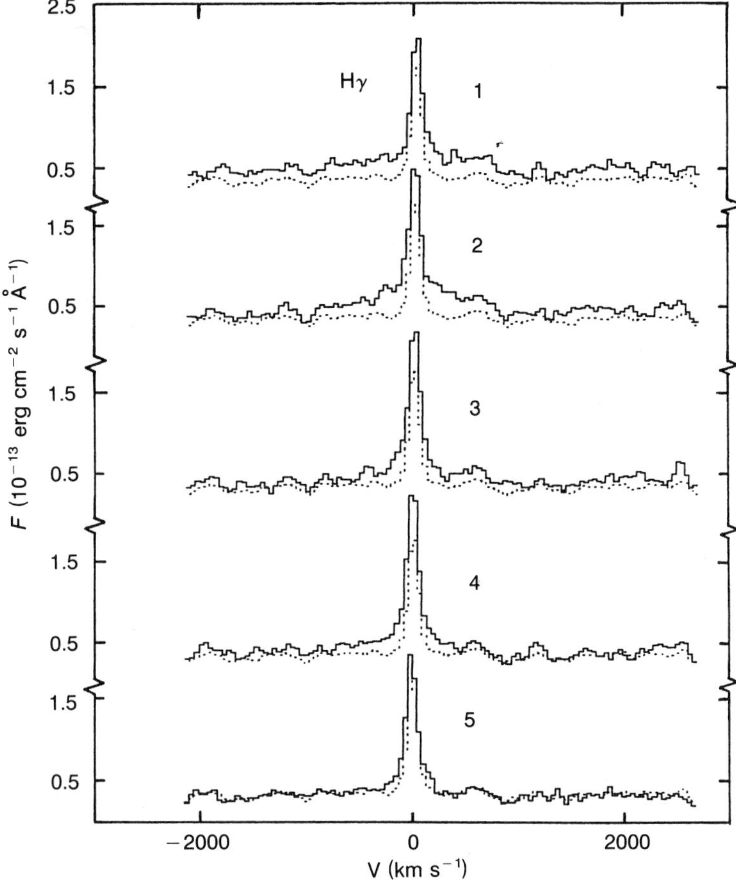

Figure 3. Broadening of the hydrogen Balmer Hγ line during a stellar flare in five consecutive spectra taken during a 5-min interval showing the broadening referred to in the text. The dotted profile is the profile of the same line in the quiescent star. *(Reprinted from Doyle et al., 1988.)*

Observations of the rotation rates of late-type stars in a sequence of stellar clusters of increasing age confirm this general picture. In particular, the recent work of John R. Stauffer has elegantly demonstrated the process. Nevertheless, Stauffer's work also shows that there are "rogue" stars that refuse to follow the general trend towards slower rotation with increasing cluster age; instead, they retain their rapid rotation.

What causes the breakdown of the braking mechanism in these objects? Binarity, with tidally enforced corotation was a favorite mechanism about a decade and a half ago. Studies at that time, especially by Bernard W. Bopp and his co-workers, showed that many of the more rapidly rotating, active, solar-neighborhood stars are single in the sense that upper limits on the masses of any companions are so low as to render them incapable of enforcing corotation.

Coupling between the stellar magnetic field, which permeates the deep convection zones of late-type stars, and the circumstellar environment almost certainly plays a considerable, if not critical, role in the removal of primeval angular momentum. On the other hand, the most active stars are characterized by closed magnetic loop structures, whose height is a fraction of a stellar radius. Hence the connectivity of such closed loops to the zone far from the stellar surface may be poor. Therefore the most active stars may be least efficient at losing their angular momentum from an early stage of their development.

Resolution of these problems is important to our understanding of active late-type stars and their evolution. But it must await further work on rotation rates in a substantial number of young star clusters.

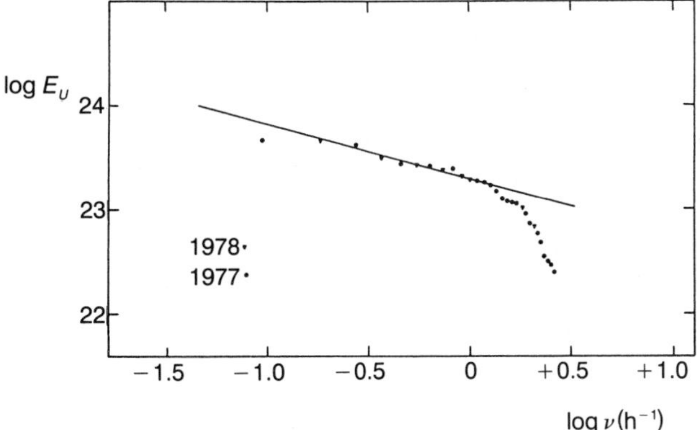

Figure 4. A cumulative frequency (log ν) diagram for a typical flare star, FL Aqr. *(Byrne and McFarland, 1980.)*

"QUIESCENCE" IN FLARE STARS

When one observes a flare star, the flare occurrence frequency depends on the energy of the weakest flares detectable. The most energetic flares occur only rarely, whereas weaker flares occur with ever-increasing frequency (Fig. 4). The weak-flare detection threshold is, apparently, only a function of the instrumentation used. Larger telescopes and more sensitive detectors result in the detection of smaller flares. Below some limit, however, we can no longer accumulate sufficient signal in a time comparable to the duration of a typical flare to be sure that a given fluctuation is not random. Many authors have therefore speculated on whether there is a component of a flare star's visible light, especially in the blue, hotter end of the spectrum, that is due solely to overlapping small flares.

Because stellar flares are extremely hot, the contrast against the background starlight is greatest the shorter the wavelength of observation. Because the radiative output of a flare is greatest in the x-ray region, we might expect to observe the effects of continuous flaring, or "microflaring," in that part of the spectrum. Attempts to observe microflaring have so far led to encouraging, but thus far inconclusive, results. John G. Doyle and C. John Butler first noted the tight correlation between the apparently quiescent coronal x-ray emission and the time-averaged energy released in flares. This could indicate simply that the mean quiescent corona had a common origin with the flares, or, on the other hand, that the apparently quiescent corona was comprised of the late stages of flares, too small to be recorded individually.

In follow-up work, Butler and Doyle and their respective coworkers used time-resolved x-ray observations of flare stars made by the *EXOSAT* satellite to show that the apparently steady corona was in fact varying continuously in exactly the way that would be expected if the hypothesis of a flare corona were true. Particularly impressive was the demonstration of Butler's group that such low amplitude x-ray variability was highly correlated with flare-like brightenings in the chromosphere, detected as impulsive enhancements of the hydrogen Balmer lines.

Carol Ambruster and associates also detected continuous, low-level variability of the coronal x-ray emission from active late-type dwarfs based on an analysis of data from the *Einstein* satellite. They disagree with the Butler et al. interpretation of these variations, arguing that they arise in the normal low-level variability of individual active regions. A resolution of this issue will probably have to await studies with the new generation of x-ray telescopes in the early 1990s. With improved sensitivity, it will hopefully be possible to detect individual microflares.

CONCLUSION

The study of flare activity in late-type stars is a vibrant field of modern astrophysics in which great advances are being made. It leans heavily on the experience of solar physicists and has resulted in a much closer association between the stellar and solar research communities. This rapid progress is likely to continue into the foreseeable future as improvements in space-satellite-borne instruments benefit the astronomer. Furthermore, as our understanding of our closest stellar neighbors improves, it will bring benefits in our understanding of the universe at large and the role played in it by these, its most abundant constituents.

Additional Reading

Ambruster, C. W., Sciortino, S., and Golub, L. (1987). Rapid, low-level x-ray variability in active late-type dwarfs. *Astrophys. J. (Suppl.)* **65** 273.

Bopp, B. W. and Fekel, Jr., F. (1977). Binary incidence among the BY Draconis variables. *Astron. J.* **82** 490.

Butler, C. J., Rodono, M., Foing, B. H., and Haisch, B. M. (1986). Coordinated Exosat and spectroscopic observations of flare stars and coronal heating. *Nature* **321** 679.

Byrne, P. B. (1983). Optical photometry of flares and flare statistics. In *Activity in Red-Dwarf Stars*, P. B. Byrne and M. Rodonò, eds. D. Reidel, Dordrecht, p. 157.

Byrne, P. B. (1989). Multi-wavelength observations of stellar flares. *Solar Phys.* **121** 61.

Byrne, P. B. and Doyle, J. G. (1987). Activity in late-type dwarfs. II. Flares and spot variations on Gl 867A (= FK Aqr) in 1989. *Astron. Ap.* **186** 268.

Byrne, P. B. and McFarland, J. (1980). Gleise 867–further observations of a multiple flare star system. *Mon. Not. R. Astron. Soc.* **193** 525.

Doyle, J. G. and Butler, C. J. (1985). Ultraviolet radiation from stellar flares and the coronal x-ray emission for dwarf-Me stars. *Nature* **313** 378.

Doyle, J. G., Butler, C. J., Byrne, P. B. and van den Oord, G. H. J. (1988). Rotational modulation and flares on RS CVn and BY Dra systems. VII. Simultaneous x-ray, radio and optical data for the dMe star YZ CMi on 4/5 March 1985. *Astron. Ap.* **193** 229.

Joy, A. H. and Humason, M. L. (1949). Observations of the faint dwarf star L726-8. *Publ. Astron. Soc. Pac.* **61** 133.

Kunkel, W. E. (1973). Activity of flare stars of the solar neighborhood. *Astrophys. J. (Suppl.)* **25** 1.

Neidig, D. F. (1989). The importance of solar white-light flares. In *Solar and Stellar Flares*, B. M. Haisch and M. Rodonò, eds. Kluwer Academic Publ., Dordrecht, p. 261.

Stauffer, J. R. and Hartmann, L. W. (1987). The distribution of rotational velocities for low-mass stars in the Pleiades. *Astrophys. J.* **318** 337.

van Maanan, A. (1940). The photographic determination of stellar parallaxes with the 60- and 100-inch reflectors. Seventeenth series. *Astrophys. J.* **91** 503.

van Maanan, A. (1945). Variable star with faint absolute magnitude. *Publ. Astron. Soc. Pac.* **57** 216.

See also **Stars, Activity and Starspots; Stars, Atmospheres, X-Ray Emission; Stars, Radio Emission; Stars, Magnetism, Theory.**

Stars, Red Supergiant

Wendy Hagen Bauer

Observations of luminous stars in the Milky Way and the Magellanic Clouds have shown that there is an upper limit to stellar luminosity (see Fig. 1 in the entry on Stars, High Luminosity). This upper limit is brighter for hot high-luminosity stars than it is for cool supergiants: The most luminous F and G hypergiants are less luminous than Of stars or B supergiants, and M supergiants have approximately the same bolometric luminosity as a typical Be star. The existence of this upper limit makes the brightest red supergiants useful as extragalactic distance indicators.

The evolutionary history of a given red giant or supergiant is not unambiguously determined by its position in the HR diagram. Stars still contracting to the main sequence, and stars with He or C or even heavier-element cores, are all found in the luminous cool-star region of the HR diagram. All cool supergiants show evidence of substantial mass loss capable of affecting their evolution. It is currently believed that stars with initial masses greater than about 40–60 solar masses will have sufficiently strong mass loss to expose a He or a C core, and therefore may never become M supergiants but would return to the hot-star portion of the HR diagram. Unfortunately, this mass limit is highly uncertain. Late M supergiants, for example, VX Sgr or VY CMa, may have had initial masses near 50–60 solar masses, but by the time they become red supergiants, their masses are probably more like 25–30 solar masses. Their luminosities are too low for them to be descended from objects like η Car.

VARIABILITY

Nearly all M giants and supergiants show at least some variability. Cool supergiants tend to show periods of hundreds to thousands of days, with low amplitudes, typically less than 1.5 mag. Most cool supergiants are classified as SRc (semiregular) or Lc (irregular) variable stars. SRc variables show smaller visible light variations than the Miras, and can show changing cycle lengths and intervals of nearly constant brightness. Some irregular stars have variations that appear to be random or chaotic as opposed to those due to pulsations. Perhaps surface irregularities such as spots or convective cells contribute more to the light variations than do pulsations or other periodic processes. The best-studied of the Lc variables, Betelgeuse (α Ori, spectral type M2 Iab), has shown random fluctuations in brightness on time scales of a year or less, superimposed on a cycle of about six years over which both brightness and radial velocity are seen to vary. The F and G hypergiants also show complex variability. Nightly variations are superimposed upon a cycle of several hundred days in length. Occasional longer-term variations are seen, which may be due to episodic mass ejection.

HYPERGIANTS

Hypergiants, the most luminous stars of a given spectral type, range from late O to early K. Among those of type F and later, the best-known galactic hypergiants are ρ Cas (F8 Ia$^+$), HR 8752 (G0–G5 Ia$^+$), and HR 5171 (G8 Ia$^+$). They all show indications of instability associated with their high luminosity. HR 8752 ranged from a spectral type of G0–K5 over a 25-yr time scale. ρ Cas, currently classified as an F star, has shown M-star characteristics, including TiO bands. Hypergiants show considerable evidence for extended envelopes and extensive mass loss.

ATMOSPHERIC STRUCTURE

The advent of space astronomy permitted ultraviolet and x-ray observations of cool stars and showed that this area of the HR diagram splits into two regions. The main-sequence stars, the giants hotter than about K0–K2, and the supergiants hotter than about G8 show evidence for coronae: ultraviolet observations indicating plasma at about 100,000 K, and x rays revealing temperatures of several million kelvins. Stars with spectral types later than these limits show no evidence for temperatures hotter than 20,000 K; they have neither transition regions nor coronae, but do show evidence for significant mass loss through some form of stellar wind. The atmospheres of these noncoronal cool stars can be divided into three regions. Innermost is the photosphere, where the visible continuum and most absorption lines form. The warmer, turbulent chromosphere extends to several stellar radii. Finally, an expanding circumstellar envelope can extend to many thousands of stellar radii.

Photospheres

Red supergiants are the largest stars in the visible HR diagram: some are as large as the orbits of Jupiter and Saturn. Despite their distances, their large sizes make them good candidates for angular diameter measurements using various interferometric techniques. Lunar occultations have also provided direct measurements of angular diameters. A few red supergiants, including VV Cephei, are members of the ζ Aurigae-type eclipsing binary systems, and their eclipse durations can provide radius determinations. Finally, radii can be estimated from an estimate of the bolometric luminosity and the effective temperature, although neither of these quantities can as yet be very precisely determined. It should be noted that even the photospheres of these objects are very extended: Angular diameter measurements vary considerably with wavelength. Diameters measured in strong absorption lines can be as much as twice as large as those measured at continuum wavelengths.

Red supergiants may have brightness variations on their surfaces. It is predicted theoretically that supergranulation occurs on a much larger scale than that of the Sun. Only a few such cells would cover the stellar surface. Some indication of nonuniform photospheric brightness for α Ori has been observed with speckle interferometry.

Chromospheres

Before space-based ultraviolet astronomy, stellar chromospheres were mainly observed through emission in the H and K lines of Ca II. Ground-based study of near-ultraviolet Fe II lines in α Ori revealed emission arising from a turbulent region extending to 1.8 $R_\odot$. These lines shared the radial velocity variations of photospheric lines, believed to be due to stellar pulsation.

The *IUE* satellite observations of α Ori detected emission from Mg II, Fe II, C II, Si II, and Al II, a fairly typical chromospheric spectrum. The relative strengths of the C II (UV 0.01) multiplet can be used to estimate chromospheric densities, and the absolute intensities of these lines can then provide an idea of the extent of the chromosphere. This method applied to α Ori indicates that matter at 6000–8000 K extends to about one stellar radius from the surface.

Further evidence for an extended chromosphere around α Ori comes from radio continuum observations, which have been modeled as due to free-free emission from partially ionized gas extending to several stellar radii.

Circumstellar Envelopes

Extended circumstellar envelopes are detected around cool giants and supergiants by a number of methods. Blueshifted absorption cores arising from the expanding envelopes are seen superimposed on strong low-excitation photospheric lines. Typical expansion velocities inferred from these lines are on the order of 10 km s^{-1}. Excess emission in the infrared arises from circumstellar dust. In oxygen-rich objects a feature at 10 μm due to silicates is seen. In the radio region, thermal emission from SiO and CO has been detected for sufficiently thick circumstellar envelopes. In the cooler objects, maser emission has been observed from SiO, H_2O, and OH.

These circumstellar envelopes are extremely extended; consider α Ori as a particularly well-studied case. Spatial heterodyne interferometry has shown that the dust shell begins at about 12 stellar radii. Polarization from circumstellar dust has been observed out to 90 arcseconds, or about 3600 stellar radii. The *Infrared Astronomical Satellite* 60-μm observations show extension at a scale of several arcminutes for α Ori and other cool supergiants, including μ Cep.

For most of the optically visible red supergiants, the condensation of silicates does not appear to be complete. Among these objects, stars with a higher dust/gas ratio (e.g., μ Cep) do not show chromospheric Ca II H and K emission (although they do show chromospheric emission of other ions in the *IUE* spectral region, but at reduced levels). These stars tend to show molecular maser emission and Balmer emission similar to that produced by shock waves in Mira variables, whereas the stars with lower dust/gas ratios (e.g., α Ori) do not. For stars with thicker circumstellar envelopes, the dust/gas ratio does appear to be constant to reflect complete condensation of silicates.

Despite extensive effort, our understanding of circumstellar envelopes does not yet permit the determination of mass loss rates to better than a factor of 2–10. Mass loss rates tend to increase for cooler and more luminous objects. Mass-loss rates for GK Ib supergiants are typically 10^{-8}–10^{-9} $M_\odot$ yr^{-1} (where $M_\odot$ is the solar mass) and those for the hypergiants are on the order of 10^{-5} $M_\odot$ yr^{-1} (although that for ρ Cas may be as high as 10^{-3} $M_\odot$ yr^{-1}). Mass-loss rates for the optically visible red supergiants tend to range from 10^{-6}–10^{-5} $M_\odot$ yr^{-1}. The mechanism of mass loss is not definitely known. Radiation pressure on dust grains probably drives the mass loss for the coolest objects, and contributes in warmer objects in which dust grains could not condense in the photosphere. For these objects, the material must be lifted far enough for grains to condense. This might be done by pulsation, turbulence, or Alfvén waves.

Additional Reading

de Jager, C. (1980). *The Brightest Stars.* D. Reidel, Dordrecht.

Humphreys, R. M. (1990). Luminous cool stars in the upper HR diagram or "stars near the edge of existence." In *Cool Stars, Stellar Systems and the Sun (Sixth Cambridge Workshop)*, G. Wallerstein, ed. Astronomical Society of the Pacific, San Francisco, p. 387. Many papers pertaining to red supergiants can be found in this and the proceedings of the previous workshops in this series.

Johnson, H. R. and Querci, F. R., eds. (1986). *The M-Type Stars.* NASA Spec. Pub. No. 492, NASA, Washington, D.C.

Kwok, S. and Pottasch, S. R., eds. (1987). *Late Stages of Stellar Evolution*. D. Reidel, Dordrecht.

See also **Binary Stars, Atmospheric Eclipses; Stars, Circumstellar Matter; Stars, High Luminosity.**

Stars, Rotation, Observed Properties

Steven D. Kawaler

It has been known for nearly four centuries that the Sun rotates with a period of about one month; the motion of sunspots across the face of the Sun was attributed to the solar rotation by David Fabricius and Galileo in the early seventeenth century. The determination of the rotation rates for other stars is necessarily more difficult because of our inability to resolve surface details. Modern determinations of stellar rotation rates rely on photometric measurements, determination of Doppler broadening of spectral lines, and less direct tracers of rotation. These techniques have allowed determination of the rotation rates for thousands of stars at all phases of stellar evolution. In turn, these observed rotation rates provide valuable information regarding stellar evolution, stellar winds and mass loss, and mixing in stellar interiors, and they place important constraints on theories of star formation.

MEASURING STELLAR ROTATION RATES

Photometric Rotation Periods

Some stars, such as young solar-type stars and others with strong magnetic fields, exhibit measurable and somewhat regular brightness variations. The phase of these variations can change with time; thus the light curves do not result from intrinsic variation of the stellar luminosity (such as pulsation). The light curves are most simply explained as the result of the rotation of the stars bringing dark starspots across the observable hemisphere of the star. Because these spots can persist for several rotation cycles, the rotation period can therefore be determined from the period of such a light curve.

Spectroscopic Rotation Velocities

Consider a star with its rotation axis perpendicular to the line of sight. If the star rotates with an angular velocity Ω, then the leading edge of the observable disk moves towards the observer with a velocity ΩR, where R is the radius of the star. The trailing edge of the disk moves away from the observer with the same velocity. Light that originates at the leading edge of the disk is Doppler shifted to shorter wavelengths, and light coming from the trailing edge is redshifted. The Doppler shift decreases with decreasing distance from the rotation axis. Thus spectral lines originating in the stellar photosphere will be broadened by the star's rotation; the geometry of this problem dictates that such a *rotationally broadened* line has a characteristic dish shape. The total width of the line in wavelength is increased by a factor $2\Omega R/c$, where c is the speed of light. Measurement of rotational broadening of lines therefore provides rotation velocities for stars.

The rotation axes of stars can point in any direction. Therefore, the observed rotational broadening includes the projection of the rotation axis on the plane of the sky, and is actually a measurement of $v \sin i$, where i is the inclination of the rotation axis away from the line of sight. Without additional information, then, rotational broadening provides only a lower limit to the true rotation velocity. Current techniques limit the measurement accuracy of $v \sin i$ to a few kilometers per second in most types of stars. Thus rotation velocities similar to that of the Sun (2 km s^{-1}) represent the slowest rotation that can be studied using this technique.

Other Indicators of Rotation

Many other techniques have been applied to the determination of stellar rotation. For example, one technique involves time-series measurements of spectral features (in particular, the strength of the Ca II H and K emission lines). These features are formed in localized active regions on the surface of stars. Thus periodic variations in spectral line strengths can indicate stellar rotation in an analogous way to photometric variations. More indirect indicators of stellar rotation rates are frequently employed. For example, empirical relationships between stellar activity (as measured by emission line strength) and rotation rates in young low-mass stars can provide estimates of rotation rates for stars with known levels of activity.

OBSERVED STELLAR ROTATION RATES

Rotation Rates of Main-Sequence Stars

Main-sequence stars of a given spectral type display a wide range of rotation velocities. For stars of intermediate mass [with $M \geq 1.5\ M_{\odot}$ (solar masses)], the distribution of rotation velocities within a given spectral type appears somewhat Maxwellian, with a peak at low velocities and a high-velocity tail reaching to nearly the equatorial breakup velocity. The mean rotation velocity of approximately 200 km s^{-1} is very nearly $\frac{1}{3}$ the equatorial breakup velocity for stars between 6 and 1.5 solar masses (i.e., between early-B and early-F spectral types). With the assumption that these stars rotate as solid bodies, this relationship corresponds to the angular momentum being proportional to the stellar mass squared; this relationship is known as the Kraft curve.

Observations depart below the Kraft curve at higher and lower masses as the consequence of angular momentum loss by stellar winds. The mean rotation velocity drops dramatically at masses below about 1.5 $M_{\odot}$, where stars have deep surface convection zones and can lose a significant fraction of their angular momentum through magnetic stellar winds during the approach to the main sequence. High-mass O and B stars produce vigorous thermal winds, which carry away significant amounts of angular momentum as well as mass.

Rotation Rates of Young Stars

Rotation rates for T Tauri stars have been measured with photometric and spectroscopic techniques. Rotation velocities range from about 10 up to over 100 km s^{-1} but the distribution of rotation velocities is strongly peaked towards low velocities. The difficulty of assigning accurate masses to these stars (many of which lie on the convective portion of the pre-main-sequence evolutionary track) precludes comparing them with their main-sequence counterparts.

However, it appears that the angular momentum in intermediate-mass protostars is within a factor of 2 of the mean angular momentum of main-sequence stars of comparable masses.

As mentioned above, low-mass main-sequence stars appear to have lost angular momentum because they fall below the Kraft curve. Observations of low-mass stars with known ages show that the rotation velocity decreases as the square root of the stellar age. This behavior is a natural consequence of angular momentum loss by a magnetic stellar wind.

Rotation Rates of Post-Main-Sequence Stars

As a star exhausts the hydrogen fuel source at its center, it leaves the main sequence on the way to becoming a red giant. In the process its radius increases, resulting in a corresponding increase in its moment of inertia. To conserve angular momentum, the rotation rate must therefore decrease substantially for post-main-sequence stars as they approach the giant branch. Indeed, subgiant stars are observed to rotate more slowly than main-sequence stars. For intermediate-mass subgiants later than mid-G spectral types, rotation velocities typically fall below 5 km s^{-1}. Rotation on the giant branch is further slowed by stellar mass loss.

Rotation Rates of Highly Evolved Stars

Numerous investigations have demonstrated that stars retain observable rotation following evolution on the giant branch. For example, some horizontal branch stars in globular clusters show rotation velocities of order 20 km s^{-1}. The rotation rate of horizontal branch stars is correlated with the blueness of the horizontal branch.

Whereas the most famous case of rotation in late stages of stellar evolution pertains to pulsars, other highly evolved objects show evidence for rotation. White dwarfs with strong magnetic fields show periodic changes in their spectra that indicate rotation, with periods ranging from 100 min to 3 days. Spectroscopic rotation velocities for normal white dwarfs can, in principle, be measured, but the lower limit is about 20 km s^{-1}. Though the statistics are still uncertain, $\frac{1}{3}$ of white dwarfs studied rotate with velocities greater than 20 km s^{-1}, indicating that they have angular momenta that are comparable to pulsars.

Additional Reading

Fukuda, I. (1982). A statistical study of rotational velocities of the stars. *Publ. Astron. Soc. Pac.* **94** 271.

Kippenhahn, R. and Weigert, A. (1990). *Stellar Structure and Evolution*. Springer-Verlag, Berlin. See Part VIII, Stellar rotation, p. 428.

Kraft, R. P. (1970). Stellar rotation. In *Spectroscopic Astrophysics*, G. H. Herbig, ed. University of California Press, Berkeley, p. 385.

Stauffer, J. R. and Hartmann, L. W. (1986). The rotational velocities of low mass stars. *Publ. Astron. Soc. Pac.* **98** 1233.

See also **Stars, Activity and Starspots; Stars, Magnetic and Chemically Peculiar; Stars, Neutron, Physical Properties and Models; Stars, White Dwarf, Observed Properties; Stars, Young, Rotation; Stellar Evolution, Rotation.**

Stars, RR Lyrae Type

James Nemec

RR Lyrae stars are 10–15-Gyr-old low-mass variable stars that pulsate radially with periods between about 0.2 and 0.9 d. They were first discovered in nearby globular clusters about a century ago. Today, ~1500 RR Lyrae stars have been identified in globular clusters, and ~6000 isolated field stars are known to be RR Lyraes. Because RR Lyrae stars are luminous and are easily recognized from their periods and from the shapes of their light curves, and because they have a small range in their luminosities, they are useful for distance estimation. They have also been used extensively for studying the structure and evolution of the Galaxy, and are currently being used as probes of other galaxies. The most distant of the known RR Lyrae stars are those discovered recently in the Local Group galaxies M31, M33, NGC 185, and NGC 147, which have distances ~700 kpc. As electronic detectors on ground-based telescopes are improved, and the Hubble Space Telescope is optically repaired, it is likely that RR Lyrae stars in galaxies beyond the Local Group will be found.

There are two basic types of RR Lyrae stars: Bailey ab-type stars, which pulsate in the fundamental mode with periods between ~0.4 and 0.9 d; and Bailey c-type stars, which pulsate in the first-overtone mode with periods between ~0.2 and 0.5 d. In general, RRab stars have cooler surface temperatures and are much more common than RRc stars. Furthermore, whereas the RRab stars typically have large amplitudes and asymmetric light curves, the RRc stars tend to have smaller amplitudes and more sinusoidal light curves. In globular clusters, the two types are easily distinguishable if the periods and amplitudes are accurately known. In that case, a plot of amplitude versus period, that is, a *P-A* diagram, shows that the RRab stars have decreasing amplitudes for increasing periods, whereas the *P-A* relationship for RRc stars is less clear.

In 1939 and 1944, P. Th. Oosterhoff discovered that the periods of RR Lyrae stars in globular clusters can be used to sort the clusters into two groups: type I systems (e.g., M3, M5), in which the mean periods of the ab- and c-type RR Lyrae stars are $\langle P_{ab} \rangle \sim 0.54$ d and $\langle P_c \rangle \sim 0.3$ d, respectively; and type II systems (e.g., M15, M92), in which $\langle P_{ab} \rangle = 0.64$ d and $\langle P_c \rangle \sim 0.37$ d. Later it was shown that the two groups can be discriminated by metal abundance. Over the past 50 years much work has gone into confirming the apparent near-absence of globular clusters with $\langle P_{ab} \rangle \sim 0.60$ d, and into explaining the Oosterhoff dichotomy.

In addition to ab- and c-type RR Lyrae stars, several other types are now recognized. Included among these are the Blazhko variables, in which the pulsations are characterized by ~20–200-d periodic modulation of the amplitude and phase of maximum light. The brightest known Blazhko variable is RR Lyrae itself, which has a fundamental-mode period of 0.567 d, and an amplitude-modulation period of 41 d. Although Blazhko variables are quite common, and tend to be found among the shortest-period RRab stars, the mechanism that produces their amplitude modulations is not well understood. Another type of RR Lyrae star, the double-mode (or d-type) RR Lyraes, of which ~40 are known, pulsate simultaneously in the fundamental and the first-overtone modes, with periods in the ratio $P_1/P_0 \sim 0.746 \pm 0.002$. These stars are exceedingly valuable sources of information because their masses can be determined using only the Petersen P_1/P_0 versus P_0 diagram, and theoretical mass-calibration curves. Masses derived in this way can be compared with less-direct mass estimates based on pulsation theory, stellar evolution theory, etc. It has been suggested that RRd stars might be in a transitional phase in which they are switching pulsation modes from the fundamental to the first-overtone mode as they evolve from ab to c type, or vice versa. One triple-mode RR Lyrae star, AC And, has been identified. It pulsates simultaneously in the fundamental, first-overtone, and second-overtone modes. There has been considerable speculation about the possibility of RR Lyrae stars pulsating in the second-overtone mode; however, such stars have yet to be identified with certainty.

It was once common to classify Cepheid variable stars with periods between about 1 and 3 d as long-period RR Lyrae stars. However, the majority of these stars are now known to be either Population II Cepheids that are more luminous than RR Lyrae stars, anomalous Cepheids (recent evidence suggests that these may be coalesced binary systems), or Population I Cepheids that are more massive and younger than RR Lyrae stars. At periods shorter than ~ 0.25 d, the period distribution of RR Lyrae stars overlaps that of young, massive Population I dwarf Cepheids (i.e., δ Scuti stars) and old Population II dwarf Cepheids (i.e., SX Phoenicis stars).

PULSATIONAL INSTABILITY

The physics of the pulsations of RR Lyrae stars is very similar to that of Cepheid variables. The observed light and radial velocity curves of both types of stars are mirror images; that is, at maximum light the radial velocity is at a minimum, and at minimum light the radial velocity is at a maximum. Originally this was interpreted as being due to line-of-sight motion in a spectroscopic binary system. However, because this hypothesis failed to provide satisfactory orbital solutions, and failed to explain a number of other observations (such as the irregular variations seen in some light curves and in some periods, the observed spectral-type changes over each light cycle, and the complete absence of spectral features that are expected for binary systems), it was soon rejected. The hypothesis that RR Lyrae stars and Cepheids are single stars that are pulsating radially was proposed as an alternative, and continues to be the most plausible explanation.

For a radially pulsating star, the observed velocity is the sum of the radial velocity arising from the projection of the space motion of the star onto the line of sight, and the contribution from the radial pulsations. Considering only the contributions arising from the pulsations, the fact that the light curve and radial velocity curve are mirror images suggests the following interpretation. At maximum luminosity, the expansion of the star is most rapid. As the star dims, the expansion continues, the diameter reaching a maximum when the radial velocity is zero. After this the star contracts, and the positive radial velocities measured by an observer are due to the apparent recession of the atmosphere. At minimum light the contraction is most rapid, at which time the higher temperatures associated with the compressed gas lead to an increased flux of radiation emerging from the star, and subsequent brightening. At maximum compression the radial velocity is again zero, and the star is once again on its way to reaching maximum light.

In an attempt to account for the observed light and radial velocity curves of Cepheids and RR Lyrae stars, Arthur S. Eddington developed a mathematical model for radial pulsations. This model, while unable to explain fully the observed phase relationships between velocity, luminosity, and temperature, did succeed in explaining the pulsation periods, and provided the foundations of modern pulsational theory. The starting point of Eddington's investigations was J. Homer Lane's pioneering 1869 study of the temperature stratification within a star, and August Ritter's 1878–1889 series of articles on the analysis of adiabatic pulsations of gaseous stars in convective equilibrium. Eddington, following Karl Schwarzschild, assumed radiative rather than convective equilibrium throughout the star. The aim of his model was to explain the pulsations, while balancing the energy continually being liberated in the star and the energy needed to excite and maintain the pulsations, as well as the obvious loss of energy from the surface of the star. A significant finding was that in RR Lyrae stars and Cepheids, it is the outer layers that are subject to the pulsational instabilities.

Two of Eddington's pulsation mechanisms, the gamma and kappa mechanisms, remain relevant today. Gamma mechanism: From Ritter's thermodynamical studies, it is known that if the ratio of the specific heats of a gaseous star, γ ($= c_p/c_v$, where $c_p =$ the specific heat at constant pressure, and $c_v =$ the specific heat at constant volume), is less than $\frac{4}{3}$ (on average), then the total energy of the star is positive, and an equilibrium configuration is not possible. Therefore, in the adiabatic central regions of stars, where the ideal gas law applies and $\gamma = \frac{5}{3}$, pulsations do not occur. On the other hand, γ can be lowered to values $\sim \frac{4}{3}$ or less, in the hydrogen and helium ionization zones near the cooler, nonadiabatic surface of an RR Lyrae star. Upon compression, because of the smaller γ, these ionization zones remain cooler than their surroundings and heat is absorbed when the temperature is high. Upon expansion, when the temperature falls, heat is given off. These conditions, which resemble the valve mechanism of a heat engine, can lead to pulsational instability if the density of material in one (or more) of these zones is sufficiently great to provide enough mechanical energy to drive the pulsations.

The κ mechanism (named for the symbol used to represent the opacity of the gas) also involves ionization zones and the conversion of radiative energy into mechanical energy. In regions where high temperatures give rise to high opacities, upon compression the temperature rises, causing the opacity to rise, thereby decreasing the flow of radiation through the star. The damming of trapped photons eventually causes the envelope to expand, thus reducing the temperature of the zone and lowering the opacity. This leads to an increased outflow of radiation. Eventually, gravity halts the expansion and the envelope collapses on a free-fall (or dynamical) time scale, and the cycle starts again.

It was not until 1953 that the second helium ionization zone was identified as the zone most likely to be responsible for exciting and maintaining the pulsations. The observed phase relationship between luminosity and radial velocity was explained in the 1960s, when nonadiabatic and nonlinear numerical models of RR Lyrae stars were computed, and accurate computer modeling of the light and radial velocity curves became possible. Today, fully hydrodynamic models are being used to calculate synthetic light and radial velocity curves for RR Lyrae stars, which have been found to provide good fits to the observations.

PHYSICAL CHARACTERISTICS

In color-magnitude diagrams, RR Lyrae stars are readily seen to be horizontal-branch stars in the Cepheid instability strip, with mean $B-V$ colors between 0.17 and 0.42 mag (corresponding to effective temperatures in the range $6000 \lesssim T_e \lesssim 9000$ K). Unlike Cepheids, which are more luminous than RR Lyrae stars and are found over a large luminosity range, RR Lyrae stars have luminosities in the narrow range $40 \lesssim L/L_\odot \lesssim 90$ (corresponding to absolute visual magnitudes in the range $+1.3 \gtrsim M_V \gtrsim +0.3$); $L_\odot$ is the solar luminosity. The RRab stars tend to be located on the low-temperature side of the strip, and the RRc stars tend to be hotter. The temperature of the red edge of the RR Lyrae instability strip, which separates the RRab stars from the red horizontal branch stars, and the temperature of the blue edge, which separates the RRc stars from blue horizontal branch stars, is determined by the depths of the narrow hydrogen and helium ionization zones, and the energy transport mechanisms in the outer layers of the star. At the red edge, the pulsation is damped by the onset of efficient convective energy transport, and at the blue edge the instability is damped because the location of the second helium ionization zone is too near the surface to be effective. Given the mean T_e and the mean L of an RR Lyrae star, its mean radius, R, can be calculated from the Stefan–Boltzmann law, $L = 4\pi R^2 \sigma T_e^4$, where σ is the Stefan–Boltzmann constant. A typical mean radius is $\sim 5\ R_\odot$ (solar radii), characteristic of a giant star. During each pulsation cycle, it is common for an RR Lyrae star to change its radius by $\delta R/R \sim 20\%$.

Stellar evolution models have shown that RR Lyrae stars are in an advanced stage of their life, deriving their energy from the nucleosynthesis of helium (via the triple-α process) in a small central core, and from hydrogen burning (via the CNO cycle) in a narrow shell outside of the core. This energy is transported by convection in the inner core, and by radiation through the outer regions of the star. Numerical models show the central temperature to be $\sim 1.2 \times 10^8$ K, and the mass density at the center of the star to be $\sim 20{,}000$ g cm^{-3}. Depending on the particular assumptions that are used in the models, estimates of the core mass range from 0.47 to 0.51 $M_\odot$, and total masses are in the range $0.50 \lesssim$

$M/M_{\odot} \lesssim 0.80$. Most RR Lyrae stars exhibit slowly changing periods. At one time it was hoped that the measured period change rates could be used to determine the time required to evolve across the instability strip; however, the period changes have a stochastic nature, which remains unexplained. Nevertheless, the observed pulsation properties of RR Lyrae stars are sufficiently well understood that they are used to test ideas about the advanced evolution of old low-mass stars.

DISTANCES AND LUMINOSITIES

The use of RR Lyrae stars as distance indicators stems from their relatively small range in luminosity (or, equivalently, absolute bolometric magnitude $M_{\rm bol}$). By measuring the mean apparent bolometric magnitude of an RR Lyrae star, $m_{\rm bol}$, and the total extinction caused by interstellar material, $A_{\rm bol}$, the distance to the star, d, follows from the distance modulus equation: $(m-M)_{\rm bol} = 5\log d - 5 + A_{\rm bol}$. In practice, RR Lyrae stars are usually observed through blue (B) and visual (V) filters, and the visual apparent magnitude m_V and the reddening E_{B-V} ($\sim 0.33 A_V$) are measured. Conversion to bolometric quantities is made by applying bolometric corrections, which are small for RR Lyrae stars.

Within a globular cluster that is rich in RR Lyrae stars, the apparent magnitude of the RR Lyrae stars usually varies by only ± 0.2 mag about the mean magnitude. For this reason, a single absolute visual magnitude, M_V, is usually assumed for all the stars. Then, because the diameter of the cluster is small relative to the distance to the cluster, the distance to the cluster can be assumed to be identical to that of the ensemble of cluster RR Lyrae stars. Given the distance to the cluster, and reddening-corrected apparent magnitudes, the luminosity level of its main-sequence turnoff can be computed, and the cluster age determined by fitting theoretical isochrones to the cluster main-sequence turnoff.

RR Lyrae stars have also proved to be useful for determining the distance to the galactic center. By identifying large numbers of RR Lyrae stars in low-extinction "windows" (directions in which the interstellar reddening is relatively small, the most famous of which is Baade's window), and making reasonable assumptions about M_V and E_{B-V}, the peak of the histogram of the derived distances to the RR Lyrae stars gives the modal distance to the stars. The distance to the galactic center then follows by assuming that the RR Lyrae stars are spherically distributed about the center, and that the modal distance of the RR Lyrae stars coincides with the distance to the center. From recent measurements (which include observations made at infrared wavelengths, where the extinction is much reduced) it has been inferred that the distance to the galactic center is 8.0 ± 0.5 kpc. This estimate agrees well with distance determinations made using non-RR Lyrae star techniques.

The greatest concern in using RR Lyrae stars as distance indicators is the uncertainty in their absolute magnitudes. It has long been known (from theoretical models) that RR Lyrae stars evolve away from their initial positions on the zero-age horizontal branch on a time scale $\sim 10^8$ yr. During this time they become more luminous and eventually ascend the red giant branch for a second time (as asymptotic giant branch stars). The RR Lyrae stars in the most advanced evolutionary states are presumed to be those that started out with the greatest masses (but not so great that they are no longer RR Lyrae stars). Thus, changes in both luminosity and effective temperature are to be expected over the lifetime of an RR Lyrae star because of this dependence on evolutionary phase. Furthermore, because L and T_e depend on the age of the star; the metal abundance [Fe/H]; the helium abundance Y; the abundances of carbon, nitrogen, and oxygen; and other parameters (such as rotation, strengths of magnetic fields, etc.), knowledge of the nature of this dependency is required before reliable estimates of M_V can be obtained. Presently, the best estimates of M_V range between $+0.2$ and $+1.0$, with metal-poor RR Lyrae stars having greater luminosities and masses than metal-rich RR Lyrae stars.

HALO AND THICK-DISK RR LYRAE STARS

The majority of the RR Lyrae stars in the Galaxy have yet to be identified. Nevertheless, from careful surveys of selected areas of the sky it is clear that the total number of galactic RR Lyrae stars is very large. The same surveys tell us that the number density of RR Lyrae stars is proportional to R^{-3}, where R is the distance from the galactic center. At $R = 0.6$ kpc, the density is ~ 4000 stars kpc^{-3}, and at $R = 1.5$ kpc the density is only ~ 260 stars kpc^{-3}. Although the number density is very low, RR Lyrae stars are still seen out to $R \sim 50$ kpc.

Studies of the chemical compositions and space motions of the RR Lyrae stars in the Galaxy have shown that they do not constitute a homogeneous stellar population, but rather are a mixture of halo and thick-disk (or old-disk) stars. The halo RR Lyrae stars have metal abundances $[\mathrm{Fe/H}] \lesssim -0.8$, and are found in metal-poor globular clusters and throughout the Galaxy (the term in brackets is defined under "Heavy Elements" in the entry on Star Clusters, Globular, Chemical Composition). They appear to have similar kinematic properties and compositions to halo subdwarfs. The entire system of halo RR Lyrae stars occupies an approximately spherical volume centered on the galactic center, and the system rotates (relative to a galactic rest frame) very slowly, if at all. Assuming that the field halo RR Lyrae stars are as old as their counterparts in globular clusters, then their ages are ~ 15 Gyr.

Thick-disk RR Lyrae stars are more metal rich than $[\mathrm{Fe/H}] \sim 0.8$, and are found only in the inner regions of the Galaxy. The spatial distribution of the system of thick-disk RR Lyrae stars is flattened, like an oblate spheroid with an axial ratio $c/a \sim 0.6$ (where c is the semiminor axis measured in the plane of the Galaxy, and a is the semimajor axis measured perpendicular to the plane). The system of thick-disk RR Lyrae stars rotates faster than the system of halo stars, but both systems lag behind the rapidly rotating stars in the thin disk of the Galaxy. It has been estimated that in the solar neighborhood there are ~ 20 thick-disk subdwarf stars for every halo subdwarf. However, only one in four of the nearby RR Lyrae stars is metal rich. The reason for this discrepancy is that metal-rich horizontal branch stars rarely evolve to hot enough temperatures to enter the instability strip, and thus they are much rarer. For the same reason, it is unusual to find RR Lyrae stars in globular clusters that only have red horizontal branches.

The question of the origin of these halo and thick-disk RR Lyrae stars is presently unsolved. Did the metal-poor stars form before the more metal-rich stars? Or, are they all approximately the same age, with the metal-rich RR Lyrae stars having formed out of clouds chemically enriched as a result of nearby supernovae explosions? Were all (or some fraction) of the field RR Lyrae stars once in globular clusters? And if so, did they enter the field as a result from the disruption of entire globular clusters, or are they isolated escapees? The answers to these questions are tied directly to the fundamental question of how the Galaxy formed. Because our present knowledge of the halo and disk stellar populations in the Galaxy is still very incomplete, RR Lyrae stars will continue to play a major role in solving the puzzle. And, when more RR Lyrae stars are discovered in other galaxies it will be of considerable interest to see if they too divide into different population types, providing valuable clues to the histories of these other systems.

Additional Reading

Baker, N. and Kippenhahn, R. (1962). The pulsations of models of δ Cephei stars. *Z. Ap.* **54** 114.

Christy, R. F. (1966). A study of pulsation in RR Lyrae models. *Astrophys. J.* **144** 108.

Cox, A. N., Hodson, S. W., and Clancy, S. P. (1983). Double-mode RR Lyrae variables in M15. *Astrophys. J.* **266** 94.

Eddington, A. S. (1930). *The Internal Constitution of the Stars.* Cambridge University Press, Cambridge.

Iben, I., Jr. (1971). Globular-cluster stars: Results of theoretical evolution and pulsation studies compared with the observations. *Publ. Astron. Soc. Pac.* **83** 697.

Oort, J. H. and Plaut, L. (1975). The distance to the galactic centre derived from RR Lyrae variables, the distribution of these variables in the Galaxy's inner region and halo, and a rediscussion of the galactic rotation constants. *Astron. Ap.* **41** 71.

Sandage, A. R. (1990). The Oosterhoff period effect: Luminosities of globular cluster zero-age horizontal branches and field RR Lyrae stars as a function of metallicity. *Astrophys. J.* **350** 631.

See also **Star Clusters, Globular, Variable Stars; Stars, Horizontal Branch; Stars, Pulsating, Overview; Stars, Pulsating, Theory; Stellar Evolution, Pulsations.**

Stars, Spectral Classification

Janet Rountree

Spectral classification is a branch of astronomy that has the goal of arranging the spectra of the stars into meaningful and self-consistent groups. Several purposes are served by this procedure: (1) a detailed study of the prototype star in each group is, in principle, sufficient for an understanding of all of the members, thus avoiding the necessity of studying each star separately; (2) the astrophysical parameters (effective temperature and luminosity, and eventually the mass and radius) of the stars can be derived through a suitable calibration of the spectral types; (3) peculiar objects can be isolated from the groups of "normal" stars for further study; (4) the distribution of stars in various groups can be used for statistical studies of stellar populations and stellar evolution.

HISTORY

The continuous spectrum of the Sun, our nearest star, was discovered by Isaac Newton in 1666, in the course of his optics experiments. But the absorption lines, on which most spectral classification systems are based, were first seen in 1802 by William H. Wollaston, and first described in detail some years later by Josef von Fraunhofer, in whose honor they are still called Fraunhofer lines. The first extensive classification program using stellar absorption spectra was carried out in the 1860s by Father Angelo Secchi, the observatory director at the Jesuits' Roman College. After making visual observations of over 4000 stars, Secchi concluded that the great majority of them fell into four main spectral types, which he called I–IV and which we would now identify as spectra dominated by (I) hydrogen lines, (II) metallic lines, (III) titanium oxide bands, and (IV) molecular carbon bands. Secchi's classification remained in general use until it was supplanted by the Harvard, or Henry Draper, system.

The system developed at Harvard College Observatory used letters of the alphabet rather than Roman numerals to denote the spectral types; originally, the stellar spectra were arranged from A to Z in order of "increasing complexity." The series was later rearranged in order of decreasing color temperature, from blue to red, and extraneous types were discarded. The types that were retained, with their principal characteristics in parentheses here, form the well-known spectral sequence: O (ionized helium lines); B (neutral helium lines); A (predominantly hydrogen lines); F (both hydrogen and metallic lines); G (strong CH band); K (similar to G, but with more prominent metallic lines and weaker hydrogen); M (titanium oxide bands); R and N (molecular carbon bands); and S (zirconium oxide bands). The position of a stellar spectrum along this sequence is indicated by a letter and a number—for example, O9, B5, G2. Annie Jump Cannon classified some 325,000 objective prism spectra on this system. Her work was published as the *Henry Draper Catalogue* and the *Henry Draper Extension*, beginning in 1918.

The Harvard spectral-type sequence is well correlated with the stars' effective temperature. But it became increasingly apparent in the 1920s and 1930s that a second dimension, correlated with stellar luminosity, was needed for a complete classification system. Work along this line was inspired partly by the development of the Hertzsprung–Russell diagram, which showed that all the common stars are restricted to several small areas in the temperature-luminosity plane. Two schools of classifiers at the Mt. Wilson Observatory and the Dominion Astrophysical Observatory attempted to graft a second dimension onto the Harvard spectral sequence by deriving an absolute magnitude for each star from a set of calibrated line ratios. But the two-dimensional system that eventually received broad acceptance was the MK classification, developed by William W. Morgan and Philip C. Keenan. In the MK system, first published in 1941, each star is assigned a spectral type (sometimes called a temperature type) in a notation similar to the Harvard system, and a "luminosity class" on a scale of I–V, where I is the most luminous. Members of the different luminosity classes are commonly called dwarfs (class V, the main sequence stars), giants (class III), and supergiants (class I). [Ia and Ib are the more luminous and less luminous supergiants, respectively; II are luminous giants and IV are subgiants.] Stars hotter than the Sun are sometimes called early-type stars, and those cooler than the Sun are late-type stars. The assignment of spectral type and luminosity class is made by comparing the spectrum of the "unknown" star with a predetermined set of standard stars, which serve as paradigms.

PRINCIPLES AND PROCEDURE

The principal difference between the MK classification and the systems that preceded it is not the use of two dimensions rather than one. The MK system introduced two unique principles into spectral classification: (1) Only information obtained from a visual inspection of the stellar spectrum is used in the classification. (2) The system is completely defined by its set of standard stars. The first principle means that knowledge of a star's color, distance, variability, cluster membership, apparent brightness, or any astrophysical parameter is not allowed to influence the classification—the spectral type is confronted with such information only after the fact. The second principle implies that the spectral type is independent of the specific criteria used to compare the "unknown" stars with the standards—all the information available in the observed spectrum is taken into account. The result of a strict adherence to these two principles is that spectral types are independent of any astrophysical calibration in terms of temperature, luminosity, mass, age, etc.—calibrations that tend to change as the state of astrophysical theory advances. Spectral types are also largely independent of the specific instrumentation used to obtain the spectra. Thus the MK system is stable and essentially invariant with time, except for continuing small refinements, and this fact has enabled it to retain its usefulness for the half-century since its inception.

In performing a spectral classification program, the first step is to obtain spectra of as many as possible of the standard stars that occupy cells in that portion of the two-dimensional array (spectral type and luminosity class) that is likely to be of interest. The program stars are then observed with exactly the same instrumentation and processing techniques as the standards, where "processing" includes development procedures for photographic plates, or mathematical manipulation for digital data. Each "unknown" is then closely compared with the three or four standards most similar to it, and is assigned to the cell containing the standard that it most closely resembles. The two spectra will rarely be identical, but if the program star differs from the standard in some important way, it may be labeled "peculiar." If many stars exhibit the same peculiarity, but to different degrees, they may

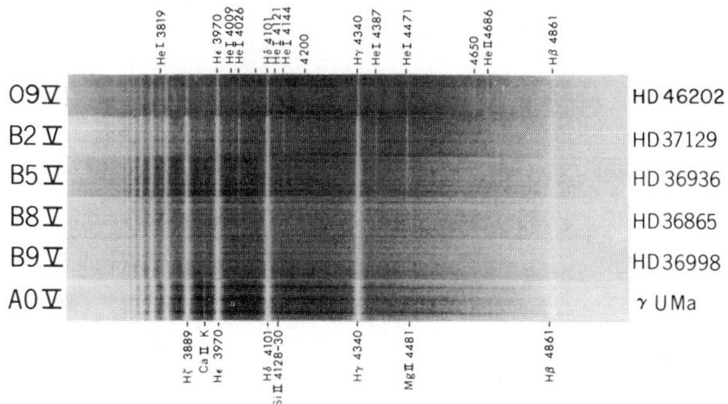

Figure 1. A series of photographic spectra of early-type, main-sequence stars recorded at an original dispersion of 128 Å mm^{-1}. The lines used as classification criteria are identified above and below the spectra. (*From* An Atlas of Low-Dispersion Grating Stellar Spectra *by H. A. Abt, A. B. Meinel, W. W. Morgan, and J. W. Tapscott, Kitt Peak National Observatory, Tucson, 1968. Reprinted by permission of the authors.*)

define a local third dimension in the classification system. Only after the classification is complete are the program stars assigned effective temperatures, absolute magnitudes, masses, radii, and so forth, which may be based on a prior calibration of the cells in the classification array. Figure 1 shows a typical set of photographic spectra taken at classification dispersion. The wavelength range covered is approximately 3800–4900 Å.

RESULTS

Over 100,000 stars have been classified on the MK system, which is the only purely *spectral* classification system in wide use today (for photometric classification systems, see the entry on Magnitude Scales and Photometric Systems). In addition, the entire *Henry Draper Catalogue* is currently being reclassified on the MK system. The classification system has been calibrated in terms of effective temperature and luminosity, from which stellar distances, masses, radii, and ages can be derived with the adoption of suitable models. Thus the global astrophysical parameters of a large number of stars are known through their spectral classification, always assuming the accuracy of the calibration and the models.

One of the first applications of the MK system was the study of the spatial distribution of the O and B stars near the Sun, leading to the discovery of at least two galactic spiral arms in the solar neighborhood. This was the first definite evidence of the spiral structure of the Galaxy. Combined with radial velocities and proper motions, distances derived from spectral types (spectroscopic parallaxes) have also been used to study the pattern of space motions of the early-type stars around the Sun, a pattern produced by a combination of differential rotation around the galactic center and local expansion.

Many types of spectrally peculiar stars have been isolated by means of spectral classification techniques. In-depth study of these groups of objects often leads to new insights in the areas of stellar atmospheres, interiors, and evolution. Among the types of stars identified in this way are the metallic-line stars (Am, mercury-manganese, and helium-weak B stars); the peculiar or "magnetic" A and B stars (having enhanced absorption lines due to silicon or rare-earth elements, and later found to exhibit strong magnetic fields); and "shell" stars, whose extended atmospheres produce absorption lines with very narrow profiles. Other groups of objects that do not fit conveniently into a two-dimensional system include emission-line stars, Wolf–Rayet stars, subdwarfs, and white dwarfs. These objects have themselves become the subject of extensive spectral classification work.

CURRENT PROBLEMS IN SPECTRAL CLASSIFICATION

As better and better photographic plate material has become available, the MK system has frequently been refined, without being basically altered. Thus new "fractional" spectral types (like B1.5) have been introduced, and luminosity classes have sometimes been split in two (e.g., B2 IV–V; G8 IIIa, IIIab, and IIIb; A0 Va and A0 Vb). But the question often arises as to whether two dimensions are sufficient to classify the spectra of all the normal stars. Because the location of a star in the *theoretical* Hertzsprung–Russell diagram is uniquely determined by mass, age, and chemical composition, there is some reason to expect that a third parameter (besides temperature and luminosity) may be involved in the empirical scheme as well. A local third dimension has been added to the MK system in some restricted areas of the HR diagram; most often, the third parameter is related to chemical composition. For stars of spectral type F8 to M, Keenan introduced a set of composition indices that describe the strengths of certain metallic absorption lines relative to those in the standard stars, which are assumed to be of solar composition. On the other hand, when classifying stars drawn from an entirely different population from that of the solar neighborhood—for example, stars in the galactic halo or in other galaxies—a new set of standards needs to be defined, and there may not be a one-to-one correspondence with the usual MK spectral types.

New observing techniques present new problems and new opportunities for the spectral classifier. Many observers now routinely use electronic detectors instead of photographic plates, even in the visual-wavelength region, because electronic detectors have greater sensitivity and linearity of response. Although it is usually possible to use the digital output of these detectors to simulate a photographic spectrum, such a procedure is cumbersome and generally unnecessary. Spectral classification can be performed directly on plots of normalized intensity versus wavelength, provided that the usual precautions are taken: The standard spectra must be obtained and processed in the same way as the program stars, and all spectra should have the same signal-to-noise ratio. The same spectral lines that have been found to be temperature or luminosity sensitive on photographic plates can be used as classification criteria on tracings as well, but the classifier must become accustomed to their appearance in the new medium. In particular, the eye tends to respond to a line's equivalent width on the photographic plate, whereas on a tracing the central depth is the most striking feature.

The development of efficient detectors for near- and far-infrared radiation, and of satellite-borne instruments, has made it possible to consider the use of spectral classification techniques in wavelength regions outside the original visual range, thus observing the stars at the wavelengths where they emit most of their energy. The vacuum ultraviolet, accessible only by spacecraft, may prove to be the most appropriate region for the classification of hot, blue stars, whereas the infrared may be useful not only for late-type stars but also for distant, reddened O and B stars.

In principle, a set of standard stars can be defined for any desired two- or n-dimensional matrix of spectral types. If the MK standards are selected, the resulting system will approximate the MK system. If not, an entirely new system may be created. In the ultraviolet, the strongest lines in the spectra of early-type stars tend to be those of highly ionized carbon, silicon, and nitrogen, which are mostly formed in the stellar wind and which therefore are likely to be governed by parameters other than effective temperature and surface gravity. However, there are enough photospheric lines even in this region to perform a normal, two-dimensional classification. Figure 2 shows a set of ultraviolet spectra observed by the *International Ultraviolet Explorer* satellite, and displayed in a form resembling photographic data; although these spectra could be used for classification, it is more practical to display the data in the form of tracings. Similarly, in the infrared, the presence of extended dust

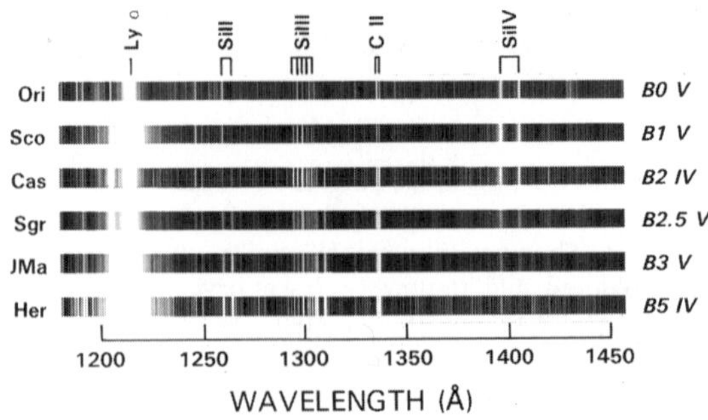

Figure 2. A sequence of B-star spectra taken by the *International Ultraviolet Explorer* satellite in the high-dispersion mode. The spectra have been resampled to a resolution of 0.25 Å and artificially widened by pixel replication. The most prominent spectral features in the 1200–1450 Å region are marked, but these are not necessarily the criteria used for spectral classification. (*From Criteria for the Spectral Classification of B Stars in the Ultraviolet, by J. Rountree and G. Sonneborn,* Astrophysical Journal, *March 1991. Reprinted by permission of the authors.*)

shells around some stars requires care in the choice of classification criteria. But the systematic study of stellar spectra at these wavelengths, which is still in its infancy, offers the possibility of gaining information about new portions of the atmospheres of normal stars.

Additional Reading

Garrison, R. F., ed. (1974). *The MK Process and Stellar Classification*. David Dunlap Observatory, Toronto.

Jaschek, C. and Jaschek, M. (1987). *The Classification of Stars*. Cambridge University Press, Cambridge.

Kaler, J. B. (1989). *Stars and Their Spectra: An Introduction to the Spectral Sequence*. Cambridge University Press, Cambridge.

McCarthy, M. F., Philip, A. G. D., and Coyne, G. V., eds. (1979). *Spectral Classification of the Future*. Vatican Observatory, Vatican City.

Morgan, W. W. and Keenan, P. C. (1973). Spectral classification. *Ann. Rev. Astron. Ap.* **11** 29.

See also **Magnitude Scales and Photometric Systems; Stars, Hertzsprung–Russell Diagram.**

Stars, Symbiotic

Scott J. Kenyon

In 1932, Paul W. Merrill and Milton L. Humason published a short note describing three *stars with combination spectra*. They noted that CI Cygni, RW Hydrae, and AX Persei displayed strong TiO absorption bands and a very intense He II 4686 Å emission line. Subsequent studies of photometric data from the Harvard College Observatory plate collection showed that CI Cyg and AX Per had undergone a 2–3 mag nova-like eruption prior to the observations of Merrill and Humason, and also revealed variations in their quiescent light levels with amplitudes of 0.5–1.0 mag and periods of several years. Roughly two dozen similar objects had been identified by the 1950s, including Z Andromedae (the original prototype), R Aquarii (an unusual Mira variable surrounded by a planetary nebula), and T Coronae Borealis (a recurrent nova). The presence of a stellar photosphere with a temperature of ~ 3000 K and a highly ionized region with a temperature of at least 100,000 K in an apparently single object was unique at that time, and Merrill suggested the term *symbiotic stars* to describe an object containing two such seemingly hostile components.

There are now about 150 known symbiotic stars in our own galaxy, and a handful have been discovered in the Magellanic Clouds. Symbiotic stars are identified by spectra covering the wavelength interval 0.1–3 μm; all of them display

1. a red continuum and absorption features of a late-type giant star (spectral type K or M), which includes Ca I, Fe I, H_2O, CO, and TiO, among others;
2. a blue continuum with bright H I and He I emission lines and either
   (a) additional emission lines from ions such as He II, [O III], [Ne V], and [Fe VII] with an equivalent width exceeding 1 Å (typical of quiescent objects and some eruptive systems),
   or
   (b) an A- or F-type continuum with additional absorption lines from H I, He I, and singly-ionized metals (typical of systems in outburst).

SYMBIOTIC STARS AS INTERACTING BINARY SYSTEMS

A binary model is an obvious explanation for a symbiotic star, because the two temperature extremes can be associated with separate stellar objects. The presence of an evolved red giant star has been demonstrated in nearly all symbiotics with near-infrared photometry and spectroscopy. Ultraviolet spectra acquired with the *International Ultraviolet Explorer* satellite have proved the existence of a hot, compact companion star and also have shown that this object ionizes a surrounding gaseous nebula. Orbital periods for symbiotic binaries are typically ~ 1–3 yr, and spectroscopic radial velocity curves suggest masses of 1–3 $M_\odot$ (solar masses) for the cool giant and 0.5–1.0 $M_\odot$ for the hot companion.

Even though radiation from the hot component is responsible for producing the spectroscopic characteristics of a symbiotic binary, recent analyses suggest that the hot object must accrete material lost by the red giant to maintain its luminosity. Thus, the rate at which the giant loses material determines whether or not a given long-period binary system can produce a symbiotic optical spectrum. This conclusion has led to the division of symbiotic binaries into semidetached systems, which transfer mass via tidal interaction, and detached systems, which transfer mass through a stellar wind. Detached symbiotics can be divided further into systems with pulsating red giants (Mira variables) that lose mass rapidly and nonpulsating red giants that lose mass more slowly. These three broad classes of symbiotic binaries are summarized in Table 1. (Infrared types S and D are defined near the end of this entry.)

Semidetached Symbiotics

The symbiotic stars that are most closely related to other types of interacting binaries contain lobe-filling red giants and a main-sequence-star companion. The prototypical example of a semidetached symbiotic binary is CI Cyg (orbital period = 855.25 d); a schematic representation of its stellar components is shown in the upper right panel of Fig. 1. The red giant radius has been estimated from its periodic eclipses of the hot companion to be roughly 200 $R_\odot$ (solar radii), which is approximately the radius of its inner Lagrangian surface. Observations obtained during the 1975 eclipse suggest that the dimensions of the hot component are 200×20–40 $R_\odot$, where the first dimension refers to the radius in the plane of the orbit and the second refers to the radius perpendicular to the

Table 1. Types of Symbiotic Binary Stars

Primary	*Secondary*	*Orbital Period*	*Infrared Type*	*Outbursts*
Red giant	Main-sequence dwarf star	1–5 yr	S	Nova-like
Red giant	White dwarf	1–5 yr	S	Nova-like
Mira variable	White dwarf	Decades	D	Nova-like

orbital plane. It is unlikely that such an object can be a star, but the geometry is consistent with that of an accretion disk. The luminosity and effective temperature of the disk suggests that the central object is a main-sequence star similar to our Sun.

An illustrative spectral energy distribution for CI Cyg is shown in the upper left panel of Fig. 1. During quiescent periods, the red giant component produces most of the energy received at wavelengths exceeding 0.5 μm (solid line), and the accretion disk is responsible for the continuum at shorter wavelengths and the intense emission lines (the emission lines have been omitted from the energy distribution for clarity). Most emission lines are produced in a region confined to the Roche volume of the hot star,

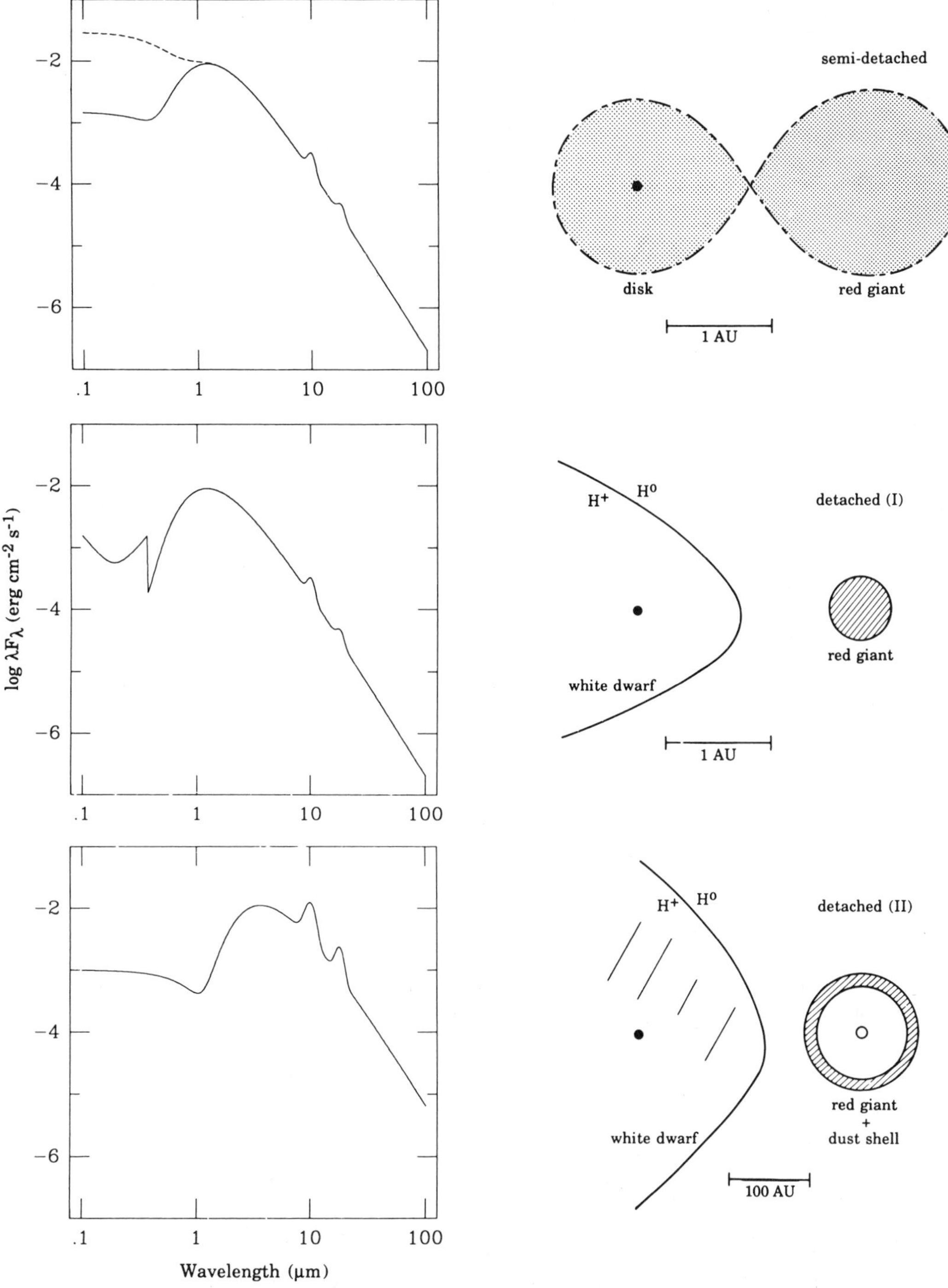

Figure 1. Representative spectral energy distributions and orbital configurations for various types of symbiotic binaries (see Table 1). The left-hand panels illustrate the energy emitted as a function of wavelength for the three classes of symbiotics (top to bottom, semi-detached, detached I, and detached II) during quiescent periods. The dashed portion of the top-left panel indicates the change in the energy distribution that occurs in an eruption. The right-hand panels illustrate the sizes of the stellar components relative to that of the orbit. A length scale in AU (astronomical units) is included with each panel. Regions of ionized hydrogen (H^+) and neutral hydrogen (H^0) are marked accordingly.

although some high-ionization lines appear to be formed in a bipolar structure at larger distances from the disk.

The accretion disk brightens dramatically during an eruption, as the mass-flow rate through the disk increases from a few $\times 10^{-5}$ $M_{\odot}$ yr^{-1} to 10^{-3} $M_{\odot}$ yr^{-1}. Aside from an obvious change in the energy distribution (indicated by the dashed line in Fig. 1), the optical spectrum evolves into a configuration resembling an A or F supergiant star. The high-ionization emission lines disappear as the system brightens, but the H I and possibly the He I features increase in intensity and develop strong central absorption cores. This high state is maintained for several months, and the system declines back to its quiescent level in 1–2 yr.

Theories for CI Cyg and related systems account for eruptions by changes in the structure of the red giant or of the disk itself. One type of model involves an increase in the mass-loss rate from the red giant star into the disk and has become known as the mass transfer instability mechanism. Hydrodynamic calculations have indicated that lobe-filling red giants are unstable to episodes of enhanced mass loss, and the duration and recurrence time scale of these events are in reasonable agreement with the observations. Another type of outburst mechanism is a disk instability. If the mass flow rate *through* the disk is smaller than the transfer rate *into* the disk, then the disk will increase in mass until it reaches a critical limit. The mass flow rate through the disk can increase dramatically at that point and give rise to an optical outburst. The evolution of the light output of the disk is similar for the two cases, and both models can reproduce observed light curves. However, there has been no attempt to model the spectroscopic development of CI Cyg and related systems through an eruption cycle.

Detached Symbiotics I

The most abundant type of symbiotic binary star contains a fairly normal red giant star that underfills its inner Lagrangian surface and loses material via a stellar wind rather than tidal overflow. Good examples of this class are RW Hya (a quiescent system; orbital period ~ 370 d) and AG Peg (an erupting system; orbital period ~ 830 d); a simple picture for these objects is shown in the middle-right panel of Fig. 1. Essentially all hot components in detached symbiotics are white dwarfs, having temperatures ranging from about 30,000–200,000 K and radii of 0.01–1 $R_{\odot}$. One system, V2116 Ophiuchi, contains a neutron star. Each hot component is surrounded by an ionized nebula which has a radius of 1–200 AU.

An example of an energy distribution for a quiescent detached symbiotic binary is presented in the middle left panel of Fig. 1. As in the semidetached systems, the red giant dominates at wavelengths greater than 0.5 μm. The Rayleigh–Jeans tail of the hot component and Balmer continuum radiation from the ionized nebula are visible at shorter wavelengths.

The eruptions of several detached symbiotic binaries closely resemble those of classical novae, and they have been called *symbiotic novae* to distinguish them from novae which occur in cataclysmic variable systems. Roughly half of the known symbiotic novae resemble A or F supergiant stars at visual maximum (as do classical novae), and radiation from the red giant is overwhelmed by its companion near maximum light. The remaining objects have spectra similar to those of planetary nebulae, and the spectrum of the giant usually is still visible near maximum. The bolometric luminosity of a symbiotic nova appears to be independent of its spectrum at visual maximum, so the temperature of the hot component is the distinguishing characteristic of the two classes.

It is currently believed that symbiotic nova eruptions are the results of thermonuclear runaways (hydrogen shell flashes) similar to those in classical novae. Hydrogen from the red giant accumulates in a thin shell on the surface of the white dwarf component until the pressure at the base of this envelope is sufficient to ignite the accreted material. If material is accreted at a fairly slow rate (less than 10^{-9} $M_{\odot}$ yr^{-1}), then the white dwarf atmosphere is degenerate at the onset of runaway and produces a cool photosphere at visual maximum. More rapid accretion results in a nondegenerate white dwarf atmosphere when the runaway begins and produces a hot photosphere at visual maximum. Thus, symbiotic novae with A or F supergiant spectra probably are the result of degenerate flashes, whereas nondegenerate flashes are responsible for other symbiotic novae.

The duration of a typical symbiotic nova eruption is far longer than the eruptions in the semidetached systems described above. It seems that 50–200 yr is a reasonable time scale for a symbiotic nova to remain in a high-luminosity state, and most of this time is spent at high effective temperatures ($\gtrsim 10^5$ K). Such long eruptions are expected in the thermonuclear model, because the time required to burn the accreted material is 100–400 yr for a typical 0.5–1.0 $M_{\odot}$ white dwarf.

Very rapid (duration of several months) thermonuclear eruptions are possible if the white dwarf mass is close to the Chandrasekhar limit (1.4 $M_{\odot}$). Theoretical calculations show that the evolution of runaways on very massive white dwarfs have some features in common with the observed outbursts of recurrent novae, such as RS Ophiuchi. Accretion-powered models can also explain some phenomena observed in recurrent novae, and additional data are required to verify if a particular type of outburst can be associated with a given recurrent nova.

Detached Symbiotics II

Roughly 20% of all known symbiotic binaries contain Mira variables instead of nonpulsating red giant stars. Good examples of this type of system are V1016 Cyg and H1-36.

There are several major differences between these symbiotics and those with nonpulsating red giant stars, as illustrated in the lower-right panel of Fig. 1. Mira symbiotics are the most prodigious radio emitters among this class of binary, and the radio flux requires a mass-loss rate of 10^{-6}–10^{-5} $M_{\odot}$ yr^{-1} for the Mira. These mass-loss rates are 1–2 orders of magnitude larger than those estimated for symbiotics containing nonpulsating red giants. The radio spectrum is sensitive to the fraction of the Mira wind that is ionized by the hot component and to the binary separation, and results obtained for the most luminous systems suggest that the wind is not completely ionized and that the orbital separations are several hundred astronomical units. Thus, Mira symbiotics have the longest orbital periods of any interacting binary.

The energy distributions for the Mira symbiotics peak at a wavelength of roughly 3 μm instead of 1 μm. It was originally thought that this unusual energy distribution was the result of *emission* by dust grains heated to temperatures of 1000–2000 K by the hot component, and these systems became known as D-type (dusty) symbiotics to distinguish them from systems with S-type (stellar) near-infrared energy distributions. It now appears that the amount of dust *absorption* along the line of sight to the Mira component in some D-type symbiotics is a factor of 5–20 *larger* than the extinction to the hot component and its ionized nebula. This extinction is local to the system and probably is caused by an optically thick dust shell that surrounds the Mira. The dust absorbs near-infrared radiation and reemits this energy at far-infrared wavelengths, and models for red giants enveloped in thick dust shells agree very well with infrared photometry obtained from the ground and from the *Infrared Astronomical Satellite*. The mass-loss rates required to produce the optically thick dust shells are consistent with those estimated from the radio data; this provides further support for this model.

Roughly half of the known symbiotic novae have erupted in Mira symbiotics, and their behavior is essentially identical to other symbiotic novae. The properties of these eruptive systems 10–20 yr after optical maximum are similar to those of other Mira symbiotics, which have not been observed to undergo an outburst in the past 100 yr. Thus, it is plausible that many of the known

Mira symbiotics are still declining from eruptions that began several hundred years ago.

Additional Reading

Allen, D. A. (1984). A catalog of symbiotic stars. *Proc. Astron. Soc. Aust.* **5** 369.

Friedjung, M. and Viotti, R., eds. (1982). *The Nature of Symbiotic Stars*. D. Reidel, Dordrecht.

Kafatos, M. and Michalitsianos, A. G. (1984). Symbiotic Stars. *Scientific American* **251** (No. 1) 84.

Kenyon, S. J. (1986). *The Symbiotic Stars*. Cambridge University Press, Cambridge.

Mikołajewska, J., Friedjung, M., Kenyon, S. J., and Viotti, R., eds. (1988). *The Symbiotic Phenomenon*. Kluwer Academic Publ., Dordrecht.

Seaquist, E. R. and Taylor, A. R. (1990). The collective radio properties of symbiotic stars. *Astrophys. J.* **349** 313.

See also **Binary Stars, Cataclysmic; Stars, Radio Emission.**

Stars, T Tauri

Gibor Basri

In the middle of the century, spectroscopic surveys turned up a new class of star, characterized by a late-type spectrum and strong Hα emission: the T Tauri stars. The stars are cool and luminous, and are associated with dark clouds and reflection nebulae. The T Tauri stars are solar-type stars in their pre–main-sequence phase. The main distinguishing characteristics of the classical T Tauri stars are excess continuum emission from the ultraviolet to the far infrared and strong line emission in selected lines. These characteristics are thought to be caused by both accretion and mass loss in the systems. Those processes in turn are currently both thought to be due to the presence of an accretion disk around the star, the final phase of star formation. Some young stars appear not to have these disks, whereas others apparently keep them for as long as a few million years. The disk accretion is accompanied by strong mass loss. During the earlier phases, this can give rise to spectacular optical jets and associated Herbig–Haro emission objects emerging perpendicular to the disk. The stars themselves are both more rapidly rotating and more convective than their main-sequence counterparts, and therefore are much more magnetically active.

The evidence that T Tauri stars are pre–main-sequence stars of typically a solar mass or less is compelling. It naturally explains their association with star-forming interstellar molecular clouds, with which they are found to share kinematics. Observations of strong lithium resonance line absorption supports their youthful status, since surface lithium is depleted on the main sequence in convective stars. One finds T Tauri stars where pre–main-sequence low-mass stars should be, on almost vertical (Hayashi) tracks on the right side of the Hertzsprung–Russell (HR) diagram. In this stage they are fully convective and shrink down while shining through deuterium burning and gravitational contraction, before becoming radiative and moving towards their higher final main-sequence temperature and smaller size.

The definition of the class began with the Hα emission, but there are a number of other spectral anomalies associated with T Tauri stars. They tend to exhibit irregular (and sometimes regular) photometric variability, which is why most of them have variable star names like the prototype itself: T Tauri. The amplitude of the variability can be from a few tenths of a magnitude to several magnitudes. In addition to Hα, the higher Balmer lines of hydrogen are often also seen in emission along with the Ca II H and K lines. In less than half of the cases there is also forbidden line emission of [O I] and [S II]. In some cases many other lines of Fe, Na, and Ca are also visible. Excess continuum emission is also quite common, particularly in the near infrared and ultraviolet. The infrared excess has a slope that is usually shallower than that of a Rayleigh–Jeans curve, likely indicating a range of emitting temperatures. The ultraviolet excess often carries into the optical range, where it "veils" the optical absorption lines, sometimes so heavily that they are only visible in high-resolution spectra. X-ray emission is observed from some T Tauri stars; it is quite likely to be found in stars with little Hα emission. The stars are typically rapidly rotating compared to their main-sequence counterparts (5–30 km s^{-1} or higher), but quite slow compared to their breakup velocities.

The original class of T Tauri stars (TTS) were identified in objective prism Hα emission surveys. These "classical" stars (CTTS) have now been supplemented by several other classes of young stars, which are often also referred to as T Tauri stars. Stars that have Hα lines too weak to have been found in the original surveys are called "weak" or "naked" T Tauri stars (WTTS or NTTS). Those with exceptionally strong veiling are often referred to as "continuum" stars. TTS that show a sudden and persistent increase of several magnitudes (or are thought to have recently done this by current evidence) are called FU Ori objects. This brightening is often accompanied by a change in apparent spectral type. TTS that show redshifted absorption features in their Balmer lines are YY Ori objects. Other related types of objects include certain infrared sources, Ae and Be stars, and other phenomena (such as outflows) associated with young stars. Today the designation "T Tauri star" can mean any of the manifestations of optically visible young solar-type stars; the differences between them will be elucidated below.

Explanations for this remarkable variety of interesting phenomena associated with young stars fall in three basic categories: magnetic (chromospheric) activity, accretion phenomena, and outflow phenomena. From the outset, the superficial similarity of the emission spectra of T Tauri stars to that of the solar chromosphere has been noted. Many of the same lines appear in the strong emission cases, namely, the strongest absorption lines in normal stars. The resemblance is only superficial; the lines do not appear in the same relative ratios, and the line profiles are typically much broader and more asymmetric than solar lines. If the stellar chromosphere occurs very deep in the atmosphere (relative to where the solar temperature rise occurs) the fluxes of some emission lines and even the continuum jumps and veiling might be reproduced. The real difficulty with the chromospheric model comes when one evaluates the total amount of excess energy present in the spectrum of an active TTS. The true stellar photospheric contribution can be estimated by studying the absorption line spectrum (and the extent of veiling thereof); it then turns out that a major fraction of the total stellar luminosity appears to be nonphotospheric. Sometimes the excess energy is even up to several times as much as the photospheric luminosity. It greatly strains credulity to imagine all this energy bypassing the photosphere to emerge in chromospheric layers.

For this reason, and because one might expect accretion to be present in very young stars, it has often been suggested that accretion might be the source of the excess energy. The problem with spherical accretion is that the optical extinction to the T Tauri systems, while substantial, is incompatible with the amount of dust needed to explain the infrared excess. As for mass loss, there are a number of strong indications of it. It can be seen directly in the bipolar radio outflows, and in the motions of the Herbig–Haro objects and optical jets seen emanating from a number of younger objects. Even in the typical T Tauri star, the Hα line (and sometimes several others) often has a blueshifted absorption component superposed on the emission. Similarly, the forbidden lines are almost completely blueward of line center. The mass-loss rates are quite uncertain, but for typical cases are thought to be in the range of 10^{-9}–10^{-7} $M_\odot$ yr^{-1}, where $M_\odot$ is the solar mass.

It now appears that there is some truth to each of these basic paradigms. The presence of strong magnetic activity is established through a number of different diagnostics. The presence of periodic variability is thought to be due to rotationally modulated dark (magnetic) spots (starspots) on some stars. The traditional chromospheric diagnostic—Ca II H and K emission—is quite strong in TTS, and in WTTS the Ca II H and K lines appear very much as they do on magnetically active main-sequence stars. The amount and characteristics of the x-ray emission in WTTS are also entirely consistent with an origin in a strong magnetic corona. This conclusion is reinforced by evidence that the x-ray luminosity is a function of the stellar rotation, accepted as a strong indicator of magnetic dynamo activity. Finally, in the WTTS it has been shown that the photospheric lines are preferentially filled in as though by emission from an atmospheric structure similar to (but more energetic than) the solar chromosphere.

In recent years a new conceptual framework for understanding the primary T Tauri phenomena in a unified way has rapidly gained support. This is the disk paradigm—that CTTS are surrounded by active accretion disks that produce most of the excess emission that distinguishes these systems. The disk can be responsible for the infrared excess and spectral shape through reprocessing of stellar light plus intrinsic accretion luminosity emitted from an optically thick dust and gas disk. The power comes from the dissipation of gravitational potential energy as material moves through the disk towards the star. The precise mechanism through which angular momentum is transferred outward in the disk (allowing accretion) is not known (Fig 1).

The spectral shape of infrared emission from for the disk arises because the disk will be hotter nearer the star, due both to the increased reprocessing of stellar radiation and the increasing gravity of the star. Indeed, the spectral shape expected is almost the same from these two sources, making it difficult to determine whether a disk is passive (pure reprocessing) or active (accreting) from the infrared spectrum alone. Only when the infrared luminosity is more than 50% of the stellar luminosity can one be certain of accretion, but in either case the presence of a disk is indicated. The spectral shape of some CTTS match simple disk models quite well, whereas others are flatter, or even rise towards the far infrared. The disks are, of course, potential sites for planetary formation, and might be modern examples (amenable to study) of the "solar nebula" which preceded our own planetary system.

As the disk material approaches the stellar surface, it must slow from typical Keplerian velocities of 200–300 km s^{-1} to a typical surface rotation velocity of 10–30 km s^{-1}. The amount of kinetic energy dissipated is equal to the gravitational potential energy dissipated in reaching the star, but now this energy must be dissipated in a small region at the star's equator instead of over a disk many astronomical units in extent. This means that the temperature of the radiating region will be far higher. Whereas temperatures in the disk range from 100 K far from the star to up to 2500 K near the star, the so-called "boundary layer" will have a temperature of 7000–11,000 K and radiate in the ultraviolet and visible part of the spectrum. This could be the source of the ultraviolet excess and optical veiling observed in many CTTS. The fact that accretion only occurs near the equator of the star avoids the problems that a strong accretion shock over the entire star might pose.

Unified steady-state disk models have now been constructed which self-consistently and simultaneously satisfy all the constraints mentioned above for individual CTTS. These involve a central star from 0.5–1.3 $M_\odot$, with a radius from 1.5–3.5 $R_\odot$ (solar radii) and a luminosity that is several times the solar luminosity. The disk is very large (tens of astronomical units), containing perhaps 0.01–1.0 $M_\odot$ of dust and gas. It is usually optically thick in dust until the dust is destroyed in the inner disk by rising temperatures, and then optically thick in gas until the boundary layer. The accretion rates seem to lie between 5×10^{-9} and 5×10^{-7} $M_\odot$ yr^{-1} for the CTTS. The overall continuum shape and observed luminosity can be accounted for by these models, as can the Balmer jump and some emission line fluxes. The currently available models are both very simplified and parametrized, and need determinations of the stellar mass and radius, the inclination of and extinction to the system, and the accretion rate and boundary-layer size in the disk. The accretion viscosity, boundary-layer geometry, and vertical structure of the disk are currently very poorly understood. In light of this disk model, previous determina-

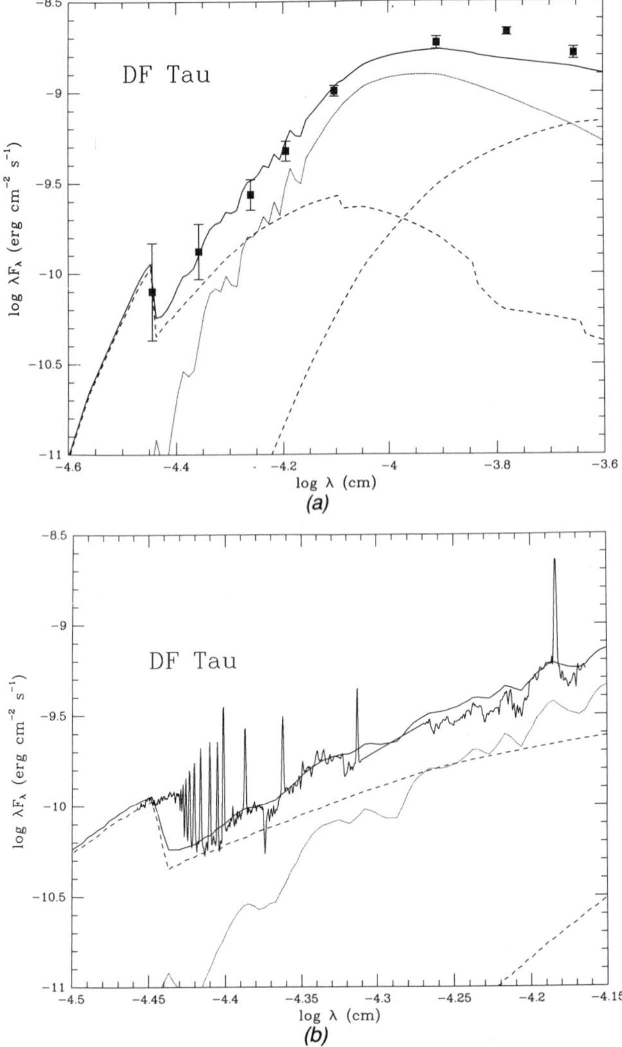

Figure 1. Confrontation of a disk model for a T Tauri star with observations. Shown is the spectrum of DF Tau, an active but not extreme example of a T Tauri star. Upper panel: the continuum distribution from the ultraviolet through the near infrared. The squares are photometric measurements and the error bars indicate the range of variability. The solid line is the model fit to the measurements. It is composed of a stellar photosphere (the thin dotted line) and a disk model (dashed lines); the dashes on the right represent emission from the disk itself and the dashes extending all the way to the left represent radiation from an optically thin boundary layer. Lower panel: detail in the optical region, showing also a spectrum at moderate resolution. Note the Balmer line emission and the Balmer continuum jump (due to the boundary layer). Because the photospheric light is substantially diluted by boundary-layer emission, the photospheric lines are veiled.

tions of the position of CTTS in the HR diagram (dependent on extinctions and stellar luminosities) must be regarded as uncertain by a factor of 2 or more. The presence of substantial accretion over the life of the disk may significantly modify the evolution of the star along pre–main-sequence tracks.

The question of how common the disks are hinges on the WTTS. These stars are well intermingled with the CTTS both spatially and temporally (meaning on the HR diagram). They share similar absorption spectra (unveiled) and narrow emission lines, so the stellar atmosphere itself is probably quite similar in the two classes. The main difference between them is the lack of any disk or strong outflow symptoms in the WTTS. It is not known whether the WTTS are CTTS that have lost their disks, or whether they are young stars that never have disks when they become optically visible. There is some evidence from far infrared observations for residual disks around some WTTS, but because they are not accreting onto the star they do not give rise to the usual T Tauri phenomena. Because searches for the WTTS are currently incomplete compared to searches for the CTTS, it is difficult to say what fraction of young stars are WTTS. The observed fraction is about one-half in the Taurus cloud and less in the ρ Oph cloud (the two nearest clouds). Of course, every CTTS eventually loses its disk, since main-sequence stars are never observed to have active disks.

Other evidence for disks has also been found. The appearance of the forbidden line emission is suggestive, because it is primarily blueshifted. The forbidden lines are thought to arise in a region many astronomical units in radius, which is expanding; one ought to be able to see the receding side of the flow as well as the approaching side, because occultation by the star is negligible. The lack of the red emission is explained if a large occulting screen (the disk) blocks our view of the far side. This also provides evidence of the size of the disk. The FU Ori stars provide kinematic evidence for disks, because the velocity broadening of their infrared lines is lower than for optical lines. This is expected because the optical lines are formed at higher temperatures near the star, where the orbital velocities are also higher. There is even a hint of the double-peaked nature expected for lines arising in a disk in some observations. The large increase in luminosity in these systems can be explained if the accretion rate rises suddenly by 1 or 2 orders of magnitude, and the change in spectral type comes from the fact that the stellar absorption lines are now swamped by the disk absorption spectrum. Finally, we are beginning to be able to image the disks directly. Although of subarcsecond size even for the nearest TTS, asymmetric infrared images have been seen by speckle techniques and the *Hubble Space Telescope* may be able to barely resolve some disks. Submillimeter interferometry is another promising technique.

The question of what produces the strong outflows also seen from the CTTS is one of the most pressing current topics. The fact that the WTTS do not exhibit this phenomenon, and are also missing the other signs of an accretion disk, strongly suggests that the disk itself is intimately connected with the mass-loss mechanism. Probably the only other safe conjecture at this point is that magnetic fields are very likely involved. The optical emission lines should in principle be telling us about the region where the mass loss originates. Hα in particular is well observed; yet it remains one of the least-understood aspects of the T Tauri phenomenon.

The broad lines come in a variety of shapes. The most common Hα lines have emission with fairly symmetric far wings extending to 200–400 km s^{-1}. There is often a blueshifted absorption feature near 100 km s^{-1} that usually does not go below the continuum, leaving twin emission peaks with the red peak usually brighter than the blue one. Less common is a more or less flat-topped emission feature with central absorption that can be unshifted or shifted to either side. Finally, there are fairly symmetric, triangular emission lines, often with little or no absorption. These shapes can be seen in the other Balmer lines (although for these the absorption often goes below the continuum) and in other strong emission lines. The weaker of these other Balmer lines seem to occur preferentially

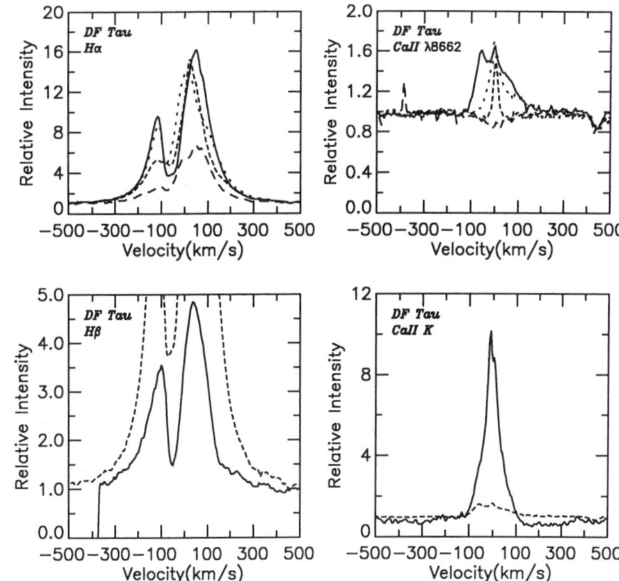

Figure 2. Emission line profiles from DF Tau. In the upper panels, profiles of Hα and a Ca II infrared triplet line are shown from several epochs. The solid lines in all panels are from October 1986. The dashed lines in the upper panels correspond to observations in December 1986, the long dashes are data from November 1987, and the short dashes are from November 1988. Note how much weaker the lines were in 1987. The triplet line shows two examples of broad emission, one of narrow emission, and one without emission at all. In the lower panels, Hβ and the Ca II K line are shown, with the Hα and the triplet lines, respectively, from the same time shown with dashes (for a scaling comparison). The velocity scale is zeroed at the photospheric velocity, and the continuum is normalized to unity.

with the triangular shape with little absorption. In extreme cases, the same line in the same star has shown all these shapes, which may mean they are different manifestations of a common underlying structure (Fig. 2).

The lines can vary in intensity and shape on time scales down to an hour, indicating that they arise (at least in part) in small regions quite near the star. The boundary layer is an obvious candidate for the base of the wind—it is small, fast moving, turbulent, energetic, and very likely entangled in the stellar magnetic field. The symmetry of the lines argues for a substantial orbital or turbulent component in the velocity broadening. Spherical wind models do not naturally lead to the typical observed profiles, but can be made to satisfy the observations to a certain extent. Exploration of conical, bipolar, ring-like, or disk-like geometries has just begun, but these look promising. Some lines in some stars are quite stable in appearance, whereas others undergo large intensity changes with relatively stable profiles, and others change their shape dramatically. The absorption components seem to be more stable, and probably arise in a larger region further from the star. The relation between the optical emission lines and other phenomena such as optical jets has not yet been established. There are many exceptions to almost anything one can say about the emission lines. It has recently become possible to monitor the lines with high resolution and signal-to-noise; hopefully this will help clarify our understanding of the emission lines of T Tauri stars in the next few years.

Additional Reading

Appenzeller, I. and Mundt, R. (1989). T Tauri stars. *Astron. Ap. Rev.* **1** 291.

Bertout, C. (1989). T Tauri stars: Wild as dust. *Ann. Rev. Astron. Ap.* **27** 351.
Cohen, M. (1988). *In Darkness Born*. Cambridge University Press, New York.
Dupree, A. K. and Lago, M. T. V. T., eds. (1988). *Formation and Evolution of Low Mass Stars*. Kluwer Academic Publ., Dordrecht.
See also **Herbig–Haro Objects and Their Exciting Stars; Planetary Systems, Formation, Observational Evidence; Pre–Main Sequence Objects: Ae Stars and Related Objects; Protostars; Protostars, Theory; Stars, Pre–Main Sequence, Winds and Outflow Phenomena; Stars, Pre–Main Sequence, X-Ray Emission; Stars, Young, Continuum Radio Observations; Stars, Young, Jets; Stars, Young, Millimeter and Submillimeter Observations; Stars, Young, Rotation; Stellar Associations, R- and T-Type.**

Stars, Temperatures and Energy Distributions

Roger A. Bell

STELLAR ENERGY DISTRIBUTIONS

The absolute fluxes and energy distributions of stars are fundamental data for the determination of stellar effective temperatures and bolometric corrections, and, in some cases, can be used to give stellar gravities and chemical abundances. These data are required over a wide range of wavelengths and the discussion of the methods used to obtain them and the results that are available can be conveniently separated into individual discussions of the ultraviolet ($\lambda < 3000$ Å), the visible ($3000 < \lambda < 12{,}000$ Å), and the infrared ($\lambda > 12{,}000$ Å). Detailed reviews of the problems of calibration have been given by Donald S. Hayes and by Arthur D. Code.

Ultraviolet

Measurements of absolute stellar fluxes in the ultraviolet have been made by five satellites—*OAO-2*, *TD1*, *ANS*, *Copernicus* and *IUE*. In addition, other data are available from *Apollo 17*, sounding rocket experiments, and Voyager.

Code described the problems of calibration in the ultraviolet. Tungsten lamps, for example, are difficult to use because of their faintness in the ultraviolet as compared to the visible and the consequent difficulties of controlling scattered light. The University of Wisconsin group used radiation from synchrotron storage rings to calibrate *OAO-2*.

In order to determine the calibration of *IUE*, Ralph Bohlin and co-workers adopted an absolute flux calibration for the star η UMa. This flux scale is based upon observations made with different satellites and is essentially the *OAO-2* calibration longward of 2000 Å and the sounding rocket scale of William H. Brune et al. at shorter wavelengths. Intercomparison of the adopted scale with that determined for other satellites indicates that the agreement is generally within 10%, with the greatest differences being less than 20%.

Absolute ultraviolet flux data for many thousands of stars are now available in the *IUE* archives, in a form that is very convenient for the determination of stellar temperatures.

Visible

The bright star Vega (HR 7001, HD 172167) has been adopted as the primary standard star for absolute fluxes in the visible. The data are presented as the apparent absolute monochromatic flux at some standard wavelength, either 5000 or 5556 Å, and the absolute energy distribution, which is the individual apparent monochromatic flux values normalized to the value of the reference wavelength. These values have been found for Vega by comparison of observations with standard lamps. The importance of a careful treatment of extinction in deriving the final results, in particular the need to differentiate between horizontal and vertical extinction, was shown by Hayes and David W. Latham.

Other stars have been carefully compared with Vega, in order that they can be used as secondary standards. These include both relatively bright stars and much fainter ones, the latter being intended to serve as flux standards for large telescopes and for the *Hubble Space Telescope*.

The use of only one star as the primary standard in spectrophotometry is rather risky, in view of the possibility of variability. This problem was reviewed by Hayes, who concluded that the evidence is not strong enough to warrant finding and observing a substitute star.

Michel Breger has published a compilation of observations of stellar fluxes by a number of observers. The work by V. Straižys and Z. Sviderskienė, which gives energy distributions as a function of spectral type, has been widely used. More recently, James E. Gunn and Linda L. Stryker have published data for 175 stars. Although originally intended as a library for the synthesis of galaxy spectra, these data are valuable for other purposes, owing to their continuous wavelength coverage over the interval 3130–10,800 Å.

Infrared

Very few observations of absolute fluxes have been made in the infrared. These observations are in the wavelength region 1–5 μm and are made at widely separated wavelengths. Russell G. Walker carried out observations of four wavelengths between 1.06 and 2.21 μm, and the Oxford University group has made observations between 1 and 5 μm. Somewhat surprisingly, the latter measurements are greater, by 11% at 5 μm and smaller amounts at shorter wavelengths, than the values predicted by the model atmospheres of Vega and the observed angular diameter. Because of the relative paucity of infrared data, some authors have calibrated their observations using the predictions of model atmospheres for either the Sun or Vega.

EFFECTIVE TEMPERATURES

The problem of determining accurate stellar effective temperatures is an important one in many areas of astronomy. These temperatures are needed for studying a wide variety of problems. One such problem is the determination of the chemical abundances of stars. Accurate temperatures are particularly needed in the studies when either molecular features or high-excitation atomic features are being used. Another similar problem is the determination of surface gravities from the comparison of the strengths of molecular features and of atomic features of the same species, for example, MgH and Mg I features.

A wide variety of methods are used to obtain effective temperatures for stars. The choice of method depends upon the temperature of the star and the data available. In a relatively few cases, sufficient data are available for the actual luminosity of the star to be found from the energy distribution measured at the Earth, when $T_{\rm eff}$ can be found by the Stefan–Boltzmann law. In the majority of cases this cannot be done, and the observational data must then be interpreted using model stellar atmospheres. The $T_{\rm eff}$ of a star is then taken to be the $T_{\rm eff}$ of the model whose predictions fit the star. Because stellar models are by no means perfect, we run the risk that the $T_{\rm eff}$ derived then becomes more of a label as to which model should be used to analyze a star, rather than a description of a fundamental property of the star itself.

The methods used when models are employed include: the infrared flux method (IRFM), which basically involves the determination of the bolometric correction for the K (or other infrared) passband; the fitting of relative absolute fluxes of stars as a function of wavelength with the corresponding model data; the comparison of stellar absolute fluxes in a particular wavelength interval with the stellar model calculations; the calibration of stellar colors in terms of T_{eff}, log g (surface gravity), and chemical composition; the comparison of computed and observed hydrogen line profiles; and the comparison of observed and computed spectral indices. In some cases, the application of these techniques required additional information about a star—that is, its chemical composition and surface gravity. This is often the case for the interpretation of colors such as $B-V$, or $b-y$ of the Strömgren system, and has caused greater interest in the use of colors where one or both passbands is in the infrared, for example, $V-R$ or $V-I$ of the Cousins system or $V-K$, originally from Johnson.

All stellar fluxes used in the determination of effective temperatures must be corrected for interstellar reddening.

For a star with an observed angular diameter ϕ and radius R at distance d from Earth, the observed absolute flux of radiation from the star measured at the Earth, l, can be converted to the stellar luminosity, L, via the equation $l = L/4\pi d^2$. Writing the luminosity $L = 4\pi R^2 \sigma T_{eff}^4$, the effective temperature can then be derived from $\sigma T_{eff}^4 = l/(\phi/2)^2$, where σ is the Stefan–Boltzmann constant.

Satellite data must be available in order to obtain the integrated flux at the Earth for stars with spectral types earlier than K. Omission of the observed flux at wavelengths less than 3000 Å causes significant errors in effective temperature even for cool stars, such as an error of about 100 K for a G0 star. The uncertainties in absolute calibration in the ultraviolet introduce errors, which are greater for the hottest stars.

Temperatures for 32 early-type stars have been found using this method by Code et al. The stars analyzed range in T_{eff} from 32,500–7460 K, with the errors ranging from 2500–140 K. Cooler stars (K0–M6) were analyzed by Stephen T. Ridgway and coworkers. The ultraviolet fluxes can be neglected for these objects and the integrated fluxes at the Earth found from broadband photometry, that is, UBVRIJKLMN. The angular diameters have been found using lunar occultations, the limb-darkening corrections coming from model atmosphere calculations. The T_{eff} values found for the sample of 31 stars range between 4950 and 2810 K, with errors between 30 and 450 K. The dominant contributor to the error in T_{eff} is the error in angular diameter.

The remaining methods, described below, do not use angular diameter data. The photometric methods rely on using ratios of fluxes.

The infrared flux method uses the ratio of the integrated flux to the flux in some reference band to determine T_{eff}. The choice of the reference band is fairly critical. In general, the K band (2.2 μm) is chosen because this gives a useful variation in the ratio. For example, the K-band ratio varies by a factor of 2.37 between model atmospheres with $T_{eff} = 5500$ and 4000, whereas the ratio for a reference band at 0.89 μm varies by only a factor of 1.15 for the same models. Reference bands at shorter wavelengths are satisfactory for hotter stars.

There are uncertainties associated with the use of the K band. First, the stellar flux decreases sharply with increasing wavelength, and so the wavelength of the K-band filter must be known accurately for the calculations of the flux ratio from the models. Allowance for the presence of CO lines in the K-band pass must be made for the coolest stars (T_{eff} approximately 4000 K). Finally, there is a vexing uncertainty for the cooler model calculations owing to uncertainty in the H^- free-free absorption coefficient. This results in an uncertainty of about 40 K for G and K stars.

This method has been used to find T_{eff} for a sample of 31 F dwarfs by M. Saxner and G. Hammarbäck, 95 G and K stars by the author and Bengt Gustafsson, and for various stars by Donald E. Blackwell and his collaborators.

The variation of flux with wavelength is a valuable tool for finding stellar temperatures. The general application of this method uses spectral scans, with a bandpass of typically 50 Å. In general, these scans are corrected for line absorption, using measurements from high dispersion spectra. The wavelength region generally used for temperature measurements is in the interval between 3650 and 8200 Å, although data at shorter wavelengths have been used for hotter stars, when the line absorption corrections are not applied. In the latter case comparisons are made with stellar fluxes computed from synthetic stellar spectra. Surface gravities can be found from the scans in many cases, by comparing fluxes above and below the Balmer limit. Robert L. Kurucz presented an informative diagram of fluxes versus wavelengths for a broad range of stellar models.

This method has been widely applied by a number of authors. In particular, it has been used for finding T_{eff} of Vega, which serves as the zero point star for synthetic color calculations. Other applications include determinations of the absolute magnitudes of Cepheid and RR Lyrae variable stars. The method appears to have fallen out of vogue in recent years, at least when using the data in the visible region of the spectrum, possibly because of the extensive observing time required to obtain the necessary data.

One of the most convenient ways to obtain stellar temperatures is by the use of dereddened colors. The problem is complicated, in some cases, by the effects of spectral lines. This problem has been tackled by the computation of synthetic colors. The colors are computed from synthetic spectra by evaluating integrals of the form

$$\int S(\lambda) F(\lambda)\, d\lambda,$$

where $S(\lambda)$ is the "sensitivity function" of one passband of a particular color system and $F(\lambda)$ is the synthetic spectrum flux. If desired, terms describing the interstellar reddening and the extinction in the Earth's atmosphere can be included in the calculation. The "sensitivity function" is primarily determined by the properties of the detector and filter being used, but is also affected, to a greater or lesser extent, by the other optical components of the photometer and telescope, for example, the reflectivity of the telescope mirrors.

The sensitivity functions of the U, R, and I passbands of the Johnson UBVRI system are a matter of some debate and have been discussed in detail in the literature.

After conversion of the integrals to a magnitude scale, the question of the zero point of the scale must be settled. This is done by identifying the colors of a particular model with the colors of a particular star. Vega is one star that can be used for this purpose, because there is an abundance of data with which the predictions of models can be compared and the effective temperature, surface gravity, and chemical composition found.

Since model atmospheres and synthetic spectra are not perfect, it is necessary to check the color calculations. The most reasonable way of doing this is to find stars where T_{eff} has been found by some other means, for example, the IRFM. It may also be necessary to know the surface gravity and chemical composition of these stars for the analysis of some colors. Such data can be found by spectroscopic means or by narrow-band photometry.

Although there is some argument as to the precise value of the solar colors, it is well known that solar models and synthetic spectra yield $B-V$ colors that are too blue. This problem is believed to result from a lack of sufficient atomic and molecular data for the synthetic spectrum calculations. Because the number of atomic lines per angstrom increases with decreasing wavelength, the solar models are too bright at shorter wavelengths. However, because the total flux from the models correspond to $T_{eff} = 5780$ K, the brighter flux at shorter wavelengths must be accompanied by the models being too faint at longer wavelengths.

Despite the solar model being too blue in $B-V$, the calculated $B-V$ colors of F dwarfs seem to match the observed colors quite well. This is shown in comparisons of the observed colors of F dwarfs of known T_{eff} with the $B-V$, T_{eff} relation for models with the solar abundance and surface gravity. The conversion of T_{eff} to

$B-V$ for F stars is an important problem in determining the ages of globular clusters, because the theoretical isochrones must be converted from the $\log L$, $\log T_{\text{eff}}$ plane (L is stellar luminosity) to the M_v, $B-V$ plane.

The author and B. Gustafsson presented tables of various colors based on observations in the V band and longer wavelengths. Their T_{eff}, $V-K$ relationship agrees reasonably well with the Ridgway et al. relationship, the latter giving temperatures about 100 K cooler at 4000 K and very nearly identical at 5000 K. The temperatures that we derived from $V-K$ also agree rather well with those deduced from the IRFM.

Temperatures can also be derived from the spectra of stars. The results have the great advantage that they are independent of interstellar reddening.

Relatively low-resolution spectra can be used to classify the spectral type and luminosity class of a star, provided that the star has approximately the solar composition. These spectral types have been calibrated in terms of T_{eff}.

The hydrogen lines can often be used as temperature indicators, although knowledge of the stellar gravity is necessary in some cases. The lines are very broad in most stellar spectra, the broadening being caused by the Stark effect. Although the strength of a line can be measured using narrow-band photometry, such as the Hβ photometry of the Strömgren system, more accurate results can probably be found by fitting the profiles of the lines. Examples of the quality of the fit that can be obtained were given by Giusa Cayrel de Strobel.

Additional Reading

Bell, R. A. (1988). Synthetic Strömgren photometry for F dwarf stars. *Astron. J.* **95** 1484.

Bell, R. A. and Gustafsson, B. (1989). The effective temperatures and colours of G and K stars. *Mon. Not. R. Astron. Soc.* **236** 653.

Bessell, M. S. (1986). On the Johnson *U* passband. *Publ. Astron. Soc. Pac.* **98** 354.

Blackwell, D. E. and Shallis, M. J. (1977). Stellar angular diameters from infrared photometry. Application to Arcturus and other stars; with effective temperatures. *Mon. Not. R. Astron. Soc.* **180** 177.

Bohlin, R. C., Holm, A. V., Savage, B. D., Snijders, M. A. J., and Sparks, W. M. (1980). Photometric calibration of the *International Ultraviolet Explorer* (*IUE*): Low dispersion. *Astron. Ap.* **85** 1.

Breger, M. (1976). Catalog of spectrophotometric scans of stars. *Astrophys. J.* (*Suppl.*) **32** 7.

Brune, W. H., Mount, G. H., and Feldman, P. D. (1979). Vacuum ultraviolet spectrophotometry and effective temperatures of hot stars. *Astrophys. J.* **227** 884.

Cayrel de Strobel, G. (1985). How precise are spectroscopic abundance determinations today? In *Calibration of Fundamental Stellar Quantities*, *IAU Symposium 111*, D. S. Hayes, L. E. Pasinetti, and A. G. D. Philip, eds. D. Reidel, Dordrecht, p. 137.

Code, A. D. (1985). The role of space observations in the calibration of fundamental stellar quantities. In *Calibration of Fundamental Stellar Quantities*, *IAU Symposium 111*, D. S. Hayes, L. E. Pasinetti, and A. G. D. Philip, eds. D. Reidel, Dordrecht, p. 209.

Code, A. D., Davis, J., Bless, R. C., and Hanbury Brown, R. (1976). Empirical effective temperatures and bolometric corrections for early-type stars. *Astrophys. J.* **203** 417.

Cousins, A. W. J. (1979). Response functions in the red and near infrared. *The Observatory* **99** 147.

Dreiling, L. A. and Bell, R. A. (1980). The chemical composition, gravity, and temperature of Vega. *Astrophys. J.* **241** 736.

Gunn, J. E. and Stryker, L. L. (1983). Stellar spectrophotometric atlas, $3130 < \lambda < 10800$ Å. *Astrophys. J.* (*Suppl.*) **52** 121.

Gustafsson, B. and Bell, R. A. (1979). The colours of G and K type giant stars. I. *Astron. Ap.* **74** 313.

Hanbury Brown, R., Davis, J., and Allen, L. R. (1974). The angular diameters of 32 stars. *Mon. Not. R. Astron. Soc.* **167** 121.

Hayes, D. S. (1970). An absolute spectrophotometric calibration of the energy distribution of twelve standard stars. *Astrophys. J.* **159** 165.

Hayes, D. S. (1985). Stellar absolute fluxes and energy distributions from 0.32 to 4.0 μm. In *Calibration of Fundamental Stellar Quantities*, *IAU Symposium 111*, D. S. Hayes, L. E. Pasinetti, and A. G. D. Philip, eds. D. Reidel, Dordrecht, p. 225.

Hayes, D. S. and Latham, D. W. (1975). A rediscussion of the atmospheric extinction and the absolute spectral-energy distribution of Vega. *Astrophys. J.* **197** 593.

Johnson, H. L. (1966). Astronomical measurements in the infrared. *Ann. Rev. Astron. Ap.* **4** 193.

Kurucz, R. L. (1979). Model atmospheres for G, F, A, B, and O stars. *Astrophys. J.* (*Suppl.*) **40** 1.

Oke, J. B. (1961). An analysis of the absolute energy distribution in the spectrum of eta Aquilae. *Astrophys. J.* **133** 90.

Oke, J. B. (1966). A spectrophotometric study of X Arietis. *Astrophys. J.* **145** 468.

Oke, J. B. and Schild, R. E. (1970). The absolute spectral energy distribution of alpha Lyrae. *Astrophys. J.* **161** 1015.

Ridgway, S. T., Joyce, R. R., White, N. M., and Wing, R. F. (1980). Effective temperatures of late-type stars: The field giants from K0 to M6. *Astrophys. J.* **235** 126.

Saxner, M. and Hammarbäck, G. (1985). An empirical temperature calibration for F dwarfs. *Astron. Ap.* **151** 372.

Stone, R. P. S. and Baldwin, J. A. (1983). Southern spectrophotometric standards for large telescopes. *Mon. Not. R. Astron. Soc.* **204** 347.

Straižys, V. and Sviderskienė, Z. (1972). Energy distribution in the stellar spectra of different spectral types and luminosities. *Bull. Vilnius Astron. Obs.* No. **35**.

Strecker, D. W., Erickson, E. F., and Witteborn, F. C. (1979). Airborne stellar spectrophotometry from 1.2 to 5.5 microns: Absolute calibration and spectra of stars earlier than M3. *Astrophys. J.* (*Suppl.*) **41** 501.

Tüg, H., White, N. M., and Lockwood, G. W. (1977). Absolute energy distributions of α Lyrae and 109 Virginis from 3295 Å to 9040 Å. *Astron. Ap.* **61** 679.

Traub, W. A. and Stier, M. T. (1976). Theoretical atmospheric transmission in the mid- and far-infrared at four altitudes. *Appl. Op.* **15** 364.

See also **Magnitude Scales and Photometric Systems; Stars, Atmospheres; Stars, Atmospheres, Radiative Transfer; Stars, Spectral Classification.**

Stars, White Dwarf, Observed Properties

Edward M. Sion

White dwarf stars are extremely dense compact stars containing a stellar mass in a volume the size of a planet. They are the final stellar remnants that result from the nonexplosive evolution of virtually all stars having initial masses up to approximately 7–8 solar masses. Thus one expects that upwards of 97% of the stars in the Milky Way galaxy should terminate their thermonuclear evolution as these burned-out cinders, compressed to complete electron degeneracy by the earlier, successively alternating stages of thermonuclear fuel ignition, fuel exhaustion, and core contraction. White dwarfs gradually cool at essentially constant radius by releasing the thermal energy of the bare atomic nuclei in their cores via the highly efficient conductive heat transport provided by the Fermi sea of degenerate electrons. These electrons provide the quantum mechanical pressure support (i.e., the exclusion energy, due to the Pauli exclusion principle that no more than two electrons with oppositely directed spins may occupy the same energy state), which prevents gravitational collapse, up to a limiting mass of 1.4 solar masses, called the Chandrasekhar limit. Stars composed of this strange matter have smaller radii the larger their mass. The electron degenerate core (usually carbon and oxygen) is surrounded by a nondegenerate (ideal gas) envelope less than 1/100 of the stellar radius, or 60–100 km thick, depending upon the total mass of the white dwarf. This envelope, having a tiny fraction of the core mass, blankets the isothermal degenerate core and transports energy by photon diffusion and by convective motions. The radiative opacity of the envelope, determined by its thermal structure and chemical constituents, regulates the rate of cooling of the star—how rapidly it loses thermal energy to space. The nondegenerate envelope is separated from the core by a partially electron-degenerate transition layer.

The spectra of the first two white dwarfs known, Sirius B and 40 Eridani B, were obtained by Walter Sydney Adams and by Williamina Paton Fleming and Edward C. Pickering, respectively. These dwarf objects had the same surface luminosity as red main-sequence dwarf stars, but their spectra and colors indicated they were much hotter (hence the term white). This implied interior densities 5–6 orders of magnitude larger than that of a main-sequence star. The physics of matter at such extremely high density was not worked out until the quantum statistical theory of the electron gas was formulated in 1926 by Enrico Fermi and Paul A. M. Dirac. Also in 1926, Sir Ralph Fowler showed that electron degeneracy pressure could support a stellar mass object against gravitational collapse. In 1939, Subrahmanyan Chandrasekhar derived the equations for the basic structure and mass–radius–core chemical composition relations that correctly combined both quantum mechanics and relativity. For this work he was awarded the Nobel Prize in Physics in 1983.

Due to their low intrinsic luminosities, their discovery is restricted to a relatively small volume of space around the Sun. The classical technique used to identify them as degenerate star candidates is to find in proper-motion lists (notably those of Willem J. Luyten and those of Henry L. Giclas and co-workers), faint stars of large proper motion whose color is bluer than those of ordinary main-sequence stars. More recently, large numbers of new degenerate stars have been identified from ultraviolet color excess surveys. White dwarfs are occasionally found as hot companions to brighter stars in binary systems, as exemplified by Sirius and Procyon.

TYPES OF WHITE DWARF STARS DEFINED BY SURFACE COMPOSITION

The white dwarfs, due to their high gravities (through gravitational diffusive separation), usually show only the lightest principal atmospheric constituent at their surface. They divide into two dominant composition sequences, those with hydrogen-rich atmospheres (denoted DA; see the following) and those with helium-rich atmospheres (non-DA). Their spectroscopic properties are determined by the complex interplay of a number of physical processes that control and/or modify the flow of elements and hence surface abundances in high-gravity atmospheres: Convective dredge-up, mixing and dilution, accretion, gravitational and thermal diffusion, radiation pressure (in the hottest white dwarfs), mass loss, residual nuclear burning, and magnetic fields. Most of these processes are still poorly understood. They manifest themselves in the variety of white dwarf surface compositions exhibited spectroscopically by absorption lines in the far-ultraviolet and optical wavelength regions.

Early systems of spectroscopic classification proposed by Gerard P. Kuiper and by Luyten were useful before the wide variety of surface compositions became known. In 1960, Jesse L. Greenstein presented an elaborated version of the earlier schemes and in 1983, the author, Jesse L. Greenstein, John D. Landstreet, James W. Liebert, Harry L. Shipman, and Gary A. Wegner introduced a similar scheme with a better description of what the spectrum actually shows and with quantitative temperature information. This system is shown in Table 1 and some basic parameters of the spectroscopic types are described below. Figure 1 displays typical examples of white dwarf optical spectra.

The DA White Dwarfs DA degenerates comprise 75%–80% of all white dwarfs hotter than 10^4 K, with the remaining 20%–25% referred to as non-DA. As a result of gravitational diffusive separation, which causes the lightest element to appear at the stellar surface, the optical spectra of DA stars generally show only the Balmer lines of H I. They have essentially pure hydrogen outer layers of uncertain thickness, and occur over a very wide range of effective temperature (T_{eff} = 6000 to 70,000–90,000 K). Below 5500 K, the Balmer lines are no longer detectable. Their hydrogen may be either primordial (i.e., remaining from earlier evolutionary stages) or accreted from the interstellar medium.

The DO White Dwarfs These carbon and helium-rich objects range from the hottest known degenerate stars ($\log g \approx 7$, $T_{eff} > 10^5$ K) with carbon, oxygen, and other metals present in their photospheres, down to the coolest known DO stars (T_{eff} = 45,000–55,000 K). In all cases, lines of ionized helium dominate their optical spectra. While their true hydrogen–helium ratios are poorly known due to their high surface temperatures, the coolest DO stars are demonstrably hydrogen-poor and several DO stars with $T_{eff} \approx 80{,}000$ K show no evidence of hydrogen based on optical data. The DO stars are thought to be the immediate precursors of the nearly pure helium DB degenerates, although this may not be invariably true. Recent abundance analyses of the hottest (PG1159-035) DO stars reveal carbon abundances as high as 55% with helium less than 25%.

Table 1. Spectral Classification and Observed Properties of White Dwarfs

Spectroscopic Types of White Dwarfs

Spectral Type	*Characteristics*
DA	Only Balmer lines; no He I or metals present
DB	He I lines; no H or metals present
DC	Continuous spectrum, no lines deeper than 5% in any part of the electromagnetic spectrum
DO	He II strong: He I or H present.
DZ	Metal lines only; no H or He
DQ	Carbon features, either atomic or molecular, in any part of the electromagnetic spectrum.

White dwarfs are further designated with a temperature index from 0–9 defined by $10 \times \theta_{eff}$ (where $\theta_{eff} = 5040/T_{eff}$), and with appropriate symbols for magnetic field/polarization (H, P), variability (V) and spectral peculiarities or unclassifiable spectra (X).

Observed Range of Physical Parameters and Statistical Properties of White Dwarf Stars

Effective temperature range	4500–150,000 K
Range of surface gravity	$7 \leq \text{Log } g \leq 9$
Range of mass	$0.4 \leq M \leq 1.4\ M_{\odot}$
Average radius	$0.0120\ R_{\odot}$
Observed range of luminosity	$-4.3 \leq \text{Log}(L/L_{\odot}) \leq 2$–3
Range of detected magnetic field strength	10^6–10^9 G
Upper limit rotation rate	$V \sin i \leq 65$ km s^{-1}
Observed Mass density	$0.002\ M_{\odot}$ pc^{-3}
Empirical white dwarf formation rate	4.5–7.5 $\times 10^{-13}$ pc^{-3} yr^{-1}
Galactic scale height of white dwarfs	250–300 pc

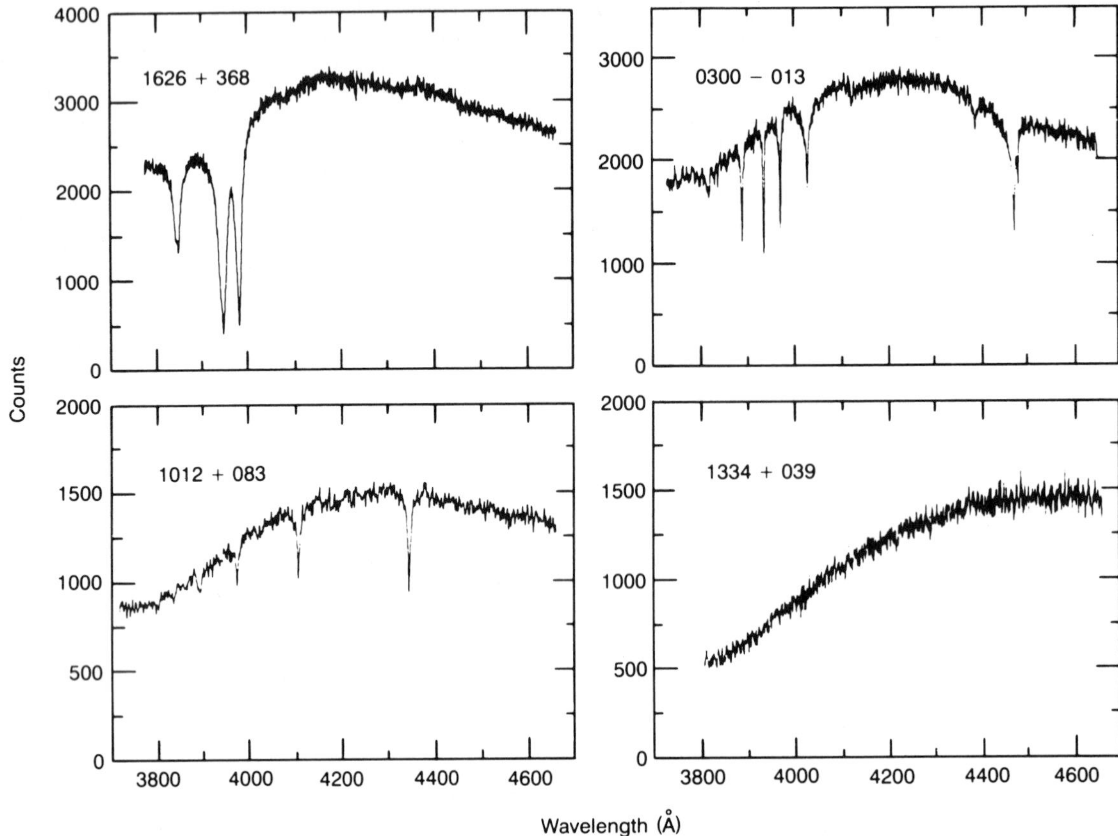

Figure 1. Optical spectra of white dwarf stars. The vertical axis is photon counts and the horizontal axis is wavelength in angstrom units. The upper-left panel displays the pressure-broadened, singly ionized calcium H and K absorption lines and a neutral magnesium absorption line of a fairly typical DZ white dwarf, WD1626 + 368 (Ross 640); The lower left panel displays the Stark-broadened Balmer absorption lines of hydrogen in the cool DA7 star, WD1012 + 083 (EG 183); The upper right panel displays the neutral helium absorption lines of the peculiar DBZ star WD0300 − 013 (GD40), but in addition the K line is present (with the H component blended with a neutral helium feature); this is one of only two DB stars known to exhibit metallic absorption. The lower-right panel displays a typical very cool DC9 white dwarf (WD1334 + 039, Wolf 489) with no evidence of any detectable absorption features. These spectra were obtained with the Multiple Mirror Telescope and MMT spectrograph, at Mount Hopkins.

The DB White Dwarfs The DB degenerates have nearly pure helium atmospheres (neutral helium lines only) with very stringent limits on the amount of hydrogen [$N(He)/N(H) > 10^4$] that can be present without showing Balmer absorption. They lie in the temperature range $12{,}000 < T_{eff} < 26{,}000$–$30{,}000$ K. Below $T_{eff} = 12{,}000$ K, He I lines disappear and DB spectra are no longer recognizable.

The DQ White Dwarfs The DQ (Formerly C_2 or $\lambda4670$ stars) degenerates generally have helium-dominated atmospheres whose optical spectra show the Swan bands of the C_2 molecule, but usually no other metals or hydrogen. They lie within the temperature range $6000 < T_{eff} < 12{,}000$ K and have carbon abundances in the range $10^{-7} < C/He < 10^{-2}$. Their far-ultraviolet spectra sometimes reveal atomic carbon (C I) lines.

The DZ White Dwarfs The DZ white dwarfs (formerly DF, DG, DK) show metallic absorption features with Ca II H and K absorption dominant, and with carbon noticeably absent or extremely weak in their usually helium-dominated atmospheres. Although DZ stars are normally too cool to show absorption lines of the dominant atmospheric constituent, either helium or hydrogen, several helium-dominated DZ stars show trace hydrogen. The hottest examples of this type have $T_{eff} \approx 10{,}500$ K and the coolest examples have $T_{eff} \approx 4500$ K. The metals appear to originate from interstellar accretion.

The DC White Dwarfs These objects are too cool to show lines of the dominant light element, be it hydrogen or helium. They exhibit featureless spectra to within 5% of the continuum. Most of these objects reveal weak carbon features when examined at high resolution on good signal-to-noise spectra. However, there are apparently still some genuine DC stars in this greatly diminished subgroup.

BASIC OBSERVED PROPERTIES

Masses of White Dwarfs

Accurate masses of white dwarfs are possible from measurements of the gravitational (Einstein) redshift of their spectral lines. The gravitational redshift of light, due to the gravitationally induced slowdown of clocks in large gravitational fields (slowing of time near a white dwarf), is one of the tests of Albert Einstein's theory of general relativity. Thus spectral lines will undergo a gravitational redshift described by

$$\Delta\lambda/\lambda = (1 - 2MG/Rc^2)^{-1/2} - 1,$$

where $\Delta\lambda = \lambda - \lambda_{inf}$, λ is the wavelength of the spectral line emitted near the gravitational radius of the white dwarf, λ_{inf} is the apparent wavelength seen by a distant observer, M and R are the mass and radius of the white dwarf, G is the universal constant of gravitation, and c is the speed of light. For velocities considerably below the speed of light, the redshift velocity is expressed with sufficient accuracy by

$$V_{rs} = GM/cR = 0.635(M/M_{\odot})(R/R_{\odot})^{-1}\ \text{km s}^{-1}.$$

Here, $M_{\odot}$ and $R_{\odot}$ are the mass and radius of the Sun, respectively. If the intrinsic wavelength shift of the spectral lines due to pressure (Stark effect) broadening is properly corrected and if the white dwarf has a well-measured line-of-sight velocity (e.g., a distant non-white dwarf binary companion of accurately known radial velocity), the gravitational redshift can be extracted and thus the mass/radius ratio of the white dwarf.

The most reliable masses of individual white dwarfs are those obtained from orbit solutions of wide (essentially noninteracting) binaries. These data points are crucial for testing the mass–radius relation for degenerate stars and for testing stellar evolution theory [e.g., the initial (parent) mass–final remnant mass relations). Among the best determinations are Sirius B ($1.053 \pm 0.028\ M_{\odot}$), Procyon B ($0.63\ M_{\odot}$), 40 Eridani B ($0.43 \pm 0.02 M_{\odot}$), and Stein 2051B ($0.50\ M_{\odot}$).

Mass estimates of white dwarfs obtained by model atmosphere analyses of colors (e.g., the gravity-sensitive portion of the Strömgren-photometry two-color diagram), energy distributions and line-profile fits yield mass values of statistical use only. The masses, radii, and gravities of large numbers of white dwarfs determined in this way, largely by Volker Weidemann and co-workers and by Harry L. Shipman have yielded statistically useful average masses and mass distributions for DA (hydrogen-rich) and non-DA (helium-rich) white dwarfs. The average mass of DA stars is $0.58 \pm 0.05\ M_{\odot}$, and the average mass of DB stars is $0.55 \pm 0.03\ M_{\odot}$.

Rotation of White Dwarfs

The most general basic conclusion about the rotation of white dwarfs to date is that they are slow rotators. This result is based upon actual determinations or inferences of slow rotation from the effects of rotation on the light curves (frequency spectrum) of the ZZ Ceti variables, from variable circular polarization due to changes in the observed magnetic field as the white dwarf rotates, and from recent analyses of the very sharp non-LTE Balmer absorption line cores of DA stars. No DA star is known with $V \sin i > 65$ km s^{-1}, where V is the rotational velocity and i is the inclination of the rotation axis to the line of sight of the rotation axis to the line of sight. For the few DB and DO stars that have been analyzed, the conclusion is the same: slow rotation. Based upon analyses of the sharp neutral helium line cores of several DB stars, all were found to have $V \sin i < 135$ km s^{-1}. The breadth of sharp metallic photospheric absorption lines in the hot DO stars implies a similar conclusion. This slow rotation of white dwarfs is unexpected if angular momentum is conserved during evolution from high-luminosity, evolved progenitors down to the white dwarf stage. Some as yet unidentified braking mechanism, possibly fast stellar-wind mass loss just prior to the white dwarf stage, may be responsible.

Pulsations of White Dwarfs

One of the most remarkable developments in the last 15 years was the discovery of three distinct classes of pulsating white dwarfs: (1) the DA pulsators, whose prototype is ZZ Ceti, with effective temperatures near 12,000 K, an instability strip of width $\delta\ T_{eff} =$ 2000 K, and multiperiodic luminosity variations in the period range 100–1200 s; (2) the DB (helium-rich) pulsators, whose prototype is GD358, with effective temperatures near 28,000 K, an instability strip of less certain width but $\delta\ T_{eff} \approx$ a few thousand degrees, and multiperiodic luminosity variations (in five of the six known pulsators) in the period range 100–1200 s; and (3) the DO pulsators, hottest of the three types of pulsators, whose prototype is GW Vir (PG1159−035), with effective temperatures between 110,000 and 150,000 K, an instability strip of approximate width $\delta T_{eff} = 40{,}000$ K and multiperiodic luminosity variations in the period range 200–2000 s. Most of the objects in all three classes have extremely complicated light curves, and all white dwarf pulsators appear to be pulsating in nonradial g modes. The ZZ Ceti mechanism appears to require very thin hydrogen layers in order to account for the high-temperature boundary (blue edge) and the observed narrow instability strip width. The observed luminosity variations are thought to be due mostly to temperature variations during a global g-mode oscillation. The oscillations are thought to be driven by partial ionization of the most abundant element in the outer layers, hydrogen in the case of the ZZ Ceti stars, helium in the case of the DB pulsators, and possibly the partial ionization of carbon and oxygen in the GW Vir variables, although the g modes may be excited by nuclear shell burning in these hottest pulsating degenerates. Extensive observational and theoretical work on these objects has been accomplished by John T. McGraw, Gilles Fontaine and Donald E. Winget.

Because the normal modes of g-mode pulsation depend upon the global properties of the star, the periods of the normal modes should change as the white dwarf evolves (cools). Since many g modes are excited in each pulsator and the observed periods are known with extreme accuracy, it is possible to measure the change in these periods in a relatively short time and thus test the cooling theory of hot degenerates (with short cooling time scales) and seismologically probe the interior regions of the white dwarfs. Exploration of this exciting seismological frontier has barely begun.

Space Density, Luminosity Function, and Space Motions of White Dwarfs

A knowledge of the space density of white dwarfs is critically important for understanding their evolutionary history by comparing their space density and formation rate with the mass density and formation rates of the types of stars which could be their progenitors, and for assessing the contribution of the white dwarfs to the local mass density of the galactic disk. The total mass density of white dwarfs in the solar neighborhood, based upon analysis of the enormous Luyten–Palomar proper-motion catalogs along with the best available trigonometric parallaxes, is $0.002\ M_{\odot}$ pc^{-3}. This result implies that white dwarfs are a relatively minor contributor to the local mass density of the galactic disk, and therefore cannot provide the so-called missing mass.

The empirical luminosity function of white dwarfs, which gives the relative numbers of stars in successive intervals of absolute magnitude within a given volume of space, provides a means of testing theoretical cooling calculations of white dwarf evolution. The simplest of these models, the Mestel cooling theory, expresses an age–luminosity law as

$$\tau_{cool} \propto \left((M/M_{\odot})^{5/7}\right)(L/L_{\odot})^{-5/7}\ \text{yr}.$$

The empirical luminosity function of white dwarfs has been constructed from observations of the Luyten proper-motion data base, extended to the intrinsically faintest local white dwarfs. Trigonometric parallaxes of high accuracy and reliable red color data are required for these faint, cool objects, and these data have been secured, principally by the U.S. Naval Observatory. Selection effects must be evaluated in correcting for incompleteness (missed stars) out to a given apparent magnitude. The resulting empirical luminosity function shows good agreement with the simple Mestel cooling theory down to luminosities as low as $\text{Log}(L/L_{\odot}) \approx -4.1$.

Below that luminosity, the empirical luminosity function exhibits a downturn, which implies a real deficiency of cool degenerate stars. In particular, despite deep intensive searches, no white dwarf has yet been found with an absolute magnitude fainter than $M_V \approx +16$. This paucity of white dwarfs fainter than $M_V = +16$ implies a maximum known cooling age of 10^{10} yr (the cooling time of a white dwarf to reach a luminosity near $\mathrm{Log}(L/L_\odot) \approx -4.5$), which may well indicate that the bulk of star formation in the disk of the Milky Way galaxy began less than 10^{10} years ago. This would imply an age for the galactic disk billions of years younger than the ages estimated for globular clusters and the galactic halo.

The space motions of the local white dwarfs indicate that they represent an admixture of stellar population subcomponents: The majority of white dwarfs belong to the old disk population subcomponent with typical total space motions of 50–60 km s^{-1} with respect to the Sun, whereas 4%–5% have total space motions greater than 150 km s^{-1}, characteristic of the halo and extreme Population II subcomponent, and several percent have motions which indicate they belong to the young disk population subcomponent and therefore are associated with young, fairly massive progenitor stars. Although the long total stellar ages of white dwarfs and the perturbative encounters they suffer during their galactic orbital motions would tend to smear out kinematical distinctions among the different types of white dwarfs (e.g., increase their velocity dispersions with age), there is evidence that the DQ (carbon-band) degenerates and the magnetic white dwarfs have higher than average and lower than average space motions, respectively.

Magnetic Fields of White Dwarfs

The magnetic white dwarfs (spectroscopically designated DH, DP, DAH, DAP, DXP; see Table 1), of which 27 are known, comprise about 2% of the total white dwarf sample and are distributed more or less uniformly throughout the magnetic field range $1 \leq B \leq 1000$ MG. Although searches for weaker fields are incomplete and detection becomes more difficult in weaker-field objects, it seems safe to conclude that nearly all white dwarfs do not have detectable magnetic fields and no more than a few percent show fields of a million gauss or higher. Like the nonmagnetic degenerates, the magnetic white dwarfs have primarily hydrogen-rich atmospheres based on their analyzable Zeeman patterns. However, the atmospheric compositions of some of the high-field examples are unknown, due to our current lack of knowledge of the Zeeman effect on atoms other than hydrogen.

From kinematical and statistical (space density) arguments, it is likely that the magnetic degenerates may be the descendants of the peculiar B and A stars on the main sequence. Some, however, could originate from extinct AM Herculis magnetic cataclysmic variables in which the cool, Roche-lobe-filling companion has been disrupted or reduced to a very low-mass (substellar) degenerate itself, thus terminating magnetically funneled mass accretion and leaving what appears to be a *single* magnetic white dwarf.

White Dwarfs in Star Clusters

White dwarfs have been identified in a number of open star clusters with the largest sample in the Hyades cluster (over a dozen DA stars and one DBA member). The mean mass of the Hyades white dwarfs is larger than the mean mass of field degenerates, thus implying more massive progenitor stars for the cluster white dwarfs. Most remarkably, white dwarfs have been identified in open clusters whose ages are so young that only the massive upper-main-sequence members of the cluster have had sufficient time to evolve toward the red giant branch. This so-called turnoff mass is 6–8 solar masses in some of these clusters, thus implying that the parents of the cluster white dwarfs were at least that massive. In addition, recent deep photometry of a few globular clusters with charge-coupled devices has led to the identification of the first white dwarf cooling sequences in globular clusters. The Hubble Space Telescope should reveal prodigious numbers of globular cluster white dwarfs and help to elucidate the effects of low metallicity and low initial parent mass on the types of white dwarf remnants that occur.

FUTURE PROSPECTS

Further advances in our understanding of key physical processes which compete with gravitational diffusion in white dwarf envelopes are inevitable in the years ahead. This progress on the theoretical front is crucial because of the looming avalanche of new spectroscopic data, an enlarged stellar statistical database, and more precise physical parameters brought about by the much higher sensitivity, better spectral resolution, and broadened wavelength coverage of the Hubble Space Telescope, other new and planned space observing missions, and the new generation of 8–10 m groundbased telescopes. The clear identification of the various evolutionary channels that lead to the formation of the different subgroups of white dwarfs remains to be elucidated. The study of white dwarfs offers the continual prospect of learning new physics, because their bizarre matter and associated physical processes cannot be studied presently in terrestrial laboratories.

Additional Reading

Greenstein, J. L. (1960). Spectra of stars below the main sequence. In *Stellar Atmospheres*, J. L. Greenstein, ed. University of Chicago Press, Chicago, p. 676.

Liebert, J. W. (1980). White dwarf stars. *Ann. Rev. Astron. Ap.* **18** 363.

Ruderman, M. A. (1971). Solid stars. *Scientific American* **224** (No. 2) 24.

Sion, E. M. (1986). Recent advances on the formation and evolution of white dwarfs. *Publ. Astron. Soc. Pac.* **98** 821.

Van Horn, H. M. (1979). The physics of white dwarfs. *Physics Today* **32** 23.

Weidemann, V. (1990). The formation and evolution of white dwarfs. *Ann. Rev. Astron. Ap.* **28**.

See also **Binary Stars, Polars (AM Herculis Type); Stars, Magnetism, Theory; Stars, Pulsating, Theory; Stars, White Dwarf, Theory; Stellar Evolution, Pulsations.**

Stars, White Dwarf, Structure and Evolution

Gilles Fontaine and Francois Wesemael

White dwarf stars are compact and dense objects that, together with neutron stars and black holes, represent the end products of stellar evolution. The study of these stellar corpses constitutes an important chapter of contemporary astrophysics, and is of interest to both the physicist and the astronomer. For instance, with stellar masses but planetary dimensions only, white dwarfs show extreme conditions of density and gravity, so extreme, in fact, that they cannot generally be reproduced on Earth. Thus, an understanding of the internal constitution of white dwarfs allows us to study the properties of matter under extreme conditions. At the same time, white dwarfs—because of their age— have the "memory" of events long past that have influenced the evolution of the stars and, indeed, the evolution of the universe itself.

It is well established that the vast majority of stars end their nuclear-burning lives in the form of white dwarfs; this is the fate awaiting our Sun, for example. After exhausting its thermonuclear fuel, the hot core of a star collapses onto itself because gravity is no longer balanced by the internal thermal pressure generated by nuclear energy sources. In most cases, the gravitational collapse stops when the stellar remnant reaches a dimension some 100 times smaller than that of the Sun. Further contraction is prevented when internal densities become so high that free electrons can no longer be packed any closer together. This phenomenon, known as electron degeneracy, is a consequence of the Pauli exclusion principle, well known in quantum mechanics. The principle "forbids" electrons from coming too close together unless they move increasingly faster. During the gravitational collapse of a burnt-out stellar remnant, the density increases, the electrons get packed closer together, and their kinetic energy increases to satisfy the Pauli principle. This increase gives rise to a quantum pressure, which is not related to the thermal motions of the particles. This pressure is called degenerate electron pressure; it ultimately prevents further contraction, and restores the condition of hydrostatic equilibrium in the now moribund star, the white dwarf.

ORIGIN AND GENERIC PROPERTIES OF WHITE DWARFS

It is generally believed that the immediate progenitors of most white dwarfs are nuclei of planetary nebulae, themselves the products of intermediate- and low-mass main-sequence evolution. Stars that begin their lives with masses less than about 7–8 $M_\odot$ (solar masses), that is, the vast majority of them, are expected to become white dwarfs. Among those which already have had the time to become white dwarfs since the formation of the Galaxy, a majority have burned hydrogen and helium in their interiors. Consequently, most of the mass of a typical white dwarf is contained in a core made of the products of helium burning, mostly carbon and oxygen. The exact proportions of C and O are unknown because of uncertainties in the nuclear rates of helium burning.

It is a remarkable fact that isolated white dwarfs show a very narrow mass distribution: $M \sim 0.6\ M_\odot$, with a dispersion of $\sigma \sim \pm 0.1\ M_\odot$. Apparently, the process of mass loss in white dwarf progenitors, which may have a wide range of initial masses, is regulated by mechanisms which are tuned finely enough to leave remnants with similar masses consistently. Also, it is currently believed that small traces of helium and hydrogen are left over after the mass-loss phases have subsided. Because of the intense gravitational field characteristic of a white dwarf ($g \sim 10^8$ cm s^{-2}, that is, roughly 4000 times larger than that of the Sun), elements rapidly segregate in the outer layers of such an object, with the lighter ones floating on top of the heavier ones. Hence, the expected structure of a typical, newly-formed white dwarf is that of a compositionally stratified object with a mass of $\sim 0.6\ M_\odot$ consisting of a C/O core surrounded by a thin, helium-rich envelope itself surrounded by a hydrogen-rich layer. Such an object has an average density of $\sim 10^6$ g cm^{-3}, a millionfold that of an ordinary star such as the Sun. The respective thicknesses of the H and He outer layers are not known a priori, and must depend on the details of the pre-white dwarf evolution. On theoretical grounds, however, it is expected that the maximum amount of He that can survive the hot planetary nebula phase is only 10^{-2} of the total mass of the star, and that the maximum fractional amount of H is about 10^{-4}. Although these outer layers are very thin, they are extremely opaque and regulate the energy outflow from the star. They consequently play an essential role in the evolution of a white dwarf. The exact masses of the H and He layers present in white dwarfs are currently at the center of a lively debate in the astronomical community.

The large opacity of the outer layers of a white dwarf implies that radiation escaping from the star originates from the outermost region—the atmosphere—which contains, typically, less than 10^{-14} of the total mass of the star. Spectroscopic observations can only probe these regions, which are usually dominated by hydrogen. Thus, a majority of white dwarfs are referred to as H-rich objects. It turns out, however, that about 25% of the white dwarfs do not possess such a hydrogen layer. These are called He-rich white dwarfs with, again, the understanding that the underlying C/O core contains essentially all of the mass, even though it is not directly observable. There is strong evidence that spectral evolution takes place among white dwarfs, that is, that H-rich stars become He-rich objects, and vice versa, during various evolutionary phases. Several physical processes, such as diffusion, convective mixing, and accretion, are clearly at work in the outer layers of white dwarfs. The investigation of the interplay of these mechanisms currently constitutes one of the most active areas of white dwarf research.

MECHANICAL STRUCTURE AND THERMAL EVOLUTION

Figures 1 and 2 summarize the results of evolutionary calculations for a typical He-rich white dwarf model. Figure 1 shows the evolutionary path followed by this model in the Hertzsprung–Russell (HR) diagram. A dying star begins the final phase of its history in the form of an extremely hot, collapsed object which can only cool off: Its nuclear energy sources are depleted, and gravita-

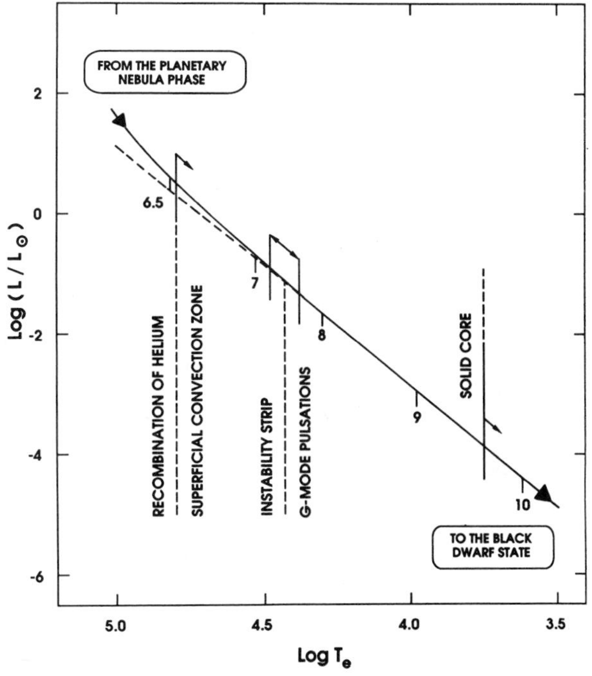

Figure 1. Hertzsprung–Russell diagram (luminosity in solar units $L/L_\odot$ versus surface or effective temperature T_e). The continuous curve shows the evolutionary track of a 0.6-$M_\odot$ white dwarf model consisting of a pure carbon core surrounded by a pure helium layer which contains 10^{-4} of the total mass of the star. The dashed line that merges with the evolutionary track corresponds to the line of constant stellar radius R through the relation $L = 4\pi R^2 \sigma T_e^4$—a straight line in this particular diagram (σ is the Stefan–Boltzmann constant). Below $L/L_\odot \sim 10^{-1}$, the white dwarf model evolves at constant radius. The numbers along the track give the logarithm of the cooling time expressed in years. The fainter the white dwarf, the longer it takes to cool further. Also indicated in the diagram are three particular phases of interest in the cooling history of this model. Around $T_e \sim 65{,}000$ K, a superficial convection zone first appears in the helium envelope. In the range $30{,}000 \gtrsim T_e \gtrsim 25{,}000$ K, the model goes through a strip in which it becomes unstable against gravity-mode pulsations. Finally, around $T_e \sim 5600$ K, carbon starts to crystallize at the center of the star.

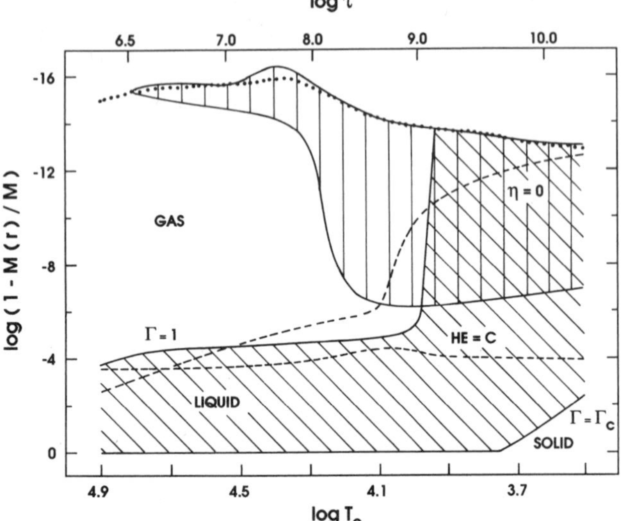

Figure 2. Fractional mass depth as a function of effective temperature (lower scale) and as a function of time expressed in years (upper scale) for the model presented in Fig. 1. Note that the choice of ordinate puts the emphasis on the outermost layers; the value $\log(1 - M(r)/M) = 0$ corresponds to the center of the star. The dotted line at the top of the graph corresponds to the position of the atmosphere; the only layers directly observable are above that line. The region defined by the vertical cross-hatching in the upper half of the diagram gives the location and the extent of the helium convection zone that develops during the evolution. The formation of this superficial convection zone begins around $T_e \sim 65{,}000$ K. With further cooling, the base of the convection zone sinks deeper into the star, and convection becomes the dominant energy transport mechanism in the outer layers of the white dwarf. The region defined by the diagonal cross-hatching corresponds to what is called a dense Coulomb fluid. The upper boundary is given by the condition $\Gamma = 1$, where Γ is the ratio of the average electrostatic potential energy to the average kinetic energy of ions in the plasma. This condition loosely defines the transition between a gas and a liquid. The lower boundary (the curve labeled $\Gamma = \Gamma_c$) corresponds to the crystallization line; below that line, the fluid solidifies. More than 99.99% of the mass of the white dwarf is in the form of a liquid during most of the evolution; only the outermost layers are in the gaseous state. After some 10^{10} years of evolution, however, more than 99% of the mass of the star has crystallized. The dashed line labeled $\eta = 0$ shows the location of the degeneracy boundary; below that line, electrons are degenerate. Clearly, most of the mass of the model is characterized by strong electron degeneracy. Finally, the dashed line labeled HE = C corresponds to the depth where the number of He ions is equal to the number of C ions; it is the dividing line between the carbon core and the helium outer layer. The line shows a certain functional dependence on time because helium and carbon diffuse with respect to each other during the evolution.

tional energy can no longer be tapped efficiently as degenerate electron pressure prevents additional contraction. Because this pressure is independent of the temperature, a white dwarf is condemned to evolve at essentially constant radius. The mechanical structure of such a star is therefore specified by the degenerate electrons. In particular, electron degeneracy is directly responsible for the curious relationship between the mass and the radius of a white dwarf: The more massive the star, the smaller its size is. Likewise, degeneracy is also responsible for the existence of a limiting mass above which a white dwarf cannot exist. This limiting mass is known as the Chandrasekhar mass, and is of the order of 1.2–1.4 $M_\odot$.

Degenerate electrons also possess another property of high relevance for white dwarfs: They are excellent conductors of heat, and thus they thermalize the internal regions of white dwarfs efficiently (a familiar illustration of this property is provided by the conduction electrons in ordinary metals). We can thus envision a white dwarf as consisting of an almost isothermal core that contains typically more than 99.99% of the mass, surrounded by a thin, opaque, insulating, nondegenerate outer layer. Typical internal temperatures in white dwarfs vary from $\sim 2\times10^7 - \sim 5\times10^6$ K for stars with surface temperatures in the range 16,000–8000 K, the most common white dwarfs. The very large temperature drop between the central regions and the surface takes place mainly in the stellar envelope. This drop usually leads to the formation of superficial convection zones, similar to that found in the Sun, which play a key role in the evolution of a white dwarf by directly affecting the cooling rate.

Largely decoupled from the electrons, the (nondegenerate) ions provide the thermal energy that slowly leaks through the outside, thereby producing the star's luminosity. In this context, the electrons do not contribute to the energy reservoir because degenerate particles cannot be "cooled" (they already occupy their states of lowest energy). As thermal energy is gradually lost from the star in the form of radiation, the ion system evolves from a gas to a fluid to a solid. Eventually, the reservoir of thermal energy becomes depleted and the star disappears from sight in the form of a cooled-off, crystallized object known as a black dwarf. The evolving structure of a typical He-rich white dwarf model during cooling is illustrated in Fig. 2.

It should be clear from this discussion that there exists an intimate relationship between dense-matter physics and the structure and evolution of white dwarfs. For instance, a detailed knowledge of the opacity and thermodynamics of strongly coupled plasmas is necessary to compute the cooling rate of a white dwarf. Indeed, this rate basically depends on how much thermal energy is stored in the interior of a star, and on how rapidly this energy is transferred from the hot core to the cold interstellar medium through the thin, opaque outer layers. Thus, a reliable description of the constitutive properties of dense plasmas is required to build a *theory* of evolving white dwarfs. By the same token, the *observed* properties of cooling white dwarfs can be used to test theories of strongly coupled plasma physics.

AREAS OF CURRENT INTEREST

As emphasized earlier, white dwarf stars hold a valuable record of the past history of the Galaxy. One property of particular interest in this respect is the fact that faint white dwarfs cool down extremely slowly (notice the logarithmic scale for the cooling age in Figs. 1 and 2). This implies that, if white dwarf formation has been going on more or less constantly over the distant past, many more faint white dwarfs than bright white dwarfs should be present in a given volume of space. This is indeed what the distribution of observed white dwarfs generally shows. However, a very important observational result of recent years has been the realization that there is a cutoff in the luminosity distribution of these stars: There is a real deficit of low-luminosity white dwarfs. The simplest model to account for this observational fact is to assume that the oldest white dwarfs in our Galaxy are still visible. In other words, the corpses of the very first generation of intermediate-mass stars in the Milky Way have not yet had the time to cool to invisibility, beyond the reach of our telescopes. By comparing the location of the observed low-luminosity cutoff in the white dwarf distribution with cooling calculations (such as those on which Fig. 1 is based), it is possible to infer the age of the white dwarf population in the Galaxy. This exciting result promises to lead to an estimate of the age of the galactic disk of unprecedented accuracy. The method is currently being refined through numerous numerical simulations of evolving white dwarfs.

Another area of current active interest is related to the presence of so-called instability strips along the cooling sequences of white dwarfs in the HR diagram. Figure 1 shows that He-rich white dwarfs go through a strip (at effective temperatures $30,000 \gtrsim T_e \gtrsim 25,000$ K) in which they become unstable against nonradial gravity-mode pulsations. There also exists an analogous instability strip for the H-rich white dwarfs, which is located at lower temperatures ($13,000 \gtrsim T_e \gtrsim 11,000$ K). In both cases, the instabilities are intimately connected to the presence of an extensive convection zone in the envelope. These instabilities manifest themselves as temperature waves at the stellar surface that cause multiperiodic luminosity variations. The importance of these two instability strips stems from the fact that they provide "windows" through which the *internal* structure of white dwarfs can be probed. Indeed, the pulsation properties of a star depend on its global structure. Thus, a comparison of the observed period structure of a pulsating white dwarf with those of models provides a unique way of inferring the internal constitution of white dwarfs, and, in particular, the run of chemical composition as a function of depth. Although the potential of this technique has barely been tapped, it is likely that asteroseismological studies of white dwarfs will soon become a major contributor to our knowledge of the internal structure of these stars.

Additional Reading

D'Antona, F. and Mazzitelli, I. (1990). Cooling of white dwarfs. *Ann. Rev. Astron. Ap.* **28** 139.

Fontaine, G. and Wesemael, F. (1985). Les naines blanches. *La Recherche* **165** 464.

Kawaler, S. D. and Winget, D. E. (1987). White dwarfs: Fossil stars. *Sky and Telescope* **74** 132.

Ruderman, M. (1971). Solid stars. *Scientific American* **224** (No. 2) 24.

Van Horn, H. M. (1977). The uncertainty principle and the structure of white dwarfs. In *The Uncertainty Principle and the Foundations of Quantum Mechanics*, W. C. Price and S. S. Chissick, eds. Wiley, London, p. 441.

Van Horn, H. M. (1979). The physics of white dwarfs. *Physics Today* **32** 23.

Weidemann, V. (1990). Masses and evolutionary status of white dwarfs and their progenitors. *Ann. Rev. Astron. Ap.* **28** 103.

Winget, D. E., Hansen, C. J., Liebert, J., Van Horn, H. M., Fontaine, G., Nather, R. E., Kepler, S. O., and Lamb, D. Q. (1987). An independent method for determining the age of the universe. *Astrophys. J. (Lett.)* **315** L77.

See also **Nebulae, Planetary, Origin and Evolution; Stars, White Dwarf, Observed Properties; Stellar Evolution, Intermediate Mass Stars; Stellar Evolution, Low Mass Stars; Stellar Evolution, Pulsations.**

Stars, Winds

Joseph P. Cassinelli

The outer atmospheres of many classes of stars undergo a continual, more or less steady outward expansion. This phenomenon of a steady radial outflow is called a "stellar wind." There is such a steady expansion of the outer atmosphere of the Sun, known as the "solar wind." The standard unit for discussing the mass of matter that is lost per unit time by a star is solar masses per year or $M_\odot$ yr^{-1}. In the case of the Sun, the solar wind carries away only 10^{-14} $M_\odot$ yr^{-1}. This mass-loss rate would be undetectable even from the distance of the nearest star. Very luminous stars, on the other hand, have much larger mass-loss rates, as high as 10^{-5} $M_\odot$ yr^{-1}. These winds are sufficiently massive to be optically thick in strong resonance lines and in certain continua. As a result the winds are detectable through observations of the emergent stellar spectra. The most direct observational evidence for the presence of a stellar wind is a so-called "P Cygni line profile." Typical lines seen in the spectra of stars are absorption lines that are symmetrical about the line center. P Cygni profiles are not symmetrical, but have an absorption component that is shifted to short wavelengths, and an emission component shifted to longer wavelengths. (See Fig. 2 of the next entry, on Stars, Wolf–Rayet Type, for the origin of P Cygni profiles.) In hot stars with very strong winds, the Balmer α line of hydrogen at a wavelength of 6563 Å and strong resonance lines, such as that of the C^{+3} ion near 1540 Å, often have P Cygni profiles. The lines are called P Cygni after an early-type supergiant that has an especially massive wind and has been intensively studied over the past century. The width of the absorption components yields information about the expansion speeds of stellar winds. The terminal wind speeds from early-type stars tend to be large, 600–3500 km s^{-1}, whereas those from the K and M stars are small, 10–100 km s^{-1}.

Figure 1 shows a Hertzsprung–Russell (or HR) diagram in the form of a plot of stellar luminosity L (in solar luminosity units) versus the stellar effective temperature, T_{eff} (in kelvins). The zero-age main sequence (ZAMS) is shown as a solid line, and regions of observationally inferred winds are shown as shaded regions. There are two dominant regions. The luminous hot stars at the upper left of the diagram have winds with large mass-loss rates (10^{-7}–10^{-5} $M_\odot$ yr^{-1}) that have wind speeds, u_∞, that are typically a factor of 1–4 times the escape speed at the surface of the star. The luminous cool giants and supergiants also have large mass-loss rates, but with wind speeds that are less than the surface escape speed. The differences are caused by the mechanisms by which the winds are driven outward, as discussed below. In the case of the luminous cool stars shown in Fig. 1, there is a cross-hatched area that

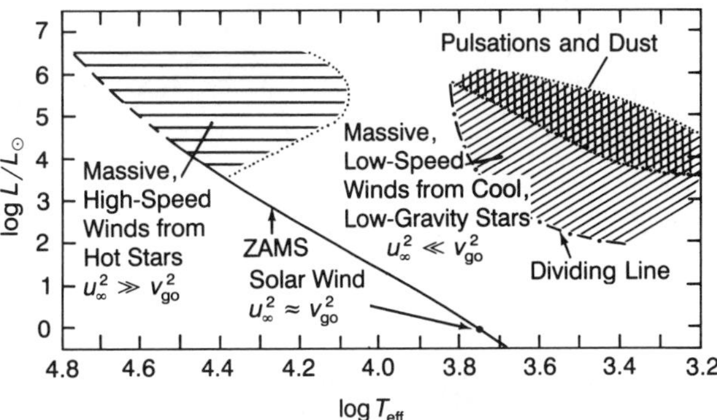

Figure 1. The Hertzsprung–Russell diagram shown in the form of a plot of stellar luminosity (L in solar units) versus stellar effective temperature (T_{eff} in kelvins). Two major zones are shown in which mass loss of the stars has been detected. At the high-temperature, high-luminosity zone, there are the fast radiation-driven winds of hot stars. For the cool luminous stars there are low-speed winds. The coolest and brightest of these are driven by a combination of pulsation and radiation pressure forces on dust that forms in the outer atmosphere. The location of the Sun on the zero-age main sequence (ZAMS), is shown by the small dot, indicated by the arrow at lower right center. (*From T. E. Holzer in NCAR / TN-306*, Proceedings of the 6th International Solar Wind Conference.)

indicates the presence of substantial atmospheric dust and regular stellar pulsations. Also shown is a dividing line that separates stars with detectable "coronae" and transition regions, but no detectable wind, from stars with no detectable corona or transition region, but with observationally inferred massive, low-speed winds. The existence of well-defined zones in the HR diagram indicates that a specific star may experience a variety of mass-loss mechanisms during its lifetime. The HR diagram is most commonly used to discuss stellar evolution. As a star ages it experiences a history of differing luminosities and surface temperatures and is found at different locations in the HR diagram.

In the following sections are presented, first, the observational evidence and diagnostics of stellar winds. A brief discussion follows of the mechanisms by which winds can be driven at various evolutionary phases of a star's life. Our focus is on single stars with more or less steady winds. Nonetheless, it should be mentioned that very interesting phenomena are associated with winds from stars in binary systems. In fact, the first strong evidence that cool giant stars have winds came from the observations of the binary star α Her. The low-speed outflow was shown to extend to several hundred stellar radii by noting that the wind absorption lines were also seen superimposed on the spectrum of the main-sequence companion star. In regards to the steady-state assumption, very careful observational studies appear to indicate that variability occurs in essentially all stellar winds.

OBSERVATIONAL EVIDENCE FOR MASS LOSS

Much of the current interest in stellar winds can be traced to the discovery of high-velocity outflows from O and B supergiants by Donald C. Morton and his co-workers. Their rocket-ultraviolet observations showed broad P Cygni profiles of lines of moderate stages of ionization, from ions such as C^{+3}, N^{+4}, Si^{+3}. Figure 2 shows an example of an ultraviolet spectral line of an early-type star. Figure 2 shows Morton's (1967) astounding observation of ζ Ori, an O9.5 Ia supergiant. Figure 3 shows the P Cygni profiles of N^{+4} (N v in spectroscopic notation) and O^{+5} (= O vi), for the very luminous O4f star ζ Pup and the much less luminous BO V main sequence star τ Sco. The ζ Ori and ζ Pup profiles clearly show the shortward-displaced absorption and longward-displaced emission components to the spectral lines. For stars like τ Sco, the wind evidence is less obvious, but the profiles that are shown here are nonetheless considered strong evidence for the presence of a stellar wind. The lines and their profiles provide powerful diagnostics of

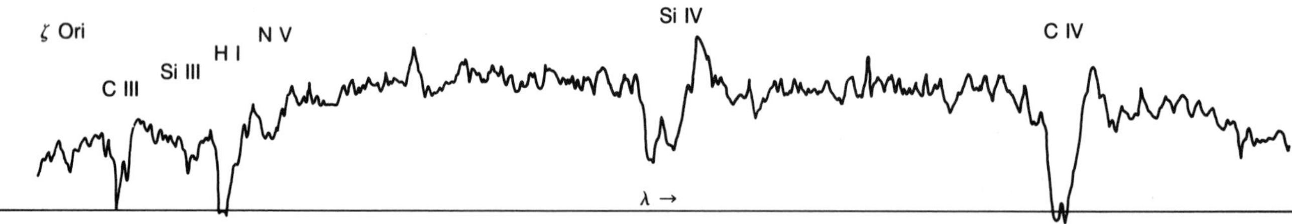

Figure 2. Donald C. Morton's early observation of the ultraviolet spectrum of ζ Ori O9.5 Ia. Wavelengths increase to the right from 1140–1630 Å. Shown are the P Cygni profiles of C iv and Si iv and shortward-displaced absorption lines of C iii, N v, and Si iii. [*From D. C. Morton (1967),* Astrophys. J. ***147*** *1017.*]

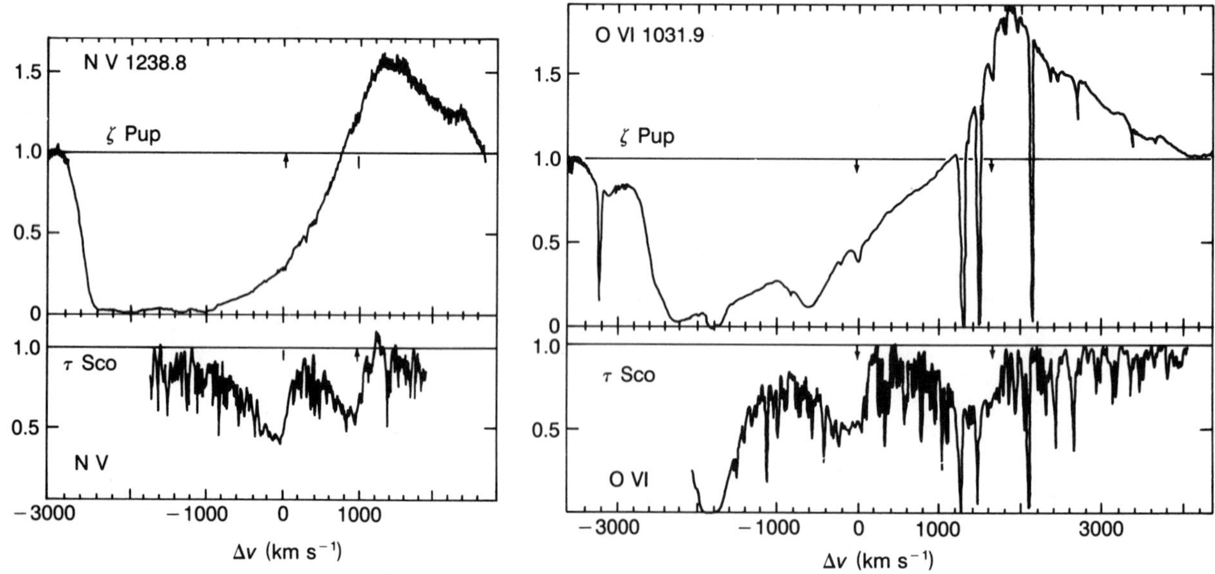

Figure 3. The spectra of ζ Pup O4f and τ Sco B0 V in the region of the resonance doublets of O vi and N v. The horizontal axes give the velocity (kilometers per second) in the frame of the star. The arrows indicate the laboratory wavelengths of the lines. In the O vi spectral region, the strong line at -1900 km s^{-1} is Ly-β. The sharp lines in the spectrum of ζ Pup are interstellar lines. [*Adapted from H. J. G. L. M. Lamers (1976),* Physique des Mouvements dans les Atmospheres Stellaires, *CNRS, Paris, p. 405.*]

the physical state of the plasma that is in the wind. For example, for a star with a sharp shortward edge of the absorption component, such as those seen in Fig. 3 for ζ Pup, one can use the Doppler formula $\Delta\lambda/\lambda = v_\infty/c$ to derive v_∞ the terminal wind speed; c is the speed of light. More detailed studies using computer models can provide the mass-loss rate and ionization state, and allow for studies of temporal changes in the wind conditions.

In regards to the winds of cool stars, the observational analyses are somewhat more complicated. There are ultraviolet lines that provide some diagnostic information about the winds from the cool luminous stars. The Mg^+ doublet (2795.5 Å, K line and 2802.7 Å, H line), appears in the ultraviolet spectra of cool stars as an extremely deep photospheric line, with a centrally reversed emission core. The emission signals the presence of a chromospheric temperature rise. If there is an expansion of the extended chromospheric region, the central reversal can be shifted to shorter wavelengths. This tends to cause the core of the Mg^+ lines to have an asymmetrical emission that can be used to diagnose the outflow.

Continuous energy distributions are also useful for studying stellar wind conditions. In the case of hot stars, one can rather easily detect excess infrared and radio emission that arises from the wind. In fact, the very best diagnostic of the wind mass-loss rate is the radio flux. Free-free opacity of winds increases towards long wavelengths. As a result, at radio wavelengths, one can only detect radiation that arises in the stellar wind at several hundred stellar radii from the star. There the density distribution is known to decrease as r^{-2}, so the radio flux is simply a measure of the radio size of the star, and this depends only on the unknown mass loss rate.

In cool stars the infrared observations often show distinctive emission peaks associated with dust that has formed in the rather cool (≤ 1000 K) expanding wind. The wavelength of the peak of the infrared emission provides an indication of the dust temperature (the hotter the dust, the shorter the wavelength of the peak emission). The magnitude of the infrared dust flux provides a measure of the amount of dust in the outflow. From this plus an estimate of the dust condensation radius and the wind speed, one can estimate the mass-loss rate. Other fascinating phenomena can be caused by the strong ambient infrared radiation. It can give rise to the pumping of OH and H_2O maser emission, the measurement of which can provide good information about the terminal velocity of the winds.

STELLAR WIND THEORY

The most basic version of stellar wind theory was developed by Eugene N. Parker in a classic article in which he predicted that the solar wind caused by the thermal pressure gradients in the hot solar corona would be supersonic. Soon after Morton's discovery of the winds of hot luminous stars, it was realized that those winds could not be explained as a consequence of thermal pressure gradients. This is because the speeds are so large (~ 2000 versus ~ 300 km s^{-1} for the Sun) that coronal temperatures of $> 10^8$ K would be required. However, the lines that were observed to have P Cygni profiles were from rather low ionization stages, such as C^{+3} and N^{+4}. These for the most part can be explained as being produced by photoionization by the hot photospheric radiation field. (The lines from higher stages of ionization, such as O VI, indicated the presence of somewhat elevated temperatures or the presence of some zones of x-ray emission). Another mechanism for driving the winds of hot stars was developed by Leon B. Lucy and Philip M. Solomon in 1970. They showed that the very strong resonance lines, such as that of C^{+3} 1550 Å, could drive off a fast wind because of its scattering and blockage of the photospheric light near 1550 Å. As the wind gas accelerates, the Doppler effect causes the C^{+3} line to block even more photospheric light and the wind is accelerated further. Lucy and Solomon could thus explain the high-speed winds. However, only one or a few lines were not sufficient to explain the large mass-loss rates ($> 10^{-6}\ M_\odot$ yr^{-1}) that were observed. John I. Castor, David C. Abbott, and Richard I. Klein later showed that there can be many lines contributing to the radiative opacity of the wind, and that essentially all of the extreme ultraviolet ($\lambda < 912$ Å) radiation of the star can be scattered. This made it possible to explain both the high wind speeds and large mass-loss rates of the hot stars. Over the subsequent years there have been a number of improvements to the basic Castor, Abbott, and Klein model, and it can now explain most classes of hot star winds. After the discovery of dust emission from cool stars, it became clear that radiative forces on dust could also explain some of the properties of cool star winds. The situation is somewhat more complicated, because it appears that another mechanism is required to provide appropriate conditions for the dust formation. Dust can form only if the temperature is less than ~ 2000 K and the density is sufficiently high. The radius at which the temperature of an extended atmosphere reduces to the dust condensation temperature can be derived from the radiative equilibrium condition. This states that the energy emitted by a dust grain equals the input energy. However, for the input energy to be low enough the grain must be at a height of several tenths of a stellar radius. However, at that location the density is too low. It has recently been noted that stellar pulsations can effectively increase the density scale height, and hence the combination of pulsation and dust formation can explain some of the cool star winds.

Not all hot star winds can be explained by the line-driven wind theory alone, and not all of the cool star winds can be explained by the combination of pulsation and dust formation. There are some classes of stars with strong winds that show no evidence of dust. Hot stars tend to be rapid rotators. The combination of rapid rotation and strong magnetic field gives rise to "magnetic rotator models." In the simplest version of centrifugal magnetic rotation theory, the field is sufficiently strong to cause solid body rotation out to the radius at which the centrifugal force equals the inward force of gravity. The mass-loss rate is determined by the centrifugal forces, but the terminal velocity is determined by the magnetic forces and line radiation forces operating in the supersonic region of the flow.

For the cool stars, acoustically driven and Alfvén-wave–driven wind models have been investigated. A major difficulty is in explaining the low speeds (≤ 100 km s^{-1}) of the winds. In regards to wind modeling, a useful general result is this: Energy or forces applied in the subsonic region of the wind increase the mass-loss rate of the star. On the other hand, heating mechanisms or forces that deposit energy or momentum in the supersonic part of a wind do not increase the mass loss but instead increase the terminal velocity of the wind. Thus the radiation force in spectral lines forms a natural mechanism for explaining the high-speed winds of hot stars, whereas the lack of high speeds in the cool stars poses severe difficulties for most wave- or magnetically driven wind models.

In summary, astronomers have developed good methods for determining the properties of stellar winds. For many, but not all of the stars, there exists a plausible wind explanation. The mass-loss rates from the winds are large enough to significantly change the total mass of a star during its lifetime. The high speeds of the winds of hot stars, coupled with their high mass-loss rates, means that these stars have a major effect on the surrounding interstellar medium. It appears that stellar winds will continue to be an active research area for many years, both because of the interest in explaining the outflow, and also because of the consequences that the winds have on stellar evolution and on the interstellar environment.

Additional Reading

Brandt, J. C. (1970). *Introduction to the Solar Wind*. W.H. Freeman, San Francisco.

Cassinelli, J. P. (1979). Stellar winds. *Ann. Rev. Astron. Ap.* **17** 275.

Cassinelli, J. P. and MacGregor, K. B. (1986). Stellar chromospheres, coronae and winds. In *Physics of the Sun*, Vol. 3, P. A. Sturrock, T. E. Holzer, D. M. Mihalas, and R. K. Ulrich, eds., D. Reidel, Dordrecht, p. 47.

Holzer, T. E. (1988). Acceleration of stellar winds. In *Proceedings of the Sixth International Solar Wind Conference*, National Center for Atmospheric Research TN 306, V. J. Pizzo, T. E. Holzer, and D. G. Sime, eds., Chap. 1.

Kudritski, R. P. and Hummer, D. G. (1990). Quantitative spectroscopy of hot stars. *Ann. Rev. Astron. Ap.* **28** 303.

Mihalas, D. M. (1978). *Stellar Atmospheres*, 2nd ed. W.H. Freeman, San Francisco, Chaps. 14 and 15.

See also **Interstellar Medium, Radio Recombination Lines; Interstellar Medium, Stellar Wind Effects; Nebulae, Cometary and Bipolar; Nebulae, Planetary, Origin and Evolution; Nebulae, Wolf-Rayet; Stars, Atmospheres, X-Ray Emission; Stars, Circumstellar Matter; Stars, Evolved, Circumstellar Masers; Stars, Pre-Main Sequence, Winds and Outflow Phenomena; Stars, R Coronae Borealis; Stars, Radio Emission; Stars, Red Supergiant; Stars, T Tauri; Stars, Wolf-Rayet Type; Stars, Young, Continuum Radio Observations; Stellar Evolution, Massive Stars; Sun, Coronal Holes and Solar Wind.**

Stars, Wolf–Rayet Type

Marc A. Azzopardi

In 1867, in the early days of stellar spectroscopy, two French astronomers from the Paris Observatory, Charles J. E. Wolf and Georges Rayet, using the 40-cm Foucault telescope with a prism spectroscope, discovered, in the constellation of Cygnus, three stars showing bright and broad lines standing out against a faint continuum. Since then, this kind of peculiar emission-line star has been called a Wolf–Rayet (WR) star. In spite of their enigmatic spectral features (normal stellar spectral features are seen in absorption, that is to say, dark lines on a relatively bright continuum) and the further detection of more objects of the same type, it was not until 1935 and the spectral classification system defined by Carlyle S. Beals and John S. Plaskett (adopted in 1938 by the International Astronomical Union Commission 29) that the study of this kind of object really began. Much work has been done since that time, but the true nature of WR stars and the part they play in stellar evolution is still the subject of controversy.

OPTICAL SPECTRAL CLASSIFICATION

The optical spectra of WR stars are dominated by emission lines of helium, carbon, nitrogen, and oxygen with occasionally some hy-

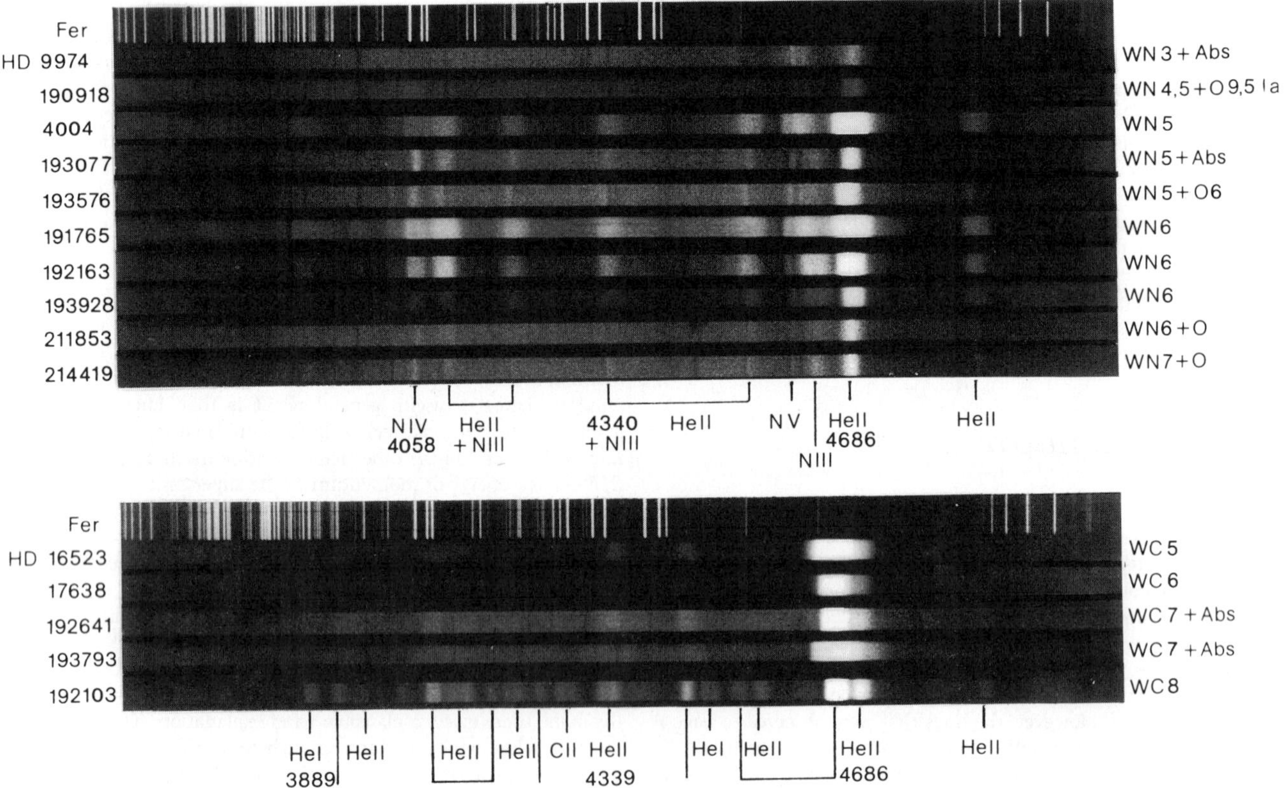

Figure 1. Typical spectra (3600–5000 Å wavelength range) of some galactic WR stars identified by their Henry Draper (HD) catalog numbers, which have been secured by Yvette Andrillat with the Haute-Provence Observatory (Centre National de la Recherche Scientifique) 120-cm telescope equipped with a grating spectrograph (65 Å mm^{-1} dispersion). The most typical spectral features, carbon (C), helium (He), and nitrogen (N) have been identified for the WN and WC subclasses, respectively. Roman numerals give the degree of ionization (i.e., I indicates the neutral element, II a once-ionized element, etc.). Wavelength calibration is provided by an iron spectrum (Fer). The WN or WC subtype is indicated for each WR star, as well as the spectral type of the companion for the binary stars; Abs means that absorption lines are also present. (*Plate from the Centre National de la Recherche Scientifique*).

drogen (Fig. 1). According to their spectral features, WR stars are set into two main subgroups: The WN subclass, containing the stars whose spectra are dominated by ionized helium and nitrogen lines (He II, N III, N IV, N V) and the WC subclass, containing the stars whose spectra exhibit ionized helium, carbon, and oxygen lines (He II, C II, C III, C IV, O III, O IV, O V, O VI). In the 1960s, a more subtle classification of WN and WC stars into different subtypes was worked out, mainly by H. J. Smith. Spectral classification criteria and new subtypes have led to the grouping of WR stars into three spectral sequences: WN2–9, WC4–10, and WO1–5, where the various subtypes depend on the ionization degree of nitrogen, carbon, and oxygen, respectively. Recently introduced by Michael J. Barlow and David G. Hummer, the WO sequence contains those WC stars that exhibit unusually strong oxygen lines. Attempts to quantify the spectral-type classification of the WN and WC sequences by measuring typical optical line strengths in galactic WR spectra brought the heterogeneity of the WN stars to light. In WC stars the line strengths are more homogeneous. This clearly shows that the present classification system of WR stars is still based on their optical spectral features (as they should be in accordance with William W. Morgan's philosophy of spectral classification) and not on the physical properties of their atmospheres, that is, effective temperature, luminosity, chemical composition, and so on.

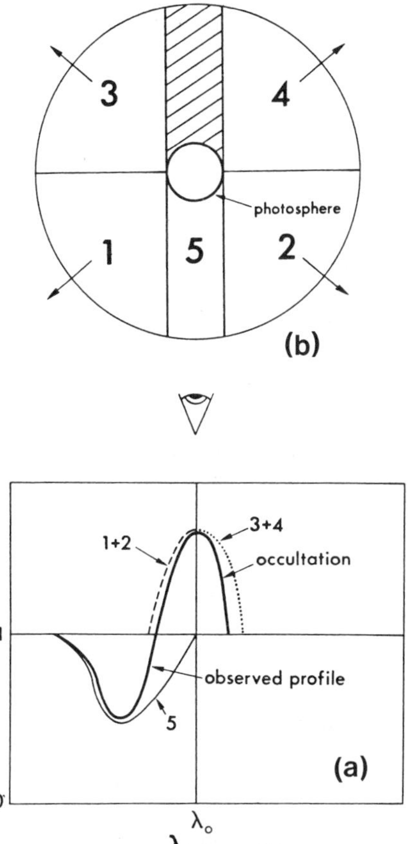

Figure 2. The so-called P Cygni line profile which characterizes mass loss in stars of high luminosity, such as WR stars, is mainly composed of a central emission component [contributions 1 + 2 and 3 + 4 in diagram (*a*)] bordered on its blue wing by an absorption component [contribution 5 in (*a*)]. The formation of such a line profile essentially results from a redistribution in wavelength (λ) of photons emitted in a radiative line transition with rest wavelength λ_0 due to Doppler effects caused by the rapid expansion of an envelope surrounding a central star [diagram (*b*)]. In this model, line photons emitted by atoms moving towards a distant observer [e.g., stellar envelope regions 1 and 2 in diagram (*b*)] give rise to the formation of a blue emission component. Similarly, the emission of photons by atoms that recede from a distant observer [e.g., regions 3 and 4 in (*b*)] accounts for the red emission wing. Finally, the absorption of the photospheric continuum of the WR star by atoms located between the stellar disk and the observer [see region 5 in (*b*)] leads to the formation of a blueshifted absorption component. All these components contribute to the observed profile. (*Graphs kindly provided by Jean Surdej, Universite de Liege, Institut d'Astrophysique.*)

THE MASS-LOSS PHENOMENON

Population I WR stars are very hot and luminous objects. There are also fainter WR stars of low masses which are the nuclei of some planetary nebulae. They will not be discussed here. Temperatures from 30,000–60,000 K have been derived, but large uncertainties exist. The luminosity of these stars, the total light radiation energy, is typically 10^5–10^6 times the luminosity of the Sun. However, what especially distinguishes WR stars from other very hot and luminous stars is that they have very extended and ionized atmospheres that account for the bright emission lines. This means that these stars lose their envelope in some way. Two possible mechanisms can explain this phenomenon: mass transfer in binaries (close stars in interaction) and mass loss by a stellar wind.

Simply, the stellar wind must be defined as stellar atmosphere material moving outward, driven by complex physical mechanisms not yet very well understood. One possible explanation is given by David C. Abbot, who imputes the stellar winds to pulsating motions arising from an instability of the deep stellar layers. Stellar winds in WR stars can reach velocities of up to 2500 km s^{-1}, and eject the stellar material into the interstellar medium, producing an expanding circumstellar envelope. NGC 6888 is a prime example of this [see Fig. 1(*a*) of the entry on Nebulae, Wolf–Rayet, p. 481]. Mass loss is revealed by the peculiar profiles of some spectral lines that are similar to those displayed by the bright star P Cygni (Fig. 2). Space observations of WR stars with the *IUE* (*International Ultraviolet Explorer*) satellite have shown that, in the far-ultraviolet region, emission lines with P Cygni profiles dominate the spectra of WR stars, confirming the presence of extremely strong stellar winds responsible for very high mass-loss rates (from 10^{-5}–10^{-4} solar mass per year), which are among the highest observed in early-type stars. Also this result is strengthened by radio observations of a sample of galactic WR stars with the Very Large Array radio telescope.

In fact, stellar winds occur over a great range of strength, in all stars of high luminosity, such as early- and late-type giants and supergiants (O, Of, OB, M stars, . . .), as well as in normal stars like the Sun. For instance, the solar wind ejects about 10^{-14} solar mass per year; although this effect is negligible for the Sun's evolution, it is, however, responsible for aurorae on the Earth.

CHEMICAL SURFACE ABUNDANCES

WR stars lose mass up to 100 times more rapidly than some stars of comparable luminosity. Consequently, most of their stellar envelope is being removed and the products of nucleosynthesis, normally confined in the stellar core, appear at the stellar surface. Because it is the longest stage in stellar evolution, core hydrogen burning is the power supply of most of the stars (90%), including the Sun. When hydrogen is exhausted, the stellar core shrinks and the central temperature becomes hotter, up to a few million kelvins. Then, core helium burning supplies the energy to almost all the remaining 10% of stars. Hence, WR stars seem to offer the wonderful opportunity to see the inner stellar material: Helium and nitrogen spectroscopically observed in the WN subtypes are the well-known products of hydrogen burning, and helium, carbon, and oxygen present in the WC subtypes are thought to be the main products of helium burning. *IUE* high-resolution spectroscopic observations, mainly by Allan J. Willis, tend to confirm the uncommon WR star surface abundance. However, a different opinion has been expressed by Anne B. Underhill, who considers that WR stars have a solar composition, but that differing electron temperatures and densities are the parameters that explain the visible spectra of WR stars. Whereas the major chemical components in the atmospheres of normal stars are hydrogen (~ 73%) and helium (~ 25%), WN stars have 98% helium, 1–2% nitrogen, less than 1% carbon, and some residual hydrogen on their surfaces, adopting Willis' point of view. As for the WC star surfaces, which contain no hydrogen at all, the composition is 10–50% carbon, a variable quantity of helium, and a few percent oxygen.

Another point of interest is the influence of WR stars on the interstellar medium in their vicinity. Owing to the very high velocity of the stellar winds and the related large amount of material ejected, part of WR star energy is converted into mechanical energy that has considerable influence on the dynamics of the surrounding interstellar environment. Furthermore, WR stars contribute significantly to the galactic enrichment in various specific elements. For instance, most of the ^{22}Ne in the Galaxy probably results from WC-star mass loss.

POPULATION AND STATISTICS

WR stars occur as extreme Population I objects that are either isolated or embedded in H II regions. WR stars are very rare and differ both in number and in distribution among the different subtypes from galaxy to galaxy.

The Galaxy

In the Galaxy, only 161 Population I WR stars have been identified, while the complete census of those objects is estimated at about 1200 by Andre Maeder and James Lequeux. Most of the known objects are listed in Karel A. van der Hucht's catalog. The spatial distribution of WR stars in the Galaxy is similar, as expected, to the massive OB star distribution, in the sense that they are concentrated toward the galactic plane, in particular in the inner spiral arms. That explains the difficulty of surveying WR stars at large distances from the Sun owing to great amounts of intervening absorbing dust, especially in the direction of the galactic center. However, one can estimate that the WR star population sample is reasonably complete within the nearest 3 kpc. In this volume-limited sample, there is a strong decrease of the number of WR stars with increasing distance from the galactic center, and statistics show that there are relatively more WC than WN types toward the galactic center.

Besides overcoming observational limitations, several other reasons led to surveys of WR stars in external galaxies. For instance, as massive stars and WR stars are closely related, the latter, which can more easily be identified through their strong emission-line features, may be used as tracers of the massive-star population. As mentioned above, one recognizes the great influence of the most massive stars on the enrichment of the interstellar medium of a galaxy in heavy elements, and therefore on the galaxy's chemical evolution. As yet, only nearby galaxies [Large Magellanic Cloud (LMC), Small Magellanic Cloud (SMC), M31, M33, NGC 6822 and IC 1613] have been intensively searched for WR stars, but the presence of this kind of object has also been reported in more distant systems.

Magellanic Clouds

Low-resolution spectroscopy, either through objective prisms or transmission gratings, is likely to be the most effective way to survey WR stars. This search method has been used for systematic WR star detection in the Magellanic Clouds, our nearest galactic neighbors. In the 1960s, several surveys of the LMC led to the identification of about 107 objects, most of them being listed in Jacques Breysacher's catalog. Systematic survey of the SMC by the author and Breysacher resulted in the identification of only eight WR stars. It is now considered that the census of the WR population in these systems is very close to complete, which therefore makes statistical studies possible. Hence, it is interesting to note that the ratio of the numbers of WN to WC stars is about 1 in the Galaxy, 4 in the LMC and 7 in the SMC. Also, the ratio of the number of red supergiants to the number of WR stars, which is about 9 and 24 for the LMC and SMC, respectively, led to the suggestion that the presence of heavy elements, because they influence the rate of mass loss, is probably an important factor governing the production of WR stars.

Other Local Group Galaxies

Spectroscopic surveys of WR stars in the nearby galaxies beyond the Magellanic Clouds are difficult, because they are relatively faint, and narrow-band filter imagery is, at present, the only efficient way to identify possible WR star candidates. Although not always possible so far, subsequent spectroscopy is an absolute requirement to confirm their WR star nature. Although more than 100 WR stars have been found in M33, mainly in OB associations by Philip Massey, at present only a few tens of these stars have been reported by Anthony F. J. Moffat and Michael Shara in M31, mostly due both to heavy obscuration and to the unfavorable orientation of this galaxy. Attempts by Taft E. Armandroff and Massey to survey NCC 6822 resulted in the identification of 12 possible WR star candidates. Subsequent slit spectroscopy allowed them to confirm the WR star nature of four of them, including the one previously known. At present only one WR star has been spectroscopically found in IC 1613, by Sandro D'Odorico and Michael Rosa.

EVOLUTIONARY SCENARIOS

Various scenarios have been suggested for WR star formation and evolution. In the 1970s, mass transfer in binaries, that is to say, the transfer of the hydrogen envelope of one star (WR progenitor) to its close companion, was the single suggested process for WR star formation. However, statistics on WR+OB galactic systems led to the conclusion that the overall percentage of WR binaries is less than 50%. Hence, although rather efficient, the mass-transfer scenario in binaries is inadequate to account for the formation of all WR stars. Other scenarios have been proposed, mainly by Peter S. Conti and Maeder, who suggest that WR stars are highly evolved remnants of massive stars. According to initial mass and mass-loss rates, a massive star may proceed toward different evolutionary stages. Owing to extremely strong mass loss, the outer stellar layers of the stars with an initial mass of 50–60 $M_\odot$ (solar masses) and more, are peeled off during the main-sequence (hydrogen-burning) and the blue-supergiant phases, and an O star becomes an Of star, then a blue-supergiant star, before becoming a WR star. If the initial mass of the stars is from 20–30 up to 50–60 $M_\odot$, the blue-supergiant phase is followed by a red-supergiant phase, as the mass loss is less important, but the stellar wind still removes the outer envelope of the red star, which finally becomes a WR star. Below about 20 $M_\odot$, the mass loss during both the blue- and the red-supergiant phases is insufficient for a star ever to reach the WR stage and the star will end as a red supergiant. Most WR stages will reach the final core collapse, possibly giving rise to a supernova explosion. Although these scenarios are commonly accepted at this time, a different interpretation of WR star formation is held by Underhill, who suspects that Population I WR stars are young massive stars still surrounded by a remnant of their natal cloud and a wind that originates in the disk.

Additional Reading

Abbot, D. C. and Conti, P. S. (1987). Wolf–Rayet stars. *Ann. Rev. Astron. Ap.* **25** 113.

Chiosi, C. and Maeder, A. (1986). The evolution of massive stars with mass loss. *Ann. Rev. Astron. Ap.* **24** 329.

Conti, P. S. et al. (1988). One perspective on O, Of, and Wolf–Rayet stars, emphasizing winds and mass loss, with remarks on environments and evolution. In *O Stars and Wolf–Rayet Stars*, NASA Spec. Pub. No. 497, P. S. Conti and A. B. Underhill, eds. NASA, Washington, DC, p. 79.

Maeder, A. (1983). Des astres qui s'evaporent: Les etoiles de Wolf–Rayet. *Recherche* **142** (14) 300.

Underhill, A. B. (1988). Another perspective on O, Of, and Wolf–Rayet stars, emphasizing model atmospheres and possibilities for atmosphere heating. In *O Stars and Wolf–Rayet Stars*, NASA Spec. Pub. No. 497, P. S. Conti and A. B. Underhill, eds. NASA, Washington, DC, p. 271.

van der Hucht, K. and Hidayat, B. (1991). *Wolf–Rayet Stars and Interrelations with Other Massive Stars in Galaxies*. Kluwer Academic Publ., Dordrecht.

See also **Galactic Structure, Optical Tracers; Nebulae, Wolf–Rayet; Stars, High Luminosity; Stars, Spectral Classification; Stars, Winds; Stellar Evolution, Massive Stars.**

Stars, Young, Continuum Radio Observations

John H. Bieging

This entry reviews the current state of knowledge about continuum radio emission from young stars. By "young stars," we mean those stars still evolving toward the main sequence in the Hertzsprung–Russell diagram, that is, toward lower luminosity and higher temperature as they complete their gravitational contraction. Excluded from the discussion will be other stars that emit radio radiation, such as the RS CVn binaries, the dMe flare stars, and the hottest (OB and Wolf–Rayet) stars with strong stellar winds. The young stars considered here have masses typically between one and two times the mass of the Sun and ages of a few hundred thousand years.

"Radio" emission here includes the wavelength range between about 20 cm and 0.35 mm, which is a limit set by the sensitivity of existing telescopes (notably the Very Large Array and new millimeter and submillimeter telescopes). The phenomena to be discussed would be observable in principle over a wider range of wavelengths, if sufficiently sensitive telescopes were available.

"Continuum" radiation is restricted to thermal and nonthermal processes not including atomic and molecular spectral lines. Thermal emission may arise from ionized gas or from cool dust particles near the star. Nonthermal emission is believed to come from relativistic electrons that are trapped in stellar magnetic fields and are radiating by the synchrotron process.

THERMAL EMISSION FROM IONIZED GAS

There is clear observational evidence that young stars eject material from their immediate vicinity—either from the stellar surface or possibly from a circumstellar disk—in a "stellar wind," which may have a velocity of 100 km s^{-1} or more. The reason such winds occur is not well understood, but evidently the wind is related to the process of star formation, because older stars (like the Sun) have only very weak winds at best.

Such winds are detectable by their radio continuum emission, if the wind material is ionized rather than neutral. It is well known that hot stars (spectral types O and B) can ionize the gas in their vicinity to produce H II regions such as the famous Orion nebula (M42). However, young stars with masses of only 1 or 2 solar masses are much cooler than O or B stars and so emit virtually no photons energetic enough to ionize hydrogen. Yet sensitive radio observations of young stars have revealed that at least some have ionized material in the form of a stellar wind. The lack of ionizing photons from these stars implies that another mechanism is responsible for the ionization of the wind. The most likely mechanism is "collisional" ionization, which taps the kinetic energy of the wind itself. Because the wind velocities are high, many times the speed of sound, shocks and turbulence in the flowing material are believed to be capable of heating the gas through dissipation of kinetic energy, and to result in ionization of at least some fraction of the stellar wind material.

An alternative source of ionization may be possible for some stars. If the density of the wind is high, a substantial number of hydrogen atoms may exist in the first excited state (principle quantum number $n=2$). These atoms can then be ionized by absorbing a Balmer-continuum photon. This mechanism obviates the need for a high-temperature source of the more energetic Lyman-continuum photons, which these relatively cool stars cannot supply.

Young stars are often found to be sources of bipolar molecular outflows, in which two lobes of gas are observed to be flowing away from the star in opposite directions. It has been argued that these bipolar flows consist of gas driven outward by a high-velocity wind from the star and collimated by some mechanism—possibly a circumstellar accretion disk, or by a more extended toroidal cloud of dense gas surrounding the star. The kinetic energy and momentum of these bipolar flows is generally found to be much larger than that in any wind from the central star, as determined by radio continuum measurements. This discrepancy has led to the suggestion that winds from young stars may be predominantly neutral, because the radio continuum observations can detect only the ionized part of the wind. This suggestion has been verified in at least a few cases, where high-velocity atomic or molecular gas is detected with sufficiently large mass-loss rates to account for the energy and momentum of the more extended bipolar flow. It is important to realize, then, that the ionized component of such stellar winds, as detected by radio continuum measurements, may represent only a small fraction of the total wind material.

The connection between stellar continuum emission and a stellar wind is suggested also by the observed association between radio emission from T Tauri stars (a class of very young stars) and the presence of Herbig–Haro (HH) objects. The HH objects are nebulous emission-line regions that apparently mark the interaction between a high-velocity stellar wind and a denser, stationary "placental" cloud out of which the star has formed. Thus, the observed correlation between the detectability of radio continuum emission from the star and the presence of associated HH objects is circumstantial evidence for a stellar wind mechanism common to both phenomena.

Another possible source of thermal radio emission in young stars is an accretion flow, where matter is still falling onto the surface of the star as part of the formation process. Such accretion is predicted to develop a specific density distribution that, if partly ionized (as discussed above) will produce a variation of radio intensity with frequency that is well matched by observations of at least some young stars. In this case, in contrast to the stellar wind, matter is flowing toward, not away from, the star.

NONTHERMAL EMISSION FROM STELLAR ACTIVITY

Some young stars produce radio continuum radiation that cannot be explained by thermal (i.e., bremsstrahlung or "free-free" radiation) emission from an ionized gas. The most important observational characteristics of this nonthermal emission include (1) rapid time variability of the observed intensity, with large variations (factor of 2) occurring in days or even hours; (2) a variation of the intensity as a function of wavelength that is incompatible with a thermal source—in particular, the intensity decreases strongly with increasing frequency; and (3) very small source sizes, implying very high source brightnesses, too high for any plausible physical temperatures of the stellar sources.

These characteristics can be explained by models in which the emission is produced by mildly relativistic electrons (with kinetic energies of a few times the electron rest mass, 0.511 MeV), trapped in magnetic structures attached to the star. The radio emission occurs by the synchrotron process (often called "gyrosynchrotron" for such moderate-energy electrons), where the particle radiates as it spirals around magnetic field lines. Such magnetic structures, in the form of loops anchored to sunspots, are known to occur on the Sun, especially at times of maximum solar activity. Energetic electrons can be released into magnetic loops by solar flares in the sunspot regions, causing solar radio "bursts." An analogous process is believed to occur on young stars, which show evidence for activity like that seen in the Sun, but of substantially greater magnitude.

Observations of x-ray emission from young stars, including x-ray flares (sudden brightening in the x-ray intensity) are one indicator of stellar activity. Another is the regular, periodic variation of the visible light from some stars, which is interpreted as the effect of large long-lived starspots (like sunspots) rotating with the star. Spots (i.e., regions of the stellar surface which are cooler by 500–1000 K than the average) are indicators of magnetic activity. Such stars tend to be rotating rapidly, and if their outer layers are pervaded by convective motions (as expected on theoretical grounds), conditions are appropriate for the creation of a dynamo within the star; this dynamo generates strong magnetic fields. Thus, the combination of convection and rotation is thought to be intimately related to stellar magnetic activity. The nonthermal radio radiation detected in such active stars is believed to occur when energetic electrons are injected into magnetic loops by a flare event, although the details of particle injection and acceleration are poorly understood, even in the case of the Sun.

Theoretical models for nonthermal emission from magnetically trapped relativistic electrons are more complicated than those for thermal emission from an ionized stellar wind. There are a large number of free parameters that are not well constrained by observations. Even so, or perhaps because of this complexity, it is possible to make theoretical models of nonthermal radio emission that can explain a number of aspects of the observations, such as the frequency dependence of the intensity. The general picture of magnetically confined energetic electrons as the source of the observed nonthermal emission is therefore likely to be correct, though a detailed picture has yet to emerge.

CONTINUUM EMISSION FROM DUST

When stars form through the gravitational collapse of a cloud of gas and dust, not all of the collapsing material can fall directly into the star, because of the conservation of angular momentum. Some of the collapsing cloud should take the form of a flattened disk that contains a modest fraction of the initial cloud mass but most of the original angular momentum. (Such disks may eventually produce planets by further condensation and collapse processes.) These circumstellar disks of gas and dust are expected on theoretical grounds, and it has recently become possible, with a new generation of radio telescopes operating at very short wavelengths of about 1 mm, to observe such disks directly by the thermal continuum radiation from the dust particles in them.

This dust component contains about 1% of the total mass of the disk. The temperature of the dust grains is expected to be around 50 K. At this temperature, the maximum thermal emission intensity occurs at a wavelength of 60 μm, in the far infrared. However, the dust should still emit at wavelengths around 1 mm, with a characteristic frequency dependence of the intensity. Recent observations in the wavelength range from 1.1–0.35 mm of T Tauri stars and related objects have detected emission at the level expected for such dusty circumstellar disks. (The observed emission is at least 100,000 times too strong to be coming from the star itself.) By comparing the observed intensities at several wavelengths with theoretical calculations, it is possible to estimate the size, temperature, and mass of the disk. Typical values obtained for stars comparable to the Sun are disk radii of about 100 AU, temperatures of between 15 and 40 K, and disk masses of a few tenths of the mass of the Sun. These values are in good agreement with the disk properties expected in the formation of solar-like stars. The amount of gas and dust left in the disk is believed to be comparable to that out of which the solar system formed, which suggests that planetary systems may be a natural consequence or by-product of the star formation process.

Additional Reading

André, P. (1987). Radio emission from young stellar objects. In *Protostars and Molecular Clouds*, T. Montmerle and C. Bertout, eds. Commissariat à l'Energie Atomique/Doc, Saclay, p. 143.

Cassinelli, J. P. (1979). Stellar winds. *Ann. Rev. Astron. Ap.* **17** 275.

Dulk, G. A. (1985). Radio emission from the Sun and stars. *Ann. Rev. Astron. Ap.* **23** 169.

Kuipers, J. (1985). Radio observable processes in stars. In *Radio Stars*, R. M. Hjellming and D. M. Gibson, eds. D. Reidel, Dordrecht, p. 3.

Lada, C. J. (1985). Cold outflows, energetic winds, and enigmatic jets around young stellar objects. *Ann. Rev. Astron. Ap.* **23** 267.

Montmerle, T. (1987). Stellar vs. solar activity: The case of pre-main sequence stars. In *Solar and Stellar Physics*, E. H. Schroeter and M. Schuessler, eds. Springer, Berlin, p. 117.

Padman, R., Lasenby, A. N., and Green, D. A. (1991). Jets in the Galaxy. In *Beams and Jets in Astrophysics*, P. A. Hughes, ed. Cambridge University Press, Cambridge, p. 484.

See also **Herbig–Haro Objects and Their Exciting Stars; Stars, Circumstellar Disks; Stars, Pre-Main Sequence, Winds and Outflow Phenomena; Stars, T Tauri; Stars, Young, Jets; Stars, Young, Millimeter and Submillimeter Observations.**

Stars, Young, Jets

William J. Zealey

Over the past decade it has become widely accepted that most young stellar objects (YSOs) are the sources of strong outflows. These often attain speeds in excess of 40 times the local speed of sound and have mass-loss rates of between 10^{-8} and 10^{-6} $M_{\odot}$ yr^{-1} (where $M_{\odot}$ is the solar mass). Such stars are usually deeply embedded in the molecular clouds from which they have recently formed. Optical jets are only one of many features associated with the impact of these supersonic winds upon their surroundings. Other optical phenomena include the faint emission nebulosities known as Herbig–Haro objects. At millimeter and radio wavelengths, high-velocity molecular gas is often observed in the emission lines of ^{12}CO.

Typical outflows exhibit some or all of the following features (Fig. 1).

1. A central stellar source embedded in a dense disk of material that may act to collimate the outflow. The visibility of the driving source depends on both the thickness of the disk and on its orientation.

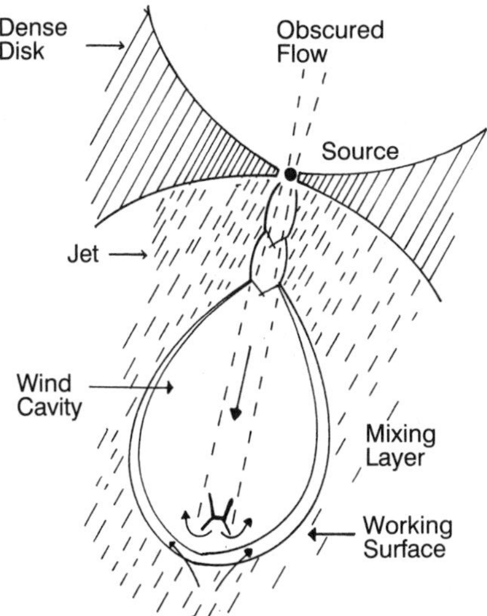

Figure 1. A diagram of a typical outflow from a young stellar object.

2. A region where the outflow appears as a highly collimated jet structure due to confinement of the outflowing material by the dense disk and ambient material.
3. A stellar wind cavity in which the collimated flow expands relatively freely. Surrounding interstellar material may be swept into the cavity. CO emission from such material has been recently observed at the edges of the outflow associated with the L1551 dark cloud.
4. A working surface at the tip of the cavity, where the collimated outflow impacts the interstellar medium. The resultant shock heating gives rise to Herbig–Haro objects.

MORPHOLOGY OF JETS

We know of more than 20 optical jets associated with molecular outflows and young stellar objects. The term "jet" is applied to any complex in which extended emission or a string of emission knots is observed along the axis of the flow. These jets may extend between 0.01 (2000 AU) and 1 pc (200,000 AU) from the source of the outflow. The most highly collimated jets are observed to have opening angles of only one or two degrees.

Well-collimated jets, the best example being HH34 (Fig. 2), often consist of strings of roughly equidistant knots. In HH34 a well-aligned jet is seen extending south from a stellar source. Like many other jets, that in HH34 terminates in a cusp of bright nebulosity. These cusps or clumps of emission nebulosity are interpreted as working surfaces where the jet finally encounters the stationary interstellar medium. In HH34, as in the majority of outflows, only a single-sided jet is observed. This is blueshifted and thus approaching the observer. Although only the southern jet is visible, a second working surface may be seen to the north of the HH34 source. This is associated with the northern, receding arm of the flow. It is likely, in this and similar cases, that the receding counterjet is obscured by the dense, tilted, circumstellar disk that surrounds the source of the flow.

Nearly all of the observed jets are associated with outflows that lie near the surfaces of dense molecular clouds. Flows that are more deeply embedded can only be observed in the infrared and radio wavelength regions. In general, searches at these wavelengths lack the necessary resolution to reveal highly collimated jets. Where suitable geometry prevails, both red- and blueshifted jets may be observed emanating from the central outflow source. In these cases there are often obvious differences in both the morphology and spectra of the opposed jets as they encounter differing density

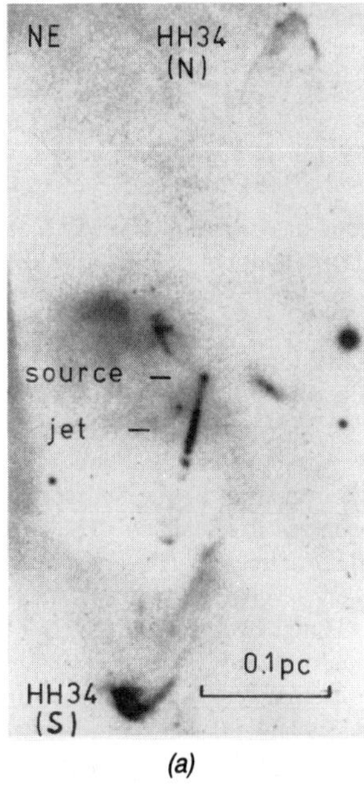

(a)

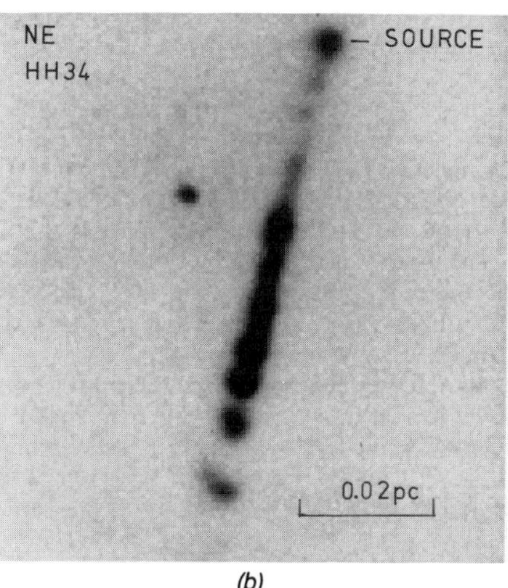

(b)

Figure 2. (*a*) An image of the HH34 outflow, obtained in the light of [S II] with a charge-coupled device. Notable are the highly collimated southern jet and the working surfaces of the flow visible both to the north and south of the source. Also visible are nebulosities near the source of the outflow, indicative of a dense circumstellar disk. (*Courtesy of Reinhard Mundt.*) (*b*) A closeup view of the HH34 jet seen in [S II] emission. (*Courtesy of Reinhard Mundt.*) Dimensions in (*a*) and (*b*) are indicated in parsecs (pc).

conditions as they expand into the surrounding material. M78/HH24 is one such outflow, where the approaching jet encounters a dense clump of stationary material 20,000 AU from the central source and ends in a bright, extended Herbig–Haro object. The redshifted jet apparently expands freely to the northwest.

Few jets are perfectly straight. Many, like HH46/47, have "wriggles" superimposed on them. An extreme case is found in HH12, where the jet appears to zigzag quite sharply. Such behavior has led several authors to consider precessing jet models of varying complexity. The suggested periods of such precession range from a few hundred years to a few thousand years.

Many of the central stars of outflow complexes have been observed at radio wavelengths. Continuum emission has been detected from several of the jets themselves. As observed in L1551

IRS5 this emission is generally elongated along the flow axis. In some cases, but not all, the spectral index is indicative of a nonthermal process.

SPECTROSCOPY OF JETS

Long-slit spectroscopy has proved a powerful method to observe the variations in velocity, velocity dispersion, electron density, and line excitation along the jets. In common with the spectra of Herbig–Haro objects, the emission spectra of the jets strongly resemble those observed from gas that has been rapidly compressed by a sudden shock wave. The cooling gas produces a unique spectrum of emission lines that may be used to derive the temperature and density of the emitting material. With the exception of some reflected light from the driving source, little optical continuum radiation is observed from the jets. The dominant emission is from the recombination lines of Hα and the forbidden emission lines ([S II], [O I], etc.). The spectrum of the knots is that of a low-excitation gas heated by a comparatively slow shock passing through the material at about 50 km s^{-1}.

The ratio of the intensity of the [S II] lines at 671.6 and 673.1 nm indicates that electron densities behind the shocks are typically 400–2000 cm^{-3}. This leads to an estimate of the density of hydrogen in the undisturbed flow of between 20 and 100 atoms cm^{-3}. In many cases the jet density is observed to decrease away from the source in a way expected of a steady wind traveling outward in an expanding jet.

Although molecular hydrogen emission has been observed in the infrared from the working surface of outflows, infrared spectroscopy of the jets is as yet in its infancy. The broad HH7–11 jet shows evidence for faint molecular hydrogen emission, broadened to widths of 200 km s^{-1}. This implies the presence of shocked material moving with velocities ranging from a few kilometers a second up to the full speed of the outflow. Such low-excitation emission from high-velocity molecules may arise at the edge of the flow in a similar way to the CO emission.

Searches for the expected strong emission from Fe II (1.67 μm) and molecular hydrogen (2.12 μm) in the more highly collimated jets of HH34, M78, and HH1/HH2 have as yet been unsuccessful.

DYNAMICS OF JETS

In all jets, material having large radial velocities is seen at visible wavelengths moving away from the central stars. These velocities are on the order of 100, but can be as high as 400 km s^{-1}. The relatively high velocity dispersions of up to 100 km s^{-1} observed are indicative of a complex structure within the jets.

Jets terminating in bright Herbig–Haro objects show evidence for deceleration near their ends. This provides further support for the view that these mark the working surfaces of the outflows.

Although a considerable amount of data on the proper motions of Herbig–Haro objects are available, little comparable information is available for the jets themselves. It has recently been shown that the knots in the L1551 jet have moved about 1 arcsec in 4 yr. When combined with the observed radial velocities, this may be interpreted as an approaching jet moving at 283 km s^{-1} and inclined at 53° to the line of sight.

The dynamic ages for the jets may be found from the lengths of the jets and their observed velocities. These ages range from a hundred to a few thousand years. This is a factor of 10 less than the age of the optical outflow phase that is estimated from the frequency of occurrence of flows in young stars.

SOURCES OF JETS

The driving sources of known optical jets are by no means all highly luminous objects. Bolometric luminosities of between 1–700 $L_{\odot}$ (solar luminosities) have been derived for sources associated with jets, a range of 3 orders of magnitude. In comparison, the driving sources of molecular CO outflows have a far wider range of luminosities of over 6 orders of magnitude.

The sources associated with jets include visible T Tauri stars (e.g., R Mon) and FU Orionis stars (L1551 IRS5). As well as often being irregular variables, the sources show highly developed P Cygni emission profiles associated with mass outflow. (P Cygni profiles are explained in Fig. 2 of the entry on Stars, Wolf–Rayet Type.) The terminal velocities of the outflowing gas derived from such profiles again lie in the range 100–400 km s^{-1}. This is similar to the radial velocities of the jets themselves. The mass-loss rates derived range from 10^{-9}–10^{-7} $M_{\odot}$ yr^{-1}.

In nearly all cases, optical or infrared reflection nebulae, indicative of a circumstellar disk, may be detected near the source. In the case of R Mon this results in a highly visible biconical or cometary nebula, the axis of which is closely aligned with the observed jets.

In a number of complexes the central source is visible only at infrared or at radio wavelengths. In these the central source may suffer up to 50 mag of visual extinction from a disk aligned closely to the observer's line of sight. One well-documented case is SSV63E, the source of the M78/HH24 outflow. Evidence for a disk is found in the form of adjacent pairs of infrared and optical reflection nebulosities that lie on both sides of source. A radio continuum source that was detected with the Very Large Array, though associated with the outflow and disk structure, is offset from the projected line of the jets.

Many outflow sources have infrared spectra in which the first-overtone bands of CO, more commonly seen in absorption, appear in emission superimposed on a slowly rising continuum spectrum. Such spectra are indicative of excited molecular gas associated either with a circumstellar disk, the outer layers of the stellar atmosphere, or with the outflow itself.

THEORETICAL MODELS OF JETS

Jet structures are not limited to young stellar objects. The discovery of jets in sources as distinct as young stellar objects, evolved stars, extragalactic objects, and quasars has led to a search for a common mechanism to explain this widespread phenomenon. The main areas requiring explanation are those of the wind-generation mechanism, the jet-collimation mechanism, and the jet-propagation mechanism.

It has been proposed by many authors that the energetic outflows associated with young stellar objects are due in some way to massive, circumstellar disks. Stellar mass loss from FU Orionis stars and other young stellar objects may be initially spherically symmetric but collimated close to the star due to the formation of de Laval nozzles. Some winds might even be intrinsically bipolar and collimated at the source. In either case subsequent collimation by a dense circumstellar disk may well result in a highly collimated flow. Direct observation of the jet-forming regions close to the stellar surface is difficult; hence the question of the formation mechanism is at present largely unanswered.

Once collimated the jets may develop instabilities and internal shock structures visible as strings of emission knots. The knots in the highly collimated astrophysical jets appear similar to the luminous shock cones seen in the exhausts of jet engines. By analogy the bright knots in astrophysical jets may be explained in terms of internal shock structures or "Mach disks" in the flow. Such internal shocks may be excited as the jet first overexpands on leaving the nozzle, while attempting to match its pressure to its surroundings. Subsequently the expansion is reversed and the jet envelope reconverges. The process may be repeated until the jet finally expands freely. The outflowing material thus encounters a series of stationary shocks or Mach disks as it moves down the jet axis. Although the observations are closely matched by this model, the proper motions and radial velocities observed in jets disagree with the stationary shock structures expected from a steady wind. The large motions observed in the knots might, however, be matched by an unsteady or periodic outflow of this kind.

An alternative explanation may lie in instabilities in the flow. Kelvin–Helmholtz instabilities occur when two fluids move across each other while in pressure equilibrium. Where the jet is denser than its surroundings, surface waves at the interface between the jet and the surrounding medium narrow the jet. The formation of internal shock waves produces regularly spaced emission knots traveling at velocities close to that of the jet.

In the case of a diffuse jet, in which the jet density is less than that of the surroundings, it is suggested that the head is embedded in a thick cocoon of backflowing gas. Perturbations in this cocoon or instabilities at its surface may excite shock waves in the jet, although these are predicted to be less regular than those for a dense jet.

CONCLUSION

A considerable amount of observational data has been gathered on young stellar objects and their associated jets. Although the overall structure of the jets and outflows is fairly clear, the detailed physics behind them is less so. The following represent just a few of the many unanswered questions relating to jets.

What physical mechanism drives the mass loss?
What is the collimation mechanism for the jets?
Are the bright knots associated with Mach disks instabilities in the flow, or are they cloudlets moving with the flow?
Do jets precess and if so, what is the precession mechanism?

I wish to thank R. Mundt and T. Ray for providing the CCD frames (charge-coupled device images) which made this paper possible.

Additional Reading

Bally, J. (1983). Bipolar gas jets in star forming regions. *Sky and Telescope* **66** 94.

Cohen, M. (1988). Circumstellar matter and winds in young stars. In *Pulsations and Mass Loss in Stars*, R. Stalio and L. A. Wilson, eds. Kluwer Academic Publ., Dordrecht, p. 83.

Mundt, R. (1984). Jets from young stars. *Sky and Telescope* **67** 130.

Mundt, R. (1984). Star birth on display. *Sky and Telescope* **67** 227.

Mundt, R. (1987). A Herbig–Haro object unmasked. *Sky and Telescope* **73** 30.

Mundt, R. (1987). Recent observations of Herbig–Haro objects, optical jets and their sources. In *Circumstellar Matter*, I. Appenzeller and C. Jordan, eds. D. Reidel, Dordrecht, p. 147.

Norman, C. A. (1987). Theory of bipolar flows and jets from young stars. In *Circumstellar Matter*, I. Appenzeller and C. Jordan, eds. Reidel, Dordrecht, p. 51.

Norman, M. L., Smarr, L., and Winkler, K.-H. A. (1985). Fluid dynamical mechanisms for knots in astrophysical jets. In *Numerical Astrophysics*, J. Centrella, J. M. LeBlanc, and R. L. Bowers, eds. Jones and Bartlett, Boston, p. 88.

Padman, R., Lasenby, A. N., and Green, D. A. (1991). *Jets in the Galaxy*. In *Beams and Jets in Astrophysics*, P. A. Hughes, ed. Cambridge University Press, Cambridge, p. 484.

Rodriguez, L. F. (1981). Searching for the energy source of the Herbig–Haro objects. *Mercury* **10** (No. 2) 34.

Rodriguez, L. F. (1988). Interstellar and circumstellar toroids. In *Galactic and Extragalactic Star Formation*, R. E. Pudritz and M. Fich, eds. Kluwer Academic Publ., Dordrecht, p. 97.

Schwartz, R. D. (1985). The nature and origin of Herbig–Haro objects. In *Protostars and Planets II*, D. M. Black and M. S. Mathews, eds. University of Arizona Press, Tucson, p. 405.

See also **Herbig–Haro Objects and Their Exciting Stars; Nebulae, Cometary and Bipolar; Stars, Pre-Main Sequence, Winds and Outflow Phenomena; Stars, Young, Continuum Radio Observations.**

Stars, Young, Masers

William J. Welch

During the more than two decades since maser emission at radio wavelengths was discovered emanating from interstellar molecules in galactic clouds, studies of this fascinating effect have given us important insights into processes in the interstellar medium (ISM) on very small spatial scales. In fact, the effect is observable in two classes of sources, red giant star envelopes and regions of high density in the ISM, but this summary will deal only with the masers in the latter regions. Two of the first molecules found in the ISM by their radio emission, OH and H_2O, exhibit maser emission, and among the approximately 50 further molecules discovered through their radio spectra in the past two decades, another five show this effect. These include SiO, CH_3OH, H_2CO, and NH_3. However, they occur in fewer sources, and, with the possible exception of SiO, have not yet provided as many new insights.

OH and H_2O remain the most important maser molecules. They show us the regions where the more massive stars are now forming. They indicate the regions of the highest densities. They allow us to determine the magnetic fields in these dense regions. The regular motions of the dense maser clumps provide information about the dynamical processes of star formation. Perhaps most important, the application of the methods of statistical parallax to the motions of the H_2O masers gives us direct distance measurements at very large distances.

SIGNPOSTS OF STAR FORMATION

Wherever in interstellar clouds there are OH and H_2O masers, there are nearby imbedded infrared sources and/or compact ionized (H II) regions. The latter contain hot O and B stars whose ultraviolet radiation ionizes the gas. The compact H II regions have lifetimes of only about 10^5–10^6 yr, 10% or less of the lifetimes of these massive stars. After this interval, most of the gas and dust surrounding the stars is dispersed by the winds and radiation of the stars. Thus, the maser process identifies the early stages in the lives of the massive stars. In those regions where there are only imbedded infrared sources, the luminosity of these sources is nevertheless 1000 times that of the Sun or more, so that the masers generally indicate the formation of the more massive stars.

OH MASERS

The OH molecule has a complex spectrum at radio wavelengths, and there is maser emission in nearly a dozen of the lowest energy states. Observations employing very long baseline interferometry (VLBI) show the regions of emission to be clusters of small bright spots, with each cluster containing several sources on the order of 10^{14} cm in size. The spots are often completely linearly or circularly polarized. The overall spread in Doppler radial velocities of the spots is typically 10–15 km s^{-1}. Densities in the clumps must be in the range 10^5–10^8 hydrogen molecules per cubic centimeter, so that the masers reveal clumpy density concentrations. The observed frequency shift between the right- and left-circularly polarized emission of a given spot is due to the Zeeman effect and typically implies high magnetic field strengths in the range of 1–10 mG (milligauss). This most interesting result suggests that the field is frozen into the gas, having increased in strength as the clump collapsed and became more dense. It is consistent with collapse from average ISM conditions of one hydrogen atom per cubic centimeter and a few microgauss of magnetic field, with the field proportional to the square root of density, as is theoretically expected. The emission from the OH masers is relatively stable, with typical fluctuations of 10% over months.

H_2O MASERS

Like the OH masers, the H_2O masers are tiny bright clusters of spots. Unlike the OH, the H_2O masers are much more intense, fluctuate wildly on time scales of weeks, show relatively little polarization, and have a wide range of radial velocities, sometimes as large as several hundred kilometers per second. Densities in the H_2O spots are higher, in the range of 10^7–10^{10} hydrogen molecules per cubic centimeter. The Zeeman splitting in this molecule is weaker than in OH, and in the few regions where polarization is observed, magnetic fields in the range of 20–50 mG have been inferred. Again the field is what is expected for such dense clumps if it were frozen in and the clump collapsed from a region of average interstellar conditions.

The most important characteristic of the H_2O masers is their wide spread in Doppler radial velocity. Proper-motion studies of these spots using VLBI reveal transverse velocities comparable to the radial velocities, showing that the radial velocities are truly kinematic. One of the intriguing features of young stellar objects is their powerful winds, and the first direct evidence of this came from a proper-motion study of water masers in Orion showing their outflow from an infrared star. (It should be noted that the distributions of radial velocity and positions of both OH and SiO masers in Orion show evidence of a disk around the central infrared star.)

EXCITATION MECHANISMS

The general principles of maser operation are understood. For some sources satisfactory detailed models have been worked out, but for many cases there is no general agreement as to the most likely detailed mechanism. The maser emission is perhaps the most stunning evidence of the lack of thermodynamic equilibrium in the ISM. The molecular gas must be in contact with thermodynamic reservoirs at different temperatures through gas collisions or interaction with radiation in order to achieve the necessary population inversion for the maser action. For the water masers, energy arguments coupled with observations indicate that the high-energy source of excitation must be contained within the small emitting region. These spots are in motion at high speed, and the most likely source of the needed energy is in the collision of the clump with stationary material. However, there is considerable debate about the details of how that energy might pump the maser.

DISTANCE MEASUREMENTS

One of the most important results of the discovery of H_2O masers has been the realization that they can be used to measure distances directly on very large scales. The water masers are very small, very intense, and moving very fast. Thus, a distant cluster of maser spots may be used to provide a direct distance measurement by the classical method of statistical parallax. Roughly, the ratio of the radial velocity dispersion to the rate of angular expansion gives the distance. This technique has been used successfully in measurements of the distances of the Orion nebula, 500 pc, and the galactic center, 7100 pc. The classical use of this technique on star clusters works only to about 50 pc. The next step, of course, is the determination of distances to those galaxies that exhibit strong water masers. Work is in progress on the nearby spiral M33, and will be extended to greater distances when the Very Long Baseline Array is complete.

Additional Reading

Cohen, R. J. (1989). Compact maser sources. *Rep. Prog. Phys.* **52** 881.

Cook, A. H. (1977). *Celestial Masers.* Cambridge University Press, Cambridge.

Elitzur, M. (1982). Physical characteristics of astronomical masers. *Rev. of Mod. Phys.* **54** 1225.

Goldreich, P. (1980). Interpretation of circumstellar masers. In *Interstellar Molecules*, B. H. Andrew, ed. D. Reidel, Dordrecht, p. 551.

Reid, M. J. (1989). The distance to the galactic center: R_0. In *The Center of the Galaxy*, M. Morris, ed. Kluwer Academic Publ., Dordrecht, p. 37.

Reid, M. J. and Moran, J. M. (1981). Masers. *Ann. Rev. Astron. Ap.* **19** 231.

See also **Masers, Interstellar and Circumstellar; Masers, Interstellar and Circumstellar, Theory; Stars, Pre-Main Sequence, Winds and Outflow Phenomena.**

Stars, Young, Millimeter and Submillimeter Observations

James P. Emerson

This entry describes observations of young stars over the decade of wavelengths between 3 and 0.3 mm corresponding to frequencies between 100 and 1000 GHz. The shortest submillimeter wavelength at which observations are possible from the driest ground-based observatories is 0.35 mm. The discussion is limited to young stars that have already become visible and hence whose spectral type may be determined optically, thus excluding protostars and other young stellar objects that are optically obscured by large amounts of dust. Young hot stars associated with H II regions are also excluded from the discussion because the complexity of such regions makes isolating the stellar emission very difficult. The young stars are less than about 10^7 years old, or some 0.1% the age of our Galaxy, have already acquired most of their mass, and are contracting toward their equilibrium configuration on the main sequence in the Hertzsprung–Russell diagram. Young stars can be surrounded by remnant material from the parent interstellar cloud, which may still be accreting onto the young star, or which may form into planets (certainly this was the case for the solar system) and the mass and dynamics of such material can be deduced from millimeter and submillimeter observations and used for comparison with theories of star and planetary formation.

STAR FORMATION

Before describing the observations it is useful to have a conceptual framework of star formation. Stars form out of clouds of interstellar gas and dust under the influence of gravity, and initially their energy output is derived entirely from the gravitational energy released as the material contracts. Such objects are known as protostars and have core temperatures too low to sustain nuclear fusion. The protostar contracts and is heated by release of gravitational potential energy. Eventually the resulting increase in density causes increases in the opacity and temperature of the core, and leads to conditions close to hydrostatic equilibrium, under which the outward pressure of the protostellar gas is approximately balanced by the inward gravitational pressure due to material in the outer layers. At about the same time the predominant energy transport mechanism in the interior changes from radiation to convection. At this stage the object becomes known as a pre-main-sequence star, with energy production still being predominantly gravitational. The pre-main-sequence star then contracts until the density and temperature in its core become high enough for fusion of hydrogen to helium to dominate the energy production. At this point the contraction is halted and the object becomes a bona fide star in which an equilibrium situation is reached where energy is derived from nuclear fusion. The star then settles onto the main sequence, where it will spend the majority of its life.

The accretion of material onto the star from the surrounding material is unlikely to take place in a spherically symmetric way, as the clouds are rotating. This makes collapse easier parallel to the rotation axis than perpendicular to it, so that the parent cloud will evolve a flattened disk-like component onto which material will fall before passing through the disk to fall onto the star. During much of this process the forming star is surrounded by a thick cocoon of obscuring dust and is optically invisible, so millimeter, submillimeter, and infrared observations must be used to study the energetics and physical properties of the star and its cocoon material. Eventually the young star loses its thick cocoon of obscuring dust and becomes optically visible. The dust may be pushed away from the star by radiation, or by a wind from the star; it may be destroyed, or it may agglomerate into structures that will cause relatively little extinction of the stellar light, by forming into a thin disk or into bodies of planetary or asteroidal size. At this stage the object is a young optically visible star of the type to be discussed below.

The star-forming material consists of molecular gas and dust. Even though the majority of the material is gaseous, the dust being typically only 0.5% by mass of the interstellar material, the dust component is very important because of its opacity to the stellar radiation field at optical and shorter wavelengths. The dust grains absorb radiation emitted from the star. This warms the dust, and then collisions between the molecules of the surrounding gas and the dust grains in turn heat the gas. Thus the presence of dust around a star strongly affects the temperature of the region around the forming star, as well as attenuating its optical flux.

Young stars are not currently *directly* detectable at millimeter and submillimeter wavelengths, as may be demonstrated by calculating the flux density radiated at 1 mm by a 4500 K blackbody star of twice the radius of the Sun at the 140-pc distance of one of the nearest known star formation regions in Taurus. This flux density is 2500 times fainter than current typical detection limits of 0.01 Jy. However, the young stars are surrounded by very large volumes of gas and dust closely associated with them, which, because this material is cool, can be best studied in the millimeter and submillimeter regions of the spectrum.

THE OBSERVATIONS

Emission Lines from Molecular Gas

The millimeter and submillimeter regions contain emission lines arising from excited rotational levels of molecules in the gaseous fraction of the star-forming material. Emission from rotationally excited CO at 2.6 mm, and many other emission lines from other molecules, are used to probe the temperature, density, and distribution of the material around star-forming regions. However, the gas studied in these observations is generally not *closely enough* associated with the star to warrant consideration as part of the young star itself, but rather represents the molecular cloud out of which the star was born.

CO observations at 2.6 mm show that certain regions have high-velocity redshifted and blueshifted molecular gas in a bipolar configuration and that the centers of these configurations coincide with infrared sources, which are assumed, and in some cases known, to be dust-embedded young stars. These bipolar outflows consist of material swept up by a wind originating at or near the young stars, and their bipolar nature led to the suggestion that the material close to the star must be arranged in an equatorial disk, either so that material could flow out unimpeded in the polar directions, or so that an intrinsically anisotropic wind off the disk itself could be responsible for the outflow, although the mechanism driving these winds is still not understood.

The hypothesis of a disk, also suggested by theoretical considerations of the collapse of a rotating cloud, and the perceived history of formation of our solar system, led to an interest in studying optically visible solar-mass stars to look for the presence of disks. The star most thoroughly studied, HL Tau, is a T Tauri star at a distance of about 140 pc. It is in one of the nearest regions of star formation, which is essential for determining the true spatial structures around young stars, which are even harder to resolve in more distant objects. The fact that HL Tau is visible allows one to estimate its spectral type and hence its mass and stellar luminosity. Using the Owens Valley Millimeter-Wave Interferometer, it was shown that the 2.7-mm ^{13}CO emission close to this star is confined to a (projected) annular configuration and that the clumps on either side of the star appear to be moving towards and away from the Earth, respectively, the velocities being consistent with material in a disk of extent about 4000 AU orbiting a 1-$M_\odot$ star, suggestive of a proto-solar-system disk.

Continuum Emission from Dust

The continuum emission comes from solid dust particles that are heated to temperatures of a few 10s kelvin by the stellar radiation field (and perhaps also by other mechanisms, such as energy generated by viscous forces during accretion) and which then radiate away their heat in the form of blackbody radiation modified by the emissivity of the dust grains, which decreases with increasing wavelength. For cool regions, which contain much of the mass, significant flux is radiated in the submillimeter region of the spectrum. Observations of the millimeter and submillimeter portions of the spectrum (along with measurements at shorter wavelengths) are used to determine the complete spectra of the regions, from which their total luminosities may be deduced; the wavelength at which the spectrum begins to fall off with increasing wavelength, which indicates the temperature of the coolest parts of the dust material; and the millimeter and submillimeter flux density, which, together with the dust temperature and emissivity, can be used to deduce the mass of material.

HL Tau and many other T Tauri stars, which have masses similar to the Sun, have broad infrared spectral distributions that can be explained as a superposition of modified blackbody emissions from material with a range of temperatures. Millimeter and submillimeter continuum observations can determine the mass of this material if the temperature and emissivity of the dust grains are known, because, at these long wavelengths, the emission should become optically thin (because the grain opacity decreases rapidly with decreasing frequency) and in the Rayleigh–Jeans limit the emission is linearly dependent on the (uncertain) dust temperature. By contrast, at shorter wavelengths, radiation is more dependent on dust temperature and the material is more likely to be optically thick, in which case only lower limits to the dust mass can be determined. Several groups made millimeter and submillimeter continuum measurements of young T Tauri stars in 1988 and 1989 and used their results to deduce the mass of material around the young stars. The observational data are in good agreement, but different groups made different assumptions about the dust emissivity at 1 mm, leading to variations in deduced masses. The masses found are nevertheless all in the range 0.001–1 $M_\odot$ (solar mass), the majority being of the same order as the minimum mass of material that is thought to have been present in the nebula from which our planetary system formed, suggesting that there is sufficient material in many of these systems to form planetary systems. The mass estimates assume the dust is mostly in the form of grains smaller than a few hundred micrometers in radius, and there could be mass contained in much larger structures that would contribute little to the observed millimeter/submillimeter continuum flux.

Various arguments suggest that for these young stars most of the nearby material is in the form of a disk: The CO observations referred to above; the fact that only blueshifted optical emission lines, formed in winds up to 100 AU from the star, are seen, suggesting that the redshifted material is obscured by an optically thick disk of radius 100 AU; near-infrared high-resolution images of scattered light; the far-infrared (0.06-mm) emission is consistent with that from a disk which is optically thick at 0.06 mm; if the mass of material deduced from the millimeter and submillimeter observations were distributed in a uniform sphere of radius 100 AU around the stars it would produce more visual extinction than observed. The material must thus be distributed so that its emission can be seen, but so that it extinguishes relatively little of the optical emission from the young star. Confining the material in a disk is a simple way to satisfy this condition. In several cases there are indications that the disks themselves radiate more luminosity than would be expected if they were thin and simply intercepting optical light and reemitting infrared photons. This can perhaps be explained by a disk whose thickness increases with distance from the young star, but, at least in some cases, another and more likely explanation is that some luminosity is being generated in the disk itself, as well as in the star. A natural source for such luminosity would be energy dissipated by viscous forces in the disk as material accretes through the disk onto the young star. However, the observations do not fit the most straightforward of such models, where the accretion is at a steady rate throughout the disk.

Further studies of millimeter and submillimeter line and continuum emission, particularly observations of nearby young stars made with high angular resolution, promise further understanding of the potential of these young stars to form associated planetary systems.

Additional Reading

Adams, F. C., Emerson, J. P., and Fuller, G. A. (1990). Submillimeter photometry and disk masses of T Tauri disk systems. *Astrophys. J.* **357** 606.

Adams, F. C., Lada, C. J., and Shu, F. H. (1988). The disks of T Tauri stars with flat infrared spectra. *Astrophys. J.* **326** 865.

Beckwith, S. V. W., Sargent, A. I., Chini, R. S., and Gusten, R. (1990). A survey for circumstellar disks around young stellar objects. *Astron. J.* **99** 924.

Emerson, J. P. (1988). Infrared emission processes. In *Formation and Evolution of Low Mass Stars*, A. K. Dupree and M. T. V. T. Lago, eds. Kluwer Academic Publ. Dordrecht, p. 21.

Lada, C. J. (1985). Cold outflows, energetic winds, and enigmatic jets around young stellar objects. *Ann. Rev. Astron. Ap.* **23** 267.

Sargent, A. I. and Beckwith, S. (1987). Kinematics of the circumstellar gas of HL Tau and R Monocerotis. *Astrophys. J.* **323** 294.

Shu, F. H., Adams, F. C., and Lizano, S. (1987). Star formation in molecular clouds: Observation and theory. *Ann. Rev. Astron. Ap.* **25** 23.

Shu, F. H., Ruden, S. P., Lada, C. J., and Lizano, S. (1991). Star formation and the nature of bipolar outflows. *Astrophys. J. (Lett.)* **370** L31.

See also **Interstellar Clouds, Molecular; Molecular Clouds and Globules, Relation to Star Formation; Protostars; Protostars, Theory; Stars, Pre-Main Sequence, Winds and Outflow Phenomena; Stars, T Tauri.**

Stars, Young, Rotation

Jérôme Bouvier

Because forming stars inherit rotation from their parental cloud, the study of stellar rotational velocities brings insight into the process of star formation. In this respect, the rotational velocities of young stars, being the least altered by stellar evolution, are especially valuable. Later, as the star ages, rotational velocity variations are caused both by structural changes in the stellar interior and by the star's interaction with its surroundings. The rotational history of young stars thus provides clues to early stellar evolution.

Two methods are commonly used to measure the surface rotation of stars. One consists of measuring the Doppler broadening of line profiles recorded on high-resolution spectrograms, which yields an estimate of the star's rotational velocity projected onto the line of sight. The other method applies to stars whose surface is covered with dark or bright spots and is somewhat similar to that first used in 1611 by Galileo Galilei to measure the rotation period of the Sun. As surface spots are carried across the stellar disk by rotation, the star's luminosity is modulated with a periodicity that directly reflects the star's rotation period, and with the knowledge of the stellar radius, the star's equatorial velocity is deduced. Because rotational modulation is not affected by projection effects, it provides much more accurate rotational velocity estimates than spectroscopic methods. Its application, however, is restricted to spotted stars, which usually have spectral types G or later.

Theoretical models predict that protostars should rotate very rapidly due to increased rotation during the protostellar collapse. Yet, because protostars are deeply embedded in dusty cocoons that are opaque to visible light, their rotational velocities cannot be directly measured. The time by which a star first becomes visible depends upon its mass. Low-mass stars appear at optical wavelengths long before they reach the main sequence, whereas high-mass stars evolve much more rapidly and already are on the main sequence when their surroundings become transparent to visible light. Hence, pre-main-sequence stars with a mass less than a few solar masses are the least-evolved objects whose rotational velocities can be measured.

At an age between 1 and 10^6 years, these stars still lie near their birthplace within large dark clouds distributed along the plane of the Galaxy. Historically, they have been divided into two broad classes according to their mass. Pre-main-sequence stars more massive than 2 solar masses, first defined as a class by George H. Herbig, have spectral types A or B and are now known as the Herbig Ae-Be stars. Less massive ones, with spectral types G or later, were first classified by Alfred H. Joy in 1945, and are referred to as T Tauri stars, after the prototype of the class. Lately, an additional distinction has been introduced between Joy's classical T Tauri stars which exhibit strong emission lines in their optical spectrum, and newly discovered weak-line T Tauri stars (also called weak or "naked" T Tauri stars), which are much less active than classical ones. Classical T Tauri stars drive powerful winds and are thought to simultaneously accrete material from a circumstellar disk that is a remnant of the star formation process, whereas weak-line T Tauri stars lack both strong winds and accretion disks.

In principle, the concentration of large numbers of young stars within dark clouds covering a small area of the sky facilitates their study. Furthermore, all stars within a given dark cloud have roughly similar ages, which offers the opportunity to perform statistical studies of stars having a well-defined evolutionary status. However, a major difficulty arises from the remote location of stellar formation regions, which mostly lie at 150 pc or more. At such a distance, stars of a solar luminosity appear as faint objects that are difficult to observe, and this explains why astronomers had to await the development of new detector technology in the early

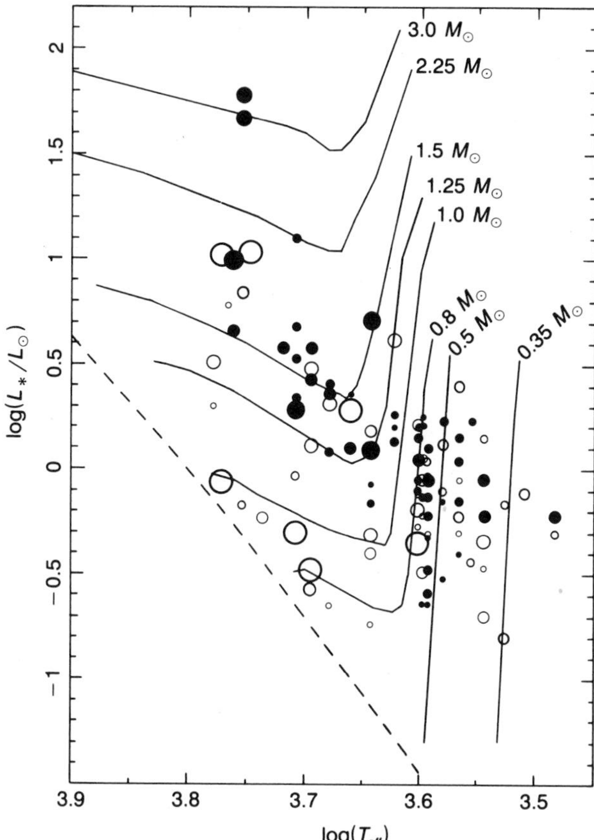

Figure 1. Rotational velocities of low-mass pre-main-sequence stars in the Hertzsprung–Russell diagram. The stellar luminosity normalized to the solar luminosity is plotted versus the stellar effective temperature. Weak-line and classical T Tauri stars are represented by open and dark circles, respectively, and the area of a circle is proportional to the stellar projected velocity. Theoretical pre-main-sequence evolutionary tracks for stars of various masses are shown as solid lines, and the zero-age main sequence is drawn as a dashed line.

1980s to be able to start large-scale studies of rotational velocities in young stars.

The first systematic study of rotation in T Tauri stars was performed in 1981 by Stuart N. Vogel and Leonard V. Kuhi. Although they were unable to measure projected velocities below 30 km s^{-1} because of instrumental problems, they found that most of the stars they observed had projected velocities below this limit. Since then, rotation rates down to a few kilometers per second have been measured for more than 100 T Tauri stars, and the results are illustrated in Fig. 1, where T Tauri stars are plotted in a theoretical Hertzsprung–Russell diagram. Open and dark circles distinguish between weak-line and classical T Tauri stars, and each circle's area is proportional to the stellar projected velocity. Solid lines represent the theoretical paths followed by pre-main-sequence stars of various masses as they contract to the main sequence.

Due to limited spectral resolution, only upper limits of 10–15 km s^{-1} are available for the projected velocities of 20% of the stars. For the remainder, projected velocities range from 6–100 km s^{-1}, with a mean velocity of the order of 25 km s^{-1}. As first made clear by the study of Vogel and Kuhi, the rotational velocities of T Tauri stars are much lower than those predicted for protostars. Comparison of T Tauri stars with more evolved stars of similar masses shows that T Tauri stars rotate much faster, on the average, than main-sequence stars of a solar mass or less, whose projected velocities rarely exceed 5 km s^{-1}. Only main-sequence stars with a mass larger than 1.2 solar masses have mean rotation rates in excess of 50 km s^{-1}, which implies that the more massive T Tauri stars must accelerate during their evolution prior to the main sequence.

The mean projected rotational velocity of pre-main-sequence stars is found to scale with stellar mass. Most T Tauri stars with a mass less than about 1.25 solar masses have projected velocities below 20 km s^{-1}, whereas more massive ones rotate twice as fast on the average; this trend continues toward Herbig's Ae-Be stars with masses larger than 3 solar masses, which usually rotate at rates between 100–200 km s^{-1}. The observed increase of rotation with mass in pre-main-sequence stars may be a direct consequence of the star formation process; that is, more massive stars would form with higher rotation rates. Alternatively, low-mass pre-main-sequence stars may already have experienced strong rotational braking as a result of magnetically channeled mass loss occurring at their surface. Indeed, this mechanism was originally proposed by Evry Schatzman in 1962 to account for the low rotation rates of low-mass main-sequence stars in the solar neighborhood. More massive stars are less affected by rotational braking because they lack deep convective envelopes where the stellar magnetic field is amplified.

The finding that weak-line and classical T Tauri stars have similar rotation rates clearly indicates that the strength of the emission-line spectrum of T Tauri stars shares no direct relation with rotational velocity. It also suggests that weak-line and classical T Tauri stars have a comparable rotational history, which is somewhat surprising in view of their widely different properties. This result may in fact indicate that weak-line T Tauri stars are evolved from classical ones, as the latter dissipate their circumstellar disks.

The next piece of information astronomers have about the rotational velocities of young stars relates to low-mass stars in open clusters. It takes approximately 3×10^7 years for a solar-mass star to complete its pre-main-sequence evolution and, at this point, it has usually moved far away from its parental cloud. In some cases, however, stars have small enough space motions to remain close to each other over much longer time scales. Then, as the gas cloud rapidly dissipates, an open star cluster results. Known open clusters have an age ranging from a few 10^7 to a few 10^8 years, which amounts to only a tiny fraction of the 10^{10} years or so a solar-mass star spends on the main sequence. Thus, low-mass stars in young open clusters, although slightly older than T Tauri stars, are still in their infancy.

Until recently, the opinion prevailed that the rotational velocities of low-mass stars steadily decrease with time from the star's birth to the end of its main-sequence evolution. This belief received support from the slow decline of rotation with age in cool main-sequence stars, first hinted at by Robert P. Kraft in 1967 and later quantified by Skumanich's law, which states that surface rotation decreases as the square root of age. The more recent finding that T Tauri stars of a solar mass or less rotate much faster than old main-sequence stars of similar masses was still consistent with this hypothesis. In the early 1980s, however, the discovery of young main-sequence stars in open clusters with rotational velocities in excess of 150 km s^{-1} implied a much more complex rotational evolution than previously envisioned.

This is illustrated in Fig. 2 where the rotational velocities of low-mass stars in three open clusters are plotted against spectral type. From top to bottom, each panel displays stars belonging to open clusters of successively greater age, thus providing a time sequence of the evolution of rotation in low-mass stars as they start their evolution onto the main sequence.

Assuming that T Tauri stars are the progenitors of low-mass stars in the Alpha Persei cluster, the existence of stars with rotation rates in excess of 20 and up to more than 150 km s^{-1} in this

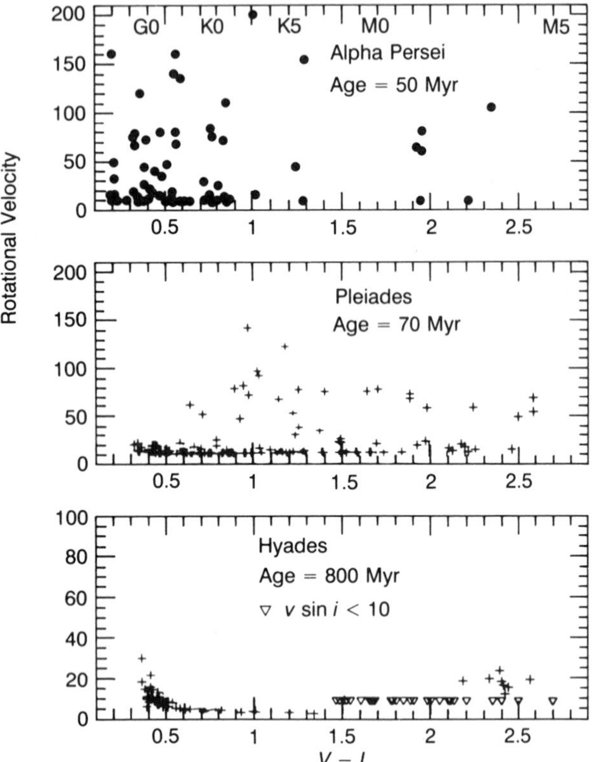

Figure 2. Rotational velocities (in km s^{-1}) of stars less massive than 1.3 solar masses in three young open clusters. The estimated cluster ages are indicated in each panel, and the correspondence between $(V - I)$ photometric index (in magnitudes) and spectral type is given in the upper panel. For the Alpha Persei and Pleiades clusters, projected velocities are shown and stars plotted at 10 km s^{-1} have only upper limits set on their projected velocity. For most of the stars with $(V - I)$ less than 1.4 mag in the Hyades cluster, true equatorial velocities were derived from rotational modulation studies, whereas only projected velocities were measured for less massive stars. Upper limits of 10 km s^{-1} on the projected velocity are shown as inverted triangles. Note the change of vertical scale in the lower panel. [*By permission from J. R. Stauffer (1987),* Cool Stars, Stellar Systems, and the Sun, *J. L. Linsky and R. E. Stencel, eds., Springer-Verlag, Berlin, p. 182.*]

cluster clearly indicates that they must be strongly accelerated during their pre-main-sequence evolution. In case no rotational braking occurs, increased rotational velocities are expected to result both from the star's contraction as it approaches the main sequence and from the development of a radiative core in the stellar interior. There is, however, an approximately equal number of stars in the Alpha Persei cluster with rotational velocities below 20 km s^{-1}. In fact, the existence of both slowly and rapidly rotating stars at any mass is one of the most puzzling aspects of the rotational velocity distribution of low-mass stars in the Alpha Persei cluster.

Comparison between the projected velocities of stars in the Alpha Persei and Pleiades clusters shows that all rapidly rotating stars with spectral type G in the Alpha Persei cluster are spun down to rotational velocities less than 20 km s^{-1} by the time they reach the age of the Pleiades cluster, that is, less than 2×10^7 years after their arrival onto the main sequence. This suggests that slowly rotating G stars in the Alpha Persei clusters may be slightly older than rapidly rotating ones, and thus have already experienced strong braking while on the main sequence. Unfortunately, determinations of the relative ages of these stars are not accurate enough to test this hypothesis. Alternatively, the assumption that T Tauri stars are the progenitors of all low-mass stars in open clusters may be questioned. Pre-main-sequence stars having much smaller rotational velocities than T Tauri stars, indeed a few kilometers per second at most, would reach the main sequence with rotation rates lower than 20 km s^{-1} in spite of pre-main-sequence spin up, and would thus account for slowly rotating stars in young clusters. However, because there is at present no observational evidence for the existence of this elusive population of very slowly rotating pre-main-sequence stars, astronomers usually prefer the more conventional view that rotational braking prevents at least some stars from spinning up during their evolution prior to the main sequence.

Figure 2 shows that the early rotational evolution of low-mass stars onto the main sequence depends upon their mass. Comparison between the rotational velocity distributions of stars in the three clusters displayed in this figure indicates that it takes a longer time to spin down the surface of less massive stars. By the age of the Pleiades cluster, only stars with spectral type G have been braked, whereas rapid rotation is still found among stars of lower mass. And in the much older Hyades cluster, only the least massive stars are left with significant rotation rates. Current models show that only the outer convective envelope of stars is braked during early main-sequence evolution, whereas the inner radiative core remains in rapid rotation. This may explain why less massive stars retain rapid surface rotation for longer time scales, as their thicker convective envelope is more difficult to brake.

As rotation measurements became widely available in the last 10 years, astronomers' knowledge of the rotational evolution of young stars has dramatically improved. It is now clear that pre-main-sequence stars have moderate rotation rates and that at least some of them must experience strong spin up in order to reach the main sequence with rotation rates in excess of 100 km s^{-1}. Once on the main sequence, the surface of low-mass stars is rapidly spun down to rotational velocities of less than 20 km s^{-1}. For a solar-mass star, this phase of strong braking lasts only a few thousandths of its total main-sequence lifetime. During the rest of the star's evolution onto the main sequence, rotational braking is still effective, but at a much lower rate, as predicted by Skumanich's law. Although the rotational evolution of young stars is roughly understood, several enigmas remain, which will undoubtedly stimulate increased efforts on the part of both theoreticians and observers in the coming decade.

Additional Reading

Bouvier, J. (1989). Stellar rotation prior to the main sequence. In *Rotation and Mixing in Stellar Interiors*, N.-J. Goupil and J.-P. Zahn, eds. Springer-Verlag, Berlin, p. 47.

Bouvier, J. (1990). Rotation in T Tauri stars. II. Clues for magnetic activity. *Astron. J.* **99** 946.

Hartmann, L. W. and Noyes, R. W. (1987). Rotation and magnetic activity in main sequence stars. *Ann. Rev. Astron. Ap.* **25** 271.

Kraft, R. P. (1970). Stellar rotation. In *Spectroscopic Astrophysics*, G. H. Herbig, ed. University of California Press, Berkeley, p. 385.

Stauffer, J. R. (1987). Rotational velocity evolution on and prior to the main sequence. In *Cool Stars, Stellar Systems, and the Sun*, J. L. Linsky and R. E. Stencel, eds. Springer-Verlag, Berlin, p. 182.

Stauffer, J. R. and Hartmann, L. W. (1986). The rotational velocities of low-mass stars. *Publ. Astron. Soc. Pac.* **98** 1233.

See also **Pre-Main Sequence Objects: Ae Stars and Related Objects; Stars, Rotation, Observed Properties; Stars, T Tauri; Stellar Evolution, Rotation.**

Stellar Associations, OB-Type

Catharine D. Garmany

OB associations are loose groupings of O- and B-type stars found in the plane of the Galaxy. They range up to several hundred parsecs in diameter, and are frequently associated with giant gas and dust clouds. In the hierarchy of physically bound groupings of stars, an OB association is the closest thing to nothing that is still something. Why, then, are such amorphous structures important to the study of stellar evolution and the interstellar energy balance?

O- and B-type stars are at the upper end of the mass and luminosity spectrum. They are prodigious sources of energy as well as continuing suppliers of matter to the interstellar medium through the action of their stellar winds. They are very short lived by stellar standards, and O-stars in particular cannot move far from their place of birth in the few million years they spend on the main sequence, unless they are assisted by some external accelerating process, such as a three-body slingshot mechanism. Thus a nest of these stars is almost like a snapshot of them at birth. With temperatures between 45,000 and 25,000 K, the OB stars emit their maximum radiation in the ultraviolet, and ionize the surrounding gas out to tens and even hundreds of parsecs. Such H II regions are seen in our own and in external galaxies, and are indicators of star formation, as well as "standard candles" for distance. The evolution of the O- and B-type stars is particularly important, as these are the precursors of supernovae, which in their explosions eject newly formed heavy elements to the interstellar medium.

STRUCTURE

The term "association" was first used in 1947 by Victor A. Ambartsumian, who also suggested that these groups might be expanding. The motions of the stars in a few nearby associations were studied by Adriaan Blaauw and others in the 1950s, and indeed the expansion was confirmed. Although the space density in associations is much too low to bind them gravitationally, the time it takes for them to be pulled apart by galactic tidal forces is comparable to the lifetimes of the most massive stars. In a comprehensive review, Blaauw discussed the associations within 1 kpc of the Sun. The question of the dimensions of associations is still best described in Blaauw's words: "For the smallest associations the dimensions are usually as well defined as those of galactic clusters. With increasing size, however, the boundaries become increasingly vague and... the association gradually merges with the general field population." Few have had the temerity to catalog the boundaries of associations. The most commonly used listing is found in *Trans. IAU Vol.* **12b** (1966). More importantly, this reference standardized the nomenclature: Cep OB1, Cep OB2, etc., to designate the associations according to constellation. Apparently overlapping associations at different distances are common, especially in the direction of galactic spiral arms. Examining the association boundaries suggests that the upper limit to the size of associations is about 200 pc, although they range down to the size of large open clusters (e.g., 10 pc).

The problem of the dimensions of associations is tied to that of membership. Although the Hertzsprung–Russell (HR) diagram of an association provides a good start, the contamination of the lower-mass stars by field stars requires knowledge of radial velocities and proper motions to resolve the question of membership. Lack of such information makes it impossible to say anything about late B-, A-, and F-type members. In many cases photometric data are insufficient to construct a reasonable HR diagram beyond the brightest stars, or else the data suggest very different distances for the stars, as happens in Cep OB2.

Associations are known to consist of subgroups of stars, spatially distinct and having different photometric ages. For example, the subgroups in the Scorpius-Centaurus Sco OB2 association have nuclear ages, computed from isochrones based on theoretical stellar evolutionary tracks, of 5×10^6, 11×10^6, and 14×10^6 years. These subgroups have been noted among the nearest associations, within 1000 pc. However the data are not sufficient to resolve groups of different ages from their isochrones in the HR diagram, there have been no attempts to look for subgroups in more distant associations. There is some evidence that the most concentrated subgroup is usually the least evolved and contains the largest proportion of early-type stars. These observations can best be understood in the context of models for star formation discussed below.

The interesting connection between associations and open clusters is not well understood. Aside from cases in which poor distance determinations of both cluster and association make them appear to coincide, many associations encompass open clusters, which can be likened to raisins in a muffin. In general the OB clusters within associations are slightly larger than the average open cluster, and they contain O-type stars, which is not true of the majority of open clusters. Whereas a lone open cluster is 5 pc or less in diameter, those within the associations are typically about 10 pc in diameter, especially if they contain O-type stars. A catalog of the most luminous members of galactic clusters and associations was prepared by Roberta M. Humphreys.

STAR FORMATION

Much recent work on associations has concentrated on the giant molecular clouds (GMCs) in which they are often embedded. These GMCs form in the galactic arms and provide the raw material for star formation. When the star forming efficiency is low (less than a percent), the stellar association left behind after the gas and dust have cleared will be unbound, and consequently will expand naturally. Star formation takes place in the dense cores of these warm molecular clouds, and *Infrared Astronomical Satellite* observations have revealed ultracompact H II regions that probably represent the stars that are not yet visible. To become visible, stars must either move out of the clouds or blow the gas away, thus creating a blister in the GMC. Stars may spend between 10 and 20% of their lives thus enshrouded. The ages of the GMCs are on the order of 10 million years, based on their crossing time of the galactic spiral arm. This places an upper limit to the period in which star formation can take place. Figure 1 is a schematic based on Anneila I. Sargent's work on Cep OB3, showing the relation of the two subgroups to the molecular cloud. Lyman-continuum radiation from an OB cluster will drive an ionization front into the adjacent molecular cloud. As this shocked layer becomes gravitationally unstable, it will collapse to form more massive stars. Thus the successive subgroups form.

But what initiates star formation? One suggestion is the passage of galactic spiral density waves, although there is no good evidence for the smooth progression in age between the stellar subgroups that would be expected were this the principal mechanism: In Sco OB2 the oldest subgroup is between the two younger ones. Another idea is that a supernova initiates the fragmentation and collapse of the cloud.

Do all of the stars form coevally? We have already discussed the age difference among the subgroups, with the implicit assumption that the stars within the subgroup formed at the same time. However, there has been considerable discussion about the star-forming history of young clusters, some of which are within OB associations. The original idea proposed by George H. Herbig in 1962 was that low- and intermediate-mass stars form first, and the birth of O-type stars terminates star birth by disrupting and dispersing the gas. Subsequent studies claimed a more quantitative relation between masses and ages of cluster members: This is usually known as sequential star formation. It is not clear whether star formation in clusters is in any way different from that in OB

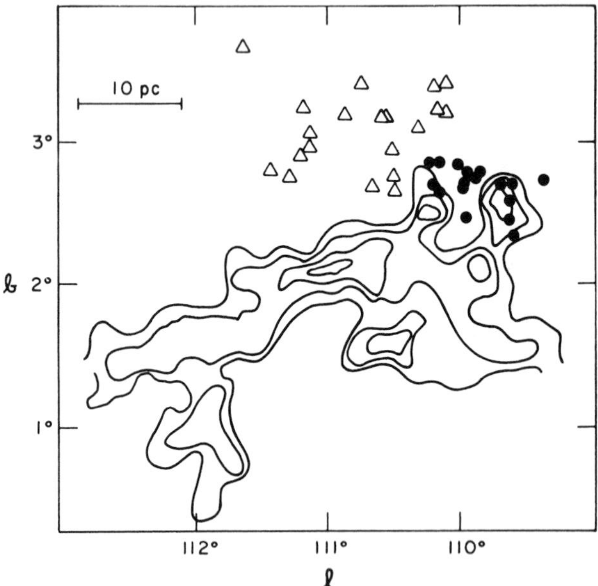

Figure 1. The association Cep OB3, showing two stellar subgroups and the isotherms representing the associated molecular cloud. Filled circles indicate the younger subgroup, triangles the older group. The horizontal and vertical coordinates are galactic longitude and latitude, respectively; the scale is indicated by the line 10 parsecs long at upper right.

associations. As already mentioned, clusters within associations seem to differ from other clusters both in size, and of course, in the presence of OB stars. In fact, a survey of all stars of spectral type O6.5 and earlier in the Galaxy shows that about 90% are clearly or probably members of associations, and of these stars, about 60% are in or very near an OB cluster. Perhaps an OB cluster is necessary for the production of early O-type stars. Richard B. Larson has suggested an evolution in the mass spectrum of clusters, with low-mass stars forming first over an extended region and higher-mass stars forming later in a more condensed core. Lower-mass stars may form under less-extreme conditions: T Tauri stars do not appear to require H II regions and ionization fronts to start their fragmentation process, and may form as a result of undisturbed cloud fragmentation.

MASS FUNCTION IN ASSOCIATIONS

There is a generally held belief that all stars form in clusters and associations. To test this, it is necessary to know how many stars, and of what mass distribution, form in a typical association. The mass of a cluster can be estimated by extrapolating the luminosity function, which is just the number of stars per magnitude interval, and converting this information to mass via the mass–luminosity relation. This method, when applied to associations, typically suggests a total stellar mass of about 500 solar masses from the stars above 5 solar masses. Associations are not known to produce stars with masses below about 2 solar masses, but this could reflect the observational limit and the fact that such stars have not yet reached the main sequence. If stars as small as 1 solar mass are produced in associations, then the total stellar mass involved would increase to about 1600 in a typical association, and the field star population could be explained. However, there is repeated evidence that the luminosity function in OB clusters and associations cuts off somewhere around 5 solar masses. Is this indeed true, and if so, why?

The initial mass function (IMF) for massive stars, defined as the probability distribution for the formation of stars of different masses, is an important clue to star formation. The theory must be able to explain the power-law distribution observed: At the same time, the observations of the IMF should produce a power-law exponent that can clearly be shown either to remain constant with stellar mass or to vary in a predictable way. The competing ideas, as applied to massive stars, pit fragmentation versus coagulation in the giant cloud from which the stars form. The derived slopes for the IMF are heavily dependent on the assumptions made in converting stellar observables to stellar mass, and on the completeness of available data. Generally, the number of stars in young clusters scales as mass to the power -1.5.

RUNAWAY AND FIELD STARS

No discussion of associations would be complete without a comment on runaway stars. These are massive O- and B-type stars found far from any OB cluster or association. Their velocities, based on proper motions and/or radial velocities, indicate that they originated in a nearby association. The first suggestion to explain their high velocities was that they were originally close binary stars whose primary exploded as a supernova, ejecting the star we now see. Another theory put forth recently proposes that runaway stars have suffered ejection following close gravitational encounters with binary systems.

There are still a significant number of massive O-type stars that appear to have no connection with known associations. It may be that some massive stars form in isolated gas clouds, although given the amorphous nature of association boundaries and uncertain distances of stars contained therein, care must be taken in deciding which can properly be labeled as field stars. Serious work on associations in nearby galaxies, in particular the Magellanic Clouds, is now underway, and the answers to many questions posed here will undoubtedly be found.

Additional Reading

Blaauw, A. (1964). The O associations in the solar neighborhood. *Ann. Rev. Astron. Ap.* **2** 213.

Boland, W. and van Woerden, H., eds. (1985). *Birth and Evolution of Massive Stars and Stellar Groups*. D. Reidel, Dordrecht.

Gies, D. (1987). The kinematical and binary properties of association and field O stars. *Astrophys. J. (Suppl.)* **64** 545.

Humphreys, R. M. (1978). Studies of luminous stars in nearby galaxies. I. Supergiants and O stars in the Milky Way. *Astrophys. J. (Suppl.)* **38** 309.

Larson, R. (1982). Mass spectra of young stars. *Mon. Not. R. Astron. Soc.* **200** 159.

Leonard, P. J. T. (1990). On the origin of the OB runaway stars. *J. Royal Astron. Soc. Canada* **84** 216.

Sargent, A. (1979). Molecular clouds and star formation. II. Star formation in Cepheus OB3 and Perseus OB2 Molecular Clouds. *Astrophys. J.* **233** 163.

See also **Galactic Structure, Optical Tracers; Star Formation, Propagating; Stars, Masses, Luminosities, and Radii; Stellar Associations, R- and T-Type; Stellar Associations and Open Clusters, Extragalactic.**

Stellar Associations, R- and T-Type

William Herbst

An association is a loose grouping of objects of the same type. Astronomers use the term primarily for three different kinds of objects, O and B stars, reflection nebulae, and T Tauri stars. We have, therefore, OB associations, R (reflection nebulae) associations, and T associations. What these objects have in common is that they are each mainly comprised of relatively young stars, that is, with ages no more than about 10 million years or so. Such objects have not had time to move far from their formation sites, so associations (of all types) are believed to be the birthplaces of stars. The study of associations is important, therefore, because it will lead eventually to an understanding of the star formation process and its role in galactic structure and evolution.

Associations may be distinguished from star clusters. The latter are much denser groupings of stars that can easily be picked out on photographs of the sky by the great concentration of stars visible in a small area. Star clusters are gravitationally bound units; associations are not. It is estimated that about 10% of all stars form in clusters, with the remaining 90% forming in associations. It should be noted that one often finds a star cluster embedded within an association.

Whereas star clusters can easily be found by simple examination of photographs, the detection of OB and T associations requires that the star types be determined first. This must be done spectroscopically—most efficiently with an objective prism. Wide-field (Schmidt) telescopes are used to record stellar spectra over tens of square degrees of sky simultaneously, from which associations of O and B or T Tauri stars can be identified. The former have strong hydrogen and helium absorption lines indicative of a high surface temperature; the latter are discovered primarily by their emission lines (usually the first line of the Balmer series of hydrogen, Hα, but more recently the Ca II H and K lines as well), which arise in an extended chromosphere or circumstellar disk or shell.

R associations can be found on direct plates without recourse to spectra. One simply searches for groups of stars with nebulous patches surrounding them. The nebulosity is starlight reflected from dust grains in the star's vicinity, as a spectrum will confirm. Reflection nebulae may be distinguished from emission nebulae by comparing their appearance on photographs taken at two different wavelengths and using the fact that the reflection nebulae are bluer than their illuminating star (shorter-wavelength light being scattered more efficiently). Most of the stars illuminating reflection nebulae have spectral types between B2 and A0. Such earlier-type (hotter) stars are able to ionize their surroundings, whereas later-type (cooler) stars are not sufficiently luminous to light up their surroundings and produce a detectable nebula.

The high luminosities of O and B stars imply short lifetimes. Such stars are also known to be massive—containing 10–50 times the mass of the Sun. The T Tauri stars are believed to be young for a variety of reasons, which include their presence in young clusters with O and B stars, their high lithium abundances, and their inferred strong surface magnetic activity. It is also widely believed now that they are surrounded by disks of matter which may, in some cases, give rise to planetary systems. T-Tauri-star masses are poorly known, but are believed to be in the range of a few tenths to about three solar masses. R-association member stars are generally believed to be young also, because of their close association with O and B stars, and because they have dense interstellar matter (out of which they presumably formed). Some also have emission lines and are known as Herbig Ae and Be stars. Some are T Tauri stars. Except for the T Tauri stars, the masses inferred from their spectral types are in the intermediate range of 3–10 solar masses, that is, between the O and early B stars and the T Tauri stars.

To summarize, then, associations are believed to be the birthplaces of 90% of all stars, with OB associations being where the most massive stars are born, R associations being the intermediate-mass birth sites, and T associations being the nurseries for the low-mass stars. Often all three kinds of associations are found in the same general region of space, for example, in Orion. Also, young clusters have both high-, intermediate-, and low-mass stars in them. Clearly, stars of different mass can form in the same general regions of space. There are also examples, however, of T associations with no corresponding OB association (e.g., in Taurus). This proves that low-mass stars can form in the absence of high-mass stars.

Whether the converse is true is still unknown. Certainly stars can form in very small associations—numbering only a few stars—and there is at least one apparently "lone" T Tauri star, TW Hya. It is unknown at present whether most solar-mass stars form in OB associations, or in T or R associations where massive stars are not present. It is, of course, also unknown whether the present-day situation in the Galaxy would apply to the time 4.5 billion years ago when the Sun formed. So we cannot (yet) argue from the data on associations about the likelihood that the Sun and solar system were born in a region with high-mass stars, like Orion, or without them, like Taurus.

R ASSOCIATIONS

Besides marking regions of recent intermediate-mass star formation, some properties of R associations make them useful to astronomers in other ways. Because most of the illuminating stars of the nebulae are fully formed, normal (main sequence) stars with well-known intrinsic properties (e.g., surface temperatures, radii, luminosities, etc.), they can be used to probe the dust and gas clouds in which they are embedded. In particular, one can determine "reddening" laws for the dust—that is, the wavelength dependence of the extinction (absorption and scattering) caused by the grains. It is found that the interstellar extinction law is often "grayer"—that is, there is less reddening for a given amount of extinction—in star forming regions than in the general interstellar medium. This indicates a relatively high abundance of larger grains or the relative absence of smaller ones. Grain growth is probably occurring in the denser portions of clouds and destruction of small grains is underway near O stars in OB associations.

The intimate association of illuminating stars and their dust and gas clouds (commonly known as molecular clouds because of the presence of many simple molecules within them) can also be exploited to provide distance determinations for the clouds. The distribution of star-forming clouds can therefore be ascertained, and shows three distinct linear groupings that are thought to be pieces of spiral arms of the Galaxy. Since line-of-sight velocities for the clouds can easily be determined from the Doppler shift of the molecular emission lines, it is also possible to probe the local kinematics of the Galaxy in this way, although OB associations, containing more luminous stars visible to greater distances, are more useful for these purposes.

The morphology of R associations and its relation to other components of a star-forming region provides additional clues about star formation. Basically, an R association marks a region of recent intermediate-mass star formation. This is generally only a portion of the whole star-forming complex or molecular cloud. In two cases, known as CMa R1 and Mon R2, the associations have a ring-like structure, suggesting that star formation has occurred at certain points along the periphery of a shell. One possibility is that a supernova produced the shells and effectively "triggered" the star formation seen. Determining the star-forming history of any region is, however, very difficult, because the stars, once they form, begin to modify their surroundings through the action of radiation pressure and stellar winds. Only by carefully accumulating data on *all* components of a star-forming region and comparing these can we hope to learn these histories and begin to address questions such as the likely early history of our solar system.

T ASSOCIATIONS

As discussed above, T associations are thought to be the sites of formation of about 90% of all stars with masses less than or equal to three times the Sun's mass. The best-studied T association, because it is one of the closest, at about 160 pc (520 ly) and also well positioned in the sky for observation from the northern hemisphere, is in Taurus. It is not part of an OB association.

Until recently the major components of this star-forming region were thought to be the T Tauri stars and the molecular cloud. Satellites have now provided infrared and x-ray views of the region and turned up additional components. In the infrared one can find deeply-embedded, very recently formed, highly-obscured stars, sometimes too reddened even to be visible optically. These are often found close to the densest portions of the clouds ("cores") as determined from molecular line studies. It is therefore believed that the dense cores are regions where stars have just formed or are about to form and that the infrared objects are the youngest stellar objects detectable, with ages of perhaps 100,000 yr.

An unanticipated new component of T associations was discovered recently by analysis of the results of x-ray surveys of star forming regions. There are a large number of stars that are strong x-ray emitters but are otherwise rather unremarkable and had not previously been noted as interesting in optical surveys. These are known as "weak" or "naked" T Tauri stars because they have some characteristics of T Tauri stars (e.g., Hα emission, but at lesser levels than their "classical" counterparts). Weak T Tauri stars are actually now believed to be the *main* stellar component of T associations. (Because they were discovered by their x-ray emission it might be time to rename T associations "X associations.")

The relationship between classical and weak T Tauri stars is a matter of considerable controversy at present. A popular view is that weak T Tauri stars lack circumstellar disks (hence the term "naked"), whereas classical T Tauri stars have them. The continuity between the groups in various properties suggests, however, that the distinction may not be quite that dramatic. It may have more to do with the degree to which the disk interacts with the star (e.g., the amount of accretion currently taking place). Curiously there is no large difference in the ages of typical classical or weak T Tauri stars. A one-million-year-old, one-solar-mass star can be found in either the classical or the weak phase, apparently. It may be that for any one star there is a progression from the classical stage to the weak stage to the fully formed, main-sequence phase, but that not all stars go through this sequence at the same rate, even if they have the same mass (i.e., a second parameter such as angular momentum or magnetic field strength may be involved). Current research is addressing questions such as these.

What is clear about the weak T Tauri stars is that they have very active stellar dynamos that generate strong magnetic fields over large areas of their surfaces. Starspots, analogous to sunspots, are found covering 30% or 40% of the star in some cases. The characteristic x-rays indicate considerable coronal gas at temperatures of 1 million degrees. Flares are often seen in the optical and in x-rays. Young stars are clearly very magnetically active stars, and there is a smooth (roughly exponential) decline of activity with time. The current, relatively low levels observed in the Sun are typical of a star of its age.

In summary, then, our current understanding is that low-mass stars such as the Sun probably form in T associations. The molecular clouds form first, and have a "stringy" appearance, with dense clumps aligned along "streamers;" these clumps can stretch over 100 ly or so. The clouds' shapes and evolution may largely be controlled by magnetic fields. Within the densest cores, stars form and become detectable first as infrared sources, and later as optically visible, classical, T Tauri stars. At this point they are probably of order one million years old. Their distinguishing spectroscopic characteristics (emission lines) and infrared and ultraviolet excesses may arise primarily in disks surrounding the stars and in an accretion boundary layer between the disk and star. When, for whatever reason, the disk or boundary layer is absent or passive, only the characteristics of the magnetically active, underlying weak T Tauri star are seen. As time goes on, even these characteristics (flares, x-rays, etc.) disappear and the star becomes a member of the general field population of the Galaxy. This occurs at about 10 million years for a one-solar-mass star. An association of stars that formed together cannot be traced for times longer than this. Only for the minority of objects that form in gravitationally bound clusters is it possible to establish a common heritage among older stars.

Additional Reading

Black, D. C. and Matthews, M. S., eds. (1985). *Protostars and Planets II*. University of Arizona Press, Tucson.

Gehrels, T., ed. (1978). *Protostars and Planets I*. University of Arizona Press, Tucson.

Herbst, W. and Assousa, G. E. (1979). Supernovas and star formation. *Scientific American* **241** (No. 2) 138.

Peimbert, M. and Jugako, J., eds. (1987). *Star Forming Regions*. D. Reidel, Boston.

Shu, F. H., Adams, F. C., and Lizano, S. (1987). Star formation in molecular clouds: Observation and theory. *Ann. Rev. Astron. Ap.* **25** 23.

See also **Pre-Main Sequence Objects: Ae Stars and Related Objects; Stars, Pre-Main Sequence, X-ray Emission; Stars, T Tauri; Stellar Associations, OB-Type; Stellar Associations and Open Clusters, Extragalactic.**

Stellar Associations and Open Clusters, Extragalactic

Edward W. Olszewski

One useful way to study and to compare the star formation histories of galaxies, and to learn the range of conditions under which groupings of stars will form, is to observe stellar associations and open clusters (and globular clusters) in galaxies other than the Milky Way. Stellar associations in nearby galaxies are useful as probes of the places where massive star formation is occurring. There is a relationship between giant molecular clouds, H II regions, and associations of hot stars in the Milky Way; studying associations in other galaxies offers the promise of teaching us how a galaxy turns its largest gas clouds into stars.

The study of open clusters is a bit more problematic. Historically, the terms globular cluster, open cluster, and stellar association referred to the appearance of such systems on a photographic plate. Later, other connotations were attached to the words: For instance, age, abundance of the elements in the stars, kinematics, and position within the Milky Way are properties that come to mind immediately. When we move to the extragalactic arena, these terms and their connotations are seen to be inadequate to describe all of the visible star clusters fully. Furthermore, a canonical Milky Way open cluster, the naked-eye "constellation" of the Pleiades, shrinks almost to insignificance at the distance of the Andromeda galaxy, which is the nearest big spiral galaxy. Luckily, in other galaxies, massive cluster formation has proceeded in a different way from that which has occurred in the Milky Way.

ASSOCIATIONS

We know remarkably little about stellar associations in other galaxies. The simple model of the connection of associations with H II regions and with molecular clouds is only beginning to be tested outside the Milky Way. With the commissioning of modern millimeter- and submillimeter-wave telescopes, this situation is expected to improve rapidly. At the moment, the state of the art is concerned with simpler questions, such as the working definition of an association, the stellar population within an association, and the total amount of massive star formation occurring within all associations in a galaxy.

DEFINITION OF AN ASSOCIATION

A simple working definition of an association in another galaxy might be "a region of enhanced density of bright blue stars." To realize this definition, and to make a list of associations, one simply examines images taken in two or more colors and circles the concentrations of blue stars. Clearly, because supporting kinematic data will undoubtedly be lacking, this definition will not necessarily divide the blue stars into physically separate groups. Furthermore, the definition of an association will be dependent upon the types of images used, and upon the distance to the galaxy in question. All things being equal, the farther away a galaxy is, the more likely it will be that the astronomer will circle ever-larger clumps of stars and call them single associations.

In fact, the available data show exactly this trend. Furthermore, the more distant galaxies have fewer total associations, which probably means that individual associations are being blurred into larger objects. For the moment, while more detailed data are lacking, we should perhaps view lists of extragalactic associations as pointers to hot stars, rather than as groups of physically associated stars.

STELLAR POPULATIONS IN ASSOCIATIONS

Until recently, there were few available facts concerning the population of stars in extragalactic associations. Almost all of this knowledge came from studies of associations in the Large Magellanic Cloud (LMC). We knew that there were many bright blue main-sequence stars, and many blue evolved stars, a few of which had been classified as O stars or as Wolf–Rayet stars. We knew that there were variable stars, a few of which were Cepheid variables, and many of which were red irregular variables (red supergiants). There was not sufficient information to derive the elemental abundances, or to deduce how and if a massive blue star evolves to the red.

Until now, astronomers were forced to shy away from stellar associations when they compiled catalogs of the most luminous and most massive stars in external galaxies. This very serious selection effect is now being removed. With the coming of the new solid-state detectors, programs aimed at detecting the most massive main-sequence stars and their later evolutionary stages can be carried out. This breakthrough was due to the ability to measure accurately stellar brightnesses and colors in crowded regions contaminated with nebulosity, and the ability to record sky-subtracted stellar spectra in the same regions of glowing gas.

30 Doradus, the largest H II region in the Large Magellanic Cloud, is now known to possess a population of hundreds of massive stars, some of which are almost hidden from sight in knots of nebulosity. Recent work on the association embedded in the Small Magellanic Cloud (SMC) H II region NGC 346 shows that it is very rich in stars more massive than 25 solar masses. The number of the hottest, high-mass, main-sequence stars in this association alone has doubled the known number of these O-type stars cataloged in the Small Magellanic Cloud. Besides identifying the bluest stars and confirming their temperature with spectroscopy, one can also design photometric passbands to allow the discovery of Wolf–Rayet stars, the evolutionary descendants of the most massive stars. Clearly, the study of stellar associations is mandatory to delineation of an unbiased sample of massive stars in a galaxy.

Finally, we have some knowledge of variable stars within the Large Magellanic Cloud associations. There are crucial questions about the comparison of stellar pulsation theory to stellar evolution theory that are best answered by studying the characteristics of the groups of stars containing Cepheid variable stars, whose periods are 3–30 d. There are also about 25 long-period red irregular variables known, which are of interest if we are to understand the evolution of massive main-sequence stars, and the effects of mass loss.

GAS AND ASSOCIATIONS

Carbon monoxide (CO) has now been detected in both Magellanic Clouds. Although the data for the CO detections, the neutral hydrogen, and for the associations themselves are not easily merged, it is clear that some of the emission regions containing CO are adjacent to stellar associations. In the published literature, no careful and comprehensive comparison of the positions of stellar associations and molecular emission has yet been done. In the case of the Andromeda galaxy, a region of significant CO and H I lies within a dark cloud near a cataloged association. An H II region lies between the stellar association and the molecular cloud, suggesting that as massive stars formed, they altered the molecular cloud from which they came. The work on the Magellanic Clouds and on the Andromeda galaxy are small first steps in extending our understanding of Milky Way molecular clouds and associations to other environments.

OPEN CLUSTERS

Open clusters are small clusters with a modest number of stars. Inasmuch as cluster formation is also an indicator of (noncluster) star formation, open clusters in external galaxies can be used to show how star formation has progressed in different regions of a galaxy. The idea here is to use the brightness of the brightest blue star in a cluster as a rough indicator of age. If the individual clusters are resolved into stars, and if color information is available, crude ages can be determined very quickly. When this work was done for 500 Large Magellanic Cloud clusters, a "movie" could be put together showing star formation. It was noted that cluster formation would occur in a given place in the LMC for a while, then stop. It would then start in another part of that galaxy. These groups of clusters fit with the simple notion that if the gas density is high enough, star formation commences, and then proceeds until the gas density gets too low. The size scale on which this group of clusters would form is about 1 kpc, which is approximately the scale of clumpiness of hydrogen gas in galaxies. Perhaps after a "resting" period, significant star formation can reoccur in that part of the galaxy.

POPULOUS CLUSTERS

The Magellanic Clouds contain a large population of star clusters that are luminous and globular shaped, but are of young or intermediate age. Some of these clusters are very blue in color, unlike the yellow-red galactic globulars; we now understand that the blue color comes from the presence of luminous main-sequence stars that finished their lifetimes long ago in the look-alike galactic globulars. The intermediate-age clusters took a bit longer to understand, as they are also yellow-red clusters. The family of these "populous" clusters includes objects in both Magellanic Clouds, a small number in the Andromeda galaxy, a good-sized population in M33, and perhaps five or so in NGC 6822. IC 1613 seems to have no globular or populous clusters, though it has a few open clusters.

The rest of this section will concentrate on Magellanic Cloud clusters, because these clusters are easy to resolve into stars, and because they have been the primary subjects of this work in the past few years.

In a stellar population, the relative number of stars seen in a certain stage of evolution reflects the relative lifetime of that stage. If a certain phase of evolution has a lifetime of 1/1000 that of the main-sequence lifetime, there will be only one star in that short-lived phase for every 1000 main-sequence stars; this fact is what makes the populous Magellanic Cloud clusters valuable, because they may have masses of 10,000 or more solar masses. Only in these very populous clusters will there be the sheer number of stars needed to explore short-lived phases of evolution. (In general, we can only know the age of stars if they occur in star clusters; field stars are very difficult to date precisely.)

As an example, the LMC populous cluster NGC 1866 is now known to contain approximately 20 Cepheid variable stars. The largest number of Cepheids in any Milky Way star cluster is three.

If we are going to understand what happens to the properties of Cepheids as they evolve, we need to study a population of Cepheids of the same age; to a reasonable degree of approximation, all stars in a star cluster are the same age.

Because the system of populous clusters in the LMC and SMC spans a wide range in age, we can use these clusters to study at what age a certain type of star appears, how the properties of that type of star change with age, and at what age that type of star will no longer be formed. The study of the carbon star population in Magellanic Cloud clusters is a good example. It was found that clusters younger than 8×10^8 years do not contain carbon stars; this fact is probably due to the effects of mass loss. Systems older than 8×10^9 years can contain less-luminous carbon stars that form in a slightly different way. In the intermediate-age range, the luminosity of carbon stars is related to the age of the cluster, with younger (more massive) carbon stars being more luminous than their older counterparts (note that individual carbon stars are in that evolutionary stage for a very short time, so we are not watching individual stars fade, but are watching stars of different mass become carbon stars). Besides providing a way to date systems containing C stars and defining an empirical relation, these observations, among others, have shown an important weakness in the details of the theory of the formation of carbon stars. The theory could make very bright carbon stars that nature does not; the interplay between theory and observation showed that the theory needed adjustment.

The populous cluster system itself can be used to derive the kinematics of a galaxy at various ages, allowing us to piece together the shapes of the Magellanic Clouds, and perhaps to find evidence for their interaction with the Milky Way. The available data show that the LMC clusters of all ages are heavily confined to a disk. This fact is different from the situation in the Milky Way, where the oldest (globular) clusters are distributed in a large halo, and the open clusters are strongly confined to a disk. From these data, we may be able to understand how each type of galaxy formed.

If we age-date their clusters, we find that the LMC and SMC have very different cluster systems. The LMC has a few old, luminous clusters. There is then a large gap, with almost no clusters with ages between 3×10^9 and 1.2×10^{10} years. Below three billion years, there are populous clusters of all ages. The SMC, which has a smaller cluster system, may have no extremely old clusters. It does, however, have some clusters spanning the entire range up to 1.2×10^{10} years. The big gap in cluster ages does not exist. The Milky Way has 100–150 very old, luminous clusters, with no populous clusters younger than 1.2–1.5×10^{10} years. As mentioned above, IC 1613, which has approximately the same luminosity as the SMC, and might be thought to have an identical cluster system, has no luminous clusters of any age.

As we look at ever more distant galaxies, we will ultimately need to depend on integrated properties of the clusters; for we will not be able to resolve individual stars in these clusters. The Magellanic Cloud clusters, with their wide range of ages and stellar content, and with our ability to study them star by star and as single entities, will probably be our major source of the information that will allow us to interpret the integrated light from the distant systems.

Additional Reading

Hodge, P. W. (1973). The recent evolutionary history of the cluster system of the Large Magellanic Cloud. *Astron. J.* **78** 807.

Hodge, P. W. (1986). *Galaxies*. Harvard University Press, Cambridge.

Hodge, P. W. (1986). Systems of stellar associations in galaxies. In *Luminous Stars and Associations in Galaxies*, C. W. H. De Loore, A. J. Willis, and P. Laskarides, eds. D. Reidel, Dordrecht, p. 369.

Israel, F. P., De Graauw, Th., Van De Stadt, H., and De Vries, C. P. (1986). Carbon monoxide in the Magellanic Clouds. *Astrophys. J.* **303** 186.

Lada, C. J., Margulis, M., Sofue, Y., Nakai, N., and Handa, T. (1988). Observations of molecular and atomic clouds in M31. *Astrophys. J.* **328** 143.

Massey, P., Parker, J. W., and Garmany, C. D. (1989). The stellar content of NGC 346: A plethora of O Stars in the SMC. *Astron. J.* **98** 1305.

Mould, J. and Aaronson, M. (1986). The formation and evolution of carbon stars. *Astrophys. J.* **303** 10.

See also **Andromeda Galaxy; Galaxies, Stellar Content; Magellanic Clouds; Star Clusters, Open; Stellar Associations, OB Type.**

Stellar Evolution

Malcolm P. Savedoff

Stellar evolution theory interprets the properties and frequency of different kinds of stars and their interrelationships. For what masses will stars eventually end as black holes, neutron stars, or white dwarfs, for example? In the latter half of the nineteenth century, interest centered on the age of the solar system. Lord Kelvin recognized that the energy released by an extended diffuse cloud contracting to form the present Sun could maintain sunlight at its present intensity for approximately 100 million years. This calculation still serves to define the Kelvin–Helmholtz time scale, which governs thermal equilibrium in stars. In these studies, our ability to understand the astronomical data appears to be limited only by our mastery of statistical, atomic, and nuclear physics and by the finite resources for computation.

A consequence of the virial theorem is that half the gravitational energy released by a contracting star is converted into radiation and the remainder into internal energy. An isolated star radiates and must replace its lost energy from its nuclear fuels, if available, or else must contract and use gravitational energy sources. Once past the *protostar* stage, it evolves very slowly as a main-sequence star. The continued conversion of hydrogen into helium leads to structural modifications, including increased central temperatures, that lead eventually to exhaustion of the hydrogen in the center of the star. The luminosity generally increases and evolution is more rapid as hydrogen now burns in a shell surrounding a contracting helium-rich core, which increases in density and temperature until halted by the burning of helium into carbon and oxygen. This cycle of core and shell burning may continue until halted by stabilization of the star by degeneracy: electrons or, conceivably, neutrons filling all the lowest single-particle energy states up to ε_F, the Fermi energy, with $\varepsilon_F \gg kT$, where k is Boltzmann's constant and T is the temperature. Otherwise, once the stablest nuclei (the iron group) are formed, further contraction is destabilized by nuclear dissociation, resulting in an implosion that, paradoxically, induces the explosion observed as a *supernova*.

By using star clusters and stellar evolution theory, binary and multiple stars, as well as OB associations and R and T associations, it is possible to make comparisons between stars of different mass but essentially the same age, initial composition, and environment, a comparison often represented through the Hertzsprung–Russell (HR) diagram. A successful model should predict the dependence of the only measurable quantities, those at the surface: the radius, luminosity (in joules per second), and surface composition of a star as a function of its age. Because the interior of a star is essentially unobservable, the theory of *stellar atmospheres* provides the connection between the radius and luminosity with observed temperature, and so forth, enabling empirical tests as discussed recently by Alvio Renzini and Flavio Fusi Pecci. To the extent that the interior materials of stars become observable through mixing or expulsion, the models predict abundance changes.

For a nonrotating, nonmagnetic, isolated star the models provide the evolution over time of the radius, temperature, pressure, density, luminosity, and composition of the star as a function of mass interior to the radius r. The effects of slow rotation, weak magnetic fields, prescribed mass loss and accretion, and, for nonisolated stars, the consequences of tidal forces and other disturbances, can then be calculated, usually in a linear approximation, and can then be compared with observations. The masses implied by stellar evolution models can be compared with masses for which the theory of pulsating stars predicts the observed periods: Any discrepancy implies a need for extension or improvement of the models. Thus the lower mass estimated for the evolved horizontal-branch RR Lyrae stars as compared with the bright end of the main sequence, a less evolved stage, is attributed to mass loss in globular clusters. Pulsation periods provide an independent test of evolution theory. For example, the rich spectrum of nonradial solar oscillations provides information on interior differential rotation, an additional constraint on the interior and evolution of the Sun.

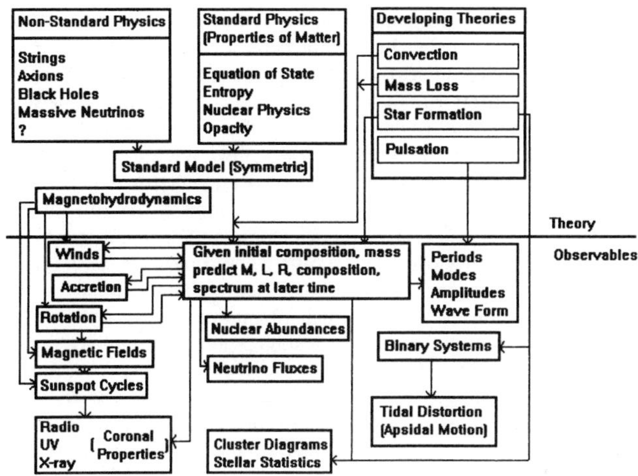

Figure 1. Stellar evolution: Relation between elements of theory and observation.

STANDARD MODEL

In the absence of perturbing fields and internal motions (contraction, convection, and rotation), the star conforms to four easily understood differential equations. (A substitute equation describes convection using the quasiempirical relations of mixing-length theory, a description of energy transport by mass motions whose efficiency must be estimated empirically.)

1. The total mass interior to radius r is the sum of the mass in each interior shell, which is the product of the shell volume and mass density.

2. A shell's mass produces an increase in the gas pressure beneath it equal to its weight per unit area, which is the product of its density, its thickness, and the acceleration of gravity. Thus, given the density at r and the mass within the radius r, the change in the pressure is known. The evaluation of the temperature of material in terms of local conditions, namely, the density, pressure, and composition of this shell, is provided by the *equation of state*. The determination of the equation of state and the three other functions described next is a continuing project exploiting advances in the physics of materials.

3. A temperature difference across a shell results in an energy flow (luminosity per unit area) controlled by the opacity of the material. The opacity is the second function of local conditions to be evaluated. When the temperature gradient consistent with the opacity is steeper than the adiabatic limit, energy is transported at least partially by convection, and the temperature difference is estimated by a more complex, approximate representation.

4. The luminosity (the energy per second radiated by a sphere of radius r) must be the sum of the energy released within the radius r by all interior shells. In the absence of motions, each shell contributes an amount proportional to its mass and the rate of energy generation per unit mass in the shell: This third function describes the nuclear physics and depends again on local conditions. For contraction or expansion, the energy released per unit mass can be calculated from the temperature and the rate of change of the entropy. Entropy is a function of local conditions whose determination is the fourth physical problem.

These four differential equations and the four functions describing the equation of state, the opacity, nuclear energy generation rate, and entropy define the calculational problem for a simple star. Many aspects of this model are described in texts by Donald D. Clayton and by Rudolf Kippenhahn and Alfred Weigert.

For static stars, these differential equations yield a unique solution if we add appropriate boundary conditions. (Mathematically, a countable set of solutions can occur, and one must always keep alert for possible multiple solutions.) Clearly, at the center of a star, the mass and luminosity interior to r should vanish. The exterior boundary conditions are approximated by requiring that the temperature and pressure vanish simultaneously. (A more exact requirement is continuity of pressure, temperature, radius, luminosity, and mass at the base of a stellar atmosphere that bounds the stellar interior model.) These four boundary conditions for a given mass and composition can be satisfied only for a unique radius and luminosity. This is known as the Russell–Vogt theorem and provides that for a stated detailed composition and mass a star should be represented by a unique point (or points) in the Hertzsprung–Russell diagram. For a given initial state near equilibrium, the time-dependent solutions approach a unique curve in the HR diagram in a Kelvin–Helmholtz time. The historical development of the realization of these solutions was described by Steven W. Stahler in 1988. Detailed models and their testing was reviewed recently by Renzini and Fusi Pecci. Don A. VandenBerg provided an example of the extended tables of models available today.

In part, discrepancies between different model sequences reflect choices in the four relations describing the properties of the stellar material noted above. Although those are better determined today, there remain several issues that are still open, where our representation of the behavior of matter reflects our ignorance of the physical relationships or an inability to carry out the necessary calculations. Often, the unique model sequences that are published depend on choices made for empirical parameters chosen to best match the observations. Figure 1 summarizes the relationships between stellar evolution theory, observations, and supporting physical investigations.

EMPIRICAL PARAMETERS

Convection is one of the open issues mentioned above. It arises in the models when application of the four equations yields unstable layers and hence mass motion. These motions can be represented by replacing the second equation discussed previously by hydrodynamic equations in two or three dimensions. The hydrodynamic equations include not only the pressure forces described in our second equation, but also the inertia and viscosity of the material. Representing convection in detail greatly increases the computational complexity of the problem by requiring more detail both in the temporal description of the gas and in its spatial description, ideally in three dimensions. Estimates are that the convection is turbulent. (The entry on Stars, Atmospheres, Turbulence and Convection discusses how this affects the outer layers of stars.) Detailed motions are chaotic and not amenable to deterministic calculation. In many stellar evolution calculations the effective relation between energy flow and temperature differences has been estimated by phenomenological theories such as mixing-length theory, in which a free parameter, the characteristic length in units of the pressure scale height, may be chosen empirically (e.g., 1.6 as in the VandenBerg tables). The highly efficient convective transport in the stellar interior makes the calculations nearly independent of this parameter, but it does influence the structure of the outer layers: In particular, estimates of the radius and surface temperature are sensitive to this parameter. Can a single value apply reliably under such diverse conditions as found in the Sun, metal-poor globular cluster stars, supergiants, and neutron stars? In addition, convective overshoot results when convective eddies coast into the stable regions, and the resulting peripheral mixed layers modify evolution by providing fuels or catalysts where none would otherwise be found. This too is a subject for contemporary exploration.

The rich phenomena observed in the outer layers of the Sun warn us that rotation and magnetic fields can generate a time-dependent response of considerable complexity. Magnetohydrodynamic effects in stars inhibit fluid motions across magnetic fields: This resists the mixing of elements and entropy in both natural and forced convection. Energy can be transferred from the radiation field into either the magnetic or rotation fields modifying the solar constant in the case of the Sun and returned in ways still being explored. The observed rapid decrease in surface rotation in low-mass stars with age can be explained by the magnetic coupling of rotation with stellar winds. Most contemporary discussions for stars are limited to one-dimensional models in which these effects can be described phenomenologically, because a direct attack with full multidimensional representation is still beyond our present competence and computer resources. Some applications beyond this have been made in solar models. Particularly intriguing is the search for the mechanism by which the observed solar constant variability is correlated with sunspots.

Stellar winds and accretion disks, as in cataclysmic binaries produce mass changes at rates that must still be empirically determined. To the extent that these processes modify a star, the "impact" on its evolutionary locus is poorly determined. On dimensional grounds, one choice used sets the rate of mass loss proportional to the product of the luminosity and radius divided by the mass (Reimer's law). Empirical discussions by Cesare S. Chiosi and Andre Maeder show no simple relationship. The final stage of low-mass evolution from red giant to white dwarf requires remarkably thin hydrogen-rich and helium-rich outer zones, suggesting that chemical composition may dominate the process. Among several sequences of models connecting white dwarfs with the main sequence, sequences based on Reimer's law seem consistent with the narrow spread of white dwarf masses that is observed.

In contrast, accretion rates are determined externally and are dependent upon the extrastellar medium and its properties. Of particular interest is the slow accretion of hydrogen-rich material from a companion star onto a white dwarf, which should result in a classical nova. Mixing of the high-angular-momentum accreted material into the catalyst-rich, low-angular-momentum white dwarf core is a complex computational problem for which a deterministic description is needed. Under what conditions will cataclysmic and noncataclysmic evolution take place? Again, empirical prescriptions may make extensions of our models possible but are inherently incomplete and disquieting.

In binary systems the evolution of two stars is coupled by their interactions. For evolved stars, tidal effects can strip material that in part may be captured by the companion star. The coupled evolution can be followed if we prescribe the transfer of mass and momentum between the stars. The situation becomes even more acute when a white dwarf moves within the envelope of its highly evolved companion, requiring gas-dynamical treatment bridging the wide range of pertinent scales. In the general case, even ignoring magnetic effects for simplicity, there seems to be no justification for requiring circular orbits, steady or symmetrical accretion, and fixed efficiency of capture. Thus again, evolutionary models are available, but they are not necessary models required by the laws of hydrodynamics and mechanics. Mass loss or exchange occurred earlier in the Sirius binary system (which has a primary of 2.6 $M_\odot$ and a white dwarf secondary of 1.05 $M_\odot$), and accounts for why the more evolved star has the lower mass. The constraint that the two stars be of essentially the same age, combined with the orbital data, helps determine which prescriptions work.

To the extent that a "standard model" exists, which has been validated observationally, one can modify this standard model to place limits on the population of exotic objects, perhaps micro–black holes or even cosmic strings, or "strange matter." In some cases, cosmologically significant limits can be placed.

Despite our theoretical and computational limitations, sufficient models exist for most known stellar objects, even for such rarities as supernovae: The success in representing the supernova neutrino and optical observations of SN 1987A in the Large Magellanic Cloud is noteworthy. In large part, processes have been identified that explain the abundances of elements both in common and unusual objects. One is still embarrassed that, for over a generation, solar models have been unable to represent the measured *solar* neutrinos. It is to some extent gratifying that our knowledge of stellar evolution is now mature enough that we cannot accommodate any arbitrary neutrino flux with minor adjustments. One anticipates resolution of this contradiction perhaps through new experiments or through independent confirmation of one of the plethora of ad hoc theoretical speculations that have been proposed.

Additional Reading

Böhm-Vitense, E. (1991). *Introduction to Stellar Astrophysics. Vol. 3. Stellar Structure and Evolution*. Cambridge University Press, Cambridge.

Chiosi, C. and Maeder, A. (1986). The evolution of massive stars with mass loss. *Ann. Rev. Astron. Ap.* **24** 329.

Clayton, D. D. (1983). *Principles of Stellar Evolution and Nucleosynthesis*. University of Chicago Press, Chicago.

Kippenhahn, R. and Weigert, A. (1990). *Stellar Structure and Evolution*. Springer-Verlag, Berlin.

Renzini, A. and Fusi Pecci, F. (1988). Tests of evolutionary sequences using color–magnitude diagrams of globular clusters. *Ann. Rev. Astron. Ap.* **26** 194.

Stahler, S. W. (1988). Understanding young stars: A history. *Pub. Astron. Soc. Pacific*, **100** 1474.

VandenBerg, D. A. (1985). Evolution of 0.7–3.0 $M_\odot$ stars having $-1.0 \leq$ [Fe/H] ≤ 0.0. *Astrophys. J. (Suppl.)* **58**, 711.

See also **Binary and Multiple Stars, Origin; Binary Stars, Cataclysmic; Black Holes, Theory; Magnetohydrodynamics, Astrophysical; Neutrinos, Solar; Neutrinos, Supernova; Protostars, Theory; Star Clusters, Globular, Mass Loss; Star Clusters, Stellar Evolution; Stars, Hertzsprung–Russell Diagram; Stars, Horizontal Branch; Stars, Interiors, Radiative Transfer; Stars, Pulsating, Theory; Stars, White Dwarf, Structure and Evolution; Stars, Young, Rotation; Stellar Associations, all entries; Stellar Evolution, all entries; Sun, Interior and Evolution; Sun, Oscillations; Supernovae, Type I, Theory and Interpretation; Supernovae, Type II, Theory and Interpretation; Supernova Remnants, Evidence for Nucleosynthesis.**

Stellar Evolution, Binary Systems

Ronald E. Taam

With the launch of astronomical satellite observatories in the 1970s and 1980s the study of binary star systems has had a resurgence of interest. The observations in the ultraviolet and x-ray spectral energy bands from above the Earth's atmosphere have led to exciting new discoveries of unexpected types of systems. Coupled with the fact that the majority of stars in the Galaxy are members of binary or multiple systems, the evolution of binary stars has become a very active field of research.

In a simplified scheme, binary star systems can be grouped according to the relative separation of their components. For a widely separated system, the individual stellar components are so far apart that their evolution is unaffected. Astronomers have always had a keen interest in studying the individual stellar components in these systems because they provide the only means by which stellar masses and radii can be reliably determined. The fundamental data obtained from the observational study of these components can then, without complications associated with the presence of a companion, be applied to single stars. On the other hand, for close binary systems the stellar components nearly touch and they cannot evolve as if they were in isolation. Hence the application of, for example, mass–radius relations deduced for single stars is not very useful for quantitative understanding of the components of such close binary systems.

The interactions between these binary star components can be manifest in terms of distortions of the stellar surfaces, evidence of gas streams and mass transfer between the components, and the effect of irradiation of light on one component by the other. Because the companion can impose a limit, by tidal effects, on the extent to which the other component expands, the relative proximity of the two stars in a close binary system influences the evolution of the individual stellar components. This circumstance can lead to an evolution that significantly departs from the evolution of a single star. Any expansion beyond the tidal limit results in either mass transfer between the components or to mass loss from the binary system (see below). This very simple fact demonstrates that stellar evolution in binary systems is dramatically affected by the presence of a close companion. There are many such close binary systems in our galaxy where mass transfer or mass loss have already played a significant role in the evolution of the system to the present state and many other binaries where such processes will govern evolution in the future. A particular example, which has played a pivotal role in the history of the development of the subject, was the observational paradox posed by the binary system Algol. In this system, the more evolved, red giant star (i.e., the large cool component) was found to be less massive than its nearly unevolved, main sequence companion. This is quite contrary to ideas about stellar evolution for single stars where it is known that the more massive star evolves faster. The apparent paradox was resolved when it was suggested in the 1950s by John A. Crawford that the progenitor of the currently more evolved star was the more massive component and that the system had undergone a rapid, large-scale mass transfer phase during which the mass ratio of the components was reversed.

During the last two decades of the Space Age considerable progress has been made in understanding the nature, origin, and evolution of these close binary systems. We will briefly outline some of the key principles of mass transfer in binary systems and discuss the recent findings that have enhanced our understanding of the evolution of these systems.

FUNDAMENTAL CONCEPTS OF MASS TRANSFER

Fundamental to the ideas of stellar evolution in binary star systems is the role played by tidal or Roche lobes. Suppose we consider a binary system that rotates at a constant angular velocity and whose orbit is circular. In the reference frame that corotates with the orbital motion of the binary system, the effective gravitational potential of the system can be regarded as the sum of the gravitational potentials from each of the two stellar components and the centrifugal potential. The curves of constant potential of the binary system in its orbital plane are displayed in Fig. 1. Upon inspection it is evident that the distortion of the equipotential surfaces becomes greater with increasing distance from the center of each star until a local maximum is reached between the two components where the effective gravity vanishes. The critical surface corresponding to this local maximum (in the shape of a horizontal figure eight), when rotated about a line joining the center of the two stars, encompasses a volume about each star known as its respective tidal or Roche lobe. For greater distances from the center of mass of the system, the centrifugal force dominates and the effective gravity vanishes at two other local maxima along the line joining the centers of the two stars. The variation of the effective potential along the line joining the centers of the two stars is illustrated in Fig. 2. The three local potential maxima where the effective gravity vanishes are known as the collinear Lagrangian points. As depicted in Fig. 2, when the initially more massive component of the binary expands to meet its respective Roche lobe,

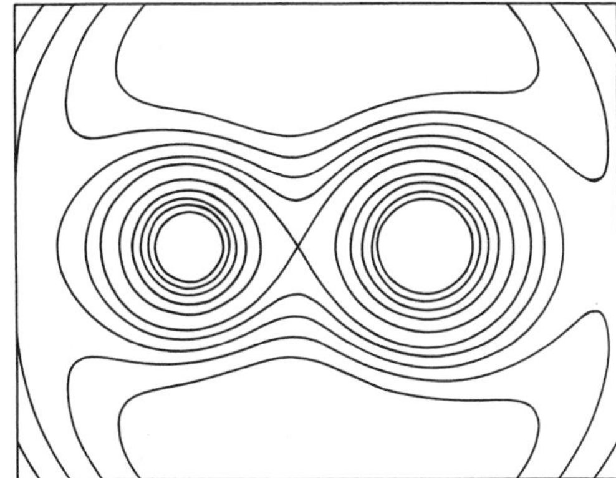

Figure 1. The Roche equipotentials in the binary orbital plane for a mass ratio equal to 1.5. The more massive component is located on the right. The largest volumes enclosing each star (rather than both stars) obtained by rotating the equipotentials about the line joining the centers of the two stars are known approximately as Roche lobes.

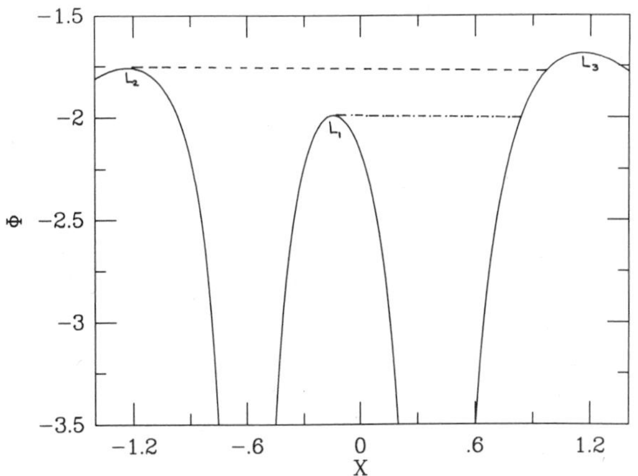

Figure 2. The variation of the Roche potential along the line joining the two stars for a binary system characterized by a mass ratio of 1.5 with the local potential maxima corresponding to the Lagrangian points marked. The inner Lagrangian point is denoted by L_1 and the outer Lagrangian points are noted as L_2 and L_3. The origin ($X = 0$) lies at the center of mass of the binary system. If the more massive star on the right fills its Roche lobe (illustrated as a dashed-dotted curve), then matter from its surface will flow down the potential well from L_1 to its companion on the left. The dashed line indicates a situation where the two stars fill a common envelope.

mass will necessarily be transferred to the less massive component because the matter at the Roche surface will flow down the potential well to its companion.

IMPORTANCE OF MASS AND ANGULAR MOMENTUM LOSS

Within the Roche model framework, the evolution of the binary system can be classified as either conservative or nonconservative depending upon whether or not the total mass and angular momentum of the system are conserved. From the observations of many different types of close binary systems, the present observed properties (viz. masses and orbital periods) of systems are difficult to reproduce in evolutionary scenarios without significant mass and angular momentum loss. To cite one generic example, consider the origin of interacting binary systems consisting of a main sequence-like star and a compact companion like a white dwarf. Such a configuration is standard for cataclysmic-variable binary systems. The difficulty encountered in evolving the progenitor systems to their present state is made evident by the fact that the compact stars were once the remnants of stars that had radii much larger than the current orbital separations. In the transformation of the progenitor system from one of long orbital period to one of short orbital period, the loss of mass and angular momentum from the system must have been substantial. Although very luminous red giant stars (whose cores are white dwarfs) can lose mass effectively via stellar winds, the same is not true for angular momentum. Constructing a theory for angular momentum loss, which is at present phenomenological with little predictive ability, is a major objective of current research in binary star evolution.

One proposed solution to the angular momentum loss problem, which is based on the Roche equipotential picture, involves the formation of a common corotating envelope that extends to the outer Lagrangian point L_2 (see Fig. 2). This can occur when the less massive star structurally readjusts, in response to mass accretion, to fill its respective Roche lobe. Instead of mass transfer between components, further expansion of the two stars leads to a filling of the common envelope. In this case, the mass loss through the outer Lagrangian point leads to a loss of angular momentum. Because the loss of angular momentum per unit mass is more than 6.7 times the mean of the system, the orbital separation shrinks, resulting in a short-period binary system. The difficulty with this approach is shown by the fact that the binary orbit shrinks so rapidly that the assumption of a corotating envelope breaks down. Thus, other mechanisms for angular momentum loss have been sought. In particular, an approach that is the main focus of current studies of stellar evolution in binaries involves the transfer of orbital angular momentum to the common envelope, which is subsequently ejected. This solution is made possible by relaxing the assumption of corotation between the spin and orbital motion. In this picture, the binary orbit decays as the two stellar cores spiral toward each other. The common envelope, which in contrast to the previous solution is not corotating, is ejected, carrying with it a large fraction of the initial angular momentum of the binary. This ejection is made possible by the hydrodynamic expansion driven by the high rate of energy deposition generated by friction or by the processes normally responsible for mass loss in red giant stars.

EVOLUTION TO THE COMMON ENVELOPE STAGE

The binary system can evolve into the common envelope stage by a variety of evolutionary paths. For the case cited above where compact white dwarf remnants are produced in cataclysmic-variable binary systems, the progenitor systems consisting of a red giant star and normal main sequence-like star are most relevant. If we consider such a system, there are circumstances in which the mass transfer process, itself, is unstable. Specifically, when the evolutionary states of the two stars in the system are so very different, the rate of mass loss from the giant is so high that the matter cannot be assimilated by its less evolved companion. In response to the high rate of mass flow, the accreting companion expands to fill its own Roche lobe after only a relatively small amount of mass ($\sim 0.01\ M_\odot$) is accreted. As remarked in the preceding section, the added mass causes the less evolved star to expand to fill the common envelope. Because the time scale to maintain uniform rotation throughout the envelope is much longer than this rapid evolutionary stage, the rotation of the giant loses synchronism with the orbital motion and the system enters into a common envelope stage.

This result can also be accomplished if the mass ratio of the system is extreme because the rotational momentum of a red giant can be comparable to the total orbital rotational momentum. In this case, tidal effects are unable to force the giant into a state of corotation because there is insufficient angular momentum available in the orbit. That is, at the onset of mass transfer, the red giant is not rotating synchronously with the orbital motion. Consequently, when the red giant expands, the companion is engulfed, leading to the result that the main sequence-like star plunges into the red giant interior.

OUTCOME OF COMMON ENVELOPE EVOLUTION

The quantitative studies of the common envelope evolutionary phase in which the two stellar cores spiral toward each other have, indeed, predicted large scale systemic mass and angular momentum loss. The orbital decay is very rapid ($\lesssim 10{,}000$ yr) and the energy released in the process is sufficient to eject the common envelope. The mass ejection process is characteristically not isotropic because matter is preferentially ejected along the equatorial plane of the binary system. The fact that the ejection is nonspherical is supported by the observational appearance of post-spiral systems (i.e., planetary nebulae with binary nuclei) with elliptical or butterfly-shaped nebulae. This nonspherical ejection follows from the fact that the energy dissipation process associated with friction in the common envelope is not distributed uniformly over the common envelope, but rather in the equatorial plane. Because the orbital-decay time scale is much greater than the mass-loss time scale during the late stages of common envelope evolution, at least for the progenitors of cataclysmic variables, it is likely that the entire common envelope will be ejected. Hence, the formation of the post-common-envelope-evolution system consisting of an evolved core of the red giant (i.e., a white dwarf) and the main sequence-like companion is highly likely from this extremely nonconservative binary evolution. Although the formation of such systems is likely, it is by no means the only possibility. For stellar components that are less evolved and/or more comparable in size, it is quite likely that the outcome of the common envelope stage is merger of the two stellar cores. In this case, a single, rapidly rotating star will emerge from the common envelope phase rather than a close binary system.

OUTLOOK

Astronomy in the next decade will be a golden age when many proposed space missions will be operating. The interrelationship between observation and theory will enable us to obtain a much better understanding of the interactions between stars in close binary systems. Stimulated by observations gathered by past space missions, a growing consensus has hypothesized that the common envelope phase of binary evolution provides the long sought for conceptual framework for understanding the transformation of long period systems into short period systems. Recent work has confirmed that the physical processes acting during this phase can lead to a very nonconservative evolution in which substantial mass and angular momentum can be lost.

Future studies will be directed toward quantitatively establishing the relationship between the type of system that emerges from the common envelope and its progenitor binary system. The outlook for further understanding of stellar evolution in binaries is bright, and future work in these areas will help clarify our understanding of this very important phase of binary star evolution.

Additional Reading

Bodenheimer, P. and Taam, R. E. (1986). Common envelope evolution. In *The Evolution of Galactic X-Ray Binaries*, J. Truemper, W. H. G. Lewin, and W. Brinkmann, eds. D. Reidel, Dordrecht, p. 13.

Eggleton, P. P. (1985). Nomenclature—the stellar zoo. In *Interacting Binary Stars*, J. E. Pringle and R. A. Wade, eds. Cambridge University Press, London, p. 21.

Kopal, Z. (1978). *Dynamics of Close Binary Systems*. D. Reidel, Dordrecht.

Livio, M. (1989). Common envelope evolution of binary stars. *Space Sci. Rev.* **50** 299.

Webbink, R. F. (1988). Late stages of close binary evolution—clues to common envelope evolution. In *Critical Observations Versus Physical Models for Close Binary Systems*, K. C. Leung, ed. Gordon and Breach, New York, p. 403.

See also **Binary and Multiple Stars, Origin; Binary Stars, Cataclysmic; Binary Stars, Contact; Binary Stars, Observations of Mass Loss and Transfer in Close Systems; Binary Stars, Semi-Detached; Binary Stars, Theory of Mass Loss and Transfer in Close Systems; Binary Stars, X-Ray, Formation and Evolution.**

Stellar Evolution, Intermediate Mass Stars

Volker Weidemann

By widely agreed definition, intermediate mass stars are those that evolve from the main sequence to central helium burning without the development of a degenerate core; thus they do not go through a central helium flash. Canonical theory sets the lower limit for intermediate mass stars at about 2.2 $M_\odot$ (solar masses); however, its value depends on composition and structural details, for example, convective overshoot, which are discussed below. The upper limit is conventionally given by the mass beyond which stars do not develop a degenerate carbon–oxygen core, but evolve instead to nondegenerate carbon ignition. Depending again on structural details, this occurs at about 8 $M_\odot$, beyond which one has the range of massive stars, which normally evolve toward supernovae of type II. Intermediate mass stars could also evolve toward supernova explosion if the carbon–oxygen core grows to degenerate carbon ignition with ensuing carbon detonation or deflagration (supernovae of type I). However, empirical studies have shown that mass loss in the preceding double-shell burning (asymptotic giant branch, AGB) phase is so strong as to stop core growth and nuclear evolution before carbon ignites: The remnant cores in general evolve via the planetary nebulae phase and become white dwarfs. Characteristic differences of intermediate mass versus low mass stars are the occurrence of blue loops in the Hertzsprung–Russell (HR) diagram away from the red giant branch during central helium burning stages (with crossing of the Cepheid variability strip) and the possibility of growing larger C–O cores with correspondingly larger luminosities on the AGB. However, as far as the existence of a thermal pulsing AGB with increasing mass loss and final evolution toward the white dwarf stage is concerned, low and intermediate mass stars behave similarly.

GENERAL REMARKS

Evolutionary model calculations for intermediate mass stars (IMS) will be considered here only under certain restrictions, excluding the influence of rotation, magnetic fields, and mass exchange between binaries. We also do not discuss Cepheid and Mira variability for which there are separate entries in this encyclopedia. Instead we will concentrate on the main features of IMS evolution and their dependence on current input physics. We shall consider structural changes, and evolution in the HR diagram as well as in the temperature–density plane through the sequential stages from the main sequence to the termination of nuclear burning at the end of the AGB.

INPUT PARAMETERS AND PHYSICS

For model calculations the following input parameters have to be specified: initial mass M_i, in general, with homogeneous composition (X, Y, Z_{CNO}, Z: hydrogen, helium, CNO, and metal abundances by mass fraction) and radiative and conductive opacities, mixing length parameter $\alpha = 1/H_p$ (H_p is the pressure scale height) for convective energy transport, overshoot parameter (measuring the extent to which convective overshoot occurs beyond the limit given by the Schwarzschild criterion), nuclear cross sections, especially for the $^{12}\text{C}(\alpha,\gamma)^{16}\text{O}$ rate, and several neutrino-producing processes. Furthermore, it is important to assume or specify a mass loss law, be it for the stellar wind (Reimers formula $\dot{M} \propto LR/M$, where L and R are the luminosity and radius, with proportionality parameter η_R) or the so-called superwind at the end of the AGB, leading to the formation of thick circumstellar shells or planetary nebulae (PN; sometimes specified by the PN efficiency parameter b). Different results obtained during the last decade are almost entirely due to different assumptions for these quantities or parameters, as will be shown. The influence of computational procedures and codes is, by comparison, minor but not negligible.

*MAIN SEQUENCE TO CENTRAL HYDROGEN EXHAUSTION (*ECHB*)*

IMS burn hydrogen in the CNO cycle and, due to its high temperature dependence, have convective cores, the outer boundary of which slowly retreat with diminishing fuel, leaving behind a somewhat He-enriched intermediate zone. Central temperatures and densities are around $\log \rho \approx 1.5$ and $\log T \approx 7.4$ (ρ and T are in grams per cubic centimeter and kelvins, respectively). The main sequence lifetime decreases with M_i from 10^9 to 4.10^7 yr from 2 to 8 $M_\odot$. It increases with convective overshoot (because there is more fuel available): for example, from 8×10^7 to 1.2×10^8 yr for $M_i = 5\ M_\odot$. The size of the luminosity and radius increase from the zero age main sequence to the ECHB phase also grows with overshoot: Therefore, the width of the observed main-sequence band has been used to estimate that overshoot is fairly strong. Changes of initial composition cause the main-sequence band to be shifted parallel to itself, but this is of minor importance as long as one restricts the models to the range of X, Y, Z for Population I, $0.20 \le Y \le 0.30$, $0.02 \le Z \le 0.03$. (Exception: Magellanic Clouds, with smaller metallicity, namely $Z \approx 0.01$ for younger clusters and $Z \approx 0.001$ for older clusters.)

ECHB TO CENTRAL HE IGNITION

After core hydrogen exhaustion, an at first thick and then thinning H shell source is established that is accompanied by core contraction and expansion of the envelope: The IMS move in the HR diagram nearly horizontally, but with slightly decreasing L to the red giant branch, thereby crossing the Hertzsprung gap in a short time scale. After IMS reach the Hayashi line the luminosity (and radius) increases again and an outer convective envelope is established that reaches farther and farther down until it enters the layers that were helium enriched during main sequence burning. In the first dredge-up phase some He is brought to the surface. In the center, temperature and density increase due to gravitational contraction until the triple alpha process is ignited at $\log T \approx 8.1$ and $\log \rho_c \approx 4$. The helium core mass at this stage is around $0.1\ M_i$.

CENTRAL HELIUM BURNING TO CORE HELIUM EXHAUSTION

For IMS this stage is characterized by the development of blue loops in the HR diagram, the maximum extent of which is reached after about half of the central helium has been burned. The core is again convective, and also develops semiconvection, which extends its fuel supply and therefore lengthens the central He-burning time —generally about one-fifth of the main sequence lifetime. Again, this depends in detail on the occurrence of convective overshoot, as does the size of the blue loops, which is also strongly dependent on Z and the detailed structure of the burning profiles. The luminosity is at first predominantly supplied by the onburning hydrogen shell, but later it is supplied more and more by the convective helium core whose radial extent increases rather than decreases as in the main sequence phase. Because He ignition stops the contraction of the core, the outer envelope shrinks and thereby causes the blueward motion in the HR diagram. As soon as the core begins to contract again toward the end of central He burning, the envelope expands and the star returns to the Hayashi line. Recently it has been found that when nearing He exhaustion the core undergoes some (3–5) "breathing" pulses that reach outward and bring fresh fuel to the center, thereby prolonging the central He-burning phase and leaving behind a larger burned-out (C–O) core mass. The final C–O core mass is larger and contains more O, because $C + \alpha$ (α is the alpha particle) reactions are strongest in the He-exhaustion phase. This also depends crucially on the cross section for this reaction, which is now taken to be three times stronger than assumed in the past, resulting in cores consisting predominantly of oxygen. The hydrogen-free core mass M_H comprises about one-fifth of the initial mass.

The central He-burning phase lasts from 10^8 to 10^7 yr for M_i between 3 and 7 $M_\odot$, about one-fourth of the main sequence lifetime. Stronger overshoot considerably reduces this ratio and, thus, also reduces the number ratio of He-burning giants to main sequence stars (in clusters).

CORE HELIUM EXHAUSTION THROUGH EARLY AGB

Following central He exhaustion the remaining helium underneath the hydrogen shell is comparatively quickly devoured by He burning in a thick shell. The H-burning shell is extinguished and the increasing luminosity is solely supplied by the outward moving He shell. In the HR diagram the stars climb the Hayashi line, on the so-called early asymptotic giant branch (EAGB). During the EAGB phase the outer convective envelope increases in depth until it reaches, for stars above 4 $M_\odot$, the He layer below the extinct H shell. Accordingly He is mixed into the envelope and brought to the surface in the second dredge-up. At the same time the hydrogen-free core mass M_H is reduced compared to the values reached at the end of core He burning.

Below the He-burning shell, the C–O core grows in mass and contracts until it reaches electron degeneracy at $\log \rho \approx 5$ and $\log T \approx 8.2$–8.4. When the outward-moving He shell approaches the hydrogen discontinuity, the envelope convection retreats and the H shell ignites again. In the interior the maximum temperature is reached off-center because neutrino emission, which increases with density, removes energy from the center. Neutrino emission continues to cool the center down to $\log T_c \approx 8.1$ at $\log \rho \approx 6$–7. Thus, a highly degenerate core is produced.

The lifetime of the EAGB phase is short, of the order of 10^6 yr, because helium has to be burned rapidly in order to cover the high energy losses. At the end, the C–O core mass has grown to between 0.5 and 1.1 $M_\odot$ depending mainly on M_i and overshoot assumptions.

DOUBLE SHELL BURNING ON THE TP-AGB

In this phase both hydrogen and helium shells are burning. However, as soon as the He shell approaches the H shell it starts to burn intermittently, using up most of its fuel in so-called shell flashes or thermal pulses (TP). During pulses a convective tongue reaches outward, mixing freshly created carbon into the intermediate layer. Because calculations show that pulses increase in strength during further evolution, it is expected that the tongue finally extends to regions into which the outer envelope penetrates (although the molecular weight discontinuity at the H/He interface constitutes a barrier) and mixing of carbon and other nuclear products (e.g., *s*-process elements) from the burning zone occurs in the third dredge-up. During the TP-AGB phase the average luminosity steadily increases according to a well established core mass–luminosity relation, which is nearly universal and only slightly dependent on composition and mass. The existence of such a relation is due to the fact that the burning takes place at the surface of a geometrically minute and highly compact degenerate core that is surrounded by a comparatively thin and very extended envelope with little back-reaction on the physical conditions in the burning region. During the shell flashes the H shell is, for a short time, extinguished. This leads to a temporary decrease of the total luminosity (after a short spike due to the flash itself).

With increasing luminosity (and radius) the stars lose more and more mass by the Reimers wind ($\dot{M} \propto LR/M$). From the core evolution one derives that the climbing rate on the TP-AGB is given by about 1 (bolometric) mag in 10^6 yr, which is also the order of the total lifetime on the TP-AGB, after which even a Reimers wind would remove the remaining hydrogen envelope. However, experience has shown that the core masses (or luminosities) do not grow to this limit, but that a superwind increases the mass loss rates to values of $10^{-4}\ M_\odot\ \mathrm{yr}^{-1}$, thus bringing the TP-AGB evolution to a nearly abrupt stop. After the remnant hydrogen layer is reduced to about $10^{-4}\ M_\odot$, the star leaves the AGB and shrinks at constant luminosity, thereby increasing its surface temperature until, at $T_{\mathrm{eff}} \approx 30{,}000$ K, its ultraviolet radiation becomes strong enough to ionize the surrounding material, which then appears as a planetary nebula. After the H shell is finally extinguished, the luminosity strongly decreases and the star moves down toward the white dwarf region.

In this context it is important to realize that the white dwarf is prefabricated already as the degenerate core on the TP-AGB, which also determines the initial conditions for post-AGB evolution.

Consistent calculations of central star or white dwarf evolution must thus start in the AGB phase without suppression of thermal pulses and with inclusion of mass loss. This has been done only recently. During the TP-AGB stage the stars are also long period variables, a fact that is probably responsible for the increased rate of mass loss due to shock formation in the outer envelope and further acceleration by radiation pressure on dust that is formed above the photosphere. Information about mass loss on the TP-AGB and thus on the initial–final mass relation can also be obtained from AGB luminosity functions and from white dwarfs in clusters with known turn-off mass: In both cases, one concludes that intermediate mass stars terminate their nuclear evolution by mass loss and not by carbon ignition and/or supernova explosion. The upper mass limit for IMS appears to coincide with the upper mass limit for white dwarf production. Its numerical value, however, is dependent on assumptions for core overshooting, and ranges from about 6–8 $M_\odot$. In this context one should mention that strong overshooting, in general, causes the evolution to proceed at larger core masses and luminosities compared to the canonical theory, thereby also causing the lower IMS limit (for nondegenerate He ignition) to be reduced from 2.2 to about 1.6 $M_\odot$.

Additional Reading

Bertelli, G., Bressan, A., Chiosi, C., and Angerer, K. (1986). Evolutionary models for low and intermediate mass stars with convective overshooting. *Astron. Ap. Suppl.* **66** 191.

Iben, I., Jr. and Renzini, A. (1983). Asymptotic giant branch evolution and beyond. *Ann. Rev. Astron. Ap.* **21** 271.

Iben, I., Jr. and Renzini, A. (1984). Single star evolution I. Massive stars and early evolution of low and intermediate mass stars. *Phys. Rep.* **105** 329.

Kippenhahn, R. and Weigert, A. (1990). *Stellar Structure and Evolution*. Springer-Verlag, Berlin.

Maeder, A. and Meynet, G. (1989). Grids of evolutionary models from 0.85 to 120 $M_\odot$: Observational test and mass limits. *Astron. Ap.* **210** 155.

Renzini, A. (1987). Some embarrassments in current treatments of convective overshooting. *Astron. Ap.* **188** 49.

Weidemann, V. (1987). The initial–final mass relation: Galactic disk and Magellanic Clouds. *Astron. Ap.* **188** 74.

See also **Nebulae, Planetary, Origin and Evolution; Stars, Carbon; Stars, Cepheid Variable; Stars, Evolved, Circumstellar Masers; Stars, Long Period Variable; Stars, White Dwarf, Structure and Evolution; Stellar Evolution.**

Stellar Evolution, Low Mass Stars

Francesca D'Antona and Italo Mazzitelli

In stellar astrophysics, we define low mass stars as all those stars for which, at the end of the central hydrogen burning phase, the electrons in the remnant helium core become degenerate. Computations show that this occurrence takes place for stars having an initial mass smaller than 2.2–2.5 $M_\odot$ (solar masses), the exact value depending on the chemical composition. Our Sun, the vast majority of the nearby stars, and even the bright star Sirius, belong to this class. According to the physical and chemical modifications occurring during the life of a low mass star, we can distinguish a number of main evolutionary phases, which we usually define as: pre-main sequence, main sequence, red giant, horizontal branch (broadly speaking), asymptotic giant branch, and, ultimately, white dwarf. Not all of the stars go through the whole evolutionary path just outlined: In fact, the lower the star's mass is, the larger the number of phases bypassed, to arrive directly at the white dwarf stage.

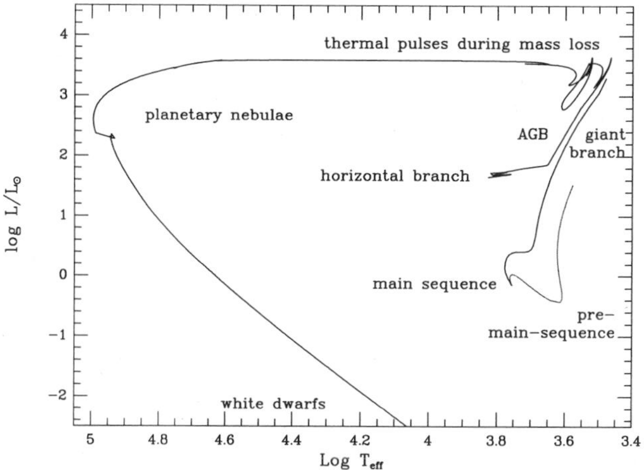

Figure 1. A complete evolutionary track in the HR diagram for a typical 1 $M_\odot$ star is shown from the pre-main sequence (dotted) to white dwarf cooling. The results are taken from sets of computations by the authors.

Our aim is to provide in this entry a qualitative and, where possible, semiquantitative description of the relevant physical features of each of the quoted evolutionary phases (see Fig. 1). In doing so, we will stick to the theory as long as possible, even if the theory is presently able to elucidate only the physiological aspects of stellar evolution, and not also the pathological ones, which are perhaps more interesting. It has often been stated that to derive information about the life of stars on the ground of the last few centuries of observations is like reaching conclusions about the life of human beings after observing a crowd for a couple of minutes. Even if this claim is not strictly correct, there is some truth in it.

PRE-MAIN SEQUENCE

The birth of a star, that is the first gathering and heating of interstellar matter until a nearly spherical configuration in nearly hydrostatic equilibrium appears, is a not yet well-understood phenomenon. Gas and dust in the dense interstellar clouds, where stars usually originate, are an obstacle to the observations, and the magnetohydrodynamics of a turbulent, rotating, supersonic plasma out of thermodynamic equilibrium is still far from the grasp of any reliable theory. However, both simplified models and observations lead to the conclusion that accidental local compressions of matter in clouds lead to the appearance of gravitationally bound regions. Matter begins to freely fall toward the mass center, and density increases until it becomes so large that the falling particles undergo frequent collisions. The kinetic energy is then transformed into thermal energy, leading to heating and ionization of gas. Matter becomes opaque to radiation, and a gradient of (gas) pressure sets in, stopping the collapse. From this point on, further contraction will be allowed only as the release of gravitational energy compensates for the energy radiated away from the surface. A protostar in hydrostatic equilibrium is born. The mass of this protostar can be only a few hundredths of solar mass; if the gravitationally bound region contains more mass, accumulation onto the surface of the star goes on for some million years at a rate of 10^{-7}–10^{-6} $M_\odot$ yr^{-1} until all the nebula is cleared.

In the meantime, slow gravitational contraction leads to progressive increase of the central temperature. When the temperature reaches $\sim 10^6$ K, deuterium, which is present in the original matter in a few parts per 100,000, begins a nuclear reaction with hydrogen, freeing energy and building ^{4}He. The energy supply from the nuclear reactions is sufficient to power the outflowing luminosity for a while; moreover, the energy flux coming from the center of the star is so large that radiation is not any more adequate to carry energy through the star, and the star becomes fully convective.

When deuterium is exhausted, gravitational contraction resumes. The central density and temperature increase until one of the two following situations occurs.

1. Density becomes so large that electron degeneracy sets in. Because a density profile gives rise to a degeneracy pressure profile even at zero temperature, contraction stops and the star cools down to invisibility in some hundreds of millions of years. Objects of this kind having a mass smaller than 0.08 $M_\odot$ are usually called brown dwarfs; they can be regarded as a class of white dwarfs, in which the internal chemical composition is still the original one, unprocessed by nuclear reactions.
2. Ignition of hydrogen nuclear fusion.

MAIN SEQUENCE

When the central temperature reaches 10–20$\times 10^6$ K, hydrogen begins nuclearly reacting, giving rise to ^{4}He. Because hydrogen is abundant (70% or more in mass), the energy output of the reactions is large (6×10^{18} erg g^{-1}) and the luminosity of the star is relatively low, this is the long-lasting phase in the life of the star. Our Sun has been on the main sequence for $\sim 4.5 \times 10^9$ yr and will remain there for an equivalent amount of time. If the central temperature is lower than 15×10^6 K, hydrogen burns via the so-called proton–proton chain, which starts from the direct fusion of two protons. At greater temperatures, the elements carbon, nitrogen, and oxygen behave as catalysts; a nucleus of ^{12}C absorbs a proton and, following a complex path of disintegrations and absorptions through several isotopes of C, N and O, gives rise to ^4He, restoring the initial ^{12}C. Because the CNO cycle steeply depends on temperature, stars of mass larger than 1.2–1.3 $M_\odot$ for which the CNO cycle is the maximum contributor, have nuclear energy generation that is strongly peaked around the center. The resulting energy flux in the central regions is so large that, again, convection sets in, this time limited to the core. The larger the star's mass is, the larger are the luminosity and temperature. When the surface temperature grows larger than $\sim$20,000 K, matter is almost fully ionized also in the external layers, and radiative transport is efficient, so that stars of $\sim 2\ M_\odot$ are radiative in the external layers (but convective in the core). Stars of lower mass, for which the surface temperature is lower and ionization incomplete, have instead convective external layers (but not convective cores), until, for masses smaller than 0.35 $M_\odot$, external convection reaches the center.

Surface luminosity varies as the cube of the star's mass. Because the total nuclear energy reservoir is only linear with mass, the larger the star's mass is, the shorter its life is on the main sequence. When hydrogen at the center is completely exhausted and a helium core makes its first appearance, hydrogen burning moves to a spherical shell around the core, and the main sequence phase is over.

RED GIANT BRANCH

The luminosity increases and the external layers of the star expand when the hydrogen burning shell approaches the surface. Due to the fast decrease of the surface gravity (and to other, not well-understood mechanisms), mass loss, a phenomenon already present on the main sequence, begins increasing, and the total mass can decrease by some tenths of $M_\odot$. If in this process the total mass goes below approximately 0.5 $M_\odot$, the hydrogen burning shell reaches very close to the surface, and is then turned off. The external layers rapidly contract and become degenerate on the top of the already degenerate helium core. The star becomes a white dwarf with an internal helium composition. In the opposite case, when the total mass of the star remains larger, the helium core grows up to 0.5 $M_\odot$. At this critical value, the temperature in the core reaches 10^8 K, and the helium fusion reactions ignite, giving rise to carbon and oxygen. The first release of nuclear energy causes an increase in temperature. Because the matter is degenerate, this has little influence upon the internal pressure profiles, and reexpansion with reabsorption of gravitational energy cannot occur, in response, to decrease the temperature and stabilize burning. Therefore nuclear burning accelerates due to the increased temperature, until a violent, although nondisruptive, explosion (the helium flash) generates enough energy to lift matter from electron degeneracy and restore equilibrium conditions with central helium burning.

HORIZONTAL BRANCH

The star is now furnished with two nuclear energy sources: helium burning in the (convective) core, and hydrogen burning in a shell around the helium core. The larger the relative contribution of the helium reactions (i.e., the star is closer to its helium burning main sequence) and the smaller the hydrogen envelope, the larger is the surface temperature of the star. Horizontal branch stars with extended *H*-envelopes and efficient *H*-burning shells tend instead to resemble more the red giants. In globular clusters (where helium burning dominates), the horizontal branch appears as a conspicuous group of hot stars, in open clusters (hydrogen shell burning dominated) the appearance of these stars remains much the same as red giants.

When helium is completely exhausted in the convective core, helium burning stops for a while, and then resumes in a shell around the now-degenerate carbon–oxygen core.

ASYMPTOTIC GIANT BRANCH

This phase can be completely skipped if the hydrogen burning shell has reached the surface during the previous phase. At the exhaustion of helium burning, the star is left without any source of nuclear energy, and it becomes a white dwarf of mass smaller than 0.53–0.55 $M_\odot$. If the mass is larger and some hydrogen envelope is then still present, a double shell burning phase follows. During this phase, helium thermal pulses occur in a very complex physical framework roughly summarized in the following.

Helium stops burning and shell hydrogen burning accumulates a thick layer of helium until, at the base of this layer, helium reactions resume with a growth rate faster than the energy transfer timescale (this has nothing to do with electron degeneracy). The local temperature increases, leading to impulsive helium burning followed by the onset of a convective shell and expansion, the hydrogen shell burning turns off, and steady helium burning turns on until almost all the helium layer is consumed. Hydrogen shell burning then resumes, and the whole process iterates up to 20–30 times, depending on the mass of the star. During the pulses, external convection can reach into the helium- and carbon-enriched region, leading to the appearance of the so-called carbon stars and also, presumably of *s*-process material.

This phase ends with the final ejection of the whole hydrogen-rich envelope, which forms a planetary nebula around a helium, carbon, and oxygen core. The processes responsible for the ejection of the planetary nebula may be related to pulsational instability, to the effect of radiation pressure on grains formed in the cool red giant envelopes, or also may be related to reaching the critical Eddington luminosity during the thermal pulse phase. In any case, none of these mechanisms has been completely investigated or is really well-understood as of today, also because of the involved hydrodynamic computations.

WHITE DWARF SEQUENCE

The surface helium- and, possibly, hydrogen-rich layers contract upon the carbon–oxygen core, and a white dwarf appears with negligible nuclear energy sources or none at all. Radiation from the surface and, at least in a first phase, neutrinos from the core, cool the structure, but electron degeneracy prevents any further contraction. Very soon, due to the high efficiency of degenerate electrons in transferring heat, the structure becomes almost isothermal.

When the internal temperature becomes low enough, coulomb interactions take over, and the ions form a lattice structure: The white dwarf crystallizes, freeing a small amount of phase transition heat. At even lower temperatures, the Debye conditions are reached, and the specific heat of the matter rapidly drops to zero. The star dies as a cold crystal in a few billion years.

Additional Reading

Böhm-Vitense, E. (1991). *Introduction to Stellar Astrophysics. Vol. 3. Stellar Structure and Evolution*. Cambridge University Press, Cambridge.

D'Antona, F. and Mazzitelli, I. (1990). Cooling of white dwarfs. *Ann. Rev. Astron. Ap.* **28** 139.

Iben, I., Jr. and Renzini, A. (1983). Asymptotic giant branch evolution and beyond. *Ann. Rev. Astron. Ap.* **21** 271.

Kaler, J. B. (1985). Planetary nebulae and their central stars. *Ann. Rev. Astron. Ap.* **23** 89.

Liebert, J. and Probst, R. G. (1980). Very low mass stars. *Ann. Rev. Astron. Ap.* **25** 473.

Trimble, V. (1986). White dwarfs: The once and future suns. *Sky and Telescope* **72** 348.

See also **Nebulae, Planetary, Origin and Evolution; Neutrinos, Solar; Stars, Carbon; Stars, Horizontal Branch; Stars, White Dwarf, Structure and Evolution; Stellar Evolution.**

Stellar Evolution, Massive Stars

Achim Weiss and James W. Truran

Massive stars are among the brightest objects in galaxies, exhibiting extremely strong stellar winds and ending their lives in spectacular fashion as supernovae. They are the sites of the nucleosynthesis of many of the heavy elements found, for example, in our own bodies, and they pose a variety of unresolved problems of basic physics and stellar evolution theory. Although they comprise only 10% of all stars, they play an extremely important role in the lives of galaxies. In this entry, we present brief discussions of many interesting features of these exciting objects.

GENERAL PROPERTIES

The mass range of "massive stars" extends from about 9 to ≈ 150 $M_{\odot}$ (1 $M_{\odot} = 1.989 \times 10^{33}$ g, the solar mass). The lower limit is defined by the requirement that the star will reach sufficiently high temperatures in the center to start nuclear burning of carbon and/or oxygen under nondegenerate conditions. The upper limit is determined by various instabilities that might inhibit the existence of more massive stellar objects for times long enough to burn hydrogen during the main-sequence phase. One such instability that was found theoretically by Paul Ledoux in the 1940s involves exponentially growing pulsations that will lead to the ejection of the stellar envelope. The creation of e^+e^- (electron–positron) pairs at temperatures higher than 10^9 K can also lead to a dynamical instability. Finally, there exists a maximum mass that can be formed out of the interstellar medium. Although there are indications that even more massive objects might exist for a short time, such theoretically proposed "supermassive stars" will not be included in this entry.

Following the approximate relation between mass (M) and luminosity (L) on the main sequence ($L \sim M^{3.5}$), the luminosities of massive stars extend from 3000 $L_{\odot}$ to several million solar luminosities (1 $L_{\odot} = 3.826 \times 10^{33}$ erg s^{-1}, the solar luminosity), which translate to absolute bolometric magnitudes M_{bol} between -4 and -12. The effective temperatures are of the order of 10,000–100,000 K, but can change during evolution to a few thousand kelvins when the stars become red supergiants. A Hertzsprung–Russell (HR) diagram of observed massive stars in the Milky Way is shown in Fig. 1. Due to their high main-sequence luminosities, massive stars are rather short-lived objects, with lifetimes extending from 3×10^7 yr (9 $M_{\odot}$) to only 3×10^6 yr (120 $M_{\odot}$), as compared to 10^{10} yr for less massive stars.

On the main sequence, massive stars burn hydrogen into helium by nuclear fusion in very extended convective cores, which include up to 80% of the total stellar mass. All subsequent burning phases take place within the former convective core. The star passes through phases of helium, carbon, neon, oxygen, and silicon burning, until finally an iron core is developed, in which no further nuclear burning can take place. This core will then collapse, and the gravitational energy released powers a supernova explosion, during which a large variety of heavy elements through iron and even beyond will be created and ejected into the interstellar medium.

While being transformed internally by nuclear processes, massive stars also lose mass by very strong stellar winds. The mass loss rates can be as high as 10^{-4} $M_{\odot}$ yr^{-1} for the most massive and luminous objects, which implies that major fractions of the stellar envelope will be lost and deeper layers that already have been processed might be uncovered.

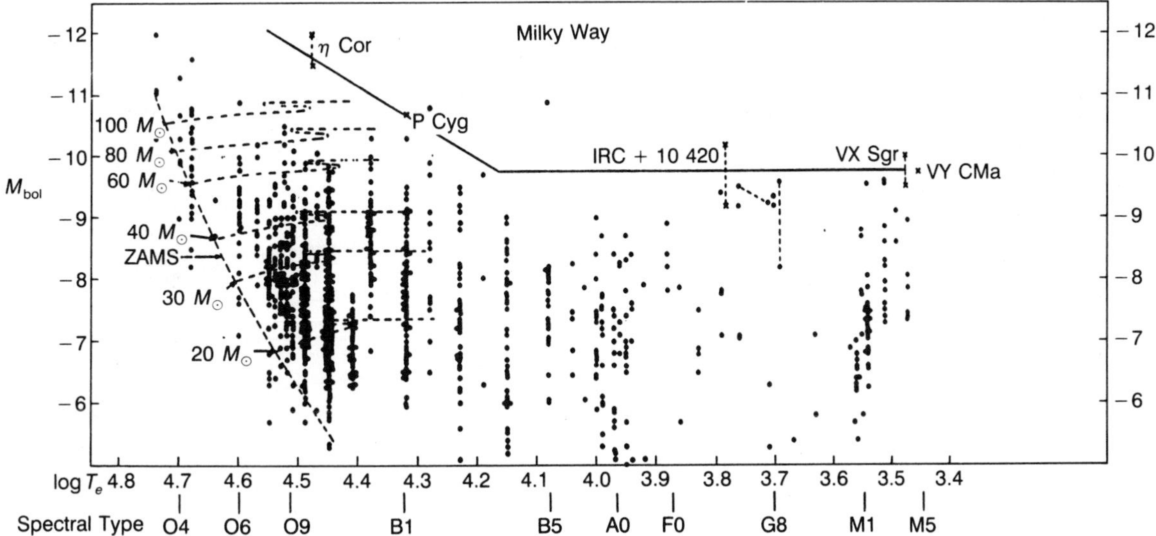

Figure 1. A Hertzsprung–Russell diagram of massive stars in the Milky Way. Some of the most massive objects along the Humphreys–Davidson limit are identified. The theoretical zero age main sequence and a few evolutionary paths up to the end of central hydrogen burning are indicated by dashed lines. [*Adapted from Humphreys and Davidson*, Astrophys. J. *232 409 (1979) with permission of the University of Chicago Press.*]

TYPES OF MASSIVE STARS

According to their effective temperature (on the main sequence) and luminosities, massive stars are of spectral types O and early B, and belong to the luminosity classes III (giants), II (bright giants), and I (supergiants).

From Fig. 1 it is also evident that massive stars occupy two regions of very distinct effective temperatures in the HR diagram, with a clear gap between them. Those on the hotter side are generally called blue supergiants, which include all stars lying still within the main-sequence band, as well as those that evolve back to higher temperatures during subsequent phases. The cooler objects are called red supergiants; they comprise stars that are in the central helium-burning phase or in later evolutionary stages. The relative numbers of these red and blue objects are observable quantities that can be used to test stellar evolution theory and are the subject of much debate.

Other types of massive stars are known by their peculiar properties. The most interesting class is the Wolf–Rayet (WR) stars. They are characterized by extremely unusual chemical compositions: strong underabundances of hydrogen with accompanying increases in their helium abundances, as well as overabundances of nitrogen (subtype WN), carbon (WC), and oxygen (WO). It is generally believed that these objects have lost their hydrogen envelopes partially or completely due to heavy mass loss, thereby uncovering regions of matter that has been processed in hydrogen burning (WN) or even helium burning (WC, WO). The masses of WR stars are typically above 30 $M_\odot$. Although they are located close to the main sequence, their internal evolution might have progressed past this stage.

Another subclass of massive stars, the OBN (or OBC) stars, show normal or somewhat depleted hydrogen abundances, with a dramatically increased N/C ratio, which is typical for CNO-processed material. It is generally assumed that this material has been brought to the stellar surface by rotation-induced mixing and/or by mass loss. Indeed, massive stars above $\approx 20\ M_\odot$ are among the fastest rotators known, with some of them rotating close to the centrifugal breakup limit.

From Fig. 1 it is also evident that above a certain luminosity limit (the Humphreys–Davidson limit), very few supergiants exist. At or near this limit, there are situated many variable supergiants, classified as luminous blue variables (LBV), that comprise the subtypes of the Hubble–Sandage, S Doradus, P Cygni, and η Carinae variables. It is believed that there exists an instability that leads to periodic or sporadic ejections of envelope material, which thereby reduces the stellar mass, and that the observed LBV are in this stage. The masses of the LBV are among the highest known; their evolutionary status is unknown, but they most likely have evolved past the core hydrogen burning phase.

EVOLUTIONARY ASPECTS

The evolution of massive stars is straightforward in the sense that their high central temperatures, which are required to provide pressure to balance gravity, permit successive phases of nuclear burning of hydrogen, helium, carbon, neon, oxygen, and silicon. The details of this evolution are, however, sensitive to a number of physical parameters. We review some interesting evolutionary features in this section.

Overshooting

As all other stars do, massive stars spend more than 75% of their lives in the hydrogen core burning phase on the main sequence. During this phase (see Fig. 2 for theoretical results), they evolve to lower effective surface temperatures and higher luminosities. The

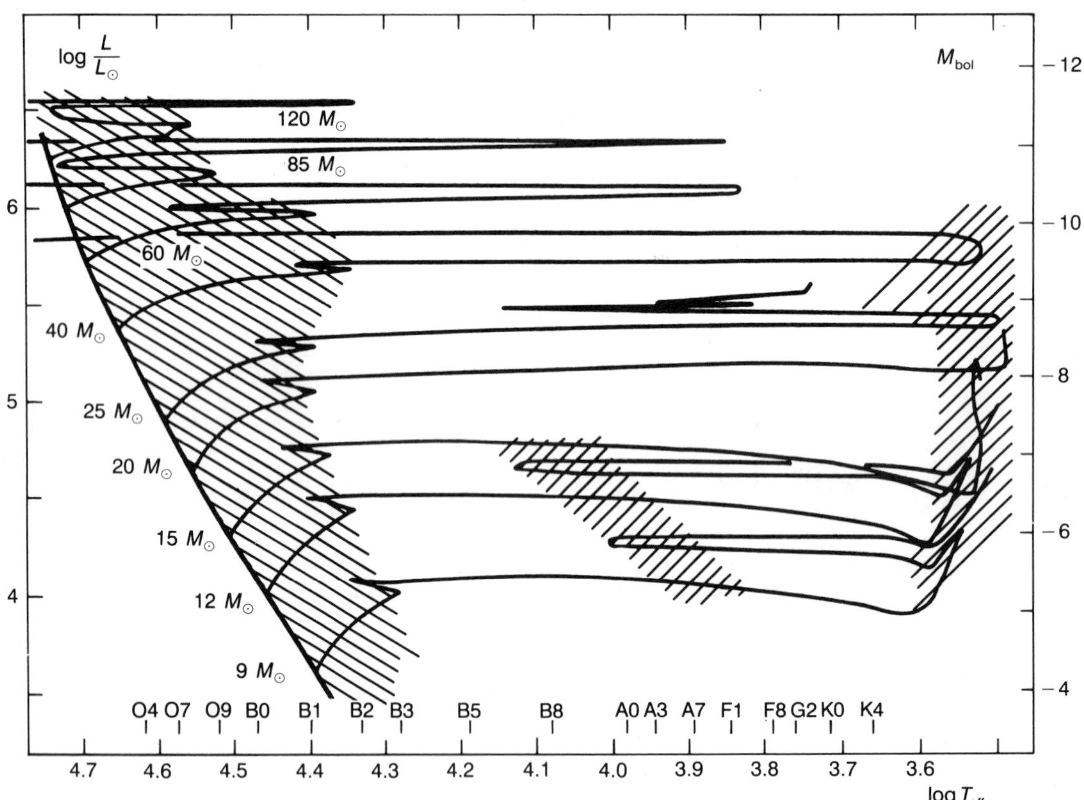

Figure 2. Theoretical evolution of massive stars with mass loss and overshooting taken into account. Hatched regions indicate phases of slow nuclear burning; the left one is identical to the main sequence. [*Adapted from Maeder and Meynet,* Astron. Ap. ***210*** *155 (1989) with permission of Springer-Verlag*.]

duration of the main-sequence phase and, therefore, the width of the main-sequence band (left hatched region in Fig. 2) depends on the amount of hydrogen available for nuclear burning. Due to the existence of large convective cores, fresh hydrogen-rich material is constantly mixed into the hottest regions of the star at the center, where the nuclear reactions are most effective and most of the energy is produced. Thus, the size of the convective core influences the main sequence lifetime. However, the boundary of a convective region is not well determined theoretically. Even if a layer inside a star is convectively stable, it might be penetrated by convective elements from the unstable region, due to the momentum of these elements. This dynamical effect, called *overshooting*, causes the convective cores to be larger than they would be if treated statically. The magnitude of convective overshoot is still under discussion, but the latest results indicate a moderate increase in core size by some 10% of a pressure scale height or roughly an additional 5% in mass. Overshooting (also known to occur in later stages of core burning) effects the entire evolution and is a major subject of current stellar evolution research.

Mass Loss

A second effect that is especially important for massive stars is mass loss. One theoretical view is that massive-star winds are driven by radiation pressure, but other models (fluctuation around thermal equilibrium; corona–wind interaction) are also under discussion. The mass loss rates vary from 10^{-11}, for a zero age 9-$M_{\odot}$ star, to as much as 10^{-4} $M_{\odot}$ yr^{-1} for a 40-$M_{\odot}$ star on the red giant branch, or an even more massive star during the main-sequence phase. All these rates are obtained from observation and are incorporated into theoretical calculations (Fig. 2) by empirical formulae. As a consequence of the high mass loss rates, the most massive stars can lose up to 20% of their mass prior to leaving the main sequence and more than 50% during their whole lives. Even a 9-$M_{\odot}$ star might lose 5%. As a result, layers of matter that were processed in nuclear reactions in earlier evolutionary phases can be uncovered; also, some stars may never evolve to red giants.

Blue–Red–Blue Evolution

After the main-sequence phase, massive stars usually evolve to red giants, igniting core helium burning either in the transition phase or upon reaching the Hayashi track, and spending most of the helium core burning time there. However, they might also experience "blue loops" during this phase. Phases of slow helium burning are indicated by the right hatched regions in Fig. 2. The extension of these loops, in terms of both effective temperature and time, depends on the two major effects previously mentioned, as well as on the characteristics of convection in inhomogeneous layers (*semiconvection*), another area of uncertainty in convection theory. (If overshooting is large enough, semiconvection might not occur.)

Final Evolution

Outside the helium burning core, a hydrogen burning shell is still active, providing a major fraction of the luminosity. Both burning regions lie inside the original convective hydrogen core. After helium exhaustion and the initiation of helium shell burning, carbon burning starts in an even smaller central region and lasts for only a few thousand years. Subsequent burning phases follow in the same manner, the changes now occurring so rapidly that the star cannot adjust its outer structure, and therefore keeps its stellar appearance. Finally, the star terminates its life as a supernova of type II (hydrogen in spectrum); WR stars, however, might account for

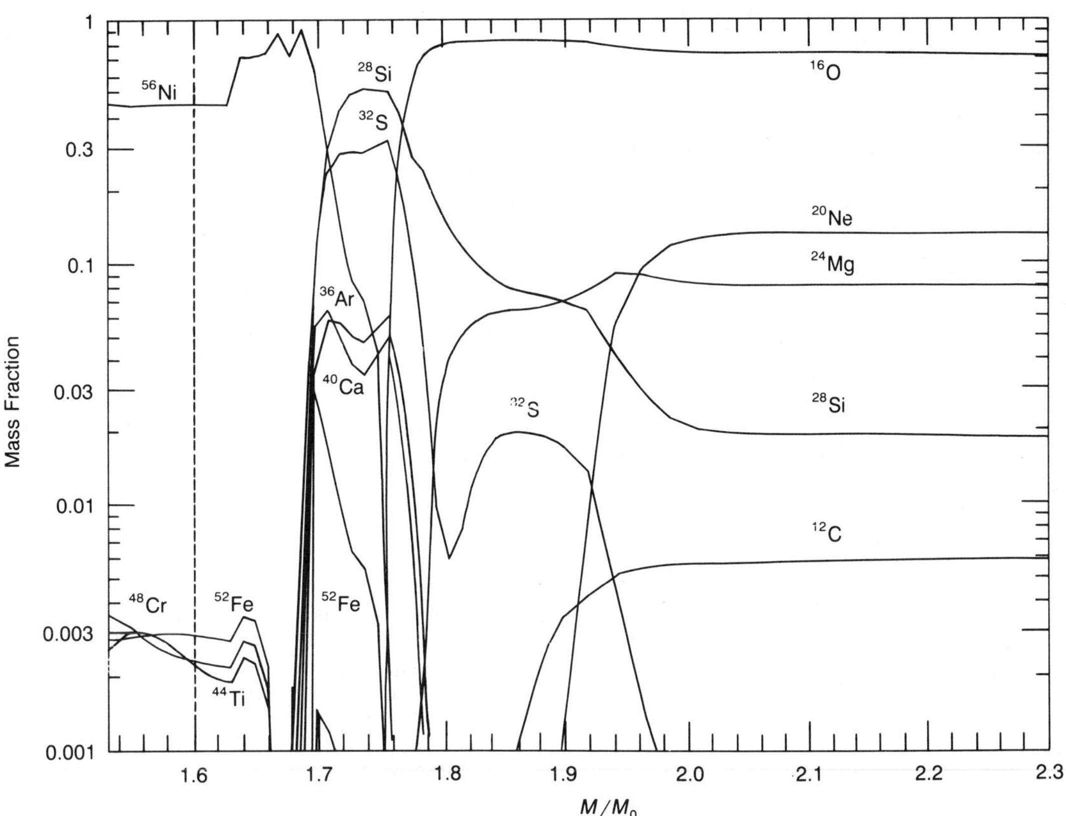

Figure 3. The inner core of a 20-$M_{\odot}$ star at the end of its stellar life and after the shock front caused by the supernova explosion has passed through it, leading to additional nuclear processes. (*Courtesy of F.-K. Thielemann.*)

some of the type I supernovae (hydrogen-free). The remnants are presumed to be neutron stars (pulsars) and possibly black holes.

Other Aspects

Due to their short lifetimes, observed massive stars belong to the youngest stellar population in the Galaxy and have high concentrations ($\approx$ 2–3%) of metals (elements heavier than helium). Theoretical work has therefore concentrated on Population I stars. As a result of Supernova 1987A in the Large Magellanic Cloud, a metal-poor system, recent research also has explored the effects of low metallicity, which seems to influence the evolution significantly as reflected in the stellar appearance. In particular, it might explain the fact that the progenitor of SN1987A was not a red supergiant, but a blue supergiant of spectral type B3. Previously, it had been assumed that most massive stars end their life as red supergiants, with the exception of the WR stars.

NUCLEOSYNTHESIS

Massive stars are believed to be extremely important contributors to the abundances of the heavy elements observed in nature. Successive stages of burning of hydrogen, helium, carbon, neon, oxygen, and silicon fuels define their presupernova evolution, leading to the creation of a layered configuration of these nuclear burning products.

The iron cores of approximately the Chandrasekhar limiting mass ($\approx 1.4\ M_{\odot}$), formed in silicon burning, lack further nuclear fuel and are compelled to contract under gravity, leading to collapse. When the ashes of prior burning epochs are subsequently subjected to high temperatures and densities accompanying supernova ejection, further nuclear processing yields elemental and isotopic abundance patterns that mimic very closely those of solar system matter. The inner structure of a star of 20 $M_{\odot}$ subsequent to shock processing in a supernova is shown in Fig. 3.

The heavy elements, ranging in mass from oxygen to nickel, that are present in galactic matter are believed to have been formed in these massive stars and associated type II supernova environments. The recent studies of SN1987A provide a strong confirmation of this theoretical picture. The consistency of the light curve with the time scales and energetics of the decays of ^{56}Ni through ^{56}Co to ^{56}Fe provides evidence for the formation of approximately 0.1 $M_{\odot}$ of iron peak nuclei, including specifically the elements chromium, manganese, iron, cobalt, and nickel. The supernova phase of evolution of these massive stars may also provide the site of production of even heavier elements through uranium and thorium.

Additional Reading

Arnett, W. D., Bahcall, J. N., Kirshner, R. P., and Woosley, S. E. (1989). Supernova 1987A. *Ann. Rev. Astron. Ap.* **27** 629.

Chiosi, C. and Maeder, A. (1986). The evolution of massive stars with mass loss. *Ann. Rev. Astron. Ap.* **24** 329.

Humphreys, R. M. and Davidson, K. (1979). Studies of luminous stars in nearby galaxies. III. Comments on the evolution of the most massive stars in the Milky Way and the Large Magellanic Cloud. *Astrophys. J.* **232** 409.

Kippenhahn, R. and Weigert, A. (1990). *Stellar Structure and Evolution*. Springer-Verlag, Berlin.

Maeder, A. and Meynet, G. (1989). Grids of evolutionary models from 0.85 to 120 $M_{\odot}$: Observational test and the mass limits. *Astron. Ap.* **210** 155.

Truran, J. W. (1984). Nucleosynthesis. *Ann. Rev. Nucl. Particle Sci.* **34** 53.

Woosley, S. E. and Weaver, T. A. (1986). The physics of supernova explosions. *Ann. Rev. Astron. Ap.* **24** 205.

See also **Stars, Carbon; Stars, High Luminosity; Stars, Red Supergiant; Stars, Wolf–Rayet Type; Stellar Evolution, Intermediate Mass Stars; Supernovae, Type I, Theory and Interpretation; Supernovae, Type II, Theory and Interpretation; Supernova Remnants, Evidence for Nucleosynthesis.**

Stellar Evolution, Pre-Main Sequence

Steven W. Stahler

The life history of a star is conveniently divided into a number of different phases. These phases can be viewed as a sequence of events initiated within the star in response to the compressive force of gravity. For most of the star's lifetime, gravity is precisely balanced by the thermal pressure generated by hydrogen fusion in the stellar core. During this main-sequence phase, the star is a static object, changing slowly in chemical composition as hydrogen is converted into helium. During the prior, pre-main-sequence phase, the star's central temperature is too low to sustain fusion. The force of gravity predominates and the star contracts. This process of overall contraction is the essence of pre-main-sequence evolution.

If a pre-main-sequence star's self-gravity were entirely unopposed, the star would be in a state of free-fall collapse. Such a rapid collapse does not occur because the force of gravity squeezing on the gas generates significant thermal pressure. This pressure would, indeed, eventually halt the contraction but for the fact that the star is continually radiating away energy from its surface. The net result is that, at any instant, the star's gravity exceeds only slightly its internal pressure. Pre-main-sequence contraction is said to be quasistatic, and it proceeds at a rate that is set by the star's surface luminosity.

In any object in which thermal pressure nearly balances gravity, the central temperature T_c is related to the mass M and the radius R by the proportionality

$$T_c \propto M/R.$$

This relation shows explicitly how the large size of pre-main-sequence stars makes them colder than their main-sequence counterparts and therefore incapable of burning hydrogen at their centers. During the course of their evolution, these stars shrink by about a factor of 10, until T_c rises to approximately 10^7 K, at which point the onset of fusion halts further contraction.

CONVECTION AND CONTRACTION

The luminosity emitted from the stellar surface is related to the radius and surface temperature by Stefan's law (where σ is the Stefan–Boltzmann constant):

$$L = 4\pi R^2 \sigma T_{\mathrm{surf}}^4.$$

For sufficiently large radii, that is, for sufficiently young stars, the surface luminosity is so large that it cannot be transported from the interior of the star by the diffusion of radiation alone. Thus, young pre-main-sequence stars must rely on another mechanism for energy transport: thermal convection. Hot, buoyant blobs of gas rise from the deep interior, emit their heat as radiation at the stellar surface, and then sink back down. The presence of convection is not, of course, unique to pre-main-sequence stars. The present-day Sun, for example, has an outer convection zone that covers a substantial fraction of its radius. The pre-main-sequence Sun, however, was *entirely* convective, at least during its early contraction.

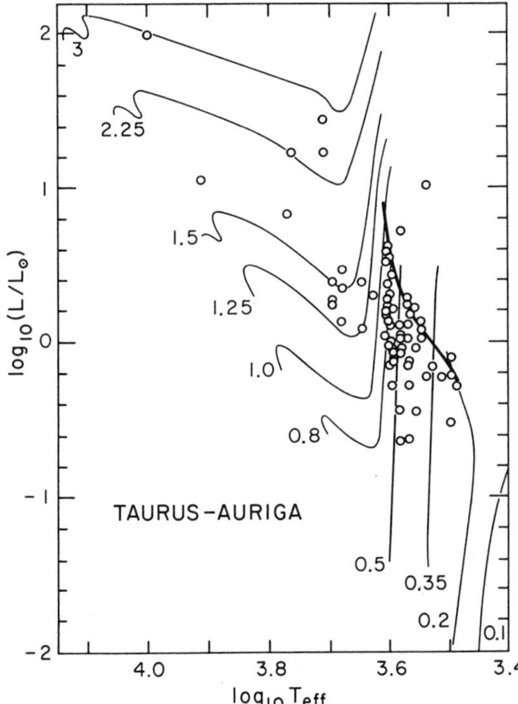

Figure 1. Hertzsprung–Russell diagram for pre-main-sequence stars. The lighter curves are theoretical evolutionary tracks; each track is labeled by the mass of the associated star in solar masses. The heavy curve is the theoretical birthline, that is, the predicted upper envelope for pre-main-sequence stars. The open circles are observations of T Tauri stars in the Taurus–Auriga molecular cloud complex. [*Reproduced from Shu, Adams, and Lizano (1987). Reproduced, with permission, from the* Annual Review of Astronomy and Astrophysics ***25*** © *Annual Reviews, Inc.*]

Historically, the crucial role of convection in young stars was slow to be realized. Such eminent astrophysicists as Karl Schwarzschild, Arthur S. Eddington, and Subrahmanyan Chandrasekhar either dismissed the possibility of stellar convection entirely or else considered only its role in the inner regions of massive main-sequence stars. The importance of outer convection came to be appreciated in the 1950s, first through the work of Donald Osterbrock on low-mass main-sequence stars. L. G. Henyey, R. LeLevier, and R. D. Levée published the first computer calculation of quasistatic contraction in young stars in 1955, but continued to ignore surface convection. Finally, Chushiro Hayashi, in 1961, pointed out that there is a minimum surface temperature for pre-main-sequence stars of a given mass. Thus, a large radius at early time necessarily implies a large luminosity, making convection inevitable.

The physical origin for Hayashi's minimum temperature lies in the nature of the opacity in the star's subphotospheric layers. At typical pre-main-sequence surface temperatures of 3000–5000 K, this opacity is provided principally by H^- ions, whose abundance is extremely sensitive to temperature. A drop in T_{surf} below the critical Hayashi value lowers the ion abundance and the opacity so much that the stellar material is transparent to the transported radiation, indicating that the true stellar photosphere cannot be at that temperature.

Hayashi went on to show that the actual surface temperature remains close to the minimum value as long as the star is fully convective. In the Hertzsprung–Russell (HR) diagram, the star follows a nearly vertical, downward path. The surface luminosity smoothly decreases, along with the radius, until L is below the maximum value that can be carried out by radiative diffusion. At this point, the shrinking star develops, at its center, a radiatively stable region. As this radiative region grows, T_{surf} increases and the star follows a more nearly horizontal track in the HR diagram. The radius of the star continues to decrease until hydrogen ignites. These features of pre-main-sequence evolution were calculated with high accuracy by Icko Iben, Jr. and by Dilhan Ezer and Alastair G. W. Cameron, in the years immediately following Hayashi's discovery.

INFLUENCE OF THE STELLAR MASS

The evolutionary time scale for pre-main-sequence contraction, that is, the time for the radius to shrink by a significant factor, is the Kelvin–Helmholtz time τ_{KH}, defined in terms of the mass, radius, luminosity, and G (the constant of gravitation) as

$$\tau_{KH} = GM^2/(RL).$$

Because the product RL drops with decreasing radius, the evolution is relatively rapid at first, but then slows down as the main sequence is approached. For the Sun, substitution of solar values into the definition of τ_{KH} gives a contraction time prior to hydrogen ignition of 3×10^7 yr, in agreement with more accurate numerical computations. This pre-main-sequence lifetime is only about 0.3% of the Sun's estimated main-sequence life of 10^{10} yr.

Stars of lower mass than the Sun reach smaller radii, and therefore lower luminosities, on the main sequence. These stars have substantially longer pre-main-sequence lifetimes. On the other hand, their main-sequence lives are proportionally even longer. On a statistical basis alone, therefore, pre-main-sequence stars must be very rare compared to their main-sequence counterparts. The T Tauri stars are believed to be pre-main-sequence objects with masses ranging from about 0.1–2 $M_{\odot}$. The observed pre-main-sequence stars of higher mass are known as Herbig Ae and Be stars. The Kelvin–Helmholtz time becomes so brief for masses greater than about 10 $M_{\odot}$ that these stars must ignite hydrogen while still surrounded by the interstellar cloud from which they originally condensed. Thus, no visible pre-main-sequence phase is possible for such massive objects.

Stellar structure theory shows that the amount of luminosity that can be carried by radiation in a star scales as $M^{11/2}R^{-1/2}$. The steep dependence on mass implies that stars of sufficiently low mass can never be radiatively stable. Thus, stars with mass less than about 0.5 $M_{\odot}$ follow vertical, convective tracks in the HR diagram right down to the main sequence. At masses lower by another factor of 10, in the "brown dwarf" regime, the rise of electron degeneracy pressure at high density prevents the contracting star's central temperature from ever achieving the critical value for hydrogen fusion.

ORIGIN OF PRE-MAIN-SEQUENCE STARS

We have already noted that, in a star of fixed mass, the contraction time scale lengthens as the radius shrinks. This fact implies that the picture of quasistatic contraction cannot hold indefinitely far into the past. For if we consider a hypothetical pre-main-sequence star of extremely large radius, its luminosity would be so high that contraction would proceed at supersonic speed, violating the assumption that internal pressure nearly balances gravity.

A self-consistent theoretical picture now exists for a dynamical phase of stellar evolution preceding pre-main-sequence. In this "protostar" phase, a primitive star gathers mass, through its gravitational attraction, from its parent cloud of gas and dust. Although the prostar is hidden at optical wavelengths by the surrounding cloud material, many of its properties have been derived by careful numerical calculations performed over the last decade. These calculations have been used to derive the locus in the

Hertzsprung–Russell diagram where pre-main-sequence stars of approximately solar mass or less should first appear. This stellar "birthline" is predicted to form the upper envelope to the distribution of such stars in the diagram; this prediction has been verified by the actual observations of T Tauri stars.

One of the most intriguing observational properties of pre-main-sequence stars is the presence of circumstellar winds. In stars that are well past their protostar phase, these winds are not energetic enough to play a major role in the star's development. In younger stars near the birthline, however, the winds are observed as massive, high-velocity gas flows streaming away from the location of the star. There is an emerging consensus among astrophysicists that it is the onset of these more powerful winds that ends the protostar phase by blowing back the ambient cloud material. However, neither the theory of protostars nor that of pre-main-sequence stars, as outlined here, gives much indication of why such winds should arise.

Finally, there has been much interest in recent years in the possibility of gaseous disks surrounding pre-main-sequence stars. Such disks would be the precursors to the planetary systems that presumably form around many stars. Their origin can readily be understood as the natural outcome of the collapse of rotating protostellar clouds. What is more puzzling is that the same explanation also indicates that the pre-main-sequence stars possessing such disks should be rapidly rotating, which they are not. It is likely that the winds emanating from these stars, when coupled to magnetic lines of force, provide efficient braking, but this picture has yet to be supported by detailed theoretical calculations.

Additional Reading

Clayton, D. D. (1983). *Principles of Stellar Evolution and Nucleosynthesis*. University of Chicago Press, Chicago.

Cohen, M. (1984). The T Tauri stars. *Phys. Rep.* **116** (No. 4) 173.

Kippenhahn, R. and Weigert, A. (1990). *Stellar Structure and Evolution*. Springer-Verlag, Berlin. See Part V, Early stellar evolution, p. 247.

Lada, C. J. (1985). Cold outflows, energetic winds, and enigmatic jets around young stellar objects. *Ann. Rev. Astron. Ap.* **23** 267.

Shu, F. H., Adams, F. C., and Lizano, S. (1987). Star formation in molecular clouds: Observation and theory. *Ann. Rev. Astron. Ap.* **25** 23.

See also **Pre-Main Sequence Objects: Herbig Ae Stars and Related Objects; Protostars, Theory; Stars, Pre-Main Sequence, Winds and Outflow Phenomena; Stars, T Tauri; Stars, Young, all entries; Stellar Evolution; Sun, Interior and Evolution.**

Stellar Evolution, Pulsations

Norman H. Baker

Stellar evolution studies have always gone hand-in-hand with the analysis of stellar pulsation. Pulsations provide a probe of the stellar interior that is quite different from other observational tests of stellar evolution, and can be used to test and refine the results of stellar structure theory. In fact, it would be accurate to say that pulsation theory is an intrinsic part of the study of stellar structure and evolution.

A classic example illustrates this point. Seventy years ago Sir Arthur S. Eddington derived an equation that describes the pulsations of Cepheid variable stars and predicts their periods. He also proposed a thermodynamic "valve" mechanism to explain their instability that requires that the outer layers of a star become more opaque during compression and that they be able to store heat for part of a cycle. This appeared to demand that an abundant element be partially ionized in these layers. This mechanism was implausible on the basis of contemporary stellar models, because it was thought that stars consisted mainly of elements heavier than carbon. Only later, after stellar atmosphere studies showed that hydrogen and helium are the most abundant elements in normal stars, did it become possible to calculate that the ionization of H and He makes the valve work just as Eddington suggested.

It might be thought that pulsation has little to do with evolution, because stellar evolution involves primarily nuclear processes deep in a star's core, whereas pulsation is mainly a surface phenomenon (because pulsation amplitudes are usually very small in the deep interior). But in fact pulsation studies can throw light on such fundamental parameters as stellar masses, luminosities, radii, and composition, and on the evolutionary status of stars, and it is the purpose of this entry to indicate what can be learned in this way for several important classes of pulsating stars.

CEPHEID VARIABLE STARS

Stars with masses greater than 2–3 $M_\odot$ (solar masses) develop nondegenerate isothermal cores at the end of main-sequence H burning, and move rapidly to the right across the Hertzsprung–Russell (HR) diagram to become red giants. All such stars cross at least once the "Cepheid instability strip," a narrow region (700–800 kelvins wide) that is nearly vertical in the HR diagram. Within this strip, nearly all stars pulsate. As can be seen in Fig. 1, at the lowest masses stars cross the strip only once, but more massive stars may cross three or five times, whereas stars above 15–20 $M_\odot$ again make only one crossing. Of course the more massive, more lumi-

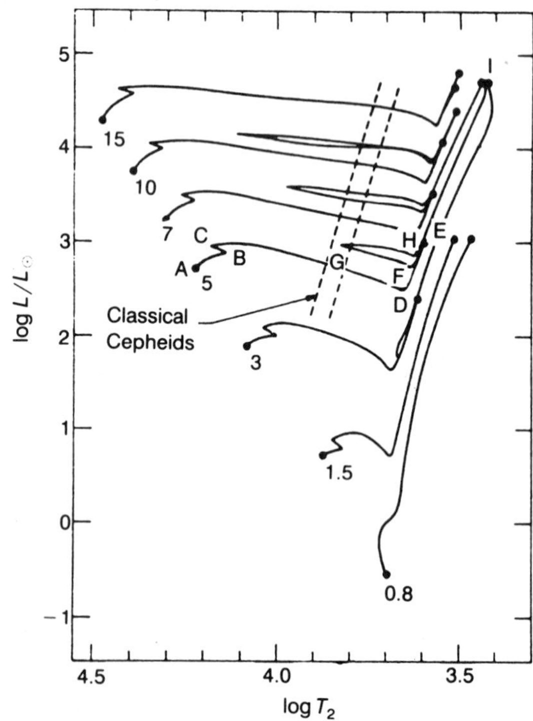

Figure 1. Post-main-sequence evolutionary tracks in the Hertzsprung–Russell diagram for Population I stellar models. The two dashed lines show the location of the Cepheid instability strip. Tracks for stars of intermediate mass make multiple crossings of the strip, whereas those of higher and lower mass cross only once. [*From J. P. Cox, Pulsating stars,* Rep. Prog. Phys. ***37*** *563 (1974); after B. Paczyński,* Acta Astron. ***20*** *47 (1970).*]

nous stars evolve more rapidly, but for any star the first crossing is quite rapid because it corresponds to a stage of thermal instability during hydrogen shell burning. The second crossing occurs during helium core burning, and is normally the slowest one, so this is the stage in which most Cepheids will be found. This circumstance makes it possible to infer a mass–luminosity relation for Cepheids (though of course there will be a few Cepheids on more rapid crossings that will not conform). The mass–luminosity relation depends on composition, especially on the heavy element abundance. Because the existence of the evolutionary loops at a given mass depends on composition, so does the frequency of Cepheids of that mass. For this reason, the average mass of a Cepheid in our galaxy is greater than it is in the Large or Small Magellanic Clouds, which have somewhat different average compositions.

It is easily shown that the pulsation period of a Cepheid varies as the inverse square root of its mean density. When this information is combined with the mass–luminosity relation and the circumstance that all Cepheids have nearly the same effective temperature, the important period–luminosity relation is readily derived. If the color of the Cepheid is known, it is possible to correct for the finite width of the instability strip.

A number of the predicted properties of Cepheids are affected by the physics that is employed in stellar models, and altering such things as nuclear reaction rates, compositions, opacities, and so forth, may produce better or worse agreement with observations. The periods themselves are the most direct diagnostic. Through the period–mean density relation they depend on luminosity, effective temperature, and mass, and because the first two of these are in principle measurable, Cepheid masses can be inferred. These "pulsation masses" are in general agreement with those inferred from evolutionary models. Indeed there are several other ways to estimate the mass and they are, with one exception, generally consistent. The exception comes from double-mode Cepheids, which are pulsating simultaneously in more than one mode, normally the fundamental and the first overtone. Pulsation theory shows that the ratio of these periods is a function of mass, and the masses implied by observed period ratios are smaller by a factor of at least 2 than those given by other methods. This may indicate the necessity of adjusting some properties of the star models, but that is hard to do without changing other properties that agree with observations. Another diagnostic is the frequency distribution of periods, because this depends on the details of the evolutionary tracks, and here the agreement is satisfactory. Pulsation theory also predicts the position of the high-temperature edge of the instability strip, which is strongly dependent on the composition in the outer part of a star, and here also theory and observation are in good agreement.

On the whole, the properties of Cepheids provide strong support for present ideas of stellar evolution. But as observations are improved, it seems certain that new discrepancies will arise that will demand refinements in our present understanding of stellar evolution.

RR LYRAE STARS

Stars less massive than about 2 $M_\odot$ evolve very differently than more massive ones. The effect of such evolution is seen most dramatically in the HR diagrams of globular clusters, which contain only low-mass stars. These clusters are among the oldest objects in our galaxy (ages 10–20 Gyr) and are often metal-deficient; the heavy-element abundance varies from nearly the solar value to as little as 10^{-4} that amount for the most metal-deficient clusters. Globular-cluster stars that are burning helium in their cores are found on the horizontal branch in the HR diagram, a region at roughly constant luminosity ($M_v \approx 0.0$–1.0) but broad in color. The horizontal branch (HB) intersects the nearly vertical instability strip, and stars found in their intersection will pulsate. These are the RR Lyrae stars, having periods of 0.1–1.0 day. Globular clusters, horizontal-branch stars, and RR Lyr variables may furnish important clues to the early history of our galaxy and to the primordial abundance of helium.

Statistical properties of RR Lyr stars are intimately connected with horizontal-branch morphology. In some clusters the HB is rather uniformly populated (the instability strip is somewhere in the middle) and as many as 30% of HB stars may be RR Lyraes. In other clusters, most of the HB stars lie either to the blue or to the red side of the instability strip, and there are fewer RR Lyraes, sometimes none at all. This morphology is correlated with abundance: In metal-rich clusters the HB stars are predominantly red, whereas a very blue HB is a sign of extreme metal deficiency. In order to reproduce the properties of the HB it is necessary to assume that the stars have a range of masses, in some cases varying by 30–40% within a cluster. This dispersion is apparently the consequence of mass loss in the preceding red-giant phase. When a star begins to burn helium in its core it settles onto the "zero-age horizontal branch." The initial evolution, depending on composition, may be either to the red or to the blue, but in the final stages of helium core burning all stars become brighter and redder. The periods of the RR Lyr stars depend on mass, luminosity, and effective temperature, and all of these may vary within a given cluster. To sort all this out it is necessary to construct not just a few evolutionary tracks, but an entire ensemble of tracks representing the stars of different mass, and then to see how the HB changes with time.

Among the RR Lyraes one finds stars pulsating in the first-overtone mode (type c) as well as the more common fundamental-mode pulsators (type ab). The type c RR Lyrae stars are found toward the blue side of the instability strip, and overlap in color little if at all with the cooler ab stars. The number ratio of type c to type ab stars increases with metal deficiency, as does the mean period of the ab stars. Furthermore it is found that, at a given effective temperature, the pulsation periods of metal-deficient stars are greater. One way to explain this is to hypothesize that HB luminosity increases with decreasing metal abundance, but this contradicts results from zero-age HB models. It now appears probable that the metal-poor clusters are simply more evolved than the others, and that their RR Lyr stars are brighter because they are nearing the state of He exhaustion in their cores. If this is so, it implies that there is an inverse correlation between age and metallicity for globular clusters. This is consistent with the inference from other data that outer halo clusters formed over a period of several billions (10^9) of years rather than rapidly during the initial collapse of the Galaxy, as has been thought. This view is by no means universally accepted at the present time, and the statistical properties of the RR Lyraes still present puzzles, but this account shows how the complex interplay between stellar evolution and pulsation studies may throw light on the structure and history of the Galaxy.

NONRADIAL PULSATIONS

Cepheids and RR Lyrae stars pulsate radially: The entire star, at any given radius, moves in and out together. Many kinds of stars undergo *nonradial* pulsation, in which the amplitude may be a function of latitude and longitude. Though harder to observe, and not yet so well studied, nonradial pulsations are potentially even more useful than radial ones for the study of stellar structure. There are two classes of nonradial modes. The p modes are like standing sound waves and are largely confined to the outer layers; the radial modes are a special case. The g modes, on the other hand, are similar to internal gravity waves and may have large amplitudes in the deep interior. Because of their small surface amplitudes they are hard to detect, but they can provide direct evidence about the structure of stellar interiors.

Nonradial oscillations have been most intensely studied in the Sun. Low-amplitude p modes are observed in the solar photosphere, having periods of around 5 min. The excitation mechanism is unknown, but probably arises in the Sun's surface convection zone. Study of these modes has already yielded details of the Sun's internal structure, such as the depth of the convection zone. Because the modes are split by rotation, it is also possible to infer the variation of rotational velocity with depth. The modes of lowest degree (those having the longest horizontal wavelengths) are hard to observe accurately, but because they penetrate deeper they may be able to provide information about the deep interior, for example, the rotational velocity of the solar core. The g modes too would give information about the core, but they appear not yet to have been observed. The field of solar seismology is a very active one at present.

Nonradial oscillations occur in several other classes of stars, most of them on or near the main sequence. Of much current interest are the solar-like oscillations of the peculiar A (Ap) stars and a few others. These are p modes with periods of 6–12 min, and in some cases a number of modes have been identified. Of course it is unlikely that we shall ever have such detailed data for other stars as we do for the Sun, but it is quite possible that information about stellar interiors will be deduced from the study of oscillations like those of Ap stars. For example, analysis of the frequency spectrum can, in principle, be used to infer a star's mass and age.

WHITE DWARFS

In the last 15–20 yr multiperiodic pulsations of white dwarfs, with periods of hundreds to thousands of seconds, have been observed. These oscillations have been identified as nonradial g modes of low degree. The most abundant variables are the DAV stars with hydrogen-rich envelopes and the DBV stars with helium-rich envelopes. The pulsations appear to be excited by the same type of mechanism that drives Cepheids, depending on H (DAV) or He (DBV) ionization.

Because the periods of white dwarfs are so short they can be obtained with great precision, and the many frequencies can be measured with some confidence. This circumstance, along with the fact that only certain modes are observed and the requirement that the excitation mechanism should work, allows a detailed picture to be made of the compositional stratification of the important outer layers. In at least one case an accurate value of the star's mass was also obtained, and this should be possible for other white dwarfs. Study of the pulsations has already helped to increase substantially the understanding of white dwarf structure, and the database continues to grow.

Period changes have also been observed. In this way the energy loss rate through neutrino emission (for hot white dwarfs) has been verified, and white dwarf cooling times can be measured directly. It has been proposed that these cooling times can help to fix the ages of the oldest white dwarfs and thus to infer the age of the galactic disk.

Oscillations of neutron stars have not yet been seen, but it would not be surprising if they were to be discovered. This would open an exciting new field of investigation.

Additional Reading

Christensen-Dalsgaard, J. and Frandsen, S., eds. (1987). *Advances in Helio- and Asteroseismology*. D. Reidel, Dordrecht.

Cox, J. P. and Giuli, R. T. (1968). *Principles of Stellar Structure*. Gordon and Breach, New York.

Kippenhahn, R. and Weigert, A. (1990). *Stellar Structure and Evolution*. Springer-Verlag, Berlin. See Part VII, Pulsating stars, p. 397.

Madore, B. F., ed. (1985). *Cepheids: Theory and Observations*. Cambridge University Press, London.

Sandage, A. R. (1986). The population concept, globular clusters, subdwarfs, ages, and the collapse of the Galaxy. *Ann. Rev. Astron. Ap.* **24** 421.

See also **Star Clusters, Globular, Variable Stars; Stars, Cepheid Variable; Stars, Cepheid Variable, Period-Luminosity Relation and Distance Scale; Stars, Hertzsprung–Russell Diagram; Stars, Horizontal Branch; Stars, Magnetic and Chemically Peculiar; Stars, Nonradial Pulsations in B-Type; Stars, Pulsating, Theory; Stars, RR Lyrae Type; Stellar Evolution, Low Mass Stars; Sun, Oscillations.**

Stellar Evolution, Rotation

Jean-Louis Tassoul

The study of stellar rotation began about 1610, when Galileo recognized sunspots as being associated with the visible surface of the Sun and measured the rotation rate of this star by observing their motions across the solar disk. Yet, it is to the Jesuit Father Christopher Scheiner that belongs the credit of showing, in 1630, that the solar photosphere does not rotate like a solid body, its period of rotation depending upon heliocentric latitude. During the seventeenth century, in the wake of these discoveries, many scientists argued that the variability in light of some stars was the direct consequence of axial rotation, the spinning bodies showing alternatively their bright (unspotted) and dark (spotted) hemispheres to the observer. (This idea was popularized in Bernard le Fontenelle's *Entretiens sur la Pluralité des Mondes* [*Conversations on the Plurality of Worlds*], a highly successful introduction to astronomy that went through many revised editions during 1686–1742.) Although this explanation for the variable stars did not withstand the passage of time, it is nevertheless worth mentioning because it shows the interest and fascination that stellar rotation has aroused since its inception.

In fact, it is not until 1909 that Frank Schlesinger found convincing observational evidence that other stars also rotate. In the 1930s, Otto Struve and his associates measured the rotation rates of many single and double stars by the broadening of their spectral lines. By that time, however, all the ideas necessary for the theory of rotating stars had already been proposed in the pioneering studies of Edward A. Milne, H. von Zeipel, and others.

GENERAL PROPERTIES

Theoretically, rotation has two main effects on the structure of a star: (i) a global expansion due to the centrifugal force, and (ii) a departure from spherical symmetry due to the nonspherical part of the effective gravity. The first effect causes a reduction in the total luminosity of the star. The second effect, which crucially depends on the angular momentum distribution within the star, causes the polar caps to appear hotter than the equatorial belt. A useful measure of the effects of rotation is the ratio of the centrifugal force to gravity at the equator, $\varepsilon = \Omega^2 R^3/GM$, where Ω is the angular velocity, R is the mean radius, G is the constant of gravitation, and M is the total mass. This parameter is actually a measure of the flattening of the star's outer surface. (In a realistic main-sequence model in uniform rotation, ε does not exceed the critical value $\varepsilon_c \approx 0.4$, at which point effective gravity vanishes at the equator so that equatorial breakup is most likely to occur.) In terms of ε, main-sequence stars are slowly rotating bodies: In early-type stars one has in general $\varepsilon \approx 0.01$–0.10, whereas in solar-type stars ε may be two or three orders of magnitude smaller. This is basically the reason why the standard theory of stellar structure and evolution, which deals with spherically symmetric models, has been so successful in explaining the major observed properties of stars. Except in a few cases (such as the Be stars which may be close to equatorial breakup), it would thus seem that the effects of rotation on stellar structure are dynamically unimportant along the main sequence.

More important are the effects of rotation during the early and late phases of stellar evolution. As is well known, its effects during the very early phases are of direct relevance to the formation of double (and multiple) stars and planetary systems. Similarly, the implosion of a star's core during the very late phases may lead—by conservation of angular momentum—to the formation of rapidly spinning neutron stars. The fact that the majority of white dwarfs are slow rotators thus indicates that these stars have gradually lost most of their angular momentum during the pre-white-dwarf phase of evolution.

If we exclude these early and late phases during which rotation plays the role of midwife, so to speak, the most interesting effect of rotation is to generate a whole spectrum of small-scale motions as well as large-scale meridional currents in stellar radiative zones. The importance of these motions lies in the fact that, under certain conditions, they may induce some degree of mixing. This is of direct relevance to stellar evolution because evolved stars frequently exhibit anomalous surface abundances, which are indicative of deep mixing of processed material. (The standard evolution theory does not account for such a mixing.) Large-scale meridional currents may also prevent the gravitational sorting of the elements in the surface layers of most (but not all) early-type stars. These currents thus provide the missing link that was needed to explain the correlation between slow rotation and abnormal spectrum in the A-type stars. This constitutes a strong argument in favor of the diffusion model for the Hg-Mn and Fm-Am stars.

BREAK NEAR SPECTRAL TYPE F5V

Because stellar rotation is basically a problem of fluid dynamics, it is important to distinguish between those stars having an outer convective zone (such as the Sun) from those having a subphotospheric envelope in which radiative equilibrium prevails. In the former case, magnetically coupled winds carrying away angular momentum cause the slow but inexorable spin down of the star as it leisurely evolves on the main sequence. In the latter case, because there is little or no surface convection, magnetic braking is virtually inoperative so that the star suffers little or no loss of angular momentum during its main-sequence lifetime. In fact, this is the most likely cause of the sharp decline in main-sequence rotational velocities in the middle *F*s because, as was originally pointed out by Evry Schatzman in 1962, the transition to small rotation rates occurs precisely at the spectral type where main-sequence stars develop an outer convective envelope. One may also argue that the abnormally low rotational velocities of the late-type stars are due to the systematic occurrence of planets (or brown dwarfs?) around these stars. If so, the retardation process should be completed at an early stage of stellar evolution. The recent observational data indicate that the mean rotational velocities of the late-type main-sequence stars decrease continuously with age. These studies are consistent with Schatzman's idea that the continuous transfer of angular momentum from the surface convective layers to outer space causes the gradual spin down of the solar-type stars (although one cannot a priori exclude the formation of planets as an occasional sink of angular momentum). In these stars, thus, the actual surface rotation rates result from intricate interactions between turbulent convection, rotation, meridional circulation, dynamo-generated magnetic fields, and mass loss.

EARLY-TYPE STARS

All theoretical speculations about the angular momentum distribution within an early-type star have their roots in von Zeipel's paradox, which states that the conditions of mechanical and radiative equilibrium are, in general, incompatible in a rotating barotrope. This result quickly led Arthur S. Eddington and Heinrich Vogt to point out, in 1925, that the small departures from spherical symmetry in a rotating star lead to unequal heating along the polar and equatorial radii; this, in turn, causes a large-scale flow of matter in meridian planes passing through the rotation axis. (Thermally driven currents also exist in a tidally distorted star, as well as in a magnetic star, because both the tidal interaction with a companion and the Lorentz force generate small departures from spherical symmetry.) In 1950, P. A. Sweet made the first detailed study of these currents in a uniformly rotating, nonmagnetic star. He found a quadrupolar circulation pattern, with rising motions at the poles and sinking motions at the equator. The typical speed of the meridional flow is of the order of $\varepsilon R/t_{KH}$, where t_{KH} is the Kelvin–Helmholtz time ($t_{KH} = GM^2/RL$, where L is the total luminosity). Unfortunately, because Sweet neglected dissipation altogether, he was unable to streamline the core-envelope interface and the outer surface so that his steady laminar solution has unwanted mathematical singularities at the boundaries. Moreover, because his procedure neglected the back reaction of the meridional flow on the driving motion, the Coriolis force acting on these steady currents remains unbalanced. For the sake of convenience, it was therefore assumed that there exists an inconspicuous magnetic field that can effectively enforce solid-body rotation in spite of the slow but inexorable transport of angular momentum by Sweet's laminar meridional flow. The existence of such a magnetic field has never been properly demonstrated, however.

Thirty years elapsed before the problem was reconsidered from the viewpoint of physical fluid dynamics. The newly proposed solution rests essentially on a dynamical linkage between eddy-like and/or wave-like motions (which may be called anisotropic turbulence because they are predominantly two-dimensional) and the mean flow (i.e., the differential rotation and concomitant meridional currents). To be more specific, because strict radiative equilibrium prevents a rotating star from being a barotrope, the main idea is that the chemically homogeneous parts in a radiative envelope are filled with small-scale transient motions that are caused by the ever-present barotropic-baroclinic instabilities. This anisotropic turbulence, in turn, generates thin thermo-viscous boundary layers so that Sweet's circulation velocities do not become infinite near the boundaries. Simultaneously, the turbulent friction acting on the differential rotation can be made to balance the transport of angular momentum by the meridional flow. Thus, by taking into account the eddy-mean flow interaction which takes place continuously in a stellar radiative zone, one can obtain a simple but adequate description of the mean state of motion in the envelope of an early-type star. (Nothing is known about the state of motion in the convective core of such a star, however, although it is customary to describe this core as a uniformly rotating barotrope in strict convective equilibrium.)

In the radiative envelope of a slowly rotating star, the meridional motion consists of a single cell extending from the core boundary to the free surface, with interior upwelling at the poles which is compensated by interior downwelling at the equator. Because of the presence of boundary layers, there are no singularities in the mean flow so that the circulation velocities remain uniformly small everywhere. The derivation of the rotation rate as a function of the coordinates is a much more intricate problem, however, because the viscous forces are directly proportional to the eddy viscosities. Because it is impossible at this time to perform a meaningful evaluation of these coefficients, there is no hope of calculating the departures from solid-body rotation with any accuracy. (A similar difficulty occurs in the theories of solar rotation; but, then, it is at least possible to adjust the theoretical rotation law to the observed surface rotation rate.) Perhaps the only sure thing is that there is no longer any reason to claim that almost-uniform rotation is forced upon a star by the mere presence of an inconspicuous axisymmetric magnetic field. Detailed calculations show that, with little or no turbulence, the inner rotation rate in a radiative envelope must necessarily tend toward a solution which has, in general, a large gradient in the angular velocity near the rotation axis. In other words, almost-uniform rotation can only be achieved with some kind of viscous action; purely magnetic action cannot do the job.

SOLAR SPIN DOWN

The recent observational data indicate that much of the Sun's radiative interior is spinning at a rate close to that of the surface equatorial belt, while the hydrogen-burning core is apparently rotating more rapidly than the chemically homogeneous parts of the core. Theoretically, the Sun's interior differs in two important respects from the radiative envelope of an early-type star: (i) the thermally-driven meridional currents are utterly negligible in the solar core because the Eddington–Sweet circulation time, $t_{ES} = t_{KH}/\varepsilon$, is much larger than the Sun's age, and (ii) angular momentum is continuously transferred away from the solar convective envelope to outer space. Accordingly, there must exist a very effective mechanism of angular momentum transport that keeps the inner and outer parts of the radiative interior rotating nearly uniformly in spite of the inexorable solar-wind torque. Again, it has been shown that a moderate amount of small-scale transient motions in the chemically homogeneous parts of the core is sufficient to reproduce the present quasisolid inner and outer rotation rates of the Sun. Moreover, the constraints on the eddy viscosities derived from the turbulent diffusion of momentum are in good agreement with those derived from the turbulent mixing of material and the solar surface abundances. Although there is no doubt that purely magnetic action will retain its attractiveness to some, these results strongly suggest that rotationally induced turbulence is a most promising avenue for further research. This is nonstandard stellar evolution theory.

Additional Reading

Kippenhahn, R. and Weigert, A. (1990). *Stellar Structure and Evolution*. Springer-Verlag, Berlin. See Part VIII, Stellar rotation, p. 427.

Michaud, G. (1988). Main sequence abundances, mass loss and meridional circulation. In *Atmospheric Diagnostics of Stellar Evolution: Chemical Peculiarity, Mass Loss, and Explosion*, K. Nomoto, ed. Springer-Verlag, Berlin, p. 3.

Smith, R. C. (1987). Rotating stellar interiors. In *Physics of Be Stars*, A. Slettebak and T. P. Snow, eds. Cambridge University Press, Cambridge, p. 123.

Stauffer, J. R. and Hartmann, L. W. (1986). The rotational velocities of low-mass stars. *Publications of the Astronomical Society of the Pacific* **98** 1233.

Tassoul, J.-L. (1978). *Theory of Rotating Stars*. Princeton University Press, Princeton.

Tassoul, J.-L. (1990). The effects of rotation on stellar structure and evolution. In *Angular Momentum and Mass Loss for Hot Stars*, L. A. Willson and R. Stalio, eds. Kluwer Academic Publ., Boston, p. 7.

Tassoul, J.-L., and Tassoul, M. (1989). The internal rotation of the Sun. *Astron. Ap.* **213** 397.

See also **Interstellar Clouds, Collaspe and Fragmentation; Stars, Magnetic and Chemically Peculiar; Stars, Neutron, Physical Properties and Models; Stars, Rotation, Observed Properties; Sun, Oscillations.**

Stellar Orbits, Galactic

Peter O. Vandervoort

In our galaxy, the Milky Way, each star moves on a galactic orbit under the influence of the gravitational forces exerted by all of the other matter and in accordance with Newton's laws of motion. Other "objects" in the Galaxy, such as star clusters and clouds of interstellar gas, similarly follow orbits in the prevailing gravitational field. The individual orbits underlie and determine the dynamical behavior of the Galaxy as a whole. Therefore, an understanding of the orbits plays a fundamental role in the interpretation of the observed structure of the Galaxy and of the observed rotation and random motions of objects in the Galaxy.

The true gravitational field acting on a star in the Galaxy is not accurately known. In practice, therefore, the theory of stellar orbits must deal with orbits in a model of the field. The models that are generally used for this purpose represent the Galaxy as a smooth, time-independent distribution of mass. Most models also incorporate two important symmetries that the Galaxy possesses approximately. The first of these, called axial symmetry, is the property that the structure and appearance of the Galaxy are those of a figure of revolution about the axis of galactic rotation. The second important symmetry, reflection symmetry, is the property that the half of the Galaxy that lies below the galactic plane is the mirror image of the half that lies above. In any particular model, the prevailing gravitational field must be specified mathematically. Theoretical orbits in the Galaxy are constructed by solving Newton's equations of motion in which the force is calculated from the adopted model of the field. For special classes of orbits and in cases where certain approximations can be justified, the equations can be solved with the aid of mathematical analysis. More generally, the equations must be solved with the aid of numerical computations.

The orbital periods of objects in the Galaxy are very long. Therefore, it is not possible to observe the orbits of individual objects or to make direct comparisons of their true orbits with theoretical orbits. The comparison of theory and observation takes the form of interpretations of the structure and kinematics of the Galaxy in terms of statistical models based on theoretical orbits. In describing the spatial distributions of the different galactic constituents we identify distinct structural features of the Galaxy such as the *disk* (which contains the spiral arms) and the *spheroidal component* (which includes the galactic halo and the galactic bulge). These different structural features exist and maintain their integrity, because they are composed of objects that move through the Galaxy on orbits of particular kinds. Thus, for example, constituents of the disk (for example, interstellar clouds, OB stars, long period variable stars, open clusters, and the common stars in the solar neighborhood) move on orbits that, for their energies, have high angular momenta around the axis of galactic rotation. In contrast, constituents of the spheroidal component (for example, globular clusters, RR Lyrae variable stars, and subdwarf stars) move on orbits that, for their energies, have low angular momenta around the axis of galactic rotation. The different populations of stars and gas form distinct subsystems of the Galaxy, each with its own rate of rotation.

STELLAR ORBITS IN THE GALACTIC DISK

The circumstance that the galactic disk is observed to be thin and in a state of rapid rotation implies that the constituents of the disk move on nearly circular orbits that remain near the galactic plane. The theory of such orbits and the use of the theory in conjunction with the observed kinematics of stars and gas in order to probe the structure of the Galaxy was developed in the 1930s along the following lines by Jan H. Oort, Bertil Lindblad, and others.

Circular Orbits and Galactic Rotation

An important class of solutions of the equations of motion describes orbits that lie in the plane of the Galaxy. An example of the simplest kind of orbit in the plane is a *circular orbit* consisting of motion at a constant speed on a circle centered on the galactic center. There is no motion toward or away from the galactic center, because the centrifugal and gravitational forces on the star just balance in the manner

$$\frac{[V_c(r)]^2}{r} = g_r(r),$$

where r is the radius of the orbit, $V_c(r)$, called the circular velocity, is the speed of the star, and $g_r(r)$ is the strength of the gravitational field in the direction of the galactic center.

The simplest model of galactic rotation in the disk assumes that objects move on circular orbits. Accordingly, the rotational velocity $V_{rot}(r)$ deduced from observations at a given distance r from the galactic center is to be identified as the circular velocity $V_c(r)$. Consequently, the observed rotation curve of the Galaxy, a plot of $V_{rot}(r)$ against r, may be used in order to probe the prevailing gravitational field in the galactic plane. The observation that the rotation curve remains flat (that is, that the rotational velocity remains relatively constant) to distances from the galactic center considerably beyond that of the solar neighborhood implies that the field is much stronger than can be accounted for in terms of the mass of the observed luminous constituents of the Galaxy. This is the basis for the conjecture that a population of "dark matter" must be a significant constituent of the Galaxy.

In the solar neighborhood, located at a distance of about 8.5 kpc from the galactic center, the circular velocity is approximately 220 km s^{-1}. A more refined model of galactic rotation considers that the stellar orbits are not strictly circular. Stars on noncircular orbits pass through a given region with velocity components in the direction of galactic rotation that are, on the average, less than the local circular velocity. Thus, the more noncircular the orbits of the stars belonging to a given subsystem of the Galaxy, the slower the rotation of that subsystem. It is observed that the different stellar subsystems of the disk follow this trend with rotational velocities ranging typically from 10–30 km s^{-1} less than the circular velocity.

Epicyclic Motions

General, noncircular orbits in the galactic plane are solutions of the classical central-field problem in dynamics. Thus, they trace out "rosette" paths of the kind illustrated in the left-hand panel of Fig. 1. It is useful to describe such an orbit in terms of an epicyclic motion illustrated in the right-hand panel of Fig. 1. The motion is referred to a point, called an *epicenter*, that moves on a circular orbit of radius r_0, say, with the circular velocity $V_c(r_0)$ appropriate to that radius. The angular velocity of the epicenter is $\Omega = V_c(r_0)/r_0$. The star moves on an *epicycle* around the moving epicenter. If the orbit is nearly circular and the epicycle accordingly small, then the epicycle is an ellipse, as illustrated in the figure, with an axis ratio

$$\frac{a^2}{b^2} = \frac{r_0}{2V_c(r_0)}\left(\frac{\partial V_c(r_0)}{\partial r_0} + \frac{V_c(r_0)}{r_0}\right),$$

where $\partial V_c(r_0)/\partial r_0$ is the rate at which the circular velocity changes with distance from the galactic center. The star moves around the epicycle with a circular frequency κ (the ordinary frequency multiplied by a factor 2π) that is given by

$$\kappa^2 = \frac{2V_c(r_0)}{r_0}\left(\frac{\partial V_c(r_0)}{\partial r_0} + \frac{V_c(r_0)}{r_0}\right).$$

It is customary to call κ the *epicyclic frequency*.

The epicyclic character of stellar motions in the disk imposes a remarkable constraint on the kinematics of the stars in a given region. The velocity of a star parallel to the galactic plane has components u and v, say, perpendicular and parallel, respectively, to the direction of galactic rotation. Let $\langle u^2\rangle$ and $\langle (v-V_{rot})^2\rangle$

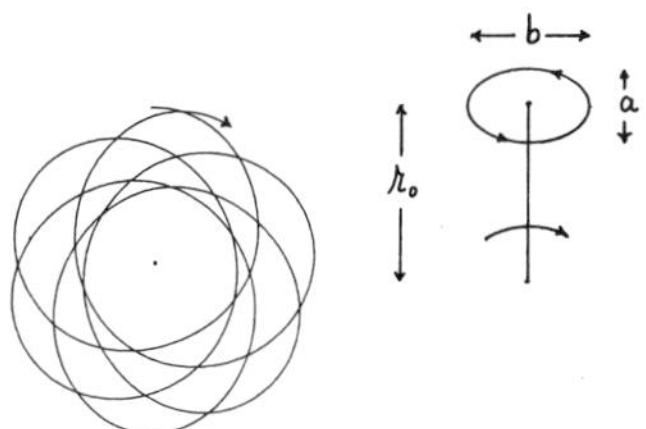

Figure 1. Orbit of a star in the galactic plane. The left-hand panel shows the rosette pattern of the orbit from the point of view of a stationary observer. The right-hand panel shows the resolution of the orbit into the circular motion of an epicenter at a distance r_0 from the galactic center and the epicyclic motion of the star on an ellipse of axes a and b. Arrows indicate the directions of the motion of the epicenter and of the epicyclic motion.

denote the average values of u^2 and $(v-V_{rot})^2$, respectively, for the stars of a given subsystem that are located in a given small region of the Galaxy. It is a theoretical consequence of the epicyclic motions of the stars that

$$\frac{\langle (v-V_{rot})^2\rangle}{\langle u^2\rangle} = \frac{r}{2V_c}\left(\frac{\partial V_c}{\partial r} + \frac{V_c}{r}\right).$$

All of the quantities in this relationship can be measured in observational studies of the law of galactic rotation and of the random motions of the stars in the solar neighborhood. When tested observationally, the relationship is found to be satisfied reasonably well. From the results of such investigations it is found that the period of the circular orbit of an epicenter in the solar neighborhood is about $2\pi/\Omega = 2.4\times10^8$ yr, whereas the period of the epicyclic motion is about $2\pi/\kappa = 1.7\times10^8$ yr.

Vertical Motions and Structure

In general, stellar orbits are not confined to the galactic plane. However, an orbit that remains near the plane can be represented as a superposition of the motion parallel to the plane that the star would have if it were strictly confined to the plane and a small "vertical" motion perpendicular to the plane. The parallel and vertical components of the motion influence each other so weakly that the two motions are approximately independent. The vertical motion is an oscillation periodic in time, because the vertical component of the gravitational field provides a restoring force that is everywhere directed toward the galactic plane. If the amplitude of the vertical motion is sufficiently small, then that motion is a simple harmonic oscillation with a circular frequency ν, called the *vertical frequency*, that is given by

$$\nu^2 = 4\pi G\rho_0,$$

where G is the constant of gravitation and ρ_0 is the mass density of the Galaxy evaluated in the galactic plane at the position of the epicenter of the orbit.

The oscillations of the stars through the galactic plane underlie the equilibrium of a given subsystem of the Galaxy (e.g., the K-type giants) in the vertical direction and thus play a fundamental role in observational determinations of the component g_z of the gravitational field of the Galaxy in the vertical direction. For, if the vertical component of the field were stronger, then the amplitudes of the vertical oscillations of the stars would be reduced, and the subsystem would be "compressed" vertically toward the plane. Likewise, if the velocities of the stars in the vertical direction were reduced, then the amplitudes of the vertical oscillations would be reduced, and the subsystem would "collapse" toward the plane. These principles, formulated quantitatively, provide the basis for a determination of the variation of g_z with distance from the galactic plane from an analysis of the spatial distribution of the stars in the vertical direction and the distribution of the vertical components of their velocities. From the derived run of g_z with vertical distance, one can calculate the vertical distribution of the mass that must be the source of the gravitational field of the Galaxy. The galactic mass density, obtained dynamically in this way, is about 0.18 $M_\odot$ pc^{-3} (solar masses per cubic parsec) in the solar neighborhood (i.e., in the galactic plane). The corresponding period of a small vertical oscillation is $2\pi/\nu = 6\times10^7$ yr. The directly observed luminous constituents of the solar neighborhood account for only about 0.11 $M_\odot$ pc^{-3}. The discrepancy is considered to provide further evidence for the presence of dark matter in the Galaxy.

FURTHER DEVELOPMENTS

The models of stellar orbits just described neglect the deflections caused by the gravitational perturbations that occur during close encounters with other objects in the Galaxy. The work of James H. Jeans early in this century and subsequent investigations in the 1940s by Subrahmanyan Chandrasekhar and others showed that the cumulative effects of the encounters of stars with other stars have been negligible since the Galaxy formed. However, encounters of stars with massive objects such as giant molecular clouds will have produced significant deflections of the stellar orbits, and, as a consequence, the subsystems of the galactic disk must have suffered slow changes of their structures and gradual reductions of their rotations. The orbits will have been further perturbed by encounters of the Galaxy with other stellar systems such as the Magellanic Clouds.

The orbit of a star in the galactic disk, as already described, is *regular* in the sense that it can be resolved into three independent periodic components, the motion of the epicenter, the epicyclic motion, and the vertical motion, with definite frequencies Ω, κ, and ν, respectively. The investigations of George Contopoulos and others since 1960 have shown that many orbits that extend well outside the disk of the Galaxy are likewise regular in the sense that the motion can be resolved into three periodic components, each with a definite frequency. However, other orbits have been found to be *chaotic* in the sense that the motion involves many (probably infinitely many) frequencies. The phenomenon of order and chaos in stellar orbits is a major area of modern research.

Spiral structure in the Galaxy is a departure from axial symmetry that can perturb stellar orbits. If spiral structure is a uniformly rotating wave in the galactic disk, as is envisaged in some modern theories, then, over most of the galactic disk, the gravitational fields of the spiral arms cause only modest distortions of the circular orbits of the epicenters of the orbits and of the epicyclic motions, and the orbits are essentially regular. However, in certain *resonance regions* of the Galaxy, where a star encounters spiral arms periodically in time with a frequency commensurate with either the angular velocity of the epicenter or the epicyclic frequency, the spiral perturbations will strongly distort and modify the motion of the epicenter and the epicyclic motion. Orbits in the resonance regions are often chaotic.

Additional Reading

Binney, J. J. and Tremaine, S. D. (1987). *Galactic Dynamics*. Princeton University Press, Princeton.

Mihalas, D. and Binney, J. J. (1981). *Galactic Astronomy*. W. H. Freeman, San Francisco.

Ogorodnikov, K. F. (1965). *Dynamics of Stellar Systems*. Pergamon Press, Oxford.

Ollongren, A. (1965). Theory of stellar orbits in the Galaxy. *Ann. Rev. Astron. Ap.* **3** 113.

Oort, J. H. (1965). Stellar dynamics. In *Galactic Structure*, A. Blaauw and M. Schmidt, eds. University of Chicago Press, Chicago, p. 455.

See also **Galactic Structure, Interstellar Clouds; Galactic Structure, Large Scale; Galactic Structure, Optical Tracers; Galactic Structure, Spiral, Observations; Galactic Structure, Stellar Kinematics; Missing Mass, Galactic.**

Sun

Jack Zirker

The Sun is the nearest star: a typical, middle-aged cool star like billions of others in our galaxy. It is the only star on which astronomers can resolve fine details or make extremely precise measurements. As a result, much of the knowledge we now have on the interiors and atmospheres of stars derives from a close study of the Sun. A distinct scientific discipline, solar physics, is devoted toward understanding, in exact mathematical and physical terms, the processes that occur on the Sun and toward extending this understanding to other types of stars. The Sun also has a powerful influence on interplanetary space and, in particular, the Earth's immediate environment. Therefore, solar research has strong ties to geophysics, magnetospheric physics, and astronomy.

Solar physics is a mature discipline. Except in a few areas (e.g., the interior of the Sun), most of the raw phenomena are well known and reasonably well described observationally. However, a detailed understanding, based on physical principles, is lacking for many of these phenomena, and accurate prediction, an important goal in solar–terrestrial research, is still far off. Many of the observable structures and events are highly complex on close examination, and often require coordinated observations, over many decades of the electromagnetic spectrum from space and Earth–based observatories. To interpret these data, solar physicists draw on the full armory of contemporary physics: particle physics, radiative transfer theory, atomic and nuclear physics, magnetohydrodynamics, and plasma physics. Synoptic observations, that trace the evolution of solar structures over months and years, are also an important aspect of solar research.

Some of the major advances of the last 150 years include the discovery of the sunspot cycle (Heinrich Schwabe, 1843), the first determination of the solar constant (John Herschel, 1847), the identification of terrestrial elements in the Sun (Gustav R. Kirchhoff, 1860), the discovery of helium (Joseph Norman Lockyer, 1868), the proof that the corona is solar (W. R. Campbell, 1868), the discovery of sunspot magnetic fields (George Ellery Hale, 1908), the invention of the coronagraph (Bernard F. Lyot, 1930), the proof that the corona is hot (Bengt Edlen, 1942), the detection of solar radio radiation (Grote Reber, 1944), and the first extreme ultraviolet spectrum of the Sun (Richard Tousey, 1946).

SOLAR INTERIOR

Until the mid-1970s, our knowledge of the physical conditions inside the Sun depended entirely on theoretical models of stellar structure and evolution. The observational constraints on these models consisted of only three data: the solar radius, age, and luminosity. With the measurement of the solar neutrino flux (by Raymond Davis and associates), a striking discovery between fact and theory appeared: three-quarters of the predicted neutrino flux was missing. The neutrinos are released as a by-product of energy generation in the core of the Sun. If their predicted flux was wrong, the whole theory of stellar evolution was in jeopardy. After a painstaking examination of all facets of the problem, researchers concluded recently that the most likely explanation lies in the "oscillations" of neutrinos between several allowed species, during their transit from the Sun to Earth. New experiments to test this explanation, as well as other contenders, are being built presently.

The interior of the Sun remained unobservable until the discovery and interpretation of the global oscillations, in about 1974. The entire Sun is vibrating as a bell vibrates, at discrete, closely spaced frequencies that are fixed by the run of temperature and composition throughout the interior. By observing and analyzing these oscillations, which are visible as pulsations of the solar surface, "helioseismologists" can explore the solar interior, much as terrestrial seismologists probe the Earth's core. This technique has been highly developed as an observational tool, along with the theory of vibrating stars that is needed to interpret and guide the observations.

One of the most important goals of this research is the mapping of convective motions and differential rotation inside the Sun. These motions combine, in ways that are not fully understood, to regenerate the solar magnetic fields that are observed in the solar atmosphere in an 11-year cycle of activity. Important advances have been made recently in measuring these internal motions. The Sun appears to rotate at its surface rate to a surprisingly great depth but may accelerate in a fast core, to perhaps twice its surface speed. The variation of rotation with latitude as well as depth, is now beginning to be known, and large-scale, long-lived convective cells are also becoming detectable.

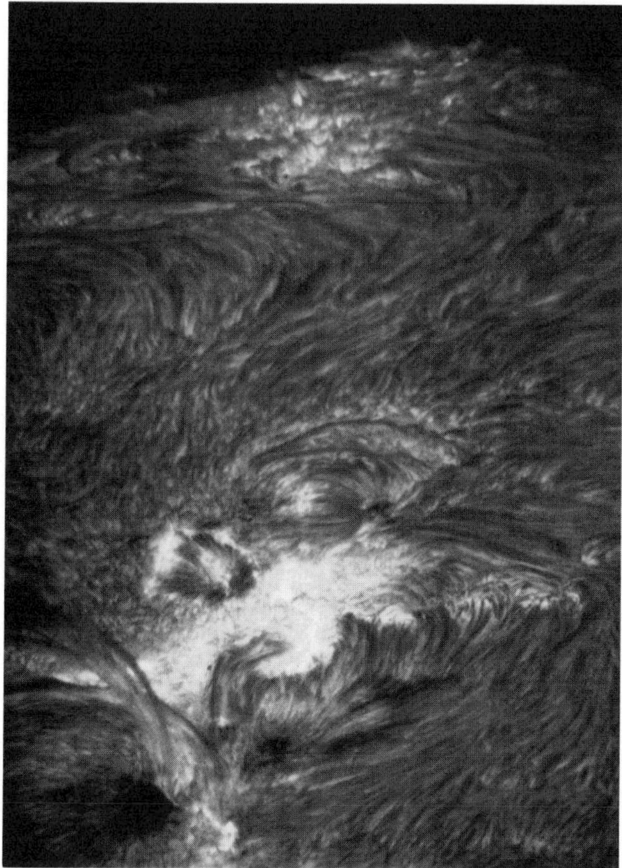

Figure 1. Chromospheric structures near an active region, photographed in the red hydrogen Fraunhofer line. The thread-like fibrils trace the magnetic fields that emerge in and around sunspots (lower left and right center). Away from spots, the fields rise high into the overlying corona. (*National Solar Observatory photograph.*)

In order to obtain the extremely precise oscillation frequencies required for their work, researchers have observed the Sun for 24 h a day at the South Pole during several weeks of the Antarctic summer. Because even longer continuous observations are required, they have joined to build a network of identical observing stations around the world, the Global Oscillation Network.

QUIET SUN

Astronomers have come to realize that all the structures they observe in the solar atmosphere are shaped, and probably heated, by magnetic fields. The fields emerge from the solar surface (the photosphere) as tiny flux tubes, with kilogauss field strengths. Mathematical simulations suggest that this improbable arrangement arises from the interaction of convective cells with emerging flux. The convective cells shuffle magnetic flux over the solar surface and store kinetic energy in the magnetic field. However, a thorough understanding of this process, an important goal of current research, awaits further observations with the highest attainable spatial resolution.

From the surface, the magnetic fields expand upward in the solar atmosphere, to form the highly filamentary structures of the chromosphere and corona (Fig. 1). We know these layers are heated principally by nonthermal energy, and that the magnetic field is crucial, but again the essential physical processes seem to lie at the limit of attainable spatial resolution. One of the most promising clues on the heating process is the recent discovery of ultraviolet "microflares"—tiny explosions in the magnetic network, that may arise from the release of energy stored in the magnetic field.

In the corona, the magnetic field creates the grand streamers, loops, and arches that are visible to the naked eye at total eclipses. Most of the coronal magnetic flux loops back to the surface, trapping hot plasma, but some field lines extend far out into interplanetary space. These "open" lines allow heated plasma to stream outward as the solar wind, at speeds reaching 800 km s^{-1}. As with the quiet corona, the precise mechanism that heats and accelerates the solar wind to such speeds is not definitely known, although reasonable possibilities have been proposed.

For a related overview, see the entry on Solar Activity.

Additional Reading

Leibacher, J. W., Noyes, R. W., Toomre, J., and Ulrich, R. K. (1985). Helioseismology. *Scientific American* **253** (No. 3) 48.

Noyes, R. W. (1982). *The Sun—Our Star*. Harvard University Press, Cambridge, Mass.

Wentzel, D. G. (1989). *The Restless Sun*. Smithsonian Institution Press, Washington, D.C.

Zirin, H. (1988). *Astrophysics of the Sun*. Cambridge University Press, London.

See also **Coronagraphs, Solar; Interplanetary Medium, Solar Cosmic Rays; Interplanetary Medium, Solar Wind; Neutrino Observatories; Neutrinos, Solar; Solar Activity, all entries; Solar Magnetographs; Solar Physics, Space Missions; Sun, all entries; Telescopes and Observatories, Solar.**

Sun, Atmosphere

Eugene H. Avrett

The solar atmosphere is the outer envelope of the Sun that emits the light that escapes into space. The Sun's heat is produced by thermonuclear reactions in its core, and this energy is transmitted outward by radiation and convection until it reaches the atmosphere and escapes as electromagnetic radiation.

The atmosphere is not uniform: it has two types of structure. The first is the basic vertical stratification: The visible surface region is called the photosphere; the chromosphere is a hotter region located higher in the atmosphere; the very hot corona is located still higher and extends into interplanetary space. The second type of structure has the form of horizontal inhomogeneities at all levels of the atmosphere and includes prominent features such as sunspots, active regions, and flares.

The Sun is a gaseous body held together by gravity. The interior is very dense and the outer regions are very tenuous; for some purposes the Sun's outer envelope may be considered to extend beyond the orbit of the Earth. The "visible surface," like the edge of a distant fog bank, is simply the deepest region from which light reaches us directly. The hot gas of the atmosphere, like the Sun as a whole, consists of about 90% hydrogen by number of atoms, 10% helium, and trace amounts of almost all other natural elements. The most abundant trace elements are oxygen (0.09%) and carbon (0.04%).

We can observe only the light (and some of the energetic particles) produced by the atmosphere, but much of the observed atmospheric structure, both vertical and horizontal, is caused by motions beneath the surface that cannot be observed directly. Below the atmosphere most of the outward flow of energy is carried by convective motions. Convection ceases at the lower boundary of the atmosphere, but the cellular granulation pattern seen at high spatial resolution is due to the uppermost vestige of convective motions. These subsurface motions also generate pressure waves that carry energy outward through the atmosphere.

Interior convection also interacts with solar rotation to generate magnetic fields. These fields extend through the atmosphere as concentrated projections of magnetic flux that also carry energy outward.

Deep in the atmosphere where the pressure exerted by the magnetic field is low compared with the gas pressure, the magnetic field is compressed into small flux tubes. Higher in the atmosphere where the gas pressure is much less, the magnetic flux tubes expand to fill most of the region. The magnetic field pattern can be seen extending up far into the corona either as closed loop structures or as open regions carrying energetic particles from the atmosphere into interplanetary space.

The energy carried outward by pressure waves and magnetic fields heats the outer atmosphere. The temperature decreases from values near 6000 K at the visible surface to a minimum value (T_{min}) of roughly 4400 K about 500 km higher in the atmosphere.

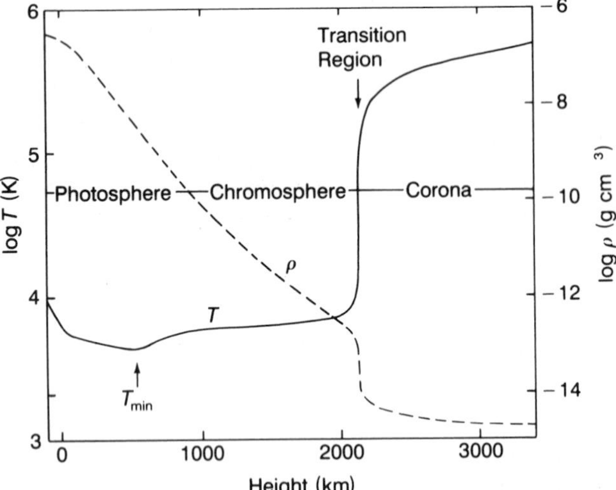

Figure 1. Temperature (left ordinate) and density (right ordinate) as functions of height in the solar atmosphere, determined from observations of average quiet regions of the Sun.

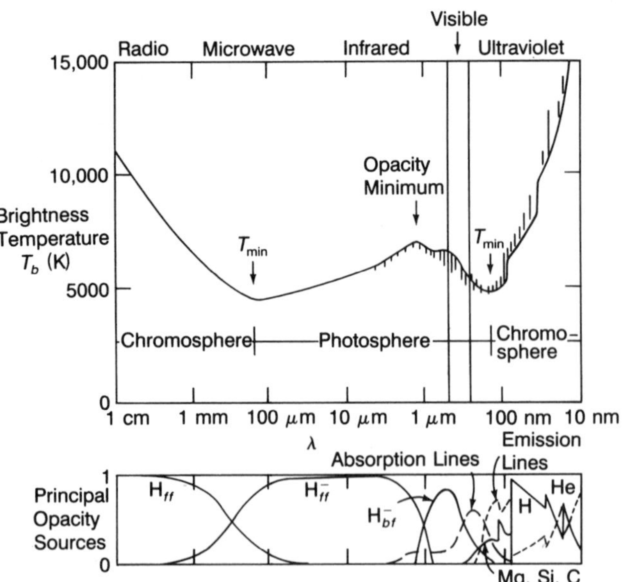

Figure 2. The approximate wavelength distribution of the solar brightness temperature from 1 cm to 10 nm (upper panel), and the principal sources of opacity (lower panel) at the depths where the emitted radiation is formed. (Symbols are defined in the text.)

Further out, the temperature increases with height, first to 6000–8000 K in the chromosphere, and then to several million degrees in the corona. Figure 1 shows the temperature and gas density as functions of height based on observations of average "quiet" regions of the Sun, that is, without prominent features or unusual solar activity. The chromosphere and corona are separated by a transition region that is less than 100 km thick.

Results such as those in Fig. 1 are determined from observations of the Sun's radiation over a wide wavelength range, from radio wavelengths to x-rays. Because the opacity of the solar atmosphere varies with wavelength, we can probe different atmospheric layers by observing radiation of different wavelengths. At wavelengths where the opacity is low we see radiation that originates deep in the atmosphere, and where the opacity is greater we see higher layers. The observed intensity of light as a function of wavelength λ can be used to determine the temperature at the depth in the atmosphere where the light of a given wavelength is emitted.

The upper panel of Fig. 2 shows the observed brightness temperature T_b of the average quiet Sun as a function of wavelength between 1 cm and 10 nm. Here $T_b(\lambda)$ is a measure of the brightness of the observed spectrum of radiation, and is defined as the temperature of an idealized source that emits the same intensity of light at the wavelength λ as that which is observed at that wavelength.

The solar atmosphere has the lowest opacity at $\lambda = 1.6\ \mu$m. The light at that infrared wavelength has a brightness temperature of 6800 K and originates in the deepest and hottest layers of the atmosphere that are directly observable. For infrared wavelengths longer than 1.6 μm, the opacity increases as λ increases so that the observed radiation is emitted from higher, cooler layers in the atmosphere and the observed brightness temperature decreases, reaching a minimum value of about 4500 K at $\lambda = 150\ \mu$m. The opacity continues to increase with increasing λ for $\lambda > 150\ \mu$m, and T_b now increases because of the increasing temperature in the chromosphere.

Now consider the wavelength region $\lambda < 1.6\ \mu$m, which includes the wavelength band 400–700 nm that is visible to the human eye. The opacity increases as λ decreases and near $\lambda = 160$ nm, T_b is about 4400 K, which seems to be the lowest temperature found anywhere in the average quiet Sun. At even lower values of λ, T_b increases rapidly because at these short wavelengths the light again originates in the hotter chromosphere. Figure 2 only indicates brightness temperatures up to 15,000 K, but million degree temperatures are found from observations at much shorter and much longer wavelengths than those in Fig. 2, and particularly from emission lines at extreme-ultraviolet and x-ray wavelengths.

The lower panel of Fig. 2 indicates the principal physical processes that absorb radiation at the depths where the observed light is emitted, that is, it indicates the principal sources of opacity. The symbols H and He refer to absorption by hydrogen and helium, H^-_{bf} and H^-_{ff} to bound–free and free–free absorption by the negative hydrogen ion, H_{ff} to hydrogen free–free absorption, and Mg, Si, and C to bound–free absorption by magnesium, silicon, and carbon. The vertical scale of this panel gives the fractional contribution of each of the processes shown.

Observations such as those indicated in Fig. 2, together with computer simulations that take the important physical processes and opacities into account, lead to models, such as that in Fig. 1, of the stratification of temperature, gas density, and other physical parameters.

Such a physical description of the atmosphere allows us to determine, for example, the amount of heating required to account for the high temperatures of the chromosphere and corona. Although almost all of the energy from the Sun escapes from the photosphere, only about 10^{-4} of the total energy output is involved in heating the chromosphere, and only about 10^{-5} and 10^{-6} of the total is required to produce the chromosphere–corona transition region and the corona, respectively. Even so, radiation and energetic particles from the corona have important effects on the Earth, particularly when solar flares occur.

We know much more about the Sun than any other star. Because we can observe the surface features in detail and can study a far greater range of phenomena, the Sun presents us with a greater number of clearly defined problems to be solved than arise in the study of other individual stars. There is much evidence that the chromosphere is heated by the dissipation of pressure waves that originate beneath the photosphere, but the types of waves that are responsible for this heating have not been identified, and the extent to which the magnetic fields interact with wave motions in the chromosphere is unclear. There is much evidence that the corona is heated by magnetic fields that penetrate into the corona from moving regions below the photosphere, but the details of this process remain highly uncertain. The reason for the abrupt increase of temperature (8000–10^6 K) in the chromosphere–corona transition region is still not understood. This phenomenon cannot be explained in terms of simple thermal conduction; a more detailed procedure is needed to properly calculate the energy transport in this region. The buildup and storage of magnetic energy and its sudden release in flares is understood only schematically. One of the goals of solar research is the successful prediction of flares hours or days in advance. Such predictions may depend on the recognition of characteristic large-scale patterns of photospheric motions that are caused by unusual changes in the subsurface magnetic field.

These and other fundamental problems in understanding the solar atmosphere need to be solved. Further observations and detailed analysis should succeed in giving us an improved description of the basic physical processes that occur in the solar atmosphere.

Additional Reading

Athay, R. G. (1988). The hot solar envelope. In *Multiwavelength Astrophysics*, F. Córdova, ed. Cambridge University Press, London, p. 7.

Cox, A. N., Livingston, W. C., and Matthews, M., eds. (1991). *The Solar Interior and Atmosphere*. University of Arizona Press, Tucson.

Stenflo, J. O., ed. (1990). *Solar Photosphere: Structure, Convection, and Magnetic Fields*. Kluwer Publishing Co., Utrecht.

Sturrock, P. A., Holzer, T. E., Mihalas, D. M., and Ulrich, R. K., eds. (1986). *Physics of the Sun, II: The Solar Atmosphere*. D. Reidel Publishing Co., Dordrecht.

See also **Heliosphere; Solar Activity; Stars, Atmospheres, all entries; Sun, Atmosphere, all entries; Sun, Coronal Holes and Solar Wind; Sun, Magnetic Field.**

Sun, Atmosphere, Chromosphere

Robert F. Stein

CHROMOSPHERE

The chromosphere is the middle part of the solar atmosphere—above the photosphere (where most of the solar radiation that reaches us on Earth comes from) and below the corona (the extremely hot outer atmosphere of the Sun). It is a region where the temperature rises slowly with height from about 4300 K at the bottom (the temperature minimum) to about 10^4 K at the top where a steep temperature rise to coronal values begins (the transition region; see Fig. 1 in the entry Sun, Atmosphere). It is best observed in the light from the hydrogen Balmer lines and the resonance lines of ionized calcium and magnesium. The crucial problem in understanding the chromosphere is understanding why its temperature increases with height.

Heat flows from hot regions to cold regions. If the temperature of the solar atmosphere were controlled only by the exchange of energy between the photons streaming outward through it and the gas in the atmosphere (radiative equilibrium), then the temperature of the atmosphere would decrease slightly from its surface value, becoming constant as the interaction with the radiation decreased with decreasing gas density. In fact, the temperature of the solar atmosphere decreases for only a short distance above the surface and then starts to rise again in the chromosphere. This means there must be some nonradiative, mechanical or magnetic energy source to heat the chromosphere.

CLUES TO ITS STRUCTURE

Clues to the structure of the chromosphere and the source of the energy to heat it are provided by both solar and stellar observations. Emission from the solar chromosphere is not uniform over the solar surface. It is concentrated into a network that outlines the boundaries of supergranulation-scale convection cells, and at bright points that appear in the cell interiors. Both of these sites are tightly correlated with magnetic flux tubes that extend upward through the surface. Thus the mechanism providing energy to the chromosphere clearly involves the magnetic field. The bright point emission typically repeats with periods of approximately 3 min and its phase is coherent over areas of order 8 Mm (megameters) in diameter, which indicates that the heating mechanism may be associated with the global solar p-mode oscillations. Stellar observations of cool stars show that the chromospheric Ca II and Mg II flux has a well defined lower limit, with a large scatter at each effective temperature above the lower limit. The minimum fluxes are a function of effective temperature, but depend only weakly on gravity. For the Sun, these minimum fluxes are found in regions of weak magnetic field in the centers of the supergranule cells in the quiet Sun. The scatter in chromospheric emission above the lower limit is found to depend on the rotation rate of the star. The coronal soft x-ray flux is well correlated with the excesses of the Ca II and Mg II fluxes over their lower limits. These results suggest that the excess flux above the lower limit depends on the magnetic field strength, but that the minimum flux is independent of the magnetic field. For the Sun, a constant energy input per unit mass of 4.5×10^9 erg g^{-1} s^{-1} is required to balance the chromospheric radiative losses between the temperatures of 6000 and 8000 K. A successful theory of chromospheric heating must reproduce this result.

SOURCES OF HEATING

Turbulent convection beneath the solar surface produces fluctuations in entropy, vorticity, and divergence of the Reynolds stress. These fluctuations in turn act as sources for the generation of acoustic waves, internal gravity waves, and, in the presence of isolated magnetic flux tubes, magnetohydrodynamic sausage, kink, and torsional tube waves. The large scale convective motions (granulation and supergranulation) also shuffle around the magnetic flux tubes that thread the surface. A small fraction of the energy transported outward by convection below the solar surface is converted into these motions instead of being converted back into thermal radiation at the surface. Rough estimates show that somewhat more energy is converted into wave and magnetic field motions than is needed to heat the chromosphere and corona.

Acoustic waves, with frequencies above the acoustic cutoff frequency $\omega = c_s/2H$, where c_s is the sound speed and H is the pressure scale height, propagate isotropically at the sound speed. Because the sound speed increases only slightly through the chromosphere they suffer little refraction there.

Internal gravity waves, with frequencies below the Brunt–Väisälä frequency $N_{BV}^2 = (g/T)[dT/dz - (dT/dz)_{adiabatic}]$, propagate primarily horizontally at speeds somewhat less than the sound speed. They are predominantly transverse and very dispersive.

There are three types of magnetic flux tube waves: a torsional, Alfvèn wave that propagates at the Alfvèn speed of the plasma inside the tube; a transverse, "kink," wave that propagates at the mean of the internal and external Alfvèn speeds; and a nearly longitudinal, "sausage," acoustic wave that propagates at the subsonic, sub-Alfvènic cusp speed $c_T = c_s c_A/(c_s^2 + c_A^2)^{1/2}$, where c_A is the Alfvèn speed. The kink and sausage modes are dispersive and have cutoff frequencies. An important property of all these tube modes is that they propagate energy along the magnetic flux tubes and are not refracted by the increase in Alfvèn and sound speeds with height.

These wave modes, as well as slow, quasisteady (DC) shuffling of the magnetic flux tubes, transport energy from the convection zone up into the solar atmosphere. However, all the motions, except the Alfvèn waves and the DC field motions, are more or less compressive, so they lose energy in the photosphere by the emission and absorption of radiation. In order to heat the chromosphere, the remaining energy must be converted to thermal energy of the plasma, that is, dissipated. All dissipation requires the transport of momentum or energy on the scale of the mean free path of the particles doing the transporting: electrons, protons, ions, or photons. In the chromosphere these scales are typically small, except for the last, which becomes so large that radiative dissipation is ineffective. Hence, dissipation requires the generation of small scale structure or fluctuations.

Because of the drop in density with height, wave amplitudes grow. Longitudinal, compressive waves steepen into shocks in which regions of very different velocity and energy are found within a mean free path of each other so that energy can be thermalized by viscosity and conduction. Internal gravity wave amplitudes also increase with height, which increases their shear, but they do not form shocks.

In nonuniform media, such as found in the chromosphere, other means of producing small scale structure also exist. One is the refraction of waves, which increases the component of the wavevector in the direction of the gradient of the propagation speed. For tube waves it can go on increasing forever, because they are not subject to total internal reflection. Hence, very small scale structures transverse to the tube develop which will dissipate by viscosity and resistivity. One can look at this process as waves traveling on adjacent field lines, with different propagation speeds, getting out of phase with each other. Another process is the resonant coupling of the wave modes. If there exists a region where another wave locally has the same propagation speed along the tube as the tube wave, then strong coupling between the modes will occur. If the other wave has a very short wavelength, as can occur for plasma waves, then finite Larmor radius effects will be important and dissipation will occur.

The slow motions of magnetic flux tube foot points will produce interactions with neighboring flux tubes above the height (in the mid-chromosphere) where the tubes have spread out and are no longer isolated. Where parallel currents exist, they attract one another. Hence, flux tubes that twist in the same sense are pulled toward one another, which squeezes out the plasma between them and produces a narrow current sheet where resistivity produces dissipation.

CONCLUSION

There is as yet no complete and self-consistent theory of chromospheric heating. We believe that we know many of the ingredients to such a theory: the convection zone dynamics generating the motions, the propagation of the waves and slow field readjustments into the atmosphere, the development of small scale structures, the dissipation of the mechanical and magnetic energy, and the radiative emission and absorption that transports and removes thermal energy. For some of these processes details remain to be calculated and a global theory of chromospheric heating, putting all the pieces together, still needs to be developed.

Additional Reading

Durrant, C. J. (1988). *The Atmosphere of the Sun*. Adam Hilger, Bristol.

Kuperus, M., Ionson, J. A., and Spicer, D. S. (1981). On the theory of coronal heating mechanisms. *Ann. Rev. Astron. Ap.* **19** 7.

Jordan, S., ed. (1981). *The Sun as a Star*. NASA SP-450, NASA, Washington, D.C.

Stein, R. F. and Leibacher, J. (1974). Waves in the solar atmosphere. *Ann. Rev. Astron. Ap.* **12** 407.

Sturrock, P. A., Holzer, T. E., Mihalas, D. M., and Ulrich, R. K., eds. (1986). *Physics of the Sun*, Vol. II. D. Reidel, Dordrecht.

Zirin, H. (1988). *Astrophysics of the Sun*. Cambridge University Press, Cambridge.

See also **Magnetohydrodynamics, Astrophysical; Solar Activity; Stars, Atmospheres, Radiative Transfer; Stars, Atmospheres, Turbulence and Convection; Sun, Atmosphere; Sun, Magnetic Field.**

Sun, Atmosphere, Corona

Neil R. Sheeley, Jr.

The corona is the outermost part of the Sun's atmosphere, extending from just above the visible surface into interplanetary space. By earthly standards, the corona is truly vacuous. Its average density falls from about 1.7×10^{-16} g cm^{-3} at 1.1 $R_\odot$ ($R_\odot$ being the solar radius) to 5×10^{-19} g cm^{-3} at 3 $R_\odot$, and still further to 2×10^{-20} g cm^{-3} at 10 $R_\odot$, corresponding to proton densities of 1×10^8, 3×10^5, and 1×10^4 cm^{-3}, respectively. Further into the heliosphere, the proton density falls more slowly as r^{-2}, reaching about 5 cm^{-3} at the 215 $R_\odot$ orbit of Earth.

We can appreciate how tenuous the corona is by comparing it with the Sun itself and with the Earth. Just inside the Sun's visible disk, the density is about 3×10^{-7} g cm^{-3}, which is roughly 10^{-4} the density of air that you are breathing at sea level on Earth (1.3×10^{-3} g cm^{-3}). The relatively higher temperature of the Sun's atmosphere (6000 K compared to 300 K on Earth) causes the discrepancy in pressures to be less. Nevertheless, the 1.3×10^{5} dyne cm^{-2} pressure just inside the Sun's disk is still only about 10% of the 1×10^{6} dyne cm^{-2} sea-level pressure on Earth. Moving into the Sun, the density rises, passing Earth's sea-level density of 10^{-3} g cm^{-3} at 0.9 $R_{\odot}$, reaching the 1 g cm^{-3} density of water at 0.5 $R_{\odot}$, and attaining the enormous density of 160 g cm^{-3} at the center of the Sun. This central density is about 8 times that of platinum, whose density of 21 g cm^{-3} is both one of the largest in the periodic table and about twice that of lead (11 g cm^{-3}). With an average density of 1.4 g cm^{-3}, about 90% of the Sun's mass is contained within the relatively small core below 0.5 $R_{\odot}$. Thus, by comparison, the corona refers to hardly anything at all.

This tenuous coronal material is relatively hot, reaching temperatures in the range $1\text{–}2\times10^{6}$ K near the Sun and 10^{5} K farther out into the heliosphere. At these high temperatures and low densities, the material is a highly ionized plasma, consisting mainly of protons and electrons with some helium nuclei and trace amounts of heavier ions whose emission lines dominate the x-ray region of the coronal spectrum. This million-degree plasma has a very low electrical resistivity, corresponding to an equivalent rate of 1 m^{2} s^{-1} at which it can diffuse across the coronal magnetic field. Consequently, on most time scales of interest, the plasma is firmly attached to the coronal magnetic field lines whose shapes it traces out.

Figure 1 is a composite of an eclipse photograph obtained at 12:50 UT on June 30, 1973 from Kenya and a soft x-ray image obtained an hour earlier from the *Skylab* space station. The x-ray image shows the corona against the Sun's disk (demagnified slightly relative to the size of the moon). Here, the intensity is contributed by the emission lines within the bandpass of the soft x-ray instrument, and it maps out the location of the corresponding highly ionized atoms. The white-light image shows the outer corona in projection against the sky. The intensity is produced by visible light that originated from the Sun's surface and has been Thomson-scattered toward the observer by electrons along the line of sight in the corona. Consequently, this white-light intensity is a measure of the number of electrons (and thus protons for this electrically neutral hydrogen plasma) along the line of sight.

Several characteristics of the inner corona are visible in the x-ray image. Bright loops of relatively dense material are visible against a fainter background, which itself seems to consist of loops. The brighter loops outline field lines joining the opposite poles of newly erupted bipolar magnetic regions on the Sun's surface. Some of these loops are visible extending beyond the limb. The very dark area that extends meridionally from the north pole into the southern hemisphere is a region of exceptionally low density. Such coronal holes map the origin of the open magnetic field lines whose other ends do not return to the Sun.

Figure 1. Composite photograph of the outer corona in visible light and the inner corona in soft x-rays (slightly demagnified insert) during the total eclipse of June 30, 1973. (*Composite of High Altitude Observatory white light image and American Science and Engineering x-ray image provided courtesy of AS & E.*)

Along the northern flanks of the coronal hole, bright emission is seen extending into the outer corona where it forms part of the large white-light density structures called helmet streamers. These streamers mark the boundary of the hole in the outer corona. At lower latitudes, other streamers are visible apparently criss-crossing in front of one another along the line of sight. All of these streamers become increasingly aligned in the radial direction at greater distances from the Sun.

This entry concerns the inner corona with its loops and holes, and the outer corona with its helmet streamers. We are really addressing the topology of the solar magnetic field outlined by these tenuous, but highly conducting, plasma structures.

INNER CORONA

Figure 2 compares an image of the inner corona (top) with a map of the line-of-sight magnetic field on the visible disk (bottom). The coronal image was obtained in the Fe XV 284-Å emission line during the *Skylab* mission in 1973, and is printed as a negative. The magnetogram represents positive and negative polarities by lighter-than-average and darker-than-average features, respectively. Coronal loops are visible connecting magnetic regions of opposite polarity, not only within individual bipolar magnetic regions, but also between adjacent regions. Time-lapse sequences of such images show that the interconnecting loops form while the new bipolar regions are growing, as if some of the field lines were changing their connections. Such field-line reconnection was found to be a normal characteristic of coronal evolution, and seemed to take place nearly continuously as new flux erupted, despite the extremely low electrical resistivity of the plasma.

A puzzling question that arose during the *Skylab* era was why coronal holes such as that shown in Fig. 1 were able to maintain their meridional shapes for several months without becoming sheared by the differential rotation of the Sun's surface. At its observed rate, differential rotation ought to shear and destroy such holes after only a few months. A popular speculation was that the holes are manifestations of rigidly rotating phenomena hidden below the Sun's surface.

This romantic idea was somewhat dashed when we later discovered that the rigid rotation of coronal holes can be explained in terms of observable properties of the Sun's atmosphere. One of these properties is the rotation of the outer corona, which is observed to be much more rigid than the rotation of the Sun's surface. This rigid rotation of the outer corona stems from the fact that it is largely free of electrical currents. In a current-free extension of the magnetic field, small-scale features are filtered out with increasing height, so that there is a limit to how tightly the coronal field can be wound.

This current-free property is also responsible for making the footpoints of open field lines corotate with their outer-coronal extensions, rather than with the flux elements on the differentially rotating surface. Otherwise, the field lines would become curled, currents would form, and unsupportable volume forces would occur. Consequently, the footpoint connections continually change, and the boundary of the coronal hole moves as a "reconnection wave" or shadow of the rigidly rotating field in the outer corona, rather than as a physical structure subject to the differential rotation of the Sun's surface.

Nevertheless, as a rigidly rotating coronal hole drifts across the surface, the hole will eventually encounter the boundary of the

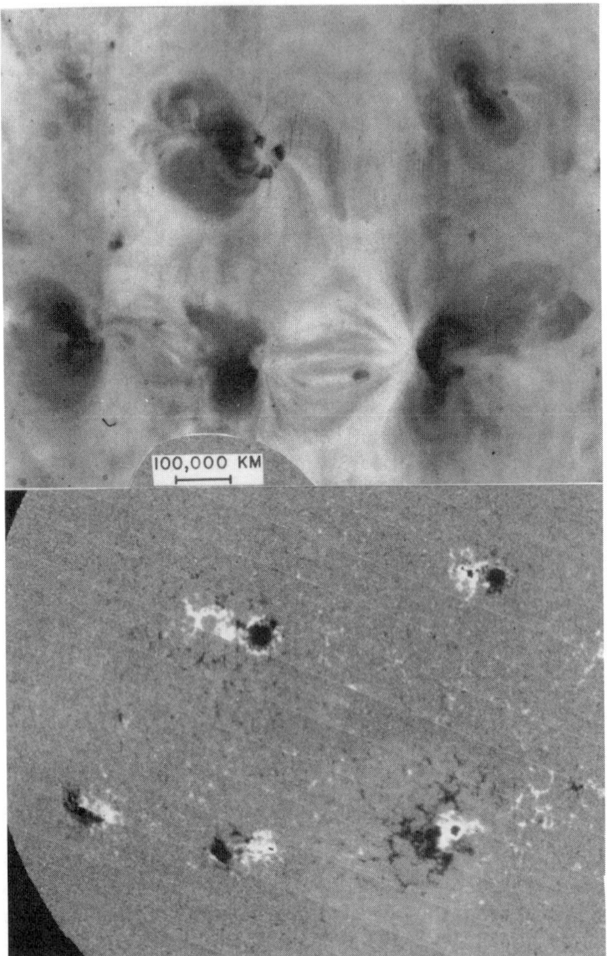

Figure 2. Comparison of an Fe xv 284-Å image (negative print, above) of the inner corona and a map of magnetic fields on the Sun's surface (below; white refers to positive polarity and black to negative). (*Courtesy of the Naval Research Laboratory and the National Solar Observatory.*)

unipolar magnetic region in which it is embedded. The hole cannot long endure the simultaneous constraints of conforming to the increasingly sheared shape of this magnetic boundary and of corotating with the outer-coronal field. Thus, soon after the encounter, the hole will begin to die, eventually leaving only the axisymmetric polar hole and a dwindling remnant near the equator.

This fatal and inexorable encounter was delayed during the *Skylab* mission because the magnetic boundary was confined to low latitudes where the shearing rate is small. This is typically the case near the time of sunspot minimum when the polar fields are strong, sunspot activity is reduced, and the belts of sunspot eruption have drifted close to the equator. However, near sunspot maximum, this divider of surface polarities extends to high latitude and is responsible for the more rapid shearing and shorter lifetimes of coronal holes.

OUTER CORONA

Electrons in the outer corona reveal their presence by scattering visible light from the Sun's surface toward the observer. The resulting intensity I depends on the product of three factors: the intensity I_e of the light incident on the scattering electrons, the electron density n_e at the point of scattering, and the scattering cross section σ. By multiplying these factors together and adding the contributions from points all along the line of sight, one obtains the observed intensity at a given location in the coronal image according to $I = \int I_e n_e \sigma \, dl$.

We can increase our depth perception somewhat by recognizing that the scattered light is polarized. For the component polarized tangentially to concentric circles in the sky plane, the cross section is $\sigma = \sigma_0/2$, where $\sigma_0 = 6.6 \times 10^{-25}$ cm^2 is the Thomson scattering cross section. Thus, for tangential polarization, the cross section is independent of the scattering angle ψ, and the density contributions are "unweighted." For the component polarized along the radii of these circles, the cross section is $\sigma = (\sigma_0/2)\cos^2\psi$. This cross section for "radial" polarization vanishes in the sky plane where $\psi = 90°$, and approaches its maximum value of $\sigma_0/2$ at large angles from the sky plane. Thus, although the intensity of the radial component is always less than that of the unweighted tangential component, the difference is greatest if the electron density is concentrated toward the sky plane where the cross section for radial polarization is least. On the other hand, if the scattering electrons were located relatively far out of the sky plane, then there would be little difference between the intensities of tangential and radial polarization.

It is easy to see that this polarization difference will depend on how fast the electron density decreases with true radial distance from the Sun. A large radial falloff of electron density will cause most of the scattering electrons to be located near the sky plane closest to the Sun. Consequently, this polarization technique for improving depth perception is less effective in the inner corona where the falloff rate is steepest.

One can estimate the coronal intensity I in terms of the intensity I_s of the Sun's surface by adopting a density profile of the form $n_e = n_0(r/R)^{-\alpha}$ [where r is the true (three-dimensional) radial distance from the Sun], and assuming that the projected (sky-plane) radial distance ρ is sufficiently large that the Sun may be regarded as a point source from positions along the line of sight. In this case, $I_e \sim I_s(r/R)^{-2}$ and $I \sim I_e n_e \sigma \rho \sim I_s \cdot \sigma_0 n_0 \rho \cdot (\rho/R)^{-(\alpha+2)}$. For a nominal base density $n_0 \sim 1 \times 10^8$ cm^{-3} and falloff rate $\alpha \sim 4$, this gives $I/I_s \sim 10^{-8}$ at a projected distance of $\rho/R = 3$. Consequently, the visible disk must be occulted in order to detect the much weaker coronal intensity. This may be accomplished naturally by the moon as in the eclipse photograph in Fig. 1 or artificially by an occulting disk as in the coronagraph images in Fig. 3.

Obtained by the *SOLWIND* coronagraph in Earth orbit, the images in Fig. 3 show the corona over an annular range of projected distances from 2.5 to about 9 $R_\odot$. The small white disk at the center of the 07:06 UT image indicates the size of the over-occulted Sun. The observations were obtained through a "tangential" polarizer everywhere except in two eccentric rings near 5 and 8 $R_\odot$, where approximately "radial" polarizers were used. The radial polarizers diminish the intensity of the background corona and of several helmet streamers, indicating that most of the scattering electrons lie near the sky plane. Of course, streamers cannot always be located in the sky plane, so that part of this effect must be due to the large radial falloff of their electron density with true radial distance from the Sun. On the other hand, diffuse clouds of material are occasionally ejected well out of the sky plane, where they fill in the gaps produced by the radial polarizers.

In addition to aiding in the depth perception of coronal features, these polarizers help to distinguish the polarized Thomson-scattered radiation from other unpolarized sources of radiation. For example, whereas the tangential polarizer does not reduce the intensity of Thomson-scattered radiation from a coronal streamer located in the plane of the sky, it halves the unpolarized intensity of light scattered from submicron-sized dust particles in the vicinity of the Sun (called the F component of the corona because it reflects the Fraunhofer absorption spectrum of the Sun's disk). Also, when relatively cool prominence material is ejected into the coronagraph's field of view, its unpolarized Hα 6563-Å line emission is transmit-

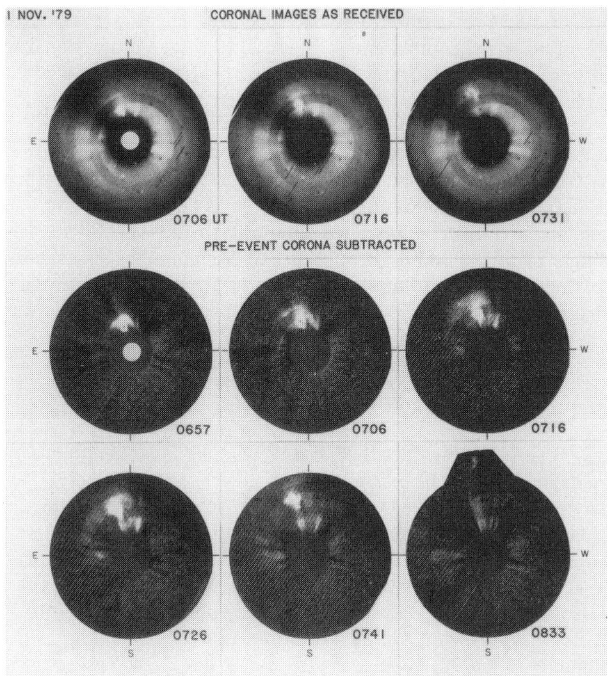

Figure 3. Images of the outer corona obtained with the *SOLWIND* coronagraph on the *P78-1* Earth-orbiting satellite. Times in Universal Time appear below and to the right of each image; 0716 means 07:16 UT. To show the coronal mass ejection more clearly, the pre-event coronal intensity has been subtracted from the lower images. (*Courtesy of the Naval Research Laboratory.*)

ted equally by the tangential and radial polarizers, thereby distinguishing it from visible light scattered from coronal electrons.

Figure 3 shows such a transient event in progress during 06:57–08:33 UT on November 1, 1979 near sunspot maximum. During this time, a cloud of coronal material is visible moving out through the field of view at a projected speed of about 1400 km s^{-1}, which is well in excess of the nominal 500 km s^{-1} magnetoacoustic speed of the corona. Following behind it, an eruptive prominence is visible, moving undiminished across the polarizing rings and escaping from the field at 08:33 UT.

The accompanying shock wave passed Venus at 22:28 UT the next day and was detected by the *Pioneer Venus* orbiter. Venus's angular location, 47° behind the east limb of the Sun as seen from Earth, was consistent with the backside origin of the erupted prominence. However, Venus's radial position, 0.73 AU (156 $R_{\odot}$) from the Sun, corresponded to a transit speed of only 765 km s^{-1}, so that the shock must have decelerated en route, as is typically observed for the fastest shocks.

The amount of ejected material was in the range 10^{15}–10^{16} g. Although this may seem large by earthly standards (250 million tons), in this volume of $(10\ R_{\odot})^3$ it amounts to a density of only 10^{-21} g cm^{-3} or 10^3 protons cm^{-3}, which is a very good vacuum indeed.

Additional Reading

Gibson, E. G. (1973). *The Quiet Sun*. NASA SP-303, U.S. Govt. Printing Office, Washington D.C.

Koutchmy, S. (1977). Solar corona. In *Illustrated Glossary For Solar and Solar–Terrestrial Physics*, A. Bruzek and C. J. Durrant, eds. D. Reidel, Dordrecht, p. 39.

Noyes, R. W. (1982). *The Sun—Our Star*. Harvard University Press, Cambridge.

Wang, Y.-M., Nash, A. G., and Sheeley, N. R., Jr. (1989). Magnetic flux transport on the sun. *Science* **245** 712.

Wentzel, D. G. (1989). *The Restless Sun*. Smithsonian Institution Press, Washington, D.C.

Wolfson, R. (1983). The active solar corona. *Scientific American* **248** (No. 2) 104.

Zirin, H. (1988). *Astrophysics of the Sun*. Cambridge University Press, Cambridge.

See also **Coronagraphs, Solar; Heliosphere; Interplanetary Magnetic Field; Interplanetary Medium, Solar Wind; Solar Activity, Coronal Mass Ejections; Solar Activity, Solar Flares; Solar Activity, Solar Flares, Theories; Stars, Atmospheres, X-Ray Emission; Sun, Coronal Holes and Solar Wind; Sun, Eclipses; Sun, Magnetic Field; Sun, Radio Emissions.**

Sun, Atmosphere, Photosphere

Gary A. Chapman

The solar photosphere is that layer of the Sun's atmosphere that is visible from the Earth. It is the source of what we call sunlight, the source of most of the heat and light striking the Earth. The photosphere, in a telescope, appears to be (and is often called) the surface of the Sun. In fact, the Sun has no surface; the photosphere is a region of rapidly changing temperature and pressure, like the Earth's atmosphere but with a much higher temperature. Furthermore, because the Sun is entirely gaseous, the atmosphere has no bottom, unlike the atmosphere of the Earth.

ONE SHOULD NEVER LOOK DIRECTLY AT THE SUN; THE PHOTOSPHERE CAN DAMAGE THE HUMAN EYE.

The photosphere is distinguished by the appearance of small grains called the solar granulation, first photographed by the French astronomer Jules Janssen, in 1885. We know that the granulation, visible in telescopes as small as 4–6 in., is the visible sign of turbulent convection, the boiling of the upper layer of the solar atmosphere. The average granule has a lifetime of about 15 min. The average size of an individual granule is approximately 1000 km and granules rise, as they cool, with a velocity of about 1 km s^{-1}. As one looks toward the edge of the solar disk (called the limb), the granulation becomes more and more difficult to see. The limb, which appears sharp, has a fuzziness of about 70 km in width, a dimension that is beyond the capability of present day solar telescopes to resolve.

More dramatic structures in the photosphere than the granulation are sunspots. Sunspots, seen in ancient times with the naked eye through heavy fog or smoke, are dark features that are now identified with strong magnetic fields. Figure 1 shows a photograph of the solar photosphere, with sunspots, as seen in that part of the spectrum to which the human eye is most sensitive.

Sunspots are more or less dark circular regions where the temperature is lower than the surrounding photosphere. The cause of this lowering of temperature is a matter of current research, but it is closely associated with the imbedded magnetic field.

Often associated with sunspots are scattered, irregularly shaped bright "patches" called faculae (a Latin term meaning little torches). Both sunspots and faculae are intimately related to magnetic fields. The faculae, which are normally seen near the limb in white light, are seen across the solar disk as the photospheric network when viewed in the light of narrow spectral intervals of most Fraunhofer absorption lines.

Variations in the photosphere, due to sunspots and faculae, are thought to be the major cause of the variations in the solar output measured by several spacecraft.

A more subtle aspect of the photosphere is the gradual decrease in intensity from the center of the disk of the Sun to the limb, called limb darkening. This effect is such as to give the solar image a "vignetted" appearance. This effect is in contrast to the appear-

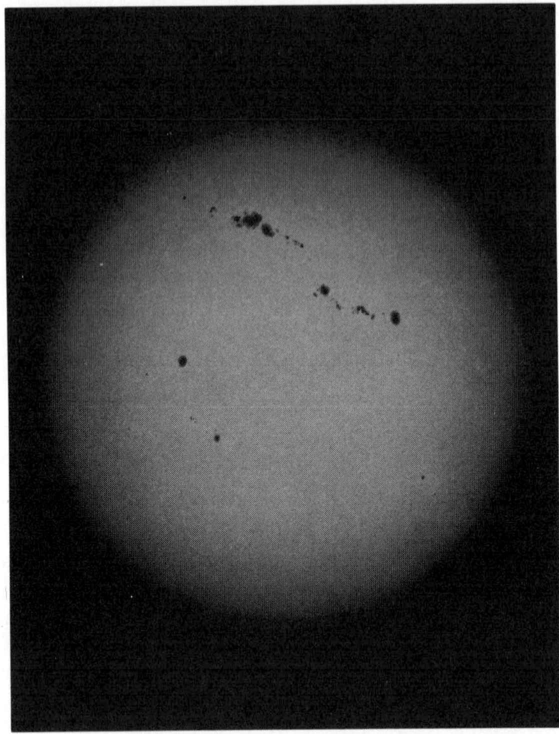

Figure 1. The solar photosphere photographed with a 15-cm telescope on March 20, 1969 at the San Fernando Observatory. At this scale the obvious features are the limb darkening and sunspots. Geocentric north and east are at the top and left, respectively. (© *San Fernando Observatory/CSUN.*)

ance of the moon, which has a more or less uniform brightness across its disk. The limb darkening is caused primarily by the decrease in the temperature with height in the Sun's atmosphere that is seen as a decrease in brightness from the center of the disk to the limb.

The visible part of the photosphere has an average temperature of approximately 5800 K. Because the temperature decreases with height, as for planetary atmospheres, deeper layers correspond to higher temperatures. Interpreting limb darkening as a change in temperature, the lower level has a temperature of about 6200 K, whereas the upper level has a temperature of about 5400 K. At greater heights, seen at ultraviolet wavelengths from balloons, rockets, or spacecraft, the temperature falls to about 4400 K. The exact value of this temperature minimum is not firmly established but depends on details of the observations and their interpretation.

The photosphere is the region of the Sun's atmosphere where many of the spectral lines are formed that are used to determine the temperature, pressure, and chemical composition of the Sun. Certain of these spectral lines are especially suited for studies of magnetic or velocity fields. Some lines are especially sensitive to the presence of magnetic fields, which split them into several components, due to the Zeeman effect. Other lines are not split in the presence of magnetic fields and are especially suited for the study of velocity fields, using the Doppler effect.

The study of short-period oscillatory velocities has now become the chief part of a separate subfield, called helioseismology. These oscillations, which have most of their power at a period of roughly 5 min, also produce an oscillatory intensity pattern in the photosphere as was seen most clearly in photographs obtained in 1985 from the Space Shuttle.

Most of our knowledge of magnetic fields in the Sun comes from measurements of the Zeeman splitting of certain Fraunhofer lines. This phenomenon (discussed in the section on "Methods" in the entry Stars, Magnetism, Observed Properties) causes a wavelength shift of polarized components of these lines that is proportional to the strength and polarity of the magnetic field at each place at the photosphere. This is the principle by which magnetic maps of the Sun are made at several solar observatories.

Observations at the highest possible resolution with present day telescopes show that the solar granulation becomes altered when magnetic fields are present. The cooler lanes between the granules have bright "filigree" scattered amongst them. The filigree are thought to correspond to the presence of discrete tubes of magnetic flux that, at lower resolution, are called faculae, but are not clearly detectable until they are at least one-half of a solar radius away from the center of the disk.

Additional Reading

Bray, R. J. and Loughhead, R. E. (1984). *The Solar Granulation*. Chapman and Hall, London.

Chapman, G. A. (1987). Variations of solar irradiance due to magnetic activity. *Ann. Rev. Astron. Ap.* **25** 633.

Durrant, C. J. (1988). *The Atmosphere of the Sun*. Adam Hilger, IOP Publishing, Ltd., Bristol.

Hudson, H. S. (1988). Observed variability of the solar luminosity. *Ann. Rev. Astron. Ap.* **26** 473.

Foukal, P. V. (1990). *Solar Astrophysics*. Wiley, New York.

Noyes, R. W. (1982). *The Sun, Our Star*. Harvard University Press, Cambridge, Mass.

Orrall, F. Q., ed. (1981). *Solar Active Regions*. Colorado Associated University Press, Boulder.

Shine, R. A., et al. (1987). White light sunspot observations from the solar optical universal polarimeter on *Spacelab-2*. *Science* **238** 1265.

Zirin, H. (1988). *Astrophysics of the Sun*. Cambridge University Press, Cambridge.

See also **Magnetographs, Solar; Solar Activity, Sunspots and Active Regions, Observed Properties; Solar Activity, Sunspots and Active Regions, Theories; Stars, Atmospheres, Turbulence and Convection; Stars, Chemical Composition; Sun, Atmosphere; Sun, Solar Constant; Sun, Oscillations.**

Sun, Coronal Holes and Solar Wind

George L. Withbroe

The earliest evidence that plasma flowed outward from the Sun was provided by studies of geomagnetic phenomena. Some geomagnetic disturbances were found to be correlated with solar activity such as flares, others were correlated with so-called "M" regions that could not be readily associated with specific solar features such as sunspots. The M regions, which are now known to be low density regions in the corona called coronal holes, were postulated as the solar sources of corpuscular radiation causing magnetic disturbances which tended to reoccur at 27 day intervals. That interval corresponds to the rotation period of the Sun as viewed from the Earth. Other early evidence for the presence of outward-flowing solar plasma was provided by the behavior of the tails of comets. Ludwig Biermann found that the accelerations observed in some comet tails required more pressure than could be supplied by solar photons alone. He showed that a likely source of this additional pressure was gas or plasma flowing outward from the Sun with a speed of about 1000 km s^{-1}. In the late 1950s theoretical studies by Eugene N. Parker and others demonstrated that the plasma in the outer solar corona is sufficiently hot that it is not bound by the Sun's gravitational field, but can flow outward into interplanetary space, forming the solar wind. In the early 1960s, near the dawn of the space age, in situ measurements made by

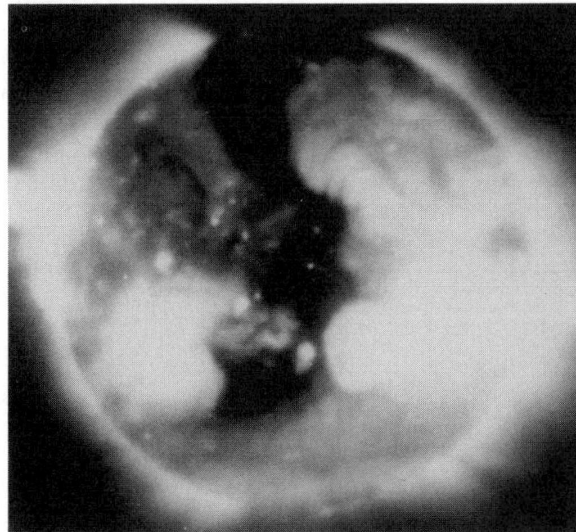

Figure 1. *Skylab* x-ray photograph showing a large coronal hole (dark region) near the central meridian of the Sun and extending from below the solar equator to the north pole. (*Courtesy of American Science and Engineering and the Harvard College Observatory.*)

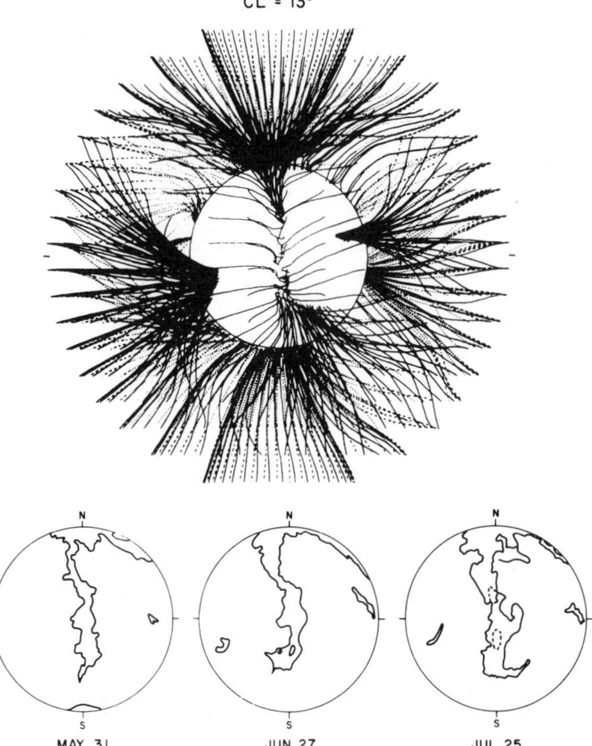

Figure 2. Comparison of calculated magnetic field configuration for field lines swept out by the solar wind (upper part of figure) and contour maps giving locations of coronal holes determined from x-ray photographs (lower part of figure).

spacecraft flown far from the Earth confirmed the existence of the solar wind, a low density plasma flowing outward from the Sun with a mean speed of several hundred kilometers per second.

Since then, the properties of the solar wind near the ecliptic have been measured in situ from as close as 0.3 AU to the Sun and out to the outer limits of the solar system. These measurements have shown that there are two basic types of steady-state solar wind: low speed flows with typical speeds of about 350 km s^{-1}, and high speed flows with typical speeds of about 700 km s^{-1}. A third type of wind consists of transient flows resulting from explosive phenomena on the Sun (solar flares and coronal mass ejections). The transient flows often have different parameters than the steady state flows, such as a high helium abundance, low electron and proton temperatures, unusually high ionization states of heavy ions, and magnetic clouds. Coronal mass ejections, which may or may not be flare-related, are the source of the transient flows and typically have speeds ranging from several hundred to over 1000 km s^{-1}.

The corona, the hot outer layer of the solar atmosphere, is highly structured. Some of this structure is shown in the x-ray photograph presented in Fig. 1. The brightest areas are regions with strong, closed magnetic field configurations. The regions of intermediate brightness are regions with weak, predominately closed magnetic configurations, whereas the dark region near the central meridian of the Sun and extending from south of the solar equator to the north pole is a large coronal hole. Coronal holes have lower coronal densities (by a factor of 3 or more) and lower temperatures (about 10^6 as compared to 1.5 to 3×10^6 K) than the surrounding regions.

Observations such as those illustrated in Fig. 1, coupled with calculations of the configuration of the coronal magnetic field, indicate that most of the solar surface, about 80%, appears to be covered with closed magnetic fields. The remainder of the surface has open magnetic configurations where only one end of each field line is attached to the surface. Several solar radii above the solar surface, the magnetically open regions expand to occupy the entire volume and the solar wind flow becomes predominately radial. Figure 2 presents a schematic view illustrating this concept. The upper part of the figure contains a plot of magnetic field lines that extend out into interplanetary space. (Field lines from magnetically closed regions are not plotted.) These field lines were calculated from measurements of the magnetic field at the solar surface using the assumption that the coronal magnetic field is potential (no electric currents above the solar surface). The contour maps in the lower part of the figure show the locations of coronal holes determined from x-ray photographs similar to Fig. 1. The most prominent magnetically open regions are large unipolar magnetic regions associated with coronal holes. Extrapolation back to the Sun of the solar wind flows in recurrent high speed streams shows that they originate in coronal holes and that the magnetic polarity associated with a given high speed solar wind stream corresponds to that in the originating coronal hole. Additional evidence relating high speed streams and coronal holes is provided by the coincidence between global maps of high-speed flows derived from radio scintillation observations of compact astronomical radio sources, and the corresponding maps defining the size and shape of polar coronal holes.

The discovery that coronal holes are the source of recurrent high speed streams resolved one outstanding problem in solar-terrestrial physics. However, a second problem remains: Why are the flow speeds in high speed streams so large? The classic Parker solar wind model assumes that the solar wind is driven by the thermal pressure of the hot coronal gas. This model can account for the observed velocity of the low speed wind, but not for the high speed wind. An additional acceleration force is needed. The most likely source is Alfvèn waves. In situ measurements of the solar wind far from the Sun show evidence for the presence of Alfvèn waves, particularly in high speed streams. Measurements of spectral line widths in the low corona are also consistent with this hypothesis, because the measured widths are larger than expected for lines that are broadened only by thermal motions. The excess width is consistent with the broadening expected if there is present in the low corona a sufficiently large flux of Alfvèn waves to accelerate the wind to the observed speeds.

The major remaining problem in the physics of coronal holes is determining the source of the plasma heating that produces the 10^6-K temperature of the coronal plasma in these regions. Because the overall magnetic configuration in coronal holes is open, there are fewer options for possible heating mechanisms than in regions with closed magnetic fields. In the latter it is possible to store energy magnetically via twisted field lines and then dissipate the stored energy by dissipation of electric currents or by annihilation of magnetic fields. In magnetically open regions, such as coronal holes, some form of wave heating appears to be responsible. Thus, it appears likely that magnetohydrodynamic (MHD) waves are responsible both for the plasma heating and for the acceleration of the solar wind in these regions. However, the detailed mechanisms for accomplishing this have not been identified. Improved empirical and theoretical work is required to establish what type(s) of waves are involved and how they deliver their energy to the plasma.

A closely related problem is that of the coronal fine structure. Observations of polar coronal holes show that they contain ray-like fine structures that extend far above the solar surface. These features, polar plumes, appear to overlie small magnetic bipoles, which are often marked by bright extreme ultraviolet (EUV) and x-ray emissions, known as coronal bright points. The coronal bright points are magnetically closed at low coronal heights whereas the overlying polar plumes appear to be magnetically open. It is possible that magnetic activity in the magnetically closed regions produces plasma heating and/or accelerates plasma magnetically and serves to drive the solar wind. The EUV emission from these regions is variable on a time scale of a few minutes. This variable emission appears to be caused by an impulsive, stochastic heating mechanism involving rapid dissipation of magnetic energy.

There is evidence for other localized transient dissipations of energy at the coronal base that could play a role in heating and acceleration of the solar wind plasma. These transient energy releases produce small-scale (10^3–10^4 km) jet-like phenomena, such as spicules and macrospicules, which have upward velocities of 25–150 km s^{-1}, and more energetic explosive events associated with motions of up to 400 km s^{-1}. These phenomena are observed best in spectral lines formed at relatively cool temperatures, 10^4–10^5 K. The required small-scale energy releases are believed to involve dissipation of magnetic energy in small magnetic structures near the coronal base. At the present time it is unclear whether the heating and acceleration of the coronal plasma in coronal holes is produced by averaging over many of these small-scale energy releases or by MHD waves as suggested earlier. It is possible that MHD waves may be associated with the generation of small-scale phenomena, in addition to heating the coronal plasma and accelerating the solar wind.

Additional Reading

Biermann, L. (1951). Kometenschweife und solare Korpuskularstrahlung. *Z. Astrophysik* **29** 274.

Brueckner, G. E. and Bartoe, D. F. (1983). Observations of high-energy jets in the corona above the quiet Sun, the heating of the solar corona, and the acceleration of the solar wind. *Astrophys. J.* **272** 329.

Kiepenheuer, K. O. (1953). Solar activity. In *The Sun*, G. P. Kuiper, ed. University of Chicago Press, Chicago, p. 322.

Kojima, M. and Kakinuma, T. (1987). Solar cycle evolution of solar wind stream structure between 1973 and 1985 observed with the interplanetary scintillation method. *J. Geophys. Res.* **92** 7269.

Parker, E. N. (1958). Dynamics of the interplanetary gas and magnetic fields. *Astrophys. J.* **128** 664.

Pizzo, V. J., Holzer, T. E., and Sime, D. G. (1988). *Proceedings of the Sixth International Solar Wind Conference*. NCAR TN-306. National Center for Atmospheric Research, Boulder.

Wentzel, D. G. (1989). *The Restless Sun*. Smithsonian Institution Press, Washington, D.C.

Withbroe, G. L. (1986). Origins of the solar wind in the corona. In *The Sun and the Heliosphere in Three Dimensions*, R. G. Marsden, ed. D. Reidel, Dordrecht, p. 19.

Zirker, J. B. (1977). *Coronal Holes and High Speed Wind Streams*. Colorado Associated University Press, Boulder.

See also **Coronagraphs, Solar; Heliosphere; Interplanetary Magnetic Field; Interplanetary Medium, Solar Wind; Magnetohydrodynamics, Astrophysical; Solar Activity, Coronal Mass Ejections; Solar Physics, Space Missions; Sun, Atmosphere, Corona.**

Sun, Eclipses

Jay M. Pasachoff

Eclipses of the Sun occur when the Moon passes between the Sun and the Earth. An eclipse is an example of a syzygy, a three-body alignment.

During a solar eclipse, the Moon's shadow falls upon the Earth. The darkest part of the shadow, from which no part of the Sun can be seen, is the *umbra*. From within the umbra, we see a *total solar eclipse*. During a total solar eclipse, the solar photosphere is hidden from our view and we can see the faint, outer layers of the solar atmosphere that are normally fainter than the sky background. These layers, the solar chromosphere and corona, remain special objects of scientific study at eclipses.

The umbra, a long conical shadow, often misses the Earth because of the inclination of the lunar orbit. Sometimes the *penumbra*, the surrounding shadow from which part of the Sun can be seen, hits the Earth, making a *partial solar eclipse*. From the Earth, the edge of the Moon is seen silhouetted against the solar surface. Though the irregularities at the Moon's edge show under high resolution, little of scientific value is carried out at partial solar eclipses in the optical part of the spectrum in contrast to the valuable optical observations made at total solar eclipses. In the radio part of the spectrum, partial solar eclipses can be used to give finer resolution of occulted active regions on the Sun than is normally available.

Total eclipses of the Sun as we see them occur because of the happy accident that the Moon is the same factor closer to the Earth than the Sun that it is smaller than the Sun. This factor is about 400. Sometimes, chiefly because of the Moon's elliptical orbit around the Earth but with some contribution from the Earth's elliptical orbit around the Sun, the angular diameter of the Moon is slightly smaller (up to about 10%) than that of the disk of the solar photosphere. At those times, an annulus (ring) of photospheric sunlight remains visible around the moon, making an *annular eclipse*. Occasionally an eclipse is annular along part of its path on the Earth and total along the rest.

About $\frac{1}{3}$ of solar eclipses are total, $\frac{1}{3}$ are annular, and $\frac{1}{3}$ are merely partial. Because they result from an exact alignment of Sun, Moon, and Earth, solar eclipses always occur at full moon.

SAROS

As has been known since it was discovered by ancient Greek astronomers, the circumstances of total eclipses repeat every 18 years $11\frac{1}{3}$ days (plus or minus a day, depending on how leap years fall), a period known as the Saros. The Saros is caused by the coincidence of several important lunar and solar periods (see Table 1). The lunar period is marked by synodic months, the period of the phases. The solar period is marked by eclipse years, returns of the Sun through one of the nodes where the inclined lunar orbit meets the ecliptic, the path of the Sun in the sky. Similarly, nodical months (also known as draconic months to honor the mythological dragon that supposedly devoured the Sun to cause an eclipse) mark another lunar period by the passages of the Moon through the node. In addition to the date repeating, the overall coverage of the Sun by the Moon and the duration of totality also repeat because of the further coincidence of an integral number of anomalistic months, the period of variation of the Moon–Earth distance,

Table 1. Causes of the Saros

223 lunar (synodic) months	6585.32 days
242 lunar (nodical) months	6585.36 days
19 eclipse years (solar)	6585.78 days
239 anomalistic months	6585.54 days

which differs from other lunar months because of the precession of the Moon's orbit.

The 38 days around each nodal passage when the Sun is sufficiently close to the node for an eclipse to occur is called an *eclipse season*. At least one solar eclipse takes place during each eclipse season, and there can be as many as three eclipse seasons in a year. Thus as many as five partial and total solar eclipses can occur in a given calendar year.

FUTURE ECLIPSES

The 150 total and annular solar eclipses during 1901–2000 are evenly divided between total and annular. Total eclipses range in duration from seconds up through about 7 min. The ellipse of the lunar shadow projected onto the Earth sweeps a path across the Earth's surface that is up to about 400 km wide and several thousands of kilometers long. Eclipses near the equator at noon have the longest durations, because then the rotational velocity of points of the Earth surface within the zone of totality is highest, most closely matching the velocity of the umbra through space.

Desirable eclipses to study are often those that last the longest, though many eclipse phenomena worthy of study occur only at the beginning or the end of the total phases, making the scientific value of an eclipse less dependent on duration. Weather forecasts are also significant determinants for optical observers.

Table 2 lists total and annular eclipses during 1991–2010. Note that, matching the Saros, the long, desirable eclipse of 11 July 1991 is repeated on 22 July 2009.

ECLIPSE PHENOMENA

The paths of total solar eclipses now draw not only professional astronomers but also many amateur astronomers and tourists. A partial eclipse and also the partial phases that precede a total eclipse must be viewed through special eye-protection filters or else projected with a telescope or pinhole camera onto a screen, because the surface brightness of the photosphere is sufficiently bright as to cause serious eye damage or blindness. These partial phases commonly last 2 h. During the last minute or so, the chiaroscuro of *shadow bands*, waves of low-contrast light and dark due to diffraction effects in the Earth's atmosphere, may be seen running across the landscape. The darkness caused by the eclipse is noticeable only in the last few minutes before totality. Also, the light then takes on an eerie quality and shadows appear especially sharp, because the visible diameter of the part of the Sun that is casting the shadows is less than that of the whole Sun.

About 15 s before totality, the thin remaining crescent of sunlight is broken into a series of *Baily's beads* by the mountains on the edge of the Moon. The last Baily's bead glows so brightly compared with other visible features, including perhaps the solar corona that forms a band visible on the rest of the lunar edge, that it is known as the *diamond-ring effect*. As the diamond ring diminishes, the solar chromosphere is visible as a pinkish band on the leading edge of the Moon; it radiates chiefly in the emission

Table 2. Total and Annular Eclipses, 1991–2010*

Date	*Type*	*Maximum Duration*	*Locale*
15 January 1991	Annular	7 m 55 s	Tasmania, New Zealand
11 July 1991	Total	6 m 54 s	Hawaii, Mexico, Central and South America
4 January 1992	Annular	11 m 43 s	partial: Hawaii; end: California
30 June 1992	Total	5 m 21 s	Uruguay, Atlantic Ocean
10 May 1994	Annular	6 m 14 s	Mexico, U.S.
3 November 1994	Total	4 m 23 s	S. America
29 April 1995	Annular	6 m 38 s	S. America
24 October 1995	Total	2 m 10 s	S. Asia
9 March 1997	Total	2 m 50 s	Mongolia, U.S.S.R., Arctic
26 February 1998	Total	4 m 09 s	Panama, Colombia, Venezuela
22 August 1998	Annular	3 m 14 s	Indonesia, Malaysia, Oceania
16 February 1999	Annular	1 m 18 s	Australia
11 August 1999	Total	2 m 23 s	Europe, S. Asia
21 June 2001	Total	4 m 57 s	Angola, Zambia, Mozambique
14 December 2001	Annular	3 m 54 s	Pacific Ocean, Central America
10 June 2002	Annular	1 m 13 s	Pacific; ends off Baja California
4 December 2002	Total	2 m 04 s	Southern Africa, W. Australia
31 May 2003	Annular	3 m 41 s	Arctic
23 November 2003	Total	1 m 57 s	Antarctica
8 April 2005	Ann/Total	0 m 42 s	Panama, Colombia, Venezuela
3 October 2005	Annular	4 m 31 s	Portugal, Spain, N. and E. Africa
29 March 2006	Total	4 m 07 s	Africa, Turkey, U.S.S.R.
22 September 2006	Annular	7 m 09 s	N.E. South America, Atlantic
7 February 2008	Annular	2 m 14 s	Antarctica
1 August 2008	Total	2 m 27 s	N. Canada, N. Greenland, Asia
26 January 2009	Annular	7 m 56 s	Indonesia
22 July 2009	Total	6 m 40 s	India, Bangladesh, China, S. Japan Is.
15 January 2010	Annular	11 m 11 s	Africa, India, Myanmar, China
11 July 2010	Total	5 m 20 s	S. Chile, S. Argentina

*Data from F. Espenak, *Fifty Year Canon of Solar Eclipses, 1986–2035*. NASA Reference Publication 1178 (revised), NASA, Greenbelt, Md.

lines of hydrogen (the Balmer series) and ionized calcium (the H and K lines in the ultraviolet). The D lines of sodium and the nearby D_3 line from which helium was discovered are also visible at this time.

After 10 s or so, the chromosphere is covered and the corona is visible. It may have been visible slightly earlier, depending on the clarity of the sky. The corona is an ionized plasma at an average temperature of 2,000,000 K, and shows streamers and plumes held in place by the solar magnetic field. It has about the same total brightness as the full moon, and thus can be looked at directly, without filters. Most of the visible light that the corona radiates is in the emission lines of fourteen-times ionized iron at 530.3 nm and ten-times ionized iron at 637.4 nm. The emission lines are known as the E corona, and fall off sharply with distance from the Sun. The corona also reflects photospheric light to Earth; the high Doppler velocities of the coronal electrons that are doing the scattering washes out the photospheric Fraunhofer lines and the resulting radiation appears as a continuum. This part of the corona is known as the K corona, from the German word for continuous. Another coronal contribution, weaker than the continuous corona within about 1 $R_\odot$ (solar radius) above the solar edge but stronger beyond, is the F corona, caused by scattering of photospheric light by interplanetary dust and thus retaining the solar Fraunhofer lines. A thermal corona, or T corona, has been reported in the infrared and results from the self-emission of interplanetary dust.

The shape of the coronal streamers varies with the solar-activity cycle. The corona is most oblong during solar minimum, when a few coronal streamers are visible at the solar equator and small coronal plumes appear at the solar poles. The corona is roundest during solar maximum, when so many solar streamers are visible that they project into the plane of the sky as a disk.

SCIENTIFIC VALUE

The corona is so seldom visible and is accessible for such brief periods that much remains of scientific interest to study at eclipses. Further, new scientific questions arise to be answered. Also, studies are made of how the corona changes over the 11-yr solar-activity cycle. Major scientific expeditions continue to be mounted by several countries to study the eclipses that are classed as major in terms of duration and favorable weather prospects.

Coronagraphs on certain high mountains on Earth can observe the inner part of the corona, chiefly in the emission lines but also through study of the polarization of the K corona. Eclipses excel for studying fainter phenomena in the chromosphere and inner corona as well as all phenomena above in the middle corona.

Coronagraphs in space have so far had to occult not only the solar photosphere but also $\frac{3}{4}$ of a solar radius (for the *Solar Maximum Mission*) or so around it. Diffraction rings around the occulting disk hide even more of the corona. Future satellites with improved coronagraphs may eventually solve this problem.

Additional Reading

Carton, W. H. C. (1989). Oppolzer's great *Canon of Eclipses*. *Sky and Telescope* **78** 475.

Codona, J. L. (1991). The enigma of shadow bands. *Sky and Telescope* **81** 482.

Espenak, F. (1987). *Fifty Year Canon of Solar Eclipses: 1986–2035*. NASA Reference Publication 1178 (revised), NASA, Greenbelt, Md.

Menzel, D. H. and Pasachoff, J. M. (1990). *A Field Guide to the Stars and Planets*, 2nd ed. Houghton Mifflin Co., Boston.

Pasachoff, J. M. (1989). *Contemporary Astronomy*, 4th ed. Saunders College Publishing, Philadelphia.

Zirin, H. (1988). *Astrophysics of the Sun*. Cambridge University Press, New York.

Zirker, J. B. (1984). *Total Eclipses of the Sun*. Van Nostrand Reinhold, New York.

See also **Coronagraphs, Solar; Sun, Atmosphere, Corona.**

Sun, High-Energy Particle Emissions

Reuven Ramaty

High-energy solar emissions (γ rays, hard x-rays, and neutrons) result from the interactions of particles accelerated in solar flares with the ambient solar atmosphere. These emissions, as well as the accelerated particles that escape from the Sun to interplanetary space, can be observed with detectors carried above the Earth's atmosphere by rockets, balloons, and spacecraft. By studying these radiations, information is obtained on the acceleration and transport of accelerated particles in solar flares, and on properties of the solar atmosphere such as its chemical composition and the structure of its magnetic fields. Particle acceleration is a widespread process in astrophysics, occurring at diverse sites ranging from planetary magnetospheres to distant objects such as supernovae, active galaxies, and quasars. But because of the proximity of the Sun and the intense high-energy emissions that it produces, the study of these emissions provides one of the best techniques for investigating accelerated charged particle phenomena in astrophysics. Furthermore, of all astrophysical sites, only in the case of the Sun can both the high-energy emissions and the particles that produce them be simultaneously observed.

OVERVIEW OF THE OBSERVATIONAL DATA

The first evidence for particle acceleration in solar flares came from the 1942 observation of particles of energies higher than several gigaelectron volts (1 GeV = 10^9 eV) from two large solar flares with ground-based detectors, which were sensitive to the secondary particles produced in the atmosphere by the primary energetic particles. Since then, accelerated particles from solar flares have been routinely observed in interplanetary space with orbiting spacecraft and interplanetary spaceprobes, as well as with instruments on the ground. The energy spectrum of the particles (i.e., the number of particles of various kinetic energies) has been measured over a broad energy range, from below 1 MeV (10^6 eV) to several gigaelectron volts. The elemental and isotopic compositions of the particles have also been observed. These compositions are highly variable, changing from flare to flare, and with time within a flare. It is thought that these variations are caused mostly by the acceleration process; however, the variations could also reflect variations in space and time of the composition of the ambient solar atmosphere from which the particles are accelerated.

High-energy electromagnetic emission from solar flares was first observed in 1959 with a detector flown on a balloon. Hard x-rays of energies around 500 keV (1 keV = 10^3 eV) were detected. Many more x-ray observations followed with instruments on rockets, balloons, and spacecraft. Characteristic γ-ray line emission, resulting from the bombardment of nuclei with fast particles, was first observed from two flares in 1972 with a detector flown on the seventh *Orbiting Solar Observatory*. Recently, very substantial progress in the study of x-rays and γ rays from solar flares has been achieved by observations with instruments on the *Solar Maximum Mission* (*SMM*). Hard x-rays were observed with a sensitive *SMM* spectrometer, capable of measuring time variations on millisecond time scales, and with an imaging instrument, capable of localizing the x-ray sources with 8 arcsec angular resolution. Gamma rays were observed with a spectrometer capable of observing line emission in the energy range from about 300 keV to 8 MeV. *SMM* was launched in 1980 and, after operating successfully for almost a decade, reentered and disintegrated in the atmosphere on December 3, 1989. X-rays and γ rays were also observed with the Japanese satellite *Hinotori*, which is no longer operational.

Neutrons, resulting mostly from the breakup of He nuclei by accelerated particles, were observed with the γ-ray spectrometer on *SMM* and with groundbased neutron monitors, which detect the nucleonic cascade produced in the atmosphere by the incoming solar neutrons. Neutron observations complement the x-ray and γ-ray observations, but because the neutrons are not stable (they decay with a half-life of 10.6 min into protons), they cannot be observed from distant astrophysical sources. The survival probability of solar neutrons in transit from the Sun to the Earth increases with increasing neutron energy. This probability is close to unity at 1 GeV, but below 100 MeV most of the neutrons decay before they reach the Earth. However, by detecting the protons resulting from their decay in interplanetary space, these low-energy neutrons can also be studied.

MODELS AND PHYSICAL PROCESSES

A large solar flare releases as much as 10^{32} erg and a significant fraction of this energy appears in the form of accelerated particles. The flare energy, which ultimately must be derived from subphotospheric mechanical energy (e.g., differential rotation and convection), is most probably stored in magnetic fields and is released by rapid reconnection of magnetic field lines, probably in the corona. Magnetic loops in the solar atmosphere almost certainly play a role in the flare process. These loops are anchored in the strong magnetic fields of the photosphere, and extend through the chromosphere into the corona. A plausible model for the production of the observed high-energy emissions involves the acceleration of electrons and ions (protons and heavier nuclei) in the coronal part of a loop or an arcade of loops. The detailed nature of the acceleration mechanism is still not completely understood. A leading candidate is the mechanism proposed by Enrico Fermi in 1949 to explain the origin of the cosmic rays. In this mechanism the particles are accelerated stochastically by colliding with randomly moving magnetized clouds; but in the more modern approach, stochastic acceleration is thought to be caused by resonant interactions of the particles with magnetohydrodynamic turbulence of wavelengths comparable to the particles' gyroradius in the magnetic field. The turbulence could be produced in the corona during the primary flare energy release and could be sustained in the ionized corona, but is expected to be damped quickly by collisions between ions and neutral atoms in the chromosphere and below. This is an important argument favoring acceleration in the corona.

The accelerated particles propagate along the magnetic field lines and produce radiation as they interact with matter and magnetic fields in the loop. Interaction of the electrons with magnetic fields produces radio and microwave emissions, which are not discussed in this entry. Interactions of the electrons with the ambient gas produce bremsstrahlung (braking radiation), which is generated as the fast electrons are slowed down in the electrostatic Coulomb potentials of the ambient nuclei and electrons. The energy spectrum of bremsstrahlung is continuous (i.e., it contains no lines) and extends up to a photon energy comparable to the initial electron energy. Interactions of the ions with the gas produce excited and radioactive nuclei, as well as neutrons and pions. These secondary products lead to a variety of γ-ray line and continuum emissions, which we will discuss in more detail. The interactions of the accelerated particles with ambient gas are thought to occur predominantly in the chromospheric portions of the loops where the gas density is much higher than in the corona.

Nuclear deexcitation lines result from the bombardment of ambient C and heavier nuclei by accelerated protons and α particles (He nuclei), and from the inverse reactions in which ambient hydrogen and helium are bombarded by accelerated carbon and heavier nuclei. The former lead to narrow lines whose widths are determined by the recoil velocities of the heavy targets, whereas the latter produce broad lines whose widths are due to the velocities of the accelerated heavy projectiles. Because of their low relative abundances, interactions between accelerated and ambient heavy nuclei are not very important. Furthermore, because H and He have no bound excited states, proton–proton and proton–α-particle interactions can be ignored in considerations of deexcitation-line production. However, interactions of α particles with ambient He produce two strong lines, at 478 keV from ^{7}Li and 429 keV from ^{7}Be.

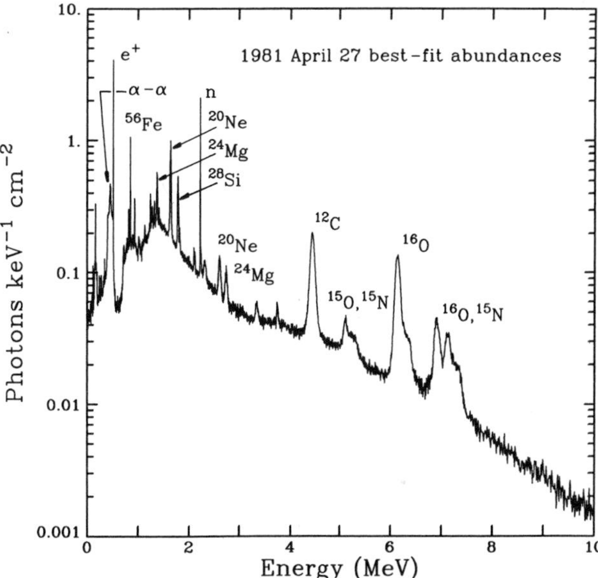

Figure 1. Calculated solar flare γ-ray spectrum, corresponding to abundances that best fit the observed spectrum of the 1981 April 27 limb flare. The observations were carried out with the γ-ray spectrometer on *SMM*.

Neutrons in solar flares are produced in proton–α-particle, α-particle–α-particle, and proton–proton interactions. Neutrons can also be produced in proton and α-particle interactions with C and heavier nuclei, but these reactions are not very important in solar flares. The neutrons can penetrate into the photosphere where the density is high enough for the neutrons to be captured before they decay. Most of the captures are on H and ^{3}He. Capture on H produces a line at 2.223 MeV, which is the strongest line from solar flares, except for flares very close to the limb of the Sun. For limb flares the 2.223-MeV line is strongly attenuated by scattering in the photosphere. The other nuclear lines originate higher in the atmosphere and are attenuated only for flares behind the solar limb.

Positrons (antielectrons) in solar flares result mainly from the decay of radioactive nuclei and charged pions. The positron, discovered in 1932 by Carl D. Anderson, annihilates with an electron to produce two γ rays of energies equal to the electron rest-mass energy, 511 keV. The 511-keV line has been observed from several flares, and its delayed nature, due to the finite lifetimes of the radioactive nuclei, has been confirmed. In many cases the positrons form short-lived positronium atoms before annihilation, which, in addition to annihilating into 511-keV line photons, also produce a characteristic continuum. Because of instrumental limitations this continuum has not yet been observed from solar flares, even though it has been seen with high-resolution γ-ray spectrometers from the direction of the galactic center.

A theoretical nuclear interaction spectrum, calculated by incorporating a large body of nuclear data into a computer code, is shown in Fig. 1. There are strong nuclear deexcitation lines at 6.129 MeV from ^{16}O, 4.438 MeV from ^{12}C, 1.779 MeV from ^{28}Si, 1.634 MeV from ^{20}Ne, 1.369 MeV from ^{24}Mg, and at 0.847 MeV from ^{56}Fe. The neutron capture line at 2.223 MeV and the positron annihilation line at 511 keV can also be seen. The feature just below the 511-keV line is due to the superposition of the ^{7}Be and ^{7}Li lines at 429 keV and 478 keV. The underlying continuum is also of nuclear origin. It results from the superposition of the broad lines due to the inverse reactions and unresolved line emission from heavy nuclei. The shape of the spectrum depends on elemen-

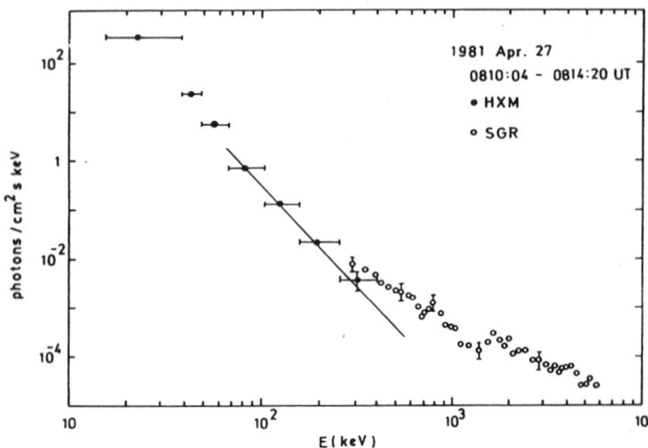

Figure 2. Extended high-energy spectrum of the 1981 April 27 flare observed with detectors on *Hinotori*. [*From M. Yoshimori et al.*, J. Phys. Soc. Japan *54 4462 (1985), courtesy of the author.*]

tal abundances in both the ambient gas and the accelerated particles, whereas the absolute normalization of the spectrum depends on the number of accelerated particles that interact at the Sun. The results shown in Fig. 1 were obtained by fitting the calculations to the observed spectrum of the 1981 April 27 flare. These data were obtained with the γ-ray spectrometer on *SMM*.

An extended high-energy spectrum of this solar flare, measured from 20 keV to 7 MeV with instruments on the *Hinotori* satellite, is shown in Fig. 2. At energies greater than about 400 keV, line emission can be seen. From energies of several tens of kiloelectron volts to about 400 keV the emission is almost certainly bremsstrahlung produced by accelerated electrons, often referred to as nonthermal bremsstrahlung. But in the energy region around 20 keV other processes may also contribute to the observed emission. At these low energies, nonthermal bremsstrahlung production is energetically inefficient, that is, the ratio of the bremsstrahlung energy loss rate to the total energy loss rate of an electron is very small ($\sim 10^{-5}$). This inefficiency has led to extremely large estimates of the total energy content in subrelativistic electrons. These estimates are so large that the energy contained in the electrons appears to be capable of accounting for the entire flare energy. Although this in itself may not be a problem, the transport of very large quantities of energy by $\sim$ 20-keV electrons involves very large electric currents, the stability of which has not yet been demonstrated.

An alternative to nonthermal bremsstrahlung is bremsstrahlung produced in a quasithermal plasma of temperature in excess of 10^7 K. This is a rapidly heated plasma in which the electrons and protons have not yet come into thermal equilibrium. The efficiency of such bremsstrahlung production is strongly model dependent, but it could be significantly higher than that of nonthermal bremsstrahlung, and this would lead to lower estimates of the total electron energy content. The plasma could be heated by protons of energies less than 1 MeV. One of the attractions of this concept is that the energetic particles that transport the energy are ions, rather than electrons. With a mean ion energy of 1 MeV, the number of ions required to produce a given amount of x-ray emission is much smaller than the required number of electrons with mean energy of only a few tens of kiloelectron volts. A reduction in the number of energy-carrying particles greatly alleviates the flare energy transport problem.

IMPLICATIONS OF THE OBSERVATIONS

Observation of the time profiles of the x-ray and γ-ray emissions from solar flares reveals the impulsive nature of these emissions, suggesting that particle acceleration in flares is closely associated with the primary energy release process. Even though the total energy in accelerated particles is somewhat uncertain, it is nevertheless clear that a significant fraction of the flare energy is contained in accelerated protons and electrons. The energy contained in protons of energies greater than several megaelectron volts, reliably determined from the observation of γ-ray lines, amounts to several percent of the total flare energy. Much more energy could be contained in protons of energies less than 1 MeV but in this energy range the protons do not produce high-energy emissions directly, and thus their presence cannot be easily inferred. The energy contained in accelerated electrons, as discussed above, could be very large, but the exact value is model dependent and somewhat uncertain. Nevertheless, the overall energetics of the accelerated particles clearly indicate that particle acceleration must play a fundamental role in the flare process.

The observations place constraints on the acceleration mechanism. For example, the observations of high-energy neutrons and pion decay radiation show that protons are accelerated to hundreds of megaelectron volts in less than 10 s and to several gigaelectron volts in less than 1 min. Fermi acceleration by magnetohydrodynamic turbulence can be rapid enough to account for the observed time scales provided that the turbulent energy density is at least several percent of the ambient magnetic field energy density (which is several hundred ergs per cubic centimeter in the corona).

The observations also place constraints on the magnetic field geometry and on the trapping of the particles. By comparing the number of protons that interact at the Sun to produce nuclear deexcitation line emission with the number observed in interplanetary space with detectors on spacecraft, it is found that the number of interacting protons is generally much larger. This implies that the protons are efficiently trapped and forced to interact at the Sun. Magnetic loops anchored in the photosphere provide a natural trapping geometry. On the other hand, there are flares for which there are more escaping than interacting protons. Here the escaping particles are probably accelerated on open field lines with ready access to interplanetary space.

Information on the structure of the magnetic field in the loops is provided by observations of continuum γ-ray emission at high energies ($\gtrsim$ 10 MeV) where the bremsstrahlung is strongly beamed along the direction of motion of the electrons. Gamma-ray emitting flares at these energies are observed from sites located predominantly near the limb of the Sun. This result can be explained by the motion of an electron in a chromospheric magnetic field whose average direction is normal to the photosphere and whose strength increases with decreasing distance from the photosphere. Such a field can reflect or mirror the charged particles. It has been shown that mirroring produces an anisotropic bremsstrahlung angular distribution, peaking at directions tangential to the photosphere. This distribution can explain the observed uneven distribution of flare locations on the Sun. The implied magnetic field gradient is such that the field increases from about 100 G in the corona to 1000 G in the photosphere.

Solar flare γ-ray observations also provide new techniques for abundance determinations in the solar atmosphere. The fitting of spectra such as shown in Fig. 1 to the data require chromospheric gas abundances that differ from photospheric abundances. The C and O abundances are reduced relative to those of Mg, Si, and Fe. The Ne abundance is not measured in the photosphere, but the chromospheric Ne-to-O ratio is enhanced relative to its value in the corona. The origin of the variations is not understood. The 2.223-MeV line resulting from neutron capture by protons in the photosphere is a probe of the photospheric ^{3}He abundance. Nonradiative capture on ^{3}He competes with capture on H, allowing the determination of the ^{3}He abundance from observations of the time-dependent flux of the 2.223-MeV line. Using this technique, an upper limit, $^3\text{He}/\text{H} < 3.5 \times 10^{-5}$, has been derived from 2.223-MeV line observations. This limit is sufficiently low to be consistent with the ^{3}He abundance expected solely from cosmological nucleosynthesis, suggesting that the contribution of turbulent diffusion from the solar interior, where ^3He is made via deuterium burning, is not very important.

CONCLUDING REMARKS

High-energy observations of the Sun have opened a new window on the study of solar activity. Particle acceleration in the dynamic, magnetized, solar atmosphere plays an essential role in this activity. Accelerated particle interactions in solar flares produce γ-ray lines, γ-ray and hard x-ray continuum emissions, and neutrons. In addition, the accelerated particles escaping from the Sun are directly observed. The observations reveal a picture in which much of the available flare energy is initially contained in accelerated particles. The observations also show that the particles are accelerated impulsively on short time scales, and that the bulk of the particles are confined and forced to interact in magnetic flare loops. In addition, the high energy solar emissions allow the study of the structure of the magnetic fields and of the composition of the solar atmosphere.

Additional Reading

Dennis, B. R. (1988). Solar flare hard x-ray observations. *Solar Phys.* **118** 49.

Forman, M. A., Ramaty, R., and Zweibel, E. G. (1986). The acceleration and propagation of solar flare energetic particles. In *The Physics of the Sun*, vol. II, P. A. Sturrock, ed. D. Reidel, Dordrecht, p. 249.

Kundu, M. R., Woodgate, B., and Schmahl, E. J., eds. (1989). *Energetic Phenomena on the Sun*. Kluwer Academic Publ., Dordrecht.

Ramaty, R. and Murphy, R. J. (1987). Nuclear processes and accelerated particles in solar flares. *Space Sci. Rev.* **45** 213.

Reames, D. V. (1990). Energetic particles from impulsive solar flares. *Astrophys. J. (Suppl.)*, **73** 235.

Rieger, E. (1987). Solar flares: High-energy radiations and particles. *Solar Phys.* **121** 323.

See also **Cosmic Rays, Origin; Stars, High-Energy Photon and Cosmic Ray Sources; Telescopes, Detectors, and Instruments, Gamma Ray.**

Sun, Interior and Evolution

Sabatino Sofia

We cannot readily see the inside of the Sun. Thus, to find out its internal structure (run of density, temperature, chemical composition, etc.) and dynamics (rotation, turbulent motions, etc.) it is necessary to use indirect techniques. Observationally, the means currently available to infer information about the solar interior are the detection of neutrinos and the observation of solar oscillations.

Neutrinos are produced in the central portion of the Sun and travel outward with little or no interaction with the stellar material. This lack of interaction causes the neutrinos to retain information on the conditions of the solar core. However, because they can only be seen when they interact with the detecting instrument, the observation of solar neutrinos is a very difficult undertaking. The results obtained to date have been so unexpected that they have created their own questions rather than elucidating the properties of the solar interior. Because there is currently a massive international effort to tackle this problem, the situation should improve greatly in the coming years.

Solar oscillations reflect the conditions (structure and rotation) of the regions in the Sun where they are excited. Because different oscillation modes are excited at different depths in the Sun, it is possible to deduce the solar internal conditions, layer by layer, by studying the appropriate oscillation modes. In practice, the required accuracy of the observations and number of modes are difficult to achieve, and although significant progress has been made (indeed, these studies have generated a new subdiscipline in solar physics called helioseismology), more definitive answers await the operation of extensive ground- and space-based observational facilities, besides the development of new analytical techniques. Again, great improvements are expected to occur in the coming years.

So far the most solidly founded understanding of the internal structure of the Sun has been obtained theoretically. In this approach, to find out the conditions of the current Sun we must start with the conditions of the Sun when it was first formed, and then compute the changes during its lifetime. This undertaking makes use of the techniques of stellar evolution.

Notice that we know the global properties of the Sun (mass, radius, luminosity, and chemical composition) far more accurately than for any other star. Similarly, we know the solar age (by geological dating of the approximately co-eval objects in the solar system) and the initial chemical composition of the solar nebula (by measuring the composition of pristine comets during their first approaches to the Sun) far better than for other stars. Finally, we know solar properties such as the surface differential rotation (the period of rotation in the equatorial region is shorter than at higher solar latitudes), magnetic (sunspot) activity cycle, and so forth, not directly detectable even for the closest solar neighbors. As these properties are incorporated into our models, these models more realistically represent the real Sun. A great deal of research effort is currently going into developing these increasingly sophisticated solar models. The main features of the Sun, however, can be understood in terms of the simpler models, and we emphasize these in the rest of this entry.

MODELLING THE SUN

To compute a stellar model we can assume that it is in hydrostatic and thermal equilibrium, that mass and energy are conserved, and we must define the efficiency of energy transport, whether radiative or convective, appropriate for the region considered. In order to implement these conditions, we require an equation of state (a relationship between pressure, density, and temperature of the stellar material), the opacity of the stellar material, and means to evaluate energy sources or sinks, whether by nuclear processes, or as a consequence of gravitational readjustments. The most simple solar models are spherically symmetric, which means that such complicating features as rotation and magnetic fields are ignored. Given the appropriate initial and boundary conditions for the problem, the resulting mathematical system is fully defined and can be solved numerically with the use of computers. We will present hereafter the results of these studies, which indicate how the Sun was formed, how it evolved to the present configuration, and its ultimate fate.

Like all other stars, the Sun formed out of a portion of an interstellar cloud that became bound by self-gravitation when a shock wave propagated through the medium. The initial composition of the cloud was primarily hydrogen (over 70% by mass), helium (approximately 25%), and a few percent of assorted heavier elements. As the protosolar cloud contracted, there was a conversion of potential energy into kinetic energy. According to the virial theorem, half of this energy was radiated away and the other half increased the temperature of the cloud. As this process accelerated, the cloud began to shine as a cool, luminous "star." At this stage, which is known as the pre-main sequence (PMS) phase, the proto-Sun was fully convective and thus thoroughly mixed. The PMS stage lasted approximately 10×10^6 yr.

STAGES OF EVOLUTION

The best way to show the stages of evolution is by means of the Hertzsprung–Russell (HR) diagram, in which we plot the surface

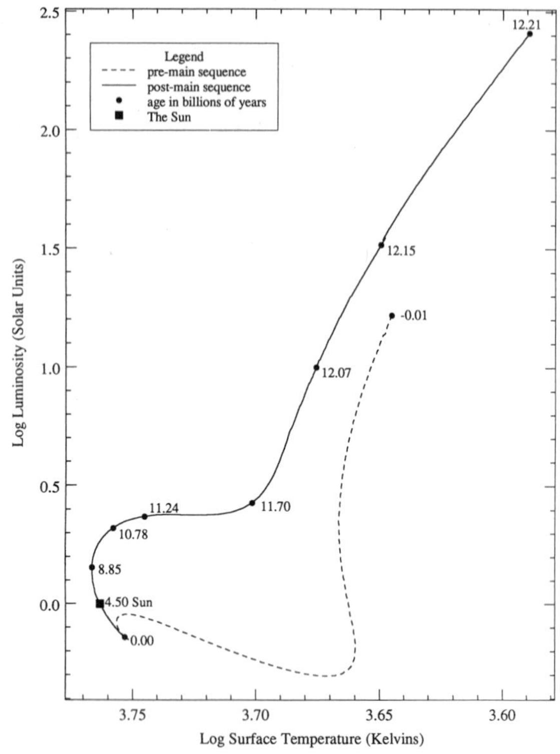

Figure 1. The HR diagram for the Sun. Age 0.00 represents the ZAMS stage, and the square represents the present Sun. The ages are given in billions of years.

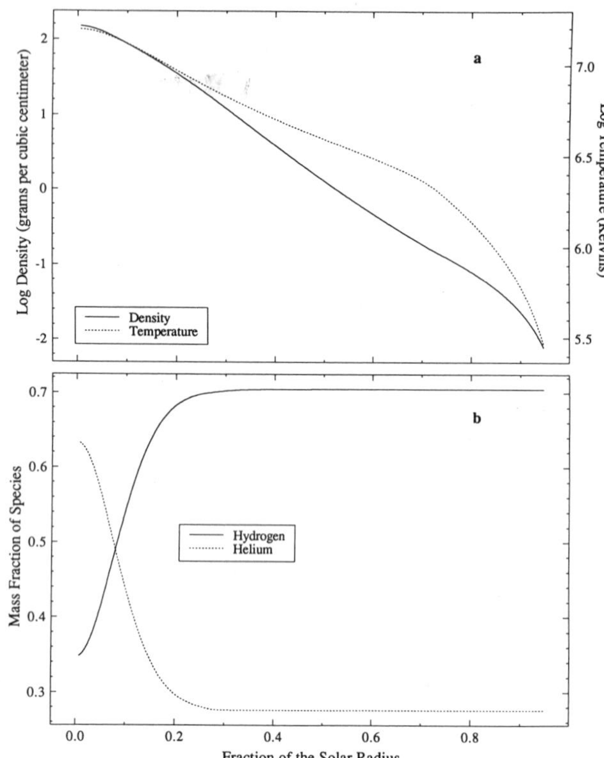

Figure 2. (*a*) Runs of density and temperature of the present Sun as a function of fractional radius. (*b*) Hydrogen and helium abundance of the present Sun as a function of fractional radius. Notice that the efficiency of nuclear processing is highest at the very center of the Sun.

temperature and the luminosity (total rate of energy output) of the object (see Fig. 1). Notice that, following astronomical custom, temperature increases to the left in this diagram.

As the contraction continued, the cloud core continued to warm. Radiative processes began to carry an increasing fraction of the energy flow, until the core material ceased to convect and became fully radiative. Eventually, as the central temperature reached about 12×10^6 K, hydrogen nuclei began fusing to form helium, and in this nuclear process generated vast amounts of energy. This energy provided sufficient pressure support to halt further contraction and to maintain the central temperature despite the energy loss to the surface layers. It was at this point, denoted as the zero-age main sequence (ZAMS), that the Sun became what we can properly call a star.

Because the nuclear process converting hydrogen into helium is extremely efficient, the stage of hydrogen core "burning," which is called the main sequence (MS) phase, is the longest lasting state in the solar lifetime. For the Sun it will last approximately 10×10^9 yr. Presently the Sun has consumed approximately one-half of its core hydrogen fuel and it is located at the point marked with a square in the HR diagram. During the entire MS phase, the Sun slowly expands and its luminosity increases at the rate of about 6% per billion years.

The run of temperature, density, and the hydrogen and helium abundance of the present Sun is shown in Fig. 2.

FATE OF THE SUN

The end of the MS stage will occur when the solar core no longer contains hydrogen. The temperature, by then close to 20×10^6 K, will be insufficient to allow helium burning, which requires about 100×10^6 K. Deprived of the means to resupply the energy lost from the core, the core material will cool down, decreasing the internal pressure support, and so the core will begin again to shrink. The gravitational energy released through the shrinkage will heat the core and supply energy for the flow to continue. The hydrogen-rich material in the shells just outside the core will become sufficiently hot to begin nuclear burning there. This total energy will exceed the ability of the solar envelope to dispose of it to the outside, so the envelope will begin to expand and cool down. The Sun will become a red giant. Over a few million years the core radius will have decreased to about one-fiftieth of the original size, and the central temperature will have increased to about 100×10^6 K. The outer solar envelope will have expanded all the way to the Earth's orbit. By that time the surface temperature will be about 3500 K. When the core reaches 100×10^6 K, helium burning will begin (in an explosive fashion), producing primarily carbon, plus some oxygen. Eventually, all the helium in the core will be used up and it will begin again to contract, and some helium burning will occur in the shells just outside the core. This will cause the Sun to expand again to become a red supergiant. At this point the Sun will be as large as the orbit of Mars. A complex series of flashes and thermal pulses will ensue, causing the ejection of all the material surrounding the core, which will include approximately one-half of the solar mass. The exposed hot core will heat the expanding material, producing a planetary nebula. Meanwhile, the leftover core will settle down as a hot, nuclearly inert, dense object, a white dwarf star. This Earth-sized object, the end result of solar evolution, will slowly cool and darken to forever disappear from the view of our galactic neighbors. This dark cinder and the even darker outer planets that have not been swallowed up by the Sun during its expanding stage, are all that will remain of our solar system.

Additional Reading

Böhm-Vitense, E. (1991). *Introduction to Stellar Astrophysics*. Cambridge University Press, Cambridge, U.K.

Kippenhahn, R. and Weigert, A. (1990). *Stellar Structure and Evolution*. Springer-Verlag, Berlin.
Stix, M. (1989). *The Sun*. Springer-Verlag, Berlin.
See also **Neutrinos, Solar; Stellar Evolution, Low Mass Stars; Sun, Oscillations.**

Sun, Magnetic Field

Herschel B. Snodgrass

The magnetic field of the Sun is not constant. Instead it undergoes an intriguing sequence of changes that repeats every ~22 yr, containing two 11-yr segments of opposite polarity, called activity cycles (or sunspot cycles). Apart from magnetic polarities, the activity cycles are similar to each other; the activity consists of the eruption of sunspots and other manifestations of the interaction between the magnetic field and the solar plasma. It is thought that the driving forces lie beneath the solar surface, where the magnetic field is generated by the large-scale motions of convection and rotation. This process is an unsolved problem of considerable interest to astrophysicists.

At all times the solar surface contains large-scale patterns of weak (~2-G) magnetic field. Sunspots, in contrast, are small regions, a few thousand kilometers in diameter, of intense (250–5000-G) magnetic field that are confined to a zone of latitudes that migrates, during each 11-yr activity cycle, from midlatitudes to the solar equator. Individual spots come and go, lasting from a few hours to a few months, but certain regions become major complexes of activity, with sunspots erupting over and over again at the same longitude. The solar activity is judged by the number and intensity of spots that are present. As one zone reaches the equator, the next is just forming; at this time the activity is at a minimum and the weak field is concentrated at the poles of the Sun, which are of opposite magnetic polarity. This field configuration is not, however, an axial dipole field, because the field strength varies roughly as the eighth power of the cosine of the colatitude.

The polar magnetic fields of the Sun reverse their polarity shortly after the phase of maximum activity, which occurs ~4 yr after minimum. The polarities of the sunspot fields are related to the polarities of the polar fields. Sunspots tend to come in bipolar pairs, with the leading spot, in the sense of the solar rotation, having the same polarity as the premaximum polar field for the hemisphere in which it is located. Thus the sunspot bipoles are oriented oppositely in different hemispheres, and the orientations reverse from one activity cycle to the next.

There is historical evidence for occasional extended periods of time during which the activity cycle seems to go away. During these periods, the Sun remains in a solar minimum-like state, with no evidence of sunspot activity. The most recent such period, known as the Maunder minimum, began in the 1630s and lasted for about 75 years. We have no knowledge of the weak solar fields during such periods.

OBSERVATIONS

The magnetic field on a remote body can be measured if the body emits light containing atomic spectra. By the Zeeman effect, the field splits spectral lines into components with distinct polarizations; splitting is proportional to field strength, and polarization is determined by field direction. Measurement is most feasible for field components that are directed along the line of sight. George Ellery Hale first made such measurements at Mount Wilson Observatory in 1910 to discover the magnetic fields in sunspots, and by 1919 he and his colleagues had found the reversing sunspot polarity pattern, which is known as the Hale–Nicholson law.

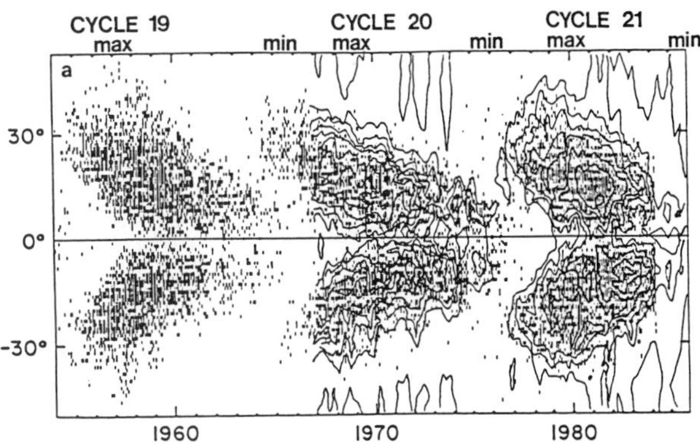

Figure 1. Sunspots and magnetic flux [plot of solar latitude distribution versus time of the averages over solar longitude of sunspots and total (positive plus negative) magnetic flux]. The presence of a dot indicates the presence of a sunspot or sunspot group, and the contour levels for magnetic flux are $3, 6, 9, 12, \ldots \times 10^{21}$ Mx. The cycle number and the times of maximum and minimum sunspot number are indicated. The coincidence of flux and sunspot zones is evident, but note that the flux maximum occurs somewhat after the maximum in sunspot number. All observations were taken at Mount Wilson Observatory; the flux contours begin in 1967, when the systematic magnetograph observation program was started.

The primary device now used for measuring solar magnetic fields is the magnetograph, invented by Horace and Harold Babcock in the 1950s. By passing alternate circular polarizations through a high-resolution ($\pm 10^{-4}$ Å) spectrograph, the magnetograph measures the photospheric magnetic flux $\int \mathbf{B} \cdot d\mathbf{A}$ (where **B** is the magnetic field strength and **A** is the area) through an area projected on the Sun's surface, determined by an entrance aperture at the telescope focus. The total flux is underestimated, both because bipolar field features smaller than this area cancel and because the instrument saturates when the field is too strong. Dividing by the area gives the mean line-of-sight projected magnetic field, which with reference to the Sun is the radial field near disk center and a tangential field (poloidal near the poles; toroidal elsewhere) near the limb.

With their magnetograph the Babcocks discovered the weak polar fields and observed their reversal after the solar maximum of 1957. Since then, several long-term magnetograph observation programs have been underway, mapping the line-of-sight projection of the global solar field. Tracking this projection as the Sun rotates, with its axis tilted by 7° relative to the Earth's orbital plane, allows some resolution of the field's tangential and radial components. Direct measurement of the full vector magnetic field is presently feasible only for the strong fields in and around active regions: A few vector magnetographs are in operation and others are being developed. These are mainly used in the study of solar flares.

High-resolution magnetograph observations reveal that the magnetic flux at the solar surface is not smoothly distributed, but instead forms a network of small bundles with strengths of ~1000 G. This may result from the field being swept into the downdrafts of the various tiers of convection cells. The total (positive plus negative) flux for the whole of the Sun's surface is $\sim 10^{23}$ Mx, which varies by a factor of ~3 over the cycle and is concentrated (99%) in the sunspot-producing zones (Fig. 1). Flux comes and goes at a rate of about 10% per day, and it is estimated that 90% of the surface magnetic flux disappears in situ, whether by submergence, evaporation, or cancellation, rather than by being transported away from its point of emergence. But poleward of active regions patches of flux are observed that appear to have broken off and are drifting poleward and backward like smoke from the stack of a moving ship.

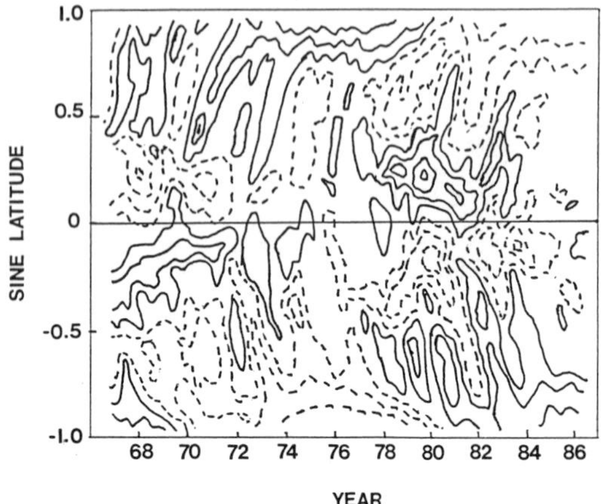

Figure 2. Latitude distribution of solar magnetic fields (averages over longitude versus time). Solid contours represent positive magnetic polarity; dashed contours negative. Contours are ± 0.5, 1, 2, and 4 G. This figure shows the transitions between the background axial dipole at solar minimum (1975–6 and 1986) and more complicated field patterns. The poleward motions of unipolar field zones, leading to the polarity reversal of the background field are evident; the clearest indication of a triple reversal is seen in the northern hemisphere during 1967–73.

When the magnetic field is averaged over all longitudes (Fig. 2), the equatorward edge of the sunspot zone is seen to be dominated by the polarity of the leader spots (the background field polarity before reversal). The poleward side of the sunspot zone has the opposite, or follower-spot polarity; the neutral line between this and the polar field zone is called the polar crown. Around solar maximum, the zone of follower-spot polarity suddenly begins to move poleward, as if carried by a meridional flow. This appears to cancel, and ultimately reverse, the polar fields. In some cycles, as seen in Fig. 2, several streams with alternating polarity move poleward, causing the polar field to oscillate before settling down to its reversed state.

The field structure above the photosphere may be seen with special filters (for Hα and the Ca K-line). These reveal the Sun's chromosphere and structures such as filaments and prominences with which one can map the neutral lines between unipolar regions of opposite polarity. One also sees dynamical phenomena such as the surges and flares associated with changes and reconnections in the field. The large-scale patterns in the field above the surface are seen in coronagraph, eclipse, and x-ray observations of the solar corona. Of special interest are the dark regions called coronal holes, which are cooler than adjacent coronal regions and lie above large unipolar areas where material flows out into the solar wind. Here the field lines close only at great distances from the Sun.

MODELS OF THE SOLAR DYNAMO

The Maxwell condition Div $\mathbf{B} = 0$ (magnetic field lines form closed loops) means that fields observed on the solar surface must be connected both above and below the surface. The subsurface field is believed to be organized into flux tubes, which are buoyant and exert tension along their length. Because a magnetic field is strongly anchored to a conducting fluid, the flux tubes are subject to the rotational and convective motions of the solar plasma. Bipolar sunspot pairs occur where the tubes loop up through the surface.

It is widely believed that the Sun's magnetic field is generated by dynamo action either in the convection zone or in an overshoot layer just beneath it. A dynamo is a mechanism that expends mechanical energy to intensify a magnetic field; in the Sun it consists of plasma motions that organize and intensify the field into flux tubes by stretching, twisting, and wrapping. In 1953, just after the weak solar-minimum polar fields were discovered, Thomas G. Cowling proposed that they were connected by poloidal field lines lying just below the solar surface, and that the Sun's differential rotation, in which the equatorial rotation rate is greater than that at the poles, acts to turn and intensify the poloidal field into a toroidal field from which the active regions arise. This mechanism is now called an omega dynamo. Eugene N. Parker then showed that the poloidal field could be maintained by the "cyclonic convection" of rising flux from the toroidal field. This process later became known as the alpha dynamo.

After observing the polar reversal, Horace Babcock proposed that the solar cycle combines Cowling's omega dynamo with a poleward drift of follower spots to cancel and reverse the polar fields, and the cancellation of leader polarities across the equator. Robert B. Leighton then modeled the surface wanderings of the flux tubes as a diffusion process, but this was contradicted by observations of the patterns of field transport (Fig. 2). Recently it has been found possible to match the observed patterns of drift, polar reversal, and equatorial cancellation by combining diffusion with meridional flow.

This picture is, however, oversimplified at best. The Cowling omega dynamo theory predicts correctly that sunspot activity should begin at midlatitudes but then concludes incorrectly that it should spread toward both the equator and the poles, and unless the resulting toroidal field is both deep and weak (< 200 G) or held down in some way, magnetic buoyancy should rapidly float it away. A similar objection has been raised to the diffusion plus meridional flow mechanism for polar reversal: If the poleward-moving field elements remain attached to the toroidal field, then magnetic tension should quickly stop them. If they detach, they should float away before they get very far.

Mathematical approaches to modeling the solar dynamo have thus far met with little success. It has been shown that convection coupled with rotation can generate a magnetic field and maintain it against dissipation, but no one has yet modeled a field with (1) the correct global properties and cycle of changes that is (2) consistent with the observed patterns of solar rotation and flow. Dynamo models are of two types: One type is called self-consistent because it yields both the differential rotation and the magnetic fields from parametrized first principles. These models predict a pattern of giant convective cells encircling the Sun at the equator like a cartridge belt, but observations have yielded no trace of such cells. They also predict a variation with depth of the differential rotation that is contradicted by observations of solar oscillations. An alternative presently under study is that the dynamo lies in an overshoot layer beneath the convection zone.

A second type of dynamo model, called kinematic, assumes forms for differential rotation and convection, and derives the magnetic field. Although none of these models has succeeded in reproducing all aspects of the cycle, they are rich in predictions. For example, with the nonlinear feedback of a Lorentz-force modification of the differential rotation, they are able to reproduce the extended solar minima.

Some researchers have suggested that the Sun's magnetic field does not arise in a turbulent dynamo process, arguing, for example, that the cycle timing is too precise. One alternative put forth is that the surface fields are bits of a primordial field expelled from the Sun's core at regular intervals by an unknown mechanism. The timing argument was based in part on patterns found in Australian varves (ancient lake deposits), which were thought to be correlated with the solar cycle. But because these varves are now believed to be of tidal origin, the argument may no longer be viable.

CONCEPT OF THE EXTENDED CYCLE

Another alternative, suggested by M. Trellis and M. Waldmeier in 1957 on the basis of coronal and solar wind observations, is that the solar cycle is a zonal phenomenon, in which toroidal zones of activity are created at the poles at ~11-yr intervals, around solar maximum, and migrate to the equator during the course of ~18 yr. This implies that during the period from solar maximum to the following minimum two such zones are simultaneously present in each hemisphere. As a zone migrates, its activity increases, changing from high-latitude coronal emission during the first ~7 yr to sunspot activity as it reaches lower latitudes.

The extended cycle idea was overshadowed by the Babcock picture, which conveyed the notion of a global solar involvement in distinct 11-yr, opposite-polarity phases. That notion has guided much theoretical work, but it is neither an essential part of the Babcock picture nor compelled by the observations. The extended cycle would have an omega dynamo generate bands of toroidal magnetic field with alternating polarities beneath successive activity zones, beginning at the time of the polar reversal. In the 1950s, Parker proposed a "dynamo wave" mechanism that could cause such field bands to propagate.

In 1980 a zone of enhanced rotational shear, called the torsional oscillation, was discovered. The shear zone starts at high latitudes and migrates to the equator in ~18 yr, and during the last 11 yr of this migration it coincides with the sunspot zone. This argues strongly for the extended cycle and leads to speculation on its mechanism: It has been proposed that the shear is a surface signature of the downflow zone between deep-lying toroidal convective rolls. This mode of convection may be favored near the poles or in the presence of strong toroidal fields. The enhanced shear is then a Coriolis effect; deep in the downflow region it would act as the omega dynamo for field bands while the downflow countered their buoyancy. Such bands would produce a "thermal shadow," which would in turn accentuate the downflow, and migration of the bands via Parker's dynamo wave would thus tend to drag the rolls along. Activity complexes would appear as the rolls broke up under field and Coriolis stresses at low latitudes.

An alternative picture of the extended cycle is one in which all migrations are poleward, as is suggested in Fig. 2. In this picture the cycle begins at solar minimum with the sunspot fields. As it progresses sunspots erupt at successively lower latitudes, which makes the activity zone appear to migrate. The polar field stems from poleward migration of the follower flux zones following solar maximum, and the polar fields persist until they are cancelled by flux-zone migrations following the next maximum. Proponents of this idea assert that the torsional oscillation could be a Lorentz-force effect of the active-region fields, and that its high latitude portion is as yet ambiguous.

Extended cycle models are just beginning to be investigated, and there is no consensus presently regarding their chances for success. But whatever the models of the future may be, the emphasis may have to shift away from approaches based on traditional deterministic dynamics. Because the magnetohydrodynamic equations are nonlinear and exceedingly complex, the system is surely chaotic. Chaos theory, a relatively new discipline, predicts that direct attempts at solution of equations of this sort, which require parametrization and numerical integration, cannot derive the actual behavior of the system. Thus the Sun itself, for now, may have to be our best teacher. But the elegance of the magnetic cycle, alternating occasionally with the extended minima, suggests that there may be two or more quasistable states for which there are entities known in chaos theory as attractors. Although for the present, the methods of this new discipline are far too primitive for a system as complex as the Sun, at some time in the future, solar research may focus on identifying these attractors, on studying their structure, and on searching for others.

Additional Reading

Noyes, R. W. (1982). *The Sun, Our Star*. Harvard University Press, Cambridge.

Parker, E. N. (1983). Magnetic fields in the cosmos. *Scientific American* **249** (No. 2) 44.

Parker, E. N. (1987). The dynamo dilemma. *Solar Physics* **110** 11.

Robinson, L. J. (1987). The sunspot cycle: Tip of the iceberg. *Sky and Telescope* **73** 589.

Wentzel, D. G. (1989). *The Restless Sun*. Smithsonian Institution Press, Washington, D.C.

Zirin, H. (1988). *Astrophysics of the Sun*. Cambridge University Press, Cambridge, U.K.

See also **Magnetohydrodynamics, Astrophysical; Solar Activity; Solar Magnetographs; Stars, Magnetism, Observed Properties.**

Sun, Oscillations

Edward J. Rhodes, Jr.

Oscillatory motions in the solar photosphere and chromosphere were first predicted to exist on purely theoretical grounds as early as 1948. More than a decade after the first of these theoretical predictions was made, the first evidence for the existence of these oscillations was obtained at the Mount Wilson Observatory during the summers of 1960 and 1961. These pioneering observations, which were made by Robert B. Leighton, Robert W. Noyes, and George W. Simon, who were then at the California Institute of Technology, showed that the solar photosphere and chromosphere are constantly moving inward and outward in a nearly radial motion, which has a period of roughly 5 min. The existence of these so-called 5-min oscillations was quickly confirmed by other observers. At the present time these 5-min oscillations have been studied so extensively that we now know them to be the surface manifestations of radial and nonradial acoustic (and possibly also gravity) modes that are trapped within the solar interior. This knowledge that the Sun is actually a very-small-amplitude variable star has opened up an exciting new window into the previously invisible solar interior: the study of helioseismology.

Helioseismology is a rapidly growing subfield of solar astronomy that combines ground- and spaced-based observations of the solar surface with computational techniques similar to those employed by geophysicists in their study of the Earth's interior to probe the thermodynamic structure and gas dynamics of the solar interior as functions of radius, latitude, and time.

HISTORICAL DEVELOPMENT OF HELIOSEISMOLOGY

The discovery of the existence of the solar 5-min oscillations did not immediately result in the establishment of helioseismology. In fact, for nearly a decade after their discovery, these oscillations were initially viewed as simply the response of the solar atmosphere to the turbulent overshooting from the solar convection zone below. Then, in 1970, the suggestion that these oscillations might have an origin deeper in the solar interior was first made by Roger K. Ulrich at the University of California, Los Angeles. Ulrich suggested that the 5-min oscillations are in reality acoustic waves that are trapped beneath the solar photosphere.

Ulrich's explanation of the 5-min oscillations as the surface manifestations of processes occurring deep beneath the photosphere was not immediately accepted. However, in a breakthrough paper Franz-Ludwig Deubner of West Germany obtained the first observational evidence for Ulrich's trapped acoustic modes. As had been the case with the original discovery of the oscillations themselves in 1960, Deubner's observations were quickly confirmed, in 1977 by the author, Ulrich, and Simon. In the first use of the 5-min oscillations as a tool for helioseismology, they inferred that the convection zone of the Sun was considerably deeper than was generally accepted at the time.

The observations mentioned thus far all consisted of time series of one- or two-dimensional images of the solar velocity fields. However, beginning in the mid-1970s observations of solar oscillation modes were also obtained by two different groups who employed instruments that provided no spatial resolution of the solar disk at all. Two of these groups (the first led by George R. Isaak and H. B. van der Raay of the University of Birmingham and the second led by Eric Fossat and Gerard Grec of the Observatory of Nice) used gas-cell resonance spectrometers in which the solar light was passed through small cells that contained heated potassium or sodium vapor clouds. Both of these instruments combined the light from the entire visible solar disk and hence observed the Sun as if it were a star. The first observations of the radial and nearly radial solar 5-min oscillations were made with one of these resonance cell instruments. Eventually, both groups obtained power spectra having such improved frequency resolution that Douglas Gough of Cambridge University and Jorgen Christensen-Dalsgaard of the University of Aarhus were able to unambiguously identify individual radial, dipole, quadrupole, and octopole acoustic modes that extended over a range of periods centered about 5 min. Identifications of additional low-degree acoustic modes were made by a group at the Stanford Solar Observatory headed by Philip H. Scherrer.

In addition to providing a probe of the thermodynamic properties of the solar interior through analysis of the frequencies of the various oscillation modes, helioseismology can also provide a probe of the rotational state of the solar interior. In 1977, the author demonstrated that the differences in the observed frequencies of the solar oscillation modes traveling in opposite directions around the solar equator could be employed to obtain a measurement of the Sun's surface rotation rate. In 1979, Deubner, Ulrich, and the author extended this idea to show that measurements of the frequency splittings of modes having different horizontal and radial structure could be used to provide a depth-dependent probe of the solar internal angular velocity. This idea was later exploited by John W. Harvey and Thomas L. Duvall, Jr. with observations they obtained at Kitt Peak National Observatory.

In addition to the use of solar Doppler shifts, solar oscillations have been observed with time series of:

1. Spatially resolved and whole-disk intensity measurements.
2. Measurements of the total solar irradiance that have been observed from above the Earth's atmosphere from instruments onboard such spacecraft as the *Solar Maximum Mission*.
3. Measurements of the limb-darkening function at a few selected angles around the solar image.

NATURE OF THE OSCILLATIONS

Oscillatory wave motions may be created within a gravitationally stratified, compressible gas such as that in the solar interior in response to perturbations by the forces of gas pressure gradients and gravity. Gradients in the gas pressure give rise to acoustic or pressure waves, also known as p modes. The action of gravity gives rise to buoyancy effects that become manifest as convective motions in the solar convection zone or as gravity waves, called g modes, where the solar gas is locally stable against convection. To date only the p modes have been observed unambiguously, although numerous recent studies have claimed to observe the surface effects of low-degree internal g modes.

In the gaseous interior of the Sun there are no solid walls in which its acoustic modes are confined. Nevertheless, the strong radial variations in the temperature and density of the solar gas allow acoustic cavities of many different effective lengths to be present simultaneously in the solar interior. Each of these cavities possesses both an inner and an outer reflecting boundary at which the direction of travel of the waves is reversed. Just below the solar surface layers the gas density increases rapidly over a short radial distance. This rapid density increase reflects outward traveling sound waves back into the deeper layers of the solar interior. The reflected sound waves then are traveling toward the center of the Sun where the gas temperature and density are both much higher than they are near the surface. This inward increase in the gas temperature makes the speed of sound in the interior also increase toward the Sun's center. In the case of all of the nonradial oscillation modes, the acoustic wavefronts are inclined in such a way that a part of each wave is located closer to the center of the Sun than is the rest of that wave. Because the sound speed is increasing inwardly, this innermost portion of the wavefront travels slightly faster than does the rest of the wavefront. This difference in speed causes the wavefront to be refracted away from its original inward direction. Eventually, the wavefront is refracted through an angle of 90°. At this point the wave can travel no farther into the Sun. The wave begins traveling back outward toward the outer reflecting layer.

The radial distance inside the Sun where an inward-moving wave is turned back outward is said to be the inner reflecting boundary of the acoustic cavity. The depth at which this inner reflecting boundary is located is different for different solar p modes. The depth of penetration of a given solar acoustic mode is related directly to its horizontal length scale at the solar surface. Because modes of vastly different horizontal scales are all present within the Sun at any given time, the Sun may be viewed as possessing acoustic cavities of many different lengths.

Within the Sun the existence of both inward- and outward-propagating waves results in standing waves within the acoustic cavities. The frequency of the standing waves depends in part upon the radial extent of the cavity in which they resonate. Therefore, the existence of acoustic cavities of various lengths within the Sun means that there are many different frequencies excited in the Sun at any one time.

A single solar acoustic mode is illustrated schematically in Fig. 1. In this computer-generated cutaway drawing of the Sun, both the surface structure and radial variations of this mode are illustrated. The motions or intensity fluctuations that would be visible at the solar surface are depicted as the alternating light and dark regions. In the case of instruments that would measure the line-of-sight Doppler shifts of the surface layers, the light regions would represent portions of the surface that would be moving inward at this moment in time. The dark regions would correspond to portions of the surface that were moving outward at the same moment. In another computer image made roughly $2\frac{1}{2}$ min after this one, the light regions would have turned dark and vice versa. After an additional $2\frac{1}{2}$ min the same pattern shown in Fig. 1 would have returned.

The surface pattern shown in Fig. 1 is just one example of a mathematical surface called a spherical harmonic. Two quantities are enough to uniquely describe every spherical harmonic function that can possibly exist on the Sun. These are called the degree l and the azimuthal order m.

The degree corresponds to the number of nodal lines in the pattern. That is, it represents the total number of circles in the surface at which the amplitude of the pattern is equal to zero. In the case of the radial modes the entire surface of the Sun will be moving with the same phase at any given moment of time. Hence, there are no nodal circles on the solar surface and $l = 0$.

All modes for which the degree is not equal to zero are referred to as nonradial modes. Low-degree modes are those having few nodal circles and large horizontal scale lengths. The lowest degree modes are the most nearly radial of the nonradial modes. For high-degree modes there are many, closely spaced nodal circles on the surface and the horizontal scale length of a given mode is small.

The azimuthal order is equal to the number of nodal circles that cross the solar equator. Nonradial modes for which $m = 0$ have no nodal circles that intersect the equator. Their nodal circles run parallel to the equator. These modes are referred to as zonal harmonics. When the degree and the azimuthal order are both equal, the nodal circles all lie at right angles to the equator. These modes are called sectoral harmonics. All the other nonradial modes in which $l \neq m$ are called tesseral harmonics.

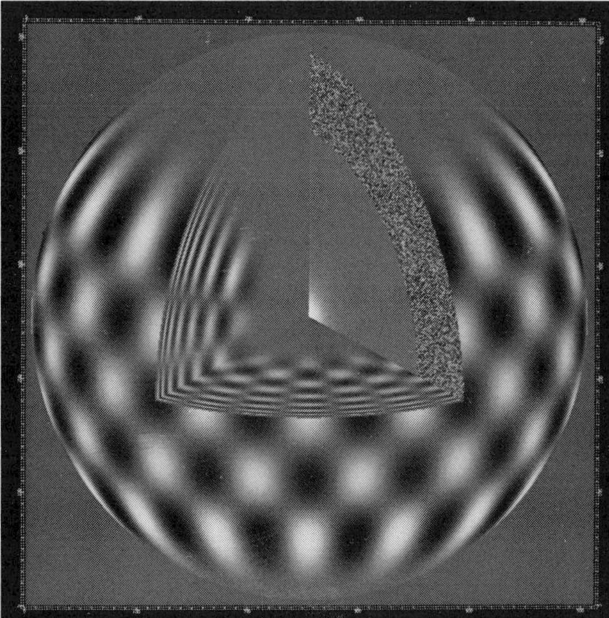

Figure 1. A computer-generated simulation of the three-dimensional structure of a single nonradial acoustic oscillation mode. The light-colored regions on the surface and along the two panels of the cutaway portion of the Sun are regions where the solar material is momentarily moving inward, away from the Earth. In the dark-colored regions the gas is moving outward at that same moment. After $2\frac{1}{2}$ min, all of the outward-moving regions would be moving inward and vice versa. The three portions of the solar interior are shown along the right-hand side of the cutaway. The convection zone is the stippled region beneath the surface. The radiative zone is beneath the convection zone and the energy-generating core is at the very center. (*Figure courtesy of John W. Harvey and John Leibacher, National Solar Observatory.*)

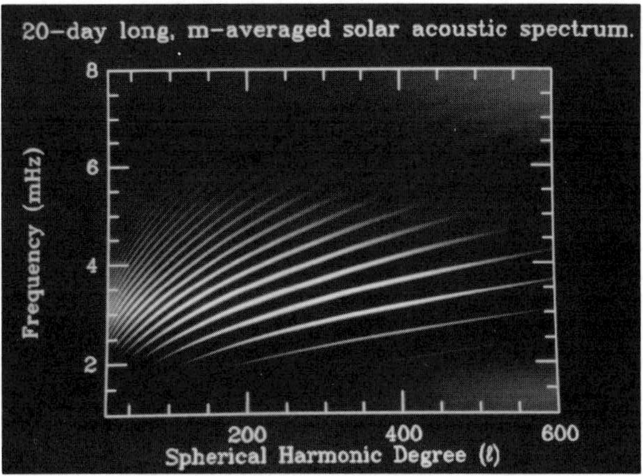

Figure 2. Two-dimensional (l, ν) velocity power spectrum of solar p-mode oscillations as obtained from observations made with a sodium magnetooptical filter over 20 consecutive days in July, 1988, at the 60-ft Solar Tower Telescope at the Mount Wilson Observatory. The degree l, of the zonal ($m = 0$), spherical harmonic surface patterns increases horizontally from 0 (for the radial modes) at the left to 600 at the right. The frequency axis increases upward. Frequencies corresponding to a period near 5 min are near 3.3 mHz, or where the observed ridges are the darkest. The solar p-mode ridges are obvious here from low degrees up through $l = 600$. The ridges have been shown to exist at higher degrees in observations having finer spatial resolution. Each p-mode ridge corresponds to a particular overtone number n. The higher ridges contain more nodes in their radial behavior than do the lower ridges. Solar internal gravity modes, if they become visible at the solar surface, will be below the p-mode ridges and be concentrated at very low degrees (i.e., near the left-hand axis).

The tesseral pattern shown in Fig. 1 represents a spatial standing wave. All such waves may also be thought of as a superposition of two oppositely directed traveling waves. In the case of the sectoral harmonics the two traveling waves of a given degree are moving in opposite directions around the solar equator. One mode is propagating in the direction of solar rotation and the second is traveling in the retrograde direction. In the case of tesseral harmonics such as that shown in Fig. 1 the two modes of a given degree are propagating at oblique angles relative to the equator.

The radial behavior of the particular mode shown in Fig. 1 is shown on the left and bottom panels of the cutaway into the solar interior. These cutaway panels illustrate the additional nodal surfaces in the solar interior that have the shape of concentric spheres. These nodal surfaces represent radial points at which the amplitude of this harmonic mode is always quite close to zero. The number of nodes between the inner and outer reflecting boundaries for a given mode is called its radial order, or overtone number n. For each unique combination of l and m, such as the one shown here, there is a series of modes each of which has a unique radial order. The lowest frequency is the fundamental mode and all of the overtone modes have higher frequencies. Each oscillating mode is then uniquely specified by the combination l, m, and n. Each of these combinations of l, m, and n corresponds to a unique frequency. The actual surface pattern that is observed at any moment in time is a superposition of many million different spherical harmonic modes, each oscillating with a unique frequency and spatial scale.

An example of the dependence of the frequencies of the solar p modes upon the degree of the modes and upon their overtone number is shown in Fig. 2. This is a two-dimensional power spectrum. Each different "ridge" of observed power corresponds to the set of zonal p modes of a given overtone number for degrees between 0 and 600.

Because of the relationship between the horizontal scale length of a mode and the distance it will travel into the Sun before it is reflected, the lower-degree modes tend to penetrate more deeply into the Sun than do the higher-degree modes. For modes of a given degree the depth of the cavity increases with increasing radial order. By contrast, the radial modes penetrate all the way to the geometric center of the Sun. The internal g modes, should be primarily confined to the deep solar interior where few of the p modes penetrate. Thus, should they be detected, these g modes will provide information about the solar interior that is complementary to that provided by the p modes.

A third class of modes is the torsional oscillation, in which gas motion is horizontal rather than radial. Torsional oscillations have been observed on the Sun, but thus far they have not been employed in any helioseismological studies. The restoring force for these modes is the Sun's toroidal magnetic field and hence these modes will be useful in constraining future dynamo models.

RECENT RESULTS

One of the principal recent results from helioseismology is the realization that our best current "standard" solar models do not predict oscillation frequencies that match the observed frequencies to the accuracy of the observations. This discrepancy is leading theoreticians to try many different refinements to the solar models. One such refinement has been a decrease in the central temperature of the models, which lowers their predicted solar neutrino fluxes. A second refinement has been the inclusion of so-called weakly interacting massive particles (or WIMPs) into the solar interior as a possible vehicle for such a lowered central temperature. Thus far, no single revised solar model has been able to "solve" the neutrino problem, include WIMPs, and match all of the observed oscillation frequencies. Other refinements currently being explored include changes to the equation of state of the solar gas and to the treatment of convection in models of the outer portion of the solar interior.

The second principal result has been the measurement of the Sun's internal angular velocity over the outer half of its radius. The picture that is emerging is that the Sun maintains the latitudinal differential rotation that is seen at its surface (in which its equator rotates faster than the higher latitudes) throughout its convection zone, but then appears to be rotating more like a solid sphere in its radiative interior beneath the convection zone. Such a rotational profile, if verified, will place severe constraints on models of the dynamo that is thought to be driving the solar activity cycle.

An additional topic of much current research concerns the method of excitation of p modes. Although the earliest theoretical work indicated that these modes might be due to the same radiative process that drives the large nonradial oscillations seen in high-amplitude variable stars, more recent theoretical and observational work has concentrated on the possibility that individual modes are stochastically excited by the motion of convective eddies in the convection zone.

Additional Reading

Bartusiak, M. (1990). Seeing into the Sun. *Mosaic* **21** (No. 1) 23.

Brown, T. M., Mihalas, B. W., and Rhodes, E. J., Jr. (1986). Solar waves and oscillations. In *Physics of the Sun*, vol. I, P. A. Sturrock, ed. D. Reidel, Dordrecht, p. 177.

Deubner, F.-L. and Gough, D. O. (1984). Helioseismology: Oscillations as a diagnostic of the solar interior. *Ann. Rev. Astron. Ap.* **22** 593.

Leibacher, J. W., Noyes, R. W., Toomre, J., and Ulrich, R. K. (1985). Helioseismology. *Scientific American* **253** (No. 3) 48.

Libbrecht, K. G. (1988). Solar and stellar seismology. *Space Sci. Rev.* **47** 275.

Rolfe, E. J., ed. (1988). *Seismology of the Sun and Sun-Like Stars*. European Space Agency Report SP-286, Paris.

Toomre, J. (1986). Properties of solar oscillations. In *Seismology of the Sun and the Distant Stars*, D. O. Gough, ed. D. Reidel, Dordrecht, p. 1.

See also **Stars, Pulsating Theory; Sun, Interior and Evolution; Sun, Magnetic Field; Sun, Solar Constant.**

Sun, Radio Emissions

George A. Dulk

Radio waves from the Sun were discovered in 1943 when they were recorded as interference on radar receivers. But much earlier it was anticipated that the Sun should be a source of radio waves. For example, in 1880 Thomas A. Edison attempted to record them with a crude receiver attached to a coil of wire wrapped around a mass of iron ore. He was unsuccessful, partly because low-frequency waves are cut off by the ionosphere, whose existence was then unknown. This was only seven years after James C. Maxwell had predicted the existence of radio waves and only three years after Heinrich Hertz had discovered them in his laboratory. Today, radio emissions from the Sun have been recorded at all wavelengths from 1 mm–30 km, corresponding to frequencies from 300 GHz–10 kHz and source distances from the Sun's surface of a few thousand kilometers to 1 AU. Because of the ionosphere, receivers in space must be used to record waves of frequency less than 10–20 MHz.

Radio waves originate from the quiet Sun, from the strong magnetic fields of active regions, and from solar flares. Most radiation of the quiet Sun and active regions is due to the high temperature of the solar atmosphere, that is, to electron-ion *bremsstrahlung*. In regions of strong magnetic field, especially near sunspots, the hot coronal electrons spiral around the lines of force and emit *gyroresonance radiation* at low harmonics of the cyclotron frequency.

At times of most solar flares, the electrons are accelerated to high energies and several other processes become important.

1. Electrons of energy greater than about 100 keV emit at frequencies of 10–100 times the gyrofrequency, giving *gyrosynchrotron radiation*. (This term often refers to magneto-bremsstrahlung in general, including gyroresonance radiation at the nonrelativistic limit and synchrotron radiation at the relativistic limit.)
2. In some plages and near some sunspots in the low corona, the magnetic field is strong enough so that the gyrofrequency $\nu_B = eB/2\pi m_e c \approx 2.8 \times 10^6 B$ is higher than the plasma frequency $\nu_p = (n_e e^2/\pi m_e)^{1/2} \approx 9000\sqrt{n_e}$. (Here, e and m_e are the charge and mass of the electron, respectively, B is the magnetic field strength, n_e is the number density of electrons, and c is the speed of light.) At those locations, whenever the electron velocity distribution is anisotropic, waves at the gyrofrequency and at frequencies perhaps twice as high can be amplified to give *cyclotron maser radiation*.
3. There are other locations on the Sun where the magnetic field is less strong so that $\nu_B \lesssim \nu_p$. There, when the electron velocity distribution is anisotropic, electrostatic (Langmuir) waves near the plasma frequency are amplified and some of their energy is converted to radio waves, producing *plasma radiation*.

Two of the emission mechanisms described above, bremsstrahlung and gyrosynchrotron, are incoherent in the sense that the electrons emit individually. The other two, cyclotron maser radiation and plasma radiation, are coherent in the sense that radiation is amplified to levels far higher than is possible if the electrons emit individually. Resonances between electrons and waves lead to a direct and efficient transfer of the free energy from an unstable electron distribution into wave energy.

In general, the lower the radio frequency, the farther from the surface of the Sun the radiation originates. This happens because emission at a given frequency ν can arise only from regions where $\nu_p \le \nu$, and because n_e and ν_p decrease with altitude in the solar atmosphere. Several kinds of radio bursts originate at frequencies near the plasma frequency or its harmonic, coming from a thin layer above the *plasma level* where $\nu_p = \nu$. The principal exceptions are bursts at frequencies above about 3 GHz that arise in the corona whereas their plasma levels are in the chromosphere or in the chromosphere-corona transition region. Often there are disturbances that travel through the atmosphere, creating radiation at the local plasma frequency, and producing a radio burst in which the emission frequency descends more or less rapidly from high to low, the rate depending on the speed of the disturbance and on whether the radiation is emitted at the fundamental or the harmonic of the plasma frequency.

RADIO EMISSION OF THE QUIET SUN

The appearance of the quiet Sun in the radio region depends on the frequency of observation. At frequencies higher than about 10 GHz, it resembles the Sun at visible wavelengths: a disk of brightness corresponding to a temperature of about 15,000 K, that is, *brightness temperature* $T_b \approx 15{,}000$ K (compared to $T_b \approx 6{,}000$ K in visible light), on which are superimposed slightly brighter areas that manifest active regions. At lower frequencies, the brightness of the disk increases slowly and that of active regions increases

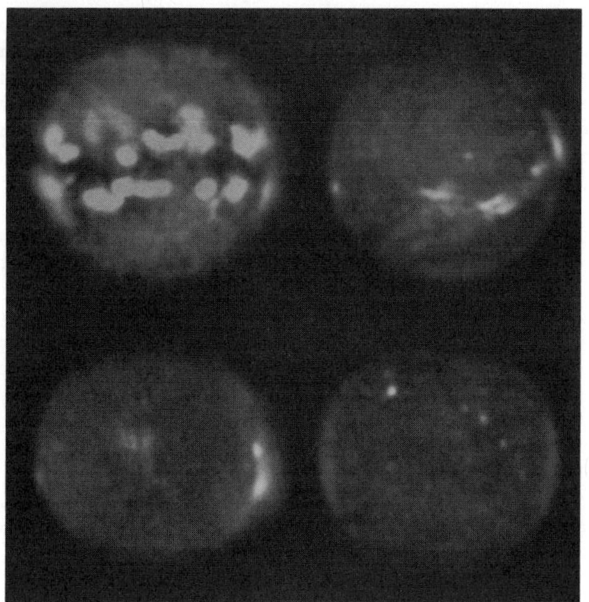

Figure 1. Images of the Sun at 1.4 GHz recorded by the Very Large Array, supplemented in some cases by the Arecibo or Green Bank telescopes: (*a*) 1981 September 26 (upper left, near sunspot maximum), (*b*) 1984 August 16 (upper right), (*c*) 1985 July 14 (lower left), (*d*) 1986 January 24 (lower right, near sunspot minimum). The images were processed using maximum entropy method reconstruction techniques by Dr. Timothy S. Bastian.

rapidly. At $\nu \approx 1$ GHz, the disk has $T_b \approx 10^5$ K, active regions have $T_b \approx 1-2\times10^6$ K, and there are dark regions, coronal holes, of $T_b \approx 5\times10^4$ K. At still lower frequencies, the brightness of the disk continues to increase while that of active regions becomes slightly lower. At $\nu \lesssim 100$ MHz, the average brightness is $T_b \approx 1\times10^6$ K, the active regions are usually not evident, coronal holes may or may not be visible, and the solar disk is larger by about 50% than that in visible light.

Figure 1 shows the appearance of the quite Sun at 1.4 GHz on four days between September 1981, when there were many sunspots and active regions, and January 1986, near sunspot minimum when there was hardly any activity. The images resemble those recorded in soft x-rays which also exhibit bright active regions and dark coronal holes. Fig. 1*a*, corresponding to a date near the maximum of the sunspot cycle, has two bands of activity parallel to the solar equator. The brightest areas have $T_b \approx 2.2\times10^6$ K. Away from active regions there are unresolved structures, possibly magnetic loops, with $T_b \approx 10^5$ K. As the sunspot cycle proceeds toward minimum (Fig. 1*b*–*d*), the number of active regions diminishes, the vestiges of activity are less bright, the average brightness of the disk diminishes considerably (to $\approx 6\times10^4$ K), and coronal holes often appear, especially close to the poles. The long, narrow, dark regions, especially evident in Fig. 1*d*, correspond to low-density coronal regions surrounding filaments.

All of the radiation of the quiet Sun and active regions in Fig. 1 probably arises from bremsstrahlung, none from gyroresonance emission. At $\nu \lesssim 2$ GHz, it seems that the magnetic field of active regions is able to contain enough electrons and ions that the region is optically thick, $\tau_\nu \gg 1$, due to bremsstrahlung. But at $\nu \gtrsim 2$ GHz, where $\tau_\nu \lesssim 1$, the low corona with its localized, strong fields is visible, and small, bright sources due to gyroresonance emission are seen near sunspots and other locations of strong magnetic field.

RADIO BURSTS ASSOCIATED WITH SOLAR FLARES

Flares are the source of large numbers of energetic particles and many kinds of waves. Figure 2 illustrates the radio bursts that are frequently created by a large flare. Not all of these bursts are produced by all flares. For example, there are many simple flares that last for only a few minutes and produce only the bursts of the *impulsive phase*, that is, the microwave impulsive bursts and those of types III and V.

Microwave Bursts

At centimeter wavelengths ($\nu \gtrsim 3$ GHz), often called microwaves, radio bursts arise in the low corona from regions where most of the magnetic field structures are closed, that is, in the form of magnetic loops. Only a few lines of force are open, along which fast electrons and ions can traverse the corona and enter into interplanetary space. There is a very close association between microwave bursts and those recorded in hard x-rays: Not only is the microwave flux approximately proportional to hard x-ray flux at $\gtrsim 50$ keV, but also the temporal variations are nearly identical, even on scales of less than one second.

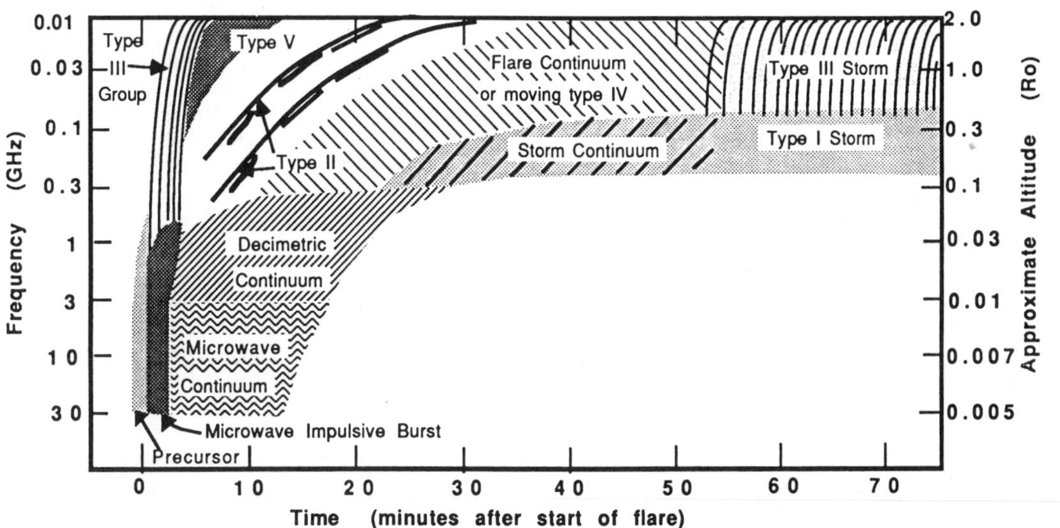

Figure 2. Schematic diagram showing the main kinds of radio bursts that accompany a major solar flare according to their dynamic spectra, the variations of emission frequency with time. The approximate altitude of the emitting sources in units of the solar radius is shown on the ordinate scale at the right.

There are several species of microwave bursts. *Microwave impulsive bursts* occur during the impulsive phase of flares when the rate of energy release is the largest. Most of the energy appears in the form of accelerated electrons of $E \sim 10$–100 keV. The more energetic of these electrons then emit microwaves by the gyrosynchrotron mechanism. Observations with the Very Large Array demonstrate that the radiation usually arises from a series of magnetic loops that arch over the neutral line that separates the two magnetic polarities. Simultaneously, the fast electrons emit hard x-rays by bremsstrahlung when they collide with ions, either in the magnetic loops or in the dense chromosphere at the footpoints of the loops.

Microwave continuum bursts, often called type IV, are energetic but rare. They occur with energetic flares and shock waves. Ordinarily this radiation starts with the impulsive phase and increases in intensity over an interval of some tens of minutes. The spectrum of the emission becomes harder and harder as electrons are accelerated to relativistic energies, several million electron volts, and emit gyrosynchrotron emission in the magnetic loops in which they are trapped.

Microwave post-bursts frequently follow the main phases of flares of large or moderate intensity. The appear as an enhancement of flux that endures for several minutes or several tens of minutes after the principal bursts. It seems that they arise from bremsstrahlung in the hot ($\sim 10^7$ K), dense plasma that has been heated by the flare and is the source of intense soft x-ray emission.

Microwave spike bursts are characterized by short durations (milliseconds) and high degrees of circular polarization (often 100%). They attain brightnesses of $T_b > 10^{13}$ K, showing that the emission is coherent. The cyclotron maser mechanism is the most plausible. Similar bursts are observed from stars and, at dekameter and longer wavelengths, from Earth, Jupiter, and other planets; these are convincingly attributed to cyclotron maser emission.

Bursts at Decimetric Wavelengths

The decimetric regime manifests what is certainly the most complicated and confusing variety of solar bursts. Some are similar to the type III bursts described below, some are *decimetric continuum* (type IV) bursts on which are superimposed several kinds of fine structures, and some are similar to microwave spike bursts. Excepting the spike bursts, most of these probably arise from a variant of plasma emission, perhaps at the upper hybrid frequency $\nu_{uh}^2 = \nu_p^2 + \nu_B^2$, where $\nu_p \gtrsim \nu_B$. There must be some form of free energy, probably a loss cone anisotropy that develops when fast electrons are partially trapped in magnetic loops. The complicated fine structures possibly arise from the fact that the gryofrequency and plasma frequency are similar, whence many kinds of waves are possible.

Bursts at Metric and Longer Wavelengths

At frequencies lower than about 300 MHz, several kinds of bursts have been identified, as illustrated in Fig. 2. Most of these were discovered by P. Wild and his colleagues in Australia after they had constructed the first dynamic spectrograph in the 1940s and 1950s.

Type III bursts are created after packets of electrons are accelerated in active regions and travel along lines of force, forming streams that traverse the corona and the solar wind at a speed of 0.1–$0.3c$. About 10% of these streams are associated with reported flares, producing groups of type III bursts during the impulsive phase as illustrated in Fig. 2.

For classical type III bursts, the starting frequency is several hundred megahertz, emitted at a height of $\approx 0.1\ R_\odot$. In about 10 s, the frequency descends to 20 MHz (the typical limit of ground-based observations), emitted at a height of about $1.0\ R_\odot$. Then in about an hour, it descends further to 30 kHz, near 1 AU. As the electrons traverse the corona and solar wind, the faster ones outpace the slower ones; hence at some distance from the acceleration region the fast electrons, intermixed with those of the ambient plasma, form a streaming, bump-in-the-tail anisotropy that is unstable to the generation of electron plasma (Langmuir) waves at the local plasma frequency. These waves attain a high level of intensity.

There are two main ideas of the subsequent development:

1. The waves may react on the fast electrons in such a way as to reduce the free energy, converting the bump to a plateau by the process of quasilinear diffusion.
2. The waves may become concentrated in isolated regions, becoming more and more intense in a smaller and smaller volume, collapsing into a soliton.

In either case, radio waves are generated both at the fundamental and harmonic of the local plasma frequency, radiation that is observed to attain $T_b \approx 10^8$–10^{12} K in the corona and up to 10^{15} K in the solar wind. An immense amount of work has been done on the theory of type III bursts, probably more than on any other plasma-physical problem in astrophysics.

Type II bursts arise from shock waves that traverse the corona at a speed of about 500–1500 km s^{-1}, sometimes to and beyond 1 AU. Electrons are accelerated in the shocks and develop some kind of anisotropy that leads to high levels of Langmuir waves. For some shocks there is evidence of fast electrons streaming from the shock, but for most there is not. The means of electron acceleration, the development of the anisotropy, and the details of the conversion to radio waves remain poorly understood.

Type IV bursts are of several varieties that have the common trait of consisting mainly of continuum radiation. The first observation involved a source that moved bodily outward several tenths of a solar radius before disappearing; others have moved several radii. But these moving type IV bursts are quite rare, there being only a few of them per year. It seems that their sources are self-contained configurations of plasma, magnetic field and fast electrons that emit either gyrosynchrotron or plasma radiation. On a few occasions they have been observed in association with large-scale ejections of material seen in visible light, the coronal transients.

Flare continuum bursts are more common. Their sources remain nearly fixed at any given frequency, but the lower the frequency, the farther from the solar surface is the source. Brightness temperatures attain values up to 10^{12} K, usually with a moderate degree of circular polarization. It seems that these bursts arise from plasma radiation when fast electrons become trapped in large magnetic loops and develop some sort of anisotropy.

Storm continuum and type I storms occur frequently in years of high solar activity. As indicated in Fig. 2, storm continuum radiation starts about a half hour after the impulsive phase and continues for some hours. Type I storms occur in association with active regions containing one or more large sunspots, sometimes starting with or being strengthened by flares. Storms are composed of continuum radiation of bandwidth ≈ 200 MHz on which are superimposed myriads of type I bursts of bandwidth about 2 MHz and duration about 1 s. The radiation of both bursts and continuum is due to plasma emission at the fundamental and is often nearly 100% circularly polarized. It remains a mystery how the required fast electrons continue to be accelerated for days or weeks, how they develop the required anisotropy, what is the form of the anisotropy, and what is the origin of the intense waves of very low frequency (perhaps ion-sound waves) that are probably necessary for the conversion of the Langmuir waves into radio waves.

Type III storms occur at frequencies lower than 50–100 MHz (see Fig. 2). In contrast to the accompanying type I storms that occur in magnetic loops within $\approx 0.5\ R_\odot$ of the Sun, type III storms are formed on magnetic field lines that are open into the interplanetary medium; the bursts are often observed at kilometric wavelengths.

CONCLUSIONS

Radio astronomy of the Sun has advanced considerably since its inception in about 1945. It is probable that all emission mechanisms are known, but most are imperfectly understood. Today the emphasis is on the study of radio emissions in combination with observations in visible, ultraviolet, and x-radiation, aiming to understand the quiet solar atmosphere and the properties of the magnetic field as well as the energetic particles and the many varieties of waves that are abundant in solar flares.

Radio astronomy of stars, on the other hand, is relatively new and the emphasis is largely on the discovery of what kinds of stars produce observable radiation, on the identification of the mechanisms of emission, and on the properties of the sources. Often the solar analogy is used in efforts to understand the stellar radiation. But that analogy must be used with care because the circumstances of the emission on the stars may differ greatly from that on the Sun.

Additional Reading

Dulk, G. A. (1985). Radio emission from the Sun and stars. *Ann. Rev. Astron. Ap.* **23** 169.

Goldman, M. V. (1989). Electron beams and instabilities during solar radio emission. In *Proc. 1988 Yosemite Conf.* American Geophysical Union.

Kundu, M. R., and Lang, K. R. (1985). The Sun and nearby stars: Microwave observations at high resolution. *Science* **228** 9.

Lecacheux, A., Steinberg, J.-L., Hoang, S., and Dulk, G. A. (1989). Characteristics of type III bursts in the solar wind from simultaneous observations from ISEE 3 and Voyager. *Astron. Ap.* **217** 237.

McLean, D. J., and Labrum, N. R., eds. (1985). *Solar Radiophysics.* Cambridge University Press, Cambridge.

Melrose, D. B. (1980). *Plasma Astrophysics: Nonthermal Processes in Diffuse Magnetized Plasmas*, **1** and **2**. Gordon and Breach, New York.

See also **Radiation, High-Energy Interaction with Matter; Radio Astronomy, Space Missions; Solar Activity, all entries; Sun, Atmosphere.**

Sun, Solar Constant

Hugh S. Hudson

The total radiant energy from the Sun incident upon the "top" of the Earth's atmosphere [about 1367 watts per square meter ($W\ m^{-2}$)] is called the *solar constant*. This input of solar energy ultimately defines the Earth's weather and climate. The long-term variations of the solar luminosity (which are not very well understood observationally, because precise observations began only recently) obviously could play an important role in environmental issues on Earth, such as the suspected global warming trend of the "greenhouse effect." This entry nevertheless discusses solar luminosity variations mainly from the astrophysical point of view.

The term "solar constant" misleadingly implies that the solar luminosity does not vary with time. In fact, we find that it has varied significantly, on all time scales sampled, since precise radiometric observations from space began in late 1978. The variations reflect the existence of several distinct physical processes continuously at work in the solar interior, and the interpretations of these processes contribute to our knowledge of stellar structure. In particular, the luminosity variations may help us to understand stellar rotation, convection, and magnetism.

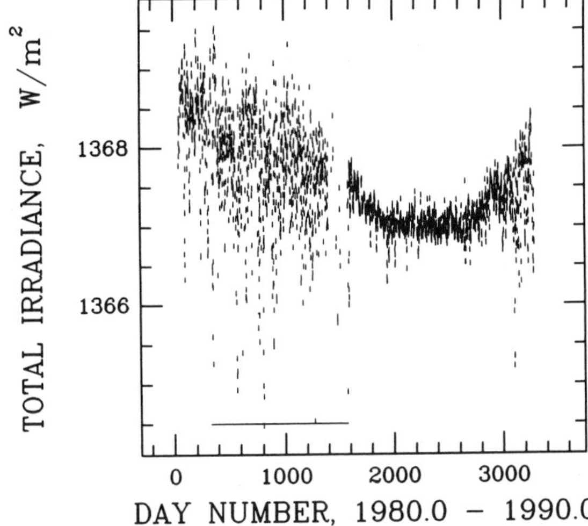

Figure 1. Time variation of the total solar irradiance, as observed by the ACRIM instrument on board the *Solar Maximum Mission* spacecraft. The horizontal line at the bottom shows an interval of degraded performance before the *Challenger* astronauts repaired the spacecraft in 1984. The "dips" due to sunspots are clearly visible, as is the increase in 1988–1989 coinciding with the increase of sunspot activity of the solar cycle.

MEASUREMENT PROCESS

The measurement of total solar luminosity has several noteworthy technical aspects. We can observe only the *solar irradiance*, the radiant energy flux emitted by the Sun in the direction of the observer, rather than the total radiant energy emitted in all directions (the *luminosity*). Some components of the solar radiation that vary on short time scales are highly directional; accordingly, the corresponding variations do not relate directly to true changes in solar luminosity. The total irradiance *does* refer to the flux integrated over wavelength, however, so that the total irradiance measurements are guides to the true bolometric luminosity of the Sun.

By a remarkable coincidence, the solar variability detected to date has an amplitude that is just at the limit of ground-based astronomical photometry: The fluctuations are on the order of 0.1% of the average irradiance. Thus all of the elaborate early attempts (Claude-Servais-Mathias Pouillet, John Herschel, Samuel P. Langley, Charles G. Abbot) to detect variations in the solar constant by ground-based techniques failed or, more precisely, they only established upper limits.

Instruments put into space, namely, equipment aboard the *Nimbus 7* satellite (which began observations in late 1978) and, especially, the active cavity radiometer irradiance monitor (ACRIM) on the *Solar Maximum Mission* satellite which operated during 1980–1989, made breakthroughs in detector stability and sampling frequency. These were essential for the detection of the full range of variations of the total solar irradiance. The ACRIM demonstrated a remarkable degree of long-term stability, within limits of a few parts per million per year.

OBSERVED VARIABILITY OF TOTAL IRRADIANCE

Figure 1 shows the time series of data from ACRIM from 1980 to 1988. Figure 2 summarizes these variations in terms of the *power spectrum*. The power spectrum displays the distribution of variance with frequency of variation or its time scale; it has some points of similarity with an ordinary radiation spectrum of intensity versus wavelength. In the solar case, Fig. 2 shows the existence of both lines and continuum. The periods of variability represented

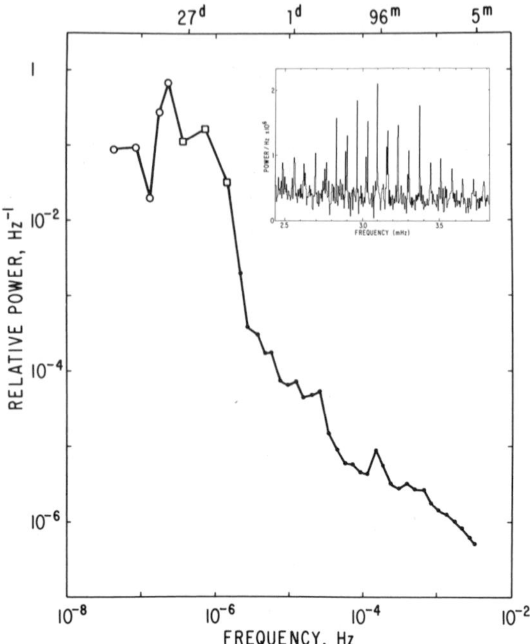

Figure 2. Power spectra of ACRIM total-irradiance observations. The distribution of variation as a function of the frequency is plotted; a continuum shows the lack of characteristic time scales over the corresponding range of frequencies. The inset shows the sharply-defined "p-mode" oscillations at about a 5-min period; each oscillation has an amplitude of less than a few parts per million of the total irradiance. (*Courtesy* Sky & Telescope. *Reproduced with permission*.)

in Fig. 2 range from a few minutes (twice the sample interval) up to several years (the lifetime of the *Solar Maximum Mission*). The continuum power results from broadband (incoherent) variability associated with solar surface features, such as sunspots and granulation, and the spectral lines result from the (coherent) global oscillations of the whole body of the Sun.

Table 1 summarizes the forms of variability detected thus far in the total solar irradiance. We fully expect to identify further contributing mechanisms of solar variability as observations continue. One known mechanism of little direct significance to the *total* irradiance is solar rotation; this does not produce a pronounced modulation at the solar rotational period of about 27 days because of the competing effects of sunspots and faculae, as described later.

INTERPRETATION OF THE VARIATIONS

Each of the variations listed in Table 1 has a different physical explanation, and the interpretation of the observables (time scales, amplitudes, degree of coherence, relationship with other kinds of data, etc.) leads us in different ways to various degrees of understanding of the physics of the solar interior. One tool for this interpretation is the power spectrum, as given in Fig. 2. In the following paragraphs we will discuss each of the listed variability

Table 1. Types of Solar Variability

Mechanism	*Time Scale*	*Amplitude*
Oscillations	5 min	Few parts per million
Granulation	Tens of minutes	Tens of parts per million
Sunspots	Few days	$\lesssim$ 0.2% peak-to-peak
Faculae	Tens of days	$\lesssim$ 0.1% peak-to-peak
Solar cycle	11 years	~ 0.1% peak-to-peak

mechanisms briefly, summarizing our present state of knowledge and suggesting how this new observational material will aid in understanding the Sun and the physical processes it presents to us.

Oscillations

The so-called five-minute oscillations have been known for some three decades. Initially, observers thought them to be rather incoherent surface waves, but we know them to be the photopheric fingerprints of resonant waves trapped in the solar interior. These standing waves or normal modes of oscillation are quite remarkable and form the basis of the new science of stellar seismology or "asteroseismology." Perhaps 30 million individual modes are continuously excited (but at very small amplitudes, with surface displacements on the order of one meter) in the Sun. The solar constant only reflects the subset of oscillation modes that have relatively simple geometries; more complex modes rapidly suffer from mutual cancellation of their brightness crests and troughs, as viewed in integrated sunlight, and thus become undetectably small to nonimaging detectors. The simple solar oscillations that are detected in solar constant measurements are particularly interesting from the stellar point of view, because these same "low-degree" modes will eventually be observable on other stars and will greatly aid our understanding of how stellar interiors differ across the range of stellar types.

The value of each oscillation frequency represents an integral over the interior volume of the Sun, weighted by the residence time (the inverse of the local sound speed). Large-scale motion, such as rotation, perturbs the resonances in much the same way that the Zeeman effect perturbs the wavelengths of magnetically sensitive emission lines in the electromagnetic spectrum. The first fruits of helioseismology, the study of the solar standing waves, exploit this property: It has been shown that the solar interior is rotating at about the same rate as the surface, and that the differential rotation seen in the photosphere extends radially inwards to the base of the convection zone (about one-third of the distance inwards from the photosphere). In other words, the dependence of rotational velocity on latitude is preserved to a substantial depth in the solar interior. These facts are crucially important for the operation of the "dynamo" thought to be the origin of solar magnetism, so we can expect rapid progress in this field.

Granulation

The higher-frequency continuum of variation shown in Fig. 2 results from small-scale structures in the photosphere, for example, granulation. The surface granulation has a close connection with the convective energy flow in the outer part of the solar interior; this eventually may offer us a new means of inferring the properties of convection in other stars, provided that stellar photometry can match the precision of this solar photometry. The power spectrum in this frequency range, interestingly enough, tends to follow the approximately $1/f$ law often noted elsewhere in diverse natural systems.

Sunspots

The sunspots cause the prominent "dips" seen in Fig. 1. Figure 3 illustrates a particularly clear example, with a large enough scale to show how the rotation of the spot group changes the irradiance according to the spot's projected area as seen against the solar disk.

Some astronomers expected that the flux deficit in sunspots would be balanced by enhanced emission in a "bright ring" surrounding a sunspot or group of spots. However, we infer, from the accuracy with which the total-irradiance deficit matches the prediction made from sunspot areas, that there is very little local reemission of the convective energy flux diverted around the spot. This establishes that the sunspots alter the solar luminosity itself, albeit

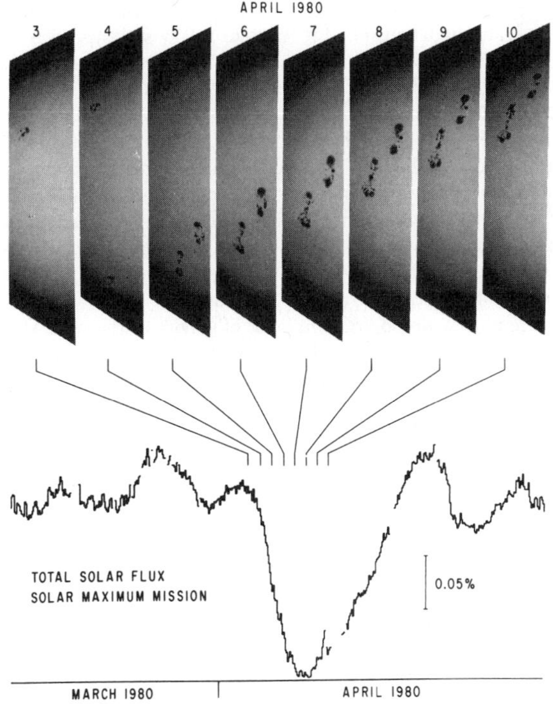

Figure 3. Prominent dip in total irradiance due to a large sunspot crossing the visible hemisphere of the Sun is revealed by ACRIM measurements. The sequence of solar images shows the development of the sunspot group as it traverses the solar disk. (*Courtesy* Sky & Telescope. *Reproduced with permission*.).

for the short times that the solar material can store this blocked energy flux. The detailed mechanism of this storage, and the time scales it works on, have not been determined.

Faculae

Faculae are bright regions that tend to congregate near sunspots to form solar *active regions*. They are composed of bright granules, which have relatively greater contrast against the photosphere when seen at the limb. Within a given active region the excess brightness of the faculae tends to compensate for the darkness of the sunspots.

Both spots and faculae represent regions of strong magnetic fields, and so it might be puzzling at first sight why the one is dark and the other bright. The "hot wall" picture is one standard explanation of facular brightness: The *magnetic pressure* in the facular flux tube depresses the gas of the photosphere, allowing the higher-temperature interior regions to radiate from the walls of the cavity thus formed. The energy supply is quickly replenished by the rapid heat conduction of the convective material just below the photosphere, so that no "dark ring" forms around the facular granule, just as no "bright ring" forms around a sunspot. Sunspots, in this picture, differ solely by having large areas, so that the wall effect becomes negligible and the flux blockage by convection predominates.

The Solar Cycle

The solar-cycle variation evident in Fig. 1 has been discovered so recently that no fully satisfactory theory exists to explain it. However, its time profile over the cycle matches fairly well with that of the so-called active network. Faculae in active regions live longer than the sunspots do and tend to disperse slowly away from the site as a result of differential rotation. Their excess brightness then is distributed in the "network" that outlines the supergranulation, the larger scale of convective cells visible via their emission-line tracers in the chromosphere. Apparently, the active network rises and falls with the solar cycle and is the direct cause of the total irradiance variation. What is unknown at present is the mechanism within the solar interior that allows this distribution of tiny facular granules to retain their organization over such a long time scale.

This recent finding that solar luminosity reaches a *maximum* during the maximum of an 11-yr sunspot cycle was surprising, because it represents the opposite sense from the short-term effects of spots (the dips). This relationship is consistent with the temporal coincidence between the "Maunder minimum," a period of several decades when there were few sunspots on the Sun, and the so-called Little Ice Age in seventeenth-century Europe, as pointed out by the solar physicist John A. Eddy.

THE SUN AS A VARIABLE STAR

If the Sun were at a distance of many light-years—a typical star—what would its variations look like? First, of course, the small amplitude of the variations described previously would be undetectable by normal Earth-based astronomical photometry. As pointed out, the Sun had not been detected as a variable star until the *Nimbus 7* and *Solar Maximum Mission* satellites were launched.

But there are many stars known that exhibit the solar types of variability in forms that are more exaggerated and thus more easily detectable. Nonradial oscillations with some similarities to those detectable in the solar constant have been found in Ap stars and other pulsators, even in certain white dwarfs. "Starspots" are frequently detectable in late-type stars, as are stellar cycles comparable in length to the solar cycle. Thus we believe that the solar luminosity variations provide a close-up view of physics that has more general applications. With the steady increase of observational power in astronomy, it will be quite interesting to find how far the solar paradigm can carry us and where new physics comes into play in our understanding of stars and their behavior.

Additional Reading

Chapman, G. A. (1987). Variations of solar irradiance due to magnetic activity. *Ann. Rev. Astron. Ap.* **25** 633.

Foukal, P. V. (1990). The variable Sun. *Scientific American* **262** (No. 2) 34.

Hudson, H. S. (1988). The observed variability of the solar luminosity. *Ann. Rev. Astron. Ap.* **26** 473.

Lean, J. (1989). Contribution of ultraviolet irradiance variations to changes in the Sun's total irradiance. *Science* **244** 197.

Leibacher, J. W., Noyes, R. W., Toomre, J., and Ulrich, R. K. (1985). Helioseismology. *Scientific American* **253** (No. 3) 48.

Willson, R. C. and Hudson, H. S. (1988). Solar luminosity variations in solar cycle 21. *Nature* **332** 810.

See also **Solar Activity; Stars, Pulsating, Overview; Sun, Atmosphere, Photosphere; Sun, Interior and Evolution; Sun, Magnetic Field; Sun, Oscillations.**

Superclusters, Dynamics and Models

Guido Chincarini, Roberto Scaramella, and Paolo Vettolani

On very large scales, gravity seems to be the dominating force so that the distribution of matter is intimately related to the motions on large scales if matter is clumped as we observe in the distribution of galaxies (see Fig. 1).

We know, on the other hand, that the observations of the Cosmic Microwave Background Radiation (CMBR) show homo-

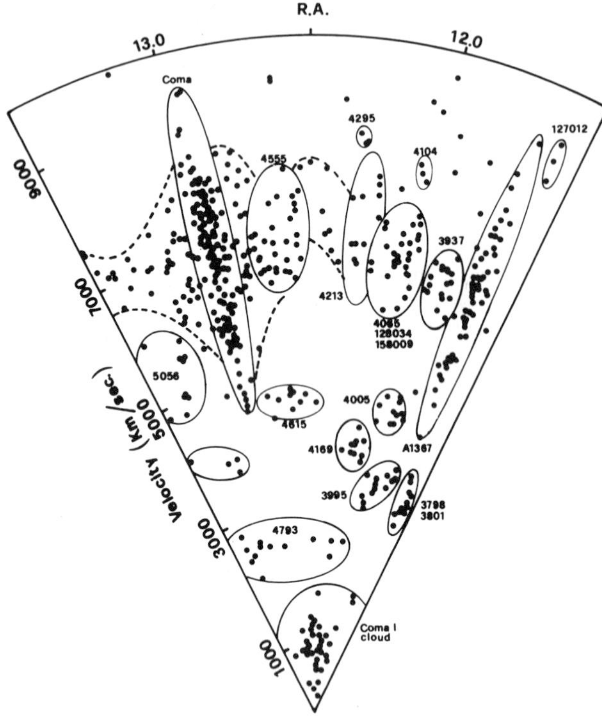

Figure 1. Part of the supercluster Coma A1367. The supercluster is located at a distance corresponding to the redshift of about 7000 km s^{-1} (this corresponds to a formal distance of 70 Mpc using H_0 = 100 km s^{-1} Mpc^{-1} or to 140 Mpc using H_0 = 50 km s^{-1} Mpc^{-1}, a more likely value).

geneity at all angular scales to a very high degree of accuracy. This implies that the universe is homogeneous on very large scales, as is assumed by the Friedmann–Lemaitre–Robertson–Walker cosmological model.

These very general considerations define to a large extent some of the questions that the current observational cosmology addresses. What is the local topology of the matter distribution in the universe, that is, how and on what scale is matter clumped? What do we mean by matter, baryons, exotic particles, and invisible matter? What are the effects that the estimated distribution of matter has on the reported large scale motions with respect to the Hubble flow and how is the determination of the Hubble constant affected by local streamings? Is it possible to devise new methods to estimate the mass content of the universe? How did the largest structures, usually called superclusters (SC), form and evolve?

A PANORAMIC VIEW

Recent redshift surveys covering very large volumes of space indicate that in spite of our efforts we are still limited by the size of the sample, and that we do not yet observe a fair sample of the universe; we must observe in a deeper and more complete way. The spatial distribution of galaxies has in recent times been described as a hierarchy of clusters, a network of filaments, and as an irregular lattice of cells and/or bubbles.

It is important to note, though, that only clusters of galaxies, and to a lesser extent, groups of galaxies, uncontroversially appear to be completely meaningful entities, that are gravitationally bound and that delimit regions of space in which localized physical processes are at work (e.g., x-ray diffuse luminosity). In this respect, SCs are a much less sharply defined class of astronomical objects, in that up to now they have in general been defined by *subjective*, albeit sensible, criteria of overdensity thresholds in the number of galaxies and/or clusters of galaxies. Obviously these procedures carry with them a large factor of uncertainty that must be well-borne in mind. As an example, SC has been used to denote fairly different systems, although in general it refers to galaxy systems with scales of tens of megaparsecs [1 Mpc $\approx 3\times10^6$ ly; we will use the value $H_0 = 100h$ km s^{-1} Mpc^{-1} for the Hubble constant].

Clearly, the actual distribution of SCs (in a broad sense) is very much related to the initial conditions of the matter distribution in our universe: On the very large scales the distances, the sizes, and the masses involved are so large that there has been not enough time since the Big Bang for them to reach a nonlinear stage of evolution. This very same fact gives us hope that we can make a direct connection between the initial conditions and the present observable large-scale structure. Moreover, a clearly established structure of SCs would argue in favor of gravity as the main driving force for the formation of today's structure and pose insurmountable difficulties to other proposed mechanisms, like the suggestions of cosmic explosions or radiation-driven instabilities as main sources for galaxy and cluster formation processes.

It is then important to note that quantitative measures of clustering, especially the two-point correlation function (a measure of deviation from a random distribution), seem to indicate the presence over a finite distance range of a scale-free distribution of the same form but with different amplitudes for both galaxies and clusters. Clusters are more spatially correlated among themselves than galaxies are, and it is still not clear if the same applies to SC with respect to clusters of galaxies.

While the detailed picture must await deeper and more extended observations, the fundamental fact persists that the distribution of observed objects is characterized by inhomogeneities even on large scales and that, assuming that gravity is at work, such irregularities of the distribution of the mass must perturb the otherwise smooth expansion of the universe. We discuss this point in the next section.

DYNAMICS

The most dense large-scale structures, SCs, consist of large, perhaps unbound, agglomerates of galaxies and clusters of galaxies. A crucial problem is to determine the mass of these overdensities. This point is directly related to dynamical and theoretical considerations, and is concerned with the problem of the behavior of the mass-to-light ratio, M/L, when the length scale increases.

Indeed, while we are able to directly measure luminosities, mass determinations are more complex and usually require a few very strong assumptions on the dynamical state of the object under study.

To have a theoretically appealing, spatially flat universe, one would need to have much more mass than that obtained by summing the mass directly seen in galaxies through their luminous component (L_B). This then translates into a very large value for the average mass-to-luminosity ratio: $\langle M/L_B\rangle \approx 1300$ (this would mean that we are able to see only the fraction 1/1300 of the mass of the universe). An experimental value smaller than this ratio would imply that the universe is open, whereas a larger value would indicate that it is closed.

This theoretical number is much larger than those usually derived from dynamical studies: For galaxies, $\langle M/L_B\rangle \approx 10$, while for clusters of galaxies, $\langle M/L_B\rangle \approx 300$, a fact which suggests that indeed the dominant part of the matter in the universe consists of yet unseen dark matter. Therefore, although by increasing the size considered the M/L ratio increases, these astronomical observations suggest an open universe.

It then becomes crucial to determine what happens on scales larger than those of clusters of galaxies and whether the rising trend of M/L continues up to the point of reaching the theoretically appealing value.

The major problem is that virial theorem mass estimates are reliable only when they are determined for well-relaxed, dynamically old systems, while this is not the case for SCs: These have crossing times which are comparable or even greater than the age of the universe, resulting in dynamically young, unrelaxed systems. Also, while the x-ray luminosity of rich clusters gives useful information on the total mass contained within the cluster radius, this method is not applicable to SCs, for which only upper limits to their diffuse x-ray luminosity are available.

Another interesting line of approach is that of taking advantage of the fact that SCs are young systems, and therefore still in a linear stage of evolution.

Indeed, in linear perturbation theory, by the continuity equation, there is a relationship among a mass overdensity, the gravitationally-induced peculiar velocity with respect to the Hubble flow, and the circumstance that the universe is spatially flat, open or closed. In other words, when a galaxy is in the neighborhood of a very large mass, this galaxy is subject to the gravitational attraction of the large mass, and this has the effect of slowing down the cosmic expansion velocity (Hubble flow) of that particular galaxy, which then has a nonzero peculiar velocity. By comparing the amplitudes of the expansion velocities of several galaxies close to a SC with those of galaxies at the same distance from us which are in more homogeneous regions and expand with the Hubble flow (i.e., they have zero peculiar velocity), in principle one should be able to measure the extent of the SC gravitational pull, and hence its mass and its M/L.

Much effort has been devoted to apply this technique to the infall of our galaxy towards the Virgo cluster, the nearest cluster of galaxies (see Fig. 2). There are many practical difficulties, one of which is trying to determine the exact amount of our peculiar velocity towards Virgo: We can estimate the total peculiar velocity of our galaxy through the measured dipole anisotropy in the CMBR. This anisotropy is not primordial, but is a sort of Doppler effect, due to our specific motion: We see the CMBR sky hotter on the direction toward which we are moving, and colder on the opposite side (an observer on a galaxy with zero peculiar velocity would not measure such a dipole effect).

The total peculiar velocity of the few galaxies that constitute with our own, the Local Group of galaxies (LG), is estimated through the dipole anisotropy to be ~ 600 km s^{-1}, but it is not directed towards Virgo. Because from linear theory one finds that the peculiar velocity is parallel to the direction of the net acceleration that is felt, it follows that there are other large mass contributors besides Virgo to the peculiar motion of the LG. Because gravity falls with the square of distance, masses farther away than Virgo must have much larger masses than the Virgo cluster to have appreciable influence on our peculiar velocity.

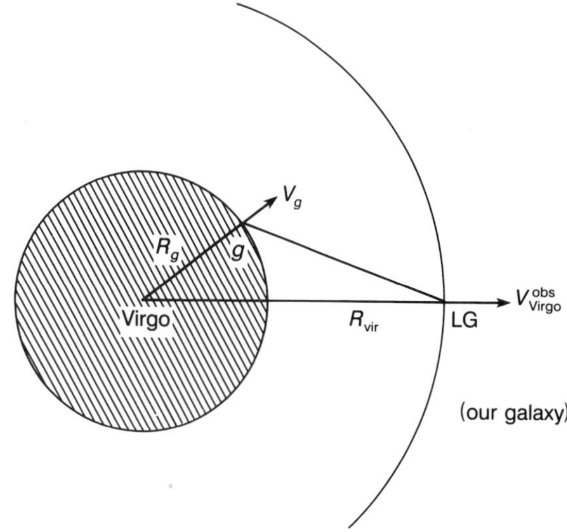

Figure 2. The expansion velocity of the galaxy g is affected by the mean density of the sphere with radius R_g. The expansion velocity of the Local Group of galaxies (LG) depends on the mean density of the sphere with radius R_{vir}.

It is also important to note that, in an expanding universe, besides the more intuitive pulls given by mass concentrations (infalls), one also has "pushes" from underdense regions: If a galaxy happens to be on the edge of a void, it will be blown away because voids tend to expand faster than average, contrary to dense regions, which tend to contract.

From various studies, the region of the Hydra–Centaurus (H–C) SC emerged recently as the most interesting one for this problem At first it was thought that the H–C SC was itself at rest, and that it was the main one responsible for the peculiar acceleration needed in addition to that due to Virgo to explain our peculiar velocity. However, recent studies of peculiar motions of galaxies within 50 Mpc h^{-1} (where $h = H_0/100$ km s^{-1} Mpc^{-1}) from the LG confirmed earlier claims of the presence of large peculiar velocities coherent on large regions of space—the Rubin–Ford effect. These studies not only pointed out distortions of the Hubble flow larger than previously thought, and hence the need for even larger mass concentrations, but also showed that the H–C SC was itself

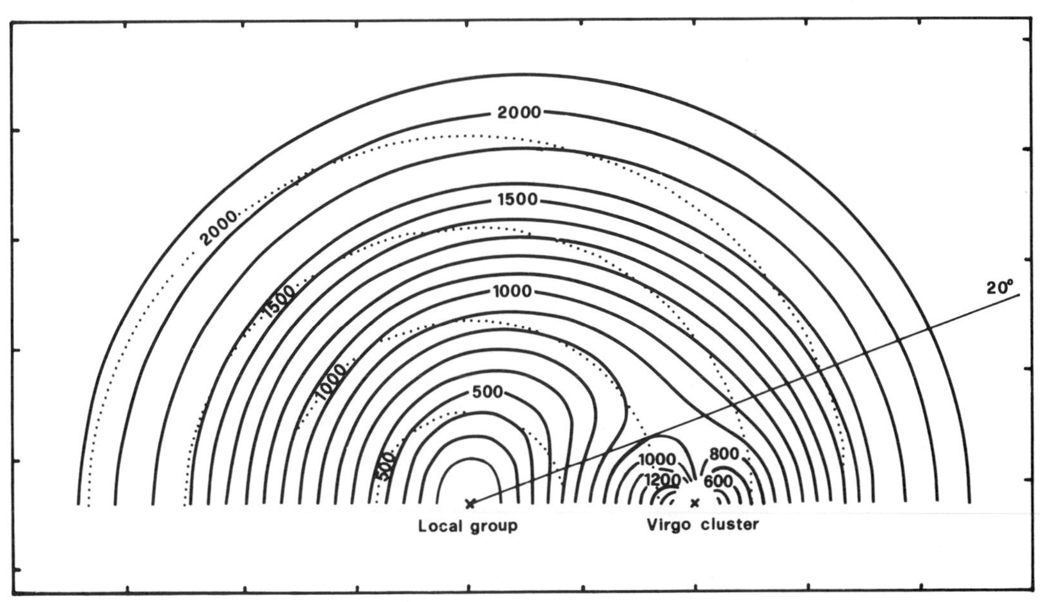

Figure 3. Deviation from the Hubble expansion due to the Virgo cluster density perturbation. The dashed lines show the unperturbed velocity field.

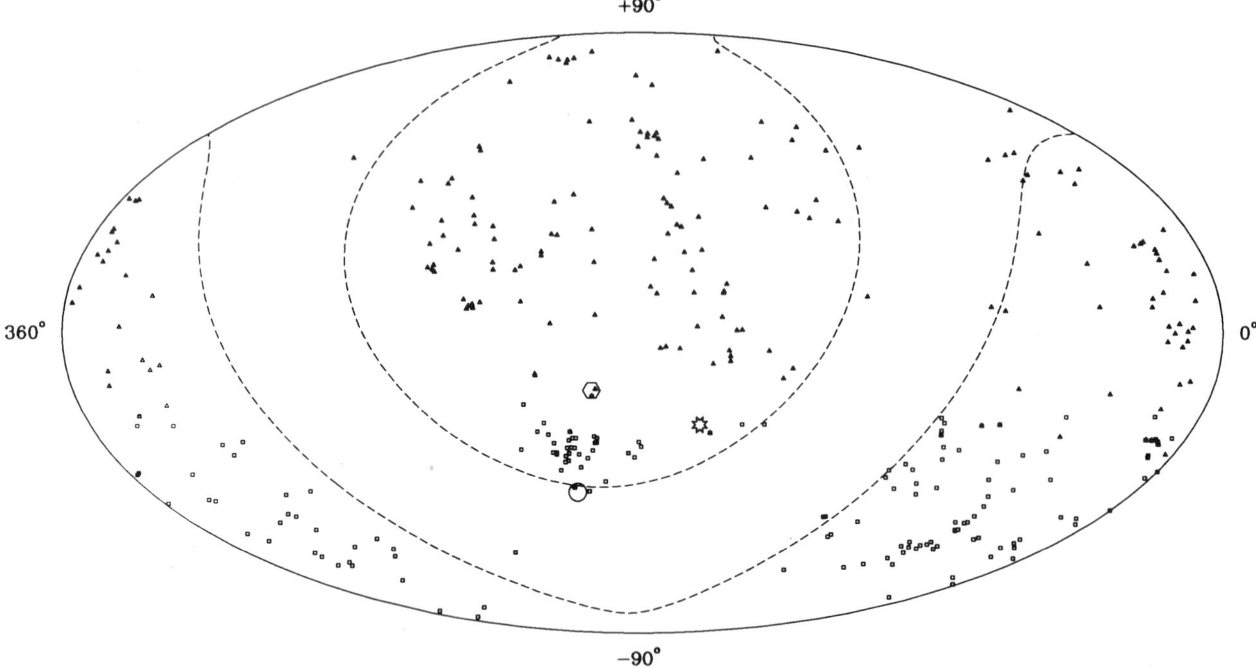

Figure 4. Equi-area projection in right ascension and declination of all the rich clusters with measured or estimated distances within distances corrresponding to a redshift of 20,000 km s^{-1}. Triangles and squares represent Abell and ACO (Abell, Corwin and Olowin) clusters, respectively (some clusters are listed in both catalogues). The star shows the direction of the cosmic microwave background radiation dipole, whereas the empty hexagon shows the direction of the peculiar flow reported in Gott III, et al. [*Ap. J.* **340** 625 (1989)]. Between the latter two symbols the cluster concentration discussed in the text is clearly visible. The dashed lines indicate the avoidance zone due to our galaxy plane ($|b| \leq 20°$).

moving with respect to the Hubble flow, and not at rest as was once assumed (see Fig. 3).

These motions on very large scales have the main component of the flow directed towards the Centaurus direction and, because the SC there itself appears to be moving, must be directed at some very large mass behind it. This hypothetical mass concentration has been nicknamed the Great Attractor (GA). Indeed, the Centaurus cluster is part of an overdensity located at ~ 35 Mpc, but is itself falling so it cannot be the GA. In the same direction, behind it there is another concentration of galaxies at ~ 45 Mpc which is likely to produce a strong attraction on Centaurus, our galaxy, and the LG, and on the galaxies which show the general motion in that direction. Therefore a very large mass is required to explain the whole effect. On the other hand, we discovered that in the very same direction, three times farther away at ~ 140 Mpc, lies the largest nearby concentration of rich clusters known so far (see Fig. 4). Is this a coincidence? Probably not. A very important consequence of this alignment is that, once determined, the fraction of the pull due to this extremely rich SC as a residual from the subtraction of those due to the more nearby structures (including the GA) that can be studied more easily, will enable us to make the *first direct measure* of the mass of a very rich SC. This in turn will allow measurement of the value M/L on the very large scales and therefore directly test if the universe is spatially open (more likely), flat (perhaps), or closed (less likely).

PRESENT VIEW

From what has been described earlier, it is clear that our picture of the universe has changed drastically during the last decade. After previously considering highly homogeneous model with a few clusters of galaxies scattered around in an otherwise uniform sea of galaxies expanding uniformly and smoothly, we have now recognized a locally high-structured distribution organized by gravity. Primordial fluctuations of the order of a few hundredths of the size (3000 Mpc) of the current horizon evolved such as to almost reach the nonlinear stage, with gravity playing the most important role. Irregularities in the distribution of mass on such scales affect the Hubble expansion, perturbing the Hubble flow. On much larger scales there must be uniformity and the whole is embedded in a bath of uniform radiation, the CMBR, rather than in a smooth sea of galaxies which has been denied by the observations: Excesses of densities alternate with regions devoid of galaxies. What the transition scale is between inhomogeneity and uniformity has yet to be determined. Whether or not there is a cellular structure of the universe will be a matter for future researchers and surely enough, nature will show us further unexpected features.

Additional Reading

Bahcall, N. A. (1988). Large-scale structure in the universe indicated by galaxy clusters. *Ann. Rev. Astron. Ap.* **26** 631.

Broadhurst, T. J., Ellis, R. S., Koo, D.C., and Szalay, A. S. (1990). Large-scale distribution of galaxies at the galactic pole. *Nature* **343** 726.

Chincarini, G. (1978). Clumpy structure of the universe and the general field. *Nature* **282** 515.

Chincarini, G., and Rood, H. J. (1980). The cosmic tapestry. *Sky and Telescope* **59** 364.

Dressler, A. (1991). The Great Attractor: Do galaxies trace the large-scale mass distribution? *Nature* **350** 391.

Dressler, A., et al. (1987). Spectroscopy and photometry of elliptical galaxies: A large-scale streaming motion in the local universe. *Astrophys. J.* (*Lett.*) **313** L37.

Gott III, J. R., et al. (1989). The topology of large-scale structure III: Analysis of observations. *Astrophys. J.* **340** 625.
Oort, J. H. (1983). Superclusters. *Ann. Rev. Astron. Ap.* **21** 373.
Postman, M., Geller, M. J., Huchra, J. P. (1988). The dynamics of the Corona Borealis supercluster. *Astron. J.* **95** 267.
Rood, H. J. (1988). Voids. *Ann. Rev. Astron. Ap.* **26** 254.
Scaramella. R., Baiesi-Pillastrini, G., Chincarini, G., Vettolani, G., and Zamorani, G. (1989). A marked concentration of galaxy clusters: Is this the origin of large-scale motions? *Nature* **338** 562.

See also **Background Radiation, Microwave; Cosmology, Clustering and Superclustering; Cosmology, Theories; Dark Matter, Cosmological; Galaxies, Local Group; Superclusters, Observed Properties; Virgo Cluster; Voids, Extragalactic.**

Superclusters, Observed Properties

Anthony P. Fairall

Superclusters are large conglomerations of clusters and groups of galaxies on a scale exceeding 100 million light years [10^{21} km or 30 megaparsecs (Mpc)]. Individual superclusters and "complexes" of superclusters interconnect to form the largest known structures in the universe: a sponge-like network of high-density regions, spaced apart by a labyrinth of voids (Fig. 1). The tendency in recent years is for ever-larger structures to be recognized, so that we may not have yet established the top of the hierarchy of clustering, that is, the largest inhomogeneities in the universe. The greatest strides have been made since the mid-1970s, as large numbers of galaxy redshifts have been obtained.

The first examinations of the large-scale distribution of galaxies were carried out in the 1930s independently by Edwin Hubble and Harlow Shapley. Hubble worked in numerous narrow selected fields. Shapley covered most of the sky using wide-angle photographs. He noted regions where the galaxy count was much higher than average and labeled these as "clouds" of galaxies. Many of his clouds are recognized today as superclusters.

Improved wide-angle photographic surveys, such as the classic National Geographic Society–Palomar Observatory Sky Survey, do reveal hundreds of thousands of galaxies, but give only a two-dimensional view of their distribution. To some extent, the third dimension (distance) can be gauged by the angular diameters of the galaxies. A nearby galaxy appears larger than a more distant one, although one may sometimes be misled by a small nearby galaxy mimicking the appearance of a larger distant galaxy. However, there would be little problem in deciding between nearby and distant clusters of galaxies. In this way, in the 1950s and 1960s, George Abell and Fritz Zwicky independently gauged the relative distances of clusters on the Palomar Sky Survey. Abell concentrated on rich clusters, whereas Zwicky's cluster boundaries lay far from the central condensations. Abell also noted that certain regions of the sky had a greater number of clusters than others. Thus, both suspected much larger entities. In a similar way, C. D. Shane, working from Lick photographs, described "superclusters" or (as he preferred to describe them, using Shapley's term) "clouds of galaxies."

The true recognition of superclusters required a three-dimensional view. When distances to relatively nearby galaxies were calibrated, Gerard de Vaucouleurs advocated the existence of a "supergalaxy," now recognized as our local supercluster.

At larger distances, the most effective way of obtaining galaxy distances is to measure redshifts. Due to the overall expansion of the universe, an observer in any galaxy would see its neighbors moving away from it. The velocity of recession increases with distance from the observer's galaxy according to Hubble's well-

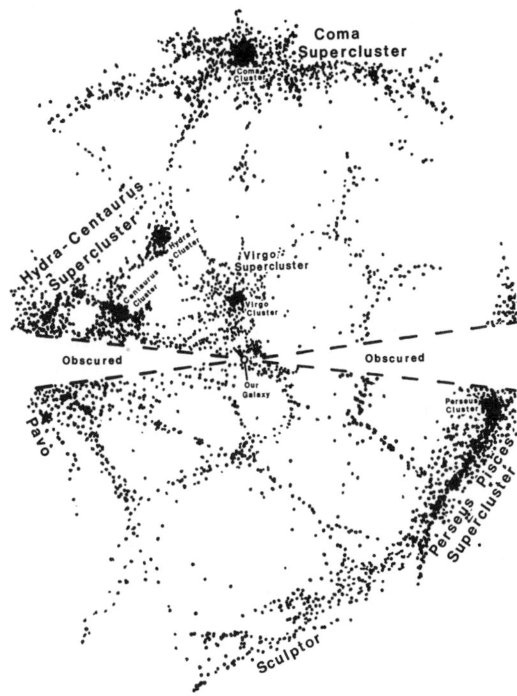

Figure 1. A schematic representation of neighboring superclusters (as prepared by the author from various published plots). Structures are shown out to approximately 300 million light years from our galaxy. Foreground obscuration in the plane of our galaxy hides some of the structure.

known relation

$$V = H_0 d,$$

where V is the velocity of recession, d is the distance, and H_0 is the Hubble constant (in the range 50–100 km s^{-1} Mpc^{-1} or 15–30 km s^{-1} per million light-years). The velocity can be measured by the Doppler shift of spectral features (redshift): either absorption or emission lines in the optical spectra, or 21-cm neutral hydrogen emission in the radio region. Knowing the velocity, the distance can be inferred.

Strictly speaking, the observed velocity is not entirely cosmological but should be expressed as

$$V_0 = H_0 d + V_s + V_p,$$

where the velocity V_0 is now corrected for our Sun's motion within the Galaxy and for the streaming motion of our galaxy, V_s is the streaming motion associated with the *observed* galaxy, and V_p is the observed galaxy's own individual motion over and above systematic streaming; V_p is significant (several hundred kilometers per second) in rich clusters. Because it is difficult to disentangle these velocities, and because of the uncertainty in the value of the Hubble constant, it is customary to plot data in "redshift space" (with dimensions shown as kilometers per second) rather than in conventional three-dimensional space. Gross structures are much the same in either space, but the peculiar velocities in rich clusters make the clusters appear stretched radially (the "Finger of God" effect) in redshift space (examples of this can be seen in Fig. 2).

In the past, obtaining the spectrum of a single galaxy called for a photographic exposure of some hours duration. The advent of electronic image intensifiers and the replacement of the photographic plate by charge-coupled devices (CCDs) and Reticon arrays has greatly accelerated the acquisition of redshifts. Around 1950 little more than 100 redshifts were known, by 1960, 1000 were known and twice that number had been collected by 10 years later. Yet, by 1980 the figure exceeded 10,000 and by 1990 it was around 40,000.

With greater numbers of redshifts available in the mid-1970s, Guido Chincarini pointed out that, even when clusters are avoided in a study, redshifts seem to favor certain values and to avoid others (for the region of sky involved). Thus superclustering was revealed in the third dimension, and three-dimensional mapping became possible. Much pioneering work was done in the region of the Coma cluster where a bridge to a neighboring cluster, Abell 1367, was discerned, with a void immediately in front of the structure.

NEARBY SUPERCLUSTERS

In order to map completely the neighboring superclusters to our own "Virgo supercluster," one would need to work out to a redshift corresponding to several thousand kilometers per second. Within such a volume of space, there are hundreds of thousands of giant galaxies. Even with present technology, it is quite impossible to obtain all their redshifts (the number to be observed would be much greater because a vast number of background galaxies would have to be candidates). Thus, it is necessary to restrict observations to a "representative" sample. The most common approach is to observe all galaxies brighter than a selected limiting apparent luminosity (apparent magnitude). Such a choice is satisfactory for the capabilities of the telescopes involved, but nearer low-luminosity galaxies are included in the sample, whereas distant high-luminosity galaxies may be excluded. The data thin with distance, and low galactic latitudes (where light from distant objects is subject to extinction by matter in the Milky Way) have to be avoided. Nevertheless, knowledge of the galaxy luminosity function (the relative numbers of galaxies versus luminosity) allows one to derive true number densities and other statistics. Difficulties could arise if the luminosity function varies with environment; such tendencies have been claimed in the literature. The alternative approach is to disregard a strict magnitude limit and to observe what appears to constitute a representative sample on the sky. This allows for initial mapping of superclusters (mainly because the intervening voids are almost completely empty), but cannot produce quantitative parameters. Whatever system of sampling is used, the outcome is plots revealing filamentary or sponge-like structures, reminiscent of aqueous media even though the data are in the form of discrete points.

Figure 1 gives an indication of neighboring superclusters. The nearest of these is the Hydra–Centaurus–Pavo supercluster. The names reflect the main constellations in which the structure is seen on the sky (although constellations are based on nearby stars in our galaxy, they conveniently represent general directions when looking far beyond those stars). Although the Hydra–Centaurus portion is seen on the sky on one side of the Milky Way and the Pavo portion on the other side, the agreement in redshift and general continuity of structure point to its being a single entity, with the main bulk lying in Centaurus or probably behind the foreground obscuration of the Milky Way. The Hydra condensation centers around the Hydra I cluster (redshift 3500 km s^{-1}) and there is only a relatively weak bridge to the Centaurus concentration. The latter is dominated by the Centaurus cluster, which shows a composite structure in redshift space, with concentrations at both 3000 and 4500 km s^{-1}. However, the weaker 4500 km s^{-1} concentration may be, to some extent, background galaxies because 4500 km s^{-1} is the dominant redshift for the bulk of the extended Centaurus superclustering. The same redshift (4500 km s^{-1}) is picked up on the Pavo side, which contains a number of weaker clusters. All these clusters contain both elliptical and spiral galax-

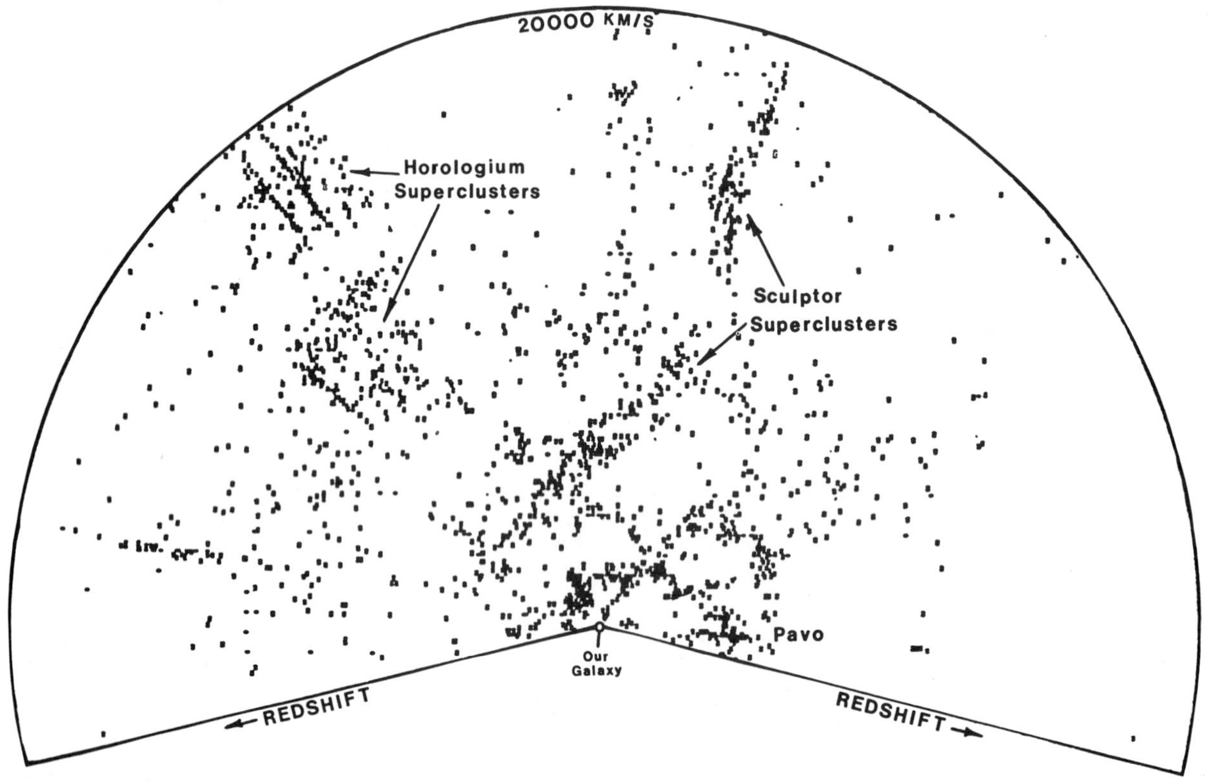

Figure 2. Some prominent features in the southern sky (southern galactic hemisphere) revealed by galaxy redshift data in a slice-like volume out to 20,000 km s^{-1}. Because this is "redshift" space, rather than "conventional" three-dimensional space, clusters of galaxies (such as in the further Horologium supercluster) appear as radial striations. The region covered is between right ascension 18^h–7^h and declination $-22\frac{1}{2}°$ to $-67\frac{1}{2}°$ (but restricted to $-47\frac{1}{2}°$ in the central portion so as not to smear the structures too much).

ies, but the cores of both Hydra I and Centaurus have greater proportions of elliptical and S0 galaxies. There seem to be some filamentary links between our own Virgo supercluster and the Centaurus concentration, almost as if our supercluster were something of an appendage. We are separated from the Hydra and Pavo concentrations by foam-like voids with only a sprinkling of galaxies around their perimeters.

The Coma supercluster, mentioned earlier, is separated from our own supercluster and Hydra–Centaurus by a number of voids. Its structure is centered on the Coma cluster, the nearest rich cluster of galaxies (composed almost entirely of elliptical and S0 galaxies). From this central concentration, extensions run out more or less perpendicular to our line of sight. Although that running (westwards) to the Abell 1367 cluster was the first discovered, present-day surveys (particularly the Harvard–Smithsonian "slices") show it extending both east, north, and southward, so much so that it has been referred to as the Great Wall, and its full extent is still being assessed.

Although the Virgo, Centaurus, and Coma superclusters dominate the northern galactic hemisphere, the other side of the obscuring band of the Milky Way is dominated by the Perseus–Pisces supercluster (which has been mapped extensively by the radio redshifts of Martha P. Haynes and Riccardo Giovanelli). Particularly interesting in this supercluster is a filamentary central condensation that is well marked by elliptical galaxies. It is some 4000 km s^{-1} long in redshift space and runs perpendicular to the line of sight. Toward one end lies the Perseus cluster. Voids again intervene between this supercluster and ours; their peripheral galaxies provide tenuous interconnections. Toward the south the supercluster continues and connects to another heavy wall-like structure, in the Sculptor region, that runs at an angle to our line of sight.

It has been remarked that these surrounding structures give a sort of "tree ring" appearance to the distribution within distances from our galaxy that correspond to redshifts of several thousand kilometers per second. More redshifts still are needed to define the nature and extent of such larger patterns. What is relevant is that, whenever a volume of space is sampled, there always seems to be structure with a dimension comparable to that of the volume surveyed. This has led to considerations of fractal structures (identical forms repeated on ever-increasing scales) occurring in the universe. If this is correct, although one would gain a geometrical interpretation to the nature of the structures, it would make a physical explanation extremely difficult.

MORE DISTANT SUPERCLUSTERS

Beyond redshifts of several thousand kilometers per second, a number of further superclusters have been mapped tentatively. The volume is incompletely sampled, although it is unlikely that very conspicuous superclusters would have been overlooked. For example, in the north there is a pair of superclusters in Hercules (at redshifts around 10,000 km s^{-1}), whereas in the south Shapley's cloud of galaxies in Horologium is resolved into two superclusters (at redshifts of 12,000 and 18,000 km s^{-1}) seen along a common line of sight (see Fig. 2).

An alternative approach for reaching out to larger distances is to assume that Abell's clusters, which mark peak number densities, flag the high points of superclusters. Thus, distant superclusters can be recognized as groupings (in redshift space) of Abell clusters, and examinations of even larger volumes of space can be carried out. On this basis, ever larger conglomerations (to scales of 30,000 km s^{-1}) have been claimed. The work recently has been extended to the southern skies.

The future holds exciting prospects. Just as human eyes scanned the galaxies on the wide-angle photographs, the finest-quality photographs of the U.K. Schmidt telescope and the new Palomar sky surveys are now being scrutinized by machine and millions of galaxies already have been detected and cataloged. From these sky densities comes evidence of possible larger superclusters. In parallel, the development of fiber-optic spectrographs that can be used to observe many galaxies simultaneously may lead to the mass determination of redshifts. We can look forward to finding even more remarkable structures in the superclustering of galaxies.

Additional Reading

Bahcall, N. A. (1988). Large-scale structure in the universe indicated by galaxy clusters. *Ann. Rev. Astron. Ap.* **26** 631.

Burns, J. O. (1986). Very large structures in the universe. *Scientific American* **255** (No. 1) 30.

Chincarini, G. and Rood, H. J. (1980). The Cosmic Tapestry. *Sky and Telescope* **59** 364.

Finkbeiner, A. K. (1990). Mapmaking on the cosmic scale. *Mosaic* **21** (No. 3) 12.

Gregory, S. A. and Thompson, L. A. (1982). Superclusters and voids in the distribution of galaxies. *Scientific American* **246** (No. 3) 88.

Oort, J. H. (1983). Superclusters. *Ann. Rev. Astron. Ap.* **21** 373.

Schwarzschild, B. (1990). Gigantic structures challenge standard view of cosmic evolution. *Physics Today* **43** (No. 6) 20.

Silk, J., Szalay, A. S., and Zel'dovich, Y. B. (1983). The large-scale structure of the universe. *Scientific American* **249** (No. 4) 56.

See also **Clusters of Galaxies; Galaxies, Local Supercluster; Superclusters, Dynamics and Models; Voids, Extragalactic.**

Supernovae, General Properties

Robert A. Fesen

A supernova (SN) is a spectacular energy outburst caused by the explosive disruption of a star, marking the end of its life. Such stellar deaths are nature's most energetic single events, creating truly impressive celestial fireworks. Supernovae have typical kinetic energy releases of 10^{51} erg and total energy releases of up to 10^{53} erg, roughly equivalent to the Sun's energy output summed over 10 billion years then multiplied by 100. During the first few weeks, a supernova can reach an absolute magnitude of -19 to -20, rivaling the combined light produced by an entire galaxy's billions of stars. Yet, the brilliant optical (visible) light we see from the explosion represents less than 1% of a supernova's total energy. Most of the energy comes out in the form of neutrinos, whereas a SN's kinetic energy output is manifested by fragments of the original star that are hurled into space with velocities as high as 20,000 km s^{-1}.

Supernovae, however, are relatively rare events, occurring only once every 25–100 yr in a typical galaxy. Analyses of ancient records from Europe, China, Japan, Korea, and Arabia indicate that there have been fewer than 10 such events "eye-witnessed" to have occurred in the Milky Way galaxy over the last two millennia. These galactic supernovae have appeared as bright "temporary" or "guest" stars with peak apparent magnitudes ranging from ≈ 0 (SN 1181 AD) to -9 (SN 1006 AD) thereby ranking them among the brightest celestial objects ever observed. The handful of these historic galactic supernovae includes the famous new stars reported by Tycho Brahe in 1572 and Johannes Kepler in 1604 as well as the 1054 AD Chinese guest star at whose position we now see the Crab nebula.

Despite their relative rarity, supernova explosions represent a very important phenomenon in astrophysics. Expanding shock waves and debris from such stellar explosions create supernova remnants (SNRs), like the Crab nebula or the Cygnus loop, that can remain visible for up to a few times 10^5 yr and that constitute some of the brightest radio, x-ray, and gamma-ray sources in the sky. Besides marking the dramatic end of some stars' lives, supernovae are also the sites for neutron star and black hole formation and are a major, if not the sole, source of cosmic rays. In addition, they are extremely effective mechanisms for dispersing into the interstellar gas the heavy elements produced by the star during its lifetime, thereby enriching the raw material from which new generations of stars are born. Finally, although brief and unpredictable events, supernovae are a million times brighter than the distance-yardstick Cepheid variable stars. This has led astronomers to try to use them as standard candles to help establish the cosmological distance scale.

DISCOVERY

Supernovae went unrecognized until the 1920s, when Edwin Hubble determined that the distances to spiral nebulae were such as to make them distant and separate star systems (galaxies) like the Milky Way. This discovery meant that the bright nova-like stellar outburst (S Andromeda) observed in 1885 ($m_{max} = 5.8$) in the Andromeda galaxy and a similar one observed in 1895 ($m_{max} = 8.0$) in the galaxy NGC 5253 were, in fact, thousands of times brighter than ordinary galactic novae. Fritz Zwicky, who coined the term "super-nova" (the hyphen was later dropped), quickly realized the potential importance of supernovae and in the early 1930s began a systematic search and study of this phenomenon. This search eventually turned up dozens of extragalactic supernovae between 1930 and 1960, which greatly helped to define the types and statistics of these stellar explosions.

Through both accidental discoveries and dedicated monitoring of large numbers of galaxies, nearly two dozen supernovae are detected each year. A supernova is designated by the year of its initial occurrence followed by a capital letter signifying the order of discovery within that year. Thus SN 1979C, which occurred in the bright spiral galaxy M 100, was the third supernova reported during that year. If more than 26 are discovered in a single year, then double lowercase letters are used; for example, 1988aa. Over 660 extragalactic supernovae have been recorded since 1885, the vast majority of which have been discovered over the last 40 years (see Fig. 1). Until very recently, most detected supernovae were discovered accidently. Today, however, systematic searches for supernovae, using small automated telescopes with sensitive electronic detectors (e.g., charge-coupled devices) such as those used by the Berkeley Supernova Search Program, have begun to produce useful numbers of supernova discoveries. Nonetheless, many supernovae are still discovered by accident and an Australian amateur astronomer, Rev. Robert O. Evans, has discovered visually over a dozen supernovae.

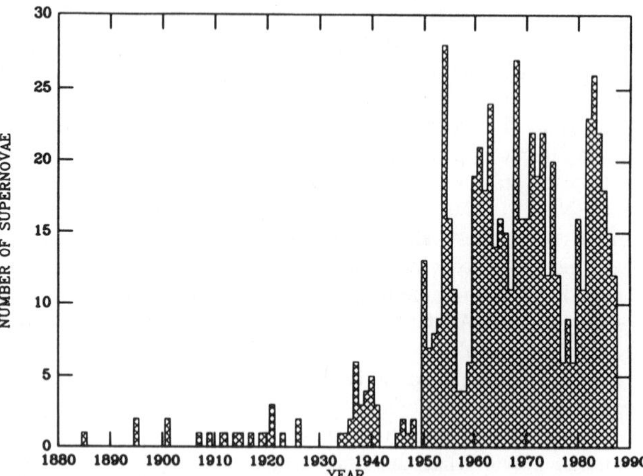

Figure 1. Supernova discoveries by year. (*By permission from* The Supernova Story *by L. A. Marschall, Plenum Press.*)

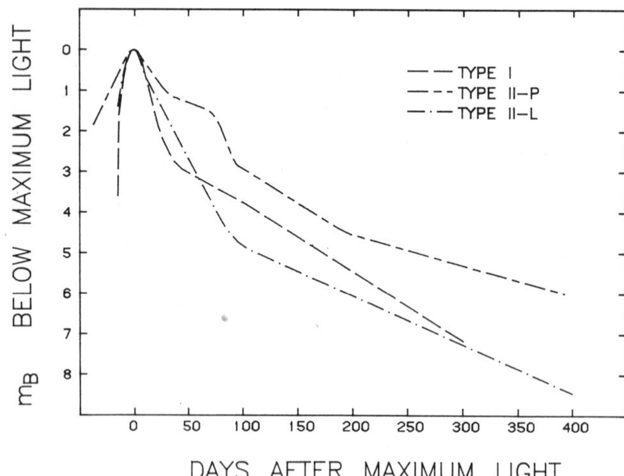

Figure 2. Typical supernova light curves. [*From J. Doggett and D. Branch (1985),* Astron. J. ***90*** *2303.*]

CLASSIFICATION TYPES

Much of our knowledge about the variety of supernova explosions comes not from the bright historical outbursts but rather from the study of the much fainter extragalactic supernovae. In the 1940s, Rudolph L. Minkowski established an empirical supernova classification scheme based upon a supernova's observed optical spectral properties near maximum light. Minkowski divided supernovae into two types: type I, whose optical spectra do not show any evidence of hydrogen in either absorption or emission, and type II, which do exhibit hydrogen. This simple scheme permits easy object classification despite severe initial problems with the interpretation of many spectral features due to the large Doppler broadening. Zwicky later established SN types III, IV, and V for three somewhat peculiar type II supernovae that were, ironically, all sighted in 1961. However, type II supernovae display such a wide intrinsic range of properties that defining new classes for every different-appearing SN was viewed by many astronomers as confusing. On the other hand, the simple and widely adopted Minkowski scheme has the added benefit of dividing stellar explosions into bins defined by the presence or absence of hydrogen and thus is sensitive to possible differences in explosion mechanism of the progenitor.

Classical type I supernovae, now commonly referred to as SN Ia, are the brightest and most uniform subclass of SN outbursts. SN Ia systematically follow a standard pattern of optical postmaximum decline and spectral changes (see Fig. 2). They are found in all kinds of galaxies and represent about 75% of all type I supernovae detected. SN Ia display a spectrum near peak brightness having a 15,000-K blackbody-type continuum with resonance absorption lines of Si II, Ca II, S II, Mg II, and O I. All SN Ia have a strong absorption feature at 6150 Å due to blueshifted Si II 6355 Å that, besides a lack of hydrogen, is their most distinctive spectral feature. Within a few weeks of maximum, permitted Fe II lines develop that later dominate the late-time spectrum. Expansion velocities differ somewhat among individual cases but typical values are around 10,000–13,000 km s^{-1} at maximum, which decrease to around 9000 km s^{-1} two months later.

Light curves for all SN Ia are also very similar, showing a rapid rise to maximum brightness at around $M_B = -18.2 + 5 \log h$, where h is the Hubble constant in units of 100 km s^{-1} Mpc^{-1}. Thus, for a Hubble constant of 50 km s^{-1} Mpc^{-1}, SN Ia are -19.7 mag at peak. SN Ia stay at maximum for only a short time, fading by 3 mag in just 30 days but thereafter decreasing much less rapidly in a nearly linear decline of 0.015 mag day^{-1} starting at about day 50. This late-time decline is believed to be caused by the radioac-

Table 1. SN Type Distribution According to Parent Galaxy Type

SN Type	*E*	*S0*	*Sa*	*Sab*	*Sb*	*Sbc*	*Sc*	*Scd*	*Sd*	*Sm*	*S*	*I*	*nc*	*Total*
I/Ipec	13	12	6	2	11	6	12	3	1	2	8	4	7	87
Ia	2	7	2	1	4	11	11	3	3	1	3	4	8	60
Ib		2			1	1	9			1		1		15
I (all)	15	21	8	3	16	18	32	6	4	4	11	9	15	162
II/IIpec		1	5	4	12	7	44	16	2	1	3	3	7	105
Unknown	29	18	30	10	61	30	76	4	2	4	62	26	42	394
Total	44	40	43	17	89	55	152	26	8	9	76	38	64	661

Source: (Adapted from The Asiago Supernova Catalogue *by R. Barbon, E. Cappellaro, and M. Turatto (1989),* Astron. Ap. Suppl. Ser. ***81*** *421.)*

tive decay energy input of ^{56}Co. Current models suggest a probably CO white dwarf progenitor that, due to mass accretion from a binary companion, grows hydrostatically unstable after exceeding the 1.4-$M_\odot$, Chandrasekhar mass limit, where $M_\odot$ is the solar mass. Although no spectra were obtained, the galactic supernovae of 1006 AD and 1572 AD (Tycho's SN) are believed to have been type Ia events.

A class of hydrogen-deficient supernovae called SN Ib was formally recognized in 1985 with the discovery of SN 1985F, although examples had been reported as early as the 1960s. This new type of SN I lacks both hydrogen lines and the SN Ia's distinctive Si II 6150-Å absorption feature. SN Ibs instead show prominent He lines. Other differences include a 1.5-mag dimmer optical maximum, site locations only in spiral galaxies (especially near active star-forming regions), quite different infrared light curves, and somewhat lower peak expansion velocities of around 10,000 km s^{-1} or less. The late-time spectrum of a SN Ib is dominated by extremely wide (Doppler-broadened) emission lines of [O I] 5577, 6300, 6363, Ca II, Na 5890, 5896, and Mg I] 4562. Recently, a few type I supernovae with neither the 6150-Å Si II line nor detectable helium lines have been reported and have been tentatively classified as SN Ic.

SN Ib light curves show a decline of 0.06 mag day^{-1} for the first 40 days after maximum light, followed by a less steep linear decline of 0.010 mag day^{-1} compared to a SN Ia's 0.015 mag day^{-1}. Again, radioactive decay of ^{56}Co is suspected to contribute to the luminosity variations. Spectroscopic interpretations and theoretical models suggest SN Ib progenitors contain severe solar masses and may involve Wolf–Rayet type post–main-sequence stars, although exploding white dwarfs have also been proposed. The youngest galactic remnant currently known, Cas A ($\approx$1680 AD), may have resulted from type Ib supernova event.

Type II supernovae are stellar explosions involving stars that, unlike type I supernovae, have retained their hydrogen envelopes. This type of SN is found only in spiral galaxies and mostly in the spiral arms. SN II comprise an inhomogeneous group with at least two distinct light curves. One shows a halt or "plateau" in its postmaximum decline (SN II-P), producing a nearly constant luminosity between days 30–80; the other type exhibits little or no plateau, having instead a linear postmaximum decline of 0.05 mag day^{-1} (SN II-L). As a group, SN II show a peak magnitude of $M_B = -17.5 + 5 \log h$. However, they display an intrinsic peak magnitude range of at least two magnitudes. They also show a wide range of expansion velocities at maximum light, anywhere from 2000 to 20,000 km s^{-1}.

Spectroscopically, SN II typically exhibit at maximum a practically continuous spectrum with only weak hydrogen and sometimes helium absorptions. Later, broad hydrogen line emission grows ever stronger while the hydrogen line absorptions weaken and vanish. Late-time features include resonance line absorptions of ions such as Ca II, Na I, Fe II, Ti II, and Sc II. Type II SN are currently believed to originate mostly in red supergiant stars of 8 $M_\odot$ or more. Yet, the recent case of SN 1987A has shown that some fainter SN II are due to blue supergiants as well. Some of the intrinsic variation among SN II probably comes from varying progenitor masses. Progenitor mass differences between SN II and SN Ib/Ic are not currently known, but at least one case (SN 1987K) underwent a spectral metamorphosis, starting out as a SN II but developing a SN-Ib–type spectrum at late times. This suggests some overlap of progenitor masses and that differences between types Ib and II may reflect an outer layer of hydrogen. Detection of neutrinos from SN 1987A clearly demonstrated the correctness of current SN II models involving core collapse, neutron star formation, and core bounce leading to a shock wave that, with the help of energy deposited from the escaping flood of neutrinos, is able finally to explode the star.

SITES AND RATES

The two basic types of supernova occur at different rates in different types of galaxies. Galaxies are divided into either elliptical or spiral galaxies, with spiral galaxies subdivided into classes based upon the tightness of the spiral arms and the size of the nucleus. Table 1 gives the distribution of classified and unclassified supernovae as a function of parent galaxy type. Note how both types are commonly seen in Sc-type galaxies, whereas only SN Ia are observed to occur in elliptical (E) galaxies. This difference is believed to be due to differences in the ages of the progenitors of SN I versus SN II. The more massive stars responsible for SN II are commonly found in spiral galaxies but are rare in ellipticals. On the other hand, SN Ia are believed to be explosions of white dwarfs in binary star systems that should be common in both elliptical and spiral galaxies.

Absolute SN rates and peak brightness statistics are difficult to determine because of the following: the large fraction of serendipitous SN detections (two-thirds of all SN detected are type Ia because they are brighter and more easily discovered); uncertainties in the extinction in highly inclined spiral galaxies; earlier photographic searches often overexposed the central portions of galaxies and thereby possibly missed some supernovae; and unusually active star-forming galaxies ("starburst galaxies"), which can affect counting statistics. Supernovae rate estimates with corrections for these effects indicate that elliptical galaxies have only about one-third as many supernovae as spiral galaxies, correcting for the number of stars. Also, the bluer the spiral galaxy, the higher the SN rate. Recent estimates on the absolute SN rates suggest a SN rate per 10^{10} $L_{B\odot}$ (where $L_{B\odot}$ is the luminosity of the Sun in the photometric B band) averaged over all galaxy types of 0.3 per century for both SN Ia and SN Ib, with 1.0 per century for SN II. For the Milky Way, with roughly 2×10^{10} $L_{B\odot}$, these values suggest a galactic rate of around 3.2 supernovae per century, suggesting that we have seen only a fraction of our galaxy's recent supernovae.

Additional Reading

Goldsmith, D. (1989). *Supernova! The Exploding Star of 1987.* St. Martin's Press, New York.

Marschall, L. A. (1988). *The Supernova Story*. Plenum Press, New York.

Murdin, P. and Murdin, L. (1985). *Supernovae*. Cambridge University Press, London.

Petschek, A. G., ed. (1990). *Supernovae*. Springer, New York.

Trimble, V. (1982). Supernovae: Part I: The events. *Rev. Modern Phys.* **54** 1183; Supernovae: Part II: The aftermath. *Rev. Modern Phys.* **55** 511.

Woosley, S. E. and Weaver, T. A. (1986). The physics of supernova explosions. *Ann. Rev. Astron. Ap.* **24** 205.

Woosley, S. E. and Weaver, T. A. (1989). The great supernova of 1987. *Scientific American*, **261** (No. 2) 32.

See also **Distance Indicators, Extragalactic; Neutrinos Supernova; Supernova, all entries; Supernova Remnants, all entries.**

Supernovae, Historical

David H. Clark

A large number of spectacular stellar outbursts were recorded before the advent of the telescope for astronomical purposes: outbursts that we now realize must have been predominantly novae and supernovae, although to the astronomers of the time they represented the transient appearance of a "new star" where none had been observable to the naked eye previously. Most pre-Renaissance observations of new stars were made in the Orient (China, Korea, and Japan); such events seem to have attracted little interest in the Occident prior to the sixteenth century.

ANCIENT ORIENTAL OBSERVATIONS

From an early period in China a very complex and rigid astrological system existed, and this remained virtually unchanged until modern times. The Emperor was regarded to be the "Son of Heaven," so that celestial phenomena, no matter how trivial, were of great concern to the throne. In order to detect celestial admonitions and take counteraction promptly, court astronomers/astrologers were appointed by the ruler to maintain an assiduous watch of the sky and record their sightings. Many of these observations were preserved in the official dynastic histories, plus other sources. Although crude by modern standards, these ancient Chinese observations, complemented later by those from Korea and Japan (whose rulers introduced systems modeled on those of China), represent an enormously rich source of almost continuous astronomical data over a 2000-yr period. In addition to new stars, the records include observations of sunspots, aurorae, planetary conjunctions, lunar and solar eclipses, daytime sightings of Venus, and so forth. Anything that attracted the attention of the imperial astronomers and was deemed useful for astrological prognostication purposes was recorded.

The oriental records list three kinds of new star. The first were called *k'o-hsing*, meaning "guests stars" or "visiting stars"—stars not previously present, but that made a transient visit to the heavens. Spectacular outbursts in 1006, 1054, 1572, and 1604 AD, events we now know to have been supernovae (and that were in fact also recorded in the Occident), were all classified as k'o-hsing. The second category of new stars were the *hui-hsing*, meaning "broom stars" or "sweeping stars," almost certainly comets with an observable tail. The final (and least frequent type) were the *po-hsing*, "rayed stars" or "bushy stars," which we suspect were usually comets without discernible tails.

By diligent researching in Oriental dynastic histories, encyclopedias, diaries, and astronomical works, a large number of guest star records have been located. The Swedish astronomer Kurt Lundmark seems to have been the first astronomer in modern times to appreciate the value of this fascinating field of research, using it in his efforts to gather historical accounts of stellar outbursts that must have been intrinsically brighter than normal novae. Much of the inspiration for more recent surveys came from the monumental work on ancient Chinese astronomy by Joseph Needham in the 1950s.

Almost all historical records of hui-hsing and po-hsing mention motion, so it is safe to assume that these were comets. This reduces the list of potential novae and supernovae from a full 2000 yr of oriental astronomical records to just 75 "guest star" events. A remarkable, although perhaps not totally unexpected, fact is that for these 75 cases the records indicate that the guest stars were seen for either fewer than 25 days or more than about 50 days. Events of intermediate duration are completely absent. Because a conspicuous supernova fades fairly slowly over many months, we can assume the short-duration stars were novae and concentrate on the longer duration events in trying to track down the records of historical supernovae.

SUPERNOVA REMNANTS AND GUEST STARS

Supernovae are expected to leave evidence of their outburst for tens, even hundreds of thousands of years. These so-called supernova remnants are most frequently detected at radio wavelengths, where interaction of the interstellar magnetic field with the shock wave generated in the supernova explosion plus ambient cosmic rays produce nonthermal radio emission. Some 140 extended radio sources in our galaxy are believed to be the radio remnants of ancient supernovae. Those supernova remnants that lie comparatively nearby and that, therefore, suffer only slight obscuration from interstellar dust, may also be observed at visible wavelengths. Some of the most spectacular optical nebulosities in the heavens, such as the lace-like Veil nebula in Cygnus and the entangled web of the Crab nebula, are supernova remnants. Nearby remnants are often also observed in x-rays. Under certain conditions the stellar remnant of a supernova might also be observable as a pulsar.

When the positions of historical guest stars are described with sufficient precision, it is possible to relate them to catalogued supernova remnants. This is feasible, with a degree of certainty, for just six guest stars over the past two millennia. These historical supernovae are listed in Table 1 and will be described briefly in turn.

185 AD The single Chinese record of this event reads "[185 AD, December 7] ... a guest star appeared within Nan-men [an asterism, the Southern Gate]. It was as large as half a mat, it was multicoloured, and it scintillated. It gradually became smaller and disappeared in the 6th month of the year after next. According to the standard prognostication, this means insurrection."

The "Southern Gate" can be uniquely associated with the stars α and β Centauri, and the term used for "within" implied that the guest star lay between these two bright southern stars. At the

Table 1. Historical Supernovae

Date (AD)	*Where Sighted*	*Duration*	*Supernova Remnant*
185	China	20 months	RCW 86
1006	China, Japan, Korea, Europe, Arab lands	Several years	PKS 1459 − 41
1054	China, Japan, Arab lands	22 months	Crab nebula
1181	China, Japan	185 days	3C 58
1572	China, Korea, Europe	16 months	Tycho's
1604	China, Korea, Europe	12 months	Kepler's

inferred position we observe today an extended supernova remnant known by its catalogue number, RCW 86. It is also observed in x-rays (from interstellar material heated by the expanding shock wave from the supernova outburst) and optical nebulosity is prominent over part of its periphery. The properties of RCW 86 are commensurate with it being of the required age, and it is almost certainly the remnant of the guest star of 185 AD.

1006 AD This is without any doubt the most spectacular guest star recorded historically. There are extensive observations from the Far and Middle East, as well as from Europe. Many of the reports make extravagant allusions to its spectacular brilliance, for example, "glittering in aspect and dazzling the eyes," "its rays on the Earth were like the rays of the Moon," "its form was like the half Moon, with pointed rays shining so brightly that one could see things clearly." At maximum brightness its apparent magnitude must have been at least −10. From a mixture of positional information, its location can be determined with considerable precision; it coincides with that of the radio supernova remnant PKS 1459−41, which is also observable in x-rays and with faint optical filaments.

1054 AD This is the best known guest star of all because of its association with the famous Crab nebula, one of the most studied objects in astronomy. The association is irrefutable; the properties of the Crab place its birth as being compatible with the arrival of the guest star announced (according to *The Essentials of Sung History*) by the Chinese court astrologer Yang Wei-te on August 27, 1054 AD. "Yang Wei-te said, 'I humbly observed that a guest star has appeared; above the star in question there is a faint glow, yellow in colour. If one carefully examines the prognostications concerning the emperor, the interpretation is as follows: The fact that the guest star does not trespass against Pi and its brightness is full means that there is a person of great worth ....' All the officials presented their congratulations."

In the case of the Crab the central stellar remnant, the pulsar, is the "power house" of the nebula (which as a consequence displays central brightening, rather than the peripheral brightening that is characteristic of most supernova remnants).

1181 AD Although not as bright as the guest star of 1054, and as a consequence not as extensively observed, the guest star of 1181 AD has also been associated with a centrally brightened supernova remnant known by its radio catalogue name of 3C 58. Although also detected (weakly) at x-ray and optical wavelengths, 3C 58 does not have a detected pulsar. The association of 3C 58 with the guest star of 1181 AD is probable, rather than certain; however, there is no other creditable remnant in the vicinity of the guest star's appearance.

1572 and 1604 AD It is fortunate that the next two galactic supernovae appeared during the lifetimes of two of the greatest pretelescope astronomers, Tycho Brahe and Johannes Kepler. Although neither discovered the supernovae that now bear their names (Tycho for the guest star of 1572 AD and his pupil Kepler for that of 1604 AD), the extent and quality of their observations ensured this distinction. Both events were extensively observed elsewhere in Europe and the Far East, and unambiguous associations have been made with supernova remnants observed in the radio, in x-rays, and optically.

From the study of supernova remnants, and where possible their association with historical guest star records, astronomers have been able to infer a great deal about the evolution of remnants and the nature of the stellar explosions that produced them.

Additional Reading

Clark, D. H. and Stephenson, F. R. (1976). *The Historical Supernovae*. Pergamon Press, New York.

Needham, J. (1959). *Science and Civilisation in China*, Vol. 3. Cambridge University Press, New York.

See also **Supernovae, General Properties; Supernova Remnants, all entries.**

Supernovae, New Types

J. Craig Wheeler

Supernovae represent the cataclysmic explosions of stars at the endpoints of their evolution. Some have been observed historically in our galaxy by naked eye observations prior to the invention of the telescope. Seven or eight have been confirmed or at least strongly suspected from historical records in the last 2000 yr. One was recorded by Tycho Brahe in 1572 and one by Johannes Kepler in 1604, but there were no other unambiguous naked eye events until Supernova 1987A was discovered on February 23, 1987 in the Large Magellanic Cloud, a satellite galaxy to our own Milky Way. In the modern era, the establishment of the cosmological distance scale and the existence of other galaxies were required before the true nature of supernovae was deduced. An ordinary "new star" or nova represents the explosion of a thin skin of matter on the surface of a compact evolved star. However, the great brightness of supernovae deduced from their huge distances in external galaxies proved that the energies involved were of the order that would destroy the whole star. Supernovae are currently discovered at the rate of about 20 or 30 per year, and several hundred have been catalogued. Other evidence for stellar explosions is manifested by the production of supernova remnants, the expanding clouds of ejected matter that plow into the interstellar matter and emit radio waves and x-rays as well as optical radiation, and by neutron stars that represent the compact remnants of stellar collapse and explosion. There are several hundred each of these extended and compact remnants of supernovae observed scattered about our galaxy and other nearby galaxies.

All modern supernovae have been discovered in other galaxies. Their true nature as exploding stars was first recognized about 50 yr ago. The first step in their study was to classify them. The fundamental classification scheme for supernovae is based on their spectra near maximum light, because that is when they are brightest and easiest to observe. The first broad categories were enumerated as type I (SN I), for those that displayed no evidence for the basic element hydrogen in their spectra, and type II (SN II), for those that did show characteristic features of hydrogen. In recent years, important subdivisions of these basic categories have been elucidated. An associated development has been the realization that spectral information at later, dimmer epochs than maximum, although more difficult to obtain, can give qualitatively different kinds of information concerning the nature of the explosion event.

There have also been parallel developments in theoretical understanding of supernova explosions. There are two basic processes by which stars explode. One involves a thermonuclear runaway and is predicted to completely disrupt the star. The other involves the development and collapse of a heavy element core to form a neutron star. Some of the gravitational energy liberated in this process is presumed to power the explosion, leaving behind the neutron star as a remnant. One of the major goals of the study of supernovae is to understand which mechanism underlies different types of observed supernovae. For a long time there was a tendency to associate thermonuclear explosions with observed type I events, and to identify core collapse with observed type II events, but recent developments have called these simple categorizations into question.

CLASSIFICATION BY MAXIMUM LIGHT SPECTRA AND LIGHT CURVES

Type II Supernovae

Although the classification scheme for supernovae has traditionally depended on the spectrum near maximum light, there are other differentiating properties, such as the temporal pattern of the light output, and the galactic environment in which the supernovae explode. Figure 1 shows a sample of spectra from near or shortly

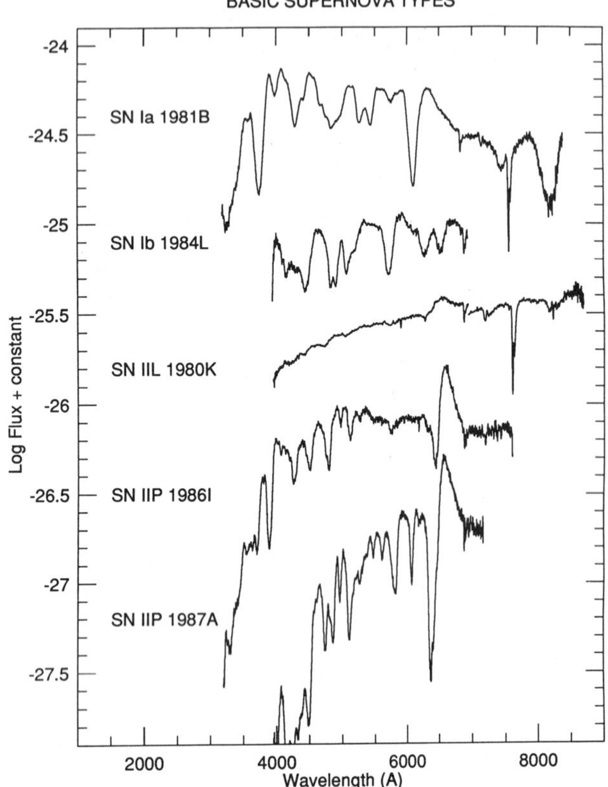

Figure 1. Characteristic spectra near maximum light are shown for the type Ia SN 1981B, and type Ib SN 1984L, and the type II-L SN 1980K. Spectra a few weeks after discovery are shown for the type II-P SN 1986I and SN 1987A. The SN Ia spectrum shows the strong characteristic absorption line of ionized silicon at 6150 Å that is not seen in the spectra of SN Ib. The SN Ib has strong absorption lines of neutral helium at about 5700 Å and other wavelengths that are not observed in SN Ia. The SN II-L event shows a nearly continuous spectrum near maximum light, with some suggestion of the presence of hydrogen in the bump near 6500 Å. The later spectra of the SN II-P and of SN 1987A show a strong hydrogen feature in both emission and blue shifted absorption. [*From Wheeler and Harkness,* Rep. Prog. Phys. *(1990)* **53** *1467.*]

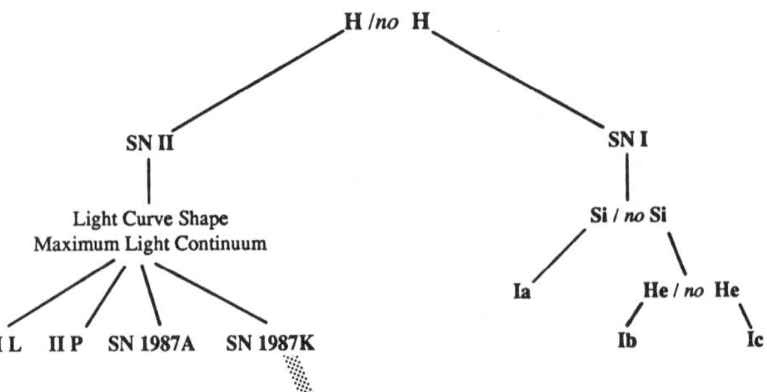

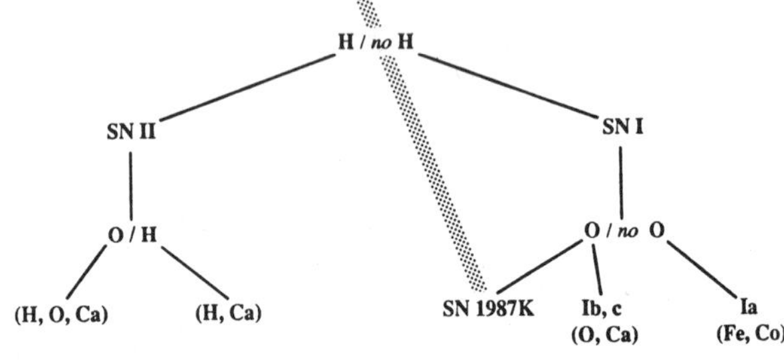

Figure 2. The schematic classification scheme for supernovae based on early and late time spectral features. [*From Harkness and Wheeler, in Supernovae, A. Petschek, ed. Springer, New York (1990)* p. 1.]

after maximum light for supernovae of various types and Fig. 2 gives a schematic representation of the current classification scheme. The basic differentiating property remains the spectral evidence for hydrogen near maximum light. SN II are further differentiated by the shape of the light curve. Those with an extended plateau lasting two or three months after maximum light are called type II plateau (SN II-P). They are thought to arise from explosions in stars with extended hydrogen-rich red giant envelopes. These envelopes are heated by the shock waves of the explosions, and the diffusion of this radiant heat from the envelope is thought to power the plateau of the light curve. Other SN II decline nearly exponentially from maximum light and are called type II linear (SN II-L) after the linear nature of their decline in logarithmic coordinates. The physical characteristics that give rise to this pattern of light emission have not been clearly elucidated. Whether SN II-P and SN II-L are distinct subclasses or extremes of a continuum is debated.

SN 1987A displayed a light curve that fell into neither category, but it is thought to represent a variation of an SN II-P. The light curve of SN 1987A was different than that of a classical SN II-P for the first three months due to the blue, relatively compact nature of the progenitor. The small size, about 20 times the radius of the Sun as opposed to several hundred solar radii for typical models of SN II-P, caused SN 1987A to lose a great deal of its original shock-induced heat by adiabatic expansion before the ejecta became sufficiently thin to radiate efficiently. This caused it to be dimmer than most SN II-P for the first two months or so. As it faded from maximum, SN 1987A did show some similarity to SN II-P as they fade from the plateau. The question of why the progenitor star was so small has not been adequately resolved. Although SN 1987A displayed some different spectral behavior, it is not clear that traditional SN II-L and P events can be differentiated from their spectra alone.

After the decline from the plateau, some SN II-P show a flatter tail in their light curves. Hypotheses for the origin of this light range from the power of a remnant neutron star that could be a rotating, magnetized, pulsar, to radioactive decay. The latter was definitely established in the case of SN 1987A by modeling of the light curve and by the detection first of x-rays from scattered γ rays produced in the decay and then by direct detection of the γ rays themselves.

The primary origin of the radioactive power is the decay of unstable ^{56}Ni. This element is produced in abundance when the shock from the supernova core impacts on surrounding layers of heavy elements and induces thermonuclear burning. Under these extreme conditions, the burning proceeds to the iron peak because iron and related elements have the highest binding energy per nucleon. The rapid shock burning takes place on the strong nuclear time scale and there is no time for the weak interactions to proceed. Because the common fuels consist of equal numbers of protons and neutrons, so must the reaction product. The nucleus with the highest binding energy per nucleon with equal numbers of protons and neutrons is ^{56}Ni. This element is unstable to decay by emission of γ rays and positrons with a time to decay by a factor of

e being 8.8 days. The decay product is ^{56}Co, which in turn decays to stable ^{56}Fe with an e-folding time of 114 days. Direct detection of this process in SN 1987A gives grounds to suspect that it occurs in SN II-P as well.

SN II are associated with the spiral arms of spiral galaxies and with other environments characterized by recent star formation. This circumstantial evidence suggests that they are short-lived stars that do not drift far from their birth site before exploding. No SN II has ever been discovered in an elliptical galaxy. Elliptical galaxies are thought (but not without some dispute) to have ceased their star formation shortly after they formed, over ten billion years ago. Because massive stars burn their fuel quickly and die, the evidence for short lifetimes for the progenitors of SN II suggests that they arise in massive stars. Various statistical arguments suggest that they arise in stars with mass in roughly the range 10–25 $M_\odot$ (solar masses).

With this evidence for the mass range of the progenitors of SN II-P, theory and other circumstantial evidence suggest that these stars must undergo core collapse, and hence produce and leave behind neutron star remnants. This basic picture has been amply supported by the observations of SN 1987A. The progenitor star is deduced to have been of about 20 $M_\odot$, and the direct detection of neutrinos confirms that gravitational collapse occurred.

Type I Supernovae

Although minor variations have been established, the majority of hydrogen-deficient SN I events have a very characteristic spectral development. A key feature near maximum light is the presence of the strong 6347-Å feature of once-ionized silicon that appears as a Doppler blue-shifted absorption feature at 6150 Å. This sort of type I event has been subclassified SN Ia. Such events are observed in elliptical galaxies and in spiral galaxies where, unlike SN II, they tend to eschew the spiral arms but are rather associated with populations of older stars between the arms. This circumstantial evidence suggests that SN Ia arise from stars that live long lives before exploding. The popular hypothesis, which is not supported by quantitative theory or direct observation, is that these events arise in binary systems in which there is an extended quiescent phase before the matter from one star is deposited on the other, triggering it to explode.

A model that agrees rather well with the data is one in which the explosion of the SN Ia is initiated in a compact white dwarf star composed of roughly equal fractions of carbon and oxygen and supported by the pressure of compacted electrons. Such cores naturally arise from the evolution of stars in the mass range from about 1–8 $M_\odot$. The assumption is that these stars lose their outer hydrogen-rich envelopes, but then later have mass added back to them from a companion star to drive them to an explosive endpoint. The hiatus before the onset of the mass accretion phase is supposed to account for the existence of these events among older populations of stars.

The explosion ensues when the total mass closely approaches the Chandrasekhar limit of about 1.4 $M_\odot$, above which degenerate electrons can no longer support the core, and the conditions of density and temperature get so extreme that carbon begins thermonuclear burning. The electron pressure supporting the star is insensitive to the added heat, so the star does not undergo a normal process of expansion, which would absorb the heat and regulate the burning. The heat does serve to accelerate the nuclear burning of the carbon and then the oxygen as well, and this unstable process is thought to cause the violent explosion. The nature of the explosion is debated. The two possibilities are a supersonic *detonation* in which the shock triggers the burning, which in turn drives the shock, and *deflagration*, a form of subsonic, but still rapid, combustion in which the "flame front" is driven by turbulent heat exchange. The former mechanism receives some support from theoretical analysis of the onset of nuclear burning, but the latter mechanism agrees better with observation. The subsonic burning is naturally quenched, leaving partially burned matter that is identified in the spectra.

Either mechanism of thermonuclear burning results in the complete disruption of the star and the production of a great deal of ^{56}Ni. The compact white dwarf progenitors envisaged lose all the original shock heat in expansion, and the decay of the ^{56}Ni to ^{56}Co and then to stable ^{56}Fe is thought to power the entire light curve in these events. The late-time light curve shows an exponential decay with a slope that is steeper than expected from the natural decay of ^{56}Co. Models show that the relatively low mass of the ejecta allows an increasing large fraction of the γ rays to escape directly. There is thus a smaller fraction of the decay energy deposited in the ejecta to power the light curve, and hence the light decreases more steeply than if all the γ-ray energy were trapped in the ejecta.

Despite a certain basic agreement with observations of the model of a SN Ia based on the thermonuclear explosion in a carbon–oxygen white dwarf, there are still major uncertainties in terms of the evolution that leads a white dwarf to an explosive endpoint, the nature of the carbon-burning runaway that is complicated by exotic neutrino loss mechanisms, and the nature of the propagation of the burning front.

There has been some discussion of the possibility that type II-L events may arise by a related thermonuclear explosion in stars in which the core is still surrounded by a hydrogen envelope. Recent evidence suggests, however, that most SN II-L events are too dim to derive their light from the large amount of ^{56}Ni decay that must accompany such explosions. The origin of SN II-L thus remains unclear.

Other hydrogen deficient events do not display the characteristic silicon feature of SN Ia (see Figs. 1 and 2). They have recently been assigned to separate subclasses. The events that fail to show the strong silicon feature near maximum light can be further differentiated by the presence or absence of strong lines of neutral helium. The events that show no Si near maximum light, but do show He are identified as type Ib. There are other events that fail to show either H or Si near maximum light, and show only weak evidence for He. Whereas the strength of the helium lines increases for a couple of months in SN Ib events, the helium lines are never strong in this other subclass. The weak helium line events have been tentatively classified as SN Ic, a nomenclature that has yet to be universally adopted. These SN Ic events are probably physically closely related to the SN Ib.

Like SN II, but unlike SN Ia events, the SN Ib and SN Ic events are found in regions of spiral galaxies characterized by ongoing star formation. This suggests that they arise in stars of at least moderate mass. Models of their light curves suggest that the ejected mass is about 5 or 6 $M_\odot$. This is too big to represent the conditions that lead to thermonuclear explosion, and many researchers have concluded that SN Ib and Ic arise from the hydrogen-denuded cores of massive stars. There are two possible means to eliminate the outer hydrogen envelope from the star. One is in a strong stellar wind driven by the radiation pressure associated with large luminosity. The other is by transfer of the mass to a binary companion. Statistical evidence suggests that SN Ib and Ic explode too often to come only from stars that are so massive that they can shed their envelopes in a wind. The remaining possibility is that they have binary companions. A reasonable hypothesis that requires further proof is that SN Ib and Ic are virtually identical to SN II-P events, except that they are the fraction of such stars that have companions in close enough orbits to induce them to lose their outer envelopes. If this is the case, they should explode by the process of core collapse and leave behind neutron star remnants. If the neutron star is left in orbit around the posited companion, the system could later evolve to become an x-ray binary, perhaps of the class with massive companions, typified by Centarus X-3.

If the progenitor stars that explode to make SN Ib are the bare cores of massive stars, they are so small in radius that they will lose all the heat deposited by the original shock by the time the ejecta are dilute enough to radiate. Even though the cores may be different in detail from those thought to give rise to SN Ia, the circumstances still suggest that there must be a separate source of heat to power the light curve. The light curves of SN Ib are similar in shape to those of SN Ia over the peak that defines maximum light. They are, however, dimmer by about a factor of 4. This is an important difference because the luminosity of SN Ib is too small to be produced by nickel decay in a thermonuclear explosion. The light curve can be reproduced by models in which about 0.1 $M_\odot$ of ^{56}Ni is ejected. This amount of nickel is representative of that expected to be produced by core collapse and is about the amount directly determined in the explosion of SN 1987A. This lends some support to the notion that SN Ib proceed from the same basic physical processes as SN II and SN 1987A. The rather sparse data on SN Ic light curves suggest that SN Ic are about as bright at maximum with similar light curve shapes as SN Ib events, so the preceding arguments probably apply to them as well. The later-time light curves of some SN Ib show an exponential decline that closely matches that expected from the decay of ^{56}Co, further suggesting that radioactive decay powers the light curve and that, unlike SN Ia, the ejecta mass is large enough to trap all the γ rays from the decay for an appreciable length of time. There is some fragmentary evidence that the light curves of some SN Ic fall off more quickly at later times. This might be a hint that they have a smaller ejecta mass than SN Ib. One hypothesis is that the stellar cores have lost most of their helium layers as well as their hydrogen envelopes. The implication for the progenitor stars is not clear.

The preceding discussion has been concerned with optical light curves. There is also some inteesting data on the light curves of SN I in the infrared. SN Ia show a uniform behavior in the infrared with a decline after maximum light to a relative minimum about 20 days after maximum and then a rise to a secondary maximum about 10 days later. This dip is thought to come from some absorptive material in the ejecta, but its nature is not understood. SN Ib and SN Ic show infrared light curves similar to one another, but do not display the post-maximum dip nor the secondary maximum of SN Ia. One suggestion is that the ejecta of SN Ia are silicon rich and may form grains or molecules that cause the minimum, whereas the outer layers of SN Ib and Ic primarily consist of helium and oxygen, which do not form such absorbing material.

No SN Ia has ever been observed to emit radio radiation associated with the explosion. Radio emission has been detected from two SN Ib events. The explanation seems to be that the ejecta of the supernova collide with the matter of a stellar wind that had occurred prior to the explosion of the star. The difference between SN Ia and SN Ib presumably argues that the progenitors of SN Ia have little or no circumstellar matter compared to SN Ib events. There is some evidence from the studies of the ultraviolet spectra that the highest velocity matter is truncated even in SN Ia events, suggesting that even they may have some circumstellar matter, however dilute.

NEW EVIDENCE FROM LATE-TIME SPECTRA

As the ejecta of supernovae expand, densities decline, and the spectra become dominated by nebular emission lines, beginning a month or so after maximum light. Most events in this phase are still recognizable as SN I or SN II as defined by the early-time spectra. Events with H lines near maximum light also show strong hydrogen-line emission in the nebular phase. Some events show strong lines of neutral oxygen and once-ionized calcium as well as of H. Others are dominated by the H and Ca emission with weaker evidence for O. The origin of this difference and its possible correlation with the SN II-L and II-P subclassification scheme is not clear.

Type Ib and Ic events have proven to be distinctly different in their nebular phases from SN Ia. In this phase, SN Ia are dominated by strong emission lines of once- and twice-ionized iron. SN Ib and Ic events show spectra in this phase that are very similar to each other, characterized by strong emission lines of neutral oxygen and magnesium and once-ionized calcium.

An interesting exception to the classification scheme of supernovae has emerged with SN 1987K. Near maximum light this event showed obvious evidence for hydrogen, and numerical models suggest there must be an appreciable amount. Four months later in the nebular phase, spectra of SN 1987K showed no evidence of H. Rather, the spectra were very similar to those of SN Ib and Ic events at this phase. This transition suggests that SN Ib are more closely related to SN II than to SN Ia. The physical conditions necessary to effect this transition in spectral type have not yet been understood.

CONCLUSIONS

Supernova 1987A, the newest naked eye supernova, was also a new type of supernova in terms of its relatively compact progenitor, dim light curve, and some special spectral characteristics. Nevertheless, it brought basic confirmation of the picture in which massive stars were expected to develop heavy element cores that collapse to form neutron stars. The other new types of supernovae, the subclasses type Ib and Ic, have given new insights into the variety of conditions under which stars explode. They are hydrogen deficient like SN Ia but various lines of evidence suggest that they represent the cores of massive stars. If so, they presumably explode by means of core collapse and leave neutron star remnants, plausibly in binary orbit with a massive star companion. The fact that the enshrouding hydrogen envelope is missing from the progenitor stars means that the explosion can reveal details of the core structure and composition that otherwise are masked from view in any event that is not as nearby, bright, and well studied as SN 1987A. The nature of the inner core is the best clue to the evolution of the progenitor and to the explosion itself. The great promise of SN 1987A and of the new subtypes of SN I is that study of them will teach us lessons that expand our knowledge of all types of stellar explosions.

Additional Reading

Harkness, R. P. and Wheeler, J. C. (1990). Classification of supernovae. In *Supernovae*, A. Petschek, ed. Springer, New York, p. 1.

Porter, A. C. and Filippenko, A. V. (1987). The observational properties of type Ib supernovae. *Astron. J.* **93** 1372.

Wheeler, J. C. and Harkness, R. P. (1986). Helium-rich supernovas. *Scientific American* **257** (No. 5) 50.

Wheeler, J. C. and Harkness, R. P. (1990). Type I supernovae. *Rep. Prog. Phys.* **53** 1467.

Woosley, S. E. and Weaver, T. A. (1986). The physics of supernova explosions. *Ann. Rev. Astron. Ap.* **24** 205.

See also **Neutrinos, Supernova; Stellar Evolution, Massive Stars; Supernovae, all entries.**

Supernovae, Type I, Theory and Interpretation

Ken'ichi Nomoto

Supernovae are stellar explosions that release energies of $\sim 10^{51}$ erg and shine as bright as a whole galaxy. Supernovae are classified into two major types: type I and type II, where type I supernovae are identified from the absence of hydrogen lines in the maximum-light spectra, in contrast to their presence in type II supernovae. Type I supernovae are further subclassified from the helium features in the spectra, namely, type Ia (no helium), type Ib (helium-rich), and type Ic (helium-poor). The lack of hydrogen lines implies that the progenitor of a type I supernova has lost its hydrogen-rich envelope before the explosion. The candidates for the progenitors of type I supernovae are white dwarfs, Wolf–Rayet stars, and helium stars in close binary systems. The currently popular models are the carbon deflagration of accreting C+O white dwarfs for type Ia supernovae and the explosion of Wolf–Rayet stars and helium stars for type Ib and Ic supernovae.

TYPE Ia SUPERNOVAE

For type Ia supernovae, accreting white dwarfs have been considered to be promising candidates for the progenitor stars. The explosion mechanism originally suggested by Fred Hoyle and William A. Fowler, that is, the thermonuclear explosion of electron-degenerate cores, basically has been confirmed by extensive numerical modeling and comparison with observations.

White Dwarf Progenitors

Isolated white dwarfs are simply cooling stars that eventually end up as invisible frigid stars. The white dwarf in a close binary system evolves differently, however, because the companion star expands and transfers matter to the white dwarf at a certain stage of its evolution. This mass accretion can *rejuvenate* the cold white dwarf.

The mass accretion onto the white dwarf releases gravitational energy at the white dwarf surface. Most of the released energy is radiated away from the shocked region as ultraviolet light and does not contribute much to heating the white dwarf's interior. The continuing accretion compresses the previously accreted matter and releases gravitational energy in the interior. A part of this energy is transported to the surface and is radiated away from the surface (radiative cooling) but the rest goes into thermal energy of the interior matter (compressional heating). Thus the interior temperature of the white dwarf is determined by the competition between compressional heating and radiative cooling; that is, the white dwarf is hotter if the mass accretion rate $\dot{M}$ is larger, and vice versa.

The scenario that possibly brings a close binary system to a type I supernova explosion is as follows (although the exact evolutionary origin is not yet understood): Initially, the close binary system consists of two intermediate-mass stars [$M < 8\ M_\odot$ (solar masses)]. As a result of Roche lobe overflow, the primary star of this system becomes a white dwarf composed of carbon and oxygen (a C+O white dwarf). When the secondary star evolves, it begins to transfer hydrogen-rich matter to the white dwarf.

When a certain amount of hydrogen is accumulated on the white dwarf surface, hydrogen shell burning is ignited. Its outcome depends on $\dot{M}$: For slow accretion ($\dot{M} < \sim 1\times10^{-8}\ M_\odot\ \mathrm{yr}^{-1}$), hydrogen shell burning is unstable and tends to "flash," which leads to the ejection of most of the accreted matter from the white dwarf; the strongest flash grows into a *nova* explosion. For these cases, the white dwarf does not become a *supernova* because its mass cannot grow. In other words, novae are *not* the precursors of supernovae.

For intermediate accretion rates ($3\times10^{-6}\ M_\odot\ \mathrm{yr}^{-1} > \dot{M} > 1\times10^{-8}\ M_\odot\ \mathrm{yr}^{-1}$), on the other hand, the hydrogen flashes and the subsequent helium flashes are of moderate strength, thereby increasing the C+O white dwarf mass toward the Chandrasekhar mass. When the white dwarf mass becomes $1.4\ M_\odot$ and the central density reaches $\sim 3\times10^{9}\ \mathrm{g\ cm}^{-3}$; explosive carbon burning starts at the white dwarf's center.

If the accretion rate is higher than $\sim 3\times10^{-6}\ M_\odot\ \mathrm{yr}^{-1}$, the accreted matter is too hot to be "swallowed" by the white dwarf. The matter forms a common envelope, which is eventually lost from the system. As a result of mass and angular-momentum losses from the system, some binaries form a pair of C+O white dwarfs. Further evolution of such a double white dwarf system is driven by gravitational-wave radiation and leads to a Roche lobe overflow of the smaller mass C+O white dwarf. The fate of these merging white dwarfs is not clear yet but would be either a type I supernova explosion or a collapse to form a single neutron star.

Carbon Deflagration

When carbon is ignited at the white dwarf's center, carbon burning is so explosive as to incinerate the material into iron-peak elements; the central temperature reaches $\sim 10^{10}$ K. The resulting shock wave is not strong enough to ignite carbon in the adjacent layer; in other words, a *detonation* wave that propagates at supersonic speed does not form. Instead, the interface between the burned and unburned layers becomes convectively unstable. As a result of mixing with the hot material, fresh carbon is ignited. In this way, a carbon-burning front propagates outward on the time scale for convective heat transport. This kind of explosive burning front that propagates at a subsonic speed is called a *convective deflagration* wave. In the standard model, the propagation speed of the convection deflagration wave is on the average about one-fifth of the sound speed. It takes about one second for the front to reach the surface region, which is significantly slower than the supersonic detonation wave. Hence, the white dwarf expands during the propagation of the deflagration wave.

Behind the deflagration wave, the material undergoes explosive nuclear burning of silicon, oxygen, neon, and carbon, depending on the peak temperatures. In the inner layer, nuclear reactions are rapid enough to incinerate the material into iron-peak elements, mostly ^{56}Ni. When the deflagration wave arrives at the outer layers, the density it encounters has already decreased due to the expansion of the white dwarf. At such low densities, the peak temperature is too low to complete silicon burning and thus only Ca, Ar, S, and Si are produced from oxygen burning. In the intermediate layers, explosive burning of carbon and neon synthesizes S, Si, and Mg. In the outermost layers, the deflagration wave dies and C+O remain unburned. The composition structure after freeze-out is shown in Fig. 1.

In the standard carbon deflagration model, the amount of ^{56}Ni produced is $M_{\mathrm{Ni}} = 0.6\ M_\odot$, and the explosion energy is E = (nuclear energy release) − (binding energy of the white dwarf) = 1.3×10^{51} erg. The nuclear energy release is large enough to disrupt the white dwarf completely and no compact star is left behind.

The outcome of carbon deflagration depends on its propagation speed, which involves a parameter such as the mixing length of

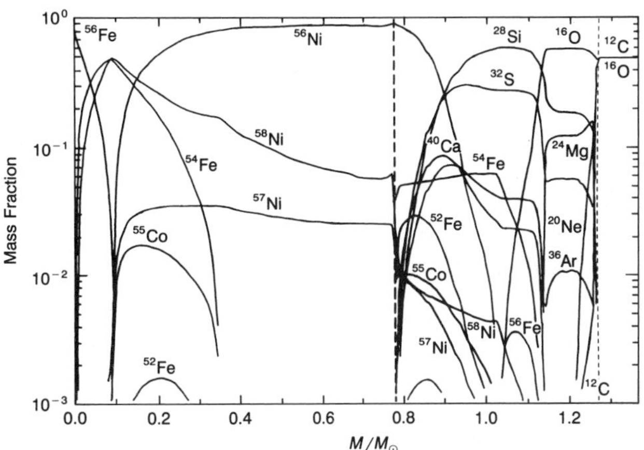

Figure 1. Composition of a carbon deflagration model for type Ia supernovae as a function of interior mass. The white dwarf undergoes incineration into iron-peak elements (mostly ^{56}Ni) at $M_r < 0.7\ M_\odot$ and partial explosive burning in the intermediate region at $0.7\ M_\odot < M_r < 1.3\ M_\odot$; in the outermost layer, carbon and oxygen remain unburned. (M_r designates the total mass interior to radius r.)

convection. The preceding standard model has been chosen because it accounts well for the observed light curves and spectra at both early and late times of type Ia supernovae.

Light Curve

The explosion energy goes into the kinetic energy of expansion, and without a late-time energy source the exploding white dwarf would not be bright. However, during the expansion phase, ^{56}Ni decays into ^{56}Co with a half-life of 6.6 days and ^{56}Co decays into ^{56}Fe with a half-life of 77 days. These radioactive decays produce gamma rays and positrons, whose energies power the light curve as follows.

Gamma rays originating from radioactive decays are degraded into x-rays by multiple Compton scatterings. The photoelectric absorption of x-rays and the collisional ionization due to energetic electrons eventually heat the expanding materials and produce the optical light.

The light curve powered by the radioactive decays reaches its peak at about 15 days after the explosion and declines because of the increasing transparency of the ejecta to gamma rays and due to the decreasing number of radioactive elements. The calculated curve is in good agreement with the observed bolometric light curves of SN 1972E and SN 1981B.

Spectra

Because type Ia supernovae do not have a thick hydrogen-rich envelope, elements newly synthesized during the explosion can be observed in the spectra; this enables us to diagnose the internal hydrodynamics and nucleosynthesis in type Ia supernovae.

Synthetic spectra are calculated based on the abundance distribution and expansion velocities of the standard model and are found to be in excellent agreement with the observed optical spectrum of SN 1981B as seen in Fig. 2. The material velocity at the photosphere near maximum light is $\sim 10{,}000$ km s^{-1} and the spectral features are identified as P-Cygni profiles of Fe, Ca, S, SI, Mg, and O.

At late times, the spectra are dominated by the emission lines of Fe and Co. The outer layers are transparent and the inner Ni–Co–Fe core is exposed. Synthetic spectra of emission lines of [Fe II] and [Co I] agree quite well with the spectra observed at such phases. The agreement implies that both explosion energy and nucleosynthesis in the carbon deflagration model are consistent with the observations of type Ia supernovae.

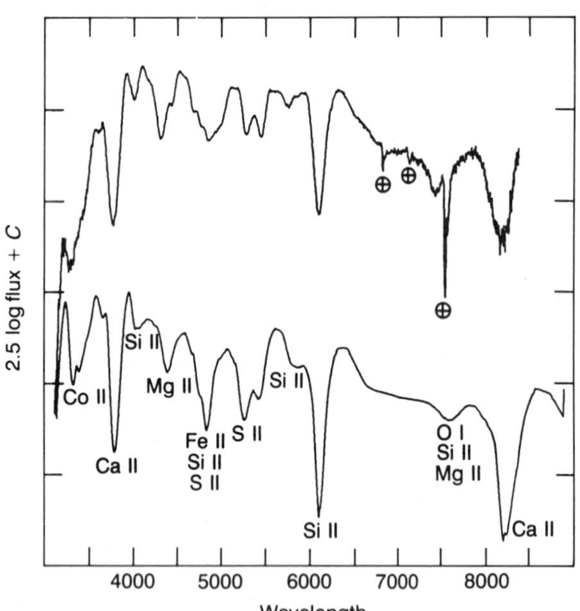

Figure 2. The maximum light spectrum of SN 1981B (top; ⊕ designates absorption lines due to the Earth's atmosphere) is compared with a synthetic spectrum (bottom) for the carbon deflagration model shown in Fig. 1.

TYPES Ib AND Ic SUPERNOVAE

The difference of the maximum-light spectra of types Ib and Ic supernovae from those of type Ia supernovae was first recognized in terms of the lack of a Si feature at 6100 Å. A more fundamental difference is the presence of a He line feature around 5800 Å in spectra of types Ib and Ic, which type Ia spectra do not have. The He feature of type Ib is strong whereas that of type Ic is fairly weak. Another important difference is found in the late-time spectra; the broad emission lines of oxygen appear in types Ib and Ic, whereas iron features dominate the type Ia spectra.

The exponential tails of the light curves imply that the decays of ^{56}Ni and ^{56}Co power the light curves of types Ib and Ic supernovae. The peak luminosities are lower than those of type Ia supernovae by a factor of roughly 4, which implies that the amount of ^{56}Ni produced is about 0.15 $M_\odot$ in types Ib and Ic.

Most type Ib and Ic supernovae are associated with star-forming regions. This fact has led to the currently popular idea that the progenitors of type Ib and Ic supernovae are helium stars more massive than $\sim 3\ M_\odot$.

Helium Star Model

Helium stars considered here are formed from stars more massive than $\sim 12\ M_\odot$ that have lost their hydrogen-rich envelope by strong wind as in Wolf–Rayet stars or by Roche lobe overflow in close binary systems. Such helium stars evolve in the same manner as helium cores in massive stars, thereby initiating a supernova explosion by the iron core collapse as in type II supernovae.

Theoretical light curves and spectra for the helium star models are basically consistent with observations of type Ib supernovae. The light curves of type Ic supernovae tend to decline faster than type Ib; this suggests that the progenitors of type Ic supernovae are somewhat less massive than those of type Ib.

Additional Reading

Branch, D. (1987). Supernovae. *Encyclopedia of Physical Science and Technology* **13** 507.

Nomoto, K. (1985). Explosive nucleosynthesis in carbon deflagration models for type I supernovae. In *Nucleosynthesis: Challenges and New Developments*, W. D. Arnett and J. W. Truran, eds. University of Chicago Press, Chicago, p. 202.

Trimble, V. (1982). Supernovae. Part I: The events. *Rev. Modern Physics* **54** 1183.

Wheeler, J. C. and Harkness, R. P. (1990). Type I supernovae. *Rep. Prog. Phys.* **53** 1467.

Woosley, S. E. and Weaver, T. A. (1986). The physics of supernova explosions. *Ann. Rev. Astron. Ap.* **24** 205.

See also **Stars, White Dwarf, Structure and Evolution; Stars, Wolf–Rayet Type; Stellar Evolution, Intermediate Mass Stars; Stellar Evolution, Massive Stars; Supernovae, all entries.**

Supernovae, Type II, Theory and Interpretation

Stanford E. Woosley

A supernova (SN) of type II is the observable event corresponding to the death of a massive star. The class is defined by spectroscopic evidence for hydrogen in the ejecta, indicating that the star that explodes is one that has not lost its envelope. Although it is possible, in principle, for an intermediate mass star of roughly 5–8 $M_\odot$ (solar masses) to ignite carbon burning while still maintaining a partially intact envelope and thus to provide a type II event with a thermonuclear power source, just as in type I events, another mechanism is generally deemed responsible for type II supernovae: gravitational collapse. This paradigm received welcome confirmation recently when a burst of neutrinos was detected at the onset of SN 1987A. Such a neutrino burst could only have been generated by the collapse of the iron core of a massive evolved star to a neutron star. From photographs of the region taken before the explosion, it has been determined that the progenitor of SN 1987A had a mass on the main sequence near 20 $M_\odot$. In theory, the lightest star that can experience core collapse to a neutron star and thus power a type II supernova is near 8 $M_\odot$, a value that is somewhat sensitive to how convection is treated in the evolutionary calculation. The highest mass star that makes a type II supernova is determined by the efficiency of the still poorly understood supernova mechanism and by the possibility that stars heavier than about 40 $M_\odot$ may lose their entire hydrogen envelope either to a pulsational instability or to a radiatively driven wind. If such stars still were to succeed in exploding, they would be designated type Ib.

THE LIVES OF MASSIVE STARS

The life of a massive star is comparatively brief because such stars are very luminous and are profligate spenders of their nuclear energy reserves. A star of 10 $M_\odot$ will live for about 30 million years, a star of 20 $M_\odot$ for about 10 million years, and a 40 $M_\odot$ star, only 5 million years. Most of the time is spent burning hydrogen to helium. The remainder is spent in five other burning stages characterized by rapidly decreasing time scales. First helium burns to carbon and oxygen, then carbon to neon and magnesium, neon to (more) oxygen and magnesium, oxygen and magnesium to silicon and sulfur, and finally silicon and sulfur burn to elements of the iron group. Along the way, traces of many other elements are made, for example, sodium and aluminum in carbon burning, phosphorus in neon burning, and chlorine and potassium in oxygen burning. Each of these burning stages must surmount increasing charge barriers to nuclear fusion and so each stage occurs at a higher temperature than the previous one. Carbon burns at about one billion degrees kelvin (10^9 K), but silicon burning requires 3.5×10^9 K. At temperatures this great, energy escapes from the core not only by the usual means of radiative diffusion and convection, but by neutrinos. The neutrinos are generated by the annihilation of electrons and positrons, themselves produced by copious gamma rays in the core. Because the neutrino losses scale as about the ninth power of the temperature and because nuclear resources are very limited, as fusion produces progressively heavier elements the time scales become shorter. Carbon may take 1000 years or more to burn, whereas oxygen burns in about one year and silicon burning takes only one week. For stars at the lower end (8–10 $M_\odot$) of the type II supernova mass range, this standard scenario and the time scales are altered somewhat after carbon burning because the electrons become degenerate and provide an additional source of pressure in the core. In the end, however, the cores of all stars in the mass range we are considering collapse to neutron stars.

At the end of silicon burning the (typical) massive star consists of an iron core of about 1.4 $M_\odot$ surrounded by layers consisting of the ashes of previous burning stages and a low-density hydrogen envelope (Fig. 1). The fraction consisting of helium and heavier elements ranges from roughly one-fourth (at 10 $M_\odot$) to one-half (at 50 $M_\odot$) of the original mass of the star. Smaller still is the fraction consisting of carbon and heavier elements. Once iron has been formed in the center, no further energy can be released by nuclear fusion. Gravitational contraction raises the temperature and density but not enough to provide the pressure needed to balance gravity. As the temperature of the iron core grows to 10×10^9 K and more and the density to 10×10^9 g cm^{-3}, electrons are squeezed into the nuclei, leading to heavier neutron-rich isotopes. Concurrently, high-energy radiation begins to tear nuclei apart into α particles. Both processes rob the core of energy and pressure and although the pressure never actually decreases, it becomes weaker compared to gravity. The collapse of the core accelerates, eventually reaching speeds of up to 70,000 km s^{-1}.

As the central density reaches and then exceeds the density of the atomic nucleus (2.7×10^{14} g cm^{-3}) new forces and pressures come into play as the repulsive component of the strong, or nuclear, force brings an abrupt halt to the collapse. Roughly half of the core, that part in sonic communication, halts as a unit. The other half runs into this inner core at supersonic speed and bounces. A shock wave is born. A rebound of the compressed inner core as well as the energy from the reflecting material itself gives energy to this shock and it moves outwards. If enough energy is provided to the shock, it can exit the core, which is now making the transition to a neutron star, while retaining enough outward momentum and energy to eject the rest of the star with a kinetic energy near 10^{51} erg.

THE EXPLOSION MECHANISM

Unfortunately, most current calculations show that the shock loses so much energy to dissipative processes (neutrino losses and the photodissociation of bound nuclei into neutrons and protons) on the way out that by the time it reaches the edge of the core it has lost all outward kinetic energy. Were nothing else to intervene, there would be no supernova, just the relentless growth of the collapsed core by accretion to a state where not even the strong force could prevail against gravity. A black hole would be formed.

However, the collapsed core, although very dense and neutron-rich, is not yet a neutron star. The binding energy of a cold neutron star (about 20% of its rest mass, or 3×10^{53} erg) must still be radiated away over the next few seconds as neutrinos. If just a tiny fraction, only a few tenths of one percent of this neutrino energy, is deposited at the outer edge of the core (just behind where it has temporarily stalled and material is accreting), a powerful explosion still may develop. Detailed calculations by one research group have demonstrated this occurrence. Heating from neutrino energy deposition causes expansion, blowing a large bubble filled with radiation and electron–positron pairs. Expansion of this bubble causes the shock to move outward again with enough energy to eject all of the star external to the core with high velocity.

The success or failure of this "delayed mechanism," so called because it takes roughly one second to develop in contrast to the 20 ms characterizing the shock crossing time of the core, depends upon a variety of nonthermal microscopic physical processes (neutrino scattering on electrons and nuclei, neutrino capture on neutrons and protons, and neutrino–neutrino annihilation, to name a few) and on macroscopic processes such as convection (and rotation and magnetic fields?), all coupled in a situation that can only be studied using sophisticated numerical codes. Not too surprisingly, no consensus has yet emerged as to the general validity of this mechanism. Almost certainly the explosion of some stars, if not all type II supernovae, occurs via neutrino energy transport, but there may well be additional physics that has yet to be modeled correctly in the codes.

Whatever the situation on the computer, the existence of pulsars and supernovae assures us that some explosion mechanism works in nature. In order that material not fall back onto the neutron star and turn it into a black hole, a strong shock wave must somehow find its way into the heavy-element layers surrounding the collapsed core and expel them with high velocity. As this shock transits the shells of silicon and oxygen just outside the core, the high temperature it produces leads to a frenzy of nuclear fusion, producing a number of heavy elements from silicon through zinc. One of the most abundant of these is the radioactive nucleus ^{56}Ni, the most tightly bound of all nuclei having equal numbers of

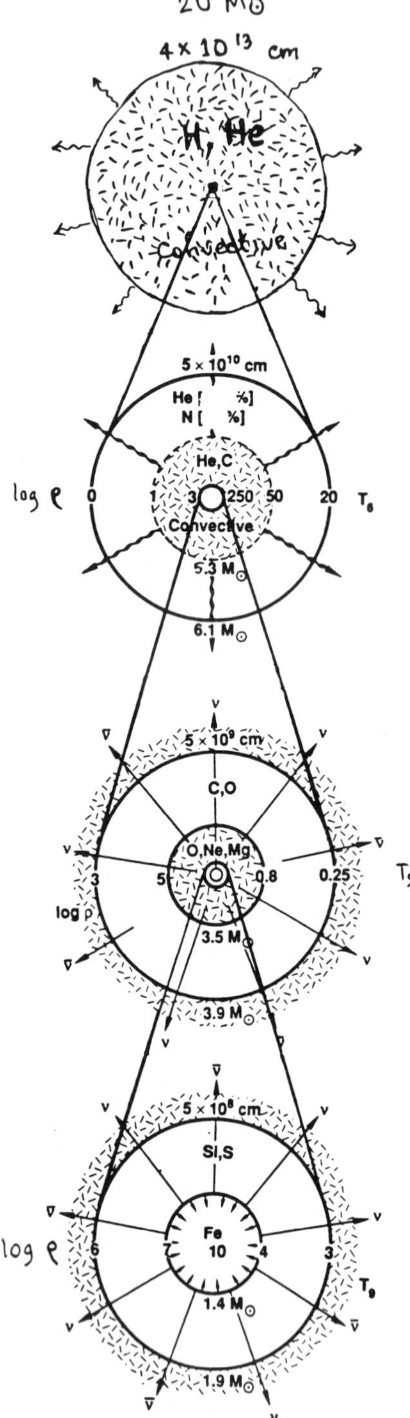

Figure 1. The structure and composition of a presupernova star of 20 $M_{\odot}$ is illustrated as a sequence of enlargements showing the core structure to radial scale. Top: The star at this point is a red supergiant. Almost all of its volume is comprised by the low-density envelope ($\rho \sim 10^{-7}$ g cm^{-3}) of hydrogen and helium where energy is being transported by convection (indicated by stippling). Magnifying the central point by a factor of 1000 reveals the helium core (second sketch from top). The total mass of the helium core and heavier elements within is 6.1 $M_{\odot}$. In a portion of this helium, fusion to carbon is providing most of the energy that comes out as radiation from the surface. Temperature here is in millions of degrees kelvin and ρ is the logarithm of the density in grams per cubic centimeter. Magnifications by two additional factors of 10 reveal the carbon–oxygen, oxygen–neon–magnesium, and silicon–sulfur shells (third sketch from top), and finally the iron core itself (bottom sketch), which, having reached a central temperature of 10^{10} K and density 10^{10} g cm^{-3}, is now collapsing. In the bottom two frames temperature is in billions of degrees kelvin and the mass at each boundary is the total contained within. A vast flood of neutrinos is carrying away energy from the hot shells of heavy elements and from the iron core. The surface luminosity in radiation at this point is 100,000 solar luminosities, but the neutrino luminosity is about 10^{16} solar luminosities and rising rapidly.

neutrons and protons (28). This nucleus decays with a half-life of 6.1 days to ^{56}Co, which in turn decays to ^{56}Fe with a half-life of 77.2 days. It is now believed that most iron in nature has been produced explosively as ^{56}Ni in both type I and type II supernovae and not as stable ^{56}Fe. Theory also suggests that from a few thousandths (10-$M_{\odot}$ supernova) to a few tenths (40-$M_{\odot}$) of a solar mass of ^{56}Ni are produced in the explosion of each type II supernova. Analysis of SN 1987A has shown that 0.07 $M_{\odot}$ of ^{56}Ni was produced in that explosion. The decay of ^{56}Co is also a very important energy source to the supernova at late times (discussed below).

THE SUPERNOVA APPEARS

The shock wave moves on through the rest of the star, exiting the helium core in about one minute and the hydrogen envelope in from two hours to one day (depending upon whether the supernova occurred in a compact blue star as did SN 1987A or in a red supergiant as have most other type II supernovae). As the shock exits, the surface of the star is heated to temperatures near one-half million degrees and is accelerated to velocities of up to 30,000 km s^{-1}. It is at this point that the supernova first becomes visible to the (nonneutrino) astronomer. Early on, the emission is chiefly in the ultraviolet, but as the surface expands and cools the spectrum shifts into the optical. Within about one week the temperature near the surface of the supernova declines to near 6000 K and hydrogen begins to recombine. This recombination removes electrons that were the chief source of opacity and releases the energy that had been deposited in the hydrogen envelope by the shock wave. Recombination propagates as a front that, although carried outward in space by the expansion of the envelope, eventually moves inward in mass until the entire envelope has recombined. During this period, which may last from one week to three months depending upon the mass and radius of the envelope at the time the star explodes, the spectrum is similar to that of a blackbody having temperature 6000 K. Thus most of the emission is in optical wavelengths. Typically, a total of about 10^{49} erg, or roughly 1% of the kinetic energy and 0.01% of the total neutrino energy, is radiated during this phase although the actual amount varies greatly from supernova to supernova.

Once the envelope has recombined, the luminosity declines precipitously until a new energy source is found. That source is usually radioactivity: specifically, the energetic photons released as ^{56}Co decays to ^{56}Fe. Initially, all the gamma rays deposit their energy by scattering with electrons deep within the supernova. Then most of the energy still comes out at optical wavelengths. As time passes, however, an increasing fraction of the gamma rays (and x-rays from partially thermalized gamma rays) escape, although enough still deposit energy to keep the supernova shining brightly. X-rays and gamma rays having energy characteristic of ^{56}Co decay have been detected from SN 1987A. For a period of several years the luminosity of the supernova tracks the exponential decay of this radioactive isotope corrected for the partial escape of the gamma rays. At such late times other energy sources may also contribute to the light. These include the decay of radioactive isotopes besides ^{56}Co (especially ^{57}Co and ^{44}Ti), energy input by a pulsar, or energy from the supernova running into circumstellar material. Because the many factors upon which the light curve depends (the mass and radius of the hydrogen envelope, the mass of ^{56}Ni synthesized in the explosion, the presence of a pulsar or circumstellar material, and the energy of the explosion itself) are likely to vary from star to star, it is not surprising that the emission of type II supernovae is far less regular than that of type I supernovae.

Spectroscopically, the emission of a type II supernova is dominated by lines of calcium, oxygen, and hydrogen. Velocities from 2000 to 30,000 km s^{-1} are inferred, with most of the mass (including the heavy elements made in the explosion) being ejected at the lower velocities, typically less than 4000 km s^{-1}. As the heavy elements expand, they cool. Providing that the density remains sufficiently high, a portion of the ejecta eventually may condense to form both molecules and dust. Emission from both has been seen in the late-time spectrum of SN 1987A. The presence of dust, as well as a decrease in the average excitation energy of collisionally excited heavy elements, results in a shift of a major fraction of the emission to the infrared at times later than about 600 days. Again, SN 1987A has provided an important example.

Additional Reading

Arnett, W. D., Bahcall, J. N., Kirshner, R. P., and Woosley, S. E. (1989). Supernova 1987A. *Ann. Rev. Astron. Ap.* **27** 629.

Behthe, H. A. and Brown, G. E. (1985). How a supernova explodes. *Scientific American* **252** (No. 5) 60.

Petschek, A., ed. (1989). *Supernovae.* D. Reidel, Dordrecht. A collection of essays on the supernova phenomenon.

Woosley, S. E. and Phillips, M. M. (1988). Supernova 1987A! *Science* **240** 750.

Woosley, S. E. and Weaver, T. A. (1986). The physics of supernova explosions. *Ann. Rev. Astron. Ap.* **24** 205.

Woosley, S. E. and Weaver, T. A. (1989). The great supernova of 1987. *Scientific American* **261** (No. 2) 32.

See also **Neutrinos, Supernova; Stellar Evolution, Massive Stars; Supernova, all entries.**

Supernova Remnants and Pulsars, Galactic Distribution

Richard N. Manchester

In 1934, not long after the discovery of the neutron, Fritz Zwicky suggested that the burnt-out core of a massive star would collapse to form a neutron star and that the energy released in this collapse would blow off the outer layers of the star to form a supernova. It was not until 34 years later that this idea was confirmed by the discovery of pulsars within the Vela and Crab supernova remnants and by the subsequent identification of pulsars as rotating neutron stars. Since then, the number of known pulsars has grown rapidly to more than 450, and the number of known supernova remnants has grown more slowly to around 150. Surprisingly, no other convincing associations between pulsars and supernova remnants were found until quite recently. Searches for periodicities in x-ray data, obtained with the *Einstein* observatory by Frederick Seward and his collaborators in the early 1980s, resulted in the detection of pulsars in G 320.4 − 1.2 and in 0540 − 693, a supernova remnant located in the Large Magellanic Cloud. More recently still, weak radio pulsars have been found in the galactic remnants CTB 80 (G69.0 + 2.7) and W 44 (G34.7 − 0.4).

Despite this rather small number of associations, it is widely agreed that most if not all neutron stars are formed in supernova events. Many of these neutron stars are potentially detectable as pulsars. One might therefore expect the distributions of supernova remnants and pulsars through the Galaxy to be similar. However, the observed distributions are quite different. Much of the difference can be attributed to the effect of observational selection in pulsar searches, but even when this is allowed for, there are significant differences between the two distributions. In the next section the observed distributions are described; the final section shows how the observed distributions are corrected for observational selection and describes the results obtained.

OBSERVED DISTRIBUTIONS

In general, pulsars and supernova remnants are both easiest to detect at radio frequencies. Consequently most of the known examples have been found in large-scale surveys using radio telescopes. Searches for pulsars exploit the unique characteristics of pulsar emissions (highly periodic trains of dispersed pulses), and the detection of pulsars is usually unequivocal. On the other hand, identifying a radio source as a supernova remnant is often difficult. The main identifying characteristics are an annular form, a nonthermal spectrum for the radio continuum (that is, flux density proportional to $\nu^{-\alpha}$ where ν is the frequency and α is a positive constant, usually in the range 0.3–0.7), the presence of linear polarization and the absence of recombination-line emission. Detection of recombination lines would indicate that the source was an H II region and not a supernova remnant. Determination of these characteristics often requires high resolution observations at several radio frequencies, and so it is not uncommon for suggested identifications to remain uncertain for many years.

All surveys are affected by selection effects. The most obvious of these is simply the region of sky searched. Apart from this, the principal selection effect is on flux density; survey instruments have a limited sensitivity and only objects brighter than this limit can be detected. For nearby and older supernova remnants, the limit is really on surface brightness, that is, flux density per unit area on the sky. As supernova remnants expand, their surface brightness decreases and eventually they merge into the general background of galactic emission. For pulsars, several other effects are important. All surveys are insensitive below a minimum pulsar period, which is related to the sampling interval used in the survey. For most of the distant pulsars, interstellar scattering and, to a lesser extent, interstellar dispersion limit the sensitivity of many surveys, especially those at lower radio frequencies. Both these effects tend to smear the pulses out and hence make them difficult to detect.

Fortunately, for both supernova remnants and pulsars, the whole celestial sphere has been searched to a reasonable limiting flux density. The distribution in galactic coordinates of known supernova remnants and pulsars is shown in Fig. 1.

The two distributions are quite different. Supernova remnants are much more concentrated in latitude along the galactic equator. Very few are at latitudes greater than 10°, whereas pulsars are found even within a few degrees of the galactic poles. Both supernova remnants and pulsars are concentrated in the inner galactic quadrants (toward the galactic center). The concentrations of pulsars near the equator from $l = 40$–$60°$ and from the center to 40° do not represent real enhancements in the density of pulsars but are simply the results of especially sensitive surveys in those directions. The first concentration is in the region of the inner galactic plane accessible to the Arecibo radio telescope and the second results from a survey at the relatively high radio frequency of 1.4 GHz of the region between latitudes ±2° using the Jodrell Bank radio telescope.

Figure 1 shows the observed galactic distribution of supernova remnants and pulsars in two coordinates, but there is of course a third coordinate necessary to specify the galactic position. Distances to supernova remnants can be estimated in several ways. In the case of the Crab nebula, measurement of the expansion veloc-

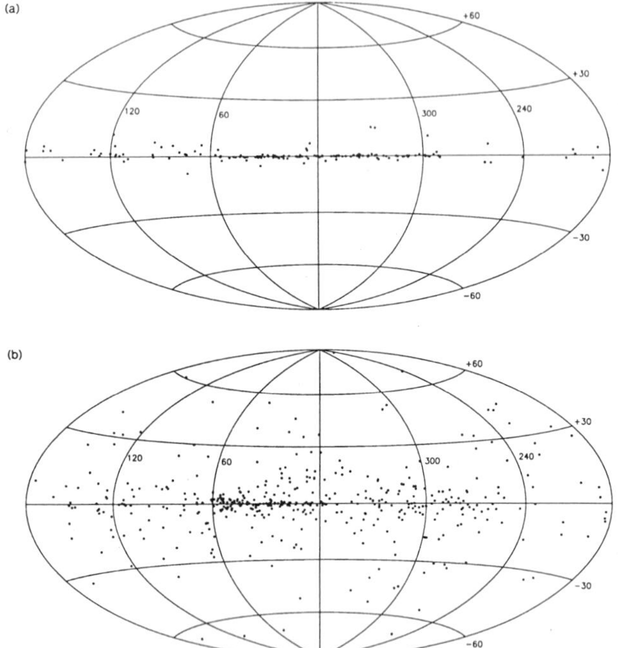

Figure 1. Distribution of the 155 known galactic supernova remnants (*a*) and the 450 known pulsars (*b*) plotted in galactic coordinates (longitude *l* and latitude *b*). In each figure the central horizontal line is the galactic equator (which runs along the Milky Way); the galactic center ($l = 0$, $b = 0$) is at the center of each figure.

ity of optical filaments in the radial direction (from the Doppler shift of spectral lines) and in the transverse direction (from the proper motion or change in position with time) has allowed a direct estimation of the remnant distance. This method cannot be applied generally, and less reliable methods must be used for most remnants. Several remnants are (or appear to be) associated with stellar associations whose distance can be estimated. An important method that has been applied to many remnants relies on the observation of interstellar absorption lines (usually of hydrogen) in their radio spectra. From the velocity of the absorbing gas relative to the Sun and a model for the differential rotation of the Galaxy, a distance to the remnant (known as the kinematic distance) can often be estimated. These methods allow estimation of the distances to about one-third of the known remnants. For the remainder we must rely on the Σ-D method. This method is based on a model for the expansion of remnants that is an oversimplification of the actual situation and hence gives distances that have an uncertainty of 50% or more. Observations of remnants with known distances suggest that the radio surface brightness Σ is related to the remnant diameter D by a relation of the form $\Sigma = AD^{\beta}$, where A and β are constants. Determinations of β differ somewhat from sample to sample, but recent results suggest that $\beta \sim -3$. Because $\Sigma \sim S\theta^{-2}$, where S is the observed flux density and θ is the angular diameter, we obtain a relation for the distance in terms of measurable quantities $d = B(S\theta)^{-1/3}$, where B is a constant.

The situation for pulsars is similar, in that for a few pulsars the distance can be estimated in a reasonably model-free way, but for the rest an approximate method must be used. Direct measurements of annual parallax using radio interferometric techniques have been made for a few pulsars. Kinematic distances have been obtained from neutral hydrogen absorption measurements on about 35 pulsars. But for the vast majority of pulsars we rely on the pulse dispersion produced by interstellar electrons to estimate the distances. The pulse dispersion is proportional to $\int n_e \, dl$ along the path to the pulsar, where n_e is the interstellar electron density and dl is an element of length. Therefore, if we have a model for the distribution of interstellar electrons in the Galaxy, we can compute the distance to a pulsar directly from its dispersion. Unfortunately, there is no independent method of estimating the electron density to the required precision, and we must rely on the pulsars with measured distances to define the model. In fact, a constant value of $n_e = 0.03$ cm^{-3} gives a distance that is almost as good as that given by the more complicated models. This technique obviously ignores small-scale fluctuations in the interstellar electron density

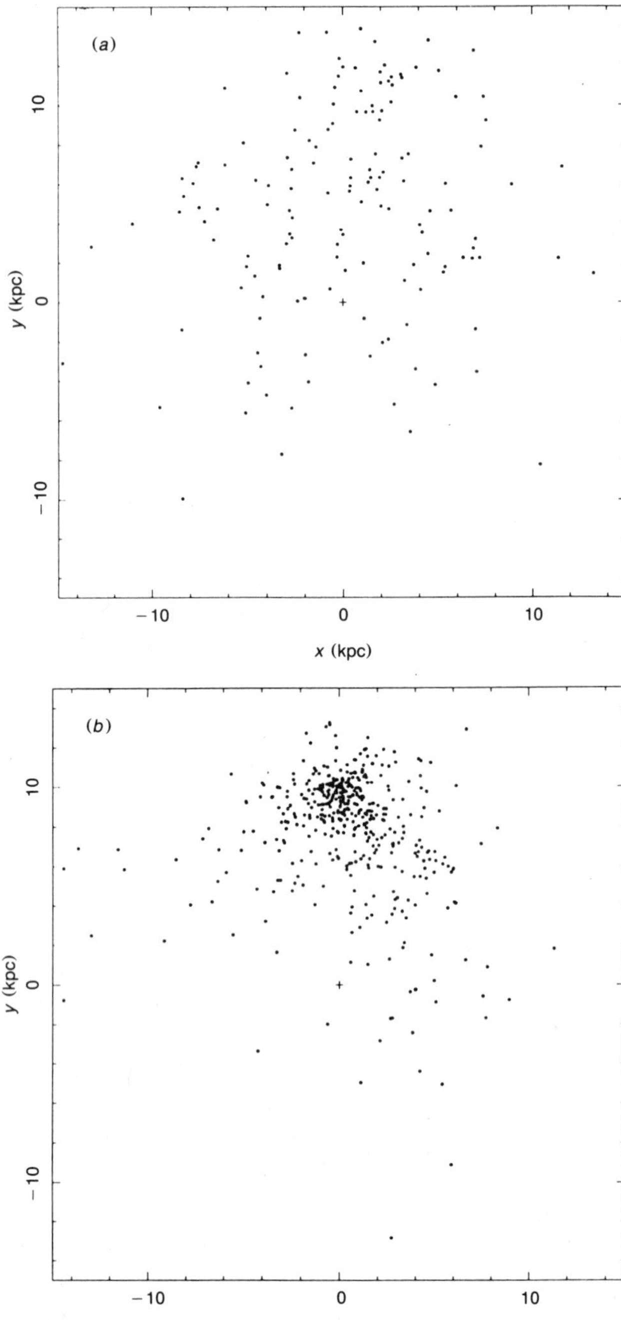

Figure 2. Distribution of known supernova remnants (*a*) and pulsars (*b*) across the galactic plane. The galactic center ($x = 0$, $y = 0$) is marked by a cross and the Sun is located at $x = 0$, $y = 10$ kpc in each figure. Known pulsars are strongly concentrated around the Sun, whereas the distribution of supernova remnants is more uniform across the Galaxy.

and hence distances calculated using it also have an uncertainty of the order of 50%.

Using distances computed in these ways, we can plot the distribution of the observed supernova remnants and pulsars across the plane of the Galaxy as shown in Fig. 2

This figure illustrates the main reason why the distributions of supernova remnants and pulsars in galactic latitude are so different. On the scale of the Galaxy, most of the known pulsars are relatively close to the Sun, whereas supernova remnants are observed over most of the galactic disk. The limited range of known pulsars is totally due to selection; most pulsars are of low radio luminosity and cannot be detected with current instruments unless they are relatively close to the Sun. The highly sensitive surveys with the Arecibo and Jodrell Bank telescopes simply observe pulsars to a greater distance and hence are able to detect a greater number in a given area.

Most of the pulsars shown in Figs. 1 and 2 are located within the galactic disk. There is, however, a small but growing group of pulsars that are being found in galactic globular clusters. These pulsars all have very short periods and have an evolutionary history different from that of the bulk of galactic pulsars. They belong to the class of "millisecond" pulsars, the first of which, PSR 1937+21, has a period of only 1.5 ms and was discovered in 1982. This pulsar and several others with similar properties are located within the galactic disk. Despite their short periods, millisecond pulsars are very old pulsars that have been given a new lease on life by being spun up as a result of accretion of matter from a binary companion.

GALACTIC DISTRIBUTIONS

Although most of the known pulsars are relatively close to the Sun, there is no reason to believe that the region around the Sun is special on a galactic scale. Indeed, we must assume that the distribution of both pulsars and supernova remnants has cylindrical symmetry about an axis through the galactic center in order to correct the observed distributions for selection effects.

The correction procedure for supernova remnants is relatively simple. Figure 2 shows that we can detect supernova remnants across most of the Galaxy. In particular, there is little selection

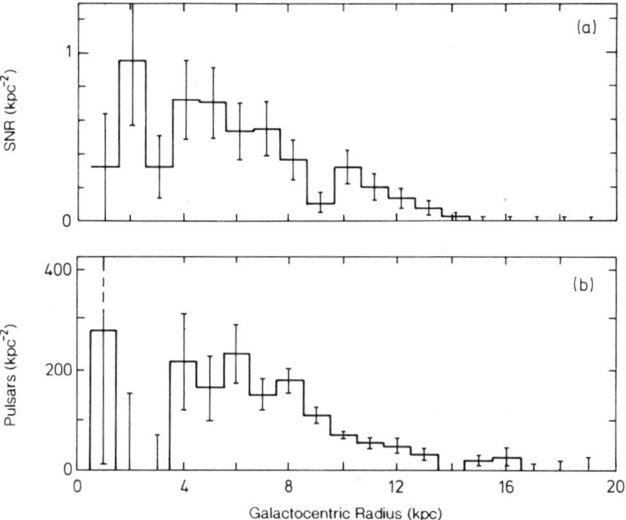

Figure 3. Distribution of supernova remnants (a) and pulsars (b) as a function of galactocentric radius R after correction for the principal selection effects. The ordinate in each case is density on the galactic plane after summing over z distance and luminosity to some limiting value. Error bars show the uncertainty in density for each bin.

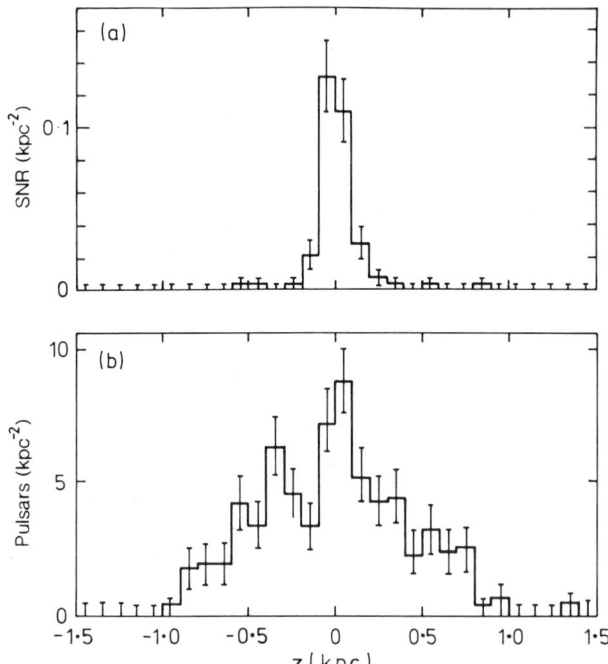

Figure 4. Distribution of supernova remnants (a) and pulsars (b) as a function of galactic z distance after correction for the principal selection effects. The ordinate in each case is the density per 100 pc bin of z distance at $R = 10$ kpc after summing to a limiting luminosity. Error bars show the uncertainty in density for each bin.

against the brighter remnants in the half of the galactic plane nearer the Sun. Because the surface brightness is related to the remnant diameter by the Σ-D relation, we can get a reasonably unbiased sample simply by considering only those remnants with $y > 0$ and diameters less than some limiting value. Figure 3a shows the distribution of such supernova remnants with linear diameter less than 50 pc as a function of galactocentric radius $R = (x^2 + y^2)^{1/2}$ and Fig. 4a shows the distribution in distance perpendicular to the galactic plane, $z = d \sin b$.

For pulsars the effects of selection are more severe and the correction procedures are more complex. We approximate the distribution of pulsars in R, z, and radio luminosity L by the product of three different independent distribution functions, $\rho_R(R)$, $\rho_z(z)$, and $\rho_L(L)$. We then solve for each of these functions using equations of the form

$$\rho_R(R) = N_R(R) \Big/ \left[\sum_z \sum_L V(R, z, L) \rho_z(z) \rho_L(L) \right],$$

where $N_R(R)$ is the observed distribution as a function of R, $V(R, z, L)$ is the volume of the Galaxy searched for pulsars of galactocentric radius, R, z-distance z, and luminosity L, and Σ_z and Σ_L are summations over z and L. Selection effects are considered in the computation of the volume V; for example, the volume of the Galaxy searched for low luminosity pulsars is very small because they are below the limiting flux density if they are too far away. Therefore the true number of these pulsars in the Galaxy is much larger than the observed number. The distributions ρ_R and ρ_z resulting from the iterative solution of these equations are shown in Figs. 3b and 4b, respectively.

Within the uncertainties, the form of the R distributions for supernova remnants and pulsars is the same. This supports the notion that these two classes of object have a common origin. However, the density of pulsars is more than 2 orders of magnitude greater than the density of supernova remnants. If we integrate over the R distributions we find that the total number of supernova remnants in the Galaxy above the luminosity or surface brightness limit is 180, whereas the total number of pulsars is about 100,000. This difference is simply a result of the much longer active lifetime of pulsars. Independent estimates of the lifetime of pulsars are of the order of 10^7 yr, whereas supernova remnants last about 10^4 yr before they merge into the galactic background. When account is taken of the fact that we can only observe about 20% of active pulsars because the beams of many do not sweep over the Earth, the derived birthrates for pulsars and supernova remnants are comparable, about one birth per 50 yr. This provides further support for a common origin.

Unlike the R distributions, the z distributions for supernova remnants and pulsars are rather different. Both peak at $z = 0$, the galactic plane, but the pulsar distribution is much wider than that for supernova remnants. The scale height for supernova remnants is about 80 pc, comparable to the scale height of the progenitor O and B stars; that of pulsars is about 400 pc. Direct measurements of pulsar proper motions show that pulsars are high velocity objects, with a median velocity of the order of 100 km s^{-1}. The observations are consistent with pulsars being born near the galactic plane and moving away from it during their lifetime.

Additional Reading

Clarke, D. H. and Caswell, J. L. (1976). A study of galactic supernova remnants, based on Molonglo–Parkes observational data. *Monthly Not. Roy. Astron. Soc.* **174** 267.

Green, D. A. (1988). A revised reference catalogue of galactic supernova remnants. *Ap. Space Sci.* **148** 3.

Lyne, A. G., Manchester, R. N., and Taylor, J. H. (1985). The galactic population of pulsars. *Monthly Not. Roy. Astron. Soc.* **213** 613.

Lyne, A. G. and Graham-Smith, F. (1990). *Pulsar Astronomy*. Cambridge University Press, Cambridge.

Manchester, R. N. and Taylor, J. H. (1976). *Pulsars*. W. H. Freeman, New York.

Taylor, J. H. and Stinebring, D. A. (1986). Recent progress in the understanding of pulsars. *Ann. Rev. Astron. Ap.* **24** 285.

Weiler, K. W. and Sramek, R. A. (1988). Supernovae and supernova remnants. *Ann. Rev. Astron. Ap.* **26** 295.

See also **Pulsars, all entries; Supernova Remnants, Observed Properties.**

Supernova Remnants, Evidence for Nucleosynthesis

Claude R. Canizares

Most of the chemical elements found on Earth were formed in the interiors of stars over the past 5–10 billion years. Just after the Big Bang, the universe contained only hydrogen and helium with tiny traces of other light elements like lithium. These early species were the simplest elements, those whose atomic nuclei consist of one or at most a few protons and neutrons. Heavier elements like carbon, oxygen, and iron, had to come later. They were, and continue to be, "synthesized" deep in the interiors of stars by a step-by-step process of nuclear fusion: so-called nucleosynthesis. Those stars that end their normal lives explosively as supernovae spew out their newly formed chemical species together with still heavier atomic nuclei that were fused during the explosion itself. In interstellar space, the heavy elements eventually mix with other material to form clouds, some of which may later collapse to make new stars and possibly planets. Thus a galaxy like our Milky Way hosts a continual cycle of chemical enrichment. Most of the time, that process is hidden from our view by the dense outer layers of the star. The enriched material may be seen only during the relatively brief interval following its ejection by the supernova but before it loses its identity in the interstellar medium. A major focus of observations of supernova remnants has been the detection and quantitative measurement of these chemical elements.

How far the process of nucleosynthesis proceeds in a star depends on its mass. Stars like the Sun normally produce elements as heavy as carbon and oxygen, but their densities and temperatures are too low to fuse these into heavier elements. More massive stars do support the required conditions to synthesize elements such as oxygen, magnesium, silicon, and iron, as well as some still heavier ones. The various species are spatially segregated within the stellar core, with the heavier elements lying closer to the center where the densities are highest. When lower-mass stars explode, as type I supernovae, additional fusion does occur, giving rise to almost a solar mass of radioactive elements that quickly decay to form iron. Massive stars exploding as type II supernovae also create some iron and other heavier elements in addition to ejecting their preexisting oxygen, magnesium, and silicon.

The tool for determining the chemical composition of material at astronomical distances is spectral analysis. Each element in the periodic table is capable of emitting radiation at a unique set of frequencies or wavelengths that serve as its signature. Precisely where these wavelengths occur in the electromagnetic spectrum depends on the detailed atomic structure of the element and, in the case of atoms that are lacking one or more of their usual complement of electrons, on their state of ionization. Thus a spectral study of the radiation from a supernova remnant will show patterns of emission at certain wavelengths, spectral emission lines, that reveal the chemical composition of the radiating material. The strength of a given set of emission lines depends on the physical processes causing the atoms to radiate at all and on the quantity of the particular element: its abundance. By studying enough lines and comparing them to detailed astrophysical calculations, it is possible to separate out the latter, thus performing a quantitative chemical analysis. Such studies are most advanced in the visible part of the spectrum, where they have been applied for many decades. But many elements and ions have no characteristic lines at visible wavelengths. Space-borne ultraviolet telescopes, primarily the *International Ultraviolet Explorer*, have helped, but even these cannot detect the bulk of the material in a supernova remnant because it is too hot: As the ejecta run into the surrounding interstellar material they are shock-heated to 10^7–10^8 K and radiate primarily x-rays. The capabilities of x-ray detectors for spectral studies have been limited, but instruments on the *Einstein*, *EXOSAT*, and *Tenma* satellites, among others, have permitted some preliminary study of elemental abundances.

ENRICHED EJECTA OF MASSIVE SUPERNOVA REMNANTS

The unusual elemental abundances in the supernova remnant Casseopeia A (Cas A) were first noted by Walter Baade and Rudolph L. Minkowski in 1954. Cas A is the ~350-yr-old remnant of a type II supernova explosion in a star of roughly 20 $M_\odot$ at a distance of about 2 kpc. The mass estimates are made from x-ray images (Fig. 1). Optical pictures of Cas A show a series of wispy, gaseous filaments containing 0.01% of a solar mass. Proper motion studies indicate that some of these clumps of material are moving at velocities of ~6000 km s^{-1}, whereas others are moving at only ~150 km s^{-1}. Spectral studies of the fast-moving knots reveal that many are composed of nearly pure oxygen, whereas others have oxygen together with elements like sulfur, argon, and calcium, which are synthesized by the nuclear burning of oxygen. The more slowly moving knots also show unusual abundances but with a predominance of lighter elements such as nitrogen and possibly helium. X-ray spectral studies indicate excess silicon, sulfur, cal-

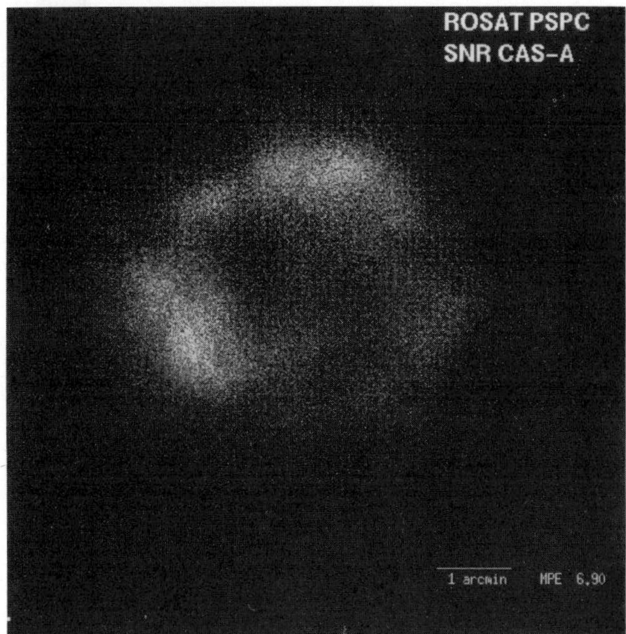

Figure 1. X-ray image of Cassiopeia A, a young supernova remnant with unusual elemental abundances. The image was obtained with the Position Sensitive Proportional Counter on the *Roentgen Satellite* (*ROSAT*), which was launched from Cape Canaveral on June 1, 1990. (*Courtesy of the ROSAT Team, Max Planck Institute for Physics and Astrophysics, Institute for Extraterrestrial Physics.*)

cium, and argon, and show that the bulk of the material is moving roughly as fast as the fast-moving knots.

All of this is fully consistent with the standard picture of nucleosynthesis, as already described. The fast-moving knots were presumably explosively ejected from the interior of the star. Their exact composition tells something about their point of origin in the core: The pure oxygen knots originated farther out than those that also contain heavier oxygen-burning products. The slower clumps, enriched with still lighter elements, are probably from the stellar envelope, which was ejected at much lower velocities during an interval prior to the supernova explosion.

There are a handful of other supernova remnants bearing similarities to Cas A. One of these, called N 132D, is located in our neighbor galaxy, the Large Magellanic Cloud. The optical and x-ray spectra of this object show a strong overabundance of oxygen, together with neon and other species. N 132D seems to have relatively less of the slightly heavier elements, like sulfur, than Cas A. This suggests that in N 132D the enrichment processes did not progress as far as in Cas A, which suggests that its progenitor star was of somewhat lower mass. For reasons not yet understood, optical pictures show that the oxygen-rich filaments in N 132D form a ring-like structure, expanding at ~ 3000 km s^{-1}.

Puppis A is a middle-aged supernova remnant, about 4000 yr old. This means that the stellar ejecta are already mixed with roughly 10 times their mass of interstellar material. Nevertheless, x-ray spectral studies indicate an overabundance of oxygen amounting to 3–5 times the mass of the Sun in this element alone. Some pure oxygen filaments have been found optically as well. To produce so much oxygen, the exploding star must have been even more massive than the progenitor of Cas A. The Puppis A supernova alone has synthesized and returned to the Milky Way enough oxygen for several hundred solar systems.

The progenitor star of the Crab nebula was probably at the low end of the high mass range, about 8 $M_{\odot}$, although some researchers believe the mass could have been up to 20 $M_{\odot}$. This remnant of an explosion noted by Chinese astronomers in 1054 AD left behind a rapidly spinning neutron star whose prodigious effusion of radiation and energetic particles overwhelms any emission from the shock-heated material. This is particularly true in the x-ray spectral region, where no emission lines have been detected, but it also confuses the interpretation of the optical spectra of the filaments. Nevertheless, there is good evidence that the Crab nebula is significantly enriched with helium and intermediate mass elements like carbon and oxygen.

REMNANTS OF LOW-MASS STARS

In contrast to the situation with high-mass type II supernova remnants, the low-mass type I remnants show scant evidence of unusual abundances. Their optical spectra are generally dominated by the emission lines of hydrogen, the most common element in the Universe. Most of the emitting hydrogen is probably not from the exploding star itself, but rather from the interstellar medium being heated by the expanding shock wave. Why the enriched material is so elusive has been something of a puzzle for the past several decades.

The youngest type I supernova remnants in the Milky Way are named Tycho and Kepler, after the astronomers who observed and studied the light from their explosions in 1572 and 1604, respectively. It is their careful records of the brightness of these supernovae that permit modern astronomers to classify them with some confidence. X-ray studies of Tycho show possible enhancements of elements such as silicon and sulfur, but there is nothing as dramatic as the pure oxygen knots of the type II remnants. Some of the knots in the Kepler remnant have overabundances of nitrogen. Like the slowly moving filaments in Cas A, this material appears to have undergone only moderate nuclear processing.

The most troubling question about type I supernova remnants is "Where is the iron?" Supernova theory has long predicted that a type I explosion should produce nearly a solar mass of iron, a hypothesis that has been bolstered by direct observations of the relevant processes in SN 1987A, a supernova in the Large Magellanic Cloud (which, however, is closer to type II than to type I). The problem is that iron emission lines are not unusually strong in type I remnants: x-ray studies find an iron abundance not too different from that found in the Sun or in typical interstellar matter, and the optical filaments show no excess iron. The resolution of this enigma seems to be that remnants like those of Tycho's and Kepler's supernovae are too young. The intense heat of the supernova explosion is quickly radiated away, so for tens to hundreds of years the expanding ejecta in a remnant are relatively cool and unobservable until they are reheated by the shock wave caused by their collision with surrounding matter. Presumably, the iron at the center of the younger type I remnants has not yet been reheated, and so it radiates no emission lines. Support for this explanation comes from observations of ultraviolet starlight that has passed through the center of the type I remnant of a supernova known from Chinese records to have exploded in the year 1006 AD. The spectral analysis shows an absorption line characteristic of excess iron at relatively low temperature.

Additional Reading

Matthews, G. J. and Cowan, J. J. (1990). New insights into the astrophysical *r*-process. *Nature* **345** 491.

Raymond, J. C. (1984). Observations of supernova remnants. *Ann. Rev. Astron. Ap.* **22** 75.

Seward, F., Gorenstein, P., and Tucker, W. L. (1985). Young supernova remnants. *Scientific American* **253** (No. 2) 88.

Shu, F. (1982). *The Physical Universe.* University Science Books, Mill Valley, Calif., pp. 102–157.

Trimble, V. (1975). The origin and abundances of the chemical elements. *Rev. Mod. Phys.* **47** 877.

Trimble, V. (1983). Supernovae. Part II: The aftermath. *Rev. Mod. Phys.* **55** 511.

See also **Stellar Evolution, Intermediate Mass Stars; Stellar Evolution, Massive Stars; Supernovae, all entries; Supernova Remnants, Evolution and Interaction with the Interstellar Medium; Supernova Remnants, Observed Properties.**

Supernova Remnants, Evolution and Interaction with the Interstellar Medium

Roger A. Chevalier

Supernovae are stellar explosions that deposit large amounts of energy in the interstellar medium. The resulting shock waves can heat and accelerate the surrounding gas and emission from this gas identifies supernova remnants. The evolution of supernova remnants depends on the properties of the surrounding medium, such as its cloud structure. The energy from supernovae can create large bubble structures and may determine the basic properties of the interstellar medium, such as its pressure and cloud velocities.

In 1946, Jan Oort identified the filamentary nebula called the Cygnus Loop (Fig. 1) as a supernova remnant, based on the observed velocities, on the momentum of the material that this implied, and on the lack of any existing stellar energy source for the nebula. This led to the "snowplow" model for the propagation of a supernova remnant, in which a cool shell sweeps up surrounding material that cools and is added to the shell. The cooling gas gives rise to the filamentary emission.

In the 1940s and 1950s, the remnants of Tycho's supernova (1572) and Kepler's supernova (1604) were identified at optical and radio wavelengths and they did not show the widespread, cooling, shock wave emission seen in the Cygnus Loop; yet they were clearly interacting strongly with the surrounding gas. In 1962, Iosif S. Shklovskii proposed that a theory developed for the expansion of nuclear explosions in the Earth's atmosphere could be applied to supernova remnants. This theory depends on the deposition of a given amount of energy in a uniform medium. The shock wave properties can be found by scaling to the parameters of interest, whether they be on an astronomical scale or one that is much smaller. The theory was eventually widely used in the interpretation of x-ray observations of supernova remnants.

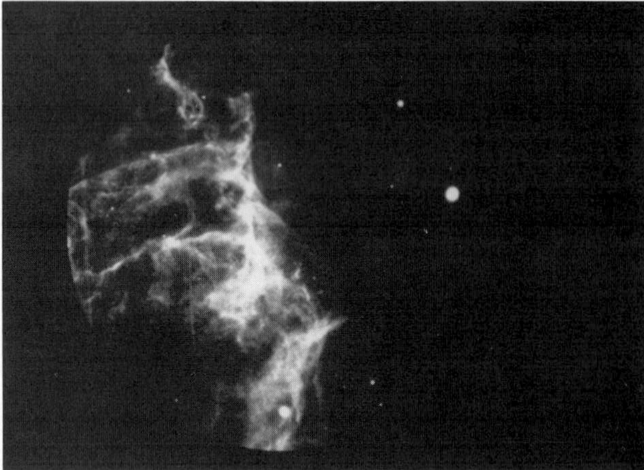

Figure 1. A portion of the Cygnus Loop, as photographed in ultraviolet light at an effective wavelength of 1550 Å on December 5, 1990, by the Ultraviolet Imaging Telescope of the *Astro-1* mission, (*Courtsey of Theodore P. Stecher, NASA/Goddard Space Flight Center*.)

In the 1950s, a strong radio source, Cassiopeia A, was identified with a clump of optical emission knots. Studies of these knots showed that they were freely expanding at high velocities away from a central point. The knots appeared to be fragments of an exploded star, the result of a supernova. The variety of appearances of supernova remnants was eventually placed into an evolutionary picture.

EVOLUTIONARY PHASES OF SUPERNOVA REMNANTS

The idealized problem of a supernova explosion in a uniform medium can be divided into four evolutionary phases. The first phase is dominated by the free expansion of the exploded star. Mass in the star is accelerated in the explosion and subsequently moves at constant velocity. Only the very outer parts of the exploded star interact with the surrounding medium and are slowed down by passing through a shock front. As the supernova expands, it interacts with more of the surrounding gas and an increasing amount of the supernova gas is shocked and decelerated. When the amount of swept-up gas is comparable to that in the exploded star, most of the exploded star is slowed and the free expansion phase has ended.

At this point, most of the supernova energy has been transferred to the surrounding medium and a blast wave is driven into that medium. The nature of the expansion is similar to that of the initial expansion of a nuclear explosion in the Earth's atmosphere. Although energy is conserved during the expansion, the shock wave slows down and the temperature of the shocked gas decreases. As this occurs, the shocked interstellar gas is able to radiate away its energy more effectively and eventually the total radiated energy becomes a substantial fraction of the initial supernova energy. The supernova remnant makes a transition from the energy-conserving phase to the radiative phase.

Because the gas density is the highest immediately behind the shock wave, most of the radiative cooling occurs there and a dense, cool shell forms behind the shock front. The shell acts as a snowplow, as described by Oort, and newly shocked gas is added to the shell. Although Oort estimated the shell expansion by the conservation of momentum in the shell, it has since become clear that the pressure of the hot, low-density gas inside the shell is significant and the expansion is thus somewhat more rapid than would be expected from momentum conservation. However, the shock front steadily decelerates and the pressure of the postshock gas decreases. Eventually the supernova remnant pressures and velocities become comparable to those in the interstellar medium and the remnant is no longer distinguishable from the interstellar medium. This fourth, and final, phase signals the end of the supernova remnant's life.

For typical interstellar conditions, the initial free expansion phase lasts several hundred years, until the supernova has expanded to a radius of about 10 light-years (ly). The remnant Cassiopeia A is at least partially in this phase, as indicated by the fast-moving knots. The energy-conserving phase can last for tens of thousands of years, until the remnant has expanded to a radius of about 100 ly. Tycho's remnant may be in this phase, whereas the filamentary emission from the Cygnus Loop indicates that it is mostly radiative. The radiative phase can last for hundreds of thousands of years.

Although this scheme provides a useful guide to supernova remnant evolution, it has become clear that there are complications. Many supernovae are thought to have massive star progenitors: stars with initial masses at least eight times that of the Sun. During their initial evolutionary phases, these are hot stars with energetic stellar winds and with large luminosities of photoionizing radiation. The effect of the winds and radiation on the surrounding interstellar medium is to create a low-density bubble with a radius of 100 ly or more. In a later evolutionary phase, massive stars become cool supergiant stars with slow, dense winds. The explosion of a massive star initially interacts with this wind; the Cassiopeia A remnant shows evidence for such interaction. Supernova remnant expansion into the low-density wind bubble can prolong the free expansion phase. Remnants may be difficult to observe in such a phase because they do not radiate effectively.

Another complication is that the interstellar medium itself is not uniform, but shows a cloud structure. This structure can blur the distinction between the second and third evolutionary phases of a remnant because the shock waves in clouds may be radiative whereas the shock waves in the intercloud medium are energy-conserving. With an inhomogeneous medium present inside supernova remnants, the dense gas is at a lower temperature because of pressure equilibrium, so that heat conduction can transfer energy between the hot and cool gas if the charged particle motions are not impeded by magnetic fields. The result can be an exchange of mass between the hot and cool media and a range of intermediate-temperature gas in the conductive interfaces. Clear evidence for the operation of heat conduction in remnants has not yet been obtained.

RELATIVISTIC PARTICLES

All supernova remnants are observed to be sources of continuous radio emission that is interpreted as synchrotron emission from relativistic electrons gyrating in a magnetic field. In the younger remnants, like that of Tycho's supernova, the radio emission is close to the site of the outer shock front and a shock wave acceleration mechanism for relativistic electrons is indicated. A theory for shock wave acceleration has been developed that is able to produce a power law energy spectrum of relativistic particles, as is implied by the observations. In such a spectrum, the number of particles of a given energy is proportional to that energy to a fixed power. In older radiative remnants, there is a strong compression of the interstellar gas behind the shock front and the compression of the interstellar cosmic ray electrons and of the interstellar magnetic field may be able to produce the observed radio emission. Direct shock acceleration may also be a factor in this case.

The shock wave acceleration mechanism is thought to be more effective for protons than for electrons, but the overall efficiency of the acceleration mechanism is not well-determined. If the efficiency is very high, the relativistic particles can have an effect on the supernova remnant dynamics, in particular, on the shock wave compression. Particle acceleration in supernova remnants is an excellent candidate for the acceleration of the cosmic-ray particles observed at the Earth and throughout the Milky Way galaxy because relativistic particles are clearly generated in remnants and because supernovae can provide the energy required for the population of cosmic rays.

EMISSION PROCESSES IN SUPERNOVA REMNANTS

Although synchrotron radiation is commonly observed from supernova remnants, the emission is from a minor dynamical component and it is the thermal gas emission that provides the most useful diagnostic information. In young remnants, the gas is shock-heated to a temperature of one million degrees kelvin (10^6 K) or more, so that the radiation is primarily in the x-ray wavelength region. Although the atomic processes involved are fairly well understood, there are uncertainties in the shock-heating because it depends on poorly understood plasma processes. The shock transition is called collisionless because ordinary charged particle collisions are not important. The degree of electron heating in such a shock wave is not known. Other complications are that the gas does not have time to come into ionization equilibrium so that the state of ionization depends on the time history of the gas and that the heavy element abundances may be determined by the supernova ejecta. High spectral and spatial resolution observations will be necessary to sort out the physical state of the gas.

If dust is mixed in with the gas, as expected for interstellar gas, the dust grains can be heated by collisions with the hot gas so that they radiate infrared emission before they are sputtered away by the collisions. Recent observations have shown that young supernova remnants are infrared sources; the infrared luminosity is an indicator of the dust density and thus of the gas density.

Finally, optical and ultraviolet emission close to shock fronts is a useful diagnostic tool. In the energy-conserving stage, the hot gas is too highly ionized to emit significant optical emission, but neutral atoms that come through the shock front can have transitions that are excited before ionization of the atoms occurs. This mechanism emphasizes hydrogen line emission, which is observed in a number of young remnants. In older remnants with cooling shock waves, the cooling gas emits a range of ultraviolet, optical, and infrared lines. The line intensities can be used to estimate the shock velocity, the preshock density, and the heavy-element abundances in the gas.

COLLECTIVE EFFECTS

The combined effects of a number of supernovae can have a substantial impact on the interstellar medium. Massive stars are known to occur in groups of tens of stars. The combined effects of the supernova remnants and stellar winds from the stars can result in a large shell hundreds of light-years in radius. Such shells have been observed and are known as superbubbles. Their size can become comparable to the typical height to which interstellar gas extends from the galactic plane. When it reaches this height, the shock wave can break out into the galactic halo, where a large volume can be shock-heated because of the low density. A hot galactic corona can be created in this way.

Even in individual supernova remnants the cooling time for the hot, low-density gas inside of the shell in the radiative phase is long. In many cases, the cooling time is more than the time for the gas to be reheated by another supernova remnant so that the gas remains hot and a hot phase of the interstellar medium is created. There is evidence for such a phase from observations of absorption by highly ionized oxygen; the implication is that at least 20% of the volume of the interstellar medium is in the hot phase. The role of supernovae in creating this medium probably also means that they determine the pressure of interstellar gas.

Another effect of supernova remnants is that they are able to produce interstellar cloud velocities, either by overrunning and accelerating a preexisting cloud or by the formation of a shell of cloud material in the radiative phase. The role of supernovae is assured because they are a dominant energy source in the interstellar medium.

Finally, supernovae are thought to be the dominant source of heavy elements in the universe. Observations of stars and the typical interstellar medium show that these elements are quite well mixed throughout the Galaxy. Much of the mixing probably takes place during supernova remnant evolution, a process that is fundamental to the evolution of stars and galaxies.

Additional Reading

Chevalier, R. A. (1977). The interaction of supernovae with the interstellar medium. *Ann. Rev. Astron. Ap.* **15** 175.

Chevalier, R. A. (1990). Interaction of supernovae with circumstellar matter. In *Supernovae*, A. G. Petschek, ed. Springer, New York, p. 91.

Falle, S. A. E. G. (1987). The evolution of older remnants. In *Supernova Remnants and the Interstellar Medium*, R.S. Roger and T. L. Landecker, eds. Cambridge University Press, London, p. 419.

McCray, R. and Snow, T. P., Jr. (1979). The violent interstellar medium. *Ann. Rev. Astron. Ap.* **17** 213.

McKee, C. F. (1987). Supernova remnant shocks in an inhomogeneous interstellar medium. In *Supernova Remnants and the Interstellar Medium*, R. S. Roger and T. L. Landecker, eds. Cambridge University Press, London, p. 205.

See also **Cosmic Rays, Acceleration; Cosmic Rays, Origin; Interstellar Medium, Galactic Corona; Interstellar Medium, Hot Phase; Shock Waves, Astrophysical; Supernova Remnants, Evidence for Nucleosynthesis; Supernova Remnants, Observed Properties.**

Supernova Remnants, Observed Properties

John R. Dickel

The explosion of a star, a supernova, releases over 10^{50} erg of energy, hurtling its innards out into the surrounding region at highly supersonic speeds, and is truly a spectacular event. Its remains can still be identified thousands of years later as a supernova remnant (SNR) that produces radiation covering nearly the entire electromagnetic spectrum, from radio waves through the infrared and visual regions to the very high-energy, short-wavelength x-rays. The radiation in each wavelength range has a different physical origin. A great variety of structures are seen in each wavelength range and from one wavelength range to another, both within an individual remnant and from remnant to remnant. Although the space around the supernova is nearly a vacuum, there is some interstellar material present. The tremendous speed of its expansion causes the remnant quickly to sweep up more matter than was ejected, so that the diversity of structure actually shows the tremendous range of conditions present in the medium surrounding the supernova. Some hints about the ejecta are also present in the early years. Because SNRs appear so different in each wavelength range, we shall discuss each kind of radiation separately while tying together common threads that give us a full picture of the interaction of the remnant with its surroundings. Figure 1 is a sample composite picture of the emission from a young SNR at two wavelengths.

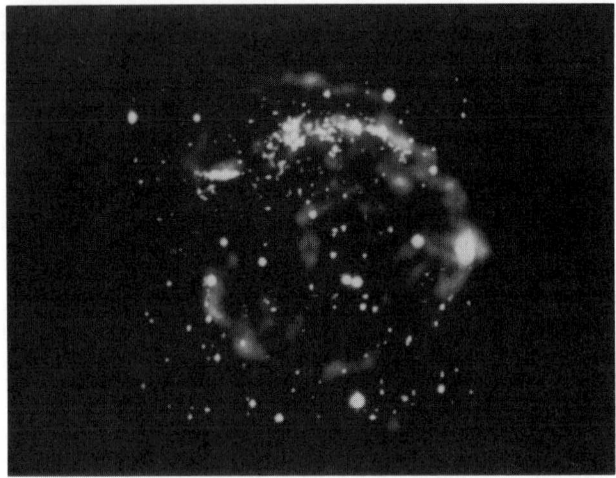

Figure 1. Visual representation of two separate images of the supernova remnant Cassiopeia A, the remains of a star that exploded about three centuries ago and hurled its material out into its surroundings at about 20 million miles per hour. The ejecta have slowed down somewhat and have the appearance of a shell that is about 10 light-years in diameter. The bright white represents the filaments seen in visible light, or what your eye would see, but greatly enhanced by a deep photograph (i.e., a long exposure) taken with a large telescope. The grey areas are regions that emit very long wavelength radio waves, which are normally invisible; this part of the image is what the object would look like if we could see radio waves. Also see Figure 1, p. 899. (*Data from J. Dickel, E. Greisen, and S. van den Bergh.*)

RADIO OBSERVATIONS OF SNRs

Supernova remnants come from evolved massive stars, which are concentrated in the plane of our galaxy (the Milky Way). This means that they are readily obscured by intervening interstellar dust, which is also concentrated in the galactic plane. Radio waves, which cover the long-wavelength part of the spectrum, can penetrate the dust better than visible light, ultraviolet light, and x-rays so that the best way to identify SNRs is by their radio emission.

This emission is produced by the synchrotron mechanism: Superenergetic electrons, moving at nearly the speed of light, are trapped and accelerated by interstellar magnetic fields. A similar process happens in laboratory particle accelerators called synchrotrons; thus, supernova remnants can be regarded as giant cosmic synchrotrons. As the relativistic electrons spiral around magnetic field lines, they produce a characteristic radio synchrotron emission. The spectrum (the intensity of this emission as a function of wavelength) depends upon the energies of the collection of radiating electrons. These energies are generally distributed over a wide range, but with the number of electrons increasing at lower energies. This means that we observe a continuous spectrum with the greatest brightness at low radio frequencies and a gradual decrease in brightness toward higher frequencies. The detailed relation can be represented by a power law such that the emission intensity I is proportional to the frequency f raised to a constant power α, often called the spectral index; in other words, we find $I = (\text{constant}) f^{\alpha}$, where α is typically -0.5. Individual remnants can vary from the average by about ± 0.2 in α. Spectra of SNRs with ages of only a few hundred years generally have a somewhat steeper slope (i.e., the absolute value of α is larger) than those with ages of thousands of years. The gradual change in spectral slope indicates that the electrons still are being accelerated to higher energies by turbulent interaction and/or by shock waves created by the interaction of the expanding material with irregularities in the surrounding interstellar and circumstellar gas. Some relativistic electrons were present as part of the cosmic-ray background but many more have been spawned by the interaction processes. Indeed, supernova remnants do not "turn on" until about 100 yr after the explosion, when sufficient particles have been accelerated. As a remnant further ages and its energy is dissipated, the acceleration processes die down and the brightness begins to decrease slowly.

The magnetic fields in SNRs (which must be present to account for the observed synchrotron radiation) were present originally, threading through interstellar space, but have been strengthened greatly by being stretched and squeezed at the unstable interfaces between clumps in the surrounding medium and the expanding blast wave. The magnetic fields direct the electrons so that they move in very prescribed paths. This results in an orientation, or polarization, of the synchrotron radiation in a direction perpendicular to the alignment of the magnetic field. Thus, by measuring the polarization we can determine both the degree of ordering of the magnetic field and its direction. The measurements generally show only a small degree of polarization, so the magnetic field must be very disordered and tangled on small scales, but the alignment is found to have a small net value in a direction stretching radially outward from the explosion center.

The other important observed characteristics of the radio emission from SNRs concern the morphology, or apparent structure of the remnants. The most obvious feature is the appearance of a hollowed-out shell with a thickness of perhaps one-fourth the radius of the remnant. The emission is brightest around the edge but there is some emission from the direction of the center that is caused by radiation from the front and back parts of the shell along the line of sight. The outline of the shell often is not perfectly round, because, as the expanding remnant sweeps up the surrounding material, clumps in the interstellar medium will retard the expansion with respect to the expansion in the regions between clumps. The older the remnant, the more irregular its outline becomes.

It should be mentioned that a few young SNRs actually appear to fill completely the volume they occupy instead of being shells. The prototype of these objects (which sometimes are called plerions) is the Crab nebula, so called because of its (rather far-fetched) resemblance to the shape of a crab, seen in an early telescopic study. The Crab nebula contains a central pulsar that is the neutron-star remains of the condensed central volume of the star that became the supernova. This rapidly spinning, extremely compact object is gradually slowing down through friction with its surroundings. The energy lost during this "spin down" powers the Crab SNR, particularly at high energies, so that it and similar SNRs have flatter than average synchrotron spectra. Some leakage of particles and field apparently produces emission from the whole volume. Indeed, at middle times, as the influence of the pulsar decreases, some remnants show a composite structure with an inner "crab-type" structure and an outer decoupled shell.

The pure shell remnants also show changes in structure with time. Not only does the outline become less round as holes, clouds, and other features of the interstellar medium are encountered in different directions, but the generally smooth shell breaks up into myriads of very thin filaments. There is some controversy as to whether these filaments are thin sheets that are seen edge-on or if they are indeed string-like. In reality, they are probably bits and pieces of sheets that have been shredded to ribbons that take on the beautiful wispy character virtually identical to that seen in visual wavelengths.

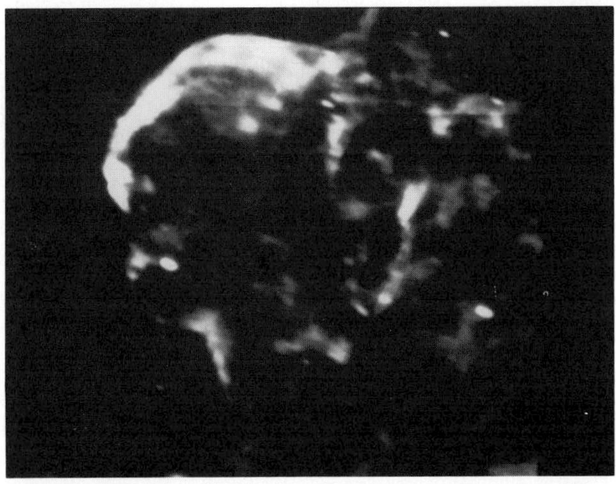

(a)

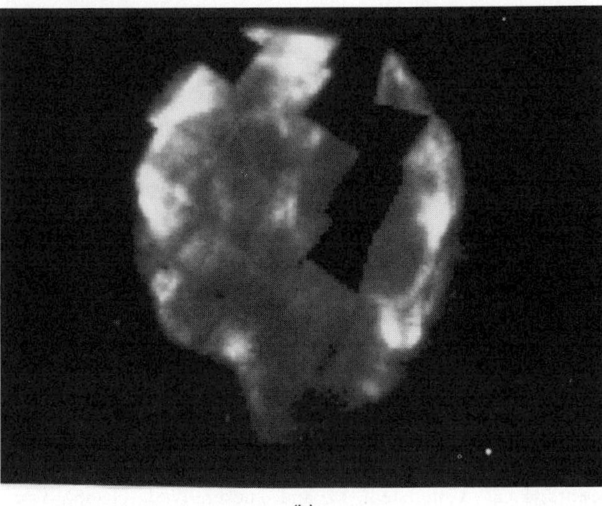

(b)

Figure 2. (*a*) An image of the infrared emission from the old supernova remnant called the Cygnus Loop, as observed at a wavelength of 60 μm by the *IRAS* satellite. This SNR is perhaps 40,000 yr old and is over 100 light-years across. There obviously is heated dust present although the outline is very irregular and some contaminating emission is visible on the upper right (*courtesy of R. Arendt*). (*b*) An energetic x-ray image of the same object taken with the *Einstein* observatory (*W. Ku and co-workers*). Also see Figure 1, p. 900.

INFRARED PROPERTIES OF SNRs

Infrared emission is generally associated with thermal radiation from cold dust in the interstellar medium and it was somewhat of a surprise when the *Infrared Astronomical Satellite* (*IRAS*) found in 1983 that energetic supernova remnants were bright in the infrared (see Fig. 2). Indeed, in all cases where comparative measurements are available, the infrared emission of a SNR is brighter than that in any other wavelength range. Circumstellar dust, either in the general interstellar medium or from preexplosion mass loss by the initial star, is heated collisionally by the very hot, shocked gas from the more tenuous regions between the clumps of dusty material or is heated by x-rays.

It is often difficult to identify the structure of a supernova remnant in the infrared because near the plane of the Milky Way there are so many clouds of interstellar material that confusion along the line of sight distorts our view. Only about one-third of all known SNRs can be identified unambiguously. When the SNR is obvious, the emission is very clumpy and irregular as individual interstellar cloudlets are engulfed by the expanding material and shock waves. In some instances, we see a dense cloud abutting the expanding remnant identified at radio wavelengths. The cloud that is retarding the expansion will also be heated by the interaction with the ejecta and outward-moving shock, producing a very sharp observable boundary to the SNR. The energetics of the interaction can be evaluated by comparing the emission in the four different infrared bands measured by *IRAS* to determine the temperatures of both the heated and the cold dust. In many cases, the cloud essentially halts the expansion but often it will be heated and will eventually evaporate.

SNRs AT VISIBLE AND ULTRAVIOLET WAVELENGTHS

At visible wavelengths the emission from supernova remnants is frequently disappointing although a number of beautiful examples of fantastic filamentary structures are observed. The problem is that interstellar gas and dust so completely scatter and absorb visible light that only a few nearby remnants can be seen at all. These few do provide us with a wealth of information, however.

The optical and ultraviolet emission of a SNR is almost entirely in spectral lines from gas that has been excited and compressed by the expanding shock wave and by thermal instabilities. Slightly denser regions will tend to cool faster and thus compress further in order to try to maintain pressure balance with material between clumps. The visible gas generally has a temperature near 10,000 K and a density of a few hundred atoms per cubic centimeter, as opposed to millions of degrees kelvin and less than one particle per cubic centimeter between the filaments. The temperature and density can be determined by spectral measurements comparing the intensities of various individual lines. For example, the atomic

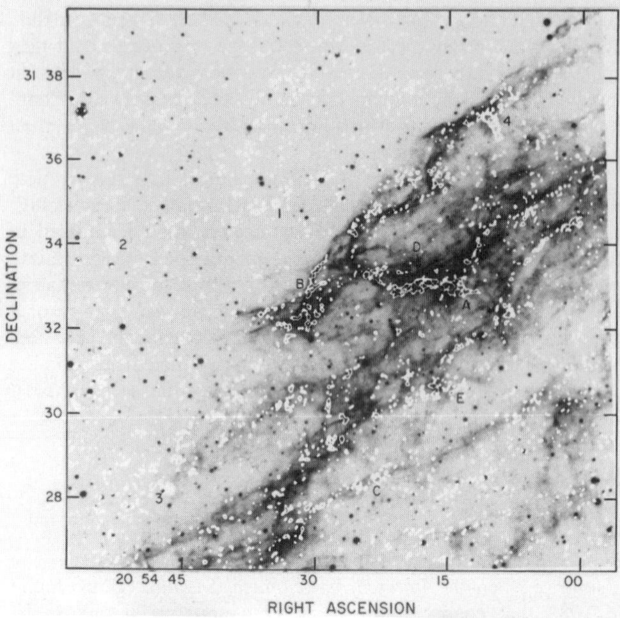

Figure 3. Part of the northeastern Cygnus Loop. The photograph shows light from a spectral line of ionized sulfur that should be very similar to that from the Balmer α line of hydrogen. The contours represent the brightness of 20-cm radio radiation. Strong radio emission correlates almost perfectly with the hydrogen glow. [*From W. Straka, J. Dickel, W. Blair, and R. Fesen,* Astrophysical Journal ***306*** *266 (1986).*]

levels that produce the Hα line of hydrogen at wavelength 656.3 nanometers (nm) are most readily excited at temperatures near 10,000 K and this line is very prominent in all SNRs. There is a pair of lines of singly ionized sulfur [S II] at wavelengths 671.7 and 673.1 nm whose relative brightnesses depend upon collisions among atoms to populate the energy levels of the atom responsible for their emission (see Fig. 3). The number of collisions depends, of course, on the density so that the ratio of these lines is a good density diagnostic. We note that in a terrestrial laboratory the density is so high that collisions tend to depopulate these levels completely and the sulfur lines are not visible at all; hence they are called forbidden lines and are visible only in the very tenuous gas of interstellar space. Another forbidden line, doubly ionized oxygen at wavelength 500.7 nm, is a good diagnostic of shock waves (which provide the energy to remove the two electrons from the atom). Thus this oxygen line is very prominent in supernova remnants. In addition, emission can be observed at both the visible and ultraviolet wavelengths from very highly ionized and excited gas, such as nine-times-ionized iron. This represents material that is either cooling from a very hot, shocked condition or being evaporated and heated from the denser, cooler clumps.

Although lines of many elements are present in the spectra of SNRs, most remnants have swept up so much interstellar material by the time they are observed, that their optical spectra generally cannot be used to diagnose the enrichment of heavy elements in the interstellar medium by the supernova process. One exception is a small class of oxygen-rich SNRs that, although still composed mostly of hydrogen, do show an overabundance of oxygen relative to most SNRs and the general interstellar medium. Much of this was probably material slowly lost from just below the atmosphere of the unstable star before the supernova explosion. It is now being encountered and excited by the rapid supernova ejecta and supersonic shocks.

Radial velocities (determined through the Doppler effect) of the various spectral lines reveal two important characteristics of SNRs. First, the observed wavelengths are significantly shifted from their rest values, indicating that the remnants are expanding very rapidly: Young SNRs expand at many thousands of kilometers per second and the older ones, which have already swept up much surrounding material, are still going at hundreds of kilometers per second. Most of the visible remnants have radiated away only a negligible amount of energy and so the expansion is guided by conservation of energy and must slow down as mass is accumulated. A second feature of the spectral lines is their large widths, which indicate a very turbulent and disturbed zone of interaction with significant random velocities for the material.

X-RAYS FROM SUPERNOVA REMNANTS

In SNRs we also find very hot gas that has been heated by passage of the shocks through both the circumstellar medium and the ejecta. This gas, at temperatures of up to 10^7 K, produces strong thermal bremsstrahlung radiation at x-ray wavelengths. It is generally fairly smoothly distributed as the shocks move radially and encompass the entire surroundings. As more material is encountered and the shock speed decreases, the temperature will go down. In addition, atoms of heavy elements (which are present in small numbers in the interstellar medium but in larger fractions in the ejecta) will be highly multiply ionized and will produce bright emission lines during the recombination process. The shock, of course, is a very transient phenomenon and so there is not a steady state. The material is very underionized compared to an equilibrium process.

SUMMARY

The observations of supernova remnants at all wavelengths demonstrate a very dynamic, ever-changing process as a tremendous amount of energy is deposited by the supernova ejecta into spatially variable surroundings. The future holds great promise as we begin to accumulate high-resolution, sensitive data at all wavelengths over a sufficient time baseline to allow us actually to observe the temporal changes resulting from the dynamic interaction of the ejecta with the material around them.

Additional Reading

Marschall, L. A. (1988). *The Supernova Story*. Plenum Press, New York.

Raymond, J. C. (1984). Observations of supernova remnants. *Ann. Rev. Astron. Ap.* **22** 75.

Reynolds, S. P. (1988). Supernova remnants. In *Galactic and Extra Galactic Radio Astronomy*, G. L. Verschuur and K. I. Kellermann, eds. Springer, Berlin, p. 439.

Roger, R. S. and Landecker, T. L., eds. (1988). *Supernova Remnants and the Interstellar Medium* Cambridge University Press, London.

Seward, F. D., Gorenstein, P., and Tucker, W. L. (1985). Young supernova remnants. *Scientific American* **253** (No. 2) 88.

See also **Radio Sources, Emission Mechanisms; Shock Waves, Astrophysical; Supernovae, General Properties; Supernovae, Historical; Supernova Remnants, all entries.**

T

Telescopes and Observatories, Solar

Oddbjørn Engvold

No two solar telescopes are alike, which suggests that the perfect telescope has not yet been conceived. Each designer weighs the various design parameters differently, depending on the scientific objectives for the instrument, characteristics of the particular site, and personal preferences. Major observatories usually have several solar telescopes designed for particular observational tasks such as high-resolution imaging, spectroscopy, polarimetry, coronal observations, and regular patrolling of solar activity.

The basic telescope design types are (1) *steerable telescopes* that point directly at the Sun, and (2) *fixed telescopes* that use large flat mirrors to reflect the light into a usually large, stationary image-forming system. Several modern solar telescopes are actually combinations of these two types. Such telescopes relay the light beam to stationary secondary or tertiary foci to allow for fixed mountings of post-focus instruments such as spectrographs and narrow-band filters (monochromators).

Steerable telescopes are necessarily compact and utilize optical elements with short focal length (f) relative to the mirror or lens diameter (D), that is, the f-ratio, $(f/D) \leq 20$. Telescope aberrations (spherical, coma, astigmatism) are nonnegligible in such telescopes and usually limit the usable field of view to less than the full disk of the Sun.

Telescopes used for observations of the very faint solar corona, so-called coronagraphs, utilize single objectives in order to minimize the internal scattering of light in the instrument itself. For observations over a wide optical wavelength range, one has to use achromatic doublet lenses such as found in the solar 50-cm refractor at Pic-du-Midi, France.

Fixed telescopes can be arranged both horizontally and vertically by using a two-mirror *coelostat* mounting, or the more compact alt-azimuth arrangement referred to as the *turret*, or "Sun seeker," system. Alternatively, the telescope light path can be oriented along the direction of the Earth's polar axis by directing the solar light beam into the telescope via one single flat *heliostat* mirror. Such systems make use of long focal length optics (f-ratios in the range 40–100). Examples of large fixed telescopes with coelostat mounts are the early Mt. Wilson Solar Tower, the Italian Arcetri Solar Tower, the modern Solar Vacuum Tower at Kitt Peak, Arizona, and the 60-cm German vacuum solar telescope at Izaña, Canary Islands. The McMath Solar Telescope of the National Solar Observatory (NSO) at Kitt Peak is the largest instrument of its kind. The unorthodox design includes a single 2-m diameter *heliostat* mirror and a long sloping tube and wind shield. (Table 1 is a list of optical solar telescopes.)

METEOROLOGICAL CONDITIONS AT SOLAR OBSERVATORY SITES

A number of successful solar observatories are located at inland lakes, taking advantage of the thermally stabilizing effect of a large body of water. The Big Bear Solar Observatory and San Fernando Observatory, both in California, are examples. Both test measurements and actual astronomical observations have shown that high-level island and coastal sites in certain latitude belts around the Earth show superior performance to inland sites both concerning daytime and nighttime *seeing*. The location of good sites is, furthermore, closely connected to the large-scale global circulation pattern, with ascending motion near the equator and descending air masses in subtropical latitudes, forming the trade wind system. High-level sites in these latitudes are generally located in semipermanent high-pressure systems above an inversion layer and are immersed in subsiding, dry and stable air masses. Both the Hawaiian and the Canary Island archipelagos fulfill these conditions and their excellent suitability for astronomical observations is demonstrated by the fact that about half a dozen telescopes have been built and are being operated successfully at each of them.

OPTICAL TURBULENCE IN THE VICINITY OF TELESCOPES

Turbulence close to the telescope aperture degrades the seeing. Metallic surfaces and other local surfaces heated by the Sun will generate thermal convection and turbulence. A telescope protruding in the wind is a turbulence generator. The latter is largely eliminated if the air can flow smoothly over the structure such as for the domeless and aerodynamic towers of the vacuum telescope at the NSO, on Sacramento Peak, New Mexico, the Japanese Hida telescope, and the Swedish telescope at La Palma, Canary Islands.

EVACUATION OF TELESCOPE LIGHT PATH

Evacuating the telescope light path eliminates *internal* seeing. Most modern solar telescopes are therefore vacuum telescopes. The major drawback of such systems is that thermal and mechanical stresses in the entrance windows give rise to optical aberration and polarization. The light path of the planned large-aperture telescope LEST will be filled with helium gas and thus require the use of a 1–2-cm-thick entrance window.

EXAMPLES OF NEW SOLAR TELESCOPE FACILITIES

The German solar telescope installations at Observatorio del Teide, Izaña, consist of the 70-cm $f/66$ Vacuum Tower Telescope (VTT) of the Kiepenheuer Institute, Freiburg, and the 45-cm $f/56$ Gregory–Coudé of Göttingen Observatory, Göttingen. The VTT has a classic coelostat configuration and a vertical tower telescope.

The Swedish solar telescope at Roque de los Muchachos, La Palma, is patterned after the vacuum tower telescope at Sacramento Peak. The combination of an optically simple and good system and a superb site has made this telescope one of the very best in the world for high-resolution studies of the Sun. Its domeless turret design is very compact and its aerodynamic shape gives a minimum of local disturbance and eliminates dome seeing completely.

NEAR FUTURE SOLAR TELESCOPE PROGRAMS

The French polarization-free solar telescope, the Télescope Héliographique pour l'Etude du Magnetism et des Instabilités Solaires (THEMIS) will operate at Izaña, on the island of Tenerife, Canary Islands.

The Large Earth-Based Solar Telescope (LEST) is an ambitious ground-based solar telescope program for high spatial resolution polarimetric observations of the Sun. This international collaboration includes the United States and several major European countries. Its aperture diameter will be 2.4 m and the diffraction limit of the modified Gregorian system will be 0.05 arcsecond at $\lambda 5000$ Å. The LEST will be placed on La Palma (2360 m).

Table 1. Optical Solar Telescopes Listed According to Aperture Diameter

Telescope and Observatory	*Mount / System*	*Aperture, Diameter (D) (cm)*	*Focal Ratio f_{eff}/D*	*Tower Height (m)*	*Altitude of Observatory (m)*
McMath Solar Telescope NSO, Kitt Peak, AZ, USA	H	M 152	60	31	2060
Solar Tower Telescope Crimea, USSR	C	M 90	56		570
Vacuum Tower Telescope NSO, Sacramento Peak, NM, USA	T	M 76	72	41.5	2810
Baikal Solar Telescope USSR	S	L 76	53	24	660
Vacuum Solar Telescope NSO, Kitt Peak, AZ, USA	C	M 70	60	23	2060
San Fernando Solar Observatory California State U., Northridge, USA	E/G–C	L 61	32.0	14	366
Meudon Solar Tower, France	C	M 60	54	23	—
Domeless Solar Tower Telescope Kwasan and Hida Observatory, Japan	A/G	M 60	53.7	23	1300
German Vacuum Tower Telescope Izaña, Tenerife, Spain	C	M 60	76	38.5	2400
Coronagraph Peak Alma Ata, USSR	E	L 53	24		3000
Swedish Solar Tower La Palma, Spain	T	L 50	45	16	2360
Solar Refractor Pic-du-Midi, France	E	L 50	70		2860
Coronagraph Sayan Observatory, USSR	E/C	L 50	24	8	2000
Horizontal Solar Telescope Ondrejov, Czechoslovakia	C	45	78	7	
Göttingen Vacuum Telescope Izaña, Tenerife, Spain	E/G–C	M 45	53	19	2400
Monte Mario Solar Tower Rome, Italy	C	L 45	62	34	143
Horizontal Solar Telescope Peak Alma Ata, USSR	C/N	M 44	40		3000
Solar Tower Telescope Nanjing University, China	C	M 43	50	21	36
Big Bear Solar Observatory Big Bear City, CA, USA	E/S	M 41–23	35–14	9	2042
Horizontal Solar Telescope Yunnan Observatory, China	C	M 40	40	—	2000
German–Spanish Solar Telescope Izaña, Tenerife, Spain	E/N	M 40	7	12.5	2387
Kanzelhöhe Solar Telescope Graz University, Austria	A/G	M 40	65	12	1526
Solar Tower Telescope Kodaikanal Observatory, India	C	L 38	95	11	2343
Huairou Telescope–Magnetograph Beijing, China	E	L 35	8	23	900
Domeless Coudé at Capri Kiepenheuer Institute, Germany	E/C	L 35	46	9	137
Mt. Wilson Solar Tower Pasadena, CA, USA	C	L 30	150	46	1740
Solar Tower Telescope Kislovodsk, Pulkovo Observatory, USSR	C	M 30	57		2070
Arcetri Solar Tower Florence, Italy	C	L 30	151	52	
Norikura Horizontal Telescope Tokyo Astronomical Observatory, Japan	E	L 25	35		2876
Ottawa River Solar Observatory, Canada	E	L 25	17	5	58

Table 1. (*continued*)

Telescope and Observatory	*Mount / System*	*Aperture, Diameter (D) (cm)*	*Focal Ratio f_{eff}/D*	*Tower Height (m)*	*Altitude of Observatory (m)*
Mees Solar Observatory Haleakala, Hawaii, USA	E	L 25	17		3050
Debrecen Coronagraph Heliographic Observatory, Hungary	E	L 25	18		127
Solar Optical Telescope National Observatory, Greece	E	L 25	20		110
Horizontal Solar Telescope Kiev, USSR	C	M 22	36		162
Arosa Coronagraph ETH, Switzerland	E	L 20	11	3	2050
Coronagraph Skalnate Pleso, Czechoslovakia	E	L 20	15		1783

Telescope mounts: H, heliostat; C, coelostat; S, siderostat; T, turret; A, alt-azimuth; E, equatorial. Optical systems: G, Gregorian; N, Newtonian; C, Coudé; G–C, Gregorian–Coudé; S, several. Aperture: L, lens; M, mirror.

SOLAR MISSIONS IN SPACE

The Solar and Heliospheric Observatory (SOHO) is a space mission for studies of the solar interior and the outer solar atmosphere. The payload will include six instruments for studies of structures and dynamics of the Sun's chromosphere and corona. SOHO is currently scheduled for launch in July 1995 and is being designed for a lifetime of two years, but it will be equipped with sufficient on-board consumables for an extra four years.

The Orbiting Solar Laboratory (OSL) of NASA is a scaled-down version of the former proposed Solar Optical Telescope (SOT) and the High-Resolution Solar Observatory (HRSO). OSL will be a free-flying, polar-orbiting facility for high spatial and temporal resolution of the Sun over a spectral range from the x-ray to the near infrared. The OSL satellite will have a 1-m-aperture telescope optimized for λλ2000–11,000 Å, which feeds a narrow-band filter, a set of fixed broadband filters, and a visible-light echelle spectrograph.

The Naval Research Laboratory's High Resolution Telescope and Spectrograph (HRTS), which covers the wavelength range λλ1175–1700 Å will, in addition to being part of the OSL payload, also be flown on rockets during the next few years.

Additional Reading

Domingo, V. and Poland, A. I. (1989). Scientific and technical aspects of the instruments. In *The SOHO Mission*, ESA SP-1104, p. 7.

Dunn, R. B. (1985). High resolution solar telescopes. *Solar Phys.* **100** 1.

Livingston, W. C., Harvey, J., Pierce, A. K., Schrage, D., Gillespie, B., Simmons, J., and Slaughter, C. (1976). The Kitt Peak 60-cm vacuum telescope. *Applied Optics* **15** 33.

Mayfield, E. B., Vrabec, D., Rogers, E., Janssen, T., and Becker, R. A. (1969). A new solar observatory in California. *Sky and Telescope* **37** 208.

Schröter, E. H., Soltau, D., and Wiehr, E. (1985). The German solar telescopes at the Observatorio del Teide. *Vistas in Astron*. **28** 519.

Wyller, A. A. and Scharmer, G. B. (1985). Sweden's solar and stellar telescopes on La Palma. *Vistas in Astron*. **28** 467.

Zirin, H. (1970). The Big Bear Solar Observatory. *Sky and Telescope* **39** 215.

See also **Coronagraphs, Solar; Filters, Tunable Optical; Solar Magnetographs.**

Telescopes, Detectors and Instruments, Gamma Ray

Edward L. Chupp

Gamma ray telescopes are used by astrophysicists to study the highest-energy processes occurring at remote sites in the universe. Gamma rays, a high-energy form of electromagnetic radiation, extend in energy from about 100 keV, the energy of typical x-ray machines, to over a trillion (10^{12}) times this energy. Whereas conventional optical and some x-ray telescopes form images using well-known reflection optical systems and a detector consisting of fine grain film, a video camera, or a solid state array, gamma ray telescopes use less familiar techniques to obtain a "picture" of the sky. To study the majority of the gamma rays, which cannot penetrate the Earth's atmosphere, instruments must be carried above most of the Earth's atmosphere in order to detect the weak intensities of gamma rays from cosmic sources. Only the most energetic gamma rays can be detected at ground level where the Earth's atmosphere itself is used as a detector. Gamma ray telescopes for use in the 21st century will be so large and complex that they may be assembled on space stations or on the Moon.

GAMMA RAY TELESCOPE PRINCIPLES

Gamma ray telescope design is based on the fundamental properties of a gamma ray, which are its energy, polarization, and direction of origin. In Fig. 1 we summarize the basic steps and components involved in the design of a gamma ray telescope. Energetic ions and electrons that enter the gamma ray telescope must be prevented from causing false signals. This is accomplished by use of a thin detector as a filter, sensitive only to charged particles. Gamma rays pass through the filter and enter the focusing or imaging stage of the telescope (see Fig. 1). The gamma ray is then converted into a charged particle(s) by interacting with matter by one of three basic mechanisms, the probability of occurrence of each depending on the gamma ray photon energy. If the gamma ray energy is less than about 500 keV, then in most materials, it will be converted into a photoelectron which carries nearly all the gamma ray's energy. For energies between 500 keV and 20 MeV, a gamma ray will usually undergo Compton scattering and produce an electron ("Compton electron") and a gamma ray of low energy. This phenomenon, first observed at x-ray energies as early as 1920, was correctly explained by Arthur H. Compton in 1922, who

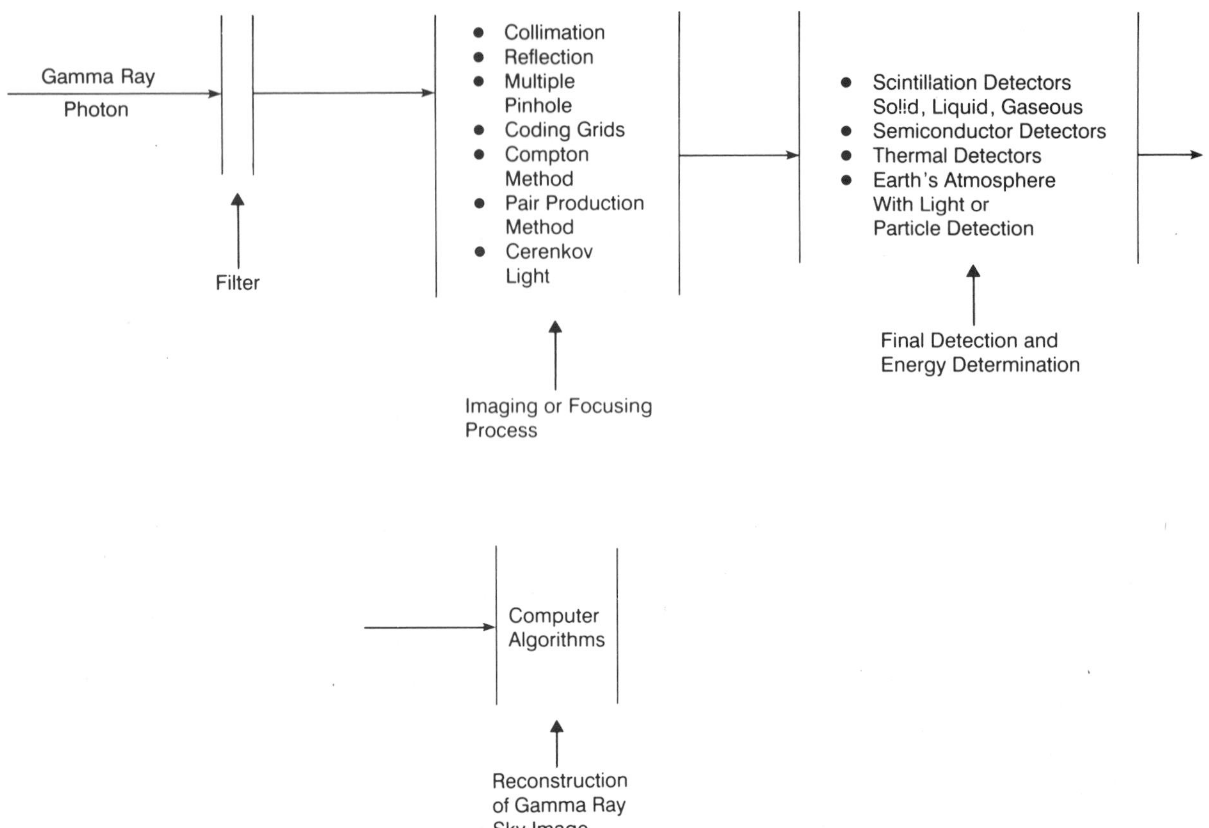

Figure 1. Design principles for gamma ray telescopes.

received a Nobel Prize in 1927 for this discovery. Finally, if the gamma ray has an energy above about 20 MeV, it will most probably convert into a pair of electrons. In all three cases the electrons are recorded to give the gamma ray energy, as described in the following text.

To obtain a true image of the sky, some technique must be devised to establish the directions of the incoming photons (see Fig. 1). The reflection of gamma ray photons, as in reflecting optical telescopes, appears to be practical only for the lower-energy gamma rays (< 500 keV). Earlier gamma ray telescopes operated at energies below 10 MeV, used collimators which only allowed gamma rays from one direction to be detected, and did not form a true image. Modern gamma ray telescopes, which image photons with energies above 1 MeV, use other techniques. To understand the imaging process, we consider each energy range separately.

LOW ENERGY (100 keV–10 MeV)

Modern imaging gamma ray telescopes that operate at low energies use the principle of Compton scattering, collimation, or the pinhole camera. A telescope based on Compton scattering (the COMPTEL) was included in NASA's *Gamma Ray Observatory* (*GRO*), launched by the space shuttle *Columbia* on April 5, 1991. Figure 2 shows a cutaway drawing of the COMPTEL at the center of the *GRO*. The incoming photon scatters and creates a "Compton electron" in the upper scintillators, which are viewed by several photomultiplier tubes. This electron can be identified and the scattered photon that leaves the upper scintillator is then detected in one of the lower scintillators. The path of the scattered photon is then detected quite accurately and with this knowledge the direction of origin of the initial photon can be located. By recording the arrival directions of several incoming photons, the sky can be imaged in gamma rays of any energy between 1–20 MeV, to an accuracy of about 1/8 of a degree.

Figure 2. The *Gamma Ray Observatory* (*GRO*) is shown with cutaway views of its four instruments, EGRET, COMPTEL, OSSE, and BATSE. The instruments are described in the text. (*Courtesy of C. J. Pellerin, NASA*).

At the left in Fig. 2 is a cutaway drawing of a second *GRO* instrument, the Oriented Scintillator Spectrometer Experiment (OSSE). This detector is not a true imaging telescope, but restricts photon arrival directions by using a passive tungsten collimator. Four large scintillators, held at different angles, record gamma ray spectra. The crystals are then reoriented (rocked) and additional spectra recorded. By comparing the spectra obtained for all known orientations of the scintillators, it is possible to determine if excess gamma rays are coming from a given region of the sky. This instrument operates over a gamma ray energy range 100 keV–20 MeV.

A third experiment on the *GRO*, called the Burst and Transient Source Experiment (BATSE), uses several simple scintillation spectrometers, which measure the gamma ray spectra (over the energy range 100 keV–10 MeV) from transient events, such as gamma ray bursts or solar flares. Two of these spectrometers are positioned at each of the four corners of the *GRO* satellite as shown in Fig. 2 at different orientations. A transient burst of gamma rays from a particular direction will be recorded differently in each of the detectors and a crude ($\sim 1° \rightarrow 10°$) burst location accuracy can be achieved.

For imaging low-energy gamma rays, the pinhole camera or coded aperture concept shows great promise. It is well known that a single, small pinhole placed above a position-sensitive photon detector (such as film for visible light) will give an excellent image as long as diffraction effects are avoided. Because cosmic gamma ray fluxes are very low, many pinholes are required to allow a sufficient number of photons to reach the detector. Images formed by the different pinholes overlap and mathematical methods must be used to "unfold" an image of the sky. The greater the number of pinholes used, the better is the quality of the image obtained. Because low-energy gamma rays are very penetrating, the pinhole array or "mask" must be made of a material of high atomic number and high density, such as lead or tungsten, and must be a few centimeters thick. The number of pinholes used for low-energy gamma ray imaging ranges from approximately 1000 to as few as 9. The positional accuracy of such telescopes depends on the distance of the mask from the detector plane and the spatial resolution of the detector (see Fig. 1). Instruments in use thus far have an accuracy of about 1/12 of a degree, although higher location accuracy is possible.

Besides recording the gamma rays that enter through the pinholes, the position-sensitive detector must also measure, with the highest accuracy, the energy spectrum of the photons. This is accomplished by using detectors made of germanium (Ge) crystals which are operated as semiconductor devices. Gamma ray interactions in the crystals produce electrons whose energy can be determined with high precision and this in turn gives an accurate measure of the gamma ray energy. An example of a low-energy gamma ray telescope, using a coded aperture mask and an array of germanium crystals, is shown in Fig. 3. The upper part of the figure

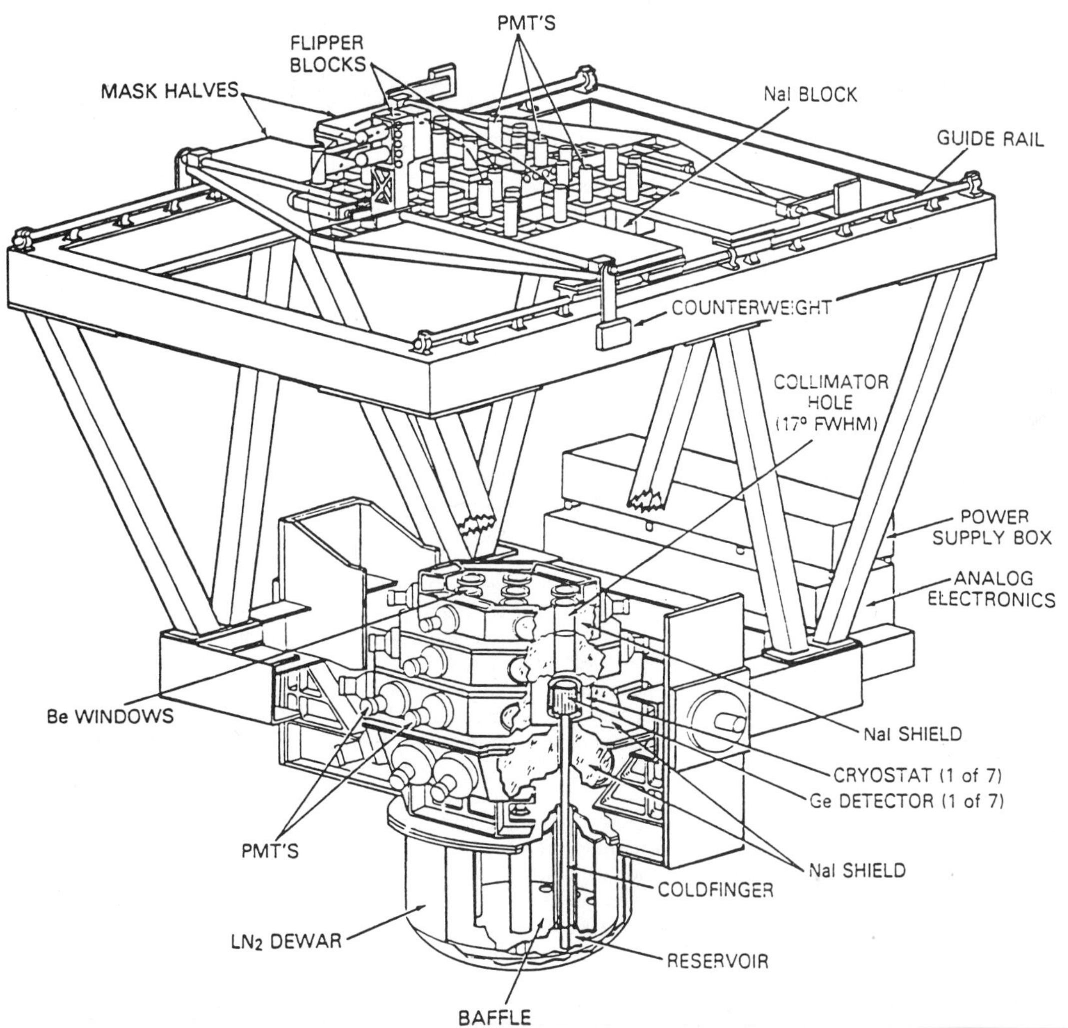

Figure 3. The Gamma Ray Imaging Spectrometer (GRIS) is shown with cutaway views of key instrument elements which are described in the text. This instrument module is placed in a balloon gondola (not shown). (*Courtesy of the GRIS Collaboration: Goddard Space Flight Center, AT & T Bell Laboratories, and Sandia National Laboratory.*)

depicts the coded mask array in which the opaque elements are made of scintillation crystals viewed by photomultiplier tubes (PMT). In the lower part of the figure, the position-sensitive detector is represented by seven germanium detectors. This array is surrounded by sodium iodide (NaI) scintillation detectors, whose purpose is to electronically shield the germanium detectors from ionizing charged particle effects. The telescope section shown in this figure is mounted in a gondola (the size of a small house trailer) and will be carried to the stratosphere by a 300-ft diameter helium-filled balloon.

MEDIUM / HIGH-ENERGY (20 MeV–30 GeV)

The most practical detection technique, for higher energies, relies on the production of pairs of electrons, the third mechanism by which a gamma ray interacts with matter. The most advanced instrument using this process is the Energetic Gamma Ray Experiment Telescope (EGRET) on the *GRO* spacecraft, shown at the far right in Fig. 2. This device consists of an electron imaging device known as a "spark chamber," surrounded by a plastic scintillator dome that electronically rejects entering charged particles from the analysis. Gamma rays convert into a pair of electrons in closely spaced, thin tantalum foils. The pair of electrons, which leave the foil in which they were produced, initiate sparks between several pairs of foils, which delineate the tracks of the electrons. By "imaging" the electrons, the direction of the incoming photons can be accurately determined. Below the spark chambers a thick sodium iodide scintillation detector records the total energy of the pair of electrons and this, in turn, gives the energy of the gamma ray.

Figure 4. The 10-m optical reflector used for very high energy gamma ray studies at the Fred Lawrence Whipple Observatory, Mt. Hopkins, AZ. (*Courtesy of Whipple Observatory.*)

VERY HIGH AND ULTRA HIGH ENERGY (> 10^{11} eV)

Above a photon energy of about 100 GeV (10^{11} eV), the extremely low fluxes of cosmic gamma rays necessitate instruments much larger than can be placed on Earth-orbiting satellites or space stations. However, to investigate this gamma ray energy range, the Earth's atmosphere is used as the detector, because a primary gamma ray in this energy range can initiate a shower of secondary particles which reach to the ground level. In the energy range from about 10^{11}–10^{13} eV (the TeV range), the secondary electrons produce a flash of visible blue light known as Čerenkov emission, which is observed by directional mirror systems that focus the light on photomultiplier tubes, much like a conventional telescope. One of the first TeV telescopes is at the Fred Lawrence Whipple Observatory on Mount Hopkins, at Amado, Arizona, and is shown in Fig. 4. This device uses 248 individual mirrors to focus light from a point source to a single focal point 7.3 m from the mirror, where it is detected by photomultipliers. Other mirror telescopes for the TeV energy range are operating in New Mexico, Utah, Australia, and France.

In the energy range above 10^{14} eV, the secondary particles are directly detected at ground level. These secondary particles are electrons, photons, protons, neutrons, and mesons, depending on whether a cosmic gamma ray or charged primary cosmic ray initiated the shower in the atmosphere. If the shower is produced by a PeV gamma ray, then the shower will contain a much smaller proportion of certain elementary particles, μ mesons, than if produced by a charged particle. Thus it is possible to determine the nature of the primary radiation whose arrival direction can be determined to an accuracy of 1/10 of a degree. The electromagnetic cascade shower, produced by the high-energy primary gamma rays, can extend over several hundred meters.

CURRENT AND FUTURE OBSERVATIONS

The field of gamma ray astronomy is still in its infancy and some instrument designs described here only went into operation recently. In the high-energy range, the EGRET on *GRO* (with major instruments supplied by NASA and the Federal Republic of Germany) began observations in 1991. (A French/Soviet high-energy spark chamber telescope, also using a coded aperture mask, known as *GAMMA-1*, was launched before *GRO*, but unfortunately is substantially inoperative due to an equipment failure.) *GRO* also carried the low-energy instruments, OSSE, COMPTEL, and BATSE. To study gamma rays at low energies, a second French/Soviet instrument known as *SIGMA* was launched on the Soviet *GRANAT* spacecraft in 1990; on October 13–14, 1990, it obtained significant evidence for a positron-annihilation source near but distinct from the galactic center. Gamma ray burst experiments will be placed on board several Soviet, ESA (European Space Agency), and U.S. spacecraft within the next 10 years.

The ultimate promise of the fledgling field of gamma ray astronomy will not be fulfilled until very large, sophisticated telescopes are assembled on space stations or on the Moon. One advanced instrument known as the *Pinhole Occulter Facility* is now under study as a potential experiment for the U.S. space station *Freedom*. Complex mathematical techniques, such as now used in medical imaging, will be increasingly applied in the future of telescope designs.

Additional Reading

Bertsch, D. L., Fichtel, C. E., and Trombka, J. I. (1988). Instrumentation for gamma-ray astronomy. *Space Sci. Rev.* **48** 113.

Chupp, E. L. (1976). *Gamma Ray Astronomy: Nuclear Transition Region*. D. Reidel, Dordrecht. (See chapter 6, Experimental considerations for nuclear γ-ray astronomy, p. 204.)

Gamma Ray Program Working Group (1988). *Gamma Ray Astrophysics to the Year 2000*. NASA, Washington, DC.

Greisen, K. (1966). Experimental gamma-ray astronomy. In *Perspectives in Modern Physics*, R. E. Marshak, ed. Wiley, New York, p. 355.

Kniffen, D. A. (1991). The Gamma Ray Observatory. *Sky and Telescope* **81** 488.

See also **Background Radiation, Gamma Rays; Gamma Ray Astronomy, Space Missions; Gamma Ray Bursts, Observed Properties and Sources.**

Telescopes, Detectors and Instruments, Infrared

Jonathan H. Elias

The infrared spectral region, for an astronomer, extends from the limit of sensitivity of conventional visible-wavelength detectors, at about 1.0 μm, to the region where radio wavelength techniques become more efficient, at about 1000 μm. Although astronomical observations in the infrared were made almost two centuries ago by William Herschel, modern infrared astronomy effectively dates from the 1960s, when technological developments pushed the sensitivity of infrared instrumentation to the point where most classes of astrophysical phenomena became observable in the infrared.

Advances in instrumentation since then have been rapid; improvements in the efficiency of infrared instrumentation over the last 25 years have been such that observations that were difficult or impossible with the largest available telescopes in the mid-1960s are now considered routine with 0.6-m telescopes. Continued advances in instrumentation are likely to lead to similar improvements over at least the next decade.

BASIC CONSTRAINTS

Despite the rapid advances in the capabilities of infrared instrumentation, there are a few basic principles of telescope and instrument design that represent permanent constraints. These are conveniently considered by comparison with the field of visible-wavelength instrumentation.

At wavelengths longer than 2.0 μm, thermal emission from ambient-temperature objects becomes comparable to or greater than airglow (mainly OH molecule) emission from the nighttime sky. What this means is that every warm object in the optical train of the instrument and telescope emits detectable photons, and thus it is desirable to cool as much of the instrument as is possible. Ground-based telescopes cannot be cooled practically, so the next best alternative is to design them with maximum efficiency and a minimum of warm emissive material in the field of view of the instrument. Rocket and balloon-borne telescopes, as well as spacecraft, can be cooled to the point where their thermal emission is negligible, and also permit observations above the warm, emissive atmosphere; these small, cooled telescopes can have sensitivities as good as that of a large, ground-based telescope.

Even with an appropriately optimized instrument and telescope, the background flux—emission from sky, telescope, and instrument—is usually much greater relative to the flux from astronomical objects than is the case at visible wavelengths, and must be measured to correspondingly greater precision if its effects are to be removed.

The Earth's atmosphere limits the wavelengths that can be observed from the ground in the infrared to a number of "windows" between atmospheric absorption bands, due principally to carbon dioxide and water vapor. In the near infrared, there are a series of these wavelength intervals, centered at 1.2, 1.6, 2.2, 3.5, 4.8, and 10 μm. These are quite transparent and permit observations to be made from any site suitable for visible-wavelength astronomy. There are additional windows accessible from dry, high-altitude sites—one centered at 20 μm and a second at 34 μm. From 40–300 μm, the atmosphere is completely opaque, and observations can be made only from high-altitude aircraft, balloons, rockets, or spacecraft. At still longer wavelengths, several windows are accessible from dry sites at wavelengths between 350 and 1000 μm. Atmospheric transmission continues to improve at millimeter radio wavelengths.

INFRARED TELESCOPES

Few large telescopes have been built deliberately optimized for infrared observations—the two largest are the 3.8-m United Kingdom Infrared Telescope (UKIRT) and the 3-m NASA Infrared Telescope Facility (IRTF), both on Mauna Kea, Hawaii. Nevertheless, most of the world's largest telescopes have been modified to permit observations in the infrared.

The main characteristic of a telescope optimized for the infrared is that thermal emission that might be detected by the instrument is minimized. Infrared instruments are typically mounted at the Cassegrain focus, and there is usually a special secondary mirror that is used for infrared observations. This mirror is undersized and, if possible, permits the sky to be viewed past its edges. This secondary mirror rather than by the primary telescope mirror thus acts as the principal defining stop of the telescope, and the edges of the stop are defined by the sky rather than by the primary mirror cell. This ensures that emission from the primary mirror cell and the secondary mirror mount do not reach the instrument.

Any astronomical measurement is in fact a measurement of the object of interest plus background emission ("sky"), which must be accurately removed if a good measurement of the object itself is to be made. The very high background fluxes present in the infrared are normally measured and removed by rapidly moving the field of view of the detector from the object to be measured to a blank region nearby in the sky and back again, usually several times per second. The difference between the fluxes measured at the two points is the net flux from the object. Because a large telescope cannot move back and forth rapidly on the sky, normal practice is to mount the telescope secondary mirror on pivots that allow it to be tilted slightly about one axis, using solenoids or electromechanical drivers. These chopping secondaries (also known as wobbling or nodding secondaries) are designed to be relatively small and lightweight (40-cm diameter or less); typical final focal ratios of infrared secondaries are thus $f/30$ or slower, and the mirrors can chop at frequencies of 20 Hz or greater.

The development of infrared detector arrays has, to a certain extent, reduced the need for fast chopping, because a single exposure provides (in principle) the measurements of the object and of the nearby blank sky. Nevertheless, correction for nonuniformities in the array is often done by slow chopping or by moving the telescope.

INFRARED INSTRUMENTS

Infrared instruments are designed and operated in a manner similar to conventional visible wavelength instruments, except for differences imposed by the need to minimize radiation from the telescope and the instrument itself. Thus the optical design of an infrared instrument will often resemble that of its visible-wavelength counterpart, but the mechanical design will be quite different, because the entire instrument must be cooled to liquid nitrogen temperature (77 K).

Single-Channel Photometers

The oldest instrument type is the single-channel photometer. Because highly efficient single-element detectors are available at wavelengths where detector arrays are not—or are less—efficient, these instruments continue to be used. In addition, the lower cost of a single-channel instrument permits one to provide an infrared capability for a small telescope.

A typical photometer is housed in a cryogenic dewar that contains the filters, entrance apertures, Fabry lens, detector, and first-stage electronics. The filters must be cooled so that the detector "sees" radiation only at wavelengths that the filter transmits. Following the filters and entrance apertures in the optical train is a Fabry lens, which reimages the telescope secondary onto the detector. The Fabry lens ensures that radiation falling on the detector is limited to light passing through the telescope or emitted from the mirrors; it also limits variations in the instrument response due to small guiding errors.

Imagers

The design of imagers ("cameras") is complicated by the need to limit thermal radiation falling on the detector array. This means that one cannot simply put an array directly at the focus of a telescope, but must use additional optics that first form an image of the telescope secondary on a cold stop and then reimage the telescope focal plane onto the detector. A typical design contains a Fabry lens or mirror, followed by the cold stop, cold filters, and a reimaging lens, and then the detector. Normal practice is to build the entire instrument in a single cryogenic dewar.

Grating Spectrometers

The optical design of infrared spectrometers is virtually the same as visible-wavelength spectrometers, except that a Fabry lens is usually placed near the entrance slit in order to image the telescope secondary on the spectrograph collimator, thus using it as a cold stop. Again, in order to minimize background radiation, it is necessary to cool the entire instrument. This in turn leads to a less modular design than is the case for visible-wavelength instruments: compactness is highly desirable, and any necessity for changing individual components, such as dispersion gratings, requires warming up the instrument to room temperature. The need for compactness and the limited sizes of infrared arrays mean that the spectral resolution of the current generation of infrared grating spectrometers is relatively limited, with maximum resolutions of 1 part in 3000 or less.

Higher spectral resolution can be achieved in a number of ways. Michelson interferometers are capable of providing very high resolution; for bright objects, spectral resolutions better than 1 part in 100,000 are possible. A small number of these have been built for astronomical infrared spectroscopy. Fabry–Perot interferometers can also be used to achieve high resolution, in conjunction with single-element detectors, imagers, or low-resolution spectrometers; these are typically used to obtain spectral resolutions ranging from roughly 1 part in 500 to 1 part in 20,000 or higher. Finally, infrared analogs of the echelle spectrograph can in principle be built to extend the resolution of infrared grating spectrographs to higher resolution. As the sizes of available infrared arrays increase, infrared echelles become increasingly attractive despite their size and complexity; a few such instruments are in existence or under construction.

Detectors

The heart of any instrument is the detector. Different types of detectors are used for different purposes and at different wavelengths. Astronomical infrared detectors operate on either of two basic principles: energy detection and photon detection.

The astronomical bolometer or energy detector is a small piece of semiconductor (usually germanium or silicon) mounted on fine wires and cooled to a temperature below 2 K. When radiation falls on the bolometer, its temperature rises and its resistance falls; measurements of the resistance can be used to measure the incident flux. Bolometers have high efficiencies and can be built to operate with high fluxes; they are typically used for photometry at wavelengths of 10 μm and beyond.

In photon detectors, incident photons excite electrons in a semiconductor from the ground state into the conduction band. Depending on the material and the design of the detector, this produces either a voltage at a diode junction (photovoltaic detector) or a change in resistance (photoconductor). Because there is normally a minimum energy required to excite the electrons, there is a corresponding cutoff wavelength, beyond which the detector is not sensitive. The most commonly used detector materials for the 1–5-μm spectral region are InSb and HgCdTe. Beyond 5 μm, a variety of detectors are currently in use.

Detector arrays are constructed in any of three ways. The simplest, least efficient way is to assemble several single-element detectors into a small array. Astronomical instruments have been constructed using up to eight individual detectors. At the other extreme are monolithic arrays, analogous to the visible-wavelength silicon charge-coupled device (CCD), where detection and readout are accomplished by a single device. The third type is the so-called hybrid array, where an array of detectors is fabricated and bonded to a CCD-like readout device.

Additional Reading

Gillett, F. C. and Houck, J. R. (1991). The decade of infrared astronomy. *Physics Today* **44** (No. 4) 32.

Low, F. J. and Rieke, G. H. (1974). The instrumentation and techniques of infrared photometry. *Methods of Experimental Phys.* **12A** 415.

Soifer, B. T. and Pipher, J. L. (1978). Instrumentation for infrared astronomy. *Ann. Rev. Astron. Ap.* **16** 335.

Wolfe, W. L. and Zissis, G. J., eds. (1978). *The Infrared Handbook*. Office of Naval Research, Washington, DC.

Wynn-Williams, C. G. and Becklin, E. E., eds. (1987). *Infrared Astronomy with Arrays*. Institute for Astronomy, University of Hawaii, Honolulu.

See also **Infrared Astronomy, Space Missions; Observatories, Airborne; Scientific Spacecraft and Missions; Sounding Rocket Experiments, Astronomical.**

Telescopes, Detectors and Instruments, Ultraviolet

Robert C. Bless

For the purposes of this entry, we will consider the astronomical ultraviolet (UV) to extend from the Earth's atmospheric cutoff at about 3200 Å shortward to the Lyman limit at 912 Å. To avoid the primary UV atmospheric absorbers—ozone and oxygen—instruments should be at least 150 km above the Earth. The UV spectrum is of astrophysical interest for several reasons. For example, hot stars radiate most of their energy in the UV; a star with an effective temperature five times that of the Sun radiates seven or eight times as much energy in this region as in the visual. Many ions of astrophysical interest have their strongest lines in the UV, such as Mg II (2796, 2800 Å), C IV (1548, 1551 Å), Si IV (1394, 1403 Å), and N V (1239, 1243 Å). These resonance lines have been of particular importance in studies of the interstellar medium and stellar winds.

Scintillation ("twinkling") and seeing effects are of course absent above the atmosphere. Thus the angular resolution that can be achieved is limited only by the quality and size of the optics and the stability of the telescope pointing system. Because of technical difficulties, this aspect of space astronomy is only now beginning to be exploited in a major way. Finally, many classes of stars have been found to be rapidly variable in light, often with very low amplitudes. The absence of scintillation "noise" above the atmosphere presents obvious attractions for investigations of these objects.

The telescopes, instruments, and techniques used in the ultraviolet are similar to those employed in the visible part of the spectrum. Indeed, this is one reason why the UV was the first spectral region to be explored from space. Because of this similarity, only a brief description will be given of some of the instruments in three payloads which, however, are representative of those most commonly used in nonsolar UV astronomy. These instruments are on the *Orbiting Astronomical Observatory-2* (*OAO-2*), the *International Ultraviolet Explorer* (*IUE*), and the *Hubble Space Telescope* (*HST*). The *OAO-2* was the first successful astronomical space observatory; it exemplifies some early photometric techniques. The *IUE* has had a remarkably long and fruitful life of spectroscopy, more than 12 years as of this writing. The *HST*, one of the most ambitious scientific satellites yet constructed by NASA, was designed to exploit the high-quality imaging possibilities attainable above the Earth's atmosphere. It was launched on April 24, 1990.

SIMPLE PHOTOMETERS AND SCANNERS

The *OAO-2* was a double-ended spacecraft with a set of telescopes looking out each end. One set was designed to make an ultraviolet survey of a significant fraction of the sky, using as detectors a then-new type of television tube. A similar modern UV-sensitive detector will be described later. The payload looking out the other end was primarily intended for photometry of point sources.

The larger portion of the photometric payload consisted of five independent telescopes, photometer heads, and associated photomultipliers (PMTs) for filter photometry. Their design was identical to conventional filter photometers used at ground-based observatories and, except for their UV-sensitive photocathodes, these instruments could have been (and in fact were) used on the ground. The primary difference was that, because of the relatively poor pointing accuracy of the *OAO* spacecraft, the most commonly used photometric aperture was large, projecting to 10 arcminutes on the sky. Photometric quality was equal to that attainable on the ground, however. Two objective plane-grating scanning spectrometers completed this payload. Spacecraft space limitations dictated the use of plane gratings as the light-collecting mirrors, otherwise the design was again conventional. Stepper motors rotated the gratings thereby causing the spectrum to be scanned past a PMT. Spectrophotometry and filter photometry of hot stars were possible to sixth and tenth visual magnitudes, respectively.

THE IUE INSTRUMENT

Though photomultipliers are admirable for many purposes, they are single-channel detectors and so are inefficient in many applications. As soon as detectors with two-dimensional formats were developed for ground observatories, they were used in space as, for example, in the *IUE*. The telescope in this satellite feeds two echelle spectrographs. The high-resolution two-dimensional echelle format is achieved by a spherical grating that cross-disperses the diffracted radiation from the echelle; the echelle orders are thereby separated and focused onto the detector. When a plane mirror is inserted in front of the echelle, only the spherical grating disperses the radiation, producing a low-resolution one-dimensional spectrum.

Each spectrograph feeds one of two redundant detectors that consists of a UV-to-visible image converter followed by a secondary electron conduction television camera. Incident UV photons liberate electrons from the image converter's cesium–telluride photocathode; these are accelerated by a 5-keV potential onto a phosphor where each electron produces about 60 blue photons. The resulting image is fiber optically coupled to the bialkali photocathode of a vidicon television camera. Blue photons cause photoelectrons to be emitted from the cathode which are accelerated to the KCl vidicon target where secondary electrons are produced and conducted away. Residual positive charge image is discharged by the read beam in a 768×768 raster scan made with 37-μm steps. The video signal is digitized by an 8-bit D/A converter and transmitted to the ground. Because the target is nonconductive, integration times of hours are possible. Fifteenth magnitude stars have been observed at low resolution.

THE HUBBLE SPACE TELESCOPE

With the *HST*, high-quality visible and ultraviolet imaging possible from above the Earth's atmosphere will be realized for the first time. This telescope has the largest astronomical mirror placed into orbit. Its primary was intended to be the most nearly perfect large astronomical mirror ever made, having a wavefront error of about $\lambda/60$ at 6328 Å. Overall telescope performance was expected to be about $\lambda/20$. The pointing control system of the spacecraft was expected to limit image motion to 0.007 arcsecond rms. If these characteristics had been achieved in orbit, a point source would have produced an image with a full width at half maximum of about 0.04 arcsecond at 4000 Å, or an order of magnitude better than the very best that can be achieved (only occasionally) from the ground. (See the section On Performance in Orbit of the entry, Hubble Space Telescope.—Ed.)

As shown in Table 1, five instruments analyze the light collected by this telescope (see Fig. 1). In addition, one of the fine guidance sensors can be used to obtain accurate astrometric data. The Faint Object Camera (FOC) is designed to exploit fully the imaging capabilities of *HST*. It has two separate cameras, one operating at $f/48$, the other operating at $f/96$. In the latter mode, 25-μm detector pixels subtend 0.022 arcsecond on the sky, thereby adequately sampling the best images produced by the telescope. The field of view is small, however, only 11×11 arcseconds. Wheels in the optical path enable 44 different bandpass, polarizers, and neutral density filters to be inserted in the beam, as well as objective prisms. In addition, a small coronographic disk can be inserted at the entrance aperture to reduce the brightness of a point source. This makes it possible to observe faint objects in the presence of a nearby bright companion.

A field of view four times in area is provided by an independent $f/48$ camera in the FOC which is similar in overall design to the $f/96$ system. In addition to its imaging capabilities, it has a long slit (10 arc-seconds) spectrographic mode with a resolving power of about 2300. Spectra of extended objects at the $f/48$ angular resolution are thus possible.

Detectors for both FOC cameras are identical—a television tube optically coupled to a three-stage image intensifier which has a gain of about 10^5. The television system operates in a pulse-counting mode with a format as large as 512×512 pixels. The speed of the pulse-counting system limits the brightest stars that can be observed to about 21 visual magnitude in its full format. Neutral density filters can extend this to brighter limits, however.

The Wide Field and Planetary Camera (WFPC) on *HST* also operates at two different focal ratios, $f/12.9$ and $f/30$, giving square fields of 2.67 and 1.15 arcminutes across, respectively. The detectors for each WFPC camera are mosaics of four change-coupled devices (CCDs) yielding a 1600×1600 pixel format. The 15-μm pixels correspond to 0.10 and 0.043 arcsecond in the wide field and planetary modes, respectively, thereby compromising somewhat the angular resolution of the telescope. To reduce dark noise and allow long exposures, the detectors are cooled to -95°C by thermoelectric coolers coupled to an external heat radiator. CCDs have excellent spectral sensitivity in the red and a thin coating of coronene provides UV response by converting blue and UV photons to visible photons. Consequently, it is possible for this

Table 1. Representative Payloads

Project	*Telescope*	*Instrument*	*Detector*[a]	*Purpose*
OAO-2	Four 20-cm Herschelian	Filter photometer	PMT	Photometry,
(12/68–2/73)	One 40-cm prime focus	Filter photometer	PMT	4200–1200 Å
	Two objective gratings	Spectrometer	PMT	Low-resolution spectroscopy
	Four 30-cm Schwarzschild	Wide angle camera	UV SEC vidicon	UV sky survey
IUE (1/78–)	One 45-cm Ritchey–Chrétien	Crossed echelle–grating spectrograph	Image converter and SEC vidicon	0.2 Å spectroscopy, 1500–3200 Å
		Grating spectrograph	Same	6 Å spectroscopy 1150–3200 Å
HST (4/90–)	2.4-m Ritchey–Chrétien	Wide-field camera	4 CCDs, with total of 1600 × 1600 pixels	0.1 arcsecond imagery, 1200–10,000 Å
		Planetary camera	Same	0.04 arcsecond imagery, 1200–10,000 Å
		Faint object camera	Image intensifier and vidicon	0.02 arcsecond imagery, 1200–6500 Å
		High-resolution spectrograph–echelle and grating	Digicon	Spectroscopy, RP 2×10^3–1×10^5, 1110–3200 Å
		Faint object spectrograph–grating; polarimetry	Digicon	Spectroscopy, RP 100–1000, 1150–7000 Å
		High-speed photometer	Image dissector–PMT	Filter photometry, sample times ≥ 10 μs

[a]CCD = change-coupled device; PMT = photomultiplier tube; RP = spectral resolving power; SEC = secondary electron conduction.

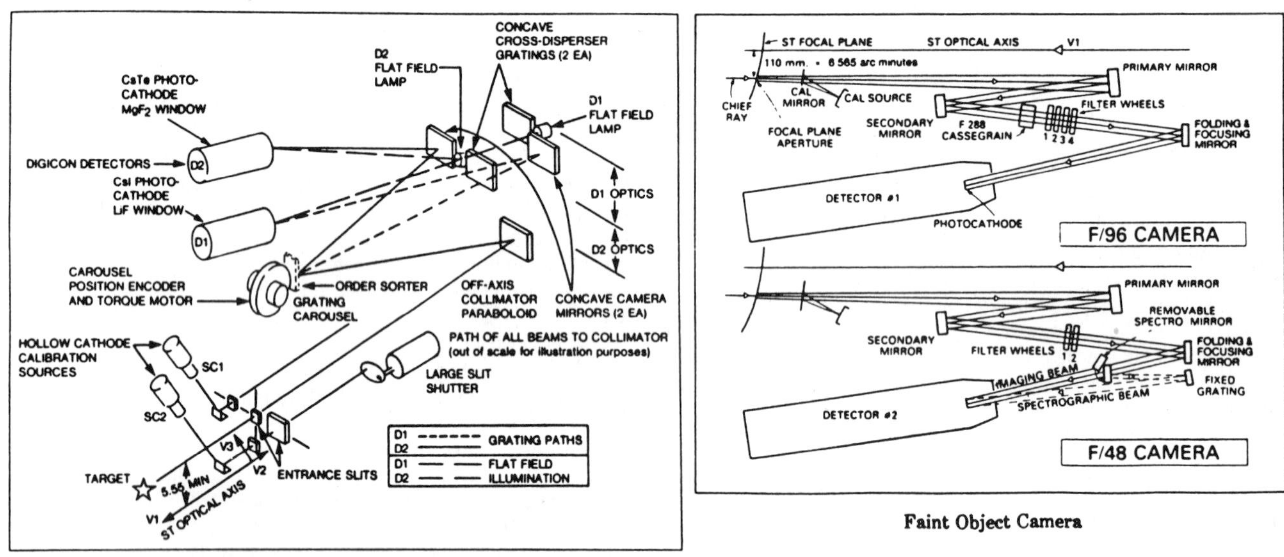

Faint Object Camera

High Resolution Spectrograph Optical Concept

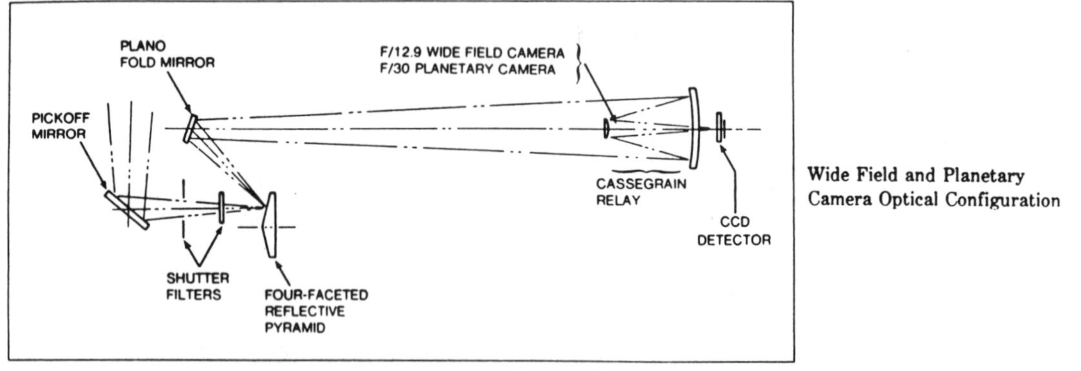

Wide Field and Planetary Camera Optical Configuration

Figure 1. Three of the scientific instruments on the *Hubble Space Telescope*.

one detector to have good response to radiation from 1200–10,000 Å.

The two spectrographs (the Faint Object Spectrograph and the Goddard High Resolution Spectrograph) use a different type of hybrid solid-state detector called Digicons. Photoelectrons emitted from the cathode are magnetically focused and accelerated onto a linear array of 512 silicon diodes, each 40 μm wide. On striking a diode, every 20–25 keV electron produces a burst of about 5500 electrons. Each diode has its own independent set of electronics which amplifies and counts the burst. Deflection coils enable the desired order of an echellogram spectrum to be positioned onto the diode array or to be stepped across the array improving the resolution, allowing interorder background measurements to be made, and so on. Using fairly conventional grating and echelle spectrographs, these two instruments provide a wide wavelength coverage with a large range in spectral resolution.

The image dissector on the High Speed Photometer is unusual in that it is the only detector mentioned here that has not been used for ground-based astronomy. Focusing coils allow only those photoelectrons from any selectable 180-μm point on the cathode to enter the photomultiplier section of the detector. This greatly reduces the dark noise. Also, by appropriate telescope pointing, radiation passing through any of about 50 filter–aperture combinations can be accessed without any moving parts in the photometer.

As this brief review has indicated, subject to limitations of weight, volume, and power, nearly any instrument used in a ground-based telescope can be adapted for use in space. There are aspects peculiar to space that play significant roles in the design and operation of payloads, however. For example, space astronomy can be done with instrumentation in near-Earth orbits, either attached to the Space Shuttle or as a free-flying satellite (*OAO-2* and *HST*) or from geosynchronous—24-hr period—orbits (*IUE*). Satellite designs for both types of orbits must take into account the thermal and charged particle environment peculiar to a given altitude. Obviously, more massive payloads can be put into a near-Earth orbit than a geosynchronous orbit and observations can be made from within the Earth's shadow. However, the Earth occults half of the sky and with the area within 40° or 50° from the Sun usually off limits to observation, a substantial portion of the sky is unavailable at any given time. In addition, only objects near the poles of the orbit can be observed continuously for more than about 35 min before data taking is interrupted for an hour or so. On average, a satellite can spend only about one-third of the time taking data, roughly the same efficiency as is achieved at observatories on Earth. Finally, real-time control of satellites near the Earth can be maintained for only a relatively small fraction of the time.

By contrast, satellites in 24-hr orbits can always be in direct contact with the ground, can achieve observing efficiencies twice as large as can satellites in near-Earth orbits, and can have much more of the sky available at any time. However, they pay a significant mass penalty, require a small propulsion system to keep the satellite properly positioned over the Earth, and are nearly always in sunlight so that the telescopes must be carefully baffled against scattered light.

Additional Reading

Bahcall, J. N. and Spitzer, L. (1982). The space telescope. *Scientific American* **247** (No. 1) 40.

Boggess, A., et al. (1978). The *IUE* spacecraft and instrumentation. *Nature* **275** 372.

Code, A. D., Houck, T. E., McNall, J. F., Bless, R. C., and Lillie, C. F. (1969). Ultraviolet photometry from a spacecraft. *Sky and Telescope* **38** 290.

Hall, D. N. B., ed. (1982). *The Space Telescope Observatory*, NASA CP-2244. NASA Scientific and Technical Information Branch.

Kondo, Y., ed. (1990). *Observatories in Earth Orbit and Beyond*. Kluwer Academic Publ., Dordrecht.

See also **Cameras and Imaging Detectors, Optical Astronomy; Hubble Space Telescope; Scientific Spacecraft and Missions; Sounding Rocket Experiments, Astronomical; Ultraviolet Astronomy, Space Missions.**

Telescopes, Detectors and Instruments, X-Ray

Martin C. Weisskopf

Instrumentation used to study cosmic x-ray emission falls into two distinct categories: The first, x-ray telescopes, serve to concentrate the x-ray emission onto a smaller area and form an image. The second are the detection devices themselves, which convert the energy carried by x-ray photons into an electronic signal measuring characteristics such as energy, position within the detector, time of occurrence, state of polarization, and so on. No single x-ray detector has to date been capable of providing all of these functions simultaneously. Devices have ranged from photographic film to sophisticated low-temperature calorimeters, currently under development in a number of laboratories.

X-RAY TELESCOPES

The field of x-ray astronomy began with the discovery in 1948 that the Sun is an x-ray emitter. This (V2) sounding-rocket experiment utilized photographic film as the detection device. The first x-ray image of the Sun was obtained in 1960, with a sounding-rocket experiment utilizing a pinhole camera with photographic film. The first detection of a celestial x-ray source, other than the Sun, was obtained in 1962 by Riccardo Giacconi and Bruno Rossi, using a Geiger counter sent above the absorbing effects of the atmosphere in a sounding rocket. As in many subsequent uses of gas-filled detectors, celestial location was obtained by means of mechanical collimation. Mechanical collimators, which precede the entrance window and restrict the field of view to, at best, about 1/2°, give nonimaging instruments their directional sensitivity. During the 1960s and 1970s, more sophisticated forms of mechanical collimation, the modulation collimators, were used by x-ray astronomers to achieve better angular resolution. These devices were arrangements of planes of wires aligned perpendicular to the viewing axis of the detector. When this axis was then either swept through, or rotated about, the position of the x-ray source, the variation in the detected x-ray signal could be correlated to the source position. More recently, and especially for observations at higher (greater than 10 keV) energies, where grazing-incidence reflection is too inefficient, "coded apertures" have come into use. These are essentially plates with multiple pinholes, casting shadows onto position-sensitive detectors. The resulting "image" can then be mathematically unscrambled, to produce an image of the portion of the sky that was viewed.

X-ray astronomy, however, benefits substantially from focusing devices capable of concentrating large amounts of x-rays onto small areas and forming a true image. This is important, not only because the image is formed, but also because the capability of seeing faint sources is significantly improved as all x-ray detectors in space are plagued with a so-called "background signal" that is proportional to the size of the detector. The background signal is a by-product of x-ray-like events produced in the detectors by cosmic rays and higher energy x-rays and gamma rays.

The development of x-ray "grazing incidence" optics is an outgrowth of x-ray astronomy, although its history dates back to Arthur H. Compton. In 1923, he showed that x-rays can be reflected from a polished surface if the angle of incidence is small. The underlying physics here is that nominally total external reflection can take place because the x-ray index of refraction in matter is less than unity. In the late 1940s and 1950s, the problem of exploiting this effect to form x-ray images was first successfully addressed by P. Kirkpatrick and Albert V. Baez and by H. Wolter. Notably, Wolter examined, theoretically, a number of geometries and showed that, in order to obtain a true image over an extended field of view, x-rays need to undergo two successive reflections from coaxial arrangements of either paraboloid–hyperboloid or paraboloid–ellipsoid combinations. The practical difficulties in actually constructing x-ray optics—the difficulty of forming the correct geometrical shape, polishing those shapes, and properly aligning the elements of the telescopes—were problems that have been solved in the context of building x-ray telescopes for x-ray astronomy.

The first use of a Wolter-type telescope for astronomical purposes occurred in 1963 on a sounding-rocket flight and utilized a telescope made from polished electroformed nickel, to record an x-ray image of the Sun onto photographic film, with an angular resolution of 1 arcminute. The first x-ray telescope to fly in an Earth-orbiting satellite was mounted in one of NASA's *High-Energy Astronomy Observatories*, *HEAO-2*, launched in 1978 and named the *Einstein* observatory. This observatory represented a major advance, not only in the development of x-ray optics, but also in the development of x-ray detectors. The *Einstein* telescope nested four paraboloid–hyperboloid mirror pairs, to obtain a large collecting area and an angular resolution of a few arcseconds.

Generally speaking, there are three factors that determine the capability of an x-ray telescope to resolve objects: figure (i.e., the degree to which the actual reflecting surfaces correspond to the theoretical shape), alignment, and surface roughness. The first two factors primarily determine the angular resolution, whereas the third is the principal factor in determining the fraction of the reflected x-rays falling within this core of the image. The *Einstein* telescope had a surface roughness at a level of tens of angstroms. Modern x-ray telescopes achieve surface roughness as small as 3 Å and an angular resolution as good as 1/2 arcsecond.

X-RAY DETECTORS

The history of the instrumentation used in x-ray astronomy has paralleled the development of x-ray telescopes. Indeed, a number of significant developments and astronomical discoveries have been made without the latter. A device that has played a major role is the gas-filled proportional counter. Developed independently of x-ray astronomy during the late 1940s, this device relies on the effects produced when an x-ray photon is photoelectrically absorbed in a gas, typically an inert one, such as argon. The majority of gas-filled detectors used in x-ray astronomy are based on sensing the charge liberated by the ionizing x-radiation. The photoelectric absorption of the incident x-ray in a proportional counter is immediately followed by the release of some of the energy through the liberation of a photoelectron. The latter loses energy, collisionally ionizing additional gas atoms, and thus creating a number of free electrons and ions. This number is proportional to the energy of the incident x-ray. In the proportional counter, this initial charge is further amplified (gas multiplication), by accelerating the charge in strong electric fields prior to its collection. This causes further ionization and, as long as the fields are not too strong, the total charge produced is proportional to the energy of the incident x-ray, hence the name. At stronger electric fields, the pulse of charge no longer retains its proportionality to the incident energy, and the device is called a Geiger counter. A more recent type of gas-filled detector used by x-ray astronomers is the gas scintillator, a proportional counter that exploits the light emitted by the gas atoms during the ionization process.

The *Einstein* observatory had four different, selectable focal-plane instruments, encompassing a number of different detector technologies, variations of which are still relevant. Two detectors were x-ray-sensitive "cameras," in that they could measure the position of an x-ray within the detector. The first of these was a position-sensitive, gas-filled proportional counter; the second, a microchannel plate x-ray detector. The latter is an outgrowth of military technology; however, its use for astronomical observations and the corresponding requirements of large areas and excellent position resolution has been influential in its development. A typical microchannel plate consists of order 10 million, close-packed channels of lead-glass tubes whose surfaces are coated with appropriate materials. Typical tubes have diameters of 10–25 μm. An x-ray striking the opening of a tube releases a photoelectron, which is accelerated toward the opposite wall by an applied electric field. More electrons are released upon impact at the opposite wall and the process continues. The total number of electrons released and exiting from the rear of the channel exceed one hundred million per incident photon. Due to the discrete nature of the tubes, this charge is localized and may be sensed to record the position to high accuracy. Such devices yield excellent spatial resolution but poor energy (spectral) resolution (the capability to distinguish photons of different x-ray energies).

The *Einstein* observatory also had three instruments specifically designed for spectroscopy—the ability to measure accurately the energy carried by the x-rays. The first was a Bragg crystal spectrometer, which exploits the energy dependence of Bragg reflection from perfect crystals. These devices offer extremely high energy resolution at the price of poor efficiency, as only one energy may be studied at a time. (Bragg reflection, and also the polarization dependence of x-rays scattering from electrons in solid materials, have been used to make x-ray polarimeters.) The second spectrometer utilized a transmission grating, located on an arm to position it immediately behind the telescope. The position-sensitive proportional counter was then used as the readout device for the dispersed energy spectrum. Gratings, used either in transmission or in reflection, have the advantage that much of the energy spectrum may be observed simultaneously and the energy resolution may be as small as a fraction of 1%.

The third *Einstein* spectrometer was a solid-state device. This type of instrumentation is based on electron-hole (absence of an electron in the conduction band) creation in solid materials (these typically are cooled and are usually silicon or, at higher energies, germanium). In x-ray astronomy, the use of solid-state detectors has been driven by the need for better energy resolution with high efficiency. Solid-state x-ray detectors consist of semiconducting material, subdivided (by impurity doping) into regions of different conductivity, within which charge collection can be accomplished. The *Einstein* detector was a silicon detector doped with lithium; it obtained an energy resolution of about 150 eV. More recently a number of different solid-state detectors are being developed for use by x-ray astronomers. These include x-ray-sensitive charge-coupled devices, which measure both position and energy, and the quantum calorimeter. The latter consists of an x-ray absorbing material in contact with a solid-state thermometer (called a "thermistor") and is sensitive to the change in temperature produced by the absorption of individual x-ray photons. To operate, these devices must be kept at extremely cold temperatures, near absolute 0. Thus far, they have achieved an energy resolution of 13 eV; however, they are theoretically limited to only a few electron-volts. The first flight of such a device may be on NASA's *Advanced X-Ray Astrophysics Facility* (*AXAF*) in the late 1990s.

Additional Reading

Aschenbach, B. (1985). X-ray telescopes. *Rep. Prog. Phys.* **48** 579.

Fraser, G. W. (1989). *X-Ray Detectors in Astronomy*. Cambridge University Press, Cambridge.

Giacconi, R. (1980). The *Einstein* X-Ray Observatory. Scientific American **242** (*No.* 2) 80.

Knoll, G. F. (1979). *Radiation Detection and Measurement*. Wiley, New York.

Kondo, Y., ed. (1990). *Observatories in Earth Orbit and Beyond*. Kluwer Academic Publ., Dordrecht.

Skinner, G. K. (1988). X-ray imaging with coded masks. *Scientific American* **259** (No. 2) 84.

Weisskopf, M. C. (1988). Astronomy and astrophysics with the *Advanced X-Ray Astrophysics Facility*. *Space Sci. Rev.* **47** 47.

See also **Scientific Spacecraft and Missions; Sounding Rocket Experiments, Astronomical; X-Ray Astronomy, Space Missions.**

Telescopes, Historical

Kevin Krisciunas

This entry reviews the development of the astronomical telescope from its conceptual gestation through the beginning of the twentieth century. It is limited to astronomy at optical wavelengths, since radio astronomy and the other new forms of astronomy began later in the century.

Optical telescopes owe their development to their ability to magnify distant objects, making the objects appear nearer. This has obvious terrestrial advantages for people as different as operagoers and military commanders. For astronomical uses the telescope has the added advantage over the eye of providing an increase in *light-gathering power*. Just as a bucket can collect more rainwater than a thimble, so does a large telescope collect more light than the eye. If one can accurately focus starlight with a telescope, the telescope plus eye can see much fainter stars than the eye alone. Long-exposure photographs or the use of modern silicon array detectors further extend the limit of light detection. Under clear, moonless skies, away from city lights, the unaided eye can detect sixth magnitude stars; with a 6-in. (15 cm) telescope the eye can reach apparent magnitude 14; with a 4-m telescope, a modern solid-state array can reach magnitude 27 in blue light.

A telescope also provides an increase in the achievable *spatial resolution*—the ability to discern greater detail in objects observed. For circular light collectors the theoretical (diffraction limited) resolution in arcseconds is given by

$$\theta = (1.22\lambda / D) \cdot 206{,}265,$$

where λ is the wavelength of light being detected and D is the diameter of the telescope objective, measured in the same units. For optical light, $3500 < \lambda < 7000$ Å (1 Å $= 10^{-10}$ m). For the dark-adapted eye, $D \approx 6$ mm, and for yellow light ($\lambda = 5500$ Å), the diffraction limit is about 23 arcseconds (though the practical limit is about 60 arcseconds). For a 6-in. optical telescope, θ is about 0.9 arcsecond. Even at a good ground-based site, it is rare to experience seeing better than 0.5 arcsecond, so for optical telescopes with $D > 0.3$ m, the diffraction limit can only be achieved by placing the telescope above the Earth's atmosphere or by "freezing out" the atmospheric turbulence using such methods as speckle interferometry.

The preceding paragraph should be further qualified by stating that it relates to the separation of close point sources of light such as double stars of comparable brightness. The eye is able to discern the presence of linear structures such as a thin wire viewed at a distance, even when the thickness of the wire subtends an angle less than 1 arcsecond. It is thus that lines of craters on Mars, viewed with the eye through a telescope, may give the impression of canals.

PRELIMINARY IDEAS

It is surprising to learn that the first description of the principles of the optics of telescopes and microscopes dates back to the year 1267, to Roger Bacon's *Opus Majus*, wherein the author describes the principles of refraction by lens and states: "And thus from an incredible distance we may read the smallest letters, and may number the smallest particles of dust and sand.... And thus a boy may appear to be a giant, and a man as big as a mountain, for as much as we may see the man under as great an angle as the mountain, and near as we please; and thus a small army may appear a very great one, and though far off, yet very near to us, and the contrary. Thus also, the sun, moon, and stars may be made to descend hither in appearance...".

Giambattista della Porta, in his *Magiae Naturalis* (1589), states: "If you do but know how to join the two [viz., the concave and the convex glasses] rightly together, you will see both remote and near objects larger than they otherwise appear, and withal very distinct."

The preface to Leonard Digges' *Pantometria* [(1591), 2nd ed.], edited by his son Thomas, contains similar words. The son says that the father "by proportional glasses, duly situate in convenient angles, not only discovered things farre [sic] off, read letters, numbered pieces of money, with the very coyne [sic] and superscription thereof, cast by some of his freends [sic] of purpose upon downs in the open fields, but also seven miles off declared what hath been doone [sic] in private places."

Nevertheless, the traditionally acknowledged inventor of the telescope is the Dutch optician Hans Lippershey, who presented a petition at The Hague on October 2, 1608, to gain recognition of his invention of an instrument for seeing at a distance. Four days later Lippershey was told that he would be awarded 900 florins for an example of his invention. On December 15, the Assembly examined a newly built telescope and commanded Lippershey to build two more for the same price each. Lippershey was not awarded the exclusive privilege to make such instruments, because, independent of him, two other Dutchmen, Zacharias Jansen and James Metius (aka Jacob Adrianzoon), laid similar claims to the invention. Metius' petition to the same Assembly is dated October 17, 1608.

Thus it was that the knowledge necessary to make a simple telescope—the *word* was not invented until 1611—was there in the minds of lens makers and the technically minded in Europe for some time.

EARLY REFRACTING TELESCOPES

Galileo Galilei (1564–1642) heard about the invention of the telescope in May of 1609 and in short order made one of his own (without ever having seen one). It consisted of a plano-convex lens as an objective and a smaller plano-concave lens, placed closer to the objective than its focus, acting as an eyepiece and giving a direct (noninverted) image. Over the course of 1609, Galileo made a number of telescopes on this model, the most powerful of which magnified 33 times. Although the objectives of his telescopes were up to 5 cm in diameter, Galileo found it necessary to mask them to about half their diameters so as to improve the image definition. Twentieth-century scholars have examined some of these very telescopes and found that they could achieve a resolution of 10–15 arcseconds, roughly twice the theoretical limit.

Johannes Kepler (1571–1630), famous for his three laws of planetary motion, suggested in his treatise on dioptrics (1611) that the refracting telescope would be improved by the use of a plano-convex eyepiece lens placed beyond the focus of the primary objective. This gives an inverted image (immaterial to viewers of celestial objects), and it affords a wider field of view. Later in the seventeenth century, the Englishman William Gascoigne (1612–1644) demonstrated that with the Keplerian model one could place a set of crosshairs in front of the eyepiece, thus greatly increasing one's ability to measure the positions of stars. Gascoigne then developed an adjustable micrometer.

Refractors of this era had two principal drawbacks. The images suffered first of all from *spherical aberration*; the lens surfaces could be made spherical, but needed to be parabolic to provide a good focus. Parabolic figures could not be ground by opticians at that time.

Figure 1. Observing with long-focus refractors at the Paris Observatory, 1705. The Marly water tower was moved to this site especially to support such telescopes.

Figure 2. Sir William Herschel's 40-ft focal length reflector, completed in 1789. It contained a speculum metal mirror 48 in. in diameter. With it Herschel discovered two new moons of Saturn, Mimas and Enceladus.

The other drawback to refractors of this era was *chromatic aberration*. Isaac Newton (1642–1727) showed in 1666 that white light is composed of all the colors of the rainbow, and that a prism bends blue light most of all, red light least of all. Because a single lens works by the same principle of refraction, the focus of the blue light is closer to the objective than the focus of the red light. Thus a single lens will always give star images that are surrounded by colored halos, owing to the nonunique focal length of the lens. Newton also stated, *incorrectly*, that the *dispersive* power of all glass (the rate of change of index of refraction with wavelength) was the same, implying that even compound lenses would exhibit chromatic aberration.

One way of diminishing spherical aberration and the blurring effect of chromatic aberration was to make very long focus telescopes. Aerial telescopes, supported on masts, became popular, especially on the Continent. A focal length of 100 ft was not uncommon, and there are reports of objectives with focal lengths in excess of 200 ft being used. Understandably, these telescopes were hard to manage, but with them astronomers such as Jean Dominique Cassini made many planetary discoveries (see Fig. 1).

EARLY REFLECTING TELESCOPES

The simplest way to eliminate chromatic aberration was to construct a reflecting telescope, because reflection is not a function of wavelength. In 1668, Newton constructed a reflector with a concave spherical mirror and a flat secondary mirror. The secondary mirror was placed closer to the primary than its focus and at a 45° angle, allowing the eyepiece to be placed at the side of the top of the tube. This is still known as the Newtonian design. Newton's first telescope, with an objective diameter of $1\frac{1}{3}$ inches, focal length of $6\frac{1}{4}$ inches and an eyepiece giving a magnification of 35, allowed him to see the moons of Jupiter and the crescent of Venus. His second telescope, a more successful example, was presented to the Royal Society of London in December 1671.

Newton's ideas had been preceded, however, by the work of the Scotsman James Gregory, who in 1663 suggested the design now known after him, the Gregorian, whereby the telescope consists of a concave parabolic primary mirror with a concave secondary mirror ground to an elliptical figure and placed beyond the focus of the primary. The primary has a cylindrical hole cut in its center, and the eyepiece is placed behind the primary, giving a noninverted image. Gregory contracted opticians in London to execute this design, but the efforts failed to produce a working telescope.

In 1672, the Frenchman Guillaume Cassegrain suggested the design named after him and which has been basically followed for the construction of all the large reflectors built in the twentieth century. Instead of Gregory's concave elliptical secondary, the Cassegrain design calls for a convex hyperboloidal secondary mirror placed closer to the primary than its focus. Compared to the Gregorian, this makes the telescope shorter by twice the focal length of the secondary. In 1779, Jesse Ramsden showed that the Cassegrain design has yet another advantage: The combination of a concave and a convex mirror tends to correct the imperfections of figure in each, whereas in the Gregorian design the aberrations are additive.

The Cassegrain design, like the Gregorian, does not appear to have been successfully executed in the seventeenth century because of the difficulty of making nonspherical surfaces. Only in 1723 did the Englishman John Hadley devise a method of making parabolic mirrors. Shortly thereafter, Hadley and the Scotsman, James Short, began manufacturing Newtonians, Cassegrains, and Gregorians.

Until the mid-nineteenth century, reflecting telescopes had mirrors made of speculum metal. A typical recipe was 2.2 parts copper to 1 part tin. These mirrors were very heavy and they tarnished rapidly (requiring frequent repolishing). However, in spite of their low reflectivity compared to the transmission of the best lenses, reflectors provided unparalleled light-gathering power. William Herschel (1738–1822), the discoverer of the planet Uranus, made many reflecting telescopes. His favorite had a primary mirror 18.8 in. in diameter and a focal length of 20 ft. His largest was 48 in. in diameter with a focal length of 40 ft. They were mounted in alt-azimuth fashion, on a rotating track for azimuth adjustments, and with ropes and pulleys for elevation adjustment. In the Herschelian design there was no secondary mirror. One looked directly with an eyepiece back down the tube by observing at the side of the top of the tube (see Fig. 2).

The largest speculum mirror telescope ever built was the 72-in. diameter reflector of William Parsons, the third earl of Rosse. The telescope was situated at Birr Castle, in Parsonstown, Ireland. It was slung between two massive walls of masonry oriented north–south. Thus one had to wait for objects to transit the celestial meridian to observe them. Completed in 1845, Lord Rosse's telescope was used for the discovery of double stars and nebulae and first revealed the spiral nature of Messier 51, the Whirlpool Nebula in Canes Venatici.

ACHROMATIC REFRACTORS

In spite of the successes of the speculum reflectors, the telescope of choice of astronomers of the eighteenth and nineteenth centuries was the achromatic refractor. The first refractor with an objective designed to reduce chromatic aberration was built by the Englishman Chester More Hall in 1733. Hall sought no credit for this development and his design was essentially unknown.

An article by the Swiss-born mathematician Leonhard Euler in 1747 rekindled interest in the subject, and another by the Swedish mathematician Klingenstierna in 1755 inspired the Englishman John Dollond to perform various optical experiments that demonstrated the incorrectness of Newton's proof of the impossibility of an achromatic objective. In 1758, Dollond made his first example and was awarded a patent.

An achromatic objective consists of two elements, a double convex lens of crown glass with a plano-concave lens of flint glass behind it. Owing to the greater dispersive power of flint glass, this combination allows one to make the focal length of the objective the same for two different colors of light. The triple achromat, consisting of two double convex crown lenses sandwiching one double concave flint lens, was developed by Dollond's son Peter in 1765. This further reduced chromatic aberration, but the larger number of optical surfaces led necessarily to a greater diminution of image brightness, owing to the transmission of glass being less than 100%. Nevertheless, these developments made the refractor superior to any reflector of equivalent objective diameter.

The largest telescope lenses successfully manufactured were produced by the Americans Alvan Clark (1804–1887) and his son Alvan Graham Clark (1832–1897). They made the 26-in. objective for the U.S. Naval Observatory (1873), followed by a 30-in. for Pulkovo Observatory (1885), the Lick Observatory 36-in. (1888), and the 40-in. for Yerkes Observatory (1897). Other large refractors included the Vienna 27-in. (1881) and Royal Observatory (Greenwich) 28-in. (1894) by Howard Grubb; the Bischoffsheim Observatory (Nice) 30.3-in. (1886) and Meudon 32.5-in. (1891) by Paul Ferdinand Gautier and Paul and Prosper Henry; and the 31.5-in. Potsdam refractor by the firm of Steinheil and Repsold (1899). A 49.2-in. objective of 197-ft focal length was made by Gautier in 1900 and was mounted as a siderostat (a plane mirror driven to follow the stars deflected the light into the horizontally mounted, stationary lens), but this telescope was not a success.

FURTHER DEVELOPMENTS

A major breakthrough for the development of astronomy was the ability to deposit silver on glass, a process developed by the German chemist Justus von Liebig in the early 1850s. The first telescopes made with silver-on-glass mirrors were constructed by Carl August von Steinheil in 1856 and Léon Foucault in 1857. Foucault also developed a simple and accurate method of testing the figure of mirrors, known to all telescope makers as the Foucault test.

Glass mirrors had the advantage of being lighter than speculum metal and easier to grind. Also, the reflectivity of silver was much higher, yielding greater light-gathering power. With the success of the 36-in. Crossley reflector at Lick (1895), the 60-in. at Mount Wilson (1908), and the 100-in. at Mount Wilson (1917), reflectors supplanted refractors as the telescope of choice.

An aluminized mirror keeps its reflectivity longer than a silvered one. With the development of vacuum technology, aluminized telescope mirrors have been used since the 1930s. Today certain

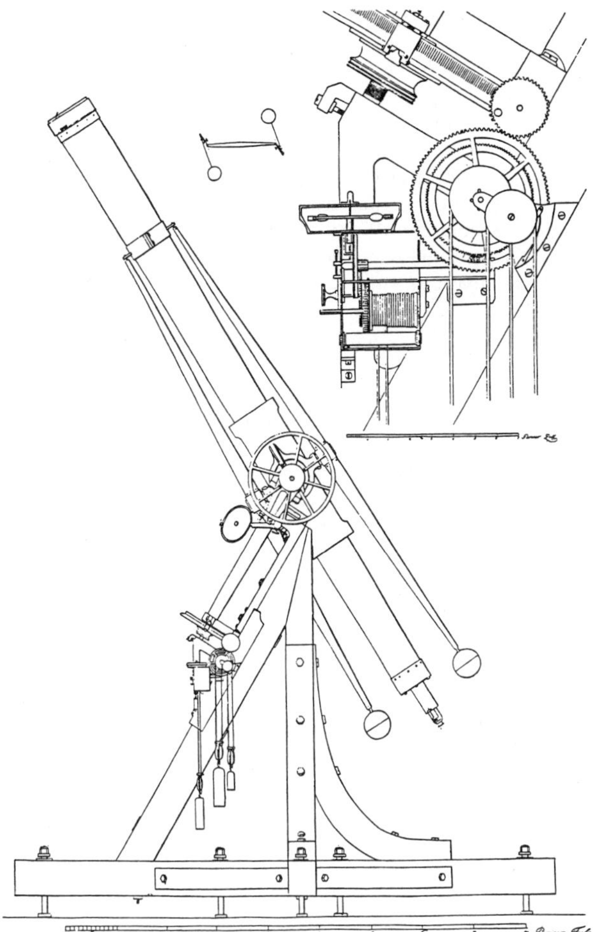

Figure 3. The 9.6-in. Dorpat Observatory refractor, built by Joseph von Fraunhofer. This telescope had the first modern equatorial mounting and was used by Wilhelm Struve for studies of double stars and the measurement of the trigonometric parallax of the star Vega.

optical coatings are available, such as magnesium fluoride, that increase the durability of aluminized mirrors.

It is one thing to have a telescope with good optics, but the telescope must also have a solid mounting and be able to track celestial objects accurately and smoothly. The first modern telescope mounting was the German equatorial of Joseph von Fraunhofer (1787–1826), constructed for the 9.6-in. refractor for the Dorpat Observatory in Estonia. It had a polar axis directed parallel to the Earth's axis of rotation, allowing the observer to position it in declination and to drive it in right ascension by means of a falling weight (see Fig. 3).

From an electronic and mechanical point of view, significant developments in telescope design of the past century include further development of motorized clock drives, servo-controlled guiding, and mirror support systems.

From an optical standpoint, one of the most important inventions was the Schmidt telescope (first made by Bernhard Schmidt in 1931) which gives good (point-like) star images over the telescope's field of view by means of a special corrector plate situated at the top of the telescope tube. This eliminates the effect called *coma*, whereby off-axis star images exhibit fan-like shapes that are most spread out at the edge of the field.

Glass making has evolved as well. The 100-in. telescope at Mount Wilson has a mirror made of plate glass. It was originally planned to cast the Palomar Observatory 200-in. mirror of fused quartz, which has a much lower coefficient of expansion than plate glass, but the research and development costs became prohibitive, and the mirror was cast of Pyrex. Today there exist materials like Cervit and Zerodur, which hardly expand or contract with temperature at all, leading to better images at the focus of the telescope.

Until a century ago the principal astronomical detector at the focus of the telescope was the human eye. Photography replaced the eye by the turn of the century for everything except planetary drawings and double star measurements. Photomultiplier tubes became common after World War II. Image tubes followed. In the 1970s, charge-coupled devices were invented, and in the 1980s, arrays were made to work at infrared wavelengths as well.

Additional Reading

Gingerich, O., ed. (1984). *Astrophysics and Twentieth-Century Astronomy, Part A*. (*The General History of Astronomy*, Vol. 4.) Cambridge University Press, Cambridge.

Howse, D. (1986). The Greenwich list of observatories. *J. History Astron.* **17** (Part 4) i–iv, 1–100.

King, H. C. (1979). *The History of the Telescope*. Dover, New York.

Krisciunas, K. (1988). *Astronomical Centers of the World*. Cambridge University Press, Cambridge.

Van Helden, A. (1977). The invention of the telescope. *Trans. Amer. Philos. Soc.* **67** (Part 4) 67.

Wright, H. (1966). *Explorer of the Universe: A Biography of George Ellery Hale*. E. P. Dutton, New York.

See also **Cameras and Imaging Detectors, Optical Astronomy; Telescopes, Large Optical; Telescopes, Wide Field.**

Telescopes, Large Optical

David L. Crawford

Most of the entries in this Encyclopedia would not have been possible without the contribution of large optical telescopes. These powerful instruments have been at the core of astronomical research for many decades and will continue to be essential tools in frontier research for many more decades to come.

WHAT IS A LARGE OPTICAL TELESCOPE?

The first question is: What is "large"? Naturally, the answer has been changing with time, as our abilities and our expectations have evolved. A few hundred years ago, any optical aid, compared to the eye alone, might well have been considered "large." Then came the major increase in instrument aperture employed by the Herschels.

The next wave was that of the large refracting telescopes, including the great refractors still operating at Lick Observatory and Yerkes Observatory. Then came the early modern large reflectors, leading from the 60-in. and the 100-in. telescopes at Mount Wilson Observatory to the 200-in. Hale Telescope at Palomar. The Soviet Union has put a 6-m (240-in.) reflector in operation, currently the largest optical telescope. The Hale Telescope, with its 5-m mirror, is the second largest in regular operation. A list of the largest telescopes [with apertures or primary mirror diameters of 3 m (120 in.) or greater] that are in regular operation as of May 1991 is given in Table 1.

Reflectors are easier to make than refractors, due to the fact that only one surface must be made optically (nearly) perfect, rather than the two or more lens surfaces in refractors; in addition, the required accurate support of the optical elements in the telescope is easier with a reflector than with a refractor, where the support can only be at the edge of the lens, rather than at the edge and behind (or within) as in the case of a telescope with a mirror. Generally, the overall size of a reflecting telescope can be considerably smaller than that of a refractor of the same aperture: The dome housing the Yerkes 40-in. refractor is about the same size as the dome housing the 200-in. reflector.

We are now seeing a new generation of telescopes of all sizes, including several much larger than any of the older-generation ones currently in operation. Of course, what is "large" depends on what type of research one is doing and what one's funding and expectations are. Clearly, a 16-in. telescope is large for the amateur astronomer, but small for most professionals. A 16-in. telescope is also a rather large solar telescope. The *Hubble Space Telescope* (*HST*) is a large optical telescope, as well as a space telescope capable of operating in the ultraviolet and infrared, even though it "only" has a 94-in. mirror. So a clear statement of what is "large" is not possible and not really needed.

The next question is: What is "optical?" As technology improves, the definition of "optical" is not as clear as it was several decades ago. The *HST* is clearly an optical telescope, though it operates a great deal of the time in the ultraviolet wavelength region of the spectrum. Almost all optical telescopes can operate in the near ultraviolet, though the eye is not sensitive to this region. Likewise, most reflective optical telescopes can work fairly effectively in the near infrared, where again the eye is not sensitive but the atmosphere is transparent.

Certainly, one of the most exciting advances in astronomy has been the progress of research from the optical spectrum into the ultraviolet, the infrared, the sub-millimeter, the millimeter, and the radio regions. But the optical still remains critically important for frontier research, and the motivation for larger ground-based telescopes will continue for the indefinite future. All of these telescopes, at all the spectral regions, are complementary, and all are needed for progress with frontier research. Indeed, "complementarity" is a splendid description of the mutual roles of telescopes in astronomy: All wavelength regions offer important information to astronomers, and we also need telescopes of smaller sizes, which can still be powerful research tools. Think of a truck fleet operator with only large tractor-trailers and no pickups to help out. It would not be an efficient or cost-effective system. The same is true for astronomy.

A FEW BASICS ABOUT A TELESCOPE

What is a telescope designed to do? It is designed to collect light coming from celestial objects, so that it can be analyzed with such instrumentation as is attached to the telescope; the photometer, imaging devices, spectrometers or spectrographs, and other powerful tools. The amount of light arriving is "astronomically" small in many cases, so that we want to have the largest collecting area possible, hence a "large" telescope. The light-collecting power of a telescope depends on the area of the primary mirror, or the square of the diameter.

The electronics age allows many other improvements now being incorporated into new designs, such as active optics (where the mirror support system is actively controlled, in real time, to help improve the quality of the optical image), automatic control, even remote control, of the telescope and the instrumentation, sometimes even from another continent.

Due to improvements incorporated into these new-generation telescopes, we are able to not only build larger and better telescopes, but to do it at considerably less cost, relatively speaking, than was possible with the older ones. In fact, there is no way that these new telescopes could be built without the new technological advances.

The following list briefly summarizes the various developments that make a telescope "new generation." Note that most of these hold no matter what the size of the telescope, and new-generation small telescopes are very efficient research tools. Coupled with the new detectors, they have, in fact, become "large."

Table 1. Optical Telescopes with Primary Mirrors of 3.0 m Diameter or Greater in Full Operation, May 1991

Name and Observatory	*Aperture and Mirror Material*	*Comments*
Bolshoi Altazimuth Telescope, Special Astrophysical Observatory	6.0 m; glass	Two $f/30$ Naysmyth foci; $f/4$ prime focus; moving weight, 650 tons; altazimuth mounting
Hale Telescope, Palomar Observatory	5.0 m; Pyrex	Cassegrain[a] ($f/16$), Coudé ($f/30$), and prime ($f/3.3$) foci; horseshoe yoke equatorial mounting
Multiple Mirror Telescope, MMT Observatory	4.5 m; fused silica	Six coaligned, 1.8-m reflectors; rotating building
William Herschel Telescope, Roque de los Muchachos Observatory	4.2 m; Cer-Vit	Cassegrain and two Naysmith foci; $f/2.5$ primary mirror; altazimuth mounting
N. U. Mayall Telescope, Kitt Peak National Observatory	4.0 m; fused silica	Cassegrain ($f/8$), Coudé ($f/190$), and prime ($f/2.7$) foci; equatorial mounting; 100-ft-high, 37-ft-diam. concrete pier
4-m Telescope, Cerro Tololo Inter-American Observatory	4.0 m; Cer-Vit	Cassegrain ($f/8$), infrared ($f/30$), and prime ($f/2.7$) foci; flipping secondary mirror; equatorial mounting; 500-ton rotating dome
Anglo-Australian Telescope, Anglo-Australian Observatory	3.9 m; Cer-Vit	Cassegrain ($f/8$), Coudé ($f/36$), infrared ($f/15$), and prime foci; equatorial mounting
Canada–France–Hawaii Telescope, Mauna Kea	3.6 m; Cer-Vit	Cassegrain ($f/8$), Coudé ($f/20$), and prime ($f/3.8$) foci; equatorial horseshoe mounting
United Kingdom Infrared Telescope, Mauna Kea	3.6 m; Cer-Vit	Cassegrain [$f/35$ (chopping secondary), $f/9$] foci; Coudé ($f/20$) focus; English yoke equatorial mounting
ESO 3.6-m Telescope, European Southern Observatory	3.6 m; fused silica	Cassegrain, Coudé, and prime ($f/3$) foci; horseshoe and fork equatorial mounting
New Technology Telescope, European Southern Observatory	3.58 m; Zerodur	$f/11$ Naysmyth focus; $f/2.2$ meniscus primary; eight-sided, air-conditioned, rotating building; altazimuth mounting, active optics
Calar Alto Telescope, German–Spanish Astronomical Center	3.5 m; Zerodur	Horseshoe equatorial mounting
NASA Infrared Telescope Facility, Mauna Kea	3.0 m; Cer-Vit	$f/35$ chopping secondary Cassegrain focus, $f/120$ Coudé focus; equatorial yoke mounting
C. D. Shane Telescope, Lick Observatory	3.0 m; Pyrex	Cassegrain, Coudé, and prime ($f/5$) foci; equatorial fork mounting

[a] "Cassegrain" is used to indicate the corresponding focus of both conventional Cassegrain (parabolic primary) and Ritchey–Chrétien (hyperbolic primary) telescopes.

1. New optical design techniques, using computers, with faster focal ratios.
2. Ability to produce larger single mirror blanks and to servo-control small mirror segments to operate as if they were one large mirror.
3. Lighter-weight optics, hence less thermal problems with the mirrors and with the air near them. Lighter-weight mountings can be used, cutting costs.
4. New types of mountings, with computer control; most are alt-azimuth rather than equatorial in design.
5. Full computer control of the telescope, the optics, and the attached instrumentation.
6. Much smaller buildings and dome, due to the smaller mountings and faster focal ratios.
7. The identification and use of better observing sites, as we are better able to understand site quality.
8. Use of computers in operating and scheduling the telescopes, thus potentially greatly improving the efficiency of use and handling of the vastly increasing amount of data being produced by the new detectors.

All these will mean much better science per dollar, whatever the size of the telescope and will also mean that astronomy has the chance of affording at least a few of the giant and very costly telescopes needed to work at the forefront of knowledge.

NEW PROJECTS

Several telescopes in the 3.5-m class are currently under development, including the nearly completed 3.5-m reflector of the Astrophysical Research Consortium at Apache Point, near Sunspot, New Mexico. They feature lightweight, short-focus primary mirrors, extensive automation, and, frequently, remote control. The same features characterize a future class of very large telescopes, with apertures ranging from 6.5 to 16 m, that is described in the entry on Telescopes, Next Generation.

PROBLEMS, AND THE FUTURE

Finally, just a few words about some of the problems, and about the future. We must assume that there will be a future, and that efforts under way to control light pollution will be successful.

There is no question that the new giant telescopes are stretching new technology. That is what must be done to make advances. But it also means that there will be problems; the schedules will go slower than many would like. But they will work very well, I am sure, when the teething days are over. They will produce excellent, exciting science. And they will lead to even larger optical telescopes in the future, both ground-based and in space. The costs of these telescopes are very large, by astronomy standards, but small when compared to the costs of other large technology developments, such as the SST and military aircraft. The astronomical telescopes will be at the cutting edge of technology, themselves, as will their instrumentation and detectors. They will be powerful tools for advancing knowledge of this marvelous and exciting universe we live in.

Additional Reading

The best references to developments relative to large optical telescopes can be found in recent issues of *Sky and Telescope* and *Astronomy*.

Beckers, J. M. et al. (1981). The Multiple Mirror Telescope. In *Telescopes for the 1980s*, G. Burbidge and A. Hewitt, eds. Annual Reviews Inc., Palo Alto, p. 63.

Fienberg, R. T. (1990). Dazzling views from Europe's NTT. *Sky and Telescope* **79** 596.

Fischer, D. (1989). A telescope for tomorrow. *Sky and Telescope* **78** 249.

Mannery, E., Siegmund, W. A., and Hull, C. L. (1989). The performance of the Apache Point Observatory 3.5 m telescope. *Ap. Space Sci.* **160** 269.

Ridpath, I. (1990). The William Herschel Telescope. *Sky and Telescope* **80** 136.

See also **Cameras and Imaging Detectors, Optical Astronomy; Hubble Space Telescope; Radio Telescopes and Radio Observatories; Telescopes and Observatories, Solar; Telescopes, all entries.**

Telescopes, Next Generation

Stephen P. Maran

Next generation telescopes (NGTs) are reflecting telescopes with primary mirrors greater than six meters (m) in diameter, the aperture of the largest existing conventional telescope. At least eight NGTs are under design or development, with one, the Keck Telescope on Mauna Kea, Hawaii, in partial operation since late 1990. The term "new technology telescopes" is often used interchangeably with "next generation telescopes," to emphasize the engineering developments that distinguish them from older instruments. However, many smaller telescopes under development or in operation also are considered new technology telescopes. Further, certain new technologies that are crucial to the design of NGTs, such as active thermal control and adaptive optics, are being installed on existing telescopes, to upgrade their performance.

The 200-in. (5-m) Hale Telescope, completed in 1948, at the Palomar Observatory was widely regarded as the largest practical telescope for astronomical observations in visible light. The fact that the 6-m telescope at Zelenchukskaya, U.S.S.R., completed in 1976, did not exceed the performance of the Hale Telescope despite much greater cost seemed to confirm this conclusion. Larger telescopes were thought to offer diminishing returns, because the blurring caused by turbulence in the atmosphere ("atmospheric seeing") was thought to be certain to prevent them from producing finer optical images (sharper spatial resolution) and to prevent the capability of detecting much fainter objects.

MODERN RATIONALE FOR GIANT TELESCOPES

In the late 1970s and 1980s, new data on the causes of blurred images in existing telescopes, as well as the development of new optical and control technologies, made it practical to begin designing telescopes with primary mirrors larger than six meters in diameter. It was found that at good mountaintop sites in Arizona, Hawaii, and Chile, much of the atmospheric seeing that blurred telescopic images was not due to the atmosphere far above the observatory as previously thought, but to air currents actually near and within the dome and even within the telescope. Experiments showed that these localized contributions to the atmospheric seeing could be reduced substantially by using exhaust fans, air conditioning domes during the warm part of the day, insulating dome floors, removing heat-generating electrical apparatus from the dome, designing the shape of the telescope building for minimal disturbance of the prevailing outside air flow, and applying specialized thermal coatings to telescope structures and domes. With the seeing now much improved, larger telescopes could provide finer resolution and thus became scientifically advantageous. However, it also was necessary to establish that they would be economically feasible.

The enormous and therefore unacceptable estimated costs of making very large telescopes of conventional design arise from the following: (1) the expense of fabricating large mirrors (both the cost of materials and the cost of figuring and polishing the optical surfaces); (2) the cost of large, heavy, precision machinery for supporting and pointing the telescopes while precisely tracking targets; (3) the cost of the very large buildings and rotating domes needed to house the telescopes (in some cases, the whole building rotates). The new technologies that substantially reduced the estimated costs of NGTs are the following: (1) methods of casting very large yet lightweight telescope mirrors that have approximately the correct shapes even before optical figuring begins; (2) methods to shape and polish telescope mirrors that are exceptionally fast (have very low focal ratios, or "*f*/numbers"); (3) methods for accurate control of large telescopes on altazimuth or similar mountings, as opposed to conventional equatorial mountings. Lightweight primary mirrors can be supported by mirror cells, telescope tubes, and telescope mountings that are much lighter and less expensive than those for heavy conventional mirrors of the same size. The faster the primary mirror, the deeper its "bowl" (or concave shape), the shorter its focal length and, accordingly, the shorter the telescope. It is often said that conventional telescopes were shaped like cannons, whereas NGTs will resemble searchlights. With shorter telescope tubes, the rotating dome or building can be smaller and therefore cheaper. A telescope of a given length on an altazimuth or similar mounting moves through a smaller overall volume when pointed over the full range of directions in the sky than does a telescope of the same length on an equatorial mounting. However, the tracking of stellar targets as the Earth turns is much simpler with an equatorial mounting; thus, nearly all large conventional telescopes built before the 1970s have equatorial mountings. The installation of the 6-m telescope in the U.S.S.R. and the Multiple Mirror Telescope (MMT) on Mount Hopkins in Amado, Arizona, in 1976 and 1979, respectively, showed that modern computer control of a large telescope on an altazimuth mounting made target tracking on such mountings readily feasible.

During the 1980s, the development of improved infrared detectors for celestial observations led to the flourishing of infrared astronomy. Visible-light (optical) observations with conventional ground-based telescopes are usually seeing-limited, meaning that a large telescope does not yield a sharper image than a somewhat smaller telescope under identical observatory conditions because atmospheric seeing blurs the typical star image obtained with each telescope to virtually the same size. However, observations made in infrared light at good sites are often diffraction-limited (depending on the wavelength of observation), meaning that the blurring is mostly due to the finite size of the primary mirror rather than to atmospheric turbulence. This occurs because infrared light has longer wavelengths than visible light and the seeing improves slightly at longer wavelengths because the individual parcels of air of different temperatures (which cause the blurring by bending light that takes different ray paths by different amounts) bend the longer-wavelength light less than the shorter-wavelength light (just as a glass prism does); also, the spatial resolution of a telescope worsens with increasing wavelength, but improves with increasing primary mirror diameter. Thus it became clear that giant telescopes, if successfully constructed, would offer even greater advantages in performance in infrared light, with respect to conventionally sized telescopes, than in visible light. These performance advantages are based on the smaller images produced by the NGTs, which concentrate the light from a given star more effectively, enabling the detection of fainter stars as well as the recording of sharper photographs.

ACTIVE OPTICS AND ADAPTIVE OPTICS

It is not sufficient to build a large telescope of high optical quality; it is also critical to maintain the precise optical shape of a telescope mirror as it is subjected to changing forces. A mirror sags under the force of gravity in different ways when the mirror is oriented at different angles to the vertical as the telescope is pointed at different directions in the sky. Also, it is deformed due to changes in its temperature or to differences in temperature between different parts of the mirrors, for example, the front versus the back. Such temperature gradients warp the mirror, however slightly, as the glass at different positions contracts or expands by different amounts. The wind also applies distorting forces to a large telescope mirror, in addition to shaking the telescope.

The technology of active optics was developed to compensate for the forces applied to telescope mirrors by gravity, temperature effects, and mechanical disturbances, all of which will be severe for the primary mirrors of NGTs, which are much thinner and less stiff than conventional mirrors of the same diameter would be. In active optics, the shape of a telescope mirror (usually the primary mirror) is adjusted through computer control of many small motorized actuators mounted on the mirror back. Also, the position and the tilt of the secondary mirror may be controlled in a similar way. In the case of a mosaic mirror that is assembled from multiple individual mirror segments, active optics also control the relative orientations and spacings of the segments.

The size of the image of a star produced by a telescope mirror in the absence of atmospheric seeing is smaller for a larger telescope mirror, providing that each mirror is polished to a suitable optical prescription. Therefore, giant telescopes offer the possibility of sharper images even in visible light, providing that the effects of atmospheric seeing can be eliminated or substantially reduced. To realize this possibility, astronomers are developing so-called adaptive optics, which correct for the effects of seeing on images. Because the seeing effects change hundreds of times per second, the corrections must be made at a similarly rapid rate. This is done by sensing the atmospheric disturbances on the image of a bright reference star and making rapid adjustments in the shape of a small, thin, deformable mirror in the light path of the telescope, through the use of computer-controlled actuators on the mirror back. This is called the rubber mirror method, although the mirror used is usually a normal optical material such as glass or metal. Experiments show that adaptive optics can make major improvements in image quality provided that the bright reference star is very close to the target of study on the sky. Because suitable reference stars will not be found near most targets, astronomers are studying the use of an "artificial reference star," consisting of laser light projected through the telescope and producing a small bright spot where it is reflected by a high-altitude layer of atmospheric sodium atoms. An artificial reference star (also called a beacon) can be produced near the direction of any celestial target at which the telescope points. In April 1991, the National Science Foundation (NSF) announced that adaptive optics and beacon technology developed by the U.S. Department of Defense had been declassified and transferred to the NSF for application to astronomical telescopes.

Some numerical examples obtained under real conditions illustrate the value of the new telescope technologies. At the European Southern Observatory's (ESO) station at Cerro La Silla, Chile, a conventional 3.6-m (142-in.) telescope, which began operating in 1976, yielded visible-light star images under excellent atmospheric conditions with diameters of about 1 arcsecond (arcsec). In contrast, the slightly smaller ESO 3.5-m New Technology Telescope (NTT), which saw "first light" on March 22–23, 1989 and which was equipped with a fast ($f/2.2$), lightweight (6 tons) primary mirror, active thermal control, active optics, and a compact rotating building whose shape was validated by wind-tunnel experiments, produces image diameters of only 0.33 arcsec under similar conditions on the same mountaintop. Yet the NTT cost only 25 million deutsche marks, about one-third the cost of the 3.6-m telescope. Then, when an experimental rubber mirror system of adaptive optics was installed on the older 3.6-m telescope in 1990, it improved the image size in infrared light at wavelength 2.2 micrometers [(μm), called the K band by astronomers] from about 0.8 arcsec to only 0.18 arcsec.

DESIGN APPROACHES FOR NEXT GENERATION TELESCOPES

Telescope Types

The giant telescopes under development can be classified as single or multiple telescopes. A single telescope has one primary mirror. The mirror can be monolithic, meaning that it was cast as a single piece, or it can be a segmented mirror, composed of individual pieces that may fit together (mosaic mirror) or not, depending on their shapes. A multiple telescope has more than one primary mirror; the separate primary mirrors can be mounted together on a single telescope structure or on independent structures, with the light collected by all primaries routed by fiber optics or small mirrors to a central point where it is combined and measured or photographed.

Primary Mirror Types

At the heart of every NGT project is a lightweight primary mirror design. Most current NGTs incorporate monolithic primary mirrors of either of two types: lightweight honeycomb mirrors or thin meniscus mirrors. A honeycomb primary mirror consists of a concave front plate, a flat back plate, and a honeycomb-patterned array of glass ribs between the two plates. Such a mirror is very light but stiff. The principal source of honeycomb mirrors is the Mirror Laboratory at the University of Arizona, in Tucson, where they are fabricated by piling glass lumps in a mold placed in a large oven. The oven is rotated at a carefully calculated rate, so that the molten glass assumes the approximate desired concave shape due to "centrifugal force." A thin meniscus primary mirror is a single curved plate of glass or glass–ceramic, concave on the front surface and convex on the back. According to the procedure employed by the principal source for such mirrors, the Schott Glass Works in Mainz, Germany, they are cast by pouring a molten glass–ceramic (tradenamed Zerodur) into a rotating mold. A meniscus mirror is not very stiff, so the telescope operator relies on an active optics system to maintain its shape.

The greater the diameter of a monolithic mirror of a given design, the thicker and heavier it must be to attain the needed stiffness. In contrast, the thickness of a large mosaic mirror does not depend on the diameter of the mirror, but only on the diameter of the largest mirror segment, because a segment sags under its own weight, not the total weight of the mosaic.

Telescope primary mirrors are figured with the use of a convex lap, or polishing tool, that rubs a liquid slurry of fine abrasive powder across the mirror blank. In parabolic or hyperbolic mirrors, as used on most large telescopes, the curvature of the mirror surface changes with distance from the center, unlike a spherical mirror. In practice, the faster and larger the mirror, the more difficult it is to figure with a conventional rigid tool. This is a severe problem for the huge, very concave primaries of the NGTs and is addressed by the development of a "stressed lap" technique at the University of Arizona Mirror Laboratory. The shape of an aluminum polishing lap is adjusted by lever arms, steel bands, and actuators that apply forces as large as 1000 lb under active computer control so that the lap is always suitable for polishing the part of the mirror blank with which it is instantaneously in contact. A related problem was foreseen in the development of the mirror segments for the mosiac primary mirror of the Keck Telescope. Unlike the primary mirror as a whole, an individual segment is not symmetrical about an axis perpendicular to the center and, therefore, is very difficult to polish with a conventional rotating tool (or with a stationary tool applied to a rotating mirror blank). To address this problem, "stressed mirror" polishing was developed: The mirror segment blank is deformed deliberately by forces applied by weights and levers during the polishing process. After polishing, the force is removed and the mirror hopefully relaxes to the desired shape.

NEXT GENERATION TELESCOPE PROJECTS

Keck Telescope

The Keck Telescope is a single telescope with a 10-m (400-in.) diameter, $f/1.75$ hyperbolic mosaic primary mirror. The primary consists of 36 1.8-m (72-in.) hexagonal glass–ceramic segments with 3-mm-wide intersegment spacings. The 94-million-dollar telescope project is in the charge of the California Association for Research in Astronomy, a corporation formed by the California Institute of Technology and the University of California. The structure that carries the primary mirror of a telescope is called the mirror cell; in the Keck Telescope, each mirror segment is mounted in its own subcell and the segments are adjusted to an accuracy of one-millionth of an inch. The telescope is on Mauna Kea, altitude 13,800 ft, with Keck Observatory headquarters at Kamuela, Hawaii. In late 1990, with nine of the mirror segments in place, the telescope saw first light, obtaining photographs of a spiral galaxy that indicated that the system that controls the positioning of the mirror segmentsworks satisfactorily. Although the Keck Telescope has twice the aperture of the Hale Telescope, it fits in a dome that is 101 ft high and 122 ft wide, versus 135 and 137 ft, respectively, for the Hale telescope. The glass in the Keck primary mirror weighs 28,800 lb, almost the same as the 29,000-lb Hale mirror, which has half the diameter. On April 26, 1991, it was announced that the W. M. Keck foundation would provide approximately $75,000,000 to fund 80% of the cost of a second telescope of equal size and similar design, the Keck II telescope. The Keck II will be located 75 m from the Keck I telescope and the pair of telescopes will be usable as an optical interferometer with a baseline of that length. Used together on a common target, the two Keck Telescopes will provide a collecting area equivalent to that of a single 14-m (550-in.) reflector. Keck I will begin research use in 1992 and Keck II will be completed in 1996.

Very Large Telescope

The Very Large Telescope (VLT), a project of the European Southern Observatory, will be a multiple telescope consisting of four separately mounted reflectors arranged on a north–south baseline. Each reflector will be furnished with an 8-m (315-in.) glass–ceramic, thin meniscus primary mirror (with a focal ratio of approximately $f/1.8$) that is only 7 inches thick. When the light of the four reflectors is combined, the VLT will provide a light-collection capability equivalent to that of a single 16-m (630-in.) telescope. The reflectors also will be usable individually, so that they can be pointed at different targets at the same time or they can observe the same target simultaneously with different scientific instruments. In addition, they can be used as an interferometer, obtaining spatial resolution equivalent to a single telescope with an aperture that equals the length of the VLT baseline, about 100 m (330 ft). The shapes of the primary mirrors will be changed from hyperbolic to parabolic as required for observations at Ritchey–Chrétien and Cassegrain foci, respectively, by means of an active optics control system. The VLT will be erected at an altitude of 2664 m (8740 ft) on Cerro Paranal, an isolated peak in the Atacama desert, about 80 mi south of Antofagasta, Chile. In 1989, the budgeted cost of the VLT was 382 million deutsche marks, equal at that time to about 210 million dollars. The first 8-m test mirror blank was cast in 1991 at the Schott Glass Works.

Columbus Project

The Columbus Project, a joint effort of Ohio State University, the University of Arizona, and Italian organizations, is developing a telescope consisting of two co-mounted reflectors, sometimes described as resembling a huge pair of binoculars. Each reflector will be furnished with an 8-m $f/1.2$ borosilicate glass honeycomb primary mirror. Together they will furnish a light-gathering power equivalent to that of a single 11.3-m (445-in.) telescope. The intended site is Emerald Peak, on 10,720-ft Mount Graham, in the Pinaleno mountains of eastern Arizona. The use of this site in the Coronado National Forest has been opposed by organizations concerned with possible impacts on the Mount Graham red squirrels, an endangered subspecies.

Multiple Mirror Telescope Conversion

The Multiple Mirror Telescope Conversion represents a planned retrofitting of the existing Multiple Mirror Telescope at Mount Hopkins, near Amado, Arizona, with a 6.5-m (260-in.) borosilicate glass honeycomb mirror. In its original form, the MMT consisted of six 1.8-m (72-in.) reflectors on a common mounting, with an equivalent aperture of 4.5 m (176 in.). The MMT is operated by the Smithsonian Astrophysical Observatory and the University of Arizona.

Japan National Large Telescope

The Japan National Large Telescope, to be erected on Mauna Kea, was long conceived as a 7.5-m thin meniscus mirror telescope. In late 1990, plans were revised slightly to make it an 8-m telescope at a projected cost of 38,000 million yen.

Magellan Project

The Magellan Project, a cooperative effort of the Carnegie Institution of Washington and the University of Arizona, will build a single 8-m telescope with an $f/1.2$ borosilicate glass honeycomb mirror, to be located at the Las Campanas Observatory in Chile. The design concept as of April 1991 is shown in Fig. 1.

Twin 8-Meter Advanced Technology Telescopes

Twin 8-Meter Advanced Technology Telescopes, previously called the National 8-Meter Telescopes or NOAO 8-Meter Telescopes and likely to be renamed again, are proposed as a joint project of the United States, the United Kingdom, and likely a third country, whose assent to the project has not yet (early 1991) been established. The project is directed by the National Optical Astronomy Observatories (NOAO) in Tucson, Arizona. The telescopes, each with an 8-m $f/1.8$ borosilicate glass honeycomb mirror, would be

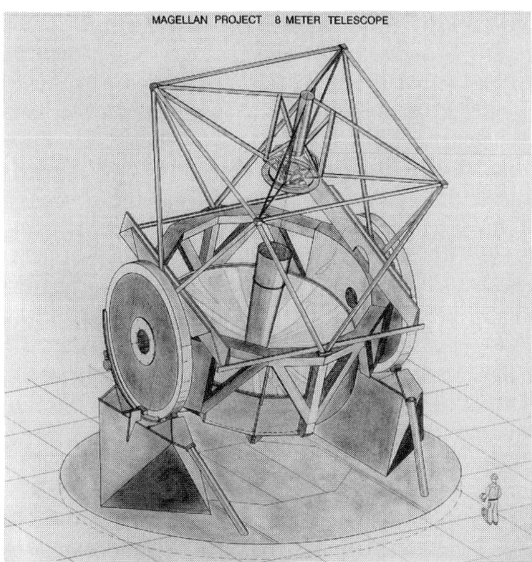

Figure 1. Artist's impression of the future 8-m telescope of the Magellan Project. (*Drawing by Steven Gunnels, courtesy of the Carnegie Institution of Washington and L & F Industries.*)

capable of observing opposite hemispheres of the sky, with the northern hemisphere telescope at Mauna Kea and the southern hemisphere telescope at Cerro Pachon, altitude 9000 ft, in the Chilean Andes. The project is planned for completion around 1998. The 8-m telescope at Mauna Kea would be the only currently planned very large telescope that would be optimized for maximum infrared performance, including adaptive optics for high infrared spatial resolution.

Spectroscopic Survey Telescope

The Spectroscopic Survey Telescope (SST), a joint project of Pennsylvania State University and the University of Texas, will have a segmented primary mirror having a light-collecting power equivalent to that of a monolithic 8-m mirror. The 85 identical, comounted 36-in.-diameter spherical mirror segments, polished by inexpensive student labor, will not fit together in a mosaic, because they are circular in outline rather than hexagonal like the Keck Telescope segments. The simple mounting of the SST will keep it always pointed at an altitude of 70°, but will rotate the telescope in azimuth to acquire targets as they cross that altitude at different points in the sky. Exposures will be limited to about one hour, due to the limited tracking range provided by a movable optical component. The spherical mirrors will not provide a high-definition image suitable for photography, but the SST will constitute a relatively economical means for collecting a great deal of light for spectroscopic analysis of astronomical objects. The SST will be erected at the McDonald Observatory, altitude 6800 ft, near Fort Davis, Texas.

Additional Reading

Astronomy and Astrophysics Survey Committee (1991). *The Decade of Discovery in Astronomy and Astrophysics*. National Academy Press, Washington, D.C. See especially Chapter 3, Existing programs, p. 55.

Fienberg, R. T. (1990). Dazzling views from Europe's NTT. *Sky & Telescope* **79** 596.

Maran, S. P. (1987). A new generation of giant eyes gets ready to probe the Universe. *Smithsonian* **18** (No. 3) 41.

Maran, S. P. (1989). New eyes on the sky. In *Science Year 1990*. World Book, Chicago, page 89.

Martin, B., Hill, J. M., and Angel, R. (1991). The new ground-based optical telescopes. *Physics Today* **44** (No. 3) 22.

Sinnott, R. W. (1990). The Keck Telescope's giant eye. *Sky & Telescope* **80** 15.

Waldrop, M. M. (1990). Keck Telescope ushers in a new era. *Science* **249** 1244.

See also **Telescopes, Large Optical.**

Telescopes, Wide Field

Colin M. Humphries

SCHMIDT OPTICAL SYSTEM

In any image-forming instrument the angular field over which images can usefully be recorded or resolved from each other is limited by the extent to which optical aberrations are present. The theory and properties of wavefront aberrations are discussed in several texts on optics, including the summary by Warren J. Smith, but for the present purpose we note that aberrations such as coma and astigmatism become more pronounced as the off-axis field angle increases and, for many telescope designs, usable fields of view without some special form of correction are typically only a few arcminutes in diameter as projected on the sky. [Of particular importance here are the so-called third-order, or Seidel, aberrations (spherical aberration, coma, astigmatism, field curvature, and field distortion), as well as the chromatic aberrations (longitudinal and transverse) and the chromatic variations of Seidel terms (such as chromatic differences of spherical aberration and field curvature). Fifth- and higher-order aberrations may also require consideration depending on their relative sizes.]

In 1931, Bernhard Schmidt, an Estonian working at the Hamburg Observatory in Bergedorf, Germany, published a new and elegant method of overcoming the problem of field limitation. Whereas it was known that, with an object at infinity, a paraboloidal mirror forms a stigmatic (point-like) on-axis image free of aberrations, Schmidt's system used a spherical mirror together with a weak aspheric lens or plate. The aspheric plate was located at the center of curvature of the sphere (Fig. 1) and its purpose was to provide a precorrection for the spherical aberration that would otherwise be obtained if the spherical mirror were used alone.

In effect, the sphere plus corrector lens is optically equivalent to an on-axis paraboloid; and, because the corrector plate acts as an aperture stop at the center of curvature of a sphere, the system has no unique axis and is therefore free of the off-axis aberrations coma and astigmatism. Thus, apart from a slight obliquity effect, the corrector plate provides the same optical processing for off-axis light as for on-axis light.

The focal surface is convex spherical toward the mirror and has a radius equal to the focal length; it lies midway between the corrector plate and the mirror and is concentric with the latter. The

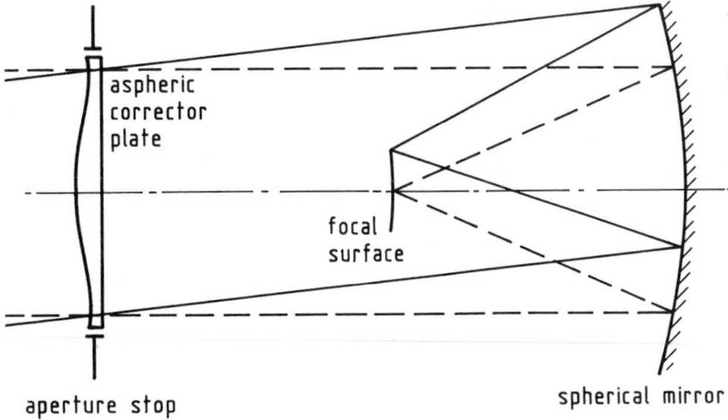

Figure 1. Schmidt camera/telescope (with aspheric profile of corrector plate shown greatly exaggerated).

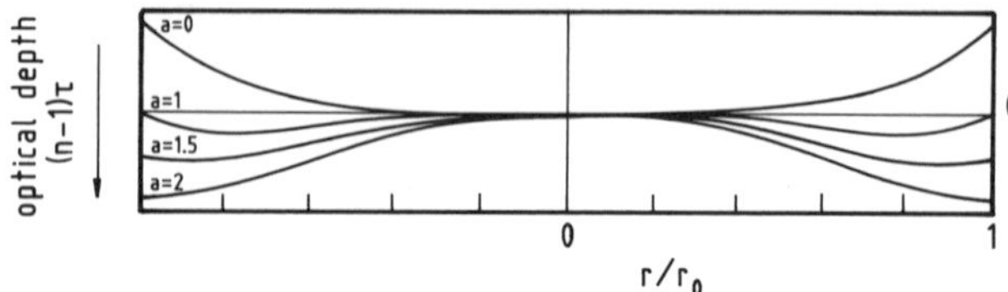

Figure 2. Corrector plate profiles for different values of parameter *a*. (*Based on a diagram given by E. H. Linfoot,* Recent Advances in Optics; *reproduced by permission of Oxford University Press.*)

spherical mirror must then have a diameter larger than that of the corrector plate so that it receives all of the off-axis radiation for the chosen field size (the aperture size quoted for a Schmidt camera or telescope is that of the corrector, not the mirror).

In this way, images of excellent quality can be obtained over fields several degrees in diameter for primary mirrors with focal aperture ratios as fast as $f/0.5$. Because the capacity of a telescope to record information from the sky depends on the square of the angular field as well as on the square of the aperture, a Schmidt system, even one of moderate aperture, is a powerful instrument.

There is no unique solution for the profile of a Schmidt corrector plate because there is an infinite number of paraboloidal surfaces of different focal lengths to which the chosen sphere may be referenced for specifying the required wavefront retardations at each zone (i.e., radial distance) of the corrector plate. Assuming that the correction is to be obtained by aspherizing at a single surface, the required depth (t) of material to be removed by optical figuring from a plane-parallel plate is given by

$$t = \frac{\left(ar_o^2 r^2 - r^4\right)}{32(n-1)f^3},$$

where r is the zone radius, r_o is the outer radius, n is the refractive index, $2f$ is the radius of curvature of the sphere, and a is a dimensionless parameter that depends on the focal aperture ratio of the system and on the focal lengths of the chosen reference paraboloid and the sphere. The parameter a is normally chosen so that chromatic aberration introduced by the corrector is minimized, leading to $a \simeq 1.5$. Some examples of profiles for different values of a are shown in Fig. 2. For very fast focal aperture ratios, it may be necessary to include additional terms of higher power in the previous expression for the corrector profile.

A Schmidt corrector plate may be used with the figured surface facing either inwards or outwards but ghost image problems are minimized in the latter arrangement. A better method of reducing ghost image problems is available at the design and manufacture stages, by performing the figuring on a slightly curved meniscus instead of a plane-parallel plate. In that case all ghost images from the corrector can be substantially enlarged so that in most cases they have a surface brightness below the sky background threshold.

Having gained a reputation as an artisan of precision optics, Schmidt produced a working 36-cm $f/1.7$ version of his coma-free camera in 1930. He performed all of the grinding and polishing work himself using only one hand, having had his right hand and forearm amputated after a childhood accident. The method that he devised for producing his aspheric corrector plate is an early example of stress polishing; the plate was polished while stressed on a ring support by a partial vacuum at the rear, so that on releasing the vacuum it sprang back to give the desired asphericity to the upper surface.

USES OF SCHMIDT SYSTEMS

With its wide-field imaging properties and fast focal ratio capability, a Schmidt system is well suited for astrophotography and for use as a sky survey telescope. In the 1950s, the first comprehensive, deep, all-sky survey of the northern hemisphere was carried out by the Palomar 1.2-m Schmidt telescope in the United States. Sky surveys in the southern hemisphere have since been performed by the 1.2-m United Kingdom Schmidt telescope in Australia and by the 1.0-m European Southern Observatory Schmidt telescope in Chile. Instead of having just a simple aspheric corrector plate of the type already described, each of these survey telescopes has an achromatic doublet corrector plate (though the original Palomar survey used only a singlet corrector). With a doublet, using glasses with different dispersions, the corrector asphericity can be distributed over more than one surface and balanced chromatic corrections may be obtained from the visible blue region of the spectrum to near infrared wavelengths. In this way a Schmidt telescope or camera may be corrected simultaneously for third-order spherical aberration, coma, astigmatism, and field distortion, while spherochromatic aberration (wavelength-dependent difference of spherical aberration) is small. The remaining uncorrected aberrations are higher-order ones at oblique angles of incidence.

Fine-grain photographic emulsions are generally used for detection with a Schmidt telescope and, coated on glass plate or film, these can be curved in a special plate holder to fit the spherical focal surface. Color filters, together with the spectral properties of photographic emulsions, are used to define passbands and, for a 1.2-m $f/2.5$ Schmidt telescope, sky-background limited exposures reaching objects as faint as the 23rd B-magnitude are obtained typically in 60–90 minutes depending on the choice of emulsion and filter; for exposures longer than this, differential atmospheric refraction elongates the images at opposite edges of the field. In conditions of good atmospheric seeing such a telescope is capable of producing images of 1 arcsecond diameter over a field of $6.5 \times 6.5°$ on photographic plates 36-cm square. A single sky-background limited exposure on a field of this size will contain up to one million images reaching as faint as the sky limit. Efficient reduction of such large quantities of data then requires the use of an automatic measuring machine.

Schmidt telescopes are particularly effective for the discovery of previously undetected objects, such as comets, asteroids, and so on; and for astronomical programs that depend on statistical sifting or evaluation of large quantities of source data; for example, extensive searches for high-redshift quasars can be made with a Schmidt telescope, either through multicolor photometry or by placing a small-angle prism in front of the corrector plate (so that on a single photographic plate many thousands of individual spectra may be studied). An alternative method of obtaining multiobject spectral observations is to couple optical fibers to the focal surface and to feed these to the slit of a floor-mounted spectrograph.

Several variations of Schmidt's original configuration have been devised: a field-flattened version (mentioned in the following discussion); an off-axis folded system; an all-reflecting version; and solid versions, with glass filling the space between either the corrector and the mirror or the focal surface and the mirror. A short-tube version will also be mentioned.

Of the many other applications that have been found for this excellent and versatile optical system, its use as a spectrograph camera is an important one in situations where a low focal ratio (giving high detection speed) and a wide field are both required. Disadvantages of a Schmidt system are fairly few but include (1) a relatively long overall length, (2) a central obstruction by the focal surface of between 5% and 25% of the incident light depending on unvignetted field size, and (3) a rather inconveniently placed, curved, focal surface.

OTHER WIDE-FIELD TELESCOPES AND CAMERAS

Catadioptric Systems

Where a combination of lenses and mirrors is used, as in the Schmidt telescope, the system is known as catadioptric. Just as spherical aberration in a Schmidt may be compensated by an aspheric plate, so a similar compensation may be obtained by using a thick meniscus lens, as in the *Bouwers* monocentric design (Fig.

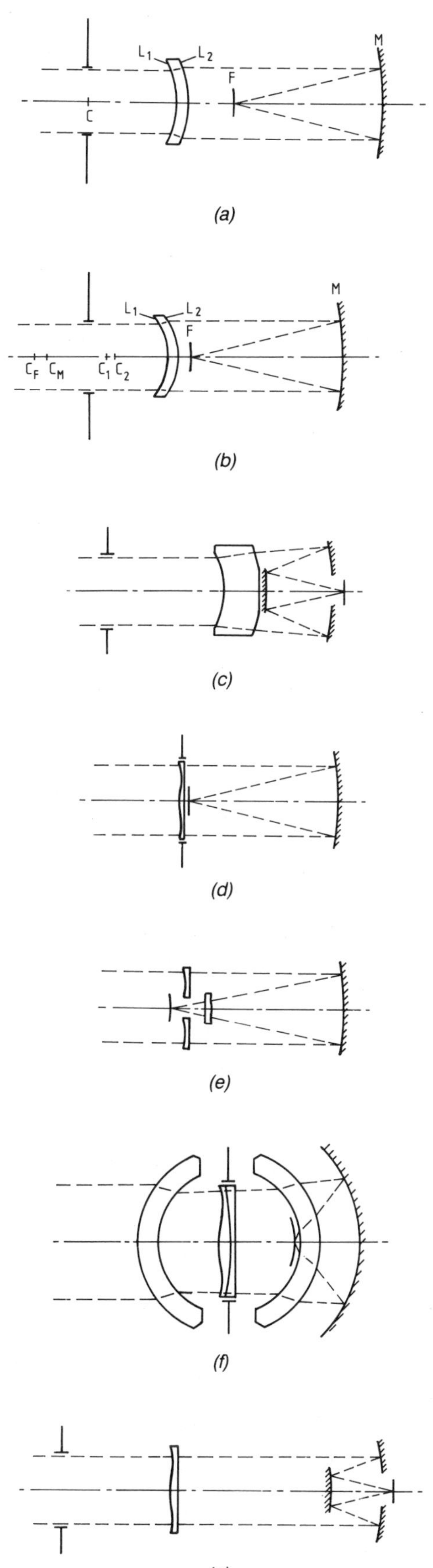

Figure 3. Other catadioptric wide-field optical systems: (*a*) Bouwers monocentric, (*b*) Maksutov, (*c*) Maksutov–Cassegrain, (*d*) Wright, (*e*) modified Wright (after D. S. Brown), (*f*) super-Schmidt, and (*g*) Schmidt–Cassegrain.

3*a*; named after Albert Bouwers of Delft, Holland, from studies performed by him in the 1940s). Here, all of the surfaces involved (L_1, L_2, M, and F) are spherical and have a common center of curvature C, at which position an aperture stop is located to remove off-axis aberrations. For a collimated beam input the optical path is greater at the edge of the meniscus than at its center, and so the meniscus corrects third-order spherical aberration. In other words, the spherical aberration introduced by the meniscus is equal and of opposite sign to that of the spherical mirror.

In the *Maksutov* design (after Dmitry Maksutov of Pulkovo, USSR, 1944) the spherical surfaces are no longer concentric (Fig. 3*b*), allowing correction of axial chromatic aberration as well as third-order spherical aberration and coma. By coating a reflecting layer in the center of the meniscus rear surface, the convergent beam can be sent to a surface behind the spherical primary mirror, as in the *Maksutov–Cassegrain* design; alternatively, the second reflecting surface may be manufactured as a separate component (Fig. 3*c*). In either case the focal surface can be flattened by choosing the meniscus thickness and position so that its astigmatism is equal and opposite to that of the mirror pair. Such an optical system can be used as a compact, wide-field spectrograph camera in which the focal plane is easily accessible for an array detector.

The tube-length and curved-field disadvantages of a classical Schmidt are avoided in the *Wright* camera (from the proposal by Franklin Wright of the United States in 1935), where the aspheric corrector plate is placed close to the focus of the primary mirror (Fig. 3*d*) to give a flat-field aplanat (i.e., a flat field essentially free of third-order spherical aberration and coma). For this short-tube modified Schmidt, the mirror is aspherized to become nearly an oblate spheroid and the figuring of the corrector is deeper than for a classical Schmidt. However, astigmatism now limits the off-axis performance and, for an $f/4$ system, the images at a field angle of only 1° are already 4 arcseconds in diameter. By adding a second, small, aspheric corrector plate close to the focal surface (Fig. 3*e*), and by making both correctors achromatic, a considerable improvement in performance can be achieved. With the inclusion of sixth-order figuring terms for each element, the geometric image spread is well under 1 arcsecond diameter over a broad spectral range and for a field of diameter 5°.

A divided meniscus is used in the *super-Schmidt* camera (Fig. 3*f*), also known as the meniscus-Schmidt. All surfaces of the meniscus lenses are spherical and concentric with that of the mirror. The lenses correct most of the spherical aberration of the mirror, whereas residual aberrations and chromatic aberration are compensated by an achromatic aspheric corrector plate at the center of curvature. Cameras such as these can be made as fast as $f/0.8$ with a field diameter of 50°, and were developed originally in the 1950s for meteor photography based on a design by James Baker of the United States. Other wide-field cameras that have been designed especially for Earth satellite tracking are the *Baker–Nunn* camera (a 0.6-m aperture $f/1$ system with a three-element achromatic correcting lens, spherical mirror, and a $5 \times 30°$ field) and the *Hewitt* camera (a 0.63-m $f/1$ field-flattened Schmidt with a 10° diameter field).

A flat-field *Schmidt–Cassegrain* configuration is shown in Fig. 3*g* and, like the Maksutov–Cassegrain, offers the advantage over a standard Schmidt of an easily accessible focal surface capable of good imaging over a reasonably large field. To produce an anastigmatic flat field (i.e., a flat field essentially free of third-order spherical aberration, coma, and astigmatism), at least one of the mirrors has to be nonspherical as well as the corrector plate, and for the system shown in Fig. 3*g* the primary mirror is slightly ellipsoidal.

The telescope systems mentioned previously have been mainly those that are capable of large aperture and have good imaging properties over fields several degrees in diameter. If the required field is only 1°–2° or less, then excellent astronomical imaging performance may be obtained by using multielement field corrector lenses at the prime or secondary foci of Cassegrain or

Table 1. Two-Mirror Telescopes Using Conic Surfaces

Type	*Primary*	*Secondary*	*Features*
Cassegrain	Paraboloid	Convex hyperboloid	Small field; limited mainly by coma
Ritchey–Chrétien	Hyperboloid	Convex hyperboloid	Aplanatic field limited by astigmatism
Dall–Kirkham	Ellipsoid	Convex spherical	Very small field; severely limited by coma
Gregorian	Paraboloid	Concave ellipsoid	Small field; limited mainly by coma
Aplanatic Gregorian	Ellipsoid	Concave ellipsoid	Aplanatic field limited by astigmatism
Couder	Hyperboloid	Concave ellipsoid	Anastigmatic; wide-field; curved focal surface

Ritchey–Chrétien telescopes. Such field corrector systems have been studied in detail by, amongst others, the British optical designer Charles G. Wynne.

All-Reflective Wide-Field Systems

Table 1 summarizes the properties of several two-mirror telescopes using concave primaries and shows in each case the combination of conics used for the primary and secondary surfaces. Sometimes higher-order figuring terms (in r^4, r^6, r^8, etc.) are added to the basic conic surfaces to refine the aberration characteristics of a particular configuration.

With two-mirror telescopes in which the focal surface is located near or behind the primary mirror (for example, Cassegrain and Gregorian systems) the focal aperture ratio of the final beam is necessarily high (typically $f/8$ or larger) if the size of the secondary mirror, and therefore the geometric obstruction, is to be kept acceptably small. Even in the aplanatic versions (Ritchey–Chrétien and Aplanatic Gregorian, respectively), where coma as well as spherical aberration is eliminated, astigmatism limits the usable fields for astronomical purposes to diameters of a few tens of arcminutes. However, regardless of the aberration characteristics of particular designs, these long focal length configurations do not lend themselves to low focal aperture ratios and therefore wide-field coverage.

In the *Couder* telescope (after André Couder of Paris, 1926) light from the concave primary mirror is intercepted before its focus by a concave secondary mirror to give a highly convergent final beam that does have good wide-field imaging properties (Fig. 4*a*). Disadvantages of the Couder system are the relatively large secondary mirror required to obtain the field coverage, the extra length necessitated by the need for a sky fog baffle tube and the position of the focal surface close to the secondary with a curvature approximately twice that of a classical Schmidt of the same focal length.

In a three-mirror telescope, assuming all of the mirrors are curved, there are enough surfaces to produce a flat-field anastigmat without the need for additional correction. A design using a paraboloidal primary mirror with spherical secondary and tertiary mirrors of equal curvature was first proposed by Maurice Paul in 1935 (Fig. 4*b*). Later, Baker showed that a flat field could be obtained by making the secondary and tertiary curvatures unequal, and the secondary mirror ellipsoidal instead of spherical, giving the system that is now known as the *Paul–Baker*. It gives very low image spread but, as for three-mirror telescopes in general if the mirrors are on the same optical axis, either the geometric obstruction ratio tends to be high or the field size has to be limited. Figure 4*c* shows a three-mirror design due to Roderick *Willstrop* in Britain that gives image spreads under 0.3 arcsecond over a field of 4° for an $f/1.5$ final beam but has an obstruction ratio of 36% of the light incident upon the primary; the perforated primary is approximately paraboloidal and all three mirrors have higher-order figuring terms.

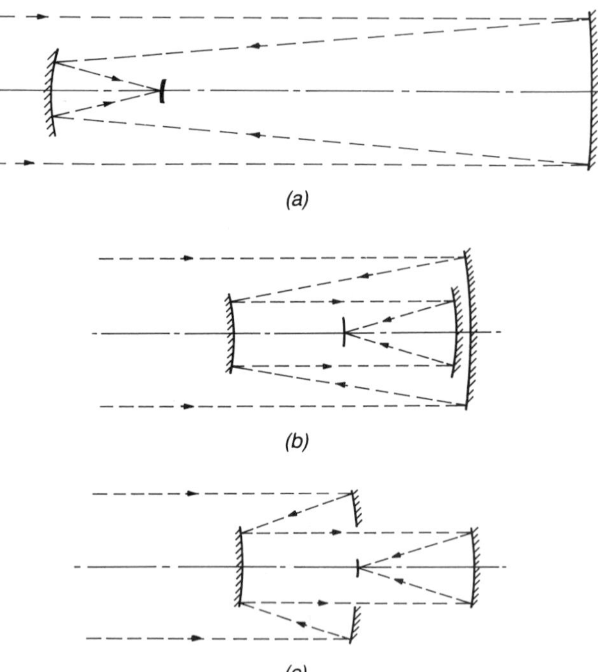

Figure 4. All-reflective wide-field telescopes: (*a*) Couder, (*b*) Paul–Baker, and (*c*) Willstrop.

Additional Reading

Bowen, I. S. (1960). Schmidt cameras. In *Telescopes*, G. P. Kuiper and B. M. Middlehurst, eds. University of Chicago Press, Chicago, p. 43.

Linfoot, E. H. (1955). *Recent Advances in Optics*. Oxford University Press, London.

Schroeder, D. J. (1987). *Astronomical Optics*, Chap. 7. Academic Press, San Diego.

Smith, W. J. (1978). *The Infrared Handbook*, W. J. Wolfe and G. J. Zissis, eds., Chap. 8. Office of Naval Research, Department of the Navy, Washington, D.C.

Wynne, C. G. (1971). Maksutov spectrograph cameras. *Monthly Notices Roy. Astron. Soc.* **153** 261.

Wynne, C. G. (1972). Field correctors for astronomical telescopes. *Progress in Optics* **10** 139.

See also **Cameras and Imaging Detectors, Optical Astronomy; Spectrographs, Astronomical; Telescopes, Large Optical.**

Time and Clocks

William J. H. Andrewes

Time is a word that eludes definition until it is given some practical application. In a scientific context, its measurement has enabled the introduction of a standard by which distance and speed can be determined and by which the operation of machines can be synchronized; in a social and economic context, it has become the primary organizer of business and daily life throughout the industrialized world. The history of time measurement and its growing dominance over human activity is directly linked to the development of devices that made its measurement possible. These can be divided into three categories: time finding, time measuring, and timekeeping. Instruments devised to find time by determining the position of the Earth's rotation in relation to the heavens include sundials, astrolabes, nocturnals, sextants, and transit instruments. Instruments designed to measure specific intervals of time include sandglasses, fire clocks, certain types of water clocks, mechanical timers, and some atomic clocks. In the third category are instruments designed to make time accessible on demand; these include water clocks, mechanical, electric, and quartz clocks and watches, as well as atomic clocks.

The measurement of time is based on both natural and artificial intervals. There are three natural intervals of significance: the solar day, the lunar cycle, and the solar year. Because the solar day has the most pronounced influence on human activity with its marked divisions of daylight and darkness, it became the standard unit of measurement for determining the length of the lunar month and the solar year. All smaller units of time were introduced purely for the convenience of dividing the day when the need for shorter intervals arose.

DIVISIONS OF THE DAY

The word "hour" comes from the Greek word *hora*, which denoted a season or a portion of the year. It was later adopted to signify a portion of the day. Drawings and texts on the inner side of Egyptian coffin lids indicate that the division of the night into 12 parts had been introduced by 2100 B.C. The adoption of 12 as the number of hours into which the day and night are divided is almost certainly derived from the Egyptian "decanal hours," which were determined by the rising of stars on the horizon. The constellations were divided into 36 "decans," each of which served as the indicator of 1 hour of the night for a period of 10 days. During the summer months, if the period of twilight is disregarded, 12 decans will be seen to rise and set during the night. Because this occurrence took place about the time of the first annual appearance of Sirius, the star which in principle marked the beginning of the Egyptian year, 12 was adopted as a standard number of divisions for each period of daylight as well as each period of darkness. Because these hours varied in length as the duration of the period of daylight and darkness changed with the seasons, they were referred to as temporal, unequal, or sometimes planetary hours. Robertus Anglicus, in his *Commentary on the Sphere of Sacrobosco* written in 1271, stated that these hours "are said to be unequal, not because the hours of the day are unequal among themselves, nor the hours of the night unequal among themselves, but rather because the hours of the day are unequal to the hours of the night." Temporal hours were adopted by the Romans, and their use, at least in astrology, continued in Europe until the mid-17th century. The introduction of the mechanical clock, which was not suited to such inconstant divisions, brought about the change to the equal-hour systems of dividing the day.

In medieval Europe, the bell became an important part of monastic and community life because it was the most efficient method of public communication for announcing the events and offices of the day and alerting the population to important occasions. Determining the appropriate time to ring the bell posed a problem in Northern Europe: Overcast skies limited the use of sundials, and freezing temperatures in winter rendered water clocks inoperable. Sandglasses were introduced in the late 13th century, but their use was restricted to time measurement: Because of their limited duration and silent operation, they could not be left unattended for any extended period. The growth of urban populations from the 11th century onwards increased the demand for a more reliable device to ensure that the bell was rung at the correct time. In the *Commentary of 1271*, Robertus Anglicus describes how makers of clocks ("artifices horologiarum") were attempting to make some form of weight-driven timekeeping device with a wheel that made exactly one revolution in a sidereal day. The need for improvement in the method of determining the time is revealed by his statement: "this device would be more useful than any astrolabe or any other astronomical instrument for determining hours."

The origins of the mechanical clock are obscure. Early references are confused by the fact that the word *horologium* (from the Greek words *hora*, an hour, and *logeo*, to tell) was a generic term applied to all instruments of time, including the sundials and water clocks that were the principal instruments used for finding and keeping time before the mechanical clock. The mechanical clock had several advantages over all these devices: Its operation was not seriously affected by climatic changes; its time and rate could be easily adjusted; in the tradition of astronomical geared devices, it was naturally suited to show any number of indications; and, most important, it could be made to ring a large bell automatically, so that an entire community would be made aware of the time. Church records at Dunstable Priory, a house of Augustinian Canons in Bedfordshire, England, describe that in 1283 a *horologium* was installed above the rood-screen, a partition between the nave and the choir. This device could not have been a sundial or other time-finding device because it was located inside the building. Its position above the rood-screen, which would have been an inconvenient place for a water clock, suggests that it had a dial to indicate the time. Five similar references occur in English church records before the end of the century. Supported by the fact that astronomical planetariums and other complicated geared devices were known at that time, this evidence indicates that mechanical clocks were being used in England during the last quarter of the 13th century. Similar documents in France and Italy from the same period reveal that knowledge of these new devices also existed on the Continent. By the second quarter of the 14th century, mechanical clocks were well known throughout Europe. When Richard of Wallingford began the description of his highly complicated astronomical clock in 1327, he referred only briefly to the timekeeping part of the mechanism, as if such knowledge was commonplace.

The mechanical clock was an important addition to urban communities in particular. By striking a bell automatically throughout the course of the day, it provided the time by which the entire population of a city organized their lives. It is not surprising, therefore, that the contemporary word for a bell, *clocca*, was adopted in some countries to describe these new timekeepers. *Clocca*, an echoic word derived from the sound produced by striking a hard object, is of Celtic origin. In the Low Countries, this new timekeeper became known as a *klokke* and in England, it became known as a *clock*. The evolution of the mechanical clock was made possible by the invention of the escapement. The escapement controls the rotation of the wheels of the timekeeper, driven by a power source such as a weight or a spring, and releases the energy of the power source at regular intervals to maintain the motion of an oscillator that governs the speed at which the mechanism operates.

With the advent of the mechanical clock, the system of dividing the day into 24 equal hours was adopted quite rapidly throughout Europe. However, the time of day that the first hour commenced and the method of counting the hours varied from one region to the next. Astronomical hours, which began at noon, were used for determining star positions and for other astronomical purposes.

Babylonian hours, which began at dawn, were useful for determining the number of hours of daylight and thereby the duration of the work day. Italian hours, which began at dusk, were employed in many towns in Northern Italy and were useful for determining the number of hours until sunset. In Nuremberg, equal hours were used to number each period of daylight and each period of darkness, and thereby combined the advantages of both the Babylonian and the Italian horary systems. Common, French, or small clock hours, which divided the day into two 12 equal-hour periods, were adopted as the standard system of counting the hours after the introduction of the pendulum clock.

FROM HOURS TO MINUTES

The division of the hour into minutes is known to have been used in the 15th century. Although no timekeepers made before 1550 with this feature are known to survive, a manuscript written in 1475 by a German monk named Paulus Almanus illustrates a clock with a dial showing the division of minutes. In the text, Almanus actually described this dial as indicating the time of sunrise, an error which would suggest that the division of the hour into minutes was most unusual at that time. Indeed, such small divisions of time would have had no application in civil or domestic life. A more precise indication of the time became necessary only when the clock was used as a scientific instrument. On January 16, 1484, the Nuremberg astronomer Bernhard Walther used a mechanical clock to measure the difference in time between the rising of Mercury and sunrise. The clock, which he claimed was well regulated by the Sun from noon on one day to noon on the next, did not possess a dial indicating the division of the hour. Therefore, to time the event, he had to count the revolutions of the hour wheel and express a fraction of a rotation by the number of gear teeth. Astronomical observations of this kind demanded a standardized and more accurate system of dividing the hour. Because smaller units of time were introduced for astronomical purposes, it is not surprising that they were based on the sexagesimal system, which had long been established in astronomy for the division of a degree. The divisions of a degree were called *minuta*, the primary 60 divisions being the *minuta prima*, the secondary 60 divisions the *minuta secunda*, and so on. These were abbreviated to minuta and secunda, and from these, the words "minutes" and "seconds" were derived. During the course of the next hundred years, clocks incorporating minutes and seconds were made, but surviving examples are rare. By 1587, the Danish astronomer Tycho Brahe claimed to have clocks that indicated both minutes and seconds, although, because he found them to be unreliable, he preferred to use angular measuring instruments for his observations. Such precise divisions of time are seldom seen on clocks made before the application of the pendulum in 1657 or on portable timekeepers made before the invention of the spiral balance spring in 1675.

FROM MINUTES TO SECONDS

The mechanical clock did not become a reliable instrument for measuring and keeping time until the pendulum was used to control its operation. The pendulum proved to be significantly more accurate than the balance wheel or foliot that were used to regulate earlier timekeepers, because it has an intrinsic frequency controlled by the force of gravity. It is known to have been used as a time-measuring device before it was applied as a regulator for clocks. Leonardo da Vinci, Benvenuto Volpaia, and others made drawings of pendulums, but there is no evidence that their schemes were ever put into practice. Galileo devised a mechanism to maintain the motion of a pendulum but this instrument had no provision for telling the time. The first person to apply the pendulum to a mechanical clock and thereby introduce a substantial improvement in the accuracy of timekeepers was the celebrated Dutch mathematician Christiaan Huygens. The first pendulum clocks, made in Holland in 1657, attracted considerable interest and, within two years, these new timekeepers were being manufactured in France and England. During the next 15 years, the accuracy of the pendulum clock was further improved by the introduction of the long pendulum and the invention of the anchor escapement.

THE EQUATION OF TIME

Through the course of the seasons, the observed motion of the Sun in the sky is not constant. This is due to two factors: the tilt of the Earth's axis relative to its path and the ellipticity of the Earth's orbit around the Sun. Therefore, apparent solar time (the time shown by a sundial) and mean solar time (the time shown by a clock) agree only four times a year (on or about April 15, June 13, September 2, and December 25) and can differ by as much as 16 minutes. The difference between solar and mean time is known as the equation of time. The pendulum clock, which required more accurate methods of regulation than earlier timekeepers, demanded a resolution of this discrepancy when a sundial was used to find the time. Tables showing the equation of time for each day of the year were published in several books on dialing. Sometimes clockmakers and watchmakers supplied these tables in the form of a broadside pasted inside clock cases or printed on a small circular piece of paper (known as a watch paper) that served as padding between the inner and outer cases of watches. The equation of time is also found engraved on sundials. Clocks designed especially to keep solar time were made both in England and France, but surviving examples are quite rate.

SIDEREAL TIME

Owing to the variations of solar time, pendulum clocks used in observatories were regulated by the observation of the transit of a star from one night to the next. This system of time measurement, which measured the rotation of the Earth on its axis relative to the stars, is known as sidereal time (from the Latin *siderius*, relating to the stars). In the course of one year, the Earth, because of its orbit around the Sun, will rotate one more time with respect to the Sun than it will with respect to the stars. In other words, in 365.2422 days, the time it takes the Earth to travel from vernal equinox to vernal equinox, sidereal time will differ from solar time by 24 hours. Therefore, a sidereal day is 3 minutes 56.4 seconds shorter than a mean solar day. Clocks that were designed specifically for keeping sidereal or mean solar time were called regulators, because they provided the standard time by which other clocks were regulated. A regulator incorporates several special features: a pendulum designed especially to compensate for temperature variation; a special escapement to provide a regular impulse to the pendulum, maintaining power to keep the clock running while it is being wound; and a dial especially designed to indicate clearly minutes and seconds. Regulators were made generally without striking mechanisms.

The ability to maintain an accurate record of the time revealed new opportunities for science. One of these related to the problem of determining longitude. As the Earth rotates on its axis once in a period of 24 hours, it follows that each degree of its rotation takes 4 minutes. Therefore, the difference in the longitude of two places can be expressed by the difference in their local times. To find this difference, however, the local time of one place (found during the day by taking equal altitudes of the Sun to establish the exact time of noon, or at night by measuring the position of the circumpolar stars) has to be compared with the local time of the other place. The problem of finding or recording the local time of the other place that serves as a reference for determining the difference in longitude led to the development of the marine timekeeper. With the growth of maritime traffic during the 17th century, the number of ill-fated voyages caused by ignorance of the longitude increased at great cost to trade, colonization, and exploration. Therefore, major seafaring countries offered substantial rewards for a solution to the problem. In order to keep time at sea, a timekeeper had to overcome numerous difficulties, such as the rocking of the ship, variations in gravity in different latitudes, and dramatic changes in temperature. Notwithstanding Isaac Newton's belief that such a device would never be produced, John Harrison devoted most of his life to the pursuit of developing an accurate and reliable marine timekeeper. The watch that he completed in 1759 qualified for the £20,000 prize offered by the British government. By 1790, the marine chronometer, as this device came to be known, had been developed into such a sophisticated machine that its basic design remained unchanged throughout World War II. The regulator and the chronometer, which set new standards of accuracy for determining measurements of speed and distance, had a major influence on astronomy, navigation, surveying, and many other branches of science.

STANDARDIZING TIME

In the days when the stagecoach was the principal form of public transport, each community lived by its own local time, determined by the position of the Sun at noon. The variation in the local time of one community to that of another was of no consequence before the advent of the railroad. The railroad substantially reduced the time that it took to travel from one community to another and thereby created a need for a uniform time by which the operation of trains could be regulated. The development of the electric telegraph enabled time to be distributed over great distances from observatories that served as the source from which each region's time was determined. Mass production of clocks and watches also aided the growing consciousness of the importance of time in the expanding industrialized society. The first public time service, based on clock beats telegraphed from the Harvard College Observatory in Cambridge, MA, was introduced in December 1851. The Royal Observatory at Greenwich introduced its public time service in August 1852. In October 1884 at the International Meridian Conference in Washington, DC, the world was divided into 24 time zones with the Prime Meridian established on the main transit instrument at the Royal Observatory at Greenwich, and Greenwich Mean Time became the official time that was gradually adopted throughout the civilized world. France, reluctant to accept the Greenwich meridian over the meridian of the Paris Observatory, finally passed a law in 1911 (which remained in force until 1978) that made GMT the legal time in France, but expressed it quite specifically as "Paris Mean Time, retarded by 9 minutes 21 seconds.'

COORDINATED UNIVERSAL TIME

In 1928, at a time when the most sophisticated electromechanical clocks were being installed in many of the leading observatories, Warren Marrison of the Bell Telephone Laboratories was investigating the properties of quartz crystal as a frequency standard. In 1939, the first quartz crystal clock was installed at the Greenwich Observatory, and within eight years, the first atomic clock had been developed in Washington, DC. The accuracy of the cesium beam atomic clock (about one second in 3000 years) confirmed that the Earth's rotation was not constant. Therefore, in 1967, the atomic second was internationally adopted as the fundamental unit of time measurement, defined as the duration of 9,192,631,770 periods of the radiation corresponding to the transition between the two hyperfine levels of the ground state of the cesium-133 atom. By 1972, the average rates of several atomic clocks in different parts of the world had determined that the Earth was slowing down by approximately one second per year. The world's time signals (referred to as UTC—Universal Time Coordinated or Coordinated Universal Time) are coordinated at present by the Bureau International de Poids et Mesures in Paris. These signals, which are based on atomic time, are adjusted by leap seconds in the middle or at the end of each year to keep them within nine-tenths of a second of the time determined by observation of the heavens (UT1—Universal Time 1). Atomic time (TAI—Temps Atomique International) is based on the average rate of approximately 150 atomic clocks in 30 countries.

Additional Reading

Andrewes, W. J. H. (1985). Time for the astronomer, 1484–1884. *Vistas in Astronomy*, **28** 69.

Brusa, G. (1990). Early mechanical horology in Italy. *Antiquarian Horology* **18** (No. 5) 485.

Gouk, P. (1988). *The Ivory Sundials of Nuremberg 1500–1700*. Whipple Museum of the History of Science, Cambridge.

Landes, D. L. (1983). *Revolution in Time*. Harvard University Press, Cambridge, MA.

Neugebauer, O. (1955). The Egyptian "decans." *Vistas in Astronomy*, **1** 205.

Thorndyke, L. (1941). Invention of the mechanical clock about 1271 A.D. *Speculum—A Journal of Medieval Studies* **16** (No. 2) 242.

See also **Astrometry, Techniques and Telescopes; Calendars; Constellations and Star Maps; Coordinates and Reference Systems.**

Three-Body Problem

Victor Szebehely

This problem deals with the motion of three bodies that attract each other according to the Newtonian law of gravity. The important characteristics of this problem are its broad ranges of applicability and its nonintegrability. The second property indicates the fundamental difficulty in obtaining generally valid solutions for long time durations.

The gravitational two-body problem is integrable, reliable long-time predictions are available, and the motion is stable for negative values of the total energy. The change of the dynamical properties, when the third body is introduced, is remarkable and it can be explained by the increase of the order of the system, or in simpler terms, by the increase of the number of variables involved when the change is made from two to three bodies. At the same time the basic physical information available does not change. Both the two- and three-body systems are energy and momentum conserving and the centers of mass of both systems move on a straight line with constant velocity. So, the essential dynamical properties are the same and, consequently, the number of equations expressing these properties remains the same. When the number of bodies increases from two to three the information to be obtained increases.

It is also important to consider the force law acting between the participating bodies. For instance, if the force is directly proportional to the distance, as is the case when linear springs act between the bodies, the problem is readily integrable for any number of participating bodies. This is not the case for gravitational forces that are inversely proportional to the square of the distance.

The nonexistence of generally valid analytic solutions, for long time durations, raises the three-body problem to the level of "unsolvable" problems, the challenge of which proved to be irresistible for the most outstanding minds in the history of astronomy, ever since Isaac Newton's attempts to establish the rules governing the motion of the Moon as influenced by the Earth and by the Sun.

Before a short historical review is offered in the next section, considering the contributions made in the field of three-body research, a frequently encountered misinterpretation of the previously mentioned concepts of unsolvability and nonintegrability of the problem should be clarified. With the use of modern high-speed computers, there is no difficulty in numerically integrating the problem of three bodies and obtaining trajectories for finite time durations. This, however, does not mean that the problem is solved in a general sense. For a given set of initial positions and velocities, the computation of specific trajectories for three bodies with known masses presents no problem. The unsolvability comes in when detailed trajectories are needed with long time validity. The limit of predictability depends on the specific starting conditions and on the masses. Even the most powerful computers cannot offer solutions for arbitrarily long times of the gravitational three-body problem. Once again, the proper meaning of this statement must be understood. The details of the trajectories (i.e., the precise locations of the bodies) are lost sooner or later, but the *qualitative* aspects of the behavior can be predicted for infinite times, even without the use of computers. These recent results will be summarized in a later section.

Besides the "unsolvable" challenges attracting researchers, the three-body problem is also recognized as one of the fundamental systems in dynamical astronomy. The reader might find it interesting to note this example of the inherent limitations of sciences. The general solution of this basic problem is unattainable because of fundamental limitations. The applications of the three-body problem in celestial mechanics, space research, and astrophysics will be summarized in a later section.

HISTORICAL REVIEW

The first scientifically important contributions to the three-body problem were made by Newton, who studied the Earth–Moon–Sun problem in 1687 and complained of headaches and sleeplessness due to the severity of the subject.

Leonhard Euler's (1760) contributions included the introduction of the concept of the restricted problem of three bodies and the use of a rotating or synodic coordinate system. The model known as the restricted problem is widely used in space dynamics. It consists of two bodies of large masses moving on circular orbits around each other. The third body with much smaller mass moves under the influence of the two large bodies and it does not affect their motion. In this way the problem is separated into a two-body problem and a restricted three-body problem where the motion of only the small body is to be determined.

It is not unusual in celestial mechanics that models introduced originally only for theoretical reasons later on turn out to be of fundamental practical importance. Newton's idea of throwing stones horizontally from the top of a mountain with higher and higher velocities is today the model for the orbiting of artificial satellites. Similarly, Euler's idea of the restricted problem is used today to establish Earth-to-Moon trajectories of spaceprobes. (In fact, the spaceprobe is always the third small body.)

Joseph Louis Lagrange, in 1772, showed the existence of five equilibrium points in the Sun–Jupiter restricted problem where the Trojan asteroids represented the third bodies.

The lunar theory was not solved by Newton (in fact, it is still not "solved" because of the nonintegrability principle) and was attacked later by George W. Hill (1878) and Brown with various analytical series approximations. Hill also introduced the idea of curves of zero velocity and established possible and forbidden regions of motion. Such qualitative results are of considerable importance for nonintegrable dynamical systems because their validity can be extended to arbitrary long time durations.

The nonintegrabilities were shown by Jules Henri Poincaré (1899) for the general as well as for the restricted three-body problems.

Another approach of considerable importance for nonintegrable systems is the study of periodic orbits. These orbits repeat themselves in equal intervals of time for arbitrary lengths of time. Such orbits form well-established families in the restricted problem as shown by Poincaré, Elis Strömgren, and George H. Darwin, and they are used for present-day space missions with small propulsion requirements for guidance.

For the general three-body problem, neither the concept of limiting zero-velocity curves nor families of periodic orbits have immediate and practical applications.

An approach to treating three-body and many-body problems when the solution is close to a well-known two-body motion is called *perturbation* theory. The complete overview of the literature along this line would be overwhelming. A few of the classical contributors in addition to those mentioned previously are Brouwer, Charlier, Encke, Gauss, Herget, Laplace, Moulton, Newcomb, Siegel, Smart, Tisserand, Von Zeipel, and Wintner.

CLASSIFICATION OF MOTIONS

The description of possible motions concerning the restricted problem is based on families of periodic orbits. One of the pertinent parameters is the value of the mass ratio of the two large bodies. This value is usually expressed by $\mu = m_2/(m_1 + m_2)$, where m_1 and m_2 are the large masses and $m_2 < m_1$ according to generally accepted conventions. When $\mu = 1/2$ the large masses are equal and when μ is 0, the problem is reduced to the two-body problem consisting of m_1 and the small "third" body. A large number of families have been established for various values of μ. The applications for solar system dynamics and space research require μ to be small. For example, for Earth-to-Moon trajectories, $\mu \cong 0.0122$; for lunar theories, when the effects of the Sun and of the Earth on the Moon are treated, $\mu \cong 3 \times 10^{-6}$; and for asteroids controlled by Jupiter and the Sun, $\mu \cong 0.00095$.

The number of families of periodic orbits increases every day because increasing the complexity of such orbits, new families can be established by means of numerical integration. The number of "basic families" of simple periodic orbits known since 1920 is about 20.

The classification of orbits in the general problem of three bodies is much more complicated because there are three nonzero participating masses influencing each other. Recent computer results allow us to offer qualitative classification of and to make predictions concerning the final outcome of the possible motions.

The total energy of the system is constant during the motion and it is given by

$$h = \frac{1}{2}\left(m_1 v_1^2 + m_2 v_2^2 + m_3 v_3^2\right) - G\left(\frac{m_1 m_2}{r_{12}} + \frac{m_2 m_3}{r_{12}} + \frac{m_3 m_1}{r_{31}}\right),$$

where G is the constant of gravity, the participating masses are m_1, m_2, and m_3, their velocities are v_1, v_2, and v_3, and the distances between the bodies are r_{12}, r_{23}, and r_{31}.

If the total energy is positive or 0, at least one of the three bodies will escape. If the total energy of the system is negative, we can have bounded or unbounded motion but because of the basic instability property of the bounded three-body solutions, the long-time outcome is usually the formation of a binary with the third body's escape. Prior to escape several possible bounded motions can be observed which are known as interplay, ejection, revolution, equilibrium, periodic orbits, and oscillatory motions. *Interplay* occurs when the three bodies perform bounded motion in a random fashion. As long as no collisions, close approaches, or external perturbations occur, the motion continues and the three bodies may stay together for a long time. A variation of this type of motion is when one of the bodies is *ejected* on a perturbed elliptic orbit and it returns to the binary left behind. If the ejection orbit is hyperbolic, we speak about an *escape* which is the outcome having the highest probability. The effect of the escaping third body on the binary becomes 0 as its distance approaches infinity and we are left with a binary configuration. Note that ejections might also result in binary formations in a galaxy because the ejected third body might be captured by other systems if it departs sufficiently far from the original two components. Another possible triple configuration is known as *revolution*, when a binary is surrounded by the orbit of the third body. Once again we might observe that if the orbit of the third body is far from the binary, the system is stable but other perturbing effects in the galaxy might take the third body away. If the separation is small, that is, if the orbit of the outer body is close to the inner binary, the revolution will become an interplay and an escape might be expected sooner or later. The *equilibrium* configurations or Lagrangian solutions for masses of the same order of magnitude are inherently unstable and in addition require special initial conditions. The same applies for *periodic orbits*. For the sake of completeness, *oscillatory motions* should be mentioned which also require special sets of initial conditions (with measure 0) and therefore have little practical significance.

The effect of the angular momentum, in addition to the total energy of the system, should be mentioned as an important parameter to determine the qualitative properties of the motion. Small values of the angular momentum result in triple close approaches, which in turn lead to escape. In conclusion, it can be stated that numerical and analytical results verify the fact that randomly selected initial conditions for a three-body system lead to its breakup into a binary system and an escaping third body.

APPLICATIONS

The model of the restricted problem of three bodies is used to describe the dynamical behavior of many aspects of planetary systems such as the motions of asteroids, comets, planetary satellites, and so on. The field of astrodynamics which is concerned with the motion of artificial bodies, such as spaceprobes, uses for all computations the model of the restricted problem. When problems of stellar dynamics emerge, the model of the general three-body problem is applied. The limits of applicability of the much simpler restricted problem are not always obvious because gravitational effects depend on varying distances and on the masses. An outstanding, famous example concerning the applicability of simple models is the motion of the Moon. The restricted model leads to a conclusion of definite stability (according to Hill), whereas the general model allows no such conclusion.

The basically unstable behavior of triple stellar systems mentioned previously indicates that the general problem of three bodies is able to offer an explanation for the overwhelming majority of binary stars in our galaxy. For triple systems having positive values of the total energy, the separation into an escaping star and a binary takes less time than for systems having negative energy. Separation takes place after a triple close approach, the accurate numerical evaluation of which presents fundamental difficulties because the equations contain singularities. The role of angular momentum is opposite to that of the energy because high values delay escape.

Usually, but not always, the body with the smallest mass escapes from the general three-body system. Note that this is not the case for the restricted problem where triple close approaches cannot occur because the distance between the bodies with large masses is constant.

The essential difference, therefore, between the restricted and the general problems is that in the restricted problem periodic orbits dominate, whereas the general problem results in escape orbits. Planetary and lunar theories are based on periodic solutions and expansions in Fourier series. Triple stellar systems, on the other hand, are analyzed by techniques based on asymptotic escape orbits.

Numerical results indicate that the time of disruption of a triple system formed by three stars having masses of the order of the Sun's mass and placed initially at relative distances of the order of parsecs is 10^{19} yr. The behavior of a triple stellar system placed in a galaxy should be about the same because the critical mechanism of escape is the triple close approach when the influence of other members of the galaxy can be ignored.

The long-time behavior of the solar system is a much more difficult problem to analyze analytically or numerically than the behavior of triple stellar systems. The presently available and reliable numerical results indicate Laplacian stability for 10^8 yr, during which time the orbits of the major planets are not expected to cross and consequently no collisions or escapes will occur.

Additional Reading

The immense literature of the three-body problem includes the names of the greatest mathematicians, astronomers, and physicists, making this problem, as Whittaker puts it "the most celebrated of all dynamical problems." The thousands of contributions might be put in chronological order or might be organized according to subjects. The crossings of these two classifications reveal the evolution of the sciences of astronomy, celestial mechanics, and space research.

It is interesting to note that not much can be found regarding three-body considerations and associated perturbation theories in the works of the great thinkers of classical astronomy and celestial mechanics, such as Aristotle, Ptolemaeus, Copernicus, Brahe, Galileo, and Kepler. After these came Newton and his work in the lunar theory, as mentioned before. Alexander Pope's description of Newton's contributions is: "Nature and nature's laws lay hid in night. God said 'Let Newton be,' and all was light." Following Newton, contributors concentrated on astronomical applications concerning the solar system and on the mathematical aspects of dynamics. The principal names and dates of publications are: Euler (1760–1772), Lagrange (1772–1788), Laplace (1773–1825), Jacobi (1836–1866), Hill (1887–1907), Tisserand (1889–1896), Poincaré (1890–1910), Brown (1896–1933), Liapunov (1896–1892), Darwin (1897–1911), Whittaker (1899–1916), Moulton (1900–1960), Charlier (1900–1907), Happel (1900–1941), Levi-Civita (1903–1920), Plummer (1903–1932), Newcomb (1909), Sundman (1912), Birkhof (1913–1950), Elis Strömgren (1913–1935), Wintner (1925–1941), Brouwer (1933–1961), and Siegel (1936–1956).

The contributors to the dynamical aspects of the exploding fields of space research, of modern celestial mechanics, and of stellar systems presented their works in thousands of articles and books that cannot be referenced here. Some of the authors of textbooks and reference volumes with dates of publications are Baker and Makemson (1967), Bate, Mueller, and White (1971), Battin (1964–1987), Chebotarev (1965), Danby (1962–1989), Duboshin (1968–1978), Fitzpatrick (1958), Hagihara (1957–1976), Herrick (1971–1972), King-Hele (1962–1964), Kovalevsky (1963–1967), McCuskey (1963), Moser (1971–1973), Pollard (1966), Roy (1978–1988), Stumpff (1956–1965), Szebehely (1967–1989), and Taff (1985).

In addition to these authors, attention is directed to the excellent historical reviews by Gautier (1817), Lovett (1911), Marcolongo (1919), Bell (1937), Koestler (1959), Lerner (1973), Hall (1981), and Mark (1987).

See also **Interplanetary Spacecraft Dynamics; *N*-Body Problem.**

U

Ultraviolet Astronomy, Space Missions

Yoji Kondo

Ultraviolet observations from above the Earth's atmosphere became possible when captured German V2 rockets became available in the United States at the end of World War II. The Sun was the first astronomical object observed in the ultraviolet from a rocket, thanks to the copious photons received from the Sun due to its proximity to Earth. By the 1960s, rocket-pointing and telescope-detector systems had advanced sufficiently to enable observations of bright stars in the ultraviolet. However, the observing time available to a rocket-borne payload above some 150 km altitude was limited typically to a few minutes per flight. Balloon-borne ultraviolet telescopes extended the observing time to several hours per flight but from a float altitude of some 40 km, which is above most of the ozone layer, only the mid-ultraviolet range longward of about 2000 Å is observable. Clearly, orbiting astronomical telescopes were needed to explore the universe more fully in the ultraviolet. Here, we shall limit our discussion to orbiting astronomical satellites designed to observe objects other than the Sun, which has been studied from orbiting telescopes beginning with a series of orbiting solar observatories in the 1960s.

NASA ORBITING ASTRONOMICAL OBSERVATORIES

The age of major ultraviolet astronomical satellite missions arrived on December 7, 1968, with the successful launching of the *Orbiting Astronomical Observatory* (*OAO-2*) by the National Aeronautics and Space Administration (NASA). As a historical note, the first *orbiting* astronomical ultraviolet instrument was actually a Hasselblad camera with an objective grating, carried aboard the *Gemini 11* and *Gemini 12* spacecraft in 1966 and operated by the astronauts. The *OAO-A1*, which was launched into orbit in 1966, did not function due to a battery failure. *OAO-2*, originally designated *OAO-A2*, was a backup for the *OAO-A1* and was identical in design to its predecessor. It consisted of the Wisconsin Experiment Package and the Smithsonian Astrophysical Observatory's Celescope. The University of Wisconsin instrument consisted of the following: four 20-cm stellar photometers, with bandpasses centered at 1430, 1550, 1910, 2460, 2980, and 3320 Å, which were used to observe all sorts of individual stars and external galaxies; two 20-cm scanning spectrophotometers, which were used to observe individual astronomical objects; and a 40-cm nebular photometer, which was used to study emission nebulae. The far-ultraviolet spectrophotometer had a resolution of 10 Å; the mid-ultraviolet spectrophotometer with a resolution of 20 Å suffered from radiation belt particle noise. The Smithsonian Celescope consisted of four 30-cm telescopes equipped with ultraviolet-sensitive television (Uvicon) tubes; it was used to survey a substantial fraction of the sky in several ultraviolet bandwidths. Among the major scientific results from *OAO-2* were improved knowledge of the interstellar grains and the excess-ultraviolet radiation from some external galaxies.

The *OAO-B*, which contained the Goddard Experiment Package, was launched on November 30, 1970, but failed to attain orbit due to a launch vehicle shroud separation system failure.

The *OAO-C* was launched on August 21, 1972. Upon attaining its orbit at an altitude of about 775 km, it became *OAO-3*. However, to commemorate the 500th anniversary of the birth of Nicolaus Copernicus in 1973, the satellite was renamed *Copernicus*. It was equipped with Princeton University's 80-cm aperture ultraviolet telescope with high-resolution spectrophotometers, and with three focusing x-ray telescopes supplied by the University of London and the University of Leicester. The far-ultraviolet spectrophotometers, called U1 and U2, employing second-order images, had resolutions of 0.05 and 0.2 Å, respectively; the spectral range covered was 700–1650 Å. However, no stellar spectrum was obtained on the short-wavelength side of 912 Å, where the Lyman continuum absorption begins. The mid-ultraviolet spectrophotometers, called V1 and V2, using first-order images, had resolutions of 0.1 and 0.4 Å, respectively. The spectral range covered was 1550–3000 Å; however, these phototubes suffered from the radiation belt particle noise. The primary scientific objective of *Copernicus* was the study of the interstellar medium. The discovery of the pervasive, hot tenuous gas within several tens of parsecs of the Sun was one of its important achievements.

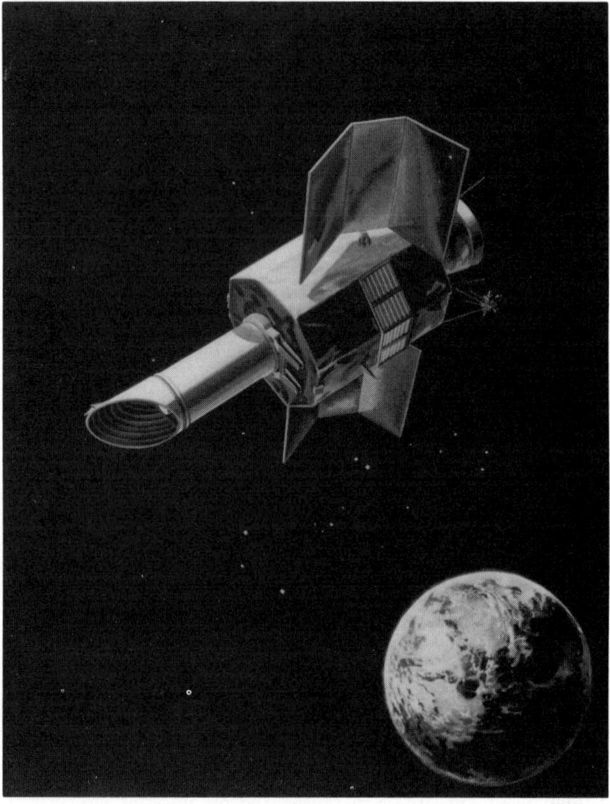

Figure 1. Artist's conception of the *International Ultraviolet Explorer* in geosynchronous orbit.

EUROPEAN SATELLITES

The European Space Agency's *TD-1* satellite, which was launched on March 12, 1972, into a 550-km orbit, carried two ultraviolet experiments. The *TD-1* was operated through 1974. The S2/68 Ultraviolet Sky Survey Telescope had a 27.5-cm diameter off-axis paraboloid. The spectral range covered was 1350–2550 Å and the dispersion was 36 Å mm^{-1}. The S59 spectrometer covered the ranges 2064–2158, 2497–2591, and 2777–2868 Å at a resolution of 1.7 Å. The primary contribution of the *TD-1* was the *Ultraviolet Bright-Star Spectrophotometric Catalogue* and its supplement.

The *Astronomical Netherlands Satellite* (*ANS*), which was launched on August 30, 1974, carried an ultraviolet experiment package consisting of a 22-cm diameter Cassegrain telescope as well as two x-ray experiments. The *ANS* continued its operation through 1976. The central wavelengths for the ultraviolet photometer bandpasses were 1550, 1800, 2200, 2500, and 3300 Å. The *ANS* orbit had an apogee of 250 km and a perigee of 1100 km. The *ANS* obtained accurate five-band ultraviolet photometry for a large number of stars, providing a data set for systematic study of ultraviolet colors of normal stars, white dwarfs, and subdwarfs, and of the interstellar extinction. High photometric accuracy enabled investigation of variable stars, such as pulsating variables, eclipsing binaries, and cataclysmic variables. The large entrance aperture of the *ANS* allowed relatively good signal-to-noise data to be collected for extended objects like globular clusters and nearby galaxies, making it possible to study their stellar population contents.

SKYLAB AND VOYAGER MISSIONS

NASA's *Skylab*, launched in 1973 and manned by three successive crews through early 1974, carried an ultraviolet sky survey instrument known as the S-019 experiment. It had a 15-cm aperture and achieved a resolution of 2, 12, and 42 Å at 1400, 2000, and 2800 Å, respectively.

The interplanetary probes *Voyager 1* and *Voyager 2* have far- and extreme-ultraviolet spectrometers, operating effectively in the 500–1400-Å range at a resolution of 15–30 Å. Despite their modest size, those spectrometers have been used to observe some relatively bright astronomical objects when the spacecraft could be spared from their primary mission of observing the planets.

SOVIET SATELLITE ASTRON

The high-apogee satellite *Astron* was launched by the Soviet Union in 1983. It had an 80-cm diameter telescope equipped with three scanning spectrometers operating in the 1100–3500-Å range.

INTERNATIONAL ULTRAVIOLET EXPLORER

A new era in space observatories was ushered in with the successful launching of the *International Ultraviolet Explorer* (*IUE*) on January 26, 1978. This satellite, which was placed in a geosynchronous orbit with a mean orbital altitude of 34,890 km, was developed under the joint auspices of NASA, the European Space Agency (ESA), and the Science and Engineering Research Council (SERC) of the United Kingdom. Astronomical observations with the *IUE* are conducted around the clock, 16 hours a day from the NASA Goddard Space Flight Center in Greenbelt, MD, near Washington, DC, and 8 hours a day from ESA's Villafranca Ground Station near Madrid, Spain. One major advantage of the *IUE* is its capability to observe continuously all day. In contrast, a low-Earth-orbit satellite observatory can observe a typical target during only about a third of its orbital period, which is in the range of some 90–100 minutes. The *IUE* is operated in *real time*; a telemetered new observation may be examined while the next one is in progress.

Each ground station is staffed with resident astronomers and telescope operators, who assist the guest observers with the planning and the conduct of their observations. This observatory made it possible for a large number of astronomers, without any experience in space observations, to observe in space. Over the first 10 years of its operations, the *IUE* was used by over 1600 scientists from all corners of the world. This is a substantial fraction of all research astronomers on Earth. One measure of the productivity of a scientific program is the number of scientific articles published. During the first 12 years, the total number of articles published in *refereed* journals using *IUE* observations was in excess of 1874. *IUE* observing time is awarded to research programs that are selected by peer review from among the proposals submitted once a year to either NASA or ESA.

The *IUE* is equipped with spectrographs that operate in the far ultraviolet (1150–2000 Å) and mid-ultraviolet (1900–3200 Å). It may be operated at a high (0.1–0.3 Å) or a low (6–7 Å) resolution. It has been used to observe a star as bright as −1.4 magnitude and as faint as magnitude 21, providing an effective dynamical range of over 20 magnitudes or a factor of one hundred million. By early 1990, some 70,000 spectral images had been obtained; these spectra are made available to the astronomical community from NASA and ESA. The *IUE* regional data analysis facilities help astronomers to analyze the data efficiently.

Highlights of the *IUE* results include the discovery of the hot gaseous halos surrounding the Milky Way and other galaxies. *IUE* observations have also shown that practically all stars, including even some white dwarf stars, lose matter in the form of stellar winds. Our understanding of the evolutionary processes in interacting binary stars, including such exotic objects as novae and x-ray binaries, have substantially improved. The first space observation of Halley's comet was obtained with the *IUE* in September 1985. Based on the observations of OH emission that were obtained, the comet was ejecting water vapor at the rate of several tons a second as it approached the Sun. The *IUE* provided continual coverage of Halley's comet at the times of the encounter of ESA's *Giotto*, the Soviet Union's *Vega 1* and *Vega 2*, Japan's *Sakigake* and *Suisei*, and the U.S. *ICE* missions with the comet. The last space observation of the comet was also made with the *IUE*.

When Supernova 1987A was discovered in the Large Magellanic Cloud on February 24, 1987, the *IUE* was used to obtain its ultraviolet spectrum within hours of the report. It was the brightest supernova since Johannes Kepler's naked-eye observation in 1604. The far-ultraviolet spectrum, which was taken after the initial ultraviolet flash subsided at the shortest wavelength, showed two stars remaining in the direction of the supernova. Comparing it with the presupernova photographic plate taken from the ground, whose analysis showed three stars in the same region, the progenitor of the supernova was identified as a blue supergiant. This was the first time that the progenitor of a supernova was positively identified.

MISSIONS FOR THE 1990s

The 2.4-m aperture *Hubble Space Telescope* (*HST*), discussed in the corresponding entry, was launched from the Space Shuttle on April 24, 1990. The *HST* is designed to attain higher spectral resolutions and reach significantly fainter astronomical objects. The *Extreme Ultraviolet Explorer* (*EUVE*) of the University of California at Berkeley is slated for launch on a Delta rocket in late 1991. The *EUVE* will extend the spectral coverage to beyond the Lyman continuum absorption that begins at 912 Å, closing the gap between the far-ultraviolet and the soft x-ray regions which begin at about 100 Å. The *Astro-1* mission of the Space Shuttle *Columbia* in December 1990 flew an attached payload consisting of the Broad Band X-Ray Telescope and three ultraviolet telescopes. They were: (1) the Hopkins Ultraviolet Telescope (HUT) for studying spectra of faint astronomical objects such as quasars, active galactic nuclei, and normal galaxies in the far ultraviolet; (2) the Ultraviolet Imaging Telescope (UIT) for imaging objects such as globular star clusters and galaxies in broad ultraviolet wavelength bands and with a wide field of view; and (3) the Wisconsin Ultraviolet Photo-Polarimeter Experiment (WUPPE) for studying the polarization of hot stars, galactic nuclei, and quasars. These three telescopes will fly again on Astro-2.

Additional Reading

Boggess, A., et al. (1978). The *IUE* spacecraft and instrumentation. *Nature* **275** 372.

Boksenberg, A., et al. (1973). The Ultraviolet Sky-Survey Telescope in the *TD-1A Satellite*. *Monthly Notices Roy. Astron. Soc.* **163** 291.

Broadfoot, A. L., et al. (1977). Ultraviolet spectrometer experiment for the *Voyager* mission. *Space Sci. Rev.* **21** 183.

Code, A. D., ed. (1971). The scientific results from the *Orbiting Astronomical Observatory (OAO-A2)*. NASA SP-310.

de Jager, C., et al. (1974). The orbiting stellar ultraviolet spectrophotometer S59 in ESRO's *TD-1A Satellite*. *Astrophys. Space Sci.* **26** 207.

Kondo, Y., ed. (1990). *Observatories in Earth Orbit and Beyond*. Kluwer Academic Publ., Dordrecht.

Kondo, Y., et al., eds. (1987). *Exploring the Universe with the IUE Satellite*. D. Reidel, Dordrecht.

Kondo, Y., Boggess, A., Maran, S. P. (1989). Astrophysical contributions of the *International Ultraviolet Explorer*. *Ann. Rev. Astron. Ap.* **27** 397.

Maran, S. P. (1991). Astro: Science in the fast lane. *Sky and Telescope* **81** 591.

Rogerson, J. B., Spitzer, L., et al. (1973). Spectrophotometric results from the *Copernicus* satellite. I. Instrumentation and performance. *Ap. J. (Lett.)* **181** L97.

Shore, L. A. (1987). *IUE*: Nine years of astronomy. *Astronomy* **15** (No. 4) 14.

van Duinen, R. J. (1975). The ultraviolet experiment on board the *Astronomical Netherlands Satellite—ANS*. *Astron. Ap.* **39** 159.

See also **Hubble Space Telescope; Observatories, Balloon; Scientific Spacecraft and Missions; Sounding Rocket Experiments, Astronomical; Telescopes, Detectors and Instruments, Ultraviolet.**

Uranus and Neptune, Atmospheres

F. Michael Flasar

Except for Pluto, Uranus and Neptune are the farthest-known planets from the Sun. They are similar in size and have radii about four times that of Earth (see Table 1). Like Jupiter and Saturn, their atmospheres are primarily hydrogen (H_2) and helium (He). Compared to Earth, the atmospheres of Uranus and Neptune are quite deep. Current theoretical models suggest that they comprise the outer 30% of the planets' radial extent and overlie an "ocean" of water, ammonia, and methane at high pressures (2×10^5 bar) and temperatures (2500 K). However, little is known about the atmospheres at these great depths, and most observations, whether from Earth or spacecraft, have probed only the outermost "skin" of the atmosphere, at altitudes where the pressure is less than 100 bar. The encounters of *Voyager 2* with Uranus (1986) and Neptune (1989) added much to our knowledge of their atmospheres.

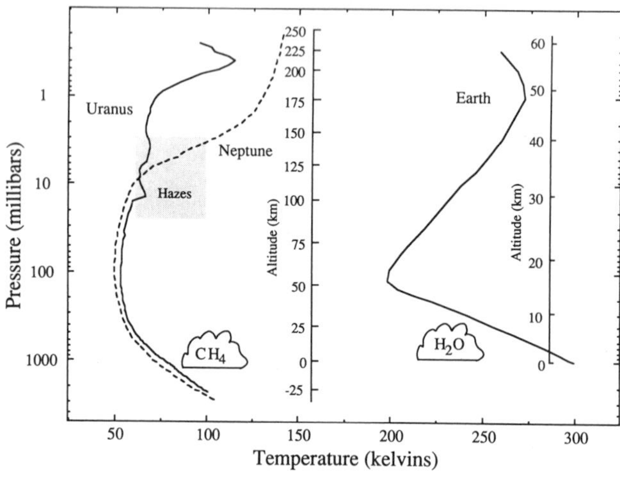

Figure 1. Vertical profiles of temperature on Uranus, Neptune, and Earth as a function of atmospheric pressure (1000 mbar = 1 bar = 0.987 atm). The temperatures of Uranus were obtained from the radio occultation experiment on the *Voyager 2* spacecraft. Those of Neptune were derived from Earth-based observations at infrared and submillimeter wavelengths and a preliminary analysis of *Voyager 2* radio occultation measurements. Clouds and hazes are indicated schematically.

MEAN VERTICAL STRUCTURE AND ENERGY BALANCE

Figure 1 displays the vertical profiles of temperature in Uranus' and Neptune's atmospheres, compared to that in Earth's. As is often done in meteorology, the vertical axis is marked in terms of atmospheric pressure instead of altitude. The two quantities are uniquely related, because the atmospheric pressure at a specified altitude is equal to the weight of the overlying atmosphere. Hence pressure decreases with increasing altitude, and "upward" in Fig. 1 is toward the top; altitude scales are included as inserts. The altitude range in the figure is much larger for Uranus and Neptune

Table 1. Atmospheres of Uranus, Neptune, and Earth

	Uranus	*Neptune*	*Earth*
Length of day (hr)	17.2	16.1	24.0
Orbital period (yr)	84	165	1
Radius (km) at 1 bar	25,600	24,800	6,378
"Surface" gravity (m s^{-2})	9.0	11.0	9.8
Distance from Sun (AU)	19.2	30.1	1.0
Obliquity of rotation axis to orbit pole (deg)	97	29	23.5
Major constituents	Dry gas:* H_2 (85%) He (15%) Condensibles: CH_4, NH_3(?) H_2O(?), H_2S(?)	Dry gas:* H_2 (> 80%) He (< 20%) Condensibles: CH_4, NH_3(?) H_2O(?), H_2S(?)	Dry air:* N_2(78.1%) O_2 (21.9%) Ar (0.9%) Condensibles: H_2O ($\leq$ 2.2%)†

*Composition (by volume) ignores presence of condensibles.
†Spatially and temporally variable.

than for Earth, because the mean molecular mass of a hydrogen–helium atmosphere is much less than that of air (cf. Table 1).

The temperatures of Uranus and Neptune in Fig. 1 were not obtained with in situ thermometers, but from remote sensing. One method consists of observing the planets' emitted electromagnetic radiation at infrared, submillimeter, and radio wavelengths (micrometers through centimeters). Because the opacity of the atmosphere varies with wavelength, the radiation observed at each wavelength originates at a different altitude. The amount of radiation emitted depends on the temperature of the atmosphere at the emitting level. Thus, from the spectrum observed at different wavelengths, one can in principle retrieve atmospheric temperature as a function of altitude. In practice, this is only possible if the abundance of the opacity source is known. Fortunately, over much of the pressure range depicted in Fig. 1, the dominant species, hydrogen, is the principal source of opacity. Another method consists of observing the atmospheric refraction of radio waves emitted by a source as it is occulted by the planet. The amount of refraction depends on the temperature profile of the atmosphere. This method was used when the *Voyager 2* spacecraft transmitted radio signals to Earth as it passed behind Uranus and Neptune.

The atmospheric temperatures on Uranus and Neptune are much colder than on Earth, because these planets are so much farther from the Sun (Table 1). At pressures greater than 100 mbar, the temperatures on Uranus and Neptune are quite similar, even though Neptune is 50% farther away. The reason for this lies in the atmospheric energy balance of both planets. Like Earth, both planets receive sunlight, reflecting some of it back to space and absorbing the remainder. The latter can heat the atmosphere, evaporate condensates, and drive atmospheric motions, but it is ultimately radiated back to space as infrared radiation, at wavelengths several hundred times longer than those of sunlight. A careful analysis of *Voyager 2* infrared measurements, obtained as the spacecraft flew by Uranus early in 1986, indicates that Uranus' total infrared emission can be explained by this simple balance. However, Earth-based observations have shown that Neptune radiates significantly more energy (2.5 times) at infrared wavelengths than it receives and absorbs from sunlight. The source of this extra energy is thought to be the cooling of its interior. Temperatures there are probably as high as 8000 K. They resulted from the compression the primordial planet experienced when it formed as a gaseous ball from the solar nebula and contracted under its own self-gravitation 4.5 billion years ago. Most of the infrared radiation from Uranus and Neptune is emitted within a scale height of the 400-mbar level. (Radiation emitted from deeper levels will be reabsorbed at higher altitudes. That from higher levels does not amount to much, because of the lower densities there.) The amount of energy radiated to space depends on the temperature. It increases as temperature increases, approximately as the fourth power of temperature. By coincidence, the amount of heat leakage from Neptune's interior, combined with its absorbed sunlight, nearly equals the amount of sunlight absorbed on Uranus. Hence Neptune and Uranus emit similar amounts of infrared radiation, and their temperatures near 400 mbar are comparable.

The giant hydrogen planets Jupiter and Saturn also radiate more energy than they receive from the Sun. What makes Uranus so different from Jupiter, Saturn, and Neptune? The transport of heat from the interior to the level of the atmosphere at which infrared radiation escapes to space can only be achieved by convection, that is, by motions in which hotter masses of gas deep in the planet rise vertically and mix with the cooler ambient environment higher up. Transport of heat from the interior by radiation or conduction is much too slow. The convective motions are thought to be highly efficient, and they establish a vertical gradient in temperature that is slightly larger than adiabatic. In other words, the decrease in temperature with altitude (or with decreasing pressure) is nearly the same as that experienced by a parcel of atmosphere displaced upward the same distance as it expands and cools without exchanging heat or mass with its environment. An atmosphere in which temperature decreases much more rapidly with altitude is so unstable to convection that heat is quickly transported upward, warming the upper altitudes and reducing the temperature gradient to close to the adiabatic gradient. Conversely, an atmosphere in which temperature decreases more slowly than the adiabat is stable to convection, and the upward heat transport is shut off.

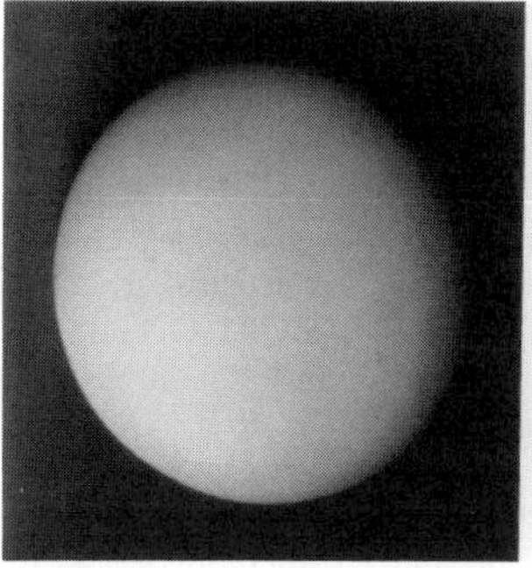
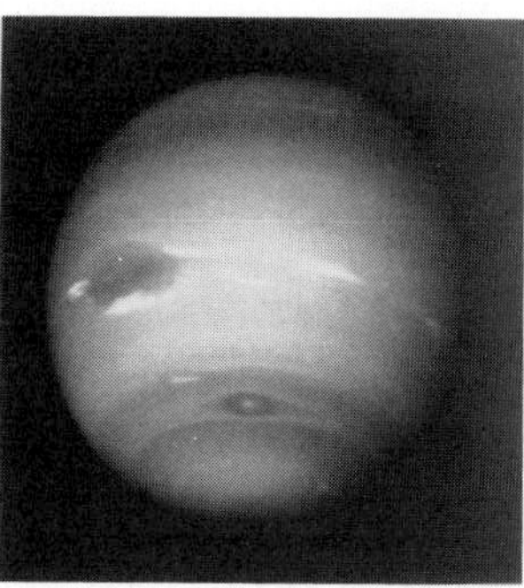

Figure 2. Uranus (left) and Neptune (right) as observed by *Voyager 2*. The view of Uranus is mostly in the southern hemisphere; the south pole is slightly to the left of and below the center of the disk. The view of Neptune is centered south of the equator. The striations and banded structure are aligned along latitude circles and the south pole is near the bottom of the disk. Uranus appears bland, whereas Neptune has several well-defined features, the largest of which is the Great Dark Spot, reminiscent in several ways of the Great Red Spot on Jupiter. The contrast in the Neptune image has been enhanced somewhat.

In this scenario the interiors of both Uranus and Neptune cooled by transporting heat convectively early in their histories. However, warming by sunlight sets a minimum temperature in the atmosphere at the level at which infrared radiation escapes to space (300–400 mbar). Being closer to the Sun, Uranus has a higher value of this minimum temperature, about 58 K. Since its formation, Uranus' interior has evidently cooled to the extent that, in the face of this minimum, it can no longer maintain a vertical gradient in temperature that is slightly greater than adiabatic. Hence convection from its interior has shut off. Neptune, with a lower minimum temperature set by the Sun in its atmosphere, has not yet reached this final state.

In addition to hydrogen and helium, both Uranus' and Neptune's atmospheres contain methane (CH_4). Selective absorption of sunlight at red wavelengths by methane accounts for the bluish appearance of both planets. In addition, methane has been identified from its characteristic spectral emission features that have been observed at infrared and visible wavelengths. The mole fraction of methane may exceed 0.02 at pressures greater than 1 bar; although not precisely determined, it is certain that the ratio of carbon to hydrogen in Uranus' and Neptune's atmospheres is much greater than the Sun's. At these high abundances methane should condense near 1 bar (cf. Fig. 1). Methane clouds have not been unambiguously observed on Uranus; in fact, the visual appearance of Uranus is rather bland (cf. Fig. 2). Neptune shows more structure; *Voyager 2* observed several bright features on Neptune, suggestive of clouds that may be composed of methane crystals. Although they have not been directly observed, ammonia (NH_3), hydrogen sulfide (H_2S), and water (H_2O) are also thought to be present and to condense to form clouds at deeper levels. Clouds of NH_3 should form at pressures of several bars. H_2S does not condense by itself, but combines with NH_3 to form ammonium hydrosulfide clouds, which are thought to form near pressures of 50 bar. Finally, H_2O clouds are expected to form even deeper. Depending on the H_2O abundance, they should form at pressures ranging from 100–2000 bar.

Uranus' and Neptune's vertical profiles of temperature, illustrated in Fig. 1, are close to adiabatic at pressures greater than 800 mbar, consistent with the discussion earlier. Temperatures have actually been retrieved from ground-based observations at infrared through millimeter wavelengths to pressures as large as 10 bar. The derived vertical profiles down to this level are also close to adiabatic. Deeper levels are probed at radio wavelengths, for which the principal opacities are probably from NH_3 and H_2O, condensibles that are not uniformly distributed. Because the spatial distributions of these opacity sources are not known, it has not been possible to unambiguously derive temperatures at these levels.

Voyager 2 also detected hazes in Uranus' and Neptune's stratospheres, at pressures less than 100 mbar. Much higher in the atmosphere, at pressures less than 1 mbar, solar radiation at ultraviolet wavelengths destroys methane, and the ensuing chemical reactions form ethane, acetylene, and other complex hydrocarbon molecules. The molecules condense at higher temperatures than methane, and as they are transported by atmospheric motions downward toward the temperature minima in Fig. 1, they can condense into clouds and hazes. These may account for the high-altitude hazes observed by *Voyager*.

Neptune becomes much warmer than Uranus at pressures less than 10 mbar. In part, this may be caused by absorption of solar radiation by the hazes just described; those on Neptune may be thicker than those on Uranus. However, gaseous methane is also an important absorber of sunlight at high altitudes. For reasons not well understood, Neptune appears to have much more stratospheric methane than Uranus. That in Uranus' stratosphere is consistent with a "cold trapping" near 100 mbar. In this case the stratosphere is supplied with methane from the deep atmosphere. Condensation and precipitation of methane where temperatures are low, near 100 mbar, limits the stratospheric abundance to a very low value. Because Neptune and Uranus have similar temperatures in this region, Neptune would be expected to have a similar amount of stratospheric methane. Instead, the methane abundance may be comparable to that in the deep atmosphere. Perhaps precipitation in methane clouds is not as effective on Neptune as on Uranus.

HORIZONTAL STRUCTURE AND WINDS

Observations from the *Voyager 2* spacecraft, of Uranus in 1986 and Neptune in 1989, have provided the most detailed information on the variation with latitude of winds and temperatures on these planets. The temperature field is remarkably bland. Unlike the 30-K decrease of temperature from equator to pole near the surface on Earth, the polar and equatorial temperatures are nearly equal on Uranus and Neptune. In part, the lack of contrast results from both planets being so far from the Sun and receiving much less sunlight than Earth. Over a complete orbit about the Sun, Uranus' poles actually receive more sunlight than the equator, because Uranus' equatorial plane is nearly perpendicular to its orbital plane. If Uranus' atmosphere simply radiated back at infrared wavelengths what it absorbed in sunlight at each location, the poles would, on the average, be 7 K warmer than the equator at the 1-bar level. Neptune's obliquity is more like Earth's (Table 1), and its equator would be warmer than its poles, by 9 K on the average. That not even these smaller contrasts are observed implies that atmospheric motions must be redistributing heat efficiently over the globe to keep the temperatures so uniform. The meridional velocities required are not large, only about 2 cm s^{-1}, well below the threshold of detectability from *Voyager*. Both planets do exhibit small depressions in temperature, amounting to 2–3 K near 1 bar, which are localized at mid-latitudes in each hemisphere. These depressions are not readily explained in terms of variations in solar heating arising from localized differences in albedo. Instead, they appear to result from adiabatic expansion and cooling of the atmosphere associated with upwelling at mid-latitudes.

Winds have been observed on Uranus and Neptune, but they are zonal; that is, they blow along latitude circles. The usual method of measuring wind velocities on the giant planets is by tracking distinct cloud features. On Earth, winds are measured relative to its surface, but these bodies lack rigid surfaces. Instead, the winds are referenced to the (presumably) uniform rotation of the planets' interiors. These are inferred from periodic modulations of radio signals from the planets and of the magnetic fields, which are believed to be corotating with their interiors. Uranus' rotation period is 17.2 hr, less than Earth's; Neptune's is shorter, 16.1 hr. Unlike Jupiter and Saturn, very few distinct features were seen on Uranus (Fig. 2), so the determination of wind speeds from cloud motions could only be accomplished at a few latitudes. Observations of the slight distortion of Uranus' figure from a nearly spherical ellipsoid were also used to infer departures of the atmosphere from the planet's rotational period. In addition, it is known from terrestrial meteorology that zonal winds are related to the gradients of temperature with latitude. The pattern of zonal winds deduced from these three methods are similar to those in Earth's atmosphere, but the winds are much stronger. At mid-latitudes the winds blow in the direction of Uranus' rotation with speeds up to 180 m s^{-1}, akin to the westerlies (which blow from the west) on Earth. At low latitudes the zonal winds blow counter to the planetary rotation with speeds up to 100 m s^{-1}, reminiscent of Earth's easterly trade winds. Neptune displayed several distinct cloud features that could be tracked. Qualitatively, the meridional pattern of zonal winds is also similar to Earth's, with winds blowing in the direction of the planetary rotation at high latitudes and counter to it near the equator. Preliminary analysis of the *Voyager 2* images indicates that the equatorial winds may be huge, as large as 600 m s^{-1}. Whether the similarity between the winds on Uranus and Neptune and those on Earth is fundamental or merely coincidental is not known. The zonal wind structure on Earth is thought to be the result of two competing processes: (1) The tendency of air parcels aloft to conserve angular momentum, that is, to "spin up" and move faster along latitude circles as they move toward higher latitudes and closer to Earth's rotation axis. (2) The strong friction exerted by the surface on the atmosphere just above it. It is not clear what plays the role of the rigid surface on Uranus and Neptune.

The similarity in the meridional structure of winds and temperatures on Uranus and Neptune was unexpected. The patterns of insolation on the two planets differ significantly, because their obliquities are so different. Furthermore, Neptune has a much stronger internal heat source than Uranus. Both processes were thought to be important in determining the general circulations of these planetary atmospheres. The *Voyager* observations suggest that this may not be the case. Together with Jupiter and Saturn, Uranus and Neptune provide the atmospheric scientist with several natural laboratories of deep atmospheres having both similar and differing characteristics. Studying them as a group, in the context of the control parameters that nature has provided (e.g., rotation rate, insolation, planetary size), may turn out to be the best way to unravel their individual mysteries.

Additional Reading

Beatty, J. K. (1990). Getting to know Neptune. *Sky and Telescope* **79** 146.

Bergstrahl, J. and Matthews, M. S., eds. (1990). *Uranus*. University of Arizona Press. Tucson.

Stone, E. C., Miner, E. D. et al. (1986). The *Voyager 2* encounter with the Uranian system. *Science* **233** 39.

Stone, E. C., et al. (1987). The *Voyager 2* encounter with Uranus. *J. Geophys. Res.* **92** 14,873.

Stone, E. C., Miner, E. D., et al. (1989). The *Voyager 2* encounter with the Neptunian system. *Science* **246** 1417.

See also **Outer Planets, Space Missions; Planetary Atmospheres, Dynamics; Planetary Atmospheres, Structure and Energy Transfer; Saturn, Atmosphere.**

Uranus and Neptune, Satellites

Carolyn Collins Petersen

The satellites in orbit around Uranus and Neptune have long been thought to be repositories for primitive materials dating back to the condensation of the solar nebula, some 4.5 billion years ago. Because they orbit far from the heat of the Sun, these worlds have undergone relatively little of the thermal modification that has changed the surfaces of the inner solar system bodies.

Aside from historical implications, these moons have turned out to be a fascinating and diverse array of objects. Their compositions are predominantly water ice and trace amounts of other volatiles, such as methane (CH_4) and ammonia (NH_3), and in some cases, a fraction of rocky material. The smaller moons may be mostly ices, whereas the larger moons of the two planets are ice and rock agglomerations. Three major processes appear to be responsible for the startling variety of surface features seen throughout the two systems, ranging from impact craters to geysers: tectonism (caused by forces acting within the moons' crusts), volcanism, and impact cratering.

The larger satellites of each system had been discovered from the Earth. *Voyager 2* detected the smaller moons during flybys of Uranus in 1986 and Neptune in 1989.

Table 2. Rings of Uranus and Neptune[a]

Ring	*Distance from Planet Center (km)*	*Albedo*
Uranus		
1986U2R	(38,000)	(0.03)
6	41,840	(0.03)
5	42,230	(0.03)
4	42,580	(0.03)
Alpha	44,720	(0.03)
Beta	45,670	(0.03)
Eta	47,190	(0.03)
Gamma	47,630	(0.03)
Delta	48,290	(0.03)
Lambda	50,020	(0.03)
Epsilon	51,140	(0.03)
Neptune		
1989N3R	41,900	(low)
1989N2R	53,200	(low)
1989N4R	53,200–59,100	(low)
1989N1R	62,930	(low)

[a]Values in parentheses are uncertain by more than 10%.

URANIAN SYSTEM

Before the January 1986 *Voyager 2* flyby, the known satellites at Uranus were Miranda, Ariel, Umbriel, Titania, and Oberon (in order of increasing distance from the planet; see Table 1). Such information as albedo (surface reflectivity), surface composition, and orbital characteristics for these "classical" moons had been derived from ground-based spectroscopic, photometric, and radiometric studies of the five satellites. Earth-based studies in 1977 revealed the existence of nine rings around Uranus (see Table 2). These are named 6, 5, 4, alpha, beta, eta, gamma, delta, and epsilon (in order of increasing distance from the planet). They are extremely difficult to see from Earth, and their existence was inferred from stellar occultation studies, notably from the Kuiper Airborne Observatory.

Table 1. Satellites of Uranus and Neptune

Moon	*Diameter*[a] *(km)*	*Mean Distance from planet (km)*	*Mass (g)*	*Mean density* *(g cm^{-3})*	*Albedo*	*Date Discovered*	*Discoverer*
Uranus							
Cordelia	(30)	49,750	?	?	0.05[b]	1986	*Voyager 2*
Ophelia	(30)	53,760	?	?	0.05[b]	1986	*Voyager 2*
Bianca	(40)	59,160	?	?	0.05[b]	1986	*Voyager 2*
Cressida	(70)	61,770	?	?	0.04[b]	1986	*Voyager 2*
Desdemona	(60)	62,660	?	?	0.04[b]	1986	*Voyager 2*
Juliet	(80)	64,360	?	?	0.06[b]	1986	*Voyager 2*
Portia	(110)	66,100	?	?	0.09[b]	1986	*Voyager 2*
Rosalind	(60)	69,930	?	?	0.04[b]	1986	*Voyager 2*
Belinda	(70)	75,260	?	?	0.05[b]	1986	*Voyager 2*
Puck	150	86,010	?	?	0.07[b]	1986	*Voyager 2*
Miranda	670	129,780	6.89×10^{22}	1.35	0.34	1948	G. Kuiper
Ariel	1160	191,240	1.26×10^{24}	1.66	0.40	1851	W. Lassell
Umbriel	1170	265,970	1.33×10^{24}	1.51	0.19	1851	W. Lassell
Titania	1580	435,840	3.48×10^{24}	1.68	0.28	1787	W. Herschel
Oberon	1520	582,600	3.03×10^{24}	1.58	0.24	1787	W. Herschel
Neptune							
Naiad	(50)	48,000	?	?	0.06	1989	*Voyager 2*
Thalassa	(80)	50,000	?	?	0.06	1989	*Voyager 2*
Despoina	(180)	52,500	?	?	0.06	1989	*Voyager 2*
Galatea	(150)	62,000	?	?	0.054	1989	*Voyager 2*
Larissa	(190)	73,600	?	?	0.052	1989	*Voyager 2*
Proteus	(400)	117,600	?	?	0.060	1989	*Voyager 2*
Triton	2700	354,800	2.14×10^{25}	2.07	0.6—0.9	1846	W. Lassell
Nereid	(340)	5,513,400	?	?	0.14	1949	G. Kuiper

[a]Values in parentheses are uncertain by more than 10%. Also see tables on pages 607, 608, 613 and 616.
[b]Geometric albedo. All other albedos given are normal reflectivity.

Voyager 2 imaged the uranian ring system and large satellites of Uranus, and found two more rings [1986U1R (lambda) and 1986U2R] and ten more satellites (see Tables 1 and 2). The rings are populated by meter-sized chunks of ice coated with dark and presumably carbonaceous materials.

The 10 newly discovered moons share the dark coloration of the ring particles, and their albedoes are 5–7%. By contrast, Earth's moon has an albedo of 12%. These bodies have been described as "black as soot" and "looking like charcoal." Their masses and densities are not reliably determined.

The five largest uranian satellites have much higher reflectivities, ranging from 20–40%—brighter than Earth's moon. Their densities range from 1.4–1.7 g cm^{-3}, slightly greater than the density of water. It is not surprising then, that water ice is present on the surfaces of the satellites. (It is important to remember that these densities imply that some fraction of each of the larger moons is made up of rock.)

However, the larger uranian moons appear grayish. This predominance of neutral and dark surface coloration of those moons is not well understood. It is thought that the moons may sweep up dark material in orbit around the planet. Alternatively, the brightness of the initially transparent methane ice mixed with the water ice may be changed by magnetospheric particle bombardment. In this process, prolonged proton bombardment of a methane–water ice mixture drives hydrogen molecules out of the ice mixture, creating more complex organic compounds that darken the ice.

These five moons are now most likely geologically inactive, frozen solid. Surface temperatures hover around 70–80 K. With densities slightly higher than water, it is likely that each of the satellites is 40–50% water ice, mixed with other ices, carbon, and silicates.

The largest moon of Uranus is Titania, with a diameter of 1580 km. Much of Titania's surface is heavily cratered, with some multiringed impact basins up to 70 km in diameter. Faults and scarps split the surface, and in some places the faults widen out to form dropped valleys called *grabens*. Voyager scientists have suggested that the faults are the result of extensional forces that split Titania's crust as the moon froze to its interior. As these cracks widened, ices oozed out and spread over the surface, creating localized relatively smooth areas.

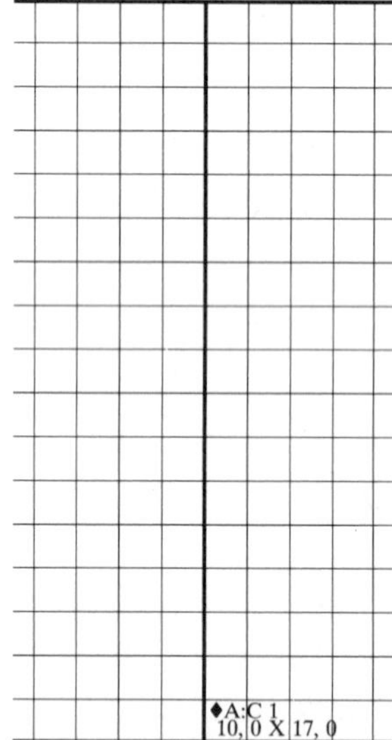

Figure 1. The surface of Miranda. Arden Corona lies on the upper right-hand quadrant of Miranda. To the left is a chevron-shaped area called Inverness Corona that is bounded on three sides by a series of faults. Verona Chasma (top, just above the chevron) is a 15-km-high ice cliff. (*Photo courtesy NASA/JPL; R. Batson U.S.G.S—Flagstaff, Arizona.*)

Oberon is the outermost known satellite of Uranus and the planet's second-largest moon. Like Titania, its surface is dominated by very large impact craters. Bright ejecta splash out in radial patterns from some of the larger craters. In the centers of some craters, dark material has oozed from the interior and frozen on the surface. Scarps and faults split the surface, and some arc-shaped features suggest remnants of impact basins, long since resurfaced by ice flows.

Umbriel is the darkest classical (pre-*Voyager*) uranian moon, with an albedo of 20%. The high number of impact scars implies an ancient surface, perhaps dating back to the early history of the solar system. The higher density (1.61 g cm^{-3}) may imply that Umbriel's ratio of ice to heavier carbon and silicate materials is somewhat lower than for the other uranian moons.

Complex geomorphic features characterize both Ariel and Miranda. Parts of Ariel's surface are older, cratered terrain, split by faults and wide sinuous grabens that might be relics of ice flows. Other, younger areas on Ariel have been resurfaced by ice flows. Craters larger than 60 km in diameter are missing.

For sheer variety of surface features, Miranda stands alone in the system (see Fig. 1). *Voyager 2* photographs show a clear global dichotomy, and mapping of Miranda based on these images has resulted in a division of the mirandan surface into two distinct units: *cratered terrain* (dominated by a variety of impact craters) and *coronae* (roughly circular- to rectangular-shaped regions that seem to dominate the surface). The cratered terrain seems to be the underlying surface over which the more striking coronae formations are overlaid. (The coronae themselves are cratered as well, although to a lesser extent than the surrounding plains.) Crater diameters on Miranda range from 500 m to 50 km.

The mirandan coronae have sparked intense speculation regarding their origin, and the formation of Miranda itself. These areas, first named "the chevron," "the ovoids," and "trapezoids" by various Voyager science team members, exhibit unusual terrain types. They show complex terrain within their boundaries and have high-contrast albedo markings, as if bright material has been juxtaposed alongside darker surface areas. Fracture zones cut across all units of terrain. The overwhelming question regarding the observed surface characteristics of Miranda is "How did this satellite form with the surface features it exhibits?"

Current theory suggests that Miranda may have broken up and re-accreted several times since its original formation. Each time the satellite reformed, it differentiated; that is, heavier rocky particles sank to the center of the moon, forming a core, while the lighter volatiles formed an icy coating. After the last accretion, Miranda lost whatever heat source it had, and partially differentiated agglomerates of ice and rock froze midway through the satellite, according to this idea. If so, what we see on the surface today is a jumbled mixture of ancient core materials and former surface units from a previous incarnation of Miranda. Much of the intense interest in this moon stems from this crazy quilt of terrain units, imaged so sharply by the *Voyager 2* cameras.

NEPTUNIAN SYSTEM

As it did for Uranus, *Voyager 2* made a number of startling satellite and ring discoveries at Neptune. Prior to the August 1989 flyby, only the satellites Triton and Nereid were known to be orbiting around the planet. Triton, the larger of the two, had been spotted from Earth in 1846; Nereid was discovered in 1949. Fragments of rings, or ring arcs, and a possible third satellite had been inferred from stellar occultation studies taken from Earth in the 1980s but were not generally accepted as proven. It remained for *Voyager 2* to image the satellites and map the ring structure.

During the encounter, *Voyager 2* discovered six additional moons inside the orbit of Triton: Naiad, Thalassa, Despoina, Galatea, Larissa, and Proteus (in order of increasing distance from the planet). It also confirmed the existence of four rings or ring areas, 1989N3R, 1989N2R, 1989N4R, and 1989N1R (also in order of distance from the planet), and a broad band of dust stretching

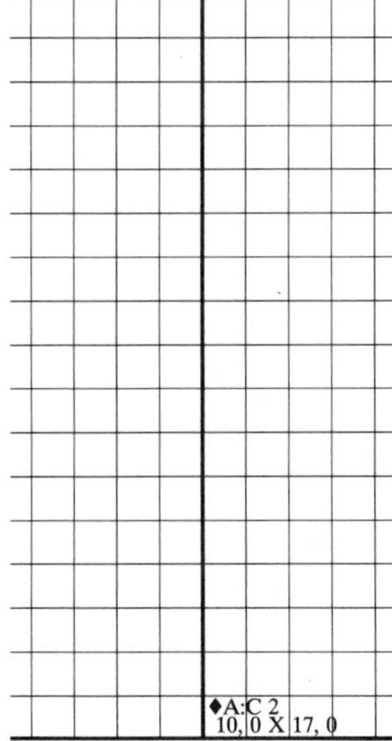

Figure 2. A mosaic of Triton images. The bright southern polar cap is at the left. Note the dark streaks deposited by wind-driven eruptive plumes scattered across the image. Below the streaks is the area of cantaloupe terrain. (*Photo courtesy NASA/JPL.*)

across the ring plane. Larissa is thought to be the "third satellite" seen from Earth during stellar occultation studies.

(Note: Upon their discovery, solar system bodies such as satellites and rings are given provisional names. The designation is usually the year of discovery, the first initial of the planet around which the body orbits, and the number of the object. At a later time, the International Astronomical Union adopts final names for the objects. As of early 1991, four of the six newly discovered neptunian satellites had final names, whereas the designations for the other two, Galatea and Larissa, were pending.)

The newly discovered satellites range in size from 50–400 km in diameter and have albedoes of only about 5–6%, making them hard to see. The moons Proteus and Larissa are dark and heavily cratered. Proteus is frozen hard, irregularly shaped, and larger than Nereid (see Table 1). However, it was never seen from Earth because it orbits too close to Neptune and is lost in the glare of the planet. Despoina and Galatea each appear to accompany the rings 1989N2R and 1989N1R, respectively. These may be examples of so-called shepherd satellites that are thought to confine particles within the ring boundaries.

The other two satellites of Neptune, Thalassa and Naiad, were discovered in close encounter search images, and do not seem to correspond to any ring regions. They are both quite small and may be cratered as well. Naiad orbits at a 4.7° inclination to Neptune's equator, whereas the other *Voyager*-discovered satellites orbit within 1° of the equatorial plane.

Outermost Nereid is the third-largest satellite in the neptunian system. It was very distant from the spacecraft during the *Voyager* encounter, and little image detail was obtained.

Triton is set apart from the other satellites by its size (2700 km diameter) and its highly inclined, retrograde orbit. Its peculiar orbital characteristics suggest that Triton did not originate in the neptunian system; rather, it has been suggested that Triton was captured by Neptune. Triton's orbit may have been elliptical at first, but over time, the tidal action of Neptune's gravity circularized the orbit, locking the moon into synchronous rotation, so that the same side always faces Neptune. The other satellites orbit prograde and they are thought to be in synchronous rotation as well.

Like Io at Jupiter, Triton has a tenuous atmosphere. The mix of gases observed consists predominantly of nitrogen, with methane found near the surface. The atmospheric pressure is about 16 microbars—70,000 times lower than the Earth. At times, atmospheric N_2 is released from sunlit regions and a westward wind drives atmospheric mixing. Wispy clouds and haze layers sometimes hover low in the 25–50-km-thick atmosphere. The surface temperature at Triton is 38 K—the lowest observed temperature on any natural body in the solar system.

The density of Triton is 2.07 g cm^{-3}, twice that of water. That figure has led to speculation that this moon is a well-differentiated body containing a rocky core about 1000 km in diameter. Overlying this putative core would be a series of frozen layers, each containing NH_3, CH_4, and water ices in various concentrations. The uppermost crust is a thin veneer of CH_4 and N_2 ices. The crust makes Triton brighter than the other satellites of Neptune, with a high albedo of between 70 and 90%. The surface appears to be mostly nitrogen ice, which acts very much like a rock at the low temperatures on the satellite. The crust exhibits color variations, ranging from whitish N_2 frost, to a reddish-looking south-polar ice cap, perhaps tinted by the effects of energetic particle bombardment.

Triton has been described as a hybrid object, displaying many terrain types seen previously on other icy satellites in the solar system (see Fig. 2). These features include a dense concentration of pits and ridges called cantaloupe terrain. Elsewhere, smooth plains and terraced ice lakes indicate that successive floods of material oozed from beneath the crust. Few craters are to be found on Triton, evidence of the constant resurfacing activities that must be erasing any impacts.

Smeared across the polar regions of Triton are windstreaks of some darker material. These are thought to be the ejecta from small volcanoes, or geysers, expelling a mixture of nitrogen gas and hydrocarbon-rich materials. Two of the observed eruptions feed plumes that reach up 8 km into the thin tritonian atmosphere. Current theory suggests that the energy to power these eruptions may come from the heat of sunlight trapped just beneath the surface by an icy equivalent of the greenhouse effect.

Again, as at Miranda, the overwhelming question is "How did Triton form the surface features it now exhibits?" Tidal forces from Neptune probably forced Triton's orbit into a circular shape. Heat generated by this process may have kept the satellite's exterior fluid during much of its early history. The moon differentiated, creating the large silicate core that exists today. Heating from the core and possible bombardment by orbiting debris melted and remelted the surface, causing the variety of terrain units and volcanic features seen during the *Voyager 2* flyby.

The *Voyager* data provide a great deal of insight into the behavior of ices under tectonic stress and under extremes of temperature and pressure. Because the moons of Uranus and Neptune are repositories of volatile materials left over from the formation of the solar system, these satellites deserve more detailed study in the future.

Additional Reading

Abell, G. O., Morrison, D., and Wolff, S. C. (1987). *Exploration of the Universe*, 5th ed. Saunders College Publishing, New York.

Beatty, J. K. and Chaikin, A., eds. (1990). *The New Solar System*, 3rd ed. Sky Publishing, Cambridge, MA.

Burns, J. A. and Matthews, M. S., eds. (1986) *Satellites*. University of Arizona Press, Tucson.

Greeley, R. (1987). *Planetary Landscapes*. Allen & Unwin, Boston.

Morrison D. and Wolff, S. C. (1990). *Frontiers of Astronomy*. Saunders College Publishing, New York.

Murray, C. D. and Thompson, R. P. (1990). Orbits of shepherd satellites deduced from the structure of the rings of Uranus. *Nature* **348** 499.

Smith, B. et al. (1986). *Voyager 2* in the uranian system: Imaging science results. *Science* **233** 43.

Stone, E. C. and Miner, E. D. (1989) The *Voyager 2* encounter with the neptunian system. *Science* **246** 1417.

See also **Outer Planets, Space Missions; Planetary and Satellite Atmospheres; Planetary Rings; Satellites, all entries.**

V

Venus, Atmosphere

Alvin Seiff

The atmosphere of Venus is the gaseous envelope of the second planet from the Sun, located at 0.72 times the Earth's distance from the Sun. Venus' atmosphere is 105 times as massive as Earth's atmosphere, per unit surface area. Its principal constituent is CO_2 (mole fraction, 0.965), with 0.035 N_2 and trace amounts of SO_2 (180 ppm), argon (70 ppm), CO (30 ppm), and O_2 (20 ppm). The surface temperature and pressure at the mean planetary radius, 6051.5 km, are 735 K and 95.0 bar (1 atm = 1.013 bar). Even though nitrogen is a minor component, its partial pressure at the surface is 4.16 times that on Earth.

Water, on the other hand, is not abundant on Venus. At the high surface temperature, liquid water, and thus oceans, cannot exist. Temperatures remain above the normal boiling point of water up to an altitude of 47 km, just 2 km below the cloud deck. The partial pressure of water at the 45-km level has been reported as 0.4 mb, corresponding to a relative humidity of only 0.026%. Measured water content drops off toward the surface, and above the clouds, it falls to a few parts per million at most. Thus, at present, Venus has no surface water, and its atmosphere is very dry. The planet's deficiency in water, massive atmosphere, and other major differences from the atmosphere of what has been called its "sister planet," Earth, have attracted much scientific interest.

Closer proximity to the Sun is clearly a factor contributing to these differences. Venus' insolation is 2 times that of Earth's, but at the present time, the clouds of Venus reflect 76% of the incident solar energy, so that Venus absorbs (and reemits) less solar heat than does Earth. This may not have been the case during the earlier evolutionary period. The massive atmosphere is consistent with the release of volatiles from the surface which is promoted by the high surface temperature. The quantity of CO_2 in the atmosphere of Venus is stored in the surface and oceans on Earth, fixed in carbonate rocks and dissolved in the oceans. It is not clear whether the increased nitrogen abundance in Venus' atmosphere can also be accounted for by the absence of other reservoirs. There are clear indications of past volcanic activity on Venus, also a source of atmospheric gases. The atmosphere comprises 10^{-4} of the mass of Venus. The corresponding fractions for Earth and Mars are 10^{-6} and less than 10^{-7}, respectively. (Note the regularity of the variation with distance from the Sun.)

To account for the lack of water on Venus, a loss process in which water is photochemically dissociated in the upper atmosphere, followed by escape of H to space and downward diffusion of O_2 to react with the surface, has been suggested. Support for this hypothesis is found in the enriched deuterium/hydrodgen ratio on Venus, which is approximately 100 times that on Earth. (Deuterium escapes to space more slowly than hydrogen.)

THERMAL STRUCTURE OF THE ATMOSPHERE

Temperature profiles with altitude, determined by experiments on the *Pioneer Venus* entry probes, orbiter, and bus, from the surface to 180 km, are shown in Fig. 1. Temperature decreases from 735 K at the surface to approximately 240 K at the cloud tops. The usual explanation for Venus' high surface temperature is "greenhouse effect" conservation of upwelling infrared radiation, with radiative transfer controlling the temperature profile. However, in green-

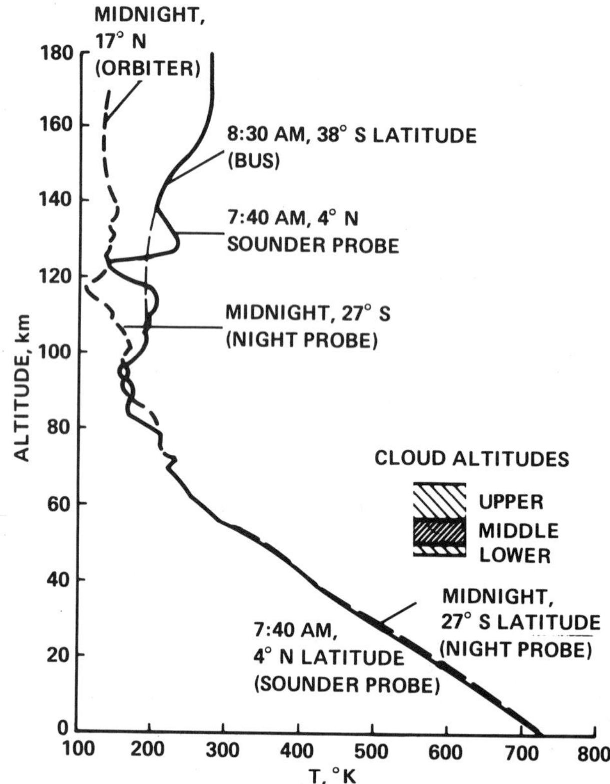

Figure 1. Temperatures in the atmosphere of Venus from the surface to 180 km, as given by *Pioneer Venus* probe and orbiter experiments.

house models that fit the observations, temperatures below altitudes of 35–50 km are not those for radiative equilibrium, but instead approximately follow an adiabat. This indicates that the temperature structure below the clouds is controlled by atmospheric dynamics.

The cloud top temperatures are determined by radiative equilibrium between infrared emission and solar absorption. The tropopause is found near the cloud tops at a pressure level of approximately 200 mb, which varies a little with latitude. It is the upper boundary of the lower atmosphere and is marked by a sharp change in the temperature lapse rate with altitude, dT/dz. Above the tropopause, the atmosphere is stably stratified and can therefore support gravity waves, which have been observed. There are stably stratified regions below the tropopause as well.

Diurnal and latitudinal temperature variations are small below the clouds (Fig. 2). These small temperature contrasts over distances comparable to the planet radius are a remarkable feature of Venus' meteorology. In the upper atmosphere, however, temperature varies spectacularly from day to night, from approximately 300 K on the day side to as cool as 130 K at night (Fig. 1). This large diurnal difference begins to develop near 100 km. Between 95 and 160 km, temperature increases by approximately 125 K on the day side, whereas it decreases by as much as 35 K on the night side.

Superimposed on the upper atmospheric mean profiles are large amplitude oscillations, which increase in amplitude with altitude to as much as 50 K, and have vertical wavelengths of about 15 km.

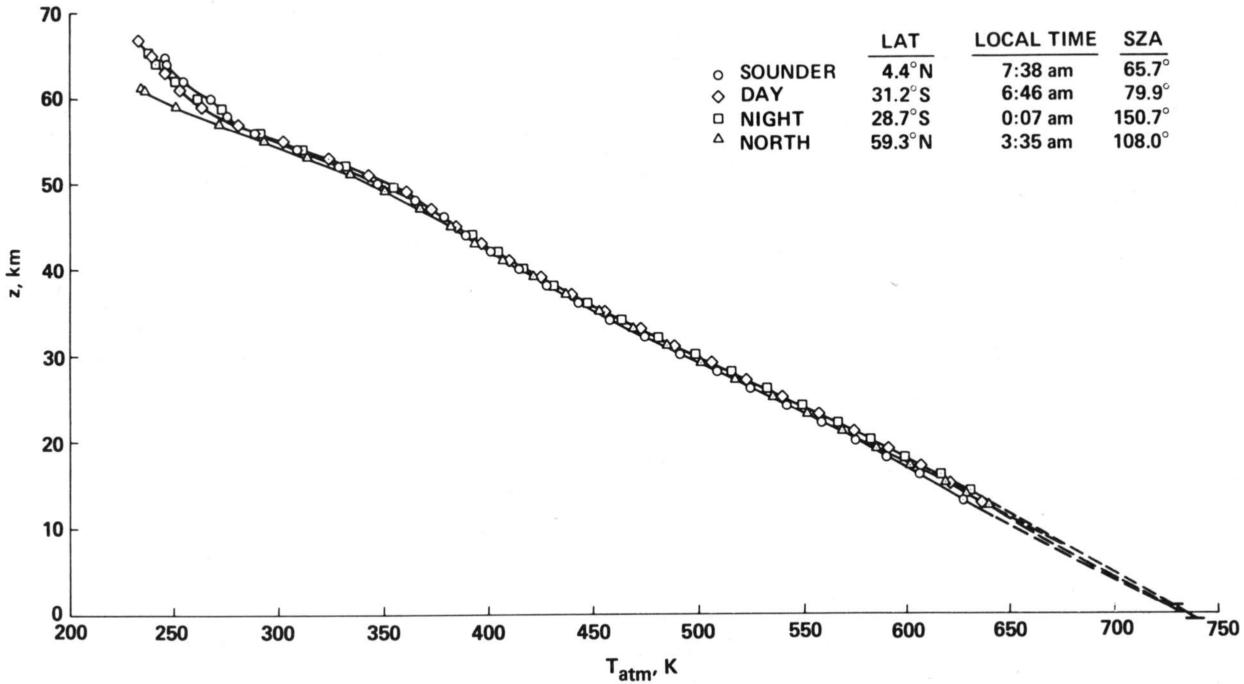

Figure 2. Temperature profiles below the clouds at four widely separated locations on the day and night sides of Venus, from the four *Pioneer Venus* probes (latitudes from 4°–59°).

One such observed oscillation is shown in Fig. 1. These are caused by wave motions with vertical displacements and a vertical component in the direction of propagation, but that are otherwise roughly analogous to surface waves on water. As fluid masses undergo vertical excursions about equilibrium levels, they are adiabatically compressed and expanded, leading to observed temperature oscillations. There are numerous other indications of wave motions in the atmosphere of Venus, on different scales. These include patterns seen in ultraviolet images of the cloud tops. A stationary wave, the solar fixed semidiurnal thermal tide, also has been observed in the Venus middle atmosphere, between the cloud tops and 100 km. It has a peak amplitude of about ± 5 K. Thus the physics and dynamics of Venus' middle and upper atmosphere are complex.

CLOUDS

Venus is continually blanketed by opaque clouds, so that the planetary surface is never visible. Measurements made at the surface have shown that only about 3% of the solar photons reach the surface in the subsolar region, after multiple scattering. Surface light levels fall off toward the terminators, for example, to 0.1% at solar zenith angle (SZA) = 85°.

The cloud levels, physical and optical depths, and light-scattering effectiveness have been examined by both in-situ and remote-sensing measurements on Soviet *Venera* and U.S. *Pioneer Venus Orbiter* spacecraft. Altitudes and physical depths of the clouds are indicated in Fig. 1. There are generally three cloud layers, between about 48 and 67 km altitude. The divisions between them are not always sharp and distinct, but optical differences among the three layers have been confirmed by discontinuities in temperature lapse rate that occur at the cloud boundaries. At some times and locations, the lower two layers appear to merge, whereas at other times, detached layers may occur lying beneath the "lower" cloud. The variability in structure is, however, far less than in the clouds of Earth.

The lower cloud, at about 48 km altitude, is about 1 km deep. The deeper middle cloud lies between 50 and 57 km. The upper cloud extends typically from 57– ~ 66 km and is more tenuous. Above this, there is a haze layer that grows increasingly tenuous with altitude, and may extend to 90 km. Total cloud optical depth has been estimated to be between 25 and 35. The opacity, although high, is much lower than was supposed before the *Pioneer Venus* mission. The thin lower cloud contributes a major part of the opacity.

Particle sizes in the clouds and hazes are from a few microns to submicron. Total number densities are less than 10^3 cm^{-3}. Thus the clouds are tenuous, and apparently do not form droplets large enough to rain. From the small cloud material mass, it has been concluded that vaporization and condensation processes are not energetically important in the atmosphere.

The primary cloud material is concentrated sulfuric acid, about 80% by weight, identified from its index of refraction and from detection of its vapor species. Theoretical models postulate the cyclic formation of the aerosols from SO_2 and water vapor as raw materials. Photochemistry plays a role at the upper levels. In the models, aerosols grow and settle to levels where they vaporize and dissociate, followed by upward diffusion of the gaseous species.

Some recent observations have indicated inhomogeneities in the clouds which permit seeing down to a level of possibly 50 km or deeper at 1.7 and 2.3 μm wavelength on the night side. The intensity of radiation seen at these wavelengths indicates it cannot be thermal emission from the cloud tops, but originates at higher temperature levels, deeper in the atmosphere. The zonal velocity of the emission features also indicates a deeper origin. These infrared "windows" indicate a significant local variation in cloud structure, the exact nature of which has not been identified.

ATMOSPHERIC CIRCULATION

Wind patterns at and below the cloud levels have been established by radio tracking of descending probes and by theoretical analysis of planet-wide variations in temperature at a given pressure, obtained by radio occultation and infrared radiometry. The first-order description is a pattern of zonal (east to west) winds, spiraling very gradually at cloud top levels toward the poles and culminating in a polar vortex. Long period waves are superimposed on these winds. The winds are generally in the direction of planetary rotation, which is sluggishly retrograde (i.e., opposite to the direction of the movement of the planets around the Sun). Whereas Venus rotates once in 243 days, the cloud top winds circle the planet in 4 days, at 100 m s^{-1}. A quantitative explanation for the generation and maintenance of these winds remains to be achieved. Below the clouds, winds decrease through a series of high and low shear intervals to a value of approximately 1 m s^{-1} near the surface. Even this low velocity represents considerable momentum, because of the high surface density of the atmosphere (0.065 times that of water).

Zonal winds measured in situ by tracking the *Pioneer Venus* probes have been shown to be in cyclostrophic balance with the pressure field. That is, the pressure gradient with latitude provides the horizontal component of centripetal force needed to support zonal flow at measured velocities. Subsequently, the assumption of cyclostrophic balance was used to calculate wind velocity from the pressure variation with latitude measured globally by remote sensing. An interesting feature that emerges is a jet stream at 45° latitude, centered at the cloud tops (~ 65 km altitude), where peak velocities are 110–120 m s^{-1}.

Above the clouds, zonal winds diminish, approaching 0 at 90 km altitude. In the upper atmosphere, above 100 km, no direct wind measurements exist, but the circulation indicated by computer modeling of observations of the variations in pressure and composition with solar zenith angle and altitude is a subsolar to antisolar flow with a zonal flow superimposed. Peak velocities reach approximately 250 m s^{-1} in flow across the terminator.

ATMOSPHERIC CONTRASTS AND STABILITY

The interaction between atmosphere structure and dynamics on Venus is intimate. Each influences the other. Horizontal variations in temperature at a given pressure, termed contrasts, are of key importance to atmospheric dynamics. Theoretically, it was predicted that they would be about 10^{-2} K in the deep atmosphere, because of thermal inertia and long radiative time constants; and approximately 0.1 K at cloud levels, limited by rapid zonal motion from day to night sides. However, contrasts measured by the four *Pioneer Venus* probes, widely separated over the Earth-facing hemisphere of the planet, were found to be about a few degrees kelvin (Fig. 2), small compared to those found on Earth (which are approximately 100 K), but large compared to the prediction.

Two phenomena are responsible for the larger than expected contrasts: an inherent, unpredicted variation in thermal structure $T(p)$ with latitude; and the waves present in the atmosphere. A systematic variation in structure with latitude begins at about 30° latitude at altitudes above 33 km. Models have been given for this latitude variation. Temperature at a given altitude decreases with increasing latitude. Simultaneously, the tropopause and the cloud top altitude are lowered. These changes establish the meridional pressure gradients that balance the zonal winds.

Another aspect of thermal structure important to dynamics is the stability of the atmosphere, $S = dT/dz - \Gamma$, where z is the altitude and Γ is the adiabatic lapse rate. In a stable atmosphere, i.e., one with $S > 0$, a fluid parcel displaced vertically will return to its equilibrium altitude. In an unstable atmosphere, $S < 0$, a small vertical displacement leads to divergence, either upward or downward. Unstable or neutrally stable atmospheres tend to become convective, whereas stable atmospheres are stratified and support gravity waves.

Venus' lower atmosphere had been expected to be convective, that is, unstable. *Pioneer Venus* probe data showed, however, that over much of its depth, the lower atmosphere is stable (Fig. 3). Hence it can no longer be assumed that the entire lower atmosphere is convective, although layers from 20–30 km at latitudes up to 30°, and the near surface layer are probably convective.

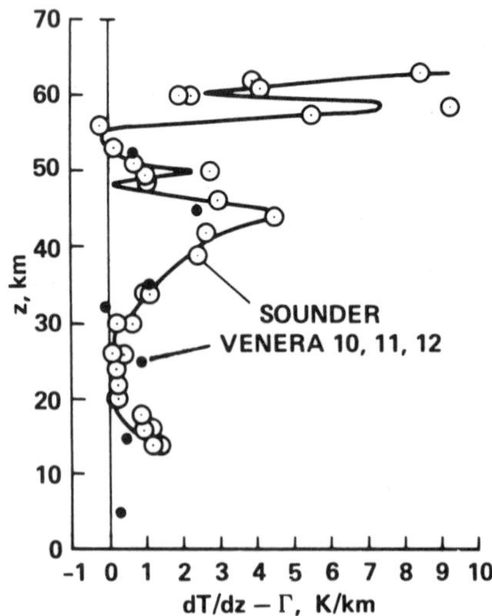

Figure 3. Stability of the lower atmosphere of Venus at the *Pioneer Venus* Large Probe (Sounder) site.

Gravity waves have been observed in the remaining stratified layers.

In the upper atmosphere, where temperature contrasts between the day and night hemispheres are large (see the previous discussion), the temperature adjustment on streamlines crossing the terminator is rapid. It occurs in a local Venus time interval (used as a measure of solar zenith angle) of 3 or 4 hours. This is possible because the predominant gas in the upper atmosphere below 150 km is CO_2, an efficient radiator in the infrared. As the atmosphere cools in crossing the terminator, its volume contracts, causing subsidence of the flow. Composition is transported downward along descending streamlines.

Additional Reading

Kliore, A. J., Moroz, V. I., and Keating, G. M., eds. (1985). *The Venus International Reference Atmosphere, Adv. in Space Res.* **5** (No. 11). Pergamon, New York.

Marov, M. Ya. (1978). Results of Venus missions. *Ann. Rev. Astron. Ap.* **16** 141.

Pioneer Venus Special Issue (1980). *J. Geophys. Res.* **80** (No. A13).

Pollack, James B. (1990). Atmospheres of the terrestrial planets. In *The New Solar System*, 3rd ed., J. K. Beatty and A. Chaikin, eds. Sky Publishing Corp., Cambridge, MA and Cambridge University Press, Cambridge, UK, p. 91.

Seiff, A. (1983). Thermal structure of the atmosphere of Venus. In *Venus*, D. M. Hunten, L. Colin, T. M. Donahue, and V. I. Moroz, eds. University of Arizona, Tucson, p. 215.

See also **Mercury and Venus, Space Missions; Observatories Balloon; Planetary Atmospheres, Clouds and Condensates; Planetary Atmospheres, Dynamics; Planetary Atmospheres, Escape processes; Planetary Atmospheres, Ionospheres; Planetary Atmospheres, Solar Activity Effects; Venus, Geology and Geophysics; Venus, Magnetic Fields.**

Venus, Geology and Geophysics

Maria T. Zuber

The geology and geophysics of Venus are of great interest because that planet's size and bulk properties are most similar to those of the Earth. As noted in Table 1, Venus and Earth differ by less than 10% in equatorial radius, mean density, and surface gravity, and by

Table 1. Comparison of Venus and Earth

Physical Property	*Venus*	*Earth*	*Venus / Earth*
Mass (10^{24} kg)	4.87	5.97	0.82
Equatorial radius (km)	6051	6371	0.95
Mean density (kg m^{-3})	5250	5515	0.95
Moment of inertia factor (I/MR^2)*	0.34	0.33	1.02
Surface gravity (m s^{-2})	8.87	9.78	0.91
Sidereal rotation period (day)	−243.01	0.997	−243.74
Magnetic field (G)	~ 0	0.5	~ 0
Surface temperature (K)	740	300	2.46
Surface pressure (MPa)	9.3	0.1	93.0

*I = moment of inertia; M = mass; R = radius.

less than 20% in mass. Because it has been recognized that the geologic evolution of the solid ("terrestrial") planets is primarily linked to size and mass, such that the larger, more massive planets undergo more active geologic histories, it has been suggested that Venus has evolved in a manner similar to Earth. However, Earth and Venus differ in several other important respects, with Venus being closer to the Sun and having a much higher surface temperature, a much slower (and retrograde) rotation, and a lack of near-surface liquid water and of an intrinsic magnetic field. Thus the nature of Venus' geologic evolution remains an open question, and one of great importance in understanding the terrestrial planets.

Despite the relatively close proximity of Venus and Earth, little was known about the nature of Venus' surface until recently because it is obscured by an optically opaque atmosphere. To view broad areas of the surface, images have been obtained using radar (i.e., microwave) techniques. Radar images of Venus as well as radar-derived topography, surface roughness, and reflectivity data have been acquired from Earth-based observatories and from orbiting spacecraft (see Fig. 1). The Arecibo and Goldstone Radar Observatories have imaged selected areas of the Venus surface at spatial resolutions as fine as 1.5 km. The U.S. *Pioneer Venus Orbiter* mapped more than 90% of the surface at an average spatial resolution of about 70 km. The Soviet *Venera* 15/16 spacecraft mapped an area of the northern hemisphere corresponding to about 25% of the total surface at a spatial resolution of 1–2 km. Other observations of the Venus surface have been obtained in local areas from nine Soviet *Venera* and *Vega* landers and include optical images and information on composition and surface properties (e.g., rock strength and density). Insight into the nature of the Venusian interior has been gained from Earth-based measurements of mass and rotation rate, and spacecraft-derived measurements of gravity, magnetics, and surface topography.

INTERNAL STRUCTURE, COMPOSITION, AND THERMAL STATE

The similarities in mean density and moment of inertia factor (I/MR^2) between Venus and Earth (Table 1) suggest that the two planets have similar bulk compositions and internal structures. For instance, the relatively large value of mean density strongly suggests that Venus contains an iron core. However, Venus evidently lacks an intrinsic magnetic field that would be strong evidence that at least part of its core exists in the liquid state, as is the case for Earth. If solidification of the liquid core drives the geodynamo and if Venus' interior temperatures are at least as high as Earth's, then Venus may lack the energy to generate a magnetic field because its central pressure is less than that at the Earth's solid inner core/molten outer core boundary. Constraints on the size of the

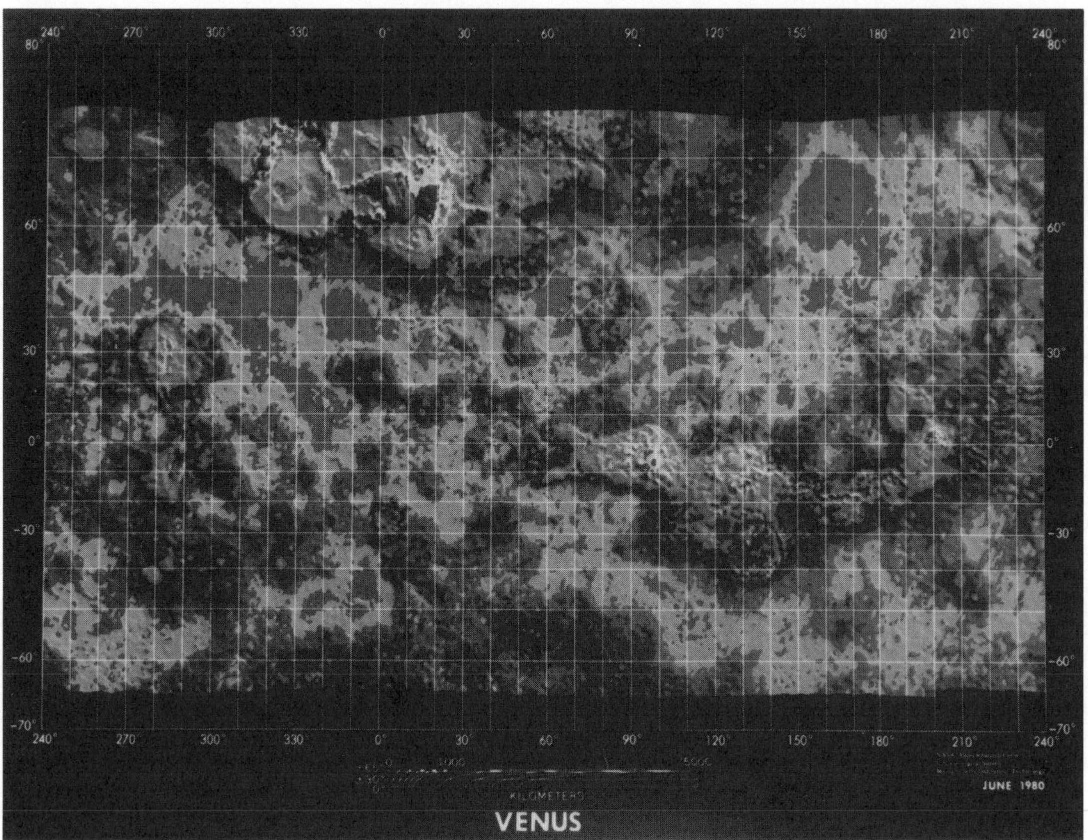

Figure 1. Surface topography of Venus as determined from the radar altimeter aboard the U.S. *Pioneer Venus Orbiter*. Elevations are referenced to a mean planetary radius of 6051 km. The spatial resolution is approximately 60–100 km and vertical resolution is approximately 0.5 km. (*By permission from G. H. Pettengill.*)

core and the compositions of the core and mantle are based primarily on geochemical and cosmochemical models of the Earth's bulk composition adapted to predicted Venusian internal temperatures and pressures. These models suggest that the interior of Venus is compositionally similar to the Earth, except for small differences in iron, sulfur, and oxygen content. Most models predict the radii of the core and mantle to be approximately 3200 and 2800 km, respectively.

Estimates of the composition of Venus' crust are based on in-situ chemical analyses of the surface. Surface compositions at the landing sites resemble those of terrestrial igneous rocks, which suggests that the crust formed due to melting of an upper mantle with an Earth-like composition. Most of the sites exhibit compositions similar to basalt, a common volcanic rock that comprises most of the Earth's seafloor. The thickness of the crust has implications for the amount of melting that has taken place in the upper mantle. Theoretical models constrained by the observed depths of impact craters and by typical spacings of ridges and rifts imply crustal thicknesses in many parts of Venus of 10–20 km, which is less than the average thickness of the Earth's continental crust (~ 45 km) but greater than the thickness of the oceanic crust (~ 6 km). Alternatively, arguments relating to venusian mantle mineralogy and high topographic elevations suggest crustal thicknesses in some areas of 100 km or more.

Lateral variations in Venus' internal density structure have been determined from gravity anomalies derived from orbital perturbations of the *Pioneer Venus Orbiter*. Gravity anomalies on Venus have magnitudes comparable to those of Earth. However, gravity and topography on Venus are strongly correlated at long wavelengths (> 500 km). This is in contrast to the Earth where long-wavelength gravity and topography are generally uncorrelated, and gravity anomalies are believed to be associated with convection in the mantle. Gravity and topography have also been used to determine the depth of isostatic compensation of various areas of Venus, which provides information on the mechanisms that support or "hold up" surface topography. Compensation depths are generally greater than on Earth, which suggests that the topography is either geologically very young or is supported by deep internal processes such as mantle upwelling.

If Venus has a similar composition and internal heat sources as the Earth, then its high surface temperature (~ 740 K) requires that the thickness of its lithosphere, or rigid outer shell, is much less than Earth's. Theoretical models constrained by laboratory experiments on the flow and fracture properties of rock extrapolated to venusian conditions predict a maximum lithosphere thickness on the order of 50 km.

Considerable debate exists concerning the primary mechanism by which Venus transports heat from its interior. Heat transport from planetary interiors to surfaces via the lithosphere is the driving mechanism responsible for global surface processes such as volcanism and tectonics (folding and faulting). On Earth, internal cooling is mainly accomplished by plate tectonics, in which heat is transported to the base of the lithosphere by the upwelling limbs of large mantle convective cells, and cold mantle material is returned to the deep interior in corresponding downwelling cells. In this process, lithospheric plates recycle; new lithospheric material is created at spreading centers (volcanic ridge systems on the ocean floor) above regions of upwelling, and old lithosphere is subducted into the mantle at areas of downwelling. Several lines of evidence suggest that heat loss on Venus may instead by accomplished largely by mantle plumes or hotspots, which are narrow regions of hot mantle upwelling. However, other possible mechanisms such as plate recycling or lithospheric conduction cannot presently be excluded as inconsequential. It is possible that a combination of heat loss mechanisms operates on Venus.

GLOBAL SURFACE CHARACTERISTICS

The global distribution of topographic elevations on Venus is unimodal, with a dynamic range (~ 13 km) similar to the Earth. However, Earth exhibits a bimodal distribution of elevations that earmarks the distinct differences in elevation and composition between the ocean basins and continents. On Venus, approximately 65% of the topography lies near the mean planetary radius and forms the rolling plains; about 27% of the topography lies approximately 1–2 km below the mean radius and corresponds to the lowlands; and approximately 8% of the topography rises as much as 11 km above the mean radius and constitutes the highlands. Thus the distribution of elevations on Venus is much closer to the mean radius than on Earth. Most of Venus' highland topography is concentrated into two provinces: Ishtar Terra, which is located in the northern hemisphere and is approximately the size of Australia, and Aphrodite Terra, which lies near the equator and is larger than Africa in surface area.

The distribution of regional (~ 100 km scale) slopes on Venus has been derived from topographic data. Although such slopes on the Earth and Venus span a similar range of values, their distributions are distinctly different, with Earth exhibiting generally lower slopes that may largely reflect erosional processes.

Measurements of global surface properties show that most of Venus is composed of dense, rocky surfaces, whereas less than a quarter of the planet is covered by unconsolidated, higher-porosity materials such as soils, which dominate the surfaces of the Moon and Mars. A very small percentage of the surface, mostly in topographically high regions, is characterized by very high radar reflectivities that may indicate the presence of highly conducting materials such as iron-bearing minerals.

VOLCANISM, TECTONISM, AND IMPACT CRATERING

On the order of 70% of the Venus surface appears similar to basaltic (i.e., lava) plains on other planetary surfaces. Radar images of the venusian plains contain many dark and bright flow-like features, some of which appear to emanate from volcanic source craters and fissure-like linear features. Volcanic landforms ranging from small, irregularly distributed clusters of conical domes to major volcanos have also been recognized. Theoretical studies suggest that despite differences in cooling history related mainly to atmospheric density and surface temperature, lava flows on Venus and Earth should develop in similar manners. However, certain styles of explosive volcanism are not likely to exist on Venus unless volatile contents (i.e., dissolved gas concentrations) of magmas are as great or greater than those observed in some very volatile-rich magmas on Earth. The rate of volcanism, as estimated from the areal density of impact craters, calculations of internal heat flow, and experiments on surface–atmosphere chemical reaction rates, suggests that currently Venus is, on average, less volcanically active than Earth. However, the spatial distribution of volcanic landforms has yet to be systematically mapped over the entire planet.

Venus exhibits numerous surface features that apparently formed due to horizontal compression (folding and mountain building) and extension (rifting) resulting from the mechanisms by which the planet transfers heat from its interior. Contrasting styles of tectonic deformation are manifested by patterns of ridges and valleys distributed in either narrow zones or wide areas, and by regions with broad, domal topography. Deformation occurs most commonly in, but is not limited to, the highlands.

One style of tectonism occurs in regions known as Maxwell, Frejya, and Akna Montes, and consists of subparallel ridges and valleys, with lengths of hundreds of kilometers and typical spacings of 5–20 km. These features bound the upland plateau of Lakshmi Planum in Ishtar Terra and have been interpreted as mountain belts formed as a consequence of horizontal compression. Other ridge and valley features are arrayed in belts that trend parallel to the ridges and valleys. The most prominent assemblage of these "ridge belts" occurs in the rolling plains east of Atalanta Planitia, the deepest regional depression on Venus, and strikes north–south with an average belt-to-belt spacing of about 300 km. The question

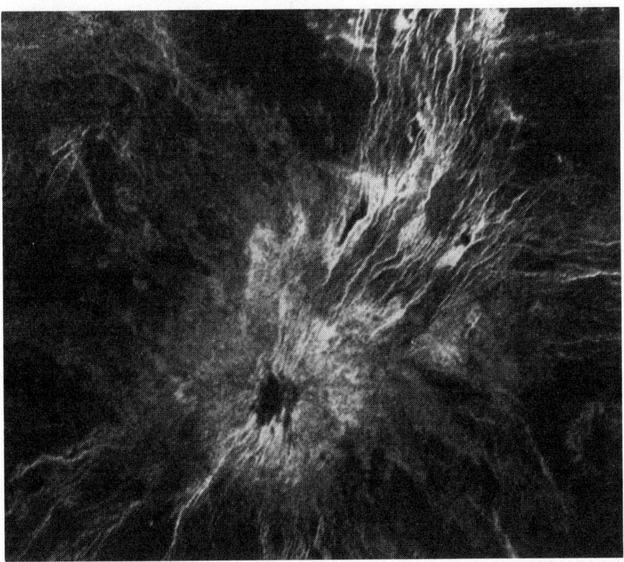

Figure 2. Radar image of part of the Beta Regio upland, obtained from the Arecibo Observatory. Relatively rough areas are bright, whereas smooth areas are dark. The diffuse bright circular feature with the dark central region is Theia Mons, which is interpreted to be a large shield volcano. The bright lineations are interpreted to be faults. The resolution of the image is 2 km. The dark quasicircular region has a diameter of about 100 km. (*By permission from D. B. Campbell.*)

of whether the ridge belts formed due to compression, extension, or some combination of forces is a matter of current debate.

A unique terrain type, known as "tessera," consists of elevated plateau-like areas with spatial extents of several thousands of kilometers. Within the tessera are ridges and grooves in intersecting, curving, and often chaotic patterns. The detailed nature of the tessera is enigmatic and may reflect multiple stages of deformation associated with gravitationally driven processes.

Venus contains a number of broad (hundreds of kilometers), dome-like features that rise up to 3–5 km above the surrounding plains and are in some cases characterized by rift zones, which are linear depressions formed where the Venus lithosphere has ruptured due to horizontal extension. The most prominent venusian rift is Beta Regio (Fig. 2). Two large shield-like volcanos, Rhea and Theia Mons, are associated with the rift. Beta Regio has been compared to the Earth's East African Rift zone on the basis of similarities in spatial scale, domal topography, and the presence of volcanic landforms. Like East Africa, the source of both the dome topography and the extensional stresses responsible for rifting is thought to be dynamic uplift due to a mantle plume.

In western Aphrodite Terra, bilateral topographic symmetry and linear discontinuities in topography and surface roughness that trend perpendicular to the long axis of the highland have been recognized. These characteristics are also observed at terrestrial spreading centers, which has led to the suggestion that Aphrodite is an area of lithospheric divergence. Alternatively, the topographic form of this area has been interpreted as the surficial expression of a mantle plume.

Another landform unique to Venus is the "coronae" or "ovoids." These are circular features, 150–600 km in diameter, that consist of an annular ring of ridges and valleys that surround a complex central region. The coronae are elevated up to 2 km above their surroundings and have raised rims and a narrow, bounding moat. More than two dozen of these features have been identified, with most located in volcanic plains surrounding the Ishtar Terra highland. The formation of the coronae may have been related to mantle plume-related uplift and subsequent gravitational relaxation of surface topography. A possibly related class of features are the "arachnoids," which are 50–200 km diameter spider-shaped systems of radial and concentric ridges that have an unknown origin.

Over 140 impact craters, with diameters ranging from 8–144 km, have been recognized on the Venus surface. The craters vary in preservation state, and thus apparent age. Measurements of the number of craters per unit area, in combination with models of the presumed flux of impacting bodies, suggest that the age of Venus' rolling plains is in the range 200 million to 1.5 billion years. This is much younger than the surfaces of the Moon and Mercury, which have ages on the order of 4 billion years, but is older than most terrestrial volcanic surfaces, such as the ocean floor, which is everywhere less than 200 million years old.

SURFACE MODIFICATION PROCESSES

The high surface temperature and predicted low erosion rates suggest that viscous relaxation of relief may be the dominant mechanism for the reduction of topography on Venus. Viscous relaxation is the horizontal spreading or flow of topography, driven by gravity. This process is believed to occur rapidly (in a geologic sense) for rocks at the high temperatures that exist near the Venus surface. Models of viscous relaxation indicate that topographically high areas on Venus are unlikely to be more than a few hundred million years in age (or else they would have flowed away), which lends additional support to the contention that the planet has been geologically active in its recent past.

Because of the dense atmosphere and present lack of liquid water on Venus, mechanical weathering is much less important than on Earth. Only two surficial mass transport processes are probable on the Venus surface. Mass wasting, the gravitationally driven downslope movement of unconsolidated material, is suggested by the presence of angular rocks at one of the lander sites, but is thought to be of only local importance. Aeolian (wind driven) transport is indicated by laboratory simulations of the Venus environment, measured wind speeds, and a variety of observations from radar images and lander sites, but likely affects only fine materials. The experiments indicate that mechanical weathering due to wind blasting may be much different on Venus than Earth. On Venus, windblown dust particles can be effectively accreted onto rocks to form thin surface films, altering the physical and chemical character of their exposed surfaces.

Chemical weathering on Venus is inferred on the basis of thermodynamic considerations that suggest that some basaltic minerals should be unstable at the surface; in particular, oxidation and sulfur-based alteration of surface rocks are possible. The near-infrared reflectance properties at some landing sites show evidence that the surface in these areas may be relatively oxidized. It appears that chemical weathering has been the primary mechanism for soil production on the surface of Venus during the past billion years.

SUMMARY

The Venus surface has apparently been extensively shaped by volcanism and tectonism and, to a more limited extent, by other processes. Evidence for these processes is temporally and spatially distributed, indicating that Venus has undergone an active, complex, and diverse geologic history. Refinement of our understanding of the structure and geologic evolution of Venus will be forthcoming with information from NASA's *Magellan* radar mapping mission, which was launched from the Kennedy Space Center on May 4, 1989, entered Venus orbit on August 10, 1990, and began radar mapping of Venus on September 15, 1990. *Magellan* is globally imaging the planet at spatial resolutions of up to 120 m and is mapping the topography, gravity, and radar surface properties at much higher resolutions than were previously available.

Additional Reading

Barsukov, V. L., et al. (1986). The geology and geomorphology of the Venus surface as revealed by radar images obtained by *Veneras 15* and *16*. *J. Geophys. Res.* **91** (No. B4) D378.

Basilevsky, A. T. and Head, J. W., III (1988). The geology of Venus. *Ann. Rev. Earth Planet. Sci.* **16** 295.

Basilevsky, A. T. (1989). The planet next door. *Sky and Telescope* 360.

Masursky, H., Eliason, E., Ford, P. G., McGill, G. E., Pettengill, G. H., Schaber, G. G., and Schubert, G. (1980). *Pioneer Venus* radar results: geology from images and altimetry. *J. Geophys. Res.* **85** 8232.

Pettengill, G. H., Eliason, E., Ford, P. G., Loriot, G. B., Masursky, H., and McGill, G. E. (1979). *Pioneer Venus* radar results: altimetry and surface properties. *J. Geophys. Res.* **85** 8261.

Phillips, R. J. and Malin, M. C. (1983). The interior of Venus and tectonic implications. In *Venus*, D. M. Hunten, L. Colin, T. M. Donohue, and V. I. Moroz, eds. University of Arizona Press, Tucson, p. 159.

Saunders, R. S. and Pettengill, G. H. (1991). *Magellan*: Mission summary. *Science* **252** 247. [See accompanying papers on geology and geophysics of Venus in *Science* **252** (No. 5003).]

See also **Mecury and Venus, Space Missions; Planetary Tectonic Processes, Terrestrial Planets; Planetary Volcanism and Surface Features; Venus, Atmosphere.**

Venus, Magnetic Fields

Christopher T. Russell and Janet G. Luhmann

Magnetic dynamos, in which currents deep inside a body generate an intrinsic magnetic field, appear to be the rule rather than the exception in the larger solar system bodies. All solar system bodies ranging in size from the Earth to the Sun generate their own magnetic fields. But the score is mixed for the smaller planets. Mercury has a small magnetic moment, Mars has at most a very weak one, and Venus has essentially none at all. Determining the size of any internally generated magnetic field is important because its size provides a clue as to the nature of the interior of the planet and such clues are few in number in the absence of missions to the surfaces of these bodies. However, deducing the precise magnitude of the intrinsic field is difficult in the presence of external magnetic fields produced by the interaction of the expanding ionized outer atmosphere, or corona, of the Sun (called the solar wind) with the planet's ionized outer atmosphere. This supersonically expanding ionized gas also has an embedded magnetic field of its own that is imposed on it at the Sun. Thus we first review the properties of the interaction of the solar wind with Venus. This will enable us to examine the procedures for determining the very stringent upper limits presently placed on the intrinsic magnetic field of Venus and to understand the often apparently conflicting evidence for a martian intrinsic magnetic field.

PLANETARY PROPERTIES AND EXPECTATIONS

Venus is very close to the Earth in size with a radius of 6052 km compared to the terrestrial radius of 6371 km. Venus rotates much more slowly than the Earth, turning once about its axis with respect to the stars every 243 Earth days. This is about one-quarter the rotation rate of Mercury. From dimensional considerations we would expect that the magnetic field in the dynamo region would be proportional to the core radius and rotation rate. The magnetic moment, which equals the magnetic field times the cube of the radius, should therefore be proportional to the product of the rotation rate and the fourth power of the rotation rate. Scaling the Mercury magnetic moment from the terrestrial moment by using this relationship gives a value within a factor of about 2 of the observed magnetic field. The predicted venusian magnetic moment would be about 2.5×10^{13} Tm^3 (and for Mars about 6×10^{14} Tm^3). Magnetic moments of these values would be quite readily detected from orbiting spacecraft.

Venus has a dense lower atmosphere but its upper atmosphere is not too unlike that of the Earth. The solar extreme ultraviolet (EUV) radiation ionizes this upper atmosphere producing an ionosphere with a peak ion density of slightly under 10^6 ions per cubic centimeter at about 140 km altitude in the subsolar region. At low altitudes the temperature of this ionized gas is a few hundred degrees Celsius but at high altitude the temperature rises to a couple of thousand degrees. The electron temperature is about twice the ion temperature.

A variety of chemical reactions occur in the ionosphere of Venus. An important one is the dissociative recombination of ionized diatomic oxygen with an electron that produces two excited atoms of monatomic oxygen which fly apart at a velocity sufficient to produce a neutral oxygen exosphere up to an altitude of 4000 km. These neutrals may then be ionized by the extreme ultraviolet radiation producing ions at high altitudes.

THE SOLAR WIND INTERACTION WITH VENUS

The ionosphere of Venus is very highly electrically conducting so that the magnetic field carried to Venus from the Sun by the solar wind is excluded from it. This causes the magnetic field of the solar wind to pile up outside of the ionosphere. Moreover, during solar maximum (when solar extreme ultraviolet radiation intensity peaks), the thermal pressure in the ionosphere is usually sufficient to balance the pressure exerted on it by the flowing solar wind at about 300 km altitude or above. The surface where these pressures balance is called the ionopause. Thus the ionosphere, with the help of the magnetic barrier built up above it, is able to deflect the solar wind around the planet. The solar wind is supersonic; that is, the

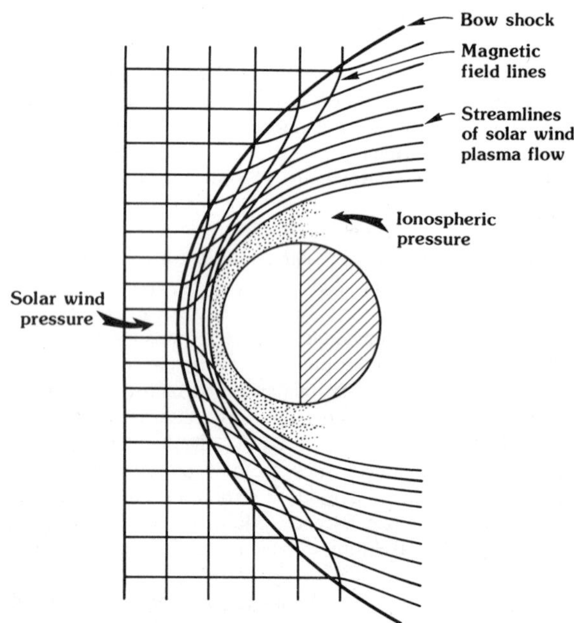

Figure 1. Schematic illustration of the solar wind interaction with Venus, showing the essential features. At the ionopause, the incident solar wind pressure is balanced by the pressure of the ionospheric plasma. The solar wind flows from the left parallel to the streamlines. The magnetic field is carried along with the flow and draped over the ionosphere.

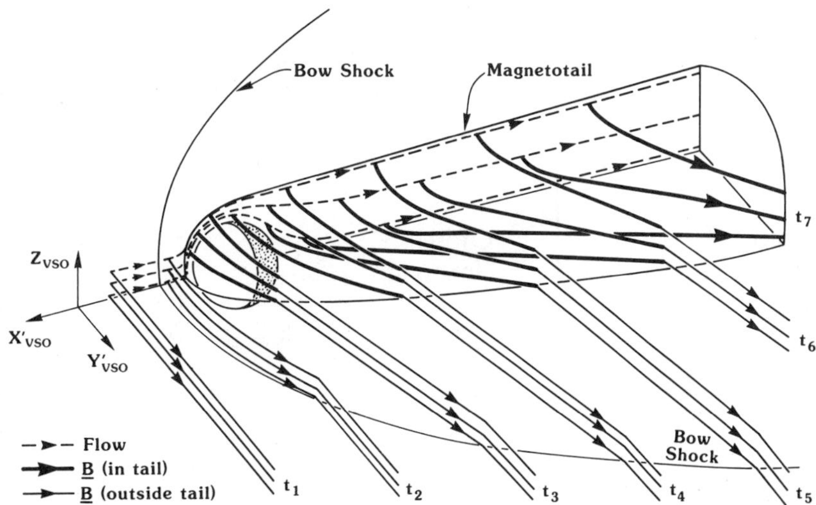

Figure 2. Illustration of the draping of the interplanetary magnetic field, which is initially embedded in the solar wind, around the Venus obstacle. The field sinks into the wake behind the planet because it is loaded down with heavy ionospheric ions. This creates a "magnetotail" with fields pointing toward or away from the Sun in accord with the interplanetary field orientation.

velocity of compressional waves in the solar wind is lower than the velocity of the solar wind relative to Venus. Thus a shock front has to form in front of Venus to slow the solar wind so that it can be deflected around the ionospheric obstacle. This shock is called a bow shock. All of the other planets also appear to have associated bow shocks. The location of the venusian bow shock varies with the solar cycle as the upper atmosphere increases and decreases in density. Figure 1 shows a sketch of this interaction.

ION PICKUP AND THE VENUS MAGNETOTAIL

As the solar wind flows past Venus, the plasma that passes closest to the planet moves more slowly than the material that passes further away. The magnetic field lines that are carried in the flow therefore become draped around the planet in a tail-like fashion as sketched in Fig. 2. Hence the name "magnetotail" is adopted for this structure, although it has a different meaning than in the case of the Earth. The amount of stretching depends on how strongly the flow near the planet is slowed. The neutral hot oxygen exosphere above the ionopause provides additional slowing when it becomes ionized. When ions are produced in the flowing solar wind plasma, they are "picked up" in the flow. Because momentum must be conserved, the additional ions from the exosphere add mass and so must slow the flow. The hot oxygen exosphere thus plays a role in determining the field draping in the Venus tail.

MAGNETIC FIELDS IN THE DAY-SIDE IONOSPHERE

Although the ionosphere usually has sufficient thermal pressure to stand off the solar wind, a magnetic field from the magnetic barrier still manages to penetrate the ionosphere in two ways. First, it diffuses into the ionosphere; second, it convects into the ionosphere in tubes.

The tubes would not be expected to sink in the ionosphere because they are buoyant. However, as these tubes approach the ionosphere, they pick up heavy ions from the photoionization of the oxygen exosphere and sink. In the process of sinking, they become twisted to resemble the helical magnetic structures or flux ropes often seen on the Sun. Venus, however, provides the opportunity to study measurements made inside flux ropes. Measurements by the *Pioneer Venus Orbiter* magnetometer show that these ropes can have field lines in the configuration illustrated in Fig. 3. The magnetic field is strong and parallel to the axis of the rope in the center of the rope but winds around the axis further out. As the distance from the center of the rope increases, the field weakens. A small number of Venus ionospheric ropes assume an almost "force-free" structure with the inward forces associated with the twist of the magnetic field balancing the outward forces associated with the magnetic pressure gradient.

Figure 3. Configuration of field lines in a "flux rope" in the Venus ionosphere as suggested by the *Pioneer Venus Orbiter* observations. The fields are straight and are strongest along the axis.

Another type of magnetic field that appears in the day-side ionosphere is a large-scale horizontal field that is oriented roughly in the direction of the overlying magnetic barrier field. Figure 4 compares the appearance of the large-scale fields with the flux rope fields as seen in altitude profiles obtained by the *Pioneer Venus Orbiter*. The large-scale field diffuses downward through the ionopause boundary and then is convected by a downward drift of the ionospheric plasma to low altitudes where it decays. The decay of the field is caused by collisions of ionized particles and neutral particles (which are numerous at low altitudes), because the collisions reduce the currents associated with ion motion relative to the electrons. Both frictional interactions and recombination damp out the currents, and, without ionospheric currents, there is no ionospheric magnetic field near the bottom of the ionosphere (below ~140 km). The strength of the large-scale field depends on the altitude of the ionopause. It is larger when the ionopause is low, because the more frequent collisions at the ionopause expedite the transport of the field across that boundary. The vertical drift also becomes faster at lower altitudes. Its highest value of 60 m s^{-1} near 190 km in fact produces the distinctive shape of the altitude profiles of the field in Fig. 4.

MAGNETIC FIELDS IN THE NIGHT IONOSPHERE

Above about 200 km we may consider the ionosphere to be produced on the day side and lost at night. In this altitude range

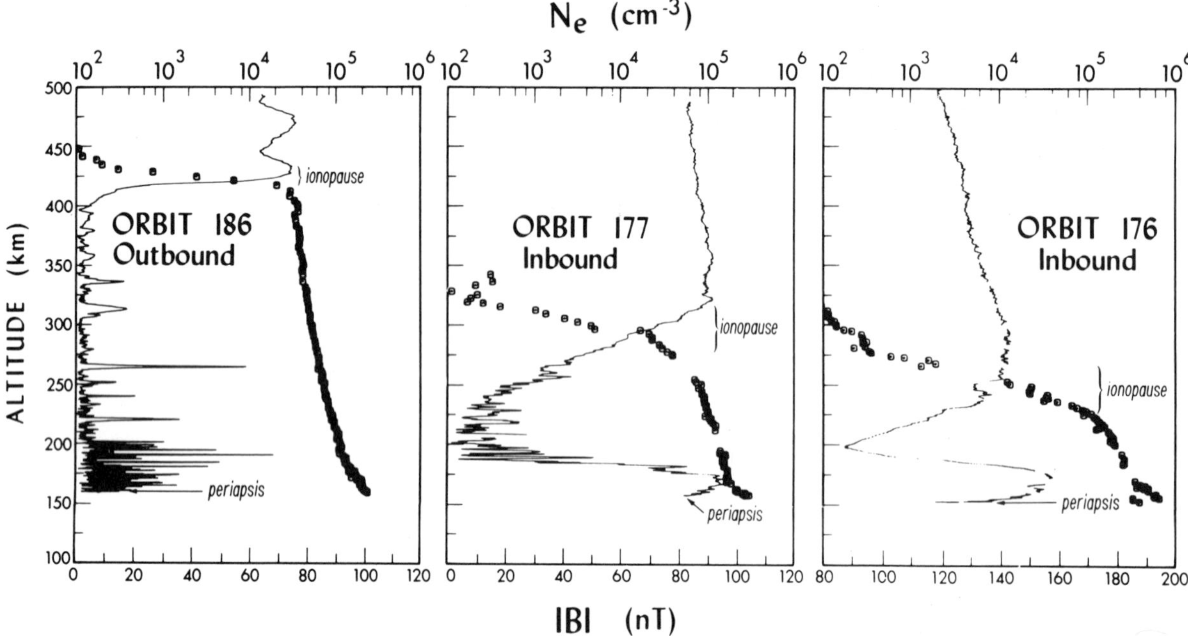

Figure 4. Examples of altitude profiles of magnetic field magnitude (lines) and electron density (points) observed in the day-side Venus ionosphere by instruments on the *Pioneer Venus Orbiter*. The example on the left shows the bottom of the magnetic barrier and the ionopause at the top, and flux ropes within the ionospheric plasma. The other two examples illustrate how flux ropes disappear and large-scale fields appear when the height of the ionopause decreases in response to increasing solar wind pressure.

there is a strong flow from day to night which carries to the night side any and all magnetic fields that managed to penetrate the highly conducting day-side ionosphere. Thus in some senses the night ionosphere is Venus' wastebasket of magnetic fields. Nevertheless, we see some order here. There are regions of strong magnetic field accompanied by low plasma density. These have been called ionosopheric holes. Surrounding these holes are regions of weak magnetic field accompanied by high plasma densities. The two regions are in pressure equilibrium so that the sum of the magnetic and thermal pressures in each are equal. When the dynamic pressure of the solar wind is high, the magnetic field in the night ionosphere is strong and horizontal.

THE INTRINSIC MAGNETIC FIELD

Because the effect of the solar wind interaction is to compress the magnetic field on the day side and stretch it out on the night side, we might expect that any weak planetary magnetic field would be more easily observed on the night side of the planet. Moreover, the ionospheric pressure is much weaker on the night side, also aiding the detection of any intrinsic magnetic field. As mentioned previously we might expect the magnetic moment of Venus to be about 2.5×10^{13} Tm^3 or about 0.3% of the terrestrial magnetic field. However, a careful search of the observations of the *Pioneer Venus Orbiter* at low altitudes reveals no evidence for a magnetic field of moment above 8×10^{10} Tm^3 which is 10^{-5} of the terrestrial magnetic field and 0.3% of the expected magnetic field strength. This observation suggests that Venus actually has no internal dynamo action at all. It probably has a conducting liquid core, but it is possible that the process that drives the dynamo in the center of the Earth, which is believed to be the solidification of the inner core, has not yet begun on Venus. Mars may be in a similar state for a different reason. The expected martian moment of 6×10^{14} Tm^3 is also not observed. However, Mars, being much smaller than the Earth and Venus, may have cooled off faster so that there solidification of the inner core is complete. Mars may have had an ancient magnetic dynamo leaving remanent magnetization on the surface, whereas Venus' interior never began dynamo action. We will never know if this latter speculation is true because the surface of Venus is too hot to preserve a remanent magnetic record. In short, Venus, Earth, and Mars may represent three stages of a planetary dynamo. At Venus, dynamo activity has not yet begun but may begin later; on Earth, the dynamo is going strong, whereas at Mars, the dynamo has ceased to function.

Additional Reading

Keating, G. M., ed. (1990). *The Venus Atmosphere*. In *Advances in Space Research*, **10** (No. 5).

Luhmann, J. G. (1986). The solar wind interaction with Venus. *Space Sci. Rev.* **44** 241.

Russell, C. T. (1987). Planetary magnetism. In *Geomagnetism*, Vol. 2, p. 457. Academic, New York.

See also **Earth, Magnetosphere and Magnetotail; Mercury and Venus, Space Missions; Mercury, Magnetosphere; Planetary Atmospheres, Solar Activity Effects; Venus, Atmosphere.**

Virgo Cluster

Bruno Binggeli

The Virgo Cluster is the closest and best-studied great cluster of galaxies, lying at a distance of approximately 20 Mpc in the constellation of Virgo. Cosmographically, the Virgo Cluster is the nucleus of the Local Supercluster of galaxies, in whose outskirts we (in the Milky Way, in the Local Group) are situated. As early as 1784, Charles Messier noted an unusual concentration of "nebulae" in Virgo; 15 out of the 109 "Messier" objects are, in fact, Virgo Cluster galaxies, the most famous of which is Messier 87,

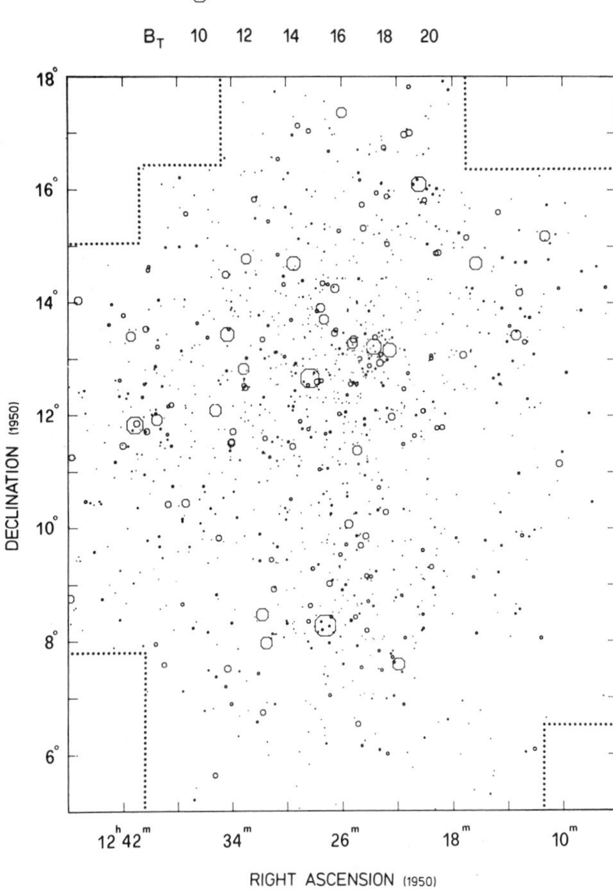

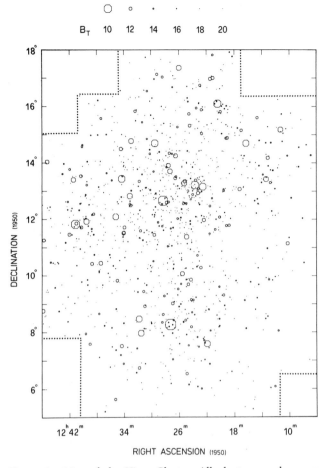

Figure 1. Map of the Virgo Cluster. All cluster members are plotted with luminosity-weighted symbols. The symbol size (area) is proportional to the luminosity of the galaxy. The magnitude scale (blue total apparent magnitudes) is given at the top of the figure. This map should be a fair representation of how the cluster appears in the sky. The two brightest galaxies, at right ascension approximately equal to $12^{\mathrm{h}}28^{\mathrm{m}}$ and declination approximately equal to 12°40′ and right ascension approximately equal to $12^{\mathrm{h}}27^{\mathrm{m}}$ and declination approximately equal to 8°17′, are M87 and M49, respectively. [*Reproduced by permission from Binggeli, Tammann, and Sandage (1987),* Astron. J. ***94*** *251.*]

the giant elliptical galaxy with the mysterious jet. After Edwin P. Hubble's 1923 discovery of Cepheids in M31, the true nature of the group of nebulae in Virgo as a self-gravitating system of hundreds of galaxies was soon realized, and the first systematic investigations of the Virgo Cluster (as it was subsequently called) were carried out by Harlow Shapley and Adelaide Ames. Ever since, the Virgo Cluster has been, and still is, of primary importance for extragalactic astronomy: Large numbers of equidistant galaxies of all types and luminosities can be observed here in great detail, rendering the cluster: (1) an ideal laboratory for the study of the systematic properties of galaxies and (2) a fundamental stepping stone for the cosmological distance scale.

MORPHOLOGY

The Virgo Cluster is a fairly poor, loosely concentrated, irregularly shaped (see Fig. 1) cluster of galaxies with a high abundance of spiral galaxies among the bright cluster members (see Table 1). It is representative of the most common class of galaxy clusters, which is characterized by these properties. Rich, dense, regularly

Table 1. Known Member Galaxies of the Virgo Cluster

Morphological Type	*Number*
Elliptical	30
S0	49
Spiral	128
Dwarf elliptical	828
Dwarf S0	30
Dwarf irregular	89
Dwarf irregular/elliptical	89
Other	34
Total	1277

shaped clusters of predominantly E and S0 galaxies are much rarer. Nevertheless, owing to its proximity, the "mediocre" Virgo Cluster could be mapped to an unsurpassed level of depth and morphological detail, rendering it presently the richest cluster of galaxies in terms of the number of known member galaxies. As Table 1 shows, dwarf galaxies, the dwarf elliptical (dE) types in particular, numerically dominate the cluster population. These stellar systems of low surface brightness are hard to detect, even in nearby Virgo. There must be thousands more extremely faint and diffuse cluster members that are still awaiting discovery.

The distribution of the presently known Virgo Cluster members is shown in Fig. 1. The cluster covers a large, roughly circular sky area of approximately 10° diameter. Several subconcentrations can be distinguished. There is a major subcluster (A) of galaxies around the giant E galaxy M87, centered on right ascension $\simeq 12^{\mathrm{h}}25^{\mathrm{m}}$ and declination $\simeq 13°$; there is a smaller, less dense subcluster (B) around the brightest cluster member M49, centered on right ascension $\simeq 12^{\mathrm{h}}27^{\mathrm{m}}$ and declination $\simeq$ 8°30′. A third, barely significant subclump (C) has been identified around M59, at right ascension approximately equal to $12^{\mathrm{h}}40^{\mathrm{m}}$ and declination approximately equal to 12°. Although M87 is most often taken as the center of the Virgo Cluster, it is off the center (density peak) of A by approximately 1° in the direction toward C. With respect to morphological type, the elliptical and S0 member galaxies are the most strongly clustered species; they constitute the "skeleton" of the cluster. The E types, in particular, are distributed preferentially (almost chain-like) along the axis A–C. Remarkably, even the jet of M87 is aligned with this fundamental cluster axis. Spiral and (dwarf) irregular galaxies, on the other hand, are scattered over the whole face of the cluster, almost without noticeable concentration.

DYNAMICS

As its irregular structure suggests, the Virgo Cluster is not in a state of dynamical equilibrium—not even in the central region, which is more surprising. There is evidence that the cluster is still in the making.

From the presently known radial velocities (redshifts) of about 350, mostly bright Virgo members, one derives a mean heliocentric, systemic velocity of the cluster of $\langle v_{\odot} \rangle \simeq 1100$ km s^{-1}. Although this mean is invariant, the velocity *distribution* differs substantially for different galaxy types. Late-type (spiral and irregular) galaxies have a broad velocity distribution with a dispersion (standard deviation from $\langle v \rangle$) of $\sigma_v \simeq 900$ km s^{-1}, whereas early-type (E, S0, dE, dS0) galaxies show a narrow distribution with $\sigma_v \simeq 550$ km s^{-1}. The late types are thus more dispersed, not only in space, but also in velocity. This has been taken as evidence that spiral and irregular galaxies have only recently (in the last few 10^9 yr) fallen, or are still in the process of falling, into the cluster from the environment: These galaxies are not yet settled down in the cluster ("dynamically relaxed") but are streaming inward and outward in the manner of a damped oscillation. Such an infall scenario is plausible, as the Local Supercluster is indeed made up of large "clouds" of spiral and irregular galaxies: One such cloud seems to be falling into the Virgo cluster at this very epoch. Likewise, the southern subcluster *B* may be falling into the main subcluster *A*.

The well-concentrated early-type galaxies of subcluster *A* must then be viewed as the oldest cluster members that formed in the densest part(s) of the cluster or fell into it very early on. However, these galaxies do not constitute a dynamically relaxed cluster core, as one would expect. Rather, the central part of the Virgo Cluster seems to consist of a small number of subclumps of galaxies, one of which is defined by M87 alone. In spite of its enormous mass of approximately 5×10^{13} $M_{\odot}$, which is indicated by its large, x-ray emitting halo of hot gas, this giant galaxy is off the cluster center in space *and* velocity ($\Delta v \approx 200$ km s^{-1}). However, as a result of "dynamical friction," the subclumps will rapidly merge. We may, in fact, be living in a very special time, shortly ($\approx 10^9$ yr) before the final formation of a relaxed cluster core in Virgo.

This is exciting but it also complicates the dynamical modeling of the Virgo Cluster. The virial theorem can no longer be applied to derive a cluster mass. Nevertheless, requiring simply that the cluster be gravitationally bound (total energy equal to 0), one gets $M_{tot} \geq 5\times10^{14}$ $M_{\odot}$, and a mass-to-light ratio of $M/L \gtrsim 450$ in solar units—which clearly indicates the dominance of dark matter.

A PROBE FOR COSMOLOGY

Bright cluster members have traditionally been used to derive the Hubble constant (expansion rate of the universe), H_0. In fact, almost all determinations of H_0 are based on the Virgo Cluster, because it is the center of a large velocity perturbation pattern that embraces the whole supercluster, including us. The velocity of the Virgo Cluster, if referred to the centroid of the Local Group (removing the motion of the Sun in the Milky Way, removing the motion of the Galaxy in the Local Group), is $\langle v_{LG}\rangle \approx 1000$ km s^{-1}; if referred to the Sun, it is $\langle v_{\odot}\rangle \approx 1100$ km s^{-1}. The Local Group is falling toward (but will not fall into!) the Virgo Cluster with 200–300 km s^{-1} (the value is debated), so the true, cosmic expansion velocity of the cluster is $\langle v_{cos}\rangle \approx 1200$–$1300$ km s^{-1}. As the distance estimates for the Virgo Cluster range from 15–22 Mpc, one arrives at a value of the Hubble constant (H_0) between 50–100 km s^{-1} Mpc^{-1}. Thus the present uncertainty in H_0 is essentially the difficulty in pinning down the distance to the Virgo Cluster.

Once this important problem is solved, attention is likely to shift back to the cluster as such, and the freed energy may be used to exploit this great galaxy mine for the sake of a better understanding of the formation and evolution of structure in the universe, rendering the Virgo Cluster a true probe for cosmology.

Additional Reading

Binggeli, B., Tammann, G. A., and Sandage, A. (1987). Studies of the Virgo Cluster. VI. Morphology and kinematics of the Virgo Cluster. *Astron. J.* **94** 251.

Jacoby, G. H., Ciardullo, R., Ford, H. C. (1990). Planetary nebulae as standard candles. V. The distance to the Virgo Cluster. *Ap. J.* **356** 332.

Richter, O.-G., and Binggeli, B., eds. (1985). *The Virgo Cluster, ESO Conference and Workshop Proceedings 20*. European Southern Observatory, Garching.

Sarazin, C. L. (1988). *X-Ray Emission from Clusters of Galaxies*. Cambridge University Press, Cambridge.

Tully, R. B. and Fisher, J. R. (1987). *Nearby Galaxies Atlas*. Cambridge University Press, Cambridge.

See also **Clusters of Galaxies, all entries; Galaxies, Local Group; Galaxies, Local Supercluster.**

Voids, Extragalactic

Valérie de Lapparent-Gurriet

Part of the progress in our understanding of the distribution of galaxies outside the Milky Way was achieved by examining galaxy catalogs that map the angular position on the sky of galaxies brighter than a chosen magnitude limit (see Fig. 1). These "projected"—onto the celestial sphere—or "two-dimensional" catalogs showed the inhomogeneity of the light-emitting component of the universe and led to the discovery of concentrations of galaxies on scales from approximately 1–100 Mpc (1 Mpc $\approx 3\times10^6$ ly), corresponding to clusters and superclusters.

Parallel to the acquisition of larger galaxy catalogs, the application of the Hubble law allowed recovery of the third coordinate, the distance of the galaxies from the Milky Way. By measuring the recessional velocities of the galaxies from their redshifts over selected regions of projected catalogs, it was then possible to obtain "three-dimensional" maps of the galaxy distribution, also called redshift surveys. These new maps confirmed that most superclusters in projected catalogs were real superclusters in three-dimensional space and also indicated that the superclusters are associated with large voids. These "extragalactic" voids are defined by regions of supercluster size that are underdense compared to the surroundings and are frequently devoid of any galaxies brighter than the catalog limit.

OBSERVATIONS

Most of the early redshift surveys contained empty regions of about 10–20 Mpc. Considering the low signal-to-noise ratio in the delineation of the structures, some of these voids could be statistical accidents. However, the discovery of a strikingly large void in a redshift survey toward the Boötes constellation suggested that voids might be real structures. The Boötes void is enclosed in a sphere with a diameter of 60 Mpc (for a Hubble constant of 100 km s^{-1} Mpc^{-1}; we use this value hereafter) and centered at 150 Mpc from the Milky Way. Given the clustering properties of galaxies and the characteristics of the sample in which the Boötes void was discovered, the probability that such a large void could be a chance fluctuation is very small (about one chance in a million).

Results from a recent redshift survey suggest a new picture of the galaxy distribution in which structures like the Boötes void are common features. This new survey has the configuration of a "slice": the galaxy catalog is a thin strip of the sky of 6° in declination by 117° in right ascension, which lies in Fig. 1 between the heavy tick marks and contains 1059 galaxies. The "pie diagram" of Fig. 2 plots the recessional velocity in kilometers per second of all the galaxies in the strip (marked as dots) as a function of right ascension (a galaxy with a velocity of 10,000 km s^{-1} is at a distance of 100 Mpc from the Milky Way, located at the apex of the cone). The declination coordinate is suppressed because of the narrow angle covered. Because the catalog is limited by a threshold in apparent magnitude (blue magnitude brighter than 15.5), the density of galaxies decreases at large velocity where only the intrinsically brightest galaxies are detected. This decrease in density defines a characteristic depth for the catalog of approximately 100 Mpc.

The concentration of points in the center of the map in Fig. 2 corresponds to the Coma cluster of galaxies (see Fig. 1). The peculiar velocities of the galaxies in the gravitational potential of the cluster cause the elongation along the line of sight, the "finger-of-god" effect. Outside clusters, the distortions due to the peculiar velocities are small (a few 100 km s^{-1}), and the map in velocity approximates the map in real space. Figure 2 suggests that the galaxy distribution is dominated by large voids with diameters ranging from 20–50 Mpc and delineated by sharp linear structures. The picture in which galaxies are distributed in thin sheet-like or

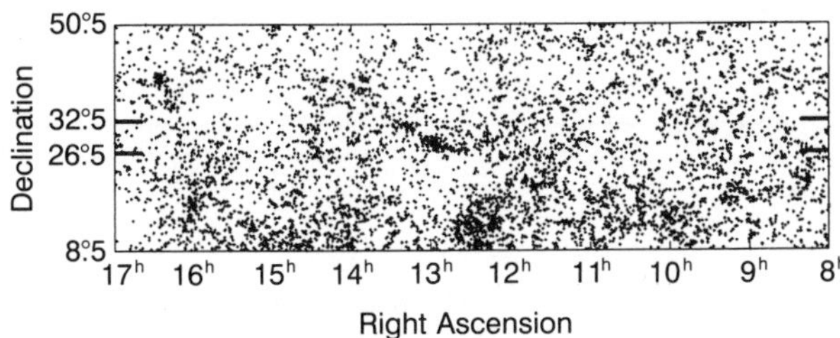

Figure 1. Distribution on the sky of the 7031 galaxies brighter than the apparent blue magnitude of 15.5 in a selected region of the catalog by Zwicky, Herzog, Wild, Karpowicz, and Kowal. The strip of Fig. 2 is indicated by the heavy tick marks.

shell-like structures surrounding vast voids provides a simple interpretation for this cellular network. The comparatively smooth distribution of galaxies in projection onto the sky (see Fig. 1) supports this interpretation. Moreover, additional data for adjacent slices to the one in Fig. 2 do show that the voids and sheets extend in declination across more than 30 Mpc.

The alternation of superclusters with voids in most previous catalogs had already suggested a cellular pattern for the galaxy distribution, but the distribution was not sampled densely enough for clear delineation of the sheets. For example, in pie diagrams of the Boötes void, the edges of the void are poorly defined because redshifts were measured for only 2% of the galaxies in the region and the separation between the galaxies in the sheets is of the order of the size of the void. In the map of Fig. 2, the separation between the galaxies is much smaller than the size of the voids. If these sheet-like structures are frequent, the filament-like superclusters observed in other surveys might be portions of shells intersected by the survey. In addition, all superclusters are connected, and they are geometrical arrangements rather than dynamical systems. The real physical units could be the voids: A smaller amount of energy would be required to move matter across the radii of the voids than across the longer span of the connected shells.

IMPLICATIONS OF VOIDS AND SHEETS

The size and the geometry of the voids and sheets challenge the measurement of the average properties of the galaxy distribution. Because the size of the largest voids in Fig. 2 is comparable with the depth of the map, this survey is *not* a fair representation of the general galaxy distribution, and the mean density of galaxies is poorly determined. Under reasonable assumptions on the spectrum of void diameters, the fluctuations in the mean number of galaxies are determined by the number of the largest voids in the sample. Therefore, a better determination of the mean matter density associated with galaxies requires the completion of redshift surveys extending to larger velocities and thus containing more of the largest voids.

Because the largest void that can be detected in a survey is limited by the depth of the survey, voids with diameters larger than the present upper limit of 50 Mpc could also be discovered in deeper redshift surveys. Such surveys, requiring use of the largest existing ground-based telescopes, would provide better observational limits on the scale above which the visible universe might become homogeneous. In addition, deeper surveys might reveal a yet undetected population of very faint objects inside the observed

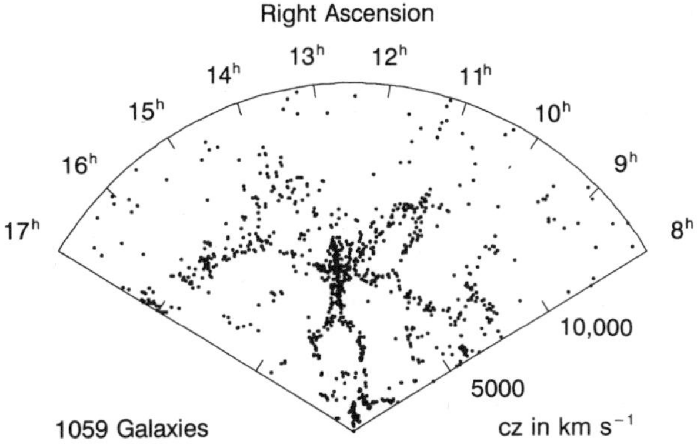

Figure 2. Distribution of 1059 galaxies in a 6° wide slice of the universe, as a function of angle across the sky and distance from the Milky Way.

voids. So far, other types of galaxies that might not have been included in the *Catalogue of Galaxies and of Clusters of Galaxies* by Fritz Zwicky and his collaborators (i.e., low-surface-brightness, emission-line, or IRAS galaxies) trace the same large-scale structures as in Fig. 2.

The strong asymmetry and contrast of the sheets delineating the voids require using statistical measures of the galaxy distribution that make no assumptions of local spherical symmetry and are poorly sensitive on the mean density. Simple statistics satisfying these requirements were applied to the map of Fig. 2 and put tight constraints on the theoretical models for the formation of large-scale structure: The fraction of the total volume occupied by galaxies (20%); the thickness of the sheets (5 Mpc); and the shape and the radius of the voids. More sophisticated statistics like the probability of finding a void of a given volume or the radius of curvature of the sheets constrain the general topology of the distribution. Some of these statistics are useful for discriminating between a bubble-like topology, where all the voids are isolated from each other by shells, and a sponge-like topology, where the high- and low-density regions are equivalent and form two interwoven networks, and further constrain the theoretical models.

Additional Reading

Burns, J. O. (1986). Very large structures in the universe. *Scientific American* **255** (No. 1) 38.

Geller, M. J. and Huchra, J. P. (1989). Mapping the universe. *Science* **246** 897.

Hubble, E. (1936). *The Realm of the Nebulae*. Yale University Press, New Haven.

Oort, J. H. (1983). Superclusters. *Ann. Rev. Astron. Ap.* **21** 373.

Peebles, P. J. E. (1980). *The Large-Scale Structure of the Universe*. Princeton University Press, Princeton.

Zeldovich, Ya. B., Einasto, J., and Shandarin, S. F. (1982). Giant voids in the universe. *Nature* **300** 407.

Zwicky, F., Herzog, E., Wild, P., Karpowicz, M., and Kowal, C. T. (1961–1968). *Catalogue of Galaxies and of Clusters of Galaxies*. Vols. 1–6. California Institute of Technology, Pasadena.

See also **Cosmology, Clustering and Superclustering; Superclusters, Dynamics and Models; Superclusters, Observed Properties.**

X-Ray Astronomy, Space Missions

Hale Bradt, Takaya Ohashi, and Kenneth A. Pounds

Observations of celestial x-ray sources must be carried out above most of the Earth's atmosphere because it is highly opaque to x-rays. Here we briefly describe the principal missions in celestial (nonsolar) x-ray astronomy and selected significant advances from each. Table 1 summarizes the orbiting missions that were launched through 1990.

EARLY YEARS (1962–1969)

The post–World War II availability of sounding rockets that reached altitudes of approximately 100 km gave astronomers their first glimpse of x-rays from the Sun in 1949 and of a more distant celestial x-ray source in 1962. The standard rocket detector was the proportional counter sensitive to relatively low-energy x-rays, typically 1–10 keV. Experiments with ultrathin plastic windows allowed studies of x-rays at about 0.25 keV. The spectral resolution was crude (~ 20% at 6 keV), as was the angular resolution which was provided by mechanical collimators, for example, bundles of tubes to define a circular field of view of about 1°. The use of "modulation collimators" (MCs) provided angular resolution of approximately 1 arcminute. (A MC consists of grids of parallel wires.) The timing resolution was excellent, better than milliseconds, but this capability was not often required. A flight lasted only a few minutes, and payloads were typically recovered via parachute for future use.

Balloons made flights of about 10 hours at altitudes of approximately 40 km. They could carry heavier payloads of larger aperture, typically crystal scintillator detectors. At these altitudes, only x-rays exceeding approximately 20 keV in energy could penetrate the residual atmosphere and reach the detectors. These higher-energy x-rays are emitted from a given source much less frequently than are lower-energy x-rays, which precludes their detection in the short period of a rocket flight. The larger detectors and longer exposures possible with balloons yielded important studies in the 20–60-keV regions, nicely complementing the lower-energy rocket x-ray studies.

The potential of x-ray astronomy was revealed in this period despite the relative paucity of observing time. Major discoveries included: (1) the discovery of Sco X-1 (the brightest x-ray source) and its identification as a blue star with an unusual emission-line spectrum; (2) the detection of a diffuse background of x-rays; (3) the detection of x-ray emission from the Crab nebula supernova remnant, and pulsed x-ray emission from the pulsar within it; (4) the discovery of a "transient" source, Cen X-2, that flared to an intensity comparable to that of Sco X-1; (5) the x-ray detection of the galaxy M87 in the Virgo cluster; and (6) the detection of an approximately 20-min flare from Sco X-1.

SMALL SATELLITES (1967–1977)

Uhuru

The first Earth-orbiting mission dedicated to celestial x-ray astronomy was the *Uhuru* satellite launched in 1970. Its detector complement was simple but powerful: a large area (0.084 m^2) of proportional counters sensitive to 2–20-keV x-rays. The satellite was in a low equatorial Earth orbit (to minimize background due to particles trapped in the radiation belts) and rotated about an axis perpendicular to the field of view, with a period typically of 12 min. It provided the first comprehensive view of the entire x-ray sky, to about 10^{-3} the intensity of the Crab nebula, with much higher sensitivity and uniformity of sky coverage than previously possible. The *Uhuru* x-ray catalog contains 339 objects which, for the most part, are binary stellar systems, supernova remnants, Seyfert galaxies, and clusters of galaxies. A map of these sources shows dramatically the clustering of galactic sources on the galactic plane and the isotropy of numerous, and heretofore unexpected, extragalactic x-ray sources.

Three *Uhuru* discoveries gave penetrating insight into the underlying nature of many x-ray sources.

1. The discovery of two eclipsing x-ray pulsars, Her X-1 and Cen X-3, with Doppler-shifting periods, demonstrated the existence of a class of binary x-ray sources consisting of a normal star and a spinning, magnetized neutron star. (Her X-1 also exhibited a 35-day period due to the precessing accretion disk.) The power source for the x-ray emission is the gravitational energy released as gas from the normal star accretes onto the neutron star. The spin (typical period about 1 s) of the magnetized neutron star results in seemingly pulsing x-rays. This, the second manifestation of neutron stars, after radio pulsars, revealed a whole new domain of binary system evolution.
2. Diffuse x-ray emission from several clusters of galaxies suggested the presence of a dilute hot gas interspersed among the galaxies. This interpretation was later verified with *Ariel V* and *OSO-8* (see the following discussion). This led to important evolutionary studies of clusters.
3. The detection of rapid aperiodic temporal variability of the intensity of Cyg X-1 led to rocket and ground-based observations that identified the optical counterpart. Measures of its optical mass function suggested that the source was a binary system containing a stellar black hole. *Uhuru* also demonstrated that variability was a common phenomenon among x-ray sources.

These discoveries and numerous other pioneering studies by *Uhuru* brought x-ray astronomy solidly into the mainstream of astronomy.

Nondedicated Missions

Beginning in 1967 (prior to *Uhuru*), x-ray experiments were flown on board a series of satellites that were not actually dedicated to x-ray astronomy, the *Orbiting Solar Observatories* (*OSO*s). *OSO-3* and *OSO-5* each carried a scintillator experiment for the study of the hard-x-ray diffuse background.

These instruments resided in the rotating "wheel" section of the satellite and thus scanned along great circles of the sky that covered the entire sky in six months of satellite operations. Spectra from 8 to 200 keV were obtained, and the isotropy of the background was established. *OSO-7* (launched 1971) carried two celestial x-ray scanning experiments: a system of multiple proportional counters with wide energy response (1–60 keV) and a high-energy ($\gtrsim$ 10 keV) crystal scintillator experiment. These instruments also scanned the entire sky from the wheel section of the satellite. A "multicolor" catalog of 184 sources and the discovery of rapid intensity variability (60% in 6 days) of x-rays from the nucleus of the galaxy Centaurus A came from this mission.

Table 1. Orbiting Missions with Celestial X-Ray Astronomy Capability through 1990

Mission	*Dates*	*X-Ray Instruments*	*Features (Spin / Point)*	*Science (Sample)*
OSO-3	1967–1968	H	S	Diffuse background
OSO-5	1969–1972	H	S	Diffuse background
Vela series	1969–1979	M	Four satellites	X-ray bursts
Uhuru	1970–1973	M	S, survey	Binaries, clusters
OSO-7	1971–1973	M, H	S, survey	Four-color catalog
Copernicus	1972–1981	M, C	P, primarily UV	SNR mapping
ANS	1974–1976	L, M, C, B	P	X-ray bursts
Ariel V	1974–1980	M, B, Pol, SM, MC	S, survey	AGN, transients
SAS-3	1975–1979	L, M, MC, C	S, P	Bursts/positions
OSO-8	1975–1978	L, M, H, B, Pol	S, low background	Fe-line emission
HEAO-1	1977–1979	L, M, H, MC	S, P, large area	All-sky catalog
Einstein (*HEAO-2*)	1978–1981	F, IPC, HRI B, SS, M, TG	S, x-ray Imaging	Clusters, QSOs, SNR
Ariel VI	1979–1981	L, M, C	Pri. cosmic ray	Spectra, timing
Hakucho	1979–1984	M, MC, SM	S	Bursts
Tenma	1983–1984	L, M, G, C, SM, MC	S, good Spectral resolution	Fe lines
EXOSAT	1983–1986	L, M, G, F, TG	P, 4-day orbit	QPOs
Ginga	1987	M, SM	P, large area	SN1987A, AGN
Röntgen/Kvant	1987–	M, H, CM, G	Space station *MIR*	SN 1987A
Granat	1989	M, H, CM	P, 4-day orbit	Timing, spectra
ROSAT	1990–	F, HRI, IPO	P	Survey, imaging
Astro-1	1990	F,SD	P, 8-day mission	Spectra

B, Bragg crystal spectroscopy; C, collector (reflecting); CM, coded-mask imaging; F, focusing optics; G, gas scintillation proportional counter; H, high-energy, (10–200 keV) crystal scintillator; HRI, high resolution imager; IPC, imaging proportional counter; L, low-energy (0.1–1 keV) proportional counter; M, medium energy (1–20 keV) proportional counter; MC modulation collimator; Pol, polarimetry; SM, sky monitor; SD, solid-state detector; SS, solid-state spectrometer; TG transmission gratings; P, pointed (= three axis stabilized); S, spinners (may carry instruments pointed along spin axis).

The third *Orbiting Astronomical Observatory* (*Copernicus*, 1972) was stabilized about three axes. It carried a cluster of relatively small proportional counters with grazing incidence reflecting collectors and was provided by a British group, beginning a pattern of international collaboration in x-ray astronomy. As a pointed instrument, it could obtain a much greater exposure to a given source than could a scanning instrument of the same aperture. Images were obtained via raster scans of the supernova remnants Puppis A and (a part of) the Cygnus loop with approximately 10-arcminute resolution at about 1 keV. A change in the x-ray intensity of the galaxy Centaurus A was first reported by *Copernicus* observers.

The classified U.S. *Vela* satellites carried x-ray detectors during the entire decade of the 1970s. They were not intended primarily for astronomical studies but did provide much useful data. *Vela 5A* and *Vela 5B* were launched in 1969, and *Vela 6A* and *Vela 6B* were launched in 1970. Each operated for about 1 year except *Vela 5B* which provided useful data until mid-1979. Major accomplishments included: (1) the detection of the bright transient source Cen X-4, monitored through its rise and decay, (2) the codiscovery of x-ray bursts in 1976, and (3) the discovery of a 300-day periodicity in Cyg X-1.

In the mid-1970s, two other shared missions with substantial capabilities were launched, *ANS* and *OSO-8*. They flew in parallel with two dedicated satellites which will also be discussed. The last half of the decade was very productive for x-ray astronomy.

The *Astronomical Netherlands Satellite* (*ANS*), launched in 1974, carried reflecting x-ray collectors and 2–40-keV proportional counters, one of which was used with a Bragg crystal spectrometer. This system was provided by a U.S. group. Two major results were the codiscovery of x-ray bursts (also seen with *Vela*) and the detection of the first x-ray flare observed from a "normal" (not compact) star other than the Sun. The bursts from a source in the globular cluster NGC 6624 were a surprising new phenomenon due to thermonuclear explosions on the surface of a neutron star.

The *OSO-8* satellite (1975) carried a complement of celestial x-ray experiments, both scanning and pointed. These included proportional counters with extremely low noise, broad energy response, and good energy resolution. The iron-line emission detected from several clusters of galaxies showed that thermal emission from a hot gas is an intrinsic property of clusters. (The first such detection was by *Ariel V*; see below.) It was also demonstrated that x-ray bursts exhibit a blackbody spectrum. That was a crucial element in our understanding that the bursts are thermonuclear flashes. This mission also carried a soft x-ray survey experiment to study the diffuse background at approximately 0.2 keV, a scintillator x-ray experiment, a Bragg crystal spectrometer, and a polarimeter. The scintillator detected a hard x-ray source near the galactic center, and the polarimeter placed the first meaningful upper limits on polarization in celestial x-ray sources (e.g., $< 0.6\%$ from Sco X-1).

The manned *Skylab* mission in 1973 carried two grazing incidence telescopes that focused an image of the Sun onto film at the focal plane. The spectacular and often-reproduced images of the Sun in x-rays demonstrated that x-ray astronomy could enter the field of imaging astronomy.

More Rockets and Balloons

Experiments with rockets and balloons continued in the 1970s with emphasis on objectives that could not be carried out with existing satellite instrumentation. Notable results included the following:

1. The detection of polarization of the x-ray flux from the Crab nebula, which showed that the x-ray emission arises from synchrotron emission.
2. The celestial position of an x-ray source, GX3 + 1, was measured with great precision, to 0.3 arcsecond, with a lunar occultation experiment.

3. X-ray emission from a white dwarf star in a binary system (SS Cygni) was found for the first time.
4. Intense flares of millisecond duration were discovered from the black hole candidate Cyg X-1.
5. Thermally broadened iron-line emission was discovered in the x-ray emission from the supernova remnant Cas A and from the x-ray binary Cyg X-3.
6. Extremely soft x-ray emission from an isolated white dwarf star, HZ 43, was detected with a rocket (and with *SAS-3*).
7. The extended x-ray emission from the galaxy M87 was studied with a two-dimensional imaging system.
8. An emission feature at 58 keV (or possibly an absorption feature at $\sim$ 35 keV) in the spectrum of the binary star Her X-1 was discovered. The feature is believed to be due to the quantized cyclotron transitions expected in the strong magnetic field of the neutron star in this system.
9. A *narrow* x-ray spectral line (of oxygen) was detected in the supernova remnant Puppis A by means of Bragg crystal spectroscopy.
10. Images of supernova remnants clearly revealed the shock waves (temperature $\sim 10^6$ K) propagating through the interstellar medium.
11. The diffuse all-sky radiation at very low x-ray energies (0.1–1 keV) was mapped with multiple rocket flights, revealing the uneven distribution of hot plasma (temperature $\sim 10^6$ K) in the solar region of the Galaxy.

The "Small" Dedicated Missions Ariel V and SAS-3

The British *Ariel V* (1974) and the American third *Small Astronomy Satellite* (*SAS-3*, 1975) carried out studies of the x-ray sky, each with new capabilities that led to additional discoveries. Each satellite weighed about 200 kg. The two satellites were dedicated to x-ray astronomy, and they could be oriented so that most of the sky was available to the principal instruments at any time of the year. The operations were controlled, in each case, by a small group of scientists with a major stake in the scientific yield. Both missions were well suited to studies of variable x-ray emitters and could respond rapidly to celestial events.

Ariel V was a spinning satellite that monitored the entire sky with scanning slat-collimated proportional counters (0.03 m^2) and "pinhole" x-ray cameras. The latter, known collectively as the All Sky Monitor, were provided by a U.S. group. A proportional counter detector also viewed along the spin axis. A Bragg crystal spectrometer and polarimeter were also on board. Major contributions included: (1) the discovery of several long-period (minutes) x-ray pulsars, (e.g., 1118-61), and the monitoring of their long-term light curves; (2) the discovery of several bright transient x-ray sources including A0620-00 which exceeded Sco X-1 in brightness (in 1975) and is now considered very likely to contain a black hole; (3) the establishment that Seyfert 1 galaxies (active galactic nuclei) are universally x-ray emitters; (4) the first report of Fe emission from an extragalactic source, the Perseus cluster; (5) a catalog of 251 detected x-ray sources.

The *SAS-3* was designed as a spinning satellite, but its spin rate was controlled by a gyroscope that could be commanded to stop the spin. Thus all its instruments could be pointed, albeit with a modest drift. This provided continuous trains of data from pulsars, bursters, and transient sources. *SAS-3* carried a proportional counter array with slat and tubular collimators ($\sim$0.03 m^2), a small collector system with thin window proportional counters for the study of approximately 0.2 keV emission, and a modulation collimator system of substantial area (0.03 m^2) to measure source positions to about 1 arcminute. The scientific yield included: (1) the discovery of a dozen x-ray burst sources including the dramatic and unique Rapid Burster, which gains its energy from accretion instabilities rather than nuclear flashes; (2) the discovery of the highly magnetic white dwarf binary system, AM Her, through its x-ray emission and the codiscovery of x-ray emission from HZ 43; (3) an all-sky survey of the soft x-ray flux; and (4) the precise celestial locations of approximately 60 x-ray sources which brought about the first identifications of bursting x-ray sources with visible stellar systems, the identification of the first quasar located through its x-ray emission, and evidence for the central location of x-ray sources in globular clusters.

THE LARGE HEAO MISSIONS (1977–1981)

Beginning in 1977, NASA launched a series of very large scientific payloads called *High Energy Astronomy Observatories* (*HEAO*). They were launched on Atlas Centaur rockets. The payloads were about 2.5 m$\times$5.8 m in size and approximately 3000 kg in mass. The telemetry rate was large, approximately 6400 bits per second compared to the less than 1 kb s^{-1} typical of earlier satellites. Two of these missions were dedicated to x-ray astronomy. *HEAO-1* was a spinning survey mission and *HEAO-2*, or *Einstein*, was a pioneering imaging mission. (*HEAO-3* was a cosmic ray and gamma ray mission.)

HEAO-1

The *HEAO-1* carried four major experiments: (1) a Large Area Sky Survey experiment (LASS), a 1.0-m^2 proportional counter array sensitive in the 1–20 keV range and designed to survey the sky for discrete sources; (2) a smaller (but still quite large) proportional counter array, the Cosmic X-ray Experiment (CXE, $\sim$0.4 m^2), designed to study the diffuse x-ray background from 0.2–60 keV; (3) a modulation collimator (MC) experiment for the determination of modestly precise ($\sim$1 arcminute) celestial positions; and (4) a high-energy experiment (A4, extending to $\sim$10 MeV gamma rays) of which a modest aperture (0.020 m^2) was devoted to 15–100-keV x-rays.

HEAO-1 was primarily a scanning mission; it rotated once every 30 min about the Earth–Sun line. In this manner, the instruments scanned a great circle in the sky that lay 90° from the Sun. As the Sun moved through the sky due to the orbital motion of the Earth, the scan circle moved around the sky at 1° day^{-1}. The instruments had fields of view of order 1°–4° (except for the 1°$\times$20° slat collimators of the high-energy experiment). Thus a given source near the ecliptic was viewed for only a few days. However, sources near the ecliptic pole were scanned nearly continuously during the entire mission. In this manner a deep and thorough survey of the sky was obtained by each instrument in a 6-month period. The sky was scanned almost three times during the mission. The satellite had a limited pointing capability that was used on occasion during its final year to obtain continuous coverage of a given source. It also had a high-telemetry-rate mode (128 kb s^{-1}) that was invoked for brief periods.

The LASS experiment yielded a comprehensive catalog of 842 x-ray sources (1–20 keV) in a systematic all-sky survey which reached to approximately 10^{-3} the intensity of the Crab nebula. It also showed that Cyg X-1 varies aperiodically on time scales down to 3 ms and made the first detection of an eclipse in a low-mass x-ray binary system.

The CXE experiment yielded: (1) numerous broadband spectra of AGN which permitted meaningful comparisons between quasars, BL Lac type objects, and so on; (2) dramatic 100-s variability of the Seyfert galaxy NGC 6814 and also pulsations in the cataclysmic variables SS Cygni and U Gem; (3) the discovery of the Cygnus "superbubble"; and (4) a definitive broadband spectrum of the diffuse x-ray background showing a remarkably thermal shape with temperature 4.6×10^8 K.

The MC positions have led to several hundred optical identifications and source classifications that permit astronomers to study the *HEAO-1* sources at all wavelengths. The identifications include active galactic nuclei (AGN), cataclysmic variables, active coronal type stars, and clusters of galaxies, a number of which have proved to have quite unusual characteristics.

The A4 experiment yielded: (1) a catalog of approximately 40 high-energy x-ray sources (> 25 keV); (2) high-energy x-ray spectra of AGN that raise fundamental questions about the origin of the diffuse background; (3) LMC X-4, the second known (after Her X-1) binary system with periodic (30 days) on–off states surely due to a precessing accretion disk; and (4) the second example of cyclotron absorption, in a binary system 0115 + 63.

Einstein

The *HEAO-2* spacecraft, launched in 1978 and named *Einstein*, made the first sustained use of grazing-incidence focusing optics for celestial x-ray astronomy. This focusing requires that each ray be reflected two times. The results are true in-focus images of extended objects and an enormous gain in sensitivity for weak point-like sources. The most sensitive detectors could reach sources of intensity about 10^{-7} the Crab nebula. The energy-dependent effective aperture was approximately 0.02 m^2, and the instruments generally covered the range 0.1–4 keV. Four instruments could be rotated, one at a time, into the focal plane: an Imaging Proportional Counter (IPC) with high sensitivity and modest (~ 1 arcminute) resolution, a High-Resolution (~ 2 arcseconds) Imager (HRI), a Solid-State Spectrometer (SSS) of moderate sensitivity and spectral resolution significantly exceeding that of proportional counters, and a Bragg Focal Plane Crystal Spectrometer (FPCS) with very high spectral resolution ($\Delta\lambda/\lambda = 0.003$ at 1 keV). *Einstein* also carried a (nonfocusing) Monitor Proportional Counter (MPC) array of 0.07 m^2 area to monitor the higher-energy x-ray emission (2–15 keV) in the view direction of the focusing telescopes. It was necessarily a pointed mission.

Einstein's imaging capability and high sensitivity to faint sources opened up new realms of x-ray astronomy.

1. Coronal x-ray emission from normal stars was found to be much stronger and pervasive than expected. The x-ray intensity from late-type (cool) stars was highly correlated with stellar rotation and strong convection, indicating that the release of energy in twisted magnetic field lines heats the x-ray emitting coronae.
2. High-resolution spectroscopy of supernova remnants by the SSS and FPCS showed evidence of the nucleosynthesis of oxygen, silicon, and sulphur in supernovae.
3. More than a dozen collapsed objects were discovered at the center of supernova remnants by virtue of their x-ray-emitting synchrotron nebulae, and studies of more than 30 supernova remnants in the Large Magellanic Cloud provided strong constraints on their evolution.
4. Numerous discrete x-ray sources were resolved in the Andromeda galaxy and the Magellanic Clouds.
5. X-ray jets were discovered emerging from the cores of the nuclei of the galaxies M87 and Centaurus A, aligned with the known radio jets.
6. A hot interstellar medium was discovered in elliptical galaxies.
7. The radial distribution and temperatures of the x-ray halo surrounding the galaxy M87 revealed an inflow of gas from the cluster in which it resides and showed the presence of dark matter.
8. The distributions of x-ray-emitting gas and galaxies in clusters of galaxies showed a clear distinction between different stages of cluster evolution.
9. It was found that all quasars are x-ray emitters and that the numbers of quasars at great distances account for a large fraction, and the possibly all, of the diffuse x-ray background.

With *Einstein*, x-ray astronomy became an equal partner with the other major branches of astronomy. The field could boast comparable positional accuracy, sensitivity, spectral resolution, and the detection of a wide range of source types, from stars to quasars. For the first time, large numbers of guest observers used an x-ray observatory.

THE 1980s

With the reentry from orbit of *SAS-3* and *HEAO-1* in 1979 and *Einstein* in 1981, the U.S. program for x-ray astronomy entered a long hiatus. *No* U.S. x-ray astronomy missions were launched in the 1980s. The field carried on with the European *EXOSAT*, the Japanese *Hakucho*, *Tenma*, and *Ginga*, and the Soviet *Kvant* missions. These were individually more modest in scale than the *HEAO*s and were directed toward in-depth studies of known phenomena. They had, however, new features that allowed continued advances. The British *Ariel VI* (1979) was primarily a cosmic ray mission; it carried two modest x-ray experiments that were not very productive because of severe electromagnetic interference from ground-based radar which precluded efficient pointing operations. A few studies (e.g., of AGN spectra and pulsar timing) were successfully carried out. The *EXOSAT*, *Ginga*, and *Kvant* missions enjoyed substantial international collaboration and/or guest observing.

EXOSAT

EXOSAT, a three-axis pointed satellite, was launched in 1983. It carried 0.16-m^2 of medium energy (ME) proportional counters and had a relatively restricted field of view (0.75° FWHM) which permitted the study of individual sources in the energy range 1–50 keV. This aperture was much larger than that of previous satellites that could point continuously at a given source. For the first time, a collecting area comparable to that of *Uhuru* (~ 0.08 m^2) could be pointed routinely for long periods toward a given source. At the same time, a comparable area could be pointed to adjacent blank sky to obtain background rates. The other major instrument on *EXOSAT* was an imaging telescope sensitive to 0.05–2 keV with 20-arcsecond on-axis resolution and peak effective area of 0.0010 m^2. Transmission gratings were used with this system for a few spectral studies. A gas scintillation proportional counter (GSPC) of modest area (0.010 m^2) provided improved spectral resolution over that of a normal proportional counter.

A unique feature of *EXOSAT* was its huge elliptical orbit with apogee 190,000 km, perigee 350 km, and orbital period 90 hr. Although this subjected *EXOSAT* to somewhat higher radiation backgrounds that were dependent on solar activity, it provided periods up to several days of uninterrupted viewing of a source. (Low Earth orbit missions are subject to Earth occultations of their targets for $\lesssim$ 30 min every ~ 95 min, except for sources near the orbital poles.)

EXOSAT made major contributions to the understanding of coronal systems, cataclysmic variables, neutron-star systems, and AGN through the study of long-term x-ray light curves. Notable were (1) the determination of the spin-luminosity evolution of the x-ray nova 2030 + 375, (2) the determination of periodicities, of five previously undeciphered low-mass x-ray binaries, (3) the prevalence of relatively "soft" x-ray emission from AGN, and (4) the finding that short term variation is a common characteristic of AGN, and that this variation lacks a characteristic time scale.

A major new *EXOSAT* discovery was that the emission from low-mass x-ray binaries exhibits transient quasiperiodicities of order 20–500 ms. The phenomenon, called quasiperiodic oscillations (QPOs), is closely related to the details of the accretion and emission processes. The frequency variations (for QPOs of ~ 15–50 Hz) with luminosity and spectral "hardness" are highly characteristic and repeatable for a given source. This phenomenon promises to become a powerful diagnostic of accretion processes. The QPOs could well indicate the presence of a hitherto undetected rapidly rotating neutron star.

Hakucho, Tenma, and Ginga

The Japanese have flown three successful x-ray astronomy satellites, beginning with *Hakucho* in 1979. They were launched into low Earth orbits of inclination about 31° from Kagoshima Space Center (KSC) in Kyushu. The data stored on board are telemetered to KSC when the satellite passes overhead. This occurs when the Earth's rotation brings Japan under the northern portion of the inclined satellite orbit, that is, during approximately 5 of the 15 satellite orbits each day. The highest telemetry rate data and the commanding opportunities occurred during those 5 orbits. The sequence of three missions is marked by a rapid growth in capability. *Tenma* and *Ginga* also carried gamma-ray-burst detectors.

Hakucho was a spinning satellite that carried proportional counters that viewed along the spin axis, some with modulation collimators. This allowed pointed continuous observations of sources, except when interrupted by Earth occultations and high background regions (e.g., the South Atlantic Anomaly). Detailed studies of burst sources served to focus attention on the disparity between burst phenomena and the standard thermonuclear burst model. The luminosities of burst sources in the region of the galactic center were found to be unreasonably large for the then-accepted 10 kpc distance to the galactic center. This suggested the galactic center may be somewhat closer than 10 kpc in accord with growing evidence at other wavelengths.

Tenma (1983) was also a spinner. Most of the instrument complement pointed along the spin axis, again providing pointed observations. It featured a relatively large (~ 0.06 m^2) area of Gas Scintillation Proportional Counters (GSPC) with broad energy range (1–60 keV) and exceptionally good spectral resolution for proportional counters (9.5% at 6 keV). Other instruments included a wide-field sky monitor which located and studied transient sources. Iron-line emission was studied in many classes of sources. For the first time, such emission was found in low-mass x-ray binaries (LMXBs) and from the galactic ridge. The high spectral resolution made possible the distinction between "cold" iron-line emission at 6.4 keV believed to arise from x-rays impinging on cold matter (found in pulsars and AGN) and the highly ionized iron (6.7 keV) believed to arise from a hot plasma (found in LMXBs and the galactic ridge). An absorption feature at 4.1 keV was discovered in spectra of x-ray bursts from several sources, a clear diagnostic of the matter through which the burst x-rays pass and possibly the gravitational field at the neutron star surface.

Ginga (1987) is a three-axis stabilized spacecraft which features an exceptionally large area (0.40 m^2) of U.K.-built proportional counters that are sensitive from 1.5–37 keV. This and the pointed capability brought exceptional sensitivity to an instrument extending to energies above a few keV. A sky monitor capability and a U.S.-built gamma burst detector were also carried. A broad range of studies of temporal and spectral phenomena in a wide variety of sources (QPOs, bursters, transients, pulsars, clusters of galaxies, and AGN) have been carried out. Individual QPO oscillations were detected; two very bright transients were discovered with the sky monitor, one similar to A0620-00, a leading black hole candidate; frequent weak transients were discovered in the galactic ridge; spectra of QSOs out to $z = 1.4$ were obtained; absorption features, probably due to cold gaseous matter in the central nucleus, were found in Seyfert 1 spectra; sensitive studies of the x-ray background fluctuations constrain the contributions of discrete sources; the x-ray spectral state was found to be correlated with radio flux for several sources (e.g., GX 17+2); cyclotron absorption lines in gamma ray bursts were detected together with two harmonics at the expected frequencies, which renders the cyclotron interpretation compelling; and cyclotron absorption features were detected from several x-ray pulsars. Finally, and most important, *Ginga* (and also *Kvant*) detected x-rays from the supernova SN1987A.

The Soviet Program

A strong program in x-ray astronomy is emerging in the Soviet Union. The *Röntgen* x-ray observatory was launched in 1987 aboard the *Kvant* module which docked to the *MIR* space station. The complement of detectors included a sensitive high-energy experiment, a "coded-mask" system which is effective for the imaging of high-energy photons not normally amenable to focusing, and gas scintillation detectors. *Röntgen* was an international endeavor with contributions from the Federal Republic of Germany, the United Kingdom, and the Netherlands. The primary result was the discovery and study of x-rays from SN1987A (mentioned previously). High-energy spectra of galactic and extragalactic sources were obtained as were timing results for the Her X-1 pulsar.

The *Granat* mission was launched in December 1989. It carries seven x-ray and gamma-ray systems. High-pressure proportional counters (~ 0.4 m^3) are sensitive over 3–150 keV (ART-S and ART-P), and one of these uses a coded mask of 6-arcminute resolution. The French SIGMA experiment (30–1300 keV) features a coded mask, and position-sensitive scintillation detectors, yielding 13-arcminute resolution. An all-sky monitor system (WATCH, provided by Denmark) is also on board and also several gamma-ray-burst experiments. The 4-day orbit permits long uninterrupted observations of sources. Early results include a high-energy map of the galactic-center region and high-energy spectra of galactic and extragalactic sources.

THE 1990s

A major resurgence of x-ray astronomy is expected in the 1990s. Two missions were launched in 1990. The Space Shuttle *ASTRO-1* mission carried a Broad Band Reflecting X-Ray Telescope (BBXRT) of large aperture, energy response up to 12 keV, excellent energy resolution (90 eV at 1 keV) and moderate angular resolution. It made ~ 100 observations of ~ 60 sources during the 8-day mission. Early spectral results indicate a rich yield of astrophysics will be forthcoming. The German *ROSAT* carries grazing incidence collectors of substantial aperture, a U.K. extreme ultraviolet survey instrument, and a U.S. imaging detector. The primary objective of *ROSAT* is the production of a deep all-sky survey of "soft" x-ray sources (0.1–2.4 keV). The survey, nearly complete at this writing (early 1991), contains at least 50,000 sources. An intensive program of pointed observations will follow.

Later in the decade, the Japanese ASTRO-D, with U.S. participation, will carry a substantial aperture of collectors that will reach to higher energies than *ROSAT*. The Soviet Spectrum-X (with European participation) will emphasize imaging and relatively high resolution spectroscopy. The Italian SAX mission will carry a collection of high-performance instruments. The U.S. X-Ray Timing Explorer with a large proportional counter area will be optimized to study temporal variability of sources on a large range of time scales. Major missions currently in preparation are the U.S. Advanced X-Ray Astrophysics Facility (AXAF) patterned after *Einstein* and stressing high-resolution imaging and spectroscopy, and the European Space Agency *XMM* mission featuring high throughput and spectroscopy. Both are large focusing instruments that offer major improvements in sensitivity and a long duration ($\gtrsim 10$ yr) in orbit.

Additional Reading

Beatty, J. K. (1990). *ROSAT* and the x-ray universe. *Sky and Telescope* **80** 128.

Bleeker, J. and Hermsen, W., eds. (1990). X-ray and gamma-ray astronomy. *Adv. in Space Res.* **10** (No. 2).

Blair, W. P. and Gull, T. R. (1990). *Astro*: Observatory in a Shuttle. *Sky and Telescope* **79** 591.

Bradt, H. and McClintock, J. (1983). The optical counterparts of compact galactic x-ray sources, *Ann. Rev. Astron. Ap.* **21** 13.
Clark, G. W. (1977). X-ray stars in globular clusters, *Scientific American* **237** (No. 4) 42.
Elvis, M., ed. (1990). *Imaging X-Ray Astronomy*. Cambridge University Press, Cambridge.
Frazer, G. (1989). *X-Ray Detectors in Astronomy*. Cambridge University Press, Cambridge.
Giacconi, R. (1980). The *Einstein* X-Ray Observatory. *Scientific American* **242** (No. 2) 80.
Holt, S. and McCray, R. (1982). Spectra of cosmic x-ray sources. *Ann. Rev. Astron. Ap.* **20** 323.
Kondo, Y., ed. (1990). *Observatories in Earth Orbit and Beyond*. Kluwer Academic Publ., Dordrecht.
Oda, M. (1987). What do we learn from space? Space science in Japan. *Physics Today* **40** (No. 12) 26.
Pallavicini, R. and White, N., eds. (1988). X-ray astronomy with *EXOSAT*. *Memoire della Societa Astronomica Italiana* **59** (No. 1–2).
Tucker, W. (1984). *The Star Splitters*, NASA SP-466.
Tucker, W. and Giacconi, R. (1985). *The X-Ray Universe*. Harvard University Press, Cambridge, MA.

See also **Background Radiation, X-Ray; Scientific Spacecraft and Missions; Sounding Rocket Experiments, Astronomical; Telescopes, Detectors and Instruments, X-Ray; X-Ray Bursters; X-Ray Sources, Galactic Distribution; X-Ray Sources, Quasiperiodic Oscillators.**

X-Ray Bursters

Walter H. G. Lewin

X-ray bursts occur in binary systems where a neutron star is accreting matter from a nearby companion star (the donor). The bursts are of two kinds (type I and type II) each with a very different origin. The source of energy of a type I burst is thermonuclear fusion, whereas that of a type II burst is gravity.

If a neutron star and a companion star (e.g., one similar to our Sun) in a binary system are close enough to each other, matter can "fall" from the companion star to the neutron star; it will reach a speed of about one-third of the speed of light as it crashes onto the surface of the neutron star. The kinetic energy of this matter will be released in the form of heat. The source of this energy is gravity; therefore, we call it gravitational potential energy.

If the mass transfer from the donor to the neutron star is about 10^{17} g s^{-1}, the energy release is so high that the surface layers of the neutron star heat to temperatures of roughly 10^7 K. At such high temperatures these layers radiate primarily x-rays with energies in the range from roughly 1–10 keV. Only about 100 such binary systems in our galaxy exhibit this behavior. (Because x-rays are heavily absorbed by the Earth's atmosphere, the observations are made from satellites.)

There is only one known system, called the Rapid Burster, where the accretion occurs in a spasmodic fashion. For reasons that are not well understood, the matter, before it reaches the neutron star, is held back by some unknown "barrier"; the continuous flow of matter "pushes" onto this barrier. As time goes on, more and more matter piles up, and finally the barrier gives in, and a blob of matter falls onto the neutron star resulting in a sudden energy release in the form of a burst of x-rays. Depending on the size of the blob, these x-ray bursts (called type II bursts) can last from about one second up to hundreds of seconds (see Fig. 1). The idea of a barrier and of increasing pressure as the matter piles up is strongly supported by the observations which show that after a given x-ray burst, the waiting time to the next burst is roughly proportional to the total number of x-rays observed in that burst (see Fig. 1). Thus the larger the blob that penetrates the barrier (the more x-rays in the burst), the larger the reduction of pressure onto the barrier; therefore, the longer it will take for the pressure to build up again to a critical level so that another burst can occur. This behavior is not unlike that of lightning; the stronger the electric discharge, the longer it will take for the next discharge to occur.

Let us return to the other neutron star x-ray binary systems where the accretion is "normal" (i.e., not spasmodic). In most cases, the surface layers of the donor star consist largely of hydrogen. This hydrogen, once accreted onto the neutron star, burns (fusion) to helium via a complicated network of thermonuclear reactions. Often, a degenerate helium layer is then formed under the freshly accreted hydrogen. When the temperature and density become high enough, helium nuclei can fuse to produce carbon and other heavy elements, and the entire helium layer can explode like a giant nuclear fusion bomb (this is called a thermonuclear flash). (Once the helium ignites, in general, some of the hydrogen above the helium layer will also be entrained in the thermonuclear reactions.) When the helium reactions start, energy is released and thus the temperature goes up, but due to the degeneracy of the helium, this will *not* cause a decrease in density (a density decrease would lower the reaction rate). Thus the thermonuclear reaction rate further increases, which in turn increases the temperature, which then increases the reaction rate, and so on, and a thermonuclear runaway (flash) becomes inevitable.

This thermonuclear flash increases the temperature of the neutron star surface which causes a sudden increase in x-ray production; this is called a type I x-ray burst. It takes typically a few minutes (but, depending on the amount of matter that "ignites," it can take as long as half an hour) for the neutron star surface to cool down to the preburst temperature. This cooling, which is not observed in type II bursts, can be clearly seen in type I bursts (see Fig. 2).

There are about 40 type I burst sources known (a dozen of them are located in globular clusters). If the amount of helium that fuses in the explosion is roughly 10^{21} g (this is not always so) and for an accretion rate of 10^{17} g s^{-1}, one would expect the bursts to recur about every 3 hr. Observed type I burst intervals typically range from about 1 to several hours, but the production of type I bursts can cease all together when the accretion rate (mass flow from the donor) surpasses a certain level. There is little doubt that in those cases the helium will still fuse to carbon (and heavier elements), but this no longer occurs in the form of a flash.

There is no obvious reason why the Rapid Burster with its unique spasmodic accretion should not be able to produce type I bursts. After all, once the blobs have fallen onto the neutron star, producing type II x-ray bursts, the same events as before could lead to a thermonuclear flash in the Rapid Burster, and type I bursts could be produced.

The Rapid Burster does, indeed, produce type I bursts, and this is of particular historical interest. Type I bursts (originally not called this) with typical intervals of several hours were discovered in 1975; type II bursts (up to 1000 bursts per day) from the Rapid Burster were discovered in early 1976. Investigators believed that all these bursts had one and the same origin. However, it was immediately clear that the frequent bursts from the Rapid Burster had to be due to accretion (gravitational potential energy); they could not possibly be due to thermonuclear flashes. If they were, one would expect to observe roughly about 100 times more energy in the form of x-rays *between* the bursts than *in* the bursts, and this was not observed (see Fig. 1). The reason for this is rather simple. As matter accretes onto a neutron star, it releases about 100 times more energy (gravitational potential energy) than when this same amount of matter fuses from helium to heavier elements (thermonuclear energy). Thus the discovery of the Rapid Burster in early 1976 showed either that not all bursts have a thermonuclear origin or that there is more than one burst mechanism; the latter did not seem plausible at the time.

A turning point came in the fall of 1977 when it was discovered that the Rapid Burster produces two very different kind of bursts: (1) the frequent bursts which we called type II, and (2) the bursts at

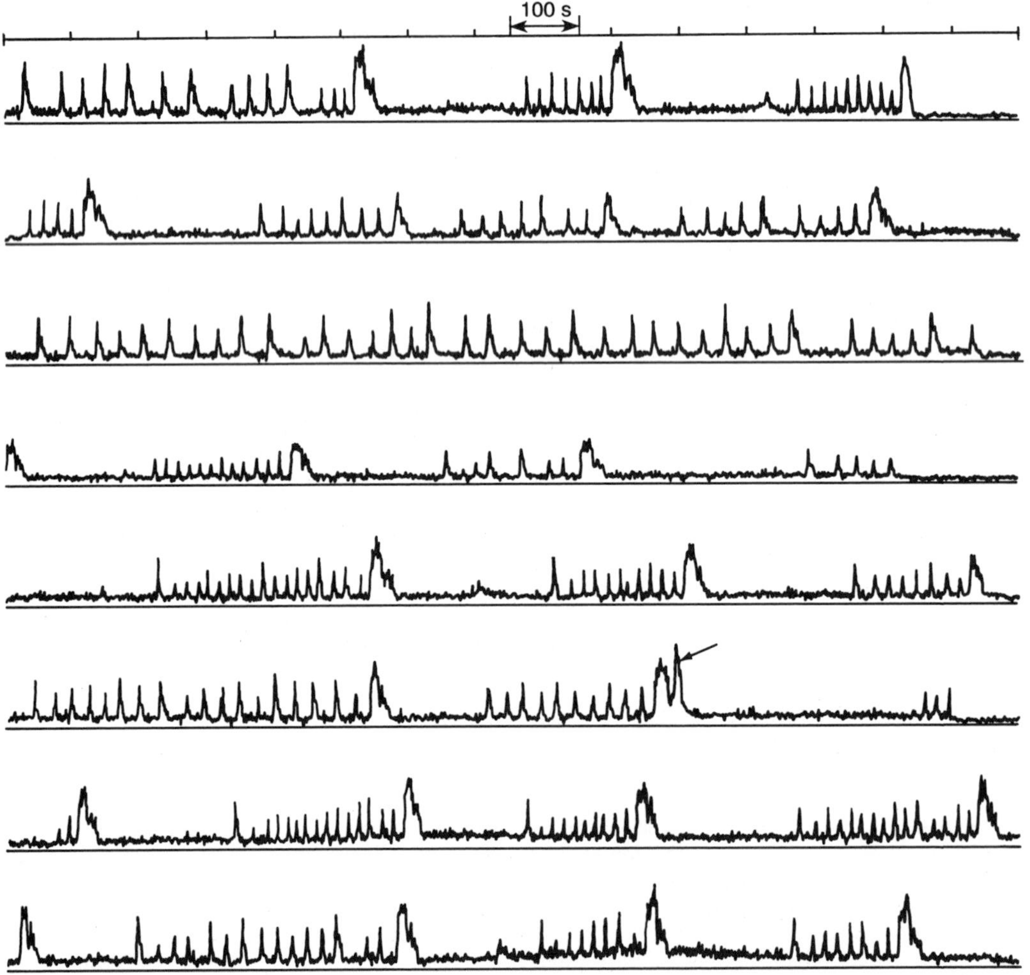

Figure 1. Type II x-ray bursts from the Rapid Burster (MXB 1730-335). These are data (24-min intervals) obtained during eight different orbits of *SAS-3* on March 2–3, 1976, at the time that this unique object was discovered. Note the strong correlation between the integrated number of x-rays in a burst and the waiting time to the following burst. These bursts are due to blobs of matter falling onto a neutron star (gravitational potential energy). If the bursts were due to thermonuclear flashes, the amount of energy (in the form of x-rays) between the bursts would be approximately 100 times as high as that in the bursts, which is clearly not the case. The majority of x-rays between the bursts are not from the Rapid Burster but from the x-ray binary MXB 1728-34 which is only about half a degree away from the Rapid Burster and which was simultaneously observed. The arrow indicates a type I burst (thermonuclear flash) from MXB 1728-34. (*From* Accretion Driven Stellar X-Ray Sources, *Walter H. G. Lewin and Edward P. J. van den Heuvel, eds. Cambridge University Press, Cambridge.*)

intervals of several hours which we called type I (Fig. 3). It was immediately clear that the first were due to accretion (as was believed all along) and the latter due to thermonuclear flashes. The observational support for this came from the fact that after a day of observing, the total energy in the form of x-rays in the rapid type II bursts was about 120 times more than that accumulated in the type I bursts. This number of 120 was the clincher (see Fig. 3)! And so it happened that the Rapid Burster, which at first spoiled the waters, became a Rosetta Stone.

This result stimulated the theorists, and detailed calculations that started in 1977 have since greatly strengthened the idea that type I bursts are the result of thermonuclear flashes on the surface of neutron stars.

Additional Reading

Belian, R. D., Conner, J. P., and Evans, W. D. (1976). The discovery of x-ray bursts from a region in the constellation Norma. *Ap. J.* (*Lett.*) **206** L135.

Grindlay, J., Gursky, H., Schnopper, H., et al. (1976). Discovery of intense x-ray bursts from the globular cluster NGC 6624. *Ap. J.* (*Lett.*) **205** L127.

Hoffman, J. A., Marshall, H., and Lewin, W. H. G. (1978). Dual character of the rapid burster and a classification of x-ray bursts. *Nature* **271** 630.

Joss, P. C. (1978). Helium-burning flashes on an accreting neutron star: A model for x-ray burst sources. *Ap. J.* (*Lett.*) **225** L123.

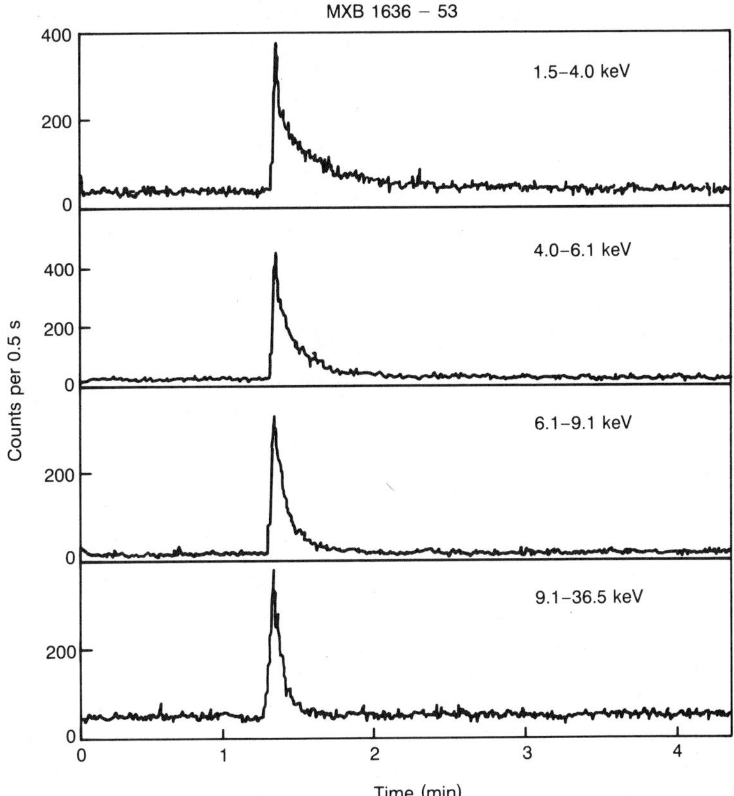

Figure 2. A type I x-ray burst from the x-ray binary MXB 1636-53. Notice that the burst lasts about 2–3 min in the lowest energy range (top panel) but less than half a minute in the highest energy range (bottom panel). This is the signature of cooling of the neutron star surface after the occurrence of a thermonuclear flash. The higher the temperature of the neutron star surface, the higher the energy of the x-rays. Thus, as the surface cools, there is first a reduction in the number of very high energy x-rays, and subsequently a reduction at lower energies. Observed burst intervals for this source range from about half an hour to about 15 hours. The steady stream of x-rays observed before and after the bursts, clearly visible in all four panels, is due to accreting matter. In a given observing time (e.g., a day), their integrated energy exceeds the energy in the bursts by a large factor as is expected because the type I bursts are due to thermonuclear flashes, and the x-rays between the bursts are due to gravitational potential energy. (*Courtesy of Professor Y. Tanaka.*)

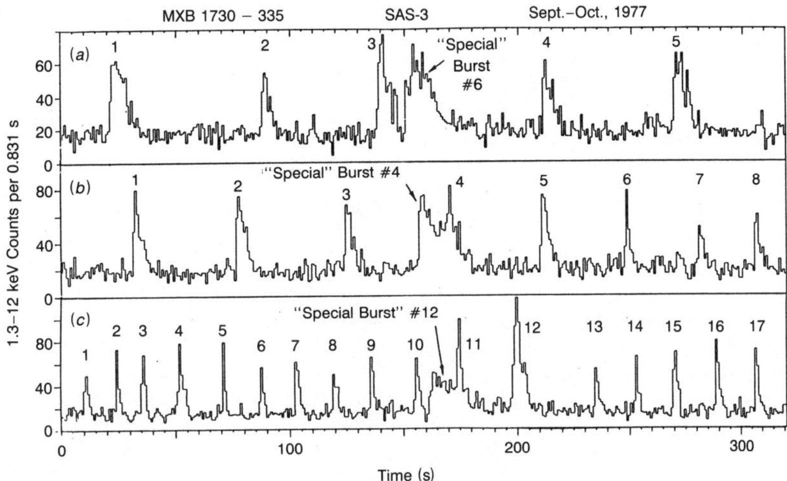

Figure 3. Discovery of type I x-ray bursts from the Rapid Burster (MXB 1730-335). The type I bursts (marked as "special") occur independently of the sequence of the rapidly repetitive type II bursts (numbered separately). In about one day of observing, the integrated energy in the type II bursts exceeded that in the type I bursts by a factor of about 120 which strongly suggested the idea that the type I bursts are due to thermonuclear flashes (the type II bursts are due to gravitational potential energy). [*From Hoffman, J. A., Marshall, H., and Lewin, W. H. G. (1978),* Nature **271** *630.*]

Lewin, W. H. G. and Joss, P. C. (1983). X-ray bursters and the x-ray sources of the galactic bulge. In *Accretion Driven Stellar X-Ray Sources*, W. H. G. Lewin and E. P. J. van den Heuvel, eds. Cambridge University Press, Cambridge, p. 41.

Lewin, W. H. G., Doty, J., Clark, G. W., et al. (1976). The discovery of rapidly repetitive x-ray bursts from a new source in Scorpius. *Ap. J.* (*Lett.*) **207** L95.

Ulmer, M. P. and Melia, F. (1990). X-ray and gamma ray bursts. In *Astrophysics and Particle Physics*, F. L. Navarria and P. G. Pelfer, eds. *Nucl. Phys. B, Proc. Suppl.* **14B**, North-Holland, Amsterdam, p. 129.

See also **Star Clusters, Globular, X-Ray Sources; X-Ray Astronomy, Space Missions; X-Ray Sources, Galactic Distribution; X-Ray Sources, Quasiperiodic Oscillators.**

X-Ray Sources, Galactic Distribution

Robert S. Warwick

The subject of galactic x-ray astronomy dates back to the serendipitous discovery in 1962 of the first cosmic x-ray source in the constellation of Scorpio. Nowadays, Sco X-1 is recognized as the brightest persistent source in the x-ray sky but is only one of approximately 150 discrete galactic x-ray sources listed in the x-ray source catalogs derived from the all-sky surveys conducted by the *Uhuru*, *Ariel V*, and *HEAO-1* satellites. Accurate x-ray positions for many of these sources have enabled their optical counterparts to be identified and hence different classes of x-ray sources to be established. Detailed timing, spectral, and imaging measurements in the x-ray regime have helped elucidate the physical processes that give rise to x-ray luminosities ranging from $10^{32}-10^{38}$ erg s^{-1} in these objects. The majority of the more luminous x-ray sources are binary star systems in which a compact secondary star, such as a neutron star or black hole, accretes matter from its companion. Other classes of x-ray sources include supernova remnants, cataclysmic variables, and stellar coronal emitters. There is also evidence for diffuse x-ray emission arising from a very hot gaseous component of the interstellar medium.

MEDIUM-ENERGY X-RAY OBSERVATIONS

The medium-energy 2–20-keV band of x-ray astronomy has been very successfully exploited over the last two decades through the use of collimated gas counter experiments flown on a series of x-ray astronomy satellites. One reason for the intense interest in the medium-energy x-ray band is that the interstellar medium is essentially transparent in this spectral regime and, with current instrumentation, it is possible to detect x-ray sources more luminous than approximately 10^{36} erg s^{-1} over the full extent of the Galaxy. However, systematic studies of the distribution of galactic x-ray sources have been hampered by two effects. The first is that in order to provide adequate sky coverage, the survey instruments have generally had rather limited angular resolution, typically worse than 1°, and as a consequence the all-sky surveys have suffered from problems of source confusion, particularly in the more crowded regions of the galactic plane. A second factor is that the essential step of identifying the optical counterpart to an x-ray source can prove to be an extremely difficult process for relatively distant sources in regions of high optical extinction, even in the ideal circumstance when an x-ray position accurate to much better than an arcminute is available.

The situation is somewhat different in the adjacent soft x-ray band in which the x-ray telescopes flown on the *Einstein* observatory and on *EXOSAT* have provided a full imaging capability. The advantages of these imaging systems over the nonimaging collimated instruments include significantly enhanced sensitivity and the ability to measure x-ray source positions with arcsecond accuracy. Soft x-ray observations have made a substantial contribution to our overall perspective of galactic x-ray astronomy and, in particular, to our knowledge of relatively nearby, low-luminosity x-ray emitters such as the stellar coronae of normal stars. However, for x-ray energies below about 2 keV photoelectric absorption in the gaseous component of the interstellar medium poses problems for studies of x-ray source distributions within the Galaxy. For example, the mean free path through the galactic plane of a 0.5 keV x-ray photon is only approximately 1 kpc, and at 0.1 keV the horizon is typically less than 100 pc. As a consequence the present review concentrates on the results obtained in the *medium-energy x-ray band*.

SPATIAL AND FLUX DISTRIBUTIONS

The most direct way of investigating the galactic x-ray source population is to consider the distribution in galactic latitude and longitude of sources detected in the all-sky x-ray surveys. From such studies it is readily established that there is a general concentration of x-ray sources toward the galactic plane and that many of the brightest sources congregate within about 20° of the galactic center. It is also evident that the galactic latitude distribution of the sources is considerably wider in the central regions of the Galaxy than elsewhere in the galactic plane. These features are illustrated in Fig. 1, which shows a map of the 2–6 keV emission for a central 100° strip of the galactic plane. The x-ray survey observations can be interpreted in terms of at least two distinct populations of galactic x-ray sources, one associated with the galactic bulge and the other with the galactic disk. The former distribution is confined within about 3 kpc of the galactic center, whereas, because many galactic x-ray sources are observed in the second and third galactic quadrants, the disk distribution must extend throughout the galactic plane out to large galactocentric radii.

In principle, information on the galactic x-ray source distribution can also be derived from studying the x-ray source counts, that is, the $\log N - \log S$ relation, at low galactic latitudes. Here the number of sources per steradian, N, brighter than a particular x-ray flux, S, is plotted as a function of S. At high fluxes N is found to vary as $S^{-0.5}$, but with a steepening to $N \propto S^{-1.0}$ at lower fluxes. The flatter slope is consistent with a one-dimensional "spiral" arm distribution whereas the steeper form of the relationship is representative of a two-dimensional disk distribution of sources. Unfortunately, it turns out that the interpretation of the available data is complicated by a variety of factors such as source variability and catalog incompleteness and reveals little new information about the spatial distribution of faint x-ray sources in the Galaxy. This situation will change, however, when deep unbiased surveys of the galactic plane in the medium-energy x-ray band become available in the late-1990s from x-ray observatory missions such as *AXAF* and *XMM*.

X-RAY SOURCE POPULATIONS

The most luminous x-ray sources in the Galaxy are the x-ray binary systems in which a compact object accretes material from a companion star. Two main categories of x-ray binaries have been distinguished both through their characteristic x-ray properties and through optical follow-up studies. These are the high-mass and the low-mass x-ray binaries, designated HMXRB and LMXRB, respectively.

HMXRB

The HMXRB contain a massive early-type primary ($M > 5\ M_\odot$ [solar masses]) that fuels matter by accretion on to a companion neutron star or black hole either via a stellar wind or through a process known as Roche lobe overflow. In terms of their x-ray

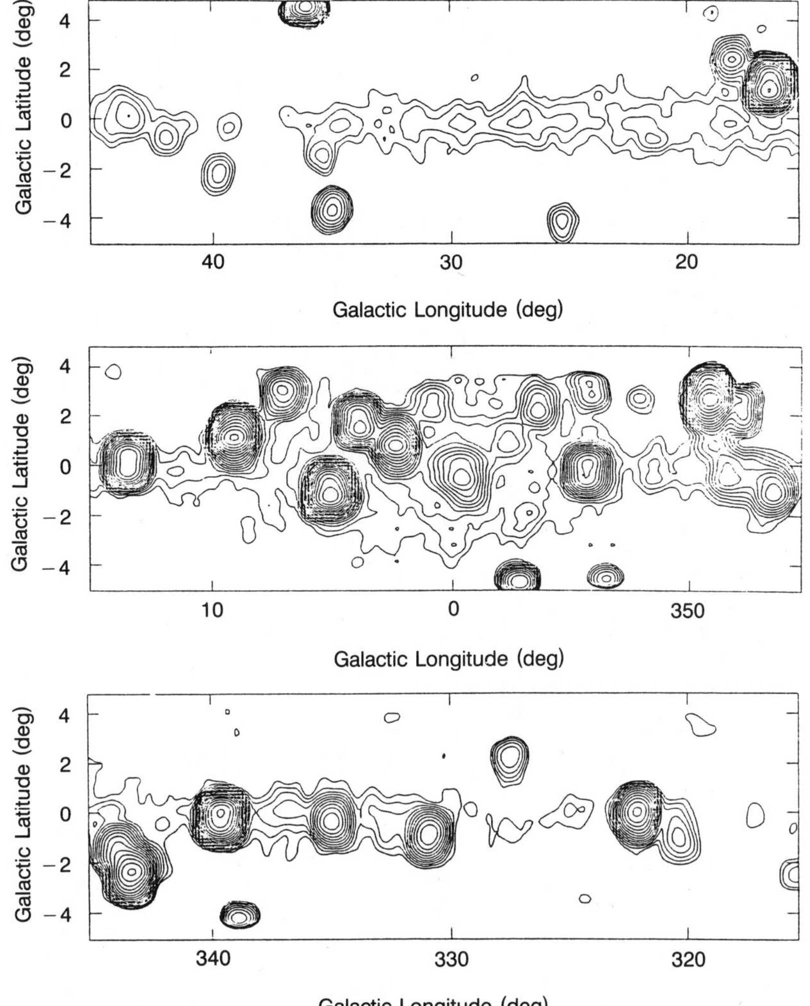

Figure 1. A map of the 2–6-keV x-ray emission from a central 100° strip in the galactic plane. The three sections of the map were produced in a series of scanning observations with the medium-energy proportional counters on *EXOSAT* in June 1984. The angular resolution is limited by the 0.75° × 0.75° field of view of the *EXOSAT* detectors. The contours are plotted on a logarithmic scale.

properties, the HMXRB are characterized by hard x-ray spectra and often exhibit coherent x-ray pulsations at the rotation period of the underlying neutron star. The HMXRB population divides into two subcategories, the OB supergiant binaries, such as Cen X-3 and Cyg X-1, and those systems containing a Be-main-sequence star. The OB systems are predominantly persistent sources with x-ray luminosities in the range 10^{36}–10^{38} erg s^{-1}, whereas the Be-star x-ray binaries are generally highly variable or transient in nature with peak luminosities similar to the OB systems but with quiescent luminosities at least two orders of magnitude lower. Be-star binaries with peak x-ray luminosities closer to 10^{32}–10^{35} erg s^{-1} have also been observed, the nearby systems X Per and γ Cas being the best known examples. The brightness of the primary star in a HMXRB means that such systems are relatively easy to detect optically and up to the present time some 30 or so HMXRB have been identified within the Galaxy. Figure 2 shows the inferred galactic distribution of those systems for which distance estimates are available. HMXRB are relatively young systems (typically $< 10^8$ yr) and should follow a spiral arm (extreme Population I) distribution in the Galaxy. This is supported by Fig. 2, if we take into account the large uncertainties associated with the distance estimates. The concentration of sources near the Sun in Fig. 2 is due to unavoidable observational biases, the most severe being the problem of optically identifying even the very brightest of stellar systems at large distances near the galactic plane. The mean distance above the galactic plane of the optically identified HMXRB is approximately 75 pc, consistent with their Population I classification.

LMXRB

The LMXRB consist of a low-mass, late-type star (typically $M < 1$ $M_{\odot}$) plus a neutron star or black hole that accretes material through Roche lobe overflow. LMXRB generally have relatively soft x-ray spectra and as a class exhibit a variety of interesting x-ray properties such as x-ray bursts, quasiperiodic oscillations, and absorption dips. About a quarter of LMXRB are designated as soft x-ray transients. For a given x-ray flux, LMXRB are typically 5–10 magnitudes fainter in the optical band than HMXRB, a fact which explains the relatively slow progress in the optical identification of LMXRB. X-ray reprocessing in an accretion disk often provides the dominant source of optical light in LMXRB and hence, even when an optical identification is established, the spectral class and absolute magnitude of the companion star and hence the distance of

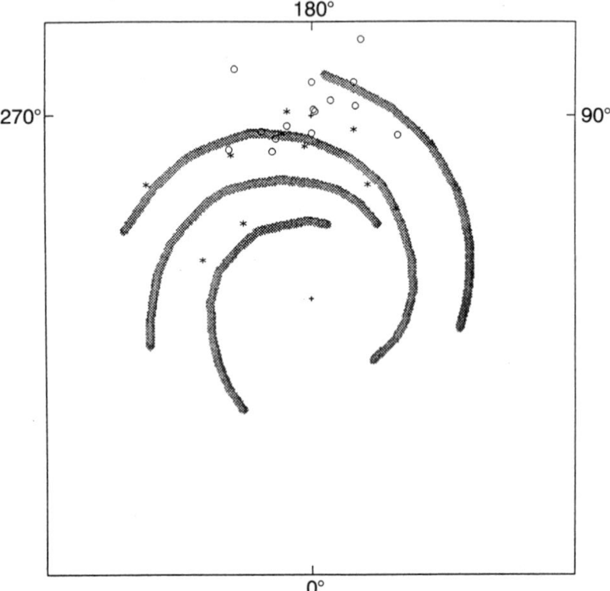

Figure 2. The distribution of HMXRB systems in the Galaxy. Be-star systems are shown as open circles, whereas OB systems are shown as stars. The crosses represent the position of the Sun and the galactic center. The shaded bands trace the spiral structure of the Galaxy as deduced from radio and optical studies. The HMXRB concentration in the solar neighborhood is due to observational selection effects.

Table 1. Galactic X-Ray Source Populations

Class of Object	*Galactic Distribution*	*Number in Galaxy*	L_X *(erg s^{-1})*
HMXRB			
OB systems	Thin disk	15	10^{36-38}
Be systems	Thin disk	25	10^{35-37}
LMXRB			
Bright bulge	Bulge	10	10^{38}
Globular cluster	Bulge + halo	12	10^{36-38}
Disk	Thick disk	100	10^{35-38}
Low L_X			
Be binaries	Thin disk	10^4	10^{32-35}
SNR	Thin disk	1000	10^{33-35}
CVs	Thick disk	10^5	10^{30-34}
RS CVn	Thin disk	10^6	10^{30-32}
Diffuse x-rays	Disk + bulge	1	10^{38}

the system are difficult to determine. Using a variety of defining criteria, the total number of galactic x-ray sources now established as LMXRB is approximately 70, with approximately half of these having reasonably firm optical identifications. The LMXRB form, in fact, a somewhat mixed class of objects, which can be divided into three subclasses. The 10 or so bright x-ray sources associated with the galactic bulge are generally considered as a distinct set of highly luminous LMXRB. Making the reasonable assumption that these sources are typically at about 10 kpc, that is, the approximate distance of the galactic center, then the observed x-ray fluxes imply x-ray luminosities of $1-3\times10^{38}$ erg s^{-1}, comparable with the Eddington luminosity of a compact solar mass object. The dozen or so bright x-ray sources located in the cores of globular clusters (star clusters found in the extended bulge and halo of the Galaxy) comprise a second category of LMXRB with x-ray luminosities ranging from $10^{36}-10^{38}$ erg s^{-1}. LMXRB are also observed outside the confines of the central regions of our galaxy in a thick disk component, although precise details of the distribution are uncertain due to the incompleteness of the sample. The identification of dwarf companions and the association of some LMXRB with globular clusters are clear pointers to LMXRB being predominantly old, Population II objects ($>10^9$ yr).

Other Classes of X-Ray Sources

Although the x-ray binaries dominate the total luminosity radiated by our galaxy in the medium-energy x-ray band, there are several classes of objects with x-ray luminosities in the range $10^{32}-10^{35}$ erg s^{-1}, which have been detected in the all-sky surveys at distances up to 10 kpc. For example, young supernova remnants (SNR) produce thermal x-rays in the shocked heated gas within the expanding SNR shell and in few instances contain nebulae that emit synchrotron radiation and that are powered by an active pulsar. Cataclysmic variables (CV) are low-mass binary systems in which the compact object is a white dwarf star rather than a neutron star or black hole and which are characterized by fairly hard x-ray spectra, although soft x-ray components dominate the output of certain categories of CV particularly during outburst states. Finally, in the medium-energy band, the harder spectral components of the RS CVn binaries produce x-ray luminosities of up to 10^{32} erg s^{-1}. Information on the galactic distribution of these lower-luminosity classes of x-ray emitter are generally based on space densities derived from optical studies. However, evidence for the presence in the Galaxy of an extensive distribution of lower-luminosity x-ray sources is available directly from x-ray observations. Away from the crowding of the central 10° of the Galaxy, a narrow ridge of emission is apparent in Fig. 1, which extends along the galactic plane on both sides of the galactic center out to approximately 40° longitude. Detailed analysis shows that the galactic x-ray ridge has an x-ray luminosity of approximately 10^{38} erg s^{-1}, a scale height of 100 pc, and a radial extent within the Galaxy of about 6.5 kpc. If this ridge feature is interpreted in terms of discrete sources, then an individual source in the underlying population must have an x-ray luminosity of less than 3×10^{35} erg s^{-1}. Furthermore, if results from galactic plane surveys conducted by *Einstein* are taken into account, this estimate must be revised downward by at least a factor of 10. SNR, CV, and RS CVn systems and also low-luminosity Be-star binaries may all make substantial contributions to the galactic x-ray ridge.

Diffuse Emission

An alternative hypothesis for the origin of the galactic ridge is in terms of diffuse x-ray emission. Recent observations with the *TENMA* and *GINGA* satellites suggest that there is an extensive distribution of very hot ($\sim10^8$ K) gas within the galactic disk, the galactic bulge region, and possibly the galactic halo. The origin of this extreme component of the interstellar medium is at present unclear.

SUMMARY

Table 1 provides details of the different classes of galactic x-ray sources including their distribution in the Galaxy, the likely total number of objects (in the Galaxy) in each category, and the observed range of x-ray luminosity. X-ray source populations that emit predominantly in the soft x-ray band, such as normal stellar coronae, are, however, excluded. The total x-ray luminosity of the Galaxy is about 3×10^{39} erg s^{-1} in the 2–20-keV band with the dominant contribution coming from LMXRB in the galactic bulge.

Additional Reading

Bradt, H. V. D. and McClintock, J. E. (1983). Optical counterparts of compact galactic x-ray sources. *Ann. Rev. Astron. Ap.* **21** 13.

Condon, J. J. and Lockman, F. J., eds. (1990). *Large-Scale Surveys of the Sky*. National Radio Astronomy Observatory, Green Bank.

Elvis, M., ed. (1990). *Imaging X-Ray Astronomy*. Cambridge University Press, Cambridge.

Lewin, W. H. G. and van den Heuvel, E. P. J., eds. (1983). *Accretion Driven Binary X-Ray Sources*. Cambridge University Press, Cambridge.

Longair, M. S. (1981). *High Energy Astrophysics*. Cambridge University Press, Cambridge.

Tucker, W. and Giaconni, R. (1985). *The X-Ray Universe*. Harvard University Press, Cambridge, MA.

See also **Background Radiation, Soft X-Ray; Binary Stars, Cataclysmic; Binary Stars, Polars (AM Herculis Type); Binary Stars, RS Canum Venaticorum Type; Binary Stars, X-Ray, Formation and Evolution; Galaxies, X-Ray Emission; Interstellar Medium, Hot Phase; Star Clusters, Globular, X-Ray Sources; Stars, Atmospheres, X-Ray Emission; Stars, Pre-Main Sequence, X-Ray Emission; Supernova Remnants and Pulsars, Galactic Distribution; X-Ray Bursters; X-Ray Sources, Quasiperiodic Oscillators.**

X-Ray Sources, Quasiperiodic Oscillators

Michiel van der Klis

Quasiperiodic oscillations (QPOs) is a term x-ray astronomers use for any type of aperiodic variability in x-ray intensity that consists of fluctuations with frequencies that are constrained to a narrow range. Such oscillations cause a peak of finite width in the power spectrum of the x-ray intensity variations (Fig. 1 bottom left; Fig. 2). Unlike the case in some other areas of physics, the use of the term QPO here is rather loose and applies to any type of aperiodic variability that produces a narrow peak. In most cases, the sources are too weak to reveal the exact nature of the signals, which can only be detected by averaging large amounts of data into a single power spectrum. QPOs are only one aspect of a broader range of stochastic variations in x-ray intensity observed from compact x-ray sources—broad power spectral components that do *not* form a peak are usually called "noise" components.

QPOs gained a foothold in x-ray astronomy as a major diagnostic in 1985, when they were discovered with the European Space Agency's x-ray astronomy satellite *EXOSAT*, in several of the brightest low-mass x-ray binaries (LMXBs). The high observed oscillation frequencies of 6–60 Hz suggested an origin close to the compact object, where the strong gravitational forces cause rapid motion. After some initial confusion, it became clear in 1987 that two entirely different types of QPOs had been discovered in the same sources in rapid succession. The properties of these two types of QPOs and of other noise components were found to strongly correlate with changes in the time-averaged x-ray spectrum of the sources. By studying this correlation, it was possible to distinguish two different classes of LMXB, called the Z sources and the atoll sources. Sources in each of these classes exhibit different states, characterized by different x-ray spectral and stochastic-variability properties. QPOs have only been observed from Z sources—in both Z and atoll sources, noise components are observed which provide a powerful additional diagnostic of source state.

One of the two types of QPOs ("horizontal-branch" QPOs) is now believed by most experts to require a rapidly spinning neutron star with a magnetic-field strength of approximately 10^9–10^{10} G, weak relative to the approximately 10^{12} G in neutron stars in massive x-ray binaries. If correct, then these QPOs are the first direct indication of the presence of weakly magnetized neutron stars in low-mass x-ray binaries, which is a prediction of the scenario in which these systems finally evolve into millisecond radio pulsars. These QPOs then provide the only presently available observational insight into the somewhat enigmatic physical process of neutron star magnetic-field decay in LMXBs and their progenitor systems, which in turn is a sensitive probe into the precise properties of the supra-nuclear-density matter in the interior of the neutron star.

The second type of QPO ("normal-flaring branch" QPOs) occurs when an x-ray source becomes so luminous that it approaches the Eddington limit, where radiation pressure matches gravity. It may therefore arise as the result of an instability in the interplay between the radiative and gravitational forces on the accreting matter. This would imply that while exhibiting this second type of QPO, these sources provide us with x-ray standard candles, "calibrated" sources of radiation whose luminosity is close to the Eddington luminosity of a neutron star ($\sim 10^{38}$ erg s^{-1}), and whose distance can consequently be calculated directly from the inverse-square law.

Maybe as significant as the presence of QPOs in the very bright Z sources is their absence in the atoll sources, LMXBs which are only slightly fainter. Maybe this absence is explained by these sources having a different evolutionary history resulting in an even lower neutron star magnetic-field strength than in QPO sources, and in lower accretion rates. There is evidence that binary orbital periods of atoll sources are shorter than those of Z sources, supporting the idea of a different evolutionary history.

A small number of sources (2 or 3) that show QPOs does not fit in (as yet) with either the Z or the atoll class; among these is the famous Rapid Burster, which is illustrated in Figure 1 of the entry on X-Ray Bursters.

THE Z SOURCES

Z sources derive their name from the roughly Z-shaped pattern they trace out in an x-ray color–color diagram (Fig. 1, top left). Such a diagram, analogous to an optical U–B, B–V diagram, shows that these sources have three x-ray spectral states, corresponding to the three branches of the Z pattern, respectively the *horizontal branch* (HB, the uppermost branch in the Z), the *normal branch* (NB, middle branch), and the *flaring branch* (FB, lower branch).

Z sources typically move along the Z in an irregular fashion, on a time scale of days. They never jump from one branch to the next but always follow the Z. For this reason, it is generally assumed that the x-ray spectral changes, of which the motion of a Z source in the color–color diagram is the expression, are caused by changes in the accretion rate $\dot{M}$. It is believed that $\dot{M}$ increases from HB via NB to FB, and induces changes in the geometry and other physical properties of the accretion flow that are responsible for the x-ray spectral changes. Arguments for this sense of $\dot{M}$ along the Z are obtained from the magnetospheric model for HB QPOs and from the observation that the ultraviolet (UV) flux, mostly x-rays reprocessed into UV in the accretion disk, increases monotonically along the Z: The disk serves as an x-ray detector with a better view of the x-ray source than we have. As a similar Z pattern is also observed in diagrams of x-ray color versus x-ray intensity, an important consequence of this interpretation is that x-ray intensity is *not* monotonically related to $\dot{M}$.

It has been proposed that the HB–NB transition can be identified with the change in the geometry of the accretion flow that occurs when the inner accretion disk is puffed up sufficiently by radiation pressure to engulf the magnetosphere of the neutron star, and that in the FB the mass transfer rates become super-Eddington and mass outflow occurs. The x-ray spectral changes in both NB and FB are consistent with their being due to changes in the Comptonization optical depth of matter surrounding the x-ray source.

Radio, optical, and UV properties of Z sourves are correlated with the x-ray states, with radio flux strongest when the accretion rate is low and optical and UV flux gradually increasing with the accretion rate. Power spectral properties, and in particular QPO characteristics, are also strongly correlated with x-ray spectral state.

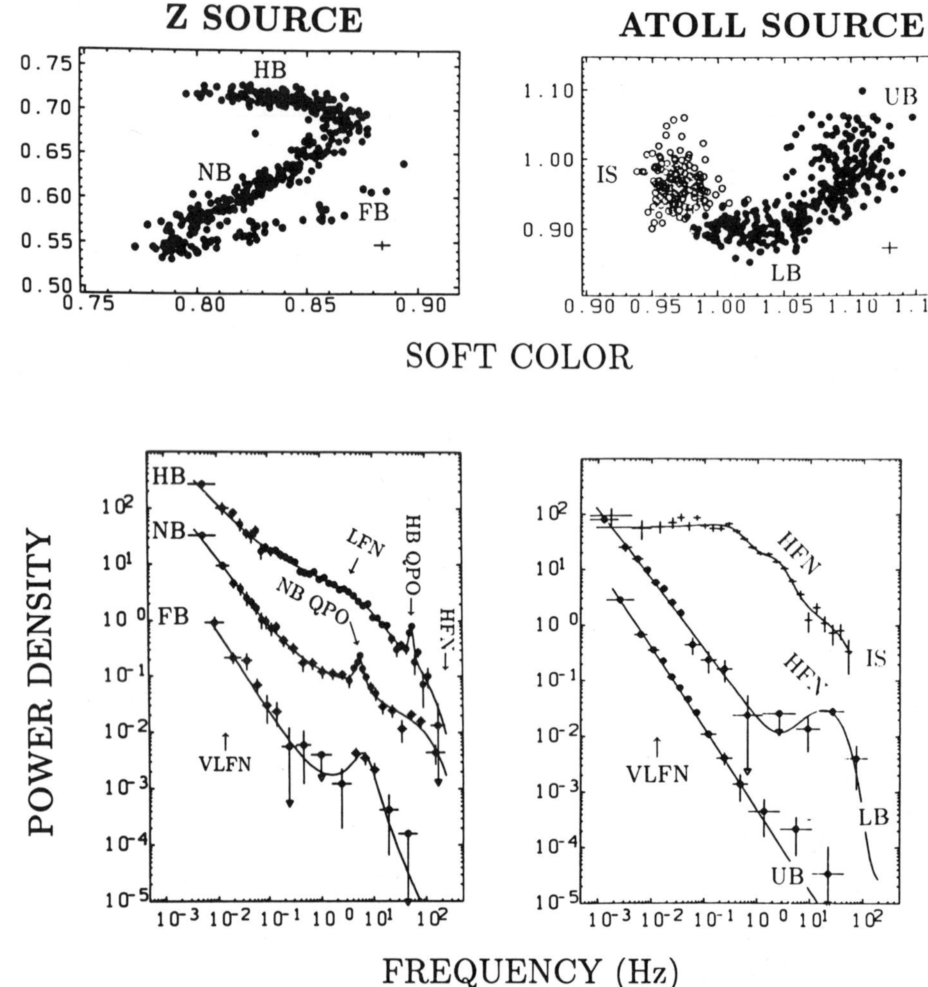

Figure 1. A compilation of x-ray color–color data (top) and power spectra (bottom) illustrating Z (left) and atoll (right) behavior. Color–color diagrams, illustrating the changes in the x-ray spectrum, exhibit a three-branched Z-shaped pattern in Z sources and "islands" and a curved "banana branch" in the atoll sources. Power spectra, displaying the amount of x-ray variability as a function of variability frequency, differ depending on the location of the source in the color–color diagram. Separate power spectra are shown for the three branches in the Z, and for the island and lower and upper banana state in an atoll source. QPO peaks are seen in the Z source power spectra, and broad noise components occur in both Z and atoll source power spectra.

HORIZONTAL-BRANCH QPOs AND LOW-FREQUENCY NOISE

In the horizontal branch, QPOs occur with a frequency ν_{QPO} that increases (Fig. 2) from approximately 15 to 60 Hz when the source moves along the branch from left to right. It is generally suspected that these QPOs are produced as the result of an interaction between the inner accretion disk and the neutron star magnetic field, leading to a quasiperiodic modulation of the accretion rate.

The "magnetospheric" radius r_m, within which the accretion disk is disrupted by magnetic forces and where this interaction is supposed to take place, is located only a few tens of kilometers above the neutron star surface. It varies with accretion rate as $r_m = C\dot{M}^{-2/7}$, where the constant C depends on the magnetic-field strength. Therefore, with Kepler's law, the frequency with which matter, in its Keplerian orbit at the inner edge of the disk, near r_m, circulates around the neutron star increases with accretion rate as $\nu_K(r_m) = C^{-3/2}\dot{M}^{3/7}$.

It is assumed that the circulating matter in the inner disk contains inhomogeneities, or "clumps," that penetrate the magnetosphere more easily at specific locations (e.g., the magnetic poles) than at others. Every time a clump passes such a location of easier entry, part of it enters the magnetosphere and accretes onto the neutron star surface. An enhancement in $\dot{M}$ and therefore in x-ray intensity results. This process continues until all of the clump has accreted. The x-ray intensity variations we observe as QPOs consist of the superposition of the signals of many individual clumps, each of which contributes a short-lived oscillation to the x-ray intensity with a duration equal to the clump lifetime.

The frequency ν_{QPO} of this oscillation is the *beat frequency* of the Keplerian disk frequency $\nu_K(r_m)$ with which the clumps at r_m circulate around the neutron star, and the star's spin frequency ν_S, because *this* is the frequency with which a clump passes a location of easier entry. So, the model predicts $\nu_{\mathrm{QPO}} = C^{-3/2}\dot{M}^{3/7} - \nu_S$. According to this "modulated-accretion beat-frequency model," HB QPOs and low-frequency noise (LFN) contain information about the magnetic-field strength B and the spin rate ν_S of the neutron star. Derived values, $B \sim 10^9$–10^{10} G and $\nu_S \sim 100$ Hz suffer from uncertainties in the theoretical predictions for the relations of x-ray intensity and r_m with $\dot{M}$, but are in good general agreement with the idea that these sources evolve into millisecond radio pulsars.

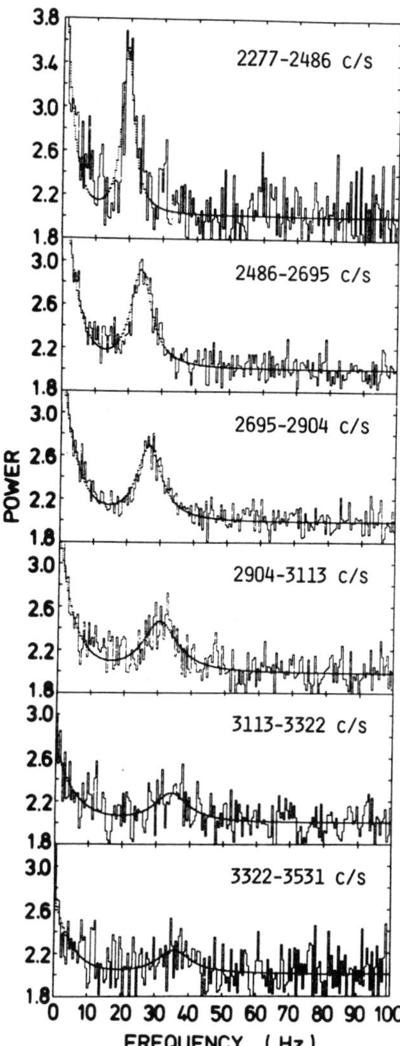

Figure 2. Power spectra of horizontal-branch QPOs and low-frequency noise in the Z-type source GX 5-1. For increasing x-ray photon count rates (displayed in each frame), the QPO peak in the power spectrum shifts to higher frequencies and becomes less clear. The LFN is clearly seen at frequencies below the QPO peak.

A natural consequence of the model is that the power spectrum of the x-ray intensity variations contains not only a QPO peak, but also additional power at lower frequencies, the LFN, which represents the irregular variability in x-ray intensity caused by the random occurrence of the clumps in the accretion flow. Such LFN is indeed observed (Fig. 1, bottom left; Fig. 2).

If, in addition to the QPO, the neutron star spin frequency could be measured directly, it would be possible to test these theoretical predictions. In LMXBs this has so far not been possible. However, QPO peaks have been found in a number of strong-field accreting neutron stars in massive x-ray binaries, where ν_S is known. Here, it may soon be possible to perform these tests.

NORMAL-FLARING BRANCH QPOs

In the middle of the NB, another type of QPO peak occurs (Fig. 1, bottom left) which differs from HB QPOs in several respects. Its frequency is about 6 Hz and hardly changes when the source moves along the branch. No associated LFN is observed. There is no doubt that this is a different phenomenon from HB QPOs, as the latter have on occasion been observed to persist (weakly) into the NB and to occur simultaneously with this 6-Hz peak. When the source moves from the NB into the FB, the frequency of the NB/FB QPO gradually increases from approximately 6–10 Hz. In the FB, frequency is correlated with position in the branch and varies from approximately 10–20 Hz. NB QPOs have been observed to be out of phase by nearly 180° when observed in x-ray bands below and above approximately 6 keV.

Judging from their distance and brightness, Z sources have x-ray luminosities of roughly 10^{38} erg s^{-1}, of the order of the Eddington limit L_{Edd} of a 1–2 $M_{\odot}$ (solar mass) neutron star. If the mass transfer rate increases from HB via NB to FB, then it is expected that L_{Edd} is approached somewhere along the NB–FB track. This means that radiative forces become important and may cause instabilities in the accretion flow. Indeed, numerical hydrodynamic calculations of Eddington-limited accretion flows suggest that an oscillation with a frequency near 6 Hz could be set up. Such oscillations would result in a variation in the Compton scattering optical depth of the flow and cause the x-ray spectrum to quasiperiodically "rock" around a pivot point near 6 keV. This, of course, agrees nicely with the approximate 180° phase lag observed in NB QPOs above and below 6 KeV.

ATOLL SOURCES

Atoll sources move much more slowly in the x-ray color–color diagram than do Z sources. Therefore, with typical x-ray observations lasting of order one day, the diagrams of these sources, composed of several different observations, attain a lumpy appearance (Fig. 1, top right), somewhat resembling a geographical map of an atoll. When their x-ray intensity (and probably accretion rate) is lowest, they all but stop moving in the color–color diagram and produce a single lump: an *island*. When they get brighter, one curved branch is traced out on a time scale of days: the *banana*. It seems likely that when $\dot{M}$ increases, a source moves successively from the island state via the left end of the banana up into the banana, which at its hard-energy (upper right) end sometimes curves back to the upper left. The power spectrum of the x-ray intensity variations varies drastically along the track (Fig. 1, bottom right). It can be described in terms of two noise components, whose strength monotonically varies with $\dot{M}$. They are called *high-frequency noise* (HFN), which dominates in the island state, and *very low frequency noise* (VLFN), which dominates in the extreme banana. Similar noise components, having the same dependence on accretion rate, are also observed in addition to QPO phenomena, in the Z sources. They may be a general feature of the disk accretion process.

No HB QPO/LFN or NB/FB QPOs have been observed from atoll sources. An explanation may be that the magnetic-field strengths of their neutron stars are lower (explaining the absence of magnetospheric HB QPO/LFN), and their accretion rates are lower as well (no Eddington-limited-flow NB/FB QPOs). The fact that atoll sources, as a group, are fainter than Z sources seems to provide an additional argument in favor of the latter idea. There are indications that the donor stars accompanying the x-ray sources are giants or subgiants in Z sources and dwarfs in atoll sources. This would explain the difference in mass transfer rate, as giants are expected to transfer more mass due to their evolutionary radius expansion. It is possible that the difference in evolutionary history leading to the presence of different companions could also have caused a difference in the decay of the neutron star magnetic field or in its initial value.

Additional Reading

Adv. in Space Res. **8** (No. 2-3) (1988). Articles by E. P. J. van den Heuvel, L. Stella, G. Hasinger, M. van der Klis, K. Mitsuda et al., A. Tennant, W. Collmar et al., F. K. Lamb, J. Shaham, and N. D. Kylafis).

Hasinger, G. and van der Klis, M. (1989). Two patterns of correlated x-ray timing and spectral behaviour in low-mass x-ray binaries. *Astron. Ap.* **225** 79.

Lamb, F. K. (1989). Accretion by magnetic neutron stars. In *Timing Neutron Stars*, H. Ögelman and E. P. J. van den Heuvel, eds. Kluwer Academic Publishers, Dordrecht, p. 649.

Lewin, W. H. G., van Paradijs, J., and van der Klis, M. (1988). A review of quasi-periodic oscillations in low-mass x-ray binaries. *Space Sci. Rev.* **46** 273

van der Klis, M. (1988). Quasi-periodic oscillations in celestial x-ray sources. *Scientific American* **256** (No. 11) 50.

van der Klis, M. (1989). Quasi-periodic oscillations and noise in low-mass x-ray binaries. *Ann. Rev. Astron. Ap.* **27** 517.

See also **Binary Stars, X-Ray, Formation and Evolution; Pulsars, Binary; Pulsars, Millisecond; Star Clusters, Globular, X-Ray Emission; X-Ray Astronomy, Space Missions; X-Ray Sources, Galactic Distribution.**

Z

Zodiacal Light and Gegenschein

René Dumont

"In February, and for a little before and a little after that month, about six in the evening, when the Twilight hath almost deserted the Horizon, you shal see a plainly discernable way of the Twilight striking up towards the Pleiades or Seven Starrs, and seeming almost to touch them." In 1661, Joshua Childrey gave (in his *Britannia Banonica*) this first precise description of the zodiacal light (Z.L.)—for its evening, western occurrence. Another similar "zodiacal cone" can be seen in the east, during the first half of autumn, before the dawn.

The phenomenon was known in the Middle Ages from middle eastern scientists. Scarce and questionable, on the other hand, were the allusions to it in Europe, prior to the 17th century (Aristotle? Seneca? Pontanus, ca. 1500). Since the first assiduous observations of Z.L. in 1683 by Giovanni Domenico (Jean Dominique) Cassini, it was for him and for other people a matter of surprise that it had so rarely been noticed before. When Bernard de Fontenelle, in the 1714 version of his famous *Pluralité des Mondes*, added a few pages commenting on Cassini's "discovery," he wrote that Z.L. "...seems to all appearances to have been visible from earliest times, and yet nobody had ever seen it. —How does a Light manage to hide? To do that, it must be unusually skilful," replied the beautiful marchioness to her astronomy teacher.

In his *Découverte de la lumière qui paroist dans le Zodiaque* (1693), Cassini ascribed the cones to a circumsolar cloud flattened upon the solar equator and reflecting sunlight increasingly, as the line of sight approaches either the Sun or the symmetry plane of the cloud. This was a correct view, except that the symmetry plane is somewhat different—linked to planets more than to the Sun, therefore closer to the ecliptic than to the solar equator. The 1706 total solar eclipse led Cassini to (rather correctly) regard the solar corona as the innermost part of the cloud, or the continuation of the cones across unobservable gaps of about 20° width in elongation.

The seasonal changes of visibility of the cones from middle latitudes are governed by the angle between ecliptic and horizon after the end and before the beginning of the twilights, as clearly explained and sketched (Fig. 1) by J. J. D. de Mairan (*Traité Physique et Historique de l'Aurore Boréale*, 1733). Adopting Cassini's interplanetary cloud, de Mairan emphasized its extension beyond the terrestrial orbit, simply because the cones can sometimes be distinguished up to solar elongations greater than 90°. Very good sights under very good conditions may follow a dimmer path of light along the whole ecliptic, with a slight reincrease in the antisolar region, which Alexander von Humboldt named the Gegenschein (see the right part of Fig. 2*a*). Visual detection of the Gegenschein, as a faint oval patch, is only possible when the antisun is far from the Milky Way, and preferably in scotopic sight, when one is watching it "out of the corner of one's eye."

The 19th century brought extensive observations along the whole ecliptic, often from the tropics, in which the position and shape of the cones were noted to change, sometimes rather quickly. The telluric origin (airglow; see the following discussion) of these evolutions has been sensed at odd times. Subjectivity in visual observations, and their discrepancies, led to confusing theories: Cassini's and de Mairan's model was challenged in 1856 by a circumterrestrial cloud hypothesis (Jones). The Gegenschein was lavishly theorized, both as a gaseous tail of the Earth which had not yet faded in the 1950s and as an accretion of matter near the external collinear libration point L3 of the Sun–Earth system, at 0.01 AU from the Earth in the antisolar direction.

Progress in quantitative measurements was slow until the photoelectric era. If, surprisingly, a correct polarization degree had been given for the cones (15%) as early as 1874 (Wright), on the other hand, the brightnesses, for instance at 60° or 90° elongation, remained uncertain by up to a factor of 2 in 1965. At the same time, the residual brightness, which had long been suspected to exist toward the ecliptic poles (Dufay, in 1928) from the fact that the cloud surrounds the Earth, remained uncertain by a factor of 5.

ZODIACAL LIGHT AND GEGENSCHEIN AS A SIGNAL OF INTERPLANETARY DUST: OBSERVATIONAL PROBLEMS AND RESULTS

Earth-Based Photopolarimetry

As in most studies of extended faint sources, mixture and disentangling problems are the main ones. In the field of an instrument (assumed to be cleared from any discretely discernible object), Z.L. coexists with (1) light from all stars; (2) light from all galaxies, which are both beyond the limiting magnitude of the instrument; and (3) diffuse galactic light due to interstellar matter. When the observation is made from the ground or from a balloon, that is, through the airglow layers, (4) a rather bright continuous emission adds to the response, even if the main lines and bands of the airglow are avoided by the appropriate filters.

The intensities of components 1 and 3 are decreased by pointing at medium galactic latitudes, but component 1 remains of importance for very small photometers with fields of several square-degrees and low limiting magnitudes. Component 2 is weak. Against component 4, which is worrying due to its temporal and directional variations, one fair disentangling method takes advantage of the brightness covariance [Barbier (1955)] between the continuum and the [O I] 557.7-nm emission line of the airglow [R. Dumont (1965)]. The covariance can be calibrated at the celestial pole, apart from an unknown intercept of extraterrestrial origin.

Our present knowledge of the distribution of the brightness and polarization of the Z.L. and Gegenschein, as seen from the Earth's orbit, comes from satellite or rocket observations and from long exacting ground-based programs. The former are free of airglow noise, but rigidities in their implementation often led to fragmentary investigations. The latter have to compromise with airglow, but they allow for reducing the starlight by reducing the field of view, and they place no impediments on a thorough celestial coverage.

Isophotes of Fig. 2 are drawn from the Landolt–Börnstein compilation of data by H. Fechtig, Ch. Leinert, and E. Grün. The photometric pattern is from the Tenerife (Spain) ground-based observations by the author, Sánchez, and Soulié, and from a rocket program (Leinert and collaborators) for the inner zodiacal light. Their mutual agreement at their junction (30° elongation) is very good. The main features are the Gegenschein, which is 30% brighter than the minimum along the ecliptic, and the high-latitude residual brightness, which is about half of that minimum. The polarimetric map is from ground-based observations at Haleakala (Hawaii) by J. L. Weinberg, Misconi, and Mann, from *OSO-5* satellite observations (J. G. Sparrow and E. Ney), from the Thisbe balloon program (Frey et al.), and from the already quoted rocket data. This map exhibits a region of rather strong partial linear polarization near 60° elongation (about 20%, slightly less in the

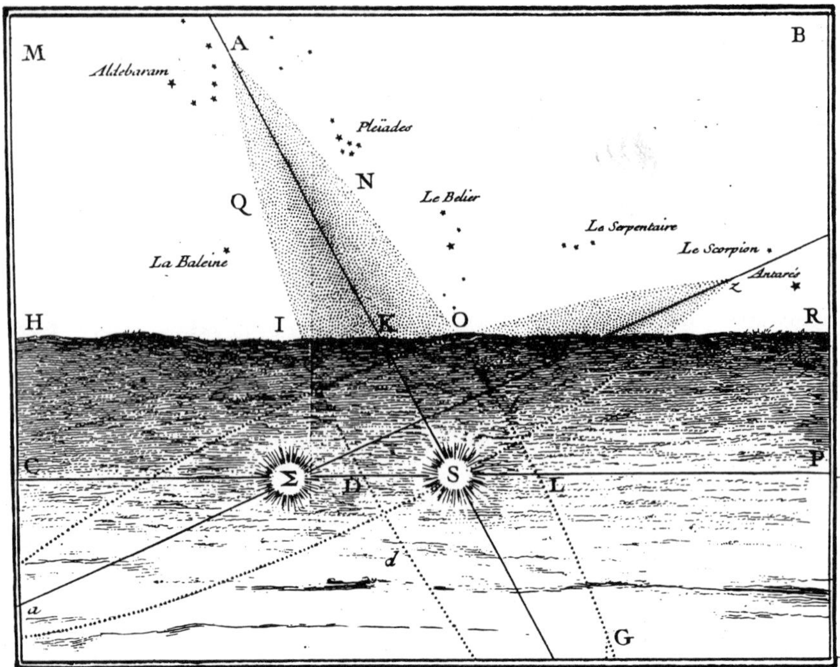

Figure 1. Visibility of zodiacal cones from northern middle latitudes. The figure is valid for the end of winter. At the end of the evening twilight, the Sun (S) is depressed by 18° and the ecliptic is tilted by more than 60°, both with respect to the horizon. The zodiacal cone, which approximately follows the ecliptic, is fairly well extricated from the unclear vicinity of the horizon. At the beginning of the dawn (Sun at Σ), the tilt is less than 30° and the cone is much more difficult to see. [*From de Mairan (1733); see text.*]

ecliptic, slightly more far from it). Around the antisun and about 10° away of it, there is a ring of negative polarization, that is, with the Fresnel vector lying in the scattering plane, instead of perpendicular to it, as in the rest of the sky.

Photopolarimetry from Space Probes

A less parochial view of the zodiacal cloud has been given, outside the Earth's orbit by *Pioneer 10* and *Pioneer 11* (Weinberg et al.); inside it by *Helios 1* and *Helios 2* (Leinert et al.). In addition to a satisfying agreement with the previous Earth-based results when they were near the Earth's orbit, these deep space probes solved important problems during their missions.

The *Pioneer* spacecraft detected Z.L. up to 3 AU heliocentric distance and gave precious data about the "astronomical" (i.e., components 1, 2, and 3, mentioned previously) components of the skylight. Analyses of the interplanetary signal gave clues toward the heliocentric-dependent properties of the dust [Schuerman (1980)]. Because scans of the ecliptic did show an antisolar peak even far from the Earth, the *Pioneer* spacecraft proved the Gegenschein to be unrelated to Earth, contrary to most of the previous interpretations. The Gegenschein is a matter of phase function of the interplanetary grains. It is not intrinsically different from the Z.L., and it is simply due to an enhanced reflectivity of the dust at backscattering angles—just like the rear reflectors on a vehicle.

Along inner orbits with 0.3 AU perihelion, the *Helios* spacecraft provided several years of surveying of the Z.L., with outstanding reliability with respect to any kind of variations. Despite severe restrictions upon the celestial coverage, especially at low ecliptic latitudes, major results were obtained. For any given pointing coordinates referred to the Sun ($\lambda-\lambda_\odot, \beta$), the same brightness and the same polarization degree are recovered when the same observing heliocentric distance, R, is restored. This shows a great stability of the cloud, as previously concluded—with somewhat lesser precision, and from $R=1$ AU only—from 12 years of ground-based survey at Tenerife. Almost irrespective also of the pointing coordinates, heliocentric dependences were found as $R^{-2.3}$ for the brightness and as $R^{0.3}$ for the polarization degree. This implies a great smoothness for the cloud; the latter dependence, however, shows that the dust properties cannot be the same everywhere.

Radiometry: The Thermal Emission

Detection and studies of the thermal emission from interplanetary dust have been rather tentative until the radiometric survey by the *Infrared Astronomical Satellite* (*IRAS*), although limited to solar differential longitudes $\lambda-\lambda_\odot$ between $\pm 60°$ and $\pm 120°$. *IRAS* showed the zodiacal emission at each of the wavelengths of 12, 25, 60, and 100 μm (Michael G. Hauser et al.) with peculiar intensity at the first two wavelengths.

Radiometric data gave access to the temperature of the dust (≈ 260 K), and their comparison with visual photometric data led to its albedo (≈ 0.08), both at 1 AU. Inversion techniques led to heliocentric dependences of these quantities, which invalidate once more the idea of a homogeneous cloud. The gradients retrieved from *IRAS* data agree with those from *ZIP* (*Zodiacal Infrared Project*; Price et al.), the other main source of radiometric data (at 10.9 and 20.9 μm). However, a discrepancy by a factor of 2 between *IRAS* and *ZIP* absolute brightnesses stresses the calibration and disentangling difficulties, in the present state of infrared observations. The coverage is wider for *ZIP* than for *IRAS*; in the antisolar region, no "radiometric Gegenschein" appears, and the contrary would have conflicted with the well-known isotropy of thermal radiation.

Other Observational Approaches

Colorimetry [i.e., the study of the sunlight scattered by the dust in the widest range of $\lambda\lambda$, from the near infrared (2.4 μm) to the ultraviolet (0.1 μm)] provides tests about the size distributions of the grains. Despite a controversial situation in the ultraviolet, it

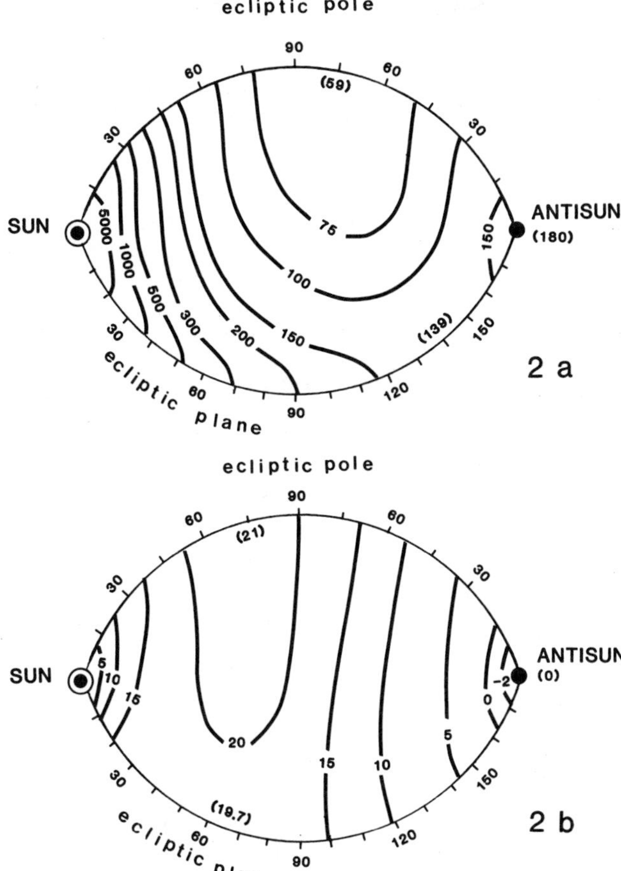

Figure 2. Zodiacal brightness and polarization over the sky. Lines of equal brightness (*a*), of equal polarization degree (*b*) over a quarter of the sky limited by the Sun, the antisun, half the ecliptic, and half the orthogonal great circle (containing an ecliptic pole). The three other quarters are the same, disregarding second-order seasonal oscillations. The brightness unit is the flux of a G2 V star of 10th visual magnitude per square degree (S_{10}). Polarization degrees are in percent. [*Interpolated from the tables of Landolt–Börnstein, VI 2a (1981); see text.*]

seems that the Z.L. spectrum departs rather little from the solar spectrum, except some possible reddening at small elongations.

Spectrometry (i.e., the study of Doppler shifts of the Fraunhofer absorption lines present in the Z.L. spectrum) provides tests about the kinematics of the grains, and shows that most of them orbit in the same sense as the planets do. Due to the weakness of their radial velocities, however, and to averaging effects along the line of sight, it is difficult to decide whether or not their orbital velocities differ significantly from the Keplerian ones.

ZODIACAL LIGHT AS A NOISE: A CUMBERSOME FOREGROUND

Scattered and emitted radiation from zodiacal dust is increasingly worrisome, as background targets become fainter and fainter.

In radiometry, thermal emission dominates other sources in the 10–25-μm range and at low and medium ecliptic latitudes. Its subtraction, according to models that are not free of assumptions or uncertainties, requires much care, in order not to affect severely the knowledge of often weaker and colder background sources.

In the optical domain, observations of background sources are in principle less seriously affected by "noise" from the Z.L.; the *Hubble Space Telescope* (*HST*), however, aims to detect objects of 28th or 29th magnitude. One figure showing that it has to compromise with Z.L. is the mean interplanetary flux (assuming an average brightness of 100 S_{10}; see Fig. 2*a*) inside the theoretical central diffraction disk of a 2.4-m perfectly stigmatic telescope at $\lambda = 0.5$ μm (area, 0.0086 square arcsecond) which is just the flux of a star of 28th magnitude. The *Instrument Handbook* of the Space Telescope Science Institute refers to the Tenerife [A. C. Levasseur-Regourd and R. Dumont (1980)] brightness table as a basis of correction for this "zodiacal veil." Figure 2*a* shows this noise to be minimized by a proper choice of the epoch—the best one being when the longitude of the Sun is that of the target ± about 135°. Until the *HST* is optically corrected, the problem is more significant.

THE ZODIACAL DUST CLOUD ACCORDING TO ITS LIGHT AND HEAT

Radiative studies are only one of the possible approaches toward the properties of the dust. Most of the optical and thermal parameters (polarization, temperature, albedo, etc.) and their gradients versus location in the solar system are mentioned in the entry on Interplanetary Dust, Remote Sensing.

The cloud appears to be steady and smooth, with little short-term variations and no solar-cycle-related or secular changes. The space density in the symmetry plane seems to be inversely proportional to the heliocentric distance, r. At a given r, the density is reduced by a factor of approximately 2 or slightly more by an inclination of 20° upon the symmetry plane. The shape of the isodensity surfaces remains controversial, but could be ellipsoidal with some central bulge. As discussed previously, converging evidence shows that the volume average properties of the dust complex are heliocentric-dependent, probably also inclination-dependent. What is changing is another matter: Size? Composition? Ratio of two populations?

The quasiabsence of seasonal oscillations of brightness near the ecliptic means that the cloud is highly rotationally symmetric. From the amplitudes and phases of the oscillations observed toward ecliptic poles, one can retrieve (after a careful account of the eccentricity of Earth's orbit) the obliquity and the ascending node of the symmetry plane upon the ecliptic. After a period of discrepancy between the previous optical and new *IRAS* results, some agreement seems to be now found near 1.6° and 90°. These figures (valid at least "for the level of the Earth"—because a warping of the symmetry surface versus r is not excluded) approximately identify the symmetry plane near 1 AU with the invariable plane of the solar system.

Additional Reading

Fechtig, H., Leinert, Ch., and Grün, E. (1981). Interplanetary dust and zodiacal light. In *Landolt-Börnstein*, Neue Serie VI 2a, p. 228. Springer-Verlag, Berlin.

Giese, R. H., Kneissel, B., and Rittich, U. (1986). Three-dimensional models of the zodiacal dust cloud: a comparative study. *Icarus* **68** 395.

Leinert, Ch. (1975). Zodiacal light—a measure of the interplanetary environment. *Space Sci. Rev.* **18** 281.

Leinert, Ch. and Grün, E. (1990). Interplanetary dust. In *Physics of the Inner Heliosphere. Physics and Chemistry in Space—Space and Solar Physics*, vol. 20, R. Schwenn and E. Marsch, eds. Springer-Verlag, Berlin, p. 207.

Levasseur-Regourd, A. C. and Dumont R. (1980). Absolute photometry of zodiacal light. *Astron. Ap.* **84** 277.

Olson, D. W. (1989). Who first saw the zodiacal light? *Sky and Telescope* **77** 146.

Weinberg, J. L. and Sparrow, J. G. (1978). Zodiacal light as an indicator of interplanetary dust. In *Cosmic Dust*, J. A. M. McDonnell, ed. Wiley, New York, p. 75.

See also **Diffuse Galactic Light; Interplanetary Dust, Collection and Analysis; Interplanetary Dust, Dynamics; Interplanetary Dust, Remote Sensing; Sun, Eclipses.**

Index